Symbols, Units, and Abbreviations

Quantity	Symbol	Unit	Abbreviation
angular frequency	ω	radian/second	rad/s
capacitance	C	farad	F
charge	Q, q	coulomb	C
current	I, i	ampere	A
electric field intensity	$\mathscr{E}$	volts/meter or newtons/coulomb	V/m N/C
energy, work	W, w	joule	J
flux density	B	tesla	T
frequency	f	hertz	Hz
impedance	$\mathbf{Z}$	ohm	Ω
inductance	L	henry	H
magnetic flux	Φ	weber	Wb
mmf	$\mathscr{F}$	amp-turns	At
period	T	seconds	s
power	P, p	watt	W
reactance	X	ohm	Ω
reluctance	$\mathfrak{R}$	amp-turns/weber	
resistance	R	ohm	Ω
time	t	second	s
voltage	V, v, E, e	volt	V

Prefixes

Prefix	Symbol	Multiple
tera	T	10^{12}
giga	G	10^{9}
mega	M	10^{6}
kilo	k	10^{3}
milli	m	10^{-3}
micro	μ	10^{-6}
nano	n	10^{-9}
pico	p	10^{-12}

Table 3–1 Resistivity of Materials, ρ

Material	Resistivity, ρ, at 20° C (Ω-m)
Silver	1.645×10^{-8}
Copper	1.723×10^{-8}
Gold	2.443×10^{-8}
Aluminum	2.825×10^{-8}
Tungsten	5.485×10^{-8}
Iron	12.30×10^{-8}
Lead	22×10^{-8}
Mercury	95.8×10^{-8}
Nichrome	99.72×10^{-8}
Carbon	3500×10^{-8}
Germanium	20–2300*
Silicon	≅500*
Wood	10^{8}–10^{14}
Glass	10^{10}–10^{14}
Mica	10^{11}–10^{15}
Hard rubber	10^{13}–10^{16}
Amber	5×10^{14}
Sulphur	1×10^{15}
Teflon	1×10^{16}

*The resistivities of these materials are dependent upon the impurities within the materials.

Conversion Factors

1 in. = 2.540 cm
1 ft = 0.3048 m
1 yd = 0.914 m
1 mi = 1.609 km
1 m = 39.37 in.
1 m = 3.281 ft
1 cm = 0.01 m = 1×10^{-2} m
1 km = 0.6215 mi
1 in.^{2} = 6.452 cm^2
1 m^2 = 10.76 ft^2
1 gal (US) = 3.7854×10^{-3} m^3
1 liter = 1×10^{-3} m^3

1 hp ≈ 746 watts
1 watt = 1 J/s
1 kWh = 3.6 M J

1 N = 0.2248 lb
1 lb = 4.448 N
1 kg = 2.2046 lb

1 tesla = 10^4 gauss
1 weber = 10^8 lines
$H_g = 7.96 \times 10^5 B_g$
1 Coulomb = 6.24×10^{18} electrons

Constants

Permeability of free space	$\mu_o = 4\pi \times 10^{-7}$ Wb/A-m
Permittivity of free space	$\epsilon_o = 8.854 \times 10^{-12}$ F/m
Charge on an electron	$Q_e = 1.602 \times 10^{-19}$ C
Velocity of light	$c = 299\ 792\ 458$ m/s

The Greek Alphabet

Alpha	A	α	Iota	I	ι	Rho	P	ρ
Beta	B	β	Kappa	K	κ	Sigma	Σ	σ
Gamma	Γ	γ	Lambda	Λ	λ	Tau	T	τ
Delta	Δ	δ	Mu	M	μ	Upsilon	Υ	υ
Epsilon	E	ϵ	Nu	N	ν	Phi	Φ	ϕ
Zeta	Z	ζ	Xi	Ξ	ξ	Chi	X	χ
Eta	H	η	Omicron	O	o	Psi	Ψ	ψ
Theta	Θ	θ	Pi	Π	π	Omega	Ω	ω

Straight Line

$y = m\,x + b$ where m = slope = $\Delta y/\Delta x$. Δy is the amount y changes when x changes by Δx.

Quadratic Equation

Roots of $ax^2 + bx + c = 0$ are

$$x_{1,2} = \frac{-b \pm \sqrt{b^2 - 4ac}}{2a}$$

If $(b^2 - 4ac) \geq 0$, roots are real; if $(b^2 - 4ac) < 0$, roots are complex conjugates.

Trigonometric Identities

$\sin(-\alpha) = -\sin\alpha$ $\qquad$ $\cos(-\alpha) = \cos\alpha$

$\sin(\omega t \pm 90°) = \pm\cos\omega t$ $\qquad$ $\sin(\omega t \pm 180°) = -\sin\omega t$

$\cos(\omega t \pm 90°) = \mp\sin\omega t$ $\qquad$ $\cos(\omega t \pm 180°) = -\cos\omega t$

$$\sin^2\omega t = \frac{1}{2}(1 - \cos 2\omega t)$$

Natural Logarithms

Given $e^x = y$. x may be found by noting that $\ln e^x = x$.
For example, if $e^{-5t} = 0.7788$, then

$$\ln e^{-5t} = \ln 0.7788$$
$$-5t = -0.25$$
$$t = 0.05 \text{ s}$$

Powers of *j*

$j = \sqrt{-1}$

$j^2 = -1$

$j^3 = -j$

$\dfrac{1}{j} = -j$

Calculus

$\dfrac{d}{dx}(ax) = a$ $\qquad$ $\dfrac{d}{dx}(\sin ax) = a\cos ax$

$\dfrac{d}{dx}(e^{ax}) = ae^{ax}$ $\qquad$ $\dfrac{d}{dx}(\cos ax) = -a\sin ax$

$\dfrac{d}{dx}(uv) = u\dfrac{dv}{dx} + v\dfrac{du}{dx}$

$\displaystyle\int x\,dx = \frac{x^2}{2}$ $\qquad$ $\displaystyle\int \sin ax\,dx = -\frac{1}{a}\cos ax$

$\displaystyle\int \frac{dx}{x} = \ln x$ $\qquad$ $\displaystyle\int \cos ax\,dx = \frac{1}{a}\sin ax$

Other

$$\frac{1}{1+x} \approx 1 - x \quad (x \ll 1)$$

Icon Summary	Meaning
(globe icon)	Visit the book's Web site
CircuitSim 01-1	Interactive circuit simulation
PSPICE	PSpice example or problem
multiSIM	Multisim example or problem
∫	Calculus-based material
✓	In-Process Learning Check
(speaker icon)	Audio clip

CIRCUIT ANALYSIS
THEORY AND PRACTICE

FIFTH EDITION

CIRCUIT ANALYSIS

THEORY AND PRACTICE

FIFTH EDITION

Allan H. Robbins

Wilhelm C. Miller

CENGAGE Learning®

Andover • Melbourne • Mexico City • Stamford, CT • Toronto • Hong Kong • New Delhi • Seoul • Singapore • Tokyo

Circuit Analysis: Theory & Practice, 5e
Allan H. Robbins and Wilhelm C. Miller

ISBN-13: 978-81-315-1902-8
ISBN-10: 81-315-1902-3

Cengage Learning India Private Limited
418, F.I.E., Patparganj
Delhi 110092
India

Tel: 91-11-43641111
Fax: 91-11-43641100
Email: asia.infoindia@cengage.com

Cengage Learning is a leading provider of customized learning solutions with office locations around the globe, including Andover, Melbourne, Mexico City, Stamford (CT), Toronto, Hong Kong, New Delhi, Seoul, Singapore, and Tokyo. Locate your local office at: www.cengage.com/global

Cengage Learning Products are represented in Canada by Nelson Education, Ltd.

For product information, visit our website at **www.cengage.co.in**

Printed in India
Reprint 2017

Contents

Preface

The Book and Who It Is For

This textbook is written primarily for students in electronics technology and electrical technology programs, engineering programs, and industrial training programs at colleges, universities, and career schools and in industry. Its goal continues to be to meet the training needs of today's students, their professors, and their instructors as it has throughout its four previous editions. An introductory circuit analysis textbook, it covers fundamentals of dc and ac circuits, methods of analysis, capacitance, inductance, magnetic circuits, basic transients, Fourier analysis, and other topics. When students successfully complete a course using this book, they will have a good working knowledge of basic circuit principles and a demonstrated ability to solve a variety of circuit-related problems.

Text Organization

The book contains 25 chapters and is divided into 5 main parts: Foundation dc Concepts, Basic dc Analysis, Capacitance and Inductance, Foundation ac Concepts, and Impedance Networks. Chapters 1 through 4 are introductory. They cover the foundation concepts of voltage, current, resistance, Ohm's law, and power. Chapters 5 through 9 focus on dc analysis methods. Included are Kirchhoff's laws, series and parallel circuits, mesh and nodal analysis, Y and Δ transformations, source transformations, Thévenin's and Norton's theorems, the maximum power transfer theorem, and so on. Chapters 10 through 14 cover capacitance, magnetism, and inductance, plus magnetic circuits and simple dc transients. Chapters 15 through 17 cover foundation ac concepts; ac voltage generation; and the basic ideas of frequency, period, phase, and so on. Phasors and the impedance concept are introduced and used to solve simple problems. Power in ac circuits is investigated and the concept of power factor and the power triangle are introduced. Chapters 18 through 25 then apply these ideas. Topics include ac versions of earlier dc techniques such as mesh and nodal analysis, Thévenin's theorem, and so on, as well as new ideas such as resonance, filters, Bode techniques, three-phase systems, transformers, and nonsinusoidal waveform analysis.

Several appendices round out the book. Appendix A provides operational instructions, reference material, and tips for PSpice and Multisim users; Appendix B is a math-and-calculator tutorial that describes typical mathematical and calculator usage in circuit analysis—including methods for solving simultaneous equations; Appendix C shows how to apply calculus to derive the maximum power transfer theorem for both dc circuits and ac circuits, while Appendix D contains answers to odd-numbered end-of-chapter problems.

Required Background

Students need a working knowledge of basic algebra and trigonometry and the ability to solve second-order linear equations such as those found in mesh analysis. They should be familiar with the SI metric system and the atomic nature of matter. Calculus is introduced gradually into later chapters for those who need it. However, calculus is not an essential prerequisite or corequisite, as all topics can be readily understood without it. Thus, students who know (or are studying) calculus can use this knowledge to enrich their understanding of circuit theory, whereas, because the calculus parts of the book can be omitted without any loss of continuity, students unfamiliar with calculus can comfortably navigate around it as they work through the chapters. (Calculus-based material is flagged by an ∫ icon to identify it as optional material for advanced learners.)

Features of the Book

- **A clearly written, easy-to-understand style** emphasizes principles and concepts.
- **More than 1,200 diagrams and photos.** Color and 3D visual effects are used to illustrate and clarify ideas and to aid visual learners.
- **Examples.** Over 380 examples, worked out in step-by-step detail, help promote understanding and guide the student in problem solving.
- **Practice Problems.** These follow the presentation of key ideas, encouraging the student to practice the skills just learned.

Each chapter opens with Key Terms, Outline, Objectives, Chapter Preview, and Putting It in Perspective.

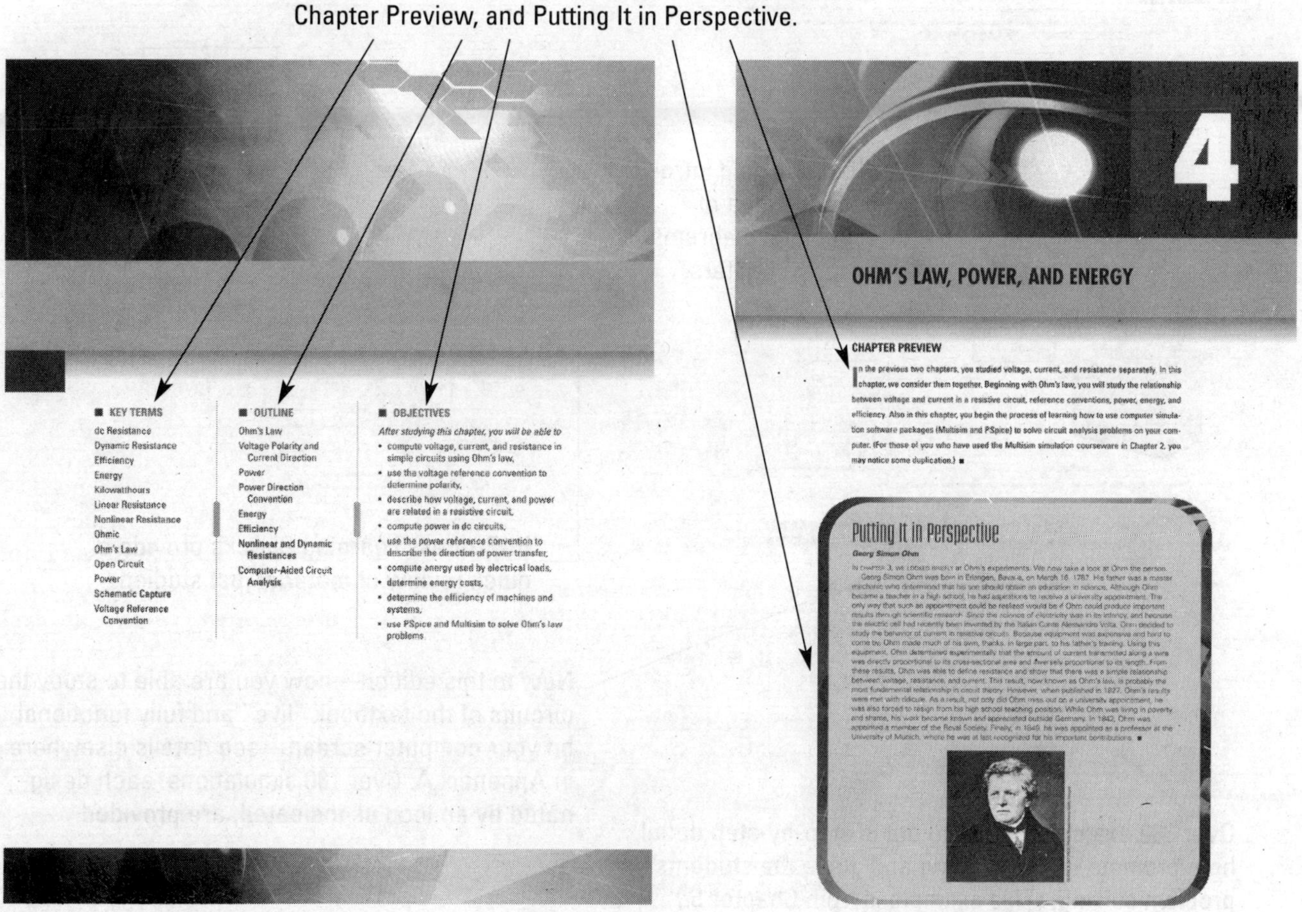

- **Key Terms** at the beginning of each chapter identify new terms to be introduced.
- **Competency-Based Objectives** define the knowledge or skill that the student is expected to gain from each chapter.
- **Chapter Previews** provide a context and a brief overview for the upcoming chapter and answer the question "Why am I learning this?"
- **Putting It in Perspective.** These are short vignettes that provide interesting background on people, events, and ideas that led to major advances or contributions in electrical science.
- **In-Process Learning Checks.** These are short, self-test quizzes that provide a quick check of the material just learned and help identify learning gaps.
- **Putting It into Practice.** These are end-of-chapter mini-projects—such as assignments that require students to do some research or to reason their way through realistic situations similar to what they might encounter in practice.
- **Margin Notes.** These include Practical Notes (which provide practical information, e.g., tips on how to use meters) plus general notes that provide additional information or add a perspective to the material being studied.
- **Audio Clips.** These clips (found on our Web site) present a more in-depth discussion of the most difficult topic for each chapter.
- **Computer Simulations.** Each chapter (beginning with Chapter 4) contains a section on computer-aided circuit analysis using Multisim and PSpice.

PSPICE MultiSim

Practice Problems build problem-solving skills and test student understanding.

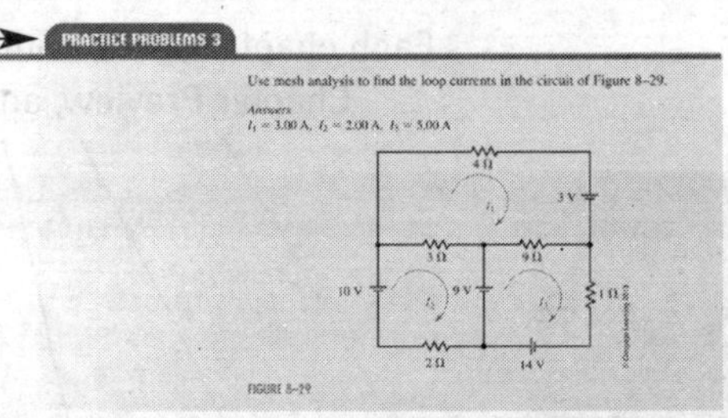

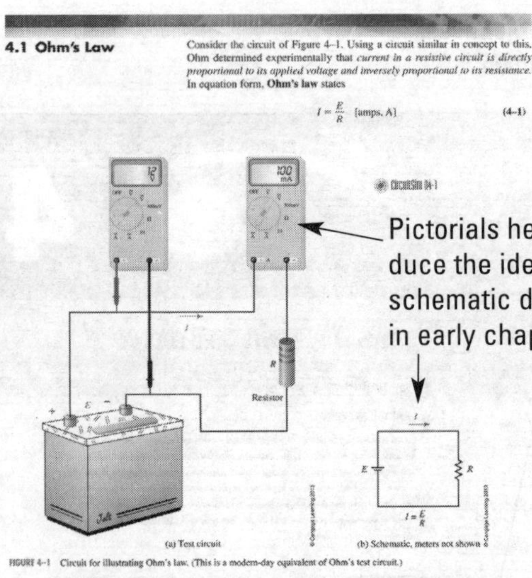

Pictorials help introduce the idea of schematic diagrams in early chapters.

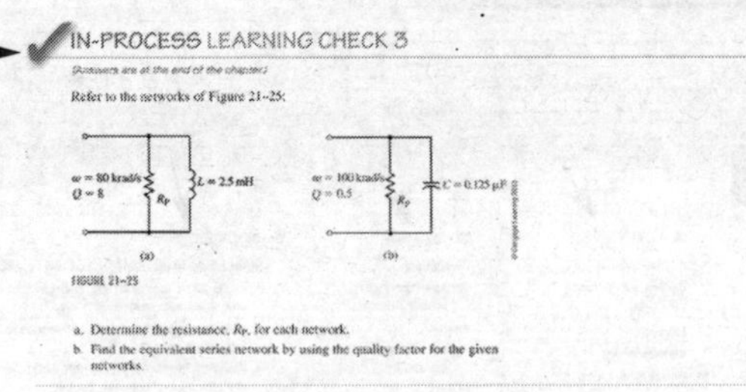

In-Process Learning Checks provide a quick review of material just studied.

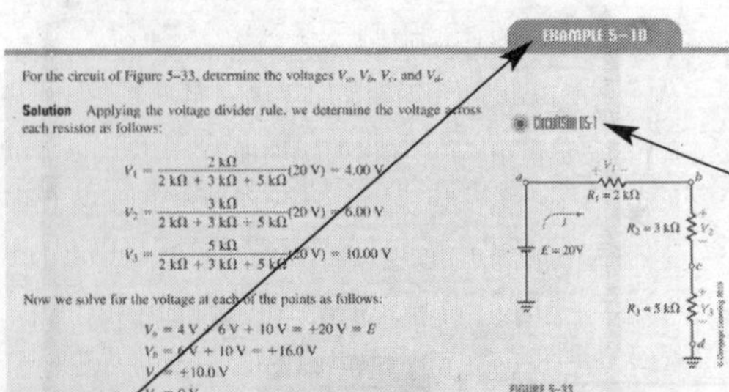

New to this edition—now you are able to study the circuits of the textbook "live" and fully functional on your computer screen—see details elsewhere in Appendix A. Over 180 simulations, each designated by an icon as indicated, are provided.

Over 380 examples, worked out in step-by-step detail, help promote understanding and guide the students in problem solving. (This example is from Chapter 5.)

These sections provide step-by-step instructions on how to build circuits on your screen, plus actual screen captures to show you what you should see when you run the simulations.

- **More than 1,600 end-of-chapter problems, Practice Problems,** and **In-Process Learning Check problems** are provided.
- **Calculators.** The use of calculators in circuit analysis has been integrated throughout the text. (Calculator techniques are also discussed in Appendix B, *Mathematics in Circuit Analysis*.)
- **Answers to Odd-Numbered Problems.** These are contained in Appendix D.

Each chapter (as applicable) includes a section showing how to use modern computer methods to solve circuits by simulation. The first example below (from Chapter 14) shows how to use Multisim to solve a transient problem, while the second example (from Chapter 23) shows how to use PSpice to solve a coupled circuit problem.

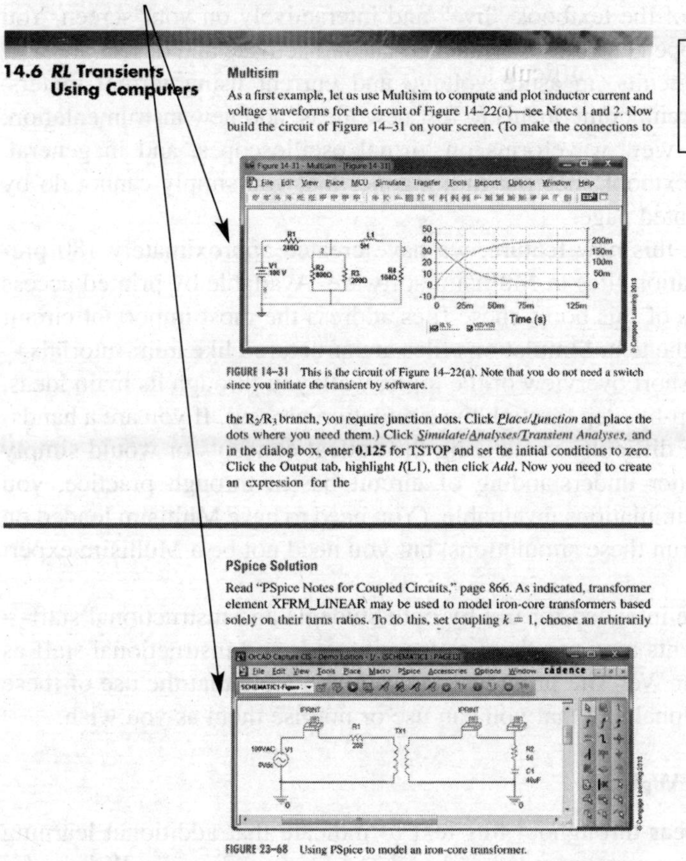

14.6 *RL* Transients Using Computers

Multisim

As a first example, let us use Multisim to compute and plot inductor current and voltage waveforms for the circuit of Figure 14–22(a)—see Notes 1 and 2. Now build the circuit of Figure 14–31 on your screen. (To make the connections to

FIGURE 14–31 This is the circuit of Figure 14–22(a). Note that you do not need a switch since you initiate the transient by software.

the R_2/R_3 branch, you require junction dots. Click *Place/Junction* and place the dots where you need them.) Click *Simulate/Analyses/Transient Analyses*, and in the dialog box, enter **0.125** for TSTOP and set the initial conditions to zero. Click the Output tab, highlight *I*(L1), then click *Add*. Now you need to create an expression for the

PSpice Solution

Read "PSpice Notes for Coupled Circuits," page 866. As indicated, transformer element XFRM_LINEAR may be used to model iron-core transformers based solely on their turns ratios. To do this, set coupling $k = 1$, choose an arbitrarily

FIGURE 23–68 Using PSpice to model an iron-core transformer.

Putting It into Practice problems are provided at the end of chapters and describe a work-related problem.

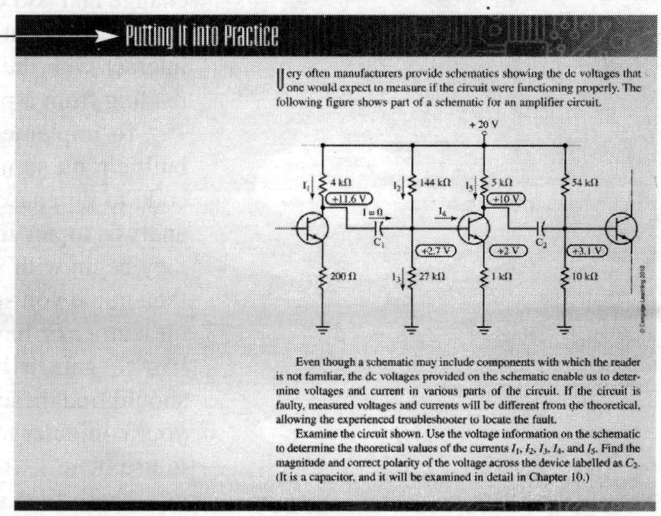

Putting It into Practice

Very often manufacturers provide schematics showing the dc voltages that one would expect to measure if the circuit were functioning properly. The following figure shows part of a schematic for an amplifier circuit.

Even though a schematic may include components with which the reader is not familiar, the dc voltages provided on the schematic enable us to determine voltages and current in various parts of the circuit. If the circuit is faulty, measured voltages and currents will be different from the theoretical, allowing the experienced troubleshooter to locate the fault.

Examine the circuit shown. Use the voltage information on the schematic to determine the theoretical values of the currents I_1, I_2, I_3, I_4, and I_5. Find the magnitude and correct polarity of the voltage across the device labelled as C_2. (It is a capacitor, and it will be examined in detail in Chapter 10.)

New and Improved Features of the Fifth Edition

Current users of a textbook generally expect that a new edition will retain the features that attracted them to the book in the first place. For such users, we have maintained the same basic structure that we used in the fourth edition, with improvements made in presentation, examples, problems, photographs, and the glossary. Additionally, the Multisim and PSpice material has been updated to the latest revisions and some new examples have been added. For users who would like to see us take advantage of new technologies such as the Internet, electronic publishing, and circuit simulation software to provide an improved learning experience, we have done that, too. Prime examples of the latter are the introduction of circuit simulation courseware (to permit interactive, hands-on learning on a personal computer), the addition of *CourseMate* (a multifaceted learning tool), and modernized and expanded Web sites for students and instructors. These are all features that we hope will appeal to new adopters as well. Let us look at each in turn.

Learning Circuit Theory Interactively on Your Computer (*New*)

While many students find it challenging to learn circuit theory simply by "reading the book," few have the resources necessary to establish their own personal laboratories where they can work their way through the learning process using real test and measurement equipment. However, the widespread use of computers and the availability of circuit simulation software now make it possible to create a hands-on learning environment on a desktop (or laptop)—in essence, a "virtual laboratory" (complete with multimeters, oscilloscopes, signal generators, and so on) where, along with our newly developed courseware, you can study the circuits of the textbook "live" and interactively on your screen. You can, for instance, open and close virtual switches to activate and deactivate simulated textbook circuits, measure voltage and current using virtual meters, change and add circuit components to test new ideas, add new instrumentation, initiate transients, view waveforms on virtual oscilloscopes, and in general, interact with the textbook material in a manner that you simply cannot do by reading from a printed page.

To implement this new feature, we have created approximately 180 prebuilt circuit simulation files in Multisim software. Available by printed access code to purchasers of this book, these files address the most important circuit analysis topics in the text. Simulations files are structured like mini-tutorials—they begin with a short overview of the topic, walk you through its main ideas, then guide you step-by-step through the simulation process. If you are a hands-on learner or have difficulty visualizing theoretical concepts or would simply like to gain a better understanding of circuit basics through practice, you should find these simulations invaluable. (You need to have Multisim loaded on your computer to run these simulations, but you need not be a Multisim expert to use them.)

These files are intended for use by both students and instructional staff—for example, students may use them as learning aids and instructional staff as teaching aids. (Our Web site provides suggestions.) Note that the use of these simulations is optional and that you can use or not use them as you wish.

Student Premium Website

Various icons appear throughout this text to indicate that additional learning tools are available at our expanded and updated *Student Premium Website for Circuit Analysis*. The printed access code at the back of this book will give students online access to the following suite of learning tools:

- **Learning Circuit Analysis Interactively on Your Computer:** This is the hands-on circuit simulation courseware described above.
- **Multisim Beginner's Tutorial:** This tutorial guides you step by step through the creation of Multisim circuit simulations. It supplements the presentation of Section 4.8 of the textbook and serves as an aid to creating your own simulations.

- **PSpice Beginner's Tutorial:** This tutorial guides you step by step through the creation of PSpice circuit simulations. It supplements the presentation of Section 4.8 of the textbook.
- **For Further Investigation (Mini-Tutorials):** This is a collection of tutorial articles that provide additional and/or advanced information to complement that found in the textbook. Included are *Advanced Studies in Transients; Matrix Representation in Circuit Analysis; Significant Digits and Numerical Accuracy*; *The Importance of Selectivity in Broadcast Radio; Bode Plots—More Examples; Fourier Series; Using Analog Multimeters; Appliance Grounding, Solving Simultaneous Equations with Complex Coefficients and How to Use Excel to Sketch Graphs.*
- **Lab Pre-Study Simulations: The Oscilloscope:** This is a group of five hands-on, Multisim-based simulations that teach basic oscilloscope usage and measurement techniques. It supplements our Lab Manual (described later).
- **Lab Pre-Study Simulation: Three-Phase Circuits:** This is a hands-on, Multisim-based simulation intended as preparation for three-phase laboratory work. (If you have no three-phase lab facilities at your institution, this simulation may serve as a useful substitute.)
- **Multisim and PSpice Solutions:** These are worked-out solutions to the PSpice and Multisim examples studied in the computer-aided circuit analysis sections of the various chapters.
- **Technical Notes Concerning Simulations:** Describes some of the practical issues involved with circuit simulations.
- **Sample Quizzes:** These sample questions give you a preview of the interactive quizzing feature available through CourseMate for Circuit Analysis.

To access the Student Premium Website, please visit www.cengagebrain.com. At the cengagebrain.com home page, enter the printed access code found inside the back cover of this book and click on the "Register" button. The Student Premium Website will be added to your CengageBrain bookshelf.

CourseMate for Circuit Analysis (*New*)

CourseMate for Circuit Analysis is an integrated, Web-based learning solution for building circuit analysis knowledge and skills. The *CourseMate* Web site includes all of the Student Premium Website assets just listed, plus the following suite of resources and study tools:

- An interactive eBook with highlighting, note taking, and search capabilities
- Interactive quizzes
- A set of student-focused PowerPoint slides
- Flashcards, crosswords, and other skill-building games

Instructors can use *CourseMate for Circuit Analysis* to access Instructor Resources and other classroom management tools.

How to Access the *CourseMate for Circuit Analysis* Site

To access these supplemental materials or see a CourseMate demo, please visit www.cengagebrain.com. At the cengagebrain.com home page, enter the ISBN of your title (from the back cover of your book) using the search box at the top of the page. This will take you to the product page where these resources can be found.

The following table lists the full array of supplemental resources available with this text and where they can be found:

	Student Premium Website	CourseMate	Reference Icon in Textbook
• Learning Circuit Analysis Interactively on Your Computer	✓	✓	CircuitSim
• Multisim Beginner's Tutorial	✓	✓	
• Capture PSpice Beginner's Tutorial	✓	✓	
• Audio Clips	✓	✓	
• For Further Investigation Mini-Tutorials	✓	✓	
• Lab Pre-Study Simulations: The Oscilloscope	✓	✓	
• Lab Pre-Study Simulation: Three-Phase Circuits	✓	✓	
• Multisim and PSpice Solutions	✓	✓	
• Technical Notes Concerning Simulations	✓	✓	
• Sample Quizzes	✓		
• Student-Focused PowerPoint Slides		✓	
• Interactive Quizzing		✓	
• Electronic Glossary		✓	
• Flashcards		✓	
• Games		✓	
• eBook		✓	

Teaching Resources

Laboratory Manual

The lab manual contains 28 hands-on labs (many with integrated computer simulation exercises), plus a comprehensive guide to equipment and laboratory measurements. Solutions to the lab manual may be found in the Instructor's Companion Website.

Instructor's Companion Website

The Instructor's Companion Website includes tools and instructional resources to enrich your classroom and make your preparation time shorter. Instructors may log in at www.cengagebrain.com and search on this textbook's ISBN to access the following resources:

- **Instructor's Guide.** Contains step-by-step solutions to all end-of-chapter (even and odd) problems and lab manual solutions, including waveforms, circuit diagrams, and more.
- **In-Process Quizzes.** These are in-depth, ready-for-the-classroom tests. Analytic in nature, patterned after—and similar in depth to—our end-of-chapter problems, these tests can be likened to chapter-based mini-exams.
- **PowerPoint® Presentation.** Slides for each chapter help you present concepts and material featuring key graphics and illustrations from the text. By importing graphics from the Image Library, you can create individualized presentations.
- **Image Library.** Contains full-color images from the book that you can use for slide shows, transparencies, and custom PowerPoint slides—or that you can simply print to use as needed.
- **Computerized Testbank.** Questions of varying levels of difficulty are provided in true/false and multiple-choice formats. These can be used to supplement the more in-depth *In-Process Quizzes* described earlier.

PSpice and Multisim Versions Used in This Book

The versions of PSpice and Multisim used throughout this text are the versions current at the time of writing—see Appendix A. Also shown in Appendix A are operational details for these products, as well as details about downloads, Web sites, helpful tutorials, and so on. [Explanatory notes or Multisim 9 versions of the Multisim material permit legacy software users to fully participate in the computer-aided circuit analysis activities of this book, as well as to use the CircuitSim circuit simulations (in sofar as the more limited features of Multisim 9 permit)—see our Web site for details.]

To the Student

Learning circuit theory should be challenging, interesting, and (we hope) fun. However, it is also hard work, because the knowledge and skills that you seek can only be gained through practice. We offer a few guidelines.

1. As you go through the material, try to gain an appreciation of where the theory comes from—that is, the basic experimental laws on which it is based. This will help you better understand the foundation ideas on which the theory is built.
2. Learn the terminology and definitions. Important new terms are introduced frequently. Learn what they mean and where they are used.
3. Study each new section carefully and be sure that you understand the basic ideas and how they are put together. Work your way through the examples with your calculator. Try the practice problems, then the end-of-chapter problems. Not every concept will be clear immediately, and most likely many will require several readings before you gain an adequate understanding.
4. When you are ready, test your understanding using the In-Process Learning Checks (self-quizzes) located in each chapter.
5. When you have mastered the material, move on to the next block. For those concepts that you are having difficulty with, consult your instructor or some other authoritative source.

To aid in this process, you may want to take advantage of the new learning support tools that we have provided, including the *CircuitSim* interactive, hands-on computer simulations, and the *CourseMate for Circuit Analysis* tools.

Calculators for Circuit Analysis and Electronics

You will need a good scientific calculator. A good calculator will permit you to more easily master the numerical aspects of problem solving, thereby leaving you more time to concentrate on the theory itself. This is especially true for ac, where complex number work dominates. There are some powerful calculators on the market that handle complex-number arithmetic almost as easily as real-number arithmetic—and there are some less expensive calculator models that also do a credible job. You should acquire an appropriate calculator (after consulting with your instructor), and learn to use it proficiently.

Acknowledgments

Many people have contributed to the development of this text. We begin by expressing our thanks to our students for providing much-needed feedback. Next, the reviewers and accuracy checkers: no textbook can be successful without the dedication and commitment of such people. We thank the following:

Reviewers

Sami Antoun, DeVry University, Columbus, OH
G. Thomas Bellarmine, Florida A & M University
Harold Broberg, Purdue University
William Conrad, IUPUI—Indiana University–Purdue University Indianapolis
Franklin David Cooper, Tarrant County College, Fort Worth, TX
Lance Crimm, Southern Polytechnic State University, Marietta, GA
David Delker, Kansas State University
Fred Dreyfuss, Pace University, White Plains, NY
Timothy Haynes, Haywood Community College
Bruce Johnson, University of Nevada, Reno, NV
Jim Pannell, DeVry University, Irving, TX
Alan Price, DeVry University, Pomona, CA
Philip Regalbuto, Trident Technical College
William Routt, Wake Tech Community College, Raleigh, NC
Carlo Sapijaszko, DeVry University, Orlando, FL
Jeffrey Schwartz, DeVry University, Long Island City, NY
John Sebeson, DeVry University, Addison, IL
Hesham Shaalan, Texas A&M University, Corpus Christi, TX
Parker Sproul, DeVry University, Phoenix, AZ
Lloyd E. Stallkamp, Montana State University
Roman Stemprok, University of Texas
Richard Sturtevant, Springfield Tech Community College, Springfield, MA

Technical Accuracy Reviewers

Chia-chi Tsui, DeVry University, Long Island City, NY
Rudy Hofer, Conestoga College, Kitchener, Ontario
Marie Sichler, Red River College, Winnipeg, Manitoba

Reviewers of the 5th Edition

Karl Huehne, Purdue University, West Lafayette, IN

George Peters, St. Clair College, Windsor, Ontario

Ken Swayne, Pellissippi State Community College, Knoxville, TN

Technical Assistance for the 5th Edition

George Foster, Red River College, Winnipeg, Manitoba

Dan Moroz, Red River College, Winnipeg, Manitoba

The following firms and individuals supplied photographs, diagrams, and other useful information:

Advance Devices Inc.
AT&T
AVX Corporation
B&K Precision Corporation
Bourns Inc.
Cadence Design Systems Inc.
Coilcraft Inc.
Fluke Corporation
Illinois Capacitor Inc.
IMRIS Inc.
JBL Professional
Kepco Inc.
National Instruments Corporation
Tektronix Inc.
United States Air Force, Nellis Air Force Base
Vansco Electronics Ltd.

We express our deep appreciation to the staff at Delmar Cengage Learning for their tireless efforts in putting this book together: To Stacy Masucci, Acquisitions Editor, for helping to define the project and smooth the way; Mary Clyne, Product Manager, for direction, encouragement, advice, and generally pulling the project together; David Arsenault, Senior Art Director, for ensuring that the new text design was done right; Barbara LeFleur, Content Project Manager, for helping to make the project work on the usual short deadline; Erica Glisson, Associate Marketing Manager, for helping get the book to its intended clientele; and Andrea Timpano, former Editorial Assistant, for providing answers and timely assistance. We also wish to thank Tania Andrabi, Associate Project Manager, CENVEO Publisher Services, and her colleagues (in particular Kevin Campbell, copyeditor) along with Joan Conlon, Freelance Developer, for guiding the book through the copyediting, page layout, and all such related tasks required for the production of this edition of the text book. To all these people and their staffs, a special thank you.

Lastly, we thank our wives and families for their support and perseverance during the preparation of this edition.

Allan H. Robbins
Wilhelm C. Miller
January 2012

About the Authors

Allan H. Robbins graduated in Electrical Engineering, earning both a Bachelor's degree and a Master's degree. After gaining industrial experience, he joined Red River College, where he served as an instructor and as head of the Department of Electrical and Computer Technology. In addition to his academic career, Allan has been a consultant and a small-business partner in the electronics/microcomputer field. He began writing as a contributing author for Osborne-McGraw-Hill in the early stages of the newly emerging microcomputer field and, in addition to this title, is joint author of several other textbooks. He has served as Section Chairman for the Institute of Electrical and Electronic Engineers and as a board member of several other organizations.

Wilhelm (Will) C. Miller obtained a diploma in Electronic Engineering Technology from Red River Community College (now Red River College) and later graduated from the University of Winnipeg with a degree in Physics and Mathematics. He worked in the communications field for ten years, including a one-year assignment with Saudi PTT in Jeddah, Saudi Arabia. Will has served as both an instructor and chair in the Electrical/Electronic Engineering Technology Department at Red River College. He currently serves as the chair of the Canadian Technology Accreditation Board, which ensures that accredited programs meet the Canadian National Technology Benchmarks. Additionally, Will provides academic consulting services, most recently in Korea and Mozambique.

FOUNDATION dc CONCEPTS

I

Circuit theory provides the tools and concepts needed to understand and analyze electrical and electronic circuits. The foundations of this theory were laid down over the past several hundred years by a number of pioneer researchers, including Count Alessandro Volta of Italy, Georg Simon Ohm of Germany, André Marie Ampère of France, James Watt of Scotland, and others. In 1800, for example, Volta developed an electric cell (battery) that provided the first source of what we now call dc voltage. Around the same time, the concept of current was evolved (even though nothing was known about the atomic structure of matter until much later). In 1826, Ohm brought the two ideas together and experimentally determined the relationship between voltage and current in a resistive circuit. This result, known as Ohm's law, set the stage for the development of modern-day circuit theory.

In Part I, we examine the foundation of this theory. We look at voltage, current, power, energy, and the relationships between them. The ideas developed here are profoundly important and constitute the fundamental ideas upon which all electrical and electronic circuit theory is built. ■

■ KEY TERMS

Application Software
Base
Block Diagram
Circuit Theory
Conversion Factor
Engineering Notation
Exponent
Horsepower (hp)
Joule
Multisim
Newton
Pictorial Diagram
Power of Ten Notation
Prefixes
PSpice
Resistance
Schematic Diagram
SI System
Watt

■ OUTLINE

Introduction
The SI System of Units
Converting Units
Power of Ten Notation
Prefixes, Engineering Notation, and Numerical Results
Circuit Diagrams
Circuit Analysis Using Computers and Calculators

■ OBJECTIVES

After studying this chapter, you will be able to

- describe the SI system of measurement,
- convert between various sets of units,
- use power of ten notation to simplify handling of large and small numbers,
- express electrical units using standard prefix notation such as μA, kV, mW, etc.,
- use a sensible number of significant digits in calculations,
- describe what block diagrams are and why they are used,
- convert a simple pictorial circuit to its schematic representation,
- describe generally how computers and calculators fit in the electrical and electronic circuit analysis picture.

1 INTRODUCTION

CHAPTER PREVIEW

An electrical circuit is a system of interconnected components such as resistors, capacitors, inductors, voltage sources, and so on. The electrical behavior of these components is described by a few basic experimental laws. These laws and the principles, concepts, mathematical relationships, and methods of analysis that have evolved from them are known as **circuit theory.**

Much of circuit theory deals with problem solving and numerical analysis. When you analyze a problem or design a circuit, for example, you are typically required to compute values for voltage, current, and power. In addition to a numerical value, your answer must include a unit. The system of units used for this purpose is the SI system (Systéme International). The SI system is a unified system of metric measurement; it encompasses not only the familiar MKS (meters, kilograms, seconds) units for length, mass, and time, but also units for electrical and magnetic quantities as well.

Quite frequently, however, the SI units yield numbers that are either too large or too small for convenient use. To handle these, engineering notation and a set of standard prefixes have been developed. Their use in representation and computation is described and illustrated.

Since circuit theory is somewhat abstract, diagrams are used to help present ideas. We look at several types—schematic, pictorial, and block diagrams—and show how to use them to represent circuits and systems.

We conclude the chapter with a brief look at computer and calculator usage in circuit analysis. Two popular software simulation packages are described, Multisim® (from National Instruments Corporation) and Orcad PSpice® (from Cadence Design Systems Inc.). These two packages are used throughout the book to illustrate ideas, to show how computers may be used to solve problems, and to help you understand the principles of circuit theory. ■

Putting It in Perspective

Hints on Problem Solving

During the analysis of electrical and electronic circuits, you will find yourself solving quite a few problems. An organized approach helps. Listed below are some useful guidelines:

1. Make a sketch (e.g., a circuit diagram), mark on it what you know, then identify what it is that you are trying to determine. Watch for "implied data" such as the phrase "the capacitor is initially uncharged." (As you will find out later, this means that the initial voltage on the capacitor is zero.) Be sure to convert all implied data to explicit data, for example, $V_o = 0V$.
2. Think through the problem to identify the principles involved, then look for relationships that tie together the unknown and known quantities.
3. Substitute the known information into the selected equation(s) and solve for the unknown. (For complex problems, the solution may require a series of steps involving several concepts. If you cannot identify the complete set of steps before you start, start anyway. As each piece of the solution emerges, you are one step closer to the answer. You may make false starts. However, even experienced people do not get it right on the first try every time. Note also that there is seldom one "right" way to solve a problem. You may therefore come up with an entirely different correct solution method than the authors do.)
4. Check the answer to see that it is sensible—that is, is it in the "right ballpark?" Does it have the correct sign? Do the units match?

1.1 Introduction

Technology has dramatically changed the way we do things; we now have Internet-connected computers and sophisticated electronic entertainment systems in our homes, electronic control systems in our cars, cell phones that can be used just about anywhere, robots that assemble products on production lines, and so on.

A first step to understanding these technologies is electric circuit theory. Circuit theory provides you with the knowledge of basic principles that you need to understand the behavior of electric and electronic devices, circuits, and systems. In this book, we develop and explore its basic ideas.

NOTES...

1. These examples are chosen to give you a concept of the scope of the application of electrical and electronic technology and its integration with other technical disciplines. To illustrate this breadth, we have chosen illustrative applications from the fields of home entertainment, health care, and industrial manufacturing processes.
2. As you go through the examples here, you will see components, devices and electrical quantities that have not yet been discussed. You will learn about these later. For the moment, just concentrate on the general ideas.

Examples of the Technology at Work

Before we begin, let us look at a few more examples of the technology at work (see Notes 1 and 2).

First, consider Figure 1–1, which shows a home theater system. This system relies on electrical and electronic circuits, magnetic circuits, and laser technology for its operation. For example, resistors, capacitors, and integrated circuits are used to control the voltages and currents that operate its motors and to amplify its audio and video signals, while laser circuitry is used to read data from the disks. The speaker system relies on magnetic circuits for its operation, while other magnetic circuits (the power transformers) drop the ac voltage from the 120-volt wall outlet voltage to the lower levels required to power the system.

iStockphoto/shippee

FIGURE 1–1 A home theater installation. This system incorporates many aspects of electrical and electronic technology.

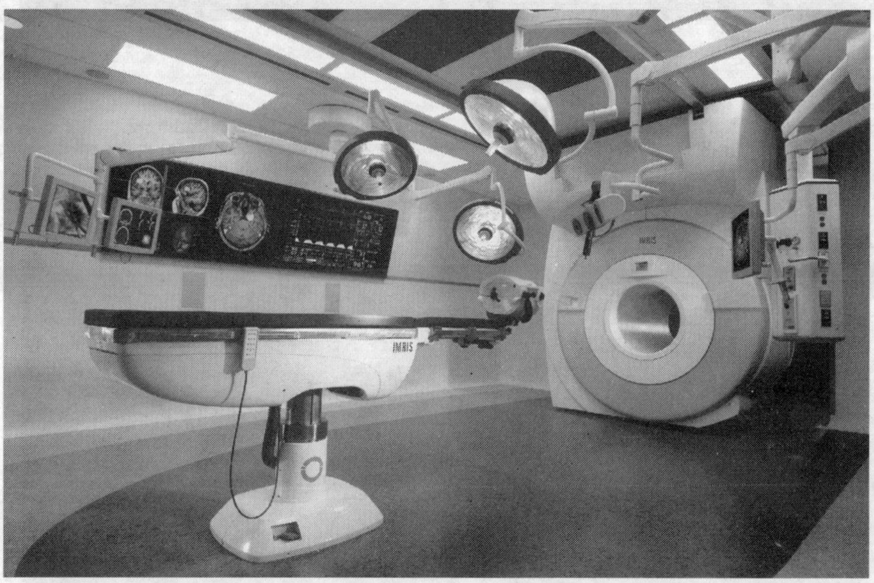

Photo courtesy of IMRIS

FIGURE 1–2 An innovative Magnetic Resonance Imaging (MRI) suite. This suite features a movable high field MR machine, a surgical information management system, a data display system, and more. The suite is engineered to permit image-guided therapy without having to move the patient between machines during the process.

Figure 1–2 shows another example, a Magnetic Resonance Imaging (MRI) suite. The heart of the suite is an MRI machine, a noninvasive medical diagnostic machine that helps physicians diagnose and treat medical conditions. The suite employs a sophisticated array of electrical, electronic, computer, mechanical, and magnetic systems to permit its use in image-guided therapy.

Figure 1–3 shows another application, a manufacturing facility where fine pitch surface-mount (SMT) components are placed on printed circuit boards at high speed using laser centering and optical verification. The bottom row of Figure 1–4 shows how small many of these components are. Computer control provides the high precision needed to accurately position parts as tiny as these.

Courtesy of Parker Vansco

FIGURE 1–3 Laser centering and optical verification in a manufacturing process. This machine is installing surface mount components of the type seen in Figure 10–30(a), Chapter 10.

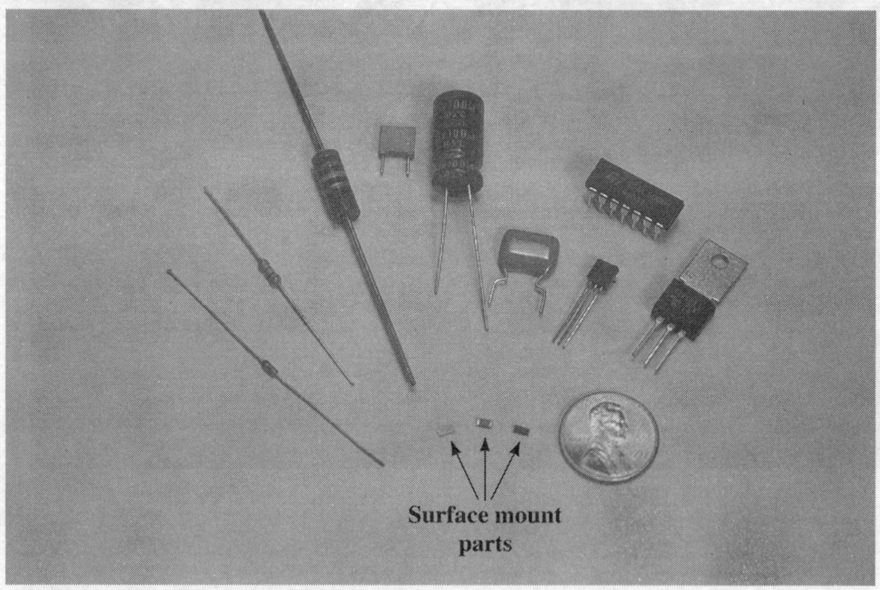

© Cengage Learning 2013

FIGURE 1–4 Some typical electronic components. The part with 7 pins visible is an integrated circuit, the small parts at the bottom are "surface mount parts," and the rest are "discrete components"—see Notes.

NOTES...

While most manufacturers now use surface mount components in their products, such parts are too small for general use in introductory electronics laboratories, so you will mostly encounter discrete components and integrated circuits in your basic studies.

1.2 The SI System of Units

The solution of technical problems requires the use of units. At present, two major systems—the English (US Customary) and the metric—are in everyday use. For scientific and technical purposes, however, the English system has been almost totally superseded. In its place the **SI system** is used. Table 1–1 shows a few frequently encountered quantities with units expressed in both systems.

The SI system combines the MKS metric units and the electrical units into one unified system: See Table 1–2 and Table 1–3. (Do not worry about the electrical units yet. We define them later, starting in Chapter 2.) Note that some symbols and abbreviations use capital letters while others use lowercase letters.

A few non-SI units are still in use. For example, electric motors are commonly rated in horsepower, and wires are frequently specified in AWG sizes (American Wire Gage, Section 3.2). On occasion, you will need to convert non-SI units to SI units. Table 1–4 may be used for this purpose.

TABLE 1–1 Common Quantities

1 meter = 100 centimeters
= 39.37 inches
1 millimeter = 39.37 mils
1 inch = 2.54 centimeters
1 foot = 0.3048 meter
1 yard = 0.9144 meter
1 mile = 1.609 kilometers
1 kilogram = 1000 grams
= 2.2 pounds
1 gallon (US) = 3.785 liters

Definition of Units

When the metric system came into being in 1792, the meter was defined as one ten-millionth of the distance from the north pole to the equator and the second as $\frac{1}{60} \times \frac{1}{60} \times \frac{1}{24}$ of the mean solar day. Later, more accurate definitions based on

TABLE 1–2 Some SI Base Units

Quantity	Symbol	Unit	Abbreviation
Length	l	meter	m
Mass	m	kilogram	kg
Time	t	second	s
Electric current	I, i	ampere	A
Temperature	T	kelvin	K

TABLE 1–3 Some SI Derived Units*

Quantity	Symbol	Unit	Abbreviation
Force	F	newton	N
Energy	W	joule	J
Power	P, p	watt	W
Voltage	V, v, E, e	volt	V
Charge	Q, q	coulomb	C
Resistance	R	ohm	Ω
Capacitance	C	farad	F
Inductance	L	henry	H
Frequency	f	hertz	Hz
Magnetic flux	Φ	weber	Wb
Magnetic flux density	B	tesla	T

*Electrical and magnetic quantities will be explained as you progress through the book. As in Table 1–2, the distinction between capitalized and lowercase letters is important.

TABLE 1–4 Conversions

	When You Know	Multiply By	To Find
Length	inches (in.)	0.0254	meters (m)
	feet (ft)	0.3048	meters (m)
	miles (mi)	1.609	kilometers (km)
Force	pounds (lb)	4.448	newtons (N)
Power	horsepower (hp)	746	watts (W)
Energy	kilowatthour (kWh)	3.6×10^6	joules† (J)
	foot-pound (ft-lb)	1.356	joules† (J)

†1 joule = 1 newton-meter.

physical laws of nature were adopted. The meter is now defined as the distance travelled by light in a vacuum in 1/299 792 458 of a second, while the second is defined in terms of the period of a cesium-based atomic clock. The definition of the kilogram is the mass of a specific platinum-iridium cylinder (the international prototype), preserved at the International Bureau of Weights and Measures in France (although this is under review at the time of writing and may be changed).

Relative Size of the Units*

To gain a feel for the SI units and their relative size, refer to Table 1–1 and Table 1–4. Note that 1 meter is equal to 39.37 inches; thus, 1 inch equals 1/39.37 = 0.0254 meter or 2.54 centimeters. A force of one pound is equal to 4.448 newtons; thus, 1 **newton** is equal to 1/4.448 = 0.225 pound of force, which is about the force required to lift a quarter-pound weight. One **joule** is the work done in moving a distance of one meter against a force of one newton. This is about equal to the work required to raise a quarter-pound weight one meter. Raising the weight one meter in one second requires about one **watt** of power.

The watt is also the SI unit for electrical power. A typical electric lamp, for example, dissipates power at the rate of 60 watts, and a toaster at a rate of about 1000 watts.

The link between electrical and mechanical units can be easily established. Consider an electrical generator. Mechanical power input produces electrical power output. If the generator were 100% efficient, then one watt of mechanical power input would yield one watt of electrical power output. This clearly ties the electrical and mechanical systems of units together.

However, just how big is a watt? While the preceding examples suggest that the watt is quite small, in terms of the rate at which a human can work, it is actually quite large. For example, a person can do manual labor at a rate of about 60 watts when averaged over an 8-hour day—just enough to power a standard 60-watt electric lamp continuously over this time! A horse can do considerably better. Based on experiment, James Watt determined that a strong dray horse could average 746 watts. From this, he defined the **horsepower (hp)** as 1 horsepower = 746 watts. This is the figure that we still use today.

1.3 Converting Units

Sometimes quantities expressed in one unit must be converted to another. For example, suppose you want to determine how many kilometers there are in 10 miles. Given that 1 mile is equal to 1.609 kilometers (Table 1–1), you can write 1 mi = 1.609 km, using the abbreviations in Table 1–4. Now multiply both sides by 10. Thus, 10 mi = 16.09 km.

This procedure is quite adequate for simple conversions. However, for complex conversions, it may be difficult to keep track of units. The procedure outlined next helps. It involves writing units into the conversion sequence, cancelling where applicable, then gathering up the remaining units to ensure that the final result has the correct units.

To get at the idea, suppose you want to convert 12 centimeters to inches. From Table 1–1, 2.54 cm = 1 in. From this, you can write

$$\frac{2.54\text{ cm}}{1\text{ in}} = 1 \quad \text{or} \quad \frac{1\text{ in}}{2.54\text{ cm}} = 1 \tag{1–1}$$

*Paraphrased from Edward C. Jordan and Keith Balmain, *Electromagnetic Waves and Radiating Systems,* Second Edition. (Englewood Cliffs, New Jersey: Prentice-Hall, Inc, 1968).

The quantities in Equation 1–1 are called **conversion factors.** As you can see, conversion factors have a value of 1, and thus you can multiply them times any expression without changing the value of that expression. For example, to complete the conversion of 12 cm to inches, choose the second ratio (so that units cancel), then multiply. Thus,

$$12 \text{ cm} = 12 \cancel{\text{cm}} \times \frac{1 \text{ in}}{2.54 \cancel{\text{cm}}} = 4.72 \text{ in}$$

When you have a chain of conversions, select factors so that all unwanted units cancel. This provides an automatic check on the final result as illustrated in part (b) of Example 1–1.

EXAMPLE 1–1

Given a speed of 60 miles per hour (mph),

a. convert it to kilometers per hour,

b. convert it to meters per second.

Solution

a. Recall, 1 mi = 1.609 km. Thus,

$$1 = \frac{1.609 \text{ km}}{1 \text{ mi}}$$

Now multiply both sides by 60 mi/h and cancel units:

$$60 \text{ mi/h} = \frac{60 \cancel{\text{mi}}}{\text{h}} \times \frac{1.609 \text{ km}}{1 \cancel{\text{mi}}} = 96.54 \text{ km/h}$$

b. Given that 1 mi = 1.609 km, 1 km = 1000 m, 1 h = 60 min, and 1 min = 60 s, choose conversion factors as follows:

$$1 = \frac{1.609 \text{ km}}{1 \text{ mi}}, \quad 1 = \frac{1000 \text{ m}}{1 \text{ km}}, \quad 1 = \frac{1 \text{ h}}{60 \text{ min}}, \quad \text{and} \quad 1 = \frac{1 \text{ min}}{60 \text{ s}}$$

Thus,

$$\frac{60 \text{ mi}}{\text{h}} = \frac{60 \cancel{\text{mi}}}{\cancel{\text{h}}} \times \frac{1.609 \cancel{\text{km}}}{1 \cancel{\text{mi}}} \times \frac{1000 \text{ m}}{1 \cancel{\text{km}}} \times \frac{1 \cancel{\text{h}}}{60 \cancel{\text{min}}} \times \frac{1 \cancel{\text{min}}}{60 \text{ s}} = 26.8 \text{ m/s}$$

You can also solve this problem by treating the numerator and denominator separately. For example, you can convert miles to meters and hours to seconds, then divide (see Example 1–2). In the final analysis, both methods are equivalent.

EXAMPLE 1–2

Do Example 1–1(b) by expanding the top and bottom separately.

Solution

$$60 \text{ mi} = 60 \cancel{\text{mi}} \times \frac{1.609 \cancel{\text{km}}}{1 \cancel{\text{mi}}} \times \frac{1000 \text{ m}}{1 \cancel{\text{km}}} = 96\,540 \text{ m}$$

$$1 \text{ h} = 1 \cancel{\text{h}} \times \frac{60 \cancel{\text{min}}}{1 \cancel{\text{h}}} \times \frac{60 \text{ s}}{1 \cancel{\text{min}}} = 3600 \text{ s}$$

Thus, velocity = 96 540 m/3600 s = 26.8 m/s as above.

PRACTICE PROBLEMS 1

1. Area $= \pi r^2$. Given $r = 8$ inches, determine area in square meters (m^2).
2. A car travels 60 feet in 2 seconds. Determine
   a. its speed in meters per second,
   b. its speed in kilometers per hour.

For part (b), use the method of Example 1–1, then check using the method of Example 1–2.

Answers
1. 0.130 m^2; 2. a. 9.14 m/s, b. 32.9 km/h

1.4 Power of Ten Notation

Electrical values vary tremendously in size. In electronic systems, for example, voltages may range from a few millionths of a volt to several thousand volts, while in power systems, voltages of up to several hundred thousand are common. To handle this large range, the **power of ten notation** (Table 1–5) is used—see Note.

To express a number in the power of ten notation, move the decimal point to where you want it, then multiply the result by the power of ten needed to restore the number to its original value. Thus, 247 000 $= 2.47 \times 10^5$. (The number 10 is called the **base,** and its power is called the **exponent.**) An easy way to determine the exponent is to count the number of places (right or left) that you moved the decimal point. Thus,

$$247\,000 = 2\,4\,7\,0\,0\,0 = 2.47 \times 10^5$$

5 4 3 2 1

Similarly, the number 0.003 69 may be expressed as 3.69×10^{-3} as illustrated below.

$$0.003\,69 = 0.0\,0\,3\,6\,9 = 3.69 \times 10^{-3}$$

1 2 3

Multiplication and Division Using Powers of Ten

To multiply numbers in power of ten notation, multiply their base numbers, then add their exponents. Thus,

$$(1.2 \times 10^3)(1.5 \times 10^4) = (1.2)(1.5) \times 10^{(3+4)} = 1.8 \times 10^7$$

For division, subtract the exponents in the denominator from those in the numerator. Thus,

$$\frac{4.5 \times 10^2}{3 \times 10^{-2}} = \frac{4.5}{3} \times 10^{2-(-2)} = 1.5 \times 10^4$$

NOTES...

The complete range of powers of ten defined in the SI system includes twenty members. However, only those powers of common interest to us in circuit theory are shown in Table 1–5.

TABLE 1–5 Common Power of Ten Multipliers

$1\,000\,000 = 10^6$	$0.000001 = 10^{-6}$
$100\,000 = 10^5$	$0.00001 = 10^{-5}$
$10\,000 = 10^4$	$0.0001 = 10^{-4}$
$1\,000 = 10^3$	$0.001 = 10^{-3}$
$100 = 10^2$	$0.01 = 10^{-2}$
$10 = 10^1$	$0.1 = 10^{-1}$
$1 = 10^0$	$1 = 10^0$

EXAMPLE 1–3

Convert the following numbers to power of ten notation, then perform the operation indicated:

a. 276×0.009,

b. 98 200/20.

Solution

a. $276 \times 0.009 = (2.76 \times 10^2)(9 \times 10^{-3}) = 24.8 \times 10^{-1} = 2.48$

b. $\dfrac{98\,200}{20} = \dfrac{9.82 \times 10^4}{2 \times 10^1} = 4.91 \times 10^3$

Addition and Subtraction Using Powers of Ten

To add or subtract, first adjust all numbers to the same power of ten. It does not matter what exponent you choose, as long as all are the same.

EXAMPLE 1–4

Add 3.25×10^2 and 5×10^3

a. using 10^2 representation,

b. using 10^3 representation.

Solution

a. $5 \times 10^3 = 50 \times 10^2$. Thus, $3.25 \times 10^2 + 50 \times 10^2 = 53.25 \times 10^2$.

b. $3.25 \times 10^2 = 0.325 \times 10^3$. Thus, $0.325 \times 10^3 + 5 \times 10^3 = 5.325 \times 10^3$, which is the same as 53.25×10^2 as found in part (a).

Powers

Raising a number to a power is a form of multiplication (or division if the exponent is negative). For example,

$$(2 \times 10^3)^2 = (2 \times 10^3)(2 \times 10^3) = 4 \times 10^6$$

In general, $(N \times 10^n)^m = N^m \times 10^{nm}$. In this notation, $(2 \times 10^3)^2 = 2^2 \times 10^{3 \times 2} = 4 \times 10^6$ as before.

Integer fractional powers represent roots. Thus, $4^{1/2} = \sqrt{4} = 2$ and $27^{1/3} = \sqrt[3]{27} = 3$.

NOTES...

Use common sense when handling numbers. With calculators, for example, it is often easier to work directly with numbers in their original form than to convert them to power of ten notation. [As an example, it is more sensible to multiply 276×0.009 directly than to convert to power of ten notation as we did in Example 1–3(a).] If the final result is needed as a power of ten, you can convert as a last step.

EXAMPLE 1–5

Expand the following:

a. $(250)^3$ b. $(0.0056)^2$ c. $(141)^{-2}$ d. $(60)^{1/3}$

Solution

a. $(250)^3 = (2.5 \times 10^2)^3 = (2.5)^3 \times 10^{2\times3} = 15.625 \times 10^6$
b. $(0.0056)^2 = (5.6 \times 10^{-3})^2 = (5.6)^2 \times 10^{-6} = 31.36 \times 10^{-6}$
c. $(141)^{-2} = (1.41 \times 10^2)^{-2} = (1.41)^{-2} \times (10^2)^{-2} = 0.503 \times 10^{-4}$
d. $(60)^{1/3} = \sqrt[3]{60} = 3.915$

PRACTICE PROBLEMS 2

Determine the following:

a. $(6.9 \times 10^5)(0.392 \times 10^{-2})$
b. $(23.9 \times 10^{11})/(8.15 \times 10^5)$
c. $14.6 \times 10^2 + 11.2 \times 10^1$ (Express in 10^2 and 10^1 notation.)
d. $(29.6)^3$
e. $(0.385)^{-2}$

Answers
a. 2.70×10^3; b. 2.93×10^6; c. $15.72 \times 10^2 = 157.2 \times 10^1$; d. 25.9×10^3; e. 6.75

1.5 Prefixes, Engineering Notation, and Numerical Results

In scientific work, it is common to find very large and very small numbers expressed in power of ten notation. However, in engineering, certain elements of style and standard practices have evolved, giving rise to what is referred to as **engineering notation.** In engineering notation, it is more common to use **prefixes** than powers of ten. The most common prefixes (along with their symbols) are listed in Table 1–6. (Note: In this notation, powers of ten must be multiples of three.) As an example, while a current of 0.0045 A (amperes) may be expressed as 4.5×10^{-3} A, it is preferable to express it as 4.5 mA or as 4.5 milliamps. Note also that there are often several equally acceptable choices. For example, a time interval of 15×10^{-5} s may be expressed as 150 μs, or as 150 microseconds, or as 0.15 ms, or as 0.15 milliseconds. Note also that it is not wrong to express the number as 15×10^{-5}; it is simply not common engineering practice. From here onward, we will use engineering notation almost exclusively.

TABLE 1–6 Engineering Prefixes

Power of 10	Prefix	Symbol
10^{12}	tera	T
10^9	giga	G
10^6	mega	M
10^3	kilo	k
10^{-3}	milli	m
10^{-6}	micro	μ
10^{-9}	nano	n
10^{-12}	pico	p

EXAMPLE 1–6

Express the following in engineering notation:

a. 10×10^4 volts b. 0.1×10^{-3} watts c. 250×10^{-7} seconds

Solution

a. $10 \times 10^4\ \text{V} = 100 \times 10^3\ \text{V} = 100$ kilovolts $= 100$ kV

b. $0.1 \times 10^{-3}\ \text{W} = 0.1$ milliwatts $= 0.1$ mW

c. $250 \times 10^{-7}\ \text{s} = 25 \times 10^{-6}\ \text{s} = 25$ microseconds $= 25\ \mu\text{s}$

EXAMPLE 1–7

Convert 0.1 MV to kilovolts (kV).

Solution

$$0.1\ \text{MV} = 0.1 \times 10^6\ \text{V} = (0.1 \times 10^3) \times 10^3\ \text{V} = 100\ \text{kV}$$

Remember that a prefix represents a power of ten, and thus the rules for power of ten computation apply. For example, when adding or subtracting, adjust to a common base, as illustrated in Example 1–8.

EXAMPLE 1–8

Compute the sum of 1 ampere (amp) and 100 milliamperes.

Solution Adjust to a common base, either amps (A) or milliamps (mA). Thus,

$$1\ \text{A} + 100\ \text{mA} = 1\ \text{A} + 100 \times 10^{-3}\ \text{A} = 1\ \text{A} + 0.1\ \text{A} = 1.1\ \text{A}$$

Alternatively, 1 A + 100 mA = 1000 mA + 100 mA
= 1100 mA = 1.1 A as above.

PRACTICE PROBLEMS 3

1. Convert 1800 kV to megavolts (MV).
2. In Chapter 4, we show that voltage is the product of current times resistance—that is, $V = I \times R$, where V is in volts, I is in amperes, and R is in ohms. Given $I = 25$ mA and $R = 4\ \text{k}\Omega$, convert these to power of ten notation, then determine V.
3. If $I_1 = 520\ \mu\text{A}$, $I_2 = 0.157$ mA, and $I_3 = 2.75 \times 10^{-4}$ A, what is $I_1 + I_2 + I_3$ in mA? In microamps?

Answers

1. 1.8 MV; 2. 100 V; 3. 0.952 mA, 952 μA

Numerical Results

While computers and calculators typically display many digits, the question is how many should you actually retain in your answer? While it is tempting to write them all down, a closer inspection shows that this may not be sensible. To see why, note that much of engineering is based on measurement. In practice, for example, voltage is measured with a voltmeter and current with an ammeter. Such measured values are approximate since it is impossible to measure anything exactly. Since measured values have an element of uncertainty, computations based on them also have an element of uncertainty—see Notes. To illustrate, suppose you want to know the area of a garden and you measure the length and width as $L = 5.76$ m and $W = 3.72$ m. Assume an uncertainty of 1 in the last digit of each—that is, $L = 5.76 \pm 0.01$ and $W = 3.72 \pm 0.01$. This means that L can be as small as 5.75 and as large as 5.77, while W can range between 3.71 and 3.73. The area can thus be as small as 5.75 m $\times$ 3.71 m = 21.3 m^2 or as large as 5.77 m $\times$ 3.73 m = 21.5 m^2. Note that we aren't even sure what the first digit is after the decimal place in this result. Now you can see why it doesn't make sense to write down all the digits that your calculator shows, since you would get 5.76 m $\times$ 3.72 m = 21.4272 m^2 on a six-digit display—and it is clearly senseless to claim that you know what the four digits following the decimal point are when in fact, you don't even know what the first digit is.

So what do we do? A practical solution is to carry all the digits throughout the solution of the problem (your calculator will probably do this anyway), then round the answer to a suitable number of digits. As a useful guideline, you should round your final answer to three digits unless it makes sense to do otherwise. For the preceding example, this would yield an answer of 21.4 m^2 which, as you can see, is the average of the high and low computed values. When performing a chain calculation, make sure that you carry enough digits all the way through—that is, do not write down three-digit intermediate answers, then reenter them to continue the calculation—this will lead to more uncertainty in the final answers.

NOTES...

1. Numbers may be exact or they may be approximate. Exact numbers are typically obtained by a process of counting or by definition—for example, 1 hour is defined as 60 minutes. Here, 60 is exact (i.e., it is not 59.99 or 60.01). Computations based solely on exact numbers contain no uncertainty. However, computations involving both exact and approximate numbers carry the uncertainty of the approximate numbers.
2. In this book, **unless otherwise noted, all numbers in examples and problems are to be taken as exact.** Thus, there will be no uncertainty in the computed answers.
3. We may show more than three digits in the answers in this book since the process of solution is usually of primary importance and it can be obscured by dropping digits too soon.
4. The topic of significant digits, numerical accuracy, and working with approximate numbers has only been touched upon here. For additional information, please visit our Web site at www.cengagebrain.com. In the search box at the top of the page, key in Robbins and Miller. (This will take you to the product page where your resources can be found.) Follow the links to *For Further Investigation,* then select *Significant Digits and Numerical Accuracy* to access it.

IN-PROCESS LEARNING CHECK 1

(Answers are at the end of the chapter.)

1. All conversion factors have a value of what?
2. Convert 14 yards to centimeters.
3. What units does the following reduce to?

$$\frac{\text{km}}{\text{h}} \times \frac{\text{m}}{\text{km}} \times \frac{\text{h}}{\text{min}} \times \frac{\text{min}}{\text{s}}$$

4. Express the following in engineering notation:

   a. 4270 ms b. 0.001 53 V c. 12.3×10^{-4} s

5. Express the result of each of the following computations as a number times 10 to the power indicated:

   a. 150×120 as a value times 10^4; as a value times 10^3.

   b. $300 \times 6/0.005$ as a value times 10^4; as a value times 10^5; as a value times 10^6.

   c. $430 + 15$ as a value times 10^2; as a value times 10^1.

   d. $(3 \times 10^{-2})^3$ as a value times 10^{-6}; as a value times 10^{-5}.

6. Express each of the following as indicated.

   a. 752 μA in mA.

   b. 0.98 mV in μV.

   c. 270 μs + 0.13 ms in μs and in ms.

1.6 Circuit Diagrams

Electrical and electronic circuits are constructed using components such as batteries, switches, resistors, capacitors, transistors, and interconnecting wires. To represent these circuits on paper, diagrams are used. In this book, we use three types: block diagrams, schematic diagrams, and pictorials.

Block Diagrams

Block diagrams describe a circuit or system in simplified form. The overall problem is broken into blocks, each representing a portion of the system or circuit. Blocks are labelled to indicate what they do or what they contain, then interconnected to show their relationship to each other. General signal flow is usually from left to right and top to bottom. Figure 1–5, for example, represents an audio amplifier. Although you have not covered any of its circuits yet, you should be able to follow the general idea quite easily—sound is picked up by

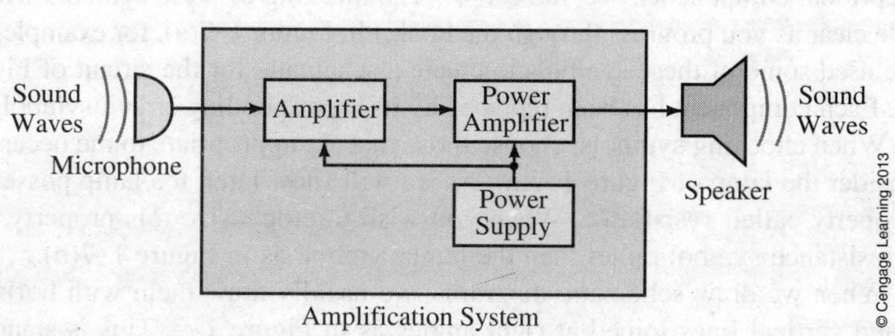

FIGURE 1–5 An example block diagram. Pictured is a simplified representation of an audio amplification system.

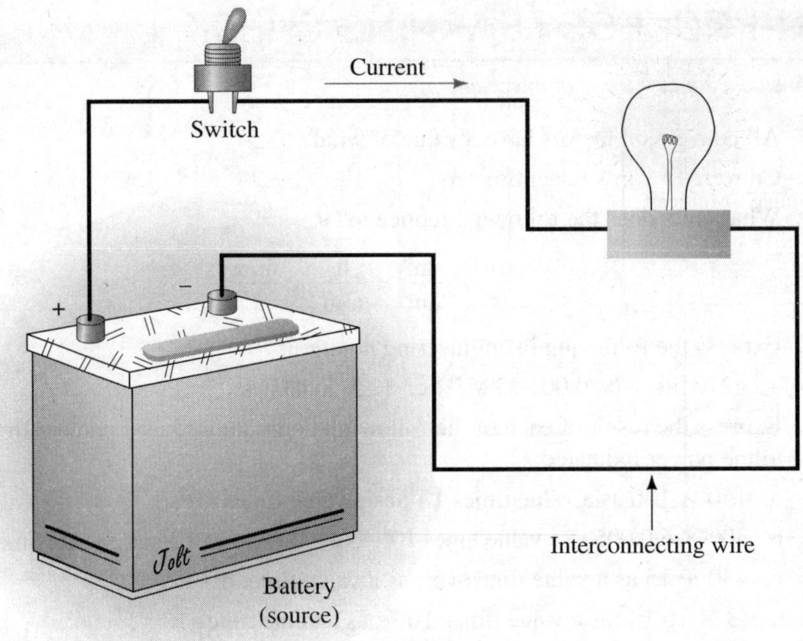

FIGURE 1–6 A pictorial diagram. The battery is referred to as a *source* while the lamp is referred to as a *load.* (The + and − on the battery are discussed in Chapter 2.)

the microphone, converted to an electrical signal, amplified by a pair of amplifiers, then output to the speaker, where it is converted back to sound. A power supply energizes the system. The advantage of a block diagram is that it gives you the overall picture and helps you understand the general nature of a problem. However, it does not provide detail.

Pictorial Diagrams

Pictorial diagrams provide detail. They help you visualize circuits and their operation by showing components as they actually look physically. For example, the circuit of Figure 1–6 consists of a battery, a switch, and an electric lamp, all interconnected by wire. Operation is easy to visualize—when the switch is closed, the battery causes current in the circuit, which lights the lamp. The battery is referred to as the source and the lamp as the load.

Schematic Diagrams

While pictorial diagrams help you visualize circuits, they are cumbersome to draw. **Schematic diagrams** get around this by using simplified, standard symbols to represent components; see Table 1–7. (The meaning of these symbols will be made clear as you progress through the book.) In Figure 1–7(a), for example, we have used some of these symbols to create a schematic for the circuit of Figure 1–6. Each component has been replaced by its corresponding circuit symbol.

When choosing symbols, choose those that are appropriate to the occasion. Consider the lamp of Figure 1–7(a). As we will show later, the lamp possesses a property called **resistance.** When you wish to emphasize this property, use the resistance symbol rather than the lamp symbol, as in Figure 1–7(b).

When we draw schematic diagrams, we usually draw them with horizontal and vertical lines joined at right angles as in Figure 1–7. This is standard practice. (At this point you should glance through some later chapters, e.g., Chapter 7, and study additional examples.)

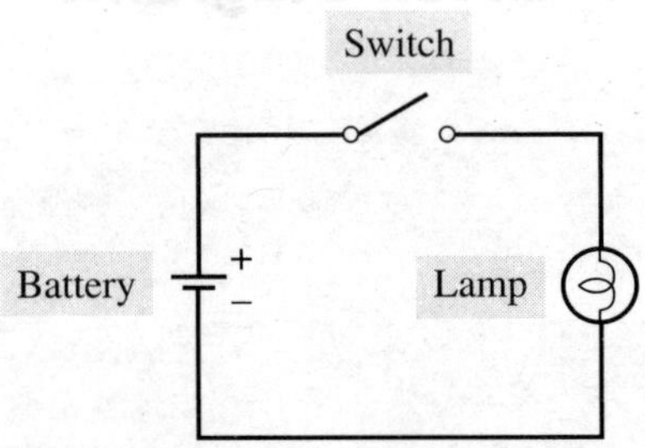

(a) Schematic using lamp symbol

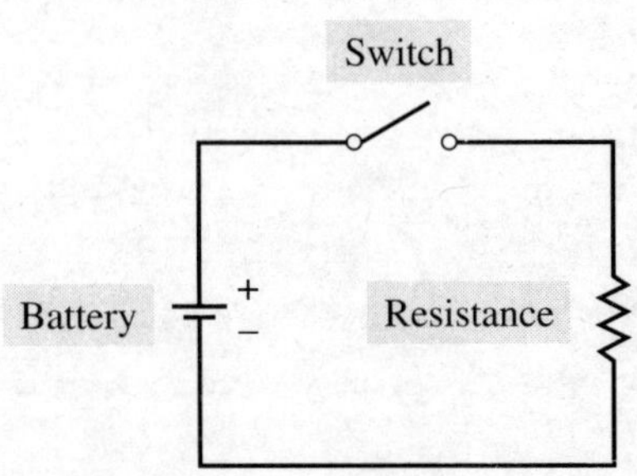

(b) Schematic using resistance symbol

FIGURE 1–7 Schematic representation of Figure 1–6. The lamp has a circuit property called resistance (discussed in Chapter 3).

TABLE 1–7 Schematic Circuit Symbols

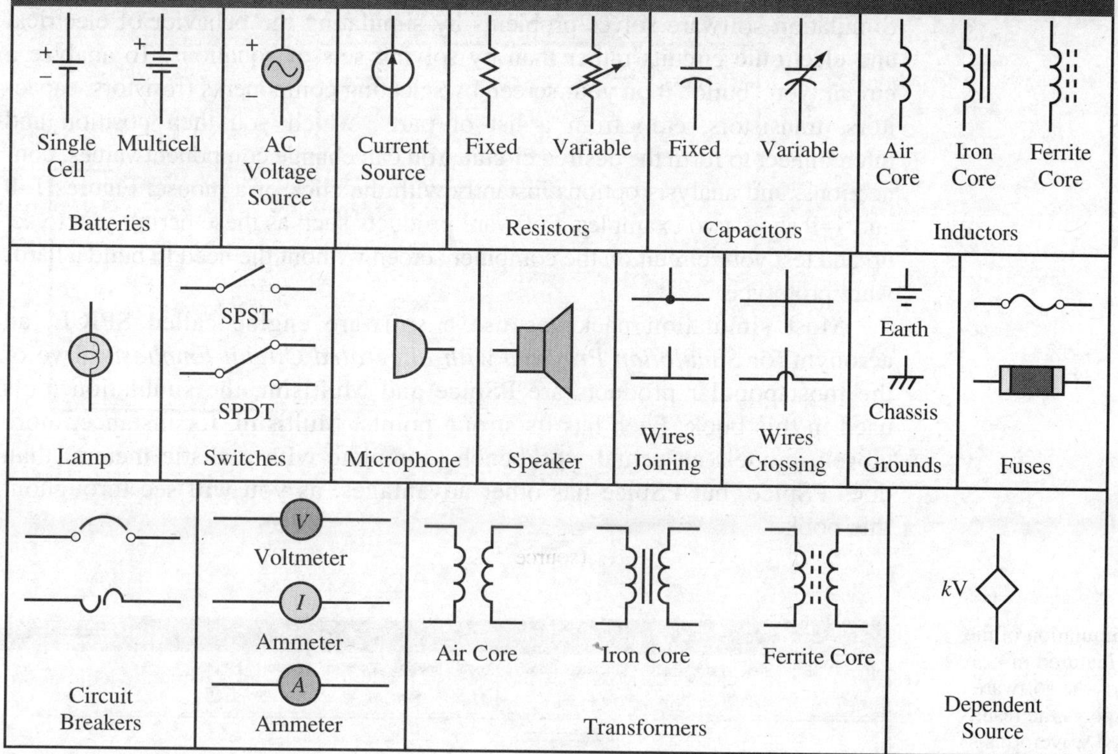

1.7 Circuit Analysis Using Computers and Calculators

Computers and calculators are widely used for circuit analysis and design. Software commonly employed for this purpose includes simulation software (such as **Multisim** and **PSpice**) and numerical analysis software such as Mathcad and Matlab—see Notes. We begin with simulation software.

NOTES...

1. Software for technical use falls into two broad categories—**application software** (such as Multisim and PSpice) and programming languages (such as Java and C++). Application software is designed to solve problems without requiring programming on the part of the user, whereas programming languages require that you write code for each type of problem to be solved. In this book, we do not consider programming languages.
2. Software tools should always be used wisely. Before you use Multisim or PSpice, for example, be sure you understand the basics of the subject that you are studying, as the uninformed use of such software can result in answers that make no sense—and you need to be able to recognize this fact. This is why you should solve many problems manually with your calculator first—to develop both an understanding of the theory and a "feel" for what is right.
3. Computer software is frequently updated, and the versions used in this book are the versions current at the time of writing—see Appendix A.
4. Multisim® is a trademark of National Instruments Company; OrCAD®, OrCAD Capture®, and PSpice® are registered trademarks of Cadence Design Systems Inc.; Matlab is a trademark of MathWorks; and Mathcad® is a trademark of Parametric Technology Corporation.

Circuit Simulation Software

Simulation software solves problems by simulating the behavior of electrical and electronic circuits rather than by solving sets of equations. To analyze a circuit, you "build" it on your screen by selecting components (resistors, capacitors, transistors, etc.) from a list of parts, which you then position and interconnect to form the desired circuit. You can change component values, connections, and analysis options instantly with the click of a mouse. Figures 1–8 and 1–9 show two examples. Software products such as these permit you to set up and test your circuit on the computer screen without the need to build a hardware prototype.

Most simulation packages use a software engine called SPICE, an acronym for *Simulation Program with Integrated Circuit Emphasis.* Two of the most popular products are PSpice and Multisim, the simulation tools used in this book. Each has its strong points. Multisim, for instance, more closely models an actual workbench (complete with realistic meters) than does PSpice, but PSpice has other advantages, as you will see throughout this book.

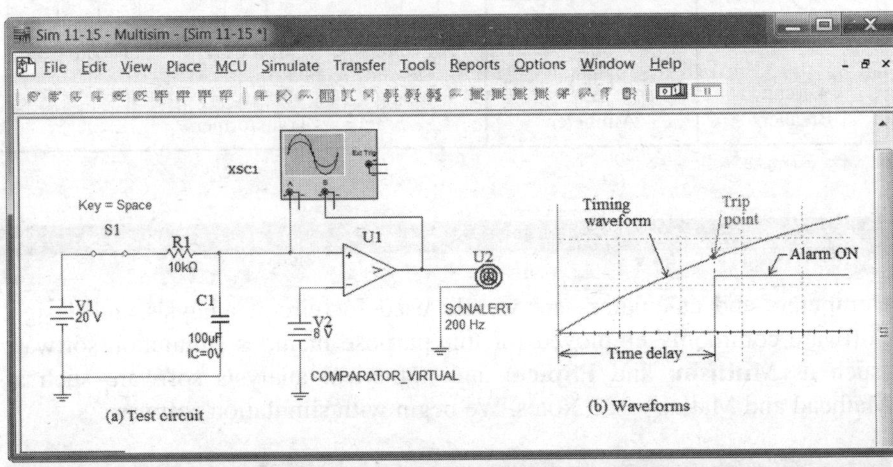

FIGURE 1–8 Multisim simulation of the time delay alarm circuit featured in Chapter 11. Not only does the software analyze the circuit, it displays the results graphically as color-coded waveforms.

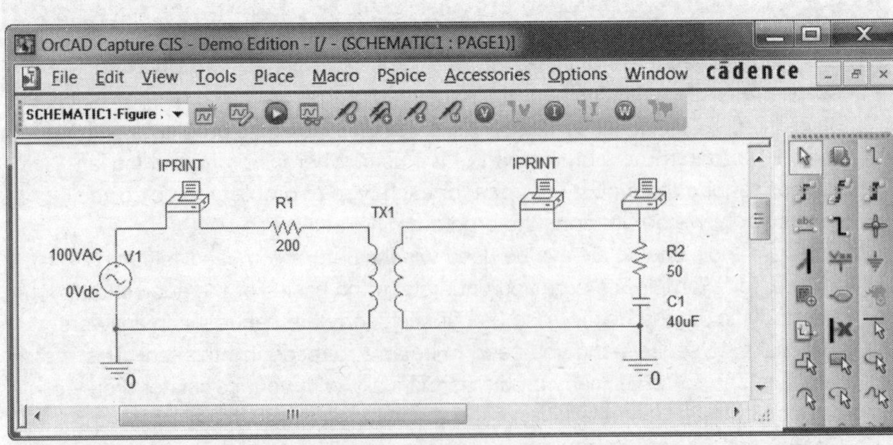

FIGURE 1–9 PSpice simulation of a transformer coupled circuit from Chapter 23. The meters measure circuit variables and show results in numerical form.

Prepackaged Math Software

Another useful category of software includes mathematics packages such as Mathcad and Matlab. These programs (which use numerical analysis techniques to solve equations, plot data, etc.) also require no programming—you simply enter the data and let the computer do the work. Generally, they use standard mathematical notation (or very close to it) and are great aids for solving simultaneous equations such as those encountered in mesh and nodal analysis, which will be discussed in Chapter 8 and Chapter 19.

Calculators in Circuit Analysis

While the preceding software programs can be very useful, a calculator will be your primary tool for learning circuit analysis. You will need one that is able to work with complex numbers in both rectangular and polar forms. Such a calculator saves a great deal of time and effort and cuts down dramatically on errors. Figure 1–10 shows a typical calculator—the Sharp EL-506W. (It is the calculator that we use for illustrative purposes throughout this book—see Notes. The EL-506W has regional variants—for example, it is sold in Canada as the EL-546W.) (Valuable tips on how to use calculators in circuit analysis are integrated throughout the text and summarized in Appendix B.)

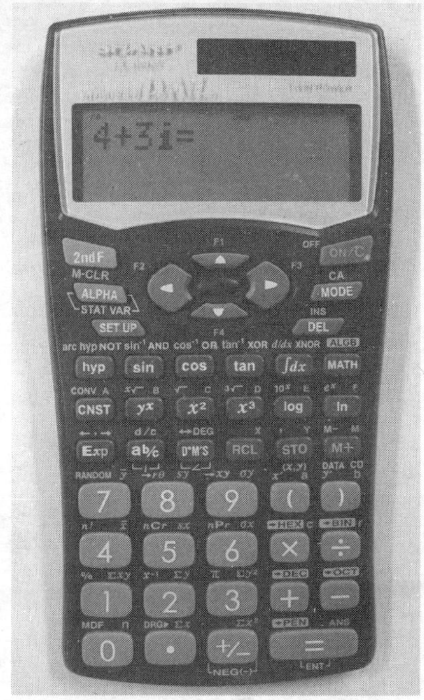

FIGURE 1–10 Calculators like this greatly simplify the work involved in circuit analysis. The photo shows a Sharp EL-506W, but check with your professor or consult your book list to see which calculator is recommended for your course.

NOTES...

A survey of current book adopters revealed a considerable variety of calculators in use, with no one calculator a clear winner—consequently, we make no recommendations about which one to buy. Be sure to check with your instructor (or your book list) before you make a choice.

Problems

NOTES...

1. Conversion factors may be found on the inside of the front cover or in the tables of Chapter 1.
2. Difficult problems have their question number printed in red.
3. Answers to odd-numbered problems are in Appendix D.

1.3 Converting Units

1. Perform the following conversions:
   a. 27 minutes to seconds
   b. 0.8 hours to seconds
   c. 2 h 3 min 47 s to s
   d. 35 horsepower to watts
   e. 1827 W to hp
   f. 23 revolutions to degrees
2. Perform the following conversions:
   a. 27 feet to meters
   b. 2.3 yd to cm
   c. 36° F to degrees C
   d. 18 (US) gallons to liters
   e. 100 sq. ft to m^2
   f. 124 sq. in. to m^2
   g. 47-pound force to newtons
3. Set up conversion factors, compute the following, and express the answer in the units indicated.
   a. The area of a plate 1.2 m by 70 cm in m^2.
   b. The area of a triangle with base 25 cm, height 0.5 m in m^2.
   c. The volume of a box 10 cm by 25 cm by 80 cm in m^3.
   d. The volume of a sphere with 10 in. radius in m^3.

4. An electric fan rotates at 300 revolutions per minute. How many degrees is this per second?
5. If the surface mount robot machine of Figure 1–3 places 15 parts every 12 s, what is its placement rate per hour?
6. If your laser printer can print 8 pages per minute, how many pages can it print in one-tenth of an hour?
7. A car gets 27 miles per US gallon. What is this in kilometers per liter?
8. The equatorial radius of the earth is 3963 miles. What is the earth's circumference in kilometers at the equator?
9. A wheel rotates 18° in 0.02 s. How many revolutions per minute is this?
10. The height of horses is sometimes measured in "hands," where 1 hand = 4 inches. How many meters tall is a 16-hand horse? How many centimeters?
11. Suppose $s = vt$ is given, where s is distance travelled, v is velocity, and t is time. If you travel at $v = 60$ mph for 500 seconds, you might make an unthinking substitution $s = vt = (60)(500) = 30{,}000$ miles. What is wrong with this calculation? What is the correct answer?
12. A round pizza has a circumference of 47 inches. How long does it take for a pizza cutter traveling at 0.12 m/s to cut diagonally across it?
13. Joe S. was asked to convert 2000 yd/h to meters per second. Here is Joe's work: velocity $= 2000 \times 0.9144 \times 60/60 = 1828.8$ m/s. Determine conversion factors, write units into the conversion, and find the correct answer.
14. The mean distance from the earth to the moon is 238 857 miles. Radio signals travel at 299 792 458 m/s. How long does it take a radio signal to reach the moon?
15. If you walk at a rate of 3 km/h for 8 minutes, 5 km/h for 1.25 h, then continue your walk at a rate of 4 km/h for 12 minutes, how far will you have walked in total?
16. Suppose you walk at a rate of 2 mph for 12 minutes, 4 mph for 0.75 h, then finish off at 5 mph for 15 minutes. How far have you walked in total?
17. You walk for 15 minutes at a rate of 2 km/h, then 18 minutes at 5 km/h, and for the remainder, your speed is 2.5 km/h. If the total distance covered is 2.85 km, how many minutes did you spend walking at 2.5 km/h?
18. You walk for 16 minutes at a rate of 1.5 mph, speed up to 3.5 mph for awhile, then slow down to 3 mph for the last 12 minutes. If the total distance covered is 1.7 miles, how long did you spend walking at 3.5 mph?
19. Your plant manager asks you to investigate two machines. The cost of electricity for operating machine #1 is 43 cents/minute, while that for machine #2 is $200.00 per 8-hour shift. The purchase price, production capacities, maintenance costs, and long-term reliability for both machines are identical. Based on this information, which machine should you purchase and why?
20. Given that 1 hp = 550 ft-lb/s, 1 ft = 0.3048 m, 1 lb = 4.448 N, 1 J = 1 N-m, and 1 W = 1 J/s, show that 1 hp = 746 W.

1.4 Power of Ten Notation

21. Express each of the following in power of ten notation with one nonzero digit to the left of the decimal point.

a. 8675
b. 0.008 72
c. 12.4×10^2
d. 37.2×10^{-2}
e. $0.003\ 48 \times 10^5$
f. $0.000\ 215 \times 10^{-3}$
g. 14.7×10^0

22. Express the answer for each of the following in power of ten notation with one nonzero digit to the left of the decimal point.
    a. $(17.6)(100)$
    b. $(1400)(27 \times 10^{-3})$
    c. $(0.15 \times 10^{6})(14 \times 10^{-4})$
    d. $1 \times 10^{-7} \times 10^{-4} \times 10.65$
    e. $(12.5)(1000)(0.01)$
    f. $(18.4 \times 10^{0})(100)(1.5 \times 10^{-5})(0.001)$
23. Repeat the directions in Question 22 for each of the following.
    a. $\dfrac{125}{1000}$
    b. $\dfrac{8 \times 10^{4}}{(0.001)}$
    c. $\dfrac{3 \times 10^{4}}{(1.5 \times 10^{6})}$
    d. $\dfrac{(16 \times 10^{-7})(21.8 \times 10^{6})}{(14.2)(12 \times 10^{-5})}$
24. Determine the answers for the following.
    a. $123.7 + 0.05 + 1259 \times 10^{-3}$
    b. $72.3 \times 10^{-2} + 1 \times 10^{-3}$
    c. $86.95 \times 10^{2} - 383$
    d. $452 \times 10^{-2} + (697)(0.01)$
25. Convert the following to power of ten notation and, without using your calculator, determine the answers.
    a. $(4 \times 10^{3})(0.05)^{2}$
    b. $(4 \times 10^{3})(-0.05)^{2}$
    c. $\dfrac{(3 \times 2 \times 10)^{2}}{(2 \times 5 \times 10^{-1})}$
    d. $\dfrac{(30 + 20)^{-2}(2.5 \times 10^{6})(6000)}{(1 \times 10^{3})(2 \times 10^{-1})^{2}}$
    e. $\dfrac{(-0.027)^{1/3}(-0.2)^{2}}{(23 + 1)^{0} \times 10^{-3}}$
26. For each of the following, convert the numbers to power of ten notation, then perform the indicated computations. Round your answer to four digits.
    a. $(452)(6.73 \times 10^{4})$
    b. $(0.009\ 85)(4700)$
    c. $(0.0892)/(0.000\ 067\ 3)$
    d. $12.40 - 236 \times 10^{-2}$
    e. $(1.27)^{3} + 47.9/(0.8)^{2}$
    f. $(-643 \times 10^{-3})^{3}$
    g. $[(0.0025)^{1/2}][1.6 \times 10^{4}]$
    h. $[(-0.027)^{1/3}]/[1.5 \times 10^{-4}]$
    i. $\dfrac{(3.5 \times 10^{4})^{-2} \times (0.0045)^{2} \times (729)^{1/3}}{[(0.00872) \times (47)^{3}] - 356}$
27. For the following,
    a. convert numbers to power of ten notation, then perform the indicated computation,
    b. perform the operation directly on your calculator without conversion. What is your conclusion?

    i. 842×0.0014 ii. $\dfrac{0.0352}{0.007\ 91}$
28. Express each of the following in conventional notation.
    a. 34.9×10^{4}
    b. 15.1×10^{0}
    c. 234.6×10^{-4}
    d. 6.97×10^{-2}
    e. $45\ 786.97 \times 10^{-1}$
    f. 6.97×10^{-5}
29. One coulomb (Chapter 2) is the amount of charge represented by 6 240 000 000 000 000 000 electrons. Express this quantity in power of ten notation.
30. The mass of an electron is 0.000 000 000 000 000 000 000 000 000 000 899 9 kg. Express as a power of ten with one nonzero digit to the left of the decimal point.
31. If 6.24×10^{18} electrons pass through a wire in 1 s, how many pass through it during a time interval of 2 h 47 min 10 s?
32. Compute the distance traveled in meters by light in a vacuum in 1.2×10^{-8} second.

33. How long does it take light to travel 3.47×10^5 km in a vacuum?
34. How far in km does light travel in one light-year?
35. While investigating a site for a hydroelectric project, you determine that the flow of water is 3.73×10^4 m^3/s. How much is this in liters/hour?
36. The gravitational force between two bodies is $F = 6.6726 \times 10^{-11} \frac{m_1 m_2}{r^2}$ N, where masses m_1 and m_2 are in kilograms and the distance r between gravitational centers is in meters. If body 1 is a sphere of radius 5000 miles and density of 25 kg/m^3, and body 2 is a sphere of diameter 20 000 km and density of 12 kg/m^3, and the distance between centers is 100 000 miles, what is the gravitational force between them?

1.5 Prefixes, Engineering Notation, and Numerical Results

37. What is the appropriate prefix and its abbreviation for each of the following multipliers?
    a. 1000
    b. 1 000 000
    c. 10^9
    d. 0.000 001
    e. 10^{-3}
    f. 10^{-12}
38. Express the following in terms of their abbreviations—for example, microwatts as μW. Pay particular attention to capitalization (e.g., V, not v, for volts).
    a. milliamperes
    b. kilovolts
    c. megawatts
    d. microseconds
    e. micrometers
    f. milliseconds
    g. nanoamps
39. Express the following in the most sensible engineering notation (e.g., 1270 μs = 1.27 ms).
    a. 0.0015 s
    b. 0.000027 s
    c. 0.00035 ms
40. Convert the following:
    a. 156 mV to volts
    b. 0.15 mV to microvolts
    c. 47 kW to watts
    d. 0.057 MW to kilowatts
    e. 3.5×10^4 volts to kilovolts
    f. 0.000 035 7 amps to microamps
41. Determine the values to be inserted in the blanks.
    a. 150 kV = ___ $\times 10^3$ V = ___ $\times 10^6$ V
    b. 330 μW = ___ $\times 10^{-3}$ W = ___ $\times 10^{-5}$ W
42. Perform the indicated operations and express the answers in the units indicated.
    a. 700 μA − 0.4 mA = ___ μA = ___ mA
    b. 600 MW + 300 $\times 10^4$ W = ___ MW
43. Perform the indicated operations and express the answers in the units indicated.
    a. 330 V + 0.15 kV + 0.2 $\times 10^3$ V = ___ V
    b. 60 W + 100 W + 2700 mW = ___ W
44. The voltage of a high-voltage transmission line is 1.15×10^5 V. What is its voltage in kV?

45. You purchase a 1500-W electric heater to heat your room. How many kW is this?

46. A portion of an electronic circuit is shown in Figure 1–11. As you will learn in Chapter 6, for this circuit, $I_4 = I_1 + I_2 + I_3$. If $I_1 = 1.25$ mA, $I_2 = 350\ \mu$A, and $I_3 = 250 \times 10^{-5}$ A, what is I_4?

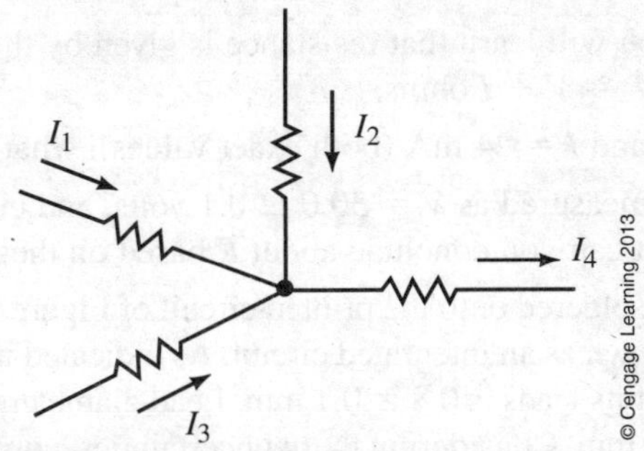

FIGURE 1–11 A nodal point in an electronic circuit. As you learn in Chapter 6, the current leaving the junction is equal to the sum of currents entering.

47. For Figure 1–12, $I_1 + I_2 - I_3 + I_4 = 0$. If $I_1 = 12$ A, $I_2 = 0.150$ kA and $I_4 = 250 \times 10^{-1}$ A, what is I_3?

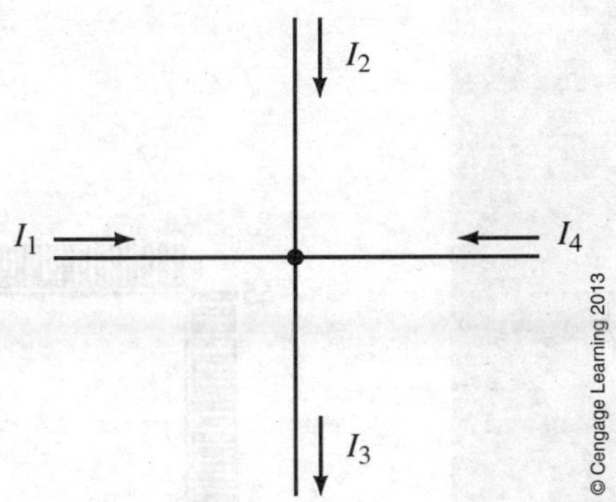

FIGURE 1–12

48. In a certain electronic circuit, $V_1 = V_2 - V_3 - V_4$. If $V_1 = 120$ mV, $V_2 = 5000\ \mu$V, and $V_3 = 20 \times 10^{-4}$ V, what is V_4?

49. While repairing an antique radio, you come across a faulty capacitor designated 39 mmfd. After a bit of research, you find that "mmfd" is an obsolete unit meaning "micromicrofarads." You need a replacement capacitor of equal value. Consulting Table 1–6, what would 39 "micromicrofarads" be equivalent to?

50. a. If 0.045 coulomb of charge (Question 29) passes through a wire in 15 ms, how many electrons is this?

    b. At the rate of 9.36×10^{19} electrons per second, how many coulombs pass a point in a wire in 20 μs?

51. A radio signal travels at 299 792.458 km/s and a telephone signal at 150 m/μs. If they originate at the same point, which arrives first at a destination 5000 km away? By how much?

52. In Chapter 4, you will learn that dc power is given by the product of voltage times current, that is, $P = V \times I$ watts.
    a. If $V = 50$ V and $I = 24$ mA (both exact values), what is P in watts?
    b. If voltage is measured with a voltmeter as $V = 50.0 \pm 0.1$ volts, and current is measured with an ammeter as $I = 24.0 \pm 0.1$ mA, what can you conclude about P based on these measured values?
53. In Chapter 4, you will learn that resistance is given by the ratio of voltage to current, that is, $R = V \div I$ ohms.
    a. If $V = 50$ V and $I = 24$ mA (both exact values), what is R?
    b. If voltage is measured as $V = 50.0 \pm 0.1$ volts, and current as $I = 24.0 \pm 0.1$ mA, what can you conclude about R based on these measured values?
54. The component soldered onto the printed circuit of Figure 1–13(a) is an electronic device known as an integrated circuit. As indicated in (b), center-to-center spacing of its leads is 0.8 ± 0.1 mm. Lead diameters can vary from 0.25 mm to 0.45 mm. Considering these uncertainties, compute the minimum and maximum distance between leads due to these manufacturing tolerances.
55. Open the mini-tutorial *Significant Digits and Numerical Accuracy* found on the Web site and do the exercise problems.

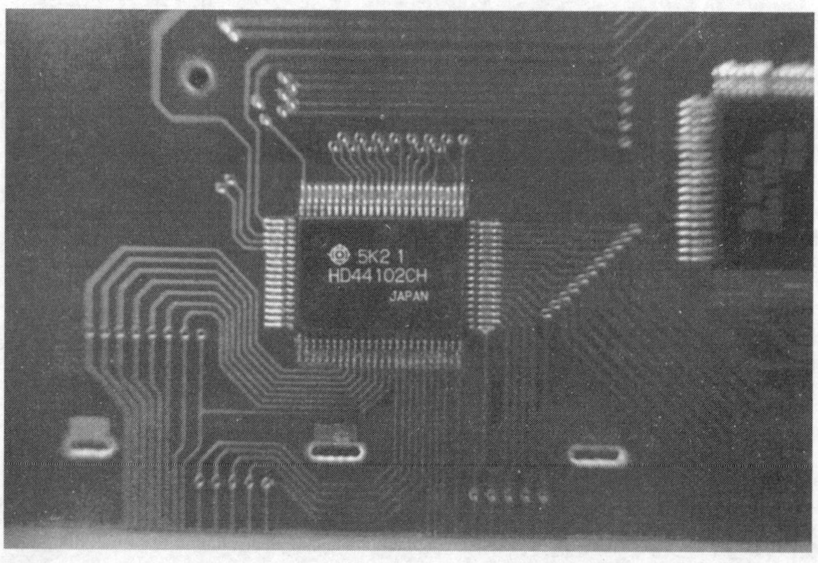

(a)

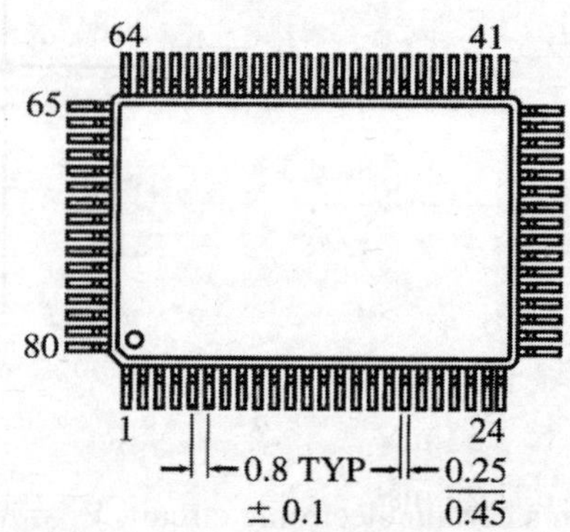

(b)

FIGURE 1–13

1.6 Circuit Diagrams

56. Consider the pictorial diagram of Figure 1–14. Using the appropriate symbols from Table 1–7, draw this in schematic form. Hint: In later chapters, there are many schematic circuits containing resistors, inductors, and capacitors. Use these as aids.

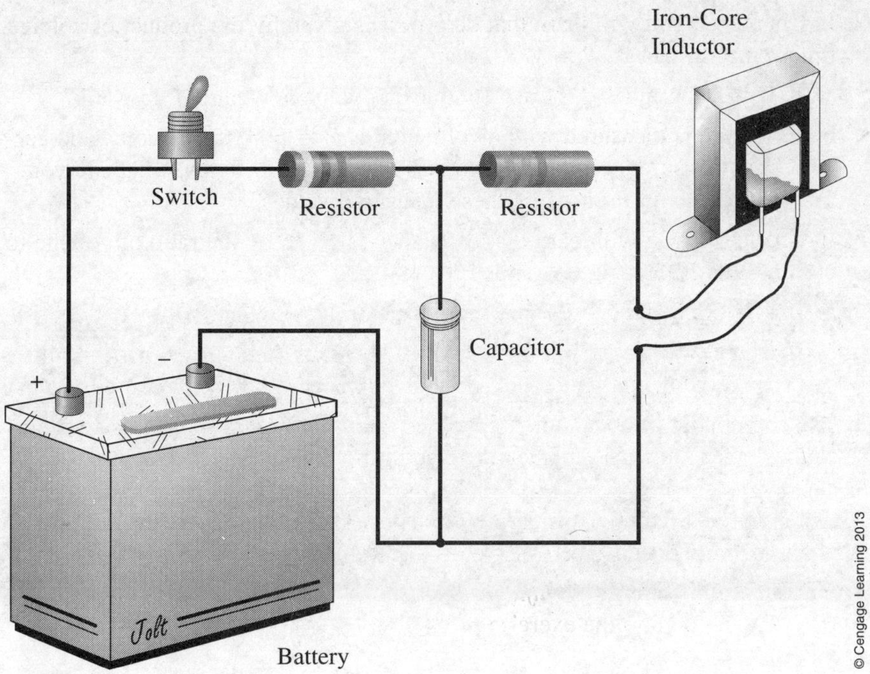

FIGURE 1–14

57. Draw the schematic diagram for a simple flashlight.

ANSWERS TO IN-PROCESS LEARNING CHECKS

IN-PROCESS LEARNING CHECK 1

1. One
2. 1280 cm
3. m/s
4. a. 4.27 s
   b. 1.53 mV
   c. 1.23 ms
5. a. $1.8 \times 10^4 = 18 \times 10^3$
   b. $36 \times 10^4 = 3.6 \times 10^5 = 0.36 \times 10^6$
   c. $4.45 \times 10^2 = 44.5 \times 10^1$
   d. $27 \times 10^{-6} = 2.7 \times 10^{-5}$
6. a. 0.752 mA
   b. 980 μV
   c. 400 μs = 0.4 ms

KEY TERMS

Ampere
Ampere-Hour
Atom
Autopolarity
Battery
Circuit Breaker
Conductor
Conventional Current
Coulomb
Coulomb's Law
Current
dc or Direct Current
dc Source
Electrical Charge
Free Electrons
Fuse
Insulator
Ion
Polarity
Potential Difference
Primary Batteries
Semiconductor
Secondary Batteries
Valence Electron
Voltage

OUTLINE

Atomic Theory Review
The Unit of Electrical Charge: The Coulomb
Voltage
Current
Practical dc Voltage Sources
Measuring Voltage and Current
Switches, Fuses, and Circuit Breakers

OBJECTIVES

After studying this chapter, you will be able to

- describe the basic makeup of an atom,
- explain the relationships between valence shells, free electrons, and conduction,
- describe the fundamental (coulomb) force within an atom and the energy required to create free electrons,
- describe what ions are and how they are created,
- describe the characteristics of conductors, insulators, and semiconductors,
- describe the coulomb as a measure of charge,
- define voltage,
- describe how a battery "creates" voltage,
- explain current as a movement of charge and how voltage causes current in a conductor,
- describe important battery types and their characteristics,
- describe how to measure voltage and current.

2

VOLTAGE AND CURRENT

CHAPTER PREVIEW

A basic electric circuit consisting of a source of electrical energy, a switch, a load, and interconnecting wire is shown in Figure 2–1. When the switch is closed, current in the circuit causes the light to come on. This circuit is representative of many common circuits found in practice, including those of flashlights and automobile headlight systems. We will use it to help develop an understanding of voltage and current—see Notes and icon *CircuitSim 02-1*.

CircuitSim 02-1

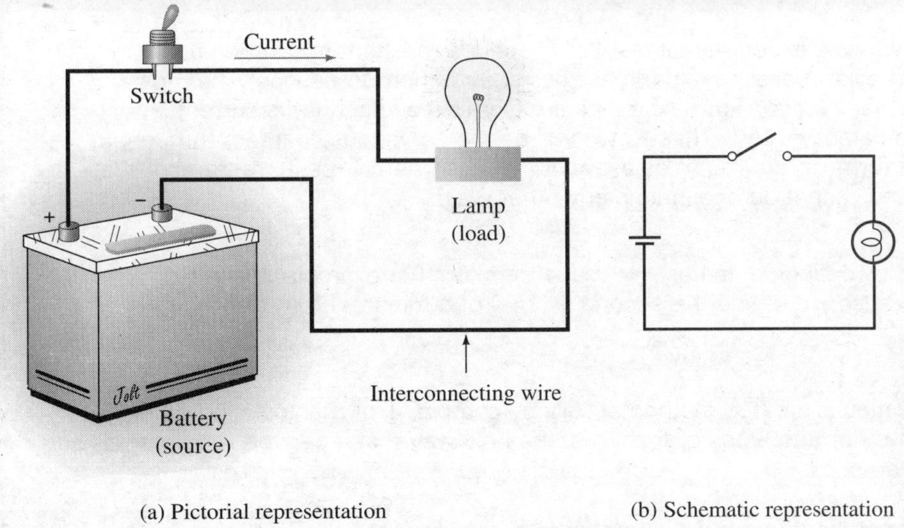

(a) Pictorial representation (b) Schematic representation

FIGURE 2–1 A basic electric circuit.

NOTES...

CircuitSim 02-1 introduces a new innovation to circuit analysis textbooks—the ability to study the circuits of the textbooks "live" and fully functional on your computer—see *Learning Circuit Theory Interactively on Your Computer* in the Preface. Each time you see such an icon in this book, it means that there is a companion simulation available.

Elementary atomic theory shows that the current in Figure 2–1 is actually a flow of charges. The cause of their movement is the "voltage" of the source. While in Figure 2–1 this source is a battery, in practice it may be any one of a number of practical sources, including generators, power supplies, solar cells, and so on.

In this chapter we look at the basic ideas of voltage and current. We begin with a discussion of atomic theory. This leads us to free electrons and the idea of current as a movement of charge. The fundamental definitions of voltage and current are then developed. Following this, we look at a number of common voltage sources. The chapter concludes with a discussion of voltmeters and ammeters and the measurement of voltage and current in practice. ■

Putting It in Perspective

The Equations of Circuit Theory

IN THIS CHAPTER you meet the first of the equations and formulas that we use to describe the relationships of circuit theory. Remembering formulas is made easier if you clearly understand the principles and concepts on which they are based. As you may recall from high school physics, formulas can come about in only one of three ways: through experiment, by definition, or by mathematical manipulation.

Experimental Formulas
Circuit theory rests on a few basic experimental results. These are results that can be proven in no other way; they are valid solely because experiment has shown them to be true. The most fundamental of these are called "laws." Four examples are Ohm's law, Kirchhoff's current law, Kirchhoff's voltage law, and Faraday's law. (These laws will be met in various chapters throughout the book.) When you see a formula referred to as a law or an experimental result, remember that it is based on experiment and cannot be obtained in any other way.

Defined Formulas
Some formulas are created by definition, that is, we make them up. For example, there are 60 seconds in a minute because we define the second as 1/60 of a minute. From this we get the formula $t_{sec} = 60 \times t_{min}$.

Derived Formulas
This type of formula or equation is created mathematically by combining or manipulating other formulas. In contrast to the other two types of formulas, the only way that a derived relationship can be obtained is by mathematics.

An awareness of where circuit theory formulas come from is important to you. This awareness not only helps you understand and remember formulas, it helps you understand the very foundations of the theory—the basic experimental premises upon which it rests, the important definitions that have been made, and the methods by which these foundation ideas have been put together. This can help enormously in understanding and remembering concepts. ■

2.1 Atomic Theory Review

The basic structure of an **atom** is shown symbolically in Figure 2–2. It consists of a nucleus of protons and neutrons surrounded by a group of orbiting electrons. As you learned in physics, the electrons are negatively charged (−), while the protons are positively charged (+). Each atom (in its normal state) has an equal number of electrons and protons, and since their charges are equal and opposite, they cancel, leaving the atom electrically neutral, that is, with zero net charge. The nucleus, however, has a net positive charge, since it consists of positively charged protons and uncharged neutrons.

The basic structure of Figure 2–2 applies to all elements, but each element has its own unique combination of electrons, protons, and neutrons. For example, the hydrogen atom, the simplest of all atoms, has 1 proton and 1 electron, while the copper atom has 29 electrons, 29 protons, and 35 neutrons. Silicon, which is important because of its use in transistors and other electronic devices, has 14 electrons, 14 protons, and 14 neutrons.

In the model of Figure 2–2, electrons that have approximately the same orbital radiuses may be thought of as forming shells. This gives us the simplified picture of Figure 2–3, where we have grouped closely spaced orbits into shells designated *K*, *L*, *M*, *N*, and so on. Only certain numbers of electrons can exist within each shell, and no electrons can exist in the space between shells. The maximum that any shell can hold is $2n^2$ where n is the shell number. Thus, there can be up to 2 electrons in the *K* shell, up to 8 in the *L* shell, up to 18 in the *M* shell, and up to 32 in the *N* shell. The number in any shell depends on the element. For instance, the copper atom, which has 29 electrons, has all 3 of its inner shells completely filled, but its outer shell (shell *N*) has only 1 electron,

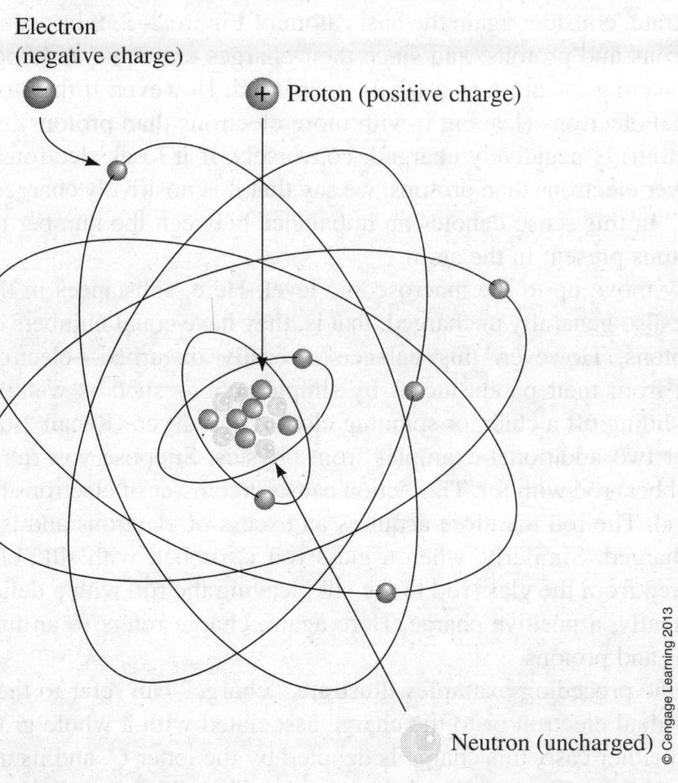

FIGURE 2–2 Bohr model of the atom. Electrons travel around the nucleus at incredible speeds, making billions of trips in a fraction of a second. The force of attraction between the electrons and the protons in the nucleus keeps them in orbit.

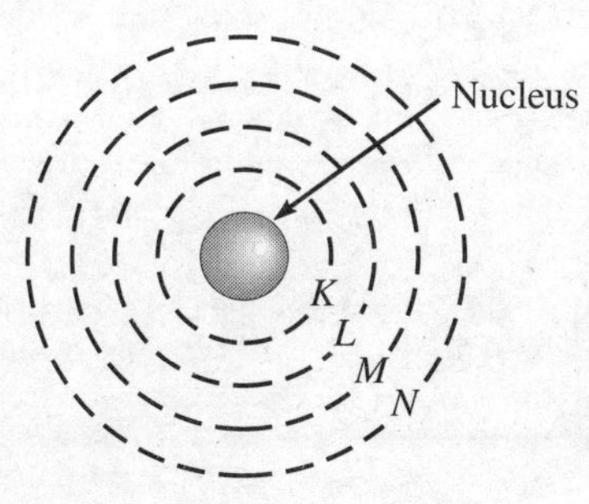

FIGURE 2–3 Simplified representation of the atom. Electrons travel in roughly spherical orbits called "shells."

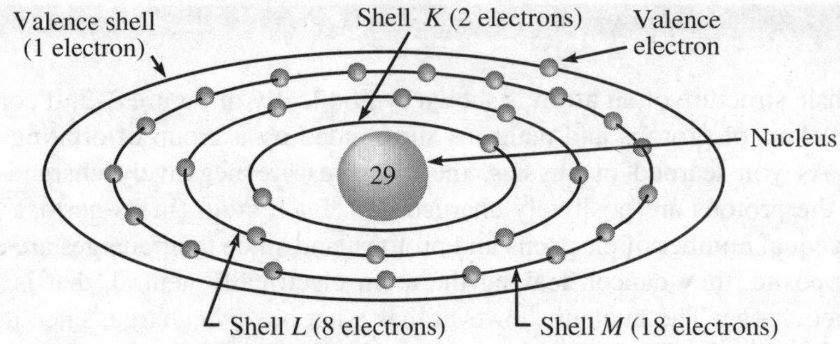

FIGURE 2–4 Copper atom. Since the valence electron is only weakly attracted to the nucleus, it is said to be "loosely bound."

as in Figure 2–4. This outermost shell is called its valence shell, and the electron in it is called its **valence electron.**

No element can have more than 8 valence electrons; when a valence shell has 8 electrons, it is filled. As we shall see, the number of valence electrons that an element has directly affects its electrical properties.

Electrical Charge

In the previous paragraphs, we mentioned the word "charge." However, we need to look at its meaning in more detail. First, we should note that **electrical charge** is an intrinsic property of electrons and protons that manifests itself in the form of forces—electrons repel other electrons but attract protons, while protons repel each other but attract electrons. It was through studying these forces that scientists determined that the charge on the electron is negative while that on the proton is positive.

However, the way in which we use the term "charge" extends beyond this. To illustrate, consider again the basic atom of Figure 2–2. It has equal numbers of electrons and protons, and since their charges are equal and opposite, they cancel, leaving the atom as a whole uncharged. However, if the atom acquires additional electrons (leaving it with more electrons than protons), we say that it (the atom) is negatively charged; conversely, if it loses electrons and is left with fewer electrons than protons, we say that it is positively charged. The term "charge" in this sense denotes an imbalance between the number of electrons and protons present in the atom.

Now move up to the macroscopic level. Here, substances in their normal state are also generally uncharged; that is, they have equal numbers of electrons and protons. However, this balance is easily disturbed—electrons can be stripped from their parent atoms by simple actions such as walking across a carpet, sliding off a chair, or spinning clothes in a dryer. (Recall "static cling.") Consider two additional examples from physics. Suppose you rub an ebonite (hard rubber) rod with fur. This action causes a transfer of electrons from the fur to the rod. The rod therefore acquires an excess of electrons and is thus negatively charged. Similarly, when a glass rod is rubbed with silk, electrons are transferred from the glass rod to the silk, leaving the rod with a deficiency and, consequently, a positive charge. Here again, charge refers to an imbalance of electrons and protons.

As the preceding examples illustrate, "charge" can refer to the charge on an individual electron or to the charge associated with a whole group of electrons. In either case, this charge is denoted by the letter Q, and its unit of measurement in the SI system is the coulomb. (The definition of the coulomb is considered in Section 2.2.) In general, the charge Q associated with a group of electrons is equal to the product of the number of electrons times the charge on each individual electron. Since charge manifests itself in the form of forces, charge is defined in terms of these forces. This is discussed next.

Coulomb's Law

The force between charges was studied by the French scientist Charles Coulomb (1736–1806). Coulomb determined experimentally that the force between two charges Q_1 and Q_2 (Figure 2–5) is directly proportional to the product of their charges and inversely proportional to the square of the distance between them. Mathematically, **Coulomb's law** states

$$F = k\frac{Q_1 Q_2}{r^2} \quad \text{[newtons, N]} \qquad \textbf{(2–1)}$$

where Q_1 and Q_2 are the charges in coulombs (to be defined in Section 2.2), r is the center-to-center spacing between them in meters, and $k = 9 \times 10^9$. Coulomb's law applies to aggregates of charges as in Figure 2–5(a) and (b), as well as to individual electrons within the atom as in (c).

As Coulomb's law indicates, force decreases inversely as the square of distance; thus, if the distance between two charges is doubled, the force decreases to $(1/2)^2 = 1/4$ (i.e., one quarter) of its original value. Because of this relationship, electrons in outer orbits are less strongly attracted to the nucleus than those in inner orbits; that is, they are less tightly bound to the nucleus than those close by. Valence electrons are the least tightly bound and will, if they acquire sufficient energy, escape from their parent atoms.

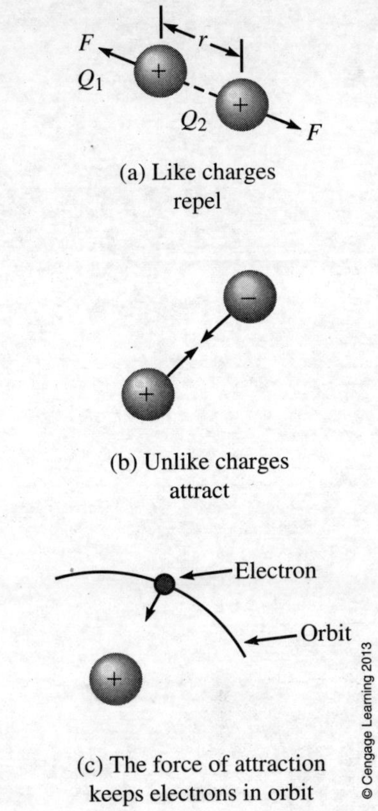

FIGURE 2–5 Coulomb law forces.

Free Electrons

The amount of energy required to escape depends on the number of electrons in the valence shell. If an atom has only a few valence electrons, there will be a relatively weak attraction between these electrons and the nucleus, and only a small amount of additional energy is needed. For example, for a metal like copper, valence electrons can gain sufficient energy from heat alone (thermal energy), even at room temperature, to escape from their parent atoms and wander from atom to atom throughout the material as depicted in Figure 2–6. (Note that these electrons do not leave the substance, they simply wander from the valence shell of one atom to the valence shell of another. The material therefore remains electrically neutral.) Such electrons are called **free electrons.** In copper, there are on the order of 10^{23} free electrons per cubic centimeter at room temperature. As we shall see, it is the presence of this large number of free electrons that makes copper such a good conductor of electric current. On the other hand, if the valence shell is full (or nearly full), valence electrons are much more tightly bound. Such materials have few (if any) free electrons.

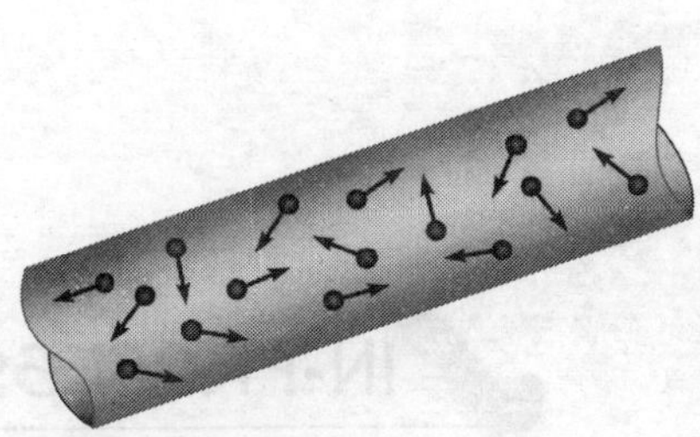

FIGURE 2–6 Random motion of free electrons in a conductor.

Ions

As noted earlier, when a previously neutral atom gains or loses an electron, it acquires a net electrical charge. The charged atom is referred to as an **ion.** If the atom loses an electron, it is called a positive ion; if it gains an electron, it is called a negative ion.

Conductors, Insulators, and Semiconductors

The atomic structure of matter affects how easily charges, that is, electrons, move through a substance and hence how it is used electrically. Electrically, materials are classified as conductors, insulators, or semiconductors.

Conductors

Materials through which charges move easily are termed **conductors.** The most familiar examples are metals. Good metal conductors have large numbers of free electrons that are able to move about easily. In particular, silver, copper, gold, and aluminum are excellent conductors. Of these, copper is the most widely used. Not only is it an excellent conductor, it is inexpensive and easily formed into wire, making it suitable for a broad spectrum of applications ranging from common house wiring to sophisticated electronic equipment. Aluminum, although it is only about 60% as good a conductor as copper, is also used, mainly in applications where light weight is important, such as in overhead power transmission lines. Silver and gold are too expensive for general use. However, gold, because it oxidizes less than other materials, is used in specialized applications; for example, critical electrical connectors in electronic equipment use it because it makes a more reliable connection than other materials.

Insulators

Materials that do not conduct (e.g., glass, porcelain, plastic, rubber, and so on) are termed **insulators.** The covering on electric lamp cords, for example, is an insulator. It is used to prevent the wires from touching and to protect us from electric shock.

Insulators do not conduct because they have full or nearly full valence shells and thus their electrons are tightly bound. However, when high enough voltage is applied, the force is so great that electrons are literally torn from their parent atoms, causing the insulation to break down and conduction to occur. In air, you see this as an arc or flashover. In solids, charred insulation usually results.

Semiconductors

Silicon and germanium (plus a few other materials) have half-filled valence shells and are thus neither good conductors nor good insulators. Known as **semiconductors,** they have unique electrical properties that make them important to the electronics industry. The most important material is silicon. It is used to make transistors, diodes, integrated circuits, and other electronic devices. Semiconductors have made possible personal computers, home theater systems, cell phones, calculators, iPods, and a host of other electronic products.

✔ IN-PROCESS LEARNING CHECK 1

(Answers are at the end of the chapter.)

1. Describe the basic structure of the atom in terms of its constituent particles: electrons, protons, and neutrons. Why is the nucleus positively charged? Why is the atom as a whole electrically neutral?
2. What are valence shells? What does the valence shell contain?
3. Describe Coulomb's law and use it to help explain why electrons far from the nucleus are loosely bound.
4. What are free electrons? Describe how they are created, using copper as an example. Explain what role thermal energy plays in the process.
5. Briefly distinguish between a normal (i.e., uncharged) atom, a positive ion, and a negative ion.

2.2 The Unit of Electrical Charge: The Coulomb

As noted in the previous section, the unit of electrical charge in the SI system is the coulomb (C). The **coulomb** is defined as the charge carried by 6.24×10^{18} electrons. Thus, if an electrically neutral (i.e., uncharged) body has 6.24×10^{18} electrons removed, it will be left with a net positive charge of 1 coulomb, that is, $Q = 1$ C. Conversely, if an uncharged body has 6.24×10^{18} electrons added, it will have a net negative charge of 1 coulomb, that is, $Q = -1$ C. Usually, however, we are more interested in the charge moving through a wire. In this regard, if 6.24×10^{18} electrons pass through a wire, we say that the charge that passed through the wire is 1 C.

We can now determine the charge on one electron. It is $Q_e = 1/(6.24 \times 10^{18}) = 1.602 \times 10^{-19}$ C.

EXAMPLE 2–1

An initially neutral body has 1.7 μC of negative charge removed. Later, 18.7×10^{11} electrons are added. What is the body's final charge?

Solution Initially the body is neutral, that is, $Q_{initial} = 0$ C. When 1.7 μC of electrons is removed, the body is left with a positive charge of 1.7 μC. Now, 18.7×10^{11} electrons are added back. This is equivalent to

$$18.7 \times 10^{11} \text{ electrons} \times \frac{1 \text{ coulomb}}{6.24 \times 10^{18} \text{ electrons}} = 0.3\ \mu\text{C}$$

of negative charge. The final charge on the body is therefore $Q_f = 1.7\ \mu\text{C} - 0.3\ \mu\text{C} = +1.4\ \mu\text{C}$.

To get an idea of how large a coulomb is, we can use Coulomb's law. If it were possible to place 2 charges of 1 coulomb each 1 meter apart, the force between them would be

$$F = (9 \times 10^9)\frac{(1 \text{ C})(1 \text{ C})}{(1 \text{ m})^2} = 9 \times 10^9 \text{ N, that is, about 1 million tons!}$$

PRACTICE PROBLEMS 1

1. Positive charges $Q_1 = 2$ μC and $Q_2 = 12$ μC are separated center to center by 10 mm. Compute the force between them. Is it attractive or repulsive?
2. Two equal charges are separated by 1 cm. If the force of repulsion between them is 9.7×10^{-2} N, what is their charge? What may the charges be, both positive, both negative, or one positive and one negative?
3. After 10.61×10^{13} electrons are added to a metal plate, it has a negative charge of 3 μC. What was its initial charge in coulombs?

Answers

1. 2160 N, repulsive; 2. 32.8 nC, both (+) or both (−); 3. 14 μC (+)

2.3 Voltage

When charges are detached from one body and transferred to another, a *potential difference* or *voltage* results between them. A familiar example is the voltage that develops when you walk across a carpet. Voltages in excess of ten thousand volts can be created in this way. (We will define the volt rigorously very shortly.) This voltage is due entirely to the separation of positive and negative charges, that is, charges that have been pulled apart.

Figure 2–7 illustrates another example. During electrical storms, electrons in thunderclouds are stripped from their parent atoms by the forces of turbulence and carried to the bottom of the cloud, leaving a deficiency of electrons (positive charge) at the top and an excess (negative charge) at the bottom. The force of repulsion then drives electrons away beneath the cloud, leaving the ground positively charged. Hundreds of millions of volts are created in this way. (This is what causes the air to break down and a lightning discharge to occur.)

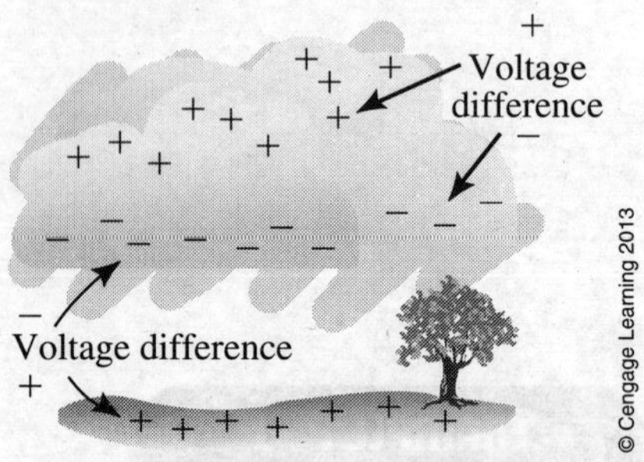

FIGURE 2–7 Voltages created by separation of charges in a thundercloud. The force of repulsion drives electrons away beneath the cloud, creating a voltage between the cloud and earth as well. If voltage becomes large enough, the air breaks down and a lightning discharge occurs.

Practical Voltage Sources

As the preceding examples show, voltage is created solely by the separation of positive and negative charges. However, static discharges and lightning strikes are not practical sources of electricity. We now look at practical sources. A common example is the **battery.** In a battery, charges are separated by chemical action. An ordinary alkaline flashlight battery, Figure 2–8, illustrates the concept. The alkaline material (a mixture of manganese dioxide, graphite, and electrolytic) and a gelled zinc powder mixture, separated by a paper barrier soaked in electrolyte, are placed in a steel can. The can is connected to the top cap to form the cathode or positive terminal, while the zinc mixture, by means of the brass pin, is connected to the bottom to form the anode or negative terminal. (The bottom is insulated from the rest of the can.) Chemical reactions result in an excess of electrons in the zinc mix and a deficiency in the manganese dioxide mix. This separation of charges creates a voltage of approximately 1.5 V, with the top end cap + and the bottom of the can −. The battery is useful as a source since its chemical action creates a continuous supply of energy that is able to do useful work, such as light a lamp or run a motor.

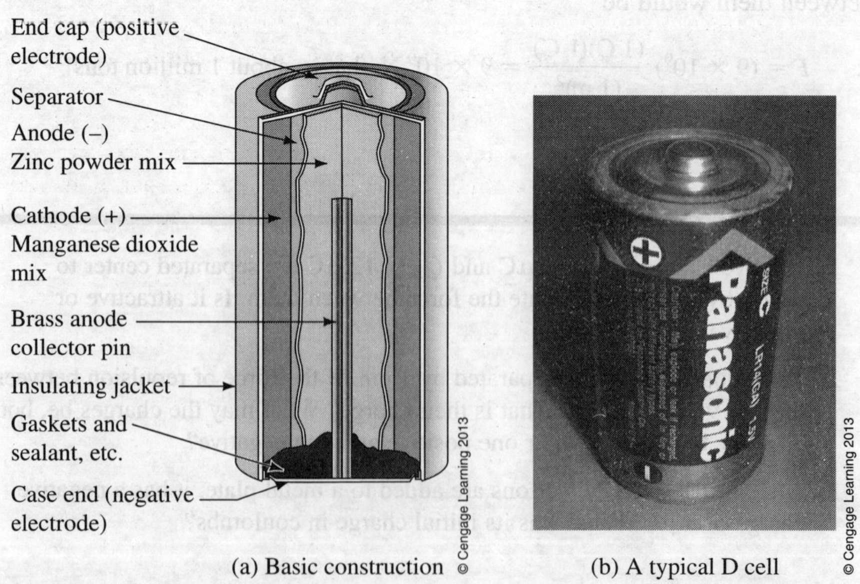

(a) Basic construction

(b) A typical D cell

FIGURE 2–8 Alkaline cell. Voltage is created by the separation of charges due to chemical action. Nominal cell voltage is 1.5 V.

NOTES...

The source of Figure 2–8 is more properly called a cell than a battery, since "cell" refers to a single cell while "battery" refers to a group of cells. However, through common usage, such cells are referred to as batteries. In what follows, we will also call them batteries.

Potential Energy

The concept of voltage is tied into the concept of potential energy. We therefore look briefly at energy.

In mechanics, potential energy is the energy that a body possesses because of its position. For example, a bag of sand hoisted by a rope over a pulley has the potential to do work when it is released. The amount of work that went into giving it this potential energy is equal to the product of force times the distance through which the bag was lifted (i.e., work equals force times distance). In the SI system, force is measured in newtons and distance in meters. Thus, work has the unit newton-meters (which we call joules, Section 1.2, Chapter 1).

In a similar fashion, work is required to move positive and negative charges apart. This gives them potential energy. To understand why, consider again the cloud of Figure 2–7, redrawn in Figure 2–9. Assume the cloud is initially uncharged. Now assume a charge of Q electrons is moved from the top of the cloud to the bottom. The positive charge left at the top of the cloud exerts a force on the electrons that tries to pull them back as they are being moved away. Since the electrons are being moved against this force, work (force times distance) is required. Since the separated charges experience a force to return to the top of the cloud, they have the potential to do work if released, that is, they possess potential energy. Similarly for the battery of Figure 2–8, the charges, which have been separated by chemical action, also possess potential energy.

Definition of Voltage: The Volt

In electrical terms, a difference in potential energy is defined as **voltage.** In general, the amount of energy required to separate charges depends on the voltage developed and the amount of charge moved. By definition, *the voltage between two points is one volt if it requires one joule of energy to move one coulomb of charge from one point to the other.* In equation form,

$$V = \frac{W}{Q} \quad \text{[volts, V]} \tag{2–2}$$

where W is energy in joules, Q is charge in coulombs, and V is the resulting voltage in volts (see Notes).

NOTES...

The discussion found here may seem a bit abstract and somewhat detached from our ordinary experience, which suggests that voltage is the "force or push" that moves electric current through a circuit. While both viewpoints are correct (we look at the latter in great detail, starting in Chapter 4), in order to establish a coherent analytic theory, we need a rigorous definition. Equation 2–2 provides that definition. Although it is a bit abstract, it gives us the foundation upon which rest many of the important circuit relationships that you will encounter shortly.

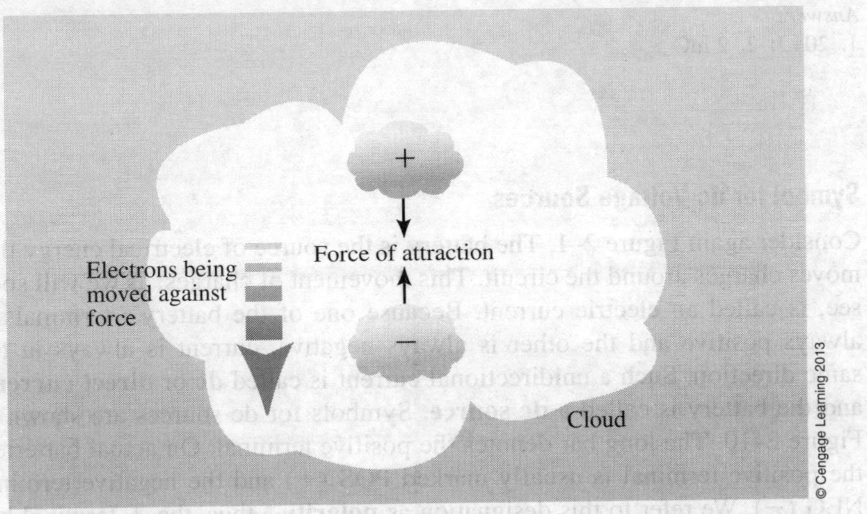

FIGURE 2–9 Work (force × distance) is required to move the charges apart.

Note carefully that voltage is defined between points. For the case of the battery, for example, voltage appears between its terminals. Thus, voltage does not exist at a point by itself; it is always determined with respect to some other point. (For this reason, voltage is also called **potential difference.** We often use the terms interchangeably.) Note also that this argument applies regardless of how you separate the charges, whether it be by chemical means as in a battery, by mechanical means as in a generator, by photoelectric means as in a solar cell, and so on.

Alternate arrangements of Equation 2–2 are useful:

$$W = QV \quad \text{[joules, J]} \tag{2–3}$$

$$Q = \frac{W}{V} \quad \text{[coulombs, C]} \tag{2–4}$$

EXAMPLE 2–2

If it takes 35 J of energy to move a charge of 5 C from one point to another, what is the voltage between the two points?

Solution

$$V = \frac{W}{Q} = \frac{35\text{ J}}{5\text{ C}} = 7\text{ J/C} = 7\text{ V}$$

PRACTICE PROBLEMS 2

1. The voltage between two points is 19 V. How much energy is required to move 67×10^{18} electrons from one point to the other?
2. The potential difference between two points is 140 mV. If 280 μJ of work are required to move a charge Q from one point to the other, what is Q?

Answers

1. 204 J; 2. 2 mC

Symbol for dc Voltage Sources

Consider again Figure 2–1. The battery is the source of electrical energy that moves charges around the circuit. This movement of charges, as we will soon see, is called an electric current. Because one of the battery's terminals is always positive and the other is always negative, current is always in the same direction. Such a unidirectional current is called **dc** or **direct current,** and the battery is called a **dc source.** Symbols for dc sources are shown in Figure 2–10. The long bar denotes the positive terminal. On actual batteries, the positive terminal is usually marked POS (+) and the negative terminal NEG (−). We refer to this designation as **polarity**—thus, the + terminal has positive polarity and the − terminal has negative polarity.

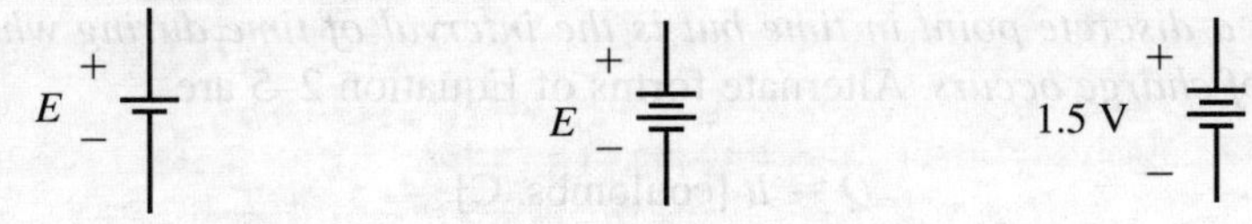

(a) Symbol for a cell (b) Symbol for a battery (c) A 1.5 volt battery

FIGURE 2–10 Battery symbol. The long bar denotes the positive terminal and the short bar the negative terminal—thus it is not necessary to put + and − signs on the diagram (although we often do). For simplicity, we use the symbol shown in (a) throughout this book for batteries as well as for cells.

2.4 Current

Earlier, you learned that there are large numbers of free electrons in metals like copper. These electrons move randomly throughout the material (Figure 2–6), but their net movement in any given direction is zero.

Assume now that a battery is connected as in Figure 2–11. Since electrons are attracted by the positive pole of the battery and repelled by the negative pole, they move around the circuit, passing through the wire, the lamp, and the battery. This movement of charge is called an electric **current.** The more electrons per second that pass through the circuit, the greater is the current. Thus, current is the *rate of flow* (or *rate of movement*) of charge.

The Ampere

Since charge is measured in coulombs, its rate of flow is coulombs per second. In the SI system, 1 coulomb per second is defined as 1 **ampere** (commonly abbreviated A). From this, we get that *1 ampere is the current in a circuit when 1 coulomb of charge passes a given point in 1 second* (Figure 2–11). The symbol for current is *I*. Expressed mathematically,

$$I = \frac{Q}{t} \quad \text{[amperes, A]} \tag{2–5}$$

where *Q* is the charge (in coulombs) and *t* is the time interval (in seconds) over which it is measured. *In Equation 2–5, it is important to note that t does not*

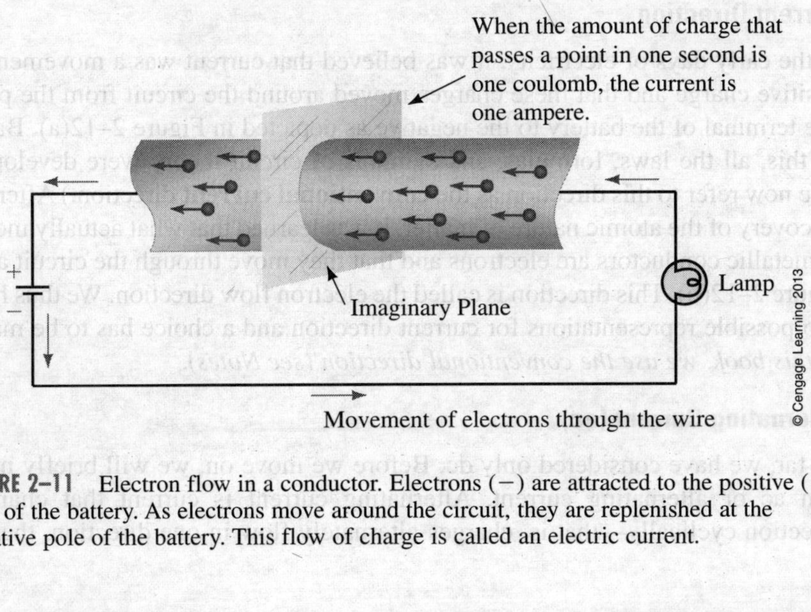

FIGURE 2–11 Electron flow in a conductor. Electrons (−) are attracted to the positive (+) pole of the battery. As electrons move around the circuit, they are replenished at the negative pole of the battery. This flow of charge is called an electric current.

represent a discrete point in time but is the interval of time during which the transfer of charge occurs. Alternate forms of Equation 2–5 are

$$Q = It \text{ [coulombs, C]} \tag{2–6}$$

and

$$t = \frac{Q}{I} \text{ [seconds, s]} \tag{2–7}$$

EXAMPLE 2–3

If 840 coulombs of charge pass through the imaginary plane of Figure 2–11 during a time interval of 2 minutes, what is the current?

Solution Convert t to seconds. Thus,

$$I = \frac{Q}{t} = \frac{840 \text{ C}}{(2 \times 60)\text{s}} = 7 \text{ C/s} = 7 \text{ A}$$

PRACTICE PROBLEMS 3

1. Between $t = 1$ ms and $t = 14$ ms, 8 μC of charge pass through a wire. What is the current?
2. After the switch of Figure 2–1 is closed, current $I = 4$ A. How much charge passes through the lamp between the time the switch is closed and the time that it is opened 3 minutes later?

Answers
1. 0.615 mA; 2. 720 C

Although Equation 2–5 is the theoretical definition of current, we never actually use it to measure current. In practice, we use an instrument called an ammeter (Section 2.6). However, it is an extremely important theoretical relationship that we will soon use to develop other more practical relationships.

NOTES...

A perfectly coherent theory can be built around either of the directions of Figure 2–12, and many circuit analysis and electronics textbooks have been written from each viewpoint. There are, however, compelling reasons for choosing the conventional direction. Among these are (1) Standard electronic symbols used in circuit diagrams and found in manufacturers' data books are based on it, (2) Computer software such as PSpice and Multisim utilizes it, and (3) All Engineering level and virtually all EET level college programs teach it.

Current Direction

In the early days of electricity, it was believed that current was a movement of positive charge and that these charges moved around the circuit from the positive terminal of the battery to the negative as depicted in Figure 2–12(a). Based on this, all the laws, formulas, and symbols of circuit theory were developed. (We now refer to this direction as the **conventional current** direction.) After the discovery of the atomic nature of matter, it was learned that what actually moves in metallic conductors are electrons and that they move through the circuit as in Figure 2–12(b). This direction is called the electron flow direction. We thus have two possible representations for current direction and a choice has to be made. *In this book, we use the conventional direction (see Notes).*

Alternating Current (ac)

So far, we have considered only dc. Before we move on, we will briefly mention ac or alternating current. Alternating current is current that changes direction cyclically—that is, charges alternately flow in one direction, then in

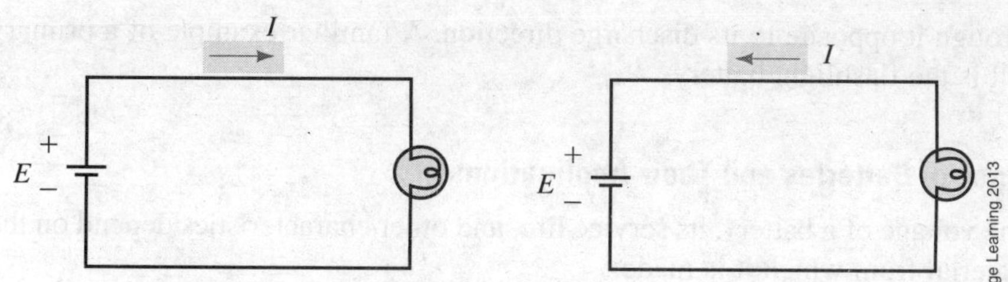

(a) Conventional current direction (b) Electron flow direction

FIGURE 2–12 Conventional current versus electron flow. In this book, we use conventional current.

the other in a circuit. The most common ac source is the commercial ac power system that supplies energy to your home. We mention it here because you will encounter it briefly in Section 2.5. It is covered in detail in Chapter 15.

IN-PROCESS LEARNING CHECK 2

(Answers are at the end of the chapter.)

1. Body *A* has a negative charge of 0.2 μC and body *B* has a charge of 0.37 μC (positive). If 87×10^{12} electrons are transferred from *A* to *B*, what are the charges in coulombs on *A* and on *B* after the transfer?
2. Briefly describe the mechanism of voltage creation using the alkaline cell of Figure 2–8 to illustrate.
3. When the switch in Figure 2–1 is open, the current is zero, yet free electrons in the copper wire are moving about. Describe their motion. Why does their movement not constitute an electric current?
4. If 12.48×10^{20} electrons pass a certain point in a circuit in 2.5 s, what is the current in amperes?
5. For Figure 2–1, assume a 12-V battery. The switch is closed for a short interval, then opened. If $I = 6$ A and the battery expends 230 040 J moving charge through the circuit, how long was the switch closed?

2.5 Practical dc Voltage Sources

Batteries

Batteries are the most common dc source. They are made in a variety of shapes, sizes, and ratings, from miniaturized button batteries capable of delivering only a few microamps to large automotive batteries capable of delivering hundreds of amps. Common sizes are the AAA, AA, C, and D. All batteries use unlike conductive electrodes immersed in an electrolyte. Chemical interaction between the electrodes and the electrolyte creates the voltage of the battery. We now look at some of the more common types—see Notes.

NOTES...

Only a brief overview of various battery types is presented here. More detailed information may be found by searching the Internet.

Primary and Secondary Batteries

Batteries eventually become "discharged." Some types of batteries, however, can be "recharged." Such batteries are called **secondary batteries.** Other types, called **primary batteries,** cannot be recharged. A familiar example of a secondary battery is the automobile battery. It can be recharged by passing current through it opposite to its discharge direction. A familiar example of a primary cell is the flashlight battery.

Types of Batteries and Their Applications

The voltage of a battery, its service life, and other characteristics depend on the material from which it is made.

Alkaline

These are popular, general-purpose batteries that are used in flashlights, portable radios, TV remote controllers, cameras, toys, and so on. With a nominal cell voltage of 1.5 V, they come in various sizes as depicted in Figure 2–13. While some alkaline batteries are designed to be recharged, most are not.

Lithium

Lithium batteries (Figure 2–14) feature small size and long life (e.g., some have shelf lives of 10 to 20 years). Applications include watches, pacemakers, cameras, and battery backup of computer memories. Several types of lithium cells are available, with voltages from of 2 V to 3.5 V and current ratings from the microampere to the ampere range.

FIGURE 2–13 Alkaline batteries. From left to right, a 9V rectangular battery, an AAA cell, a D cell, an AA cell, and a C cell.

FIGURE 2–14 An assortment of lithium batteries. The battery on the computer motherboard is for memory backup.

Nickel-Cadmium (Ni-Cad)

Ni-Cads are general-purpose, rechargeable batteries that, although once popular, are gradually being superseded by NiMH and lithium ion batteries, which have better performance characteristics and higher energy densities. Uses include cordless power tools and home entertainment systems.

Nickel-Metal Hydride (NiMH)

NiMH batteries are rechargeable batteries that can have two to three times the capacity of an equivalent-sized Ni-Cad battery. They are used for powering electric vehicles, as well as in consumer electronics, such as cameras.

Lead-Acid

This is the familiar automotive battery. Its basic cell voltage is about 2 volts, but typically, six cells are connected internally to provide 12 volts at its terminals. Lead-acid batteries are capable of delivering large current (in excess of 100 A) for short periods as required, for example, to start an automobile.

Battery Capacity

Batteries run down under use. However, an estimate of their useful life can be determined from their capacity, that is, their **ampere-hour** rating. (The ampere-hour rating of a battery is equal to the product of its current drain times the length of time that you can expect to draw the specified current before the battery becomes unusable.) For example, a battery rated at 200 Ah can theoretically supply 20 A for 10 h, or 5 A for 40 h, and so on. The relationship between capacity, life, and current drain is

$$\text{life} = \frac{\text{capacity}}{\text{current drain}} \qquad \textbf{(2–8)}$$

The capacity of batteries is not a fixed value as suggested above but is affected by discharge rates, operating schedules, temperature, and other factors. (For example, a battery discharged at a high rate will have a lower capacity than the same battery discharged at a lower rate.) At best, therefore, capacity is an estimate of expected life under certain conditions. To illustrate, consider Figure 2–15. It shows a typical variation of capacity of a Ni-Cad battery with changes in temperature.

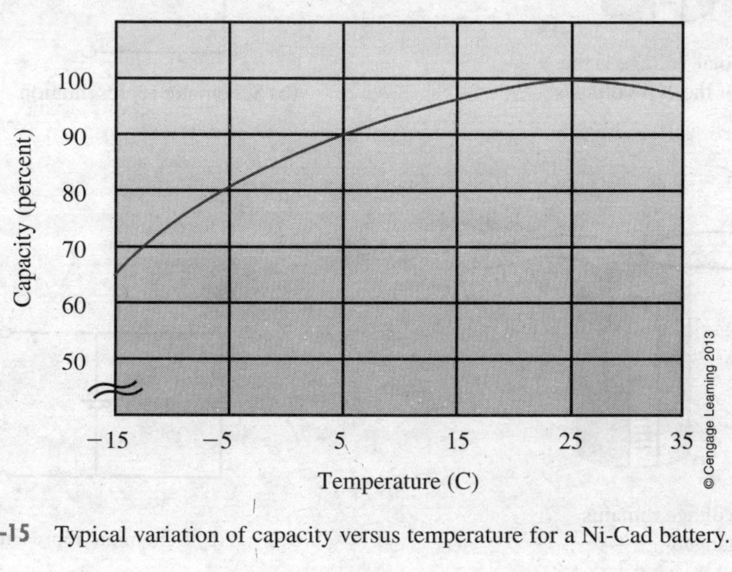

FIGURE 2–15 Typical variation of capacity versus temperature for a Ni-Cad battery.

Other Characteristics

Because batteries are not perfect, their terminal voltage drops as the amount of current drawn from them increases. (This issue is considered in Chapter 5.) In addition, battery voltage is affected by temperature and other factors that affect their chemical activity. However, these factors are not considered in this book.

EXAMPLE 2–4

Assume the battery of Figure 2–15 has a capacity of 240 Ah at 25° C. What is its capacity at −15° C?

Solution From the graph, capacity at –15° C is down to 65% of its value at 25° C. Thus, capacity = 0.65 × 240 = 156 Ah.

Cells in Series and Parallel

Cells may be connected as in Figure 2–16 and Figure 2–17 to increase their voltage and current capabilities. This is discussed in later chapters.

FIGURE 2–16 Cells connected in series to increase the available voltage.

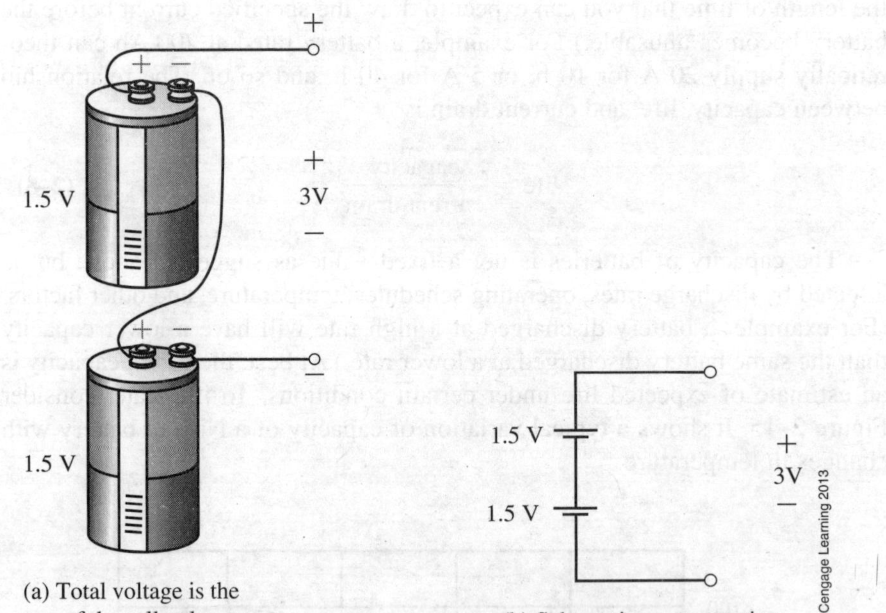

(a) Total voltage is the sum of the cell voltages

(b) Schematic representation

FIGURE 2–17 Cells connected in parallel to increase the available current. (Both must have the same voltage.) Do not do this for extended periods of time.

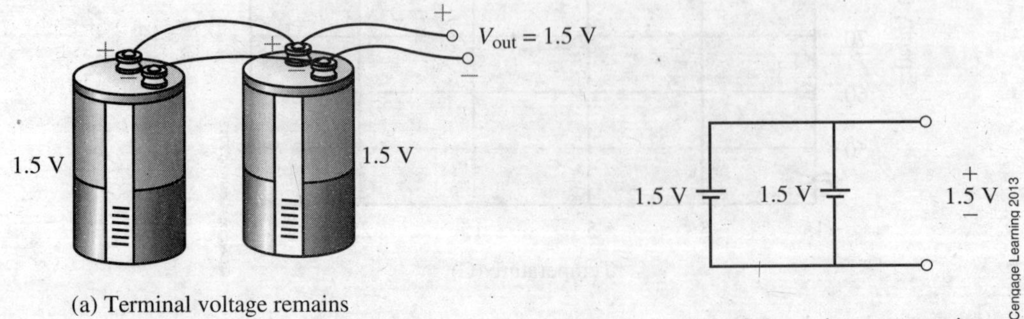

(a) Terminal voltage remains unchanged

(b) Schematic representation

Electronic Power Supplies

Electronic systems such as TV sets, Blu-ray players, computers, and so on, require dc for their internal circuit operation. Except for portable units, which use batteries, they obtain their power from the commercial ac power lines by means of built-in power supplies. Figure 2–18 shows a fixed voltage supply built for such purposes. In other applications, variable voltage may be needed, as for example during prototype development and circuit testing. Figure 2–19 shows a remotely programmable, variable dc supply whose voltage and current can be set remotely by the user.

Solar Cells

Solar cells convert light energy to electrical energy using photovoltaic means. The basic cell consists of two layers of semiconductor material. When light strikes the cell, many electrons gain enough energy to cross from one layer to the other to create a dc voltage.

Solar energy has a number of practical applications. Large solar farms, for example (some covering nearly 1000 acres or 400 ha of land), use solar panels to create dc which is then converted to ac and fed into the adjacent commercial electrical power grid. Figure 2–20 shows a typical installation, built to serve the needs of an Air Force base. Other, much smaller installations are found on house rooftops in residential areas or on the roofs of commercial buildings. In

Photo courtesy of Kepco Inc.

FIGURE 2–18 General-purpose 300 W open frame power supply.

Photo courtesy of Kepco Inc.

FIGURE 2–19 Programmable power supply. Voltage and current values may be set locally or by remote control.

FIGURE 2–20 Solar panel installation at Nellis Air Force Base in Nevada. This array of panels (which employs an advanced sun tracking system) supplies a significant percentage of the base's electrical power needs.

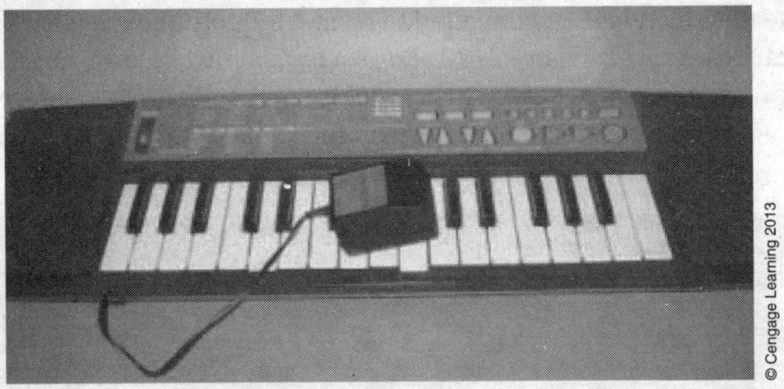

FIGURE 2–21 AC adapters are used as a source of dc for many electronic devices.

remote areas, solar panels are used to power communications systems and irrigation pumps. In space, they are used to power satellites. In everyday life, they are used to power hand-held calculators.

AC Adapter

Many electronic devices—including laptop computers, answering machines, and modems—utilize an ac adapter in order to provide dc for powering their circuits. The adapter plugs into any standard 120 V ac outlet; converts ac to dc; and uses this dc to power the desired device (such as the keyboard of Figure 2–21).

2.6 Measuring Voltage and Current

Voltage and current are usually measured in practice using a multimeter, an instrument that combines voltage, current, and resistance (and sometimes other) measuring functions in a single unit. While both digital and analog versions are available, the hand-held digital multimeter (DMM), is by far the most popular choice these days—see Notes. In terms of product choice, DMMs range from

NOTES...

In this book, we consider only digital multimeters. However, a mini-tutorial describing analog measurements may be found on our Web site at www.cengagebrain.com. Log in as per the instructions in the Preface and follow the links to *For Further Investigation* and select *Using Analog Multimeters.*

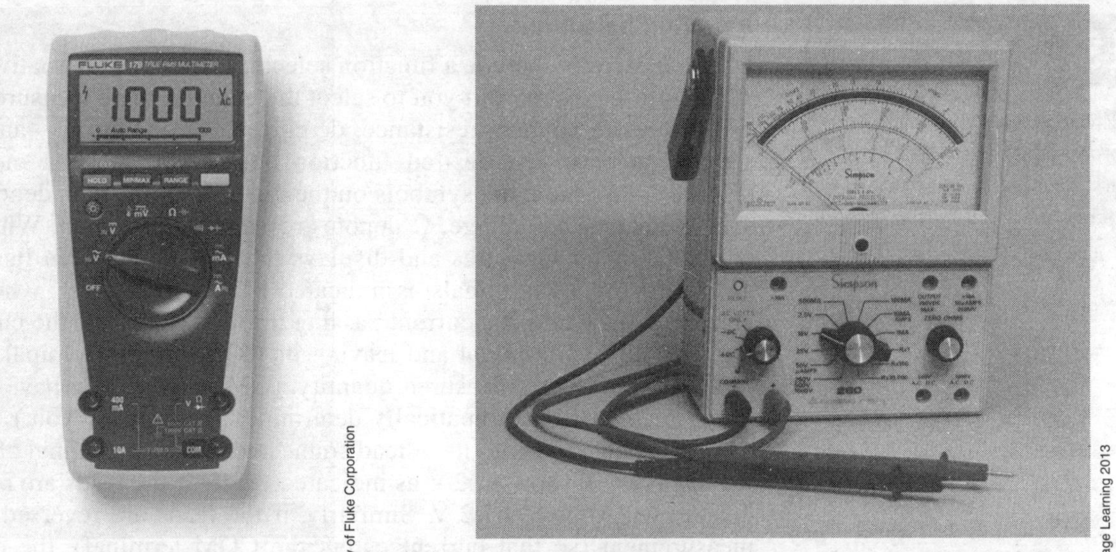

(a) Hand-held digital multimeter (DMM)

(b) Analog multimeter (VOM). Analog multimeters have been largely superseded by DMMs.

FIGURE 2–22 Multimeters are multipurpose test instruments that combine the functions of a voltmeter, ammeter, and ohmmeter (plus some secondary functions) into a single unit. Some instruments use terminal markings of VΩ and COM, while others use a + and − designation. Black and red color-coded test leads are industry standard.

inexpensive units such as those commonly used by home hobbyists to professional grade devices such as that shown in Figure 2–22.

Terminal Designations

Multimeters typically have a set of terminals marked VΩ, A, and COM that are color coded red and black as in Figure 2–22(a). Terminal VΩ is the terminal to use to measure voltage and resistance, while terminal A is used for current measurement. The terminal marked COM is the common terminal for all measurements. (Some multimeters combine the VΩ and A terminals into one terminal marked VΩA.) On some instruments the VΩ terminal is called the + terminal, and the COM terminal is called the − terminal. In the interests of generality, we use the +, − designation in this book (Figure 2–23).

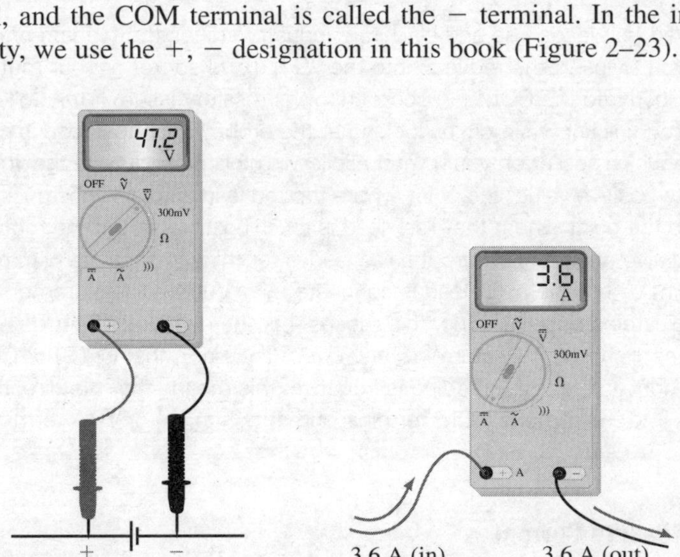

(a) Set selector to $\overline{\overline{V}}$ to measure dc voltage

(b) Set selector to $\overline{\overline{A}}$ to measure dc current

FIGURE 2–23 Measuring voltage and current with a multimeter. Be sure to set the selector switch to the correct function before you energize the circuit and connect the red lead to the VΩ (+) terminal and the black lead to the COM (−) terminal.

NOTES...

DMMs as Learning Tools

Voltage and current as presented earlier in this chapter are rather abstract concepts involving energy, charge, and charge movement. Voltmeters and ammeters are introduced at this point to help present the ideas in more physically meaningful terms. In particular, we concentrate on DMMs. Experience has shown them to be powerful learning tools. For example, when dealing with the sometimes difficult topics of voltage polarity conventions, current direction conventions, and so on (as in later chapters), the use of DMMs showing readings complete with signs for voltage polarity and current direction provides clarity and aids understanding in a way that simply drawing arrows and putting numbers on diagrams does not. You will find that in the first few chapters of this book DMMs are used for this purpose quite frequently.

NOTES...

Most DMMs have internal circuitry that automatically selects the correct range for voltage measurement. Such instruments are called "autoranging" or "autoscaling" devices.

Function Selection

DMMs generally include a function selector switch (or alternatively, a set of push buttons) that permit you to select the quantity to be measured—such as dc voltage, ac voltage, resistance, dc current, or ac current—and you must set the meter to the desired function before you make a measurement, (Figure 2–23). Note the symbols on the dial. The symbol $\overline{\mathrm{V}}$ denotes dc voltage, $\tilde{\mathrm{A}}$ denotes ac voltage, Ω denotes resistance, and so on. When set to dc volts, the meter measures and displays the voltage between its VΩ (or +) and COM (or −) terminals as indicated in Figure 2–23(a); when set to dc current, it measures the current passing through it, that is, the current entering its A (or +) terminal and leaving its COM (or −) terminal. Be sure to note the sign of the measured quantity. (DMMs generally have an **autopolarity** feature that automatically determines the sign for you.) Thus, if the meter is connected with its + lead connected to the + terminal of the source, the display will show 47.2 V as indicated, while if the leads are reversed, the display will show −47.2 V. Similarly, if the leads are reversed for current measurement (so that current enters the COM terminal), the display will show −3.6 A. Be sure to observe the standard color convention for test lead connections—see Practical Notes.

How to Measure Voltage

CircuitSim 02-2

Since voltage is the potential difference between two points, you measure voltage by placing the voltmeter leads *across* the component whose voltage you wish to determine as we saw in Figure 2–23(a). Figure 2–24 shows another example. To measure the voltage across the lamp, place one lead on each side of the lamp as shown. If the meter is not autoscale and you have no idea how large the voltage is, set the meter to its highest range, then work your way down to avoid damage to the instrument.

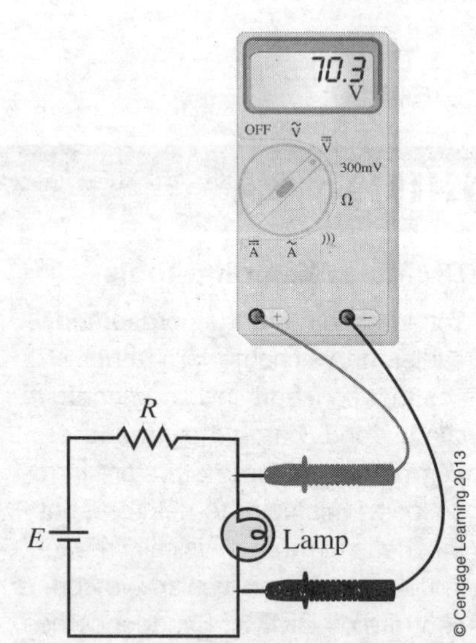

FIGURE 2–24 To measure voltage, place the voltmeter leads across the component whose voltage you wish to determine. If the voltmeter reading is positive, the point where the red lead is connected is positive with respect to the point where the black lead is connected.

PRACTICAL NOTES...

Color-coded test leads (red and black) are industry standard. Standard practice calls for you to insert the red lead into the VΩ (i.e., +) socket of your meter, and the black lead into the COM (−) socket. (This is a safety issue. If you follow this practice you will know simply by looking at the probe which test lead is connected to which socket on your meter.) Following this procedure, if the voltmeter indicates a positive value, the point where the red lead is touching is positive with respect to the point where the black lead is touching; inversely, if the meter indicates negative, the point where the red lead is touching is negative with respect to the point where the black lead is connected. For current measurements, if the meter indicates a positive value, this means that the direction of current is into the red, that is, (+) or VΩA terminal and out of the black, that is, (−) or COM terminal; conversely, if the reading is negative, this means that the direction of current is into the meter's COM terminal and out of its (+) or VΩA terminal.

How to Measure Current

As indicated by Figure 2–23(b), the current that you wish to measure must pass *through* the meter. Consider Figure 2–25(a). To measure this current, open the circuit as in (b) and insert the ammeter. The sign of the reading will be positive if current enters the A or (+) terminal or negative if it enters the COM (or −) terminal as described in the Practical Notes.

CircuitSim 02-3

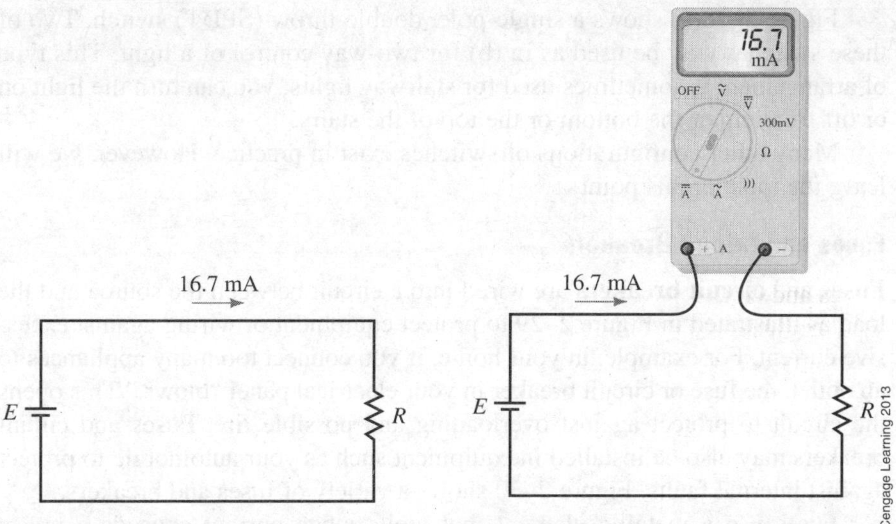

(a) Current to be measured (b) Ammeter correctly inserted

FIGURE 2–25 To measure current, insert the ammeter into the circuit so that the current you wish to measure passes through the instrument. The reading is positive here because current enters the + (A) terminal.

Meter Symbols

So far, we have shown meters pictorially. Usually, however, they are shown schematically. The schematic symbol for a voltmeter is a circle with the letter *V*, while the symbol for an ammeter is a circle with the letter *I*. The circuits of Figures 2–24 and 2–25 have been redrawn (Figure 2–26) to indicate this.

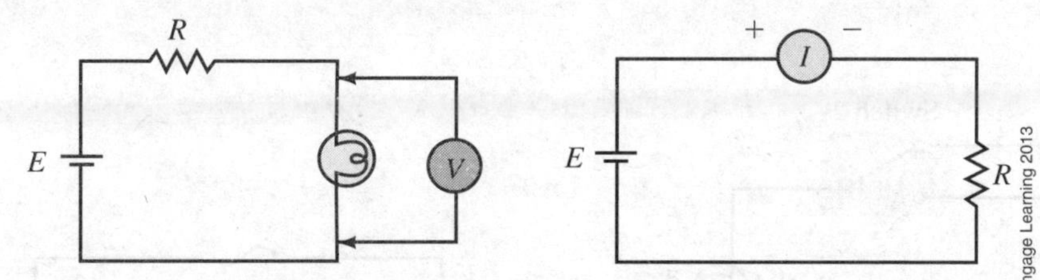

(a) Voltmeter (b) Ammeter

FIGURE 2–26 Schematic symbols for voltmeter and ammeter.

PRACTICAL NOTES...

1. One sometimes hears statements such as ". . . the voltage through a resistor" or ". . . the current across a resistor." These statements are incorrect. Voltage does not pass through anything; voltage is a potential difference and appears across things. This is why we connect a voltmeter *across* components to measure their voltage. Similarly, current does not appear across anything; current is a flow of charge that passes *through* circuit elements. This is why we put the ammeter in the current path—to measure the current in it. Thus, the correct statements are ". . . voltage across the resistor . . ." and ". . . current through the resistor. . . ."
2. Do *not* connect ammeters directly across a voltage source. Ammeters have nearly zero resistance and damage will probably result.

2.7 Switches, Fuses, and Circuit Breakers

Switches

The most basic switch is a single-pole, single-throw (SPST) switch as shown in Figure 2–27. With the switch open, the current path is broken and the lamp is off; with it closed, the lamp is on. This type of switch is used, for example, for light switches in homes.

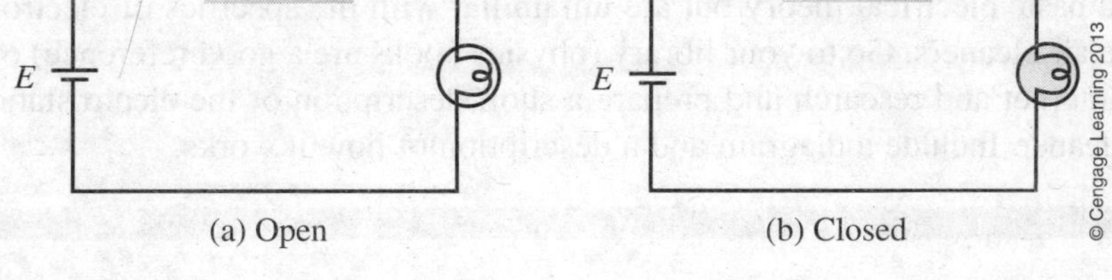

(a) Open (b) Closed

FIGURE 2–27 Single-pole, single-throw (SPST) switch.

Figure 2–28(a) shows a single-pole, double-throw (SPDT) switch. Two of these switches may be used as in (b) for two-way control of a light. This type of arrangement is sometimes used for stairway lights; you can turn the light on or off from either the bottom or the top of the stairs.

Many other configurations of switches exist in practice. However, we will leave the topic at this point.

Fuses and Circuit Breakers

Fuses and **circuit breakers** are wired into a circuit between the source and the load as illustrated in Figure 2–29 to protect equipment or wiring against excessive current. For example, in your home, if you connect too many appliances to an outlet, the fuse or circuit breaker in your electrical panel "blows." This opens the circuit to protect against overloading and possible fire. Fuses and circuit breakers may also be installed in equipment such as your automobile to protect against internal faults. Figure 2–30 shows a variety of fuses and breakers.

Fuses use a metallic element that melts when current exceeds a preset value. Thus, if a fuse is rated at 3 A, it will "blow" if more than 3 amps passes through it. Fuses are made as fast-blow and slow-blow types. Fast-blow fuses are very fast; typically, they blow in a fraction of a second. Slow-blow fuses, on the other hand, react more slowly so that they do not blow on small, momentary overloads.

CircuitSim 02-4

Circuit breakers work on a different principle. When the current exceeds the rated value of a breaker, the magnetic field produced by the excessive current operates a mechanism that trips open a switch. After the fault or overload condition has been cleared, the breaker can be reset and used again. Since they are mechanical devices, their operation is slower than that of a fuse; thus, they do not "pop" on momentary overloads as, for example, when you start a motor.

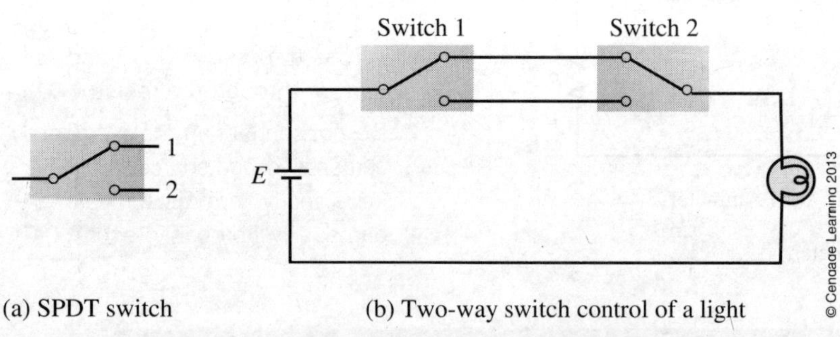

(a) SPDT switch (b) Two-way switch control of a light

FIGURE 2–28 Single-pole, single-throw (SPDT) switch.

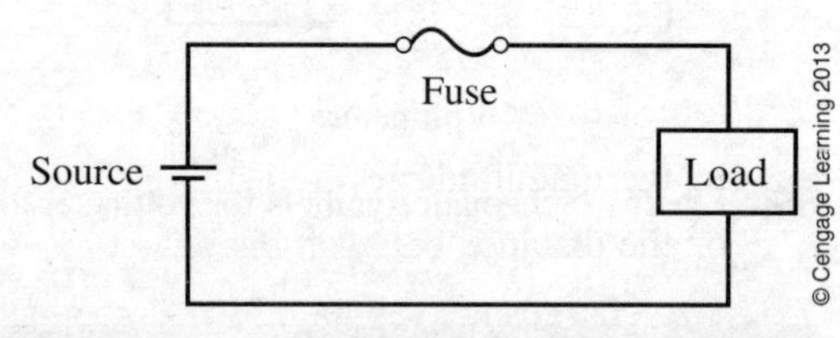

FIGURE 2–29 Using a fuse to protect a circuit.

Putting It into Practice

Your company is considering the purchase of an electrostatic air cleaner system for one of its facilities and your supervisor has asked you to prepare a short presentation for the Board of Directors. Members of the Board understand basic electrical theory but are unfamiliar with the specifics of electrostatic air cleaners. Go to your library (physics books are a good reference) or the Internet and research and prepare a short description of the electrostatic air cleaner. Include a diagram and a description of how it works.

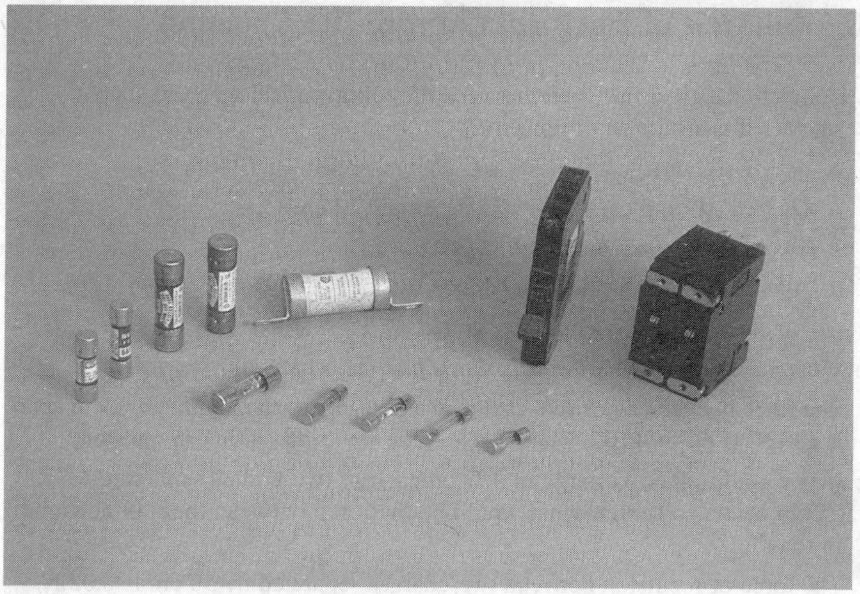

(a) A variety of fuses and circuit breakers

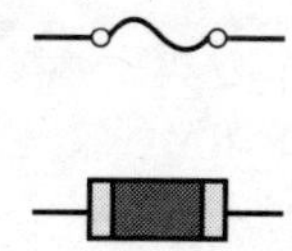

(b) Fuse symbols

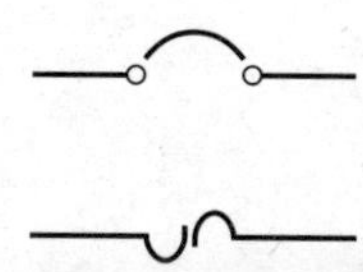

(c) Circuit breaker symbols

FIGURE 2–30 Fuses and circuit breakers.

Problems

2.1 Atomic Theory Review

1. How many free electrons are there in the following at room temperature?
   a. 1 cubic meter of copper
   b. a 5 m length of copper wire whose diameter is 0.163 cm
2. Two charges are separated by a certain distance, Figure 2–31. How is the force between them affected if
   a. the magnitudes of both charges are doubled?
   b. the distance between the charges is tripled?

FIGURE 2–31

3. Two charges are separated by a certain distance. If the magnitude of one charge is doubled and the other tripled and the distance between them halved, how is the force affected?
4. A certain material has 4 electrons in its valence shell and a second material has 1. Which is the better conductor?
5. a. What makes a material a good conductor? (In your answer, consider valence shells and free electrons.)
   b. Besides being a good conductor, list two other reasons why copper is so widely used.
   c. What makes a material a good insulator?
   d. Normally air is an insulator. However, during lightning discharges, conduction occurs. Briefly discuss the mechanism of charge flow in this discharge.
6. a. Although gold is very expensive, it is sometimes used in electronics as a plating on contacts. Why?
   b. Why is aluminum sometimes used when its conductivity is only about 60% as good as that of copper?

NOTES...

All spacings in these questions are center-to-center.

2.2 The Unit of Electrical Charge: The Coulomb

7. Compute the electrical force between the following charges and state whether it is attractive or repulsive.
   a. A + 1 μC charge and a +7 μC charge, separated 10 mm
   b. $Q_1 = 8$ μC and $Q_2 = -4$ μC, separated 12 cm
   c. Two electrons separated by 12×10^{-8} m
   d. An electron and a proton separated by 5.3×10^{-11} m
   e. An electron and a neutron separated by 5.7×10^{-11} m
8. What do we mean when we say that a body is "charged?"
9. The force between a positive charge and a negative charge that are 2 cm apart is 180 N. If $Q_1 = 4$ μC, what is Q_2? Is the force attraction or repulsion?
10. If you could place a charge of 1 C on each of two bodies separated 25 cm center to center, what would be the force between them in newtons? In tons?
11. The force of repulsion between two charges separated by 50 cm is 0.02 N. If $Q_2 = 5Q_1$, determine the charges and their possible signs.
12. How many electrons does a charge of 1.63 μC represent?
13. Determine the charge possessed by 19×10^{13} electrons.
14. An electrically neutral metal plate acquires a negative charge of 47 μC. How many electrons were added to it?
15. A metal plate has 14.6×10^{13} electrons added. Later, 1.3 μC of charge is added. If the final charge on the plate is 5.6 μC, what was its initial charge?

2.3 Voltage

16. Sliding off a chair and touching someone can result in a shock. Explain why.
17. If 360 joules of energy are required to transfer 15 C of charge through the lamp of Figure 2–1, what is the voltage of the battery?
18. If 600 J of energy are required to move 9.36×10^{19} electrons from one point to the other, what is the potential difference between the two points?
19. If 1.2 kJ of energy are required to move 500 mC from one point to another, what is the voltage between the two points?
20. How much energy is required to move 20 mC of charge through the lamp of Figure 2–24?
21. How much energy is gained by a charge of 0.5 μC as it moves through a potential difference of 8.5 kV?
22. If the voltage between two points is 100 V, how much energy is required to move an electron between the two points?
23. Given a voltage of 12 V for the battery in Figure 2–1, how much charge is moved through the lamp if it takes 57 J of energy to move it?

2.4 Current

24. For the circuit of Figure 2–1, if 27 C pass through the lamp in 9 seconds, what is the current in amperes?
25. If 250 μC pass through the ammeter of Figure 2–32 in 5 ms, what will the meter read?

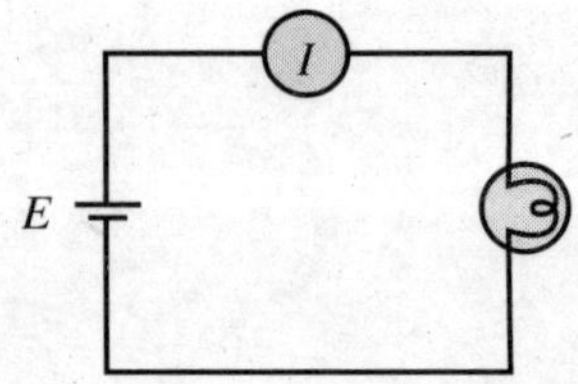

FIGURE 2–32

26. If the current $I = 4$ A in Figure 2–1, how many coulombs pass through the lamp in 7 ms?
27. How much charge passes through the circuit of Figure 2–25 in 20 ms?
28. How long does it take for 100 μC to pass a point if the current is 25 mA?
29. If 93.6×10^{12} electrons pass through a lamp in 5 ms, what is the current?
30. The charge passing through a wire is given by $q = 10t + 4$, where q is in coulombs and t in seconds,
    a. How much charge has passed at $t = 5$ s?
    b. How much charge has passed at $t = 8$ s?
    c. What is the current in amps?
31. The charge passing through a wire is $q = (80t + 20)$ C. What is the current? Hint: Choose two arbitrary values of time and proceed as in Question 30, or use calculus if you know how.
32. How long does it take 312×10^{19} electrons to pass through the circuit of Figure 2–32 if the ammeter reads 8 A?
33. If 1353.6 J are required to move 47×10^{19} electrons through the lamp of Figure 2–32 in 1.3 min, what are E and I?

2.5 Practical dc Voltage Sources

34. What do we mean by dc? By ac?
35. Consider three batteries connected as in Figure 2–33.
    a. If $E_1 = 1.47$ V, $E_2 = 1.61$ V, and $E_3 = 1.58$ V, what is E_T?
    b. If the connection to source 3 is reversed, what is E_T?
36. How do you charge a secondary battery? Make a sketch. Can you charge a primary battery?
37. A battery rated 1400 mAh supplies 28 mA to a load. How long can it be expected to last?
38. The battery of Figure 2–15 is expected to last 17 h at a current drain of 1.5 A at 25° C. How long do you expect it to last at 5° C at a current drain of 0.8 A?
39. The battery of Figure 2–15 is rated at 81 Ah at 5° C. What is the expected life (in hours) at a current draw of 5 A at −15° C?

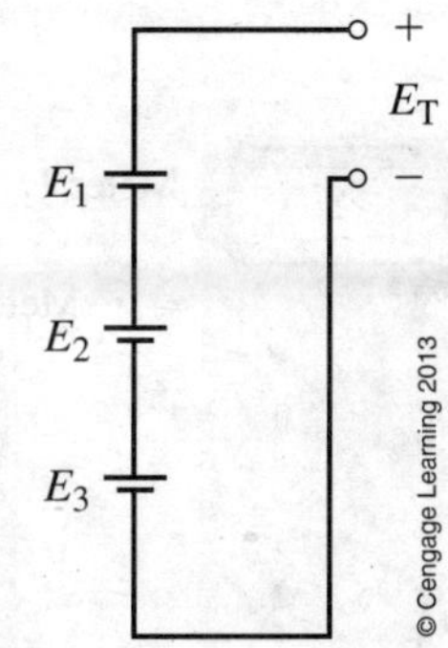

FIGURE 2–33

2.6 Measuring Voltage and Current

40. The digital voltmeters of Figure 2–34 have autopolarity. For each case, determine the readings.

CircuitSim 02-5

41. The current in the circuit of Figure 2–35 is 9.17 mA. Which ammeter correctly indicates the current? (a) Meter 1, (b) Meter 2, (c) both.
42. Redraw the circuit of Figure 2–35 using schematic symbols for ammeters instead of multimeter pictorials.
43. Current in the circuit of Figure 2–36 is 7 A. What are the expected readings for Meter 1 and Meter 2?
44. What is wrong with the statement that the voltage through the lamp of Figure 2–24 is 70.3 V?
45. What is wrong with the metering scheme shown in Figure 2–37? Fix it.

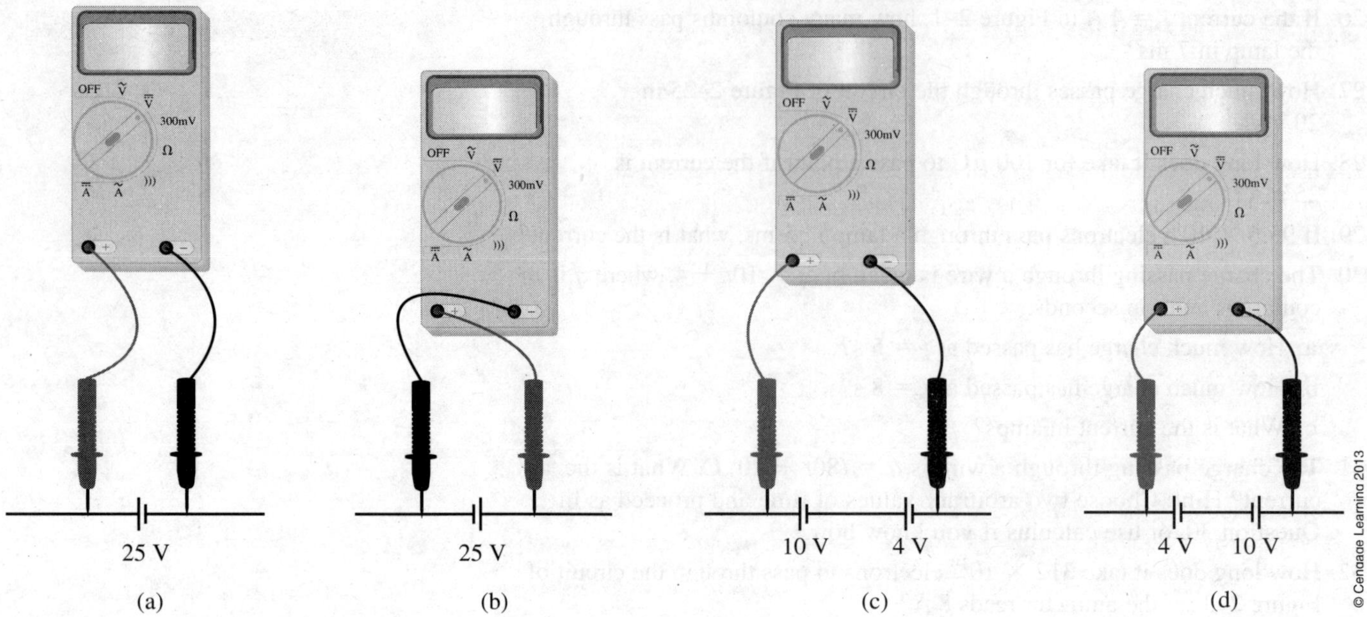

FIGURE 2–34

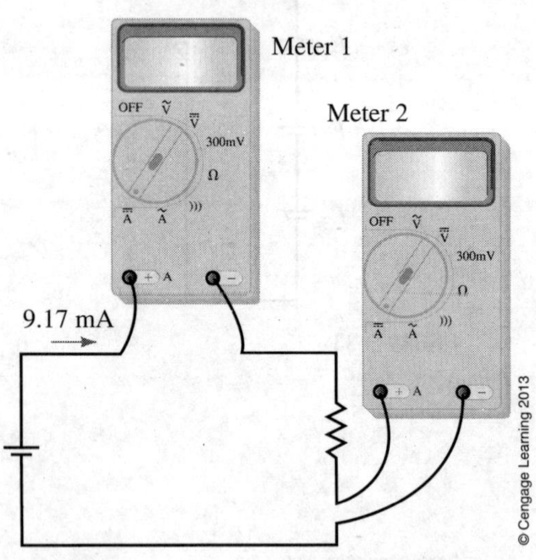

FIGURE 2–35

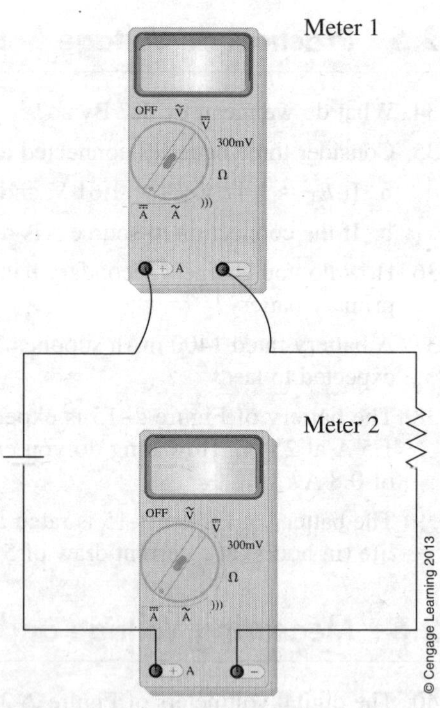

FIGURE 2–36

CircuitSim 02-6

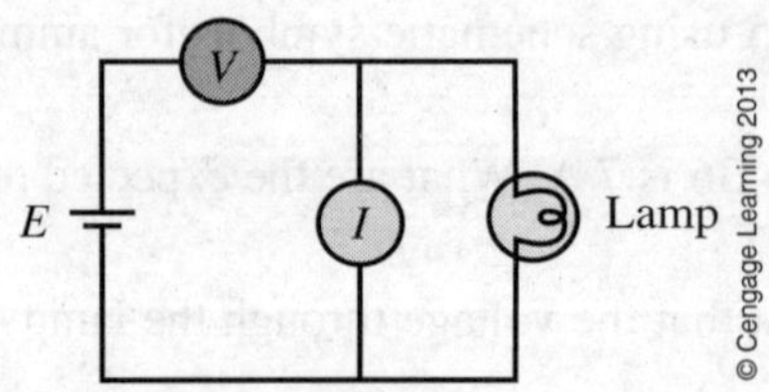

FIGURE 2–37 What is wrong here?

2.7 Switches, Fuses, and Circuit Breakers

46. We wish to control a light using two switches as indicated in Table 2–1. Draw the required circuit.
47. Fuses have a current rating so that you can select the proper size to protect a circuit against overcurrent. They also have a voltage rating. Why? Hint: Read the section on insulators, Section 2.1.

TABLE 2–1

Switch 1	Switch 2	Lamp
Open	Open	Off
Open	Closed	On
Closed	Open	On
Closed	Closed	On

ANSWERS TO IN-PROCESS LEARNING CHECKS

IN-PROCESS LEARNING CHECK 1

1. An atom consists of a nucleus of protons and neutrons orbited by electrons. The nucleus is positive because protons are positive, but the atom is neutral because it contains the same number of electrons as protons, and their charges cancel.
2. The valence shell is the outermost shell. It contains the atom's valence electrons. The number of electrons in this shell determines the properties of the material with regard to whether it is a conductor, insulator, or semiconductor.
3. The force between charged particles is proportional to the product of their charges and inversely proportional to the square of their spacing. Since force decreases as the square of the spacing, electrons far from the nucleus experience little force of attraction.
4. If a loosely bound electron gains sufficient energy, it may break free from its parent atom and wander throughout the material. Such an electron is called a free electron. For materials like copper, heat (thermal energy) can give an electron enough energy to dislodge it from its parent atom.
5. A normal atom is neutral because it has the same number of electrons as protons and their charges cancel. An atom that has lost an electron is called a positive ion, while an atom that has gained an electron is called a negative ion.

IN-PROCESS LEARNING CHECK 2

1. $Q_A = 13.7\ \mu\text{C}$ (pos.), $Q_B = 13.6\ \mu\text{C}$ (neg.)
2. Chemical action creates an excess of electrons in the zinc mixture and a deficiency in the manganese dioxide mix. This separation of charges results in a voltage of 1.5 V. The zinc mixture is connected to the top end cap by the steel can (making it the positive electrode) and the manganese dioxide mixture is connected to the bottom of the can by the brass pin (making it the negative electrode).
3. Motion is random. Since the net movement in all directions is zero, current is zero.
4. 80 A
5. 3195 s

KEY TERMS

American Wire Gauge (AWG)
Conductance
Diode
Ohmmeter
Open Circuit
Photocell
Resistance
Resistivity
Short Circuit
Superconductor
Temperature Coefficient
Thermistor
Varistor

OUTLINE

Resistance of Conductors
Electrical Wire Tables
Resistance of Wires—Circular Mils
Temperature Effects
Types of Resistors
Color Coding of Resistors
Measuring Resistance—The Ohmmeter
Thermistors
Photoconductive Cells
Nonlinear Resistance
Conductance
Superconductors

OBJECTIVES

After studying this chapter, you will be able to

- calculate the resistance of a section of a conductor, given its cross-sectional area and length,
- convert between areas measured in square mils, square meters, and circular mils,
- use tables of wire data to obtain the cross-sectional dimensions of various gauges of wire and predict the allowable current for a particular gauge of wire,
- use the temperature coefficient of a material to calculate the change in resistance as the temperature of the sample changes,
- use resistor color codes to determine the resistance and tolerance of a given fixed-composition resistor,
- demonstrate the procedure for using an ohmmeter to determine circuit continuity and to measure the resistance of both an isolated component and one that is located in a circuit,
- develop an understanding of various ohmic devices such as thermistors and photocells,
- develop an understanding of the resistance of nonlinear devices such as varistors and diodes,
- calculate the conductance of any resistive component.

3

RESISTANCE

CHAPTER PREVIEW

You have been introduced to the concepts of voltage and current in previous chapters and have found that current involves the movement of charge. In a conductor, the charge carriers are the free electrons that are moved due to the voltage of an externally applied source. As these electrons move through the material, they constantly collide with atoms and other electrons within the conductor. In a process similar to friction, the moving electrons give up some of their energy in the form of heat. These collisions represent an opposition to charge movement that is called **resistance.** The greater the opposition (i.e., the greater the resistance), the smaller will be the current for a given applied voltage.

Circuit components (called resistors) are specifically designed to possess resistance and are used in almost all electronic and electrical circuits. Although the resistor is the simplest component in any circuit, its effect is very important in determining the operation of a circuit.

Resistance is represented by the symbol R (Figure 3–1) and is measured in units of ohms (after Georg Simon Ohm). The symbol for ohms is the capital Greek letter omega (Ω).

In this chapter, we examine resistance in its various forms. Beginning with metallic conductors, we study the factors that affect resistance in conductors. Following this, we look at commercial resistors, including both fixed and variable types. We then discuss important nonlinear resistance devices and conclude with an overview of superconductivity and its potential impact and use. ■

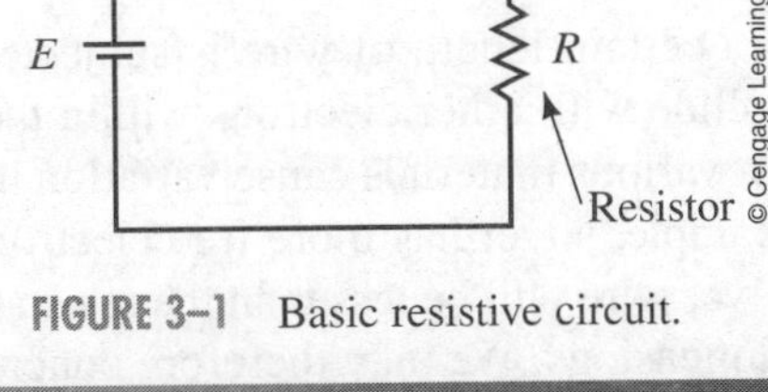

FIGURE 3–1 Basic resistive circuit.

Putting It in Perspective

Georg Simon Ohm and Resistance

ONE OF THE FUNDAMENTAL RELATIONSHIPS of circuit theory is that between voltage, current, and resistance. This relationship and the properties of resistance were investigated by the German physicist Georg Simon Ohm (1787–1854) using a circuit similar to that of Figure 3–1. Working with Volta's recently developed battery and wires of different materials, lengths, and thicknesses, Ohm found that current depended on both voltage and resistance. For a fixed resistance, he found that doubling the voltage doubled the current, tripling the voltage tripled the current, and so on. Also, for a fixed voltage, Ohm found that the opposition to current was directly proportional to the length of the wire and inversely proportional to its cross-sectional area. From this, he was able to define the resistance of a wire and show that current was inversely proportional to this resistance; that is, when he doubled the resistance, he found that the current decreased to half of its former value.

These two results when combined form what is known as Ohm's law. (You will study Ohm's law in great detail in Chapter 4.) Ohm's results are of such fundamental importance that they represent the real beginnings of what we now call electrical circuit analysis. ■

Hulton Archive/Getty Images

3.1 Resistance of Conductors

As mentioned in the chapter preview, conductors are materials that permit the flow of charge. However, conductors do not all behave the same way. Rather, we find that the resistance of a material is dependent upon several factors:

- Type of material
- Length of the conductor
- Cross-sectional area
- Temperature

If a certain length of wire is subjected to a current, the moving electrons will collide with other electrons within the material. Differences at the atomic level of various materials cause variation in how the collisions affect resistance. For example, silver has more free electrons than copper, and so the resistance of a silver wire will be less than the resistance of a copper wire having the identical dimensions. We may therefore conclude the following:

The resistance of a conductor is dependent upon the type of material.

If we were to double the length of the wire, we can expect that the number of collisions over the length of the wire would double, thereby causing the resistance to also double. This effect may be summarized as follows:

The resistance of a metallic conductor is directly proportional to the length of the conductor.

A somewhat less intuitive property of a conductor is the effect of cross-sectional area on the resistance. As the cross-sectional area is increased, the moving electrons are able to move more freely through the conductor, just as water moves more freely through a large-diameter pipe than a small-diameter pipe. If the cross-sectional area is doubled, the electrons would be involved in half as many collisions over the length of the wire. We may summarize this effect as follows:

The resistance of a metallic conductor is inversely proportional to the cross-sectional area of the conductor.

The factors governing the resistance of a conductor at a given temperature may be summarized mathematically as follows:

$$R = \frac{\rho \ell}{A} \quad [\text{ohms}, \Omega] \tag{3–1}$$

where

ρ = resistivity, in ohm-meters (Ω-m)
ℓ = length, in meters (m)
A = cross-sectional area, in square meters (m^2).

In the previous equation the lowercase Greek letter rho (ρ) is the constant of proportionality and is called the **resistivity** of the material. Resistivity is a physical property of a material and is measured in ohm-meters (Ω-m) in the SI system. Table 3–1 lists the resistivities of various materials at a temperature of 20° C. The effects on resistance due to changes in temperature will be examined in Section 3.4.

Since most conductors are circular, as shown in Figure 3–2, we may determine the cross-sectional area from either the radius or the diameter as follows:

$$A = \pi r^2 = \pi\left(\frac{d}{2}\right)^2 = \frac{\pi d^2}{4} \tag{3–2}$$

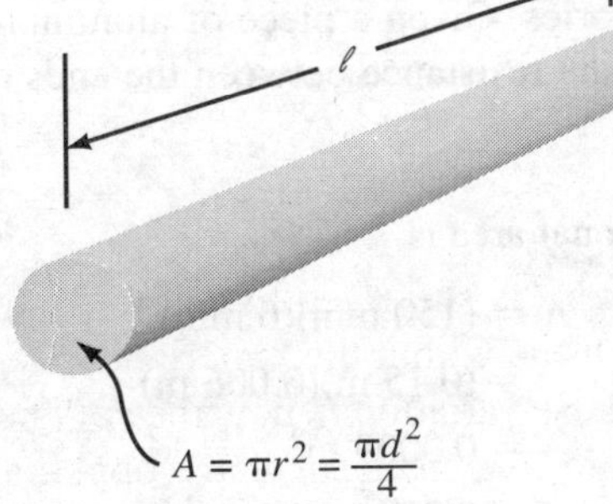

FIGURE 3–2 Conductor with a circular cross section.

TABLE 3–1 Resistivity of Materials, ρ

Material	Resistivity, ρ, at 20° C (Ω-m)
Silver	1.645×10^{-8}
Copper	1.723×10^{-8}
Gold	2.443×10^{-8}
Aluminum	2.825×10^{-8}
Tungsten	5.485×10^{-8}
Iron	12.30×10^{-8}
Lead	22×10^{-8}
Mercury	95.8×10^{-8}
Nichrome	99.72×10^{-8}
Carbon	3500×10^{-8}
Germanium	20–2300*
Silicon	≅500*
Wood	10^{8}–10^{14}
Glass	10^{10}–10^{14}
Mica	10^{11}–10^{15}
Hard rubber	10^{13}–10^{16}
Amber	5×10^{14}
Sulphur	1×10^{15}
Teflon	1×10^{16}

*The resistivities of these materials are dependent upon the impurities within the materials.

EXAMPLE 3–1

Most homes use solid copper wire having a diameter of 1.63 mm to provide electrical distribution to outlets and light sockets. Determine the resistance of 75 meters of a solid copper wire having the above diameter.

Solution We will first calculate the cross-sectional area of the wire using Equation 3–2.

$$\begin{aligned} A &= \frac{\pi d^2}{4} \\ &= \frac{\pi(1.63 \times 10^{-3}\text{ m})^2}{4} \\ &= 2.09 \times 10^{-6}\text{ m}^2 \end{aligned}$$

Now, using Table 3–1, the resistance of the length of wire is found as

$$\begin{aligned} R &= \frac{\rho \ell}{A} \\ &= \frac{(1.723 \times 10^{-8}\ \Omega\text{-m})(75\text{ m})}{2.09 \times 10^{-6}\text{ m}^2} \\ &= 0.619\ \Omega \end{aligned}$$

PRACTICE PROBLEMS 1

Find the resistance of a 100-m long tungsten wire that has a circular cross section with a diameter of 0.1 mm ($T = 20°$ C).

Answer
698 Ω

EXAMPLE 3–2

Bus bars are bare solid conductors (usually rectangular) used to carry large currents within buildings such as power generating stations, telephone exchanges, and large factories. Given a piece of aluminum bus bar as shown in Figure 3–3, determine the resistance between the ends of this bar at a temperature of 20° C.

Solution The cross-sectional area is

$$\begin{aligned} A &= (150\text{ mm})(6\text{ mm}) \\ &= (0.15\text{ m})(0.006\text{ m}) \\ &= 0.0009\text{ m}^2 \\ &= 9.00 \times 10^{-4}\text{ m}^2 \end{aligned}$$

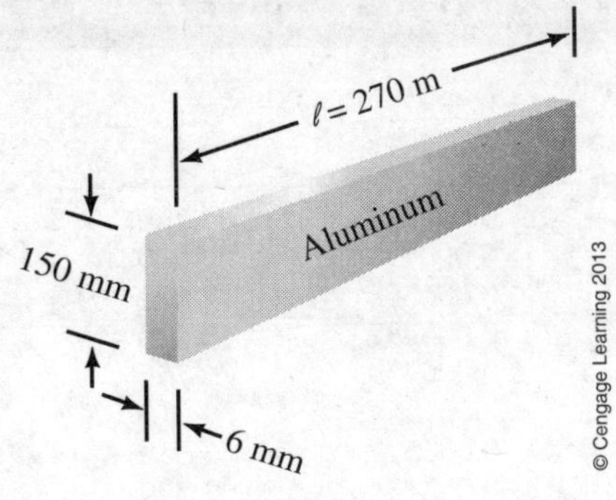

FIGURE 3–3 Conductor with a rectangular cross section.

The resistance between the ends of the bus bar is determined as

$$R = \frac{\rho \ell}{A}$$

$$= \frac{(2.825 \times 10^{-8}\ \Omega\text{-m})(270\ \text{m})}{9.00 \times 10^{-4}\ \text{m}^2}$$

$$= 8.48 \times 10^{-3}\ \Omega = 8.48\ \text{m}\Omega$$

IN-PROCESS LEARNING CHECK 1

(Answers are at the end of the chapter.)

1. Given two lengths of wire having identical dimensions, if one wire is made of copper and the other is made of iron, which wire will have the greater resistance? How much greater will the resistance be?
2. Given two pieces of copper wire that have the same cross-sectional area, determine the relative resistance of the one that is twice as long as the other.
3. Given two pieces of copper wire that have the same length, determine the relative resistance of the one that has twice the diameter of the other.

3.2 Electrical Wire Tables

Although the SI system is the standard measurement for electrical and other physical quantities, the English system is still used extensively in the United States and to a lesser degree throughout the rest of the English-speaking world. One area that has been slow to convert to the SI system is the designation of cables and wires, where the **American Wire Gauge (AWG)** is the primary system used to denote wire diameters. In this system, each wire diameter is assigned a wire gauge number. As Figure 3–4 shows, the higher the AWG number, the smaller the diameter of the cable or wire, for example, AWG 22-gauge wire is a smaller diameter than AWG 14 gauge. Since resistance is inversely proportional to the square of the diameter, a given length of 22-gauge wire will have more resistance than an equal length of 14-gauge wire. Because of the difference in resistance, we can intuitively deduce that large-diameter cables will be able to handle more current than smaller-diameter cables. Table 3–2 provides a listing of data for standard bare copper wire.

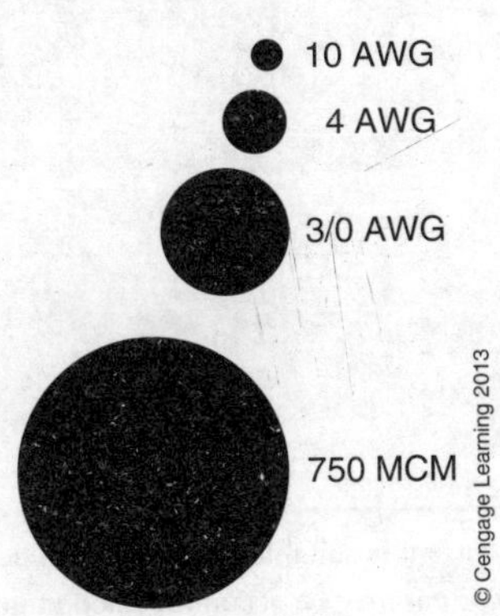

FIGURE 3–4 Typical conductor cross-sections (actual size).

Even though Table 3–2 provides data for solid conductors up to AWG 4/0, most applications do not use solid conductor sizes beyond AWG 10. Solid conductors are difficult to bend and are easily damaged by mechanical flexing.

TABLE 3–2 Standard Solid Copper Wire at 20° C

Size (AWG)	Diameter (inches)	Diameter (mm)	Area (CM)	Area (mm^2)	Resistance (Ω/1000 ft)	Current Capacity (A)
56	0.0005	0.012	0.240	0.000122	43 200	
54	0.0006	0.016	0.384	0.000195	27 000	
52	0.0008	0.020	0.608	0.000308	17 000	
50	0.0010	0.025	0.980	0.000497	10 600	
48	0.0013	0.032	1.54	0.000779	6 750	
46	0.0016	0.040	2.46	0.00125	4 210	
45	0.0019	0.047	3.10	0.00157	3 350	
44	0.0020	0.051	4.00	0.00243	2 590	
43	0.0022	0.056	4.84	0.00245	2 140	
42	0.0025	0.064	6.25	0.00317	1 660	
41	0.0028	0.071	7.84	0.00397	1 320	
40	0.0031	0.079	9.61	0.00487	1 080	
39	0.0035	0.089	12.2	0.00621	847	
38	0.0040	0.102	16.0	0.00811	648	
37	0.0045	0.114	20.2	0.0103	521	
36	0.0050	0.127	25.0	0.0127	415	
35	0.0056	0.142	31.4	0.0159	331	
34	0.0063	0.160	39.7	0.0201	261	
33	0.0071	0.180	50.4	0.0255	206	
32	0.0080	0.203	64.0	0.0324	162	
31	0.0089	0.226	79.2	0.0401	131	
30	0.0100	0.254	100	0.0507	104	
29	0.0113	0.287	128	0.0647	81.2	
28	0.0126	0.320	159	0.0804	65.3	
27	0.0142	0.361	202	0.102	51.4	
26	0.0159	0.404	253	0.128	41.0	0.75*
25	0.0179	0.455	320	0.162	32.4	
24	0.0201	0.511	404	0.205	25.7	1.3*
23	0.0226	0.574	511	0.259	20.3	
22	0.0253	0.643	640	0.324	16.2	2.0*
21	0.0285	0.724	812	0.412	12.8	
20	0.0320	0.813	1 020	0.519	10.1	3.0*
19	0.0359	0.912	1 290	0.653	8.05	
18	0.0403	1.02	1 620	0.823	6.39	5.0†
17	0.0453	1.15	2 050	1.04	5.05	
16	0.0508	1.29	2 580	1.31	4.02	10.0†
15	0.0571	1.45	3 260	1.65	3.18	
14	0.0641	1.63	4 110	2.08	2.52	15.0†
13	0.0720	1.83	5 180	2.63	2.00	
12	0.0808	2.05	6 530	3.31	1.59	20.0†
11	0.0907	2.30	8 230	4.17	1.26	
10	0.1019	2.588	10 380	5.261	0.998 8	30.0†
9	0.1144	2.906	13 090	6.632	0.792 5	
8	0.1285	3.264	16 510	8.367	0.628 1	
7	0.1443	3.665	20 820	10.55	0.498 1	
6	0.1620	4.115	26 240	13.30	0.395 2	
5	0.1819	4.620	33 090	16.77	0.313 4	
4	0.2043	5.189	41 740	21.15	0.248 5	
3	0.2294	5.827	52 620	26.67	0.197 1	
2	0.2576	6.543	66 360	33.62	0.156 3	
1	0.2893	7.348	83 690	42.41	0.123 9	
1/0	0.3249	8.252	105 600	53.49	0.098 25	
2/0	0.3648	9.266	133 100	67.43	0.077 93	
3/0	0.4096	10.40	167 800	85.01	0.061 82	
4/0	0.4600	11.68	211 600	107.2	0.049 01	

*This current is suitable for single conductors and surface or loose wiring.

†This current may be accommodated in up to three wires in a sheathed cable. For four to six wires, the current in each wire must be reduced to 80% of the indicated value. For seven to nine wires, the current in each wire must be reduced to 70% of the indicated value.

For this reason, large-diameter cables are nearly always stranded rather than solid. Stranded wires and cables use anywhere from seven strands, as shown in Figure 3–5, to in excess of a hundred strands.

As one might expect, stranded wire uses the same AWG notation as solid wire. Consequently, AWG 10 stranded wire will have the same cross-sectional conductor area as AWG 10 solid wire. However, due to the additional space lost between the conductors, the stranded wire will have a larger overall diameter than the solid wire. Also, because the individual strands are coiled as a helix, the overall strand length will be slightly longer than the cable length.

Wire tables similar to Table 3–2 are available for stranded copper cables and for cables constructed of other materials (notably aluminum).

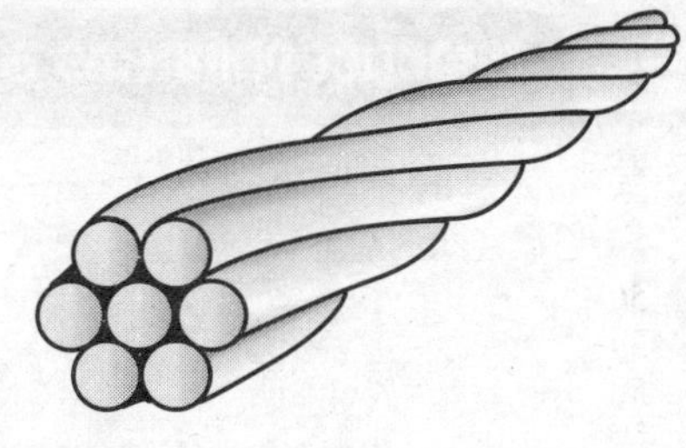

FIGURE 3–5 Stranded wire (7 strands).

EXAMPLE 3–3

Calculate the resistance of 200 feet of AWG 16 solid copper wire at 20° C.

Solution From Table 3–2, we see that AWG 16 wire has a resistance of 4.02 Ω per 1000 feet. Since we are given a length of only 200 feet, the resistance will be determined as

$$R = \left(\frac{4.02\ \Omega}{1000\ \text{ft}}\right)(200\ \text{ft}) = 0.804\ \Omega$$

By examining Table 3–2, several important points may be observed:

- If the wire size increases by three gauge sizes, the cross-sectional area will approximately double. Since resistance is inversely proportional to cross-sectional area, a given length of larger-diameter cable will have a resistance that is approximately half as large as the resistance of a similar length of the smaller-diameter cable.
- If there is a difference of three gauge sizes between cables, then the larger-diameter cable will be able to handle approximately twice as much current as the smaller-diameter cable. The amount of current that a conductor can safely handle is directly proportional to the cross-sectional area.
- If the wire size increases by ten gauge sizes, the cross-sectional area will increase by a factor of about ten. Due to the inverse relationship between resistance and cross-sectional area, the larger-diameter cable will have about one-tenth the resistance of a similar length of the smaller-diameter cable.
- For a 10-gauge difference in cable sizes, the larger-diameter cable will have ten times the cross-sectional area of the smaller-diameter cable and so it will be able to handle approximately ten times more current.

EXAMPLE 3–4

If AWG 14 solid copper wire is able to handle 15 A of current, determine the expected current capacity of AWG 24 and AWG 8 copper wire at 20° C.

Solution Since AWG 24 is ten sizes smaller than AWG 14, the smaller cable will be able to handle about one-tenth the capacity of the larger-diameter cable.

AWG 24 will be able to handle approximately 1.5 A of current.

AWG 8 is six sizes larger than AWG 14. Since current capacity doubles for an increase of three sizes, AWG 11 would be able to handle 30 A and AWG 8 will be able to handle 60 A.

PRACTICE PROBLEMS 2

1. From Table 3–2 find the diameters in millimeters and the cross-sectional areas in square millimeters of AWG 19 and AWG 30 solid wire.
2. By using the cross-sectional areas for AWG 19 and AWG 30, approximate the areas that AWG 16 and AWG 40 should have.
3. Compare the actual cross-sectional areas as listed in Table 3–2 to the areas found in Problems 1 and 2. (You will find a slight variation between your calculated values and the actual areas. This is because the actual diameters of the wires have been adjusted to provide optimum sizes for manufacturing.)

Answers

1. $d_{AWG19} = 0.912$ mm $A_{AWG19} = 0.653$ mm^2
   $d_{AWG30} = 0.254$ mm $A_{AWG30} = 0.0507$ mm^2
2. $A_{AWG16} \cong 1.31$ mm^2 $A_{AWG40} \cong 0.0051$ mm^2
3. $A_{AWG16} = 1.31$ mm^2 $A_{AWG40} = 0.00487$ mm^2

IN-PROCESS LEARNING CHECK 2

(Answers are at the end of the chapter.)

1. AWG 12-gauge wire is able to safely handle 20 amps of current. How much current should an AWG 2-gauge cable be able to handle?
2. The electrical code actually permits up to 120 A for the cable in Problem 1. How does the actual value compare to your theoretical value? Why do you think there is a difference?

3.3 Resistance of Wires—Circular Mils

The American Wire Gauge system for specifying wire diameters was developed using a unit called the circular mil (CM), which is defined as the area contained within a circle having a diameter of 1 mil (1 mil = 0.001 inch). A square mil is defined as the area contained in a square having side dimensions of 1 mil. From Figure 3–6, it is apparent that the area of a circular mil is smaller than the area of a square mil.

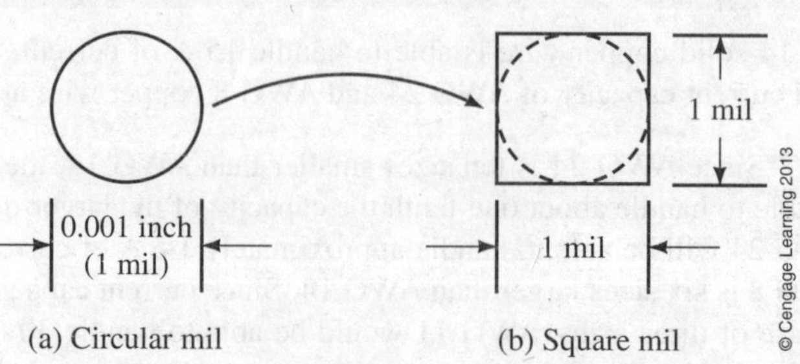

FIGURE 3–6

Because not all conductors have circular cross sections, it is occasionally necessary to convert areas expressed in square mils into circular mils. We will now determine the relationship between the circular mil and the square mil.

Suppose that a wire has the circular cross section shown in Figure 3–6(a). By applying Equation 3–2, the area, in square mils, of the circular cross section is determined as follows:

$$A = \frac{\pi d^2}{4}$$

$$= \frac{\pi(1\text{ mil})^2}{4}$$

$$= \frac{\pi}{4}\text{ sq. mil}$$

From the above derivation the following relations must apply:

$$1\text{ CM} = \frac{\pi}{4}\text{ sq. mil} \qquad (3\text{–}3)$$

$$1\text{ sq. mil} = \frac{4}{\pi}\text{ CM} \qquad (3\text{–}4)$$

The greatest advantage of using the circular mil to express areas of wires is the simplicity with which calculations may be made. Unlike previous area calculations which involved the use of π, area calculations may be reduced to simply finding the square of the diameter.

If we are given a circular cross section with a diameter d (in mils), the area of this cross section is determined as

$$A = \frac{\pi d^2}{4} \quad \text{[square mils]}$$

Using Equation 3–4, we convert the area from square mils to circular mils. Consequently, if the diameter of a circular conductor is given in mils, we determine the area in circular mils as

$$A_{CM} = d_{mil}^{\ 2} \quad \text{[circular mils, CM]} \qquad (3\text{–}5)$$

EXAMPLE 3–5

Determine the cross-sectional area in circular mils of a wire having the following diameters:

a. 0.0159 inch (AWG 26 wire)
b. 0.500 inch

Solution

a. $d = 0.0159$ inch
$= (0.0159\text{ inch})(1000\text{ mils/inch})$
$= 15.9$ mils

Now, using Equation 3–5, we obtain

$$A_{CM} = (15.9)^2 = 253\text{ CM}$$

From Table 3–2, we see that the previous result is precisely the area given for AWG 26 wire.

b. $d = 0.500$ inch
$= (0.500 \text{ inch})(1000 \text{ mils/inch})$
$= 500$ mils

$$A_{CM} = (500)^2 = 250\ 000 \text{ CM}$$

In Example 3–5(b) we see that the cross-sectional area of a cable may be a large number when it is expressed in circular mils. In order to simplify the units for area, the Roman numeral M is often used to represent 1000. If a wire has a cross-sectional area of 250 000 CM, it is more easily written as 250 MCM.

Clearly, this is a departure from the SI system, where M is used to represent one million. Since there is no simple way to overcome this conflict, the student working with cable areas expressed in MCM will need to remember that the M stands for one thousand and not for one million.

EXAMPLE 3–6

a. Determine the cross-sectional area in square mils and in circular mils of a copper bus bar having cross-sectional dimensions of 0.250 inch × 6.00 inch.
b. If this copper bus bar were to be replaced by AWG 2/0 cables, how many cables would be required?

Solution

a. $A_{\text{sq. mil}} = (250 \text{ mils})(6000 \text{ mils})$
$= 1\ 500\ 000$ sq. mils

The area in circular mils is found by applying Equation 3–4, and this will be

$$\begin{aligned} A_{CM} &= (250 \text{ mils})(6000 \text{ mils}) \\ &= (1\ 500\ 000 \text{ sq. mils})\left(\frac{4}{\pi} \text{ CM/sq. mil}\right) \\ &= 1\ 910\ 000 \text{ CM} \\ &= 1910 \text{ MCM} \end{aligned}$$

b. From Table 3–2, we see that AWG 2/0 cable has a cross-sectional area of 133.1 MCM (133 100 CM), and so the bus bar is equivalent to the following number of cables:

$$n = \frac{1910 \text{ MCM}}{133.1 \text{ MCM}} = 14.4$$

This example illustrates that 15 cables would need to be installed to be equivalent to a single 6-inch by 0.25-inch bus bar. Due to the expense and awkwardness of using this many cables, we see the economy of using solid bus bar. The main disadvantage of using bus bar is that the conductor is not covered with an insulation, and so the bus bar does not offer the same protection as cable. However, since bus bar is generally used in locations where only experienced technicians are permitted access, this disadvantage is a minor one.

As we have seen in Section 3.1, the resistance of a conductor was determined to be

$$R = \frac{\rho \ell}{A}$$

Although the original equation used SI units, the equation will also apply if the units are expressed in any other convenient system. If cable length is generally expressed in feet and the area in circular mils, then the resistivity must be expressed in the appropriate units. Table 3–3 gives the resistivities of some conductors represented in circular mil-ohms per foot.

The following example illustrates how Table 3–3 may be used to determine the resistance of a given section of wire.

TABLE 3–3 Resistivity of Conductors, ρ

Material	Resistivity, ρ, at 20° C (CM-Ω/ft)
Silver	9.90
Copper	10.36
Gold	14.7
Aluminum	17.0
Tungsten	33.0
Iron	74.0
Lead	132.
Mercury	576.
Nichrome	600.

EXAMPLE 3–7

Determine the resistance of an AWG 16 copper wire at 20° C if the wire has a diameter of 0.0508 inch and a length of 400 feet.

Solution The diameter in mils is found as

$$d = 0.0508 \text{ inch} = 50.8 \text{ mils}$$

Therefore the cross-sectional area (in circular mils) of AWG 16 is

$$A_{CM} = 50.8^2 = 2580 \text{ CM}$$

Now, by applying Equation 3–1 and using the appropriate units, we obtain the following:

$$R = \frac{\rho \ell}{A_{CM}}$$

$$= \frac{\left(10.36 \dfrac{\text{CM-}\Omega}{\text{ft}}\right)(400 \text{ ft})}{2580 \text{ CM}}$$

$$= 1.61 \ \Omega$$

PRACTICE PROBLEMS 3

1. Determine the resistance of 1 mile (5280 feet) of AWG 19 copper wire at 20° C, if the cross-sectional area is 1290 CM.
2. Compare the above result with the value that would be obtained by using the resistance (in ohms per thousand feet) given in Table 3–2.
3. An aluminum conductor having a cross-sectional area of 1843 MCM is used to transmit power from a high-voltage dc (HVDC) generating station to a large urban center. If the city is 900 km from the generating station, determine the resistance of the conductor at a temperature of 20° C. (Use 1 ft ≡ 0.3048 m.)

Answers

1. 42.4 Ω; 2. 42.5 Ω; 3. 27.2 Ω

IN-PROCESS LEARNING CHECK 3

(Answers are at the end of the chapter.)

A conductor has a cross-sectional area of 50 square mils. Determine the cross-sectional area in circular mils, square meters, and square millimeters.

3.4 Temperature Effects

Section 3.1 indicated that the resistance of a conductor will not be constant at all temperatures. As temperature increases, more electrons will escape their orbits, causing additional collisions within the conductor. For most conducting materials, the increase in the number of collisions translates into a relatively linear increase in resistance, as shown in Figure 3–7.

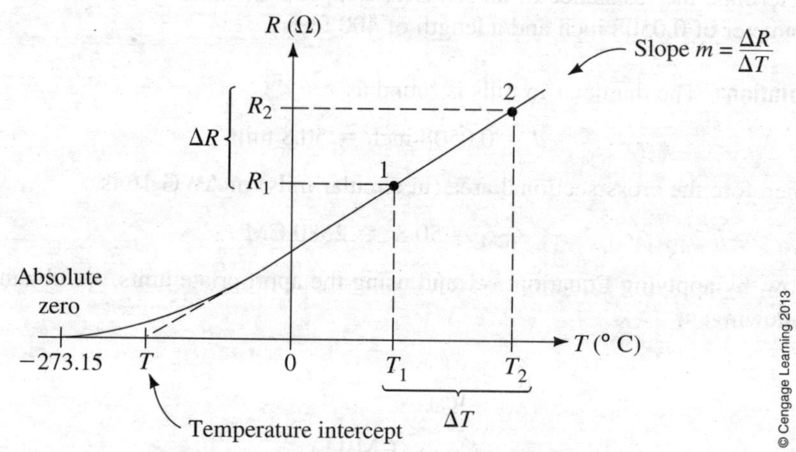

FIGURE 3–7 Temperature effects on resistance of a conductor.

The rate at which the resistance of a material changes with a variation in temperature is dependent on the **temperature coefficient** of the material, which is assigned the Greek letter alpha (α). Some materials have only very slight changes in resistance, while other materials demonstrate dramatic changes in resistance with a change in temperature.

Any material for which resistance increases as temperature increases is said to have a positive temperature coefficient.

For semiconductor materials such as carbon, germanium, and silicon, increases in temperature allow electrons to escape their usually stable orbits and become free to move within the material. Although additional collisions do occur within the semiconductor, the effect of the collisions is minimal when compared with the contribution of the extra electrons to the overall flow of charge. As the temperature increases, the number of charge electrons increases, resulting in more current. Therefore, an increase in temperature results in a decrease in resistance. Consequently, these materials are referred to as having negative temperature coefficients.

Table 3–4 gives the temperature coefficients, α per degree Celsius, of various materials at 20° C and at 0° C.

TABLE 3–4 Temperature Intercepts and Coefficients for Common Materials

	T (° C)	α (° C)$^{-1}$ at 20° C	α (° C)$^{-1}$ at 0° C
Silver	−243	0.003 8	0.004 12
Copper	−234.5	0.003 93	0.004 27
Aluminum	−236	0.003 91	0.004 24
Tungsten	−202	0.004 50	0.004 95
Iron	−162	0.005 5	0.006 18
Lead	−224	0.004 26	0.004 66
Nichrome	−2270	0.000 44	0.000 44
Brass	−480	0.002 00	0.002 08
Platinum	−310	0.003 03	0.003 23
Carbon		−0.000 5	
Germanium		−0.048	
Silicon		−0.075	

If we consider that Figure 3–7 illustrates how the resistance of copper changes with temperature, we observe an almost linear increase in resistance as the temperature increases. Further, we see that as the temperature is decreased to absolute zero ($T = -273.15°$ C), the resistance approaches zero.

In Figure 3–7, the point at which the linear portion of the line is extrapolated to cross the abscissa (temperature axis) is referred to as the temperature intercept or the inferred absolute temperature T of the material.

By examining the straight-line portion of the graph, we see that we have two similar triangles, one with the apex at point 1 and the other with the apex at point 2. The following relationship applies for these similar triangles.

$$\frac{R_2}{T_2 - T} = \frac{R_1}{T_1 - T}$$

This expression may be rewritten to solve for the resistance R_2 at any temperature T_2 as follows:

$$R_2 = \frac{T_2 - T}{T_1 - T} R_1 \qquad \text{(3–6)}$$

An alternate method of determining the resistance R_2 of a conductor at a temperature T_2 is to use the temperature coefficient α of the material. Examining Table 3–4, we see that the temperature coefficient is not a constant for all temperatures, but rather is dependent upon the temperature of the material. The temperature coefficient for any material is defined as

$$\alpha = \frac{m}{R_1} \qquad \text{(3–7)}$$

The value of α is typically given in chemical handbooks. In the preceding expression, α is measured in (° C)$^{-1}$, R_1 is the resistance in ohms at a temperature T_1, and m is the slope of the linear portion of the curve ($m = \Delta R/\Delta T$). It is left as an end-of-chapter problem for the student to use Equations 3–6 and 3–7 to derive the following expression from Figure 3–6.

$$R_2 = R_1[1 + \alpha_1(T_2 - T_1)] \qquad \text{(3–8)}$$

EXAMPLE 3–8

An aluminum wire has a resistance of 20 Ω at room temperature (20° C). Calculate the resistance of the same wire at temperatures of −40° C, 100° C, and 200° C.

Solution From Table 3–4, we see that aluminum has a temperature intercept of –236° C.

At $T = -40°$ C:
The resistance at −40° C is determined using Equation 3–6.

$$R_{-40° \text{ C}} = \left[\frac{-40° \text{ C} - (-236° \text{ C})}{20° \text{ C} - (-236° \text{ C})}\right] 20\ \Omega = \left(\frac{196° \text{ C}}{256° \text{ C}}\right) 20\ \Omega = 15.3\ \Omega$$

At $T = 100°$ C:

$$R_{100° \text{ C}} = \left[\frac{100° \text{ C} - (-236° \text{ C})}{20° \text{ C} - (-236° \text{ C})}\right] 20\ \Omega = \left(\frac{336° \text{ C}}{256° \text{ C}}\right) 20\ \Omega = 26.3\ \Omega$$

At $T = 200°$ C:

$$R_{200° \text{ C}} = \left[\frac{200° \text{ C} - (-236° \text{ C})}{20° \text{ C} - (-236° \text{ C})}\right] 20\ \Omega = \left(\frac{436° \text{ C}}{256° \text{ C}}\right) 20\ \Omega = 34.1\ \Omega$$

This phenomenon indicates that the resistance of conductors changes quite dramatically with changes in temperature. For this reason manufacturers generally specify the range of temperatures over which a conductor may operate safely.

EXAMPLE 3–9

Tungsten wire is used as filaments in incandescent light bulbs. Current in the wire causes the wire to reach extremely high temperatures. Determine the temperature of the filament of a 100-W light bulb if the resistance at room temperature is measured to be 11.7 Ω and when the light is on, the resistance is determined to be 144 Ω.

Solution If we rewrite Equation 3–6, we are able to solve for the temperature T_2 as follows:

$$\begin{aligned} T_2 &= (T_1 - T)\frac{R_2}{R_1} + T \\ &= [20° \text{ C} - (-202° \text{ C})]\frac{144\ \Omega}{11.7\ \Omega} + (-202° \text{ C}) \\ &= 2530° \text{ C} \end{aligned}$$

PRACTICE PROBLEMS 4

An HVDC (high-voltage dc) transmission line must be able to operate over a wide temperature range. Calculate the resistance of 900 km of 1843 MCM aluminum conductor at temperatures of −40° C and +40° C.

Answers
20.8 Ω, 29.3 Ω

IN-PROCESS LEARNING CHECK 4

(Answers are at the end of the chapter.)

Explain what is meant by the terms *positive temperature coefficient* and *negative temperature coefficient.* To which category does aluminum belong?

3.5 Types of Resistors

Virtually all electric and electronic circuits involve the control of voltage and/or current. The best way to provide such control is by inserting appropriate values of resistance into the circuit. Although various types and sizes of resistors are used in electrical and electronic applications, all resistors fall into two main categories: fixed resistors and variable resistors.

Fixed Resistors

As the name implies, fixed resistors are resistors having resistance values that are essentially constant. There are numerous types of fixed resistors, ranging in size from almost microscopic (as in integrated circuits) to high-power resistors that are capable of dissipating many watts of power. Figure 3–8 illustrates the basic structure of a molded carbon composition resistor.

As shown in Figure 3–8, the molded carbon composition resistor consists of a carbon core mixed with an insulating filler. The ratio of carbon to filler determines the resistance value of the component: the higher the proportion of carbon, the lower the resistance. Metal leads are inserted into the carbon core, and then the entire resistor is encapsulated with an insulated coating. Carbon composition resistors are available in resistances from less than 1 Ω to 100 MΩ and typically have power ratings from ⅛ W to 2 W. Figure 3–9 shows various sizes of resistors, with the larger resistors being able to dissipate more power than the smaller resistors.

Although carbon-core resistors have the advantages of being inexpensive and easy to produce, they tend to have wide tolerances and are susceptible to large changes in resistance due to temperature variation. As shown in Figure 3–10, the resistance of a carbon composition resistor may change by as much as 5% when temperature is changed by 100° C.

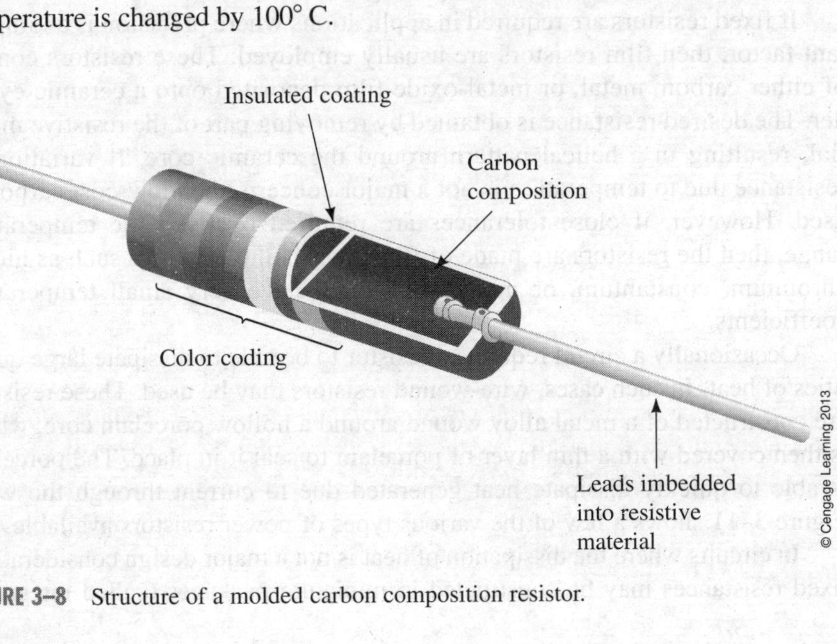

FIGURE 3–8 Structure of a molded carbon composition resistor.

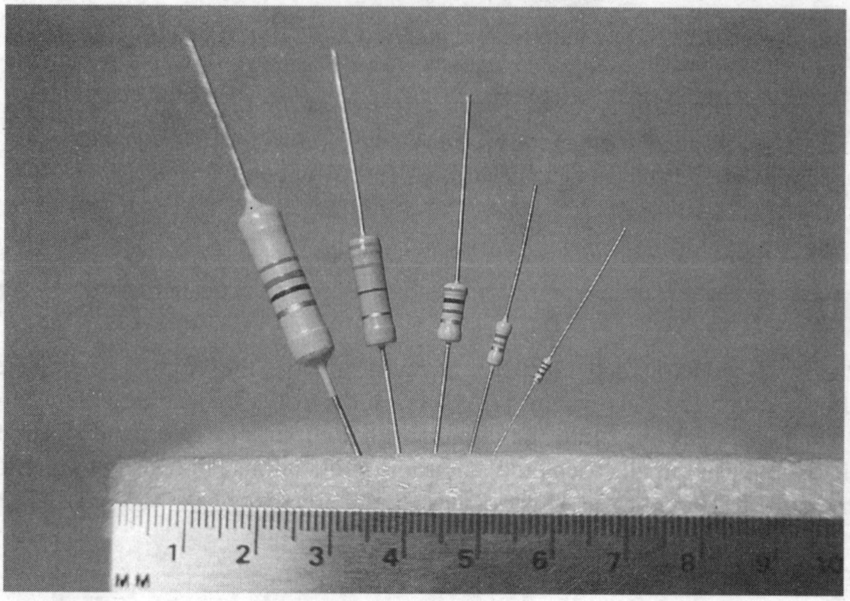

FIGURE 3–9 Actual size of carbon resistors (2 W, 1 W, ½ W, ¼ W, ⅛ W).

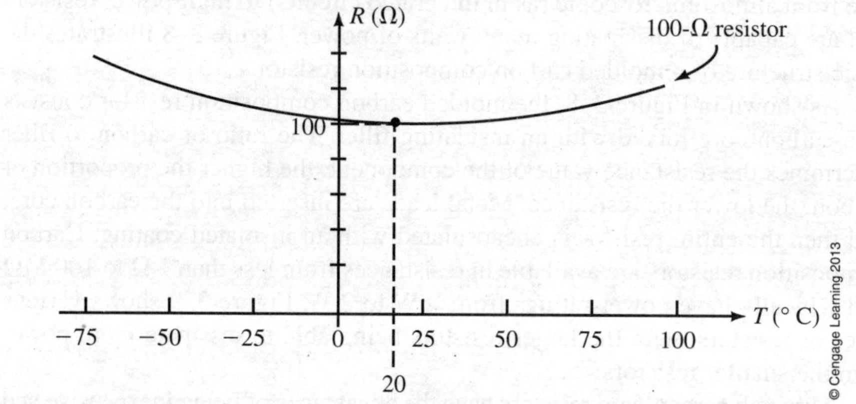

FIGURE 3–10 Variation in resistance of a carbon composition fixed resistor.

Other types of fixed resistors include carbon film, metal film, metal oxide, wire-wound, and integrated circuit packages.

If fixed resistors are required in applications where precision is an important factor, then film resistors are usually employed. These resistors consist of either carbon, metal, or metal-oxide film deposited onto a ceramic cylinder. The desired resistance is obtained by removing part of the resistive material, resulting in a helical pattern around the ceramic core. If variation of resistance due to temperature is not a major concern, then low-cost carbon is used. However, if close tolerances are required over a wide temperature range, then the resistors are made of films consisting of alloys such as nickel chromium, constantum, or manganin, which have very small temperature coefficients.

Occasionally a circuit requires a resistor to be able to dissipate large quantities of heat. In such cases, wire-wound resistors may be used. These resistors are constructed of a metal alloy wound around a hollow porcelain core, which is then covered with a thin layer of porcelain to seal it in place. The porcelain is able to quickly dissipate heat generated due to current through the wire. Figure 3–11 shows a few of the various types of power resistors available.

In circuits where the dissipation of heat is not a major design consideration, fixed resistances may be constructed in miniature packages (called integrated

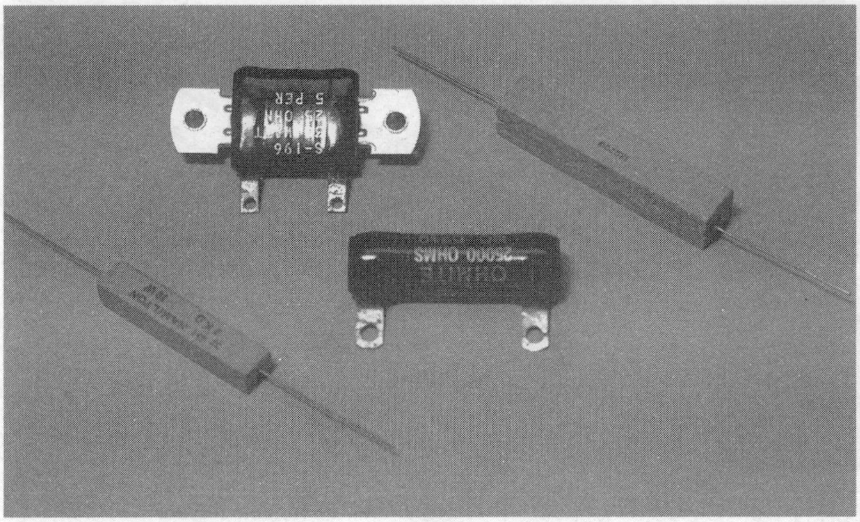

© Cengage Learning 2013

FIGURE 3–11 Power resistors.

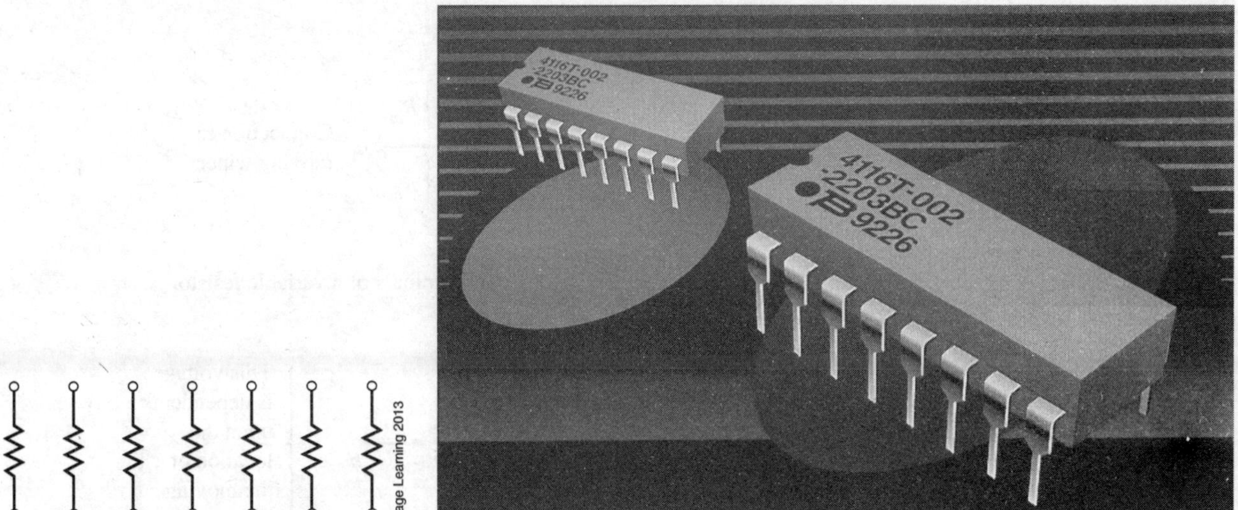

© Cengage Learning 2013

(a) Internal resistor arrangement

(b) Integrated resistor network

Courtesy of Bourns, Inc.

FIGURE 3–12

circuits or ICs) capable of containing many individual resistors. The obvious advantage of such packages is their ability to conserve space on a circuit board. Figure 3–12 illustrates a typical resistor IC package.

Variable Resistors

Variable resistors provide indispensable functions that we use in one form or another almost daily. These components are used to adjust the volume of our radios, set the level of lighting in our homes, and adjust the heat of our stoves and furnaces. Figure 3–13 shows the internal and the external view of typical variable resistors.

In Figure 3–14, we see that variable resistors have three terminals, two of which are fixed to the ends of the resistive material. The central terminal is connected to a wiper that moves over the resistive material when the shaft is rotated with either a knob or a screwdriver. The resistance between the two outermost terminals will remain constant, while the resistance between the

(a) External view of variable resistors

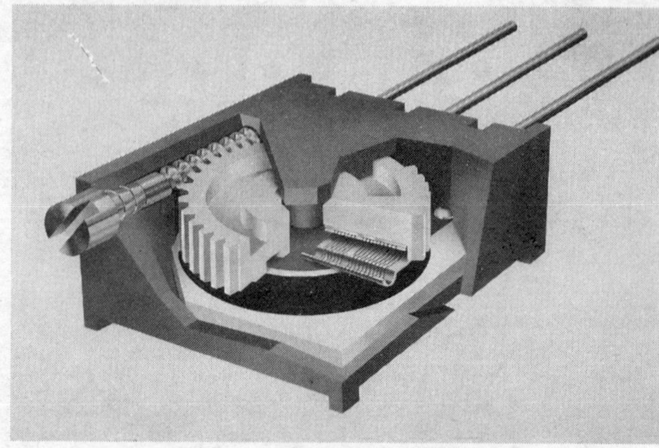

(b) Internal view of variable resistor

FIGURE 3–13 Variable resistors.

(a) Variable resistors

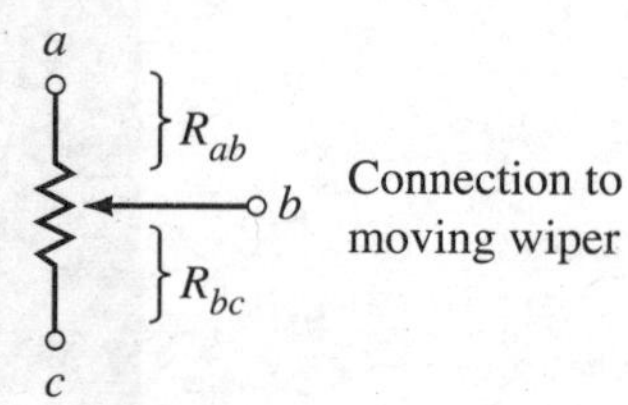

(b) Terminals of a variable resistor

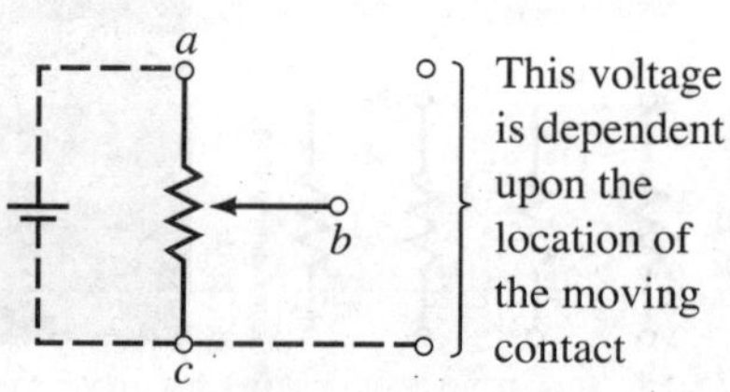

(c) Variable resistor used as a potentiometer

FIGURE 3–14

central terminal and either terminal will change according to the position of the wiper.

If we examine the schematic of a variable resistor as shown in Figure 3–14(b), we see that the following relationship must apply:

$$R_{ac} = R_{ab} + R_{bc} \tag{3–9}$$

Variable resistors are used for two principal functions. Potentiometers, shown in Figure 3–14(c), are used to adjust the amount of potential (voltage) provided to a circuit. Rheostats, the connections and schematic of which are shown in Figure 3–15, are used to adjust the amount of current within a circuit. Applications of potentiometers and rheostats will be covered in later chapters.

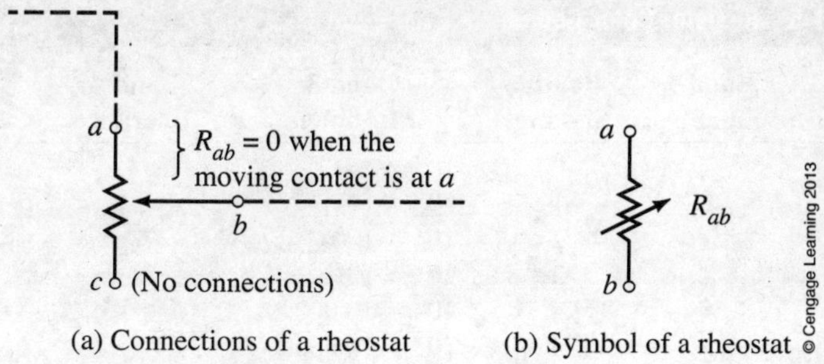

FIGURE 3–15

3.6 Color Coding of Resistors

Large resistors such as the wire-wound resistors or the ceramic-encased power resistors have their resistor values and tolerances printed on their cases. Smaller resistors, whether constructed of a molded carbon composition or a metal film, may be too small to have their values printed on the component. Instead, these smaller resistors are usually covered by an epoxy or similar insulating coating over which several colored bands are printed radially as shown in Figure 3–16.

The colored bands provide a quickly recognizable code for determining the value of resistance, the tolerance (in percentage), and occasionally the expected reliability of the resistor. The colored bands are always read from left to right, left being defined as the side of the resistor with the band nearest to it.

The first two bands represent the first and second digits of the resistance value. The third band is called the multiplier band and represents the number of zeros following the first two digits; it is usually given as a power of ten. The fourth band indicates the tolerance of the resistor, and the fifth band (if present) is an indication of the expected reliability of the component. The reliability is a statistical indication of the expected number of components that will no longer have the indicated resistance value after 1000 hours of use. For example, if a particular resistor has a reliability of 1%, it is expected that after 1000 hours of use, no more than 1 resistor in 100 is likely to be outside the specified range of resistance as indicated in the first four bands of the color codes. Table 3–5 shows the colors of the various bands and the corresponding values.

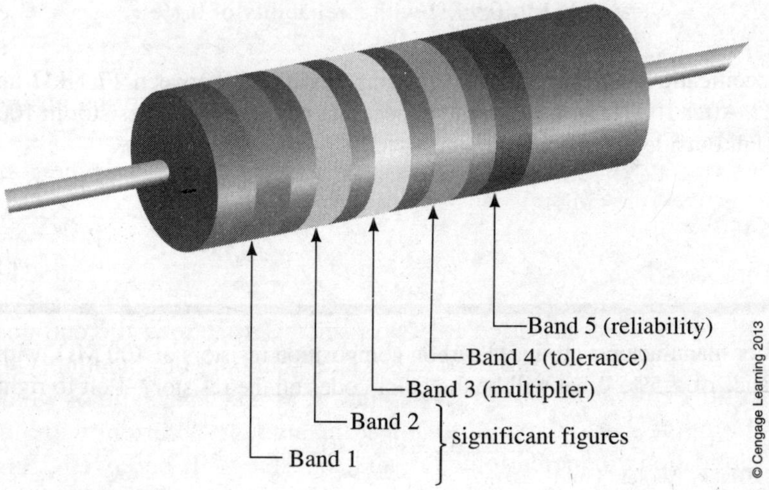

FIGURE 3–16 Resistor color codes.

TABLE 3–5 Resistor Color Codes

Color	Band 1 Sig. Fig.	Band 2 Sig. Fig.	Band 3 Multiplier	Band 4 Tolerance	Band 5 Reliability
Black		0	$10^0 = 1$		
Brown	1	1	$10^1 = 10$		1%
Red	2	2	$10^2 = 100$		0.1%
Orange	3	3	$10^3 = 1\ 000$		0.01%
Yellow	4	4	$10^4 = 10\ 000$		0.001%
Green	5	5	$10^5 = 100\ 000$		
Blue	6	6	$10^6 = 1\ 000\ 000$		
Violet	7	7	$10^7 = 10\ 000\ 000$		
Gray	8	8			
White	9	9			
Gold			0.1	5%	
Silver			0.01	10%	
No color				20%	

EXAMPLE 3–10

Determine the resistance of a carbon film resistor having the color codes shown in Figure 3–17.

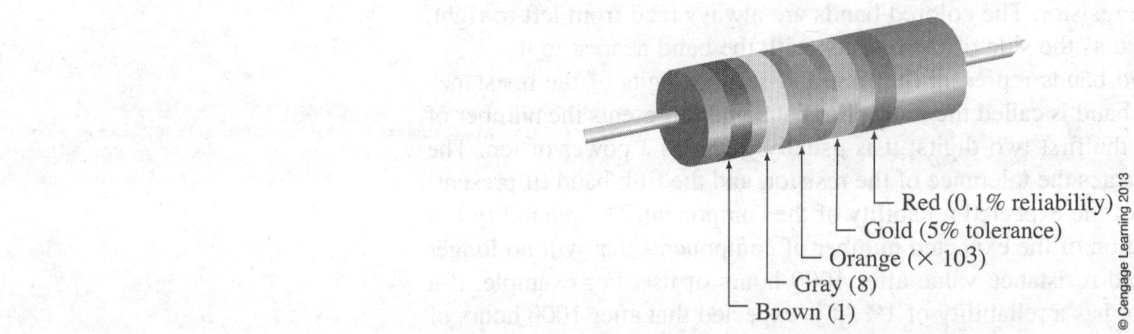

FIGURE 3–17

Solution From Table 3–5, we see that the resistor will have a value determined as

$$R = 18 \times 10^3\ \Omega \pm 5\%$$
$$= 18\ \text{k}\Omega \pm 0.9\ \text{k}\Omega \text{ with a reliability of } 0.1\%$$

This specification indicates that the resistance will fall between 17.1 kΩ and 18.9 kΩ. After 1000 hours, we would expect that no more than 1 resistor in 1000 would fall outside the specified range.

PRACTICE PROBLEMS 5

A resistor manufacturer produces carbon composition resistors of 100 MΩ, with a tolerance of ±5%. What will be the color codes on the resistor? (Left to right)

Answer
Brown, Black, Violet, Gold

3.7 Measuring Resistance—The Ohmmeter

The **ohmmeter** is an instrument that is generally part of a multimeter (usually including a voltmeter and an ammeter) and is used to measure the resistance of a component. Although it has limitations, the ohmmeter is used almost daily in service shops and laboratories to measure resistance of components and also to determine whether a circuit is faulty. In addition, the ohmmeter may also be used to determine the condition of semiconductor devices such as diodes and transistors. Figure 3–18 shows a typical digital multimeter with an ohms function.

In order to measure the resistance of an isolated component or circuit, the ohmmeter is placed across the component under test, as shown in Figure 3–19. The resistance is then simply read from the meter display.

When using an ohmmeter to measure the resistance of a component that is located in an operating circuit, the following steps should be observed:

1. As shown in Figure 3–20(a), remove all power supplies from the circuit or component to be tested. If this step is not followed, the ohmmeter reading will, at best, be meaningless, and the ohmmeter may be severely damaged.
2. If you wish to measure the resistance of a particular component, it is necessary to isolate the component from the rest of the circuit. This is done by disconnecting at least one terminal of the component from the balance of the circuit as shown in Figure 3–20(b). If this step is not followed, in all likelihood the resistance reading indicated by the ohmmeter will not be the resistance of the desired resistor, but rather the resistance of the combination.

Courtesy of Fluke Corporation

FIGURE 3–18 Digital multimeter. To measure ohms, set the dial to Ω.

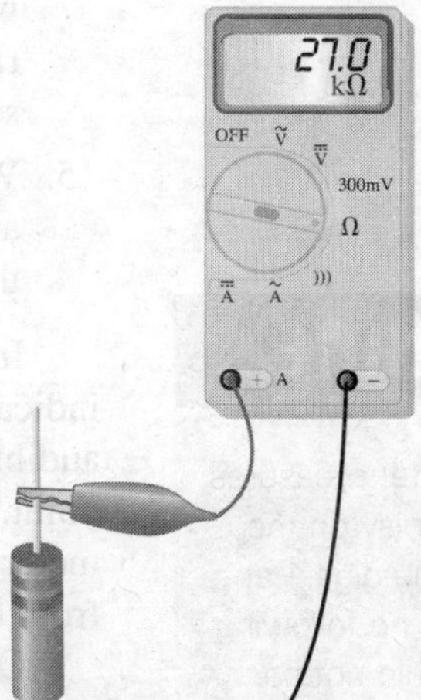

© Cengage Learning 2013

FIGURE 3–19 Ohmmeter used to measure an isolated component.

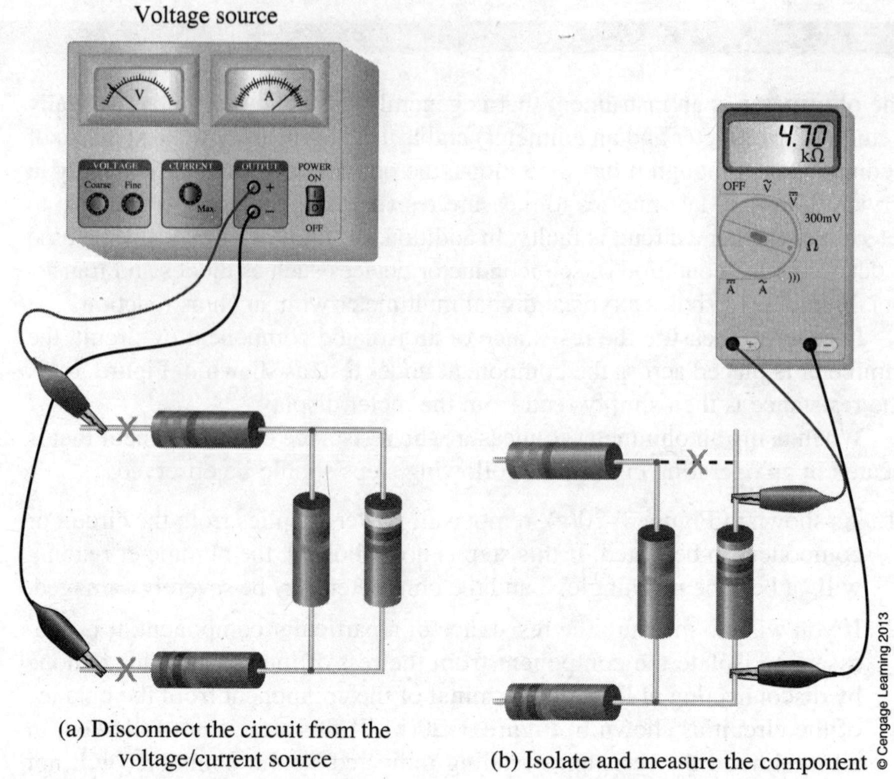

FIGURE 3–20 Using an ohmmeter to measure resistance in a circuit.

3. As shown in Figure 3–20(b), connect the two probes of the ohmmeter across the component to be measured. The black and red leads of the ohmmeter may be interchanged when measuring resistors. When measuring resistance of other components, however, the measured resistance will be dependent upon the direction of the sensing current. Such devices are covered briefly in a later section of this chapter.
4. Ensure that the ohmmeter is on the correct range to provide the most accurate reading. For example, although a digital multimeter (DMM) can measure a reading for a 1.2-kΩ resistor on the 2-MΩ range, the same ohmmeter will provide additional significant digits (hence more precision) when it is switched to the 2-kΩ range. For analog meters, the best accuracy is obtained when the needle is approximately in the center of the scale.
5. When you are finished, turn the ohmmeter off. Because the ohmmeter uses an internal battery to provide a small sensing current, it is possible to drain the battery if the probes accidently connect together for an extended period.

In addition to measuring resistance, the ohmmeter may also be used to indicate the continuity of a circuit. Many modern digital ohmmeters have an audible tone to indicate that a circuit is unbroken from one point to another point. As demonstrated in Figure 3–21(a), the audible tone of a digital ohmmeter allows the user to determine continuity without having to look away from the circuit under test.

Ohmmeters are particularly useful instruments in determining whether a given circuit has been short circuited or open circuited.

A **short circuit** occurs when a low-resistance conductor such as a piece of wire or any other conductor is connected between two points in a circuit. Due to the very low resistance of the short circuit, current will bypass the rest of the

PRACTICAL NOTE...

When a digital ohmmeter measures an open circuit, the display on the meter will usually be the digit 1 at the left-hand side, with no following digits. This reading should not be confused with a reading of 1 Ω, 1 kΩ, or 1 MΩ, which would appear on the right-hand side of the display.

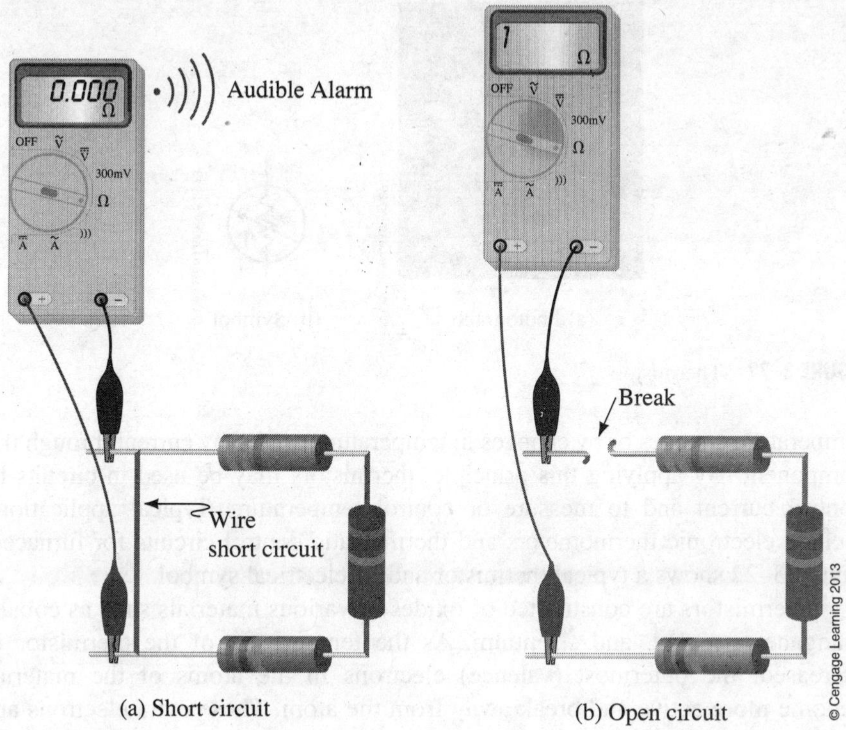

FIGURE 3–21

circuit and go through the short. An ohmmeter will indicate a very low (theoretically zero) resistance when used to measure across a short circuit.

An **open circuit** occurs when a conductor is broken between the points under test. An ohmmeter will indicate infinite resistance when used to measure the resistance of a circuit having an open circuit.

Figure 3–21 illustrates circuits having a short circuit and an open circuit.

PRACTICE PROBLEMS 6

An ohmmeter is used to measure across the terminals of a switch.

a. What will the ohmmeter indicate when the switch is closed?

b. What will the ohmmeter indicate when the switch is opened?

Answers

a. 0 Ω (short circuit)

b. ∞ (open circuit)

3.8 Thermistors

In Section 3.4 we saw how resistance changes with changes in temperature. While this effect is generally undesirable in resistors, there are many applications that use electronic components having characteristics that vary according to changes in temperature. Any device or component that causes an electrical change due to a physical change is referred to as a transducer.

A **thermistor** is a two-terminal transducer in which resistance changes significantly with changes in temperature (hence a thermistor is a "thermal resistor"). The resistance of thermistors may be changed either by external

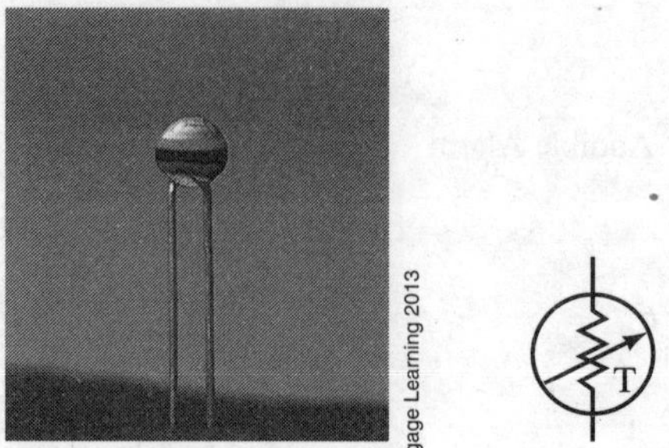

(a) Photograph (b) Symbol

FIGURE 3–22 Thermistors.

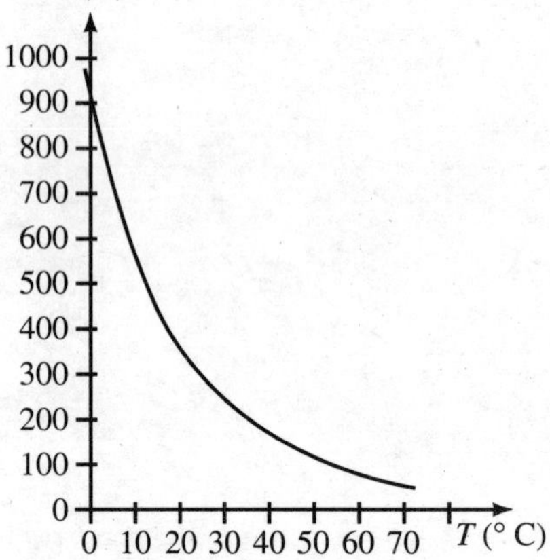

FIGURE 3–23 Thermistor resistance as a function of temperature.

temperature changes or by changes in temperature caused by current through the component. By applying this principle, thermistors may be used in circuits to control current and to measure or control temperature. Typical applications include electronic thermometers and thermostatic control circuits for furnaces. Figure 3–22 shows a typical thermistor and its electrical symbol.

Thermistors are constructed of oxides of various materials such as cobalt, manganese, nickel, and strontium. As the temperature of the thermistor is increased, the outermost (valence) electrons in the atoms of the material become more active and break away from the atom. These extra electrons are now free to move within the circuit, thereby causing a reduction in the resistance of the component (negative temperature coefficient). Figure 3–23 shows how resistance of a thermistor varies with temperature effects.

PRACTICE PROBLEMS 7

Referring to Figure 3–22, determine the approximate resistance of a thermistor at each of the following temperatures:

a. 10° C
b. 30° C
c. 50° C

Answers
a. 550 Ω; b. 250 Ω; c. 120 Ω

3.9 Photoconductive Cells

Photoconductive cells or **photocells** are two-terminal transducers that have a resistance determined by the amount of light falling on the cell. Most photocells are constructed of either cadmium sulfide (CdS) or cadmium selenide (CdSe) and are sensitive to light having wavelengths between 4000 Å (blue light) and 10 000 Å (infrared). The angstrom (Å) is a unit commonly used to measure the wavelength of light and has a dimension given as $1 \text{ Å} = 1 \times 10^{-10}$ m. Light, which is a form of energy, strikes the material of the photocell and causes the release of valence electrons, thereby reducing the resistance of the component. Figure 3–24 shows the structure, symbol, and resistance characteristics of a typical photocell.

Photocells may be used to measure light intensity and/or to control lighting. They are typically used as part of a security system.

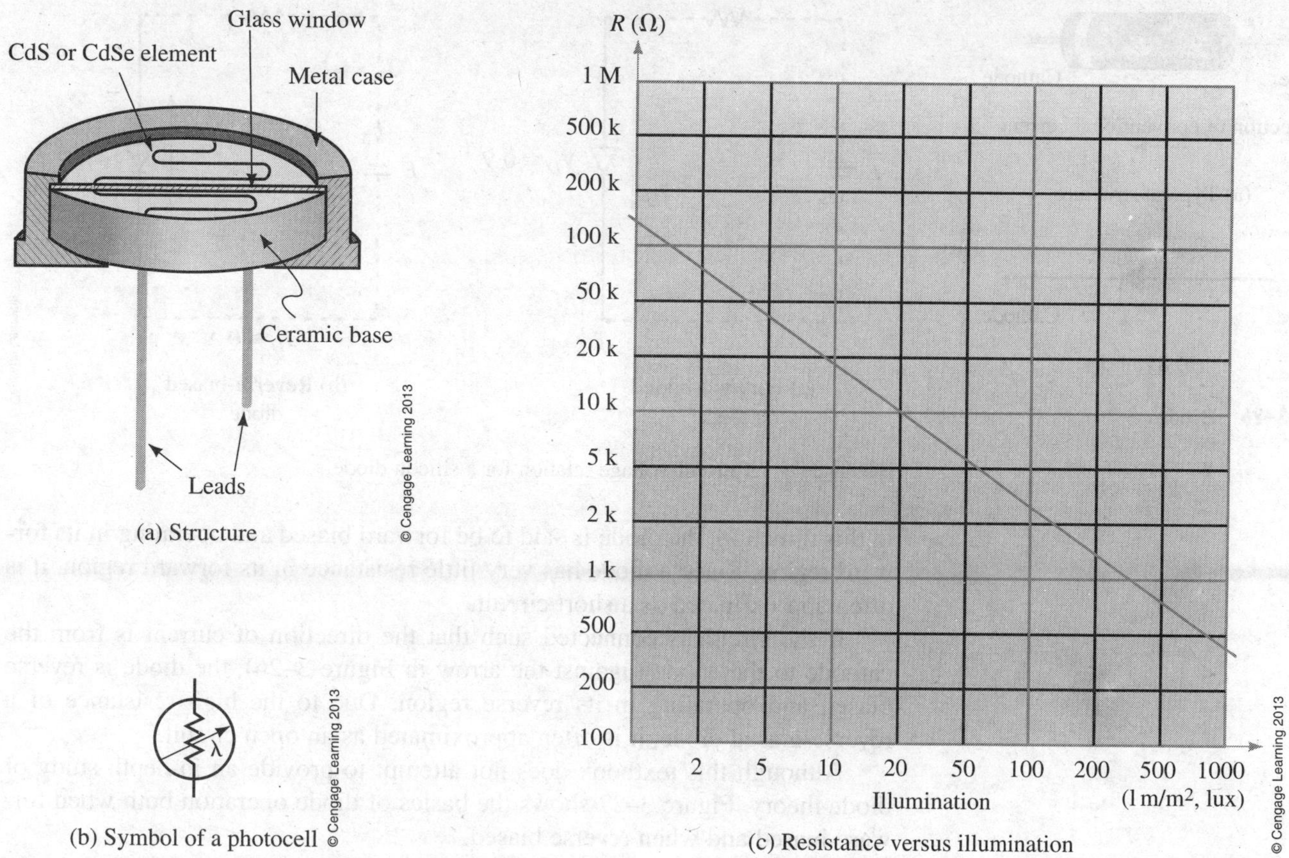

FIGURE 3–24 Photocell.

3.10 Nonlinear Resistance

Up to this point, the components we have examined have had values of resistance that were essentially constant for a given temperature (or, in the case of a photocell, for a given amount of light). If we were to examine the current versus voltage relationship for these components, we would find that the relationship is linear, as shown in Figure 3–25.

If a device has a linear (straight-line) current-voltage relation then it is referred to as an ohmic device. (The linear current-voltage relationship will be will be covered in greater detail in the next chapter.) Often in electronics, we use components that do not have a linear current-voltage relationship; these devices are referred to as nonohmic devices. On the other hand, some components, such as the thermistor, can be shown to have both an ohmic region and a nonohmic region. For large current through the thermistor, the component will get hotter. This increase in temperature will result in a decrease of resistance. Consequently, for large currents, the thermistor is a nonohmic device.

We will now briefly examine two common nonohmic devices.

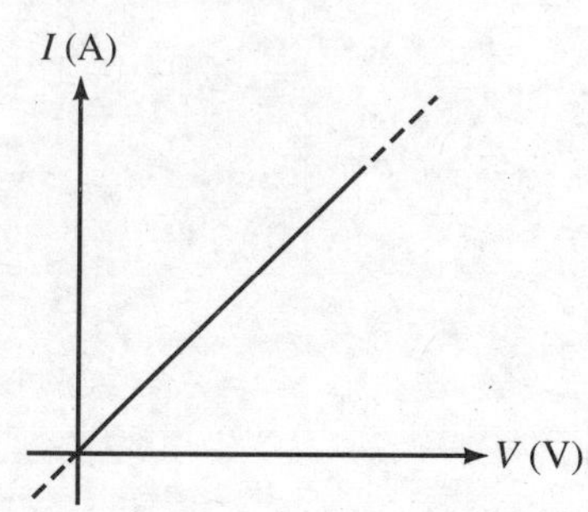

FIGURE 3–25 Linear current-voltage relationship.

Diodes

The **diode** is a semiconductor device that permits charge to flow in only one direction. Figure 3–26 illustrates the appearance and the symbol of a typical diode.

Conventional current through a diode is in the direction from the anode toward the cathode (the end with the line around the circumference). When current is

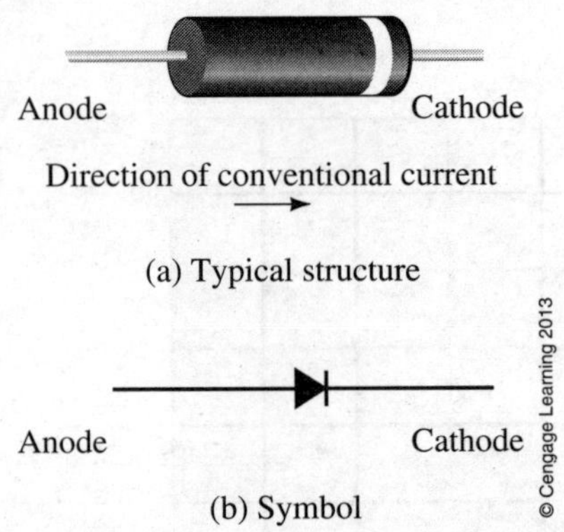

FIGURE 3–26 Diode.

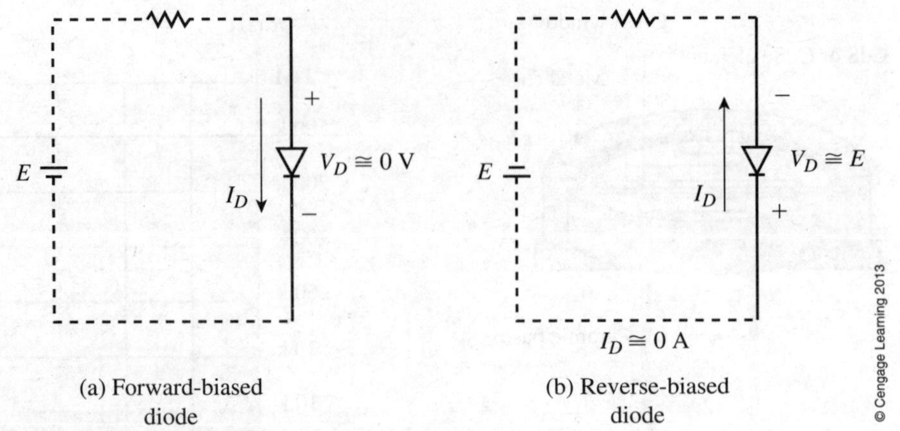

FIGURE 3–27 Current-voltage relation for a silicon diode.

in this direction, the diode is said to be forward biased and operating in its forward region. Since a diode has very little resistance in its forward region, it is often approximated as a short circuit.

If the circuit is connected such that the direction of current is from the cathode to the anode (against the arrow in Figure 3–26), the diode is reverse biased and operating in its reverse region. Due to the high resistance of a reverse-biased diode, it is often approximated as an open circuit.

Although this textbook does not attempt to provide an in-depth study of diode theory, Figure 3–27 shows the basics of diode operation both when forward biased and when reverse biased.

PRACTICAL NOTES...

Because an ohmmeter uses an internal voltage source to generate a small sensing current, the instrument may easily be used to determine the terminals (and hence the direction of conventional flow) of a diode. (See Figure 3–28.)

FIGURE 3–28 Determining diode terminals with an ohmmeter.

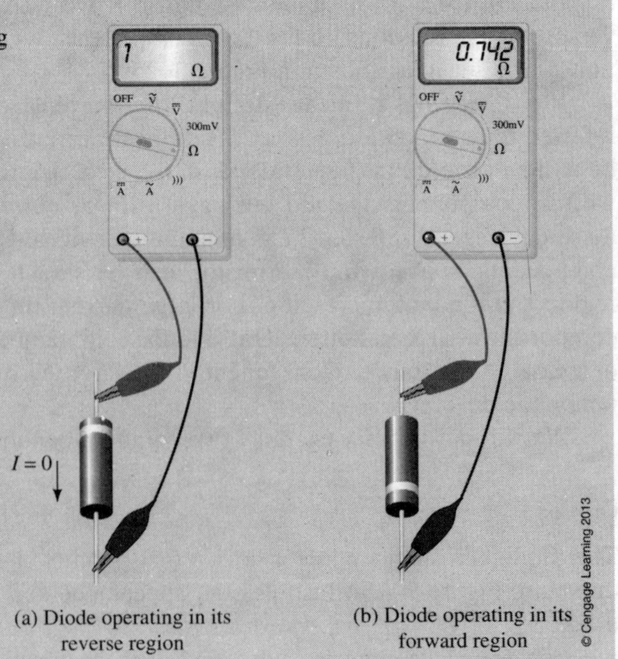

If we measure the resistance of the diode in both directions, we will find that the resistance will be low when the positive terminal of the ohmmeter is connected to the anode of the diode. When the positive terminal is connected to the cathode, virtually no current will occur in the diode and so the indication on the ohmmeter will be infinite resistance normally shown by the numeral 1 on the left.

Varistors

Varistors, as shown in Figure 3–29, are semiconductor devices that have very high resistances when the voltage across the varistors is below the breakdown value. However, when the voltage across a varistor (either polarity) exceeds the rated value, the resistance of the device suddenly becomes very small, allowing charge to flow. Figure 3–30 shows the current-voltage relation for varistors.

Varistors are used in sensitive circuits, such as those in computers, to ensure that if the voltage suddenly exceeds a predetermined value, the varistor will effectively become a short circuit to the unwanted signal, thereby protecting the rest of the circuit from excessive voltage.

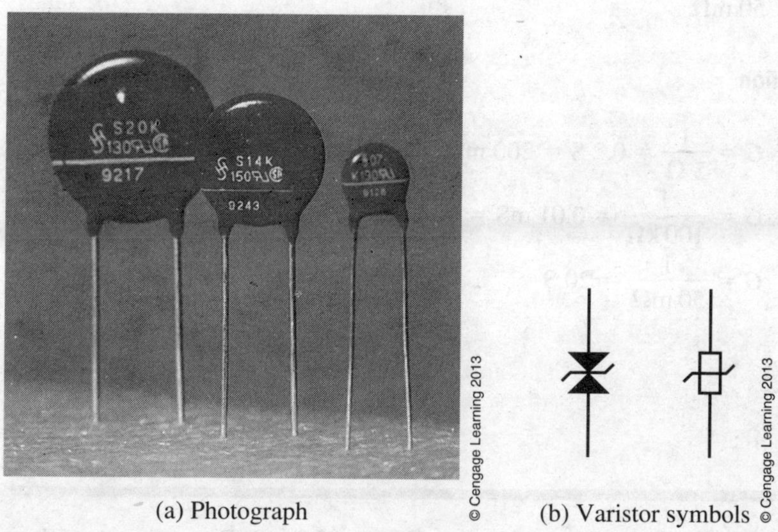

(a) Photograph (b) Varistor symbols

FIGURE 3–29 Varistors.

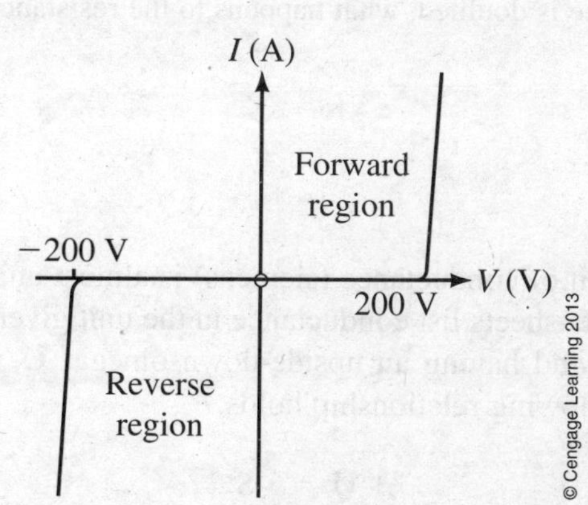

FIGURE 3–30 Current-voltage relation of a 200-V (peak) varistor.

3.11 Conductance

Conductance, *G*, is defined as the measure of a material's ability to allow the flow of charge and is assigned the SI unit the siemens (S). A large conductance indicates that a material is able to conduct current well, whereas a low value of conductance indicates that a material does not readily permit the flow of charge. Mathematically, conductance is defined as the reciprocal of resistance. Thus

$$G = \frac{1}{R} \quad \text{[siemens, S]} \tag{3–10}$$

where R is resistance, in ohms (Ω).

EXAMPLE 3–11

Determine the conductance of the following resistors:

a. 5 Ω

b. 100 kΩ

c. 50 mΩ

Solution

a. $G = \dfrac{1}{5\ \Omega} = 0.2\ \text{S} = 200\ \text{mS}$

b. $G = \dfrac{1}{100\ \text{k}\Omega} = 0.01\ \text{mS} = 10\ \mu\text{S}$

c. $G = \dfrac{1}{50\ \text{m}\Omega} = 20\ \text{S}$

PRACTICE PROBLEMS 8

1. A given cable has a conductance given as 5.0 mS. Determine the value of the resistance, in ohms.

2. If the conductance is doubled, what happens to the resistance?

Answers
1. 200 Ω; 2. It halves

Although the SI unit of conductance (siemens) is almost universally accepted, older books and data sheets list conductance in the unit given as the mho (ohm spelled backward) and having an upside-down omega, ℧, as the symbol. In such a case, the following relationship holds:

$$1\ ℧ = 1\ \text{S} \tag{3–11}$$

PRACTICE PROBLEMS 9

A specification sheet for a radar transmitter indicates that one of the components has a conductance of 5 $\mu\mu\mho$.

a. Express the conductance in the proper SI prefix and unit.

b. Determine the resistance of the component, in ohms.

Answers

a. 5 pS; b. 2×10^{11} Ω

3.12 Superconductors

As you have seen, all power lines and distribution networks have internal resistance, which results in energy loss due to heat as charge flows through the conductor. If there was some way of eliminating the resistance of the conductors, electricity could be transmitted farther and more economically. The idea that energy could be transmitted without losses along a "**superconductor**" transmission line was formerly a distant goal. However, recent discoveries in high-temperature superconductivity promise the almost magical ability to transmit and store energy with no loss in energy.

In 1911, the Dutch physicist Heike Kamerlingh Onnes discovered the phenomenon of superconductivity. Studies of mercury, tin, and lead verified that the resistance of these materials decreases to no more than one ten-billionth of the room temperature resistance when subjected to temperatures of 4.6 K, 3.7 K, and 6 K, respectively. Recall that the relationship between kelvins and degrees Celsius is as follows:

$$T_K = T_{(°C)} + 273.15° \qquad \textbf{(3–12)}$$

The temperature at which a material becomes a superconductor is referred to as the critical temperature, T_C, of the material. Figure 3–31 shows how the resistance of a sample of mercury changes with temperature. Notice how the resistance suddenly drops to zero at a temperature of 4.6 K.

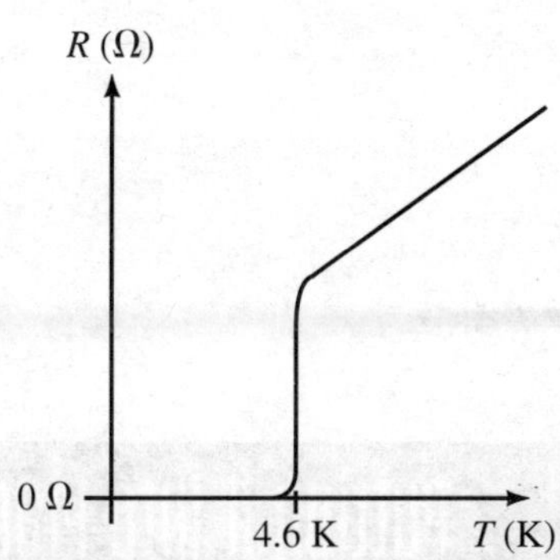

FIGURE 3–31 Critical temperature of mercury.

Experiments with currents in supercooled loops of superconducting wire have determined that the induced currents will remain undiminished for many years within the conductor provided that the temperature is maintained below the critical temperature of the conductor.

A peculiar, seemingly magical property of superconductors occurs when a permanent magnet is placed above the superconductor. The magnet will float above the surface of the conductor as if it is defying the law of gravity, as shown in Figure 3–32.

FIGURE 3–32 The Meissner effect: A magnetic cube hovers above a disk of ceramic superconductor. The disk is kept below its critical temperature in a bath of liquid nitrogen.

This principle, which is referred to as the *Meissner effect* (named after Walther Meissner), may be simply stated as follows:

> When a superconductor is cooled below its critical temperature, magnetic fields may surround but not enter the superconductor.

The principle of superconductivity is explained in the behavior of electrons within the superconductor. Unlike conductors, which have electrons moving randomly through the conductor and colliding with other electrons [Figure 3–33(a)], the electrons in superconductors form pairs which move

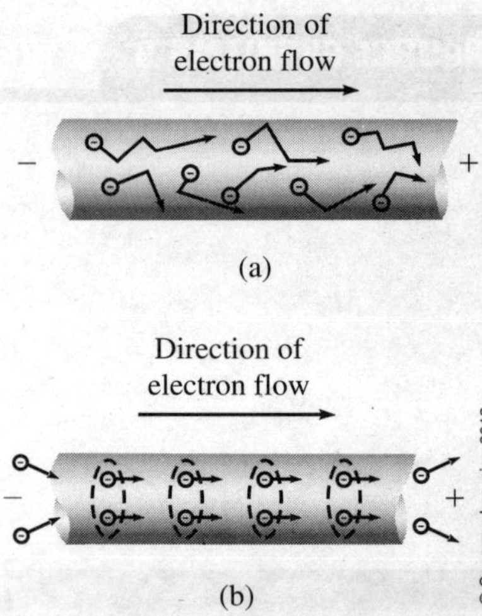

FIGURE 3–33 (a) In conductors, electrons are free to move in any direction through the conductor. Energy is lost due to collisions with atoms and other electrons, giving rise to the resistance of the conductor. (b) In superconductors, electrons are bound in pairs and travel through the conductor in step, avoiding all collisions. Since there is no energy loss, the conductor has no resistance.

through the material in a manner similar to a band marching in a parade. The orderly motion of electrons in a superconductor, shown in Figure 3–33(b), results in an ideal conductor, since the electrons no longer collide.

The economy of having a high critical temperature has led to the search for high-temperature superconductors. In the late 1980s and early 1990s, research at the IBM Zurich Research Laboratory in Switzerland and the University of Houston in Texas yielded superconducting materials that are able to operate at temperatures as high as 98 K (−175° C). While this temperature is still very low, it means that superconductivity can now be achieved by using the readily available liquid nitrogen rather than the much more expensive and rare liquid helium.

Superconductivity has been found in such seemingly unlikely materials as ceramics consisting of barium, lanthanum, copper, and oxygen. Research is now centered on developing new materials that become superconductors at ever higher temperatures and that are able to overcome the disadvantages of the early ceramic superconductors. While hopes were high that the new superconducting materials would result in significant improvements in transportation, medical treatment, and energy storage and transmission, the potential has gone largely unrealized.

Very expensive, low-temperature superconductivity continues to be used in some giant particle accelerators and, to a limited degree, in electronic components (such as superfast Josephson junctions and SQUIDs, i.e., superconducting quantum interference devices, which are used to detect very small magnetic fields). By far, the most important application of superconductors is to generate the very strong magnetic fields needed in magnetic resonance imagers (MRI). Once research produces commercially viable, high-temperature superconductors, however, the possibilities of the applications will be virtually limitless. High-temperature superconductivity promises to yield improvements in transportation, energy storage and transmission, computers, and medical treatment and research.

Putting It into Practice

You are a troubleshooting specialist working for a small telephone company. One day, word comes in that an entire subdivision is without telephone service. Everyone suspects that a cable was cut by one of several backhoe operators working on a waterline project near the subdivision. However, no one is certain exactly where the cut occurred. You remember that the resistance of a length of wire is determined by several factors, including the length. This gives you an idea for determining the distance between the telephone central office and location of the cut.

First you go to the telephone cable records, which show that the subdivision is served by 26-gauge copper wire. Then, since each customer's telephone is connected to the central office with a pair of wires, you measure the resistance of several loops from the central office. As expected, some of the measurements indicate open circuits. However, several pairs of the wire were shorted by the backhoe, and each of these pairs indicates a total resistance of 338 Ω. How far from the central office did the cut occur?

Problems

3.1 Resistance of Conductors

1. Determine the resistance, at 20° C, of 100 m of solid aluminum wire having the following radii:

    a. 0.5 mm c. 0.005 mm

    b. 1.0 mm d. 0.5 cm

2. Determine the resistance, at 20° C, of 200 feet of iron conductors having the following cross sections:

    a. 0.25 inch by 0.25 inch square

    b. 0.125 inch diameter round

    c. 0.125 inch by 4.0 inch rectangle

3. A 250-foot length of solid copper bus bar, shown in Figure 3–34, is used to connect a voltage source to a distribution panel. If the bar is to have a resistance of 0.02 Ω at 20° C, calculate the required height of the bus bar (in inches).

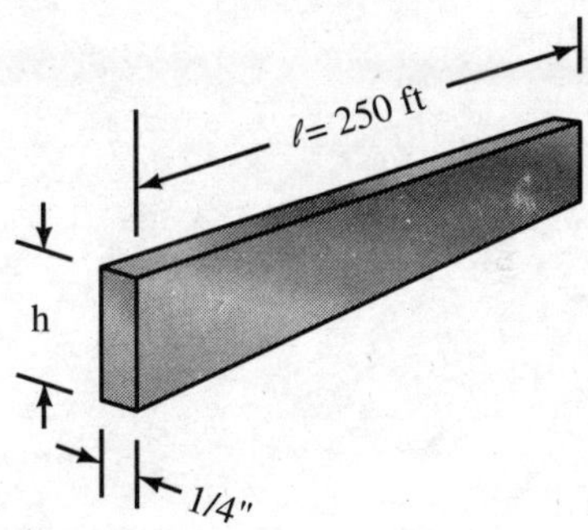

FIGURE 3–34

4. Nichrome wire is used to construct heating elements. Determine the length of 1.0-mm diameter Nichrome wire needed to produce a heating element that has a resistance of 2.0 Ω at a temperature of 20° C.

5. A copper wire having a diameter of 0.80 mm is measured to have a resistance of 10.3 Ω at 20° C. How long is this wire in meters? How long is the wire in feet?

6. A piece of aluminum wire has a resistance, at 20° C, of 20 Ω. If this wire is melted down and used to produce a second wire having a length four times the original length, what will be the resistance of the new wire at 20° C? (Hint: The volume of the wire has not changed.)

7. Determine the resistivity (in ohmmeters) of a carbon-based graphite cylinder having a length of 6.00 cm, a diameter of 0.50 mm, and a measured resistance of 3.0 Ω at 20° C. How does this value compare with the resistivity given for carbon?

8. A solid circular wire of length 200 m and diameter of 0.4 mm has a resistance measured to be 357 Ω at 20° C. Of what material is the wire constructed?

9. A 2500-m section of alloy wire has a resistance of 32 Ω. If the wire has a diameter of 1.5 mm, determine the resistivity of the material in ohm-meters. Is this alloy a better conductor than copper?

10. A section of iron wire having a diameter of 0.030 inch is measured to have a resistance of 2500 Ω (at a temperature of 20° C).

    a. Determine the cross-sectional area in square meters and in square millimeters. (Note: 1 inch = 2.54 cm = 25.4 mm.)

    b. Calculate the length of the wire in meters.

3.2 Electrical Wire Tables

11. Use Table 3–2 to determine the resistance of 300 feet of AWG 22 and AWG 19 solid copper conductors. Compare the diameters and the resistances of the wires.

12. Use Table 3–2 to find the resistance of 250 m of AWG 8 and AWG 2 solid copper conductors. Compare the diameters and the resistances of the wires.
13. Determine the maximum current that could be handled by AWG 19 wire and by AWG 30 wire.
14. If AWG 8 is rated at a maximum of 40 A, how much current could AWG 2 handle safely?
15. A spool of AWG 36 copper transformer wire is measured to have a resistance of 550 Ω at a temperature of 20° C. How long is this wire in meters?
16. How much current should AWG 36 copper wire be able to handle?

3.3 Resistance of Wires—Circular Mils

17. Determine the area in circular mils of the following conductors ($T = 20°$ C):
    a. Circular wire having a diameter of 0.016 inch
    b. Circular wire having a diameter of 2.0 mm
    c. Rectangular bus bar having dimensions 0.25 inch by 6.0 inch
18. Express the cross-sectional areas of the conductors of Problem 17 in square mils and in square millimeters.
19. Calculate the resistance, at 20° C, of 400 feet of copper conductors having the cross-sectional areas given in Problem 17.
20. Determine the diameter in inches and in millimeters of circular cables having cross-sectional areas as given below. (Assume the cables to be solid conductors.)
    a. 250 CM
    b. 1000 CM
    c. 250 MCM
    d. 750 MCM
21. A 200-foot length of solid copper wire is measured to have a resistance of 0.500 Ω.
    a. Determine the cross-sectional area of the wire in both square mils and circular mils.
    b. Determine the diameter of the wire in mils and in inches.
22. Repeat Problem 21 if the wire had been made of Nichrome.
23. A spool of solid copper wire having a diameter of 0.040 inch is measured to have a resistance of 12.5 Ω (at a temperature of 20° C).
    a. Determine the cross-sectional area in both square mils and circular mils.
    b. Calculate the length of the wire in feet.
24. Iron wire having a diameter of 30 mils was occasionally used for telegraph transmission. A technician measures a section of telegraph line to have a resistance of 2500 Ω (at a temperature of 20° C).
    a. Determine the cross-sectional area in both square mils and circular mils.
    b. Calculate the length of the wire in feet and in meters. (Note: 1 ft = 0.3048 m.) Compare your answer to the answer obtained in Problem 10.

3.4 Temperature Effects

25. An aluminum conductor has a resistance of 50 Ω at room temperature. Find the resistance of the same conductor at −30° C, 0° C, and at 200° C.

26. AWG 14 solid copper house wire is designed to operate within a temperature range of −40° C to +90° C. Calculate the resistance of 200 circuit feet of wire at both temperatures. Note: A circuit foot is the length of cable needed for a current to travel to and from a load.

27. A given material has a resistance of 20 Ω at room temperature (20° C) and 25 Ω at a temperature of 85° C.
    a. Does the material have a positive or a negative temperature coefficient? Explain briefly.
    b. Determine the value of the temperature coefficient α at 20° C.
    c. Assuming the resistance-versus-temperature function to be linear, determine the expected resistance of the material at 0° C (the freezing point of water) and at 100° C (the boiling point of water).

28. A given material has a resistance of 100 Ω at room temperature (20° C) and 150 Ω at a temperature of −25° C.
    a. Does the material have a positive or a negative temperature coefficient? Explain briefly.
    b. Determine the value of the temperature coefficient α at 20° C.
    c. Assuming the resistance-versus-temperature function to be linear, determine the expected resistance of the material at 0° C (the freezing point of water) and at −40° C.

29. An electric heater is made of Nichrome wire. The wire has a resistance of 15.2 Ω at a temperature of 20° C. Determine the resistance of the Nichrome wire when the temperature of the wire is increased to 260° C.

30. A silicon diode is measured to have a resistance of 500 Ω at 20° C. Determine the resistance of the diode if the temperature of the component is increased with a soldering iron to 30° C. (Assume that the resistance-versus-temperature function is linear.)

31. An electrical device has a linear temperature response. The device has a resistance of 120 Ω at a temperature of −20° C and a resistance of 190 Ω at a temperature of 120° C.
    a. Calculate the resistance at a temperature of 0° C.
    b. Calculate the resistance at a temperature of 80° C.
    c. Determine the temperature intercept of the material.

32. Derive the expression of Equation 3–8.

3.5 Types of Resistors

33. A 10-kΩ variable resistor has its wiper (movable terminal *b*) initially at the bottom terminal, *c*. Determine the resistance R_{ab} between terminals *a* and *b* and the resistance R_{bc} between terminals *b* and *c* under the following conditions:
    a. The wiper is at *c*.
    b. The wiper is one-fifth of the way around the resistive surface.
    c. The wiper is four-fifths of the way around the resistive surface.
    d. The wiper is at *a*.

34. The resistance between wiper terminal *b* and bottom terminal *c* of a 200-kΩ variable resistor is measured to be 50 kΩ. Determine the resistance that would be measured between the top terminal *a* and the wiper terminal *b*.

3.6 Color Coding of Resistors

35. Given resistors having the following color codes (as read from left to right), determine the resistance, tolerance, and reliability of each component. Express the uncertainty in both percentage and ohms.

a. Brown	Green	Yellow	Silver	
b. Red	Gray	Gold	Gold	Yellow
c. Yellow	Violet	Blue	Gold	
d. Orange	White	Black	Gold	Red

36. Determine the color codes required if you need the following resistors for a project:
    a. 33 kΩ ± 5%, 0.1% reliability
    b. 820 Ω ± 10%
    c. 15 Ω ± 20%
    d. 2.7 MΩ ± 5%

3.7 Measuring Resistance—The Ohmmeter

37. Explain how an ohmmeter may be used to determine whether a light bulb is burned out.
38. If an ohmmeter were placed across the terminal of a switch, what resistance would you expect to measure when the contacts of the switch are closed? What resistance would you expect to measure when the contacts are opened?
39. Explain how you could use an ohmmeter to determine approximately how much wire is left on a spool of AWG 24 copper wire.
40. An analog ohmmeter is used to measure the resistance of a two-terminal component. The ohmmeter indicates a resistance of 1.5 kΩ. When the leads of the ohmmeter are reversed, the meter indicates that the resistance of the component is an open circuit. Is the component faulty? If not, what kind of component is being tested?

3.8 Thermistors

41. A thermistor has the characteristics shown in Figure 3–23.
    a. Determine the resistance of the device at room temperature, 20° C.
    b. Determine the resistance of the device at a temperature of 40° C.
    c. Does the thermistor have a positive or a negative temperature coefficient? Explain.

3.9 Photoconductive Cells

42. For the photocell having the characteristics shown in Figure 3–24(c), determine the resistance
    a. in a dimly lit basement having an illuminance of 10 lux
    b. in a home having an illuminance of 50 lux
    c. in a classroom having an illuminance of 500 lux

3.11 Conductance

43. Calculate the conductance of the following resistances:
    a. 0.25 Ω
    b. 500 Ω
    c. 250 kΩ
    d. 12.5 MΩ
44. Determine the resistance of components having the following conductances:
    a. 62.5 μS
    b. 2500 mS
    c. 5.75 mS
    d. 25.0 S
45. Determine the conductance of 1000 m of AWG 30 solid copper wire at a temperature of 20° C.
46. Determine the conductance of 200 feet of aluminum bus bar (at a temperature of 20° C) which has a cross-sectional dimension of 4.0 inches by 0.25 inch. If the temperature were to increase, what would happen to the conductance of the bus bar?

ANSWERS TO IN-PROCESS LEARNING CHECKS

IN-PROCESS LEARNING CHECK 1

1. Iron wire will have approximately seven times more resistance than copper.
2. The longer wire will have twice the resistance of the shorter wire.
3. The wire having the greater diameter will have one quarter the resistance of the small-diameter wire.

IN-PROCESS LEARNING CHECK 2

1. 200 A
2. The actual value is less than the theoretical value. Since only the surface of the cable is able to dissipate heat, the current must be decreased to prevent heat buildup.

IN-PROCESS LEARNING CHECK 3

$A = 63.7$ CM
$A = 3.23 \times 10^{-8}\ \text{m}^2 = 0.0323\ \text{mm}^2$

IN-PROCESS LEARNING CHECK 4

Positive temperature coefficient means that resistance of a material increases as temperature increases. Negative temperature coefficient means that the resistance of a material decreases as the temperature increases. Aluminum has a positive temperature coefficient.

KEY TERMS

dc Resistance
Dynamic Resistance
Efficiency
Energy
Kilowatthours
Linear Resistance
Nonlinear Resistance
Ohmic
Ohm's Law
Open Circuit
Power
Schematic Capture
Voltage Reference Convention

OUTLINE

Ohm's Law
Voltage Polarity and Current Direction
Power
Power Direction Convention
Energy
Efficiency
Nonlinear and Dynamic Resistances
Computer-Aided Circuit Analysis

OBJECTIVES

After studying this chapter, you will be able to

- compute voltage, current, and resistance in simple circuits using Ohm's law,
- use the voltage reference convention to determine polarity,
- describe how voltage, current, and power are related in a resistive circuit,
- compute power in dc circuits,
- use the power reference convention to describe the direction of power transfer,
- compute energy used by electrical loads,
- determine energy costs,
- determine the efficiency of machines and systems,
- use PSpice and Multisim to solve Ohm's law problems.

4

OHM'S LAW, POWER, AND ENERGY

CHAPTER PREVIEW

In the previous two chapters, you studied voltage, current, and resistance separately. In this chapter, we consider them together. Beginning with Ohm's law, you will study the relationship between voltage and current in a resistive circuit, reference conventions, power, energy, and efficiency. Also in this chapter, you begin the process of learning how to use computer simulation software packages (Multisim and PSpice) to solve circuit analysis problems on your computer. (For those of you who have used the Multisim simulation courseware in Chapter 2, you may notice some duplication.) ■

Putting It in Perspective

Georg Simon Ohm

IN CHAPTER 3, WE LOOKED BRIEFLY at Ohm's experiments. We now take a look at Ohm the person.

Georg Simon Ohm was born in Erlangen, Bavaria, on March 16, 1787. His father was a master mechanic who determined that his son should obtain an education in science. Although Ohm became a teacher in a high school, he had aspirations to receive a university appointment. The only way that such an appointment could be realized would be if Ohm could produce important results through scientific research. Since the science of electricity was in its infancy, and because the electric cell had recently been invented by the Italian Conte Alessandro Volta, Ohm decided to study the behavior of current in resistive circuits. Because equipment was expensive and hard to come by, Ohm made much of his own, thanks, in large part, to his father's training. Using this equipment, Ohm determined experimentally that the amount of current transmitted along a wire was directly proportional to its cross-sectional area and inversely proportional to its length. From these results, Ohm was able to define resistance and show that there was a simple relationship between voltage, resistance, and current. This result, now known as Ohm's law, is probably the most fundamental relationship in circuit theory. However, when published in 1827, Ohm's results were met with ridicule. As a result, not only did Ohm miss out on a university appointment, he was also forced to resign from his high school teaching position. While Ohm was living in poverty and shame, his work became known and appreciated outside Germany. In 1842, Ohm was appointed a member of the Royal Society. Finally, in 1849, he was appointed as a professor at the University of Munich, where he was at last recognized for his important contributions. ■

4.1 Ohm's Law

Consider the circuit of Figure 4–1. Using a circuit similar in concept to this, Ohm determined experimentally that *current in a resistive circuit is directly proportional to its applied voltage and inversely proportional to its resistance.* In equation form, **Ohm's law** states

$$I = \frac{E}{R} \quad \text{[amps, A]} \tag{4–1}$$

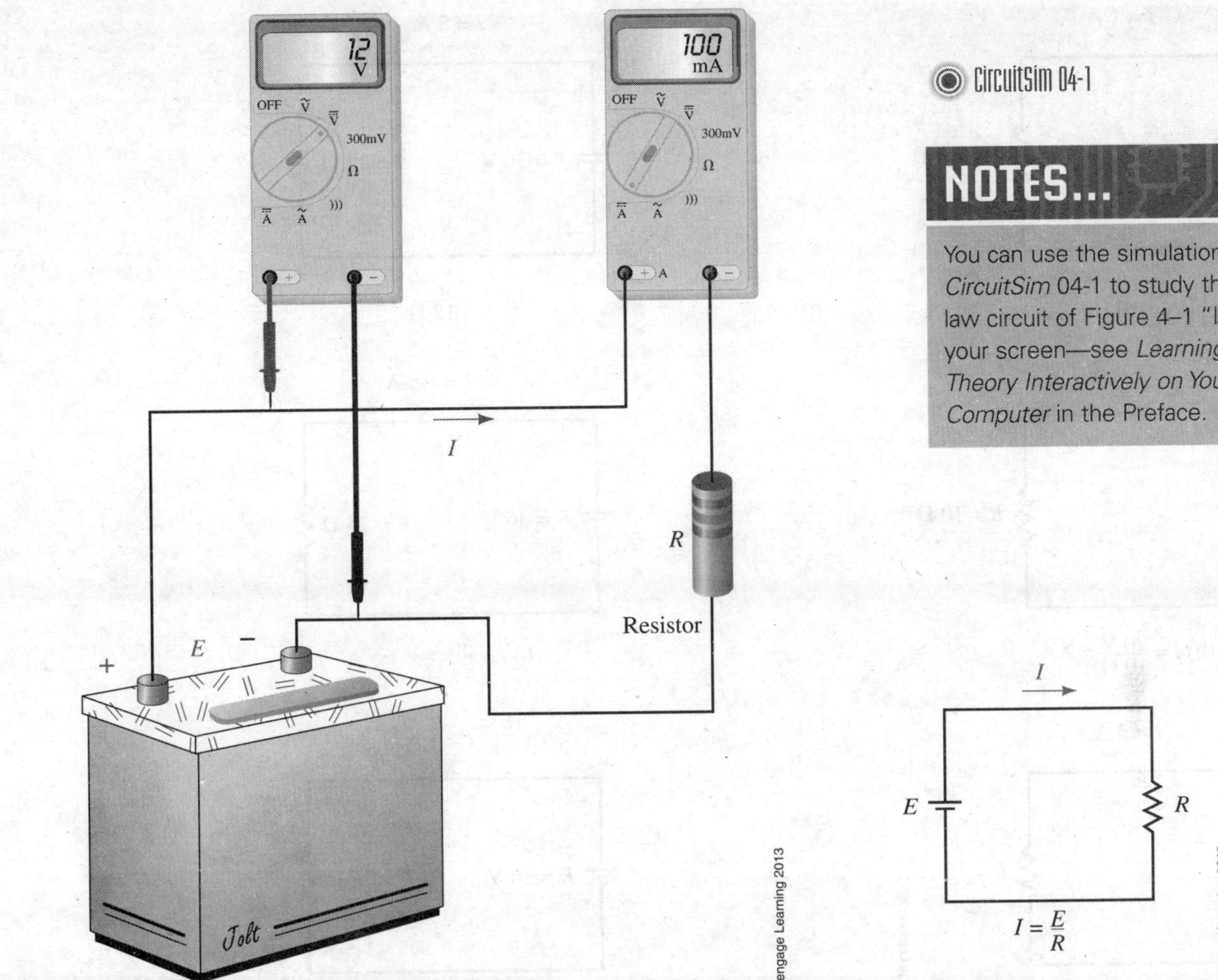

(a) Test circuit

(b) Schematic, meters not shown

FIGURE 4–1 Circuit for illustrating Ohm's law. (This is a modern-day equivalent of Ohm's test circuit.)

◉ CircuitSim 04-1

NOTES...

You can use the simulation of *CircuitSim* 04-1 to study the Ohm's law circuit of Figure 4–1 "live" on your screen—see *Learning Circuit Theory Interactively on Your Computer* in the Preface.

where

E is the applied voltage in volts,

R is the resistance in ohms,

I is the current in amperes.

From this you can see that the larger the applied voltage, the larger the current, while the larger the resistance, the smaller the current.

The proportional relationship between voltage and current described by Equation 4–1 may be demonstrated by direct substitution as indicated in Figure 4–2. For a fixed resistance, doubling the voltage as shown in (b) doubles the current, while tripling the voltage as shown in (c) triples the current, and so on.

The inverse relationship between resistance and current is demonstrated in Figure 4–3. For a fixed voltage, doubling the resistance as shown in (b) halves the current, while tripling the resistance as shown in (c) reduces the current to one-third of its original value, and so on.

Ohm's law may also be expressed in the following forms by rearrangement of Equation 4–1:

$$E = IR \text{ [volts, V]} \qquad (4\text{–}2)$$

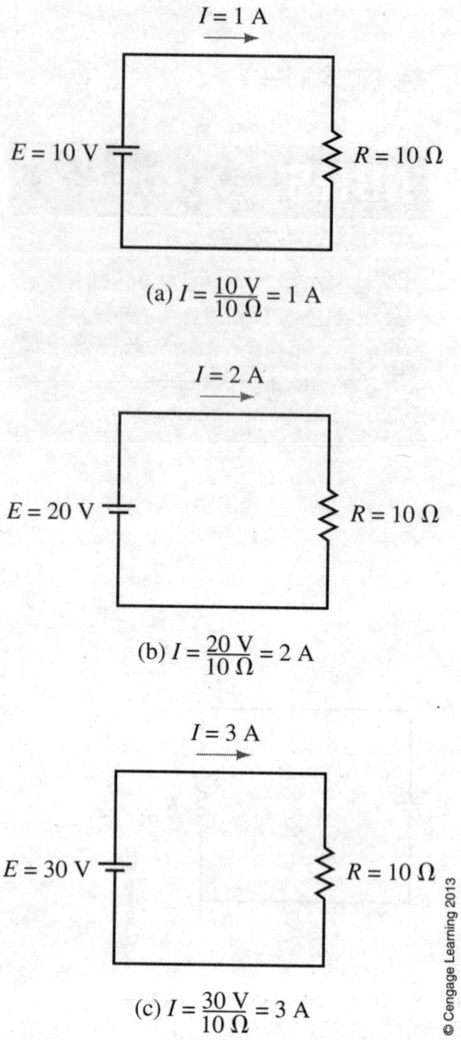

FIGURE 4–2 For a fixed resistance, current is directly proportional to voltage; thus, doubling the voltage as in (b) doubles the current, while tripling the voltage as in (c) triples the current, and so on.

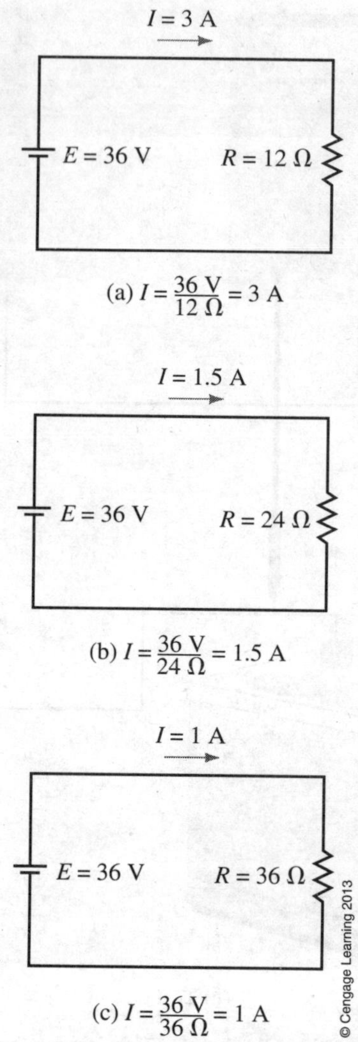

FIGURE 4–3 For a fixed voltage, current is inversely proportional to resistance; thus, doubling the resistance as in (b) halves the current, while tripling the resistance as in (c) results in one-third the current, and so on.

and

$$R = \frac{E}{I} \quad \text{[ohms, } \Omega\text{]} \tag{4–3}$$

When using Ohm's law, be sure to express all quantities in base units of volts, ohms, and amps as in Examples 4–1 to 4–3, or utilize the relationships between prefixes as in Example 4–4.

EXAMPLE 4–1

A 27-Ω resistor is connected to a 12-V battery. What is the current?

Solution Substituting the resistance and voltage values into Ohm's law yields

$$I = \frac{E}{R} = \frac{12 \text{ V}}{27 \ \Omega} = 0.444 \text{ A}$$

EXAMPLE 4–2

The lamp of Figure 4–4 draws 25 mA when connected to a 6-V battery. What is its resistance?

Solution Using Equation 4–3,

$$R = \frac{E}{I} = \frac{6\text{ V}}{25 \times 10^{-3}\text{ A}} = 240\ \Omega$$

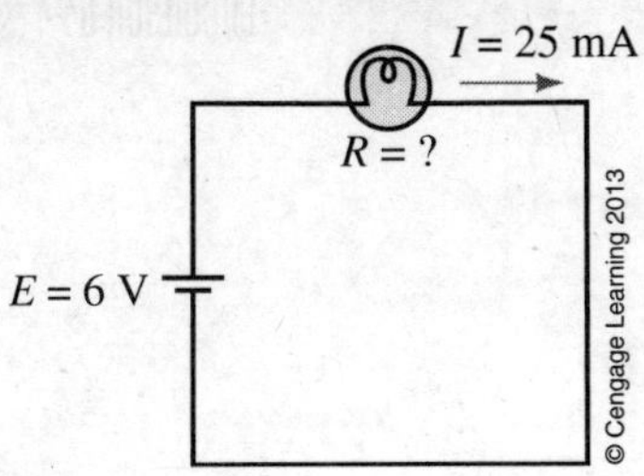

FIGURE 4–4

EXAMPLE 4–3

If 125 μA is the current in a resistor with color bands red, red, yellow, what is the voltage across the resistor?

Solution Using the color code of Chapter 3, $R = 220\text{ k}\Omega$. From Ohm's law, $E = IR = (125 \times 10^{-6}\text{ A})(220 \times 10^{3}\ \Omega) = 27.5\text{ V}$.

EXAMPLE 4–4

A resistor with the color code brown, red, yellow is connected to a 30-V source. What is I?

Solution When E is in volts and R is in kΩ, the answer comes out directly in mA. From the color code, $R = 120\text{ k}\Omega$. Thus,

$$I = \frac{E}{R} = \frac{30\text{ V}}{120\text{ k}\Omega} = 0.25\text{ mA}$$

CircuitSim 04-2

Traditionally, circuits are drawn with the source on the left and the load on the right as indicated in Figures 4–1 to 4–3. However, you will also encounter circuits with other orientations. For these, the same principles apply; as we did in Figure 4–4, simply draw the current arrow pointing out from the positive end of the source and apply Ohm's law in the usual manner. More examples are shown in Figure 4–5.

CircuitSim 04-3

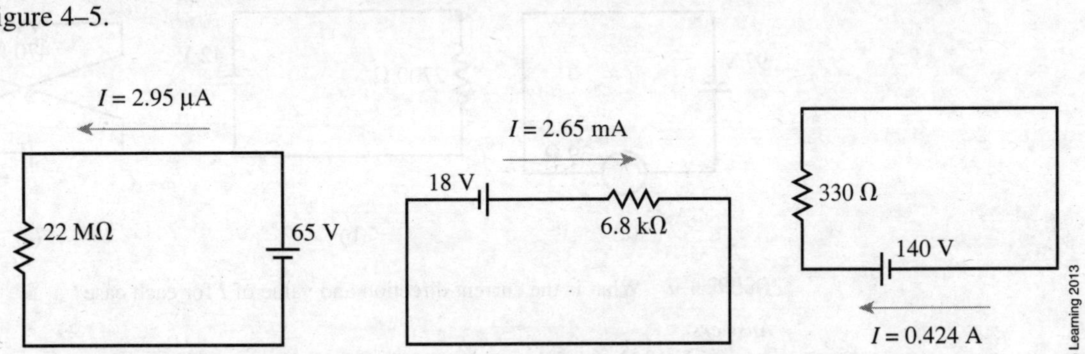

(a) $I = \dfrac{65\text{ V}}{22\text{ M}\Omega} = 2.95\ \mu\text{A}$ (b) $I = \dfrac{18\text{ V}}{6.8\text{ k}\Omega} = 2.65\text{ mA}$ (c) $I = \dfrac{140\text{ V}}{330\ \Omega} = 0.424\text{ A}$

FIGURE 4–5

Ohm's Law in Graphical Form

CircuitSim 04-4

The relationship between current and voltage described by Equation 4–1 may be shown graphically as in Figure 4–6. The graphs, which are straight lines, show clearly that the relationship between voltage and current is linear, that is, that current is directly proportional to voltage.

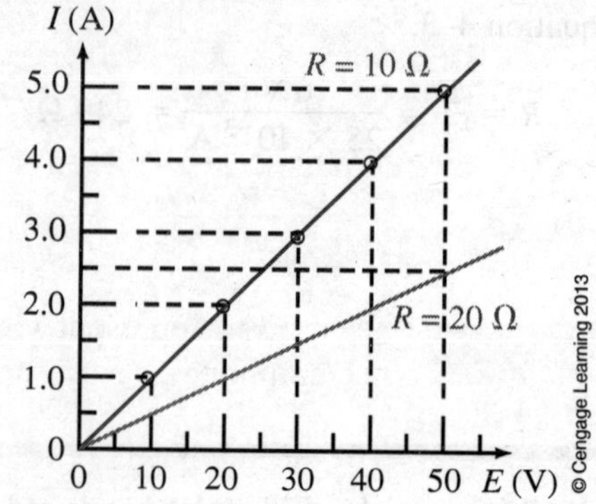

FIGURE 4–6 Graphical representation of Ohm's law. The red plot is for a 10-Ω resistor while the green plot is for a 20-Ω resistor.

PRACTICE PROBLEMS 1

1. a. For the circuit of Figure 4–2(a), show that halving the voltage halves the current.
   b. For the circuit of Figure 4–3(a), show that halving the resistance doubles the current.
   c. Are these results consistent with the verbal statement of Ohm's law?
2. For each of the following, draw the circuit with values marked, then solve for the unknown.
   a. A 10 000-milliohm resistor is connected to a 24-V battery. What is the resistor current?
   b. How many volts are required to establish a current of 20 μA in a 100-kΩ resistor?
   c. If 125 V is applied to a resistor and 5 mA results, what is the resistance?
3. For each circuit of Figure 4–7 determine the current, including its direction (i.e., the direction that the current arrow should point).
4. Verify all points on Figure 4–6 for $R = 10\ \Omega$.

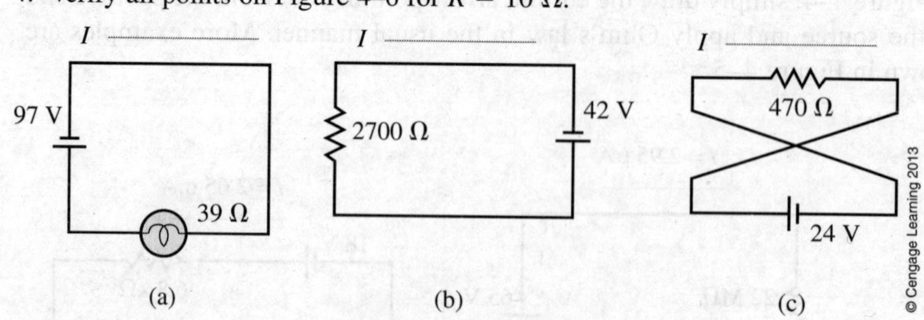

FIGURE 4–7 What is the current direction and value of *I* for each case?

Answers

1. a. 5 V/10 Ω = 0.5 A b. 36 V/6 Ω = 6 A c. Yes
2. a. 2.4 A b. 2.0 V c. 25 kΩ
3. a. 2.49 A, left b. 15.6 mA, right c. 51.1 mA, left

Open Circuits

Current can only exist where there is a conductive path (e.g., a length of wire). For the circuit of Figure 4–8, I equals zero since there is no conductor between points a and b. We refer to this as an **open circuit**, Since $I = 0$, substitution of this into Equation 4–3 yields

$$R = \frac{E}{I} = \frac{E}{0} \Rightarrow \infty \text{ ohms}$$

Thus, an open circuit has infinite resistance.

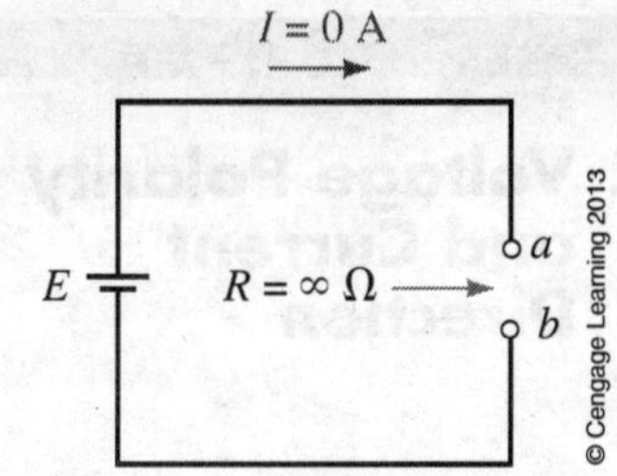

FIGURE 4–8 An open circuit has infinite resistance.

Voltage Symbols

Two different symbols are commonly used to represent voltage. For sources, use uppercase E; for loads (and other components), use uppercase V. This is illustrated in Figure 4–9.

Using the symbol V, Ohm's law may be rewritten in its several forms as

$$I = \frac{V}{R} \quad \text{[amps]} \tag{4–4}$$

$$V = IR \quad \text{[volts]} \tag{4–5}$$

$$R = \frac{V}{I} \quad \text{[ohms]} \tag{4–6}$$

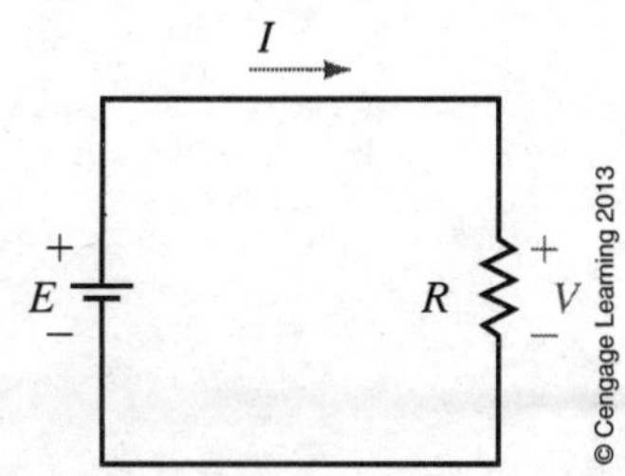

FIGURE 4–9 Symbols used to represent voltages. E is used for source voltages while V is used for voltage across circuit components such as resistors.

These relationships hold for every resistor in a circuit, no matter how complex the circuit. Since $V = IR$, these voltages are often referred to as *IR drops.*

CircuitSim 04-5

EXAMPLE 4–5

The current through each resistor of Figure 4–10 is $I = 0.5$ A (see Note). Compute V_1 and V_2.

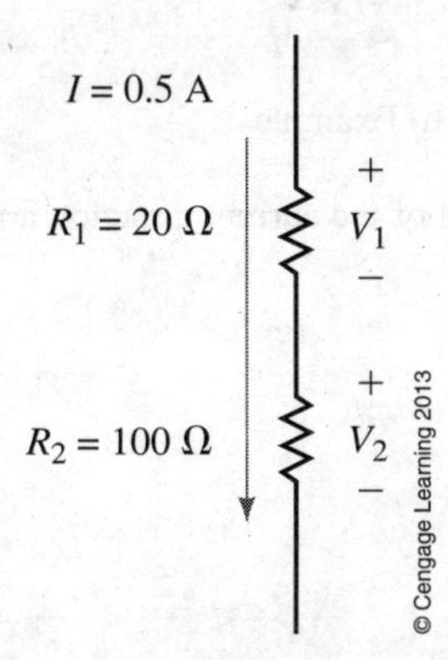

FIGURE 4–10 Ohm's law applies to each resistor.

NOTES...

In the interest of brevity (as in Figure 4–10), we sometimes draw only a portion of a circuit with the rest of the circuit implied rather than shown explicitly.

Thus, this circuit contains a source and connecting wire, even though they are not shown here.

Solution $V_1 = IR_1 = (0.5 \text{ A})(20\ \Omega) = 10$ V. Note, I is also the current through R_2. Thus, $V_2 = IR_2 = (0.5 \text{ A})(100\ \Omega) = 50$ V.

CircuitSim 04-6

4.2 Voltage Polarity and Current Direction

So far, we have paid little attention to the polarity of voltages across resistors—that is, at which end should we place the + sign for voltage and at which end should we place the − sign? However, the issue is of extreme importance. Fortunately, there is a simple relationship between current direction and voltage polarity. To get at the idea, consider Figure 4–11(a). Here the polarity of V is obvious since the resistor is connected directly to the source. This makes the top end of the resistor positive with respect to the bottom end, and $V = E = 12$ V as indicated by the meters—remember, they have autopolarity.

Now consider current. The direction of I is from top to bottom through the resistor as indicated by the current direction arrow. Examining voltage polarity, we see that the plus sign for V is at the tail of this arrow. This observation turns out to be true in general and gives us a convention (the **voltage reference convention**) for marking voltage polarity on circuit diagrams. *For voltage across a resistor, always place the plus sign at the tail of the current reference arrow.* Several additional examples are shown in Figures 4–11(b) and (c).

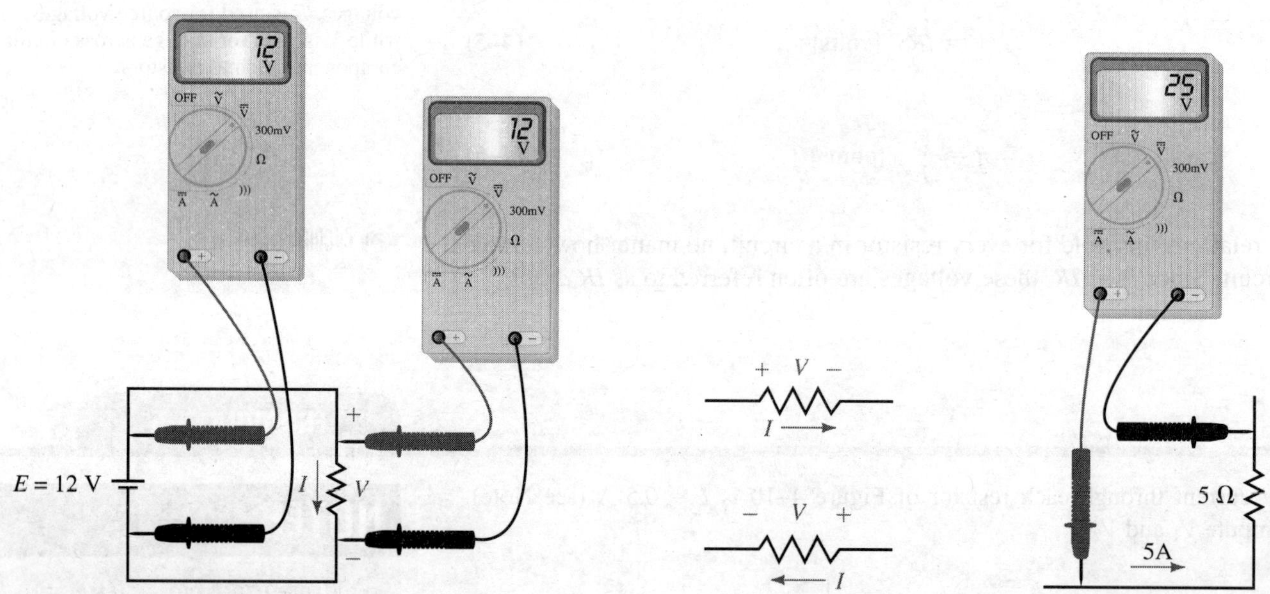

(a) Defining convention (b) Examples (c) Another example

FIGURE 4–11 Convention for voltage polarity. Place the + sign for V at the tail of the current direction arrow.

CircuitSim 04-7

PRACTICE PROBLEMS 2

For each resistor of Figure 4–12, compute V and show its polarity.

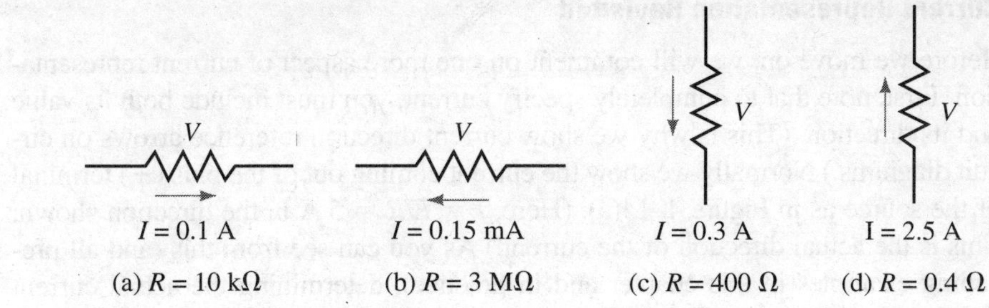

FIGURE 4–12

Answers

a. 1000 V, + at left; b. 450 V, + at right; c. 120 V, + at top; d. 1 V, + at bottom

IN-PROCESS LEARNING CHECK 1

(Answers are at the end of the chapter.)

1. A resistor has color bands brown, black, and red and a current of 25 mA. Determine the voltage across it.
2. For a resistive circuit, what is I if $E = 500$ V and R is open circuited? Will the current change if the voltage is doubled?
3. A certain resistive circuit has voltage E and resistance R. If $I = 2.5$ A, what will be the current if:
   a. E remains unchanged but R is doubled?
   b. E remains unchanged but R is quadrupled?
   c. E remains unchanged but R is reduced to 20% of its original value?
   d. R is doubled and E is quadrupled?
4. The voltmeters of Figure 4–13 have autopolarity. Determine the reading of each meter, its magnitude and sign.

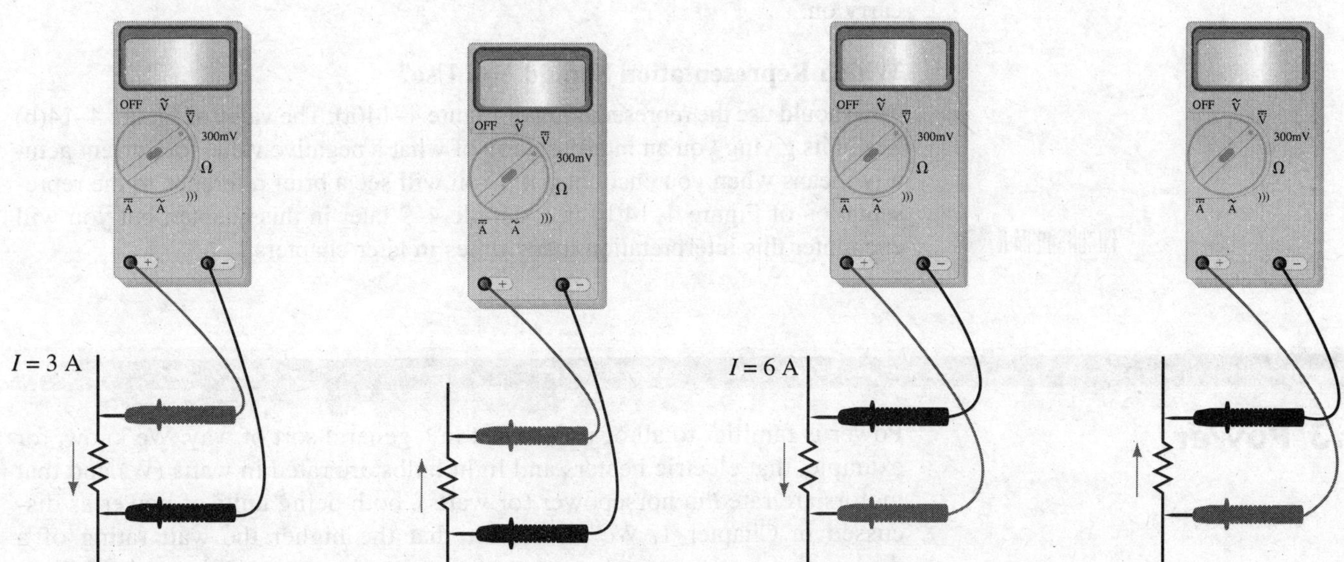

FIGURE 4–13

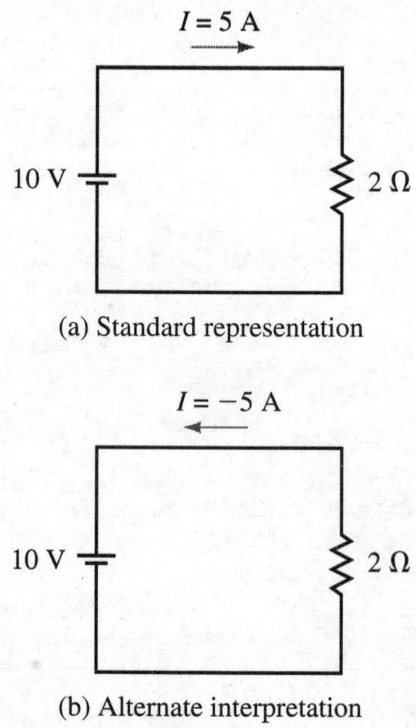

FIGURE 4–14 Two representations of the same current. While we generally focus on the representation of (a), there are practical reasons why you should be familiar with (b) as well—see text.

Current Representation Revisited

Before we move on, we will comment on one more aspect of current representation. First, note that to completely specify current, you must include both its value and its direction. (This is why we show current direction reference arrows on circuit diagrams.) Normally we show the current coming out of the plus (+) terminal of the source as in Figure 4–14(a). (Here, $I = E/R = 5$ A in the direction shown. This is the actual direction of the current.) As you can see from this (and all preceding examples in this chapter and in practice), determining the actual current direction in single-source networks is easy. However, when analyzing complex circuits (such as those with multiple sources as in later chapters and in practice), it is not always easy to tell in advance in what direction all currents will be. As a result, when you solve such problems (or measure their currents), you may find that some currents have negative values. What does this mean?

To get at the answer, consider both parts of Figure 4–14. In (a), current is shown in the usual direction, while in (b), it is shown in the opposite direction. To compensate for the reversed direction, we have changed the sign of the value of *I*. The interpretation placed on this is that a positive current in one direction is the same as a negative current in the opposite direction. Therefore, (a) and (b) are two representations of the same current. Thus, if during the solution of a problem (or in the course of a lab measurement) you obtain a positive value for current, this means that its actual direction is the same as the reference arrow; conversely, if you obtain a negative value, its direction is opposite to the reference arrow.

Why Is This Important?

In a practical sense, knowing this interpretation is important to you. For example, suppose you are in the lab and one of your current measurements is negative. What should you do—should you de-energize the circuit, reverse the ammeter leads, and power up again just to get a positive reading? Or should you carry on knowing that, even though the reading is negative, it still tells you all you need to know about the current, namely its magnitude and its direction? Most experienced lab personnel would simply log the results and carry on.

Which Representation Should You Use?

You should use the representation of Figure 4–14(a). The value of Figure 4–14(b) lies in its giving you an interpretation of what a negative value for current actually means when you encounter it. [You will see a brief reference to the representation of Figure 4–14(b) in Example 4–9 later in this chapter, but you will encounter this interpretation many times in later chapters.]

CircuitSim 04-8

4.3 Power

Power is familiar to all of us, at least in a general sort of way. We know, for example, that electric heaters and light bulbs are rated in watts (W) and that motors are rated in horsepower (or watts), both being units of power as discussed in Chapter 1. We also know that the higher the watt rating of a device, the more energy we can get out of it per unit time. Figure 4–15 illustrates the idea. In (a), the greater the power rating of the light, the more light energy it can produce per second. In (b), the greater the power rating of the heater, the more heat energy it can produce per second. In (c), the larger the power rating of the motor, the more mechanical work it can do per second.

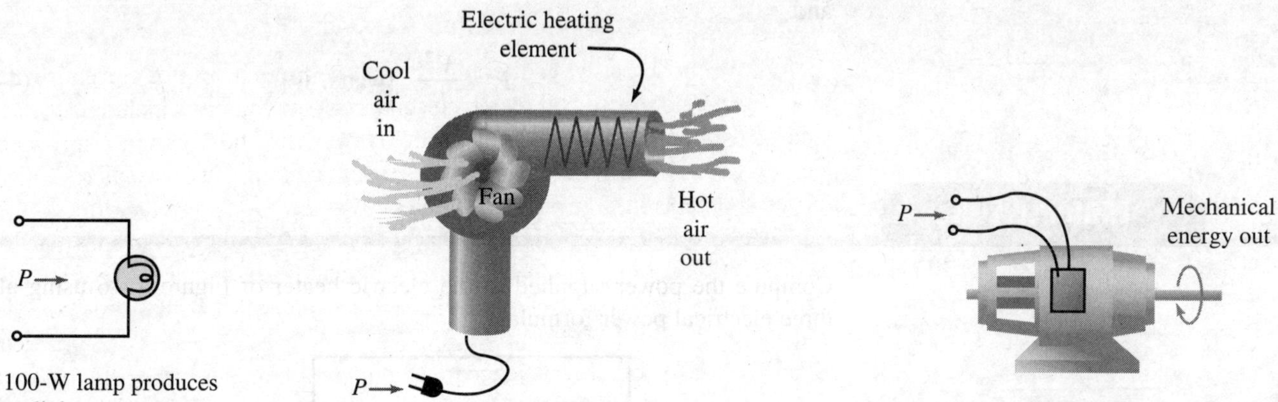

(a) A 100-W lamp produces more light energy per second than a 40-W lamp

(b) Hair dryer

(c) A 10-hp motor can do more work in a given time than a ½-hp motor

FIGURE 4–15 Energy conversion. Power P is a measure of the rate of energy conversion.

As you can see, power is related to energy, which is the capacity to do work. Formally, **power** is defined as the rate of doing work or, equivalently, as the rate of transfer of energy. The symbol for power is P. By definition,

$$P = \frac{W}{t} \quad \text{[watts, W]} \tag{4–7}$$

where W is the work (or energy) in joules and t is the corresponding time interval of t seconds (see Notes).

The SI unit of power is the watt. From Equation 4–7, we see that P also has units of joules per second. If you substitute $W = 1$ J and $t = 1$ s, you get $P = 1$ J/1 s $= 1$ W. From this, you can see that *one watt equals one joule per second.* Occasionally, you also need power in horsepower. To convert, recall that 1 hp $=$ 746 watts.

Power in Electrical and Electronic Systems

Since our interest is in electrical power, we need expressions for P in terms of electrical quantities. Recall from Chapter 2 that voltage is defined as work per unit charge and current as the rate of transfer of charge,

$$V = \frac{W}{Q} \tag{4–8}$$

and

$$I = \frac{Q}{t} \tag{4–9}$$

From Equation 4–8, $W = QV$. Substituting this into Equation 4–7 yields $P = W/t = (QV)/t = V(Q/t)$. Replacing Q/t with I, we get

$$P = VI \quad \text{[watts, W]} \tag{4–10}$$

and, for a source,

$$P = EI \quad \text{[watts, W]} \tag{4–11}$$

Additional relationships are obtained by substituting $V = IR$ and $I = V/R$ into Equation 4–10:

$$P = I^2R \quad \text{[watts, W]} \tag{4–12}$$

NOTES...

1. Since the denominator of Equation 4–7 represents an interval of time, it would be mathematically more appropriate to use Δt instead of t. Similarly, since the numerator represents an amount of work done during this interval, it would be more proper to use ΔW. In this notation, power would be expressed as

$$P = \Delta W/\Delta t \tag{4–7a}$$

However, in introductory circuit analysis texts such as this, it is common to show power as we do in Equation 4–7, and we will continue to do so here. However, in later chapters where it becomes appropriate to do so, we will change to the form represented by Equation 4–7a.

2. The symbol for energy is W and the abbreviation for watts is W. Multiple use of symbols is common in technology. In such cases, you have to look at the context in which a symbol is used to determine its meaning.

and

$$P = \frac{V^2}{R} \quad \text{[watts, W]} \tag{4–13}$$

EXAMPLE 4–6

Compute the power supplied to the electric heater of Figure 4–16 using all three electrical power formulas.

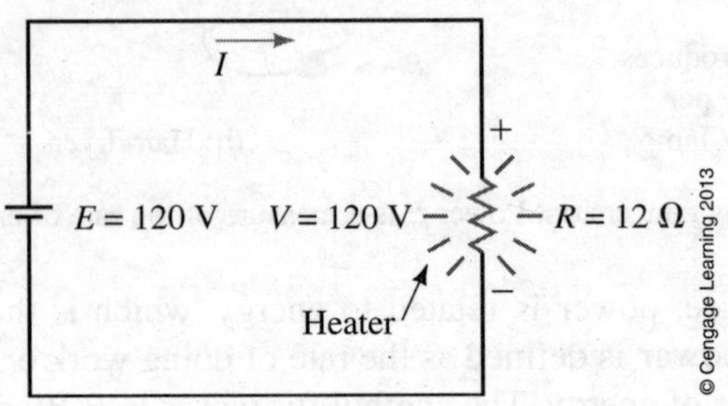

FIGURE 4–16 Power to the load (i.e., the heater) can be computed from any of the power formulas.

Solution $I = V/R = 120\ \text{V}/12\ \Omega = 10\ \text{A}$. Thus, the power may be calculated as follows:

a. $P = VI = (120\ \text{V})(10\ \text{A}) = 1200\ \text{W}$
b. $P = I^2R = (10\ \text{A})^2(12\ \Omega) = 1200\ \text{W}$
c. $P = V^2/R = (120\ \text{V})^2/12\ \Omega = 1200\ \text{W}$

Note that all give the same answer, as they must.

EXAMPLE 4–7

Compute the power to each resistor in Figure 4–17 using Equation 4–13.

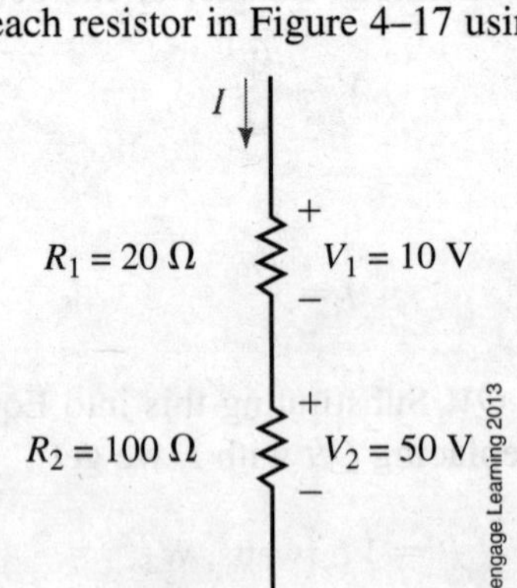

FIGURE 4–17

Solution You must use the appropriate voltage in the power equation. For resistor R_1, use V_1; for resistor R_2, use V_2.

a. $P_1 = V_1^2/R_1 = (10\ \text{V})^2/20\ \Omega = 5\ \text{W}$
b. $P_2 = V_2^2/R_2 = (50\ \text{V})^2/100\ \Omega = 25\ \text{W}$

EXAMPLE 4–8

If the dc motor of Figure 4–15(c) draws 6 A from a 120-V source,

a. Compute its power input in watts.
b. Assuming the motor is 100% efficient (i.e., that all electrical power supplied to it is output as mechanical power), compute its power output in horsepower.

Solution

a. $P_{in} = VI = (120\text{ V})(6\text{ A}) = 720\text{ W}$
b. $P_{out} = P_{in} = 720$ W. Converting to horsepower, $P_{out} = (720\text{ W})/(746\text{ W/hp}) = 0.965$ hp.

PRACTICE PROBLEMS 3

a. Show that $I = \sqrt{\dfrac{P}{R}}$
b. Show that $V = \sqrt{PR}$
c. A 100-Ω resistor dissipates 169 W. What is its current?
d. A 3-Ω resistor dissipates 243 W. What is the voltage across it?
e. For Figure 4–17, $I = 0.5$ A. Use Equations 4–10 and 4–12 to compute power to each resistor. Compare your answers to the answers of Example 4–7.

Answers
c. 1.3 A; d. 27 V; e. $P_1 = 5$ W, $P_2 = 25$ W (same)

CircuitSim 04-9

Power Rating of Resistors

Resistors must be able to safely dissipate their heat without damage. For this reason, resistors are rated in watts. (For example, common resistors of the type used in electronics are made with standard ratings of ⅛, ¼, ½, 1, and 2 W, as you saw in Figure 3–8.) To provide a safety margin, it is customary to select a resistor that is capable of dissipating two or more times its computed power. A resistor that is overrated will run a little cooler.

PRACTICAL NOTES...

A properly chosen resistor is able to dissipate its heat safely without becoming excessively hot. However, if through bad design or subsequent component failure its current becomes excessive, it will overheat, and damage may result, as shown in Figure 4–18. One of the symptoms of overheating is that the resistor becomes noticeably hotter than other resistors in the circuit. (Be careful, however, as you might get burned if you try to check by touch.) Component failure may also be detected by smell. Burned components have a characteristic odor that you will soon come to recognize. If you detect any of these symptoms, turn the equipment off and look for the source of the problem. Note, however, an overheated component is often the symptom of a problem, rather than its cause.

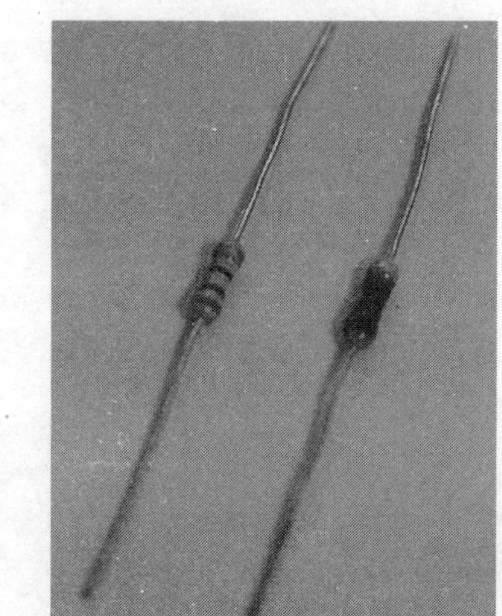

FIGURE 4–18 The resistor on the right has been damaged by overheating.

Measuring Power

Power can be measured using a device called a wattmeter. However, since wattmeters are used primarily for ac power measurement, we will hold off their consideration until Chapter 17. (You seldom need a wattmeter for dc circuits since you can determine power directly as the product of voltage times current, and V and I are easy to measure.)

4.4 Power Direction Convention

For circuits with one source and one load, energy flows from the source to the load and the direction of power transfer is obvious. For circuits with multiple sources and loads, however, the direction of energy flow in some parts of the network may not be at all apparent. We therefore need to establish a clearly defined power transfer direction convention.

A resistive load (Figure 4–19) may be used to illustrate the idea. Since the direction of power flow can only be into a resistor, never out of it (since resistors do not produce energy), we define the positive direction of power transfer as from the source to the load as in (a) and indicate this by means of an arrow: $P \rightarrow$. We then adopt the convention that, *for the relative voltage polarities and current and power directions shown in Figure 4–19(a), when power transfer is in the direction of the arrow, it is positive, whereas when it is in the direction opposite to the arrow, it is negative.*

To help interpret the convention, consider Figure 4–19(b), which highlights the source end. From this we see that *power out of a source is positive when both the current and power arrows point out from the source, both I and P have positive values, and the source voltage has the polarity indicated.*

Now consider Figure 4–19(c), which highlights the load end. Note the relative polarity of the load voltage and the direction of the current and power arrows. From this, we see that *power to a load is positive when both the current and power direction arrows point into the load, both have positive values, and the load voltage has the polarity indicated.*

In Figure 4–19(d), we have generalized the concept. The box may contain either a source or a load. If P has a positive value, power transfer is into the box; if P has a negative value, its direction is out.

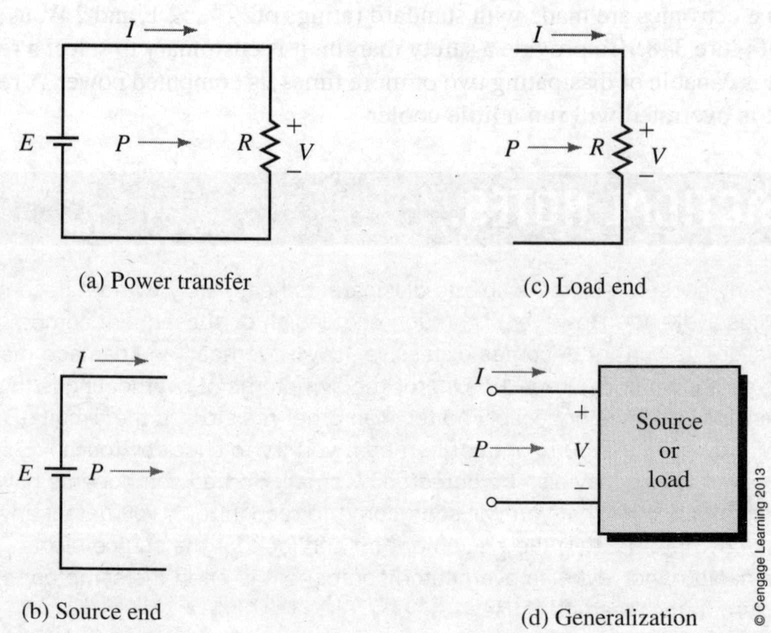

FIGURE 4–19 Reference convention for power.

EXAMPLE 4–9

Use the power direction convention to describe power transfer for the electric vehicle of Figure 4–20.

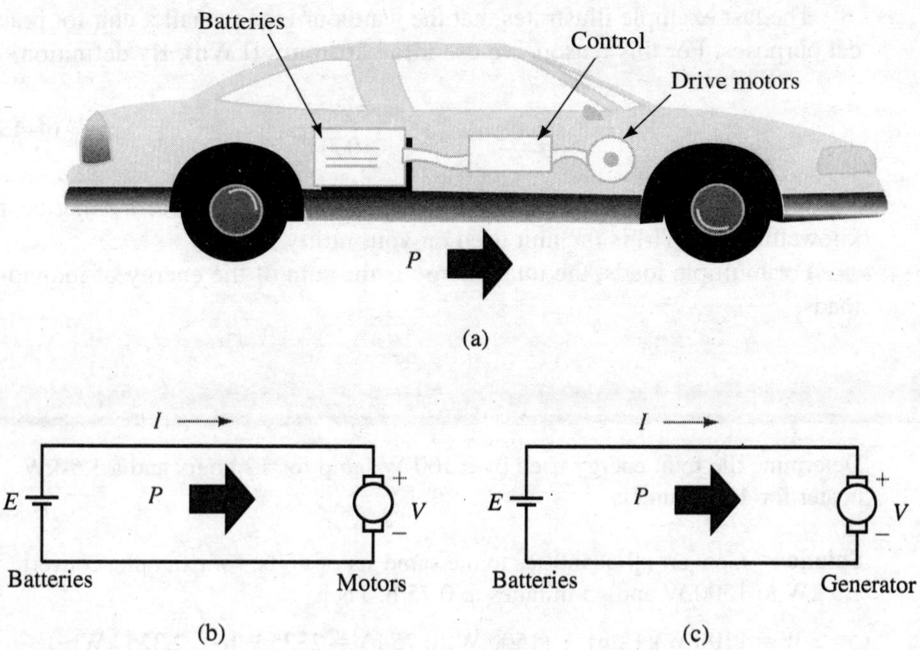

FIGURE 4–20

Solution During normal operation, the batteries supply power to the motors, and current and power are both positive, Figure 4–20(b). However, when the vehicle is going downhill, its motors are driven by the weight of the car and they act as generators—as in (c). Since the motors now act as the source and the batteries as the load, the actual current is opposite in direction to the reference arrow shown and is thus negative (recall Figure 4–14). Thus, $P = VI$ is negative. The interpretation is, therefore, that power transfer is in the direction opposite to the power reference arrow. For example, if $V = 48$ volts and $I = -10$ A, then $P = VI = (48\text{ V})(-10\text{ A}) = -480$ W. This is consistent with what is happening, since minus 480 W *into* the motors is the same as plus 480 W *out.* This 480 W of power flows from the motors to the batteries, helping to charge them as the car goes downhill.

4.5 Energy

Earlier (Equation 4–7), we defined power as the rate of doing work. When you transpose this equation, you get the formula for **energy:**

$$W = Pt \qquad (4\text{–}14)$$

If t is measured in seconds, W has units of watt-seconds (i.e., joules, J), while if t is measured in hours, W has units of watthours (Wh). Note that in Equation 4–14, P must be constant over the time interval under consideration. If it is not, apply Equation 4–14 to each interval over which P is constant as described later in this section. (For the more general case, you need calculus, which we don't consider here.)

The most familiar example of energy usage is the energy that we use in our homes and pay for on our utility bills. This energy is the energy used by the lights and electrical appliances in our homes. For example, if you run a 100-W lamp for 1 hour, the energy consumed is $W = Pt = (100\text{ W})(1\text{ h}) = 100\text{ Wh}$, while if you run a 1500-W electric heater for 12 hours, the energy consumed is $W = (1500\text{ W})(12\text{ h}) = 18\,000\text{ Wh}$.

The last example illustrates that the watthour is too small a unit for practical purposes. For this reason, we use **kilowatthours** (kWh). By definition,

$$\text{energy}_{(\text{kWh})} = \frac{\text{energy}_{(\text{Wh})}}{1000} \qquad \textbf{(4–15)}$$

Thus, for the preceding example, W = 18 kWh. In most of North America, the kilowatthour (kWh) is the unit used on your utility bill.

For multiple loads, the total energy is the sum of the energy of individual loads.

EXAMPLE 4–10

Determine the total energy used by a 100-W lamp for 12 hours and a 1.5-kW heater for 45 minutes.

Solution Convert all quantities to the same set of units, for example, convert 1.5 kW to 1500 W and 45 minutes to 0.75 h. Then,

$$W = (100\text{ W})(12\text{ h}) + (1500\text{ W})(0.75\text{ h}) = 2325\text{ Wh} = 2.325\text{ kWh}$$

Alternatively, convert all power to kilowatts first. Thus,

$$W = (0.1\text{ kW})(12\text{ h}) + (1.5\text{ kW})(0.75\text{ h}) = 2.325\text{ kWh}$$

EXAMPLE 4–11

Suppose you use the following electrical appliances: a 1.5-kW heater for 7½ hours; a 3.6-kW broiler for 17 minutes; three 100-W lamps for 4 hours; and a 900-W toaster for 6 minutes. At $0.11 per kilowatthour, how much will this cost you?

Solution Convert time in minutes to hours. Thus,

$$W = (1500)(7\tfrac{1}{2}) + (3600)\left(\frac{17}{60}\right) + (3)(100)(4) + (900)\left(\frac{6}{60}\right)$$

$$= 13\,560\text{ Wh} = 13.56\text{ kWh}$$

$$\text{cost} = (13.56\text{ kWh})(\$0.11/\text{kWh}) = \$1.49$$

In those areas of the world where in everyday life the SI system dominates, the megajoule (MJ) is sometimes used instead of the kWh (as the kWh is not an SI unit). The relationship is 1 kWh = 3.6 MJ.

Watthour Meter

In practice, energy is measured by watthour meters, many of which are electromechanical devices that incorporate a small electric motor whose speed is proportional to power to the load. This motor drives a set of dials through a gear train (Figure 4–21). Since the angle through which the dials rotate depends on

FIGURE 4–21 Watthour meter. This meter uses a gear train to drive the dials. Newer meters use electronic sensing and digital displays.

the speed of rotation (i.e., power consumed) and the length of time that this power flows, the dial position indicates energy used. Note, however, that electromechanical devices are now giving way to electronic meters, which perform this function electronically and display the result on digital readouts.

Law of Conservation of Energy

Before leaving this section, we consider the law of conservation of energy. It states that energy can neither be created nor destroyed, but is instead converted from one form to another. You have seen examples of this—for example, the conversion of electrical energy into heat energy by a resistor, and the conversion of electrical energy into mechanical energy by a motor. In fact, several types of energy may be produced simultaneously. For example, electrical energy is converted to mechanical energy by a motor, but some heat is also produced. This results in a lowering of efficiency, a topic we consider next.

4.6 Efficiency

Poor efficiency results in wasted energy and higher costs. For example, an inefficient motor costs more to run than an efficient one for the same output. An inefficient piece of electronic gear generates more heat than an efficient one, and this heat must be removed, resulting in increased costs for fans, heat sinks, and the like.

Efficiency can be expressed in terms of either energy or power. Power is generally easier to measure, so we usually use power. The **efficiency** of a device or system (Figure 4–22) is defined as the ratio of power output P_{out} to power input P_{in}, and it is usually expressed in percent and denoted by the Greek letter η (eta). Thus,

$$\eta = \frac{P_{out}}{P_{in}} \times 100\% \tag{4–16}$$

In terms of energy,

$$\eta = \frac{W_{out}}{W_{in}} \times 100\% \tag{4–17}$$

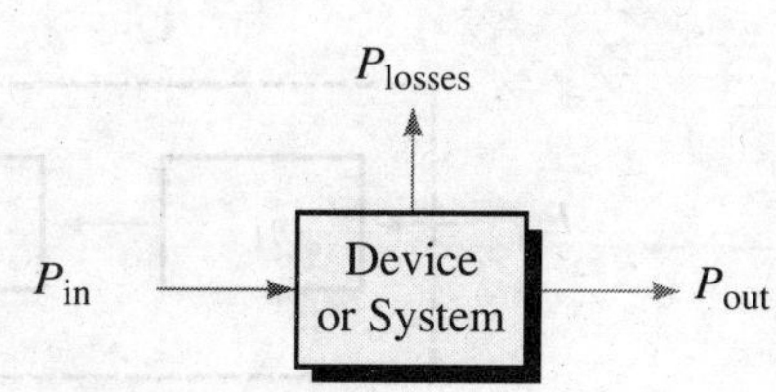

FIGURE 4–22 Input power equals output power plus losses.

Since $P_{in} = P_{out} + P_{losses}$, efficiency can also be expressed as

$$\eta = \frac{P_{out}}{P_{out} + P_{losses}} \times 100\% = \frac{1}{1 + \dfrac{P_{losses}}{P_{out}}} \times 100\% \qquad \textbf{(4–18)}$$

The efficiency of equipment and machines varies greatly. Large power transformers, for example, have efficiencies of 98% or better, while many electronic amplifiers have efficiencies lower than 50%. Note that efficiency will always be less than 100%.

EXAMPLE 4–12

A 120-V dc motor draws 12 A and develops an output power of 1.6 hp.

a. What is its efficiency?
b. How much power is wasted?

Solution

a. $P_{in} = EI = (120\text{ V})(12\text{ A}) = 1440\text{ W}$, and $P_{out} = 1.6\text{ hp} \times 746\text{ W/hp} = 1194\text{ W}$. Thus,

$$\eta = \frac{P_{out}}{P_{in}} = \frac{1194\text{ W}}{1440\text{ W}} \times 100 = 82.9\%$$

b. $P_{losses} = P_{in} - P_{out} = 1440 - 1194 = 246\text{ W}$

EXAMPLE 4–13

The efficiency of a power amplifier is the ratio of the power delivered to the load (e.g., speakers) to the power drawn from the power supply. Generally, this efficiency is not very high. For example, suppose a power amplifier delivers 400 W to its speaker system. If the power loss is 509 W, what is its efficiency?

Solution

$$P_{in} = P_{out} + P_{losses} = 400\text{ W} + 509\text{ W} = 909\text{ W}$$

$$\eta = \frac{P_{out}}{P_{in}} \times 100\% = \frac{400\text{ W}}{909\text{ W}} \times 100\% = 44\%$$

For systems with subsystems or components in cascade (Figure 4–23), overall efficiency is the product of the efficiencies of each individual part, where efficiencies are expressed in decimal form. Thus,

$$\eta_T = \eta_1 \times \eta_2 \times \eta_3 \times \cdots \times \eta_n \qquad \textbf{(4–19)}$$

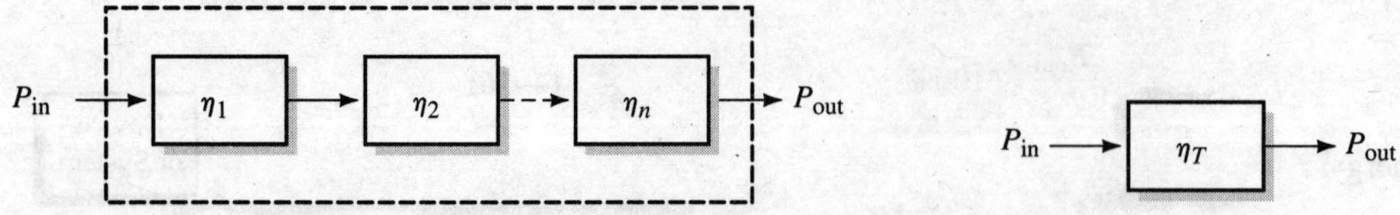

(a) Cascaded system (b) Equivalent of (a)

FIGURE 4–23 For systems in cascade, the resultant efficiency is the product of the efficiencies of the individual stages.

EXAMPLE 4–14

a. For a certain system, $\eta_1 = 95\%$, $\eta_2 = 85\%$, and $\eta_3 = 75\%$. What is η_T?
b. If $\eta_T = 65\%$, $\eta_2 = 80\%$, and $\eta_3 = 90\%$, what is η_1?

Solution

a. Convert all efficiencies to a decimal value, then multiply. Thus, $\eta_T = \eta_1\eta_2\eta_3 = (0.95)(0.85)(0.75) = 0.61$ or 61%.
b. $\eta_1 = \eta_T/(\eta_2\eta_3) = (0.65)/(0.80 \times 0.90) = 0.903$ or 90.3%

EXAMPLE 4–15

A motor drives a pump through a gearbox (Figure 4–24). Power input to the motor is 1200 W. How many horsepower are delivered to the pump?

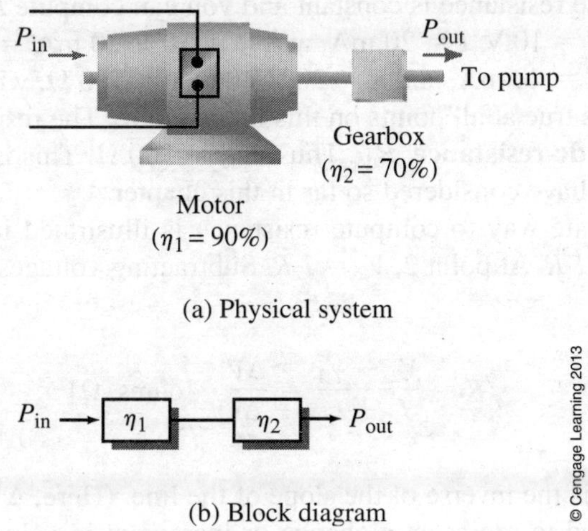

FIGURE 4–24 Motor driving pump through a gearbox.

Solution The efficiency of the motor/gearbox combination is $\eta_T = \eta_1 \times \eta_2 = (0.90)(0.70) = 0.63$. The output of the gearbox (and hence the input to the pump) is $P_{out} = \eta_T \times P_{in} = (0.63)(1200\text{ W}) = 756\text{ W}$. Converting to horsepower, $P_{out} = (756\text{ W})/(746\text{ W/hp}) = 1.01\text{ hp}$.

EXAMPLE 4–16

The motor of Figure 4–24 is operated from 9:00 A.M. to 12:00 noon and from 1:00 P.M. to 5:00 P.M. each day, for 5 days a week, outputting 7 hp to a load. At \$0.085/kWh, it costs \$22.19 per week for electricity. What is the efficiency of the motor/gearbox combination?

Solution

$$W_{in} = \frac{\$22.19/\text{wk}}{\$0.085/\text{kWh}} = 261.1\text{ kWh/wk}$$

The motor operates 35 h/wk. Thus,

$$P_{in} = \frac{W_{in}}{t} = \frac{261.1\text{ kWh/wk}}{35\text{ h/wk}} = 7460\text{ W}$$

$$\eta_T = \frac{P_{out}}{P_{in}} = \frac{(7\text{ hp} \times 746\text{ W/hp})}{7460\text{ W}} = 0.7$$

Thus, the motor/gearbox efficiency is 70%.

4.7 Nonlinear and Dynamic Resistances

All resistors considered so far have constant values that do not change with voltage or current. Such resistors are termed **linear** or **ohmic** since their current-voltage (*I-V*) plot is a straight line. However, the resistance of some materials changes with voltage or current as you saw in Chapter 3, Section 3.10. These materials are termed **nonlinear** because their *I-V* plot is curved (Figure 4–25).

Since the resistance of all materials changes with temperature, all resistors are to some extent nonlinear, since they all produce heat and this heat changes their resistance. For most resistors, however, this effect is small over their normal operating range, and such resistors are considered to be linear. (The commercial resistors shown in Figure 3–8 and most others that you will encounter in this book are linear.)

Since an *I-V* plot is a graph of Ohm's law, resistance can be computed from the ratio *V/I*. First, consider the linear plot of Figure 4–25. Because the slope is constant, the resistance is constant and you can compute *R* at any point. For example, at $V = 10$ V, $I = 20$ mA, and $R = 10$ V/20 mA $= 500\ \Omega$. Similarly, at $V = 20$ V, $I = 40$ mA, and $R = 20$ V/40 mA $= 500\ \Omega$, which is the same as before. This is true at all points on this linear curve. The resulting resistance is referred to as **dc resistance**, R_{dc}. Thus, $R_{dc} = 500\ \Omega$. This is the type of resistance that we have considered so far in this chapter.

An alternate way to compute resistance is illustrated in Figure 4–26. At point 1, $V_1 = I_1R$. At point 2, $V_2 = I_2R$. Subtracting voltages and solving for *R* yields

$$R = \frac{V_2 - V_1}{I_2 - I_1} = \frac{\Delta V}{\Delta I} \quad [\text{ohms, } \Omega] \qquad (4\text{–}20)$$

where $\Delta V/\Delta I$ is the inverse of the slope of the line. (Here, Δ is the Greek letter delta. It is used to represent a change or increment in value.) To illustrate, if you select ΔV to be 20 V, you find that the corresponding ΔI from Figure 4–26 is 40 mA. Thus, $R = \Delta V/\Delta I = 20$ V/40 mA $= 500\ \Omega$ as before. Resistance

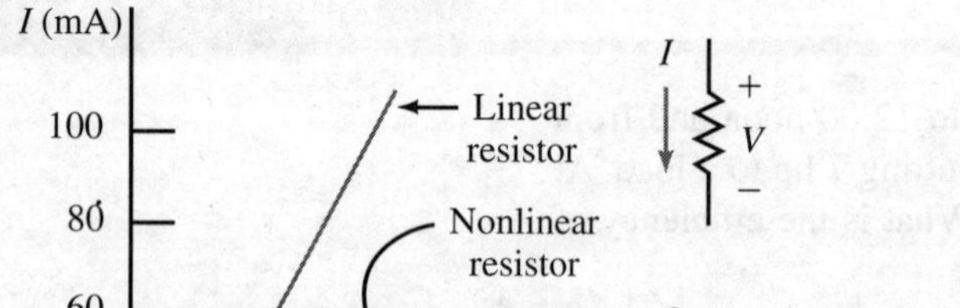

FIGURE 4–25 Linear and nonlinear resistance characteristics.

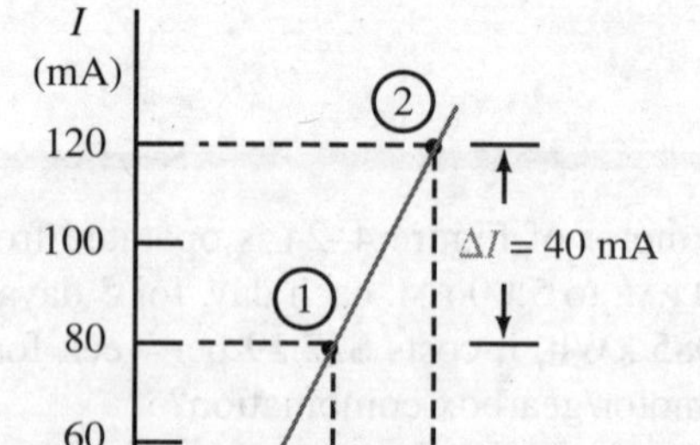

FIGURE 4–26 $R = \Delta V/\Delta I = 20$ V/40 mA $= 500\ \Omega$.

calculated as in Figure 4–26 is called **differential** or **dynamic resistance.** For linear resistors, $R_{\text{dynamic}} = R_{\text{dc}}$.

Now consider the nonlinear resistance plot of Figure 4–25. At $V = 20$ V, $I = 20$ mA. Therefore, $R_{\text{dc}} = 20$ V/20 mA $= 1.0$ kΩ; at $V = 120$ V, $I = 60$ mA, and $R_{\text{dc}} = 120$ V/60 mA $= 2.0$ kΩ. This resistance therefore increases with applied voltage. However, for small variations about a fixed point on the curve, the dynamic resistance will be constant. This is an important concept that you will utilize later in your study of electronics. We leave the topic for now.

4.8 Computer-Aided Circuit Analysis

MultiSIM

PSPICE

We end our introduction to Ohm's law by solving several simple problems using Multisim and PSpice. Note, however, that although both software packages are described here, most users will select one or the other. (Regardless of which one you choose, you should install its corresponding shortcut icon on your screen when you install the software. A shortcut icon permits you to launch your software with a click of the mouse. If you do not have the shortcut, you can still launch from the Windows Start button menu as detailed in Appendix A.)

You will create a number of files throughout this course, so an organized filing system helps. First, create a folder on your computer called Multisim Files (if you are a Multisim user), or PSpice Files (if you are a PSpice user). (This is a Windows file creation process and is not detailed here. If you need help, see Appendix A.) In this folder, create a subfolder called Ch 4 in which to save the files for Chapter 4. As you go through the book, add a subfolder for each new chapter.

To begin the analysis, you must first build the circuit on your screen, a process referred to as **schematic capture.** Let us use the circuit of Figure 4–27 to get started. Because this is your first look at circuit simulation, considerable detail is included. (Although the procedures may seem complex, they become quite intuitive with a little practice.) When setting up a problem for analysis, there are generally several ways that you can proceed, and with experience you will learn shortcuts. In the meantime, to get started, use the methods suggested here.

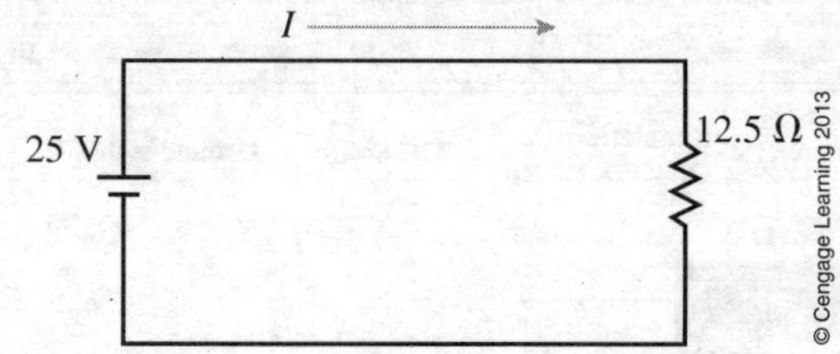

FIGURE 4–27 Simple circuit to illustrate computer analysis.

Multisim

For the most part, Multisim usage is straightforward. The example below shows you how to get started—or alternatively, proceed as per Note 1.

NOTES...

Multisim Operational Notes

1. We have put together a "hands-on" Multisim Beginners Tutorial that you might find helpful. Using annotated screen shots and illustrated instructions, it walks you step by step through the process of creating and running the simulation of Figure 4–28 on your screen. Visit our Web site at cengagebrain.com and follow the links to *Multisim Beginners Tutorial*.
2. Unless directed otherwise, use the left button for all operations.
3. Multisim circuits require a ground.
4. To select a component already placed on the screen, click it. A dashed box appears, indicating selection.
5. To deselect a component, place the cursor outside the dashed box and click the left button.
6. To reposition a component, select it, then, holding the left button down, drag it to the new position and release.
7. To delete a component, select it, then press the delete key.
8. We frequently use a shorthand designation to represent sequential actions. For example, if you are told to click *View/Toolbars*, this is a shorthand way of saying, "click *View* on the menu bar, then select *Toolbars* on the drop-down list that appears."
9. To rotate a component, select it—then, while holding down the *Ctrl* key, press the *R* key. (In shorthand notation, this may be written as *Ctrl/R*. Each *Ctrl/R* rotates the component by 90°.)
10. To place a connector dot, select *Place* on the menu bar, then click *Junction*. Position the dot on the screen and click to place it.
11. A maximum of four wires may be connected to any point (junction).

Getting Started

Double-click the Multisim shortcut icon on your computer. A screen similar to that of Figure 4–28 (except that it is blank) opens. Place your mouse pointer on *View* and left-click. If either *Spreadsheet View* or *Description Box* is selected, click the selected item to deselect it.

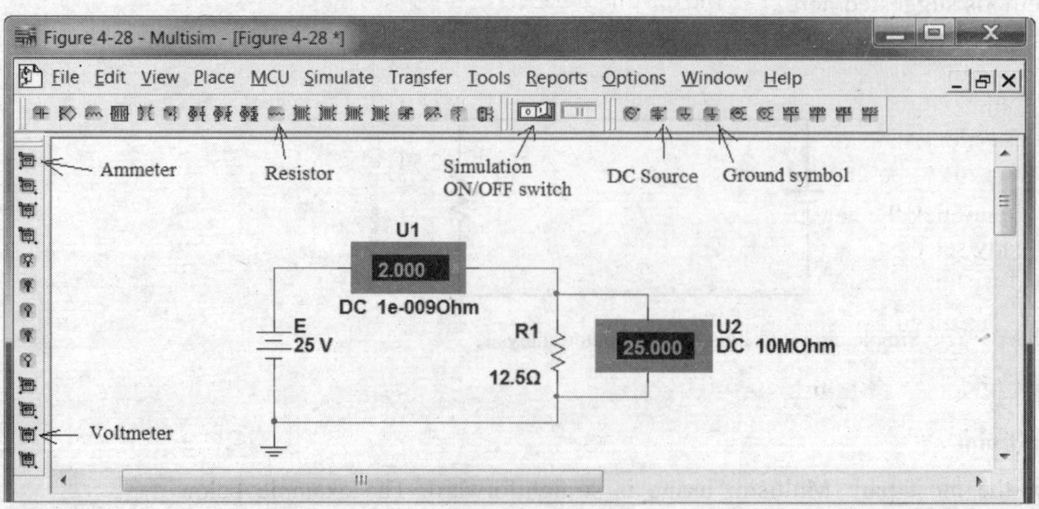

FIGURE 4–28 Multisim simulation of circuit of Figure 4–27.

To ensure that you have the toolbars selected that you need for this problem, right-click View, then confirm that the following are selected: *Basic, Measurement Components, Power Source Component,* and *Simulation Switch.* (For legacy software users, select *Simulation* instead of *Simulation Switch.*) Click any that are not selected to add them to your screen. (Don't worry if others are selected—you may need them in later examples.) Now compare your screen to that of Figure 4–28. You will probably find that the screen layouts are different, but that all items pointed to by arrows on Figure 4–28 are found somewhere on your screen.

To gain familiarity with your workspace screen, experiment a bit. For example, locate each of the items indicated by arrows on Figure 4–28 and place your mouse pointer on them. A small banner opens identifying each. (Familiarity is important because, in what follows, we refer to items by name and you need to be able to locate them.)

Creating Your First Multisim Circuit

First, briefly study the *Multisim Operational Notes*, then proceed to build the circuit as follows. (If you have utilized the *CircuitSim* interactive study companion files in this and previous chapters, you will be somewhat familiar with some of what follows.)

- With your mouse pointer, locate the dc power source icon (the dc voltage source), left-click it and move your mouse pointer out into the workspace screen. You will see that it has a battery symbol attached to it. (This symbol is referred to as a "ghost.")
- Position the ghost on your workspace as in Figure 4–28, then click the left button to place it. (If the ghost remains after the source has been placed, right-click to cancel.)
- Repeat the procedure to select and place the ammeter.
- Click the resistor symbol and move it to the workspace. Note that it is horizontal. Before placing, press Ctrl/R (see Note 9), then click to place the rotated resistor. (If you place it before you rotate it, select it and rotate as per Notes 4 and 9.)
- Similarly, select and place the ground symbol and the voltmeter symbol. Reposition anything that needs to be moved by dragging it with the mouse.
- "Wire" the circuit as follows: Move the mouse pointer to the tip of the battery symbol and it will change to a crosshair. Click to start the connection and a wire follows. Move to the destination pin (in this case, the left ammeter pin) then click again. The wire attaches. Repeat to attach a wire from the battery to the ground.
- Continue until the wiring is finished. (When placing a wire, you can click at points along its path to nail it in place, e.g., at a corner.) Reposition any wire that needs to be moved by dragging it.
- Double-click the battery symbol. In its dialog box, click the Value tab (if not already selected), change the voltage to **25,** then click OK. Similarly, change the resistor value to 12.5, select Ohms, then click OK.
- On your menu bar, click *File/Save As*. A dialog box opens. At the far right of the box labelled *Save In*, click the small arrow, and navigate to your newly created folder (i.e., Multisim Files/Ch 4) and select it. Now locate the *File Name* box near the bottom of the dialog window, type **Figure 4–28**, then click *Save*.
- Activate the circuit by clicking the simulation ON/OFF switch.

After simulation is complete, the meters will display circuit current (2 A) and resistor voltage (25 V) as depicted in Figure 4–28. To close Multisim, click its Windows close button.

PRACTICE PROBLEMS 4

Relaunch Multisim and rebuild the circuit of Figure 4–28 from scratch, except rotate the source symbol by 180° so that the source voltage is −25 V. Click *Save As* and save as Practice P4. Run the simulation and reconcile your results with the voltage polarity and current direction conventions discussed in this chapter and in Chapter 2.

PSpice

PSpice provides you with a variety of analysis options. For this example, we will use bias point analysis. While this option has limited capability, it is easy to use and is a good choice for dc circuits with fixed sources.

Getting Started

Double-click the PSpice shortcut icon—see Notes 1 and 2. On the screen that opens, click *File,* select *New*, then *Project.* In the *New Project* box, type **Figure 4–29** into the *Name* box, ensure that *Analog or Mixed A/D* is selected, then click the *Browse* button. Navigate your way through the directory tree until you locate *Ch 4*, (the subfolder that you just created), double-click, then click *OK* to close the box.

Click *OK* again. This opens the *Create PSpice Project* box. Select *Create a blank project,* then click *OK*. On the next screen, you will see *Schematic 1: Page1*. (This is the screen on which you will build your circuit. Click its *Maximize* button. It should look much like Figure 4–29 of the textbook except that it will be blank, and the *Place Part* box on the right-hand side will not be open.) Note the buttons, tools, and other items identified on Figure 4–29. Locate these on your screen. Poke around with your mouse and spend a bit of time studying the screen to get familiar with it. (Familiarity is important as we refer to items by name and you need to be able to locate them.) Note that a number of items are inactive, but they will become active when needed.

NOTES...

1. PSpice is available as a free download from Cadence Design Systems Inc.; see Appendix A to get your copy. (PSpice is available as a full-featured commercial version or as a demo version. It is the demo version that is free, but it is completely adequate for all examples in this book.)
2. The icon will likely be labelled ORCAD Capture CIS Demo on your screen.

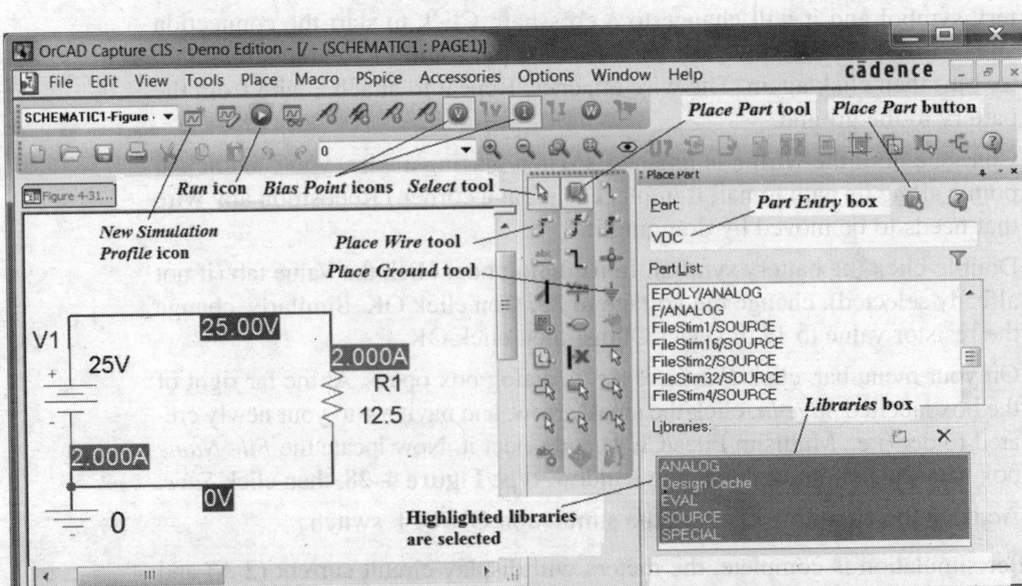

FIGURE 4–29 PSpice simulation of Figure 4–27 using the bias point analysis technique.

Building Your First PSpice Circuit

You are now ready to build the circuit of Figure 4–27. It should look like Figure 4–29 when completed. Review the *PSpice Operational Notes*, refer to the icon identifications in Figure 4–29 for guidance, then proceed as follows:

- Click the *Place Part* tool and the *Place Part* box opens. In the *Libraries* box, make sure the following are present: *ANALOG, EVAL, SOURCE,* and *SPECIAL.* If any are missing, add them. (For example, to add *EVAL,* click the *Add Library* button [which is not labelled, but looks like a dotted square to the left of the X button], select *EVAL* in the library list, then click *Open.*)
- Ensure that all entries in the libraries box are highlighted as in Figure 4–29. If they aren't, click the first entry (to highlight it, press and hold the shift key on your keyboard, then click the last entry in the list. This should highlight all entries as indicated in Figure 4–29. This means that all are selected.)
- If you know the name of the part you want, type it into the *Part Entry* box. For example, to get the dc source, type **VDC**, then click the *Place Part* button.
- A battery symbol will appear as a "ghost" attached to your cursor. Position it as in Figure 4–29, then click once. To end placement, move your pointer to an empty spot, right-click, then left-click *End Mode*—see *PSpice Operational Notes*, Note 2. This gets back your normal mouse pointer.
- Drag your mouse over entry *VDC* in the *Parts Entry* box (to highlight it), type **R** then click the *Place Part* button. A horizontal resistor appears attached to your pointer. Rotate it by pressing the *R* key on your keyboard, place it using the left button, then end placement as per *PSpice Operational Note 2*.

NOTES...

PSpice Operational Notes

1. Unless directed otherwise, use the left mouse button for all operations.
2. To end the placement of a part, click the right mouse button, then left-click *End Mode.* (Alternatively, you can press the *Esc* key on your keyboard twice.)
3. To select a particular component on your schematic, place the mouse pointer on it and click. The selected component changes color.
4. To deselect a selected component, move the pointer to an empty place on the screen and click.
5. To delete a component from your schematic, select it, then press the *Delete* key.
6. To drag a component, place the pointer over it, press and hold the left button, drag the component to where you want it, and then release.
7. To change a default value, place the pointer over the numerical value (not the component symbol) and double-click. Change the appropriate value.
8. To rotate a component, select it, click the right button, then click *Rotate.* Alternately, press the *R* key on your keyboard.
9. There must be no space between a value and its unit. Thus, for example, enter 25V, not 25 V. Alternatively just enter the value without the unit as, for instance, 25.
10. All PSpice circuits must have a ground.
11. When placing a wire, you can click at points along its path to nail it in place, as for example, at a corner. You can click on top of another wire to connect two wires together. To improve the looks of the circuit, reposition wires or components as desired by dragging them.
12. Not all buttons in the workspace are explicitly labelled. Use your mouse pointer to explore and identify unlabeled buttons.

- Click the *Place Ground* tool, select either *0/CAPSYM* or *0/SOURCE*, click *OK*, place the symbol, then end placement.
- To begin wiring, click the *Place Wire* tool—see Figure 4–29. The cursor changes to a +. Position the cursor in the little box at the bottom end of the battery symbol, and click. This attaches a wire. Now move the cursor to the top of the ground symbol, pulling the wire with you. Click and it attaches. Wire the rest of the circuit in this manner, then end wire placement by right-clicking, then left-click *End Wire*.
- Click the *Select Tool* button if necessary to get your normal mouse pointer back.
- You need to update component values. To change the resistor value, double-click its 1k default value (not its symbol), type **12.5** into the *Value* box, then click *OK*. Similarly, change the source voltage to 25V. (See *Operational Note 9*.)
- Click the *New Simulation Profile* icon, enter your file name **Figure 4–29** in the *Name* box, leave *None* in the *Inherit From* box, then click *Create*. A *Simulations Settings* box opens. (It will likely open behind your workspace screen, so you may have to temporarily minimize your workspace screen to see it.) Click the *Analysis* tab, from *Analysis type,* select *Bias Point,* then click *OK*.
- Enlarge your workspace screen again, then click the bias point voltage and bias point current icons to their ON states.
- Click the *Save Document* icon, then the *Run* icon. If an *Undo Warning* box opens, click *Yes*.
- Voltage and current values will appear on the screen as indicated in Figure 4–29. (A grayed-out window may open. Just move it out of the way.)
- To exit from PSpice, click its Windows close button. Click *Yes All* to save changes.

PSpice Voltmeters and Ammeters

As an alternative to the bias point method, you can use PSpice voltmeters and ammeters to display results. However, the bias point method is much simpler to use and is completely adequate for all dc problems in this book. For ac analysis however (Chapter 15 and beyond), you will find that the bias point method does not work—it cannot handle ac. At that point, we show how to use ac meters.

Plotting Ohm's Law

As a final example for this chapter, use PSpice to compute and graph current versus voltage. By varying the source voltage of Figure 4–27 and plotting *I*, for example, you can obtain a plot of Ohm's law. Proceed as follows.

- Launch PSpice, click *File/New/Project*, and in the *New Project Name* box, type **Figure 4–31**. Ensure that the *Analog or Mixed A/D* is selected. Use the *Browse* button and navigate to the target folder (i.e., *Ch 4*), double-click, then click *OK*.
- Click *OK*. This opens the *Create PSpice Project* box. Select *Create a blank project,* then click *OK*.
- This opens the *Schematic 1: Page 1* screen. Now rebuild the circuit of Figure 4–27. Set *R* to 10 ohms. Leave the source voltage at its default value of 0 volts.
- Click the *New Simulation Profile* icon, type **Figure 4–31** as your file name, leave *None* in the *Inherit From* box, then click *Create*. A *Simulations Setting* box opens. (You may have to minimize your workspace screen to see it.) Click the *Analysis* tab, select *DC Sweep* from the *Analysis Type* list and *Primary Sweep* from the *Options* list. For *Sweep Variable*, choose *Voltage Source*, and in the *Name* box type the name of your voltage source. (It will be V1.) Choose

Linear as the sweep type, key **0** into the *Start Value* box, **100** into the *End Value* box, **5** into the *Increment* box, then click *OK*. (This sets PSpice up to sweep the voltage from 0 V to 100 V in 5 V steps.)

- Click the *Current Marker* icon and position the marker as shown in Figure 4–30. (It will be gray, not green.) Press *Esc* to get your normal mouse pointer back.
- Click the *Save Document* icon, then the *Run* icon. If an *Undo Warning* box opens, click *Yes*.
- The Ohm's law plot of Figure 4–31 will appear on your screen (except that it may have a black background). You can either keep the black background or change it as follows: Click *Tools, Options, Color Settings, Background,* select white, then click *OK*.
- Using the cursor (see Appendix A), read values from the graph. Verify each reading with your calculator.

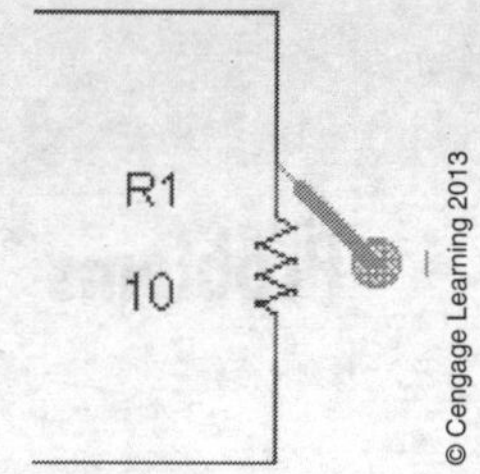

FIGURE 4–30 Using a current marker to display current. (The marker will be gray when you place it, but it takes on the color of the trace that it creates when you run the simulation.)

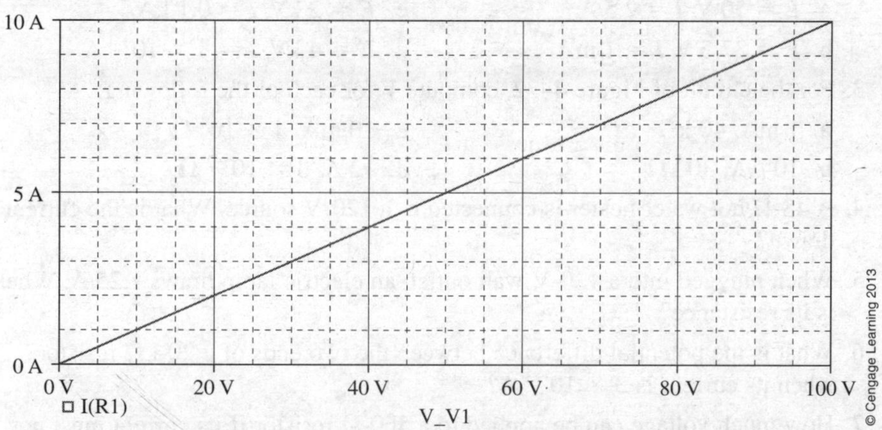

FIGURE 4–31 Ohm's law plot for the circuit of Figure 4–29 with $R = 10\ \Omega$.

Putting It into Practice

You have been assigned to perform a design review and cost analysis of an existing product. The product uses a 12-V ±5% power supply. After a catalog search, you find a new power supply with a specification of 12 V ±2%, which, because of its newer technology, costs only five dollars more than the one you are currently using. Although it provides improved performance for the product, your supervisor won't approve it because of the extra cost. However, the supervisor agrees that if you can bring the cost differential down to $3.00 or less, you can use the new supply. You then look at the schematic and discover that a number of precision 1% resistors are used because of the wide tolerance of the old supply. (As a specific example, a ±1% tolerance 220-kΩ resistor is used instead of a standard 5% tolerance resistor because the loose tolerance of the original power supply results in current through the resistor that is out of spec. Its current must be between 50 mA and 60 mA when the 12-V supply is connected across it. Several other such instances are present in the product, a total of 15 in all.) You begin to wonder whether you could replace the precision resistors (which cost $0.24 each) with standard 5% resistors (which cost $0.03 each) and save enough money to satisfy your supervisor. Perform an analysis to determine if your hunch is correct.

Problems

4.1 Ohm's Law

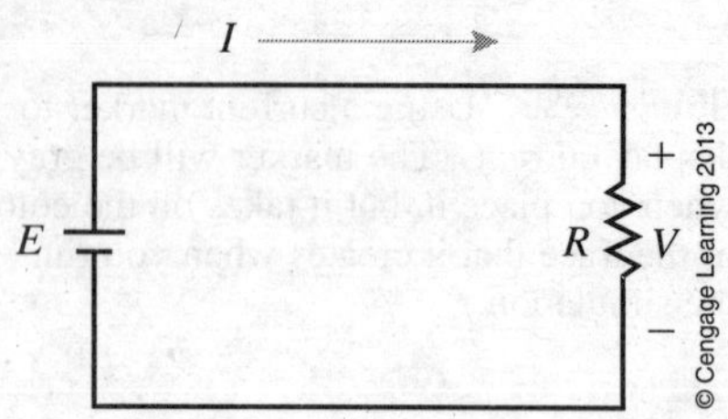

FIGURE 4–32

1. For the circuit of Figure 4–32, determine the current I for each of the following. Express your answer in the most appropriate unit: amps, milliamps, microamps, and so on.
   a. $E = 40$ V, $R = 20\ \Omega$
   b. $E = 35$ mV, $R = 5$ mΩ
   c. $V = 200$ V, $R = 40$ kΩ
   d. $E = 10$ V, $R = 2.5$ MΩ
   e. $E = 7.5$ V, $R = 2.5 \times 10^3\ \Omega$
   f. $V = 12$ kV, $R = 2$ MΩ
2. Determine R for each of the following. Express your answer in the most appropriate unit: ohms, kilohms, megohms, and so on.
   a. $E = 50$ V, $I = 2.5$ A
   b. $E = 37.5$ V, $I = 1$ mA
   c. $E = 2$ kV, $I = 0.1$ kA
   d. $E = 4$ kV, $I = 8 \times 10^{-4}$ A
3. For the circuit of Figure 4–32, compute V for each of the following:
   a. 1 mA, 40 kΩ
   b. 10 μA, 30 kΩ
   c. 10 mA, $4 \times 10^4\ \Omega$
   d. 12 A, $3 \times 10^{-2}\ \Omega$
4. A 48-Ω hot water heater is connected to a 120-V source. What is the current drawn?
5. When plugged into a 120-V wall outlet, an electric lamp draws 1.25 A. What is its resistance?
6. What is the potential difference between the two ends of a 20-kΩ resistor when its current is 3×10^{-3} A?
7. How much voltage can be applied to a 560-Ω resistor if its current must not exceed 50 mA?
8. A relay with a coil resistance of 240 Ω requires a minimum of 50 mA to operate. What is the minimum voltage that will cause it to operate?
9. For Figure 4–32, if $E = 30$ V and the conductance of the resistor is 0.2 S, what is I? Hint: See Section 3.11, Chapter 3.
10. If $I = 36$ mA when $E = 12$ V, what is I if the 12-V source is
    a. replaced by an 18-V source?
    b. replaced by a 4-V source?
11. Current through a resistor is 15 mA. If the voltage drop across the resistor is 33 V, what is its color code?
12. For the circuit of Figure 4–33,
    a. If $E = 28$ V, what does the meter indicate?
    b. If $E = 312$ V, what does the meter indicate?
13. For the circuit of Figure 4–33, if the resistor is replaced with one with red, red, black color bands, at what voltage do you expect the fuse to blow?
14. A 20-V source is applied to a resistor with color bands brown, black, red, and silver
    a. Compute the nominal current in the circuit.
    b. Compute the minimum and maximum currents based on the tolerance of the resistor.
15. An electromagnet is wound with AWG 30 copper wire. The coil has 800 turns and the average length of each turn is 3 inches. When connected to a 48-V dc source, what is the current (assuming voltage is constant)
    a. at 20° C?
    b. at 40° C?

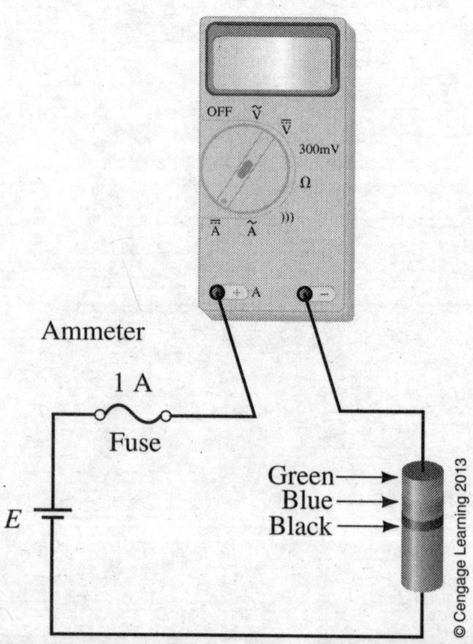

FIGURE 4–33

16. You are to build an electromagnet with 0.643-mm diameter copper wire. To create the required magnetic field, the current in the coil must be 1.75 A at 20° C. The electromagnet is powered from a 9.6-V dc source. How many meters of wire do you need to wind the coil?
17. A resistive circuit element is made from 100 m of 0.5-mm diameter aluminum wire. If the current at 20° C is 200 mA, what is the applied voltage?
18. Prepare an Ohm's law graph similar to Figure 4–7 for a 2.5-kΩ and a 5-kΩ resistor. Compute and plot points every 5 V from $E = 0$ V to 25 V. Reading values from the graph, find current at $E = 14$ V.
19. Figure 4–34 represents the *I-V* graph for the circuit of Figure 4–32. What is *R*?
20. For a resistive circuit, *E* is quadrupled and *R* is halved. If the new current is 24 A, what was the original current?
21. For a resistive circuit, $E = 100$ V. If *R* is doubled and *E* is changed so that the new current is double the original current, what is the new value of *E*?
22. You need to measure the resistance of an electric heater element, but have only a 12-V battery and an ammeter. Describe how you would determine its resistance. Include a sketch.
23. If 25 m of 0.1-mm diameter Nichrome wire is connected to a 12-V battery, what is the current at 20° C?
24. If the current is 0.5 A when a length of AWG 40 copper wire is connected to 48 V, what is the length of the wire in meters? Assume 20° C.

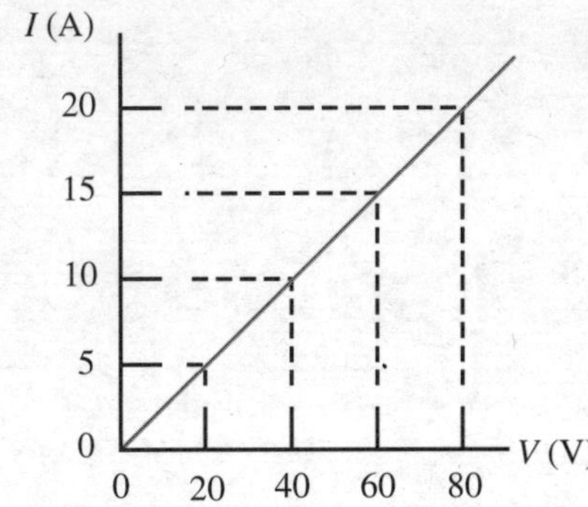

FIGURE 4–34

4.2 Voltage Polarity and Current Direction

25. For each resistor of Figure 4–35, determine voltage *V* and its polarity or current *I* and its direction as applicable.
26. The ammeters of Figure 4–36 have autopolarity. Determine their readings, magnitude, and polarity.

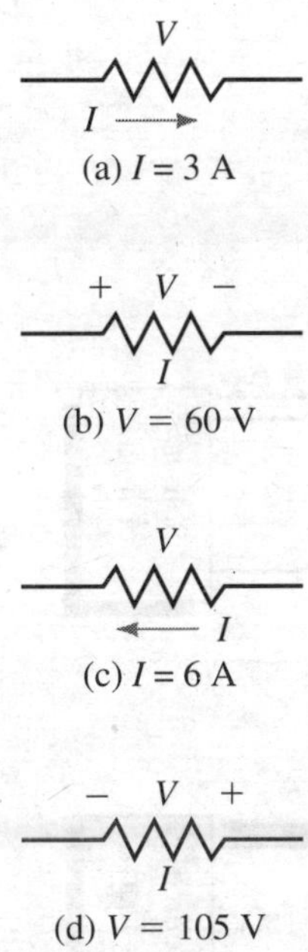

FIGURE 4–35 All resistors are 15 Ω.

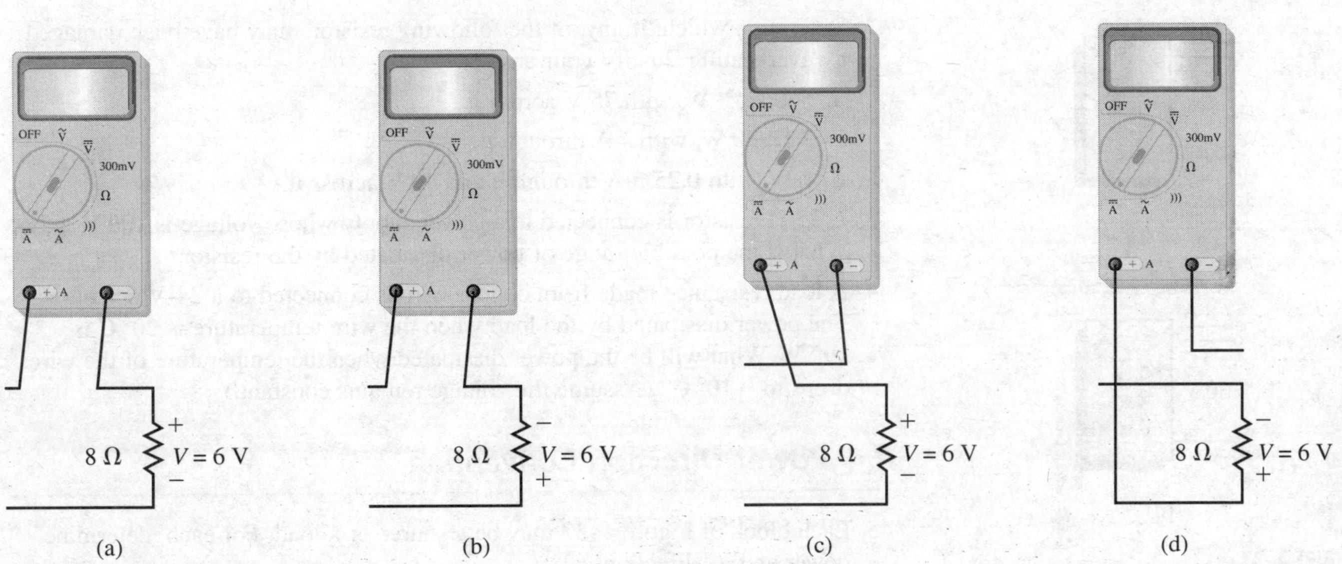

FIGURE 4–36

4.3 Power

27. A resistor dissipates 723 joules of energy in 3 minutes and 47 seconds. Calculate the rate at which energy is being transferred to this resistor in joules per second. What is its power dissipation in watts?
28. How long does it take for a 100-W soldering iron to dissipate 1470 J?
29. A resistor draws 3 A from a 12-V battery. How much power does the battery deliver to the resistor?
30. A 120-V electric coffee maker is rated 960 W. Determine its resistance and rated current.
31. A 1.2-kW electric heater has a resistance of 6 Ω. How much current does it draw?
32. A warning light draws 125 mA when dissipating 15 W. What is its resistance?
33. How many volts must be applied to a 3-Ω resistor to result in a power dissipation of 752 W?
34. What *IR* drop occurs when 90 W is dissipated by a 10-Ω resistor?
35. A resistor with color bands brown, black, and orange dissipates 0.25 W. Compute its voltage and current.
36. A 2.2-kΩ resistor with a tolerance of ±5% is connected to a 12-V dc source. What is the possible range of power dissipated by the resistor?
37. A portable radio transmitter has an input power of 0.455 kW. How much current does it draw from a 12-V battery?
38. For a resistive circuit, $E = 12$ V.
    a. If the load dissipates 8 W, what is the current in the circuit?
    b. If the load dissipates 36 W, what is the load resistance?
39. A motor delivers 3.56 hp to a load. How many watts is this?
40. The load on a 120-V circuit consists of six 100-W lamps, a 1.2-kW electric heater, and an electric motor drawing 1500 W. If the circuit is fused at 30 A, what happens when a 900-W toaster is plugged in? Justify your answer.
41. A 0.27-kΩ resistor is rated 2 W. Compute the maximum voltage that can be applied and the maximum current that it can carry without exceeding its rating.
42. Determine which, if any, of the following resistors may have been damaged by overheating. Justify your answer.
    a. 560 Ω, ½ W, with 75 V across it.
    b. 3 Ω, 20 W, with 4 A through it.
    c. ¼ W, with 0.25 mA through it and 40 V across it.
43. A 25-Ω resistor is connected to a power supply whose voltage is 100 V ±5%. What is the possible range of power dissipated by the resistor?
44. A load resistance made from copper wire is connected to a 24-V dc source. The power dissipated by the load when the wire temperature is 20° C is 192 W. What will be the power dissipated when the temperature of the wire drops to −10° C? (Assume the voltage remains constant.)

4.4 Power Direction Convention

45. Each block of Figure 4–37 may be a source or a load. For each, determine power and its direction.

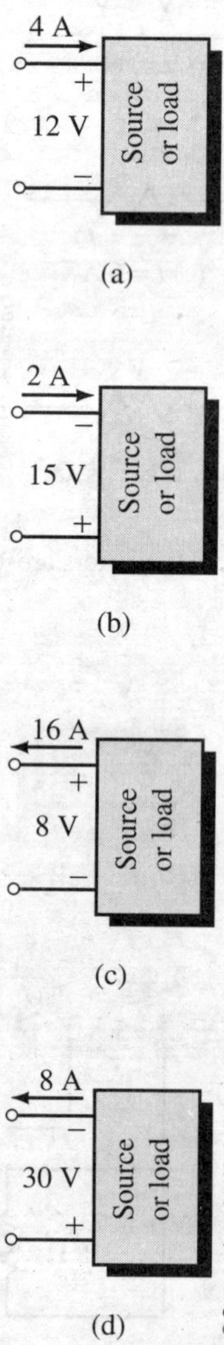

FIGURE 4–37

46. The 12-V battery of Figure 4–38 is being "charged" by a battery charger. The current is 4.5 A as indicated.
    a. What is the direction of current?
    b. What is the direction of power flow?
    c. What is the power to the battery?

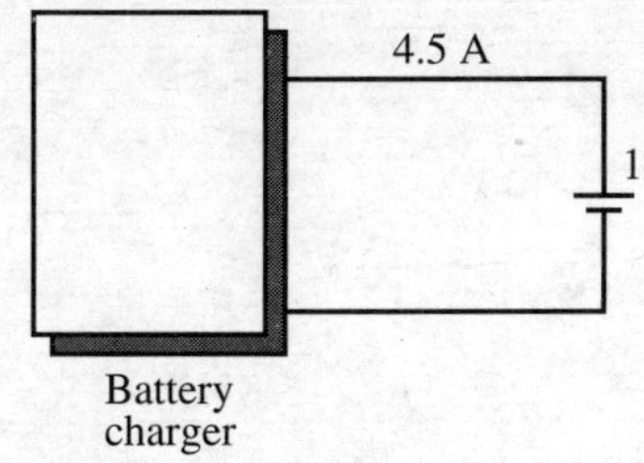

FIGURE 4–38

4.5 Energy

47. A 40-W night safety light burns for 9 hours.
    a. Determine the energy used in joules.
    b. Determine the energy used in watthours.
    c. At $0.08/kWh, how much does it cost to run this light for 9 hours?
48. An indicator light on a control panel operates continuously, drawing 20 mA from a 120-V supply. At $0.09 per kilowatthour, how much does it cost per year to operate the light?
49. Determine the total cost of using the following at $0.11 per kWh:
    a. a 900-W toaster for 5 minutes,
    b. a 120-V, 8-A heater for 1.7 hours,
    c. an 1100-W dishwasher for 36 minutes,
    d. a 120-V, 288-Ω soldering iron for 24 minutes.
50. An electric device with a cycle time of 1 hour operates at full power (400 W) for 15 minutes, at half power for 30 minutes, then cuts off for the remainder of the hour. The cycle repeats continuously. At $0.10/kWh, determine the yearly cost of operating this device.
51. While the device of Problem 50 is operating, two other loads as follows are also operating:
    a. a 4-kW heater, continuously,
    b. a 3.6-kW heater, on for 12 hours per day.

    Calculate the yearly cost of running all loads.
52. At $0.08 per kilowatthour, it costs $1.20 to run a heater for 50 hours from a 120-V source. How much current does the heater draw?
53. If there are 24 slices in a loaf of bread and you have a two-slice, 1100-W toaster that takes 1 minute and 45 seconds to toast a pair of slices, at $0.13/kWh, how much does it cost to toast a loaf?

4.6 Efficiency

54. The power input to a motor with an efficiency of 85% is 690 W. What is its power output?
    a. in watts
    b. in hp
55. The power output of a transformer with $\eta = 97\%$ is 50 kW. What is its power in?
56. For a certain device, $\eta = 94\%$. If losses are 18 W, what are P_{in} and P_{out}?
57. The power input to a device is 1100 W. If the power lost due to various inefficiencies is 190 W, what is the efficiency of the device?
58. A 240-V, 4.5-A water heater produces heat energy at the rate of 3.6 MJ per hour. Compute
    a. the efficiency of the water heater,
    b. the annual cost of operation at $0.09/kWh if the heater is on for 6 h/day.

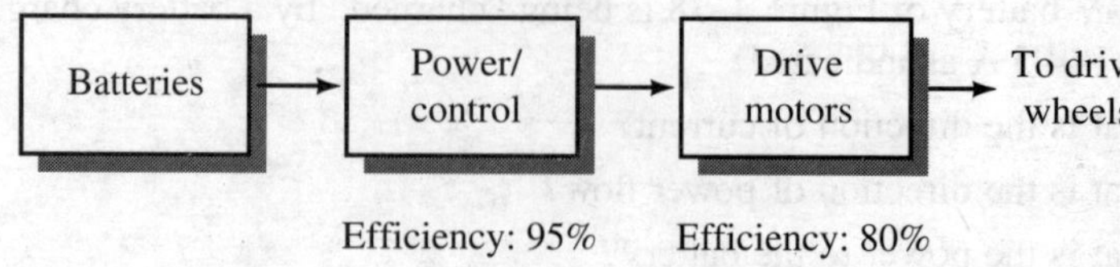

FIGURE 4–39

59. A 120-V dc motor with an efficiency of 89% draws 15 A from the source. What is its output horsepower?
60. A 120-V dc motor develops an output of 3.8 hp. If its efficiency is 87%, how much current does it draw?
61. The power/control system of an electric car consists of a 48-V onboard battery pack, electronic control/drive unit, and motor (Figure 4–39). If 180 A are drawn from the batteries, how many horsepower are delivered to the drive wheels?
62. Show that the efficiency of n devices or systems in cascade is the product of their individual efficiencies, that is, that $\eta_T = \eta_1 \times \eta_2 \times \cdots \times \eta_n$.
63. A 120-V dc motor drives a pump through a gearbox (Figure 4–24). If the power input to the pump is 1100 W, the gearbox has an efficiency of 75%, and the power input to the motor is 1600 W, determine the horsepower output of the motor.
64. If the motor of Problem 63 is protected by a 15-A circuit breaker, will it open? Compute the current to find out.
65. If the overall efficiency of a radio transmitting station is 55% and it transmits at 35 kW for 24 h/day, compute the cost of energy used per day at \$0.09/kWh.
66. In a factory, two machines, each delivering 27 kW are used on an average for 8.7 h/day, 320 days/year. If the efficiency of the newer machine is 87% and that of the older machine is 72%, compute the difference in cost per year of operating them at \$0.10 per kilowatthour.

4.7 Nonlinear and Dynamic Resistances

67. A voltage-dependent resistor has the I-V characteristic of Figure 4–40.
    a. At $V = 25$ V, what is I? What is R_{dc}?
    b. At $V = 60$ V, what is I? What is R_{dc}?
    c. Why are the two values different?

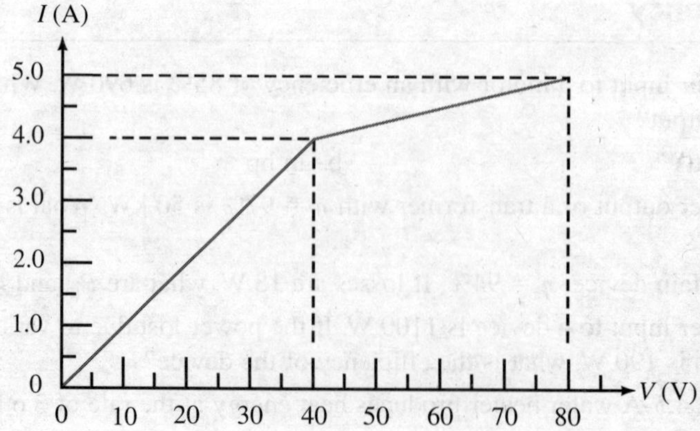

FIGURE 4–40

68. For the resistor of Figure 4–40:

a. Determine $R_{dynamic}$ for V between 0 and 40 V.

b. Determine $R_{dynamic}$ for V greater than 40 V.

c. If V changes from 20 V to 30 V, how much does I change?

d. If V changes from 50 V to 70 V, how much does I change?

4.8 Computer-Aided Circuit Analysis

69. Set up the circuit of Figure 4–32 and solve for currents for the voltage/resistance pairs of Problem 1a, 1c, 1d, and 1e. Multisim

70. A battery charger with a voltage of 12.9 V is used to charge a battery, Figure 4–41(a). The internal resistance of the charger is 0.12 Ω and the voltage of the partially run-down battery is 11.6 V. The equivalent circuit for the charger/battery combination is shown in (b). You reason that, since the two voltages are in opposition, net voltage for the circuit will be 12.9 V − 11.6 V = 1.3 V and, thus, the charging current I will be 1.3 V/0.12 Ω = 10.8 A. Set up the circuit of (b) and use Multisim to verify your conclusion. Multisim

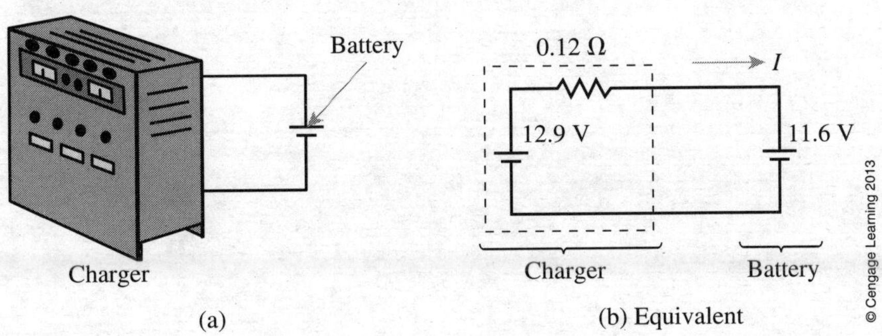

FIGURE 4–41

71. Click *Place* on the menu bar; click *Component*; from the *Group* list, select *Basic*; scroll down and click *Switch*; select *SPDT*, then click *OK*. Place on the screen, then close the *Select a Component* box. Double-click the switch symbol, and when its dialog box opens, select the *Value* tab. In the *Key for Toggle* box, select *A*, then click *OK*. (This relabels the switch as [A]. Press the A key on the keyboard several times and note that the switch opens and closes.) Add a second switch and label it (B). Add a 12-V dc source, ground, and lamp. (To add the lamp, click *Place/Component*, and from the *Group* box, select *Indicators/Virtual_lamp*.) Set up the circuit of Figure 2–28, Chapter 2. Operate the switches and determine whether you have successfully achieved two-way control. Multisim

72. Repeat Problem 69 using PSpice. PSPICE

73. Repeat Problem 70 using PSpice. PSPICE

74. Using PSpice, repeat the Ohm's law analysis (Figure 4–31) except use R = 25 Ω and sweep the source voltage from −10 V to +10 V in 1-V increments. On this graph, what does the negative value for current mean? PSPICE

PSPICE

75. The cursor may be used to read values from PSpice graphs. Get the graph from Problem 74 on the screen and:
    a. Click *Trace* on the menu bar, select *Cursor*, click *Display*. A display box opens. Position the cursor on the graph, and click again. The cursor reading is displayed on your screen.
    b. The cursor may be positioned using the mouse or the left and right arrow keys. Position the cursor at 2 V and read the current. Verify by Ohm's law. Repeat at a number of other points, both positive and negative.

ANSWERS TO IN-PROCESS LEARNING CHECKS

IN-PROCESS LEARNING CHECK 1

1. 25 V
2. 0 A, No
3. a. 1.25 A; b. 0.625 A; c. 12.5 A; d. 5 A
4. a. 30 V; b. −72 V; c. −90 V; d. 160 V

BASIC dc ANALYSIS

II

Part I provided the foundation upon which proficient circuit analysis is built. The terms, units, and definitions used in the previous chapters have developed the necessary vocabulary that is used throughout electrical and electronics technology.

In Part II, we build on this foundation, applying the concepts developed in Part I to initially analyze series dc circuits. Several very important rules, laws, and theorems will be developed in the following chapters, providing additional tools needed to extend circuit theory to the analysis of parallel and series-parallel circuits. Circuit theorems will be used to simplify the most complex circuit into an equivalent circuit that is represented as a single source and a single resistor. Although most circuits are much more complex than the circuits used in this text, all circuits, even the most complex, follow the same laws.

Several different techniques are often available to analyze a given circuit. In order to give a broad overview, this textbook uses the various methods for illustrative purposes. You are encouraged to use different techniques where possible, so that you develop the skills to become proficient in circuit analysis. ■

KEY TERMS

Chassis Ground
Earth Ground
Electric Circuit
Ground
Kirchhoff's Voltage Law (KVL)
Loading Effect (Ammeter)
Meter Loading
Point Sources
Series Circuit
Terminal Voltage
Total Equivalent Resistance
Voltage Divider Rule
Voltage Subscripts

OUTLINE

Series Circuits
Kirchhoff's Voltage Law
Resistors in Series
Voltage Sources in Series
Interchanging Series Components
The Voltage Divider Rule
Circuit Ground
Voltage Subscripts
Internal Resistance of Voltage Sources
Ammeter Loading Effects
Circuit Analysis Using Computers

OBJECTIVES

After studying this chapter, you will be able to

- determine the total resistance in a series circuit and calculate circuit current,
- use Ohm's law and the voltage divider rule to solve for the voltage across all resistors in the circuit,
- express Kirchhoff's voltage law and use it to analyze a given circuit,
- solve for the power dissipated by any resistor in a series circuit and show that the total power dissipated is exactly equal to the power delivered by the voltage source,
- solve for the voltage between any two points in a series or parallel circuit,
- calculate the loading effect of an ammeter in a circuit,
- use computers to assist in the analysis of simple series circuits.

5

SERIES CIRCUITS

CHAPTER PREVIEW

In the previous chapter we examined the interrelation of current, voltage, resistance, and power in a single resistor circuit. In this chapter we will expand on these basic concepts to examine the behavior of circuits having several resistors in series.

We will use Ohm's law to derive the voltage divider rule and to verify Kirchhoff's voltage law. A good understanding of these important principles provides an important base upon which further circuit analysis techniques are built. Kirchhoff's voltage law and Kirchhoff's current law, which will be covered in the next chapter, are fundamental in understanding *all* electrical and electronic circuits.

After developing the basic framework of series circuit analysis, we will apply the ideas to analyze and design simple voltmeters and ohmmeters. While meters are usually covered in a separate instruments or measurements course, we examine these circuits merely as an application of the concepts of circuit analysis.

Similarly, we will observe how circuit principles are used to explain the loading effect of an ammeter placed in series with a circuit. ■

Putting It in Perspective

Gustav Robert Kirchhoff

KIRCHHOFF WAS A GERMAN PHYSICIST born on March 12, 1824, in Königsberg, Prussia. His first research was on the conduction of electricity, which led to his presentation of the laws of closed electric circuits in 1845. Kirchhoff's current law and Kirchhoff's voltage law apply to all electrical circuits and therefore are fundamentally important in understanding circuit operation. Kirchhoff was the first to verify that an electrical impulse travelled at the speed of light.

Although these discoveries have immortalized Kirchhoff's name in electrical science, he is better known for his work with R. W. Bunsen in which he made major contributions in the study of spectroscopy and advanced the research into blackbody radiation.

Kirchhoff died in Berlin on October 17, 1887. ■

5.1 Series Circuits

An **electric circuit** is the combination of any number of sources and loads connected in any manner that allows charge to flow. The electric circuit may be simple, such as a circuit consisting of a battery and a light bulb. Or the circuit may be very complex, such as the circuits contained within a television set, microwave oven, or computer. However, no matter how complicated, each circuit follows fairly simple rules in a predictable manner. Once these rules are understood, any circuit may be analyzed to determine the operation under various conditions.

All electric circuits obtain their energy either from a direct current (dc) source or from an alternating current (ac) source. In the next few chapters, we examine the operation of circuits supplied by dc sources. Although ac circuits have fundamental differences when compared with dc circuits, the laws, theorems, and rules that you learn in dc circuits apply directly to ac circuits as well.

In the previous chapter, you were introduced to a simple dc circuit consisting of a single voltage source (such as a chemical battery) and a single load resistance. The schematic representation of such a simple circuit was covered in Chapter 4 and is shown again in Figure 5–1.

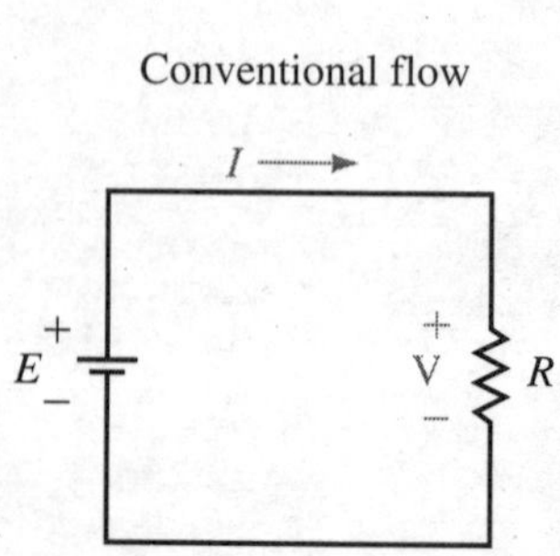

FIGURE 5–1

While the circuit of Figure 5–1 is useful in deriving some important concepts, very few practical circuits are this simple. However, we will find that even the most complicated dc circuits can generally be simplified to the circuit shown.

We begin by examining the most simple connection, the series connection. In Figure 5–2, we have two resistors, R_1 and R_2, connected at a single point in what is said to be a series connection.

Two elements are said to be in series if they are connected at a single point and if there are no other current-carrying connections at this point.

A **series circuit** is constructed by combining various elements in series, as shown in Figure 5–3. Current will leave the positive terminal of the voltage source, move through the resistors, and return to the negative terminal of the source.

In the circuit of Figure 5–3, we see that the voltage source, E, is in series with R_1, R_1 is in series with R_2, and R_2 is in series with E. By examining this circuit, another important characteristic of a series circuit becomes evident. In an analogy similar to water flowing in a pipe, current entering an element must be the same as the current leaving the element. Now, since current does not leave at any of the connections, we conclude that the following must be true:

The current is the same everywhere in a series circuit.

While the preceding statement seems self-evident, we will find that this will help to explain many of the other characteristics of a series circuit.

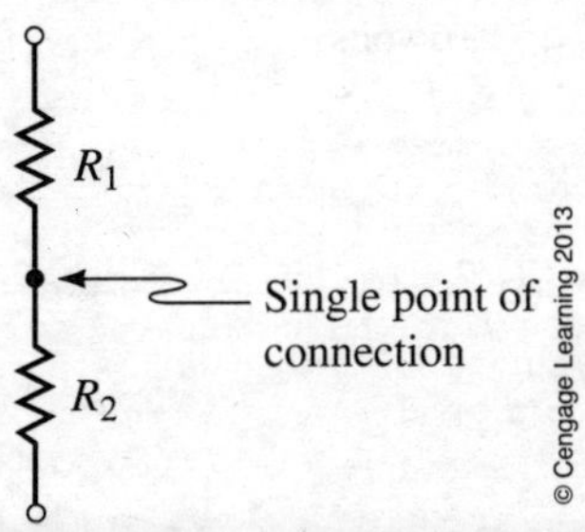

FIGURE 5–2 Resistors in series.

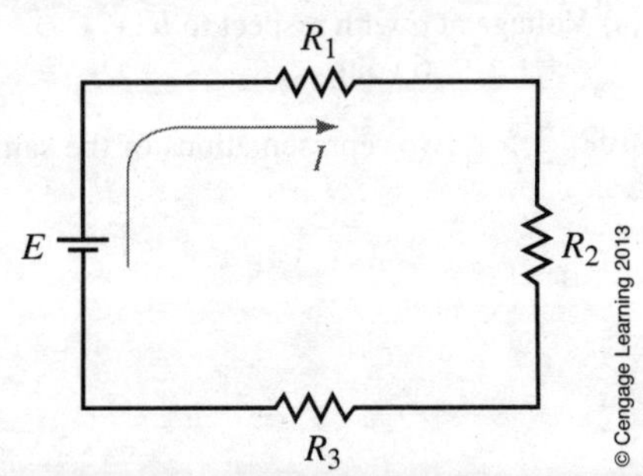

FIGURE 5–3 Series circuit.

IN-PROCESS LEARNING CHECK 1

(Answers are at the end of the chapter.)

What two conditions determine whether two elements are connected in series?

The Voltage Polarity Convention Revisited

The +, − sign convention of Figure 5–4(a) has a deeper meaning than so far considered. *Voltage exists between points, and when we place a + at one point and a − at another point, we define this to mean that we are looking at the voltage at the point marked + with respect to the point marked −.* Thus, in Figure 5–4, $V = 6$ volts means that point a is 6 V positive with respect to point b. Since the red lead of the meter is placed at point a and the black at point b, the meter will indicate +6 V.

Now consider Figure 5–5(b). [Part (a) has been repeated from Figure 5–4(b) for reference.] Here, we have placed the plus sign at point b, meaning that you are looking at the voltage at b with respect to a. Since point b is 6 V negative with respect to point a, V will have a value of minus 6 volts, that is, $V = -6$ volts. Note also that the meter indicates −6 volts since we have reversed its leads and the red lead is now at point b and the black at point a. *It is important to realize here that the voltage across R has not changed. What has changed is how we are looking at it and how we have connected the meter to measure it.* Thus, since the actual voltage is the same in both cases, (a) and (b) are equivalent representations.

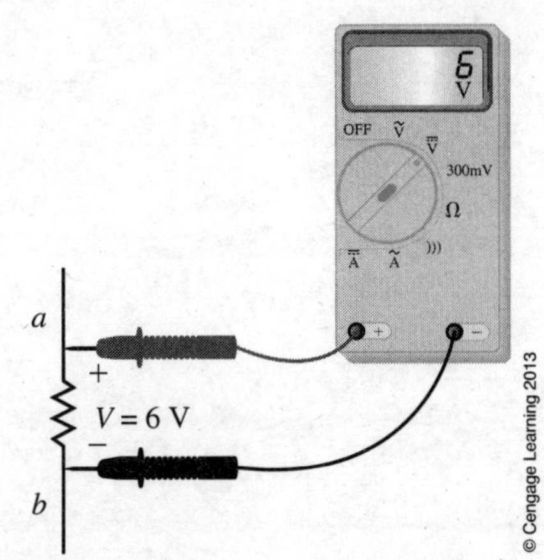

FIGURE 5–4 The +, − symbology means that you are looking at the voltage at the point marked + with respect to the point marked −.

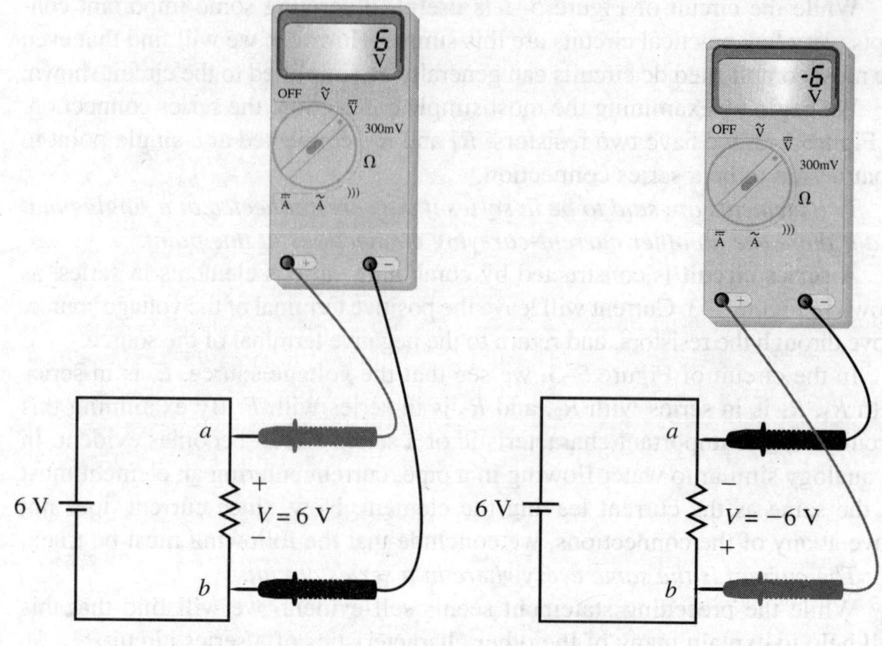

(a) Voltage at *a* with respect to *b* is 6 volts

(b) Voltage at *b* with respect to *a* is −6 volts

FIGURE 5–5 Two representations of the same voltage.

EXAMPLE 5–1

Consider Figure 5–6. In (a), $I = 3$ A, while in (b), $I = -3$ A. Using the voltage polarity convention, determine the voltages across the two resistors and show that they are equal.

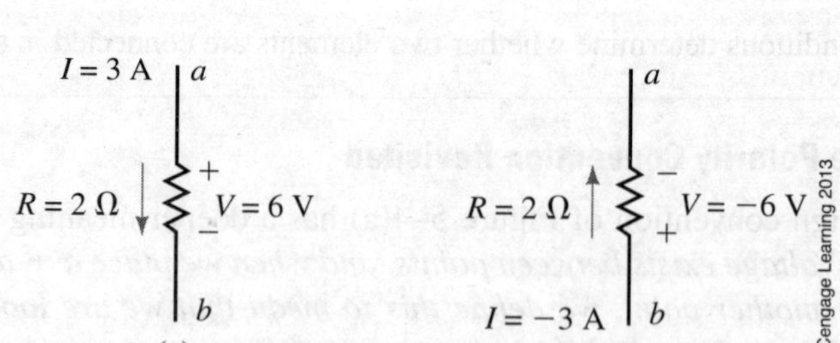

(a) (b)

FIGURE 5–6

Solution In each case, place the plus sign at the tail of the current direction arrow. Then, for (a), you are looking at the polarity of *a* with respect to *b* and you get $V = IR = (3\ \text{A})(2\ \Omega) = 6$ V as expected. Now consider (b). The polarity markings mean that you are looking at the polarity of *b* with respect to *a* and you get $V = IR = (-3\ \text{A})\ (2\ \Omega) = -6$ V. This means that point *b* is 6 V negative with respect to *a*, or equivalently, *a* is 6 V positive with respect to *b*. Thus, the two voltages are equal.

5.2 Kirchhoff's Voltage Law

Next to Ohm's law, one of the most important laws of electricity is **Kirchhoff's voltage law (KVL),** which states the following:

The summation of voltage rises and voltage drops around a closed loop is equal to zero. Symbolically, this may be stated as follows:

$$\Sigma V = 0 \quad \text{for a closed loop} \tag{5–1}$$

In the preceding symbolic representation, the uppercase Greek letter sigma (Σ) stands for summation and V stands for voltage rises and drops. A **closed loop** is defined as any path that originates at a point, travels around a circuit, and returns to the original point without retracing any segments.

An alternate way of stating Kirchhoff's voltage law is as follows:

The summation of voltage rises is equal to the summation of voltage drops around a closed loop.

$$\Sigma E_{\text{rises}} = \Sigma V_{\text{drops}} \quad \text{for a closed loop} \tag{5–2}$$

If we consider the circuit of Figure 5–7, we may begin at point a in the lower left-hand corner. By arbitrarily following the direction of the current, I, we move through the voltage source, which represents a rise in potential from point a to point b. Next, in moving from point b to point c, we pass through resistor R_1, which presents a potential drop of V_1. Continuing through resistors R_2 and R_3, we have additional drops of V_2 and V_3, respectively. By applying Kirchhoff's voltage law around the closed loop, we arrive at the following mathematical statement for the given circuit:

$$E - V_1 - V_2 - V_3 = 0$$

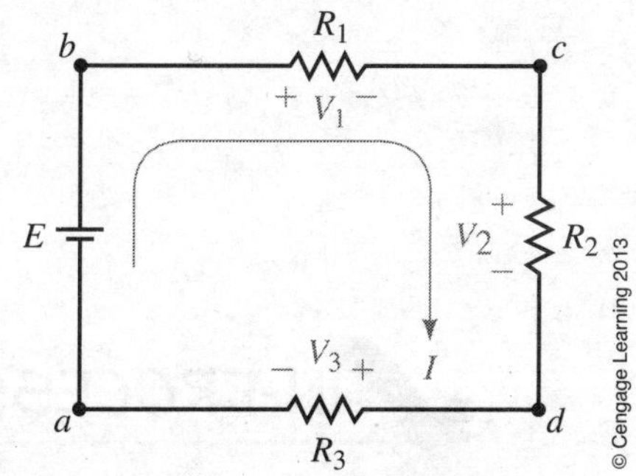

FIGURE 5–7 Kirchoff's voltage law.

Although we chose to follow the direction of current in writing Kirchhoff's voltage law equation, it would be just as correct to move around the circuit in the opposite direction. In this case the equation would appear as follows:

$$V_3 + V_2 + V_1 - E = 0$$

By simple manipulation, it is quite easy to show that the two equations are identical.

EXAMPLE 5–2

Verify Kirchhoff's voltage law for the circuit of Figure 5–8.

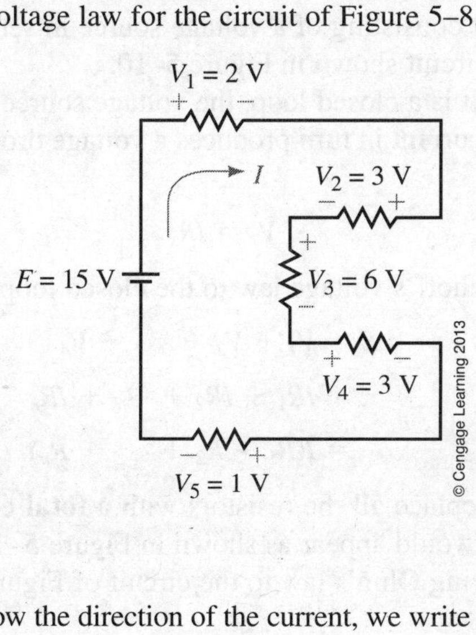

FIGURE 5–8

Solution If we follow the direction of the current, we write the loop equation as

$$15\text{ V} - 2\text{ V} - 3\text{ V} - 6\text{ V} - 3\text{ V} - 1\text{ V} = 0$$

PRACTICE PROBLEMS 1

Verify Kirchhoff's voltage law for the circuit of Figure 5–9.

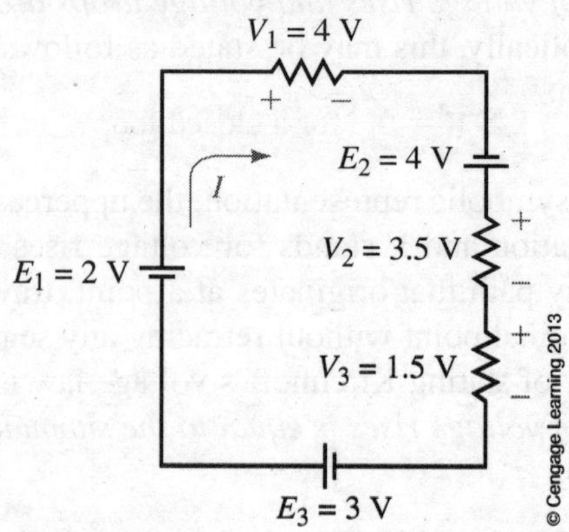

FIGURE 5–9

Answer
2 V − 4 V + 4 V − 3.5 V − 1.5 V + 3 V = 0

IN-PROCESS LEARNING CHECK 2

(Answers are at the end of the chapter.)

Define Kirchhoff's voltage law.

5.3 Resistors in Series

Almost all complicated circuits can be simplified. We will now examine how to simplify a circuit consisting of a voltage source in series with several resistors. Consider the circuit shown in Figure 5–10.

Since the circuit is a closed loop, the voltage source will cause a current I in the circuit. This current in turn produces a voltage drop across each resistor, where

$$V_x = IR_x$$

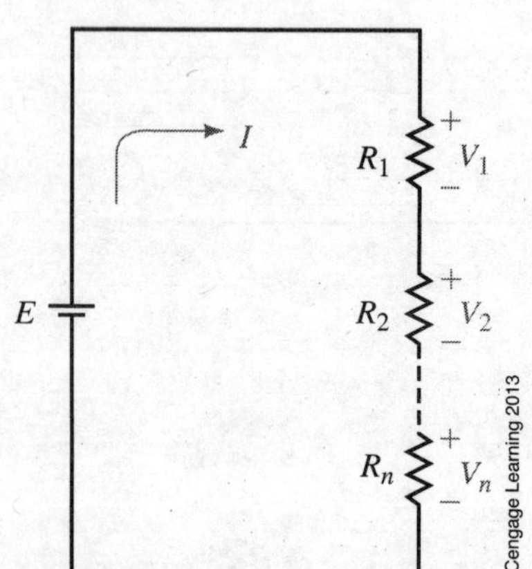

FIGURE 5–10

Applying Kirchhoff's voltage law to the closed loop gives

$$E = V_1 + V_2 + \cdots + V_n$$
$$= IR_1 + IR_2 + \cdots + IR_n$$
$$= I(R_1 + R_2 + \cdots + R_n)$$

If we were to replace all the resistors with a **total equivalent resistance**, R_T, then the circuit would appear as shown in Figure 5–11.

However, applying Ohm's law to the circuit of Figure 5–11 gives

$$E = IR_T \tag{5–3}$$

Since the circuit of Figure 5–11 is equivalent to the circuit of Figure 5–10, we conclude that this can only occur if the total equivalent resistance of the n series resistors is given as

$$R_T = R_1 + R_2 + \cdots + R_n \quad \text{[ohms, } \Omega] \tag{5–4}$$

If each of the n resistors has the same value, then the total equivalent resistance is determined as

$$R_T = nR \quad \text{[ohms, } \Omega] \tag{5–5}$$

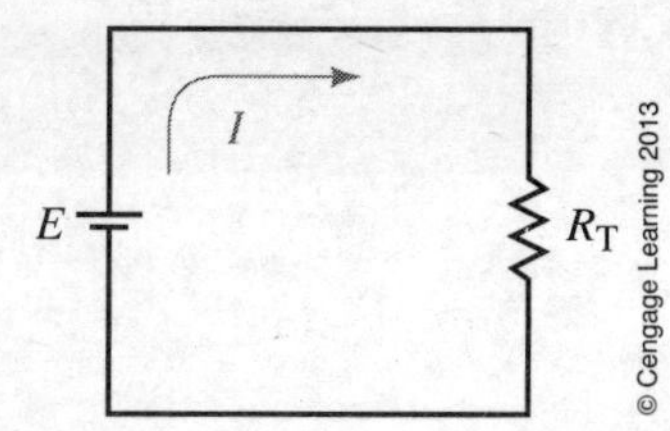

FIGURE 5–11

EXAMPLE 5–3

Determine the total resistance for each of the networks shown in Figure 5–12.

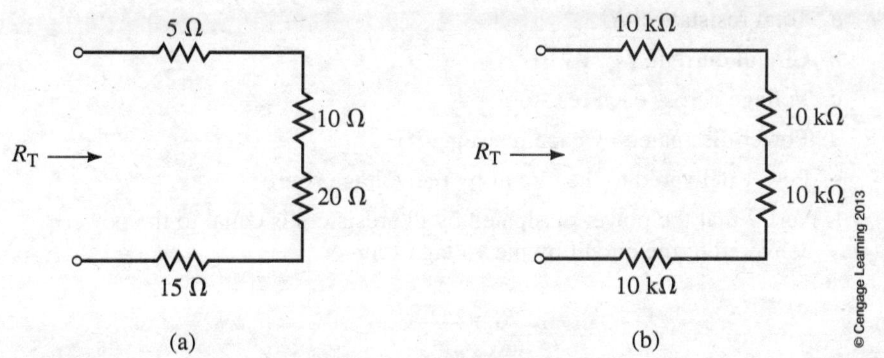

FIGURE 5–12

Solution

a. $R_T = 5\ \Omega + 10\ \Omega + 20\ \Omega + 15\ \Omega = 50.0\ \Omega$

b. $R_T = 4(10\ \text{k}\Omega) = 40.0\ \text{k}\Omega$

Any voltage source connected to the terminals of a network of series resistors will provide the same current as if a single resistance, having a value of R_T, were connected between the open terminals. From Ohm's law we get

$$I = \frac{E}{R_T} \quad \text{[amps, A]} \tag{5–6}$$

The power dissipated by each resistor is determined as

$$\begin{aligned} P_1 &= V_1 I = \frac{V_1^2}{R_1} = I^2 R_1 \quad \text{[watts, W]} \\ P_2 &= V_2 I = \frac{V_2^2}{R_2} = I^2 R_2 \quad \text{[watts, W]} \\ &\vdots \\ P_n &= V_n I = \frac{V_n^2}{R_n} = I^2 R_n \quad \text{[watts, W]} \end{aligned} \tag{5–7}$$

In Chapter 4, we showed that the power delivered by a voltage source to a circuit is given as

$$P_T = EI \quad \text{[watts, W]} \tag{5–8}$$

Since energy must be conserved, the power delivered by the voltage source is equal to the total power dissipated by all the resistors. Hence

$$P_T = P_1 + P_2 + \cdots + P_n \quad \text{[watts, W]} \tag{5–9}$$

EXAMPLE 5–4

For the series circuit shown in Figure 5–13, find the following quantities:

a. Total resistance, R_T.

b. Circuit current, I.

c. Voltage across each resistor.

d. Power dissipated by each resistor.

e. Power delivered to the circuit by the voltage source.

f. Verify that the power dissipated by the resistors is equal to the power delivered to the circuit by the voltage source.

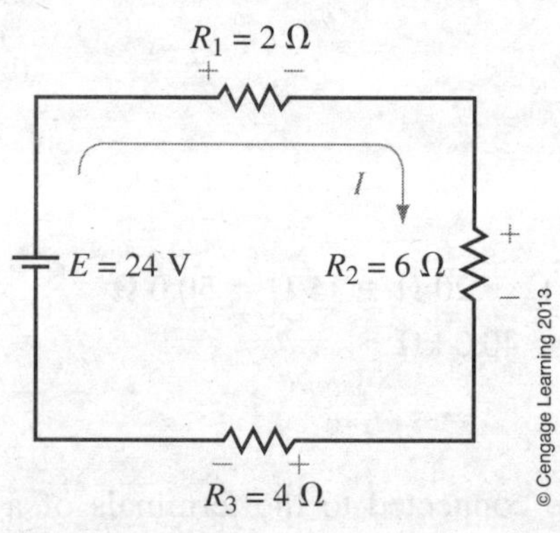

FIGURE 5–13

Solution

a. $R_T = 2\ \Omega + 6\ \Omega + 4\ \Omega = 12.0\ \Omega$

b. $I = (24\ \text{V})/(12\ \Omega) = 2.00\ \text{A}$

c. $V_1 = (2\ \text{A})(2\ \Omega) = 4.00\ \text{V}$

$V_2 = (2\ \text{A})(6\ \Omega) = 12.0\ \text{V}$

$V_3 = (2\ \text{A})(4\ \Omega) = 8.00\ \text{V}$

d. $P_1 = (2\ \text{A})^2(2\ \Omega) = 8.00\ \text{W}$

$P_2 = (2\ \text{A})^2(6\ \Omega) = 24.0\ \text{W}$

$P_3 = (2\ \text{A})^2(4\ \Omega) = 16.0\ \text{W}$

e. $P_T = (24\ \text{V})(2\ \text{A}) = 48.0\ \text{W}$

f. $P_T = 8\ \text{W} + 24\ \text{W} + 16\ \text{W} = 48.0\ \text{W}$

PRACTICE PROBLEMS 2

For the series circuit shown in Figure 5–14, find the following quantities:

a. Total resistance, R_T.
b. The direction and magnitude of the current, I.
c. Polarity and magnitude of the voltage across each resistor.
d. Power dissipated by each resistor.
e. Power delivered to the circuit by the voltage source.
f. Show that the power dissipated is equal to the power delivered.

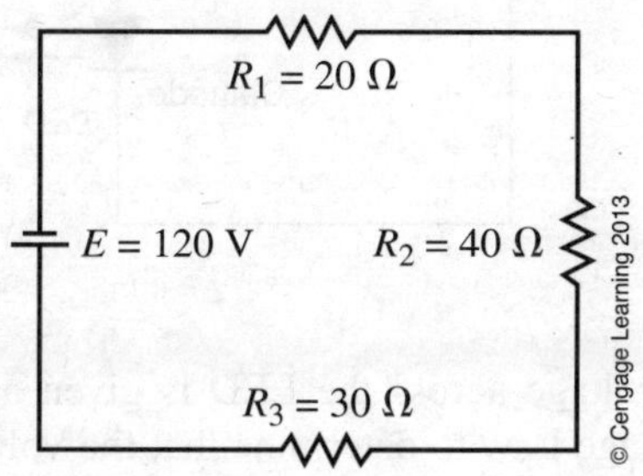

FIGURE 5–14

Answers
a. 90.0 Ω
b. 1.33 A counterclockwise
c. $V_1 = 26.7$ V, $V_2 = 53.3$ V, $V_3 = 40.0$ V
d. $P_1 = 35.6$ W, $P_2 = 71.1$ W, $P_3 = 53.3$ W
e. $P_T = 160.$ W
f. $P_1 + P_2 + P_3 = 160.\ \text{W} = P_T$

EXAMPLE 5–5

Light-emitting diodes (LEDs) are used everywhere from flashlights to backlit television sets. Due to their simplicity and low energy requirement, these ubiquitous components are being used increasingly as inexpensive light sources. In this example you will learn how simple it is to design a circuit using an LED.

While it is not necessary to know exactly how the devices work, there are a few simple characteristics that help you to understand their basic operation, which is a bit different than what you already know about other light sources. First, it is important to know that LEDs permit current in only one direction, namely that conventional current is from *anode* to *cathode*, as indicated by the direction of the arrow of the device symbol. When this is done, the LED is said to be *forward biased*. Voltage sources should not be connected against the direction of the arrow, since doing so may damage the LED. Such a diode would be said to be *reverse biased*. (Refer to Section 3.10 for a more detailed description of diode operation.)

Next, there are two important electrical characteristics that are used in the design of an LED circuit. The typical forward current of the device is given as I_F, while the nominal voltage drop across the LED when it is forward biased is given as V_F. While the actual values depend on several physical properties of the LED, the latter two values will normally be between 10 mA and 100 mA and 3–5 V, respectively.

Given that a white LED has a forward current, $I_F = 20$ mA, and a forward voltage, $V_F = 3.6$ V, determine the series resistor R that is required to meet the given conditions.

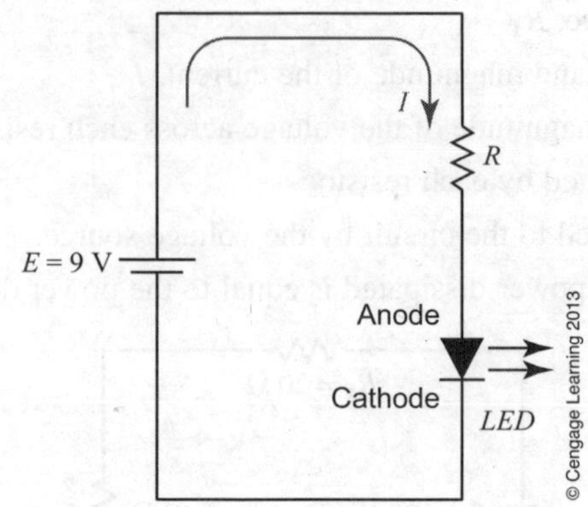

FIGURE 5–15

Solution Since the voltage across the LED is given as $V_F = 3.6$ V, we can apply Kirchhoff's Voltage Law to determine that the voltage across the resistor must be

$$V_R = 9.0\text{ V} - 3.6\text{ V} = 5.4\text{ V}$$

Now, since we have a series circuit, the current through the resistor must be the same as the current through the LED, namely $I_F = 20$ mA. This means that the value of the series resistor must be

$$R = \frac{5.4\text{ V}}{20\text{ mA}} = 270\ \Omega$$

IN-PROCESS LEARNING CHECK 3

(Answers are at the end of the chapter.)

Three resistors, R_1, R_2, and R_3, are in series. Determine the value of each resistor if $R_T = 42\text{ k}\Omega$, $R_2 = 3R_1$, and $R_3 = 2R_2$.

5.4 Voltage Sources in Series

If a circuit has more than one voltage source in series, then the voltage sources may effectively be replaced by a single source having a value that is the sum or difference of the individual sources. Since the sources may have different polarities, it is necessary to consider polarities in determining the resulting magnitude and polarity of the equivalent voltage source.

If the polarities of all the voltage sources are such that the sources appear as voltage rises in a given direction, then the resultant source is determined by simple addition, as shown in Figure 5–16.

If the polarities of the voltage sources do not result in voltage rises in the same direction, then we must compare the rises in one direction to the rises in the other direction. The magnitude of the resultant source will be the sum of the rises in one direction minus the sum of the rises in the opposite direction. The polarity of the equivalent voltage source will be the same as the polarity of whichever direction has the greater rise. Consider the voltage sources shown in Figure 5–17.

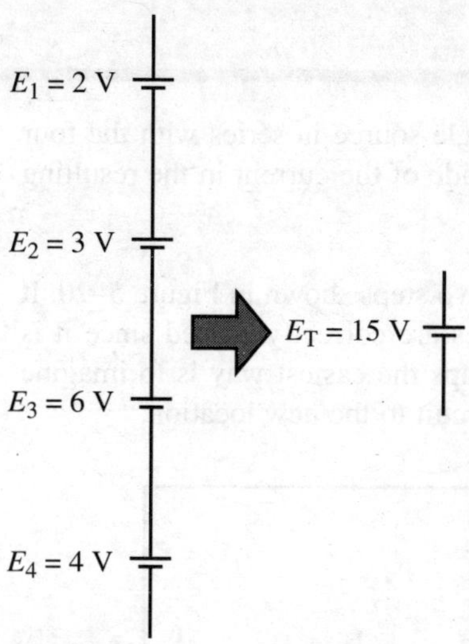

FIGURE 5–16

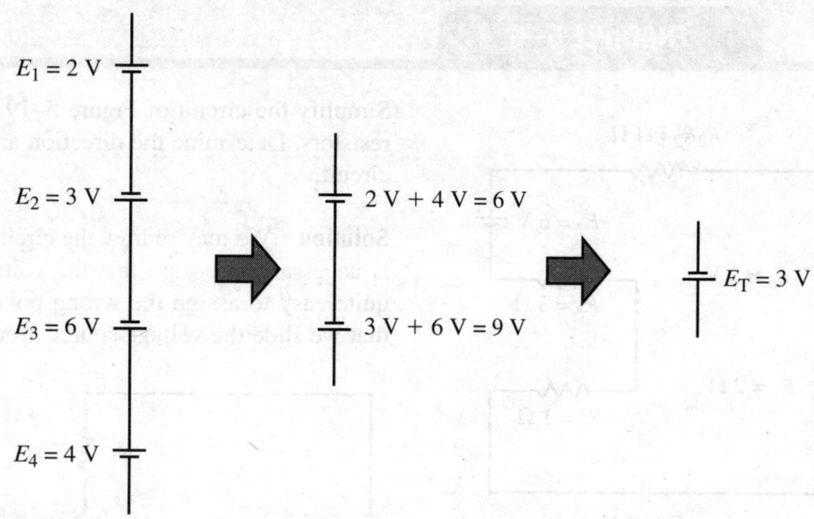

FIGURE 5–17

If the rises in one direction were equal to the rises in the opposite direction, then the resultant voltage source would be equal to zero.

IN-PROCESS LEARNING CHECK 4

(Answers are at the end of the chapter.)

A typical lead-acid automobile battery consists of six cells connected in series. If the voltage between the battery terminals is measured to be 13.06 V, what is the average voltage of each cell within the battery?

5.5 Interchanging Series Components

The order of series components may be changed without affecting the operation of the circuit.

The two circuits in Figure 5–18 are equivalent.

Very often, once the circuits have been redrawn it becomes easier to visualize the circuit operation. Therefore, we will regularly use the technique of interchanging components to simplify circuits before we analyze them.

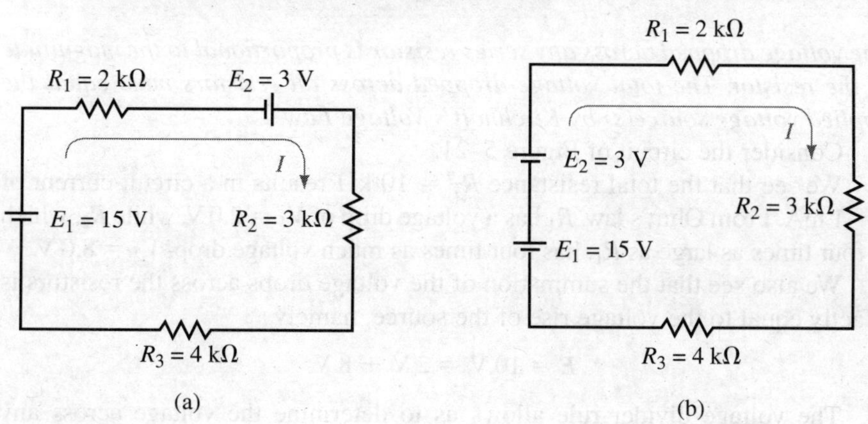

FIGURE 5–18

EXAMPLE 5–6

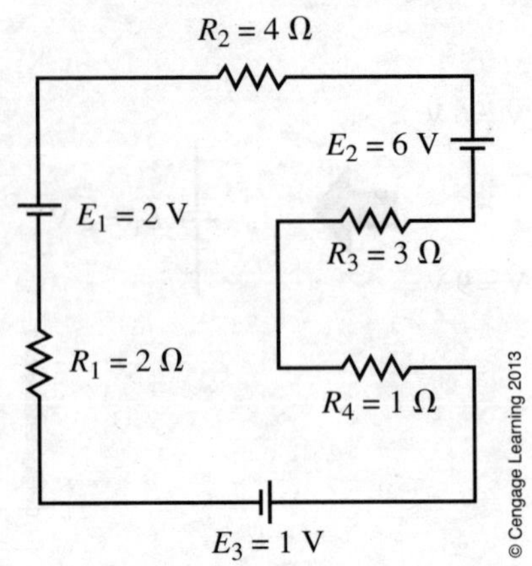

FIGURE 5–19

Simplify the circuit of Figure 5–19 into a single source in series with the four resistors. Determine the direction and magnitude of the current in the resulting circuit.

Solution We may redraw the circuit by the two steps shown in Figure 5–20. It is necessary to ensure that the voltage sources are correctly moved since it is quite easy to assign the wrong polarity. Perhaps the easiest way is to imagine that we slide the voltage source around the circuit to the new location.

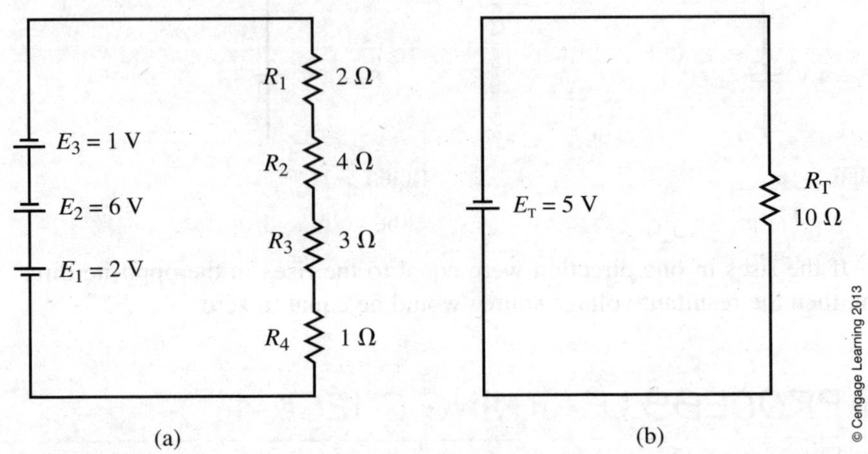

FIGURE 5–20

The current in the resulting circuit will be in a counterclockwise direction around the circuit and will have a magnitude determined as

$$I = \frac{E_T}{R_T} = \frac{6\text{ V} + 1\text{ V} - 2\text{ V}}{2\ \Omega + 4\ \Omega + 3\ \Omega + 1\ \Omega} = \frac{5\text{ V}}{10\ \Omega} = 0.500\text{ A}$$

Because the circuits are in fact equivalent, the current direction determined in Figure 5–20 also represents the direction for the current in the circuit of Figure 5–19.

5.6 The Voltage Divider Rule

The voltage dropped across any series resistor is proportional to the magnitude of the resistor. The total voltage dropped across all resistors must equal the applied voltage source(s) by Kirchhoff's Voltage Law.

Consider the circuit of Figure 5–21.

We see that the total resistance $R_T = 10\text{ k}\Omega$ results in a circuit current of $I = 1\text{ mA}$. From Ohm's law, R_1 has a voltage drop of $V_1 = 2.0\text{ V}$, while R_2, which is four times as large as R_1, has four times as much voltage drop, $V_2 = 8.0\text{ V}$.

We also see that the summation of the voltage drops across the resistors is exactly equal to the voltage rise of the source, namely,

$$E = 10\text{ V} = 2\text{ V} + 8\text{ V}$$

The voltage divider rule allows us to determine the voltage across any series resistance in a single step, without first calculating the current. We have seen that for any number of resistors in series the current in the circuit is determined by Ohm's law as

$$I = \frac{E}{R_T} \quad \text{[Amps, A]} \qquad \textbf{(5–10)}$$

where the two resistors in Figure 5–21 result in a total resistance of

$$R_T = R_1 + R_2$$

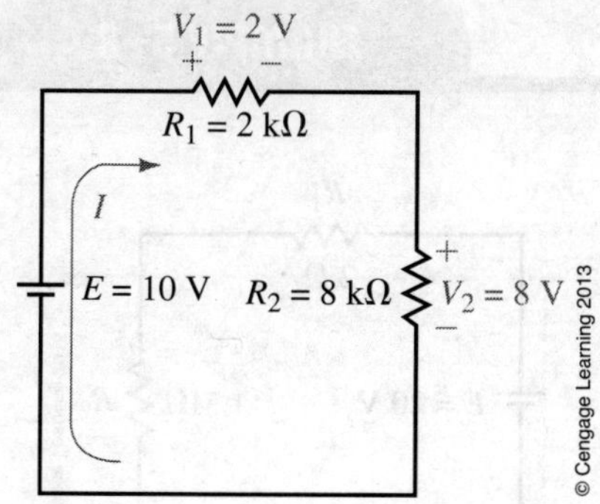

FIGURE 5–21

By again applying Ohm's law, the voltage drop across any resistor in the series circuit is calculated as

$$V_x = IR_x$$

Now, by substituting Equation 5–4 into the preceding equation we write the **voltage divider rule** for two resistors as a simple equation:

$$V_x = \frac{R_x}{R_T}E = \frac{R_x}{R_1 + R_2}E$$

In general, for any number of resistors the voltage drop across any resistor may be found as

$$V_x = \frac{R_x}{R_T}E \qquad \textbf{(5–11)}$$

EXAMPLE 5–7

Use the voltage divider rule to determine the voltage across each of the resistors in the circuit shown in Figure 5–22. Show that the summation of voltage drops is equal to the applied voltage rise in the circuit.

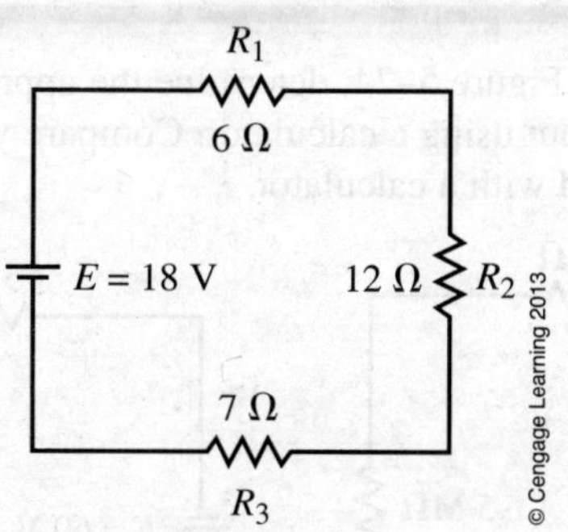

FIGURE 5–22

Solution

$$R_T = 6\ \Omega + 12\ \Omega + 7\ \Omega = 25.0\ \Omega$$

$$V_1 = \left(\frac{6\ \Omega}{25\ \Omega}\right)(18\ \text{V}) = 4.32\ \text{V}$$

$$V_2 = \left(\frac{12\ \Omega}{25\ \Omega}\right)(18\ \text{V}) = 8.64\ \text{V}$$

$$V_3 = \left(\frac{7\ \Omega}{25\ \Omega}\right)(18\ \text{V}) = 5.04\ \text{V}$$

The total voltage drop is the summation

$$V_T = 4.32\ \text{V} + 8.64\ \text{V} + 5.04\ \text{V} = 18.0\ \text{V} = E$$

EXAMPLE 5–8

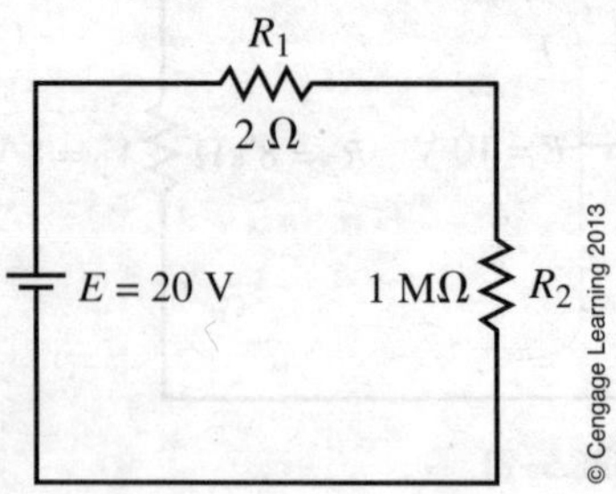

FIGURE 5–23

Using the voltage divider rule, determine the voltage across each of the resistors of the circuit shown in Figure 5–23.

Solution

$$R_T = 2\ \Omega + 1\ 000\ 000\ \Omega = 1\ 000\ 002\ \Omega$$

$$V_1 = \left(\frac{2\ \Omega}{1\ 000\ 002\ \Omega}\right)(20\ \text{V}) \approx 40\ \mu\text{V}$$

$$V_2 = \left(\frac{1.0\ \text{M}\Omega}{1.000\ 002\ \text{M}\Omega}\right)(20\ \text{V}) = 19.999\ 86\ \text{V}$$

$$\approx 20.0\ \text{V}$$

The previous example illustrates two important points that occur regularly in electronic circuits. If a single series resistance is very large in comparison with the other series resistances, then the voltage across that resistor will be essentially the total applied voltage. On the other hand, if a single resistance is very small in comparison with the other series resistances, then the voltage drop across the small resistor will be essentially zero. As a general rule, if a series resistor is more than 100 times larger than another series resistor, then the effect of the smaller resistor(s) may be effectively neglected.

PRACTICE PROBLEMS 3

For the circuits shown in Figure 5–24, determine the approximate voltage drop across each resistor without using a calculator. Compare your approximations to the actual values obtained with a calculator.

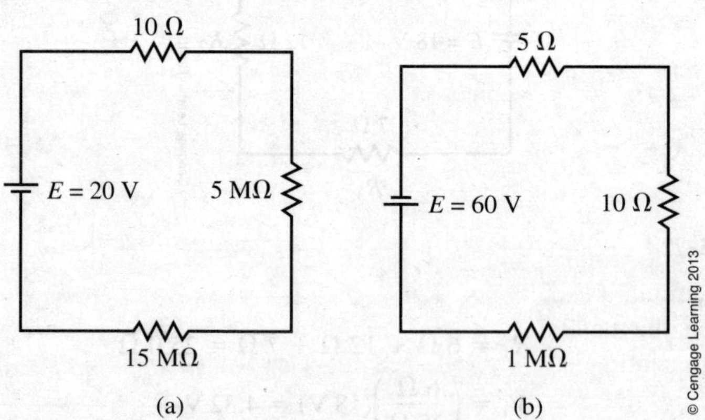

FIGURE 5–24

Answers

a. $V_{10\text{-}\Omega} \cong 0$ | $V_{5\text{-M}\Omega} \cong 5.00$ V | $V_{10\text{-M}\Omega} \cong 15.0$ V

$V_{10\text{-}\Omega} = 10.0\ \mu$V | $V_{5\text{-M}\Omega} = 5.00$ V | $V_{10\text{-M}\Omega} = 15.0$ V

b. $V_{5\text{-}\Omega} \cong 0$ | $V_{10\text{-}\Omega} \cong 0$ | $V_{1\text{-M}\Omega} \cong 60.0$ V

$V_{5\text{-}\Omega} = 0.300$ mV | $V_{10\text{-}\Omega} = 0.600$ mV | $V_{1\text{-M}\Omega} = 60.0$ V

IN-PROCESS LEARNING CHECK 5

(Answers are at the end of the chapter.)

The voltage drops across three resistors are measured to be 10.0 V, 15.0 V, and 25.0 V. If the largest resistor is 47.0 kΩ, determine the sizes of the other two resistors.

5.7 Circuit Ground

Perhaps one of the most misunderstood concepts in electronics is that of ground. This misunderstanding leads to many problems when circuits are designed and analyzed. The standard symbol for circuit ground is shown in Figure 5–25(a), while the symbol for chassis ground is shown in Figure 5–25(b).

In its most simple definition, **ground** is simply an "arbitrary electrical point of reference" or "common point" in a circuit. Using the ground symbol in this manner usually allows the circuit to be sketched more simply. When the ground symbol is used arbitrarily to designate a point of reference, it would be just as correct to redraw the circuit schematic showing all the ground points connected together or indeed to redraw the circuit using an entirely different point of reference. The circuits shown in Figure 5–26 are exactly equivalent circuits even though the circuits of Figure 5–26(a) and 5–26(c) use different points of reference.

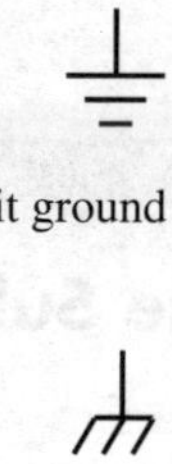

(a) Circuit ground or reference

(b) Chassis ground

FIGURE 5–25

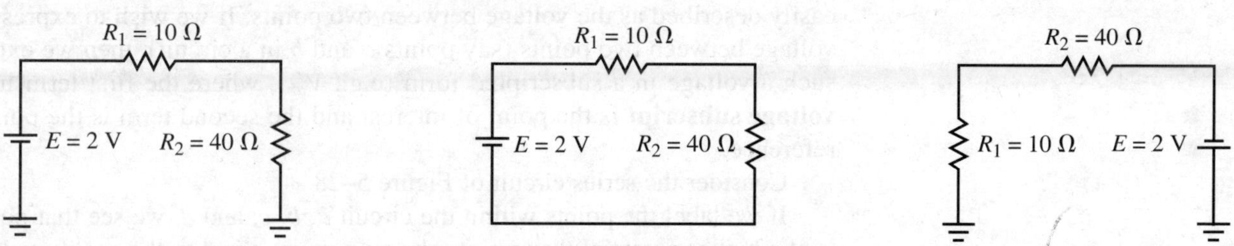

(a) (b) (c)

FIGURE 5–26

While the ground symbol is used to designate a common point of reference within a circuit, it usually has a greater meaning to the technologist or engineer. Very often, the metal chassis of an appliance is connected to the circuit ground. Such a connection is referred to as a **chassis ground** and is usually designated as shown in Figure 5–25(b).

In order to help prevent electrocution, the chassis ground is usually further connected to the **earth ground** through a connection provided at the electrical outlet box. In the event of a failure within the circuit, the chassis would redirect current to ground (tripping a breaker or fuse), rather than presenting a hazard to an unsuspecting operator.

As the name implies, the earth ground is a connection that is bonded to the earth, either through water pipes or by a connection to ground rods. Everyone is familiar with the typical 120-Vac electrical outlet shown in Figure 5–27. The rounded terminal of the outlet is always the ground terminal and is used not

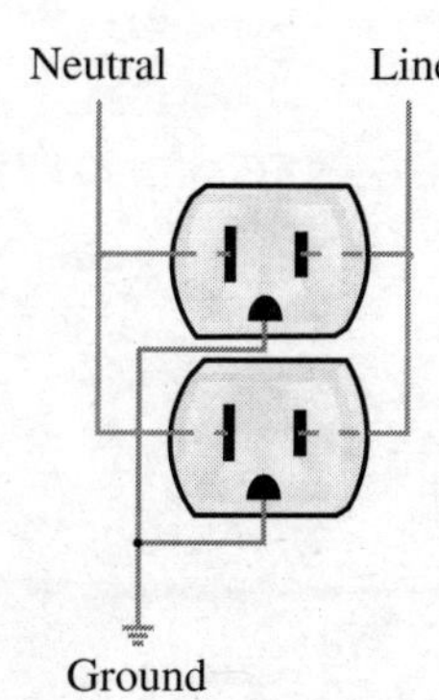

FIGURE 5–27 Ground connection in a typical 120-Vac outlet.

only in ac circuits but may also be used to provide a common point for dc circuits. When a circuit is bonded to the earth through the ground terminal the ground symbol no longer represents an arbitrary connection, but rather represents a very specific type of connection.

IN-PROCESS LEARNING CHECK 6

(Answers are at the end of the chapter.)

If you measure the resistance between the ground terminal of the 120-Vac plug and the metal chassis of a microwave oven to be zero ohms, what does this tell you about the chassis of the oven?

5.8 Voltage Subscripts

Double Subscripts

As you have already seen, voltages are always expressed as the potential difference between two points. In a 9-V battery, there is a 9-volt rise in potential from the negative terminal to the positive terminal. Current through a resistor results in a voltage drop across the resistor such that the terminal from which charge leaves is at a lower potential than the terminal into which charge enters. We now examine how voltages within any circuit may be easily described as the voltage between two points. If we wish to express the voltage between two points (say points *a* and *b* in a circuit), then we express such a voltage in a subscripted form (e.g., V_{ab}), where the first term in the **voltage subscript** is the point of interest and the second term is the point of reference.

Consider the series circuit of Figure 5–28.

If we label the points within the circuit *a*, *b*, *c*, and *d*, we see that point *b* is at a higher potential than point *a* by an amount equal to the supply voltage. We may write this mathematically as $V_{ba} = +50$ V. Although the plus sign is redundant, we show it here to indicate that point *b* is at a higher potential than point *a*. If we examine the voltage at point *a* with respect to point *b*, we see that *a* is at a lower potential than *b*. This may be written mathematically as $V_{ab} = -50$ V.

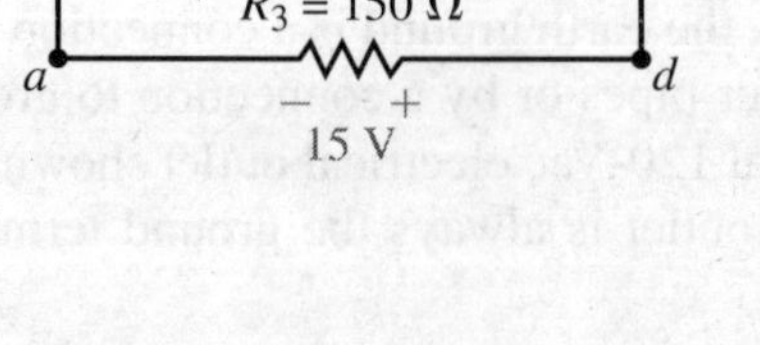

FIGURE 5–28

From the preceding illustration, we make the following general statement:

$$V_{ab} = -V_{ba}$$

for any two points a and b within a circuit.

Current through the circuit results in voltage drops across the resistors as shown in Figure 5–28. If we determine the voltage drops on all resistors and show the correct polarities, then we see that the following must also apply:

$$V_{bc} = +10\text{ V} \qquad V_{cb} = -10\text{ V}$$
$$V_{cd} = +25\text{ V} \qquad V_{dc} = -25\text{ V}$$
$$V_{da} = +15\text{ V} \qquad V_{ad} = -15\text{ V}$$

If we wish to determine the voltage between any other two points within the circuit, it is a simple matter of adding all the voltages between the two points, taking into account the polarities of the voltages. The voltage between points b and d would be determined as follows:

$$V_{bd} = V_{bc} + V_{cd} = 10\text{ V} + 25\text{ V} = +35\text{ V}$$

Similarly, the voltage between points b and a could be determined by using the voltage drops of the resistors:

$$V_{ba} = V_{bc} + V_{cd} + V_{da} = 10\text{ V} + 25\text{ V} + 15\text{ V} = +50\text{ V}$$

Notice that this result is precisely the same as when we determined V_{ba} using only the voltage source. This result indicates that the voltage between two points is not dependent upon the path taken.

EXAMPLE 5–9

For the circuit of Figure 5–29, find the voltages V_{ac}, V_{ad}, V_{cf}, and V_{eb}.

Solution First, we determine that the equivalent supply voltage for the circuit is

$$E_T = 3\text{ V} + 4\text{ V} - 2\text{ V} = 5.0\text{ V}$$

with a polarity such that current will move in a counterclockwise direction within the circuit.

Next, we determine the voltages on all resistors by using the voltage divider rule and assigning polarities based on the direction of the current.

$$V_1 = \frac{R_1}{R_T}E_T = \left(\frac{10\ \Omega}{10\ \Omega + 30\ \Omega + 40\ \Omega}\right)(5.0\text{ V}) = 0.625\text{ V}$$

$$V_2 = \frac{R_2}{R_T}E_T = \left(\frac{30\ \Omega}{10\ \Omega + 30\ \Omega + 40\ \Omega}\right)(5.0\text{ V}) = 1.875\text{ V}$$

$$V_3 = \frac{R_3}{R_T}E_T = \left(\frac{40\ \Omega}{10\ \Omega + 30\ \Omega + 40\ \Omega}\right)(5.0\text{ V}) = 2.50\text{ V}$$

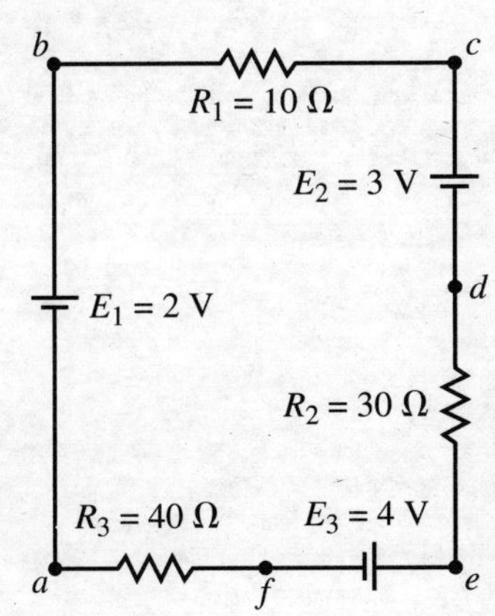

FIGURE 5–29

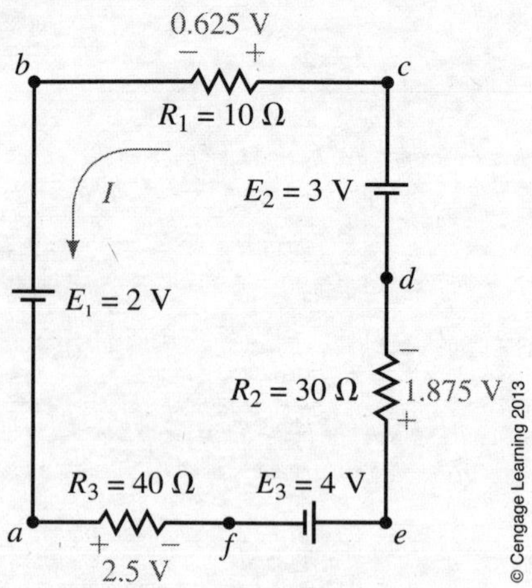

FIGURE 5–30

The voltages appearing across the resistors are as shown in Figure 5–30.

Finally, we solve for the voltages between the indicated points:

$$V_{ac} = -2.0\text{ V} - 0.625\text{ V} = -2.625\text{ V}$$

$$V_{ad} = -2.0\text{ V} - 0.625\text{ V} + 3.0\text{ V} = +0.375\text{ V}$$

$$V_{cf} = +3.0\text{ V} - 1.875\text{ V} + 4.0\text{ V} = +5.125\text{ V}$$

$$V_{eb} = +1.875\text{ V} - 3.0\text{ V} + 0.625\text{ V} = -0.500\text{ V}$$

Or, selecting the opposite path, we get

$$V_{eb} = +4.0\text{ V} - 2.5\text{ V} - 2.0\text{ V} = -0.500\text{ V}$$

PRACTICAL NOTES...

Since most students initially find it difficult to determine the correct polarity for the voltage between two points, we present a simplified method to correctly determine the polarity and voltage between any two points within a circuit.

1. Determine the circuit current. Calculate the voltage drop across all components.
2. Polarize all resistors based upon the direction of the current. The terminal at which the current enters is assigned to be positive, while the terminal at which the current leaves is assigned to be negative.
3. In order to determine the voltage at point *a* with respect to point *b*, start at point *b*. Refer to Figure 5–31. Now, imagine that you walk around the circuit to point *a*.

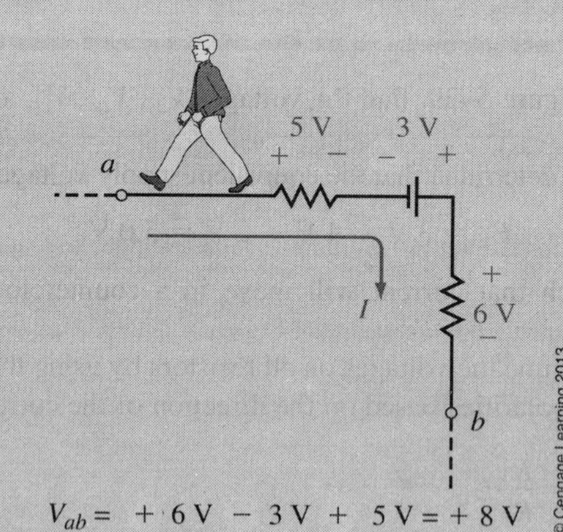

FIGURE 5–31

4. As you "walk" around the circuit, add the voltage drops and rises as you get to them. The assigned polarity of the voltage at any component (whether it is a source or a resistor) is positive if the voltage rises as you "walk" through the component and is negative if the voltage decreases as you pass through the component.
5. The resulting voltage, V_{ab}, is the numerical sum of all the voltages between *a* and *b*.

For Figure 5–31, the voltage V_{ab} is determined as

$$V_{ab} = 6\text{ V} - 3\text{ V} + 5\text{ V} = 8\text{ V}$$

PRACTICE PROBLEMS 4

Find the voltage V_{ab} in the circuit of Figure 5–32.

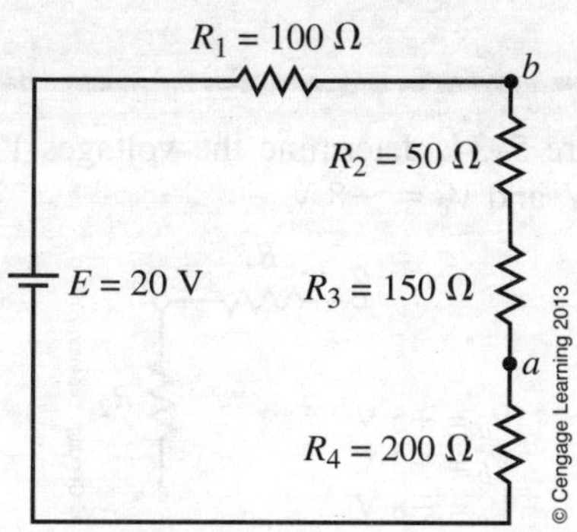

FIGURE 5–32

Answer
$V_{ab} = -8.00$ V

Single Subscripts

In a circuit which has a reference point (or ground point), most voltages will be expressed with respect to the reference point. In such a case it is no longer necessary to express a voltage using a dual subscript. Rather if we wish to express the voltage at point *a* with respect to ground, we simply refer to this as V_a. Similarly, the voltage at point *b* would be referred to as V_b. Therefore, any voltage that has only a single subscript is always referenced to the ground point of the circuit.

EXAMPLE 5–10

For the circuit of Figure 5–33, determine the voltages V_a, V_b, V_c, and V_d.

CircuitSim 05-1

Solution Applying the voltage divider rule, we determine the voltage across each resistor as follows:

$$V_1 = \frac{2\text{ k}\Omega}{2\text{ k}\Omega + 3\text{ k}\Omega + 5\text{ k}\Omega}(20\text{ V}) = 4.00\text{ V}$$

$$V_2 = \frac{3\text{ k}\Omega}{2\text{ k}\Omega + 3\text{ k}\Omega + 5\text{ k}\Omega}(20\text{ V}) = 6.00\text{ V}$$

$$V_3 = \frac{5\text{ k}\Omega}{2\text{ k}\Omega + 3\text{ k}\Omega + 5\text{ k}\Omega}(20\text{ V}) = 10.00\text{ V}$$

Now we solve for the voltage at each of the points as follows:

$$V_a = 4\text{ V} + 6\text{ V} + 10\text{ V} = +20\text{ V} = E$$

$$V_b = 6\text{ V} + 10\text{ V} = +16.0\text{ V}$$

$$V_c = +10.0\text{ V}$$

$$V_d = 0\text{ V}$$

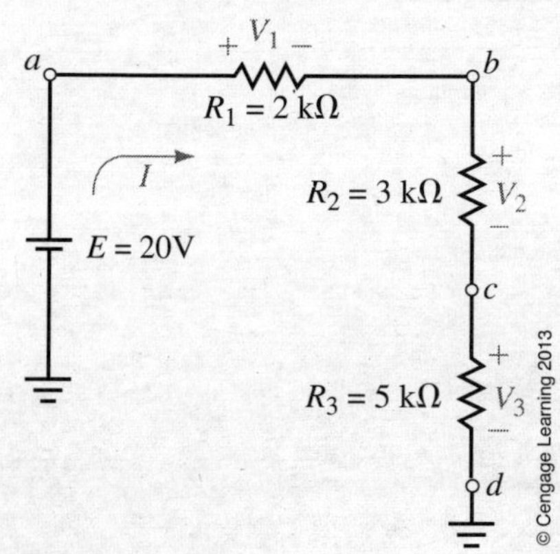

FIGURE 5–33

If the voltage at various points in a circuit is known with respect to ground, then the voltage between the points may be easily determined as follows:

$$V_{ab} = V_a - V_b \quad \text{[volts, V]} \tag{5–12}$$

EXAMPLE 5–11

For the circuit of Figure 5–34, determine the voltages V_{ab} and V_{cb} given that $V_a = +5$ V, $V_b = +3$ V, and $V_c = -8$ V.

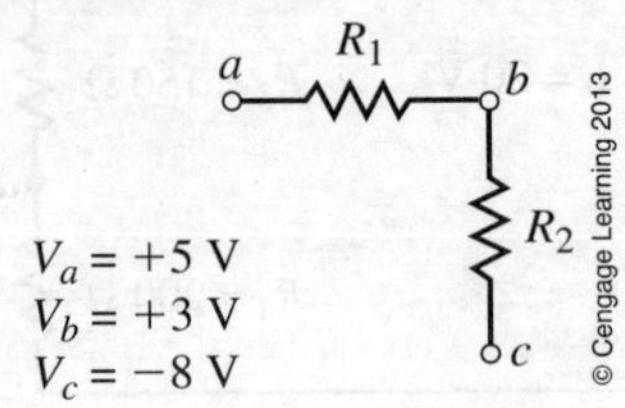

FIGURE 5–34

Solution

$$V_{ab} = +5\text{ V} - (+3\text{ V}) = +2\text{ V}$$
$$V_{cb} = -8\text{ V} - (+3\text{ V}) = -11\text{ V}$$

Point Sources

The idea of voltages with respect to ground is easily extended to include voltage sources. When a voltage source is given with respect to ground, it may be simplified in the circuit as a **point source** as shown in Figure 5–35.

Point sources are often used to simplify the representation of circuits. We need to remember that in all such cases the corresponding points always represent voltages with respect to ground (even if ground is not shown).

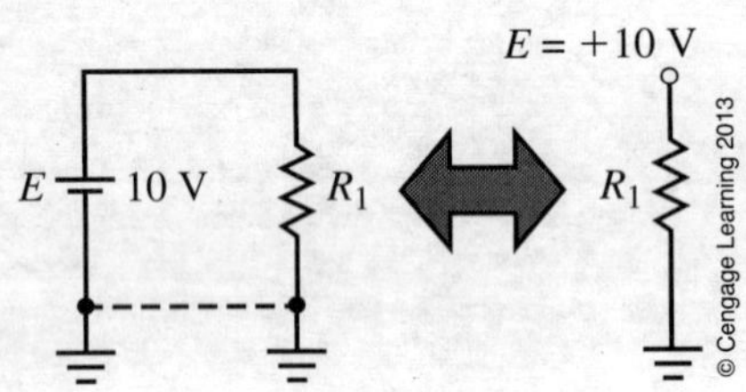

FIGURE 5–35

EXAMPLE 5–12

Determine the current and direction in the circuit of Figure 5–36.

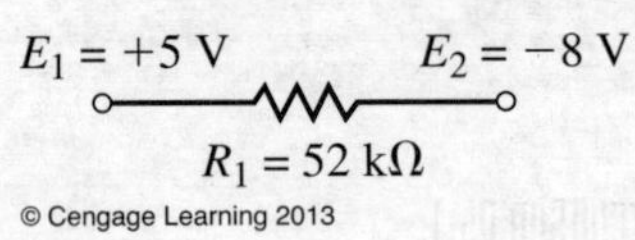

FIGURE 5–36

Solution The circuit may be redrawn showing the reference point and converting the voltage point sources into the more common schematic representation. The resulting circuit is shown in Figure 5–37.

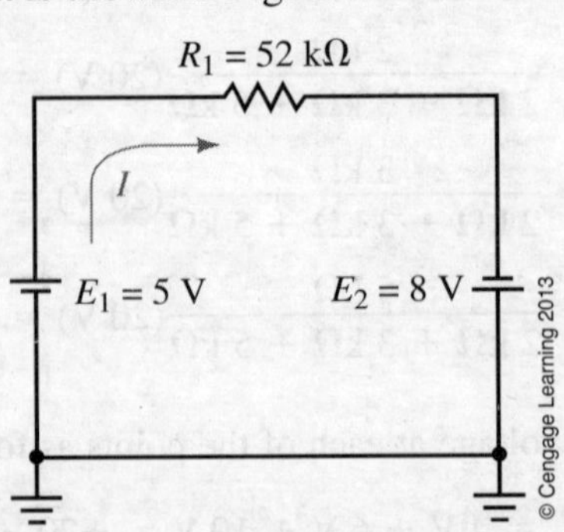

FIGURE 5–37

Now, we easily calculate the current in the circuit as

$$I = \frac{E_T}{R_1} = \frac{5\text{ V} + 8\text{ V}}{52\text{ k}\Omega} = 0.250\text{ mA}$$

IN-PROCESS LEARNING CHECK 7

(Answers are at the end of the chapter.)

Voltage measurements are taken at three locations in a circuit. They are $V_a = +5.00$ V, $V_b = -2.50$ V, and $V_c = -5.00$ V. Determine the voltages V_{ab}, V_{ca}, and V_{bc}.

5.9 Internal Resistance of Voltage Sources

So far we have worked only with ideal voltage sources, which maintain constant voltages regardless of the loads connected across the terminals. Consider a typical lead-acid automobile battery, which has a voltage of approximately 12 V. Similarly, four C-cell batteries, when connected in series, have a combined voltage of 12 V. Why then can we not use four C-cell batteries to operate the car? The answer, in part, is because the lead-acid battery has a much lower internal resistance than the low-energy C-cells. In practice, all voltage sources contain some internal resistance that will reduce the efficiency of the voltage source. We may symbolize any voltage source schematically as an ideal voltage source in series with an internal resistance. Figure 5–38 shows both an ideal voltage source and a practical or actual voltage source.

The voltage that appears between the positive and negative terminals is called the **terminal voltage.** In an ideal voltage source, the terminal voltage will remain constant regardless of the load connected. An ideal voltage source will be able to provide as much current as the circuit demands. However, in a real voltage source, the terminal voltage is dependent upon the value of the load connected across the voltage source. As expected, the real voltage source sometimes is not able to provide as much current as the load demands. Rather the current in the circuit is limited by the combination of the internal resistance and the load resistance.

Under a no-load condition ($R_L = \infty\ \Omega$), there is no current in the circuit, and so the terminal voltage will be equal to the voltage appearing across the ideal voltage source. If the output terminals are shorted together ($R_L = 0\ \Omega$), the current in the circuit will be a maximum and the terminal voltage will be equal to approximately zero. In such a situation, the voltage dropped across the internal resistance will be equal to the voltage of the ideal source.

The following example helps to illustrate the preceding principles.

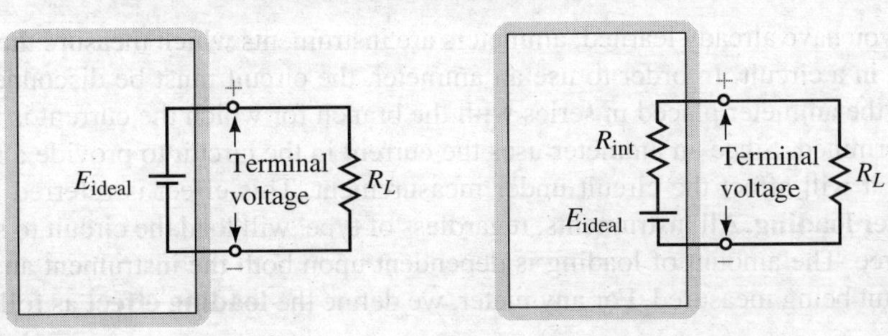

(a) Ideal voltage source (b) Real voltage source

FIGURE 5–38

EXAMPLE 5–13

Two batteries having an open-terminal voltage of 12 V are used to provide current to the starter of a car having a resistance of 0.10 Ω. If one battery has an internal resistance of 0.02 Ω and the second battery has an internal resistance of 100 Ω, calculate the current through the load and the resulting terminal voltage for each of the batteries.

Solution The circuit for each of the batteries is shown in Figure 5–39.

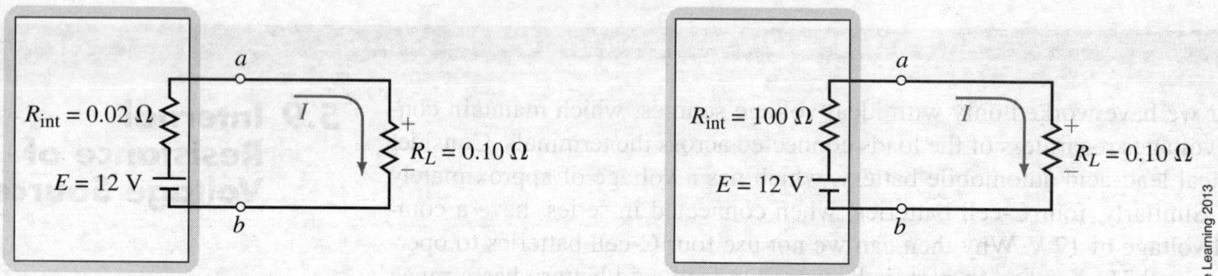

(a) Low internal resistance

(b) High internal resistance

FIGURE 5–39

$R_{int} = 0.02\ \Omega$:

$$I = \frac{12\text{ V}}{0.02\ \Omega + 0.10\ \Omega} = 100.\text{ A}$$

$$V_{ab} = (100\text{ A})(0.10\ \Omega) = 10.0\text{ V}$$

$R_{int} = 100\ \Omega$:

$$I = \frac{12\text{ V}}{100\ \Omega + 0.10\ \Omega} = 0.120\text{ A}$$

$$V_{ab} = (0.120\text{ A})(0.10\ \Omega) = 0.0120\text{ V}$$

This simple example helps to illustrate why a 12-V automobile battery (which is actually 14.4 V) is able to start a car while eight 1.5-V flashlight batteries connected in series will have virtually no measurable effect when connected to the same circuit.

5.10 Ammeter Loading Effects

As you have already learned, ammeters are instruments which measure the current in a circuit. In order to use an ammeter, the circuit must be disconnected and the ammeter placed in series with the branch for which the current is to be determined. Since an ammeter uses the current in the circuit to provide a reading, it will affect the circuit under measurement. This effect is referred to as **meter loading.** All instruments, regardless of type, will load the circuit to some degree. The amount of loading is dependent upon both the instrument and the circuit being measured. For any meter, we define the **loading effect** as follows:

$$\text{loading effect} = \frac{\text{theoretical value} - \text{measured value}}{\text{theoretical value}} \times 100\% \qquad \textbf{(5–13)}$$

EXAMPLE 5–14

For the series circuits of Figure 5–40, determine the current in each circuit. If an ammeter having an internal resistance of 250 Ω is used to measure the current in the circuits, determine the current through the ammeter and calculate the loading effect for each circuit.

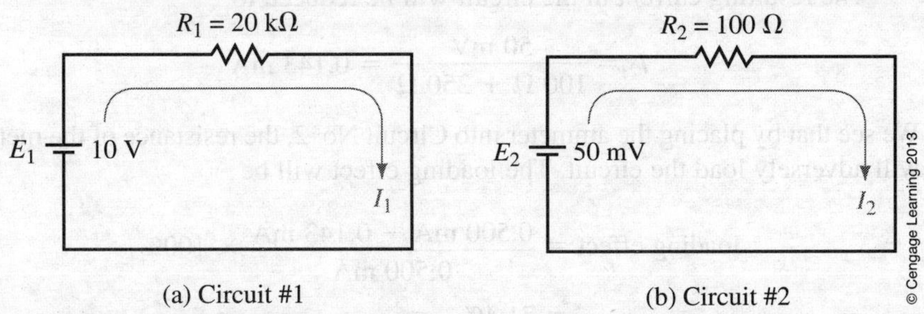

FIGURE 5–40

Solution Circuit No. 1: The current in the circuit is

$$I_1 = \frac{10\text{ V}}{20\text{ k}\Omega} = 0.500\text{ mA}$$

Now, by placing the ammeter into the circuit as shown in Figure 5–41(a) the resistance of the ammeter will slightly affect the operation of the circuit.

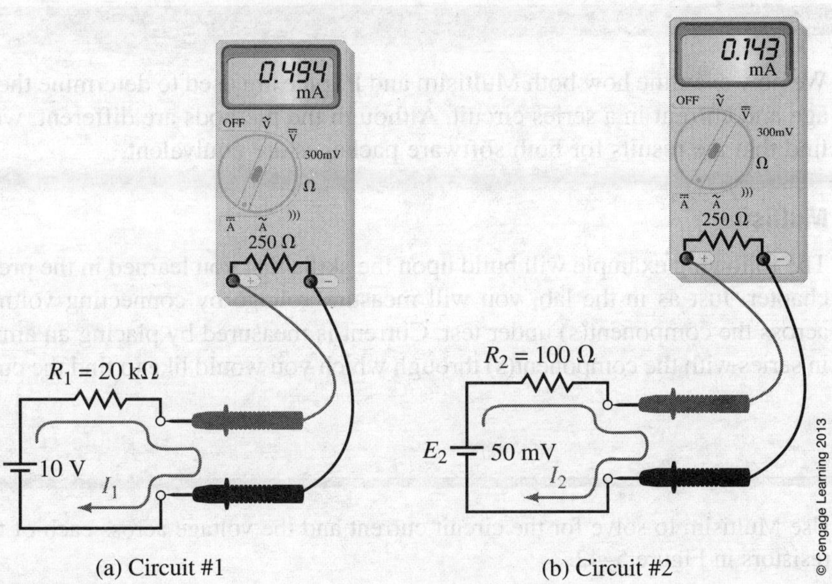

FIGURE 5–41

The resulting current in the circuit will be reduced to

$$I_1 = \frac{10\text{ V}}{20\text{ k}\Omega + 0.25\text{ k}\Omega} = 0.494\text{ mA}$$

Circuit No. 1: We see that by placing the ammeter into Circuit No. 1, the resistance of the meter slightly affects the operation of the circuit. Applying Equation 5–16 gives the loading effect as

$$\text{loading effect} = \frac{0.500\text{ mA} - 0.494\text{ mA}}{0.500\text{ mA}} \times 100\%$$

$$= 1.23\%$$

Circuit No. 2: The current in the circuit is also found as

$$I_2 = \frac{50\text{ mV}}{100\ \Omega} = 0.500\text{ mA}$$

Now, by placing the ammeter into the circuit as shown in Figure 5–41(b), the resistance of the ammeter will greatly affect the operation of the circuit.

The resulting current in the circuit will be reduced to

$$I_2 = \frac{50\text{ mV}}{100\ \Omega + 250\ \Omega} = 0.143\text{ mA}$$

We see that by placing the ammeter into Circuit No. 2, the resistance of the meter will adversely load the circuit. The loading effect will be

$$\text{loading effect} = \frac{0.500\text{ mA} - 0.143\text{ mA}}{0.500\text{ mA}} \times 100\%$$
$$= 71.4\%$$

The results of this example indicate that an ammeter, which usually has fairly low resistance, will not significantly load a circuit having a resistance of several thousand ohms. However, if the same meter is used to measure current in a circuit having low values of resistance, then the loading effect will be substantial.

5.11 Circuit Analysis Using Computers

We now examine how both Multisim and PSpice are used to determine the voltage and current in a series circuit. Although the methods are different, we will find that the results for both software packages are equivalent.

Multisim

Multisim

The following example will build upon the skills that you learned in the previous chapter. Just as in the lab, you will measure voltage by connecting voltmeters across the component(s) under test. Current is measured by placing an ammeter in series with the component(s) through which you would like to find the current.

EXAMPLE 5–15

Use Multisim to solve for the circuit current and the voltage across each of the resistors in Figure 5–42.

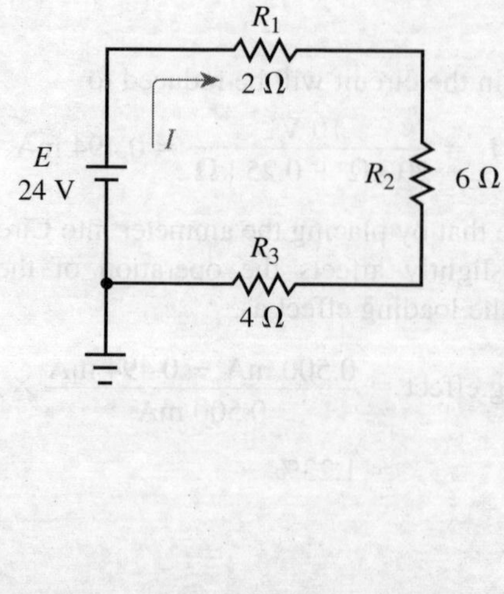

FIGURE 5–42

Solution Open Multisim and construct the circuit in Figure 5–42. If necessary, review the steps as outlined in the previous chapter. Remember that your circuit will need to have a circuit ground—found by clicking on the *Power Source Family* button. Open a new subfolder in your Multisim folder and title the project as Figure 5–42. Once your circuit resembles the circuit shown in Figure 5–42, insert the ammeters and the voltmeters into the circuit as shown in Figure 5–43. Create a new subfolder titled Ch 5 within your Multisim folder. Save this file as Example 5–15 or some other suitable name in this subfolder.

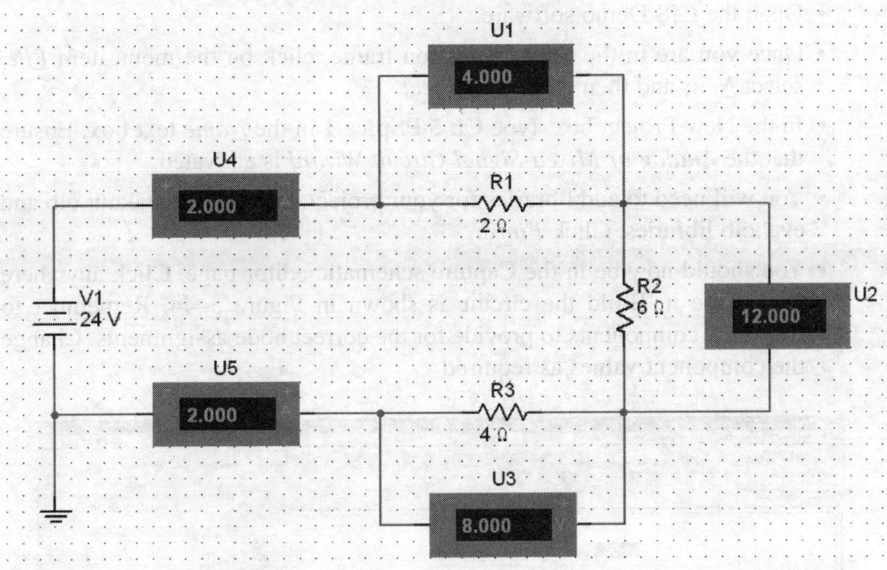

FIGURE 5–43

CircuitSim 05-2

Notice that an extra ammeter is placed into the circuit. The only reason for this is to show that the current is the same everywhere in a series circuit.

Once all ammeters and voltmeters are inserted with the correct polarities, you may run the simulator by moving the toggle switch to the ON position. Your indicators should show the same readings as the values shown in Figure 5–43. If any of the values indicated by the meters are negative, you will need to disconnect the meter(s) and reverse the terminals by using the Ctrl-R function.

Although this example is very simple, it illustrates some very important points that you will find useful when simulating circuit operation.

1. All voltmeters are connected across the components for which we are trying to measure the voltage drop.
2. All ammeters are connected in series with the components through which we are trying to find the current.
3. A ground symbol (or reference point) is required by all circuits that are to be simulated by Multisim.

PSpice

PSPICE

While PSpice has some differences compared to Multisim, we find that there are also many similarities. The following example shows how to use PSpice to analyze the previous circuit. If necessary, refer to Appendix A to find the Voltage Level marker (which indicates voltage with respect to ground) and the Current Into Pin marker.

EXAMPLE 5–16

Use PSpice to solve for the circuit current and the voltage across each of the resistors in Figure 5–42.

Solution This example lists some of the more important steps that you will need to follow. For more detail, refer to Appendix A and the PSpice example in Chapter 4.

- Open the CIS Demo software.
- Once you are in the Capture session frame, click on the menu item *File,* select *New,* and then click on *Project.*
- In the New Project box, type **Ch 5 PSpice 1** in the *Name* text box. Ensure that the *Analog or Mixed-Signal Circuit Wizard* is activated.
- You will need to add libraries for your project. Select the breakout.olb and eval.olb libraries. Click *Finish.*
- You should now be in the Capture schematic editor page. Click anywhere to activate it. Build the circuit as shown in Figure 5–44. Remember to rotate the components to provide for the correct node assignments. Change the component values as required.

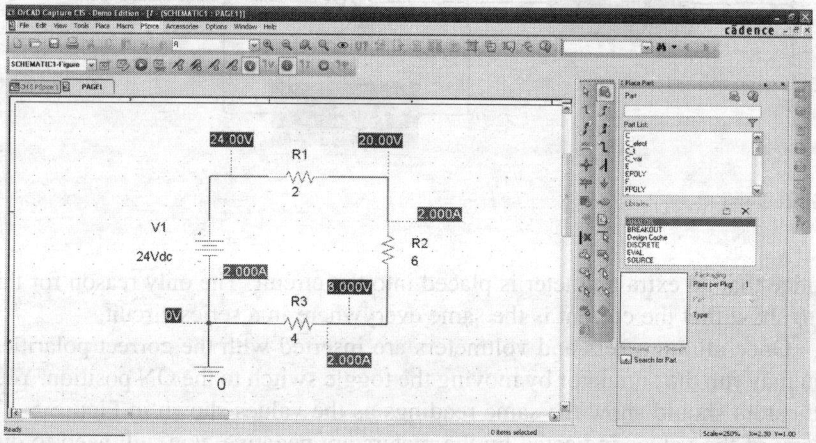

FIGURE 5–44

- Click the New Simulation Profile icon and enter a name (e.g., **Figure 5–44**) in the *Name* text box. You will need to enter the appropriate settings for this project in the *Simulation Setting* box. Click the *Analysis* tab, and select Bias Point from the *Analysis type* list. Click OK and save the document.
- Click on the *Run* icon. You will observe the A/D Demo screen. When you close this screen, you will observe the bias voltages and currents. From these results, we have the following:

$$V_1 = 24\text{ V} - 20\text{ V} = 4\text{ V}$$

$$V_2 = 20\text{ V} - 8\text{ V} = 12\text{ V}$$

$$V_3 = 8\text{ V}$$

For the supply voltage of 24 V, the current is 2.00 A. Clearly, these results are consistent with the theoretical calculations and the results obtained using Electronics Workbench.

- Save your project and exit from PSpice.

Putting It into Practice

You are part of a research team in the electrical metering department of a chemical processing plant. As part of your work, you regularly measure voltages between 200 V and 600 V. The only voltmeter available to you today has voltage ranges of 20 V, 50 V, and 100 V. Clearly, you cannot safely use the voltmeter to measure the expected voltages. However, you recall from your electrical course that you can use a voltage divider network to predictably reduce voltages. In order to keep current levels to safe values, you decide to use resistors in the megohm range. Without changing any internal circuitry of the voltmeter, show how you can use large-value resistors to change the 100-V range of the voltmeter to effectively measure a maximum of 1000 V. (Naturally, you would take extra precautions when measuring these voltages.) Show the schematic of the design, including the location of your voltmeter.

Problems

5.1 Series Circuits

1. The voltmeters of Figure 5–45 have autopolarity. Determine the reading of each meter, giving the correct magnitude and sign.
2. The voltmeters of Figure 5–46 have autopolarity. Determine the reading of each meter, giving the correct magnitude and sign.

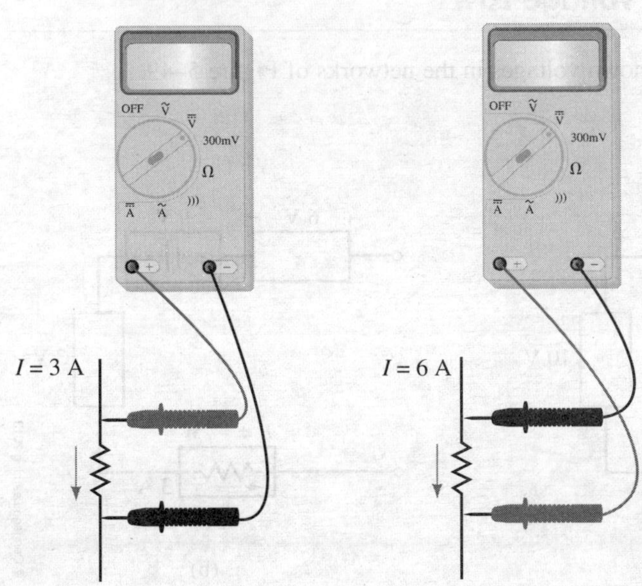

(a) $R = 10\ \Omega$ (b) $R = 15\ \Omega$

FIGURE 5–45

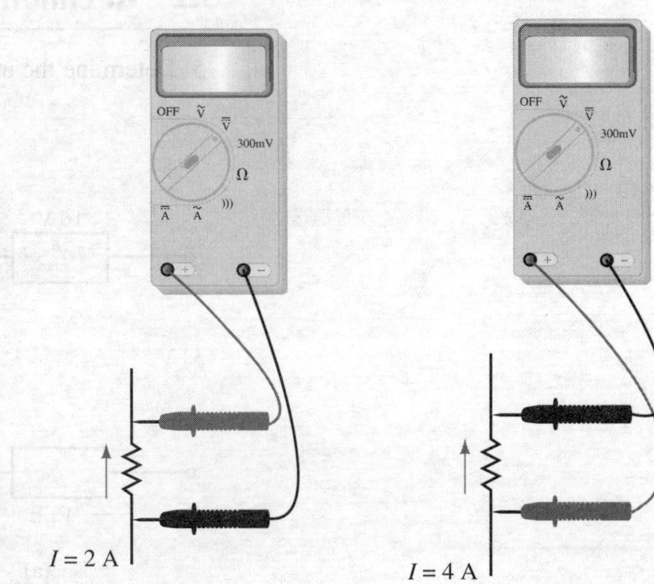

(a) $R = 36\ \Omega$ (b) $R = 40\ \Omega$

FIGURE 5–46

3. All resistors in Figure 5–47 are 15 Ω. For each case, determine the magnitude and polarity of voltage *V*.
4. The ammeters of Figure 5–48 have autopolarity. Determine their readings, giving the correct magnitude and sign.

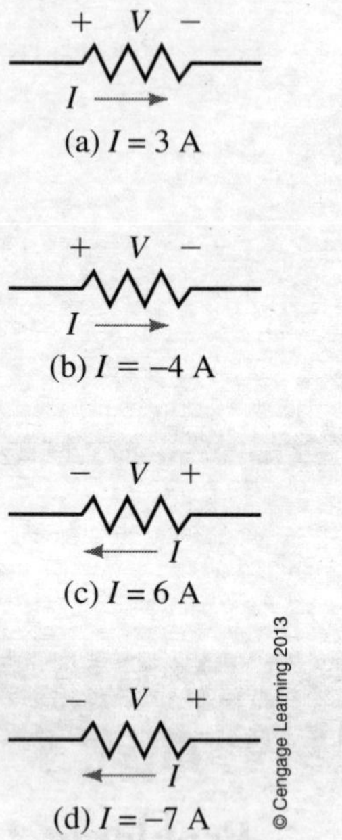

FIGURE 5–47 All resistors are 15 Ω.

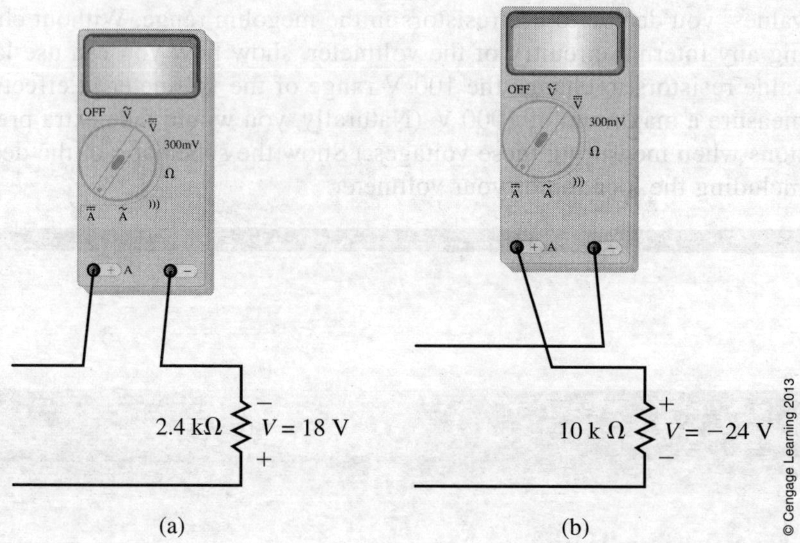

FIGURE 5–48

5.2 Kirchhoff's Voltage Law

5. Determine the unknown voltages in the networks of Figure 5–49.

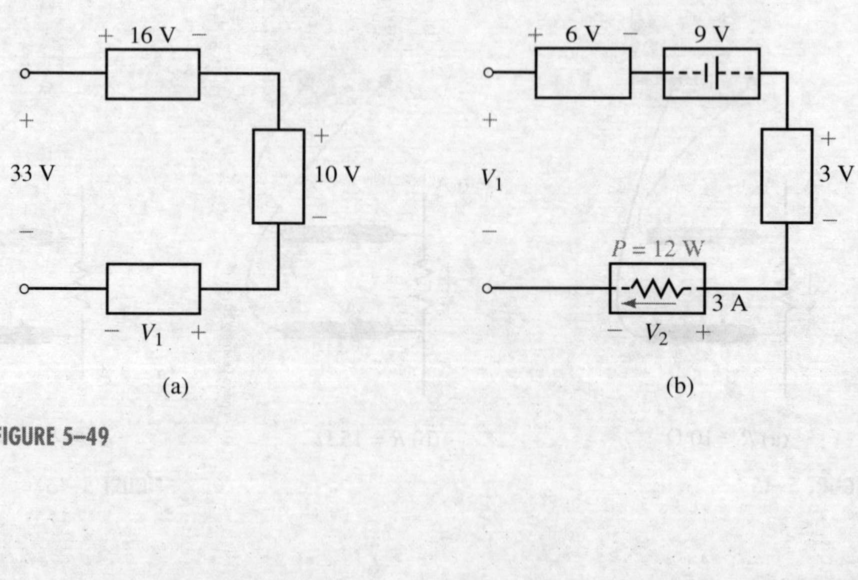

FIGURE 5–49

6. Determine the unknown voltages in the networks of Figure 5–50.

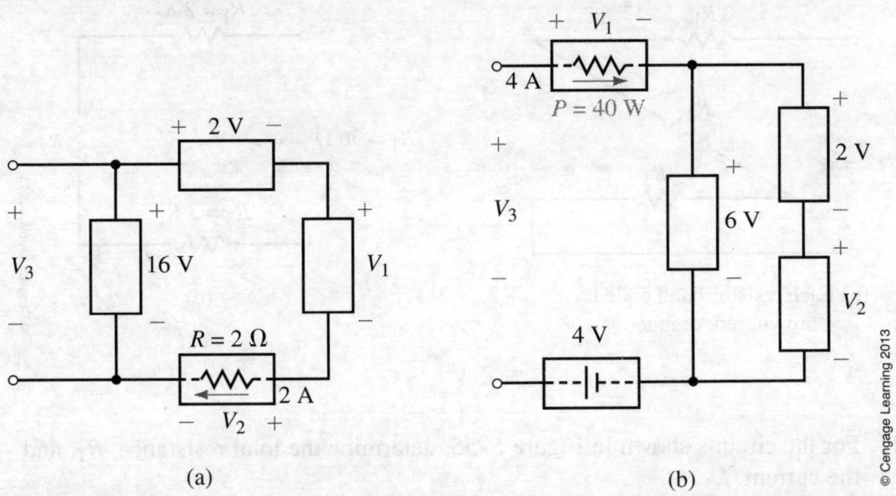

FIGURE 5–50

7. Solve for the unknown voltages in the circuit of Figure 5–51.

8. Solve for the unknown voltages in the circuit of Figure 5–52.

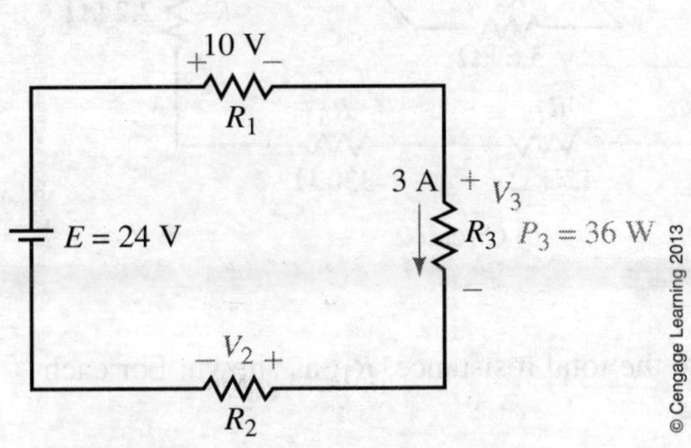

FIGURE 5–51

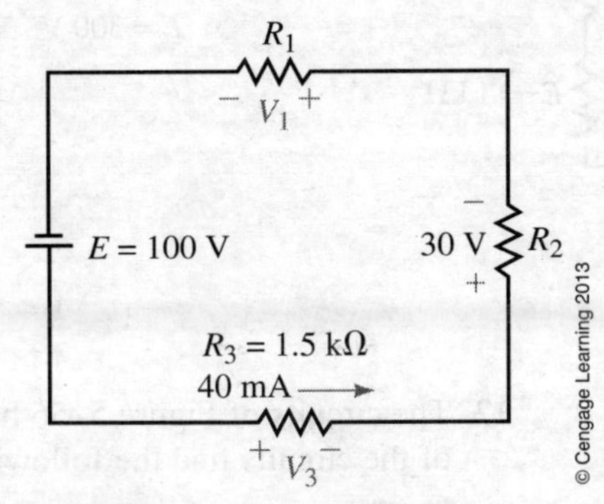

FIGURE 5–52

5.3 Resistors in Series

9. Determine the total resistance of the networks shown in Figure 5–53.

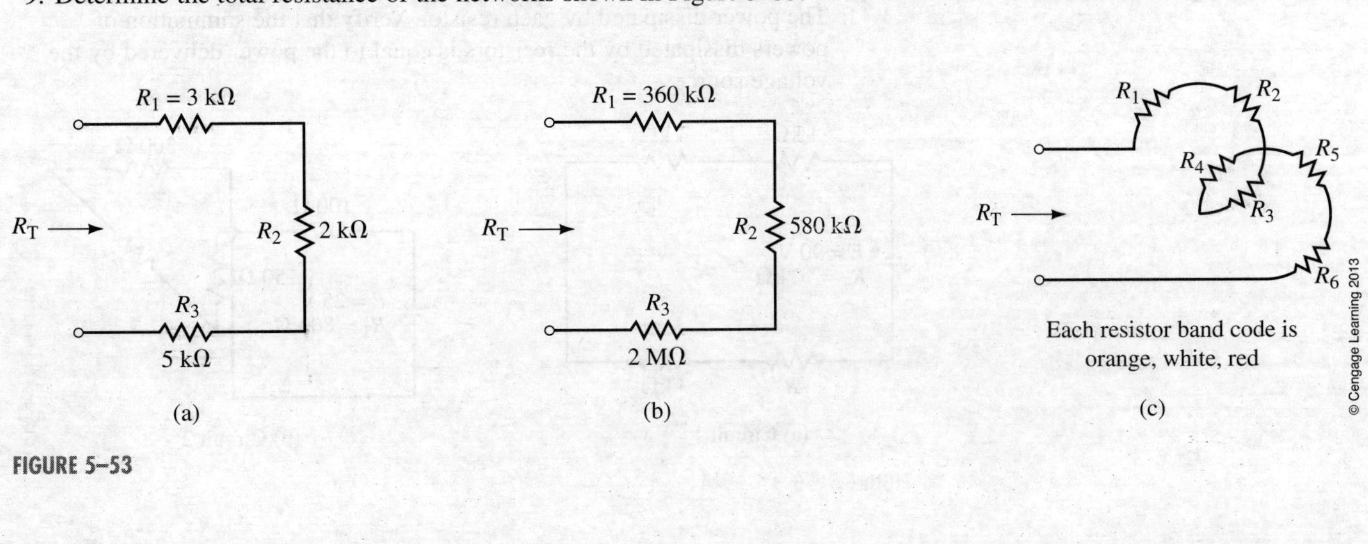

FIGURE 5–53

10. Determine the unknown resistance in each of the networks in Figure 5–54.

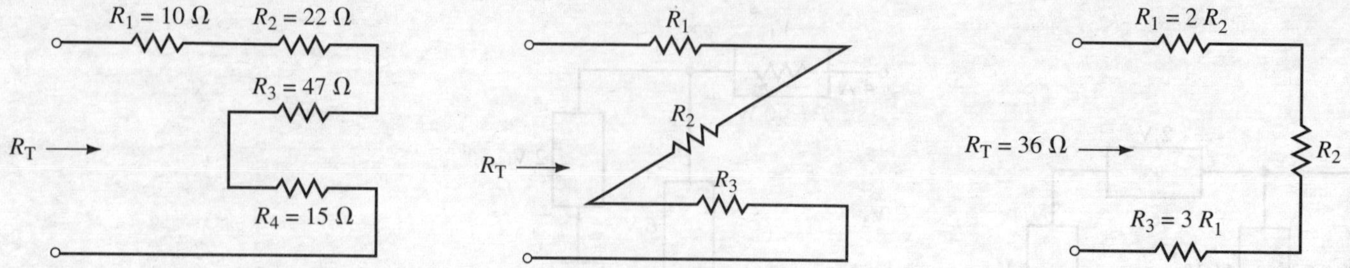

(a) (b) Each resistor band code is brown, red, orange (c)

FIGURE 5–54

11. For the circuits shown in Figure 5–55, determine the total resistance, R_T, and the current, I.

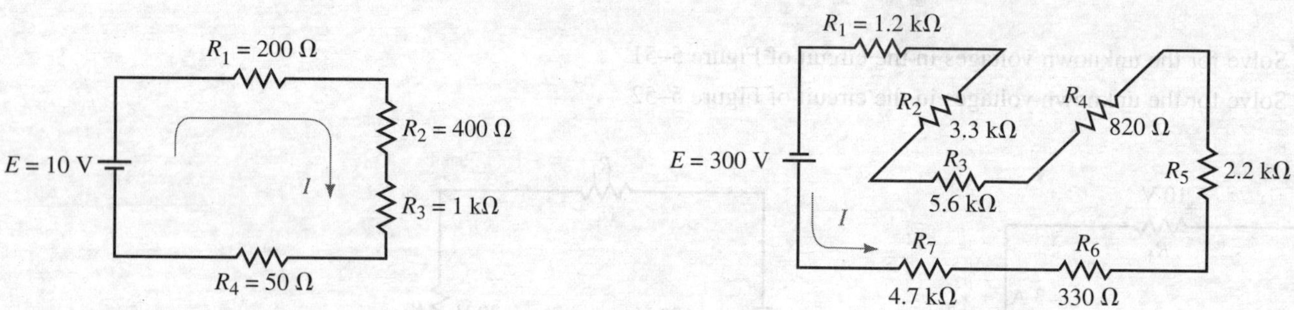

(a) Circuit 1 (b) Circuit 2

FIGURE 5–55

12. The circuits of Figure 5–56 have the total resistance, R_T, as shown. For each of the circuits find the following:
    a. The magnitude of current in the circuit.
    b. The total power delivered by the voltage source.
    c. The direction of current through each resistor in the circuit.
    d. The value of the unknown resistance, R.
    e. The voltage drop across each resistor.
    f. The power dissipated by each resistor. Verify that the summation of powers dissipated by the resistors is equal to the power delivered by the voltage source.

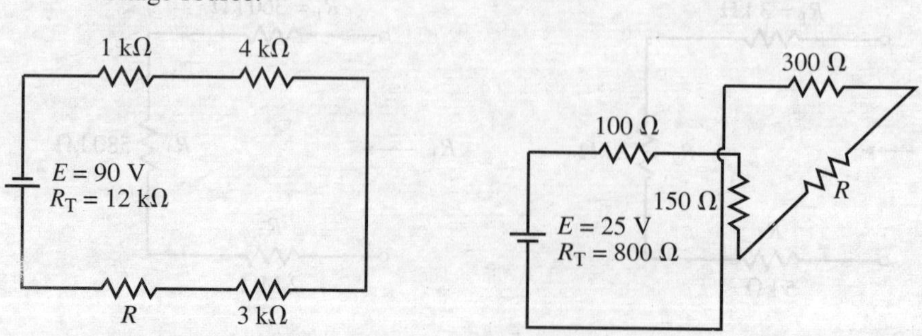

(a) Circuit 1 (b) Circuit 2

FIGURE 5–56

13. For the circuit of Figure 5–57, find the following quantities:
    a. The circuit current.
    b. The total resistance of the circuit.
    c. The value of the unknown resistance, R.
    d. The voltage drop across all resistors in the circuit.
    e. The power dissipated by all resistors.

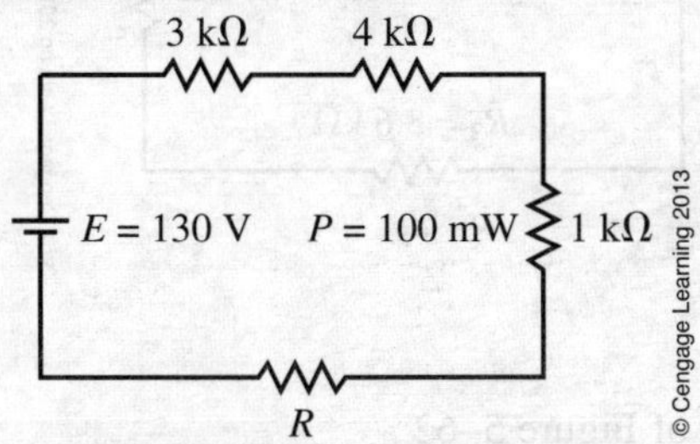

FIGURE 5–57

14. The circuit of Figure 5–58 has a current of 2.5 mA. Find the following quantities:
    a. The total resistance of the circuit.
    b. The value of the unknown resistance, R_2.
    c. The voltage drop across each resistor in the circuit.
    d. The power dissipated by each resistor in the circuit.

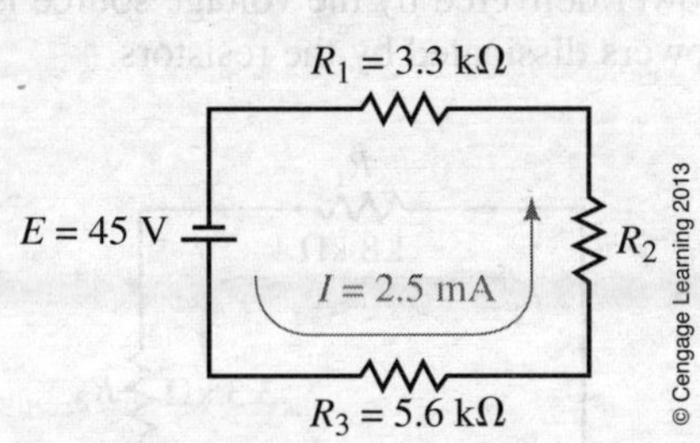

FIGURE 5–58

15. For the circuit of Figure 5–59, find the following quantities:
    a. The current, I.
    b. The voltage drop across each resistor.
    c. The voltage across the open terminals a and b.
16. Refer to the circuit of Figure 5–60:
    a. Use Kirchhoff's voltage law to find the voltage drops across R_2 and R_3.
    b. Determine the magnitude of the current, I.
    c. Solve for the unknown resistance, R_1.

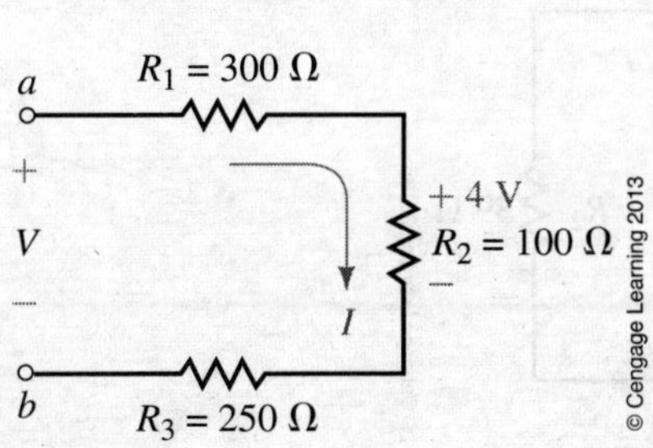

FIGURE 5–59

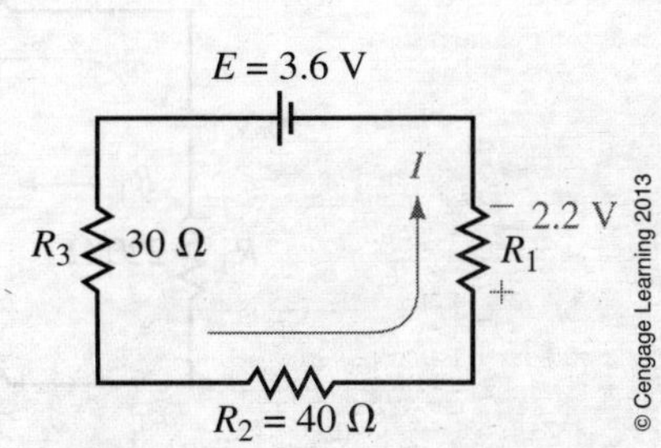

FIGURE 5–60

17. Repeat Problem 16 for the circuit of Figure 5–61.

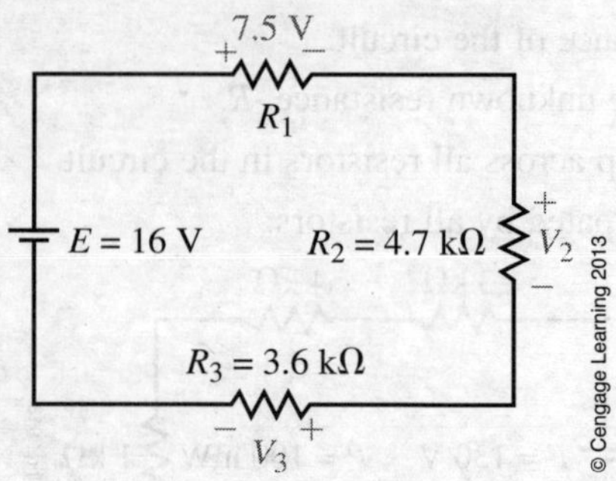

FIGURE 5–61

18. Refer to the circuit of Figure 5–62:
   a. Find R_T.
   b. Solve for the current, I.
   c. Determine the voltage drop across each resistor.
   d. Verify Kirchhoff's voltage law around the closed loop.
   e. Find the power dissipated by each resistor.
   f. Determine the minimum power rating of each resistor, if resistors are available with the following power ratings: ⅛ W, ¼ W, ½ W, 1 W, and 2 W.
   g. Show that the power delivered by the voltage source is equal to the summation of the powers dissipated by the resistors.

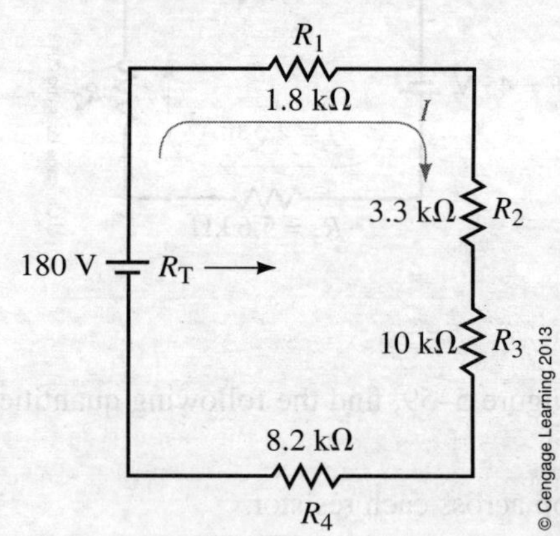

FIGURE 5–62

19. Repeat Problem 18 for the circuit of Figure 5–63.

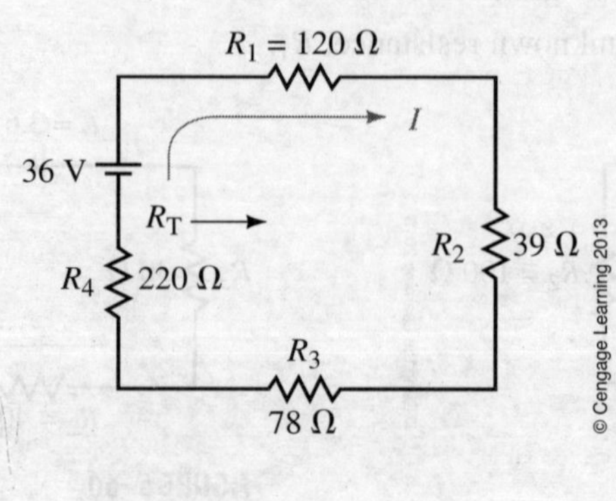

FIGURE 5–63

20. Refer to the circuit of Figure 5–64.
    a. Calculate the voltage across each resistor.
    b. Determine the values of the resistors R_1 and R_2.
    c. Solve for the power dissipated by each of the resistors.

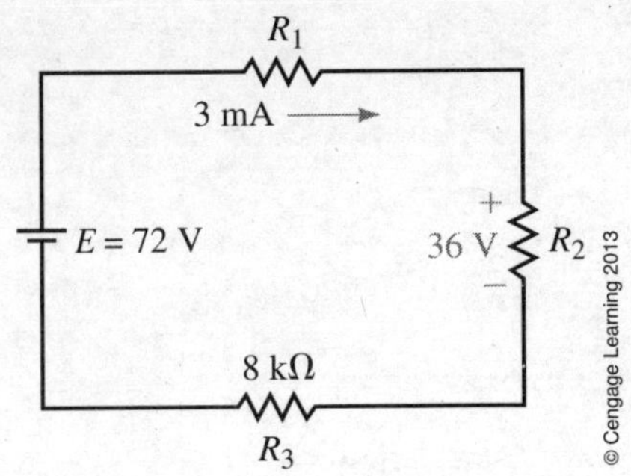

FIGURE 5–64

5.5 Interchanging Series Components

21. Refer to Figure 5–15. Determine the value of the series resistance given the following circuit parameters:
    $E = 24$ V
    $I_F = 10$ mA
    $V_F = 3.8$ V
22. Repeat Problem 21 given the following circuit parameters:
    $E = 12$ V
    $I_F = 15$ mA
    $V_F = 4.2$ V
23. Redraw the circuits of Figure 5–65, showing a single voltage source for each circuit. Solve for the current in each circuit.

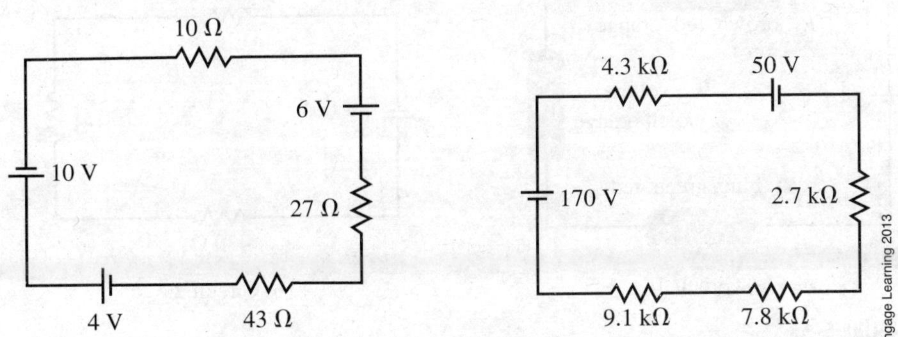

(a) Circuit 1 (b) Circuit 2

FIGURE 5–65

24. Use the information given to determine the polarity and magnitude of the unknown voltage source in each of the circuits of Figure 5–66.

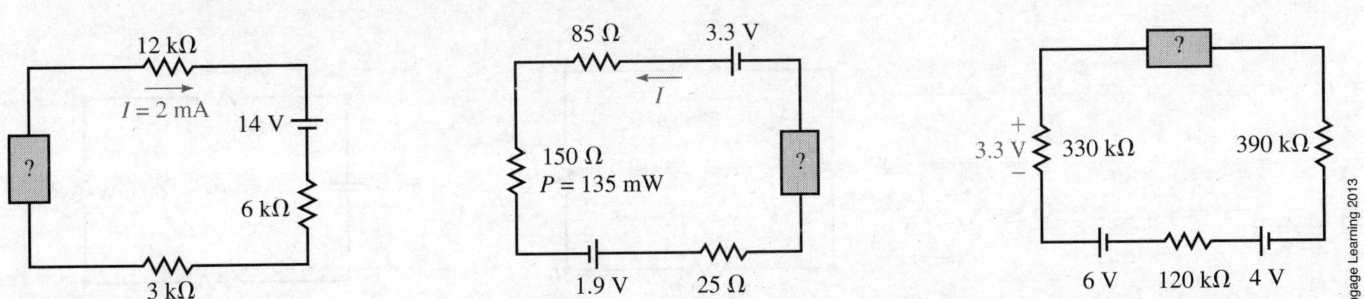

(a) (b) (c)

FIGURE 5–66

5.6 The Voltage Divider Rule

25. Use the voltage divider rule to determine the voltage across each resistor in the circuits of Figure 5–67. Use your results to verify Kirchhoff's voltage law for each circuit.

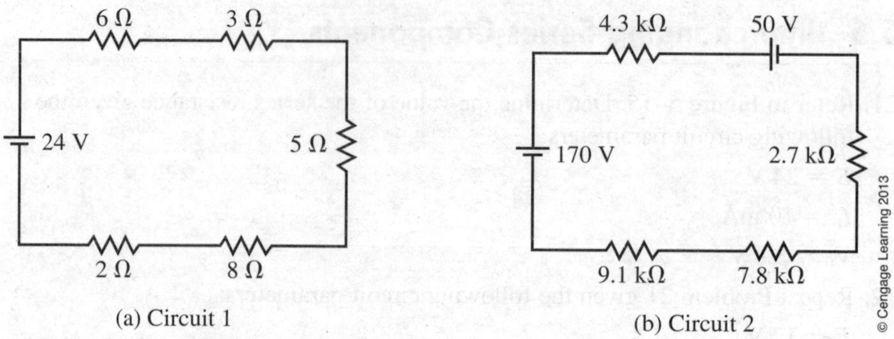

FIGURE 5–67

26. Repeat Problem 23 for the circuits of Figure 5–68.

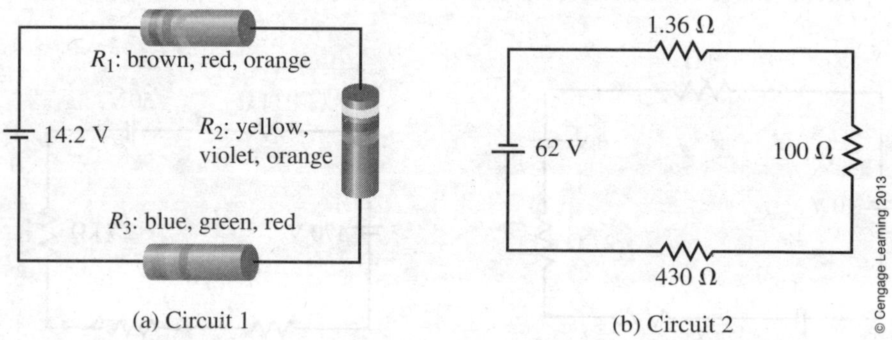

FIGURE 5–68

27. Refer to the circuits of Figure 5–69:
   a. Find the values of the unknown resistors.
   b. Calculate the voltage across each resistor.
   c. Determine the power dissipated by each resistor.

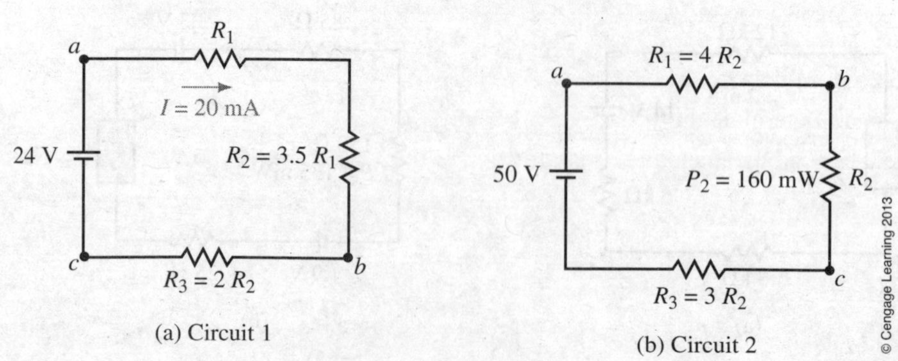

FIGURE 5–69

28. Refer to the circuits of Figure 5–70:
   a. Find the values of the unknown resistors using the voltage divider rule.
   b. Calculate the voltage across R_1 and R_3.
   c. Determine the power dissipated by each resistor.

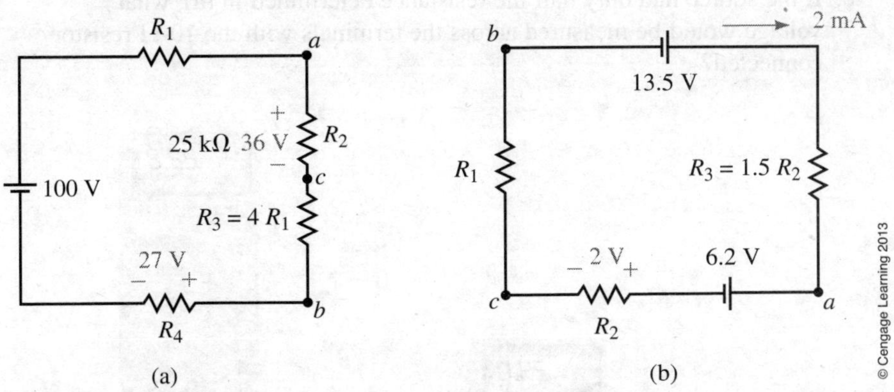

FIGURE 5–70

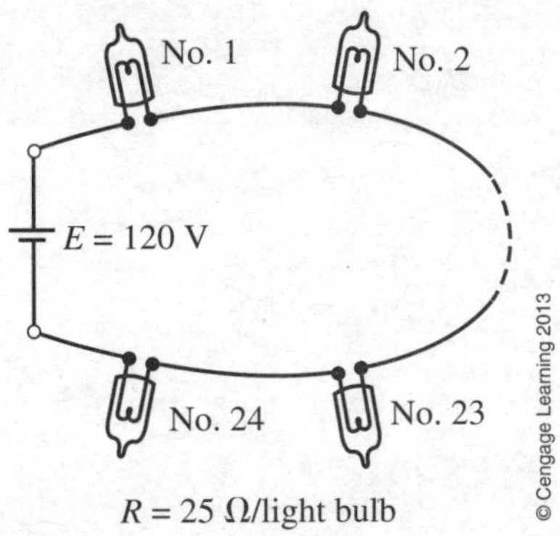

FIGURE 5–71

29. A string of 24 series light bulbs is connected to a 120-V supply as shown in Figure 5–71.
   a. Solve for the current in the circuit.
   b. Use the voltage divider rule to find the voltage across each light bulb.
   c. Calculate the power dissipated by each bulb.
   d. If a single light bulb were to become an open circuit, the entire string would stop working. To prevent this from occurring, each light bulb has a small metal strip which shorts the light bulb when the filament fails. If two bulbs in the string were to burn out, repeat Steps (a) through (c).
   e. Based on your calculations of Step (d), what do you think would happen to the life expectancy of the remaining light bulbs if the two faulty bulbs were not replaced?

30. Repeat Problem 29 for a string consisting of 36 light bulbs.

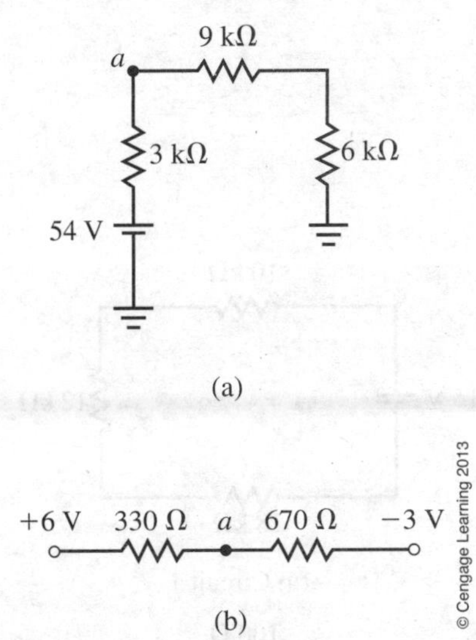

FIGURE 5–72

5.8 Voltage Subscripts

31. Solve for the voltages V_{ab} and V_{bc} in the circuits of Figure 5–69.
32. Repeat Problem 31 for the circuits of Figure 5–70.
33. For the circuits of Figure 5–72, determine the voltage across each resistor and calculate the voltage V_a.
34. Given the circuits of Figure 5–73:
   a. Determine the voltage across each resistor.
   b. Find the magnitude and direction of the current in the 180-kΩ resistor.
   c. Solve for the voltage V_a.

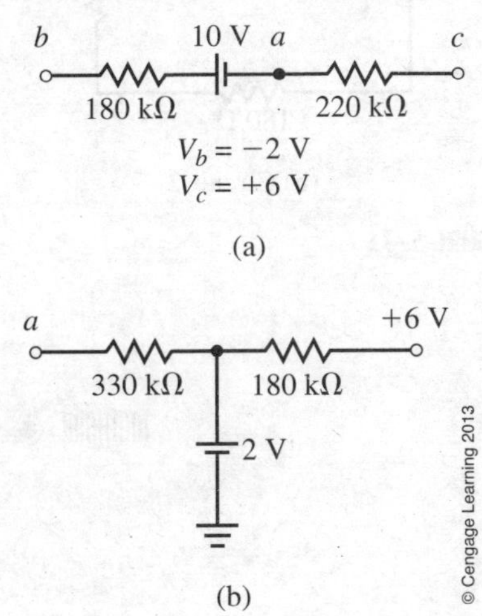

FIGURE 5–73

5.9 Internal Resistance of Voltage Sources

35. A battery is measured to have an open-terminal voltage of 14.2 V. When this voltage is connected to a 100-Ω load, the voltage measured between the terminals of the battery drops to 6.8 V.
   a. Determine the internal resistance of the battery.
   b. If the 100-Ω load were replaced with a 200-Ω load, what voltage would be measured across the terminals of the battery?

36. The voltage source shown in Figure 5–74 is measured to have an open-circuit voltage of 24 V. When a 10-Ω load is connected across the terminals, the voltage measured with a voltmeter drops to 22.8 V.
    a. Determine the internal resistance of the voltage source.
    b. If the source had only half the resistance determined in (a), what voltage would be measured across the terminals with the 10-Ω resistor connected?

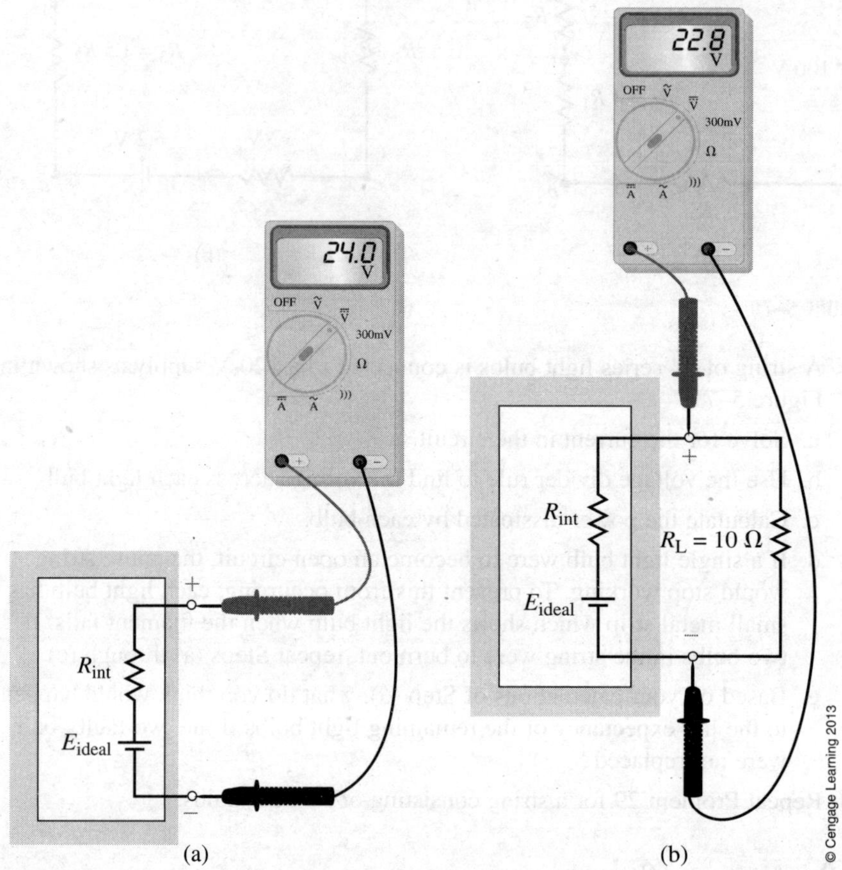

FIGURE 5–74

5.10 Ammeter Loading Effects

37. For the series circuits of Figure 5–75, determine the current in each circuit. If an ammeter having an internal resistance of 50 Ω is used to measure the current in the circuits, determine the current through the ammeter and calculate the loading effect for each circuit.
38. Repeat Problem 37 if the ammeter has a resistance of 10 Ω.

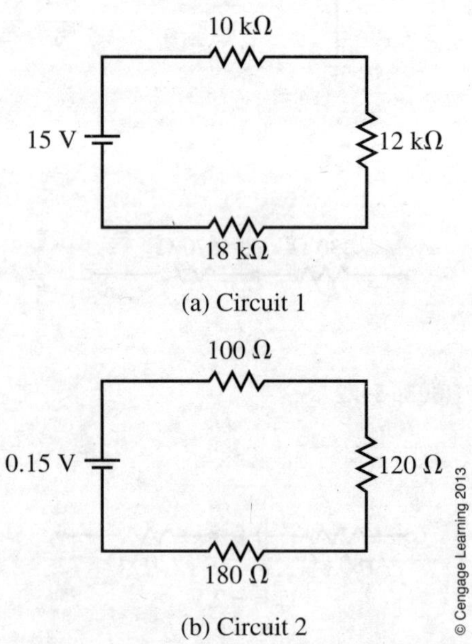

FIGURE 5–75

5.11 Circuit Analysis Using Computers

Multisim

39. Refer to the circuits of Figure 5–67. Use Multisim to find the following:
    a. The current in each circuit.
    b. The voltage across each resistor in the circuit.

40. Given the circuit of Figure 5–76, use Multisim to determine the following: Multisim

a. The current through the voltage source, *I*.

b. The voltage across each resistor.

c. The voltage between terminals *a* and *b*.

d. The voltage, with respect to ground, at terminal *c*.

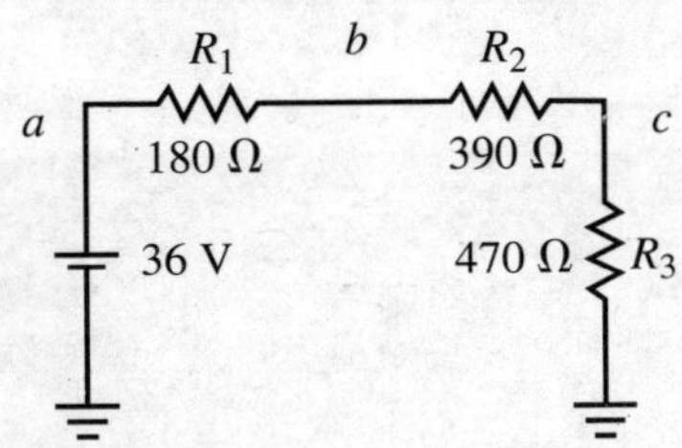

FIGURE 5–76

Multisim

41. Refer to the circuit of Figure 5–63. Use PSpice to find the following: PSPICE

a. The current in the circuit.

b. The voltage across each resistor in the circuit.

42. Refer to the circuit of Figure 5–62. Use PSpice to find the following: PSPICE

a. The current in the circuit.

b. The voltage across each resistor in the circuit.

ANSWERS TO IN-PROCESS LEARNING CHECKS

IN-PROCESS LEARNING CHECK 1

1. Two elements are connected at only one node.
2. No current-carrying element is connected to the common node.

IN-PROCESS LEARNING CHECK 2

The summation of voltage drops and rises around any closed loop is equal to zero; or the summation of voltage rises is equal to the summation of voltage drops around a closed loop.

IN-PROCESS LEARNING CHECK 3

$R_1 = 4.2\ \text{k}\Omega$
$R_2 = 12.6\ \text{k}\Omega$
$R_3 = 25.2\ \text{k}\Omega$

IN-PROCESS LEARNING CHECK 4

$E_{\text{CELL}} = 2.18\ \text{V}$

IN-PROCESS LEARNING CHECK 5

$R_1 = 18.8\ \text{k}\Omega$
$R_2 = 28.21\ \text{k}\Omega$

IN-PROCESS LEARNING CHECK 6

The chassis of the oven is grounded when it is connected to the electrical outlet.

IN-PROCESS LEARNING CHECK 7

$V_{ab} = 7.50\ \text{V}$
$V_{ca} = -10.0\ \text{V}$
$V_{bc} = 2.5\ \text{V}$

KEY TERMS

Current Divider Rule (CDR)
Kirchhoff's Current Law
Loading Effect (Voltmeter)
Nodes
Parallel (Or Shunt) Circuits
Total Conductance
Total Equivalent Resistance

OUTLINE

Parallel Circuits
Kirchhoff's Current Law
Resistors in Parallel
Voltage Sources in Parallel
Current Divider Rule
Analysis of Parallel Circuits
Voltmeter Loading Effects
Computer Analysis

OBJECTIVES

After studying this chapter you will be able to

- recognize which elements and branches in a given circuit are connected in parallel and which are connected in series,
- calculate the total resistance and conductance of a network of parallel resistances,
- determine the current in any resistor in a parallel circuit,
- solve for the voltage across any parallel combinations of resistors,
- apply Kirchhoff's current law to solve for unknown currents in a circuit,
- explain why voltage sources of different magnitudes must never be connected in parallel,
- use the current divider rule to solve for the current through any resistor of a parallel combination,
- identify and calculate the loading effects of a voltmeter connected into a circuit,
- use Multisim to observe loading effects of a voltmeter,
- use PSpice to evaluate voltage and current in a parallel circuit.

6

PARALLEL CIRCUITS

CHAPTER PREVIEW

Two fundamental circuits form the basis of all electrical circuits. They are the series circuit and the parallel circuit. The previous chapter examined the principles and rules that apply to series circuits. In this chapter we study the **parallel** (or **shunt**) **circuit** and examine the rules governing the operation of these circuits.

Figure 6–1 illustrates a simple example of several light bulbs connected in parallel with one another and a battery supplying voltage to the bulbs.

This illustration shows one of the important differences between the series circuit and the parallel circuit. The parallel circuit will continue to operate even though one of the light bulbs may have a defective (open) filament. Only the defective light bulb will no longer glow. If a circuit were made up of several light bulbs in series, however, the defective light bulb would prevent any current in the circuit, and so all the light bulbs would be off. ■

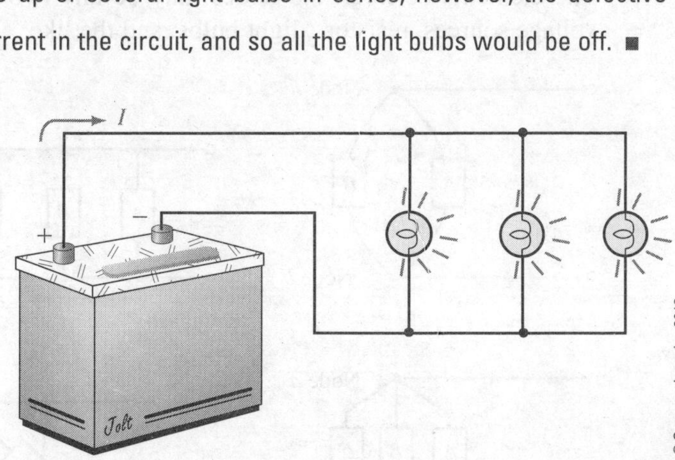

FIGURE 6–1 Simple parallel circuit.

Putting It in Perspective

Luigi Galvani and the Discovery of Nerve Excitation

LUIGI GALVANI WAS BORN IN BOLOGNA, Italy, on September 9, 1737.

Galvani's main expertise was in anatomy, a subject in which he was appointed lecturer at the university in Bologna.

Galvani discovered that when the nerves of frogs were connected to sources of electricity, the muscles twitched. Although he was unable to determine where the electrical pulses originated within the animal, Galvani's work was significant and helped to open further discoveries in nerve impulses.

Galvani's name has been adopted for the instrument called the galvanometer, which is used for detecting very small currents.

Luigi Galvani died in Bologna on December 4, 1798. Although he made many contributions to science, Galvani died poor, shrouded in controversy due to his refusal to swear allegiance to Napolean. ■

6.1 Parallel Circuits

Figure 6–1 shows that one terminal of each light bulb is connected to the positive terminal of the battery and that the other terminal of the light bulb is connected to the negative terminal of the battery. These points of connection are often referred to as **nodes.**

Elements or branches are said to be in a parallel connection when they have exactly two nodes in common. Additionally, these parallel elements or branches will have the same voltage across them.

Figure 6–2 shows several different ways of sketching parallel elements. The elements between the nodes may be any two-terminal devices such as voltage sources, resistors, light bulbs, and the like.

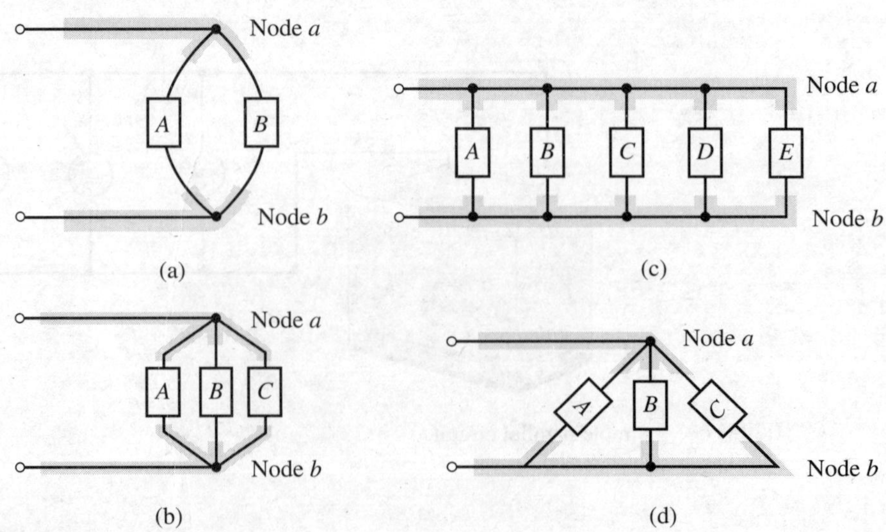

FIGURE 6–2 Parallel elements.

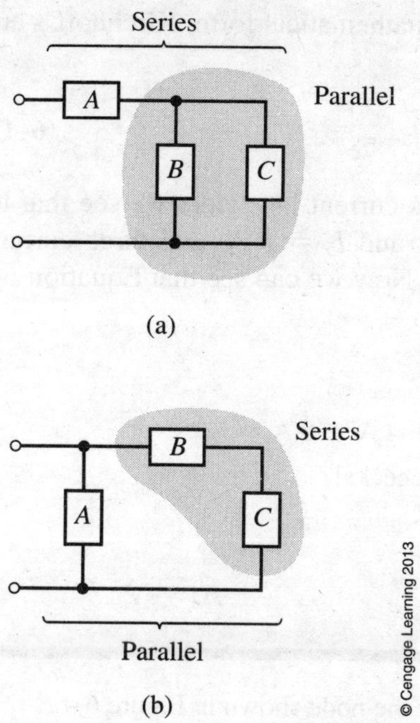

FIGURE 6–3 Series-parallel combinations.

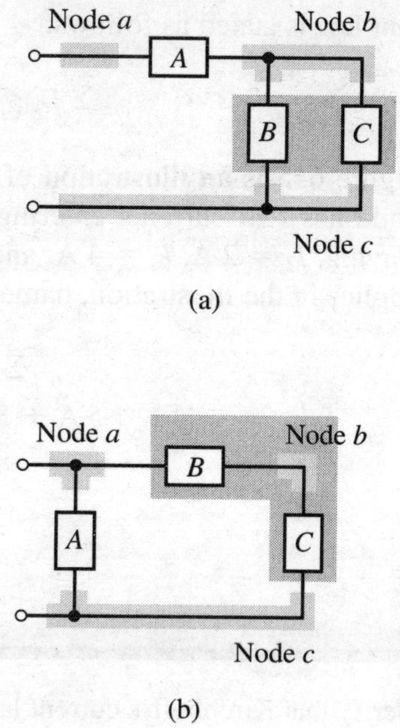

FIGURE 6–4

In Figure 6–2, notice that every element has two terminals and that each of the terminals is connected to one of the two nodes.

Very often, circuits contain a combination of series and parallel components. Although we will study these circuits in greater depth in later chapters, it is important at this point to be able to recognize the various connections in a given network. Consider the networks shown in Figure 6–3.

When analyzing a particular circuit, it is usually easiest to first designate the nodes (we will use lowercase letters) and then to identify the types of connections. Figure 6–4 shows the nodes for the networks of Figure 6–3.

In the circuit of Figure 6–4(a), we see that element B is in parallel with element C since they each have nodes b and c in common. This parallel combination is now seen to be in series with element A.

In the circuit of Figure 6–4(b), element B is in series with element C since these elements have a single common node: node b. The branch consisting of the series combination of elements B and C is then determined to be in parallel with element A.

6.2 Kirchhoff's Current Law

Recall that Kirchhoff's voltage law was extremely useful in understanding the operation of the series circuit. In a similar manner, Kirchhoff's current law is the underlying principle that is used to explain the operation of a parallel circuit. **Kirchhoff's current law** states the following:

The summation of currents entering a node is equal to the summation of currents leaving the node.

An analogy that helps us understand the principle of Kirchhoff's current law is the flow of water. When water flows in a closed pipe, the amount of water entering a particular point in the pipe is exactly equal to the amount of water leaving, since there is no loss. In mathematical form, Kirchhoff's current law is stated as follows:

$$\Sigma I_{\text{entering node}} = \Sigma I_{\text{leaving node}} \tag{6–1}$$

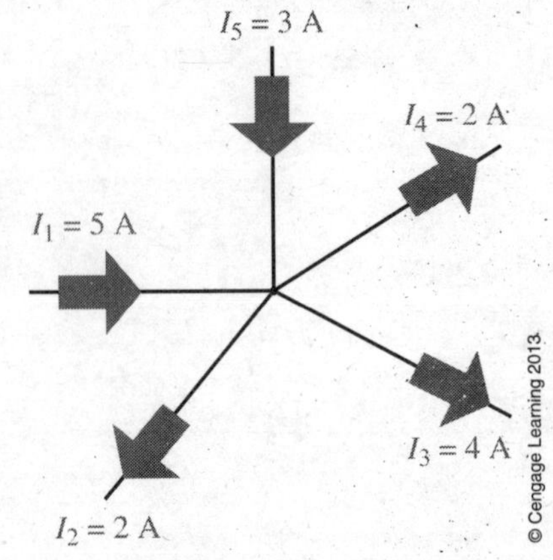

FIGURE 6–5 Kirchhoff's current law.

Figure 6–5 is an illustration of Kirchhoff's current law. Here we see that the node has two currents entering, $I_1 = 5$ A and $I_5 = 3$ A, and three currents leaving, $I_2 = 2$ A, $I_3 = 4$ A, and $I_4 = 2$ A. Now we can see that Equation 6–1 applies in the illustration, namely,

$$\Sigma I_{\text{in}} = \Sigma I_{\text{out}}$$
$$5\text{ A} + 3\text{ A} = 2\text{ A} + 4\text{ A} + 8\text{ A}$$
$$8\text{ A} = 8\text{ A (checks!)}$$

PRACTICE PROBLEMS 1

Verify that Kirchhoff's current law applies at the node shown in Figure 6–6.

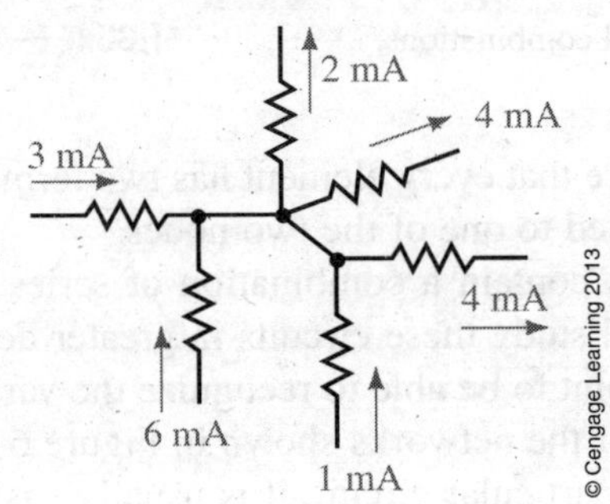

FIGURE 6–6

Answer
3 mA + 6 mA + 1 mA = 2 mA + 4 mA + 4 mA

Quite often, when we analyze a given circuit, we are unsure of the direction of current through a particular element within the circuit. In such cases, we assume a reference direction and base further calculations on this assumption. If our assumption is incorrect, calculations will show that the current has a negative sign. The negative sign simply indicates that the current is in fact opposite to the direction selected as the reference. The following example illustrates this very important concept.

EXAMPLE 6–1

Determine the magnitude and correct direction of the currents I_3 and I_5 for the network of Figure 6–7.

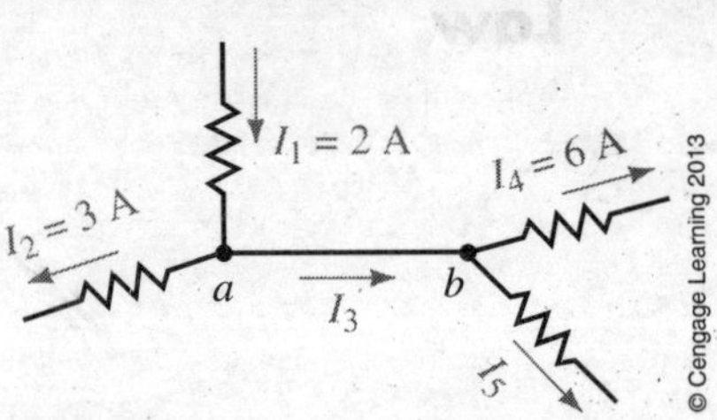

FIGURE 6–7

Solution Although points *a* and *b* are in fact the same node, we treat the points as two separate nodes with 0 Ω resistance between them.

Since Kirchhoff's current law must be valid at point *a*, we have the following expression for this node:

$$I_1 = I_2 + I_3$$

and so

$$I_3 = I_1 - I_2$$
$$= 2\text{ A} - 3\text{ A} = -1\text{ A}$$

Notice that the reference direction of current I_3 was taken to be from a to b, while the negative sign indicates that the current is in fact from b to a.

Similarly, using Kirchhoff's current law at point b gives

$$I_3 = I_4 + I_5$$

which gives current I_5 as

$$I_5 = I_3 - I_4$$
$$= -1\text{ A} - 6\text{ A} = -7\text{ A}$$

The negative sign indicates that the current I_5 is actually toward node b rather than away from the node. The actual directions and magnitudes of the currents are illustrated in Figure 6–8.

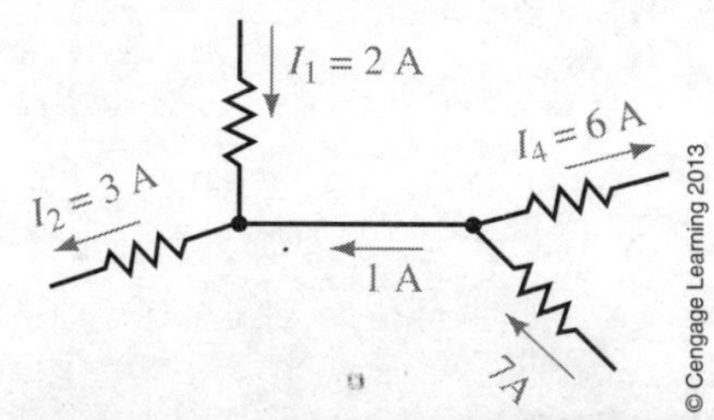

FIGURE 6–8

EXAMPLE 6–2

Find the magnitudes of the unknown currents for the circuit of Figure 6–9.

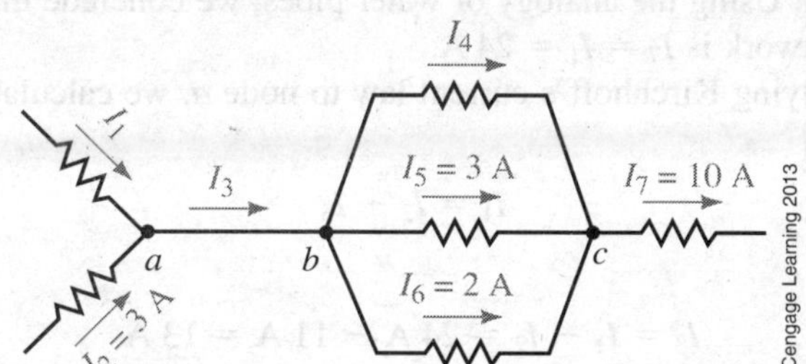

FIGURE 6–9

Solution If we consider point a, we see that there are two unknown currents, I_1 and I_3. Since there is no way to solve for these values, we examine the currents at point b, where we again have two unknown currents, I_3 and I_4. Finally, we observe that at point c there is only one unknown, I_4. Using Kirchhoff's current law we solve for the unknown current as follows:

$$I_4 + 3\text{ A} + 2\text{ A} = 10\text{ A}$$

Therefore,

$$I_4 = 10\text{ A} - 3\text{ A} - 2\text{ A} = 5\text{ A}$$

Now we can see that at point b the current entering is

$$I_3 = 5\text{ A} + 3\text{ A} + 2\text{ A} = 10\text{ A}$$

And finally, by applying Kirchhoff's current law at point a, we determine that the current I_1 is

$$I_1 = 10\text{ A} - 3\text{ A} = 7\text{ A}$$

EXAMPLE 6–3

Determine the unknown currents in the network of Figure 6–10.

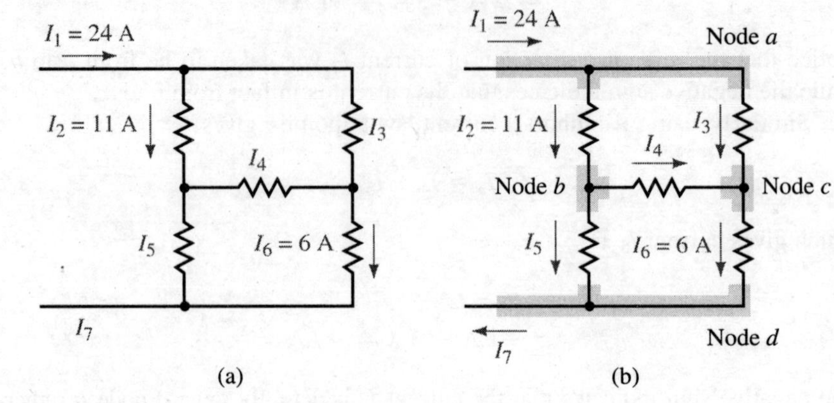

FIGURE 6–10

Solution We first assume reference directions for the unknown currents in the network.

Since we may use the analogy of water moving through conduits, we can easily assign directions for the currents I_3, I_5, and I_7. However, the direction for the current I_4 is not as easily determined, so we arbitrarily assume that its direction is to the right. Figure 6–10(b) shows the various nodes and the assumed current directions.

By examining the network, we see that there is only a single source of current $I_1 = 24$ A. Using the analogy of water pipes, we conclude that the current leaving the network is $I_7 = I_1 = 24$ A.

Now, applying Kirchhoff's current law to node a, we calculate the current I_3 as follows:

$$I_1 = I_2 + I_3$$

Therefore,

$$I_3 = I_1 - I_2 = 24\text{ A} - 11\text{ A} = 13\text{ A}$$

Similarly, at node c, we have

$$I_3 + I_4 = I_6$$

Therefore,

$$I_4 = I_6 - I_3 = 6\text{ A} - 13\text{ A} = -7\text{ A}$$

Although the current I_4 is opposite to the assumed reference direction, we do not change its direction for further calculations. We use the original direction together with the negative sign; otherwise the calculations would be needlessly complicated.

Applying Kirchhoff's current law at node b, we get

$$I_2 = I_4 + I_5$$

which gives

$$I_5 = I_2 - I_4 = 11\text{ A} - (-7\text{ A}) = 18\text{ A}$$

Finally, applying Kirchhoff's current law at node d gives

$$I_5 + I_6 = I_7$$

resulting in

$$I_7 = I_5 + I_6 = 18\text{ A} + 6\text{ A} = 24\text{ A}$$

PRACTICE PROBLEMS 2

Determine the unknown currents in the network of Figure 6–11.

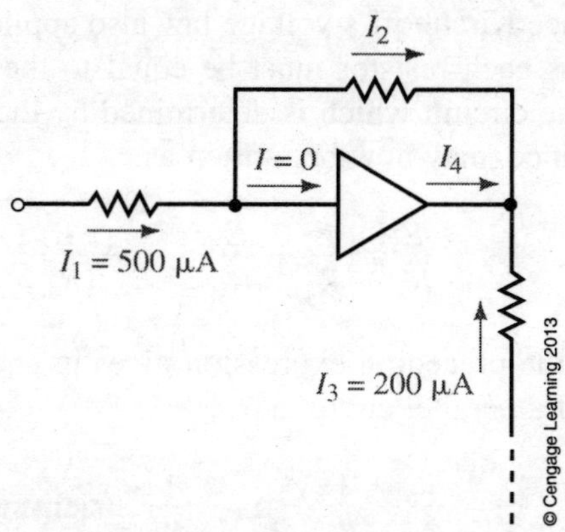

FIGURE 6–11

Answers

$I_2 = 500\ \mu A,\ I_4 = -700\ \mu A$

6.3 Resistors in Parallel

A simple parallel circuit is constructed by combining a voltage source with several resistors as shown in Figure 6–12.

The voltage source will result in current from the positive terminal of the source toward node *a*. At this point the current will split between the various resistors and then recombine at node *b* before continuing to the negative terminal of the voltage source.

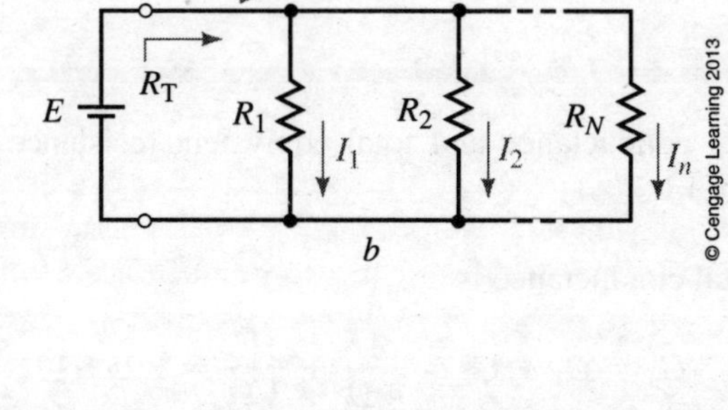

FIGURE 6–12

This circuit illustrates a very important concept of parallel circuits. If we were to apply Kirchhoff's voltage law around each closed loop in the parallel circuit of Figure 6–12, we would find that the voltage across all parallel resistors is exactly equal, namely $V_{R_1} = V_{R_2} = V_{R_3} = E$. Therefore, by applying Kirchhoff's voltage law, we make the following statement:

The voltage across all parallel elements in a circuit will be the same.

This principle allows us to determine the equivalent resistance, R_T, of any number of resistors connected in parallel. The equivalent resistance, R_T, is the effective resistance "seen" by the source and determines the total current, I_T, provided to the circuit. Applying Kirchhoff's current law to the circuit of Figure 6–11, we have the following expression:

$$I_T = I_1 + I_2 + \cdots + I_n$$

However, since Kirchhoff's voltage law also applies to the parallel circuit, the voltage across each resistor must be equal to the supply voltage, E. The total current in the circuit, which is determined by the supply voltage and the equivalent resistance, may now be written as

$$\frac{E}{R_T} = \frac{E}{R_1} + \frac{E}{R_2} + \cdots + \frac{E}{R_n}$$

Simplifying the preceding expression gives us the general expression for total resistance of a parallel circuit as

$$\frac{1}{R_T} = \frac{1}{R_1} + \frac{1}{R_2} + \cdots + \frac{1}{R_n} \quad \text{(siemens, S)} \tag{6–2}$$

Since conductance was defined as the reciprocal of resistance, we may write the preceding equation in terms of conductance, namely,

$$G_T = G_1 + G_2 + \cdots + G_n \quad \text{(S)} \tag{6–3}$$

Whereas series resistors had a total resistance determined by the summation of the particular resistances, we see that any number of parallel resistors have a **total conductance** determined by the summation of the individual conductances.

The **total equivalent resistance** of n parallel resistors may be determined in one step as follows:

$$R_T = \frac{1}{\frac{1}{R_1} + \frac{1}{R_2} + \cdots + \frac{1}{R_n}} \quad (\Omega) \tag{6–4}$$

An important effect of combining parallel resistors is that the resultant resistance will always be smaller than the smallest resistor in the combination.

EXAMPLE 6–4

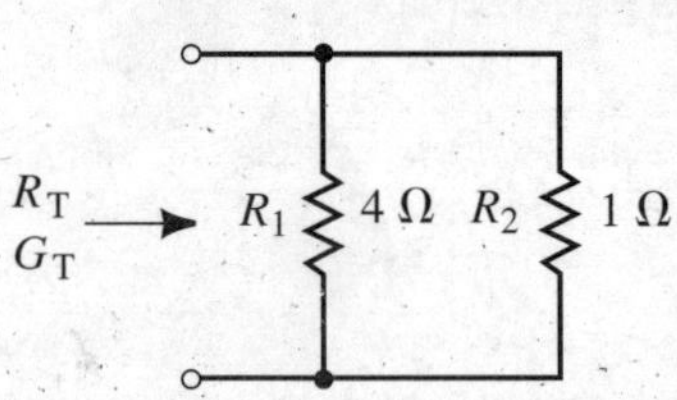

FIGURE 6–13

Solve for the total conductance and total equivalent resistance of the circuit shown in Figure 6–13.

Solution The total conductance is

$$G_T = G_1 + G_2 = \frac{1}{4\ \Omega} + \frac{1}{1\ \Omega} = 1.25\ \text{S}$$

The total equivalent resistance of the circuit is

$$R_T = \frac{1}{G_T} = \frac{1}{1.25\ \text{S}} = 0.800\ \Omega$$

Notice that the equivalent resistance of the parallel resistors is indeed less than the value of each resistor.

EXAMPLE 6–5

Determine the conductance and resistance of the network of Figure 6–14.

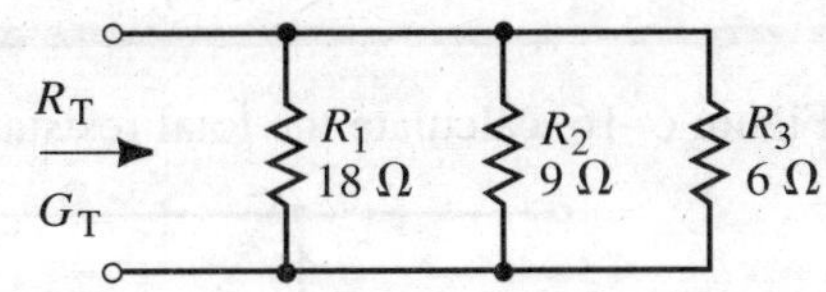

FIGURE 6–14

Solution The total conductance is

$$G_T = G_1 + G_2 + G_3$$
$$= \frac{1}{18\ \Omega} + \frac{1}{9\ \Omega} + \frac{1}{6\ \Omega}$$
$$= 0.0\overline{5}\ \text{S} + 0.\overline{1}\overline{1}\ \text{S} + 0.1\overline{6}\ \text{S}$$
$$= 0.3\overline{3}\ \text{S}$$

where the overbar indicates that the number under it is repeated infinitely to the right.

The total resistance is

$$R_T = \frac{1}{0.3\overline{3}\ \text{S}} = 3.00\ \Omega$$

Calculator Hint: A simplified method that may be used to solve for resistors in parallel is by using the x^{-1} key that is available on all scientific calculators. For some calculators, it may be necessary to use the function (Shift) key and a second key. Locate the x^{-1} key for your particular calculator. To solve for the equivalent resistance of the network in Figure 6–14, perform the following operations:

18 x^{-1} **+** **9** x^{-1} **+** **6** x^{-1} **=** x^{-1}

Depending on your calculator, your display will appear similar to the following:

```
18-1+9-1+6-1
                .333333333333
Ans-1
                            3
```

PRACTICE PROBLEMS 3

For the parallel network of resistors shown in Figure 6–15, find the total conductance G_T and the total resistance R_T.

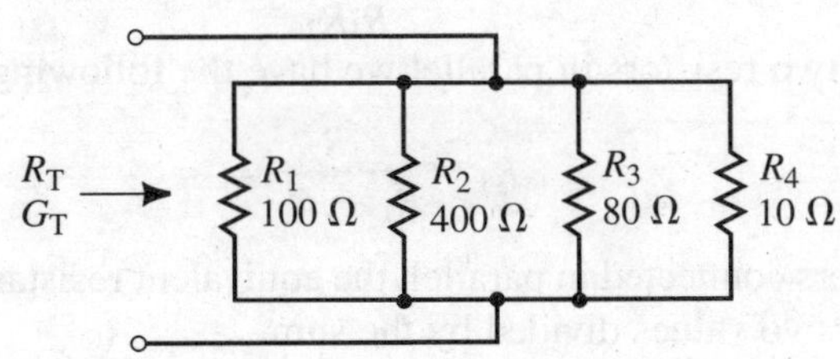

FIGURE 6–15

Answers
$G_T = 0.125$ S, $R_T = 8.00\ \Omega$

n Equal Resistors in Parallel

If we have *n* equal resistors in parallel, each resistor, *R*, has the same conductance, *G*. By applying Equation 6–3, the total conductance is found:

$$G_T = nG$$

The total resistance is now easily determined as

$$R_T = \frac{1}{G_T} = \frac{1}{nG} = \frac{R}{n} \tag{6–5}$$

EXAMPLE 6–6

For the networks of Figure 6–16, calculate the total resistance.

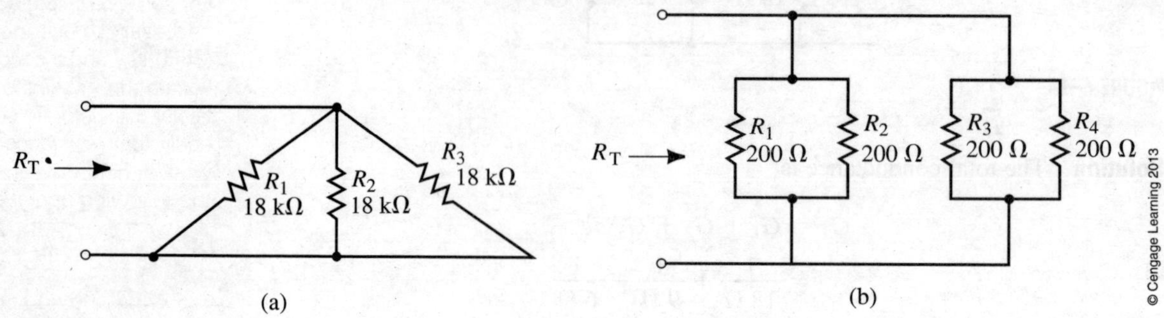

FIGURE 6–16

Solution

a. $R_T = \frac{18\ \text{k}\Omega}{3} = 6\ \text{k}\Omega$ b. $R_T = \frac{200\ \Omega}{4} = 50\ \Omega$

Two Resistors in Parallel

Very often circuits have only two resistors in parallel. In such a case, the total resistance of the combination may be determined without the necessity of determining the conductance.

For two resistors, Equation 6–4 is written as

$$R_T = \frac{1}{\frac{1}{R_1} + \frac{1}{R_2}}$$

By cross multiplying the terms in the denominator, the expression becomes

$$R_T = \frac{1}{\frac{R_1 + R_2}{R_1R_2}}$$

Thus, for two resistors in parallel we have the following expression:

$$R_T = \frac{R_1R_2}{R_1 + R_2} \tag{6–6}$$

For two resistors connected in parallel, the equivalent resistance is found by the product of the two values divided by the sum.

EXAMPLE 6–7

Determine the total resistance of the resistor combinations of Figure 6–17.

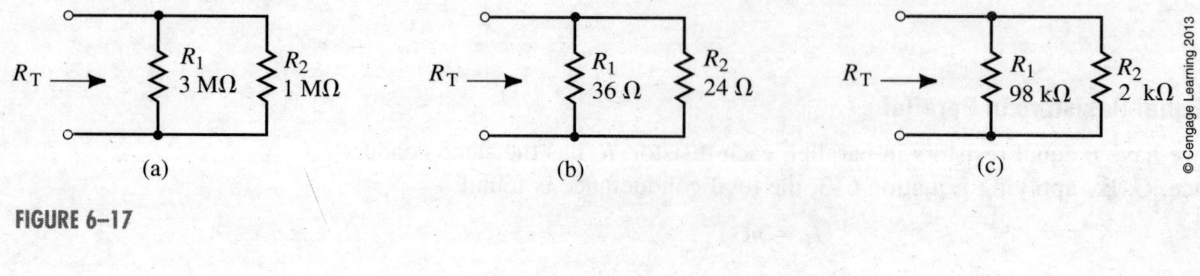

FIGURE 6–17

Solution

a. $R_T = \dfrac{(3\ \text{M}\Omega)(1\ \text{M}\Omega)}{3\ \text{M}\Omega + 1\ \text{M}\Omega} = 0.75\ \text{M}\Omega = 750\ \text{k}\Omega$

b. $R_T = \dfrac{(36\ \Omega)(24\ \Omega)}{36\ \Omega + 24\ \Omega} = 14.4\ \Omega$

c. $R_T = \dfrac{(98\ \text{k}\Omega)(2\ \text{k}\Omega)}{98\ \text{k}\Omega + 2\ \text{k}\Omega} = 1.96\ \text{k}\Omega$

Although Equation 6–6 is intended primarily to solve for two resistors in parallel, the approach may also be used to solve for any number of resistors by examining only two resistors at a time.

EXAMPLE 6–8

Calculate the total resistance of the resistor combination of Figure 6–18.

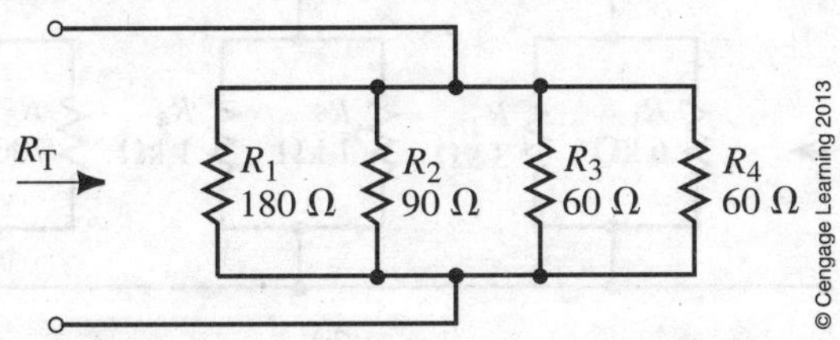

FIGURE 6–18

Solution By grouping the resistors into combinations of two, the circuit may be simplified as shown in Figure 6–19.

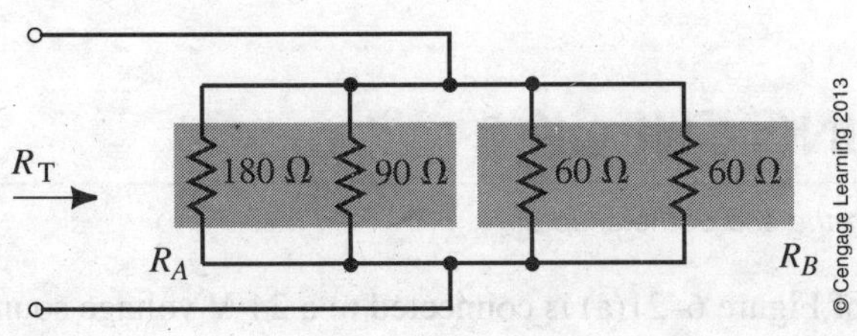

FIGURE 6–19

The equivalent resistance of each of the indicated combinations is determined as follows:

$$R_A = \frac{(180\ \Omega)(90\ \Omega)}{180\ \Omega + 90\ \Omega} = 60\ \Omega$$

$$R_B = \frac{(60\ \Omega)(60\ \Omega)}{60\ \Omega + 60\ \Omega} = 30\ \Omega$$

The circuit can be further simplified as a combination of two resistors shown in Figure 6–20.

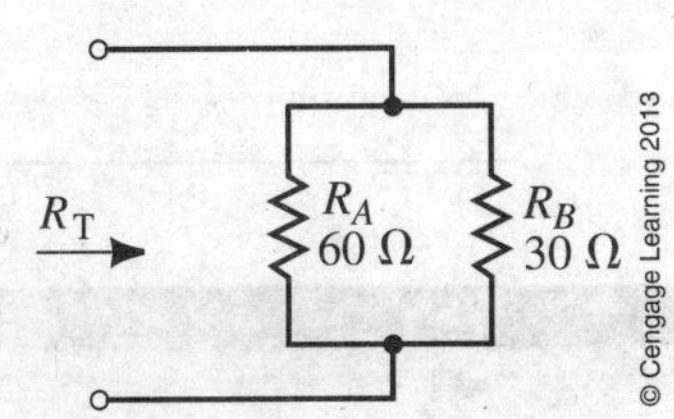

FIGURE 6–20

The resultant equivalent resistance is

$$R_T = \frac{(60\ \Omega)(30\ \Omega)}{60\ \Omega + 30\ \Omega} = 20\ \Omega$$

Three Resistors in Parallel

Using an approach similar to the derivation of Equation 6–6, we may arrive at an equation that solves for three resistors in parallel. Indeed, it is possible to write a general equation to solve for four resistors, five resistors, and more. Although such an equation is certainly useful, students are discouraged from memorizing such lengthy expressions. You will generally find that it is much more efficient to remember the principles upon which the equation is constructed. Consequently, the derivation of Equation 6–7 is left up to the student.

$$R_T = \frac{R_1R_2R_3}{R_1R_2 + R_1R_3 + R_2R_3} \qquad (6\text{–}7)$$

PRACTICE PROBLEMS 4

Find the total equivalent resistance for each network in Figure 6–21.

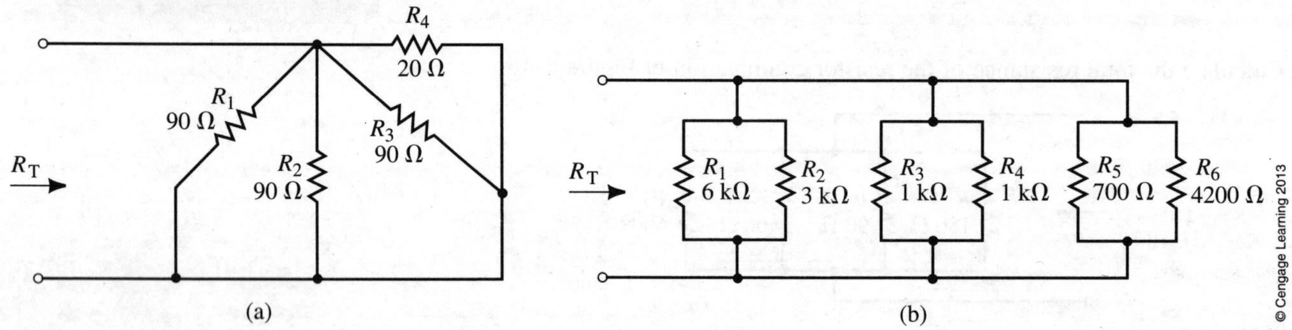

FIGURE 6–21

Answers
a. 12 Ω; b. 240 Ω

IN-PROCESS LEARNING CHECK 1

(Answers are at the end of the chapter.)

If the circuit of Figure 6–21(a) is connected to a 24-V voltage source, determine the following quantities:

a. The total current provided by the voltage source.

b. The current through each resistor of the network.

c. Verify Kirchhoff's current law at one of the voltage source terminals.

6.4 Voltage Sources in Parallel

Voltage sources of different potentials should never be connected in parallel, since to do so would contradict Kirchhoff's voltage law. However, when two equal potential sources are connected in parallel, each source will deliver half the required circuit current. For this reason automobile batteries are sometimes connected in parallel to assist in starting a car with a "weak" battery. Figure 6–22 illustrates this principle.

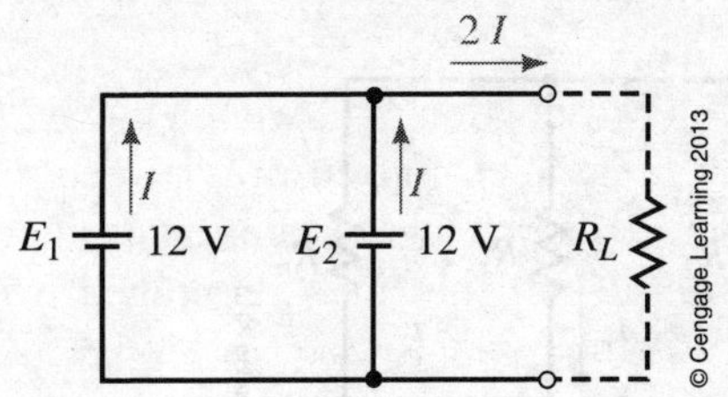

FIGURE 6–22 Voltage sources in parallel.

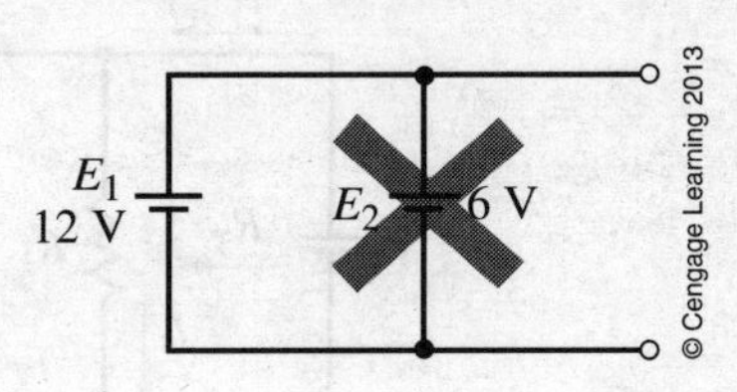

FIGURE 6–23 Voltage sources of different voltages must never be placed in parallel.

Figure 6–23 shows that if voltage sources of two different potentials are placed in parallel, Kirchhoff's voltage law will be violated around the closed loop. In practice, if voltage sources of different potentials are placed in parallel, the resulting closed loop can have a very large current. The current will occur even though there may not be a load connected across the sources. Example 6–9 illustrates the large currents that can occur when two parallel batteries of different potential are connected.

EXAMPLE 6–9

A 12-V battery and a 6-V battery (each having an internal resistance of 0.05 V) are inadvertently placed in parallel as shown in Figure 6–24. Determine the current through the batteries.

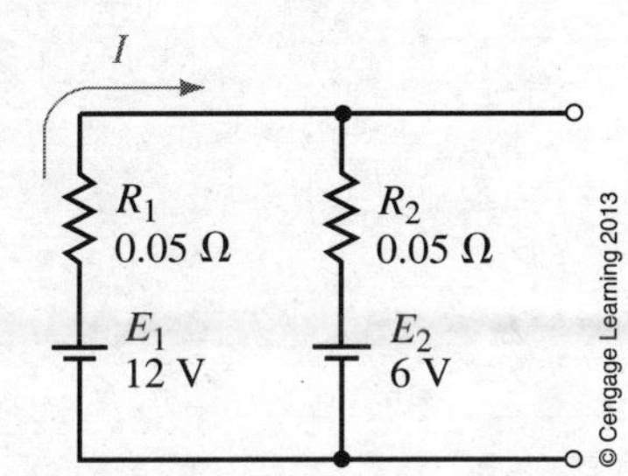

FIGURE 6–24

Solution From Ohm's law,

$$I = \frac{E_T}{R_T} = \frac{12\text{ V} - 6\text{ V}}{0.05\ \Omega + 0.05\ \Omega} = 60\text{ A}$$

This example illustrates why batteries of different potential must never be connected in parallel. Tremendous currents will occur within the sources, resulting in the possibility of a fire or explosion.

6.5 Current Divider Rule

When we examined series circuits we determined that the current in the series circuit was the same everywhere in the circuit, whereas the voltages across the series elements were typically different. The voltage divider rule (VDR) was used to determine the voltage across all resistors within a series network.

In parallel networks, the voltage across all parallel elements is the same. However, the currents through the various elements are typically different. The **current divider rule (CDR)** is used to determine how current entering a node is split between the various parallel resistors connected to the node.

Consider the network of parallel resistors shown in Figure 6–25.

If this network of resistors is supplied by a voltage source, the total current in the circuit is

$$I_T = \frac{E}{R_T} \tag{6–8}$$

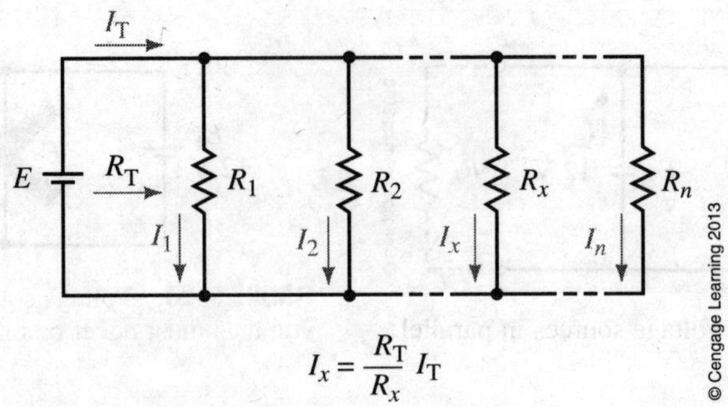

FIGURE 6–25 Current divider rule.

Since each of the n parallel resistors has the same voltage, E, across its terminals, the current through any resistor in the network is given as

$$I_x = \frac{E}{R_x} \tag{6–9}$$

By rewriting Equation 6–8 as $E = I_T R_T$ and then substituting this into Equation 6–9, we obtain the current divider rule as follows:

$$I_x = \frac{R_T}{R_x} I_T \tag{6–10}$$

An alternate way of writing the current divider rule is to express it in terms of conductance. Equation 6–10 may be modified as follows:

$$I_x = \frac{G_x}{G_T} I_T \tag{6–11}$$

The current divider rule allows us to calculate the current in any resistor of a parallel network if we know the total current entering the network. Notice the similarity between the voltage divider rule (for series components) and the current divider rule (for parallel components). The main difference is that the current divider rule of Equation 6–11 uses circuit conductance rather than resistance. While this equation is useful, it is generally easier to use resistance to calculate current.

If the network consists of only two parallel resistors, then the current through each resistor may be found in a slightly different way. Recall that for two resistors in parallel, the total parallel resistance is given as

$$R_T = \frac{R_1 R_2}{R_1 + R_2}$$

Now, by substituting this expression for total resistance into Equation 6–10, we obtain

$$I_1 = \frac{I_T R_T}{R_1}$$

$$= \frac{I_T\left(\frac{R_1 R_2}{R_1 + R_2}\right)}{R_1}$$

which simplifies to

$$I_1 = \frac{R_2}{R_1 + R_2} I_T \tag{6–12}$$

Similarly,

$$I_2 = \frac{R_1}{R_1 + R_2} I_T \qquad \textbf{(6–13)}$$

Several other important characteristics of parallel networks become evident. *If current enters a parallel network consisting of any number of equal resistors, then the current entering the network will split equally between all of the resistors.*

If current enters a parallel network consisting of several values of resistance, then the smallest value of resistor in the network will have the largest amount of current. Inversely, the largest value of resistance will have the smallest amount of current.

This characteristic may be simplified by saying that *most of the current will follow the path of least resistance.*

EXAMPLE 6–10

For the network of Figure 6–26, determine the currents I_1, I_2, and I_3.

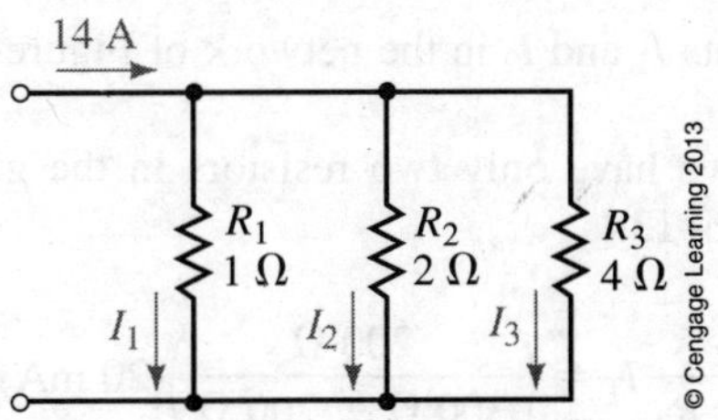

FIGURE 6–26

Solution First, we calculate the total conductance of the network.

$$G_T = \frac{1}{1\ \Omega} + \frac{1}{2\ \Omega} + \frac{1}{4\ \Omega} = 1.75\ \text{S}$$

Now the currents may be evaluated as follows:

$$I_1 = \frac{G_1}{G_T} I_T = \left(\frac{1\ \text{S}}{1.75\ \text{S}}\right) 14\ \text{A} = 8.00\ \text{A}$$

$$I_2 = \frac{G_2}{G_T} I_T = \left(\frac{0.5\ \text{S}}{1.75\ \text{S}}\right) 14\ \text{A} = 4.00\ \text{A}$$

$$I_3 = \frac{G_3}{G_T} I_T = \left(\frac{0.25\ \text{S}}{1.75\ \text{S}}\right) 14\ \text{A} = 2.00\ \text{A}$$

An alternate approach is to use circuit resistance, rather than conductance.

$$R_T = \frac{1}{G_T} = \frac{1}{1.75\ \text{S}} = 0.571\ \Omega$$

$$I_1 = \frac{R_T}{R_1} I_T = \left(\frac{0.571\ \Omega}{1\ \Omega}\right) 14\ \text{A} = 8.00\ \text{A}$$

$$I_2 = \frac{R_T}{R_2} I_T = \left(\frac{0.571\ \Omega}{2\ \Omega}\right) 14\ \text{A} = 4.00\ \text{A}$$

$$I_3 = \frac{R_T}{R_3} I_T = \left(\frac{0.571\ \Omega}{5\ \Omega}\right) 14\ \text{A} = 2.00\ \text{A}$$

EXAMPLE 6–11

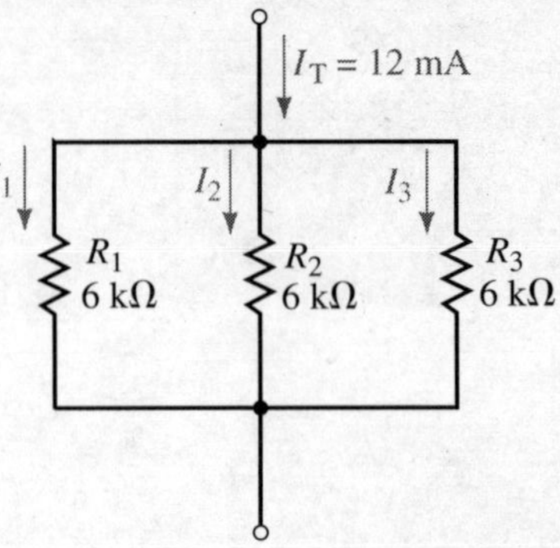

FIGURE 6–27

For the network of Figure 6–27, determine the currents I_1, I_2, and I_3.

Solution Since all the resistors have the same value, the incoming current will split equally between the resistances. Therefore,

$$I_1 = I_2 = I_3 = \frac{12\text{ mA}}{3} = 4.00\text{ mA}$$

EXAMPLE 6–12

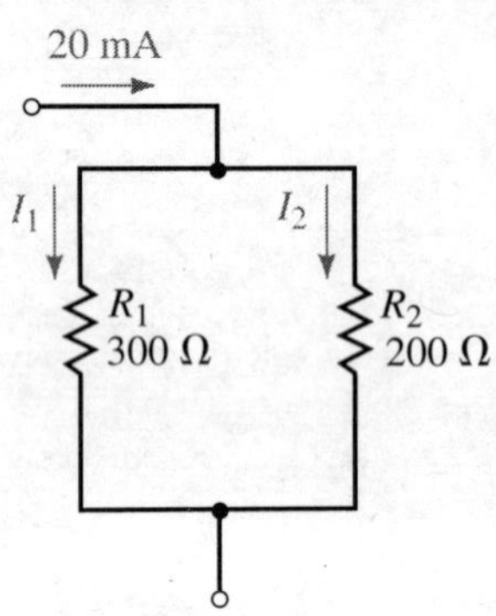

FIGURE 6–28

Determine the currents I_1 and I_2 in the network of Figure 6–28.

Solution Because we have only two resistors in the given network, we use Equations 6–12 and 6–13:

$$I_1 = \frac{R_2}{R_1 + R_2} I_T = \left(\frac{200\ \Omega}{300\ \Omega + 200\ \Omega}\right)(20\text{ mA}) = 8.00\text{ mA}$$

$$I_2 = \frac{R_1}{R_1 + R_2} I_T = \left(\frac{300\ \Omega}{300\ \Omega + 200\ \Omega}\right)(20\text{ mA}) = 12.0\text{ mA}$$

EXAMPLE 6–13

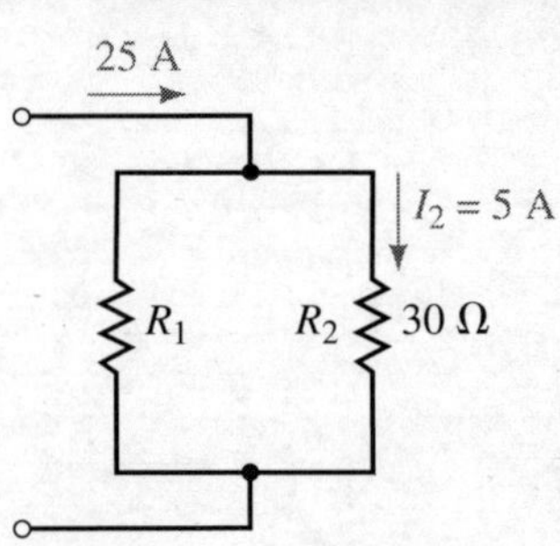

FIGURE 6–29

Determine the resistance R_1 so that current will divide as shown in the network of Figure 6–29.

Solution There are several methods that may be used to solve this problem. We will examine only two of the possibilities.

Method I: Since we have two resistors in parallel, we may use Equation 6–13 to solve for the unknown resistor:

$$I_2 = \frac{R_1}{R_1 + R_2} I_T$$

$$5\text{ A} = \left(\frac{R_1}{R_1 + 30\ \Omega}\right)(25\text{ A})$$

Using algebra, we get

$$(5\text{ A})R_1 + (5\text{ A})(30\ \Omega) = (25\text{ A})R_1$$
$$(20\text{ A})R_1 = 150\text{ V}$$
$$R_1 = \frac{150\text{ V}}{20\text{ A}} = 7.50\ \Omega$$

Method II: By applying Kirchhoff's current law, we see that the current in R_1 must be

$$I_1 = 25\text{ A} - 5\text{ A} = 20\text{ A}$$

Now, since elements in parallel must have the same voltage across their terminals, the voltage across R_1 must be exactly the same as the voltage across R_2. By Ohm's law, the voltage across R_2 is

$$V_2 = (5\text{ A})(30\ \Omega) = 150\text{ V}$$

And so

$$R_1 = \frac{150\text{ V}}{20\text{ A}} = 7.50\ \Omega$$

As expected, the results are identical. This example illustrates that there is usually more than one method for solving a given problem. Although the methods are equally correct, we see that the second method in this example is less involved.

PRACTICE PROBLEMS 5

Use the current divider rule to calculate the unknown currents for the networks of Figure 6–30.

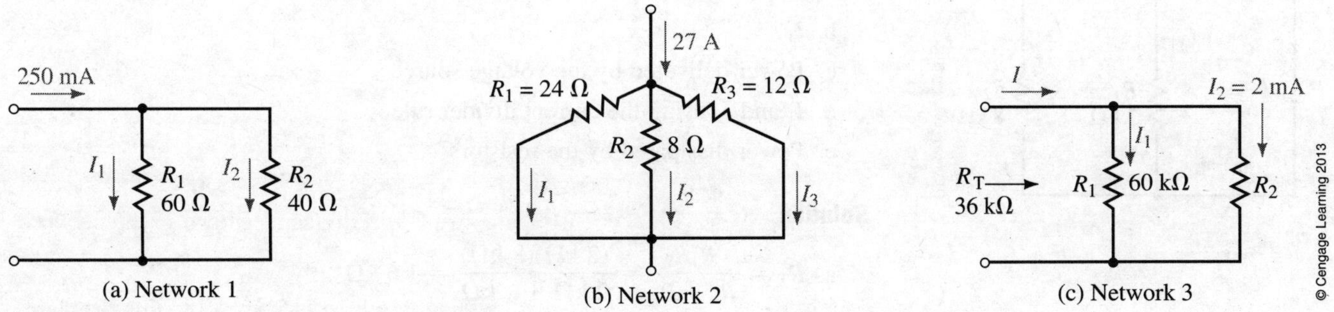

FIGURE 6–30

Answers

Network 1: $I_1 = 100\text{ mA}$, $I_2 = 150\text{ mA}$

Network 2: $I_1 = 4.50\text{ A}$, $I_2 = 13.5\text{ A}$, $I_3 = 9.00\text{ A}$

Network 3: $I_1 = 3.00\text{ mA}$, $I = 5.00\text{ mA}$

IN-PROCESS LEARNING CHECK 2

(Answers are at the end of the chapter.)

Four resistors are connected in parallel. The values of the resistors are 1 Ω, 3 Ω, 4 Ω, and 5 Ω.

a. Using only a pencil and a piece of paper (no calculator), determine the current through each resistor if the current through the 5-Ω resistor is 6 A.

b. Again, without a calculator, solve for the total current applied to the parallel combination.

c. Use a calculator to determine the total parallel resistance of the four resistors. Use the current divider rule and the total current obtained in part (b) to calculate the current through each resistor.

6.6 Analysis of Parallel Circuits

We will now examine how to use the principles developed in this chapter when analyzing parallel circuits. In the examples to follow, we find that the laws of conservation of energy apply equally well to parallel circuits as to series circuits. Although we choose to analyze circuits a certain way, remember that there is usually more than one way to arrive at the correct answer. As you become more proficient at circuit analysis you will generally use the most efficient method. For now, however, use the method with which you feel most comfortable.

EXAMPLE 6–14

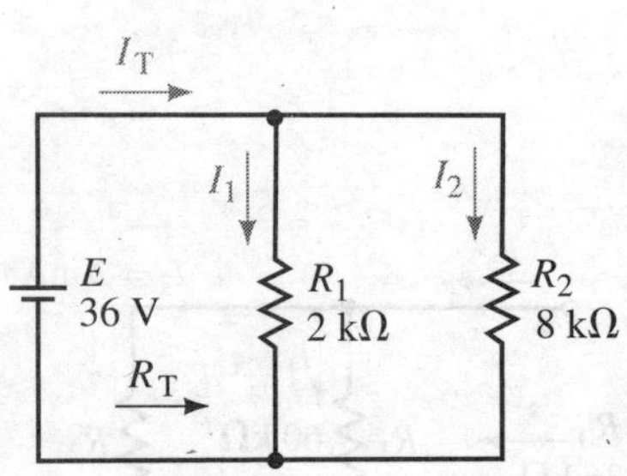

FIGURE 6–31

For the circuit of Figure 6–31, determine the following quantities:

a. R_T

b. I_T

c. Power delivered by the voltage source

d. I_1 and I_2 using the current divider rule

e. Power dissipated by the resistors

Solution

a. $R_T = \dfrac{R_1 R_2}{R_1 + R_2} = \dfrac{(2\text{ k}\Omega)(8\text{ k}\Omega)}{2\text{ k}\Omega + 8\text{ k}\Omega} = 1.6\text{ k}\Omega$

b. $I_T = \dfrac{E}{R_T} = \dfrac{36\text{ V}}{1.6\text{ k}\Omega} = 22.5\text{ mA}$

c. $P_T = EI_T = (36\text{ V})(22.5\text{ mA}) = 810\text{ mW}$

d. $I_2 = \dfrac{R_1}{R_1 + R_2} I_T = \left(\dfrac{2\text{ k}\Omega}{2\text{ k}\Omega + 8\text{ k}\Omega}\right)(22.5\text{ mA}) = 4.5\text{ mA}$

$I_1 = \dfrac{R_2}{R_1 + R_2} I_T = \left(\dfrac{8\text{ k}\Omega}{2\text{ k}\Omega + 8\text{ k}\Omega}\right)(22.5\text{ mA}) = 18.0\text{ mA}$

e. Since we know the voltage across each of the parallel resistors must be 36 V, we use this voltage to determine the power dissipated by each resistor. It would be equally correct to use the current through each resistor to calculate the power. However, it is generally best to use given information rather than calculated values to perform further calculations since it is then less likely that an error is carried through.

$$P_1 = \frac{E^2}{R_1} = \frac{(36\text{ V})^2}{2\text{ k}\Omega} = 648\text{ mW}$$

$$P_2 = \frac{E^2}{R_2} = \frac{(36\text{ V})^2}{8\text{ k}\Omega} = 162\text{ mW}$$

Notice that the power delivered by the voltage source is exactly equal to the total power dissipated by the resistors, namely $P_T = P_1 + P_2$.

EXAMPLE 6–15

Refer to the circuit of Figure 6–32:

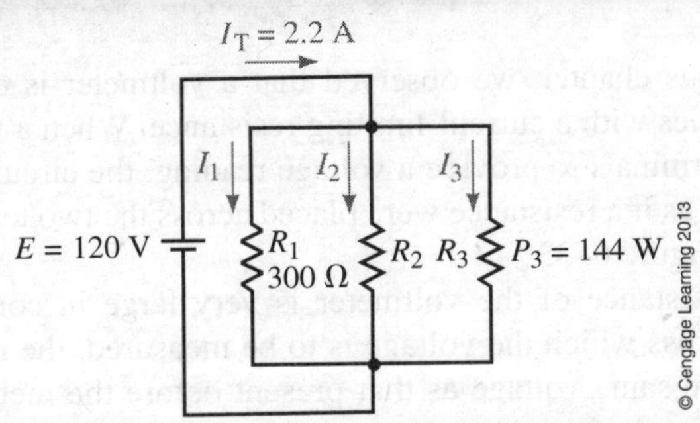

FIGURE 6–32

a. Solve for the total power delivered by the voltage source.

b. Find the currents I_1, I_2, and I_3.

c. Determine the values of the unknown resistors R_2 and R_3.

d. Calculate the power dissipated by each resistor.

e. Verify that the power dissipated is equal to the power delivered by the voltage source.

Solution

a. $P_T = EI_T = (120\text{ V})(2.2\text{ A}) = 264\text{ W}$

b. Since the three resistors of the circuit are in parallel, we know that the voltage across all resistors must be equal to $E = 120$ V.

$$I_1 = \frac{V_1}{R_1} = \frac{120\text{ V}}{300\ \Omega} = 0.4\text{ A}$$

$$I_3 = \frac{P_3}{V_3} = \frac{144\text{ W}}{120\text{ V}} = 1.2\text{ A}$$

Because KCL must be maintained at each node, we determine the current I_2 as

$$I_2 = I_T - I_1 - I_3$$
$$= 2.2\text{ A} - 0.4\text{ A} - 1.2\text{ A} = 0.6\text{ A}$$

c. $R_2 = \dfrac{V_2}{I_2} = \dfrac{120\text{ V}}{0.6\text{ A}} = 200\ \Omega$

Although we could use the calculated current I_3 to determine the resistance, it is best to use the given data in calculations rather than calculated values.

$$R_3 = \frac{V_3^2}{P_3} = \frac{(120\text{ V})^2}{144\text{ W}} = 100\ \Omega$$

d. $P_1 = \frac{V_1^2}{R_1} = \frac{(120\text{ V})^2}{300\ \Omega} = 48\text{ W}$

$P_2 = I_2E_2 = (0.6\text{ A})(120\text{ V}) = 72\text{ W}$

e. $P_{in} = P_{out}$

$264\text{ W} = P_1 + P_2 + P_3$

$264\text{ W} = 48\text{ W} + 72\text{ W} + 144\text{ W}$

$264\text{ W} = 264\text{ W}$ (checks!)

6.7 Voltmeter Loading Effects

In the previous chapter, we observed that a voltmeter is essentially a meter display in series with a current-limiting resistance. When a voltmeter is placed across two terminals to provide a voltage reading, the circuit is affected in the same manner as if a resistance were placed across the two terminals. The effect is shown in Figure 6–33.

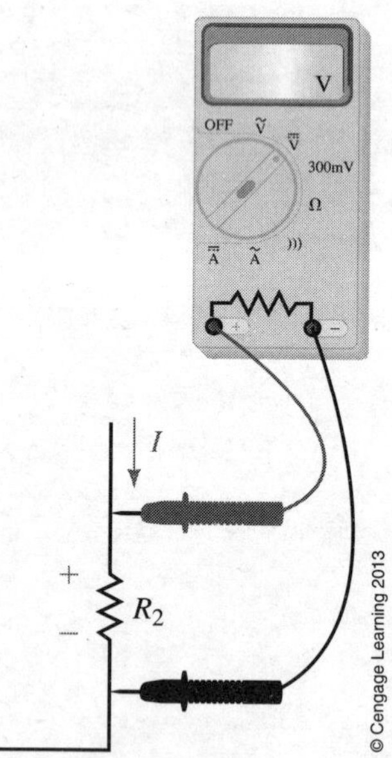

FIGURE 6–33

If the resistance of the voltmeter is very large in comparison with the resistance across which the voltage is to be measured, the meter will indicate essentially the same voltage as that present before the meter was connected. On the other hand, if the meter has an internal resistance which is near in value to the resistance across which the measurement is taken, then the meter will adversely load the circuit, resulting in an erroneous reading. Generally, if the meter resistance is more than ten times larger than the resistance across which the voltage is taken, then the **voltmeter loading effect** is considered negligible and may be ignored.

In the circuit of Figure 6–34, there is no current in the circuit since the terminals a and b are open circuited. The voltage appearing between the open terminals must be $V_{ab} = 10$ V. Now, if we place a voltmeter having an internal resistance of 200 kΩ between the terminals, the circuit is closed, resulting in a small current. The complete circuit appears as shown in Figure 6–35.

The reading indicated on the face of the meter is the voltage that occurs across the internal resistance of the meter. Applying Kirchhoff's voltage law to the circuit, this voltage is

$$V_{ab} = \frac{200\text{ k}\Omega}{200\text{ k}\Omega + 100\ \Omega}(10\text{ V}) = 9.995\text{ V}$$

Clearly, the reading on the face of the meter is essentially equal to the expected value of 10 V. Recall from the previous chapter that we defined the loading effect of a meter as follows:

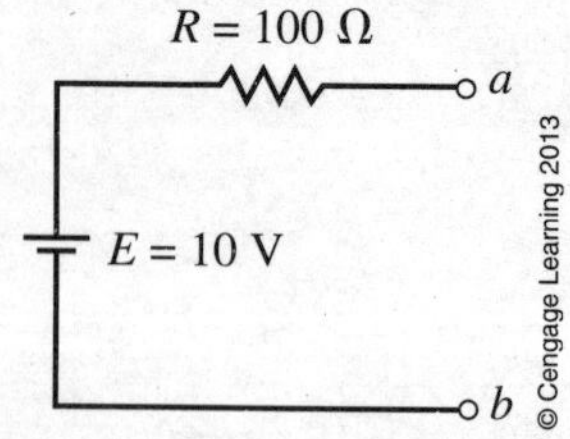

FIGURE 6–34

$$\text{loading effect} = \frac{\text{actual value} - \text{reading}}{\text{actual value}} \times 100\%$$

For the circuit of Figure 6–35, the voltmeter has a loading effect of

$$\text{loading effect} = \frac{10\text{ V} - 9.995\text{ V}}{10\text{ V}} \times 100\% = 0.05\%$$

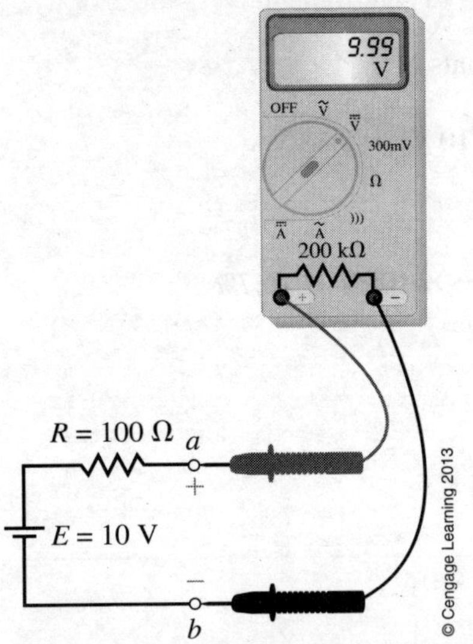

FIGURE 6–35

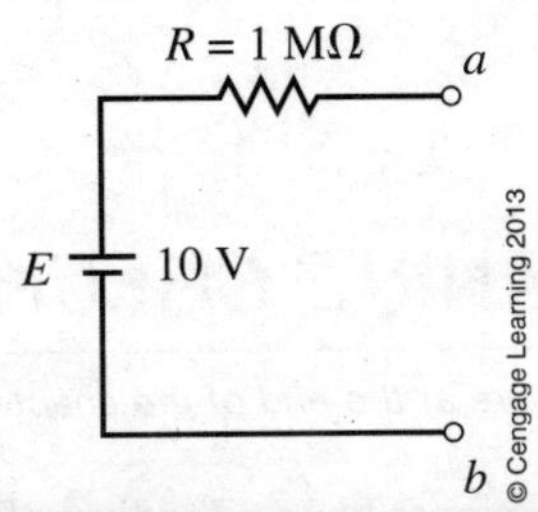

FIGURE 6–36

This loading error is virtually undetectable for the circuit given. The same would not be true if we had a circuit as shown in Figure 6–36 and used the same voltmeter to provide a reading.

Again, if the circuit were left open circuited, we would expect that $V_{ab} = 10$ V.

By connecting the 200-kΩ voltmeter between the terminals, as shown in Figure 6–37, we see that the voltage detected between terminals *a* and *b* will no longer be the desired voltage; rather,

$$V_{ab} = \frac{200\ \text{k}\Omega}{200\ \text{k}\Omega + 1\ \text{M}\Omega}(10\ \text{V}) = 1.667\ \text{V}$$

The loading effect of the meter in this circuit is

$$\text{loading effect} = \frac{10\ \text{V} - 1.667\ \text{V}}{10\ \text{V}} \times 100\% = 83.33\%$$

The previous illustration is an example of a problem that can occur when taking measurements in electronic circuits. When an inexperienced technician or technologist obtains an unforeseen result, he or she assumes that something is wrong with either the circuit or the instrument. In fact, both the circuit and the instrument are behaving in a perfectly predictable manner. The tech merely forgot to take into account the meter's loading effect. All instruments have limitations, and we must always be aware of these limitations.

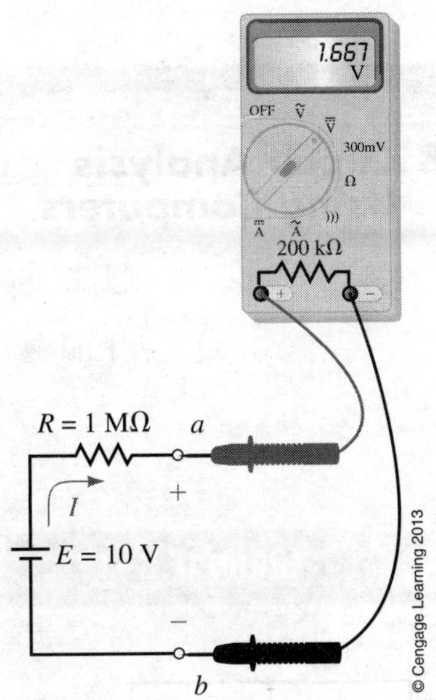

FIGURE 6–37

EXAMPLE 6–16

A digital voltmeter having an internal resistance of 5 MΩ is used to measure the voltage across terminals *a* and *b* in the circuit of Figure 6–37.

a. Determine the reading on the meter.
b. Calculate the loading effect of the meter.

Solution

a. The voltage applied to the meter terminals is

$$V_{ab} = \left(\frac{5\ \text{M}\Omega}{1\ \text{M}\Omega + 5\ \text{M}\Omega}\right)(10\ \text{V}) = 8.33\ \text{V}$$

b. The loading effect is

$$\text{loading error} = \frac{10\ \text{V} - 8.33\ \text{V}}{10\ \text{V}} \times 100\% = 16.7\%$$

IN-PROCESS LEARNING CHECK 3

(Answers are at the end of the chapter.)

All instruments have a loading effect on the circuit in which a measurement is taken. If you were given two voltmeters, one with an internal resistance of 200 kΩ and another with an internal resistance of 1 MΩ, which meter would load a circuit more? Explain.

6.8 Circuit Analysis Using Computers

Multisim PSPICE

As you have already seen, computer simulation is useful in providing a visualization of the skills you have learned. We will use both Multisim and PSpice to "measure" voltage and current in parallel circuits. One of the most useful features of Multisim is the program's ability to accurately simulate the operation of a real circuit. In this section, you will learn how to change the settings of the multimeter to observe meter loading in a circuit.

Multisim

EXAMPLE 6–17

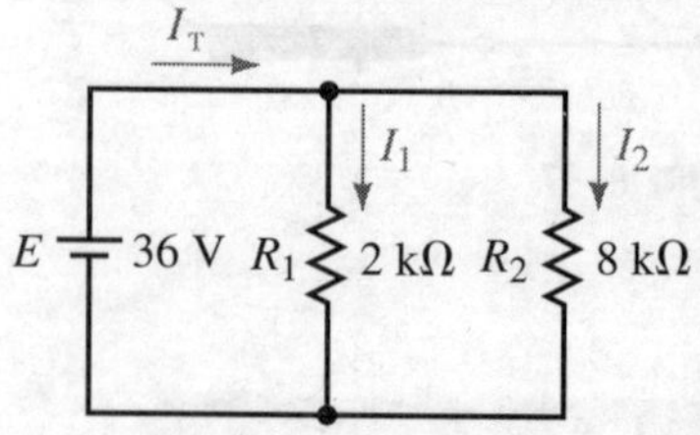

FIGURE 6–38

Use Multisim to determine the currents I_T, I_1, and I_2 in the circuit of Figure 6–38. This circuit was analyzed previously in Example 6–14.

Solution After opening the Circuit window:

- Select the components for the circuit from the *Parts bin* toolbars. You need to select the battery and ground symbol from the *Sources* toolbar. The resistors are obtained from the *Basic* toolbar.
- Once the circuit is completely wired, place an ammeter into the circuit to measure the current provided by the voltage source. Place Measurement Probes (found in the *Instruments* toolbar) to measure current in each branch.
- Simulate the circuit by clicking on the power switch. You should see results similar to those shown in Figure 6–39.

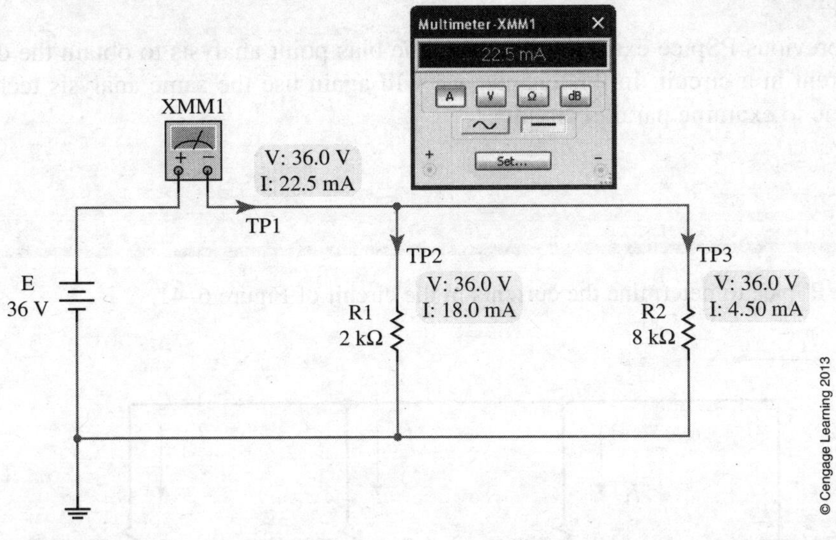

CircuitSim 06-1

FIGURE 6–39

Notice that these results are consistent with those found in Example 6–14.

EXAMPLE 6–18

Use Multisim to demonstrate the loading effect of the voltmeter used in Figure 6–37. The voltmeter is to have internal resistance of 200 kΩ.

Solution After opening the Circuit window:

- Construct the circuit by placing the battery, resistor, and ground as shown in Figure 6–37.
- Select the multimeter from the *Instruments* toolbar.
- Enlarge the multimeter by double-clicking on the symbol.
- Click on the *Settings* button on the multimeter face.
- Change the voltmeter resistance to 200 kΩ. Accept the new value by clicking on OK.
- Run the simulation by clicking on the power switch. The resulting display is shown in Figure 6–40.

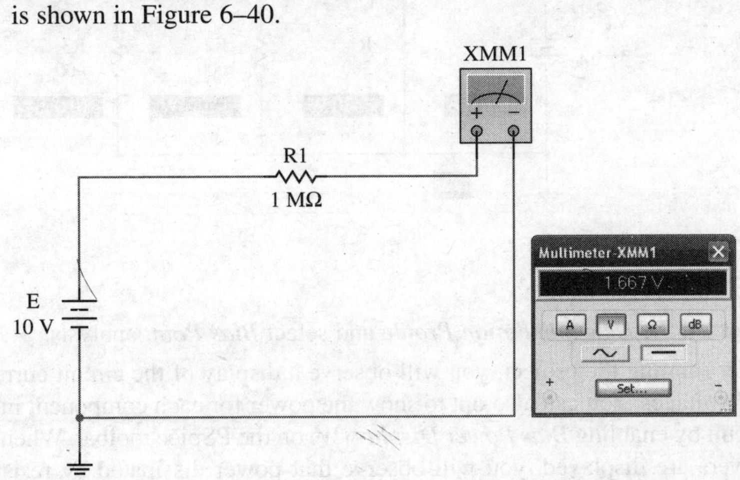

CircuitSim 06-2

FIGURE 6–40

PSpice

In previous PSpice examples, we used the bias point analysis to obtain the dc current in a circuit. In this chapter we will again use the same analysis technique to examine parallel circuits.

EXAMPLE 6–19

Use PSpice to determine the currents in the circuit of Figure 6–41.

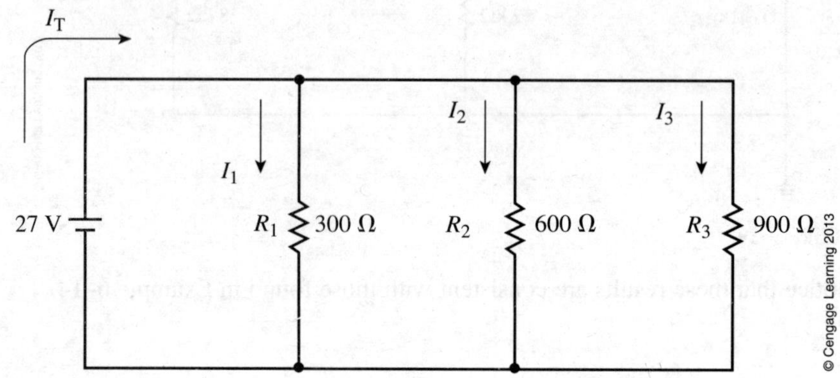

FIGURE 6–41

Solution

- Open the CIS Demo software and construct the circuit as illustrated in Figure 6–42.

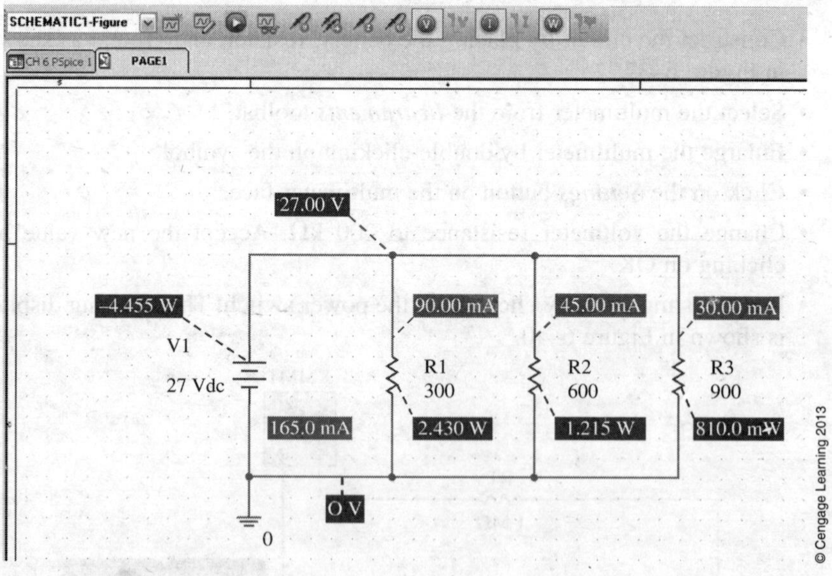

FIGURE 6–42

- Click on the *New Simulation Profile* and select *Bias Point* analysis.
- After running the project, you will observe a display of the circuit currents and voltages. You can also opt to show the power for each component in the circuit by enabling *Bias Power Display (W)* on the PSpice toolbar. When the powers are displayed, you will observe that power dissipated by resistors is positive, while power delivered by the voltage source is negative.

The currents are $I(\text{R1}) = 90$ mA, $I(\text{R2}) = 45$ mA, $I(\text{R3}) = 30$ mA, and $I(\text{V1}) = 165$ mA. Notice that the power delivered by the voltage source is 4.455 W, which is exactly equal to the summation of power dissipated by all the resistors.

PRACTICE PROBLEMS 6

Use Multisim to determine the current in each resistor of the circuit of Figure 6–21(a) if a 24-V voltage source is connected across the terminals of the resistor network.

Answers
$I_1 = I_2 = I_3 = 0.267$ A, $I_4 = 1.20$ A

PRACTICE PROBLEMS 7

Use PSpice to determine the current in each resistor of the circuit shown in Figure 6–43.

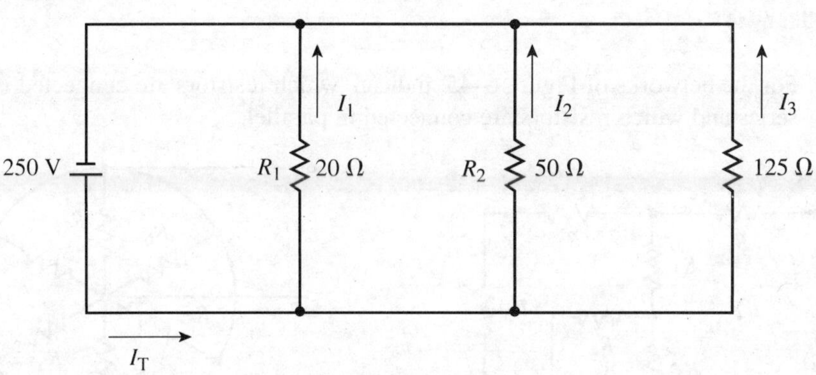

FIGURE 6–43

Answers
$I_1 = 12.5$ A, $I_2 = 5.00$ A, $I_3 = 2.00$ A, $I_T = 19.5$ A

Putting It into Practice

You have been hired as a consultant to a heating company. One of your jobs is to determine the number of 1000-W heaters that can be safely handled by an electrical circuit. All of the heaters in any circuit are connected in parallel. Each circuit operates at a voltage of 240 V and is rated for a maximum of 20 A. The normal operating current of the circuit should not exceed 80% of the maximum rated current. How many heaters can be safely installed in each circuit? If a room requires 5000 W of heaters to provide adequate heat during the coldest weather, how many circuits must be installed in this room?

Problems

6.1 Parallel Circuits

1. Indicate which of the elements in Figure 6–44 are connected in parallel and which elements are connected in series.

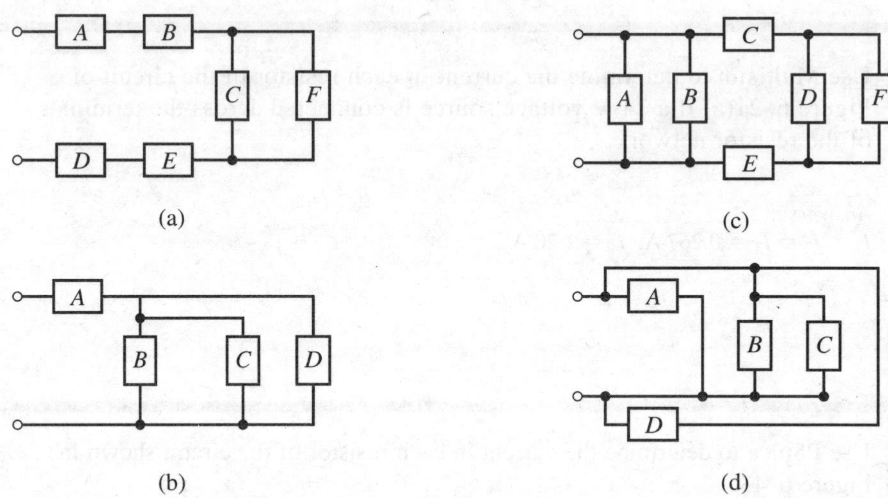

FIGURE 6–44

2. For the networks of Figure 6–45, indicate which resistors are connected in series and which resistors are connected in parallel.

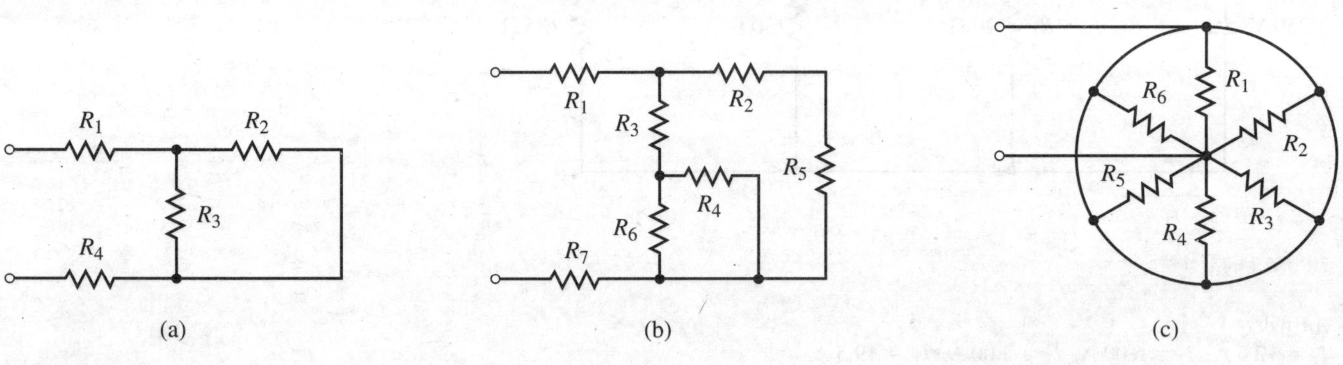

FIGURE 6–45

3. Without changing the component positions, show at least one way of connecting all the elements of Figure 6–46 in parallel.
4. Repeat Problem 3 for the elements shown in Figure 6–47.

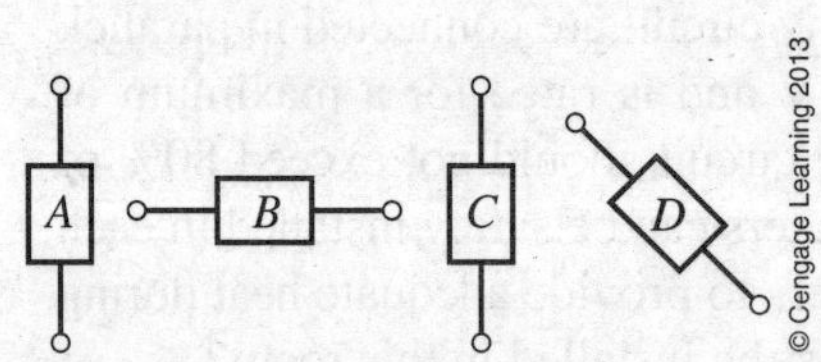

FIGURE 6–46

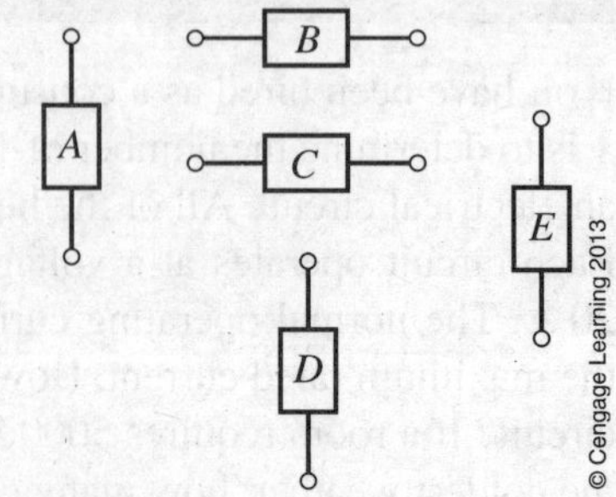

FIGURE 6–47

6.2 Kirchhoff's Current Law

5. Use Kirchhoff's current law to determine the magnitudes and directions of the indicated currents in each of the networks shown in Figure 6–48.

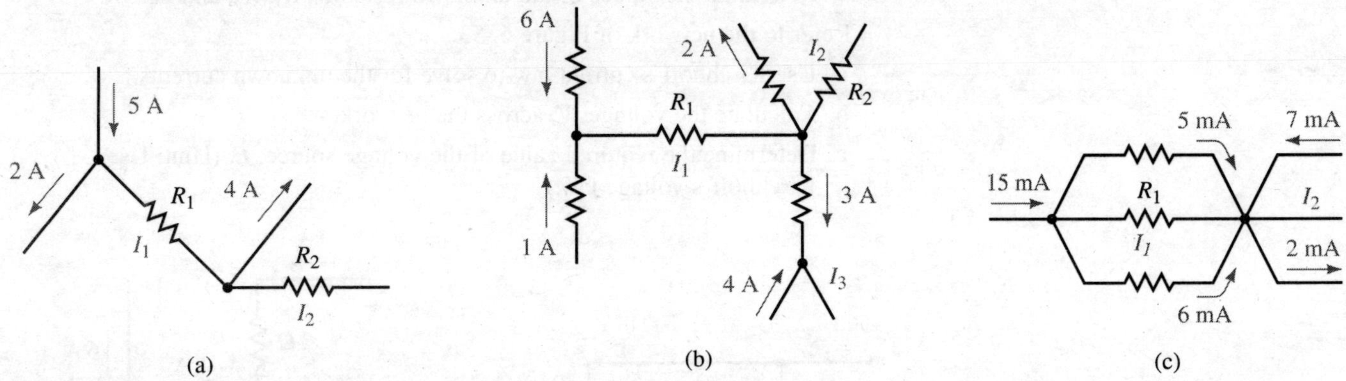

FIGURE 6–48

6. For the circuit of Figure 6–49, determine the magnitude and direction of each of the indicated currents.

7. Consider the network of Figure 6–50:

   a. Calculate the currents I_1, I_2, I_3, and I_4.

   b. Determine the value of the resistance R_3.

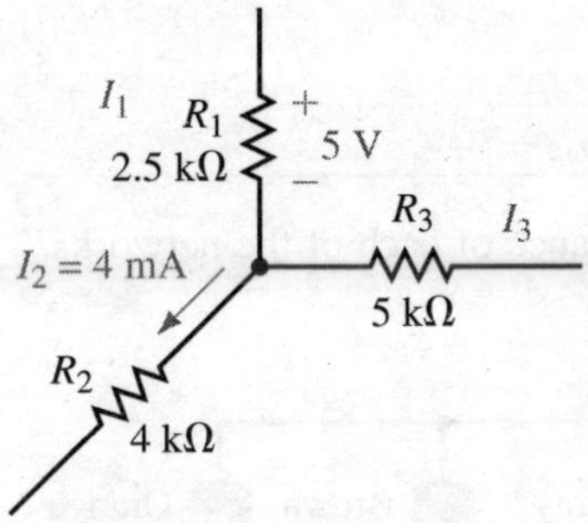

FIGURE 6–49

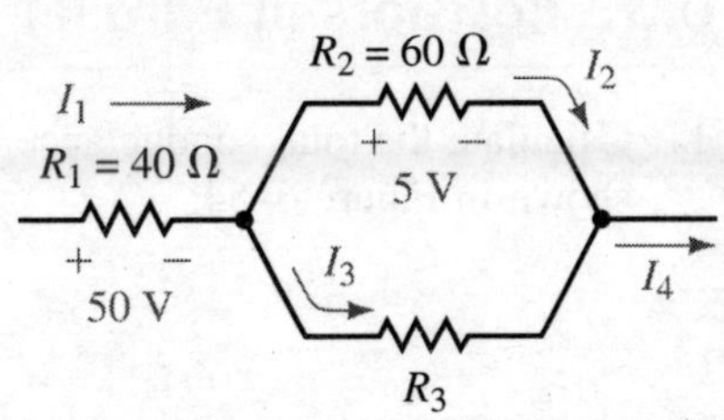

FIGURE 6–50

8. Find each of the unknown currents in the networks of Figure 6–51.

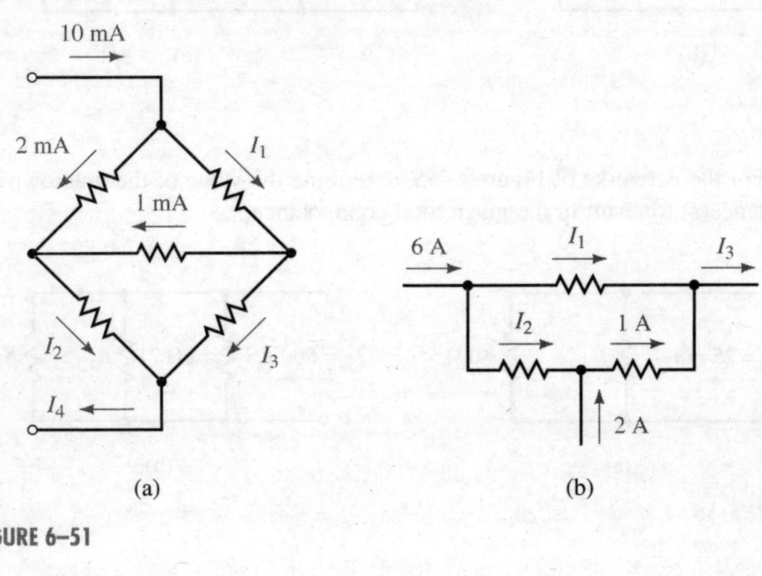

FIGURE 6–51

9. Refer to the network of Figure 6–52:
   a. Use Kirchhoff's current law to solve for the unknown currents, I_1, I_2, I_3, and I_4.
   b. Calculate the voltage, V, across the network.
   c. Determine the values of the unknown resistors, R_1, R_3, and R_4.
10. Refer to the network of Figure 6–53:
   a. Use Kirchhoff's current law to solve for the unknown currents.
   b. Calculate the voltage, V, across the network.
   c. Determine the required value of the voltage source, E. (Hint: Use Kirchhoff's voltage law.)

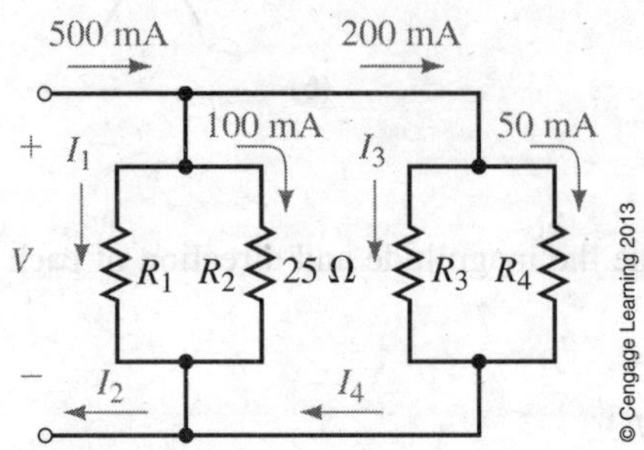

FIGURE 6–52

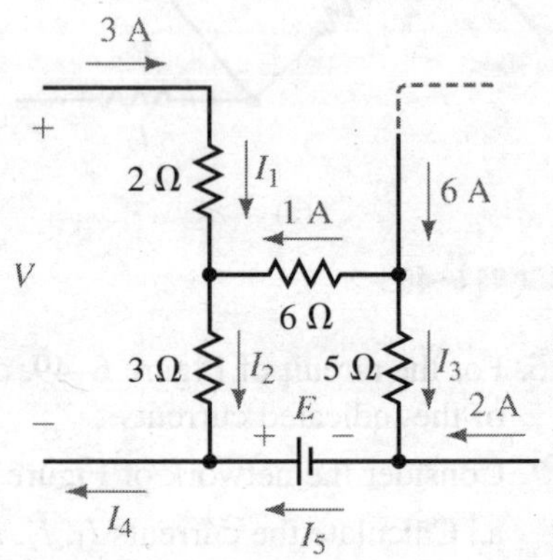

FIGURE 6–53

6.3 Resistors in Parallel

11. Calculate the total conductance and total resistance of each of the networks shown in Figure 6–54.

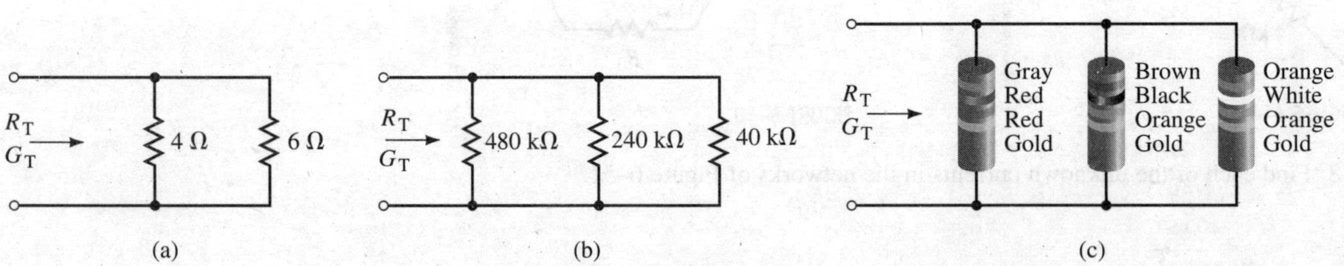

FIGURE 6–54

12. For the networks of Figure 6–55, determine the value of the unknown resistance(s) to result in the given total conductance.

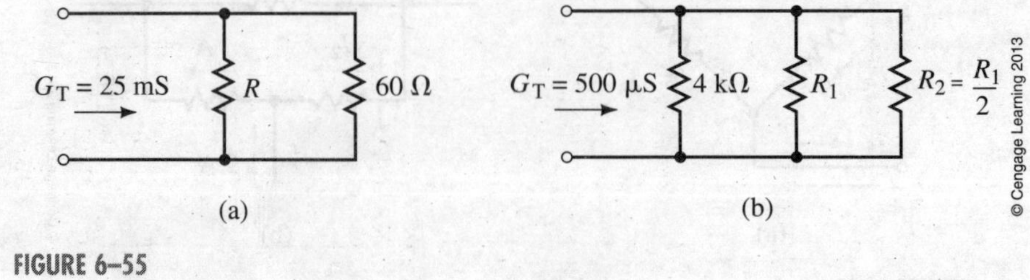

FIGURE 6–55

13. For the networks of Figure 6–56, determine the value of the unknown resistance(s) to result in the total resistances given.

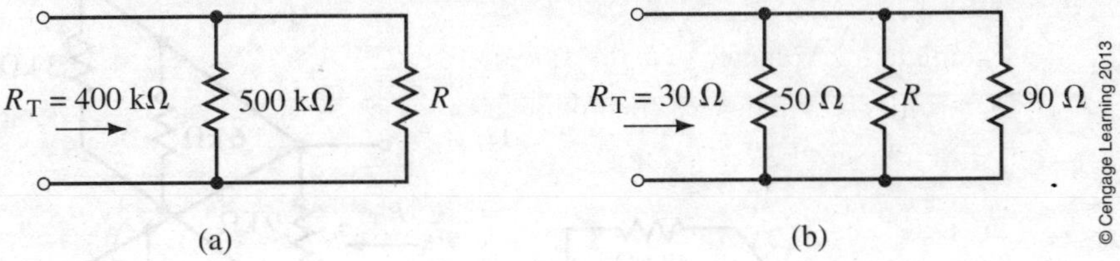

FIGURE 6–56

14. Determine the value of each unknown resistor in the network of Figure 6–57, so that the total resistance is 100 kΩ.

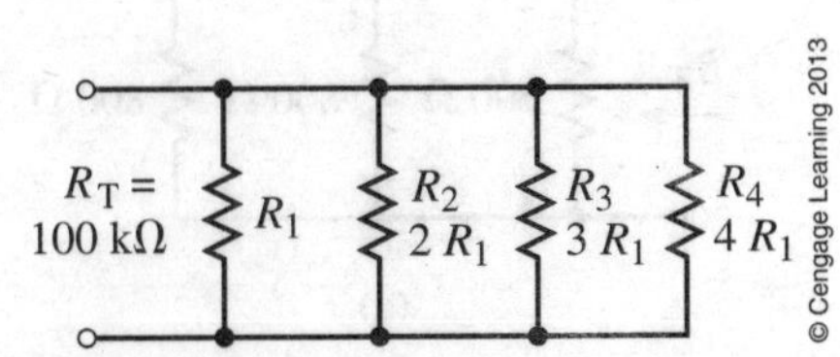

FIGURE 6–57

15. Refer to the network of Figure 6–58:
    a. Calculate the values of R_1, R_2, and R_3 so that the total resistance of the network is 200 Ω.
    b. If R_3 has a current of 2 A, determine the current through each of the other resistors.
    c. How much current must be applied to the entire network?
16. Refer to the network of Figure 6–59:
    a. Calculate the values of R_1, R_2, R_3, and R_4 so that the total resistance of the network is 100 kΩ.
    b. If R_4 has a current of 2 mA, determine the current through each of the other resistors.
    c. How much current must be applied to the entire network?

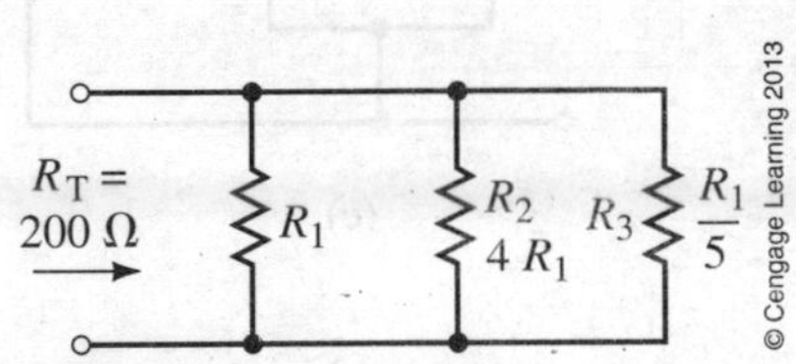

FIGURE 6–58

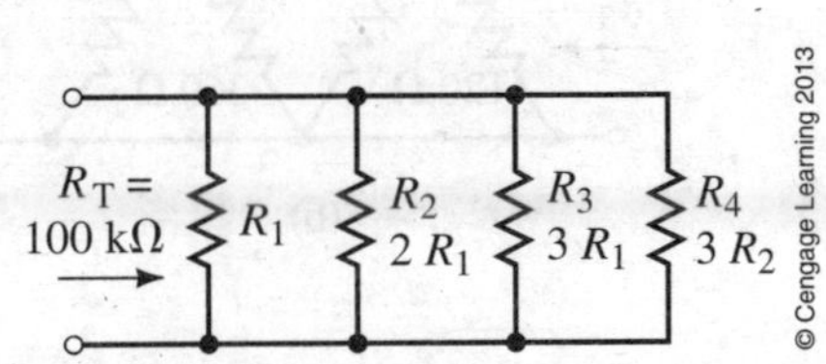

FIGURE 6–59

17. Refer to the network of Figure 6–60:
    a. Find the voltages across R_1 and R_2.
    b. Determine the current I_2.
18. Refer to the network of Figure 6–61:
    a. Find the voltages across R_1, R_2, and R_3.
    b. Calculate the current I_2.
    c. Calculate the current I_3.

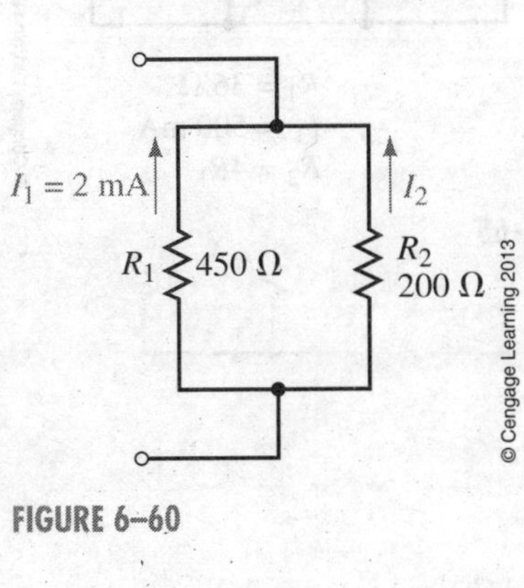

FIGURE 6–60

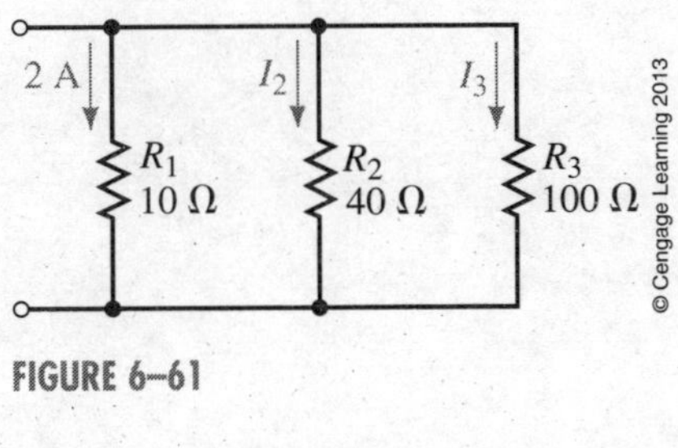

FIGURE 6–61

19. Determine the total resistance of each network of Figure 6–62.

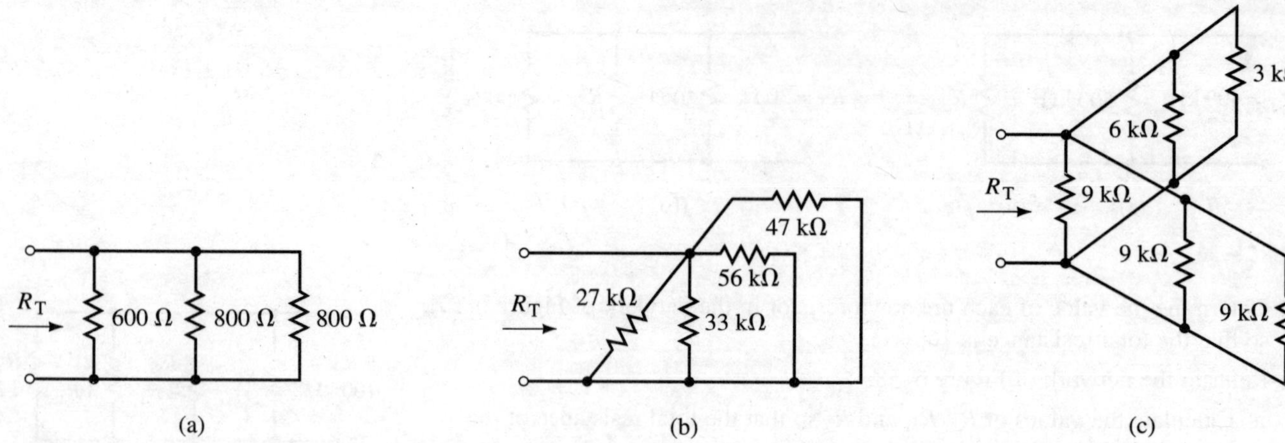

FIGURE 6–62

20. Determine the total resistance of each network of Figure 6–63.

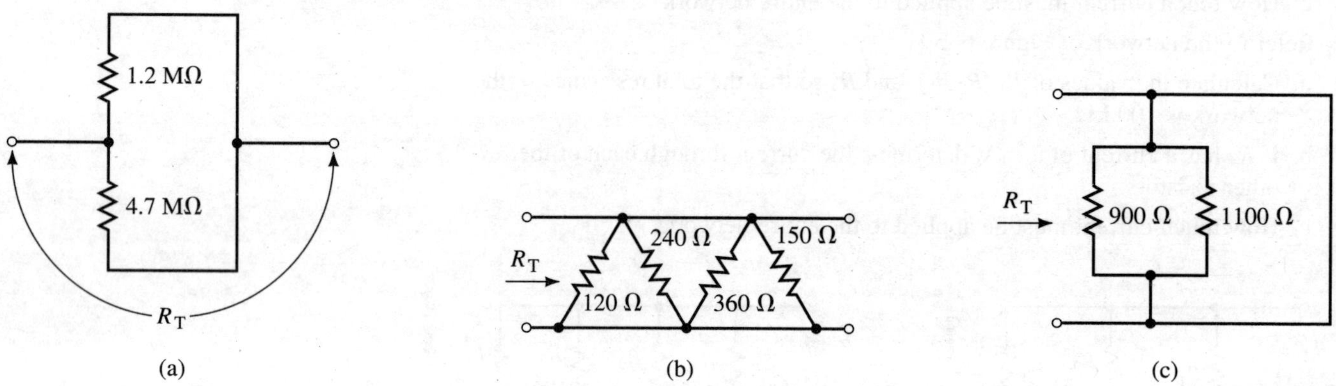

FIGURE 6–63

21. Determine the values of the resistors in the circuit of Figure 6–64, given the indicated conditions.

22. Given the indicated conditions, calculate all currents and determine all resistor values for the circuit of Figure 6–65.

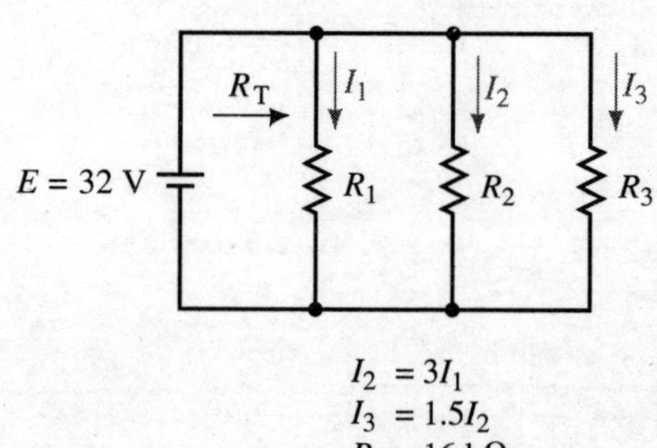

FIGURE 6–64

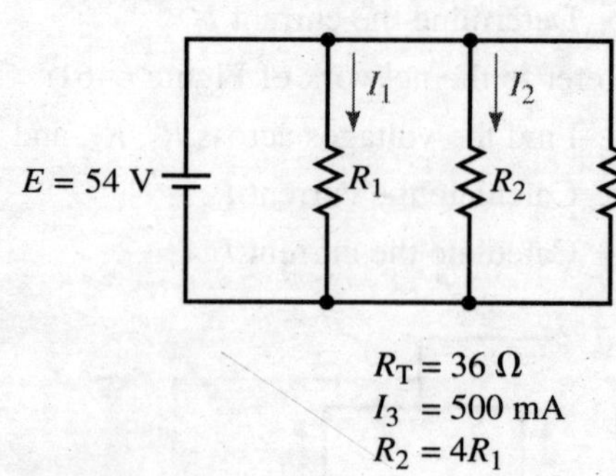

FIGURE 6–65

23. Without using a pencil, paper, or a calculator, determine the resistance of each network of Figure 6–66.

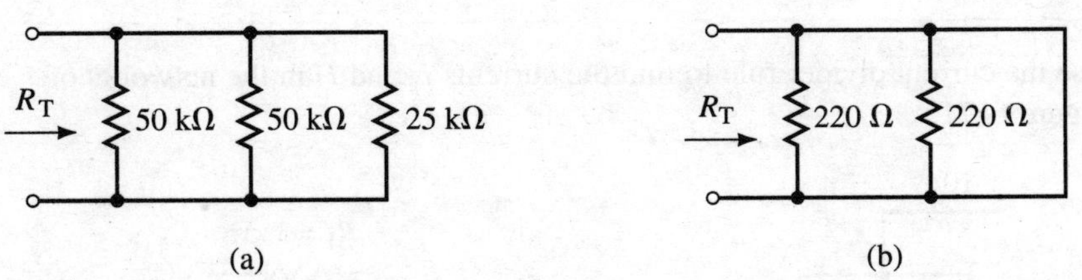

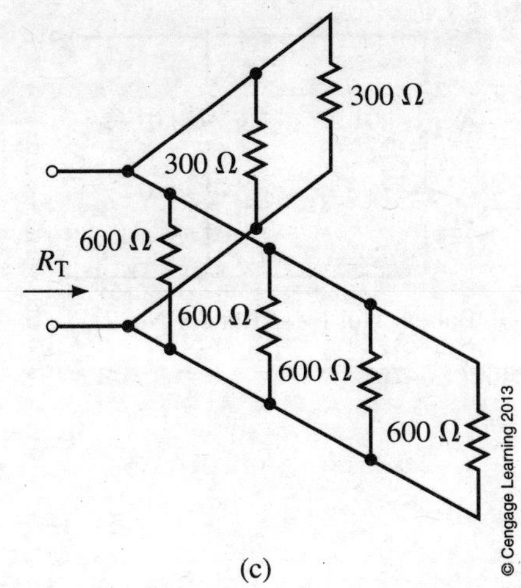

FIGURE 6–66

24. Without using a pencil, paper, or calculator, determine the approximate resistance of the network of Figure 6–67.

25. Without using a pencil, paper, or a calculator, approximate the total resistance of the network of Figure 6–68.

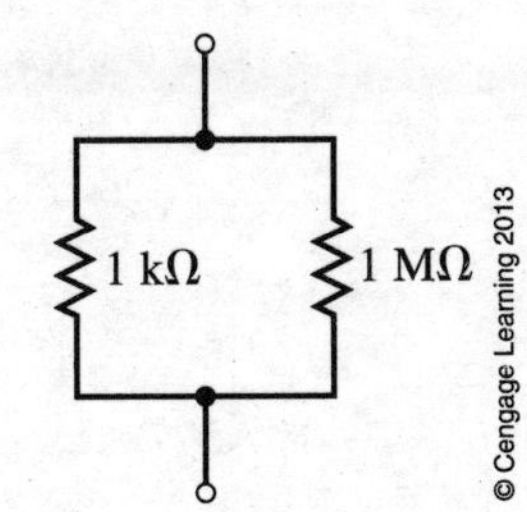

FIGURE 6–67

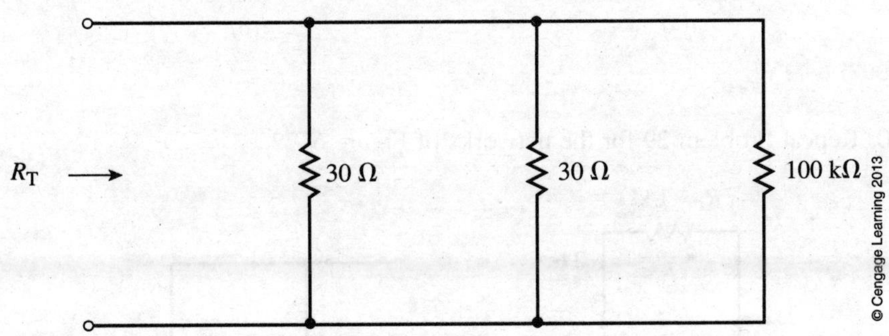

FIGURE 6–68

26. Derive Equation 6–7, which is used to calculate the total resistance of three parallel resistors.

6.4 Voltage Sources in Parallel

27. Two 20-V batteries are connected in parallel to provide current to a 100-V load as shown in Figure 6–69. Determine the current in the load and the current in each battery.

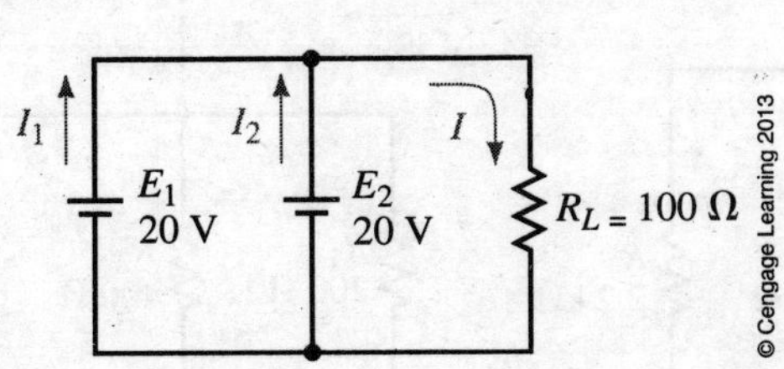

FIGURE 6–69

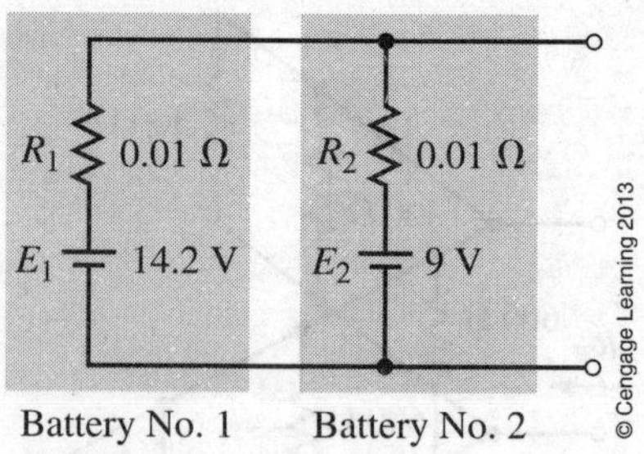

FIGURE 6–70

28. Two lead-acid automobile batteries are connected in parallel, as shown in Figure 6–70, to provide additional starting current. One of the batteries is fully charged at 14.2 V and the other battery has discharged to 9 V. If the internal resistance of each battery is 0.01 Ω, determine the current in the batteries. If each battery is intended to provide a maximum current of 150 A, should this method be used to start a car?

6.5 Current Divider Rule

29. Use the current divider rule to find the currents I_1 and I_2 in the networks of Figure 6–71.

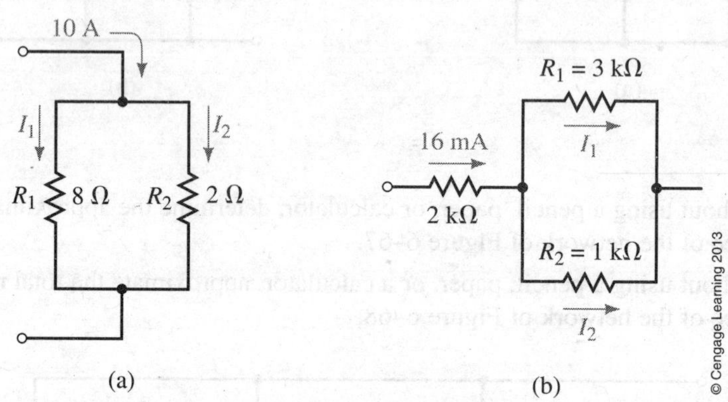

FIGURE 6–71

30. Repeat Problem 29 for the networks of Figure 6–72.

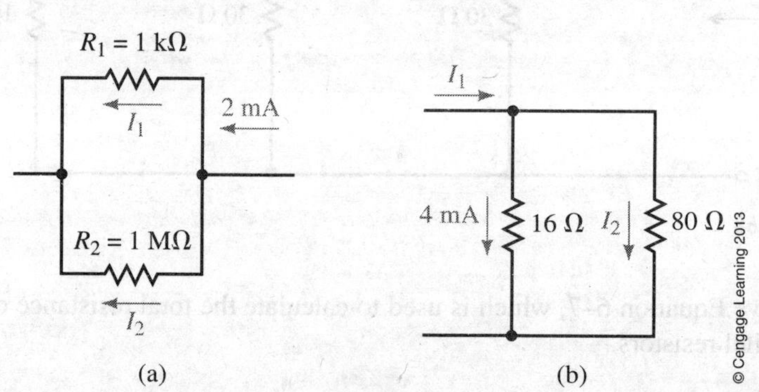

FIGURE 6–72

31. Use the current divider rule to determine all unknown currents for the networks of Figure 6–73.

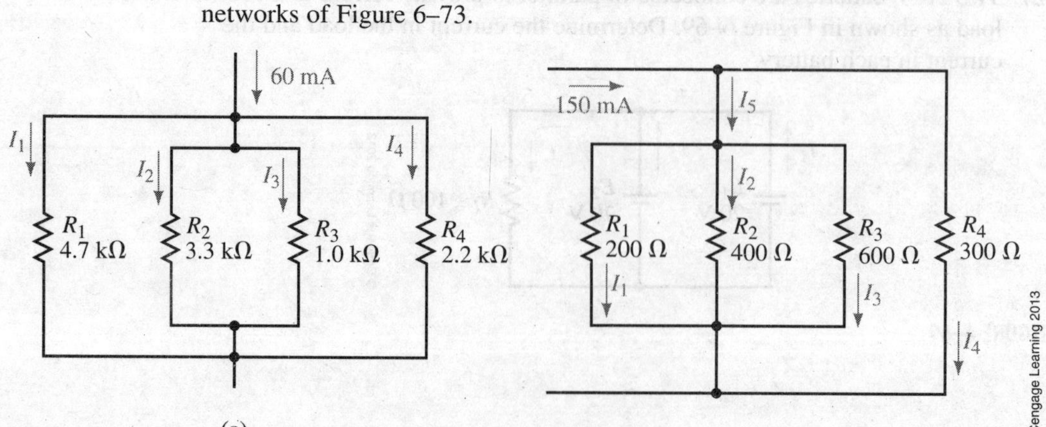

FIGURE 6–73

32. Repeat Problem 31 for the networks of Figure 6–74.

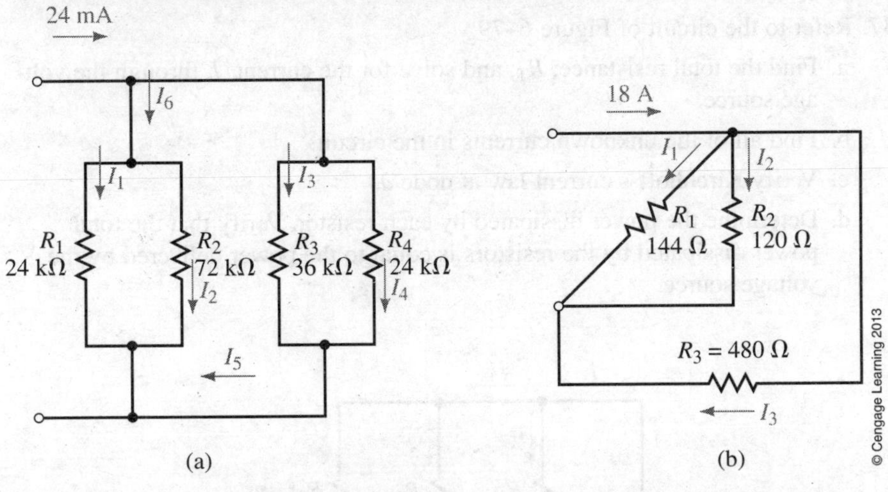

FIGURE 6–74

33. Use the current divider rule to determine the unknown resistance in the network of Figure 6–75.

34. Use the current divider rule to determine the unknown resistance in the network of Figure 6–76.

35. Refer to the circuit of Figure 6–77:

   a. Determine the equivalent resistance, R_T, of the circuit.

   b. Solve for the current I.

   c. Use the current divider rule to determine the current in each resistor.

   d. Verify Kirchhoff's current law at node *a*.

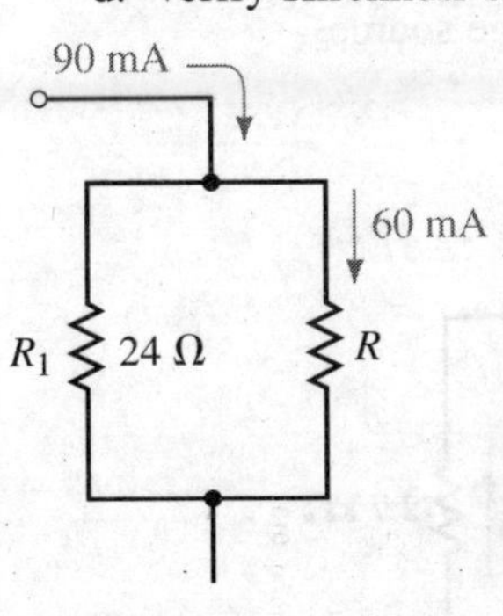

FIGURE 6–75

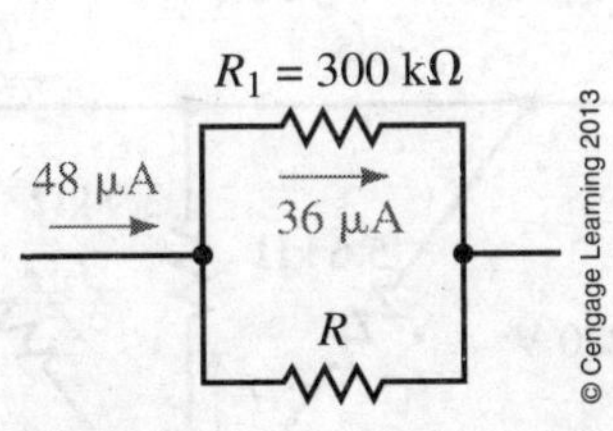

FIGURE 6–76

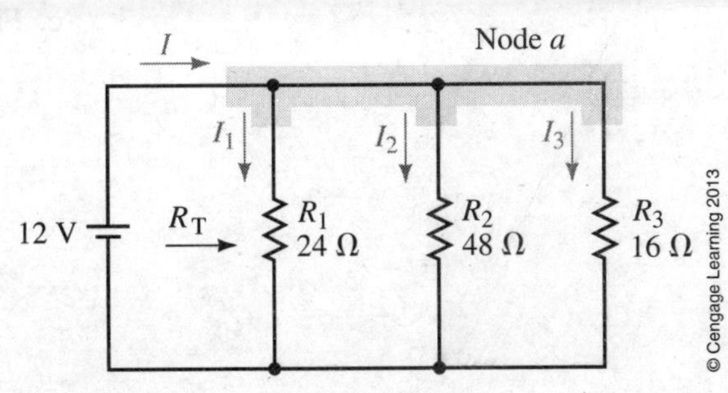

FIGURE 6–77

36. Repeat Problem 35 for the circuit of Figure 6–78.

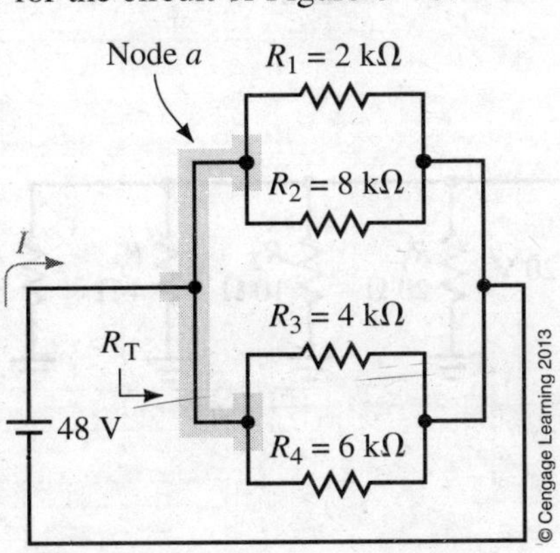

FIGURE 6–78

6.6 Analysis of Parallel Circuits

37. Refer to the circuit of Figure 6–79:

    a. Find the total resistance, R_T, and solve for the current, I, through the voltage source.

    b. Find all of the unknown currents in the circuit.

    c. Verify Kirchhoff's current law at node *a*.

    d. Determine the power dissipated by each resistor. Verify that the total power dissipated by the resistors is equal to the power delivered by the voltage source.

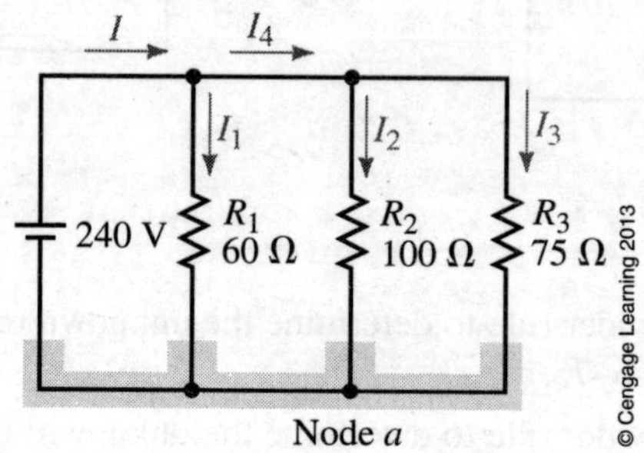

FIGURE 6–79

38. Repeat Problem 37 for the circuit of Figure 6–80.

39. Refer to the circuit of Figure 6–81:

    a. Calculate the current through each resistor in the circuit.

    b. Determine the total current supplied by the voltage source.

    c. Find the power dissipated by each resistor.

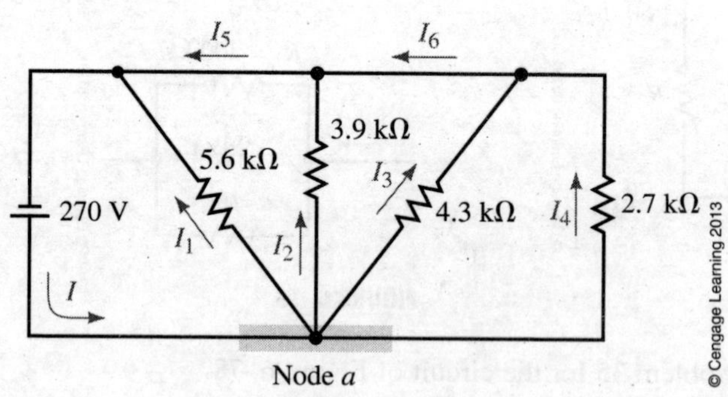

FIGURE 6–80

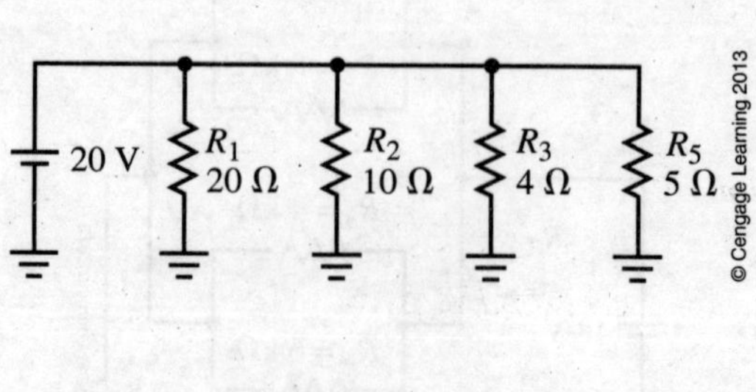

FIGURE 6–81

40. Refer to the circuit of Figure 6–82.
    a. Solve for the indicated currents.
    b. Find the power dissipated by each resistor.
    c. Verify that the power delivered by the voltage source is equal to the total power dissipated by the resistors.

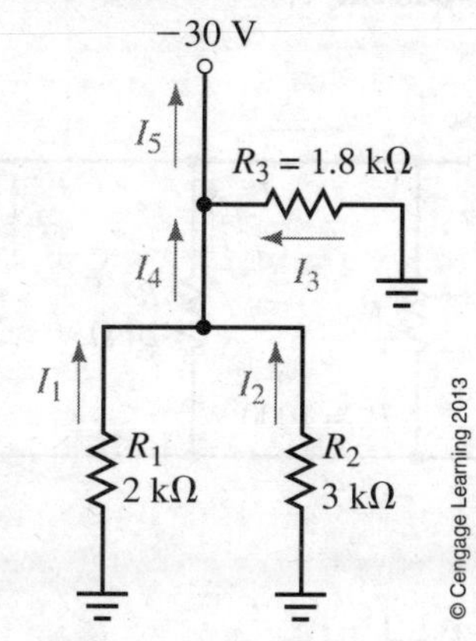

FIGURE 6–82

41. Given the circuit of Figure 6–83:
    a. Determine the values of all resistors.
    b. Calculate the currents through R_1, R_2, and R_4.
    c. Find the currents I_1 and I_2.
    d. Find the power dissipated by resistors R_2, R_3, and R_4.

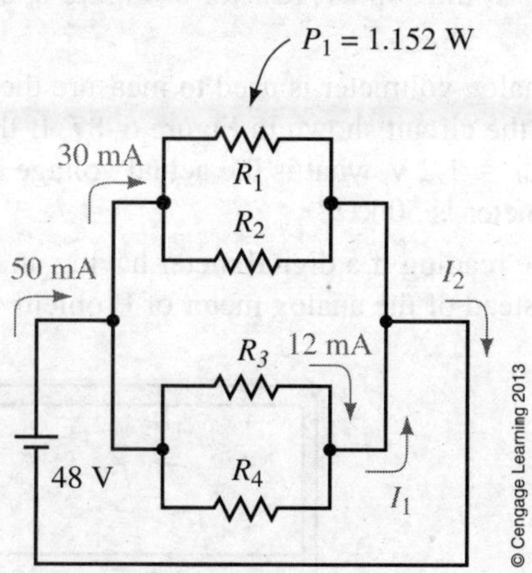

FIGURE 6–83

42. A circuit consists of four resistors connected in parallel and connected to a 20-V source as shown in Figure 6–84. Determine the minimum power rating of each resistor if resistors are available with the following power ratings: ⅛ W, ¼ W, ½ W, 1 W, and 2 W.

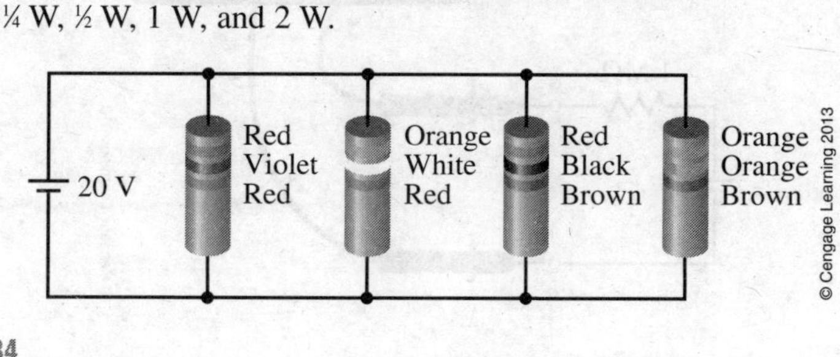

FIGURE 6–84

43. For the circuit of Figure 6–85, determine each of the indicated currents. If the circuit has a 15-A fuse as shown, is the current enough to cause the fuse to open?

44. a. For the circuit of Figure 6–85, calculate the value of R_3 that will result in a circuit current of exactly $I_T = 15$ A.
    b. If the value of R_3 is increased above the value found in part (a), what will happen to the circuit current, I_T?

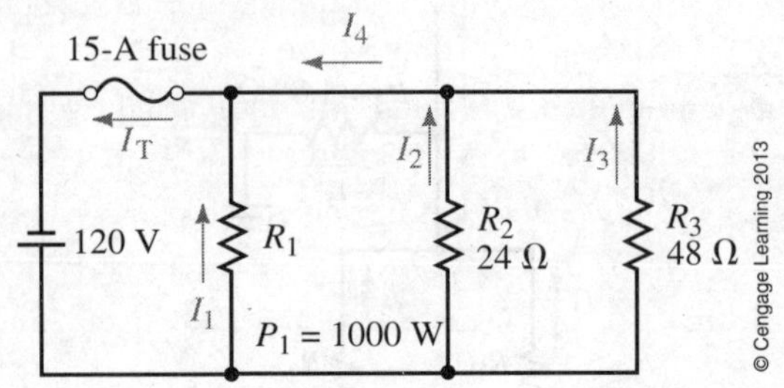

FIGURE 6–85

6.7 Voltmeter Loading Effects

45. A voltmeter having a 1-MΩ internal resistance is used to measure the indicated voltage in the circuit shown in Figure 6–86.
    a. Determine the voltage reading which will be indicated by the meter.
    b. Calculate the voltmeter's loading effect when used to measure the indicated voltage.

46. Repeat Problem 45 if the 500-kΩ resistor of Figure 6–86 is replaced with a 2-MΩ resistor.

47. An inexpensive analog voltmeter is used to measure the voltage across terminals *a* and *b* of the circuit shown in Figure 6–87. If the voltmeter indicates that the voltage $V_{ab} = 1.2$ V, what is the actual voltage of the source if the resistance of the meter is 50 kΩ?

48. What would be the reading if a digital meter having an internal resistance of 10 MΩ is used instead of the analog meter of Problem 47?

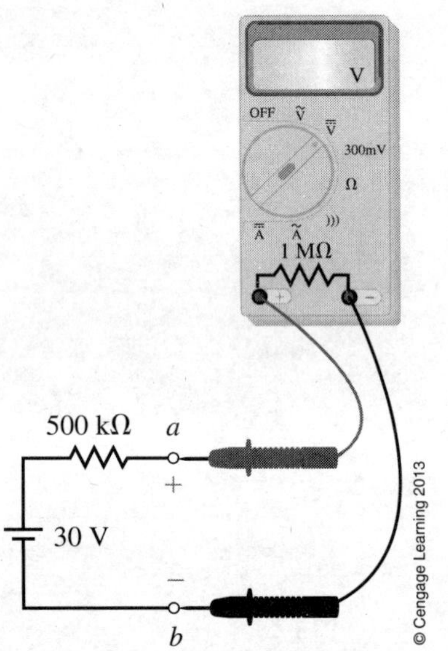

FIGURE 6–86

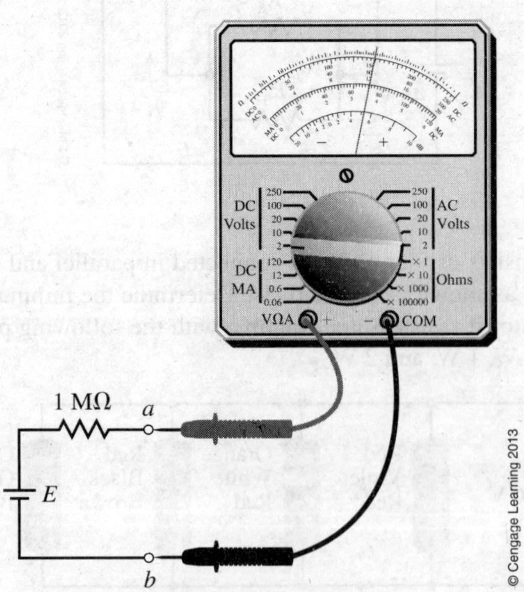

FIGURE 6–87

6.8 Circuit Analysis Using Computers

49. Use Multisim to solve for the current through each resistor in the circuit of Figure 6–79. MULTISIM
50. Use Multisim to solve for the current through each resistor in the circuit of Figure 6–80. MULTISIM
51. Use Multisim to simulate a voltmeter with an internal resistance of 1 MΩ used to measure voltage as shown in Figure 6–86. MULTISIM
52. Use Multisim to simulate a voltmeter with an internal resistance of 500 kΩ used to measure voltage as shown in Figure 6–86. MULTISIM
53. Use PSpice to solve for the current through each resistor in the circuit of Figure 6–79. PSPICE
54. Use PSpice to solve for the current through each resistor in the circuit of Figure 6–80. PSPICE

ANSWERS TO IN-PROCESS LEARNING CHECKS

IN-PROCESS LEARNING CHECK 1

a. $I = 2.00$ A
b. $I_1 = I_2 = I_3 = 0.267$ A, $I_4 = 1.200$ A
c. 3(0.267 A) + 1.200 A = 2.00 A (as required)

IN-PROCESS LEARNING CHECK 2

a. $I_{1\Omega} = 30.0$ A, $I_{3\Omega} = 10.0$ A, $I_{4\Omega} = 7.50$ A
b. $I_T = 53.5$ A
c. $R_T = 0.561\ \Omega$. [The currents are the same as those determined in part (a).]

IN-PROCESS LEARNING CHECK 3

The voltmeter with the smaller internal resistance would load the circuit more, since more of the circuit current would enter the instrument.

KEY TERMS

Branch
Branch Currents
Bridge Circuit
Loaded Output
Loading Effect
Parallel Branches
Series-Parallel Circuits
Transistor
Transistor Bias
Universal Bias
Unloaded Output
Voltage Regulation
Zener Diode

OUTLINE

The Series-Parallel Network
Analysis of Series-Parallel Circuits
Applications of Series-Parallel Circuits
Potentiometers
Loading Effects of Instruments
Circuit Analysis Using Computers

OBJECTIVES

After studying this chapter, you will be able to

- find the total resistance of a network consisting of resistors connected in various series-parallel configurations,
- solve for the current through any branch or component of a series-parallel circuit,
- determine the difference in potential between any two points in a series-parallel circuit,
- calculate the voltage drop across a resistor connected to a potentiometer,
- analyze how the size of a load resistor connected to a potentiometer affects the output voltage,
- calculate the loading effects of a voltmeter or ammeter when used to measure the voltage or current in any circuit,
- use PSpice to solve for voltages and currents in series-parallel circuits,
- use Multisim to solve for voltages and currents in series-parallel circuits.

7

SERIES-PARALLEL CIRCUITS

CHAPTER PREVIEW

Most circuits encountered in electronics are neither simple series circuits nor simple parallel circuits, but rather a combination of the two. Although **series-parallel circuits** appear to be more complicated than either of the previous types of circuits analyzed to this point, we find that the same principles apply.

This chapter examines how Kirchhoff's voltage and current laws are applied to the analysis of series-parallel circuits. We will also observe that voltage and current divider rules apply to the more complex circuits. In the analysis of series-parallel circuits, we often simplify the given circuit to enable us to more clearly see how the rules and laws of circuit analysis apply. Students are encouraged to redraw circuits whenever the solution of a problem is not immediately apparent. This technique is used by even the most experienced engineers, technologists, and technicians.

In this chapter, we begin by examining simple resistor circuits. The principles of analysis are then applied to more practical circuits such as those containing zener diodes and transistors. The same principles are then applied to determine the loading effects of voltmeters and ammeters in more complex circuits.

After analyzing a complex circuit, we want to know whether the solutions are in fact correct. As you have already seen, electrical circuits usually may be studied in more than one way to arrive at a solution. Once currents and voltages for a circuit have been found, it is very easy to determine whether the resultant solution verifies the law of conservation of energy, Kirchhoff's current law, and Kirchhoff's voltage law. If there is any discrepancy (other than rounding error), there is an error in the calculation! ■

Putting It in Perspective

Benjamin Franklin

BENJAMIN FRANKLIN WAS BORN IN BOSTON, Massachusetts, in 1706. Although Franklin is best known as a great statesman and diplomat, he also furthered the cause of science with his experiments in electricity. This particularly includes his work with the Leyden jar, which was used to store electric charge. In his famous experiment of 1752, he used a kite to demonstrate that lightning is an electrical event. It was Franklin who postulated that positive and negative electricity are in fact a single "fluid."

Although Franklin's major accomplishments came as a result of his work in achieving independence of the Thirteen Colonies, he was nonetheless a notable scientist.

Benjamin Franklin died in his Philadelphia home on February 12, 1790, at the age of eighty-four. ■

7.1 The Series-Parallel Network

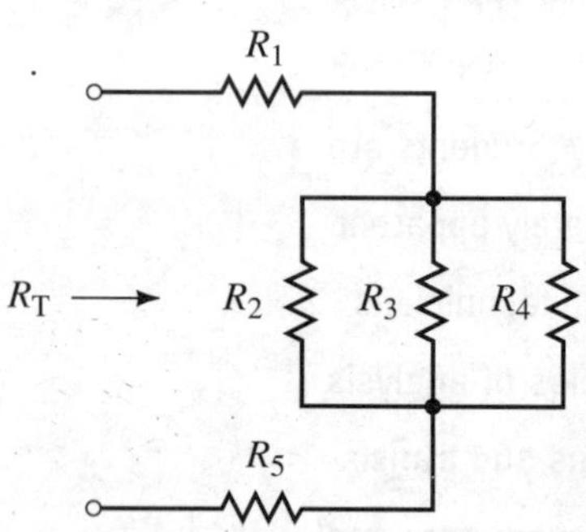

FIGURE 7–1

In electric circuits, we define a **branch** as any portion of a circuit that can be simplified as having two terminals. The components between the two terminals may be any combination of resistors, voltage sources, or other elements. Many complex circuits may be separated into a combination of both series and/or parallel elements, while other circuits consist of even more elaborate combinations that are neither series nor parallel.

In order to analyze a complicated circuit, it is important to be able to recognize which elements are in series and which elements or branches are in parallel. Consider the network of resistors shown in Figure 7–1.

We immediately recognize that the resistors R_2, R_3, and R_4 are in parallel. This parallel combination is in series with the resistors R_1 and R_5. The total resistance may now be written as follows:

$$R_T = R_1 + (R_2 \| R_3 \| R_4) + R_5$$

EXAMPLE 7–1

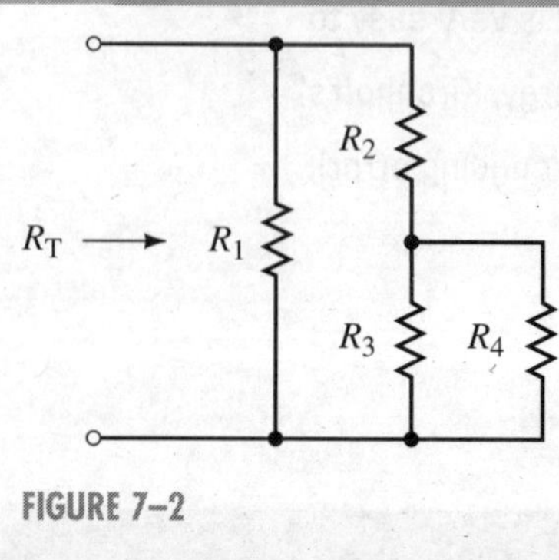

FIGURE 7–2

For the network of Figure 7–2, determine which resistors and branches are in series and which are in parallel. Write an expression for the total equivalent resistance, R_T.

Solution First, we recognize that the resistors R_3 and R_4 are in parallel: $(R_3 \| R_4)$.

Next, we see that this combination is in series with the resistor R_2: $[R_2 + (R_3 \| R_4)]$.

Finally, the entire combination is in parallel with the resistor R_1. The total resistance of the circuit may now be written as follows:

$$R_T = R_1 \| [R_2 + (R_3 \| R_4)]$$

PRACTICE PROBLEMS 1

For the network of Figure 7–3, determine which resistors and branches are in series and which are in parallel. Write an expression for the total resistance, R_T.

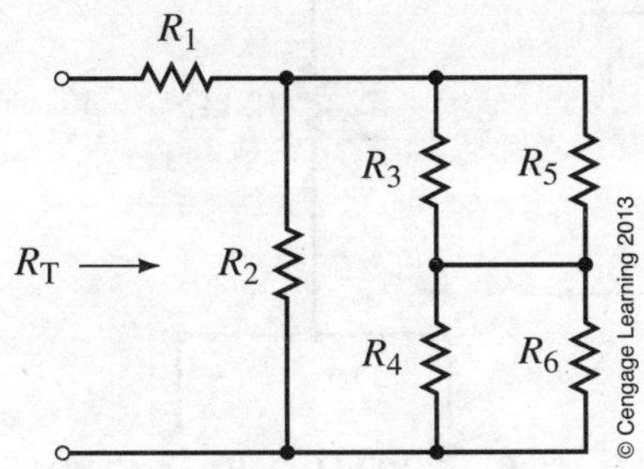

FIGURE 7–3

Answer
$R_T = R_1 + R_2 \| [(R_3\|R_5) + (R_4\|R_6)]$

7.2 Analysis of Series-Parallel Circuits

Series-parallel networks are often difficult to analyze because they initially appear confusing. However, the analysis of even the most complex circuit is simplified by following some fairly basic steps. By practicing (not memorizing) the techniques outlined in this section, you will find that most circuits can be reduced to groupings of series and parallel combinations. In analyzing such circuits, it is imperative to remember that the rules for analyzing series and parallel elements still apply.

The same current occurs through all series elements.

The same voltage occurs across all parallel elements.

In addition, remember that Kirchhoff's voltage law and Kirchhoff's current law apply for all circuits regardless of whether the circuits are series, parallel, or series-parallel. The following steps will help to simplify the analysis of series-parallel circuits:

1. Whenever necessary, redraw complicated circuits showing the source connection at the left-hand side. All nodes should be labelled to ensure that the new circuit is equivalent to the original circuit. You will find that as you become more experienced at analyzing circuits, this step will no longer be as important and may therefore be omitted.
2. Examine the circuit to determine the strategy that will work best in analyzing the circuit for the required quantities. You will usually find it best to begin the analysis of the circuit at the components most distant from the source.
3. Simplify recognizable combinations of components wherever possible, redrawing the resulting circuit as often as necessary. Keep the same labels for corresponding nodes.
4. Determine the equivalent circuit resistance, R_T.
5. Solve for the total circuit current. Indicate the directions of all currents and label the correct polarities of the voltage drops on all components.
6. Calculate how currents and voltages split between the elements of the circuit.
7. Since there are usually several possible ways at arriving at solutions, verify the answers by using a different approach. The extra time taken in this step will usually ensure that the correct answer has been found.

EXAMPLE 7–2

Consider the circuit of Figure 7–4.

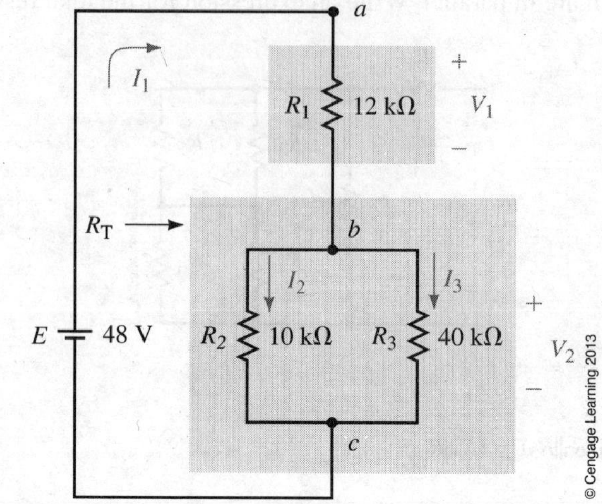

FIGURE 7–4

a. Find R_T.

b. Calculate I_1, I_2, and I_3.

c. Determine the voltages V_1 and V_2.

Solution By examining the circuit of Figure 7–4, we see that resistors R_2 and R_3 are in parallel. This parallel combination is in series with the resistor R_1.

The combination of resistors may be represented by a simple series network shown in Figure 7–5. Notice that the nodes have been labelled using the same notation.

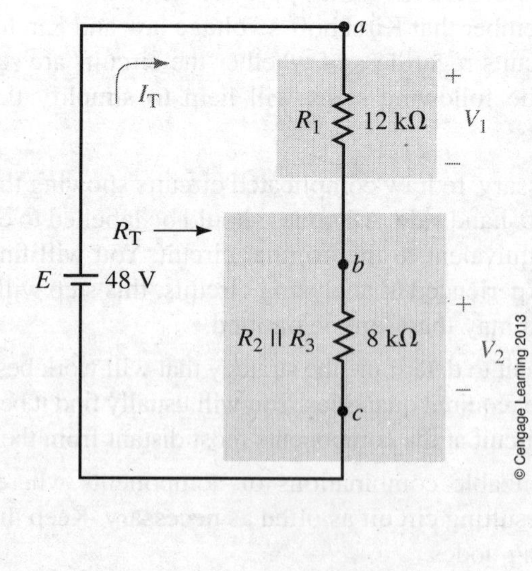

FIGURE 7–5

a. The total resistance of the circuit may be determined from the combination

$$R_T = R_1 + R_2 \| R_3$$

$$R_T = 12\text{ k}\Omega + \frac{(10\text{ k}\Omega)(40\text{ k}\Omega)}{10\text{ k}\Omega + 40\text{ k}\Omega}$$

$$= 12\text{ k}\Omega + 8\text{ k}\Omega = 20\text{ k}\Omega$$

b. From Ohm's law, the total current is

$$I_T = I_1 = \frac{48\text{ V}}{20\text{ k}\Omega} = 2.4\text{ mA}$$

The current I_1 will enter node b and then split between the two resistors R_2 and R_3. This current divider may be simplified as shown in the partial circuit of Figure 7–6. Applying the current divider rule to these two resistors gives

$$I_2 = \frac{(40\text{ k}\Omega)(2.4\text{ mA})}{10\text{ k}\Omega + 40\text{ k}\Omega} = 1.92\text{ mA}$$

$$I_3 = \frac{(10\text{ k}\Omega)(2.4\text{ mA})}{10\text{ k}\Omega + 40\text{ k}\Omega} = 0.48\text{ A}$$

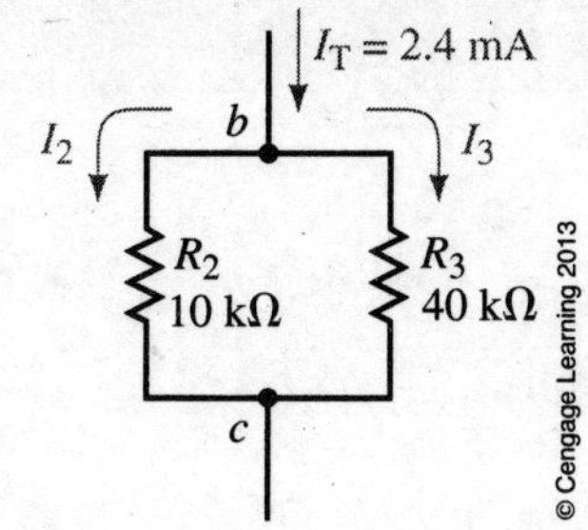

FIGURE 7–6

c. Using the preceding currents and Ohm's law, we determine the voltages:

$$V_1 = (2.4\text{ mA})(12\text{ k}\Omega) = 28.8\text{ V}$$
$$V_3 = (0.48\text{ mA})(40\text{ k}\Omega) = 19.2\text{ V} = V_2$$

In order to check the answers, we may simply apply Kirchhoff's voltage law around any closed loop which includes the voltage source:

$$\begin{aligned}\sum V &= E - V_1 - V_3 \\ &= 48\text{ V} - 28.8\text{ V} - 19.2\text{ V} \\ &= 0\text{ V (checks!)}\end{aligned}$$

The solution may be verified by ensuring that the power delivered by the voltage source is equal to the summation of powers dissipated by the resistors.

IN-PROCESS LEARNING CHECK 1

(Answers are at the end of the chapter.)

Use the results of Example 7–2 to verify that the law of conservation of energy applies to the circuit of Figure 7–4 by showing that the voltage source delivers the same power as the total power dissipated by all resistors.

EXAMPLE 7–3

Find the voltage V_{ab} for the circuit of Figure 7–7.

Solution We begin by redrawing the circuit in a more simple representation as shown in Figure 7–8. From Figure 7–8, we see that the original circuit consists of two **parallel branches**, where each branch is a series combination of two resistors.

If we take a moment to examine the circuit, we see that the voltage V_{ab} may be determined from the combination of voltages across R_1 and R_2. Alternatively, the voltage may be found from the combination of voltages across R_3 and R_4.

As usual, several methods of analysis are possible. Because the two branches are in parallel, the voltage across each branch must be 40 V. Using the voltage

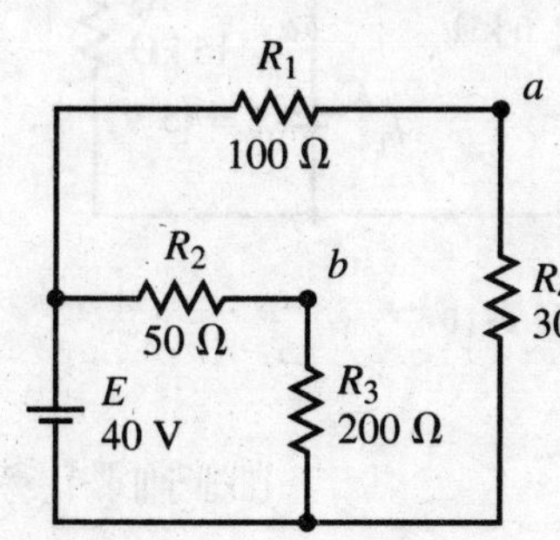

FIGURE 7–7

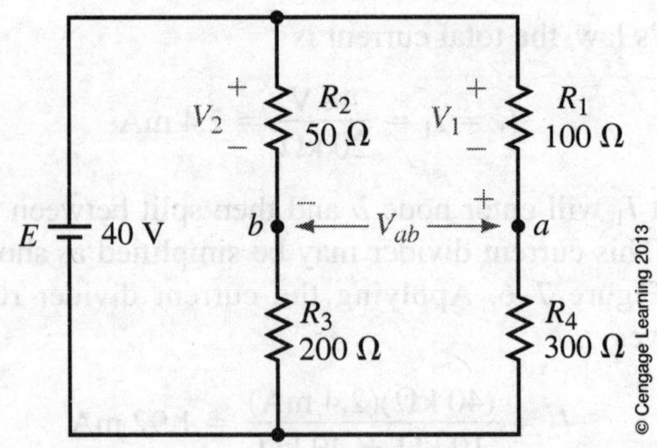

FIGURE 7–8

divider rules allows us to quickly calculate the voltage across each resistor. Although equally correct, other methods of calculating the voltages would be more lengthy.

$$V_2 = \frac{R_2}{R_2 + R_3}E$$
$$= \left(\frac{50\ \Omega}{50\ \Omega + 200\ \Omega}\right)(40\ \text{V}) = 8.0\ \text{V}$$

$$V_1 = \frac{R_1}{R_1 + R_4}E$$
$$= \left(\frac{100\ \Omega}{100\ \Omega + 300\ \Omega}\right)(40\ \text{V}) = 10.0\ \text{V}$$

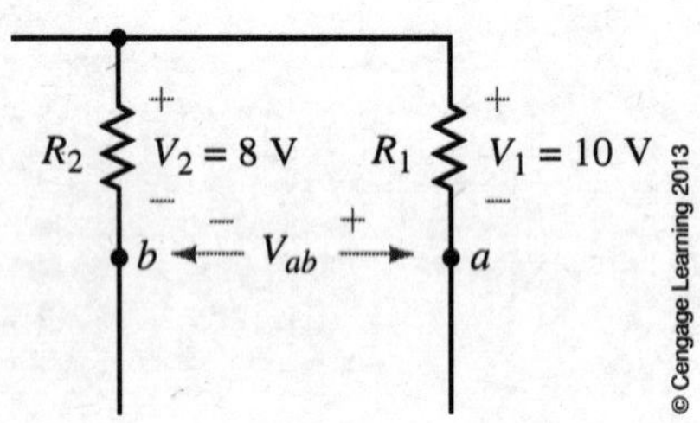

FIGURE 7–9

As shown in Figure 7–9, we apply Kirchhoff's voltage law to determine the voltage between terminals a and b.

$$V_{ab} = -10.0\ \text{V} + 8.0\ \text{V} = -2.0\ \text{V}$$

EXAMPLE 7–4

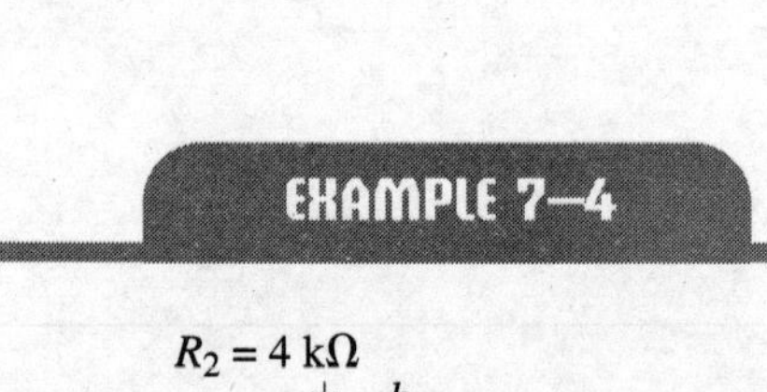

FIGURE 7–10

CircuitSim 07-1

Consider the circuit of Figure 7–10:

a. Find the total resistance R_T "seen" by the source E.

b. Calculate I_T, I_1, and I_2.

c. Determine the voltages V_2 and V_4.

Solution We begin the analysis by redrawing the circuit. Since we generally like to see the source on the left-hand side, one possible way of redrawing the resultant circuit is shown in Figure 7–11. Notice that the polarities of voltages across all resistors have been shown.

a. From the redrawn circuit, the total resistance of the circuit is

$$R_T = R_3 + [(R_1 + R_2)\|R_4]$$
$$= 3\ \text{k}\Omega + \frac{(4\ \text{k}\Omega + 6\ \text{k}\Omega)(15\ \text{k}\Omega)}{(4\ \text{k}\Omega + 6\ \text{k}\Omega) + 15\ \text{k}\Omega}$$
$$= 3\ \text{k}\Omega + 6\ \text{k}\Omega = 9.00\ \text{k}\Omega$$

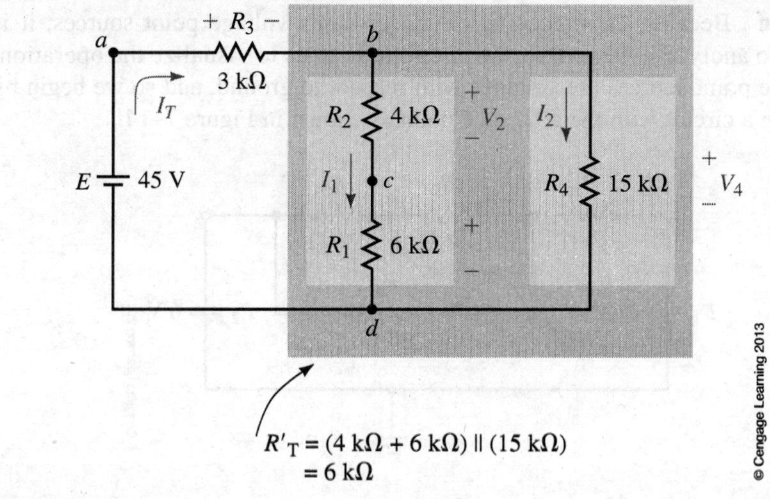

FIGURE 7–11

b. The current supplied by the voltage source is

$$I_T = \frac{E}{R_T} = \frac{45\ \text{V}}{9\ \text{k}\Omega} = 5.00\ \text{mA}$$

We see that the supply current divides between the parallel branches as shown in Figure 7–12.

Applying the current divider rule, we calculate the **branch currents** as

$$I_1 = I_T\frac{R'_T}{(R_1 + R_2)} = \frac{(5\ \text{mA})(6\ \text{k}\Omega)}{4\ \text{k}\Omega + 6\ \text{k}\Omega} = 3.00\ \text{mA}$$

$$I_2 = I_T\frac{R'_T}{R_4} = \frac{(5\ \text{mA})(6\ \text{k}\Omega)}{15\ \text{k}\Omega} = 2.00\ \text{mA}$$

Notice: When determining the branch currents, the resistance R'_T is used in the calculations rather than the total circuit resistance. This is because the current $I_T = 5$ mA splits between the two branches of R'_T and the split is not affected by the value of R_3.

c. The voltages V_2 and V_4 are now easily calculated by using Ohm's law:

$$V_2 = I_1R_2 = (3\ \text{mA})(4\ \text{k}\Omega) = 12.0\ \text{V}$$
$$V_4 = I_2R_4 = (2\ \text{mA})(15\ \text{k}\Omega) = 30.0\ \text{V}$$

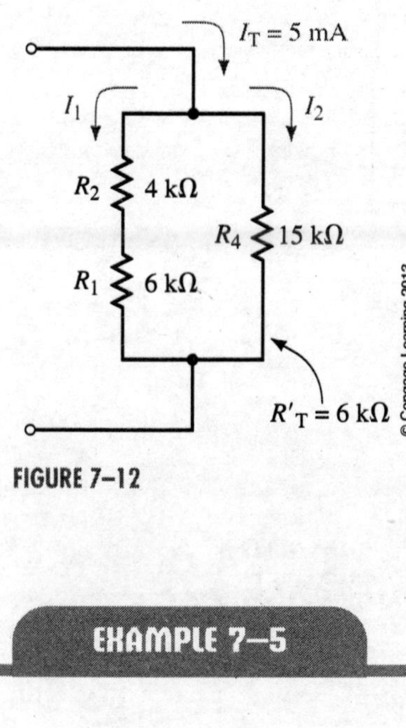

FIGURE 7–12

EXAMPLE 7–5

For the circuit of Figure 7–13, find the indicated currents and voltages.

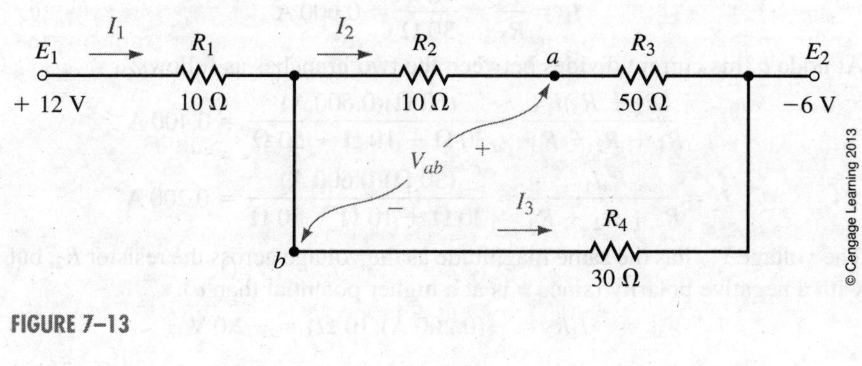

FIGURE 7–13

CircuitSim 07-2

Solution Because the preceding circuit contains voltage point sources, it is easier to analyze if we redraw the circuit to help us to visualize the operation.

The point sources are voltages with respect to ground, and so we begin by drawing a circuit with the reference point as shown in Figure 7–14.

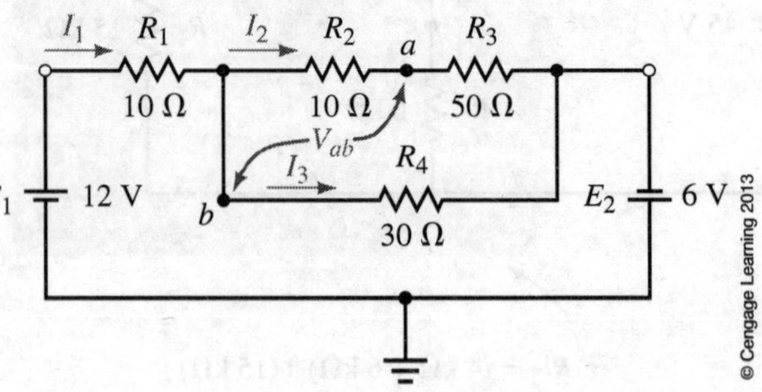

FIGURE 7–14

Now, we can see that the circuit may be further simplified by combining the voltage sources ($E = E_1 + E_2$) and by showing the resistors in a more suitable location. The simplified circuit is shown in Figure 7–15.

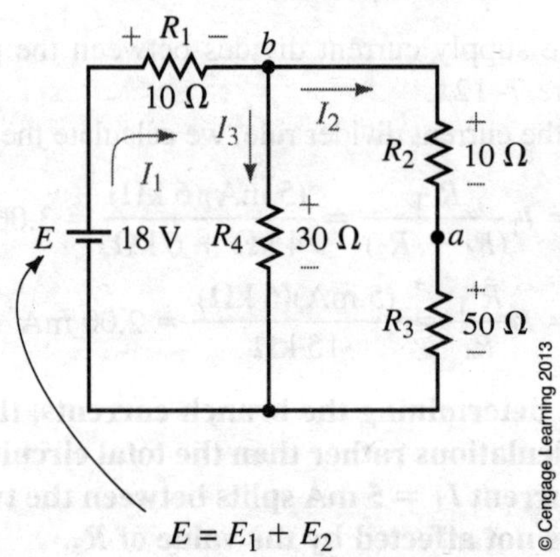

FIGURE 7–15

The total resistance "seen" by the equivalent voltage source is

$$R_T = R_1 + [R_4 \| (R_2 + R_3)]$$

$$= 10\ \Omega + \frac{(30\ \Omega)(10\ \Omega + 50\ \Omega)}{30\ \Omega + (10\ \Omega + 50\ \Omega)} = 30.0\ \Omega$$

And so the total current provided into the circuit is

$$I_1 = \frac{E}{R_T} = \frac{18\text{ V}}{30\ \Omega} = 0.600\text{ A}$$

At node b this current divides between the two branches as follows:

$$I_3 = \frac{(R_2 + R_3)I_1}{R_4 + R_2 + R_3} = \frac{(60\ \Omega)(0.600\text{ A})}{30\ \Omega + 10\ \Omega + 50\ \Omega} = 0.400\text{ A}$$

$$I_2 = \frac{R_4 I_1}{R_4 + R_2 + R_3} = \frac{(30\ \Omega)(0.600\text{ A})}{30\ \Omega + 10\ \Omega + 50\ \Omega} = 0.200\text{ A}$$

The voltage V_{ab} has the same magnitude as the voltage across the resistor R_2, but with a negative polarity (since b is at a higher potential than a):

$$V_{ab} = -I_2R_2 = -(0.200\text{ A})(10\ \Omega) = -2.0\text{ V}$$

PRACTICE PROBLEMS 2

Consider the circuit of Figure 7–16:

a. Find the total circuit resistance, R_T.
b. Determine the current I_T through the voltage sources.
c. Solve for the currents I_1 and I_2.
d. Calculate the voltage V_{ab}.

Answers
a. $R_T = 7.20\ \text{k}\Omega$; b. $I_T = 1.11$ mA; c. $I_1 = 0.133$ mA, $I_2 = 0.444$ mA;
d. $V_{ab} = -0.800$ V

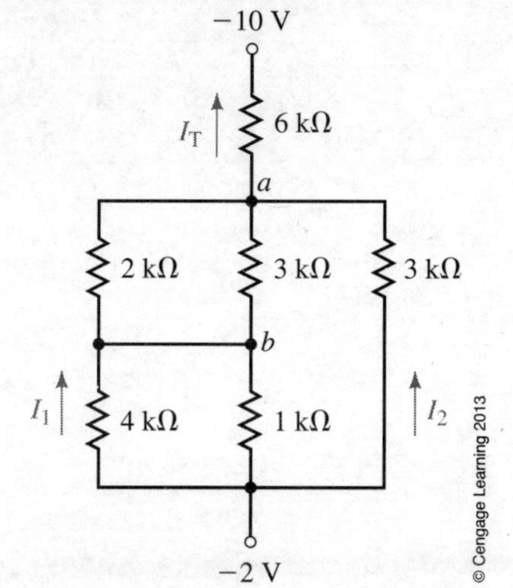

FIGURE 7–16

7.3 Applications of Series-Parallel Circuits

We now examine how the methods developed in the first two sections of this chapter are applied when analyzing practical circuits. You may find that some of the circuits introduce you to unfamiliar devices. For now, you do not need to know precisely how these devices operate, simply that the voltages and currents in the circuits follow the same rules and laws that you have used up to now.

EXAMPLE 7–6

The circuit of Figure 7–17 is referred to as a **bridge circuit** and is used extensively in electronic and scientific instruments.

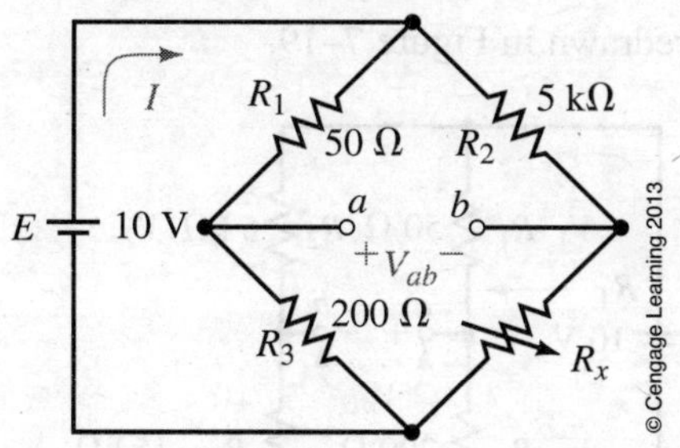

FIGURE 7–17

CircuitSim 07-3

Calculate the current I and the voltage V_{ab} when

a. $R_x = 0\ \Omega$ (short circuit)
b. $R_x = 15\ \text{k}\Omega$
c. $R_x = \infty$ (open circuit)

Solution

a. $R_x = 0\ \Omega$:

The circuit is redrawn as shown in Figure 7–18.

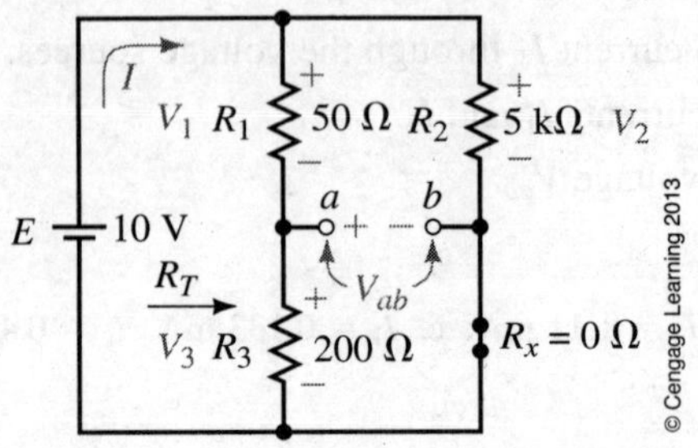

FIGURE 7–18

The voltage source "sees" a total resistance of

$$R_T = (R_1 + R_3)\|R_2 = 250\ \Omega\|5000\ \Omega = 238\ \Omega$$

resulting in a source current of

$$I = \frac{10\ \text{V}}{238\ \Omega} = 0.042\ \text{A} = 42.2\ \text{mA}$$

The voltage V_{ab} may be determined by solving for voltage across R_1 and R_2.

The voltage across R_1 will be constant regardless of the value of the variable resistor R_x. Hence

$$V_1 = \left(\frac{50\ \Omega}{50\ \Omega + 200\ \Omega}\right)(10\ \text{V}) = 2.00\ \text{V}$$

Now, since the variable resistor is a short circuit, the entire source voltage will appear across the resistor R_2, giving

$$V_2 = 10.0\ \text{V}$$

And so

$$V_{ab} = -V_1 + V_2 = -2.00\ \text{V} + 10.0\ \text{V} = +8.00\ \text{V}$$

b. $R_x = 15\ \text{k}\Omega$:

The circuit is redrawn in Figure 7–19.

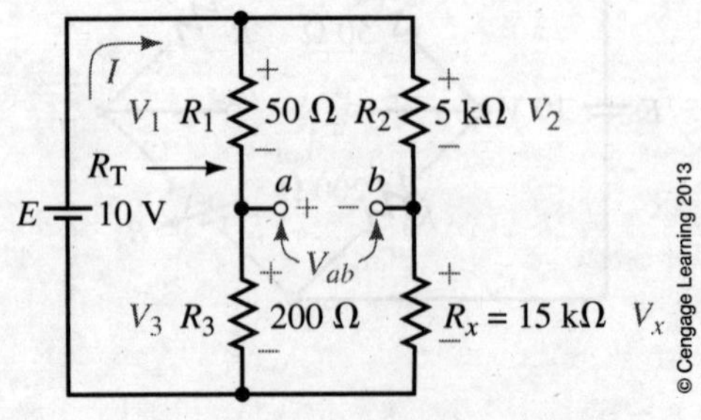

FIGURE 7–19

The voltage source "sees" a circuit resistance of

$$\begin{aligned} R_T &= (R_1 + R_3)\|(R_2 + R_x) \\ &= 250\ \Omega\|20\ \text{k}\Omega = 247\ \Omega \end{aligned}$$

which results in a source current of

$$I = \frac{10\text{ V}}{247\ \Omega} = 0.0405\text{ A} = 40.5\text{ mA}$$

The voltages across R_1 and R_2 are

$$V_1 = 2.00\text{ V} \quad \text{(as before)}$$

$$V_2 = \frac{R_2}{R_2 + R_x}E$$

$$= \left(\frac{5\text{ k}\Omega}{5\text{ k}\Omega + 15\text{ k}\Omega}\right)(10\text{ V}) = 2.50\text{ V}$$

Now the voltage between terminals a and b is found as

$$V_{ab} = -V_1 + V_2$$
$$= -2.0\text{ V} + 2.5\text{ V} = +0.500\text{ V}$$

c. $R_x = \infty$:
The circuit is redrawn in Figure 7–20. Because the second branch is an open circuit due to the resistor R_x, the total resistance "seen" by the source is

$$R_T = R_1 + R_3 = 250\ \Omega$$

resulting in a source current of

$$I = \frac{10\text{ V}}{250\ \Omega} = 0.040\text{ A} = 40.0\text{ mA}$$

The voltages across R_1 and R_2 are

$$V_1 = 2.00\text{ V} \quad \text{(as before)}$$
$$V_2 = 0\text{ V} \quad \text{(since the branch is open)}$$

And so the resulting voltage between terminals a and b is

$$V_{ab} = -V_1 + V_2$$
$$= -2.0\text{ V} + 0\text{ V} = -2.00\text{ V}$$

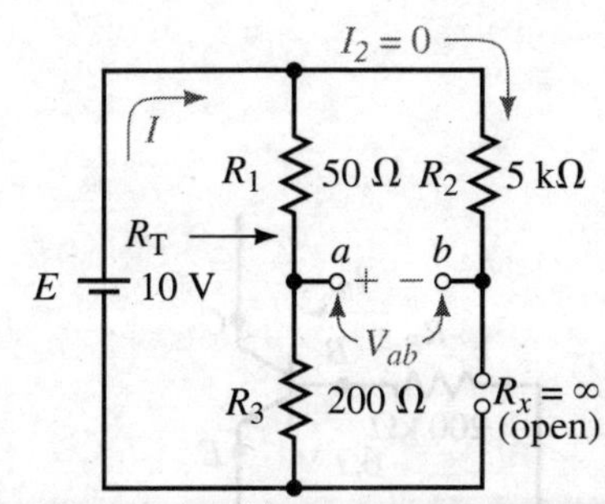

FIGURE 7–20

The previous example illustrates how voltages and currents within a circuit are affected by changes elsewhere in the circuit. In the example, we saw that the voltage V_{ab} varied from -2 V to $+8$ V, while the total circuit current varied from a minimum value of 40 mA to a maximum value of 42 mA. These changes occurred even though the resistor R_x varied from 0 Ω to ∞.

A **transistor** is a three-terminal device that may be used to amplify small signals. In order for the transistor to operate as an amplifier, however, certain dc conditions must be met. These conditions set the "bias point" of the transistor. The bias current of a transistor circuit—the **transistor bias**—is determined by a dc voltage source and several resistors. Although the operation of the transistor is outside the scope of this chapter, we can analyze the bias circuit of a transistor using elementary circuit theory.

EXAMPLE 7–7

Use the given conditions to determine I_C, I_E, V_{CE}, and V_B for the transistor circuit of Figure 7–21.

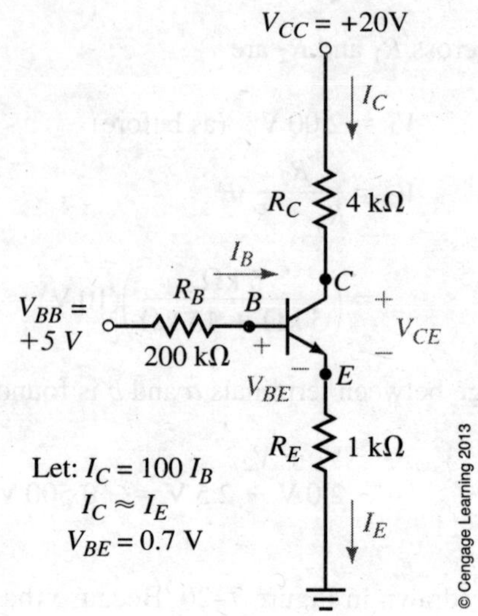

FIGURE 7–21

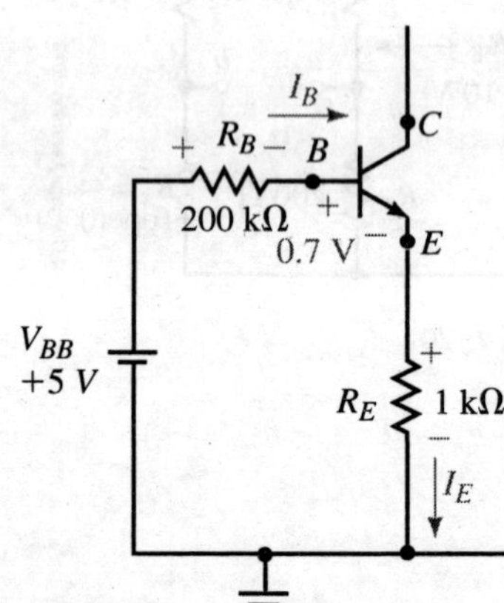

FIGURE 7–22

Solution In order to simplify the work, the circuit of Figure 7–21 is separated into two circuits: one circuit containing the known voltage V_{BE} and the other containing the unknown voltage V_{CE}.

Since we always start with the given information, we redraw the circuit containing the known voltage V_{BE} as illustrated in Figure 7–22.

Although the circuit of Figure 7–22 initially appears to be a series circuit, we see that this cannot be the case, since we are given that $I_E \cong I_C = 100 I_B$. We know that the current everywhere in a series circuit must be the same. However, Kirchhoff's voltage law still applies around the closed loop, resulting in the following:

$$V_{BB} = R_B I_B + V_{BE} + R_E I_E$$

The previous expression contains two unknowns, I_B and I_E (V_{BE} is given). From the given information we have the current $I_E \cong 100 I_B$, which allows us to write

$$V_{BB} = R_B I_B + V_{BE} + R_E(100 I_B)$$

Solving for the unknown current I_B, we have

$$5.0\text{ V} = (200\text{ k}\Omega)I_B + 0.7\text{ V} + (1\text{ k}\Omega)(100 I_B)$$

$$(300\text{ k}\Omega)I_B = 5.0\text{ V} - 0.7\text{ V} = 4.3\text{ V}$$

$$I_B = \frac{4.3\text{ V}}{300\text{ k}\Omega} = 14.3\ \mu\text{A}$$

The current $I_E \cong I_C = 100 I_B = 1.43$ mA.

As mentioned previously, the circuit can be redrawn as two separate circuits. The circuit containing the unknown voltage V_{CE} is illustrated in Figure 7–23. Notice that the resistor R_E appears in both Figure 7–22 and Figure 7–23. Applying Kirchhoff's voltage law around the closed loop of Figure 7–23, we have the following:

$$V_{R_C} + V_{CE} + V_{R_E} = V_{CC}$$

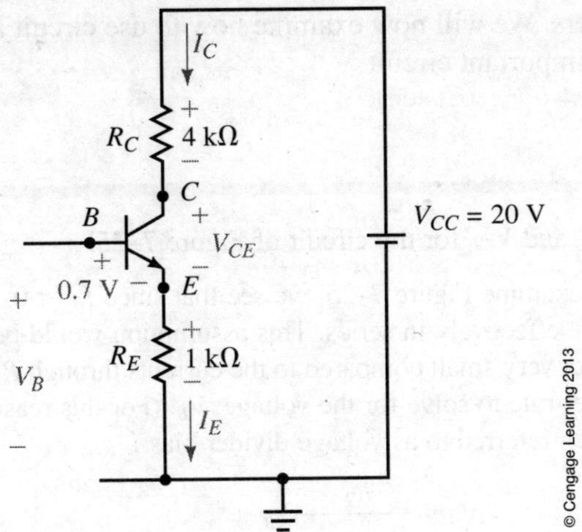

FIGURE 7–23

The voltage V_{CE} is found as

$$\begin{aligned} V_{CE} &= V_{CC} - V_{R_C} - V_{R_E} \\ &= V_{CC} - R_C I_C - R_E I_E \\ &= 20.0\text{ V} - (4\text{ k}\Omega)(1.43\text{ mA}) - (1\text{ k}\Omega)(1.43\text{ mA}) \\ &= 20.0\text{ V} - 5.73\text{ V} - 1.43\text{ V} = 12.8\text{ V} \end{aligned}$$

Finally, applying Kirchhoff's voltage law from B to ground, we have

$$\begin{aligned} V_B &= V_{BE} + V_{R_E} \\ &= 0.7\text{ V} + 1.43\text{ V} \\ &= 2.13\text{ V} \end{aligned}$$

PRACTICE PROBLEMS 3

Use the given information to find V_G, I_D, and V_{DS} for the circuit of Figure 7–24.

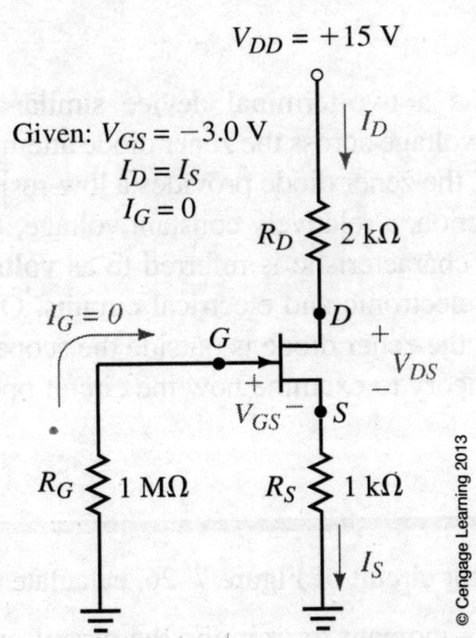

FIGURE 7–24

Answers
$V_G = 0$, $I_D = 3.00$ mA, $V_{DS} = 6.00$ V

The **universal bias** circuit is one of the most common transistor circuits used in amplifiers. We will now examine how to use circuit analysis principles to analyze this important circuit.

EXAMPLE 7–8

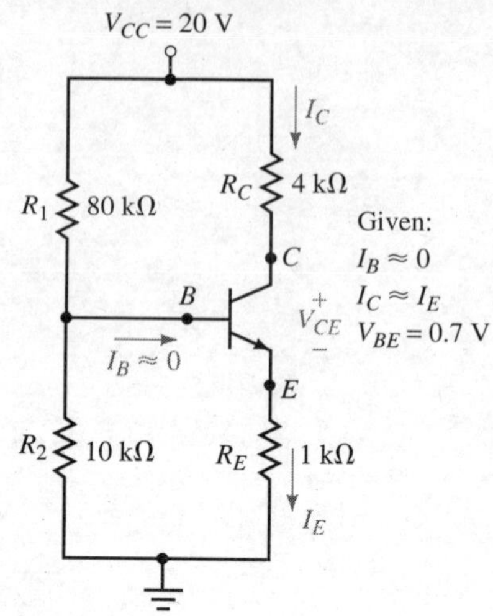

FIGURE 7–25

Determine the I_C and V_{CE} for the circuit of Figure 7–25.

Solution If we examine Figure 7–25, we see that since $I_B \approx 0$, we may assume that R_1 and R_2 are effectively in series. This assumption would be incorrect if the current I_B were not very small compared to the currents through R_1 and R_2. We use the voltage divider rule to solve for the voltage, V_B. (For this reason, the universal bias circuit is often referred to as voltage divider bias.)

$$V_B = \frac{R_2}{R_1 + R_2} V_{CC}$$
$$= \left[\frac{10\text{ k}\Omega}{80\text{ k}\Omega + 10\text{ k}\Omega}\right](20\text{ V})$$
$$= 2.22\text{ V}$$

Next, we use the value for V_B and Kirchhoff's voltage law to determine the voltage across R_E.

$$V_{RE} = 2.22\text{ V} - 0.7\text{ V} = 1.52\text{ V}$$

Applying Ohm's law, we now determine the current I_E.

$$I_E = \frac{1.52\text{ V}}{1\text{ k}\Omega} = 1.52\text{ mA} \cong I_C$$

Finally, applying Kirchhoff's voltage law and Ohm's law, we determine V_{CE} as follows:

$$V_{CC} = V_{RC} + V_{CE} + V_{RE}$$
$$V_{CE} = V_{CC} - V_{RC} - V_{RE}$$
$$= 20\text{ V} - (1.52\text{ mA})(4\text{ k}\Omega) - (1.52\text{ mA})(1\text{ k}\Omega)$$
$$= 12.4\text{ V}$$

A **zener diode** is a two-terminal device similar to a varistor (refer to Chapter 3). When the voltage across the zener diode attempts to go above the rated voltage for the device, the zener diode provides a low-resistance path for the extra current. Due to this action, a relatively constant voltage, V_Z, is maintained across the zener diode. This characteristic is referred to as **voltage regulation** and has many applications in electronic and electrical circuits. Once again, although the theory of operation of the zener diode is outside the scope of this chapter, we can apply simple circuit theory to examine how the circuit operates.

EXAMPLE 7–9

For the voltage regulator circuit of Figure 7–26, calculate I_Z, I_1, I_2, and P_Z.

Solution If we take a moment to examine the circuit, we see that the zener diode is placed in parallel with the resistor R_2. This parallel combination is in series with the resistor R_1 and the voltage source, E.

In order for the zener diode to operate as a regulator, the voltage across the diode would have to be above the zener voltage without the diode present. If we removed the zener diode, the circuit would appear as shown in Figure 7–27.

From Figure 7–27, we may determine the voltage V_2 which would be present without the zener diode in the circuit. Since the circuit is a simple series circuit, the voltage divider rule may be used to determine V_2:

$$V_2 = \frac{R_2}{R_1 + R_2}E = \left(\frac{10\ \text{k}\Omega}{5\ \text{k}\Omega + 10\ \text{k}\Omega}\right)(15\ \text{V}) = 10.0\ \text{V}$$

When the zener diode is placed across the resistor R_2 the device will operate to limit the voltage to $V_Z = 5$ V.

Because the zener diode is operating as a voltage regulator, the voltage across both the diode and the resistor R_2 must be the same, namely 5 V. The parallel combination of D_Z and R_2 is in series with the resistor R_1, and so the voltage across R_1 is easily determined from Kirchhoff's voltage law as

$$V_1 = E - V_Z = 15\ \text{V} - 5\ \text{V} = 10\ \text{V}$$

Now, from Ohm's law, the currents I_1 and I_2 are easily found to be

$$I_2 = \frac{V_2}{R_2} = \frac{5\ \text{V}}{10\ \text{k}\Omega} = 0.5\ \text{mA}$$

$$I_1 = \frac{V_1}{R_1} = \frac{10\ \text{V}}{5\ \text{k}\Omega} = 2.0\ \text{mA}$$

Applying Kirchoff's current law at node a, we get the zener diode current as

$$I_Z = I_1 - I_2 = 2.0\ \text{mA} - 0.5\ \text{mA} = 1.5\ \text{mA}$$

Finally, the power dissipated by the zener diode must be

$$P_Z = V_Z I_Z = (5\ \text{V})(1.5\ \text{mA}) = 7.5\ \text{mW}$$

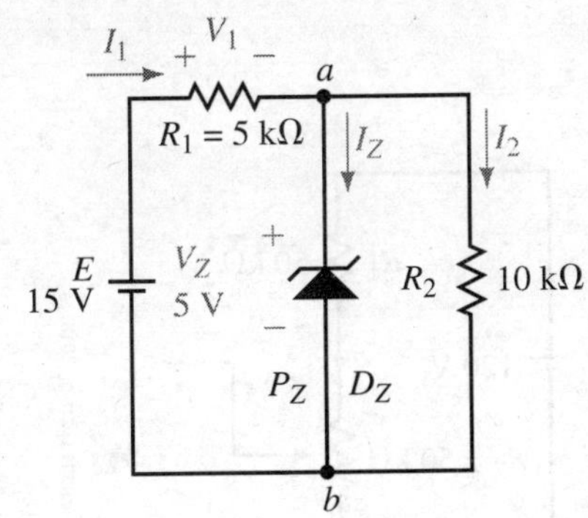

FIGURE 7–26

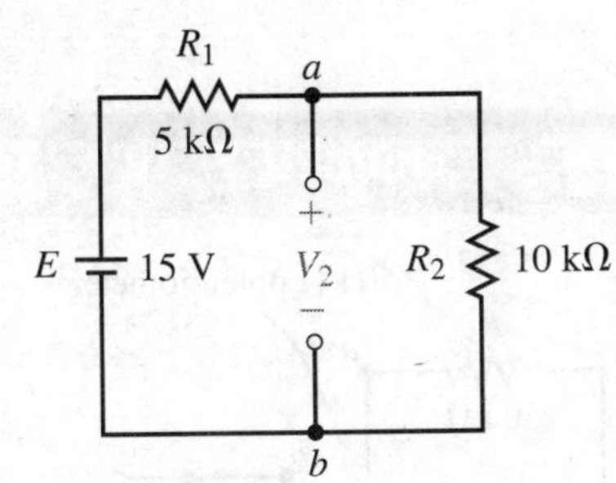

FIGURE 7–27

IN-PROCESS LEARNING CHECK 2

(Answers are at the end of the chapter.)

Use the results of Example 7–9 to show that the power delivered to the circuit by the voltage source of Figure 7–26, is equal to the total power dissipated by the resistors and the zener diode.

PRACTICE PROBLEMS 4

1. Determine I_1, I_Z, and I_2 for the circuit of Figure 7–26 if the resistor R_1 is increased to 10 kΩ.
2. Repeat Problem 1 if R_1 is increased to 30 kΩ.

Answers

1. $I_1 = 1.00$ mA, $I_2 = 0.500$ mA, $I_Z = 0.500$ mA
2. $I_1 = I_2 = 0.375$ mA, $I_Z = 0$ mA. (The voltage across the zener diode is not sufficient for the device to come on.)

7.4 Potentiometers

As mentioned in Chapter 3, variable resistors may be used as potentiometers as shown in Figure 7–28 to control voltage into another circuit.

The volume control on a receiver or amplifier is an example of a variable resistor used as a potentiometer. When the movable terminal is at the uppermost

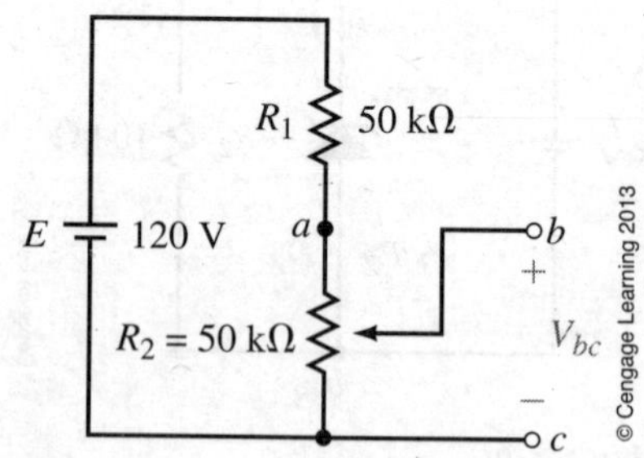

FIGURE 7–28

position, the voltage appearing between terminals b and c is simply calculated by using the voltage divider rule as

$$V_{bc} = \left(\frac{50\text{ k}\Omega}{50\text{ k}\Omega + 50\text{ k}\Omega}\right)(120\text{ V}) = 60\text{ V}$$

Alternatively, when the movable terminal is at the lowermost position, the voltage between terminals b and c is $V_{bc} = 0$ V, since the two terminals are effectively shorted and the voltage across a short circuit is always zero.

The circuit of Figure 7–28 represents a potentiometer having an output voltage which is adjustable between 0 V and 60 V. This output is referred to as the **unloaded output,** since there is no load resistance connected between the terminals b and c. If a load resistance were connected between these terminals, the output voltage, called the **loaded output,** would no longer be the same. The following example is an illustration of circuit loading.

EXAMPLE 7–10

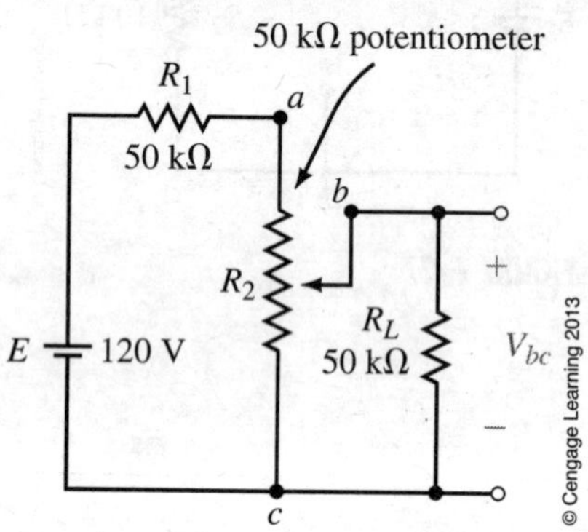

FIGURE 7–29

For the circuit of Figure 7–29, determine the range of the voltage V_{bc} as the potentiometer varies between its minimum and maximum values.

Solution The minimum voltage between terminals b and c will occur when the movable contact is at the lowermost contact of the variable resistor. In this position, the voltage $V_{bc} = 0$ V, since the terminals b and c are shorted.

The maximum voltage V_{bc} occurs when the movable contact is at the uppermost contact of the variable resistor. In this position, the circuit may be represented as shown in Figure 7–30.

In Figure 7–30, we see that the resistance R_2 is in parallel with the load resistor R_L. The voltage between terminals b and c is easily determined from the voltage divider rule, as follows:

$$V_{bc} = \frac{R_2\|R_L}{(R_2\|R_L) + R_1}E$$

$$= \left(\frac{25\text{ k}\Omega}{25\text{ k}\Omega + 50\text{ k}\Omega}\right)(120\text{ V}) = 40\text{ V}$$

We conclude that the voltage at the output of the potentiometer is adjustable from 0 V to 40 V for a load resistance of $R_L = 50$ kΩ.

By inspection, we see that an unloaded potentiometer in the circuit of Figure 7–29 would have an output voltage of 0 V to 60 V.

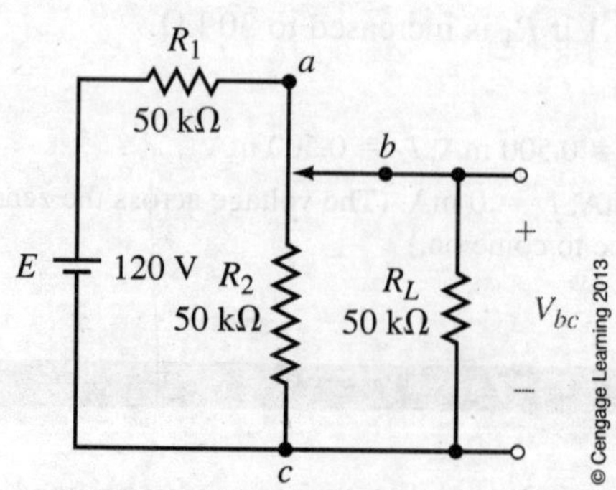

FIGURE 7–30

PRACTICE PROBLEMS 5

Refer to the circuit of Figure 7–29.

a. Determine the output voltage range of the potentiometer if the load resistor is $R_L = 5\ \text{k}\Omega$.
b. Repeat (a) if the load resistor is $R_L = 500\ \text{k}\Omega$.
c. What conclusion may be made about the output voltage of a potentiometer when the load resistance is large in comparison with the potentiometer resistance?

Answers

a. 0 V to 10 V
b. 0 V to 57.1 V
c. When R_L is large in comparison to the potentiometer resistance, the output voltage will better approximate the unloaded voltage. (In this example, the unloaded voltage is 0 V to 60 V.)

IN-PROCESS LEARNING CHECK 3

(Answers are at the end of the chapter.)

A 20-kΩ potentiometer is connected across a voltage source with a 2-kΩ load resistor connected between the wiper (center terminal) and the negative terminal of the voltage source.

a. What percentage of the source voltage will appear across the load when the wiper is one-fourth of the way from the bottom?
b. Determine the percentage of the source voltage which appears across the load when the wiper is one-half and three-fourths of the way from the bottom.
c. Repeat the calculations of (a) and (b) for a load resistor of 200 kΩ.
d. From the preceding results, what conclusion can you make about the effect of placing a large load across a potentiometer?

7.5 Loading Effects of Instruments

In Chapters 5 and 6 we examined how ammeters and voltmeters affect the operation of simple series circuits. The degree to which the circuits are affected is called the **loading effect** of the instrument. Recall that, in order for an instrument to provide an accurate indication of how a circuit operates, the loading effect should ideally be zero. In practice, it is impossible for any instrument to have zero loading effect, since all instruments absorb some energy from the circuit under test, thereby affecting circuit operation.

In this section, we will determine how instrument loading affects more complex circuits.

EXAMPLE 7–11

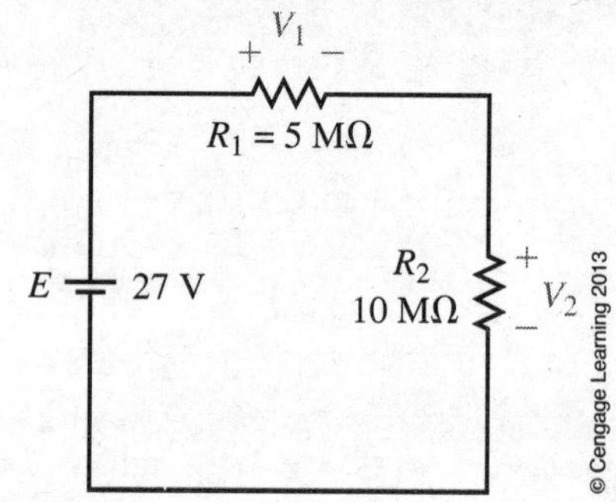

FIGURE 7–31

CircuitSim 07-4

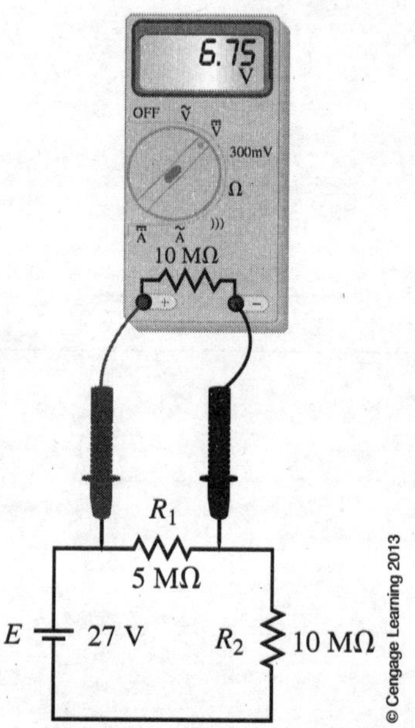

FIGURE 7–32

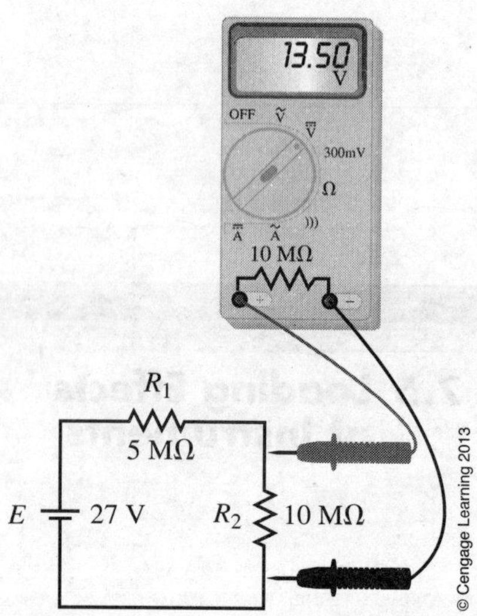

FIGURE 7–33

Calculate the loading effects if a digital multimeter, having an internal resistance of 10 MΩ, is used to measure V_1 and V_2 in the circuit of Figure 7–31.

Solution In order to determine the loading effect for a particular reading, we need to calculate both the unloaded voltage and the loaded voltage.

For the circuit given in Figure 7–31, the unloaded voltage across each resistor is

$$V_1 = \left(\frac{5\ \text{M}\Omega}{5\ \text{M}\Omega + 10\ \text{M}\Omega}\right)(27\ \text{V}) = 9.0\ \text{V}$$

$$V_2 = \left(\frac{10\ \text{M}\Omega}{5\ \text{M}\Omega + 10\ \text{M}\Omega}\right)(27\ \text{V}) = 18.0\ \text{V}$$

When the voltmeter is used to measure V_1, the result is equivalent to connecting a 10-MΩ resistor across resistor R_1, as shown in Figure 7–32.

The voltage appearing across the parallel combination of R_1 and resistance of the voltmeter is calculated as

$$V_1 = \left(\frac{5\ \text{M}\Omega \| 10\ \text{M}\Omega}{(5\ \text{M}\Omega \| 10\ \text{M}\Omega) + 10\ \text{M}\Omega}\right)(27\ \text{V})$$

$$= \left(\frac{3.33\ \text{M}\Omega}{13.3\ \text{M}\Omega}\right)(27\ \text{V})$$

$$= 6.75\ \text{V}$$

Notice that the measured voltage is significantly less than the 9 V that we had expected to measure.

With the voltmeter connected across resistor R_2, the circuit appears as shown in Figure 7–33.

The voltage appearing across the parallel combination of R_2 and resistance of the voltmeter is calculated as

$$V_2 = \left(\frac{10\ \text{M}\Omega \| 10\ \text{M}\Omega}{5\ \text{M}\Omega + (10\ \text{M}\Omega \| 10\ \text{M}\Omega)}\right)(27\ \text{V})$$

$$= \left(\frac{5.0\ \text{M}\Omega}{10.0\ \text{M}\Omega}\right)(27\ \text{V})$$

$$= 13.5\ \text{V}$$

Again, we notice that the measured voltage is quite a bit less than the 18 V that we had expected.

Now the loading effects are calculated as follows:

When measuring V_1:

$$\text{loading effect} = \frac{9.0\ \text{V} - 6.75\ \text{V}}{9.0\ \text{V}} \times 100\%$$

$$= 25\%$$

When measuring V_2:

$$\text{loading effect} = \frac{18.0\ \text{V} - 13.5\ \text{V}}{18.0\ \text{V}} \times 100\%$$

$$= 25\%$$

This example clearly illustrates a problem that novices often make when they are taking voltage measurements in high-resistance circuits. If the measured voltages $V_1 = 6.75$ V and $V_2 = 13.50$ V are used to verify Kirchhoff's voltage law, the novice would say that this represents a contradiction of the law (since $6.75 \text{ V} + 13.50 \text{ V} \neq 27.0 \text{ V}$). In fact, we see that the circuit is behaving exactly as predicted by circuit theory. The problem occurs when instrument limitations are not considered.

PRACTICAL NOTES...

Whenever an instrument is used to measure a quantity, the operator must always consider the loading effects of the instrument.

PRACTICE PROBLEMS 6

Calculate the loading effects if an analog voltmeter, having an internal resistance of 200 kΩ, is used to measure V_1 and V_2 in the circuit of Figure 7–31.

Answers

V_1: loading effect = 94.3%

V_2: loading effect = 94.3%

EXAMPLE 7–12

For the circuit of Figure 7–34, calculate the loading effect if a 5.00-Ω ammeter is used to measure the currents I_T, I_1, and I_2.

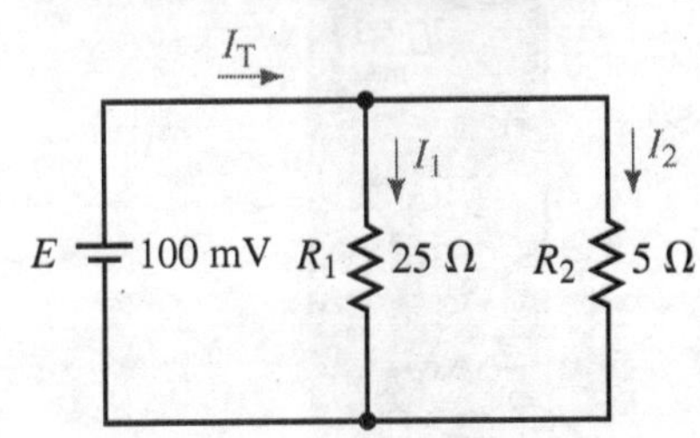

FIGURE 7–34

Solution We begin by determining the unloaded currents in the circuit. Using Ohm's law, we solve for the currents I_1 and I_2:

$$I_1 = \frac{100 \text{ mV}}{25 \ \Omega} = 4.0 \text{ mA}$$

$$I_2 = \frac{100 \text{ mV}}{5 \ \Omega} = 20.0 \text{ mA}$$

Now, by Kirchhoff's current law,

$$I_T = 4.0 \text{ mA} + 20.0 \text{ mA} = 24.0 \text{ mA}$$

If we were to insert the ammeter into the branch with resistor R_1, the circuit would appear as shown in Figure 7–35.

The current through the ammeter would be

$$I_1 = \frac{100 \text{ mV}}{25 \ \Omega + 5 \ \Omega} = 3.33 \text{ mA}$$

CircuitSim 07-5

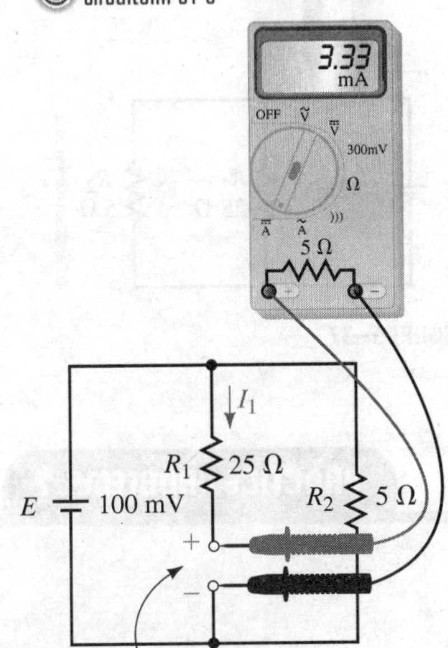

Circuit must be broken in order to measure current.

FIGURE 7–35

If we were to insert the ammeter into the branch with resistor R_2, the circuit would appear as shown in Figure 7–36.

The current through the ammeter would be

$$I_2 = \frac{100 \text{ mV}}{5 \ \Omega + 5 \ \Omega} = 10.0 \text{ mA}$$

If the ammeter were inserted into the circuit to measure the current I_T, the equivalent circuit would appear as shown in Figure 7–37.

The total resistance of the circuit would be

$$R_T = 5 \ \Omega + 25 \ \Omega \| 5 \ \Omega = 9.17 \ \Omega$$

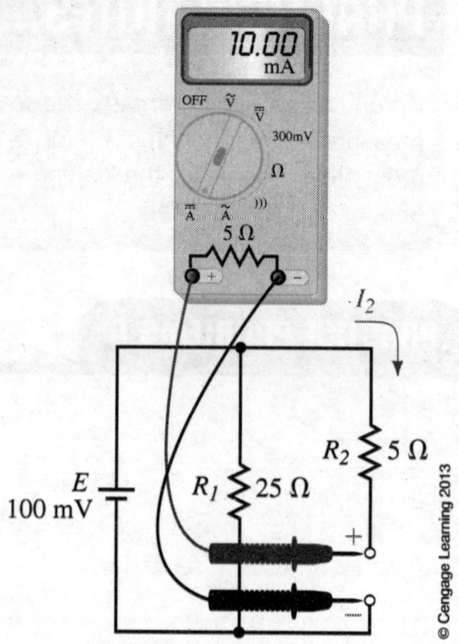

FIGURE 7–36

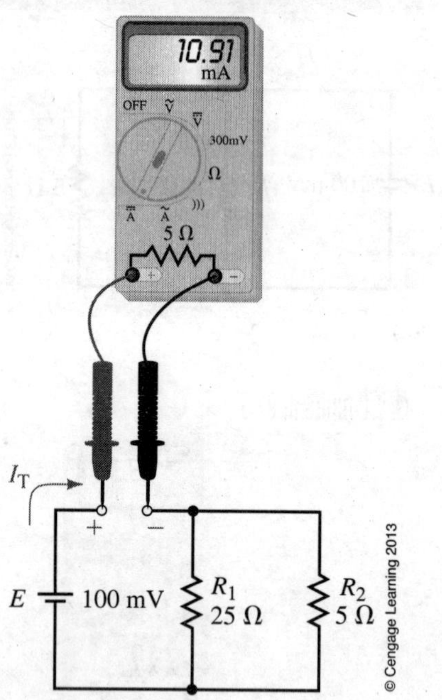

FIGURE 7–37

This would result in a current I_T determined by Ohm's law as

$$I_T = \frac{100 \text{ mV}}{9.17 \ \Omega} = 10.9 \text{ mA}$$

The loading effects for the various current measurements are as follows: When measuring I_1,

$$\text{loading effect} = \frac{4.0 \text{ mA} - 3.33 \text{ mA}}{4.0 \text{ mA}} \times 100\% = 16.7\%$$

When measuring I_2,

$$\text{loading effect} = \frac{20 \text{ mA} - 10 \text{ mA}}{20 \text{ mA}} \times 100\% = 50\%$$

When measuring I_T,

$$\text{loading effect} = \frac{24 \text{ mA} - 10.9 \text{ mA}}{24 \text{ mA}} \times 100\% = 54.5\%$$

Notice that the loading effect for an ammeter is most pronounced when it is used to measure current in a branch having a resistance in the same order of magnitude as the meter.

You will also notice that if this ammeter were used in a circuit to verify the correctness of Kirchhoff's current law (KCL), the loading effect of the meter would produce an apparent contradiction. From KCL,

$$I_T = I_1 + I_2$$

By substituting the measured values of current into the preceding equation, we have

$$10.91 \text{ mA} = 3.33 \text{ mA} + 10.0 \text{ mA}$$
$$10.91 \text{ mA} \neq 13.33 \text{ mA (contradiction)}$$

This example illustrates that the loading effect of a meter may severely affect the current in a circuit, giving results that seem to contradict the laws of circuit theory. Therefore, whenever an instrument is used to measure a particular quantity, we must always take into account the limitations of the instrument and question the validity of a resulting reading.

PRACTICE PROBLEMS 7

Calculate the readings and the loading error if an ammeter having an internal resistance of 1 Ω is used to measure the currents in the circuit of Figure 7–34.

Answers

$I_{T(RDG)}$ = 19.4 mA; loading error = 19.4%

$I_{1(RDG)}$ = 3.85 mA; loading error = 3.85%

$I_{T(RDG)}$ = 16.7 mA; loading error = 16.7%

7.6 Circuit Analysis Using Computers

Multisim

PSPICE

Multisim

The analysis of series-parallel circuits using Multisim is almost identical to the methods used in analyzing series and parallel circuits in previous chapters. The following example illustrates that Multisim produces the same solutions as those obtained in Example 7–4.

EXAMPLE 7–13

Given the circuit of Figure 7–38, use Multisim to find the following quantities:

a. Total resistance, R_T
b. Voltages V_2 and V_4
c. Currents I_T, I_1, and I_2

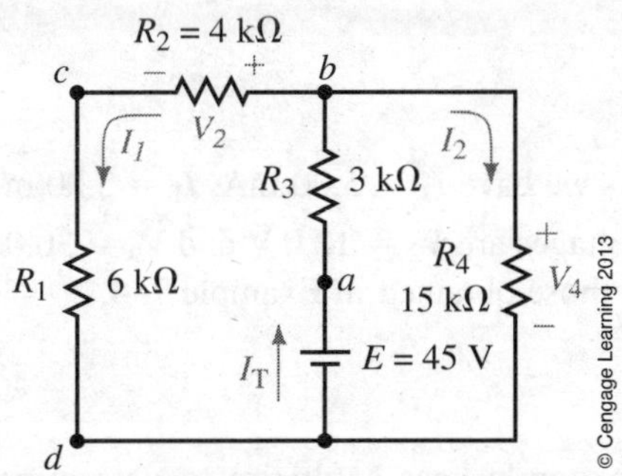

FIGURE 7–38

CircuitSim 07-6

Solution

a. We begin by constructing the circuit as shown in Figure 7–39. This circuit is identical to that shown in Figure 7–38, except that the voltage source has been omitted and a multimeter (from the Instruments button on the Parts bin toolbar) inserted in its place. The ohmmeter function is then selected and the power switch turned on. The resistance is found to be $R_T = 9.00\ \text{k}\Omega$.

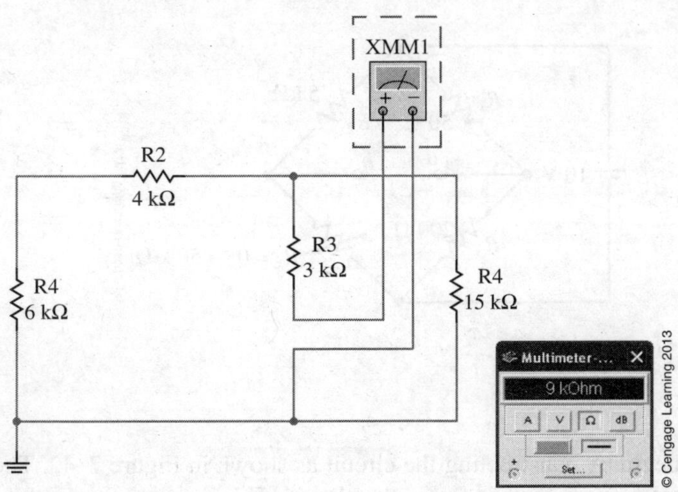

FIGURE 7–39

b. Next, we remove the multimeter and insert the 45-V source. Ammeters and voltmeters are inserted, as shown in Figure 7–40.

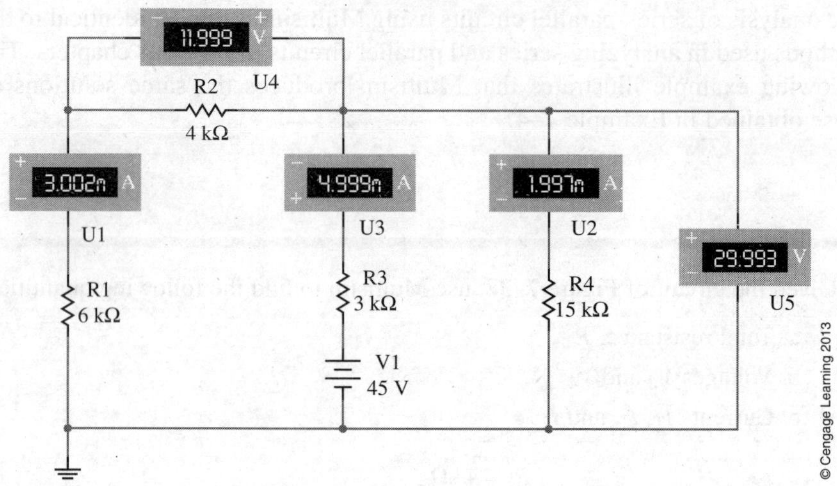

FIGURE 7–40

From the results we have $I_T = 5.00$ mA, $I_1 = 3.00$ mA, $I_2 = 2.00$ mA.

c. The required voltages are $V_2 = 12.0$ V and $V_4 = 30.0$ V. These results are consistent with those obtained in Example 7–4.

The following example uses Multisim to determine the voltage across a bridge circuit. The example uses a potentiometer to provide a variable resistance in the circuit. Multisim can provide a display of voltage (on a multimeter) as the resistance is changed.

EXAMPLE 7–14

Given the circuit of Figure 7–41, use Multisim to determine the values of I and V_{ab} when $R_x = 0\ \Omega$, 15 kΩ, and 50 kΩ.

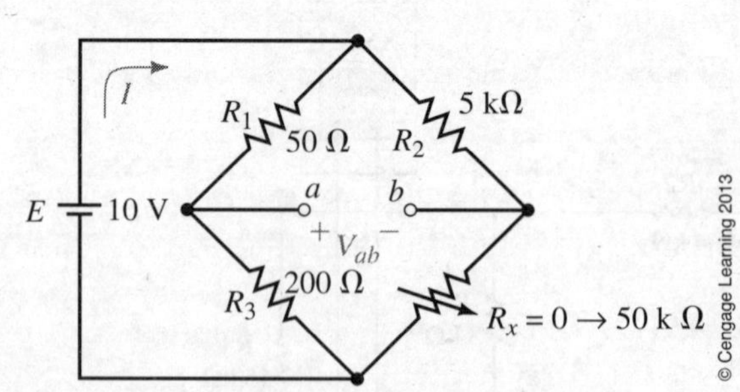

FIGURE 7–41

Solution

1. We begin by constructing the circuit as shown in Figure 7–42. The potentiometer is selected from the Basic Family group on the Virtual Components toolbar. Ensure that the potentiometer is inserted as illustrated.

CircuitSim 07-7

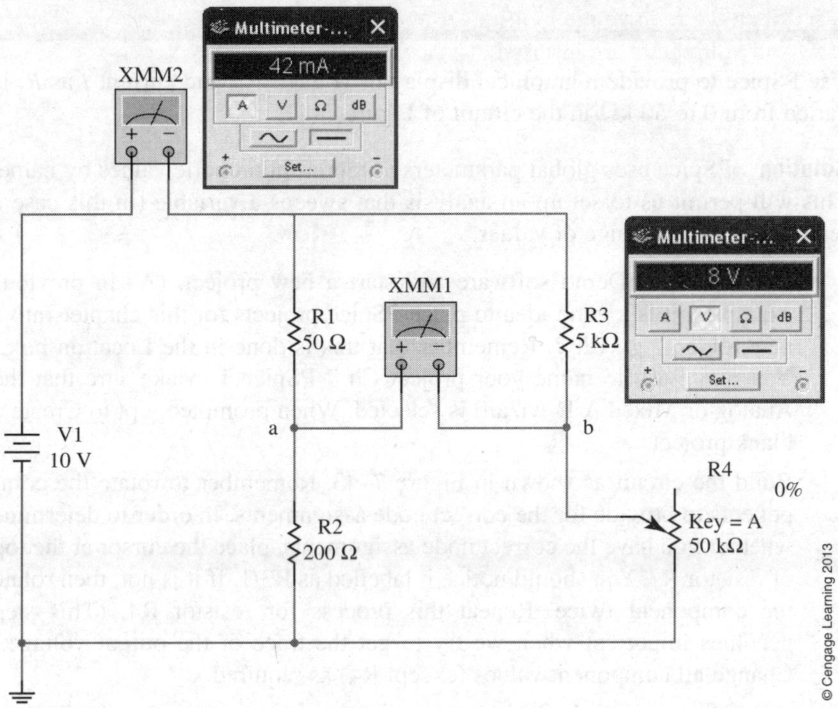

FIGURE 7–42

2. Double-click on the potentiometer symbol and change its value to 50 kΩ. Notice that the increment value is set for 5%. Adjust this value so that it is at 10%. We will use this in a following step.
3. Once the circuit is completely built, the power switch is turned on. Notice that the resistor value is at 50%. This means that the potentiometer is adjusted so that its value is 25 kΩ. The following step will change the value of the potentiometer to 0 Ω.
4. The value of the potentiometer is now easily changed by either Shift-**A** (to incrementally decrease the value) or **A** (to incrementally increase the value). As you change the value of the potentiometer, you should observe that the voltage displayed on the multimeter is also changing. It takes a few seconds for the display to stabilize. As shown in Figure 7–42, you should observe that $I = 42.0$ mA and $V_{ab} = 8.00$ V when $R_x = 0\ \Omega$.
5. Finally, after adjusting the value of the potentiometer, we obtain the following readings:

$$R_x = 15\text{ k}\Omega\ (30\%):\quad I = 40.5\text{ mA and } V_{ab} = 0.500\text{ V}$$

$$R_x = 50\text{ k}\Omega\ (100\%):\quad I = 40.18\text{ mA and } V_{ab} = -1.091\text{ V}$$

PSpice

PSpice is somewhat different from Multisim in the way it handles potentiometers. In order to place a variable resistor into a circuit, we must set the parameters of the circuit to sweep through a range of values. The following example illustrates the method used to provide a graphical display of output voltage and source current for a range of resistance. Although the method is different, the results are consistent with those of the previous example.

EXAMPLE 7–15

Use PSpice to provide a graphical display of voltage V_{ab} and current I as R_x is varied from 0 to 50 kΩ in the circuit of Figure 7–41.

Solution PSpice uses global parameters to represent numeric values by name. This will permit us to set up an analysis that sweeps a variable (in this case a resistor) through a range of values.

- Open the CIS Demo software and start a new project. (As in previous examples, it is a good idea to place PSpice projects for this chapter into a new folder, e.g., Ch 7. Remember that this is done in the Location box.) You may wish to name your project Ch 7 PSpice 1. Make sure that the Analog or Mixed A/D wizard is selected. When prompted, opt to Create a blank project.
- Build the circuit as shown in Figure 7–43. Remember to rotate the components to provide for the correct node assignments. In order to determine whether you have the correct node assignments, place the cursor at the top of resistor R3. You should notice it labelled as R3/1. If it is not, then rotate the component twice. Repeat this process for resistor R4. (This step becomes important when we try to get the trace of the output voltage.) Change all component values (except R4) as required.

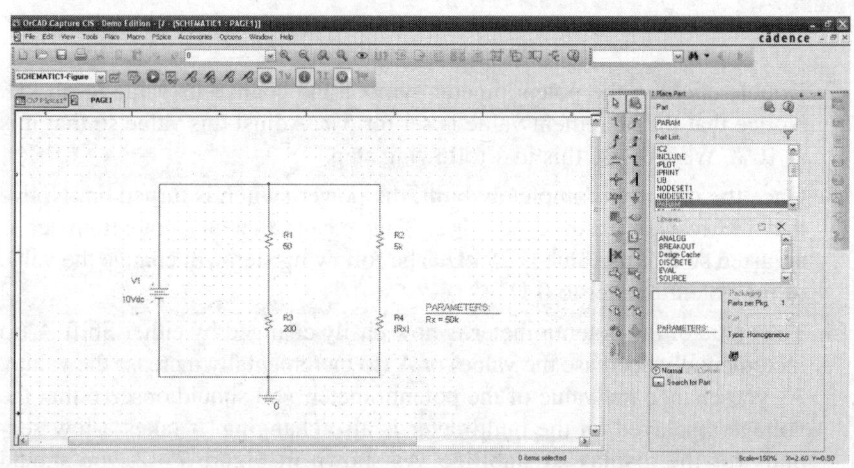

FIGURE 7–43

- Double-click on the component value for R4. Enter **{Rx}** in the *Value* text box of the *Display Properties* for this resistor. The curly braces tell PSpice to evaluate the parameter and use its value.
- Click on the *Place part* tool. Select the Special library and click on the *PARAM* part. Place the PARAM part adjacent to resistor R4.
- Double-click on *PARAMETERS*. Click on the *New Column* button. A dialog box will appear. Enter **Rx** as the name of the new column and assign a value of **50k** as the default value of this resistor. Click on *Apply*. Since we are adding only one column to this parameter, exit by clicking on Cancel. Highlight the column that you just created, click on the *Display* button, and activate the *Name* and *Value* option in the *Display* properties.

Exit the Property Editor. You should observe the circuit as it appears in Figure 7–43.

- Click on the *New Simulation* icon and give the simulation a name such as **Figure 7–43.**
- In the *Simulation Settings,* select the *Analysis Tab*. The *Analysis type* is DC Sweep. Select *Primary Sweep* from the *Options* listing. In the Sweep variable box select *Global Parameter*. Type **Rx** in the *Parameter name* text box. Select the Sweep type to be linear and set the limits as follows:

  Start value: **100**

  End value: **50k**

  Increment: **100.**

  These settings will change the resistor value from 100 Ω to 50 kΩ in 100-Ω increments. Click *OK*.
- Click on the *Run* icon. You will see a blank screen with the abscissa (horizontal axis) showing R_x scaled from 0 to 50 kΩ.
- PSpice is able to plot most circuit variables as a function of R_x. In order to request a plot of V_{ab}, click on *Add Trace*. Enter **V(R3:1) – V(R4:1)** in the *Trace Expression* text box. The voltage V_{ab} is the voltage between node 1 of R_3 and node 1 of R_4. Click *OK*.
- Finally, to obtain a plot of the circuit current *I* (current through the voltage source), we need to first add an extra axis. Click on *Plot* and then click *Add Y Axis*. To obtain a plot of the current, click on *Trace* and *Add Trace*. Select **I(V1).** The resulting display on the monitor is shown in Figure 7–44.

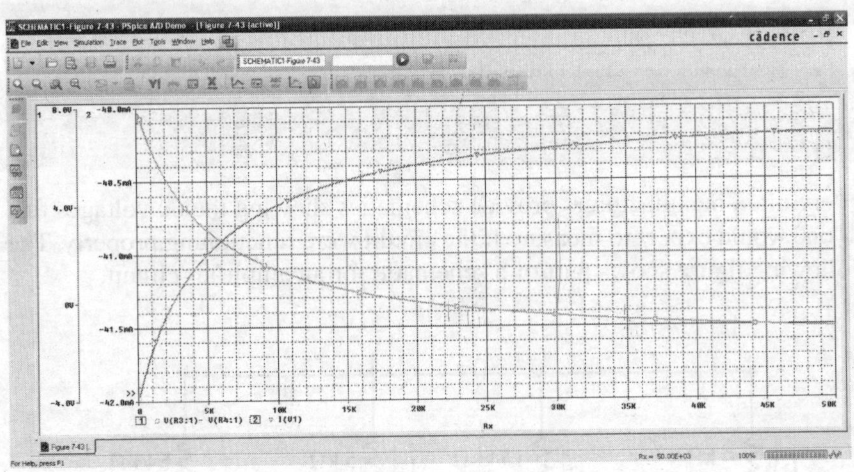

FIGURE 7–44

Notice that the current shown is negative. This is because PSpice sets the reference direction through a voltage source from the positive terminal to the negative terminal. One way of removing the negative sign is to request the current as −I(V1).

PRACTICE PROBLEMS 8

Given the circuit of Figure 7–45, use Multisim to solve for V_{ab}, I, and I_L when $R_L = 100\ \Omega$, $500\ \Omega$, and $1000\ \Omega$.

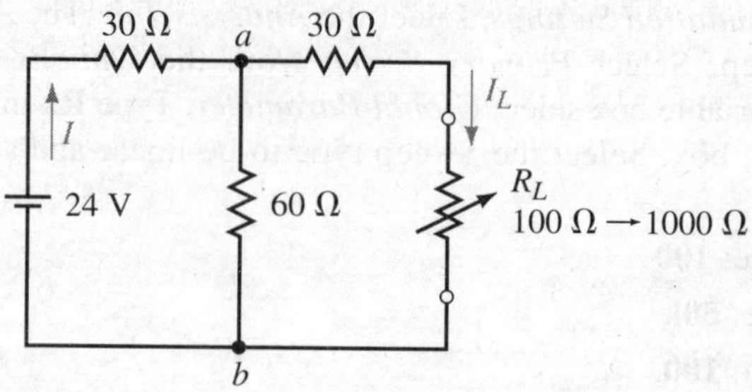

FIGURE 7–45

Answers

$R_L = 100\ \Omega$: $I = 169$ mA, $V_{ab} = 6.93$ V, $I_L = 53.3$ mA

$R_L = 500\ \Omega$: $I = 143$ mA, $V_{ab} = 7.71$ V, $I_L = 14.5$ mA

$R_L = 1000\ \Omega$: $I = 138$ mA, $V_{ab} = 7.85$ V, $I_L = 7.62$ mA

PRACTICE PROBLEMS 9

Use PSpice to input a file for the circuit of Figure 7–45. The output shall display the source current, I, the load current, I_L, and the voltage V_{ab} as the resistor R_L is varied in 100-Ω increments between 100 Ω and 1000 Ω.

Putting It into Practice

Very often manufacturers provide schematics showing the dc voltages that one would expect to measure if the circuit were functioning properly. The following figure shows part of a schematic for an amplifier circuit.

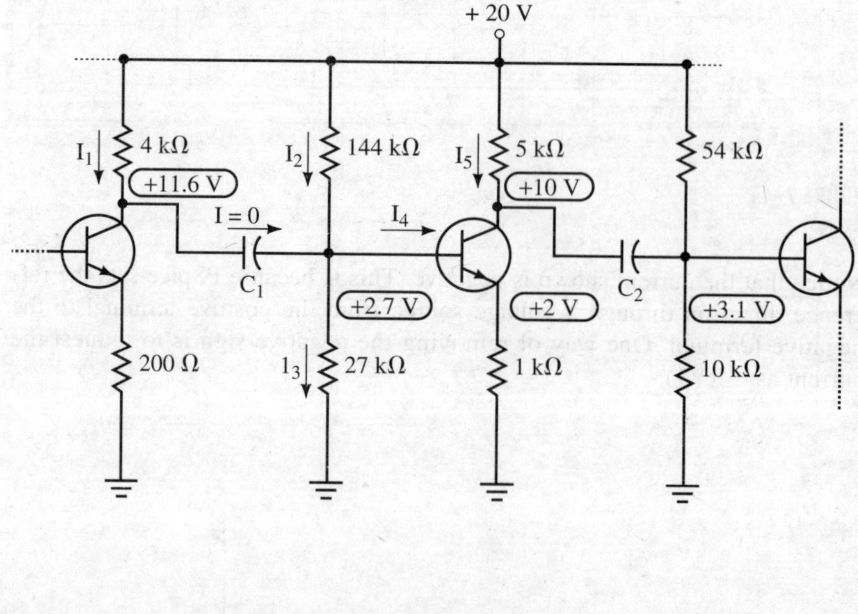

Even though a schematic may include components with which the reader is not familiar, the dc voltages provided on the schematic enable us to determine voltages and current in various parts of the circuit. If the circuit is faulty, measured voltages and currents will be different from the theoretical, allowing the experienced troubleshooter to locate the fault.

Examine the circuit shown. Use the voltage information on the schematic to determine the theoretical values of the currents I_1, I_2, I_3, I_4, and I_5. Find the magnitude and correct polarity of the voltage across the device labelled as C_2. (It is a capacitor, and it will be examined in detail in Chapter 10.)

Problems

7.1 The Series-Parallel Network

1. For the networks of Figure 7–46, determine which resistors and branches are in series and which are in parallel. Write an expression for the total resistance, R_T.
2. For each of the networks of Figure 7–47, write an expression for the total resistance, R_T.

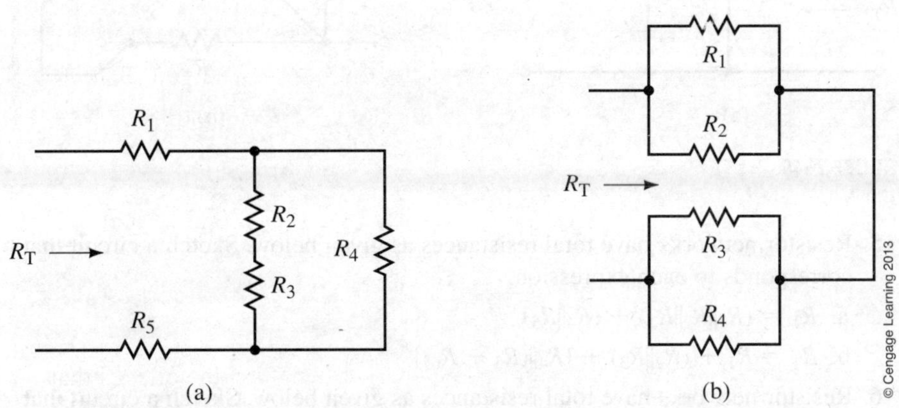

FIGURE 7–46

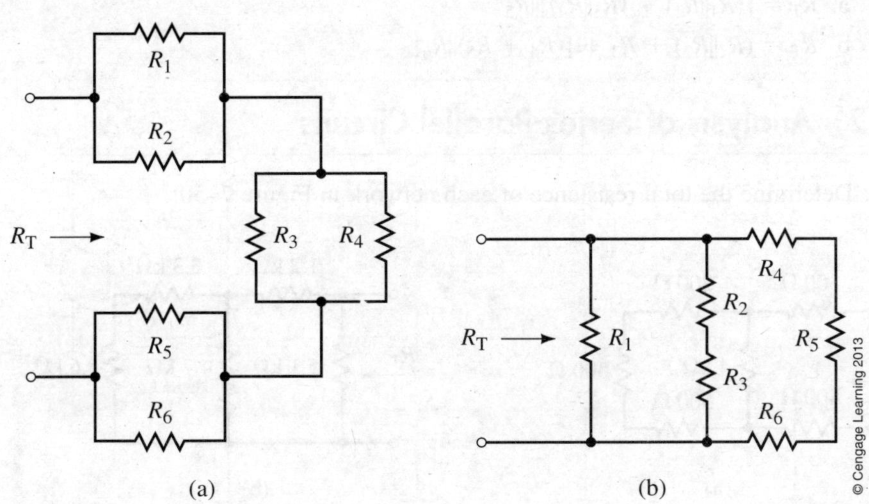

FIGURE 7–47

3. Write an expression for both R_{T_1} and R_{T_2} for the networks of Figure 7–48.
4. Write an expression for both R_{T_1} and R_{T_2} for the networks of Figure 7–49.

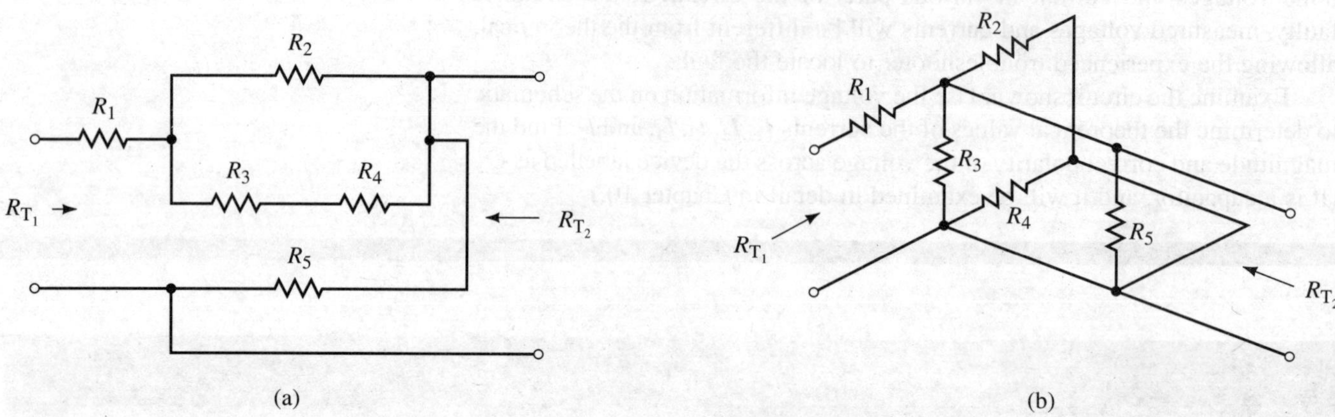

FIGURE 7–48

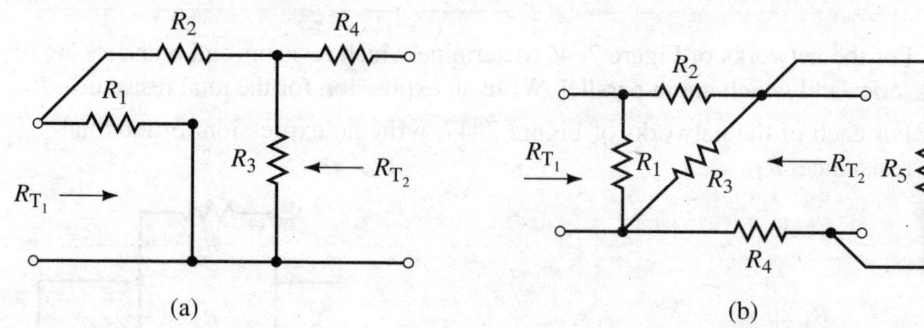

FIGURE 7–49

5. Resistor networks have total resistances as given below. Sketch a circuit that corresponds to each expression.
   a. $R_T = (R_1\|R_2\|R_3) + (R_4\|R_5)$
   b. $R_T = R_1 + (R_2\|R_3) + [R_4\|(R_5 + R_6)]$
6. Resistor networks have total resistances as given below. Sketch a circuit that corresponds to each expression.
   a. $R_T = [(R_1\|R_2) + (R_3\|R_4)]\|R_5$
   b. $R_T = (R_1\|R_2) + R_3 + [(R_4 + R_5)\|R_6]$

7.2 Analysis of Series-Parallel Circuits

7. Determine the total resistance of each network in Figure 7–50.

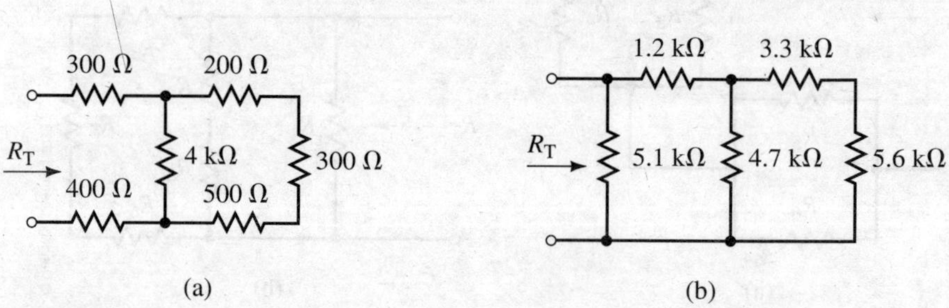

FIGURE 7–50

8. Determine the total resistance of each network in Figure 7–51.

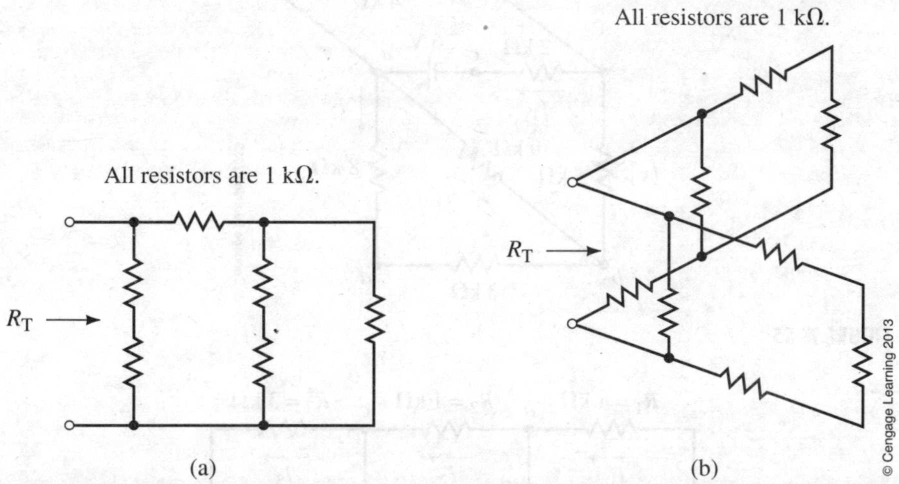

FIGURE 7–51

9. Calculate the resistances R_{ab} and R_{cd} in the circuit of Figure 7–52.
10. Calculate the resistances R_{ab} and R_{bc} in the circuit of Figure 7–53.

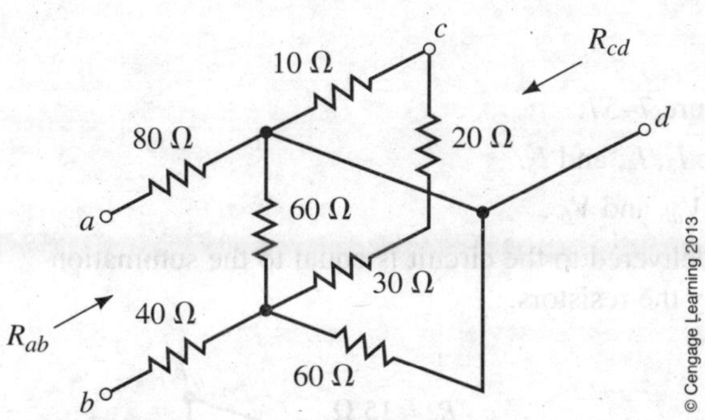

FIGURE 7–52

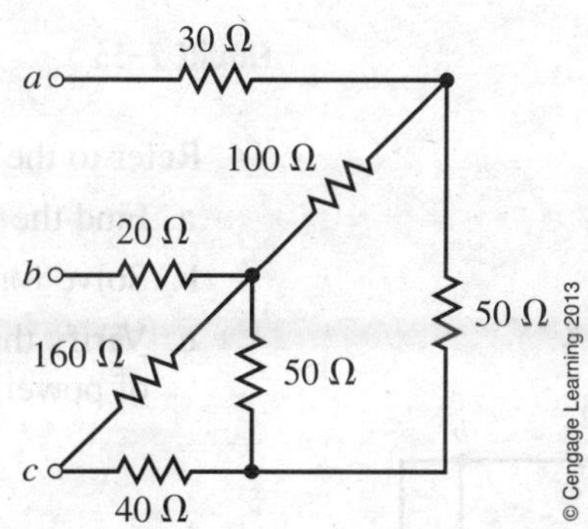

FIGURE 7–53

11. Refer to the circuit of Figure 7–54.
 Find the following quantities:
 a. R_T
 b. I_T, I_1, I_2, I_3, I_4
 c. V_{ab}, V_{bc}
12. Refer to the circuit of Figure 7–55.
 Find the following quantities:
 a. R_T (equivalent resistance "seen" by the voltage source)
 b. I_T, I_1, I_2, I_3, I_4
 c. V_{ab}, V_{bc}, V_{cd}
13. Refer to the circuit of Figure 7–56.
 a. Find the currents I_1, I_2, I_3, I_4, I_5, and I_6.
 b. Solve for the voltages V_{ab} and V_{cd}.
 c. Verify that the power delivered to the circuit is equal to the summation of powers dissipated by the resistors.

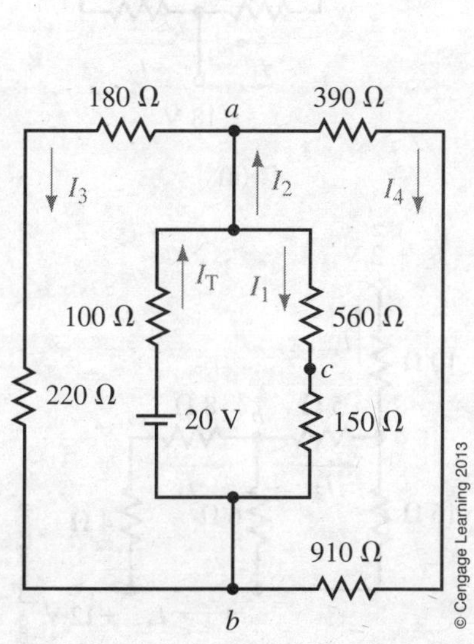

FIGURE 7–54

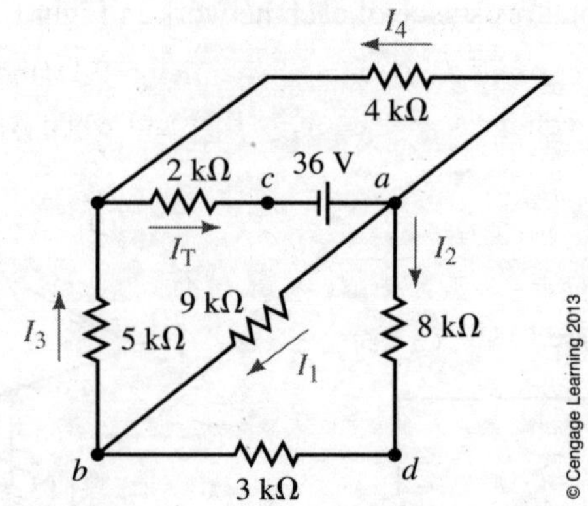

FIGURE 7–55

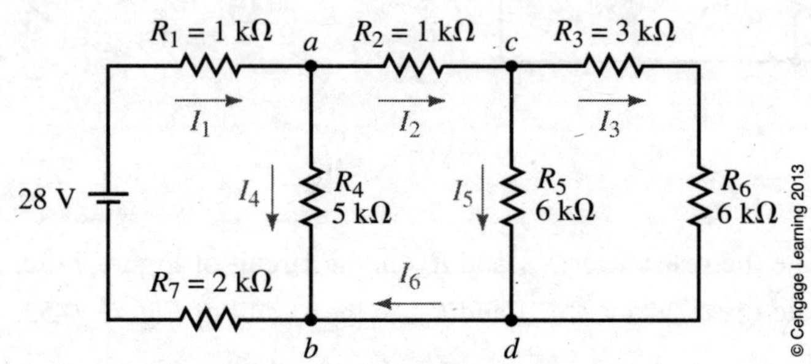

FIGURE 7–56

14. Refer to the circuit of Figure 7–57:
    a. Find the currents I_1, I_2, I_3, I_4, and I_5.
    b. Solve for the voltages V_{ab} and V_{bc}.
    c. Verify that the power delivered to the circuit is equal to the summation of powers dissipated by the resistors.

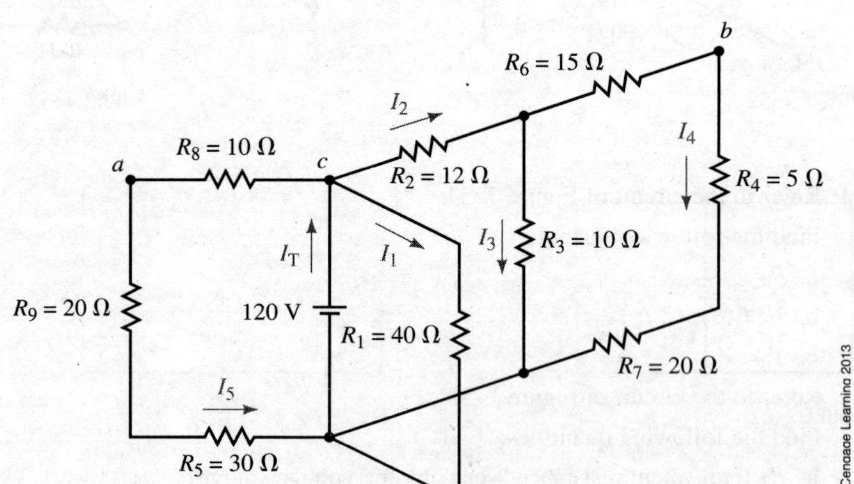

FIGURE 7–57

15. Refer to the circuits of Figure 7–58:
    a. Find the indicated currents.
    b. Solve for the voltage V_{ab}.
    c. Verify that the power delivered to the circuit is equal to the summation of powers dissipated by the resistors.

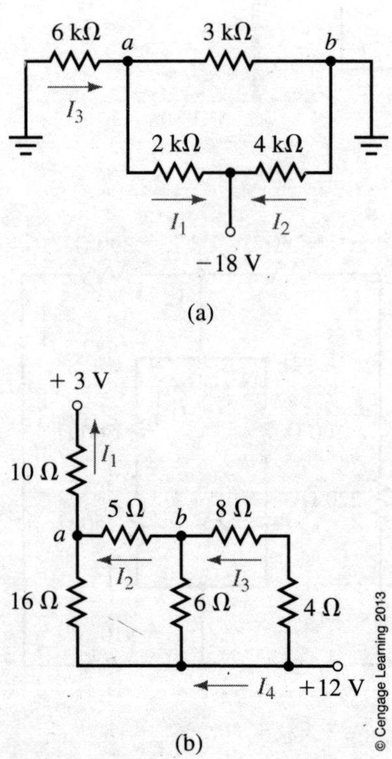

FIGURE 7–58

16. Refer to the circuit of Figure 7–59:
    a. Solve for the currents I_1, I_2, and I_3 when $R_x = 0\ \Omega$ and when $R_x = 5\ \text{k}\Omega$.
    b. Calculate the voltage V_{ab} when $R_x = 0\ \Omega$ and when $R_x = 5\ \text{k}\Omega$.

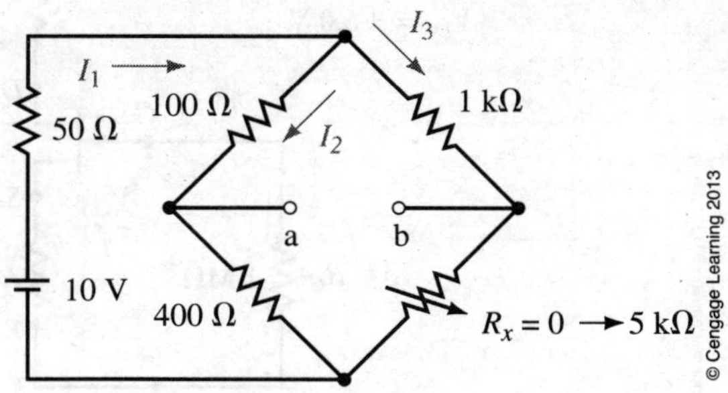

FIGURE 7–59

7.3 Applications of Series-Parallel Circuits

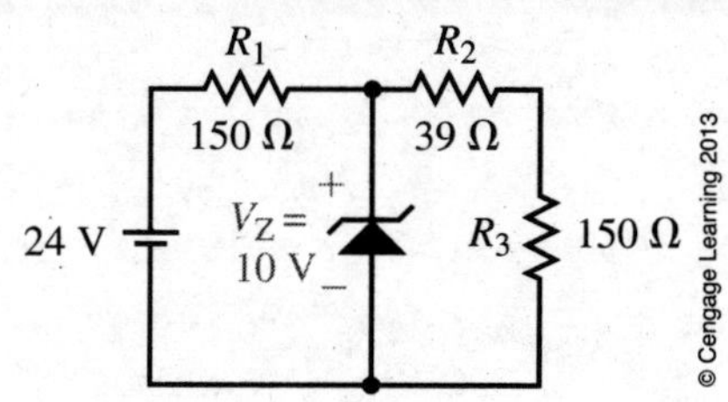

FIGURE 7–60

17. Solve for all currents and voltage drops in the circuit of Figure 7–60. Verify that the power delivered by the voltage source is equal to the power dissipated by the resistors and the zener diode.
18. Refer to the circuit of Figure 7–61:
    a. Determine the power dissipated by the 6.2-V zener diode. If the zener diode is rated for a maximum power of ¼ W, is it likely to be destroyed?
    b. Repeat part (a) if the resistance R_1 is doubled.

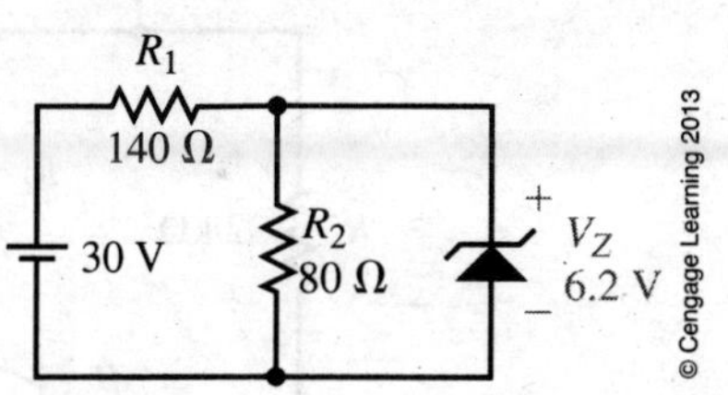

FIGURE 7–61

19. Given the circuit of Figure 7–62, determine the range of R (maximum and minimum values) that will ensure that the output voltage $V_L = 5.6$ V while the maximum power rating of the zener diode is not exceeded.
20. Given the circuit of Figure 7–63, determine the range of R (maximum and minimum values) that will ensure that the output voltage $V_L = 5.6$ V while the maximum power rating of the zener diode is not exceeded.

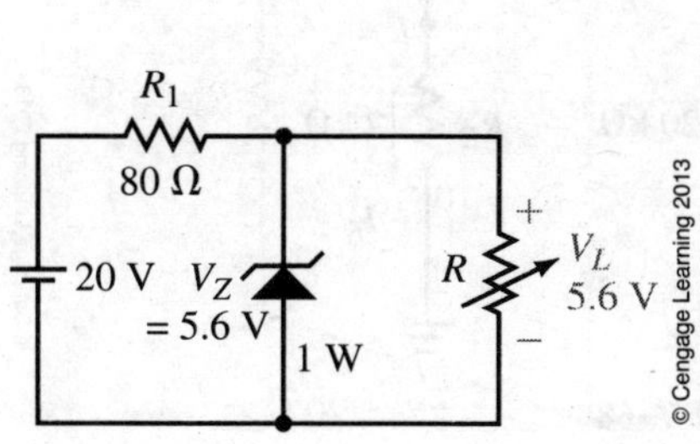

FIGURE 7–62

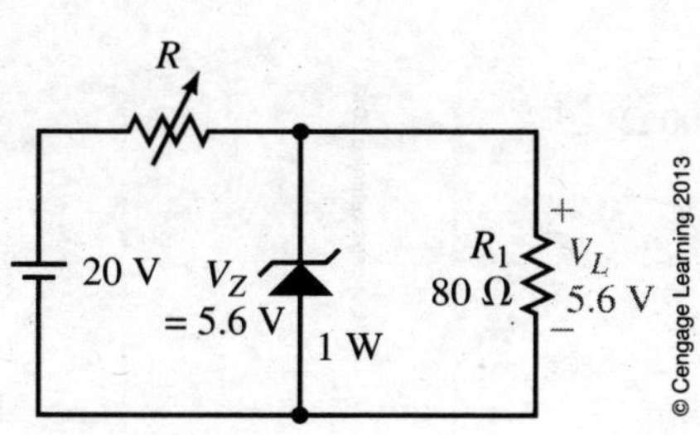

FIGURE 7–63

21. Given the circuit of Figure 7–64, determine V_B, I_C, and V_{CE}.
22. Repeat Problem 21 if R_B is increased to 10 kΩ. (All other quantities remain unchanged.)
23. Consider the circuit of Figure 7–65 and the indicated values:
    a. Determine I_D.
    b. Calculate the required value for R_S.
    c. Solve for V_{DS}.

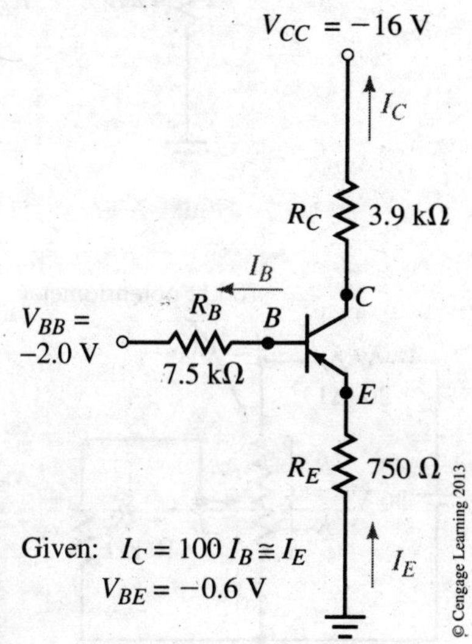

FIGURE 7–64

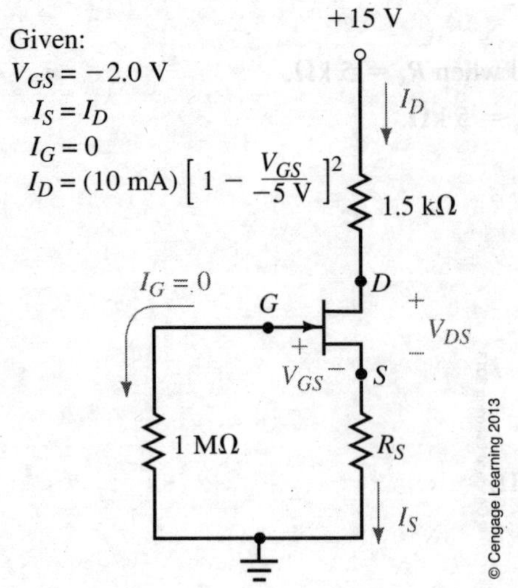

FIGURE 7–65

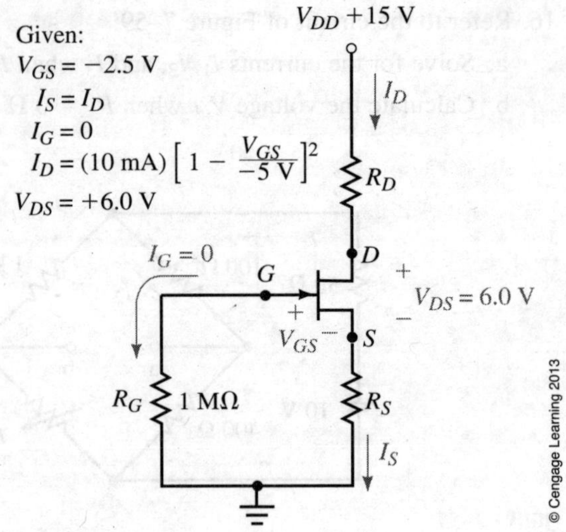

FIGURE 7–66

24. Consider the circuit of Figure 7–66 and the indicated values:
    a. Determine I_D and V_G.
    b. Design the required values for R_S and R_D.
25. Calculate I_C and V_{CE} for the circuit of Figure 7–67.
26. Calculate I_C and V_{CE} for the circuit of Figure 7–68.

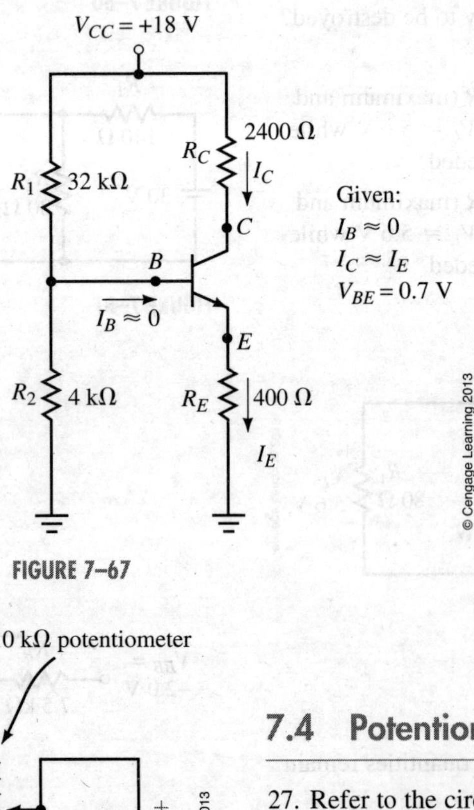

FIGURE 7–67

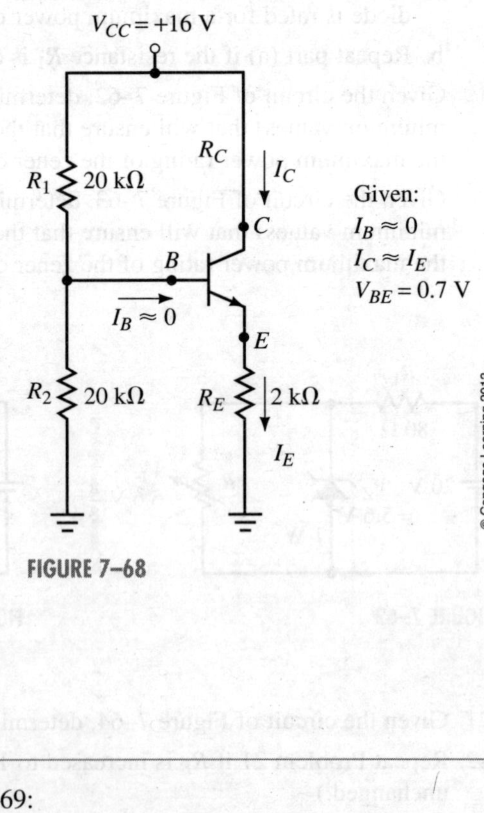

FIGURE 7–68

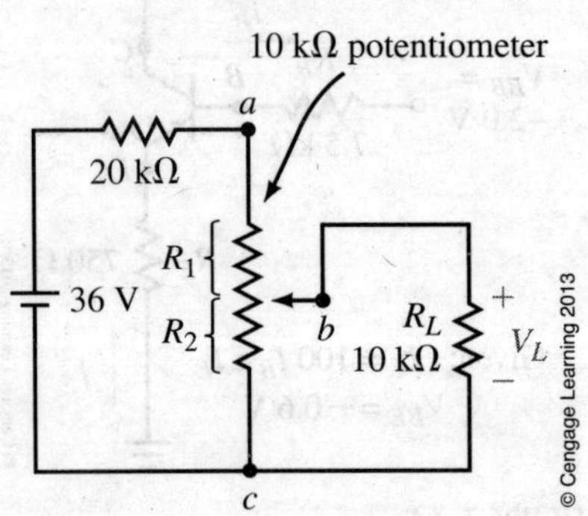

FIGURE 7–69

7.4 Potentiometers

27. Refer to the circuit of Figure 7–69:
    a. Determine the range of voltages that will appear across R_L as the potentiometer is varied between its minimum and maximum values.
    b. If R_2 is adjusted to be 2.5 kΩ, what will be the voltage V_L? If the load resistor is now removed, what voltage would appear between terminals *a* and *b*?
28. Repeat Problem 27 using the load resistor $R_L = 30$ kΩ.

29. If the potentiometer of Figure 7–70 is adjusted so that $R_2 = 200\ \Omega$, determine the voltages V_{ab} and V_{bc}.
30. Calculate the values of R_1 and R_2 required in the potentiometer of Figure 7–70 if the voltage V_L across the 50-Ω load resistor is to be 6.0 V.
31. Refer to the circuit of Figure 7–71:
    a. Determine the range of output voltage (minimum to maximum) that can be expected as the potentiometer is adjusted from minimum to maximum.
    b. Calculate R_2 when $V_{out} = 20$ V.
32. In the circuit of Figure 7–71, what value of R_2 results in an output voltage of 40 V?
33. Given the circuit of Figure 7–72, calculate the output voltage V_{out} when $R_L = 0\ \Omega$, 250 Ω, and 500 Ω.
34. Given the circuit of Figure 7–73, calculate the output voltage V_{out} when $R_L = 0\ \Omega$, 500 Ω, and 1000 Ω.

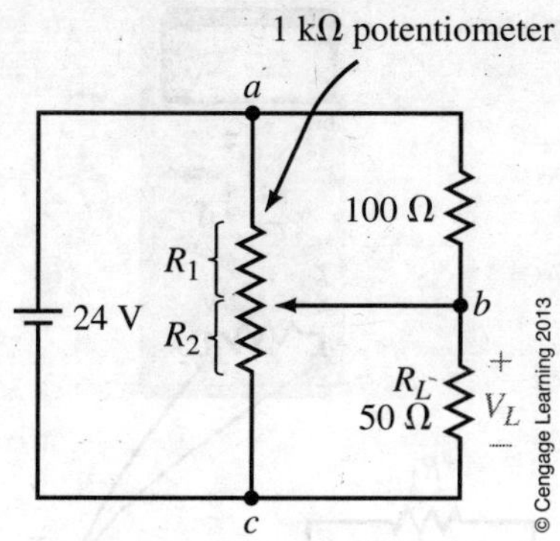

FIGURE 7–70

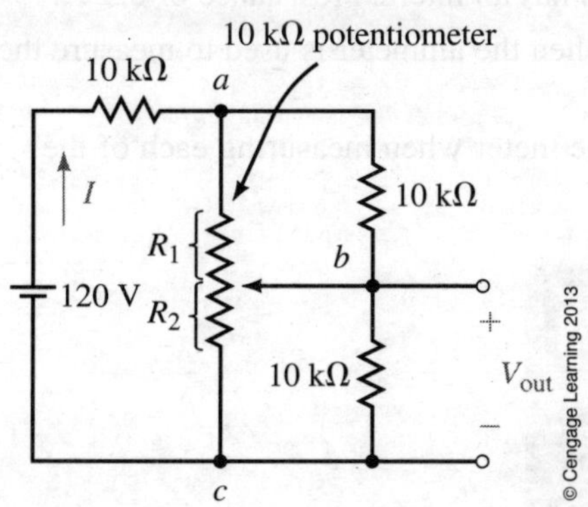

FIGURE 7–71

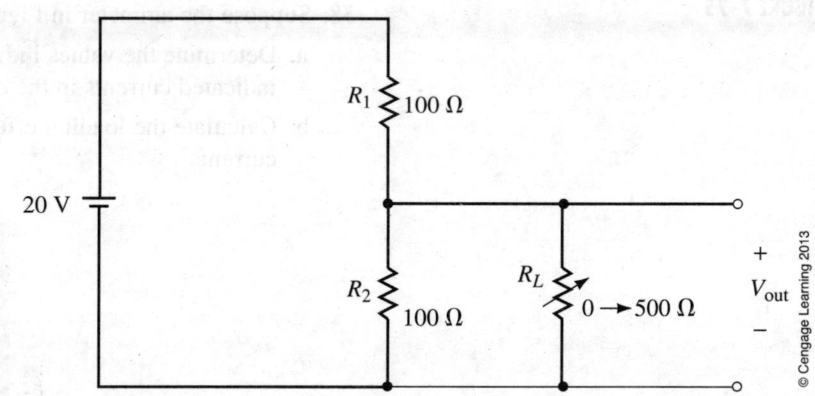

FIGURE 7–72

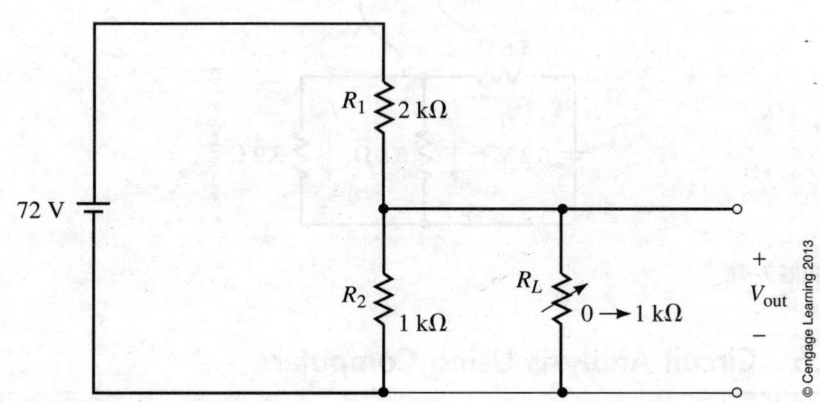

FIGURE 7–73

7.5 Loading Effects of Instruments

35. A voltmeter having a sensitivity $S = 20$ kΩ/V is used on the 10-V range (200-kΩ total internal resistance) to measure voltage across the 750-kΩ resistor of Figure 7–74. The voltage indicated by the meter is 5.00 V.
    a. Determine the value of the supply voltage, E.
    b. What voltage will be present across the 750-kΩ resistor when the voltmeter is removed from the circuit?

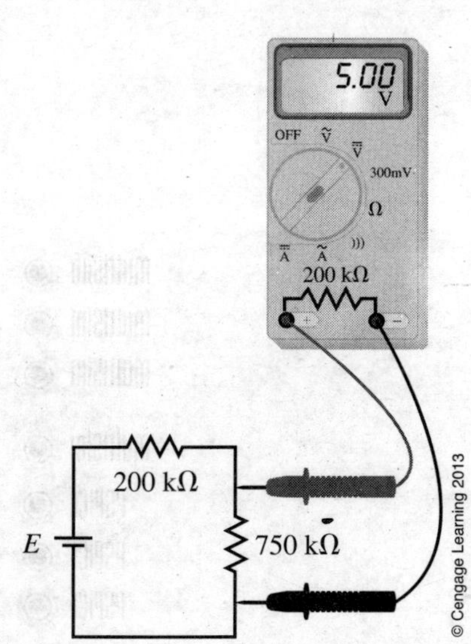

FIGURE 7–74

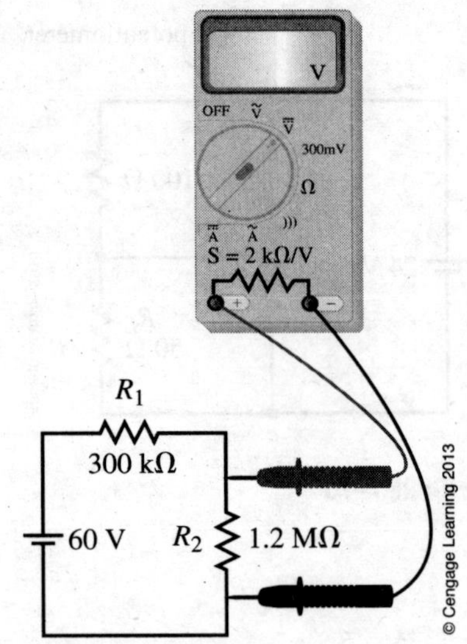

FIGURE 7–75

c. Calculate the loading effect of the meter when used as shown.

d. If the same voltmeter is used to measure the voltage across the 200-kΩ resistor, what voltage will be indicated?

36. The voltmeter of Figure 7–75 has a sensitivity $S = 2\ \text{k}\Omega/\text{V}$.

a. If the meter is used on its 50-V range ($R = 100\ \text{k}\Omega$) to measure the voltage across R_2, what will be the meter reading and the loading error?

b. If the meter is changed to its 20-V range ($R = 40\ \text{k}\Omega$) determine the reading on this range and the loading error. Will the meter be damaged on this range? Will the loading error be less or more than the error in part (a)?

37. An ammeter is used to measure current in the circuit shown in Figure 7–76.

a. Explain how to correctly connect the ammeter to measure the current I_1.

b. Determine the values indicated when the ammeter is used to measure each of the indicated currents in the circuit.

c. Calculate the loading effect of the meter when measuring each of the currents.

38. Suppose the ammeter in Figure 7–76 has an internal resistance of 0.5 Ω:

a. Determine the values indicated when the ammeter is used to measure the indicated currents in the circuit.

b. Calculate the loading effect of the meter when measuring each of the currents.

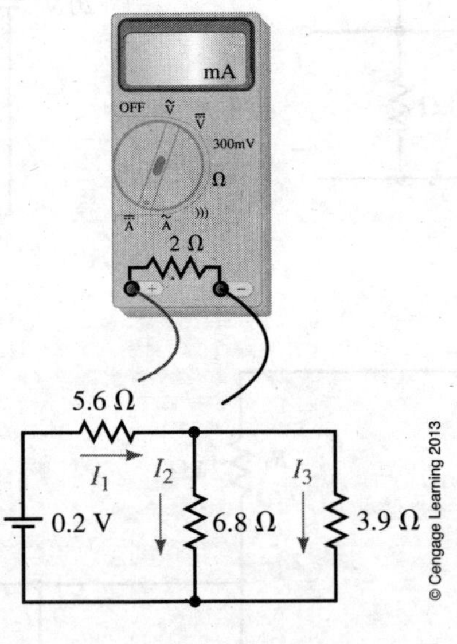

FIGURE 7–76

7.6 Circuit Analysis Using Computers

Multisim 39. Use Multisim to solve for V_2, V_4, I_T, I_1, and I_2 in the circuit of Figure 7–10.

Multisim 40. Use Multisim to solve for V_{ab}, I_1, I_2, and I_3 in the circuit of Figure 7–13.

Multisim 41. Use Multisim to solve for the meter reading in the circuit of Figure 7–75 if the meter is used on its 50-V range.

Multisim 42. Repeat Problem 41 if the meter is used on its 20-V range.

PSPICE 43. Use PSpice to solve for V_2, V_4, I_T, I_1, and I_2 in the circuit of Figure 7–10.

PSPICE 44. Use PSpice to solve for V_{ab}, I_1, I_2, and I_3 in the circuit of Figure 7–13.

PSPICE 45. Use PSpice to obtain a display of V_{ab} and I_1 in the circuit of Figure 7–59. Let R_x change from 500 Ω to 5 kΩ using 100-Ω increments.

ANSWERS TO IN-PROCESS LEARNING CHECKS

IN-PROCESS LEARNING CHECK 1

$P_T = 115.2$ mW, $P_1 = 69.1$ mW, $P_2 = 36.9$ mW, $P_3 = 9.2$ mW
$P_1 + P_2 + P_3 = 115.2$ mW as required.

IN-PROCESS LEARNING CHECK 2

$P_T = 30.0$ mW, $P_{R1} = 20.0$ mW, $P_{R2} = 2.50$ mW,
$P_Z = 7.5$ mW
$P_{R1} + P_{R2} + P_Z = 30$ mW as required.

IN-PROCESS LEARNING CHECK 3

$R_L = 2\ \text{k}\Omega$:

a. $N = ¼$: $V_L = 8.7\%$ of V_{in}

b. $N = ½$: $V_L = 14.3\%$ of V_{in}

$N = ¾$: $V_L = 26.1\%$ of V_{in}

$R_L = 200\ \text{k}\Omega$:

c. $N = ¼$: $V_L = 24.5\%$ of V_{in}

$N = ½$: $V_L = 48.8\%$ of V_{in}

$N = ¾$: $V_L = 73.6\%$ of V_{in}

d. If $RL \gg R_1$, the loading effect is minimal.

KEY TERMS

Balanced Bridge
Branch-Current Analysis
Bridge Networks
Constant-Current Sources
Current Source
Linear Bilateral Networks
Mesh (Loop) Analysis
Mutual Conductance
Nodal Analysis

OUTLINE

Constant-Current Sources
Source Conversions
Current Sources in Parallel and Series
Branch-Current Analysis
Mesh (Loop) Analysis
Nodal Analysis
Delta-Wye (Pi-Tee) Conversion
Bridge Networks
Circuit Analysis Using Computers

OBJECTIVES

After studying this chapter, you will be able to

- convert a voltage source into an equivalent current source,
- convert a current source into an equivalent voltage source,
- analyze circuits having two or more current sources in parallel,
- write and solve branch equations for a network,
- write and solve mesh equations for a network,
- write and solve nodal equations for a network,
- convert a resistive delta to an equivalent wye circuit or a wye to its equivalent delta circuit and solve the resulting simplified circuit,
- determine the voltage across or current through any portion of a bridge network,
- use PSpice to analyze multiloop circuits,
- use Multisim to analyze multiloop circuits.

8

METHODS OF ANALYSIS

CHAPTER PREVIEW

The networks you have worked with so far have generally had a single voltage source and could be easily analyzed using techniques such as Kirchhoff's voltage law and Kirchhoff's current law. In this chapter, you will examine circuits that have more than one voltage source or that cannot be easily analyzed using techniques studied in previous chapters.

The methods used in determining the operation of complex networks will include branch-current analysis, mesh (or loop) analysis, and nodal analysis. Although any of the above methods may be used, you will find that certain circuits are more easily analyzed using one particular approach. The advantages of each method will be discussed in the appropriate section.

In using the techniques outlined here, we assume that the networks are **linear bilateral networks.** The term *linear* indicates that the components used in the circuit have voltage-current characteristics that follow a straight line. Refer to Figure 8–1.

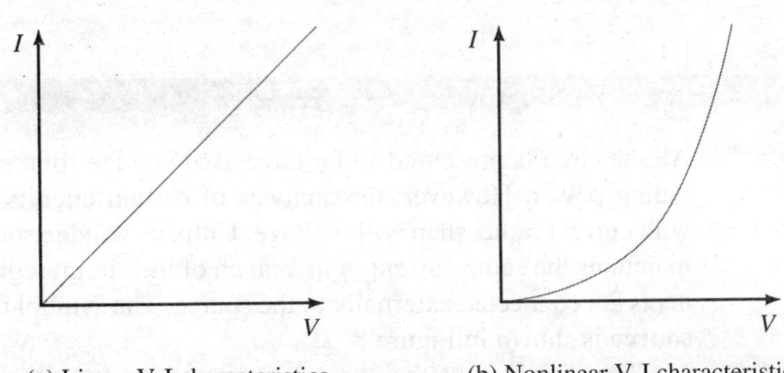

(a) Linear V-I characteristics (b) Nonlinear V-I characteristics

FIGURE 8–1

The term *bilateral* indicates that the components in the network will have characteristics that are independent of the direction of the current through the element or the voltage across the element. A resistor is an example of a linear bilateral component since the voltage across a resistor is directly proportional to the current through it and the operation of the resistor is the same regardless of the direction of the current.

In this chapter you will be introduced to the conversion of a network from a delta (Δ) configuration to an equivalent wye (Y) configuration. Conversely, we will examine the transformation from a Y configuration to an equivalent Δ configuration. You will use these conversions to examine the operation of an unbalanced bridge network. ■

Putting It in Perspective

Sir Charles Wheatstone

CHARLES WHEATSTONE WAS BORN IN GLOUCESTER, England, on February 6, 1802. Wheatstone's original interest was in the study of acoustics and musical instruments. However, he gained fame and a knighthood as a result of inventing the telegraph and improving the electric generator.

Although he did not invent the bridge circuit, Wheatstone used one for measuring resistance very precisely. He found that when the currents in the Wheatstone bridge are exactly balanced, the unknown resistance can be compared to a known standard.

Sir Charles died in Paris, France, on October 19, 1875. ■

8.1 Constant-Current Sources

All the circuits presented so far have used voltage sources as the means of providing power. However, the analysis of certain circuits is easier if you work with current rather than with voltage. Unlike a voltage source, a **current source** maintains the same current in its branch of the circuit regardless of how components are connected externally to the source. The symbol for a **constant-current source** is shown in Figure 8–2.

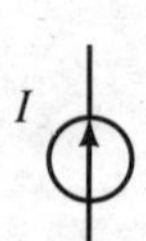

FIGURE 8–2 Ideal constant-current source.

The direction of the current source arrow indicates the direction of conventional current in the branch. In previous chapters you learned that the magnitude and the direction of current through a voltage source vary according to the size of the circuit resistances and how other voltage sources are connected in the circuit. For current sources, the voltage across the current source depends on how the other components are connected.

EXAMPLE 8–1

Refer to the circuit of Figure 8–3:

a. Calculate the voltage V_S across the current source if the resistor is 100 Ω.
b. Calculate the voltage if the resistor is 2 kΩ.

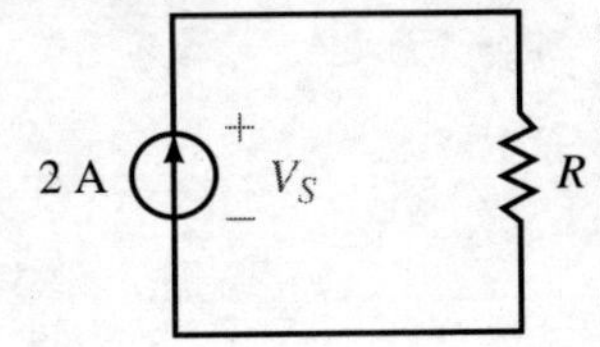

FIGURE 8–3

Solution The current source maintains a constant current of 2 A through the circuit. Therefore,

a. $V_S = V_R = (2\text{ A})(100\ \Omega) = 200\text{ V}$
b. $V_S = V_R = (2\text{ A})(2\text{ k}\Omega) = 4000\text{ V}$

If the current source is the only source in the circuit, then the polarity of voltage across the source will be as shown in Figure 8–3. This, however, may not be the case if there is more than one source. The following example illustrates this principle.

EXAMPLE 8–2

Determine the voltages V_1, V_2, and V_S and the current I_S for the circuit of Figure 8–4.

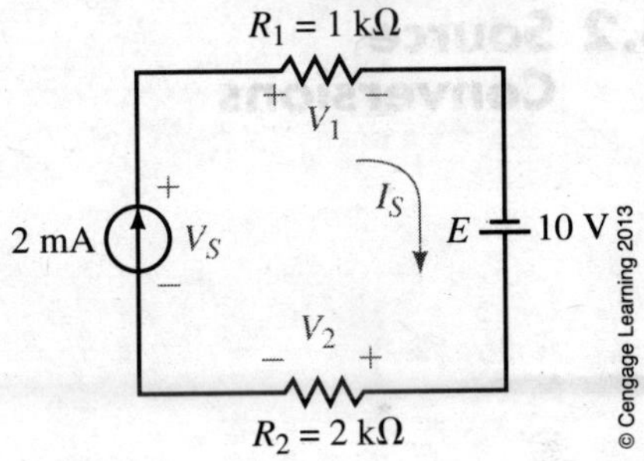

FIGURE 8–4

Solution Since the given circuit is a series circuit, the current everywhere in the circuit must be the same, namely

$$I_S = 2\text{ mA}$$

Using Ohm's law,

$$V_1 = (2\text{ mA})(1\text{ k}\Omega) = 2.00\text{ V}$$
$$V_2 = (2\text{ mA})(2\text{ k}\Omega) = 4.00\text{ V}$$

Applying Kirchhoff's voltage law around the closed loop,

$$\sum V = V_S - V_1 - V_2 + E = 0$$
$$V_S = V_1 + V_2 - E$$
$$= 2\text{ V} + 4\text{ V} - 10\text{ V} = -4.00\text{ V}$$

From the preceding result, you see that the actual polarity of V_S is opposite to that assumed.

EXAMPLE 8–3

Calculate the currents I_1 and I_2 and the voltage V_S for the circuit of Figure 8–5.

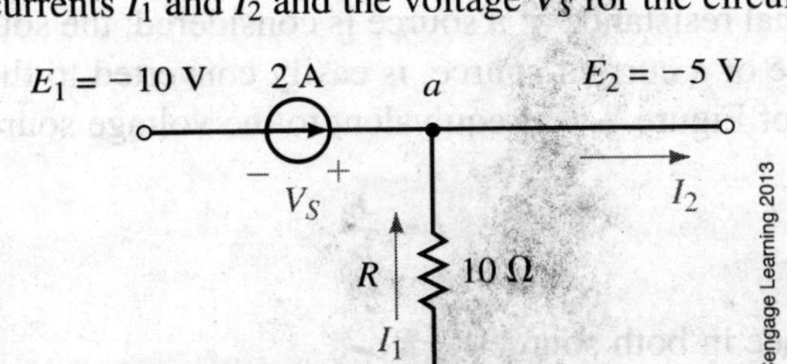

FIGURE 8–5

Solution Because the 5-V supply is effectively across the load resistor,

$$I_1 = \frac{5\ \text{V}}{10\ \Omega} = 0.5\ \text{A} \quad \text{(in the direction assumed)}$$

Applying Kirchhoff's current law at point *a*,

$$I_2 = 0.5\ \text{A} + 2.0\ \text{A} = 2.5\ \text{A}$$

From Kirchhoff's voltage law,

$$\sum V = -10\ \text{V} + V_S + 5\ \text{V} = 0\ \text{V}$$
$$V_S = 10\ \text{V} - 5\ \text{V} = +5\ \text{V}$$

By examining the previous examples, we can conclude the following about current sources:

The constant-current source determines the current in its branch of the circuit.

The magnitude and polarity of voltage appearing across a constant-current source are dependent upon the network in which the source is connected.

8.2 Source Conversions

In the previous section you were introduced to the ideal constant-current source. This is a source that has no internal resistance included as part of the circuit. As you recall, voltage sources always have some series resistance, although in some cases this resistance is so small in comparison with other circuit resistance that it may effectively be ignored when determining the operation of the circuit. Similarly, a constant-current source will always have some shunt (or parallel) resistance. If this resistance is very large in comparison with the other circuit resistance, the internal resistance of the source may once again be ignored. *An ideal current source has an infinite shunt resistance.*

Figure 8–6 shows equivalent voltage and current sources.

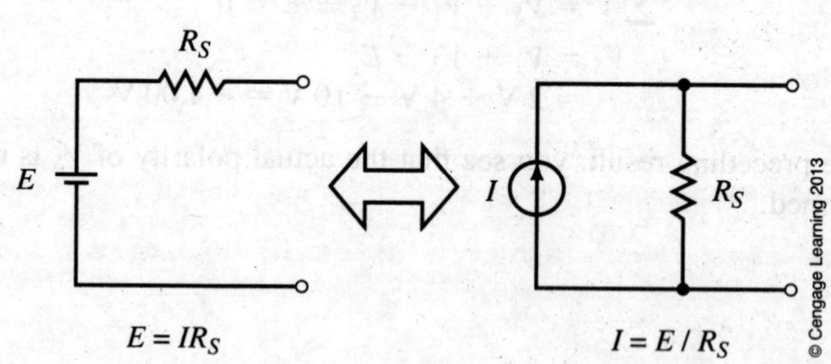

FIGURE 8–6

If the internal resistance of a source is considered, the source, whether it is a voltage source or a current source, is easily converted to the other type. The current source of Figure 8–6 is equivalent to the voltage source if

$$I = \frac{E}{R_S} \tag{8–1}$$

and the resistance in both sources is R_S.

Similarly, a current source may be converted to an equivalent voltage source by letting

$$E = IR_S \tag{8–2}$$

These results may be easily verified by connecting an external resistance, R_L, across each source. The sources can be equivalent only if the voltage across R_L is the same for both sources. Similarly, the sources are equivalent only if the current through R_L is the same when connected to either source.

Consider the circuit shown in Figure 8–7. The voltage across the load resistor is given as

$$V_L = \frac{R_L}{R_L + R_S}E \tag{8–3}$$

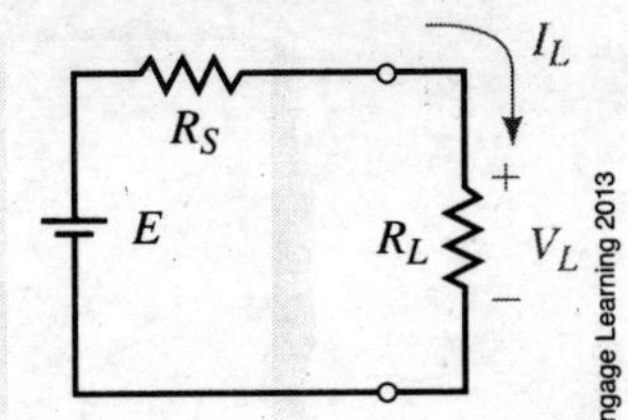

FIGURE 8–7

The current through the resistor R_L is given as

$$I_L = \frac{E}{R_L + R_S} \tag{8–4}$$

Next, consider an equivalent current source connected to the same load as shown in Figure 8–8. The current through the resistor R_L is given by

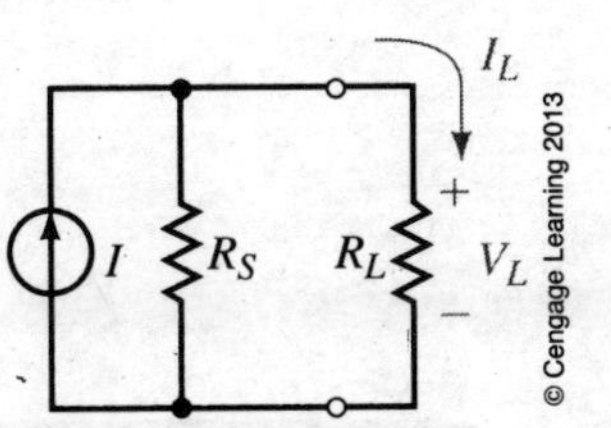

FIGURE 8–8

$$I_L = \frac{R_S}{R_S + R_L}I$$

But, when converting the source, we get

$$I = \frac{E}{R_S}$$

And so

$$I_L = \left(\frac{R_S}{R_S + R_L}\right)\left(\frac{E}{R_S}\right)$$

This result is equivalent to the current obtained in Equation 8–4. The voltage across the resistor is given as

$$V_L = I_L R_L$$

$$= \left(\frac{E}{R_S + R_L}\right)R_L$$

The voltage across the resistor is precisely the same as the result obtained in Equation 8–3. We therefore conclude that the load current and voltage drop are the same whether the source is a voltage source or an equivalent current source.

NOTES...

Although the sources are equivalent, currents and voltages within the sources may no longer be the same. The sources are only equivalent with respect to elements connected external to the terminals.

EXAMPLE 8–4

Convert the voltage source of Figure 8–9(a) into a current source and verify that the current, I_L, through the load is the same for each source.

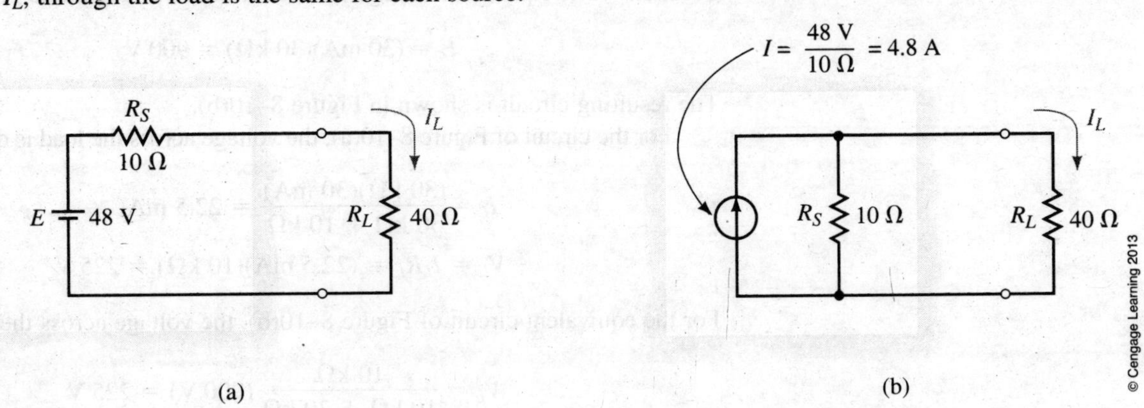

FIGURE 8–9

Solution The equivalent current source will have a current magnitude given as

$$I = \frac{48\text{ V}}{10\ \Omega} = 4.8\text{ A}$$

The resulting circuit is shown in Figure 8–9(b).

For the circuit of Figure 8–9(a), the current through the load is found as

$$I_L = \frac{48\text{ V}}{10\ \Omega + 40\ \Omega} = 0.96\text{ A}$$

For the equivalent circuit of Figure 8–9(b), the current through the load is

$$I_L = \frac{(4.8\text{ A})(10\ \Omega)}{10\ \Omega + 40\ \Omega} = 0.96\text{ A}$$

Clearly the results are the same.

EXAMPLE 8–5

Convert the current source of Figure 8–10(a) into a voltage source and verify that the voltage, V_L, across the load is the same for each source.

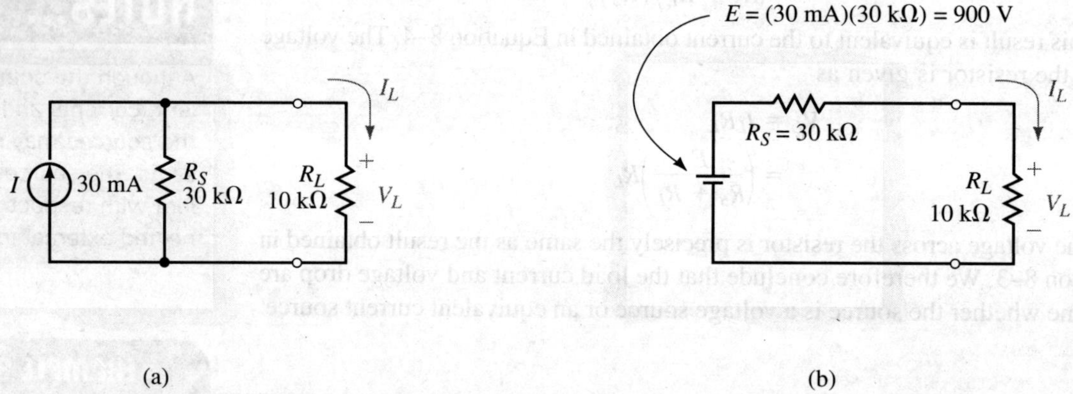

FIGURE 8–10

Solution The equivalent voltage source will have a magnitude given as

$$E = (30\text{ mA})(30\text{ k}\Omega) = 900\text{ V}$$

The resulting circuit is shown in Figure 8–10(b).

For the circuit of Figure 8–10(a), the voltage across the load is determined as

$$I_L = \frac{(30\text{ k}\Omega)(30\text{ mA})}{30\text{ k}\Omega + 10\text{ k}\Omega} = 22.5\text{ mA}$$

$$V_L = I_L R_L = (22.5\text{ mA})(10\text{ k}\Omega) = 225\text{ V}$$

For the equivalent circuit of Figure 8–10(b), the voltage across the load is

$$V_L = \frac{10\text{ k}\Omega}{10\text{ k}\Omega + 30\text{ k}\Omega}(900\text{ V}) = 225\text{ V}$$

Once again, we see that the circuits are equivalent.

PRACTICE PROBLEMS 1

1. Convert the voltage sources of Figure 8–11 into equivalent current sources.

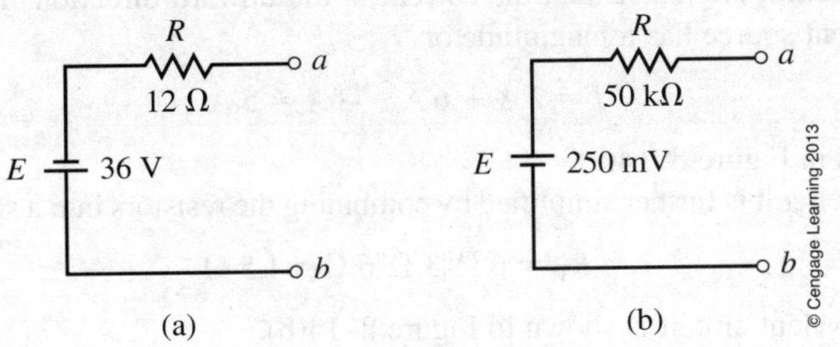

FIGURE 8–11

2. Convert the current sources of Figure 8–12 into equivalent voltage sources.

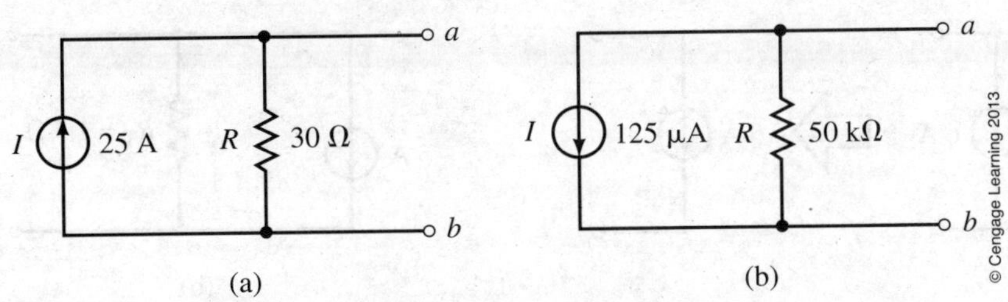

FIGURE 8–12

Answers

1. a. $I = 3.00$ A (downward) in parallel with $R = 12\ \Omega$
   b. $I = 5.00\ \mu$A (upward) in parallel with $R = 50$ kΩ
2. a. $E = V_{ab} = 750$ V in series with $R = 30\ \Omega$
   b. $E = V_{ab} = -6.25$ V in series with $R = 50$ kΩ

8.3 Current Sources in Parallel and Series

When several current sources are placed in parallel, the circuit may be simplified by combining the current sources into a single current source. The magnitude and direction of this resultant source is determined by adding the currents in one direction and then subtracting the currents in the opposite direction.

EXAMPLE 8–6

Simplify the circuit of Figure 8–13 and determine the voltage V_{ab}.

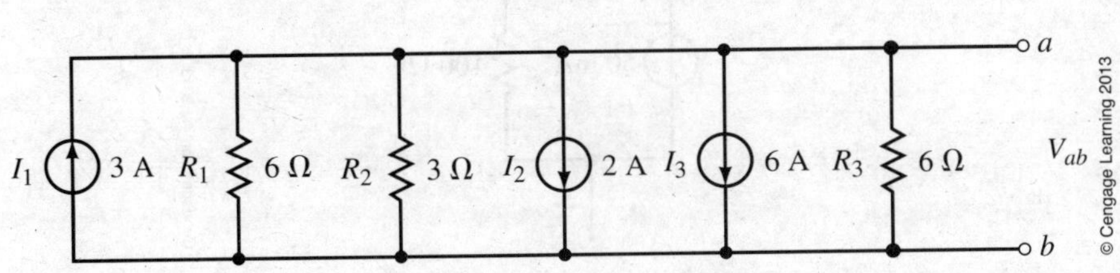

FIGURE 8–13

Solution Since all of the current sources are in parallel, they can be replaced by a single current source. The equivalent current source will have a direction that is the same as both I_2 and I_3, since the magnitude of current in the downward direction is greater than the current in the upward direction. The equivalent current source has a magnitude of

$$I = 2\text{ A} + 6\text{ A} - 3\text{ A} = 5\text{ A}$$

as shown in Figure 8–14(a).

The circuit is further simplified by combining the resistors into a single value:

$$R_T = 6\ \Omega \| 3\ \Omega \| 6\ \Omega = 1.5\ \Omega$$

The equivalent circuit is shown in Figure 8–14(b).

The voltage V_{ab} is found as

$$V_{ab} = -(5\text{ A})(1.5\ \Omega) = -7.5\text{ V}$$

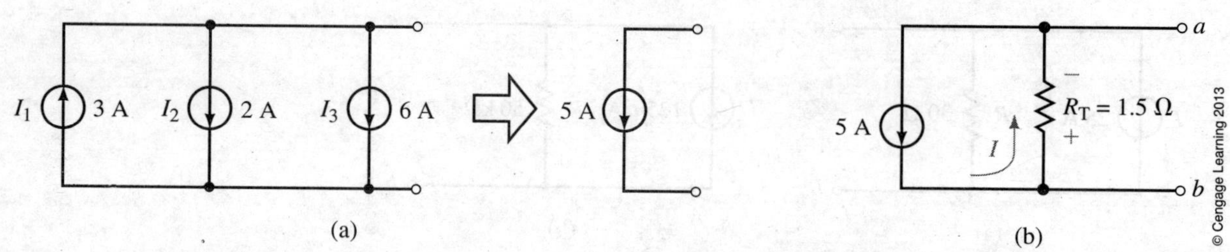

FIGURE 8–14

EXAMPLE 8–7

Reduce the circuit of Figure 8–15 into a single current source and solve for the current through the resistor R_L.

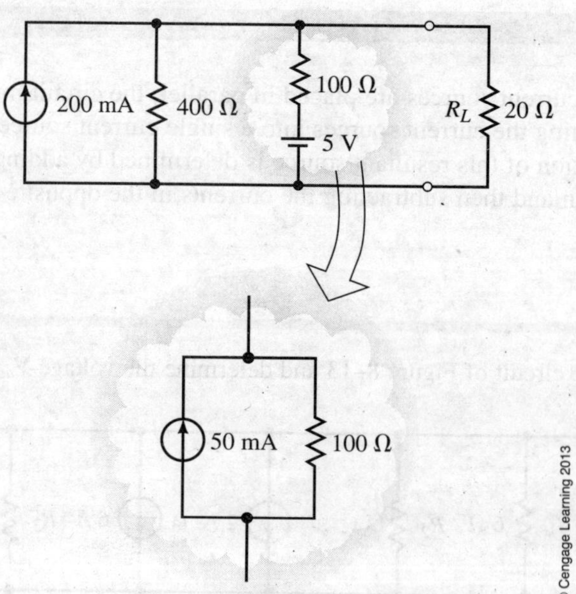

FIGURE 8–15

Solution The voltage source in this circuit is converted to an equivalent current source as shown. The resulting circuit may then be simpified to a single current source where

$$I_S = 200 \text{ mA} + 50 \text{ mA} = 250 \text{ mA}$$

and

$$R_S = 400\ \Omega \| 100\ \Omega = 80\ \Omega$$

The simplified circuit is shown in Figure 8–16.

The current through R_L is now easily calculated as

$$I_L = \left(\frac{80\ \Omega}{80\ \Omega + 20\ \Omega}\right)(250 \text{ mA}) = 200 \text{ mA}$$

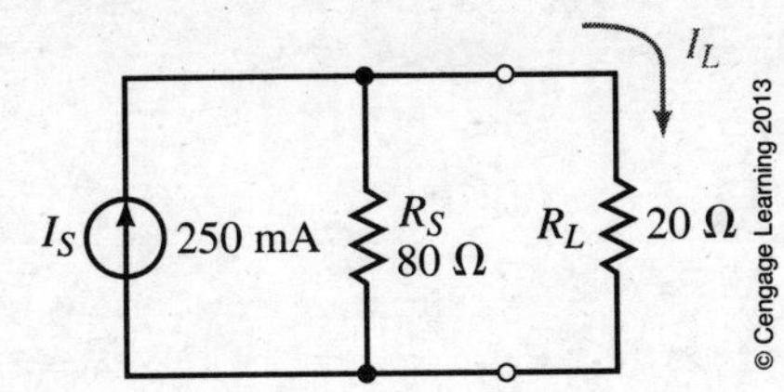

FIGURE 8–16

Current sources should never be placed in series. If a node is chosen between the current sources, it becomes immediately apparent that the current entering the node is not the same as the current leaving the node. Clearly, this cannot occur since there would then be a violation of Kirchhoff's current law (see Figure 8–17).

NOTES...

Current sources of different values are never placed in series.

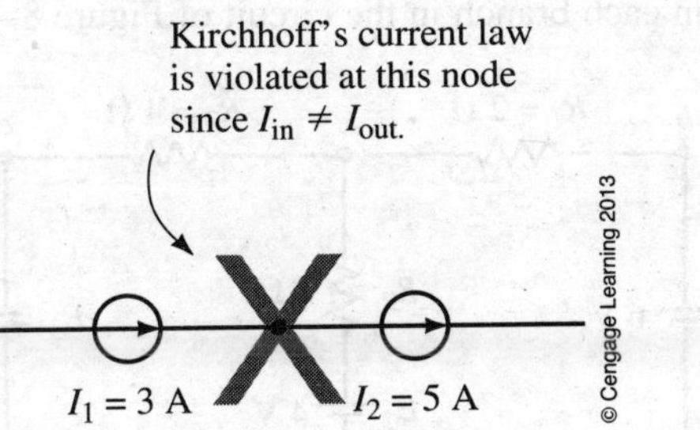

FIGURE 8–17

IN-PROCESS LEARNING CHECK 1

(Answers are at the end of the chapter.)

1. Briefly explain the procedure for converting a voltage source into an equivalent current source.
2. What is the most important rule determining how current sources are connected into a circuit?

8.4 Branch-Current Analysis

In previous chapters we used Kirchhoff's current law (KCL) and Kirchhoff's voltage law (KVL) to solve equations for circuits having a single voltage source. In this section, you will use these powerful tools to analyze circuits having more than one source.

Branch-current analysis allows us to directly calculate the current in each branch of a circuit. Since the method involves the analysis of several simultaneous linear equations, you may find that a review of determinants is in order.

Appendix B has been included to provide a review of the mechanics of solving simultaneous linear equations.

When applying branch-current analysis, you will find the following technique useful.

1. Arbitrarily assign current directions to each branch in the network. If a particular branch has a current source, then this step is not necessary since you already know the magnitude and direction of the current in this branch.
2. Using the assigned currents, label the polarities of the voltage drops across all resistors in the circuit.
3. Apply Kirchhoff's voltage law around each of the closed loops. Write just enough equations to include all branches in the loop equations. If a branch has only a current source and no series resistance, it is not necessary to include it in the KVL equations.
4. Apply Kirchhoff's current law at enough nodes to ensure that all branch currents have been included. In the event that a branch has only a current source, it will need to be included in this step.
5. Solve the resulting simultaneous linear equations.

EXAMPLE 8–8

Find the current in each branch in the circuit of Figure 8–18.

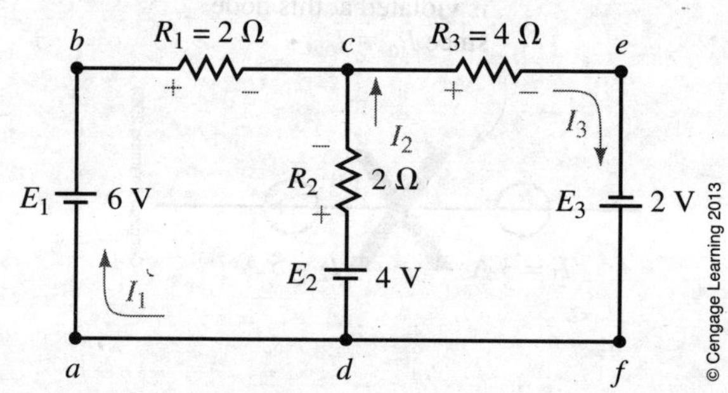

FIGURE 8–18

Solution

Step 1: Assign currents as shown in Figure 8–18.

Step 2: Indicate the polarities of the voltage drops on all resistors in the circuit, using the assumed current directions.

Step 3: Write the Kirchhoff voltage law equations.

Loop *abcda:* $6\text{ V} - (2\ \Omega)I_1 + (2\ \Omega)I_2 - 4\text{ V} = 0\text{ V}$

Notice that the circuit still has one branch which has not been included in the KVL equations, namely the branch *cefd.* This branch would be included if a loop equation for *cefdc* or for *abcefda* were written. There is no reason for choosing one loop over another, since the overall result will remain unchanged even though the intermediate steps will not give the same results.

Loop *cefdc:* $4\text{ V} - (2\ \Omega)I_2 - (4\ \Omega)I_3 + 2\text{ V} = 0\text{ V}$

Now that all branches have been included in the loop equations, there is no need to write any more. Although more loops exist, writing more loop equations would needlessly complicate the calculations.

Step 4: Write the Kirchhoff current law equation(s).

By applying KCL at node c, all branch currents in the network are included.

Node c: $$I_3 = I_1 + I_2$$

To simplify the solution of the simultaneous linear equations we write them as follows:

$$2I_1 - 2I_2 + 0I_3 = 2$$
$$0I_1 - 2I_2 - 4I_3 = -6$$
$$1I_1 + 1I_2 - 1I_3 = 0$$

The principles of linear algebra (Appendix B) allow us to solve for the determinant of the denominator as follows:

$$D = \begin{vmatrix} 2 & -2 & 0 \\ 0 & -2 & -4 \\ 1 & 0 & -1 \end{vmatrix}$$
$$= 2\begin{vmatrix} -2 & -4 \\ 1 & -1 \end{vmatrix} - 0\begin{vmatrix} -2 & 0 \\ 1 & -1 \end{vmatrix} + 1\begin{vmatrix} -2 & 0 \\ -2 & -4 \end{vmatrix}$$
$$= 2(2 + 4) - 0 + 1(8) = 20$$

Now, solving for the currents, we have the following:

$$I_1 = \frac{\begin{vmatrix} 2 & -2 & 0 \\ -6 & -2 & -4 \\ 0 & 1 & -1 \end{vmatrix}}{D}$$
$$= \frac{2\begin{vmatrix} -2 & -4 \\ 1 & -1 \end{vmatrix} - (-6)\begin{vmatrix} -2 & 0 \\ 1 & -1 \end{vmatrix} + 0\begin{vmatrix} -2 & 0 \\ -2 & -4 \end{vmatrix}}{20}$$
$$= \frac{2(2 + 4) + 6(2) + 0}{20} = \frac{24}{20} = 1.200 \text{ A}$$

Similarly,

$$I_2 = \frac{\begin{vmatrix} 2 & 2 & 0 \\ 0 & -6 & -4 \\ 1 & 0 & -1 \end{vmatrix}}{D}$$
$$= \frac{4}{20} = 0.200 \text{ A}$$

and

$$I_3 = \frac{\begin{vmatrix} 2 & -2 & 2 \\ 0 & -2 & -6 \\ 1 & 1 & 0 \end{vmatrix}}{D}$$
$$= \frac{28}{20} = 1.400 \text{ A}$$

Calculator Hint: The use of calculators and computer programs have made it very easy to find the solutions of simultaneous linear equations. If we were to use a calculator (such as the Sharp EL-546W) to solve for the three unknowns in this example, we would need to first indicate that this is a third-order (three-variable) system and then enter the coefficients a_{11}, a_{12}, a_{13}, and the constant b_1 of the first equation as follows:

```
a1,1x1…a1,3x3=b1
 a1,1=2
 a1,2=-2
 a1,3=0
 b1=2
```

Similarly, the coefficients and constants for the remaining two equations would be entered in order to result in a solution displayed as

```
x1=1.2
x2=.2
x3=1.4
```

These results are consistent with those found using determinants, namely that I_1 = 1.2 A, I_2 = 0.2 A, and I_3 = 1.4 A.

EXAMPLE 8–9

Find the currents in each branch of the circuit shown in Figure 8–19. Solve for the voltage V_{ab}.

CircuitSim 08-1

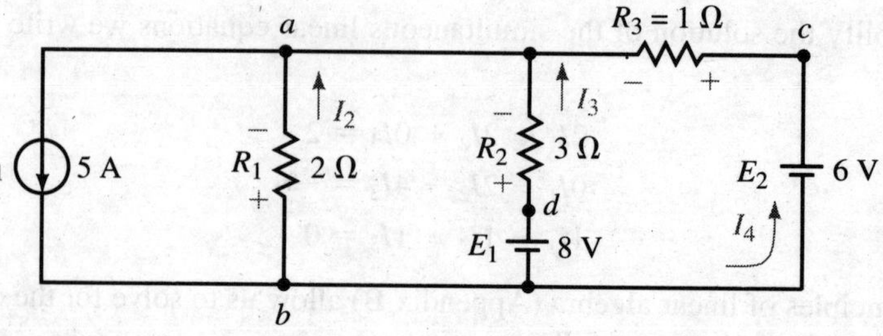

FIGURE 8–19

Solution Notice that although Figure 8–19 has four currents, there are only three *unknown* currents: I_2, I_3, and I_4. The current I_1 is given by the value of the constant-current source. In order to solve this network we will need three linear equations. As before, the equations are determined by Kirchhoff's voltage and current laws.

Step 1: The currents are indicated in the given circuit.

Step 2: The polarities of the voltages across all resistors are shown.

Step 3: Kirchhoff's voltage law is applied at the indicated loops:

Loop *badb*: $-(2\ \Omega)(I_2) + (3\ \Omega)(I_3) - 8\text{ V} = 0\text{ V}$

Loop *bacb*: $-(2\ \Omega)(I_2) + (1\ \Omega)(I_4) - 6\text{ V} = 0\text{ V}$

Step 4: Kirchhoff's current law is applied as follows:

Node *a*: $I_2 + I_3 + I_4 = 5\text{ A}$

Rewriting the linear equations,

$$-2I_2 + 3I_3 + 0I_4 = 8$$
$$-2I_2 + 0I_3 + 1I_4 = 6$$
$$1I_2 + 1I_3 + 1I_4 = 5$$

The determinant of the denominator is evaluated as

$$D = \begin{vmatrix} -2 & 3 & 0 \\ -2 & 0 & 1 \\ 1 & 1 & 1 \end{vmatrix} = 11$$

Now solving for the currents, we have

$$I_2 = \frac{\begin{vmatrix} 8 & 3 & 0 \\ 6 & 0 & 1 \\ 5 & 1 & 1 \end{vmatrix}}{D} = \frac{-11}{11} = -1.00\text{ A}$$

$$I_3 = \frac{\begin{vmatrix} -2 & 8 & 0 \\ -2 & 6 & 1 \\ 1 & 5 & 1 \end{vmatrix}}{D} = \frac{22}{11} = 2.00\text{ A}$$

$$I_4 = \frac{\begin{vmatrix} -2 & 3 & 8 \\ -2 & 0 & 6 \\ 1 & 1 & 5 \end{vmatrix}}{D} = \frac{44}{11} = 4.00\text{ A}$$

The current I_2 is negative, which simply means that the actual direction of the current is opposite to the chosen direction.

Although the network may be further analyzed using the assumed current directions, it is easier to understand the circuit operation by showing the actual current directions as in Figure 8–20.

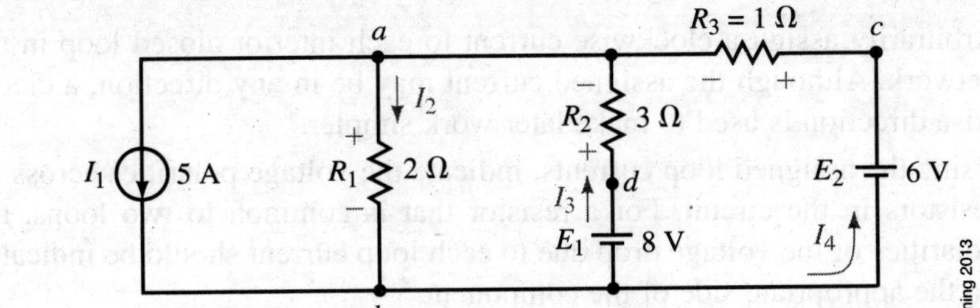

$I_2 = 1.00$ A
$I_3 = 2.00$ A
$I_4 = 4.00$ A

FIGURE 8–20

Using the actual direction for I_2,

$$V_{ab} = +(2\ \Omega)(1\ \text{A}) = +2.00\text{V}$$

PRACTICE PROBLEMS 2

Use branch-current analysis to solve for the indicated currents in the circuit of Figure 8–21.

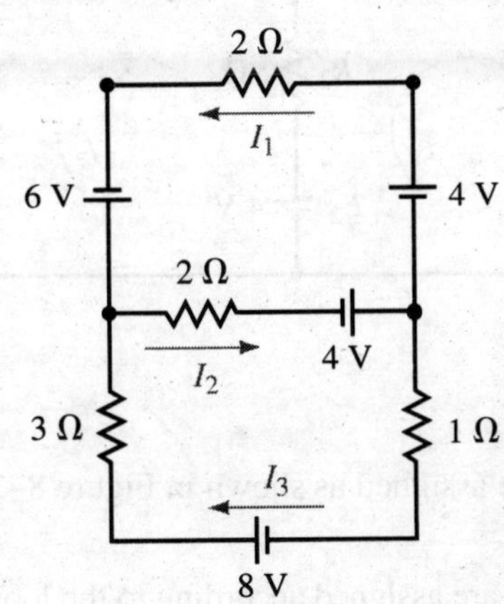

FIGURE 8–21

Answers
$I_1 = 3.00$ A, $I_2 = 4.00$ A, $I_3 = 1.00$ A

8.5 Mesh (Loop) Analysis

In the previous section you used Kirchhoff's laws to solve for the current in each branch of a given network. While the methods used were relatively simple, branch-current analysis is awkward to use because it generally involves solving several simultaneous linear equations. It is not difficult to see that the number of equations may be prohibitively large even for a relatively simple circuit.

A better approach and one that is used extensively in analyzing linear bilateral networks is called **mesh** (or **loop**) **analysis.** While the technique is similar to branch-current analysis, the number of simultaneous linear equations tends to be less. The principal difference between mesh analysis and branch-current analysis is that we simply need to apply Kirchhoff's voltage law around closed loops without the need for applying Kirchhoff's current law.

The steps used in solving a circuit using mesh analysis are as follows:

1. Arbitrarily assign a clockwise current to each interior closed loop in the network. Although the assigned current may be in any direction, a clockwise direction is used to make later work simpler.
2. Using the assigned loop currents, indicate the voltage polarities across all resistors in the circuit. For a resistor that is common to two loops, the polarities of the voltage drop due to each loop current should be indicated on the appropriate side of the component.
3. Applying Kirchhoff's voltage law, write the loop equations for each loop in the network. Do not forget that resistors that are common to two loops will have two voltage drops, one due to each loop.
4. Solve the resultant simultaneous linear equations.
5. Branch currents are determined by algebraically combining the loop currents that are common to the branch.

EXAMPLE 8–10

Find the current in each branch for the circuit of Figure 8–22.

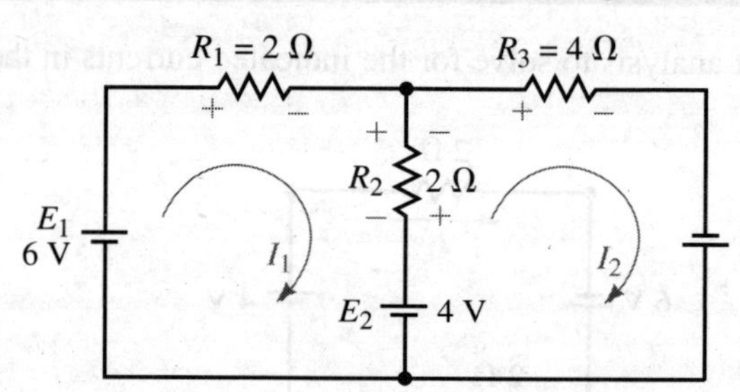

FIGURE 8–22

Solution

Step 1: Loop currents are assigned as shown in Figure 8–22. These currents are designated I_1 and I_2.

Step 2: Voltage polarities are assigned according to the loop currents. Notice that the resistor R_2 has two different voltage polarities due to the different loop currents.

Step 3: The loop equations are written by applying Kirchhoff's voltage law in each of the loops. The equations are as follows:

Loop 1: $6\ \text{V} - (2\ \Omega)I_1 - (2\ \Omega)I_1 + (2\ \Omega)I_2 - 4\ \text{V} = 0$

Loop 2: $4\ \text{V} - (2\ \Omega)I_2 + (2\ \Omega)I_1 - (4\ \Omega)I_2 + 2\ \text{V} = 0$

Note that the voltage across R_2 due to the currents I_1 and I_2 is indicated as two separated terms, where one term represents a voltage drop in the direction of I_1 and the other term represents a voltage rise in the same direction. The magnitude and polarity of the voltage across R_2 are determined by the actual size and directions of the loop currents. The preceding loop equations may be simplified as follows:

Loop 1: $(4\ \Omega)I_1 - (2\ \Omega)I_2 = 2\ \text{V}$

Loop 2: $-(2\ \Omega)I_1 + (6\ \Omega)I_2 = 6\ \text{V}$

Using determinants, the loop equations are easily solved as

$$I_1 = \frac{\begin{vmatrix} 2 & -2 \\ 6 & 6 \end{vmatrix}}{\begin{vmatrix} 4 & -2 \\ -2 & 6 \end{vmatrix}} = \frac{12 + 12}{24 - 4} = \frac{24}{20} = 1.20 \text{ A}$$

and

$$I_2 = \frac{\begin{vmatrix} 4 & 2 \\ -2 & 6 \end{vmatrix}}{\begin{vmatrix} 4 & -2 \\ -2 & 6 \end{vmatrix}} = \frac{24 + 4}{24 - 4} = \frac{28}{20} = 1.40 \text{ A}$$

From these results, we see that the currents through resistors R_1 and R_3 are I_1 and I_2, respectively.

The branch current for R_2 is found by combining the loop currents through this resistor:

$$I_{R_2} = 1.40 \text{ A} - 1.20 \text{ A} = 0.20 \text{ A (upward)}$$

The results obtained by using mesh analysis are exactly the same as those obtained by branch-current analysis. Whereas branch-current analysis required three equations, this approach requires the solution of only two simultaneous linear equations. Mesh analysis also requires that only Kirchhoff's voltage law be applied and clearly illustrates why mesh analysis is preferred to branch-current analysis.

If the circuit being analyzed contains current sources, the procedure is a bit more complicated. The circuit may be simplified by converting the current source(s) to voltage sources and then solving the resulting network using the procedure shown in the previous example. Alternatively, you may not wish to alter the circuit, in which case the current source will provide one of the loop currents.

EXAMPLE 8–11

Determine the current through the 8-V battery for the circuit shown in Figure 8–23.

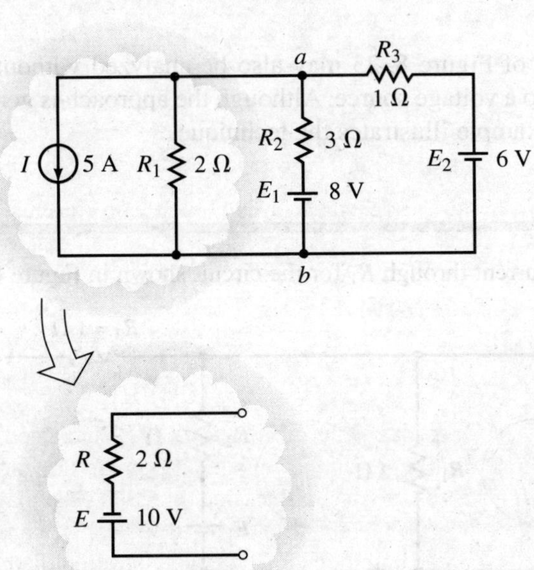

FIGURE 8–23

Solution Convert the current source into an equivalent voltage source. The equivalent circuit may now be analyzed by using the loop currents shown in Figure 8–24.

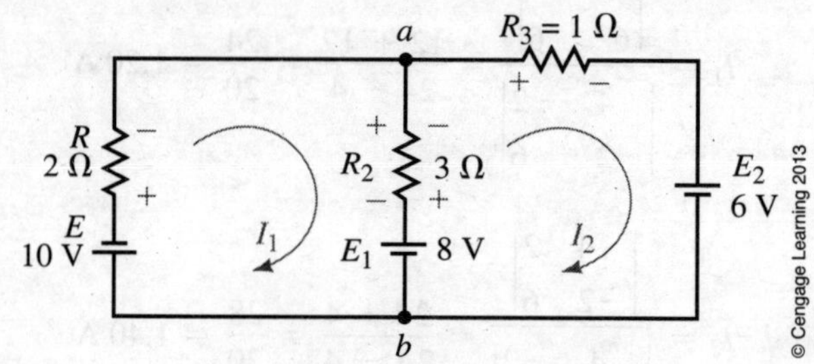

FIGURE 8–24

$$\text{Loop 1:}\quad -10\text{ V} - (2\ \Omega)I_1 - (3\ \Omega)I_1 + (3\ \Omega)I_2 - 8\text{ V} = 0$$
$$\text{Loop 2:}\quad 8\text{ V} - (3\ \Omega)I_2 + (3\ \Omega)I_1 - (1\ \Omega)I_2 - 6\text{ V} = 0$$

Rewriting the linear equations, you get the following:

$$\text{Loop 1:}\quad (5\ \Omega)I_1 - (3\ \Omega)I_2 = -18\text{ V}$$
$$\text{Loop 2:}\quad -(3\ \Omega)I_1 + (4\ \Omega)I_2 = 2\text{ V}$$

Solving the equations using determinants, we have the following:

$$I_1 = \frac{\begin{vmatrix} -18 & 3 \\ 2 & 4 \end{vmatrix}}{\begin{vmatrix} 5 & -3 \\ -3 & 4 \end{vmatrix}} = -\frac{66}{11} = -6.00\text{ A}$$

$$I_2 = \frac{\begin{vmatrix} 5 & -18 \\ -3 & 2 \end{vmatrix}}{11} = -\frac{44}{11} = -4.00\text{ A}$$

If the assumed direction of current in the 8-V battery is taken to be I_2, then

$$I = I_2 - I_1 = -4.00\text{ A} - (-6.00\text{ A}) = 2.00\text{ A}$$

The direction of the resultant current is the same as I_2 (upward).

The circuit of Figure 8–23 may also be analyzed without converting the current source to a voltage source. Although the approach is generally not used, the following example illustrates the technique.

EXAMPLE 8–12

Determine the current through R_1 for the circuit shown in Figure 8–25.

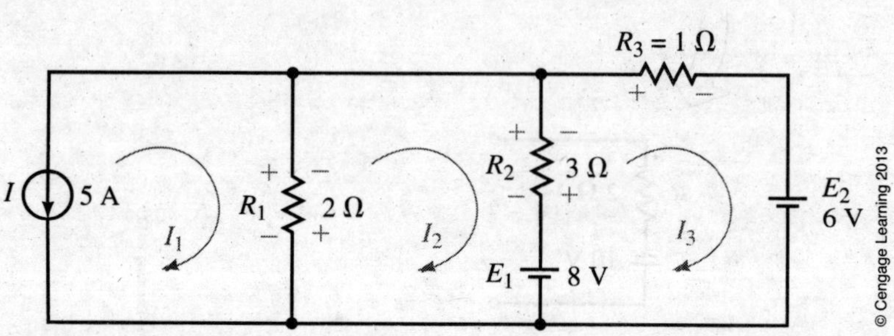

FIGURE 8–25

Solution By inspection, we see that the loop current $I_1 = -5$ A. The mesh equations for the other two loops are as follows:

$$\text{Loop 2:}\quad -(2\ \Omega)I_2 + (2\ \Omega)I_1 - (3\ \Omega)I_2 + (3\ \Omega)I_3 - 8\ \text{V} = 0$$
$$\text{Loop 3:}\quad 8\ \text{V} - (3\ \Omega)I_3 + (3\ \Omega)I_2 - (1\ \Omega)I_3 - 6\ \text{V} = 0$$

Although it is possible to analyze the circuit by solving three linear equations, it is easier to substitute the known value $I_1 = -5$ A into the mesh equation for loop 2, which may now be written as

$$\text{Loop 2:}\quad -(2\ \Omega)I_2 - 10\ \text{V} - (3\ \Omega)I_2 + (3\ \Omega)I_3 - 8\ \text{V} = 0$$

The loop equations may now be simplified as

$$\text{Loop 2:}\quad (5\ \Omega)I_2 - (3\ \Omega)I_3 = -18\ \text{V}$$
$$\text{Loop 3:}\quad -(3\ \Omega)I_2 + (4\ \Omega)I_3 = 2\ \text{V}$$

The simultaneous linear equations are solved as follows:

$$I_2 = \frac{\begin{vmatrix} -18 & -3 \\ 2 & 4 \end{vmatrix}}{\begin{vmatrix} 5 & -3 \\ -3 & 4 \end{vmatrix}} = -\frac{66}{11} = -6.00\ \text{A}$$

$$I_3 = \frac{\begin{vmatrix} 5 & -18 \\ -3 & 2 \end{vmatrix}}{11} = -\frac{44}{11} = -4.00\ \text{A}$$

The calculated values of the assumed reference currents allow us to determine the actual current through the various resistors as follows:

$$I_{R_1} = I_1 - I_2 = -5\ \text{A} - (-6\ \text{A}) = 1.00\ \text{A} \quad \text{(downward)}$$
$$I_{R_2} = I_3 - I_2 = -4\ \text{A} - (-6\ \text{A}) = 2.00\ \text{A} \quad \text{(upward)}$$
$$I_{R_3} = -I_3 = 4.00\ \text{A} \quad \text{(left)}$$

These results are consistent with those obtained in Example 8–9.

Format Approach for Mesh Analysis

A very simple technique may be used to write the mesh equations for any linear bilateral network. When this format approach is used, the simultaneous linear equations for a network having n independent loops will appear as follows:

$$R_{11}I_1 - R_{12}I_2 - R_{13}I_3 - \cdots - R_{1n}I_n = E_1$$
$$-R_{21}I_1 + R_{22}I_2 - R_{23}I_3 - \cdots - R_{2n}I_n = E_2$$
$$\vdots$$
$$-R_{n1}I_1 - R_{n2}I_2 - R_{n3}I_3 - \cdots + R_{nn}I_n = E_n$$

The terms $R_{11}, R_{22}, R_{33}, \ldots, R_{nn}$ represent the total resistance in each loop and are found by simply adding all the resistances in a particular loop. The remaining resistance terms are called the mutual resistance terms. These resistances represent resistance which is shared between two loops. For example, the mutual resistance R_{12} is the resistance in loop 1, which is located in the branch between loop 1 and loop 2. If there is no resistance between two loops, this term will be zero.

The terms containing $R_{11}, R_{22}, R_{33}, \ldots, R_{nn}$ are positive, and all of the mutual resistance terms are negative. This characteristic occurs because all currents are assumed to be clockwise.

If the linear equations are correctly written, you will find that the coefficients along the principal diagonal ($R_{11}, R_{22}, R_{33}, \ldots, R_{nn}$) will be positive. All other coefficients will be negative. Also, if the equations are correctly written, the terms will be symmetrical about the principal diagonal—for example, $R_{12} = R_{21}$.

The terms $E_1, E_2, E_3, \ldots, E_n$ are the summation of the voltage rises in the direction of the loop currents. If a voltage source appears in the branch shared by two loops, it will be included in the calculation of the voltage rise for each loop.

The method used in applying the format approach of mesh analysis is as follows:

1. Convert current sources into equivalent voltage sources.
2. Assign clockwise currents to each independent closed loop in the network.
3. Write the simultaneous linear equations in the format outlined.
4. Solve the resulting simultaneous linear equations.

EXAMPLE 8–13

Solve for the currents through R_2 and R_3 in the circuit of Figure 8–26.

Solution

Step 1: Although we see that the circuit has a current source, it may not be immediately evident how the source can be converted into an equivalent voltage source. Redrawing the circuit into a more recognizable form, as shown in Figure 8–27, we see that the 2-mA current source is in parallel with a 6-kΩ resistor. The source conversion is also illustrated in Figure 8–27.

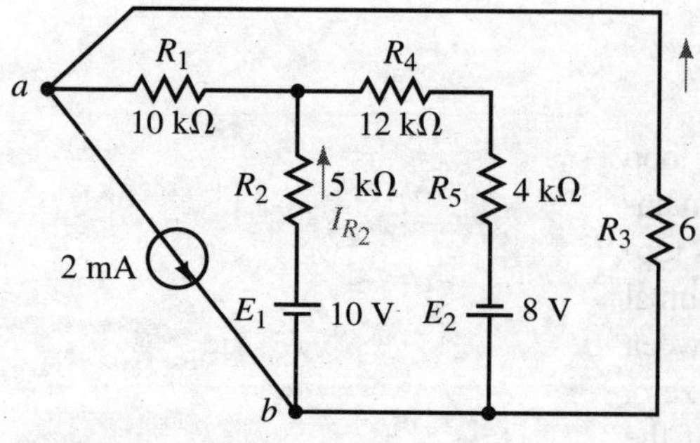

FIGURE 8–26

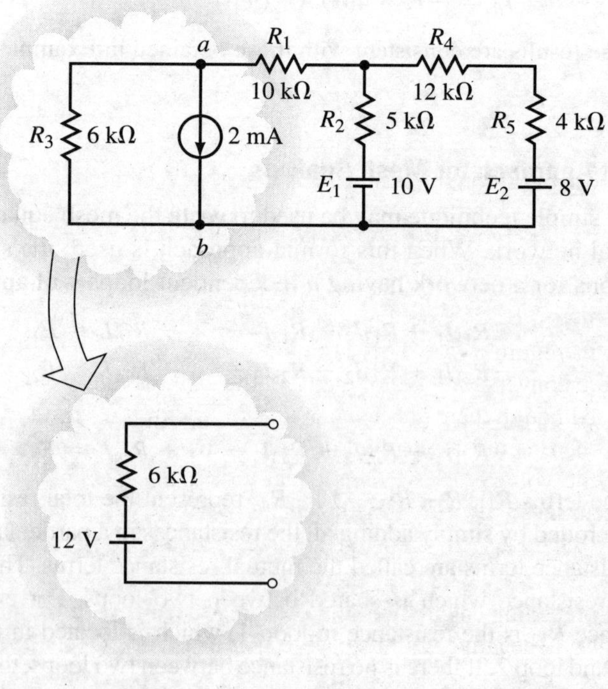

FIGURE 8–27

Step 2: Redrawing the circuit is further simplified by labelling some of the nodes, in this case a and b. After performing a source conversion, we have the two-loop circuit shown in Figure 8–28. The current directions for I_1 and I_2 are also illustrated.

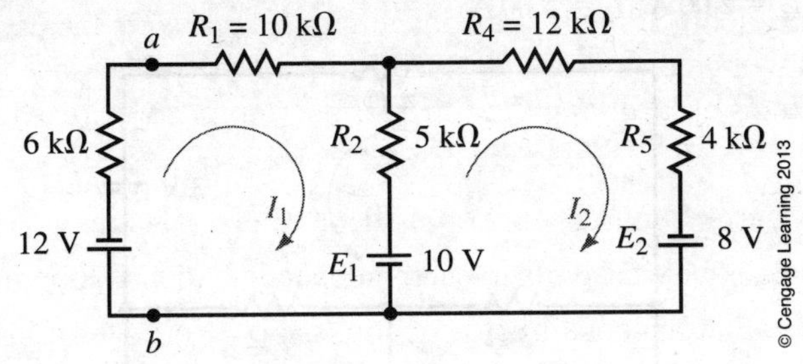

FIGURE 8–28

Step 3: The loop equations are

$$\text{Loop 1:} \quad (6\text{ k}\Omega + 10\text{ k}\Omega + 5\text{ k}\Omega)I_1 - (5\text{ k}\Omega)I_2 = -12\text{ V} - 10\text{ V}$$
$$\text{Loop 2:} \quad -(5\text{ k}\Omega)I_1 + (5\text{ k}\Omega + 12\text{ k}\Omega + 4\text{ k}\Omega)I_2 = 10\text{ V} + 8\text{ V}$$

In loop 1, both voltages are negative since they appear as voltage drops when following the direction of the loop current.

These equations are rewritten as

$$(21\text{ k}\Omega)I_1 - (5\text{ k}\Omega)I_2 = -22\text{ V}$$
$$-(5\text{ k}\Omega)I_1 + (21\text{ k}\Omega)I_2 = 18\text{ V}$$

Step 4: In order to simplify the solution of the previous linear equations, we may eliminate the units (kΩ and V) from our calculations. By inspection, we see that the units for current must be in milliamps. Solving for the currents I_1 and I_2 results in:

$$I_1 = -0.894\text{ mA}$$

and

$$I_2 = 0.644\text{ mA}$$

The current through resistor R_2 is easily determined to be

$$I_2 - I_1 = 0.644\text{ mA} - (-0.894\text{ mA}) = 1.54\text{ mA} \quad \text{(upward)}$$

The current through R_3 is not found as easily. A common mistake is to say that the current in R_3 is the same as the current through the 6-kΩ resistor of the circuit in Figure 8–28. **This is not the case.** Since this resistor was part of the source conversion it is no longer in the same location as in the original circuit.

Although there are several ways of finding the required current, the method used here is the application of Ohm's law. If we examine Figure 8–26, we see that the voltage across R_3 is equal to V_{ab}. From Figure 8–28, we see that we determine V_{ab} by using the calculated value of I_1.

$$V_{ab} = -(6\text{ k}\Omega)I_1 - 12\text{ V} = -(6\text{ k}\Omega)(-0.894\text{ mA}) - 12\text{ V} = -6.64\text{ V}$$

The preceding calculation indicates that the current through R_3 is upward (since point a is negative with respect to point b). The current has a value of

$$I_{R3} = \frac{6.64\text{ V}}{6\text{ k}\Omega} = 1.11\text{ mA}$$

PRACTICE PROBLEMS 3

Use mesh analysis to find the loop currents in the circuit of Figure 8–29.

Answers

$I_1 = 3.00$ A, $I_2 = 2.00$ A, $I_3 = 5.00$ A

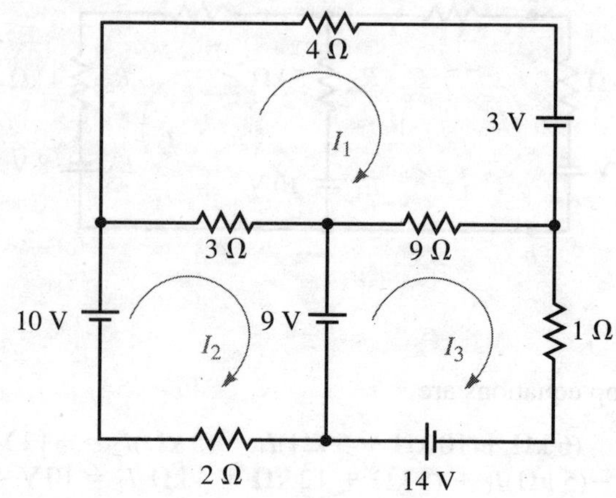

FIGURE 8–29

8.6 Nodal Analysis

In the previous section we applied Kirchhoff's voltage law to arrive at loop currents in a network. In this section we will apply Kirchhoff's current law to determine the potential difference (voltage) at any node with respect to some arbitrary reference point in a network. Once the potentials of all nodes are known, it is a simple matter to determine other quantities such as current and power within the network.

The steps used in solving a circuit using **nodal analysis** are as follows:

1. Arbitrarily assign a reference node within the circuit and indicate this node as *ground.* The reference node is usually located at the bottom of the circuit, although it may be located anywhere.
2. Convert each voltage source in the network to its equivalent current source. This step, although not absolutely necessary, makes further calculations easier to understand.
3. Arbitrarily assign voltages ($V_1, V_2, \ldots, V_n$) to the remaining nodes in the circuit. (Remember that you have already assigned a reference node, so these voltages will all be with respect to the chosen reference.)
4. Arbitrarily assign a current direction to each branch in which there is no current source. Using the assigned current directions, indicate the corresponding polarities of the voltage drops on all resistors.
5. With the exception of the reference node (ground), apply Kirchhoff's current law at each of the nodes. If a circuit has a total of $n + 1$ nodes (including the reference node), there will be n simultaneous linear equations.
6. Rewrite each of the arbitrarily assigned currents in terms of the potential difference across a known resistance.
7. Solve the resulting simultaneous linear equations for the voltages (V_1, $V_2, \ldots, V_n$).

EXAMPLE 8–14

Given the circuit of Figure 8–30, use nodal analysis to solve for the voltage V_{ab}.

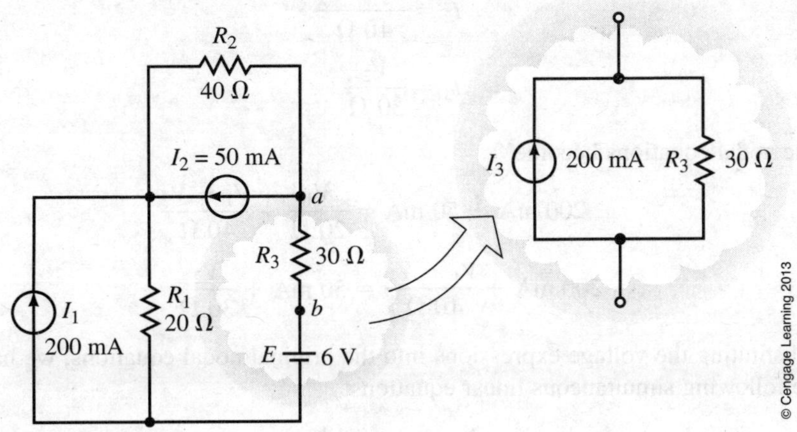

FIGURE 8–30

Solution

Step 1: Select a convenient reference node.

Step 2: Convert the voltage sources into equivalent current sources. The equivalent circuit is shown in Figure 8–31.

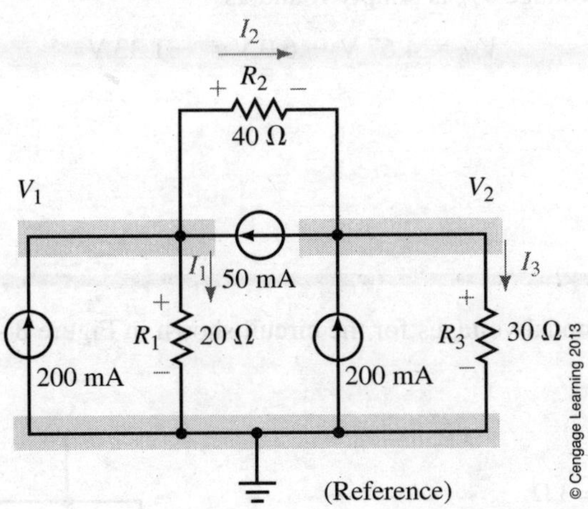

FIGURE 8–31

Steps 3 and 4: Arbitrarily assign node voltages and branch currents. Indicate the voltage polarities across all resistors according to the assumed current directions.

Step 5: We now apply Kirchhoff's current law at the nodes labelled as V_1 and V_2:

Node V_1:
$$\sum I_{\text{entering}} = \sum I_{\text{leaving}}$$
$$200 \text{ mA} + 50 \text{ mA} = I_1 + I_2$$

Node V_2:
$$\sum I_{\text{entering}} = \sum I_{\text{leaving}}$$
$$200 \text{ mA} + I_2 = 50 \text{ mA} + I_3$$

Step 6: The currents are rewritten in terms of the voltages across the resistors as follows:

$$I_1 = \frac{V_1}{20\ \Omega}$$

$$I_2 = \frac{V_1 - V_2}{40\ \Omega}$$

$$I_3 = \frac{V_2}{30\ \Omega}$$

The nodal equations become

$$200\text{ mA} + 50\text{ mA} = \frac{V_1}{20\ \Omega} + \frac{V_1 - V_2}{40\ \Omega}$$

$$200\text{ mA} + \frac{V_1 - V_2}{40\ \Omega} = 50\text{ mA} + \frac{V_2}{30\ \Omega}$$

Substituting the voltage expressions into the original nodal equations, we have the following simultaneous linear equations:

$$\left(\frac{1}{20\ \Omega} + \frac{1}{40\ \Omega}\right)V_1 - \left(\frac{1}{40\ \Omega}\right)V_2 = 0.25\text{ A}$$

$$-\left(\frac{1}{40\ \Omega}\right)V_1 + \left(\frac{1}{30\ \Omega} + \frac{1}{40\ \Omega}\right)V_2 = 0.15\text{ A}$$

If we go back to the original circuit of Figure 8–30, we see that the voltage V_2 is the same as the voltage V_a, namely

$$V_a = 4.67\text{ V} = 6.0\text{ V} + V_{ab}$$

Therefore, the voltage V_{ab} is simply found as

$$V_{ab} = 4.67\text{ V} - 6.0\text{ V} = -1.33\text{ V}$$

Calculator Hint: Most scientific calculators are able to easily work with the linear equations in this example without the need to perform intermediate calculations. The following step shows that by using the inverse function (x^{-1}), the coefficients of the first equation can be entered directly into the calculator.

```
a1,1x1+a1,2x2=b1
 a1,1=20⁻¹+40⁻¹
 a1,2=-40⁻¹
 b1=0.25
```

After entering the coefficients of the second equation, we have the solution as follows:

```
x1=4.88888888889
x2=4.66666666667
```

Therefore the solutions of the given linear equations are, V_1 = 4.89 V and V_2 = 4.67 V.

EXAMPLE 8–15

Determine the nodal voltages for the circuit shown in Figure 8–32.

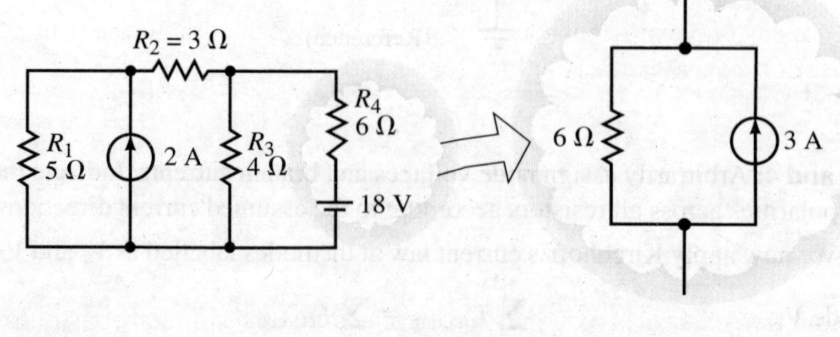

FIGURE 8–32

Solution By following the steps outlined, the circuit may be redrawn as shown in Figure 8–33.

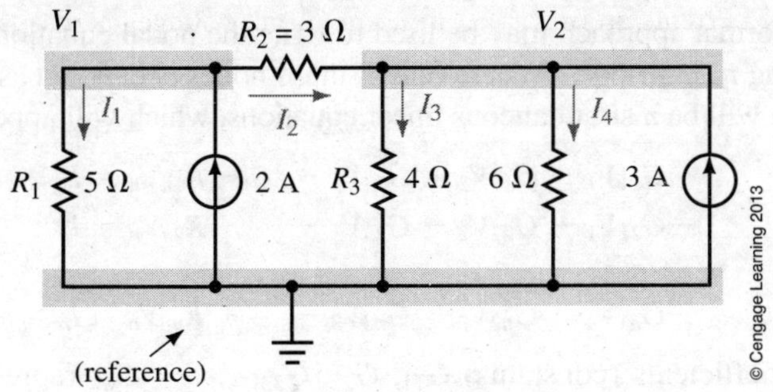

FIGURE 8–33

Applying Kirchhoff's current law to the nodes corresponding to V_1 and V_2, the following nodal equations are obtained:

$$\sum I_{\text{leaving}} = \sum I_{\text{entering}}$$

Node V_1: $\quad I_1 + I_2 = 2\text{ A}$

Node V_2: $\quad I_3 + I_4 = I_2 + 3\text{ A}$

The currents may once again be written in terms of the voltages across the resistors:

$$I_1 = \frac{V_1}{5\ \Omega}$$

$$I_2 = \frac{V_1 - V_2}{3\ \Omega}$$

$$I_3 = \frac{V_2}{4\ \Omega}$$

$$I_4 = \frac{V_2}{6\ \Omega}$$

The nodal equations become

Node V_1: $\quad \dfrac{V_1}{5\ \Omega} + \dfrac{(V_1 - V_2)}{3\ \Omega} = 2\text{ A}$

Node V_2: $\quad \dfrac{V_2}{4\ \Omega} + \dfrac{V_2}{6\ \Omega} = \dfrac{(V_1 - V_2)}{3\ \Omega} + 3\text{ A}$

These equations may now be simplified as

Node V_1: $\quad \left(\dfrac{1}{5\ \Omega} + \dfrac{1}{3\ \Omega}\right)V_1 - \left(\dfrac{1}{3\ \Omega}\right)V_2 = 2\text{ A}$

Node V_2: $\quad -\left(\dfrac{1}{3\ \Omega}\right)V_1 + \left(\dfrac{1}{4\ \Omega} + \dfrac{1}{6\ \Omega} + \dfrac{1}{3\ \Omega}\right)V_2 = 3\text{ A}$

We may use any of the previously used methods to solve the simultaneous linear equations. Consequently, the solutions for V_1 and V_2 are:

$$V_1 = 8.65\text{ V}$$

and

$$V_2 = 7.85\text{ V}$$

In the previous two examples, you may have noticed that the simultaneous linear equations have a format similar to that developed for mesh analysis. When we wrote the nodal equation for node V_1 the coefficient for the variable V_1 was positive, and it had a magnitude given by the summation of the conductance attached to this node. The coefficient for the variable V_2 was negative and had a magnitude given by the mutual conductance between nodes V_1 and V_2.

Format Approach

A simple format approach may be used to write the nodal equations for any network having $n + 1$ nodes. Where one of these nodes is denoted as the reference node, there will be n simultaneous linear equations, which will appear as follows:

$$
\begin{aligned}
G_{11}V_1 - G_{12}V_2 - G_{13}V_3 - \cdots - R_{1n}V_n &= I_1 \\
-G_{21}V_1 + G_{22}V_2 - G_{23}V_3 - \cdots - R_{2n}V_n &= I_2 \\
&\vdots \\
-G_{n1}V_1 - G_{n2}V_2 - G_{n3}V_3 - \cdots + R_{nn}V_n &= I_n
\end{aligned}
$$

The coefficients (constants) G_{11}, G_{22}, G_{33}, . . . , G_{nn} represent the summation of the conductances attached to the particular node. The remaining coefficients are called the **mutual conductance** terms. For example, the mutual conductance G_{23} is the conductance attached to node V_2, which is common to node V_3. If there is no conductance that is common to two nodes, then this term would be zero. Notice that the terms G_{11}, G_{22}, G_{33}, . . . , G_{nn} are positive and that the mutual conductance terms are negative. Further, if the equations are written correctly, then the terms will be symmetrical about the principal diagonal—for example, $G_{23} = G_{32}$.

The terms V_1, V_2, . . . , V_n are the unknown node voltages. Each voltage represents the potential difference between the node in question and the reference node.

The terms I_1, I_2, . . . , I_n are the summation of current sources entering the node. If a current source has a current such that it is leaving the node, then the current is simply assigned as negative. If a particular current source is shared between two nodes, then this current must be included in both nodal equations.

The method used in applying the format approach of nodal analysis is as follows:

1. Convert voltage sources into equivalent current sources.
2. Label the reference node as ⏚. Label the remaining nodes as V_1, V_2, . . . , V_n.
3. Write the linear equation for each node using the format outlined.
4. Solve the resulting simultaneous linear equations for V_1, V_2, . . . , V_n.

The next examples illustrate how the format approach is used to solve circuit problems.

EXAMPLE 8–16

Determine the nodal voltages for the circuit shown in Figure 8–34.

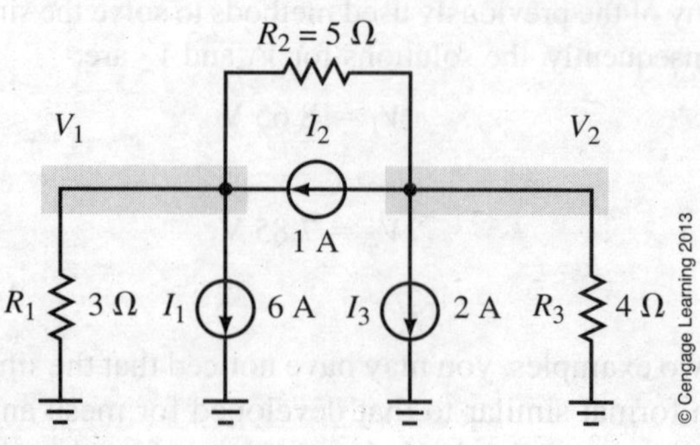

FIGURE 8–34

Solution The circuit has a total of three nodes: the reference node (at a potential of zero volts) and two other nodes, V_1 and V_2.

By applying the format approach for writing the nodal equations, we get two equations:

Node V_1: $$\left(\frac{1}{3\ \Omega}+\frac{1}{5\ \Omega}\right)V_1-\left(\frac{1}{5\ \Omega}\right)V_2=-6\ \text{A}+1\ \text{A}$$

Node V_2: $$-\left(\frac{1}{5\ \Omega}\right)V_1+\left(\frac{1}{5\ \Omega}+\frac{1}{4\ \Omega}\right)V_2=-1\ \text{A}-2\ \text{A}$$

On the right-hand sides of the preceding equations, those currents that are leaving the nodes are given a negative sign.

The solution of the preceding linear equations gives:

$$V_1=-14.25\ \text{V}$$

and

$$V_2=-13.00\ \text{V}$$

EXAMPLE 8–17

Use nodal analysis to find the nodal voltages for the circuit of Figure 8–35. Use the answers to solve for the current through R_1.

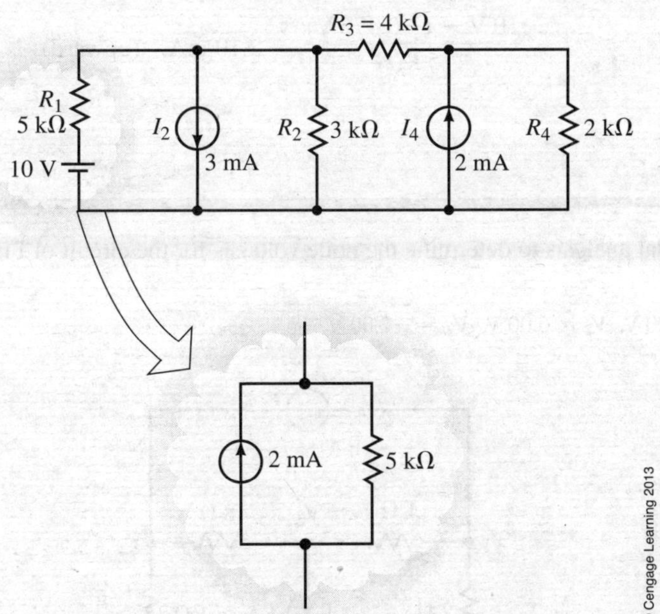

CircuitSim 08-2

FIGURE 8–35

Solution In order to apply nodal analysis, we must first convert the voltage source into its equivalent current source. The resulting circuit is shown in Figure 8–36.

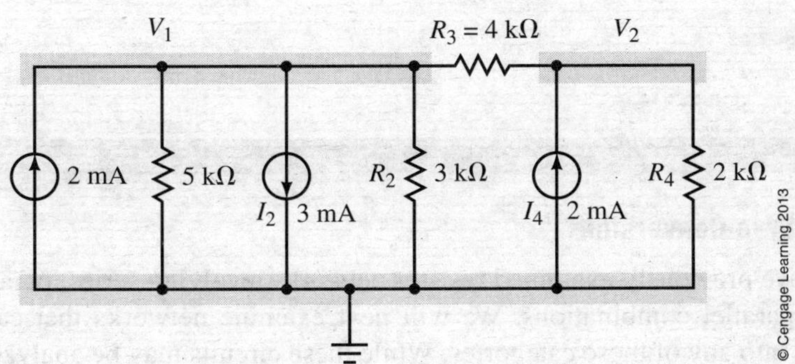

FIGURE 8–36

NOTES...

A common mistake is that the current is determined by using the equivalent circuit rather than the original circuit. You must remember that the circuits are only equivalent external to the conversion.

Labelling the nodes and writing the nodal equations, we obtain the following:

$$\text{Node } V_1: \quad \left(\frac{1}{5\text{ k}\Omega} + \frac{1}{3\text{ k}\Omega} + \frac{1}{4\text{ k}\Omega}\right)V_1 - \left(\frac{1}{4\text{ k}\Omega}\right)V_2 = 2\text{ mA} - 3\text{ mA}$$

$$\text{Node } V_2: \quad -\left(\frac{1}{4\text{ k}\Omega}\right)V_1 + \left(\frac{1}{4\text{ k}\Omega} + \frac{1}{2\text{ k}\Omega}\right)V_2 = 2\text{ mA}$$

Because it is inconvenient to use kilohms and milliamps throughout our calculations, we may eliminate these units in our calculations. You have already seen that any voltage obtained by using these quantities will result in the units being "volts." Therefore, when entering the values into a calculator, we can neglect the units.

The solutions are as follows:

$$V_1 = -0.476\text{ V}$$

and

$$V_2 = 2.51\text{ V}$$

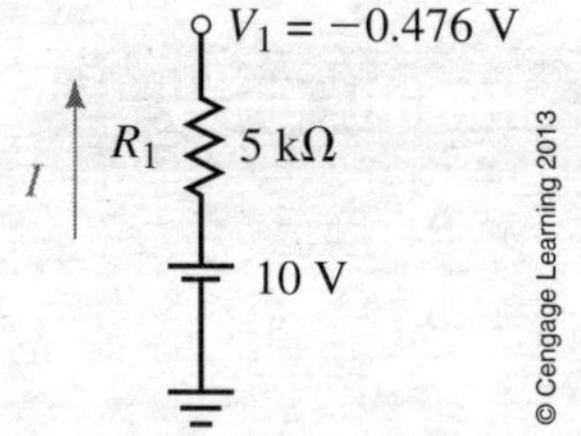

FIGURE 8–37

Using the values derived for the nodal voltages, it is now possible to solve for any other quantities in the circuit. To determine the current through resistor $R_1 = 5\text{ k}\Omega$, we first reassemble the circuit as it appeared originally. Since the node voltage V_1 is the same in both circuits, we use it in determining the desired current. The resistor may be isolated as shown in Figure 8–37.

The current is easily found as

$$I = \frac{10\text{ V} - (-0.476\text{ V})}{5\text{ k}\Omega} = 2.10\text{ mA} \quad \text{(upward)}$$

PRACTICE PROBLEMS 4

Use nodal analysis to determine the node voltages for the circuit of Figure 8–38.

Answers

$V_1 = 3.00\text{ V},\ V_2 = 6.00\text{ V},\ V_3 = -2.00\text{ V}$

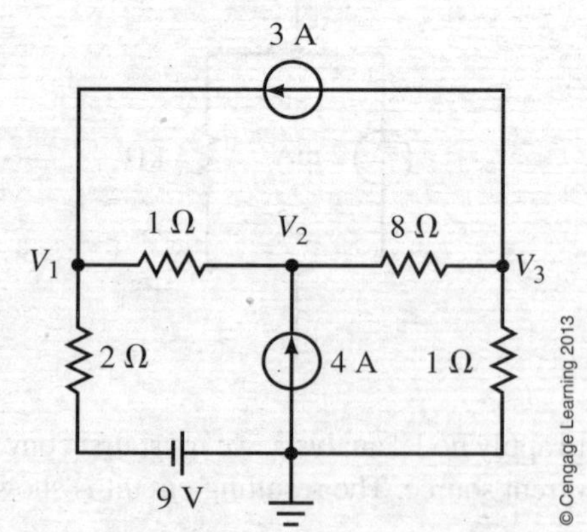

FIGURE 8–38

8.7 Delta-Wye (Pi-Tee) Conversion

Delta-Wye Conversion

You have previously examined resistor networks involving series, parallel, and series-parallel combinations. We will next examine networks that cannot be placed into any of these categories. While these circuits may be analyzed using techniques developed earlier in this chapter, there is an easier approach. For example, consider the circuit shown in Figure 8–39.

This circuit could be analyzed using mesh analysis. However, you see that the analysis would involve solving four simultaneous linear equations, since there are four separate loops in the circuit. If we were to use nodal analysis, the solution would require determining three node voltages, since there are three nodes in addition to a reference node. Unless a computer is used, both techniques are very time-consuming and prone to error.

As you have already seen, it is occasionally easier to examine a circuit after it has been converted to some equivalent form. We will now develop a technique for converting a circuit from a **delta** (or **pi**) into an equivalent **wye** (or **tee**) **circuit.** Consider the circuits shown in Figure 8–40. We start by making the assumption that the networks shown in Figure 8–40(a) are equivalent to those shown in Figure 8–40(b). Then, using this assumption, we will determine the mathematical relationships between the various resistors in the equivalent circuits.

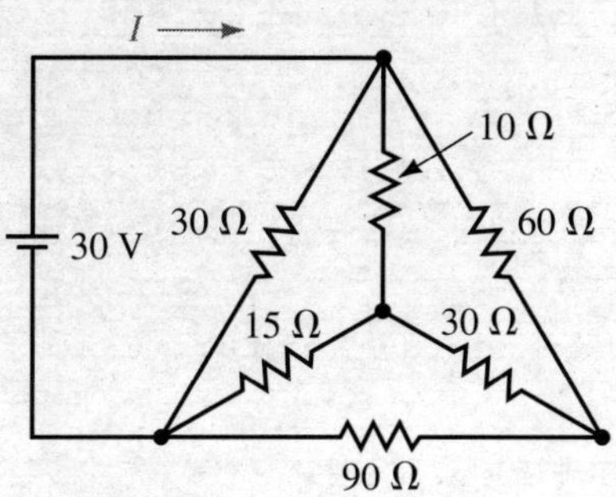

FIGURE 8–39

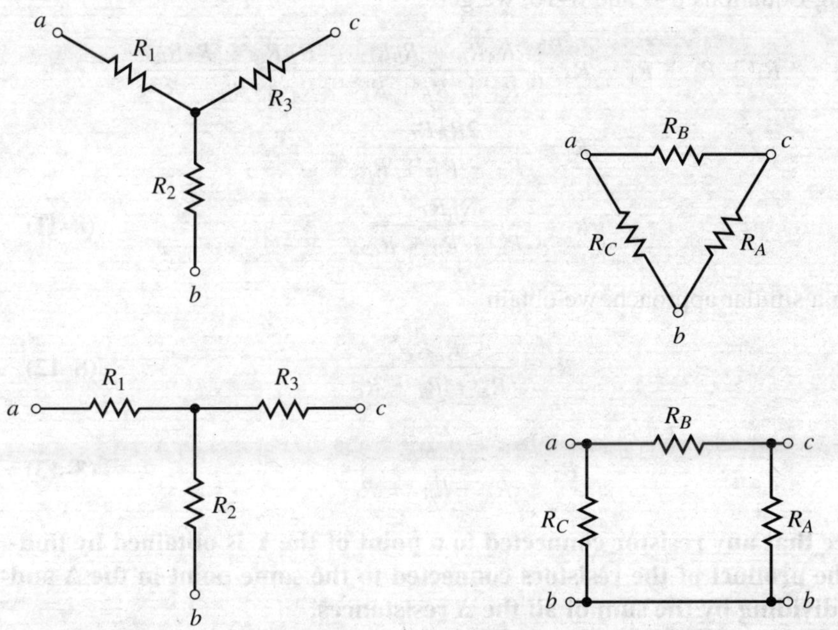

(a) Wye (Y) or Tee (T) network (b) Delta (Δ) or Pi (Π) network

FIGURE 8–40

The circuit of Figure 8–40(a) can be equivalent to the circuit of Figure 8–40(b) only if the resistance "seen" between any two terminals is exactly the same. If we were to connect a source between terminals *a* and *b* of the Y, the resistance between the terminals would be

$$R_{ab} = R_1 + R_2 \tag{8–5}$$

But the resistance between terminals *a* and *b* of the Δ is

$$R_{ab} = R_C \| (R_A + R_B) \tag{8–6}$$

Combining Equations 8–5 and 8–6, we get

$$R_1 + R_2 = \frac{R_C(R_A + R_B)}{R_A + R_B + R_C}$$

$$R_1 + R_2 = \frac{R_A R_C + R_B R_C}{R_A + R_B + R_C} \tag{8–7}$$

Using a similar approach between terminals b and c, we get

$$R_2 + R_3 = \frac{R_A R_B + R_A R_C}{R_A + R_B + R_C} \tag{8–8}$$

and between terminals c and a we get

$$R_1 + R_3 = \frac{R_A R_B + R_B R_C}{R_A + R_B + R_C} \tag{8–9}$$

If Equation 8–8 is subtracted from Equation 8–7, then

$$\begin{aligned} R_1 + R_2 - (R_2 + R_3) &= \frac{R_A R_C + R_B R_C}{R_A + R_B + R_C} - \frac{R_A R_B + R_A R_C}{R_A + R_B + R_C} \\ R_1 - R_3 &= \frac{R_B R_C - R_A R_B}{R_A + R_B + R_C} \end{aligned} \tag{8–10}$$

Adding Equations 8–9 and 8–10, we get

$$\begin{aligned} R_1 + R_3 + R_1 - R_3 &= \frac{R_A R_B + R_B R_C}{R_A + R_B + R_C} + \frac{R_B R_C - R_A R_B}{R_A + R_B + R_C} \\ 2R_1 &= \frac{2R_B R_C}{R_A + R_B + R_C} \\ R_1 &= \frac{R_B R_C}{R_A + R_B + R_C} \end{aligned} \tag{8–11}$$

Using a similar approach, we obtain

$$R_2 = \frac{R_A R_C}{R_A + R_B + R_C} \tag{8–12}$$

$$R_3 = \frac{R_A R_B}{R_A + R_B + R_C} \tag{8–13}$$

Notice that any resistor connected to a point of the Y is obtained by finding the product of the resistors connected to the same point in the Δ and then dividing by the sum of all the Δ resistances.

If all the resistors in a Δ circuit have the same value, R_Δ, then the resulting resistors in the equivalent Y network will also be equal and will have a value given as

$$R_Y = \frac{R_\Delta}{3} \tag{8–14}$$

EXAMPLE 8–18

Find the equivalent Y circuit for the Δ circuit shown in Figure 8–41.

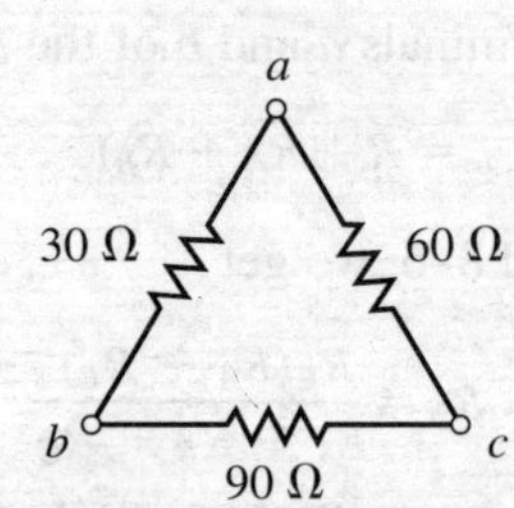

FIGURE 8–41

Solution From the circuit of Figure 8–41, we see that we have the following resistor values:

$$R_A = 90\ \Omega$$
$$R_B = 60\ \Omega$$
$$R_C = 30\ \Omega$$

Applying Equations 8–11 through 8–13 we have the following equivalent Y resistor values:

$$R_1 = \frac{(30\ \Omega)(60\ \Omega)}{30\ \Omega + 60\ \Omega + 90\ \Omega} = \frac{1800\ \Omega}{180} = 10\ \Omega$$

$$R_2 = \frac{(30\ \Omega)(90\ \Omega)}{30\ \Omega + 60\ \Omega + 90\ \Omega} = \frac{2700\ \Omega}{180} = 15\ \Omega$$

$$R_3 = \frac{(60\ \Omega)(90\ \Omega)}{30\ \Omega + 60\ \Omega + 90\ \Omega} = \frac{5400\ \Omega}{180} = 30\ \Omega$$

The resulting circuit is shown in Figure 8–42.

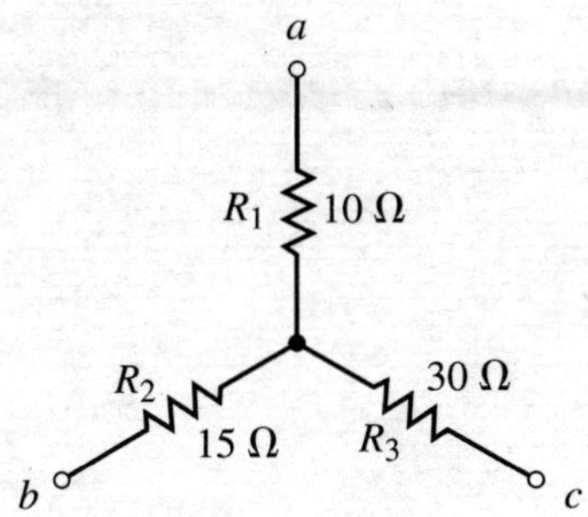

FIGURE 8–42

Wye-Delta Conversion

By using Equations 8–11 to 8–13, it is possible to derive another set of equations that allow the conversion from a Y into an equivalent Δ. Examining Equations 8–11 through 8–13, we see that the following must be true:

$$R_A + R_B + R_C = \frac{R_A R_B}{R_3} = \frac{R_A R_C}{R_2} = \frac{R_B R_C}{R_1}$$

From the preceding expression we may write the following two equations:

$$R_B = \frac{R_A R_1}{R_2} \tag{8–15}$$

$$R_C = \frac{R_A R_1}{R_3} \tag{8–16}$$

Now, substituting Equations 8–15 and 8–16 into Equation 8–11, we have the following:

$$R_1 = \frac{\left(\frac{R_A R_1}{R_2}\right)\left(\frac{R_A R_1}{R_3}\right)}{R_A + \left(\frac{R_A R_1}{R_2}\right) + \left(\frac{R_A R_1}{R_3}\right)}$$

By factoring R_A out of each term in the denominator, we are able to arrive at

$$R_1 = \frac{\left(\frac{R_A R_1}{R_2}\right)\left(\frac{R_A R_1}{R_3}\right)}{R_A\left[1 + \left(\frac{R_1}{R_2}\right) + \left(\frac{R_1}{R_3}\right)\right]}$$

$$R_1 = \frac{\left(\frac{R_A R_1 R_1}{R_2 R_3}\right)}{\left[1 + \left(\frac{R_1}{R_2}\right) + \left(\frac{R_1}{R_3}\right)\right]}$$

$$= \frac{\left(\frac{R_A R_1 R_1}{R_2 R_3}\right)}{\left(\frac{R_1 R_2 + R_1 R_3 + R_2 R_3}{R_2 R_3}\right)}$$

$$= \frac{R_A R_1 R_1}{R_1 R_2 + R_1 R_3 + R_2 R_3}$$

Rewriting the preceding expression gives

$$R_A = \frac{R_1 R_2 + R_1 R_3 + R_2 R_3}{R_1} \quad \text{(8–17)}$$

Similarly,

$$R_B = \frac{R_1 R_2 + R_1 R_3 + R_2 R_3}{R_2} \quad \text{(8–18)}$$

and

$$R_C = \frac{R_1 R_2 + R_1 R_3 + R_2 R_3}{R_3} \quad \text{(8–19)}$$

In general, we see that the resistor in any side of a Δ is found by taking the sum of all two-product combinations of Y resistor values and then dividing by the resistance in the Y, which is located directly opposite to the resistor being calculated.

If the resistors in a Y network are all equal, then the resultant resistors in the equivalent Δ circuit will also be equal and will be given as

$$R_\Delta = 3R_Y \quad \text{(8–20)}$$

EXAMPLE 8–19

Find the Δ network equivalent of the Y network shown in Figure 8–43.

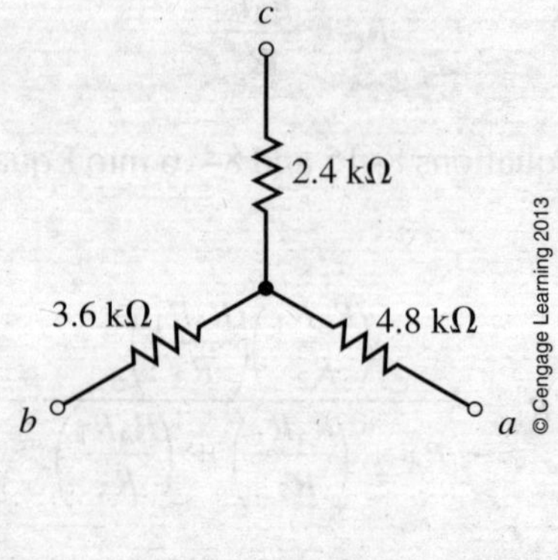

FIGURE 8–43

Solution The equivalent Δ network is shown in Figure 8–44.

The values of the resistors are determined as follows:

$$R_A = \frac{(4.8\text{ k}\Omega)(2.4\text{ k}\Omega) + (4.8\text{ k}\Omega)(3.6\text{ k}\Omega) + (2.4\text{ k}\Omega)(3.6\text{ k}\Omega)}{4.8\text{ k}\Omega}$$

$$= 7.8\text{ k}\Omega$$

$$R_B = \frac{(4.8\text{ k}\Omega)(2.4\text{ k}\Omega) + (4.8\text{ k}\Omega)(3.6\text{ k}\Omega) + (2.4\text{ k}\Omega)(3.6\text{ k}\Omega)}{3.6\text{ k}\Omega}$$

$$= 10.4\text{ k}\Omega$$

$$R_C = \frac{(4.8\text{ k}\Omega)(2.4\text{ k}\Omega) + (4.8\text{ k}\Omega)(3.6\text{ k}\Omega) + (2.4\text{ k}\Omega)(3.6\text{ k}\Omega)}{2.4\text{ k}\Omega}$$

$$= 15.6\text{ k}\Omega$$

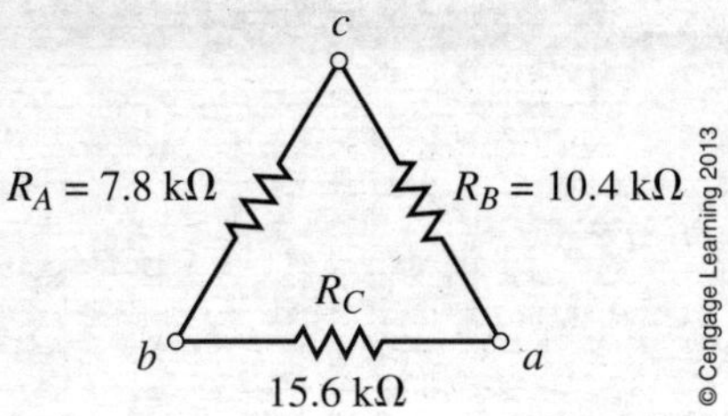

FIGURE 8–44

EXAMPLE 8–20

Given the circuit of Figure 8–45, find the total resistance, R_T, and the total current, *I*.

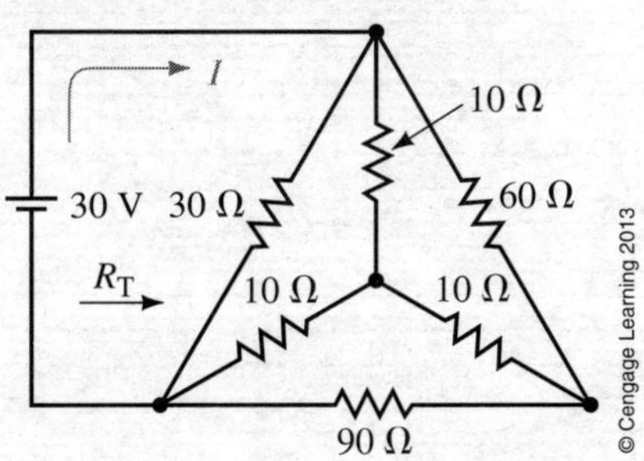

FIGURE 8–45

Solution As is often the case, the given circuit may be solved in one of two ways. We may convert the Δ into its equivalent Y and solve the circuit by placing the resultant branches in parallel, or we may convert the Y into its equivalent Δ. We choose to use the latter conversion since the resistors in the Y have the same value. The equivalent Δ will have all resistors given as

$$R_\Delta = 3(10\ \Omega) = 30\ \Omega$$

The resulting circuit is shown in Figure 8–46(a).

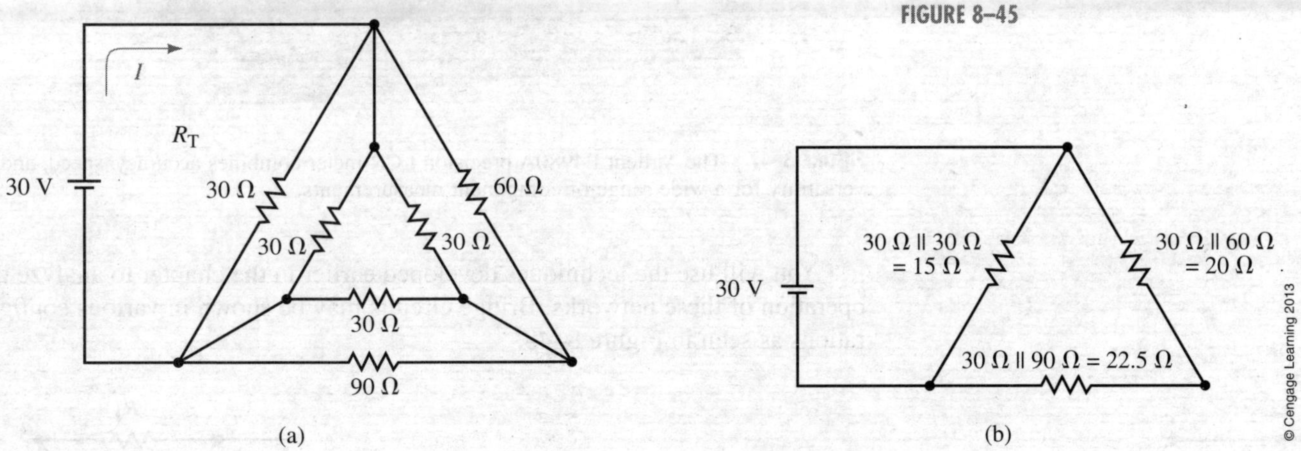

FIGURE 8–46

We see that the sides of the resulting Δ are in parallel, which allows us to simplify the circuit even further as shown in Figure 8–46(b). The total resistance of the circuit is now easily determined as

$$R_T = 15\ \Omega \| (20\ \Omega + 22.5\ \Omega)$$
$$= 11.09\ \Omega$$

This results in a circuit current of

$$I = \frac{30\text{ V}}{11.09\ \Omega} = 2.706\text{ A}$$

PRACTICE PROBLEMS 5

Convert the Δ network of Figure 8–44 into an equivalent Y network. Verify that the result you obtain is the same as that found in Figure 8–43.

Answers
$R_1 = 4.8\ \text{k}\Omega,\ R_2 = 3.6\ \text{k}\Omega,\ R_3 = 2.4\ \text{k}\Omega$

8.8 Bridge Networks

In this section you will be introduced to the **bridge network.** Bridge networks are used in electronic measuring equipment to precisely measure resistance in dc circuits and similar quantities in ac circuits. The bridge circuit was originally used by Sir Charles Wheatstone in the mid-nineteenth century to measure resistance by balancing small currents. The Wheatstone bridge circuit is still used to measure resistance very precisely. The digital bridge, shown in Figure 8–47, is an example of one such instrument.

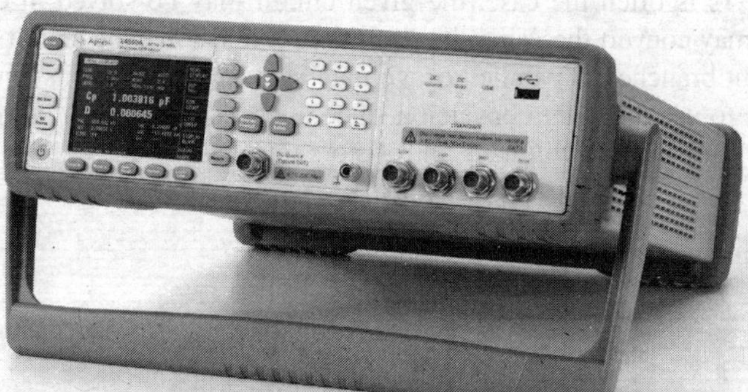

FIGURE 8–47 The Agilent E4980A precision LCR meter combines accuracy, speed, and versatility for a wide range of component measurements.

You will use the techniques developed earlier in the chapter to analyze the operation of these networks. Bridge circuits may be shown in various configurations as seen in Figure 8–48.

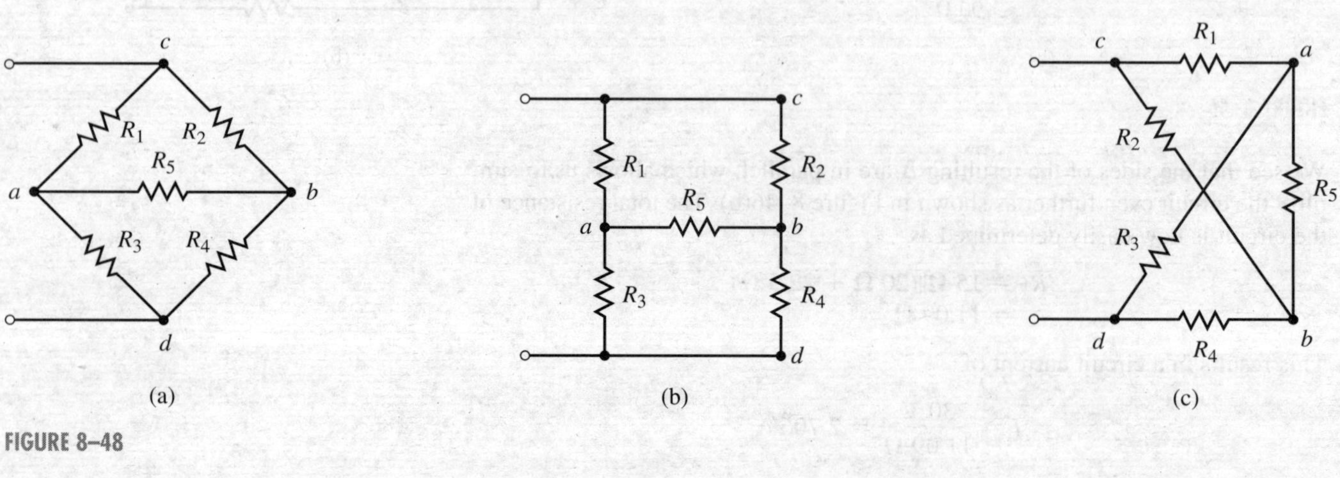

FIGURE 8–48

Although a bridge circuit may appear in one of three forms, you can see that they are equivalent. There are, however, two different states of bridges: the balanced bridge and the unbalanced bridge.

A **balanced bridge** is one in which the current through the resistance R_5 is equal to zero. In practical circuits, R_5 is generally a variable resistor in series with a sensitive galvanometer. When the current through R_5 is zero, then it follows that

$$V_{ab} = (R_5)(0\text{ A}) = 0\text{ V}$$

$$I_{R_1} = I_{R_3} = \frac{V_{cd}}{R_1 + R_3}$$

$$I_{R_2} = I_{R_4} = \frac{V_{cd}}{R_2 + R_4}$$

But the voltage V_{ab} is found as

$$V_{ab} = V_{ad} - V_{bd} = 0$$

Therefore, $V_{ad} = V_{bd}$ and

$$R_3 I_{R_3} = R_4 I_{R_4}$$

$$R_3\left(\frac{V_{cd}}{R_1 + R_3}\right) = R_4\left(\frac{V_{cd}}{R_2 + R_4}\right),$$

which simplifies to

$$\frac{R_3}{R_1 + R_3} = \frac{R_4}{R_2 + R_4}$$

Now, if we invert both sides of the equation and simplify, we get the following:

$$\frac{R_1 + R_3}{R_3} = \frac{R_2 + R_4}{R_4}$$

$$\frac{R_1}{R_3} + 1 = \frac{R_2}{R_4} + 1$$

Finally, by subtracting 1 from each side, we obtain the following ratio for a balanced bridge:

$$\frac{R_1}{R_3} = \frac{R_2}{R_4} \qquad \textbf{(8–21)}$$

From Equation 8–21, we notice that a bridge network is balanced whenever the ratios of the resistors in the two arms are the same.

An unbalanced bridge is one in which the current through R_5 is not zero, and so the preceding ratio does not apply to an unbalanced bridge network. Figure 8–49 illustrates each condition of a bridge network.

If a balanced bridge appears as part of a complete circuit, its analysis is very simple since the resistor R_5 may be removed and replaced with either a short (since $V_{R_5} = 0$) or an open (since $I_{R_5} = 0$).

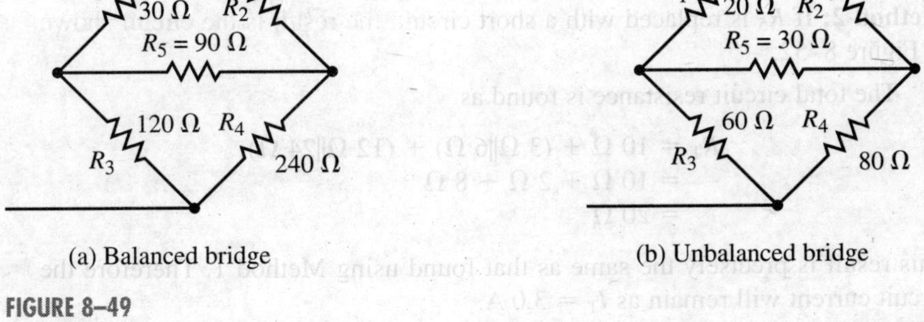

(a) Balanced bridge (b) Unbalanced bridge

FIGURE 8–49

However, if a circuit contains an unbalanced bridge, the analysis is more complicated. In such cases, it is possible to determine currents and voltages by using mesh analysis, nodal analysis, or by using Δ to Y conversion. The following examples illustrate how bridges may be analyzed.

EXAMPLE 8–21

Solve for the currents through R_1 and R_4 in the circuit of Figure 8–50.

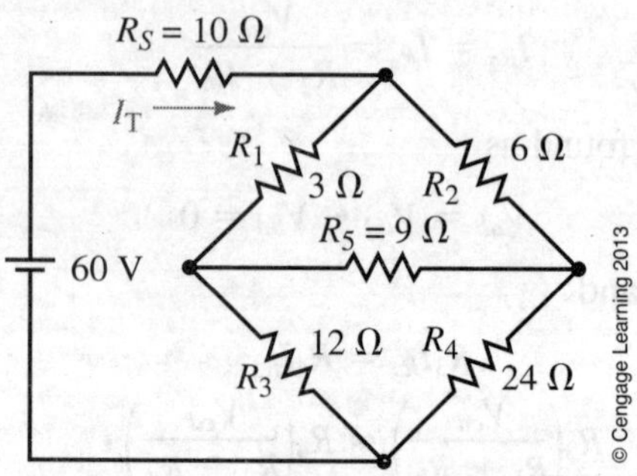

FIGURE 8–50

Solution We see that the bridge of the circuit in Figure 8-50 is balanced (since $R_1/R_3 = R_2/R_4$). Because the circuit is balanced, we may remove R_5 and replace it with either a short circuit (since the voltage across a short circuit is zero) or an open circuit (since the current through an open circuit is zero). The remaining circuit is then solved by one of the methods developed in previous chapters. Both methods will be illustrated to show that the results are exactly the same.

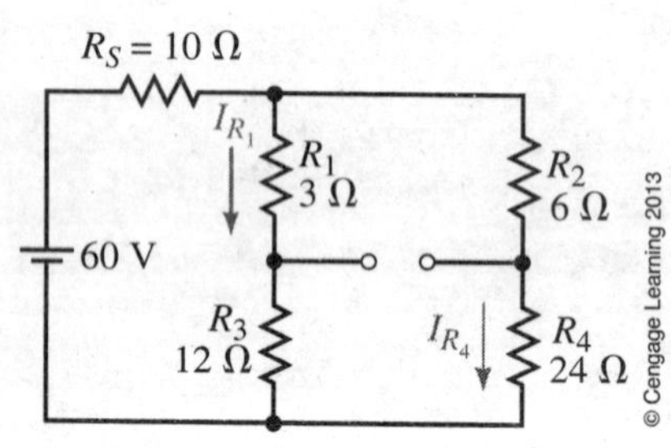

FIGURE 8–51

Method 1: If R_5 is replaced by an open, the result is the circuit shown in Figure 8–51.

The total circuit resistance is found as

$$\begin{aligned} R_T &= 10\ \Omega + (3\ \Omega + 12\ \Omega)\|(6\ \Omega + 24\ \Omega) \\ &= 10\ \Omega + 15\ \Omega\|30\ \Omega \\ &= 20\ \Omega \end{aligned}$$

The circuit current is

$$I_T = \frac{60\ \text{V}}{20\ \Omega} = 3.0\ \text{A}$$

The current in each branch is then found by using the current divider rule:

$$I_{R1} = \left(\frac{30\ \Omega}{30\ \Omega + 15\ \Omega}\right)(3.0\ \text{A}) = 2.0\ \text{A}$$

$$I_{R4} = \frac{10\ \Omega}{24\ \Omega + 6\ \Omega}(3.0\ \text{A}) = 1.0\ \text{A}$$

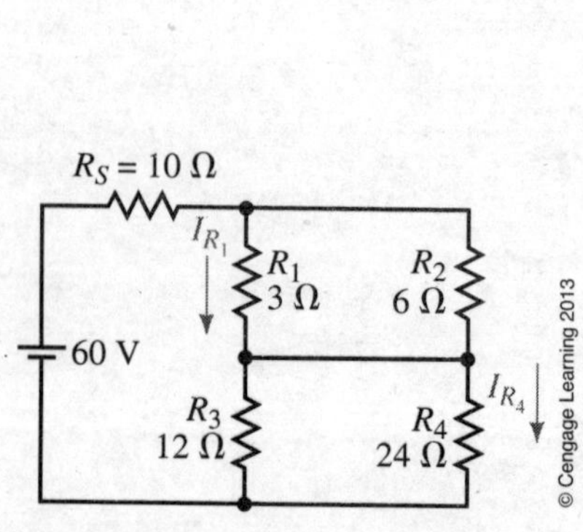

FIGURE 8–52

Method 2: If R_5 is replaced with a short circuit, the result is the circuit shown in Figure 8–52.

The total circuit resistance is found as

$$\begin{aligned} R_T &= 10\ \Omega + (3\ \Omega\|6\ \Omega) + (12\ \Omega\|24\ \Omega) \\ &= 10\ \Omega + 2\ \Omega + 8\ \Omega \\ &= 20\ \Omega \end{aligned}$$

This result is precisely the same as that found using Method 1. Therefore the circuit current will remain as $I_T = 3.0$ A.

The currents through R_1 and R_4 may be found by the current divider rule as

$$I_{R_1} = \left(\frac{6\ \Omega}{6\ \Omega + 3\ \Omega}\right)(3.0\ \text{A}) = 2.0\ \text{A}$$

and

$$I_{R_4} = \left(\frac{12\ \Omega}{12\ \Omega + 24\ \Omega}\right)(3.0\ \text{A}) = 1.0\ \text{A}$$

Clearly, these results are precisely those obtained in Method 1, illustrating that the methods are equivalent. Remember, though, R_5 can be replaced with a short circuit or an open circuit only when the bridge is balanced.

EXAMPLE 8–22

Use mesh analysis to find the currents through R_1 and R_5 in the unbalanced bridge circuit of Figure 8–53.

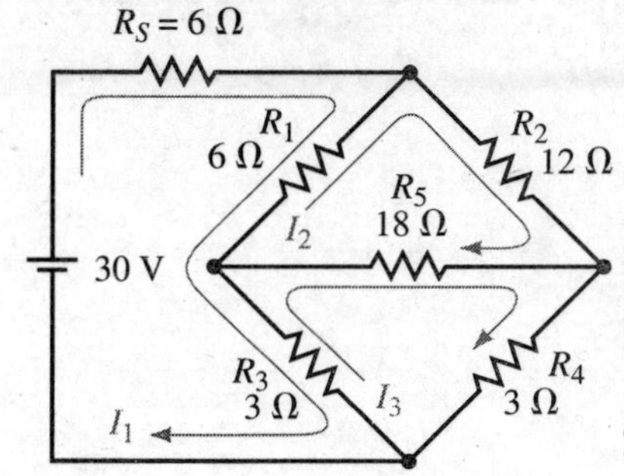

FIGURE 8–53

Solution After assigning loop currents as shown, we write the loop equations as

Loop 1: $(15\ \Omega)I_1 - (6\ \Omega)I_2 - (3\ \Omega)I_3 = 30\ \text{V}$

Loop 2: $-(6\ \Omega)I_1 + (36\ \Omega)I_2 - (18\ \Omega)I_3 = 0$

Loop 3: $-(3\ \Omega)I_1 - (18\ \Omega)I_2 + (24\ \Omega)I_3 = 0$

The loop currents are evaluated as

$$I_1 = 2.586\ \text{A}$$

$$I_2 = 0.948\ \text{A}$$

and

$$I_3 = 1.034\ \text{A}$$

The current through R_1 is found as

$$I_{R_1} = I_1 - I_2 = 2.586\ \text{A} - 0.948\ \text{A} = 1.638\ \text{A}$$

The current through R_5 is found as

$$I_{R_5} = I_3 - I_2 = 1.034\ \text{A} - 0.948\ \text{A}$$
$$= 0.086\ \text{A} \quad \text{(to the right)}$$

The previous example illustrates that if the bridge is not balanced, there will always be some current through resistor R_5. The unbalanced circuit may also be easily analyzed using nodal analysis, as in the following example.

EXAMPLE 8–23

Determine the node voltages and the voltage V_{R_5} for the circuit of Figure 8–54.

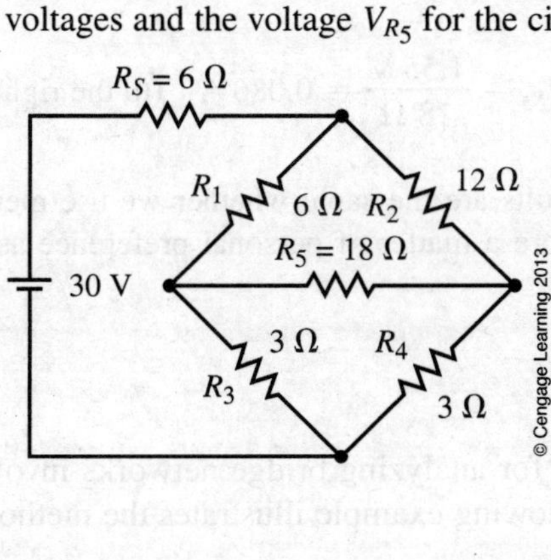

FIGURE 8–54

Solution By converting the voltage source into an equivalent current source, we obtain the circuit shown in Figure 8–55.

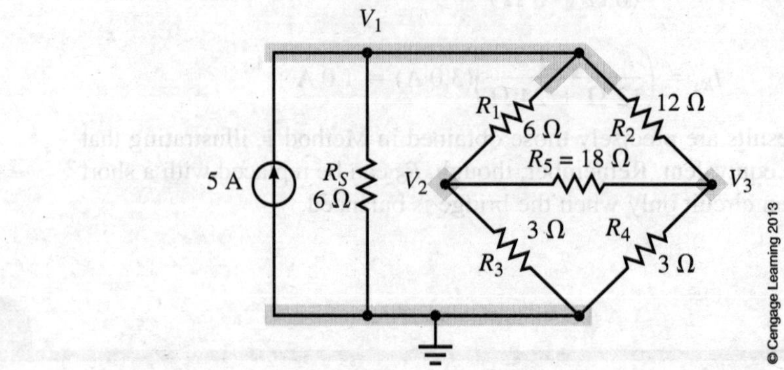

FIGURE 8–55

The nodal equations for the circuit are as follows:

Node 1: $$\left(\frac{1}{6\ \Omega}+\frac{1}{6\ \Omega}+\frac{1}{12\ \Omega}\right)V_1-\left(\frac{1}{6\ \Omega}\right)V_2-\left(\frac{1}{12\ \Omega}\right)V_3=5\ \text{A}$$

Node 2: $$-\left(\frac{1}{6\ \Omega}\right)V_1+\left(\frac{1}{6\ \Omega}+\frac{1}{3\ \Omega}+\frac{1}{18\ \Omega}\right)V_2-\left(\frac{1}{18\ \Omega}\right)V_3=0\ \text{A}$$

Node 3: $$-\left(\frac{1}{12\ \Omega}\right)V_1-\left(\frac{1}{18\ \Omega}\right)V_2+\left(\frac{1}{3\ \Omega}+\frac{1}{12\ \Omega}+\frac{1}{18\ \Omega}\right)V_3=0\ \text{A}$$

Again notice that the elements on the principal diagonal are positive and that the determinant is symmetrical about the principal diagonal.

The node voltages are calculated to be

$$V_1 = 14.48\ \text{A}$$
$$V_2 = 4.66\ \text{V}$$

and

$$V_3 = 3.10\ \text{V}$$

Using the preceding results, we find the voltage across R_5:

$$V_{R_5} = V_2 - V_3 = 4.655\ \text{V} - 3.103\ \text{V} = 1.55\ \text{V}$$

and the current through R_5 is

$$I_{R_5} = \frac{1.55\ \text{V}}{18\ \Omega} = 0.086\ \text{A} \quad \text{(to the right)}$$

As expected, the results are the same whether we use mesh analysis or nodal analysis. It is therefore a matter of personal preference as to which approach should be used.

A final method for analyzing bridge networks involves the use of Δ to Y conversion. The following example illustrates the method used.

EXAMPLE 8–24

Find the current through R_5 for the circuit shown in Figure 8–56.

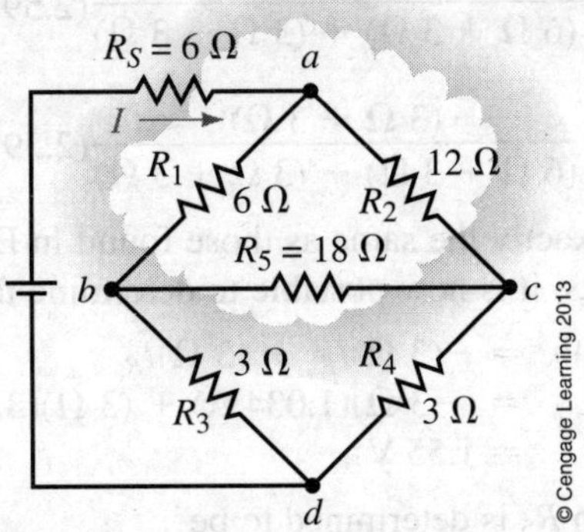

FIGURE 8–56

Solution By inspection we see that this circuit is not balanced, since

$$\frac{R_1}{R_3} \neq \frac{R_2}{R_4}$$

Therefore, the current through R_5 cannot be zero. Notice, also, that the circuit contains two possible Δ configurations. If we choose to convert the top Δ to its equivalent Y, we get the circuit shown in Figure 8–57.

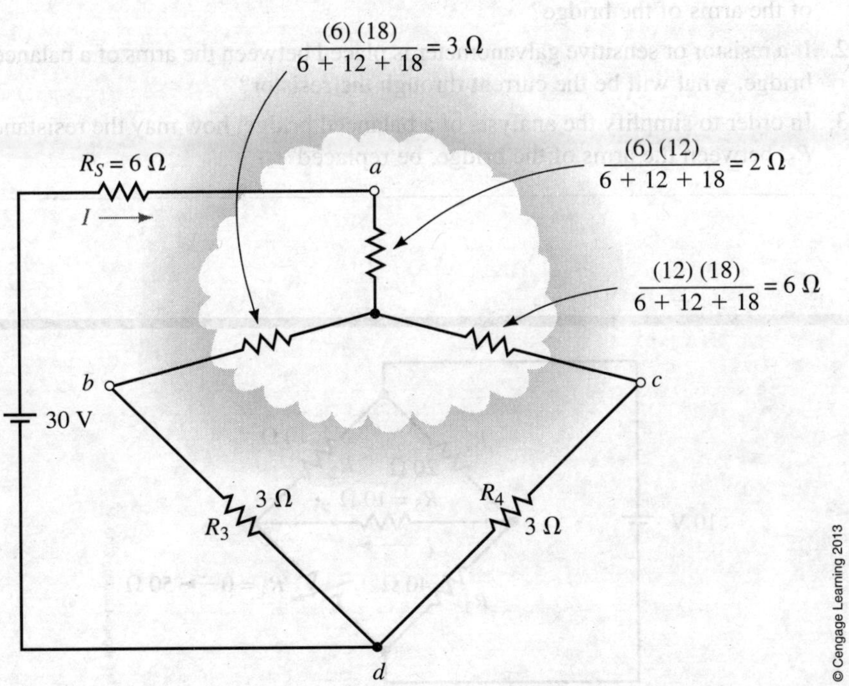

FIGURE 8–57

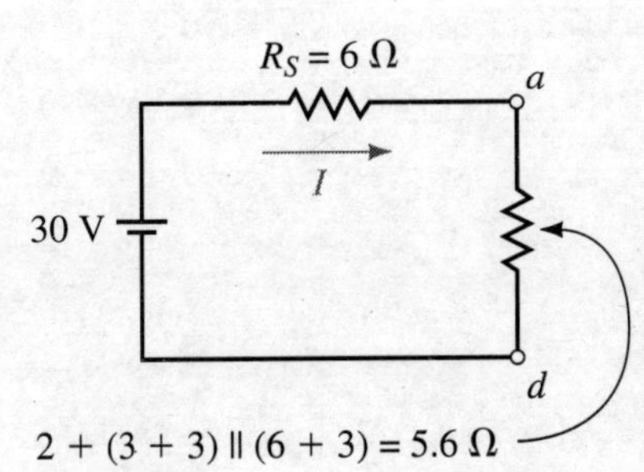

FIGURE 8–58

By combining resistors, it is possible to reduce the complicated circuit to the simple series circuit shown in Figure 8–58.

The circuit of Figure 8–58 is easily analyzed to give a total circuit current of

$$I = \frac{30\ \text{V}}{6\ \Omega + 2\ \Omega + 3.6\ \Omega} = 2.59\ \text{A}$$

Using the calculated current, it is possible to work back to the original circuit. The currents in the resistors R_3 and R_4 are found by using the current divider rule for the corresponding resistor branches, as shown in Figure 8–57.

$$I_{R_3} = \frac{(6\ \Omega + 3\ \Omega)}{(6\ \Omega + 3\ \Omega) + (3\ \Omega + 3\ \Omega)}(2.59\ \text{A}) = 1.55\ \text{A}$$

$$I_{R_4} = \frac{(3\ \Omega + 3\ \Omega)}{(6\ \Omega + 3\ \Omega) + (3\ \Omega + 3\ \Omega)}(2.59\ \text{A}) = 1.03\ \text{A}$$

These results are exactly the same as those found in Examples 8–21 and 8–22. Using these currents, it is now possible to determine the voltage V_{bc} as

$$\begin{aligned} V_{bc} &= -(3\ \Omega)I_{R_4} + (3\ \Omega)I_{R_3} \\ &= (-3\ \Omega)(1.034\ \text{A}) + (3\ \Omega)(3.103\ \text{A}) \\ &= 1.55\ \text{V} \end{aligned}$$

The current through R_5 is determined to be

$$I_{R_5} = \frac{1.55\ \text{V}}{18\ \Omega} = 0.086\ \text{A} \quad \text{(to the right)}$$

IN-PROCESS LEARNING CHECK 2

(Answers are at the end of the chapter.)

1. For a balanced bridge, what will be the value of voltage between the midpoints of the arms of the bridge?
2. If a resistor or sensitive galvanometer is placed between the arms of a balanced bridge, what will be the current through the resistor?
3. In order to simplify the analysis of a balanced bridge, how may the resistance R_5, between the arms of the bridge, be replaced?

PRACTICE PROBLEMS 6

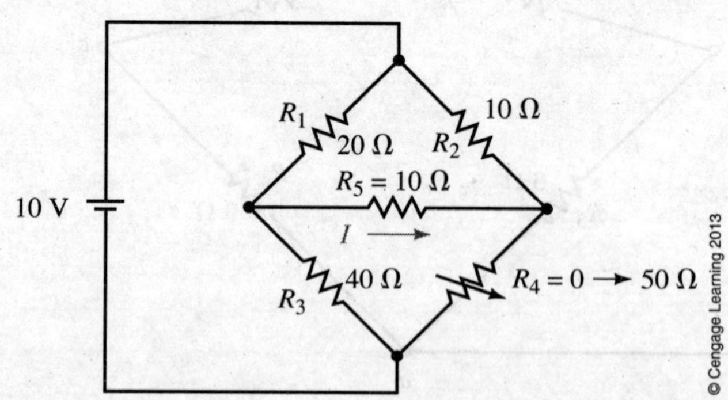

FIGURE 8–59

1. For the circuit shown in Figure 8–59, what value of R_4 will ensure that the bridge is balanced?
2. Determine the current I through R_5 in Figure 8–59 when $R_4 = 0\ \Omega$ and when $R_4 = 50\ \Omega$.

Answers

1. 20 V; 2. 286 mA, −52.6 mA

8.9 Circuit Analysis Using Computers

Multisim

PSPICE

Multisim and PSpice are able to analyze a circuit without the need to convert between voltage and current sources or having to write lengthy linear equations. It is possible to have the program output the value of voltage across or current through any element in a given circuit. The following examples were previously analyzed using several other methods throughout this chapter.

EXAMPLE 8–25

Given the circuit of Figure 8–60, use Multisim to find the voltage V_{ab} and the current through each resistor.

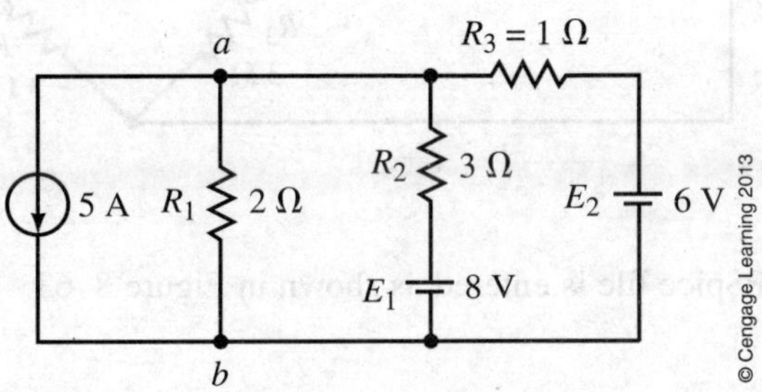

FIGURE 8–60

Solution The circuit is entered as shown in Figure 8–61. The current source is obtained by clicking on the *Signal source* button in the *Virtual Components* toolbar. As before, it is necessary to include a ground symbol in the schematic although the original circuit of Figure 8–60 did not have one. Make sure that all values are changed from the default values to the required circuit values. In this example, we use the Measurement Probe from the Instruments toolbar. Notice that, for simplicity, all unnecessary quantities have been removed from the Properties of the probe.

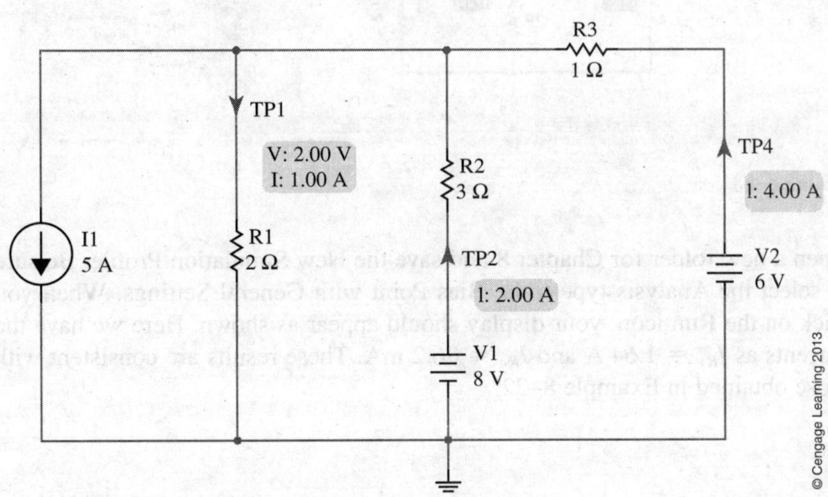

FIGURE 8–61

From the preceding results, we have the following values:

$$V_{ab} = 2.00 \text{ V}$$
$$I_{R_1} = 1.00 \text{ A} \quad \text{(downward)}$$
$$I_{R_2} = 2.00 \text{ A} \quad \text{(upward)}$$
$$I_{R_3} = 4.00 \text{ A} \quad \text{(to the left)}$$

PSpice

EXAMPLE 8–26

Use PSpice to find currents through R_1 and R_5 in the circuit of Figure 8–62.

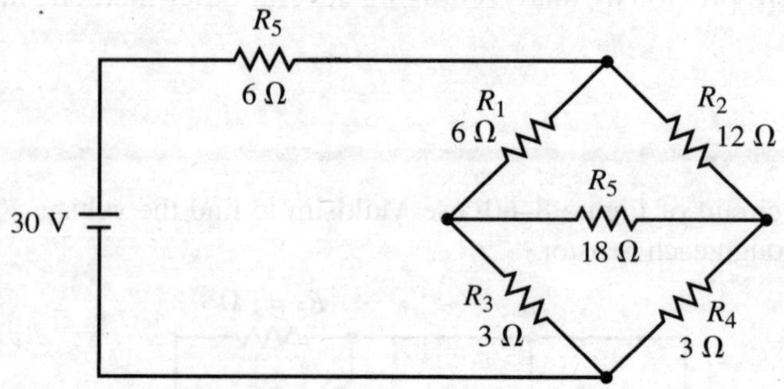

FIGURE 8–62

Solution The PSpice file is entered as shown in Figure 8–63.

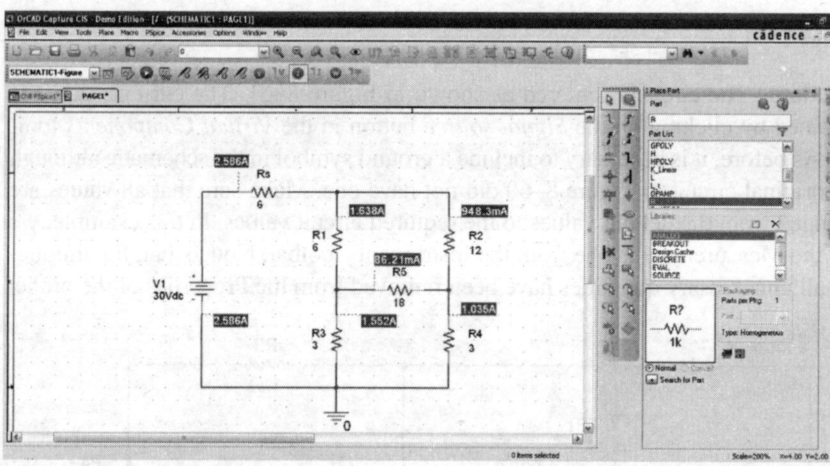

FIGURE 8–63

Open a new folder for Chapter 8 and save the New Simulation Profile. Be sure to select the Analysis type to be Bias Point with General Settings. When you click on the Run icon, your display should appear as shown. Here we have the currents as $I_{R_1} = 1.64$ A and $I_{R_5} = 86.2$ mA. These results are consistent with those obtained in Example 8–22.

PRACTICE PROBLEMS 7

Use Multisim to determine the currents I_T, I_{R_1}, and I_{R_4} in the circuit of Figure 8–50. Compare your results to those obtained in Example 8–21.

Answers

$I_T = 3.00$ A, $I_{R_1} = 2.00$ A, and $I_{R_4} = 1.00$ A

PRACTICE PROBLEMS 8

Use PSpice to input the circuit of Figure 8–50. Determine the value of I_{R_5} when $R_4 = 0\ \Omega$ and when $R_4 = 48\ \Omega$.

Note: Since PSpice cannot let $R_4 = 0\ \Omega$, you will need to let it equal some very small value such as 1 $\mu\Omega$ (1e-6).

Answers

$I_{R_5} = 1.08$ A when $R_4 = 0\ \Omega$ and $I_{R_5} = 0.172$ A when $R_4 = 48\ \Omega$

PRACTICE PROBLEMS 9

Use PSpice to input the circuit of Figure 8–54, so that the output file will provide the currents through R_1, R_2, and R_5. Compare your results to those obtained in Example 8–23.

Putting It into Practice

Strain gauges are manufactured from very fine wire mounted on insulated surfaces that are then glued to large metal structures. These instruments are used by civil engineers to measure the movement and mass of large objects such as bridges and buildings. When the very fine wire of a strain gauge is subjected to stress, its effective length is increased (due to stretching) or decreased (due to compression). This change in length results in a corresponding minute change in resistance. By placing one or more strain gauges into a bridge circuit, it is possible to detect variation in resistance, ΔR. This change in resistance can be calibrated to correspond to an applied force. Consequently, it is possible to use such a bridge as a means of measuring very large masses. Consider that you have two strain gauges mounted in a bridge as shown in the accompanying figure.

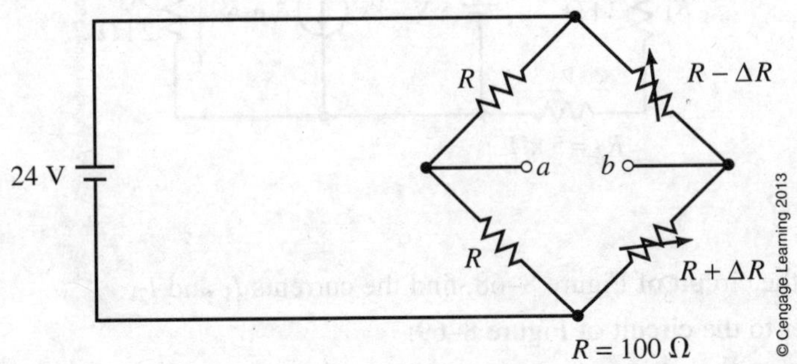

Strain gauge bridge.

The variable resistors R_2 and R_4 are strain gauges that are mounted on opposite sides of a steel girder used to measure very large masses. When a mass is applied to the girder, the strain gauge on one side of the girder will compress, reducing the resistance. The strain gauge on the other side of the girder will stretch, increasing the resistance. When no mass is applied, there will be neither compression nor stretching and so the bridge will be balanced, resulting in a voltage $V_{ab} = 0$ V.

Write an expression for ΔR as a function of V_{ab}. Assume that the scale is calibrated so that resistance variation of $\Delta R = 0.02\ \Omega$ corresponds to a mass of 5000 kg. Determine the measured mass if $V_{ab} = -4.20$ mV.

Problems

8.1 Constant-Current Sources

1. Find the voltage V_S for the circuit shown in Figure 8–64.
2. Find the voltage V_S for the circuit shown in Figure 8–65.

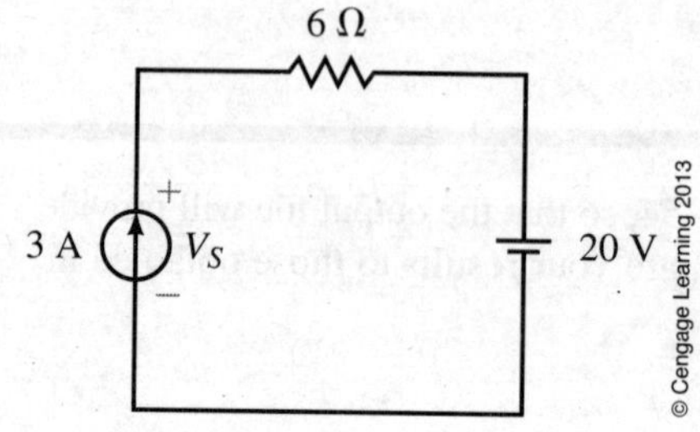

FIGURE 8–64

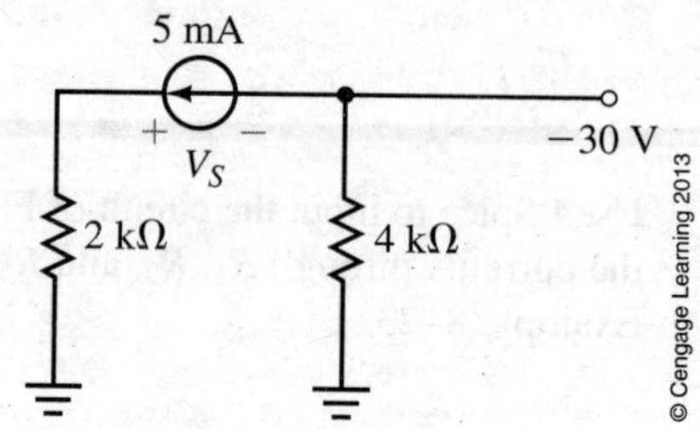

FIGURE 8–65

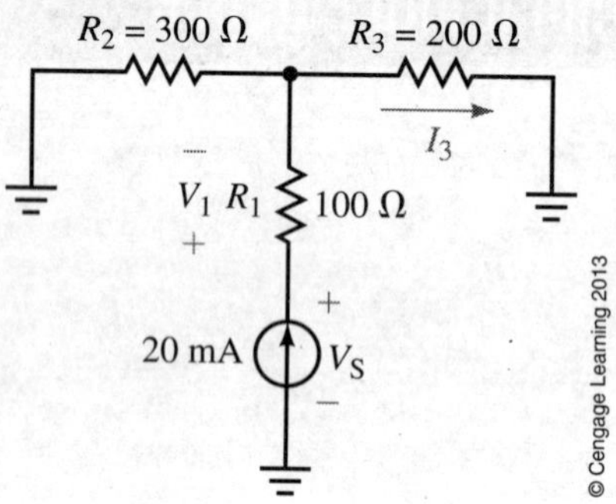

FIGURE 8–66

3. Refer to the circuit of Figure 8–66:
   a. Find the current I_3.
   b. Determine the voltages V_S and V_1.
4. Consider the circuit of Figure 8–67:
   a. Calculate the voltages V_2 and V_S.
   b. Find the currents I and I_3.

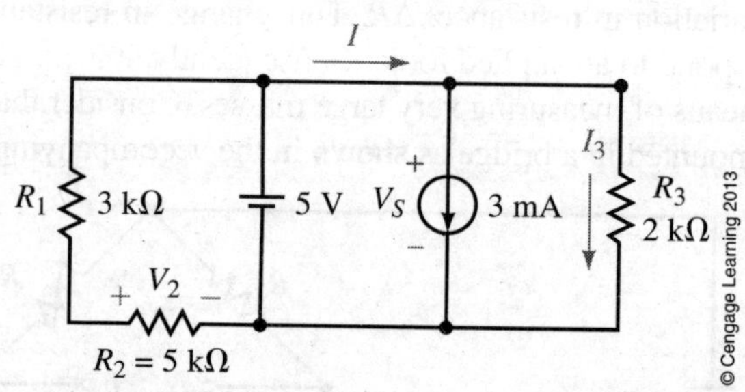

FIGURE 8–67

5. For the circuit of Figure 8–68, find the currents I_1 and I_2.
6. Refer to the circuit of Figure 8–69:
   a. Find the voltages V_S and V_2.
   b. Determine the current I_4.

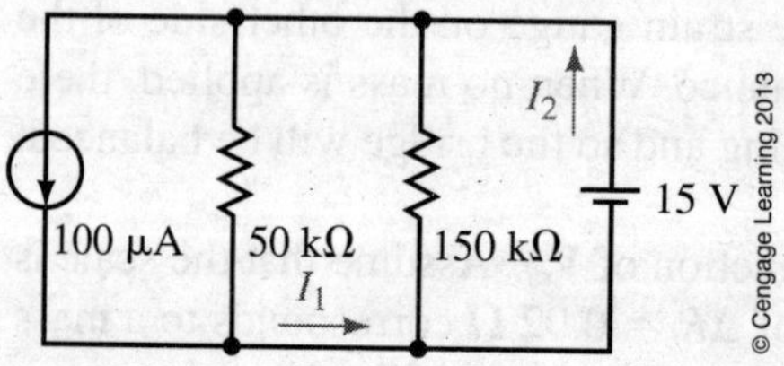

FIGURE 8–68

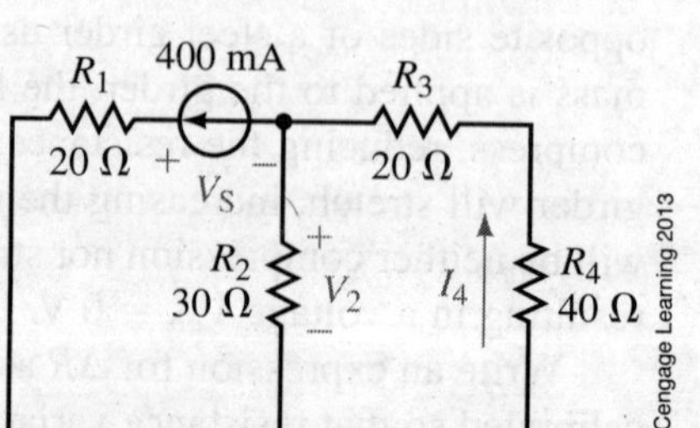

FIGURE 8–69

7. Verify that the power supplied by the sources is equal to the summation of the powers dissipated by the resistors in the circuit of Figure 8–68.
8. Verify that the power supplied by the source in the circuit of Figure 8–69 is equal to the summation of the powers dissipated by the resistors.

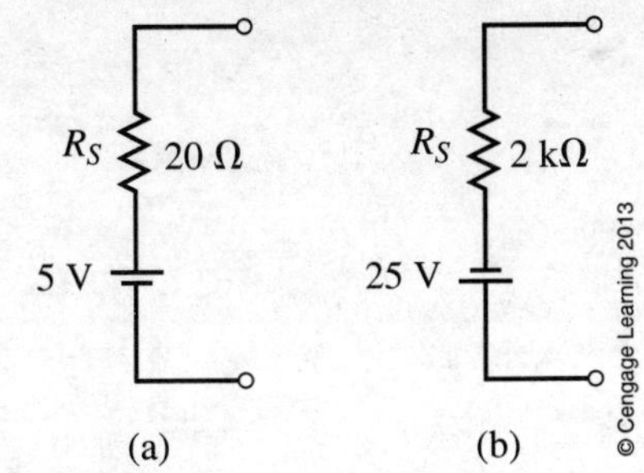

FIGURE 8–70

8.2 Source Conversions

9. Convert each of the voltage sources of Figure 8–70 into its equivalent current source.
10. Convert each of the current sources of Figure 8–71 into its equivalent voltage source.

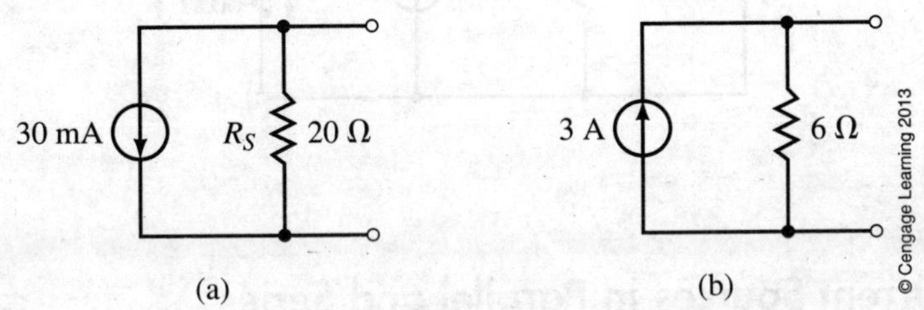

FIGURE 8–71

11. Refer to the circuit of Figure 8–72:
    a. Solve for the current through the load resistor using the current divider rule.
    b. Convert the current source into its equivalent voltage source and again determine the current through the load.
12. Find V_{ab} and I_2 for the network of Figure 8–73.

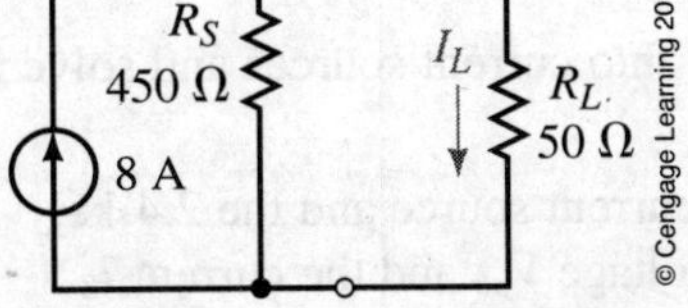

FIGURE 8–72

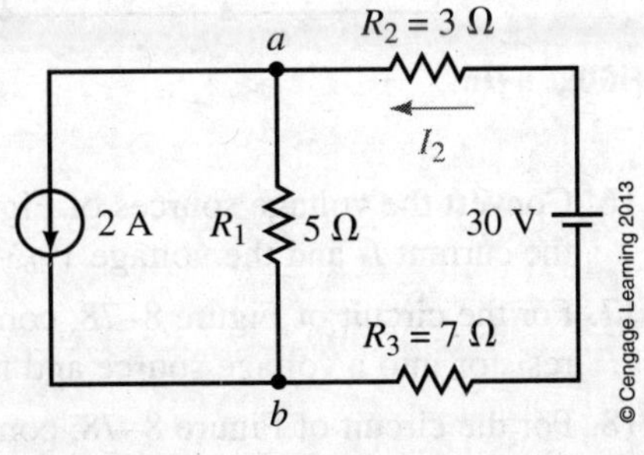

FIGURE 8–73

13. Refer to the circuit of Figure 8–74:
    a. Convert the current source and the 330-Ω resistor into an equivalent voltage source.
    b. Solve for the current I through R_L.
    c. Determine the voltage V_{ab}.

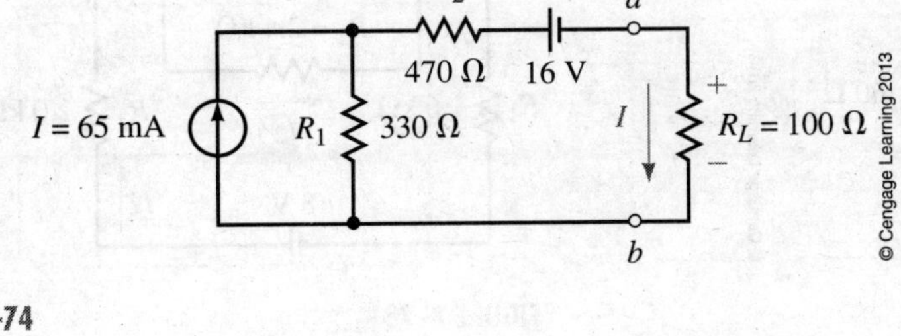

FIGURE 8–74

14. Refer to the circuit of Figure 8–75:
    a. Convert the voltage source and the 36-Ω resistor into an equivalent current source.
    b. Solve for the current I through R_L.
    c. Determine the voltage V_{ab}.

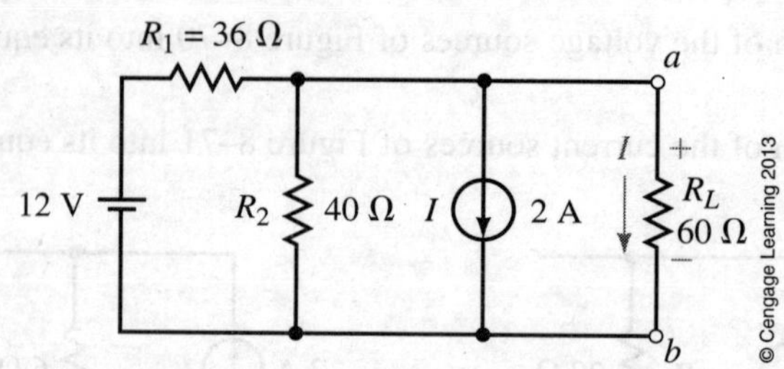

FIGURE 8–75

8.3 Current Sources in Parallel and Series

15. Find the voltage V_2 and the current I_1 for the circuit of Figure 8–76.

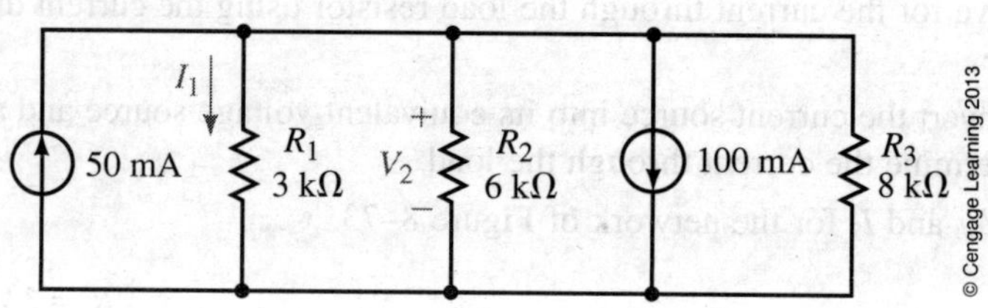

FIGURE 8–76

16. Convert the voltage sources of Figure 8–77 into current sources and solve for the current I_1 and the voltage V_{ab}.
17. For the circuit of Figure 8–78, convert the current source and the 2.4-kΩ resistor into a voltage source and find the voltage V_{ab} and the current I_3.
18. For the circuit of Figure 8–78, convert the voltage source and the series resistors into an equivalent current source.
    a. Determine the current I_2.
    b. Solve for the voltage V_{ab}.

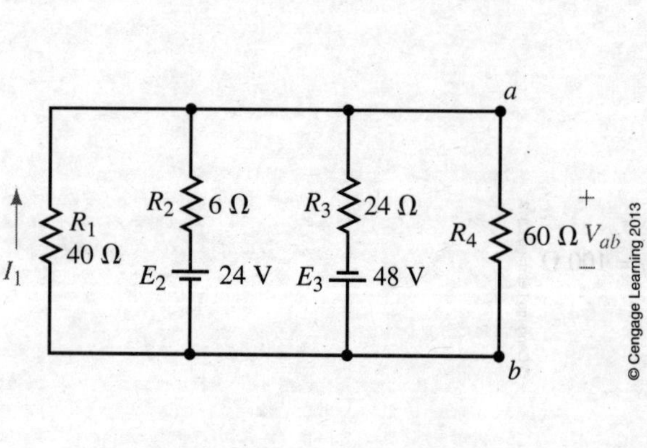

FIGURE 8–77

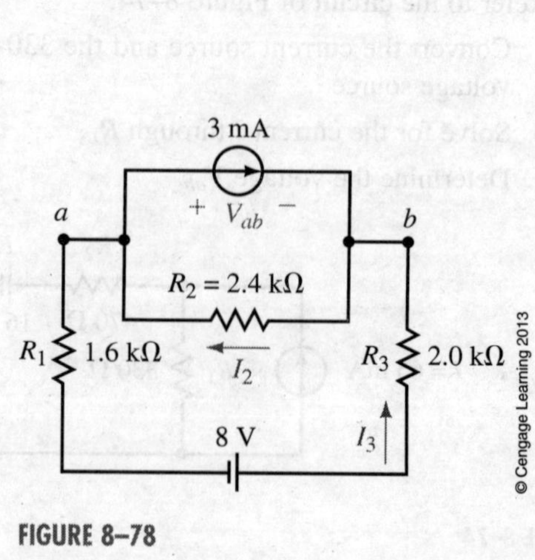

FIGURE 8–78

8.4 Branch-Current Analysis

19. Write the branch-current equations for the circuit shown in Figure 8–79 and solve for the branch currents using determinants.
20. Refer to the circuit of Figure 8–80:
    a. Solve for the current I_1 using branch-current analysis.
    b. Determine the voltage V_{ab}.

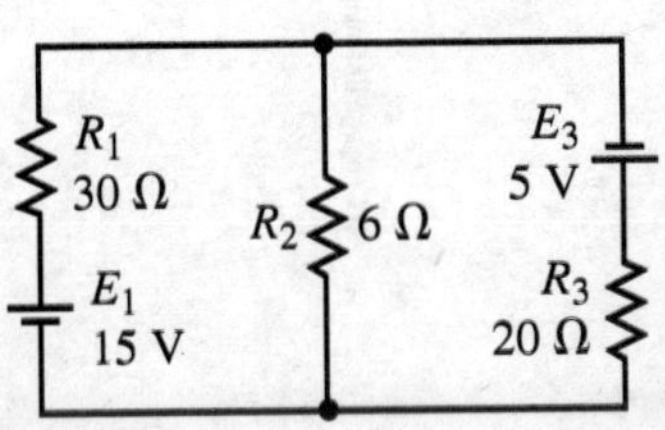

FIGURE 8–79

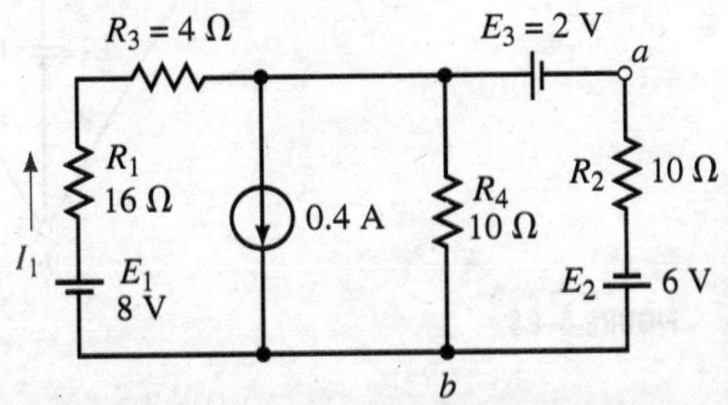

FIGURE 8–80

21. Write the branch-current equations for the circuit shown in Figure 8–81 and solve for the current I_2.

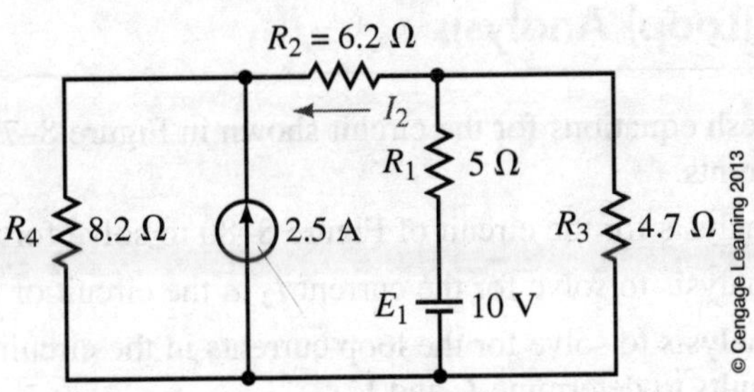

FIGURE 8–81

22. Refer to the circuit shown in Figure 8–82:
    a. Write the branch-current equations.
    b. Solve for the currents I_1 and I_2.
    c. Determine the voltage V_{ab}.
23. Refer to the circuit shown in Figure 8–83:
    a. Write the branch-current equations.
    b. Solve for the current I_2.
    c. Determine the voltage V_{ab}.

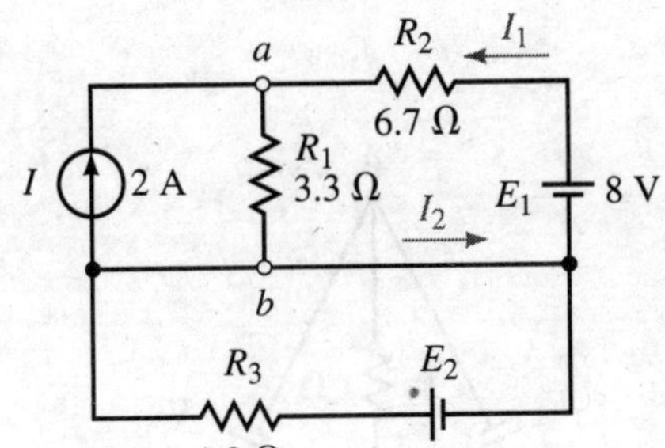

FIGURE 8–82

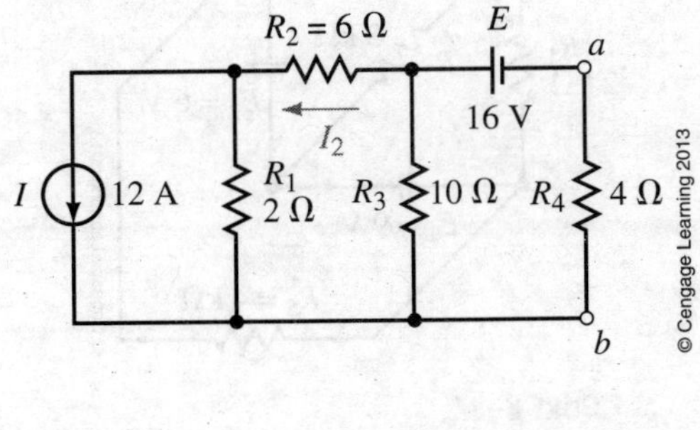

FIGURE 8–83

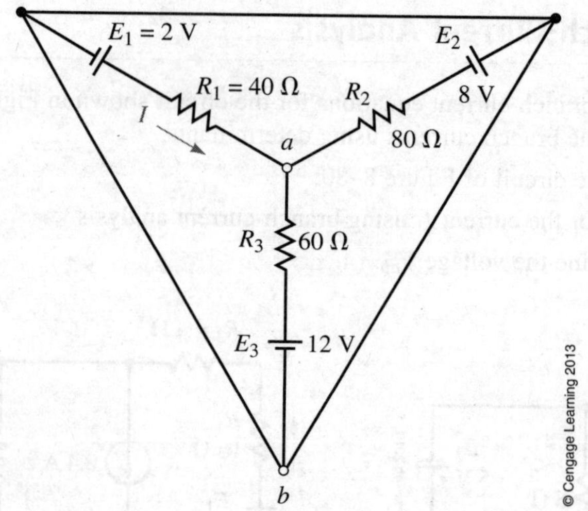

FIGURE 8–84

24. Refer to the circuit shown in Figure 8–84:
    a. Write the branch-current equations.
    b. Solve for the current I.
    c. Determine the voltage V_{ab}.

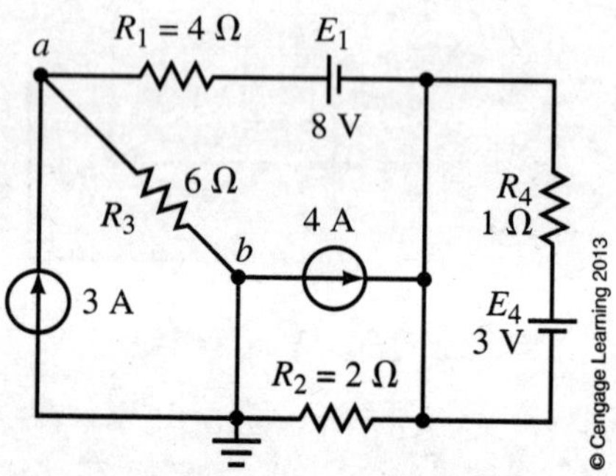

FIGURE 8–85

8.5 Mesh (Loop) Analysis

25. Write the mesh equations for the circuit shown in Figure 8–79 and solve for the loop currents.
26. Use mesh analysis for the circuit of Figure 8–80 to solve for the current I_1.
27. Use mesh analysis to solve for the current I_2 in the circuit of Figure 8–81.
28. Use mesh analysis to solve for the loop currents in the circuit of Figure 8–83. Use your results to determine I_2 and V_{ab}.
29. Use mesh analysis to solve for the loop currents in the circuit of Figure 8–84. Use your results to determine I and V_{ab}.
30. Using mesh analysis, determine the current through the 6-Ω resistor in the circuit of Figure 8–85.
31. Write the mesh equations for the network in Figure 8–86. Solve for the loop currents using determinants.
32. Repeat Problem 31 for the network in Figure 8–87.

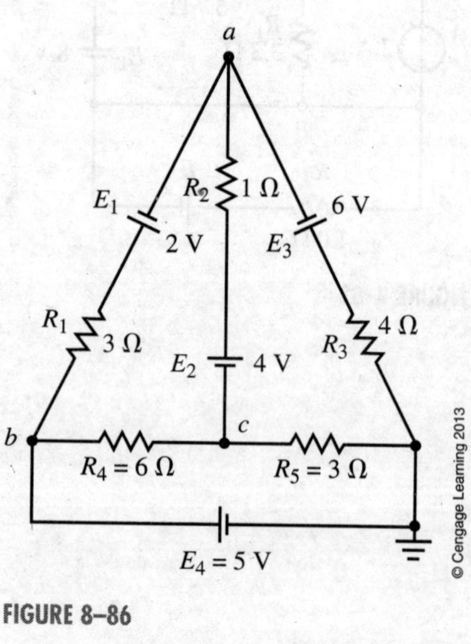

FIGURE 8–86

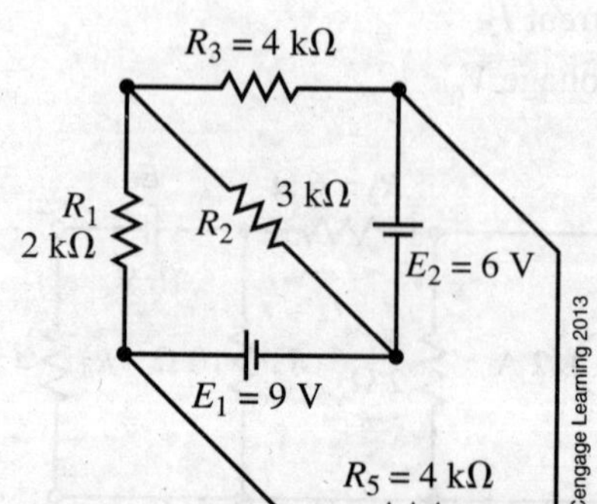

FIGURE 8–87

8.6 Nodal Analysis

33. Write the nodal equations for the circuit of Figure 8–88 and solve for the nodal voltages.
34. Write the nodal equations for the circuit of Figure 8–89 and determine the voltage V_{ab}.
35. Repeat Problem 33 for the circuit of Figure 8–90.

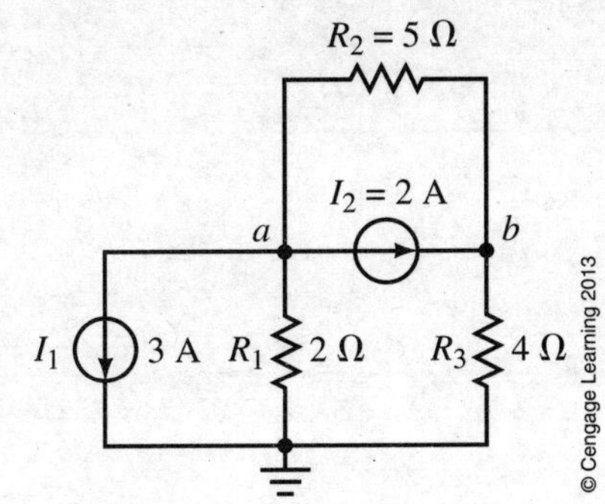

FIGURE 8–88

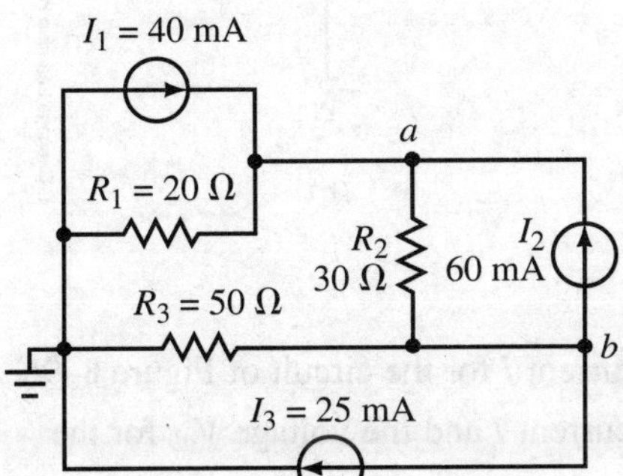

FIGURE 8–89

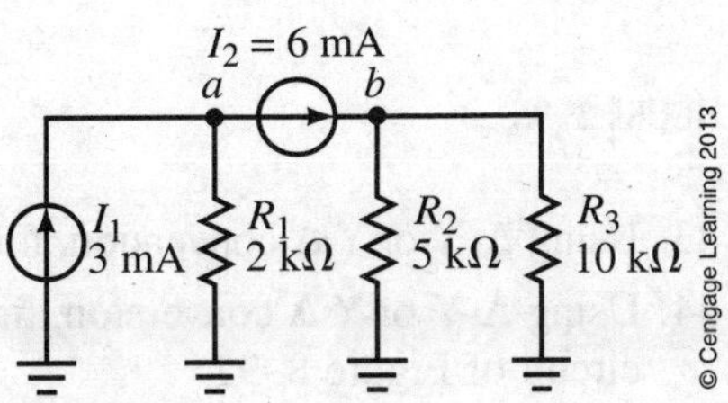

FIGURE 8–90

36. Repeat Problem 34 for the circuit of Figure 8–91.
37. Write the nodal equations for the circuit of Figure 8–86 and solve for $V_{6\,\Omega}$.
38. Write the nodal equations for the circuit of Figure 8–85 and solve for $V_{6\,\Omega}$.

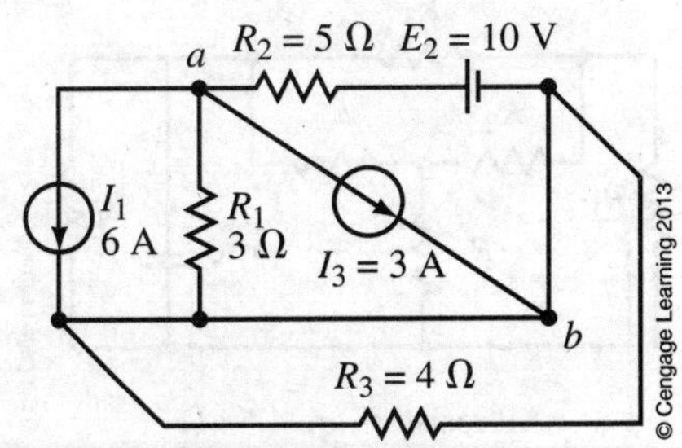

FIGURE 8–91

8.7 Delta-Wye (Pi-Tee) Conversion

39. Convert each of the Δ networks of Figure 8–92 into its equivalent Y configuration.
40. Convert each of the Δ networks of Figure 8–93 into its equivalent Y configuration.

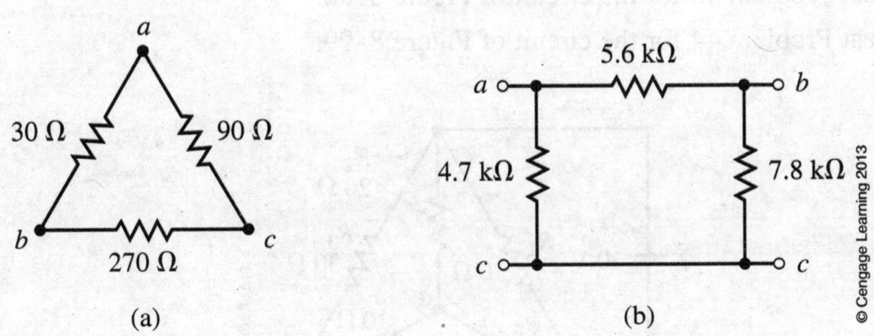

FIGURE 8–92

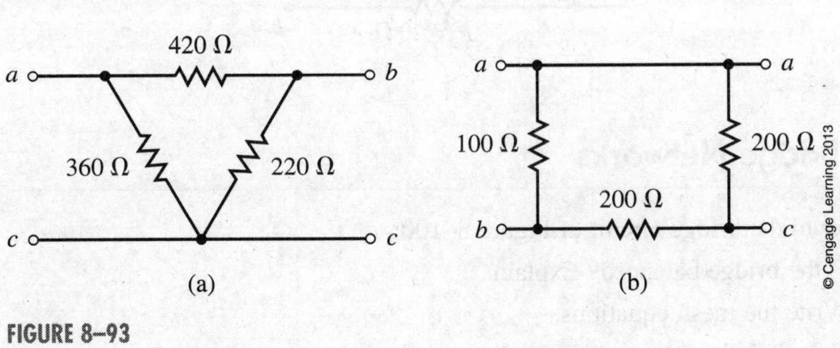

FIGURE 8–93

41. Convert each of the Y networks of Figure 8–94 into its equivalent Δ configuration.

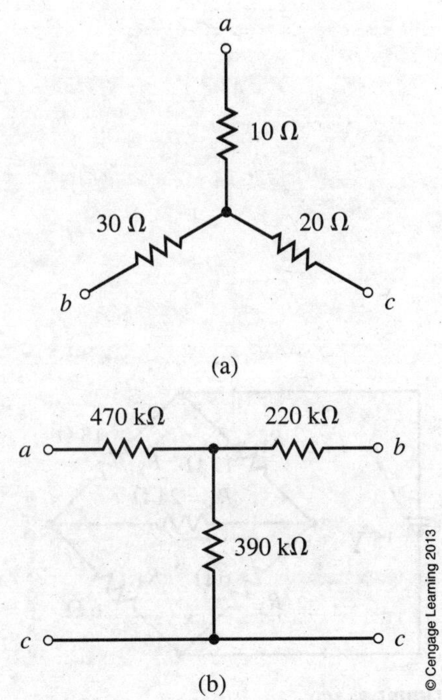

FIGURE 8–94

42. Convert each of the Y networks of Figure 8–95 into its equivalent Δ configuration.

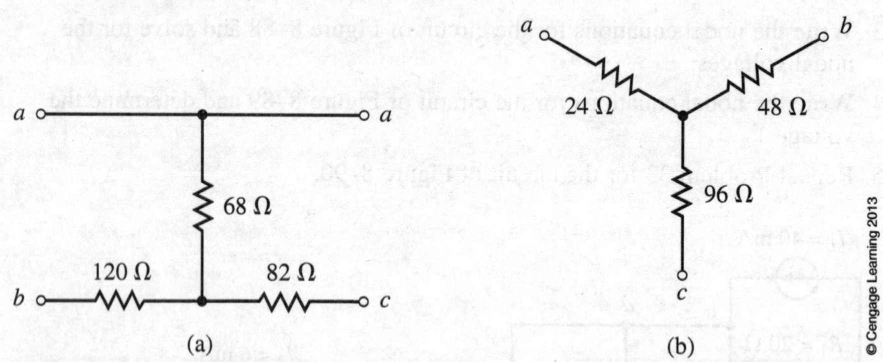

FIGURE 8–95

43. Using Δ-Y or Y-Δ conversion, find the current I for the circuit of Figure 8–96.
44. Using Δ-Y or Y-Δ conversion, find the current I and the voltage V_{ab} for the circuit of Figure 8–97.

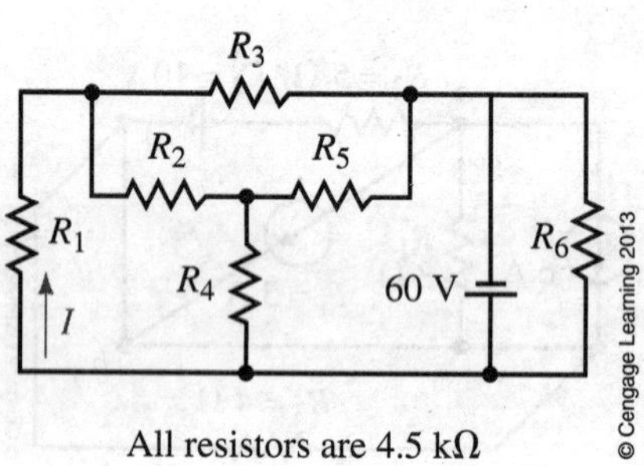

FIGURE 8–96

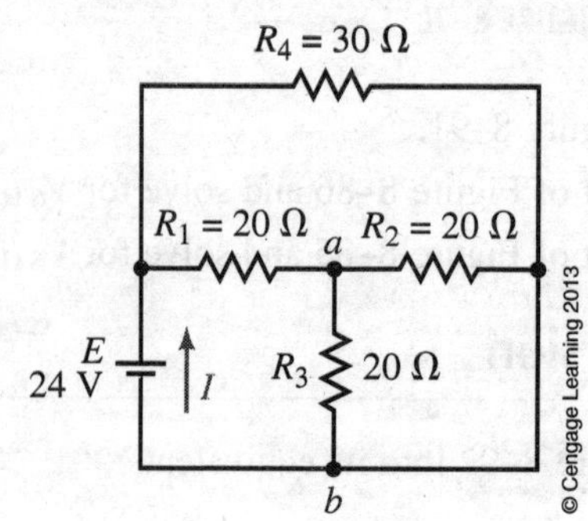

FIGURE 8–97

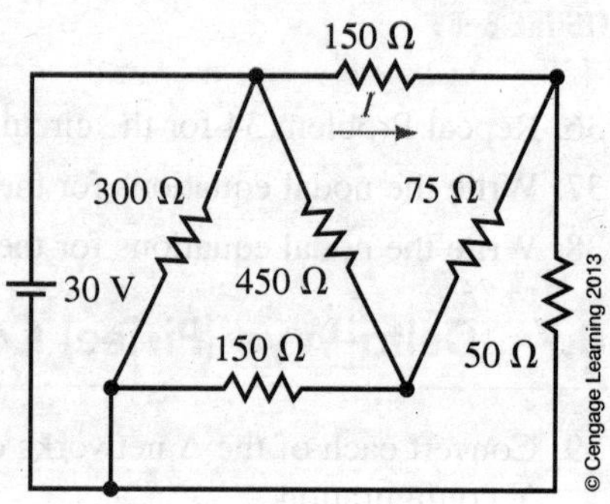

FIGURE 8–98

45. Repeat Problem 43 for the circuit of Figure 8–98.
46. Repeat Problem 44 for the circuit of Figure 8–99.

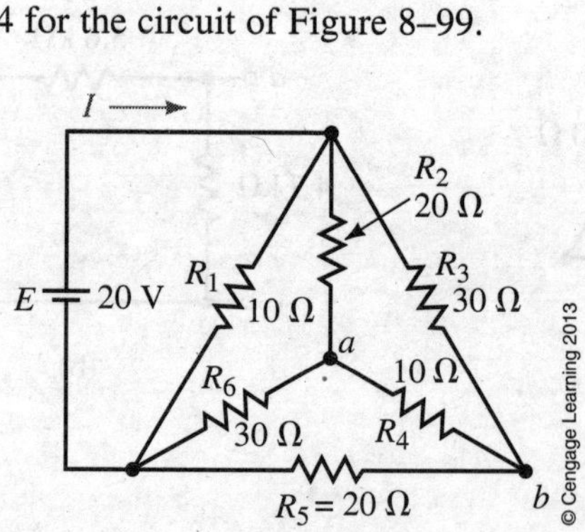

FIGURE 8–99

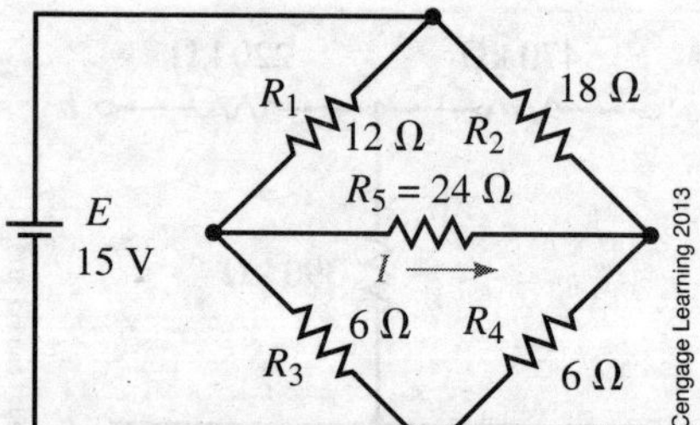

FIGURE 8–100

8.8 Bridge Networks

47. Refer to the bridge circuit of Figure 8–100:
    a. Is the bridge balanced? Explain.
    b. Write the mesh equations.
    c. Calculate the current through R_5.
    d. Determine the voltage across R_5.

48. Consider the bridge circuit of Figure 8–101:
    a. Is the bridge balanced? Explain.
    b. Write the mesh equations.
    c. Determine the current through R_5.
    d. Calculate the voltage across R_5.
49. Given the bridge circuit of Figure 8–102, find the current through each resistor.
50. Refer to the bridge circuit of Figure 8–103:
    a. Determine the value of resistance R_x such that the bridge is balanced.
    b. Calculate the current through R_5 when $R_x = 0\ \Omega$ and when $R_x = 10\ \text{k}\Omega$.

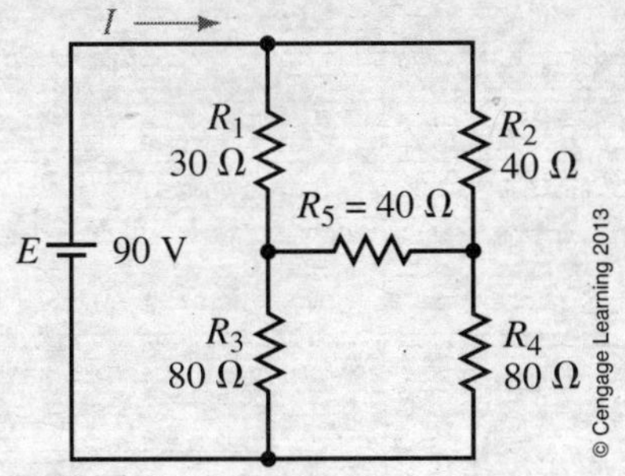

FIGURE 8–101

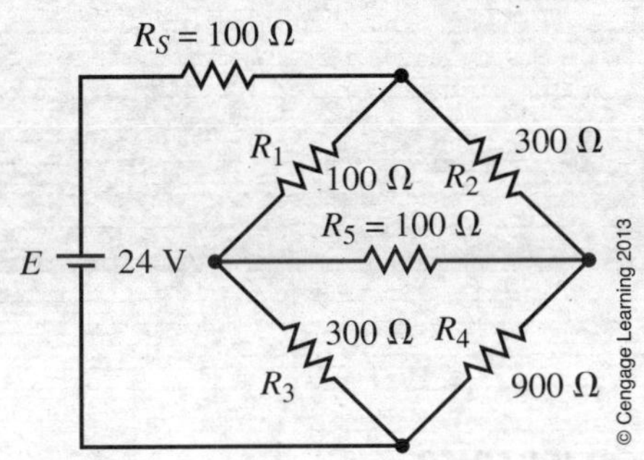

FIGURE 8–102

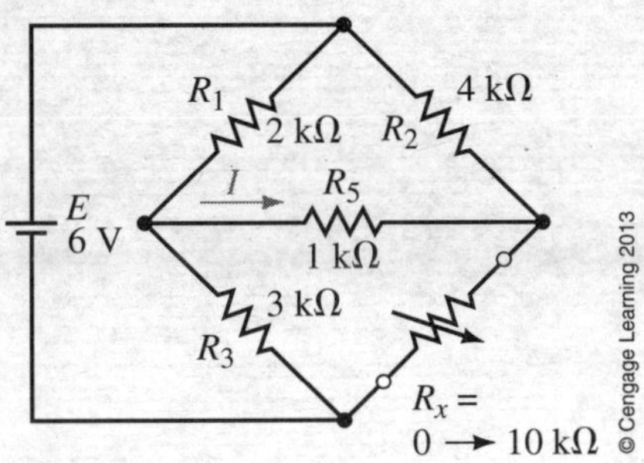

FIGURE 8–103

8.9 Circuit Analysis Using Computers

51. Use Multisim to solve for the currents through all resistors of the circuit shown in Figure 8–86. Multisim
52. Use Multisim to solve for the voltage across the 5-kΩ resistor in the circuit of Figure 8–87. Multisim
53. Use PSpice to solve for the currents through all resistors in the circuit of Figure 8–96. PSPICE
54. Use PSpice to solve for the currents through all resistors in the circuit of Figure 8–97. PSPICE

ANSWERS TO IN-PROCESS LEARNING CHECKS

IN-PROCESS LEARNING CHECK 1

1. A voltage source E in series with a resistor R is equivalent to a current source having an ideal current source $I = E/R$ in parallel with the same resistance, R.
2. Current sources are never connected in series.

IN-PROCESS LEARNING CHECK 2

1. Voltage is zero.
2. Current is zero.
3. R_5 can be replaced with either a short circuit or an open circuit.

KEY TERMS

Maximum Power Transfer Theorem

Millman's Theorem

Norton's Theorem

Reciprocity Theorem

Substitution Theorem

Superposition Theorem

Thévenin's Theorem

OUTLINE

Superposition Theorem

Thévenin's Theorem

Norton's Theorem

Maximum Power Transfer Theorem

Substitution Theorem

Millman's Theorem

Reciprocity Theorem

Circuit Analysis Using Computers

OBJECTIVES

After studying this chapter you will be able to

- apply the superposition theorem to determine the current through or voltage across any resistance in a given network,
- state Thévenin's theorem and determine the Thévenin equivalent circuit of any resistive network,
- state Norton's theorem and determine the Norton equivalent circuit of any resistive network,
- determine the required load resistance of any circuit to ensure that the load receives maximum power from the circuit,
- apply Millman's theorem to determine the current through or voltage across any resistor supplied by any number of sources in parallel,
- state the reciprocity theorem and demonstrate that it applies for a given single-source circuit,
- state the substitution theorem and apply the theorem in simplifying the operation of a given circuit.

9

NETWORK THEOREMS

CHAPTER PREVIEW

In this chapter you will learn how to use some basic theorems which will allow the analysis of even the most complex resistive networks. The theorems that are most useful in analyzing networks are the superposition, Thévenin, Norton, and maximum power transfer theorems.

You will also be introduced to other theorems which, while useful for providing a well-rounded appreciation of circuit analysis, have limited use in the analysis of circuits. These theorems, which apply to specific types of circuits, are the substitution, reciprocity, and Millman theorems. Your instructor may choose to omit the latter theorems without any loss in continuity. ■

Putting It in Perspective

André Marie Ampère

Hulton Archive/Stringer/Hulton Archive/Getty Images

ANDRÉ MARIE AMPÈRE WAS BORN in Polémieux, Rhône, near Lyon, France, on January 22, 1775. As a youth, Ampère was a brilliant mathematician who was able to master advanced mathematics by the age of twelve. However, the French Revolution, and the ensuing anarchy that swept through France from 1789 to 1799, did not exclude the Ampère family. Ampère's father, who was a prominent merchant and city official in Lyon, was executed under the guillotine in 1793. Young André suffered a nervous breakdown from which he never fully recovered. His suffering was further compounded in 1804, when after only five years of marriage, Ampère's wife died.

Even so, Ampère was able to make profound contributions to the field of mathematics, chemistry, and physics. As a young man, Ampère was appointed as professor of chemistry and physics in Bourg. Napoleon was a great supporter of Ampère's work, though Ampère had a reputation as an "absent-minded professor." Later he moved to Paris, where he taught mathematics.

Ampère showed that two current-carrying wires were attracted to one another when the current in the wires was in the same direction. When the current in the wires was in the opposite direction, the wires repelled. This work set the stage for the discovery of the principles of electric and magnetic field theory. Ampère was the first scientist to use electromagnetic principles to measure current in a wire. In recognition of his contribution to the study of electricity, current is measured in the unit of amperes.

Despite his personal suffering, Ampère remained a popular, friendly human being. He died of pneumonia in Marseille on June 10, 1836 after a brief illness. ■

9.1 Superposition Theorem

The **superposition theorem** is a method that allows us to determine the current through or the voltage across any resistor or branch in a network. The advantage of using this approach instead of mesh analysis or nodal analysis is that it is not necessary to use determinants or matrix algebra to analyze a given circuit. The theorem states the following:

The total current through or voltage across a resistor or branch may be determined by summing the effects due to each independent source.

In order to apply the superposition theorem it is necessary to remove all sources other than the one being examined. In order to "zero" a voltage source, we *replace it with a short circuit,* since the voltage across a short circuit is zero volts. A current source is zeroed by *replacing it with an open circuit,* since the current through an open circuit is zero amps.

If we wish to determine the power dissipated by any resistor, we must first find either the voltage across the resistor or the current through the resistor:

$$P = I^2R = \frac{V^2}{R}$$

NOTES...

The superposition theorem does not apply to power, since power is not a linear quantity, but rather is found as the square of either current or voltage.

EXAMPLE 9–1

Consider the circuit of Figure 9–1:

a. Determine the current in the load resistor, R_L.
b. Verify that the superposition theorem does not apply to power.

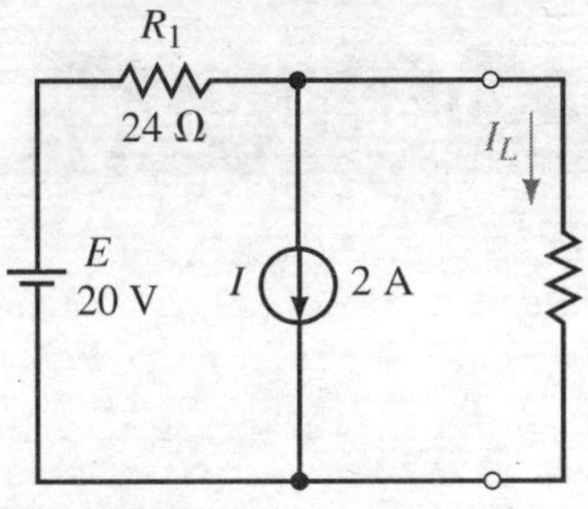

FIGURE 9–1

Solution

a. We first determine the current through R_L due to the voltage source by removing the current source and replacing in with an open circuit (zero amps) as shown in Figure 9–2.

The resulting current through R_L is determined from Ohm's law as

$$I_{L(1)} = \frac{20\text{ V}}{16\ \Omega + 24\ \Omega} = 0.500\text{ A}$$

Next, we determine the current through R_L due to the current source by removing the voltage source and replacing it with a short circuit (zero volts) as shown in Figure 9–3.

The resulting current through R_L is found with the current divider rule as

$$I_{L(2)} = -\left(\frac{24\ \Omega}{24\ \Omega + 16\ \Omega}\right)(2\text{ A}) = -1.20\text{ A}$$

The resultant current through R_L is found by applying the superposition theorem:

$$I_L = 0.5\text{ A} - 1.2\text{ A} = -0.700\text{ A}$$

The negative sign indicates that the current through R_L is opposite to the assumed reference direction. Consequently, the current through R_L will, in fact, be upward with a magnitude of 0.7 A.

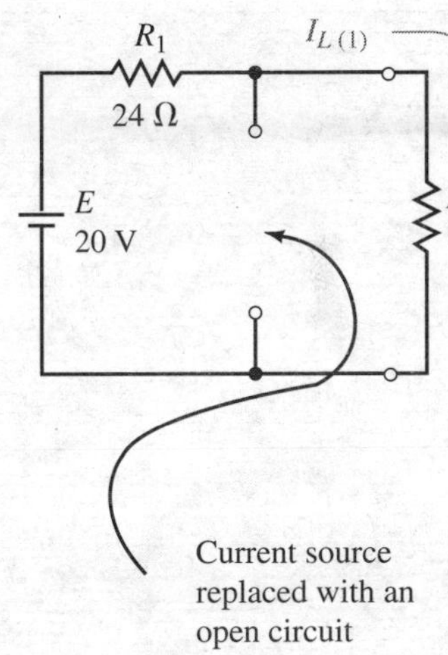

FIGURE 9–2

b. If we assume (incorrectly) that the superposition theorem applies for power, we will have the power due the first source given as

$$P_1 = I_{L(1)}^2 R_L = (0.5\text{ A})^2(16\ \Omega) = 4.0\text{ W}$$

and the power due the second source as

$$P_2 = I_{L(2)}^2 R_L = (1.2\text{ A})^2(16\ \Omega) = 23.04\text{ W}$$

The total power, if superposition applies, will be

$$P_\text{T} = P_1 + P_2 = 4.0\text{ W} + 23.04\text{ W} = 27.04\text{ W}$$

Clearly, this result is wrong, since the actual power dissipated by the load resistor is correctly given as

$$P_L = I_L^2 R_L = (0.7\text{ A})^2(16\ \Omega) = 7.84\text{ W}$$

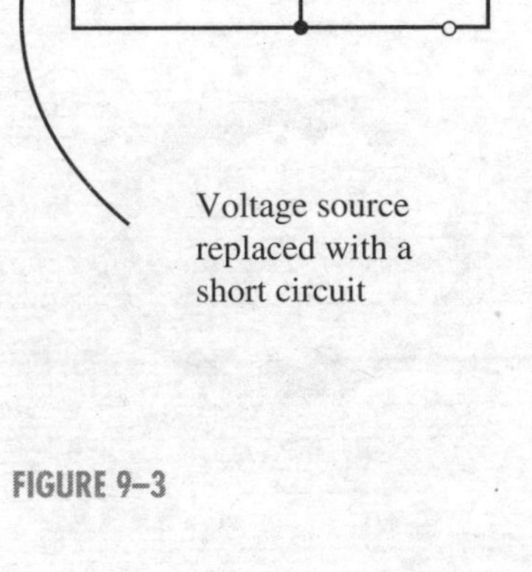

FIGURE 9–3

The superposition theorem may also be used to determine the voltage across any component or branch within the circuit.

EXAMPLE 9–2

Determine the voltage drop across the resistor R_2 of the circuit shown in Figure 9–4.

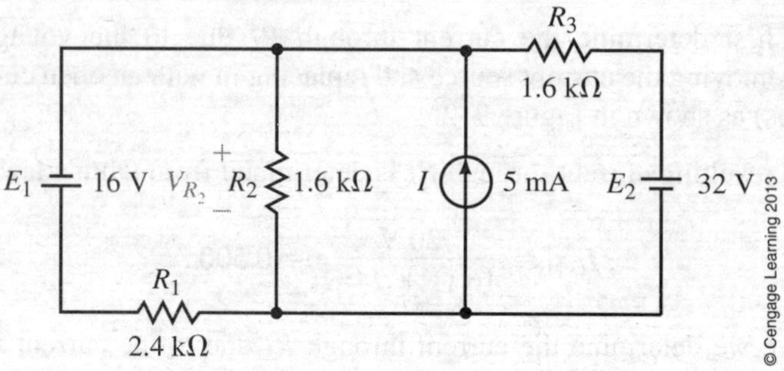

FIGURE 9–4

Solution Since this circuit has three separate sources, it is necessary to determine the voltage across R_2 due to each individual source.

First, we consider the voltage across R_2 due to the 16-V source as shown in Figure 9–5.

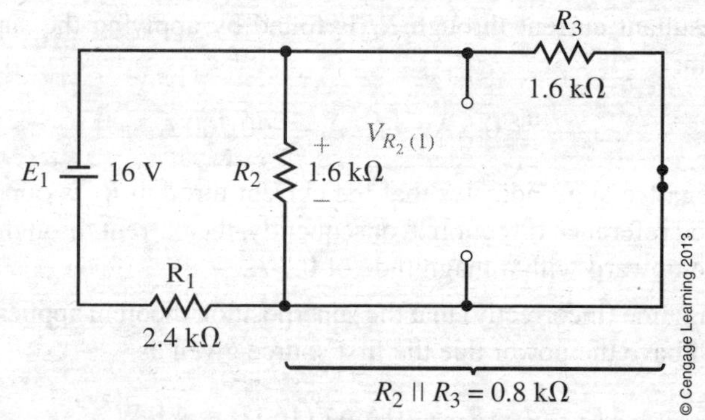

FIGURE 9–5

The voltage across R_2 will be the same as the voltage across the parallel combination of $R_2 \| R_3 = 0.8\ \text{k}\Omega$. Therefore,

$$V_{R2(1)} = -\left(\frac{0.8\ \text{k}\Omega}{0.8\ \text{k}\Omega + 2.4\ \text{k}\Omega}\right)(16\ \text{V}) = -4.00\ \text{V}$$

The negative sign in the preceding calculation simply indicates that the voltage across the resistor due to the first source is opposite to the assumed reference polarity.

Next, we consider the current source. The resulting circuit is shown in Figure 9–6. From this circuit, you can observe that the total resistance "seen" by the current source is

$$R_T = R_1 \| R_2 \| R_3 = 0.6\ \text{k}\Omega$$

The resulting voltage across R_2 is

$$V_{R_2(2)} = (0.6\ \text{k}\Omega)(5\ \text{mA}) = 3.00\ \text{V}$$

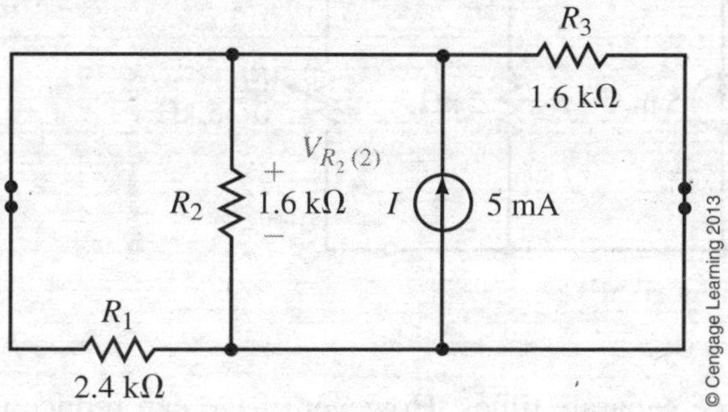

FIGURE 9–6

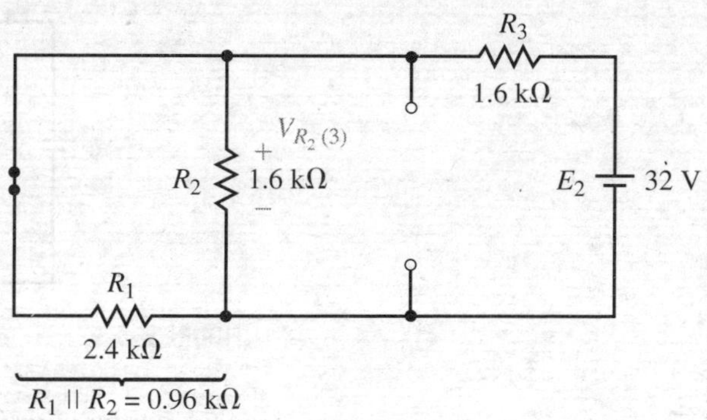

FIGURE 9–7

Finally, the voltage due to the 32-V source is found by analyzing the circuit of Figure 9–7. The voltage across R_2 is

$$V_{R2(3)} = \left(\frac{0.96\ \text{k}\Omega}{0.96\ \text{k}\Omega + 1.6\ \text{k}\Omega}\right)(32\ \text{V}) = 12.0\ \text{V}$$

By superposition, the resulting voltage is

$$V_{R_2} = -4.0\ \text{V} + 3.0\ \text{V} + 12.0\ \text{V} = 11.0\ \text{V}$$

PRACTICE PROBLEMS 1

Use the superposition theorem to determine the voltage across R_1 and R_3 in the circuit of Figure 9–4.

Answers

$V_{R_1} = 27.0$ V, $V_{R_3} = 21.0$ V

IN-PROCESS LEARNING CHECK 1

(Answers are at the end of the chapter.)

Use the final results of Example 9–2 and Practice Problem 1 to determine the power dissipated by the resistors in the circuit of Figure 9–4. Verify that the superposition theorem does not apply to power.

9.2 Thévenin's Theorem

In this section, we will apply one of the most important theorems of electric circuits. **Thévenin's theorem** allows even the most complicated circuit to be reduced to a single voltage source and a single resistance. The importance of such a theorem becomes evident when we try to analyze a circuit as shown in Figure 9–8.

If we wanted to find the current through the variable load resistor when $R_L = 0$, $R_L = 2$ kΩ, and $R_L = 5$ kΩ using existing methods, we would need to

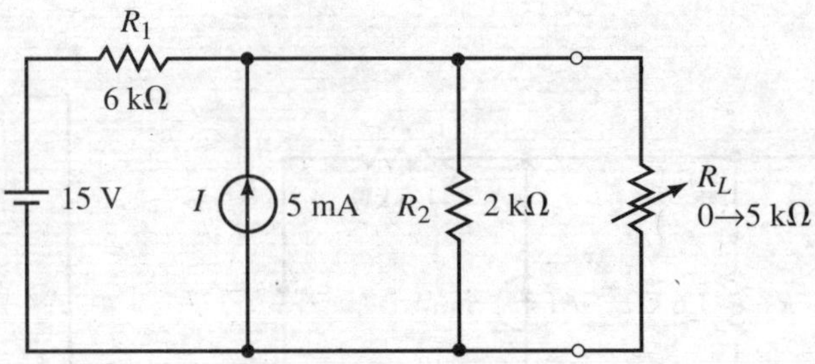

FIGURE 9–8

analyze the entire circuit three separate times. However, if we can reduce the entire circuit external to the load resistor to a single voltage source in series with a resistor, the solution becomes very easy.

Thévenin's theorem is a circuit analysis technique that reduces any linear bilateral network to an equivalent circuit having only one voltage source and one series resistor. The resulting two-terminal circuit is equivalent to the original circuit when connected to any external branch or component. In summary, Thévenin's theorem is simplified as follows:

Any linear bilateral network may be reduced to a simplified two-terminal circuit consisting of a single voltage source in series with a single resistor as shown in Figure 9–9.

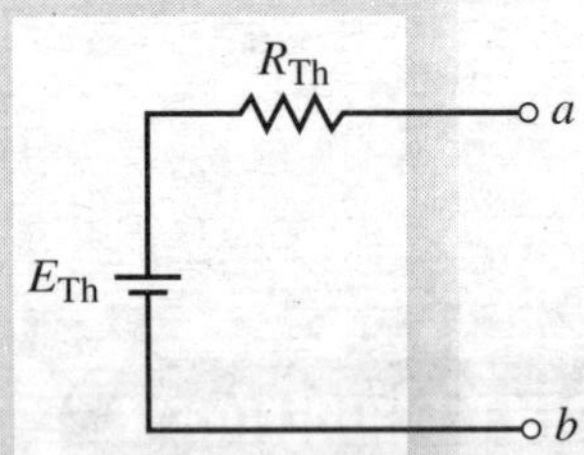

FIGURE 9–9

A linear network, remember, is any network that consists of components having a linear (straight-line) relationship between voltage and current. A resistor is a good example of a linear component since the voltage across a resistor increases proportionally to an increase in current through the resistor. Voltage and current sources are also linear components. In the case of a voltage source, the voltage remains constant, although current through the source may change.

A bilateral network is any network that operates in the same manner regardless of the direction of current in the network. Again, a resistor is a good example of a bilateral component, since the magnitude of current through the resistor is not dependent upon the polarity of voltage across the component. (A diode is not a bilateral component, since the magnitude of current through the device is dependent upon the polarity of the voltage applied across the diode.)

The following steps provide a technique that converts any circuit into its Thévenin equivalent:

1. Identify and remove the load from the circuit.
2. Label the resulting two terminals. We will label them as *a* and *b*, although any notation may be used.
3. Set all sources in the circuit to zero.

   Voltage sources are set to zero by replacing them with short circuits (zero volts).

   Current sources are set to zero by replacing them with open circuits (zero amps).
4. Determine the Thévenin equivalent resistance, R_{Th}, by calculating the resistance "seen" between terminals *a* and *b*. It may be necessary to redraw the circuit to simplify this step.
5. Replace the sources removed in Step 3, and determine the open-circuit voltage between the terminals. If the circuit has more than one source, it may be necessary to use the superposition theorem. In that case, it will be

necessary to determine the open-circuit voltage due to each source separately and then determine the combined effect. The resulting open-circuit voltage will be the value of the Thévenin voltage, E_{Th}.

6. Draw the Thévenin equivalent circuit using the resistance determined in Step 4 and the voltage calculated in Step 5. As part of the resulting circuit, include that portion of the network removed in Step 1.

EXAMPLE 9–3

Determine the Thévenin equivalent circuit external to the resistor R_L for the circuit of Figure 9–10. Use the Thévenin equivalent circuit to calculate the current through R_L.

Solution

Steps 1 and 2: Removing the load resistor from the circuit and labelling the remaining terminals, we obtain the circuit shown in Figure 9–11.

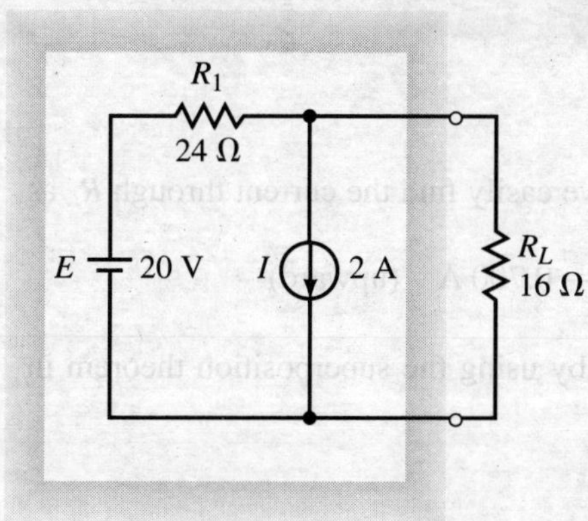

FIGURE 9–10

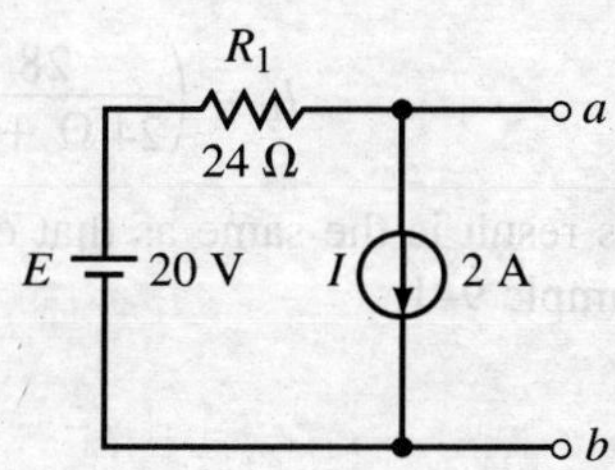

FIGURE 9–11

Step 3: Setting the sources to zero, we have the circuit shown in Figure 9–12.

Step 4: The Thévenin resistance between the terminals is $R_{Th} = 24\ \Omega$.

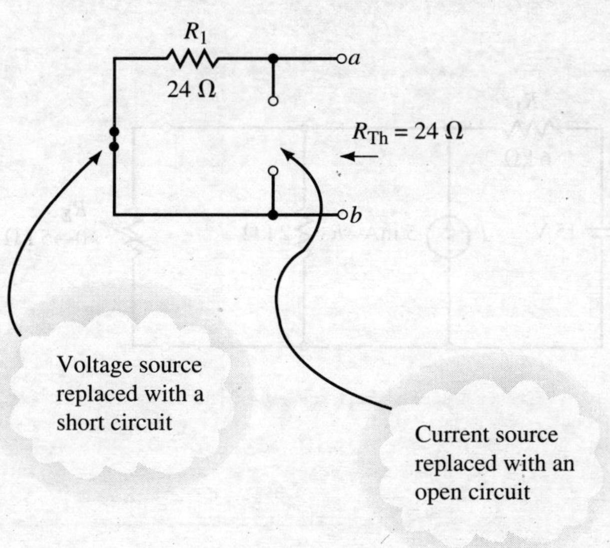

FIGURE 9–12

Step 5: From Figure 9–11, the open-circuit voltage between terminals *a* and *b* is found as

$$V_{ab} = 20\text{ V} - (24\ \Omega)(2\text{ A}) = -28.0\text{ V}$$

Step 6: The resulting Thévenin equivalent circuit is shown in Figure 9–13.

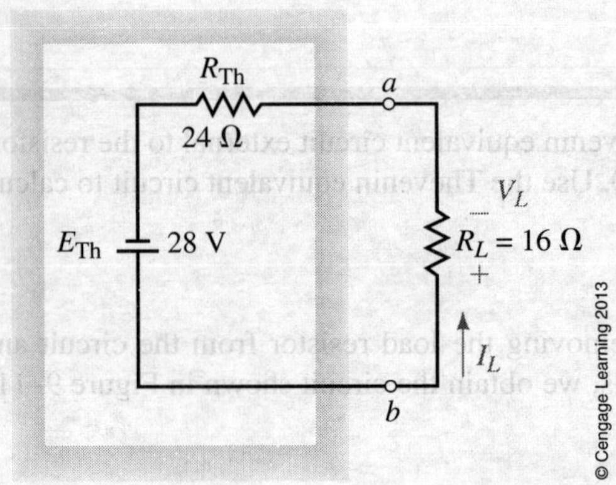

FIGURE 9–13

Using this Thévenin equivalent circuit, we easily find the current through R_L as

$$I_L = \left(\frac{28\text{ V}}{24\ \Omega + 16\ \Omega}\right) = 0.700\text{ A} \quad \text{(upward)}$$

This result is the same as that obtained by using the superposition theorem in Example 9–1.

EXAMPLE 9–4

Find the Thévenin equivalent circuit of the indicated area in Figure 9–14. Using the equivalent circuit, determine the current through the load resistor when $R_L = 0$, $R_L = 2\text{ k}\Omega$, and $R_L = 5\text{ k}\Omega$.

CircuitSim 09-1

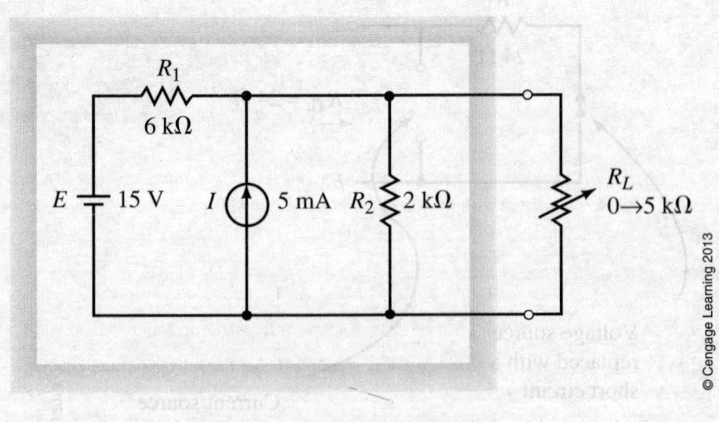

FIGURE 9–14

Solution

Steps 1, 2, and 3: After removing the load, labelling the terminals, and setting the sources to zero, we have the circuit shown in Figure 9–15.

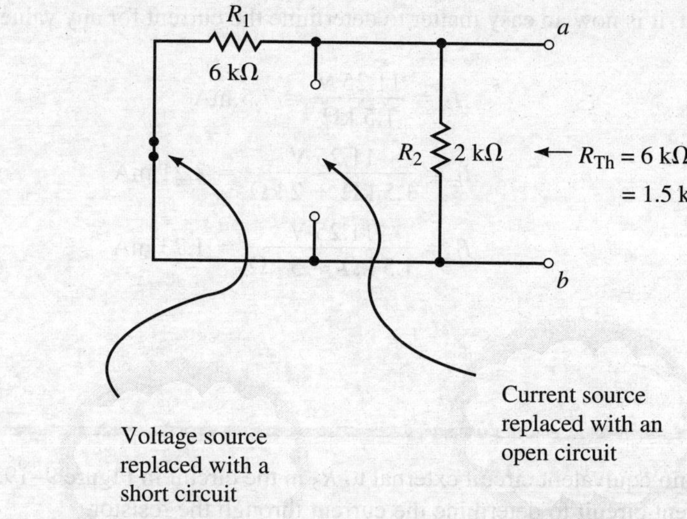

FIGURE 9–15

Step 4: The Thévenin resistance of the circuit is

$$R_{Th} = 6\ k\Omega \| 2\ k\Omega = 1.5\ k\Omega$$

Step 5: Although several methods are possible, we will use the superposition theorem to find the open-circuit voltage V_{ab}. Figure 9–16 shows the circuit for determining the contribution due to the 15-V source.

$$V_{ab(1)} = \left(\frac{2\ k\Omega}{2\ k\Omega + 6\ k\Omega}\right)(15\ V) = +3.75\ V$$

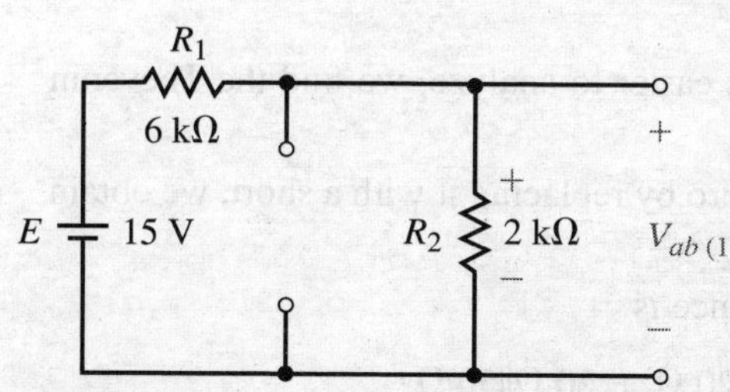

FIGURE 9–16

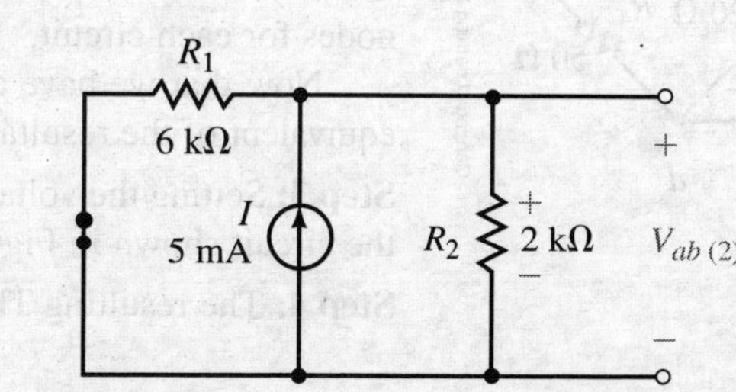

FIGURE 9–17

Figure 9–17 shows the circuit for determining the contribution due to the 5-mA source.

$$V_{ab(2)} = \left(\frac{(2\ k\Omega)(6\ k\Omega)}{2\ k\Omega + 6\ k\Omega}\right)(5\ mA) = +7.5\ V$$

The Thévenin equivalent voltage is

$$E_{Th} = V_{ab(1)} + V_{ab(2)} = +3.75\ V + 7.5\ V = 11.25\ V$$

Step 6: The resulting Thévenin equivalent circuit is shown in Figure 9–18.

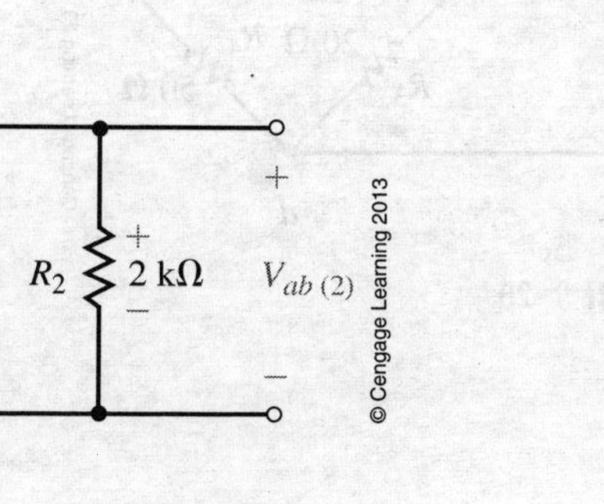

FIGURE 9–18

From this circuit, it is now an easy matter to determine the current for any value of load resistor:

$R_L = 0\ \Omega$: $\quad I_L = \dfrac{11.25\ \text{V}}{1.5\ \text{k}\Omega} = 7.5\ \text{mA}$

$R_L = 2\ \text{k}\Omega$: $\quad I_L = \dfrac{11.25\ \text{V}}{1.5\ \text{k}\Omega + 2\ \text{k}\Omega} = 3.21\ \text{mA}$

$R_L = 5\ \text{k}\Omega$: $\quad I_L = \dfrac{11.25\ \text{V}}{1.5\ \text{k}\Omega + 5\ \text{k}\Omega} = 1.73\ \text{mA}$

EXAMPLE 9–5

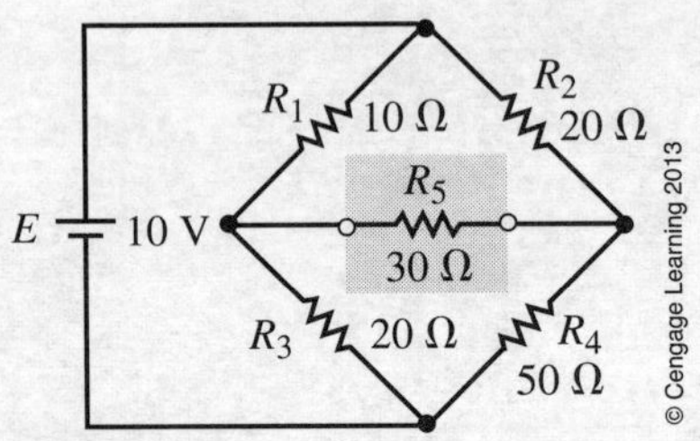

FIGURE 9–19

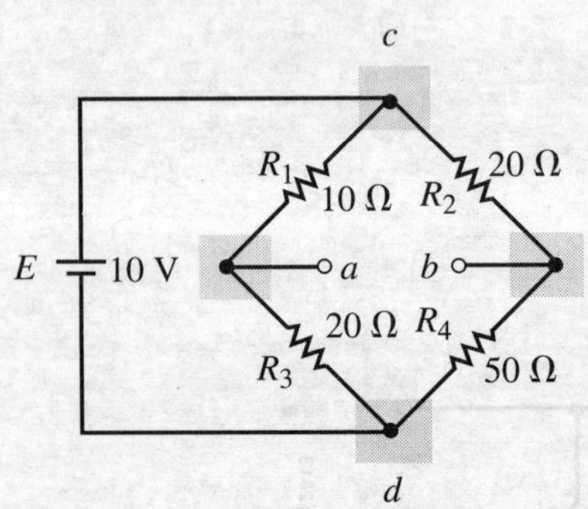

FIGURE 9–20

Find the Thévenin equivalent circuit external to R_5 in the circuit in Figure 9–19. Use the equivalent circuit to determine the current through the resistor.

Solution Notice that the circuit is an unbalanced bridge circuit. If the techniques of the previous chapter had to be used, we would need to solve either three mesh equations or three nodal equations.

Steps 1 and 2: Removing the resistor R_5 from the circuit and labelling the two terminals a and b, we obtain the circuit shown in Figure 9–20.

By examining the circuit shown in Figure 9–20, we see that it is no simple task to determine the equivalent circuit between terminals a and b. The process is simplified by redrawing the circuit as illustrated in Figure 9–21.

Notice that the circuit of Figure 9–21 has nodes a and b conveniently shown at the top and bottom of the circuit. Additional nodes (node c and node d) are added to simplify the task of correctly placing resistors between the nodes.

After simplifying a circuit, it is always a good idea to ensure that the resulting circuit is indeed an equivalent circuit. You may verify the equivalence of the two circuits by confirming that each component is connected between the same nodes for each circuit.

Now that we have a circuit that is easier to analyze, we find the Thévenin equivalent of the resultant.

Step 3: Setting the voltage source to zero by replacing it with a short, we obtain the circuit shown in Figure 9–22.

Step 4: The resulting Thévenin resistance is

$$\begin{aligned} R_{\text{Th}} &= 10\ \Omega \| 20\ \Omega + 20\ \Omega \| 50\ \Omega \\ &= 6.67\ \Omega + 14.29\ \Omega = 20.95\ \Omega \end{aligned}$$

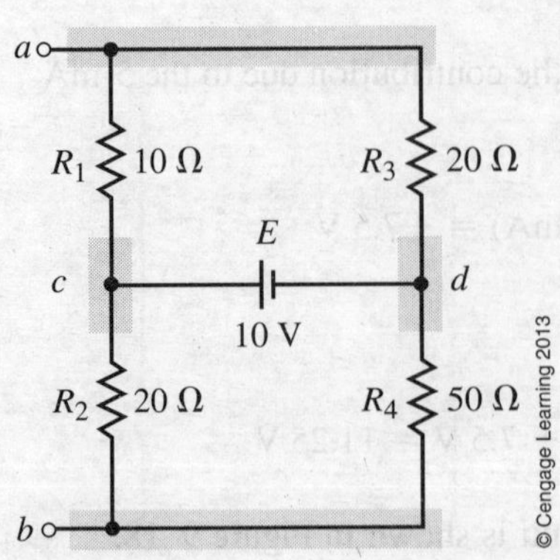

FIGURE 9–21

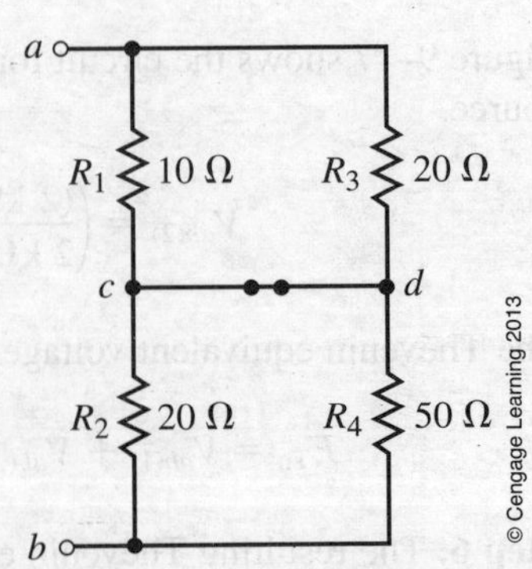

FIGURE 9–22

Step 5: The open-circuit voltage between terminals a and b is found by first indicating the loop currents I_1 and I_2 in the circuit of Figure 9–23.

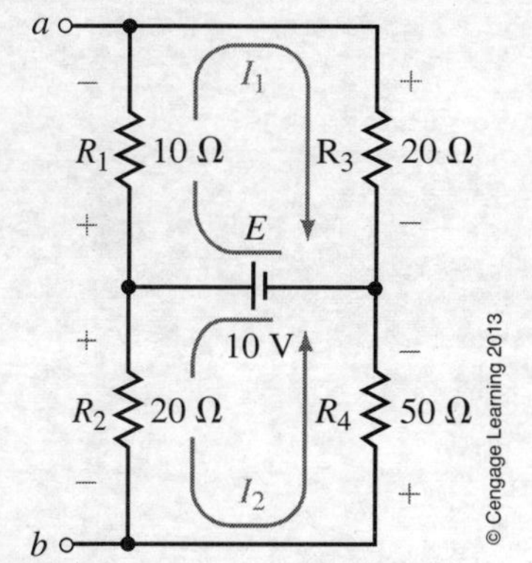

FIGURE 9–23

Because the voltage source E provides a constant voltage across the resistor combinations R_1-R_3 and R_2-R_4, we simply use the voltage divider rule to determine the voltage across the various components:

$$V_{ab} = -V_{R_1} + V_{R_2}$$

$$= -\frac{(10\ \Omega)(10\ \text{V})}{30\ \Omega} + \frac{(20\ \Omega)(10\ \text{V})}{70\ \Omega}$$

$$= -0.476\ \text{V}$$

Note: The preceding technique could not be used if the source had some series resistance, since then the voltage provided to resistor combinations R_1-R_3 and R_2-R_4 would no longer be the entire supply voltage but rather would be dependent upon the value of the series resistance of the source.

Step 6: The resulting Thévenin circuit is shown in Figure 9–24.

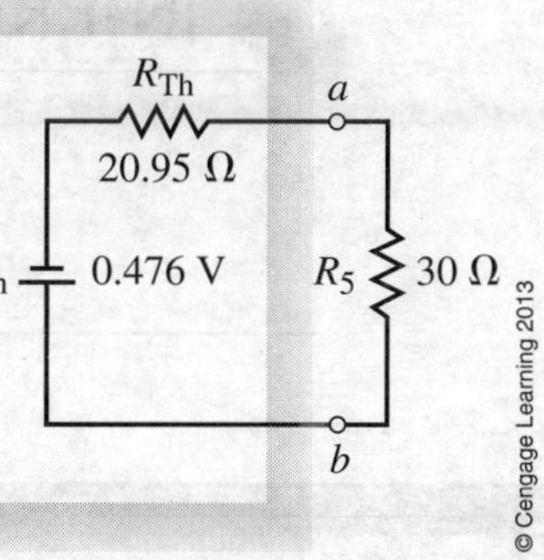

FIGURE 9–24

From the circuit of Figure 9–24, it is now possible to calculate the current through the resistor R_5 as

$$I = \frac{0.476\ \text{V}}{20.95\ \Omega + 30\ \Omega} = 9.34\ \text{mA} \quad (\text{from } b \text{ to } a)$$

This example illustrates the importance of labelling the terminals that remain after a component or branch is removed. If we had not labelled the terminals and drawn an equivalent circuit, the current through R_5 would not have been found as easily.

PRACTICE PROBLEMS 2

Find the Thévenin equivalent circuit external to resistor R_1 in the circuit of Figure 9–1.

Answer
$R_{Th} = 16\ \Omega$, $E_{Th} = 52\ \text{V}$

PRACTICE PROBLEMS 3

Use Thévenin's theorem to determine the current through load resistor R_L for the circuit of Figure 9–25.

Answer
$I_L = 10.0\ \text{mA}$ (upward)

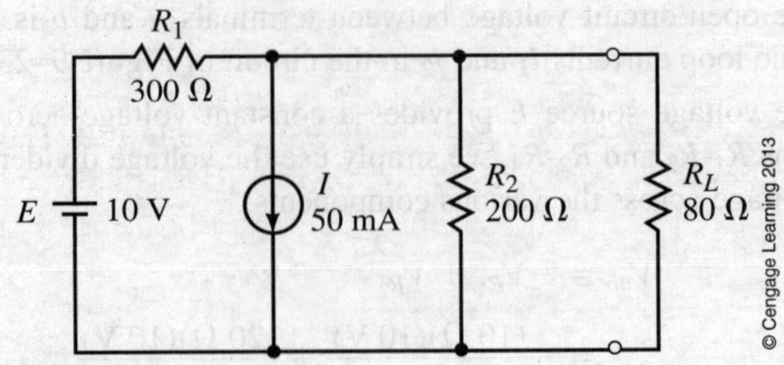

FIGURE 9–25

IN-PROCESS LEARNING CHECK 2

(Answers are at the end of the chapter.)

In the circuit of Figure 9–25, what would the value of R_1 need to be for the Thévenin resistance to be equal to $R_L = 80\ \Omega$?

9.3 Norton's Theorem

Norton's theorem is a circuit analysis technique that is similar to Thévenin's theorem. By using this theorem we reduce the circuit to a single current source and one parallel resistor. As with the Thévenin equivalent circuit, the resulting two-terminal circuit is equivalent to the original circuit when connected to any external branch or component. In summary, **Norton's theorem** may be simplified as follows:

Any linear bilateral network may be reduced to a simplified two-terminal circuit consisting of a single current source and a single shunt resistor as shown in Figure 9–26.

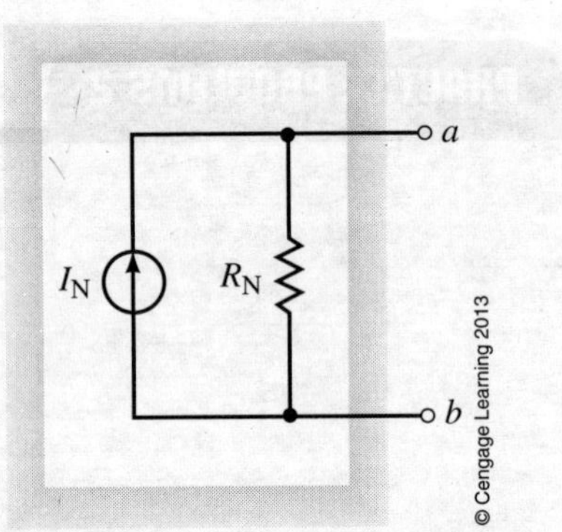

FIGURE 9–26

The following steps provide a technique that allows the conversion of any circuit into its Norton equivalent:

1. Identify and remove the load from the circuit.
2. Label the resulting two terminals. We will label them as *a* and *b*, although any notation may be used.
3. Set all sources to zero. As before, voltage sources are set to zero by replacing them with short circuits, and current sources are set to zero by replacing them with open circuits.
4. Determine the Norton equivalent resistance, R_N, by calculating the resistance seen between terminals *a* and *b*. It may be necessary to redraw the circuit to simplify this step.
5. Replace the sources removed in Step 3, and determine the current that would occur in a short if the short were connected between terminals *a* and *b*. If the original circuit has more than one source, it may be necessary to use the superposition theorem. In this case, it will be necessary to determine the short-circuit current due to each source separately and then determine the combined effect. The resulting short-circuit current will be the value of the Norton current I_N.

6. Sketch the Norton equivalent circuit using the resistance determined in Step 4 and the current calculated in Step 5. As part of the resulting circuit, include that portion of the network removed in Step 1.

The Norton equivalent circuit may also be determined directly from the Thévenin equivalent circuit by using the source conversion technique developed in Chapter 8. As a result, the Thévenin and Norton circuits shown in Figure 9–27 are equivalent.

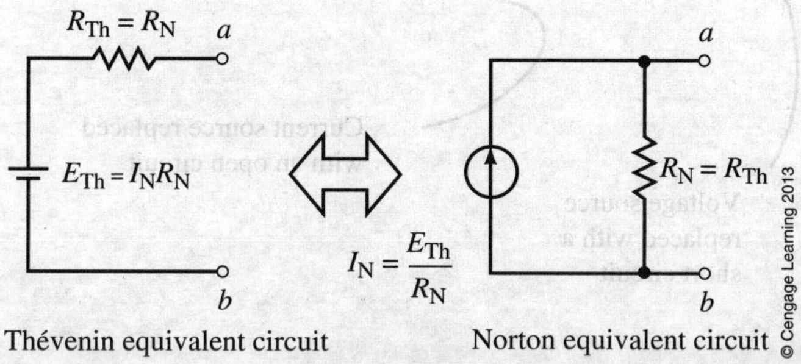

FIGURE 9–27

From Figure 9–27 we see that the relationship between the circuits is as follows:

$$E_{Th} = I_N R_N \tag{9–1}$$

$$I_N = \frac{E_{Th}}{R_{Th}} \tag{9–2}$$

EXAMPLE 9–6

Determine the Norton equivalent circuit external to the resistor R_L for the circuit of Figure 9–28. Use the Norton equivalent circuit to calculate the current through R_L. Compare the results to those obtained using Thévenin's theorem in Example 9–3.

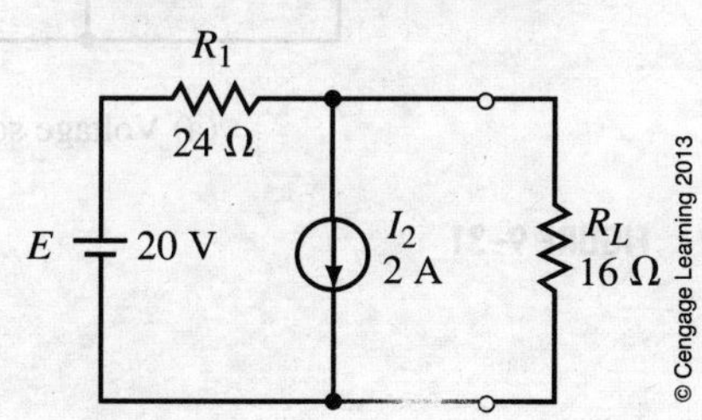

FIGURE 9–28

Solution

Steps 1 and 2: Remove load resistor R_L from the circuit and label the remaining terminals as *a* and *b*. The resulting circuit is shown in Figure 9–29.

Step 3: Zero the voltage and current sources as shown in the circuit of Figure 9–30.

Step 4: The resulting Norton resistance between the terminals is

$$R_N = R_{ab} = 24\ \Omega$$

Step 5: The short-circuit current is determined by first calculating the current through the short due to each source. The circuit for each calculation is illustrated in Figure 9–31.

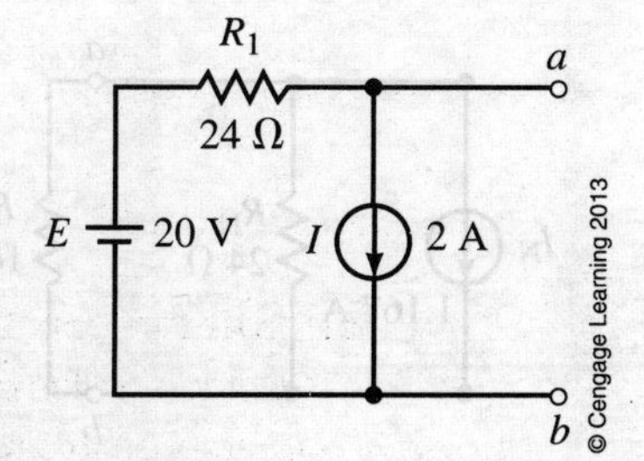

FIGURE 9–29

Voltage Source, E: The current in the short between terminals *a* and *b* [Figure 9–31(a)] is found from Ohm's law as

$$I_{ab(1)} = \frac{20\ \text{V}}{24\ \Omega} = 0.833\ \text{A}$$

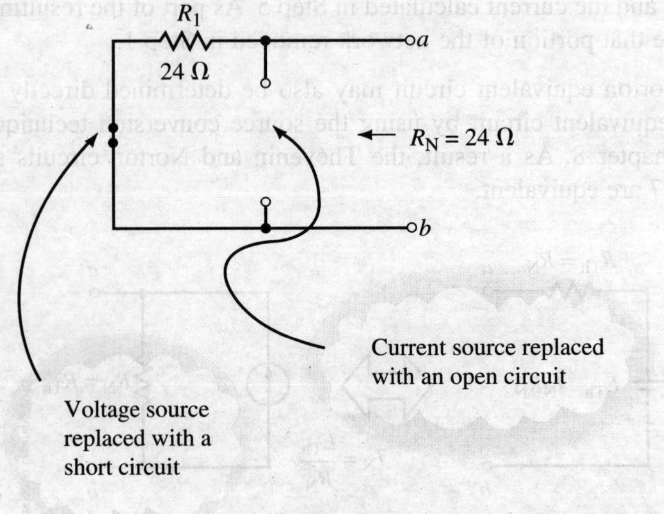

FIGURE 9–30

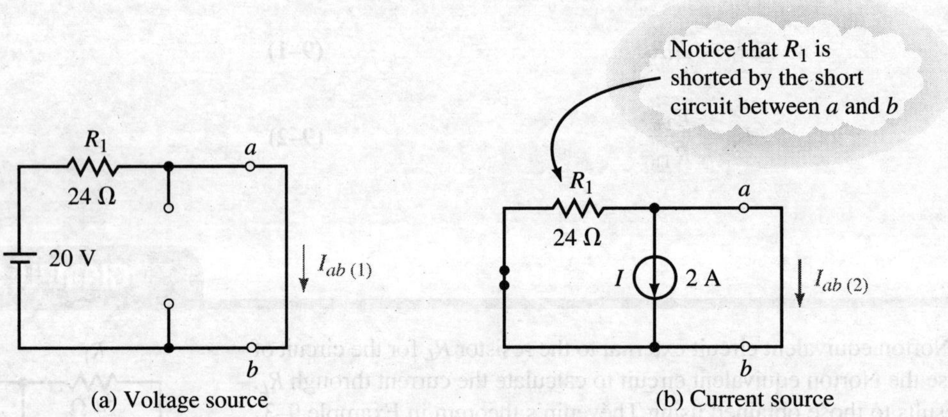

FIGURE 9–31

Current Source, I: By examining the circuit for the current source [Figure 9–31(b)] we see that the short circuit between terminals a and b effectively removes R_1 from the circuit. Therefore, the current through the short will be

$$I_{ab(2)} = -2.00 \text{ A}$$

Notice that the current I_{ab} is indicated as being a negative quantity. As we have seen before, this result merely indicates that the actual current is opposite to the assumed reference direction.

Now, applying the superposition theorem, we find the Norton current as

$$I_N = I_{ab(1)} + I_{ab(2)} = 0.833 \text{ A} - 2.0 \text{ A} = -1.167 \text{ A}$$

As before, the negative sign indicates that the short-circuit current is actually from terminal b toward terminal a.

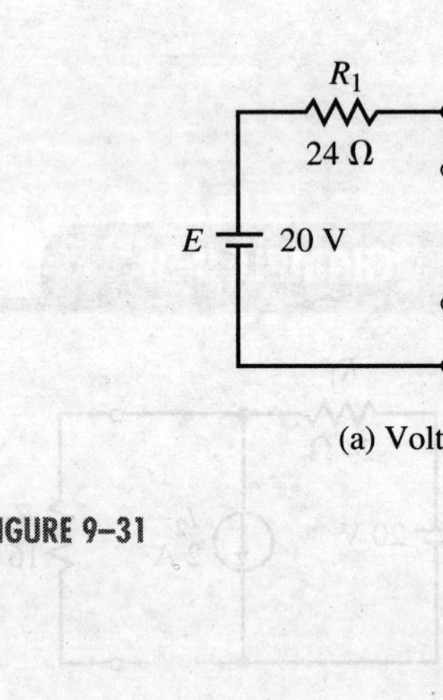

FIGURE 9–32

Step 6: The resultant Norton equivalent circuit is shown in Figure 9–32. Now we can easily find the current through load resistor R_L by using the current divider rule:

$$I_L = \left(\frac{24 \ \Omega}{24 \ \Omega + 16 \ \Omega}\right)(1.167 \text{ A}) = 0.700 \text{ A} \quad \text{(upward)}$$

By referring to Example 9–3, we see that the same result was obtained by finding the Thévenin equivalent circuit. An alternate method of finding the Norton

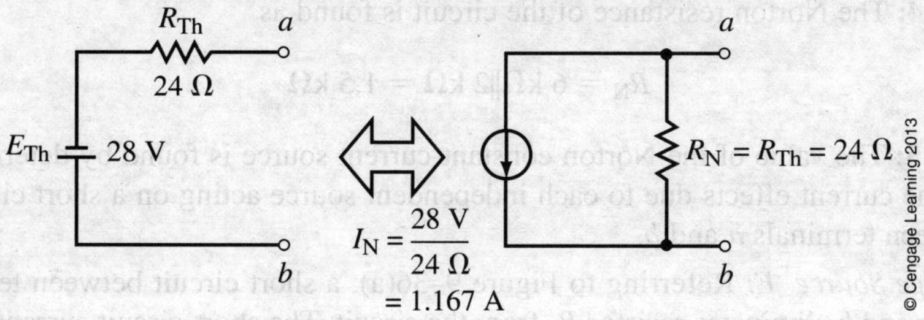

FIGURE 9–33

equivalent circuit is to convert the Thévenin circuit found in Example 9–3 into its equivalent Norton circuit shown in Figure 9–33.

EXAMPLE 9–7

Find the Norton equivalent of the circuit external to resistor R_L in the circuit in Figure 9–34. Use the equivalent circuit to determine the load current I_L when $R_L = 0$, 2 kΩ, and 5 kΩ.

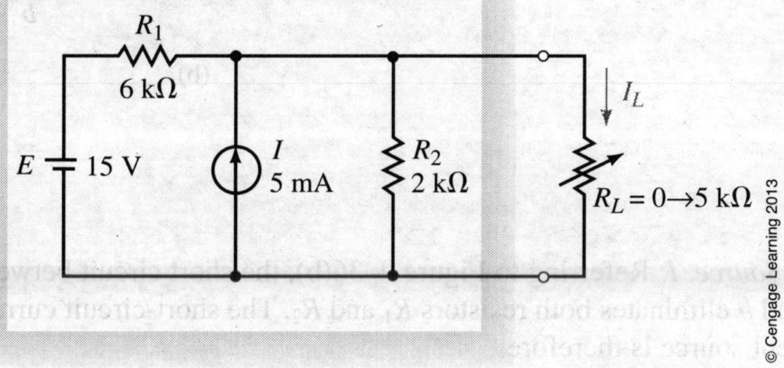

FIGURE 9–34

Solution

Steps 1, 2, and 3: After removing the load resistor, labelling the remaining two terminals a and b, and setting the sources to zero, we have the circuit of Figure 9–35.

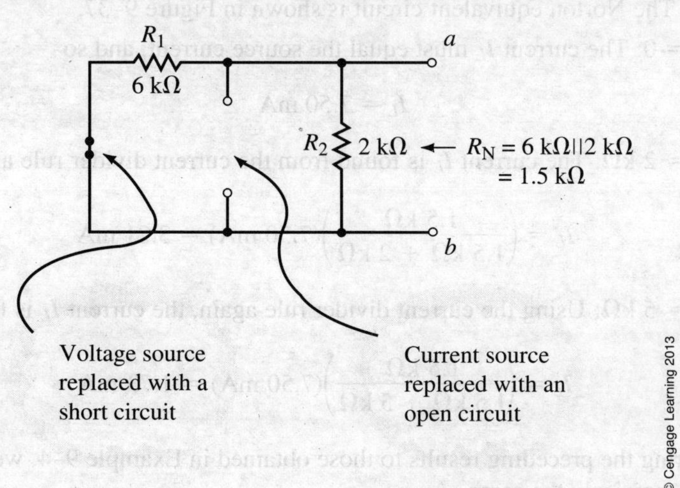

FIGURE 9–35

Step 4: The Norton resistance of the circuit is found as

$$R_N = 6\ k\Omega \| 2\ k\Omega = 1.5\ k\Omega$$

Step 5: The value of the Norton constant-current source is found by determining the current effects due to each independent source acting on a short circuit between terminals *a* and *b*.

Voltage Source, E: Referring to Figure 9–36(a), a short circuit between terminals *a* and *b* eliminates resistor R_2 from the circuit. The short-circuit current due to the voltage source is

$$I_{ab(1)} = \frac{15\ V}{6\ k\Omega} = 2.50\ mA$$

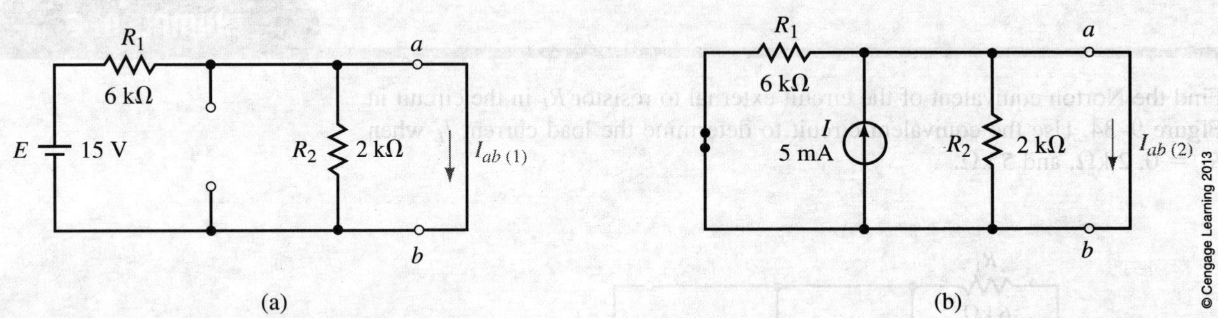

FIGURE 9–36

Current Source, I: Referring to Figure 9–36(b), the short circuit between terminals *a* and *b* eliminates both resistors R_1 and R_2. The short-circuit current due to the current source is therefore

$$I_{ab(2)} = 5.00\ mA$$

The resultant Norton current is found from superposition as

$$I_N = I_{ab(1)} + I_{ab(2)} = 2.50\ mA + 5.00\ mA = 7.50\ mA$$

Step 6: The Norton equivalent circuit is shown in Figure 9–37.

Let $R_L = 0$: The current I_L must equal the source current, and so

$$I_L = 7.50\ mA$$

Let $R_L = 2\ k\Omega$: The current I_L is found from the current divider rule as

$$I_L = \left(\frac{1.5\ k\Omega}{1.5\ k\Omega + 2\ k\Omega}\right)(7.50\ mA) = 3.21\ mA$$

Let $R_L = 5\ k\Omega$: Using the current divider rule again, the current I_L is found as

$$I_L = \left(\frac{1.5\ k\Omega}{1.5\ k\Omega + 5\ k\Omega}\right)(7.50\ mA) = 1.73\ mA$$

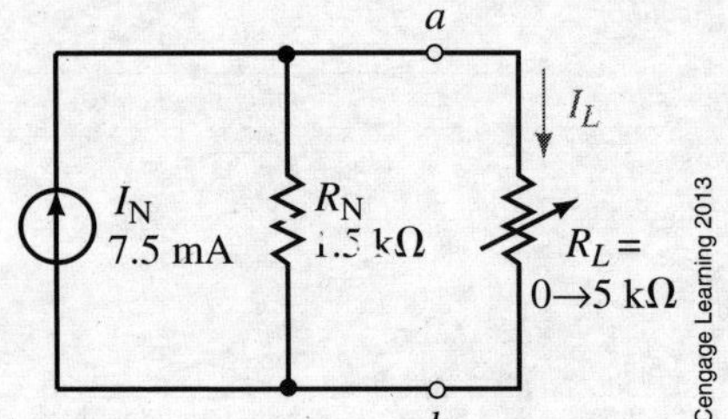

FIGURE 9–37

Comparing the preceding results to those obtained in Example 9–4, we see that they are precisely the same.

EXAMPLE 9–8

Consider the circuit of Figure 9–38:

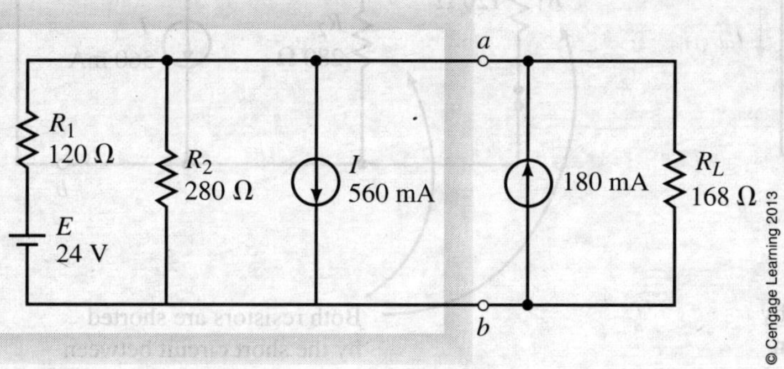

FIGURE 9–38

a. Find the Norton equivalent circuit external to terminals *a* and *b*.
b. Determine the current through R_L.

Solution

a. **Steps 1 and 2:** After removing the load, (which consists of a current source in parallel with a resistor), we have the circuit of Figure 9–39.

Step 3: After zeroing the sources, we have the network shown in Figure 9–40.

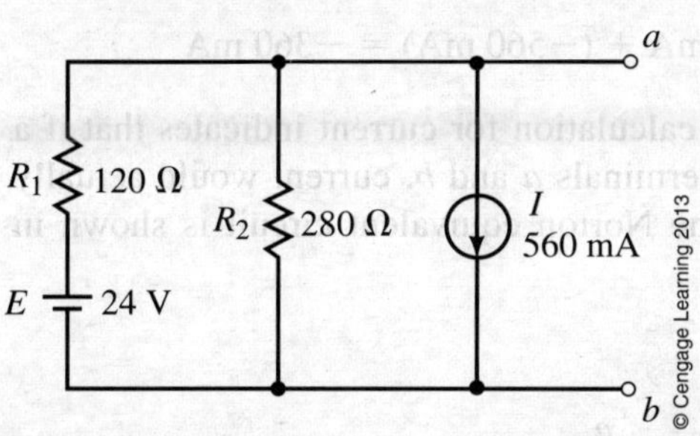

FIGURE 9–39

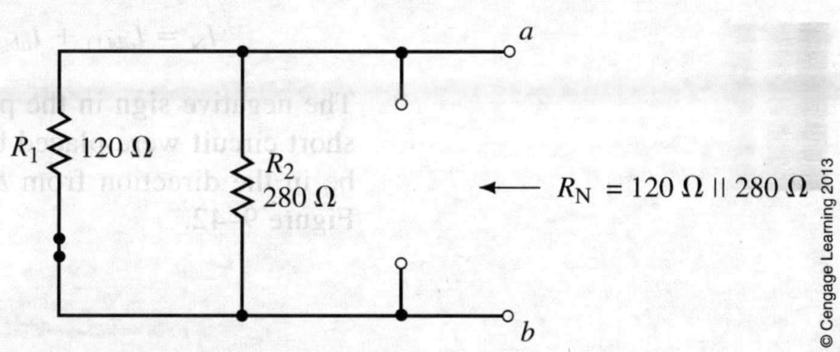

FIGURE 9–40

Step 4: The Norton equivalent resistance is found as

$$R_N = 120\ \Omega \| 280\ \Omega = 84\ \Omega$$

Step 5: In order to determine the Norton current we must again determine the short-circuit current due to each source separately and then combine the results using the superposition theorem.

Voltage Source, E: Referring to Figure 9–41(a), notice that the resistor R_2 is shorted by the short circuit between terminals *a* and *b*, and so the current in the short circuit is

$$I_{ab(1)} = \frac{24\text{ V}}{120\ \Omega} = 0.2\text{ A} = 200\text{ mA}$$

Current Source, I: Referring to Figure 9–41(b), the short circuit between terminals *a* and *b* will now eliminate both resistors. The current through the short will

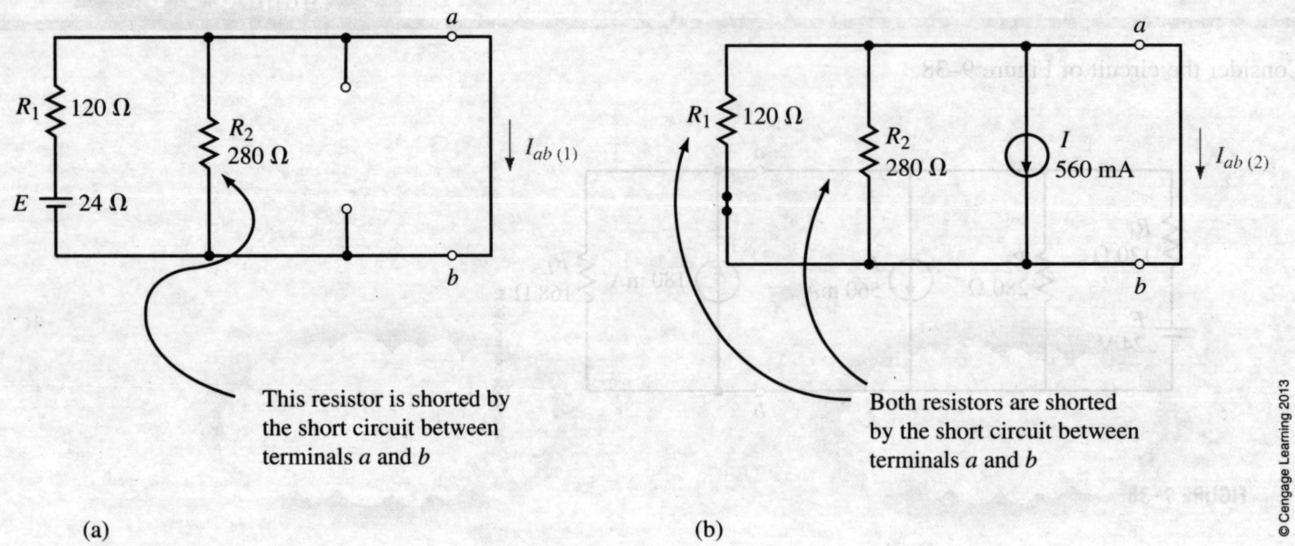

FIGURE 9–41

simply be the source current. However, since the current will not be from *a* to *b* but rather in the opposite direction, we write

$$I_{ab(2)} = -560 \text{ mA}$$

Now the Norton current is found as the summation of the short-circuit currents due to each source:

$$I_{\text{N}} = I_{ab(1)} + I_{ab(2)} = 200 \text{ mA} + (-560 \text{ mA}) = -360 \text{ mA}$$

The negative sign in the preceding calculation for current indicates that if a short circuit were placed between terminals *a* and *b*, current would actually be in the direction from *b* to *a*. The Norton equivalent circuit is shown in Figure 9–42.

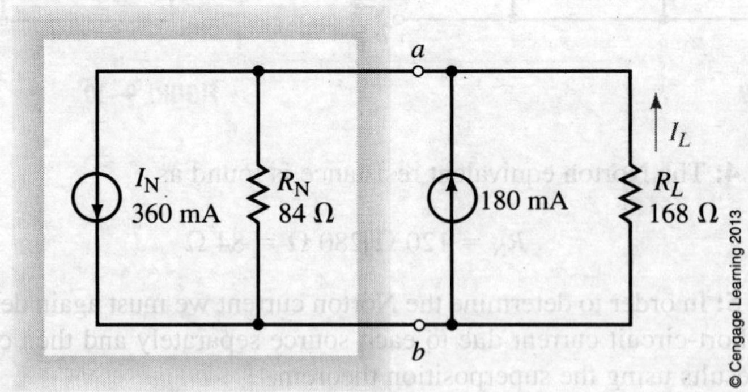

FIGURE 9–42

b. The current through the load resistor is found by applying the current divider rule:

$$I_L = \left(\frac{84 \ \Omega}{84 \ \Omega + 168 \ \Omega}\right)(360 \text{ mA} - 180 \text{ mA}) = 60 \text{ mA} \quad \text{(upward)}$$

PRACTICE PROBLEMS 4

Find the Norton equivalent of the circuit shown in Figure 9–43. Use the source conversion technique to determine the Thévenin equivalent of the circuit between points *a* and *b*.

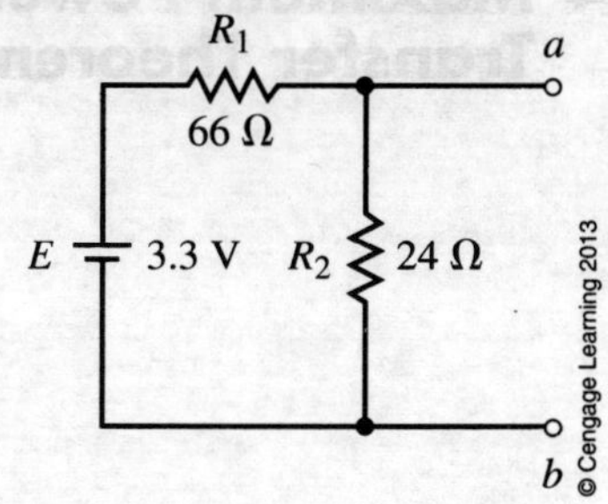

FIGURE 9–43

Answers
$R_N = R_{Th} = 17.6\ \Omega$, $I_N = 0.05$ A, $E_{Th} = 0.88$ V

PRACTICE PROBLEMS 5

Find the Norton equivalent external to R_L in the circuit of Figure 9–44. Solve for the current I_L when $R_L = 0$, 10 kΩ, 50 kΩ, and 100 kΩ.

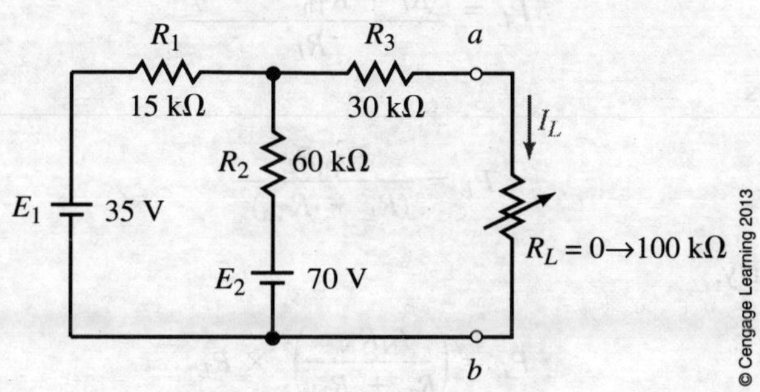

FIGURE 9–44

Answers
$R_N = 42$ kΩ, $I_N = 1.00$ mA
For $R_L = 0$: $I_L = 1.00$ mA
For $R_L = 10$ kΩ: $I_L = 0.808$ mA
For $R_L = 50$ kΩ: $I_L = 0.457$ mA
For $R_L = 100$ kΩ: $I_L = 0.296$ mA

IN-PROCESS LEARNING CHECK 3

(Answers are at the end of the chapter.)

1. Show the relationship between the Thévenin equivalent circuit and the Norton equivalent circuit. Sketch each circuit.
2. If a Thévenin equivalent circuit has $E_{Th} = 100$ mV and $R_{Th} = 500\ \Omega$, draw the corresponding Norton equivalent circuit.
3. If a Norton equivalent circuit has $I_N = 10\ \mu$A and $R_N = 20$ kΩ, draw the corresponding Thévenin equivalent circuit.

9.4 Maximum Power Transfer Theorem

In amplifiers and in most communication circuits such as radio receivers and transmitters, it is often desired that the load receive the maximum amount of power from a source.

The **maximum power transfer theorem** states the following:

A load resistance will receive maximum power from a circuit when the resistance of the load is exactly the same as the Thévenin (Norton) resistance looking back at the circuit.

The proof for the maximum power transfer theorem is determined from the Thévenin equivalent circuit and involves the use of calculus. This theorem is proved in Appendix C.

From Figure 9–45 we see that once the network has been simplified using either Thévenin's or Norton's theorem, maximum power will occur when

$$R_L = R_{Th} = R_N \tag{9–3}$$

Examining the equivalent circuits of Figure 9–45, shows that the following equations determine the power delivered to the load:

$$P_L = \frac{\left(\dfrac{R_L}{R_L + R_{Th}} \times E_{Th}\right)^2}{R_L}$$

which gives

$$P_L = \frac{E_{Th}^2 R_L}{(R_L + R_{Th})^2} \tag{9–4}$$

Similarly,

$$P_L = \left(\frac{I_N R_N}{R_L + R_N}\right)^2 \times R_L \tag{9–5}$$

Under maximum power conditions ($R_L = R_{Th} = R_N$), the preceding equations may be used to determine the maximum power delivered to the load and may therefore be written as

$$P_{max} = \frac{E_{Th}^2}{4R_{Th}} \tag{9–6}$$

$$P_{max} = \frac{I_N^2 R_N}{4} \tag{9–7}$$

NOTES...

Optional problems and derivations using calculus are marked by a ∫ icon. These may be omitted without loss of continuity by those who do not require calculus.

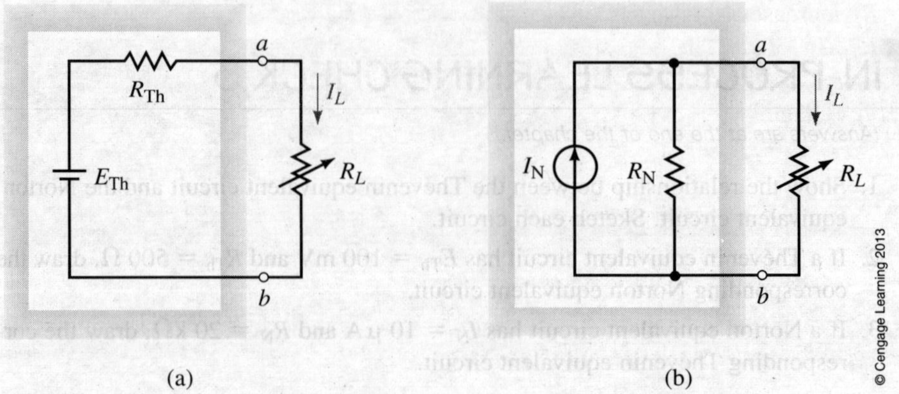

FIGURE 9–45

EXAMPLE 9–9

For the circuit of Figure 9–46, sketch graphs of V_L, I_L, and P_L as functions of R_L.

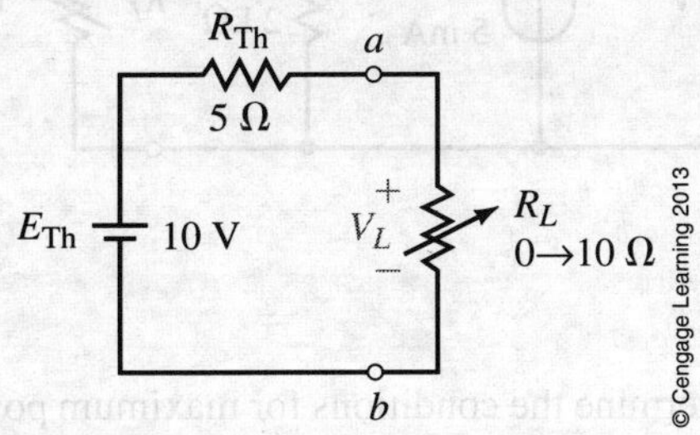

FIGURE 9–46

CircuitSim 09-2

TABLE 9–1

R_L (Ω)	V_L (V)	I_L (A)	P_L (W)
0	0	2.000	0
1	1.667	1.667	2.778
2	2.857	1.429	4.082
3	3.750	1.250	4.688
4	4.444	1.111	4.938
5	5.000	1.000	5.000
6	5.455	0.909	4.959
7	5.833	0.833	4.861
8	6.154	0.769	4.734
9	6.429	0.714	4.592
10	6.667	0.667	4.444

Solution We may first set up a table of data for various values of resistance, R_L. See Table 9–1. Voltage and current values are determined by using the voltage divider rule and Ohm's law, respectively. The power P_L for each value of resistance is determined by finding the product $P_L = V_L I_L$, or by using Equation 9–4.

If the data from Table 9–1 are plotted on linear graphs, the graphs will appear as shown in Figures 9–47, 9–48, and 9–49.

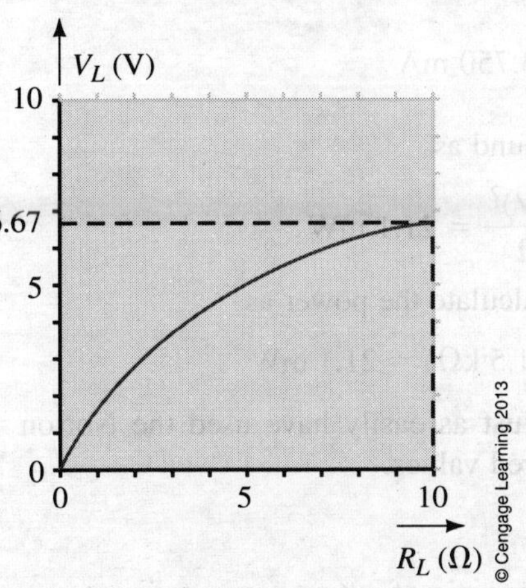

FIGURE 9–47 Voltage versus R_L.

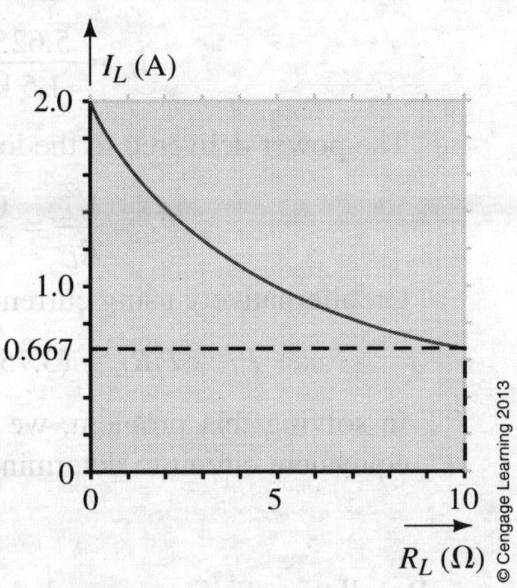

FIGURE 9–48 Current versus R_L.

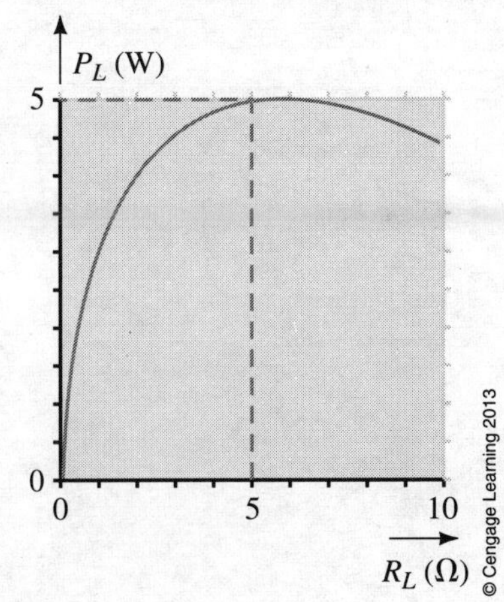

FIGURE 9–49 Power versus R_L.

Notice in the graphs that although voltage across the load increases as R_L increases, the power delivered to the load will be a maximum when $R_L = R_{Th} = 5\ \Omega$. The reason for this apparent contradiction is because, as R_L increases, the reduction in current more than offsets the corresponding increase in voltage.

EXAMPLE 9–10

Consider the circuit of Figure 9–50:

a. Determine the value of load resistance required to ensure that maximum power is transferred to the load.

b. Find V_L, I_L, and P_L when maximum power is delivered to the load.

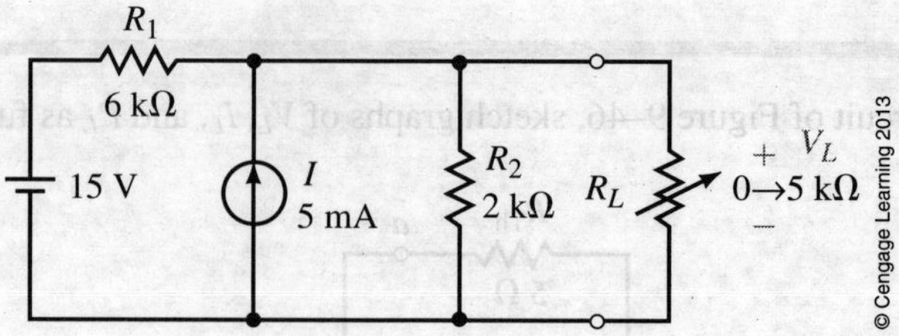

FIGURE 9–50

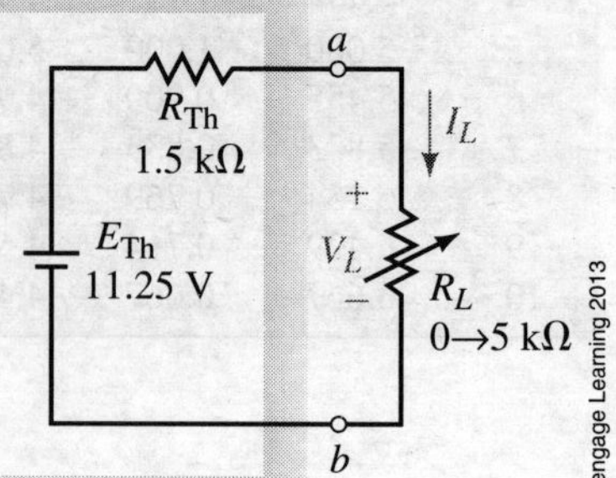

FIGURE 9–51

Solution

a. In order to determine the conditions for maximum power transfer, it is first necessary to determine the equivalent circuit external to the load. We may determine either the Thévenin equivalent circuit or the Norton equivalent circuit. This circuit was analyzed in Example 9–4 using Thévenin's theorem, and we determined the equivalent circuit to be as shown in Figure 9–51.

Maximum power will be transferred to the load when $R_L = 1.5\ \text{k}\Omega$.

b. Letting $R_L = 1.5\ \text{k}\Omega$, we see that half of the Thévenin voltage will appear across the load resistor and half will appear across the Thévenin resistance. So, at maximum power,

$$V_L = \frac{E_{Th}}{2} = \frac{11.25\ \text{V}}{2} = 5.625\ \text{V}$$

$$I_L = \frac{5.625\ \text{V}}{1.5\ \text{k}\Omega} = 3.750\ \text{mA}$$

The power delivered to the load is found as

$$P_L = \frac{V_L^2}{R_L} = \frac{(5.625\ \text{V})^2}{1.5\ \text{k}\Omega} = 21.1\ \text{mW}$$

Or, alternatively using current, we calculate the power as

$$P_L = I_L^2 R_L = (3.75\ \text{mA})^2(1.5\ \text{k}\Omega) = 21.1\ \text{mW}$$

In solving this problem, we could just as easily have used the Norton equivalent circuit to determine required values.

Recall that efficiency was defined as the ratio of output power to input power:

$$\eta = \frac{P_{out}}{P_{in}}$$

or as a percentage:

$$\eta = \frac{P_{out}}{P_{in}} \times 100\%$$

By using the maximum power transfer theorem, we see that under the condition of maximum power the efficiency of the circuit is

$$\eta = \frac{P_{out}}{P_{in}} \times 100\% = \frac{\dfrac{E_{Th}^2}{4R_{Th}}}{\dfrac{E_{Th}^2}{2R_{Th}}} \times 100\% = 0.500 \times 100\% = 50\% \qquad \text{(9–8)}$$

For communication circuits and for many amplifier circuits, 50% represents the maximum possible efficiency. At this efficiency level, the voltage presented to the following stage would only be half of the maximum terminal voltage.

In power transmission such as the 115-Vac, 60-Hz power in your home, the condition of maximum power is not a requirement. Under the condition of maximum power transfer, the voltage across the load will be reduced to half of the maximum available terminal voltage. Clearly, if we are working with power supplies, we would like to ensure that efficiency is brought as close to 100% as possible. In such cases, load resistance R_L is kept much larger than the internal resistance of the voltage source (typically $R_L \geq 10R_{int}$), ensuring that the voltage appearing across the load will be very nearly equal to the maximum terminal voltage of the voltage source.

EXAMPLE 9–11

Refer to the circuit of Figure 9–52, which represents a typical dc power supply.

a. Determine the value of R_L needed for maximum power transfer.
b. Determine terminal voltage V_L and the efficiency when the value of the load resistor is $R_L = 50\ \Omega$.
c. Determine terminal voltage V_L and the efficiency when the value of the load resistor is $R_L = 100\ \Omega$.

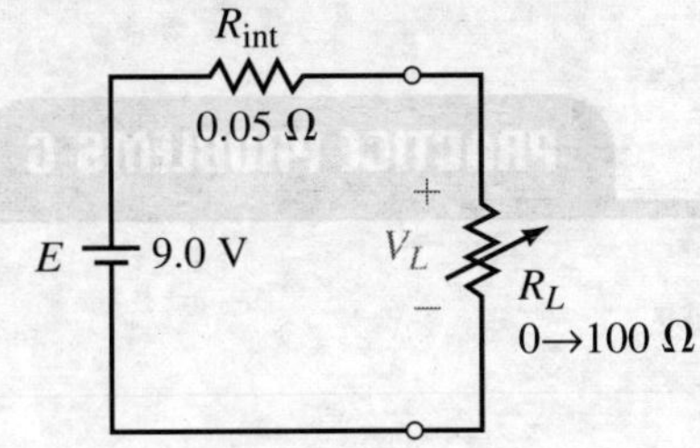

FIGURE 9–52

Solution

a. For maximum power transfer, the load resistor will be given as $R_L = 0.05\ \Omega$. At this value of load resistance, the efficiency will be only 50%.

b. For $R_L = 50\ \Omega$, the voltage appearing across the output terminals of the voltage source is

$$V_L = \left(\frac{50\ \Omega}{50\ \Omega + 0.05\ \Omega}\right)(9.0\ \text{V}) = 8.99\ \text{V}$$

The efficiency is

$$\eta = \frac{P_{out}}{P_{in}} \times 100\%$$

$$= \frac{\frac{(8.99\ \text{V})^2}{50\ \Omega}}{\frac{(9.0\ \text{V})^2}{50.05\ \Omega}} \times 100\%$$

$$= \frac{1.6168\ \text{W}}{1.6184\ \text{W}} \times 100\% = 99.90\%$$

c. For $R_L = 100\ \Omega$, the voltage appearing across the output terminals of the voltage source is

$$V_L = \left(\frac{100\ \Omega}{100\ \Omega + 0.05\ \Omega}\right)(9.0\ \text{V}) = 8.995\ 50\ \text{V}$$

The efficiency is

$$\eta = \frac{P_{out}}{P_{in}} \times 100\%$$

$$= \frac{\dfrac{(8.9955\ \text{V})^2}{100\ \Omega}}{\dfrac{(9.0\ \text{V})^2}{100.05\ \Omega}} \times 100\%$$

$$= \frac{1.6168\ \text{W}}{1.6184\ \text{W}} \times 100\% = 99.95\%$$

From this example, we see that if efficiency is important, as it is in power transmission, then the load resistance should be much larger than the resistance of the source (typically $R_L \geq 10R_{int}$). If, on the other hand, it is more important to ensure maximum power transfer, then the load resistance should equal the source resistance ($R_L = R_{int}$).

PRACTICE PROBLEMS 6

Refer to the circuit of Figure 9–44. For what value of R_L will the load receive maximum power? Determine the power when $R_L = R_N$, when $R_L = 25\ \text{k}\Omega$, and when $R_L = 50\ \text{k}\Omega$.

Answers

$R_L = 42\ \text{k}\Omega$: $P_L = 10.5\ \text{mW}$

$R_L = 25\ \text{k}\Omega$: $P_L = 9.82\ \text{mW}$

$R_L = 50\ \text{k}\Omega$: $P_L = 10.42\ \text{mW}$

IN-PROCESS LEARNING CHECK 4

(Answers are at the end of the chapter.)

A Thévenin equivalent circuit consists of $E_{Th} = 10\ \text{V}$ and $R_{Th} = 2\ \text{k}\Omega$. Determine the efficiency of the circuit when

a. $R_L = R_{Th}$

b. $R_L = 0.5R_{Th}$

c. $R_L = 2R_{Th}$

IN-PROCESS LEARNING CHECK 5

(Answers are at the end of the chapter.)

1. In what instances is maximum power transfer a desirable characteristic of a circuit?
2. In what instances is maximum power transfer an undesirable characteristic of a circuit?

9.5 Substitution Theorem

The **substitution theorem** states the following:

Any branch within a circuit may be replaced by an equivalent branch, provided the replacement branch has the same current through it and voltage across it as the original branch.

This theorem is best illustrated by examining the operation of a circuit. Consider the circuit of Figure 9–53.

The voltage V_{ab} and the current I in the circuit of Figure 9–53 are given as

$$V_{ab} = \left(\frac{6\ \text{k}\Omega}{4\ \text{k}\Omega + 6\ \text{k}\Omega}\right)(10\ \text{V}) = +6.0\ \text{V}$$

and

$$I = \frac{10\ \text{V}}{4\ \text{k}\Omega + 6\ \text{k}\Omega} = 1\ \text{mA}$$

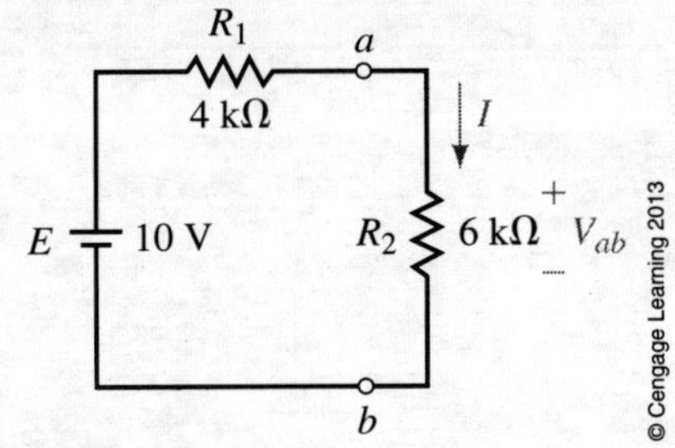

FIGURE 9–53

The resistor R_2 may be replaced with any combination of components, provided that the resulting components maintain the preceding conditions. We see that the branches of Figure 9–54 are each equivalent to the original branch between terminals a and b of the circuit in Figure 9–53.

Although each of the branches in Figure 9–54 is different, the current entering or leaving each branch will be same as that in the original branch. Similarly, the voltage across each branch will be the same. If any of these branches is substituted into the original circuit, the balance of the circuit will operate in the same way as the original. It is left as an exercise for the student to verify that each circuit behaves the same as the original.

This theorem allows us to replace any branch within a given circuit with an equivalent branch, thereby simplifying the analysis of the remaining circuit.

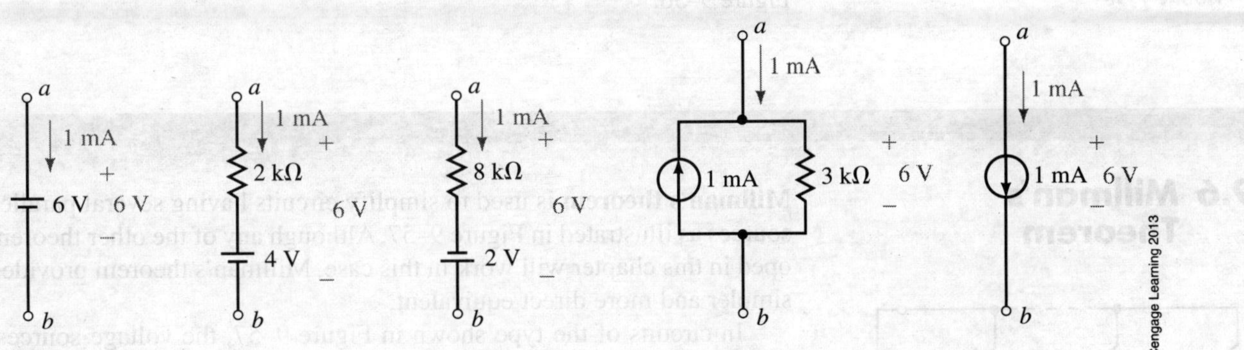

FIGURE 9–54

EXAMPLE 9–12

If the indicated portion in the circuit of Figure 9–55 is to be replaced with a current source and a 240-Ω shunt resistor, determine the magnitude and direction of the required current source.

Solution The voltage across the branch in the original circuit is

$$V_{ab} = \left(\frac{40\ \Omega \| 60\ \Omega}{16\ \Omega + (40\ \Omega \| 60\ \Omega)}\right)(20\ \text{V}) = \left(\frac{24\ \Omega}{16\ \Omega + 24\ \Omega}\right)(20\ \text{V}) = 12.0\ \text{V}$$

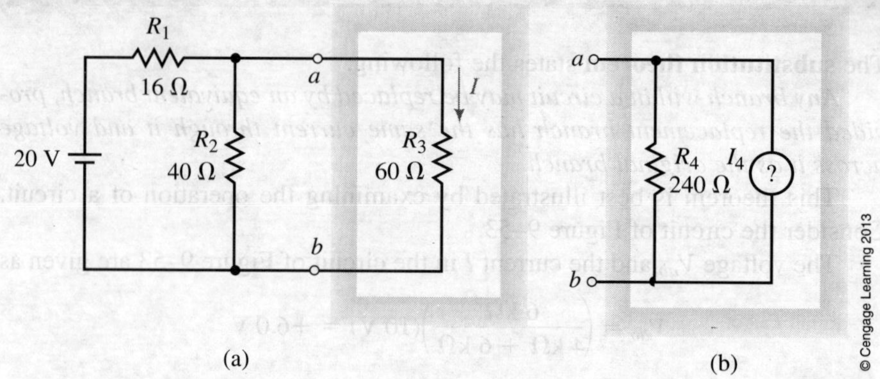

FIGURE 9–55

which results in a current of

$$I = \frac{12.0\text{ V}}{60\ \Omega} = 0.200\text{ A} = 200\text{ mA}$$

In order to maintain the same terminal voltage, V_{ab} = 12.0 V, the current through resistor R_4 = 240 Ω must be

$$I_{R4} = \frac{12.0\text{ V}}{240\ \Omega} = 0.050\text{ A} = 50\text{ mA}$$

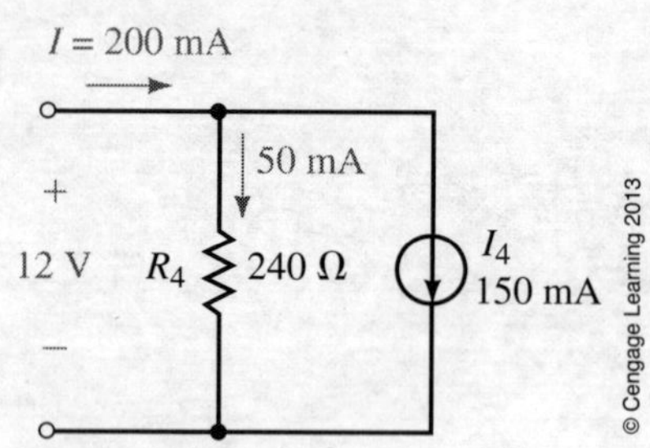

FIGURE 9–56

Finally, we know that the current entering terminal *a* is *I* = 200 mA. In order for Kirchhoff's current law to be satisfied at this node, the current source must have a magnitude of 150 mA and the direction must be downward, as shown in Figure 9–56.

9.6 Millman's Theorem

Millman's theorem is used to simplify circuits having several parallel voltage sources as illustrated in Figure 9–57. Although any of the other theorems developed in this chapter will work in this case, Millman's theorem provides a much simpler and more direct equivalent.

In circuits of the type shown in Figure 9–57, the voltage sources may be replaced with a single equivalent source as shown in Figure 9–58.

To find the values of the equivalent voltage source E_{eq} and series resistance R_{eq}, we need to convert each of the voltage sources of Figure 9–57 into its equivalent current source using the technique developed in Chapter 8. The value of each current source would be determined by using Ohm's law (i.e., $I_1 = E_1/R_1$, $I_2 = E_2/R_2$, etc.). After the source conversions are completed, the circuit appears as shown in Figure 9–59.

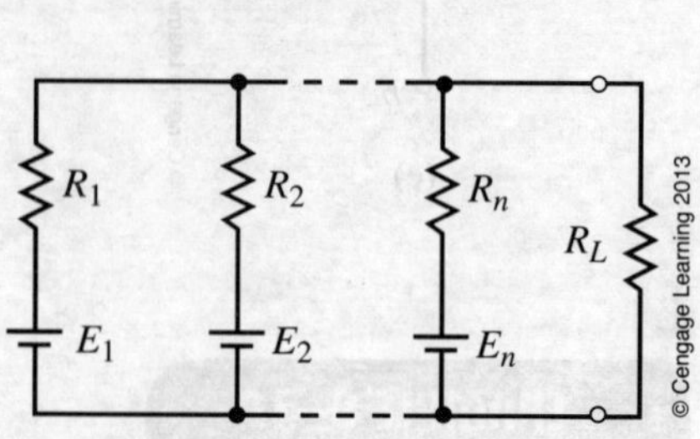

FIGURE 9–57

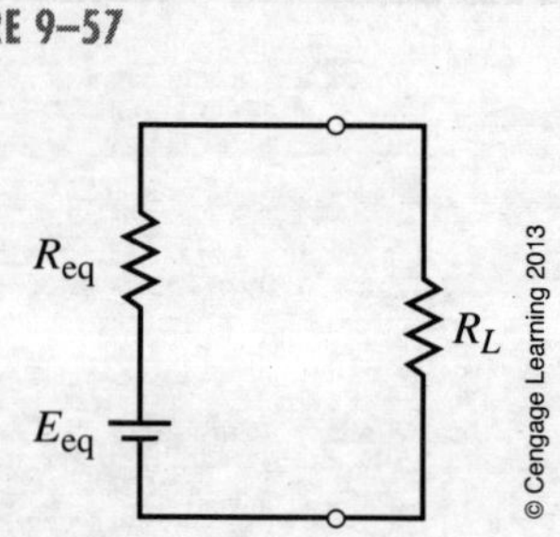

FIGURE 9–58

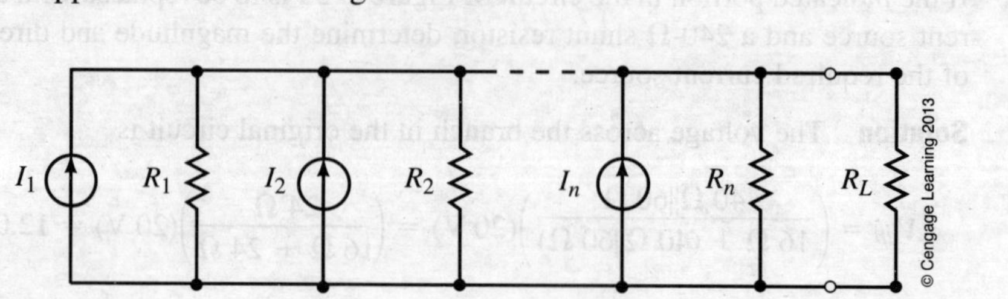

FIGURE 9–59

From the circuit of Figure 9–59 we see that all of the current sources have the same direction. Clearly, this will not always be the case since the direction of each current source will be determined by the initial polarity of the corresponding voltage source.

It is now possible to replace the *n* current sources with a single current source having a magnitude given as

$$I_{eq} = \sum_{x=0}^{n} I_x = I_1 + I_2 + I_3 + \cdots + I_n \tag{9–9}$$

which may be written as

$$I_{eq} = \frac{E_1}{R_1} + \frac{E_2}{R_2} + \frac{E_3}{R_3} + \cdots + \frac{E_n}{R_n} \tag{9–10}$$

If the direction of any current source is opposite to the direction shown, then the corresponding magnitude will be subtracted, rather than added. From Figure 9–59, we see that removing the current sources results in an equivalent resistance given as

$$R_{eq} = R_1 \| R_2 \| R_3 \| \cdots \| R_n \tag{9–11}$$

which may be determined as

$$R_{eq} = \frac{1}{G_{eq}} = \frac{1}{\frac{1}{R_1} + \frac{1}{R_2} + \frac{1}{R_3} + \cdots + \frac{1}{R_n}} \tag{9–12}$$

The general expression for the equivalent voltage is

$$E_{eq} = I_{eq}R_{eq} = \frac{\frac{E_1}{R_1} + \frac{E_2}{R_2} + \frac{E_3}{R_3} + \cdots + \frac{E_n}{R_n}}{\frac{1}{R_1} + \frac{1}{R_2} + \frac{1}{R_3} + \cdots + \frac{1}{P_n}} \tag{9–13}$$

EXAMPLE 9–13

Use Millman's theorem to simplify the circuit of Figure 9–60 so that it has only a single source. Use the simplified circuit to find the current in the load resistor, R_L.

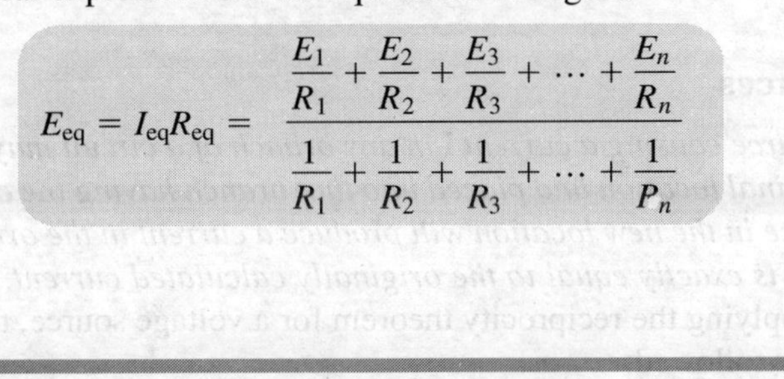

FIGURE 9–60

Solution From Equation 9–13, we express the equivalent voltage source as

$$V_{ab} = E_{eq} = \frac{\frac{-96\text{ V}}{240\ \Omega} + \frac{40\text{ V}}{200\ \Omega} + \frac{-80\text{ V}}{800\ \Omega}}{\frac{1}{240\ \Omega} + \frac{1}{200\ \Omega} + \frac{1}{800\ \Omega}}$$

$$V_{ab} = \frac{-0.300}{10.42\text{ mS}} = -28.8\text{ V}$$

The equivalent resistance is

$$R_{eq} = \frac{1}{\dfrac{1}{240\ \Omega} + \dfrac{1}{200\ \Omega} + \dfrac{1}{800\ \Omega}} = \frac{1}{10.42\ \text{mS}} = 96\ \Omega$$

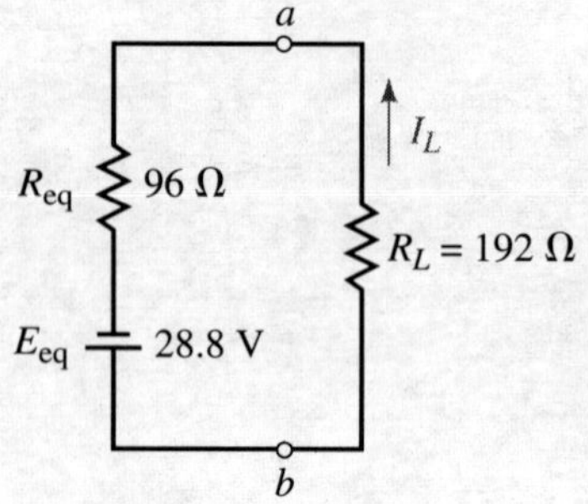

FIGURE 9–61

The equivalent circuit using Millman's theorem is shown in Figure 9–61. Notice that the equivalent voltage source has a polarity that is opposite to the originally assumed polarity. This is because the voltage sources E_1 and E_3 have magnitudes which overcome the polarity and magnitude of the source E_2.

From the equivalent circuit of Figure 9–61, it is a simple matter to determine the current through the load resistor:

$$I_L = \frac{28.8\ \text{V}}{96\ \Omega + 192\ \Omega} = 0.100\ \text{A} = 100\ \text{mA} \quad \text{(upward)}$$

9.7 Reciprocity Theorem

The **reciprocity theorem** is a theorem that can only be used with single-source circuits. This theorem, however, may be applied to either voltage sources or current sources. The theorem states the following:

Voltage Sources

A voltage source causing a current I *in any branch of a circuit may be removed from the original location and placed into that branch having the current* I. *The voltage source in the new location will produce a current in the original source location that is exactly equal to the originally calculated current,* I.

When applying the reciprocity theorem for a voltage source, the following steps must be followed:

1. The voltage source is replaced by a short circuit in the original location.
2. The polarity of the source in the new location is such that the current direction in that branch remains unchanged.

Current Sources

A current source causing a voltage V *at any node of a circuit may be removed from the original location and connected to that node. The current source in the new location will produce a voltage in the original source location that is exactly equal to the originally calculated voltage,* V.

When applying the reciprocity theorem for a current source, the following conditions must be met:

1. The current source is replaced by an open circuit in the original location.
2. The direction of the source in the new location is such that the polarity of the voltage at the node to which the current source is now connected remains unchanged.

The following examples illustrate how the reciprocity theorem is used within a circuit.

EXAMPLE 9–14

Consider the circuit of Figure 9–62:

a. Calculate the current I.
b. Remove voltage source E and place it into the branch with R_3. Show that the current through the branch which formerly had E is now the same as the current I.

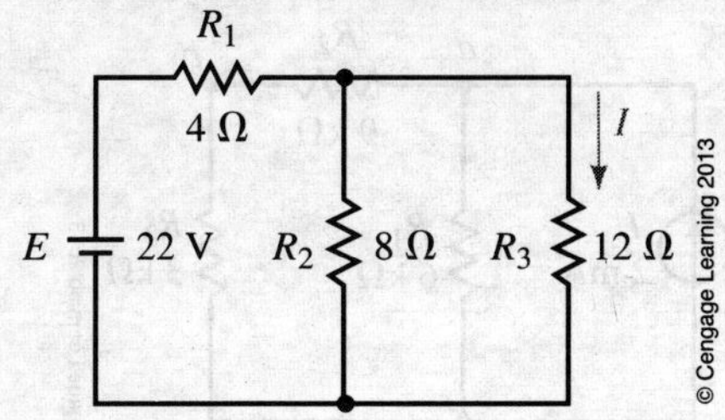

FIGURE 9–62

Solution

a. $$V_{12\,\Omega} = \left(\frac{8\ \Omega\|12\ \Omega}{4\ \Omega + (8\ \Omega\|12\ \Omega)}\right)(22\ \text{V}) = \left(\frac{4.8}{8.8}\right)(22\ \text{V}) = 12.0\ \text{V}$$

$$I = \frac{V_{12\,\Omega}}{12\ \Omega} = \frac{12.0\ \text{V}}{12\ \Omega} = 1.00\ \text{A}$$

b. Now removing the voltage source from its original location and moving it into the branch containing the current I, we obtain the circuit shown in Figure 9–63.

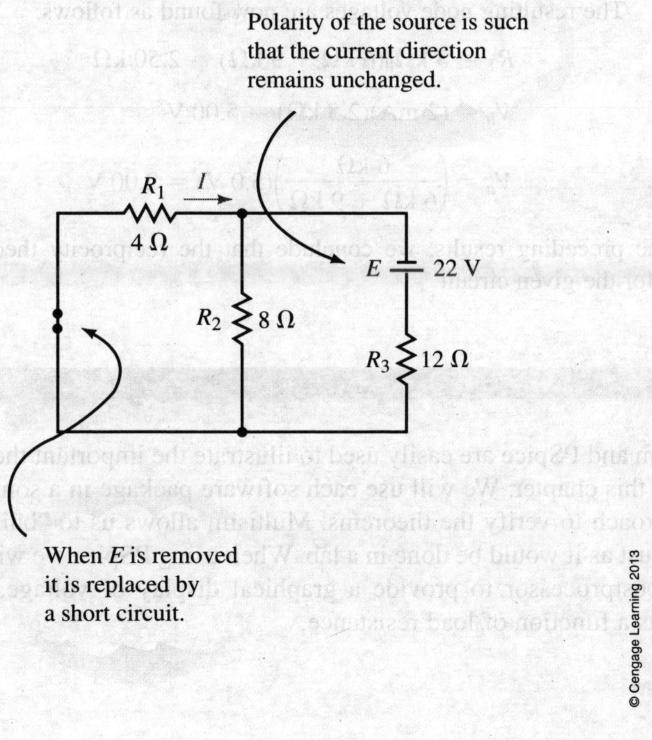

FIGURE 9–63

For the circuit of Figure 9–63, we determine the current I as follows:

$$V_{4\,\Omega} = \left(\frac{4\ \Omega\|8\ \Omega}{12\ \Omega + (4\ \Omega\|8\ \Omega)}\right)(22\ \text{V}) = \left(\frac{2.6}{14.6}\right)(22\ \text{V}) = 4.00\ \text{V}$$

$$I = \frac{V_{4\,\Omega}}{4\ \Omega} = \frac{4.00\ \text{V}}{4\ \Omega} = 1.00\ \text{A}$$

From this example, we see that the reciprocity theorem does indeed apply.

EXAMPLE 9–15

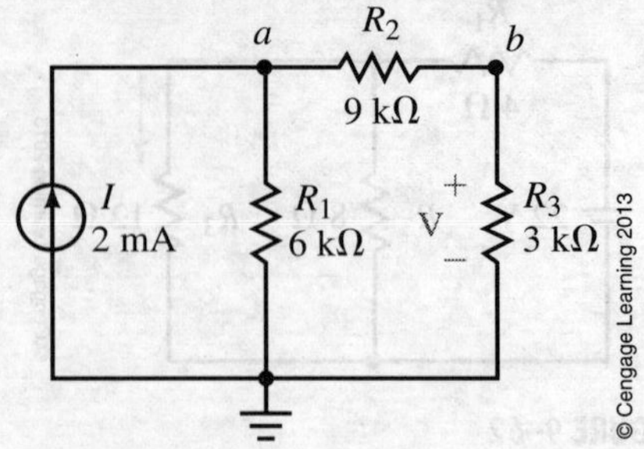

FIGURE 9–64

Consider the circuit shown in Figure 9–64:

a. Determine the voltage V across resistor R_3.

b. Remove the current source I and place it between node b and the reference node. Show that the voltage across the former location of the current source (node a) is now the same as the voltage V.

Solution

a. The node voltages for the circuit of Figure 9–64 are determined as follows:

$$R_T = 6\text{ k}\Omega \| (9\text{ k}\Omega + 3\text{ k}\Omega) = 4\text{ k}\Omega$$

$$V_a = (2\text{ mA})(4\text{ k}\Omega) = 8.00\text{ V}$$

$$V_b = \left(\frac{3\text{ k}\Omega}{3\text{ k}\Omega + 9\text{ k}\Omega}\right)(8.0\text{ V}) = 2.00\text{ V}$$

b. After relocating the current source from the original location, and connecting it between node b and ground, we obtain the circuit shown in Figure 9–65.

The resulting node voltages are now found as follows:

$$R_T = 3\text{ k}\Omega \| (6\text{ k}\Omega + 9\text{ k}\Omega) = 2.50\text{ k}\Omega$$

$$V_b = (2\text{ mA})(2.5\text{ k}\Omega) = 5.00\text{ V}$$

$$V_a = \left(\frac{6\text{ k}\Omega}{6\text{ k}\Omega + 9\text{ k}\Omega}\right)(5.0\text{ V}) = 2.00\text{ V}$$

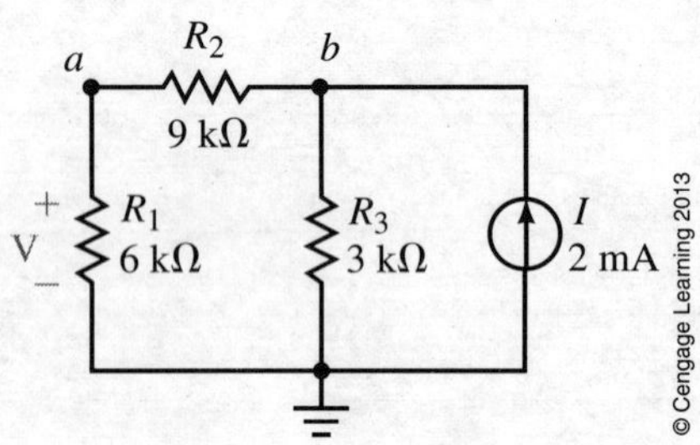

FIGURE 9–65

From the preceding results, we conclude that the reciprocity theorem again applies for the given circuit.

9.8 Circuit Analysis Using Computers

MULTISIM

PSPICE

Multisim and PSpice are easily used to illustrate the important theorems developed in this chapter. We will use each software package in a somewhat different approach to verify the theorems. Multisim allows us to "build and test" a circuit just as it would be done in a lab. When using PSpice we will activate the Probe postprocessor to provide a graphical display of voltage, current, and power as a function of load resistance.

Multisim

EXAMPLE 9–16

CircuitSim 09-3

Use Multisim to find both the Thévenin and the Norton equivalent circuits external to the load resistor in the circuit of Figure 9–66:

Solution

1. Using Multisim, construct the circuit as shown in Figure 9–67.
2. Just as in a lab, we will use a multimeter to find the open circuit (Thévenin) voltage and the short-circuit (Norton) current. As well, the multimeter is used to measure the Thévenin (Norton) resistance.

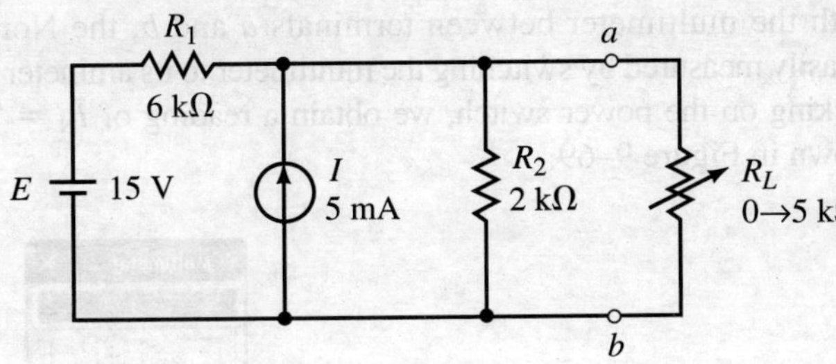

FIGURE 9–66

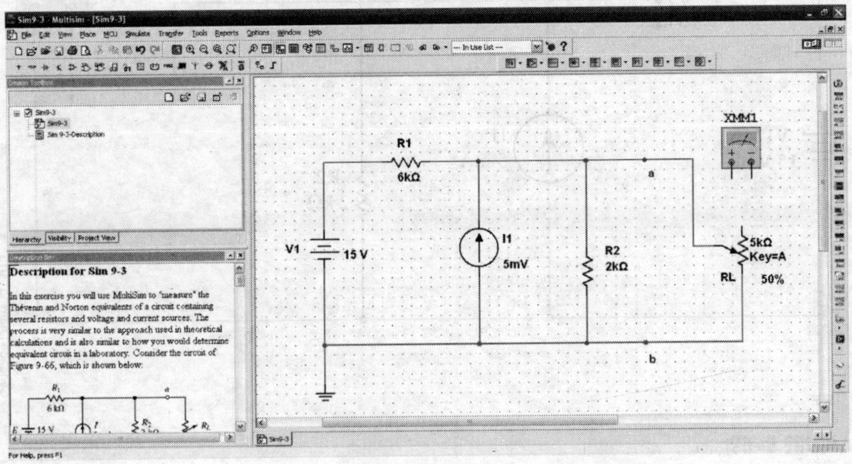

FIGURE 9–67

The steps in these measurements are essentially the same as those used to theoretically determine the equivalent circuits.

a. We begin by removing the load resistor R_L from the circuit (using the *Edit* menu and the *Cut/Paste* menu items). While not absolutely necessary, the remaining terminals are labelled as *a* and *b* for clarity.

b. The Thévenin voltage is measured by simply connecting the multimeter between terminals *a* and *b*. After clicking on the power switch, we obtain a reading of $E_{Th} = 11.25$ V as shown in Figure 9–68.

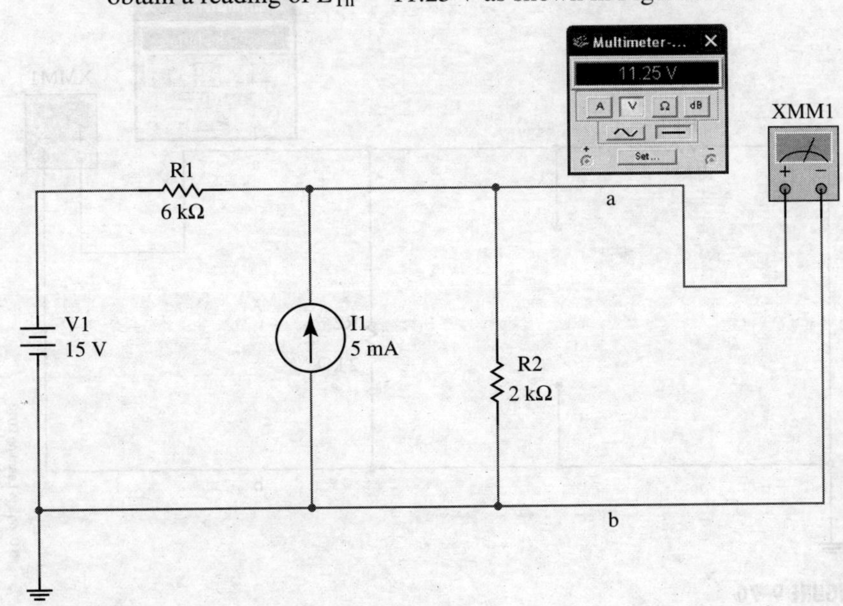

FIGURE 9–68

c. With the multimeter between terminals a and b, the Norton current is easily measured by switching the multimeter to its ammeter range. After clicking on the power switch, we obtain a reading of $I_N = 7.50$ mA as shown in Figure 9–69.

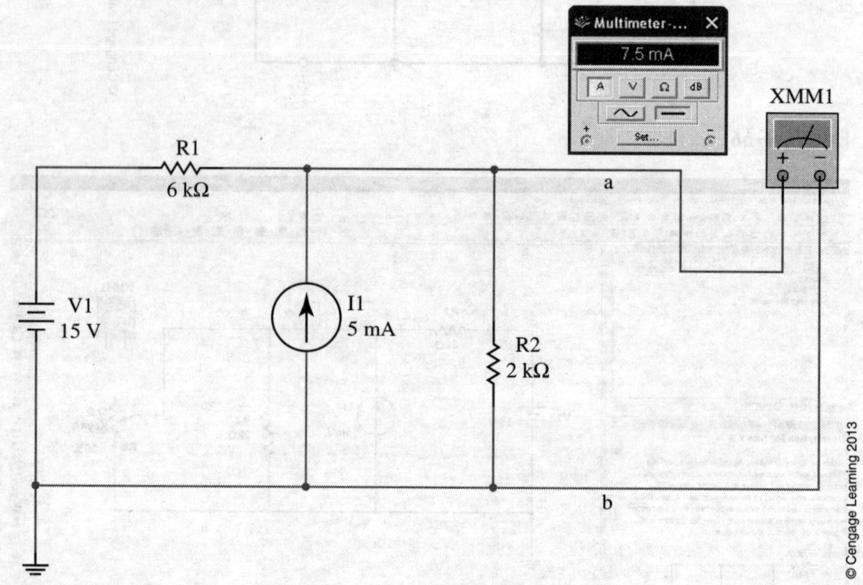

FIGURE 9–69

d. In order to measure the Thévenin resistance, the voltage source is removed and replaced by a short circuit (wire) and the current source is removed and replaced by an open circuit. Now, with the multimeter connected between terminals a and b, it is set to measure resistance (by clicking on the Ω button). After clicking on the power switch, we have the display shown in Figure 9–70. The multimeter provides the Thévenin resistance as $R_{Th} = 1.5\ \text{k}\Omega$.

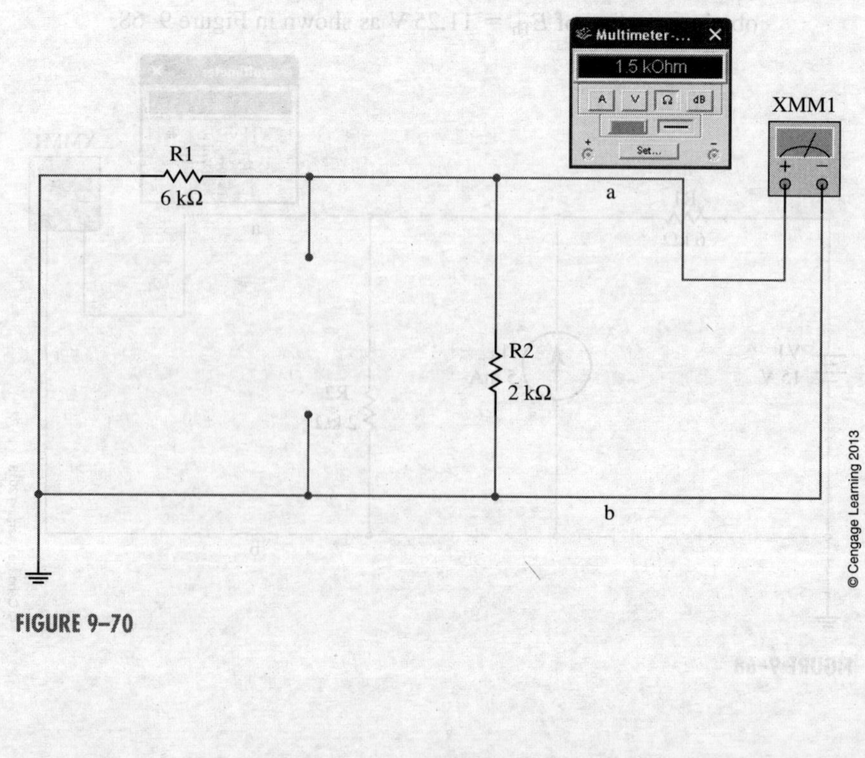

FIGURE 9–70

3. Using the measured results, we are able to sketch both the Thévenin equivalent and the Norton equivalent circuit as illustrated in Figure 9–71.

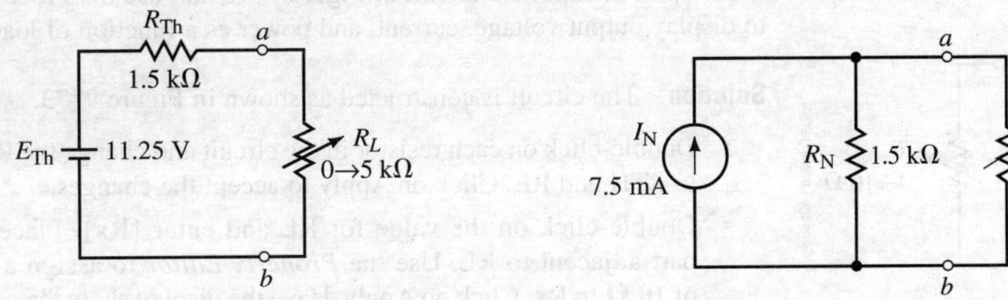

(a) Thévenin equivalent circuit (b) Norton equivalent circuit

FIGURE 9–71

These results are consistent with those obtained in Example 9–4.

Note: From the previous example we see that it is not necessary to directly measure the Thévenin (Norton) resistance since the value can be easily calculated from the Thévenin voltage and the Norton current. The following equation is an application of Ohm's law and always applies when finding an equivalent circuit.

$$R_{Th} = R_N = \frac{E_{Th}}{I_N} \quad \textbf{(9–14)}$$

Applying Equation 9–14 to the measurements of Example 9–16, we get

$$R_{Th} = R_N = \frac{11.25\text{ V}}{7.50\text{ mA}} = 1.50\text{ k}\Omega$$

Clearly, this is the same result as that obtained when we went through the extra step of removing the voltage and current sources. This approach is the most practical method and is commonly used when actually measuring the Thévenin (Norton) resistance of a circuit.

PRACTICE PROBLEMS 7

Use Multisim to find both the Thévenin and the Norton equivalent circuits external to the load resistor in the circuit of Figure 9–25.

Answers
$E_{Th} = 2.00$ V, $I_N = 16.67$ mA, $R_{Th} = R_N = 120\ \Omega$

PSpice

PSPICE

As we have already seen, PSpice has an additional postprocessor called Probe, which is able to provide a graphical display of numerous variables. The following example uses PSpice to illustrate the maximum power transfer theorem.

EXAMPLE 9–17

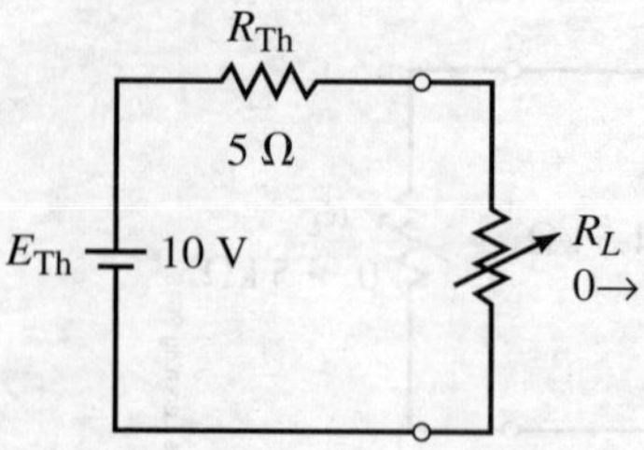

FIGURE 9–72

PSPICE

Use PSpice to input the circuit of Figure 9–72 and use the Probe postprocessor to display output voltage, current, and power as a function of load resistance.

Solution The circuit is constructed as shown in Figure 9–73.

- Double-click on each resistor in the circuit and change the Reference cells to RTH and RL. Click on Apply to accept the changes.
- Double-click on the value for RL and enter **{Rx}.** Place the PARAM part adjacent to RL. Use the *Property Editor* to assign a default value of **10 Ω** to Rx. Click on Apply. Have the display show the name and value and then exit the *Property Editor*. If necessary, refer to Example 7–15 for the complete procedure on how to assign parameters to a circuit component.

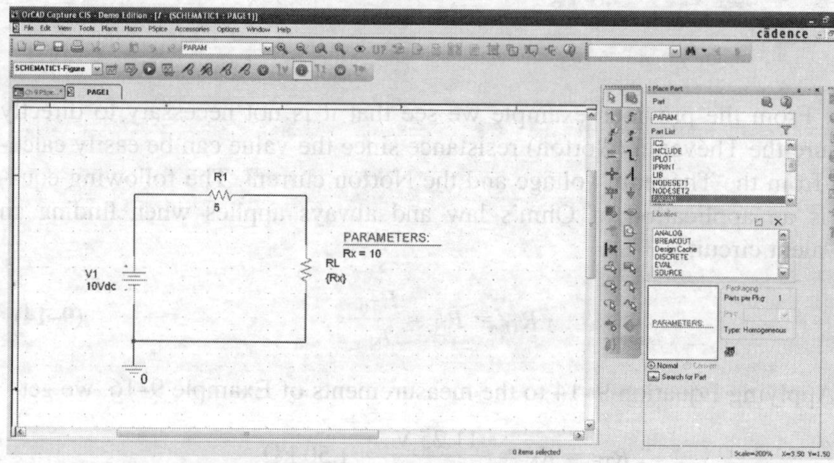

FIGURE 9–73

- Adjust the *Simulation Settings* to result in a dc sweep of the load resistor from **0.1 Ω** to **10 Ω** in **0.1 Ω** increments.
- Click on the *Run* icon once the circuit is complete.
- Once the design is simulated, you will see a blank screen with the abscissa (horizontal axis) showing RX scaled from 0 to 10 Ω.
- Since we would like to have a simultaneous display of voltage, current, and power, it is necessary to do the following:

  To display V_L: Click *Trace* and then *Add Trace*. Select **V(RL:1).** Click *OK* and the load voltage will appear as a function of load resistance.

  To display I_L: First add another axis by clicking on *Plot* and *Add Y Axis*. Next, click *Trace* and then *Add Trace*. Select **I(RL).** Click OK and the load current will appear as a function of load resistance.

  To display P_L: Add another Y axis. Click *Trace* and *Add Trace*. One method of showing power is to enter **W(RL)** and then click *OK*. The display will appear as shown in Figure 9–74. The color and line thickness of a trace can be changed by right-clicking anywhere on the desired trace and selecting *Trace Property*.

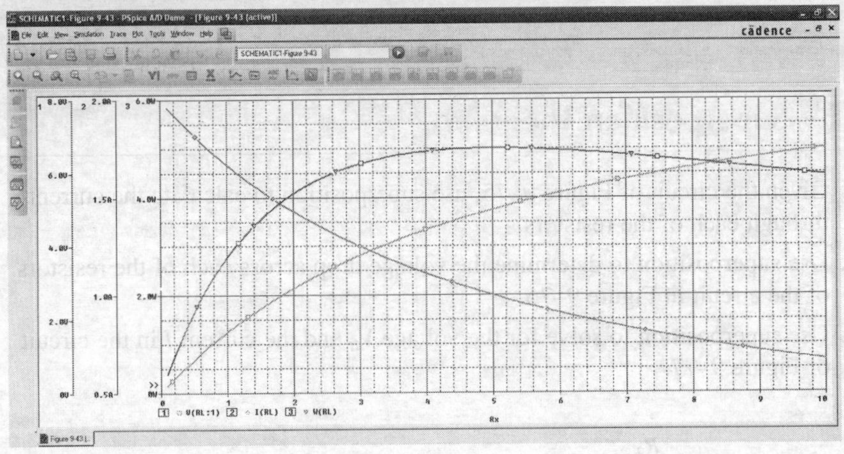

FIGURE 9–74

PRACTICE PROBLEMS 8

Use PSpice to input the circuit of Figure 9–66. Use the Probe postprocessor to obtain voltage, current, and power for the load resistor as it is varied from 0 to 5 kΩ.

Putting It into Practice

A simple battery cell (such as a D cell) can be represented as a Thévenin equivalent circuit as shown in the accompanying figure.

The Thévenin voltage represents the open-circuit (or unloaded) voltage of the battery cell, while the Thévenin resistance is the internal resistance of the battery. When a load resistance is connected across the terminals of the battery, the voltage V_{ab} will decrease due to the voltage drop across the internal resistor. By taking two measurements, it is possible to find the Thévenin equivalent circuit of the battery.

When no load is connected between the terminals of the battery, the terminal voltage is found to be $V_{ab} = 1.493$ V. When a resistance of $R_L = 10.6\ \Omega$ is connected across the terminals, the voltage is measured to be $V_{ab} = 1.430$ V. Determine the Thévenin equivalent circuit of the battery. Use the measurements to determine the efficiency of the battery for the given load.

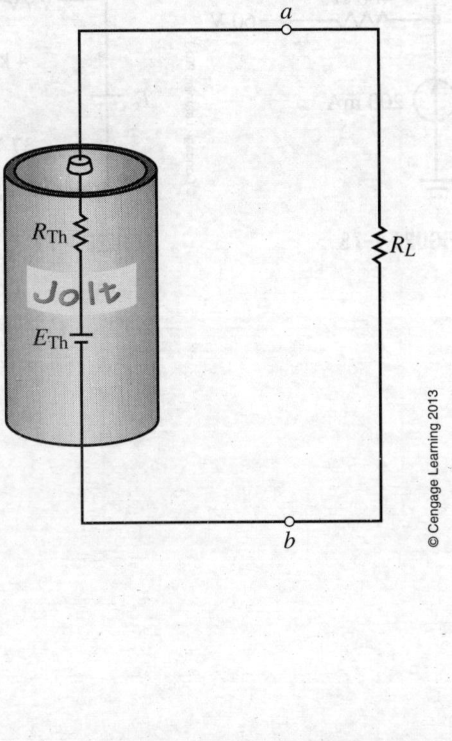

Problems

9.1 Superposition Theorem

1. Given the circuit of Figure 9–75, use superposition to calculate the current through each of the resistors.
2. Use superposition to determine the voltage drop across each of the resistors of the circuit in Figure 9–76.
3. Use superposition to solve for the voltage V_a and the current I in the circuit of Figure 9–77.

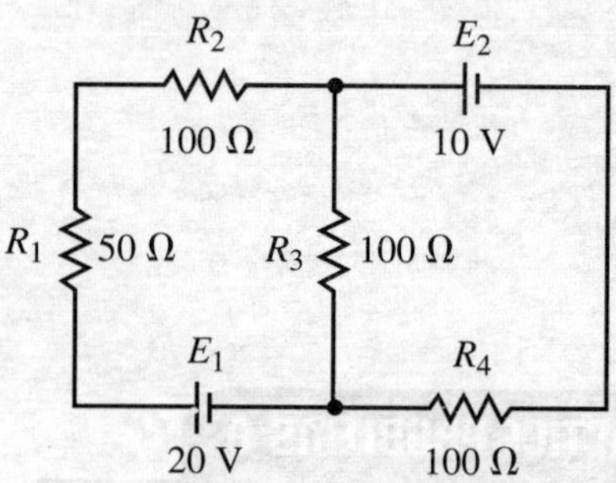

FIGURE 9–75

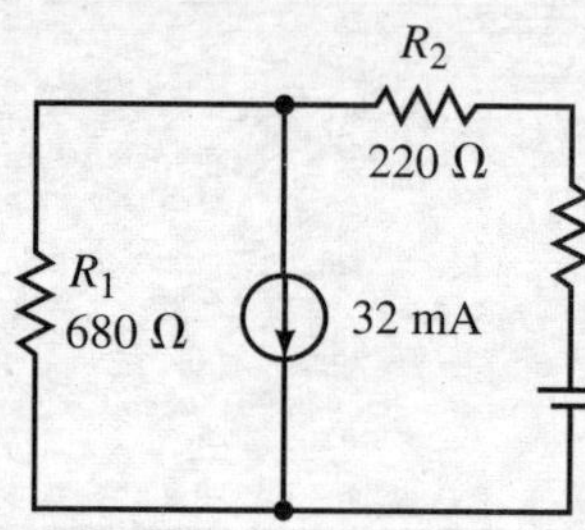

FIGURE 9–76

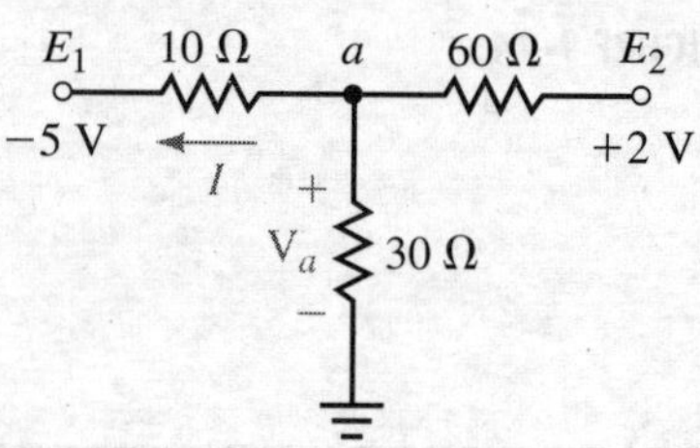

FIGURE 9–77

4. Using superposition, find the current through the 480-Ω resistor in the circuit of Figure 9–78.
5. Given the circuit of Figure 9–79, what must be the value of the unknown voltage source to ensure that the current through the load is $I_L = 5$ mA as shown? Verify the results using superposition.
6. If the load resistor in the circuit of Figure 9–80 is to dissipate 120 W, determine the value of the unknown voltage source. Verify the results using superposition.

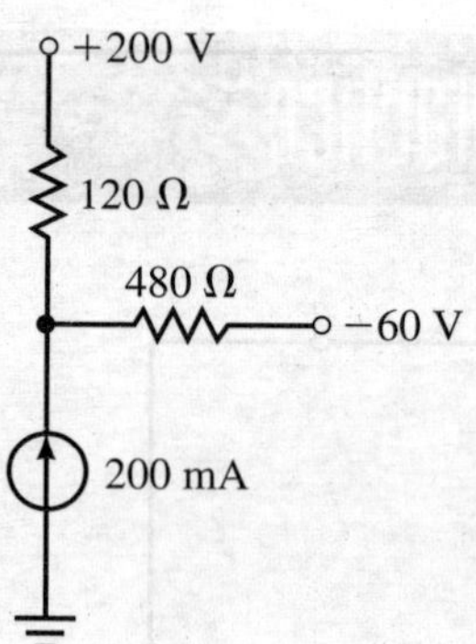

FIGURE 9–78

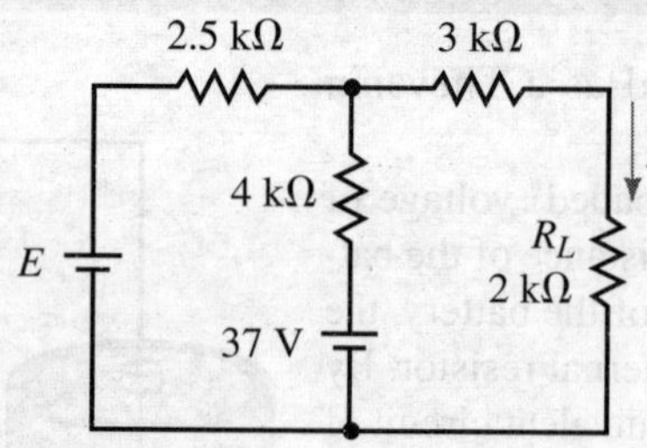

FIGURE 9–79

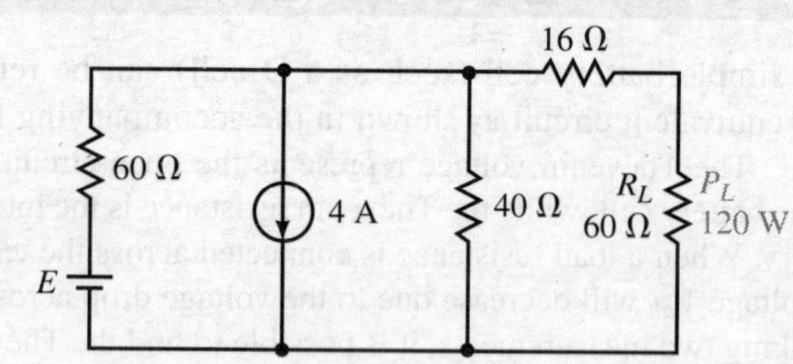

FIGURE 9–80

9.2 Thévenin's Theorem

7. Find the Thévenin equivalent external to R_L in the circuit of Figure 9–81. Use the equivalent circuit to find V_{ab}.
8. Repeat Problem 7 for the circuit of Figure 9–82.
9. Repeat Problem 7 for the circuit of Figure 9–83.
10. Repeat Problem 7 for the circuit of Figure 9–84.
11. Refer to the circuit of Figure 9–85:
    a. Find the Thévenin equivalent circuit external to R_L.
    b. Use the equivalent circuit to determine V_{ab} when $R_L = 20\ \Omega$ and when $R_L = 50\ \Omega$.

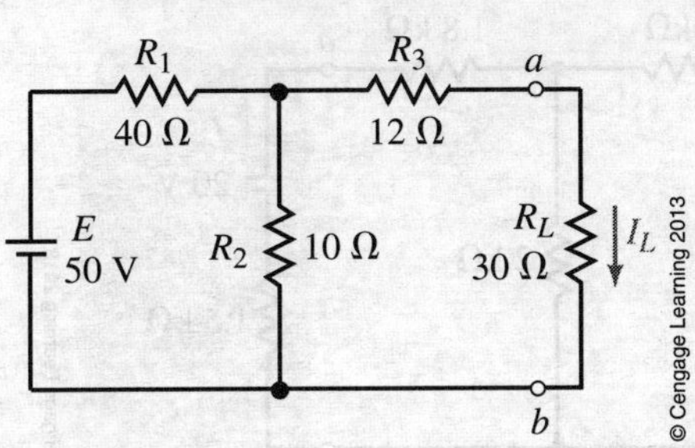

FIGURE 9–81

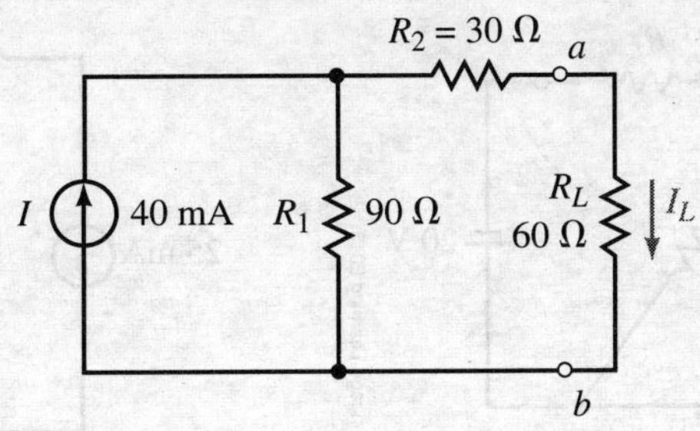

FIGURE 9–82

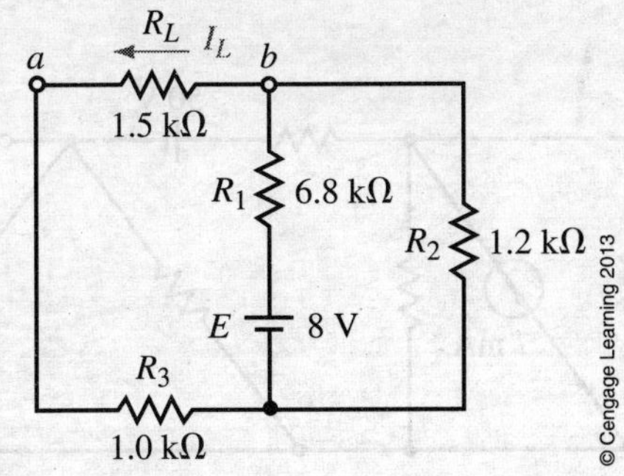

FIGURE 9–83

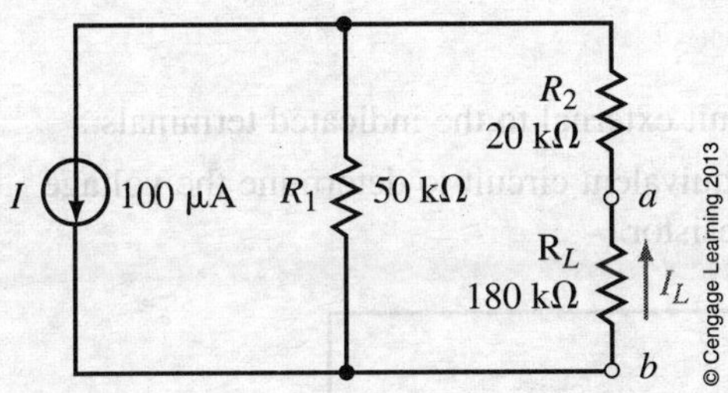

FIGURE 9–84

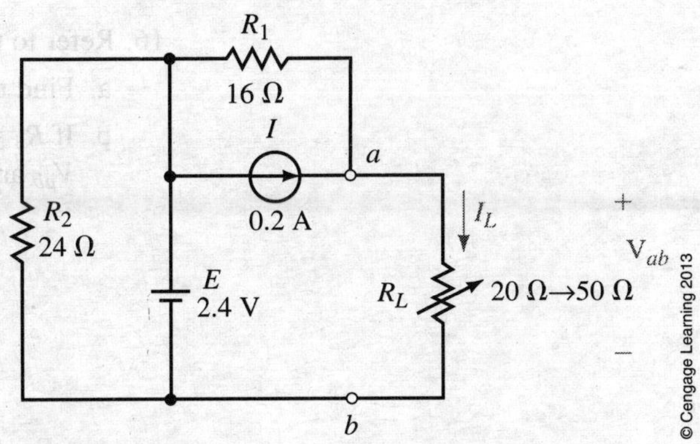

FIGURE 9–85

12. Refer to the circuit of Figure 9–86:
   a. Find the Thévenin equivalent circuit external to R_L.
   b. Use the equivalent circuit to determine V_{ab} when $R_L = 10\ \text{k}\Omega$ and when $R_L = 20\ \text{k}\Omega$.
13. Refer to the circuit of Figure 9–87:
   a. Find the Thévenin equivalent circuit external to the indicated terminals.
   b. Use the Thévenin equivalent circuit to determine the current through the indicated branch.

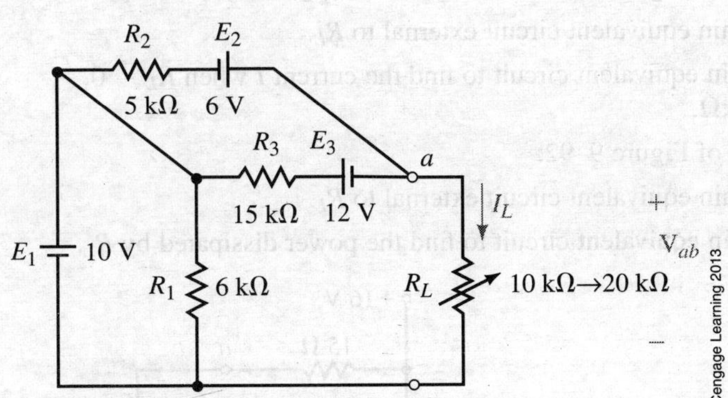

FIGURE 9–86

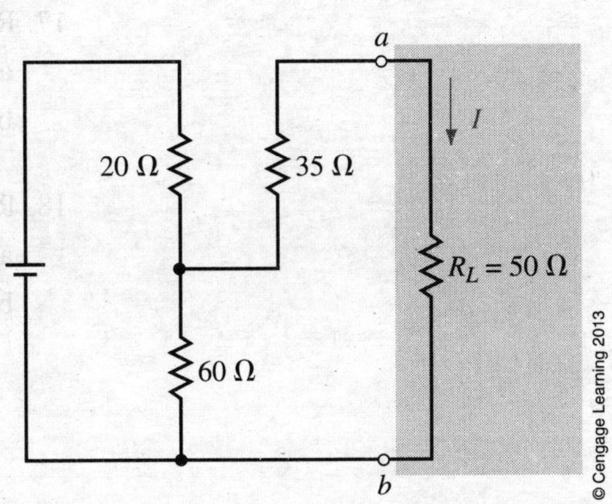

FIGURE 9–87

14. Refer to the circuit of Figure 9–88:
   a. Find the Thévenin equivalent circuit external to R_L.
   b. Use the Thévenin equivalent circuit to find V_L.
15. Refer to the circuit of Figure 9–89:
   a. Find the Thévenin equivalent circuit external to the indicated terminals.
   b. Use the Thévenin equivalent circuit to determine the current through the indicated branch.

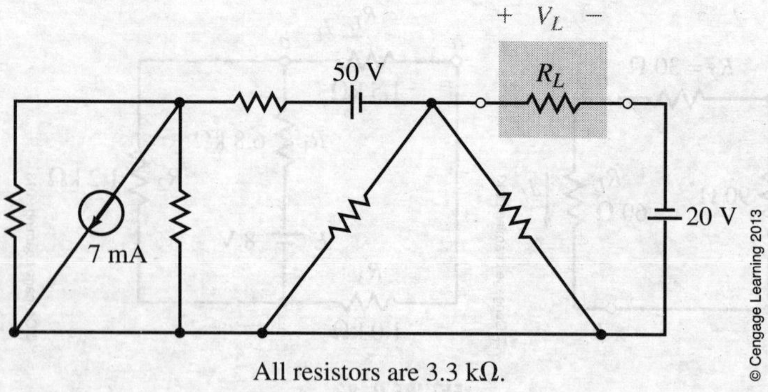

FIGURE 9–88

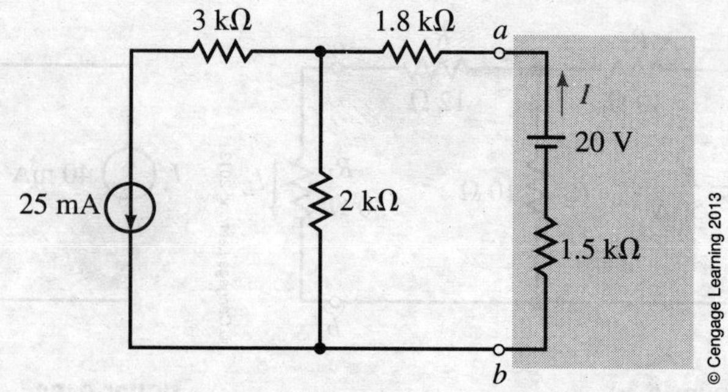

FIGURE 9–89

16. Refer to the circuit of Figure 9–90:
    a. Find the Thévenin equivalent circuit external to the indicated terminals.
    b. If $R_5 = 1\ \text{k}\Omega$, use the Thévenin equivalent circuit to determine the voltage V_{ab} and the current through this resistor.

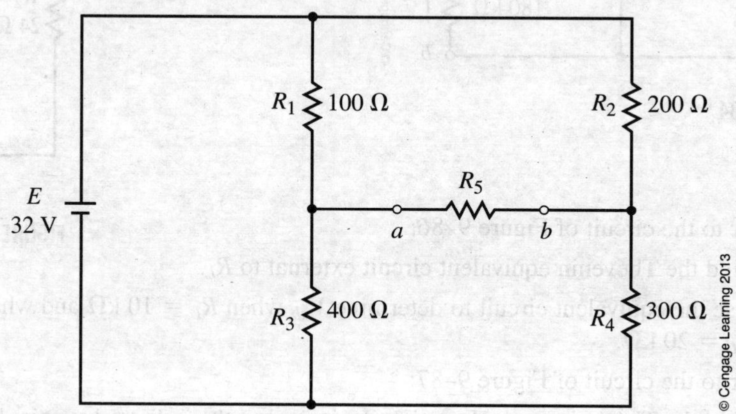

FIGURE 9–90

17. Refer to the circuit of Figure 9–91:
    a. Find the Thévenin equivalent circuit external to R_L.
    b. Use the Thévenin equivalent circuit to find the current I when $R_L = 0$, 10 kΩ, and 50 kΩ.
18. Refer to the circuit of Figure 9–92:
    a. Find the Thévenin equivalent circuit external to R_L.
    b. Use the Thévenin equivalent circuit to find the power dissipated by R_L.

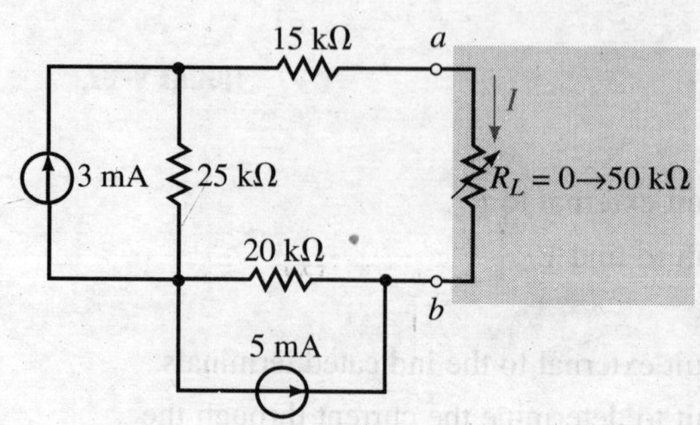

FIGURE 9–91

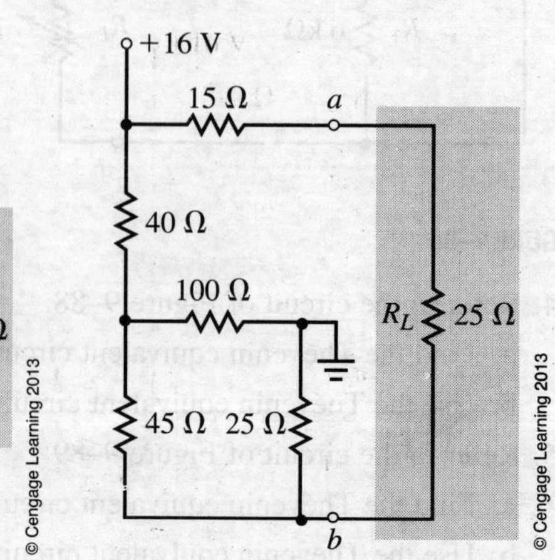

FIGURE 9–92

19. Repeat Problem 17 for the circuit of Figure 9–93.
20. Repeat Problem 17 for the circuit of Figure 9–94.

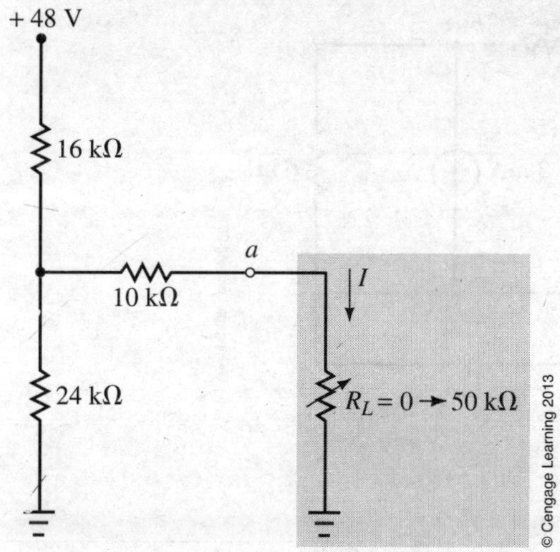

FIGURE 9–93

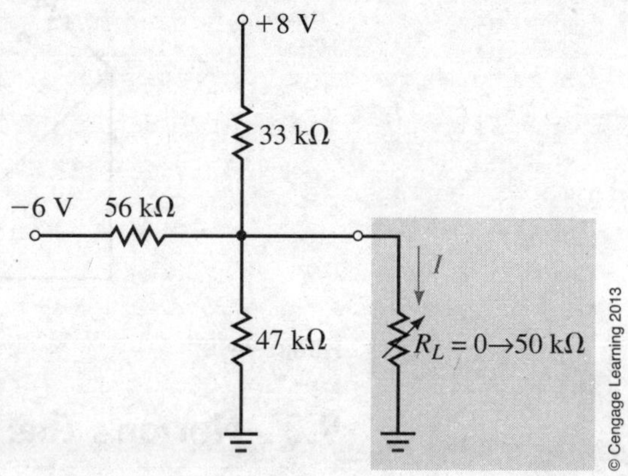

FIGURE 9–94

21. Find the Thévenin equivalent circuit of the network external to the indicated branch as shown in Figure 9–95.
22. Refer to the circuit of Figure 9–96.
    a. Find the Thévenin equivalent circuit external to the indicated terminals.
    b. Use the Thévenin equivalent circuit to determine the current through the indicated branch.
23. Repeat Problem 22 for the circuit of Figure 9–97.

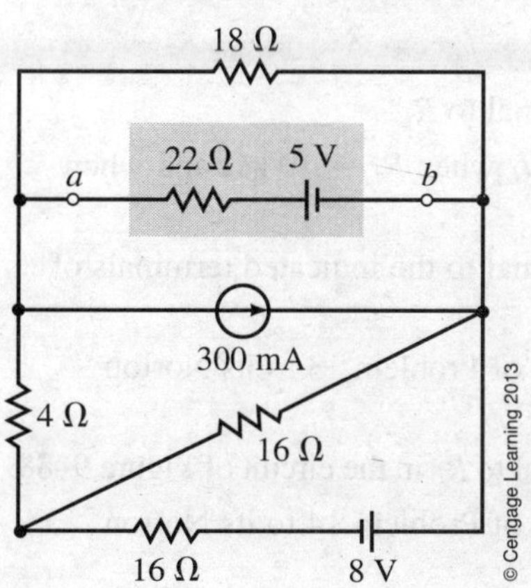

FIGURE 9–95

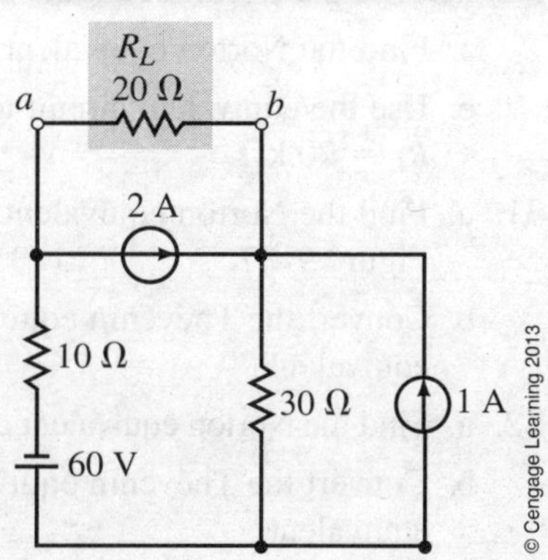

FIGURE 9–96

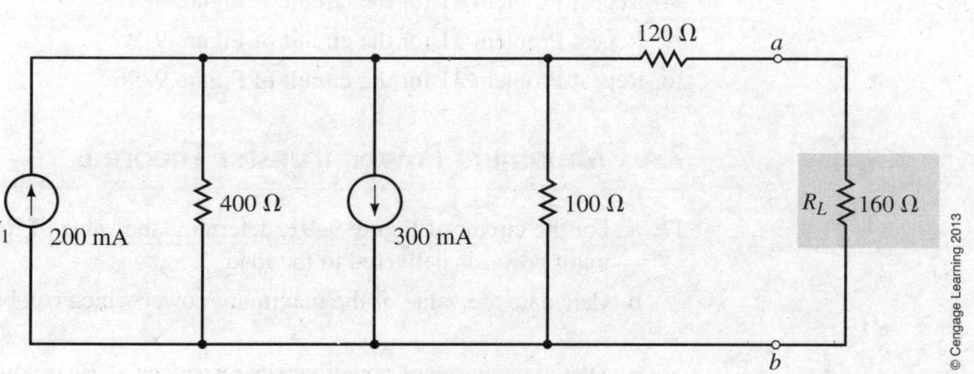

FIGURE 9–97

24. Repeat Problem 22 for the circuit of Figure 9–98.

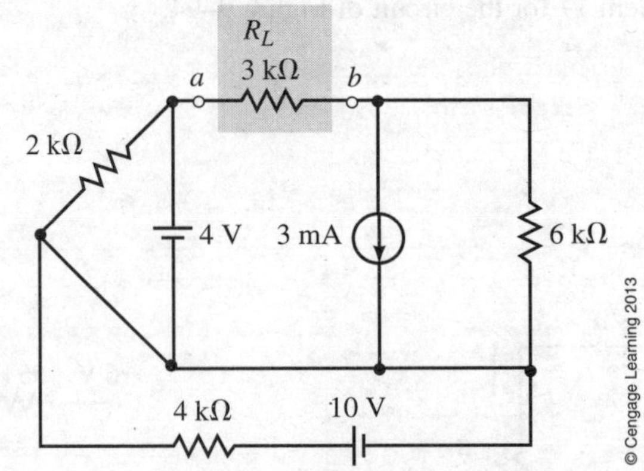

FIGURE 9–98

9.3 Norton's Theorem

25. Find the Norton equivalent circuit external to R_L in the circuit of Figure 9–81. Use the equivalent circuit to find I_L for the circuit.
26. Repeat Problem 25 for the circuit of Figure 9–82.
27. Repeat Problem 25 for the circuit of Figure 9–83.
28. Repeat Problem 25 for the circuit of Figure 9–84.
29. Refer to the circuit of Figure 9–85:
    a. Find the Norton equivalent circuit external to R_L.
    b. Use the equivalent circuit to determine I_L when $R_L = 20\ \Omega$ and when $R_L = 50\ \Omega$.
30. Refer to the circuit of Figure 9–86:
    a. Find the Norton equivalent circuit external to R_L.
    b. Use the equivalent circuit to determine I_L when $R_L = 10\ \text{k}\Omega$ and when $R_L = 20\ \text{k}\Omega$.
31. a. Find the Norton equivalent circuit external to the indicated terminals of Figure 9–87.
    b. Convert the Thévenin equivalent circuit of Problem 13 to its Norton equivalent.
32. a. Find the Norton equivalent circuit external to R_L in the circuit of Figure 9–88.
    b. Convert the Thévenin equivalent circuit of Problem 14 to its Norton equivalent.
33. Repeat Problem 31 for the circuit of Figure 9–91.
34. Repeat Problem 31 for the circuit of Figure 9–92.
35. Repeat Problem 31 for the circuit of Figure 9–95.
36. Repeat Problem 31 for the circuit of Figure 9–96.

9.4 Maximum Power Transfer Theorem

37. a. For the circuit of Figure 9–91, determine the value of R_L so that maximum power is delivered to the load.
    b. Calculate the value of the maximum power which can be delivered to the load.
    c. Sketch the curve of power versus resistance as R_L is adjusted from 0 Ω to 50 kΩ in increments of 5 kΩ.

38. Repeat Problem 37 for the circuit of Figure 9–94.
39. a. For the circuit of Figure 9–99, find the value of R so that $R_L = R_{Th}$.
    b. Calculate the maximum power dissipated by R_L.
40. Repeat Problem 39 for the circuit of Figure 9–100.

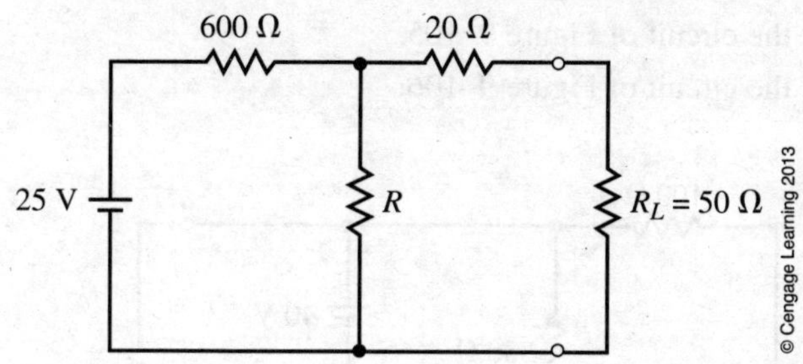

FIGURE 9–99

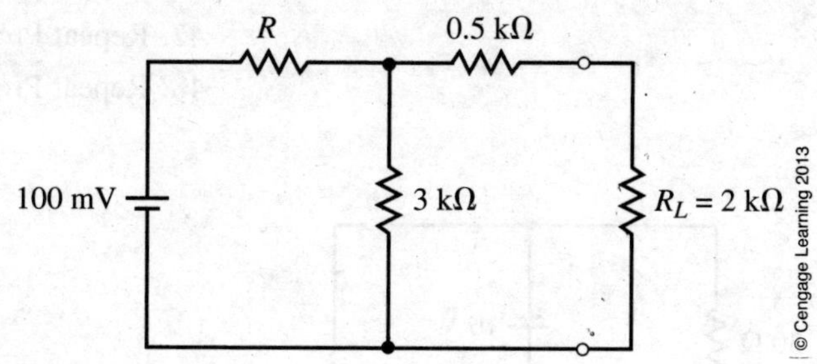

FIGURE 9–100

41. a. For the circuit of Figure 9–101, determine the values of R_1 and R_2 so that the 32-kΩ load receives maximum power.
    b. Calculate the maximum power delivered to R_L.

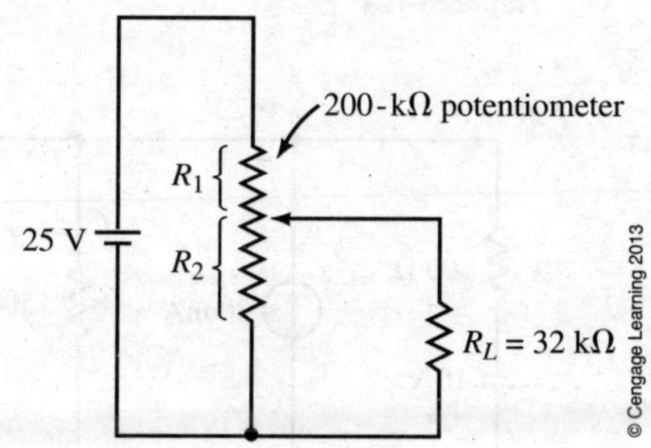

FIGURE 9–101

42. Repeat Problem 41 if the load resistor has a value of $R_L = 25$ kΩ.

9.5 Substitution Theorem

43. If the indicated portion of the circuit in Figure 9–102 is to be replaced with a voltage source and a 50-Ω series resistor, determine the magnitude and polarity of the resulting voltage source.
44. If the indicated portion of the circuit in Figure 9–102 is to be replaced with a current source and a 200-Ω shunt resistor, determine the magnitude and direction of the resulting current source.

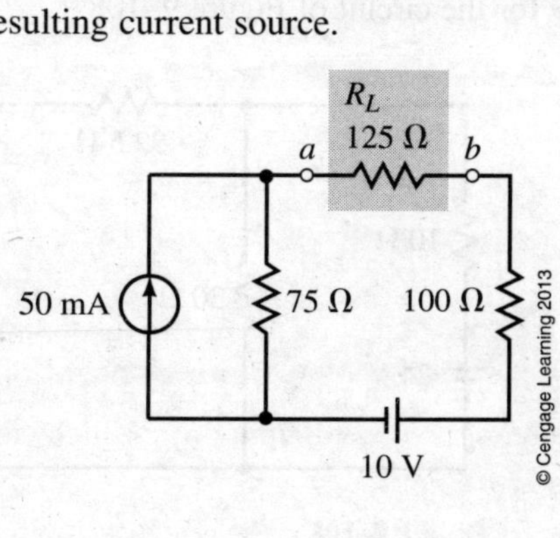

FIGURE 9–102

9.6 Millman's Theorem

45. Use Millman's theorem to find the current through and the power dissipated by R_L in the circuit of Figure 9–103.
46. Repeat Problem 45 for the circuit of Figure 9–104.
47. Repeat Problem 45 for the circuit of Figure 9–105.
48. Repeat Problem 45 for the circuit of Figure 9–106.

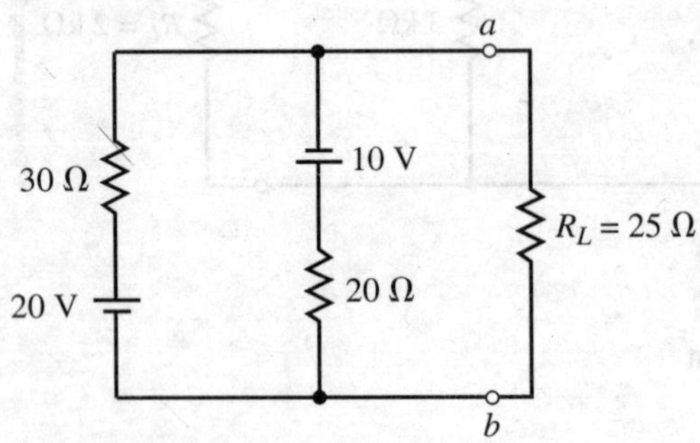

FIGURE 9–103

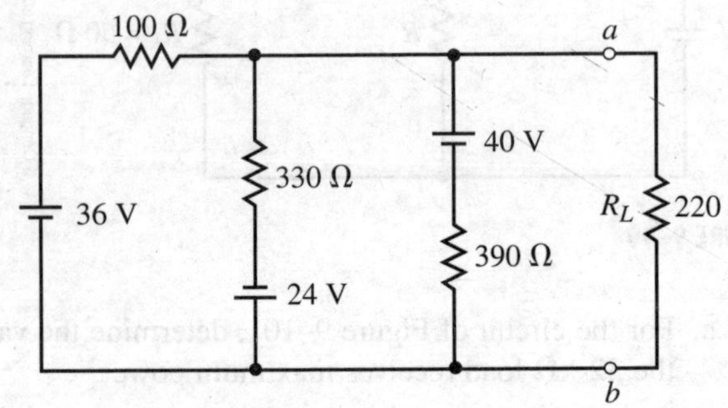

FIGURE 9–104

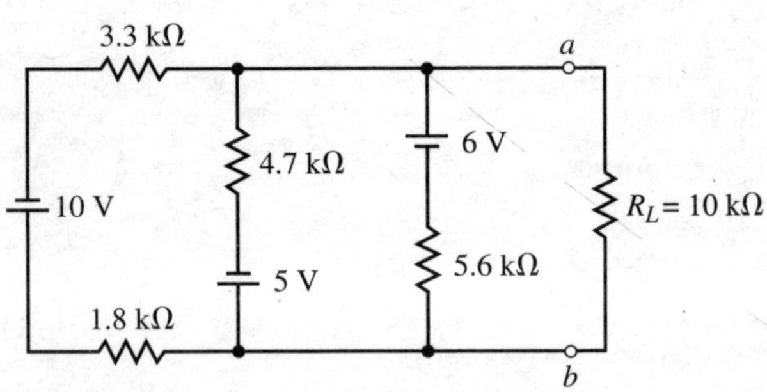

FIGURE 9–105

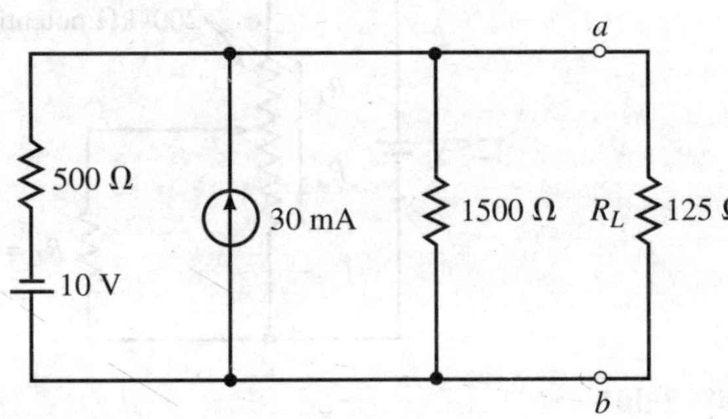

FIGURE 9–106

9.7 Reciprocity Theorem

49. a. Determine the current I in the circuit of Figure 9–107.
    b. Show that reciprocity applies for the given circuit.
50. Repeat Problem 49 for the circuit of Figure 9–108.

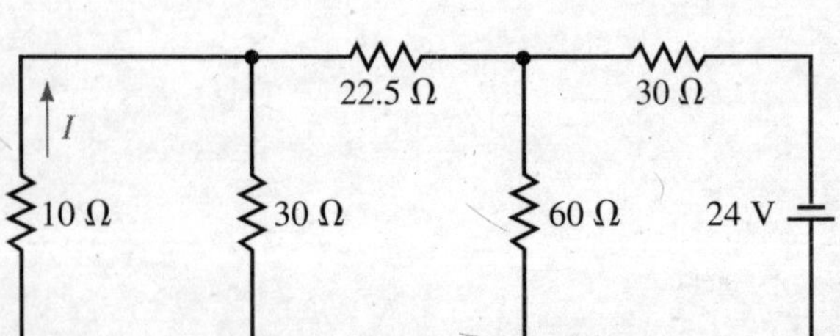

FIGURE 9–107

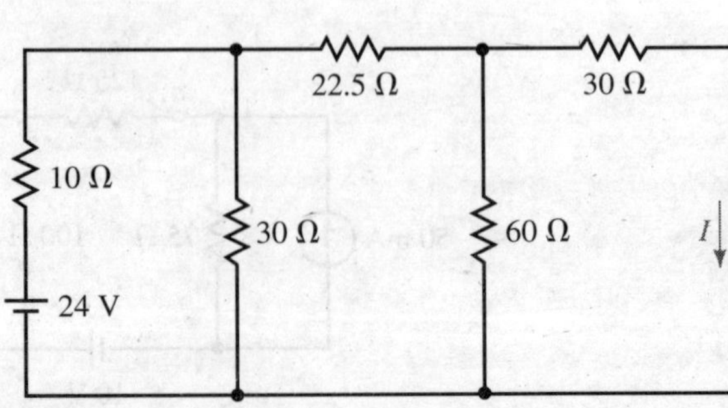

FIGURE 9–108

51. a. Determine the voltage V in the circuit of Figure 9–109.
 b. Show that reciprocity applies for the given circuit.
52. Repeat Problem 51 for the circuit of Figure 9–110.

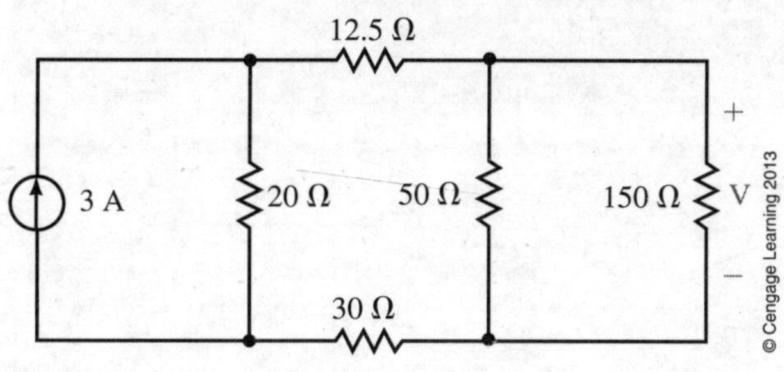

FIGURE 9–109

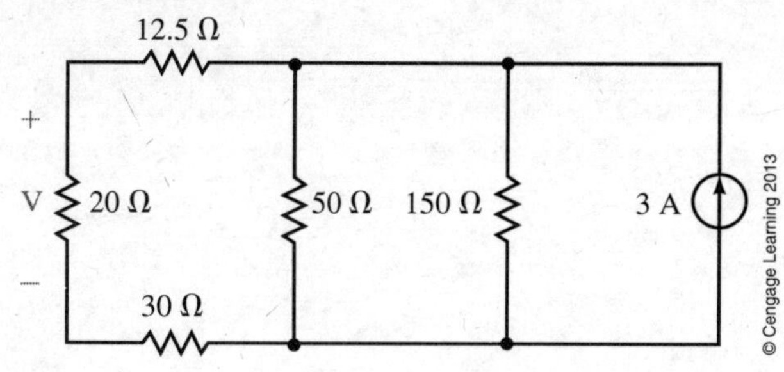

FIGURE 9–110

9.8 Circuit Analysis Using Computers

53. Use Multisim to find both the Thévenin and the Norton equivalent circuits external to the load resistor in the circuit of Figure 9–81. Multisim
54. Repeat Problem 53 for the circuit of Figure 9–82. Multisim
55. Use the schematic editor of PSpice to input the circuit of Figure 9–83 and use the Probe postprocessor to display output voltage, current, and power as a function of load resistance. Use the cursor in the Probe postprocessor to determine the value of load resistance for which the load will receive maximum power. Let the load resistance vary from 100 Ω to 4000 Ω in increments of 100 Ω. PSPICE
56. Repeat Problem 55 for the circuit of Figure 9–84. Let the load resistance vary from 1 kΩ to 100 kΩ in increments of 1 kΩ. PSPICE

ANSWERS TO IN-PROCESS LEARNING CHECKS

IN-PROCESS LEARNING CHECK 1

$P_{R_1} = 304$ mW, $P_{R_2} = 76$ mW, $P_{R_3} = 276$ mW
Assuming that superposition applies for power:
$P_{R_1(1)} = 60$ mW, $P_{R_1(2)} = 3.75$ mW, $P_{R_1(3)} = 60$ mW
But $P_{R_1} = 304$ mW · 123.75 mW

IN-PROCESS LEARNING CHECK 2

$R_1 = 133\ \Omega$

IN-PROCESS LEARNING CHECK 3

1. Refer to Figure 9–27.
2. $I_N = 200\ \mu A$, $R_N = 500\ \Omega$
3. $E_{Th} = 0.2$ V, $R_{Th} = 20$ kΩ

IN-PROCESS LEARNING CHECK 4

a. $\eta = 50\%$
b. $\eta = 33.3\%$
c. $\eta = 66.7\%$

IN-PROCESS LEARNING CHECK 5

1. In communication circuits and some amplifiers, maximum power transfer is a desirable characteristic.
2. In power transmission and dc voltage sources, maximum power transfer is not desirable.

III

CAPACITANCE AND INDUCTANCE

Resistance, inductance, and capacitance are the three basic circuit properties that we use to control voltages and currents in electrical and electronic circuits. However, each behaves in a fundamentally different way. Resistance, for example (as you learned in earlier chapters), opposes current, while (as you will soon see) inductance opposes any change in current, and capacitance opposes any change in voltage. In addition, resistance dissipates energy, while inductance and capacitance both store energy—inductance in its magnetic field and capacitance in its electric field.

Circuit elements that are built to possess capacitance are called capacitors, while elements built to possess inductance are called inductors. In Part III of this book, we explore these elements, their properties, and their behavior in electric circuits.

Note

Part III of this book deals with the basics of capacitors, inductors, magnetic circuits, and simple transients. However, some colleges and universities cover some of these topics in an order different than presented here, for example, by teaching ac before transients. To provide flexibility, Part III has been organized so that all or parts of Chapters 11, 12, and 14 may be delayed if desired (without loss of continuity) to fit individual curriculums. ■

KEY TERMS

Capacitance
Capacitor
Charged
Dielectric
Dielectric Absorption
Dielectric Breakdown
Dielectric Constant
Dielectric Strength
Electric Constant
Electric Field
Electric Field Intensity
Electric Flux Density
Electrolytic Capacitors
Equivalent Series Resistance (ESR)
Farad
Flux
Instantaneous Quantities
Leakage
Permittivity
Polarized
Relative Permittivity
Supercapacitor
Surface Mount Devices
Temperature Coefficient
Working Voltage

OUTLINE

Capacitance
Factors Affecting Capacitance
Electric Fields
Dielectrics
Nonideal Effects
Types of Capacitors
Capacitors in Parallel and Series
Capacitor Current and Voltage During Charging
Energy Stored by a Capacitor
Capacitor Failures and Troubleshooting

OBJECTIVES

After studying this chapter, you will be able to

- describe the basic construction of capacitors,
- explain how capacitors store charge,
- define capacitance,
- describe what factors affect capacitance and in what way,
- describe the electric field of a capacitor,
- compute the breakdown voltages of various materials,
- describe various types of commercial capacitors,
- compute the capacitance of capacitors in series and in parallel combinations,
- compute capacitor voltage and current for simple time-varying waveforms,
- determine stored energy,
- describe capacitor faults and the basic troubleshooting of capacitors.

10

CAPACITORS AND CAPACITANCE

CHAPTER PREVIEW

A capacitor is a circuit component designed to store electrical charge. If you connect a dc voltage source to a capacitor, for example, the capacitor will "charge" to the voltage of the source. If you then disconnect the source, the capacitor will remain charged, that is, its voltage will remain constant at the value that it attained while connected to the source (assuming no leakage). Because of this tendency to hold voltage, *a capacitor opposes changes in voltage.* It is this characteristic that gives capacitors their unique properties.

Capacitors are widely used in electrical and electronic applications. They are used in electronic systems for signal conditioning and timing, for example, in speaker system crossover networks to separate high- and low-frequency signals, in cameras to store the charge that fires the photoflash, on pump and refrigeration motors to increase starting torque, in electric power systems to increase operating efficiency, and so on. A photo of some typical capacitors is shown on the right.

Capacitance is the electrical property of capacitors: it is a measure of how much charge a capacitor can hold. In this chapter, we look at capacitance and its basic properties. In Chapter 11, we look at capacitors in dc and pulse circuits; in later chapters, we look at capacitors in ac applications. ■

Typical capacitors.

Putting It in Perspective

Michael Faraday and the Field Concept

THE UNIT OF CAPACITANCE, the farad, is named after Michael Faraday (1791–1867). Born in England to a working class family, Faraday received limited education. Nonetheless, he was responsible for many of the fundamental discoveries of electricity and magnetism. Lacking mathematical skills, he used his intuitive ability rather than mathematical models to develop conceptual pictures of basic phenomena. It was his development of the field concept, for example, that made it possible to map out the fields that exist around magnetic poles and electrical charges.

To get at this idea, recall from Chapter 2 that unlike charges attract and like charges repel, that is, a force exists between electrical charges. We call the region where this force acts an electric field. To visualize this field, we use Faraday's field concept and draw lines of force (or flux lines) that show at every point in space the magnitude and direction of the force. Now, rather than supposing that one charge exerts a force on another, we instead visualize that the original charges create a field in space and that other charges introduced into this field experience a force due to the field. This concept is helpful in studying certain aspects of capacitors, as you will see in this chapter.

The development of the field concept had a significant impact on science. We now picture several important phenomena in terms of fields, including electric fields, gravitation, and magnetism. When Faraday published his theory in 1844, however, it was not taken seriously, much like Ohm's work two decades earlier. It is also interesting to note that the development of the field concept grew out of Faraday's research into magnetism, not electric charge. ■

10.1 Capacitance

A **capacitor** consists of two conductors separated by an insulator. One of its basic forms is the parallel-plate capacitor shown in Figure 10–1. It consists of two metal plates separated by a nonconducting material (i.e., an insulator) called a **dielectric.** The dielectric may be air, oil, mica, plastic, ceramic, or other suitable insulating material.

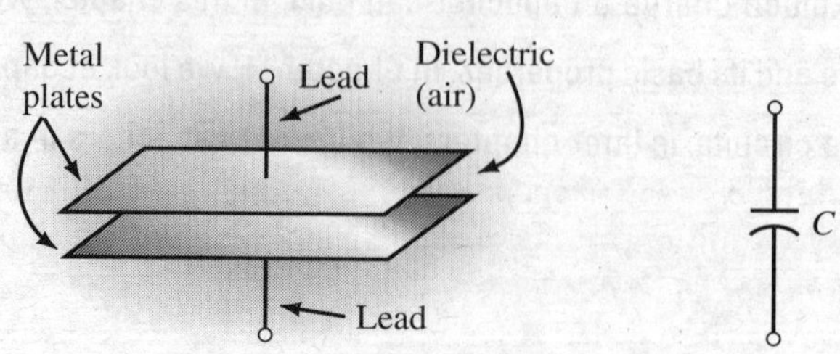

FIGURE 10–1 Parallel-plate capacitor.

Since the plates of the capacitor are metal, they contain huge numbers of free electrons. In their normal state, however, they are uncharged, that is, there is no excess or deficiency of electrons on either plate. If a dc source is now connected (Figure 10–2), electrons are pulled from the top plate by the positive potential of the battery and the same number deposited on the bottom plate. This leaves the top plate with a deficiency of electrons (i.e., positive charge) and the bottom plate with an excess (i.e., negative charge). In this state, the capacitor is said to be **charged.** (Note that no current can pass through the dielectric between the plates—thus, the movement of electrons illustrated in Figure 10–2 will cease when the capacitor reaches full charge.)

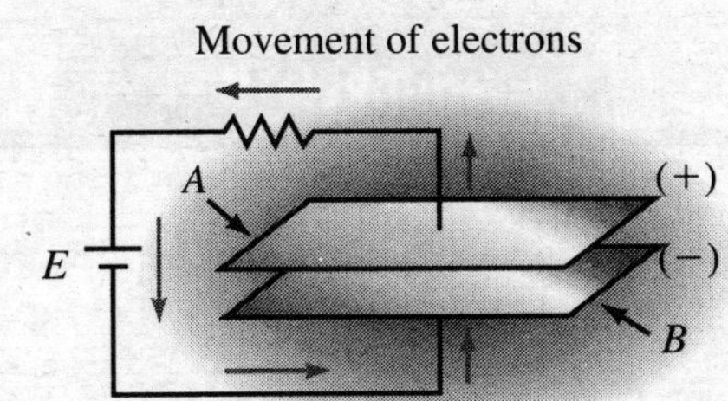

FIGURE 10–2 Capacitor during charging. At the instant that you connect the source, there is a momentary surge of current as electrons are pulled from plate A and an equal number deposited on plate B. This leaves the top plate positively charged and the bottom plate negatively charged. When charging is complete, there is no further movement of electrons, and the current is thus zero.

CircuitSim 10-1

If Q coulombs of electrons are moved during the charging process (leaving the top plate with a deficiency of Q electrons and the bottom with an excess of Q), we say that the capacitor has a charge of Q.

If we now disconnect the source (Figure 10–3), the excess electrons that were moved to the bottom plate remain trapped as they have no way to return to the top plate. The capacitor therefore remains charged with voltage E across it even though no source is present. Because of this, we say that *a capacitor can store charge.* Capacitors with little leakage (Section 10.5) can hold their charge for a considerable time—sometimes for years.

Large-value capacitors charged to high voltages contain a great deal of energy and can store a potentially lethal charge; you must therefore be very careful with them. Even high-value, low-voltage units can contain enough energy to vaporize a screwdriver and spray metal in your eyes if you attempt to short the leads. If you intend to handle such capacitors, or work on them, always discharge them after power has been removed. You can do this by connecting a high-wattage resistor (sometimes referred to as a bleeder resistor) of about 50 Ω/V of the rated voltage of the capacitor across the capacitor's terminals until it is discharged, but be careful to protect yourself from shock. Note also that some voltage may spring back due to dielectric absorption after the discharge resistor is removed—see Section 10.5.

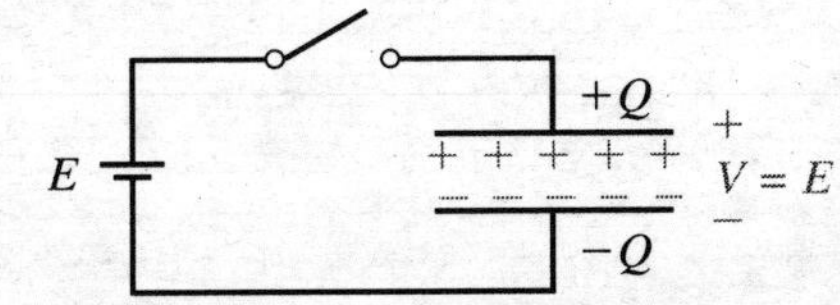

FIGURE 10–3 Capacitor after charging. When you disconnect the source, electrons are trapped on the bottom plate and cannot return—thus, charge is stored.

Definition of Capacitance

The amount of charge Q that a capacitor can store depends on the applied voltage. Experiments show that for a given capacitor, Q is proportional to voltage. Let the constant of proportionality be C. Then,

$$Q = CV \tag{10–1}$$

Rearranging terms yields

$$C = \frac{Q}{V} \quad \text{(farads, F)} \tag{10–2}$$

The term C is defined as the capacitance of the capacitor. As indicated, its unit is the **farad.** By definition, *the* ***capacitance*** *of a capacitor is one farad if it stores one coulomb of charge when the voltage across its terminals is one volt.* The farad, however, is a very large unit. Most practical capacitors (except supercapacitors, Section 10.6) range in size from picofarads (pF or 10^{-12} F) to microfarads (μF or 10^{-6} F). The larger the value of C, the more charge that the capacitor can hold for a given voltage.

EXAMPLE 10–1

a. How much charge is stored on a 10-μF capacitor when it is connected to a 24-volt source?

b. The charge on a 20-nF capacitor is 1.7 μC. What is its voltage?

Solution

a. From Equation 10–1, $Q = CV$. Thus, $Q = (10 \times 10^{-6}\text{ F})(24\text{ V}) = 240\ \mu\text{C}$.

b. Rearranging Equation 10–1, $V = Q/C = (1.7 \times 10^{-6}\text{ C})/(20 \times 10^{-9}\text{ F}) = 85\text{ V}$.

10.2 Factors Affecting Capacitance

Effect of Area

As shown by Equation 10–2, capacitance is directly proportional to charge. This means that the more charge you can put on a capacitor's plates for a given voltage, the greater will be its capacitance. Consider Figure 10–4. The capacitor of (b) has four times the area of (a). Since it has the same number of free electrons per unit area, it has four times the total charge and hence four times the capacitance. This turns out to be true in general, that is, *capacitance is directly proportional to plate area.*

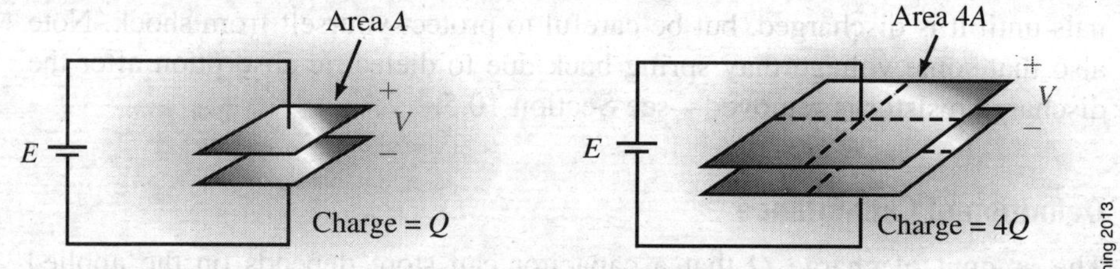

(a) Capacitor with area A and charge Q

(b) Plates with four times the area have four times the charge and therefore, four times the capacitance

FIGURE 10–4 For a fixed separation, capacitance is proportional to plate area.

Effect of Spacing

Now consider Figure 10–5. Since the top plate has a deficiency of electrons and the bottom plate an excess, a force of attraction exists across the gap. For a fixed spacing as in (a), the charges are in equilibrium. Now move the plates closer together as in (b). As spacing decreases, the force of attraction increases, pulling more electrons from within the bulk material of plate B to its top surface. This creates a deficiency of electrons throughout B. To replenish these, the source moves additional electrons around the circuit, leaving A with an even greater deficiency and B with an even greater excess. The charge on the plates therefore increases and hence, according to Equation 10–2, so does the capacitance. We therefore conclude that decreasing spacing increases capacitance, and vice versa. In fact, as we will show later, *capacitance is inversely proportional to plate spacing.*

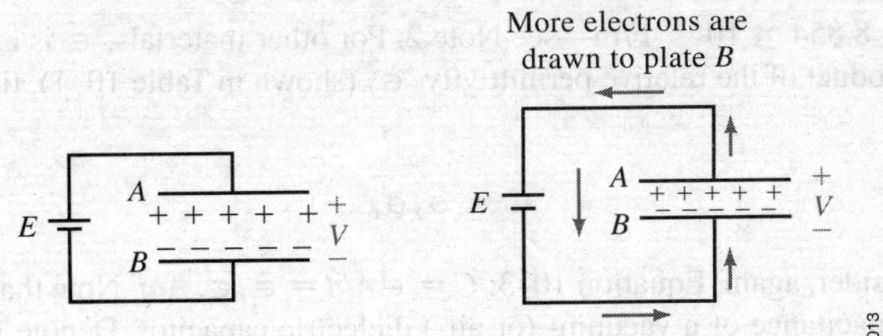

(a) Charges in equilibrium

(b) As spacing decreases, more electrons move from A to B, increasing the charge on the plates and hence, their capacitance

FIGURE 10–5 Decreasing spacing increases capacitance.

TABLE 10–1 Relative Permittivity of Various Materials

Material	ϵ_r (Nominal Values)
Vacuum	1
Air	1.0006
Ceramic	30–7500
Mica	5.5
Mylar (Note 3)	3
Oil	4
Paper (dry)	2.2
Polystyrene	2.6
Teflon (Note 3)	2.1

Effect of Dielectric

Capacitance also depends on the dielectric. Consider Figure 10–6(a), which shows an air-dielectric capacitor. If you substitute different materials for air, the capacitance increases. Table 10–1 shows the factor by which capacitance increases for a number of different materials. For example, if Teflon® is used instead of air, capacitance is increased by a factor of 2.1. This factor is called the **relative permittivity** of the material (or in older terminology, its **dielectric constant**). (**Permittivity** is a measure of how easy it is to establish electric flux in a material. We revisit it below.) Note that high-permittivity ceramic increases capacitance by as much as 7500, as indicated in Figure 10–6(b).

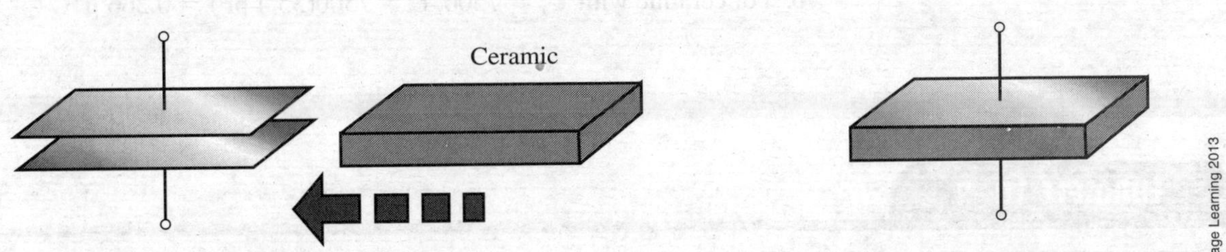

(a) $C = 200$ pF with air dielectric

(b) $C = 1.5$ μF with high-permittivity ceramic dielectric

FIGURE 10–6 The factor by which a dielectric causes capacitance to increase is termed its *relative permittivity.* The ceramic used here has a value of 7500.

Capacitance of a Parallel-Plate Capacitor

From the preceding observations, we see that capacitance is directly proportional to plate area, inversely proportional to plate separation, and dependent on the dielectric. In equation form,

$$C = \epsilon \frac{A}{d} \quad \text{(F)} \tag{10–3}$$

where area A is in square meters and spacing d is in meters—see Note 1.

Permittivity (Revisited)

The constant ϵ in Equation 10–3 is the absolute permittivity of the insulating material. Its units are farads per meter (F/m). For air or vacuum, ϵ has a value

NOTES...

1. Actually, Equation 10–3 is true in general, that is, it holds for other shapes as well as for the parallel-plate configuration. It is just harder to determine the effective area and spacing for other geometries.
2. In modern terminology, ϵ_0 (the absolute permittivity of free space) is also called the **electric constant.**
3. Teflon is a trademark of DuPont. Mylar is a trademark of Dupont Teijin Films.

of $\epsilon_o = 8.854 \times 10^{-12}$ F/m—see Note 2. For other materials, ϵ is expressed as the product of the relative permittivity, ϵ_r (shown in Table 10–1), times ϵ_o. That is,

$$\epsilon = \epsilon_r \epsilon_o \tag{10–4}$$

Consider, again, Equation 10–3: $C = \epsilon A/d = \epsilon_r \epsilon_o A/d$. Note that $\epsilon_o A/d$ is the capacitance of a vacuum- (or air-) dielectric capacitor. Denote it by C_o. Then, for any other dielectric,

$$C = \epsilon_r C_o \tag{10–5}$$

EXAMPLE 10–2

Compute the capacitance of a parallel-plate capacitor with plates 10 cm by 20 cm, separation of 5 mm, and

a. an air dielectric,

b. a ceramic dielectric with a relative permittivity of 7500.

Solution Convert all dimensions to meters. Thus, A = (0.1 m)(0.2 m) = 0.02 m^2, and $d = 5 \times 10^{-3}$ m.

a. For air, $C = \epsilon_o A/d = (8.854 \times 10^{-12})(2 \times 10^{-2})/(5 \times 10^{-3}) = 35.4 \times 10^{-12}$ F = 35.4 pF.

b. For ceramic with $\epsilon_r = 7500$, $C = 7500(35.4 \text{ pF}) = 0.266\ \mu\text{F}$.

EXAMPLE 10–3

A parallel-plate capacitor with air dielectric has a value of $C = 12$ pF. What is the capacitance of a capacitor that has the following:

a. The same separation and dielectric but five times the plate area?

b. The same dielectric but four times the area and one-fifth the plate spacing?

c. A dry paper dielectric, six times the plate area, and twice the plate spacing?

Solution

a. Since the plate area has increased by a factor of five and everything else remains the same, C increases by a factor of five. Thus, $C = 5(12 \text{ pF}) = 60$ pF.

b. With four times the plate area, C increases by a factor of four. With one-fifth the plate spacing, C increases by a factor of five. Thus, $C = (4)(5)(12 \text{ pF}) = 240$ pF.

c. Dry paper increases C by a factor of 2.2. The increase in plate area increases C by a factor of six. Doubling the plate spacing reduces C by one-half. Thus, $C = (2.2)(6)(\frac{1}{2})(12 \text{ pF}) = 79.2$ pF.

IN-PROCESS LEARNING CHECK 1

(Answers are at the end of the chapter.)

1. A capacitor with plates 7.5 cm × 8 cm and plate separation of 0.1 mm has an oil dielectric:
   a. Compute its capacitance;
   b. If the charge on this capacitor is 0.424 μC, what is the voltage across its plates?
2. For a parallel-plate capacitor, if you triple the plate area and halve the plate spacing, how does capacitance change?
3. For the capacitor of Figure 10–6, if you use mica instead of ceramic, what will be the capacitance?
4. What is the dielectric for the capacitor of Figure 10–7(b)?

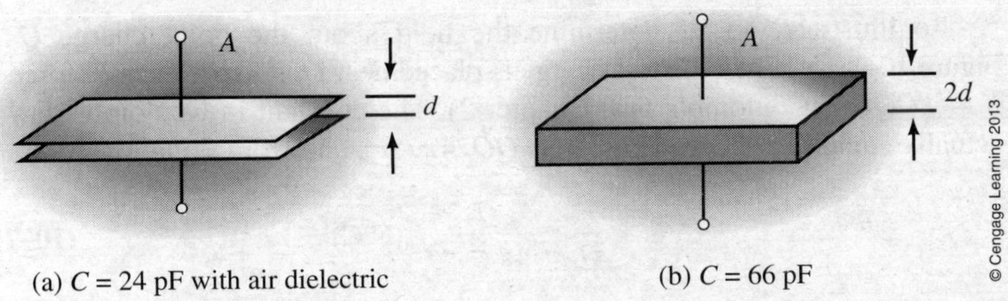

(a) $C = 24$ pF with air dielectric (b) $C = 66$ pF

FIGURE 10–7

10.3 Electric Fields

Electric Flux

In Chapter 2, we learned that unlike charges attract while like charges repel; that is, a force exists between them. The region where this force exists is called an **electric field.** To visualize this field, Faraday introduced the idea of using a small positive test charge Q_t to map the forces. He defined the direction of a field as the direction of force on the test charge, and the strength of the field as the force per unit charge (considered below). He then mapped the field as lines of force where the direction of the lines represented the direction of the force, and the density of lines represented its strength. To illustrate, consider Figure 10–8(a). Because like charges repel, the force on the positive test charge is directed outward, resulting in a radial field as indicated. Now consider (b).

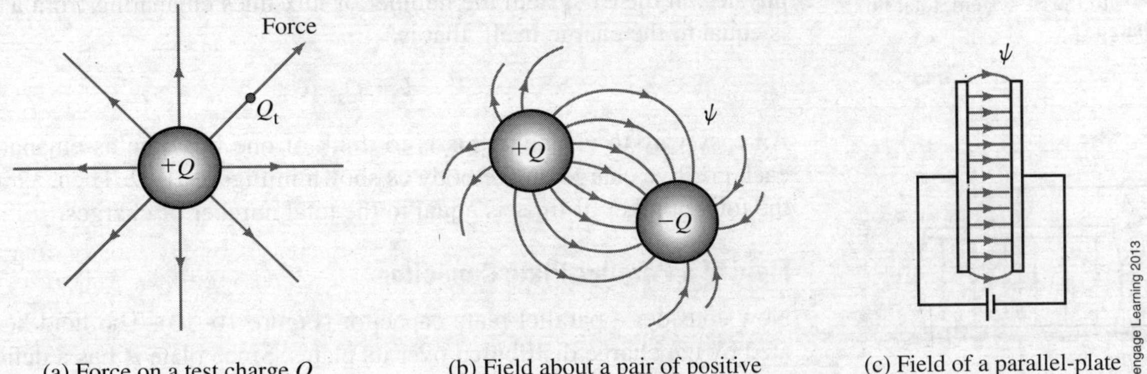

(a) Force on a test charge Q_t (b) Field about a pair of positive and negative charges (c) Field of a parallel-plate capacitor

FIGURE 10–8 Electric fields.

The test charge will be repelled by charge $+Q$ and attracted by charge $-Q$, resulting in the curved field shown. Faraday determined that lines never cross and that the stronger the field, the more dense the lines. Figure 10–8(c) shows another example: the field of a parallel-plate capacitor. In this case, the field is uniform across the gap with some fringing near its edges. Electric field lines (also called **flux** lines) are denoted by the Greek letter ψ (psi).

Electric Field Intensity $\mathscr{E}$

As noted previously, the strength of an electric field (more usually called its intensity) is defined as the force per unit charge that the field exerts on a small positive test charge Q_t. Let this **electric field intensity** be denoted by the script letter $\mathscr{E}$. Then by definition,

$$\mathscr{E} = F/Q_t \quad \text{(newtons/coulomb, N/C)} \tag{10–6}$$

To illustrate, let us determine the field about the point charge Q of Figure 10–8(a). When the test charge is placed near Q, it experiences a force of $F = kQQ_t/r^2$ (Coulomb's law, Chapter 2). The constant in Coulomb's law is actually equal to $1/4\pi\in$. Thus, $F = QQ_t/4\pi\in r^2$, and from Equation 10–6,

$$\mathscr{E} = \frac{F}{Q_t} = \frac{Q}{4\pi\in r^2} \quad \text{(N/C)} \tag{10–7}$$

Electric Flux Density

Because of the presence of $\in$ in Equation 10–7, the electric field intensity depends on the medium in which the charge is located. Let us define a new quantity, D, that is independent of the medium. Let

$$D = \in\mathscr{E} \tag{10–8}$$

D is known as **electric flux density.** As is shown in physics, D represents the density of flux lines in space, that is,

$$D = \frac{\text{total flux}}{\text{area}} = \frac{\psi}{A} \tag{10–9}$$

where ψ is the flux passing through area A.

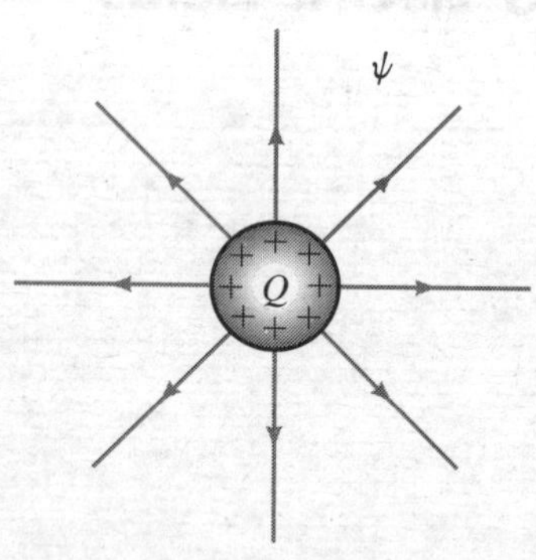

FIGURE 10–9 In the SI system, total flux ψ equals charge Q.

Electric Flux (Revisited)

Consider Figure 10–9. Flux ψ is due to the charge Q. Again, as shown in physics, in the SI system the number of flux lines emanating from a charge Q is equal to the charge itself, that is,

$$\psi = Q \quad \text{(C)} \tag{10–10}$$

An easy way to visualize this is to think of one flux line as emanating from each positive charge on the body as shown in Figure 10–9. Then, as indicated, the total number of lines is equal to the total number of charges.

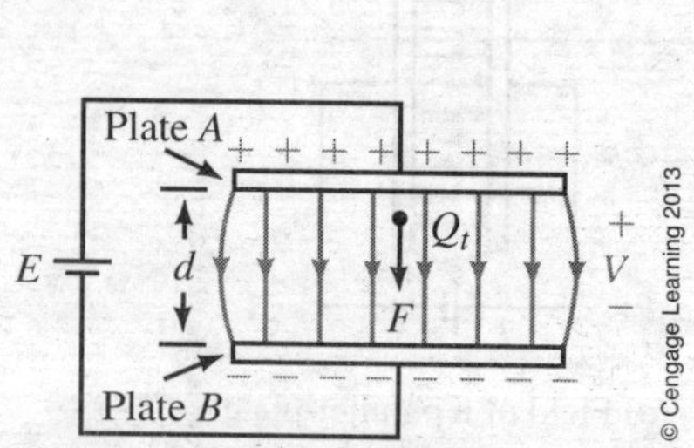

FIGURE 10–10 Work moving test charge Q_t is force times distance.

Field of a Parallel-Plate Capacitor

Now consider a parallel-plate capacitor (Figure 10–10). The field here is created by the charge distributed over its plates. Since plate A has a deficiency of electrons, it looks like a sheet of positive charge, while plate B looks like a sheet of negative charge. A positive test charge Q_t between these sheets is therefore repelled by the positive sheet and attracted by the negative sheet.

Now move the test charge from plate B to plate A. The work W required to move the charge against the force F is force times distance. Thus,

$$W = Fd \quad \text{(J)} \tag{10–11}$$

In Chapter 2, we defined voltage as work divided by charge, that is, $V = W/Q$. Since the charge here is the test charge, Q_t, the voltage between plates A and B is

$$V = W/Q_t = (Fd)/Q_t \quad \text{(V)} \tag{10–12}$$

Now divide both sides by d. This yields $V/d = F/Q_t$. But $F/Q_t = \mathscr{E}$, from Equation 10–6. Thus,

$$\mathscr{E} = V/d \quad \text{(V/m)} \tag{10–13}$$

Equation 10–13 shows that the electric field intensity between capacitor plates is equal to the voltage across the plates divided by the distance between them. Thus, if $V = 30$ V and $d = 10$ mm, $\mathscr{E} = 3000$ V/m.

EXAMPLE 10–4

Suppose that the electric field intensity between the plates of a capacitor is 50 000 V/m when 80 V is applied:

a. What is the plate spacing if the dielectric is air? If the dielectric is ceramic?
b. What is $\mathscr{E}$ if the plate spacing is halved?

Solution

a. $\mathscr{E} = V/d$, independent of dielectric. Thus,

$$d = \frac{V}{\mathscr{E}} = \frac{80 \text{ V}}{50 \times 10^3 \text{ V/m}} = 1.6 \times 10^{-3} \text{ m}$$

b. Since $\mathscr{E} = V/d$, $\mathscr{E}$ will double to 100 000 V/m.

PRACTICE PROBLEMS 1

1. What happens to the electric field intensity of a capacitor if you do the following:
   a. Double the applied voltage?
   b. Triple the applied voltage and double the plate spacing?
2. If the electric field intensity of a capacitor with polystyrene dielectric and plate size 2 cm by 4 cm is 100 kV/m when 50 V is applied, what is its capacitance?

Answers

1. a. Doubles, b. Increases by a factor of 1.5; 2. 36.8 pF

Capacitance (Revisited)

With the preceding background, we can examine capacitance a bit more rigorously. Recall, $C = Q/V$. Using the above relationships yields

$$C = \frac{Q}{V} = \frac{\psi}{V} = \frac{AD}{\mathscr{E}d} = \frac{D}{\mathscr{E}}\left(\frac{A}{d}\right) = \epsilon\frac{A}{d}$$

This confirms Equation 10–3 that we developed intuitively in Section 10.2.

10.4 Dielectrics

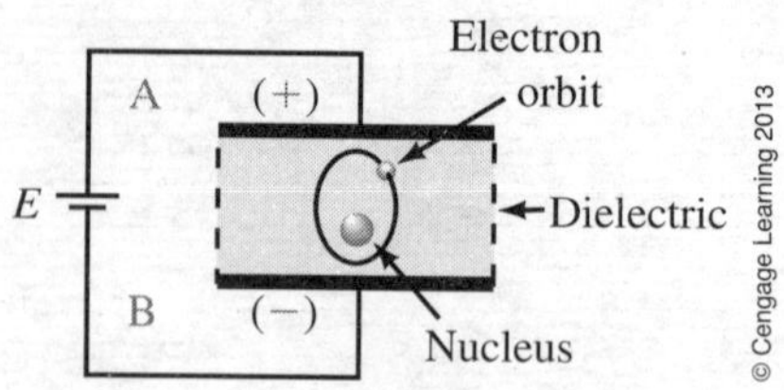

FIGURE 10–11 Effect of the capacitor's electric field on an atom of its dielectric.

As you saw in Figure 10–6, a dielectric increases capacitance. We now examine why. Consider Figure 10–11. For a charged capacitor, electron orbits (which are normally nearly circular) become elliptical as electrons are attracted toward the positive (+) plate and repelled from the negative (−) plate. This makes the end of the atom nearest the positive plate appear negative while its other end appears positive. Such atoms are **polarized.** Throughout the bulk of the dielectric, the negative end of a polarized atom is adjacent to the positive end of another polarized atom, and the effects cancel. However, at the surfaces of the dielectric, there are no atoms to cancel, and the net effect is as if a layer of negative charge exists on the surface of the dielectric at the positive plate and a layer of positive charge at the negative plate. This makes the plates appear closer together, thus increasing capacitance. Materials for which the effect is largest result in the greatest increase in capacitance.

Dielectric Breakdown

TABLE 10–2 Dielectric Strength*

Material	kV/mm
Air	3
Ceramic (high ϵ_r)	3
Mica	40
Mylar	16
Oil	15
Polystyrene	24
Rubber	18
Teflon®	60

*Values depend on the composition of the material. These are the values we use in this book.

If the voltage of Figure 10–11 is increased beyond a critical value, the force on the electrons is so great that they are literally torn from orbit. This is called **dielectric breakdown,** and the electric field intensity (Equation 10–13) at breakdown is called the **dielectric strength** of the material. For air, breakdown occurs when the voltage gradient reaches 3 kV/mm. The breakdown strengths for other materials are shown in Table 10–2. Since the quality of a dielectric depends on many factors, dielectric strength varies from sample to sample.

Breakdown is not limited to capacitors; it can occur with any type of electrical apparatus whose insulation is stressed beyond safe limits. (For example, air breaks down and flashovers occur on high-voltage transmission lines when they are struck by lightning.) The shape of conductors also affects breakdown voltage. Breakdown occurs at lower voltages at sharp points than at blunt points. This effect is made use of in lightning arresters.

EXAMPLE 10–5

A capacitor with plate dimensions of 2.5 cm by 2.5 cm and a ceramic dielectric with $\epsilon_r = 7500$ experiences breakdown at 2400 V. What is C?

Solution From Table 10–2 dielectric strength = 3 kV/mm. Thus, $d = 2400\ \text{V}/3000\ \text{V/mm} = 0.8\ \text{mm} = 8 \times 10^{-4}\ \text{m}$. So

$$\begin{aligned} C &= \epsilon_r \epsilon_o A/d \\ &= (7500)(8.854 \times 10^{-12})(0.025\ \text{m})^2/(8 \times 10^{-4}\ \text{m}) \\ &= 51.9\ \text{nF} \end{aligned}$$

PRACTICAL NOTES...

Breakdown in solid dielectrics is generally destructive. Typically, a carbonized pinhole results, and since this is a conductive path, the dielectric is no longer a useful insulator. Air, vacuum, and insulating liquids like oil, on the other hand, recover from the flashover once it is extinguished.

PRACTICE PROBLEMS 2

1. At what voltage will breakdown occur for a Mylar dielectric capacitor with plate spacing of 0.25 cm?
2. An air-dielectric capacitor breaks down at 500 V. If the plate spacing is doubled and the capacitor is filled with oil, at what voltage will breakdown occur?

Answers
1. 40 kV; 2. 5 kV

Capacitor Voltage Rating

Because of dielectric breakdown, capacitors are rated for maximum operating voltage (called **working voltage**) by their manufacturer (indicated on the capacitor as WVDC or working voltage dc). If you operate a capacitor beyond its working voltage, you may damage it.

10.5 Nonideal Effects

So far, we have assumed ideal capacitors. However, real capacitors have several nonideal characteristics.

Leakage Current

CircuitSim 10-2

When a charged capacitor is disconnected from its source, it will eventually discharge. This is because no insulator is perfect, and a small amount of charge "leaks" through the dielectric. Similarly, a small leakage current will pass through its dielectric when a capacitor is connected to a source.

The effect of **leakage** is modeled by a resistor in Figure 10–12. Except for electrolytic capacitors (see Section 10.6), leakage is very small and R is very large, typically hundreds of megohms. The larger R is, the longer a capacitor can hold its charge. For most applications, leakage can be neglected, and you can treat the capacitor as ideal.

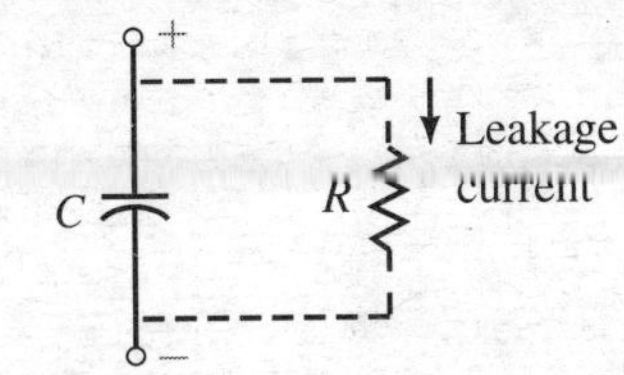

FIGURE 10–12 Leakage current.

Equivalent Series Resistance (ESR)

Real capacitors have a small amount of resistance due to the resistance of their leads, the contact connections between leads and plates, and ac losses in the dielectric. This resistance, known as **equivalent series resistance** or **ESR,** can be modeled as a resistance in series with the capacitor. Note that ESR is of concern only with ac operation—it does not affect *dc* operation. For many types of capacitors, ESR is so small that it can be neglected, but in others, ESR is a serious concern, largely because it is responsible for heat generation inside the capacitor, and heat is often the cause of capacitor failure. (ESR is mainly a problem with electrolytic capacitors.) Typically, ESR values are less than 0.1 Ω for high-frequency capacitors, but they may be 10 ohms or more for electrolytic capacitors. Note that ESR cannot be measured with an ohmmeter, but it must be measured using specialized test instruments of the type depicted in Section 10.10.

Dielectric Absorption

When you discharge a capacitor by temporarily placing a resistor across its terminals, it should have zero (or nearly zero) volts across itself after the resistor is removed. However, considerable residual voltage may spring back. To see why, recall polarization, Figure 10–11. In some dielectrics, when source voltage is removed, the relaxation of atoms to their nonpolarized state may take considerable time, and it is the residual charge from these remaining polarized atoms that creates the residual voltage. This effect is known as **dielectric absorption.** In electronic circuits, the voltage due to dielectric absorption can upset circuit voltage levels; in old CRT-type TV tubes and with electrical power apparatus, it can result in large and potentially dangerous voltages. You may have to put the shorting resistor back on to complete the discharge.

Temperature Coefficient

Because dielectrics are affected by temperature, capacitance may change with temperature. If capacitance increases with increasing temperature, the capacitor is said to have a positive temperature coefficient; if it decreases, the capacitor has a negative temperature coefficient; if it remains essentially constant, the capacitor has a zero temperature coefficient.

The **temperature coefficient** is specified as a change in capacitance in parts per million (ppm) per degree Celsius. Consider a 1-μF capacitor. Since 1 μF = 1 million pF, 1 ppm is 1 pF. Thus, a 1-μF capacitor with a temperature coefficient of 200 ppm/° C could change as much as 200 pF per degree Celsius. If the circuit you are designing is sensitive to a capacitor's value, you may have to use a capacitor with a small temperature coefficient.

10.6 Types of Capacitors

Since no single capacitor type suits all applications, capacitors are made in a variety of types and sizes. Among these are fixed and variable types with differing dielectrics and recommended areas of application. (In the following, we talk in general about recommended areas of application, but many manufacturers typically recommend where each type that they make should be used.)

Fixed Capacitors

Fixed capacitors are often identified by their dielectric. Common dielectric materials include ceramic, plastic, and mica, plus, for electrolytic capacitors, aluminum and tantalum oxide. Design variations include tubular and interleaved plates. The interleave design (Figure 10–13) uses multiple plates to increase effective plate area. A layer of insulation separates plates, and alternate plates are connected together. The tubular design (Figure 10–14) uses sheets of metal foil separated by an insulator such as plastic film. Fixed capacitors are encapsulated

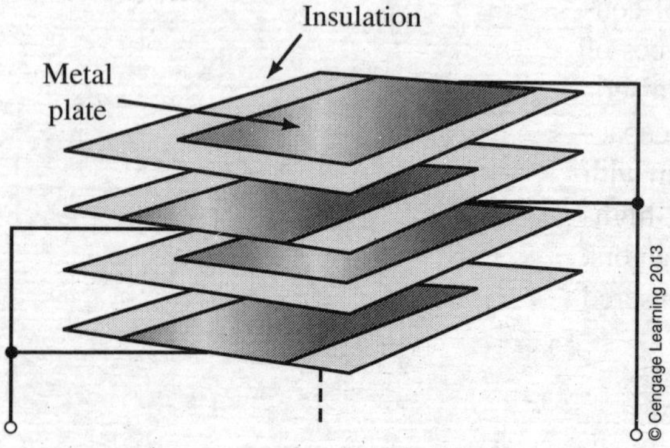

FIGURE 10–13 Stacked capacitor construction. The stack is compressed, leads attached, and the unit coated with epoxy resin or other insulating material.

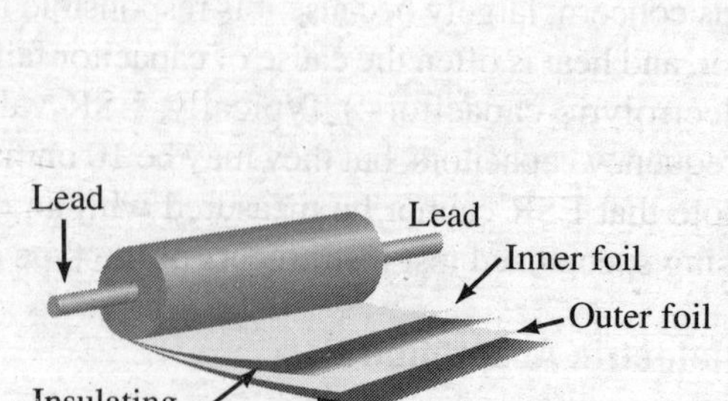

FIGURE 10–14 Tubular capacitor with axial leads.

in plastic, epoxy resin, or other insulating material and identified with value, tolerance, and other appropriate data either via body markings or color coding. Electrical characteristics and physical size depend on the dielectric used.

Ceramic Capacitors

First, consider ceramic. The permittivity of ceramic varies widely (as indicated in Table 10–1). At one end are ceramics with extremely high permittivity. These permit packaging a great deal of capacitance in a small space, but they yield capacitors whose characteristics vary widely with temperature and operating voltage. However, they are popular in applications where temperature variations are modest and small size and cost are important. At the other end are ceramics with highly stable characteristics. They yield capacitors whose values change little with temperature, voltage, or aging. However, since their dielectric constants are relatively low (typically 30 to 80), these capacitors are physically larger than equivalent capacitors made using high-permittivity ceramic. Many surface mount capacitors (considered later in this section) use ceramic dielectrics, as do disc capacitors of the type illustrated in Figure 10–15. Capacitors as in Figure 10–15 are used in bypass, filtering, and dc blocking applications (tasks that you will learn about in your electronics courses).

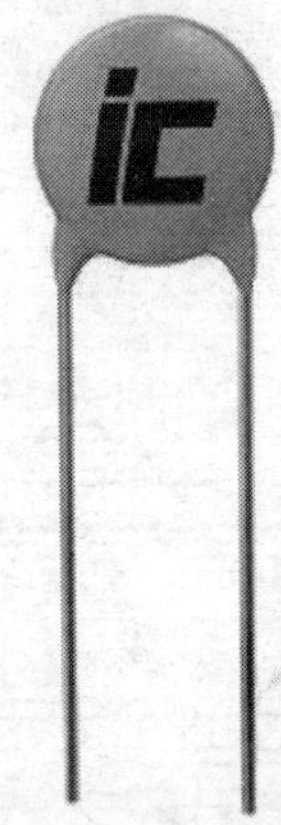

FIGURE 10–15 Ceramic disc capacitor.

Plastic Film Capacitors

Plastic-film capacitors are of two basic types: film/foil or metallized film. Film/foil capacitors use metal foil separated by plastic film as in Figure 10–14, while metallized-film capacitors have their foil material vacuum-deposited directly onto plastic film. Film/foil capacitors are generally larger than metallized-foil units, but have better capacitance stability and higher insulation resistance. Typical film materials are polyester, Mylar, polypropylene, and polycarbonate. Figure 10–16 shows a selection of plastic-film capacitors.

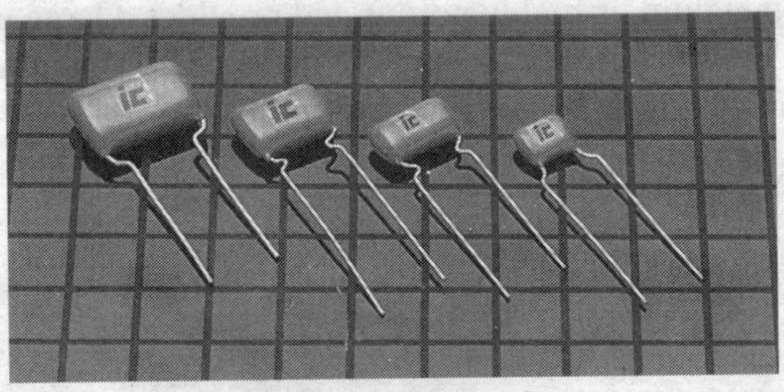

FIGURE 10–16 Radial lead film capacitors.

Metallized-film capacitors are self-healing. Thus, if voltage stress at an imperfection exceeds breakdown, an arc occurs that evaporates the metallized area around the fault, isolating the defect. (Film/foil capacitors are not self-healing.)

Silver Mica Capacitors

Silver mica capacitors provide high accuracy, low leakage, and good stability. Available values range from a few picofarads to about 0.1 μF.

Electrolytic Capacitors

Electrolytic capacitors provide relatively large capacitance (i.e., up to several thousand microfarads) at a relatively low cost. (Their capacitance is large because they have a very thin layer of oxide as their dielectric.) However, their leakage is relatively high and breakdown voltage relatively low. Electrolytics have either aluminum or tantalum as their plate material. Tantalum devices are smaller than aluminum devices, have less leakage, and are more stable.

The basic aluminum electrolytic capacitor construction is similar to that of Figure 10–14, with strips of aluminum foil separated by gauze saturated with an electrolyte. During manufacture, chemical action creates a thin oxide layer that acts as the dielectric. This layer must be maintained during use. For this reason, electrolytic capacitors are polarized (marked with a + and − sign), and the plus (+) terminal must always be kept positive with respect to the minus (−) terminal. Aluminum electrolytic capacitors have a shelf life; that is, if they are not used for an extended period, they may fail when powered up again. Figure 10–17(a) shows a selection of typical electrolytic capacitors, while Figure 10–17(b) shows their graphic symbol as used on circuit schematics. These capacitors are typically used in power supplies to smooth rectified waveforms.

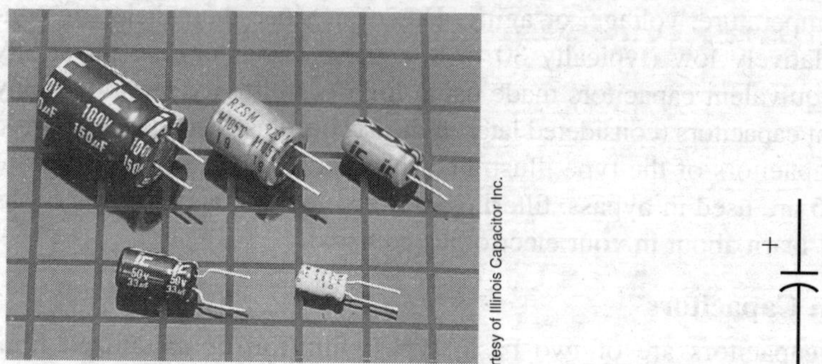

(a) Radial lead aluminium electrolytic capacitors

(b) Schematic symbol for an electrolytic capacitor

FIGURE 10–17 Electrolytic capacitors.

Tantalum capacitors come in two basic types: wet slug and solid dielectric. Figure 10–18 shows a cutaway view of a solid tantalum unit. The slug, made from powdered tantalum, is highly porous and provides a large internal surface area that is coated with an oxide to form the dielectric. Tantalum capacitors are polarized and must be inserted into a circuit properly.

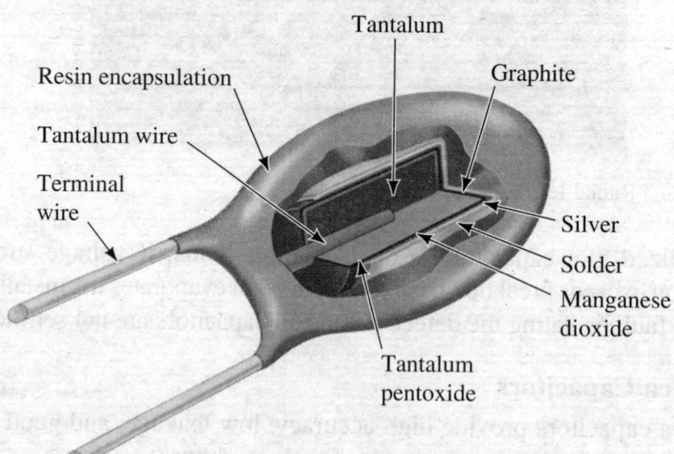

FIGURE 10–18 Cutaway view of a solid tantalum capacitor.

Surface Mount Capacitors

Most electronic products now use **surface mount devices** (SMDs). (SMDs do not have connection leads, but are soldered directly onto printed circuit boards by automated assembly machines.) Figure 10–19 shows a surface mount, ceramic chip capacitor. Such devices are extremely small and provide high packaging density as can be seen in Figure 10–30 later in this chapter.

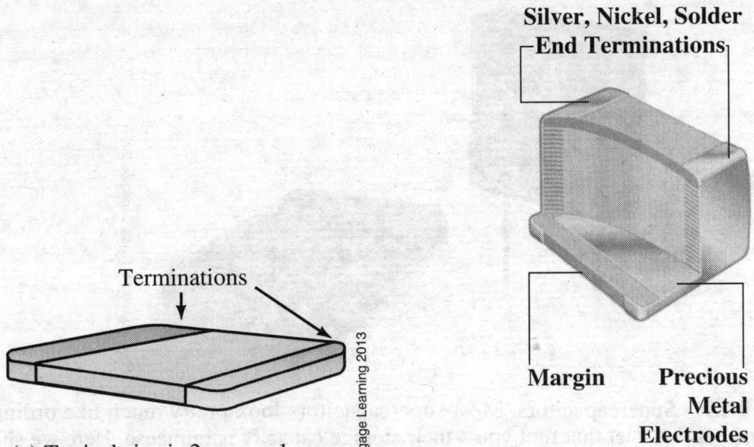

(a) Typical size is 2 mm × 1 mm
(See Figure 1–4)

(b) Cutaway view

FIGURE 10–19 Surface mount, ceramic chip capacitor.

Variable Capacitors

Several types of variable capacitors are still in use. One type (used for many years in radios to tune in stations) has a set of stationary plates and a set of movable plates that are ganged together and mounted on a shaft. As the user rotates the shaft, the movable plates mesh with the stationary plates, changing the effective surface area (and hence the capacitance).

Another adjustable type is the trimmer or padder capacitor, which is used for fine adjustments, usually over a very small range. A trimmer is usually set to its required value, then never touched again.

The symbol for a variable capacitor may be found in the table of schematic symbols, Table 1–7 of Chapter 1.

Standard Capacitor Values

Capacitors are manufactured only in specific standard value sizes, specifically 0.1 μF, 0.22 μF, 0.47 μF, and so on, plus multiples and submultiples thereof such as 0.1 pF, 0.22 pF, and 0.47 pF.

Supercapacitors

A relatively recent entry among capacitor offerings are **supercapacitors.** These devices, also known as ultracapacitors, are available with enormous capacitance values—values that extend up into the hundreds of farads range and beyond. (By comparison, the largest electrolytic capacitors usually do not exceed a few thousand microfarads.) The areas of application of supercapacitors are somewhat different than ordinary capacitors, however—typically, they are used in power source roles (roles normally serviced by batteries)—in GPS tracking systems, automotive regenerative braking systems, medical equipment, PDAs, and alarm and security systems. A common setup utilizes a supercapacitor in conjunction with a battery, wherein the battery supplies the steady bulk power demands of the load while the supercapacitor supplies the high-energy bursts that are required from time to time.

Most major capacitor manufacturers now have supercapacitors in their product lines. As indicated in Figure 10–20, from the outside, the capacitor looks like an ordinary capacitor. It achieves its massive capacitance through its internal construction, which creates an enormous amount of surface area with a very small separation between "plates." However, their voltage rating is quite low, typically only a few volts.

FIGURE 10–20 Supercapacitors. Most supercapacitors look pretty much like ordinary capacitors, but don't let that fool you—their storage capacity is immense. Here we show a pair of flat pack supercapacitors.

PRACTICE PROBLEMS 3

A 2.5-μF capacitor has a tolerance of +80% and −20%. Determine what its maximum and minimum values could be.

Answer
4.5 μF and 2 μF

10.7 Capacitors in Parallel and Series

Capacitors in Parallel

For capacitors in parallel, the effective plate area is the sum of the individual plate areas; thus, the total capacitance is the sum of the individual capacitances. This is easily shown mathematically. Consider Figure 10–21. The charge on each capacitor is given by Equation 10–1, that is, $Q_1 = C_1V$ and $Q_2 = C_2V$. Thus, $Q_T = Q_1 + Q_2 = C_1V + C_2V = (C_1 + C_2)\,V$. But $Q_T = C_TV$. Thus, $C_T = C_1 + C_2$. For more than two capacitors,

$$C_T = C_1 + C_2 + \cdots + C_N \tag{10–14}$$

That is, *the total capacitance of capacitors in parallel is the sum of their individual capacitances.*

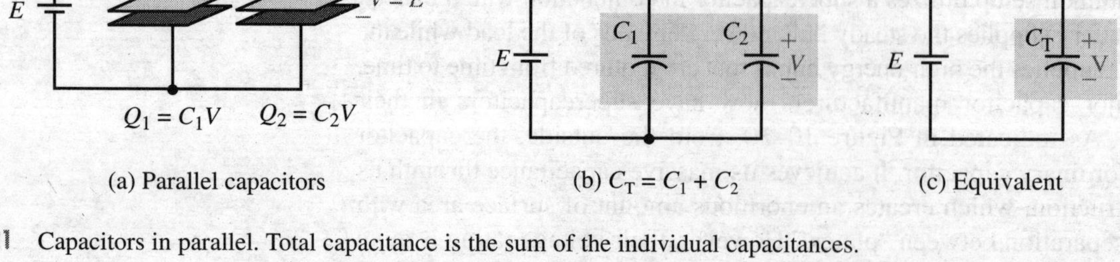

(a) Parallel capacitors (b) $C_T = C_1 + C_2$ (c) Equivalent

FIGURE 10–21 Capacitors in parallel. Total capacitance is the sum of the individual capacitances.

EXAMPLE 10–6

A 10-μF, a 15-μF, and a 100-μF capacitor are connected in parallel across a 50-V source. Determine the following:

a. Total capacitance.
b. Total charge stored.
c. Charge on each capacitor.

Solution

a. $C_T = C_1 + C_2 + C_3 = 10\ \mu F + 15\ \mu F + 100\ \mu F = 125\ \mu F$
b. $Q_T = C_T V = (125\ \mu F)(50\ V) = 6.25\ mC$
c. $Q_1 = C_1 V = (10\ \mu F)(50\ V) = 0.5\ mC$
$Q_2 = C_2 V = (15\ \mu F)(50\ V) = 0.75\ mC$
$Q_3 = C_3 V = (100\ \mu F)(50\ V) = 5.0\ mC$
Check: $Q_T = Q_1 + Q_2 + Q_3 = (0.5 + 0.75 + 5.0)\ mC = 6.25\ mC$.

PRACTICE PROBLEMS 4

1. Three capacitors are connected in parallel. If $C_1 = 20\ \mu F$, $C_2 = 10\ \mu F$, and $C_T = 32.2\ \mu F$, what is C_3?
2. Three capacitors are paralleled across an 80-V source, with $Q_T = 0.12$ C. If $C_1 = 200\ \mu F$ and $C_2 = 300\ \mu F$, what is C_3?
3. Three capacitors are paralleled. If the value of the second capacitor is twice that of the first and the value of the third is one-quarter that of the second and the total capacitance is 70 μF, what are the values of each capacitor?

Answers
1. 2.2 μF; 2. 1000 μF; 3. 20 μF, 40 μF, and 10 μF

Capacitors in Series

For capacitors in series (Figure 10–22), the same charge appears on each. Thus, $Q = C_1V_1$, $Q = C_2V_2$, and so on. Solving for voltages yields $V_1 = Q/C_1$, $V_2 = Q/C_2$, and so on. Applying KVL, we get $V = V_1 + V_2 + \cdots + V_N$. Therefore,

$$V = \frac{Q}{C_1} + \frac{Q}{C_2} + \cdots + \frac{Q}{C_N} = Q\left(\frac{1}{C_1} + \frac{1}{C_2} + \cdots + \frac{1}{C_N}\right)$$

But $V = Q/C_T$. Equating this with the right side and cancelling Q yields

$$\frac{1}{C_T} = \frac{1}{C_1} + \frac{1}{C_2} + \cdots + \frac{1}{C_N} \qquad (10\text{–}15)$$

For two capacitors in series, this reduces to

$$C_T = \frac{C_1 C_2}{C_1 + C_2} \qquad (10\text{–}16)$$

NOTES...

1. For capacitors in parallel, total capacitance is always larger than the largest capacitance, while for capacitors in series, total capacitance is always smaller than the smallest capacitance.
2. The formula for capacitors in parallel is similar to the formula for resistors in series, while the formula for capacitors in series is similar to the formula for resistors in parallel.

For N equal capacitors in series, Equation 10–15 yields $C_T = C/N$.

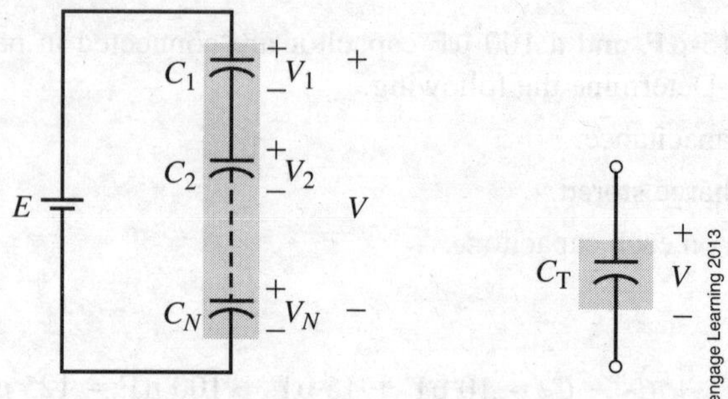

(a) Series connection (b) Equivalent

FIGURE 10–22 Capacitors in series: $\frac{1}{C_T} = \frac{1}{C_1} + \frac{1}{C_2} + \cdots + \frac{1}{C_N}$

EXAMPLE 10–7

Calculator Hint: You can also use your calculator to evaluate Equation 10–15 directly. For the capacitors of Example 10–7, the solution for C_T would look similar to the following, depending on the make and model of your calculator.

```
30⁻¹+60⁻¹+20⁻¹
                .1
Ans⁻¹
                10
```

Thus, $C_T = 10\ \mu F$.

Refer to Figure 10–23(a):

a. Determine C_T.

b. If 50 V is applied across the capacitors, determine Q.

c. Determine the voltage on each capacitor.

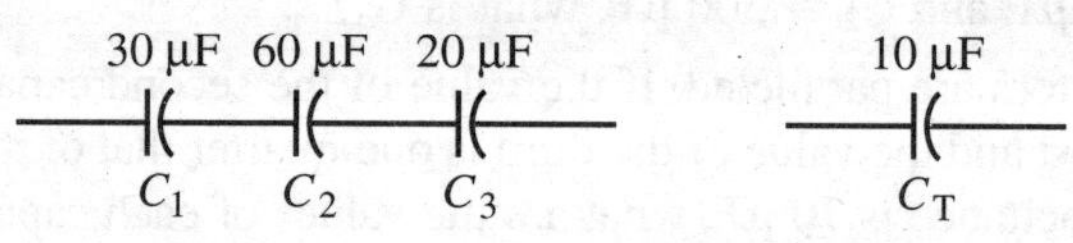

(a) (b)

FIGURE 10–23

Solution

a. $$\frac{1}{C_T} = \frac{1}{C_1} + \frac{1}{C_2} + \frac{1}{C_3} = \frac{1}{30\ \mu F} + \frac{1}{60\ \mu F} + \frac{1}{20\ \mu F}$$

$$= 0.0333 \times 10^6 + 0.0167 \times 10^6 + 0.05 \times 10^6 = 0.1 \times 10^6$$

Therefore as indicated in (b),

$$C_T = \frac{1}{0.1 \times 10^6} = 10\ \mu F$$

b. $Q = C_T V = (10 \times 10^{-6}\ F)(50\ V) = 0.5\ mC$

c. $V_1 = Q/C_1 = (0.5 \times 10^{-3}\ C)/(30 \times 10^{-6}\ F) = 16.7\ V$

$V_2 = Q/C_2 = (0.5 \times 10^{-3}\ C)/(60 \times 10^{-6}\ F) = 8.3\ V$

$V_3 = Q/C_3 = (0.5 \times 10^{-3}\ C)/(20 \times 10^{-6}\ F) = 25.0\ V$

Check: $V_1 + V_2 + V_3 = 16.7 + 8.3 + 25 = 50\ V$.

EXAMPLE 10–8

For the circuit of Figure 10–24(a), determine C_T.

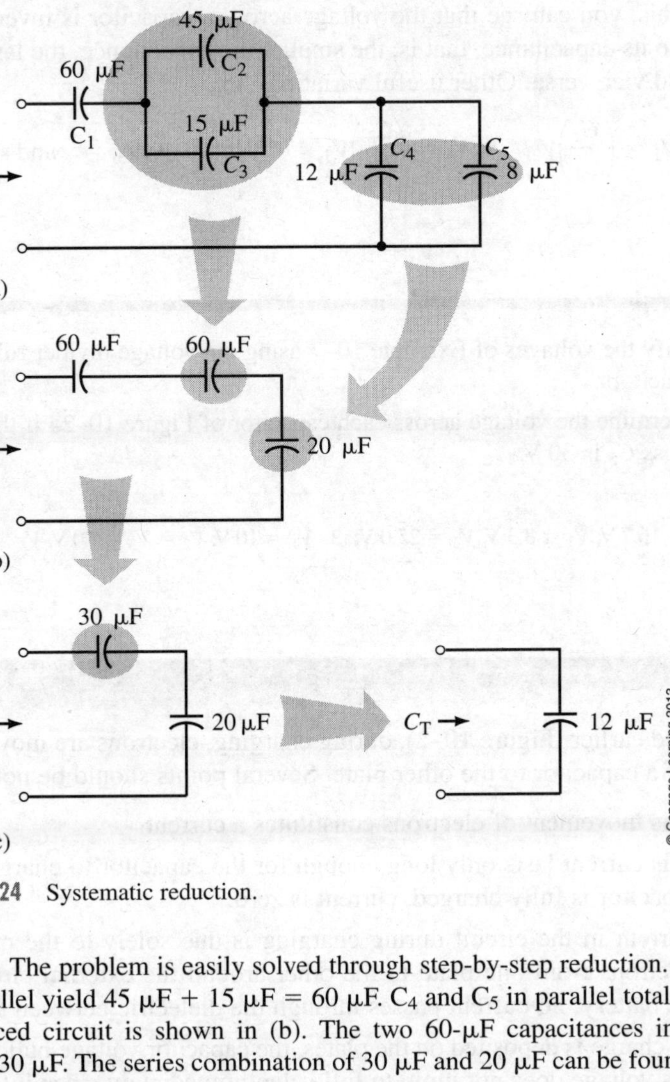

FIGURE 10–24 Systematic reduction.

Solution The problem is easily solved through step-by-step reduction. C_2 and C_3 in parallel yield 45 μF + 15 μF = 60 μF. C_4 and C_5 in parallel total 20 μF. The reduced circuit is shown in (b). The two 60-μF capacitances in series reduce to 30 μF. The series combination of 30 μF and 20 μF can be found from Equation 10–16. Thus,

$$C_T = \frac{30\ \mu\text{F} \times 20\ \mu\text{F}}{30\ \mu\text{F} + 20\ \mu\text{F}} = 12\ \mu\text{F}$$

Alternately, use your calculator to reduce (b) as described in Example 10–7.

Voltage Divider Rule for Series Capacitors

For capacitors in series (Figure 10–25) a simple voltage divider rule can be developed. Recall, for individual capacitors, $Q_1 = C_1V_1$, $Q_2 = C_2V_2$, and so on, and for the complete string, $Q_T = C_TV_T$. As noted earlier, $Q_1 = Q_2 = \cdots = Q_T$. Thus, $C_1V_1 = C_TV_T$. Solving for V_1 yields

$$V_1 = \left(\frac{C_T}{C_1}\right)V_T$$

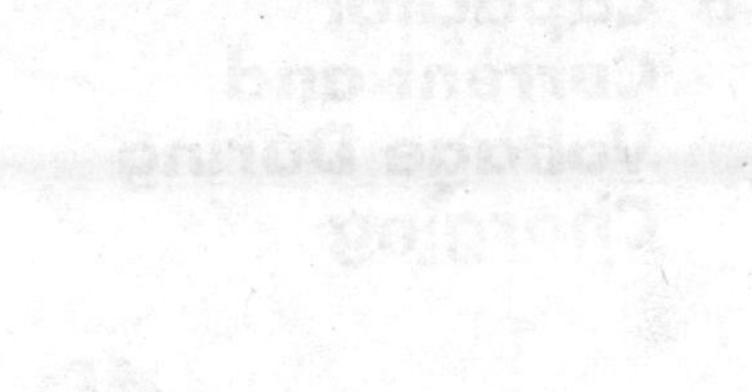

FIGURE 10–25 Capacitive voltage divider.

This type of relationship holds for all capacitors. Thus,

$$V_x = \left(\frac{C_T}{C_x}\right)V_T \tag{10–17}$$

From this, you can see that the voltage across a capacitor is inversely proportional to its capacitance, that is, the smaller the capacitance, the larger the voltage, and vice versa. Other useful variations are

$$V_1 = \left(\frac{C_2}{C_1}\right)V_2, \quad V_1 = \left(\frac{C_3}{C_1}\right)V_3, \quad V_2 = \left(\frac{C_3}{C_2}\right)V_3, \quad \text{and so on.}$$

PRACTICE PROBLEMS 5

1. Verify the voltages of Example 10–7 using the voltage divider rule for capacitors.
2. Determine the voltage across each capacitor of Figure 10–24 if the voltage across C_5 is 30 V.

Answers
1. $V_1 = 16.7$ V, $V_2 = 8.3$ V, $V_3 = 25.0$ V; 2. $V_1 = 10$ V, $V_2 = V_3 = 10$ V, $V_4 = V_5 = 30$ V

10.8 Capacitor Current and Voltage During Charging

As noted earlier (Figure 10–2), during charging, electrons are moved from one plate of a capacitor to the other plate. Several points should be noted.

1. This movement of electrons constitutes a current.
2. This current lasts only long enough for the capacitor to charge. When the capacitor is fully charged, current is zero.
3. Current in the circuit during charging is due solely to the movement of electrons from one plate to the other around the external circuit through the battery; no current passes through the dielectric between the plates.
4. As charge is deposited on the plates, the capacitor voltage builds. However, this voltage does not jump to full value immediately since it takes time to move electrons from one plate to the other. (Billions of electrons must be moved.)
5. Since voltage builds up as charging progresses, the difference in voltage between the source and the capacitor decreases and hence the rate of movement of electrons (i.e., the current) decreases as the capacitor approaches full charge.

Figure 10–26 shows what the voltage and current look like during the charging process. As indicated, the current starts out with an initial surge, then decays to zero while the capacitor voltage gradually climbs from zero to full voltage. The charging time typically ranges from nanoseconds to milliseconds, depending on the resistance and capacitance of the circuit. (We study these relationships in detail in Chapter 11.) A similar surge (but in the opposite direction) occurs during discharge.

As Figure 10–26 indicates, current exists only while the capacitor voltage is changing. This observation turns out to be true in general, that is, *current in a capacitor exists only while capacitor voltage is changing.* The reason is not

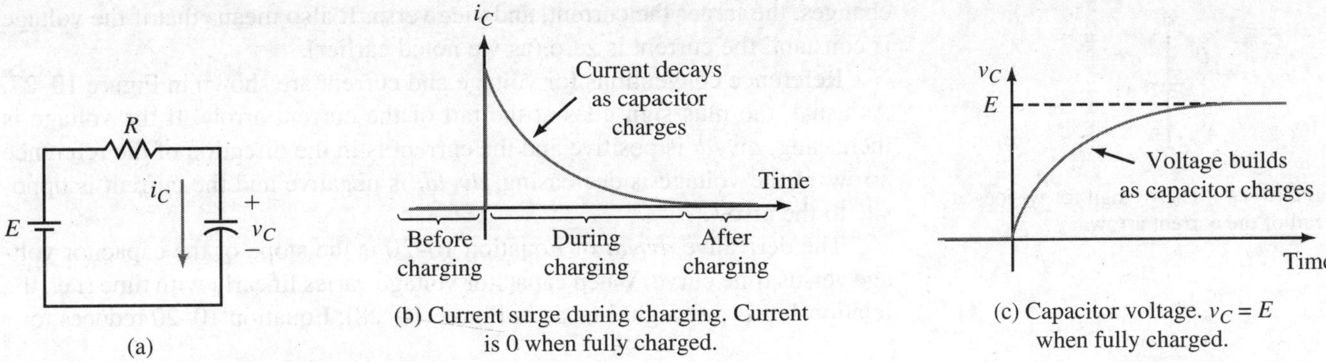

FIGURE 10–26 The capacitor does not charge instantaneously, as a finite amount of time is required to move electrons around the circuit.

hard to understand. As you saw before, a capacitor's dielectric is an insulator, and consequently no current can pass through it (assuming zero leakage). The only charges that can move, therefore, are the free electrons that exist on the capacitor's plates. When capacitor voltage is constant, these charges are in equilibrium, no net movement of charge occurs, and the current is thus zero. However, if the source voltage is increased, additional electrons are pulled from the positive plate; inversely, if the source voltage is decreased, excess electrons on the negative plate are returned to the positive plate. Thus, in both cases, capacitor current results when capacitor voltage is changed. As we show next, this current is proportional to the rate of change of voltage. Before we do this, however, we need to look at symbols.

CircuitSim 10-3

Symbols for Time-Varying Voltages and Currents

Quantities that vary with time are called **instantaneous quantities.** *Standard industry practice requires that we use lowercase letters for time-varying quantities, rather than capital letters as for dc.* Thus, we use v_C and i_C to represent changing capacitor voltage and current rather than V_C and I_C. (Often we drop the subscripts and just use v and i.) Since these quantities are functions of time, they may also be shown as $v_C(t)$ and $i_C(t)$.

Capacitor *v-i* Relationship

The relationship between charge and voltage for a capacitor is given by Equation 10–1. For the time-varying case, it is

$$q = Cv_C \quad \textbf{(10–18)}$$

But current is the rate of movement of charge. In calculus notation, this is $i_C = dq/dt$. Differentiating Equation 10–18 yields

$$i_C = \frac{dq}{dt} = \frac{d}{dt}(Cv_C) \quad \textbf{(10–19)}$$

Since C is constant, we get

$$i_C = C\frac{dv_C}{dt} \text{ (A)} \quad \textbf{(10–20)}$$

Equation 10–20 shows that *current through a capacitor is equal to C times the rate of change of voltage across it.* This means that the faster the voltage

NOTES...

Calculus is introduced at this point to aid in the development of ideas and to help explain concepts. However, not everyone who uses this book knows calculus. Therefore, the material is presented in such a manner that it never relies entirely on mathematics; thus, where calculus is used, intuitive explanations accompany it. However, to provide the enrichment that calculus offers, optional derivations and problems are included, but they are marked with a ▯ icon so that they may be omitted if desired.

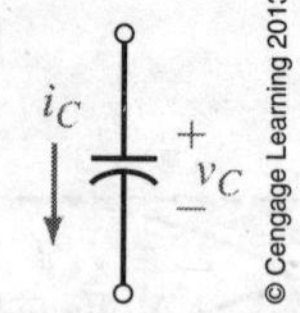

FIGURE 10–27 The + sign for v_C goes at the tail of the current arrow.

changes, the larger the current, and vice versa. It also means that if the voltage is constant, the current is zero (as we noted earlier).

Reference conventions for voltage and current are shown in Figure 10–27. As usual, the plus sign goes at the tail of the current arrow. If the voltage is increasing, dv_C/dt is positive and the current is in the direction of the reference arrow; if the voltage is decreasing, dv_C/dt is negative and the current is opposite to the arrow.

The derivative dv_C/dt of Equation 10–20 is the slope of the capacitor voltage versus time curve. When capacitor voltage varies linearly with time (i.e., the relationship is a straight line as in Figure 10–28), Equation 10–20 reduces to

$$i_C = C\frac{\Delta v_C}{\Delta t} = C\frac{\text{rise}}{\text{run}} = C \times \text{slope of the line} \qquad \textbf{(10–21)}$$

EXAMPLE 10–9

A signal generator applies voltage to a 5-μF capacitor with a waveform as in Figure 10–28(a). The voltage rises linearly from 0 V to 10 V in 1 ms, falls linearly to −10 V at t = 3 ms, remains constant until t = 4 ms, rises to 10 V at t = 5 ms, and remains constant thereafter.

a. Determine the slope of v_C in each time interval.
b. Determine the current and sketch its graph.

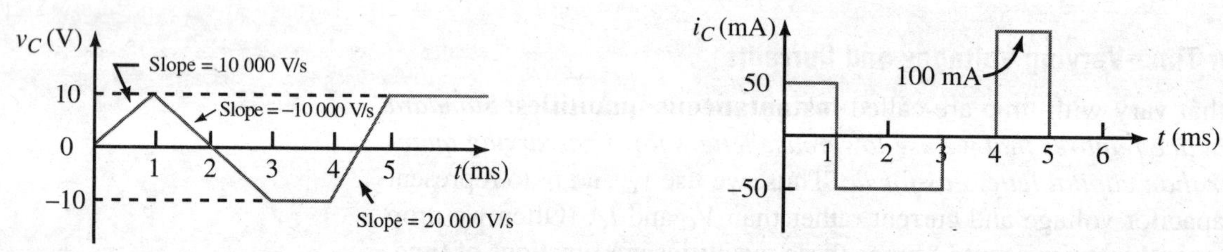

FIGURE 10–28

Solution

a. We need the slope of v_C during each time interval where slope = rise/run = $\Delta v/\Delta t$.

0 ms to 1 ms: Δv = 10 V; Δt = 1 ms; therefore, slope = 10 V/1 ms = 10 000 V/s.

1 ms to 3 ms: Slope = −20 V/2 ms = −10 000 V/s.

3 ms to 4 ms: Slope = 0 V/s.

4 ms to 5 ms: Slope = 20 V/1 ms = 20 000 V/s.

CircuitSim 10-4

b. $i_C = C dv_C/dt = C$ times slope. Thus,

0 ms to 1 ms: $i = (5 \times 10^{-6}\text{ F})(10\,000\text{ V/s}) = 50\text{ mA}$.

1 ms to 3 ms: $i = -(5 \times 10^{-6}\text{ F})(10\,000\text{ V/s}) = -50\text{ mA}$.

3 ms to 4 ms: $i = (5 \times 10^{-6}\text{ F})(0\text{ V/s}) = 0\text{ A}$.

4 ms to 5 ms: $i = (5 \times 10^{-6}\text{ F})(20\,000\text{ V/s}) = 100\text{ mA}$.

The current is plotted in Figure 10-28(b).

EXAMPLE 10–10

CircuitSim 10-5

The voltage across a 20-μF capacitor is $v_C = 100\,t\,e^{-t}$ V. Determine current i_C.

Solution Differentiation by parts using $\frac{d(uv)}{dt} = u\frac{dv}{dt} + v\frac{du}{dt}$ with $u = 100\,t$ and $v = e^{-t}$ yields

$$i_C = C\frac{d}{dt}(100\;t\,e^{-t}) = 100\,C\frac{d}{dt}(t\,e^{-t}) = 100\,C\left(t\frac{d}{dt}(e^{-t}) + e^{-t}\frac{dt}{dt}\right)$$

$$= 2000 \times 10^{-6}(-t\,e^{-t} + e^{-t})\text{ A} = 2.0\,(1 - t)e^{-t}\text{ mA}$$

CircuitSim 10-6

10.9 Energy Stored by a Capacitor

An ideal capacitor does not dissipate power. When power is transferred to a capacitor, all of it is stored as energy in the capacitor's electric field. When the capacitor is discharged, this stored energy is returned to the circuit. You will find that this is an important observation when you study ac in Chapter 15.

To determine the stored energy, consider Figure 10–29. Power is given by $p = vi$ watts. Using calculus (see ∫), it can be shown that the stored energy is given by

$$W = \frac{1}{2}CV^2 \text{ (J)} \qquad (10\text{–}22)$$

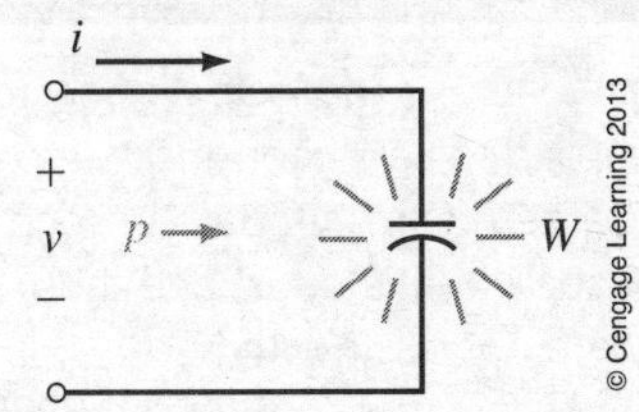

FIGURE 10–29 Storing energy in a capacitor.

where V is the voltage across the capacitor. This means that the energy at any time depends on the value of the capacitor's voltage at that time. We will leave this topic at this time, but return to it again in Chapter 17 and later chapters where you encounter it in an ac context.

Deriving Equation 10–22

Power to the capacitor (Figure 10–29) is given by $p = vi$, where $i = Cdv/dt$. Therefore, $p = Cvdv/dt$. However, $p = dW/dt$. Equate the two values of p, then after a bit of manipulation, you can integrate. Thus,

$$W = \int_0^t pdt = C\int_0^t v\frac{dv}{dt}\,dt = C\int_0^V vdv = \frac{1}{2}CV^2$$

10.10 Capacitor Failures and Troubleshooting

Although capacitors are quite reliable, they may fail because of misapplication, excessive voltage, current, temperature, or simply because they age. They can short internally, leads may become open, dielectrics may become excessively leaky, and they may fail catastrophically due to incorrect use. (If an electrolytic capacitor is connected with its polarity reversed, for example, it may explode.) Capacitors should be used well within their rating limits. Excessive voltage can lead to dielectric puncture creating pinholes that short the plates together. High temperatures may cause an increase in leakage and/or a permanent shift in capacitance. High temperatures may be caused by inadequate heat removal, excessive current, lossy dielectrics, or an operating frequency beyond the capacitor's rated limit. Generally, capacitors are so inexpensive that you simply replace them if you suspect they are faulty. To help locate faulty capacitors, you can sometimes use an ohmmeter or a capacitor tester, as described next.

Basic Testing with an Ohmmeter

Some basic (out-of-circuit) tests can be made with an analog ohmmeter. The ohmmeter can detect opens and shorts and, to a certain extent, leaky dielectrics. First, ensure that the capacitor is discharged, then set the ohmmeter to its highest range and connect it to the capacitor. (For electrolytic devices, ensure that the plus [+] side of the ohmmeter is connected to the plus [+] side of the capacitor.)

Initially, the ohmmeter reading should be low, then for a good capacitor gradually increase to infinity as the capacitor charges through the ohmmeter circuit. (Or at least a very high value, since most good capacitors, except electrolytics, have a resistance of hundreds of megohms.) (For small capacitors, this test is probably useless, as the time for charging may be too short to yield useful results.)

Faulty capacitors respond differently. If a capacitor is shorted, the meter resistance reading will stay low. If it is leaky, the reading will be lower than normal. If it is open circuited, the meter will indicate infinity immediately, without dipping to zero when first connected.

NOTES...

Quality factor and dissipation factor as applied to capacitor quality are specialized topics that are beyond the scope of this book, and thus they are not dealt with here. Manuals (such as that for the 889B as found on the manufacturer's Web site) are often good sources of information. However, wait until you have studied Chapter 21 of this book before you try to understand the concepts.

Capacitor Testers

Ohmmeter testing is of limited usefulness as it tells you nothing about capacitor quality issues such as ESR. For investigating these, more sophisticated tester/analyzers are needed. Figure 10–30 shows two. The *Smart Tweezers*™ device of (a) is designed primarily for testing surface mount components (such as that shown in Figure 10–19). It is a fully featured LCR tester that can test capacitors, resistors, and inductors. For capacitor testing, it determines capacitance, ESR, and dissipation factor, and also checks for shorts. As indicated, values are displayed directly in standard units. The *BK Precision 889B* bench mount instrument of (b) is an LCR/ESR component tester designed to test inductors, capacitors, and resistors. In terms of capacitor testing, it measures capacitance, determines leakage, dielectric absorption, ESR, quality factor Q and more—see Note.

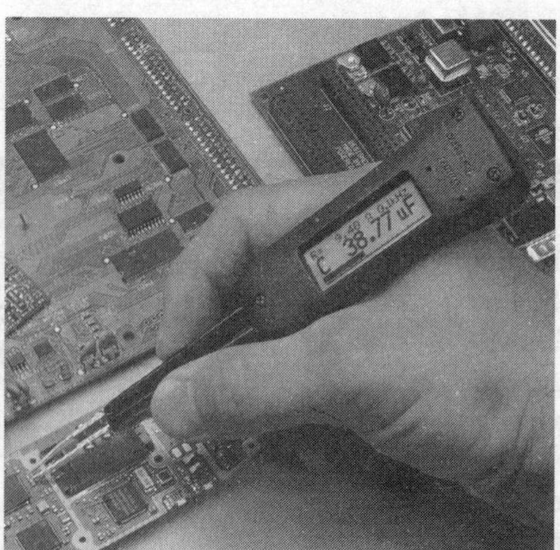

(a) These Smart Tweezers™ measure capacitance, inductance, and resistance plus secondary parameters such as capacitor ESR.

Photo courtesy of Advance Devices, Inc.

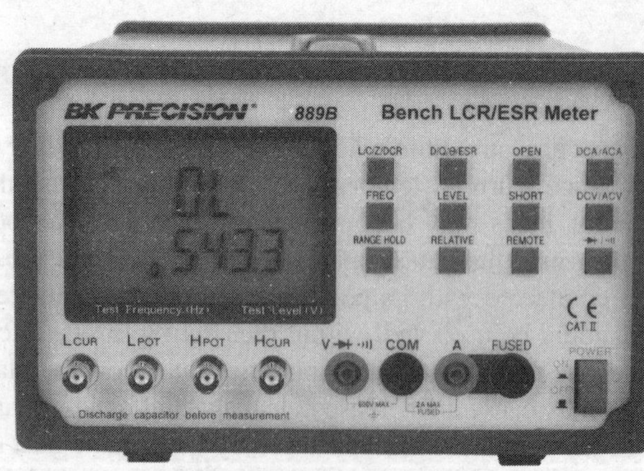

(b) This LCR/ESR tester/analyzer features a USB interface for connection to a computer.

Courtesy of B & K Precision

FIGURE 10–30 Instruments for capacitor testing and troubleshooting.

Problems

10.1 Capacitance

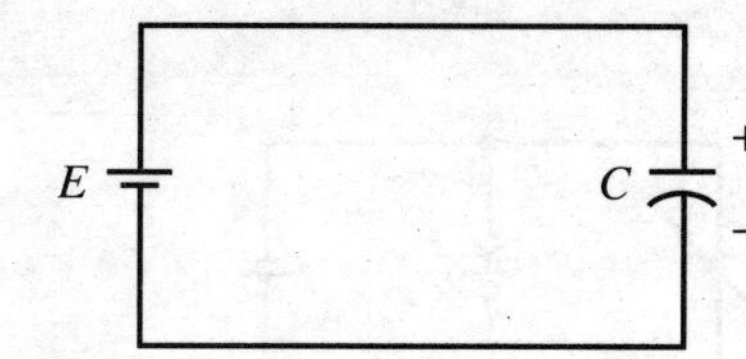

FIGURE 10–31

1. For Figure 10–31, determine the charge on the capacitor, its capacitance, or the voltage across it as applicable for each of the following.
   a. $E = 40$ V, $C = 20$ μF
   b. $V = 500$ V, $Q = 1000$ μC
   c. $V = 200$ V, $C = 500$ nF
   d. $Q = 3 \times 10^{-4}$ C, $C = 10 \times 10^{-6}$ F
   e. $Q = 6$ mC, $C = 40$ μF
   f. $V = 1200$ V, $Q = 1.8$ mC
2. Repeat Question 1 for the following:
   a. $V = 2.5$ kV, $Q = 375$ μC
   b. $V = 1.5$ kV, $C = 0.04 \times 10^{-4}$ F
   c. $V = 150$ V, $Q = 6 \times 10^{-5}$ C
   d. $Q = 10$ μC, $C = 400$ nF
   e. $V = 150$ V, $C = 40 \times 10^{-5}$ F
   f. $Q = 6 \times 10^{-9}$ C, $C = 800$ pF
3. The charge on a 50-μF capacitor is 10×10^{-3} C. What is the potential difference between its terminals?
4. When 10 μC of charge is placed on a capacitor, its voltage is 25 V. What is the capacitance?
5. You charge a 5-μF capacitor to 150 V. Your lab partner then momentarily places a resistor across its terminals and bleeds off enough charge that its voltage falls to 84 V. What is the final charge on the capacitor?

10.2 Factors Affecting Capacitance

6. A capacitor with circular plates 0.1 m in diameter and an air dielectric has 0.1 mm spacing between its plates. What is its capacitance?
7. A parallel-plate capacitor with a mica dielectric has dimensions of 1 cm × 1.5 cm and separation of 0.1 mm. What is its capacitance?
8. For the capacitor of Problem 7, if the mica is removed, what is its new capacitance?
9. The capacitance of an oil-filled capacitor is 200 pF. If the separation between its plates is 0.1 mm, what is the area of its plates?
10. A 0.01-μF capacitor has ceramic with a dielectric constant of 7500. If the ceramic is removed, the plate separation doubled, and the spacing between plates filled with oil, what is the new value for C?
11. A capacitor with a Teflon dielectric has a capacitance of 33 μF. A second capacitor with identical physical dimensions but with a Mylar dielectric carries a charge of 55×10^{-4} C. What is its voltage?
12. The plate area of a capacitor is 4.5 in.^{2} and the plate separation is 5 mils. If the relative permittivity of the dielectric is 80, what is C?

10.3 Electric Fields

13. a. What is the electric field strength $\mathscr{E}$ at a distance of 1 cm from a 100-mC charge in transformer oil?
    b. What is $\mathscr{E}$ at twice the distance?
14. Suppose that 150 V is applied across a 100-pF parallel-plate capacitor whose plates are separated by 1 mm. What is the electric field intensity $\mathscr{E}$ between the plates?

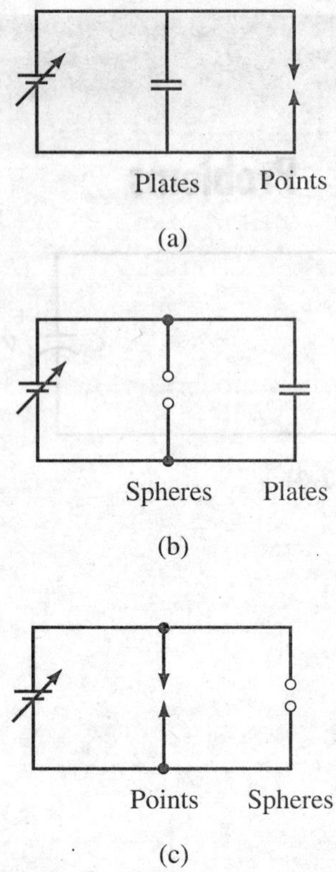

FIGURE 10–32 Source voltage is increased until one of the gaps breaks down. (The source has high internal resistance to limit current following breakdown.)

10.4 Dielectrics

15. An air-dielectric capacitor has plate spacing of 1.5 mm. How much voltage can be applied before breakdown occurs?
16. Repeat Problem 15 if the dielectric is mica and the spacing is 2 mils.
17. A mica-dielectric capacitor breaks down when *E* volts is applied. The mica is removed and the spacing between plates doubled. If breakdown now occurs at 500 V, what is *E*?
18. Determine at what voltage the dielectric of a 200 nF Mylar capacitor with a plate area of 0.625 m^2 will break down.
19. Figure 10–32 shows several gaps, including a parallel-plate capacitor, a set of small spherical points, and a pair of sharp points. The spacing is the same for each. As the voltage is increased, which gap breaks down for each case?
20. If you continue to increase the source voltage of Figures 10–32(a), (b), and (c) after a gap breaks down, will the second gap also break down? Justify your answer.

10.5 Nonideal Effects

21. A 25-μF capacitor has a negative temperature coefficient of 175 ppm/° C. By how much and in what direction might it vary if the temperature rises by 50° C? What would be its new value?
22. If a 4.7-μF capacitor changes to 4.8 μF when the temperature rises 40° C, what is its temperature coefficient?

10.7 Capacitors in Parallel and Series

23. What is the equivalent capacitance of 10 μF, 12 μF, 22 μF, and 33 μF connected in parallel?
24. What is the equivalent capacitance of 0.10 μF, 220 nF, and 4.7×10^{-7} F connected in parallel?
25. Repeat Problem 23 if the capacitors are connected in series.
26. Repeat Problem 24 if the capacitors are connected in series.
27. Determine C_T for each circuit of Figure 10–33.
28. Determine total capacitance looking at the terminals for each circuit of Figure 10–34.
29. A 30-μF capacitor is connected in parallel with a 60-μF capacitor, and a 10-μF capacitor is connected in series with the parallel combination. What is C_T?
30. For Figure 10–35, determine C_x.
31. For Figure 10–36, determine C_3 and C_4.
32. For Figure 10–37, determine C_T.
33. You have capacitors of 22 μF, 47 μF, 2.2 μF, and 10 μF. Connecting these any way you want, what is the largest equivalent capacitance you can get? The smallest?
34. A 10-μF and a 4.7-μF capacitor are connected in parallel. After a third capacitor is added to the circuit, $C_T = 2.695$ μF. What is the value of the third capacitor? How is it connected?
35. Consider capacitors of 1 μF, 1.5 μF, and 10 μF. If $C_T = 10.6$ μF, how are the capacitors connected?

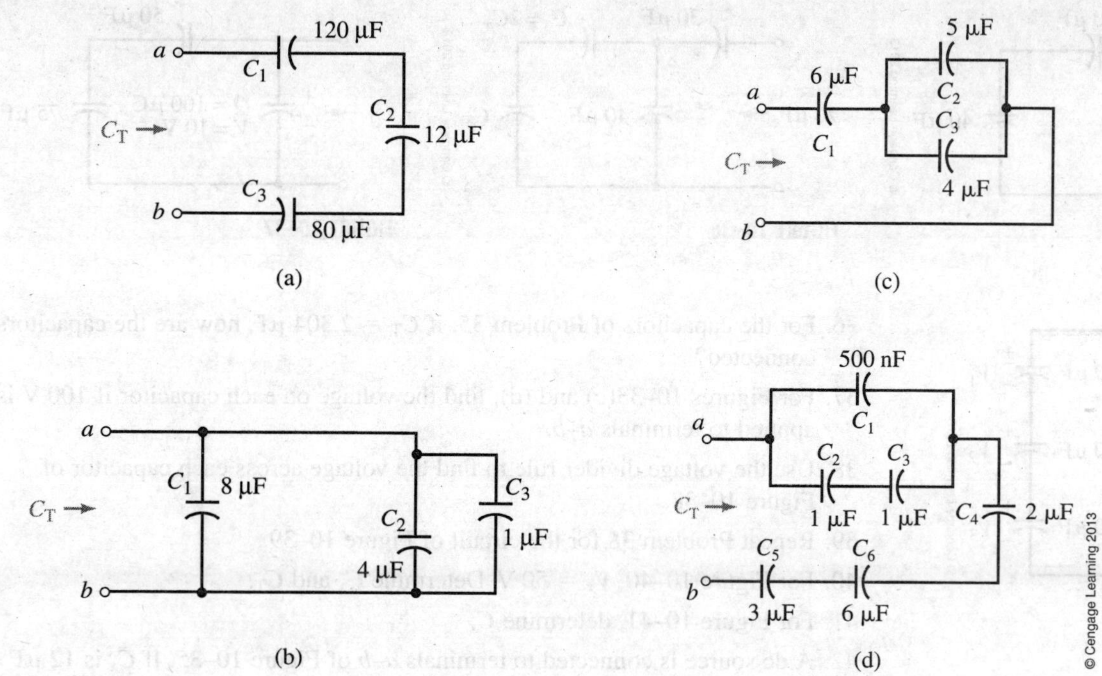

FIGURE 10–33

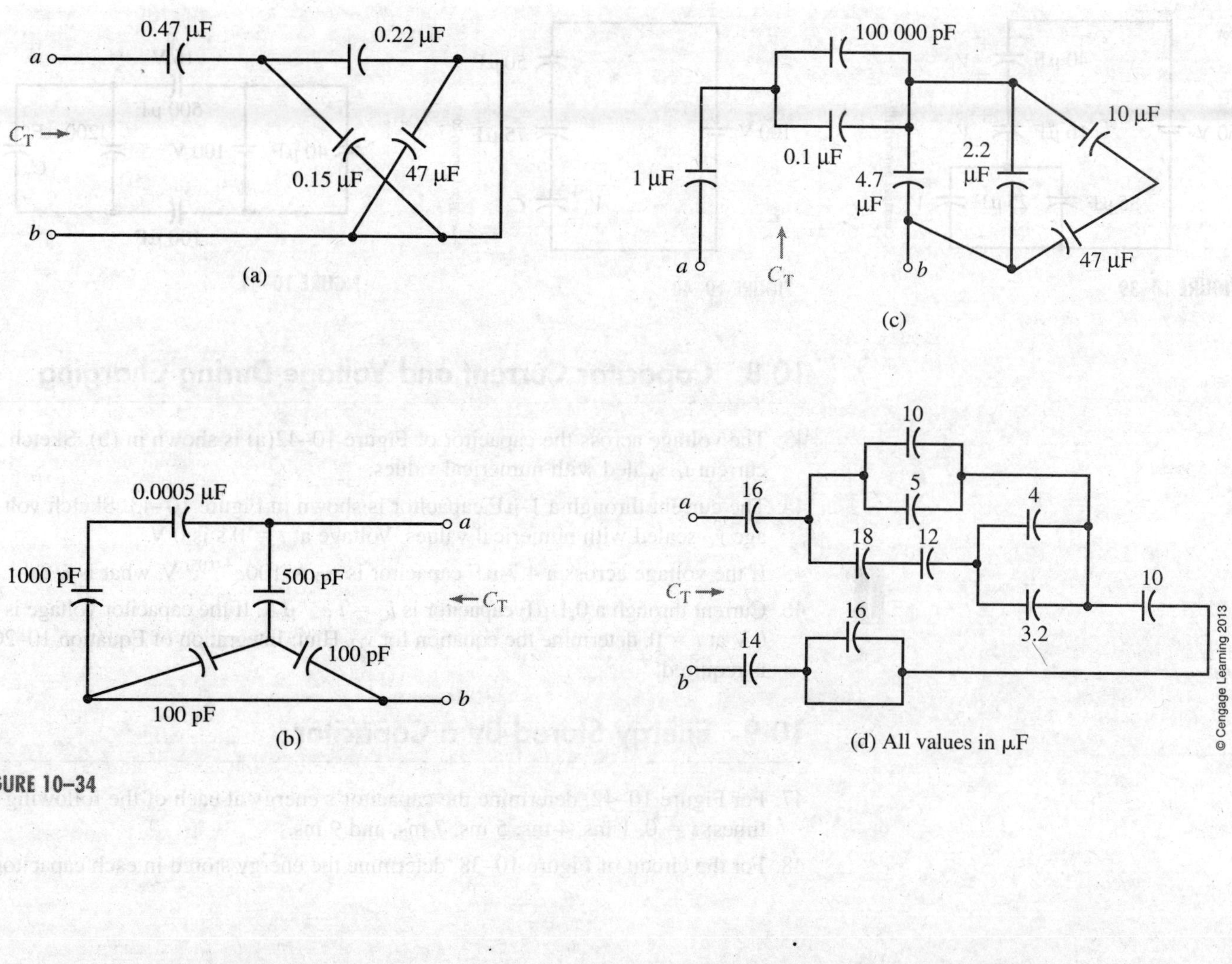

FIGURE 10–34

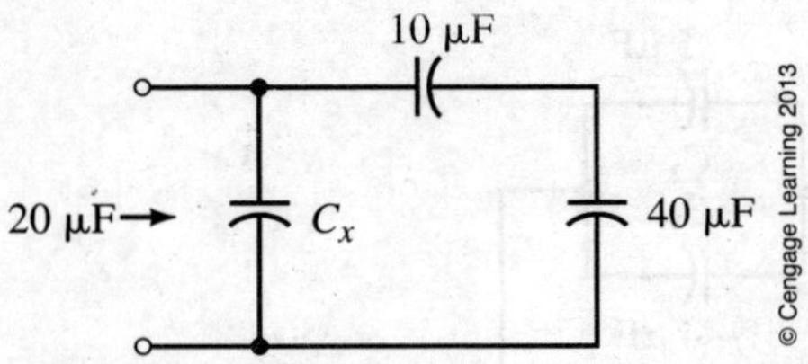

FIGURE 10–35

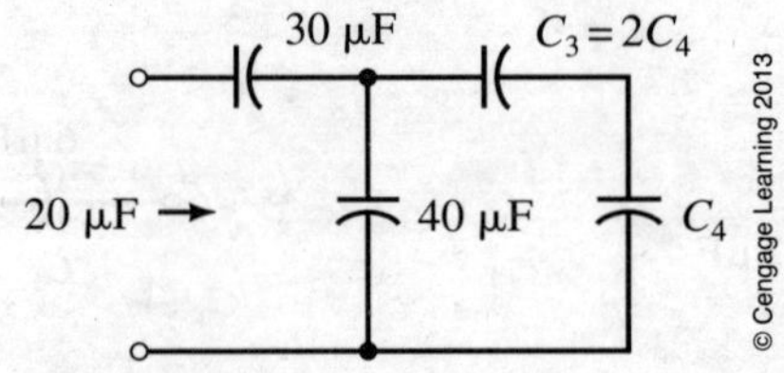

FIGURE 10–36

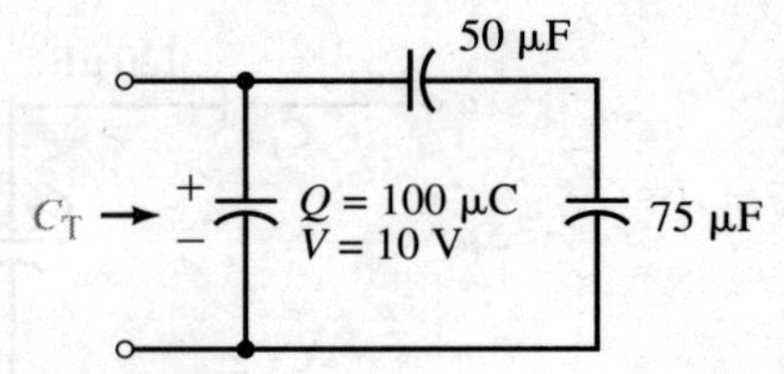

FIGURE 10–37

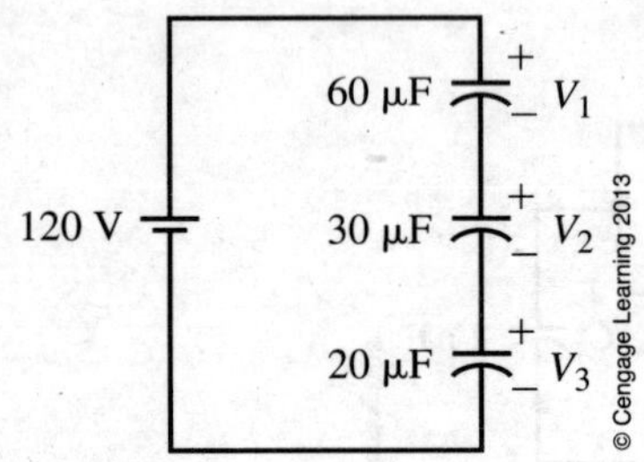

FIGURE 10–38

36. For the capacitors of Problem 35, if $C_T = 2.304\ \mu F$, how are the capacitors connected?
37. For Figures 10–33(c) and (d), find the voltage on each capacitor if 100 V is applied to terminals *a–b*.
38. Use the voltage divider rule to find the voltage across each capacitor of Figure 10–38.
39. Repeat Problem 38 for the circuit of Figure 10–39.
40. For Figure 10–40, $V_x = 50$ V. Determine C_x and C_T.
41. For Figure 10–41, determine C_x.
42. A dc source is connected to terminals *a–b* of Figure 10–35. If C_x is 12 μF and the voltage across the 40-μF capacitor is 80 V,
    a. What is the source voltage?
    b. What is the total charge on the capacitors?
    c. What is the charge on each individual capacitor?

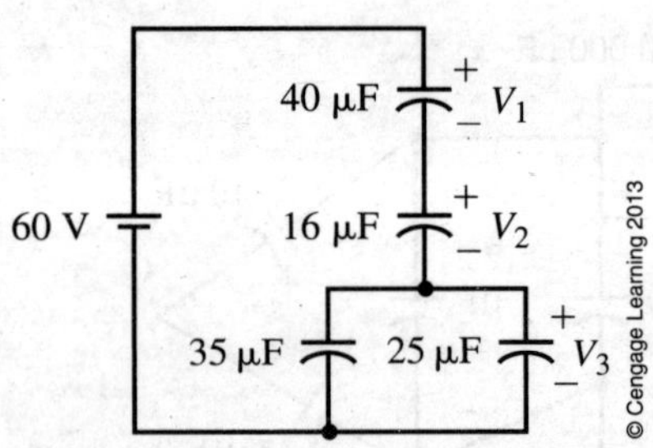

FIGURE 10–39

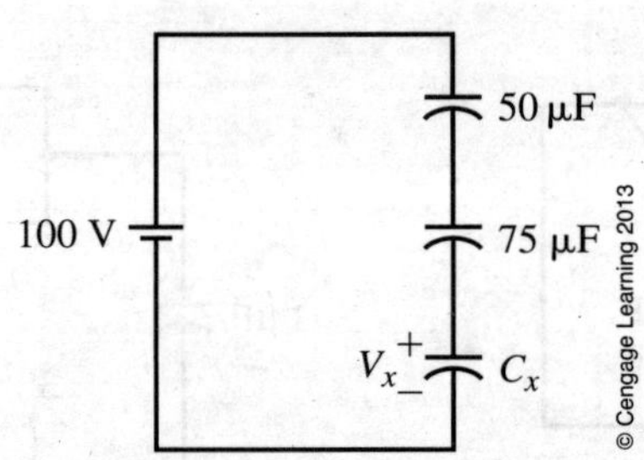

FIGURE 10–40

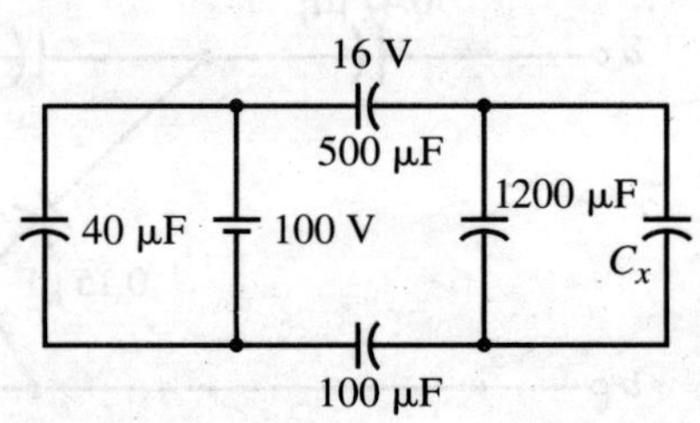

FIGURE 10–41

10.8 Capacitor Current and Voltage During Charging

43. The voltage across the capacitor of Figure 10–42(a) is shown in (b). Sketch current i_C scaled with numerical values.
44. The current through a 1-μF capacitor is shown in Figure 10–43. Sketch voltage v_C scaled with numerical values. Voltage at $t = 0$ s is 0 V.
45. If the voltage across a 4.7-μF capacitor is $v_C = 100e^{-0.05t}$ V, what is i_C?
46. Current through a 0.1-μF capacitor is $i_C = t\,e^{-t}$ μA. If the capacitor voltage is 0 V at $t = 0$, determine the equation for v_C. Hint: Integration of Equation 10–20 is required.

10.9 Energy Stored by a Capacitor

47. For Figure 10–42, determine the capacitor's energy at each of the following times: $t = 0$, 1 ms, 4 ms, 5 ms, 7 ms, and 9 ms.
48. For the circuit of Figure 10–38, determine the energy stored in each capacitor.

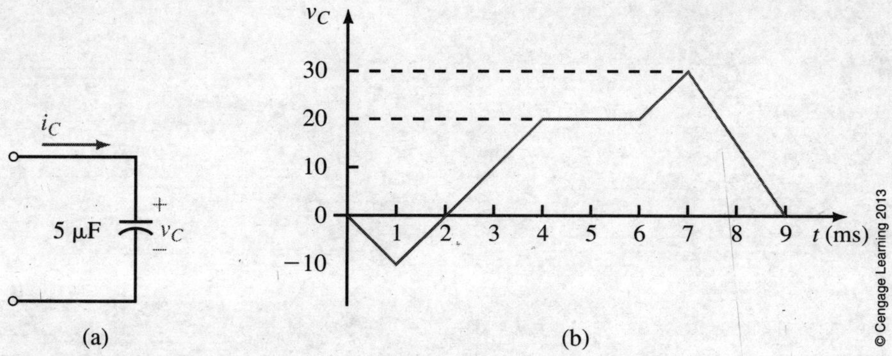

FIGURE 10–42

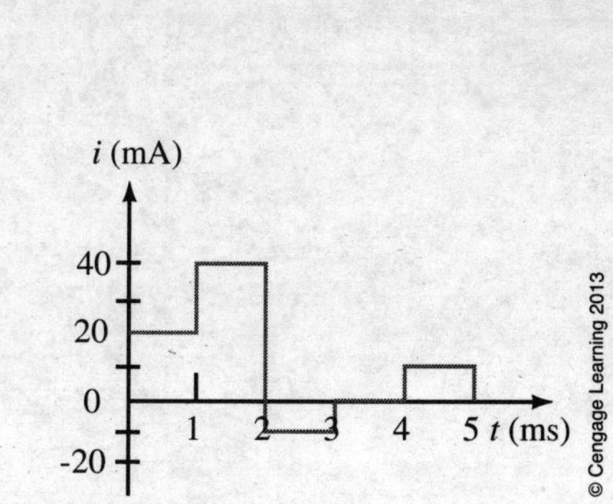

FIGURE 10–43

10.10 Capacitor Failures and Troubleshooting

49. The capacitor tester in Figure 10–44 measures and displays capacitance. For each case, determine the likely fault.

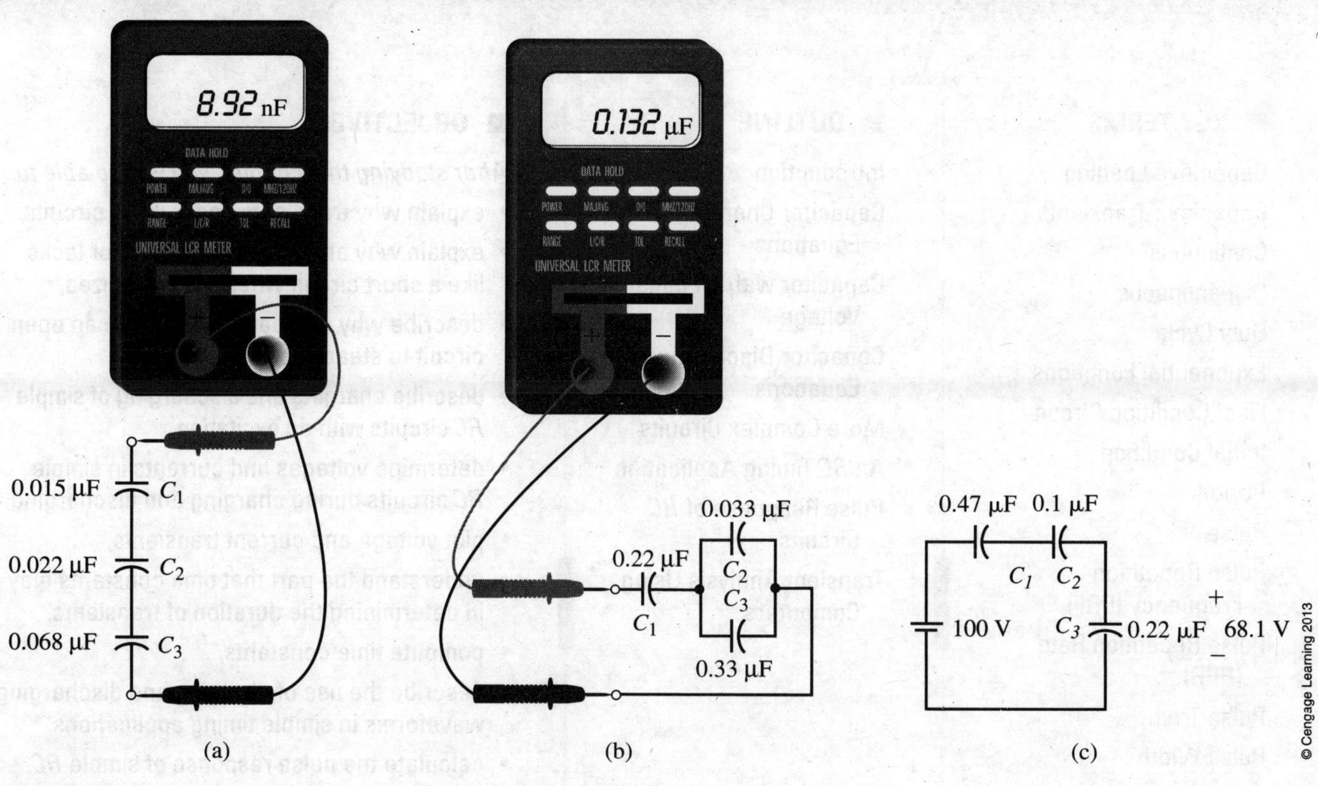

FIGURE 10–44

ANSWERS TO IN-PROCESS LEARNING CHECKS

IN-PROCESS LEARNING CHECK 1

1. a. 2.12 nF
   b. 200 V
2. It becomes six times larger.
3. 1.1 nF
4. Mica

KEY TERMS

Capacitive Loading
Capacitive Transients
Continuous
Discontinuous
Duty Cycle
Exponential Functions
Final Condition Circuit
Initial Condition
Period
Pulse
Pulse Repetition Frequency (PRF)
Pulse Repetition Rate (PRR)
Pulse Train
Pulse Width
Rise and Fall Times
Square Wave
Steady State
Stray Capacitance
Time Constant
Transient
Transient Duration
Universal Time Constant Curve

OUTLINE

Introduction
Capacitor Charging Equations
Capacitor with an Initial Voltage
Capacitor Discharging Equations
More Complex Circuits
An *RC* Timing Application
Pulse Response of *RC* Circuits
Transient Analysis Using Computers

OBJECTIVES

After studying this chapter, you will be able to

- explain why transients occur in *RC* circuits,
- explain why an uncharged capacitor looks like a short circuit when first energized,
- describe why a capacitor looks like an open circuit to steady state dc,
- describe charging and discharging of simple *RC* circuits with dc excitation,
- determine voltages and currents in simple *RC* circuits during charging and discharging,
- plot voltage and current transients,
- understand the part that time constants play in determining the duration of transients,
- compute time constants,
- describe the use of charging and discharging waveforms in simple timing applications,
- calculate the pulse response of simple *RC* circuits,
- solve simple *RC* transient problems using PSpice and Multisim.

11

CAPACITOR CHARGING, DISCHARGING, AND SIMPLE WAVESHAPING CIRCUITS

CHAPTER PREVIEW

As you saw in Chapter 10, Figure 10–26, a capacitor does not charge instantaneously; instead, voltages and currents take time to reach their new values. As we show in this chapter, this time depends on the capacitance of the circuit and the resistance through which it charges—the larger the resistance and capacitance, the longer it takes. (Similar comments hold for discharge.) Since the voltages and currents that exist during charging and discharging are transitory in nature, they are called **transients.** Transients do not last very long, typically only a fraction of a second. However, they are important to us for a number of reasons, some of which you will learn in this chapter.

Transients occur in both capacitive and inductive circuits. In capacitive circuits, they occur because capacitor voltage cannot change instantaneously; in inductive circuits, they occur because inductor current cannot change instantaneously. In this chapter, we look at **capacitive transients;** in Chapter 14, we look at inductive transients. As you will see, many of the basic principles are the same. ■

Putting It in Perspective

Desirable and Undesirable Transients

TRANSIENTS OCCUR IN CAPACITIVE AND inductive circuits whenever circuit conditions are changed, for example, by the sudden application of a voltage, the switching in or out of a circuit element, or the malfunctioning of a circuit component. Some transients are desirable and useful; others occur under abnormal conditions and are potentially destructive in nature.

An example of the latter is the transient that results when lightning strikes a power line. Following a strike, the line voltage, which may have been only a few thousand volts before the strike, momentarily rises to many hundreds of thousands of volts or higher, then rapidly decays, while the current, which may have been only a few hundred amps, suddenly rises to many times its normal value. Although these transients do not last very long, they can cause serious damage. While this is a rather severe example of a transient, it nonetheless illustrates that during transient conditions, many of a circuit's or system's most difficult problems may arise.

Some transient effects, on the other hand, are useful. For example, many electronic devices and circuits (such as oscillators and timers) utilize transient effects due to capacitor charging and discharging as the basis for their operation. You will see an example of such a usage in Section 11.6 of this chapter. ■

11.1 Introduction

Capacitor Charging

Capacitor charging and discharging may be studied using the simple circuit of Figure 11–1. We will begin with charging. First, assume the capacitor is uncharged and that the switch is open—see Note 1. Now move the switch to the charge position, Figure 11–2(a). At the instant the switch is closed the current jumps to E/R amps, then decays to zero, while the voltage, which is zero at the instant the switch is closed, gradually climbs to E volts. This is shown in (b) and (c). The shapes of these curves are easily explained.

First, consider voltage. In order to change capacitor voltage, electrons must be moved from one plate to the other. Even for a relatively small capacitor, billions of electrons must be moved. This takes time. Consequently, *capacitor voltage cannot*

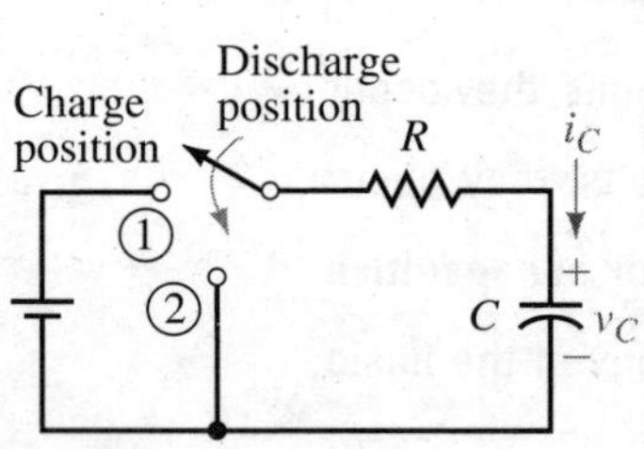

FIGURE 11–1 Circuit for studying capacitor charging and discharging. Transient voltages and currents result when the circuit is switched.

NOTES...

1. In textbooks, it is traditional to use switches to initiate transients. Although transients in real life result from a variety of causes, the principles that govern their behavior are the same no matter how they are created. In a learning environment, the simplest way to create a transient is by means of a switch—so that is what we do here.
2. For those with a calculus background, recall,

$$i_C = C\frac{dv_C}{dt}$$

If voltage is not continuous, the derivative yields an infinite current spike. Since infinite current spikes are not possible, we conclude that voltage cannot be discontinuous.

change instantaneously, that is, it cannot jump abruptly from one value to another. Instead, it climbs gradually and smoothly as illustrated in Figure 11–2(b). Stated another way, capacitor voltage must be **continuous** at all times—see Note 2.

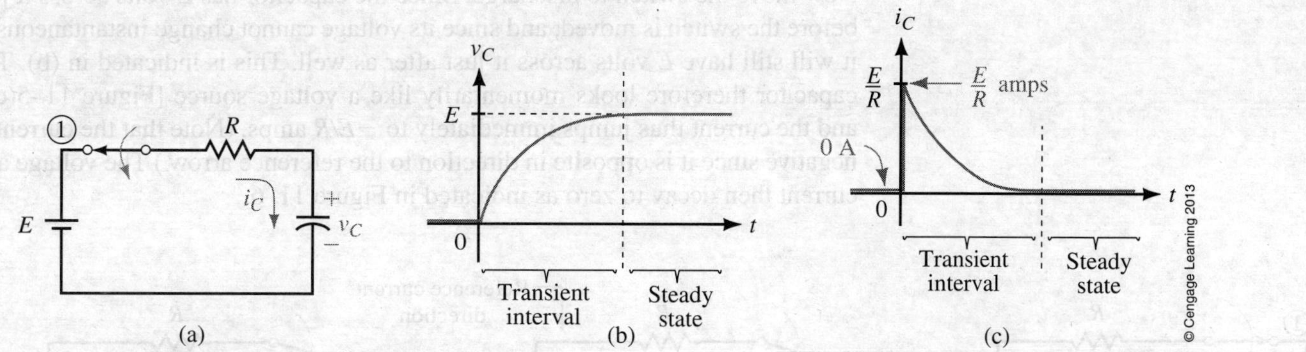

FIGURE 11–2 Capacitor voltage and current during charging. Time $t = 0$ s is defined as the instant the switch is moved to the charge position. The capacitor is initially uncharged.

Now consider current. The movement of electrons noted previously is a current. As indicated in Figure 11–2(c), this current jumps abruptly from 0 to E/R amps, that is, the current is **discontinuous.** To understand why, consider Figure 11–3(a). Since capacitor voltage cannot change instantaneously, its value just after the switch is closed will be the same as it was just before the switch is closed, namely 0 V. Since the voltage across the capacitor just after the switch is closed is zero (even though there is current through it), *the capacitor looks momentarily like a short circuit.* This is indicated in (b). This is an important observation and is true in general, that is, *an uncharged capacitor looks like a short circuit at the instant of switching.* Applying Ohm's law yields $i_C = E/R$ amps. This agrees with what we indicated in Figure 11–2(c).

CircuitSim 11-1

Finally, note the trailing end of the current curve Figure 11–2(c). Since the dielectric between the capacitor plates is an insulator, no current can pass through it. This means that the current in the circuit, which is due entirely to the movement of electrons from one plate to the other through the battery, must decay to zero as the capacitor charges.

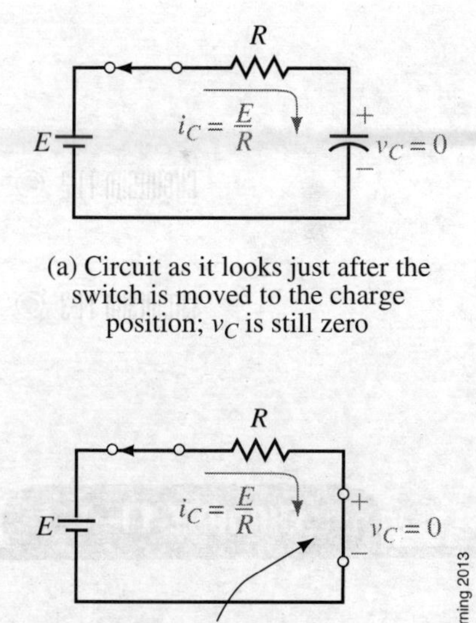

(a) Circuit as it looks just after the switch is moved to the charge position; v_C is still zero

(b) Since $v_C = 0$, $i_C = E/R$

FIGURE 11–3 An uncharged capacitor looks like a short circuit at the instant of switching.

Steady State Conditions

When the capacitor voltage and current reach their final values and stop changing [Figure 11–2(b) and (c)], the circuit is said to be in **steady state.** Figure 11–4(a) shows the circuit after it has reached steady state. Note that $v_C = E$ and $i_C = 0$. Since the capacitor has voltage across it but no current through it, it looks like an open circuit as indicated in (b). This is also an important observation and one that is true in general, that is, *a capacitor looks like an open circuit while in steady state dc.* [The circuit of Figure 11–4(b) is referred to as a **final condition circuit**.]

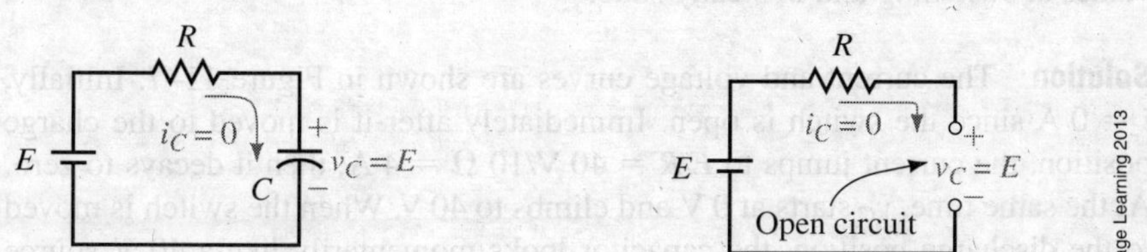

(a) $v_C = E$ and $i_C = 0$

(b) Equivalent circuit for the capacitor

FIGURE 11–4 Charging circuit after it has reached steady state. Since the capacitor has voltage across it but no current, it looks like an open circuit in steady state dc.

Capacitor Discharging

Now consider the discharge case, Figures 11–5 and 11–6. First, assume the capacitor is charged to E volts prior to switching as indicated in Figure 11–5(a). Now move the switch to discharge. Since the capacitor has E volts across it just before the switch is moved, and since its voltage cannot change instantaneously, it will still have E volts across it just after as well. This is indicated in (b). The capacitor therefore looks momentarily like a voltage source [Figure 11–5(c)], and the current thus jumps immediately to $-E/R$ amps. (Note that the current is negative since it is opposite in direction to the reference arrow.) The voltage and current then decay to zero as indicated in Figure 11–6.

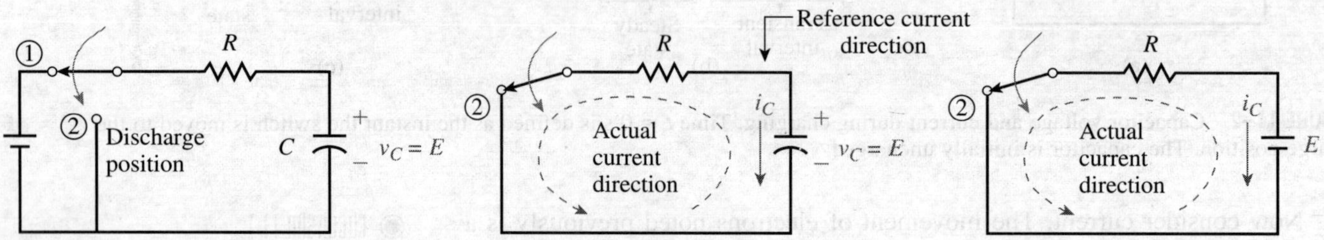

(a) Voltage $v_C = E$ just before the switch is moved to discharge position.

(b) Immediately after the switch is moved, v_C still equals E.

(c) The capacitor therefore momentarily looks like a voltage source. Ohm's law yields $i_C = -E/R$.

FIGURE 11–5 According to the voltage/current reference convention that you learned in Chapter 4, the current reference arrow must be drawn such that the + sign for v_C is at its tail end. Since the polarity of v_C is already known from (a), this establishes the reference direction for i_C as in (b). Since the actual current is opposite in direction to this reference, i_C is negative as indicated in (c).

CircuitSim 11-2

CircuitSim 11-3

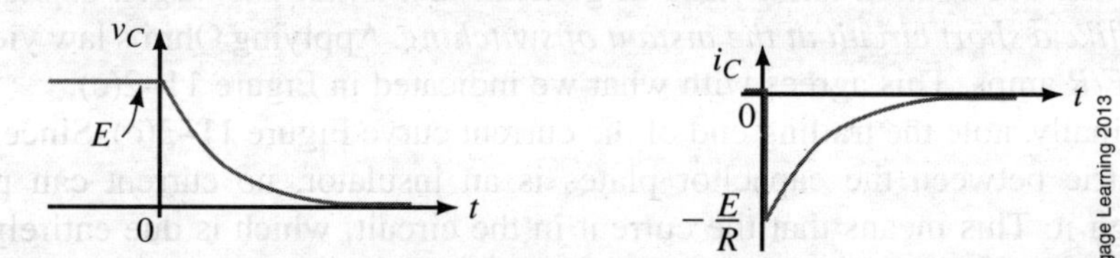

(a) (b)

FIGURE 11–6 Voltage and current during discharge. Time $t = 0$ s is defined as the instant the switch is moved to the discharge position.

EXAMPLE 11–1

For Figure 11–1, $E = 40$ V, $R = 10\ \Omega$, and the capacitor is initially uncharged. The switch is moved to the charge position and the capacitor allowed to charge fully. Then the switch is moved to the discharge position and the capacitor allowed to discharge fully. Sketch the voltages and currents and determine the values at switching and in steady state.

Solution The current and voltage curves are shown in Figure 11–7. Initially, $i = 0$ A since the switch is open. Immediately after it is moved to the charge position, the current jumps to $E/R = 40\text{ V}/10\ \Omega = 4$ A; then it decays to zero. At the same time, v_C starts at 0 V and climbs to 40 V. When the switch is moved to the discharge position, the capacitor looks momentarily like a 40-V source and the current jumps to a negative value of $-40\text{ V}/10\ \Omega = -4$ A; then it decays to zero. At the same time, v_C also decays to zero.

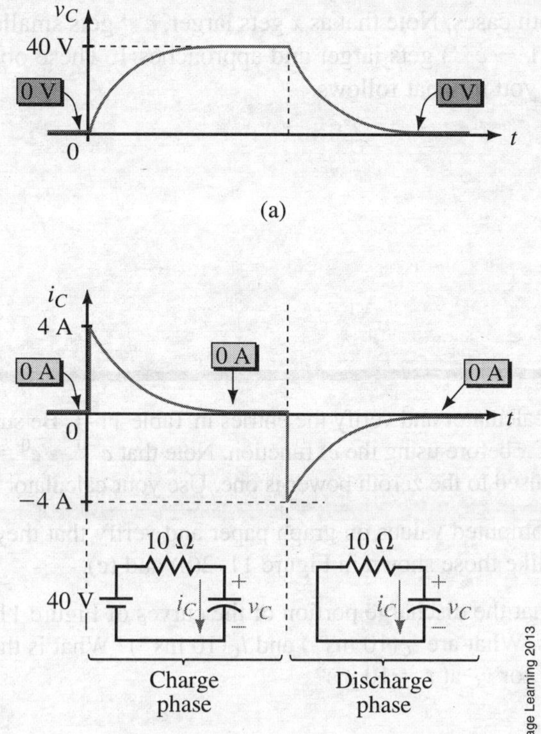

(b) The insets show what the circuit looks like during each time interval.

FIGURE 11–7 A charge/discharge example.

CircuitSim 11-4

CircuitSim 11-5

The Meaning of Time in Transient Analysis

Time t used in transient analysis is measured from the instant of switching. Thus, $t = 0$ in Figure 11–2 is defined as the instant the switch is moved to charge, while in Figure 11–6, it is defined as the instant the switch is moved to discharge. Following switching, voltages and currents may be represented by functions of time as $v_C(t)$ and $i_C(t)$. For example, the voltage across a capacitor at $t = 0$ s may be denoted as $v_C(0)$, while the voltage at $t = 10$ ms may be denoted as $v_C(10 \text{ ms})$, and so on. If you know numeric values, they can be included. Thus, if a capacitor has a voltage of 9 V at $t = 5$ ms and a current of 3 mA, these can be represented as $v_C(5 \text{ ms}) = 9$ V and $i_C(5 \text{ ms}) = 3$ mA, respectively.

A problem seems to arise when a quantity is discontinuous as is the current of Figure 11–2(c). Since its value is changing at $t = 0$ s, $i_C(0)$ cannot be defined. To get around this problem, we define two values for 0 s. We define $t = 0^-$ s as $t = 0$ s just prior to switching and $t = 0^+$ s as $t = 0$ s just after switching. In Figure 11–2(c), therefore, $i_C(0^-) = 0$ A while $i_C(0^+) = E/R$ amps. For Figure 11–6, $i_C(0^-) = 0$ A and $i_C(0^+) = -E/R$ amps. Note that capacitor voltage is continuous—thus, $v_C(0^+) = v_C(0^-)$—see Note.

NOTES...

That $v_C(0^+)$ equals $v_C(0^-)$ is actually a statement of continuity for capacitor voltage. If switching occurs at some time other than $t = 0$ s, a similar statement holds. For example, if switching occurs at $t = 20$ s, and voltage has a value of 40 volts at that instant, then the statement of continuity becomes $v_C(20^+) = v_C(20^-) = 40$ V.

Exponential Functions

As we will soon show, the waveforms of Figures 11–2 and 11–6 are exponential and vary according to e^{-x} or $(1 - e^{-x})$, where e is the base of the natural logarithm. Fortunately, **exponential functions** are easy to evaluate with

TABLE 11–1 Table of Exponentials

x	e^{-x}	$1 - e^{-x}$
0	1	0
1	0.3679	0.6321
2	0.1353	0.8647
3	0.0498	0.9502
4	0.0183	0.9817
5	0.0067	0.9933

modern calculators using their e^x function. You need to be able to evaluate both e^{-x} and $(1 - e^{-x})$ for any value of x. Table 11–1 shows a tabulation of values for both cases. Note that as x gets larger, e^{-x} gets smaller and approaches zero, while $(1 - e^{-x})$ gets larger and approaches 1. These observations will be important to you in what follows.

PRACTICE PROBLEMS 1

1. Use your calculator and verify the entries in Table 11–1. Be sure to change the sign of x before using the e^x function. Note that $e^{-0} = e^0 = 1$ since any quantity raised to the zeroth power is one. Use your calculator to confirm this.
2. Plot the computed values on graph paper and verify that they yield curves that look like those shown in Figure 11–2(b) and (c).
3. Assume that the discharge portion of the curves of Figure 11–7 begin at $t = 10$ ms. What are $i_C(10\text{ ms}^-)$ and $i_C(10\text{ ms}^+)$? What is the statement of continuity for v_C at $t = 10$ ms?

Answers

3. $i_C(10\text{ ms}^-) = 0$ A; $i_C(10\text{ ms}^+) = -4$ A; $v_C(10\text{ ms}^+) = v_C(10\text{ ms}^-) = 40$ V

11.2 Capacitor Charging Equations

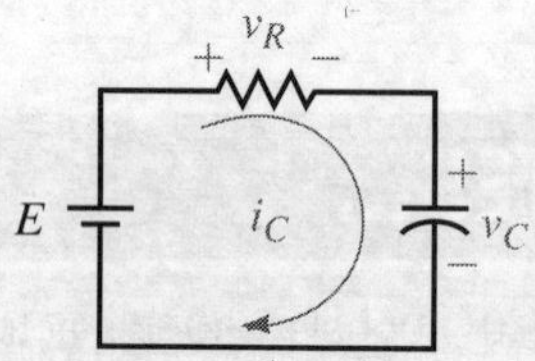

FIGURE 11–8 Circuit for the charging case. Capacitor is initially uncharged.

We will now develop equations for voltages and current during charging. Consider Figure 11–8. Kirchhoff's voltage law (KVL) yields

$$v_R + v_C = E \tag{11–1}$$

But $v_R = Ri_C$ and $i_C = Cdv_C/dt$ (Equation 10–20). Thus, $v_R = RCdv_C/dt$. Substituting this into Equation 11–1 yields

$$RC\frac{dv_C}{dt} + v_C = E \tag{11–2}$$

Equation 11–2 can be solved for v_C using basic calculus (see ∫). The result is

$$v_C = E(1 - e^{-t/RC}) \tag{11–3}$$

where R is in ohms, C is in farads, t is in seconds, and $e^{-t/RC}$ is the exponential function discussed earlier. The product RC has units of seconds. (This is left as an exercise for the student to show.)

Solving Equation 11–2 (Optional Derivation—See Notes)

First, rearrange Equation 11–2:

$$\frac{dv_C}{dt} = \frac{1}{RC}(E - v_C)$$

Rearrange again:

$$\frac{dv_C}{E - v_C} = \frac{dt}{RC}$$

Now multiply both sides by −1 and integrate.

$$\int_0^{v_C} \frac{dv_C}{v_C - E} = -\frac{1}{RC}\int_0^t dt$$

$$\Big[\ln(v_C - E)\Big]_0^{v_C} = \left[-\frac{t}{RC}\right]_0^t$$

Next, substitute integration limits,

$$\ln(v_C - E) - \ln(-E) = -\frac{t}{RC}$$

$$\ln\left(\frac{v_C - E}{-E}\right) = -\frac{t}{RC}$$

Finally, take the inverse log of both sides. Thus,

$$\frac{v_C - E}{-E} = e^{-t/RC}$$

When you rearrange this, you get Equation 11–3. That is,

$$v_C = E(1 - e^{-t/RC})$$

Now consider the resistor voltage. From Equation 11–1, $v_R = E - v_C$. Substituting v_C from Equation 11–3 yields $v_R = E - E(1 - e^{-t/RC}) = E - E + Ee^{-t/RC}$. After cancellation, you get

$$v_R = Ee^{-t/RC} \qquad (11\text{–}4)$$

Now divide both sides by R. Since $i_C = i_R = v_R/R$, this yields

$$i_C = \frac{E}{R}e^{-t/RC} \qquad (11\text{–}5)$$

The waveforms are shown in Figure 11–9. Values at any time may be determined by substitution.

NOTES...

Optional problems and derivations using calculus are marked by a ∫ icon. These may be omitted without loss of continuity by those who do not require calculus.

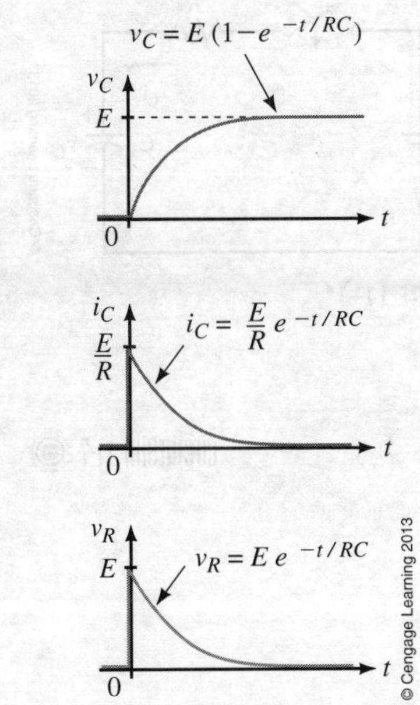

FIGURE 11–9 Curves for the circuit of Figure 11–8.

EXAMPLE 11–2

Suppose $E = 100$ V, $R = 10$ kΩ, and $C = 10$ μF:

a. Determine the expression for v_C.

b. Determine the expression for i_C.

c. Compute the capacitor voltage at $t = 150$ ms.

d. Compute the capacitor current at $t = 150$ ms.

e. Locate the computed points on the curves.

Solution

a. $RC = (10 \times 10^3\ \Omega)(10 \times 10^{-6}\ \text{F}) = 0.1$ s. From Equation 11–3, $v_C = E(1 - e^{-t/RC}) = 100(1 - e^{-t/0.1}) = 100(1 - e^{-10t})$ V.

b. From Equation 11–5, $i_C = (E/R)e^{-t/RC} = (100\ \text{V}/10\ \text{k}\Omega)e^{-10t} = 10e^{-10t}$ mA.

c. At $t = 0.15$ s, $v_C = 100(1 - e^{-10t}) = 100(1 - e^{-10(0.15)}) = 100(1 - e^{-1.5}) = 100(1 - 0.223) = 77.7$ V.

d. $i_C = 10e^{-10t}$ mA $= 10e^{-10(0.15)}$ mA $= 10e^{-1.5}$ mA $= 2.23$ mA.

e. The corresponding points are shown in Figure 11–10.

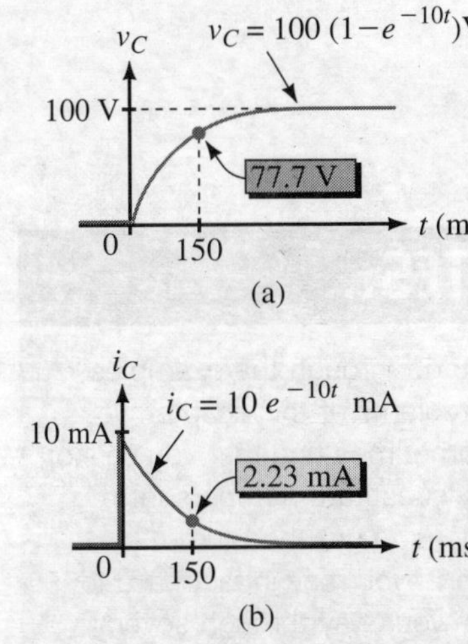

FIGURE 11–10 The computed points plotted on the v_C and i_C curves.

CircuitSim 11-6

In the preceding example, we expressed voltage as $v_C = 100(1 - e^{-t/0.1})$ and as $100(1 - e^{-10t})$ V. Similarly, current can be expressed as $i_C = 10e^{-t/0.1}$ or as $10e^{-10t}$ mA. Although some people prefer one notation over the other, both are correct and you can use them interchangeably.

PRACTICE PROBLEMS 2

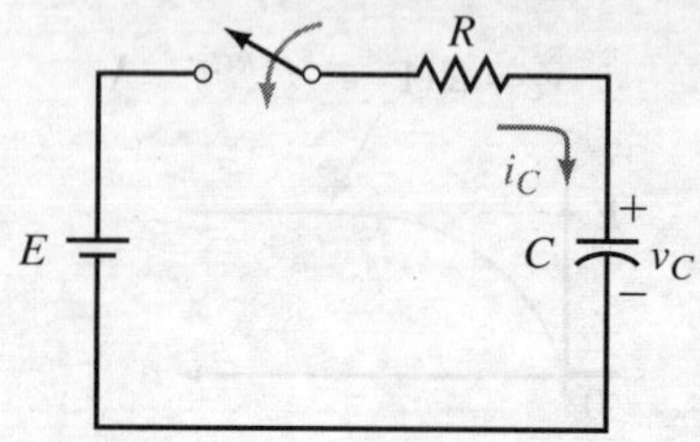

FIGURE 11–11

1. Determine additional voltage and current points for Figure 11–10 by computing values of v_C and i_C at values of time from $t = 0$ s to $t = 500$ ms at 100-ms intervals. Plot the results.
2. The switch of Figure 11–11 is closed at $t = 0$ s. If $E = 80$ V, $R = 4$ kΩ, and $C = 5$ μF, determine expressions for v_C and i_C. Plot the results from $t = 0$ s to $t = 100$ ms at 20-ms intervals. Note that charging takes less time here than for Problem 1.

Answers

CircuitSim 11-7

1.

t(ms)	v_C(V)	i_C(mA)
0	0	10
100	63.2	3.68
200	86.5	1.35
300	95.0	0.498
400	98.2	0.183
500	99.3	0.067

2. $80(1 - e^{-50t})$V $20e^{-50t}$ mA

t(ms)	v_C(V)	i_c(mA)
0	0	20
20	50.6	7.36
40	69.2	2.70
60	76.0	0.996
80	78.6	0.366
100	79.4	0.135

EXAMPLE 11–3

For the circuit of Figure 11–11, $E = 60$ V, $R = 2$ kΩ, and $C = 25$ μF. The switch is closed at $t = 0$ s, opened 40 ms later and left open. Determine equations for capacitor voltage and current and plot.

Solution $RC = (2\ \text{k}\Omega)(25\ \mu\text{F}) = 50$ ms. As long as the switch is closed (i.e., from $t = 0$ s to 40 ms), the following equations hold:

$$v_C = E\,(1 - e^{-t/RC}) = 60(1 - e^{-t/50\text{ ms}})\ \text{V}$$

$$i_C = (E/R)e^{-t/RC} = 30e^{-t/50\text{ ms}}\ \text{mA}$$

Voltage starts at 0 V and rises exponentially. At $t = 40$ ms, the switch is opened, interrupting charging. At this instant, $v_C = 60(1 - e^{-(40/50)}) = 60(1 - e^{-0.8}) =$ 33.0 V. Since the switch is left open, the voltage remains constant at 33 V thereafter as indicated in Figure 11–12. (The dotted curve shows how the voltage would have kept rising if the switch had remained closed.)

Now consider current. The current starts at 30 mA and decays to $i_C = 30e^{-(40/50)}$ mA = 13.5 mA at $t = 40$ ms. At this point, the switch is opened, and the current drops instantly to zero. (The dotted line shows how the current would have decayed if the switch had not been opened.)

NOTES...

As you go through the examples and problems of this chapter, you will sometimes find that you cannot simply substitute numbers into equations to get the correct solution—you may in fact have to reason your way through several steps. This may be seen in Example 11–3, where early switching prematurely stops charging.

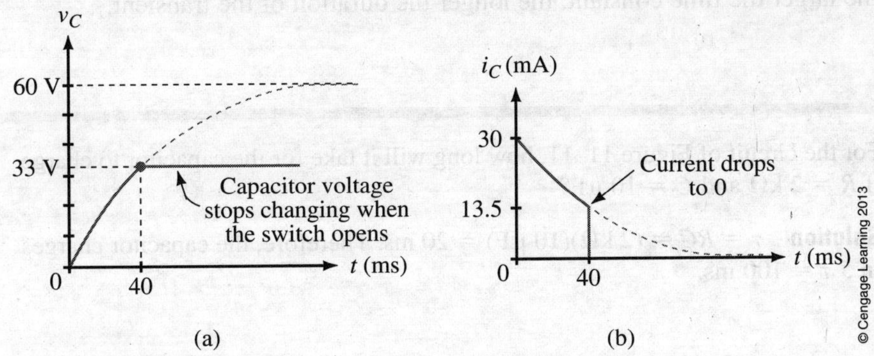

FIGURE 11–12 Interrupted charging. The switch of Figure 11–11 was opened at $t = 40$ ms, causing charging to cease.

CircuitSim 11-8

The Time Constant

The rate at which a capacitor charges depends on the product of R and C. This product is known as the **time constant** of the circuit and is given the symbol τ (the Greek letter tau). As noted earlier, RC has units of seconds. Thus,

$$\tau = RC \text{ (seconds, s)} \tag{11–6}$$

Using τ, Equations 11–3 to 11–5 can be written as

$$v_C = E(1 - e^{-t/\tau}) \tag{11–7}$$

$$i_C = \frac{E}{R}e^{-t/\tau} \tag{11–8}$$

and

$$v_R = Ee^{-t/\tau} \tag{11–9}$$

Duration of a Transient

The length of time that a transient lasts (i.e., the **transient duration**) depends on the exponential function $e^{-t/\tau}$. As t increases, $e^{-t/\tau}$ decreases, and when it reaches zero, the transient is gone. Theoretically, this takes infinite time. In practice, however, over 99% of the transition takes place during the first five time constants (i.e., transients are within 1% of their final value at $t = 5\,\tau$). This can be verified by direct substitution. At $t = 5\,\tau$, $v_C = E(1 - e^{-t/\tau}) = E(1 - e^{-5}) = E(1 - 0.0067) = 0.993E$, meaning that the transient has achieved 99.3% of its final value. Similarly, the current falls to within 1% of its final value in five time constants. Thus, *for all practical purposes, transients can be considered to last for only five time constants* (Figure 11–13). Figure 11–14 summarizes how

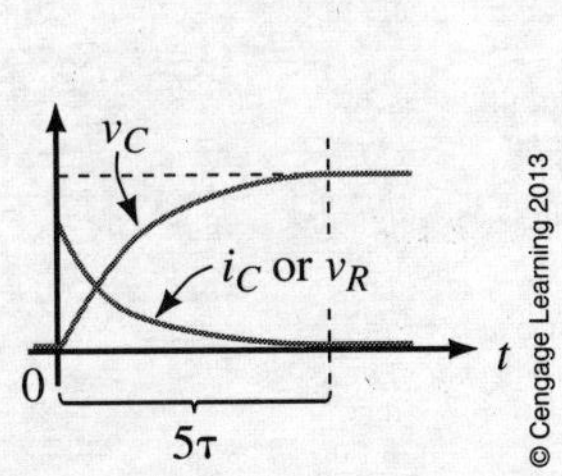

FIGURE 11–13 Transients last five time constants.

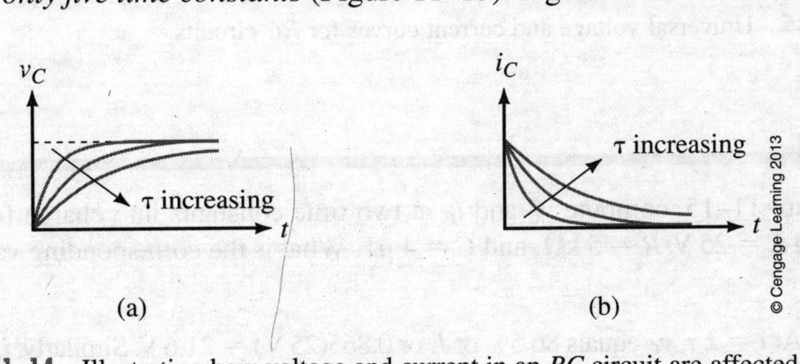

FIGURE 11–14 Illustrating how voltage and current in an RC circuit are affected by its time constant. The larger the time constant, the longer the capacitor takes to charge.

transient voltages and currents are affected by the time constant of a circuit—the larger the time constant, the longer the duration of the transient.

EXAMPLE 11–4

For the circuit of Figure 11–11, how long will it take for the capacitor to charge if $R = 2\ \text{k}\Omega$ and $C = 10\ \mu\text{F}$?

Solution $\tau = RC = (2\ \text{k}\Omega)(10\ \mu\text{F}) = 20$ ms. Therefore, the capacitor charges in $5\ \tau = 100$ ms.

EXAMPLE 11–5

The transient in a circuit with $C = 40\ \mu\text{F}$ lasts 0.5 s. What is R?

Solution $5\ \tau = 0.5$ s. Thus, $\tau = 0.1$ s and $R = \tau/C = 0.1\ \text{s}/(40 \times 10^{-6}\ \text{F}) = 2.5\ \text{k}\Omega$.

Universal Time Constant Curves

Let us now plot capacitor voltage and current with their time axes scaled as multiples of τ and their vertical axes scaled in percent. The results, Figure 11–15, are **universal time constant curves.**

Points are computed from $v_C = 100(1 - e^{-t/\tau})$ and $i_C = 100e^{-t/\tau}$. For example, at $t = \tau$, $v_C = 100(1 - e^{-t/\tau}) = 100(1 - e^{-\tau/\tau}) = 100(1 - e^{-1}) = 63.2$ V, that is, 63.2%, and $i_C = 100e^{-\tau/\tau} = 100e^{-1} = 36.8$ A, which is 36.8%, and so on. These curves provide an easy method to determine voltages and currents with a minimum of computation.

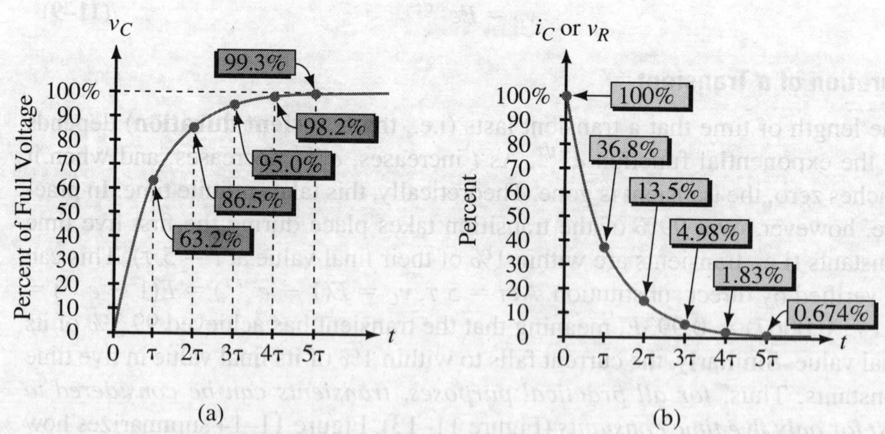

FIGURE 11–15 Universal voltage and current curves for RC circuits.

EXAMPLE 11–6

Using Figure 11–15, compute v_C and i_C at two time constants into charge for a circuit with $E = 25$ V, $R = 5\ \text{k}\Omega$, and $C = 4\ \mu\text{F}$. What is the corresponding value of time?

Solution At $t = 2\ \tau$, v_C equals 86.5% of E or $0.865(25\ \text{V}) = 21.6$ V. Similarly, $i_C = 0.135I_0 = 0.135\ (E/R) = 0.675$ mA. These values occur at $t = 2\ \tau = 2RC = 40$ ms.

IN-PROCESS LEARNING CHECK 1

(Answers are at the end of the chapter.)

1. If the capacitor of Figure 11–16 is uncharged, what is the current immediately after closing the switch?
2. Given $i_C = 50e^{-20t}$ mA.
   a. What is τ?
   b. Compute the current at $t = 0^+$ s, 25 ms, 50 ms, 75 ms, 100 ms, and 250 ms and sketch it. Verify answers using the universal time constant curves.
3. Given $v_C = 100(1 - e^{-50t})$ V, compute v_C at the same time points as in Problem 2 and sketch.
4. For Figure 11–16, determine expressions for v_C and i_C. Compute capacitor voltage and current at $t = 0.6$ s. Verify answers using the universal time constant curves.
5. Refer to Figure 11–10:
   a. What are $v_C(0^-)$ and $v_C(0^+)$?
   b. What are $i_C(0^-)$ and $i_C(0^+)$?
   c. What are the steady state voltage and current?
6. For the circuit of Figure 11–11, the current just after the switch is closed is 2 mA. The transient lasts 40 ms and the capacitor charges to 80 V. Determine E, R, and C.

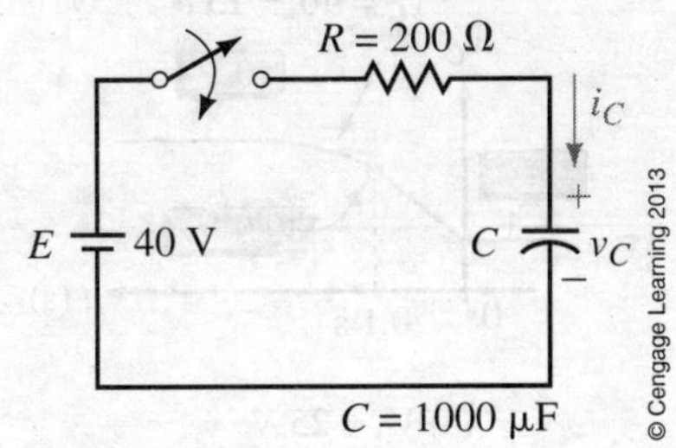

FIGURE 11–16

11.3 Capacitor with an Initial Voltage

Suppose a previously charged capacitor has not been discharged and thus still has voltage on it. Let this voltage be denoted as V_0. (V_0 is referred to as an **initial condition.**) If the capacitor is now placed in a circuit like that in Figure 11–16, the voltage and current during charging will be affected by the initial voltage. In this case, Equations 11–7 and 11–8 become

$$v_C = E + (V_0 - E)e^{-t/\tau} \quad \text{(11–10)}$$

$$i_C = \frac{E - V_0}{R}e^{-t/\tau} \quad \text{(11–11)}$$

A few comments are in order about these equations. Consider Equation 11–10. When you set $t = 0$, you get $v_C = E + (V_0 - E) = V_0$. This agrees with our assertion that the capacitor was initially charged to V_0. If you now set $t = \infty$, you get $v_C = E$, which confirms that the capacitor charges to E volts as expected. Consider Equation 11–11. When you set $t = 0$, you get $i_C = (E - V_0)/R$. Recalling that an initially charged capacitor looks like a voltage source [see Figure 11–5(c)], you can see that if you replace C in Figure 11–16 with a source V_0, the current at $t = 0$ will be $(E - V_0)/R$ as noted. Note also that these revert to Equation 11–7 and Equation 11–8 when you set $V_0 = 0$ V.

EXAMPLE 11–7

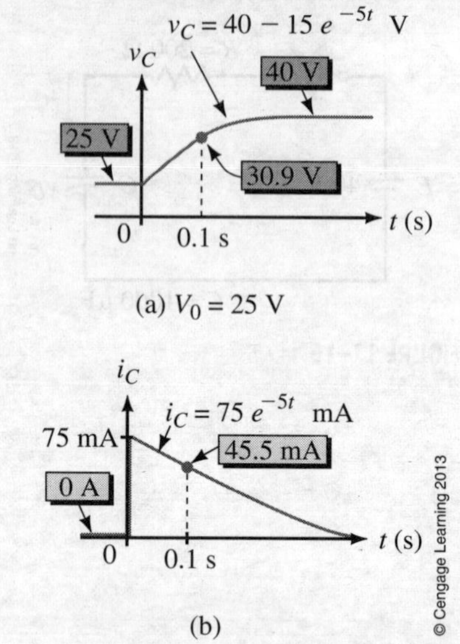

FIGURE 11–17 Capacitor with an initial voltage.

CircuitSim 11-9

Suppose the capacitor of Figure 11–16 has 25 volts on it with polarity shown at the time the switch is closed.

a. Determine the expression for v_C.

b. Determine the expression for i_C.

c. Compute v_C and i_C at $t = 0.1$ s.

d. Sketch v_C and i_C.

Solution $\tau = RC = (200\ \Omega)(1000\ \mu\text{F}) = 0.2$ s

a. From Equation 11–10,

$$v_C = E + (V_0 - E)e^{-t/\tau} = 40 + (25 - 40)e^{-t/0.2} = 40 - 15e^{-5t}\text{ V}$$

b. From Equation 11–11,

$$i_C = \frac{E - V_0}{R}e^{-t/\tau} = \frac{40 - 25}{200}e^{-5t} = 75e^{-5t}\text{ mA}$$

c. At $t = 0.1$ s,

$$v_C = 40 - 15e^{-5t} = 40 - 15e^{-0.5} = 30.9\text{ V}$$
$$i_C = 75e^{-5t}\text{ mA} = 75e^{-0.5}\text{ mA} = 45.5\text{ mA}$$

d. The waveforms are shown in Figure 11–17 with the preceding points plotted.

PRACTICE PROBLEMS 3

1. For Example 11–7, compute voltage and current at $t = 0.25$ s.
2. Repeat Example 11–7 for the circuit of Figure 11–16 if $V_0 = -150$ V.

Answers

1. 35.7 V, 21.5 mA
2. a. $40 - 190e^{-5t}$ V
   b. $0.95e^{-5t}$ A
   c. −75.2 V; 0.576 A
   d. Curves are similar to Figure 11–17 except that v_C starts at −150 V and rises to 40 V while i_C starts at 0.95 A and decays to zero.

11.4 Capacitor Discharging Equations

To determine the discharge equations, move the switch to the discharge position (Figure 11–18). (Note carefully the reference direction for current i_C.) KVL yields $v_R + v_C = 0$. Substituting $v_R = RCdv_C/dt$ from Section 11.2 yields

$$RC\frac{dv_C}{dt} + v_C = 0 \qquad \textbf{(11–12)}$$

This can be solved for v_C using basic calculus. The result is

$$v_C = V_0e^{-t/RC} \qquad \textbf{(11–13)}$$

where V_0 is the voltage on the capacitor at the instant the switch is moved to discharge. Now consider the resistor voltage. Since $v_R + v_C = 0$, $v_R = -v_C$ and

$$v_R = -V_0e^{-t/RC} \quad \textbf{(11–14)}$$

Now divide both sides by R. Since $i_C = i_R = v_R/R$,

$$i_C = -\frac{V_0}{R}e^{-t/RC} \quad \textbf{(11–15)}$$

Note that this is negative, since, during discharge, the actual current is opposite in direction to the reference arrow of Figure 11–18. Voltage v_C and current i_C are shown in Figure 11–19. As in the charging case, *discharge transients last five time constants.* You can also write these equations in terms of τ, for example, $v_C = V_0e^{-t/\tau}$, and so forth.

In Equations 11–13 to 11–15, V_0 represents the voltage on the capacitor at the instant the switch is moved to the discharge position. If the switch has been in the charge position long enough for the capacitor to fully charge, $V_0 = E$, and Equations 11–13 and 11–15 become $v_C = Ee^{-t/RC}$ and $i_C = -(E/R)e^{-t/RC}$, respectively.

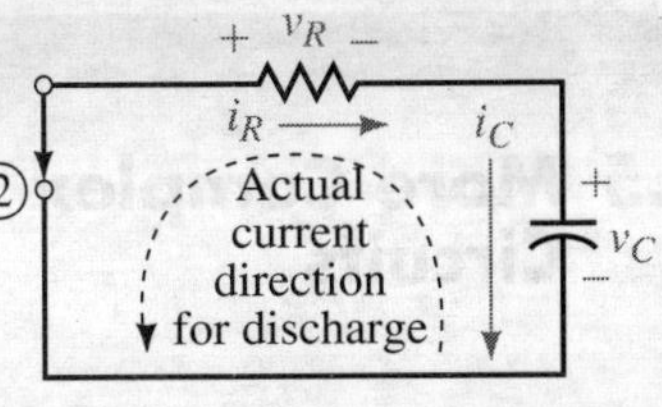

FIGURE 11–18 Discharge case. Initial capacitor voltage is V_0. Note the reference for i_C. (To conform to the standard voltage/current reference convention, i_C must be drawn in this direction so that the + sign for v_C is at the tail of the current arrow.) Since the actual current direction is opposite to the reference direction, i_C will be negative. This is indicated in Figure 11–19(b).

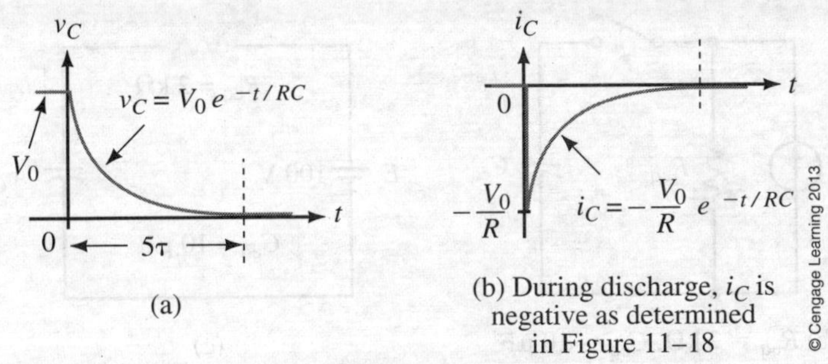

FIGURE 11–19 Capacitor voltage and current for the discharge case.

CircuitSim 11-10

EXAMPLE 11–8

For the circuit of Figure 11–18, assume the capacitor is charged to 100 V before the switch is moved to the discharge position. Suppose $R = 5\ \text{k}\Omega$ and $C = 25\ \mu\text{F}$. After the switch is moved to discharge,

a. Determine the expression for v_C.
b. Determine the expression for i_C.
c. Compute the voltage and current at 0.375 s.

Solution $RC = (5\ \text{k}\Omega)(25\ \mu\text{F}) = 0.125$ s and $V_0 = 100$ V. Therefore,

a. $v_C = V_0e^{-t/RC} = 100e^{-t/0.125} = 100e^{-8t}$ V.
b. $i_C = -(V_0/R)e^{-t/RC} = -20e^{-8t}$ mA.
c. At $t = 0.375$ s,

$$v_C = 100e^{-8t} = 100e^{-3} = 4.98\text{ V}$$

$$i_C = -20e^{-8t}\text{ mA} = -20e^{-3}\text{ mA} = -0.996\text{ mA}$$

Let us verify the answers of Example 11–8 using the appropriate universal time constant curve. As noted, $\tau = 0.125$ s, thus 0.375 s = 3τ. From Figure 11–15(b), we see that capacitor voltage has fallen to 4.98% of E at 3τ. This is (0.0498)(100 V) = 4.98 V as we computed earlier. Current can be verified similarly. I will leave this for you to do.

NOTES...

The universal time constant curve of Figure 11–15(b) may also be used for discharge problems since it has the same shape as the discharge waveforms.

11.5 More Complex Circuits

The charge and discharge equations and the universal time constant curves apply only to circuits of the forms shown in Figures 11–2 and 11–18. Fortunately, many circuits can be reduced to these forms using standard circuit reduction techniques such as series and parallel combinations, source conversions, Thévenin's theorem, and so on. Once a circuit has been reduced to its series equivalent, you can use any of the techniques that we have developed so far.

EXAMPLE 11–9

For the circuit of Figure 11–20(a), determine expressions for v_C and i_C. Capacitors are initially uncharged.

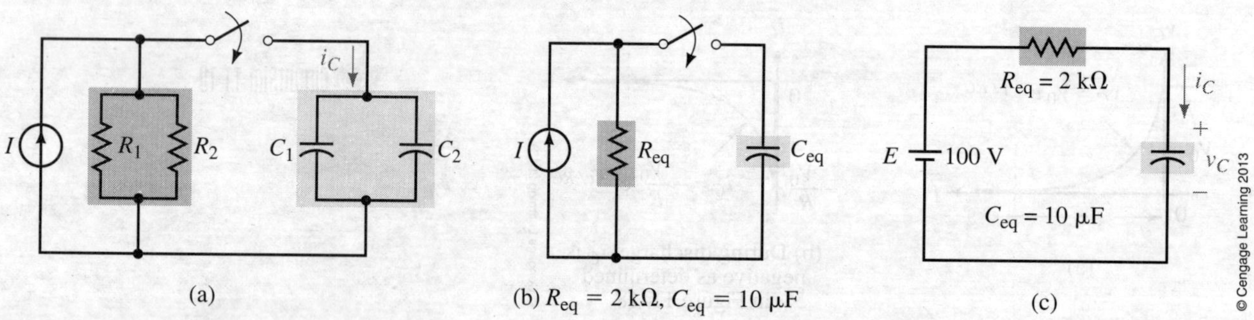

FIGURE 11–20 $I = 50$ mA, $R_1 = 3$ kΩ, $R_2 = 6$ kΩ, $C_1 = 8$ μF, $C_2 = 2$ μF.

Solution $R_{eq} = 3\text{ k}\Omega \| 6\text{ k}\Omega = 2\text{ k}\Omega$, $C_{eq} = 8\ \mu\text{F} \| 2\ \mu\text{F} = 10\ \mu\text{F}$. The reduced circuit is shown in (b). Converting to a voltage source representation yields (c).

$$R_{eq}C_{eq} = (2\text{ k}\Omega)(10 \times 10^{-6}\text{ F}) = 0.020\text{ s}$$

Thus,

$$v_C = E(1 - e^{-t/R_{eq}C_{eq}}) = 100(1 - e^{-t/0.02}) = 100(1 - e^{-50t})\text{ V}$$

$$i_C = \frac{E}{R_{eq}}e^{-t/R_{eq}C_{eq}} = \frac{100}{2000}e^{-t/0.02} = 50\ e^{-50t}\text{ mA}$$

EXAMPLE 11–10

The capacitor of Figure 11–21 is initially uncharged. Close the switch at $t = 0$ s.

a. Determine the expression for v_C.
b. Determine the expression for i_C.
c. Determine capacitor current and voltage at $t = 5$ ms and $t = 10$ ms.

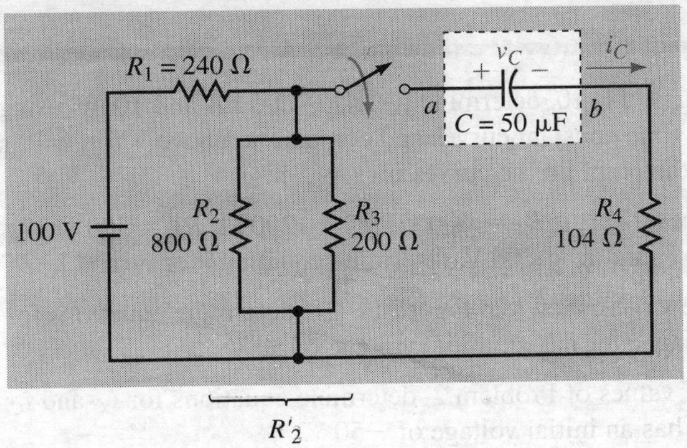

FIGURE 11–21

CircuitSim 11-11

Solution Reduce the circuit to its series equivalent using Thévenin's theorem:

$$R'_2 = R_2 \| R_3 = 160\ \Omega$$

From Figure 11–22(a),

$$R_{Th} = R_1 \| R'_2 + R_4 = 240 \| 160 + 104 = 96 + 104 = 200\ \Omega$$

From Figure 11–22(b),

$$V'_2 = \left(\frac{R'_2}{R_1 + R'_2}\right) E = \left(\frac{160}{240 + 160}\right) \times 100\text{ V} = 40\text{ V}$$

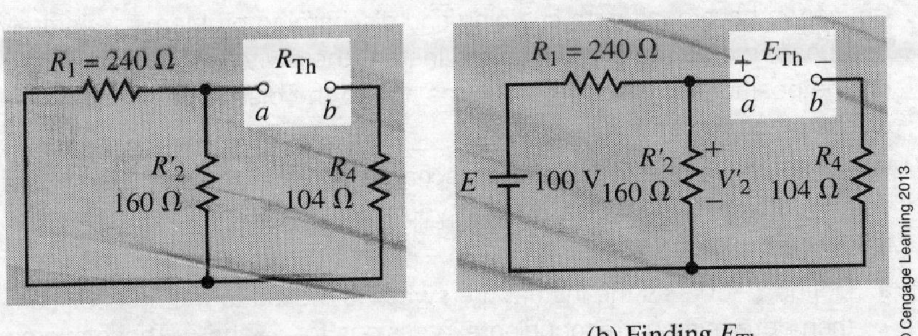

(a) Finding R_{Th} (b) Finding E_{Th}

FIGURE 11–22 Determining the Thévenin equivalent of Figure 11–21 following switch closure.

From KVL, $E_{Th} = V'_2 = 40$ V. The resultant equivalent circuit is shown in Figure 11–23.

$$\tau = R_{Th}C = (200\ \Omega)(50\ \mu\text{F}) = 10\text{ ms}$$

a. $v_C = E_{Th}(1 - e^{-t/\tau}) = 40(1 - e^{-100t})$ V

b. $i_C = \dfrac{E_{Th}}{R_{Th}} e^{-t/\tau} = \dfrac{40}{200} e^{-t/0.01} = 200e^{-100t}$ mA

c. At $t = 5$ ms, $i_C = 200e^{-100(5\text{ ms})} = 121$ mA. Similarly, $v_C = 15.7$ V. Similarly, at 10 ms, $i_C = 73.6$ mA and $v_C = 25.3$ V.

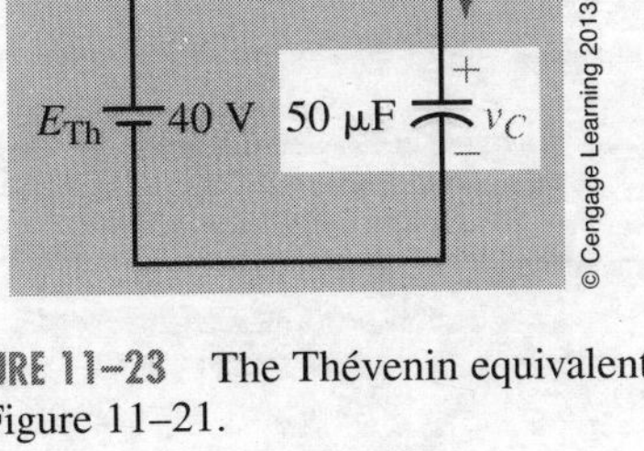

FIGURE 11–23 The Thévenin equivalent of Figure 11–21.

PRACTICE PROBLEMS 4

1. For Example 11–10, determine v_C and i_C at 5 ms and 10 ms using the universal time constant curves and compare to above. (You will have to estimate the point on the curves for $t = 5$ ms.)
2. For Figure 11–21, if $R_1 = 400\ \Omega$, $R_2 = 1200\ \Omega$, $R_3 = 300\ \Omega$, $R_4 = 50\ \Omega$, $C = 20\ \mu\text{F}$, and $E = 200$ V, determine equations for v_C and i_C.
3. Using the values shown in Figure 11–21, determine equations for v_C and i_C if the capacitor has an initial voltage of 60 V.
4. Using the values of Problem 2, determine equations for v_C and i_C if the capacitor has an initial voltage of −50 V.

Answers

1. 1.57 V, 121 mA; 25.3 V, 73.6 mA
2. $75(1 - e^{-250t})$ V; $0.375e^{-250t}$ A
3. $40 + 20e^{-100t}$ V; $-0.1e^{-100t}$ A
4. $75 - 125e^{-250t}$ V; $0.625e^{-250t}$ A

PRACTICAL NOTES...

Notes about Time References and Time Constants

1. So far, we have dealt with charging and discharging problems separately. For these, we define $t = 0$ s as the instant the switch is moved to the charge position for charging problems and to the discharge position for discharging problems.
2. When you have both charge and discharge cases in the same example, you need to establish clearly what you mean by "time." We use the following procedure:
   a. Define $t = 0$ s as the instant the switch is moved to the first position, then determine corresponding expressions for v_C and i_C. These expressions and the corresponding time scale are valid until the switch is moved to its new position.
   b. When the switch is moved to its new position, shift the time reference and make $t = 0$ s the time at which the switch is moved to its new position, determine new expressions for v_C and i_C, then use continuity considerations to link the two solutions together. (These new expressions are only valid from the new $t = 0$ s reference point. The old expressions are not valid on the new time scale.)
   c. We now have two time scales for the same graph. However, we generally only show the first scale explicitly; the second scale is implied rather than shown. Computations based on the new equations must use the new scale.
   d. Use τ_c to represent the time constant for charging and τ_d to represent the time constant for discharging. Since the equivalent resistance and capacitance for discharging may be different than for charging, the time constants may be different for the two cases.

EXAMPLE 11–11

The capacitor of Figure 11–24(a) is uncharged. The switch is moved to position 1 for 10 ms, then to position 2, where it remains.

a. Determine v_C during charge.
b. Determine i_C during charge.
c. Determine v_C during discharge.
d. Determine i_C during discharge.
e. Sketch the charge and discharge waveforms.

NOTES...

When solving a transient problem with multiple switch operations, you should always draw the circuit as it looks during each time interval of interest. (It doesn't take long to do this, and it helps clarify just what it is you need to look at for each part of the solution.) This is illustrated in Example 11–11. Here, we have drawn the circuit in Figure 11–24(b) as it looks during charging and in (c) as it looks during discharging. These diagrams make clear which components are relevant to the charging phase and which are relevant to the discharging phase.

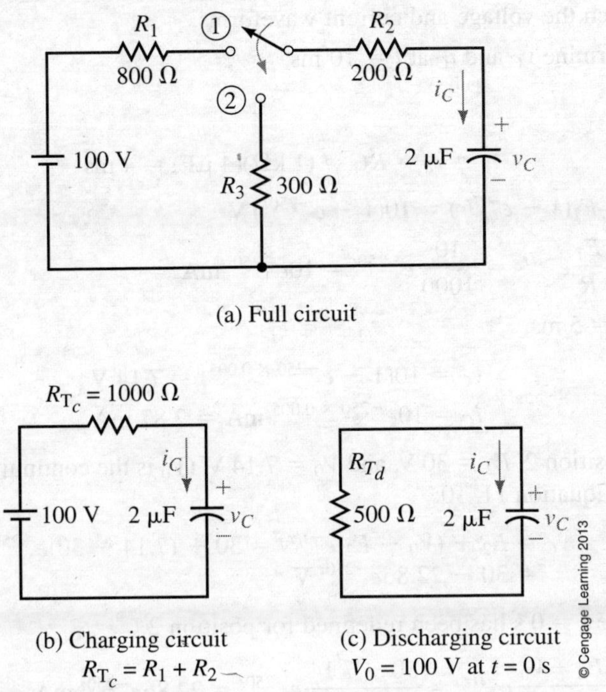

FIGURE 11–24 R_{T_c} is the total resistance of the charge circuit, while R_{T_d} is the total resistance of the discharge circuit.

Solution Figure 11–24(b) shows the equivalent charging circuit. Here,

$$\tau_C = (R_1 + R_2)C = (1\ \text{k}\Omega)(2\ \mu\text{F}) = 2.0\ \text{ms}.$$

a. $v_C = E(1 - e^{-t/\tau_c}) = 100(1 - e^{-500t})$ V

b. $i_C = \dfrac{E}{R_{T_c}}e^{-t/\tau_c} = \dfrac{100}{1000}e^{-500t} = 100e^{-500t}$ mA

Since $5\tau_c = 10$ ms, charging is complete by the time the switch is moved to discharge. Thus, $V_0 = 100$ V when discharging begins. (This is the continuity linkage.)

c. Figure 11–24(c) shows the equivalent discharge circuit.

$$\tau_d = (500\ \Omega)(2\ \mu\text{F}) = 1.0\ \text{ms}$$
$$v_C = V_0e^{-t/\tau_d} = 100e^{-1000t}\ \text{V}$$

where $t = 0$ s has been redefined for discharge as noted earlier.

d. $i_C = -\dfrac{V_0}{R_2 + R_3}e^{-t/\tau_d} = -\dfrac{100}{500}e^{-1000t} = -200e^{-1000t}$ mA

e. See Figure 11–25. Note that discharge is more rapid than charge since $\tau_d < \tau_c$.

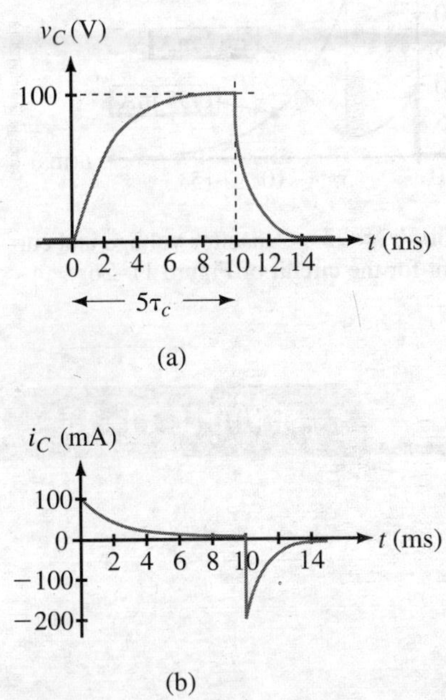

FIGURE 11–25 Waveforms for the circuit of Figure 11–24. Note that only the first time scale is shown explicitly.

CircuitSim 11-12

EXAMPLE 11–12

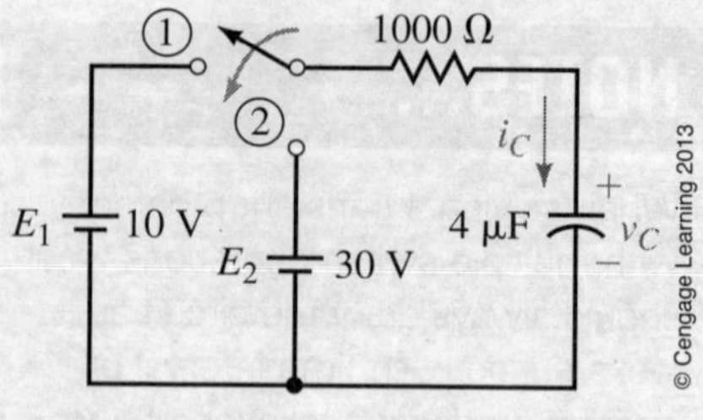

FIGURE 11–26

The capacitor of Figure 11–26 is uncharged. The switch is moved to position 1 for 5 ms, then to position 2 and left there.

a. Determine v_C while the switch is in position 1.
b. Determine i_C while the switch is in position 1.
c. Compute v_C and i_C at $t = 5$ ms.
d. Determine v_C while the switch is in position 2.
e. Determine i_C while the switch is in position 2.
f. Sketch the voltage and current waveforms.
g. Determine v_C and i_C at $t = 10$ ms.

CircuitSim 11-13

Solution

$$\tau_c = \tau_d = RC = (1 \text{ k}\Omega)(4\ \mu\text{F}) = 4 \text{ ms}$$

a. $v_C = E_1(1 - e^{-t/\tau_c}) = 10(1 - e^{-250t})$ V

b. $i_C = \dfrac{E_1}{R}e^{-t/\tau_c} = \dfrac{10}{1000}e^{-250t} = 10e^{-250t}$ mA

c. At $t = 5$ ms,

$$v_C = 10(1 - e^{-250 \times 0.005}) = 7.14 \text{ V}$$
$$i_C = 10e^{-250 \times 0.005} \text{ mA} = 2.87 \text{ mA}$$

d. In position 2, $E_2 = 30$ V, and $V_0 = 7.14$ V. (V_0 is the continuity linkage.) Use Equation 11-10:

$$v_C = E_2 + (V_0 - E_2)e^{-t/\tau_d} = 30 + (7.14 - 30)e^{-250t}$$
$$= 30 - 22.86e^{-250t} \text{ V}$$

where $t = 0$ s has been redefined for position 2.

e. $i_C = \dfrac{E_2 - V_0}{R}e^{-t/\tau_d} = \dfrac{30 - 7.14}{1000}e^{-250t} = 22.86e^{-250t}$ mA

f. See Figure 11–27.

g. $t = 10$ ms is 5 ms into the new time scale. Thus, $v_C = 30 - 22.86e^{-250(5 \text{ ms})} = 23.5$ V and $i_C = 22.86e^{-250(5 \text{ ms})} = 6.55$ mA. Values are plotted on the graph.

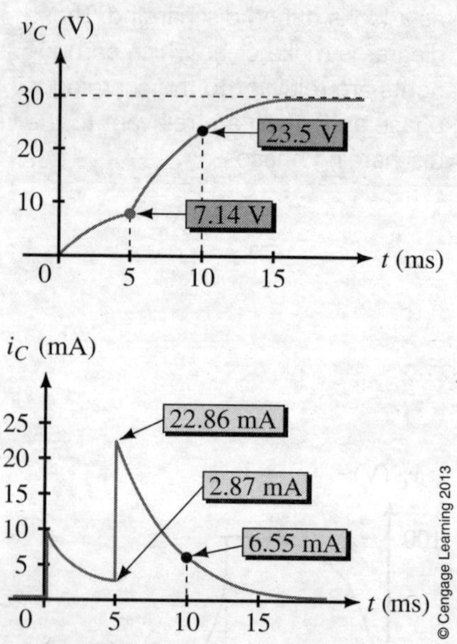

FIGURE 11–27 Capacitor voltage and current for the circuit of Figure 11–26.

EXAMPLE 11–13

In Figure 11–28(a), the capacitor is initially uncharged. The switch is moved to the charge position, then to the discharge position, yielding the current shown in (b). The capacitor takes 1.75 ms to discharge. Determine the following:

a. E b. R_1 c. C

Solution

a. Since the capacitor charges fully, it has a value of E volts when switched to discharge. The discharge current spike is therefore

$$-\frac{E}{10\ \Omega + 25\ \Omega} = -3 \text{ A}$$

Thus, $E = 105$ V.

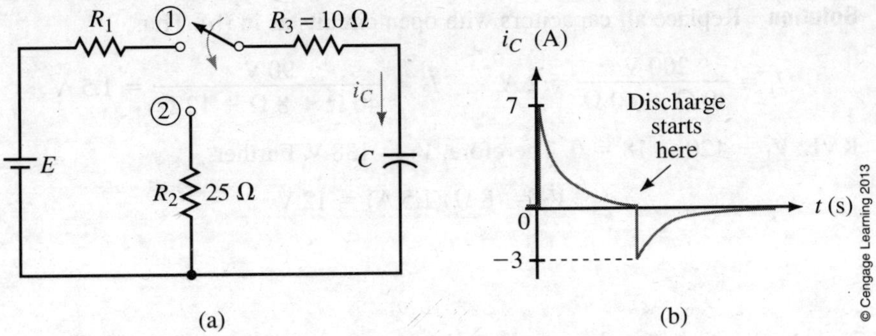

FIGURE 11–28

b. The charging current spike has a value of

$$\frac{E}{10\ \Omega + R_1} = 7\ \text{A}$$

Since $E = 105$ V, this yields $R_1 = 5\ \Omega$.

c. $5\tau_d = 1.75$ ms. Therefore, $\tau_d = 350\ \mu\text{s}$. But $\tau_d = (R_2 + R_3)C$. Thus, $C = 350\ \mu\text{s}/35\ \Omega = 10\ \mu\text{F}$.

RC Circuits in Steady State dc

When an *RC* circuit reaches steady state dc, its capacitors look like open circuits and a transient analysis is not needed—see Note.

EXAMPLE 11–14

The circuit of Figure 11–29(a) has reached steady state. Determine the capacitor voltages.

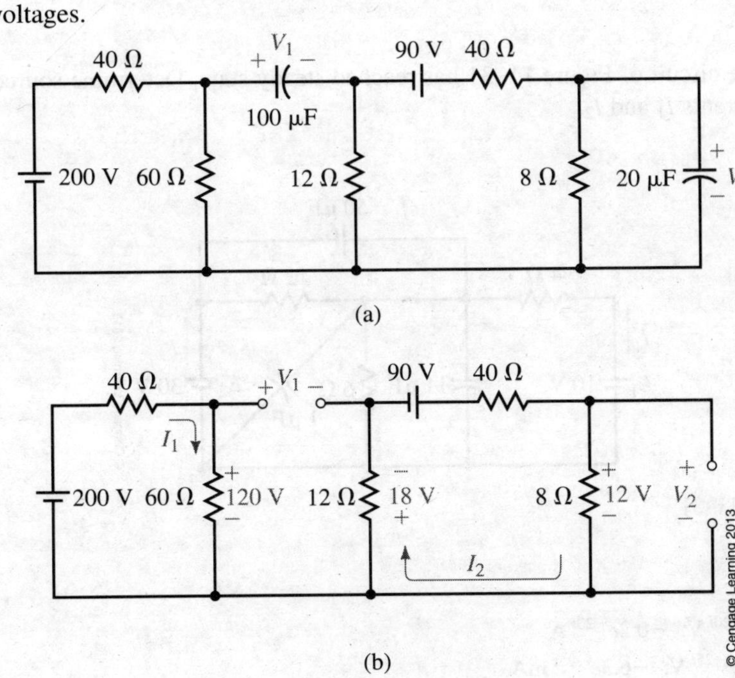

FIGURE 11–29

NOTES...

Since a capacitor consists of conducting plates separated by an insulator, there is no conductive path from terminal to terminal through the capacitor. Therefore, when a capacitor is placed across a dc source, except for a brief transient surge, its current is zero. Thus, as we concluded earlier, a capacitor looks like an open circuit to steady state dc.

CircuitSim 11-14

Solution Replace all capacitors with open circuits as in (b). Thus,

$$I_1 = \frac{200\text{ V}}{40\ \Omega + 60\ \Omega} = 2\text{ A}, \qquad I_2 = \frac{90\text{ V}}{40\ \Omega + 8\ \Omega + 12\ \Omega} = 1.5\text{ A}$$

KVL: $V_1 - 120 - 18 = 0$. Therefore, $V_1 = 138$ V. Further,

$$V_2 = (8\ \Omega)(1.5\text{ A}) = 12\text{ V}$$

PRACTICE PROBLEMS 5

1. The capacitor of Figure 11–30(a) is initially uncharged. At $t = 0$ s, the switch is moved to position 1 and 100 ms later, to position 2. Determine expressions for v_C and i_C for position 2.
2. Repeat for Figure 11–30(b). Hint: Use Thévenin's theorem.

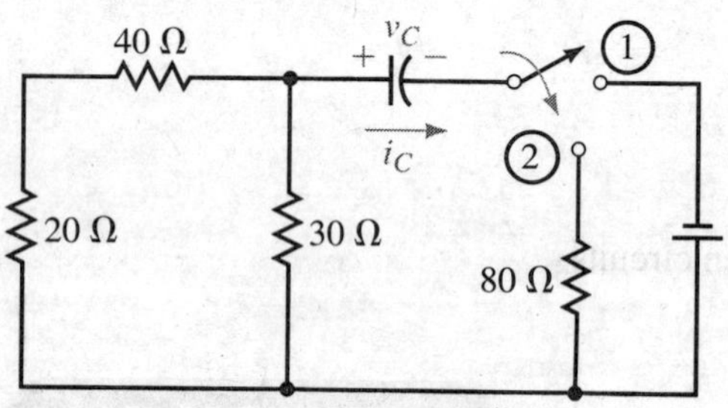

(a) $C = 500\ \mu$F

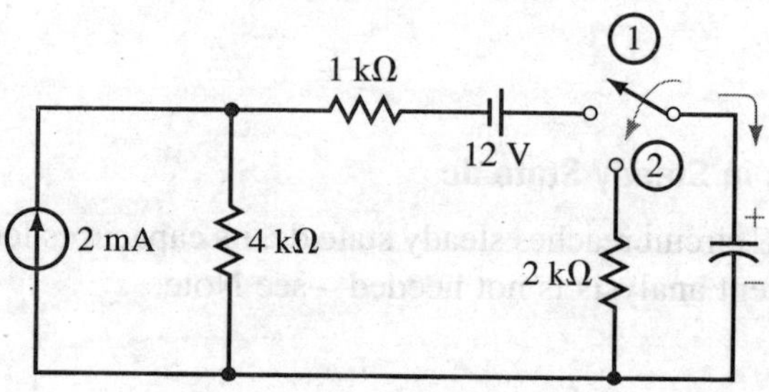

(b) $C = 20\ \mu$F

FIGURE 11–30

3. The circuit of Figure 11–31 has reached steady state. Determine source currents I_1 and I_2.

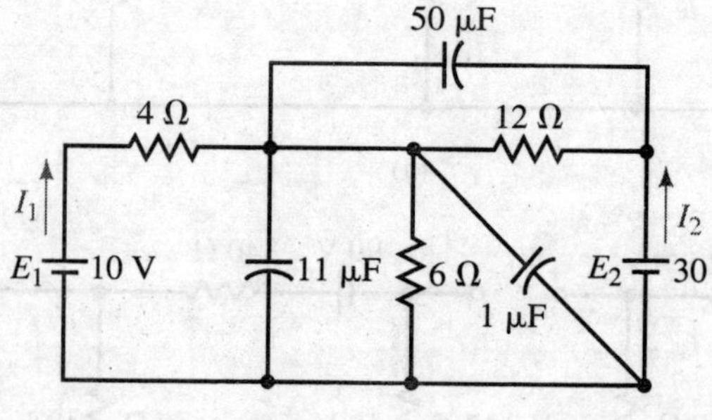

FIGURE 11–31

Answers

1. $20e^{-20t}$ V; $-0.2e^{-20t}$ A
2. $12.6e^{-25t}$ V; $-6.3e^{-25t}$ mA
3. 0 A; 1.67 A

11.6 An *RC* Timing Application

RC circuits are used to create delays for alarm, motor control, and timing applications. Figure 11–32 shows an alarm application. The alarm unit contains a threshold detector, and when the input to this detector exceeds a preset value, the alarm is turned on.

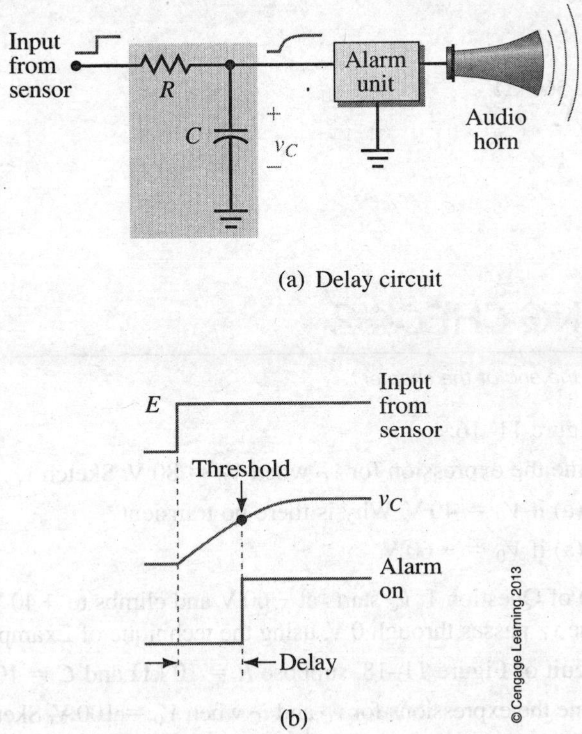

FIGURE 11–32 Creating a time delay with an *RC* circuit. Assume the alarm unit does not load the *RC* circuit.

CircuitSim 11-15

EXAMPLE 11–15

The circuit of Figure 11–32 is part of a building security system. When an armed door is opened, you have a specified number of seconds to disarm the system before the alarm goes off. If $E = 20$ V, $C = 40\ \mu$F, the alarm is activated when v_C reaches 16 V, and you want a delay of at least 25 s, what value of R is needed?

Solution $v_C = E(1 - e^{-t/RC})$. After a bit of manipulation, you get

$$e^{-t/RC} = \frac{E - v_C}{E}$$

Taking the natural log of both sides yields (see Notes)

$$-\frac{t}{RC} = \ln\left(\frac{E - v_C}{E}\right)$$

At $t = 25$ s, $v_C = 16$ V. Thus,

$$-\frac{t}{RC} = \ln\left(\frac{20 - 16}{20}\right) = \ln 0.2 = -1.6094$$

Substituting $t = 25$ s and $C = 40\ \mu$F yields

$$R = \frac{t}{1.6094C} = \frac{25\text{ s}}{1.6094 \times 40 \times 10^{-6}} = 388\text{ k}\Omega$$

Choose the next higher standard value, namely 390 kΩ.

NOTES...

The relationship between the exponential function and natural logs is discussed briefly in Appendix B.

PRACTICE PROBLEMS 6

1. Suppose you want to increase the disarm time of Example 11–15 to at least 35 s. Compute the new value of R.
2. If, in Example 11–15, the threshold is 15 V and $R = 1\ \text{M}\Omega$, what is the disarm time?

Answers

1. 544 kΩ. Use 560 kΩ.
2. 55.5 s

IN-PROCESS LEARNING CHECK 2

(Answers are at the end of the chapter.)

1. Refer to Figure 11–16:
   a. Determine the expression for v_C when $V_0 = 80$ V. Sketch v_C.
   b. Repeat (a) if $V_0 = 40$ V. Why is there no transient?
   c. Repeat (a) if $V_0 = -60$ V.
2. For part (c) of Question 1, v_C starts at -60 V and climbs to $+40$ V. Determine at what time v_C passes through 0 V, using the technique of Example 11–15.
3. For the circuit of Figure 11–18, suppose $R = 10\ \text{k}\Omega$ and $C = 10\ \mu\text{F}$:
   a. Determine the expressions for v_C and i_C when $V_0 = 100$ V. Sketch v_C and i_C.
   b. Repeat (a) if $V_0 = -100$ V.
4. Repeat Example 11–12 if voltage source 2 is reversed, i.e., $E_2 = -30$ V.
5. The switch of Figure 11–33(a) is closed at $t = 0$ s. The Norton equivalent of the circuit in the box is shown in (b). Determine expressions for v_C and i_C. The capacitor is initially uncharged.

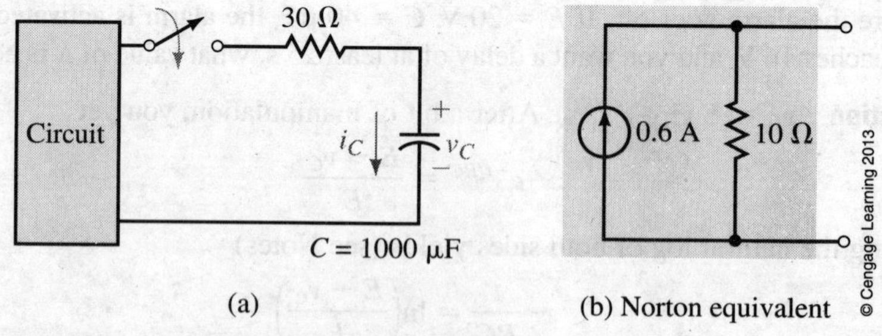

(a) (b) Norton equivalent

FIGURE 11–33 Hint: Use a source transformation.

11.7 Pulse Response of *RC* Circuits

In previous sections, we looked at the response of *RC* circuits to switched dc inputs. In this section, we consider the effect that *RC* circuits have on pulse waveforms. Since many electronic devices and systems utilize pulse or rectangular waveforms, including computers, communications systems, and motor control circuits, these are important considerations.

Pulse Basics

A **pulse** is a voltage or current that changes from one level to the other and back again as in Figure 11–34(a) and (b). A **pulse train** is a repetitive stream of pulses, as in (c). If a waveform's high time equals its low time, as in (d), it is called a **square wave.**

The length of each cycle of a pulse train is termed its **period,** *T*, and the number of pulses per second is defined as its **pulse repetition rate (PRR)** or **pulse repetition frequency (PRF).** For example, in (e), there are two complete cycles in 1 second; therefore, the PRR = 2 pulses/s. With two cycles every second, the time for one cycle is $T = ½$ s. Note that this is 1/PRR. This is true in general. That is,

$$T = \frac{1}{\text{PRR}} \text{ s} \qquad \textbf{(11–16)}$$

The width, t_p, of a pulse relative to its period [Figure 11–34(c)] is its **duty cycle.** Thus,

$$\text{percent duty cycle} = \frac{t_p}{T} \times 100\% \qquad \textbf{(11–17)}$$

A square wave [Figure 11–34(d)] therefore has a 50% duty cycle, while a waveform with $t_p = 1.5$ μs and a period of 10 μs has a duty cycle of 15%.

In practice, waveforms are not ideal, that is, they do not change from low to high or high to low instantaneously. Instead, they have finite **rise** and **fall times.** Rise and fall times are denoted as t_r and t_f and are measured between the 10% and 90% points as indicated in Figure 11–35(a). **Pulse width** is measured at the 50% point. The difference between a real waveform and an ideal waveform is often slight. For example, rise and fall times of real pulses may be only a few nanoseconds and when viewed on an oscilloscope, as in Figure 11–35(b),

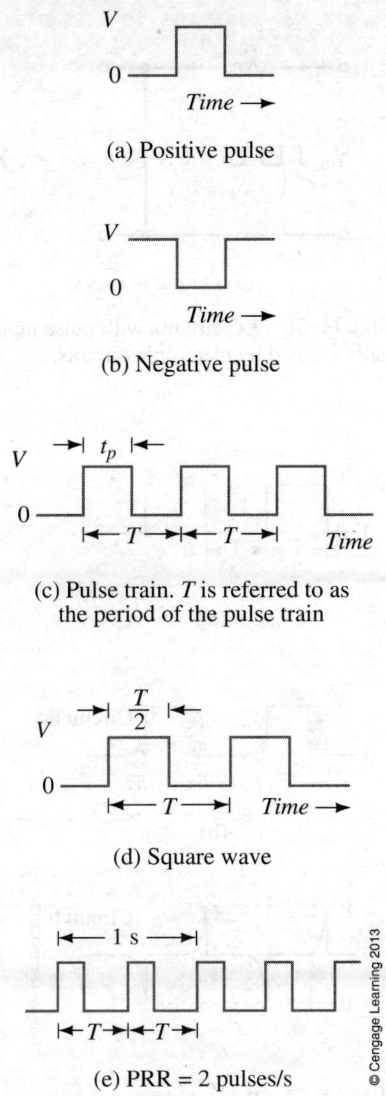

FIGURE 11–34 Ideal pulses and pulse waveforms.

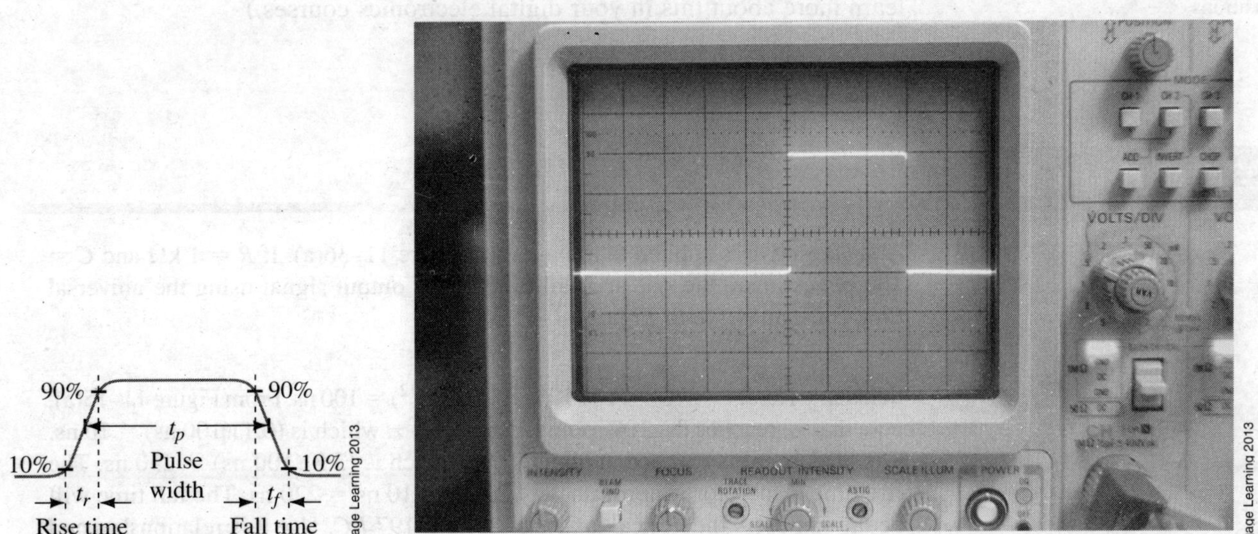

(a) Pulse definitions

(b) Pulse waveform viewed on an oscilloscope

FIGURE 11–35 Practical pulse waveforms.

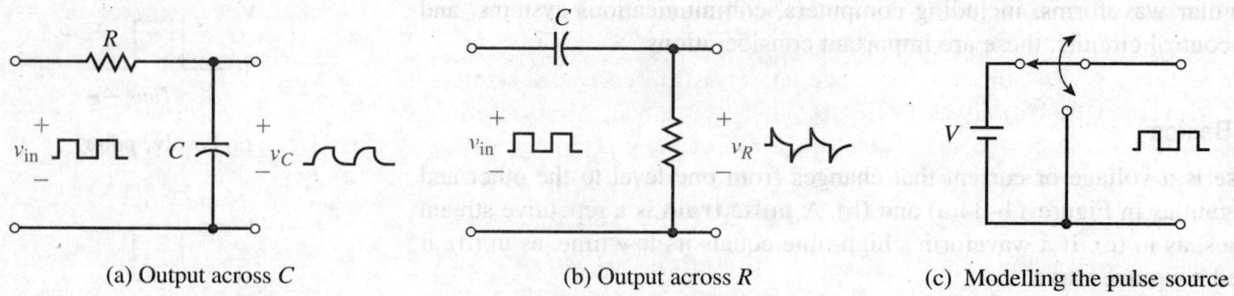

FIGURE 11–36 *RC* circuits with pulse input. Although we have modelled the source here as a battery and a switch, in practice, pulses are usually created by electronic circuits.

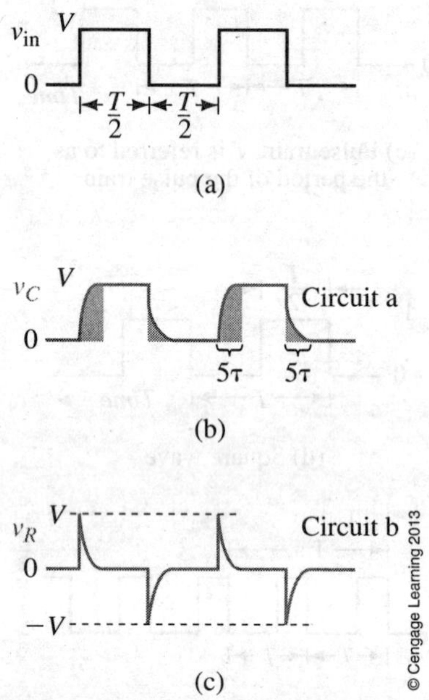

FIGURE 11–37 Pulse width much greater than 5τ. Note that the shaded areas indicate where the capacitor is charging and discharging. Spikes occur on the input voltage transitions.

appear to be ideal. While in what follows, we assume ideal waveforms, in more advanced electronics courses, you often have to take rise and fall times into account.

The Effect of Pulse Width

The width of a pulse relative to a circuit's time constant determines how it is affected by an *RC* circuit. Consider Figure 11–36. In (a), the circuit has been drawn to focus on the voltage across *C;* in (b), it has been drawn to focus on the voltage across *R*. (Otherwise, the circuits are identical.) An easy way to visualize the operation of these circuits is to assume that the pulse is generated by a switch that is moved rapidly back and forth between *V* and common as in (c). This alternately creates a charge and discharge circuit, and thus all of the ideas developed in this chapter apply directly.

Pulse Width $t_p \gg 5\tau$

First, consider the output of circuit (a). When the pulse width and time between pulses are very long compared with the circuit time constant, the capacitor charges and discharges fully, Figure 11–37(b). (This case is similar to what we have already seen in this chapter.) Note that charging and discharging occur at the transitions of the pulse. The transients therefore increase the rise and fall times of the output. In high-speed circuits, this may be a problem. (You will learn more about this in your digital electronics courses.)

EXAMPLE 11–16

A square wave is applied to the input of Figure 11–36(a). If $R = 1\ \text{k}\Omega$ and $C = 100\ \text{pF}$, estimate the rise and fall time of the output signal using the universal time constant curve of Figure 11–15(a).

CircuitSim 11-16

Solution Here, $\tau = RC = (1 \times 10^3)(100 \times 10^{-12}) = 100\ \text{ns}$. From Figure 11–15(a), note that v_C reaches the 10% point at about $0.1\ \tau$, which is (0.1)(100 ns) = 10 ns. The 90% point is reached at about $2.3\ \tau$, which is (2.3)(100 ns) = 230 ns. The rise time is therefore approximately 230 ns – 10 ns = 220 ns. The fall time will be the same. (A theoretical analysis yields 2.197 *RC*. Use this relationship and compare results.)

Now consider the circuit in Figure 11–36(b). Here, current i_C will be similar to that of Figure 11–7(b), except that the pulse widths will be narrower. Since voltage $v_R = Ri_C$, the output will be a series of short, sharp spikes that occur at input transitions as in Figure 11–37(c). Under the conditions here (i.e., pulse width much greater than the circuit time constant), v_R is an approximation to the derivative of v_{in} and the circuit is called a differentiator circuit. Such circuits have important practical uses.

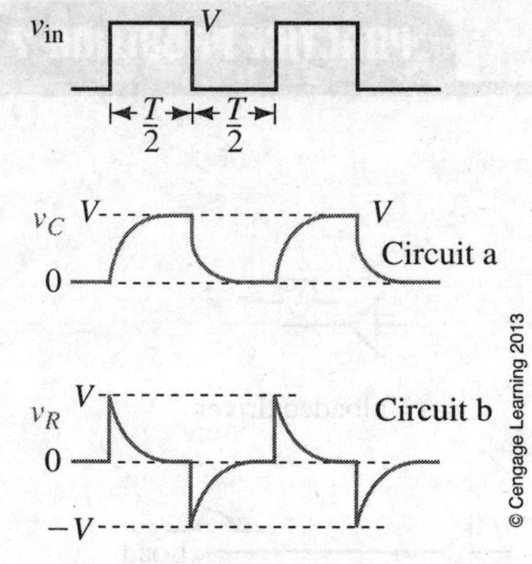

FIGURE 11–38 Pulse width equal to 5τ. These waveforms are the same as Figure 11–37, except that the transitions last relatively longer.

Pulse Width $t_p = 5\tau$

These waveforms are shown in Figure 11–38. Since the pulse width is 5τ, the capacitor fully charges and discharges during each pulse. Thus, waveforms here are again similar to what we have seen throughout this chapter.

Pulse Width $t_p \ll 5\tau$

This case differs from what we have seen so far in this chapter only in that the capacitor does not have time to charge and discharge significantly between pulses. The result is that switching occurs on the early (nearly straight line) part of the charging and discharging curves and thus, v_C is roughly triangular in shape, Figure 11–39(a). It has an average value of $V/2$. Under the conditions here, v_C is the approximate integral of v_{in} and the circuit is called an integrator circuit.

It should be noted that v_C does not reach the steady state shown in Figure 11–39 immediately. Instead, it works its way up over a period of five time constants (Figure 11–40). To illustrate, assume an input square wave of 5 V with a pulse width of 0.1 s and $\tau = 0.1$ s. Proceed as follows:

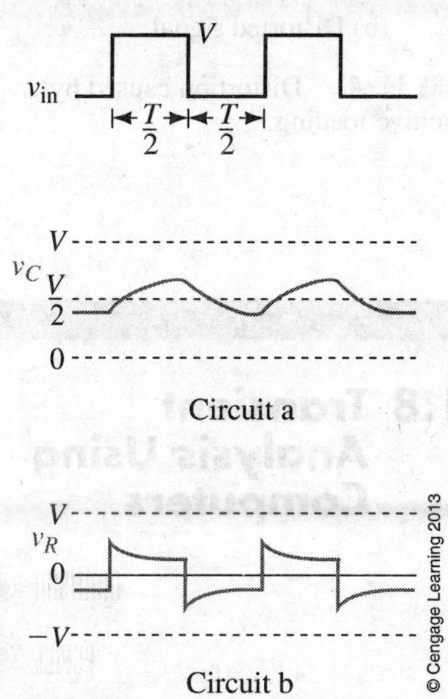

FIGURE 11–39 Pulse width much less than 5τ. The circuit does not have time to charge or discharge substantially.

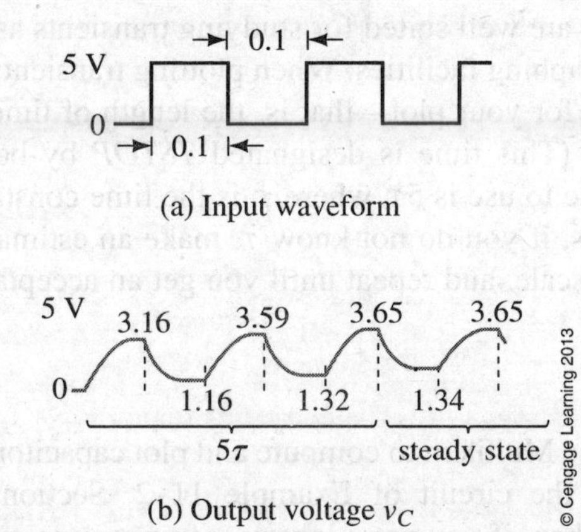

FIGURE 11–40 Notice how the solution takes five time constants to reach its steady state, after which it cycles between 1.34 and 3.65 V. You can easily see this using Multisim or PSpice.

CircuitSim 11-17

Pulse 1 $v_C = E(1 - e^{-t/\tau})$. At the end of the first pulse ($t = 0.1$ s), v_C has climbed to $v_C = 5(1 - e^{-0.1/0.1}) = 5(1 - e^{-1}) = 3.16$ V. From the end of pulse 1 to the beginning of pulse 2 (i.e., over an interval of 0.1 s), v_C decays from 3.16 V to $3.16e^{-0.1/0.1} = 3.16e^{-1} = 1.16$ V.

Pulse 2 v_C starts at 1.16 V and 0.1 s later has a value of $v_C = E + (V_0 - E)e^{-t/\tau} = 5 + (1.16 - 5)e^{-0.1/0.1} = 5 - 3.84e^{-1} = 3.59$ V. It then decays to $3.59e^{-1} = 1.32$ V over the next 0.1 s.

Continuing in this manner, the remaining values for Figure 11–40(b) are determined. After 5τ, v_C cycles between 1.34 and 3.65 V, with an average of $(1.34 + 3.65)/2 = 2.5$ V, or half the input pulse amplitude.

PRACTICE PROBLEMS 7

Verify the remaining points of Figure 11–40(b).

Capacitive Loading

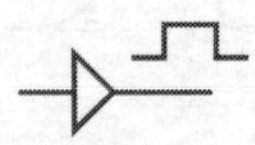

(a) Unloaded driver

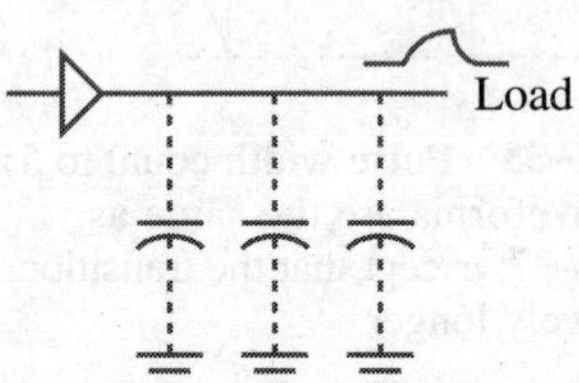

(b) Distorted signal

FIGURE 11–41 Distortion caused by capacitive loading.

Capacitance occurs whenever conductors are separated by insulating material. This means that capacitance exists between wires in cables, between traces on printed circuit boards, and so on. In general, this capacitance is undesirable, but it cannot be avoided. It is called **stray capacitance.** Fortunately, stray capacitance is often so small that it can be neglected. However, in high-speed circuits, it may cause problems.

To illustrate, consider Figure 11–41. The electronic driver of (a) produces square pulses. However, when it drives a long line as in (b), stray capacitance loads it and increases the signal's rise and fall times (since capacitance takes time to charge and discharge). This effect is referred to as **capacitive loading.** If the rise and fall times become excessively long, the signal reaching the load may be so degraded that the system malfunctions. (Capacitive loading is a serious issue, but we will leave it for future courses to deal with.)

11.8 Transient Analysis Using Computers

Multisim

PSPICE

Multisim and PSpice are well suited for studying transients as they both incorporate easy-to-use graphing facilities. When plotting transients, you must specify the time duration for your plot—that is, the length of time that you expect the transient to last. (This time is designated *TSTOP* by both Multisim and PSpice.) A good value to use is 5τ where τ is the time constant of the circuit. (For complex circuits, if you do not know τ, make an estimate, run a simulation, adjust the time scale, and repeat until you get an acceptable plot.)

Multisim

As a first example, use Multisim to compute and plot capacitor voltage and current waveforms for the circuit of Example 11–2, Section 11.2. Read the Multisim notes and then, if you are a legacy software user, go to our Web site for instructions; otherwise, proceed as follows:

- Create the circuit of Figure 11–42 on your screen. (Use a virtual capacitor and rotate it once so that its "1" end is at the top as described in Appendix A.)
- Click *Options/Sheet Properties* and under *Net Names*, click *Show All*, then *OK*. Node numbers will appear on your schematic—see Note 5. In what follows, we use the node numbers shown in Figure 11–42 in our discussion.
- Click *Simulate/Analysis* and select *Transient Analysis.* Compute the duration of the transient (it is 0.5 s) and enter this value as *TSTOP*. From the *Initial Conditions* box, choose *Set to zero,* then click the *Output* tab.
- Click *V*(2) to highlight it, then click *Add.* (This is capacitor voltage.)
- Click *I*(C1) to highlight it, then click *Add.* (This is capacitor current.)
- Click *Simulate.*

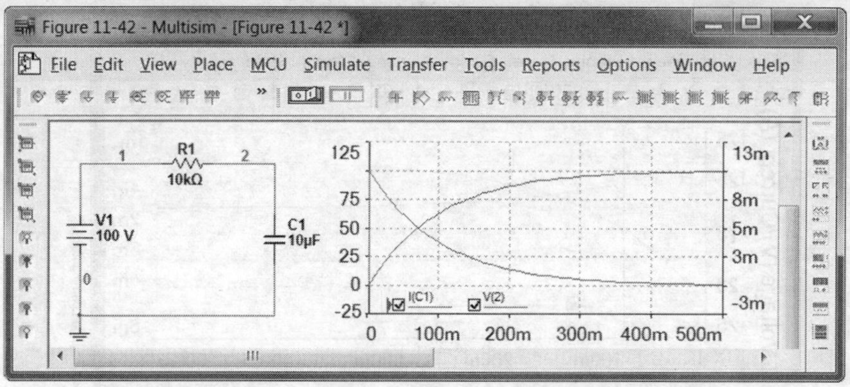

FIGURE 11–42 Multisim solution of Example 11–2. (The left-hand graph scale is for the voltage waveform.) The transient is initiated by software, so no switch is needed in the simulation setup.

- *Grapher View* opens. Click *View* and ensure that *Toolbar* is selected. Find the icon that identifies itself as *Black Background*, then click. Now find the *Show Grid* icon and click. Right-click the screen, find *Show Select Marks*, and click.

Note that the current waveform appears as a flat line along the zero axis. You need to expand it so that you can see its detail. Proceed as follows (or see Appendix A, *Displaying Multiple Traces*).

- On *Grapher,* click *Trace/Trace Properties*. In the dialog box that opens, select the *Traces* tab at the top of the box. The item marked *Trace Label* should indicate *I*(C1). If it doesn't, use the *Trace ID* up/down arrows to change it.
- Near the bottom of the box (under *Y-vertical axis*), click the *Right axis* button, then click *Apply*. Select the *Right Axis* tab at the top of the box, click the *Axis Enabled box,* click *Auto-range,* click *Apply* then *OK*. (You should have the voltage and current waveforms of Figure 11–42 on your screen, although trace colors may not match.)
- Activate the cursors by clicking the *Show Cursors* icon.
- With your mouse pointer, drag #1 cursor into the display area. Place your mouse pointer on it, right click, select *Set X Value,* type in **150m,** then click *OK*. The cursor moves to the 150 ms point and the *Cursor* display box shows the results. Your display should show current = 2.23 mA and voltage = 77.69 volts at t = 150 ms. Compare these to the answers indicated in Figure 11–10, Example 11–2.

Use your cursor to measure values at 100 ms intervals, then compare to the answers from Practice Problems 2, Question 1.

Initial Conditions in Multisim

Let us change the preceding problem to include an initial voltage of 150 V on the capacitor. Get the circuit of Figure 11–42 back on the screen, double-click the capacitor symbol, and in the dialog box under *Initial conditions*, type in **150,** ensure that units are set to V, and click *OK*. Select *Transient Analysis;* from the *Initial Conditions* box, choose *User-defined.* Run the simulation and note that the transient starts at 150 V and decays to its steady state value of 100 V in five time constants as expected.

Now expand the current trace scale as we did above. You should have the waveforms of Figure 11–43 on your screen. With your cursor, measure values at t = 100 ms and verify using Equations 11–10 and 11–11.

NOTES...

Multisim Operational Notes

1. If you are a Multisim 9 or 10 user, go to our Web site at cengagebrain.com and follow the links to *Special Instructions for Multisim 9 and 10 Users.*
2. You don't need a switch to initiate a transient. Simply build the circuit without a switch and select transient analysis. Multisim then performs the transient simulation and plots the results.
3. These examples were prepared using Multisim 11, the version current at the time of writing. Since software is updated regularly, this may not be the version that you are using.
4. To place your cursor at an exact time point on a Multisim waveform, for example, at t = 150 ms, place your mouse pointer on the cursor and right-click. Select *Set X Value,* type in **150m,** then click OK.
5. Multisim assigns node numbers according to the order in which you wire your circuit. Thus, your node numbers may be different than the ones we show.
6. Multisim computes node voltages with respect to ground. Thus, in Figure 11–42, the voltage at Node 2 is the voltage across the capacitor. Multisim 11 refers to it as V(2).

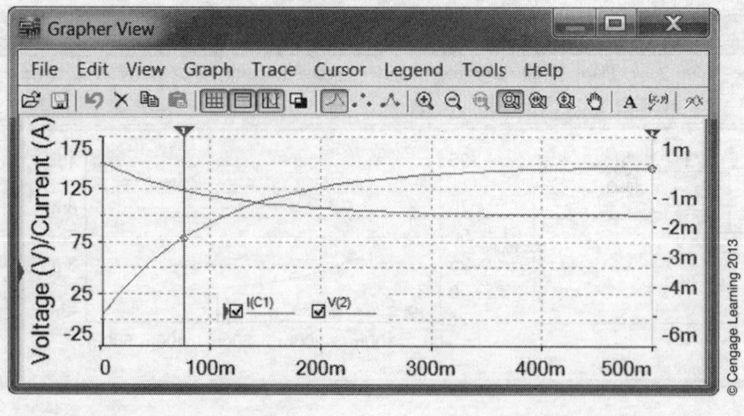

FIGURE 11–43 Solution for the circuit of Figure 11–42 with $V_0 = 150$ V.

PRACTICE PROBLEMS 8

For the circuit of Figure 11–42, set the capacitor initial condition to –50 V, then repeat the preceding simulation. Verify answers using Equations 11–10 and 11–11.

Another Example

Right-click *View* and ensure that the Signal Source Components toolbar is selected. In its toolbin, locate and click *Place Clock Voltage Source,* then build the circuit of Figure 11–44(a) on your screen. (The clock, with its default settings, produces a square wave that cycles between 0 V and 5 V with a cycle length of $T = 1$ ms.) This means that its *on* time t_p is $T/2 = 500$ μs. Since the time constant of Figure 11–44 is $\tau = RC = 50$ μs, t_p is greater than 5τ, and a waveform similar to that of Figure 11–37(c) should result. To verify, follow the procedure of the previous example, except set *End Time (TSTOP)* to 0.0025 in the *Analysis/Transient* dialog box, choose *Initial Conditions set to zero,* then select V(1) and V(2) for display. (Legacy users, select nodes 1 and 2 instead.) Click *Simulate* and the waveforms of Figure 11–44(b) appear. Note that output spikes occur on the transitions of the input waveform as predicted.

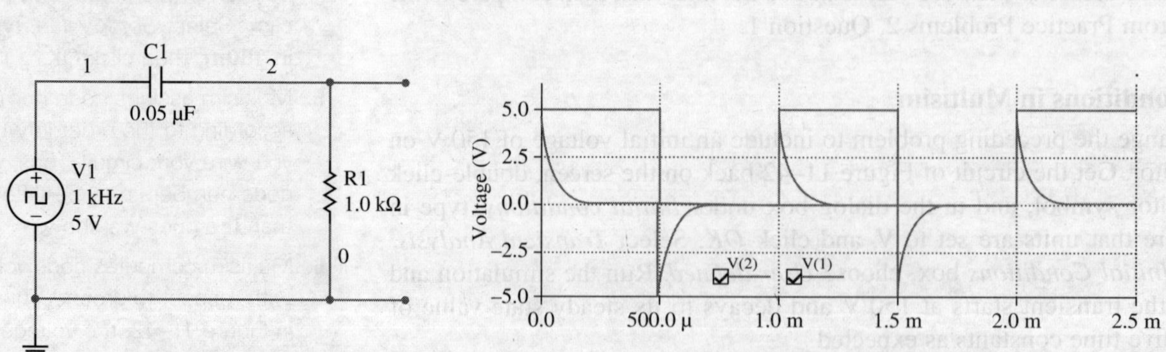

(a) (b) The red waveform is the output voltage.

FIGURE 11–44 Multisim analysis of the circuit of Figure 11–36(b). Compare (b) to Figure 11–37(c).

PRACTICE PROBLEMS 9

For the circuit of Figure 11–44, run simulations for $R = 500\ \Omega$, $2\ \text{k}\Omega$, and $20\ \text{k}\Omega$. For each case, compute τ, then note how the resulting waveform fits with the predicted waveforms of Figures 11–37 to 11–39.

PSpice

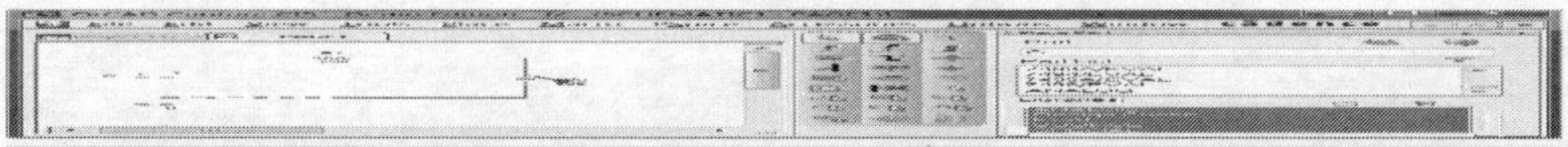

PSpice Operational Notes

1. Do not use a space between a value and its unit. Thus, use 50ms, not 50 ms, and so forth.
2. When instructed to enter data via a Property editor (e.g., initial conditions), first click the Parts tab at the bottom of the screen; scroll right until you find the cell that you want, type in its value, then click *Apply*. In the warning box, click Yes, then click the Windows close button (the x beside the word *cadence*) to close the Property Editor. (Don't click the large red boxed close button above it because that will close PSpice.)
3. For transient problems you must specify an initial condition (IC) for each capacitor and inductor, even if they are zero. The procedure is described in the examples.
4. For most simple transient analysis problems, you don't need a switch—simply specify transient analysis at run time.
5. To select the waveform that you want the cursor to apply to, click the applicable symbol at the bottom of the screen.

As a first example, consider Figure 11–2 with $R = 200\ \Omega$, $C = 50\ \mu\text{F}$, and $E = 40$ V. Let the capacitor be initially uncharged (i.e., $V_0 = 0$ V). First, read the PSpice Operational Notes, then proceed as follows:

- Create the circuit on the screen as in Figure 11–45—see Note 4. Remember to rotate the capacitor as discussed in Appendix A, then set its initial condition

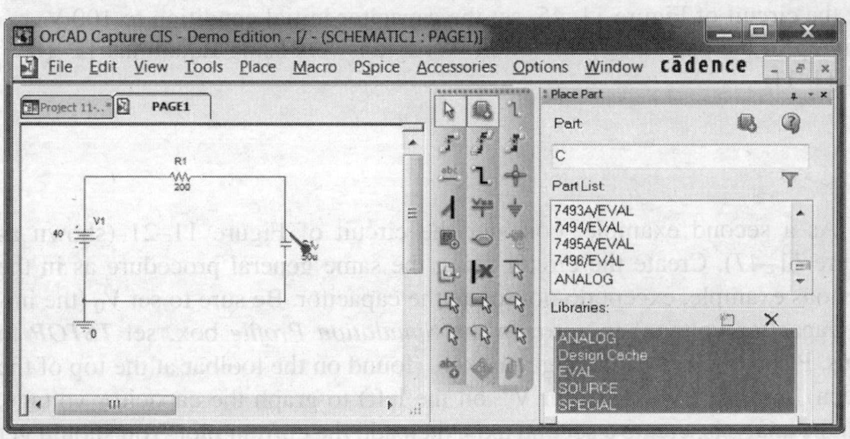

FIGURE 11–45 PSpice example. The voltage marker displays voltage with respect to ground, which, in this case, is the voltage across C_1.

(IC) to zero. To do this, double-click the capacitor symbol, type **0V** into the Property editor cell labeled IC—see Note 2—click *Apply*, then close the editor. Click the *New Simulation Profile* icon, enter a name (e.g., **Figure 11–45**) then click *Create*. In the *Simulation Settings* box, click the *Analysis* tab, select *Time Domain (Transient)* and in *Options,* select *General Settings*. Set the duration of the transient *(TSTOP)* to 50ms, then click *OK*. Find the voltage marker on the toolbar and place as shown. (It will be gray, but will change color after the simulation is run.)

- Click the *Run* icon. A trace of capacitor voltage appears. Click *Plot,* then *Add Y Axis*. Click *Trace/Add Trace,* then double-click *I(C1)*. This adds the current trace, Figure 11–46.

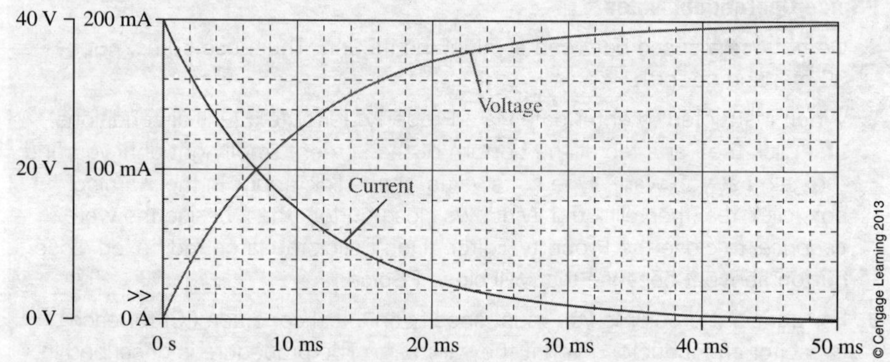

FIGURE 11–46 Waveforms for the circuits of Figures 11–45 and 11–47.

Analysis of Results

Click the *Toggle cursor* icon, then use the cursor to determine values from the screen. For example, at t = 5 ms, you should find v_C = 15.7 V and i_C = 121 mA as determined earlier in Example 11–10, part (c).

PRACTICE PROBLEMS 10

For the circuit of Figure 11–45, set the capacitor initial condition to 100 V, then repeat the preceding simulation. Verify answers using Equations 11–10 and 11–11.

As a second example, consider the circuit of Figure 11–21 (shown as Figure 11–47). Create the circuit using the same general procedure as in the previous example, except do not rotate the capacitor. Be sure to set V_0 (the initial capacitor voltage) to zero. In the *Simulation Profile* box, set *TSTOP* to 50ms. Place differential voltage markers (found on the toolbar at the top of the screen) across C (with marker V^+ on the left) to graph the capacitor voltage. Run the analysis, create a second axis, then add the current plot. You should get the same graph (i.e., Figure 11–46) as you got for the previous example, since its circuit is the Thévenin equivalent of this one.

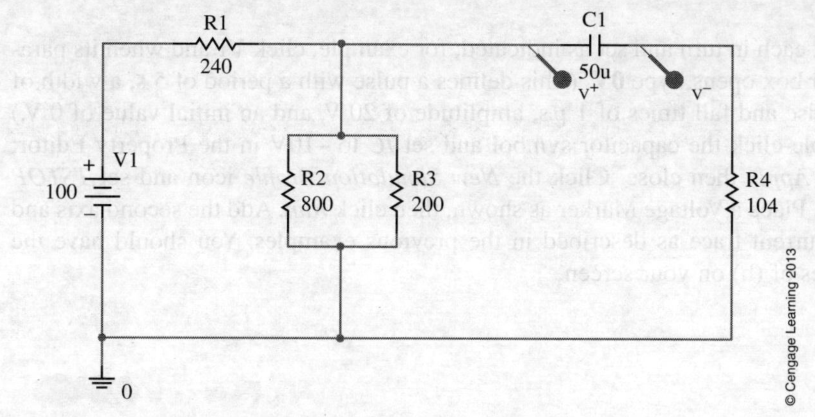

FIGURE 11–47 The differential markers display the voltage across C_1.

As a final example, consider Figure 11–48(a), which shows double switching action.

EXAMPLE 11–17

The capacitor of Figure 11–48(a) has an initial voltage of −10 V. The switch is moved to the charge position for 1 s, then to the discharge position where it remains. Determine curves for v_C and i_C.

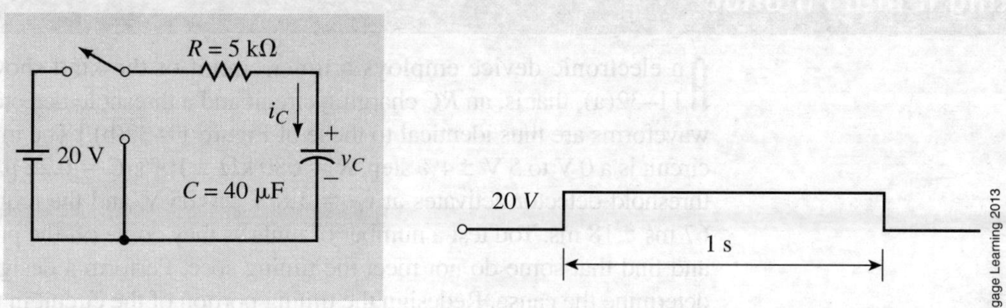

(a) Circuit to be modelled (b) The applied pulse

FIGURE 11–48 Creating a charge/discharge waveform using PSpice.

Solution PSpice has no switch that implements the switching sequence of Figure 11–48. However, moving the switch first to charge then to discharge is equivalent to placing 20 V across the *RC* combination for the charge time, then 0 V thereafter as indicated in (b). You can do this with a pulse source (VPULSE) as indicated in Figure 11–49(a). Note the parameters listed beside the symbol.

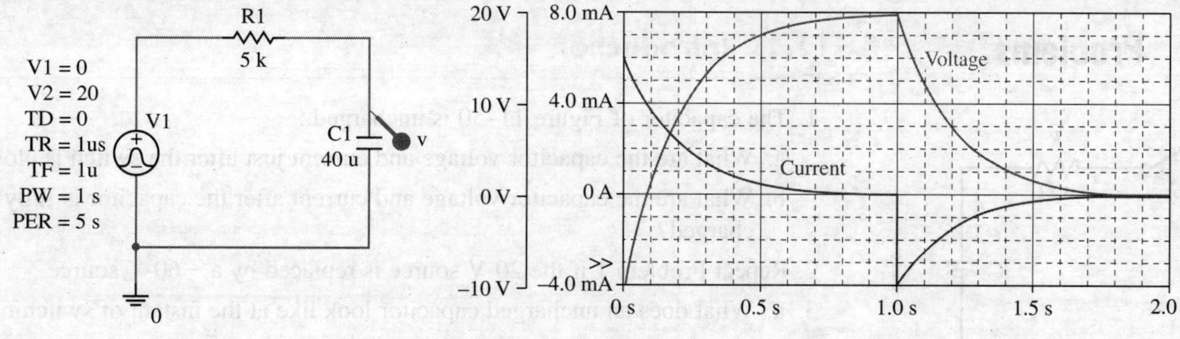

(a) (b)

FIGURE 11–49 Modelling switching action using a pulse source.

Click each in turn and set as indicated; for example, click *V1* and when its parameter box opens, type **0V**. (This defines a pulse with a period of 5 s, a width of 1 s, rise and fall times of 1 μs, amplitude of 20 V, and an initial value of 0 V.) Double-click the capacitor symbol and set *IC* to –10V in the Property Editor; click *Apply* then close. Click the *New Simulation Profile* icon and set *TSTOP* to 2s. Place a Voltage Marker as shown, then click *Run*. Add the second axis and the current trace as described in the previous examples. You should have the curves of (b) on your screen.

Note that voltage starts at −10 V and climbs to 20 V while current starts at $(E - V_0)/R$ = 30 V/5 kΩ = 6 mA and decays to zero. When the switch is turned to the discharge position, the current drops from 0 A to −20 V/5 kΩ = −4 mA and then decays to zero while the voltage decays from 20 V to zero. Thus, the solution checks.

Putting It into Practice

An electronic device employs a timer circuit of the kind shown in Figure 11–32(a), that is, an *RC* charging circuit and a threshold detector. [Its timing waveforms are thus identical to those of Figure 11–32(b).] The input to the *RC* circuit is a 0 V to 5 V ±4% step, *R* = 680 kΩ ±10%, *C* = 0.22 μF ±10%, the threshold detector activates at v_C = 1.8 V ±0.05 V, and the required delay is 67 ms ±18 ms. You test a number of units as they come off the production line and find that some do not meet the timing spec. Perform a design review and determine the cause. Redesign the timing portion of the circuit in the most economical way possible.

Problems

11.1 Introduction

1. The capacitor of Figure 11–50 is uncharged.
   a. What are the capacitor voltage and current just after the switch is closed?
   b. What are the capacitor voltage and current after the capacitor is fully charged?
2. Repeat Problem 1 if the 20-V source is replaced by a −60-V source.
3. a. What does an uncharged capacitor look like at the instant of switching?
   b. What does a charged capacitor look like at the instant of switching?
   c. What does a capacitor look like to steady state dc?
   d. What do we mean by $i(0^-)$? By $i(0^+)$?

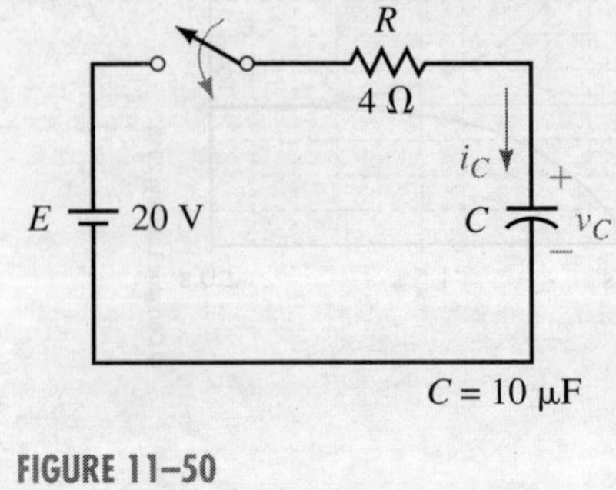

FIGURE 11–50

4. For a charging circuit, $E = 25$ V, $R = 2.2$ kΩ, and the capacitor is initially uncharged. The switch is closed at $t = 0$. What is $i(0^+)$?

5. For a charging circuit, $R = 5.6$ kΩ and $v_C(0^-) = 0$ V. If $i(0^+) = 2.7$ mA, what is E?

11.2 Capacitor Charging Equations

6. The switch of Figure 11–50 is closed at $t = 0$ s. The capacitor is initially uncharged.
   a. Determine the equation for charging voltage v_C.
   b. Determine the equation for charging current i_C.
   c. By direct substitution, compute v_C and i_C at $t = 0^+$ s, 40 μs, 80 μs, 120 μs, 160 μs, and 200 μs.
   d. Plot v_C and i_C on graph paper using the results of (c). Hint: See Example 11–2.
7. Repeat Problem 6 if $R = 500\ \Omega$, $C = 25\ \mu$F, and $E = 45$ V, except compute and plot values at $t = 0^+$ s, 20 ms, 40 ms, 60 ms, 80 ms, and 100 ms.
8. The switch of Figure 11–51 is closed at $t = 0$ s. Determine the equations for capacitor voltage and current. Compute v_C and i_C at $t = 50$ ms.
9. Repeat Problem 8 for the circuit of Figure 11–52.
10. The capacitor of Figure 11–2 is uncharged at the instant the switch is closed. If $E = 80$ V, $C = 10\ \mu$F, and $i_C(0^+) = 20$ mA, determine the equations for v_C and i_C.
11. Determine the time constant for the circuit of Figure 11–50. How long (in seconds) will it take for the capacitor to charge?
12. A capacitor takes 200 ms to charge. If $R = 5$ kΩ, what is C?
13. For Figure 11–50, the capacitor voltage with the switch open is 0 V. Close the switch at $t = 0$ and determine capacitor voltage and current at $t = 0^+$, 40 μs, 80 μs, 120 μs, 160 μs, and 200 μs using the universal time constant curves.
14. If $i_C = 25e^{-40t}$ A, what is the time constant τ and how long will the transient last?
15. For Figure 11–2, the current jumps to 3 mA when the switch is closed. The capacitor takes 1 s to charge. If $E = 75$ V, determine R and C.
16. For Figure 11–2, if $v_C = 100(1 - e^{-50t})$ V and $i_C = 25e^{-50t}$ mA, what are E, R, and C?
17. For Figure 11–2, determine E, R, and C if the capacitor takes 5 ms to charge, the current at 1 time constant after the switch is closed is 3.679 mA, and the capacitor charges to 45 volts steady state.
18. For Figure 11–2, $v_C(\tau) = 41.08$ V and $i_C(2\tau) = 219.4$ mA. Determine E and R.

11.3 Capacitor with an Initial Voltage

19. The capacitor of Figure 11–50 has an initial voltage. If $V_0 = 10$ V, what is the current just after the switch is closed?
20. Repeat Problem 19 if $V_0 = -10$ V.
21. For the capacitor of Figure 11–51, $V_0 = 30$ V.
   a. Determine the expression for charging voltage v_C.
   b. Determine the expression for current i_C.
   c. Sketch v_C and i_C.
22. Repeat Problem 21 if $V_0 = -5$ V.

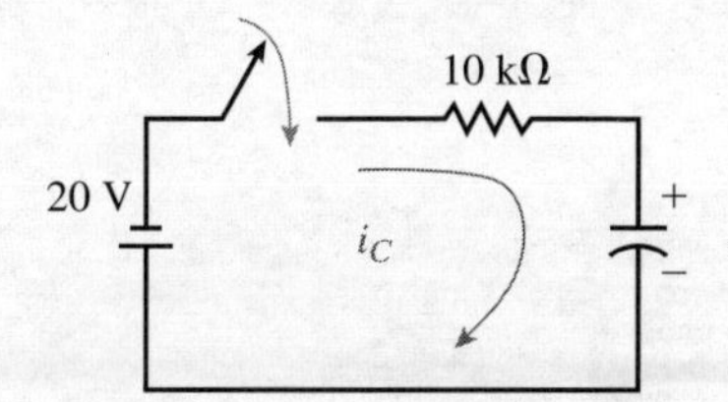

FIGURE 11–51 $V_0 = 0$ V, $C = 10\ \mu$F.

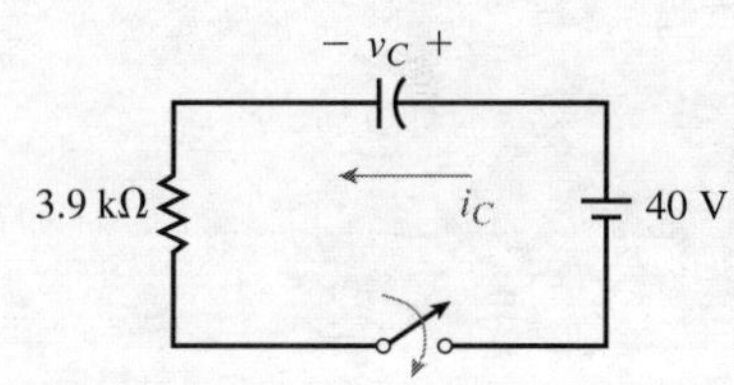

FIGURE 11–52 $C = 10\ \mu$F, $V_0 = 0$ V.

11.4 Capacitor Discharging Equations

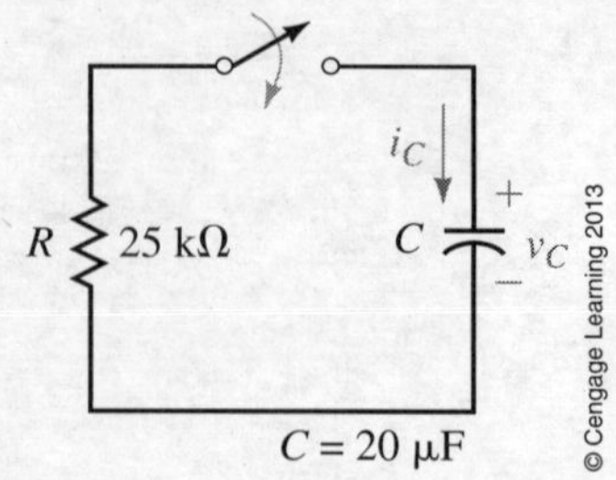

FIGURE 11–53

23. For the circuit of Figure 11–53, assume the capacitor is charged to 50 V before the switch is closed.
    a. Determine the equation for discharge voltage v_C.
    b. Determine the equation for discharge current i_C.
    c. Determine the time constant of the circuit.
    d. Compute v_C and i_C at $t = 0^+$ s, $t = \tau$, 2τ, 3τ, 4τ, and 5τ.
    e. Plot the results of (d) with the time axis scaled in seconds and time constants.
24. The initial voltage on the capacitor of Figure 11–53 is 55 V. The switch is closed at $t = 0$. Determine capacitor voltage and current at $t = 0^+$, 0.5 s, 1 s, 1.5 s, 2 s, and 2.5 s using the universal time constant curves.
25. A 4.7-μF capacitor is charged to 43 volts. If a 39-kΩ resistor is then connected across the capacitor, what is its voltage 200 ms after the resistor is connected?
26. The initial voltage on the capacitor of Figure 11–53 is 55 V. The switch is closed at $t = 0$ s and opened 1 s later. Sketch v_C. What is the capacitor's voltage at $t = 3.25$ s?
27. For Figure 11–54, let $E = 200$ V, $R_2 = 1$ kΩ, and $C = 0.5$ μF. After the capacitor has fully charged in position 1, the switch is moved to position 2.
    a. What is the capacitor voltage immediately after the switch is moved to position 2? What is its current?
    b. What is the discharge time constant?
    c. Determine discharge equations for v_C and i_C.
28. For Figure 11–54, C is fully charged before the switch is moved to discharge. Current just after it is moved is $i_C = -4$ mA and C takes 20 ms to discharge. If $E = 80$ V, what are R_2 and C?

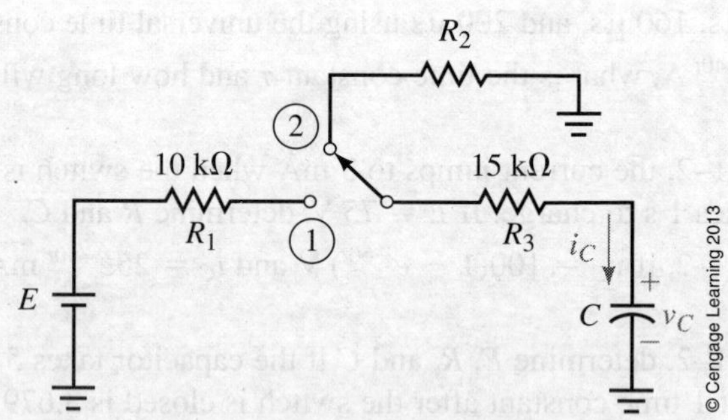

FIGURE 11–54

11.5 More Complex Circuits

29. The capacitors of Figure 11–55 are uncharged. The switch is closed at $t = 0$. Determine the equation for v_C. Compute v_C at one time constant using the equation and the universal time constant curve. Compare answers.
30. For Figure 11–56, the switch is closed at $t = 0$. Given $V_0 = 0$ V.
    a. Determine the equations for v_C and i_C.
    b. Compute the capacitor voltage at $t = 0^+$ 2, 4, 6, 8, 10, and 12 ms.
    c. Repeat (b) for the capacitor current.
    d. Why does 225 V/30 Ω also yield $i(0^+)$?

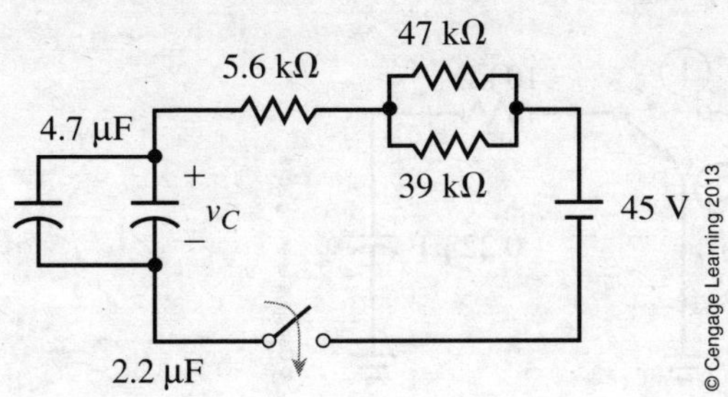

FIGURE 11–55

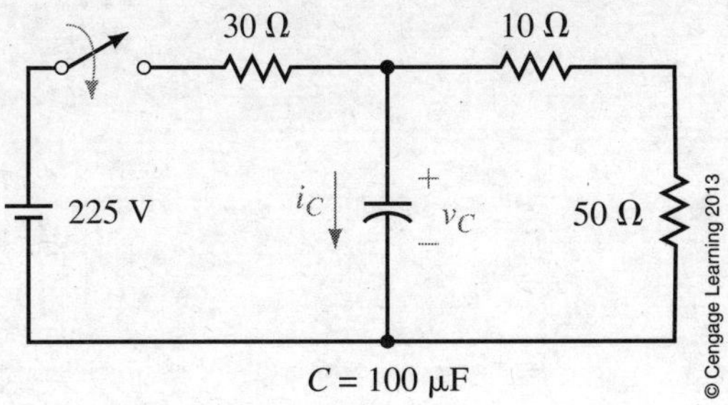

FIGURE 11–56

31. Repeat Problem 30, parts (a) to (c) for the circuit of Figure 11–57.

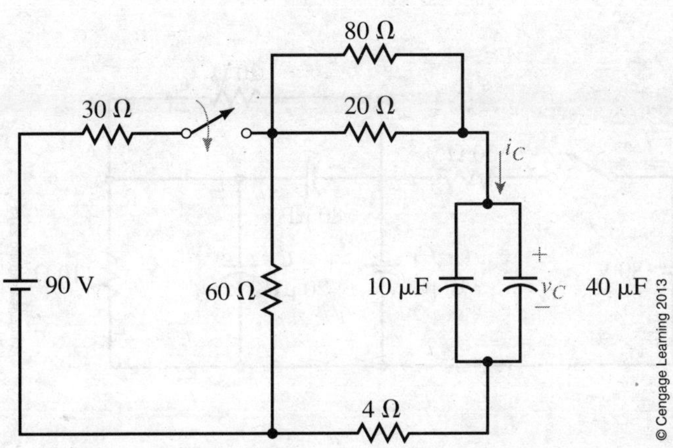

FIGURE 11–57

32. Consider again Figure 11–54. Suppose $E = 80$ V, $R_2 = 25$ kΩ, and $C = 0.5$ μF:
   a. What is the charge time constant?
   b. What is the discharge time constant?
   c. With the capacitor initially discharged, move the switch to position 1 and determine equations for v_C and i_C during charge.
   d. Move the switch to the discharge position. How long does it take for the capacitor to discharge?
   e. Sketch v_C and i_C from the time the switch is placed in charge to the time that the capacitor is fully discharged. Assume the switch is in the charge position for 80 ms.
33. For the circuit of Figure 11–54, the capacitor is initially uncharged. The switch is first moved to charge, then to discharge, yielding the current shown in Figure 11–58. The capacitor fully charges in 12.5 s. Determine E, R_2, and C.

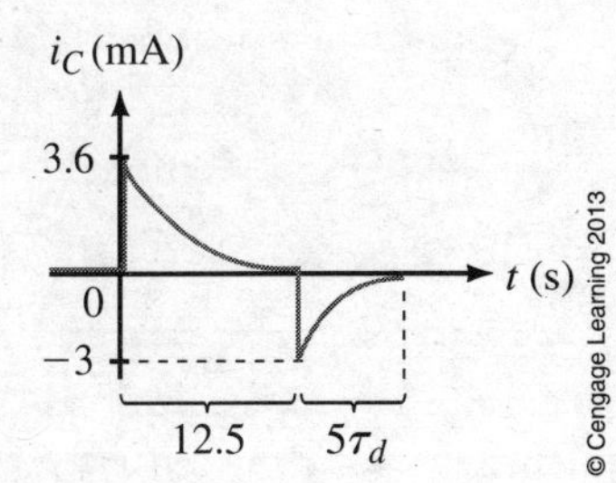

FIGURE 11–58

34. Refer to the circuit of Figure 11–59:
   a. What is the charge time constant?
   b. What is the discharge time constant?
   c. The switch is in position 2 and the capacitor is uncharged. Move the switch to position 1 and determine equations for v_C and i_C.
   d. After the capacitor has charged for two time constants, move the switch to position 2 and determine equations for v_C and i_C during discharge.
   e. Sketch v_C and i_C.

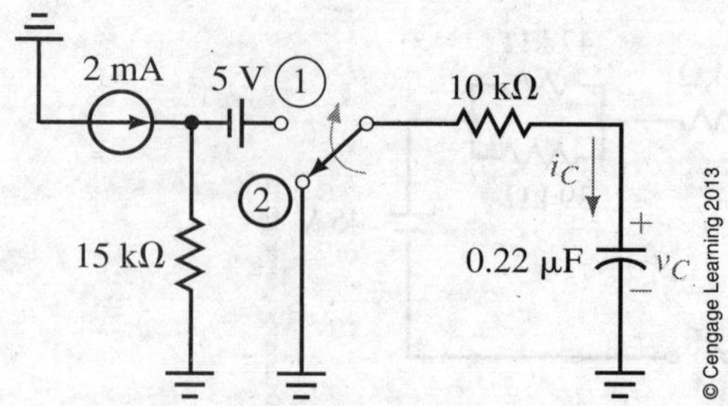

FIGURE 11–59

35. Determine the capacitor voltages and the source current for the circuit of Figure 11–60 after it has reached steady state.

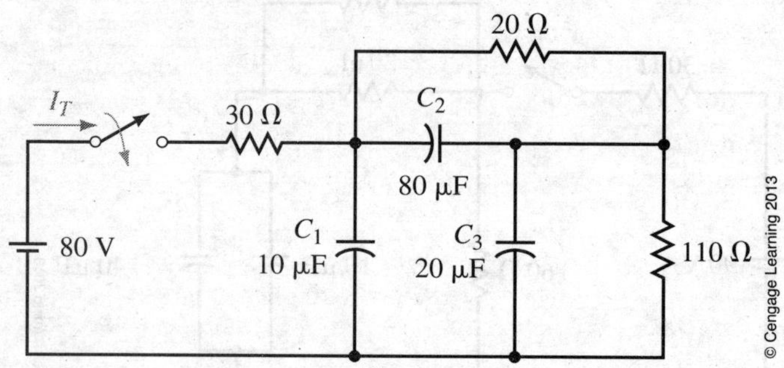

FIGURE 11–60

36. A black box containing dc sources and resistors has open-circuit voltage of 45 volts as in Figure 11–61(a). When the output is shorted as in (b), the short-circuit current is 1.5 mA. A switch and an uncharged 500-μF capacitor are connected as in (c). Determine the capacitor voltage and current 25 s after the switch is closed.

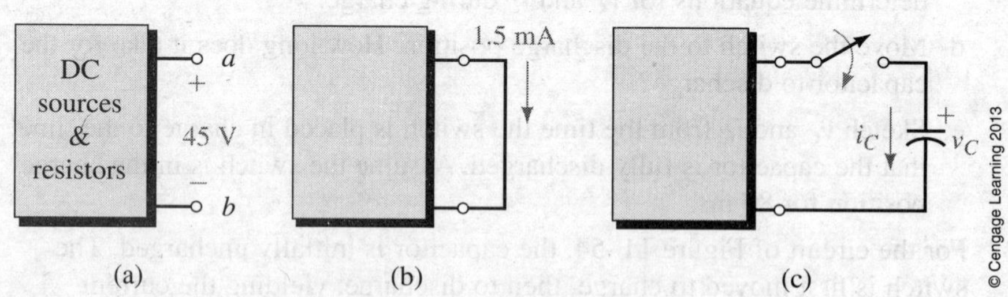

FIGURE 11–61

11.6 An *RC* Timing Application

37. For the alarm circuit of Figure 11–32, if the input from the sensor is 5 V, $R = 750\ \text{k}\Omega$, and the alarm is activated at 15 s when $v_C = 3.8$ V, what is C?
38. For the alarm circuit of Figure 11–32, the input from the sensor is 5 V, $C = 47\ \mu\text{F}$, and the alarm is activated when $v_C = 4.2$ V. Choose the nearest standard resistor value to achieve a delay of at least 37 s.

11.7 Pulse Response of *RC* Circuits

39. Consider the waveform of Figure 11–62.
    a. What is the period?
    b. What is the duty cycle?
    c. What is the PRR?
40. Repeat Problem 39 for the waveform of Figure 11–63.

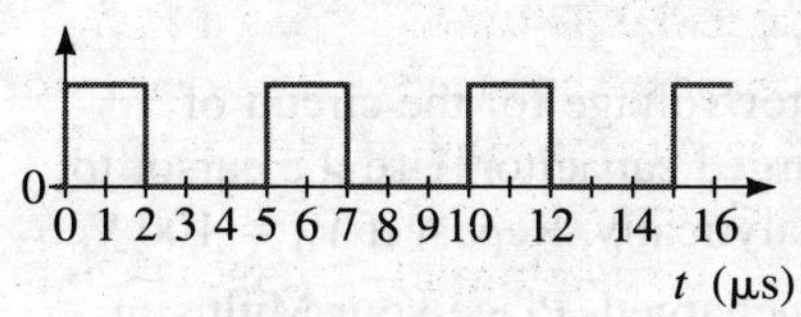

FIGURE 11–62

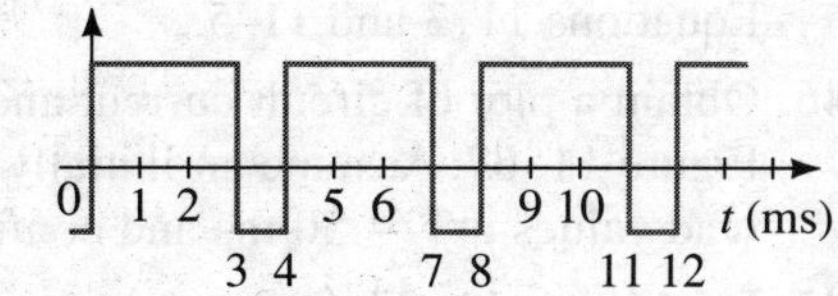

FIGURE 11–63

41. Determine the rise time, fall time, and pulse width for the pulse in Figure 11–64.
42. A single pulse is input to the circuit of Figure 11–65. Assuming that the capacitor is initially uncharged, sketch the output for each of the following sets of values:
    a. $R = 2\ \text{k}\Omega$, $C = 1\ \mu\text{F}$.
    b. $R = 2\ \text{k}\Omega$, $C = 0.1\ \mu\text{F}$.

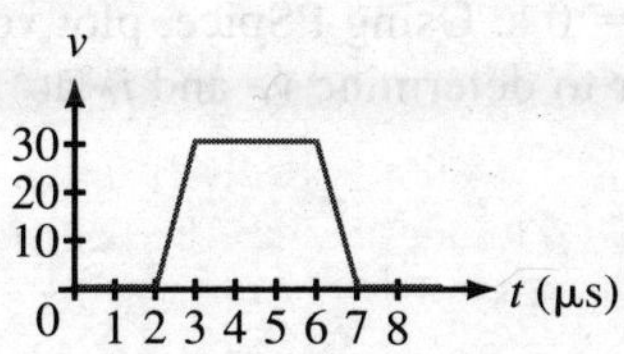

FIGURE 11–64

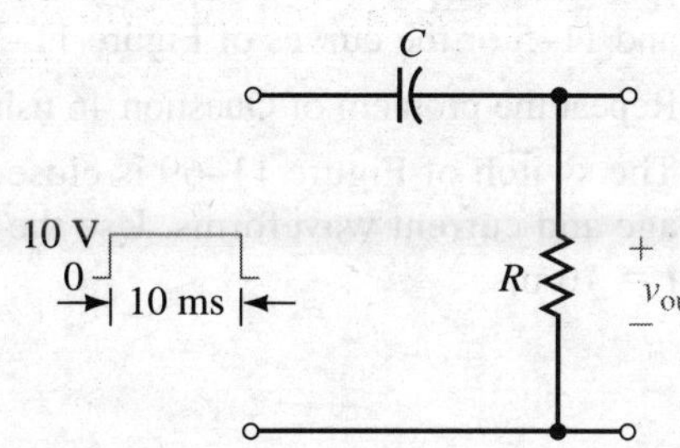

FIGURE 11–65

43. A switch (not shown) is closed to apply *E* volts to the circuit of Figure 11–66. If $R = 150\ \Omega$ and $C = 20$ pF, estimate the rise time of the output voltage.
44. A pulse train is input to the circuit of Figure 11–66. Assuming that the capacitor is initially uncharged, sketch the output for each of the following sets of values after the circuit has reached steady state:
    a. $R = 2\ \text{k}\Omega$, $C = 0.1\ \mu\text{F}$.
    b. $R = 20\ \text{k}\Omega$, $C = 1.0\ \mu\text{F}$.

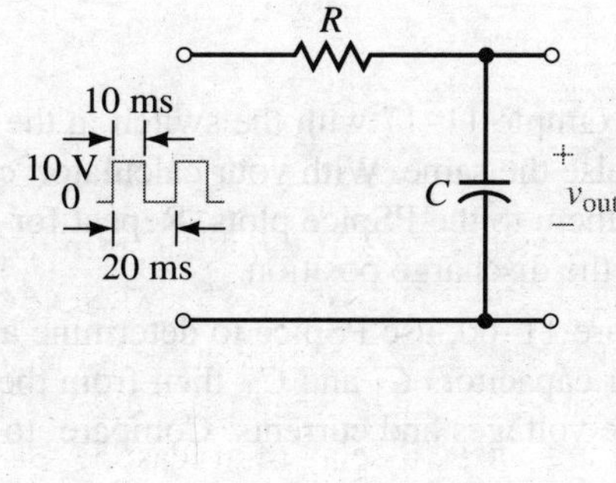

FIGURE 11–66

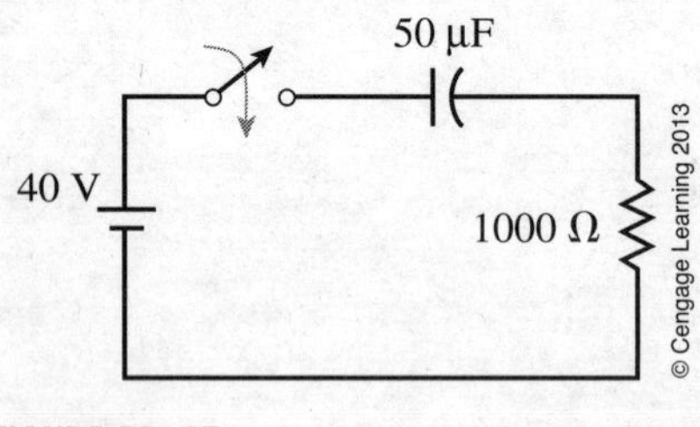

FIGURE 11–67

11.8 Transient Analysis Using Computers

Read the Notes, then use Multisim or PSpice to solve the following problems. (Note 4 applies only to Multisim users.) For Multisim current waveforms, you will need to add a new scale to expand the trace as you did in the examples of Section 11.8.

MULTISIM

45. Graph capacitor voltage and current for the circuit of Figure 11–2 with $E = -25$ V, $R = 40\ \Omega$, $V_0 = 0$ V, and $C = 400\ \mu$F. Scale values from the plot at $t = 20$ ms intervals using the cursor. Confirm your results using Equations 11–3 and 11–5.

MULTISIM

46. Obtain a plot of circuit current and resistor voltage for the circuit of Figure 11–67. Assume an initially uncharged capacitor. Use the cursor to read values at $t = 50$ ms and confirm analytically. Repeat if $V_0 = 100$ V.

MULTISIM

47. For Figure 11–57, both capacitors are uncharged. Place your Multisim ground at the bottom end of the capacitors and:
    a. Plot capacitor voltage and find v_C at $t = 4$ ms.
    b. Determine the current in the 4-Ω resistor at $t = 3.5$ ms.

MULTISIM

48. Using Multisim, construct Figure 11–68 on your screen. (The source is found in the Sources Group. Click *Place/Component* to find it.) Repeat Example 11–17 and plot capacitor voltage and current. You should get the curves of Figure 11–49(b) as your solution.

PSPICE

49. Using PSpice, graph capacitor voltage and current for a charging circuit with $E = -25$ V, $R = 40\ \Omega$, $V_0 = 0$ V, and $C = 400\ \mu$F. Scale values from the plot using the cursor. Compare to the results you get using Equations 11–3 and 11–5 or the curves of Figure 11–15.

PSPICE

50. Repeat the problem of Question 46 using PSpice.

PSPICE

51. The switch of Figure 11–69 is closed at $t = 0$ s. Using PSpice, plot voltage and current waveforms. Use the cursor to determine v_C and i_C at $t = 10$ ms.

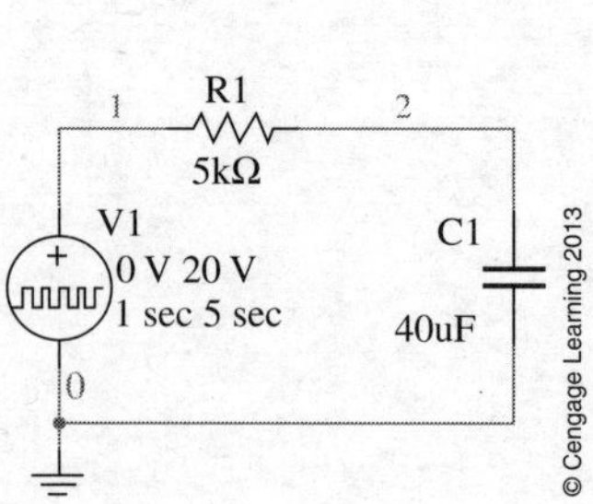

FIGURE 11–68

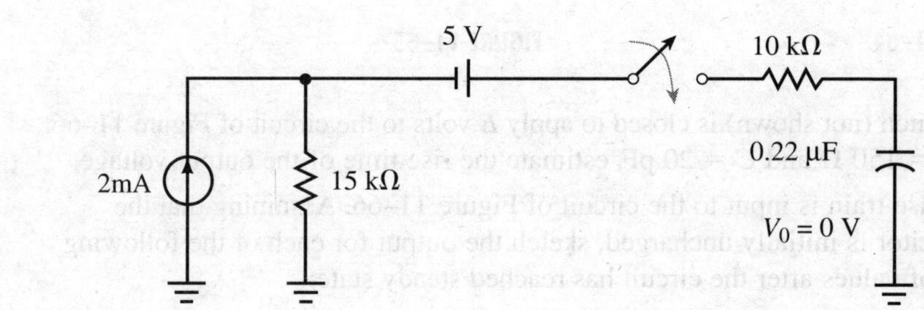

FIGURE 11–69

PSPICE

52. Using PSpice, redo Example 11–17 with the switch in the charge position for 0.5 s and everything else the same. With your calculator, compute v_C and i_C at 0.5 s and compare them to the PSpice plots. Repeat for i_C just after moving the switch to the discharge position.

PSPICE

53. For the circuit of Figure 11–60, use PSpice to determine and plot voltage and current waveforms for capacitors C_1 and C_3, then from the waveforms, determine their steady state voltages and currents. Compare to the answers of Problem 35.

ANSWERS TO IN-PROCESS LEARNING CHECKS

IN-PROCESS LEARNING CHECK 1

1. 0.2 A
2. a. 50 ms

   b.

t(ms)	i_c(mA)
0	50
25	30.3
50	18.4
75	11.2
100	6.8
250	0.337

3.

t(ms)	v_c(V)
0	0
25	71.3
50	91.8
75	97.7
100	99.3
250	100

4. $40(1 - e^{-5t})$ V, $200e^{-5t}$ mA, 38.0 V, 9.96 mA
5. a. $v_C(0^+) = v_C(0^-) = 0$; b. $i_C(0^-) = 0$; $i_C(0^+) = 10$ mA; c. 100 V, 0 A
6. 80 V, 40 kΩ, 0.2 μF

IN-PROCESS LEARNING CHECK 2

1. a. $40 + 40e^{-5t}$ V. v_C starts at 80 V and decays exponentially to 40 V.

   b. There is no transient since initial value = final value.

   c. $40 - 100e^{-5t}$ V. v_C starts at −60 V and climbs exponentially to 40 V.
2. 0.1833 s
3. a. $100e^{-10t}$ V; $-10e^{-10t}$ mA; v_C starts at 100 V and decays to 0 in 0.5 s (i.e., 5 time constants); i_C starts at −10 mA and decays to 0 in 0.5 s.

   b. $-100e^{-10t}$ V; $10e^{-10t}$ mA; v_C starts at −100 V and decays to 0 in 0.5 s (i.e., 5 time constants); i_C starts at 10 mA and decays to 0 in 0.5 s.
4. a., b., and c. Same as Example 11–12.

   d. $-30 + 37.14e^{-250t}$ V e. $-37.14e^{-250t}$ mA

   f.

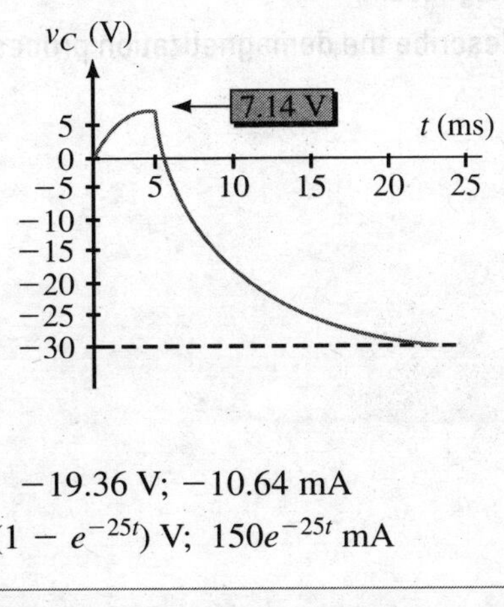

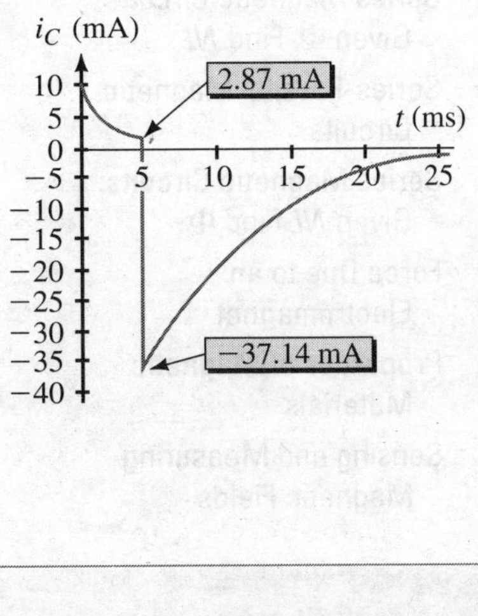

   g. −19.36 V; −10.64 mA
5. $6(1 - e^{-25t})$ V; $150e^{-25t}$ mA

NOTES...

1. Note the icon beside each problem, and use the corresponding simulation software, Multisim or PSpice for its solution.
2. Since current is discontinuous at the instant of switching, you get a meaningless value if you try to use a cursor to measure current at $t = 0$.
3. Switches are generally shown in the problem sets to denote transients, but since you use software (i.e., the Run button), to initiate Multisim and PSpice transients, switches are not usually needed.
4. Multisim examples in this book were created using Multisim 11 (the version current at the time of writing). If you are using legacy software, you may find it awkward to determine currents. In this case, simulate voltage waveforms only.

■ KEY TERMS

Ampere's Circuital Law
Ampere-Turns
B-H Curves
Domains
Ferromagnetic
Flux Density (B)
Hysteresis Loop
Magnetic Circuit
Magnetic Field
Magnetic Field Intensity (H)
Magnetic Flux
Magnetomotive Force (mmf)
Permeability
Reluctance ($\Re$)
Residual Magnetism
Saturation
Tesla (T)
Weber (Wb)

■ OUTLINE

The Nature of a Magnetic Field
Electromagnetism
Magnetic Flux and Flux Density
Magnetic Circuits and Their Applications
Air Gaps, Fringing, and Laminated Cores
Series Elements and Parallel Elements
Magnetic Circuits with dc Excitation
Magnetic Field Intensity and Magnetization Curves
Ampere's Circuital Law
Series Magnetic Circuits: Given Φ, Find NI
Series-Parallel Magnetic Circuits
Series Magnetic Circuits: Given NI, Find Φ
Force Due to an Electromagnet
Properties of Magnetic Materials
Sensing and Measuring Magnetic Fields

■ OBJECTIVES

After studying this chapter, you will be able to

- represent magnetic fields using Faraday's flux concept,
- describe magnetic fields quantitatively in terms of flux and flux density,
- explain what magnetic circuits are and why they are used,
- determine magnetic field intensity or magnetic flux density from a *B-H* curve,
- solve series magnetic circuits,
- solve series-parallel magnetic circuits,
- compute the attractive force of an electromagnet,
- explain the domain theory of magnetism,
- describe the demagnetization process.

12

MAGNETISM AND MAGNETIC CIRCUITS

CHAPTER PREVIEW

Many common devices rely on magnetism. Familiar examples include computer disk drives, audio system components such as headphones and speakers, transformers, generators, motors, and medical devices such as MRI machines. To understand their operation, you need a knowledge of magnetism and magnetic circuit principles. In this chapter, we look at fundamentals of magnetism, relationships between electrical and magnetic quantities, magnetic circuit concepts, and methods of analysis. In Chapter 13, we look at electromagnetic induction and inductance, and in Chapter 24, we apply magnetic principles to the study of transformers. ■

Putting It in Perspective

Magnetism and Electromagnetism

Photo Researchers/Photo Researchers/Getty Images

WHILE THE BASIC FACTS about magnetism have been known since ancient times, it was not until the early 1800s that the connection between electricity and magnetism was made and the foundations of modern electromagnetic theory laid down.

In 1819, Hans Christian Oersted, a Danish scientist, demonstrated that electricity and magnetism were related when he showed that a compass needle was deflected by a current-carrying conductor. The following year, Andre Ampere (1775–1836) showed that current-carrying conductors attract or repel each other just like magnets. However, it was Michael Faraday (recall Chapter 10) who developed our present concept of the magnetic field as a collection of flux lines in space that conceptually represent both the intensity and the direction of the field. It was this concept that led to an understanding of magnetism and the development of important practical devices such as the transformer and the electric generator.

In 1873, James Clerk Maxwell (see photo), a Scottish scientist, tied the then known theoretical and experimental concepts together and developed a unified theory of electromagnetism that predicted the existence of radio waves. Some 30 years later, Heinrich Hertz, a German physicist, showed experimentally that such waves existed, thus verifying Maxwell's theories and paving the way for modern radio and television. ■

12.1 The Nature of a Magnetic Field

Magnetism refers to the force that acts between magnets and magnetic materials. We know, for example, that magnets attract pieces of iron, deflect compass needles, attract or repel other magnets, and so on. This force acts at a distance and without the need for direct physical contact. The region where the force is felt is called the "field of the magnet" or simply, its **magnetic field.** Thus, *a magnetic field is a force field.*

NOTES...

Flux is perhaps an unfortunate name to apply to a magnetic field. Flux suggests a flow, but in a magnetic field, nothing actually flows; a magnetic field is simply a condition of space, that is, a region in which magnetic force exists. Nonetheless, the concept of flux is enormously helpful as an aid to visualizing magnetic phenomena, and we will continue to use it for that purpose.

Magnetic Flux

Faraday's flux concept (recall Putting It in Perspective, Chapter 10) helps us visualize this field. Using Faraday's representation, magnetic fields are drawn as lines in space. These lines, called **magnetic flux** lines or lines of force, show the direction and intensity of the field at all points. This is illustrated in Figure 12–1 for the field of a bar magnet. As indicated, the field is strongest at the poles of the magnet (where flux lines are most dense), its direction is from north (N) to south (S) external to the magnet, and flux lines never cross. The symbol for magnetic flux (Figure 12–1) is the Greek letter Φ (phi).

Figure 12–2 shows what happens when two magnets are brought close together. In (a), unlike poles attract, and flux lines pass from one magnet to the other. In (b), like poles repel, and the flux lines are pushed back as indicated by the flattening of the field between the two magnets.

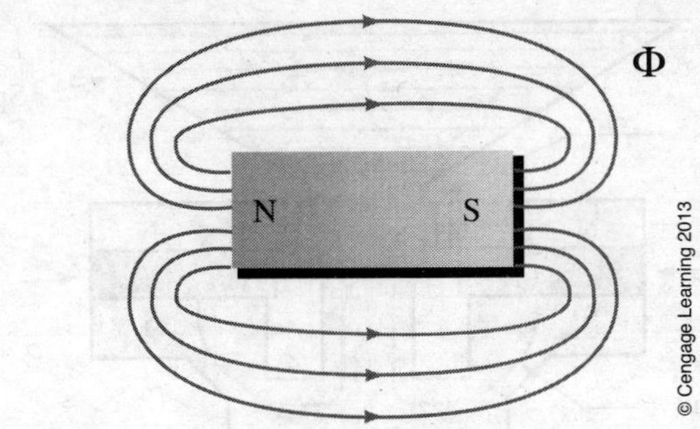

FIGURE 12–1 Field of a bar magnet. Magnetic flux is denoted by the symbol Φ.

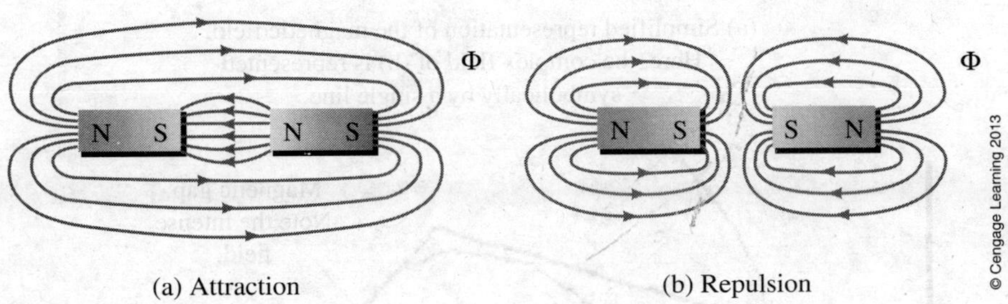

FIGURE 12–2 Field patterns due to attraction and repulsion.

Ferromagnetic Materials

Magnetic materials (materials that are attracted by magnets such as iron, nickel, cobalt, and their alloys) are called **ferromagnetic** materials. Ferromagnetic materials provide an easy path for magnetic flux. This is illustrated in Figure 12–3 where the flux lines take the longer (but easier) path through the soft iron, rather than the shorter path (of Figure 12–1) that they would normally take. Note, however, that nonmagnetic materials (plastic, wood, glass, and so on) have no effect on the field.

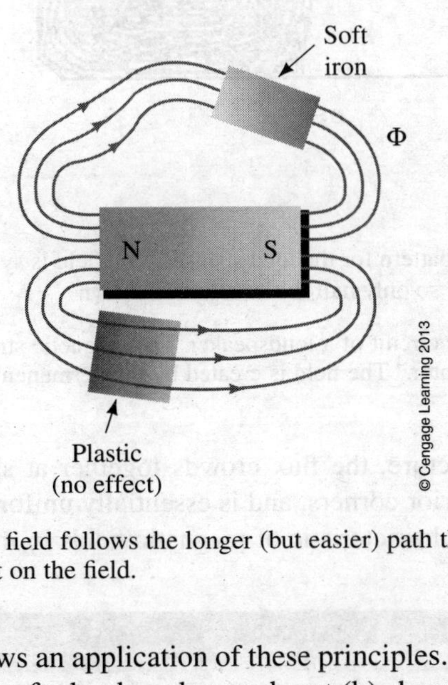

FIGURE 12–3 Magnetic field follows the longer (but easier) path through the iron. The plastic has no effect on the field.

Figure 12–4 shows an application of these principles. Part (a) shows a simplified representation of a loudspeaker, and part (b) shows expanded details of its magnetic field. The field is created by the permanent magnet, and the iron pole pieces guide the field and concentrate it in the gap where the speaker coil is placed. (For a description of how the speaker works, see Section 12.4.)

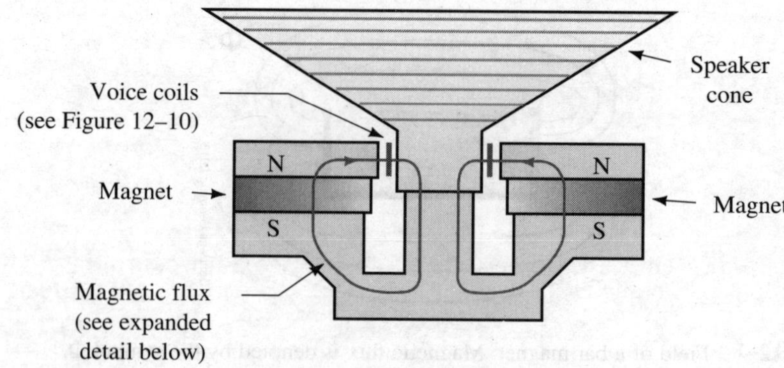

(a) Simplified representation of the magnetic field. Here, the complex field of (b) is represented symbolically by a single line.

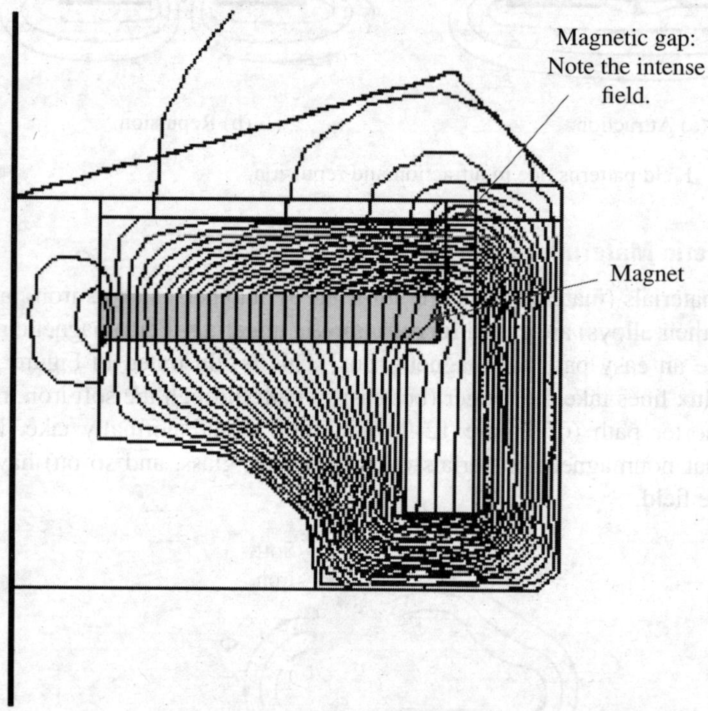

(b) Magnetic field pattern for the loud speaker. The field is symmetrical so only half the structure is shown.

FIGURE 12–4 Magnetic circuit of a loudspeaker. The magnetic structure and voice coil are called a "speaker motor." The field is created by the permanent magnet.

Within the iron structure, the flux crowds together at sharp interior corners, spreads apart at exterior corners, and is essentially uniform elsewhere, as indicated in (b). This is characteristic of magnetic fields in iron.

12.2 Electromagnetism

Most applications of magnetism involve magnetic effects due to electric currents. We look first at some basic principles. Consider Figure 12–5. The current, *I*, creates a magnetic field that is concentric about the conductor, uniform

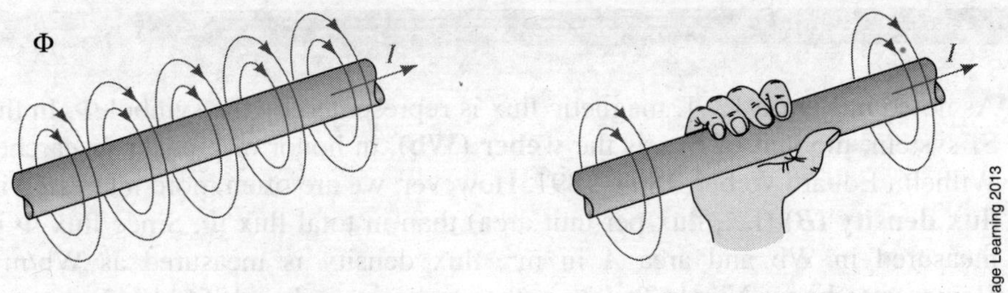

(a) Magnetic field produced by current. Field is proportional to I.

(b) Right-hand rule

FIGURE 12–5 Field about a current-carrying conductor. If current is reversed, the field remains concentric but the direction of the flux lines reverses.

along its length, and whose strength is directly proportional to I. Note the direction of the field. It may be remembered with the aid of the right-hand rule. As indicated in (b), imagine placing your right hand around the conductor with your thumb pointing in the direction of current. Your fingers then point in the direction of the field. If you reverse the direction of the current, the direction of the field reverses. If the conductor is wound into a coil, the fields of its individual turns combine, producing a resultant field as in Figure 12–6. The direction of the coil flux can also be remembered by means of a simple rule: curl the fingers of your right hand around the coil in the direction of the current and your thumb will point in the direction of the field. If the direction of the current is reversed, the field also reverses. Provided no ferromagnetic material is present, the strength of the coil's field is directly proportional to its current.

If the coil is wound on a ferromagnetic core as in Figure 12–7 (transformers are built this way), almost all flux is confined to the core, although a small amount (called stray or leakage flux) passes through the surrounding air. However, now that ferromagnetic material is present, the core flux is no longer proportional to current. The reason for this is discussed in Section 12.14.

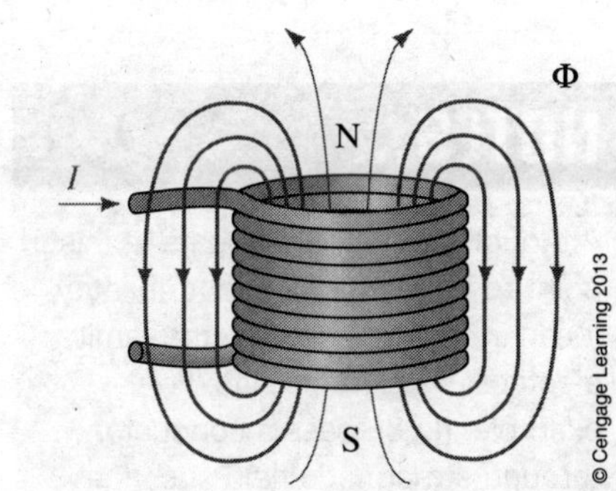

FIGURE 12–6 Field produced by a coil.

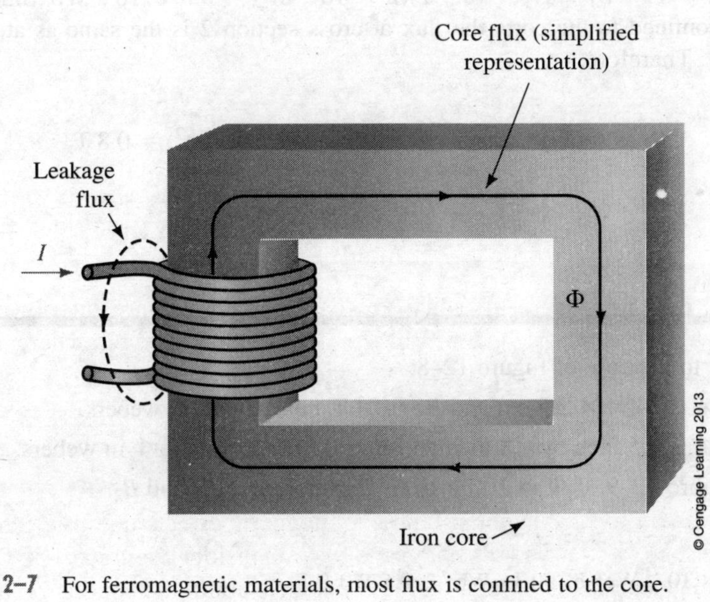

FIGURE 12–7 For ferromagnetic materials, most flux is confined to the core.

12.3 Magnetic Flux and Flux Density

As noted in Figure 12–1, magnetic flux is represented by the symbol Φ. In the SI system, the unit of flux is the **weber (Wb),** in honor of pioneer researcher Wilhelm Eduard Weber, 1804–1891. However, we are often more interested in **flux density (*B*)** (i.e., flux per unit area) than in total flux Φ. Since flux Φ is measured in Wb and area A in m^2, flux density is measured as Wb/m^2. However, to honor Nikola Tesla (another early researcher, 1856–1943) the unit of flux density is called the **tesla (T)** where $1\ T = 1\ Wb/m^2$. Flux density is found by dividing the total flux passing perpendicularly through an area by the size of the area, Figure 12–8. That is,

$$B = \frac{\Phi}{A} \quad \text{(tesla, T)} \qquad \textbf{(12–1)}$$

Thus, if $\Phi = 600\ \mu Wb$ of flux pass perpendicularly through an area $A = 20 \times 10^{-4}\ m^2$, the flux density is $B = (600 \times 10^{-6}\ Wb)/(20 \times 10^{-4}\ m^2) = 0.3\ T$. The greater the flux density, the stronger the field.

NOTES...

Although the weber appears at this point to be just an abstract quantity, it can in fact be linked to the familiar electrical system of units. For example, if you pass a conductor through a magnetic field such that the conductor cuts the flux lines at the rate of 1 Wb per second, the voltage induced is 1 V.

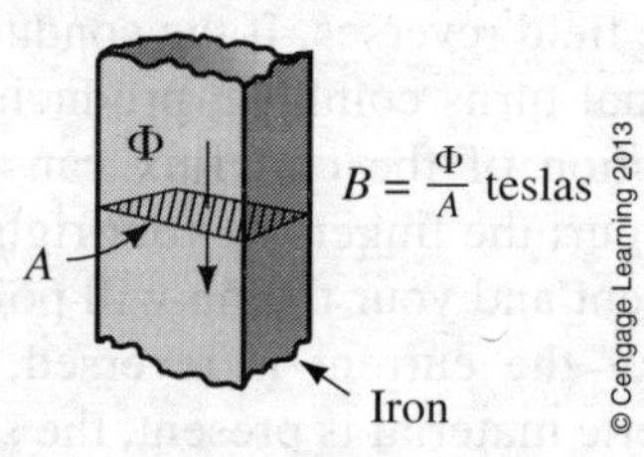

FIGURE 12–8 Concept of flux density. $1\ T = 1\ Wb/m^2$.

EXAMPLE 12–1

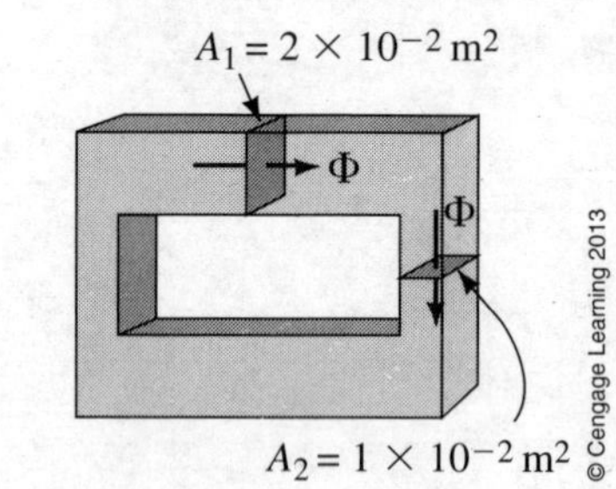

FIGURE 12–9

For the magnetic core of Figure 12–9, the flux density at cross section 1 is $B_1 = 0.4\ T$. Determine B_2.

Solution $\Phi = B_1 \times A_1 = (0.4\ T)(2 \times 10^{-2}\ m^2) = 0.8 \times 10^{-2}\ Wb$. Since all flux is confined to the core, the flux at cross section 2 is the same as at cross section 1. Therefore,

$$B_2 = \Phi/A_2 = (0.8 \times 10^{-2}\ Wb)/(1 \times 10^{-2}\ m^2) = 0.8\ T$$

PRACTICE PROBLEMS 1

1. Refer to the core of Figure 12–8:
   a. If A is 2 cm × 2.5 cm and $B = 0.4\ T$, compute Φ in webers.
   b. If A is 0.5 inch by 0.8 inch and $B = 0.35\ T$, compute Φ in webers.
2. In Figure 12–9, if $\Phi = 100 \times 10^{-4}\ Wb$, compute B_1 and B_2.

Answers

1. a. 2×10^{-4} Wb; b. 90.3 μWb; 2. 0.5 T; 1.0 T

To gain a feeling for the size of magnetic units, note that the strength of the earth's field is approximately 50 μT near the earth's surface, the field of a large generator or motor is on the order of 1 T or 2 T, and the largest fields yet produced (using superconducting magnets) are on the order of 45 T.

Other systems of units (now largely outdated) are the CGS system and the English systems. In the CGS system, flux is measured in maxwells and flux density in gauss. In the English system, flux is measured in lines and flux density in lines per square inch. Conversion factors are given in Table 12–1. We use only the SI system in this book.

TABLE 12–1 Magnetic Units Conversion Table

System	Flux (Φ)	Flux Density (*B*)
SI	webers (Wb)	teslas (T) 1 T = 1 Wb/m^2
English	lines 1 Wb = 10^8 lines	lines/in^2 1 T = 6.452 × 10^4 lines/in^2
CGS	maxwells 1 Wb = 10^8 maxwells	gauss 1 gauss = 1 maxwell/cm^2 1 T = 10^4 gauss

IN-PROCESS LEARNING CHECK 1

(Answers are at the end of the chapter.)

1. A magnetic field is a ______ field.
2. With Faraday's flux concept, the density of lines represents the ______ of the field and their direction represents the ______ of the field.
3. Three ferromagnetic materials are ______, ______, and ______.
4. The direction of a magnetic field is from ______ to ______ outside a magnet.
5. For Figures 12–5 and 12–6, if the direction of current is reversed, sketch what the fields look like.
6. If the core shown in Figure 12–7 is plastic, sketch what the field will look like.
7. Flux density B is defined as the ratio Φ/A, where A is the area ______ [parallel, perpendicular] to Φ.
8. For Figure 12–9, if A_1 is 2 cm × 2.5 cm, B_1 is 0.5 T, and B_2 = 0.25 T, what is A_2?

12.4 Magnetic Circuits and Their Applications

Many practical applications of magnetism use magnetic structures to guide and shape magnetic fields by providing well-defined paths for flux. Such structures are called **magnetic circuits.** Magnetic circuits are found in motors, generators, audio speakers, transformers, and so on. While most magnetic circuit applications involve magnetic fields created by electric currents, some involve a combination of fields created by electric currents and fixed (permanent) magnets.

The speaker structure of Figure 12–4 illustrates the latter. It uses a powerful magnet to create flux and an iron structure to guide this flux to the air gap to provide the intense field in which the voice coil is suspended. Figure 12–10 illustrates how the speaker actually works. First note that a source of audio

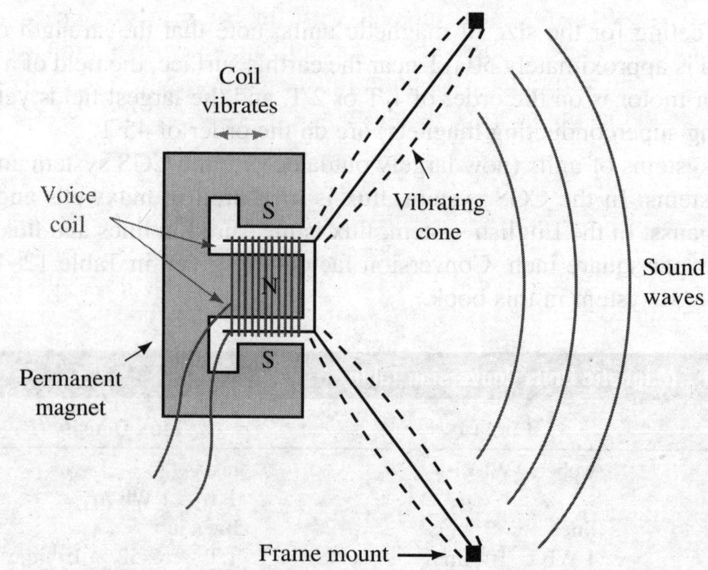

FIGURE 12–10 Audio speaker, conceptual representation.

(radio, sound reinforcement systems, etc.—recall Figure 1–5 of Chapter 1) outputs a current (proportional to the audio sound) to the speaker. The speaker utilizes a flexible cone to reproduce this sound. A coil of fine wire, wound around the cylindrical apex of this cone, is suspended in the fixed field of the air gap. Current from the amplifier passes through this coil, creating a varying magnetic field that interacts with the fixed field, creating varying forces that vibrate the cone. Since the current output by the amplifier is an amplified audio signal, the cone's vibrations correspond to this signal and thus, reproduce the original sound (in amplified form).

Figure 12–11 shows another example. Bank ATM cards, credit cards, library cards, travel reward cards, and a host of others feature a magnetic stripe that contains information encoded in the form of magnetic patterns. As the user swipes the card through the reader (past the read head), the varying magnetic field induces a voltage in the pickup winding. This tiny voltage is amplified, then sent to decoding circuitry where the data is recovered. As indicated, the pickup part of this system is a magnetic circuit.

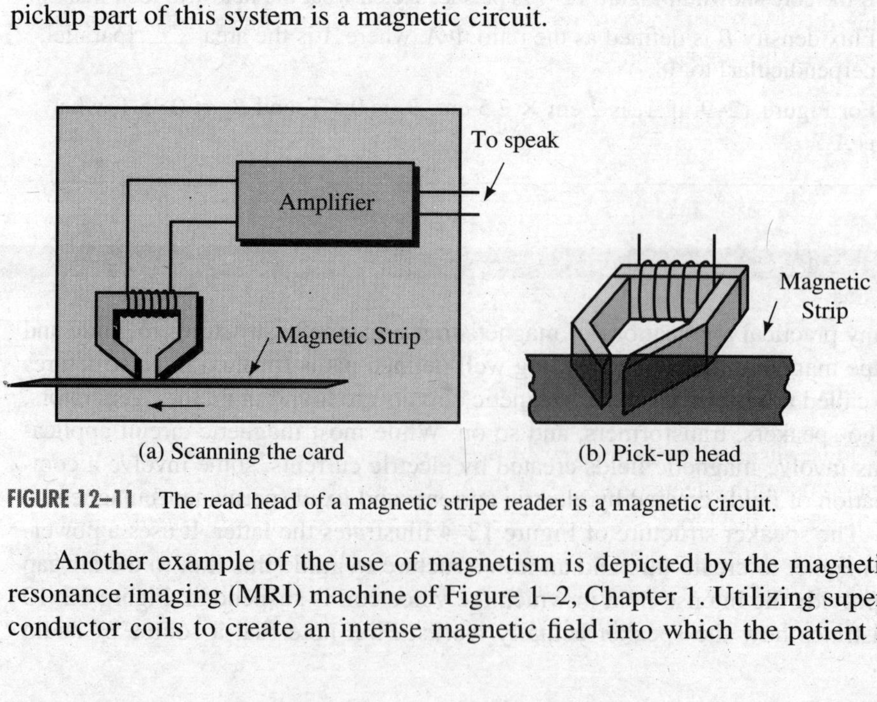

(a) Scanning the card (b) Pick-up head

FIGURE 12–11 The read head of a magnetic stripe reader is a magnetic circuit.

Another example of the use of magnetism is depicted by the magnetic resonance imaging (MRI) machine of Figure 1–2, Chapter 1. Utilizing superconductor coils to create an intense magnetic field into which the patient is placed, it uses radio waves that interact with this field to produce detailed images of internal organs of the body. The result is noninvasive, radiation-free body scanning. The particular MRI suite illustrated in Figure 1–2 utilizes a 3 T strength magnet as the heart of the system.

12.5 Air Gaps, Fringing, and Laminated Cores

Consider again Figures 12–10 and 12–11. Note that both structures have gaps in their magnetic flux paths. This is typical—most practical magnetic circuits have air gaps that are essential to their operation. However, at such gaps, fringing occurs. This fringing results in a slight weakening of the field in the gap as depicted in Figure 12–12(a). For short gaps, this weakening can often be neglected. Alternatively, you can estimate the effect of fringing in your calculations by increasing each cross-sectional dimension of the core by the length of the gap as illustrated in Example 12–2.

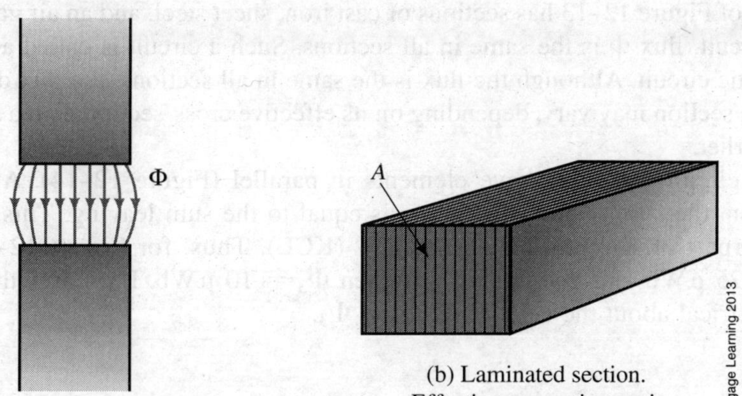

(a) Fringing at gap

(b) Laminated section. Effective magnetic area is less than the physical area.

FIGURE 12–12 Fringing and laminations.

EXAMPLE 12–2

A core with cross-sectional dimensions of 2.5 cm by 3 cm has a 0.1-mm gap. If flux density $B = 0.86$ T in the iron, what is the approximate uncorrected and corrected flux density in the gap?

Solution Neglecting fringing, gap area is the same as the core area. Thus, $B_g \approx 0.86$ T. Correcting for fringing yields

$$\Phi = BA = (0.86 \text{ T})(2.5 \times 10^{-2} \text{ m})(3 \times 10^{-2} \text{ m}) = 0.645 \text{ mWb}$$

$$A_g \simeq (2.51 \times 10^{-2} \text{ m})(3.01 \times 10^{-2} \text{ m}) = 7.555 \times 10^{-4} \text{ m}^2$$

Thus,

$$B_g \simeq 0.645 \text{ mWb}/7.555 \times 10^{-4} \text{ m}^2 = 0.854 \text{ T}$$

Now consider laminations. Many practical magnetic circuits (such as transformers) use thin sheets of stacked iron or steel as in Figure 12–12(b). Since the core is not a solid block, its effective cross-sectional area (i.e., the actual area of iron) is less than the core's physical cross-sectional area. A stacking factor, defined as the ratio of the actual area of ferrous material to the physical area of the core cross section, permits you to determine the core's effective area.

PRACTICE PROBLEMS 2

A product data sheet states that a laminated section of core has cross-sectional dimensions of 0.03 m by 0.05 m and a stacking factor of 0.9.

a. What is the effective area of the core?
b. Given $\Phi = 1.4 \times 10^{-3}$ Wb, what is the flux density, B?

Answers
a. 1.35×10^{-3} m^2; b. 1.04 T

12.6 Series Elements and Parallel Elements

Magnetic circuits may have sections of different materials. For example, the circuit of Figure 12–13 has sections of cast iron, sheet steel, and an air gap. For this circuit, flux Φ is the same in all sections. Such a circuit is called a series magnetic circuit. Although the flux is the same in all sections, the flux density in each section may vary, depending on its effective cross-sectional area as you saw earlier.

A circuit may also have elements in parallel (Figure 12–14). At each junction, the sum of fluxes entering is equal to the sum leaving. This is the counterpart of Kirchhoff's current law (KCL). Thus, for Figure 12–14, if $\Phi_1 = 25\ \mu$Wb and $\Phi_2 = 15\ \mu$Wb, then $\Phi_3 = 10\ \mu$Wb. For cores that are symmetrical about the center leg, $\Phi_2 = \Phi_3$.

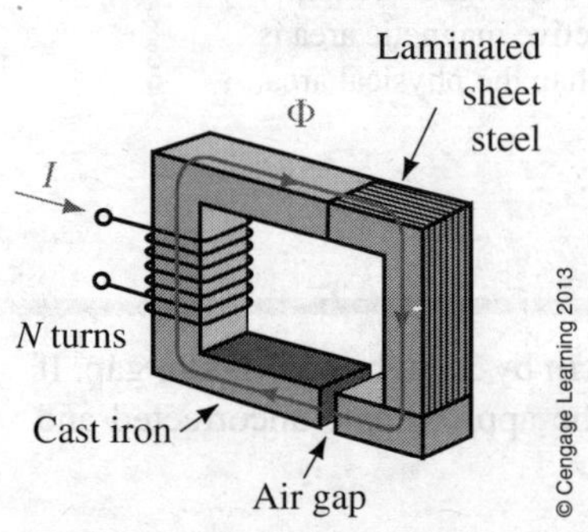

FIGURE 12–13 Series magnetic circuit. Flux Φ is the same throughout.

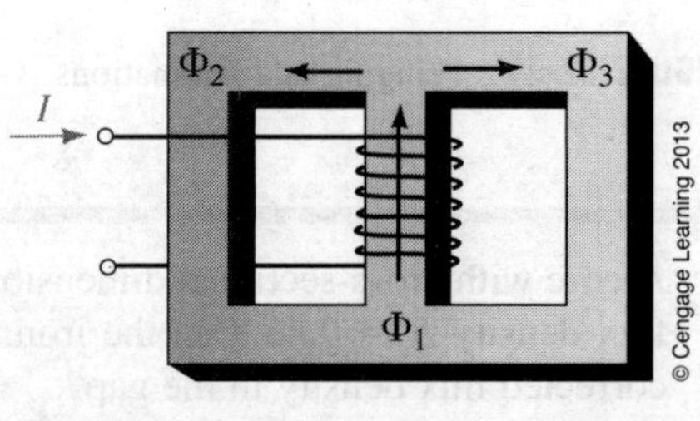

FIGURE 12–14 The sum of the flux entering a junction equals the sum leaving. Here, $\Phi_1 = \Phi_2 + \Phi_3$.

IN-PROCESS LEARNING CHECK 2

(Answers are at the end of the chapter.)

1. Why is the flux density in each section of Figure 12–13 different?
2. For Figure 12–13, $\Phi = 1.32$ mWb, the cross section of the core is 3 cm by 4 cm, the laminated section has a stacking factor of 0.8, and the gap is 1 mm. Determine the flux density in each section, taking fringing into account.
3. If the core of Figure 12–14 is symmetrical about its center leg, $B_1 = 0.4$ T, and the cross-sectional area of the center leg is 25 cm^2, what are Φ_2 and Φ_3?

12.7 Magnetic Circuits with dc Excitation

We now look at the analysis of magnetic circuits with dc excitation. There are two basic problems to consider: (1) given the flux, to determine the current required to produce it and (2) given the current, to compute the flux produced. To help visualize how to solve such problems, we first establish an analogy between magnetic circuits and electric circuits.

MMF: The Source of Magnetic Flux

Current through a coil creates magnetic flux. The greater the current or the greater the number of turns, the greater will be the flux. This flux-producing ability of a coil is called its **magnetomotive force (mmf)** and is measured in **ampere-turns.** Given the symbol $\mathcal{F}$, it is defined as

$$\mathcal{F} = NI \quad \text{(ampere-turns, At)} \qquad \textbf{(12–2)}$$

Thus, a coil with 100 turns and 2.5 amps will have an mmf of 250 ampere-turns, while a coil with 500 turns and 4 amps will have an mmf of 2000 ampere-turns.

Reluctance, $\mathfrak{R}$: Opposition to Magnetic Flux

Flux in a magnetic circuit also depends on the opposition that the circuit presents to it. Termed **reluctance ($\mathfrak{R}$),** this opposition depends on the dimensions of the core and the material of which it is made. Like the resistance of a wire, reluctance is directly proportional to length and inversely proportional to cross-sectional area. In equation form,

$$\mathfrak{R} = \frac{\ell}{\mu A} \quad \text{(At/Wb)} \qquad \textbf{(12–3)}$$

where μ is a property of the core material called its **permeability** (discussed in Section 12.8). Permeability is a measure of how easy it is to establish flux in a material. Ferromagnetic materials have high permeability and hence low $\mathfrak{R}$, while nonmagnetic materials have low permeability and high $\mathfrak{R}$.

Ohm's Law for Magnetic Circuits

The relationship between flux, mmf, and reluctance is

$$\Phi = \mathcal{F}/\mathfrak{R} \quad \text{(Wb)} \qquad \textbf{(12–4)}$$

This relationship is similar to Ohm's law and is depicted symbolically in Figure 12–15. (Remember however that flux, unlike electric current, does not flow—see note in Section 12.1.)

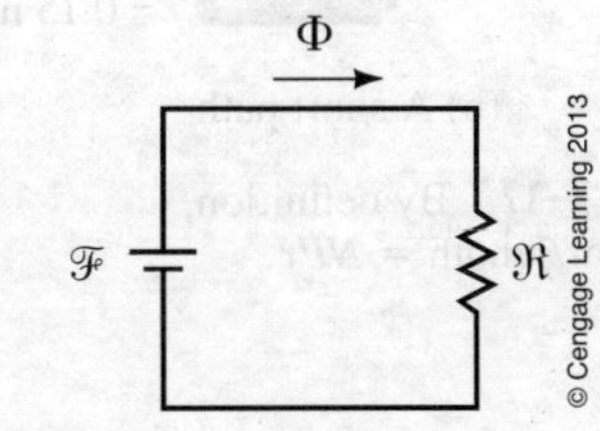

FIGURE 12–15 Electric circuit analogy of a magnetic circuit. $\Phi = \mathcal{F}/\mathfrak{R}$.

EXAMPLE 12–3

For Figure 12–16, if the reluctance of the magnetic circuit is $\mathfrak{R} = 12 \times 10^4$ At/Wb, what is the flux in the circuit?

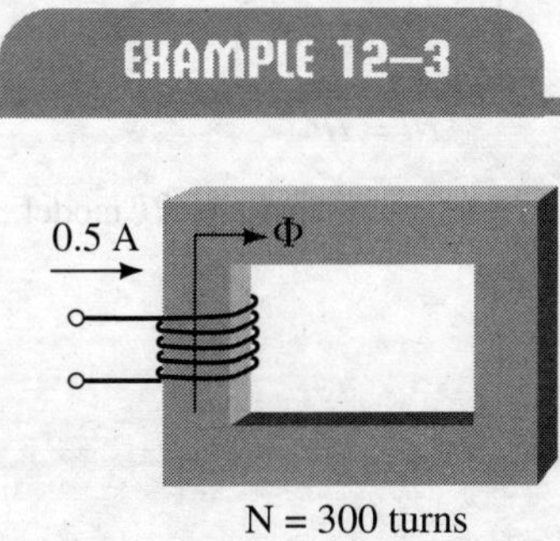

FIGURE 12–16

Solution

$$\mathcal{F} = NI = (300)(0.5\text{ A}) = 150\text{ At}$$

$$\Phi = \mathcal{F}/\mathfrak{R} = (150\text{ At})/(12 \times 10^4\text{ At/Wb}) = 12.5 \times 10^{-4}\text{ Wb}$$

In Example 12–3, we assumed that the reluctance of the core was constant. This is only approximately true under certain conditions. In general, it is not true, since $\mathfrak{R}$ is a function of flux density. Thus Equation 12–4 is not really very useful, since for ferromagnetic material, $\mathfrak{R}$ depends on flux, the very quantity that you are trying to find. The main use of Equations 12–3 and 12–4 is to provide an analogy between electric and magnetic circuit analysis.

12.8 Magnetic Field Intensity and Magnetization Curves

We now look at a more practical approach to analyzing magnetic circuits. First, we require a quantity called **magnetic field intensity, *H*** (also known as magnetizing force). It is a measure of the mmf per unit length of a circuit.

To get at the idea, suppose you apply the same mmf (say 600 At) to two circuits with different path lengths (Figure 12–17). In (a), you have 600 ampere-turns of mmf to "drive" flux through 0.6 m of core; in (b), you have the same mmf but it is spread across only 0.15 m of path length. Thus, the mmf per unit length in the second case is more intense. Based on this idea, one can define magnetic field intensity as the ratio of applied mmf to the length of path that it acts over. Thus,

$$H = \mathcal{F}/\ell = NI/\ell \quad \text{(At/m)} \tag{12–5}$$

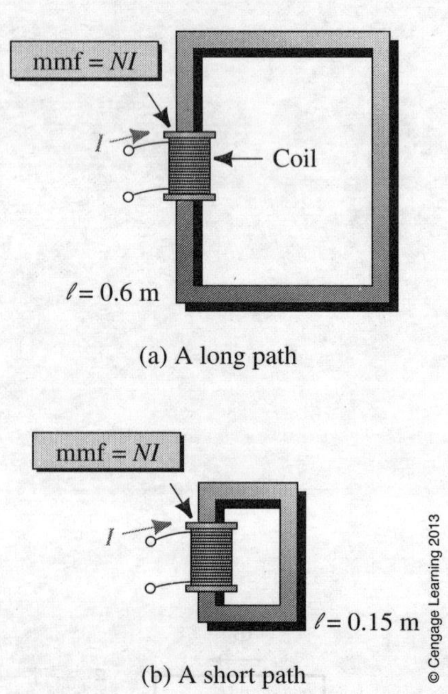

FIGURE 12–17 By definition, $H = \text{mmf/length} = NI/\ell$.

For the circuit of Figure 12–17(a), H = 600 At/0.6 m = 1000 At/m, while for the circuit of (b), H = 600 At/0.15 m = 4000 At/m. Thus, in (a) you have 1000 ampere-turns of "driving force" per meter of length to establish flux in the core, whereas in (b) you have four times as much. (However, you won't get four times as much flux, since the opposition to flux varies with the density of the flux.)

Rearranging Equation 12–5 yields an important result:

$$NI = H\ell \quad \text{(At)} \tag{12–6}$$

In an analogy with electric circuits (Figure 12–18), the NI product is an mmf source, while the $H\ell$ product is an mmf drop.

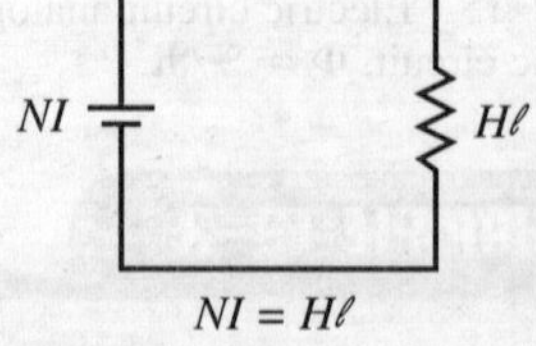

FIGURE 12–18 Circuit analogy, $H\ell$ model.

The Relationship between *B* and *H*

From Equation 12–5, you can see that magnetizing force, H, is a measure of the flux-producing ability of the coil (since it depends on NI). You also know that B is a measure of the resulting flux (since $B = \Phi/A$). Thus, B and H are related. The relationship is

$$B = \mu H \tag{12–7}$$

where μ is the permeability of the core (recall Equation 12–3).

It was stated earlier that permeability is a measure of how easy it is to establish flux in a material. To see why, note from Equation 12–7 that the larger the value of μ, the larger the flux density for a given H. However, H is proportional to current; therefore, the larger the value of μ, the larger the flux density for a given magnetizing current. From this, it follows that the larger the permeability, the more flux you get for a given magnetizing current.

In the SI system, μ has units of webers per ampere-turn-meter. The permeability of free space is $\mu_0 = 4\pi \times 10^{-7}$. For all practical purposes, the permeability of air and other nonmagnetic materials (e.g., plastic) is the same as for a vacuum. Thus, in air gaps,

$$B_g = \mu_0 H_g = 4\pi \times 10^{-7} \times H_g \tag{12–8}$$

Rearranging Equation 12–8 yields

$$H_g = \frac{B_g}{4\pi \times 10^{-7}} = 7.96 \times 10^5 B_g \quad \text{(At/m)} \qquad \textbf{(12–9)}$$

PRACTICE PROBLEMS 3

For Figure 12–16, the core cross section is 0.05 m × 0.08 m. If a gap is cut in the core and H in the gap is 3.6×10^5 At/m, what is the flux Φ in the core? Neglect fringing.

Answer
1.81 mWb

B-H Curves

For ferromagnetic materials, μ is not constant but varies with flux density, and there is no easy way to compute it. In reality, however, it isn't μ that you are interested in: What you really want to know is, given B, what is H, and vice versa. A set of curves, called ***B-H* curves,** provides this information. (These curves, also called magnetization curves, are obtained experimentally and are available in handbooks. A separate curve is required for each material.) Figure 12–19 shows typical curves for cast iron, cast steel, and sheet steel.

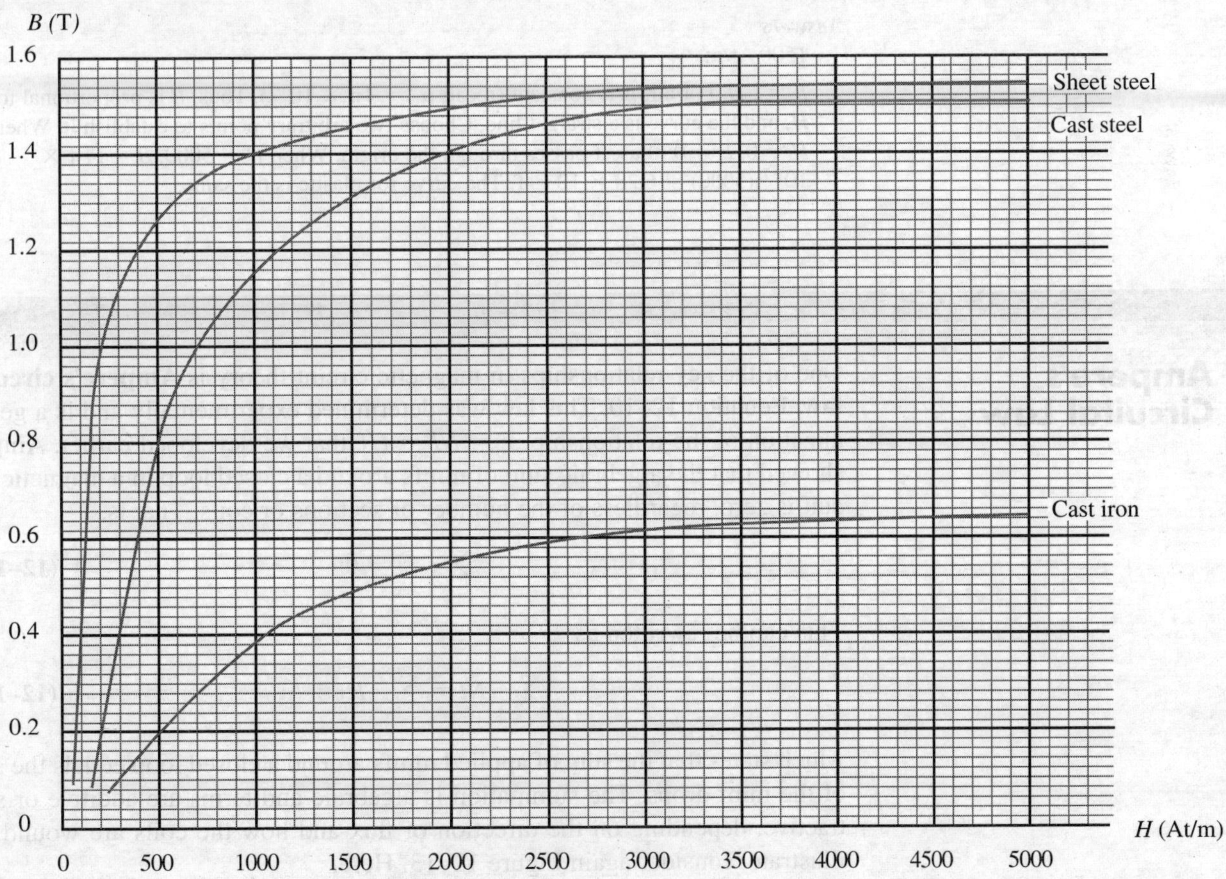

FIGURE 12–19 *B-H* curves for selected materials.

EXAMPLE 12–4

If $B = 1.4$ T for sheet steel, what is H?

Solution Enter Figure 12–19 on the axis at $B = 1.4$ T, continue across until you encounter the curve for sheet steel, then read the corresponding value for H as indicated in Figure 12–20: $H = 1000$ At/m.

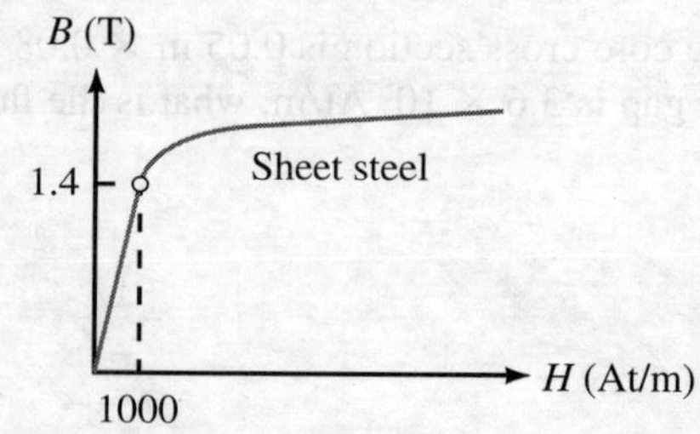

FIGURE 12–20 For sheet steel, $H = 1000$ At/m when $B = 1.4$ T.

PRACTICE PROBLEMS 4

1. The cross section of a sheet steel core is 0.1 m × 0.1 m and its stacking factor is 0.93. If $\Phi = 13.5$ mWb, what is H?
2. Sketch the B-H curves for both air and plastic.

Answers

1. 1500 At/m
2. $B = \mu H$. For air, μ is constant (recall $\mu_o = 4\pi \times 10^{-7}$). Thus, B is proportional to H, and the curve is a straight line. Choose two arbitrary points to establish it. When $H = 0$, $B = 0$; thus, it passes through the origin. When $H = 5000$, $B = (4\pi \times 10^{-7})(5000) = 6.28 \times 10^{-3}$ T. The curve for plastic is the same.

12.9 Ampere's Circuital Law

One of the key relationships in magnetic circuit theory is **Ampere's circuital law,** Equation 12–10. This law was determined experimentally and is a generalization of the relationship $\mathscr{F} = NI = H\ell$ that we developed earlier. Ampere showed that the algebraic sum of mmfs around a closed loop in a magnetic circuit is zero, regardless of the number of sections or coils. That is,

$$\sum_{\bigcirc} \mathscr{F} = 0 \qquad \textbf{(12–10)}$$

This can be rewritten as

$$\sum_{\bigcirc} NI = \sum_{\bigcirc} H\ell \quad \text{At} \qquad \textbf{(12–11)}$$

which states that the sum of applied mmfs around a closed loop equals the sum of the mmf drops. The summation is algebraic and terms are additive or subtractive, depending on the direction of flux and how the coils are wound. To illustrate, consider again Figure 12–13. Here,

$$NI - H_{\text{iron}}\ell_{\text{iron}} - H_{\text{steel}}\ell_{\text{steel}} - H_g\ell_g = 0$$

Thus,

$$\underbrace{NI}_{\text{Applied mmf}} = \underbrace{H_{iron}\ell_{iron} + H_{steel}\ell_{steel} + H_g\ell_g}_{\text{sum of mmf drops}}$$

The path to use for the $H\ell$ terms is the mean (average) path.

You now have two magnetic circuit models (Figure 12–21). While the reluctance model (a) is not very useful for solving problems, it helps relate magnetic circuit problems to familiar electrical circuit concepts. The Ampere's law model, on the other hand, permits us to solve practical problems. We look at how to do this in the next section.

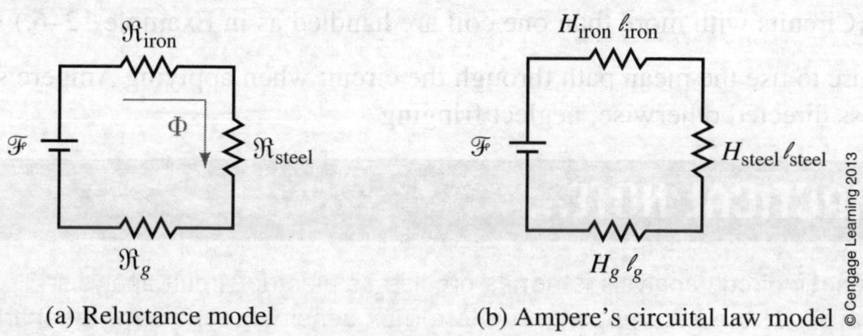

FIGURE 12–21 Two models for the magnetic circuit of Figure 12–13.

IN-PROCESS LEARNING CHECK 3

(Answers are at the end of the chapter.)

1. If the mmf of a 200-turn coil is 700 At, the current in the coil is ____ amps.
2. For Figure 12–17, if $H = 3500$ At/m and $N = 1000$ turns, then for (a), I is ____ A, while for (b), I is ____ A.
3. For cast iron, if $B = 0.5$ T, then $H =$ ____ At/m.
4. A series circuit consists of one coil, a section of iron, a section of steel, and two air gaps (of different sizes). Draw the Ampere's law model.
5. Which is the correct answer for the circuit of Figure 12–22?
   a. Ampere's law around loop 1 yields ($NI = H_1\ell_1 + H_2\ell_2$, or $NI = H_1\ell_1 - H_2\ell_2$).
   b. Ampere's law around loop 2 yields ($0 = H_2\ell_2 + H_3\ell_3$, or $0 = H_2\ell_2 - H_3\ell_3$).
6. For the circuit of Figure 12–23, the length ℓ to use in Ampere's law is ____ [0.36 m, 0.32 m, 0.28 m]. Why?

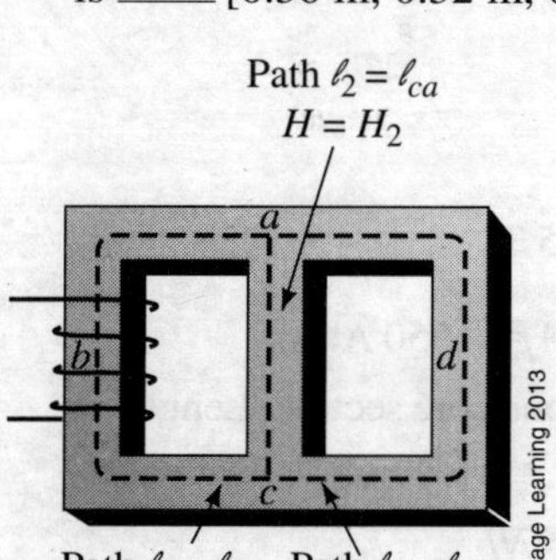

FIGURE 12–22

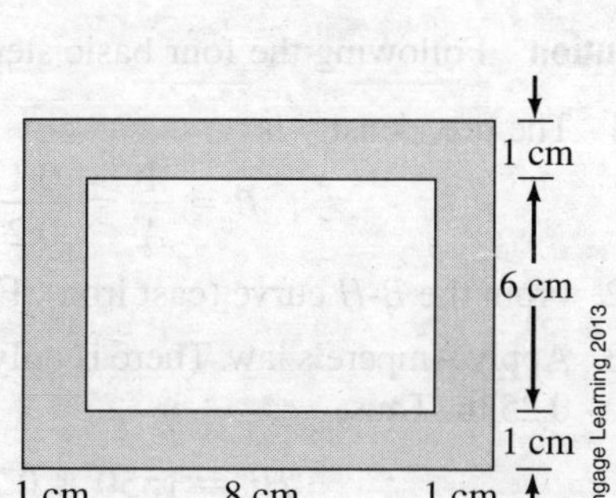

FIGURE 12–23

12.10 Series Magnetic Circuits: Given Φ, Find NI

You now have the tools needed to solve basic magnetic circuit problems. We will begin with series circuits where Φ is known and we want to find the excitation to produce it. Problems of this type can be solved using four basic steps:

1. Compute B for each section using $B = \Phi/A$.
2. Determine H for each magnetic section from the B-H curves. Use $H_g = 7.96 \times 10^5 B_g$ for air gaps.
3. Compute NI using Ampere's circuital law.
4. Use the computed NI to determine coil current or turns as required. (Circuits with more than one coil are handled as in Example 12–6.)

Be sure to use the mean path through the circuit when applying Ampere's law. Unless directed otherwise, neglect fringing.

PRACTICAL NOTE...

Magnetic circuit analysis is not as precise as electric circuit analysis because (1) the assumption of uniform flux density breaks down at sharp corners as you saw in Figure 12–4, and (2) the *B-H* curve is a mean curve and has considerable uncertainty as discussed later (Section 12–14).

Although the answers are approximate, they are adequate for many purposes.

EXAMPLE 12–5

If the core of Figure 12–24 is cast iron and $\Phi = 0.1 \times 10^{-3}$ Wb, what is the coil current?

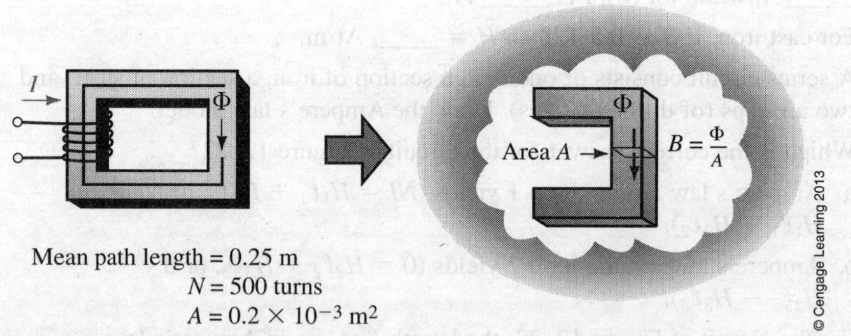

Mean path length = 0.25 m
$N = 500$ turns
$A = 0.2 \times 10^{-3}$ m²

FIGURE 12–24

Solution Following the four basic steps:

1. The flux density is

$$B = \frac{\Phi}{A} = \frac{0.1 \times 10^{-3}}{0.2 \times 10^{-3}} = 0.5 \text{ T}$$

2. From the *B-H* curve (cast iron), Figure 12–19, H = 1550 At/m.
3. Apply Ampere's law. There is only one coil and one core section. Length = 0.25 m. Thus,

$$H\ell = 1550 \times 0.25 = 388 \text{ At} = NI$$

4. Solve for I:

$$I = H\ell/N = 388/500 = 0.78 \text{ amps}$$

EXAMPLE 12–6

A second coil is added as shown in Figure 12–25. If $\Phi = 0.1 \times 10^{-3}$ Wb as before, but $I_1 = 1.5$ amps, what is I_2?

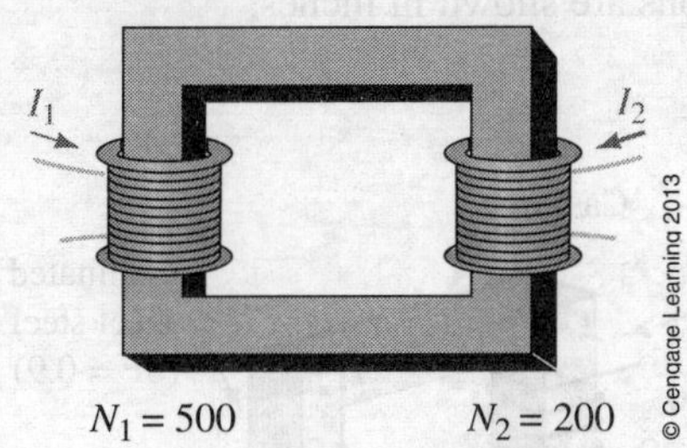

FIGURE 12–25

NOTES...

Since magnetic circuits are nonlinear, you cannot use superposition; that is, you cannot consider each coil of Figure 12–25 by itself, then sum the results. You must consider them simultaneously as we do in this example.

Solution From the previous example, you know that a current of 0.78 amps in coil 1 produces $\Phi = 0.1 \times 10^{-3}$ Wb. But you already have 1.5 amps in coil 1. Thus, coil 2 must be wound in opposition so that its mmf is subtractive. Applying Ampere's law yields $N_1I_1 - N_2I_2 = \mathrm{H}\ell$. Hence,

$$(500)(1.5\,A) - 200I_2 = 388 \text{ At}$$

and so $I_2 = 1.8$ amps.

More Examples

If a magnetic circuit contains an air gap, add another element to the conceptual models (recall Figure 12–21). Since air represents a poor magnetic path, its reluctance will be high compared with that of iron. Recalling our analogy to electric circuits, this suggests that the mmf drop across the gap will be large compared with that of the iron. You can see this in the following example.

EXAMPLE 12–7

The core of Figure 12–24 has a 0.008-m gap cut as shown in Figure 12–26. Determine how much the current must increase to maintain the original core flux. Neglect fringing.

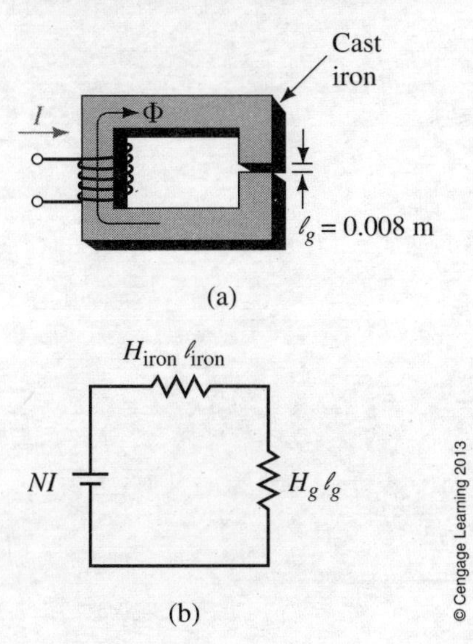

FIGURE 12–26

Solution

Iron

$\ell_{iron} = 0.25 - 0.008 = 0.242$ m. Since Φ does not change, B and H will be the same as before. Thus, $B_{iron} = 0.5$ T and $H_{iron} = 1550$ At/m.

Air Gap

B_g is the same as B_{iron}. Thus, $B_g = 0.5$ T and $H_g = 7.96 \times 10^5 B_g = 3.98 \times 10^5$ At/m.

Ampere's Law

$NI = H_{iron}\ell_{iron} + H_g\ell_g = (1550)(0.242) + (3.98 \times 10^5)(0.008) = 375 + 3184 = 3559$ At. Thus, $I = 3559/500 = 7.1$ amps. Note that the current had to increase from 0.78 amp to 7.1 amps in order to maintain the same flux, over a ninefold increase.

EXAMPLE 12–8

The laminated sheet steel section of Figure 12–27 has a stacking factor of 0.9. Compute the current required to establish a flux of $\Phi = 1.4 \times 10^{-4}$ Wb. Neglect fringing. All dimensions are shown in inches.

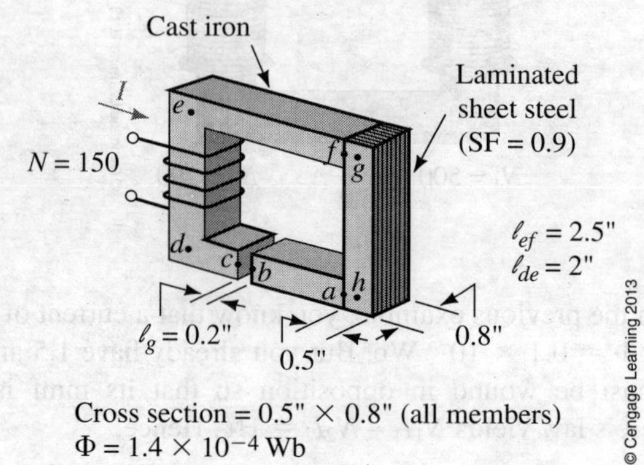

FIGURE 12–27

Solution Convert all dimensions to metric.

Cast Iron

$$\ell_{iron} = \ell_{adef} - \ell_g = 2.5 + 2 + 2.5 - 0.2 = 6.8 \text{ in} = 0.173 \text{ m}$$
$$A_{iron} = (0.5 \text{ in})(0.8 \text{ in}) = 0.4 \text{ in}^2 = 0.258 \times 10^{-3} \text{ m}^2$$
$$B_{iron} = \Phi/A_{iron} = (1.4 \times 10^{-4})/(0.258 \times 10^{-3}) = 0.54 \text{ T}$$
$$H_{iron} = 1850 \text{ At/m} \quad \text{(from Figure 12–19)}$$

Sheet Steel

$$\ell_{steel} = \ell_{fg} + \ell_{gh} + \ell_{ha} = 0.25 + 2 + 0.25 = 2.5 \text{ in} = 6.35 \times 10^{-2} \text{ m}$$
$$A_{steel} = (0.9)(0.258 \times 10^{-3}) = 0.232 \times 10^{-3} \text{ m}^2$$
$$B_{steel} = \Phi/A_{steel} = (1.4 \times 10^{-4})/(0.232 \times 10^{-3}) = 0.60 \text{ T}$$
$$H_{steel} = 125 \text{ At/m} \quad \text{(from Figure 12–19)}$$

Air Gap

$$\ell_g = 0.2 \text{ in} = 5.08 \times 10^{-3} \text{ m}$$
$$B_g = B_{iron} = 0.54 \text{ T}$$
$$H_g = (7.96 \times 10^5)(0.54) = 4.3 \times 10^5 \text{ At/m}$$

Ampere's Law

$$NI = H_{iron}\ell_{iron} + H_{steel}\ell_{steel} + H_g\ell_g$$
$$= (1850)(0.173) + (125)(6.35 \times 10^{-2}) + (4.3 \times 10^5)(5.08 \times 10^{-3})$$
$$= 320 + 7.9 + 2184 = 2512 \text{ At}$$
$$I = 2512/N = 2512/150 = 16.7 \text{ amps}$$

EXAMPLE 12–9

Figure 12–28 shows a portion of a solenoid. Flux $\Phi = 4 \times 10^{-4}$ Wb when $I = 2.5$ amps. Find the number of turns on the coil.

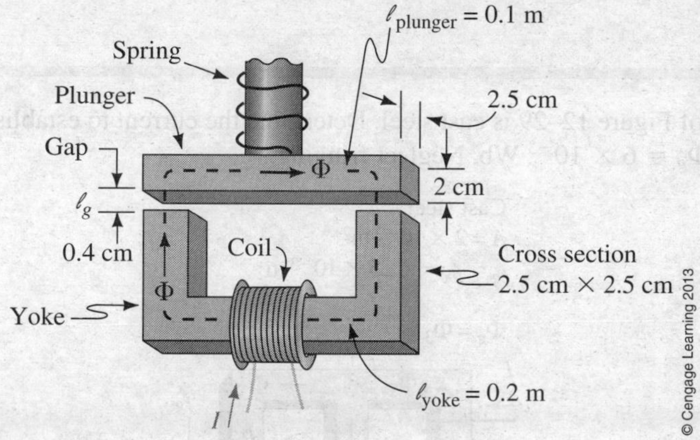

FIGURE 12–28 Solenoid. All parts are cast steel.

Solution

Yoke

$$A_{\text{yoke}} = 2.5 \text{ cm} \times 2.5 \text{ cm} = 6.25 \text{ cm}^2 = 6.25 \times 10^{-4} \text{ m}^2$$

$$B_{\text{yoke}} = \frac{\Phi}{A_{\text{yoke}}} = \frac{4 \times 10^{-4}}{6.25 \times 10^{-4}} = 0.64 \text{ T}$$

$$H_{\text{yoke}} = 410 \text{ At/m} \quad \text{(from Figure 12–19)}$$

Plunger

$$A_{\text{plunger}} = 2.0 \text{ cm} \times 2.5 \text{ cm} = 5.0 \text{ cm}^2 = 5.0 \times 10^{-4} \text{ m}^2$$

$$B_{\text{plunger}} = \frac{\Phi}{A_{\text{plunger}}} = \frac{4 \times 10^{-4}}{5.0 \times 10^{-4}} = 0.8 \text{ T}$$

$$H_{\text{plunger}} = 500 \text{ At/m} \quad \text{(from Figure 12–19)}$$

Air Gap

There are two identical gaps. For each,

$$B_g = B_{\text{yoke}} = 0.64 \text{ T}$$

Thus,

$$H_g = (7.96 \times 10^5)(0.64) = 5.09 \times 10^5 \text{ At/m}$$

The results are summarized in Table 12–2.

Ampere's Law

$$NI = H_{\text{yoke}}\ell_{\text{yoke}} + H_{\text{plunger}}\ell_{\text{plunger}} + 2H_g\ell_g = 82 + 50 + 2(2036) = 4204 \text{ At}$$

$$N = 4204/2.5 = 1682 \text{ turns}$$

TABLE 12–2

Material	Section	Length (m)	A (m²)	B (T)	H (At/m)	$H\ell$ (At)
Cast steel	yoke	0.2	6.25×10^{-4}	0.64	410	82
Cast steel	plunger	0.1	5×10^{-4}	0.8	500	50
Air	gap	0.4×10^{-2}	6.25×10^{-4}	0.64	5.09×10^5	2036

12.11 Series-Parallel Magnetic Circuits

Series-parallel magnetic circuits are handled using the sum of fluxes principle (Figure 12–14) and Ampere's law.

EXAMPLE 12–10

The core of Figure 12–29 is cast steel. Determine the current to establish an air-gap flux $\Phi_g = 6 \times 10^{-3}$ Wb. Neglect fringing.

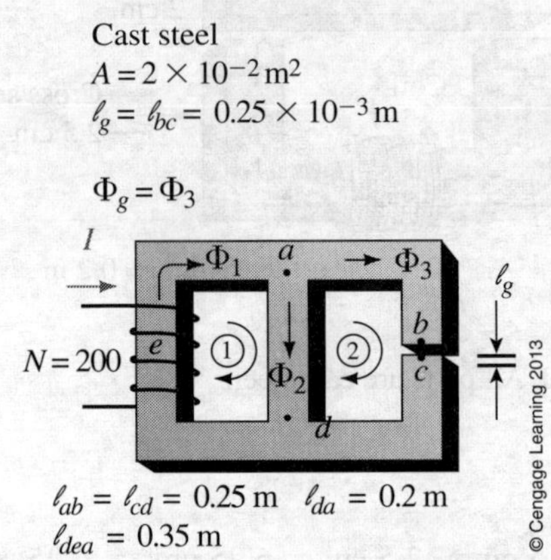

FIGURE 12–29

Solution Consider each section in turn.

Air Gap

$$B_g = \Phi_g/A_g = (6 \times 10^{-3})/(2 \times 10^{-2}) = 0.3 \text{ T}$$
$$H_g = (7.96 \times 10^5)(0.3) = 2.388 \times 10^5 \text{ At/m}$$

Sections ab and cd

$$B_{ab} = B_{cd} = B_g = 0.3 \text{ T}$$
$$H_{ab} = H_{cd} \doteq 250 \text{ At/m (from Figure 12–19)}$$

Ampere's Law (Loop 2)

$\sum_{\bigcirc} NI = \sum_{\bigcirc} H\ell$. Since you are going opposite to flux in leg *da*, the corresponding term (i.e., $H_{da}\ell_{da}$) will be subtractive. Also, $NI = 0$ for loop 2. Thus,

$$0 = \sum_{\bigcirc \text{ loop2}} H\ell$$

$$0 = H_{ab}\ell_{ab} + H_g\ell_g + H_{cd}\ell_{cd} - H_{da}\ell_{da}$$
$$= (250)(0.25) + (2.388 \times 10^5)(0.25 \times 10^{-3}) + (250)(0.25) - 0.2H_{da}$$
$$= 62.5 + 59.7 + 62.5 - 0.2H_{da} = 184.7 - 0.2H_{da}$$

Thus, $0.2H_{da} = 184.7$ and $H_{da} = 925$ At/m. From Figure 12–19, $B_{da} = 1.12$ T.

$$\Phi_2 = B_{da}A = 1.12 \times 0.02 = 2.24 \times 10^{-2} \text{ Wb}$$
$$\Phi_1 = \Phi_2 + \Phi_3 = 2.84 \times 10^{-2} \text{ Wb.}$$
$$B_{dea} = \Phi_1/A = (2.84 \times 10^{-2})/0.02 = 1.42 \text{ T}$$
$$H_{dea} = 2125 \text{ At/m} \quad \text{(from Figure 12–19)}$$

Ampere's Law (Loop 1)

$$NI = H_{dea}\ell_{dea} + H_{ad}\ell_{ad} = (2125)(0.35) + 184.7 = 929 \text{ At}$$
$$I = 929/200 = 4.65 \text{ A}$$

PRACTICE PROBLEMS 5

The cast-iron core of Figure 12–30 is symmetrical. Determine current I. Hint: To find NI, you can write Ampere's law around either loop. Be sure to make use of symmetry.

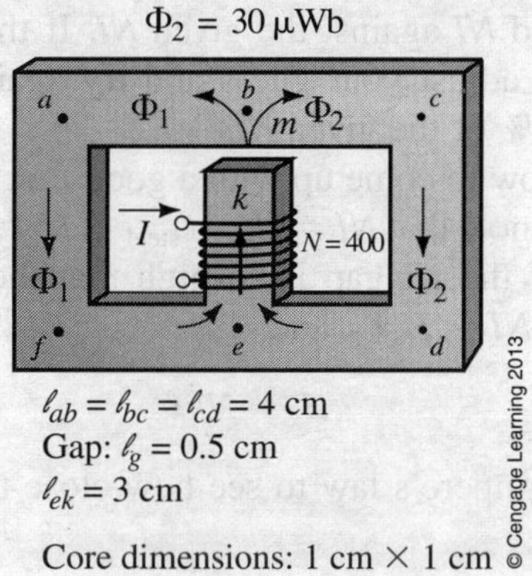

FIGURE 12–30

Answer
6.5 A

12.12 Series Magnetic Circuits: Given *NI*, Find Φ

In previous problems, you were given the flux and asked to find the current. We now look at the converse problem: given NI, find the resultant flux. For the special case of a core of one material and constant cross section (Example 12–11) this is straightforward. For all other cases, trial and error must be used.

EXAMPLE 12–11

For the circuit of Figure 12–31, $NI = 250$ At. Determine Φ.

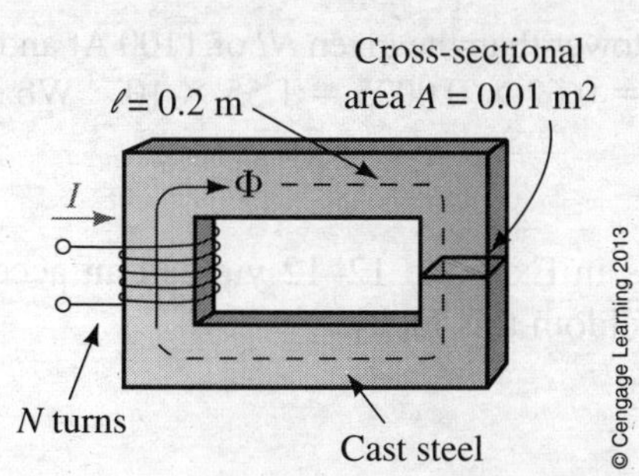

FIGURE 12–31

Solution $H\ell = NI$. Thus, $H = NI/\ell = 250/0.2 = 1250$ At/m. From the *B-H* curve of Figure 12–19, $B = 1.24$ T. Therefore, $\Phi = BA = 1.24 \times 0.01 = 1.24 \times 10^{-2}$ Wb.

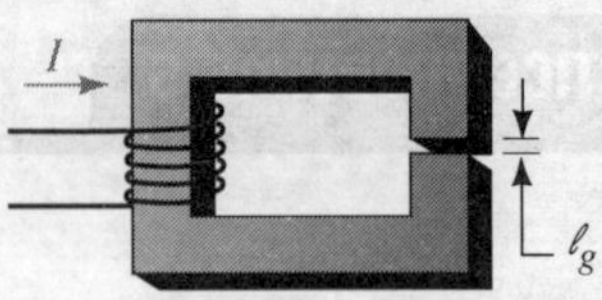

$\mathcal{F} = H_{steel}\ell_{steel} + H_g\ell_g$

$\simeq H_g\ell_g$ if $H_g\ell_g >> H_{steel}\ell_{steel}$

FIGURE 12–32

For circuits with two or more sections, the process is not so simple. Before you can find H in any section, for example, you need to know the flux density. However, in order to determine flux density, you need to know H. Thus, neither Φ nor H can be found without knowing the other first.

To get around this problem, use trial and error. First, take a guess at the value for flux, compute NI using the 4-step procedure of Section 12.10, then compare the computed NI against the given NI. If they agree, the problem is solved. If they don't, adjust your guess and try again. Repeat the procedure until you are within 5% of the given NI.

The problem is how to come up with a good first guess. For circuits of the type of Figure 12–32, note that $NI = H_{steel}\ell_{steel} + H_g\ell_g$. As a first guess, assume that the reluctance of the air gap is so high that the full mmf drop appears across the gap. Thus, $NI \simeq H_g\ell_g$, and

$$H_g \simeq NI/\ell_g \tag{12–12}$$

You can now apply Ampere's law to see how close to the given NI your trial guess is (see Notes).

EXAMPLE 12–12

The core of Figure 12–32 is cast steel, $NI = 1100$ At, the cross-sectional area everywhere is $0.0025\ \text{m}^2$, $\ell_g = 0.002$ m, and $\ell_{steel} = 0.2$ m. Determine the flux in the core.

NOTES...

Since you know that some of the mmf drop appears across the steel, you can start at less than 100% for the gap. Some basic knowledge and a bit of experience help. The relative size of the mmf drops depends on the core material. For cast iron, the percentage drop across the iron is larger than the percentage across a similar piece of sheet steel or cast steel. This is illustrated in Examples 12–12 and 12–13.

Solution

Initial Guess

Assume that 90% of the mmf appears across the gap. The applied mmf is 1100 At. Ninety percent of this is 990 At. Thus, $H_g \simeq 0.9NI/\ell = 990/0.002 = 4.95 \times 10^5$ At/m and $B_g = \mu_0 H_g = (4\pi \times 10^{-7})(4.95 \times 10^5) = 0.62$ T.

Trial 1

Since the area of the steel is the same as that of the gap, the flux density is the same, neglecting fringing. Thus, $B_{steel} = B_g = 0.62$ T. From the B-H curve, $H_{steel} = 400$ At/m. Now apply Ampere's law:

$$NI = H_{steel}\ell_{steel} + H_g\ell_g = (400)(0.2) + (4.95 \times 10^5)(0.002)$$
$$= 80 + 990 = 1070 \text{ At}$$

This answer is 2.7% lower than the given NI of 1100 At and is therefore acceptable. Thus, $\Phi = BA = 0.62 \times 0.0025 = 1.55 \times 10^{-3}$ Wb.

The initial guess in Example 12–12 yielded an acceptable answer on the first trial. (You are seldom this lucky.)

EXAMPLE 12–13

If the core of Example 12–12 is cast iron instead of steel, compute Φ.

Solution Because cast iron has a larger H for a given flux density (Figure 12–19), it will have a larger $H\ell$ drop, and less will appear across the gap. Assume 75% across the gap.

Initial Guess

$H_g \simeq 0.75\ NI/\ell = (0.75)(1100)/0.002 = 4.125 \times 10^5$ At/m.

$B_g = \mu_0 H_g = (4\mu \times 10^{-7})(4.125 \times 10^5) = 0.52$ T.

Trial 1

$B_{iron} = B_g$. Thus, $B_{iron} = 0.52$ T. From the *B-H* curve, $H_{iron} = 1700$ At/m.

Ampere's Law

$$NI = H_{iron}\ell_{iron} + H_g\ell_g = (1700)(0.2) + (4.125 \times 10^5)(0.002)$$
$$= 340 + 825 = 1165 \text{ At} \quad \text{(high by 5.9\%)}$$

Trial 2

Reduce the guess by 5.9% to $B_{iron} = 0.49$ T. Thus, $H_{iron} = 1500$ At/m (from the *B-H* curve) and $H_g = 7.96 \times 10^5$ $B_g = 3.90 \times 10^5$ At/m.

Ampere's Law

$$NI = H_{iron}\ell_{iron} + H_g\ell_g = (1500)(0.2) + (3.90 \times 10^5)(0.002)$$
$$= 300 + 780 = 1080 \text{ At}$$

The error is now 1.82%, which is excellent. Thus, $\Phi = BA = (0.49)(2.5 \times 10^{-3}) = 1.23 \times 10^{-3}$ Wb. If the error had been larger than 5%, a third trial would have been needed.

12.13 Force Due to an Electromagnet

Electromagnets are used in relays, door bells, lifting magnets, and so on. For an electromagnetic relay as in Figure 12–33, it can be shown that the force created by the magnetic field is

$$F = \frac{B_g^2 A_g}{2\mu_0} \qquad (12\text{–}13)$$

where B_g is flux density in the gap in teslas, A_g is gap area in square meters, and F is force in newtons.

EXAMPLE 12–14

Figure 12–33 shows a typical relay. The force due to the current-carrying coil pulls the pivoted arm against spring tension to close the contacts and energize the load. If the pole face is ¼ inch square and $\Phi = 0.5 \times 10^{-4}$ Wb, what is the pull on the armature in pounds?

FIGURE 12–33 A typical relay.

Solution Convert to metric units.

$$A_g = (0.25 \text{ in})(0.25 \text{ in}) = 0.0625 \text{ in}^2 = 0.403 \times 10^{-4} \text{ m}^2$$

$$B_g = \Phi/A_g = (0.5 \times 10^{-4})/(0.403 \times 10^{-4}) = 1.24 \text{ T}$$

Thus,

$$F = \frac{B_g^2 A}{2\mu_0} = \frac{(1.24)^2(0.403 \times 10^{-4})}{2(4\pi \times 10^{-7})} = 24.66 \text{ N} = 5.54 \text{ lb}$$

Figure 12–34 shows how a relay may be used in practice. When the switch is closed, the energized coil pulls the armature down. This closes the contacts and energizes the load. When the switch is opened, the spring pulls the contacts open again. Schemes like this use relatively small currents to control large loads. In addition, they permit remote control, as the relay and load may be a considerable distance from the actuating switch.

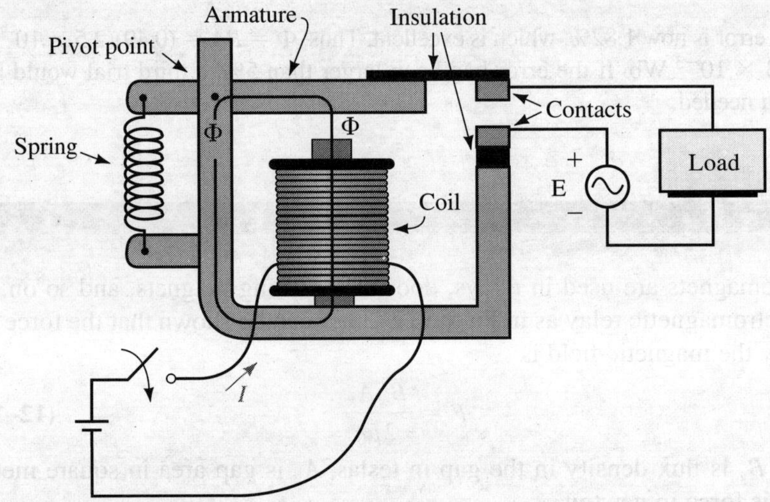

FIGURE 12–34 Controlling a load with a relay.

12.14 Properties of Magnetic Materials

Magnetic properties are related to atomic structure. Each atom of a substance, for example, produces a tiny atomic-level magnetic field because its moving (i.e., orbiting) electrons constitute an atomic-level current, and currents create magnetic fields. For nonmagnetic materials, these fields are randomly oriented and cancel. However, for ferromagnetic materials, the fields in small regions, called **domains** (Figure 12–35), do not cancel. (Domains are of microscopic size, but are large enough to hold from 10^{17} to 10^{21} atoms.) If the domain fields in a ferromagnetic material line up, the material is magnetized; if they are randomly oriented, the material is not magnetized.

Magnetizing a Specimen

A nonmagnetized specimen can be magnetized by making its domain fields line up. Figure 12–36 shows how this can be done. As current through the

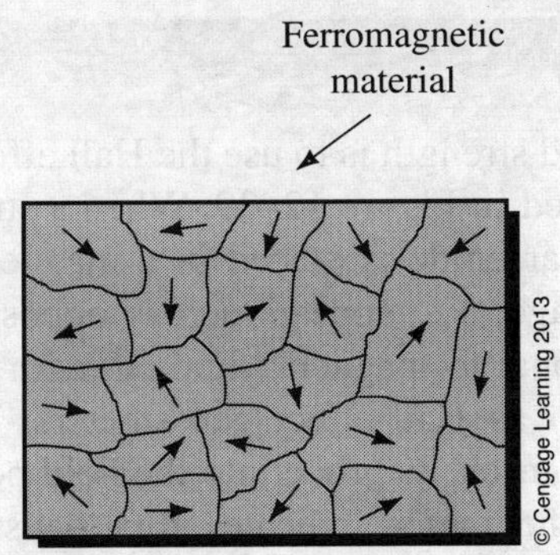

FIGURE 12–35 Random orientation of microscopic fields in a nonmagnetized ferromagnetic material. The small regions are called domains.

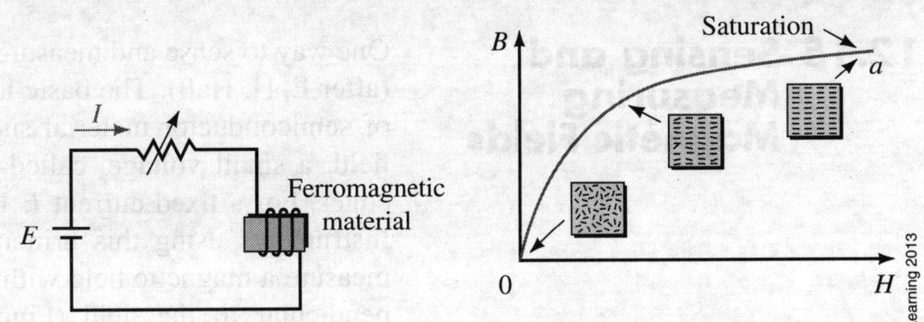

(a) The magnetizing circuit

(b) Progressive change in the domain orientations as the field is increased. H is proportional to current I.

FIGURE 12–36 The magnetization process.

coil is increased, the field strength increases and more and more domains align themselves in the direction of the field. If the field is made strong enough, almost all domain fields line up, and the material is said to be in **saturation** (the almost flat portion of the *B-H* curve). In saturation, the flux density increases slowly as magnetization intensity increases. This means that once the material is in saturation, you cannot magnetize it much further no matter how hard you try. Path 0-*a* of Figure 12–36 (traced from the nonmagnetized state to the saturated state) is termed the dc curve or normal magnetization curve. (This is the *B-H* curve that you used earlier when you solved magnetic circuit problems.)

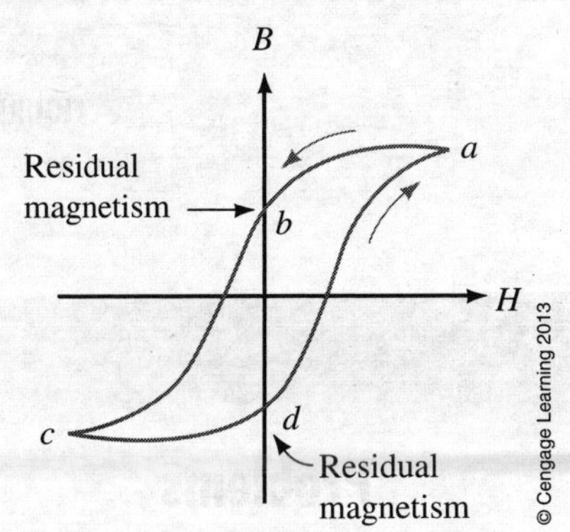

FIGURE 12–37 Hysteresis loop.

Hysteresis

If you now reduce the current to zero, you will find that the curve does not track back to zero. Instead, it follows a different path (Figure 12–37), with the result that the specimen retains some magnetism (called **residual magnetism**) as indicated by point *b*. If now you reverse the current, the flux reverses and you trace the path to *c*. If you again decrease the current to zero, you trace the path to *d*. Reversing the current and increasing it takes you back to *a*. The result is called a **hysteresis loop.** A major source of uncertainty in magnetic circuit behavior should now be apparent: As you can see, flux density depends not just on current; it also depends on which arm of the curve the sample is magnetized on, that is, it depends on the circuit's past history. For this reason, *B-H* curves are the average of the two arms of the hysteresis loop, as in the curve of Figure 12–36.

The Demagnetization Process

As indicated previously, simply turning the current off does not demagnetize ferromagnetic material. To demagnetize it, you must successively decrease its hysteresis loop to zero as in Figure 12–38. You can place the specimen inside a coil that is driven by a variable ac source and gradually decrease the coil current to zero, or you can use a fixed ac supply and gradually withdraw the specimen from the field. Such procedures are referred to as "degaussing." ("Degauss" is a term that dates back to the earlier use of gauss as the unit of magnetic flux density—recall Table 12–1.)

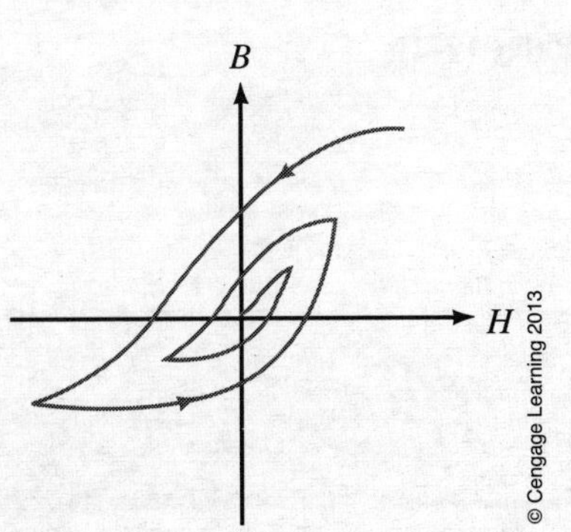

FIGURE 12–38 Demagnetization by successively shrinking the hysteresis loop.

12.15 Sensing and Measuring Magnetic Fields

One way to sense and measure magnetic field strength is to use the Hall effect (after E. H. Hall). The basic idea is illustrated in Figure 12–39. When a strip of semiconductor material such as indium arsenide is placed in a magnetic field, a small voltage, called the Hall voltage, V_H, appears across opposite edges. For a fixed current I, V_H is proportional to magnetic field strength B. Instruments using this principle are known as Hall-effect gaussmeters. To measure a magnetic field with such a meter, insert its probe into the field perpendicular to the field (Figure 12–40). The meter indicates flux density directly in teslas.

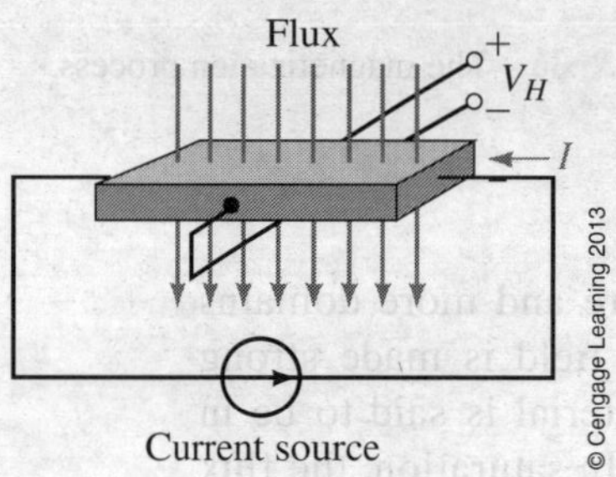

FIGURE 12–39 The Hall effect.

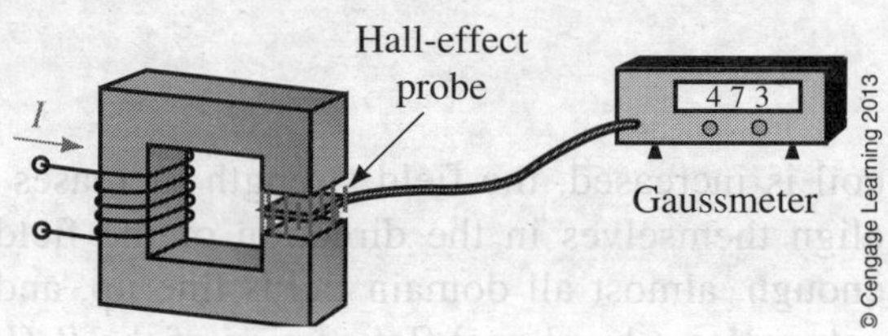

FIGURE 12–40 Magnetic field measurement.

Problems

12.3 Magnetic Flux and Flux Density

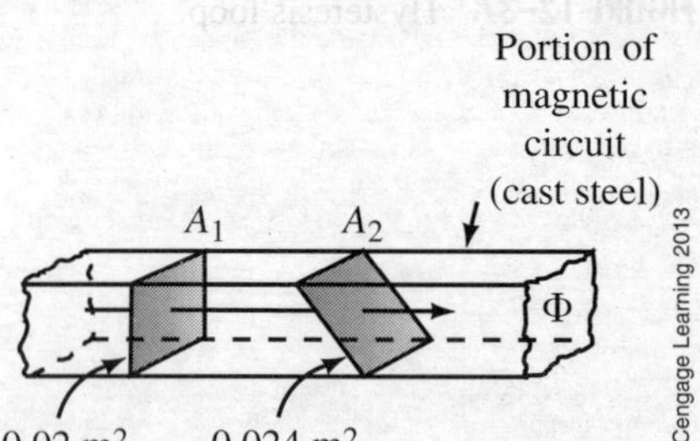

FIGURE 12–41

1. Refer to Figure 12–41:
   a. Which area, A_1 or A_2, do you use to calculate flux density?
   b. If $\Phi = 28$ mWb, what is flux density in teslas?
2. For Figure 12–41, if $\Phi = 250\ \mu$Wb, $A_1 = 1.25$ in^2, and $A_2 = 2.0$ in^2, what is the flux density in the English system of units?
3. The toroid of Figure 12–42 has a circular cross section and $\Phi = 628\ \mu$Wb. If $r_1 = 8$ cm and $r_2 = 12$ cm, what is the flux density in teslas?
4. If r_1 of Figure 12–42 is 3.5 inches and r_2 is 4.5 inches, what is the flux density in the English system of units if $\Phi = 628\ \mu$Wb?

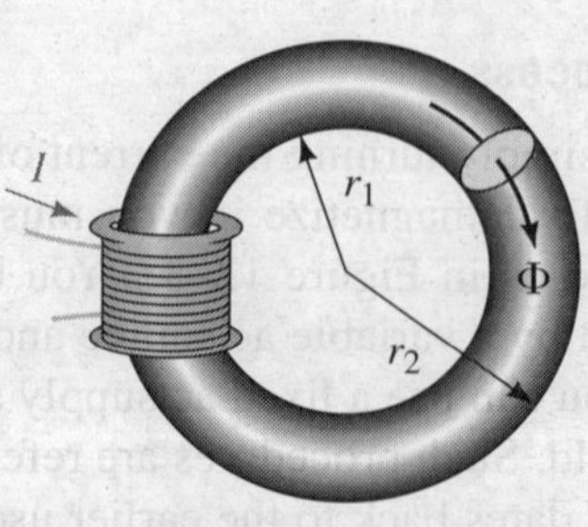

FIGURE 12–42

12.5 Air Gaps, Fringing, and Laminated Cores

5. If the section of core in Figure 12–43 is 0.025 m by 0.04 m, has a stacking factor of 0.85, and $B = 1.45$ T, what is Φ in webers?

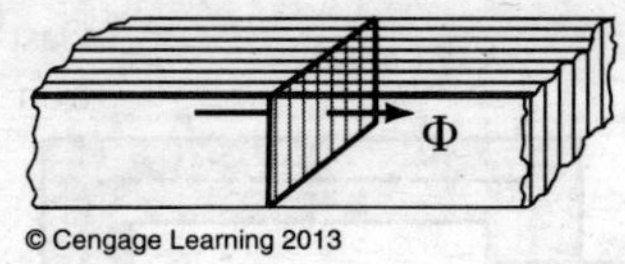

FIGURE 12–43

12.6 Series Elements and Parallel Elements

6. For the iron core of Figure 12–44, flux density $B_2 = 0.6$ T. Compute B_1 and B_3.
7. For the section of iron core of Figure 12–45, if $\Phi_1 = 12$ mWb and $\Phi_3 =$ 2 mWb, what is B_2?
8. For the section of iron core of Figure 12–45, if $B_1 = 0.8$ T and $B_2 = 0.6$ T, what is B_3?

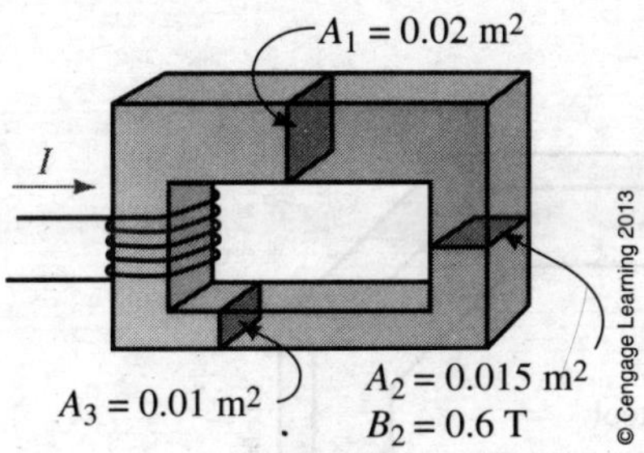

FIGURE 12–44

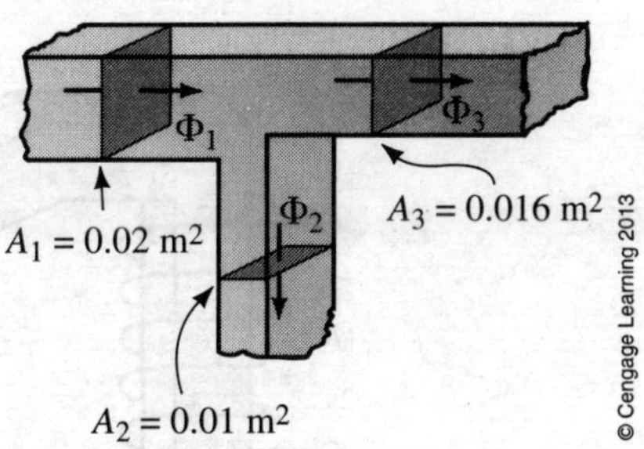

FIGURE 12–45

12.8 Magnetic Field Intensity and Magnetization Curves

9. A core with dimensions 2 cm × 3 cm has a magnetic intensity of 1200 At/m. What is Φ if the core is cast iron? If it is cast steel? If it is sheet steel with stacking factor = 0.94?
10. Figure 12–46 shows the two electric circuit equivalents for magnetic circuits. Show that μ in $\Re = \ell/\mu A$ is the same as μ in $B = \mu H$.
11. Consider again Figure 12–42. If $I = 10$ A, $N = 40$ turns, $r_1 = 5$ cm, and $r_2 =$ 7 cm, what is H in ampere-turns per meter?

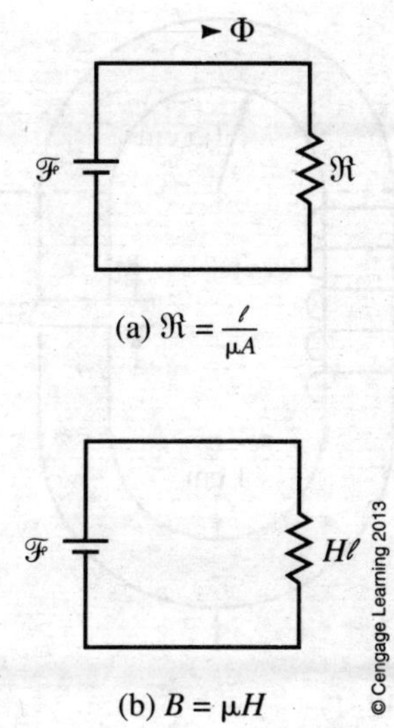

FIGURE 12–46 $\mathcal{F} = NI$.

12.9 Ampere's Circuital Law

12. Let H_1 and ℓ_1 be the magnetizing force and path length, respectively, where flux Φ_1 exists in Figure 12–47 and similarly for Φ_2 and Φ_3. Write Ampere's law around each of the windows.
13. Assume that a coil N_2 carrying current I_2 is added on leg 3 of the core shown in Figure 12–47 and that it produces flux directed upward. Assume, however, that the net flux in leg 3 is still downward. Write the Ampere's law equations for this case.
14. Repeat Problem 13 if the net flux in leg 3 is upward but the directions of Φ_1 and Φ_2 remain as in Figure 12–47.

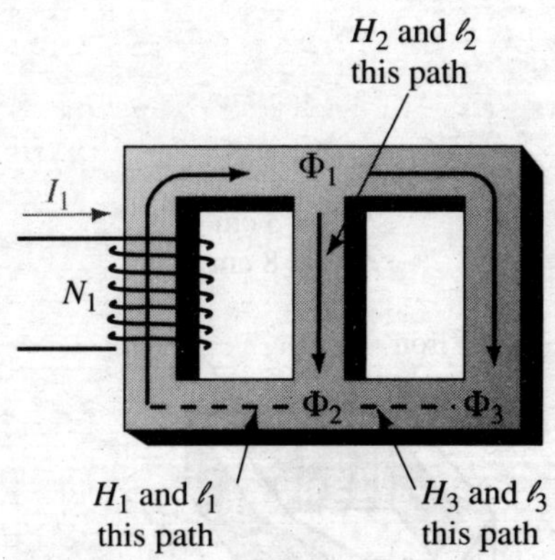

FIGURE 12–47

12.10 Series Magnetic Circuits: Given Φ, Find NI

15. Find the current I in Figure 12–48 if $\Phi = 0.16$ mWb.
16. Let everything be the same as in Problem 15 except that the cast steel portion is replaced with laminated sheet steel with a stacking factor of 0.85.

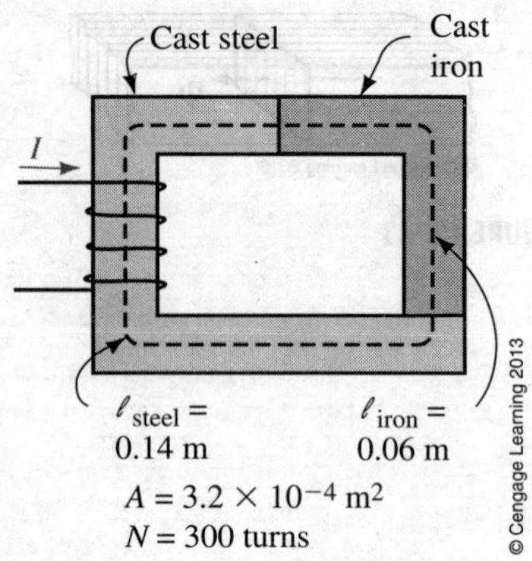

FIGURE 12–48

17. A gap of 0.5 mm is cut in the cast steel portion of the core in Figure 12–48. Find the current for $\Phi = 0.128$ mWb. Neglect fringing.
18. Two gaps, each 1 mm, are cut in the circuit of Figure 12–48, one in the cast steel portion and the other in the cast iron portion. Determine current for $\Phi = 0.128$ mWb. Neglect fringing.
19. The cast iron core of Figure 12–49 measures 1 cm × 1.5 cm, $\ell_g = 0.3$ mm, the air-gap flux density is 0.426 T, and $N = 600$ turns. The end pieces are half circles. Taking into account fringing, find current I.
20. For the circuit of Figure 12–50, $\Phi = 141\ \mu$Wb and $N = 400$ turns. The bottom member is sheet steel with a stacking factor of 0.94, while the remainder is cast steel. All pieces are 1 cm × 1 cm. The length of the cast steel path is 16 cm. Find current I.

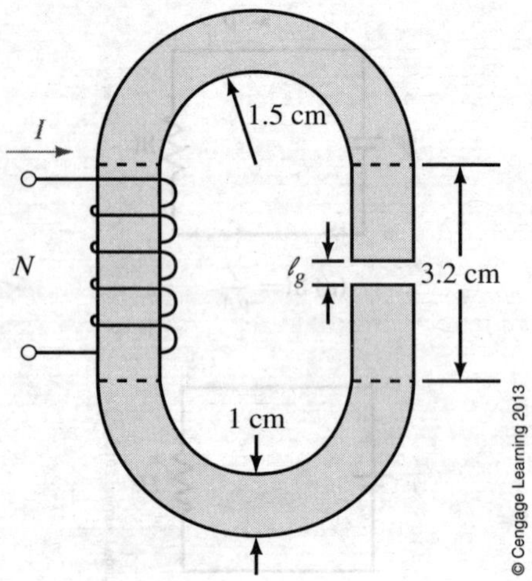

FIGURE 12–49

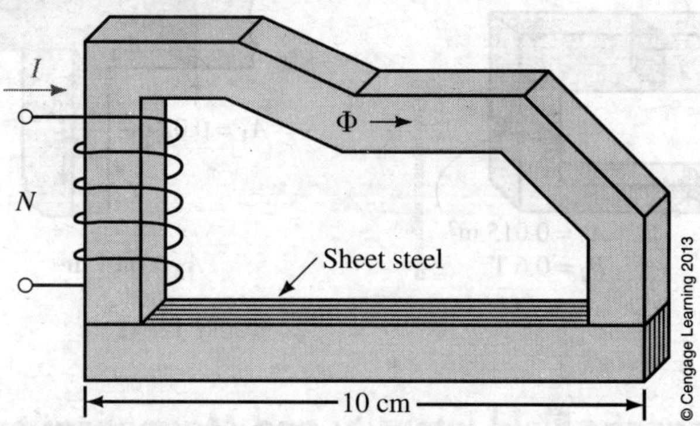

FIGURE 12–50

21. For the circuit of Figure 12–51, $\Phi = 30\ \mu$Wb and $N = 2000$ turns. Neglecting fringing, find current I.
22. For the circuit of Figure 12–52, $\Phi = 25{,}000$ lines. The stacking factor for the sheet steel portion is 0.95. Find current I.
23. A second coil of 450 turns with $I_2 = 4$ amps is wound on the cast steel portion of Figure 12–52. Its flux is in opposition to the flux produced by the original coil. The resulting flux is 35,000 lines in the counterclockwise direction. Find the current I_1.

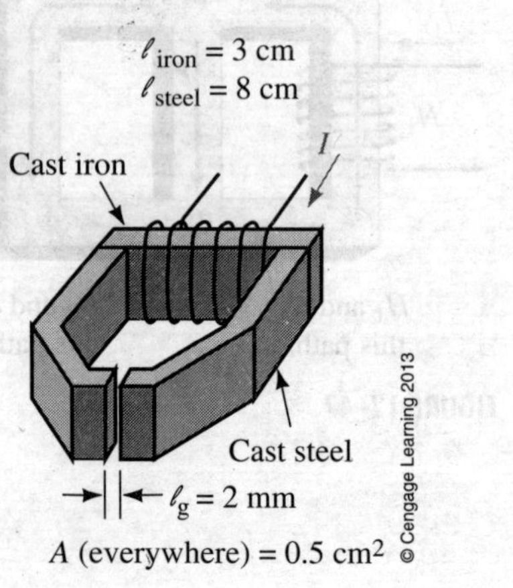

FIGURE 12–51

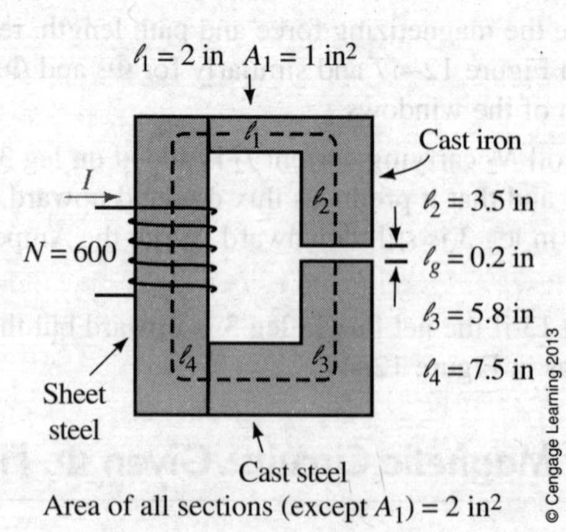

FIGURE 12–52

12.11 Series-Parallel Magnetic Circuits

24. For Figure 12–53, if $\Phi_g = 80\ \mu\text{Wb}$, find I.
25. If the circuit of Figure 12–53 has no gap and $\Phi_3 = 0.2$ mWb, find I.

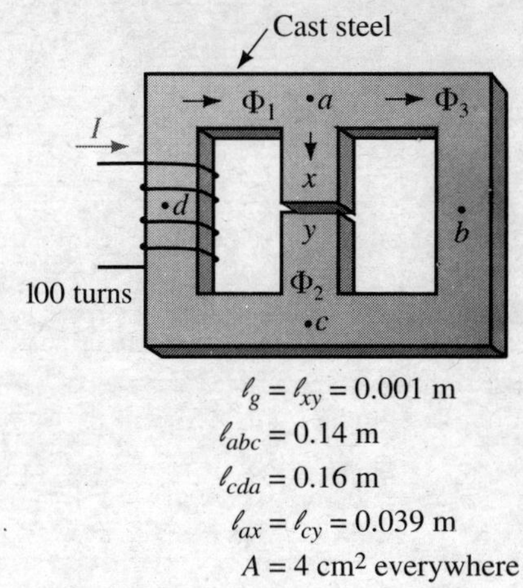

FIGURE 12–53

12.12 Series Magnetic Circuits: Given NI, Find Φ

26. A cast steel magnetic circuit with $N = 2500$ turns, $I = 200$ mA, and a cross-sectional area of 0.02 m^2 has an air gap of 0.00254 m. Assuming 90% of the mmf appears across the gap, estimate the flux in the core.
27. If $NI = 644$ At for the cast steel core of Figure 12–54, find the flux, Φ.
28. A gap $\ell = 0.004$ m is cut in the core of Figure 12–54. Everything else remains the same. Find the flux, Φ.

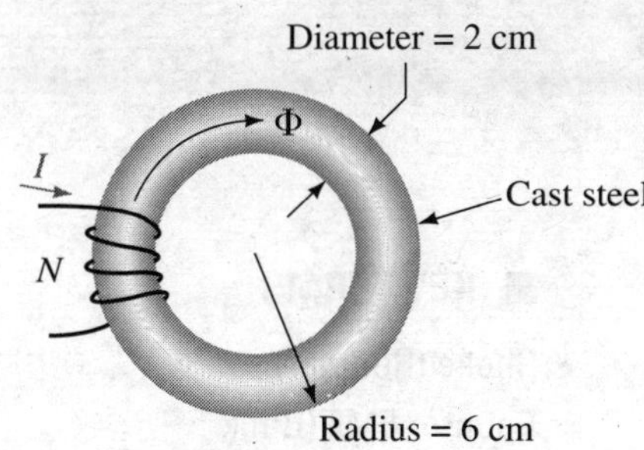

FIGURE 12–54

12.13 Force Due to an Electromagnet

29. For the relay of Figure 12–34, if the pole face is 2 cm by 2.5 cm and a force of 2 pounds is required to close the gap, what flux (in webers) is needed?
30. For the solenoid of Figure 12–28, $\Phi = 4 \times 10^{-4}$ Wb. Find the force of attraction on the plunger in newtons and in pounds.

12.15 Sensing and Measuring Magnetic Fields

31. Computers store information on hard disk drives using magnetic means. This data is read back by read heads. One of the latest read-head technologies is magnetoresistance technology. The earliest versions of this technology were known as AMR (sometimes shortened to MR). A more advanced version is known as GMR. Go to the Internet and research these read-head technologies.

ANSWERS TO IN-PROCESS LEARNING CHECKS

IN-PROCESS LEARNING CHECK 1

1. Force
2. Strength, direction
3. Iron, nickel, cobalt
4. North, south
5. Same except direction of flux reversed
6. Same as Figure 12–6, since plastic does not affect the field.
7. Perpendicular
8. 10 cm^2

IN-PROCESS LEARNING CHECK 2

1. While flux is the same throughout, the effective area of each section differs.
2. $B_{\text{iron}} = 1.1$ T; $B_{\text{steel}} = 1.38$ T; $B_g = 1.04$ T
3. $\Phi_2 = \Phi_3 = 0.5$ mWb

IN-PROCESS LEARNING CHECK 3

1. 3.5 A
2. a. 2.1 A
   b. 0.525 A
3. 1550 At/m
4. Same as Figure 12–21(b) except add $H_{g_2}\ell_{g_2}$.
5. a. $NI = H_1\ell_1 + H_2\ell_2$
   b. $0 = H_2\ell_2 - H_3\ell_3$
6. 0.32 m; use the mean path length.

■ KEY TERMS

Choke (Reactor)
Counter EMF (Back Voltage)
Faraday's Law
Flux Linkage
Henry (H)
Induced Voltage
Inductance
Inductor
Lenz's Law
Self-Inductance

■ OUTLINE

Electromagnetic Induction
Induced Voltage and Inductance
Self-Inductance
Computing Induced Voltage
Inductances in Series and Parallel
Practical Considerations
Inductance and Steady State dc
Energy Stored by an Inductance
Inductor Troubleshooting Hints

■ OBJECTIVES

After studying this chapter, you will be able to

- describe what an inductor is and what its effect on circuit operation is,
- explain Faraday's law and Lenz's law,
- compute induced voltage using Faraday's law,
- define inductance,
- compute voltage across an inductance,
- compute inductance for series and parallel configurations,
- compute inductor voltages and currents for steady state dc excitation,
- compute energy stored in an inductance,
- describe common inductor problems and how to test for them.

13

INDUCTANCE AND INDUCTORS

CHAPTER PREVIEW

In this chapter, we look at self-inductance (usually just called inductance) and inductors. To get at the idea, recall that when current flows through a conductor, it creates a magnetic field; as you will see in this chapter, this field affects circuit operation. To describe this effect, we introduce a circuit parameter called **inductance.** Inductance is due entirely to the magnetic field created by the current, and its effect is to slow the buildup and collapse of the current and—in general—oppose its change. Thus, in a sense, inductance can be likened to inertia in a mechanical system. The advantage of using inductance in our analyses is that we can dispense with all considerations of magnetism and magnetic fields—and just concentrate on familiar circuit quantities, voltage, current, and the newly introduced circuit parameter, inductance.

A circuit element built to possess inductance is called an **inductor.** In its simplest form an inductor is simply a coil of wire, Figure 13–1(a). Ideally, inductors have only inductance. However, since they are made of wire, practical inductors also have some resistance. Initially, however, we assume that this resistance is negligible and treat inductors as ideal (i.e., we assume that they have no property other than inductance). (Coil resistance is considered in Sections 13.6 and 13.7.) In practice, inductors are also referred to as **chokes** (because they try to limit or "choke" current change) or as **reactors** (for reasons to be discussed in Chapter 16). In this chapter, we refer to them mainly as inductors.

On circuit diagrams and in equations, inductance is represented by the letter *L.* Its circuit symbol is a coil as shown in Figure 13–1(b). The unit of inductance is the henry.

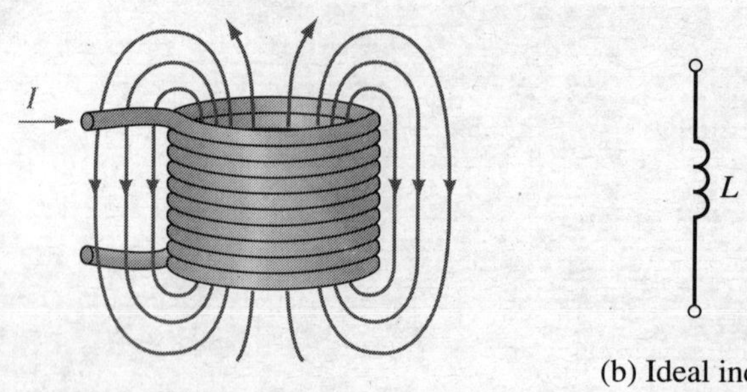

(a) A basic inductor

(b) Ideal inductor symbol

FIGURE 13–1 Inductance is due to the magnetic field created by an electric current.

Inductors are used in many places. In electronics, for example, they are used in switched-mode power supplies as energy storage devices, in electrical power systems they are used to help control short circuit currents during fault conditions, and in telecommunication systems they are used for electrical noise reduction in communication circuits. ■

Putting It in Perspective

The Discovery of Electromagnetic Induction

MOST OF OUR IDEAS CONCERNING INDUCTANCE and induced voltages are due to Michael Faraday (recall Chapter 12) and Joseph Henry (1797–1878). Working independently (Faraday in England and Henry—shown at left—in the USA), they discovered, almost simultaneously, the fundamental laws governing electromagnetic induction.

While experimenting with magnetic fields, Faraday developed the transformer. He wound two coils on an iron ring and energized one of them from a battery. As he closed the switch energizing the first coil, Faraday noticed that a momentary voltage was induced in the second coil, and when he opened the switch, he found that a momentary voltage was again induced but with opposite polarity. When the current was steady, no voltage was produced at all.

Faraday explained this effect in terms of his magnetic lines of the flux concept. When current was first turned on, he visualized the lines as springing outward into space; when it was turned off, he visualized the lines as collapsing inward. He then visualized that voltage was produced by these lines as they cut across circuit conductors. Companion experiments showed that voltage was also produced when a magnet was passed through a coil or when a conductor was moved through a magnetic field. Again, he visualized these voltages in terms of flux cutting a conductor.

Working independently in the United States, Henry discovered essentially the same results. In fact, Henry's work preceded Faraday's by a few months, but because he did not publish them first, credit was given to Faraday. However, Henry is credited with the discovery of self-induction, and in honor of his work the unit of inductance was named the henry. ■

13.1 Electromagnetic Induction

Inductance depends on **induced voltage.** Thus, we begin with a review of electromagnetic induction. First, we look at Faraday's and Henry's results. Consider Figure 13–2. In (a), a magnet is moved through a coil of wire, and this action induces a voltage in the coil. When the magnet is thrust into the coil, the meter deflects upscale; when it is withdrawn, the meter deflects downscale, indicating that polarity has changed. The voltage magnitude is proportional to how fast the magnet is moved. In (b), when the conductor is moved through the field, voltage is induced. If the conductor is moved to the right, its far end is positive; if it is moved to the left, the polarity reverses and its far end becomes negative. Again, the voltage magnitude is proportional to how fast the wire is moved. In (c), voltage is induced in coil 2 due to the magnetic field created by the current in coil 1. At the instant the switch is closed, the meter kicks upscale; at the instant it is opened, the meter kicks downscale. In (d) voltage is induced in a coil by its own current. At the instant the switch is closed, the top end of the coil becomes positive, while at the instant it is opened, the polarity reverses

NOTES...

Since we work with time-varying flux linkages in this chapter, we use ϕ rather than Φ for flux (as we did in Chapter 12). This is in keeping with the standard practice of using lowercase symbols for time-varying quantities and uppercase symbols for dc quantities.

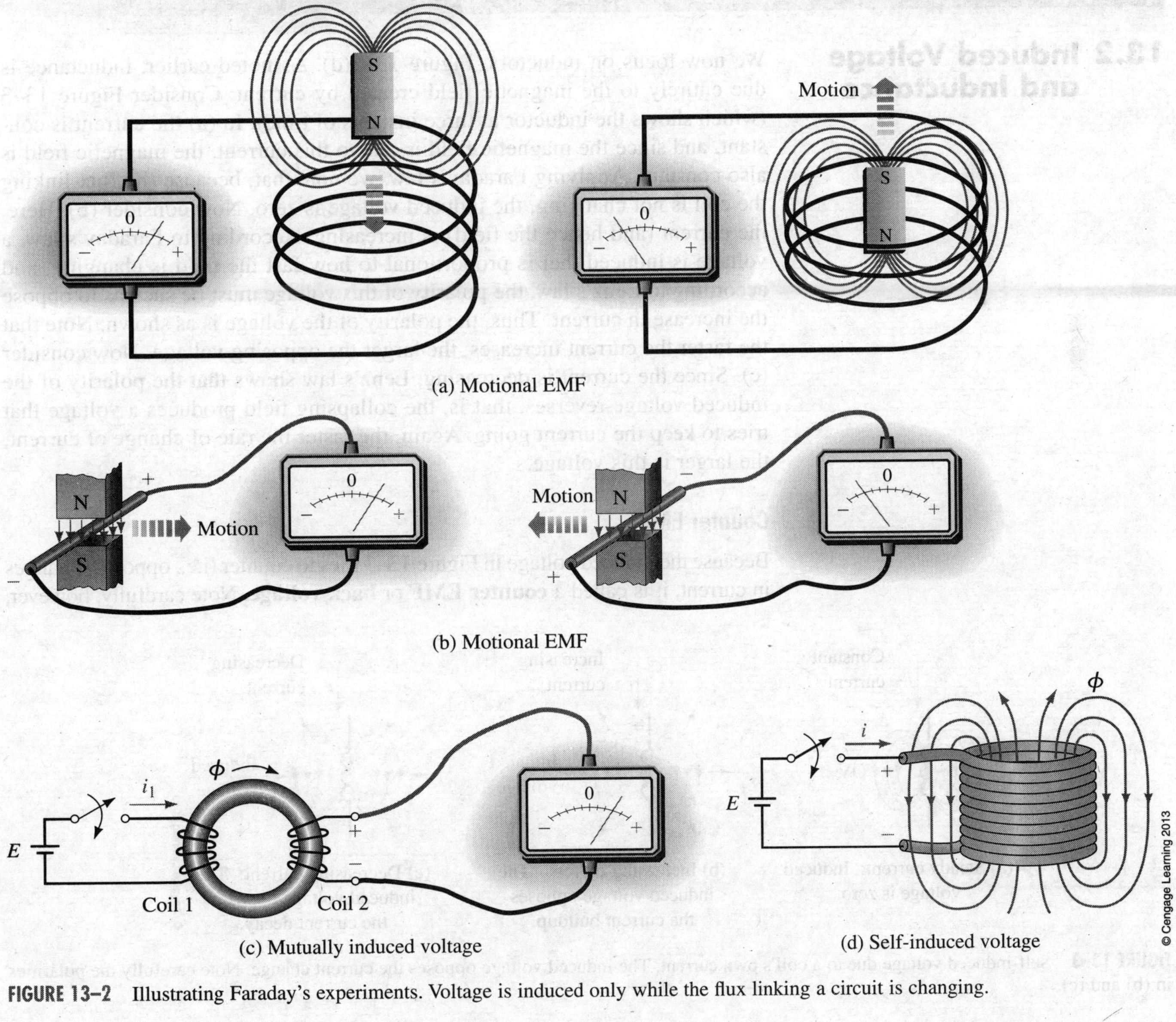

FIGURE 13–2 Illustrating Faraday's experiments. Voltage is induced only while the flux linking a circuit is changing.

and the top end becomes negative. Note that no voltage is induced in any of these cases when the flux linking the circuit is unchanging—that is, when the magnet is stationary in (a), when the wire is unmoving in (b), or when the current has reached a steady state in (c) and (d).

Faraday's Law

Based on these observations, Faraday concluded that *voltage is induced in a circuit whenever the flux linking* (i.e., passing through) *the circuit is changing and that the magnitude of the voltage is proportional to the rate of change of the flux linkages.* This result, known as **Faraday's law,** is also sometimes stated in terms of the rate of cutting flux lines. We look at this viewpoint in Chapter 15.

Lenz's Law

Heinrich Lenz (a Russian physicist, 1804–1865) determined a companion result. He showed that *the polarity of the induced voltage is such as to oppose the cause producing it.* This result is known as **Lenz's law.**

13.2 Induced Voltage and Inductance

We now focus on inductors, Figure 13–2(d). As noted earlier, inductance is due entirely to the magnetic field created by current. Consider Figure 13–3 (which shows the inductor at three instants of time). In (a) the current is constant, and since the magnetic field is due to this current, the magnetic field is also constant. Applying Faraday's law, we note that, because the flux linking the coil is not changing, the induced voltage is zero. Now consider (b). Here, the current (and hence the field) is increasing. According to Faraday's law, a voltage is induced that is proportional to how fast the field is changing, and according to Lenz's law, the polarity of this voltage must be such as to oppose the increase in current. Thus, the polarity of the voltage is as shown. Note that the faster the current increases, the larger the opposing voltage. Now consider (c). Since the current is decreasing, Lenz's law shows that the polarity of the induced voltage reverses, that is, the collapsing field produces a voltage that tries to keep the current going. Again, the faster the rate of change of current, the larger is this voltage.

Counter EMF

Because the induced voltage in Figure 13–3 tries to counter (i.e., opposes) changes in current, it is called a **counter EMF** or **back voltage.** Note carefully, however,

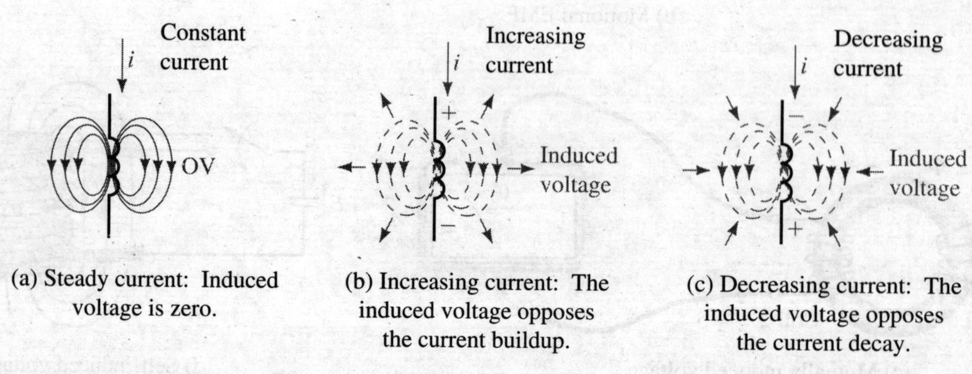

(a) Steady current: Induced voltage is zero.

(b) Increasing current: The induced voltage opposes the current buildup.

(c) Decreasing current: The induced voltage opposes the current decay.

FIGURE 13–3 Self-induced voltage due to a coil's own current. The induced voltage opposes the current change. Note carefully the polarities in (b) and (c).

that this voltage does not oppose current, it opposes only changes in current. It also does not prevent the current from changing; it only prevents it from changing abruptly. The result is that current in an inductor changes gradually and smoothly from one value to another as indicated in Figure 13–4(b). The effect of inductance is thus similar to the effect of inertia in a mechanical system. The flywheel used on an engine, for example, prevents abrupt changes in engine speed but does not prevent the engine from gradually changing from one speed to another.

(a) Current cannot jump from one value to another like this

(b) Current must change smoothly with no abrupt jumps

FIGURE 13–4 Current in inductance.

Iron-Core and Air-Core Inductors

As Faraday discovered, the voltage induced in a coil depends on flux linkages, and flux linkages depend on core materials. Coils with ferromagnetic cores (called iron-core coils) have their flux almost entirely confined to their cores, while coils wound on nonferromagnetic materials do not. (The latter are sometimes called air-core coils because all nonmagnetic core materials have the same permeability as air and thus behave magnetically the same as air.)

First, consider the iron-core case, Figure 13–5. Ideally, all flux lines are confined to the core and hence pass through (link) all turns of the winding. The product of flux times the number of turns that it passes through is defined as the **flux linkage** of the coil. For Figure 13–5, ϕ lines pass through N turns yielding a flux linkage of $N\phi$. By Faraday's law, the induced voltage is proportional to the rate of change of $N\phi$. In the SI system, the constant of proportionality is 1, and Faraday's law for this case may therefore be stated as

$$e = N \times \text{the rate of change of } \phi \quad \textbf{(13–1)}$$

In calculus notation,

$$e = N\frac{d\phi}{dt} \quad \text{(volts, V)} \quad \textbf{(13–2)}$$

where ϕ is in webers, t in seconds, and e in volts. Thus, if the flux changes at the rate of 1 Wb/s in a 1-turn coil, the voltage induced is 1 volt.

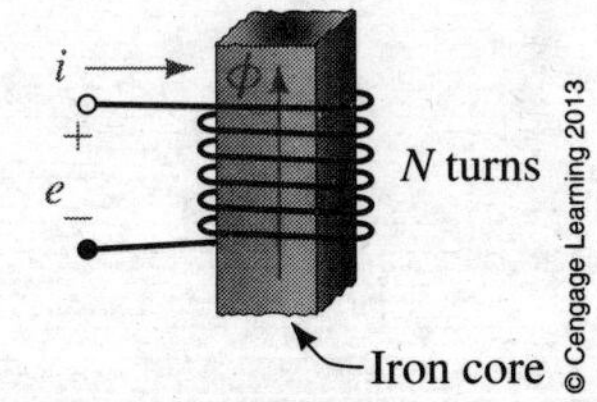

FIGURE 13–5 When flux ϕ passes through all N turns, the flux linking the coil is $N\phi$.

NOTES...

Equation 13–2 is sometimes shown with a minus sign. However, the minus sign is unnecessary. In circuit theory, we use Equation 13–2 to determine the magnitude of the induced voltage and Lenz's law to determine its polarity.

EXAMPLE 13–1

If the flux through a 200-turn coil changes steadily from 1 Wb to 4 Wb in 1 second, what is the voltage induced?

Solution The flux changes by 3 Wb in 1 second. Thus, its rate of change is 3 Wb/s.

$$\begin{aligned} e &= N \times \text{rate of change of flux} \\ &= (200 \text{ turns})(3 \text{ Wb/s}) = 600 \text{ volts} \end{aligned}$$

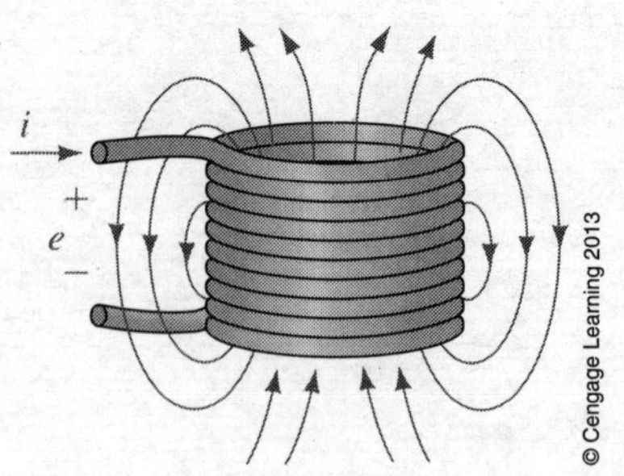

FIGURE 13–6 The flux linking the coil is proportional to current. Flux linkage is LI.

Now consider an air-core inductor (Figure 13–6). Since not all flux lines pass through all windings, you cannot determine flux linkages as above. However, (since no ferromagnetic material is present) flux is directly proportional to current. In this case, then, since induced voltage is proportional to the rate of change of flux, and since flux is proportional to current, induced voltage will be proportional to the rate of change of current. Let the constant of proportionality be L. Thus,

$$e = L \times \text{rate of change of current} \qquad (13\text{–}3)$$

In calculus notation, this can be written as

$$e = L\frac{di}{dt} \quad \text{(volts, V)} \qquad (13\text{–}4)$$

L is called the **self-inductance** (or usually, just inductance) of the coil, and in the SI system its unit is the henry. (This is discussed in more detail in Section 13.3.)

We now have two equations for coil voltage. Equation 13–4 is the more useful form for this chapter, while Equation 13–2 is needed for some of the circuits of Chapter 24. We look at Equation 13–4 in the next section.

IN-PROCESS LEARNING CHECK 1

(Answers are at the end of the chapter.)

1. Which of the current graphs shown in Figure 13–7 cannot be the current in an inductor? Why?

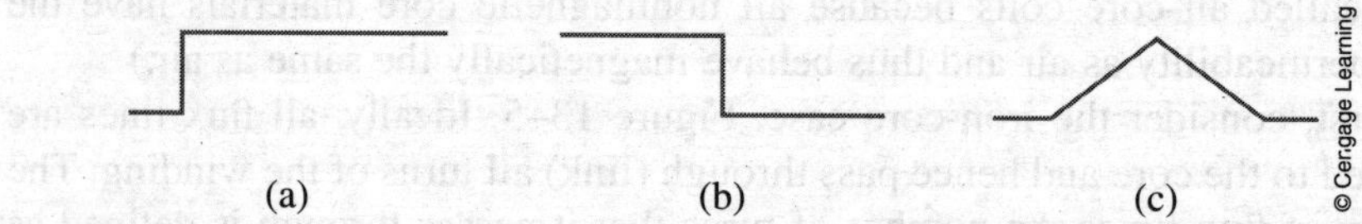

FIGURE 13–7

2. Compute the flux linkage for the coil of Figure 13–5, given $\phi = 500$ mWb and $N = 1200$ turns.
3. If the flux ϕ of Question 2 changes steadily from 500 mWb to 525 mWb in 1 s, what is the voltage induced in the coil?
4. If the flux ϕ of Question 2 changes steadily from 500 mWb to 475 mWb in 100 ms, what is the voltage induced?
5. If the flux for the 200-turn coil of Figure 13–5 is given by $\phi = 25\, t\, e^{-t}$ mWb, what is the equation for its induced voltage?

13.3 Self-Inductance

In the preceding section, we showed that the voltage induced in a coil is $e = Ldi/dt$, where L is the self-inductance of the coil (usually referred to simply as inductance as noted previously) and di/dt is the rate of change of its current. In the SI system, L is measured in henries. As can be seen from Equation 13–4, it is the ratio of voltage induced in a coil to the rate of change of current producing it. From this, we get the definition of the **henry (H)**. By definition, *the inductance of a coil is one henry if the voltage created by its changing current is one volt when its current changes at the rate of one ampere per second.*

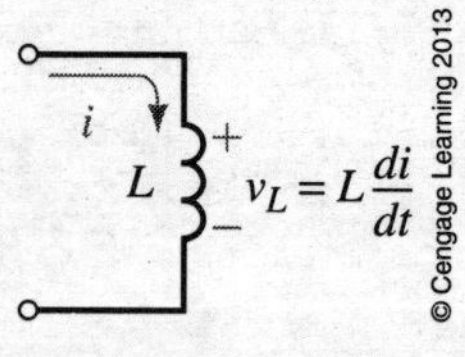

FIGURE 13–8 Voltage-current reference convention. As usual, the plus sign for voltage goes at the tail of the current arrow.

In practice, the voltage across an inductance is usually denoted by v_L rather than by e (Figure 13–8). Thus,

$$v_L = L\frac{di}{dt} \quad \text{(V)} \qquad (13\text{–}5)$$

EXAMPLE 13–2

If the current through a 5-mH inductance changes at the rate of 1000 A/s, what is the voltage induced?

Solution

$$v_L = L \times \text{rate of change of current}$$
$$= (5 \times 10^{-3}\ \text{H})(1000\ \text{A/s}) = 5\ \text{volts}$$

PRACTICE PROBLEMS 1

1. The voltage across an inductance is 250 V when its current changes at the rate of 10 mA/μs. What is L?
2. If the voltage across a 2-mH inductance is 50 volts, how fast is the current changing?

Answers
1. 25 mH; 2. 25×10^3 A/s

Inductance Formulas

Inductance for some simple shapes can be determined using the principles of Chapter 12. For example, the approximate inductance of the air core coil of Figure 13–9 can be shown to be

$$L = \frac{\mu N^2 A}{\ell} \quad \text{(H)} \tag{13–6}$$

where ℓ is in meters, A is in square meters, N is the number of turns, and μ is the permeability of the core. (Details can be found in many physics books.)

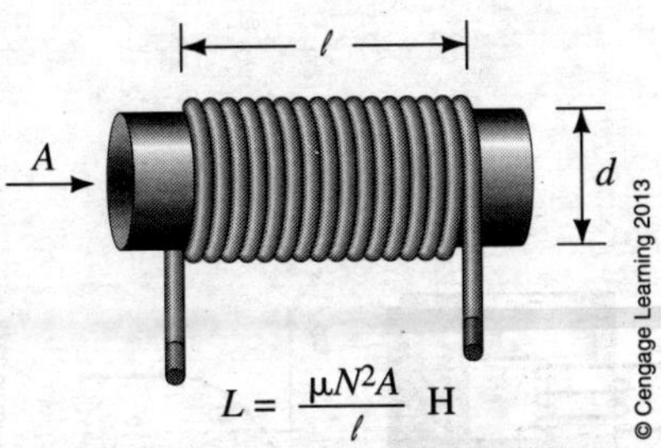

FIGURE 13–9 Approximate inductance formula for a single-layer air coil.

EXAMPLE 13–3

A 0.15-m-long air-core coil has a radius of 0.006 m and 120 turns. Compute its inductance.

Solution

$$A = \pi r^2 = 1.131 \times 10^{-4}\ \text{m}^2$$
$$\mu = \mu_0 = 4\pi \times 10^{-7}$$

Thus,

$$L = 4\pi \times 10^{-7}\,(120)^2\,(1.131 \times 10^{-4})/0.15 = 13.6\ \mu\text{H}$$

The accuracy of Equation 13–6 breaks down for small ℓ/d ratios. (If ℓ/d is greater than 10, the error is less than 4%.) Improved formulas may be found on the Internet or in design handbooks, such as the *Radio Amateur's Handbook* published by the American Radio Relay League (ARRL).

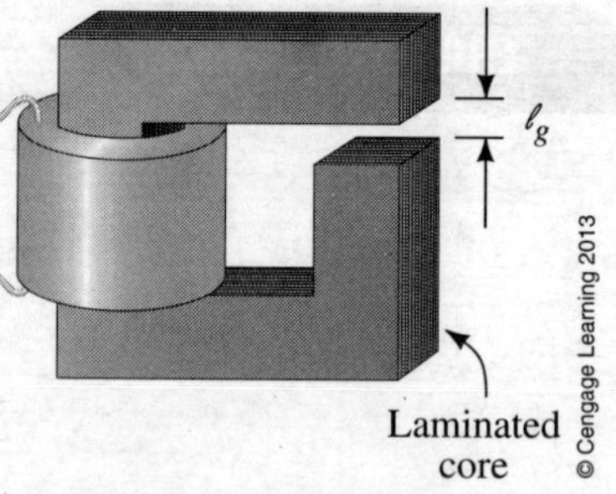

FIGURE 13–10 Iron-core coil with air gap. The gap keeps the core from going into saturation.

To provide greater inductance in smaller spaces, iron cores are sometimes used. Unless the core flux is kept below saturation, however, permeability varies and inductance is not constant. To get relatively constant inductance, an air gap may be used (Figure 13–10). If the gap is wide enough to dominate, coil inductance is approximately

$$L \approx \frac{\mu_0 N^2 A_g}{\ell_g} \quad \text{(H)} \tag{13–7}$$

where μ_0 is the permeability of air, A_g is the area of the air gap, and ℓ_g is its length. (See end-of-chapter Problem 11.) Another way to increase inductance is to use a ferrite core (Section 13.6).

EXAMPLE 13–4

The inductor of Figure 13–10 has 1000 turns, a 5-mm gap, and a cross-sectional area at the gap of 5×10^{-4} m^2. What is its inductance?

Solution

$$L \approx (4\pi \times 10^{-7})(1000)^2(5 \times 10^{-4})/(5 \times 10^{-3}) = 0.126 \text{ H}$$

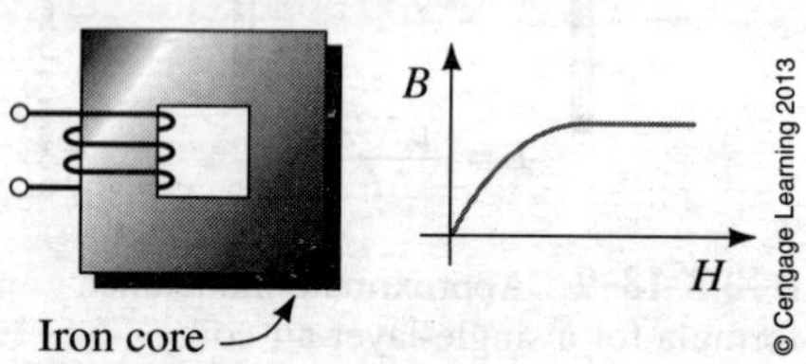

FIGURE 13–11 This coil does not have a fixed inductance because its flux is not proportional to its current.

PRACTICAL NOTES...

1. Since inductance is due to a conductor's magnetic field, it depends on the same factors that the magnetic field depends on. The stronger the field for a given current, the greater the inductance. Thus, a coil of many turns will have more inductance than a coil of a few turns (L is proportional to N^2) and a coil wound on a magnetic core will have greater inductance than a coil wound on a nonmagnetic form.
2. However, if a coil is wound on a magnetic core, the core's permeability μ may change with flux density. Since flux density depends on current, L becomes a function of current. For example, the inductor of Figure 13–11 has a nonlinear inductance due to core saturation. All inductors encountered in this book are assumed to be linear, that is, of constant value.

IN-PROCESS LEARNING CHECK 2

(Answers are at the end of the chapter.)

1. The voltage across an inductance whose current changes uniformly by 10 mA in 4 μs is 70 volts. What is its inductance?
2. If you triple the number of turns in the inductor of Figure 13–10, but everything else remains the same, by what factor does the inductance increase?

13.4 Computing Induced Voltage

Earlier, we determined that the voltage across an inductance is given by $v_L = Ldi/dt$, where the voltage and current references are shown in Figure 13–8. Note that the polarity of v_L depends on whether the current is increasing or decreasing. For example, if the current is increasing, di/dt is positive and so v_L is positive, while if the current is decreasing, di/dt is negative and v_L is negative.

To compute voltage, we need to determine di/dt. In general, this requires calculus. However, since di/dt is slope, you can determine voltage without calculus for currents that can be described by straight lines, as in Figure 13–12. For any Δt segment, slope $= \Delta i/\Delta t$, where Δi is the amount that the current changes during time interval Δt.

EXAMPLE 13–5

Figure 13–12 is the current through a 10-mH inductance. Determine voltage v_L and sketch it.

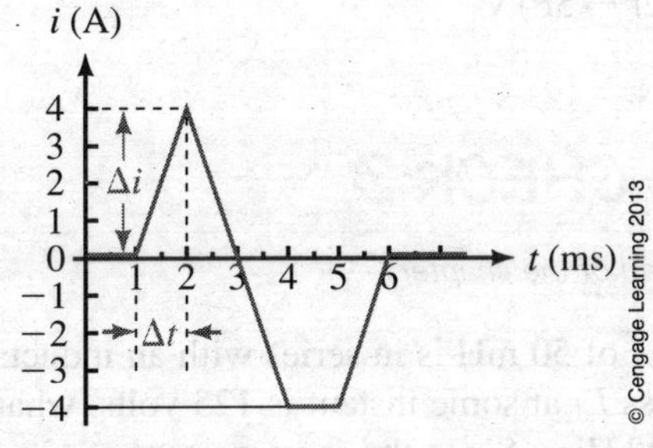

FIGURE 13–12

Solution Break the problem into intervals over which the slope is constant, determine the slope for each segment, then compute voltage using $v_L = L \times$ slope for that interval:

0 to 1 ms:	Slope = 0. Thus, $v_L = 0$ V.
1 ms to 2 ms:	Slope $= \Delta i/\Delta t = 4\text{ A}/(1 \times 10^{-3}\text{ s}) = 4 \times 10^3$ A/s. Thus, $v_L = L\Delta i/\Delta t = (0.010\text{ H})(4 \times 10^3\text{ A/s}) = 40$ V.
2 ms to 4 ms:	Slope $= \Delta i/\Delta t = -8\text{ A}/(2 \times 10^{-3}\text{ s}) = -4 \times 10^3$ A/s. Thus, $v_L = L\Delta i/\Delta t = (0.010\text{ H})(-4 \times 10^3\text{ A/s}) = -40$ V.
4 ms to 5 ms:	Slope = 0. Thus, $v_L = 0$ V.
5 ms to 6 ms:	Same slope as from 1 ms to 2 ms. Thus, $v_L = 40$ V.

The voltage waveform is shown in Figure 13–13.

CircuitSim 13-1

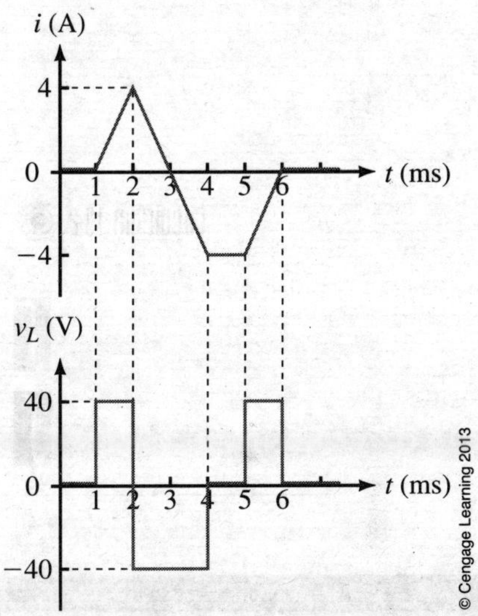

FIGURE 13–13

For currents that are not linear functions of time, you need to use calculus as illustrated in the following example.

EXAMPLE 13–6

What is the equation for the voltage across a 12.5-H inductance whose current is $i = te^{-t}$ amps?

Solution Differentiate by parts using

$$\frac{d(uv)}{dt} = u\frac{dv}{dt} + v\frac{du}{dt} \text{ with } u = t \text{ and } v = e^{-t}$$

Thus,

$$v_L = L\frac{di}{dt} = L\frac{d}{dt}(te^{-t}) = L[t(-e^{-t}) + e^{-t}] = 12.5e^{-t}(1 - t) \text{ volts}$$

PRACTICE PROBLEMS 2

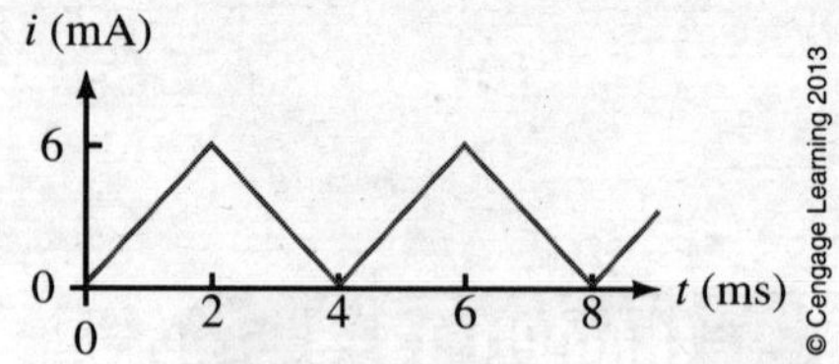

FIGURE 13–14

1. Figure 13–14 shows the current through a 5-H inductance. Determine voltage v_L and sketch it.
2. If the current of Figure 13–12 is applied to an unknown inductance and the voltage from 1 ms to 2 ms is 28 volts, what is L?
3. The current in a 4-H inductance is $i = t^2e^{-5t}$ A. What is voltage v_L?

Answers

1. v_L is a square wave. Between 0 ms and 2 ms, its value is 15 V; between 2 ms and 4 ms, its value is −15 V, and so on.
2. 7 mH; 3. $4e^{-5t}(2t - 5t^2)$ V

IN-PROCESS LEARNING CHECK 3

(Answers are at the end of the chapter.)

1. An inductance L_1 of 50 mH is in series with an inductance L_2 of 35 mH. If the voltage across L_1 at some instant is 125 volts, what is the voltage across L_2 at that instant? Hint: Since the same current passes through both inductances, the rate of change of current is the same for both.
2. Current through a 5-H inductance changes linearly from 10 A to 12 A in 0.5 s. Suppose now the current changes linearly from 2 mA to 6 mA in 1 ms. Although the currents are significantly different, the induced voltage is the same in both cases. Why? Compute the voltage.

CircuitSim 13-2

3. If the current for a 5-H coil is given by $i = 4t^3e^{-2t} + 4$ A, what is the equation for voltage v_L?
4. Repeat Question 3 if $i = 50 \sin 1000\,t$ mA.

13.5 Inductances in Series and Parallel

For inductances in series or parallel, the equivalent inductance is found by using the same rules that you used for resistance. For the series case (Figure 13–15) the total inductance is the sum of the individual inductances—see Practice Problems 3, Question 2. Thus,

$$L_T = L_1 + L_2 + L_3 + \cdots + L_N \tag{13–8}$$

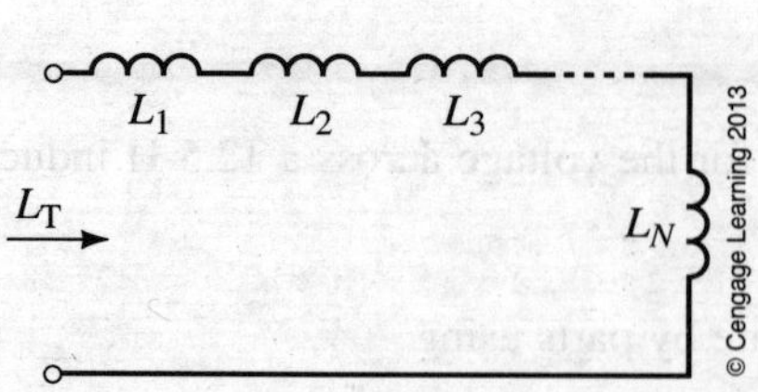

FIGURE 13–15 $L_T = L_1 + L_2 + \cdots + L_N$.

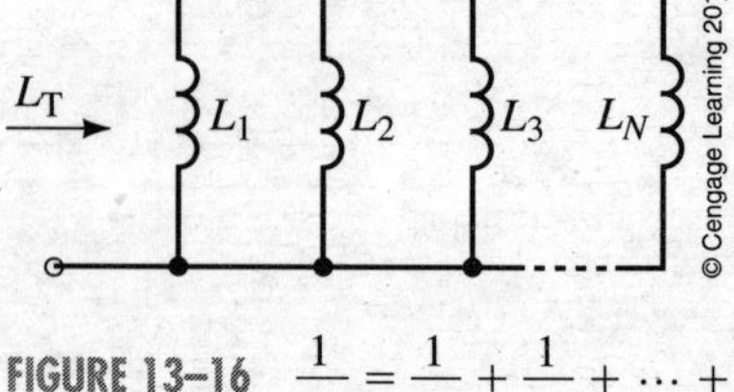

FIGURE 13–16 $\frac{1}{L_T} = \frac{1}{L_1} + \frac{1}{L_2} + \cdots + \frac{1}{L_N}$.

For the parallel case (Figure 13–16),

$$\frac{1}{L_T} = \frac{1}{L_1} + \frac{1}{L_2} + \frac{1}{L_3} + \cdots + \frac{1}{L_N} \tag{13–9}$$

For two inductances, Equation 13–9 reduces to

$$L_T = \frac{L_1 L_2}{L_1 + L_2} \tag{13–10}$$

EXAMPLE 13–7

Find L_T for the circuit of Figure 13–17.

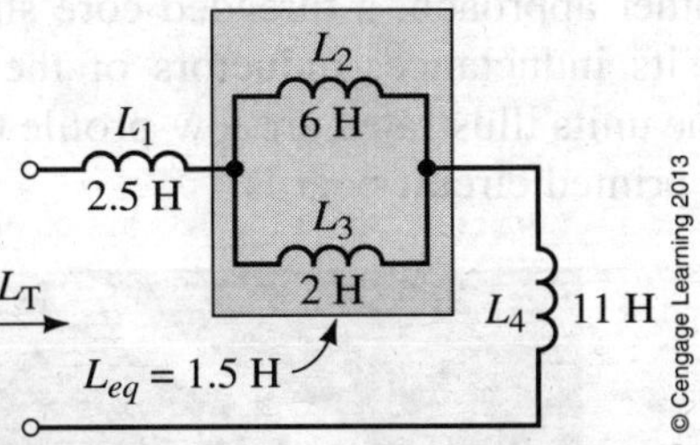

FIGURE 13–17

Solution The parallel combination of L_2 and L_3 is

$$L_{eq} = \frac{L_2 L_3}{L_2 + L_3} = \frac{6 \times 2}{6 + 2} = 1.5 \text{ H}$$

This is in series with L_1 and L_4. Thus, $L_T = 2.5 + 1.5 + 11 = 15$ H.

Calculator Hint: Using the inverse key on your calculator, you can determine the parallel equivalent of inductances using the same procedure you used for parallel resistances. Typically, your calculator steps would look something like the following (depending on the particular calculator you are using):

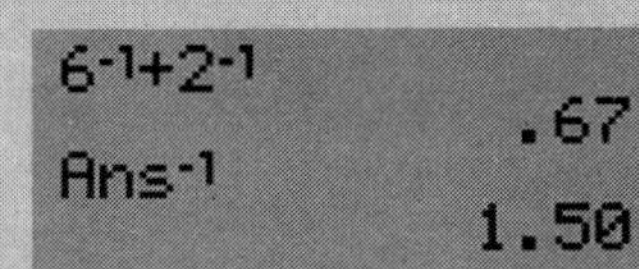

PRACTICE PROBLEMS 3

1. For Figure 13–18, $L_T = 2.25$ H. Determine L_x.
2. For Figure 13–15, the current is the same in each inductance and $v_1 = L_1 di/dt$, $v_2 = L_2 di/dt$, and so on. Apply Kirchhoff's voltage law (KVL) and show that $L_T = L_1 + L_2 + L_3 + \cdots + L_N$.

Answer
1. 3 H

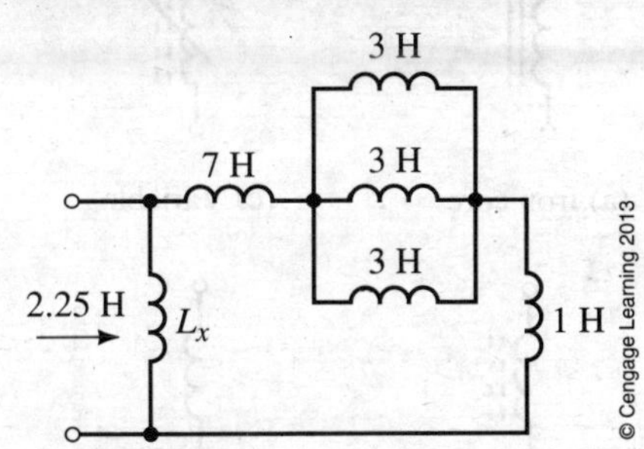

FIGURE 13–18

13.6 Practical Considerations

Core Types

The type of core used in an inductor depends to a great extent on its intended use and frequency range. (Although you have not studied frequency yet, you can get a feel for frequency by noting that the electrical power system operates at low frequency [60 cycles per second, called 60 hertz], while radio and TV systems operate at high frequency [hundreds of megahertz].) Inductors used in audio or power supply applications generally have laminated iron cores (because they need large inductance values). (Laminations reduce core loss—see Chapter 23, Section 23.6.) Inductors used for radio-frequency circuits generally have air or ferrite (or similar) cores. (Ferrite is a mixture of iron oxide in

NOTES...

Plastic or similar screwdrivers should be used so as not to disturb the coil's magnetic field and upset its value while making the adjustments.

a ceramic binder. It has characteristics that make it suitable for high-frequency work.) Iron cores cannot be used in high-frequency applications, however, since they have large power losses at high frequencies (for reasons discussed in Chapter 17).

Variable Inductors

Inductors can be made so that their inductance is variable. In one approach, inductance is varied by changing coil spacing with a screwdriver adjustment—see Note. In another approach, a threaded core slug is screwed in or out of the coil to vary its inductance. Inductors of the latter type are shown in Figure 13–19. The units illustrated are low-profile tunable inductors designed for mounting on printed circuit boards.

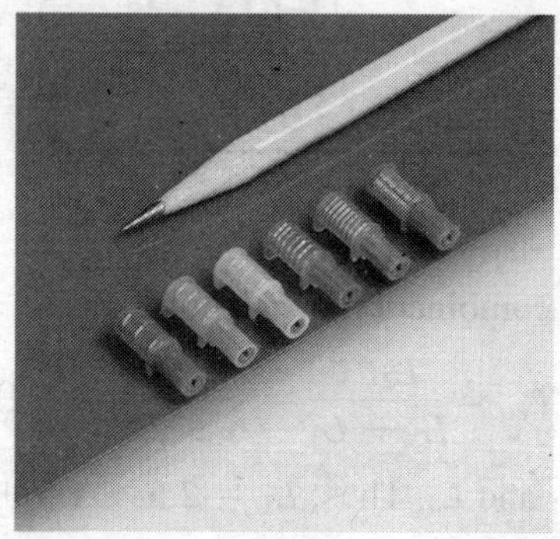

Photo courtesy of Coilcraft Inc.

FIGURE 13–19 A selection of color-coded, tunable, low-profile, high-frequency inductors. A threaded core slug is used to adjust inductance.

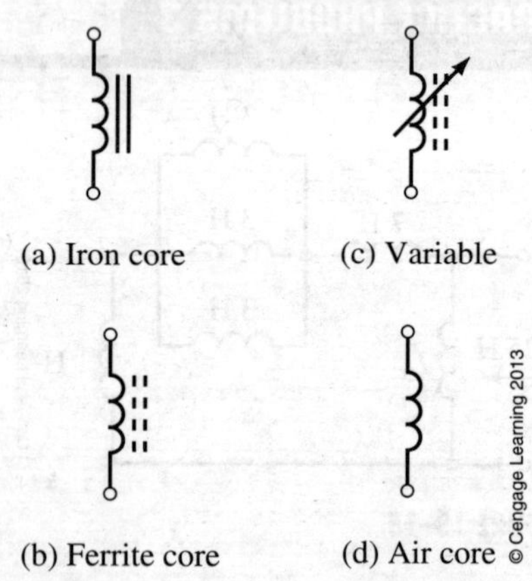

FIGURE 13–20 Circuit symbols for inductors.

Circuit Symbols

Figure 13–20 shows inductor symbols. Iron cores are identified by double solid lines, while dashed lines denote a ferrite core. Air-core inductors have no core symbol. An arrow indicates a variable inductor.

Coil Resistance

Ideally, inductors have only inductance. However, since inductors are made of imperfect conductors (e.g., copper wire), they also have resistance. [We can view this resistance as being in series with the coil's inductance as indicated in Figure 13–21(a). Also shown is stray capacitance, considered next.] Although coil resistance is generally small, it cannot always be ignored, and thus it must

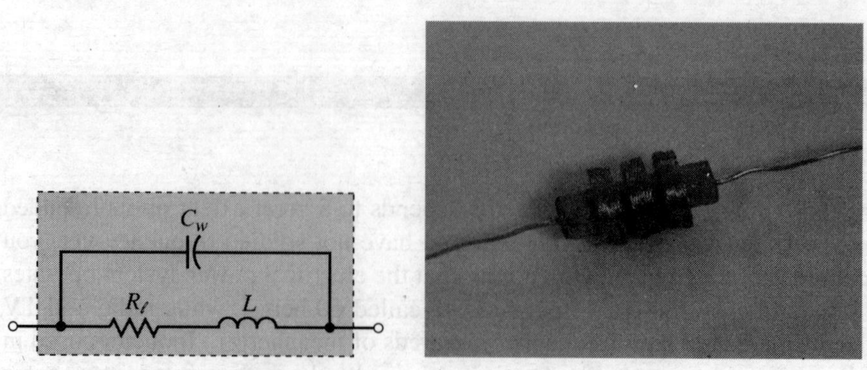

(a) Real inductors have stray capacitance and winding resistance

(b) Separating coil into sections helps reduce stray capacitance

FIGURE 13–21 A ferrite-core choke.

sometimes be included in the analysis of a circuit. In Section 13.7, we show how this resistance is taken into account in dc analysis; in later chapters, you will learn how to take it into account in ac analysis.

Stray Capacitance

Because the turns of an inductor are separated from each other by insulation, a small amount of capacitance exists from winding to winding. This capacitance is called stray or parasitic capacitance. Although this capacitance is distributed from turn to turn, its effect can be approximated by lumping it as in Figure 13–21(a). The effect of stray capacitance depends on frequency. At low frequencies, it can usually be neglected; at high frequencies, it may have to be taken into account as you will see in later courses. Some coils are wound in multiple sections, as in Figure 13–21(b), to reduce stray capacitance.

Stray Inductance

Because inductance is due entirely to the magnetic effects of electric current, all current-carrying conductors have inductance. This means that leads on circuit components such as resistors, capacitors, transistors, and so on, all have inductance, as do traces on printed circuit boards and wires in cables. We call this inductance "stray inductance." Fortunately, in many cases, the stray inductance is so small that it can be neglected (see Practical Notes).

PRACTICAL NOTES...

Although stray inductance is small, it is not always negligible. In general, stray inductance will not be a problem for short wires at low to moderate frequencies. However, even a short piece of wire can be a problem at high frequencies, or a long piece of wire at low frequencies. For example, the inductance of just a few centimeters of conductor in a high-speed logic system may be nonnegligible because the current through it changes at such a high rate.

13.7 Inductance and Steady State dc

We now look at inductive circuits with constant dc current. Consider Figure 13–22. The voltage across an ideal inductance with constant dc current is zero because the rate of change of current is zero. This is indicated in (a). Since the inductor has current through it but no voltage across it, it looks like a short circuit, (b). This is true in general, that is, *an ideal inductor looks like a short circuit in steady state dc.* (This should not be surprising since it is just a piece of wire to dc.) For a nonideal inductor, its dc equivalent is its coil resistance (Figure 13–23). For steady state dc, problems can be solved using simple dc circuit analysis techniques.

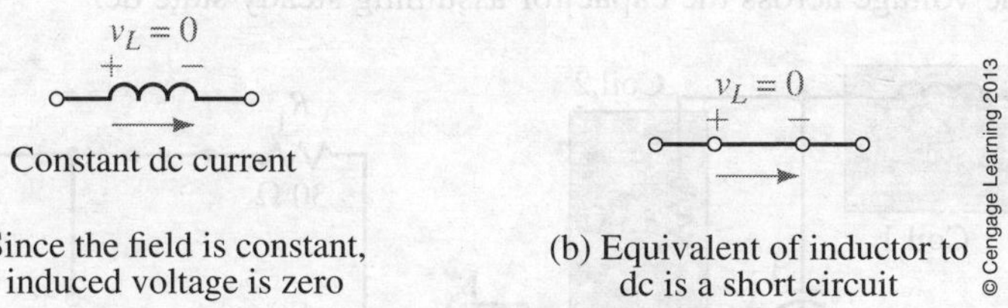

FIGURE 13–22 Inductance looks like a short circuit to steady state dc.

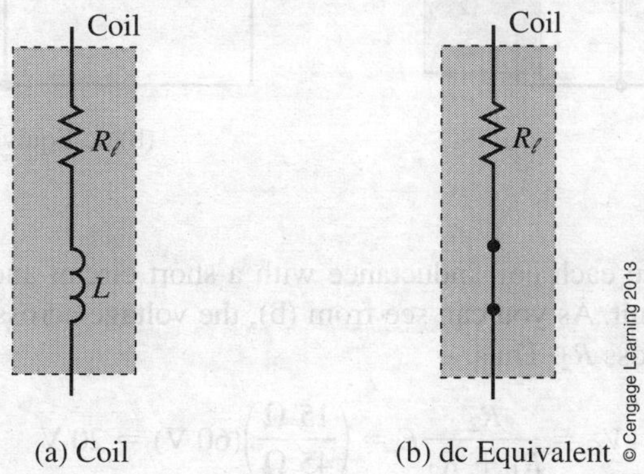

FIGURE 13–23 Steady state dc equivalent of a coil with winding resistance.

EXAMPLE 13–8

In Figure 13–24(a), the coil resistance is 14.4 Ω. What is the steady state current I?

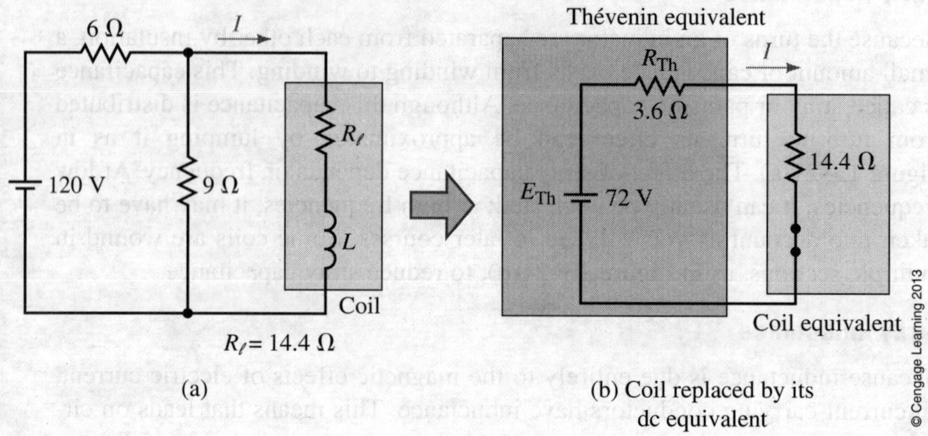

FIGURE 13–24

Solution Reduce the circuit as in (b).

$$E_{Th} = (9/15)(120) = 72 \text{ V}$$

$$R_{Th} = 6\ \Omega \| 9\ \Omega = 3.6\ \Omega$$

Now replace the coil with its dc equivalent circuit as in (b). Thus,

$$I = E_{Th}/R_T = 72/(3.6 + 14.4) = 4\text{A}$$

EXAMPLE 13–9

The resistance of coil 1 in Figure 13–25(a) is 30 Ω and that of coil 2 is 15 Ω. Find the voltage across the capacitor assuming steady state dc.

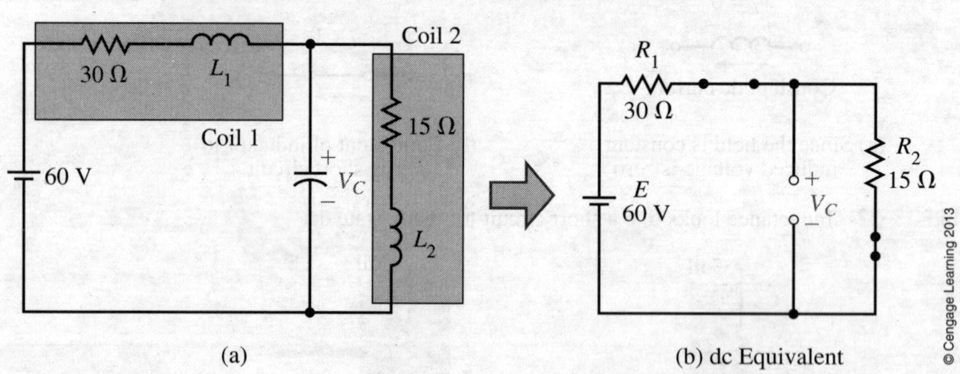

FIGURE 13–25

Solution Replace each coil inductance with a short circuit and the capacitor with an open circuit. As you can see from (b), the voltage across C is the same as the voltage across R_2. Thus,

CircuitSim 13-3

$$V_C = \frac{R_2}{R_1 + R_2}E = \left(\frac{15\ \Omega}{45\ \Omega}\right)(60 \text{ V}) = 20 \text{ V}$$

PRACTICE PROBLEMS 4

For Figure 13–26, find I, V_{C_1}, and V_{C_2} in the steady state.

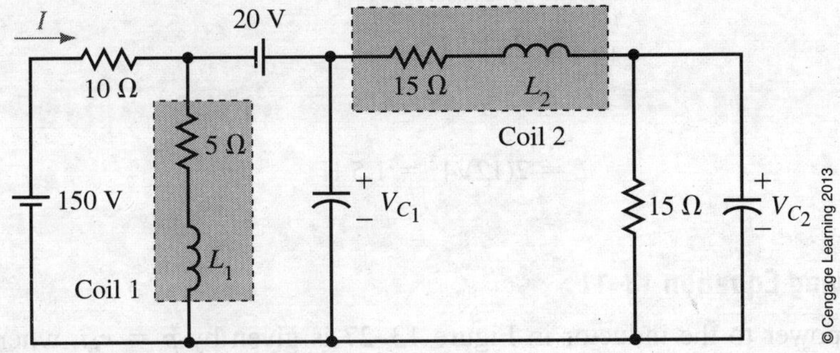

FIGURE 13–26

CircuitSim 13-4

Answers
10.7 A; 63 V; 31.5 V

13.8 Energy Stored by an Inductance

When power flows into an inductor, energy is stored in its magnetic field. When the field collapses, this energy is returned to the circuit. For an ideal inductor, $R_\ell = 0$ ohm and hence no power is dissipated; thus, an ideal inductor has zero power loss.

To determine the energy stored by an ideal inductor, consider Figure 13–27. Power to the inductor is given by $p = v_L i$ watts, where $v_L = L di/dt$. By summing this power (see next ▮), the energy is found to be

$$W = \frac{1}{2}Li^2 \quad \text{(J)} \tag{13–11}$$

where i is the instantaneous value of current. When current reaches its steady state value I, $W = ½LI^2$ J. This energy remains stored in the field as long as the current continues. When the current goes to zero, the field collapses and the energy is returned to the circuit. (We look at this idea in more detail in Chapter 17 when we consider ac.)

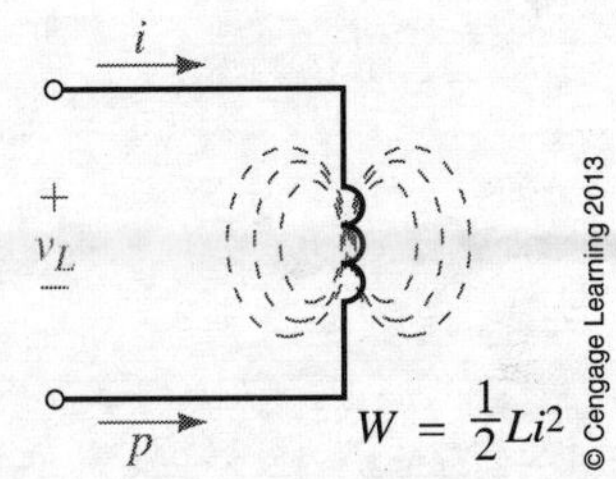

FIGURE 13–27 Energy is stored in the magnetic field of an inductor.

EXAMPLE 13–10

The coil of Figure 13–28(a) has a resistance of 15 Ω. When the current reaches its steady state value, the energy stored is 12 J. What is the inductance of the coil?

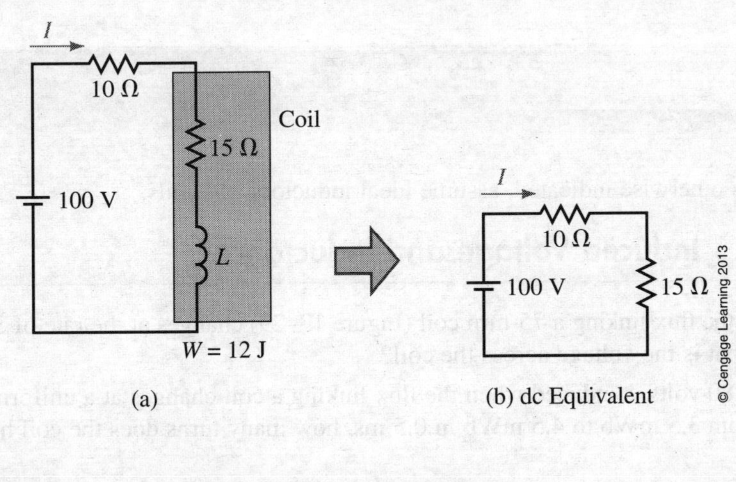

FIGURE 13–28

Solution From (b),

$$I = 100 \text{ V}/25\ \Omega = 4 \text{ A}$$

$$W = \frac{1}{2}LI^2 \text{ J}$$

$$12 \text{ J} = \frac{1}{2}L(4 \text{ A})^2$$

Thus,

$$L = 2(12)/4^2 = 1.5 \text{ H}$$

Deriving Equation 13–11

The power to the inductor in Figure 13–27 is given by $p = v_L i$, where $v_L = Ldi/dt$. Therefore, $p = Lidi/dt$. However, $p = dW/dt$. Equating the two expressions for power, then integrating both sides yields

$$W = \int_0^t pdt = \int_0^t Li\frac{di}{dt}dt = L\int_0^i idi = \frac{1}{2}Li^2$$

13.9 Inductor Troubleshooting Hints

Inductors may fail by either opening or shorting. Failures may be caused by misuse, defects in manufacturing, or faulty installation.

Open Coil

Opens can be the result of poor solder joints or broken connections. First, make a visual inspection. If nothing wrong is found, disconnect the inductor and check it with an ohmmeter. An open-circuited coil has infinite resistance.

Shorts

Shorts can occur between windings or between the coil and its core (for an iron-core unit). A short may result in excessive current and overheating. Again, check visually. Look for burned insulation, discolored components, an acrid odor, and other evidence of overheating. An ohmmeter can be used to check for shorts between windings and the core. However, checking coil resistance for shorted turns is often of little value, especially if only a few turns are shorted. This is because the shorting of a few windings may not change the overall resistance enough to be measurable. The LCR testers of Chapter 10 may also prove useful. Sometimes, however, the only conclusive test is to substitute a known good inductor for the suspected bad one.

Problems

Unless otherwise indicated, assume ideal inductors and coils.

13.2 Induced Voltage and Inductance

1. If the flux linking a 75-turn coil (Figure 13–29) changes at the rate of 3 Wb/s, what is the voltage across the coil?
2. If 80 volts is induced when the flux linking a coil changes at a uniform rate from 3.5 mWb to 4.5 mWb in 0.5 ms, how many turns does the coil have?

3. Flux changing at a uniform rate for 1 ms induces 60 V in a coil. What is the induced voltage if the same flux change takes place in 0.01 s?

13.3 Self-Inductance

4. The current in a 0.4-H inductor (Figure 13–30) is changing at the rate of 200 A/s. What is the voltage across it?
5. The current in a 75-mH inductor (Figure 13–30) changes uniformly by 200 μA in 0.1 ms. What is the voltage across it?
6. The voltage across an inductance is 25 volts when the current changes at 5 A/s. What is L?
7. The voltage induced when current changes uniformly from 3 amps to 5 amps in a 10-H inductor is 180 volts. How long did it take for the current to change from 3 amps to 5 amps?
8. Current changing at a uniform rate for 1 ms induces 45 V in a coil. What is the induced voltage if the same current change takes place in 100 μs?
9. Compute the inductance of the air-core coil of Figure 13–31, given $\ell = 20$ cm, $N = 200$ turns, and $d = 2$ cm.
10. The iron-core inductor of Figure 13–32 has 2000 turns, a cross section of 1.5 × 1.2 inches, and an air gap of 0.2 inch. Compute its inductance.

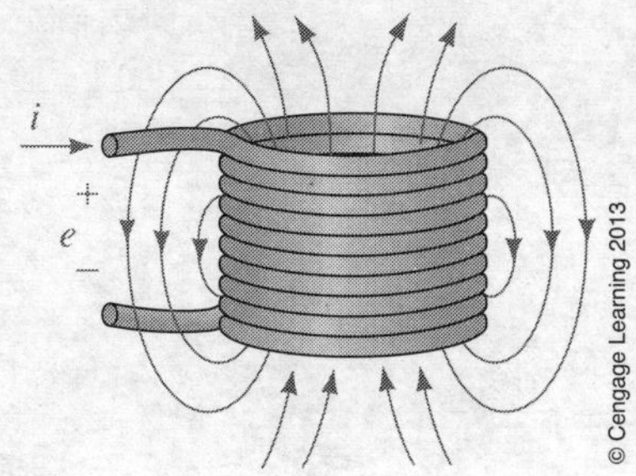

FIGURE 13–29

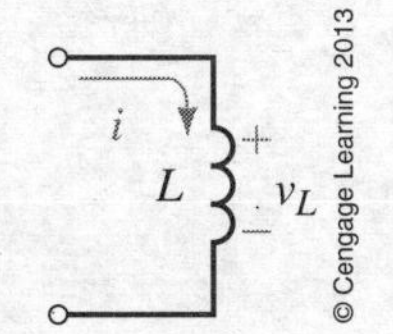

FIGURE 13–30

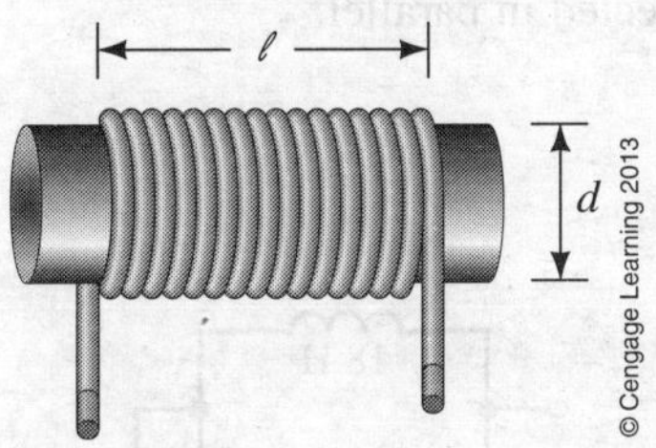

FIGURE 13–31

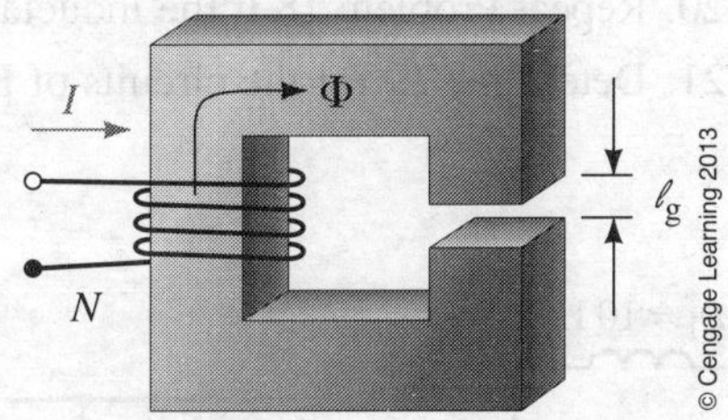

FIGURE 13–32

11. The iron-core inductor of Figure 13–32 has a high-permeability core. Therefore, by Ampere's law, $NI \simeq H_g \ell_g$. Because the air gap dominates, saturation does not occur, and the core flux is proportional to the current, that is, the flux linkage equals LI. In addition, since all flux passes through the coil, the flux linkage equals $N\Phi$. By equating the two values of flux linkage and using ideas from Chapter 12, show that the inductance of the coil is

$$L = \frac{\mu_0 N^2 A_g}{\ell_g}$$

13.4 Computing Induced Voltage

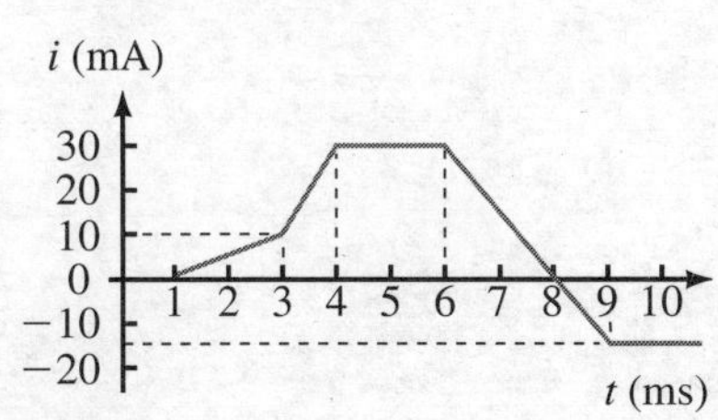

FIGURE 13–33

12. Figure 13–33 shows the current in a 0.75-H inductor. Determine v_L and plot its waveform.
13. Figure 13–34 shows the current in a coil. If the voltage from 0 ms to 2 ms is 100 volts, what is L?
14. Why is Figure 13–35 not a valid inductor current? Sketch the voltage across L to show why. Pay particular attention to $t = 10$ ms.
15. Figure 13–36 shows the graph of the voltage across an inductance. The current changes from 4 A to 5 A during the time interval from 4 s to 5 s.
    a. What is L?
    b. Determine the current waveform and plot it.
    c. What is the current at $t = 10$ s?

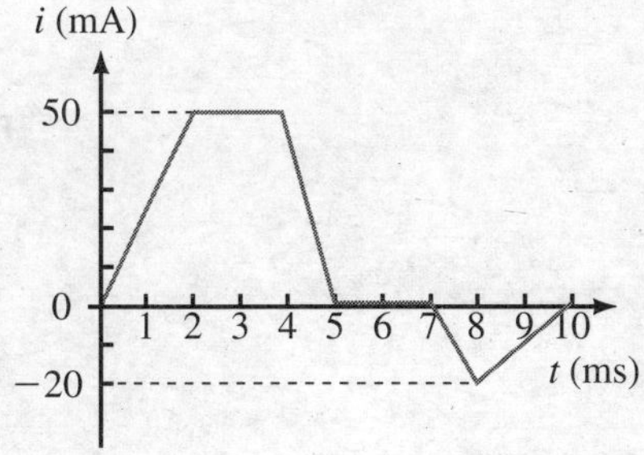

FIGURE 13–34

FIGURE 13–35

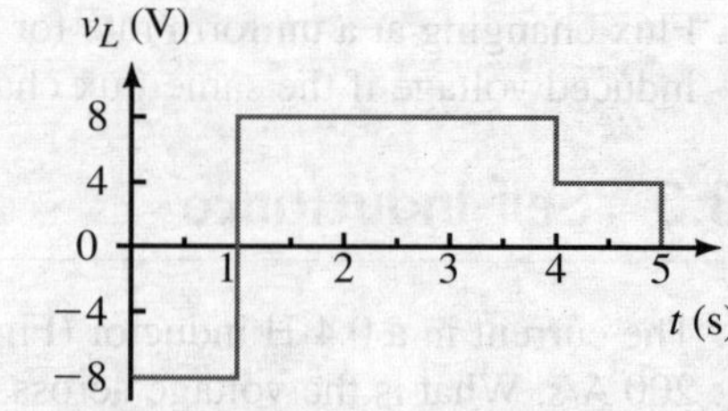

FIGURE 13–36

16. If the current in a 25-H inductance is $i_L = 20e^{-12t}$ mA, what is v_L?

13.5 Inductances in Series and Parallel

17. What is the equivalent inductance of 12 mH, 14 mH, 22 mH, and 36 mH connected in series?
18. What is the equivalent inductance of 0.010 H, 22 mH, 86×10^{-3} H, and 12 000 μH connected in series?
19. Repeat Problem 17 if the inductances are connected in parallel.
20. Repeat Problem 18 if the inductances are connected in parallel.
21. Determine L_T for the circuits of Figure 13–37.

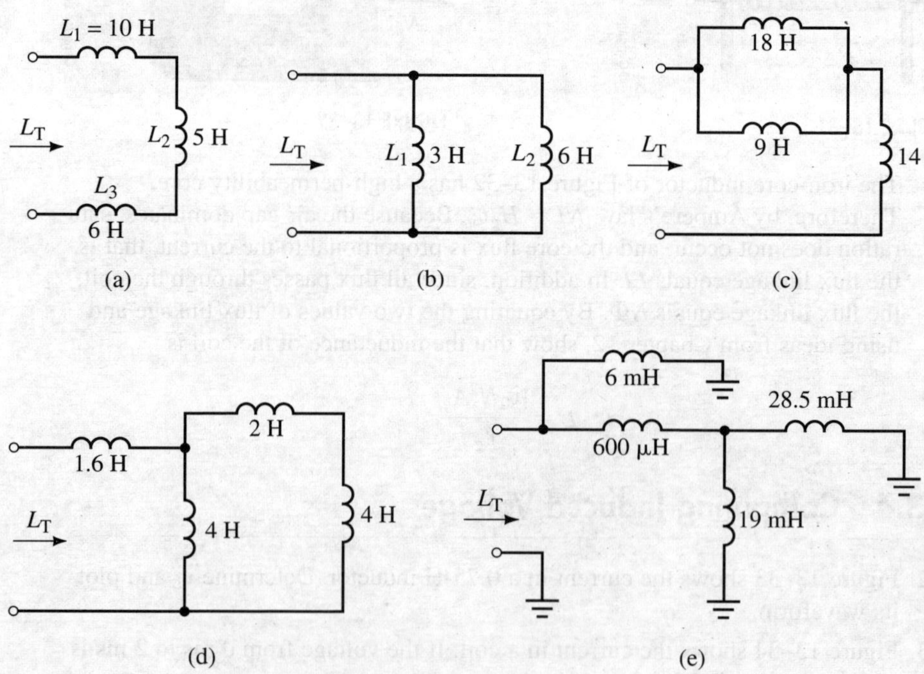

FIGURE 13–37

22. Determine L_T for the circuits of Figure 13–38.
23. A 30-μH inductance is connected in series with a 60-μH inductance, and a 10-μH inductance is connected in parallel with the series combination. What is L_T?

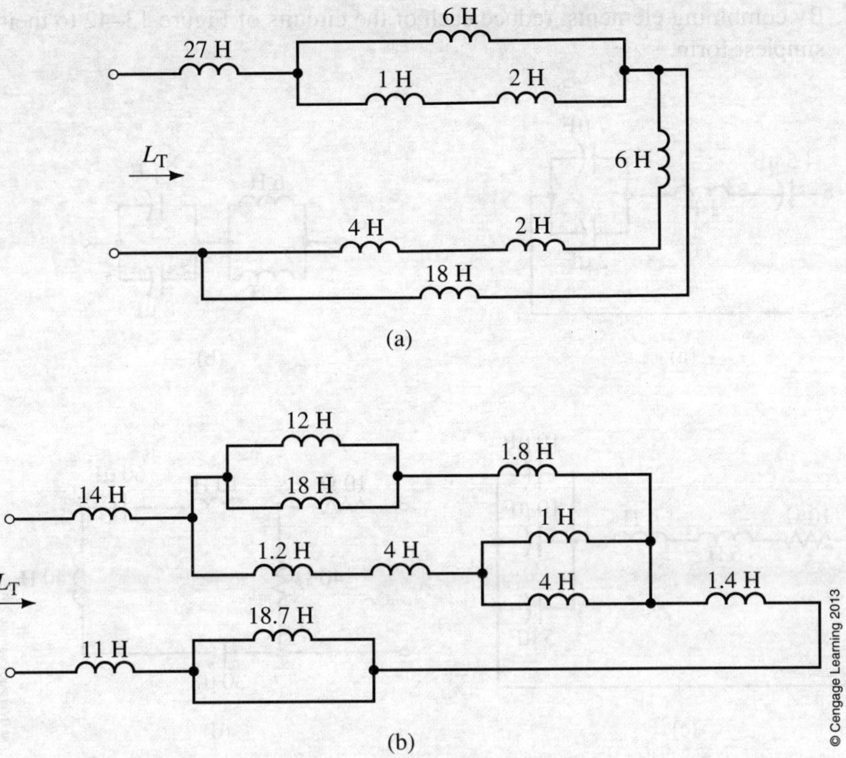

FIGURE 13–38

24. For Figure 13–39, determine L_x.
25. For the circuits of Figure 13–40, determine L_3 and L_4.
26. You have inductances of 24 mH, 36 mH, 22 mH, and 10 mH. Connecting these any way you want, what is the largest equivalent inductance you can get? The smallest?

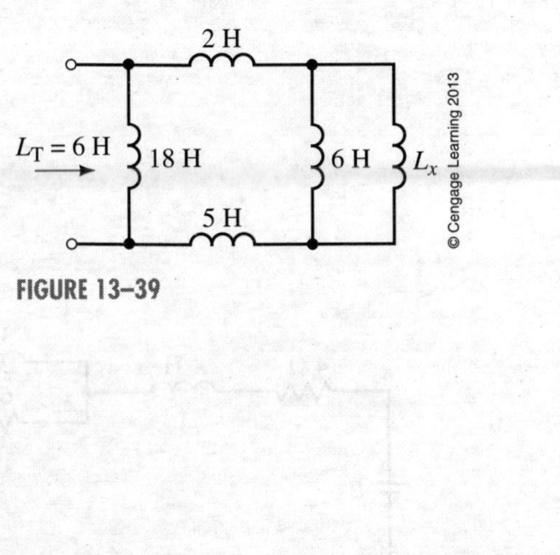

FIGURE 13–39

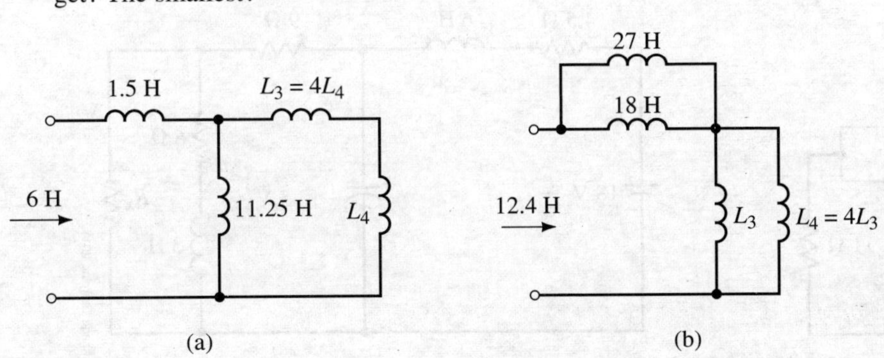

FIGURE 13–40

27. A 6-H and a 4-H inductance are connected in parallel. After a third inductance is added, $L_T = 4$ H. What is the value of the third inductance and how was it connected?
28. Inductances of 2 H, 4 H, and 9 H are connected in a circuit. If $L_T = 3.6$ H, how are the inductors connected?
29. Inductances of 8 H, 12 H, and 1.2 H are connected in a circuit. If $L_T = 6$ H, how are the inductors connected?
30. For inductors in parallel (Figure 13–41), the same voltage appears across each. Thus, $v = L_1 di_1/dt$, $v = L_2 di_2/dt$, and so on. Apply KCL and show that $1/L_T = 1/L_1 + 1/L_2 + \cdots + 1/L_N$.

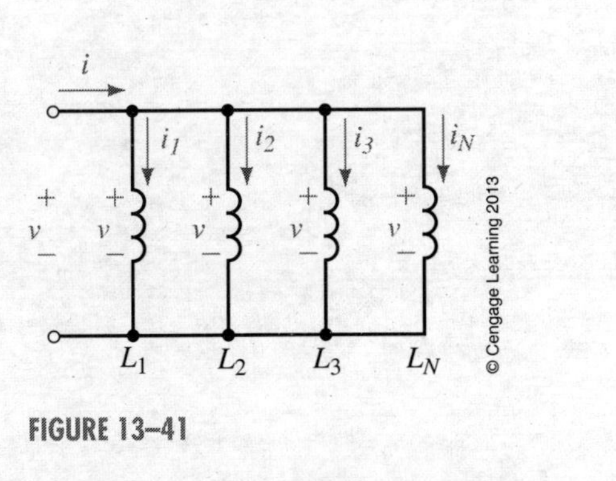

FIGURE 13–41

31. By combining elements, reduce each of the circuits of Figure 13–42 to their simplest form.

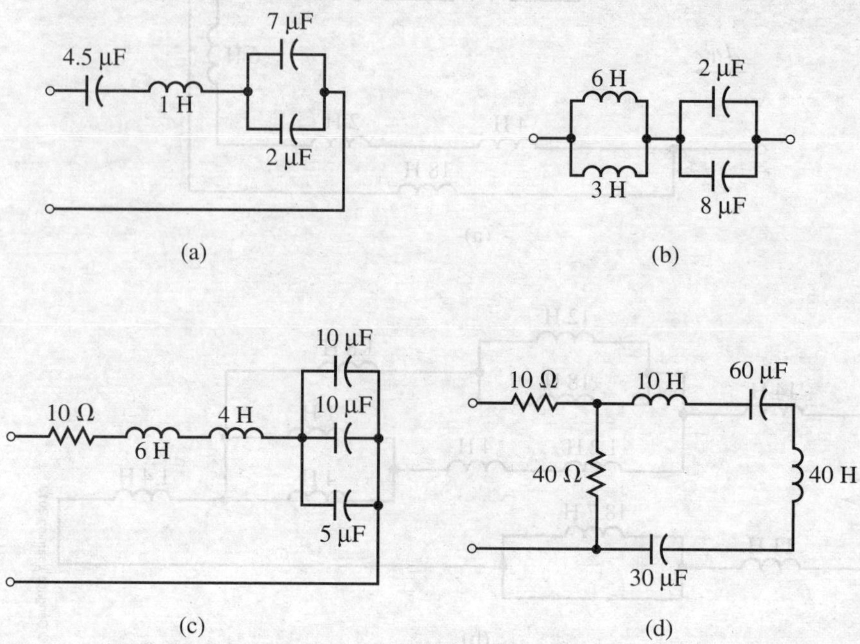

FIGURE 13–42

13.7 Inductance and Steady State dc

32. For each of the circuits of Figure 13–43, the voltages and currents have reached their final (steady state) values. Solve for the quantities indicated.

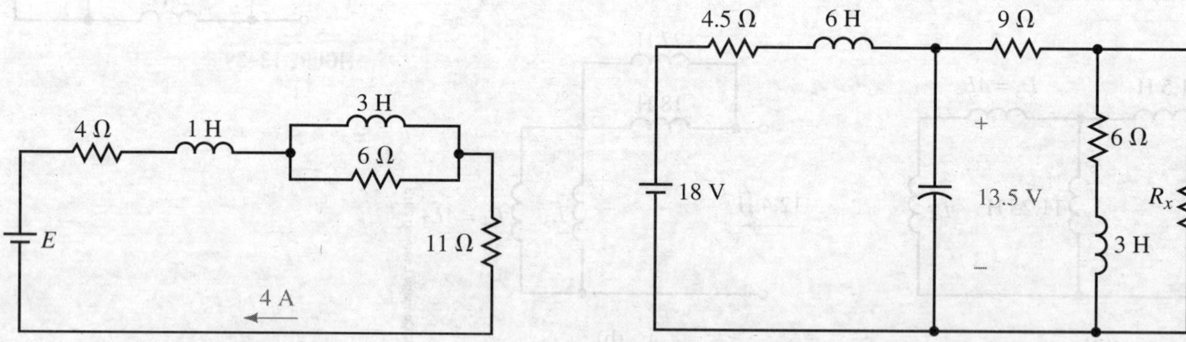

(a) Find E (b) Find R_x

FIGURE 13–43

13.8 Energy Stored by an Inductance

33. Find the energy stored in the inductor of Figure 13–44.

34. In Figure 13–45, $L_1 = 2L_2$. The total energy stored is $W_T = 75$ J. Find L_1 and L_2.

13.9 Inductor Troubleshooting Hints

35. In Figure 13–46, an inductance meter measures 7 H. What is the likely fault?

36. Referring to Figure 13–47, an inductance meter measures $L_T = 8$ mH. What is the likely fault?

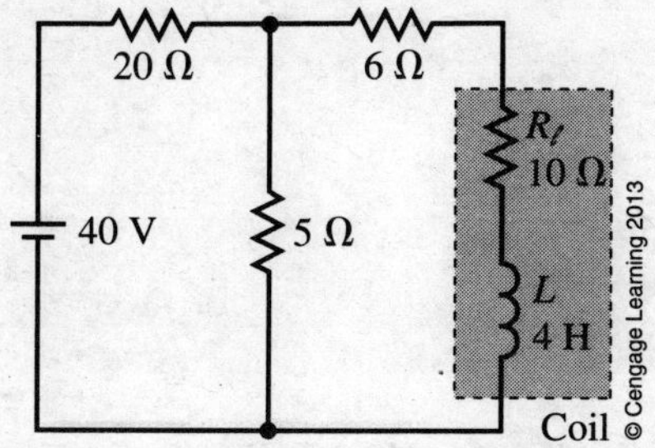

FIGURE 13–44

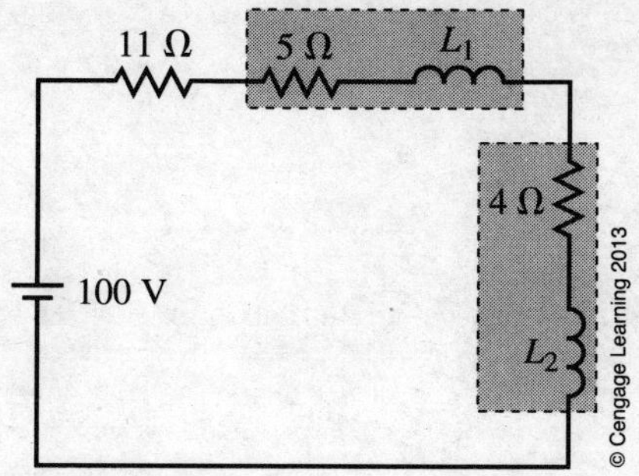

FIGURE 13–45

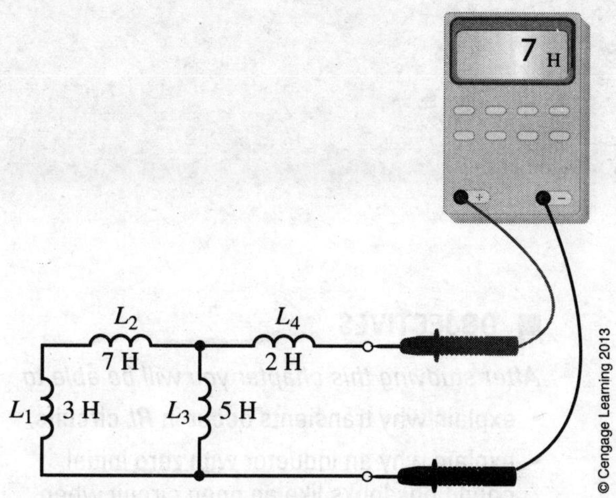

FIGURE 13–46

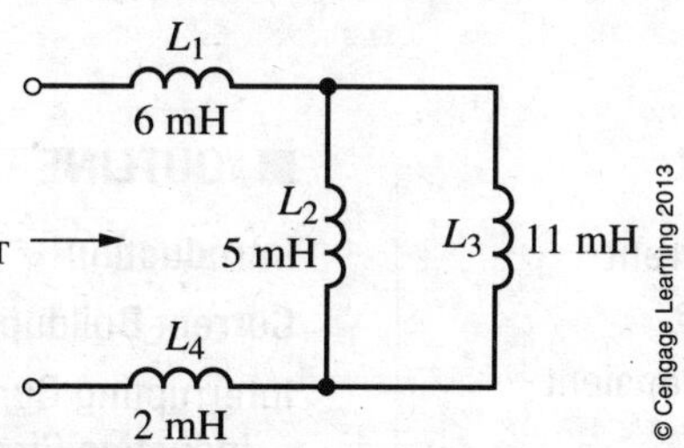

FIGURE 13–47

ANSWERS TO IN-PROCESS LEARNING CHECKS

IN-PROCESS LEARNING CHECK 1

1. Both a and b. Current cannot change instantaneously.
2. 600 Wb turns
3. 30 V
4. −300 V
5. $5e^{-t}(1 - t)$ V

IN-PROCESS LEARNING CHECK 2

1. 28 mH
2. 9 times

IN-PROCESS LEARNING CHECK 3

1. 87.5 V
2. Rate of change of current is the same; 20 V
3. $20t^2 e^{-2t}(3 - 2t)$ V
4. 250 cos 1000t

KEY TERMS

Continuity of Current for Inductance

De-Energizing Transient

Inductive Kick

Inductive Transients

Initial Condition Circuit

OUTLINE

Introduction

Current Buildup Transients

Interrupting Current in an Inductive Circuit

De-Energizing Transients

More Complex Circuits

RL Transients Using Computers

OBJECTIVES

After studying this chapter you will be able to

- explain why transients occur in *RL* circuits,
- explain why an inductor with zero initial conditions looks like an open circuit when first energized,
- compute time constants for *RL* circuits,
- compute voltage and current transients in *RL* circuits during the current buildup phase,
- compute voltage and current transients in *RL* circuits during the current decay phase,
- explain why an inductor with nonzero initial conditions looks like a current source when disturbed,
- solve moderately complex *RL* transient problems using circuit simplification techniques,
- solve *RL* transient problems using Multisim and PSpice.

14

INDUCTIVE TRANSIENTS

CHAPTER PREVIEW

In Chapter 11, you learned that transients occur in capacitive circuits because capacitor voltage cannot change instantaneously. In this chapter, you will learn that transients occur in inductive circuits because inductor current cannot change instantaneously. Although the details differ, you will find that many of the basic ideas are the same.

Inductive transients (also called *RL* transients) result when circuits containing inductance are disturbed. More so than capacitive transients, inductive transients are potentially destructive and dangerous. For example, if you break the current in an inductive circuit, a voltage spike of a few hundred or more volts may result. Such a spike can easily injure personnel or damage sensitive electronic components if proper precautions are not taken. (Current turn-on transients do not create such spikes.)

In this chapter, you study basic *RL* transients. You look at transients during current buildup and decay and learn how to calculate the voltages and currents that result. In later studies, you will learn how to both control and safeguard circuits from the effects of those transients that might be potentially destructive. ■

Putting It in Perspective

Inductance, the Dual of Capacitance

INDUCTANCE IS THE DUAL of capacitance. This means that the effect that inductance has on circuit operation is identical with that of capacitance if you interchange the term current for voltage, open circuit for short circuit, and so on. For example, for simple dc transients, current in an *RL* circuit has the same form as voltage in an *RC* circuit: they both rise to their final value in an exponential fashion according to $1 - e^{-t/\tau}$. Similarly, voltage across inductance decays in the same manner as current through capacitance, that is, according to $e^{-t/\tau}$. In fact, as you will see, there is complete duality between all the equations that describe transient voltage and current behavior in capacitive and inductive circuits.

Duality applies to steady state and initial condition representations as well. To steady state dc, for example, a capacitor looks like an open circuit, while an inductor looks like a short circuit. Similarly, the dual of a capacitor that looks like a short circuit at the instant of switching is an inductor that looks like an open circuit. Finally, the dual of a capacitor that has an initial condition of V_0 volts is an inductance with an initial condition of I_0 amps.

The principle of duality is helpful in circuit analysis as it lets you transfer the principles and concepts learned in one area directly into another. You will find, for example, that many of the ideas you learned in Chapter 11 reappear here in their dual form. ■

14.1 Introduction

As you saw in Chapter 11, when a circuit containing capacitance is disturbed, voltages and currents do not change to their new values immediately, but instead pass through a transitional phase as the circuit capacitance charges or discharges. The voltages and currents during this transitional interval are called transients. In a dual fashion, transients occur when circuits containing inductances are disturbed. In this case, however, transients occur because current in inductance cannot change instantaneously.

To get at the idea, consider Figure 14–1. In (a), we see a purely resistive circuit. At the instant the switch is closed, current jumps from 0 to *E*/*R* as required by Ohm's law. Thus, no transient (i.e., transitional phase) occurs because current reaches its final value immediately. Now consider (b). Here, we have added inductance. At the instant the switch is closed, a counter EMF appears across the inductance. This voltage attempts to stop the current from changing and consequently slows its rise. Current thus does not jump to *E*/*R* immediately as in (a), but instead climbs gradually and smoothly as in (b). The larger the inductance, the longer the transition takes.

NOTES...

The continuity statement for inductor current has a sound mathematical basis. Recall, induced voltage is proportional to the rate of change of current. In calculus notation,

$$v_L = L\frac{di}{dt}$$

This means that the faster the current changes, the larger the induced voltage. If inductor current could change from one value to another instantaneously as in Figure 14–1(a), its rate of change (i.e., *di*/*dt*) would be infinite and hence the induced voltage would be infinite. But infinite voltage is not possible. Thus, we conclude that inductor current cannot change instantaneously.

Continuity of Current

As Figure 14–1(b) illustrates, *current through an inductance cannot change instantaneously, that is, it cannot jump abruptly from one value to another, but must be continuous at all values of time.* This observation is known as the statement of **continuity of current for inductance** (see Notes). You will find this statement of great value when analyzing circuits containing inductance. We will use it numerous times in what follows.

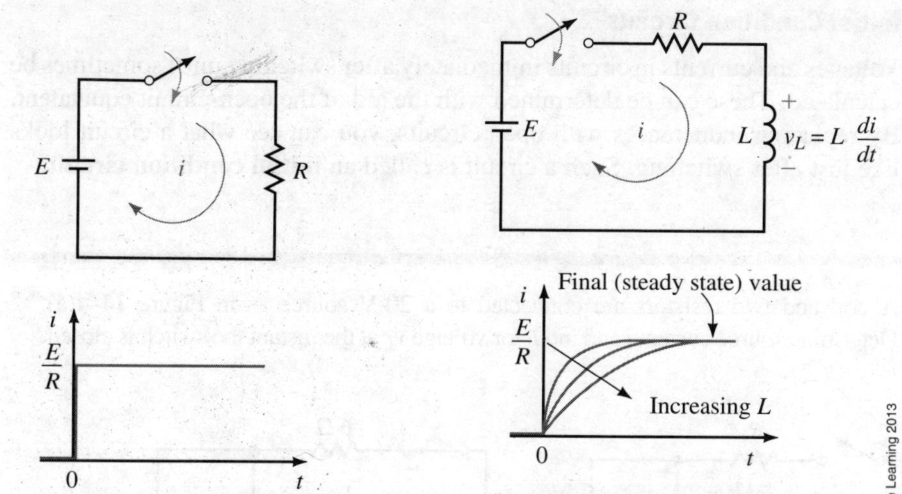

(a) No transient occurs in a purely resistive circuit

(b) Adding inductance causes a transient to appear. R is held constant here.

FIGURE 14–1 Transient due to inductance. Adding inductance to a resistive circuit as in (b) slows the current rise and fall, thus creating a transient.

CircuitSim 14-1

Inductor Voltage

Now consider inductor voltage. With the switch open as in Figure 14–2(a), the current in the circuit and voltage across L are both zero. Now close the switch. Immediately after the switch is closed, the current is still zero (since it cannot change instantaneously). Since $v_R = Ri$, the voltage across R is also zero and thus the full source voltage appears across L as shown in (b). The inductor voltage therefore jumps from 0 V just before the switch is closed to E volts just after. It then decays to zero, since, as we saw in Chapter 13, the voltage across inductance is zero for steady state dc. This is indicated in (c).

CircuitSim 14-2

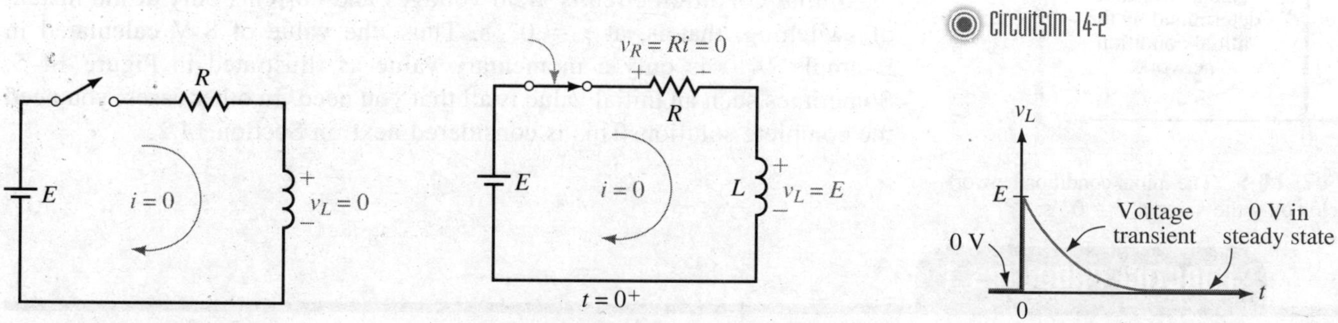

(a) Circuit before switch is closed. Current $i = 0$

(b) Circuit just after the switch has been closed. Current is still equal to zero. Thus, $v_L = E$.

(c) Voltage across L

FIGURE 14–2 Voltage across L.

Open-Circuit Equivalent of an Inductance

Consider again Figure 14–2(b). Note that just after the switch is closed, the inductor has voltage across it but no current through it. It therefore momentarily looks like an open circuit. This is indicated in Figure 14–3. This observation is true in general, that is, *an inductor with zero initial current looks like an open circuit at the instant of switching.* (In Section 14.3, we extend this statement to include inductors with nonzero initial currents.)

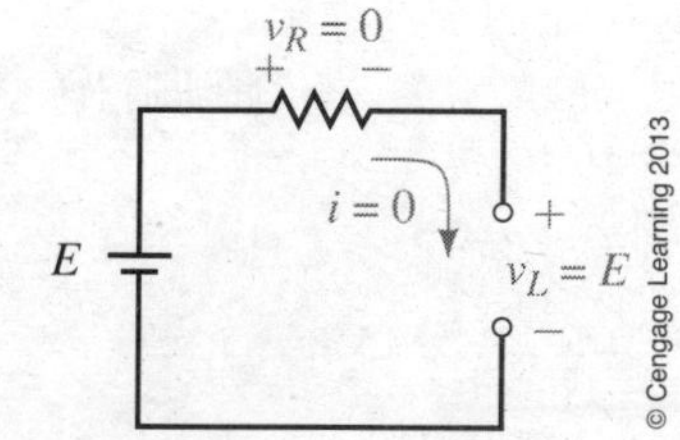

FIGURE 14–3 Inductor with zero initial current looks like an open circuit at the instant the switch is closed.

Initial Condition Circuits

Voltages and currents in circuits immediately after switching must sometimes be calculated. These can be determined with the aid of the open-circuit equivalent. By replacing inductances with open circuits, you can see what a circuit looks like just after switching. Such a circuit is called an **initial condition circuit.**

EXAMPLE 14–1

A coil and two resistors are connected to a 20-V source as in Figure 14–4(a). Determine source current i and inductor voltage v_L at the instant the switch is closed.

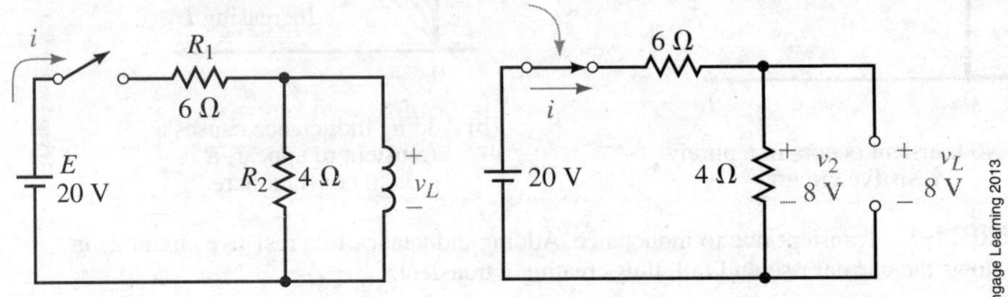

(a) Original circuit (b) Initial condition network

FIGURE 14–4

Solution Replace the inductance with an open circuit. This yields the network shown in (b). Thus, $i = E/R_T = 20\ \text{V}/10\ \Omega = 2\ \text{A}$, and the voltage across R_2 is $v_2 = (2\ \text{A})(4\ \Omega) = 8\ \text{V}$. Since $v_L = v_2$, $v_L = 8$ volts as well.

CircuitSim 14-3

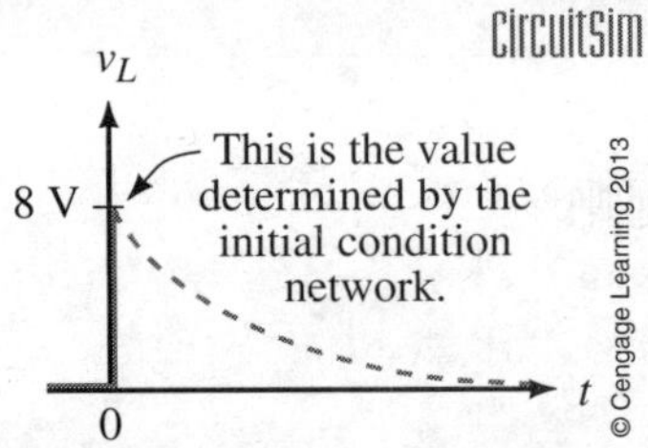

FIGURE 14–5 The initial condition network yields only the value at $t = 0^+$ s.

Initial condition circuits yield voltages and currents only at the instant of switching, that is, at $t = 0^+$ s. Thus, the value of 8 V calculated in Example 14–1 is only a momentary value as illustrated in Figure 14–5. Sometimes such an initial value is all that you need. In other cases, you need the complete solution. This is considered next, in Section 14.2.

PRACTICE PROBLEMS 1

Determine all voltages and currents in the circuit of Figure 14–6 immediately after the switch is closed and in steady state.

CircuitSim 14-4

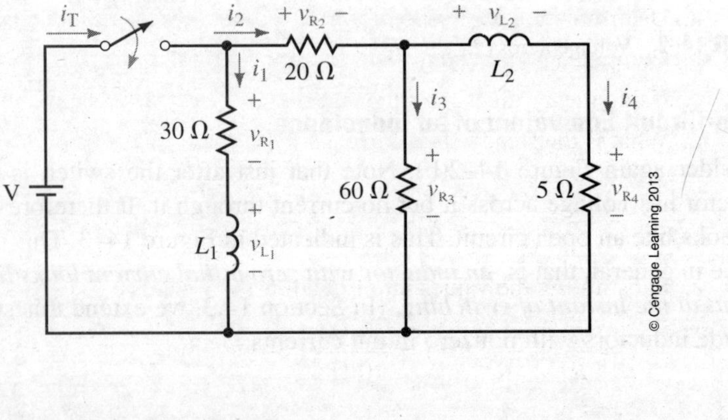

FIGURE 14–6

Answers

Initial: $v_{R_1} = 0$ V; $v_{R_2} = 40$ V; $v_{R_3} = 120$ V; $v_{R_4} = 0$ V; $v_{L_1} = 160$ V; $v_{L_2} = 120$ V; $i_T = 2$ A; $i_1 = 0$ A; $i_2 = 2$ A; $i_3 = 2$ A; $i_4 = 0$ A

Steady State: $v_{R_1} = 160$ V; $v_{R_2} = 130$ V; $v_{R_3} = v_{R_4} = 30$ V; $v_{L_1} = v_{L_2} = 0$ V; $i_T = 11.83$ A; $i_1 = 5.33$ A; $i_2 = 6.5$ A; $i_3 = 0.5$ A; $i_4 = 6.0$ A

14.2 Current Buildup Transients

Current

We will now develop equations to describe voltages and current during energization. Let us start with current. Consider Figure 14–7. Kirchhoff's voltage law (KVL) yields

$$v_L + v_R = E \qquad (14\text{–}1)$$

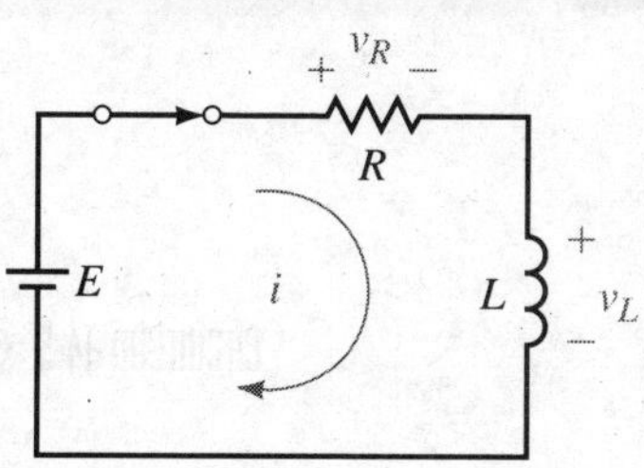

FIGURE 14–7 KVL yields $v_L + v_R = E$.

Substituting $v_L = L di/dt$ and $v_R = Ri$ into Equation 14–1 yields

$$L\frac{di}{dt} + Ri = E \qquad (14\text{–}2)$$

Equation 14–2 can be solved using basic calculus in a manner similar to what we did for *RC* circuits in Chapter 11. The result is

$$i = \frac{E}{R}(1 - e^{-Rt/L}) \text{ (A)} \qquad (14\text{–}3)$$

where *R* is in ohms, *L* is in henries, and *t* is in seconds. Equation 14–3 describes current buildup. Values of current at any point in time can be found by direct substitution as we illustrate next. Note that *E/R* is the final (steady state) current since the inductor looks like a short circuit to steady state dc (recall Section 13.7).

EXAMPLE 14–2

For the circuit of Figure 14–7, suppose $E = 50$ V, $R = 10\ \Omega$, and $L = 2$ H:

a. Determine the expression for *i*.

b. Compute and tabulate values of *i* at $t = 0^+$, 0.2, 0.4, 0.6, 0.8, and 1.0 s.

c. Using these values, plot the current.

d. What is the steady state current?

Solution

a. Substituting the values into Equation 14–3 yields

$$i = \frac{E}{R}(1 - e^{-Rt/L}) = \frac{50\text{ V}}{10\ \Omega}(1 - e^{-10t/2}) = 5(1 - e^{-5t}) \text{ amps}$$

TABLE 14–1

Time	Current
0	0
0.2	3.16
0.4	4.32
0.6	4.75
0.8	4.91
1.0	4.97

b. At $t = 0^+$ s, $i = 5(1 - e^{-5t}) = 5(1 - e^0) = 5(1 - 1) = 0$ A.
At $t = 0.2$ s, $i = 5(1 - e^{-5(0.2)}) = 5(1 - e^{-1}) = 3.16$ A.
At $t = 0.4$ s, $i = 5(1 - e^{-5(04)}) = 5(1 - e^{-2}) = 4.32$ A.
Continuing in this manner, you get Table 14–1.

c. Values are plotted in Figure 14–8. Note that this curve looks exactly like the curves we determined intuitively in Figure 14–1(b).

d. Steady state current is $E/R = 50 \text{ V}/10\ \Omega = 5$ A. This agrees with the current plot of Figure 14–8.

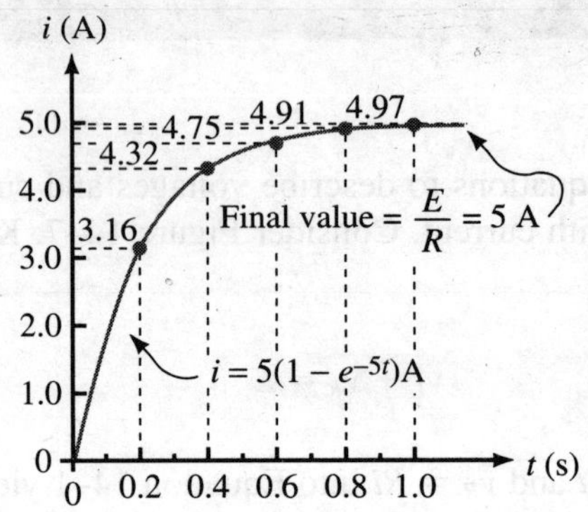

CircuitSim 14-5

FIGURE 14–8 Current buildup transient.

Circuit Voltages

With i known, circuit voltages can be determined. Consider voltage v_R. Since $v_R = Ri$, when you multiply R times Equation 14–3, you get

$$v_R = E(1 - e^{-Rt/L}) \quad \text{(V)} \tag{14–4}$$

Note that v_R has exactly the same shape as the current. Now consider v_L. Voltage v_L can be found by subtracting v_R from E as per Equation 14–1:

$$v_L = E - v_R = E - E(1 - e^{-Rt/L}) = E - E + Ee^{-Rt/L}$$

Thus,

$$v_L = Ee^{-Rt/L} \tag{14–5}$$

An examination of Equation 14–5 shows that v_L has an initial value of E at $t = 0^+$ s and then decays exponentially to zero. This agrees with our earlier observation in Figure 14–2(c).

EXAMPLE 14–3

Repeat Example 14–2 for voltage v_L.

Solution

a. From Equation 14–5,

$$v_L = Ee^{-Rt/L} = 50e^{-5t} \text{ volts}$$

b. At $t = 0^+$ s, $v_L = 50e^{-5t} = 50e^0 = 50(1) = 50$ V.
At $t = 0.2$ s, $v_L = 50e^{-5(0.2)} = 50e^{-1} = 18.4$ V.

At $t = 0.4$ s, $v_L = 50e^{-5(0.4)} = 50e^{-2} = 6.77$ V.

Continuing in this manner, you get Table 14–2.

c. The waveform is shown in Figure 14–9.

d. Steady state voltage is 0 V, as you can see in Figure 14–9.

TABLE 14–2

Time (s)	Voltage (V)
0	50.0
0.2	18.4
0.4	6.77
0.6	2.49
0.8	0.916
1.0	0.337

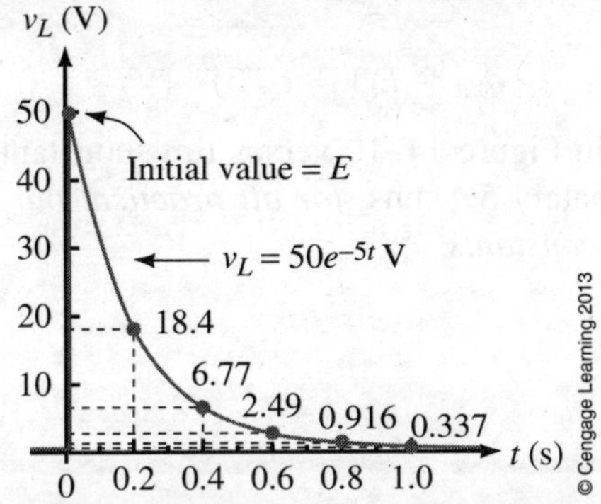

FIGURE 14–9 Inductor voltage transient.

CircuitSim 14-6

PRACTICE PROBLEMS 2

For the circuit of Figure 14–7, with $E = 80$ V, $R = 5$ kΩ, and $L = 2.5$ mH:

a. Determine expressions for i, v_L, and v_R.

b. Compute and tabulate values at $t = 0^+$, 0.5, 1.0, 1.5, 2.0, and 2.5 μs.

c. At each point in time, does $v_L + v_R = E$?

d. Plot i, v_L, and v_R using the values computed in (b).

Answers

a. $i = 16(1 - e^{-2 \times 10^6 t})$ mA; $v_L = 80e^{-2 \times 10^6 t}$ V; $v_R = 80(1 - e^{-2 \times 10^6 t})$

b.

t (μs)	v_L (V)	i_L (mA)	v_R (V)
0	80	0	0
0.5	29.4	10.1	50.6
1.0	10.8	13.8	69.2
1.5	3.98	15.2	76.0
2.0	1.47	15.7	78.5
2.5	0.539	15.9	79.5

c. Yes

d. i and v_R have the shape shown in Figure 14–8, while v_L has the shape shown in Figure 14–9, with values according to the table shown in b.

Time Constant

In Equations 14–3 to 14–5 L/R is the time constant of the circuit.

$$\tau = \frac{L}{R} \text{ (s)} \tag{14–6}$$

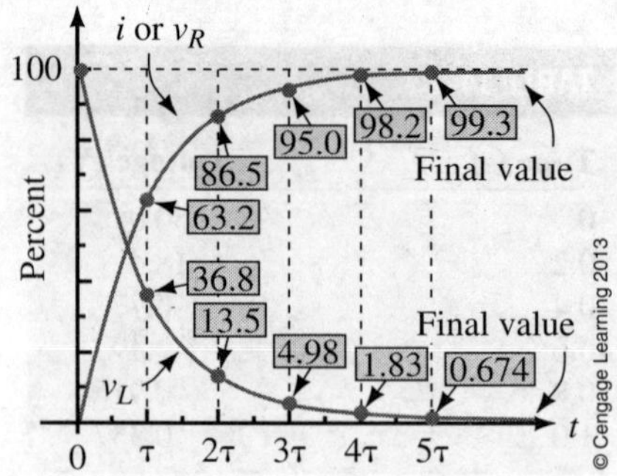

FIGURE 14–10 Universal time constant curves for the *RL* circuit.

Note that τ has units of seconds. (This is left as an exercise for the student.) Equations 14–3, 14–4, and 14–5 may now be written as

$$i = \frac{E}{R}(1 - e^{-t/\tau}) \quad \text{(A)} \tag{14–7}$$

$$v_L = Ee^{-t/\tau} \quad \text{(V)} \tag{14–8}$$

$$v_R = E(1 - e^{-t/\tau}) \quad \text{(V)} \tag{14–9}$$

Curves are plotted in Figure 14–10 versus time constant. As expected, transitions take approximately 5τ; thus, *for all practical purposes, inductive transients last five time constants.*

EXAMPLE 14–4

In a circuit where $L = 2$ mH, transients last 50 μs. What is *R*?

Solution Transients last five time constants. Thus, $\tau = 50\ \mu\text{s}/5 = 10\ \mu\text{s}$. Now $\tau = L/R$. Therefore, $R = L/\tau = 2\ \text{mH}/10\ \mu\text{s} = 200\ \Omega$.

EXAMPLE 14–5

For an *RL* circuit, $i = 40(1 - e^{-5t})$ A and $v_L = 100e^{-5t}$ V.

a. What are *E* and τ?

b. What is *R*?

c. Determine *L*.

Solution

a. From Equation 14–8, $v_L = Ee^{-t/\tau} = 100e^{-5t}$. Therefore, $E = 100$ V and $\tau = \frac{1}{5} = 0.2$ s.

b. From Equation 14–7,

$$i = \frac{E}{R}(1 - e^{-t/\tau}) = 40(1 - e^{-5t}).$$

Therefore, $E/R = 40$ A and $R = E/40\ \text{A} = 100\ \text{V}/40\ \text{A} = 2.5\ \Omega$.

c. $\tau = L/R$. Therefore, $L = R\tau = (2.5)(0.2) = 0.5$ H.

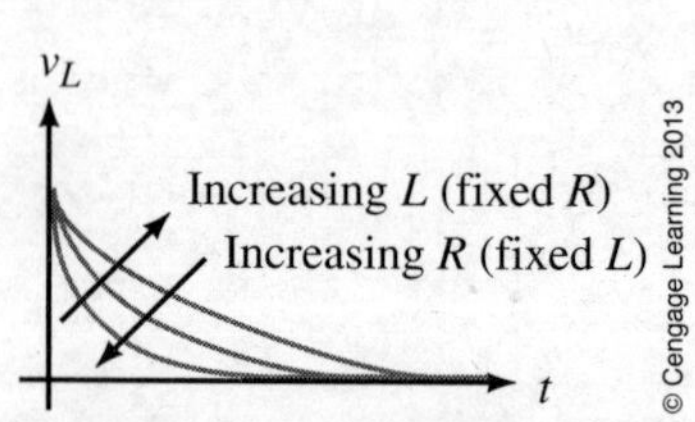

FIGURE 14–11 Effect of *R* and *L* on transient duration.

You can use the universal time constant curves to solve problems. (Be sure to convert curve percentages to a decimal value first, e.g., 63.2% to 0.632.) To illustrate, consider the problem of Examples 14–2 and 14–3. From Figure 14–10 at $t = \tau = 0.2$ s, $i = 0.632E/R$ and $v_L = 0.368E$. Thus, $i = 0.632(5\ \text{A}) = 3.16$ A and $v_L = 0.368(50\ \text{V}) = 18.4$ V as we found earlier.

The effect of inductance and resistance on transient duration is shown in Figure 14–11. The larger the inductance, the longer the transient for a given resistance. Resistance has the opposite effect: for a fixed inductance, the larger the resistance, the shorter the transient. [This is not hard to understand. As R increases, the circuit looks more and more resistive. If you get to a point where inductance is negligible compared with resistance, the circuit looks purely resistive, as in Figure 14–1(a), and virtually no transient occurs.]

IN-PROCESS LEARNING CHECK 1

(Answers are at the end of the chapter.)

1. For the circuit of Figure 14–12, the switch is closed at $t = 0$ s.
   a. Determine expressions for v_L and i.
   b. Compute v_L and i at $t = 0^+$, 10 μs, 20 μs, 30 μs, 40 μs, and 50 μs.
   c. Plot curves for v_L and i.
2. For the circuit of Figure 14–7, $E = 85$ V, $R = 50\ \Omega$, and $L = 0.5$ H. Use the universal time constant curves to determine v_L and i at $t = 20$ ms.
3. For a certain RL circuit, transients last 25 s. If $L = 10$ H and steady state current is 2 A, what is E?
4. An RL circuit has $E = 50$ V and $R = 10\ \Omega$. The switch is closed at $t = 0$ s. What is the current at the end of 1.5 time constants?

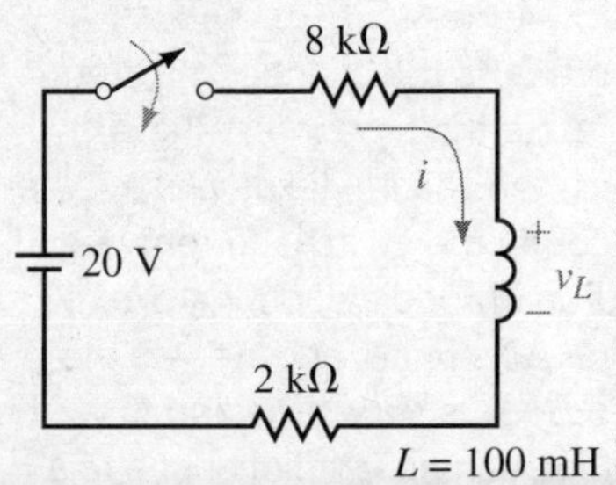

FIGURE 14–12

14.3 Interrupting Current in an Inductive Circuit

We now look at what happens when inductor current is interrupted. Consider Figure 14–13. At the instant the switch is opened, the field begins to collapse, which induces a voltage in the coil. If inductance is large and current is high, a great deal of energy is released in a very short time, creating a huge voltage that may damage equipment and create a shock hazard. (This induced voltage is referred to as an **inductive kick.**) For example, abruptly breaking the current through a large inductor (such as a motor or generator field coil) can create voltage spikes up to several thousand volts, a value large enough to draw long arcs as indicated in Figure 14–13. Even moderate-sized inductances in electronic systems can create enough voltage to cause damage if protective circuitry is not used.

CircuitSim 14-7

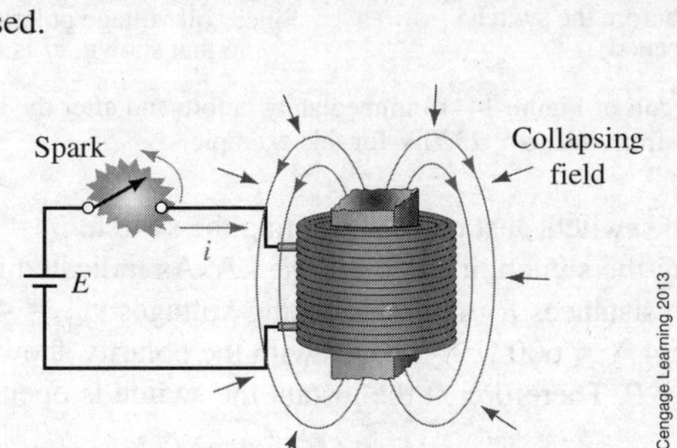

FIGURE 14–13 The sudden collapse of the magnetic field when the switch is opened causes a large induced voltage across the coil. (Several thousand volts may result.) The switch arcs over due to this voltage.

The dynamics of the switch flashover are not hard to understand. When the field collapses, the voltage across the coil rises rapidly. Part of this voltage appears across the switch. As the switch voltage rises, it quickly exceeds the breakdown strength of air, causing a flashover between its contacts. Once struck,

NOTES...

The intuitive explanation here has a sound mathematical basis. Recall, back EMF (induced voltage) across a coil is given by

$$v_L = L\frac{di}{dt} \approx L\frac{\Delta i}{\Delta t}$$

where Δi is the change in current and Δt is the time interval over which the change takes place. When you open the switch, current begins to drop immediately toward zero. Since Δi is finite, and $\Delta t \to 0$ the ratio $\Delta i/\Delta t$ is very large and thus, the voltage across L rapidly rises to a very large value, causing a flashover to occur. After the flashover, current has a path through which to decay and thus Δt, although small, no longer approaches zero. The result is a large but finite voltage spike across L.

the arc is easily maintained, as it creates ionized gases that provide a relatively low-resistance path for conduction. As the contacts spread apart, the arc elongates and eventually extinguishes as the coil energy is dissipated and coil voltage drops below that required to sustain the arc—see Notes.

There are several important points to note here:

1. Flashovers, as in Figure 14–13, are generally undesirable. However, they can be controlled through proper engineering design. (One way is to use a discharge resistor, as in the next example; another way is to use a diode or a varistor, as you will see in your electronics course.)
2. On the other hand, the large voltages created by breaking inductive currents have their uses. One is in the ignition system of automobiles, where current in the primary winding of a transformer coil is interrupted at the appropriate time by a control circuit to create the spark needed to fire the engine.
3. It is not possible for us to rigorously analyze the circuit of Figure 14–13 because the resistance of the arc changes as the switch opens. However, the main ideas can be established by studying circuits using fixed resistors. (Here, we assume the switch does not arc over.)

The Basic Ideas

We begin with the circuit of Figure 14–14. Assume the switch is closed and the circuit is in steady state. Since the inductance looks like a short circuit [Figure 14–15(a)], its pre-switching current is i_L = 120 V/30 Ω = 4 A.

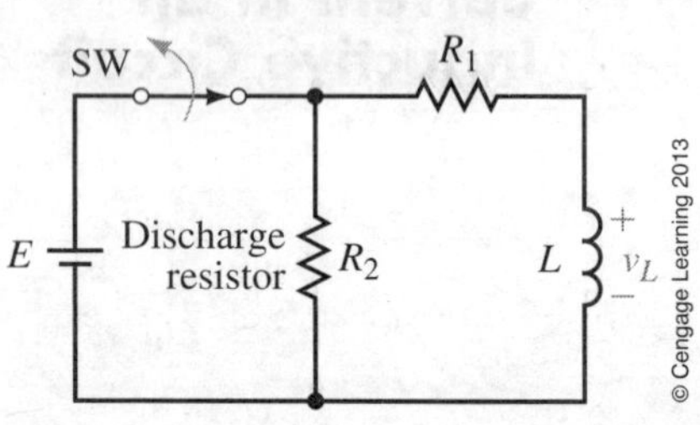

FIGURE 14–14 Discharge resistor R_2 helps limit the size of the induced voltage.

CircuitSim 14-8

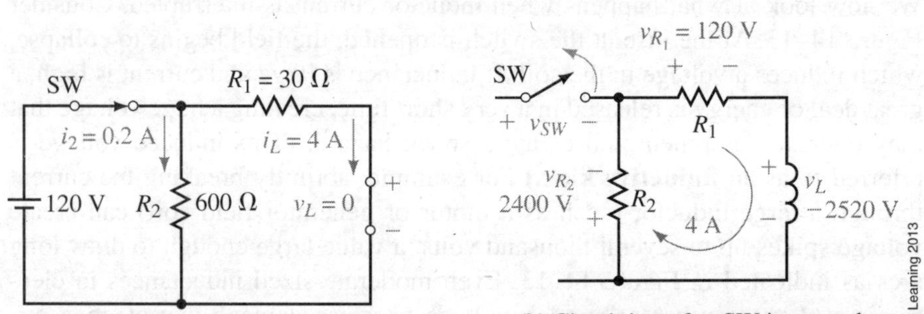

(a) Circuit just before the switch is opened

(b) Circuit just after SW is opened. Since coil voltage polarity is opposite to that shown, v_L is negative.

FIGURE 14–15 Circuit of Figure 14–14 immediately before and after the switch is opened. Coil voltage jumps from 0 V to −2520 V for this example.

Now open the switch. Just prior to opening the switch, i_L = 4 A; therefore, just after opening the switch, it must still be 4 A. As indicated in (b), this 4 A passes through resistances R_1 and R_2, creating voltages v_{R_1} = 4 A × 30 Ω = 120 V and v_{R_2} = 4 A × 600 Ω = 2400 V with the polarity shown. From KVL, $v_L + v_{R_1} + v_{R_2} = 0$. Therefore, at the instant the switch is opened,

$$v_L = -(v_{R_1} + v_{R_2}) = -2520 \text{ volts}$$

appears across the coil, yielding a negative voltage spike as in Figure 14–16. Note that this spike is more than 20 times larger than the source voltage. As we see in the next section, the size of this spike depends on the ratio of R_2 to R_1; the larger the ratio, the larger the voltage.

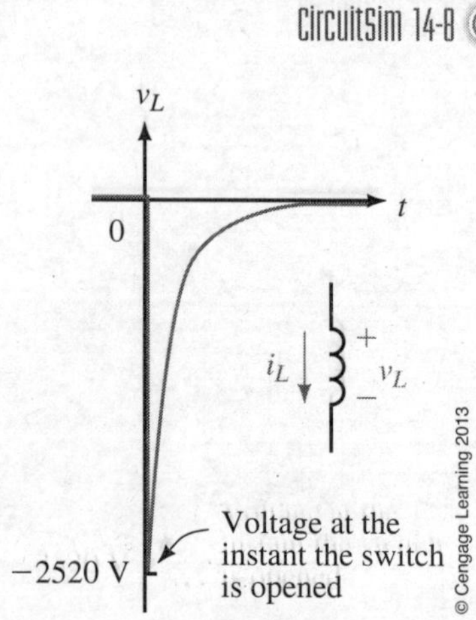

FIGURE 14–16 Voltage spike for the circuit of Figure 14–14. This voltage is more than 20 times larger than the source voltage.

Consider again Figure 14–15. Note that current i_2 changes abruptly from 0.2 A just prior to switching to −4 A just after. This is permissible, however, since i_2 does not pass through the inductor, and only currents through inductance cannot change abruptly.

PRACTICE PROBLEMS 3

Figure 14–16 shows the voltage across the coil of Figure 14–14. Make a similar sketch for the voltage across the switch and across resistor R_2. Hint: Use KVL to find v_{SW} and v_{R_2}.

Answer

v_{SW}: With the switch closed, $v_{SW} = 0$ V; When the switch is opened, v_{SW} jumps to 2520 V, then decays to 120 V. v_{R_2}: Identical in shape to Figure 14–16 except that v_{R_2} spikes at -2400 V instead of -2520 V.

Inductor Equivalent at Switching

Figure 14–17 shows the current through L of Figure 14–15. Because the current is the same immediately after switching as it is immediately before, it is constant over the interval from $t = 0^-$ s to $t = 0^+$ s. Since this is true in general, we see that *an inductance with an initial current looks like a current source at the instant of switching*. Its value is the value of the current at switching, Figure 14–18. As indicated in Figure 14–17, this is $I_0 = i_L(0^+)$. In any given problem, you can use either representation, but depending on what you are focusing on, you might choose one over the other.

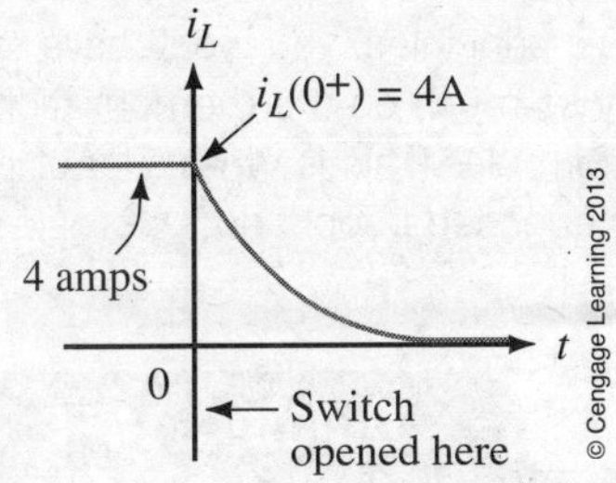

FIGURE 14–17 Inductor current for the circuit of Figure 14–15. Note that initial current $i_L(0^+)$ is often represented by the symbol I_0. In this notation, $I_0 = 4$ A.

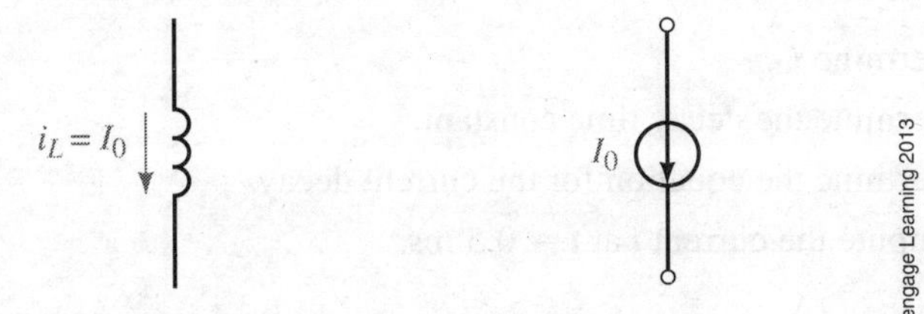

(a) Current at switching (b) Current source equivalent

FIGURE 14–18 An inductor carrying current looks like a current source at the instant of switching. You may see either representation (a) or representation (b) used in practice.

14.4 De-Energizing Transients

We now look at equations for the **de-energizing transient** voltages and currents described in the previous section.

Consider Figure 14–19(a). Let the current in the inductor before the switch is opened (the initial current) be denoted as I_0 amps. Now open the switch as in (b). KVL yields $v_L + v_{R_1} + v_{R_2} = 0$. Substituting $v_L = L di/dt$, $v_{R_1} = R_1 i$, and

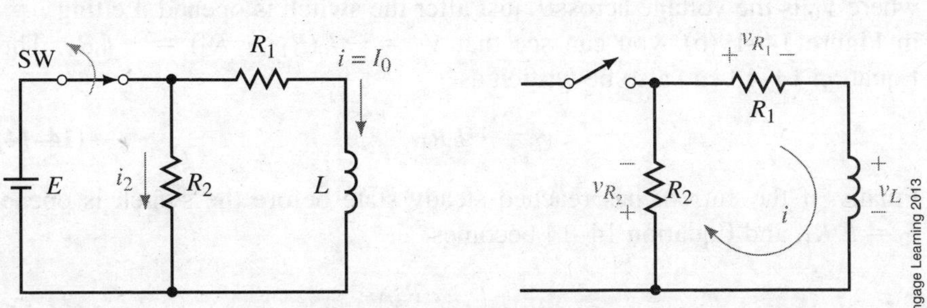

(a) Immediately before the switch is opened (b) Decay circuit

FIGURE 14–19 Circuit for studying decay transients.

NOTES...

As an aid to help you understand and solve the problems of this chapter, note that the transient portions of inductive transients (like capacitive transients) are always exponential in form, varying according to either $e^{-t/\tau}$ or $1 - e^{-t/\tau}$, and have the general shape of either Figure 14–9 or Figure 14–8. If you remember this and understand basic principles, you won't have to rely so much on the memorization of formulas. This is what a lot of experienced people do.

$v_{R_2} = R_2 i$ yields $L di/dt + (R_1 + R_2)i = 0$. Using calculus, it can be shown that the solution is

$$i = I_0 e^{-t/\tau'} \text{ (A)} \tag{14–10}$$

where

$$\tau' = \frac{L}{R_T} = \frac{L}{R_1 + R_2} \text{ (s)} \tag{14–11}$$

is the time constant of the discharge circuit. If the circuit is in steady state before the switch is opened, initial current $I_0 = E/R_1$ and Equation 14–10 becomes

$$i = \frac{E}{R_1} e^{-t/\tau'} \text{ (A)} \tag{14–12}$$

EXAMPLE 14–6

For Figure 14–19(a), assume the current has reached steady state with the switch closed. Suppose that $E = 120$ V, $R_1 = 30\ \Omega$, $R_2 = 600\ \Omega$, and $L = 126$ mH:

CircuitSim 14-9

a. Determine I_0.
b. Determine the decay time constant.
c. Determine the equation for the current decay.
d. Compute the current i at $t = 0.5$ ms.

Solution

a. Consider Figure 14–19(a). Since the circuit is in a steady state, the inductor looks like a short circuit to dc. Thus, $I_0 = E/R_1 = 4$ A.
b. Consider Figure 14–19(b). $\tau' = L/(R_1 + R_2) = 126\text{ mH}/630\ \Omega = 0.2$ ms.
c. $i = I_0 e^{-t/\tau'} = 4e^{-t/0.2\text{ ms}}$ A.
d. At $t = 0.5$ ms, $i = 4e^{-0.5\text{ ms}/0.2\text{ ms}} = 4e^{-2.5} = 0.328$ A.

Now consider voltage v_L. Using calculus, it can be shown that

$$v_L = V_0 e^{-t/\tau'} \tag{14–13}$$

where V_0 is the voltage across L just after the switch is opened. Letting $i = I_0$ in Figure 14–19(b), you can see that $V_0 = -I_0(R_1 + R_2) = -I_0 R_T$. Thus, Equation 14–13 can also be written as

$$v_L = -I_0 R_T e^{-t/\tau'} \tag{14–14}$$

Finally, if the current has reached steady state before the switch is opened, $I_0 = E/R_1$, and Equation 14–14 becomes

$$v_L = -E\left(1 + \frac{R_2}{R_1}\right)e^{-t/\tau'} \tag{14–15}$$

Note that v_L starts at V_0 volts (which is negative) and decays to zero as shown in Figure 14–20.

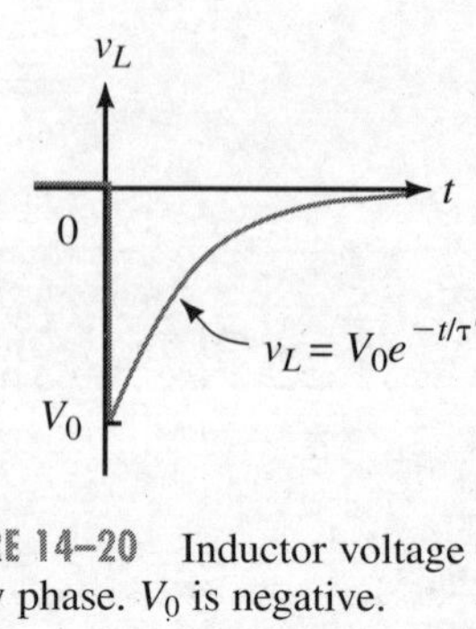

FIGURE 14–20 Inductor voltage during decay phase. V_0 is negative.

Now consider the resistor voltages. Each is the product of resistance times current (Equation 14–10). Thus,

$$v_{R_1} = R_1 I_0 e^{-t/\tau'} \quad (14\text{–}16)$$

and

$$v_{R_2} = R_2 I_0 e^{-t/\tau'} \quad (14\text{–}17)$$

CircuitSim 14-10

If current has reached steady state before switching, these become

$$v_{R_1} = Ee^{-t/\tau'} \quad (14\text{–}18)$$

and

$$v_{R_2} = \frac{R_2}{R_1} Ee^{-t/\tau'} \quad (14\text{–}19)$$

Substituting the values of Example 14–6 into these equations, we get for the circuit of Figure 14–19 $v_L = -2520e^{-t/0.2\text{ ms}}$ V, $v_{R_1} = 120e^{-t/0.2\text{ ms}}$ V and $v_{R_2} = 2400e^{-t/0.2\text{ ms}}$ V. These can also be written as $v_L = -2520e^{-5000t}$ V and so on, if desired.

Decay problems can also be solved using the decay portion of the universal time constant curves shown in Figure 14–10.

EXAMPLE 14–7

The circuit of Figure 14–21 is in steady state with the switch closed. Use Figure 14–10 to find i_L and v_L at $t = 2\ \tau$ after the switch is opened.

Solution Before the switch is opened, the inductor looks like a short circuit and its current is thus $E/R_1 = 120\text{ V}/40\ \Omega = 3$ A. Current just after the switch is opened will be the same. Therefore, $I_0 = 3$ A. At $t = 2\ \tau$, current will have decayed to 13.5% of this initial value. Therefore, $i_L = 0.135I_0 = 0.405$ A and $v_L = -(R_1 + R_1)i = -(60\ \Omega)(0.405\text{ A}) = -24.3$ V. [Alternately, $V_0 = -(3\text{ A})(60\ \Omega) = -180$ V. At $t = 2\ \tau$, this has decayed to 13.5%. Therefore, $v_L = 0.135(-180\text{ V}) = -24.3$ V as above.]

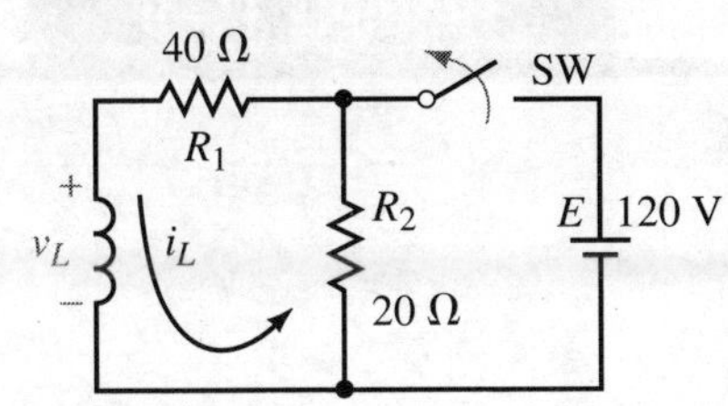

FIGURE 14–21

CircuitSim 14-11

14.5 More Complex Circuits

The equations developed so far apply only to circuits of the forms of Figures 14–7 or 14–19. Fortunately, many circuits can be reduced to these forms using circuit reduction techniques such as series and parallel combinations, source conversions, Thévenin's theorem, and so on.

EXAMPLE 14–8

Determine the expression for i_L for the circuit of Figure 14–22(a) if $L = 5$ H. Compute current at $t = 10$ ms and 30 ms.

Solution The circuit can be reduced to its Thévenin equivalent (b) as you saw in Chapter 11 (Section 11.5). For this circuit, $\tau = L/R_{\text{Th}} = 5\text{ H}/200\ \Omega = 25$ ms. Now apply Equation 14–7. Thus,

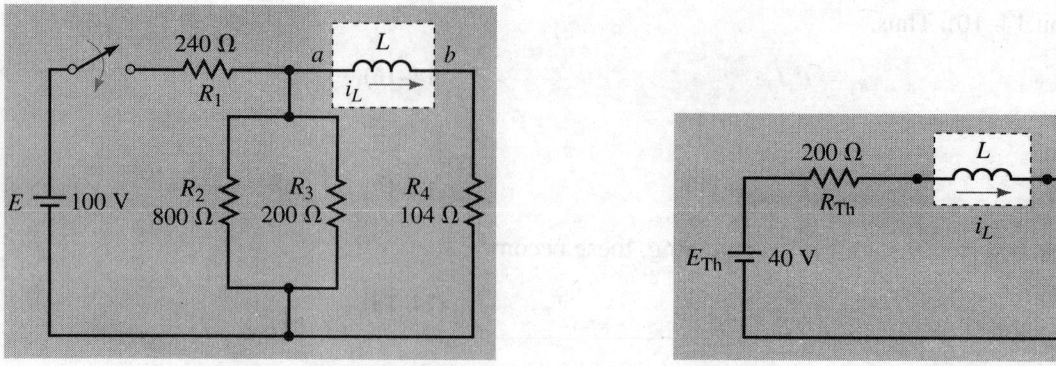

(a) Circuit (b) Thévenin equivalent

FIGURE 14–22

CircuitSim 14-12

$$i_L = \frac{E_{Th}}{R_{Th}}(1 - e^{-t/\tau}) = \frac{40}{200}(1 - e^{-t/25\text{ ms}}) = 0.2\,(1 - e^{-40t}) \quad \text{(A)}$$

At $t = 10$ ms, $i_L = 0.2(1 - e^{-(40)(0.010)}) = 0.2(1 - e^{-0.4}) = 65.9$ mA.

Similarly, at $t = 30$ ms, $i_L = 139.8$ mA.

EXAMPLE 14–9

For the circuit of Example 14–8, at what time does current reach 0.12 amps?

Solution

$$i_L = 0.2(1 - e^{-40t}) \quad \text{(A)}$$

Thus,

$$0.12 = 0.2(1 - e^{-40t}) \quad \text{(Figure 14–23)}$$
$$0.6 = 1 - e^{-40t}$$
$$e^{-40t} = 0.4$$

Taking the natural log of both sides,

$$\ln e^{-40t} = \ln 0.4$$
$$-40t = -0.916$$
$$t = 22.9 \text{ ms}$$

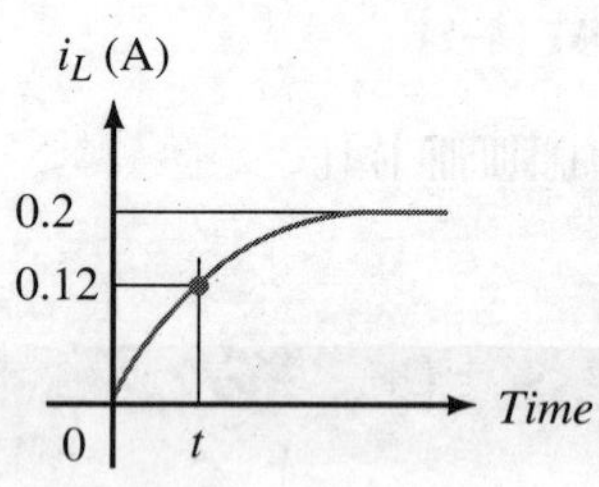

FIGURE 14–23

PRACTICE PROBLEMS 4

1. For the circuit of Figure 14–22, compute i_L and v_{R4} at $t = 50$ ms.
2. For the circuit of Figure 14–22, let $E = 120$ V, $R_1 = 600\ \Omega$, $R_2 = 3\ \text{k}\Omega$, $R_3 = 2\ \text{k}\Omega$, $R_4 = 100\ \Omega$, and $L = 0.25$ H:
   a. Determine i_L and sketch it.
   b. Determine v_L and sketch it.

3. Let everything be as in Problem 1 except L. If $i_L = 0.12$ A at $t = 20$ ms, what is L?

Answers

1. 0.173 A, 17.99 V ≈ 18.0 V
2. a. $160(1 - e^{-2000t})$ mA; b. $80e^{-2000t}$ V. i_L climbs from 0 to 160 mA with the waveshape of Figure 14–1(b), reaching steady state in 2.5 ms. v_L looks like Figure 14–2(c). It starts at 80 V and decays to 0 V in 2.5 ms.
3. 7.21 H

CircuitSim 14-13

A Note about Time Scales

Until now, we have considered energization and de-energization phases separately. When both occur in the same problem, we must clearly define what we mean by time. One way to handle this problem (as we did with RC circuits) is to define $t = 0$ s as the beginning of the first phase and solve for voltages and currents in the usual manner, then shift the time axis to the beginning of the second phase, redefine $t = 0$ s, solve the second part, and then use continuity considerations to link the two solutions together. This is illustrated in Example 14–10. Note that only the first time scale is shown explicitly on the final waveform plot.

EXAMPLE 14–10

Refer to the circuit of Figure 14–24:

a. Close the switch at $t = 0$ and determine equations for i_L and v_L.
b. At $t = 300$ ms, open the switch and determine equations for i_L and v_L during the decay phase.
c. Determine voltage and current at $t = 100$ ms and at $t = 350$ ms.
d. Sketch i_L and v_L. Mark the points from (c) on the sketch.

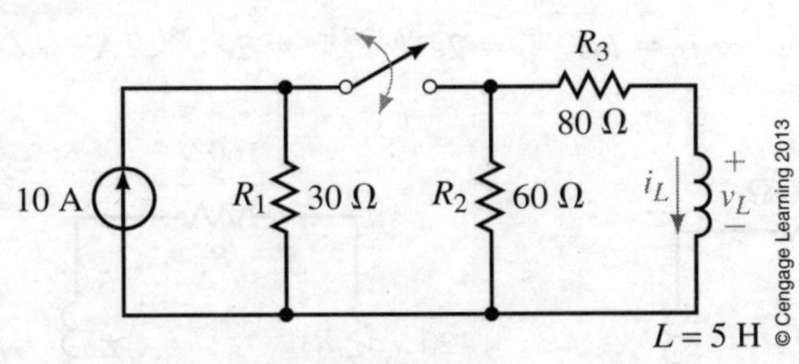

FIGURE 14–24

CircuitSim 14-14

Solution

a. Convert the circuit to the left of L to its Thévenin equivalent. As indicated in Figure 14–25(a), $R_{Th} = 60\|30 + 80 = 100\ \Omega$. From (b), $E_{Th} = V_2$, where

$$V_2 = (10\text{ A})(20\ \Omega) = 200\text{ V}$$

The Thévenin equivalent circuit is shown in Figure 14–26(a). $\tau = L/R_{Th} = 5\text{ H}/100\ \Omega = 50$ ms. Thus, during current buildup,

$$i_L = \frac{E_{Th}}{R_{Th}}(1 - e^{-t/\tau}) = \frac{200}{100}(1 - e^{-t/50\text{ ms}}) = 2\,(1 - e^{-20t})\ \text{A}$$

$$v_L = E_{Th}e^{-t/\tau} = 200\,e^{-20t}\ \ \text{V}$$

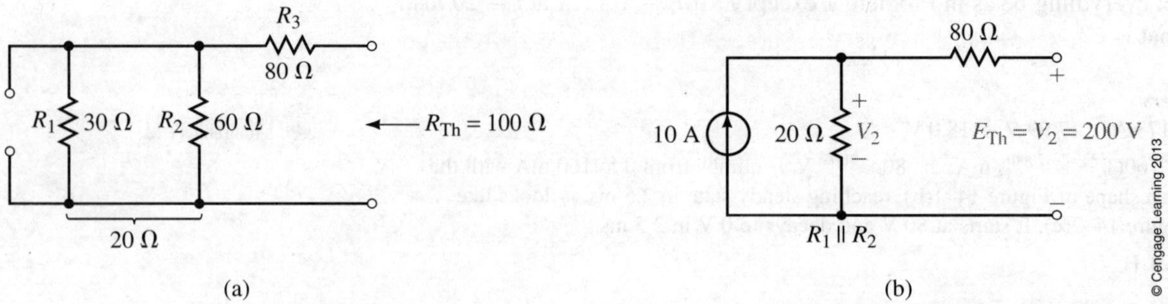

FIGURE 14–25

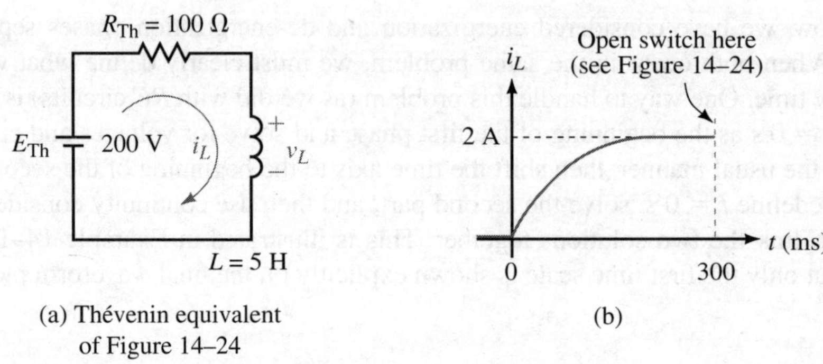

(a) Thévenin equivalent of Figure 14–24

(b)

FIGURE 14–26 Circuit and current during the buildup phase.

b. Current buildup is sketched in Figure 14–26(b). Since $5\tau = 250$ ms, current is in steady state when the switch is opened at 300 ms. Thus, $I_0 = 2$ A. When the switch is opened, current decays to zero through a resistance of 60 + 80 = 140 Ω as shown in Figure 14–27. Thus, $\tau' = 5\text{ H}/140\ \Omega = 35.7$ ms. If $t = 0$ s is redefined as the instant the switch is opened, the equation for the decay is

$$i_L = I_0 e^{-t/\tau'} = 2e^{-t/35.7\text{ ms}} = 2e^{-28t} \quad \text{A}$$

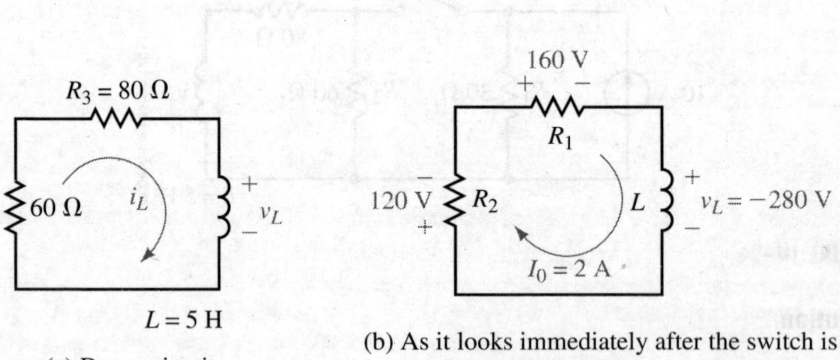

(a) Decay circuit

(b) As it looks immediately after the switch is opened. KVL yields $v_L = -280$ V

FIGURE 14–27 The circuit of Figure 14–24 as it looks during the decay phase.

Now consider voltage. As indicated in Figure 14–27(b), the voltage across L just after the switch is open is $V_0 = -280$ V. Thus,

$$v_L = V_0 e^{-t/\tau'} = -280e^{-28t} \quad \text{V}$$

c. You can use the universal time constant curves at $t = 100$ ms since 100 ms represents 2τ. At 2τ, current has reached 86.5% of its final value. Thus, $i_L = 0.865(2\text{ A}) = 1.73$ A. Voltage has fallen to 13.5%. Thus, $v_L = 0.135(200\text{ V}) = 27.0$ V. Now consider $t = 350$ ms: Note that this is 50 ms into the decay portion of the curve. However, since 50 ms is not a multiple of τ', it is difficult to use the curves. Therefore, use the equations. Thus,

$$i_L = 2\text{ A } e^{-28(50\text{ ms})} = 2\text{ A } e^{-1.4} = 0.493\text{ A}$$
$$v_L = (-280\text{ V})e^{-28(50\text{ ms})} = (-280\text{ V})e^{-1.4} = -69.0\text{ V}$$

d. The preceding points are plotted on the waveforms of Figure 14–28.

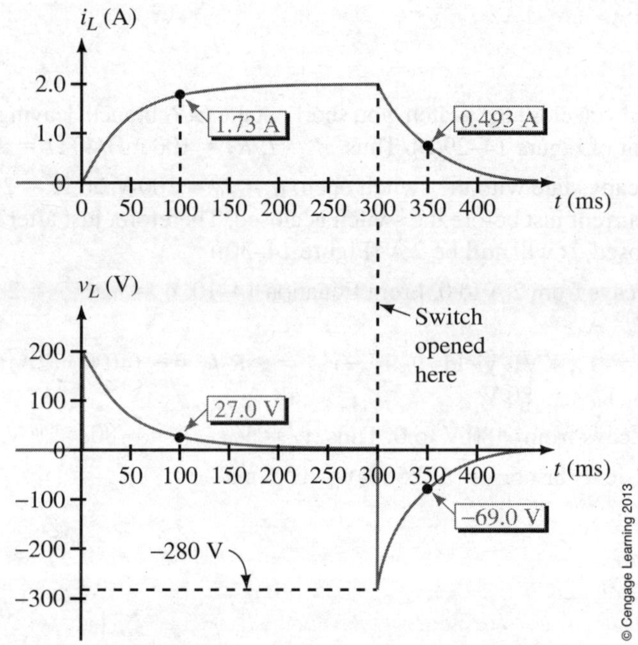

FIGURE 14–28

The basic principles that we have developed in this chapter permit us to solve problems that do not correspond exactly to the circuits of Figure 14–7 and 14–19. This is illustrated in the following example.

EXAMPLE 14–11

The circuit of Figure 14–29(a) is in steady state with the switch open. At $t = 0$ s, the switch is closed.

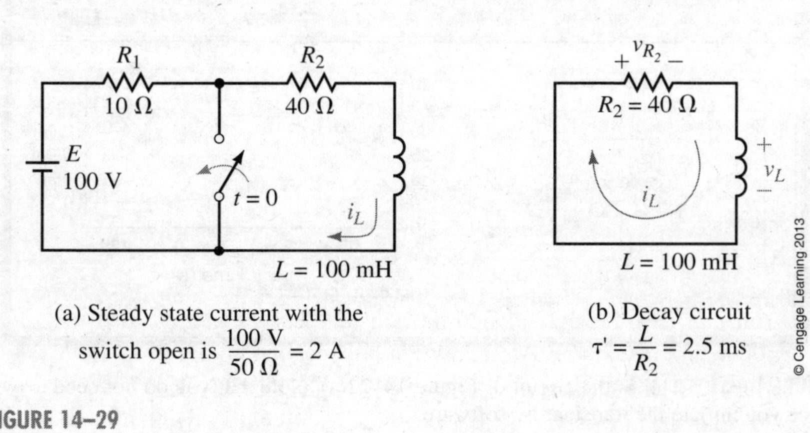

CircuitSim 14-15

FIGURE 14–29

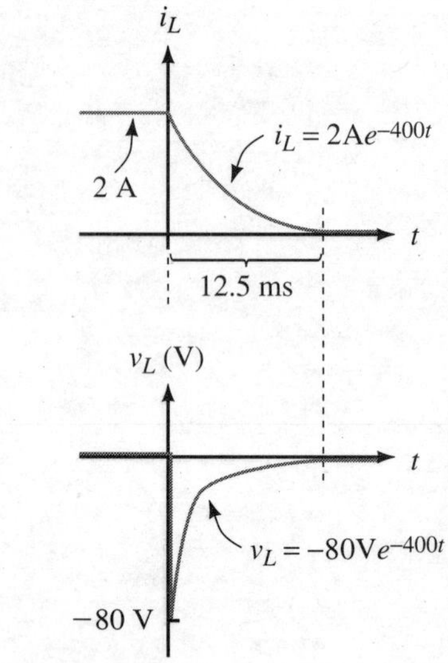

FIGURE 14–30

a. Sketch the circuit as it looks after the switch is closed and determine τ'.
b. Determine current i_L at $t = 0^+$ s.
c. Determine the expression for i_L.
d. Determine v_L at $t = 0^+$ s.
e. Determine the expression for v_L.
f. How long does the transient last?
g. Sketch i_L and v_L.

Solution

a. When you close the switch, you short out the E-R_1 branch, leaving the decay circuit of Figure 14–29(b). Thus, $\tau' = L/R_2 = 100$ mH/40 Ω = 2.5 ms.
b. In steady state with the switch open, $i_L = I_0 = 100$ V/50 Ω = 2 A. This is the current just before the switch is closed. Therefore, just after the switch is closed, i_L will still be 2 A (Figure 14–30).
c. i_L decays from 2 A to 0. From Equation 14–10, $i_L = I_0e^{-t/\tau'} = 2e^{-t/2.5 \text{ ms}} = 2e^{-400t}$ A.
d. At $t = 0^+$, KVL yields $v_L = -v_{R_2} = -R_2I_0 = -(40\ \Omega)\ (2\text{A}) = -80$ V. Thus, $V_0 = -80$ V.
e. v_L decays from −80 V to 0. Thus, $v_L = V_0e^{-t/\tau'} = -80e^{-400t}$ V.
f. Transients last $5\tau' = 5(2.5 \text{ ms}) = 12.5$ ms.

NOTES...

For Further Investigation

Additional examples may be found on our Web site. Go to cengagebrain.com and follow the links to *For Further Investigation* and select *Advanced Studies in Transients.*

14.6 *RL* Transients Using Computers

Multisim

As a first example, let us use Multisim to compute and plot inductor current and voltage waveforms for the circuit of Figure 14–22(a)—see Notes 1 and 2. Now build the circuit of Figure 14–31 on your screen. (To make the connections to

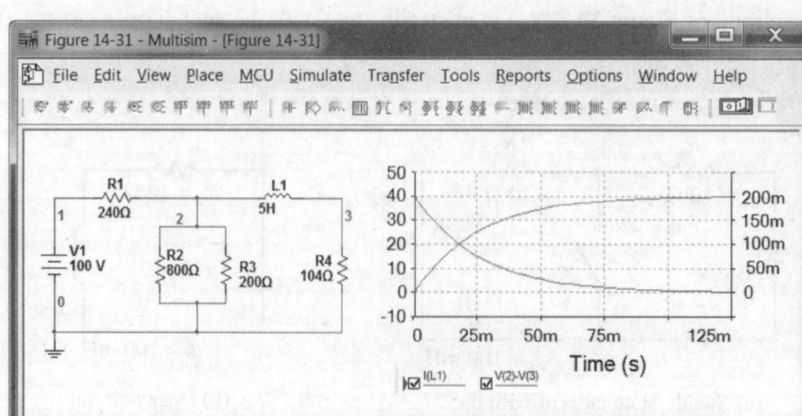

FIGURE 14–31 This is the circuit of Figure 14–22(a). Note that you do not need a switch since you initiate the transient by software.

NOTES...

Multisim

1. Because you use software to initiate the transient, you do not need the switch shown in Figure 14–22(a)—thus, it has been omitted from Figure 14–31.
2. If you are a Multisim 9 or 10 user, go to our Web site at cengagebrain.com and follow the links to *Special Instructions for Multisim 9 and 10 Users*.
3. Only basic steps are given for the following computer examples, as procedures here are similar to those of Chapter 11. If you need help, refer back to Chapter 11 or to Appendix A.
4. Multisim plots usually appear as colored traces on a black background. You can reverse the color as you did in Chapter 11.
5. Multisim scopes are grounded automatically, so an external ground is not necessary.
6. For Figure 14–32(a), to find the switch, click *Place/Components,* and from the Basic group, select *Switch/SPST,* then click *OK.*
7. To find the oscilloscope, right-click *View* and make sure *Instruments* is selected. Use your mouse to find the scope icon in the resulting tool bin on your screen.
8. If switch S1 freezes, click an empty spot on your workspace screen to unfreeze it.
9. If the ON/OFF switch is not shown, right-click *View,* then select *Simulation Switch* (or *Simulation* if you are a legacy software user).

the R_2/R_3 branch, you require junction dots. Click *Place/Junction* and place the dots where you need them.) Click *Simulate/Analyses/Transient Analyses,* and in the dialog box, enter **0.125** for TSTOP and set the initial conditions to zero. Click the Output tab, highlight *I*(L1), then click *Add*. Now you need to create an expression for the voltage across the inductor. Click *Add* Expression, double-click *V*(2), press the minus key on your keyboard, double-click *V*(3), then click *OK*. This creates the expression *V*(2)-*V*(3). (You should recognize this as the voltage across inductor L1.) Click *Simulate*. Voltage and current waveforms appear. Following the procedure that you learned in Chapter 11, add a new scale for the current waveform. Your waveforms should now look like those of Figure 14–31. Click the current curve (to select it), activate the cursors (click the cursor icon), then drag the red cursor into the display area and right-click it. Select *Set Y Value,* type in **0.12,** then click *OK*. From the cursor display, you should see that the corresponding time is 22.9 ms (which agrees with our analytic solution to Example 14–9). Click the voltage curve, then use the cursor to measure voltage at $t = 20$ ms. Compare with the answer of Practice Problems 5.

PRACTICE PROBLEMS 5

Using the circuit of Figure 14–22(b), analytically determine the equation for v_L. Compute voltage at $t = 20$ ms.

Using the Multisim Oscilloscope

Create the circuit of Figure 14–32. (Use a switch—see Note 6.) Add the oscilloscope (see Notes 5 and 7). After you have wired it in, double-click the oscilloscope, and using your mouse, activate the scrolling arrows (as indicated

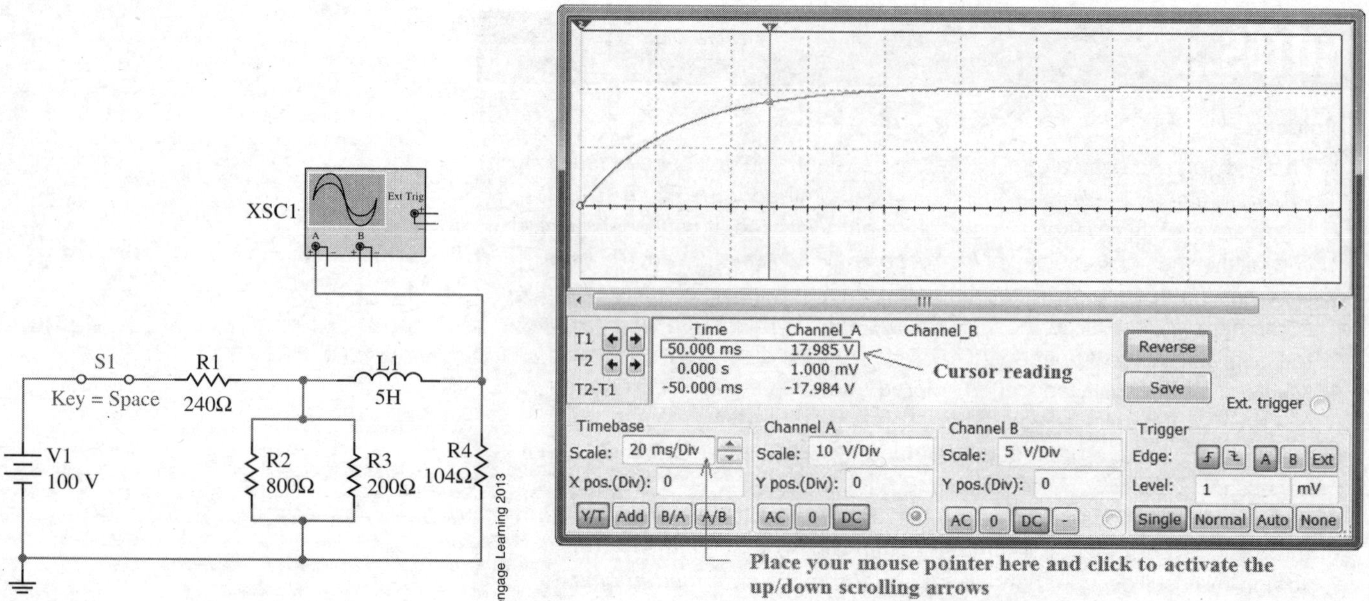

(a) Test circuit

(b) Multisim oscilloscope screen

FIGURE 14–32 This is again the circuit of Figure 14–22(a). Here we have used a switch to initiate the transient.

in Figure 14–32), then set the time base to 20 ms/Div. Similarly, set Channel A to 10 V/Div and Level to 1 mV. Select Trigger Edge buttons as indicated. Select *Single,* then on your workspace screen, click the ON/OFF simulate switch to ON—see Note 9. Press the spacebar to close switch S1—see Note 8. Before the trace reaches the edge of the screen, click the simulate switch to OFF. Position the cursor at $t = 50$ ms and read voltage. You should get 17.99 V as we did in Practice Problems 4, Question 1.

PSpice

As a first example, use PSpice to solve for inductor voltage and current for the circuit of Figure 14–45, end-of-chapter Problem 21.

- Build the circuit on the screen as in Figure 14–33 (see Notes 1 and 2). Click the *New Profile* icon and name the file. In the *Simulations Settings* box, select transient analysis, set *TSTOP* to **90ms,** then click *OK.*

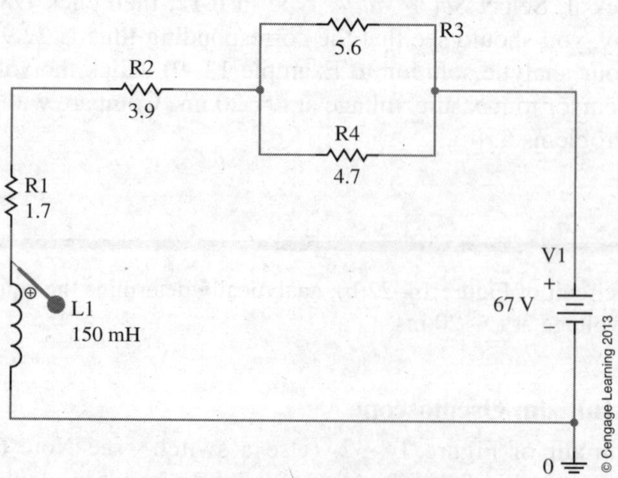

FIGURE 14–33 The voltage marker probe automatically creates the voltage trace of Figure 14–34. As is detailed in the text, you add the current trace later.

NOTES...

PSpice

1. Since the process here is similar to that of Chapter 11, only abbreviated instructions are given.
2. Because you use software to initiate the transient, you don't need the switch that is shown in Figure 14–45 of Problem 21.

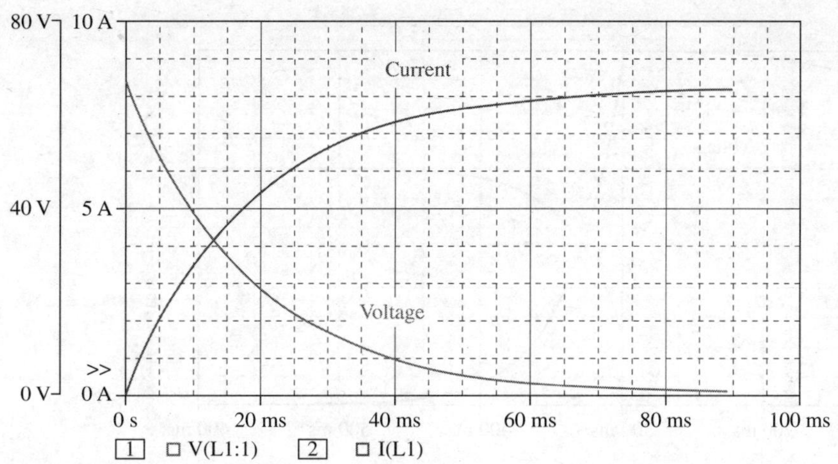

FIGURE 14–34 Inductor voltage and current for the circuit of Figure 14–33.

- Click the *Run* icon. When simulation is complete, you should have the voltage curve of Figure 14–34 on your screen.
- To add the current curve, click *Plot, Add Y Axis, Trace, Add Trace,* then double-click *I(L1)*.
- Using a cursor, measure current at t = 18.4 ms. (You should get an answer of 5.19 A.)
- Measure voltage at t = 18.4 ms. Confirm your result using Equation 14–13.

EXAMPLE 14–12

Consider the circuit of Figure 14–24, Example 14–10. The switch is closed at $t = 0$ and opened 300 ms later. Prepare a PSpice analysis of this problem and determine v_L and i_L at t = 100 ms and at t = 350 ms.

Solution PSpice doesn't have a switch that both opens and closes. However, you can simulate such a switch by using two switches as in Figure 14–35 (see Note 1). Create the circuit on your screen using IDC for the current source. Double-click *TOPEN* of switch U2 and set it to **300ms.** Click the *New Profile* icon and name the file. In the *Simulation Settings* box, select transient analysis then type in a value of **0.6** for *TSTOP*. Run the simulation—see Note 2 and Figure 14–36—then compare to the current curve of Figure 14–28. Next, create a second Y-axis, then add the voltage trace V(L1:1). Using the cursor, read values at t = 100 ms, and 350 ms. You should get approximately 27 V and 1.73 A at t = 100 ms, and −69 V and 493 mA at t = 350 ms, as in Example 14–10.

NOTES...

1. For the circuit of Figure 14–35, you need to use switches in your simulation circuit in order to simulate switching at two distinct times.
2. If you get an error message, "Sw_tClose used by X_U1 is undefined" (or similar), go to Appendix A for instructions on how to deal with it.

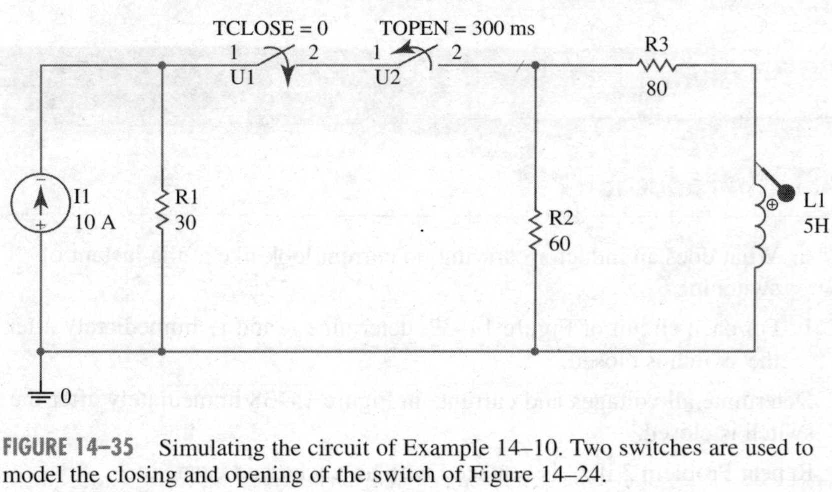

FIGURE 14–35 Simulating the circuit of Example 14–10. Two switches are used to model the closing and opening of the switch of Figure 14–24.

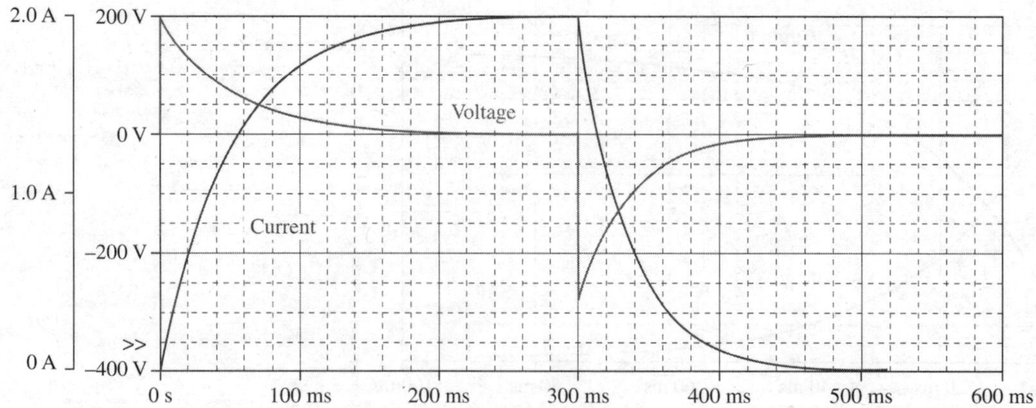

FIGURE 14–36 Inductor voltage and current for the circuit of Figure 14–35.

Putting It into Practice

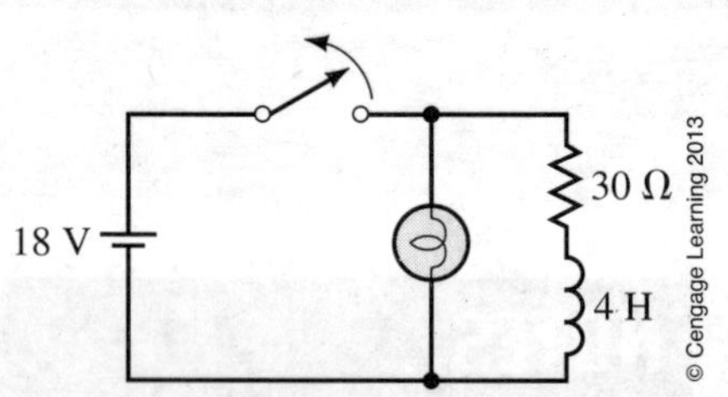

The first sample of a new product that your company has designed has an indicator light that fails. (Symptom: When you turn a new unit on, the indicator light comes on as it should. However, when you turn the power off and back on, the lamp does not come on again.) You have been asked to investigate the problem and design a fix. You acquire a copy of the schematic and study the portion of the circuit where the indicator lamp is located. As shown in the accompanying figure, the lamp is used to indicate the status of the coil; the light is to be on when the coil is energized and off when it is not. Immediately, you see the problem, solder in one component and the problem is fixed. Write a short note to your supervisor outlining the nature of the problem, explaining why the lamp burned out and why your design modification fixed the problem. Note also that your modification did not result in any substantial increase in power consumption (i.e., you did not use a resistor). Note: This problem requires a diode. If you have not had an introduction to electronics, you may not be able to do this problem.

Problems

14.1 Introduction

1. a. What does an inductor carrying no current look like at the instant of switching?
   b. For each circuit of Figure 14–37, determine i_S and v_L immediately after the switch is closed.
2. Determine all voltages and currents in Figure 14–38 immediately after the switch is closed.
3. Repeat Problem 2 if L_1 is replaced with an uncharged capacitor.

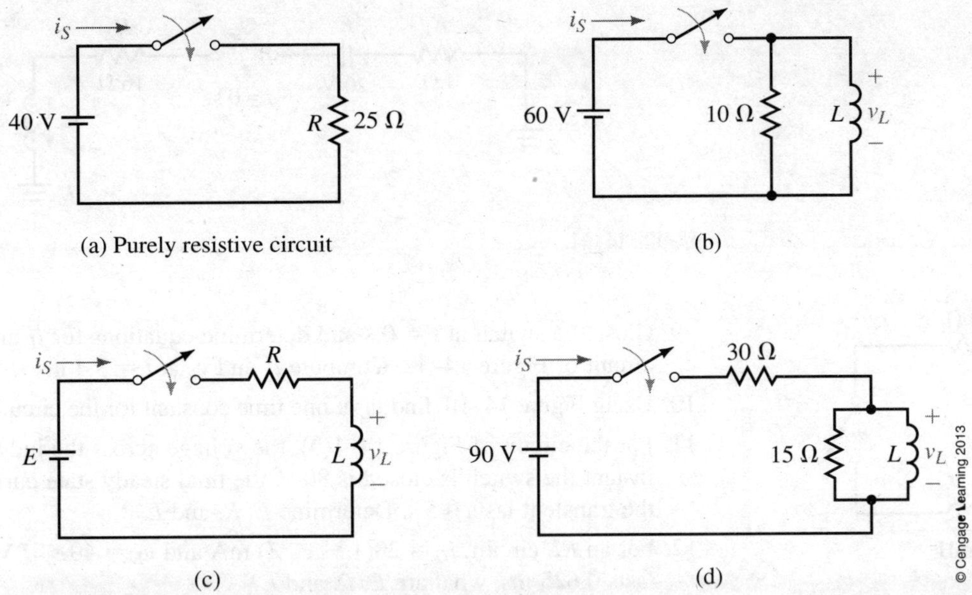

FIGURE 14–37 No value is needed for L here as it does not affect the solution.

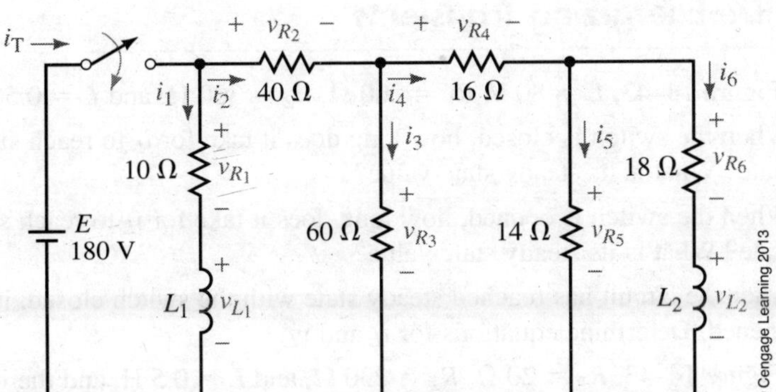

FIGURE 14–38

14.2 Current Buildup Transients

4. a. If $i_L = 8(1 - e^{-500t})$ A, what is the current at $t = 6$ ms?
   b. If $v_L = 125e^{-500t}$ V, what is the voltage v_L at $t = 5$ ms?
5. The switch of Figure 14–39 is closed at $t = 0$ s.
   a. What is the time constant of the circuit?
   b. How long is it until current reaches its steady value?
   c. Determine the equations for i_L and v_L.
   d. Compute values for i_L and v_L at intervals of one time constant from $t = 0$ to 5 τ.
   e. Sketch i_L and v_L. Label the axis in τ and in seconds.
6. Close the switch at $t = 0$ s and determine equations for i_L and v_L for the circuit of Figure 14–40. Compute i_L and v_L at $t = 1.8$ ms.
7. Repeat Problem 5 for the circuit of Figure 14–41 with $L = 4$ H.
8. For the circuit of Figure 14–39, determine inductor voltage and current at $t = 50$ ms using the universal time constant curve of Figure 14–10.

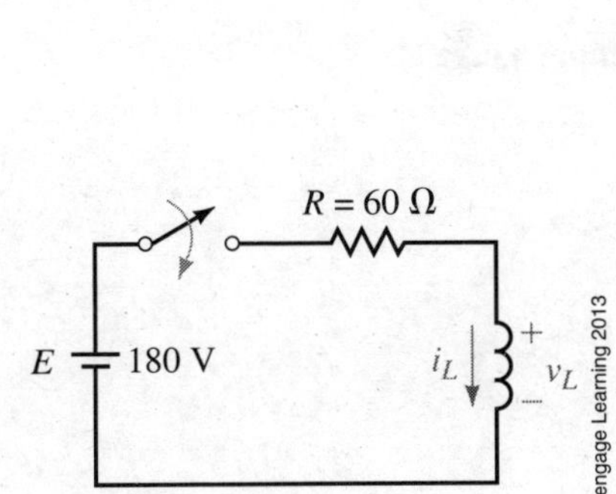

FIGURE 14–39

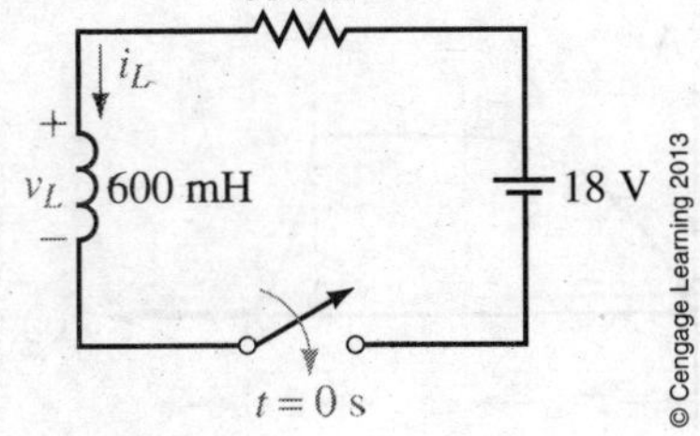

FIGURE 14–40

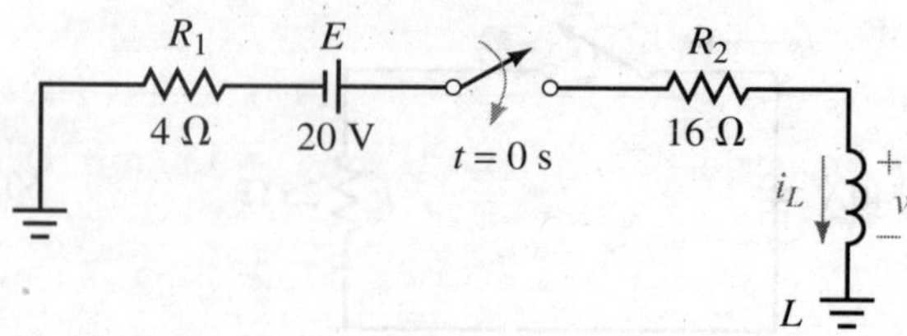

FIGURE 14–41

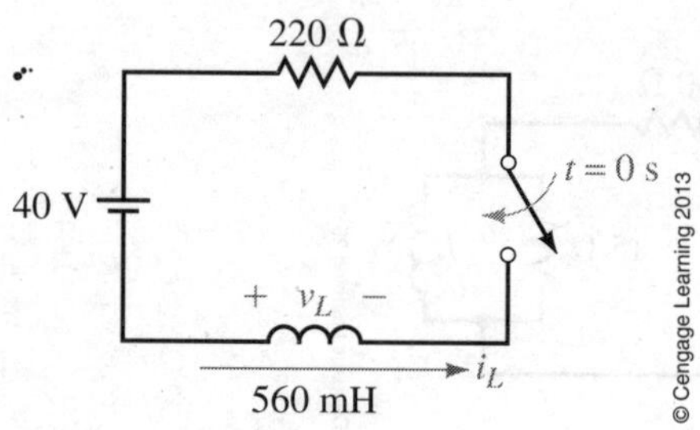

FIGURE 14–42

9. Close the switch at $t = 0$ s and determine equations for i_L and v_L for the circuit of Figure 14–42. Compute i_L and v_L at $t = 3.4$ ms.
10. Using Figure 14–10, find v_L at one time constant for the circuit of Figure 14–42.
11. For the circuit of Figure 14–1(b), the voltage across the inductance at the instant the switch is closed is 80 V, the final steady state current is 4 A, and the transient lasts 0.5 s. Determine E, R, and L.
12. For an RL circuit, $i_L = 20(1 - e^{-t/\tau})$ mA and $v_L = 40e^{-t/\tau}$ V. If the transient lasts 0.625 ms, what are E, R, and L?
13. For Figure 14–1(b), if $v_L = 40e^{-2000t}$ V and the steady state current is 10 mA, what are E, R, and L?

14.4 De-Energizing Transients

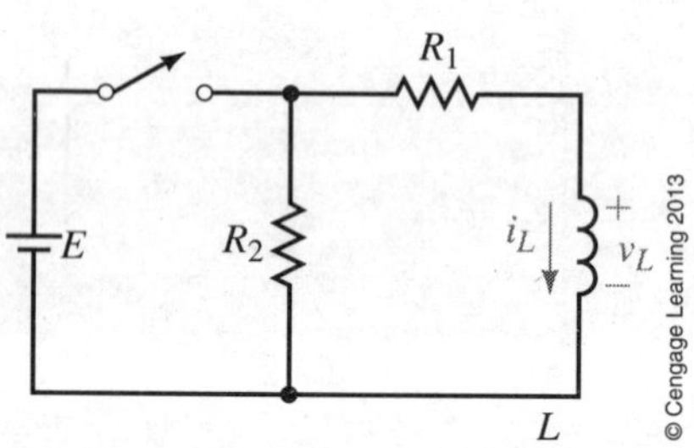

FIGURE 14–43

14. For Figure 14–43, $E = 80$ V, $R_1 = 200\ \Omega$, $R_2 = 300\ \Omega$, and $L = 0.5$ H.
    a. When the switch is closed, how long does it take for i_L to reach steady state? What is its steady state value?
    b. When the switch is opened, how long does it take for i_L to reach steady state? What is its steady state value?
    c. After the circuit has reached steady state with the switch closed, it is opened. Determine equations for i_L and v_L.
15. For Figure 14–43, $R_1 = 20\ \Omega$, $R_2 = 230\ \Omega$, and $L = 0.5$ H, and the inductor current has reached a steady value of 5 A with the switch closed. At $t = 0$ s, the switch is opened.
    a. What is the decay time constant?
    b. Determine equations for i_L and v_L.
    c. Compute values for i_L and v_L at intervals of one time constant from $t = 0$ to $5\ \tau$.
    d. Sketch i_L and v_L. Label the axis in τ and in seconds.
16. Using the values from Problem 15, determine inductor voltage and current at $t = 3\tau$ using the universal time constant curves shown in Figure 14–10.
17. Given $v_L = -2700\ \text{V}e^{-100t}$. Using the universal time constant curve, find v_L at $t = 20$ ms.
18. For Figure 14–43, the inductor voltage at the instant the switch is closed is 150 V and $i_L = 0$ A. After the circuit has reached steady state, the switch is opened. At the instant the switch is opened, $i_L = 3$ A and v_L jumps to -750 V. The decay transient lasts 5 ms. Determine E, R_1, R_2, and L.
19. For Figure 14–43, $L = 20$ H. The current during buildup and decay is shown in Figure 14–44. Determine R_1 and R_2.
20. For Figure 14–43, when the switch is moved to energization, $i_L = 2$ A $(1 - e^{-10t})$. Now open the switch after the circuit has reached steady state and redefine $t = 0$ s as the instant the switch is opened. For this case, $v_L = -400\ \text{V}e^{-25t}$. Determine E, R_1, R_2, and L.

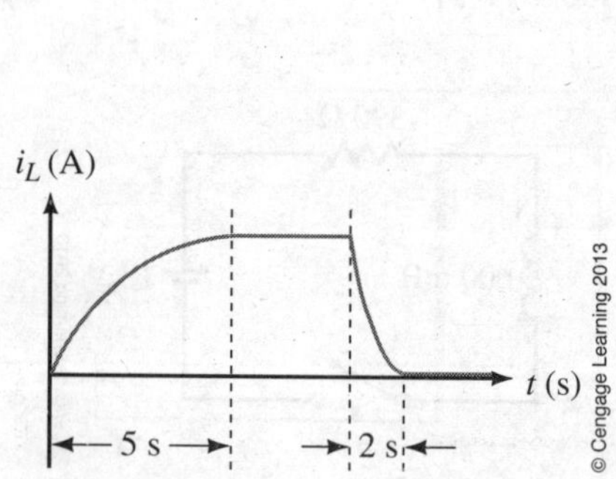

FIGURE 14–44

14.5 More Complex Circuits

21. For the coil of Figure 14–45, $R_\ell = 1.7\ \Omega$ and $L = 150$ mH. Determine coil current at $t = 18.4$ ms.

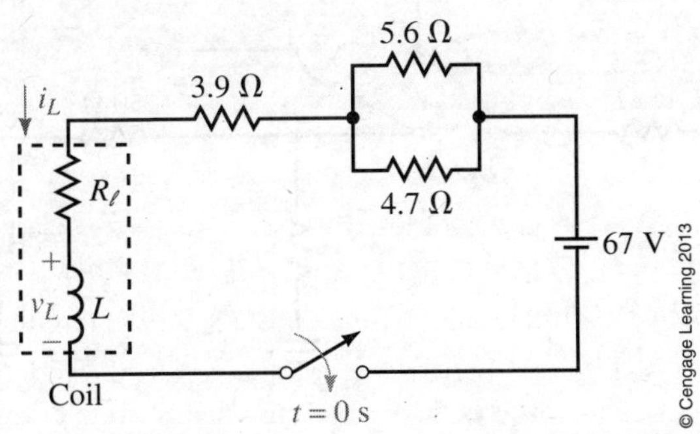

FIGURE 14–45

22. Refer to Figure 14–46:
    a. What is the energizing circuit time constant?
    b. Close the switch and determine the equation for i_L and v_L during current buildup.
    c. What is the voltage across the inductor and the current through it at $t = 20\ \mu$s?
23. For Figure 14–46, the circuit has reached steady state with the switch closed. Now open the switch.
    a. Determine the de-energizing circuit time constant.
    b. Determine the equations for i_L and v_L.
    c. Find the voltage across the inductor and current through it at $t = 17.8\ \mu$s using the equations determined previously.
24. Repeat part (c) of Problem 23 using the universal time constant curves shown in Figure 14–10.
25. a. Repeat Problem 22, parts (a) and (b) for the circuit of Figure 14–47.
    b. What are i_L and v_L at $t = 25$ ms?

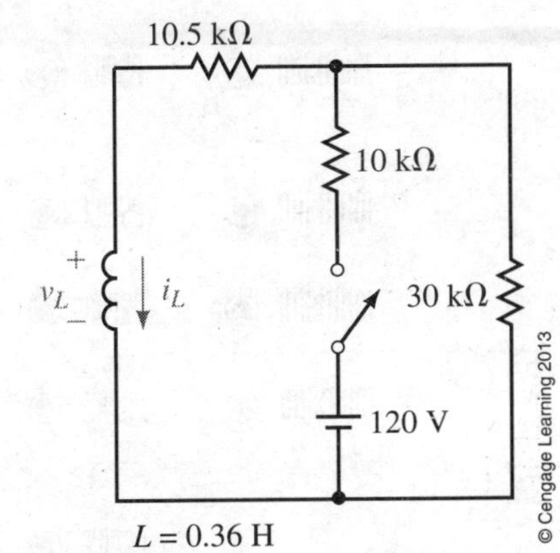

FIGURE 14–46

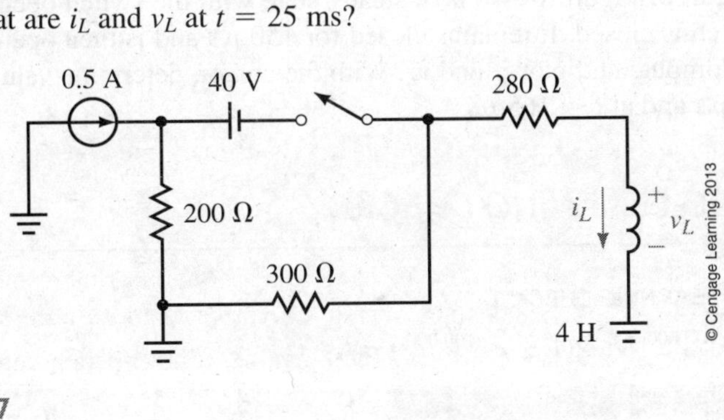

FIGURE 14–47

26. Repeat Problem 23 for the circuit of Figure 14–47, except find v_L and i_L at $t = 13.8$ ms.
27. An unknown circuit containing dc sources and resistors has an open-circuit voltage of 45 volts. When its output terminals are shorted, the short-circuit current is 0.15 A. A switch, resistor, and inductance are connected (Figure 14–48). Determine the inductor current and voltage 2.5 ms after the switch is closed.
28. The circuit of Figure 14–49 is in steady state with the switch in position 1. At $t = 0$, it is moved to position 2, where it remains for 1.0 s. It is then moved to position 3, where it remains. Sketch curves for i_L and v_L from

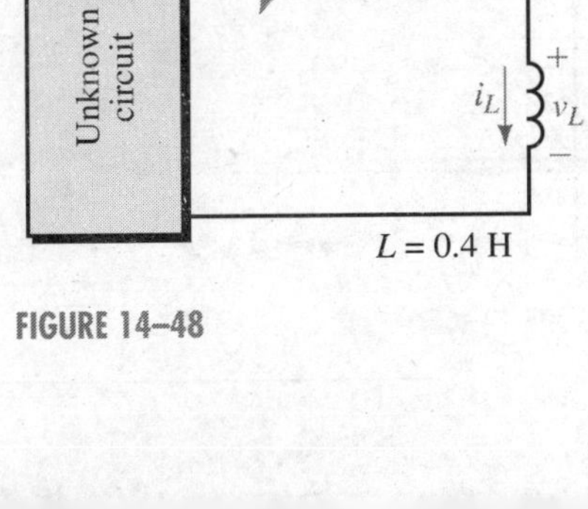

FIGURE 14–48

$t = 0^-$ until the circuit reaches steady state in position 3. Compute the inductor voltage and current at $t = 0.1$ s and at $t = 1.1$ s.

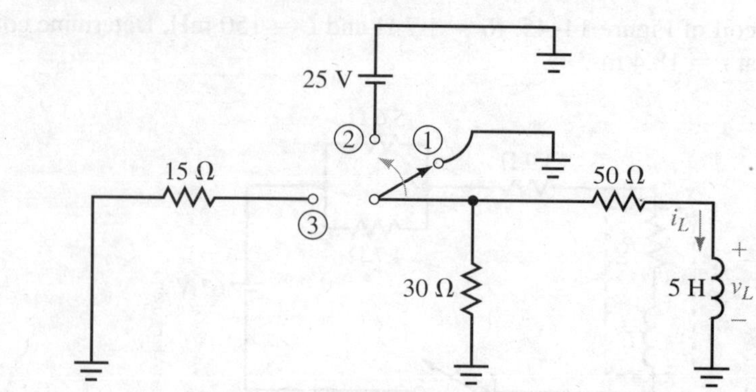

FIGURE 14–49

14.6 *RL* Transients Using Computers

Multisim PSPICE 29. The circuit of Figure 14–46 is in steady state with the switch open. At $t = 0$, it is closed and remains closed. Graph the voltage across L and find v_L at 20 μs using the cursor.

Multisim PSPICE 30. For the circuit of Figure 14–47, close the switch at $t = 0$ and find v_L at $t = 10$ ms. (For PSpice, use current source IDC.)

Multisim PSPICE 31. For Figure 14–6, let $L_1 = 30$ mH and $L_2 = 90$ mH. Close the switch at $t = 0$ and find the current in the 30 Ω resistor at $t = 2$ ms.

Multisim PSPICE 32. For Figure 14–41, let $L = 4$ H. Solve for v_L and i_L. Using the cursor, measure values at $t = 200$ ms and 500 ms. (Multisim users: You may have to create a separate vertical axis for the current waveform.)

PSPICE 33. We solved the circuit of Figure 14–22(a) by reducing it to its Thévenin equivalent. Using PSpice, analyze the circuit in its original form and plot the inductor current. Check a few points on the curve by computing values according to the solution of Example 14–8 and compare to values obtained from screen.

PSPICE 34. The circuit of Figure 14–46 is in steady state with the switch open. At $t = 0$, the switch is closed. It remains closed for 150 μs and is then opened and left open. Compute and plot i_L and v_L. With the cursor, determine values at $t = 60$ μs and at $t = 165$ μs.

ANSWERS TO IN-PROCESS LEARNING CHECKS

IN-PROCESS LEARNING CHECK 1

1. a. $20e^{-100\,000t}$ V; $2(1 - e^{-100\,000t})$ mA

b.

t(μs)	v_L(V)	i_L(mA)
0	20	0
10	7.36	1.26
20	2.71	1.73
30	0.996	1.90
40	0.366	1.96
50	0.135	1.99

c.

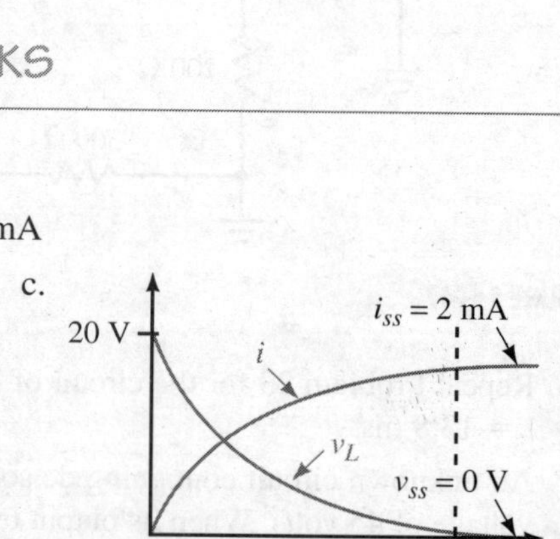

2. 11.5 V; 1.47 A
3. 4 V
4. 3.88 A

FOUNDATION ac CONCEPTS

IV

In previous chapters, we concentrated mostly on dc. We now turn our attention to ac (alternating current).

AC is important to us for a number of reasons. First, it is the basis of the electrical power system that supplies our homes and businesses with electrical energy. AC is used instead of dc because it has several important advantages, the chief one being that ac power can be transmitted easily and efficiently over long distances. However, the importance of ac extends far beyond its use in the electrical power industry. You will find, for example, that an understanding of ac concepts is necessary in virtually every branch of electronics that you will encounter, be it in the field of audio systems, communications systems, control systems, or any number of other areas. In fact, nearly every electrical and electronic device that we use in our daily lives operates from or involves the use of ac in some way.

We begin Part IV of this book with a look at fundamental ac concepts. We examine ways to generate ac voltages, methods used to represent ac voltages and currents, relationships between ac quantities in resistive, inductive, and capacitive circuits, and, finally, the meaning and representation of power in ac systems. This sets the stage for succeeding chapters that deal with ac circuit analysis techniques, including ac versions of the various methods that you have used for dc circuits throughout previous chapters of this book. ■

■ KEY TERMS

ac Current
ac Voltage
Alternating Current (ac)
Amplitude
Average Value
Cycle
Effective Value
Frequency
Hertz (Hz)
Instantaneous Value
Lag
Lead
Oscilloscope
Peak-to-Peak
Peak Value
Period (T)
Periodic Waveform
Phase Difference
Phase Shifts
Phasor
Radian Frequency
RMS (Root Mean Square) Values
Sine Wave
Waveform

■ OUTLINE

Introduction
Generating ac Voltages
Voltage and Current Conventions for ac
Frequency, Period, Amplitude, and Peak Value
Angular and Graphic Relationships for Sine Waves
Voltages and Currents as Functions of Time
Introduction to Phasors
ac Waveforms and Average Value
Effective (RMS) Values
Rate of Change of a Sine Wave (Derivative)
ac Voltage and Current Measurement
Circuit Analysis Using Computers

■ OBJECTIVES

After studying this chapter, you will be able to

- explain how ac voltages and currents differ from dc,
- draw waveforms for ac voltage and currents and explain what they mean,
- explain the voltage polarity and current direction conventions used for ac,
- describe the basic ac generator and explain how ac voltage is generated,
- define and compute frequency, period, amplitude, and peak-to-peak values,
- compute instantaneous sinusoidal voltage or current at any instant in time,
- define the relationships between ω, T, and f for a sine wave,
- define and compute phase differences between waveforms,
- use phasors to represent sinusoidal voltages and currents,
- determine phase relationships between waveforms using phasors,
- define and compute average values for time-varying waveforms,
- define and compute effective (rms) values for time-varying waveforms,
- use Multisim and PSpice to study ac waveforms.

15

ac FUNDAMENTALS

CHAPTER PREVIEW

Alternating currents (ac) are currents that alternate in direction (usually many times per second), passing first in one direction, then in the other through a circuit. Such currents are produced by voltage sources whose polarities alternate between positive and negative (rather than being fixed as with dc sources). By convention, alternating currents are called **ac currents** and alternating voltages are called **ac voltages.**

The variation of an ac voltage or current versus time is called its **waveform.** Since waveforms vary with time, they are designated by lowercase letters $v(t)$, $i(t)$, $e(t)$, and so on, rather than by uppercase letters V, I, and E as for dc. Often we drop the functional notation [e.g., $v(t)$] and just use the simpler notations v, i, and e instead.

While many waveforms are important to us, the most fundamental is the sine wave (also called sinusoidal ac). In fact, the sine wave is of such importance that many people associate the term ac with sinusoidal, even though ac refers to any quantity that alternates with time.

In this chapter, we look at basic ac principles, including the generation of ac voltages and ways to represent and manipulate ac quantities. These ideas are then used throughout the remainder of the book to develop methods of analysis for ac circuits. ■

Putting It in Perspective

Thomas Alva Edison

NOWADAYS WE TAKE IT FOR GRANTED that our electrical power systems are ac. (This is reinforced every time you see a piece of equipment rated "60 hertz ac.") However, this was not always the case. In the late 1800s, a fierce battle—the so-called "war of the currents"—raged in the emerging electrical power industry. The forces favoring the use of dc were led by Thomas Alva Edison, and those favoring the use of ac were led by George Westinghouse (Chapter 23) and Nikola Tesla (Chapter 24).

Edison, a prolific inventor who gave us the electric light, the phonograph, and many other great inventions as well, fought vigorously for dc. He had spent a considerable amount of time and money on the development of dc power and had a lot at stake, in terms of both money and prestige. So determined was Edison in this battle that he first persuaded the state of New York to adopt ac for its newly devised execution device, the electric chair, and then pointed at it with horror as an example of how deadly ac was. Ultimately, however, the combination of ac's advantages over dc and the stout opposition of Tesla and Westinghouse won the day for ac.

Edison was born in 1847 in Milan, Ohio. Most of his work was done at two sites in New Jersey—first at a laboratory in Menlo Park, and later at a much larger laboratory in West Orange, where his staff at one time numbered around 5000. He received patents as inventor or co-inventor on nearly 1300 inventions—an astonishing feat that made him probably the greatest inventor of all time.

Thomas Edison died at the age of 84 on October 18, 1931. ■

15.1 Introduction

Previously you learned that dc sources have fixed polarities and constant magnitudes and thus produce currents with constant value and unchanging direction, as illustrated in Figure 15–1. In contrast, the voltages of ac sources alternate in polarity and vary in magnitude and thus produce currents that vary in magnitude and alternate in direction.

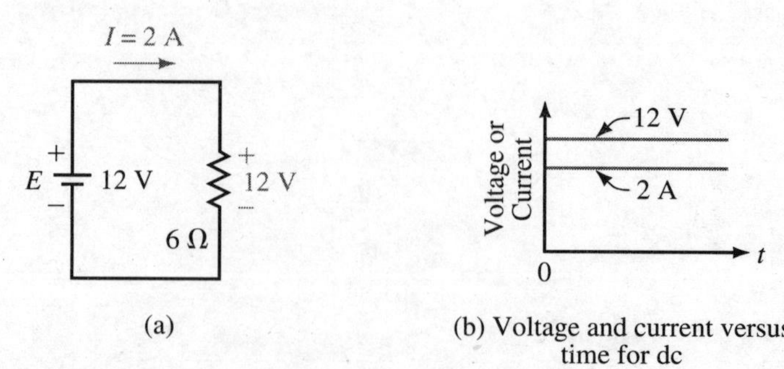

FIGURE 15–1 In a dc circuit, voltage polarities and current directions do not change.

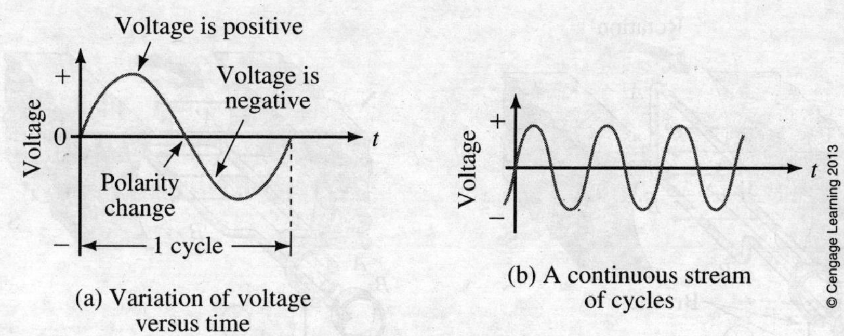

FIGURE 15–2 Sinusoidal ac waveforms. Values above the axis are positive while values below are negative.

CircuitSim 15-1

Sinusoidal ac Voltage

To illustrate, consider the voltage at the wall outlet in your home. Called a **sine wave** or **sinusoidal ac waveform** (for reasons discussed in Section 15.5), this voltage has the shape shown in Figure 15–2. Starting at zero, the voltage increases to a positive peak amplitude, decreases to zero, changes polarity, increases to a negative peak amplitude, then returns again to zero. One complete variation is referred to as a **cycle.** Since the waveform repeats itself at regular intervals as in (b), it is called a **periodic waveform.**

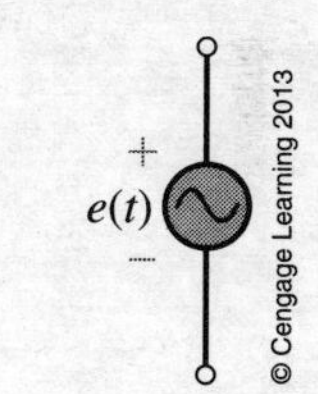

FIGURE 15–3 Symbol for a sinusoidal voltage source. Lowercase letter *e* is used to indicate that the voltage varies with time.

Symbol for an ac Voltage Source

The symbol for a sinusoidal voltage source is shown in Figure 15–3. Note that a lowercase *e* is used to represent voltage rather than *E*, since it is a function of time. Polarity marks are also shown although, since the polarity of the source varies, their meaning has yet to be established.

Sinusoidal ac Current

Figure 15–4 shows a resistor connected to an ac source. During the first half-cycle, the source voltage is positive; therefore, the current is in the clockwise direction. During the second half-cycle, the voltage polarity reverses; therefore, the current is in the counterclockwise direction. Since current is proportional to voltage, its shape is also sinusoidal (Figure 15–5).

CircuitSim 15-2

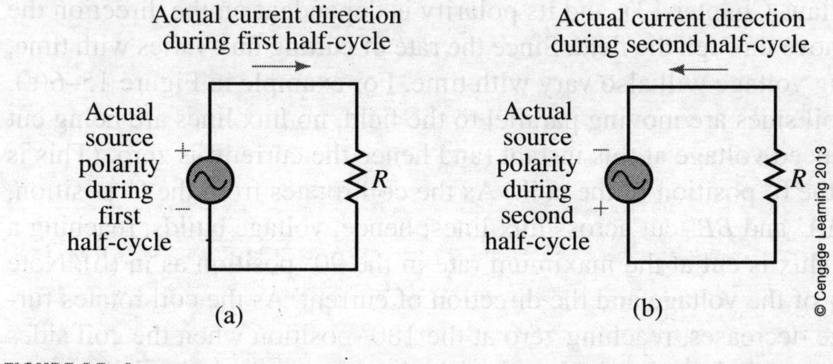

FIGURE 15–4 Current direction reverses when the source polarity reverses.

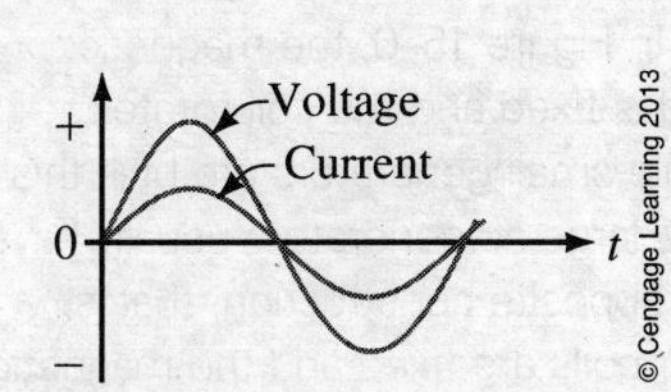

FIGURE 15–5 Current has the same waveshape as voltage.

15.2 Generating ac Voltages

One way to generate an ac voltage is to rotate a coil of wire at constant angular velocity in a fixed magnetic field—see Note. As indicated in Figure 15–6, slip rings and brushes connect the coil to the load. The magnitude of

NOTES...

There are some good animations on the Internet that show the process of ac generation. Search for *Generator Animation*. (Some animations are better than others—look until you find a good one.)

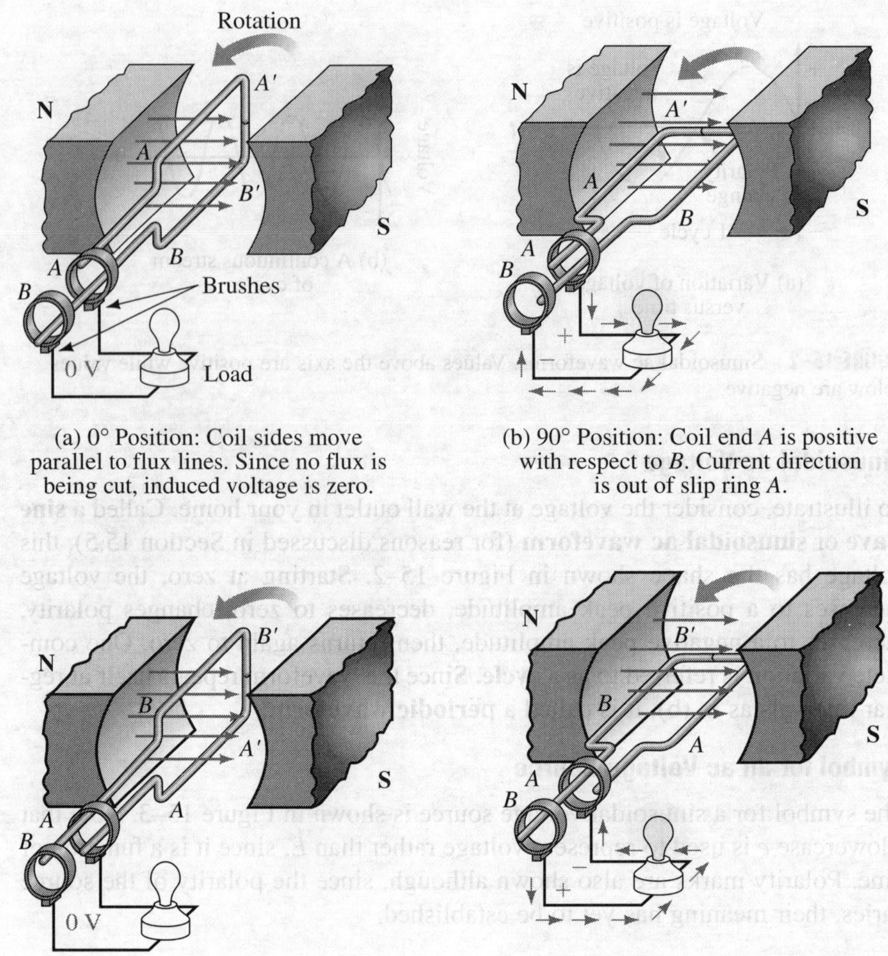

(a) 0° Position: Coil sides move parallel to flux lines. Since no flux is being cut, induced voltage is zero.

(b) 90° Position: Coil end *A* is positive with respect to *B*. Current direction is out of slip ring *A*.

(c) 180° Position: Coil again cutting no flux. Induced voltage is zero.

(d) 270° Position: Voltage polarity has reversed, therefore, current direction has also reversed.

FIGURE 15–6 Generating an ac voltage. The 0° position of the coil is defined as in (a) where the coil sides move parallel to the flux lines. In practice, the rotations are so fast that the light does not have time to go out and so appears to be continuously on.

PRACTICAL NOTES...

In practice, the coil of Figure 15–6 consists of many turns wound on an iron core. The coil, core, and slip rings rotate as a unit.

In Figure 15–6, the magnetic field is fixed and the coil rotates. While small generators are built this way, large ac generators usually have the opposite construction, that is, their coils are fixed and the magnetic field is rotated instead. In addition, large ac generators are usually made as three-phase machines with three sets of coils instead of one. This is covered in Chapter 24. However, although its details are oversimplified, the generator of Figure 15–6 gives a true picture of the voltage produced by a real ac generator.

the resulting voltage is proportional to the rate at which flux lines are cut (Faraday's law, Chapter 13), and its polarity is dependent on the direction the coil sides move through the field. Since the rate of cutting flux varies with time, the resulting voltage will also vary with time. For example in Figure 15–6(a), since the coil sides are moving parallel to the field, no flux lines are being cut and the induced voltage at this instant (and hence the current) is zero. (This is defined as the 0° position of the coil.) As the coil rotates from the 0° position, coil sides AA' and BB' cut across flux lines; hence, voltage builds, reaching a peak when flux is cut at the maximum rate in the 90° position as in (b). Note the polarity of the voltage and the direction of current. As the coil rotates further, voltage decreases, reaching zero at the 180° position when the coil sides again move parallel to the field as in (c). At this point, the coil has gone through a half-revolution.

During the second half-revolution, coil sides cut flux in directions opposite to that which they did in the first half-revolution; hence, the polarity of the induced voltage reverses. As indicated in Figure 15–6(d), voltage reaches a peak at the 270° point, and, since the polarity of the voltage has changed, so has the direction of current. When the coil reaches the 360° position, voltage is again zero and the cycle starts over. Figure 15–7 shows one cycle of

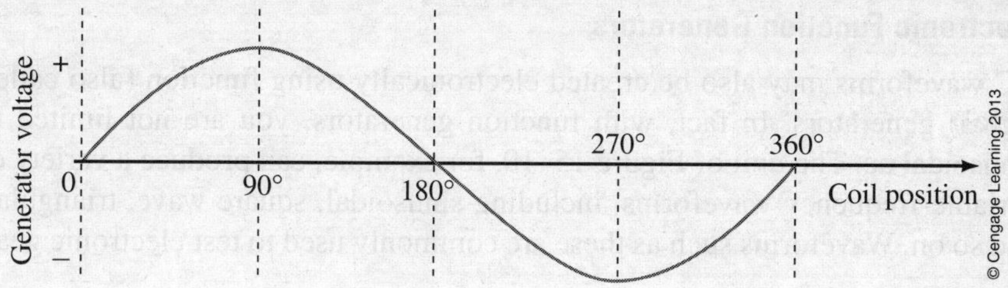

FIGURE 15–7 Coil voltage versus angular position.

the resulting waveform. Since the coil rotates continuously, the voltage produced will be a repetitive, periodic waveform as you saw in Figure 15–2(b). Current will be periodic also.

Time Scales

The horizontal axis of Figure 15–7 is scaled in degrees. Often we need it scaled in time. The length of time required to generate one cycle depends on the velocity of rotation. To illustrate, assume that the coil rotates at 600 rpm (revolutions per minute). Six hundred revolutions in 1 minute equals 600 rev/60 s = 10 revolutions in 1 second. At 10 revolutions per second, the time for 1 revolution is one-tenth of a second, that is, 100 ms. Since one cycle is 100 ms, a half-cycle is 50 ms, a quarter-cycle is 25 ms, and so on. Figure 15–8 shows the waveform rescaled in time.

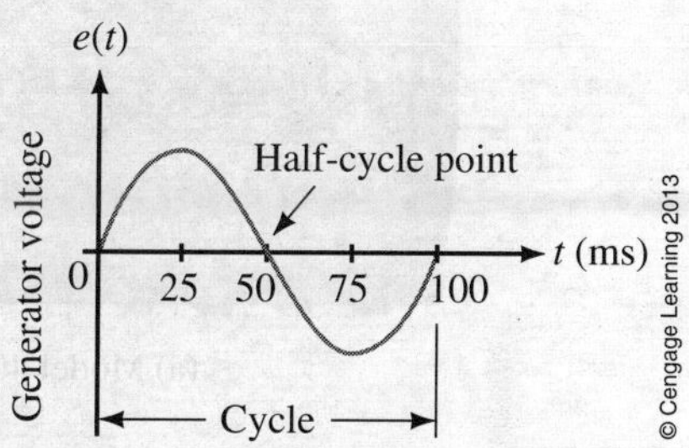

FIGURE 15–8 Cycle scaled in time. At 600 rpm, the cycle length is 100 ms.

Instantaneous Value

As Figure 15–8 shows, the coil voltage changes from instant to instant. The value of voltage at any point on the waveform is referred to as its **instantaneous value.** This is illustrated in Figure 15–9. Figure 15–9(a) shows a photograph of an actual waveform, and (b) shows it redrawn, with values scaled from the photo. For this example, the voltage has a peak value of 40 volts and a cycle time of 6 ms. From the graph, we see that at $t = 0$ ms, the voltage is zero. At $t = 0.5$ ms, it is 20 V. At $t = 2$ ms, it is 35 V. At $t = 3.5$ ms, it is −20 V, and so on.

CircuitSim 15-3

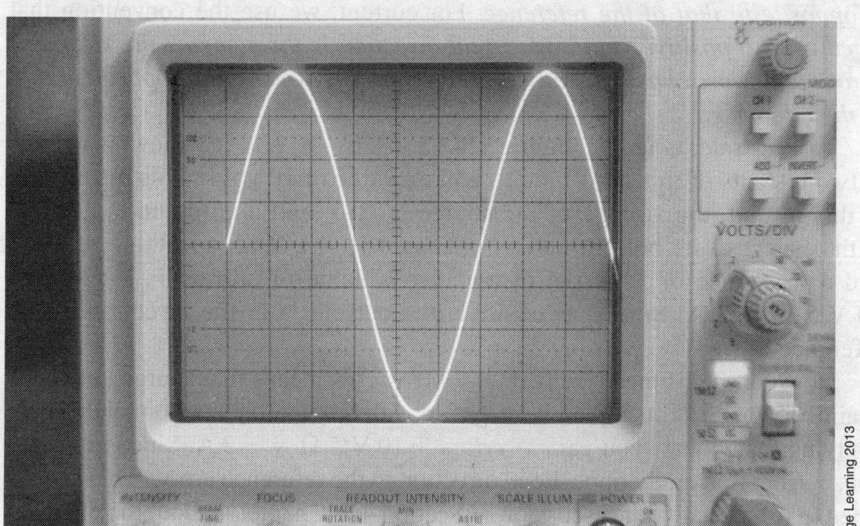

(a) Sinusoidal voltage

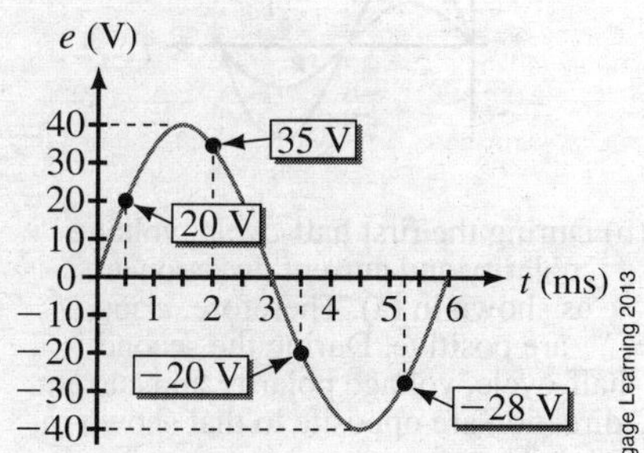

(b) Values scaled from the photograph

FIGURE 15–9 Instantaneous values.

Electronic Function Generators

AC waveforms may also be created electronically using function (also called signal) generators. In fact, with function generators, you are not limited to sinusoidal ac. The unit of Figure 15–10, for example, can produce a variety of variable-frequency waveforms, including sinusoidal, square wave, triangular, and so on. Waveforms such as these are commonly used to test electronic gear.

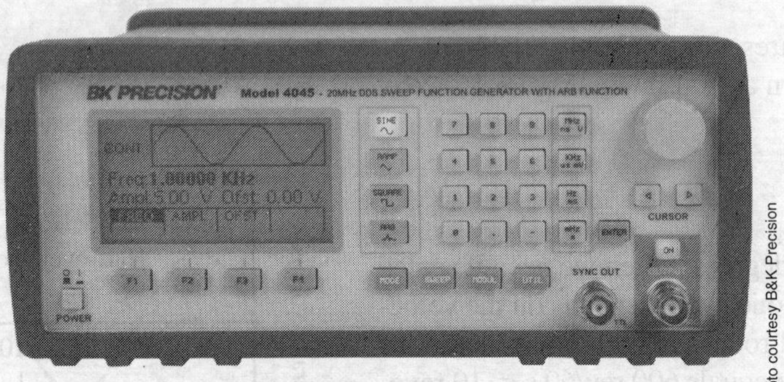

(a) Model 4045 Function Generator

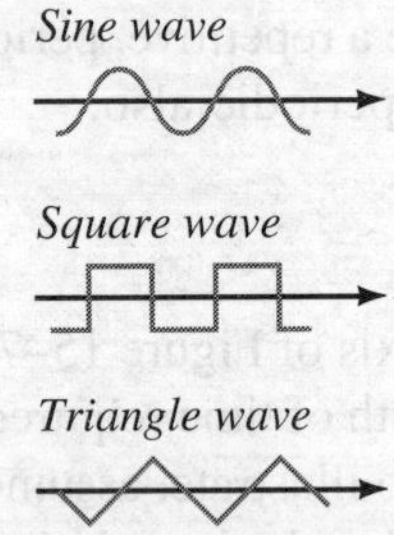

(b) Sample waveforms

FIGURE 15–10 Electronic function generators provide a variety of variable-frequency, variable-amplitude waveforms.

15.3 Voltage and Current Conventions for ac

In Section 15.1, we looked briefly at voltage polarities and current directions. At that time, we used separate diagrams for each half-cycle (Figure 15–4). However, this is unnecessary; one diagram and one set of references is all that is required. This is illustrated in Figure 15–11. First, we assign reference polarities for the source and a reference direction for the current. We then use the convention that, *when* e *has a positive value, its actual polarity is the same as the reference polarity, and when* e *has a negative value, its actual polarity is opposite to that of the reference.* For current, we use the convention that *when* i *has a positive value, its actual direction is the same as the reference arrow, and when* i *has a negative value, its actual direction is opposite to that of the reference.*

To illustrate, consider Figure 15–12. [Parts (b) and (c) show snapshots at two instants of time.] At time t_1, e has a value of 10 volts. This means that at this instant, the voltage of the source is 10 V and its top end is positive with respect to its bottom end as indicated in (b). With a voltage of 10 V and a resistance of 5 Ω, the instantaneous value of current is $i = e/R = 10\text{ V}/5\ \Omega = 2\text{ A}$. Since i is positive, the current is in the direction of the reference arrow.

Now consider time t_2. Here, $e = -10$ V. This means that source voltage is again 10 V, but now its top end is negative with respect to its bottom end. Again applying Ohm's law, you get $i = e/R = -10\text{ V}/5\ \Omega = -2\text{ A}$. Since i is negative, current is actually opposite in direction to the reference arrow. This is indicated in (c). (Recall our discussion of Section 4.2, Chapter 4 about the meaning of negative currents.)

The preceding concept is valid for any ac signal, regardless of waveshape. Let us illustrate.

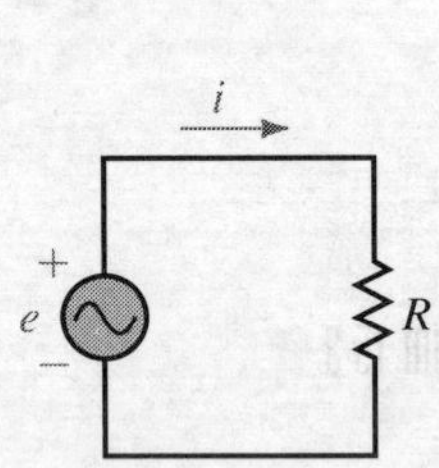

(a) References for voltage and current

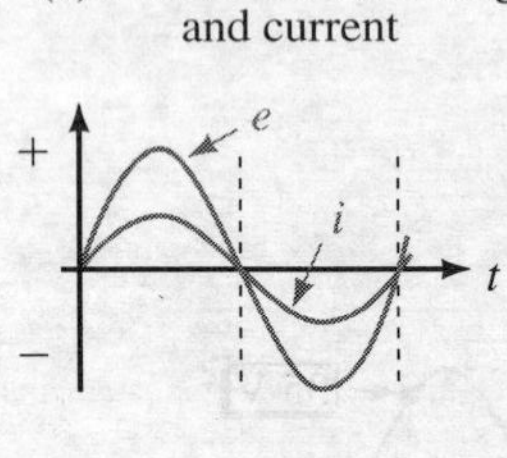

(b) During the first half-cycle, voltage polarity and current direction are as shown in (a). Therefore, e and i are positive. During the second half-cycle, voltage polarity and current direction are opposite to that shown in (a). Therefore, e and i are negative.

FIGURE 15–11 ac Voltage and current reference conventions.

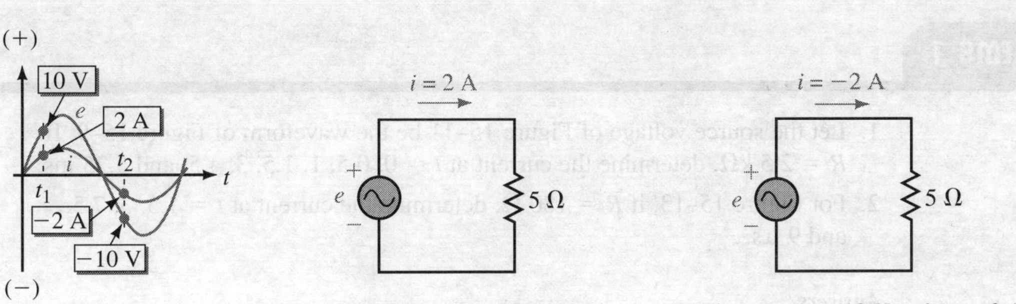

(a)

(b) Time t_1: $e = 10$ V and $i = 2$ A. Thus, voltage and current have the polarity and direction indicated.

(c) Time t_2: $e = -10$ V and $i = -2$ A. Thus, voltage polarity is opposite to that indicated and current direction is opposite to the arrow direction.

FIGURE 15–12 Illustrating the ac voltage and current convention.

CircuitSim 15-4

EXAMPLE 15–1

Figure 15–13(b) shows one cycle of a triangular voltage wave. Determine the current and its direction at t = 0, 1, 2, 3, 4, 5, 6, 7, 8, 9, 10, 11, and 12 μs and sketch.

CircuitSim 15-5

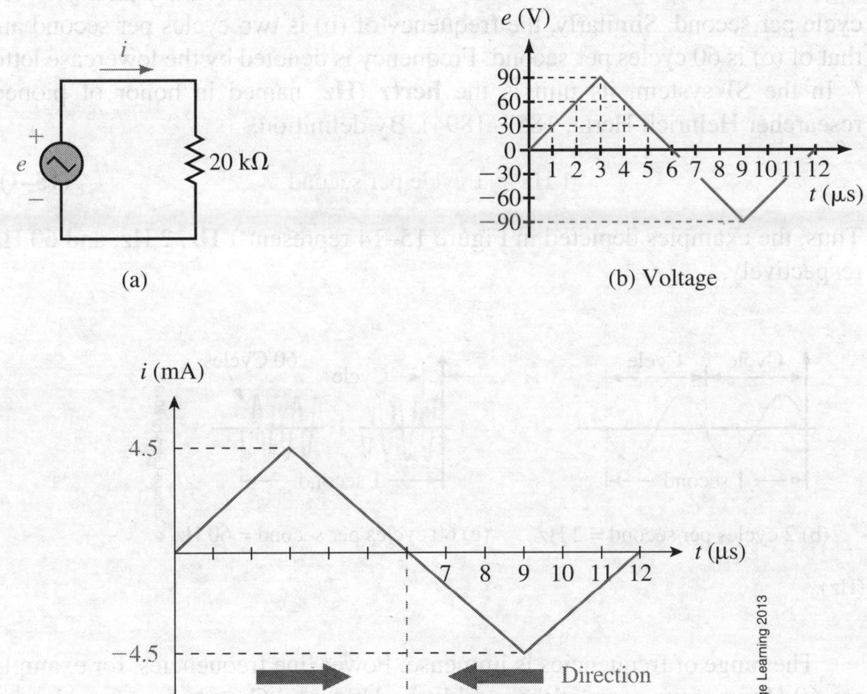

(a)

(b) Voltage

(c) Current

FIGURE 15–13

Solution Apply Ohm's law at each point in time. At t = 0 μs, e = 0 V, so $i = e/R$ = 0 V/20 kΩ = 0 mA. At t = 1 μs, e = 30 V. Thus, $i = e/R$ = 30 V/20 kΩ = 1.5 mA. At t = 2 μs, e = 60 V. Thus, $i = e/R$ = 60 V/20 kΩ = 3 mA. Continuing in this manner, you get the values shown in Table 15–1. The waveform is plotted as Figure 15–13(c).

TABLE 15–1 Values for Example 15–1

t (μs)	e (V)	i (mA)
0	0	0
1	30	1.5
2	60	3.0
3	90	4.5
4	60	3.0
5	30	1.5
6	0	0
7	−30	−1.5
8	−60	−3.0
9	−90	−4.5
10	−60	−3.0
11	−30	−1.5
12	0	0

PRACTICE PROBLEMS 1

1. Let the source voltage of Figure 15–11 be the waveform of Figure 15–9. If $R = 2.5\ \text{k}\Omega$, determine the current at $t = 0, 0.5, 1, 1.5, 3, 4.5$, and 5.25 ms.
2. For Figure 15–13, if $R = 180\ \Omega$, determine the current at $t = 1.5, 3, 7.5$, and $9\ \mu$s.

Answers
1. 0, 8, 14, 16, 0, −16, −11.2 (all mA)
2. 0.25, 0.5, −0.25, −0.5 (all A)

15.4 Frequency, Period, Amplitude, and Peak Value

Periodic waveforms (i.e., waveforms that repeat at regular intervals), regardless of their waveshape, may be described by a group of attributes such as frequency, period, amplitude, peak value, and so on.

Frequency

The number of cycles per second of a waveform is defined as its **frequency.** In Figure 15–14(a), one cycle occurs in one second; thus its frequency is one cycle per second. Similarly, the frequency of (b) is two cycles per second and that of (c) is 60 cycles per second. Frequency is denoted by the lowercase letter *f*. In the SI system, its unit is the **hertz** (**Hz**, named in honor of pioneer researcher Heinrich Hertz, 1857–1894). By definition,

$$1\ \text{Hz} = 1\ \text{cycle per second} \tag{15–1}$$

Thus, the examples depicted in Figure 15–14 represent 1 Hz, 2 Hz, and 60 Hz, respectively.

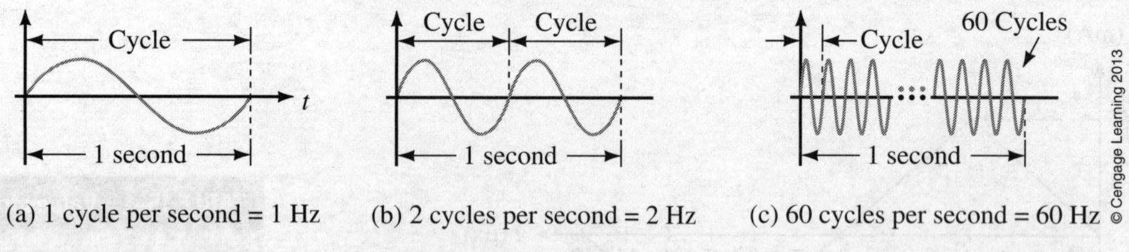

FIGURE 15–14 Frequency is measured in hertz (Hz).

The range of frequencies is immense. Power line frequencies, for example, are 60 Hz in many parts of the world (the USA and Canada for example) and 50 Hz in others. Audible sound frequencies range from about 20 Hz to about 20 kHz. The standard AM radio band occupies from 550 kHz to 1.6 MHz, while the FM band extends from 88 MHz to 108 MHz. TV transmissions occupy several bands in the 54-MHz to 890-MHz range. Above 300 GHz are optical and X-ray frequencies.

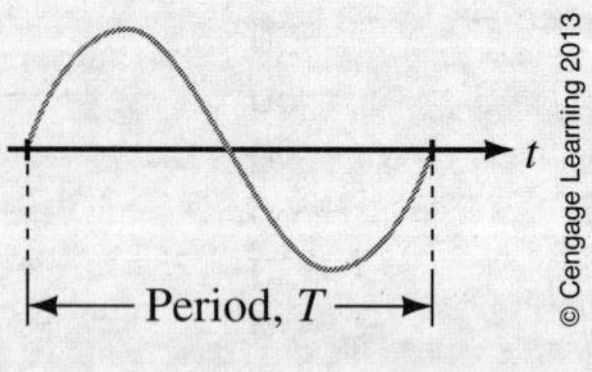

FIGURE 15–15 Period, *T* is the duration of one cycle, measured in seconds.

Period

The **period,** ***T,*** of a waveform (Figure 15–15) is the duration of one cycle. It is the inverse of frequency. To illustrate, consider again Figure 15–14. In (a), the frequency is 1 cycle per second; thus, the duration of each cycle is $T = 1$ s. In (b), the frequency is two cycles per second; thus, the duration of each cycle is $T = 1/2$ s, and so on. In general,

$$T = \frac{1}{f} \quad \text{(s)} \tag{15–2}$$

and

$$f = \frac{1}{T} \quad \text{(Hz)} \tag{15–3}$$

Note that these definitions are independent of waveshape.

EXAMPLE 15–2

a. What is the period of a 50-Hz voltage?

b. What is the period of a 1-MHz current?

Solution

a. $T = \frac{1}{f} = \frac{1}{50 \text{ Hz}} = 20 \text{ ms}$

b. $T = \frac{1}{f} = \frac{1}{1 \times 10^6 \text{ Hz}} = 1 \text{ μs}$

EXAMPLE 15–3

Figure 15–16 shows an **oscilloscope** trace of a square wave. Controls have been set such that each horizontal division represents 50 μs. Determine the waveform's frequency.

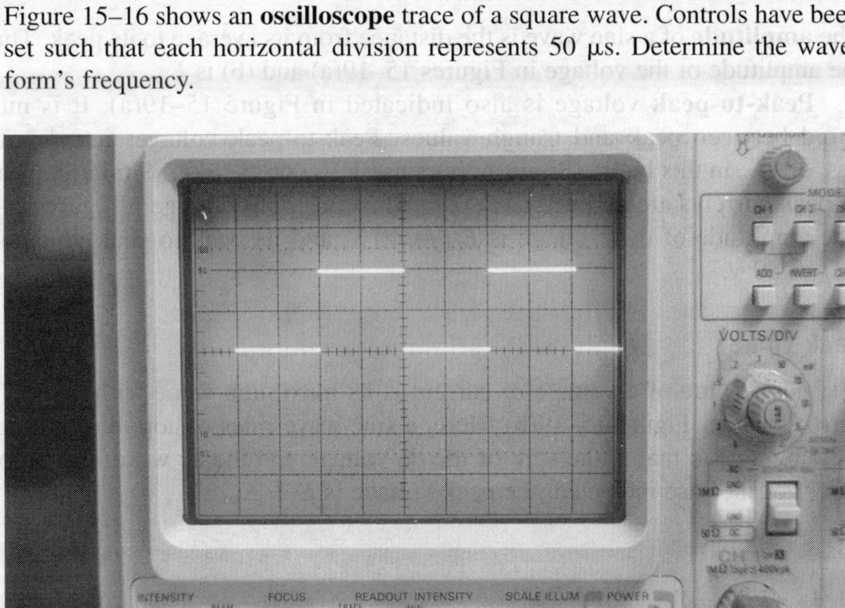

FIGURE 15–16 The concepts of frequency and period apply also to nonsinusoidal waveforms. Here, $T = 4 \text{ div} \times 50 \text{ μs/div} = 200 \text{ μs}$.

Solution Since the wave repeats itself every 200 μs, its period is 200 μs and

$$f = \frac{1}{200 \times 10^{-6} \text{ s}} = 5 \text{ kHz}$$

The period of a waveform can be measured between any two corresponding points (Figure 15–17). Often it is measured between zero points or peak points because they are easy to establish on an oscilloscope trace.

CircuitSim 15-6

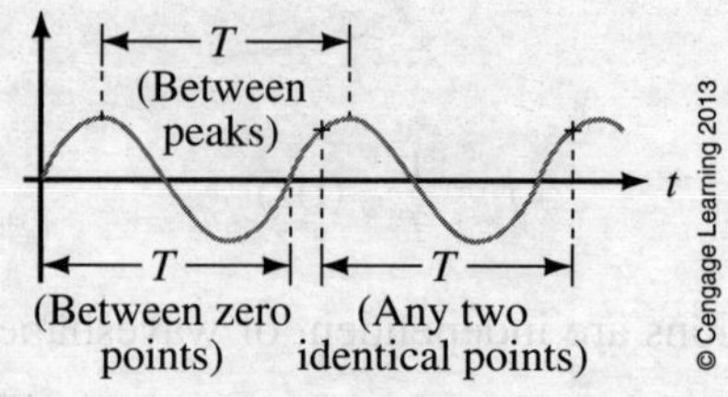

FIGURE 15–17 Period may be measured between any two corresponding points.

EXAMPLE 15–4

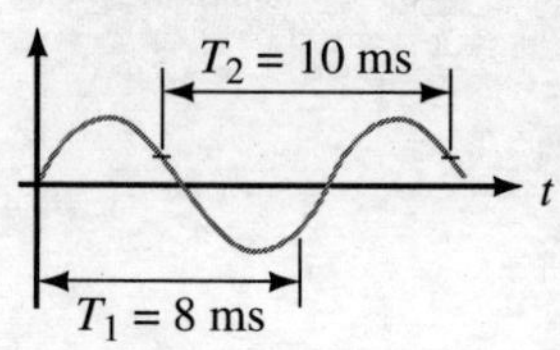

FIGURE 15–18

Determine the period and frequency of the waveform of Figure 15–18.

Solution Time interval T_1 does not represent a period as it is not measured between corresponding points. Interval T_2, however, is. Thus, $T = 10$ ms and

$$f = \frac{1}{T} = \frac{1}{10 \times 10^{-3}\ \text{s}} = 100\ \text{Hz}$$

Amplitude and Peak-to-Peak Value

The **amplitude** of a sine wave is the distance from its average to its peak. Thus, the amplitude of the voltage in Figures 15–19(a) and (b) is E_m.

Peak-to-peak voltage is also indicated in Figure 15–19(a). It is measured between peak and trough values. Peak-to-peak voltages are denoted $E_{p\text{-}p}$ or $V_{p\text{-}p}$ in this book. (Some authors use $V_{pk\text{-}pk}$ or the like.) Similarly, peak-to-peak currents are denoted as $I_{p\text{-}p}$. To illustrate, consider again Figure 15–9. The amplitude of this voltage is $E_m = 40$ V, and its peak-to-peak voltage is $E_{p\text{-}p} = 80$ V.

Peak Value

The **peak value** of a voltage or current is its maximum value with respect to zero. Consider Figure 15–19(b). Here, a sine wave rides on top of a dc value, yielding a peak that is the sum of the dc voltage and the ac waveform amplitude. For the case indicated, the peak voltage is $E + E_m$.

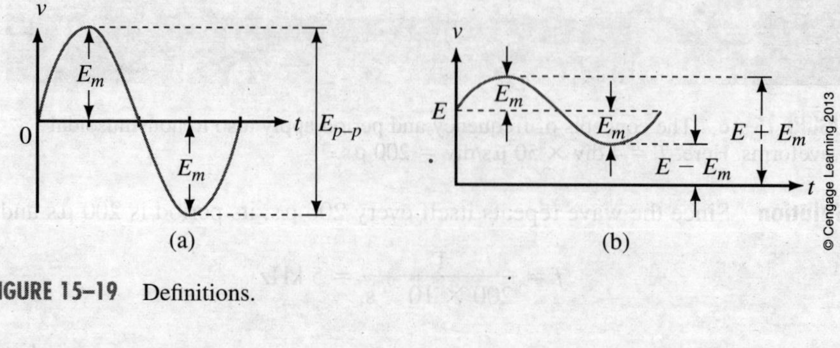

FIGURE 15–19 Definitions.

IN-PROCESS LEARNING CHECK 1

(Answers are at the end of the chapter.)

1. What is the period of an ac power system whose frequency is 60 Hz?
2. If you double the rotational speed of an ac generator, what happens to the frequency and period of the waveform?
3. If the generator of Figure 15–6 rotates at 3000 rpm, what is the period and frequency of the resulting voltage? Sketch four cycles and scale the horizontal axis in units of time.
4. For the waveform of Figure 15–9, list all values of time at which $e = 20$ V and $e = -35$ V. Hint: Sine waves are symmetrical.
5. Which of the waveform pairs of Figure 15–20 are valid combinations? Why?

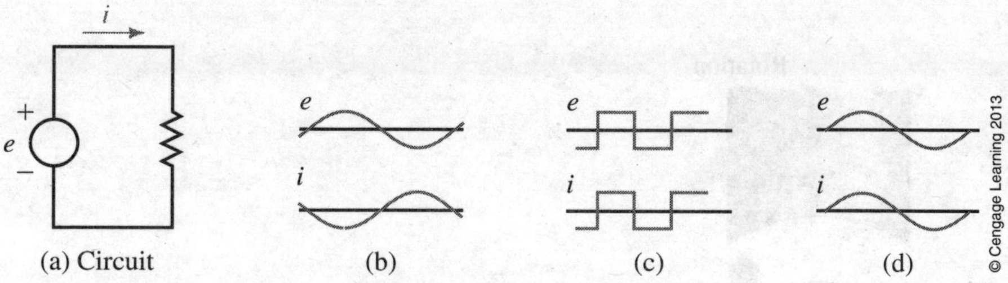

FIGURE 15–20 Which waveform pairs are valid?

6. For the waveform in Figure 15–21, determine the frequency.

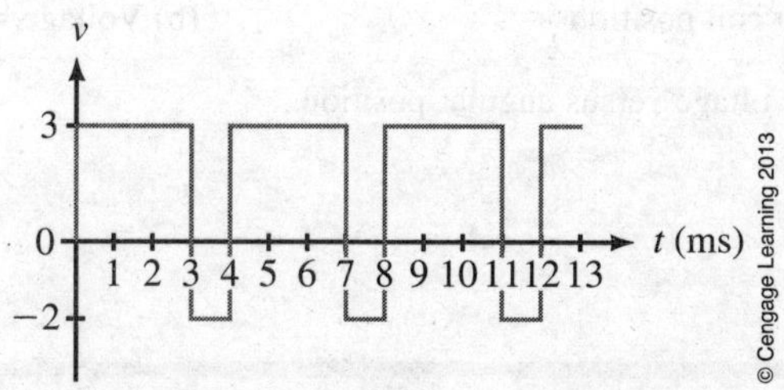

FIGURE 15–21

7. Two waveforms have periods of $T_1 = 10$ ms and $T_2 = 30$ ms. Which has the higher frequency? Compute the frequencies of both waveforms.
8. Two sources have frequencies f_1 and f_2, respectively. If $f_2 = 20f_1$, and T_2 is 1 μs, what is f_1? What is f_2?
9. Consider Figure 15–22. What is the frequency of the waveform?
10. For Figure 15–11, if $f = 20$ Hz, what is the current direction at $t = 12$ ms, 37 ms, and 60 ms? Hint: Sketch the waveform and scale the horizontal axis in ms. The answers should be apparent.
11. A 10-Hz sinusoidal current has a value of 5 amps at $t = 25$ ms. What is its value at $t = 75$ ms? See Hint in Problem 10.

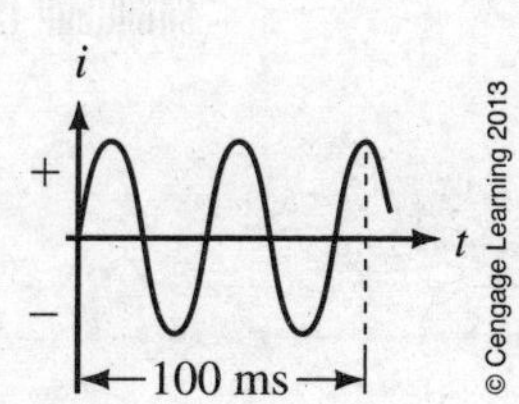

FIGURE 15–22

15.5 Angular and Graphic Relationships for Sine Waves

The Basic Sine Wave Equation

Consider again the generator of Figure 15–6, reoriented and redrawn in end view as Figure 15–23. As the coil rotates, the voltage produced is

$$e = E_m \sin \alpha \quad \text{(V)} \qquad \textbf{(15–4)}$$

NOTES...

The derivation of Equation 15–4 may be found in many advanced physics books.

where E_m is the maximum coil voltage and α is the instantaneous angular position of the coil—see Note. (For a given generator and rotational velocity, E_m is constant.) Note that $\alpha = 0°$ represents the horizontal position of the coil and that one complete cycle corresponds to 360°. Equation 15–4 states that the voltage at any point on the sine wave may be found by multiplying E_m times the sine of the angle at that point.

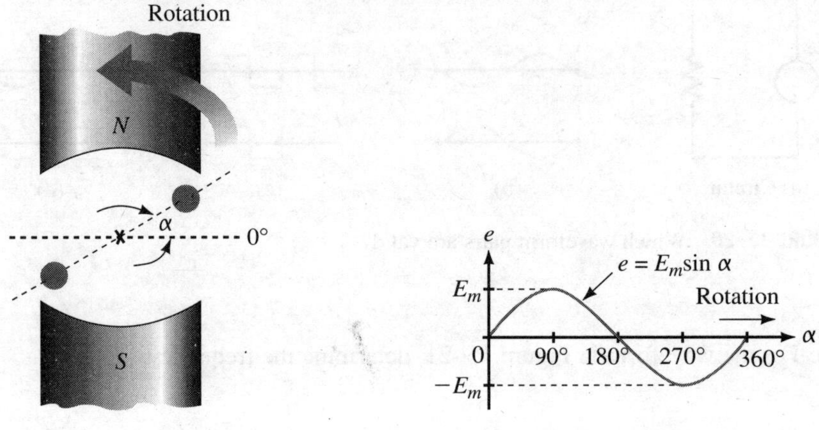

(a) End view showing coil position (b) Voltage waveform

FIGURE 15–23 Coil voltage versus angular position.

EXAMPLE 15–5

If the amplitude of the waveform of Figure 15–23(b) is $E_m = 100$ V, determine the coil voltage at 30° and 330°.

Solution At $\alpha = 30°$, $e = E_m \sin \alpha = 100 \sin 30° = 50$ V. At 330°, $e = 100 \sin 330° = -50$ V. These are shown on the graph of Figure 15–24.

CircuitSim 15-7

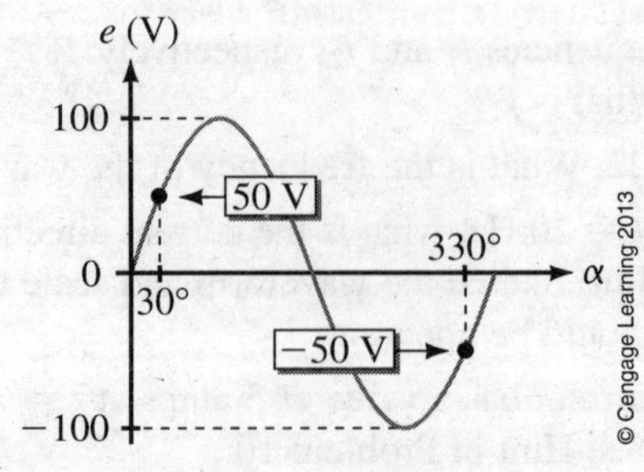

FIGURE 15–24

PRACTICE PROBLEMS 2

Table 15–2 is a tabulation of voltage versus angle computed from $e = 100 \sin \alpha$. Use your calculator to verify each value, then plot the result on graph paper. The resulting waveshape should look like Figure 15–24.

TABLE 15–2 Data for Plotting $e = 100 \sin \alpha$

Angle α	Voltage e
0	0
30	50
60	86.6
90	100
120	86.6
150	50
180	0
210	−50
240	−86.6
270	−100
300	−86.6
330	−50
360	0

Angular Velocity, ω

The rate at which the generator coil rotates is called its angular velocity. If the coil rotates through an angle of 30° in 1 second, for example, its angular velocity is 30° per second. Angular velocity is denoted by the Greek letter ω (omega). For the case cited, $\omega = 30°/\text{s}$. (Normally angular velocity is expressed in radians per second instead of degrees per second. We will make this change shortly—see Note.) When you know the angular velocity of a coil and the length of time that it has rotated, you can compute the angle through which it has turned. For example, a coil rotating at 30°/s rotates through an angle of 30° in 1 second, 60° in 2 seconds, 90° in 3 seconds, and so on. In general,

$$\alpha = \omega t \tag{15–5}$$

Expressions for t and ω can now be found. They are

$$t = \frac{\alpha}{\omega} \quad \text{(s)} \tag{15–6}$$

$$\omega = \frac{\alpha}{t} \tag{15–7}$$

NOTES...

Note as well that Equation 15–5 to Equation 15–7 also hold when α is expressed in radians and ω in radians/second. In fact, it is mostly with radian measure that we use these equations as you will soon see.

EXAMPLE 15–6

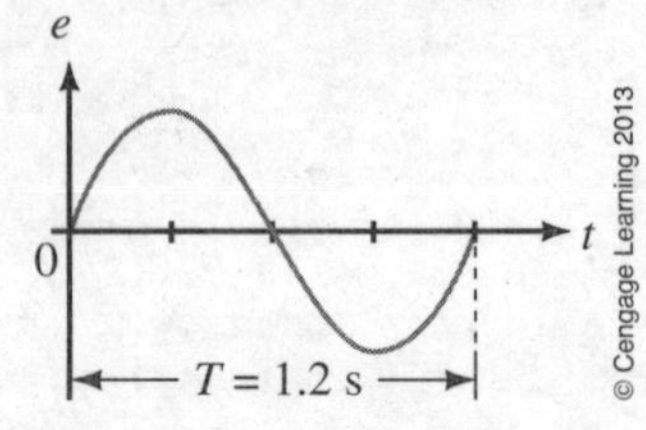

FIGURE 15–25

If the coil of Figure 15–23 rotates at $\omega = 300°/\text{s}$, how long does it take to complete one revolution?

Solution One revolution is 360°. Thus,

$$t = \frac{\alpha}{\omega} = \frac{360 \text{ degrees}}{300 \dfrac{\text{degrees}}{\text{s}}} = 1.2 \text{ s}$$

Since this is one period, we should use the symbol T. Thus, $T = 1.2$ s, as in Figure 15–25.

PRACTICE PROBLEMS 3

If the coil of Figure 15–23 rotates at 3600 rpm, determine its angular velocity, ω, in degrees per second.

Answer
21 600 deg/s

Radian Measure

In practice, ω is usually expressed in radians per second, where radians and degrees are related by the identity

$$2\pi \text{ radians} = 360° \tag{15–8}$$

One radian therefore equals $360°/2\pi = 57.296°$. A full circle, as shown in Figure 15–26(a), can be designated as either 360° or 2π radians. Likewise, the cycle length of a sinusoid, shown in Figure 15–26(b), can be stated as either 360° or 2π radians; a half-cycle as 180° or π radians, and so on.

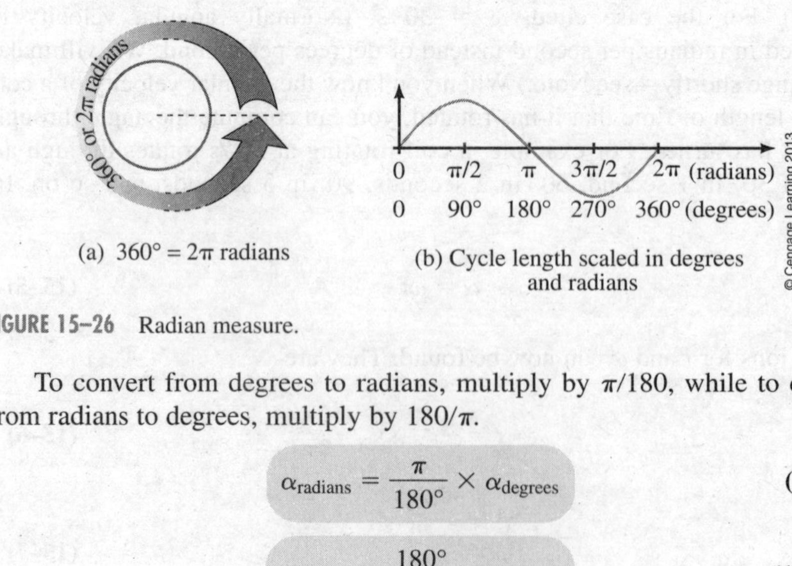

FIGURE 15–26 Radian measure.

To convert from degrees to radians, multiply by $\pi/180$, while to convert from radians to degrees, multiply by $180/\pi$.

$$\alpha_{\text{radians}} = \frac{\pi}{180°} \times \alpha_{\text{degrees}} \tag{15–9}$$

$$\alpha_{\text{degrees}} = \frac{180°}{\pi} \times \alpha_{\text{radians}} \tag{15–10}$$

Table 15–3 shows selected angles in both measures.

Scientific calculators can perform these conversions directly; use yours to confirm the answers of Example 15–7.

TABLE 15–3 Selected Angles in Degrees and Radians

Degrees	Radians
30	$\pi/6$
45	$\pi/4$
60	$\pi/3$
90	$\pi/2$
180	π
270	$3\pi/2$
360	2π

EXAMPLE 15–7

a. Convert 315° to radians.

b. Convert $5\pi/4$ radians to degrees.

Solution

a. $\alpha_{\text{radians}} = (\pi/180°)(315°) = 5.5$ rad

b. $\alpha_{\text{degrees}} = (180°/\pi)(5\pi/4) = 225°$

Graphing Sine Waves

A sinusoidal waveform can be graphed with its horizontal axis scaled in degrees, radians, or time. When scaled in degrees or radians, one cycle is always 360° or 2π radians (Figure 15–27); when scaled in time, it is frequency dependent, since the length of a cycle depends on the coil's velocity of rotation as we saw in Figure 15–8. However, if scaled in terms of period T instead of in seconds, the waveform is also frequency independent, since one cycle is always T, as shown in Figure 15–27(c).

When graphing a sine wave, you don't need many points to get a good sketch: Values every 45° (one-eighth of a cycle) are generally adequate, Table 15–4. Often, you can simply "eyeball" the curve as illustrated next in Example 15–8.

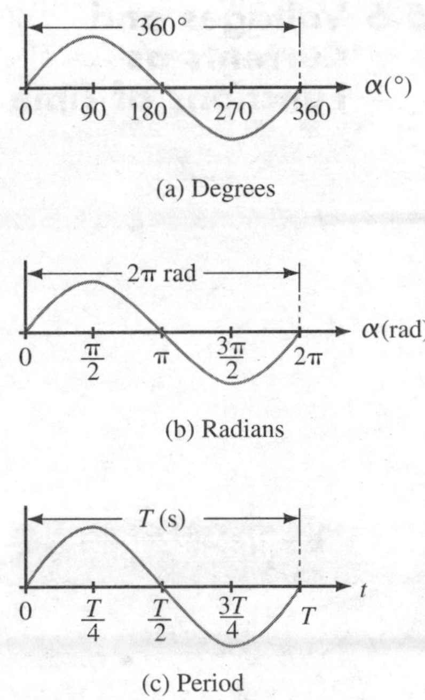

FIGURE 15–27 Comparison of various horizontal scales. Cycle length may be scaled in degrees, radians, or period. Each of these is independent of frequency.

TABLE 15–4 Values for Rapid Sketching

α (deg)	α (rad)	t (T)	Value of sin α
0	0	0	0.0
45	$\pi/4$	$T/8$	0.707
90	$\pi/2$	$T/4$	1.0
135	$3\pi/4$	$3T/8$	0.707
180	π	$T/2$	0.0
225	$5\pi/4$	$5T/8$	−0.707
270	$3\pi/2$	$3T/4$	−1.0
315	$7\pi/4$	$7T/8$	−0.707
360	2π	T	0.0

EXAMPLE 15–8

Sketch the waveform for a 25-kHz sinusoidal current that has an amplitude of 4 mA. Scale the axis in seconds.

Solution For this waveform, $T = 1/25$ kHz $= 40$ μs. Thus,

1. Lay out the time axis with the end of the cycle marked as 40 μs, the half-cycle point as 20 μs, the quarter-cycle point as 10 μs, and so on (Figure 15–28).
2. The peak value (i.e., 4 mA) occurs at the quarter-cycle point, which is 10 μs on the waveform. Likewise, −4 mA occurs at 30 μs. Now sketch.
3. Values at other time points can be determined easily if needed. For example, the value at 5 μs can be calculated by noting that 5 μs is one-eighth of a cycle, or 45°. Thus, $i = 4 \sin 45°$ mA $= 2.83$ mA. Alternately, from Table 15–4, at $T/8$, $i = (4\text{ mA})(0.707) = 2.83$ mA. As many points as you need can be computed and plotted in this manner.

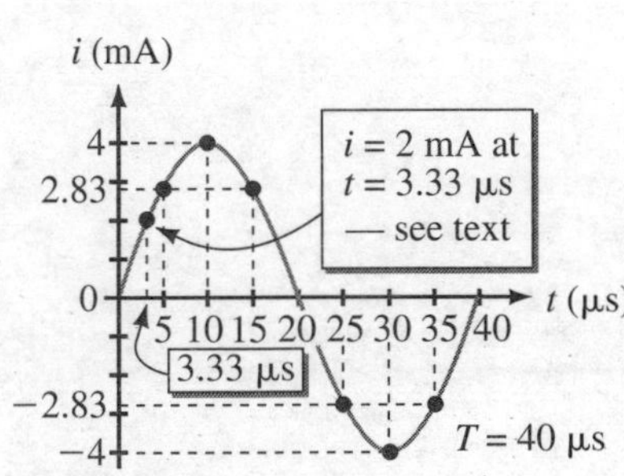

FIGURE 15–28

4. Values at particular angles can also be located easily. For instance, if you want a value at 30°, the required value is $i = 4 \sin 30°$ mA $= 2.0$ mA. To locate this point on the graph, note that 30° is one-twelfth of a cycle or $T/12 = (40\ \mu s)/12 = 3.33\ \mu s$. The point is shown on Figure 15–28.

Note that what we have done here is probably overkill, as you seldom need this much detail—it is often adequate to "eyeball" the curve as in Steps 1 and 2.

15.6 Voltages and Currents as Functions of Time

Relationship between ω, T, and f

Earlier you learned that one cycle of a sine wave may be represented as either $\alpha = 2\pi$ rads or $t = T$ s, Figure 15–27. Substituting these into $\alpha = \omega t$ (Equation 15–5), you get $2\pi = \omega T$. Transposing yields

$$\omega T = 2\pi \text{ (rad)} \quad \textbf{(15–11)}$$

Thus,

$$\omega = \frac{2\pi}{T} \text{ (rad/s)} \quad \textbf{(15–12)}$$

Recall that, $f = 1/T$ Hz. Substituting this into Equation 15–12 you get

$$\omega = 2\pi f \text{ (rad/s)} \quad \textbf{(15–13)}$$

Note that ω is referred to as **radian frequency.**

EXAMPLE 15–9

In some parts of the world, the power system frequency is 60 Hz; in other parts, it is 50 Hz. Determine ω for each.

Solution For 60 Hz, $\omega = 2\pi f = 2\pi(60) = 377$ rad/s. For 50 Hz, $\omega = 2\pi f = 2\pi(50) = 314.2$ rad/s.

PRACTICE PROBLEMS 4

1. If $\omega = 240$ rad/s, what are T and f? How many cycles occur in 27 s?
2. If 56 000 cycles occur in 3.5 s, what is ω?

Answers
1. 26.18 ms, 38.2 Hz, 1031 cycles
2. 100.5×10^3 rad/s

Sinusoidal Voltages and Currents as Functions of Time

Recall from Equation 15–4, $e = E_m \sin \alpha$, and from Equation 15–5, $\alpha = \omega t$. Combining these equations yields

$$e = E_m \sin \omega t \quad \textbf{(15–14a)}$$

Similarly,

$$v = V_m \sin \omega t \quad \textbf{(15–14b)}$$

$$i = I_m \sin \omega t \quad \textbf{(15–14c)}$$

EXAMPLE 15–10

A 100-Hz sinusoidal voltage source has an amplitude of 150 volts. Write the equation for e as a function of time.

Solution $\omega = 2\pi f = 2\pi(100) = 628$ rad/s and $E_m = 150$ V. Thus, $e = E_m \sin \omega t = 150 \sin 628t$ V.

Equations 15–14 may be used to compute voltages or currents at any instant in time. Usually, ω is in radians per second, and thus ωt is in radians. You can work directly in radians or you can convert to degrees. For example, suppose you want to know the voltage at $t = 1.25$ ms for $e = 150 \sin 628t$ V.

Working in Rads. With your calculator in the RAD mode, $e = 150 \sin(628)(1.25 \times 10^{-3}) = 150 \sin 0.785 \text{ rad} = 106$ V.

CircuitSim 15-8

Working in Degrees. 0.785 rad = 45°. Thus, $e = 150 \sin 45° = 106$ V as before.

EXAMPLE 15–11

For $v = 170 \sin 2450t$, determine v at $t = 3.65$ ms and show the point on the v waveform.

Solution $\omega = 2450$ rad/s. Therefore, $\omega t = (2450)(3.65 \times 10^{-3}) = 8.943$ rad = 512.4°. Thus, $v = 170 \sin 512.4° = 78.8$ V. Alternatively, $v = 170 \sin 8.943$ rad = 78.8 V. The point is plotted on the waveform in Figure 15–29.

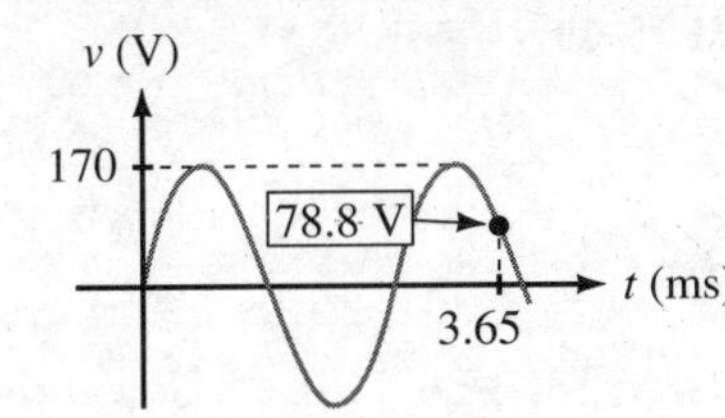

FIGURE 15–29

PRACTICE PROBLEMS 5

A sinusoidal current has a peak amplitude of 10 amps and a period of 120 ms.

a. Determine its equation as a function of time using Equation 15–14c.
b. Using this equation, compute a table of values at 10-ms intervals and plot one cycle of the waveform scaled in seconds.
c. Sketch one cycle of the waveform using the procedure of Example 15–8. (Note how much less work this is.)

Answers

a. $i = 10 \sin 52.36t$ A

c. Mark the end of the cycle as 120 ms, ½ cycle as 60 ms, ¼ cycle as 30 ms, and so on. Draw the sine wave so that it is zero at $t = 0$, 10 A at 30 ms, 0 A at 60 ms, −10 A at 90 ms, and ends at $t = 120$ ms. (See Figure 15–30.)

Determining When a Particular Value Occurs

Sometimes you need to know when a particular value of voltage or current occurs. Given $v = V_m \sin \alpha$, rewrite this as $\sin \alpha = v/V_m$. Then,

$$\alpha = \sin^{-1} \frac{v}{V_m} \qquad (15\text{–}15)$$

Compute the angle α at which the desired value occurs using the inverse sine function of your calculator, then determine the time from

$$t = \alpha/\omega$$

EXAMPLE 15–12

A sinusoidal current has an amplitude of 10 A and a period of 0.120 s. Determine the times at which

a. $i = 5.0$ A,
b. $i = -5$ A.

Solution

a. Consider Figure 15–30. As you can see, there are two points on the waveform where $i = 5$ A. Let these be denoted t_1 and t_2, respectively. First, determine ω:

$$\omega = \frac{2\pi}{T} = \frac{2\pi}{0.120\text{ s}} = 52.36 \text{ rad/s}$$

Let $i = 10 \sin \alpha$ A. Now, find the angle α_1 at which $i = 5$ A:

$$\alpha_1 = \sin^{-1}\frac{i}{I_m} = \sin^{-1}\frac{5\text{ A}}{10\text{ A}} = \sin^{-1}0.5 = 30° = 0.5236 \text{ rad}$$

Thus, $t_1 = \alpha_1/\omega = (0.5236 \text{ rad})/(52.36 \text{ rad/s}) = 0.01 \text{ s} = 10$ ms. This is indicated in Figure 15–30. Now consider t_2. Because of symmetry, t_2 is the same distance back from the half-cycle point as t_1 is in from the beginning of the cycle. Thus, $t_2 = 60 \text{ ms} - 10 \text{ ms} = 50$ ms.

b. Similarly, t_3 (the first point at which $i = -5$ A occurs) is 10 ms past midpoint, while t_4 is 10 ms back from the end of the cycle. Thus, $t_3 = 70$ ms and $t_4 = 110$ ms.

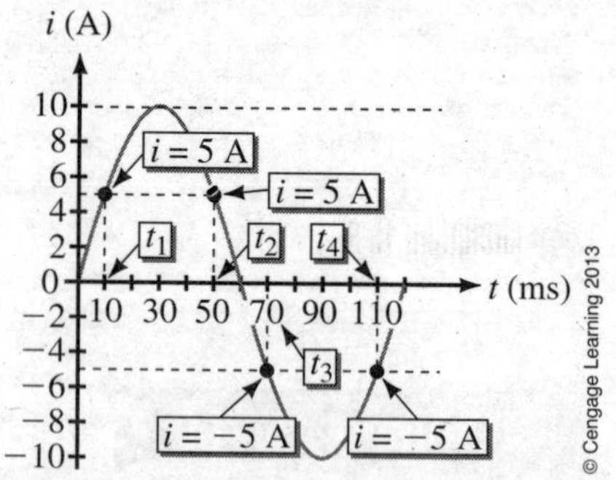

FIGURE 15–30

PRACTICE PROBLEMS 6

Given $v = 10 \sin 52.36t$, determine both occurrences of $v = -8.66$ V.

Answer
80 ms, 100 ms

Voltages and Currents with Phase Shifts

CircuitSim 15-9

If a sine wave does not pass through zero at $t = 0$ s as in Figure 15–30, it has a **phase shift.** Waveforms may be shifted to the left or to the right (see Figure 15–31). For a waveform shifted left as in (a),

$$v = V_m \sin(\omega t + \theta) \tag{15–16a}$$

while, for a waveform shifted right as in (b),

$$v = V_m \sin(\omega t - \theta) \tag{15–16b}$$

NOTES...

With equations such as 15–16(a) and (b), it is customary to express ωt in radians and θ in degrees, yielding mixed angular units (as indicated in the following examples). Although this is acceptable when the equations are written in symbolic form, you must convert both angles to the same unit (either degrees or radians) before you make numerical computations. This is illustrated in Example 15–14.

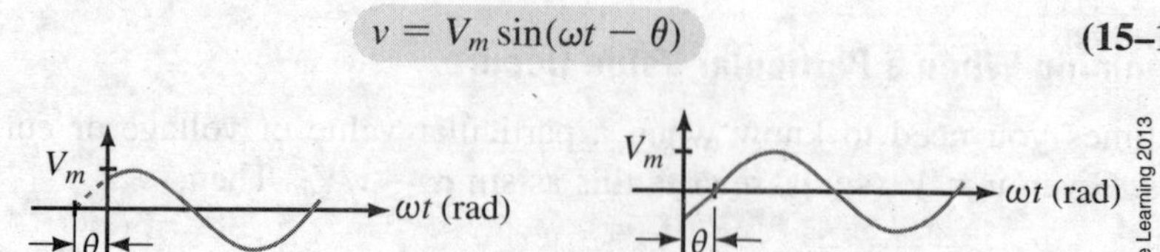

(a) $v = V_m \sin(\omega t + \theta)$ (b) $v = V_m \sin(\omega t - \theta)$

FIGURE 15–31 Waveforms with phase shifts. Angle θ is normally measured in degrees, yielding mixed angular units. (See Notes.)

EXAMPLE 15–13

Demonstrate that $v = 20 \sin(\omega t - 60°)$, where $\omega = \pi/6$ rad/s (i.e, = 30°/s), yields the shifted waveform shown in Figure 15–32.

Solution

1. Since ωt and 60° are both angles, $(\omega t - 60°)$ is also an angle. Let us define it as x. Then $v = 20 \sin x$, which means that the shifted wave is also sinusoidal.
2. Consider $v = \sin(\omega t - 60°)$. At $t = 0$ s, $v = 20 \sin(0 - 60°) = 20 \sin(-60°) = -17.3$ V as indicated in Figure 15–32.
3. Since $\omega = 30°/\text{s}$, it takes 2 s for ωt to reach 60°. Thus, at $t = 2$ s, $v = 20 \sin(60° - 60°) = 20 \sin 0° = 0$ V, and the waveform passes through zero at $t = 2$ s as indicated.

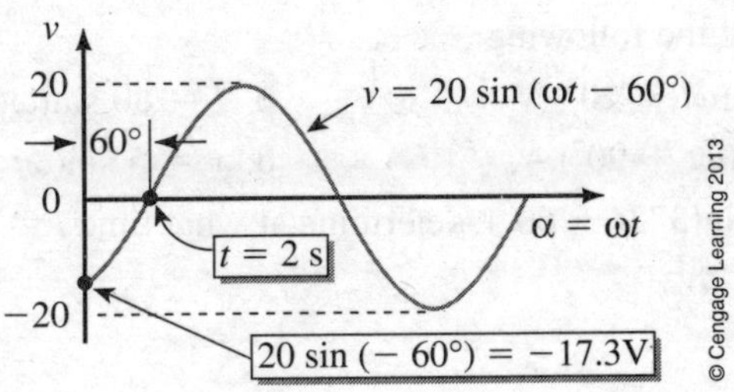

FIGURE 15–32

CircuitSim 15-10

As you can see, this example confirms that $\sin(\omega t - \theta)$ describes the wave-shape of Figure 15–31(b).

EXAMPLE 15–14

a. Determine the equation for the waveform of Figure 15–33(a), given $f = 60$ Hz. Compute current at $t = 4$ ms.

b. Repeat (a) for Figure 15–33(b).

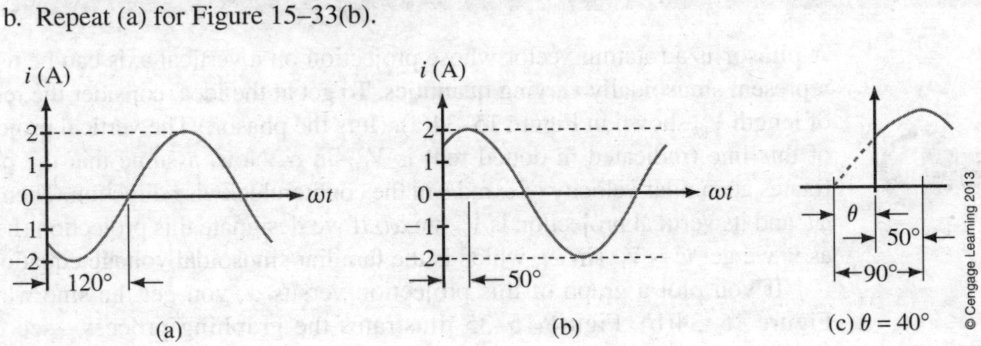

FIGURE 15–33

Solution

a. $I_m = 2$ A and $\omega = 2\pi(60) = 377$ rad/s. This waveform corresponds to Figure 15–31(b) with $\theta = 120°$. Therefore,

$$i = I_m \sin(\omega t - \theta) = 2 \sin(377t - 120°) \text{ A}$$

At $t = 4$ ms, current is

$$\begin{aligned} i &= 2 \sin(377 \times 4 \text{ ms} - 120°) = 2 \sin(1.508 \text{ rad} - 120°) \\ &= 2 \sin(86.4° - 120°) = 2 \sin(-33.64°) = -1.11 \text{ A.} \end{aligned}$$

b. This waveform matches Figure 15–31(a) as you can see if you extend the waveform back 90° from its peak as in (c). Note that $\theta = 40°$. Thus,

$$i = 2 \sin(377t + 40°) \text{ A}$$

At $t = 4$ ms, current is

$$i = 2 \sin(377 \times 4 \text{ ms} + 40°) = 2 \sin(126.4°)$$
$$= 1.61 \text{ A.}$$

PRACTICE PROBLEMS 7

CircuitSim 15-11

1. Given $i = 2 \sin(377t + 60°)$, compute the current at $t = 3$ ms.
2. Sketch each of the following:
   a. $v = 10 \sin(\omega t + 20°)$ V. b. $i = 80 \sin(\omega t - 50°)$ A.
   c. $i = 50 \sin(\omega t + 90°)$ A. d. $v = 5 \sin(\omega t + 180°)$ V.
3. Given $i = 2 \sin(377t + 60°)$, determine at what time $i = 1.8$ A.

Answers
1. 1.64 A
2. a. Same as Figure 15–31(a) with $V_m = 10$ V, $\theta = 20°$.
   b. Same as Figure 15–31(b) with $I_m = 80$ A, $\theta = 50°$.
   c. Same as Figure 15–39(b) except use $I_m = 50$ A instead of V_m.
   d. A negative sine wave with magnitude of 5 V.
3. 0.193 ms

Probably the easiest way to deal with shifted waveforms is to use phasors. We introduce the idea next.

15.7 Introduction to Phasors

A **phasor** is a rotating vector whose projection on a vertical axis can be used to represent sinusoidally varying quantities. To get at the idea, consider the red line of length V_m shown in Figure 15–34(a). (It is the phasor.) The vertical projection of this line (indicated in dotted red) is $V_m \sin \alpha$. Now, assume that the phasor rotates at angular velocity of ω rad/s in the counterclockwise direction. Then, $\alpha = \omega t$, and its vertical projection is $V_m \sin \omega t$. If we designate this projection (height) as v, we get $v = V_m \sin \omega t$, which is the familiar sinusoidal voltage equation.

If you plot a graph of this projection versus α, you get the sine wave of Figure 15–34(b). Figure 15–35 illustrates the graphing process—see Note.

NOTES...

There are some good animations on the Internet that show the process described here. Search for *Phasor Animation.* (Some animations are better than others. In addition, look for an animation that shows the generation of a sine wave, not a cosine wave.)

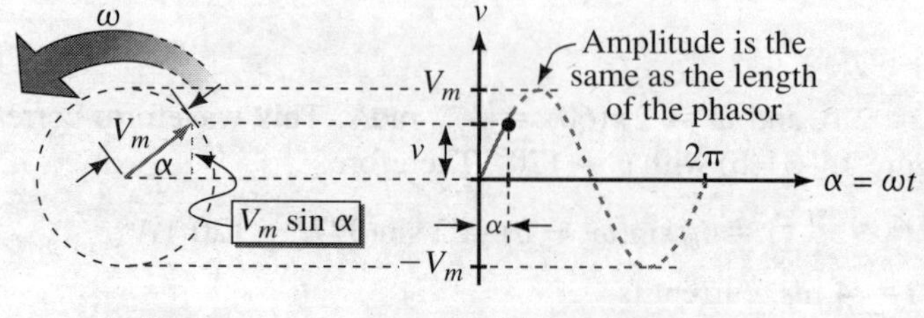

FIGURE 15–34 As the phasor rotates about the origin, its vertical projection creates a sine wave. (Figure 15–35 illustrates the process.)

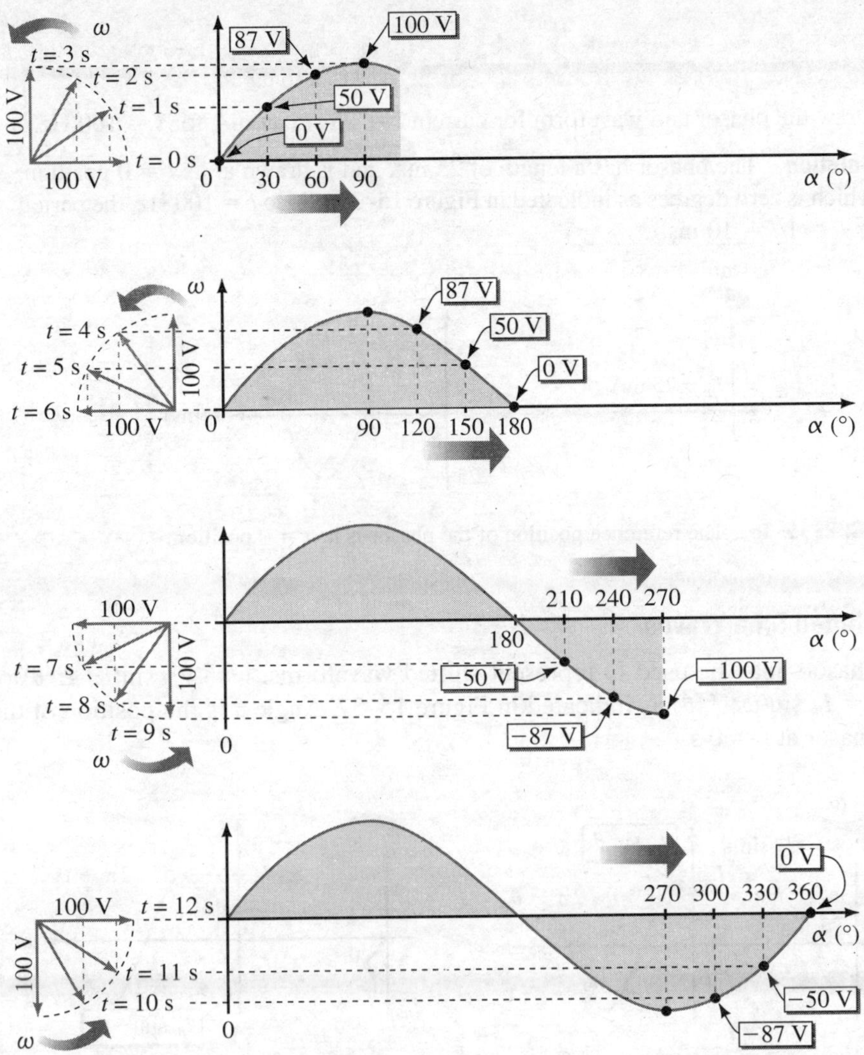

FIGURE 15–35 Evolution of the sine wave of Figure 15–34.

NOTES...

1. Although we have indicated phasor rotation in Figure 15–35 by a series of "snapshots," this is too cumbersome; in practice, we show only the phasor at its $t = 0$ s (reference) position and imply rotation rather than show it explicitly.
2. Although we are using maximum values (E_m and I_m) here, phasors are normally drawn in terms of effective (rms) values (considered in Section 15.9). For the moment, we will continue to use maximum values. We make the change to rms in Chapter 16.
3. Phasors apply only to sinusoidal waveforms.
4. In this discussion (for the sake of familiarity), we specify ω in degrees per second. Generally, however, you will use radians per second. We will make this change shortly.

It shows snapshots of the phasor and the evolving waveform at various instants of time for a phasor of magnitude $V_m = 100$ V rotating at $\omega = 30°$/s. For example, compute and plot voltage values at times $t = 0$, 1, 2, and 3 s:

1. At $t = 0$ s, $\alpha = 0$, the phasor is at its 0° position, and its vertical projection is $v = V_m \sin \omega t = 100 \sin 0° = 0$ V. The point is thus, at the origin.
2. At $t = 1$ s, the phasor has rotated 30° and its vertical projection is $v = 100 \sin 30° = 50$ V. This point is plotted at $\alpha = 30°$ on the horizontal axis.
3. At $t = 2$ s, $\alpha = 60°$, and $v = 100 \sin 60° = 87$ V, which is plotted at $\alpha = 60°$ on the horizontal axis. Similarly, at $t = 3$ s, $\alpha = 90°$, and $v = 100$ V. Continuing in this manner, the complete waveform is evolved.

From the foregoing, we conclude that *a sinusoidal waveform can be created by plotting the vertical projection of a phasor that rotates in the counterclockwise direction at constant angular velocity* ω. *If the phasor has a length of* V_m, *the waveform represents voltage; if the phasor has a length of* I_m, it represents current. Note carefully: **Phasors apply only to sinusoidal waveforms.**

EXAMPLE 15–15

Draw the phasor and waveform for current $i = 25 \sin \omega t$ mA for $f = 100$ Hz.

Solution The phasor has a length of 25 mA and is drawn at its $t = 0$ position, which is zero degrees as indicated in Figure 15–36. Since $f = 100$ Hz, the period is $T = 1/f = 10$ ms.

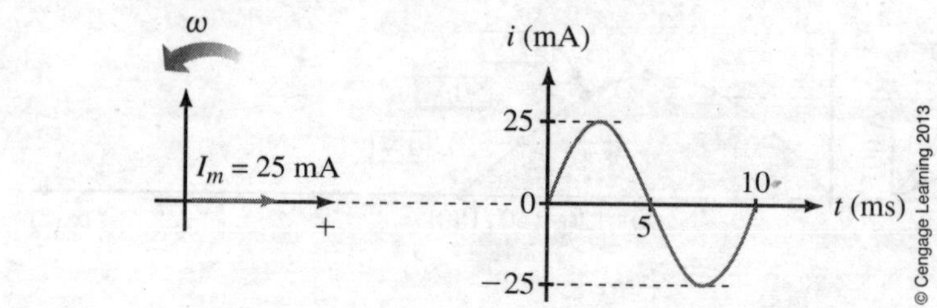

FIGURE 15–36 The reference position of the phasor is its $t = 0$ position.

Shifted Sine Waves

Phasors may be used to represent shifted waveforms, $v = V_m \sin(\omega t \pm \theta)$ or $i = I_m \sin(\omega t \pm \theta)$ as indicated in Figure 15–37. Angle θ is the position of the phasor at $t = 0$ s.

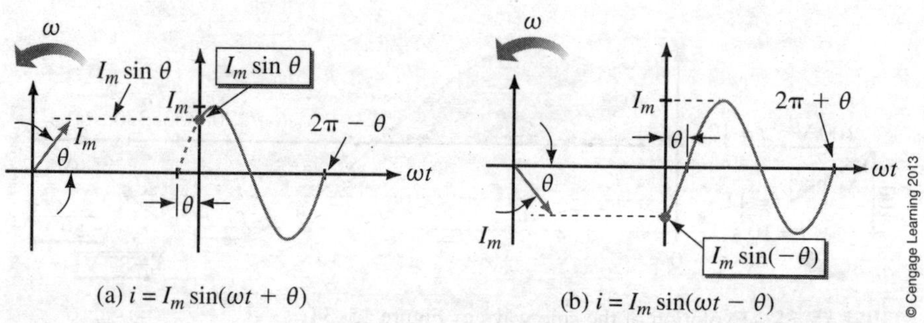

FIGURE 15–37 Phasors for shifted waveforms. Angle θ is the position of the phasor at $t = 0$ s.

EXAMPLE 15–16

Consider $v = 20 \sin(\omega t - 60°)$, where $\omega = \pi/6$ rad/s (i.e., 30°/s). Show that the phasor of Figure 15–38(a) represents this waveform.

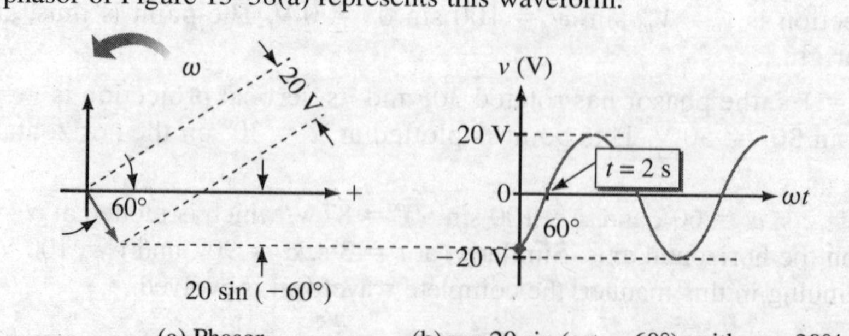

FIGURE 15–38

Solution The phasor has length 20 V and at time $t = 0$ is at –60° as indicated in (a). Now, as the phasor rotates, it generates a sinusoidal waveform, oscillating between ±20 V as indicated in (b). Note that the zero crossover point occurs at $t = 2$ s, since it takes 2 seconds for the phasor to rotate from –60° to 0° at 30 degrees per second. Now compare the waveform of (b) to the waveform of Figure 15–32, Example 15–13. They are identical. Thus, the phasor of (a) represents the shifted waveform $v = 20 \sin(\omega t - 60°)$.

EXAMPLE 15–17

With the aid of a phasor, sketch the waveform for $v = V_m \sin(\omega t + 90°)$.

Solution Place the phasor at 90° as in Figure 15–39(a). Note that the resultant waveform (b) is a cosine waveform, that is, $v = V_m \cos \omega t$. From this, we conclude that

$$\sin(\omega t + 90°) = \cos \omega t$$

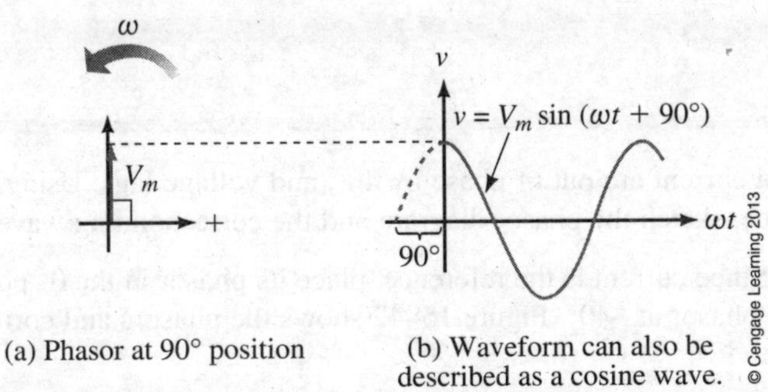

FIGURE 15–39 Demonstrating that $\sin(\omega t + 90°) = \cos \omega t$.

PRACTICE PROBLEMS 8

With the aid of phasors, show that

a. $\sin(\omega t - 90°) = -\cos \omega t$,
b. $\sin(\omega t \pm 180°) = -\sin \omega t$

Phase Difference

Phase difference refers to the angular displacement between different waveforms of the same frequency. Consider Figure 15–40. If the angular displacement is 0° as in (a), the waveforms are said to be in phase; otherwise, they are out of phase. When describing a phase difference, select one waveform as reference. Other waveforms then lead, lag, or are in phase with this reference. For example, in (b), for reasons to be discussed in the next paragraph, the current waveform is said to lead the voltage waveform, while in (c) the current waveform is said to lag.

CircuitSim 15-12

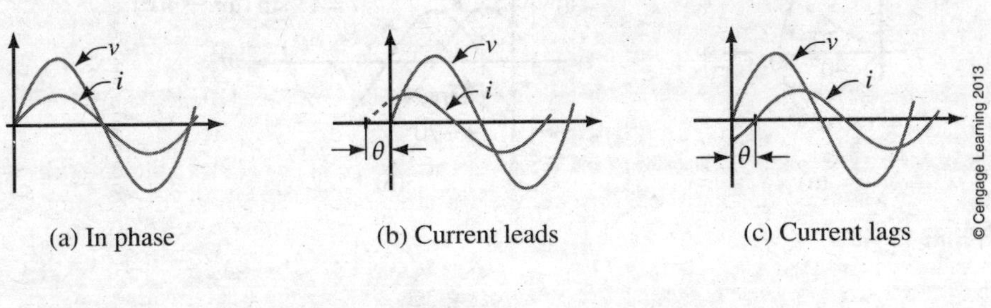

FIGURE 15–40 Illustrating phase difference. In these examples, voltage is taken as reference.

> **NOTES...**
>
> 1. To determine which waveform leads and which lags, make a quick sketch of their phasors, and the answer will be apparent. Note also that the terms *lead* and *lag* are relative. In Figure 15–41, we said that current leads voltage; you can just as correctly say that voltage lags current.
> 2. When you plot two phasors as in Figure 15–43(a), the angle between them is their phase difference.

The terms **lead** and **lag** can be understood in terms of phasors. If you observe phasors rotating as in Figure 15–41(a), the one that you see passing first is leading and the other is lagging. By definition, *the waveform generated by the leading phasor leads the waveform generated by the lagging phasor and vice versa.* In Figure 15–41, phasor I_m leads phasor V_m; thus current $i(t)$ leads voltage $v(t)$.

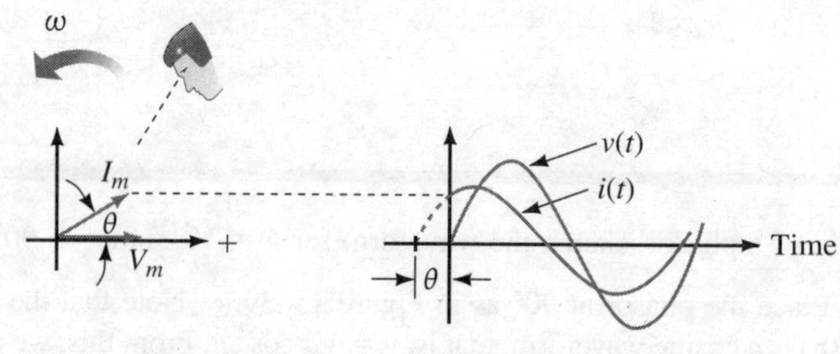

(a) I_m leads V_m (b) Therefore, $i(t)$ leads $v(t)$

FIGURE 15–41 Defining lead and lag.

EXAMPLE 15–18

Voltage and current are out of phase by 40°, and voltage lags. Using current as the reference, sketch the phasor diagram and the corresponding waveforms.

Solution Since current is the reference, place its phasor in the 0° position and the voltage phasor at –40°. Figure 15–42 shows the phasors and corresponding waveforms.

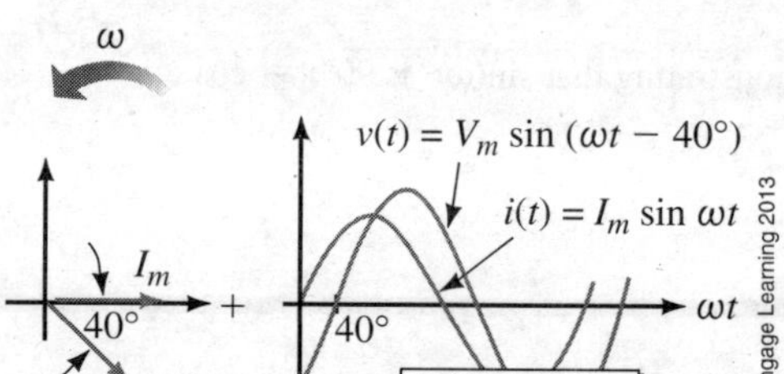

FIGURE 15–42

EXAMPLE 15–19

> **NOTES...**
>
> In all phasor diagrams to this point, we have used a stylized arrow to imply phasor rotation. In practice, however, this is not usually done—thus, from this point onward, we will omit it.

Given $v = 20 \sin(\omega t + 30°)$ and $i = 18 \sin(\omega t - 40°)$, draw the phasor diagram, determine phase relationships, and sketch the waveforms—see Note.

Solution The phasors are shown in Figure 15–43(a). From these, you can see that v leads i by 70°. The waveforms are shown in (b).

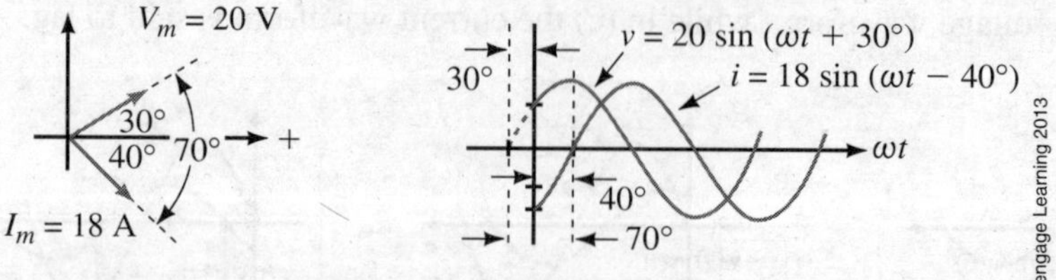

(a) (b)

CircuitSim 15-13

FIGURE 15–43

EXAMPLE 15–20

Figure 15–44 shows a pair of waveforms v_1 and v_2 on an oscilloscope. Each major vertical division represents 20 V and each major division on the horizontal (time) scale represents 20 μs. Voltage v_1 leads. Prepare a phasor diagram using v_1 as reference. Determine equations for both voltages.

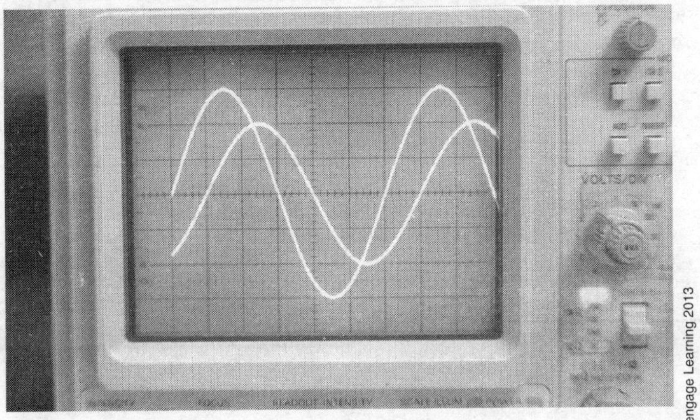

(a)

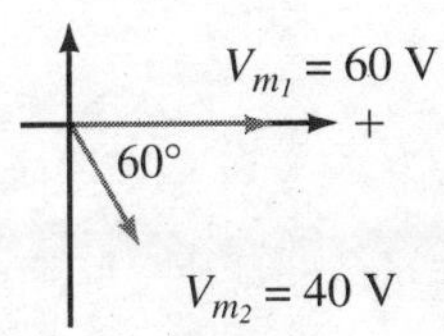

(b)

FIGURE 15–44

Solution From the photograph, the magnitude of v_1 is $V_{m_1} = 3$ div $\times$ 20 V/div = 60 V. Similarly, $V_{m_2} = 40$ V. Cycle length is $T = 6 \times 20$ μs = 120 μs, and the displacement between waveforms is 20 μs, which is ⅙ of a cycle (i.e., 60°). Selecting v_1 as reference and noting that v_2 lags yields the phasors shown in (b). Angular frequency $\omega = 2\pi/T = 2\pi/(120 \times 10^{-6}\text{ s}) = 52.36 \times 10^3$ rad/s. Thus, $v_1 = V_{m_1} \sin \omega t = 60 \sin(52.36 \times 10^3 t)$ V and $v_2 = 40 \sin(52.36 \times 10^3 t - 60°)$ V.

Sometimes voltages and currents are expressed in terms of cos ωt rather than sin ωt. As Example 15–17 shows, a cosine wave is a sine wave shifted by +90°, or alternatively, a sine wave is a cosine wave shifted by −90°. For sines or cosines with an angle, the following formulas apply.

$$\cos(\omega t + \theta) = \sin(\omega t + \theta + 90°) \quad \textbf{(15–17a)}$$

$$\sin(\omega t + \theta) = \cos(\omega t + \theta - 90°) \quad \textbf{(15–17b)}$$

To illustrate, consider $\cos(\omega t + 30°)$. From Equation 15–17a, $\cos(\omega t + 30°) = \sin(\omega t + 30° + 90°) = \sin(\omega t + 120°)$. Figure 15–45 illustrates this relationship graphically. The red phasor in (a) generates cos ωt as was shown in Example 15–17. Therefore, the green phasor generates a waveform that leads it by 30°, namely $\cos(\omega t + 30°)$. For (b), the red phasor generates sin ωt, and the green phasor generates a waveform that leads it by 120°, that is, $\sin(\omega t + 120°)$. Since the green phasor is the same in both cases, you can see that $\cos(\omega t + 30°) = \sin(\omega t + 120°)$. You may find this process easier to apply than trying to remember Equations 15–17(a) and (b).

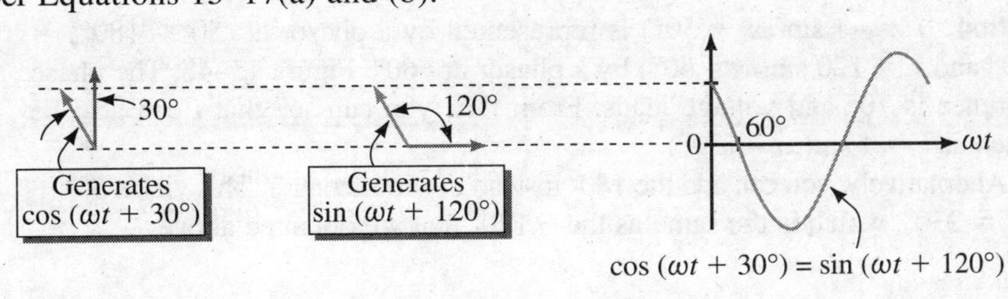

FIGURE 15–45 Using phasors to show that $\cos(\omega t + 30°) = \sin(\omega t + 120°)$.

EXAMPLE 15–21

Determine the phase angle between $v = 30 \cos(\omega t + 20°)$ and $i = 25 \sin(\omega t + 70°)$.

Solution $i = 25 \sin(\omega t + 70°)$ may be represented by a phasor at 70°, and $v = 30 \cos(\omega t + 20°)$ by a phasor at $(90° + 20°) = 110°$, Figure 15–46(a). Thus, v leads i by 40°. Waveforms are shown in (b).

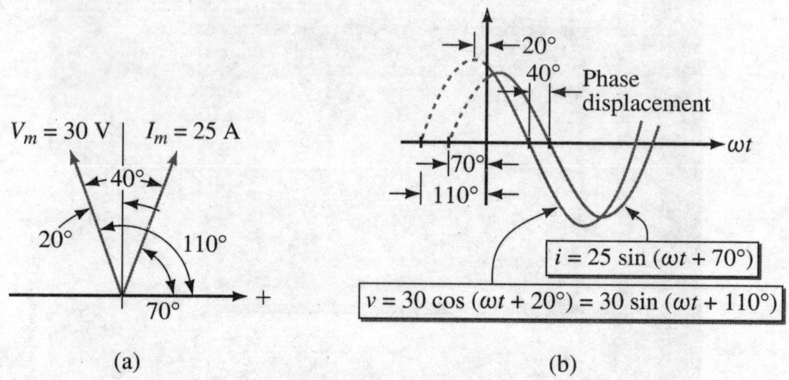

FIGURE 15–46

Sometimes you encounter negative waveforms such as $i = -I_m \sin \omega t$. To see how to handle these, refer back to Figure 15–36, which shows the waveform and phasor for $i = I_m \sin \omega t$. If you multiply this waveform by -1, you get the inverted waveform $-I_m \sin \omega t$ of Figure 15–47(a) with corresponding phasor (b). Note that the phasor is the same as the original phasor except that it is rotated by 180°. This is always true—thus, if you multiply a waveform by -1, the phasor for the new waveform is 180° rotated from the original phasor, regardless of the angle of the original phasor.

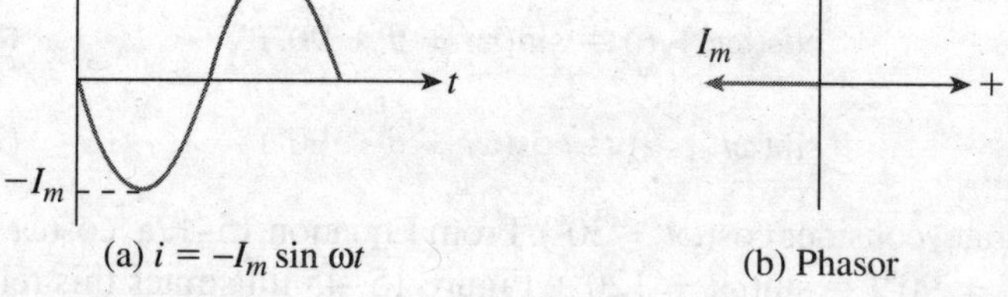

FIGURE 15–47 For a negative sine wave, the phasor is at 180°.

EXAMPLE 15–22

Find the phase relationship between $i = -4 \sin(\omega t + 50°)$ and $v = 120 \sin(\omega t - 60°)$.

Solution $i = -4 \sin(\omega t + 50°)$ is represented by a phasor at $(50° - 180°) = -130°$ and $v = 120 \sin(\omega t - 60°)$ by a phasor at $-60°$, Figure 15–48. The phase difference is 70° and voltage leads. From this, you can see that i can also be written as $i = 4 \sin(\omega t - 130°)$.

Alternatively, you can add the 180° instead of subtracting it. This yields $50° + 180° = 230°$, which is the same as the $-130°$ that we obtained above.

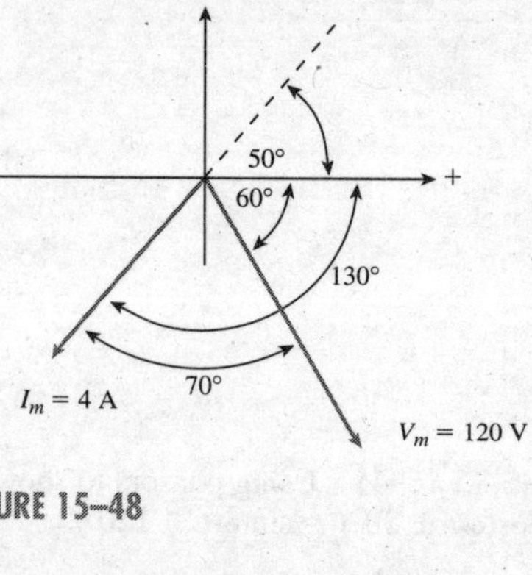

FIGURE 15–48

The importance of phasors to ac circuit analysis cannot be overstated—you will find that they are one of your main tools for representing ideas and for solving problems in later chapters and in practice. (By the time you reach Section 18.2 of Chapter 18, analysis is done exclusively with phasors.) We will leave them for the moment, but pick them up again in Chapter 16.

IN-PROCESS LEARNING CHECK 2

(Answers are at the end of the chapter.)

1. If $i = 15 \sin \alpha$ mA, compute the current at $\alpha = 0°, 45°, 90°, 135°, 180°, 225°, 270°, 315°$, and $360°$.
2. Convert the following angles to radians:
   a. 20° c. 120°
   b. 50° d. 250°
3. If a coil rotates at $\omega = \pi/60$ radians per millisecond, how many degrees does it rotate through in 10 ms? In 40 ms? In 150 ms?
4. A current has an amplitude of 50 mA and $\omega = 0.2\pi$ rad/s. Sketch the waveform with the horizontal axis scaled in
   a. degrees b. radians c. seconds
5. If 2400 cycles of a waveform occur in 10 ms, what is ω in radians per second?
6. A sinusoidal current has a period of 40 ms and an amplitude of 8 A. Write its equation in the form of $i = I_m \sin \omega t$, with numerical values for I_m and ω.
7. A current $i = I_m \sin \omega t$ has a period of 90 ms. If $i = 3$ A at $t = 7.5$ ms, what is its equation?
8. Write equations for each of the waveforms in Figure 15–49 with the phase angle θ expressed in degrees and ω in rad/s.
9. Given $i = 10 \sin \omega t$, where $f = 50$ Hz, find all occurrences of
   a. $i = 8$ A between $t = 0$ and $t = 40$ ms
   b. $i = -5$ A between $t = 0$ and $t = 40$ ms
10. Sketch the following waveforms with the horizontal axis scaled in degrees:
    a. $v_1 = 80 \sin(\omega t + 45°)$ V c. $i_1 = 10 \cos \omega t$ mA
    b. $v_2 = 40 \sin(\omega t - 80°)$ V d. $i_2 = 5 \cos(\omega t - 20°)$ mA
11. Given $\omega = \pi/3$ rad/s, determine when voltage first crosses through 0 for
    a. $v_1 = 80 \sin(\omega t + 45°)$ V b. $v_2 = 40 \sin(\omega t - 80°)$ V
12. Consider the voltages of Question 10:
    a. Sketch phasors for v_1 and v_2.
    b. What is the phase difference between v_1 and v_2?
    c. Determine which voltage leads and which lags.
13. Repeat Question 12 for the currents of Question 10.

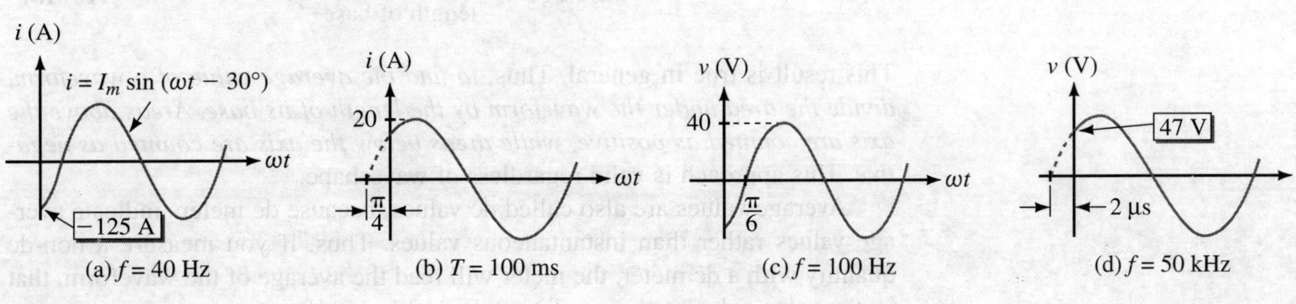

FIGURE 15–49

15.8 ac Waveforms and Average Value

While we can describe ac quantities in terms of frequency, period, instantaneous value, and other attributes, we do not yet have any way to give a meaningful value to an ac current or voltage in the same sense that we can say of a car battery that it has a voltage of 12 volts. This is because ac quantities constantly change, and thus there is no single numerical value that truly represents a waveform over its complete cycle. For this reason, ac quantities are generally described by a group of characteristics, including instantaneous, peak, average, and effective values. The first two of these we have already seen. In this section, we look at **average values;** in Section 15.9, we consider effective values.

Average Values

Many quantities are measured by their average, for instance, test and examination scores. To find the average of a set of marks for example, you add them, then divide by the number of items summed. For waveforms, the process is conceptually the same. For example, to find the average of a waveform, you can sum the instantaneous values over a full cycle, then divide by the number of points used. The trouble with this approach is that waveforms do not consist of discrete values.

Average in Terms of the Area under a Curve

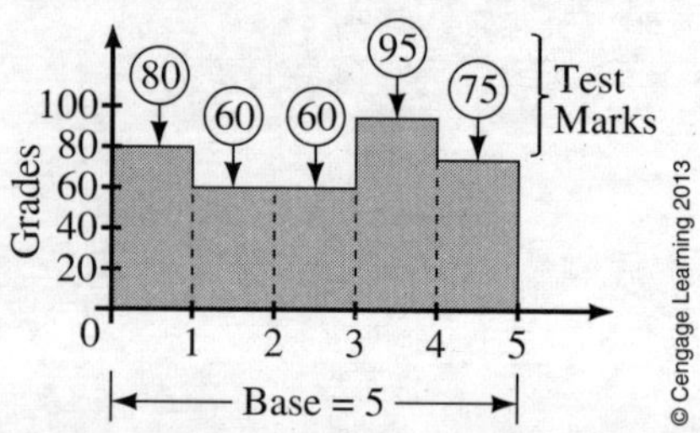

FIGURE 15–50 Determining average by area.

An approach more suitable for use with waveforms is to find the area under the curve, then divide by the baseline of the curve. To get at the idea, we can use an analogy. Consider again the technique of computing the average for a set of numbers. Assume that you earn marks of 80, 60, 60, 95, and 75 on a group of tests. Your average mark is therefore

$$\text{average} = (80 + 60 + 60 + 95 + 75)/5 = 74$$

An alternate way to view these marks is graphically as in Figure 15–50. The area under this curve can be computed as

$$\text{area} = (80 \times 1) + (60 \times 2) + (95 \times 1) + (75 \times 1)$$

Now divide this by the length of the base, namely 5. Thus,

$$\frac{(80 \times 1) + (60 \times 2) + (95 \times 1) + (75 \times 1)}{5} = 74$$

which is exactly the answer obtained above. That is,

$$\text{average} = \frac{\text{area under curve}}{\text{length of base}} \qquad \textbf{(15–18)}$$

This result is true in general. Thus, *to find the average value of a waveform, divide the area under the waveform by the length of its base. Areas above the axis are counted as positive, while areas below the axis are counted as negative.* This approach is valid regardless of waveshape.

Average values are also called dc values, because dc meters indicate average values rather than instantaneous values. Thus, if you measure a non-dc quantity with a dc meter, the meter will read the average of the waveform, that is, the value calculated according to Equation 15–18.

EXAMPLE 15–23

a. Compute the average for the current waveform of Figure 15–51.
b. If the negative portion of Figure 15–51 is −3 A instead of −1.5 A, what is the average?
c. If the current is measured by a dc ammeter, what will the ammeter indicate for each case?

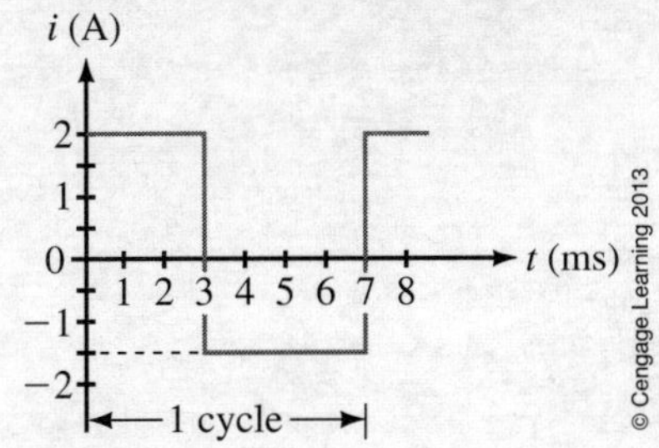

FIGURE 15–51

Solution

a. The waveform repeats itself after 7 ms. Thus, $T = 7$ ms and the average is

$$I_{avg} = \frac{(2\text{ A} \times 3\text{ ms}) - (1.5\text{ A} \times 4\text{ ms})}{7\text{ ms}} = \frac{6 - 6}{7} = 0\text{ A}$$

b. $I_{avg} = \dfrac{(2\text{ A} \times 3\text{ ms}) - (3\text{ A} \times 4\text{ ms})}{7\text{ ms}} = \dfrac{-6\text{ A}}{7} = -0.857\text{ A}$

c. A dc ammeter measuring (a) will indicate zero, while for (b) it will indicate −0.857 A.

EXAMPLE 15–24

Compute the average value for the waveforms of Figures 15–52(a) and (c). Sketch the averages for each.

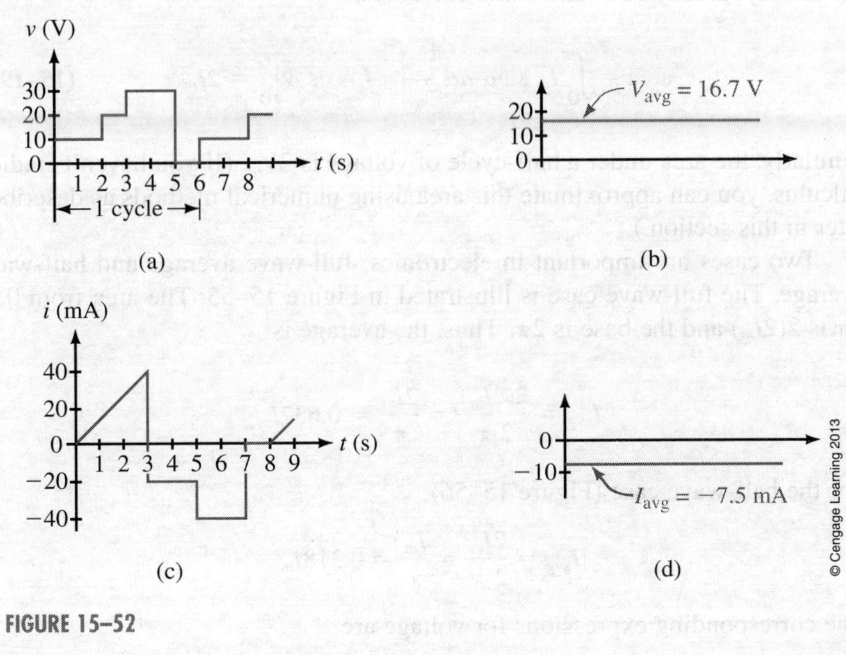

FIGURE 15–52

Solution For the waveform of (a), $T = 6$ s. Thus,

$$V_{avg} = \frac{(10\text{V} \times 2\text{s}) + (20\text{V} \times 1\text{s}) + (30\text{V} \times 2\text{s}) + (0\text{V} \times 1\text{s})}{6\text{s}} = \frac{100\text{ V-s}}{6\text{s}} = 16.7\text{V}$$

CircuitSim 15-14

The average is shown as (b). For the waveform of (c), $T = 8$ s and

$$I_{avg} = \frac{\frac{1}{2}(40\text{mA} \times 3\text{s}) - (20\text{mA} \times 2\text{s}) - (40\text{mA} \times 2\text{s})}{8\text{s}} = \frac{-60}{8}\text{mA} = -7.5\text{ mA}$$

PRACTICE PROBLEMS 9

Determine the averages for Figures 15–53(a) and (b).

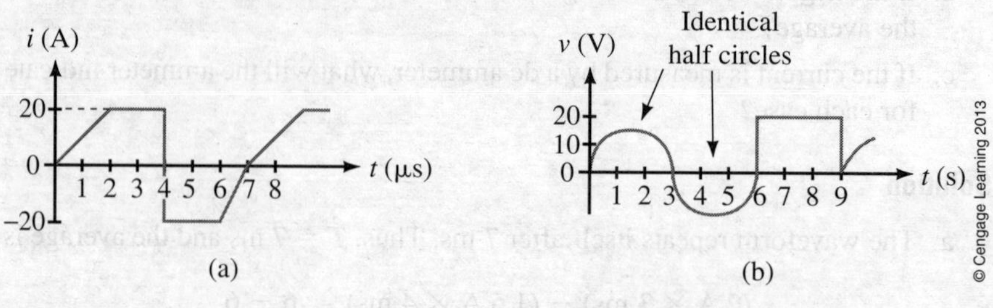

FIGURE 15–53

Answers
a. 1.43 A; b. 6.67 V

Sine Wave Averages

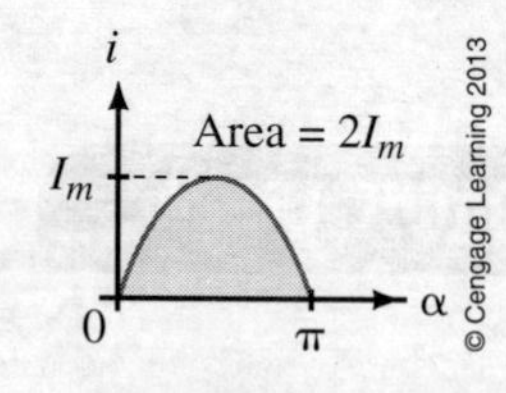

FIGURE 15–54 Area under a half-cycle.

Because a sine wave is symmetrical, its area below the horizontal axis is the same as its area above the axis; thus, over a full cycle its net area is zero, independent of frequency and phase angle. Thus, the average of sin ωt, sin$(\omega t \pm \theta)$, sin $2\omega t$, cos ωt, cos$(\omega t \pm \theta)$, cos $2\omega t$, and so on are each zero. The average of half a sine wave, however, is not zero. Consider Figure 15–54. The area under the half-cycle may be found using calculus as

$$\text{area} = \int_0^{\pi} I_m \sin\alpha \, d\alpha = \Big[-I_m \cos\alpha\Big]_0^{\pi} = 2I_m \qquad \textbf{(15–19)}$$

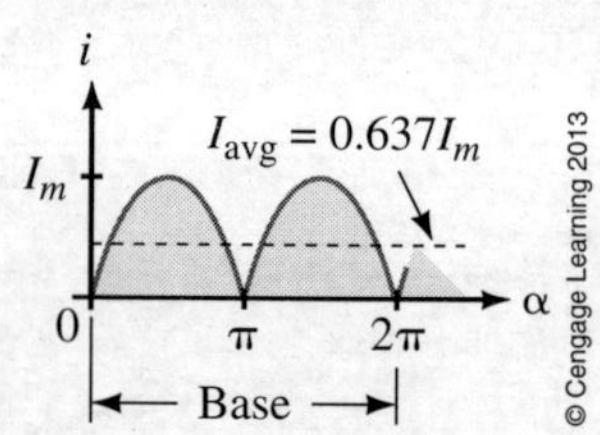

FIGURE 15–55 Full-wave average.

Similarly, the area under a half-cycle of voltage is $2V_m$. (If you haven't studied calculus, you can approximate this area using numerical methods as described later in this section.)

Two cases are important in electronics; full-wave average and half-wave average. The full-wave case is illustrated in Figure 15–55. The area from 0 to 2π is $2(2I_m)$ and the base is 2π. Thus, the average is

$$I_{avg} = \frac{2(2I_m)}{2\,\pi} = \frac{2I_m}{\pi} = 0.637I_m$$

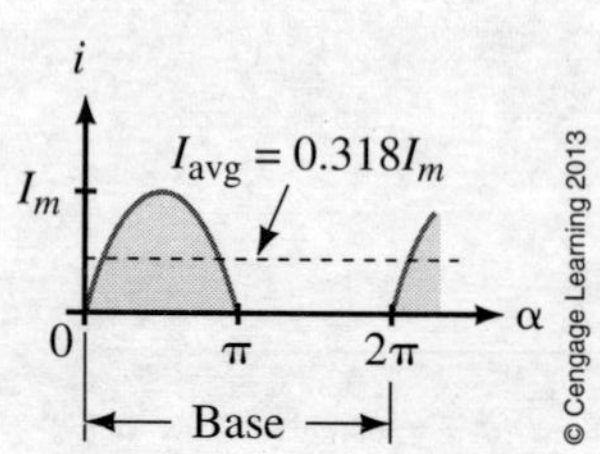

FIGURE 15–56 Half-wave average.

For the half-wave case (Figure 15–56),

$$I_{avg} = \frac{2I_m}{2\,\pi} = \frac{I_m}{\pi} = 0.318I_m$$

The corresponding expressions for voltage are

$$V_{avg} = 0.637V_m \text{ (full-wave)}$$

$$V_{avg} = 0.318V_m \text{ (half-wave)}$$

Numerical Methods

If the area under a curve cannot be computed exactly, it can be approximated. One method is to approximate the curve by straight line segments as in Figure 15–57. (If the straight lines closely fit the curve, the accuracy is very good.) Each element of area is a trapezoid (b) whose area is its average height

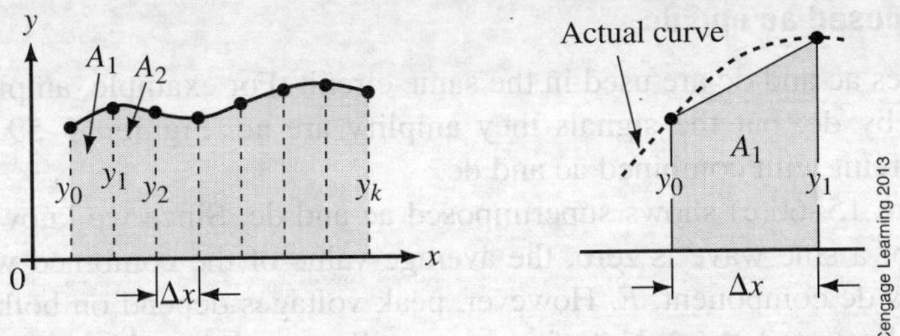

(a) Approximating the curve (b) Element of area

FIGURE 15–57 Determining area using the trapezoidal rule.

times its base. Thus, $A_1 = ½(y_0 + y_1)\Delta x$, $A_2 = ½(y_1 + y_2)\Delta x$, and so on. Total area is $A_1 + A_2 + \cdots + A_k = [½(y_0 + y_1) + ½(y_1 + y_2) + \cdots]\Delta x$ and so on. Combining terms yields

$$\text{area} = \left(\frac{y_0}{2} + y_1 + y_2 + \cdots + y_{k-1} + \frac{y_k}{2}\right)\Delta x \tag{15–20}$$

This result is known as the trapezoidal rule. Example 15–25 illustrates its use.

EXAMPLE 15–25

Approximate the area under $y = \sin(\omega t - 30°)$, Figure 15–58. Use an increment size of $\pi/6$ rad, that is, 30°.

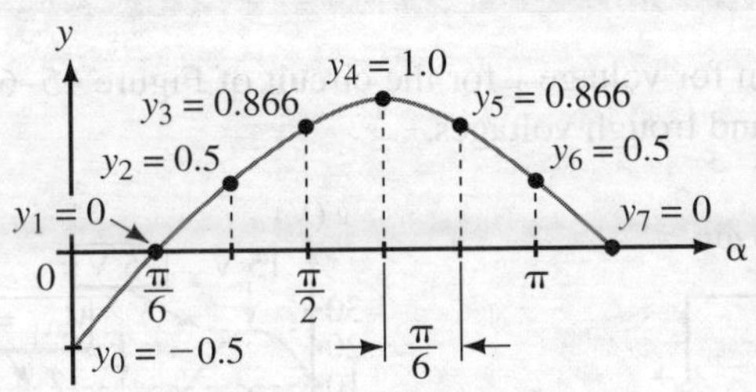

FIGURE 15–58

Solution Points on the curve $\sin(\omega t - 30°)$ have been computed by calculator and plotted as Figure 15–58. Substituting these values into Equation 15–20 yields

$$\text{area} = \left(\frac{1}{2}(-0.5) + 0 + 0.5 + 0.866 + 1.0 + 0.866 + 0.5 + \frac{1}{2}(0)\right)\left(\frac{\pi}{6}\right) = 1.823$$

The exact area (found using calculus) is 1.866; thus, the approximation of Example 15–25 is in error by 2.3%.

PRACTICE PROBLEMS 10

1. Repeat Example 15–25 using an increment size of $\pi/12$ rad. What is the percent error?
2. Approximate the area under $v = 50\sin(\omega t + 30°)$ from $\omega t = 0°$ to $\omega t = 210°$. Use an increment size of $\pi/12$ rad.

Answers

1. 1.855; 0.59%
2. 67.9 (exact 68.3; error = 0.6%)

Superimposed ac and dc

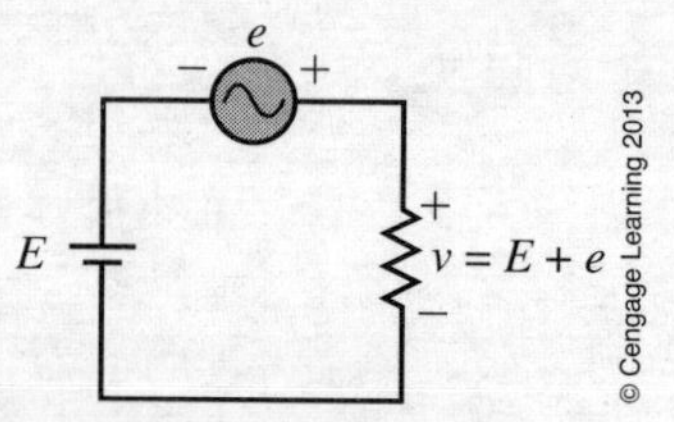

FIGURE 15–59

Sometimes ac and dc are used in the same circuit. For example, amplifiers are powered by dc, but the signals they amplify are ac. Figure 15–59 shows a simple circuit with combined ac and dc.

Figure 15–60(c) shows superimposed ac and dc. Since we know that the average of a sine wave is zero, the average value of the combined waveform will be its dc component, *E*. However, peak voltages depend on both components as illustrated in (c). Note for the case illustrated that although the waveform varies sinusoidally, it does not alternate in polarity since it never changes polarity to become negative.

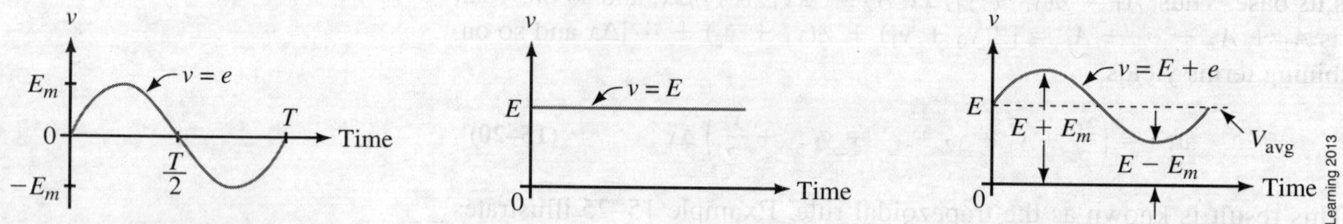

(a) AC alone. $E = 0$ V. $V_{avg} = 0$ V

(b) DC alone. $e = 0$ V. $V_{avg} = E$

(c) Superimposed ac and dc. $V_{avg} = E$

FIGURE 15–60 Superimposed dc and ac.

EXAMPLE 15–26

Draw the waveform for voltage *v* for the circuit of Figure 15–61(a). Determine its average, peak, and trough voltages.

CircuitSim 15-15

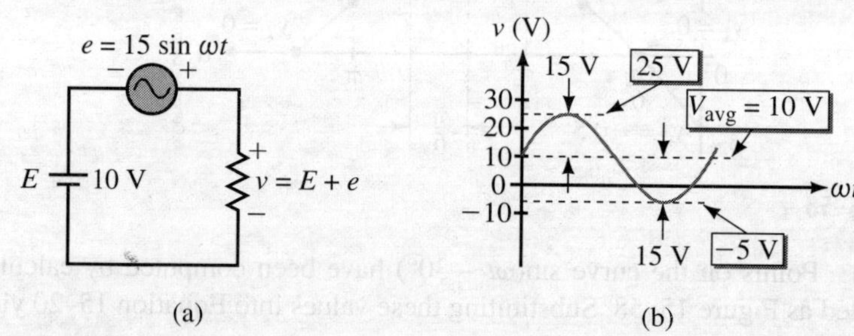

FIGURE 15–61 $v = 10 + 15 \sin \omega t$.

Solution The waveform consists of a 10-V dc value with 15-V ac riding on top of it. The average is the dc value, $V_{avg} = 10$ V. The peak voltage is $10 + 15 = 25$ V, while the trough voltage is $10 - 15 = -5$ V. This waveform alternates in polarity, although not symmetrically (as is the case when there is no dc component).

PRACTICE PROBLEMS 11

Repeat Example 15–26 if the dc source of Figure 15–61 is $E = -5$ V.

Answers

$V_{avg} = -5$ V; positive peak $= 10$ V; negative trough $= -20$ V

15.9 Effective (RMS) Values

While instantaneous, peak, and average values provide useful information about a waveform, none of them truly represents the ability of the waveform to do useful work. In this section, we look at a representation that does. It is called the waveform's **effective value.** The concept of effective value is an important one; in practice, most ac voltages and currents are expressed as effective values. Effective values are also called **rms values** for reasons discussed shortly.

What Is an Effective Value?

An effective value is an equivalent dc value: it tells you how many volts or amps of dc that a time-varying waveform is equal to in terms of its ability to produce average power. Effective values depend on the waveform. A familiar example of such a value is the value of the voltage at the wall outlet in your home. In North America its value is 120 Vac. This means that the sinusoidal voltage at the wall outlets of your home is capable of producing the same average power as 120 volts of steady dc.

Effective Values for Sine Waves

The effective value of a waveform can be determined using the circuits of Figure 15–62. Consider a sinusoidally varying current, $i(t)$. By definition, the effective value of i is that value of dc current that produces the same average power. Consider (b). Let the dc source be adjusted until its average power is the same as the average power in (a). The resulting dc current is then the effective value of the current of (a). To determine this value, determine the average power for both cases, then equate them.

First, consider the dc case. Since current is constant, power is constant, and average power is

$$P_{\text{avg}} = P = I^2R \qquad \textbf{(15–21)}$$

Now consider the ac case. Power to the resistor at any value of time is $p(t) = i^2R$, where i is the instantaneous value of current. A sketch of $p(t)$ is shown in Figure 15–62(a), obtained by squaring values of current at various points along the axis, then multiplying by R. Average power is the average of $p(t)$. Since $i = I_m \sin \omega t$,

$$\begin{aligned} p(t) &= i^2R \\ &= (I_m \sin \omega t)^2R = I_m{}^2R \sin^2\omega t \qquad \textbf{(15–22)} \\ &= I_m{}^2R\left[\frac{1}{2}(1 - \cos 2\omega t)\right] \end{aligned}$$

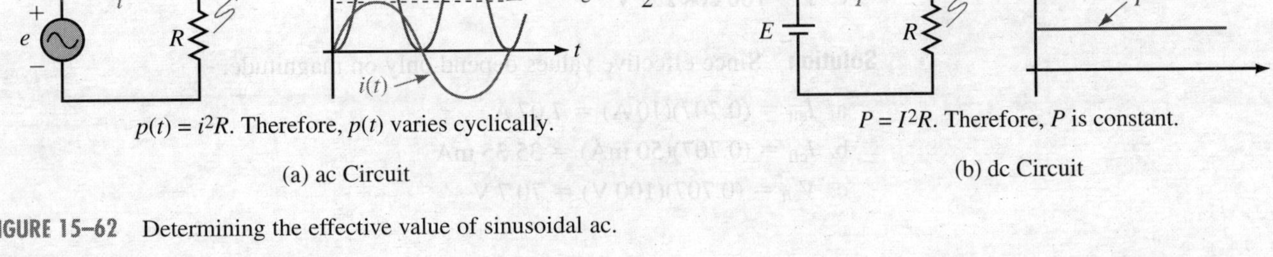

FIGURE 15–62 Determining the effective value of sinusoidal ac.

NOTES...

Because ac currents alternate in direction, you might expect average power to be zero, with power during the negative half-cycle being equal and opposite to power during the positive half-cycle and hence cancelling. However this is not true since, as Equation 15–22 shows, current is squared, and hence power is never negative. This is consistent with the idea that insofar as power dissipation is concerned, the direction of current through a resistor does not matter (Figure 15–63).

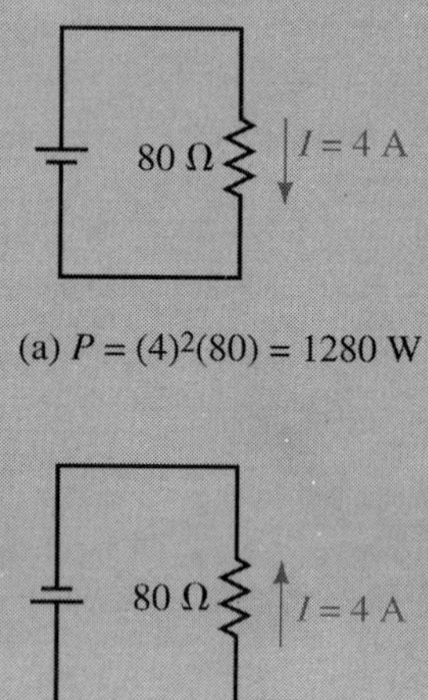

(a) $P = (4)^2(80) = 1280$ W

(b) $P = (4)^2(80) = 1280$ W

FIGURE 15–63 Since power depends only on current magnitude, it is the same for both current directions.

CircuitSim 15-16

where we have used the trigonometric identity $\sin^2 \omega t = ½(1 - \cos 2\omega t)$, from the mathematics tables inside the front cover of this book. Thus,

$$p(t) = \frac{I_m{}^2R}{2} - \frac{I_m{}^2R}{2}\cos 2\omega t \tag{15–23}$$

To get the average of $p(t)$, note that the average of $\cos 2\omega t$ is zero and thus the last term of Equation 15–23 drops off, leaving

$$P_{avg} = \text{average of } p(t) = \frac{I_m{}^2R}{2} \tag{15–24}$$

Now equate Equations 15–21 and 15–24, then cancel R.

$$I^2 = \frac{I_m{}^2}{2}$$

Now take the square root of both sides. Thus,

$$I = \sqrt{\frac{I_m{}^2}{2}} = \frac{I_m}{\sqrt{2}} = 0.707I_m$$

Current I is the value that we are looking for; it is the effective value of current i. To emphasize that it is an effective value, we will initially use subscripted notation I_{eff}. Thus,

$$I_{eff} = \frac{I_m}{\sqrt{2}} = 0.707I_m \tag{15–25}$$

Effective values for voltage are found in the same way:

$$E_{eff} = \frac{E_m}{\sqrt{2}} = 0.707E_m \tag{15–26a}$$

$$V_{eff} = \frac{V_m}{\sqrt{2}} = 0.707V_m \tag{15–26b}$$

As you can see, *effective values for sinusoidal waveforms depend only on magnitude.*

EXAMPLE 15–27

Determine the effective values of

a. $i = 10 \sin \omega t$ A
b. $i = 50 \sin(\omega t + 20°)$ mA
c. $v = 100 \cos 2\omega t$ V

Solution Since effective values depend only on magnitude,

a. $I_{eff} = (0.707)(10\text{ A}) = 7.07$ A
b. $I_{eff} = (0.707)(50\text{ mA}) = 35.35$ mA
c. $V_{eff} = (0.707)(100\text{ V}) = 70.7$ V

To obtain peak values from effective values, rewrite Equations 15–25 and 15–26. Thus,

$$I_m = \sqrt{2}I_{\text{eff}} = 1.414I_{\text{eff}} \tag{15–27}$$

$$E_m = \sqrt{2}E_{\text{eff}} = 1.414V_{\text{eff}} \tag{15–28a}$$

$$V_m = \sqrt{2}V_{\text{eff}} = 1.414V_{\text{eff}} \tag{15–28b}$$

It is important to note that these relationships hold only for sinusoidal waveforms. However, the concept of effective value applies to all waveforms, as we shall soon see.

Consider again the ac voltage at the wall outlet in your home. Since $E_{\text{eff}} =$ 120 V, $E_m = (\sqrt{2})(120 \text{ V}) = 170$ V. This means that a sinusoidal voltage alternating between ±170 V produces the same average power in a resistive circuit as 120 V of steady dc (Figure 15–64).

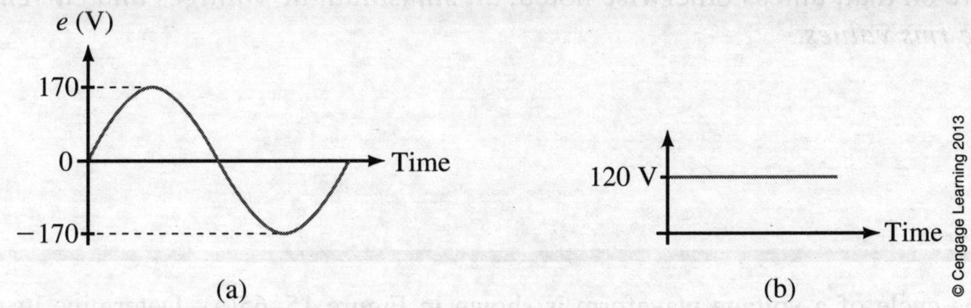

FIGURE 15–64 120 V of steady dc is capable of producing the same average power as sinusoidal ac with $E_m = 170$ V.

General Equation for Effective Values

The $\sqrt{2}$ relationship holds only for sinusoidal waveforms. For other waveforms, you need a more general formula. Using calculus, it can be shown that for any waveform

$$I_{\text{eff}} = \sqrt{\frac{1}{T}\int_0^T i^2 dt} \tag{15–29}$$

with a similar equation for voltage. This equation can be used to compute effective values for any waveform, including sinusoidal. In addition, it leads to a graphic approach to finding effective values. In Equation 15–29, the integral of i^2 represents the area under the i^2 waveform. Thus,

$$I_{\text{eff}} = \sqrt{\frac{\text{area under the } i^2 \text{ curve}}{\text{base}}} \tag{15–30}$$

To compute effective values using this equation, do the following:

Step 1: Square the current (or voltage) curve.

Step 2: Find the area under the squared curve.

Step 3: Divide the area by the length of the curve.

Step 4: Find the square root of the value from Step 3.

This process is easily carried out for rectangular-shaped waveforms since the area under their squared curves is easy to compute. For other waveforms, you have to use calculus or approximate the area using numerical methods as in end-of-chapter Problem 62. For the special case of superimposed ac and dc (Figure 15–60), Equation 15–29 leads to the following formula:

$$I_{\text{eff}} = \sqrt{I_{\text{dc}}^2 + I_{\text{ac}}^2} \qquad \textbf{(15–31)}$$

where I_{dc} is the dc current value, I_{ac} is the effective value of the ac component, and I_{eff} is the effective value of the combined ac and dc currents. Equations 15–30 and 15–31 also hold for voltage when V is substituted for I.

RMS Values

Consider again Equation 15–30. To use this equation, we compute the root of the mean square to obtain the effective value. For this reason, effective values are called **root mean square** or **rms values,** and **the terms *effective* and *rms* are synonymous.** Since, in practice, sinusoidal ac quantities are almost always expressed and measured as rms values, we shall assume from here on that, unless otherwise noted, *all sinusoidal ac voltages and currents are rms values.*

EXAMPLE 15–28

One cycle of a voltage waveform is shown in Figure 15–65(a). Determine its effective (rms) value.

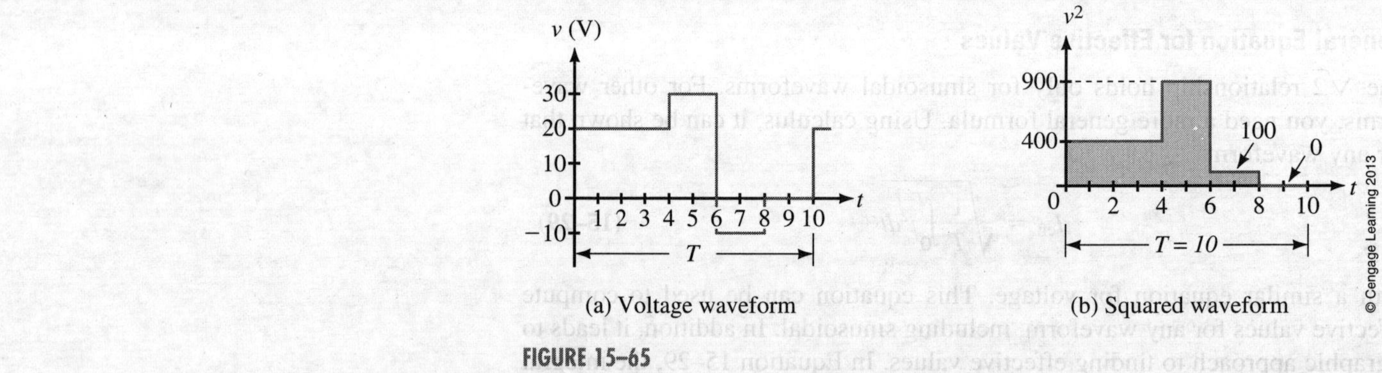

FIGURE 15–65

Solution Square the voltage waveform point by point, and plot it as in (b). Apply Equation 15–30:

$$V_{\text{eff}} = \sqrt{\frac{(400 \times 4) + (900 \times 2) + (100 \times 2) + (0 \times 2)}{10}}$$

$$= \sqrt{\frac{3600}{10}} = 19.0 \text{ V}$$

Thus, the waveform of Figure 15–65(a) has the same effective value as 19.0 V of steady dc.

EXAMPLE 15–29

Determine the effective (rms) value of the waveform of Figure 15–66(a).

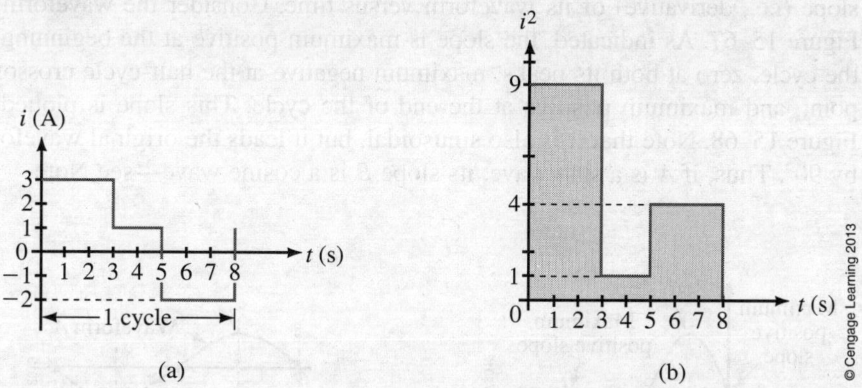

FIGURE 15–66

Solution Square the curve, then apply Equation 15–30. Thus,

$$I_{\text{eff}} = \sqrt{\frac{(9 \times 3) + (1 \times 2) + (4 \times 3)}{8}}$$

$$= \sqrt{\frac{41}{8}} = 2.26 \text{ A}$$

EXAMPLE 15–30

Compute the rms value of the waveform of Figure 15–61(b).

Solution Use Equation 15–31 (with I replaced by V). First, compute the rms value of the ac component. $V_{\text{ac}} = 0.707 \times 15 = 10.61$ V. Now substitute this into Equation 15–31. Thus,

$$V_{\text{rms}} = \sqrt{V_{\text{dc}}^2 + V_{\text{ac}}^2} = \sqrt{(10)^2 + (10.61)^2} = 14.6 \text{ V}$$

PRACTICE PROBLEMS 12

1. Determine the rms value of the current of Figure 15–51.
2. Repeat for the voltage graphed in Figure 15–52(a).

Answers

1. 1.73 A; 2. 20 V

Dropping the *eff* and *rms* Notation

The subscripts *eff* and *rms* are not used in practice. Once the concept is familiar, we drop them. From now on, they are implied rather than stated; thus, we use I instead of I_{eff} and V instead of V_{eff}.

15.10 Rate of Change of a Sine Wave (Derivative)

As you will see in Chapter 16, several important circuit effects depend on the rate of change of sinusoidal quantities. The rate of change of a quantity is the slope (i.e., derivative) of its waveform versus time. Consider the waveform of Figure 15–67. As indicated, the slope is maximum positive at the beginning of the cycle, zero at both its peaks, maximum negative at the half-cycle crossover point, and maximum positive at the end of the cycle. This slope is plotted in Figure 15–68. Note that it is also sinusoidal, but it leads the original waveform by 90°. Thus, if A is a sine wave, its slope B is a cosine wave—see Note.

NOTES...

The Derivative of a Sine Wave

The result developed intuitively here can be proven easily using calculus. To illustrate, consider the waveform sin ωt shown in Figure 15–67. The slope of this function is its derivative. Thus,

$$\text{Slope} = \frac{d}{dt}\sin\omega t = \omega\cos\omega t$$

Therefore, the slope of a sine wave is a cosine wave as depicted in Figure 15–68.

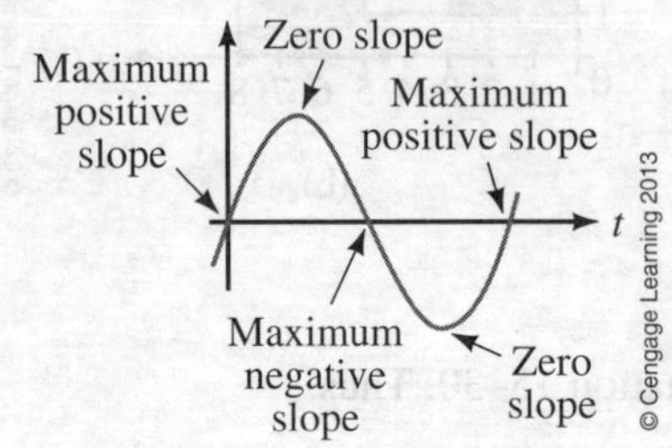

FIGURE 15–67 Slope at various places for a sine wave.

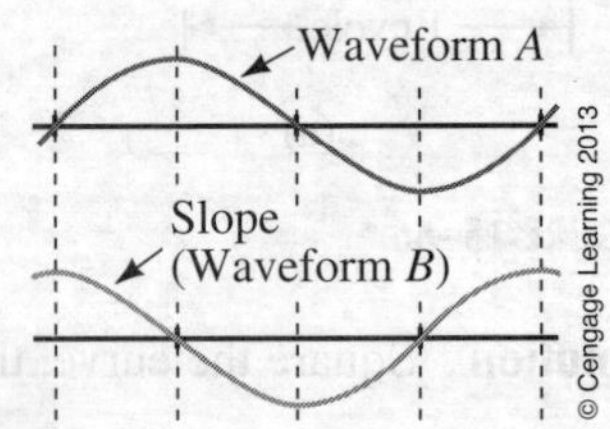

FIGURE 15–68 Showing the 90° phase shift.

CircuitSim 15-17

15.11 ac Voltage and Current Measurement

Two of the most important instruments for measuring ac quantities are the multimeter and the oscilloscope. Multimeters read the magnitude of ac voltage and current, and sometimes frequency. Oscilloscopes show waveshape and period and permit determination of frequency, phase difference, and so on.

Meters for Voltage and Current Measurement

There are two basic classes of ac meters: one measures rms correctly for sinusoidal waveforms only (called "average responding" instruments); the other measures rms correctly regardless of waveform (called "true rms" meters). Most common meters are average responding meters.

Average Responding Meters

Average responding meters use a rectifier circuit to convert incoming ac to dc. They then respond to the average value of the rectified input. Internal circuitry then rescales the measured value and displays it as an rms value. However, since these meters are calibrated for sinusoidal ac only, their readings are meaningless for all other waveforms. Most low-cost multimeters are average responding meters.

True RMS Measurement

If you need to measure the rms value of a nonsinusoidal waveform, you need a true rms meter. A true rms meter indicates true rms voltages and currents regardless of waveform. For example, for the waveform of Figure 15–64(a), any ac meter will correctly read 120 V (since it is a sine wave). For the waveform considered in Example 15–30, a true rms meter (such as that of Figure 15–69) will correctly read 14.6 V, but an average responding meter will yield only a meaningless value.

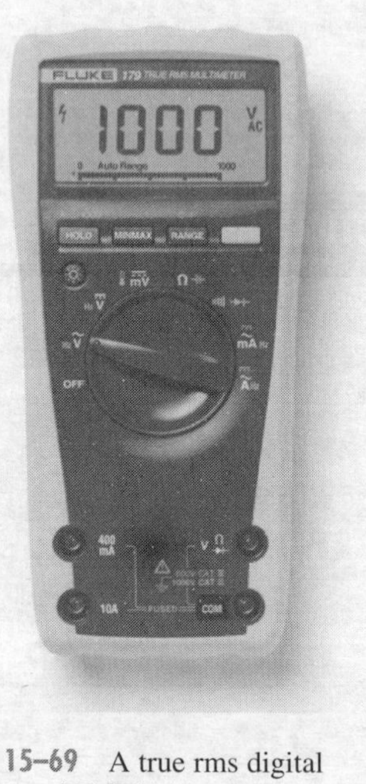

Courtesy of Fluke Corporation

FIGURE 15–69 A true rms digital multimeter (DMM).

Connecting ac Meters to Measure Voltage and Current

Unlike dc meters, ac meters are not polarity sensitive—that is, you do not need to be concerned about polarity when connecting an ac meter. For example, suppose you connect an ac voltmeter to read voltage, then disconnect it, reverse its leads, reconnect, then read voltage again. Both connections will yield a positive reading.

Oscilloscopes

Oscilloscopes (frequently referred to as scopes) are used for time domain measurement, that is, waveshape, frequency, period, phase difference, and so on, as illustrated earlier in Figures 15–16 and 15–44. As indicated, scope screens are scaled in volts vertically and in time horizontally to permit visual study of waveforms and (with the aid of cursors), measurement of waveform values—see Note.

Most modern oscilloscopes use digital technology (rather than the bulky CRT designs of a few years ago) and have many more features. The scope of Figure 15–70 for example, is a multichannel digital instrument with multicolored traces that permits the observation and analysis of many signals simultaneously.

CircuitSim 15-18

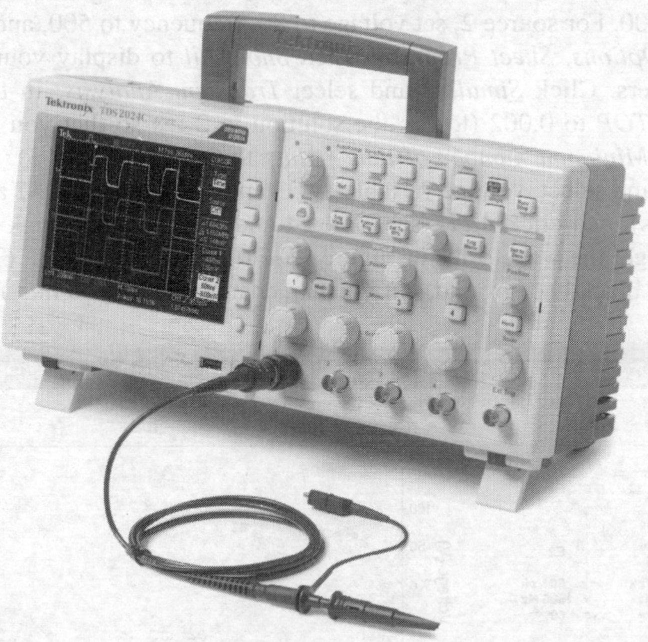

FIGURE 15–70 An oscilloscope displaying multiple waveforms simultaneously.

Oscilloscopes measure voltage. To measure current, you need to convert it to an equivalent voltage. The most inexpensive way to do this is to place a resistor in the circuit (small enough so as to not appreciably disturb the circuit), measure the voltage across this resistor with the oscilloscope, then use Ohm's law to convert the waveform to current. Alternatively, you can use a current probe that clips over the wire and monitors its magnetic field. One type of current probe uses a Hall effect sensor and can measure a wide range of currents over a considerable frequency range, including dc. However, these are quite expensive.

NOTES...

A series of hands-on, computer-based oscilloscope tutorials using circuit simulation techniques may be found on our Web site. Go to the *Student Premium Website* at cengagebrain.com and follow the links to *Lab Pre-Study Simulations—The Oscilloscope.* The simulations use Multisim.

Meter Frequency Considerations

AC meters measure voltage and current only over a limited frequency range, typically up to a few hundred kilohertz. Note, however, that accuracy may become poorer as frequency gets higher—check your manual. Oscilloscopes, on the other hand, measure very high frequencies; even moderately priced oscilloscopes work at frequencies up to hundreds of MHz.

15.12 Circuit Analysis Using Computers

MULTISIM

PSPICE

Multisim and PSpice both provide a convenient way to study phase relationships and other aspects of ac waveforms as they both incorporate easy-to-use graphing facilities. To illustrate, let us graph $e_1 = 100 \sin \omega t$ V and $e_2 = 80 \sin(\omega t + 60°)$ V and confirm the phase difference by measurement. (Since the measurement of phase difference is independent of frequency, we can choose any convenient frequency we want for this simulation. Let us use 500 Hz.)

Multisim

Select an ac signal voltage source and place it on your screen as in Figure 15–71 (see Note 1). Click Place, select *Junction*, and add a junction dot above the source. Repeat for the second source, add grounds, then wire the circuits as shown. Read Note 2, double-click source 1, then set voltage to 100 and frequency to 500. For source 2, set voltage to 80, frequency to 500, and phase to 60. Click *Options, Sheet Properties* then *Show All* to display your circuit's node numbers. Click *Simulate* and select *Transient Analysis;* in the dialog box, set *TSTOP* to 0.002 (to run the solution to 2 ms so that you see a full cycle) and *Minimum number of time points* to 200 (see Note 3). Click the *Output* tab and select voltage variables V(1) and V(2) (or nodes \$1 and \$2 for legacy users), and click *Add*—see Notes 4 and 5. When you click *Simulate,* you should get the waveforms of Figure 15–71(b) on your screen. (They may appear on a black background, but you can change that if you like.)

NOTES...

1. For this example, use the ac signal source from the Signal Sources component bin. To access it, right-click *View* and ensure that *Signal Source Components* is selected. Use your mouse to locate its icon, then click and position it on your screen.
2. To ensure that your source phase angles are set up correctly, click *Options/Global Preferences* then select the *Simulation* tab (or the *Parts* tab if you are using legacy software). Look for the small sinusoidal waveform. Ensure that the *Shift Right* button is selected.
3. Multisim generates time steps automatically when it plots waveforms. Sometimes, however, it does not generate enough points, and you get a jagged curve. If this happens, click *Minimum number of time points* and type in a larger number (say 200).

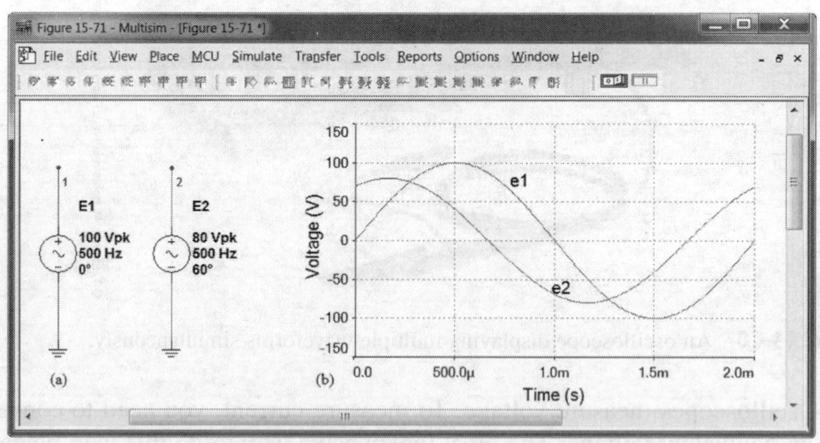

FIGURE 15–71 Studying phase relationships using Multisim.

You can verify the angle between the waveforms using cursors. First, note that the period $T = 2$ ms $= 2000$ μs. (This corresponds to 360°.) Click the *Show Grid* icon, then the *Show Cursors* icon. Using the cursors, measure the time between crossover points as indicated in Figure 15–72. You should get 333 μs. This yields an angular displacement of

$$\theta = \frac{333\ \mu\text{s}}{2000\ \mu\text{s}} \times 360° = 60°$$

as expected.

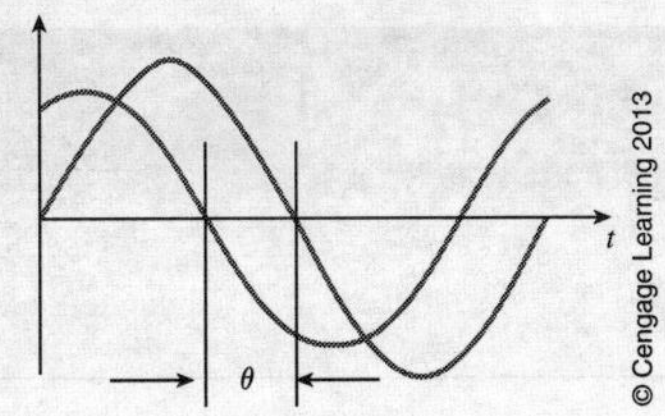

FIGURE 15–72

4. Node numbers in Figure 15–71 depend on the order in which you connect wire—thus, your node numbers may be different from those shown. However, for purposes of discussion, we will use the node numbers as illustrated.
5. Node numbers help you identify variables—for example, V(1) is the voltage (with respect to ground) at Node 1, V(2) is the voltage at Node 2, and so on. [For legacy software users, you need to use $1 instead of V(1), etc.]

PSpice

Click the *Place Part* icon, ensure that the SOURCE library is selected, type **VSIN** in the Place Part box, then click the *Place Part* button. Build the circuit as shown in Figure 15–73 on your screen. Note the empty parameter boxes beside each source. Double-click each in turn and enter 0V for the offset, 100V for the amplitude, and 500Hz for the frequency for Source 1. Similarly enter values for Source 2. Now double-click the V2 source symbol and in its Property editor window, ensure that the *Parts* tab is selected, scroll until you find a cell labelled *PHASE*, type in **60,** click *Apply,* then close. (You don't need to do this for Source 1 because PSpice will automatically use the default value of zero degrees.) Click the *New Simulations* icon and enter the file name. When the dialog box opens, select *Time Domain*, set *TSTOP* to 2 ms (to display a full cycle), then *OK*. Place voltage markers as shown to automatically plot the traces, then click *Run*. When the simulation is complete, the waveforms shown appear. (They may be on a black background, but you can change that if you like.)

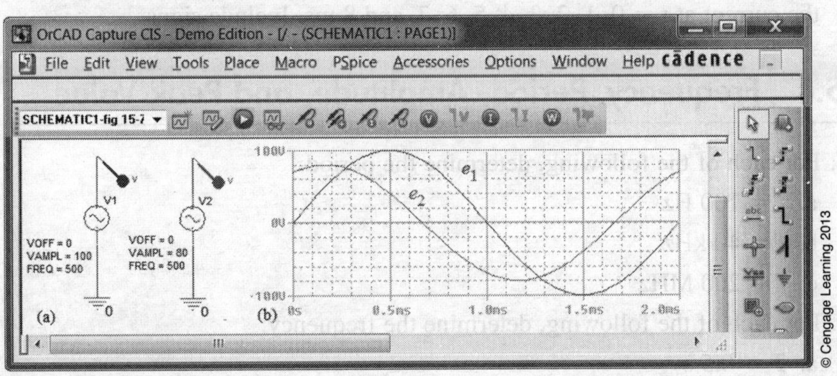

FIGURE 15–73 Studying phase relationships using PSpice.

You can verify the angle between the waveforms using cursors. First, note that the period $T = 2$ ms $= 2000$ μs. (This corresponds to 360°.) Now using the cursors, measure the time between crossover points as indicated in Figure 15–72. You should get 333 μs. This yields an angular displacement of

$$\theta = \frac{333\ \mu s}{2000\ \mu s} \times 360° = 60°$$

which agrees with the given sources.

Problems

15.1 Introduction

1. What do we mean by "ac voltage?" By "ac current?"

15.2 Generating ac Voltages

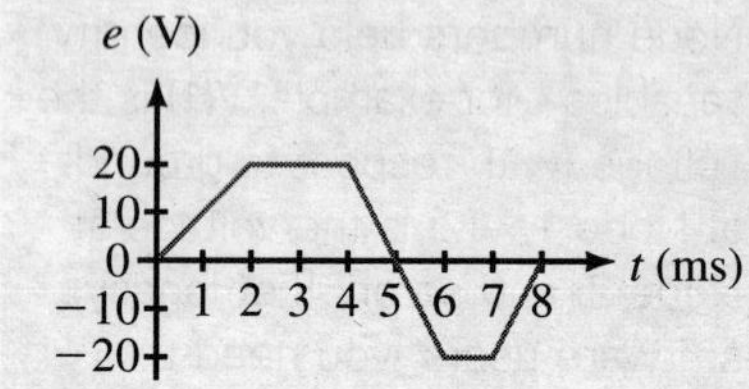

FIGURE 15–74

2. The waveform of Figure 15–8 is created by a 600-rpm generator. If the speed of the generator changes so that its cycle time is 50 ms, what is its new speed?
3. a. What do we mean by instantaneous value?
   b. For Figure 15–74, determine instantaneous voltages at $t = 0, 1, 2, 3, 4, 5, 6, 7,$ and 8 ms.

15.3 Voltage and Current Conventions for ac

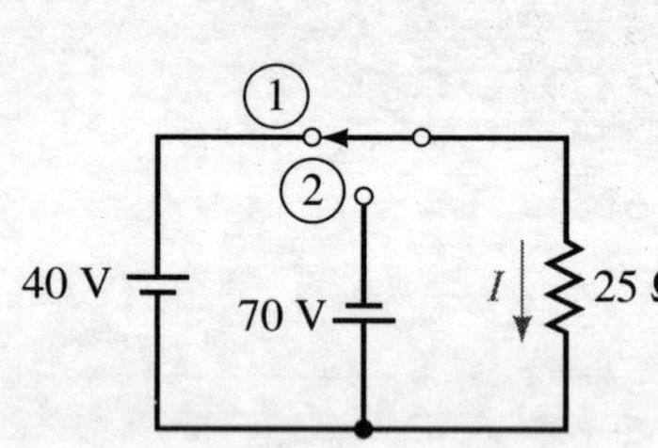

FIGURE 15–75

4. Consider the circuit of Figure 15–75. By moving the switch back and forth between positions 1 and 2, you can create a rectangular ac waveform. Suppose you let the switch remain in each position for 1 second,
   a. Sketch the voltage waveform across the load, complete with numeric values. Don't forget to scale the time axis.
   b. Sketch the load current, complete with numerical values.
5. The source of Figure 15–76 has the waveform of Figure 15–74. Determine the current at $t = 0, 1, 2, 3, 4, 5, 6, 7,$ and 8 ms. Include signs.

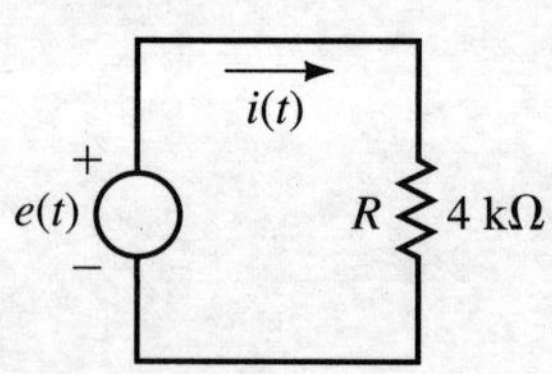

FIGURE 15–76

15.4 Frequency, Period, Amplitude, and Peak Value

6. For each of the following, determine the period:
   a. $f = 100$ Hz
   b. $f = 40$ kHz
   c. $f = 200$ MHz
7. For each of the following, determine the frequency:
   a. $T = 0.5$ s
   b. $T = 100$ ms
   c. $5T = 80$ μs
8. For a triangular wave, $f = 1.25$ MHz. What is its period? How long does it take to go through 8×10^7 cycles?
9. Determine the period and frequency for the waveform of Figure 15–77.
10. Determine the period and frequency for the waveform of Figure 15–78. How many cycles are shown?

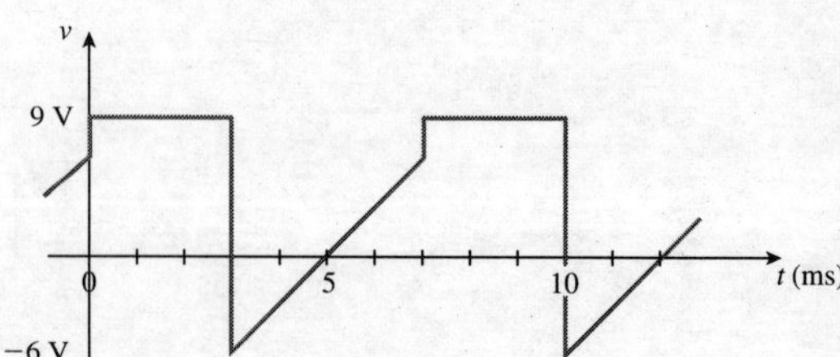

FIGURE 15–77

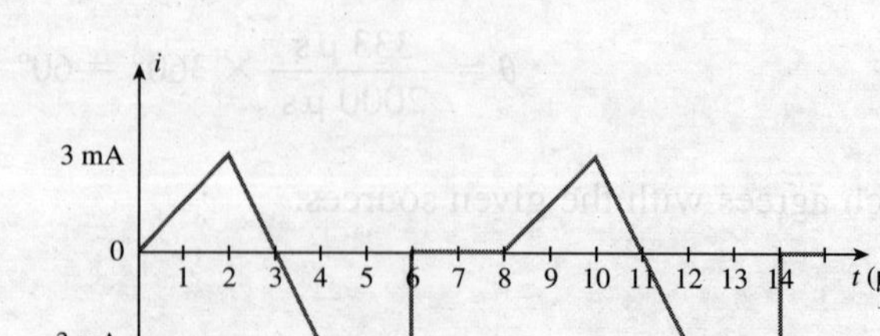

FIGURE 15–78

11. What is the peak-to-peak voltage for Figure 15–77? What is the peak-to-peak current of Figure 15–78?
12. For a certain waveform, $625T = 12.5$ ms. What is the waveform's period and frequency?
13. A square wave with a frequency of 847 Hz goes through how many cycles in 2 minutes and 57 seconds?
14. For the waveform of Figure 15–79, determine
    a. period b. frequency c. peak-to-peak value
15. Two waveforms have periods of T_1 and T_2. If $T_1 = 0.25\ T_2$ and $f_1 = 10$ kHz, what are T_1, T_2, and f_2?
16. Two waveforms have frequencies f_1 and f_2. If $T_1 = 4\ T_2$ and waveform 1 is as shown in Figure 15–77, what is f_2?

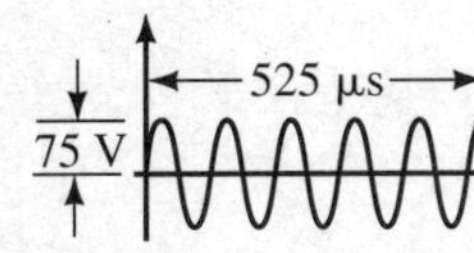

FIGURE 15–79

15.5 Angular and Graphic Relationships for Sine Waves

17. Given voltage $v = V_m \sin \alpha$. If $V_m = 240$ V, what is v at $\alpha = 37°$?
18. For the sinusoidal waveform of Figure 15–80,
    a. Determine the equation for i.
    b. Determine current at all points marked.
19. A sinusoidal voltage has a value of 50 V at $\alpha = 150°$. What is V_m?
20. Convert the following angles from radians to degrees:
    a. $\pi/12$ d. 1.43
    b. $\pi/1.5$ e. 17
    c. $3\pi/2$ f. 32π
21. Convert the following angles from degrees to radians:
    a. 10° d. 150°
    b. 25° e. 350°
    c. 80° f. 620°
22. A 50-kHz sine wave has an amplitude of 150 V. Sketch the waveform with its axis scaled in microseconds.
23. If the period of the waveform in Figure 15–80 is 180 ms, compute current at $t = 30$, 75, 140, and 315 ms.
24. A sinusoidal waveform has a period of 60 μs and $V_m = 80$ V. Sketch the waveform. What is its voltage at 4 μs?
25. A 20-kHz sine wave has a value of 50 volts at $t = 5$ μs. Determine V_m and sketch the waveform.
26. For the waveform of Figure 15–81, determine v_2.

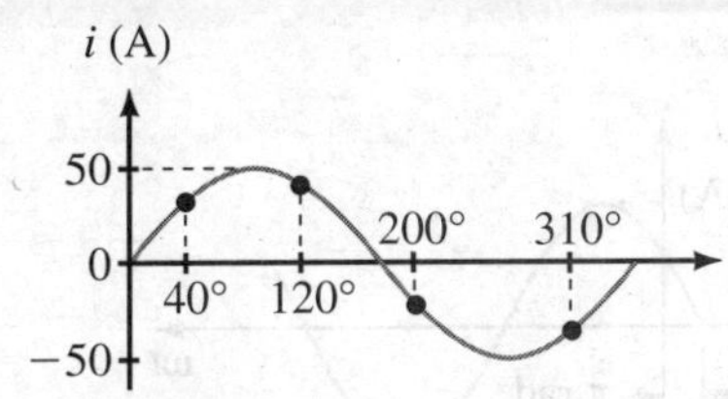

FIGURE 15–80

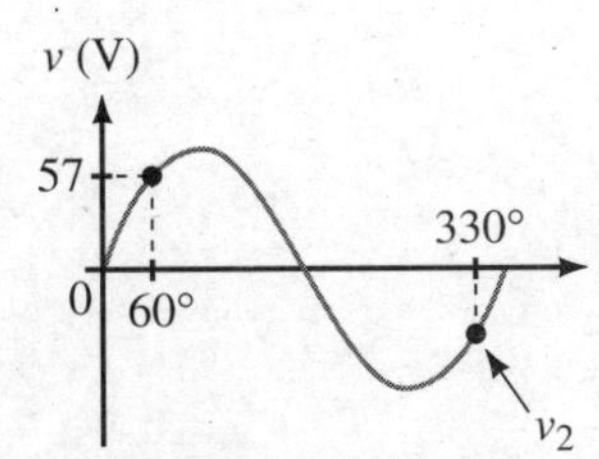

FIGURE 15–81

15.6 Voltages and Currents as Functions of Time

27. Calculate ω in radians per second for each of the following:
    a. $T = 100$ ns d. period = 20 ms
    b. $f = 30$ Hz e. 5 periods in 20 ms
    c. 100 cycles in 4 s
28. For each of the following values of ω, compute f and T:
    a. 100 rad/s b. 40 rad in 20 ms c. 34×10^3 rad/s
29. Determine equations for sine waves with the following:
    a. $V_m = 170$ V, $f = 60$ Hz c. $T = 120$ μs, $v = 10$ V at $t = 12$ μs
    b. $I_m = 40$ μA, $T = 10$ ms

30. Determine f, T, and amplitude for each of the following:
    a. $v = 75 \sin 200\pi t$ b. $i = 8 \sin 300t$
31. A sine wave has a peak-to-peak voltage of 40 V and $T = 50$ ms. Determine its equation.
32. Sketch the following waveforms with the horizontal axis scaled in degrees, radians, and seconds:
    a. $v = 100 \sin 200\pi t$ V
    b. $i = 90 \sin \omega t$ mA, $T = 80\ \mu$s
33. Given $i = 47 \sin 8260t$ mA, determine current at $t = 0$ s, 80 μs, 410 μs, and 1200 μs.
34. Given $v = 100 \sin \alpha$. Sketch one cycle.
    a. Determine at which two angles $v = 86.6$ V.
    b. If $\omega = 100\pi/60$ rad/s, at which times do these occur?
35. Write equations for the waveforms of Figure 15–82. Express the phase angle in degrees.

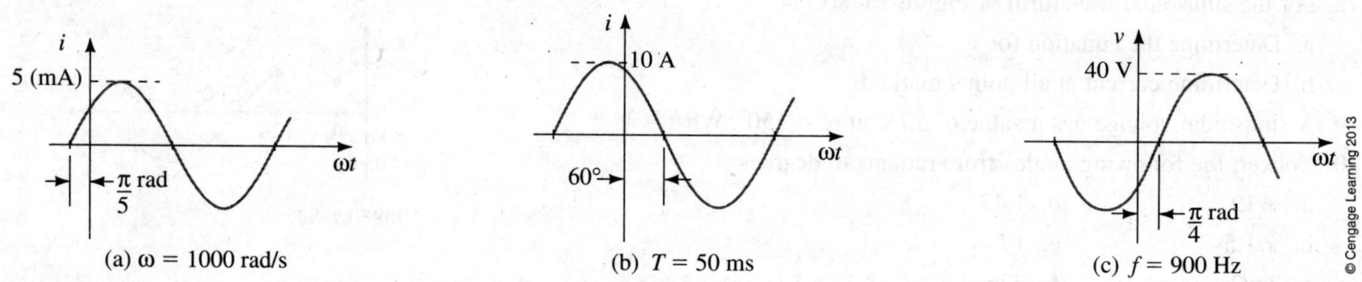

FIGURE 15–82

36. Sketch the following waveforms with the horizontal axis scaled in degrees and seconds:
    a. $v = 100 \sin(232.7t + 40°)$ V
    b. $i = 20 \sin(\omega t - 60°)$ mA, $f = 200$ Hz
37. Given $v = 5 \sin(\omega t + 45°)$. If $\omega = 20\pi$ rad/s, what is v at $t = 20$, 75, and 90 ms?
38. Repeat Problem 35 for the waveforms of Figure 15–83.

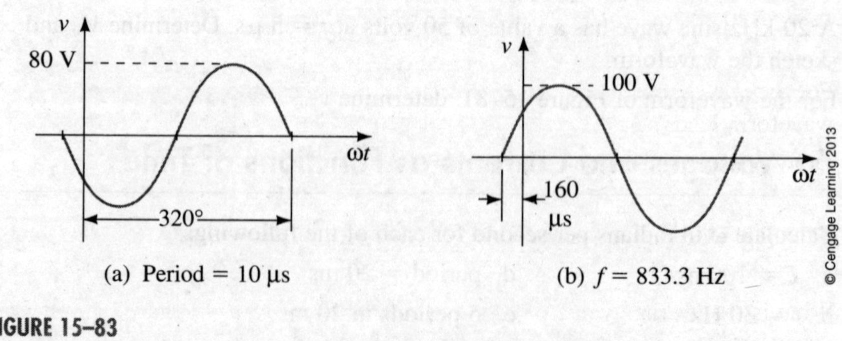

FIGURE 15–83

39. Determine the equation for the waveform shown in Figure 15–84.
40. For the waveform of Figure 15–85, determine i_2.
41. Given $v = 30 \sin(\omega t - 45°)$ where $\omega = 40\pi$ rad/s. Sketch the waveform. At what time does v reach 0 V? At what time does it reach 23 V and −23 V?

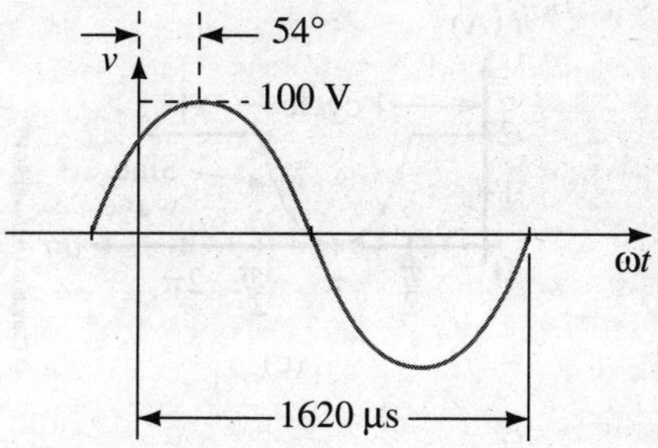

FIGURE 15–84

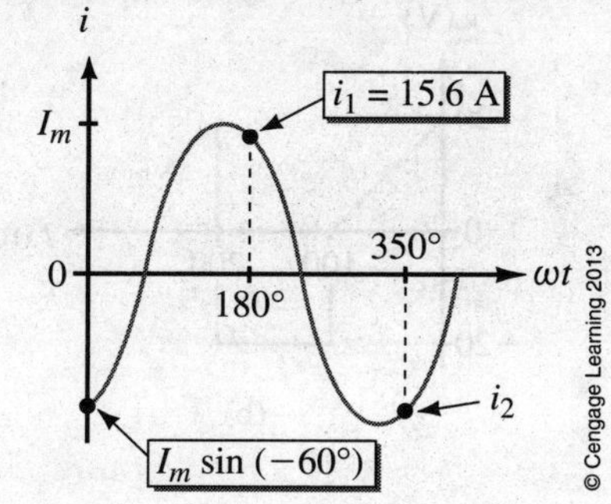

FIGURE 15–85

15.7 Introduction to Phasors

42. For each of the phasors of Figure 15–86, determine the equation for $v(t)$ or $i(t)$ as applicable, and sketch the waveform.

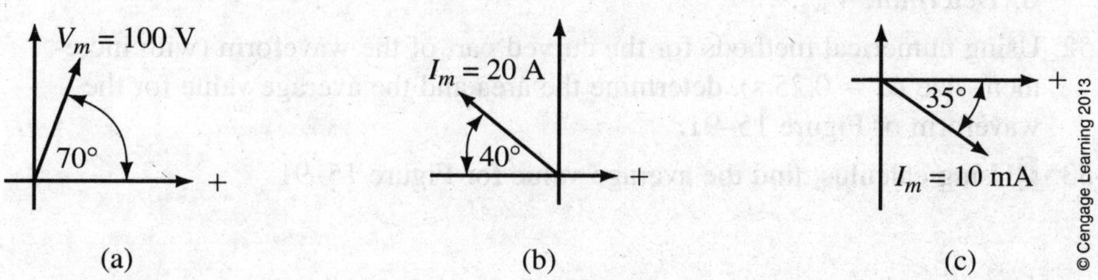

FIGURE 15–86

43. With the aid of phasors, sketch the waveforms for each of the following pairs and determine the phase difference and which waveform leads:

a. $v = 100 \sin \omega t$
$i = 80 \sin(\omega t + 20°)$

b. $v_1 = 200 \sin(\omega t - 30°)$
$v_2 = 150 \sin(\omega t - 30°)$

c. $i_1 = 40 \sin(\omega t + 30°)$
$i_2 = 50 \sin(\omega t - 20°)$

d. $v = 100 \sin(\omega t + 140°)$
$i = 80 \sin(\omega t - 160°)$

44. Repeat Problem 43 for the following:

a. $i = 40 \sin(\omega t + 80°)$
$v = -30 \sin(\omega t - 70°)$

b. $v = 20 \cos(\omega t + 10°)$
$i = 15 \sin(\omega t - 10°)$

c. $v = 20 \cos(\omega t + 10°)$
$i = 15 \sin(\omega t + 120°)$

d. $v = 80 \cos(\omega t + 30°)$
$i = 10 \cos(\omega t - 15°)$

45. For the waveforms in Figure 15–87, determine the phase differences. Which waveform leads?

46. Draw phasors for the waveforms of Figure 15–87.

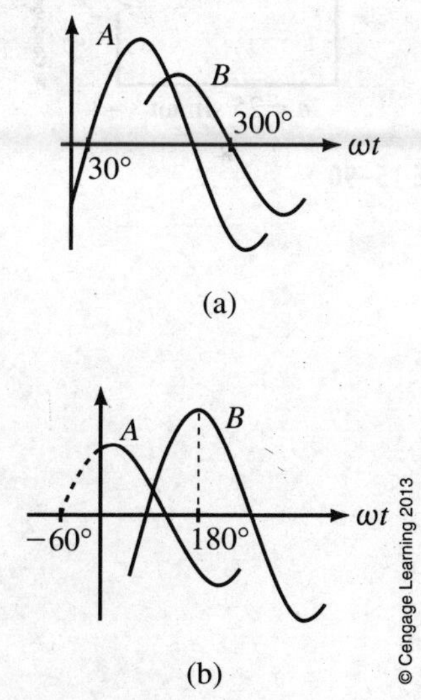

FIGURE 15–87

15.8 ac Waveforms and Average Value

47. What is the average value of each of the following over an integral number of cycles?

a. $i = 5 \sin \omega t$

b. $i = 40 \cos \omega t$

c. $v = 400 \sin(\omega t + 30°)$

d. $v = 20 \cos 2\omega t$

48. Using Equation 15–20, compute the area under the half-cycle of Figure 15–54 using increments of $\pi/12$ rad.

49. Compute I_{avg} or V_{avg} for the waveforms of Figure 15–88.

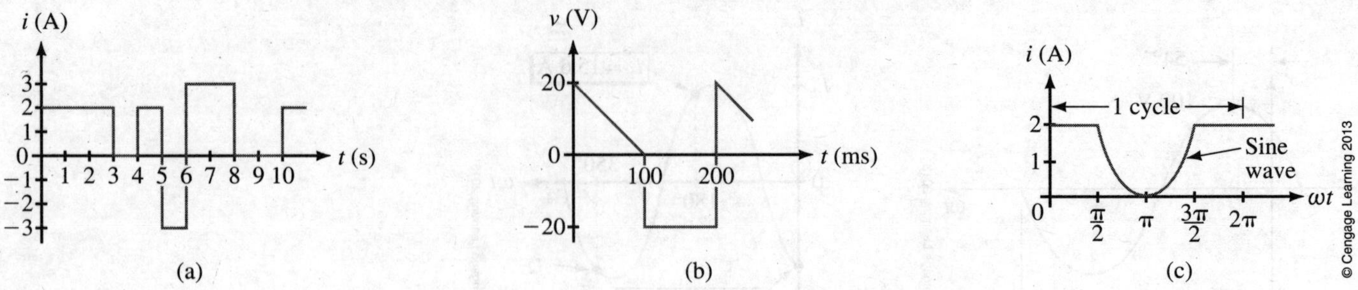

FIGURE 15–88

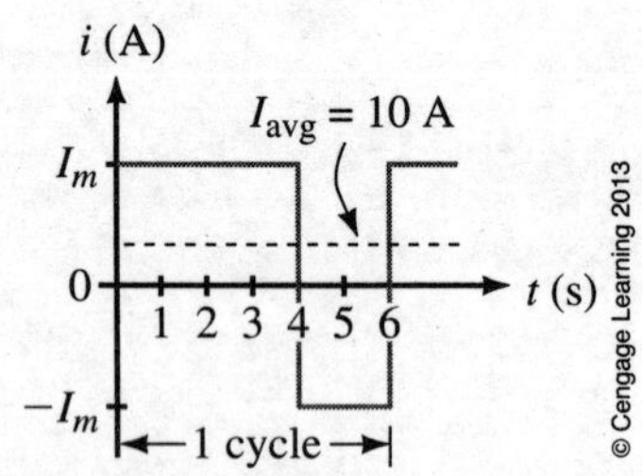

FIGURE 15–89

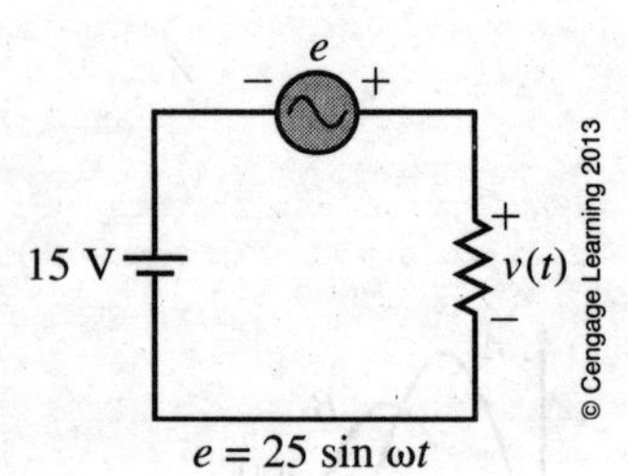

FIGURE 15–90

50. For the waveform of Figure 15–89, compute I_m.
51. For the circuit of Figure 15–90, $e = 25 \sin \omega t$ V and period $T = 120$ ms.
    a. Sketch voltage $v(t)$ with the axis scaled in milliseconds.
    b. Determine the peak and minimum voltages.
    c. Compute v at $t = 10$, 20, 70, and 100 ms.
    d. Determine V_{avg}.
52. Using numerical methods for the curved part of the waveform (with increment size $\Delta t = 0.25$ s), determine the area and the average value for the waveform of Figure 15–91.
53. Using calculus, find the average value for Figure 15–91.

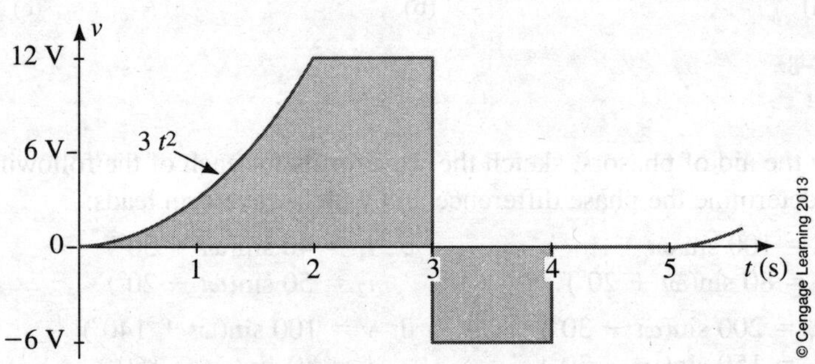

FIGURE 15–91

15.9 Effective (RMS) Values

54. Determine the effective values of each of the following:
    a. $v = 100 \sin \omega t$ V
    b. $i = 8 \sin 377t$ A
    c. $v = 40 \sin(\omega t + 40°)$ V
    d. $i = 120 \cos \omega t$ mA
55. Determine the rms values of each for the following:
    a. A 12-V battery
    b. $-24 \sin(\omega t + 73°)$ mA
    c. $10 + 24 \sin \omega t$ V
    d. $45 - 27 \cos 2\omega t$ V
56. For a sine wave, $V_{eff} = 9$ V. What is its amplitude?
57. Determine the root mean square values for
    a. $i = 3 + \sqrt{2}(4) \sin(\omega t + 44°)$ mA
    b. Voltage v of Figure 15–92 with $e = 25 \sin \omega t$ V
58. Compute the rms values for Figure 15–88(a) and Figure 15–89. For Figure 15–89, use $I_m = 30$ A.
59. Compute the rms values for the waveforms of Figure 15–92.

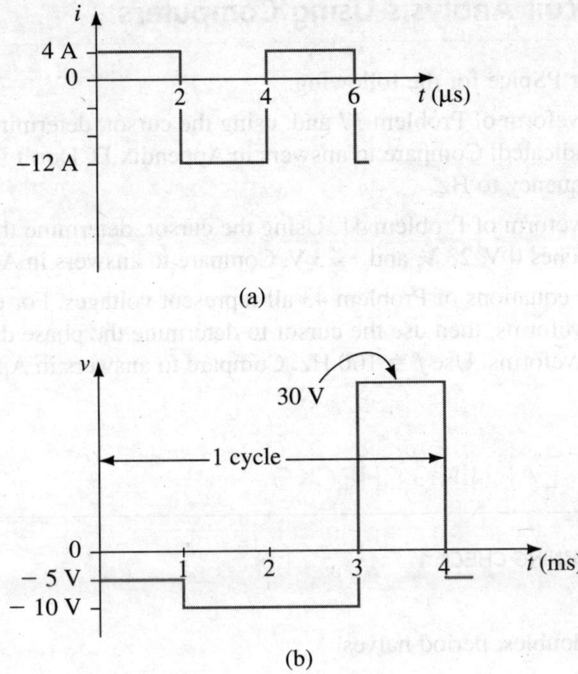

FIGURE 15–92

60. Compute the effective value for Figure 15–93.

61. Determine the rms value of the waveform of Figure 15–94. Why is it the same as that of a 24-V battery?

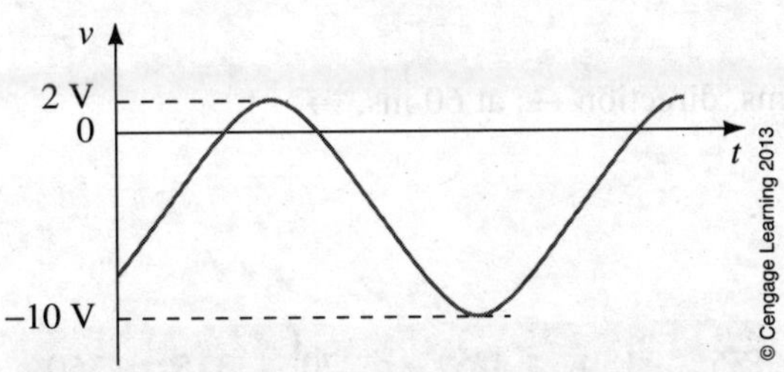

FIGURE 15–93

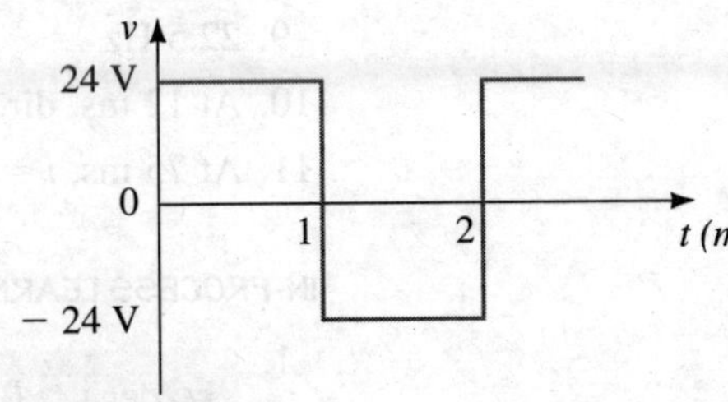

FIGURE 15–94

62. Compute the rms value of the waveform of Figure 15–52(c). To handle the triangular portion, use Equation 15–20. Use a time interval $\Delta t = 1$ s.

63. Repeat Problem 62, using calculus to handle the triangular portion.

15.11 ac Voltage and Current Measurement

64. Determine the reading of an average responding ac meter for each of the following cases. (Note: Meaningless is a valid answer if applicable.) Assume the frequency is within the range of the instrument.

   a. $v = 153 \sin \omega t$ V
   b. $v = \sqrt{2}(120) \sin(\omega t + 30°)$ V
   c. The waveform of Figure 15–61
   d. $v = 597 \cos \omega t$ V

65. Repeat Problem 64 using a true rms meter.

15.12 Circuit Analysis Using Computers

Use Multisim or PSpice for the following:

Multisim PSPICE 66. Plot the waveform of Problem 37 and, using the cursor, determine voltage at the times indicated. Compare to answers in Appendix D. Don't forget to convert the frequency to Hz.

Multisim PSPICE 67. Plot the waveform of Problem 41. Using the cursor, determine the times at which v reaches 0 V, 23 V, and −23 V. Compare to answers in Appendix D.

Multisim PSPICE 68. Assume the equations of Problem 43 all represent voltages. For each case, plot the waveforms, then use the cursor to determine the phase difference between waveforms. Use $f = 100$ Hz. Compare to answers in Appendix D.

ANSWERS TO IN-PROCESS LEARNING CHECKS

IN-PROCESS LEARNING CHECK 1

1. 16.7 ms
2. Frequency doubles, period halves
3. 20 ms; 50 Hz
4. 20 V; 0.5 ms and 2.5 ms; −35 V: 4 ms and 5 ms
5. (c) and (d); Since current is directly proportional to the voltage, it will have the same waveshape.
6. 250 Hz
7. $f_1 = 100$ Hz; $f_2 = 33.3$ Hz
8. 50 kHz and 1 MHz
9. 22.5 Hz
10. At 12 ms, direction →; at 37 ms, direction ←; at 60 ms, →
11. At 75 ms, $i = -5$ A

IN-PROCESS LEARNING CHECK 2

1.

α (deg)	0	45	90	135	180	225	270	315	360
i (mA)	0	10.6	15	10.6	0	−10.6	−15	−10.6	0

2. a. 0.349 c. 2.09
   b. 0.873 d. 4.36
3. 30°; 120°; 450°
4. Same as Figure 15–27 with $T = 10$ s and amplitude = 50 mA.
5. 1.508×10^6 rad/s
6. $i = 8 \sin 157t$ A
7. $i = 6 \sin 69.81t$ A
8. a. $i = 250 \sin(251t - 30°)$ A
   b. $i = 20 \sin(62.8t + 45°)$ A
   c. $v = 40 \sin(628t - 30°)$ V
   d. $v = 80 \sin(314 \times 10^3 t + 36°)$ V
9. a. 2.95 ms; 7.05 ms; 22.95 ms; 27.05 ms
   b. 11.67 ms; 18.33 ms; 31.67 ms; 38.33 ms

10.

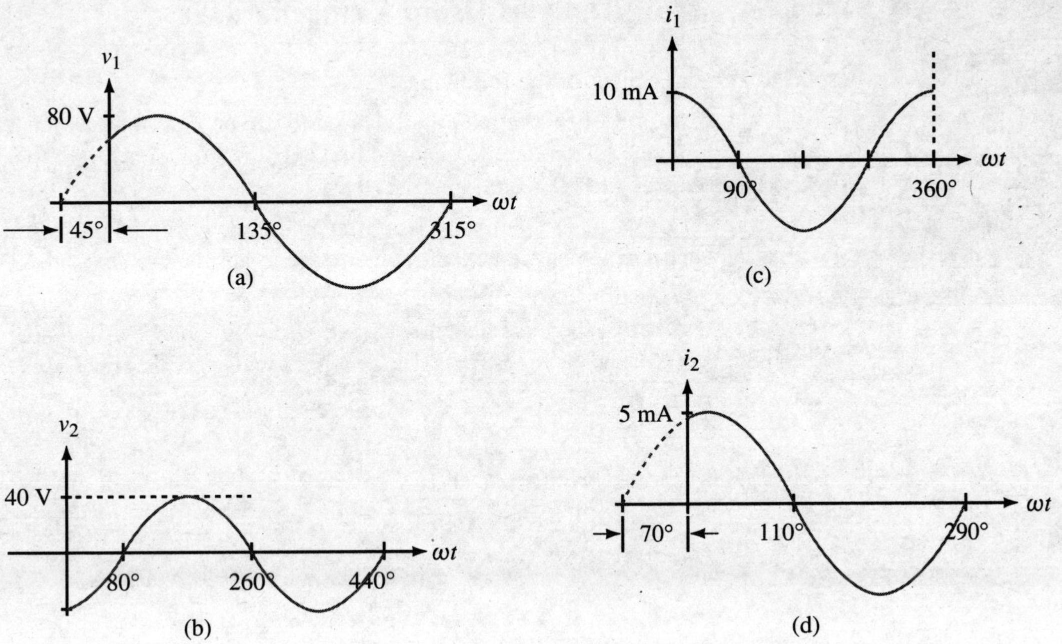

11. a. 2.25 s b. 1.33 s

12. a.

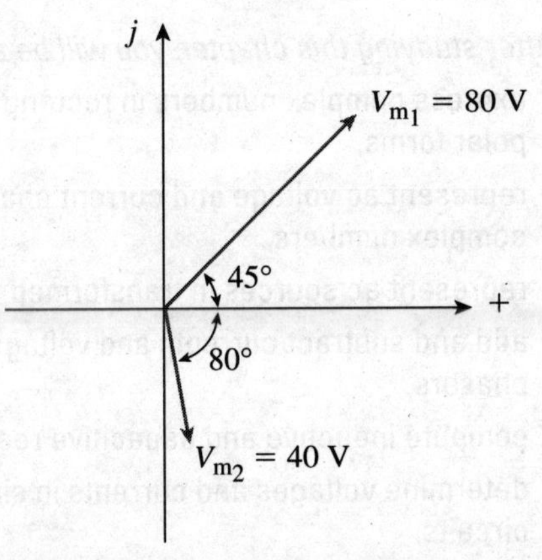

b. 125°

c. v_1 leads

13. a.

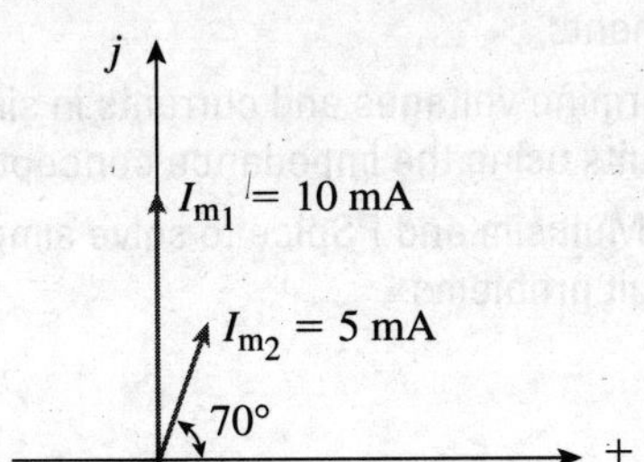

b. 20°

c. i_1 leads

■ KEY TERMS

Capacitive Reactance
Complex Number
Impedance
Inductive Reactance
Phasor Domain
Polar Form
Rectangular Form
Time Domain

■ OUTLINE

Complex Number Review
Complex Numbers in ac Analysis
R, L, and *C* Circuits with Sinusoidal Excitation
Resistance and Sinusoidal ac
Inductance and Sinusoidal ac
Capacitance and Sinusoidal ac
The Impedance Concept
Computer Analysis of ac Circuits

■ OBJECTIVES

After studying this chapter, you will be able to

- express complex numbers in rectangular and polar forms,
- represent ac voltage and current phasors as complex numbers,
- represent ac sources in transformed form,
- add and subtract currents and voltages using phasors,
- compute inductive and capacitive reactance,
- determine voltages and currents in simple ac circuits,
- explain the impedance concept,
- determine impedance for *R, L,* and *C* circuit elements,
- determine voltages and currents in simple ac circuits using the impedance concept,
- use Multisim and PSpice to solve simple ac circuit problems.

16

R, L, AND *C* ELEMENTS AND THE IMPEDANCE CONCEPT

CHAPTER PREVIEW

In Chapter 15, you learned how to analyze a few simple ac circuits in the time domain using voltages and currents expressed as functions of time. However, this is not a very practical approach. A more practical approach is to represent ac voltages and currents as phasors and circuit elements as impedances, and to analyze circuits in the phasor domain using complex algebra. With this approach, ac circuit analysis is handled much like dc circuit analysis, and all basic relationships and theorems—Ohm's law, Kirchhoff's laws, mesh and nodal analysis, superposition, and so on—apply. The major difference is that ac quantities are complex numbers, rather than real numbers, as with dc. While this complicates computational details, it does not alter basic circuit principles. This is the approach used in practice. The basic ideas are developed in this chapter.

Since phasor analysis and the impedance concept require a familiarity with complex numbers, we begin with a short review. ■

Putting It in Perspective

Charles Proteus Steinmetz

CHARLES STEINMETZ WAS BORN IN BRESLAU, Germany, in 1865 and emigrated to the United States in 1889. In 1892, he began working for the General Electric Company in Schenectady, New York, where he stayed until his death in 1923, and it was there that his work revolutionized ac circuit analysis. Prior to his time, this analysis had to be carried out using calculus, a difficult and time-consuming process. By 1893, however, Steinmetz had reduced the very complex alternating-current theory to, in his words, "a simple problem in algebra." The key concept in this simplification was the phasor—a representation based on complex numbers. By representing voltages and currents as phasors, Steinmetz was able to define a quantity called impedance and then use it to determine voltage and current magnitude and phase relationships in one algebraic operation.

Steinmetz wrote the seminal textbook on ac analysis based on his method, but at the time he introduced it, it was said that he was practically the only person who understood it. Now, however, it is common knowledge, and the methods devised by Steinmetz form the basis for nearly all ac circuit analysis techniques in use today. In this chapter, we learn the method and illustrate its application to the solution of basic ac circuit problems.

In addition to his work for GE, Charles Steinmetz was a professor of electrical engineering (1902–1913) and electrophysics (1913–1923) at Union College in Schenectady. ■

16.1 Complex Number Review

A **complex number** is a number of the form $\mathbf{C} = a + jb$, where a and b are real numbers and $j = \sqrt{-1}$. The number a is called the **real** part of **C** and b is called its **imaginary** part. (In circuit theory, j is used to denote the imaginary component rather than i to avoid confusion with current i.)

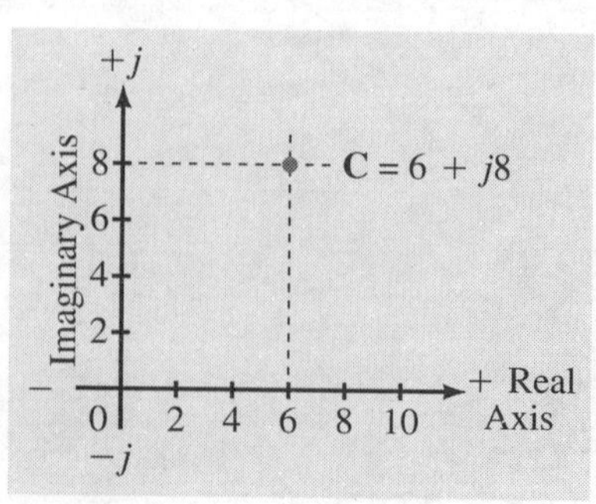

FIGURE 16–1 A complex number in rectangular form.

Geometrical Representation

Complex numbers may be represented geometrically, either in rectangular form or in polar form as points on a two-dimensional plane called the complex plane (Figure 16–1). The complex number $\mathbf{C} = 6 + j8$, for example, represents a point whose coordinate on the real axis is 6 and whose coordinate on the imaginary axis is 8. This form of representation is called the **rectangular form.**

Complex numbers may also be represented in **polar form** by magnitude and angle. Thus, $\mathbf{C} = 10 \angle 53.13°$ (Figure 16–2) is a complex number with magnitude 10 and angle 53.13°. This magnitude and angle representation is just an alternate way of specifying the location of the point represented by $\mathbf{C} = a + jb$.

Conversion between Rectangular and Polar Forms

To convert between forms, note from Figure 16–3 that

$$\mathbf{C} = a + jb \quad \text{(rectangular form)} \tag{16–1}$$

$$\mathbf{C} = C\angle\theta \quad \text{(polar form)} \tag{16–2}$$

where C is the magnitude of **C**. From the geometry of the triangle,

$$a = C\cos\theta \tag{16–3a}$$

$$b = C\sin\theta \tag{16–3b}$$

where

$$C = \sqrt{a^2 + b^2} \tag{16–4a}$$

and

$$\theta = \tan^{-1}\frac{b}{a} \tag{16–4b}$$

While Equations 16–3 and 16–4 permit conversion between forms, they are not used much in computations anymore because modern calculators do the job automatically. However, we will do one example using them to illustrate the principles involved.

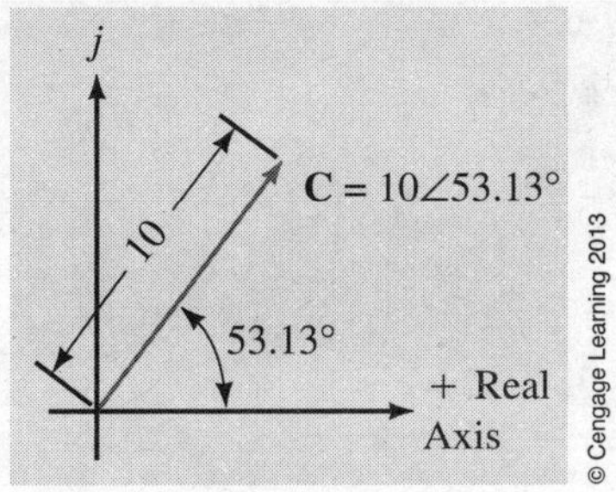

FIGURE 16–2 A complex number in polar form.

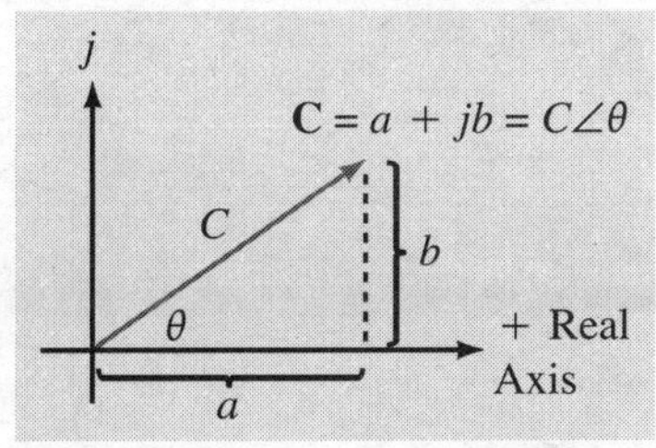

FIGURE 16–3 Polar and rectangular equivalence.

EXAMPLE 16–1

Determine rectangular and polar forms for the complex numbers **C**, **D**, **V**, and **W** of Figure 16–4(a).

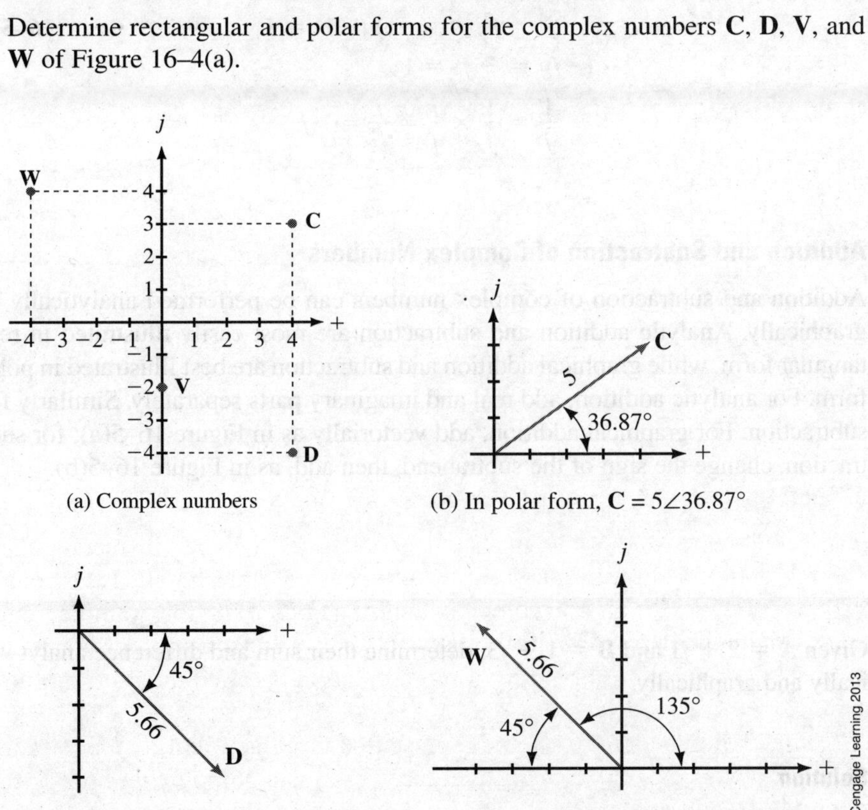

FIGURE 16–4

Solution *Point C:* Real part = 4; imaginary part = 3. Thus, C = 4 + *j*3. In polar form, $C = \sqrt{4^2 + 3^2} = 5$ and $\theta_C = \tan^{-1}(3/4) = 36.87°$. Thus, C = 5∠36.87° as indicated in (b).

Point D: In rectangular form, **D** = 4 − *j*4. Thus, $D = \sqrt{4^2 + 4^2} = 5.66$ and $\theta_D = \tan^{-1}(-4/4) = -45°$. Therefore, **D** = 5.66∠−45°, as shown in (c).

Point V: In rectangular form, **V** = −*j*2. In polar form, **V** = 2∠−90°.

Point W: In rectangular form, **W** = −4 + *j*4. Thus, $W = \sqrt{4^2 + 4^2} = 5.66$ and $\tan^{-1}(-4/4) = -45°$. Inspection of Figure 16–4(d) shows, however, that this 45° angle is the supplementary angle. The actual angle (measured from the positive horizontal axis) is 135°. Thus, **W** = 5.66∠135°.

As discussed later in this section, inexpensive calculators are available that perform such conversions directly—you simply enter the complex number components and press the conversion key. With these, the problem of determining angles for numbers such as **W** in Example 16–1 does not occur; you just enter −4 + *j*4 and the calculator returns 5.66∠135°.

Powers of *j*

Powers of *j* are frequently required in calculations. Here are some useful powers:

$$
\begin{aligned}
j^2 &= (\sqrt{-1})(\sqrt{-1}) = -1 \\
j^3 &= j^2 j = -j \\
j^4 &= j^2 j^2 = (-1)(-1) = 1 \\
(-j)j &= -j^2 = 1 \\
\frac{1}{j} &= \frac{1}{j} \times \frac{j}{j} = \frac{j}{j^2} = -j
\end{aligned}
\qquad \textbf{(16–5)}
$$

Addition and Subtraction of Complex Numbers

Addition and subtraction of complex numbers can be performed analytically or graphically. Analytic addition and subtraction are most easily illustrated in rectangular form, while graphical addition and subtraction are best illustrated in polar form. For analytic addition, add real and imaginary parts separately. Similarly for subtraction. For graphical addition, add vectorially as in Figure 16–5(a); for subtraction, change the sign of the subtrahend, then add, as in Figure 16–5(b).

EXAMPLE 16–2

Given **A** = 2 + *j*1 and **B** = 1 + *j*3, determine their sum and difference analytically and graphically.

Solution

$$\mathbf{A} + \mathbf{B} = (2 + j1) + (1 + j3) = (2 + 1) + j(1 + 3) = 3 + j4.$$

$$\mathbf{A} - \mathbf{B} = (2 + j1) - (1 + j3) = (2 - 1) + j(1 - 3) = 1 - j2.$$

Graphical addition and subtraction are shown in Figure 16–5.

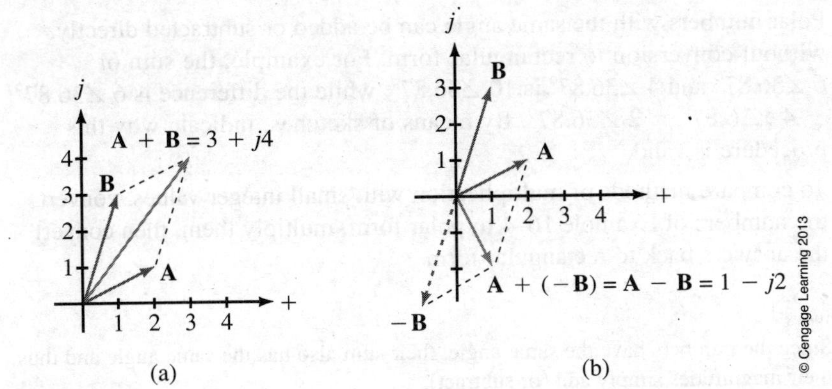

FIGURE 16–5

Multiplication and Division of Complex Numbers

These operations are best illustrated using polar representation. For multiplication, multiply magnitudes and add angles algebraically. For division, divide the magnitude of the denominator into the magnitude of the numerator, then subtract algebraically the angle of the denominator from that of the numerator. Thus, given $\mathbf{A} = A\angle\theta_A$ and $\mathbf{B} = B\angle\theta_B$,

$$\mathbf{A}\cdot\mathbf{B} = AB\underline{/\theta_A + \theta_B} \tag{16–6}$$

$$\mathbf{A/B} = A/B\underline{/\theta_A - \theta_B} \tag{16–7}$$

EXAMPLE 16–3

Given $\mathbf{A} = 3\angle 35°$ and $\mathbf{B} = 2\angle -20°$, determine the product $\mathbf{A}\cdot\mathbf{B}$ and the quotient **A/B**.

Solution

$$\mathbf{A}\cdot\mathbf{B} = (3\angle 35°)(2\angle -20°) = (3)(2)\underline{/35° - 20°} = 6\angle 15°$$

$$\frac{\mathbf{A}}{\mathbf{B}} = \frac{(3\angle 35°)}{(2\angle -20°)} = \frac{3}{2}\underline{/35° - (-20°)} = 1.5\angle 55°$$

EXAMPLE 16–4

For computations involving purely real, purely imaginary, or small integer numbers, it is sometimes more convenient to multiply directly in rectangular form than it is to convert to polar. Compute the following directly:

a. $(-j3)(2 + j4)$
b. $(2 + j3)(1 + j5)$

Solution

a. $(-j3)(2 + j4) = (-j3)(2) + (-j3)(j4) = -j6 - j^2 12 = 12 - j6$

b. $(2 + j3)(1 + j5) = (2)(1) + (2)(j5) + (j3)(1) + (j3)(j5)$
$= 2 + j10 + j3 + j^2 15 = 2 + j13 - 15 = -13 + j13$

PRACTICE PROBLEMS 1

1. Polar numbers with the same angle can be added or subtracted directly without conversion to rectangular form. For example, the sum of $6\angle 36.87°$ and $4\angle 36.87°$ is $10\angle 36.87°$, while the difference is $6\angle 36.87° - 4\angle 36.87° = 2\angle 36.87°$. By means of sketches, indicate why this procedure is valid.
2. To compare methods of multiplication with small integer values, convert the numbers of Example 16–4 to polar form, multiply them, then convert the answers back to rectangular form.

Answers

1. Since the numbers have the same angle, their sum also has the same angle and thus, their magnitudes simply add (or subtract).

NOTES...

A common mistake when determining reciprocals in rectangular form is to write the reciprocal of $a + jb$ as $\frac{1}{a+jb} = \frac{1}{a} + \frac{1}{jb}$. This is not correct. To illustrate, consider $\mathbf{C} = 3 + j4$. To find its reciprocal, enter $3 + j4$ into your calculator (see section entitled *Calculators for ac Analysis*), then press the inverse key. The result is $\frac{1}{\mathbf{C}} = \frac{1}{3+j4} = 0.12 - j0.16$. Clearly, this is not equal to $\frac{1}{3} + \frac{1}{j4} = 0.333 - j0.25$.

Reciprocals

In polar form, the reciprocal of a complex number $\mathbf{C} = C\angle\theta$ is

$$\frac{1}{C\angle\theta} = \frac{1}{C}\angle-\theta \qquad \textbf{(16–8)}$$

Thus,

$$\frac{1}{20\angle 30°} = 0.05\angle-30°$$

When you work in rectangular form, you must be somewhat careful—see the Note.

Complex Conjugates

The conjugate of a complex number (denoted by an asterisk *) is a complex number with the same real part but the opposite imaginary part. Thus, the conjugate of $\mathbf{C} = C\angle\theta = a + jb$ is $\mathbf{C}^* = C\angle-\theta = a - jb$. For example, if $\mathbf{C} = 3 + j4 = 5\angle 53.13°$, then $\mathbf{C}^* = 3 - j4 = 5\angle-53.13°$.

Calculators for ac Analysis

The analysis of ac circuits involves a considerable amount of complex number arithmetic, and you will need a calculator that works easily in both rectangular and polar forms, and (preferably) displays results in standard (or near-standard) notation as in Figure 16–6. (There are a number of calculators on the market, so be sure to check with your professors or instructors before buying to determine what make and model is recommended for your school. For purposes of illustration, we have selected the Sharp EL-506W—see Note. If you have a different calculator, consult its instruction manual, then follow through the examples. Although details differ, the concepts will generally be the same.) Computations that you will need to do in circuit analysis range from simple conversions between forms (as illustrated in Example 16–5) to rather complex computations (as illustrated in Example 16–6).

NOTES...

While preparing this edition, we surveyed current users, but found no consensus about the choice of calculator. From among the various choices, we arbitrarily selected the Sharp EL-506W for illustrative purposes.

EXAMPLE 16–5

Using your calculator,

a. determine the polar form of $\mathbf{C} = 3 + j\,4$,

b. determine the inverse of $3 - j\,4$ with the answer expressed in rectangular form.

Solution Consult your manual, then set your calculator for complex number operation with angles expressed in degrees. Using the following as general guidelines, proceed as shown:

a. Key in 3, press the + key, key in 4, press the imaginary operator key "*i*." Typically, your display will show $3 + 4i$. Now convert to polar form using the $\rightarrow r\theta$ procedure detailed in your manual. Depending on your calculator, the complete answer $5\angle 53.13°$ may appear on your screen, or you may have to extract the angle part as a second step—check your manual for details.

Sharp EL-506W Procedure: To select the complex mode, press the following keys: [Mode 3]. Key in 3, press the +, key in 4, press the "imaginary" key, press key 2nd F, press $\rightarrow r\theta$ (i.e., key 8), then the = key. The magnitude 5 appears, Figure 16–6(a). To display the angle, press 2nd F, then ←, → (i.e., the Exp key). The angle part of the answer appears, Figure 16–6(b). Thus, $3 + j4 = 5\angle 53.13°$ (rounded to a sensible number of digits).

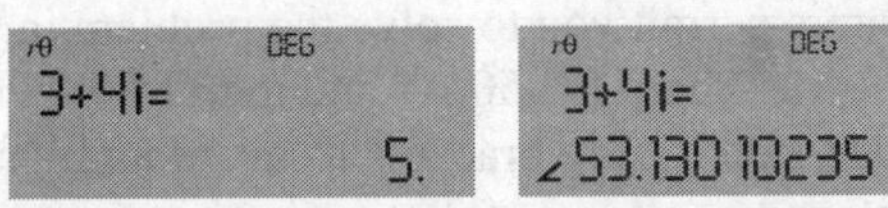

© Cengage Learning 2013

(a) (b)

FIGURE 16–6 Converting rectangular to polar form on the Sharp EL-506W. The magnitude and angle parts of the answer must be retrieved separately. (a) Showing the magnitude part of answer (b) Showing the angle part of answer.

b. Key in a left bracket, key in 3, press the minus key (not the ± key), key in 4, press the "imaginary" key, then add a right bracket. You should see $(3 - 4i)$ on your display. Now invoke the inverse operation. Your calculator should show $(3 - 4i)^{-1}$ (or something similar). Press the = key to get the answer $0.12 + 0.16i$.

Sharp EL-506W Procedure: After you enter $(3 - 4i)$, press 2nd F, then x^{-1} (i.e., key 2). This yields $(3 - 4i)^{-1}$. Press the = key. If your calculator is set to rectangular, the real part of the answer, 0.12, will appear on your display. Press 2nd F, then the ←, → key to view the imaginary part, $0.16i$. If your calculator is set to polar, you will get $0.2\angle 53.13°$. To convert this to rectangular, press 2nd F, then $\rightarrow xy$.

While all calculators with complex number capabilities can easily perform basic calculations such as those of Example 16–5, there is a great variation in how easily they handle complex calculations. To illustrate, consider the expression $(9 + j2) + 10\angle 30°$. On some calculators, you need intermediate steps, for example, you need to convert $10\angle 30°$ to rectangular in order to perform the addition. On other calculators, you can freely mix rectangular and polar numbers on input. This is illustrated in Figure 16–7. The answer will be displayed in either rectangular or polar, depending on what mode you have set.

© Cengage Learning 2013

FIGURE 16–7 On this calculator, rectangular and polar forms can be mixed freely on input. (Only the magnitude part of the answer is shown, as the angle part must be retrieved separately—recall Figure 16–6.)

(In polar form, the answer is 18.997 ∠21.62°. Note that only the magnitude of the answer is shown in Figure 16–7. A second step is required to extract the angle.)

Now consider Example 16–6.

EXAMPLE 16–6

Reduce the following to polar form.

$$(6 + j5) + \frac{(3 - j4)(10\angle 40°)}{6 + 30\angle 53.13°}$$

Solution

a. Using a calculator with basic capabilities requires a number of intermediate steps, some of which are shown below.

$$\text{answer} = (6 + j5) + \frac{(5\angle -53.13)(10\angle 40)}{6 + (18 + j24)}$$

$$= (6 + j5) + \frac{(5\angle -53.13)(10\angle 40)}{24 + j24}$$

$$= (6 + j5) + \frac{(5\angle -53.13)(10\angle 40)}{33.94\angle 45}$$

$$= (6 + j5) + 1.473\angle -58.13 = (6 + j5) + (0.778 - j1.251)$$

$$= 6.778 + j3.749 = 7.75\angle 28.95°$$

b. While all calculators permit you to solve the problem in small steps as in (a), others (such as the Sharp EL-506W and the TI-89) permit you to solve the problem in one step. Using brackets to avoid ambiguity, you can, for example, key the expression in as follows:

$$(6 + 5i) + ((3 - 4i)(10\angle 40)) \div (6 + 30\angle 53.13)$$

When you press the = key (or whatever key your calculator manual specifies), the expression is evaluated and the answer displayed. If your calculator is set to polar mode, you will get an answer of 7.75∠28.95°; if your calculator is set to rectangular mode, you will get 6.778 + 3.749*i*.

16.2 Complex Numbers in ac Analysis

Representing ac Voltages and Currents by Complex Numbers

As you learned in Chapter 15, ac voltages and currents can be represented as phasors (see Note 1). Since phasors have magnitude and angle, they can be viewed as complex numbers. To get at the idea, consider the voltage source of Figure 16–8(a). Its phasor equivalent (b) has magnitude E_m and angle θ. It therefore can be viewed as the complex number

$$\mathbf{E} = E_m\angle\theta \qquad (16\text{–}9)$$

NOTES...

1. As you saw in Chapter 15, the basic definition of a phasor is a rotating radius vector with a length equal to the amplitude (E_m or I_m) of the voltage or current that it represents. In practice, however, phasor voltages and phasor currents are always

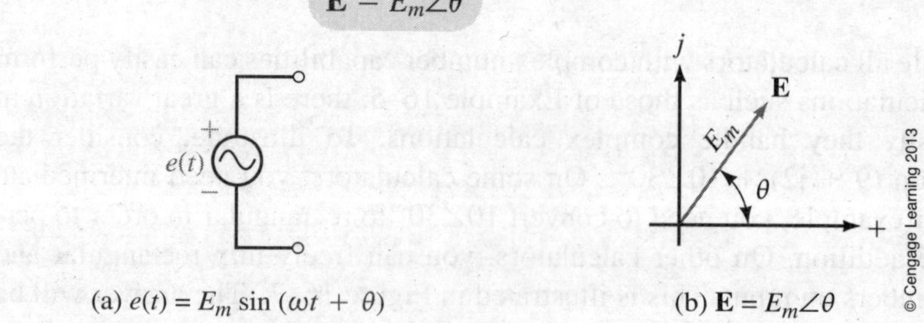

(a) $e(t) = E_m \sin(\omega t + \theta)$ (b) $\mathbf{E} = E_m\angle\theta$

FIGURE 16–8 Representation of a sinusoidal source voltage as a complex number.

From this point of view, the sinusoidal voltage $e(t) = 200 \sin(\omega t + 40°)$ of Figure 16–9(a) and (b) can be represented by its phasor equivalent, $\mathbf{E} = 200\text{ V}\angle 40°$, as in (c).

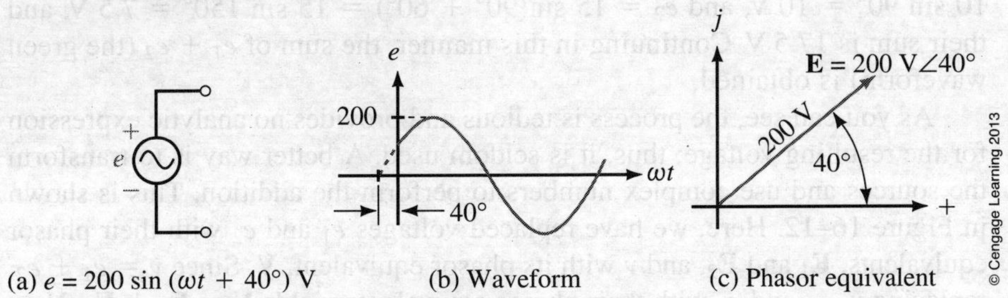

(a) $e = 200 \sin(\omega t + 40°)$ V (b) Waveform (c) Phasor equivalent

FIGURE 16–9 Transforming $e = 200 \sin(\omega t + 40°)$ V to $\mathbf{E} = 200\text{ V}\angle 40°$.

We can take advantage of this equivalence. *Rather than show a source as a time-varying voltage* e(t) *that we subsequently convert to a phasor, we can represent the source by its phasor equivalent right from the start.* This viewpoint is illustrated in Figure 16–10. Since $\mathbf{E} = 200\text{ V}\angle 40°$, this representation retains all the original information of Figure 16–9 since the sinusoidal time variation and its associated angle as illustrated in Figure 16–9(b) are implicit in the definition of the phasor.

The idea illustrated in Figure 16–10 is of fundamental importance to circuit theory. *By replacing the time function* e(t) *with its phasor equivalent* **E**, *we have transformed the source from the time domain to the phasor domain*—see Notes 2 and 3. (This is one of Steinmetz's great contributions to electrical science.)

expressed as rms values. We will make this change shortly; in the meantime, there are a few ideas that we need to explore using the phasor's fundamental definition.

2. Quantities expressed as time functions are said to be in the *time domain,* while quantities expressed as phasors are said to be in the *phasor domain.* Thus, $e = 200 \sin(\omega t + 40°)$ is in the time domain, while its phasor equivalent $\mathbf{E} = 200\text{ V}\angle 40°$ is in the phasor domain.

3. The phasor domain is sometimes referred to as the *frequency domain.*

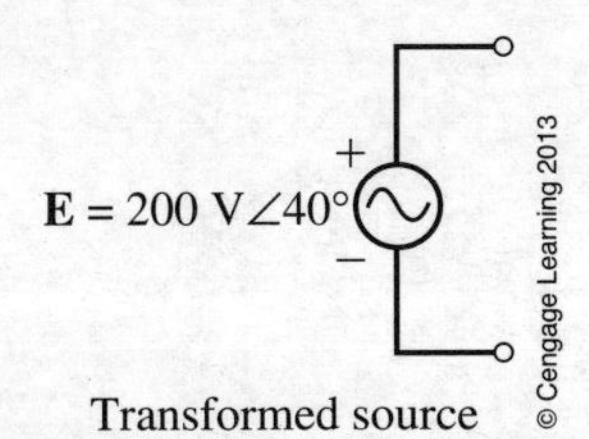

Transformed source

FIGURE 16–10 Direct transformation of the source.

Circuit Relationships in the Phasor Domain

Before we move on, we should note that both Kirchhoff's voltage law (KVL) and Kirchhoff's current law (KCL) apply in the **time domain** and in the **phasor domain**—see Notes 2 and 3. For example, $e = v_1 + v_2$ in the time domain can be transformed to $\mathbf{E} = \mathbf{V}_1 + \mathbf{V}_2$ in the phasor domain and vice versa. Similarly for currents. (Although we have stated the preceding result without proof, it can be proven rigorously—as is done in some advanced circuit theory textbooks.) The value of this approach is illustrated next.

Summing ac Voltages and Currents

Sinusoidal quantities must sometimes be added or subtracted as in Figure 16–11. Here, we want the sum of e_1 and e_2, where $e_1 = 10 \sin \omega t$ and $e_2 = 15 \sin(\omega t + 60°)$. First, let us illustrate the process in the time domain. The sum of

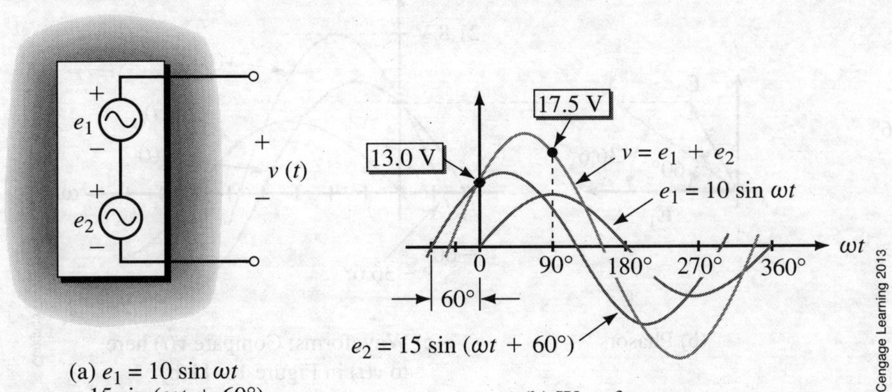

(a) $e_1 = 10 \sin \omega t$
$e_2 = 15 \sin(\omega t + 60°)$

(b) Waveforms

FIGURE 16–11 Summing waveforms point by point.

e_1 and e_2 can be found by adding waveforms point by point as in Figure 16–11(b). For example, at $\omega t = 0°$, $e_1 = 10 \sin 0° = 0$, and $e_2 = 15 \sin(0° + 60°) = 13$ V, and their sum is 13 V. Similarly, at $\omega t = 90°$, $e_1 = 10 \sin 90° = 10$ V, and $e_2 = 15 \sin(90° + 60°) = 15 \sin 150° = 7.5$ V, and their sum is 17.5 V. Continuing in this manner, the sum of $e_1 + e_2$ (the green waveform) is obtained.

As you can see, the process is tedious and provides no analytic expression for the resulting voltage; thus, it is seldom used. A better way is to transform the sources and use complex numbers to perform the addition. This is shown in Figure 16–12. Here, we have replaced voltages e_1 and e_2 with their phasor equivalents, $\mathbf{E}_1$ and $\mathbf{E}_2$, and v with its phasor equivalent, $\mathbf{V}$. Since $v = e_1 + e_2$, replacing v, e_1, and e_2 with their phasor equivalents yields $\mathbf{V} = \mathbf{E}_1 + \mathbf{E}_2$. Now $\mathbf{V}$ can be found by adding $\mathbf{E}_1$ and $\mathbf{E}_2$ as complex numbers. Once $\mathbf{V}$ is known, its corresponding time equation and companion waveform can be determined. This is the procedure used in practice.

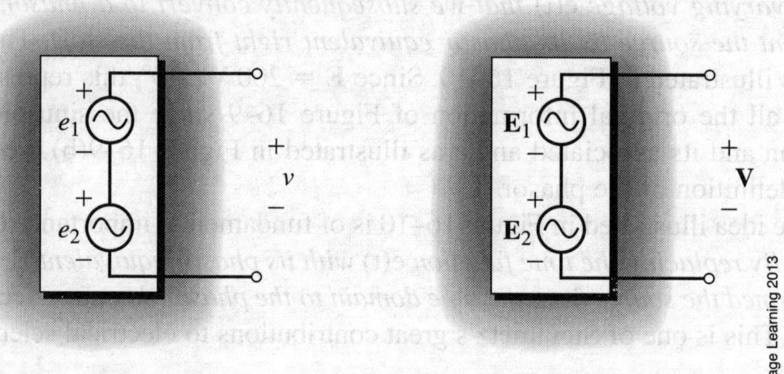

(a) Original network. $v(t) = e_1(t) + e_2(t)$

(b) Transformed network. $\mathbf{V} = \mathbf{E}_1 + \mathbf{E}_2$

FIGURE 16–12 Transformed circuit. This is one of the key ideas of sinusoidal circuit analysis.

EXAMPLE 16–7

Given $e_1 = 10 \sin \omega t$ V and $e_2 = 15 \sin(\omega t + 60°)$ V as before, determine v and sketch it.

CircuitSim 16-1

Solution

$$e_1 = 10 \sin \omega t \text{ V. Thus, } \mathbf{E}_1 = 10 \text{ V}\angle 0°.$$
$$e_2 = 15 \sin(\omega t + 60°) \text{ V. Thus, } \mathbf{E}_2 = 15 \text{ V}\angle 60°.$$

Transformed sources are shown in Figure 16–13(a) and phasors in (b).

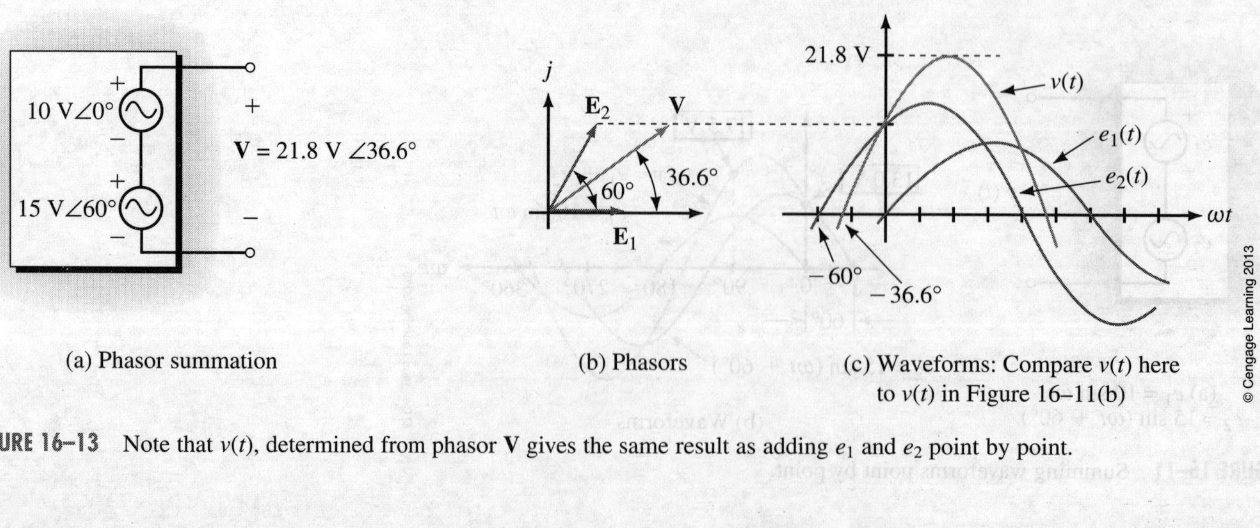

(a) Phasor summation

(b) Phasors

(c) Waveforms: Compare $v(t)$ here to $v(t)$ in Figure 16–11(b)

FIGURE 16–13 Note that $v(t)$, determined from phasor $\mathbf{V}$ gives the same result as adding e_1 and e_2 point by point.

$$\mathbf{V} = \mathbf{E}_1 + \mathbf{E}_2 = 10\angle 0° + 15\angle 60° = 21.8 \text{ V}\angle 36.6°$$

Thus, $v = 21.8 \sin(\omega t + 36.6°)$ V.

Waveforms are shown in (c). (To verify that this produces the same result as adding the waveforms point by point, see Practice Problems 2 or the Multisim simulation CircuitSim 16–1.)

PRACTICE PROBLEMS 2

Verify by direct substitution that $v = 21.8 \sin(\omega t + 36.6°)$ V, as in Figure 16–13, is the sum of e_1 and e_2. To do this, compute e_1 and e_2 at a point, add them, then compare the sum to $21.8 \sin(\omega t + 36.6°)$ V computed at the same point. Perform this computation at $\omega t = 30°$ intervals over the complete cycle to satisfy yourself that the result is true everywhere. [For example, at $\omega t = 0°$, $v = 21.8 \sin(\omega t + 36.6°) = 21.8 \sin(36.6°) = 13$ V, as we saw in Figure 16–11.]

Answers

Here are the points on the graph at 30° intervals:

ωt	0°	30°	60°	90°	120°	150°	180°	210°	240°	270°	300°	330°	360°
v	13	20	21.7	17.5	8.66	−2.5	−13	−20	−21.7	−17.5	−8.66	2.5	13

Representing Phasors as RMS Values

To this point, we have used peak values such as E_m, V_m, and I_m to represent the magnitudes of phasor voltages and currents, as this has been most convenient for our purposes. However, in practice, phasors are always expressed as rms values. Accordingly, we will now make the change. Thus, from now on, a phasor such as $\mathbf{V} = 120 \text{ V}\angle 0°$ will be taken to mean 120 V rms at an angle of zero degrees. If you need to convert this to a time function, first multiply the rms value by $\sqrt{2}$, then follow the usual procedure. Thus, the phasor $\mathbf{V} = 120 \text{ V}\angle 0°$, when converted to the time domain, yields $v(t) = \sqrt{2}(120 \text{ V}) \sin \omega t = 170 \sin \omega t$ V. See Note 1.

NOTES...

1. Although we use phasors to represent sinusoidal waveforms, it should be noted that sine waves and phasors are not the same thing. Sinusoidal voltages and currents are real—they are the actual quantities that you measure with meters and whose waveforms you see on oscilloscopes. *Phasors, on the other hand, are mathematical abstractions that we use to help visualize relationships and solve problems.*
2. To add or subtract sinusoidal voltages or currents, follow the three steps outlined in Example 16–7. That is,
   - convert sine waves to phasors and express them in complex number form,
   - add or subtract the complex numbers,
   - convert back to time functions if desired.

EXAMPLE 16–8

Express the voltage and current of Figure 16–14 in both the time and the phasor domains. (Remember to use rms values in the phasor domain.)

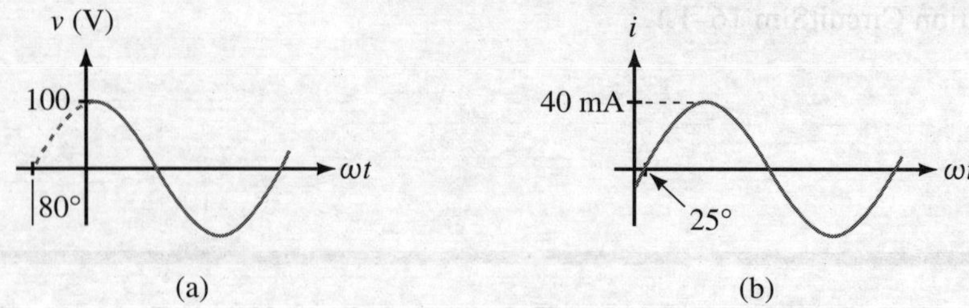

FIGURE 16–14

Solution

a. Time domain: $v = 100 \sin(\omega t + 80°)$ volts.
   Phasor domain: $\mathbf{V} = (0.707)(100\text{ V}\angle 80°) = 70.7\text{ V}\angle 80°$ (rms).

b. Time domain: $i = 40 \sin(\omega t - 25°)$ mA.
   Phasor domain: $\mathbf{I} = (0.707)(40\text{ mA}\angle -25°) = 28.3\text{ mA}\angle -25°$ (rms).

EXAMPLE 16–9

Consider two currents *i*1 and *i*2. If $i_1 = 14.14 \sin(\omega t - 55°)$ A and $i_2 = 4 \sin(\omega t + 15°)$ A:

CircuitSim 16-2

a. Determine their sum, *i*. Work with rms values.

b. If the currents are measured with ammeters, what will be their readings?

Solution

a. $\mathbf{I}_1 = (0.707)(14.14\text{ A})\angle -55° = 10\text{ A}\angle -55°$

$\mathbf{I}_2 = (0.707)(4\text{ A})\angle 15° = 2.828\text{ A}\angle 15°$

$= \mathbf{I}_1 + \mathbf{I}_2 = 10\text{ A}\angle -55° + 2.828\text{ A}\angle 15°$

$= 11.3\text{ A}\angle -41.4$

$i(t) = \sqrt{2}(11.3) \sin(\omega t - 41.4°) = 16 \sin(\omega t - 41.4°)$ A

b. 10 A, 2.828 A, and 11.3 A.

Example 16–9 illustrates a primary reason why we use rms—it ties in with standard practice in the sense that all voltmeters and ammeters display their measurements in rms. Additionally, in real life (after you have gotten through the learning process), you seldom need to deal explicitly with time functions; all you really need is magnitude and angle. Thus, very soon, we will stop working in the time domain entirely and work only with phasors. At that point, the solution will be complete when we have the answer in the form $\mathbf{I} = 11.3\text{A}\angle -41.4°$. (To help focus on rms, voltages and currents in the next two examples are expressed as an rms value times $\sqrt{2}$.)

EXAMPLE 16–10

For Figure 16–15, $v_1 = \sqrt{2}(16) \sin \omega t$ V, $v_2 = \sqrt{2}(24) \sin(\omega t + 90°)$ and $v_3 = \sqrt{2}(15) \sin(\omega t - 90°)$ V. Determine source voltage e. If measurements are made with voltmeters, what will the meters indicate?

Solution The answer can be obtained by KVL. First, convert to phasors. Thus, $\mathbf{V}_1 = 16\text{ V}\angle 0°$, $\mathbf{V}_2 = 24\text{ V}\angle 90°$, and $\mathbf{V}_3 = 15\text{ V}\angle -90°$. KVL yields $\mathbf{E} = \mathbf{V}_1 + \mathbf{V}_2 + \mathbf{V}_3 = 16\text{ V}\angle 0° + 24\text{ V}\angle 90° + 15\text{ V}\angle -90° = 18.4\text{ V}\angle 29.4°$. Converting back to a function of time yields $e = \sqrt{2}(18.4) \sin(\omega t + 29.4°)$ V. Voltmeter readings will be 16 V, 24 V, and 15 V for the voltage drops and 18.4 V for the source.

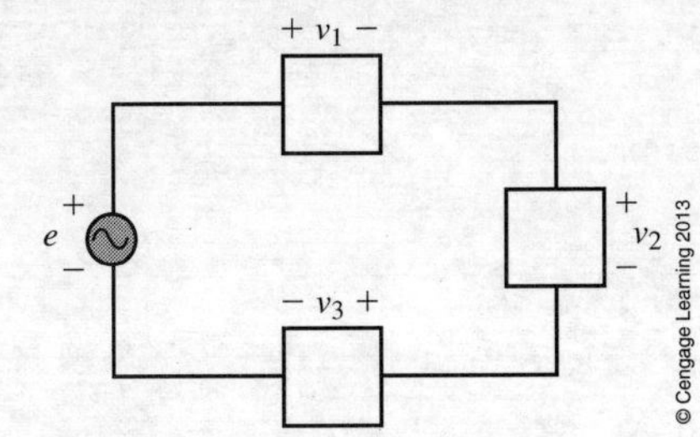

FIGURE 16–15

EXAMPLE 16–11

For Figure 16–16, $i_1 = \sqrt{2}(23) \sin \omega t$ mA, $i_2 = \sqrt{2}(0.29) \sin(\omega t + 63°)$ A, and $i_3 = \sqrt{2}(127) \times 10^{-3} \sin(\omega t - 72°)$ A. Determine current i_T. If currents are measured with ammeters, what will they indicate?

Solution Convert to phasors. Thus, $\mathbf{I}_1 = 23\text{ mA}\angle 0°$, $\mathbf{I}_2 = 0.29\text{ A}\angle 63°$, and $\mathbf{I}_3 = 127 \times 10^{-3}\text{ A}\angle -72°$. KCL yields $\mathbf{I}_T = \mathbf{I}_1 + \mathbf{I}_2 + \mathbf{I}_3 = 23\text{ mA}\angle 0° + 290\text{ mA}\angle 63° + 127\text{ mA}\angle -72° = 238\text{ mA}\angle 35.4°$. Converting back to a function of time yields $i_T = \sqrt{2}(238) \sin(\omega t + 35.4°)$ mA. Ammeters would read 23 mA, 290 mA, and 127 mA for the branch currents and 238 mA for the total current.

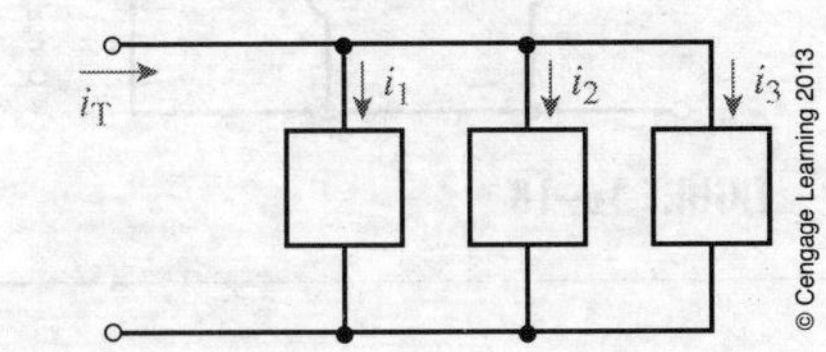

FIGURE 16–16

PRACTICE PROBLEMS 3

1. Convert the following to time functions. Values are rms.
   a. $\mathbf{E} = 500\text{ mV}\angle -20°$ b. $\mathbf{I} = 80\text{ A}\angle 40°$
2. For the circuit of Figure 16–17, using phasors, determine voltage e_1. If voltages are measured with voltmeters, what will the meters indicate?

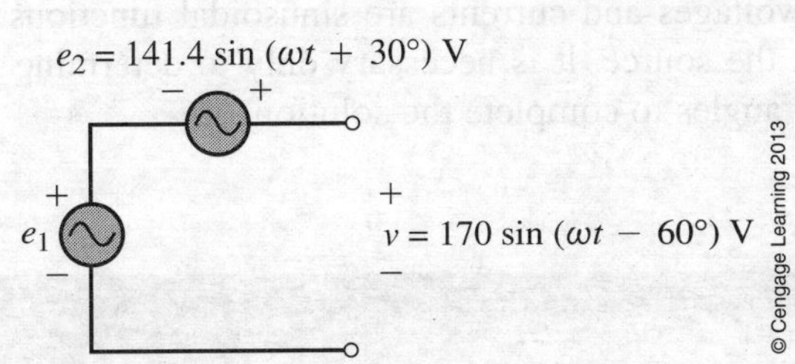

FIGURE 16–17

Answers

1. a. $e = 707 \sin(\omega t - 20°)$ mV b. $i = 113 \sin(\omega t + 40°)$ A
2. $e_1 = 221 \sin(\omega t - 99.8°)$ V, 156.3 V, 100 V for the sources and 120.2 for their sum.

IN-PROCESS LEARNING CHECK 1

(Answers are at the end of the chapter.)

1. Convert the following to polar form:
   a. $j6$ b. $-j4$ c. $3 + j3$ d. $4 - j6$
   e. $-5 + j8$ f. $1 - j2$ g. $-2 - j3$
2. Convert the following to rectangular form:
   a. $4\angle 90°$ b. $3\angle 0°$ c. $2\angle -90°$ d. $5\angle 40°$
   e. $6\angle 120°$ f. $2.5\angle -20°$ g. $1.75\angle -160°$
3. If $-\mathbf{C} = 12\angle -140°$, what is **C**?
4. Given: $\mathbf{C}_1 = 36 + j4$ and $\mathbf{C}_2 = 52 - j11$. Determine $\mathbf{C}_1 + \mathbf{C}_2$, $\mathbf{C}_1 - \mathbf{C}_2$, $1/(\mathbf{C}_1 + \mathbf{C}_2)$, and $1/(\mathbf{C}_1 - \mathbf{C}_2)$. Express in rectangular form.
5. Given: $\mathbf{C}_1 = 24\angle 25°$ and $\mathbf{C}_2 = 12\angle -125°$. Determine $\mathbf{C}_1 \cdot \mathbf{C}_2$ and $\mathbf{C}_1/\mathbf{C}_2$.
6. Compute the following and express answers in rectangular form:
   a. $\dfrac{6 + j4}{10\angle 20°} + (14 + j2)$ b. $(1 + j6) + \left[2 + \dfrac{(12\angle 0°)(14 + j2)}{6 - (10\angle 20°)(2\angle -10°)}\right]$
7. For Figure 16–18, determine i_T where $i_1 = 10 \sin \omega t$, $i_2 = 20 \sin(\omega t - 90°)$, and $i_3 = 5 \sin(\omega t + 90°)$.
8. If meters are used to measure the currents of Figure 16–18, what will be their readings?

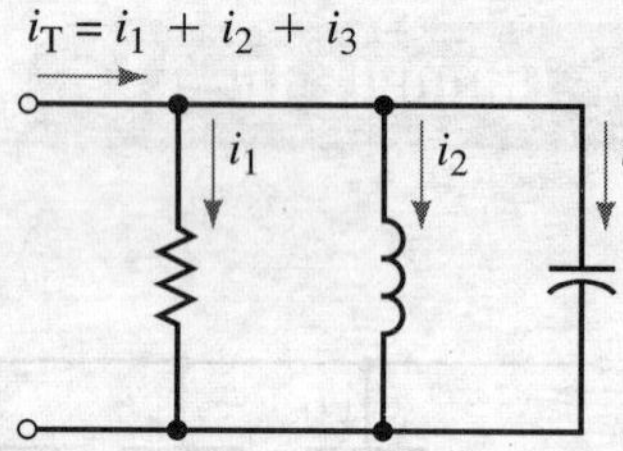

FIGURE 16–18

16.3 *R, L,* and *C* Circuits with Sinusoidal Excitation

R, L, and *C* circuit elements each have quite different electrical properties. Resistance, for example, opposes current, while inductance opposes changes in current, and capacitance opposes changes in voltage. These differences result in quite different voltage-current relationships as you saw earlier. We now investigate these relationships for the case of sinusoidal ac. Sine waves have several important characteristics that you will discover from this investigation:

1. When a circuit consisting of linear circuit elements *R, L,* and *C* is connected to a sinusoidal source, all currents and voltages in the circuit will be sinusoidal and of the same frequency as the source.
2. Since all voltages and currents are sinusoidal functions of the same frequency as the source, it is necessary only to determine their magnitudes and phase angles to complete the solution.

16.4 Resistance and Sinusoidal ac

We begin with a purely resistive circuit. Here, Ohm's law applies, and thus current is directly proportional to voltage. Current variations therefore follow voltage variations, reaching their peak when voltage reaches its peak, changing direction when voltage changes polarity, and so on (Figure 16–19). From this, we conclude that *for a purely resistive circuit, current and voltage are in phase.* Since voltage and current waveforms coincide, their phasors also coincide (Figure 16–20).

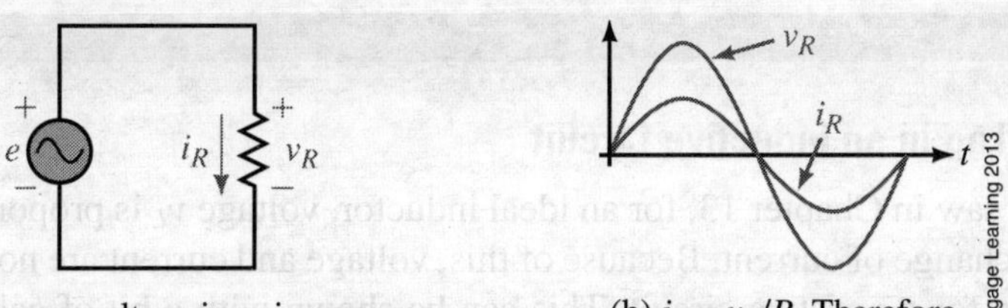

(a) Source voltage is a sine wave. Therefore, v_R is a sine wave.

(b) $i_R = v_R/R$. Therefore i_R is a sine wave also.

FIGURE 16–19 Ohm's law applies. Note that current and voltage are in phase.

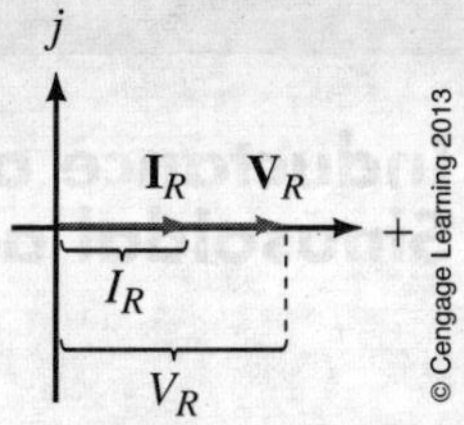

FIGURE 16–20 For a resistor, voltage and current phasors are in phase.

The relationship illustrated in Figure 16–19 may be stated mathematically as

$$i_R = \frac{v_R}{R} = \frac{V_m \sin \omega t}{R} = \frac{V_m}{R} \sin \omega t = I_m \sin \omega t \qquad \textbf{(16–10)}$$

where

$$I_m = V_m/R \qquad \textbf{(16–11)}$$

Transposing,

$$V_m = I_m R \qquad \textbf{(16–12)}$$

The in-phase relationship is true regardless of reference. Thus, if $v_R = V_m \sin(\omega t + \theta)$, then $i_R = I_m \sin(\omega t + \theta)$.

EXAMPLE 16–12

For the circuit of Figure 16–19(a), if $R = 5\ \Omega$ and $i_R = 12 \sin(\omega t - 18°)$ A, determine v_R.

Solution $v_R = Ri_R = 5 \times 12 \sin(\omega t - 18°) = 60 \sin(\omega t - 18°)$ V. The waveforms are shown in Figure 16–21.

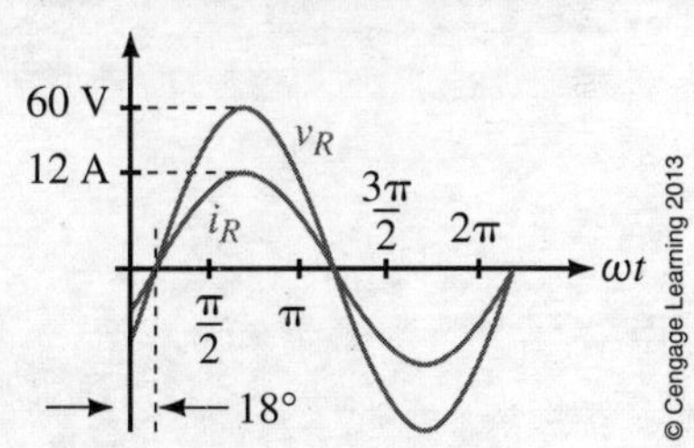

FIGURE 16–21

PRACTICE PROBLEMS 4

1. If $v_R = 150 \cos \omega t$ V and $R = 25\ \text{k}\Omega$, determine i_R and sketch both waveforms.
2. If $v_R = 100 \sin(\omega t + 30°)$ V and $R = 0.2\ \text{M}\Omega$, determine i_R and sketch both waveforms.

Answers

1. $i_R = 6 \cos \omega t$ mA. v_R and i_R are in phase.
2. $i_R = 0.5 \sin(\omega t + 30°)$ mA. v_R and i_R are in phase.

16.5 Inductance and Sinusoidal ac

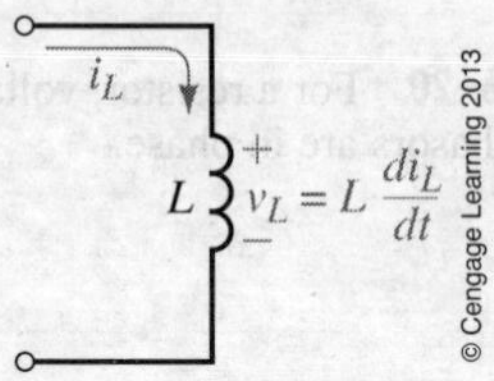

FIGURE 16–22 Voltage v_L is proportional to the rate of change of current i_L.

Phase Lag in an Inductive Circuit

As you saw in Chapter 13, for an ideal inductor, voltage v_L is proportional to the rate of change of current. Because of this, voltage and current are not in phase as they are for a resistive circuit. This can be shown with a bit of calculus. From Figure 16–22, $v_L = Ldi_L/dt$. For a sine wave of current, you get the following when you differentiate

$$v_L = L\frac{di_L}{dt} = L\frac{d}{dt}(I_m \sin \omega t) = \omega L I_m \cos \omega t = V_m \cos \omega t$$

Utilizing the trigonometric identity $\cos \omega t = \sin(\omega t + 90°)$, you can write this as

$$v_L = V_m \sin(\omega t + 90°) \tag{16–13}$$

where

$$V_m = \omega L I_m \tag{16–14}$$

Voltage and current waveforms are shown in Figure 16–23, and phasors in Figure 16–24. As you can see, *for a purely inductive circuit, current lags voltage by 90°* (i.e., ¼ cycle). Alternatively you can say that voltage leads current by 90°.

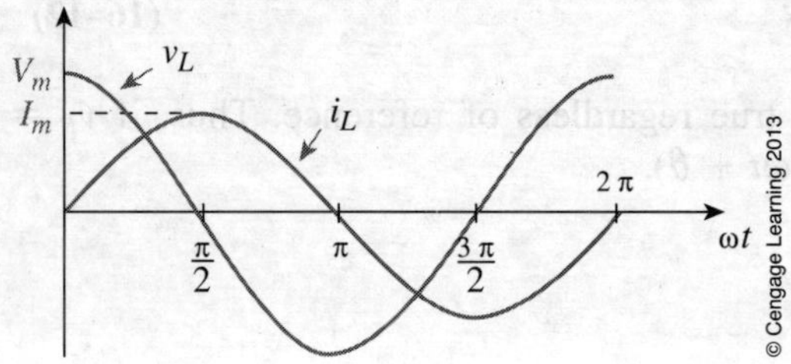

FIGURE 16–23 For inductance, current lags voltage by 90°. Here, i_L is reference.

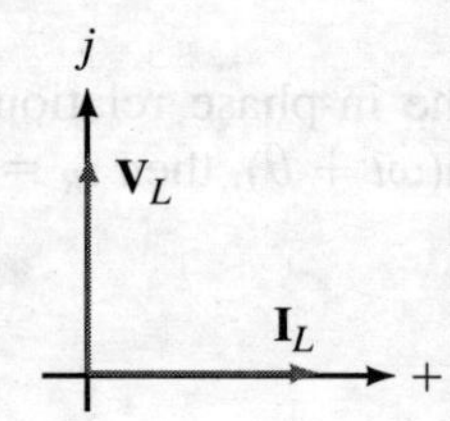

FIGURE 16–24 Phasors for the waveforms of Figure 16–23 showing the 90° lag of current.

Although we have shown that current lags voltage by 90° for the case of Figure 16–23, this relationship is true in general, that is, current always lags voltage by 90° regardless of the choice of reference. This is illustrated in Figure 16–25. Here, $\mathbf{V}_L$ is at 0° and $\mathbf{I}_L$ at −90°. Thus, voltage v_L will be a sine wave and current i_L a negative cosine wave, that is, $i_L = -I_m \cos \omega t$. Since i_L is a negative cosine wave, it can also be expressed as $i_L = I_m \sin(\omega t - 90°)$. The waveforms are shown in (b).

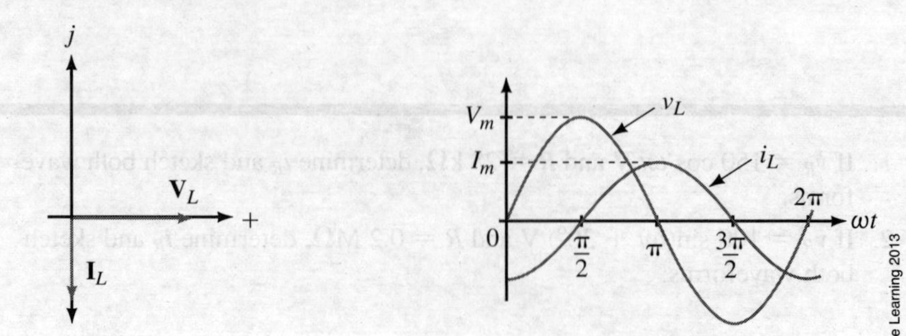

(a) Current $\mathbf{I}_L$ always lags voltage $\mathbf{V}_L$ by 90°

(b) Waveforms

FIGURE 16–25 Phasors and waveforms when $\mathbf{V}_L$ is used as reference.

Since current always lags voltage by 90° for a pure inductance, you can, if you know the phase of the voltage, determine the phase of the current, and vice versa. Thus, if v_L is known, i_L must lag it by 90°, while if i_L is known, v_L must lead it by 90°.

Inductive Reactance

From Equation 16–14, we see that the ratio V_m to I_m is

$$\frac{V_m}{I_m} = \omega L \tag{16–15}$$

The ratio of V_m to I_m is defined as **inductive reactance** and is given the symbol X_L. Since the ratio of volts to amps is ohms, reactance has units of ohms. Thus,

$$X_L = \frac{V_m}{I_m} \quad (\Omega) \tag{16–16}$$

Combining Equations 16–15 and 16–16 yields

$$X_L = \omega L \ (\Omega) \tag{16–17}$$

where ω is in radians per second and L is in henries. *Reactance* X_L *represents the opposition that inductance presents to current for the sinusoidal ac case.*

We now have everything that we need to solve simple inductive circuits with sinusoidal excitation; that is, we know that current lags voltage by 90° and that their amplitudes are related by

$$I_m = \frac{V_m}{X_L} \tag{16–18}$$

and

$$V_m = I_m X_L \tag{16–19}$$

EXAMPLE 16–13

The voltage across a 0.2-H inductance is $v_L = 100 \sin(400t + 70°)$ V. Determine i_L and sketch it.

Solution $\omega = 400$ rad/s. Therefore, $X_L = \omega L = (400)(0.2) = 80\ \Omega$.

$$I_m = \frac{V_m}{X_L} = \frac{100 \text{ V}}{80\ \Omega} = 1.25 \text{ A}$$

The current lags the voltage by 90°. Therefore $i_L = 1.25 \sin(400t - 20°)$ A as indicated in Figure 16–26.

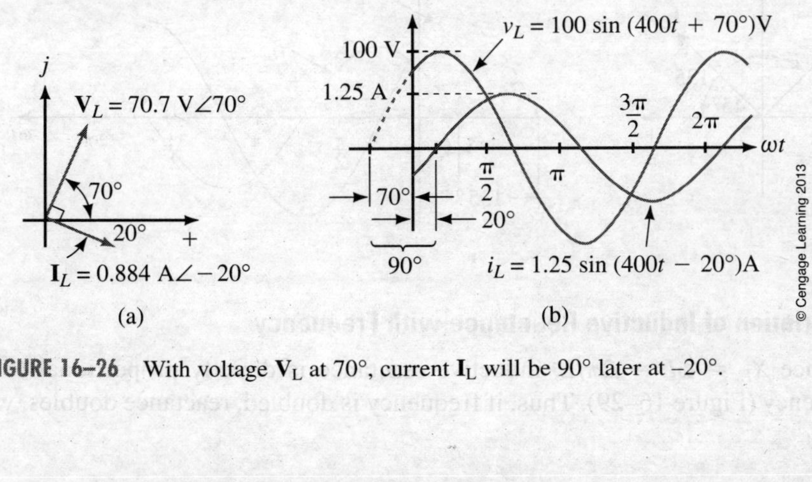

FIGURE 16–26 With voltage $\mathbf{V}_L$ at 70°, current $\mathbf{I}_L$ will be 90° later at −20°.

NOTES...

Remember to show phasors as rms values. However (by convention), we draw waveforms using peak amplitudes—V_m and I_m, and so on.

EXAMPLE 16–14

The current through a 0.01-H inductance is $i_L = 20 \sin(\omega t - 50°)$ A and $f = 60$ Hz. Determine v_L.

Solution

$$\omega = 2\pi f = 2\pi(60) = 377 \text{ rad/s}$$

$$X_L = \omega L = (377)(0.01) = 3.77\ \Omega$$

$$V_m = I_m X_L = (20\text{ A})(3.77\ \Omega) = 75.4\text{ V}$$

Voltage leads current by 90°. Thus, $v_L = 75.4 \sin(377t + 40°)$ V as shown in Figure 16–27.

CircuitSim 16-3

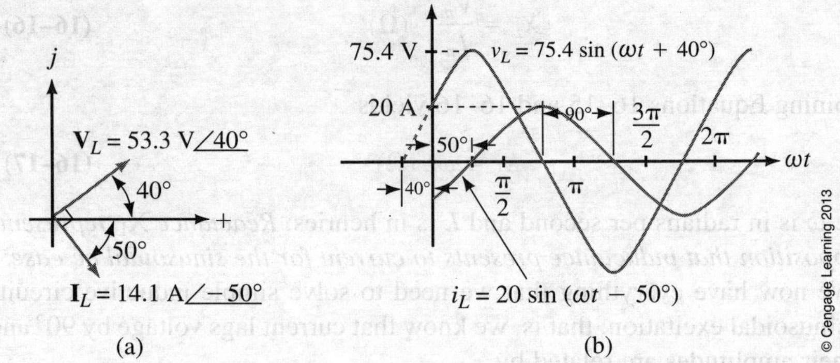

FIGURE 16–27

PRACTICE PROBLEMS 5

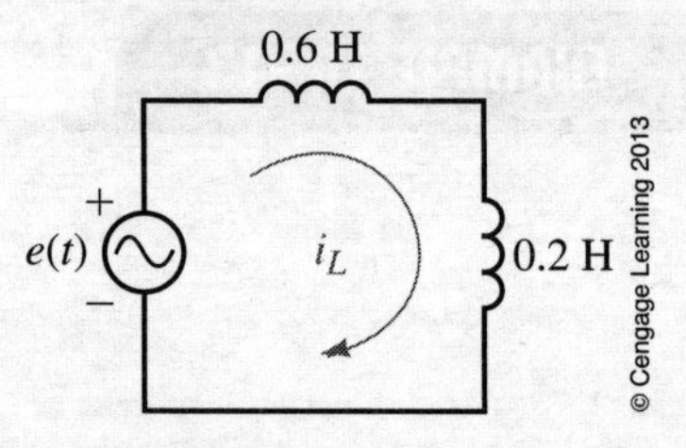

FIGURE 16–28

1. Two inductances are connected in series (Figure 16–28). If $e = 100 \sin \omega t$ and $f = 10$ kHz, determine the current. Sketch voltage and current waveforms.
2. The current through a 0.5-H inductance is $i_L = 100 \sin(2400t + 45°)$ mA. Determine v_L and sketch voltage and current phasors and waveforms.

Answers

1. $i_L = 1.99 \sin(\omega t - 90°)$ mA. Waveforms same as Figure 16–25.
2. $v_L = 120 \sin(2400t + 135°\text{ V})$. See the following figure for waveforms.

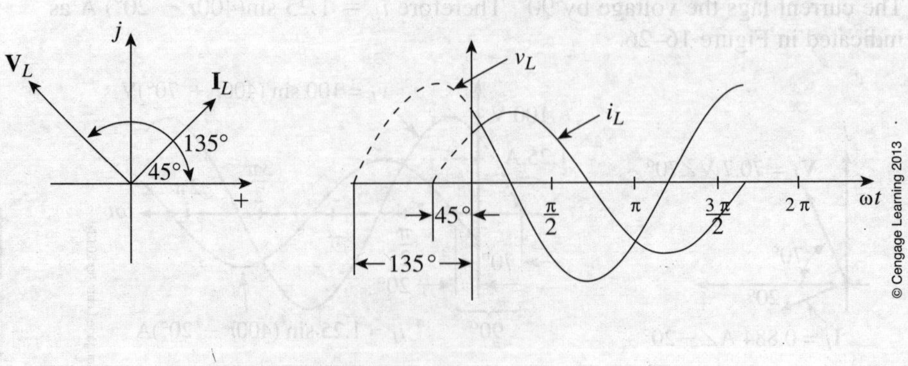

Variation of Inductive Reactance with Frequency

Since $X_L = \omega L = 2\pi f L$, inductive reactance is directly proportional to frequency (Figure 16–29). Thus, if frequency is doubled, reactance doubles, while

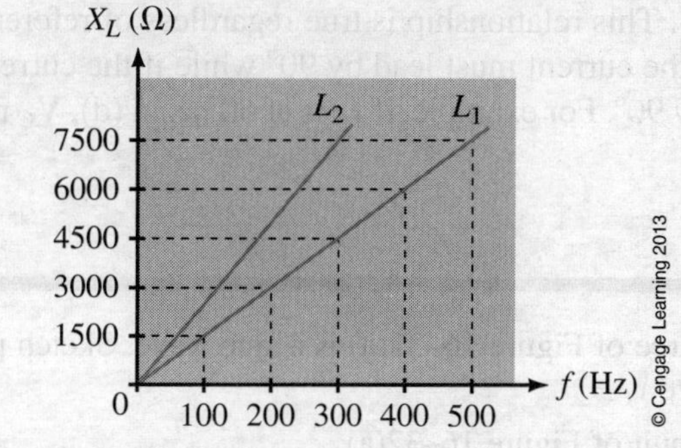

FIGURE 16–29 Variation of X_L with frequency. Note that $L_2 > L_1$.

if frequency is halved, reactance halves, and so on. In addition, X_L is directly proportional to inductance. Thus, if inductance is doubled, X_L is doubled, and so on. Note also that at $f = 0$, $X_L = 0\ \Omega$. This means that inductance looks like a short circuit to dc. (We already concluded this earlier in Chapter 13.)

PRACTICE PROBLEMS 6

A coil has 50 ohms inductive reactance. If both the inductance and the frequency are doubled, what is the new X_L?

Answer
200 Ω

16.6 Capacitance and Sinusoidal ac

Phase Lead in a Capacitive Circuit

For capacitance, current is proportional to the rate of change of voltage, that is, $i_C = C\, dv_C/dt$ [Figure 16–30(a)]. Thus, if v_C is a sine wave, you get upon substitution

$$i_C = C\frac{dv_C}{dt} = C\frac{d}{dt}(V_m \sin \omega t) = \omega C V_m \cos \omega t = I_m \cos \omega t$$

Using the appropriate trigonometric identity, this can be written as

$$i_C = I_m \sin(\omega t + 90°) \qquad \textbf{(16–20)}$$

where

$$I_m = \omega C V_m \qquad \textbf{(16–21)}$$

Waveforms are shown in Figure 16–30(b) and phasors in (c). As indicated, *for a purely capacitive circuit, current leads voltage by 90°*, or alternatively, voltage

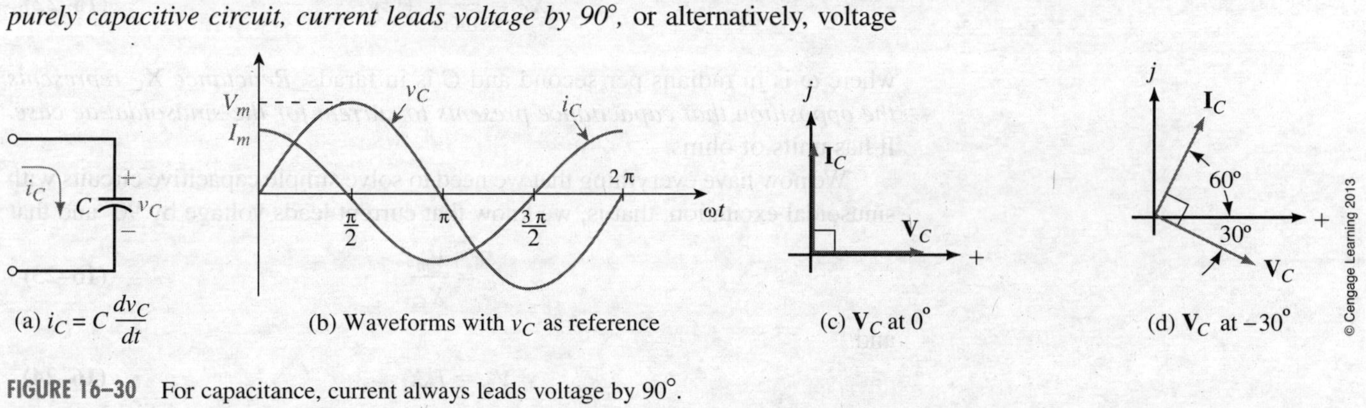

FIGURE 16–30 For capacitance, current always leads voltage by 90°.

lags current by 90°. This relationship is true regardless of reference. Thus, if the voltage is known, the current must lead by 90° while if the current is known, the voltage must lag by 90°. For example, if $\mathbf{I}_C$ is at 60° as in (d), $\mathbf{V}_C$ must be at −30°.

PRACTICE PROBLEMS 7

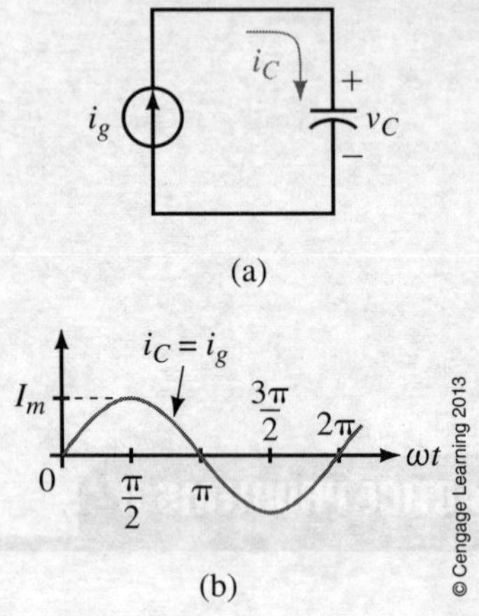

FIGURE 16–31

1. The current source of Figure 16–31(a) is a sine wave. Sketch phasors and capacitor voltage v_C.
2. Refer to the circuit of Figure 16–32(a):
   a. Sketch the phasors.
   b. Sketch capacitor current i_C.

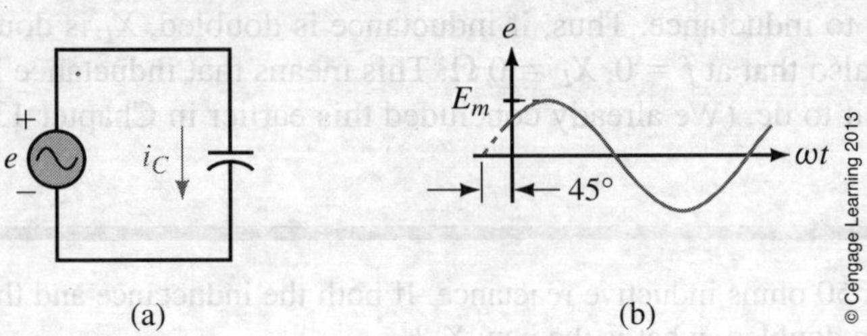

FIGURE 16–32

Answers

1. $\mathbf{I}_C$ is at 0°; $\mathbf{V}_C$ is at −90°; v_C is a negative cosine wave.
2. a. $\mathbf{V}_C$ is at 45° and $\mathbf{I}_C$ is at 135°.
   b. Waveforms are the same as for Problem 2, Practice Problems 5, except that voltage and current waveforms are interchanged.

Capacitive Reactance

Now consider the relationship between maximum capacitor voltage and current magnitudes. As we saw in Equation 16–21, they are related by $I_m = \omega C V_m$. Rearranging, we get $V_m/I_m = 1/\omega C$. The ratio of V_m to I_m is defined as **capacitive reactance** and is given the symbol X_C. That is,

$$X_C = \frac{V_m}{I_m} \quad (\Omega)$$

Since $V_m/I_m = 1/\omega C$, we also get

$$X_C = \frac{1}{\omega C} \quad (\Omega) \tag{16–22}$$

where ω is in radians per second and C is in farads. *Reactance* X_C *represents the opposition that capacitance presents to current for the sinusoidal ac case.* It has units of ohms.

We now have everything that we need to solve simple capacitive circuits with sinusoidal excitation, that is, we know that current leads voltage by 90° and that

$$I_m = \frac{V_m}{X_C} \tag{16–23}$$

and

$$V_m = I_m X_C \tag{16–24}$$

EXAMPLE 16–15

The voltage across a 10-μF capacitance is $v_C = 100 \sin(\omega t - 40°)$ V and $f = 1000$ Hz. Determine i_C and sketch its waveform.

Solution

$$\omega = 2\pi f = 2\pi(1000 \text{ Hz}) = 6283 \text{ rad/s}$$

$$X_C = \frac{1}{\omega C} = \frac{1}{(6283)(10 \times 10^{-6})} = 15.92\ \Omega$$

$$I_m = \frac{V_m}{X_C} = \frac{100 \text{ V}}{15.92\ \Omega} = 6.28 \text{ A}$$

Since current leads voltage by 90°, $i_C = 6.28 \sin(6283t + 50°)$ A as indicated in Figure 16–33.

CircuitSim 16-4

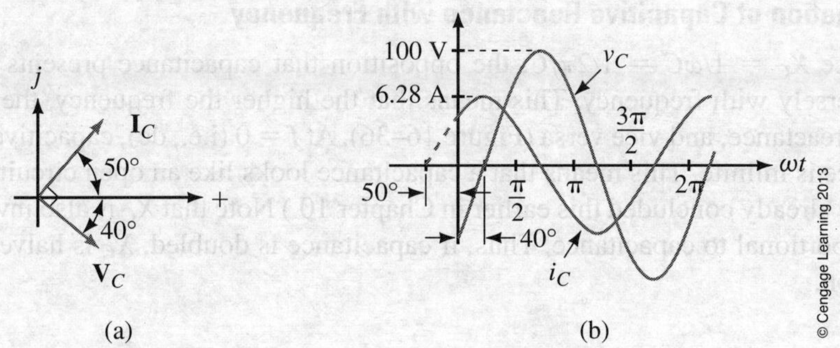

FIGURE 16–33 Phasors are not to scale with waveform.

EXAMPLE 16–16

The current through a 0.1-μF capacitance is $i_C = 5 \sin(1000t + 120°)$ mA. Determine v_C.

Solution

$$X_C = \frac{1}{\omega C} = \frac{1}{(1000 \text{ rad/s})(0.1 \times 10^{-6}\ F)} = 10\ \text{k}\Omega$$

Thus, $V_m = I_m X_C = (5 \text{ mA})(10\ \text{k}\Omega) = 50$ V. Since voltage lags current by 90°, $v_C = 50 \sin(1000t + 30°)$ V. Waveforms and phasors are shown in Figure 16–34.

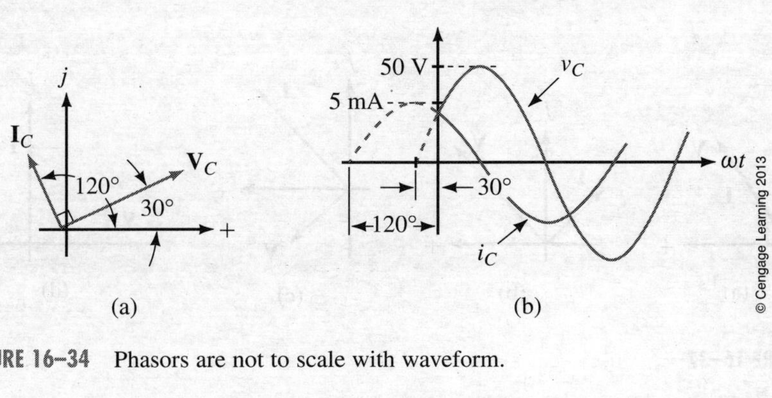

FIGURE 16–34 Phasors are not to scale with waveform.

PRACTICE PROBLEMS 8

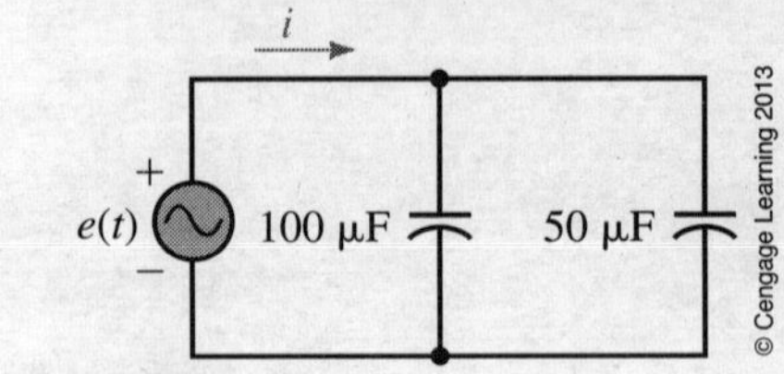

FIGURE 16–35

Two capacitances are connected in parallel (Figure 16–35). If $e = 100 \sin \omega t$ V and $f = 10$ Hz, determine the source current. Sketch current and voltage phasors and waveforms.

Answer: $i = 0.942 \sin(62.8t + 90°) = 0.942 \cos 62.8t$ A

See Figure 16–30(b) and (c).

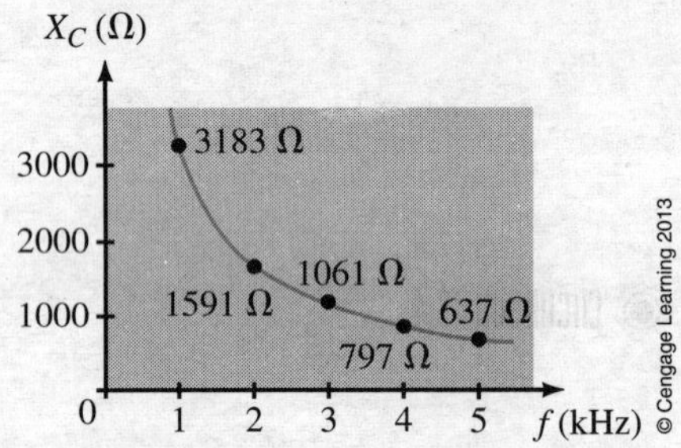

FIGURE 16–36 X_C varies inversely with frequency. Values shown are for $C = 0.05$ μF.

Variation of Capacitive Reactance with Frequency

Since $X_C = 1/\omega C = 1/2\pi fC$, the opposition that capacitance presents varies inversely with frequency. This means that the higher the frequency, the lower the reactance, and vice versa (Figure 16–36). At $f = 0$ (i.e., dc), capacitive reactance is infinite. This means that a capacitance looks like an open circuit to dc. (We already concluded this earlier in Chapter 10.) Note that X_C is also inversely proportional to capacitance. Thus, if capacitance is doubled, X_C is halved, and so on.

IN-PROCESS LEARNING CHECK 2

(Answers are at the end of the chapter.)

1. For a pure resistance, $v_R = 100 \sin(\omega t + 30°)$ V. If $R = 2\ \Omega$, what is the expression for i_R?
2. For a pure inductance, $v_L = 100 \sin(\omega t + 30°)$ V. If $X_L = 2\ \Omega$, what is the expression for i_L?
3. For a pure capacitance, $v_C = 100 \sin(\omega t + 30°)$ V. If $X_C = 2\Omega$, what is the expression for i_C?
4. If $f = 100$ Hz and $X_L = 400\ \Omega$, what is L?
5. If $f = 100$ Hz and $X_C = 400\ \Omega$, what is C?
6. For each of the phasor sets of Figure 16–37, identify whether the circuit is resistive, inductive, or capacitive. Justify your answers.

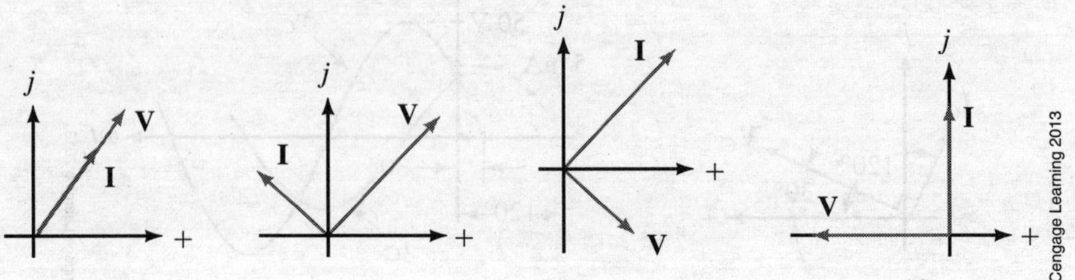

FIGURE 16–37

16.7 The Impedance Concept

During the learning process of Sections 16.5 and 16.6, we handled magnitude and phase analysis separately. However, this is not the way it is done in practice. In practice, we represent circuit elements by their impedance, and we determine magnitude and phase relationships in one step. Before we do this, however, we need to learn how to represent circuit elements as impedances.

Impedance

The opposition that a circuit element presents to current in the phasor domain is defined as its **impedance.** The impedance of the element of Figure 16–38, for example, is the ratio of its voltage phasor to its current phasor. Impedance is denoted by the boldface, uppercase letter **Z**. Thus,

$$\mathbf{Z} = \frac{\mathbf{V}}{\mathbf{I}} \quad \text{(ohms)} \qquad \textbf{(16–25)}$$

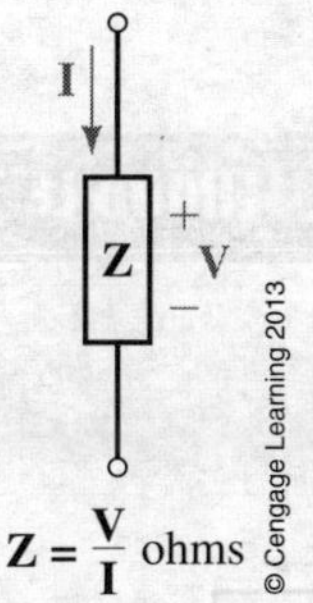

FIGURE 16–38 Impedance concept.

(This equation is sometimes referred to as Ohm's law for ac circuits.)

Since phasor voltages and currents are complex, **Z** is also complex. That is,

$$\mathbf{Z} = \frac{\mathbf{V}}{\mathbf{I}} = \frac{V}{I}\angle\theta \qquad \textbf{(16–26)}$$

where V and I are the rms magnitudes of **V** and **I,** respectively, and θ is the angle between them. From Equation 16–26,

$$\mathbf{Z} = Z\angle\theta \qquad \textbf{(16–27)}$$

where $Z = V/I$. Since $V = 0.707V_m$ and $I = 0.707I_m$, Z can also be expressed as V_m/I_m. Once the impedance of a circuit is known, the current and voltage can be determined using

$$\mathbf{I} = \frac{\mathbf{V}}{\mathbf{Z}} \qquad \textbf{(16–28)}$$

and

$$\mathbf{V} = \mathbf{IZ} \qquad \textbf{(16–29)}$$

Let us now determine impedance for the basic circuit elements R, L, and C.

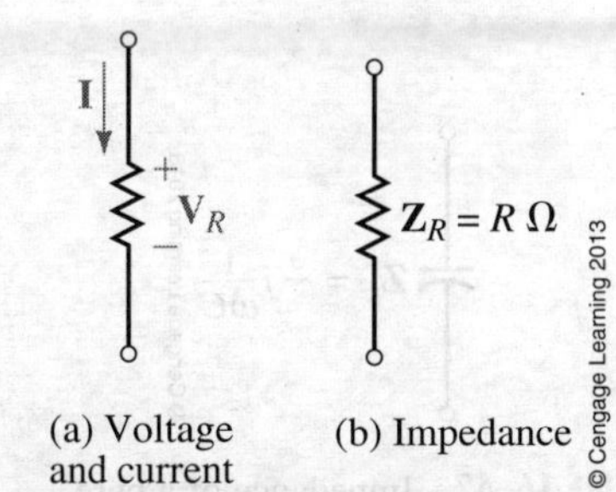

FIGURE 16–39 Impedance of a pure resistance.

Resistance

For a pure resistance (Figure 16–39), voltage and current are in phase. Thus, if voltage has an angle θ, current will have the same angle. For example, if $\mathbf{V}_R = V_R\angle\theta$, then $\mathbf{I} = I\angle\theta$. Substituting into Equation 16–25 yields:

$$\mathbf{Z}_R = \frac{\mathbf{V}_R}{\mathbf{I}} = \frac{V_R\angle\theta}{I\angle\theta} = \frac{V_R}{I}\angle 0° = R\angle 0° = R$$

Thus, the impedance of a resistor is just its resistance. That is,

$$\mathbf{Z}_R = R \qquad \textbf{(16–30)}$$

This agrees with what we know about resistive circuits, that the ratio of voltage to current is R and that the angle between them is 0°.

NOTES...

Although **Z** is a complex number, it is not a phasor. (Phasors are complex numbers that are used to represent sinusoidally varying quantities such as voltage and current). However, **Z** does not represent anything that varies with time—thus it is not a phasor.

Inductance

For a pure inductance, current lags voltage by 90°. Assuming a 0° angle for voltage (we can assume any reference we want because we are interested only

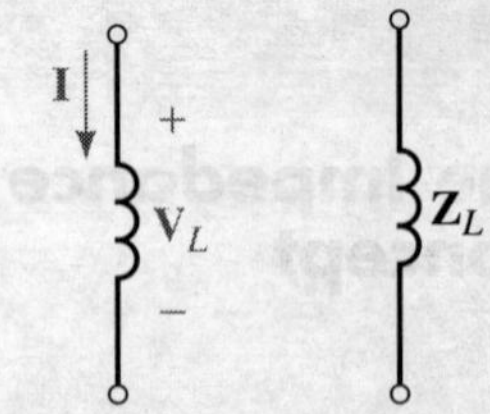

(a) Voltage and current (b) Impedance

FIGURE 16–40 Impedance of a pure inductance.

in the angle between $\mathbf{V}_L$ and $\mathbf{I}$), we can write $\mathbf{V}_L = V_L\angle 0°$ and $\mathbf{I} = I\angle -90°$. The impedance of a pure inductance (Figure 16–40) is therefore

$$\mathbf{Z}_L = \frac{\mathbf{V}_L}{\mathbf{I}} = \frac{V_L\angle 0°}{I\angle -90°} = \frac{V_L}{I}\angle 90° = \omega L\angle 90° = j\omega L$$

where we have used the fact that $V_L/I_L = \omega L$. Thus,

$$\mathbf{Z}_L = j\omega L = jX_L \tag{16–31}$$

since ωL is equal to X_L.

EXAMPLE 16–17

Consider again Example 16–13. Given $v_L = 100\ \sin(400t + 70°)$ V and $L =$ 0.2 H, determine i_L using the impedance concept.

Solution See Figure 16–41. Use rms values. Thus,

$$\mathbf{V}_L = 70.7\ \text{V}\angle 70° \quad \text{and} \quad \omega = 400\ \text{rad/s}$$

$$\mathbf{Z}_L = j\omega L = j(400)(0.2) = j80\ \Omega$$

$$\mathbf{I}_L = \frac{\mathbf{V}_L}{\mathbf{Z}_L} = \frac{70.7\angle 70°}{j80} = \frac{70.7\angle 70°}{80\angle 90°} = 0.884\ \text{A}\angle -20°$$

In the time domain, $i_L = \sqrt{2}(0.884)\ \sin(400t - 20°) = 1.25\ \sin(400t - 20°)$ A, which agrees with our previous solution, Example 16–13.

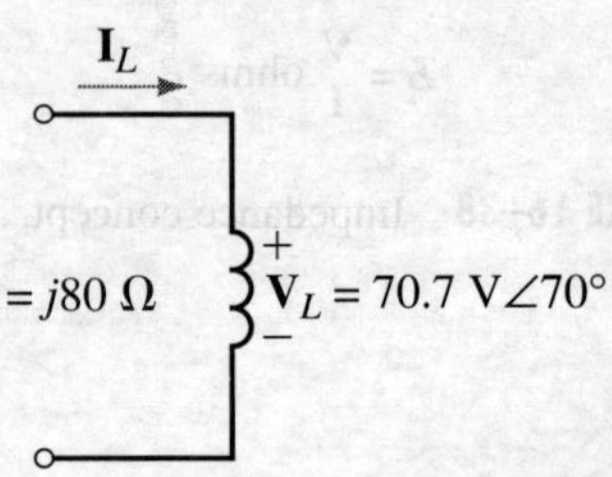

FIGURE 16–41

Capacitance

For a pure capacitance, current leads voltage by 90°. Its impedance (Figure 16–42) is therefore

$$\mathbf{Z}_C = \frac{\mathbf{V}_C}{\mathbf{I}} = \frac{V_C\angle 0°}{I\angle 90°} = \frac{V_C}{I}\angle -90° = \frac{1}{\omega C}\angle -90° = -j\frac{1}{\omega C} \quad \text{(ohms)}$$

Thus,

$$\mathbf{Z}_C = -j\frac{1}{\omega C} = -jX_C \quad \text{(ohms)} \tag{16–32}$$

since $1/\omega C$ is equal to X_C.

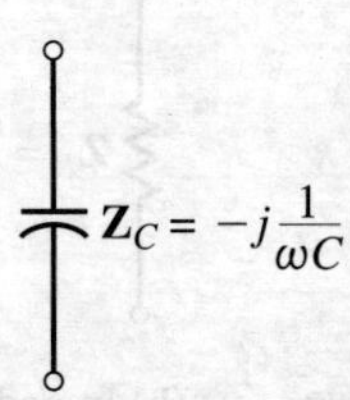

FIGURE 16–42 Impedance of a pure capacitance.

EXAMPLE 16–18

Given $v_C = 100\ \sin(\omega t - 40°)$ V, $f = 1000$ Hz, and $C = 10\ \mu$F, determine i_C in Figure 16–43.

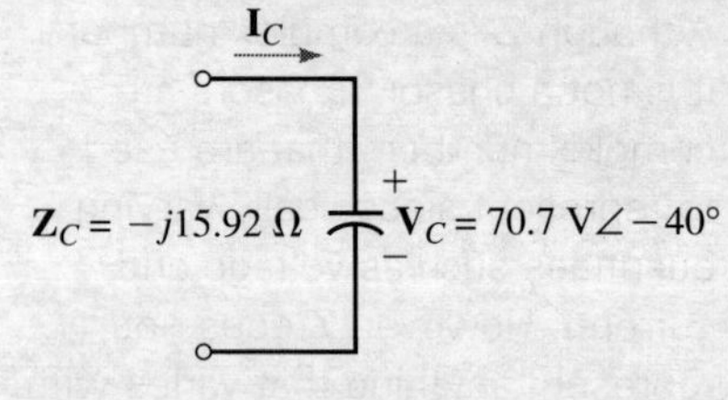

FIGURE 16–43

Solution

$$\omega = 2\pi f = 2\pi(1000\ \text{Hz}) = 6283\ \text{rads/s}$$

$$\mathbf{V}_C = 70.7\ \text{V}\angle -40°$$

$$\mathbf{Z}_C = -j\frac{1}{\omega C} = -j\left(\frac{1}{6283 \times 10 \times 10^{-6}}\right) = -j15.92\ \Omega$$

$$\mathbf{I}_C = \frac{\mathbf{V}_C}{\mathbf{Z}_C} = \frac{70.7\angle -40^\circ}{-j15.92} = \frac{70.7\angle -40^\circ}{15.92\angle -90^\circ} = 4.442\ \text{A}\angle 50^\circ$$

In the time domain, $i_C = \sqrt{2}(4.442)\sin(6283t + 50^\circ) = 6.28\sin(6283t + 50^\circ)$ A, which agrees with our previous solution, in Example 16–15.

PRACTICE PROBLEMS 9

1. If $\mathbf{I}_L = 5\ \text{mA}\angle -60^\circ$, $L = 2$ mH, and $f = 10$ kHz, what is $\mathbf{V}_L$?
2. A capacitor has a reactance of 50 Ω at 1200 Hz. If $v_C = 80 \sin 800t$ V, what is i_C?

Answers

1. 628 mV$\angle 30^\circ$
2. $0.170 \sin(800t + 90^\circ)$ A

A Look Ahead

The real power of the impedance method becomes apparent when you consider complex circuits with elements in series, parallel, and so on. This we do later, beginning in Chapter 18. Before we do this, however, there are some ideas on power that you need to know. These are considered in Chapter 17.

16.8 Computer Analysis of ac Circuits

Multisim

PSPICE

Let us now use Multisim and PSpice to verify the assertion of Practice Problems 2 (Section 16.2) that $v(t) = 21.8 \sin(\omega t + 36.6^\circ)$ is the sum of $e_1(t) = 10 \sin\omega t$ and $e_2(t) = 15 \sin(\omega t + 60^\circ)$. To do this, we will build two circuits on the screen, one to generate the sum $e_1(t) + e_2(t)$, and the other to generate the equivalent voltage $v(t) = 21.8 \sin(\omega t + 36.6^\circ)$, then compare the waveforms. Since the process is independent of frequency, we can choose any convenient frequency, say 500 Hz.

Multisim

Review Multisim Notes 1, 2, and 4, and then create the circuits of Figure 16–44 on your screen. (If you want, you can change the source designations to E1, E2, and E3 to match the original problem symbols. Double-click a source, select the *Label* tab, then change the designation.) Next, place junction dots to create nodes at the wire ends, then click *Options/Sheet Properties/Show All/OK* to display node numbers. Set source values, click *Simulate/Analyses/Transient Analysis*, set *TSTOP* to 0.002 and the *Minimum number of time points* to 200. Click the *Output* tab, add variables V(1), V(2), and V(2) − V(1) for display (see Notes 3 and 5), then click *Simulate.* Three waveforms appear. Add grid lines, then (optionally) turn off the select marks and change the background color to white.

NOTES...

Multisim

1. As noted in Chapter 15, Multisim source phase angles must be set up correctly. If you are unsure, click *Options/Global Preferences* then select the *Simulation* tab (or the *Parts* tab if you are a legacy user). Ensure that the *Shift Right* button is selected.
2. Multisim assigns node numbers according to the order in which you wire up a circuit; thus, the node numbers for your circuit may be different from those shown in Figure 16–44. For purposes of discussion, however, we will assume node numbers as in Figure 16–44.
3. Node numbers help you identify variables—for example, V(1) is the voltage (with respect to ground) at Node 1, V(2) is the voltage at Node 2, and so on. [Legacy users use $1 instead of V(1), etc.]
4. As described in Chapter 15, Multisim has two ac voltage source types. Be sure to use Signal_Voltage_Source sources in this example.
5. To create V(2)-V(1), use the *Add Expression* procedure found in the Output dialog box. If you have forgotten how, see Appendix A.

Results: Compare the results here to those of Figure 16–11 from earlier in the chapter. So far, the only waveforms plotted in Figure 16–44 are due to circuit (a). Click the *Show Legend* icon and note the legend at the bottom of the screen. Voltage V(1) represents the voltage at Node 1, which, as you can see, is source voltage e_2. Voltage V(2) represents the voltage at Node 2, which is the sum of e_1 and e_2. Thus, voltage V(2) − V(1), the difference between the two node voltages, is source voltage e_1. Waveforms are identical to those shown in Figure 16–11 (except that Multisim cannot plot the pre $t = 0$ portion of the traces—however, see Practice Problems 10).

Now add the waveform from circuit (b) as follows: click *Simulate/Analyses, then Transient Analysis,* select the *Output* tab, select V(3), click *Add,* then *Simulate.* At first glance, it appears that nothing has happened. However, your legend now shows that V(3), the voltage at Node 3, has been added. A closer look shows that the color of the "sum" trace has changed. What has happened is that since V(3) is identical to V(2), Multisim has simply graphed over the previously plotted trace. Since the two traces coincide perfectly, we conclude that the waveforms are identical, thus verifying our assertion of Practice Problems 2.

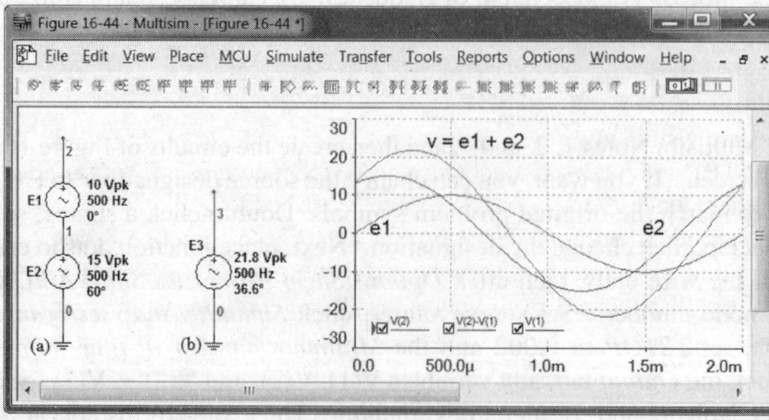

FIGURE 16–44 Using Multisim to demonstrate that 10 sin ωt + 15 sin (ωt + 60°) = 21.8 sin (ωt + 36.6°).

PRACTICE PROBLEMS 10

In the waveforms that you get from the preceding example, you cannot see the waveforms before $t = 0$. What if you want to see them? **Hint:** You can cheat a bit here. Note that the waveforms start to repeat at $t = 2$ ms on your plot. Mentally redefine this as $t = 0$, then rerun the simulation with *TSTART* set to 0.0015 and *TSTOP* to 0.004. Redefining the time scale in your mind as above, you can now see the pre $t = 0$ part of the solution.

PSpice

Create the circuit of Figure 16–45 using source *VSIN*. Double-click each parameter box in turn and key in values for voltage and frequency as indicated. Double-click the *V2* source symbol, and in its Property Editor, scroll to *PHASE* and type in **60**, click *Apply,* then close the Property Editor. Similarly, set the phase angle for source 3 to **36.6**. Click the *New Simulation Profile* icon, choose *Transient,* and set *TSTOP* to **2ms**, then click *OK*. Add differential voltage markers (probes) across source V1 and ordinary voltage markers elsewhere as indicated in Figure 16–45. (The differential markers are the ones with the +, – designations.) All markers will be gray, but they will turn color after simulation. Click the *Run* icon.

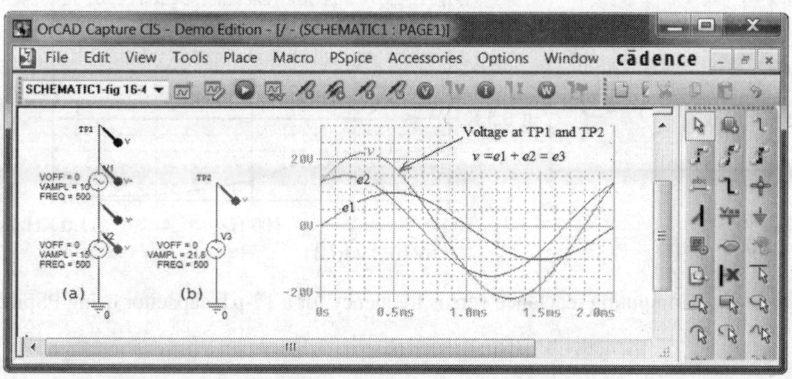

FIGURE 16–45 Using PSpice to demonstrate that $10 \sin \omega t + 15 \sin (\omega t + 60°) = 21.8 \sin (\omega t + 36.6°)$.

Results: Your graph colors may be different from those shown, but the curves should be the same. Compare waveforms to those of Figure 16–11 from earlier in the chapter. (Use the legend at the bottom of the screen to identify waveforms on your screen.) Note that results are identical. However, only three waveforms are visible in spite of the fact that we plotted four. The reason is that voltages at TP1 and TP2 are identical, and PSpice simply graphed one trace over the other. Since the two traces coincide perfectly, we conclude that the waveforms are identical, thus verifying our assertion of Practice Problems 2. (To confirm this, click the voltage probe at TP2 and delete it. You will see that voltage V3 has disappeared from your graph legend and the underlying trace, V2, remains.)

Another Example

PSpice makes it easy to study the response of circuits over a range of frequencies. This is illustrated in Example 16–19.

EXAMPLE 16–19

Compute and plot the reactance of a 12-μF capacitor over the range 10 Hz to 1000 Hz.

Solution PSpice has no command to compute reactance; however, you can calculate voltage and current over the desired frequency range, then plot their ratio. This gives reactance. Procedure: Create the circuit of Figure 16–46 on your screen. (Use source VAC here as it is the source to use for phasor analyses.) Note its default of 0V. Double-click the default value (not the symbol) and in the dialog box, enter **120V,** then click OK. Click the *New Simulations Profile* icon, enter a name, and then in the dialog box that opens, select *AC Sweep/Noise*. For the *Start Frequency,* key in **10Hz;** for the *End Frequency*, key **1kHz;** set AC Sweep type to *Logarithmic,* select *Decade,* and type **100** into the *Pts/Decade* (points per decade) box. Run the simulation and a set of empty axes appears. Click *Trace, Add Trace* and in the dialog box, click ***V1(C1),*** press the / key on the keyboard, then click ***I(C1)*** to yield the ratio V1(C1)/I(C1) (which is the capacitor's reactance). Click *OK* and PSpice will compute and plot the capacitor's reactance versus frequency. Compare its shape to Figure 16–36. Use the cursor to scale some values off the screen and verify each point using your calculator and $X_C = 1/\omega C$.

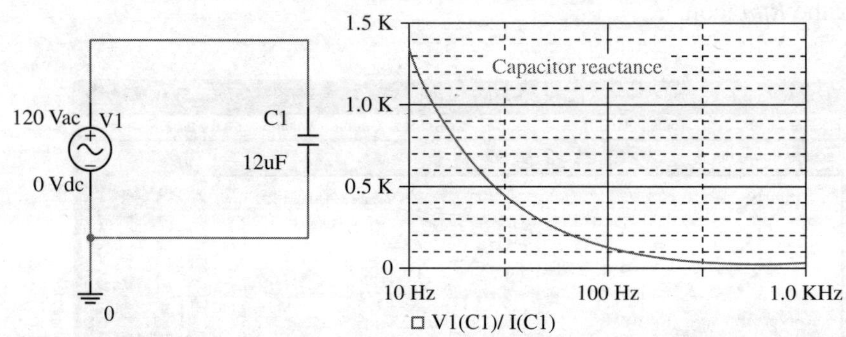

FIGURE 16–46 Computing reactance versus frequency for a 12-μF capacitor using PSpice.

Phasor Analysis

As a last example, we will show how to use PSpice to perform phasor analysis—that is, to solve problems with voltages and currents expressed in phasor form. To illustrate, consider again Example 16–18. Recall that $\mathbf{V}_C = 70.7\text{ V}\angle -40°$, $C = 10\ \mu\text{F}$, and $f = 1000$ Hz. Procedure: Create the circuit on the screen (Figure 16–47) using source *VAC* and component *IPRINT* (Note 1). Double-click the *VAC* symbol and in the Property Editor, set *ACMAG* to **70.7V** and *ACPHASE* to **−40,** click *Apply,* then close the editor—see Note 2. Double-click *IPRINT* and in the Property Editor, type **yes** into cells *AC, MAG,* and *PHASE.* Click *Apply* and close the editor. Click the *New Simulation Profile* icon, enter a name, select *AC Sweep/Noise,* set both *Start Frequency* and *End Frequency* to **1000Hz**, select *Linear,* then set *Total Points* to **1.** Run the simulation. When the simulation window opens, click *View, Output File,* then scroll to near the end of the file where you find the answers (see Note 3). The first number is the frequency (1000 Hz), the second number (*IM*) is the magnitude of the current (4.442 A), and the third (*IP*) is its phase (50 degrees). Thus, $\mathbf{I}_C = 4.442\text{ A}\angle 50°$ as we determined earlier in Example 16–18.

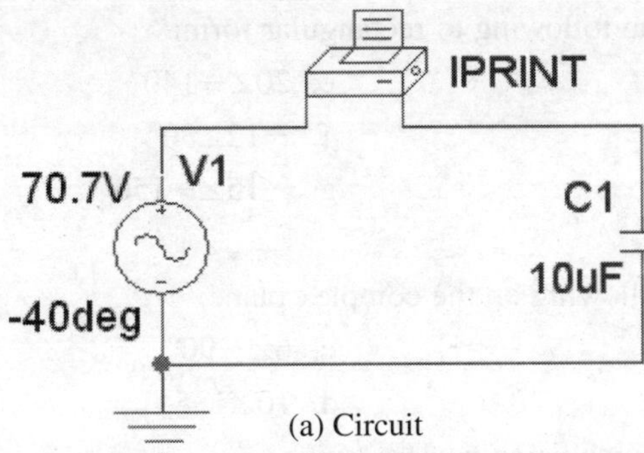

(a) Circuit

FREQ	IM(V_PRINT1)	IP(V_PRINT1)
1.000E+03	4.442E+00	5.000E+01

(b) Ammeter reading

FIGURE 16–47 Phasor analysis using PSpice. Component IPRINT is a software ammeter.

NOTES...

1. Component *IPRINT* is a software ammeter, found in the SPECIAL parts library. In this example, we configure it to display ac current in magnitude and phase angle format. Make sure that it is connected as shown in Figure 16–47, since if it is reversed, the phase angle of the measured current will be in error by 180°.
2. If you want to display the phase of the source voltage on the schematic as in Figure 16–47, double-click the source symbol and in the Property Editor click *ACPHASE, Display,* then select *Value Only.*
3. The results displayed by *IPRINT* are expressed in exponential format. Thus, frequency is shown as 1.000E+03, which is $1.000 \times 10^3 = 1000$ Hz, and so on.

PRACTICE PROBLEMS 11

Modify Example 16–19 to plot both capacitor current and reactance on the same graph. You will need to add a second Y-axis for the capacitor current. (See Appendix A if you need help.)

Problems

16.1 Complex Number Review

1. Convert each of the following to polar form:

a. $5 + j12$
b. $9 - j6$
c. $-8 + j15$
d. $-10 - j4$

2. Convert each of the following to rectangular form:
   a. $6\angle 30°$
   b. $14\angle 90°$
   c. $16\angle 0°$
   d. $6\angle 150°$
   e. $20\angle -140°$
   f. $-12\angle 30°$
   g. $-15\angle -150°$
3. Plot each of the following on the complex plane:
   a. $4 + j6$
   b. $j4$
   c. $6\angle -90°$
   d. $10\angle 135°$
4. Simplify the following using powers of *j*:
   a. $j(1 - j1)$
   b. $(-j)(2 + j5)$
   c. $j[j(1 + j6)]$
   d. $(j4)(-j2 + 4)$
   e. $(2 + j3)(3 - j4)$
5. Express your answer in rectangular form.
   a. $(4 + j8) + (3 - j2)$
   b. $(4 + j8) - (3 - j2)$
   c. $(4.1 - j7.6) + 12\angle 20°$
   d. $2.9\angle 25° - 7.3\angle -5°$
   e. $9.2\angle -120° - (2.6 + j4.1)$
   f. $\dfrac{1}{3+j4} + \dfrac{1}{8-j6}$
6. Express your answer in polar form.
   a. $(37 + j9.8)(3.6 - j12.3)$
   b. $(41.9\angle -80°)(16 + j2)$
   c. $\dfrac{42 + j18.6}{19.1 - j4.8}$
   d. $\dfrac{42.6 + j187.5}{11.2\angle 38°}$
7. Reduce each of the following to polar form:
   a. $15 - j6 - \left[\dfrac{18\angle 40° + (12 + j8)}{11 + j11}\right]$
   b. $\dfrac{21\angle 20° - j41}{36\angle 0° + (1 + j12) - 11\angle 40°}$
   c. $\dfrac{18\angle 40° - 18\angle -40°}{7 + j12} - \dfrac{16 + j17 + 21\angle -60°}{4}$

16.2 Complex Numbers in ac Analysis

NOTES...

The answers given to Problems 8 through 11 assume that you are not using rms values since we do not start using rms until later in the chapter.

8. In the manner of Figure 16–10, represent each of the following as transformed sources—see Notes.
   a. $e = 100 \sin(\omega t + 30°)$ V
   b. $e = 15 \sin(\omega t - 20°)$ V
   c. $e = 50 \sin(\omega t + 90°)$ V
   d. $e = 50 \cos \omega t$ V
   e. $e = 40 \sin(\omega t + 120°)$ V
   f. $e = 80 \sin(\omega t - 70°)$ V
9. Determine the sinusoidal equivalent for each of the transformed sources of Figure 16–48—see Notes.

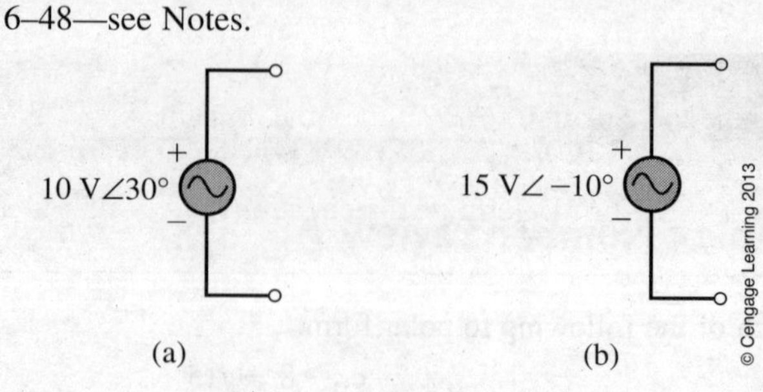

FIGURE 16–48

10. Given: $e_1 = 10 \sin(\omega t + 30°)$ V and $e_2 = 15 \sin(\omega t - 20°)$ V. Determine their sum $v = e_1 + e_2$ in the manner of Example 16–7, that is,
    a. Convert e_1 and e_2 to phasor form.
    b. Determine $\mathbf{V} = \mathbf{E}_1 + \mathbf{E}_2$.
    c. Convert $\mathbf{V}$ to the time domain.
    d. Sketch e_1, e_2, and v as per Figure 16–13.
11. Repeat Problem 10 for $v = e_1 - e_2$.

Note: For the remaining problems and throughout the remainder of the book, express phasor quantities as rms values rather than as peak values.

12. Express the voltages and currents of Figure 16–49 as time domain and phasor domain quantities.

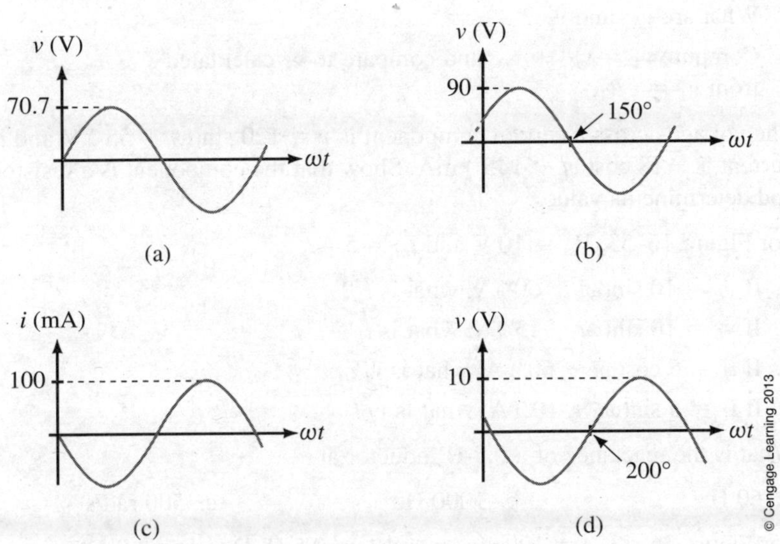

FIGURE 16–49

13. For Figure 16–50, $i_1 = 25 \sin(\omega t + 36°)$ mA and $i_2 = 40 \cos(\omega t - 10°)$ mA.
    a. Determine phasors $\mathbf{I}_1$, $\mathbf{I}_2$, and $\mathbf{I}_T$.
    b. Determine the equation for i_T in the time domain.
14. For Figure 16–50, $i_T = 50 \sin(\omega t + 60°)$ A and $i_2 = 20 \sin(\omega t - 30°)$ A.
    a. Determine phasors $\mathbf{I}_T$ and $\mathbf{I}_2$.
    b. Determine $\mathbf{I}_1$.
    c. From (b), determine the equation for i_1.
15. For Figure 16–18, $i_1 = 7 \sin \omega t$ mA, $i_2 = 4 \sin(\omega t - 90°)$ mA, and $i_3 = 6 \sin(\omega t + 90°)$ mA.
    a. Determine phasors $\mathbf{I}_1$, $\mathbf{I}_2$, $\mathbf{I}_3$ and $\mathbf{I}_T$.
    b. Determine the equation for i_T in the time domain.
16. For Figure 16–51, $i_T = 38.08 \sin(\omega t - 21.8°)$ A, $i_1 = 35.36 \sin \omega t$ A, and $i_3 = 28.28 \sin(\omega t - 90°)$ A. Determine the equation for i_2.

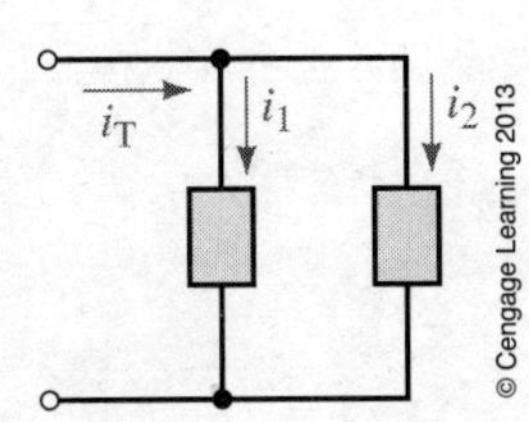

FIGURE 16–50

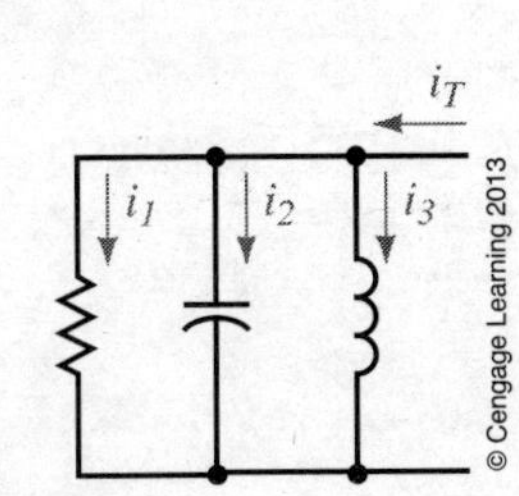

FIGURE 16–51

16.4 to 16.6

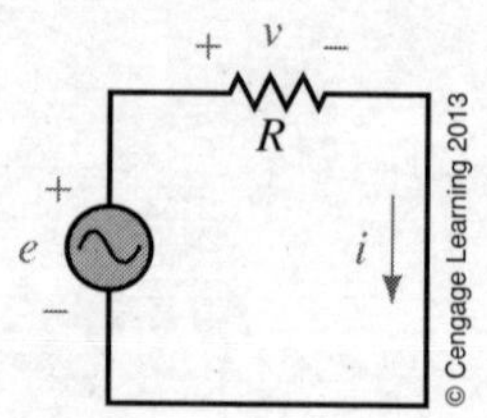

FIGURE 16–52

17. For Figure 16–52, $R = 12\ \Omega$. For each of the following, determine the unknown current or voltage and sketch.
    a. If $v = 120 \sin \omega t$ V, find i.
    b. If $v = 120 \sin(\omega t + 27°)$ V, find i.
    c. If $i = 17 \sin(\omega t - 56°)$ mA, find v.
    d. If $i = -17 \cos(\omega t - 67°)$ μA, find v.
18. Given $v = 120 \sin(\omega t + 52°)$ V and $i = 15 \sin(\omega t + 52°)$ mA, what is R?
19. Two resistors $R_1 = 10$ kΩ and $R_2 = 12.5$ kΩ are in series. If $i = 14.7 \sin(\omega t + 39°)$ mA,
    a. What are v_{R_1} and v_{R_2}?
    b. Compute $v_T = v_{R_1} + v_{R_2}$ and compare to v_T calculated from $v_T = i\, R_T$.
20. The voltage across a certain component is $v = 120 \sin(\omega t + 55°)$ V and its current is $-18 \cos(\omega t + 145°)$ mA. Show that the component is a resistor, and determine its value.

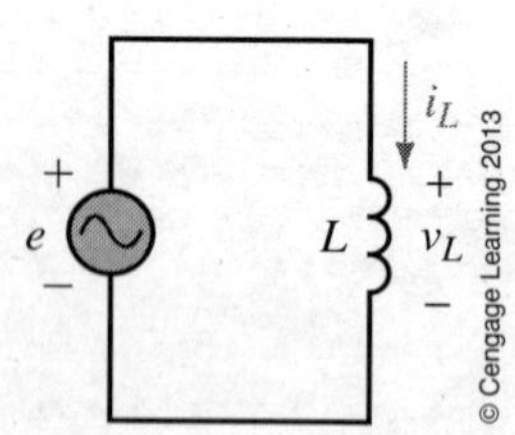

FIGURE 16–53

21. For Figure 16–53, $V_m = 10$ V and $I_m = 5$ A.
    a. If $v_L = 10 \sin(\omega t + 60°)$ V, what is i_L?
    b. If $v_L = 10 \sin(\omega t - 15°)$ V, what is i_L?
    c. If $i_L = 5 \cos(\omega t - 60°)$ A, what is v_L?
    d. If $i_L = 5 \sin(\omega t + 10°)$ A, what is v_L?
22. What is the reactance of a 0.5-H inductor at
    a. 60 Hz b. 1000 Hz c. 500 rad/s
23. For Figure 16–53, $e = 100 \sin \omega t$ and $L = 0.5$ H. Determine i_L at
    a. 60 Hz b. 1000 Hz c. 500 rad/s
24. For Figure 16–53, let $L = 200$ mH.
    a. If $v_L = 100 \sin 377t$ V, what is i_L?
    b. If $i_L = 10 \sin(2\pi \times 400t - 60°)$ mA, what is v_L?
25. For Figure 16–53, if
    a. $v_L = 40 \sin(\omega t + 30°)$ V, $i_L = 364 \sin(\omega t - 60°)$ mA, and $L = 2$ mH, what is f?
    b. $i_L = 250 \sin(\omega t + 40°)$ μA, $v_L = 40 \sin(\omega t + \theta)$ V, and $f = 500$ kHz, what are L and θ?
26. Repeat Problem 21 if the given voltages and currents are for a capacitor instead of an inductor.
27. What is the reactance of a 5-μF capacitor at
    a. 60 Hz b. 1000 Hz c. 500 rad/s
28. For Figure 16–54, $e = 100 \sin \omega t$ and $C = 5$ μF. Determine i_C at
    a. 60 Hz b. 1000 Hz c. 500 rad/s

29. For Figure 16–54, let $C = 50\ \mu F$.

a. If $v_C = 100 \sin 377t$ V, what is i_C?

b. If $i_C = 10 \sin(2\pi \times 400t - 60°)$ mA, what is v_C?

30. For Figure 16–54, if

a. $v_C = 362 \sin(\omega t - 33°)$ V, $i_C = 94 \sin(\omega t + 57°)$ mA, and $C = 2.2\ \mu F$, what is f?

b. $i_C = 350 \sin(\omega t + 40°)$ mA, $v_C = 3.6 \sin(\omega t + \theta)$ V, and $f = 12$ kHz, what are C and θ?

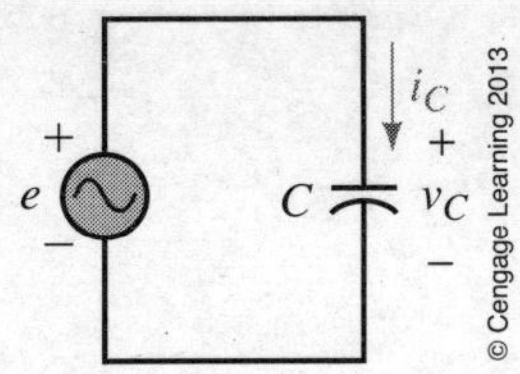

FIGURE 16–54

16.7 The Impedance Concept

31. Determine the impedance of each circuit element of Figure 16–55.

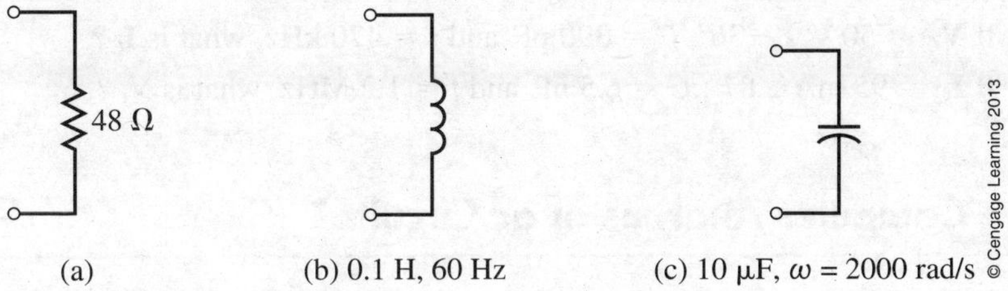

FIGURE 16–55

32. If $\mathbf{E} = 100\ \text{V}\angle 0°$ is applied across each of the circuit elements of Figure 16–56:

a. Determine each current in phasor form.

b. Express each current in time domain form.

33. If the current through each circuit element of Figure 16–56 is $0.5\ \text{A}\angle 0°$:

a. Determine each voltage in phasor form.

b. Express each voltage in time domain form.

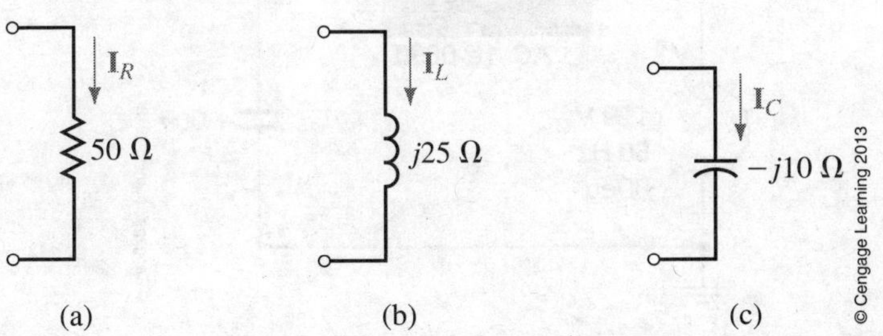

FIGURE 16–56

34. For each of the following, determine the impedance of the circuit element and state whether it is resistive, inductive, or capacitive.

a. $\mathbf{V} = 240\ \text{V}\angle{-30°}$, $\mathbf{I} = 4\ \text{A}\angle{-30°}$.

b. $\mathbf{V} = 40\ \text{V}\angle 30°$, $\mathbf{I} = 4\ \text{A}\angle{-60°}$.

c. $\mathbf{V} = 60\ \text{V}\angle{-30°}$, $\mathbf{I} = 4\ \text{A}\angle 60°$.

d. $\mathbf{V} = 140\ \text{V}\angle{-30°}$, $\mathbf{I} = 14\ \text{mA}\angle{-120°}$.

35. For each circuit of Figure 16–57, determine the unknown.

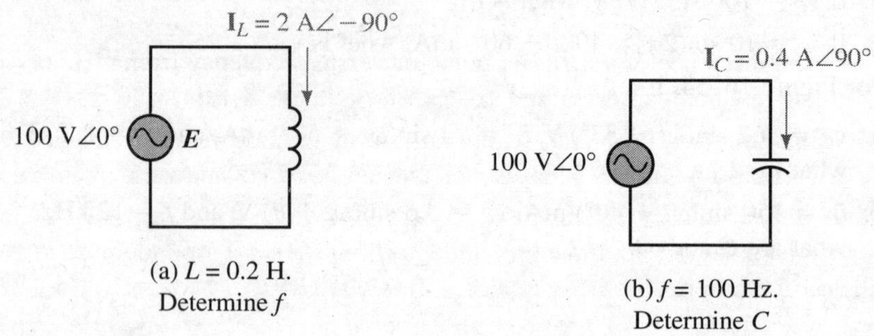

FIGURE 16–57

36. a. If $\mathbf{V}_L = 120\ \text{V}\angle 67°$, $L = 600\ \mu\text{H}$, and $f = 10$ kHz, what is $\mathbf{I}_L$?
    b. If $\mathbf{I}_L = 48\ \text{mA}\angle -43°$, $L = 550$ mH, and $f = 700$ Hz, what is $\mathbf{V}_L$?
    c. If $\mathbf{V}_C = 50\ \text{V}\angle -36°$, $C = 390$ pF, and $f = 470$ kHz, what is $\mathbf{I}_C$?
    d. If $\mathbf{I}_C = 95\ \text{mA}\angle 87°$, $C = 6.5$ nF, and $f = 1.2$ MHz, what is $\mathbf{V}_C$?

16.8 Computer Analysis of ac Circuits

Multisim

37. Create the circuit of Figure 16–58 on your screen. (To access the power ac source, right-click *View* and ensure that *Power Source Components* is selected. Use your mouse to locate its icon on your screen, click and the source attaches to your pointer. Move it to your workspace screen and click to place. To locate the ammeter, click *View* and ensure that *Measurement Components* is selected. Use your mouse to locate its icon and place it on your screen.) Double-click the ammeter symbol and set Mode to *AC*. Click the ON/OFF power switch to energize the circuit. Compare the measured reading against the theoretical value.

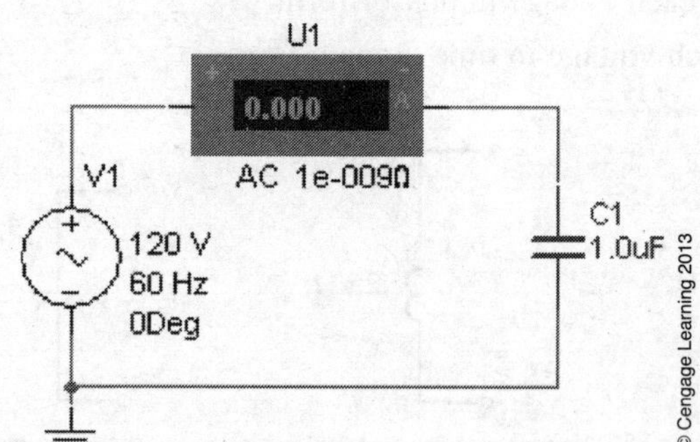

FIGURE 16–58 The source to use here is the power voltage source.

Multisim

38. Multisim can be used to plot capacitive reactance versus frequency. Create the Multisim equivalent of the PSpice circuit of Figure 16–46. Click *Simulate, Analyses, AC Analysis,* set *FSTART* to **10 Hz** and *FSTOP* to **1 kHz.** Set *Sweep type* as *Decade* and *Vertical scale* as *Linear*. Click *Output*, create the expression **V(1)/I(C1),** then click *Simulate*. You should get a magnitude graph similar to Figure 16–46 on your screen.

39. Create the circuit of Figure 16–53 on your screen. Use a source of 100 V$\angle 0°$, $L = 0.2$ H, and $f = 50$ Hz. Solve for current $\mathbf{I}_L$ (magnitude and angle). See note below. PSPICE

40. Plot the reactance of a 2.387-H inductor versus frequency from 1 Hz to 500 Hz and compare to Figure 16–29. Use a linear X-axis scale. PSPICE

41. For the circuit of Problem 39, plot current magnitude versus frequency from $f = 1$ Hz to $f = 20$ Hz. Measure the current at 10 Hz and verify with your calculator. PSPICE

Note: PSpice does not permit source/inductor loops. To get around this, add a very small resistor in series, for example, $R = 0.00001\ \Omega$.

ANSWERS TO IN-PROCESS LEARNING CHECKS

IN-PROCESS LEARNING CHECK 1

1. a. $6\angle 90°$ e. $9.43\angle 122.0°$
   b. $4\angle -90°$ f. $2.24\angle -63.4°$
   c. $4.24\angle 45°$ g. $3.61\angle -123.7°$
   d. $7.21\angle -56.3°$
2. a. $j4$ e. $-3 + j5.20$
   b. $3 + j0$ f. $2.35 - j0.855$
   c. $-j2$ g. $-1.64 - j0.599$
   d. $3.83 + j3.21$
3. $12\angle 40°$
4. $88 - j7$; $-16 + j15$; $0.0113 + j\,0.0009$; $-0.0333 - j0.0312$
5. $288\angle -100°$; $2\angle 150°$
6. a. $14.70 + j2.17$
   b. $-8.94 + j7.28$
7. $18.0 \sin(\omega t - 56.3°)$
8. 7.07 A, 14.14 A, and 3.54 A for the branch currents, 12.7 A for total current.

IN-PROCESS LEARNING CHECK 2

1. $50 \sin(\omega t + 30°)$ A
2. $50 \sin(\omega t - 60°)$ A
3. $50 \sin(\omega t + 120°)$ A
4. 0.637 H
5. 3.98 μF
6. a. Voltage and current are in phase. Therefore, *R*.
   b. Current leads by 90°. Therefore, *C*.
   c. Current leads by 90°. Therefore, *C*.
   d. Current lags by 90°. Therefore, *L*.

■ KEY TERMS

Active Power
Apparent Power
Average Power
Eddy Currents
Effective Resistance
Instantaneous Power
Power Factor
Power Factor Correction
Power Triangle
Reactive Power
Real Power
Skin Effect
VAR (Volt-Amps Reactive)
Volt-Amperes (VA)
Wattmeter

■ OUTLINE

Introduction
Power to a Resistive Load
Power to an Inductive Load
Power to a Capacitive Load
Power in More Complex Circuits
Apparent Power
The Relationship between *P, Q,* and *S*
Power Factor
ac Power Measurement
Effective Resistance
Energy Relationships for ac
Circuit Analysis Using Computers

■ OBJECTIVES

After studying this chapter, you will be able to

- explain what is meant by active, reactive, and apparent power,
- compute the active power to a load,
- compute the reactive power to a load,
- compute the apparent power to a load,
- construct and use the power triangle to analyze power to complex loads,
- compute power factor,
- explain why equipment is rated in VA instead of watts,
- measure power in single-phase circuits,
- describe why effective resistance differs from geometric resistance,
- describe energy relations in ac circuits,
- use Multisim and PSpice to study instantaneous power.

17

POWER IN ac CIRCUITS

CHAPTER PREVIEW

In Chapter 4, you studied power in dc circuits. In this chapter, we turn our attention to power in ac circuits. In ac circuits, there are additional considerations that are not present with dc. In dc circuits, for example, the only power relationship you encounter is $P = VI$ watts or its alternate forms $P = I^2R$ and $P = V^2/R$. This power is referred to as *real power* or *active power* and is the power that does useful work such as light a lamp, power a heater, run an electric motor, and so on.

In ac circuits, you also encounter active power, but for ac circuits that contain reactive elements (i.e., inductance or capacitance), a second component of power also exists. This component, termed *reactive power,* represents energy that oscillates back and forth throughout the system. For example, during the buildup of current in an inductance, energy flows from the power source to the inductance to expand its magnetic field. When the magnetic field collapses, this energy is returned to the circuit. This movement of energy in and out of the inductance constitutes a flow of power. However, since it flows first in one direction, then in the other, it contributes nothing to the average flow of power from the source to the load. For this reason, reactive power is sometimes referred to as *wattless power.* (A similar situation exists regarding power flow to and from the electric field of a capacitor.)

For a circuit that contains resistive as well as reactive elements, some energy is dissipated while the remainder is shuttled back and forth between the reactive elements as described above; thus, both active and reactive components of power are present. This combination of real and reactive power is termed apparent power—see Notes.

NOTES...

Note that real power and reactive power are not separate "types" of power, they are merely different components of the total power.

In this chapter, we look at all three components of power. New ideas that emerge include the concept of power factor, the power triangle, the measurement of power in ac circuits, and the concept of effective resistance. ■

Putting It in Perspective

Henry Cavendish

CAVENDISH, AN ENGLISH CHEMIST and physicist born in 1731, is included here, not for what he did for the emerging electrical field, but for what he didn't do. A brilliant man, Cavendish was 50 years ahead of his time, and his experiments in electricity preceded and anticipated almost all the major discoveries that came about over the next half century (e.g., he discovered Coulomb's law before Coulomb did). However, Cavendish was interested in research and knowledge purely for its own sake and never bothered to publish most of what he learned, in effect depriving the world of his findings and holding back the development of the field of electricity by many years. Cavendish's work lay unknown for nearly a century before another great scientist, James Clerk Maxwell, had it published. Nowadays, Cavendish is better known for his work in the gravitational field than for his work in the electrical field. One of the amazing things he did was to determine the mass of the earth using the rather primitive technology of his day. ■

17.1 Introduction

At any given instant, the power to a load is equal to the product of voltage times current (Figure 17–1). This means that if voltage and current vary with time, so will power. This time-varying power is referred to as **instantaneous power** and is given the symbol $p(t)$ or just p. Thus,

$$p = vi \quad \text{(watts)} \tag{17–1}$$

Now consider the case of sinusoidal ac. Since voltage and current are positive at various times during their cycle and negative at others, instantaneous power may also be positive at some times and negative at others. This is illustrated in Figure 17–2, where we have multiplied voltage times current point by point to get the power waveform. For example, from $t = 0$ s to $t = t_1$, v and i are both positive; therefore, power is positive. At $t = t_1$, $v = 0$ V and thus $p = 0$ W. From t_1 to t_2, i is positive and v is negative; therefore, p is negative. From t_2 to t_3, both v and i are negative; therefore power is positive, and so on. As discussed in Chapter 4, Section 4.4, a positive value for p means that power transfer is in the direction of the reference arrow, while a negative value means that it is in the opposite direction. Thus, during positive parts of the power cycle, power flows from the source to the load, while during negative parts, it flows out of the load back into the circuit.

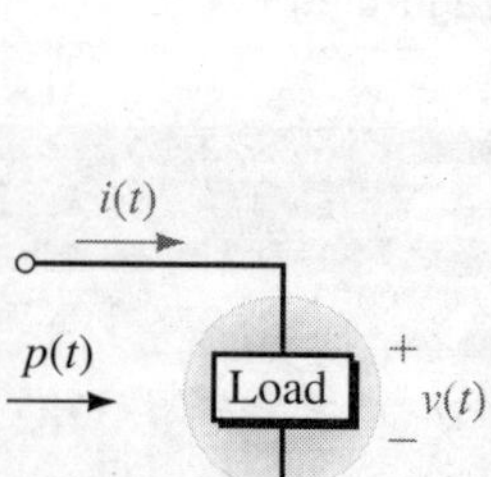

FIGURE 17–1 Voltage, current, and power references. When p is positive, power is in the direction of the reference arrow.

The waveform $p(t)$ of Figure 17–2 is the actual power waveform. We will now show that the key aspects of power flow embodied in this waveform can be described in terms of active power, reactive power, and apparent power.

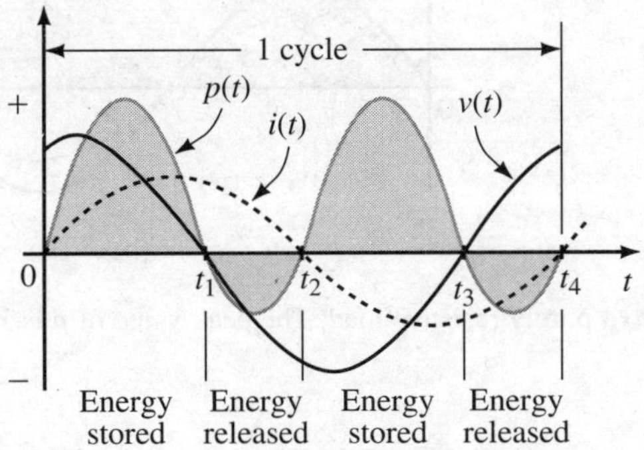

FIGURE 17–2 Instantaneous power in an ac circuit. Positive p represents power to the load; negative p represents power returned from the load.

Active Power

Since p in Figure 17–2 represents the power flowing to the load, its average will be the **average power** to the load. Denote this average by the letter P. If P is positive, then, on average, more power flows to the load than is returned from it. (If P is zero, all power sent to the load is returned.) Thus, if P has a positive value, it represents the power that is really dissipated by the load. For this reason, P is called **real power.** In modern terminology, real power is also called **active power.** Thus, *active power is the average value of the instantaneous power, and the terms real power, active power, and average power mean the same thing.* (We usually refer to it simply as power.) In this book, we use the terms interchangeably.

Reactive Power

Consider again Figure 17–2. During the intervals that p is negative, power is being returned from the load. (This can only happen if the load contains reactive elements: L or C.) The portion of power that flows into the load then back out is called **reactive power.** Since it first flows one way then the other, *its average value is zero;* thus, reactive power contributes nothing to the average power to the load.

Although reactive power does no useful work, it cannot be ignored. Extra current is required to create reactive power, and this current must be supplied by the source; this also means that conductors, circuit breakers, switches, transformers, and other equipment must be made physically larger to handle the extra current. This increases the cost of a system. (This is one of the reasons that reactive power is a major concern of power system engineers.)

As noted, the waveform of Figure 17–2 embodies both the real and reactive aspects of power. In this chapter, we learn how to separate them for purposes of analysis and measurement. We begin by looking at power to resistive, inductive, and capacitive circuit elements separately.

17.2 Power to a Resistive Load

First consider power to a purely resistive load (Figure 17–3). Here, current is in phase with voltage. Assume $i = I_m\sin \omega t$ and $v = V_m\sin \omega t$. Then,

$$p = vi = (V_m\sin \omega t)(I_m\sin \omega t) = V_mI_m\sin^2 \omega t$$

CircuitSim 17-1

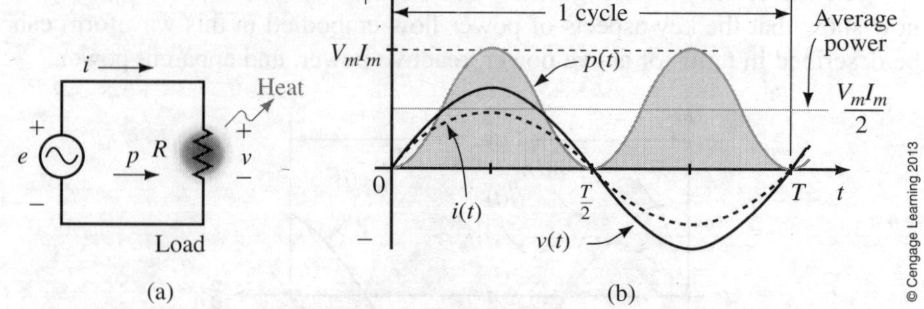

FIGURE 17–3 Power to a purely resistive load. The peak value of p is V_mI_m.

Therefore,

$$p = \frac{V_mI_m}{2}(1 - \cos 2\,\omega t) \tag{17–2}$$

where we have used the trigonometric identity $\sin^2 \omega t = ½\,(1 - \cos 2\,\omega t)$ from inside the front cover of the book.

A sketch of p versus time is shown in Figure 17–3(b). Note that p is always positive (except where it is momentarily zero). This means that power flows only from the source to the load. Since none is ever returned, all power delivered by the source is absorbed by the load. We therefore conclude that *power to a pure resistance consists of active power only.* Note also that the frequency of the power waveform is double that of the voltage and current waveforms. (This is confirmed by the double-frequency component 2ω in Equation 17–2.)

Average Power

Inspection of the power waveform of Figure 17–3 shows that its average value lies halfway between zero and its peak value of V_mI_m. That is,

$$P = V_mI_m/2$$

(You can also get the same result by averaging Equation 17–2 as we did in Chapter 15.) Since V (the magnitude of the rms value of voltage) is $V_m/\sqrt{2}$ and I (the magnitude of the rms value of current) is $I_m/\sqrt{2}$, this can be written as $P = VI$. Thus, average power to a purely resistive load is

$$P = VI \quad \text{(watts)} \tag{17–3}$$

Alternate forms are obtained by substituting $V = IR$ and $I = V/R$ into Equation 17–3. They are

$$P = I^2R \quad \text{(watts)} \tag{17–4}$$

$$= V^2/R \quad \text{(watts)} \tag{17–5}$$

Thus, the active power relationships for resistive circuits are the same for ac as for dc.

17.3 Power to an Inductive Load

For a purely inductive load as in Figure 17–4(a), current lags voltage by 90°. If we select current as reference, $i = I_m\sin \omega t$ and $v = V_m\sin(\omega t + 90°)$. A sketch of p versus time (obtained by multiplying v times i) then looks as

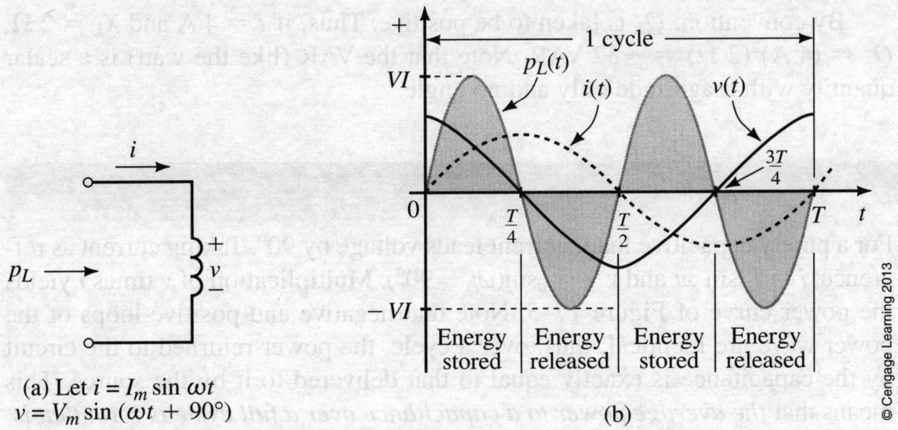

FIGURE 17–4 Power to a purely inductive load. Energy stored during each quarter-cycle is returned during the next quarter-cycle. Average power is zero.

shown in (b). Note that during the first quarter-cycle, p is positive and hence power flows to the inductance, while during the second quarter-cycle, p is negative and all power transferred to the inductance during the first quarter-cycle flows back out. Similarly for the third and fourth quarter-cycles. Thus, *the average power to an inductance over a full cycle is zero, that is, there are no power losses associated with a pure inductance.* Consequently, $P_L = 0$ W, and the only power flowing in the circuit is reactive power. This is true in general, that is, *the power that flows into and out of a pure inductance is reactive power only.*

To determine this power, consider again Equation 17–1. With $v = V_m\sin(\omega t + 90°)$ and $i = I_m\sin \omega t$, $p_L = vi$ becomes

$$p_L = V_m I_m \sin(\omega t + 90°)\sin \omega t$$

After some trigonometric manipulation, this reduces to

$$p_L = VI \sin 2\,\omega t \qquad (17\text{–}6)$$

where V and I are the magnitudes of the rms values of the voltage and current, respectively.

The product VI in Equation 17–6 is defined as reactive power and is given the symbol Q_L. Because it represents "power" that alternately flows into, then out of the inductance, Q_L contributes nothing to the average power to the load and, as noted earlier, is sometimes referred to as wattless power. As you will see in Section 17.8, however, reactive power is of major concern in the design and operation of electrical power systems.

Since Q_L is the product of voltage times current, its unit is the volt-amp (VA). To indicate that Q_L represents reactive volt-amps, an "R" is appended to yield a new unit, the **VAR (volt-amps reactive).** Thus,

$$Q_L = VI \quad \text{(VAR)} \qquad (17\text{–}7)$$

Substituting $V = IX_L$ and $I = V/X_L$ yields the following alternate forms:

$$Q_L = I^2X_L = \frac{V^2}{X_L} \quad \text{(VAR)} \qquad (17\text{–}8)$$

CircuitSim 17-2

By convention, Q_L is taken to be positive. Thus, if $I = 4$ A and $X_L = 2\ \Omega$, $Q_L = (4\text{ A})^2(2\ \Omega) = +32$ VAR. Note that the VAR (like the watt) is a scalar quantity with magnitude only and no angle.

17.4 Power to a Capacitive Load

For a purely capacitive load, current leads voltage by 90°. Taking current as reference, $i = I_m \sin \omega t$ and $v = V_m \sin(\omega t - 90°)$. Multiplication of v times i yields the power curve of Figure 17–5. Note that negative and positive loops of the power wave are identical; thus, over a cycle, the power returned to the circuit by the capacitance is exactly equal to that delivered to it by the source. This means that *the average power to a capacitance over a full cycle is zero, that is, there are no power losses associated with a pure capacitance.* Consequently, $P_C = 0$ W, and the only power flowing in the circuit is reactive power. This is true in general, that is, *the power that flows into and out of a pure capacitance is reactive power only.* This reactive power is given by

$$p_C = vi = V_m I_m \sin \omega t \sin(\omega t - 90°)$$

which reduces to

$$p_C = -VI \sin 2\omega t \qquad \textbf{(17–9)}$$

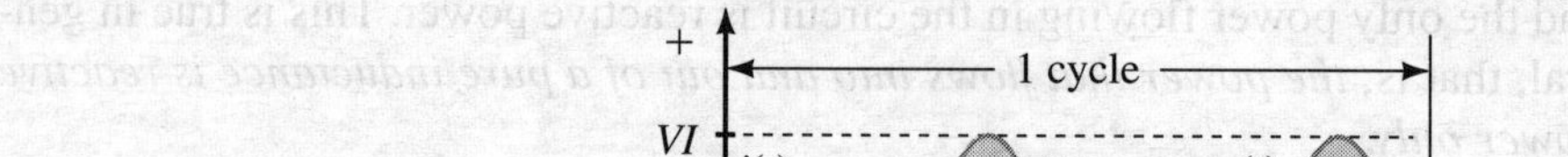

FIGURE 17–5 Power to a purely capacitive load. Average power is zero.

CircuitSim 17-3

where V and I are the magnitudes of the rms values of the voltage and current respectively. Now define the product VI as Q_C. This product represents reactive power. That is,

$$Q_C = VI \quad \text{(VAR)} \qquad \textbf{(17–10)}$$

CircuitSim 17-4

Since $V = IX_C$ and $I = V/X_C$, Q_C can also be expressed as

$$Q_C = I^2 X_C = \frac{V^2}{X_C} \quad \text{(VAR)} \qquad \textbf{(17–11)}$$

By convention, reactive power to capacitance is defined as negative. Thus, if $I = 4$ A and $X_C = 2\ \Omega$, then $I^2X_C = (4\text{ A})^2(2\ \Omega) = 32$ VAR. We can either explicitly include the minus sign as $Q_C = -32$ VAR or imply it by stating that Q represents capacitive VARs, that is, $Q_C = 32$ VAR (cap.).

EXAMPLE 17–1

For each circuit of Figure 17–6, determine real and reactive power.

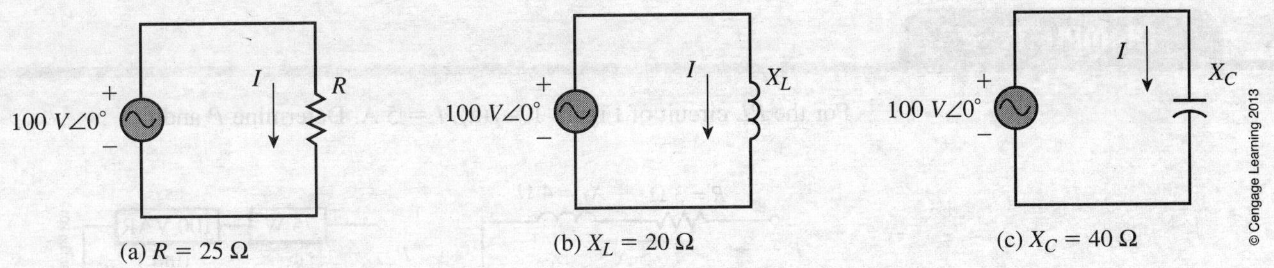

FIGURE 17–6

Solution Only voltage and current magnitudes are needed.

a. $I = 100\text{ V}/25\ \Omega = 4\text{ A}$. $P = VI = (100\text{ V})(4\text{ A}) = 400\text{ W}$. $Q = 0\text{ VAR}$.

b. $I = 100\text{ V}/20\ \Omega = 5\text{ A}$. $Q = VI = (100\text{ V})(5\text{ A}) = 500\text{ VAR}$ (ind.). $P = 0\text{ W}$.

c. $I = 100\text{ V}/40\ \Omega = 2.5\text{ A}$. $Q = VI = (100\text{ V})(2.5\text{ A}) = 250\text{ VAR}$ (cap.). $P = 0\text{ W}$.

The answer for (c) can also be expressed as $Q = -250\text{ VAR}$.

PRACTICE PROBLEMS 1

1. If the power at some instant in Figure 17–1 is $p = -27\text{ W}$, in what direction is the power at that instant?
2. For a purely resistive load, v and i are in phase. Given $v = 10 \sin \omega t$ V and $i = 5 \sin \omega t$ A. Using graph paper, carefully plot v and i at 30° intervals. Now multiply the values of v and i at these points and plot the power. [The result should look like Figure 17–3(b).]
   a. From the graph, determine the peak power and average power.
   b. Compute power using $P = VI$ and compare to the average value determined in (a).
3. Repeat Example 17–1 using Equations 17–4, 17–5, 17–8, and 17–11.

Answers

1. From the load to the source.
2. a. 50 W; 25 W
   b. Same

17.5 Power in More Complex Circuits

The relationships described previously were developed using the load of Figure 17–1. However, they hold true for every element in a circuit, no matter how complex the circuit or how its elements are interconnected. Further, in any circuit, total real power P_T is found by summing real power to all circuit elements, while total reactive power Q_T is found by summing reactive power, taking into account that inductive Q is positive and capacitive Q is negative.

It is sometimes convenient to show power to circuit elements symbolically as illustrated in the next example.

EXAMPLE 17–2

For the *RL* circuit of Figure 17–7(a), $I = 5$ A. Determine *P* and *Q*.

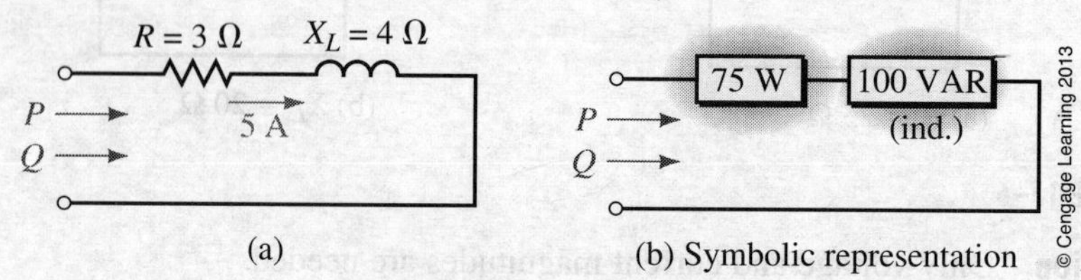

FIGURE 17–7 From the terminals, *P* and *Q* are the same for both (a) and (b).

Solution

$$P = I^2R = (5\text{ A})^2(3\ \Omega) = 75\text{ W}$$
$$Q = Q_L = I^2X_L = (5\text{ A})^2(4\ \Omega) = 100\text{ VAR (ind.)}$$

These can be represented symbolically as in Figure 17–7(b).

EXAMPLE 17–3

For the *RC* circuit of Figure 17–8(a), determine *P* and *Q*.

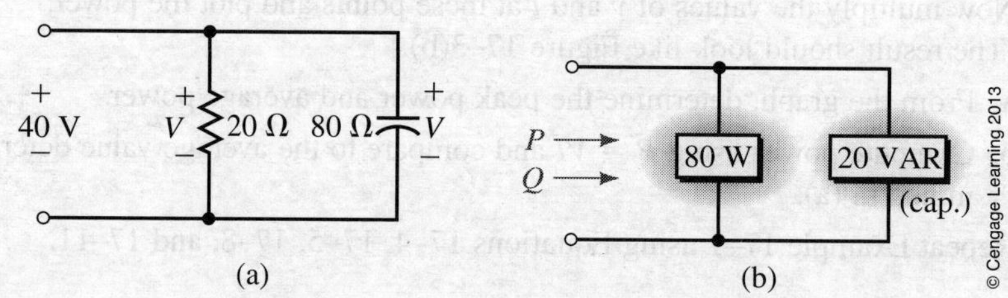

FIGURE 17–8 From the terminals, *P* and *Q* are the same for both (a) and (b).

Solution

$$P = V^2/R = (40\text{ V})^2/(20\ \Omega) = 80\text{ W}$$
$$Q = Q_C = V^2/X_C = (40\text{ V})^2/(80\ \Omega) = 20\text{ VAR (cap.)}$$

These can be represented symbolically as in Figure 17–8(b).

In terms of determining total *P* and *Q*, it does not matter how the circuit or system is connected or what electrical elements it contains. Elements can be connected in series, in parallel, or in series-parallel, for example, and the system can contain electric motors and the like, and total *P* is still found by summing the power to individual elements, while total *Q* is found by algebraically summing their reactive powers.

EXAMPLE 17–4

a. For Figure 17–9(a), compute P_T and Q_T.
b. Reduce the circuit to its simplest form.

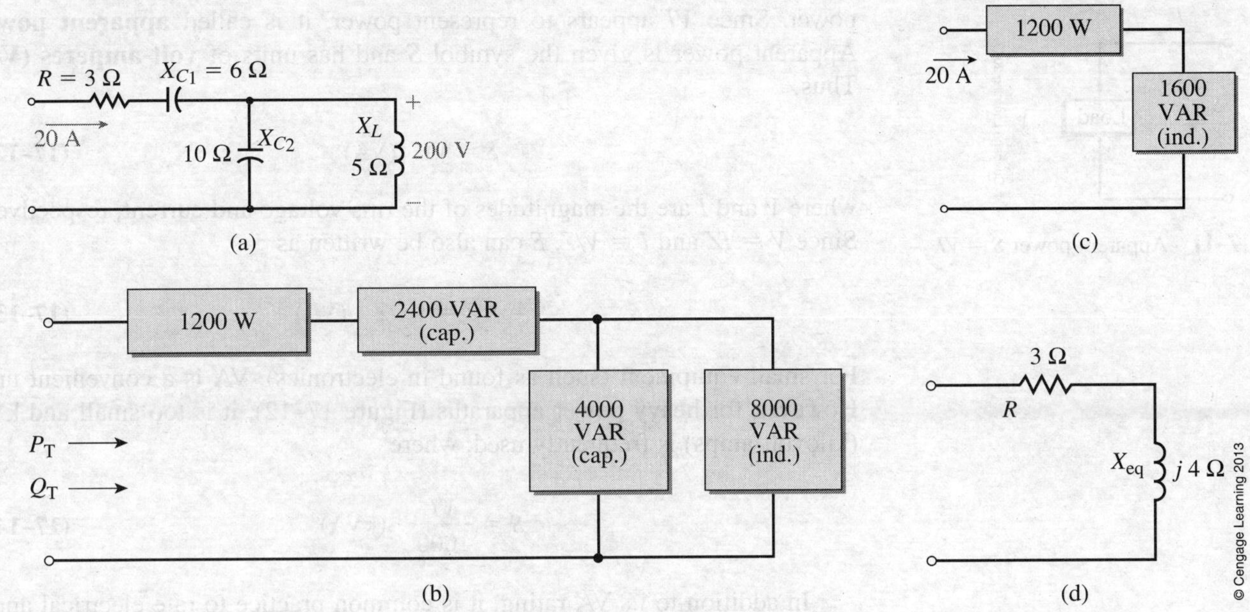

FIGURE 17–9

Solution

a. $P = I^2R = (20\text{ A})^2(3\ \Omega) = 1200\text{ W}$

$Q_{C_1} = I^2X_{C_1} = (20\text{ A})^2(6\ \Omega) = 2400\text{ VAR (cap.)}$

$$Q_{C_2} = \frac{V_2^2}{X_{C_2}} = \frac{(200\text{ V})^2}{(10\ \Omega)} = 4000\text{ VAR (cap.)}$$

$$Q_L = \frac{V_2^2}{X_L} = \frac{(200\text{ V})^2}{5\ \Omega} = 8000\text{ VAR (ind.)}$$

CircuitSim 17-5

These are represented symbolically in part (b). $P_T = 1200$ W and $Q_T = -2400\text{ VAR} - 4000\text{ VAR} + 8000\text{ VAR} = 1600\text{ VAR}$. Thus, the load is net inductive as shown in (c).

b. $Q_T = I^2X_{eq}$. Thus, $X_{eq} = Q_T/I^2 = (1600\text{ VAR})/(20\text{ A})^2 = 4\ \Omega$. Circuit resistance remains unchanged. Thus, the equivalent is as shown in (d).

PRACTICE PROBLEMS 2

For the circuit of Figure 17–10, $P_T = 1.9$ kW and $Q_T = 900$ VAR (ind.). Determine P_2 and Q_2.

Answer
300 W, 400 VAR (cap.)

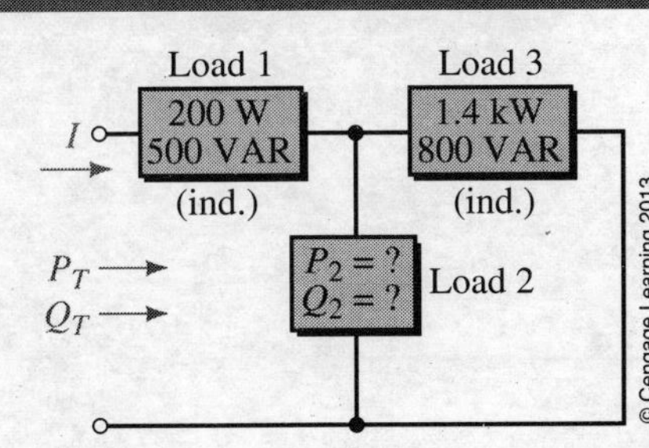

FIGURE 17–10

17.6 Apparent Power

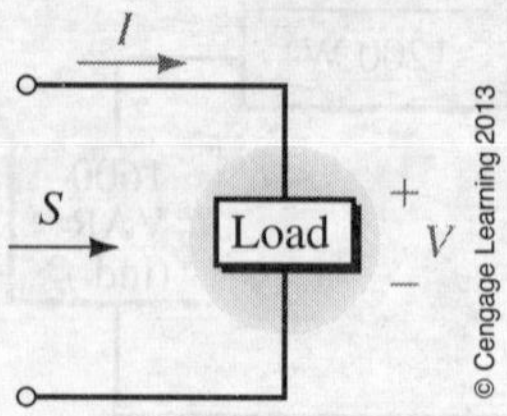

FIGURE 17–11 Apparent power $S = VI$.

When a load has voltage V across it and current I through it as in Figure 17–11, the power that appears to flow to it is VI. However, if the load contains both resistance and reactance, this product represents neither real power nor reactive power. Since VI appears to represent power, it is called **apparent power.** Apparent power is given the symbol S and has units of **volt-amperes (VA).** Thus,

$$S = VI \quad \text{(VA)} \tag{17–12}$$

where V and I are the magnitudes of the rms voltage and current, respectively. Since $V = IZ$ and $I = V/Z$, S can also be written as

$$S = I^2Z = V^2/Z \quad \text{(VA)} \tag{17–13}$$

For small equipment (such as found in electronics), VA is a convenient unit. However, for heavy power apparatus (Figure 17–12), it is too small and kVA (kilovolt-amps) is frequently used, where

$$S = \frac{VI}{1000} \quad \text{(kVA)} \tag{17–14}$$

In addition to its VA rating, it is common practice to rate electrical apparatus in terms of its operating voltage. Once you know these two, it is easy to determine rated current. For example, a piece of equipment rated 250 kVA, 4.16 kV has a rated current of $I = S/V = (250 \times 10^3 \text{ VA})/(4.16 \times 10^3 \text{ V}) = 60.1$ A.

FIGURE 17–12 Power apparatus is rated in apparent power. The transformer shown is a 167-kVA unit.

17.7 The Relationship between *P*, *Q*, and *S*

Until now, we have treated real, reactive, and apparent power separately. However, they are related by a very simple relationship through the power triangle.

The Power Triangle

Consider the series circuit of Figure 17–13(a). Let the current through the circuit be $\mathbf{I} = I\angle 0°$, with phasor representation (b). The voltages across the resistor and inductance are $\mathbf{V}_R$ and $\mathbf{V}_L$, respectively. As noted in Chapter 16, $\mathbf{V}_R$ is in phase with $\mathbf{I}$, while $\mathbf{V}_L$ leads it by 90°. Kirchhoff's voltage law applies for ac voltages in phasor form. Thus, $\mathbf{V} = \mathbf{V}_R + \mathbf{V}_L$ as indicated in (c).

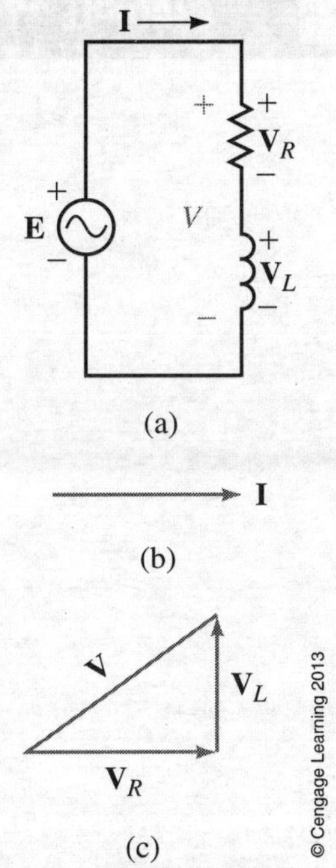

FIGURE 17–13 Steps in the development of the power triangle—continued in Figure 17–14.

The voltage triangle of (c) may be redrawn as in Figure 17–14(a) with magnitudes of V_R and V_L replaced by IR and IX_L, respectively. Now multiply all quantities by I. This yields sides of I^2R, I^2X_L, and hypotenuse VI as indicated in (b). Note that these represent P, Q, and S, respectively, as indicated in (c). This is called the **power triangle.** From the geometry of this triangle, you can see that

$$S = \sqrt{P^2 + Q_L^2} \tag{17–15}$$

Alternatively, the relationship between P, Q, and S may be expressed as a complex number:

$$\mathbf{S} = P + jQ_L \tag{17–16a}$$

or

$$\mathbf{S} = S\angle\theta \tag{17–16b}$$

If the circuit is capacitive instead of inductive, Equation 17–16a becomes

$$\mathbf{S} = P - jQ_C \tag{17–17}$$

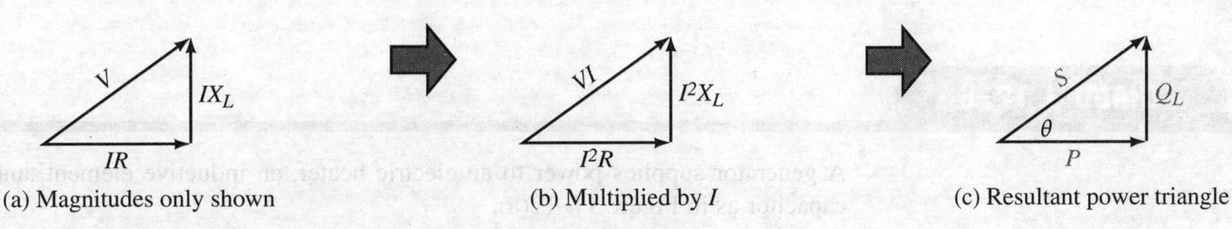

FIGURE 17–14 Steps in the development of the power triangle continued.

The power triangle in this case has a negative imaginary part as indicated in Figure 17–15.

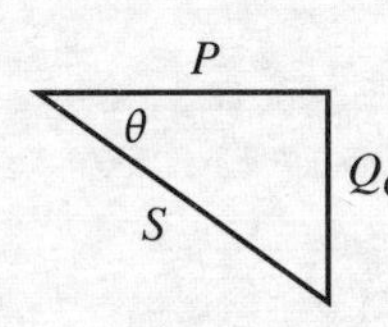

FIGURE 17–15 Power triangle for capacitive case.

The power relationships may be written in generalized forms as

$$\mathbf{S} = \mathbf{P} + \mathbf{Q} \tag{17–18}$$

and

$$\mathbf{S} = \mathbf{VI}^* \tag{17–19}$$

where $\mathbf{P} = P\angle 0°$, $\mathbf{Q}_L = jQ_L$, $\mathbf{Q}_C = -jQ_C$, and $\mathbf{I}^*$ is the conjugate of current $\mathbf{I}$—see Note. These relationships hold true for all networks regardless of what they contain or how they are configured.

NOTES...

See Chapter 16, Section 16.1 for a discussion of complex conjugates.

When solving problems involving power, remember that P values can be added to get P_T, and Q values to get Q_T (where Q is positive for inductive elements and negative for capacitive). However, apparent power values cannot be added to get S_T, that is, $S_T \neq S_1 + S_2 + \cdots + S_N$. Instead, you must determine P_T and Q_T separately, then use the power triangle to obtain S_T—see Practice Problems 3, Problem 2.

EXAMPLE 17–5

The P and Q values for a circuit are shown in Figure 17–16(a).

a. Determine the power triangle.

b. Determine the magnitude of the current supplied by the source.

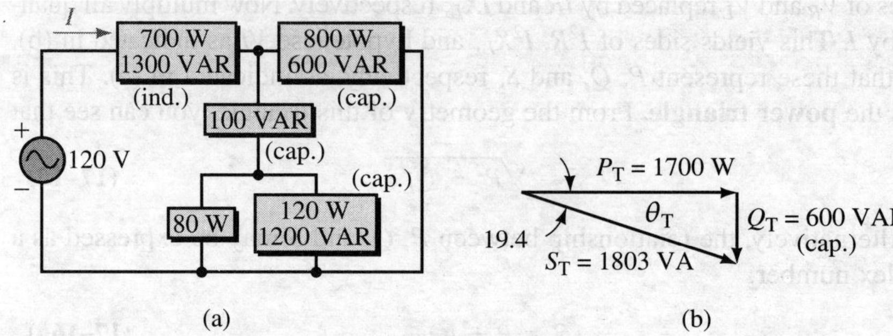

FIGURE 17–16

Solution

a. $P_T = 700 + 800 + 80 + 120 = 1700\text{ W}$

$Q_T = 1300 - 600 - 100 - 1200 = -600\text{ VAR} = 600\text{ VAR (cap.)}$

$\mathbf{S}_T = P_T + jQ_T = 1700 - j600 = 1803\angle -19.4°\text{ VA}$

The power triangle is as shown. The load is net capacitive.

b. $I = S_T/E = 1803\text{ VA}/120\text{ V} = 15.0\text{ A}$.

EXAMPLE 17–6

A generator supplies power to an electric heater, an inductive element, and a capacitor as in Figure 17–17(a).

CircuitSim 17-6

a. Find P and Q for each load.

b. Find total active and reactive power supplied by the generator.

c. Draw the power triangle for the combined loads and determine total apparent power.

d. Find the current supplied by the generator.

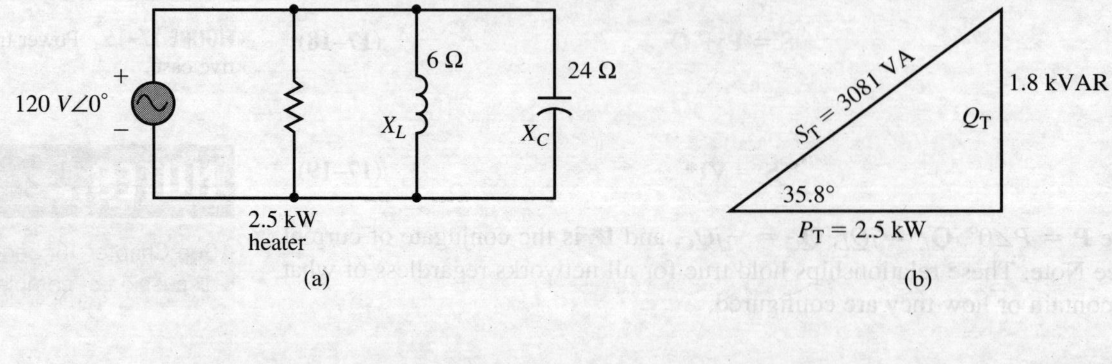

FIGURE 17–17

Solution

a. The components of power are as follows:

Heater: $P_H = 2.5 \text{ kW} \quad Q_H = 0 \text{ VAR}$

Inductor: $P_L = 0 \text{ W} \quad Q_L = \dfrac{V^2}{X_L} = \dfrac{(120 \text{ V})^2}{6\ \Omega} = 2.4 \text{ kVAR (ind.)}$

Capacitor: $P_C = 0 \text{ W} \quad Q_C = \dfrac{V^2}{X_C} = \dfrac{(120 \text{ V})^2}{24\ \Omega} = 600 \text{ VAR (cap.)}$

b. $P_T = 2.5 \text{ kW} + 0 \text{ W} + 0 \text{ W} = 2.5 \text{ kW}$

$Q_T = 0 \text{ VAR} + 2.4 \text{ kVAR} - 600 \text{ VAR} = 1.8 \text{ kVAR (ind.)}$

c. The power triangle is sketched as Figure 17–7(b). Both the hypotenuse and the angle can be obtained easily using rectangular to polar conversion. $\mathbf{S}_T = P_T + jQ_T = 2500 + j1800 = 3081\angle 35.8°$. Thus, apparent power is $S_T = 3081$ VA.

d. Generator current is $I = \dfrac{S_T}{E} = \dfrac{3081 \text{ VA}}{120 \text{ V}} = 25.7 \text{ A}$

Real and Reactive Power Equations

An examination of the power triangle of Figures 17–14 and 17–15 shows that *P* and *Q* may be expressed as

$$P = VI\cos\theta = S\cos\theta \quad \text{(W)} \tag{17–20}$$

and

$$Q = VI\sin\theta = S\sin\theta \quad \text{(VAR)} \tag{17–21}$$

where *V* and *I* are the magnitudes of the rms values of the voltage and current, respectively, and θ is the angle between them. *P* is always positive, while *Q* is positive for inductive circuits and negative for capacitive circuits. Thus, if $V = 120$ volts, $I = 50$ A, and $\theta = 30°$, $P = (120)(50)\cos 30° = 5196$ W and $Q = (120)(50)\sin 30° = 3000$ VAR.

PRACTICE PROBLEMS 3

1. A 208-V generator supplies power to a group of three loads. Load 1 has an apparent power of 500 VA with $\theta = 36.87°$ (i.e., it is net inductive). Load 2 has an apparent power of 1000 VA and is net capacitive with a power triangle angle of $-53.13°$. Load 3 is purely resistive with power $P_3 = 200$ W. Determine the power triangle for the combined loads and the generator current.
2. Using numerical values as indicated with Figure 17–10,
   a. Determine apparent power for each block, and the total apparent power to the circuit.
   b. Sum the apparent power for all blocks and show that it does not equal the actual apparent power to the circuit.

CircuitSim 17-7

Answers

1. $S_T = 1300$ VA, $\theta_T = -22.6°$, $I = 6.25$ A
2. a. 538.5 VA, 500 VA, 1612.5 VA, 2.10 KVA
   b. 2.651 kVA (Not the same.)

17.8 Power Factor

The quantity $\cos\theta$ in Equation 17–20 is defined as **power factor** and is given the symbol F_p. Thus,

$$F_p = \cos\theta \tag{17–22}$$

From Equation 17–20, we see that F_p may be computed as the ratio of real power to apparent power. Thus,

$$\cos\theta = P/S \tag{17–23}$$

Power factor is expressed as a number or as a percent. From Equation 17–23, it is apparent that the power factor cannot exceed 1.0 (or 100% if expressed in percent).

CircuitSim 17-8

The power factor angle θ is of interest. It can be found as

$$\theta = \cos^{-1}(P/S) \tag{17–24}$$

Angle θ is the angle between voltage and current. For a pure resistance, therefore, $\theta = 0°$. For a pure inductance, $\theta = 90°$; for a pure capacitance, $\theta = -90°$. For a circuit containing both resistance and inductance, θ will be somewhere between 0° and 90°; for a circuit containing both resistance and capacitance, θ will be somewhere between 0° and −90°.

Unity, Lagging, and Leading Power Factor

As indicated by Equation 17–23, a load's power factor shows how much of its apparent power is actually real power. For example, for a purely resistive circuit, $\theta = 0°$ and $F_p = \cos 0° = 1.0$. Therefore, $P = VI$ (watts) and all the load's apparent power is real power. This case ($F_p = 1$) is referred to as *unity* power factor.

For a load containing only resistance and inductance, the load current lags voltage. The power factor in this case is described as *lagging*. On the other hand, for a load containing only resistance and capacitance, current leads voltage and the power factor is described as *leading*. Thus, *an inductive circuit has a lagging power factor, while a capacitive circuit has a leading power factor.*

A load with a very poor power factor can draw excessive current. This is discussed next.

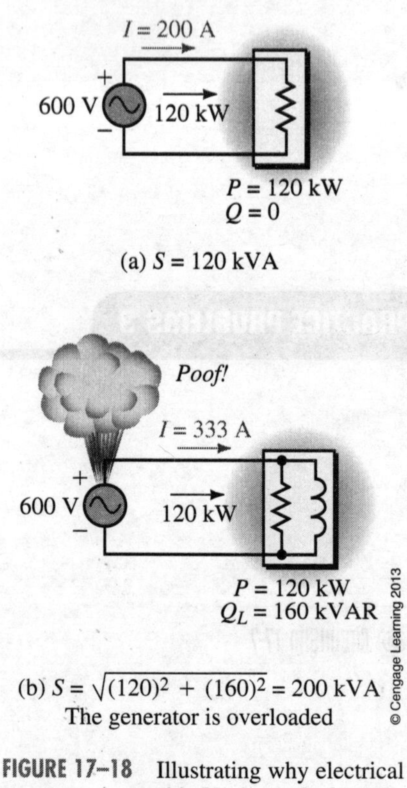

FIGURE 17–18 Illustrating why electrical apparatus is rated in VA instead of watts. Both loads dissipate 120 kW, but the current rating of generator (b) is exceeded because of the power factor of its load.

Why Equipment Is Rated in VA

As noted earlier, equipment is rated in terms of VA instead of watts. We now show why. Consider Figure 17–18. Assume that the generator is rated at 600 V, 120 kVA. This means that it is capable of supplying I = 120 kVA/600 V = 200 A. In (a), the generator is supplying a purely resistive load with 120 kW. Since $S = P$ for a purely resistive load, S = 120 kVA and the generator is supplying its rated current. In (b), the generator is supplying a load with P = 120 kW as before, but Q = 160 kVAR. Its apparent power is therefore S = 200 kVA, which means that the generator current is I = 200 kVA/600 V = 333.3 A. Even though it is supplying the same power as in (a), the generator is now greatly overloaded, and damage may result as indicated in (b).

This example illustrates clearly that rating a load or device in terms of power is a poor choice, as its current-carrying capability can be greatly exceeded (even though its power rating is not). Thus, *the size of electrical apparatus (generators, interconnecting wires, transformers, etc.) required to supply a load is governed, not by the load's power requirements, but rather by its VA requirements.*

Power Factor Correction

The problem shown in Figure 17–18 can be alleviated by cancelling some or all of the reactive component of power by adding reactance of the opposite type to the circuit. This is referred to as **power factor correction.** If you completely cancel the reactive component, the power factor angle is 0° and $F_p = 1$. This is referred to as unity power factor correction.

In practice, almost all loads, whether they be residential, industrial, or commercial, are inductive (due to the presence of motors, and other inductive elements); thus, you will probably never encounter a capacitive load that needs correcting. The upshot of this is that, in practice, virtually all power factor correction consists of adding a capacitor in order to cancel inductive effects. As is illustrated next, this capacitance is placed across the load and as near to it as practical.

EXAMPLE 17–7

For the overloaded generator problem of Figure 17–18(b), a capacitance with Q_C = 160 kVAR is added in parallel with the load as in Figure 17–19(a). Determine generator current I.

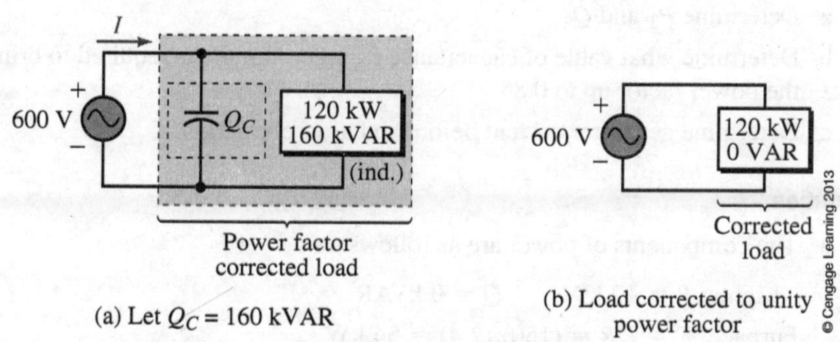

FIGURE 17–19 Power factor correction. The parallel capacitor greatly reduces source current.

Solution Q_T = 160 kVAR − 160 kVAR = 0. Therefore, S_T = 120 kW + j0 kVAR. Thus, S_T = 120 kVA, and current drops from 333 A to I = 120 kVA/600 V = 200 A. Thus, the generator is no longer overloaded—see Notes.

NOTES...

Note carefully that, what has changed here is how much current the source has to supply and consequently, how much current the transmission and distribution lines between the source and capacitor carry. Since the load itself has not changed, it requires the same current as before and the wires between the capacitor and the load must carry this larger current. (This is why power system engineers place corrective capacitors as close to the load as practical.)

CircuitSim 17-9

Residential customers are not charged directly for VARs—that is, they pay their electrical bills based solely on the number of the kilowatt-hours they use. This is because all residential customers have essentially the same power factor, and the power factor effect is simply built into the rates that they pay. Industrial customers, on the other hand, often have widely differing power factors, and the utility may have to monitor their VARs (or their power factor), as well as their watts, in order to determine a suitable charge.

CircuitSim 17-10

To illustrate, assume that the loads of Figures 17–18(a) and (b) are two small industrial plants. If the utility based its charge solely on power, both customers would pay the same amount. However, it costs the utility more to supply customer (b) since larger conductors, larger transformers, larger switchgear, and so on are required to handle the larger current. For this reason, industrial customers may pay a penalty if their power factor drops below a prescribed value as set by the utility.

EXAMPLE 17–8

Suppose an industrial client is charged a penalty if the plant power factor drops below 0.85. The equivalent plant loads are as shown in Figure 17–20. The frequency is 60 Hz—see Note 3.

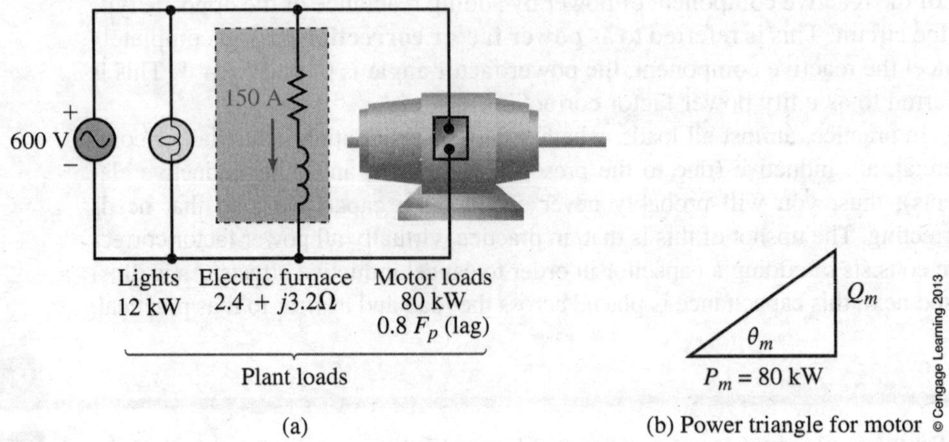

FIGURE 17–20

a. Determine P_T and Q_T.
b. Determine what value of capacitance (in microfarads) is required to bring the power factor up to 0.85.
c. Determine generator current before and after correction.

Solution

a. The components of power are as follows:

$$\text{Lights: } P = 12\text{ kW}, \qquad Q = 0\text{ kVAR}$$
$$\text{Furnace: } P = I^2R = (150)^2(2.4) = 54\text{ kW}$$
$$Q = I^2X = (150)^2(3.2) = 72\text{ kVAR (ind.)}$$

Motor: $\theta_m = \cos^{-1}(0.8) = 36.9°$. Thus, from the motor power triangle,

$$Q_m = P_m \tan\theta_m = 80 \tan 36.9° = 60\text{ kVAR (ind.)}$$
$$\text{Total: } P_T = 12\text{ kW} + 54\text{ kW} + 80\text{ kW} = 146\text{ kW}$$
$$Q_T = 0 + 72\text{ kVAR} + 60\text{ kVAR} = 132\text{ kVAR (ind.)}$$

b. The power triangle for the plant is shown in Figure 17–21(a). However, we must correct the power factor to 0.85. Thus, we need $\theta' = \cos^{-1}(0.85) = 31.8°$, where θ' is the power factor angle of the corrected load as indicated in Figure 17–21(b). The maximum reactive power that we can tolerate is thus $Q'_T = P_T \tan\theta' = 146 \tan 31.8° = 90.5$ kVAR.

Now consider Figure 17–22. $Q'_T = Q_C + 132$ kVAR, where $Q'_T = 90.5$ kVAR. Therefore, $Q_C = -41.5$ kVAR $= 41.5$ kVAR (cap.). But $Q_C = V^2/X_C$. Therefore, $X_C = V^2/Q_C = (600)^2/41.5\text{ kVAR} = 8.67\ \Omega$. But $X_C = 1/\omega C$. Thus, a capacitor of

$$C = \frac{1}{\omega X_C} = \frac{1}{(2\pi)(60)(8.67)} = 306\ \mu\text{F}$$

will provide the required correction—see Notes.

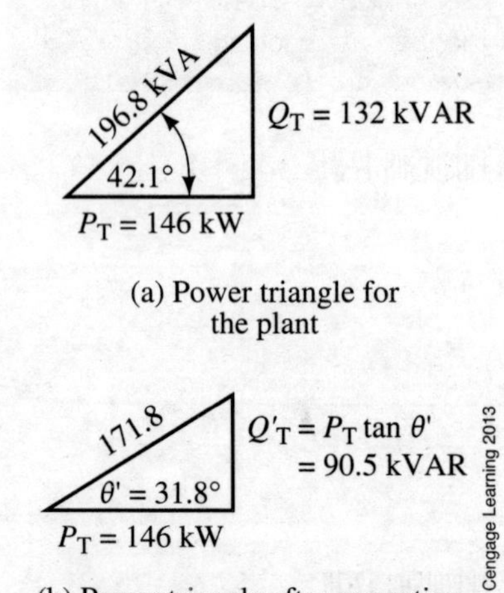

FIGURE 17–21 Initial and final power triangles. Note that P_T does not change when we correct the power factor, since for the capacitor, $P = 0$ W.

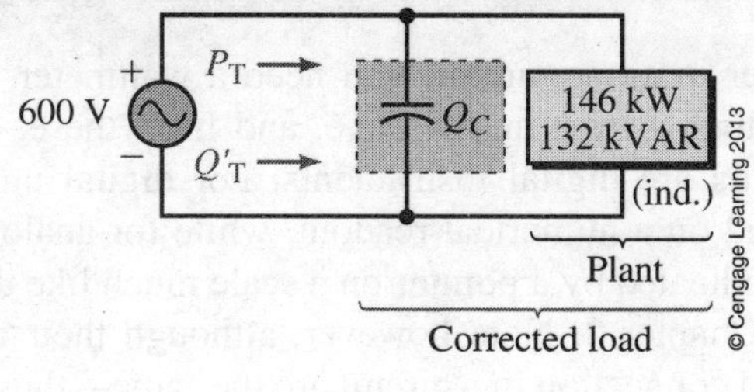

FIGURE 17–22

c. For the original circuit Figure 17–21(a), $S_T = 196.8$ kVA. Thus,

$$I = \frac{S_T}{E} = \frac{196.8 \text{ kVA}}{600 \text{ V}} = 328 \text{ A}$$

For the corrected circuit of Figure 17–21(b), $S'_T = 171.8$ kVA and

$$I = \frac{171.8 \text{ kVA}}{600 \text{ V}} = 286 \text{ A}$$

Thus, power factor correction has dropped the source current by 42 A.

NOTES...

1. Electric power companies use special utility grade capacitors for this job. The capacitor must be sized both for its value of capacitance and for its kVA rating.
2. Power factor correction capacitors are available in fixed values, or as switched banks with automatic controls that switch capacitors in or out, depending on load requirements.
3. Computer simulations for this example are found in Section 17.12.

PRACTICE PROBLEMS 4

1. By how many amps has power factor correction reduced the motor current of Figure 17–20?
2. Repeat Example 17–8, except correct the power factor to unity.
3. Due to plant expansion, 102 kW of purely resistive load is added to the plant of Figure 17–20. Determine whether power factor correction is needed to correct the expanded plant to 0.85 F_p, or better.

Answers

1. Zero—there is no change in motor current. It is the source current that is reduced.
2. 973 μF, 243 A. Other answers remain unchanged.
3. $F_p = 0.88$. No correction needed.

IN-PROCESS LEARNING CHECK 1

(Answers are at the end of the chapter.)

1. Sketch the power triangle for Figure 17–9(c). Using this triangle, determine the magnitude of the applied voltage.
2. For Figure 17–10, assume a source of $E = 240$ volts, $P_2 = 300$ W, and $Q_2 = 400$ VAR (cap.). What is the magnitude of the source current I?
3. What is the power factor of each of the circuits of Figures 17–7, 17–8, and 17–9? Indicate whether they are leading or lagging.
4. Consider the circuit of Figure 17–18(b). If $P = 100$ kW and $Q_L = 80$ kVAR, is the source overloaded, assuming it is capable of handling a 120-kVA load?

17.9 ac Power Measurement

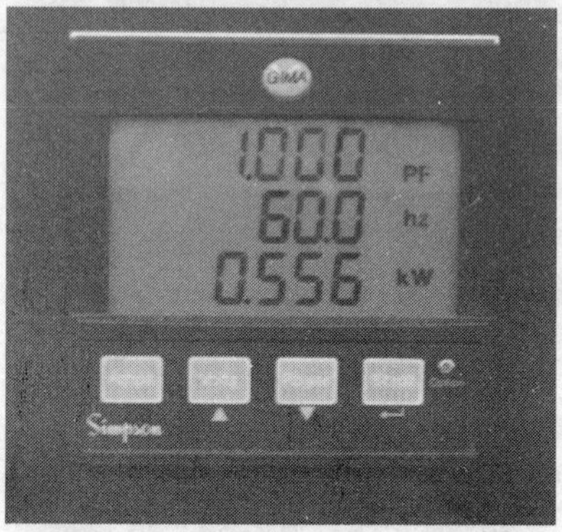

FIGURE 17–23 Multifunction power/energy meter. It can measure active power (W), reactive power (VARs), apparent power (VA), power factor, energy, and more.

To measure power in an ac circuit, you need a wattmeter. A **wattmeter** is a device that monitors current and voltage, and from these, determines power. Most modern units are digital instruments. For digital units, Figure 17–23, power is displayed on a numerical readout, while for analog instruments (see Note), power is indicated by a pointer on a scale much like the analog meter of Figure 2–22(b), Chapter 2. Note however, although their details differ, their method of use and connection in a circuit are the same—thus, the measurement techniques described herein apply to both.

To strengthen your understanding of the concept of power measurement, reexamine Figures 17–1, 17–2, and 17–3. Instantaneous load power is the product of load voltage times load current, and average power is the average of this product. One way to implement power measurement is therefore to create a meter with a current sensing circuit, a voltage sensing circuit, a multiplier circuit, and an averaging circuit. Figure 17–24 shows a simplified symbolic representation of such an instrument. Current is passed through its current coil (CC) to create a magnetic field proportional to the current, and a solid state sensor circuit connected across the load voltage reacts with this field to produce an output voltage proportional to the product of instantaneous voltage and current (i.e., proportional to instantaneous power). An averaging circuit averages this voltage and drives a display to indicate average power. (The scheme used by the meter of Figure 17–23 is actually considerably more sophisticated than this because it measures many things besides power—e.g., it measures VARs, VA, energy, and more. However, the basic idea is conceptually correct.)

NOTES...

Older wattmeters, called electrodynamometers, may still be found in practice. They are electromechanical devices that utilize both a stationary coil connected in series with the load (the current coil) and a pivoted coil (the potential coil) connected across the load. The interaction of the magnetic fields of these two coils creates a torque, and the pointer, attached to the pivoted coil, takes up a position on the scale corresponding to average power—see Practice Problems 5, Problem 2.

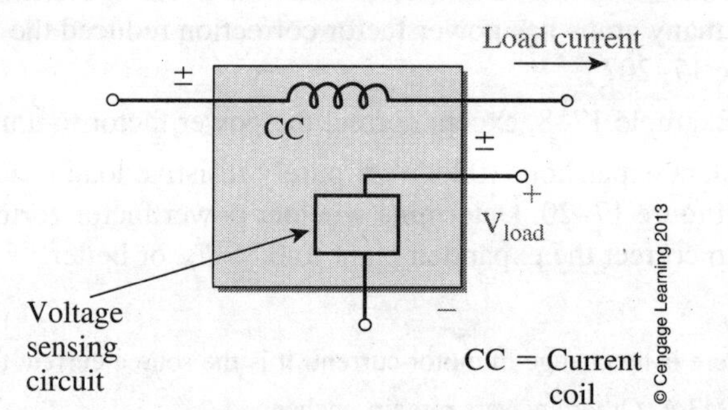

FIGURE 17–24 Conceptual representation of an electronic wattmeter.

Figure 17–25 shows how to connect a wattmeter into a circuit. Load current passes through its current coil circuit, and load voltage is impressed across its voltage sensing circuit. With this connection, the wattmeter computes and displays the product of the magnitude of the load voltage, the magnitude of the load current, and the cosine of the angle between them, that is, $V_{load} \cdot I_{load} \cdot \cos \theta_{load}$—see Notes. Thus, it measures load power. Note the ± marking on the terminals.

NOTES...

You need to be very clear on what angle to use in determining a wattmeter's reading. In terms of Figure 17–24, the angle to use is the angle between the voltage labelled V_{load} and the current labelled *Load current,* using the reference polarity for voltage and the reference direction for current shown. The diagram of Figure 17–25 helps clarify this.

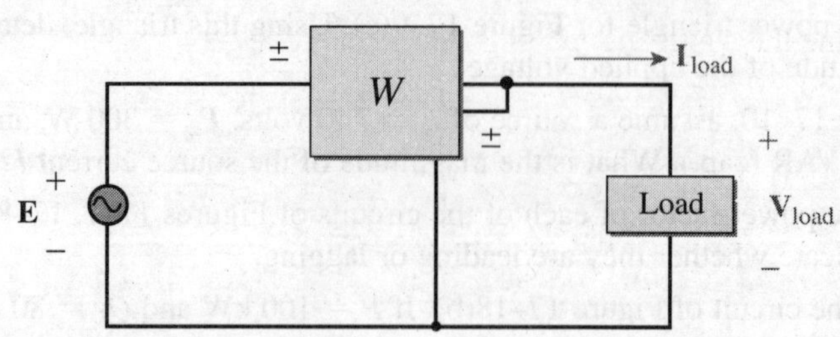

FIGURE 7–25 Connection of wattmeter.

The meter is connected so that load current enters the ± current terminal and the higher potential end of the load is connected to the ± voltage terminal. (On many meters, the ± voltage terminal is internally connected so that only three terminals are brought out as in Figure 17–26.)

When power is to be measured in a low power factor circuit, a low power factor wattmeter must be used. This is because, for low power factor loads, currents can be very high, even though the power is low. Thus, you can easily exceed the current rating of a standard wattmeter and damage it, even though the power indication on the meter is small.

EXAMPLE 17–9

For the circuit of Figure 17–25, what does the wattmeter indicate if

a. $\mathbf{V}_{load} = 100\text{ V}\angle 0°$ and $\mathbf{I}_{load} = 15\text{ A}\angle 60°$,

b. $\mathbf{V}_{load} = 100\text{ V}\angle 10°$ and $\mathbf{I}_{load} = 15\text{ A}\angle 30°$?

Solution

a. $\theta_{load} = 60°$. Thus, $P = (100)(15)\cos 60° = 750\text{ W}$,

b. $\theta_{load} = 10° - 30° = -20°$. Thus, $P = (100)(15)\cos(-20°) = 1410\text{ W}$.

Note: For (b), since $\cos(-20°) = \cos(+20°)$, it does not matter whether we include the minus sign in the calculation.

EXAMPLE 17–10

For Figure 17–26, determine the wattmeter reading.

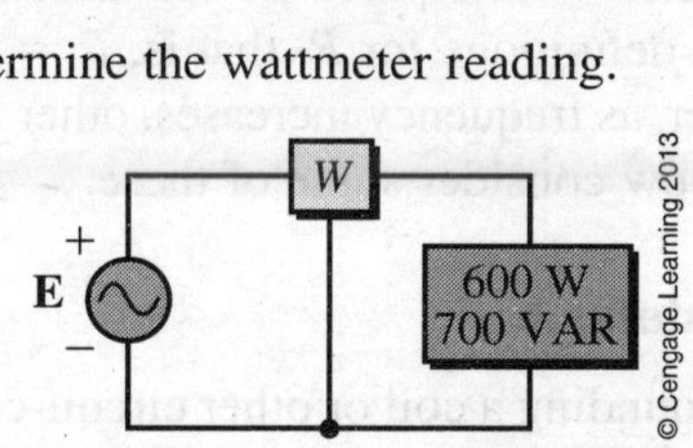

FIGURE 17–26 This wattmeter has its voltage side ± terminals connected internally.

Solution A wattmeter reads only active power. Thus, it indicates 600 W.

It should be noted that the wattmeter reads power only for circuit elements on the load side of the meter. In addition, if the load consists of several elements, it reads the total power.

CircuitSim 17-11

PRACTICE PROBLEMS 5

1. Determine the wattmeter reading for Figure 17–27.

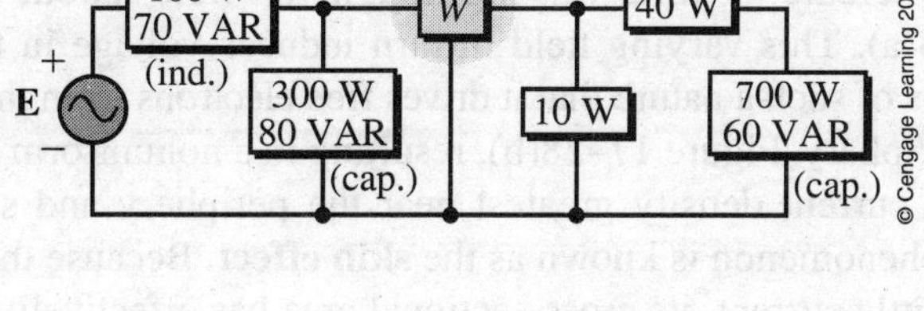

FIGURE 17–27

2. (Optional problem) Research the electrodynamometer wattmeter and prepare a short description complete with diagram showing its coils, pivoted pointer, scale, and so on. The Internet is a good source of information.

Answer
1. 750 W

17.10 Effective Resistance

Up to now, we have assumed that resistance is constant, independent of frequency. However, this is not entirely true. For a number of reasons, the resistance of a circuit to ac is greater than its resistance to dc. While this effect is small at low frequencies, it is very pronounced at high frequencies. AC resistance is known as **effective resistance.**

Before looking at why ac resistance is greater than dc resistance, we need to reexamine the concept of resistance itself. Recall from Chapter 3 that resistance was originally defined in the dc case as opposition to current, that is, $R = V/I$. (This is ohmic resistance.) Building on this, you learned in Chapter 4 that $P = I^2R$. It is this latter viewpoint that allows us to give meaning to ac resistance. That is, we define ac or effective resistance as

$$R_{\text{eff}} = \frac{P}{I^2} \quad (\Omega) \tag{17–25}$$

where P is dissipated power (as determined by a wattmeter). From this, you can see that anything that affects dissipated power affects resistance. For dc and low-frequency ac, both definitions for R, that is, $R = V/I$ and $R = P/I^2$, yield the same value. However, as frequency increases, other factors cause an increase in resistance. We will now consider some of these.

Eddy Currents and Hysteresis

The magnetic field surrounding a coil or other circuit-carrying ac current varies with time and thus induces voltages in nearby conductive material such as metal equipment cabinets, transformer cores, and so on. The resulting currents (called **eddy currents** because they flow in circular patterns like eddies in a brook) are unwanted and create power losses called eddy current losses. Since additional power must be supplied by the source to make up for these losses, P in Equation 17–25 increases, increasing the effective resistance of the coil.

If ferromagnetic material is also present, an additional power loss occurs due to hysteresis effects caused by the magnetic field alternately magnetizing the material in one direction, then the other. Hysteresis and eddy current losses are important even at low frequencies, such as the 60-Hz power system frequency. This is discussed in Chapter 23.

Skin Effect

AC currents create a time-varying magnetic field about a conductor, Figure 17–28(a). This varying field in turn induces voltage in the conductor. This voltage is of such a nature that it drives free electrons from the center of the wire to its periphery, Figure 17–28(b), resulting in a nonuniform distribution of current, with current density greatest near the periphery and smallest in the center. This phenomenon is known as the **skin effect.** Because the center of the wire carries little current, its cross-sectional area has effectively been reduced, thus increasing resistance. While skin effect is generally negligible at power line frequencies (except for conductors larger than several hundred thousand circular mils), it is so pronounced at microwave frequencies that the center of a wire carries almost no current. For this reason, hollow conductors are often used at very high frequencies instead of solid wires, as shown in Figure 17–28(c).

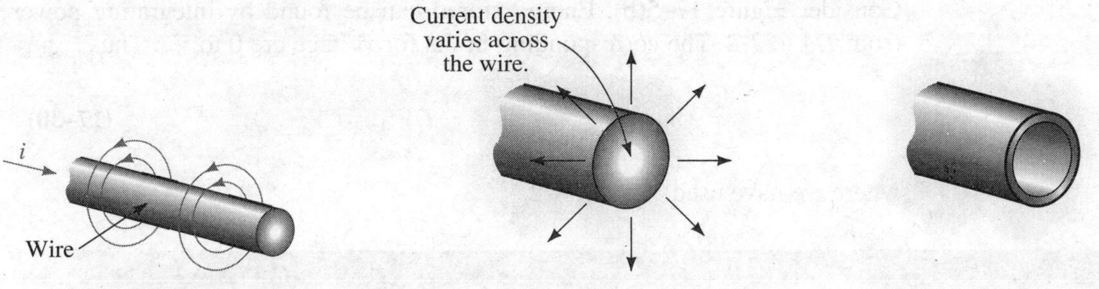

(a) Varying magnetic field induces voltage within the conductor.

(b) This voltage drives free electrons toward the periphery, leaving few electrons in the center.

(c) At high frequencies, the effect is so pronounced that hollow conductors may be used.

FIGURE 17–28 Skin effect in ac circuits.

Radiation Resistance

At high frequencies some of the energy supplied to a circuit escapes as radiated energy. For example, a radio transmitter supplies power to an antenna, where it is converted into radio waves and radiated into space. The resistance effect here is known as radiation resistance. This resistance is much higher than simple dc resistance. For example, a TV transmitting antenna may have a resistance of a fraction of an ohm to dc but several hundred ohms effective resistance at its operating frequency.

NOTES...

1. The resistance measured by an ohmmeter is dc resistance.
2. Many of the effects noted here will be treated in detail in your various electronics courses. We will not pursue them further.

17.11 Energy Relationships for ac

Recall that power and energy are related by the equation $p = dw/dt$. Thus, energy can be found by integration as

$$W = \int pdt = \int vidt \qquad (17\text{–}26)$$

Inductance

For an inductance, $v = Ldi/dt$. Substituting this into Equation 17–26, cancelling dt, and rearranging terms yields

$$W_L = \int \left(L\frac{di}{dt} \right) idt = L\int idi \qquad (17\text{–}27)$$

Recall from Figure 17–4(b) that energy flows into an inductor during time interval 0 to $T/4$ and is released during time interval $T/4$ to $T/2$. The process then repeats itself. The energy stored (and subsequently released) can thus be found by integrating power from $t = 0$ to $t = T/4$. Current at $t = 0$ is 0 and current at $t = T/4$ is I_m. Using these as our limits of integration, we find (see Note)

$$W_L = L\int_0^{I_m} idi = \frac{1}{2}LI_m^2 = LI^2 \quad \text{(J)} \qquad (17\text{–}28)$$

where we have used $I = I_m/\sqrt{2}$ to express energy in terms of effective current.

NOTES...

The idea of stored energy has been encountered before—for example, Equation 17–28 appears in Chapter 13 as Equation 13–11 and Equation 17–30 appears in Chapter 10 as Equation 10–22.

Capacitance

For a capacitance, $i = Cdv/dt$. Substituting this into Equation 17–26 yields

$$W_C = \int v\left(C\frac{dv}{dt}\right)dt = C\int v\,dv \tag{17–29}$$

Consider Figure 17–5(b). Energy stored can be found by integrating power from $T/4$ to $T/2$. The corresponding limits for voltage are 0 to V_m. Thus,

$$W_C = C\int_0^{V_m} v\,dv = \frac{1}{2}CV_m^2 = CV^2 \quad \text{(J)} \tag{17–30}$$

where we have used $V = V_m/\sqrt{2}$.

17.12 Circuit Analysis Using Computers

The power relationships described in this chapter can be investigated easily using Multisim and PSpice. To illustrate, consider Examples 17–11 and 17–12.

EXAMPLE 17–11

For the circuit of Figure 17–1 (Section 17.1), assume an *RL* load. Typical waveforms are then as shown in Figure 17–2. Since current is used as reference, we will use a current source in our simulation. Additionally, because the analysis is independent of amplitude and frequency, we have arbitrarily chosen 5 A, 100 Hz for this source.

Multisim Solution

Read Note 1, then build the circuit of Figure 17–29(a) on your screen. Ensure that nodes are displayed—see Note 2. Select transient analysis, set *TSTOP* to **0.01**, select the *Output* tab, then add *I(L1)* and *V(1)* to the output display list—see Notes 3 and 4. To create the instantaneous power waveform, click *Add Expression*, double-click *V(1)*, double-click *, double-click *I(L1)*, click *OK*, then *Simulate*. The curves of (b) should appear—see Note 5. Compare the curves to those of Figure 17–2. Using a cursor, measure peak and trough values for the power waveform. Record values (for use in Extended Example 17–11).

NOTES...

1. To select the signal current source, right-click *Place* and ensure that *Signal Source Components* is selected. From its parts bin on your screen, select *Place ac Current Source.*
2. To display nodes, click *Options* then *Sheet Properties*. Under *Net names*, click *Show all.*
3. Node numbers depend on the order in which you wire up your circuit. For our discussion, we assume node numbers as shown in Figure 17–29(a).
4. If you have legacy software, use \$1 instead of V(1), and I1#branch instead of I(L1).
5. If your display is jagged, increase the number of time points.

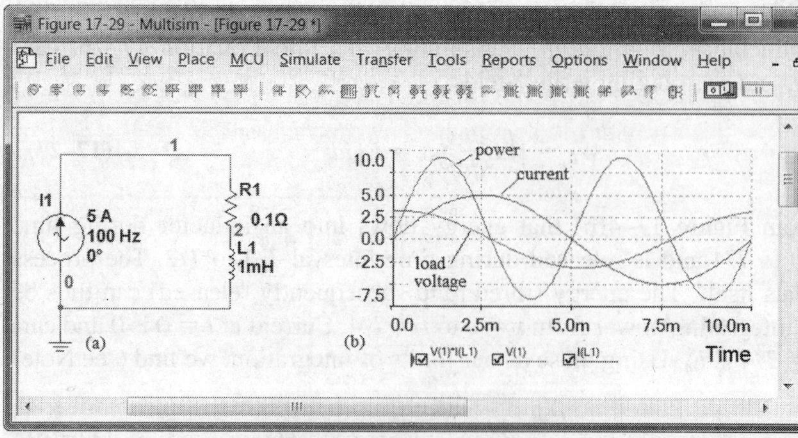

FIGURE 17–29 Investigating instantaneous power using Multisim.

Extended Example 17–11 Add a wattmeter to the circuit as in Figure 17–30. To locate the wattmeter, right-click *Place* and ensure that *Instruments* is selected. Use your mouse pointer to locate the wattmeter in the tool bin, position it on your screen as in Figure 17–30, then wire it in. Click the simulation ON/OFF switch to the ON position, double-click the wattmeter to open it, then read its value.

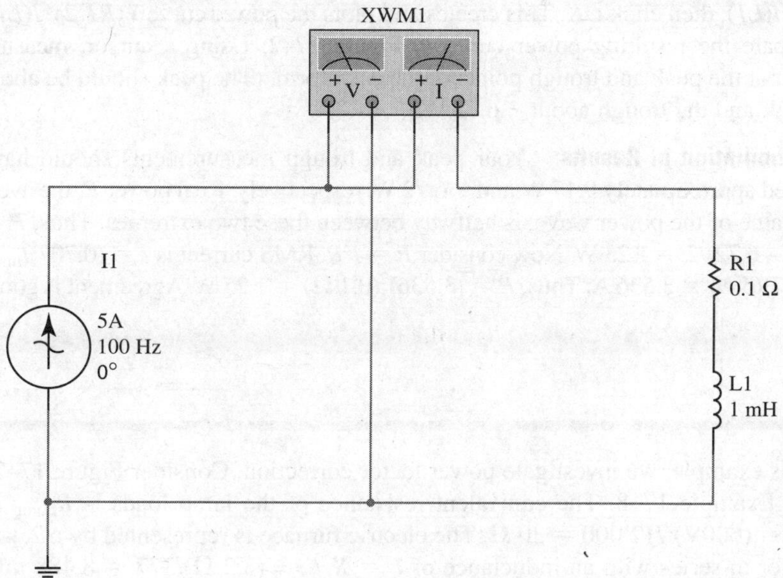

FIGURE 17–30 Verifying average power using a wattmeter.

Reconciliation of Results: First, consider the instantaneous power waveform. In the previous test, you should have measured peak and trough power values as 9.20 W and –6.70 W, respectively. Real power P, the average value of the power wave, is halfway between these two extremes. Thus, $P = (9.20 - 6.70)/2 = 1.25$ W. This agrees with the wattmeter reading. Now consider $P = I^2R$. RMS current is $I = (0.707)I_m = (0.707)(5\text{ A}) = 3.536$ A. This yields $P = (3.536)^2(0.1\Omega) = 1.25$ W. Thus, all results agree with theory.

PSpice Solution

Create the circuit on your screen as in Figure 17–31(a). Use current source *ISIN*—see Note 1. Double-click each small parameter box and enter values as

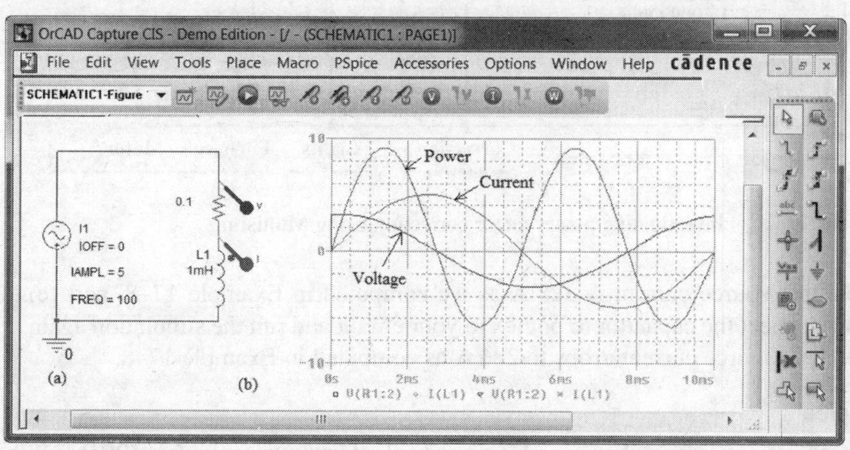

FIGURE 17–31 Investigating instantaneous power using PSpice.

NOTES...

1. PSpice represents current into devices, not out. For this reason, the source in Figure 17–31 must be rotated before positioning so that its end marked + is at the bottom.
2. Your trace names may be different. For purposes of this discussion, we use the names indicated in Figure 17–31. When you run this simulation, substitute your trace names.

shown. Now click the *New Simulation Profile* icon, type a name, choose *Transient,* set *TSTOP* to **10ms,** then click *OK*. Add voltage and current markers as shown, then run the simulation. Voltage and current waveforms as in Figure 17–31(b) appear. Note the legend below the trace—see Note 2. Voltage is identified as *V(R1:2)* and current as *I(L1)*. To plot instantaneous power, click *Trace, Add Trace,* and when the dialog box opens, click *V(R1:2)*, click the * symbol, click *I(L1)*, then click *OK*. This creates and plots the power curve *V(R1:2)*I(L1)*. Compare the resulting power curve to Figure 17–2. Using a cursor, measure power at the peak and trough points of the waveform. (The peak should be about 9.17 W and the trough about −6.72 W.)

Reconciliation of Results: Your peak and trough measurements should have yielded approximately 9.17 W and −6.72 W, respectively. Real power *P*, the average value of the power wave, is halfway between these two extremes. Thus, $P = (9.17 - 6.72)/2 = 1.23$ W. Now consider $P = I^2R$. RMS current is $I = (0.707)I_m = (0.707)(5\text{ A}) = 3.536$ A. Thus, $P = (3.536)^2(0.1\ \Omega) = 1.25$ W. Agreement is good.

EXAMPLE 17–12

In this example, we investigate power factor correction. Consider Figure 17–20 from Example 17–8. The equivalent resistance of the lamp loads is $R_{\text{Lamp}} = V^2/P = (600\text{V})^2/12\,000 = 30\ \Omega$. The electric furnace is represented by a 2.4-Ω resistor in series with an inductance of $L = X_L/\omega = (3.2\ \Omega)/377 = 8.488$ mH. We also need the equivalent circuit for the motor loads. Based on the information given, it can be shown that the equivalent circuit is a resistance of $R_m = 2.88\ \Omega$ in series with an inductance of $L_m = 5.73$ mH. (You can check this out if you like.)

Multisim Solution

Build the circuit on your screen as shown as Figure 17–32. Use the ac power source (from the *Power Source Components* toolbar). To add the ammeter, right-click Place, ensure that *Measurement Components* is selected, then locate *Place Ammeter (Horizontal)* and position it on your screen. Double-click the ammeter and set its mode to *AC,* then click the Simulate ON/OFF switch to ON.

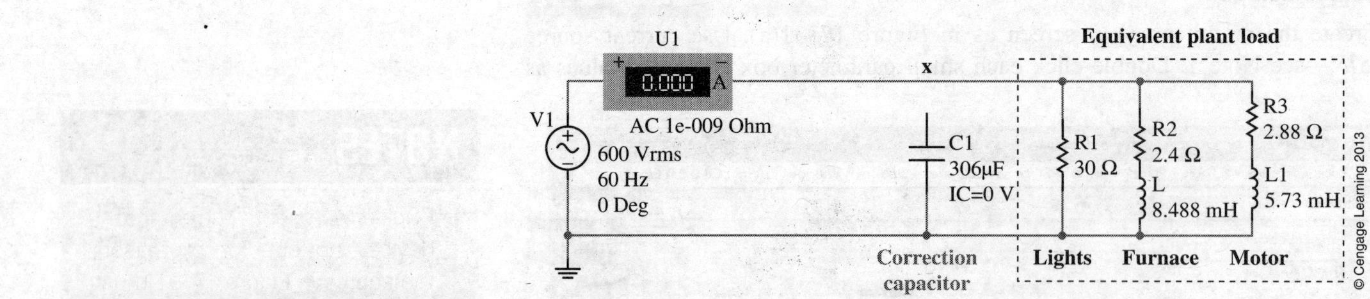

FIGURE 17–32 Investigating power factor correction using Multisim.

Note that source current is 328 A as we computed in Example 17–8, part (c). Now connect the capacitor to point *x* in your circuit and run the simulation again. Note that source current drops to 286 A as computed in Example 17–8.

PSpice

The PSpice circuit is shown as Figure 17–33. (Configure ammeter IPRINT as you did in Chapter 16.) Use power source *VAC*. Set simulation to *AC Sweep/Noise*

Analysis, set *Start Frequency* and *End Frequency* to **60** Hz, select *Linear,* set *Total Points* to **1,** click *OK,* then click the *Run* icon. When the simulation window opens, click *View/Output File,* then scroll until you find the answers indicated in Figure 17–33. As indicated, the magnitude of the current, IM(V_PRINT1), is $3.280 \times 10^2 = 328$ A, the same as we computed in Example 17–8. Now connect a 306-μF capacitor across the load between points *x-x,* then run the simulation again. Note that source current drops to 286 A as predicted in Example 17–8.

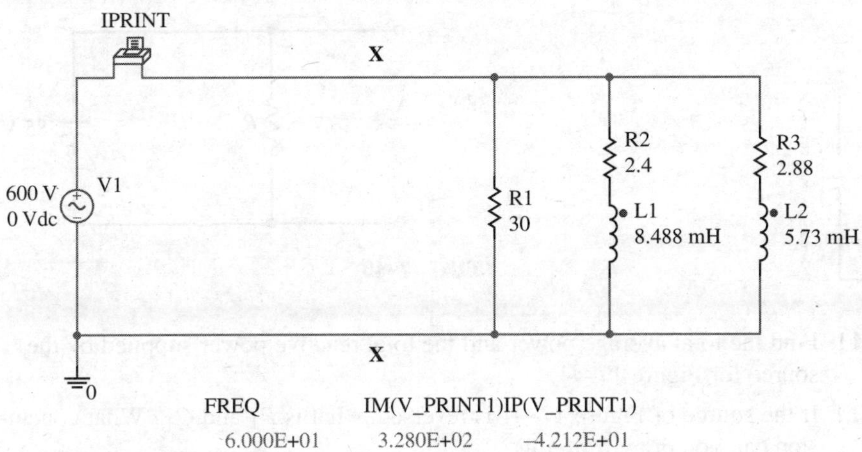

FIGURE 17–33 Investigating power factor correction using PSpice.

17.1–17.5 Problems

1. Note that the power curve of Figure 17–4 is sometimes positive and sometimes negative. What is the significance of this? Between $t = T/4$ and $t = T/2$, what is the direction of power flow?
2. What is real power? What is reactive power? Which power, real or reactive, has an average value of zero?
3. A pair of electric heating elements is shown in Figure 17–34.
   a. Determine the active and reactive power to each.
   b. Determine the active and reactive power delivered by the source.
4. For the circuit of Figure 17–35, determine the active and reactive power to the inductor.
5. If the inductor of Figure 17–35 is replaced by a 40-μF capacitor and source frequency is 60 Hz, what is Q_C?
6. Find R and X_L for Figure 17–36.
7. For the circuit of Figure 17–37, $f = 100$ Hz. Find
   a. R b. X_C c. C
8. For the circuit of Figure 17–38, $f = 10$ Hz. Find
   a. P b. X_L c. L
9. For Figure 17–39, find X_C.
10. For Figure 17–40, $X_C = 42.5\ \Omega$. Find R, P, and Q.

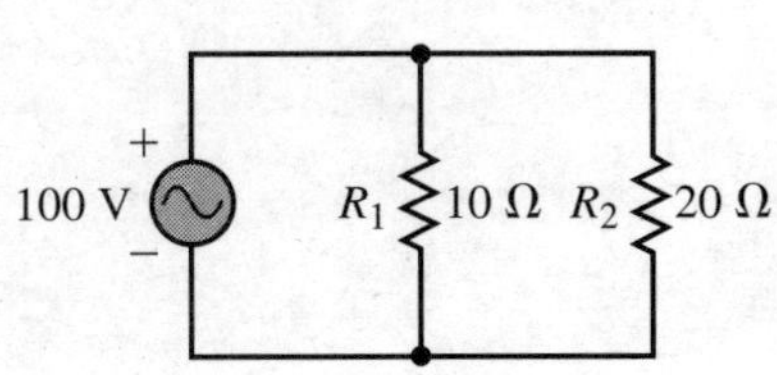

FIGURE 17–34

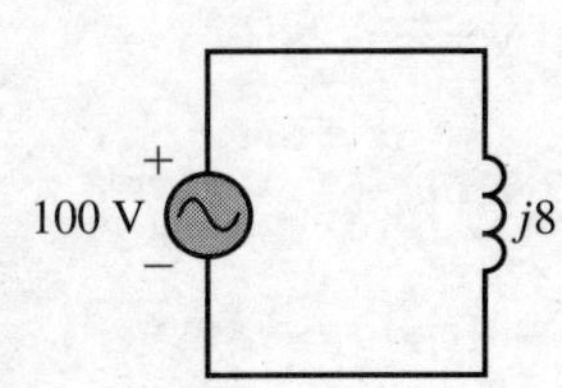

FIGURE 17–35

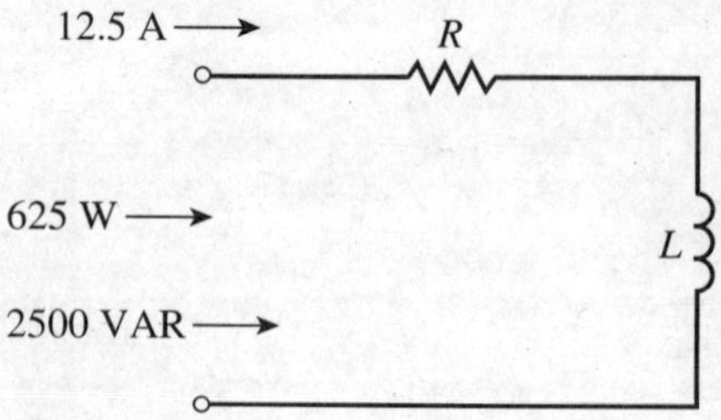

FIGURE 17–36

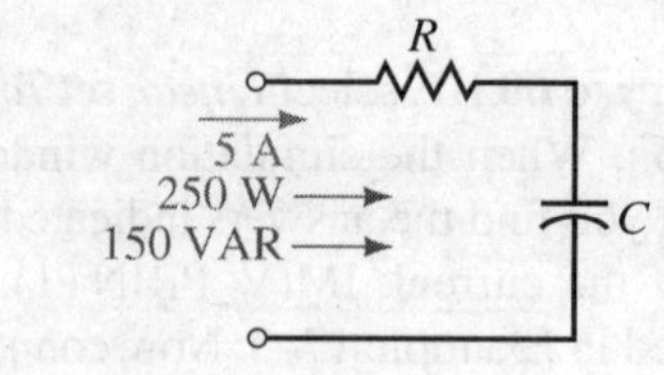

FIGURE 17–37

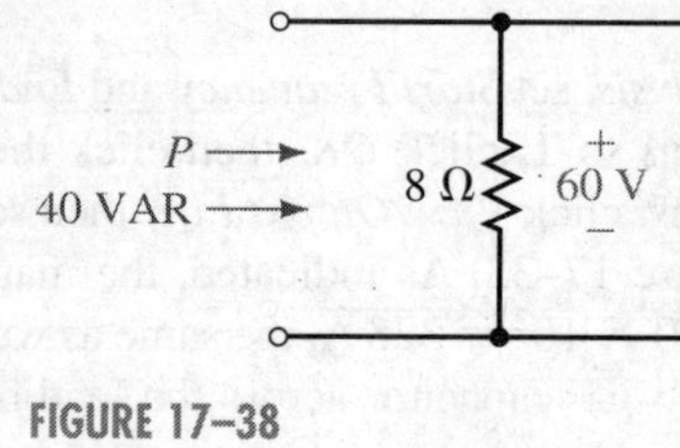

FIGURE 17–38

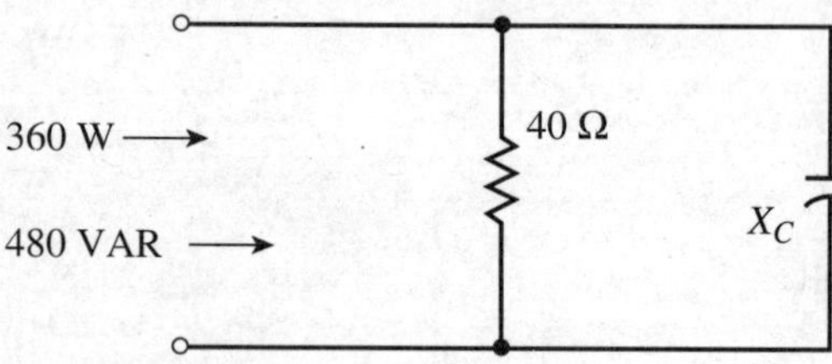

FIGURE 17–39

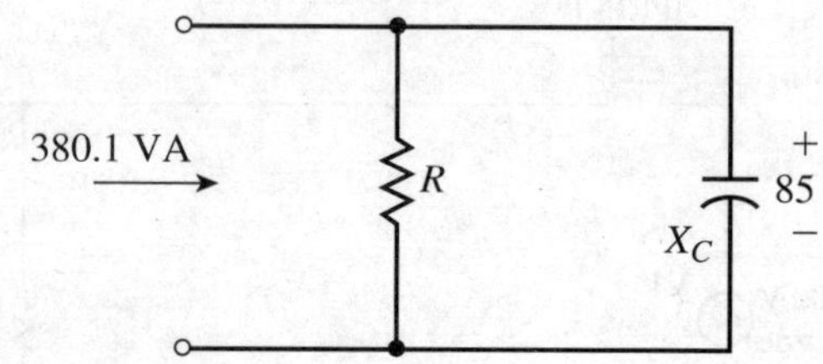

FIGURE 17–40

11. Find the total average power and the total reactive power supplied by the source for Figure 17–41.
12. If the source of Figure 17–41 is reversed, what is P_T and Q_T? What conclusion can you draw from this?

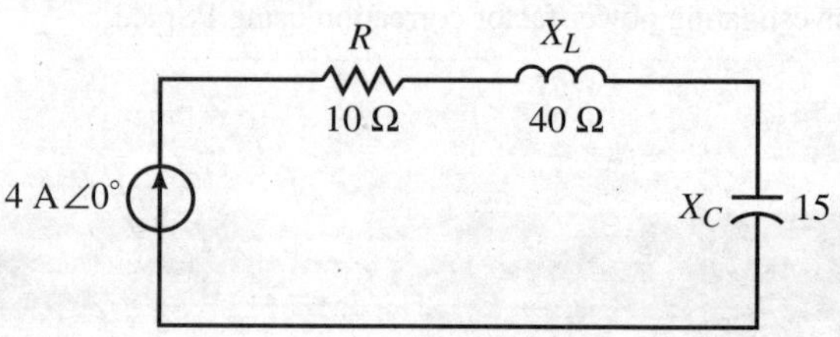

FIGURE 17–41

13. Refer to Figure 17–42. Find P_2 and Q_3. Is the element in Load 3 inductive or capacitive?

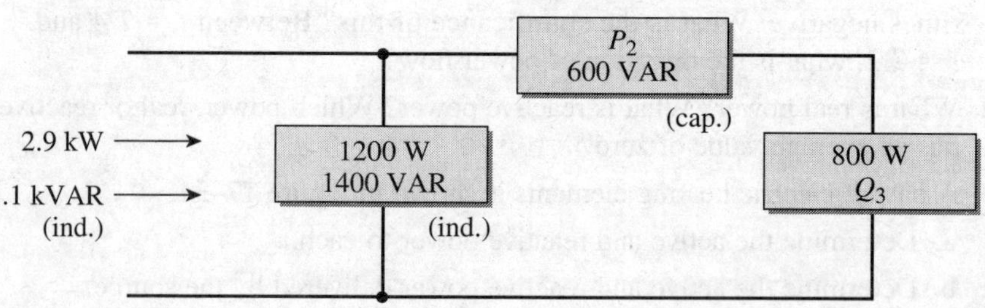

FIGURE 17–42

14. For Figure 17–43, determine P_T and Q_T.

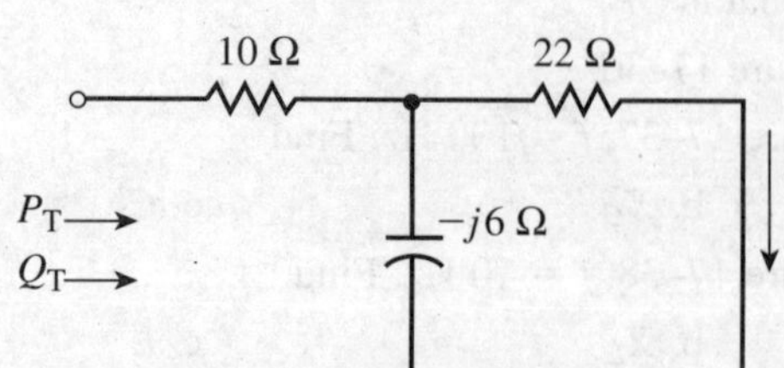

FIGURE 17–43

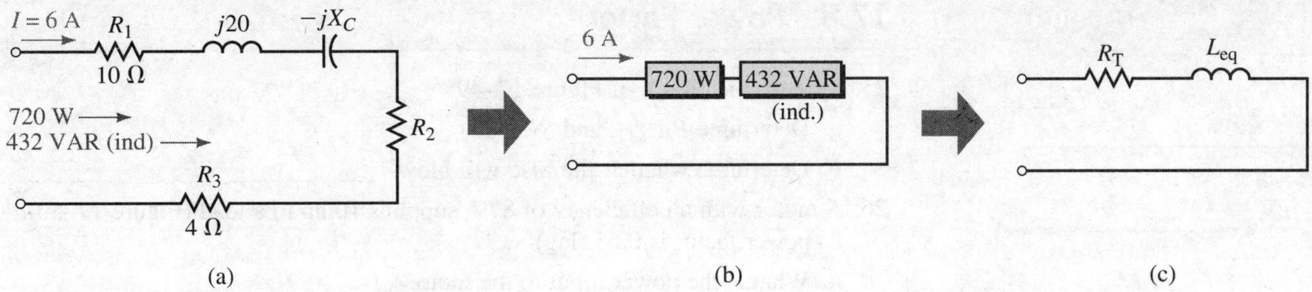

FIGURE 17–44

15. For Figure 17–44, ω = 10 rad/s. Determine

    a. R_T b. R_2 c. X_C d. L_{eq}

16. For Figure 17–45, determine the total P_T and Q_T.

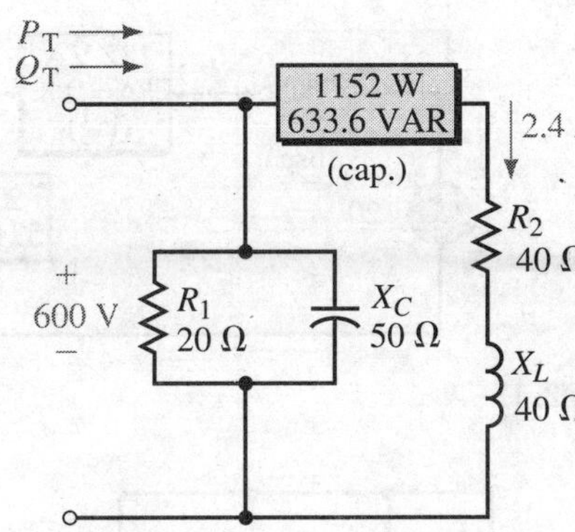

FIGURE 17–45

17.7 The Relationship between *P*, *Q*, and *S*

17. For the circuit of Figure 17–7, draw the power triangle and determine the apparent power.
18. Repeat Problem 17 for Figure 17–8.
19. Ignoring the wattmeter of Figure 17–27, determine the power triangle for the circuit as seen by the source.
20. For the circuit of Figure 17–46, what is the source current?
21. For Figure 17–47, the generator supplies 30 A. What is *R*?

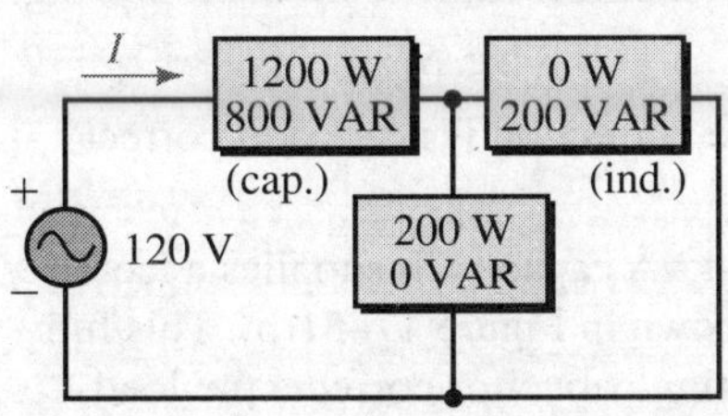

FIGURE 17–46

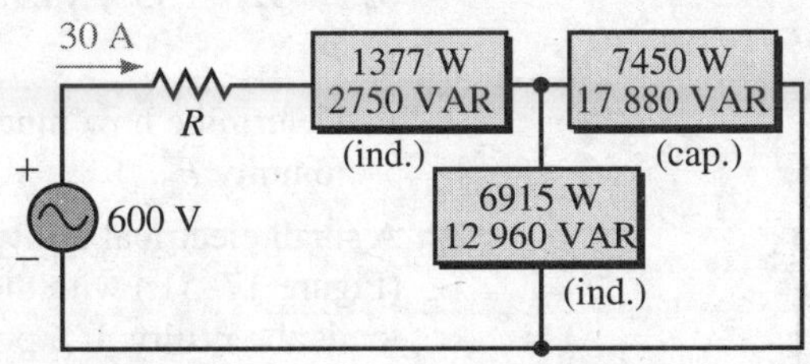

FIGURE 17–47

22. Suppose **V** = 100 V∠60° and **I** = 10A∠40°:

    a. What is θ, the angle between **V** and **I**?

    b. Determine *P* from $P = VI \cos \theta$.

    c. Determine *Q* from $Q = VI \sin \theta$.

    d. Sketch the power triangle and from it, determine **S**.

    e. Show that **S** = **VI*** gives the same answer as (d).

23. For Figure 17–48, S_{gen} = 4835 VA. What is *R*?
24. Refer to the circuit of Figure 17–16:

    a. Determine the apparent power for each box.

    b. Sum the apparent powers that you just computed. Why does the sum not equal S_T = 1803 VA as obtained in Example 17–5?

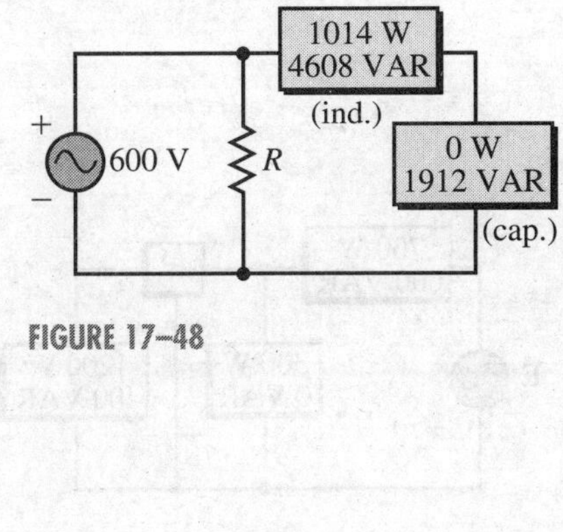

FIGURE 17–48

17.8 Power Factor

25. Refer to the circuit of Figure 17–49:
    a. Determine P_T, Q_T, and S_T.
    b. Determine whether the fuse will blow.
26. A motor with an efficiency of 87% supplies 10 hp to a load (Figure 17–50). Its power factor is 0.65 (lag).
    a. What is the power input to the motor?
    b. What is the reactive power to the motor?
    c. Draw the motor power triangle. What is the apparent power to the motor?

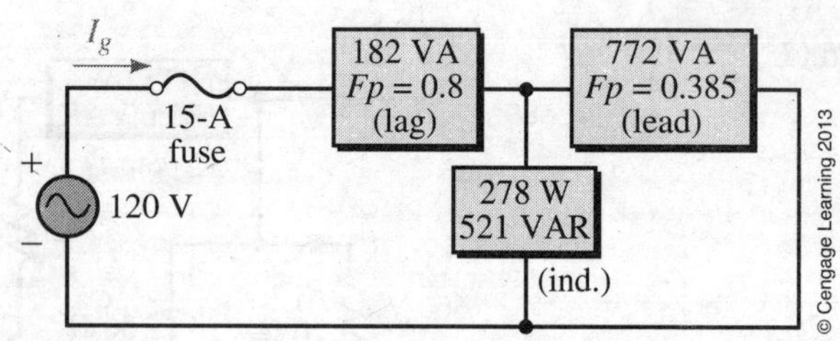

FIGURE 17–49

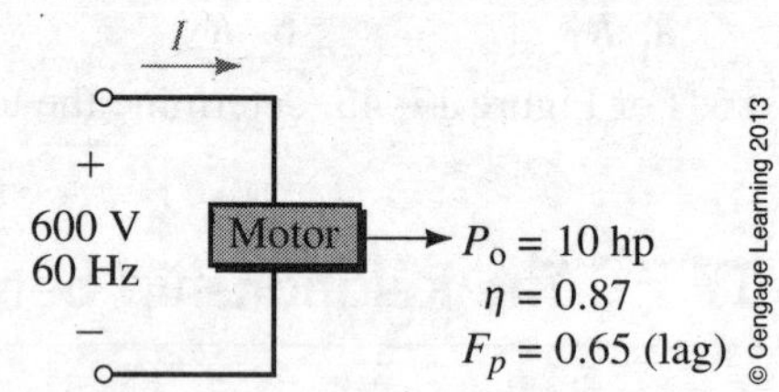

FIGURE 17–50

27. To correct the circuit power factor of Figure 17–50 to unity, a power factor correction capacitor is added.
    a. Show where the capacitor is connected.
    b. Determine its value in microfarads.
28. Consider Figure 17–20. The motor is replaced with a new unit requiring $\mathbf{S}_m = (120 + j35)$ kVA. Everything else remains the same. Find the following:
    a. P_T b. Q_T c. S_T
    d. Determine how much kVAR capacitive correction is needed to correct to unity F_p.
29. A small electrical utility has a 600-V, 300-kVA capacity. It supplies a factory (Figure 17–51.) with the power triangle shown in Figure 17–51(b). This fully loads the utility. If a power factor correcting capacitor corrects the load to unity power factor, how much more power (at unity power factor) can the utility sell to other customers?

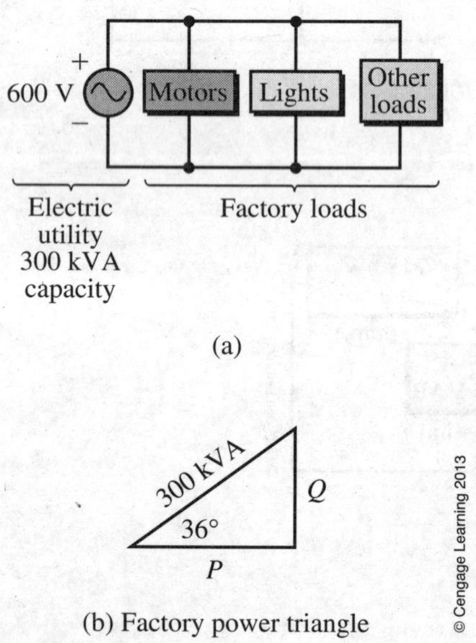

FIGURE 17–51

17.9 ac Power Measurement

30. a. Why does the wattmeter of Figure 17–52 indicate only 1200 watts?
    b. Where would the wattmeter have to be placed to measure power delivered by the source? Sketch the modified circuit.
    c. What would the wattmeter indicate in (b)?
31. Determine the wattmeter reading for Figure 17–53.

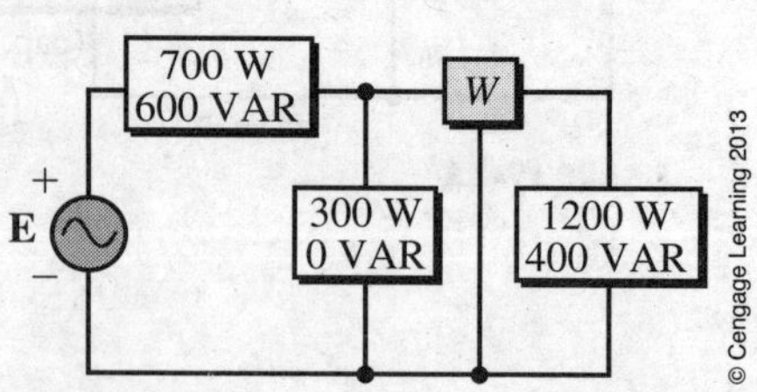

FIGURE 17–52

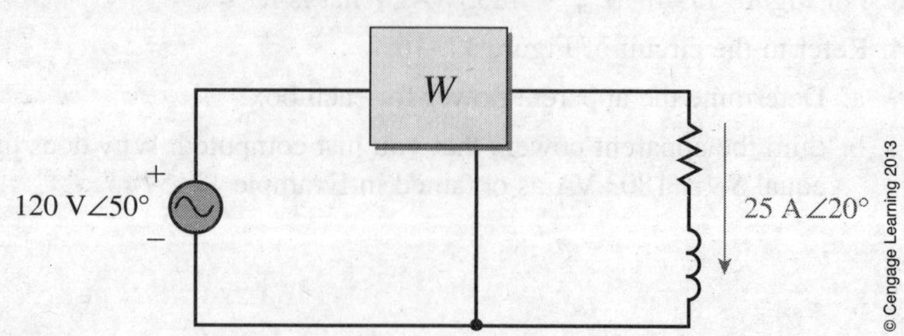

FIGURE 17–53

32. Determine the wattmeter reading for Figure 17–54.

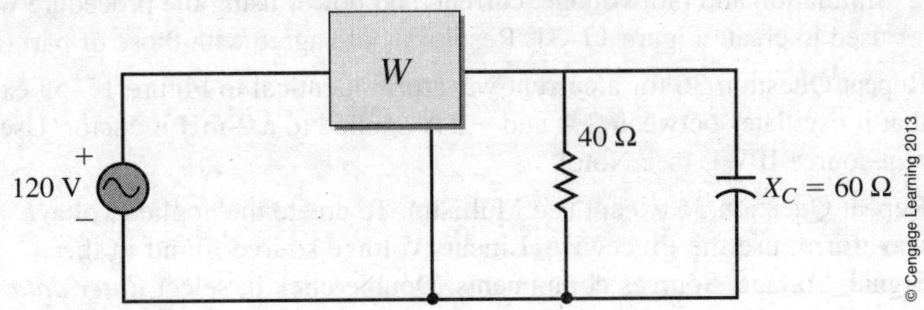

FIGURE 17–54

17.10 Effective Resistance

33. Measurements on an iron-core solenoid coil yield the following values: $V = 80$ V, $I = 400$ mA, $P = 25.6$ W, and $R = 140\ \Omega$. (The last measurement was taken with an ohmmeter.) What is the ac resistance of the solenoid coil?

17.12 Circuit Analysis Using Computers

34. Using Multisim, create the circuit of Figure 17–17 on your screen. (For the source, use the voltage source from Power Components.) Define the parallel *RLC* combination as the "load." Add an ammeter and a wattmeter to measure load current and load power. Using your calculator, compute the resistance of the heater, the inductance of the coil, and the capacitance of the capacitor, then set these values on your Multisim circuit. Run a simulation and measure load current, power, and power factor. Using the measured values, compute Q_T. Using the applied voltage and the measured current, compute apparent power. Compare these values of Q_T and S_T to the values determined in Example 17–6. Multisim

35. A 10-μF capacitor has voltage $v = 10 \sin(\omega t - 90°)$ V. Given $f = 1000$ Hz, use Multisim or PSpice to investigate the power waveform and compare to Figure 17–5. (PSpice users, use VSIN.) Multisim PSPICE

36. The voltage waveform of Figure 17–55 is applied to a 200-μF capacitor. PSPICE
    a. Using your calculator and the principles of Chapter 10, determine the current through the capacitor and sketch. (Also sketch the voltage waveform on your graph.) Multiply the two waveforms to obtain a plot of $p(t)$. Compute power at its max and min points.

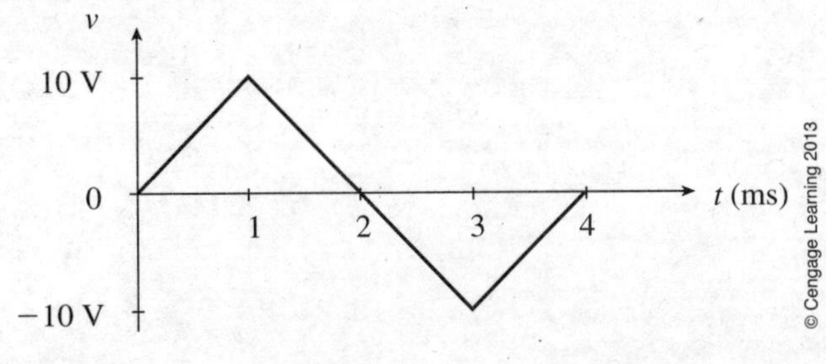

FIGURE 17–55

    b. Use PSpice to verify the results. Use voltage source VPWL. You have to describe the waveform to the source. It has a value of 0 V at $t = 0$, 10 V at $t = 1$ ms, −10 V at $t = 3$ ms, and 0 V at $t = 4$ ms. To set these, double-click the source symbol and enter values via the Property Editor as

NOTES...

PSpice represents current into devices. Thus, when you double-click a current source symbol (ISIN, IPWL, etc.) and specify a current waveform, you are specifying the current *into* the source. You must take this into consideration when you set up your current sources.

follows: **0** for T1, **0V** for V1, **1ms** for T2, **10V** for V2, and so on. Run the simulation and plot voltage, current, and power using the procedure we used to create Figure 17–31. Results should agree with those of part (a).

PSPICE 37. Repeat Question 36 for a current waveform identical to Figure 17–55 except that it oscillates between 2 A and −2 A applied to a 2-mH inductor. Use current source IPWL (see Note).

Multisim 38. Repeat Question 36 except use Multisim. To create the applied voltage waveform, use the Piecewise_Linear_Voltage source found in the Signal_Voltage_Sources components. Double-click it, select *Enter data points in table* and enter values to describe the waveform. Click *Repeat data during simulation,* then click *OK*. Choose transient analysis, set *TSTOP* to 0.008, click the *Output* tab, set items up for display, then click *Simulate.*

ANSWERS TO IN-PROCESS LEARNING CHECKS

IN-PROCESS LEARNING CHECK 1

1. 100 V

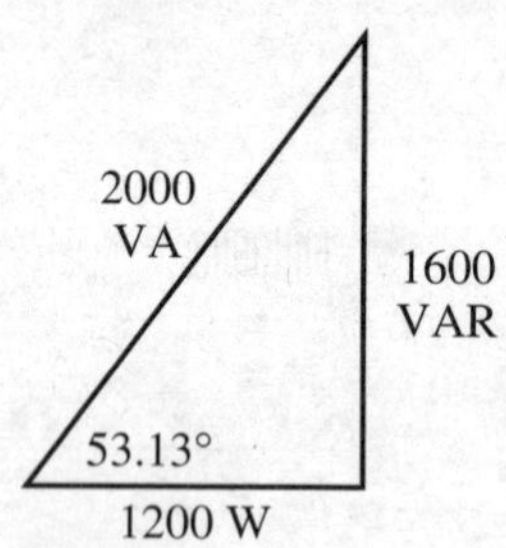

FIGURE 17–56

2. 8.76 A
3. Figure 17–7: 0.6 (lag); Figure 17–8: 0.97 (lead); Figure 17–9: 0.6 (lag)
4. Yes. (S = 128 kVA)

IMPEDANCE NETWORKS

V

As you have already observed, the impedance of an inductor or a capacitor is dependent upon the frequency of the signal applied to the element. When capacitors and inductors are combined with resistors and voltage or current sources, the circuit will behave in a predictable manner for all frequencies.

This part of the textbook examines how circuits consisting of various combinations of impedances and sources behave under specific conditions. In particular, we find that all the laws, rules, and theorems developed previously apply to even the most complicated impedance network.

Ohm's law and Kirchhoff's voltage and current laws are easily modified to give the framework for developing methods of network analysis. Just as in dc circuits, Thévenin's and Norton's theorems will allow us to simplify a complicated circuit to a single source and corresponding impedance.

The theorems and methods of analysis are applied to numerous types of circuits which are commonly encountered throughout electrical and electronics technology. Resonant circuits and filter circuits are commonly used to restrict the range of output frequencies for a given range of input frequencies.

The study of three-phase systems and transformers is particularly useful for anyone interested in commercial power distribution. These topics deal with practical applications and the drawbacks of using various types of circuits.

Finally, we examine how a circuit reacts to nonsinusoidal alternating voltages. This topic involves complex signals which are processed by impedance networks resulting in outputs which are often dramatically different from the input. ■

KEY TERMS

Admittance
Admittance Diagram
Capacitive Impedance
Cutoff (or Corner) Frequency
Impedance Diagram
Inductive Impedance
Leading and Lagging Power Factor
Resistive Circuit
Susceptance

OUTLINE

Ohm's Law for ac Circuits
ac Series Circuits
Kirchhoff's Voltage Law and the Voltage Divider Rule
ac Parallel Circuits
Kirchhoff's Current Law and the Current Divider Rule
Series-Parallel Circuits
Frequency Effects
Applications
Circuit Analysis Using Computers

OBJECTIVES

After studying this chapter, you will be able to

- apply Ohm's law to analyze simple series circuits,
- apply the voltage divider rule to determine the voltage across any element in a series circuit,
- apply Kirchhoff's voltage law to verify that the summation of voltages around a closed loop is equal to zero,
- apply Kirchhoff's current law to verify that the summation of currents entering a node is equal to the summation of currents leaving the same node,
- determine unknown voltage, current, and power for any series/parallel circuit,
- determine the series or parallel equivalent of any network consisting of a combination of resistors, inductors, and capacitors.

18

ac SERIES-PARALLEL CIRCUITS

CHAPTER PREVIEW

In this chapter we examine how simple circuits containing resistors, inductors, and capacitors behave when subjected to sinusoidal voltages and currents. Principally, we find that the rules and laws which were developed for dc circuits will apply equally well for ac circuits. The major difference between solving dc and ac circuits is that analysis of ac circuits requires using vector algebra.

In order to proceed successfully, it is suggested that the student spend time reviewing the important topics covered in dc analysis. These include Ohm's law, the voltage divider rule, Kirchhoff's voltage law, Kirchhoff's current law, and the current divider rule.

You will also find that a brief review of vector algebra will make your understanding of this chapter more productive. In particular, you should be able to add and subtract any number of vector quantities. ■

Putting It in Perspective

Heinrich Rudolph Hertz

Hulton Archive/Stringer/Hulton Archive/Getty Images

HEINRICH HERTZ WAS BORN IN HAMBURG, Germany, on February 22, 1857. He is known mainly for his research into the transmission of electromagnetic waves.

Hertz began his career as an assistant to Hermann von Helmholtz in the Berlin Institute physics laboratory. In 1885, he was appointed Professor of Physics at Karlsruhe Polytechnic, where he did much to verify James Clerk Maxwell's theories of electromagnetic waves.

In one of his experiments, Hertz discharged an induction coil with a rectangular loop of wire having a very small gap. When the coil discharged, a spark jumped across the gap. He then placed a second, identical coil close to the first, but with no electrical connection. When the spark jumped across the gap of the first coil, a smaller spark was also induced across the second coil. Today, more elaborate antennas use similar principles to transmit radio signals over vast distances. Through further research, Hertz was able to prove that electromagnetic waves have many of the characteristics of light: they have the same speed as light; they travel in straight lines; they can be reflected and refracted; and they can be polarized.

Hertz's experiments ultimately led to the development of radio communication by such electrical engineers as Guglielmo Marconi and Reginald Fessenden.

Heinrich Hertz died at the age of 36 on January 1, 1894. ■

18.1 Ohm's Law for ac Circuits

This section is a brief review of the relationship between voltage and current for resistors, inductors, and capacitors. Unlike in Chapter 16, all phasors are given as rms rather than as peak values. As you saw in Chapter 17, this approach simplifies the calculation of power.

Resistors

In Chapter 16, we saw that when a resistor is subjected to a sinusoidal voltage as shown in Figure 18–1, the resulting current is also sinusoidal and in phase with the voltage.

The sinusoidal voltage $v = V_m\sin(\omega t + \theta)$ may be written in phasor form as $\mathbf{V} = V\angle\theta$. Whereas the sinusoidal expression gives the instantaneous value of voltage for a waveform having an amplitude of V_m (volts peak), the phasor form has a magnitude which is the effective (or rms) value. The relationship between the magnitude of the phasor and the peak of the sinusoidal voltage is given as

$$V = \frac{V_m}{\sqrt{2}}$$

Because the resistance vector may be expressed as $\mathbf{Z}_R = R\angle 0°$, we evaluate the current phasor as follows:

$$\mathbf{I} = \frac{\mathbf{V}}{\mathbf{Z}_R} = \frac{V\angle\theta}{R\angle 0°} = \frac{V}{R}\angle\theta = I\angle\theta$$

NOTES...

Although currents and voltages may be shown in either time domain (as sinusoidal quantities) or in phasor domain (as vectors), resistance and reactance are never shown as sinusoidal quantities. The reason for this is that whereas currents and voltages vary as functions of time, resistance and reactance do not.

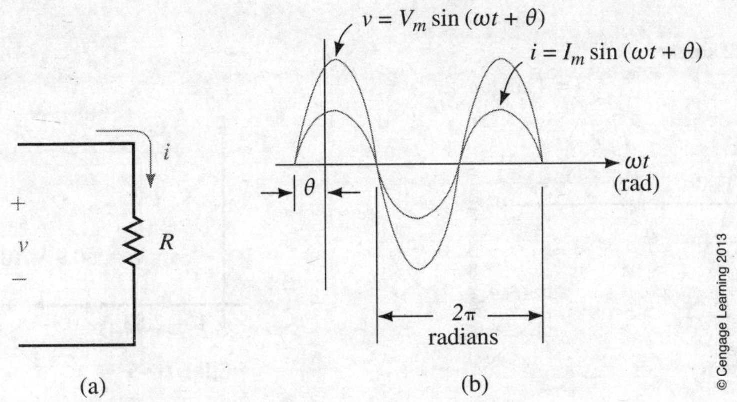

FIGURE 18–1 Sinusoidal voltage and current for a resistor.

If we wish to convert the current from phasor form to its sinusoidal equivalent in the time domain, we would have $i = I_m\sin(\omega t + \theta)$. Again, the relationship between the magnitude of the phasor and the peak value of the sinusoidal equivalent is given as

$$I = \frac{I_m}{\sqrt{2}}$$

The voltage and current phasors may be shown on a phasor diagram as in Figure 18–2.

Because one phasor is a current and the other is a voltage, the relative lengths of these phasors are purely arbitrary. Regardless of the angle θ, we see that the voltage across and the current through a resistor will always be in phase.

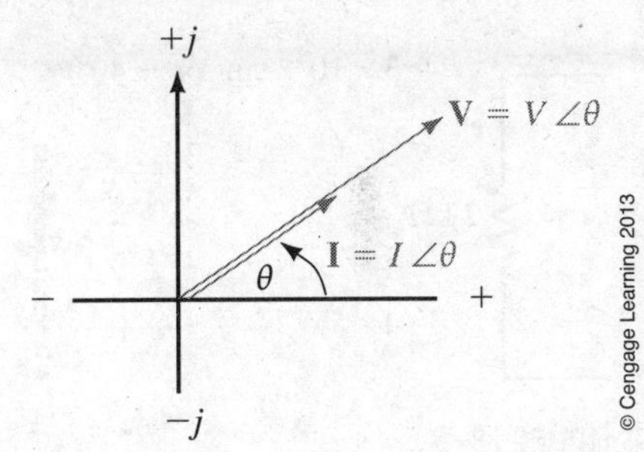

FIGURE 18–2 Voltage and current phasors for a resistor.

EXAMPLE 18–1

Refer to the resistor shown in Figure 18–3:

a. Find the sinusoidal current i using phasors.

b. Sketch the sinusoidal waveforms for v and i.

c. Sketch the phasor diagram of **V** and **I**.

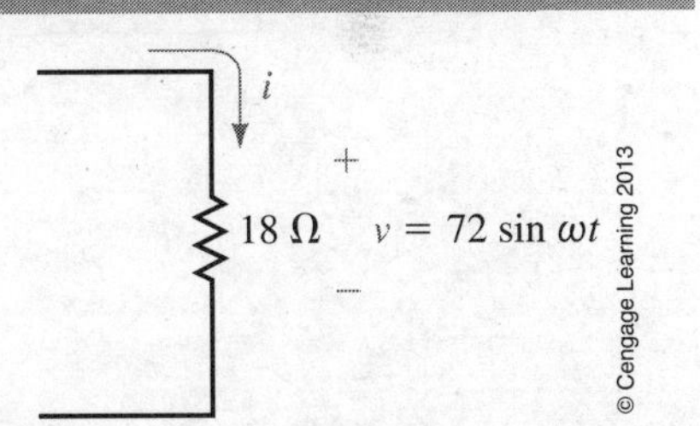

FIGURE 18–3

Solution

a. The phasor form of the voltage is determined as follows:

$$v = 72 \sin \omega t \Leftrightarrow \mathbf{V} = 50.9 \text{ V}\angle 0°$$

From Ohm's law, the current phasor is determined to be

$$\mathbf{I} = \frac{\mathbf{V}}{\mathbf{Z}_R} = \frac{50.9 \text{ V}\angle 0°}{18 \ \Omega\angle 0°} = 2.83 \text{ A}\angle 0°$$

which results in the sinusoidal current waveform having an amplitude of

$$I_m = (\sqrt{2})(2.83 \text{ A}) = 4.0 \text{ A}$$

Therefore, the current i will be written as

$$i = 4 \sin \omega t$$

b. The voltage and current waveforms are shown in Figure 18–4.

c. Figure 18–5 shows the voltage and current phasors.

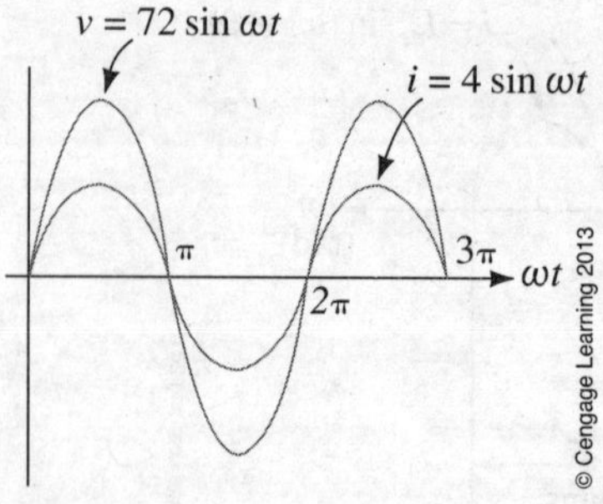

FIGURE 18–4

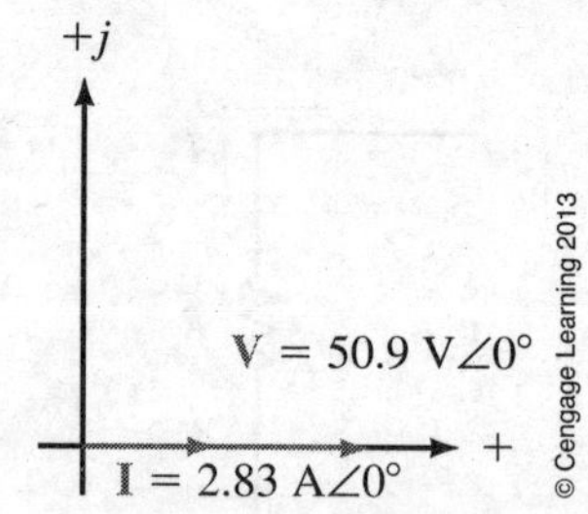

FIGURE 18–5

EXAMPLE 18–2

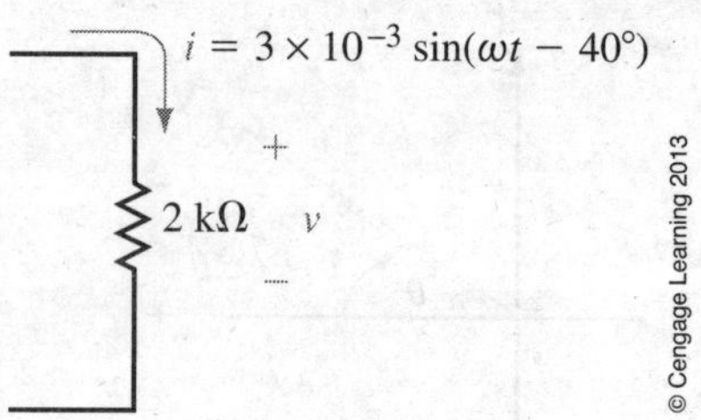

FIGURE 18–6

Refer to the resistor of Figure 18–6:

a. Use phasor algebra to find the sinusoidal voltage, v.
b. Sketch the sinusoidal waveforms for v and i.
c. Sketch a phasor diagram showing **V** and **I**.

Solution

a. The sinusoidal current has a phasor form as follows:

$$i = 3 \times 10^{-3} \sin(\omega t - 40°) \Leftrightarrow \mathbf{I} = 2.12 \text{ mA}\angle -40°$$

From Ohm's law, the voltage across the 2-kΩ resistor is determined as the phasor product

$$\begin{aligned}\mathbf{V} &= \mathbf{IZ}_R \\ &= (2.12 \text{ mA}\angle -40°)(2 \text{ k}\Omega\angle 0°) \\ &= 4.24 \text{ V}\angle -40°\end{aligned}$$

The amplitude of the sinusoidal voltage is

$$V_m = (\sqrt{2})(4.24 \text{ V}) = 6.0 \text{ V}$$

The voltage may now be written as

$$v = 6.0 \sin(\omega t - 40°)$$

b. Figure 18–7 shows the sinusoidal waveforms for v and i.

c. The corresponding phasors for the voltage and current are shown in Figure 18–8.

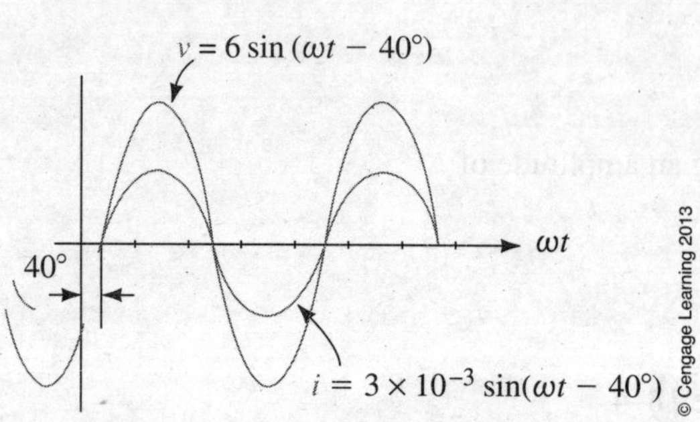

FIGURE 18–7

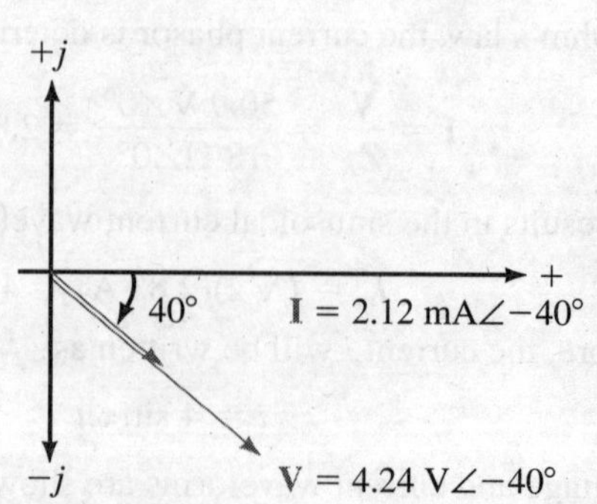

FIGURE 18–8

Inductors

When an inductor is subjected to a sinusoidal current, a sinusoidal voltage is induced across the inductor such that the voltage across the inductor leads the current waveform by exactly 90°. If we know the reactance of an inductor, then from Ohm's law the current in the inductor may be expressed in phasor form as

$$\mathbf{I} = \frac{\mathbf{V}}{\mathbf{Z}_L} = \frac{V\angle\theta}{X_L\angle 90^\circ} = \frac{V}{X_L}\angle(\theta - 90^\circ)$$

In vector form, the reactance of the inductor is given as

$$\mathbf{Z}_L = X_L\angle 90^\circ$$

where $X_L = \omega L = 2\pi f L$.

EXAMPLE 18–3

Consider the inductor shown in Figure 18–9:

a. Determine the sinusoidal expression for the current i using phasors.

b. Sketch the sinusoidal waveforms for v and i.

c. Sketch the phasor diagram showing **V** and **I**.

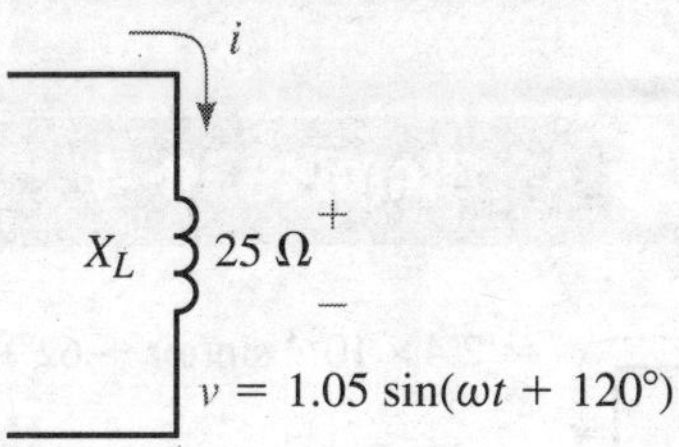

FIGURE 18–9

Solution

a. The phasor form of the voltage is determined as follows:

$$v = 1.05 \sin(\omega t + 120^\circ) \Leftrightarrow \mathbf{V} = 0.742 \text{ V } \angle 120^\circ$$

From Ohm's law, the current phasor is determined to be

$$\mathbf{I} = \frac{\mathbf{V}}{\mathbf{Z}_L} = \frac{0.742 \text{ V}\angle 120^\circ}{25\ \Omega\angle 90^\circ} = 29.7 \text{ mA}\angle 30^\circ$$

The amplitude of the sinusoidal current is

$$I_m = (\sqrt{2})(29.7 \text{ mA}) = 42 \text{ mA}$$

The current i is now written as

$$i = 0.042 \sin(\omega t + 30^\circ)$$

b. Figure 18–10 shows the sinusoidal waveforms of the voltage and current.

c. The voltage and current phasors are shown in Figure 18–11.

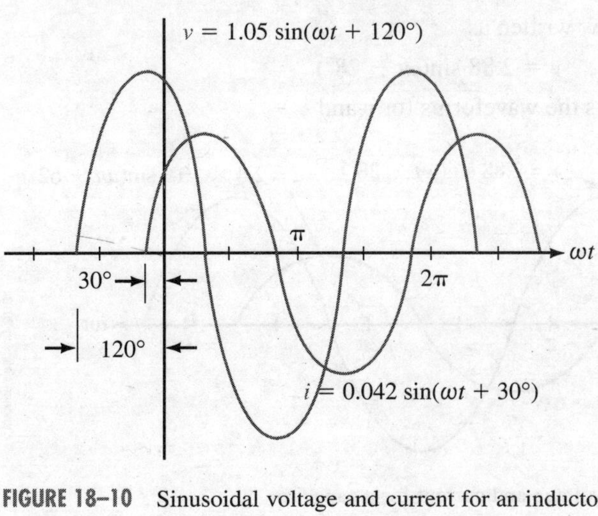

FIGURE 18–10 Sinusoidal voltage and current for an inductor.

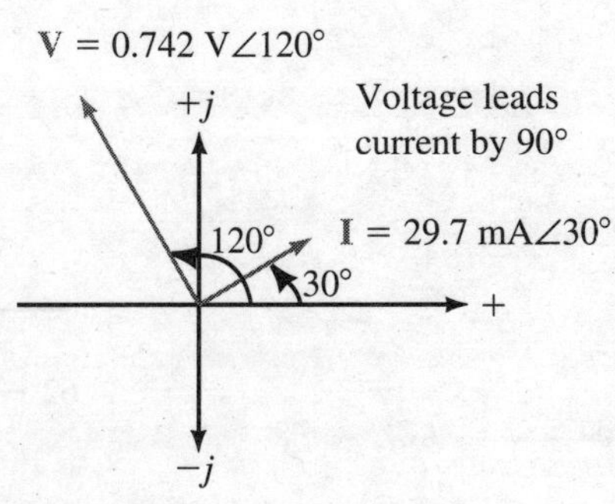

FIGURE 18–11 Voltage and current phasors for an inductor.

Capacitors

When a capacitor is subjected to a sinusoidal voltage, a sinusoidal current results. The current through the capacitor leads the voltage by exactly 90°. If we know the reactance of a capacitor, then from Ohm's law the current in the capacitor expressed in phasor form is

$$\mathbf{I} = \frac{\mathbf{V}}{\mathbf{Z}_C} = \frac{V\angle\theta}{X_C\angle-90°} = \frac{V}{X_C}\angle(\theta + 90°)$$

In vector form, the reactance of the capacitor is given as

$$\mathbf{Z}_C = X_C\angle-90°$$

where

$$X_C = \frac{1}{\omega C} = \frac{1}{2\pi fC}$$

EXAMPLE 18–4

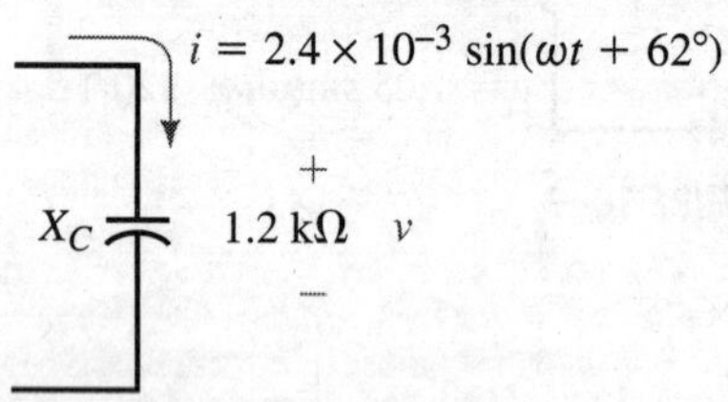

FIGURE 18–12

Consider the capacitor of Figure 18–12.

a. Find the voltage v across the capacitor.

b. Sketch the sinusoidal waveforms for v and i.

c. Sketch the phasor diagram showing **V** and **I**.

Solution

a. Converting the sinusoidal current into its equivalent phasor form gives

$$i = 2.4 \times 10^{-3}\sin(\omega t + 62°) \Leftrightarrow \mathbf{I} = 1.70\text{ mA}\angle 62°$$

From Ohm's law, the phasor voltage across the capacitor must be

$$\begin{aligned}\mathbf{V} &= \mathbf{I}\mathbf{Z}_C \\ &= (1.70\text{ mA}\angle 62°)(1.2\text{ k}\Omega\angle-90°) \\ &= 2.04\text{ V}\angle-28°\end{aligned}$$

The amplitude of the sinusoidal voltage is

$$V_m = (\sqrt{2})(2.04\text{ V}) = 2.88\text{ V}$$

The voltage v is now written as

$$v = 2.88\sin(\omega t - 28°)$$

b. Figure 18–13 shows the waveforms for v and i.

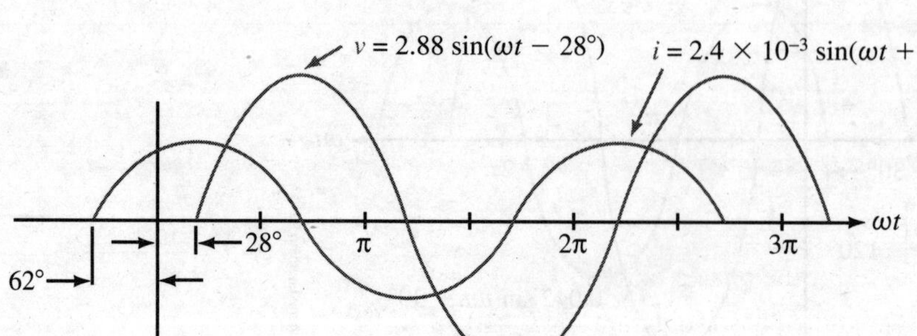

FIGURE 18–13 Sinusoidal voltage and current for a capacitor.

c. The corresponding phasor diagram for **V** and **I** is shown in Figure 18–14.

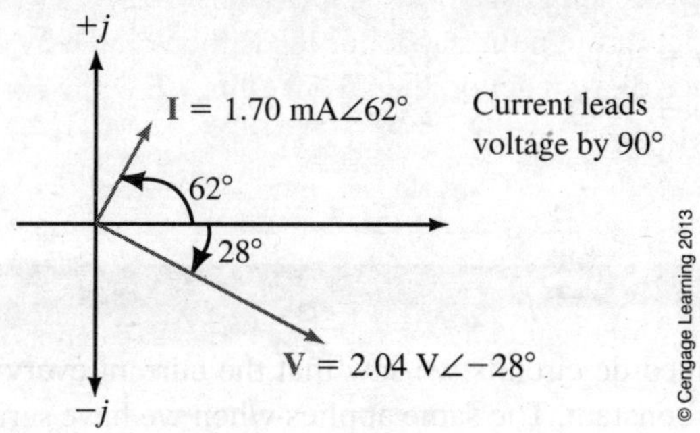

FIGURE 18–14 Voltage and current phasors for a capacitor.

The relationships between voltage and current, as illustrated in the previous three examples, will always hold for resistors, inductors, and capacitors.

IN-PROCESS LEARNING CHECK 1

(Answers are at the end of the chapter.)

1. What is the phase relationship between current and voltage for a resistor?
2. What is the phase relationship between current and voltage for a capacitor?
3. What is the phase relationship between current and voltage for an inductor?

PRACTICE PROBLEMS 1

A voltage source, $\mathbf{E} = 10\text{ V}\angle 30°$, is applied to an inductive impedance of 50 Ω.

a. Solve for the phasor current, **I**.
b. Sketch the phasor diagram for **E** and **I**.
c. Write the sinusoidal expressions for e and i.
d. Sketch the sinusoidal expressions for e and i.

Answers

a. $\mathbf{I} = 0.2\text{ A}\angle -60°$
c. $e = 14.1 \sin(\omega t + 30°)$
$i = 0.283 \sin(\omega t - 60°)$

PRACTICE PROBLEMS 2

A voltage source, $\mathbf{E} = 10\text{ V}\angle 30°$, is applied to a capacitive impedance of 20 Ω.

a. Solve for the phasor current, **I**.
b. Sketch the phasor diagram for **E** and **I**.
c. Write the sinusoidal expressions for e and i.
d. Sketch the sinusoidal expressions for e and i.

Answers
a. $\mathbf{I} = 0.5\ \text{A}\angle 120°$
c. $e = 14.1 \sin(\omega t + 30°)$
$i = 0.707 \sin(\omega t + 120°)$

18.2 ac Series Circuits

When we examined dc circuits we saw that the current everywhere in a series circuit is always constant. The same applies when we have series elements with an ac source. Further, we had seen that the total resistance of a dc series circuit consisting of n resistors was determined as the summation

$$R_T = R_1 + R_2 + \cdots + R_n$$

When working with ac circuits we no longer work with only resistance but also with capacitive and inductive reactance. *Impedance is a term used to collectively determine how the resistance, capacitance, and inductance "impede" the current in a circuit.* The symbol for impedance is the letter Z and the unit is the ohm (Ω). Because impedance may be made up of any combination of resistances and reactances, it is written as a vector quantity $\mathbf{Z}$, where

$$\mathbf{Z} = Z\angle\theta \quad (\Omega)$$

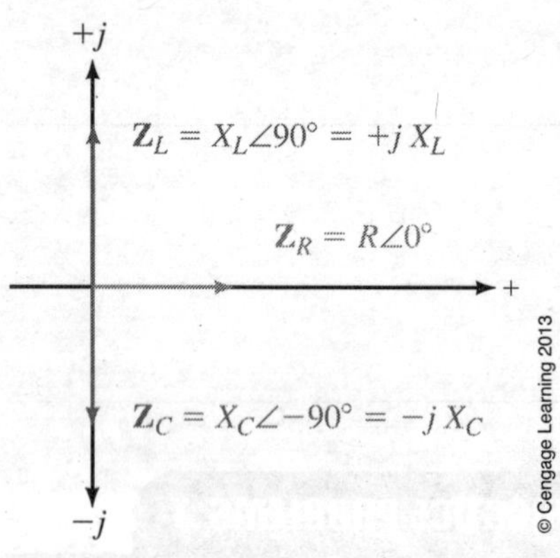

FIGURE 18–15

Each impedance may be represented as a vector on the complex plane, such that the length of the vector is representative of the magnitude of the impedance. The diagram showing one or more impedances is referred to as an **impedance diagram.**

Resistive impedance $\mathbf{Z}_R$ is a vector having a magnitude of R along the positive real axis. Purely inductive impedance $\mathbf{Z}_L$ is a vector having a magnitude of X_L along the positive imaginary axis, while the purely capacitive impedance $\mathbf{Z}_C$ is a vector having a magnitude of X_C along the negative imaginary axis. Mathematically, each of the vector impedances is written as follows:

$$\mathbf{Z}_R = R\angle 0° = R + j0 = R$$
$$\mathbf{Z}_L = X_L\angle 90° = 0 + jX_L = jX_L$$
$$\mathbf{Z}_C = X_C\angle 90° = 0 - jX_C = -jX_C$$

An impedance diagram showing each of the preceding impedances is shown in Figure 18–15.

All impedance vectors will appear in either the first or the fourth quadrant, since the resistive impedance vector is always positive.

For a series ac circuit consisting of n impedances, as shown in Figure 18–16, the total impedance of the circuit is found as the vector sum

$$\mathbf{Z}_T = \mathbf{Z}_1 + \mathbf{Z}_2 + \cdots + \mathbf{Z}_n \qquad \textbf{(18–1)}$$

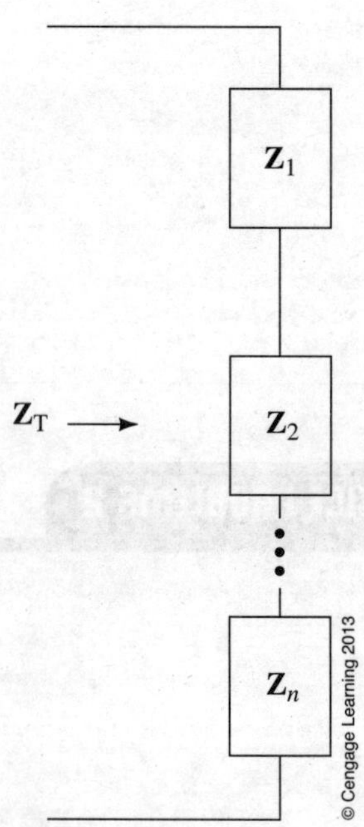

FIGURE 18–16

Consider the branch of Figure 18–17.

By applying Equation 18–1, we may determine the total impedance of the circuit as

$$\mathbf{Z}_T = (3\ \Omega + j0) + (0 + j4\ \Omega) = 3\ \Omega + j4\ \Omega$$
$$= 5\ \Omega\angle 53.13°$$

The preceding quantities are shown on an impedance diagram as in Figure 18–18.

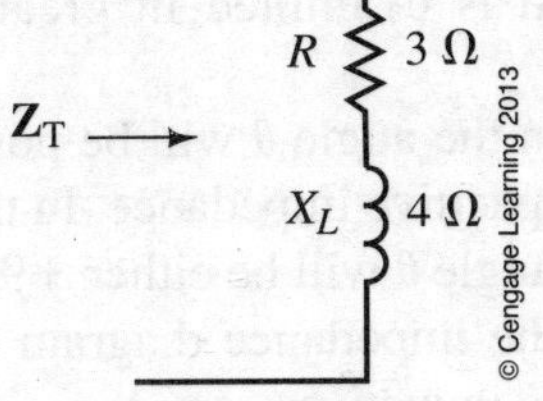

FIGURE 18–17

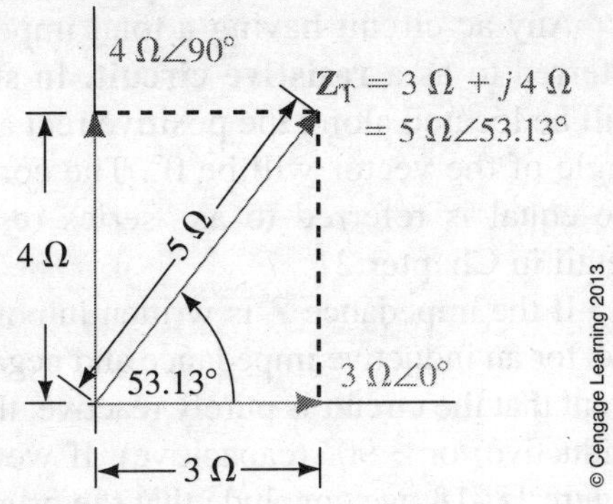

FIGURE 18–18

From Figure 18–18 we see that the total impedance of the series elements consists of a real component and an imaginary component. The corresponding total impedance vector may be written in either polar or rectangular form.

The rectangular form of an impedance is written as

$$\mathbf{Z} = R \pm jX$$

If we are given the polar form of the impedance, then we may determine the equivalent rectangular expression from

$$R = Z\cos\theta \tag{18–2}$$

and

$$X = Z\sin\theta \tag{18–3}$$

In the rectangular representation for impedance, the resistance term, R, is the total of all resistance looking into the network. The reactance term, X, is the difference between the total capacitive and inductive reactances. The sign for the imaginary term will be positive if the inductive reactance is greater than the capacitive reactance. In such a case, the impedance vector will appear in the first quadrant of the impedance diagram and is referred to as being an **inductive impedance.** If the capacitive reactance is larger, then the sign for the imaginary term will be negative. In such a case, the impedance vector will appear in the fourth quadrant of the impedance diagram, and the impedance is said to be **capacitive**.

The polar form of any impedance will be written in the form

$$\mathbf{Z} = Z\angle\theta$$

The value Z is the magnitude (in ohms) of the impedance vector **Z** and is determined as follows:

$$Z = \sqrt{R^2 + X^2} \quad (\Omega) \tag{18–4}$$

The corresponding angle of the impedance vector is determined as

$$\theta = \pm\tan^{-1}\left(\frac{X}{R}\right) \tag{18–5}$$

Whenever a capacitor and an inductor having equal reactances are placed in series, as shown in Figure 18–19, the equivalent circuit of the two components is a short circuit since the inductive reactance will be exactly balanced by the capacitive reactance.

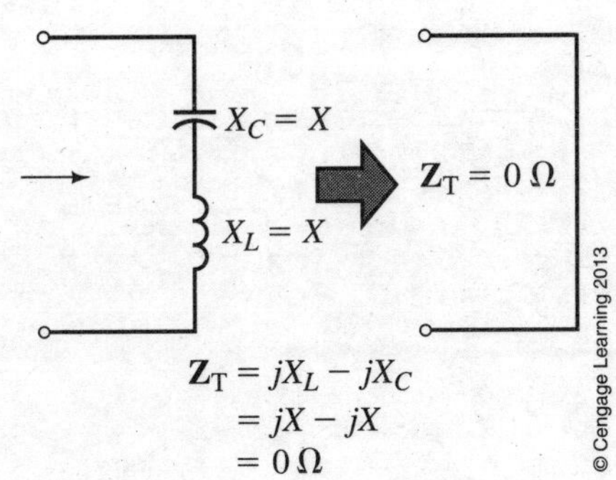

$$\begin{aligned}\mathbf{Z}_T &= jX_L - jX_C\\ &= jX - jX\\ &= 0\ \Omega\end{aligned}$$

FIGURE 18–19

Any ac circuit having a total impedance with only a real component is referred to as a **resistive circuit.** In such a case, the impedance vector $\mathbf{Z}_T$ will be located along the positive real axis of the impedance diagram and the angle of the vector will be 0°. The condition under which series reactances are equal is referred to as "series resonance" and is examined in greater detail in Chapter 21.

If the impedance **Z** is written in polar form, then the angle θ will be positive for an inductive impedance and negative for a capacitive impedance. In the event that the circuit is purely reactive, the resulting angle θ will be either +90° (inductive) or −90° (capacitive). If we reexamine the impedance diagram of Figure 18–18, we conclude that the original circuit is inductive.

EXAMPLE 18–5

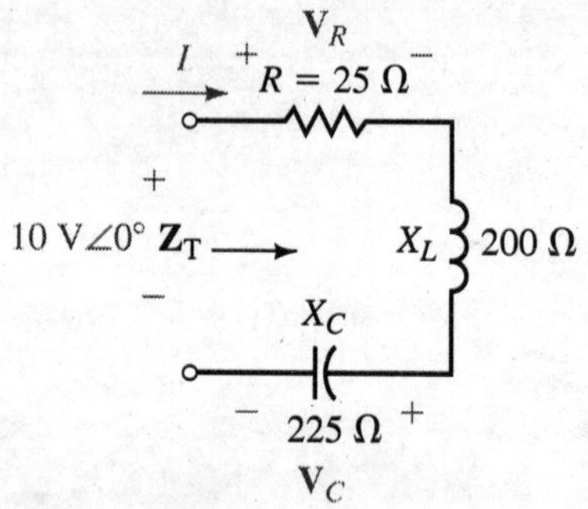

FIGURE 18–20

Consider the network of Figure 18–20.

a. Find $\mathbf{Z}_T$.

b. Sketch the impedance diagram for the network and indicate whether the total impedance of the circuit is inductive, capacitive, or resistive.

c. Use Ohm's law to determine **I**, $\mathbf{V}_R$, and $\mathbf{V}_C$.

Solution

a. The total impedance is the vector sum

$$\mathbf{Z}_T = 25\ \Omega + j200\ \Omega + (-j225\ \Omega)$$
$$= 25\ \Omega - j25\ \Omega$$
$$= 35.36\ \Omega\angle -45°$$

b. The corresponding impedance diagram is shown in Figure 18–21.

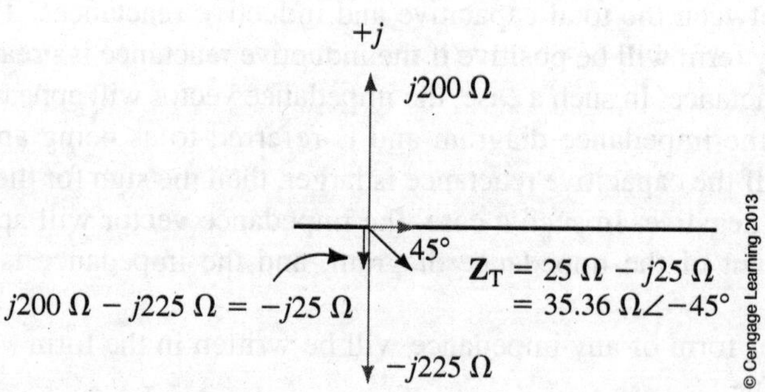

FIGURE 18–21

Because the total impedance has a negative reactance term (−j25 Ω), $\mathbf{Z}_T$ is capacitive.

c. $$\mathbf{I} = \frac{10\ \text{V}\angle 0°}{35.36\ \Omega\angle -45°} = 0.283\ \text{A}\angle 45°$$

$$\mathbf{V}_R = (282.8\ \text{mA}\angle 45°)(25\ \Omega\angle 0°) = 7.07\ \text{V}\angle 45°$$

$$\mathbf{V}_C = (282.8\ \text{mA}\angle 45°)(225\ \Omega\angle -90°) = 63.6\ \text{V}\angle -45°$$

Notice that the magnitude of the voltage across the capacitor is many times larger than the source voltage applied to the circuit. This example illustrates that the voltages across reactive elements must be calculated to ensure that maximum ratings for the components are not exceeded.

EXAMPLE 18–6

Determine the impedance **Z** which must be within the indicated block of Figure 18–22 if the total impedance of the network is $13\ \Omega\angle 22.62°$.

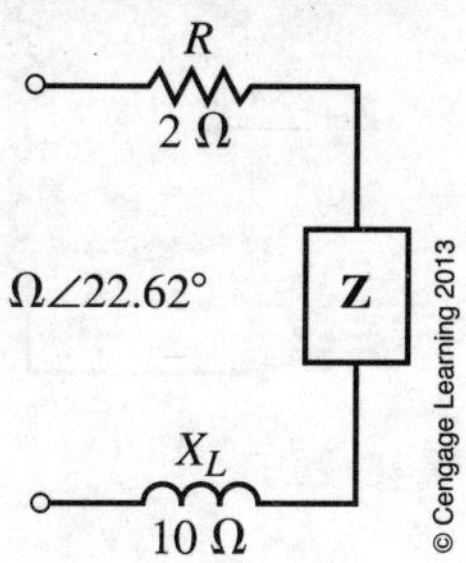

FIGURE 18–22

Solution Converting the total impedance from polar to rectangular form, we get

$$\mathbf{Z}_T = 13\ \Omega\angle 22.62° \Leftrightarrow 12\ \Omega + j5\ \Omega$$

Now, we know that the total impedance is determined from the summation of the individual impedance vectors, namely

$$\mathbf{Z}_T = 2\ \Omega + j10\ \Omega + \mathbf{Z} = 12\ \Omega + j5\ \Omega$$

Therefore, the impedance **Z** is found as

$$\begin{aligned}\mathbf{Z} &= 12\ \Omega + j5\ \Omega - (2\ \Omega + j10\ \Omega)\\ &= 10\ \Omega - j5\ \Omega\\ &= 11.18\ \Omega\angle -26.57°\end{aligned}$$

In its most simple form, the impedance **Z** will consist of a series combination of a 10-Ω resistor and a capacitor having a reactance of 5 Ω. Figure 18–23 shows the elements that may be contained within **Z** to satisfy the given conditions.

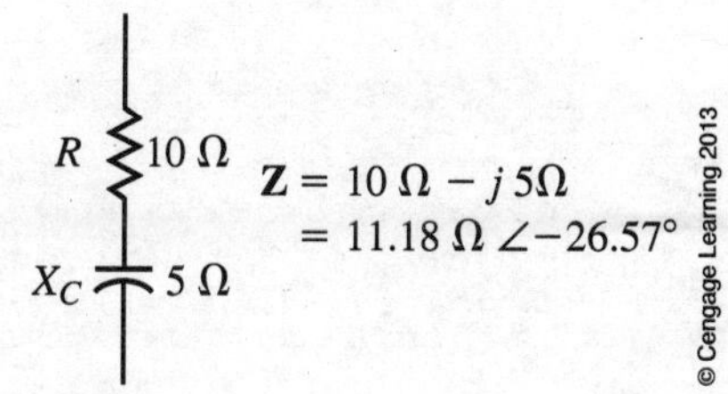

FIGURE 18–23

EXAMPLE 18–7

Find the total impedance for the network of Figure 18–24. Sketch the impedance diagram showing $\mathbf{Z}_1$, $\mathbf{Z}_2$, and $\mathbf{Z}_T$.

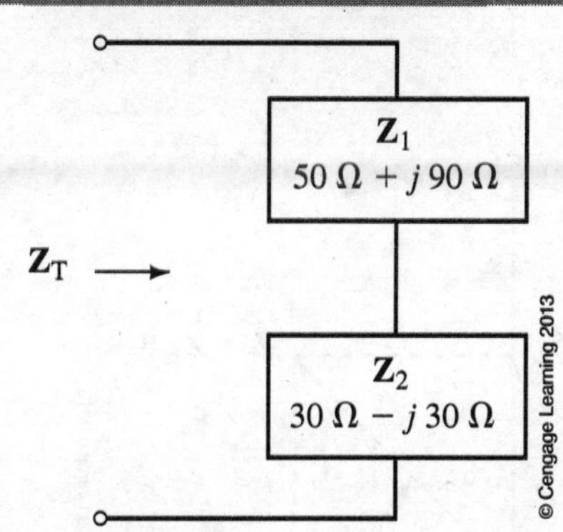

FIGURE 18–24

Solution

$$\begin{aligned}\mathbf{Z}_T &= \mathbf{Z}_1 + \mathbf{Z}_2\\ &= (50\ \Omega + j90\ \Omega) + (30\ \Omega - j30\ \Omega)\\ &= (80\ \Omega + j60\ \Omega) = 100\ \Omega\angle 36.87°\end{aligned}$$

The polar forms of the vectors $\mathbf{Z}_1$ and $\mathbf{Z}_2$ are as follows:

$$\mathbf{Z}_1 = 50\ \Omega + j90\ \Omega = 102.96\ \Omega\angle 60.95°$$

$$\mathbf{Z}_2 = 30\ \Omega - j30\ \Omega = 42.43\ \Omega\angle -45°$$

The resulting impedance diagram is shown in Figure 18–25.

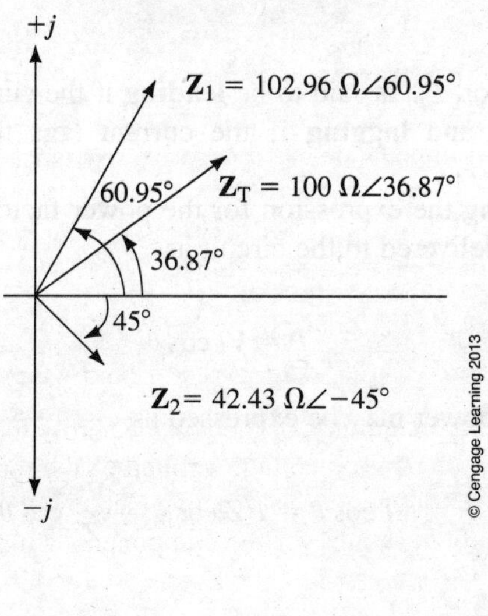

FIGURE 18–25

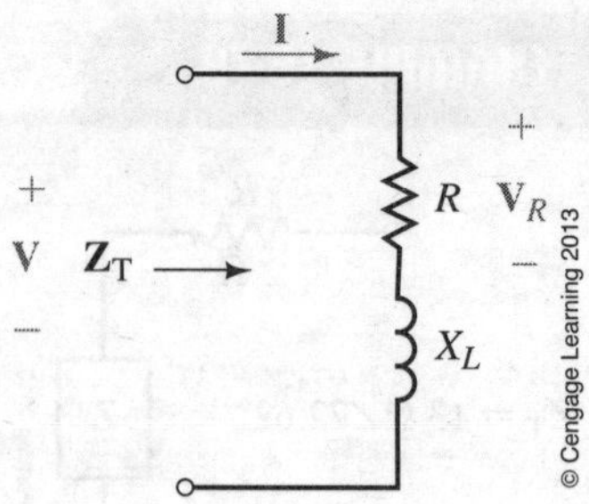

FIGURE 18–26

The phase angle θ for the impedance vector $\mathbf{Z} = Z\angle\theta$ provides the phase angle between the voltage **V** across **Z** and the current **I** through the impedance. For an inductive impedance the voltage will lead the current by θ. If the impedance is capacitive, then the voltage will lag the current by an amount equal to the magnitude of θ.

The phase angle θ is also useful for determining the average power dissipated by the circuit. In the simple series circuit shown in Figure 18–26, we know that only the resistor will dissipate power.

The average power dissipated by the resistor may be determined as follows:

$$P = V_R I = \frac{V_R^2}{R} = I^2R \qquad \textbf{(18–6)}$$

Notice that Equation 18–6 uses only the magnitudes of the voltage, current, and impedance vectors. *Power is never determined by using phasor products.*

Ohm's law provides the magnitude of the current phasor as

$$I = \frac{V}{Z}$$

Substituting this expression into Equation 18–6, we obtain the expression for power as

$$P = \frac{V^2}{Z^2}R = \frac{V^2}{Z}\left(\frac{R}{Z}\right) \qquad \textbf{(18–7)}$$

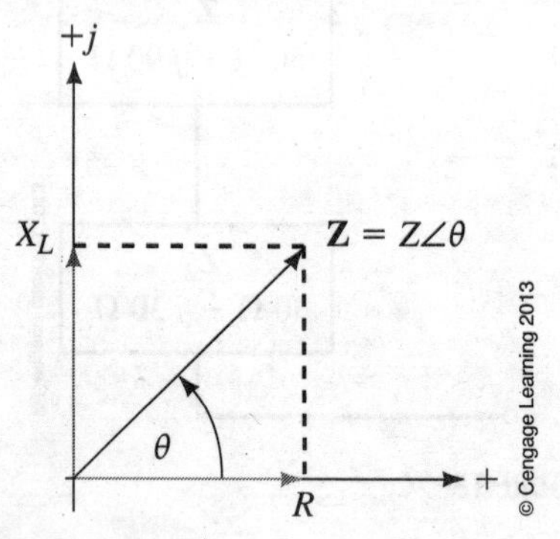

FIGURE 18–27

From the impedance diagram of Figure 18–27, we see that

$$\cos\theta = \frac{R}{Z}$$

The previous chapter had defined the power factor as $F_p = \cos\theta$, where θ is the angle between the voltage and current phasors. We now see that for a series circuit, the power factor of the circuit can be determined from the magnitudes of resistance and total impedance.

$$F_p = \cos\theta = \frac{R}{Z} \qquad \textbf{(18–8)}$$

The power factor, F_p, is said to be **leading** if the current leads the voltage (capacitive circuit) and **lagging** if the current lags the voltage (inductive circuit).

Now substituting the expression for the power factor into Equation 18–7, we express power delivered to the circuit as

$$P = VI\cos\theta$$

Since $V = IZ$, power may be expressed as

$$P = VI\cos\theta = I^2Z\cos\theta = \frac{V^2}{Z}\cos\theta \qquad \textbf{(18–9)}$$

EXAMPLE 18–8

Refer to the circuit of Figure 18–28.

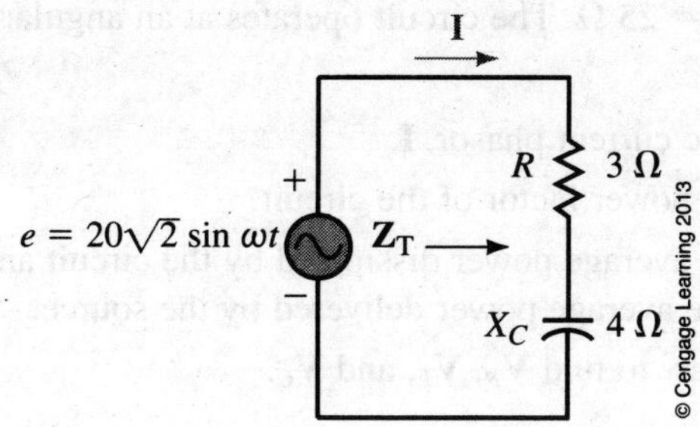

FIGURE 18–28

CircuitSim 18-1

CircuitSim 18-1A

a. Find the impedance $\mathbf{Z}_T$.
b. Calculate the power factor of the circuit.
c. Determine **I**.
d. Sketch the phasor diagram for **E** and **I**.
e. Find the average power delivered to the circuit by the voltage source.
f. Calculate the average power dissipated by both the resistor and the capacitor.

Solution

a. $\mathbf{Z}_T = 3\ \Omega - j4\ \Omega = 5\ \Omega\angle -53.13°$
b. $F_p = \cos\theta = 3\ \Omega/5\ \Omega = 0.6$ (leading)
c. The phasor form of the applied voltage is

$$\mathbf{E} = \frac{(\sqrt{2})(20\text{ V})}{\sqrt{2}}\angle 0° = 20\text{ V}\angle 0°$$

which gives a current of

$$\mathbf{I} = \frac{20\text{ V}\angle 0°}{5\ \Omega\angle -53.13°} = 4.0\text{ A}\angle 53.13°$$

d. The phasor diagram is shown in Figure 18–29.

From this phasor diagram, we see that the current phasor for the capacitive circuit leads the voltage phasor by 53.13°.

e. The average power delivered to the circuit by the voltage source is

$$P = (20\text{ V})(4\text{ A})\cos 53.13° = 48.0\text{ W}$$

f. The average power dissipated by the resistor and capacitor will be

$$P_R = (4\text{ A})^2(3\ \Omega)\cos 0° = 48\text{ W}$$
$$P_C = (4\text{ A})^2(4\ \Omega)\cos 90° = 0\text{ W} \quad \text{(as expected!)}$$

Notice that the power factor used in determining the power dissipated by each of the elements is the power factor for that element and not the total power factor for the circuit.

As expected, the summation of powers dissipated by the resistor and capacitor is equal to the total power delivered by the voltage source.

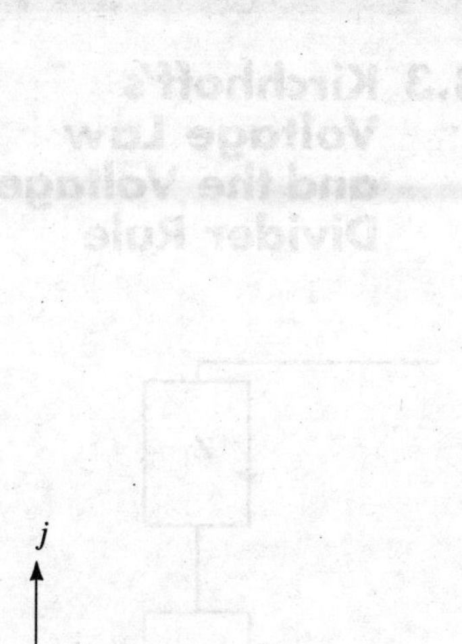

FIGURE 18–29

PRACTICE PROBLEMS 3

A circuit consists of a voltage source $\mathbf{E} = 50\text{ V}\angle 25°$ in series with $L = 20$ mH, $C = 50\ \mu$F, and $R = 25\ \Omega$. The circuit operates at an angular frequency of 2 krad/s.

a. Determine the current phasor, **I**.
b. Solve for the power factor of the circuit.
c. Calculate the average power dissipated by the circuit and verify that this is equal to the average power delivered by the source.
d. Use Ohm's law to find $\mathbf{V}_R$, $\mathbf{V}_L$, and $\mathbf{V}_C$.

Answers

a. $\mathbf{I} = 1.28\text{ A}\angle -25.19°$
b. $F_p = 0.6402$ lagging
c. $P = 41.0$ W
d. $\mathbf{V}_R = 32.0\text{ V}\angle -25.19°$
$\mathbf{V}_C = 12.8\text{ V}\angle -115.19°$
$\mathbf{V}_L = 51.2\text{ V}\angle 64.81°$

18.3 Kirchhoff's Voltage Law and the Voltage Divider Rule

When a voltage is applied to impedances in series, as shown in Figure 18–30, Ohm's law may be used to determine the voltage across any impedance as

$$\mathbf{V}_x = \mathbf{I}\mathbf{Z}_x$$

The current in the circuit is

$$\mathbf{I} = \frac{\mathbf{E}}{\mathbf{Z}_T}$$

Now, by substitution we arrive at the voltage divider rule for any series combination of elements as

$$\mathbf{V}_x = \frac{\mathbf{Z}_x}{\mathbf{Z}_T}\mathbf{E} \qquad \textbf{(18–10)}$$

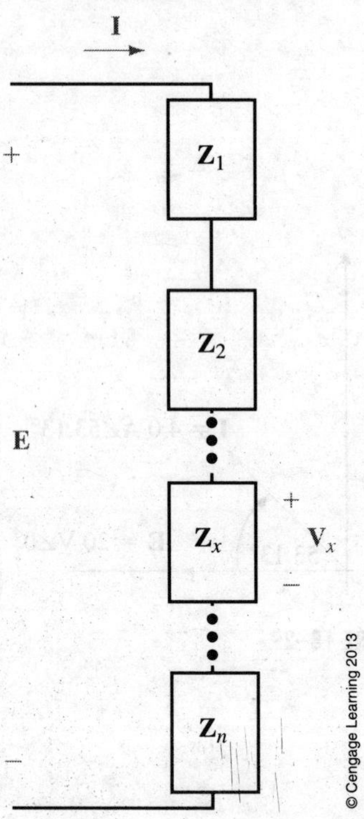

FIGURE 18–30

Equation 18–10 is very similar to the equation for the voltage divider rule in dc circuits. The fundamental differences in solving ac circuits are that we use impedances rather than resistances and that the voltages found are phasors. Because the voltage divider rule involves solving products and quotients of phasors, we generally use the polar form rather than the rectangular form of phasors.

Kirchhoff's voltage law must apply for all circuits whether they are dc or ac circuits. However, because ac circuits have voltages expressed in either sinusoidal or phasor form, Kirchhoff's voltage law for ac circuits may be stated as follows:

> *The phasor sum of voltage drops and voltage rises around a closed loop is equal to zero.*

When adding phasor voltages, we find that the summation is generally done more easily in rectangular form than in the polar form.

EXAMPLE 18–9

Consider the circuit of Figure 18–31.

a. Find $\mathbf{Z}_T$.

b. Determine the voltages $\mathbf{V}_R$ and $\mathbf{V}_L$ using the voltage divider rule.

c. Verify Kirchhoff's voltage law around the closed loop.

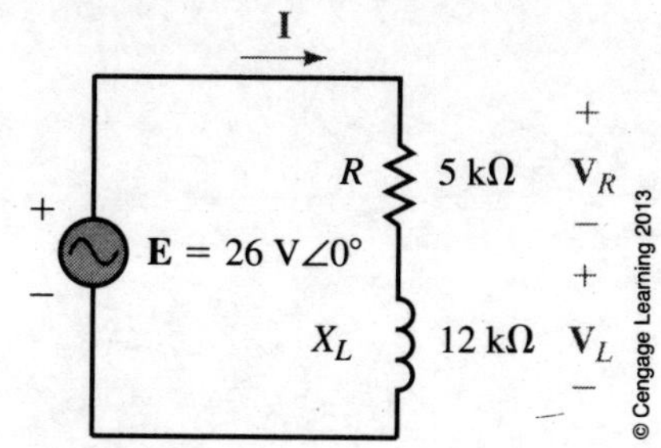

FIGURE 18–31

Solution

a. $\mathbf{Z}_T = 5\ \text{k}\Omega + j12\ \text{k}\Omega = 13\ \text{k}\Omega\angle 67.38°$

b. $$\mathbf{V}_R = \left(\frac{5\ \text{k}\Omega\angle 0°}{13\ \text{k}\Omega\angle 67.38°}\right)(26\ \text{V}\angle 0°) = 10\ \text{V}\angle -67.38°$$

$$\mathbf{V}_L = \left(\frac{12\ \text{k}\Omega\angle 90°}{13\ \text{k}\Omega\angle 67.38°}\right)(26\ \text{V}\angle 0°) = 24\ \text{V}\angle 22.62°$$

c. Kirchhoff's voltage law around the closed loop will give

$$26\ \text{V}\angle 0° - 10\ \text{V}\angle -67.38° - 24\ \text{V}\angle 22.62° = 0$$
$$(26 + j0) - (3.846 - j9.231) - (22.154 + j9.231) = 0$$
$$(26 - 3.846 - 22.154) + j(0 + 9.231 - 9.231) = 0$$
$$0 + j0 = 0$$

EXAMPLE 18–10

Consider the circuit of Figure 18–32:

a. Calculate the sinusoidal voltages v_1 and v_2 using phasors and the voltage divider rule.

b. Sketch the phasor diagram showing $\mathbf{E}$, $\mathbf{V}_1$, and $\mathbf{V}_2$.

c. Sketch the sinusoidal waveforms of e, v_1, and v_2.

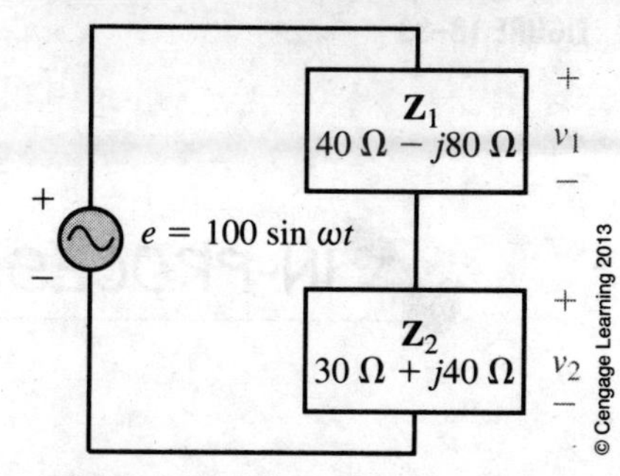

FIGURE 18–32

Solution

a. The phasor form of the voltage source is determined as

$$e = 100 \sin \omega t \Leftrightarrow \mathbf{E} = 70.71\ \angle \text{V}\ 0°$$

Applying VDR, we get

$$\mathbf{V}_1 = \left(\frac{40\ \Omega - j80\ \Omega}{(40\ \Omega - j80\ \Omega) + (30\ \Omega + j40\ \Omega)}\right)(70.71\ \text{V}\angle 0°)$$

$$= \left(\frac{89.44\ \Omega\angle -63.43°}{80.62\ \Omega\angle -29.74°}\right)(70.71\ \text{V}\angle 0°)$$

$$= 78.4\ \text{V}\angle -33.69°$$

and

$$\mathbf{V}_2 = \left(\frac{30\ \Omega + j40\ \Omega}{(40\ \Omega - j80\ \Omega) + (30\ \Omega + j40\ \Omega)}\right)(70.71\ \text{V}\angle 0°)$$

$$= \left(\frac{50.00\ \Omega\angle 53.13°}{80.62\ \Omega\angle -29.74°}\right)(70.71\ \text{V}\angle 0°)$$

$$= 43.9\ \text{V}\angle 82.87°$$

The sinusoidal voltages are determined to be

$$v_1 = (\sqrt{2})(78.4)\sin(\omega t - 33.69°)$$
$$= 111 \sin(\omega t - 33.69°)$$

and

$$v_2 = (\sqrt{2})(43.9)\sin(\omega t + 82.87°)$$
$$= 62.0 \sin(\omega t + 82.87°)$$

b. The phasor diagram is shown in Figure 18–33.

c. The corresponding sinusoidal voltages are shown in Figure 18–34.

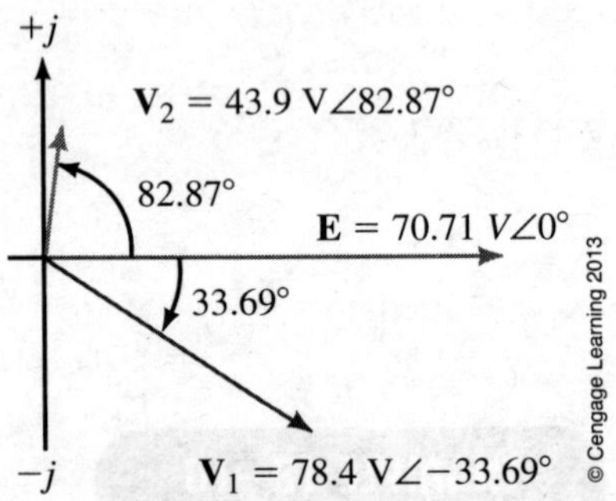

FIGURE 18–33

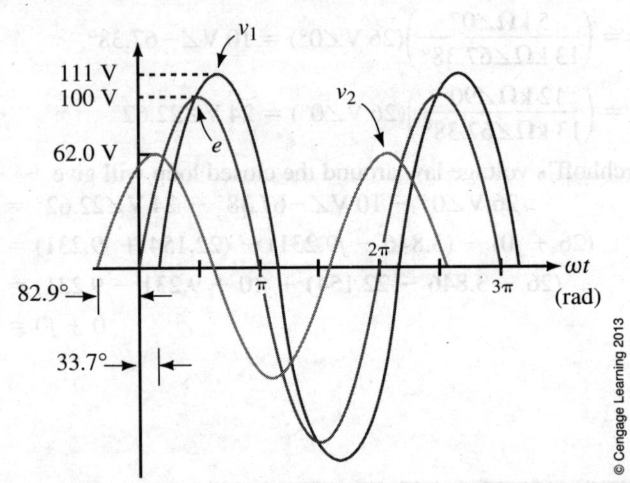

FIGURE 18–34

IN-PROCESS LEARNING CHECK 2

(Answers are at the end of the chapter.)

1. Express Kirchhoff's voltage law as it applies to ac circuits.
2. What is the fundamental difference between how Kirchhoff's voltage law is used in ac circuits and how it is used in dc circuits?

PRACTICE PROBLEMS 4

A circuit consists of a voltage source **E** = 50 V∠25° in series with L = 20 mH, C = 50 μF, and R = 25 Ω. The circuit operates at an angular frequency of 2 krad/s.

a. Use the voltage divider rule to determine the voltage across each element in the circuit.

b. Verify that Kirchhoff's voltage law applies for the circuit.

Answers

a. $\mathbf{V}_L$ = 51.2 V∠64.81°, $\mathbf{V}_C$ = 12.8 V∠ −115.19°

$\mathbf{V}_R$ = 32.0 V∠−25.19°

b. 51.2 V∠64.81° + 12.8 V∠−115.19° + 32.0 V∠−25.19° = 50 V∠25°

18.4 ac Parallel Circuits

The **admittance Y** of any impedance is defined as a vector quantity which is the reciprocal of the impedance **Z**.

Mathematically, admittance is expressed as

$$\mathbf{Y}_T = \frac{1}{\mathbf{Z}_T} = \frac{1}{Z_T\angle\theta} = \left(\frac{1}{Z_T}\right)\angle-\theta = Y_T\angle-\theta \quad \text{(S)} \tag{18–11}$$

where the unit of admittance is the siemens (S).

In particular, we have seen that the admittance of a resistor R is called conductance and is given the symbol $\mathbf{Y}_R$. If we consider resistance as a vector quantity, then the corresponding vector form of the conductance is

$$\mathbf{Y}_R = \frac{1}{R\angle0^\circ} = \frac{1}{R}\angle0^\circ = G\angle0^\circ = G + j0 \quad \text{(S)} \tag{18–12}$$

If we determine the admittance of a purely reactive component X, the resultant admittance is called the **susceptance** of the component and is assigned the symbol B. The unit for susceptance is siemens (S). In order to distinguish between inductive susceptance and capacitive susceptance, we use the subscripts L and C, respectively. The vector forms of reactive admittance are given as follows:

$$\mathbf{Y}_L = \frac{1}{X_L\angle90^\circ} = \frac{1}{X_L}\angle-90^\circ = B_L\angle-90^\circ = 0 - jB_L \quad \text{(S)} \tag{18–13}$$

$$\mathbf{Y}_C = \frac{1}{X_C\angle-90^\circ} = \frac{1}{X_C}\angle90^\circ = B_C\angle90^\circ = 0 + jB_C \quad \text{(S)} \tag{18–14}$$

In a manner similar to impedances, admittances may be represented on the complex plane in an **admittance diagram** as shown in Figure 18–35.

The lengths of the various vectors are proportional to the magnitudes of the corresponding admittances. The resistive admittance vector **G** is shown on the positive real axis, whereas the inductive and capacitive admittance vectors $\mathbf{Y}_L$ and $\mathbf{Y}_C$ are shown on the negative and positive imaginary axes, respectively.

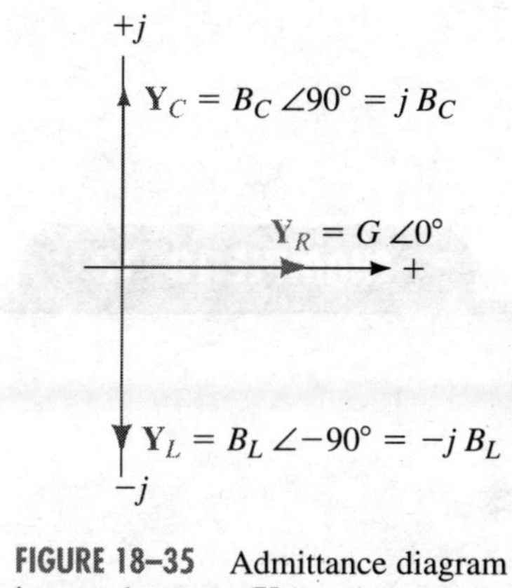

FIGURE 18–35 Admittance diagram showing conductance ($\mathbf{Y}_R$) and susceptance ($\mathbf{Y}_L$ and $\mathbf{Y}_C$).

EXAMPLE 18–11

Determine the admittances of the following impedances. Sketch the corresponding admittance diagram.

a. $R = 10\ \Omega$

b. $X_L = 20\ \Omega$

c. $X_C = 40\ \Omega$

Solution

a. $\mathbf{Y}_R = \frac{1}{R} = \frac{1}{10\ \Omega\angle0^\circ} = 100\text{ mS}\angle0^\circ$

b. $\mathbf{Y}_L = \frac{1}{X_L} = \frac{1}{20\ \Omega\angle90^\circ} = 50\text{ mS}\angle-90^\circ$

c. $\mathbf{Y}_C = \frac{1}{X_C} = \frac{1}{40\ \Omega\angle-90^\circ} = 25\text{ mS}\angle90^\circ$

The admittance diagram is shown in Figure 18–36.

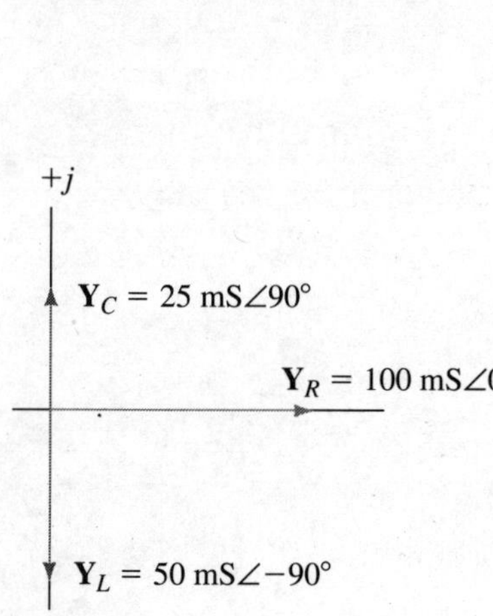

FIGURE 18–36

For any network of n admittances as shown in Figure 18–37, the total admittance is the vector sum of the admittances of the network. Mathematically, the total admittance of a network is given as

$$\mathbf{Y}_T = \mathbf{Y}_1 + \mathbf{Y}_2 + \cdots + \mathbf{Y}_n \quad \text{(S)} \qquad \textbf{(18–15)}$$

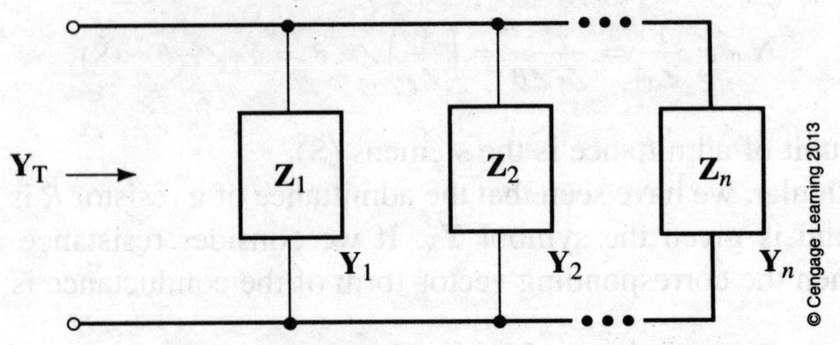

FIGURE 18–37

The resultant impedance of a parallel network of n impedances is determined to be

$$\mathbf{Z}_T = \frac{1}{\mathbf{Y}_T} = \frac{1}{\mathbf{Y}_1 + \mathbf{Y}_2 + \cdots + \mathbf{Y}_n}$$

$$\mathbf{Z}_T = \frac{1}{\frac{1}{\mathbf{Z}_1} + \frac{1}{\mathbf{Z}_2} + \cdots + \frac{1}{\mathbf{Z}_n}} \quad (\Omega) \qquad \textbf{(18–16)}$$

EXAMPLE 18–12

Find the equivalent admittance and impedance of the network of Figure 18–38. Sketch the admittance diagram.

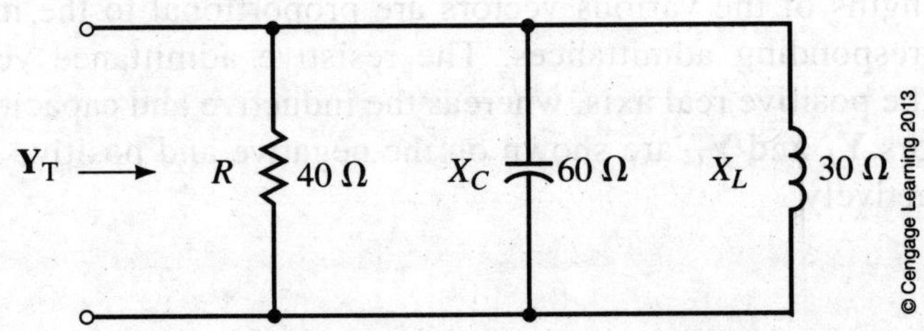

FIGURE 18–38

Solution The admittances of the various parallel elements are

$$\mathbf{Y}_1 = \frac{1}{40\ \Omega\angle 0°} = 25.0\ \text{mS}\angle 0° = 25.0\ \text{mS} + j0$$

$$\mathbf{Y}_2 = \frac{1}{60\ \Omega\angle -90°} = 16.\overline{6}\ \text{mS}\angle 90° = 0 + j16.\overline{6}\ \text{mS}$$

$$\mathbf{Y}_3 = \frac{1}{30\ \Omega\angle 90°} = 33.\overline{3}\ \text{mS}\angle -90° = 0 - j33.\overline{3}\ \text{mS}$$

The total admittance is determined as

$$\begin{aligned}\mathbf{Y}_T &= \mathbf{Y}_1 + \mathbf{Y}_2 + \mathbf{Y}_3 \\ &= 25.0\ \text{mS} + j16.\overline{6}\ \text{mS} + (-j33.\overline{3}\ \text{mS}) \\ &= 25.0\ \text{mS} - j16.\overline{6}\ \text{mS} \\ &= 30.0\ \text{mS}\angle -33.69°\end{aligned}$$

This results in a total impedance for the network of

$$\mathbf{Z}_T = \frac{1}{\mathbf{Y}_T} = \frac{1}{30.0 \text{ mS}\angle -33.69^\circ} = 33.3\ \Omega\angle 33.69^\circ$$

The admittance diagram is shown in Figure 18–39.

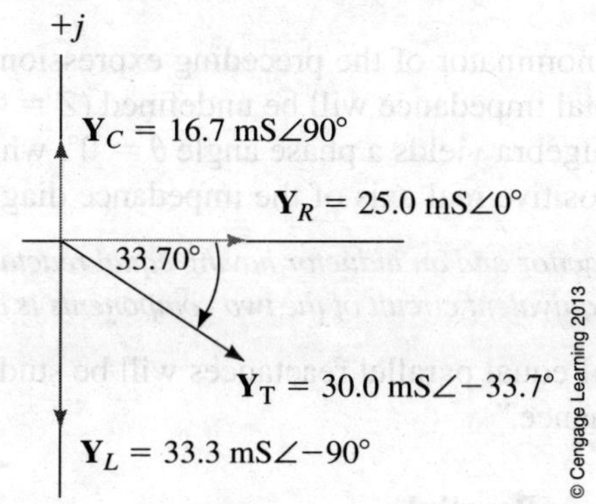

FIGURE 18–39

Two Impedances in Parallel

By applying Equation 18–14 for two impedances, we determine the equivalent impedance of two impedances as

$$\mathbf{Z}_T = \frac{\mathbf{Z}_1\mathbf{Z}_2}{\mathbf{Z}_1 + \mathbf{Z}_2} \quad (\Omega) \tag{18–17}$$

From the preceding expression, we see that for two impedances in parallel, the equivalent impedance is determined as the product of the impedances over the sum. Although the expression for two impedances is very similar to the expression for two resistors in parallel, the difference is that the calculation of impedance involves the use of complex algebra.

EXAMPLE 18–13

Find the total impedance for the network shown in Figure 18–40.

Solution

$$\mathbf{Z}_T = \frac{(200\ \Omega\angle -90^\circ)(250\ \Omega\angle 90^\circ)}{-j200\ \Omega + j250\ \Omega} = \frac{50 \text{ k}\Omega\angle 0^\circ}{50\angle 90^\circ} = 1 \text{ k}\Omega\angle -90^\circ$$

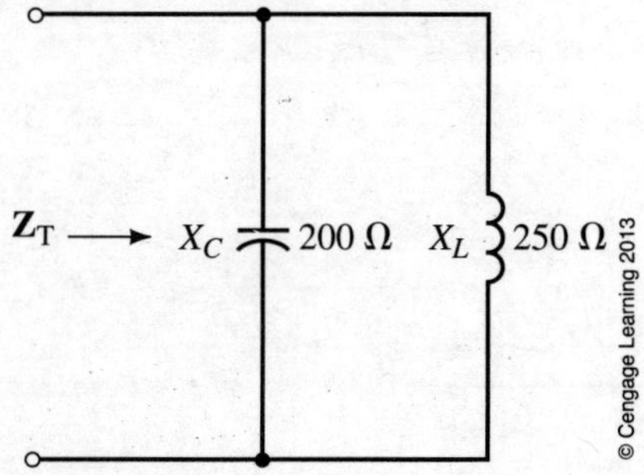

FIGURE 18–40

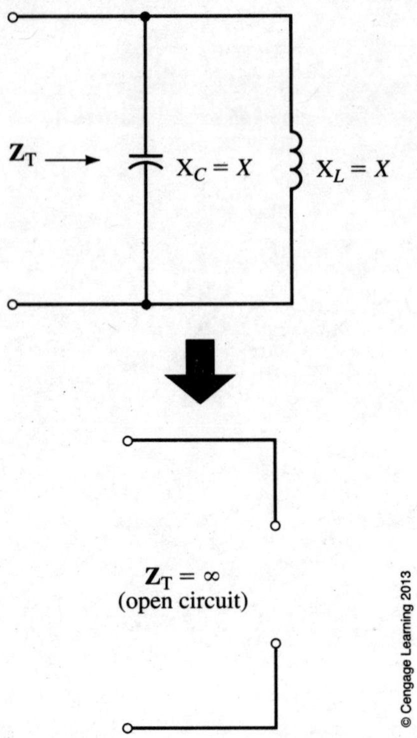

FIGURE 18–41

The previous example illustrates that unlike total parallel resistance, the total impedance of a combination of parallel reactances may be much larger than either of the individual impedances. Indeed, if we are given a parallel combination of equal inductive and capacitive reactances, the total impedance of the combination is equal to infinity (namely an open circuit). Consider the network of Figure 18–41.

The total impedance $\mathbf{Z}_T$ is found as

$$\mathbf{Z}_T = \frac{(X_L\angle 90^\circ)(X_C\angle -90^\circ)}{jX_L - jX_C} = \frac{X^2\angle 0^\circ}{0\angle 0^\circ} = \infty\angle 0^\circ$$

Because the denominator of the preceding expression is equal to zero, the magnitude of the total impedance will be undefined ($Z = \infty$). The magnitude is undefined and the algebra yields a phase angle $\theta = 0^\circ$, which indicates that the vector lies on the positive real axis of the impedance diagram.

Whenever a capacitor and an inductor having equal reactances are placed in parallel, the equivalent circuit of the two components is an open circuit.

The principle of equal parallel reactances will be studied in a later chapter dealing with "resonance."

Three Impedances in Parallel

Equation 18–16 may be solved for three impedances to give the equivalent impedance as

$$\mathbf{Z}_T = \frac{\mathbf{Z}_1\mathbf{Z}_2\mathbf{Z}_3}{\mathbf{Z}_1\mathbf{Z}_2 + \mathbf{Z}_1\mathbf{Z}_3 + \mathbf{Z}_2\mathbf{Z}_3} \quad (\Omega) \tag{18–18}$$

although this is less useful than the general equation.

EXAMPLE 18–14

Find the equivalent impedance of the network of Figure 18–42.

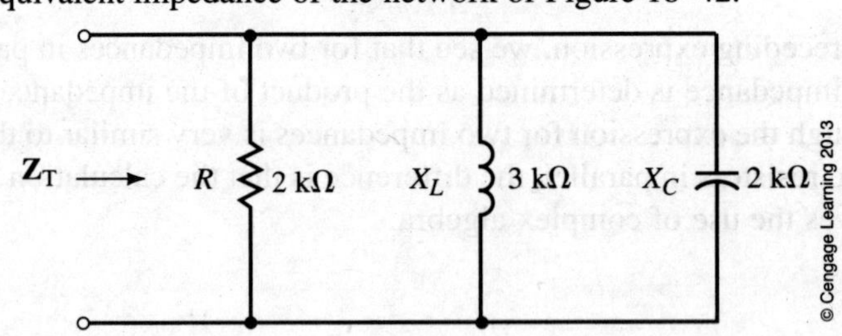

FIGURE 18–42

Solution

$$\mathbf{Z}_T = \frac{(2\text{ k}\Omega\angle 0^\circ)(3\text{ k}\Omega\angle 90^\circ)(2\text{ k}\Omega\angle -90^\circ)}{(2\text{ k}\Omega\angle 0^\circ)(3\text{ k}\Omega\angle 90^\circ) + (2\text{ k}\Omega\angle 0^\circ)(2\text{ k}\Omega\angle -90^\circ) + (3\text{ k}\Omega\angle 90^\circ)(2\text{ k}\Omega\angle -90^\circ)}$$

$$= \frac{12 \times 10^9\ \Omega\angle 0^\circ}{6 \times 10^6\angle 90^\circ + 4 \times 10^6\angle -90^\circ + 6 \times 10^6\angle 0^\circ}$$

$$= \frac{12 \times 10^9\ \Omega\angle 0^\circ}{6 \times 10^6 + j2 \times 10^6} = \frac{12 \times 10^9\ \Omega\angle 0^\circ}{6.325 \times 10^6\angle 18.43^\circ}$$

$$= 1.90\text{ k}\Omega\angle -18.43^\circ$$

And so the equivalent impedance of the network is

$$\mathbf{Z}_T = 1.80\text{ k}\Omega - j0.6\text{ k}\Omega$$

Calculator Hint: In Chapter 5, it was shown that the equivalent value of several resistors in parallel can be found using the x^{-1} key found on all scientific calculators. If your calculator is able to perform operations on complex numbers, a similar method can be used for solving the equivalent impedance of any circuit, even one with reactive elements such as in Figure 18–42. The following keystrokes show how $\mathbf{Z}_T$ is determined for a typical scientific calculator. Recognizing that all impedances are in kΩ, we can avoid needless keystrokes by simply entering the numerical values as follows:

[2] [x^{-1}] [+] [(] [i] [3] [)] [x^{-1}] [+]
[(] [+/−] [i] [2] [)] [x^{-1}] [ENTER] [x^{-1}] [ENTER]

Depending on your calculator, your display will appear similar to the following:

PRACTICE PROBLEMS 5

A circuit consists of a current source, $i = 0.030 \sin 500t$, in parallel with $L = 20$ mH, $C = 50$ μF, and $R = 25$ Ω.

a. Determine the voltage **V** across the circuit.
b. Solve for the power factor of the circuit.
c. Calculate the average power dissipated by the circuit and verify that this is equal to the power delivered by the source.
d. Use Ohm's law to find the phasor quantities, $\mathbf{I}_R$, $\mathbf{I}_L$, and $\mathbf{I}_C$.

Answers
a. $\mathbf{V} = 0.250\ \text{V}\angle 61.93°$
b. $F_p = 0.4705$ lagging
c. $P_R = 2.49\ \text{mW} = P_T$
d. $\mathbf{I}_R = 9.98\ \text{mA}\angle 61.98°$
$\mathbf{I}_C = 6.24\ \text{mA}\angle 151.93°$
$\mathbf{I}_L = 25.0\ \text{mA}\angle -28.07°$

PRACTICE PROBLEMS 6

A circuit consists of a 2.5-A_{rms} current source connected in parallel with a resistor, an inductor, and a capacitor. The resistor has a value of 10 Ω and dissipates 40 W of power.

a. Calculate the values of X_L and X_C if $X_L = 3X_C$.
b. Determine the magnitudes of current through the inductor and the capacitor.

Answers

a. $X_L = 26.7\ \Omega,\ X_C = 8.89\ \Omega$

b. $I_L = 0.75\ \text{A},\ I_C = 2.25\ \text{A}$

18.5 Kirchhoff's Current Law and the Current Divider Rule

The current divider rule for ac circuits has the same form as for dc circuits with the notable exception that currents are expressed as phasors. For a parallel network as shown in Figure 18–43, the current in any branch of the network may be determined using either admittance or impedance.

$$\mathbf{I}_x = \frac{\mathbf{Y}_x}{\mathbf{Y}_\text{T}}\mathbf{I} \quad \text{or} \quad \mathbf{I}_x = \frac{\mathbf{Z}_\text{T}}{\mathbf{Z}_x}\mathbf{I} \tag{18–19}$$

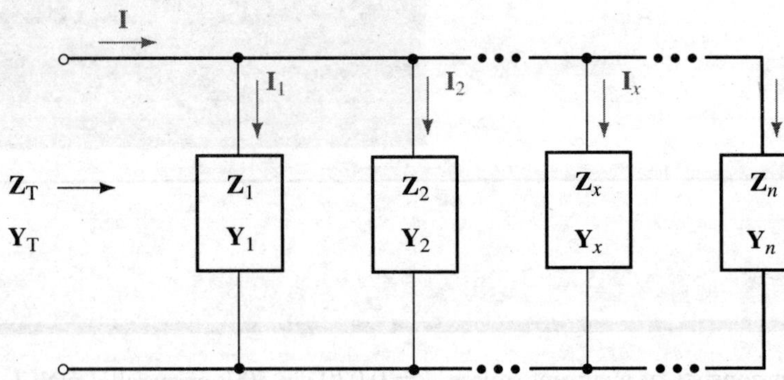

FIGURE 18–43

For two branches in parallel the current in either branch is determined from the impedances as

$$\mathbf{I}_1 = \frac{\mathbf{Z}_2}{\mathbf{Z}_1 + \mathbf{Z}_2}\mathbf{I} \tag{18–20}$$

Also, as one would expect, Kirchhoff's current law (KCL) must apply to any node within an ac circuit. For such circuits, KCL may be stated as follows:

The summation of current phasors entering and leaving a node is equal to zero.

EXAMPLE 18–15

Calculate the current in each of the branches in the network of Figure 18–44.

Solution

$$\mathbf{I}_1 = \left(\frac{250\ \Omega\angle{-90°}}{j200\ \Omega - j250\ \Omega}\right)(2\ \text{A}\angle 0°)$$

$$= \left(\frac{250\ \Omega\angle{-90°}}{50\ \Omega\angle{-90°}}\right)(2\ \text{A}\angle 0°) = 10\ \text{A}\angle 0°$$

and

$$\mathbf{I}_2 = \left(\frac{200\ \Omega\angle 90°}{j200\ \Omega - j250\ \Omega}\right)(2\ \text{A}\angle 0°)$$

$$= \left(\frac{200\ \Omega\angle 90°}{50\ \Omega\angle -90°}\right)(2\ \text{A}\angle 0°) = 8\ \text{A}\angle 180°$$

The preceding results illustrate that the currents in parallel reactive components may be significantly larger than the applied current. If the current through the component exceeds the maximum current rating of the element, severe damage may occur.

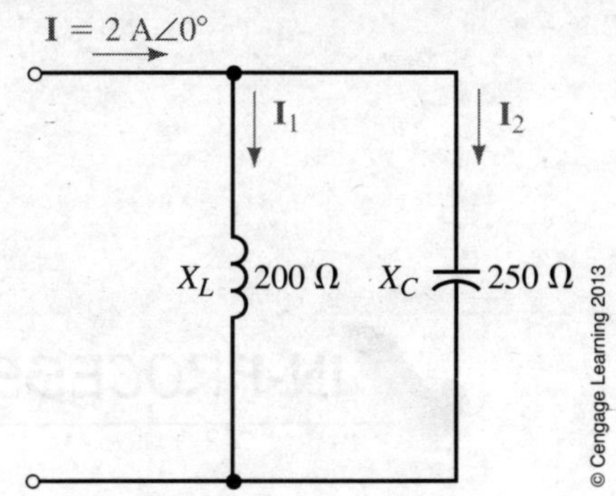

FIGURE 18–44

EXAMPLE 18–16

Refer to the circuit of Figure 18–45:

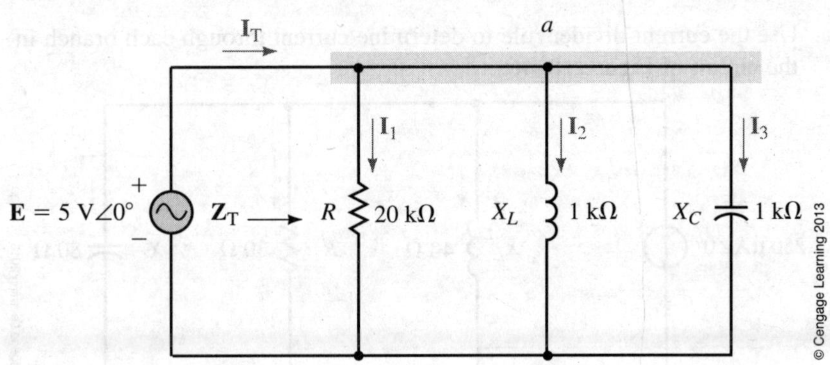

FIGURE 18–45

CircuitSim 18-2

a. Find the total impedance, $\mathbf{Z}_T$.

b. Determine the supply current, $\mathbf{I}_T$.

c. Calculate $\mathbf{I}_1$, $\mathbf{I}_2$, and $\mathbf{I}_3$ using the current divider rule.

d. Verify Kirchhoff's current law at node *a*.

Solution

a. Because the inductive and capacitive reactances are in parallel and have the same value, we may replace the combination with an open circuit. Consequently, only the resistor *R* needs to be considered. As a result

$$\mathbf{Z}_T = 20\ \text{k}\Omega\angle 0°$$

b. $$\mathbf{I}_T = \frac{5\ \text{V}\angle 0°}{20\ \text{k}\Omega\angle 0°} = 250\ \mu\text{A}\angle 0°$$

c. $$\mathbf{I}_1 = \left(\frac{20\ \text{k}\Omega\angle 0°}{20\ \text{k}\Omega\angle 0°}\right)(250\ \mu\text{A}\angle 0°) = 250\ \mu\text{A}\angle 0°$$

$$\mathbf{I}_2 = \left(\frac{20\ \text{k}\Omega\angle 0°}{1\ \text{k}\Omega\angle 90°}\right)(250\ \mu\text{A}\angle 0°) = 5.0\ \text{mA}\angle -90°$$

$$\mathbf{I}_3 = \left(\frac{20\ \text{k}\Omega\angle 0°}{1\ \text{k}\Omega\angle -90°}\right)(250\ \mu\text{A}\angle 0°) = 5.0\ \text{mA}\angle 90°$$

d. Notice that the currents through the inductor and capacitor are 180° out of phase. By adding the current phasors in rectangular form, we have

$$\mathbf{I}_T = 250\ \mu\text{A} - j5.0\ \text{mA} + j5.0\ \text{mA} = 250\ \mu\text{A} + j0 = 250\ \mu\text{A}\angle 0^\circ$$

This result satisfies Kirchhoff's current law at the node.

IN-PROCESS LEARNING CHECK 3

(Answers are at the end of the chapter.)

1. Express Kirchhoff's current law as it applies to ac circuits.
2. What is the fundamental difference between how Kirchhoff's current law is applied to ac circuits and how it is applied to dc circuits?

PRACTICE PROBLEMS 7

a. Use the current divider rule to determine current through each branch in the circuit of Figure 18–46.

CircuitSim 18-3

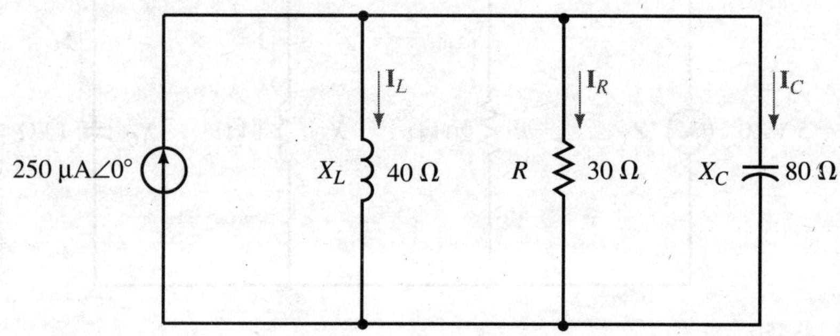

FIGURE 18–46

b. Verify that Kirchhoff's current law applies to the circuit of Figure 18–46.

Answers

a. $\mathbf{I}_L = 176\ \mu\text{A}\angle -69.44^\circ$

$\mathbf{I}_R = 234\ \mu\text{A}\angle 20.56^\circ$

$\mathbf{I}_C = 86.8\ \mu\text{A}\angle 110.56^\circ$

b. $\Sigma\mathbf{I}_{\text{out}} = \Sigma\mathbf{I}_{\text{in}} = 250\ \mu\text{A}$

18.6 Series-Parallel Circuits

We may now apply the analysis techniques of series and parallel circuits in solving more complicated circuits. As in dc circuits, the analysis of such circuits is simplified by starting with easily recognized combinations. If necessary, the original circuit may be redrawn to make further simplification more apparent. Regardless of the complexity of the circuits, we find that the fundamental rules and laws of circuit analysis must apply in all cases.

Consider the network of Figure 18–47.

We see that the impedances $\mathbf{Z}_2$ and $\mathbf{Z}_3$ are in series. The branch containing this combination is then seen to be in parallel with the impedance $\mathbf{Z}_1$.

The total impedance of the network is expressed as

$$\mathbf{Z}_T = \mathbf{Z}_1 \parallel (\mathbf{Z}_2 + \mathbf{Z}_3)$$

Solving for $\mathbf{Z}_T$ gives the following:

$$\begin{aligned}\mathbf{Z}_T &= (2\ \Omega - j8\ \Omega)\|(2\ \Omega - j5\ \Omega + 6\ \Omega + j7\ \Omega)\\ &= (2\ \Omega - j8\ \Omega) \parallel (8\ \Omega + j2\ \Omega)\\ &= \frac{(2\ \Omega - j8\ \Omega)(8\ \Omega + j2\ \Omega)}{2\ \Omega - j8\ \Omega + 8\ \Omega + j2\ \Omega)}\\ &= \frac{(8.246\ \Omega\angle-75.96^\circ)(8.246\ \Omega\angle 14.04^\circ)}{11.66\ \Omega\angle-30.96^\circ}\\ &= 5.832\ \Omega\angle-30.96^\circ = 5.0\ \Omega - j3.0\ \Omega\end{aligned}$$

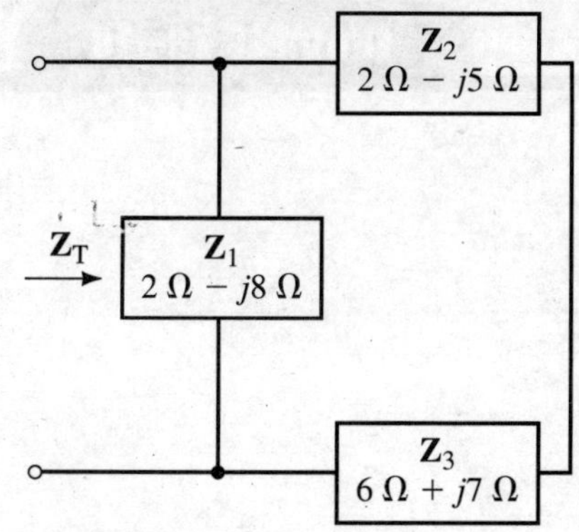

FIGURE 18–47

EXAMPLE 18–17

Determine the total impedance of the network of Figure 18–48. Express the impedance in both polar form and rectangular form.

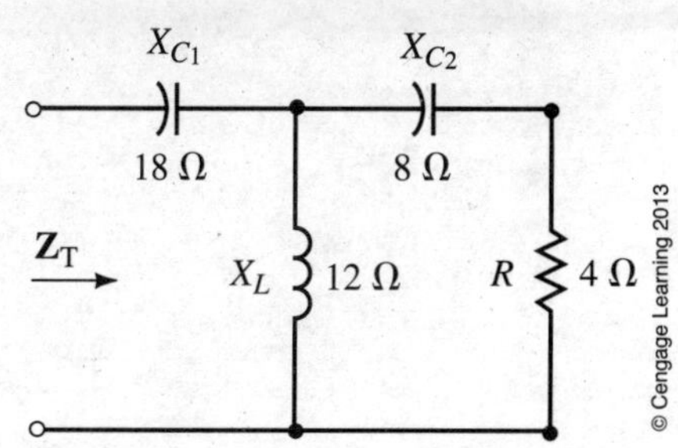

FIGURE 18–48

Solution After redrawing and labelling the given circuit, we have the circuit shown in Figure 18–49.

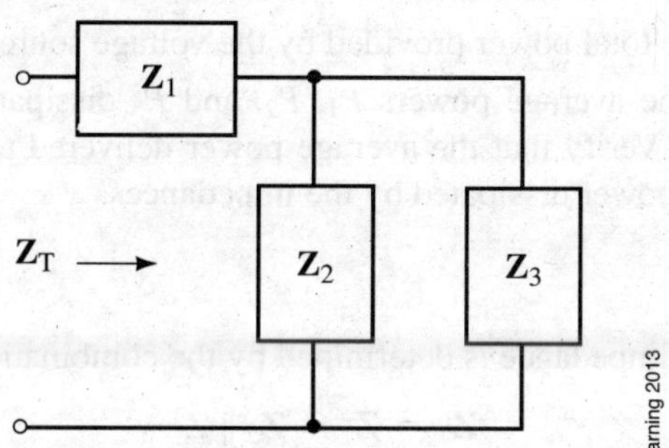

$$\begin{aligned}\mathbf{Z}_1 &= -j18\ \Omega\\ \mathbf{Z}_2 &= +j12\ \Omega = 12\ \Omega\angle 90^\circ\\ \mathbf{Z}_3 &= 4\ \Omega - j8\ \Omega = 8.94\ \Omega\angle-63.43^\circ\end{aligned}$$

FIGURE 18–49

The total impedance is given as

$$\mathbf{Z}_T = \mathbf{Z}_1 + \mathbf{Z}_2 \parallel \mathbf{Z}_3$$

where

$$\begin{aligned}\mathbf{Z}_1 &= -j18\ \Omega = 18\Omega\angle-90^\circ\\ \mathbf{Z}_2 &= j12\ \Omega = 12\Omega\angle 90^\circ\\ \mathbf{Z}_3 &= 4\ \Omega - j8\ \Omega = 8.94\ \Omega\angle-63.43^\circ\end{aligned}$$

We determine the total impedance as

$$\begin{aligned}\mathbf{Z}_T &= -j18\ \Omega + \left[\frac{(12\ \Omega\angle 90^\circ)(8.94\ \Omega\angle-63.43^\circ)}{j12\ \Omega + 4\ \Omega - j8\ \Omega}\right]\\ &= -j18\ \Omega + \left(\frac{107.3\ \Omega\angle 26.57^\circ}{5.66\angle 45^\circ}\right)\\ &= -j18\ \Omega + 19.0\ \Omega\angle-18.43^\circ\\ &= -j18\ \Omega + 18\ \Omega - j6\ \Omega\\ &= 18\ \Omega - j24\ \Omega = 30\ \Omega\angle-53.13^\circ\end{aligned}$$

EXAMPLE 18–18

Consider the circuit of Figure 18–50:

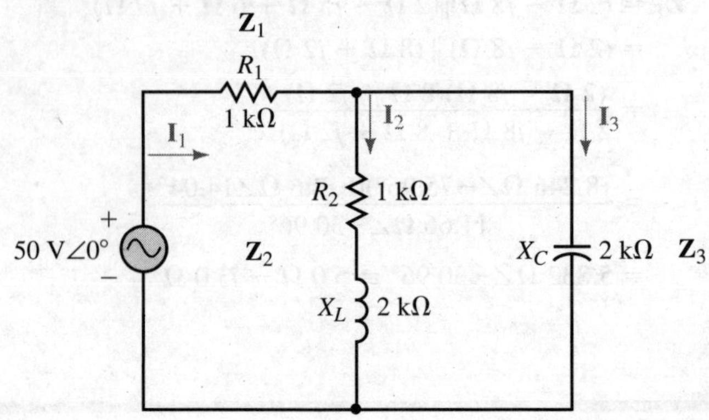

FIGURE 18–50

a. Find $\mathbf{Z}_T$.

b. Determine the currents $\mathbf{I}_1$, $\mathbf{I}_2$, and $\mathbf{I}_3$.

c. Calculate the total power provided by the voltage source.

d. Determine the average powers P_1, P_2, and P_3 dissipated by each of the impedances. Verify that the average power delivered to the circuit is the same as the power dissipated by the impedances.

Solution

a. The total impedance is determined by the combination

$$\mathbf{Z}_T = \mathbf{Z}_1 + \mathbf{Z}_2 \parallel \mathbf{Z}_3$$

For the parallel combination we have

$$\mathbf{Z}_2 \parallel \mathbf{Z}_3 = \frac{(1\ \text{k}\Omega + j2\ \text{k}\Omega)(-j2\ \text{k}\Omega)}{1\ \text{k}\Omega + j2\ \text{k}\Omega\ - j2\ \text{k}\Omega}$$

$$= \frac{(2.236\ \text{k}\Omega\angle 63.43^\circ)(2\ \text{k}\Omega\angle -90^\circ)}{1\ \text{k}\Omega\angle 0^\circ}$$

$$= 4.472\ \text{k}\Omega\angle -26.57^\circ = 4.0\ \text{k}\Omega\ - j2.0\ \text{k}\Omega$$

b. And so the total impedance is

$$\mathbf{Z}_T = 5\ \text{k}\Omega\ - j2\ \text{k}\Omega = 5.385\ \text{k}\Omega\angle -21.80^\circ$$

$$\mathbf{I}_1 = \frac{50\ \text{V}\angle 0^\circ}{5.385\ \text{k}\Omega\angle -21.80^\circ}$$

$$= 9.285\ \text{mA}\angle 21.80^\circ$$

Applying the current divider rule, we get

$$\mathbf{I}_2 = \frac{(2\ \text{k}\Omega\angle -90^\circ)(9.285\ \text{mA}\angle 21.80^\circ)}{1\ \text{k}\Omega + j2\ \text{k}\Omega\ - j2\ \text{k}\Omega}$$

$$= 18.57\ \text{mA}\angle -68.20^\circ$$

and

$$\mathbf{I}_3 = \frac{(1\ \text{k}\Omega + j2\ \text{k}\Omega)(9.285\ \text{mA}\angle 21.80^\circ)}{1\ \text{k}\Omega + j2\ \text{k}\Omega - j2\ \text{k}\Omega}$$

$$= \frac{(2.236\ \text{k}\Omega\angle 63.43^\circ)(9.285\ \text{mA}\angle 21.80^\circ)}{1\ \text{k}\Omega\angle 0^\circ}$$

$$= 20.761\ \text{mA}\angle 85.23^\circ$$

c. $$P_T = (50\ \text{V})(9.285\ \text{mA})\cos 21.80^\circ$$

$$= 431.0\ \text{mW}$$

d. Because only the resistors will dissipate power, we may use $P = I^2R$:

$$P_1 = (9.285\ \text{mA})^2(1\ \text{k}\Omega) = 86.2\ \text{mW}$$

$$P_2 = (18.57\ \text{mA})^2(1\ \text{k}\Omega) = 344.8\ \text{mW}$$

Alternatively, the power dissipated by $\mathbf{Z}_2$ may have been determined as $P = I^2Z\cos\theta$:

$$P_2 = (18.57\ \text{mA})^2(2.236\ \text{k}\Omega)\cos 63.43^\circ = 344.8\ \text{mW}$$

Since $\mathbf{Z}_3$ is purely capacitive, it will not dissipate any power:

$$P_3 = 0$$

By combining these powers, the total power dissipated is found:

$$P_T = 86.2\ \text{mW} + 344.8\ \text{mW} + 0 = 431.0\ \text{mW} \quad \text{(checks!)}$$

PRACTICE PROBLEMS 8

Refer to the circuit of Figure 18–51:

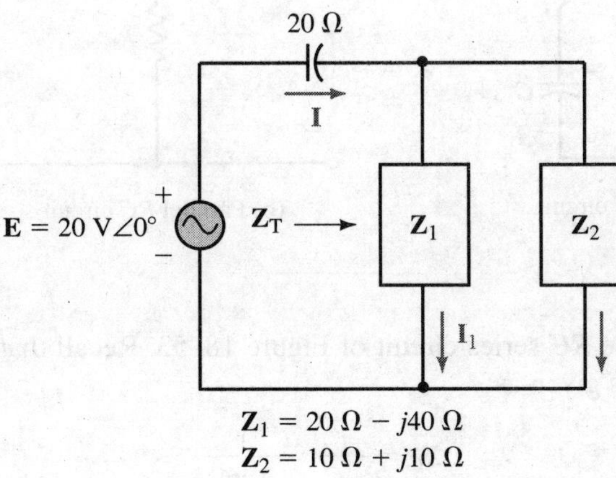

FIGURE 18–51

a. Calculate the total impedance, $\mathbf{Z}_T$.

b. Find the current **I**.

c. Use the current divider rule to find $\mathbf{I}_1$ and $\mathbf{I}_2$.

d. Determine the power factor for each impedance, $\mathbf{Z}_1$ and $\mathbf{Z}_2$.

e. Determine the power factor for the circuit.

f. Verify that the total power dissipated by impedances $\mathbf{Z}_1$ and $\mathbf{Z}_2$ is equal to the power delivered by the voltage source.

Answers

a. $\mathbf{Z}_T = 18.9\ \Omega\angle-45°$

b. $\mathbf{I} = 1.06\ \text{A}\angle45°$

c. $\mathbf{I}_1 = 0.354\ \text{A}\angle135°$, $\mathbf{I}_2 = 1.12\ \text{A}\angle26.57°$

d. $F_{P(1)} = 0.4472$ leading, $F_{P(2)} = 0.7071$ lagging

e. $F_P = 0.7071$ leading

f. $P_T = 15.0\ \text{W}$, $P_1 = 2.50\ \text{W}$, $P_2 = 12.5\ \text{W}$

$P_1 + P_2 = 15.0\ \text{W} = P_T$

18.7 Frequency Effects

As we have already seen, the reactance of inductors and capacitors depends on frequency. Consequently, the total impedance of any network having reactive elements is also frequency dependent. Any such circuit would need to be analyzed separately at each frequency of interest. We will examine several fairly simple combinations of resistors, capacitors, and inductors to see how the various circuits operate at different frequencies. Some of the more important combinations will be examined in greater detail in later chapters that deal with resonance and filters.

RC Circuits

As the name implies, *RC* circuits consist of a resistor and a capacitor. The components of an *RC* circuit may be connected either in series or in parallel as shown in Figure 18–52.

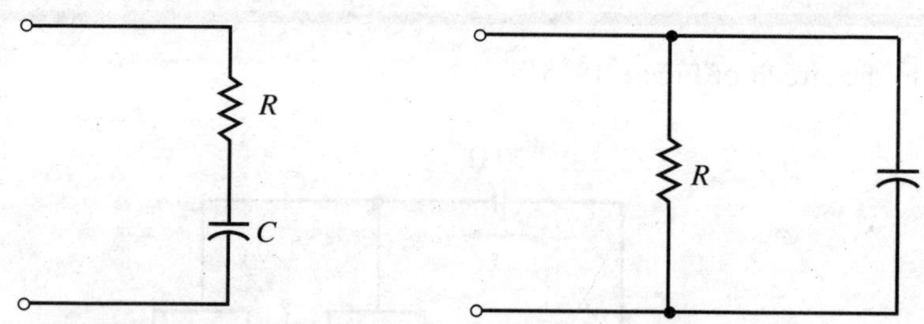

(a) Series *RC* circuit (b) Parallel *RC* circuit

FIGURE 18–52

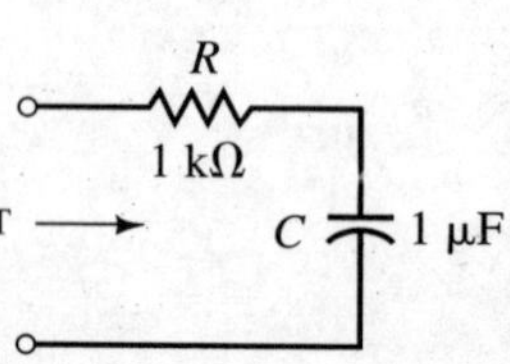

FIGURE 18–53

Consider the *RC* series circuit of Figure 18–53. Recall that the capacitive reactance, X_C, is given as

$$X_C = \frac{1}{\omega C} = \frac{1}{2\pi f C}$$

The total impedance of the circuit is a vector quantity expressed as

$$\mathbf{Z}_T = R - j\frac{1}{\omega C} = R + \frac{1}{j\omega C}$$

$$\mathbf{Z}_T = \frac{1 + j\omega RC}{j\omega C} \tag{18–21}$$

If we define the **cutoff** or **corner frequency** for an *RC* circuit as

$$\omega_c = \frac{1}{RC} = \frac{1}{\tau}\ \text{(rad/s)} \tag{18–22}$$

or equivalently as

$$f_c = \frac{1}{2\pi RC} \quad \text{(Hz)} \tag{18–23}$$

then several important points become evident.

For $\omega \leq \omega_c/10$ (or $f \leq f_c/10$) Equation 18–21 can be expressed as

$$\mathbf{Z}_\mathrm{T} \simeq \frac{1 + j0}{j\omega C} = \frac{1}{j\omega C}$$

and for $\omega \geq 10\omega_c$, the expression of Equation 18–21 can be simplified as

$$\mathbf{Z}_\mathrm{T} \simeq \frac{0 + j\omega RC}{j\omega C} = R$$

Solving for the magnitude of the impedance at several angular frequencies, we have the results shown in Table 18–1.

TABLE 18–1

Angular Frequency, ω (rad/s)	X_C (Ω)	Z_T (Ω)
0	∞	∞
1	1 M	1 M
10	100 k	100 k
100	10 k	10.05 k
200	5 k	5.099 k
500	2 k	2.236 k
1000	1 k	1.414 k
2000	500	1118
5000	200	1019
10 k	100	1005
100 k	10	1000

If the magnitude of the impedance $\mathbf{Z}_\mathrm{T}$ is plotted as a function of angular frequency ω, we get the graph of Figure 18–54. Notice that the abscissa and ordinate of the graph are not scaled linearly, but rather logarithmically. This allows for the display of results over a wide range of frequencies.

FIGURE 18–54 Impedance versus angular frequency for the network of Figure 18–53.

The graph illustrates that the reactance of a capacitor is very high (effectively an open circuit) at low frequencies. Consequently, the total impedance of the series circuit will also be very high at low frequencies. Second, we notice that as the frequency increases, the reactance decreases. Therefore, as the frequency gets higher, the capacitive reactance has a diminished effect in the circuit. At very high frequencies (typically for $\omega \geq 10\omega_c$), the impedance of the circuit will effectively be $R = 1\ \mathrm{k\Omega}$.

Consider the parallel RC circuit of Figure 18–55. The total impedance, $\mathbf{Z}_\mathrm{T}$, of the circuit is determined as

$$\mathbf{Z}_\mathrm{T} = \frac{\mathbf{Z}_R\mathbf{Z}_C}{\mathbf{Z}_R + \mathbf{Z}_C} = \frac{R\left(\dfrac{1}{j\omega C}\right)}{R + \dfrac{1}{j\omega C}}$$

$$= \frac{\dfrac{R}{j\omega C}}{\dfrac{1 + j\omega RC}{j\omega C}}$$

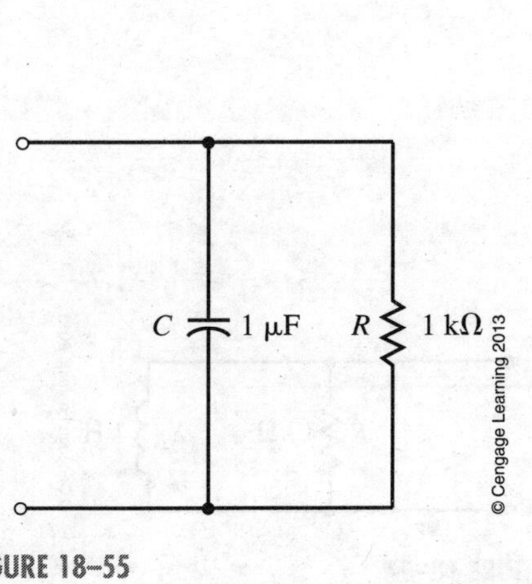

FIGURE 18–55

TABLE 18–2

Angular Frequency, ω (rad/s)	X_C (Ω)	Z_T (Ω)
0	∞	1000
1	1 M	1000
10	100 k	1000
100	10 k	995
200	5 k	981
500	2 k	894
1 k	1 k	707
2 k	500	447
5 k	200	196
10 k	100	99.5
100 k	10	10

which may be simplified as

$$\mathbf{Z}_T = \frac{R}{1 + j\omega RC} \tag{18–24}$$

As before, the cutoff frequency is given by Equation 18–22. Now, by examining the expression of Equation 18–24 for $\omega \leq \omega_c/10$, we have the following result:

$$\mathbf{Z}_T \cong \frac{R}{1 + j0} = R$$

For $\omega \geq 10\omega_c$, we have

$$\mathbf{Z}_T \cong \frac{R}{0 + j\omega RC} = \frac{1}{j\omega C}$$

If we solve for the impedance of the circuit in Figure 18–55 at various angular frequencies, we obtain the results of Table 18–2.

Plotting the magnitude of the impedance $\mathbf{Z}_T$ as a function of angular frequency ω, we get the graph of Figure 18–56. Notice that the abscissa and ordinate of the graph are again scaled logarithmically, allowing for the display of results over a wide range of frequencies.

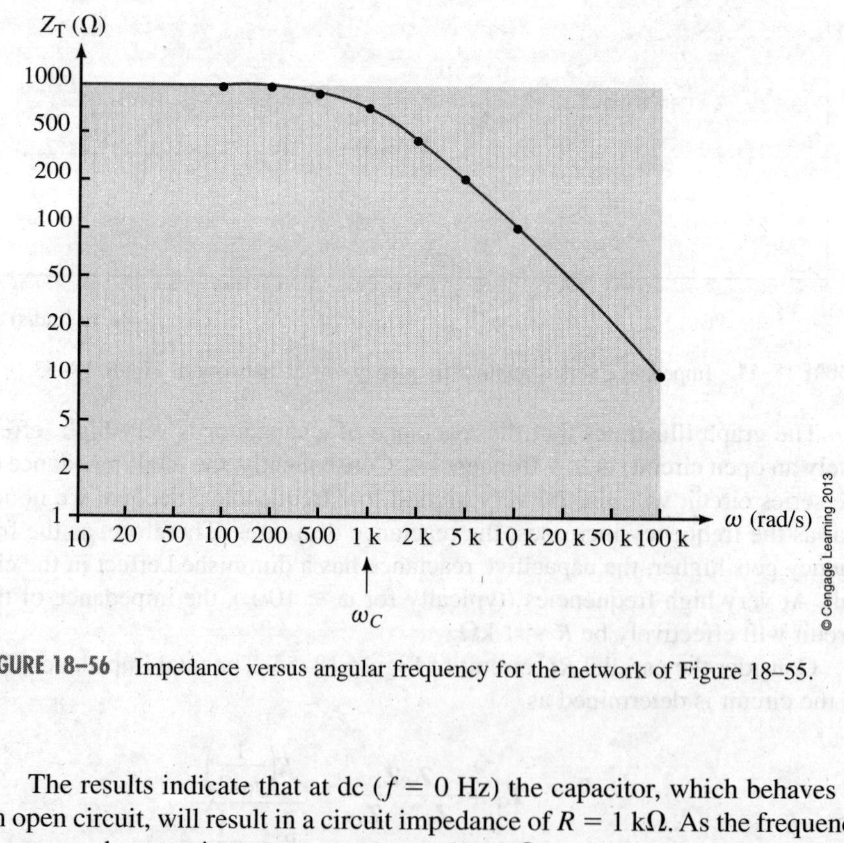

FIGURE 18–56 Impedance versus angular frequency for the network of Figure 18–55.

The results indicate that at dc ($f = 0$ Hz) the capacitor, which behaves as an open circuit, will result in a circuit impedance of $R = 1$ kΩ. As the frequency increases, the capacitor reactance approaches 0 Ω, resulting in a corresponding decrease in circuit impedance.

RL Circuits

RL circuits may be analyzed in a manner similar to the analysis of *RC* circuits. Consider the parallel *RL* circuit of Figure 18–57.

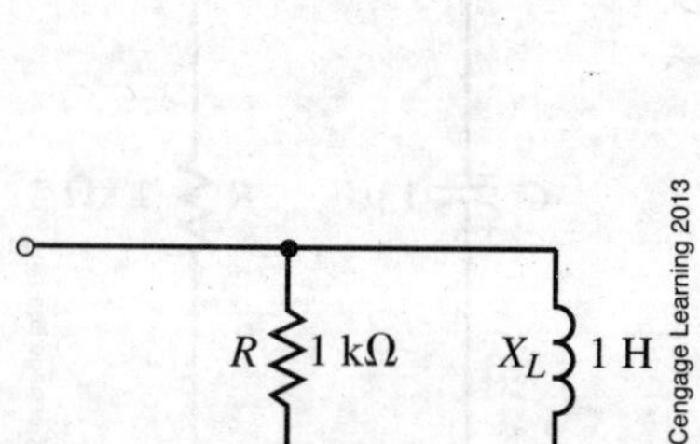

FIGURE 18–57

The total impedance of the parallel circuit is found as follows:

$$\mathbf{Z}_T = \frac{\mathbf{Z}_R \mathbf{Z}_L}{\mathbf{Z}_R + \mathbf{Z}_L} = \frac{R(j\omega L)}{R + j\omega L} \qquad \textbf{(18–25)}$$

$$\mathbf{Z}_T = \frac{j\omega L}{1 + j\omega \frac{L}{R}}$$

If we define the *cutoff* or *corner frequency* for an *RL* circuit as

$$\omega_c = \frac{R}{L} = \frac{1}{\tau} \quad \text{(rad/s)} \qquad \textbf{(18–26)}$$

or equivalently as

$$f_c = \frac{R}{2\pi L} \quad \text{(Hz)} \qquad \textbf{(18–27)}$$

then several important points become evident.

For $\omega \le \omega_c/10$ (or $f \le f_c/10$) Equation 18–25 can be expressed as

$$\mathbf{Z}_T \simeq \frac{j\omega L}{1 + j0} = j\omega L$$

The previous result indicates that for low frequencies, the inductor has a very small reactance, resulting in a total impedance which is essentially equal to the inductive reactance.

For $\omega \ge 10\omega_c$, the expression of Equation 18–25 can be simplified as

$$\mathbf{Z}_T \simeq \frac{j\omega L}{0 + j\omega \frac{L}{R}} = R$$

The preceding results indicate that for high frequencies, the impedance of the circuit is essentially equal to the resistance, due to the very high impedance of the inductor.

Evaluating the impedance at several angular frequencies, we have the results of Table 18–3.

When the magnitude of the impedance $\mathbf{Z}_T$ is plotted as a function of angular frequency ω, we get the graph of Figure 18–58.

TABLE 18–3

Angular Frequency, ω (rad/s)	X_L (Ω)	Z_T (Ω)
0	0	0
1	1	1
10	10	10
100	100	99.5
200	200	196
500	500	447
1 k	1 k	707
2 k	2 k	894
5 k	5 k	981
10 k	10 k	995
100 k	100 k	1000

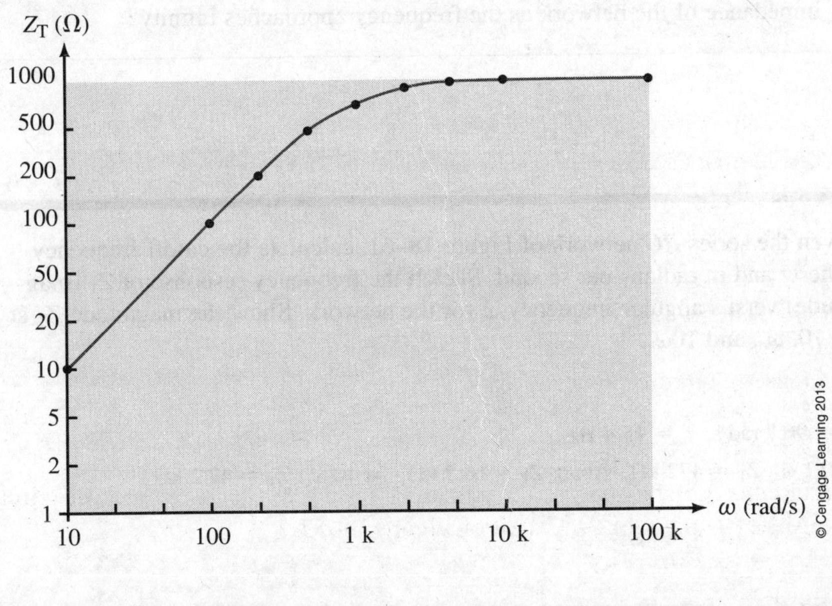

FIGURE 18–58 Impedance versus angular frequency for the network of Figure 18–57.

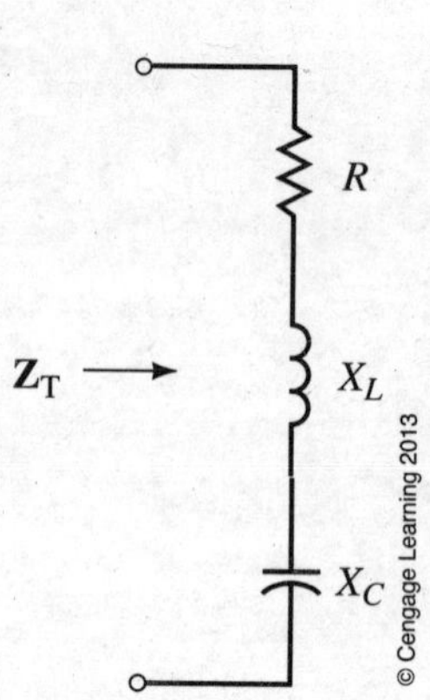

FIGURE 18–59

RLC Circuits

When numerous capacitive and inductive components are combined with resistors in series-parallel circuits, the total impedance $\mathbf{Z}_T$ of the circuit may rise and fall several times over the full range of frequencies. The analysis of such complex circuits is outside the scope of this textbook. However, for illustrative purposes we examine the simple series *RLC* circuit of Figure 18–59.

The impedance $\mathbf{Z}_T$ at any frequency will be determined as

$$\mathbf{Z}_T = R + jX_L - jX_C$$
$$= R + j(X_L - X_C)$$

At very low frequencies, the inductor will appear as a very low impedance (effectively a short circuit), while the capacitor will appear as a very high impedance (effectively an open circuit). Because the capacitive reactance will be much larger than the inductive reactance, the circuit will have a very large capacitive reactance. This results in a very high circuit impedance, $\mathbf{Z}_T$.

As the frequency increases, the inductive reactance increases, while the capacitive reactance decreases. At some frequency, f_0, the inductor and the capacitor will have the same magnitude of reactance. At this frequency, the reactances cancel, resulting in a circuit impedance which is equal to the resistance value.

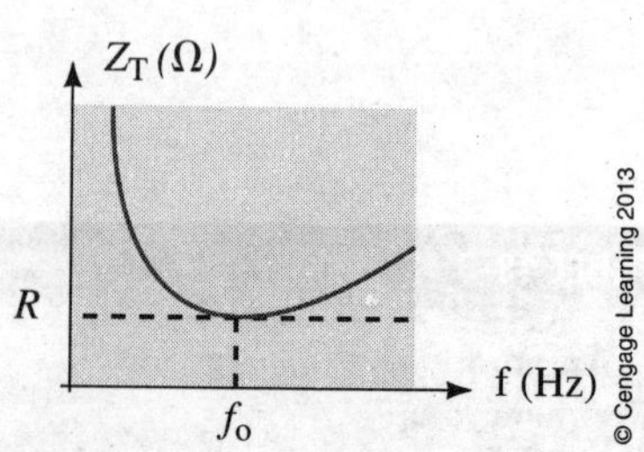

FIGURE 18–60

As the frequency increases still further, the inductive reactance becomes larger than the capacitive reactance. The circuit becomes inductive and the magnitude of the total impedance of the circuit again rises. Figure 18–60 shows how the impedance of a series *RLC* circuit varies with frequency.

The complete analysis of the series *RLC* circuit and the parallel *RLC* circuit is left until we examine the principle of resonance in a later chapter.

IN-PROCESS LEARNING CHECK 4

(Answers are at the end of the chapter.)

1. For a series network consisting of a resistor and a capacitor, what will be the impedance of the network at a frequency of 0 Hz (dc)? What will be the impedance of the network as the frequency approaches infinity?
2. For a parallel network consisting of a resistor and an inductor, what will be the impedance of the network at a frequency of 0 Hz (dc)? What will be the impedance of the network as the frequency approaches infinity?

PRACTICE PROBLEMS 9

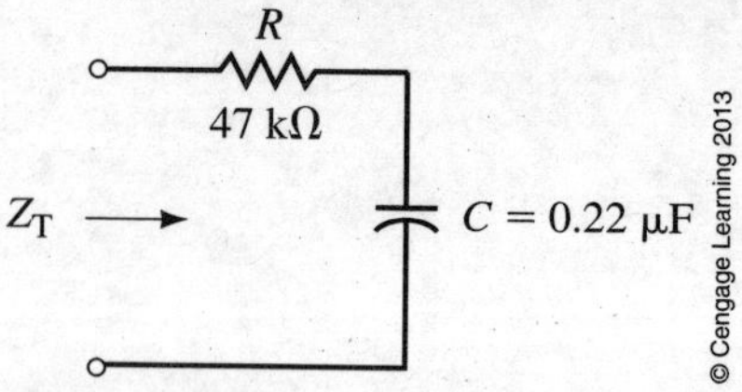

FIGURE 18–61

Given the series *RC* network of Figure 18–61, calculate the cutoff frequency in hertz and in radians per second. Sketch the frequency response of Z_T (magnitude) versus angular frequency ω for the network. Show the magnitude Z_T at $\omega_c/10$, ω_c, and $10\omega_c$.

Answers

$\omega_c = 96.7$ rad/s, $f_c = 15.4$ Hz

At 0.1 ω_c: $Z_T = 472$ kΩ, At ω_c: $Z_T = 66.5$ kΩ, At 10 ω_c: $Z_T = 47.2$ kΩ

18.8 Applications

As we have seen, we may determine the impedance of any ac circuit as a vector $\mathbf{Z} = R \pm jX$. This means that any ac circuit may now be simplified as a series circuit having a resistance and a reactance, as shown in Figure 18–62.

Additionally, an ac circuit may be represented as an equivalent parallel circuit consisting of a single resistor and a single reactance as shown in Figure 18–63. *Any equivalent circuit will be valid only at the given frequency of operation.*

We will now examine the technique used to convert any series impedance into its parallel equivalent. Suppose that the two circuits of Figure 18–62 and Figure 18–63 are exactly equivalent at some frequency. These circuits can be equivalent only if both circuits have the same total impedance, $\mathbf{Z}_T$, and the same total admittance, $\mathbf{Y}_T$.

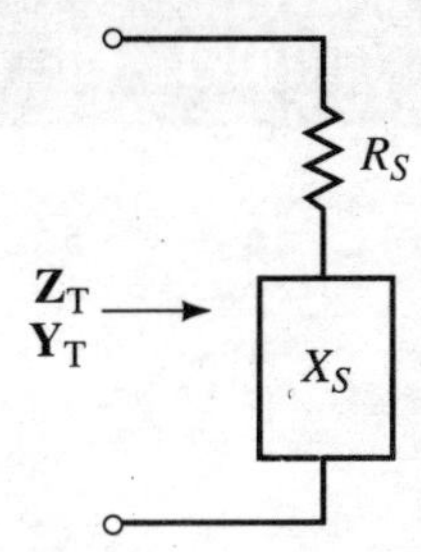

FIGURE 18–62

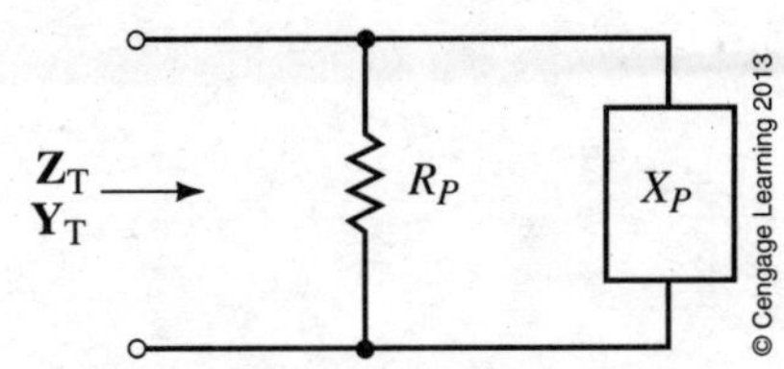

FIGURE 18–63

From the circuit of Figure 18–62, the total impedance is written as

$$\mathbf{Z}_T = R_S \pm jX_S$$

Therefore, the total admittance of the circuit is

$$\mathbf{Y}_T = \frac{1}{\mathbf{Z}_T} = \frac{1}{R_S \pm jX_S}$$

Multiplying the numerator and denominator by the complex conjugate, we obtain the following:

$$\mathbf{Y}_T = \frac{R_S \mp jX_S}{(R_S \pm jX_S)(R_S \mp jX_S)}$$

$$= \frac{R_S \mp jX_S}{R_S^2 + X_S^2} \quad (18\text{–}28)$$

$$\mathbf{Y}_T = \frac{R_S}{R_S^2 + X_S^2} \mp j\frac{X_S}{R_S^2 + X_S^2}$$

Now, from the circuit of Figure 18–63, the total admittance of the parallel circuit may be found from the parallel combination of R_P and X_P as

$$\mathbf{Y}_T = \frac{1}{R_P} + \frac{1}{\pm jX_P}$$

which gives

$$\mathbf{Y}_T = \frac{1}{R_P} \mp j\frac{1}{X_P} \quad (18\text{–}29)$$

Two vectors can only be equal if both the real components are equal and the imaginary components are equal. Therefore the circuits of Figure 18–62 and Figure 18–63 can only be equivalent if the following conditions are met:

$$R_P = \frac{R_S^2 + X_S^2}{R_S} \quad (18\text{–}30)$$

and

$$X_P = \frac{R_S^2 + X_S^2}{X_S} \quad (18\text{–}31)$$

In a similar manner, we have the following conversion from a parallel circuit to an equivalent series circuit:

$$R_S = \frac{R_P X_P^2}{R_P^2 + X_P^2} \quad (18\text{–}32)$$

and

$$X_S = \frac{R_P^2 X_P}{R_P^2 + X_P^2} \qquad (18\text{–}33)$$

EXAMPLE 18–19

A circuit has a total impedance of $\mathbf{Z}_T = 10\ \Omega + j50\ \Omega$. Sketch the equivalent series and parallel circuits.

Solution The series circuit will be an inductive circuit having $R_S = 10\ \Omega$ and $X_{LS} = 50\ \Omega$.

The equivalent parallel circuit will also be an inductive circuit having the following values:

$$R_P = \frac{(10\ \Omega)^2 + (50\ \Omega)^2}{10\ \Omega} = 260\ \Omega$$

$$X_{LP} = \frac{(10\ \Omega)^2 + (50\ \Omega)^2}{50\ \Omega} = 52\ \Omega$$

The equivalent series and parallel circuits are shown in Figure 18–64.

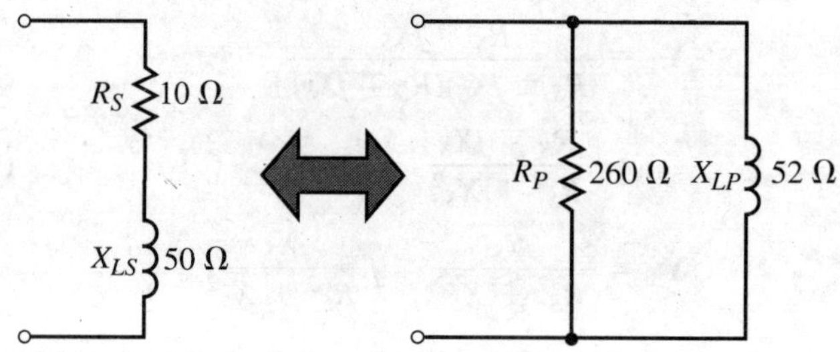

FIGURE 18–64

EXAMPLE 18–20

A circuit has a total admittance of $\mathbf{Y}_T = 0.559\ \text{mS}\angle 63.43°$. Sketch the equivalent series and parallel circuits.

Solution Because the admittance is written in polar form, we first convert to the rectangular form of the admittance.

$$G_P = (0.559\ \text{mS}) \cos 63.43° = 0.250\ \text{mS} \quad \Leftrightarrow \quad R_P = 4.0\ \text{k}\Omega$$

$$B_{CP} = (0.559\ \text{mS}) \sin 63.43° = 0.500\ \text{mS} \quad \Leftrightarrow \quad X_{CP} = 2.0\ \text{k}\Omega$$

The equivalent series circuit is found as

$$R_S = \frac{(4\ \text{k}\Omega)(2\ \text{k}\Omega)^2}{(4\ \text{k}\Omega)^2 + (2\ \text{k}\Omega)^2} = 0.8\ \text{k}\Omega$$

and

$$X_{CS} = \frac{(4\ \text{k}\Omega)^2(2\ \text{k}\Omega)}{(4\ \text{k}\Omega)^2 + (2\ \text{k}\Omega)^2} = 1.6\ \text{k}\Omega$$

The equivalent circuits are shown in Figure 18–65.

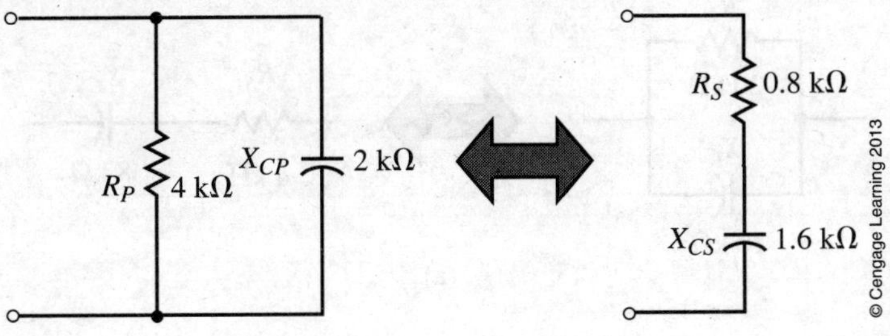

FIGURE 18–65

EXAMPLE 18–21

Refer to the circuit of Figure 18–66.

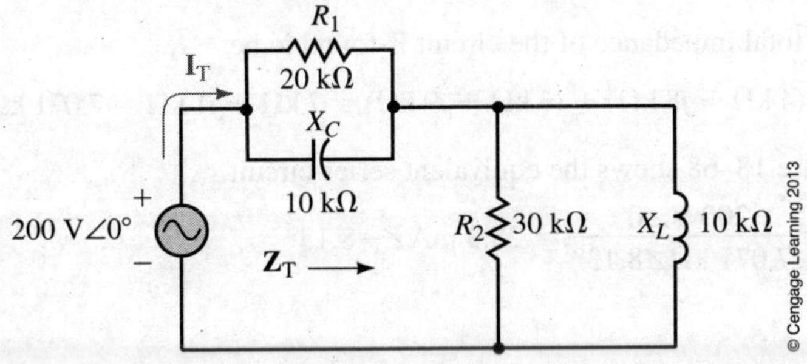

FIGURE 18–66

a. Find $\mathbf{Z}_T$.

b. Sketch the equivalent series circuit.

c. Determine $\mathbf{I}_T$.

Solution

a. The circuit consists of two parallel networks in series. We apply Equations 18–32 and 18–33 to arrive at equivalent series elements for each of the parallel networks as follows:

$$R_{S_1} = \frac{(20\text{ k}\Omega)(10\text{ k}\Omega)^2}{(20\text{ k}\Omega)^2 + (10\text{ k}\Omega)^2} = 4\text{ k}\Omega$$

$$X_{CS} = \frac{(20\text{ k}\Omega)^2(10\text{ k}\Omega)}{(20\text{ k}\Omega)^2 + (10\text{ k}\Omega)^2} = 8\text{ k}\Omega$$

and

$$R_{S_2} = \frac{(30\text{ k}\Omega)(10\text{ k}\Omega)^2}{(30\text{ k}\Omega)^2 + (10\text{ k}\Omega)^2} = 3\text{ k}\Omega$$

$$X_{LS} = \frac{(30\text{ k}\Omega)^2(10\text{ k}\Omega)}{(30\text{ k}\Omega)^2 + (10\text{ k}\Omega)^2} = 9\text{ k}\Omega$$

The equivalent circuits are shown in Figure 18–67.

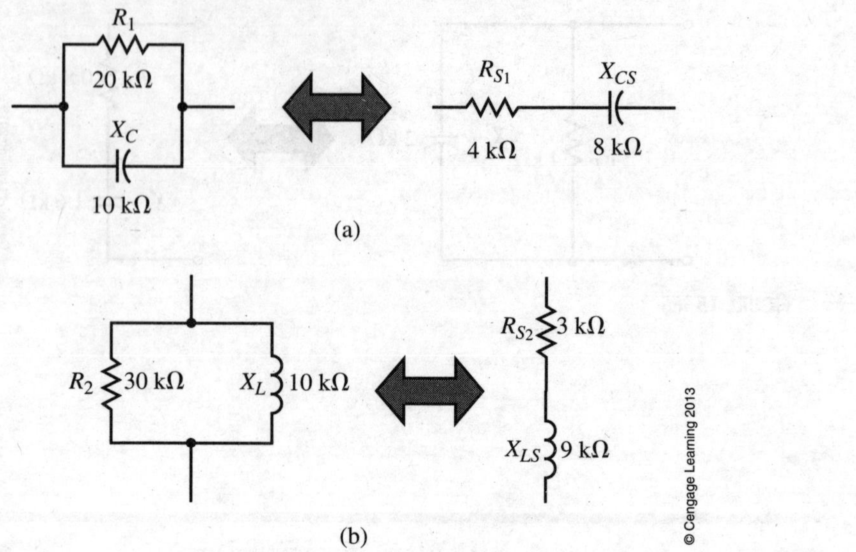

FIGURE 18–67

The total impedance of the circuit is found to be

$$\mathbf{Z}_T = (4\text{ k}\Omega - j8\text{ k}\Omega) + (3\text{ k}\Omega + j9\text{ k}\Omega) = 7\text{ k}\Omega + j1\text{ k}\Omega = 7.071\text{ k}\Omega\angle 8.13^\circ$$

b. Figure 18–68 shows the equivalent series circuit.

c. $\mathbf{I}_T = \dfrac{200\text{ V}\angle 0^\circ}{7.071\text{ k}\Omega\angle 8.13^\circ} = 28.3\text{ mA}\angle -8.13^\circ$

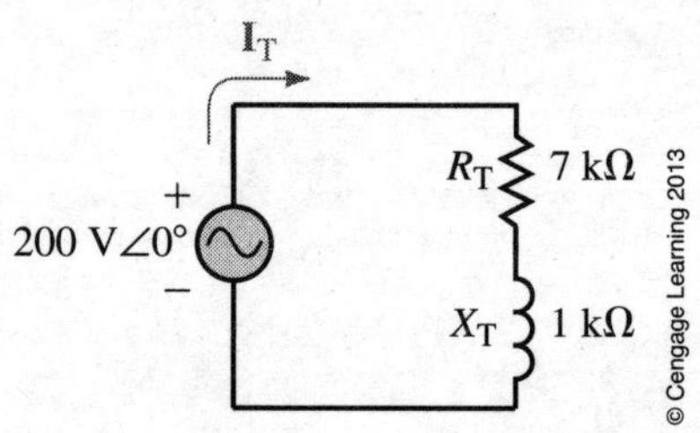

FIGURE 18–68

IN-PROCESS LEARNING CHECK 5

(Answers are at the end of the chapter.)

An inductor of 10 mH has a series resistance of 5 Ω.

a. Determine the parallel equivalent of the inductor at a frequency of 1 kHz. Sketch the equivalent showing the values of L_P (in henries) and R_P.

b. Determine the parallel equivalent of the inductor at a frequency of 1 MHz. Sketch the equivalent showing the values of L_P (in henries) and R_P.

c. If the frequency were increased still further, predict what would happen to the values of L_P and R_P.

PRACTICE PROBLEMS 10

A network has an impedance of $\mathbf{Z}_T = 50\text{ k}\Omega\angle 75^\circ$ at a frequency of 5 kHz.

a. Determine the most simple equivalent series circuit (L and R).

b. Determine the most simple equivalent parallel circuit.

Answers

a. $R_S = 12.9\text{ k}\Omega$, $L_S = 1.54\text{ H}$

b. $R_P = 193\text{ k}\Omega$, $L_P = 1.65\text{ H}$

18.9 Circuit Analysis Using Computers

Multisim

PSPICE

Multisim

In this section we will use Multisim to simulate how sinusoidal ac measurements are taken with an oscilloscope. The "measurements" are then interpreted to verify the ac operation of circuits. You will use some of the display features of the software to simplify your work. The following example provides a guide through each step of the procedure.

EXAMPLE 18–22

CircuitSim 18-4

Given the circuit of Figure 18–69.

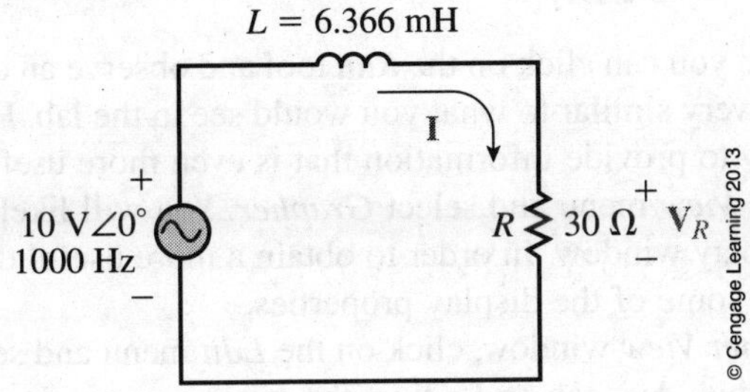

FIGURE 18–69

a. Determine the current **I** and the voltage $\mathbf{V}_R$.
b. Use Multisim to display the resistor voltage v_R and the source voltage e. Use the results to verify the results of part (a).

Solution

a. $X_L = 2\pi(1000\text{ Hz})(6.366 \times 10^{-3}\text{ H}) = 40\ \Omega$

$$\mathbf{Z} = 30\ \Omega + j40\ \Omega = 50\ \Omega\angle 53.13^\circ$$

$$\mathbf{I} = \frac{10\text{ V}\angle 0^\circ}{50\ \Omega\angle 53.13^\circ} = 0.200\text{ A}\angle -53.13^\circ$$

$$\mathbf{V}_R = (0.200\text{ A}\angle -53.13^\circ)(30\ \Omega) = 6.00\text{ V}\angle -53.13^\circ$$

b. Use the schematic editor to input the circuit shown in Figure 18–70.

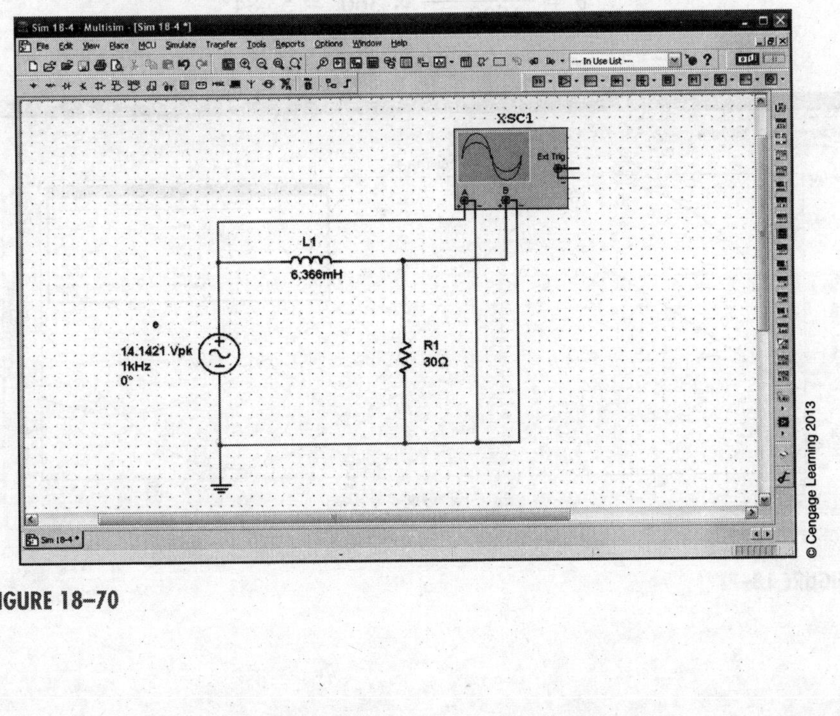

FIGURE 18–70

The ac voltage source is obtained from the Signal Source Family. The properties of the voltage source are changed by double-clicking on the symbol and then selecting the *Value* tab. Change the values as follows:

Voltage (Pk):	**14.14 V**
Frequency:	**1 kHz**
Phase:	**0 Deg**

After double-clicking on the oscilloscope (XSC1), change the settings as follows:

Time base:	**200μs/Div**
Channel A:	**5V/Div**
Channel B:	**5V/Div**

At this point, you can click on the *Run* tool and observe an oscilloscope display that will be very similar to what you would see in the lab. However, we can refine the display to provide information that is even more useful.

Click on the *View* menu and select *Grapher*. You will likely observe many cycles in the display window. In order to obtain a more useful display, it is necessary to change some of the display properties.

In the *Grapher View* window, click on the *Edit* menu and select *Properties*. Click on the *Bottom Axis* tab and adjust the range to provide a display of one cycle ($T = 1.00$ ms). A suitable range will be from 0.020 to 0.021 s, although any other 1-ms display will be equally useful.

Since we would like to be able to measure the phase angle, it is necessary to provide a grid and cursors on the graph. This is done by clicking on both the *Show Grid* tool and on the *Show Cursors* tool. After moving the cursors, it is possible to determine the phase relationship between $\mathbf{V}_R$ and **E** directly from the table of data. The display will appear similar to that shown in Figure 18–71.

After positioning the cursors on the peaks of the voltage waveforms and using the display window, we are able to obtain various measurements for the circuit. The phase angle of the resistor voltage with respect to the source voltage is found by using the difference between the cursors in the bottom axis, $dx =$ 149.5554 μs. Now we have

$$\theta = \frac{149.55\ \mu\text{s}}{1000\ \mu\text{s}} \times 360° = 53.84°$$

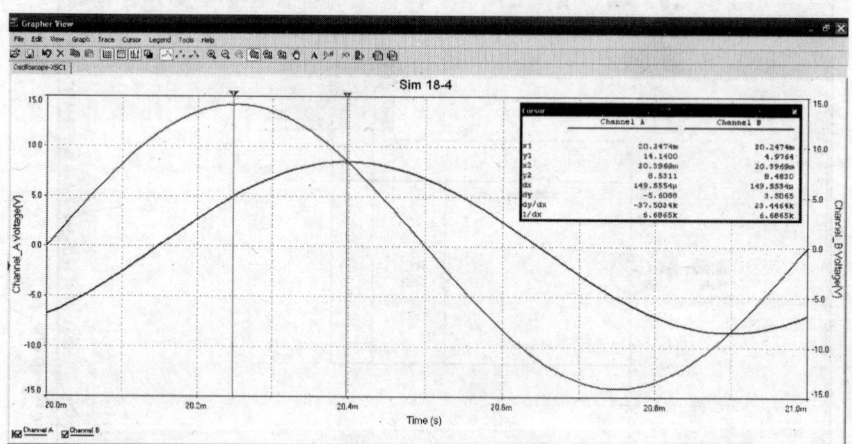

FIGURE 18–71

The amplitude of the resistor voltage is 8.483 V, which results in an rms value of 6.00 V. As a result of the measurements we have

$$\mathbf{V}_R = 6.00\ \text{V}\angle -53.84°$$

and

$$\mathbf{I} = \frac{6\ \text{V}\angle -53.84°}{30\ \Omega} = 0.200\ \text{A}\angle -53.84°$$

These values correspond very closely to the theoretical results calculated in part (a) of the example.

PSpice

In the following example we will use the Probe postprocessor of PSpice to show how the impedance of an *RC* circuit changes as a function of frequency. The Probe output will provide a graphical result that is very similar to the frequency responses determined in previous sections of this chapter.

EXAMPLE 18–23

Refer to the network of Figure 18–72. Use PSpice to input the circuit. Run the Probe postprocessor to provide a graphical display of network impedance as a function of frequency from 50 Hz to 500 Hz.

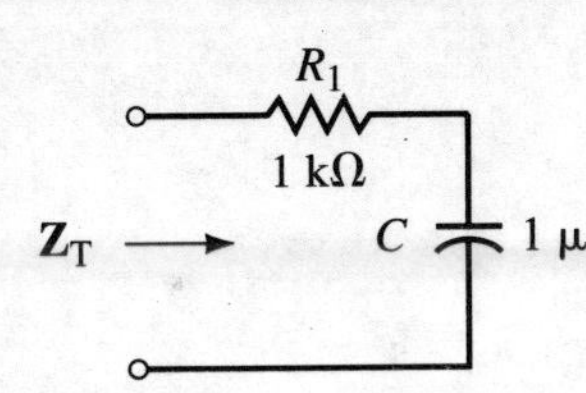

FIGURE 18–72

Solution Since PSpice is unable to analyze an incomplete circuit, it is necessary to provide a voltage source (and ground) for the circuit of Figure 18–72. The input impedance is not dependent on the actual voltage used, and so we may use any ac voltage source. In this example we arbitrarily select a voltage of 10 V.

- Open the CIS Demo software.
- Open a new project and call it Ch 18 PSpice 1. Ensure that the Analog or Mixed-Signal Circuit Wizard is activated.
- Enter the circuit as shown in Figure 18–73. Simply click on the voltage value and change its value to **10V** from the default value of **1V**. (There must be no spaces between the magnitude and the units.)
- Click on *PSpice, New Simulation Profile* and give a name such as Example 18–23 to the new simulation. The *Simulation Settings* dialog box will open.
- Click on the *Analysis* tab and select *AC Sweep/Noise* as the *Analysis* type. Select *General Settings* from the *Options* box.
- Select *Linear AC Sweep Type* (quite often, logarithmic frequency sweeps such as Decade are used). Enter the following values into the appropriate dialog boxes. *Start Frequency*; **50**, *End Frequency*: **500**, *Total Points*: **1001**. Click OK.
- Click on *PSpice* and *Run*. The Probe postprocessor will appear on the screen.

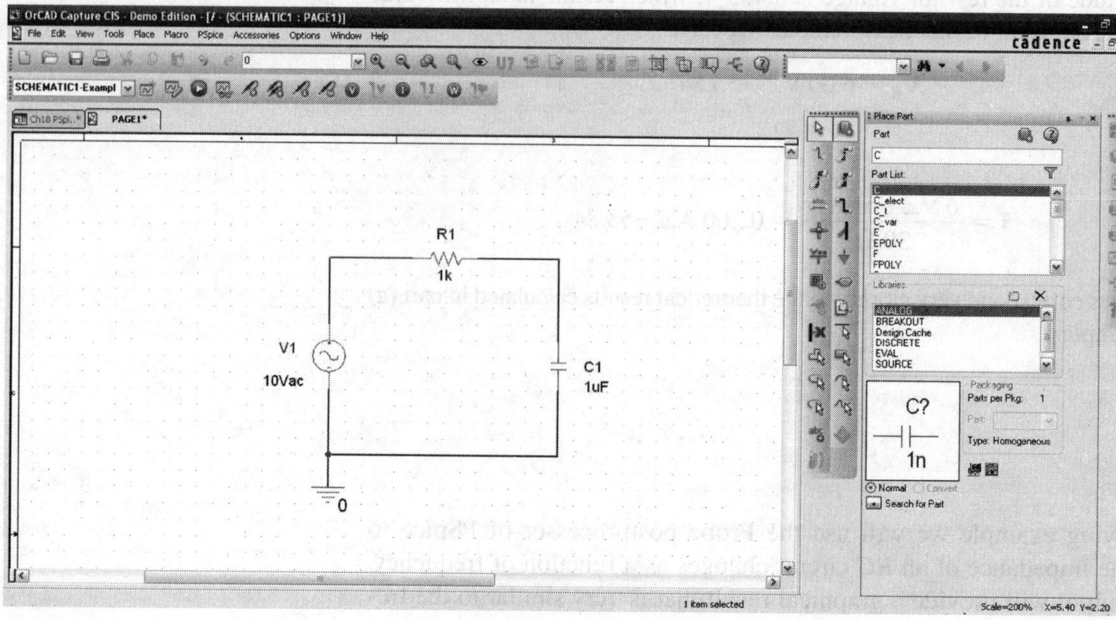

FIGURE 18–73

- Click on *Trace* and *Add Trace*. You may simply click on the appropriate values from the list of variables and use the division symbol to result in impedance. Enter the following expression into the *Trace Expression* box:

V(R1:1)/I(R1)

Notice that the preceding expression is nothing other than an application of Ohm's law. The resulting display is shown in Figure 18–74.

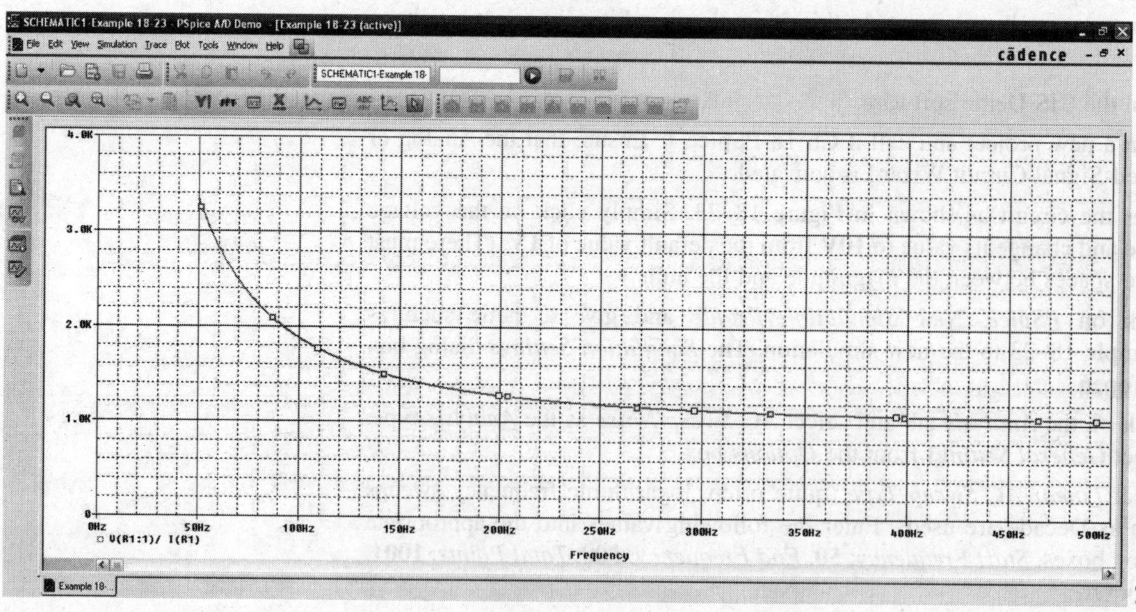

FIGURE 18–74

PRACTICE PROBLEMS 11

Given the circuit of Figure 18–75.

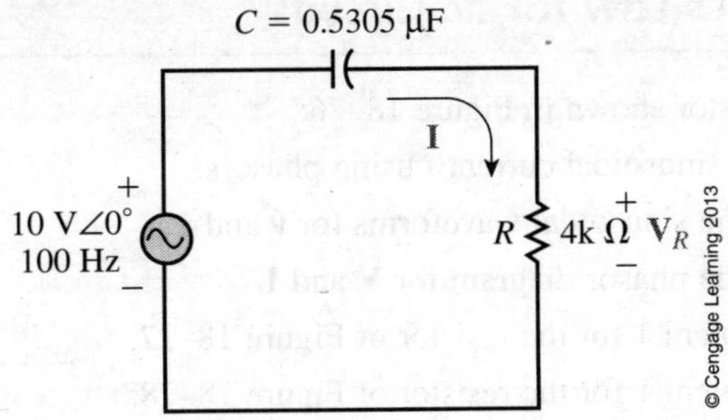

FIGURE 18–75

a. Determine the current **I** and the voltage $\mathbf{V}_R$.

b. Use Multisim to display the resistor voltage v_R and the source voltage, *e*. Use the results to verify the results of part (a).

Answers

a. $\mathbf{I} = 2.00\ \text{mA}\angle 36.87°$, $\mathbf{V}_R = 8.00\ \text{V}\angle 36.87°$

b. $v_R = 11.3 \sin(\omega t + 36.87°)$, $e = 14.1 \sin \omega t$

PRACTICE PROBLEMS 12

Use PSpice to input the circuit of Figure 18–55. Use the Probe postprocessor to obtain a graphical display of the network impedance as a function of frequency from 100 Hz to 2000 Hz.

Putting It into Practice

You are working in a small industrial plant where several small motors are powered by a 60-Hz ac line voltage of 120 Vac. Your supervisor tells you that one of the 2-hp motors that was recently installed draws too much current when the motor is under full load. You take a current reading and find that the current is 14.4 A. After doing some calculations, you determine that even if the motor is under full load, it shouldn't require that much current.

However, you have an idea. You remember that a motor can be represented as a resistor in series with an inductor. If you could reduce the effect of the inductive reactance of the motor by placing a capacitor across the motor, you should be able to reduce the current since the capacitive reactance will cancel the inductive reactance.

While keeping the motor under load, you place a capacitor into the circuit. Just as you suspected, the current goes down. After using several different values, you observe that the current goes to a minimum of 12.4 A. It is at this value that you have determined that the reactive impedances are exactly balanced. Sketch the complete circuit and determine the value of capacitance that was added into the circuit. Use the information to determine the value of the motor's inductance. (Assume that the motor has an efficiency of 100%.)

Problems

18.1 Ohm's Law for ac Circuits

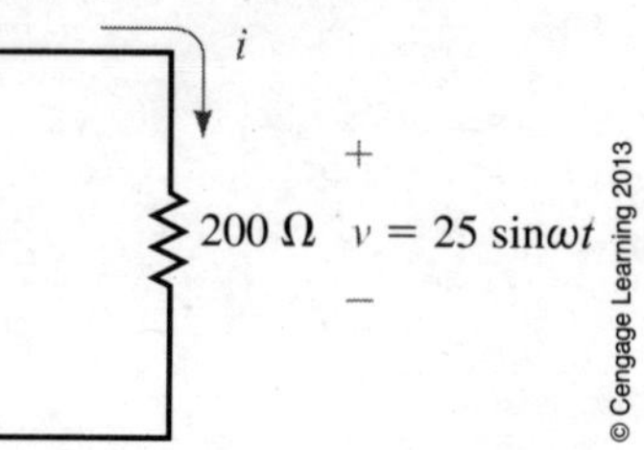

FIGURE 18–76

1. For the resistor shown in Figure 18–76:
   a. Find the sinusoidal current i using phasors.
   b. Sketch the sinusoidal waveforms for v and i.
   c. Sketch the phasor diagram for **V** and **I**.
2. Repeat Problem 1 for the resistor of Figure 18–77.
3. Repeat Problem 1 for the resistor of Figure 18–78.
4. Repeat Problem 1 for the resistor of Figure 18–79.

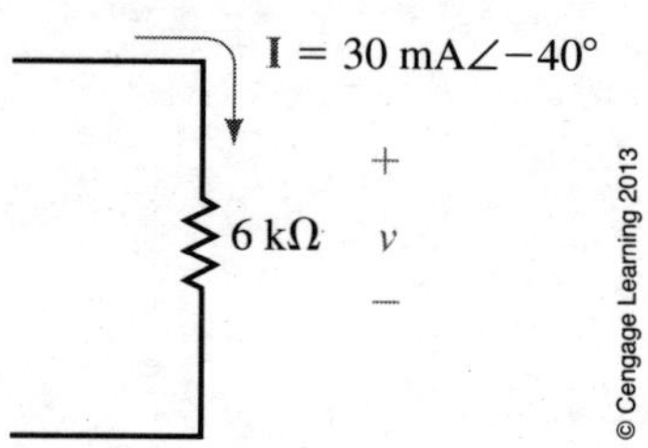

FIGURE 18–77

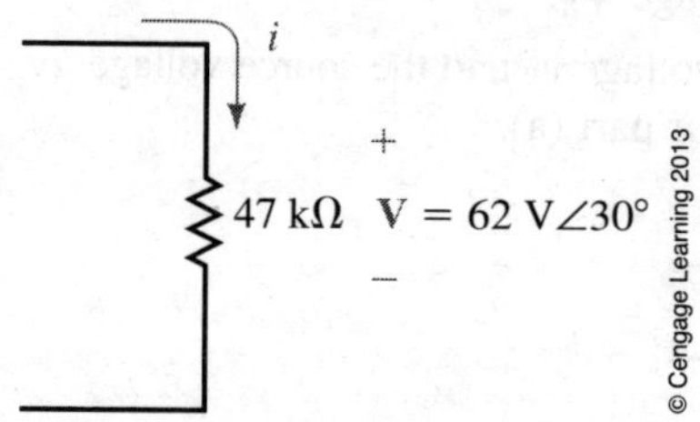

FIGURE 18–78

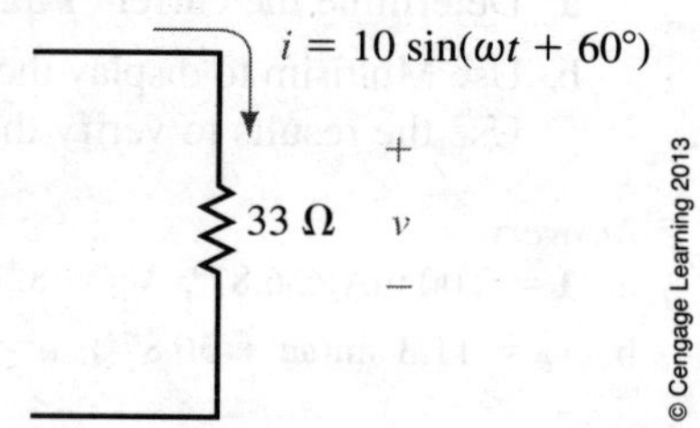

FIGURE 18–79

5. For the component shown in Figure 18–80:
   a. Find the sinusoidal voltage v using phasors.
   b. Sketch the sinusoidal waveforms for v and i.
   c. Sketch the phasor diagram for **V** and **I**.
6. For the component shown in Figure 18–81:
   a. Find the sinusoidal current i using phasors.
   b. Sketch the sinusoidal voltages for v and i.
   c. Sketch the phasor diagram for **V** and **I**.
7. For the component shown in Figure 18–82:
   a. Find the sinusoidal voltage v using phasors.
   b. Sketch the sinusoidal voltages for v and i.
   c. Sketch the phasor diagram for **V** and **I**.
8. For the component shown in Figure 18–83:
   a. Find the sinusoidal current i using phasors.
   b. Sketch the sinusoidal voltages for v and i.
   c. Sketch the phasor diagram for **V** and **I**.

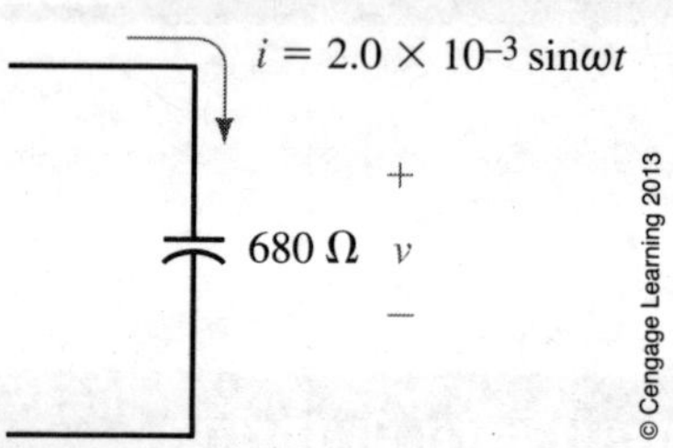

FIGURE 18–80

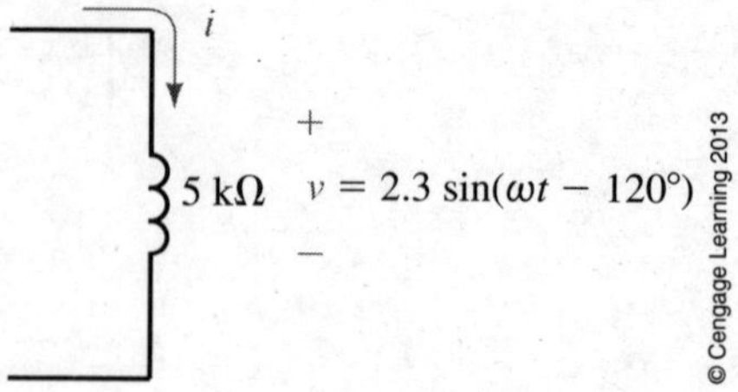

FIGURE 18–81

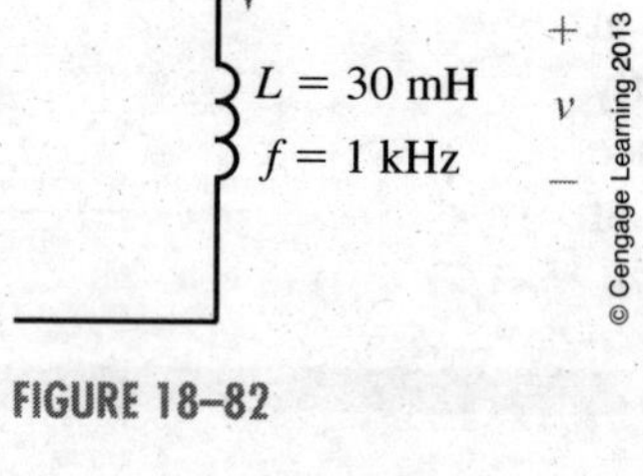

FIGURE 18–82

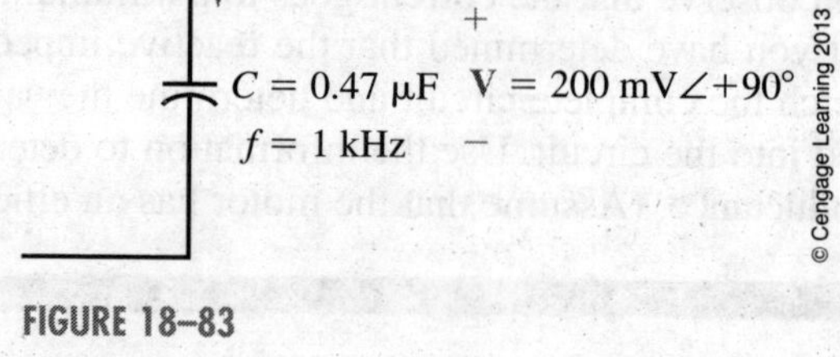

FIGURE 18–83

9. For the component shown in Figure 18–84:
   a. Find the sinusoidal voltage v using phasors.
   b. Sketch the sinusoidal voltages for v and i.
   c. Sketch the phasor diagram for **V** and **I**.
10. Repeat Problem 5 for the component shown in Figure 18–85.

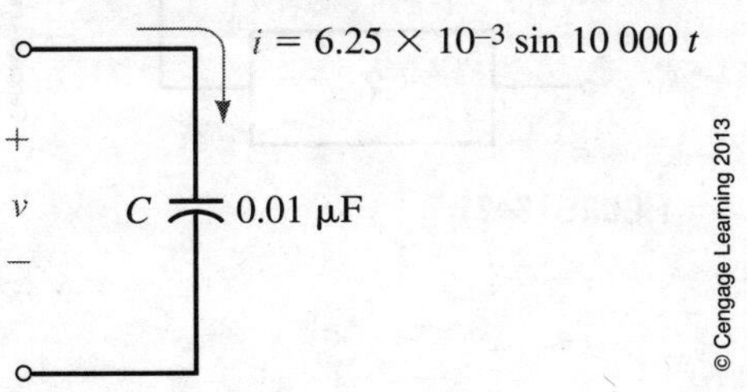

FIGURE 18–84

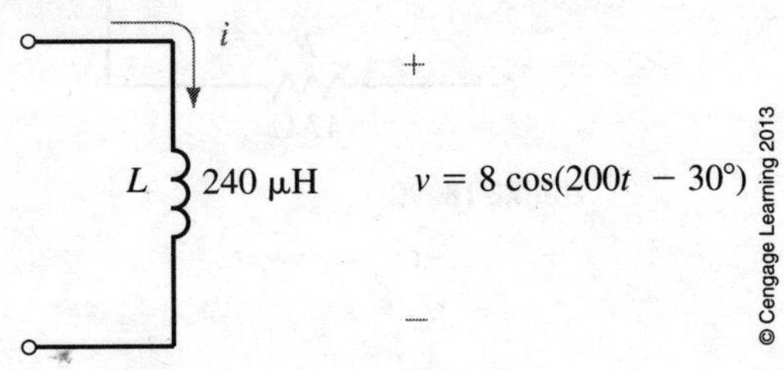

FIGURE 18–85

11. Repeat Problem 5 for the component shown in Figure 18–86.
12. Repeat Problem 5 for the component shown in Figure 18–87.

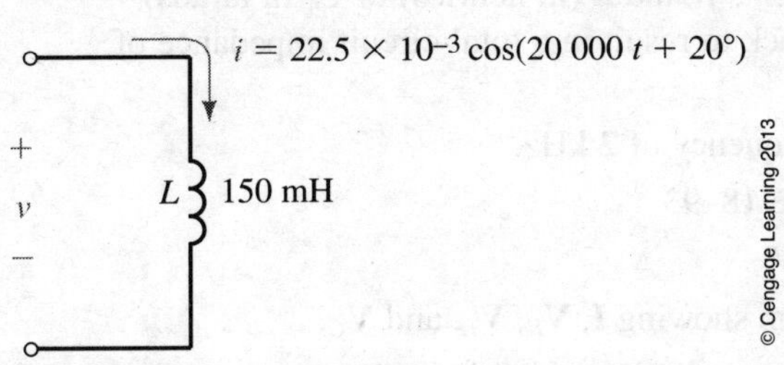

FIGURE 18–86

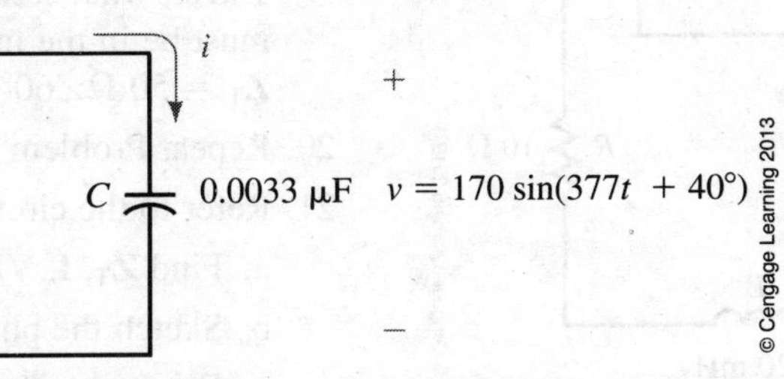

FIGURE 18–87

18.2 ac Series Circuits

13. Find the total impedance of each of the networks shown in Figure 18–88.
14. Repeat Problem 13 for the networks of Figure 18–89.

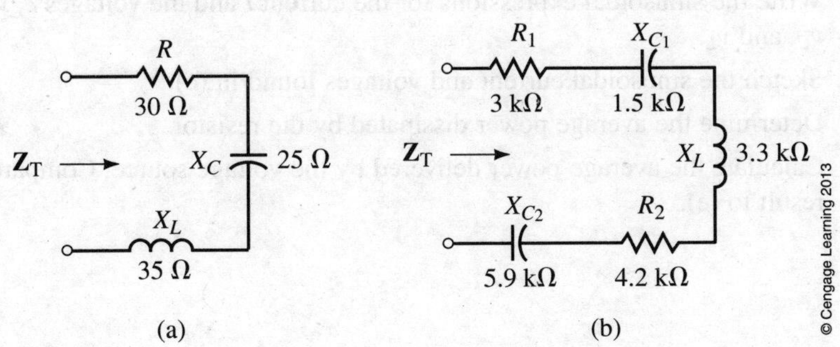

FIGURE 18–88

15. Refer to the network of Figure 18–90.
    a. Determine the series impedance **Z** that will result in the given total impedance, $\mathbf{Z}_T$. Express your answer in rectangular and polar forms.
    b. Sketch an impedance diagram showing $\mathbf{Z}_T$ and **Z**.
16. Repeat Problem 15 for the network of Figure 18–91.
17. A circuit consisting of two elements has a total impedance of $\mathbf{Z}_T = 2\ \text{k}\Omega\angle 15°$ at a frequency of 18 kHz. Determine the values in ohms, henries, or farads of the unknown elements.

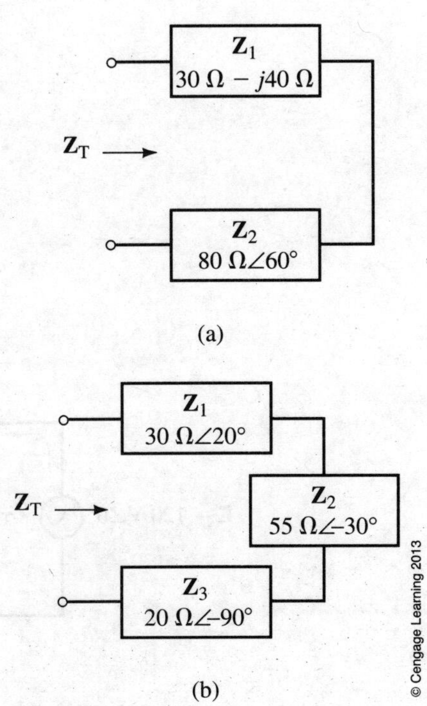

FIGURE 18–89

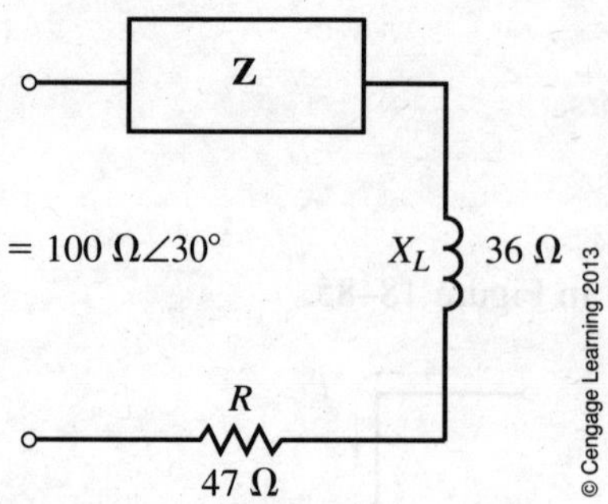

FIGURE 18–90

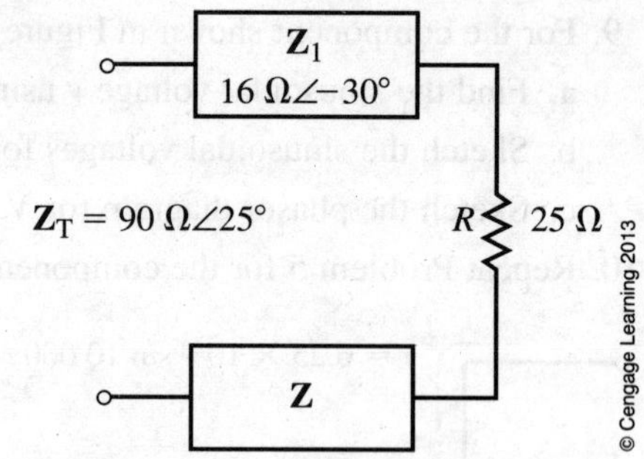

FIGURE 18–91

18. A network has a total impedance of $\mathbf{Z}_T = 24.0\ \text{k}\Omega\angle-30°$ at a frequency of 2 kHz. If the network consists of two series elements, determine the values in ohms, henries, or farads of the unknown elements.
19. Given that the network of Figure 18–92 is to operate at a frequency of 1 kHz, what series components R and L (in henries) or C (in farads) must be in the indicated block to result in a total circuit impedance of $\mathbf{Z}_T = 50\ \Omega\angle 60°$?
20. Repeat Problem 19 for a frequency of 2 kHz.
21. Refer to the circuit of Figure 18–93.
    a. Find $\mathbf{Z}_T$, $\mathbf{I}$, $\mathbf{V}_R$, $\mathbf{V}_L$, and $\mathbf{V}_C$.
    b. Sketch the phasor diagram showing $\mathbf{I}$, $\mathbf{V}_R$, $\mathbf{V}_L$, and $\mathbf{V}_C$.
    c. Determine the average power dissipated by the resistor.
    d. Calculate the average power delivered by the voltage source. Compare the result to (c).
22. Consider the circuit of Figure 18–94.
    a. Find $\mathbf{Z}_T$, $\mathbf{I}$, $\mathbf{V}_R$, $\mathbf{V}_L$, and $\mathbf{V}_C$.
    b. Sketch the phasor diagram showing $\mathbf{I}$, $\mathbf{V}_R$, $\mathbf{V}_L$, and $\mathbf{V}_C$.
    c. Write the sinusoidal expressions for the current i and the voltages e, v_R, v_C, and v_L.
    d. Sketch the sinusoidal current and voltages found in (c).
    e. Determine the average power dissipated by the resistor.
    f. Calculate the average power delivered by the voltage source. Compare the result to (e).

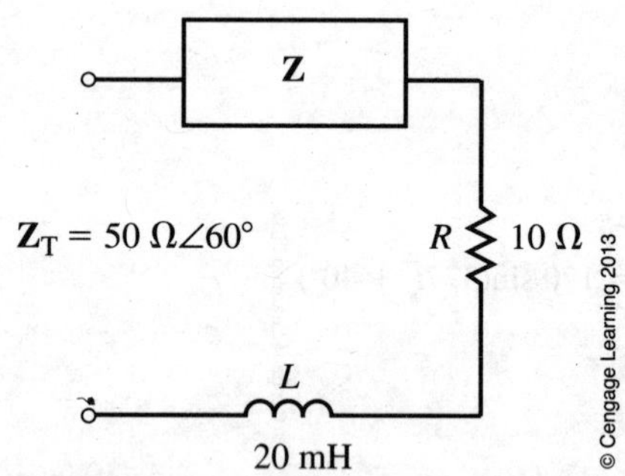

FIGURE 18–92

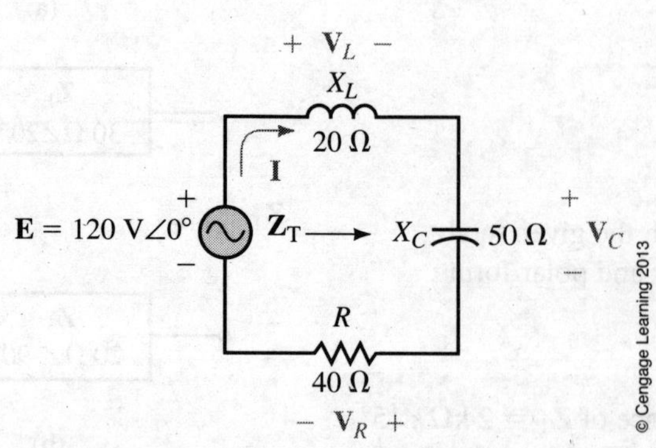

FIGURE 18–93

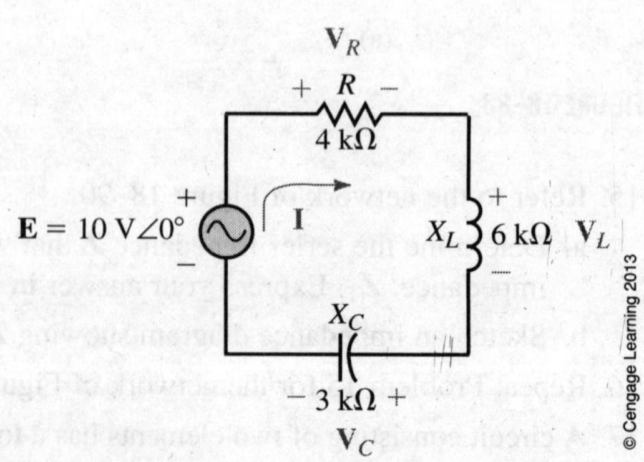

FIGURE 18–94

23. Refer to the circuit of Figure 18–95.

a. Determine the circuit impedance, $\mathbf{Z}_T$.

b. Use phasors to solve for i, v_R, v_C, and v_L.

c. Sketch the phasor diagram showing $\mathbf{I}$, $\mathbf{V}_R$, $\mathbf{V}_L$, and $\mathbf{V}_C$.

d. Sketch the sinusoidal expressions for the current and voltages found in (b).

e. Determine the average power dissipated by the resistor.

f. Calculate the average power delivered by the voltage source. Compare the result to (e).

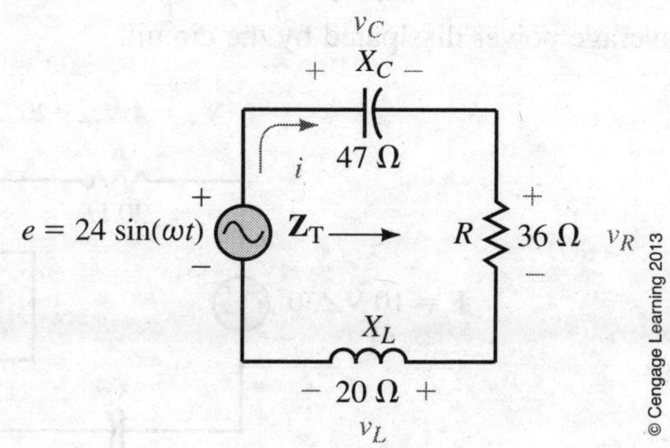

FIGURE 18–95

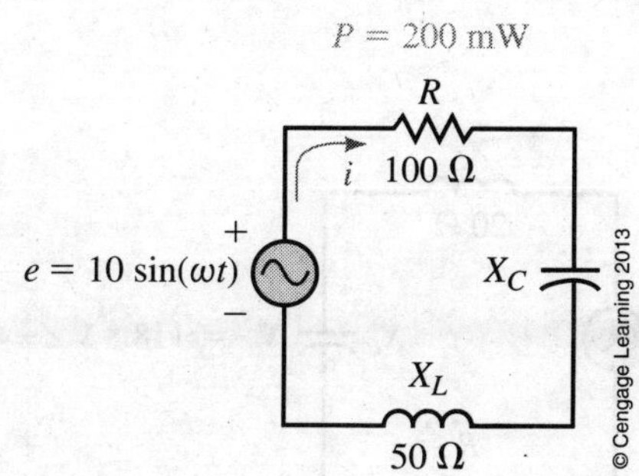

FIGURE 18–96

24. Refer to the circuit of Figure 18–96.

a. Determine the value of the capacitor reactance, X_C, needed so that the resistor in the circuit dissipates a power of 200 mW.

b. Using the value of X_C from (a), determine the sinusoidal expression for the current i in the circuit.

18.3 Kirchhoff's Voltage Law and the Voltage Divider Rule

25. a. Suppose a voltage of $10\text{ V}\angle 0°$ is applied across the network in Figure 18–88(a). Use the voltage divider rule to find the voltage appearing across each element.

b. Verify Kirchhoff's voltage law for the network.

26. a. Suppose a voltage of $240\text{ V}\angle 30°$ is applied across the network in Figure 18–89(a). Use the voltage divider rule to find the voltage appearing across each impedance.

b. Verify Kirchhoff's voltage law for each network.

27. Given the circuit of Figure 18–97:

a. Find the voltages $\mathbf{V}_C$ and $\mathbf{V}_L$.

b. Determine the value of R.

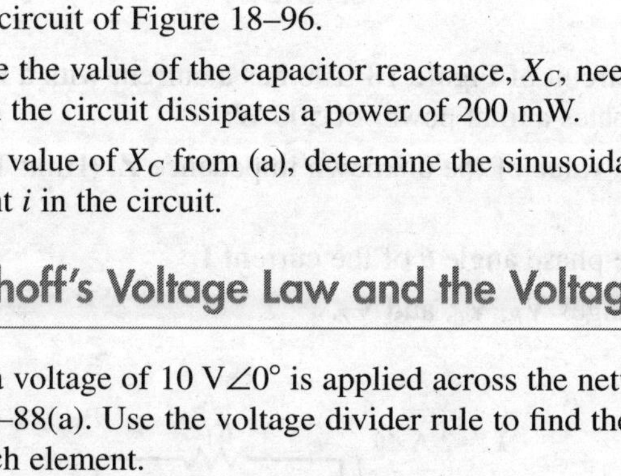

FIGURE 18–97

28. Refer to the circuit of Figure 18–98.
    a. Find the voltages $\mathbf{V}_R$ and $\mathbf{V}_L$.
    b. Determine the value of X_C.
29. Refer to the circuit of Figure 18–99:
    a. Find the voltage across X_C.
    b. Use Kirchhoff's voltage law to find the voltage across the unknown impedance.
    c. Calculate the value of the unknown impedance **Z**.
    d. Determine the average power dissipated by the circuit.

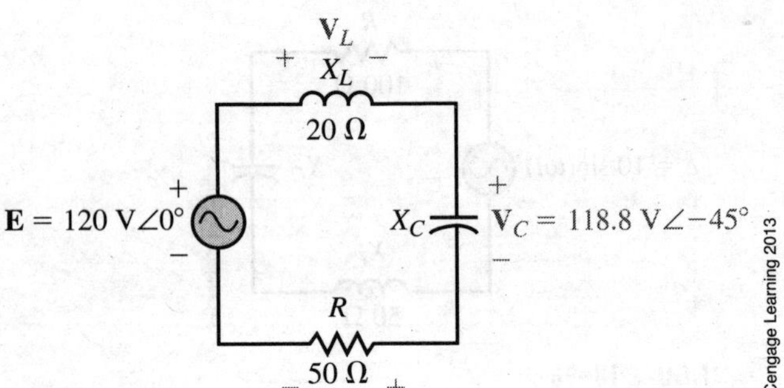

FIGURE 18–98

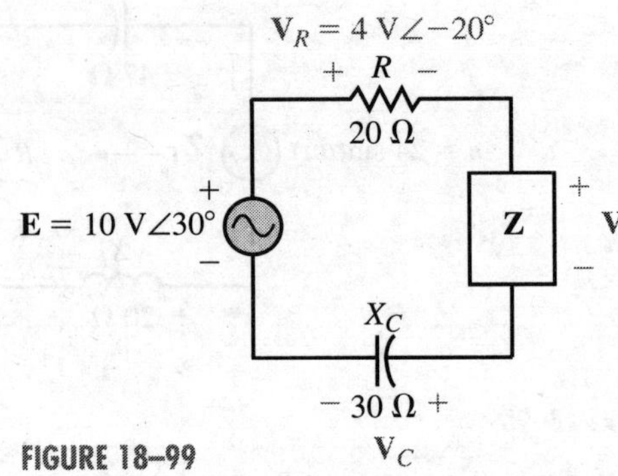

FIGURE 18–99

30. Given that the circuit of Figure 18–100 has a current with a magnitude of 2.0 A and dissipates a total power of 500 W:
    a. Calculate the value of the unknown impedance **Z**. (Hint: Two solutions are possible.)
    b. Calculate the phase angle θ of the current **I**.
    c. Find the voltages $\mathbf{V}_R$, $\mathbf{V}_L$, and $\mathbf{V}_Z$.

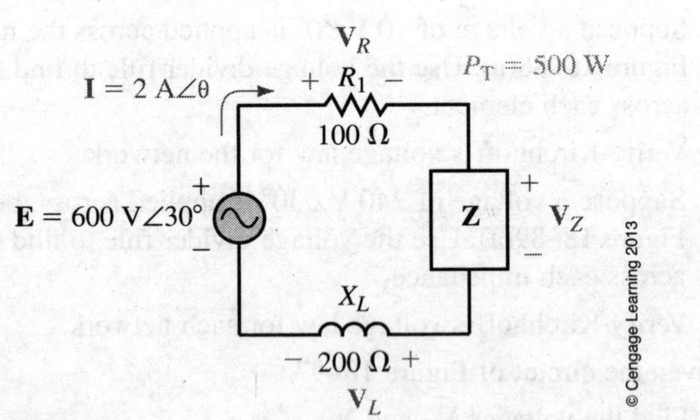

FIGURE 18–100

18.4 ac Parallel Circuits

31. Determine the input impedance, $\mathbf{Z}_T$, for each of the networks of Figure 18–101.
32. Repeat Problem 31 for Figure 18–102.
33. Given the circuit of Figure 18–103.
    a. Find $\mathbf{Z}_T$, $\mathbf{I}_T$, $\mathbf{I}_1$, $\mathbf{I}_2$, and $\mathbf{I}_3$.
    b. Sketch the admittance diagram showing each of the admittances.
    c. Sketch the phasor diagram showing **E**, $\mathbf{I}_T$, $\mathbf{I}_1$, $\mathbf{I}_2$, and $\mathbf{I}_3$.
    d. Determine the average power dissipated by the resistor.
    e. Find the power factor of the circuit and calculate the average power delivered by the voltage source. Compare the answer to the result obtained in (d).

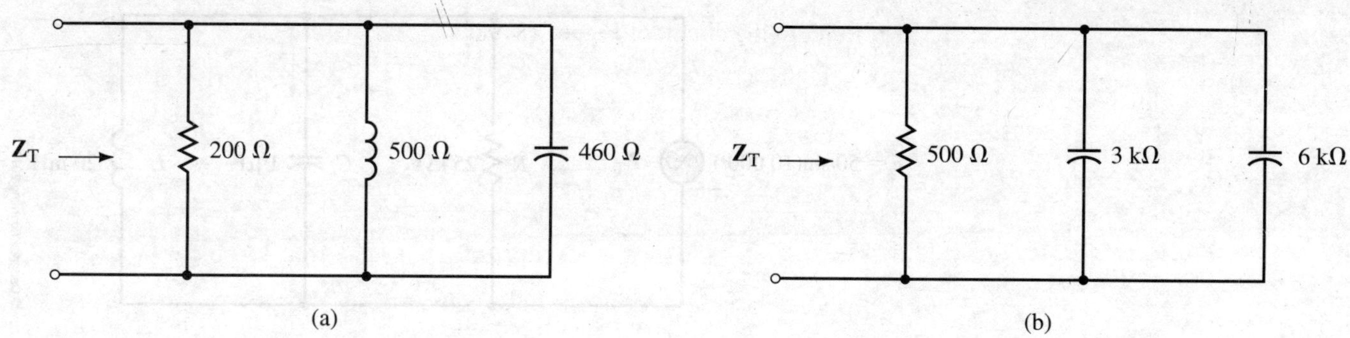

FIGURE 18–101

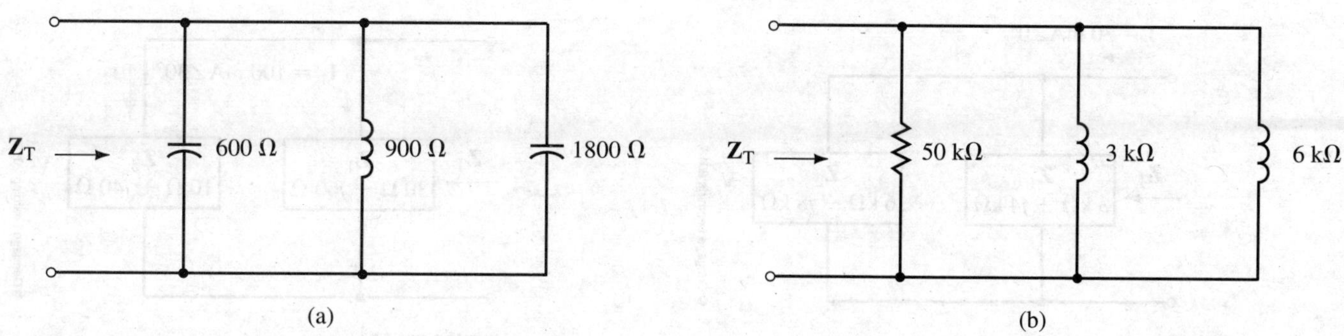

FIGURE 18–102

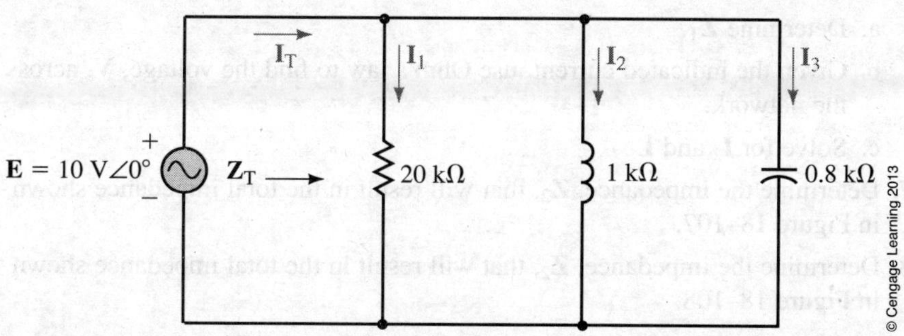

FIGURE 18–103

34. Refer to the circuit of Figure 18–104.
    a. Find $\mathbf{Z}_T$, $\mathbf{I}_T$, $\mathbf{I}_1$, $\mathbf{I}_2$, and $\mathbf{I}_3$.
    b. Sketch the admittance diagram for each of the admittances.
    c. Sketch the phasor diagram showing $\mathbf{E}$, $\mathbf{I}_T$, $\mathbf{I}_1$, $\mathbf{I}_2$, and $\mathbf{I}_3$.
    d. Determine the expressions for the sinusoidal currents i_T, i_1, i_2, and i_3.
    e. Sketch the sinusoidal voltage e and current i_T.
    f. Determine the average power dissipated by the resistor.
    g. Find the power factor of the circuit and calculate the average power delivered by the voltage source. Compare the answer to the result obtained in (f).
35. Refer to the network of Figure 18–105.
    a. Determine $\mathbf{Z}_T$.
    b. Given the indicated current, use Ohm's law to find the voltage, $\mathbf{V}$, across the network.

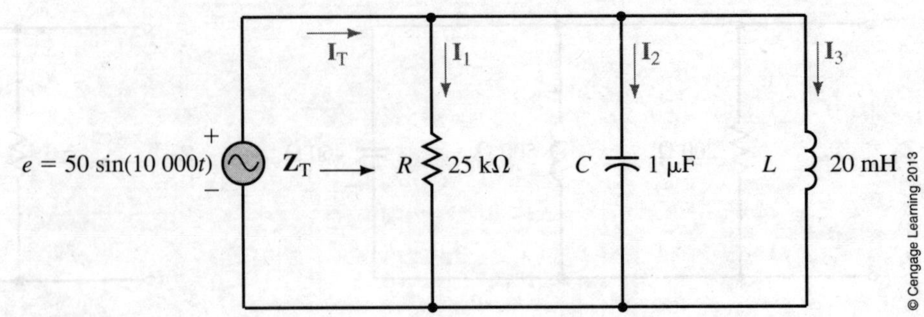

FIGURE 18–104

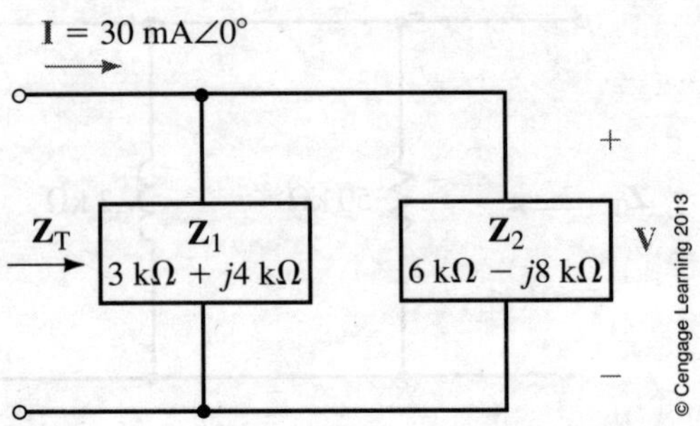

FIGURE 18–105

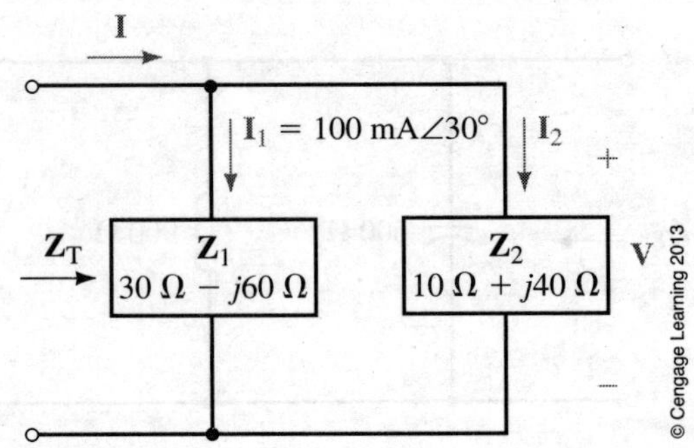

FIGURE 18–106

36. Consider the network of Figure 18–106.
    a. Determine $\mathbf{Z}_T$.
    b. Given the indicated current, use Ohm's law to find the voltage, **V**, across the network.
    c. Solve for $\mathbf{I}_2$ and **I**.
37. Determine the impedance, $\mathbf{Z}_2$, that will result in the total impedance shown in Figure 18–107.
38. Determine the impedance, $\mathbf{Z}_2$, that will result in the total impedance shown in Figure 18–108.

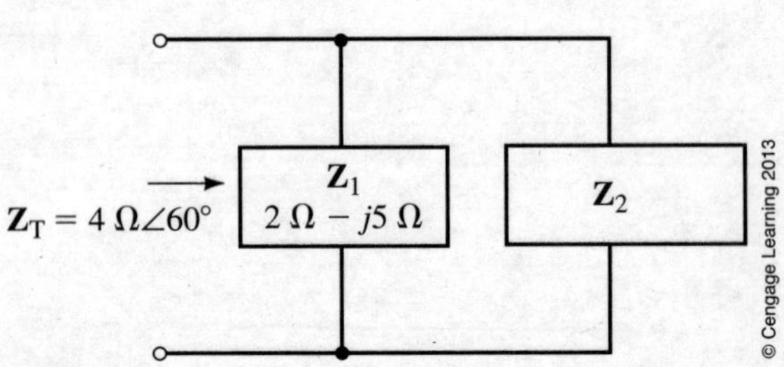

FIGURE 18–107

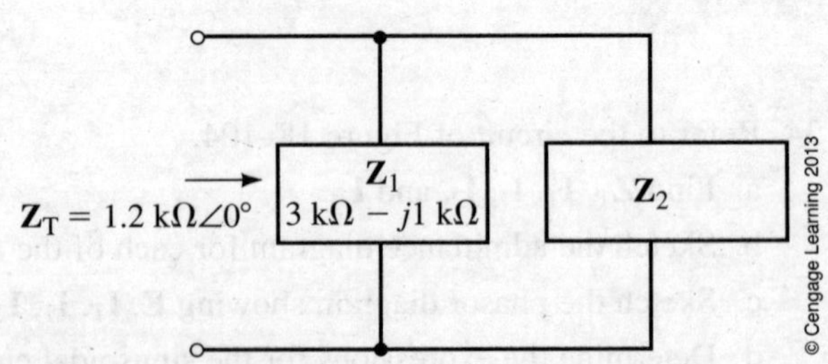

FIGURE 18–108

18.5 Kirchhoff's Current Law and the Current Divider Rule

39. Solve for the current in each element of the networks in Figure 18–101 if the current applied to each network is $10\text{ mA}\angle -30°$.
40. Repeat Problem 39 for Figure 18–102.
41. Use the current divider rule to find the current in each of the elements in Figure 18–109. Verify that Kirchhoff's current law applies.

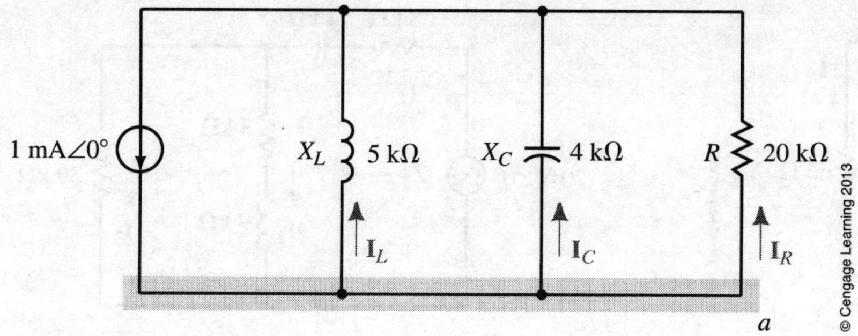

FIGURE 18–109

42. Given that $\mathbf{I}_L = 4\ \text{A}\angle 30°$ in the circuit of Figure 18–110, find the currents **I**, $\mathbf{I}_C$, and $\mathbf{I}_R$. Verify that Kirchhoff's current law applies for this circuit.
43. Suppose that the circuit of Figure 18–111 has a current **I** with a magnitude of 8 A:
    a. Determine the current $\mathbf{I}_R$ through the resistor.
    b. Calculate the value of resistance, *R*.
    c. What is the phase angle of the current **I**?

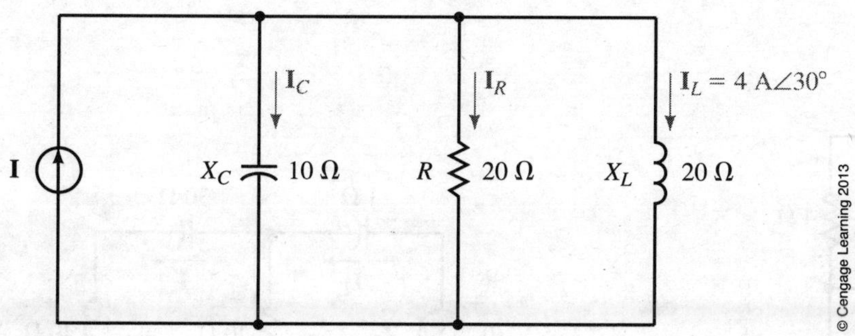

FIGURE 18–110

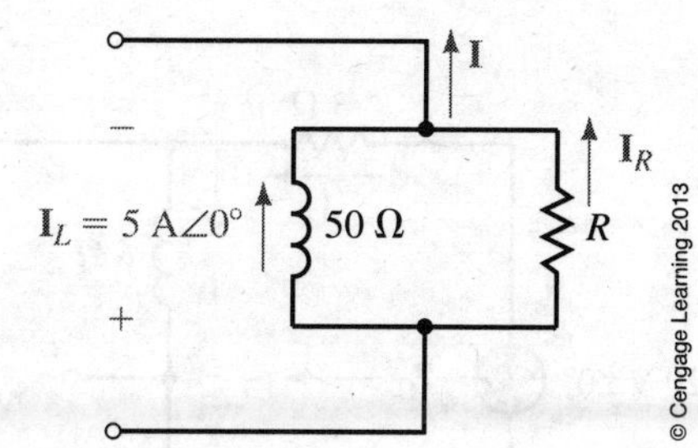

FIGURE 18–111

44. Assume that the circuit of Figure 18–112 has a current **I** with a magnitude of 3 A:
    a. Determine the current $\mathbf{I}_R$ through the resistor.
    b. Calculate the value of capacitive reactance X_C.
    c. What is the phase angle of the current **I**?

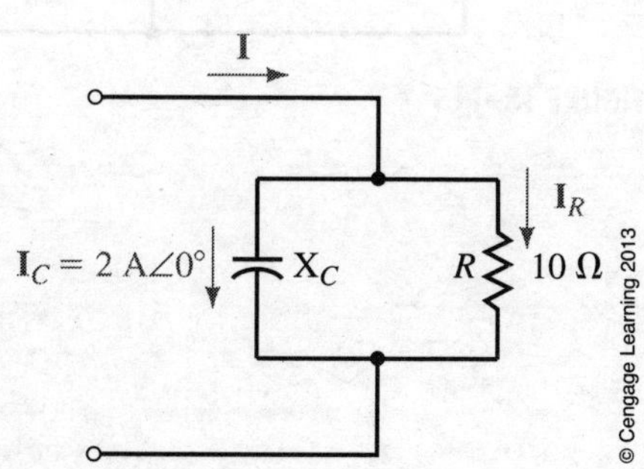

FIGURE 18–112

18.6 Series-Parallel Circuits

45. Refer to the circuit of Figure 18–113.
    a. Find $\mathbf{Z}_T$, $\mathbf{I}_L$, $\mathbf{I}_C$, and $\mathbf{I}_R$.
    b. Sketch the phasor diagram showing **E**, $\mathbf{I}_L$, $\mathbf{I}_C$, and $\mathbf{I}_R$.
    c. Calculate the average power dissipated by the resistor.
    d. Use the circuit power factor to calculate the average power delivered by the voltage source. Compare the answer to the results obtained in (c).
46. Refer to the circuit of Figure 18–114.
    a. Find $\mathbf{Z}_T$, $\mathbf{I}_1$, $\mathbf{I}_2$, and $\mathbf{I}_3$.
    b. Sketch the phasor diagram showing **E**, $\mathbf{I}_1$, $\mathbf{I}_2$, and $\mathbf{I}_3$.
    c. Calculate the average power dissipated by each of the resistors.
    d. Use the circuit power factor to calculate the average power delivered by the voltage source. Compare the answer to the results obtained in (c).

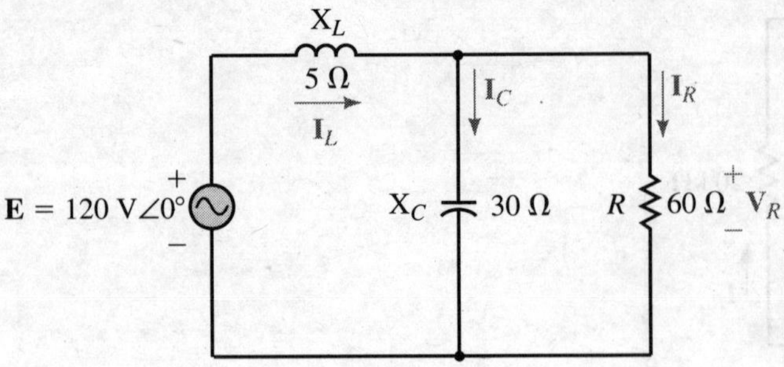

FIGURE 18–113

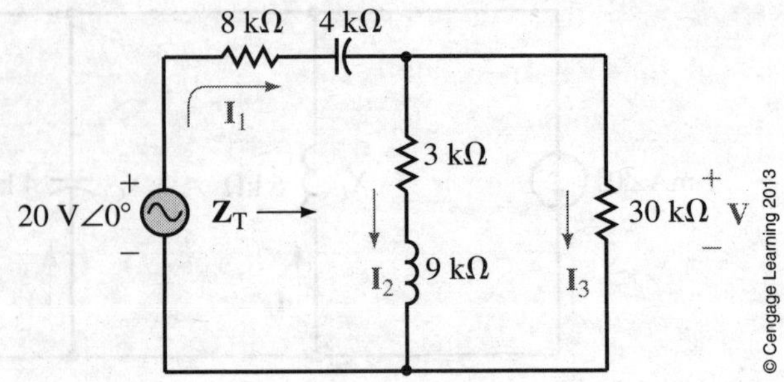

FIGURE 18–114

47. Refer to the circuit of Figure 18–115.
    a. Find $\mathbf{Z}_T$, $\mathbf{I}_T$, $\mathbf{I}_1$, and $\mathbf{I}_2$.
    b. Determine the voltage $\mathbf{V}_{ab}$.
48. Consider the circuit of Figure 18–116.
    a. Find $\mathbf{Z}_T$, $\mathbf{I}_T$, $\mathbf{I}_1$, and $\mathbf{I}_2$.
    b. Determine the voltage **V**.

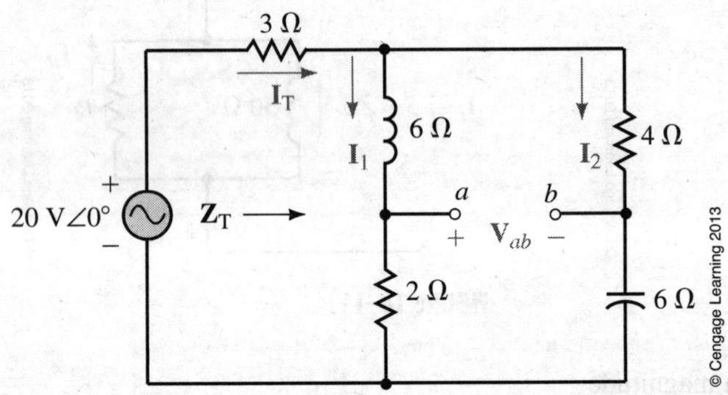

FIGURE 18–115

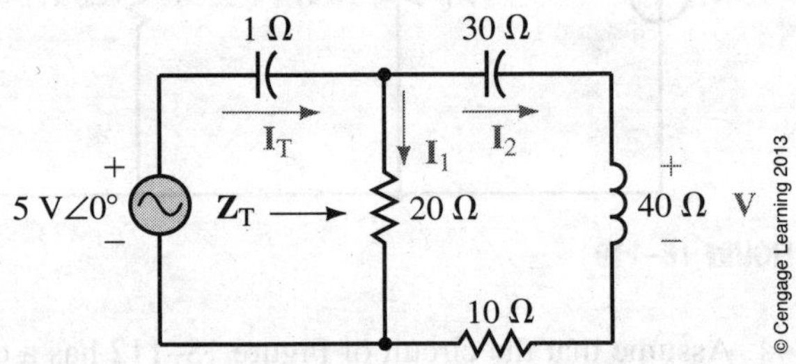

FIGURE 18–116

49. Refer to the circuit of Figure 18–117:
    a. Find $\mathbf{Z}_T$, $\mathbf{I}_1$, $\mathbf{I}_2$, and $\mathbf{I}_3$.
    b. Determine the voltage **V**.

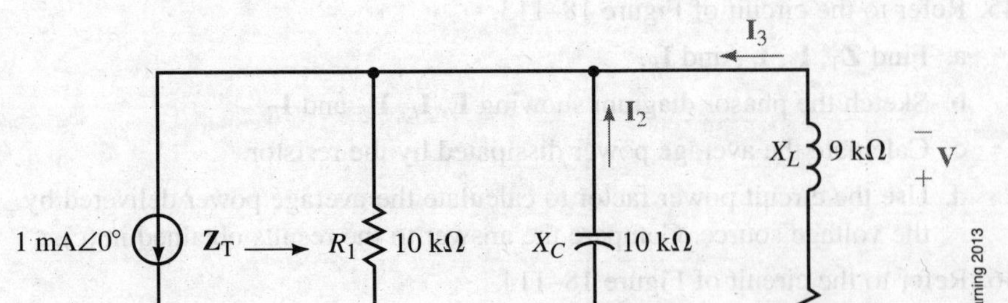

FIGURE 18–117

50. Refer to the circuit of Figure 18–118:
   a. Find $\mathbf{Z}_T$, $\mathbf{I}_1$, $\mathbf{I}_2$, and $\mathbf{I}_3$.
   b. Determine the voltage **V**.

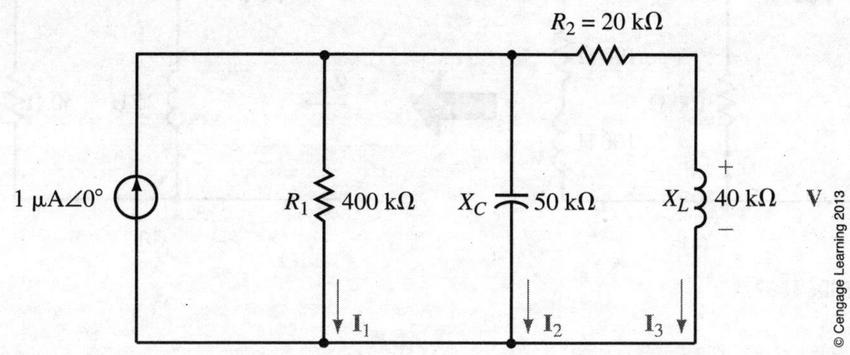

FIGURE 18–118

18.7 Frequency Effects

51. A 50-kΩ resistor is placed in series with a 0.01-μF capacitor. Determine the cutoff frequency ω_C (in rad/s) and sketch the frequency response (Z_T vs. ω) of the network.
52. A 2-mH inductor is placed in parallel with a 2-kΩ resistor. Determine the cutoff frequency ω_C (in rad/s) and sketch the frequency response (Z_T vs. ω) of the network.
53. A 100-kΩ resistor is placed in parallel with a 0.47-μF capacitor. Determine the cutoff frequency f_C (in Hz) and sketch the frequency response (Z_T vs. f) of the network.
54. A 2.7-kΩ resistor is placed in parallel with a 20-mH inductor. Determine the cutoff frequency f_C (in Hz) and sketch the frequency response (Z_T vs. f) of the network.

18.8 Applications

55. Convert each of the networks of Figure 18–119 into an equivalent series network consisting of two elements.
56. Convert each of the networks of Figure 18–119 into an equivalent parallel network consisting of two elements.
57. Show that the networks of Figure 18–120 have the same input impedance at frequencies of 1 krad/s and 10 krad/s. (It can be shown that these networks are equivalent at all frequencies.)

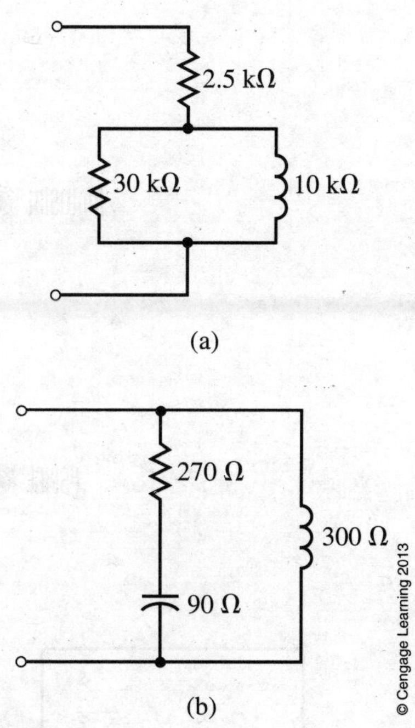

FIGURE 18–119

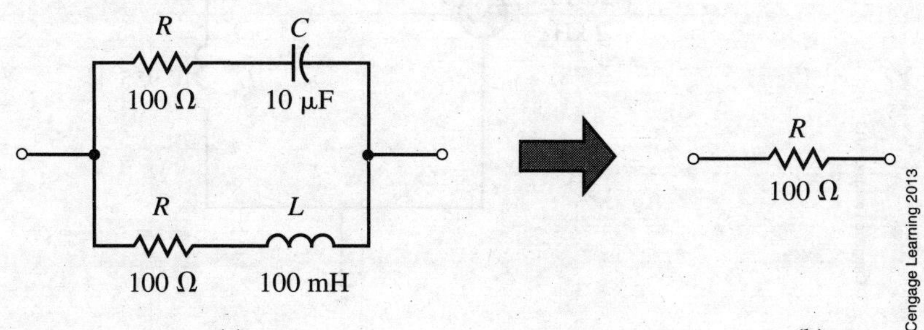

FIGURE 18–120

58. Show that the networks of Figure 18–121 have the same input impedance at frequencies of 5 rad/s and 10 rad/s.

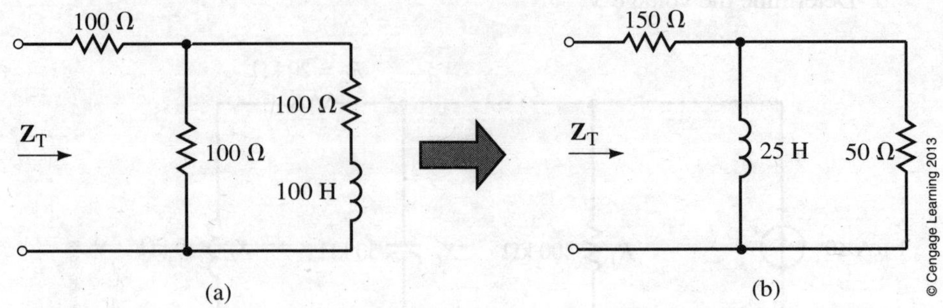

FIGURE 18–121

18.9 Circuit Analysis Using Computers

Multisim

59. Given the circuit of Figure 18–122:
    a. Use Multisim to simultaneously display the capacitor voltage v_C and e. Record your results and determine the phasor voltage $\mathbf{V}_C$.
    b. Interchange the positions of the resistor and the capacitor relative to the ground. Use Multisim to simultaneously display the resistor voltage v_R and e. Record your results and determine the phasor voltage $\mathbf{V}_R$.
    c. Compare your results to those obtained in Example 18–8.

Multisim

60. Given the circuit of Figure 18–123:
    a. Use Multisim to display the inductor voltage v_L and e. Record your results and determine the phasor voltage $\mathbf{V}_L$.
    b. Interchange the positions of the resistor and the inductor relative to the ground. Use Multisim to display the inductor voltage v_R and e. Record your results and determine the phasor voltage $\mathbf{V}_R$.
    c. Compare your results to those obtained in Example 18–9.

PSPICE

61. A 50-kΩ resistor is placed in series with a 0.01-μF capacitor. Use PSpice to input these components into a circuit. Run the Probe postprocessor to provide a graphical display of the network impedance as a function of frequency from 50 Hz to 500 Hz. Let the frequency sweep logarithmically in octaves.

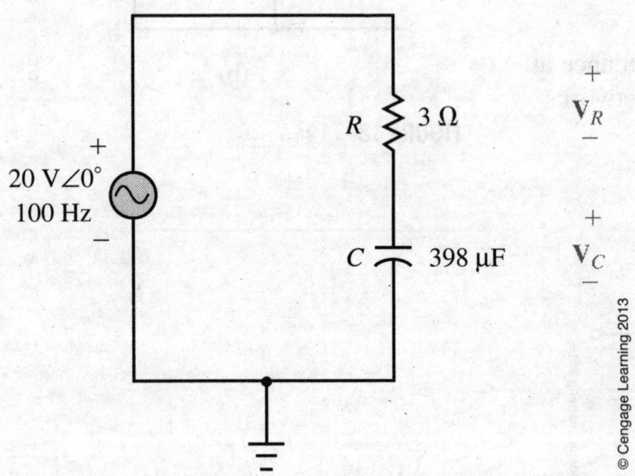

FIGURE 18–122

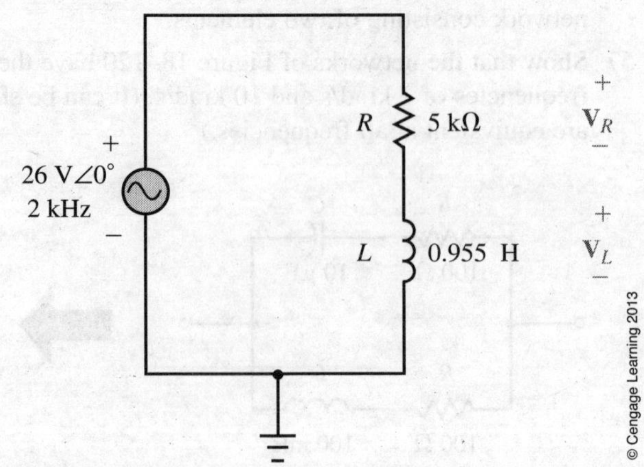

FIGURE 18–123

62. A 2-mH inductor is placed in parallel with a 2-kΩ resistor. Use PSpice to input these components into a circuit. Run the Probe postprocessor to provide a graphical display of the network impedance as a function of frequency from 50 kHz to 500 kHz. Let the frequency sweep logarithmically in octaves. PSPICE

63. A 100-kΩ resistor is placed in parallel with a 0.47-μF capacitor. Use PSpice to input these components into a circuit. Run the Probe postprocessor to provide a graphical display of the network impedance as a function of frequency from 0.1 Hz to 10 Hz. Let the frequency sweep logarithmically in octaves. PSPICE

64. A 2.7-kΩ resistor is placed in parallel with a 20-mH inductor. Use PSpice to input these components into a circuit. Run the Probe postprocessor to provide a graphical display of the network impedance as a function of frequency from 100 kHz to 1 MHz. Let the frequency sweep logarithmically in octaves. PSPICE

ANSWERS TO IN-PROCESS LEARNING CHECKS

IN-PROCESS LEARNING CHECK 1

1. Current and voltage are in phase.
2. Current leads voltage by 90°.
3. Voltage leads current by 90°.

IN-PROCESS LEARNING CHECK 2

1. The phasor sum of voltage drops and rises around a closed loop is equal to zero.
2. All voltages must be expressed as phasors, that is, $\mathbf{V} = V\angle\theta = V\cos\theta + jV\sin\theta$.

IN-PROCESS LEARNING CHECK 3

1. The phasor sum of currents entering a node is equal to the phasor sum of currents leaving the same node.
2. All currents must be expressed as phasors, that is, $\mathbf{I} = I\angle\theta = I\cos\theta + jI\sin\theta$.

IN-PROCESS LEARNING CHECK 4

1. At $f = 0$ Hz, $Z = \infty$ (open circuit). As $f \rightarrow \infty$, $Z = R$.
2. At $f = 0$ Hz, $Z = 0$ (short circuit). As $f \rightarrow \infty$, $Z = R$.

IN-PROCESS LEARNING CHECK 5

a. $R_P = 795\ \Omega$, $L_P = 10.1$ mH
b. $R_P = 790\ \text{M}\Omega$, $L_P = 10.0$ mH
c. As f is increased, R_P gets larger and L_P remains constant at 10 mH.

■ KEY TERMS

Balanced Bridges
Dependent Sources
Hay Bridge
Independent Sources
Maxwell Bridge
Schering Bridge

■ OUTLINE

Dependent Sources
Source Conversion
Mesh (Loop) Analysis
Nodal Analysis
Delta-to-Wye and Wye-to-Delta Conversions
Bridge Networks
Circuit Analysis Using Computers

■ OBJECTIVES

After studying this chapter, you will be able to

- convert an ac voltage source into its equivalent current source, and conversely, convert a current source into an equivalent voltage source,
- solve for the current or voltage in a circuit having either a dependent current source or a dependent voltage source,
- set up simultaneous linear equations to solve an ac circuit using mesh analysis,
- use complex determinants to find the solutions for a given set of linear equations,
- set up simultaneous linear equations to solve an ac circuit using nodal analysis,
- perform delta-to-wye and wye-to-delta conversions for circuits having reactive elements,
- solve for the balanced condition in a given ac bridge circuit. In particular, you will examine the Maxwell, Hay, and Schering bridges,
- use Multisim to analyze bridge circuits,
- use PSpice to calculate current and voltage in an ac circuit.

19

METHODS OF ac ANALYSIS

CHAPTER PREVIEW

To this point, we have examined only circuits having a single ac source. In this chapter, we continue our study by analyzing multisource circuits and bridge networks. You will find that most of the techniques used in analyzing ac circuits parallel those of dc circuit analysis. Consequently, a review of Chapter 8 will help you understand the topics of this chapter.

Near the end of the chapter, we examine how computer techniques are used to analyze even the most complex ac circuits. It must be emphasized that although computer techniques are much simpler than using a pencil and calculator, there is virtually no knowledge to be gained by mindlessly entering data into a computer. We use the computer merely as a tool to verify our results and to provide a greater dimension to the analysis of circuits. ■

Putting It in Perspective

Hermann Ludwig Ferdinand von Helmholtz

© CORBIS

HERMANN HELMHOLTZ WAS BORN IN POTSDAM (near Berlin, Germany) on August 31, 1821. Helmholtz was a leading scientist of the nineteenth century, whose legacy includes contributions in the fields of acoustics, chemistry, mathematics, magnetism, electricity, mechanics, optics, and physiology.

Helmholtz graduated from the Medical Institute in Berlin in 1843 and practiced medicine for five years as a surgeon in the Prussian army. From 1849 to 1871, he served as a professor of physiology at universities in Königsberg, Bonn, and Heidelberg. In 1871, Helmholtz was appointed Professor of Physics at the University of Berlin.

Helmholtz's greatest contributions were as a mathematical physicist, where his work in theoretical and practical physics led to the proof of the Law of Conservation of Energy in his paper "*Über die Erhaltung der Kraft,*" published in 1847. He showed that mechanics, heat, light, electricity, and magnetism were simply manifestations of the same force. His work led to the understanding of electrodynamics (the motion of charge in conductors), and his theory of the electromagnetic properties of light set the groundwork for later scientists to understand how radio waves are propagated.

For his work, the German emperor Kaiser Wilhelm I made Helmholtz a noble in 1883. This great scientist died on September 8, 1894, at the age of 73. ■

19.1 Dependent Sources

The voltage and current sources we have worked with up to now have been **independent sources,** meaning that the voltage or current of the supply was not in any way dependent upon any voltage or current elsewhere in the circuit. In many amplifier circuits, particularly those involving transistors, it is possible to explain the operation of the circuits by replacing the device with an equivalent electronic model. These models often use voltage and current sources which have values dependent upon some internal voltage or current. Such supplies are called **dependent sources.** Figure 19–1 compares the symbols for both independent and dependent sources.

Although the diamond is the accepted symbol for representing dependent sources, many articles and textbooks still use a circle. In this textbook we use both forms of the dependent source to familiarize the student with the various notations. The dependent source has a magnitude and phase angle determined by voltage or current at some internal element multiplied by a constant, *k*. The magnitude of the constant is determined by parameters within the particular model, and the units of the constant correspond to the required quantities in the equation.

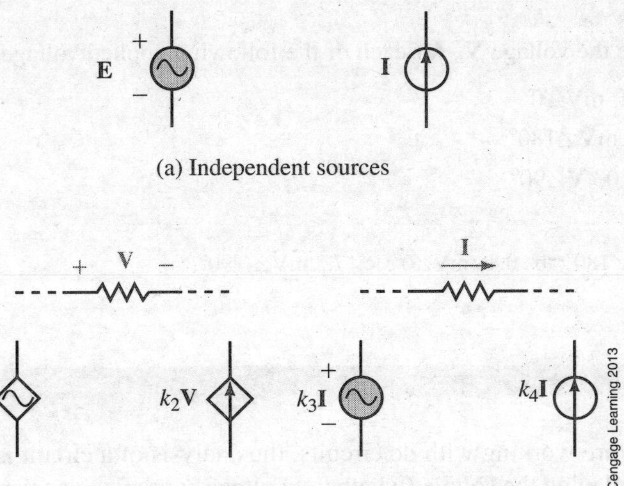

(a) Independent sources

(b) Dependent sources

FIGURE 19–1

EXAMPLE 19–1

Refer to the resistor shown in Figure 19–2. Determine the voltage $\mathbf{V}_R$ across the resistor given that the controlling voltage has the following values:

a. $\mathbf{V} = 0$ V

b. $\mathbf{V} = 5\text{ V}\angle 30°$

c. $\mathbf{V} = 3\text{ V}\angle -150°$

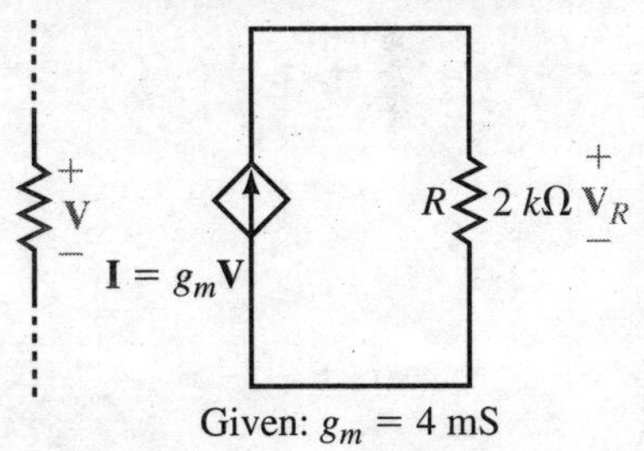

FIGURE 19–2

Solution Notice that the dependent source of this example has a constant, g_m, called the transconductance. Here, $g_m = 4$ mS.

a. $\mathbf{I} = (4\text{ mS})(0\text{ V}) = 0$
$\mathbf{V}_R = 0\text{ V}$

b. $\mathbf{I} = (4\text{ mS})(5\text{ V}\angle 30°) = 20\text{ mA}\angle 30°$
$\mathbf{V}_R = (20\text{ mA}\angle 30°)(2\text{ k}\Omega) = 40\text{ V}\angle 30°$

c. $\mathbf{I} = (4\text{ mS})(3\text{ V}\angle -150°) = 12\text{ mA}\angle -150°$
$\mathbf{V}_R = (12\text{ mA}\angle -150°)(2\text{ k}\Omega) = 24\text{ V}\angle -150°$

PRACTICE PROBLEMS 1

The circuit of Figure 19–3 represents a simplified model of a transistor amplifier.

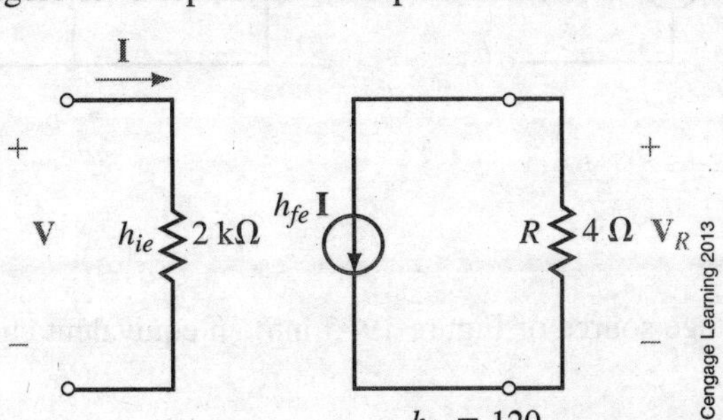

FIGURE 19–3

Determine the voltage $\mathbf{V}_R$ for each of the following applied voltages:

a. $\mathbf{V} = 10\ \text{mV}\angle 0°$

b. $\mathbf{V} = 2\ \text{mV}\angle 180°$

c. $\mathbf{V} = 0.03\ \text{V}\angle 90°$

Answers
a. 2.4 mV∠180°; b. 0.48 mV∠0°; c. 7.2 mV∠−90°

19.2 Source Conversion

When we are working with dc circuits, the analysis of a circuit is often simplified by replacing the source (whether a voltage source or a current source) with its equivalent. The conversion of any ac source is similar to the method used in dc circuit analysis.

A voltage source **E** in series with an impedance **Z** is equivalent to a current source **I** having the same impedance **Z** in parallel. Figure 19–4 shows the equivalent sources.

From Ohm's law, we perform the source conversion as follows:

$$\mathbf{I} = \frac{\mathbf{E}}{\mathbf{Z}}$$

and

$$\mathbf{E} = \mathbf{IZ}$$

It is important to realize that the two circuits of Figure 19–4 are equivalent between points *a* and *b*. This means that any network connected to points *a* and *b* will behave exactly the same regardless of which type of source is used. However, the voltages or currents within the sources will seldom be the same. In order to determine the current through or the voltage across the source impedance, the circuit must be returned to its original state.

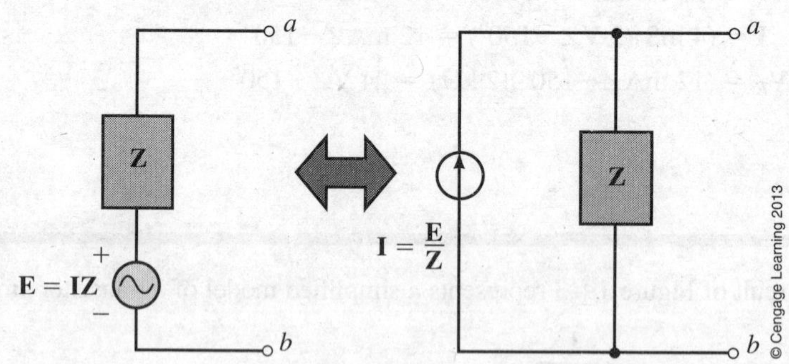

FIGURE 19–4

EXAMPLE 19–2

Convert the voltage source of Figure 19–5 into an equivalent current source.

Solution

$$\mathbf{Z}_T = 3\ \Omega + j4\ \Omega = 5\ \Omega\angle 53.13°$$

$$\mathbf{I} = \frac{10\ \text{V}\angle 0°}{5\ \Omega\angle 53.13°} = 2\ \text{A}\angle -53.13°$$

The equivalent current source is shown in Figure 19–6.

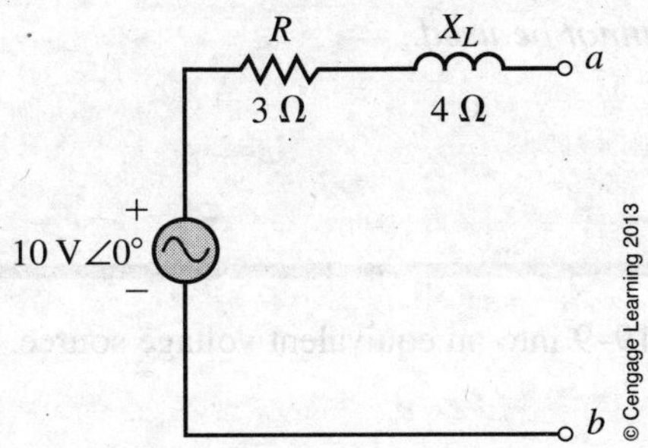

FIGURE 19–5

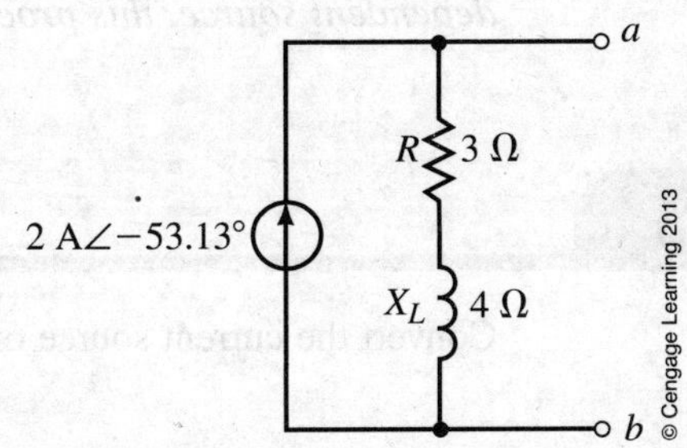

FIGURE 19–6

EXAMPLE 19–3

Convert the current source of Figure 19–7 into an equivalent voltage source.

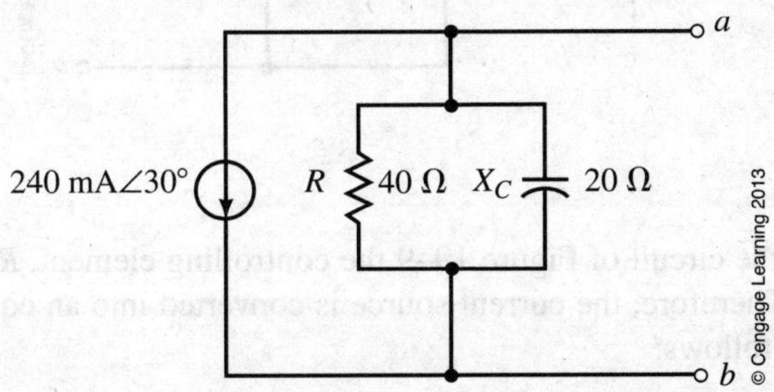

FIGURE 19–7

Solution The impedance of the parallel combination is determined to be

$$\mathbf{Z} = \frac{(40\ \Omega\angle 0°)(20\ \Omega\ \angle -90°)}{40\ \Omega - j20\ \Omega}$$

$$= \frac{800\ \Omega\ \angle -90°}{44.72\angle -26.57°}$$

$$= 17.89\ \Omega\angle -63.43° = 8\ \Omega - j16\ \Omega$$

and so

$$\mathbf{E} = (240\ \text{mA}\angle 30°)(17.89\ \Omega\angle -63.43°)$$

$$= 4.29\ \text{V}\angle -33.43°$$

The resulting equivalent circuit is shown in Figure 19–8.

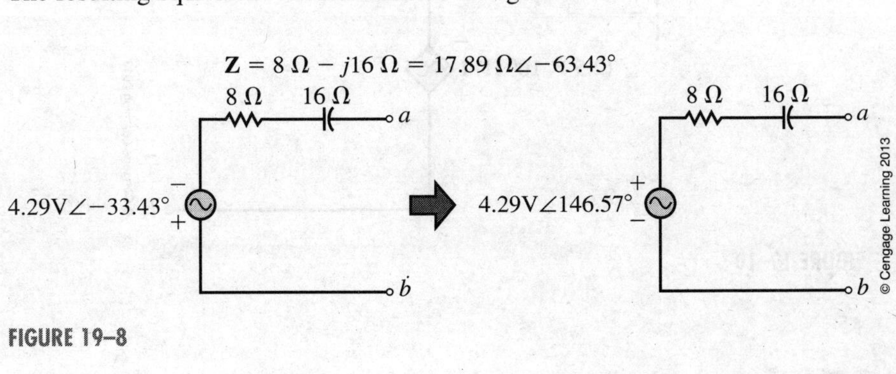

FIGURE 19–8

It is possible to use the same procedure to convert a dependent source into its equivalent provided that the controlling element is external to the circuit in which the source appears. *If the controlling element is in the same circuit as the dependent source, this procedure cannot be used.*

EXAMPLE 19–4

Convert the current source of Figure 19–9 into an equivalent voltage source.

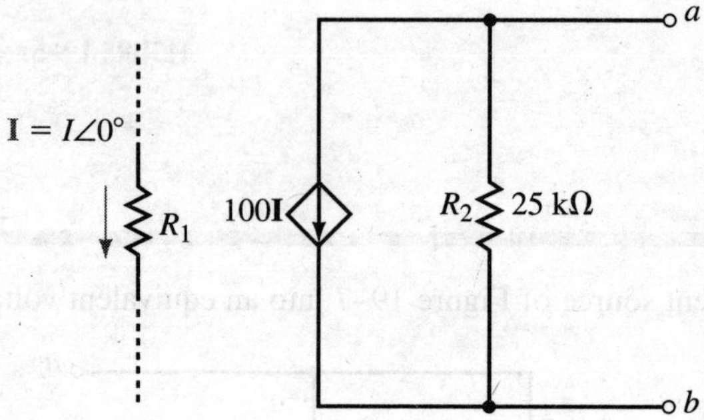

FIGURE 19–9

Solution In the circuit of Figure 19–9 the controlling element, R_1, is in a separate circuit. Therefore, the current source is converted into an equivalent voltage source as follows:

$$\mathbf{E} = (100\mathbf{I}_1)(\mathbf{Z})$$
$$= (100\mathbf{I}_1)(25\ \text{k}\Omega\angle 0^\circ)$$
$$= (2.5 \times 10^6\ \Omega)\mathbf{I}_1$$

The resulting voltage source is shown in Figure 19–10. Notice that the equivalent voltage source is dependent on the current, **I**, just as the original current source.

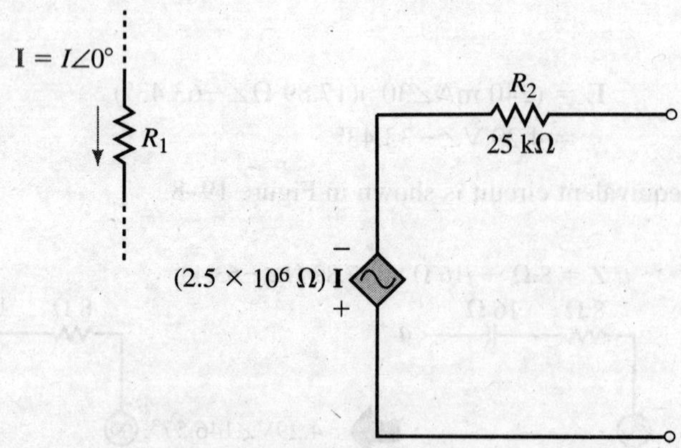

FIGURE 19–10

PRACTICE PROBLEMS 2

Convert the voltage sources of Figure 19–11 into equivalent current sources.

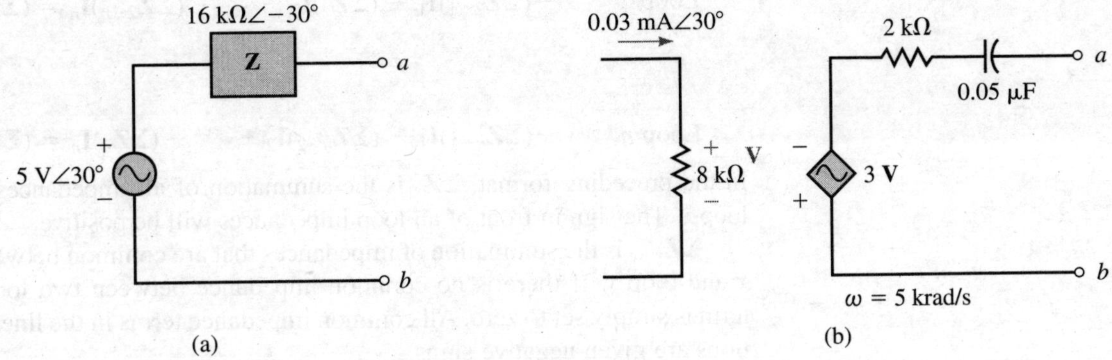

FIGURE 19–11

Answers

a. **I** = 0.3125 mA∠60° (from *b* to *a*) in parallel with **Z** = 16 kΩ∠−30°

b. **I** = 0.161 mA∠93.43° (from *a* to *b*) in parallel with **Z** = 2 kΩ − *j*4 kΩ

IN-PROCESS LEARNING CHECK 1

(Answers are at the end of the chapter.)

Given a 40-mA∠0° current source in parallel with an impedance, **Z**. Determine the equivalent voltage source for each of the following impedances:

a. **Z** = 25 kΩ∠30°

b. **Z** = 100 Ω ∠−90°

c. **Z** = 20 kΩ − *j*16 kΩ

19.3 Mesh (Loop) Analysis

Mesh analysis allows us to determine each loop current within a circuit, regardless of the number of sources within the circuit. The following steps provide a format which simplifies the process of using mesh analysis:

1. Convert all sinusoidal expressions into equivalent phasor notation. Where necessary, convert current sources into equivalent voltage sources.
2. Redraw the given circuit, simplifying the given impedances wherever possible and labelling the impedances ($\mathbf{Z}_1$, $\mathbf{Z}_2$, etc.).
3. Arbitrarily assign clockwise loop currents to each interior closed loop within a circuit. Show the polarities of all impedances using the assumed current directions. If an impedance is common to two loops, it may be thought to have two simultaneous currents. Although in fact two currents will not occur simultaneously, this maneuver makes the algebraic calculations fairly simple. The actual current through a common impedance is the vector sum of the individual loop currents.
4. Apply Kirchhoff's voltage law to each closed loop in the circuit, writing each equation as follows:

$$\Sigma\,(\mathbf{ZI}) = \Sigma\,\mathbf{E}$$

If the current directions are originally assigned in a clockwise direction, then the resulting linear equations may be simplified to the following format:

$$\begin{aligned}
&\text{Loop 1:} && +(\Sigma\mathbf{Z}_1)\mathbf{I}_1 - (\Sigma\mathbf{Z}_{1-2})\mathbf{I}_2 \cdots - (\Sigma\mathbf{Z}_{1-n})\mathbf{I}_n = (\Sigma\mathbf{E}_1)\\
&\text{Loop 2:} && -(\Sigma\mathbf{Z}_{2-1})\mathbf{I}_1 + (\Sigma\mathbf{Z}_2)\mathbf{I}_2 - \cdots - (\Sigma\mathbf{Z}_{2-n})\mathbf{I}_n = (\Sigma\mathbf{E}_2)\\
& && \vdots \\
&\text{Loop } n\text{:} && -(\Sigma\mathbf{Z}_{n-1})\mathbf{I}_1 - (\Sigma\mathbf{Z}_{n-2})\mathbf{I}_2 - \cdots + (\Sigma\mathbf{Z}_n)\mathbf{I}_n = (\Sigma\mathbf{E}_n)
\end{aligned}$$

In the preceding format, $\Sigma\mathbf{Z}_x$ is the summation of all impedances around loop x. The sign in front of all loop impedances will be positive.

$\Sigma\mathbf{Z}_{x-y}$ is the summation of impedances that are common between loop x and loop y. If there is no common impedance between two loops, this term is simply set to zero. All common impedance terms in the linear equations are given negative signs.

$\Sigma\mathbf{E}_x$ is the summation of voltage rises in the direction of the assumed current $\mathbf{I}_x$. If a voltage source has a polarity such that it appears as a voltage drop in the assumed current direction, then the voltage is given a negative sign.

5. Solve the resulting simultaneous linear equations using substitution or determinants. If necessary, refer to Appendix B for a review of solving simultaneous linear equations.

EXAMPLE 19–5

Solve for the loop equations in the circuit of Figure 19–12.

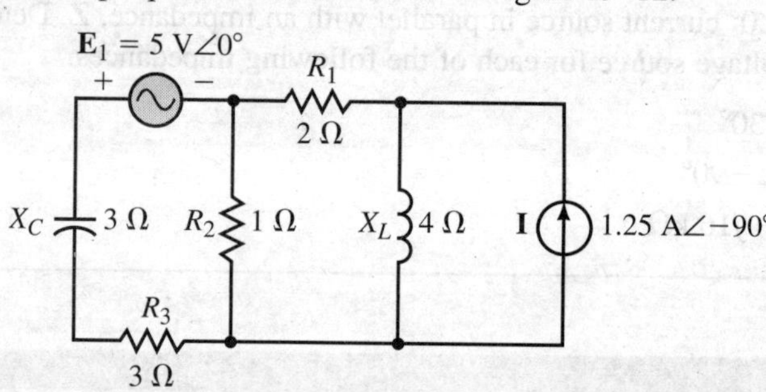

FIGURE 19–12

Solution

Step 1: The current source is first converted into an equivalent voltage source as shown in Figure 19–13.

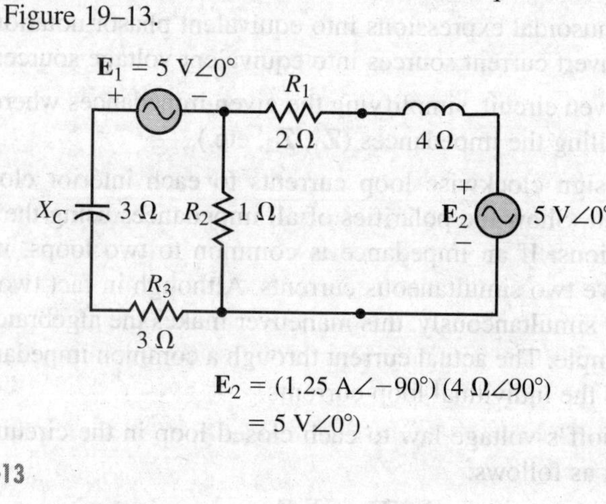

FIGURE 19–13

Steps 2 and 3: Next the circuit is redrawn as shown in Figure 19–14. The impedances have been simplified and the loop currents are drawn in a clockwise direction.

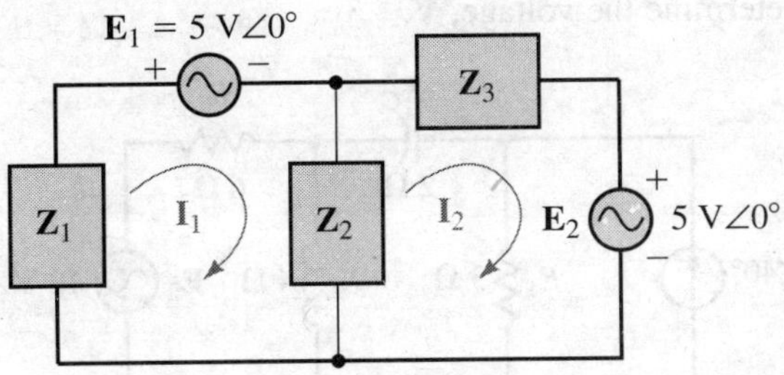

$Z_1 = 3\ \Omega - j3\ \Omega$
$Z_2 = 1\ \Omega + j0$
$Z_3 = 2\ \Omega + j4\ \Omega$

FIGURE 19–14

Step 4: The loop equations are written as

$$\text{Loop 1:} \quad (\mathbf{Z}_1 + \mathbf{Z}_2)\mathbf{I}_1 - (\mathbf{Z}_2)\mathbf{I}_2 = -\mathbf{E}_1$$

$$\text{Loop 2:} \quad -(\mathbf{Z}_2)\mathbf{I}_1 + (\mathbf{Z}_2 + \mathbf{Z}_3)\mathbf{I}_2 = -\mathbf{E}_2$$

When the values of $\mathbf{Z}_1$, $\mathbf{Z}_2$, $\mathbf{Z}_3$, $\mathbf{E}_1$, and $\mathbf{E}_2$ are entered, the simplified linear equations may be written as:

$$\text{Loop 1:} \quad (4 - j3)\mathbf{I}_1 - (1 + j0)\mathbf{I}_2 = -5\angle 0°$$

$$\text{Loop 2:} \quad -(1 + j0)\mathbf{I}_1 + (3 + j4)\mathbf{I}_2 = -5\angle 0°$$

Using determinants, we can solve for these equations as follows:

$$\mathbf{I}_1 = \frac{-(5)(3 + j4) - (5)(1)}{(3 - j3)(1) + (3 - j3)(2 + j4) + (1)(2 + j4)}$$

$$= \frac{(-15 - j20) - 5}{(3 - j3) + (6 + j6 - j^2 12) + (2 + j4)}$$

$$= \frac{-20 - j20}{23 + j7}$$

$$= \frac{28.28\angle -135°}{24.04\angle 16.93°}$$

$$= 1.18\text{ A}\angle -151.93°$$

and

$$\mathbf{I}_2 = \frac{-(5)(4 - j3) - (5)(1)}{23 + j7}$$

$$= \frac{(-20 + j15) - (5)}{23 + j7}$$

$$= \frac{-25 + j15}{23 + j7}$$

$$= \frac{29.15\angle 149.04°}{24.04\angle 16.93°}$$

$$= 1.21\text{ A}\angle 132.11°$$

Calculator Hint: A scientific calculator such as the Sharp EL-506W permits you to enter the complex number calculations in either polar or rectangular format.

EXAMPLE 19–6

Given the circuit of Figure 19–15, write the loop equations and solve for the loop currents. Determine the voltage, **V**.

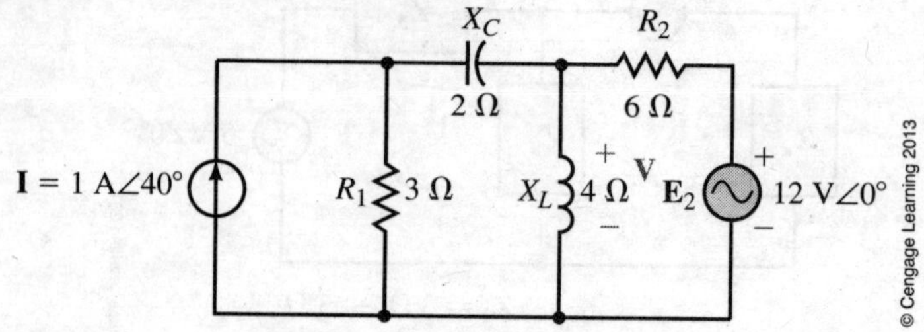

FIGURE 19–15

Solution

Step 1: Converting the current source into an equivalent voltage source gives us the circuit of Figure 19–16.

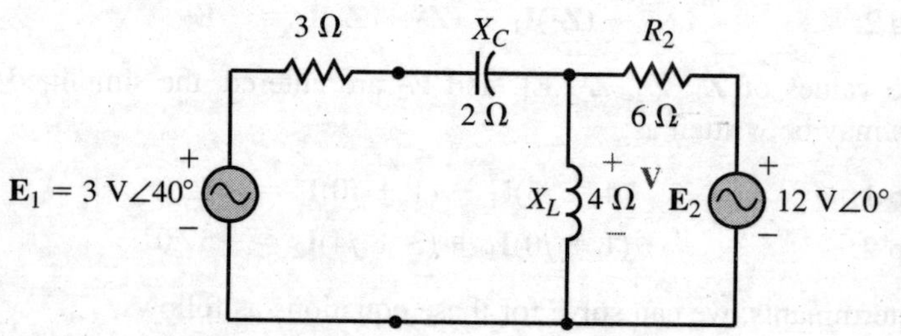

FIGURE 19–16

Steps 2 and 3: After simplifying the impedances and assigning clockwise loop currents, we have the circuit of Figure 19–17.

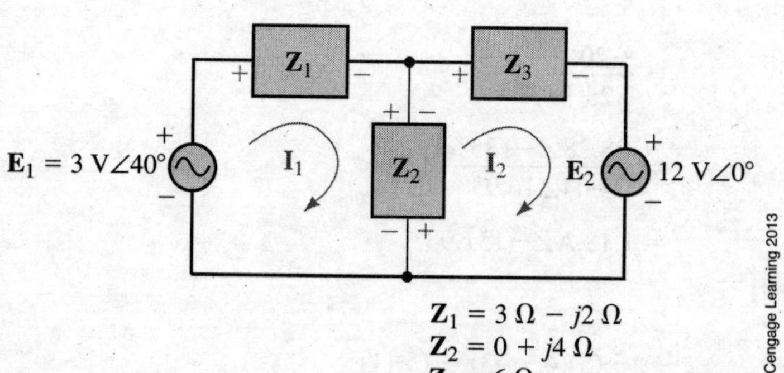

FIGURE 19–17

Step 4: The loop equations for the circuit of Figure 19–17 are as follows:

Loop 1: $(\mathbf{Z}_1 + \mathbf{Z}_2)\mathbf{I}_1 - (\mathbf{Z}_2)\mathbf{I}_2 = \mathbf{E}_1$

Loop 2: $-(\mathbf{Z}_2)\mathbf{I}_1 + (\mathbf{Z}_2 + \mathbf{Z}_3)\mathbf{I}_2 = -\mathbf{E}_2$

which, after substituting the impedance values into the expressions, become

Loop 1: $(3\ \Omega + j2\ \Omega)\mathbf{I}_1 - (j4\ \Omega)\mathbf{I}_2 = 3\ \text{V}\angle 40°$

Loop 2: $-(j4\ \Omega)\mathbf{I}_2 + (6\ \Omega + j4\ \Omega)\mathbf{I}_2 = -12\ \text{V}\angle 0°$

Step 5: Solve for the currents.

The solutions are determined as

$$\mathbf{I}_1 = 0.7887\text{ A}\angle -120.14°$$

and

$$\mathbf{I}_2 = 1.304\text{ A}\angle 156.00°$$

The current through the 4-Ω inductive reactance (in a downward direction) is

$$\begin{aligned}\mathbf{I} &= \mathbf{I}_1 - \mathbf{I}_2\\ &= (0.7887\text{ A}\angle -120.14°) - (1.304\text{ A}\angle 156.00°)\\ &= (-0.3960\text{ A} - j0.6821\text{ A}) - (-1.1913\text{ A} + j0.5304\text{ A})\\ &= 0.795\text{ A} - j1.213\text{ A} = 1.45\text{ A}\angle -56.75°\end{aligned}$$

The voltage is now easily found from Ohm's law as

$$\begin{aligned}\mathbf{V} &= \mathbf{IZ}_L\\ &= (1.45\text{ A}\angle -56.75°)(4\ \Omega\angle 90°) = 5.80\text{ V}\angle 33.25°\end{aligned}$$

EXAMPLE 19–7

Given the circuit of Figure 19–18, write the loop equations and solve for the loop currents.

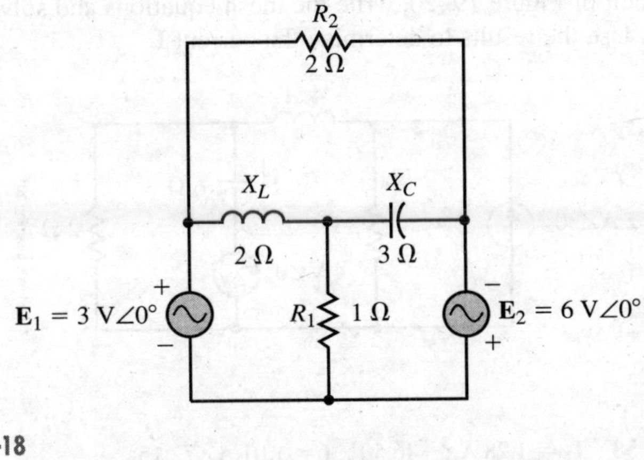

FIGURE 19–18

Solution The circuit is redrawn in Figure 19–19, showing loop currents and impedances together with the appropriate voltage polarities.

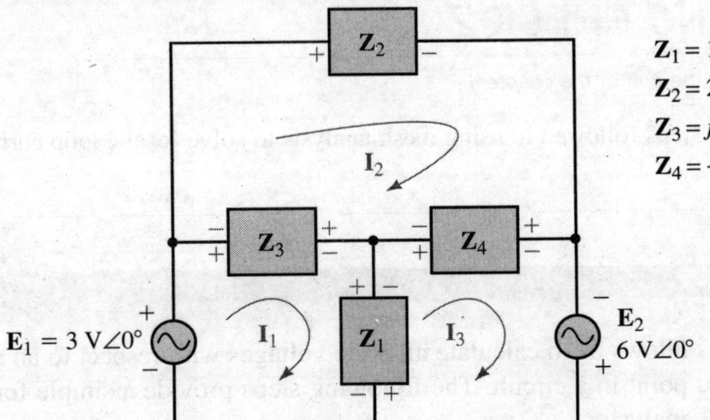

FIGURE 19–19

The loop equations may now be written as

Loop 1: $(\mathbf{Z}_1 + \mathbf{Z}_3)\mathbf{I}_1 - (\mathbf{Z}_3)\mathbf{I}_2 - (\mathbf{Z}_1)\mathbf{I}_3 = \mathbf{E}_1$

Loop 2: $-(\mathbf{Z}_3)\mathbf{I}_1 + (\mathbf{Z}_2 + \mathbf{Z}_3 + \mathbf{Z}_4)\mathbf{I}_2 - (\mathbf{Z}_4)\mathbf{I}_3 = 0$

Loop 3: $-(\mathbf{Z}_1)\mathbf{I}_1 - (\mathbf{Z}_4)\mathbf{I}_2 + (\mathbf{Z}_1 + \mathbf{Z}_4)\mathbf{I}_3 = \mathbf{E}_2$

Using the given impedance values, we have

Loop 1: $(1\ \Omega + j2\ \Omega)\mathbf{I}_1 - (j2\ \Omega)\mathbf{I}_2 - (1\ \Omega)\mathbf{I}_3 = 3\text{ V}$

Loop 2: $-(j2\ \Omega)\mathbf{I}_1 + (2\ \Omega - j1\ \Omega)\mathbf{I}_2 - (-j3\ \Omega)\mathbf{I}_3 = 0$

Loop 3: $-(1\ \Omega)\mathbf{I}_1 - (-j3\ \Omega)\mathbf{I}_2 + (1\ \Omega - j3\ \Omega)\mathbf{I}_3 = 6\text{ V}$

Notice that in the preceding equations, the phase angles ($\theta = 0°$) for the voltages have been omitted. This is because $3\text{ V}\angle 0° = 3\text{ V} + j0\text{ V} = 3\text{ V}$.

Using a calculator such as the Sharp EL-506W, we have:

$$\mathbf{I}_1 = 5.26\text{A}\ \angle -41.25°$$
$$\mathbf{I}_2 = 3.18\text{A}\ \angle -45°$$
$$\mathbf{I}_3 = 3.39\text{A}\ \angle -1.03°$$

PRACTICE PROBLEMS 3

Given the circuit of Figure 19–20, write the mesh equations and solve for the loop currents. Use the results to determine the current **I**.

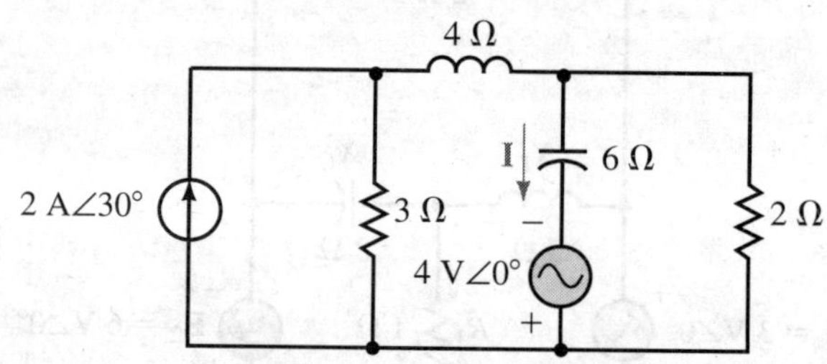

FIGURE 19–20

Answers

$\mathbf{I}_1 = 1.19\text{ A}\angle 1.58°$, $\mathbf{I}_2 = 1.28\text{ A}\angle -46.50°$, $\mathbf{I} = 1.01\text{ A}\angle 72.15°$

IN-PROCESS LEARNING CHECK 2

(Answers are at the end of the chapter.)

Briefly list the steps followed in using mesh analysis to solve for the loop currents of a circuit.

19.4 Nodal Analysis

Nodal analysis allows us to calculate all node voltages with respect to an arbitrary reference point in a circuit. The following steps provide a simple format to apply nodal analysis.

1. Convert all sinusoidal expressions into equivalent phasor notation. If necessary, convert voltage sources into equivalent current sources.

2. Redraw the given circuit, simplifying the given impedances wherever possible and relabelling the impedances as admittances ($\mathbf{Y}_1$, $\mathbf{Y}_2$, etc.).
3. Select and label an appropriate reference node. Arbitrarily assign subscripted voltages ($\mathbf{V}_1$, $\mathbf{V}_2$, etc.) to each of the remaining n nodes within the circuit.
4. Indicate assumed current directions through all admittances in the circuit. If an admittance is common to two nodes, it is considered in each of the two node equations.
5. Apply Kirchhoff's current law (KCL) to each of the n nodes in the circuit, writing each equation as follows:

$$\Sigma(\mathbf{YV}) = \Sigma\mathbf{I}_{\text{sources}}$$

The resulting linear equations may be simplified to the following format:

$$\text{Node 1:}\quad +(\Sigma\mathbf{Y}_1)\mathbf{V}_1 - (\Sigma\mathbf{Y}_{1-2})\mathbf{V}_2 - \cdots - (\Sigma\mathbf{Y}_{1-n})\mathbf{V}_n = (\Sigma\mathbf{I}_1)$$
$$\text{Node 2:}\quad -(\Sigma\mathbf{Y}_{2-1})\mathbf{V}_1 + (\Sigma\mathbf{Y}_2)\mathbf{V}_2 - \cdots - (\Sigma\mathbf{Y}_{2-n})\mathbf{V}_n = (\Sigma\mathbf{I}_2)$$
$$\vdots \qquad\qquad \vdots$$
$$\text{Node } n\text{:}\quad -(\Sigma\mathbf{Y}_{n-1})\mathbf{V}_1 - (\Sigma\mathbf{Y}_{n-2})\mathbf{V}_2 - \cdots + (\Sigma\mathbf{Y}_n)\mathbf{V}_n = (\Sigma\mathbf{I}_n)$$

In this format, $\Sigma\mathbf{Y}_x$ is the summation of all admittances connected to node x. The sign in front of all node admittances will be positive.

$\Sigma\mathbf{Y}_{x-y}$ is the summation of common admittances between node x and node y. If there are no common admittances between two nodes, this term is simply set to zero. All common admittance terms in the linear equations will have a negative sign.

$\Sigma\mathbf{I}_x$ is the summation of current sources entering node x. If a current source leaves the node, the current is given a negative sign.

6. Solve the resulting simultaneous linear equations using substitution or determinants.

EXAMPLE 19–8

Given the circuit of Figure 19–21, write the nodal equations and solve for the node voltages.

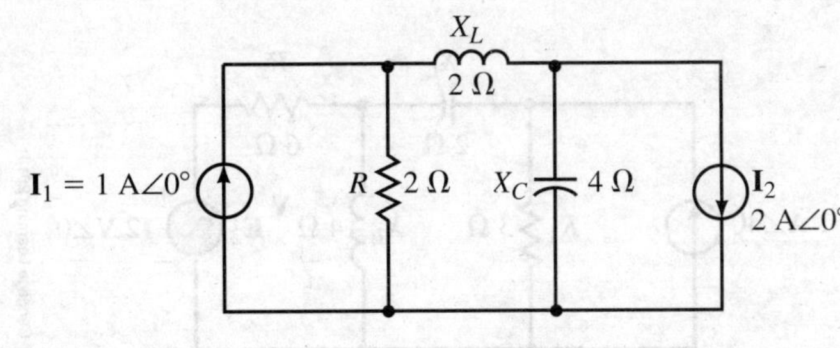

FIGURE 19–21

Solution The circuit is redrawn in Figure 19–22, showing the nodes and a simplified representation of the admittances.

The nodal equations are written as

$$\text{Node 1:}\qquad (\mathbf{Y}_1 + \mathbf{Y}_2)\mathbf{V}_1 - (\mathbf{Y}_2)\mathbf{V}_2 = \mathbf{I}_1$$
$$\text{Node 2:}\qquad -(\mathbf{Y}_2)\mathbf{V}_1 + (\mathbf{Y}_2 + \mathbf{Y}_3)\mathbf{V}_2 = -\mathbf{I}_2$$

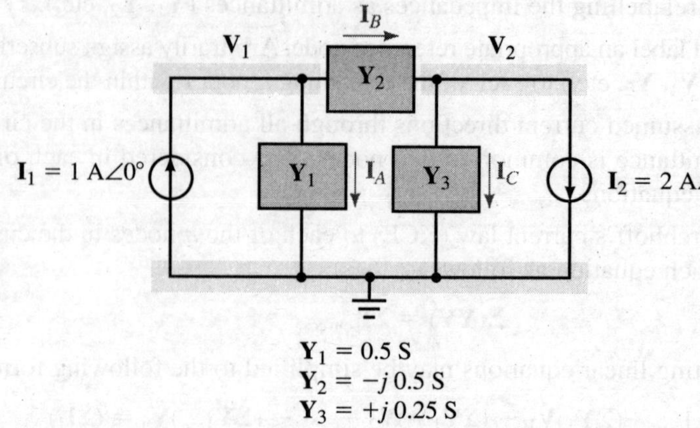

FIGURE 19–22

Now, substituting the values of admittance into the nodal equations, we get

$$\text{Node 1:} \quad (0.5 - j0.5)\mathbf{V}_1 - (-j0.5)\mathbf{V}_2 = 1\angle 0°$$

$$\text{Node 2:} \quad -(-j0.5)\mathbf{V}_1 + (-j0.25)\mathbf{V}_2 = -2\angle 0°$$

which, when solved, results in:

$$\mathbf{V}_1 = 4.243\ \text{V}\angle 135°$$

and

$$\mathbf{V}_2 = 6.324\ \text{V}\angle -161.57°$$

EXAMPLE 19–9

Use nodal analysis to determine the voltage **V** for the circuit of Figure 19–23. Compare the results to those obtained when the circuit was analyzed using mesh analysis in Example 19–6.

CircuitSim 19-1

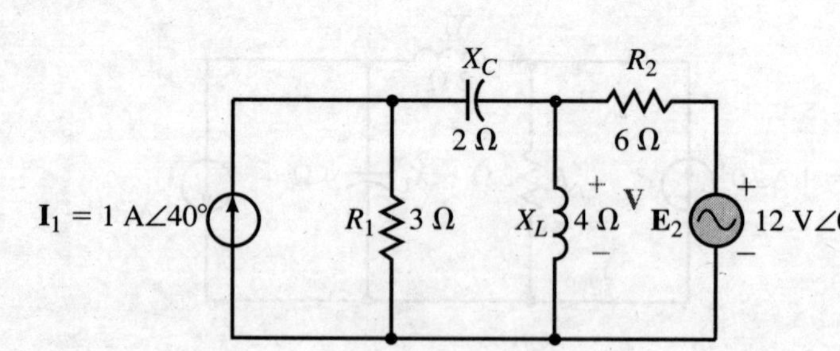

FIGURE 19–23

Solution

Step 1: Convert the voltage source into an equivalent current source as illustrated in Figure 19–24.

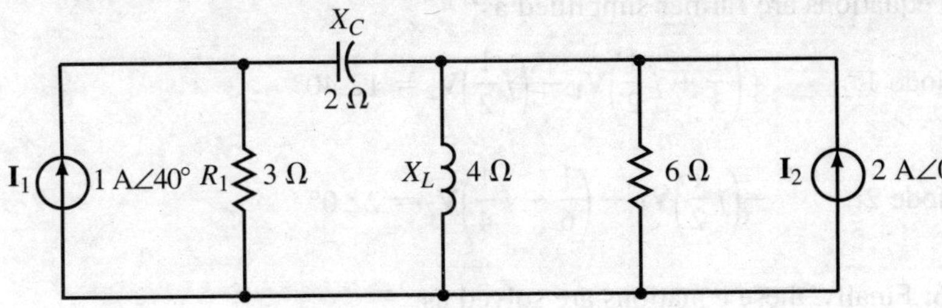

$$\mathbf{I}_2 = \frac{\mathbf{E}_2}{6\ \Omega} = \frac{12\ \text{V}\angle 0^\circ}{6\ \Omega} = 2\ \text{A}\angle 0^\circ$$

FIGURE 19–24

Steps 2, 3, and 4: The reference node is selected to be at the bottom of the circuit and the admittances are simplified as shown in Figure 19–25.

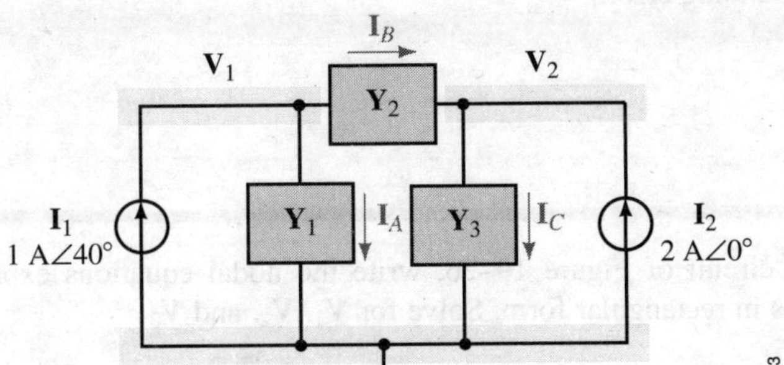

$$\mathbf{Y}_1 = \tfrac{1}{3}\ \text{S} + j\,0$$
$$\mathbf{Y}_2 = 0 + j\tfrac{1}{2}\ \text{S}$$
$$\mathbf{Y}_3 = \tfrac{1}{6}\ \text{S} - j\tfrac{1}{4}\ \text{S}$$

FIGURE 19–25

Step 5: Applying Kirchhoff's current law to each node, we have the following:

Node 1:
$$\mathbf{I}_A + \mathbf{I}_B = \mathbf{I}_1$$
$$\mathbf{Y}_1\mathbf{V}_1 + \mathbf{Y}_2(\mathbf{V}_1 - \mathbf{V}_2) = \mathbf{I}_1$$
$$(\mathbf{Y}_1 + \mathbf{Y}_2)\mathbf{V}_1 - \mathbf{Y}_2\mathbf{V}_2 = \mathbf{I}_1$$

Node 2:
$$\mathbf{I}_C = \mathbf{I}_B + \mathbf{I}_2$$
$$\mathbf{Y}_3\mathbf{V}_2 = \mathbf{Y}_2(\mathbf{V}_1 - \mathbf{V}_2) + \mathbf{I}_2$$
$$-\mathbf{Y}_2\mathbf{V}_1 + (\mathbf{Y}_2 + \mathbf{Y}_3)\mathbf{V}_2 = \mathbf{I}_2$$

Now, substituting the values of admittance into the nodal equations, we get

Node 1:
$$\left(\frac{1}{3} + \frac{1}{-j2}\right)\mathbf{V}_1 - \left(\frac{1}{-j2}\right)\mathbf{V}_2 = 1\angle 40^\circ$$

Node 2:
$$-\left(\frac{1}{-j2}\right)\mathbf{V}_1 + \left(\frac{1}{-j2} + \frac{1}{j4} + \frac{1}{6}\right)\mathbf{V}_2 = 2\angle 0^\circ$$

These equations are further simplified as:

Node 1: $$\left(\frac{1}{3}+j\frac{1}{2}\right)\mathbf{V}_1-\left(j\frac{1}{2}\right)\mathbf{V}_2=1\angle 40^\circ$$

Node 2: $$-\left(j\frac{1}{2}\right)\mathbf{V}_1+\left(\frac{1}{6}+j\frac{1}{4}\right)\mathbf{V}_2=2\angle 0^\circ$$

Step 6: Finally, these equations are solved as

$$\mathbf{V}_1 = 5.29\ \text{V}\angle 48.75^\circ$$

and

$$\mathbf{V}_2 = 5.80\ \text{V}\angle 33.27^\circ$$

Examining the circuit of Figure 19–23, we see that the voltage $\mathbf{V}$ is the same as the node voltage $\mathbf{V}_2$. Therefore $\mathbf{V} = 5.80\ \text{V}\angle 33.27^\circ$, which is the same result obtained in Example 19–6. (The slight difference in phase angle is the result of rounding error.)

EXAMPLE 19–10

Given the circuit of Figure 19–26, write the nodal equations expressing all coefficients in rectangular form. Solve for $\mathbf{V}_1$, $\mathbf{V}_2$, and $\mathbf{V}_3$.

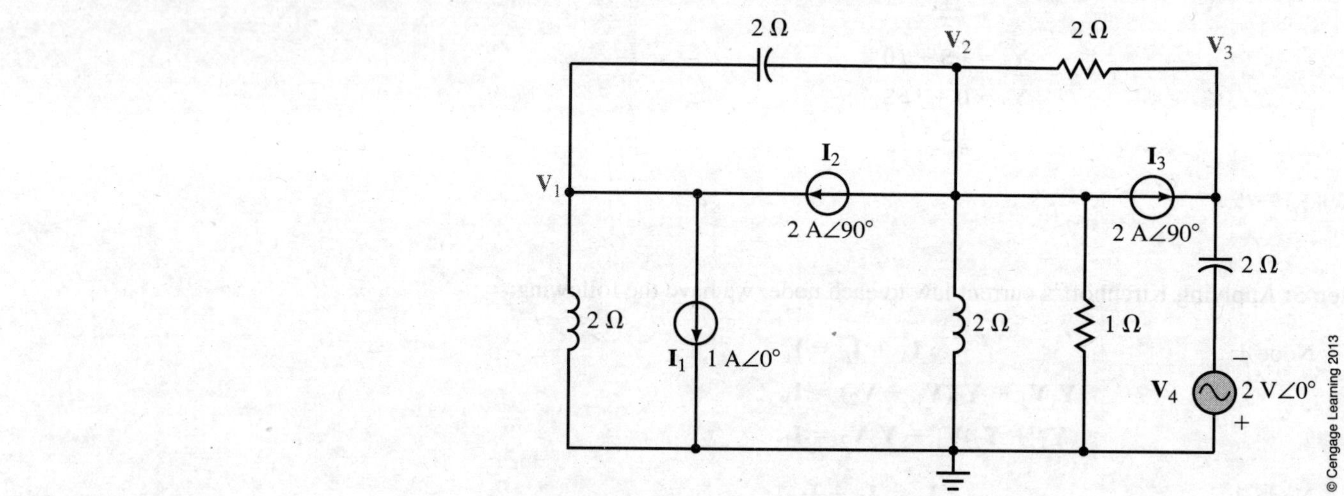

FIGURE 19–26

Solution As in the previous example, we need to first convert the voltage source into its equivalent current source. The current source will be a phasor $\mathbf{I}_4$, where

$$\mathbf{I}_4=\frac{\mathbf{V}_4}{\mathbf{X}_C}=\frac{2\ \text{V}\angle 0^\circ}{2\ \Omega\angle -90^\circ}=1.0\ \text{A}\angle 90^\circ$$

Figure 19–27 shows the circuit as it appears after the source conversion. Notice that the direction of the current source is downward to correspond with the polarity of the voltage source $\mathbf{V}_4$.

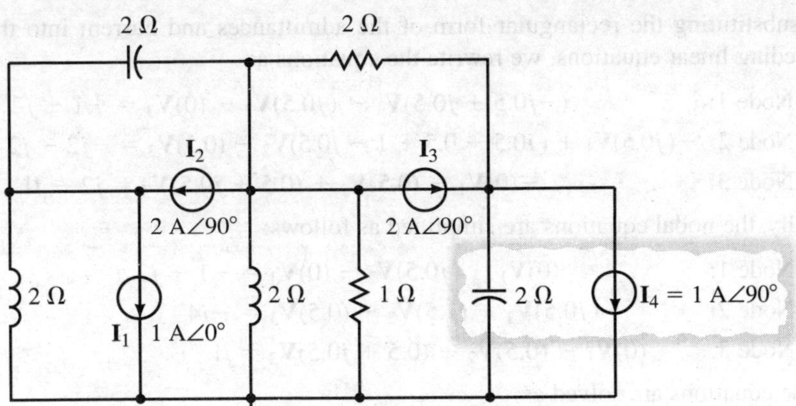

$$I_4 = \frac{2\ V\angle 0^\circ}{2\ \Omega\angle -90^\circ} = 1\ A\angle 90^\circ$$

FIGURE 19–27

Now, by labelling the nodes and admittances, the circuit may be simplified as shown in Figure 19–28.

$$\mathbf{Y}_1 = 0 - j\tfrac{1}{2}\,S \quad \mathbf{Y}_3 = \tfrac{1}{2}\,S + j0 \quad \mathbf{Y}_5 = 1\,S - j\tfrac{1}{2}\,S$$

$$\mathbf{Y}_2 = 0 + j\tfrac{1}{2}\,S \quad \mathbf{Y}_4 = 0 + j\tfrac{1}{2}\,S$$

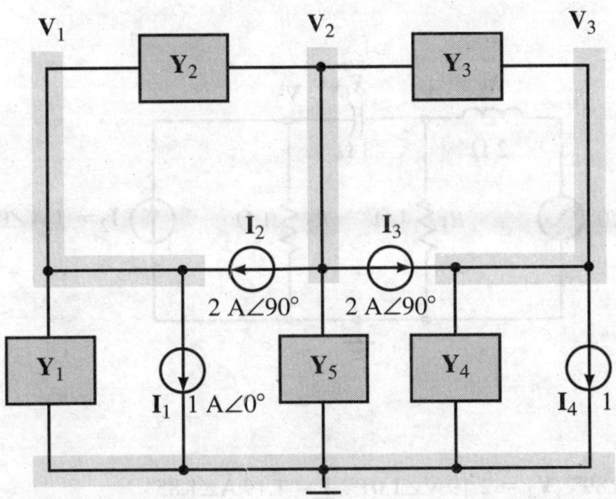

FIGURE 19–28

The admittances of Figure 19–28 are as follows:

$$\mathbf{Y}_1 = 0 - j0.5\ S$$
$$\mathbf{Y}_2 = 0 + j0.5\ S$$
$$\mathbf{Y}_3 = 0.5\ S + j0$$
$$\mathbf{Y}_4 = 0 + j0.5\ S$$
$$\mathbf{Y}_5 = 1.0\ S - j0.5\ S$$

Using the assigned admittances, the nodal equations are written as follows:

Node 1: $(\mathbf{Y}_1 + \mathbf{Y}_2)\mathbf{V}_1 - (\mathbf{Y}_2)\mathbf{V}_2 - (0)\mathbf{V}_3 = -\mathbf{I}_1 + \mathbf{I}_2$

Node 2: $-(\mathbf{Y}_2)\mathbf{V}_1 + (\mathbf{Y}_2 + \mathbf{Y}_3 + \mathbf{Y}_5)\mathbf{V}_2 - (\mathbf{Y}_3)\mathbf{V}_3 = -\mathbf{I}_2 - \mathbf{I}_3$

Node 3: $-(0)\mathbf{V}_1 - (\mathbf{Y}_3)\mathbf{V}_2 + (\mathbf{Y}_3 + \mathbf{Y}_4)\mathbf{V}_3 = \mathbf{I}_3 - \mathbf{I}_4$

By substituting the rectangular form of the admittances and current into the preceding linear equations, we rewrite the equations as

Node 1: $(-j0.5 + j0.5)\mathbf{V}_1 - (j0.5)\mathbf{V}_2 - (0)\mathbf{V}_3 = -1 + j2$

Node 2: $-(j0.5)\mathbf{V}_1 + (j0.5 + 0.5 + 1 - j0.5)\mathbf{V}_2 - (0.5)\mathbf{V}_3 = -j2 - j2$

Node 3: $-(0)\mathbf{V}_1 - (0.5)\mathbf{V}_2 + (0.5 + j0.5)\mathbf{V}_3 = j2 - j1$

Finally, the nodal equations are simplified as follows:

Node 1: $(0)\mathbf{V}_1 - (j0.5)\mathbf{V}_2 - (0)\mathbf{V}_3 = -1 + j2$

Node 2: $-(j0.5)\mathbf{V}_1 + (1.5)\mathbf{V}_2 - (0.5)\mathbf{V}_3 = -j4$

Node 3: $(0)\mathbf{V}_1 - (0.5)\mathbf{V}_2 + (0.5 + j0.5)\mathbf{V}_3 = j1$

These equations are solved as

$$\mathbf{V}_1 = 10.00\text{V}\angle 90°$$

$$\mathbf{V}_2 = 4.47\text{V}\angle -153.44°$$

$$\mathbf{V}_3 = 2.83\text{V}\angle 135°$$

PRACTICE PROBLEMS 4

Given the circuit of Figure 19–29, use nodal analysis to find the voltages $\mathbf{V}_1$ and $\mathbf{V}_2$. Use your results to find the current $\mathbf{I}$.

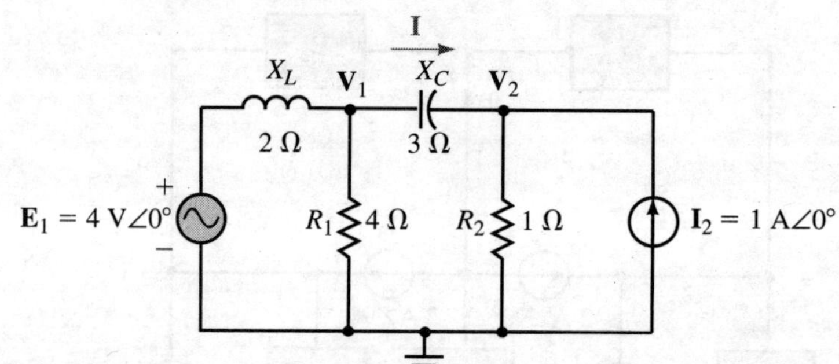

FIGURE 19–29

Answers

$\mathbf{V}_1 = 4.22\ \text{V}\angle -56.89°$, $\mathbf{V}_2 = 2.19\ \text{V}\angle 1.01°$, $\mathbf{I} = 1.19\ \text{A}\angle 1.85°$

IN-PROCESS LEARNING CHECK 3

(Answers are at the end of the chapter.)

Briefly list the steps followed in using nodal analysis to solve for the node voltages of a circuit.

19.5 Delta-to-Wye and Wye-to-Delta Conversions

In Chapter 8, we derived the relationships showing the equivalence of "delta" (or "pi") connected resistance to a "wye" (or "tee") configuration.

In a similar manner, impedances connected in a Δ configuration are equivalent to a unique Y configuration. Figure 19–30 shows the equivalent circuits.

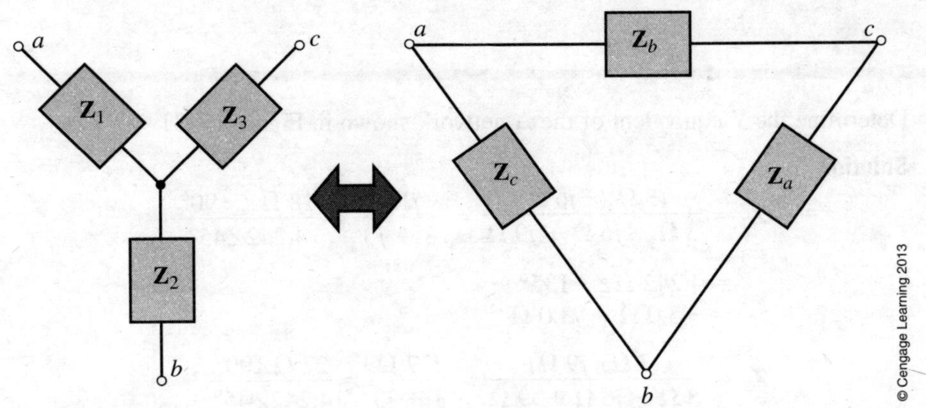

FIGURE 19–30 Delta-wye equivalence.

A Δ configuration is converted to a Y equivalent by using the following:

$$\mathbf{Z}_1 = \frac{\mathbf{Z}_b\mathbf{Z}_c}{\mathbf{Z}_a + \mathbf{Z}_b + \mathbf{Z}_c} \tag{19–1}$$

$$\mathbf{Z}_2 = \frac{\mathbf{Z}_a\mathbf{Z}_c}{\mathbf{Z}_a + \mathbf{Z}_b + \mathbf{Z}_c} \tag{19–2}$$

$$\mathbf{Z}_3 = \frac{\mathbf{Z}_a\mathbf{Z}_b}{\mathbf{Z}_a + \mathbf{Z}_b + \mathbf{Z}_c} \tag{19–3}$$

The preceding conversion indicates that the impedance in any arm of a Y circuit is determined by taking the product of the two adjacent Δ impedances at this arm and dividing by the summation of the Δ impedances.

If the impedances of the sides of the Δ network are all equal (magnitude and phase angle), the equivalent Y network will have identical impedances, where each impedance is determined as

$$\mathbf{Z}_Y = \frac{\mathbf{Z}_\Delta}{3} \tag{19–4}$$

A Y configuration is converted to a Δ equivalent by using the following:

$$\mathbf{Z}_a = \frac{\mathbf{Z}_1\mathbf{Z}_2 + \mathbf{Z}_1\mathbf{Z}_3 + \mathbf{Z}_2\mathbf{Z}_3}{\mathbf{Z}_1} \tag{19–5}$$

$$\mathbf{Z}_b = \frac{\mathbf{Z}_1\mathbf{Z}_2 + \mathbf{Z}_1\mathbf{Z}_3 + \mathbf{Z}_2\mathbf{Z}_3}{\mathbf{Z}_2} \tag{19–6}$$

$$\mathbf{Z}_c = \frac{\mathbf{Z}_1\mathbf{Z}_2 + \mathbf{Z}_1\mathbf{Z}_3 + \mathbf{Z}_2\mathbf{Z}_3}{\mathbf{Z}_3} \tag{19–7}$$

Any impedance in a Δ is determined by summing the possible two-impedance product combinations of the Y and then dividing by the impedance found in the opposite branch of the Y.

If the arms of a Y have identical impedances, the equivalent Δ will have impedances given as

$$\mathbf{Z}_\Delta = 3\mathbf{Z}_Y \tag{19–8}$$

EXAMPLE 19–11

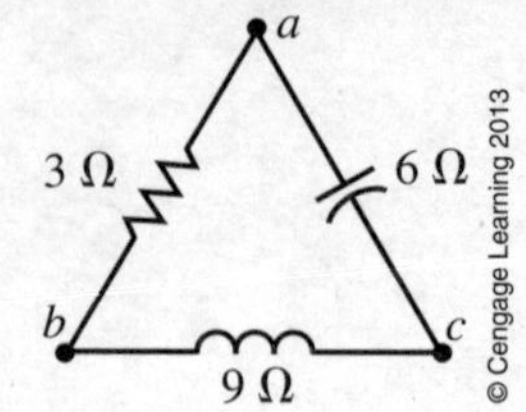

FIGURE 19–31

Determine the Y equivalent of the Δ network shown in Figure 19–31.

Solution

$$\mathbf{Z}_1 = \frac{(3\ \Omega)(-j6\ \Omega)}{3\ \Omega - j6\ \Omega + j9\ \Omega} = \frac{-j18\ \Omega}{3 + j3} = \frac{18\ \Omega\angle -90^\circ}{4.242\angle 45^\circ}$$
$$= 4.242\ \Omega\angle -135^\circ$$
$$= -3.0\ \Omega - j3.0\ \Omega$$

$$\mathbf{Z}_2 = \frac{(3\ \Omega)(j9\ \Omega)}{3\ \Omega - j6\ \Omega + j9\ \Omega} = \frac{j27\ \Omega}{3 + j3} = \frac{27\ \Omega\angle 90^\circ}{4.242\angle 45^\circ}$$
$$= 6.364\ \Omega\angle 45^\circ$$
$$= 4.5\ \Omega + j4.5\ \Omega$$

$$\mathbf{Z}_3 = \frac{(j9\ \Omega)(-j6\ \Omega)}{3\ \Omega - j6\ \Omega + j9\ \Omega} = \frac{54\ \Omega}{3 + j3} = \frac{54\ \Omega\angle 0^\circ}{4.242\angle 45^\circ}$$
$$= 12.73\ \Omega\angle -45^\circ$$
$$= 9.0\ \Omega - j9.0\ \Omega$$

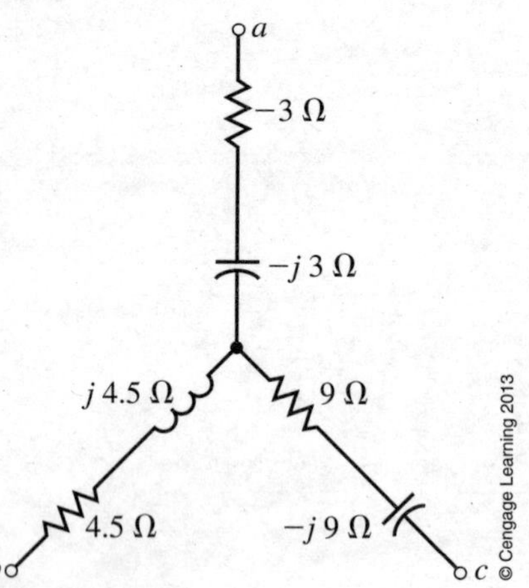

FIGURE 19–32

In the preceding solution, we see that the given Δ network has an equivalent Y network with one arm having a negative resistance. This result indicates that although the Δ circuit has an equivalent Y circuit, the Y circuit cannot actually be constructed from real components since *negative resistors* do not exist (although some active components may demonstrate negative resistance characteristics). If the given conversion is used to simplify a circuit we would treat the impedance $\mathbf{Z}_1 = -3\ \Omega - j3\ \Omega$ as if the resistance actually were a negative value. Figure 19–32 shows the equivalent Y circuit.

It is left to the student to show that the Y of Figure 19–32 is equivalent to the Δ of Figure 19–31.

EXAMPLE 19–12

Find the total impedance of the network in Figure 19–33.

Solution If we take a moment to examine this network, we see that the circuit contains both a Δ and a Y. In calculating the total impedance, the solution is easier when we convert the Y to a Δ.

The conversion is shown in Figure 19–34.

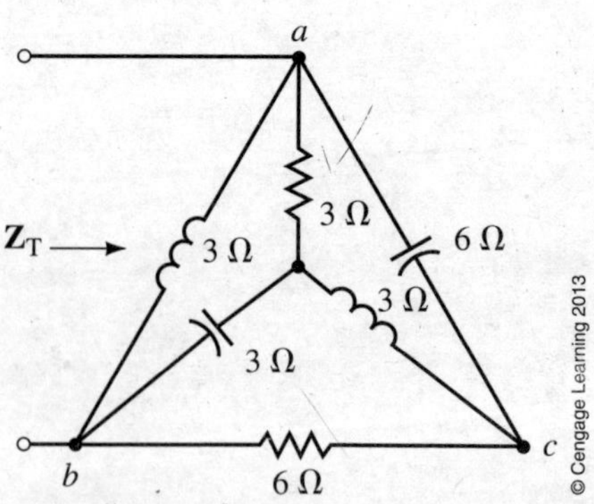

FIGURE 19–33

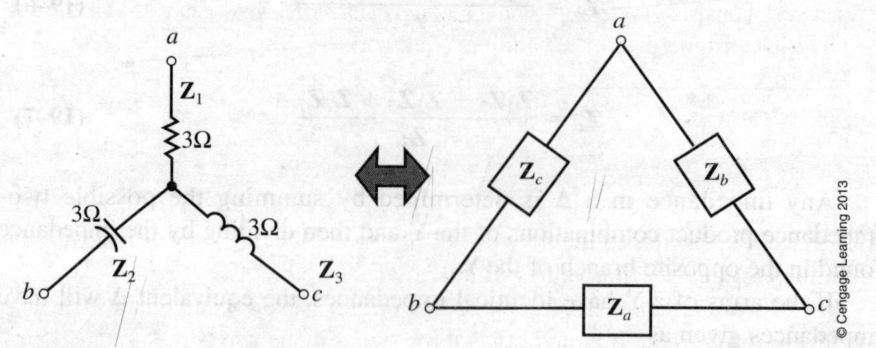

FIGURE 19–34

$$\mathbf{Z}_a = \frac{\mathbf{Z}_1\mathbf{Z}_2 + \mathbf{Z}_1\mathbf{Z}_3 + \mathbf{Z}_2\mathbf{Z}_3}{\mathbf{Z}_1}$$

$$= \frac{(3\ \Omega)(j3\ \Omega) + (3\ \Omega)(-j3\ \Omega) + (j3\ \Omega)(-j3\ \Omega)}{3\ \Omega}$$

$$= \frac{-j^2 9\ \Omega}{3} = 3\ \Omega$$

$$\mathbf{Z}_b = \frac{\mathbf{Z}_1\mathbf{Z}_2 + \mathbf{Z}_1\mathbf{Z}_3 + \mathbf{Z}_2\mathbf{Z}_3}{\mathbf{Z}_2} = \frac{9\ \Omega}{-j3} = j3\ \Omega$$

$$\mathbf{Z}_c = \frac{\mathbf{Z}_1\mathbf{Z}_2 + \mathbf{Z}_1\mathbf{Z}_3 + \mathbf{Z}_2\mathbf{Z}_3}{\mathbf{Z}_3} = \frac{9\ \Omega}{j3} = -j3\ \Omega$$

Now, substituting the equivalent Δ into the original network, we have the revised network of Figure 19–35.

The network of Figure 19–35 shows that the corresponding sides of the Δ are parallel. Because the inductor and the capacitor in the left side of the Δ have the same values, we may replace the parallel combination of these two components with an open circuit. The resulting impedance of the network is now easily determined as

$$\mathbf{Z}_T = 3\ \Omega \| 6\ \Omega + (j3\ \Omega) \| (-j6\ \Omega) = 2\ \Omega + j6\ \Omega$$

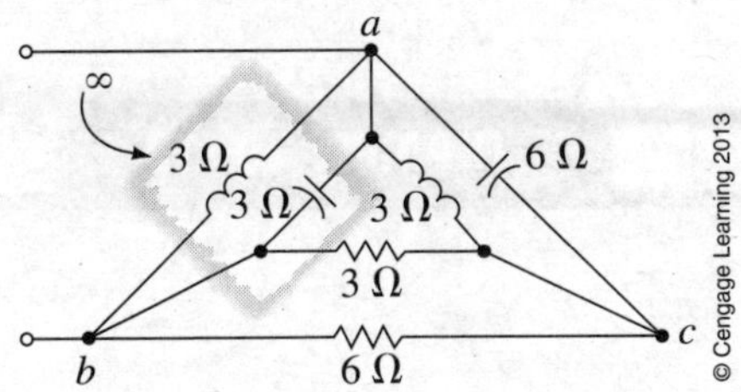

FIGURE 19–35

PRACTICE PROBLEMS 5

A Y network consists of a 60-Ω capacitor, a 180-Ω inductor, and a 540-Ω resistor. Determine the corresponding Δ network.

Answers

$\mathbf{Z}_a = -1080\ \Omega + j180\ \Omega$, $\mathbf{Z}_b = 20\ \Omega + j120\ \Omega$, $\mathbf{Z}_c = 360\ \Omega - j60\ \Omega$

IN-PROCESS LEARNING CHECK 4

(Answers are at the end of the chapter.)

A Δ network consists of a resistor, inductor, and capacitor, each having an impedance of 150 Ω. Determine the corresponding Y network.

19.6 Bridge Networks

Bridge circuits, similar to the network of Figure 19–36, are used extensively in electronics to measure the values of unknown components.

Recall from Chapter 8 that any bridge circuit is said to be balanced when the current through the branch between the two arms is zero. In a practical circuit, component values of very precise resistors are adjusted until the current through the central element (usually a sensitive galvanometer) is exactly equal

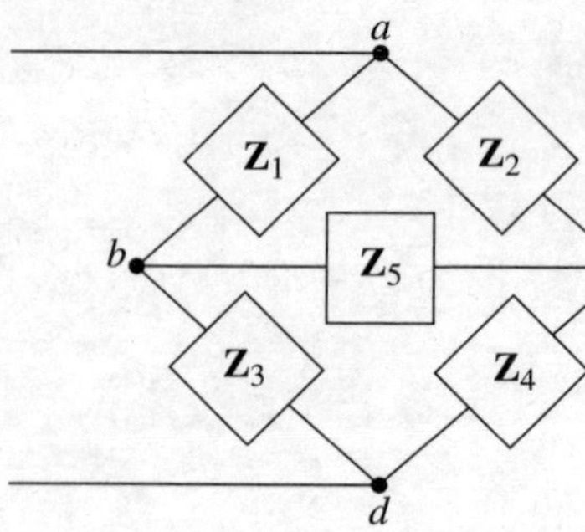

FIGURE 19–36

to zero. For ac circuits, the condition of a **balanced bridge** occurs when the impedance vectors of the various arms satisfy the following condition:

$$\frac{\mathbf{Z}_1}{\mathbf{Z}_3} = \frac{\mathbf{Z}_2}{\mathbf{Z}_4} \tag{19–9}$$

When a balanced bridge occurs in a circuit, the equivalent impedance of the bridge network is easily determined by removing the central impedance and replacing it with either an open or a short circuit. The resulting impedance of the bridge circuit is then found as either of the following:

$$\mathbf{Z}_T = \mathbf{Z}_1 \| \mathbf{Z}_2 + \mathbf{Z}_3 \| \mathbf{Z}_4$$

or

$$\mathbf{Z}_T = (\mathbf{Z}_1 + \mathbf{Z}_3) \| (\mathbf{Z}_2 + \mathbf{Z}_4)$$

If, on the other hand, the bridge is not balanced, then the total impedance must be determined by performing a Δ-to-Y conversion. Alternatively, the circuit may be analyzed by using either mesh analysis or nodal analysis.

EXAMPLE 19–13

Given that the circuit of Figure 19–37 is a balanced bridge.

a. Calculate the unknown impedance, $\mathbf{Z}_x$.
b. Determine the values of L_x and R_x if the circuit operates at a frequency of 1 kHz.

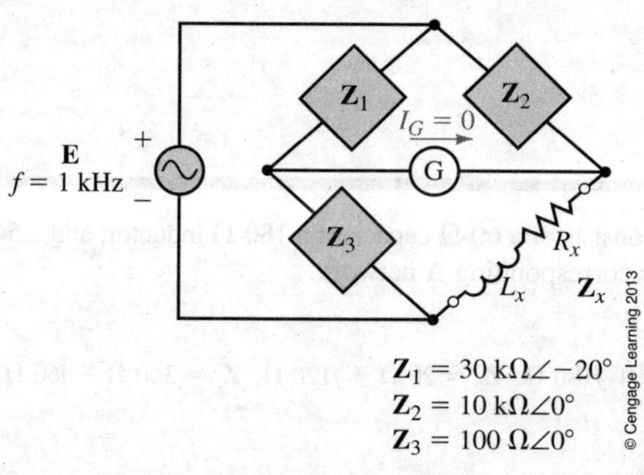

FIGURE 19–37

Solution

a. The expression for the unknown impedance is determined from Equation 19–9 as

$$\mathbf{Z}_x = \frac{\mathbf{Z}_2\mathbf{Z}_3}{\mathbf{Z}_1}$$

$$= \frac{(10\ \text{k}\Omega)(100\ \Omega)}{30\ \text{k}\Omega\angle -20°}$$

$$= 33.3\ \Omega\angle 20°$$

$$= 31.3 + j11.4\ \Omega$$

b. From the preceding result, we have

$$R_x = 31.3\ \Omega$$

and

$$L_x = \frac{X_L}{2\pi f} = \frac{11.4\ \Omega}{2\pi(1000\ \text{Hz})} = 1.81\ \text{mH}$$

We will now consider various forms of bridge circuits that are used in electronic circuits to determine the values of unknown inductors and capacitors. As in resistor bridges, the circuits use variable resistors together with very sensitive galvanometer movements to ensure a balanced condition for the bridge. However, rather than using a dc source to provide current in the circuit, the bridge circuits use ac sources operating at a known frequency (usually 1 kHz). Once the bridge is balanced, the value of unknown inductance or capacitance may be easily determined by obtaining the reading directly from the instrument. Most instruments using bridge circuitry will incorporate several different bridges to enable the measurement of various types of unknown impedances.

Maxwell Bridge

The **Maxwell bridge,** shown in Figure 19–38, is used to determine the inductance and series resistance of an inductor having a relatively large series resistance (in comparison to $X_L = \omega L$).

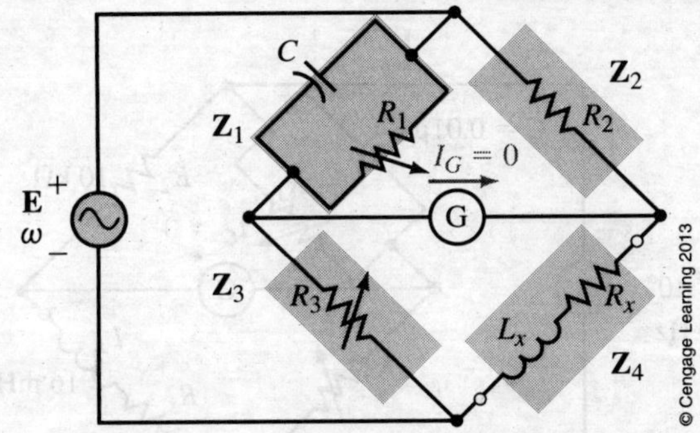

FIGURE 19–38 Maxwell bridge.

Resistors R_1 and R_3 are adjusted to provide the balanced condition (when the current through the galvanometer is zero: $I_G = 0$).

When the bridge is balanced, we know that the following condition must apply:

$$\frac{\mathbf{Z}_1}{\mathbf{Z}_2} = \frac{\mathbf{Z}_3}{\mathbf{Z}_4}$$

If we write the impedances using the rectangular forms, we obtain

$$\frac{\left[\dfrac{(R_1)\left(-j\dfrac{1}{\omega C}\right)}{R_1 - j\dfrac{1}{\omega C}}\right]}{R_2} = \frac{R_3}{R_x + j\omega L_x}$$

$$\frac{\left(-j\dfrac{R_1}{\omega C}\right)}{\left(\dfrac{\omega R_1 C - j1}{\omega C}\right)} = \frac{R_2 R_3}{R_x + j\omega L_x}$$

$$\frac{-jR_1}{\omega CR_1 - j} = \frac{R_2 R_3}{R_x + j\omega L_x}$$

$$(-jR_1)(R_x + j\omega L_x) = R_2 R_3(\omega C R_1 - j)$$

$$\omega L_x R_1 - jR_1 R_x = \omega R_1 R_2 R_3 C - jR_2 R_3$$

Now, since two complex numbers can be equal only if their real parts are equal and if their imaginary parts are equal, we must have the following:

$$\omega L_x R_1 = \omega R_1 R_2 R_3 C$$

and

$$R_1 R_x = R_2 R_3$$

Simplifying these expressions, we get the following equations for a Maxwell bridge:

$$L_x = R_2 R_3 C \qquad \textbf{(19–10)}$$

and

$$R_x = \frac{R_2 R_3}{R_1} \qquad \textbf{(19–11)}$$

EXAMPLE 19–14

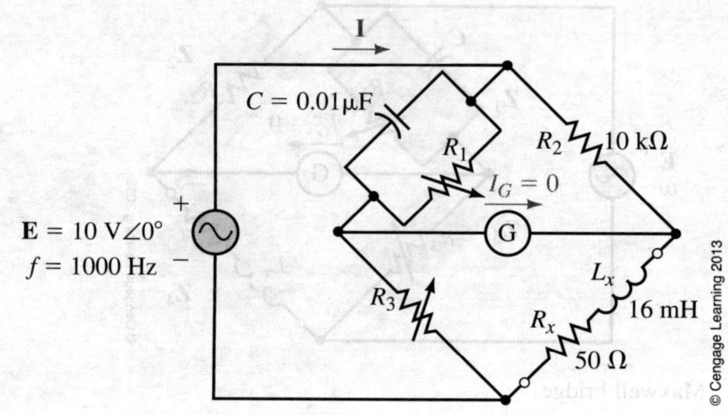

FIGURE 19–39

a. Determine the values of R_1 and R_3 so that the bridge of Figure 19–39 is balanced.
b. Calculate the current **I** when the bridge is balanced.

Solution

a. Rewriting Equations 19–10 and 19–11 and solving for the unknowns, we have

$$R_3 = \frac{L_x}{R_2 C} = \frac{16\ \text{mH}}{(10\ \text{k}\Omega)(0.01\ \mu\text{F})} = 160\ \Omega$$

and

$$R_1 = \frac{R_2 R_3}{R_x} = \frac{(10\ \text{k}\Omega)(160\ \Omega)}{50\ \Omega} = 32\ \text{k}\Omega$$

b. The total impedance is found as

$$\mathbf{Z}_T = (\mathbf{Z}_C \| \mathbf{R}_1 \| \mathbf{R}_2) + [\mathbf{R}_3 \| (\mathbf{R}_x + \mathbf{Z}_{Lx})]$$

$$\mathbf{Z}_T = (-j15.915\ \text{k}\Omega) \| 32\ \text{k}\Omega \| 10\ \text{k}\Omega + [160\ \Omega \| (50\ \Omega + j100.5\ \Omega)]$$

$$= 6.87\ \text{k}\Omega\angle{-25.6°} + 77.2\ \Omega\angle 38.0°$$

$$= 6.91\ \text{k}\Omega\angle{-25.0°}$$

The resulting circuit current is

$$\mathbf{I} = \frac{10\ \text{V}\angle 0°}{6.91\ \text{k}\Omega\angle{-25°}} = 1.45\ \text{mA}\angle 25.0°$$

Hay Bridge

In order to measure the inductance and series resistance of an inductor having a small series resistance, a **Hay bridge** is generally used. The Hay bridge is shown in Figure 19–40.

By applying a method similar to that used to determine the values of the unknown inductance and resistance of the Maxwell bridge, we can show that the following equations for the Hay bridge apply:

$$L_x = \frac{R_2R_3C}{\omega^2{R_1}^2C^2 + 1} \quad (19\text{–}12)$$

and

$$R_x = \frac{\omega^2R_1R_2R_3C^2}{\omega^2{R_1}^2C^2 + 1} \quad (19\text{–}13)$$

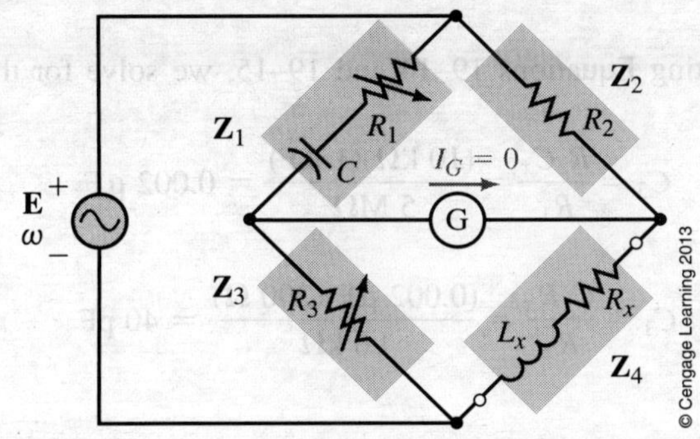

FIGURE 19–40 Hay bridge.

Schering Bridge

The **Schering bridge,** shown in Figure 19–41, is a circuit used to determine the value of unknown capacitance.

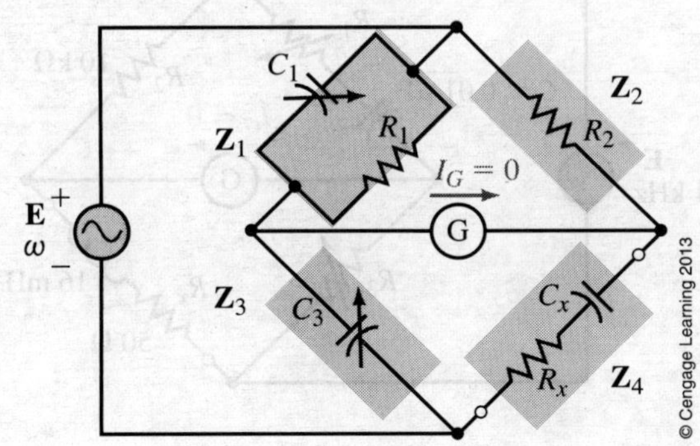

FIGURE 19–41 Schering bridge.

By solving for the balanced bridge condition, we have the following equations for the unknown quantities of the circuit:

$$C_x = \frac{R_1C_3}{R_2} \quad (19\text{–}14)$$

$$R_x = \frac{C_1R_2}{C_3} \quad (19\text{–}15)$$

EXAMPLE 19–15

Determine the values of C_1 and C_3 that will result in a balanced bridge for the circuit of Figure 19–42.

CircuitSim 19-2

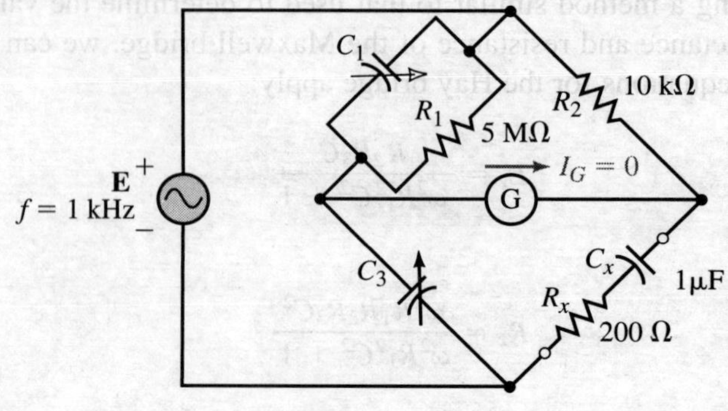

FIGURE 19–42

Solution Rewriting Equations 19–14 and 19–15, we solve for the unknown capacitances as

$$C_3 = \frac{R_2 C_x}{R_1} = \frac{(10\ \text{k}\Omega)(1\ \mu\text{F})}{5\ \text{M}\Omega} = 0.002\ \mu\text{F}$$

and

$$C_1 = \frac{C_3 R_x}{R_2} = \frac{(0.002\ \mu\text{F})(200\ \Omega)}{10\ \text{k}\Omega} = 40\ \text{pF}$$

PRACTICE PROBLEMS 6

Determine the values of R_1 and R_3 so that the bridge of Figure 19–43 is balanced.

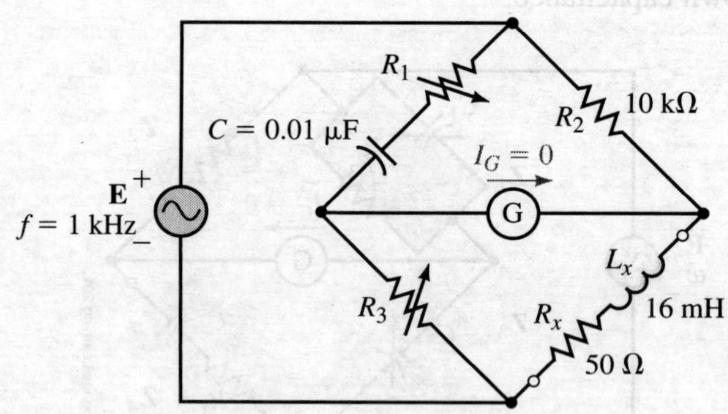

FIGURE 19–43

Answers
$R_1 = 7916\ \Omega$, $R_3 = 199.6\ \Omega$

19.7 Circuit Analysis Using Computers

In some of the examples in this chapter, we analyzed circuits that resulted in as many as three simultaneous linear equations. You have no doubt wondered if there is a less complicated way to solve these circuits without the need for using complex algebra. Computer programs are particularly useful for solving such ac circuits. Both Multisim and PSpice have individual strengths in the solution of ac circuits. As in previous examples, Multisim provides an excellent simulation of how measurements are taken in a lab. PSpice, on the other hand, provides voltage and current readings, complete with magnitude and phase angle. The following examples show how these programs are useful for examining the circuits in this chapter.

MULTISIM

PSPICE

EXAMPLE 19–16

Use Multisim to show that the bridge circuit of Figure 19–44 is balanced.

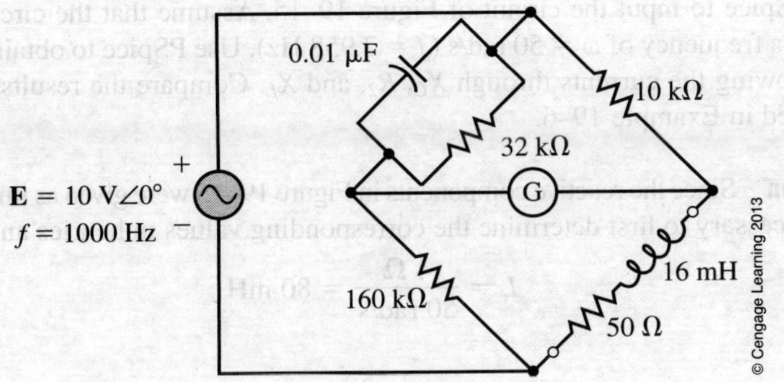

FIGURE 19–44

Solution Recall that a bridge circuit is balanced when the current through the branch between the two arms of the bridge is equal to zero. In this example, we will use a multimeter set on its ac ammeter range to verify the condition of the circuit. The ammeter is selected by clicking on **A,** and it is set to its ac range by clicking on the sinusoidal button. Figure 19–45 shows the circuit connections and the ammeter reading. The results correspond to the conditions that were previously analyzed in Example 19–14. (Note: When we use Multisim, the ammeter may not show exactly zero current in the balanced condition. This is due to the way the program does the calculations. Any current less than 5 µA is considered to be effectively zero.)

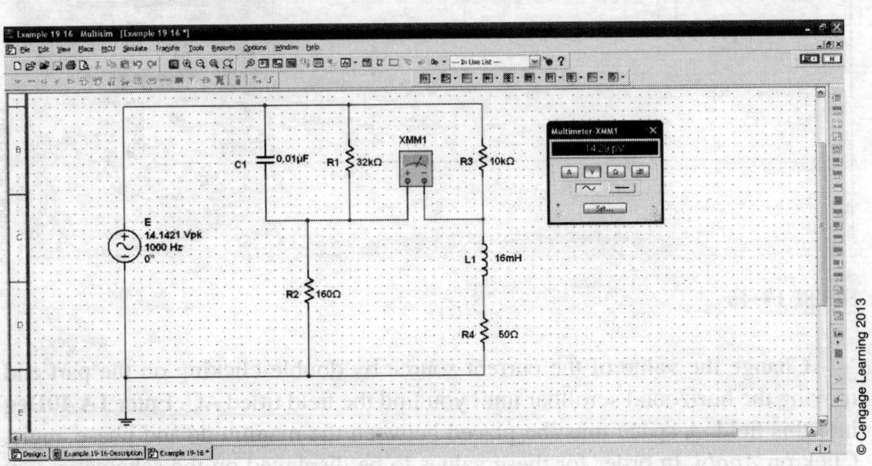

FIGURE 19–45

PRACTICE PROBLEMS 7

Use Multisim to verify that the results obtained in Example 19–15 result in a balanced bridge circuit. (Assume that the bridge is balanced if the galvanometer current is less than 5 μA.)

PSpice

EXAMPLE 19–17

Use PSpice to input the circuit of Figure 19–15. Assume that the circuit operates at a frequency of $\omega = 50$ rad/s ($f = 7.958$ Hz). Use PSpice to obtain a printout showing the currents through X_C, R_2, and X_L. Compare the results to those obtained in Example 19–6.

Solution Since the reactive components in Figure 19–15 were given as impedance, it is necessary to first determine the corresponding values in henries and farads.

$$L = \frac{4\ \Omega}{50\ \text{rad/s}} = 80\ \text{mH}$$

and

$$C = \frac{1}{(2\ \Omega)(50\ \text{rad/s})} = 10\ \text{mF}$$

Now we are ready to use OrCAD Capture CIS to input the circuit as shown in Figure 19–46. The basic steps are reviewed for you. Use the ac current source, ISRC from the SOURCE library and place one IPRINT part from the SPECIAL library. The resistor, inductor, and capacitor are selected from the ANALOG library, and the ground symbol is selected by using the Place ground tool.

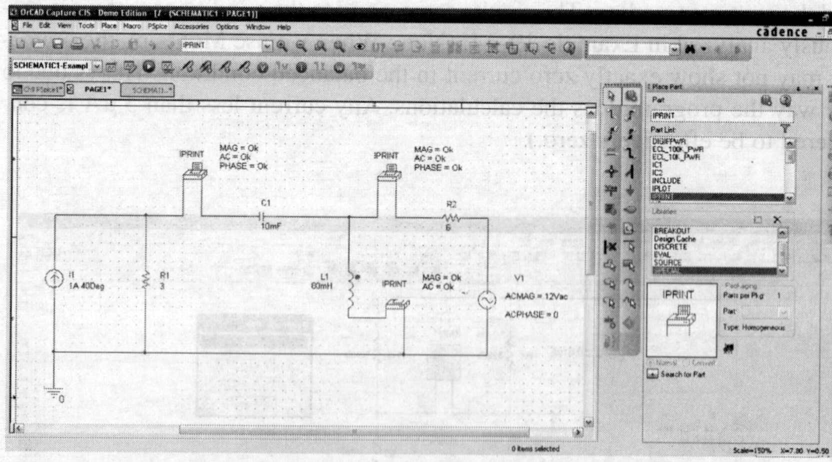

FIGURE 19–46

Change the value of the current source by double-clicking on the part and moving the horizontal scrollbar until you find the field titled AC. Enter **1A 40Deg** into this field. A space must be placed between the magnitude and phase angle. Click on *Apply*. In order for these values to be displayed on the schematic, you must click on the *Display* button and then *Value Only*. Click on OK to return to the properties editor and then close the editor by clicking on X.

The IPRINT part is similar to an ammeter and provides a printout of the current magnitude and phase angle. The properties of the IPRINT part are changed by double-clicking on the part and scrolling across to show the appropriate fields. Type **OK** in the AC, MAG, and PHASE fields. In order to display the selected fields on the schematic, you must click on the *Display* button and then select *Name* and *Value* after changing each field. Since we need to measure three currents in the circuit, we could follow this procedure two more times. However, an easier method is to click on the IPRINT part and copy the part by using <Ctrl><C> and <Ctrl><V>. Each IPRINT will then have the same properties.

Once the rest of the circuit is completed and wired, click on the *New Simulation Profile* tool. Give the simulation a name (such as **ac Branch Currents**). Click on the *Analysis* tab and select *AC Sweep/ Noise* as the analysis type. Type the following values:

Start Frequency: **7.958Hz**
End Frequency: **7.958Hz**
Total Points: **1**

Since we do not need the Probe postprocessor to run, it is disabled by selecting the *Probe Window* tab (from the *Simulation Settings* dialog box). Click on *Display Probe window* and exit the simulation settings by clicking on *OK*.

Click on the *Run* tool. Once PSpice has successfully run, click on the *View* menu and select the *Output File* menu item. Scroll through the file until the currents are shown as follows:

```
FREQ          IM(V_PRINT1)IP(V_PRINT1)
7.958E+00    7.887E-01    -1.201E+02
FREQ          IM(V_PRINT2)IP(V_PRINT2)
7.958E+00    1.304E+00      1.560E+02
FREQ          IM(V_PRINT3)IP(V_PRINT3)
7.958E+00    1.450E+00    -5.673E+01
```

The above printout provides: $\mathbf{I}_1 = 0.7887\text{ A}\angle -120.1°$, $\mathbf{I}_2 = 1.304\text{ A}\angle 156.0°$, and $\mathbf{I}_3 = 1.450\text{ A}\angle -56.73°$. These results are consistent with those calculated in Example 19–6.

PRACTICE PROBLEMS 8

Use PSpice to evaluate the node voltages for the circuit of Figure 19–23. Assume that the circuit operates at an angular frequency of $\omega = 1000$ rad/s ($f = 159.15$ Hz).

Putting It into Practice

The Schering bridge of Figure 19–74 is balanced. In this chapter, you have learned several methods that allow you to find the current anywhere in a circuit. Using any method, determine the current through the galvanometer if the value of $C_x = 0.07\ \mu\text{F}$. (All other values remain unchanged.) Repeat the calculations for a value of $C_x = 0.09\ \mu\text{F}$. Can you make a general statement for current through the galvanometer if C_x is smaller than the value required to balance the bridge? What general statement can be made if the value of C_x is larger than the value in the balanced bridge?

Problems

19.1 Dependent Sources

1. Refer to the circuit of Figure 19–47.

   Find V when the controlling current I is the following:
   a. $20\ \mu A\angle 0°$
   b. $50\ \mu A\angle -180°$
   c. $60\ \mu A\angle 60°$

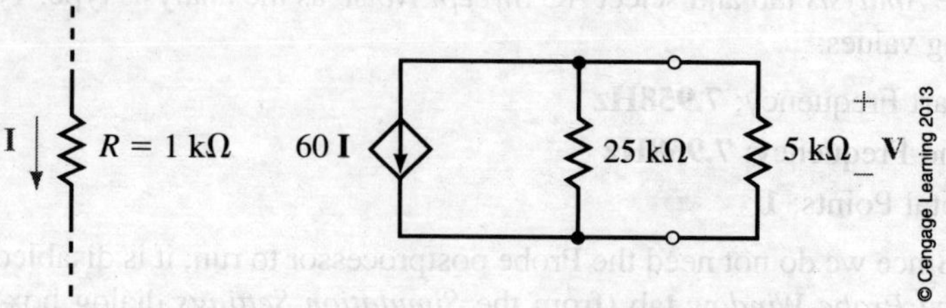

FIGURE 19–47

2. Refer to the circuit of Figure 19–48.
   Find I when the controlling voltage, V, is the following:
   a. $30\ mV\angle 0°$
   b. $60\ mV\angle -180°$
   c. $100\ mV\angle -30°$

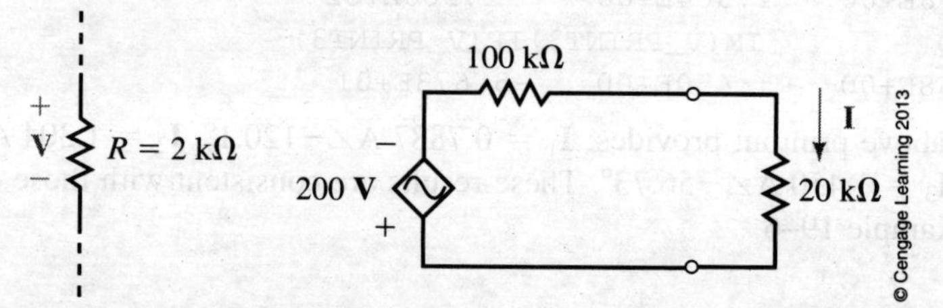

FIGURE 19–48

3. Repeat Problem 1 for the circuit of Figure 19–49.

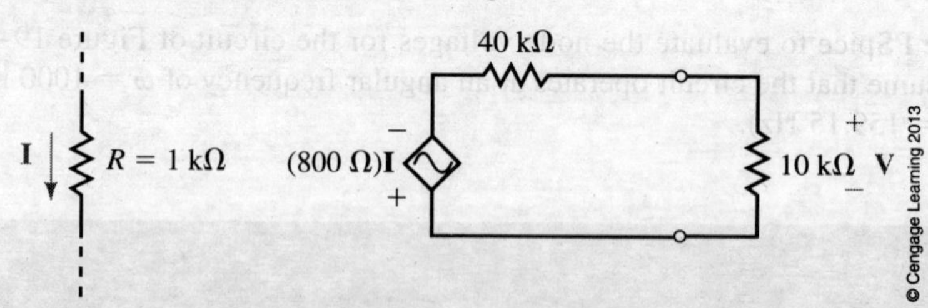

FIGURE 19–49

4. Repeat Problem 2 for the circuit of Figure 19–50.

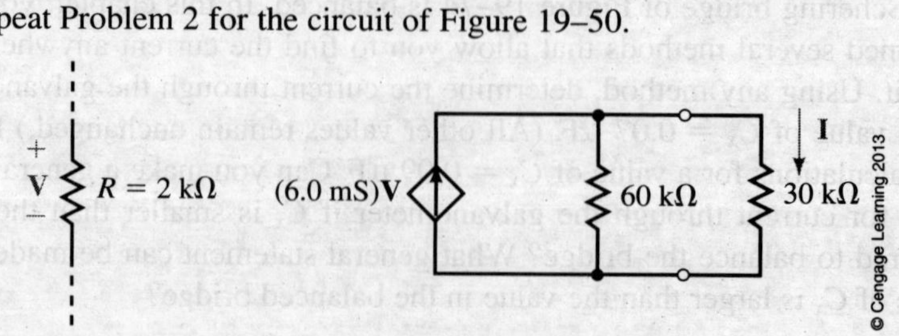

FIGURE 19–50

5. Find the output voltage, $\mathbf{V}_{out}$, for the circuit of Figure 19–51.

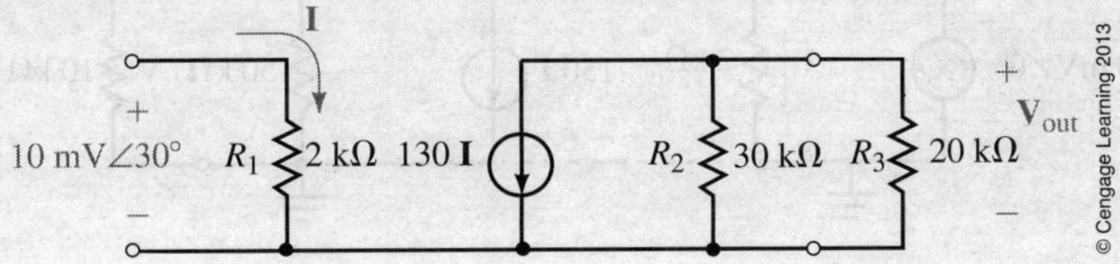

FIGURE 19–51

6. Repeat Problem 5 for the circuit of Figure 19–52.

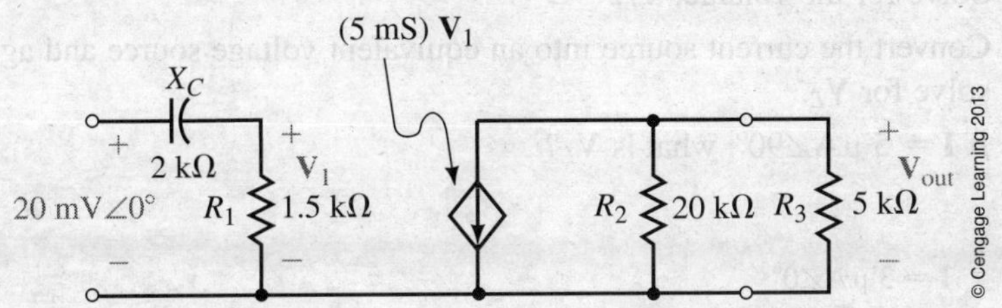

FIGURE 19–52

19.2 Source Conversion

7. Given the circuits of Figure 19–53, convert each of the current sources into an equivalent voltage source. Use the resulting circuit to find $\mathbf{V}_L$.

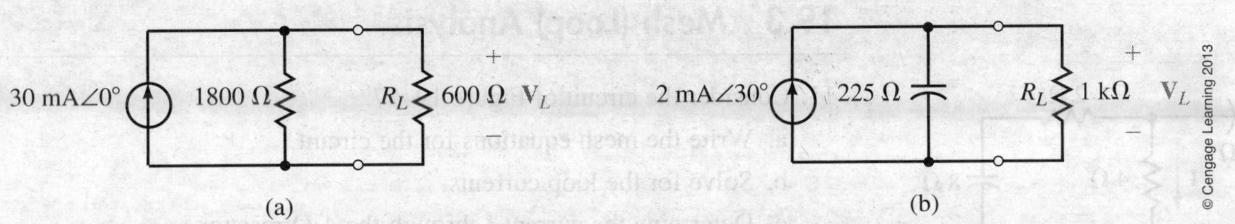

FIGURE 19–53

8. Convert each voltage source of Figure 19–54 into an equivalent current source.

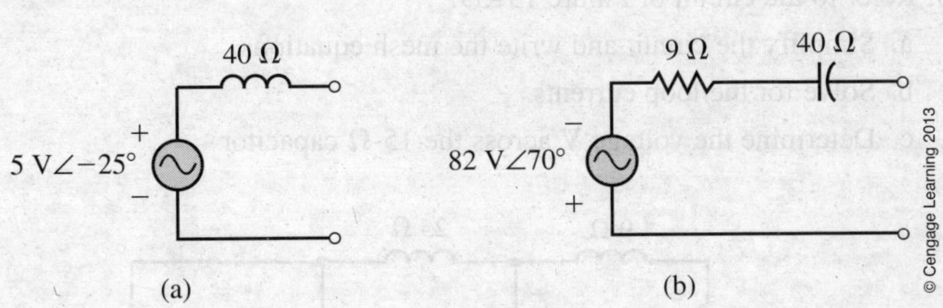

FIGURE 19–54

9. Refer to the circuit of Figure 19–55.
   a. Solve for the voltage, **V**.
   b. Convert the current source into an equivalent voltage source and again solve for **V**. Compare to the result obtained in (a).

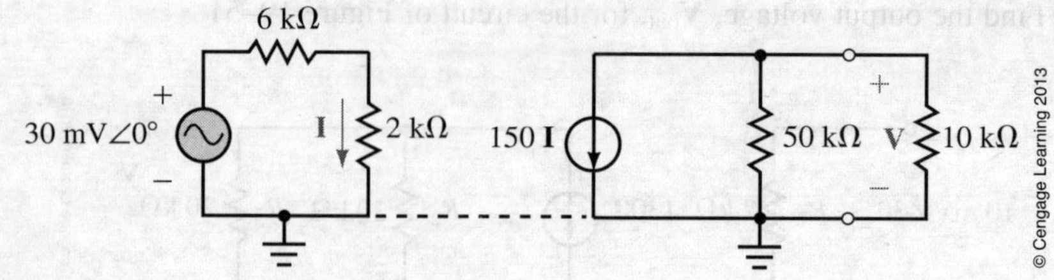

FIGURE 19–55

10. Refer to the circuit of Figure 19–56.
    a. Solve for the voltage, $\mathbf{V}_L$.
    b. Convert the current source into an equivalent voltage source and again solve for $\mathbf{V}_L$.
    c. If $\mathbf{I} = 5\ \mu\text{A}\angle 90°$, what is $\mathbf{V}_L$?

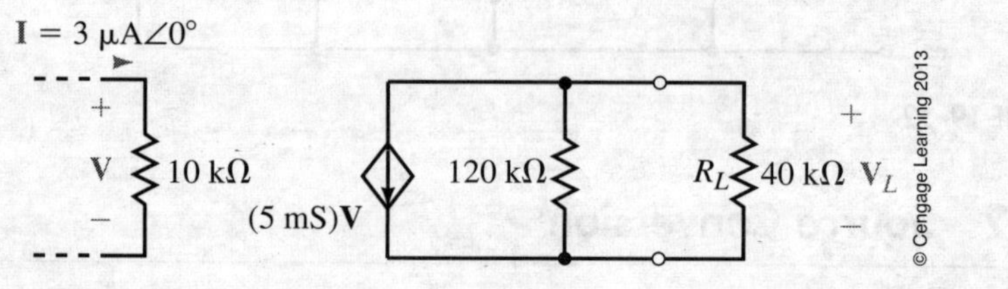

FIGURE 19–56

19.3 Mesh (Loop) Analysis

11. Consider the circuit of Figure 19–57.
    a. Write the mesh equations for the circuit.
    b. Solve for the loop currents.
    c. Determine the current **I** through the 4-Ω resistor.
12. Refer to the circuit of Figure 19–58.
    a. Write the mesh equations for the circuit.
    b. Solve for the loop currents.
    c. Determine the current through the 25-Ω inductor.
13. Refer to the circuit of Figure 19–59.
    a. Simplify the circuit and write the mesh equations.
    b. Solve for the loop currents.
    c. Determine the voltage **V** across the 15-Ω capacitor.

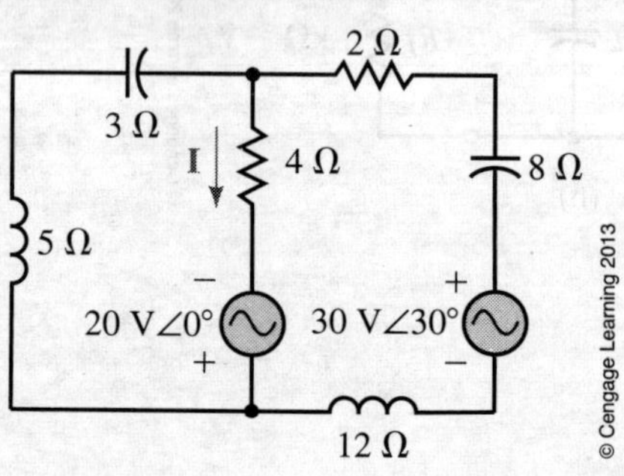

FIGURE 19–57

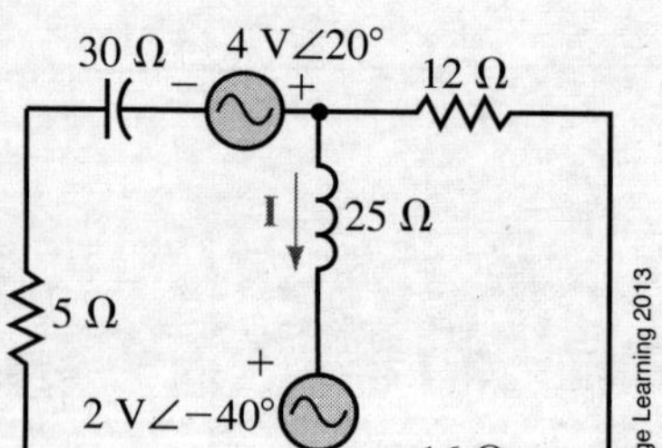

FIGURE 19–58

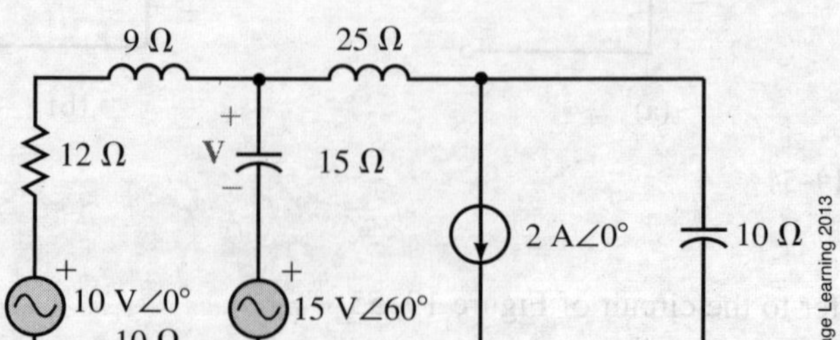

FIGURE 19–59

14. Consider the circuit of Figure 19–60.
    a. Simplify the circuit and write the mesh equations.
    b. Solve for the loop currents.
    c. Determine the voltage **V** across the 2-Ω resistor.

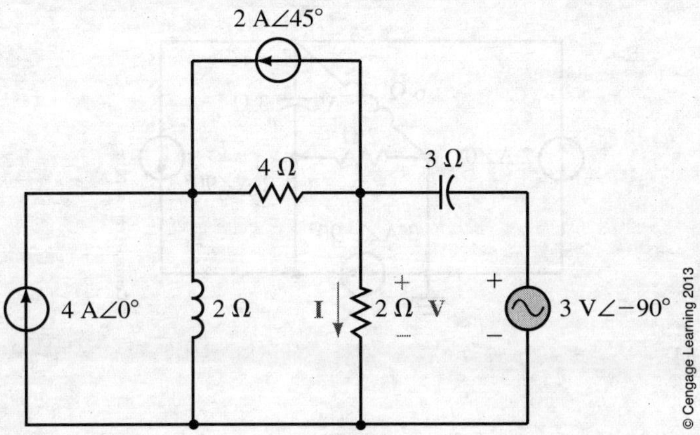

FIGURE 19–60

15. Use mesh analysis to find the current **I** and the voltage **V** in the circuit of Figure 19–61.
16. Repeat Problem 15 for the circuit of Figure 19–62.

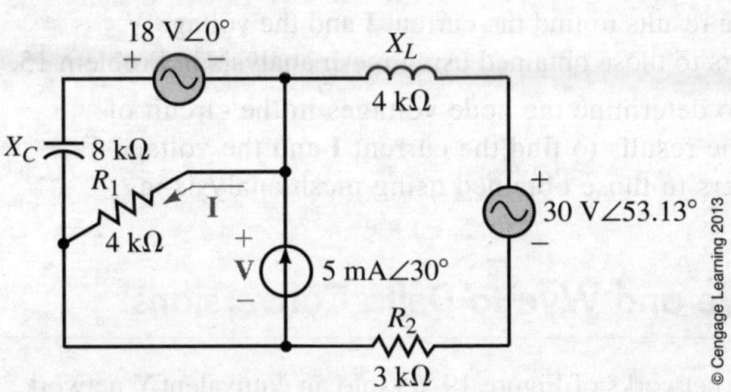

FIGURE 19–61

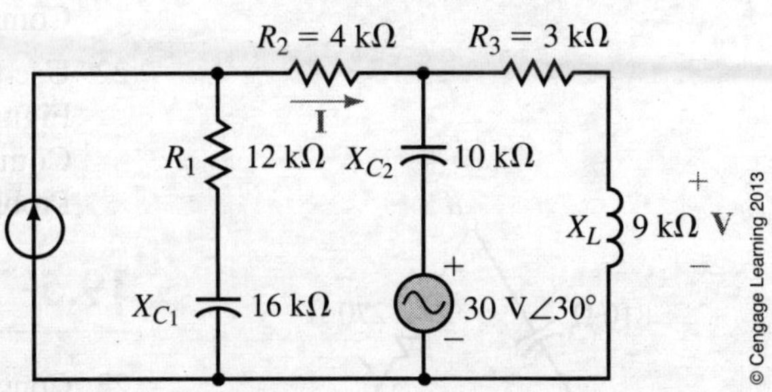

FIGURE 19–62

19.4 Nodal Analysis

17. Consider the circuit of Figure 19–63.
    a. Write the nodal equations.
    b. Solve for the node voltages.
    c. Determine the current **I** through the 4-Ω capacitor.

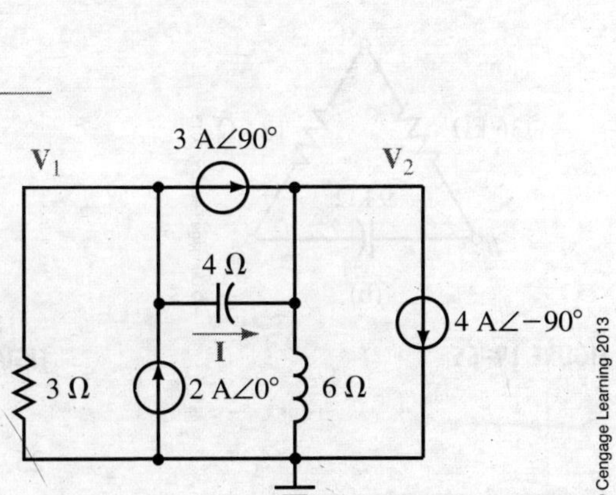

FIGURE 19–63

18. Refer to the circuit of Figure 19–64.
    a. Write the nodal equations.
    b. Solve for the node voltages.
    c. Determine the voltage **V** across the 3-Ω capacitor.

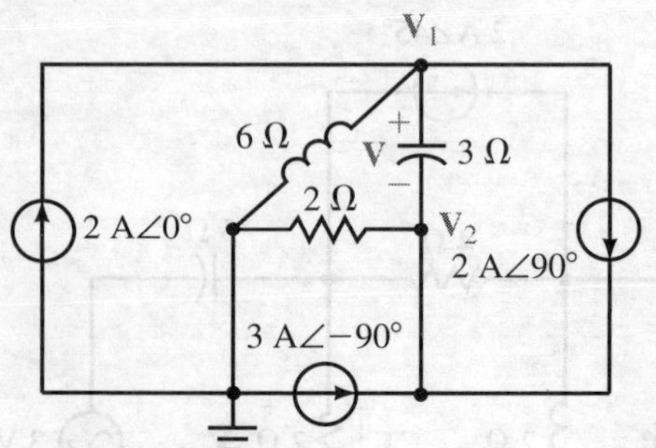

FIGURE 19–64

19. a. Simplify the circuit of Figure 19–59, and write the nodal equations.
    b. Solve for the node voltages.
    c. Determine the voltage across the 15-Ω capacitor.
20. a. Simplify the circuit of Figure 19–60, and write the nodal equations.
    b. Solve for the node voltages.
    c. Determine the current through the 2-Ω resistor.
21. Use nodal analysis to determine the node voltages in the circuit of Figure 19–61. Use the results to find the current **I** and the voltage **V**. Compare your answers to those obtained using mesh analysis in Problem 15.
22. Use nodal analysis to determine the node voltages in the circuit of Figure 19–62. Use the results to find the current **I** and the voltage **V**. Compare your answers to those obtained using mesh analysis in Problem 16.

19.5 Delta-to-Wye and Wye-to-Delta Conversions

23. Convert each of the Δ networks of Figure 19–65 into an equivalent Y network.
24. Convert each of the Y networks of Figure 19–66 into an equivalent Δ network.

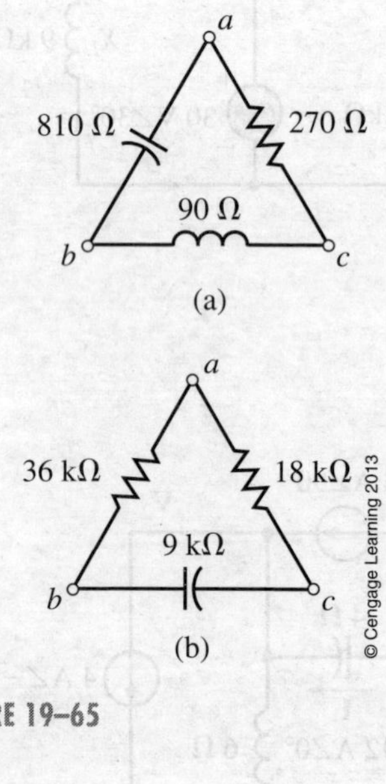

FIGURE 19–65

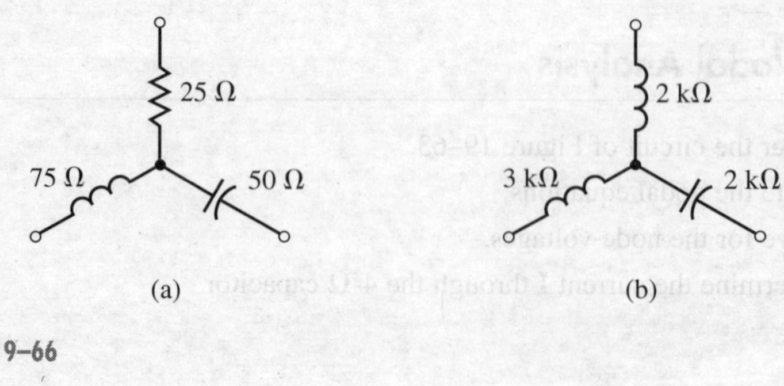

FIGURE 19–66

25. Using Δ → Y or Y → Δ conversion, calculate **I** for the circuit of Figure 19–67.
26. Using Δ → Y or Y → Δ conversion, calculate **I** for the circuit of Figure 19–68.

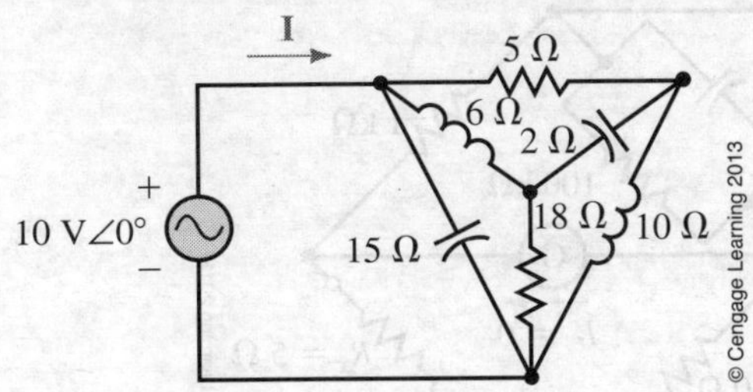

FIGURE 19–67

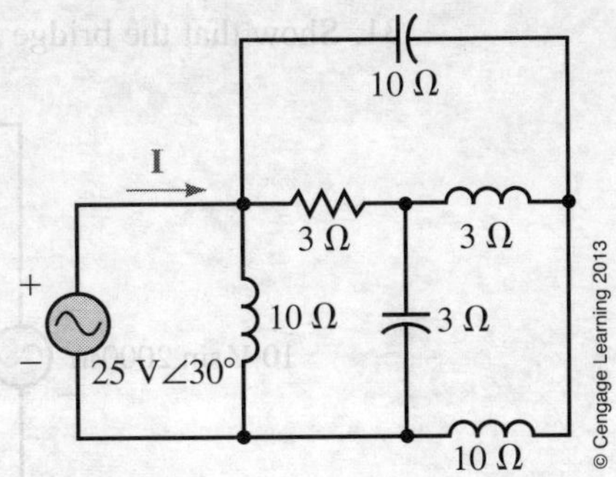

FIGURE 19–68

27. Refer to the circuit of Figure 19–69:
    a. Determine the equivalent impedance, $\mathbf{Z}_T$, of the circuit.
    b. Find the currents **I** and $\mathbf{I}_1$.
28. Refer to the circuit of Figure 19–70:
    a. Determine the equivalent impedance, $\mathbf{Z}_T$, of the circuit.
    b. Find the voltages **V** and $\mathbf{V}_1$.

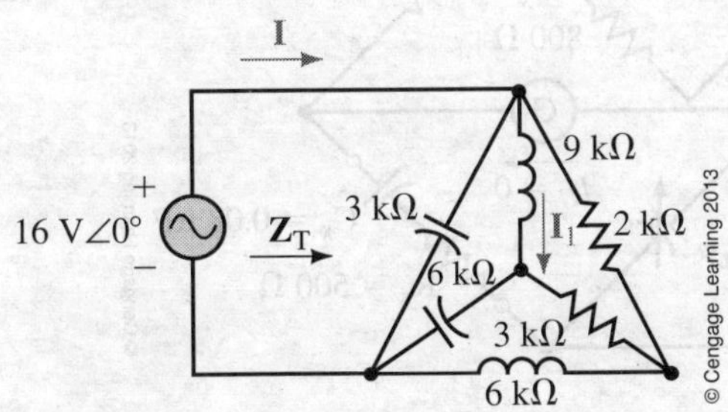

FIGURE 19–69

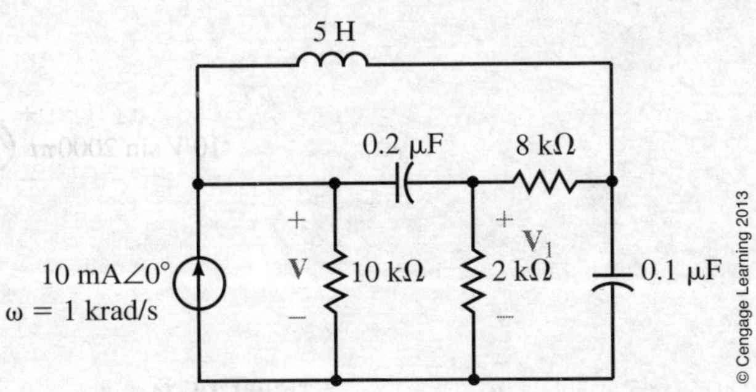

FIGURE 19–70

19.6 Bridge Networks

29. Given that the bridge circuit of Figure 19–71 is balanced:
    a. Determine the value of the unknown impedance.
    b. Solve for the current **I**.
30. Given that the bridge circuit of Figure 19–72 is balanced:
    a. Determine the value of the unknown impedance.
    b. Solve for the current **I**.

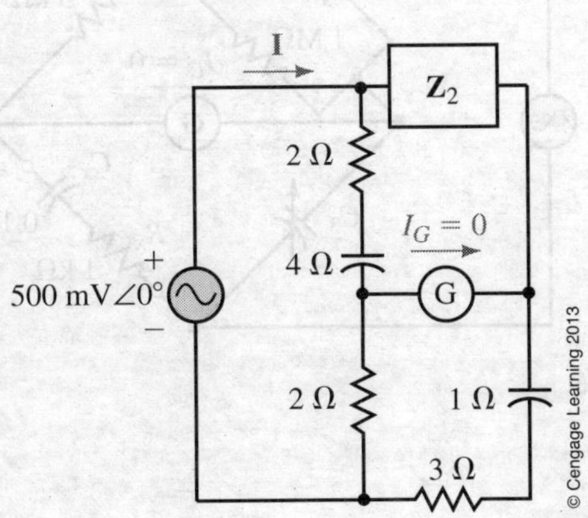

FIGURE 19–71

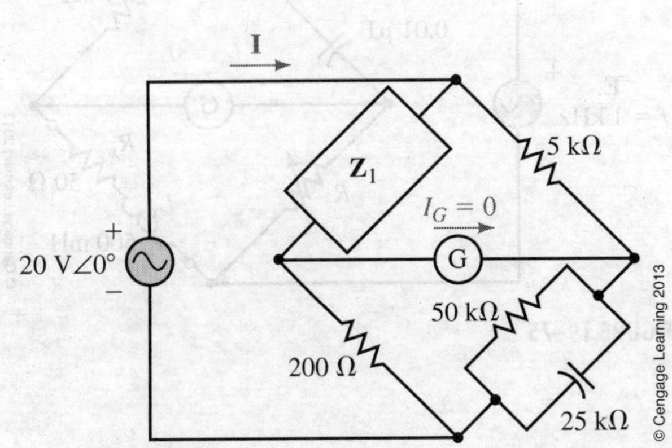

FIGURE 19–72

31. Show that the bridge circuit of Figure 19–73 is balanced.

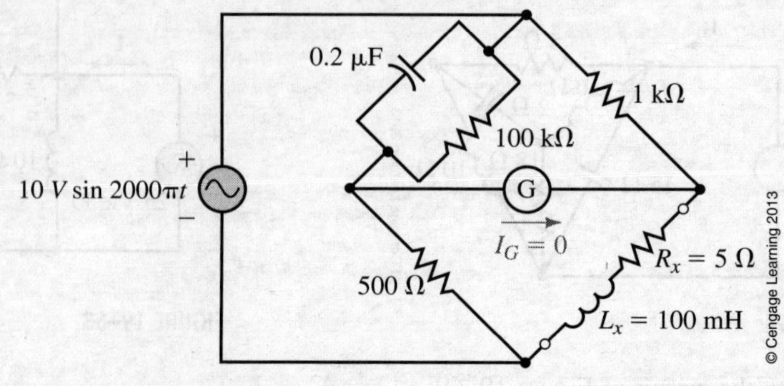

FIGURE 19–73

32. Show that the bridge circuit of Figure 19–74 is balanced.

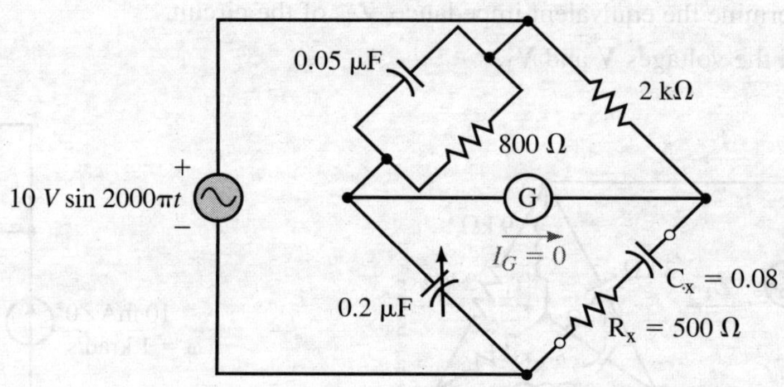

FIGURE 19–74

33. Derive Equations 19–14 and 19–15 for the balanced Schering bridge.
34. Derive Equations 19–12 and 19–13 for the balanced Hay bridge.
35. Determine the values of the unknown resistors that will result in a balanced bridge for the circuit of Figure 19–75.
36. Determine the values of the unknown capacitors which will result in a balanced bridge for the circuit of Figure 19–76.

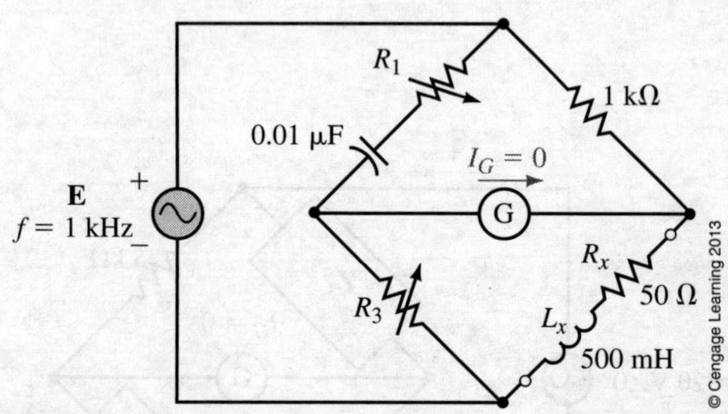

FIGURE 19–75

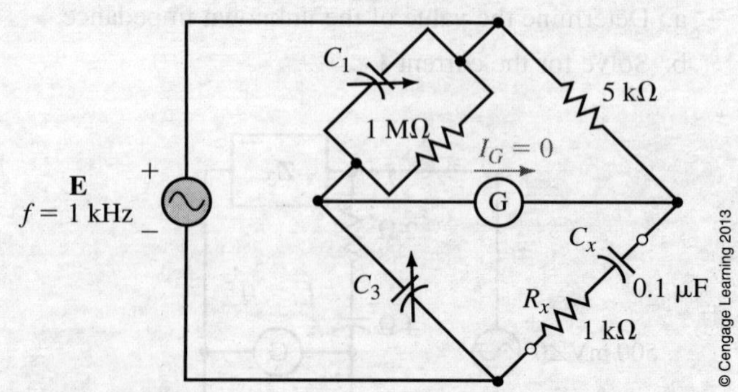

FIGURE 19–76

19.7 Circuit Analysis Using Computers

37. Use Multisim to show that the bridge circuit of Figure 19–73 is balanced. (Assume that the bridge is balanced if the galvanometer current is less than 5 μA.) Multisim
38. Repeat Problem 37 for the bridge circuit of Figure 19–74. Multisim
39. Use PSpice to input the file for the circuit of Figure 19–21. Assume that the circuit operates at a frequency $\omega = 2$ krad/s. Use IPRINT and VPRINT to obtain a printout of the node voltages and the current through each element of the circuit. PSPICE
40. Use PSpice to input the file for the circuit of Figure 19–29. Assume that the circuit operates at a frequency $\omega = 1$ krad/s. Use IPRINT and VPRINT to obtain a printout of the node voltages and the current through each element of the circuit. PSPICE
41. Use PSpice to input the file for the circuit of Figure 19–68. Assume that the circuit operates at a frequency $\omega = 20$ rad/s. Use IPRINT to obtain a printout of the current **I**. PSPICE
42. Use PSpice to input the file for the circuit of Figure 19–69. Assume that the circuit operates at a frequency $\omega = 3$ krad/s. Use IPRINT to obtain a printout of the current **I**. PSPICE

ANSWERS TO IN-PROCESS LEARNING CHECKS

IN-PROCESS LEARNING CHECK 1

a. $\mathbf{E} = 1000\ \text{V}\angle 30^\circ$ b. $\mathbf{E} = 4\ \text{V}\angle -90^\circ$
c. $\mathbf{E} = 1024\ \text{V}\angle -38.66^\circ$

IN-PROCESS LEARNING CHECK 2

1. Convert current sources to voltage sources.
2. Redraw the circuit.
3. Assign a clockwise current to each loop.
4. Write loop equations using Kirchhoff's voltage law.
5. Solve the resulting simultaneous linear equations to find the loop currents.

IN-PROCESS LEARNING CHECK 3

1. Convert voltage sources to current sources.
2. Redraw the circuit.
3. Label all nodes, including the reference node.
4. Write nodal equations using Kirchhoff's current law.
5. Solve the resulting simultaneous linear equations to find the node voltages.

IN-PROCESS LEARNING CHECK 4

$\mathbf{Z}_1 = 150\ \Omega\angle 90^\circ$, $\mathbf{Z}_2 = 150\ \Omega\angle -90^\circ$, $\mathbf{Z}_3 = 150\ \Omega\angle 0^\circ$

■ KEY TERMS

Absolute Maximum Power

Gain

Relative Maximum Power

■ OUTLINE

Superposition Theorem—Independent Sources

Superposition Theorem—Dependent Sources

Thévenin's Theorem—Independent Sources

Norton's Theorem—Independent Sources

Thévenin's and Norton's Theorems for Dependent Sources

Maximum Power Transfer Theorem

Circuit Analysis Using Computers

■ OBJECTIVES

After studying this chapter you will be able to

- apply the superposition theorem to determine the voltage across or current through any component in a given circuit,
- determine the Thévenin equivalent of circuits having independent and/or dependent sources,
- determine the Norton equivalent of circuits having independent and/or dependent sources,
- apply the maximum power transfer theorem to determine the load impedance for which maximum power is transferred to the load from a given circuit,
- use PSpice to find the Thévenin and Norton equivalents of circuits having either independent or dependent sources,
- use Multisim to verify the operation of ac circuits.

20

ac NETWORK THEOREMS

CHAPTER PREVIEW

In this chapter we apply the superposition, Thévenin, Norton, and maximum power transfer theorems in the analysis of ac circuits. Although the Millman and reciprocity theorems apply to ac circuits as well as to dc circuits, they are omitted since the applications are virtually identical with those used in analyzing dc circuits.

Many of the techniques used in this chapter are similar to those used in Chapter 9, and as a result, most students will find a brief review of dc theorems useful.

This chapter examines the application of the network theorems by considering both independent and dependent sources. In order to show the distinctions between the methods used in analyzing the various types of sources, the sections are labelled according to the types of sources involved.

An understanding of dependent sources is particularly useful when working with transistor circuits and operational amplifiers. Sections 20.2 and 20.5 are intended to provide the background for analyzing the operation of feedback amplifiers. Your instructor may find that these topics are best left until you cover this topic in a course dealing with such amplifiers. Consequently, the omission of Sections 20.2 and 20.5 will not in any way detract from the continuity of the important ideas presented in this chapter. ■

Putting It in Perspective

William Bradford Shockley

Alfred Eisenstaedt/Time & Life Pictures/Getty Images

SHOCKLEY WAS BORN THE SON of a mining engineer in London, England, on February 13, 1910. After graduating from the California Institute of Technology and Massachusetts Institute of Technology, Shockley joined Bell Telephone Laboratories.

With his co-workers John Bardeen and Walter Brattain, Shockley developed an improved solid-state rectifier using a germanium crystal which had been injected with minute amounts of impurities. Unlike vacuum tubes, the resulting diodes were able to operate at much lower voltages without the need for inefficient heater elements.

In 1948, Shockley combined three layers of germanium to produce a device that was able not only to rectify a signal but to amplify it. Thus was developed the first transistor. Since its humble beginning, the transistor has been improved and decreased in size to the point where now a circuit containing thousands of transistors can easily fit into an area not much bigger than the head of a pin.

The advent of the transistor has permitted the construction of elaborate spacecraft, unprecedented communication, and new forms of energy generation.

Shockley, Bardeen, and Brattain received the 1956 Nobel Prize in Physics for their discovery of the transistor. ■

20.1 Superposition Theorem—Independent Sources

The superposition theorem states the following:

The voltage across (or current through) an element is determined by summing the voltage (or current) due to each independent source.

In order to apply this theorem, all sources other than the one being considered are eliminated. As in dc circuits, this is done by replacing current sources with open circuits and by replacing voltage sources with short circuits. The process is repeated until the effects due to all sources have been determined.

Although we generally work with circuits having all sources at the same frequency, occasionally a circuit may operate at more than one frequency at a time. This is particularly true in diode and transistor circuits which use a dc source to set a "bias" (or operating) point and an ac source to provide the signal to be conditioned or amplified. In such cases, the resulting voltages or currents are still determined by applying the superposition theorem. The topic of how to solve circuits operating at several different frequencies simultaneously is covered in Chapter 25.

NOTES...

As in dc circuits, the superposition theorem can be applied only to voltage and current; it cannot be used to solve for the total power dissipated by an element. This is because power is not a linear quantity, but rather follows a square-law relationship ($P = V^2/R = I^2R$).

EXAMPLE 20–1

Determine the current **I** in Figure 20–1 by using the superposition theorem.

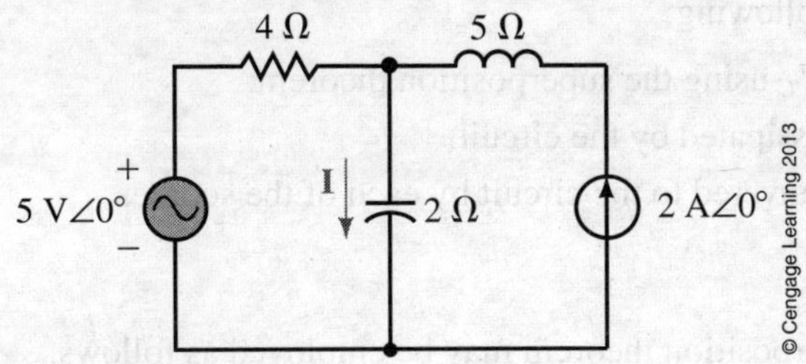

FIGURE 20–1

Solution

Current due to the 5 V∠0° voltage source: Eliminating the current source, we obtain the circuit shown in Figure 20–2.

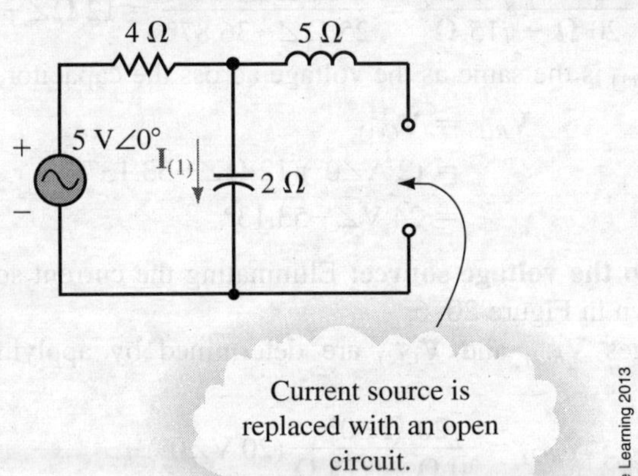

FIGURE 20–2

Applying Ohm's law, we have

$$\mathbf{I}_{(1)} = \frac{5\text{V}\angle 0°}{4 - j2\ \Omega} = \frac{5\text{V}\angle 0°}{4.472\ \Omega\angle -26.57°}$$
$$= 1.118\ \text{A}\angle 26.57°$$

Current due to the 2 A∠0° current source: Eliminating the voltage source, we obtain the circuit shown in Figure 20–3.

The current $\mathbf{I}_{(2)}$ due to this source is determined by applying the current divider rule:

$$\mathbf{I}_{(2)} = (2\ \text{A}\angle 0°)\frac{4\ \Omega\angle 0°}{4\ \Omega - j2\ \Omega}$$
$$= \frac{8\ \text{V}\angle 0°}{4.472\ \Omega\angle -26.57°}$$
$$= 1.789\ \text{A}\angle 26.57°$$

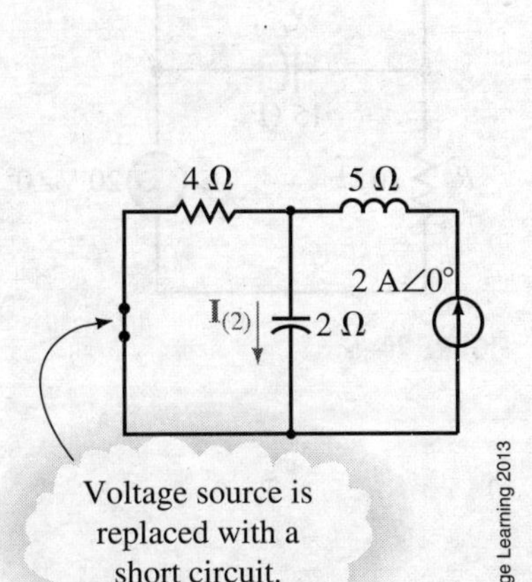

FIGURE 20–3

The total current is determined as the summation of currents $\mathbf{I}_{(1)}$ and $\mathbf{I}_{(2)}$:

$$\mathbf{I} = \mathbf{I}_{(1)} + \mathbf{I}_{(2)}$$
$$= 1.118\ \text{A}\angle 26.57° + 1.789\ \text{A}\angle 26.57°$$
$$= (1.0\ \text{A} + j0.5\ \text{A}) + (1.6\ \text{A} + j0.8\ \text{A})$$
$$= 2.6 + j1.3\ \text{A}$$
$$= 2.91\ \text{A}\angle 26.57°$$

EXAMPLE 20–2

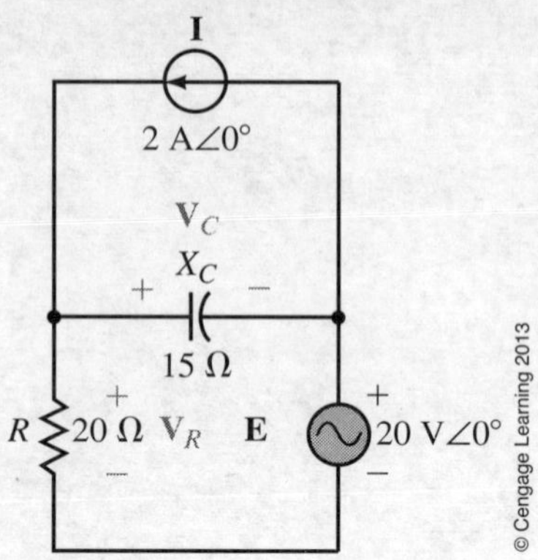

FIGURE 20–4

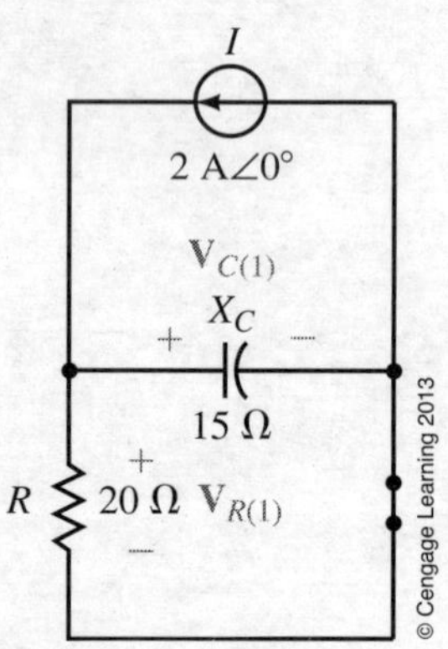

FIGURE 20–5

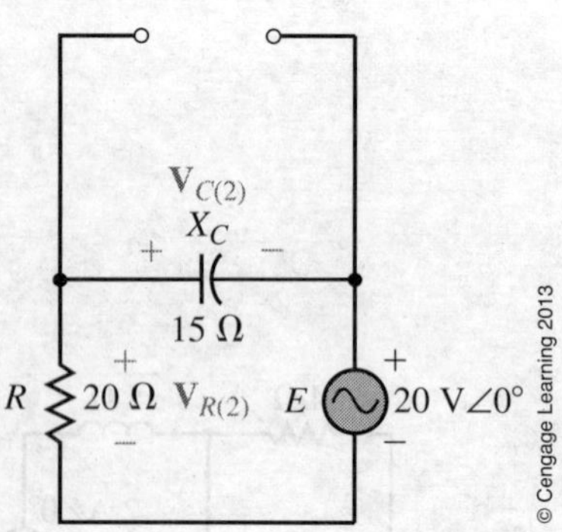

FIGURE 20–6

Consider the circuit of Figure 20–4:

Find the following:

a. $\mathbf{V}_R$ and $\mathbf{V}_C$ using the superposition theorem.

b. Power dissipated by the circuit.

c. Power delivered to the circuit by each of the sources.

Solution

a. The superposition theorem may be employed as follows:

Voltages due to the current source: Eliminating the voltage source, we obtain the circuit shown in Figure 20–5.

The impedance "seen" by the current source will be the parallel combination of $\mathbf{R}\|\mathbf{Z}_C$.

$$\mathbf{Z}_1 = \frac{(20\ \Omega)(-j15\ \Omega)}{20\ \Omega - j15\ \Omega} = \frac{300\ \Omega\angle -90^\circ}{25\ \Omega\angle -36.87^\circ} = 12\ \Omega\angle -53.13^\circ$$

The voltage $\mathbf{V}_{R(1)}$ is the same as the voltage across the capacitor, $\mathbf{V}_{C(1)}$. Hence,

$$\begin{aligned}\mathbf{V}_{R(1)} &= \mathbf{V}_{C(1)}\\ &= (2\ \text{A}\angle 0^\circ)(12\ \Omega\angle -53.13^\circ)\\ &= 24\ \text{V}\angle -53.13^\circ\end{aligned}$$

Voltages due to the voltage source: Eliminating the current source, we have the circuit shown in Figure 20–6.

The voltages $\mathbf{V}_{R(2)}$ and $\mathbf{V}_{C(2)}$ are determined by applying the voltage divider rule,

$$\begin{aligned}\mathbf{V}_{R(2)} &= \frac{20\ \Omega\angle 0^\circ}{20\ \Omega - j15\ \Omega}(20\ \text{V}\angle 0^\circ)\\ &= \frac{400\ \text{V}\angle 0^\circ}{25\ \angle -36.87^\circ} = 16\ \text{V}\angle +36.87^\circ\end{aligned}$$

and

$$\begin{aligned}\mathbf{V}_{C(2)} &= \frac{-15\ \Omega\angle -90^\circ}{20\ \Omega - j15\ \Omega}(20\ \text{V}\angle 0^\circ)\\ &= \frac{300\ \text{V}\angle 90^\circ}{25\ \angle -36.87^\circ} = 12\ \text{V}\angle 126.87^\circ\end{aligned}$$

Notice that $\mathbf{V}_{C(2)}$ is assigned to be negative relative to the originally assumed polarity. The negative sign is eliminated from the calculation by adding (or subtracting) 180° from the corresponding calculation.

By applying superposition, we get

$$\begin{aligned}\mathbf{V}_R &= \mathbf{V}_{R(1)} + \mathbf{V}_{R(2)}\\ &= 24\ \text{V}\angle -53.13^\circ + 16\ \text{V}\angle 36.87^\circ\\ &= (14.4\ \text{V} - j19.2\ \text{V}) + (12.8\ \text{V} + j9.6\ \text{V})\\ &= 27.2\ \text{V} - j9.6\ \text{V}\\ &= 28.84\ \text{V}\angle -19.44^\circ\end{aligned}$$

and

$$\begin{aligned}\mathbf{V}_C &= \mathbf{V}_{C(1)} + \mathbf{V}_{C(2)}\\ &= 24\ \text{V}\angle -53.13^\circ + 12\ \text{V}\angle 126.87^\circ\\ &= (14.4\ \text{V} - j19.2\ \text{V}) + (-7.2\ \text{V} + j9.6\ \text{V})\\ &= 7.2\ \text{V} - j9.6\ \text{V}\\ &= 12\ \text{V}\angle -53.13^\circ\end{aligned}$$

b. Since only the resistor will dissipate power, the total power dissipated by the circuit is found as

$$P_T = \frac{(28.84\ V)^2}{20\ \Omega} = 41.60\ W$$

c. The power delivered to the circuit by the current source is

$$P_1 = V_1 I \cos \theta_1$$

where $\mathbf{V}_1 = \mathbf{V}_C = 12\ V\angle -53.13°$ is the voltage across the current source and θ_1 is the phase angle between $\mathbf{V}_1$ and $\mathbf{I}$.

The power delivered by the current source is

$$P_1 = (12\ V)(2\ A) \cos 53.13° = 14.4\ W$$

The power delivered to the circuit by the voltage source is similarly determined as

$$P_2 = EI_2 \cos \theta_2$$

where $\mathbf{I}_2$ is the current through the voltage source and θ_2 is the phase angle between $\mathbf{E}$ and $\mathbf{I}_2$.

$$P_2 = (20\ V)\left(\frac{28.84\ V}{20\ \Omega}\right) \cos 19.44° = 27.2\ W$$

As expected, the total power delivered to the circuit must be the summation

$$P_T = P_1 + P_2 = 41.6\ W$$

PRACTICE PROBLEMS 1

Use superposition to find **V** and **I** for the circuit of Figure 20–7.

Answers
$\mathbf{I} = 2.52\ A\angle -25.41°$, $\mathbf{V} = 4.45\ V\angle 104.18°$

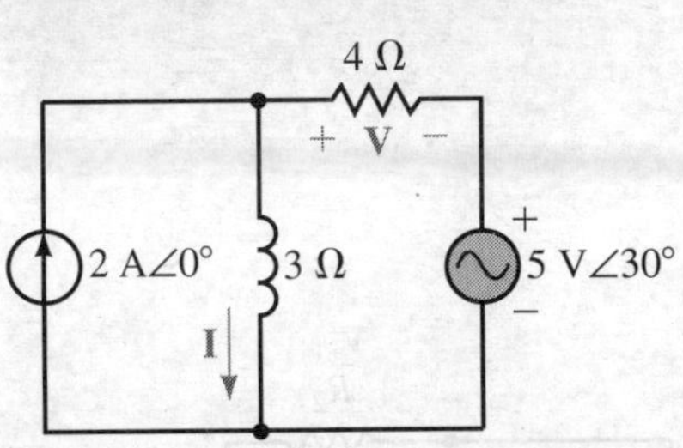

FIGURE 20–7

IN-PROCESS LEARNING CHECK 1

(Answers are at the end of the chapter.)

A 20-Ω resistor is in a circuit having three sinusoidal sources. After analyzing the circuit, it is found that the current through the resistor due to each of the sources is as follows:

$$I_1 = 1.5\ A\angle 20°$$
$$I_2 = 1.0\ A\angle 110°$$
$$I_3 = 2.0\ A\angle 0°$$

a. Use superposition to calculate the resultant current through the resistor.

b. Calculate the power dissipated by the resistor.

c. Show that the power dissipated by the resistor cannot be found by applying superposition, namely, $P_T \neq I_1^2R + I_2^2R + I_3^2R$.

20.2 Superposition Theorem—Dependent Sources

Chapter 19 introduced the concept of dependent sources. We now examine ac circuits that contain dependent sources. In order to analyze such circuits, it is first necessary to determine whether the dependent source is conditional upon a controlling element in its own circuit or whether the controlling element is located in some other circuit.

If the controlling element is external to the circuit under consideration, the method of analysis is the same as for an independent source. However, if the controlling element is in the same circuit, the analysis follows a slightly different strategy. The next two examples show the techniques used to analyze circuits having dependent sources.

EXAMPLE 20–3

Consider the circuit of Figure 20–8.

a. Determine the general expression for **V** in terms of **I**.
b. Calculate **V** if $\mathbf{I} = 1.0\ \text{A}\angle 0°$.
c. Calculate **V** if $\mathbf{I} = 0.3\ \text{A}\angle 90°$.

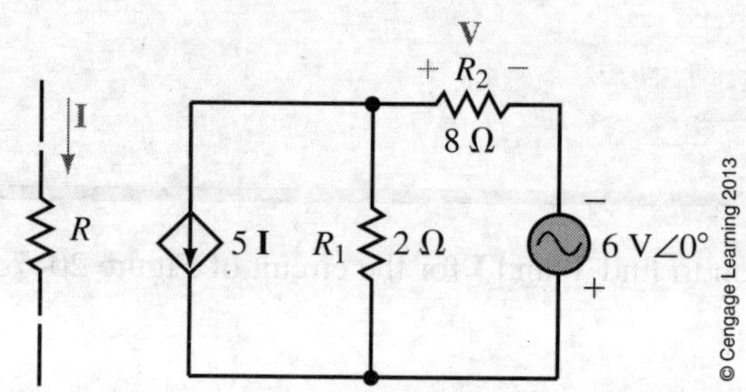

FIGURE 20–8

Solution

a. Since the current source in the circuit is dependent on current through an element that is located outside of the circuit of interest, the circuit may be analyzed in the same manner as for independent sources.

Voltage due to the voltage source: Eliminating the current source, we obtain the circuit shown in Figure 20–9.

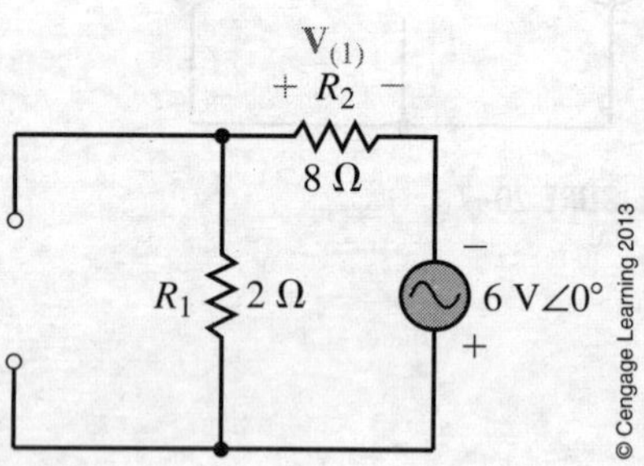

FIGURE 20–9

$$\mathbf{V}_{(1)} = \frac{8\ \Omega}{10\ \Omega}(6\ \text{V}\angle 0°) = 4.8\ \text{V}\angle 0°$$

Voltage due to the current source: Eliminating the voltage source, we have the circuit shown in Figure 20–10.

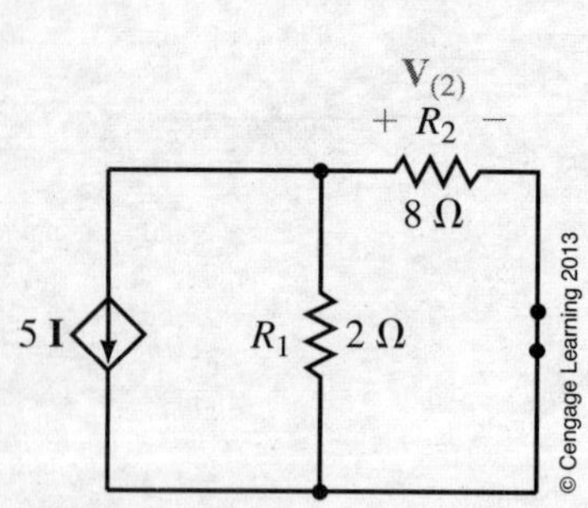

FIGURE 20–10

$$\mathbf{Z}_\text{T} = 2\ \Omega \| 8\ \Omega = 1.6\ \Omega\angle 0°$$

$$\mathbf{V}_{(2)} = \mathbf{V}_{Z_\text{T}} = -(5\mathbf{I})(1.6\ \Omega\angle 0°) = -8.0\ \Omega\mathbf{I}$$

From superposition, the general expression for voltage is determined to be

$$\mathbf{V} = \mathbf{V}_{(1)} + \mathbf{V}_{(2)}$$
$$= 4.8\ \text{V}\angle 0° - 8.0\ \Omega\mathbf{I}$$

b. If $\mathbf{I} = 1.0\ \text{A}\angle 0°$,

$$\mathbf{V} = 4.8\ \text{V}\angle 0° - (8.0\ \Omega)(1.0\ \text{A}\angle 0°) = -3.2\ \text{V}$$
$$= 3.2\ \text{V}\angle 180°$$

c. If $\mathbf{I} = 0.3\ \text{A}\angle 90°$,

$$\mathbf{V} = 4.8\ \text{V}\angle 0° - (8.0\ \Omega)(0.3\ \text{A}\angle 90°) = 4.8\ \text{V} - j2.4\ \text{V}$$
$$= 5.367\ \text{V}\angle -26.57°$$

EXAMPLE 20–4

Given the circuit of Figure 20–11, calculate the voltage across the 40-Ω resistor.

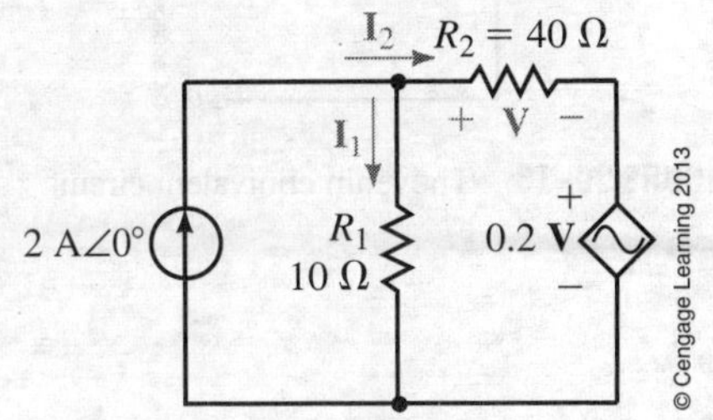

FIGURE 20–11

Solution In the circuit of Figure 20–11, the dependent source is controlled by an element located in the circuit. Unlike the sources in the previous examples, the dependent source cannot be eliminated from the circuit since doing so would contradict Kirchhoff's voltage law and/or Kirchhoff's current law.

The circuit must be analyzed by considering all effects simultaneously. Applying Kirchhoff's current law, we have

$$\mathbf{I}_1 + \mathbf{I}_2 = 2\ \text{A}\angle 0°$$

From Kirchhoff's voltage law, we have

$$(10\ \Omega)\ \mathbf{I}_1 = \mathbf{V} + 0.2\ \mathbf{V} = 1.2\ \mathbf{V}$$
$$\mathbf{I}_1 = 0.12\ \mathbf{V}$$

and,

$$\mathbf{I}_2 = \frac{\mathbf{V}}{40\ \Omega} = 0.025\ \mathbf{V}$$

Combining these expressions, we have

$$0.12\ \mathbf{V} + 0.025\ \mathbf{V} = 2.0\ \text{A}\angle 0°$$
$$0.145\ \mathbf{V} = 2.0\ \text{A}\angle 0°$$
$$\mathbf{V} = 13.79\ \text{V}\angle 0°$$

PRACTICE PROBLEMS 2

Determine the voltage **V** in the circuit of Figure 20–12.

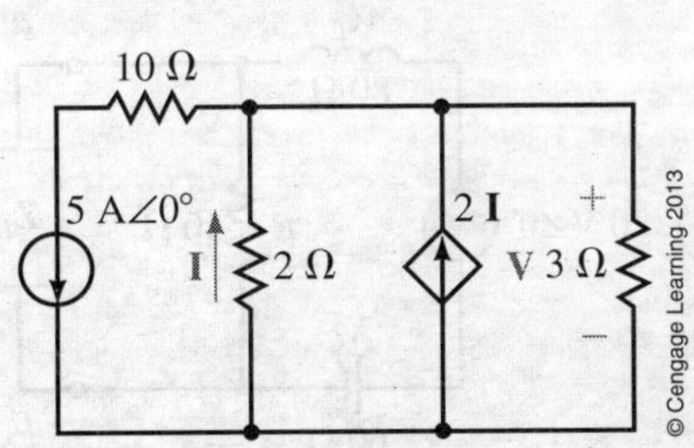

FIGURE 20–12

Answer
$\mathbf{V} = 2.73\ \text{V}\angle 180°$

20.3 Thévenin's Theorem—Independent Sources

Thévenin's theorem is a method that converts any linear bilateral ac circuit into a single ac voltage source in series with an equivalent impedance as shown in Figure 20–13.

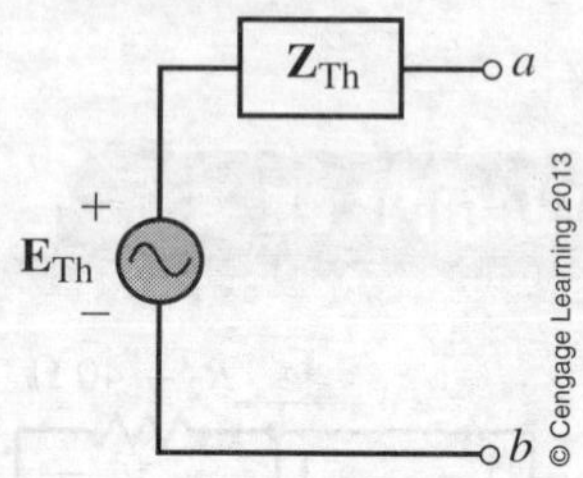

FIGURE 20–13 Thévenin equivalent circuit.

The resulting two-terminal network will be equivalent when it is connected to any external branch or component. If the original circuit contains reactive elements, the Thévenin equivalent circuit will be valid only at the frequency at which the reactances were determined. The following method may be used to determine the Thévenin equivalent of an ac circuit having either independent sources or sources which are dependent upon voltage or current in some other circuit. The outlined method may not be used in circuits having dependent sources controlled by voltage or current in the same circuit.

1. Remove the branch across which the Thévenin equivalent circuit is to be found. Label the resulting two terminals. Although any designation will do, we will use the notations *a* and *b*.
2. Set all sources to zero. As in dc circuits, this is achieved by replacing voltage sources with short circuits and current sources with open circuits.
3. Determine the Thévenin equivalent impedance $\mathbf{Z}_{Th}$ by calculating the impedance seen between the open terminals *a* and *b*. Occasionally it may be necessary to redraw the circuit to simplify this process.
4. Replace the sources removed in Step 3 and determine the open-circuit voltage across the terminals *a* and *b*. If any of the sources are expressed in sinusoidal form, it is first necessary to convert these sources into an equivalent phasor form. For circuits having more than one source, it may be necessary to apply the superposition theorem to calculate the open-circuit voltage. Since all voltages will be phasors, the resultant is found by using vector algebra. The open-circuit voltage is the Thévenin voltage, $\mathbf{E}_{Th}$.
5. Sketch the resulting Thévenin equivalent circuit by including that portion of the circuit removed in Step 1.

EXAMPLE 20–5

Find the Thévenin equivalent circuit external to $\mathbf{Z}_L$ for the circuit of Figure 20–14.

FIGURE 20–14

Solution

Steps 1 and 2: Removing the load impedance $\mathbf{Z}_L$ and setting the voltage source to zero, we have the circuit of Figure 20–15.

Step 3: The Thévenin impedance between terminals a and b is found as

$$\begin{aligned}\mathbf{Z}_{\text{Th}} &= \mathbf{R}\|(\mathbf{Z}_L + \mathbf{Z}_C)\\ &= \frac{(40\ \Omega\angle 0^\circ)(20\ \Omega\angle 90^\circ)}{40\ \Omega + j20\ \Omega}\\ &= \frac{800\ \Omega\angle 90^\circ}{44.72\ \Omega\angle 26.57^\circ}\\ &= 17.89\ \Omega\angle 63.43^\circ\\ &= 8\ \Omega + j16\ \Omega\end{aligned}$$

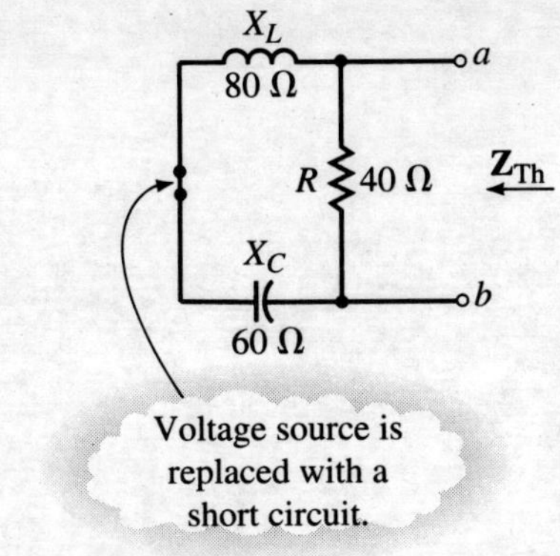

FIGURE 20–15

Step 4: The Thévenin voltage is found by using the voltage divider rule as shown in the circuit of Figure 20–16.

$$\begin{aligned}\mathbf{E}_{\text{Th}} = \mathbf{V}_{ab} &= \frac{40\ \Omega\angle 0^\circ}{40\ \Omega + j80\ \Omega - j60\ \Omega}(20\ \text{V}\angle 0^\circ)\\ &= \frac{800\ \text{V}\angle 0^\circ}{44.72\ \Omega\angle 26.57^\circ}\\ &= 17.89\ \text{V}\angle -26.57^\circ\end{aligned}$$

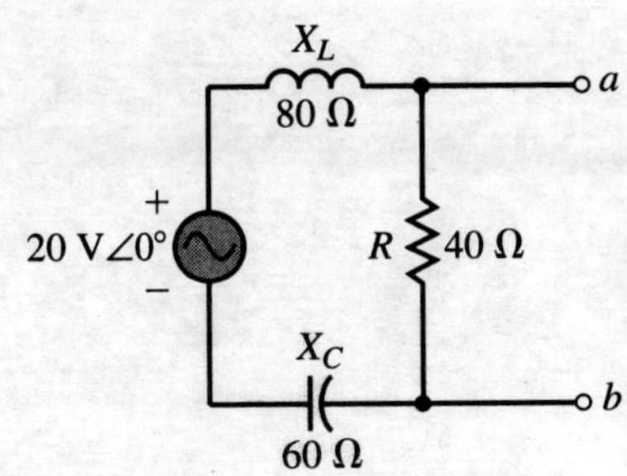

FIGURE 20–16

Step 5: The resultant Thévenin equivalent circuit is shown in Figure 20–17.

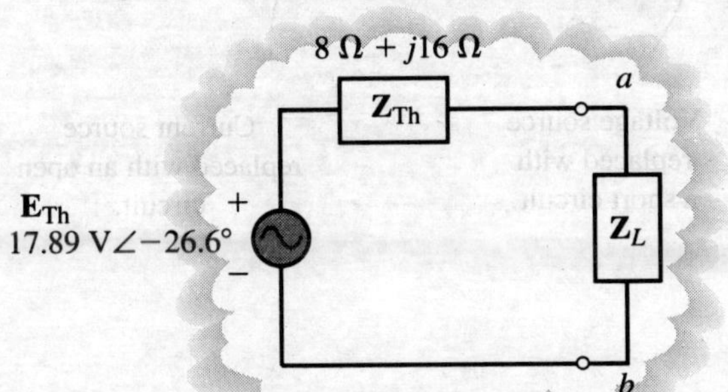

FIGURE 20–17

EXAMPLE 20–6

Determine the Thévenin equivalent circuit external to $\mathbf{Z}_L$ in the circuit in Figure 20–18.

CircuitSim 20-1

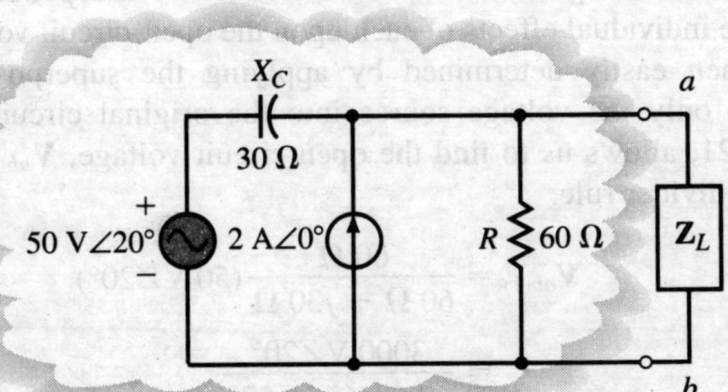

FIGURE 20–18

Solution

Step 1: Removing the branch containing $\mathbf{Z}_L$, we have the circuit of Figure 20–19.

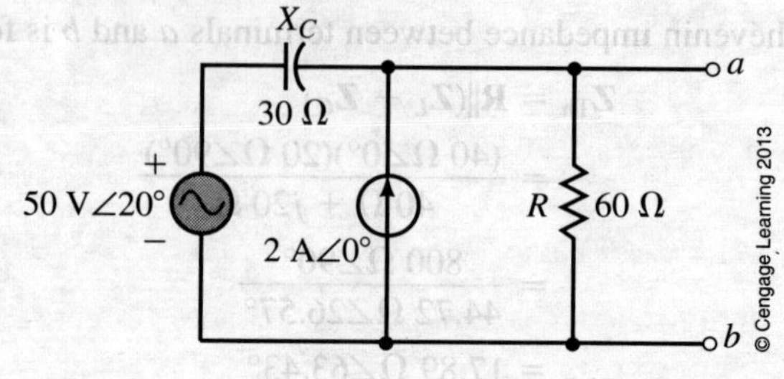

FIGURE 20–19

Step 2: After setting the voltage and current sources to zero, we have the circuit of Figure 20–20.

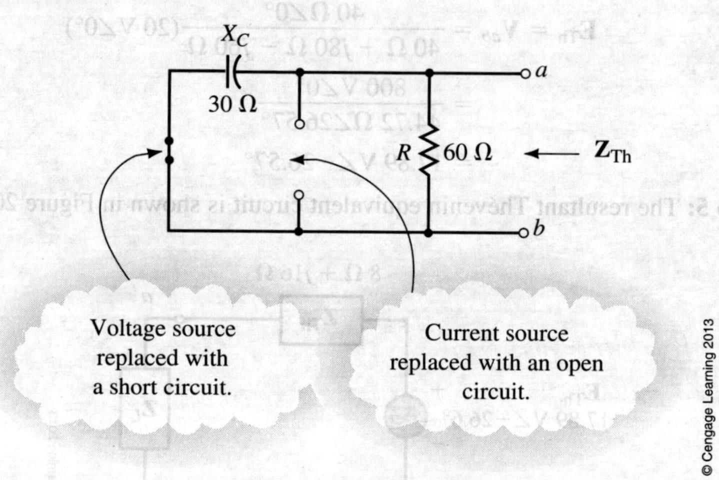

FIGURE 20–20

Step 3: The Thévenin impedance is determined as

$$\begin{aligned}\mathbf{Z}_{Th} &= \mathbf{Z}_C \| \mathbf{Z}_R \\ &= \frac{(30\ \Omega\angle -90°)(60\ \Omega\angle 0°)}{60\ \Omega - j30\ \Omega} \\ &= \frac{1800\ \Omega\angle -90°}{67.08\ \Omega\angle -26.57°} \\ &= 26.83\ \Omega\angle -63.43°\end{aligned}$$

Step 4: Because the given network consists of two independent sources, we consider the individual effects of each upon the open-circuit voltage. The total effect is then easily determined by applying the superposition theorem. Reinserting only the voltage source into the original circuit, as shown in Figure 20–21, allows us to find the open-circuit voltage, $\mathbf{V}_{ab(1)}$, by applying the voltage divider rule:

$$\begin{aligned}\mathbf{V}_{ab(1)} &= \frac{60\ \Omega}{60\ \Omega - j30\ \Omega}(50\ \text{V}\angle 20°) \\ &= \frac{3000\ \text{V}\angle 20°}{67.08\angle -26.57°} \\ &= 44.72\ \text{V}\angle 46.57°\end{aligned}$$

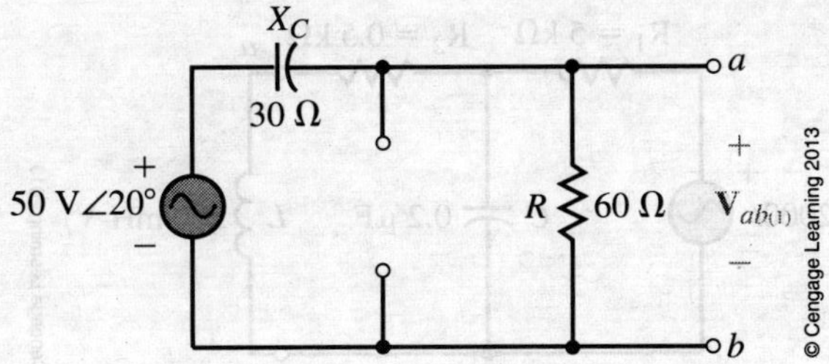

FIGURE 20–21

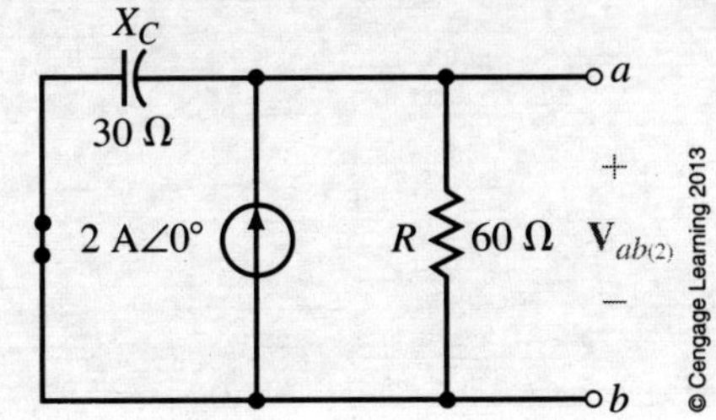

FIGURE 20–22

Now, considering only the current source as shown in Figure 20–22, we determine $\mathbf{V}_{ab(2)}$ by Ohm's law:

$$\mathbf{V}_{ab(2)} = \frac{(2\text{ A}\angle 0°)(30\ \Omega\angle -90°)(60\ \Omega\angle 0°)}{60\ \Omega - j30\ \Omega}$$
$$= (2\text{ A}\angle 0°)(26.83\ \Omega\angle -63.43°)$$
$$= 53.67\text{ V}\angle -63.43°$$

From the superposition theorem, the Thévenin voltage is determined as

$$\mathbf{E}_{\text{Th}} = \mathbf{V}_{ab(1)} + \mathbf{V}_{ab(2)}$$
$$= 44.72\text{ V}\angle 46.57° + 53.67\text{ V}\angle -63.43°$$
$$= (30.74\text{ V} + j32.48\text{ V}) + (24.00\text{ V} - j48.00\text{ V})$$
$$= (54.74\text{ V} - j15.52\text{ V}) = 56.90\text{ V}\angle -15.83°$$

Step 5: The resulting Thévenin equivalent circuit is shown in Figure 20–23.

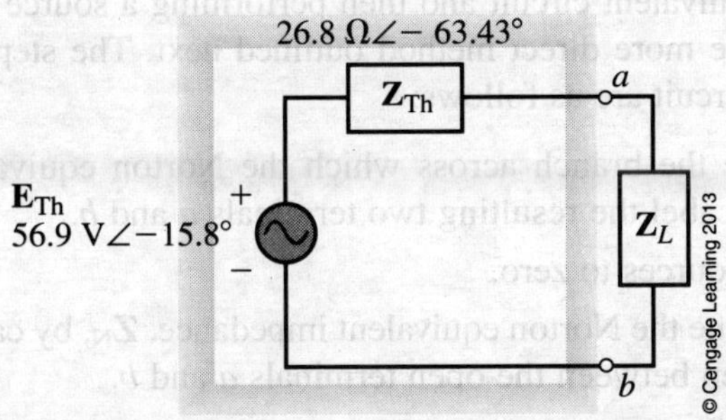

FIGURE 20–23

IN-PROCESS LEARNING CHECK 2

(Answers are at the end of the chapter.)

Refer to the circuit shown in Figure 20–24 of Practice Problems 3. List the steps that you would use to find the Thévenin equivalent circuit.

PRACTICE PROBLEMS 3

a. Find the Thévenin equivalent circuit external to the inductor in the circuit in Figure 20–24. (Notice that the voltage source is shown as sinusoidal.)

b. Use the Thévenin equivalent circuit to find the phasor output voltage, $\mathbf{V}_L$.

c. Convert the answer of (b) into the equivalent sinusoidal voltage.

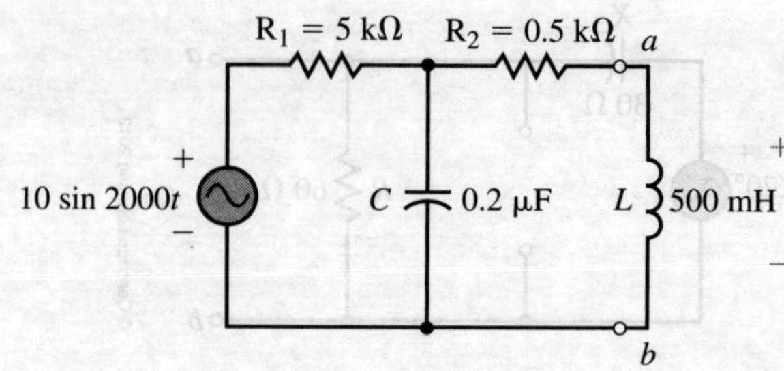

FIGURE 20–24

Answers

a. $\mathbf{Z}_{Th} = 1.5\ k\Omega - j2.0\ k\Omega = 2.5\ k\Omega\angle -53.13°$, $\mathbf{E}_{Th} = 3.16\ V\angle -63.43°$

b. $\mathbf{V}_L = 1.75\ V\angle 60.26°$

c. $v_L = 2.48 \sin(2000t + 60.26°)$

20.4 Norton's Theorem—Independent Sources

Norton's theorem converts any linear bilateral network into an equivalent circuit consisting of a single current source and a parallel impedance as shown in Figure 20–25.

Although Norton's equivalent circuit may be determined by first finding the Thévenin equivalent circuit and then performing a source conversion, we generally use the more direct method outlined next. The steps to find the Norton equivalent circuit are as follows:

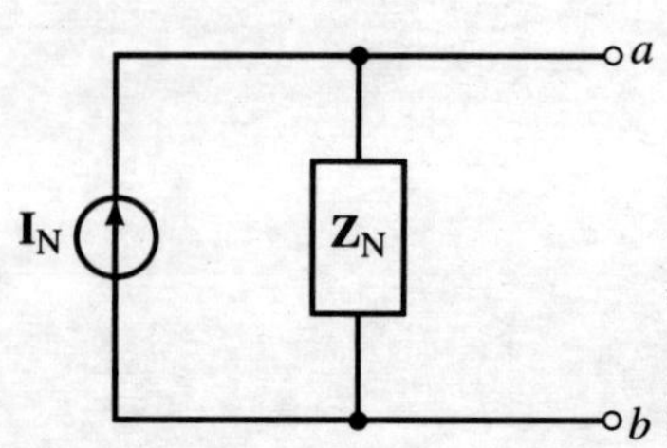

FIGURE 20–25 Norton equivalent circuit.

1. Remove the branch across which the Norton equivalent circuit is to be found. Label the resulting two terminals *a* and *b*.
2. Set all sources to zero.
3. Determine the Norton equivalent impedance, $\mathbf{Z}_N$, by calculating the impedance seen between the open terminals *a* and *b*.

   NOTE: Since the previous steps are identical to those followed for finding the Thévenin equivalent circuit, we conclude that the Norton impedance must be the same as the Thévenin impedance.
4. Replace the sources removed in Step 3 and determine the current that would occur between terminals *a* and *b* if these terminals were shorted. Any voltages and currents that are given in sinusoidal notation must first be expressed in equivalent phasor notation. If the circuit has more than one source it may be necessary to apply the superposition theorem to calculate the total short-circuit current. Since all currents will be in phasor form, any addition must be done using vector algebra. The resulting current is the Norton current, I_N.
5. Sketch the resulting Norton equivalent circuit by inserting that portion of the circuit removed in Step 1.

As mentioned previously, it is possible to find the Norton equivalent circuit from the Thévenin equivalent by simply performing a source conversion. We have already determined that both the Thévenin and Norton impedances are determined in the same way. Consequently, the impedances must be equivalent, and so we have

$$\mathbf{Z}_N = \mathbf{Z}_{Th} \quad (20\text{–}1)$$

Now, applying Ohm's law, we determine the Norton current source from the Thévenin voltage and impedance, namely,

$$\mathbf{I}_{\mathrm{N}} = \frac{\mathbf{E}_{\mathrm{Th}}}{\mathbf{Z}_{\mathrm{Th}}} \tag{20–2}$$

Figure 20–26 shows the equivalent circuits.

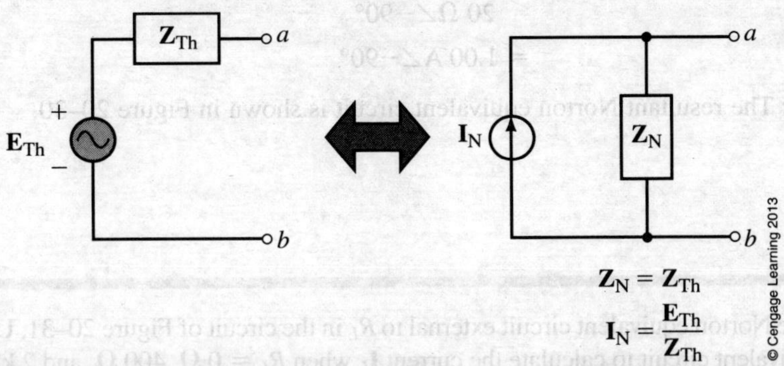

FIGURE 20–26

EXAMPLE 20–7

Given the circuit of Figure 20–27, find the Norton equivalent.

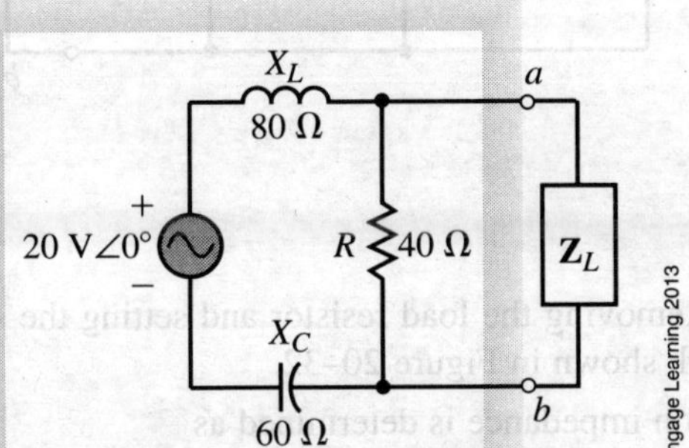

FIGURE 20–27

Solution

Steps 1 and 2: By removing the load impedance, $\mathbf{Z}_L$, and setting the voltage source to zero, we have the network of Figure 20–28.

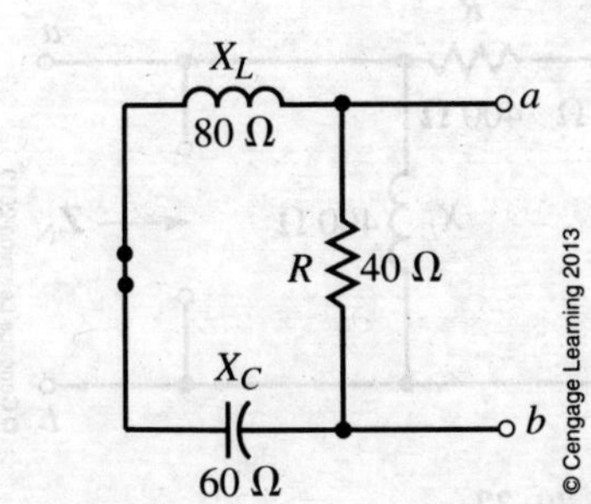

FIGURE 20–28

Step 3: The Norton impedance may now be determined by evaluating the impedance between terminals *a* and *b*. Hence, we have

$$\begin{aligned}\mathbf{Z}_{\mathrm{N}} &= \frac{(40\ \Omega\angle 0^\circ)(20\ \Omega\angle 90^\circ)}{40\ \Omega + j20\ \Omega}\\ &= \frac{800\ \Omega\angle 90^\circ}{44.72\angle 26.57^\circ}\\ &= 17.89\ \Omega\angle 63.43^\circ\\ &= 8\ \Omega + j16\ \Omega\end{aligned}$$

Step 4: Reinserting the voltage source, as in Figure 20–29, we find the Norton current by calculating the current between the shorted terminals, *a* and *b*.

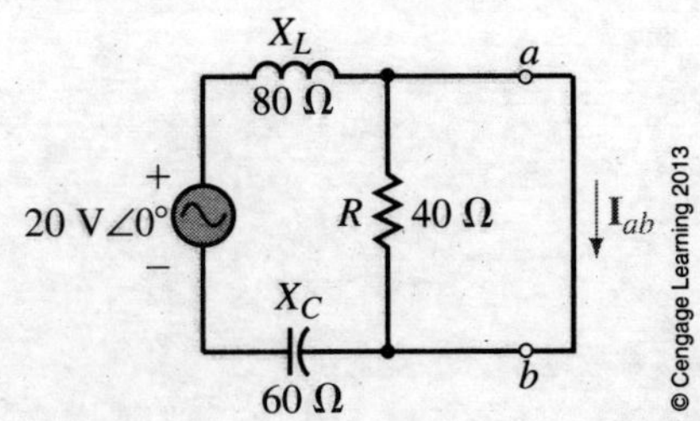

FIGURE 20–29

$Z_N = 8\ \Omega + j16\ \Omega = 17.89\ \Omega\angle 63.43°$

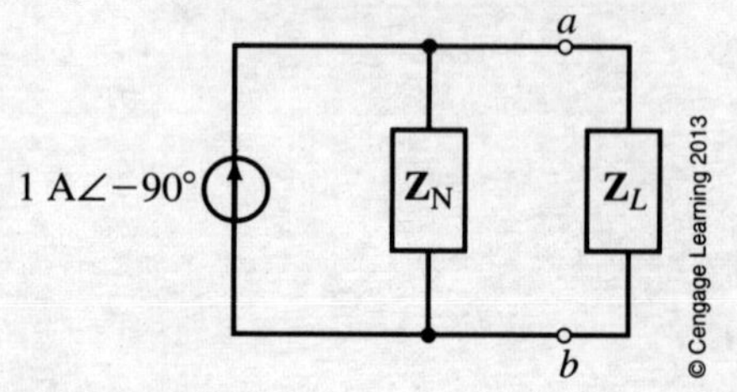

FIGURE 20–30

Because the resistor $R = 40\ \Omega$ is shorted, the current is determined by the impedances X_L and X_C as

$$\mathbf{I}_N = \mathbf{I}_{ab} = \frac{20\ \text{V}\angle 0°}{j80\ \Omega - j60\ \Omega} = \frac{20\ \text{V}\angle 0°}{20\ \Omega\angle -90°} = 1.00\ \text{A}\angle -90°$$

Step 5: The resultant Norton equivalent circuit is shown in Figure 20–30.

EXAMPLE 20–8

Find the Norton equivalent circuit external to R_L in the circuit of Figure 20–31. Use the equivalent circuit to calculate the current $\mathbf{I}_L$ when $R_L = 0\ \Omega$, $400\ \Omega$, and $2\ \text{k}\Omega$.

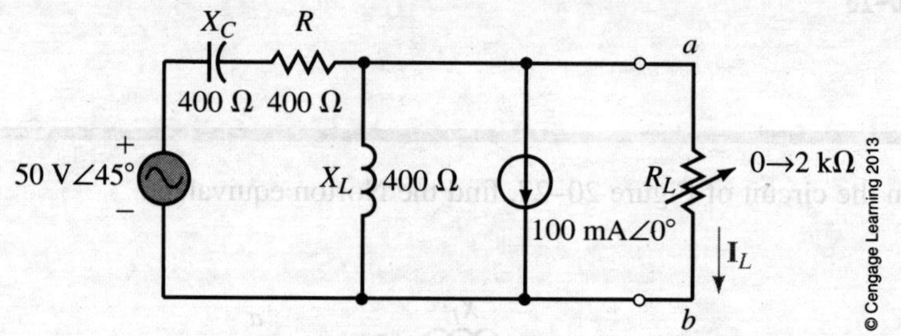

FIGURE 20–31

Solution

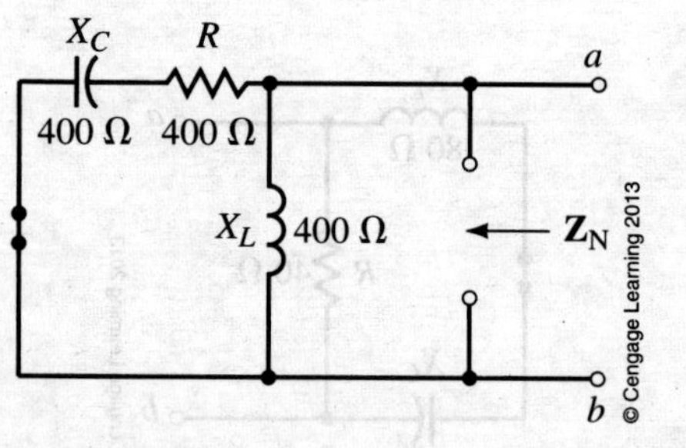

FIGURE 20–32

Steps 1 and 2: Removing the load resistor and setting the sources to zero, we obtain the network shown in Figure 20–32.

Step 3: The Norton impedance is determined as

$$\mathbf{Z}_N = \frac{(400\ \Omega\angle 90°)(400\ \Omega - j400\ \Omega)}{j400\ \Omega + 400\ \Omega - j400\ \Omega}$$

$$= \frac{(400\ \Omega\angle 90°)(565.69\ \Omega\angle -45°)}{400\ \Omega\angle 0°}$$

$$= 565.69\ \Omega\angle +45°$$

Step 4: Because the network consists of two sources, we determine the effects due to each source separately and then apply superposition to evaluate the Norton current source.

Reinserting the voltage source into the original network, we see from Figure 20–33 that the short-circuit current between terminals a and b is easily found by using Ohm's law.

$$\mathbf{I}_{ab(1)} = \frac{50\ \text{V}\angle 45°}{400\ \Omega - j400\ \Omega} = \frac{50\ \text{V}\angle 45°}{565.69\ \Omega\angle -45°} = 88.4\ \text{mA}\angle 90°$$

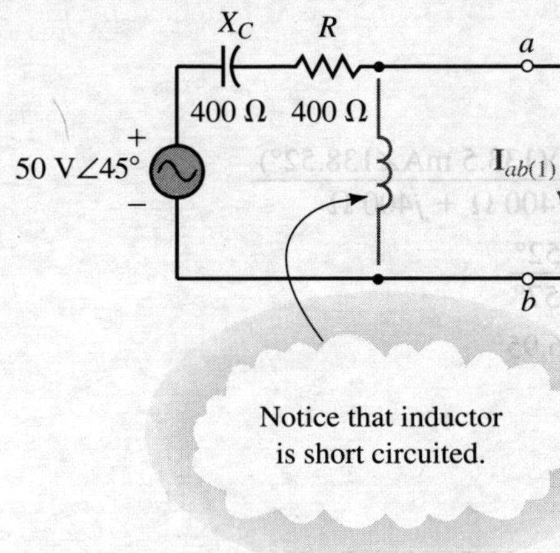

FIGURE 20–33

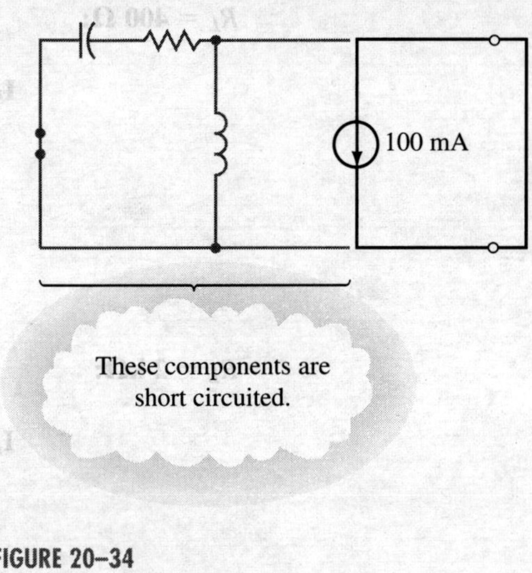

FIGURE 20–34

Since short circuiting the current source effectively removes all impedances, as illustrated in Figure 20–34, the short-circuit current between terminals a and b is given as follows:

$$\mathbf{I}_{ab(2)} = -100 \text{ mA}\angle 0^\circ$$
$$= 100 \text{ mA}\angle 180^\circ$$

Now applying the superposition theorem, the Norton current is determined as the summation

$$\mathbf{I}_\text{N} = \mathbf{I}_{ab(1)} + \mathbf{I}_{ab(2)}$$
$$= 88.4 \text{ mA}\angle 90^\circ + 100 \text{ mA}\angle 180^\circ$$
$$= -100 \text{ mA} + j88.4 \text{ mA}$$
$$= 133.5 \text{ mA}\angle 138.52^\circ$$

Step 5: The resulting Norton equivalent circuit is shown in Figure 20–35.

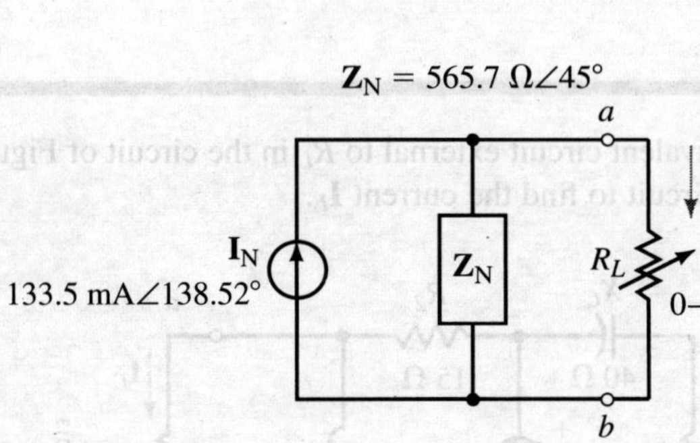

FIGURE 20–35

From the preceding circuit, we express the current through the load, $\mathbf{I}_L$, as

$$\mathbf{I}_L = \frac{\mathbf{Z}_\text{N}}{R_L + \mathbf{Z}_\text{N}} \mathbf{I}_\text{N}$$

$R_L = \mathbf{0}\ \boldsymbol{\Omega}$:

$$\mathbf{I}_L = \mathbf{I}_\text{N} = 133.5 \text{ mA}\angle 138.52^\circ$$

$R_L = 400\ \Omega$:

$$\mathbf{I}_L = \frac{\mathbf{Z}_N}{R_L + \mathbf{Z}_N}\mathbf{I}_N$$
$$= \frac{(565.7\ \Omega\angle 45°)(133.5\ \text{mA}\angle 138.52°)}{400\ \Omega + 400\ \Omega + j400\ \Omega}$$
$$= \frac{75.24\ \text{V}\angle 183.52°}{894.43\ \Omega\angle 26.57°}$$
$$= 84.12\ \text{mA}\angle 156.95°$$

$R_L = 2\ \text{k}\Omega$:

$$\mathbf{I}_L = \frac{\mathbf{Z}_N}{R_L + \mathbf{Z}_N}\mathbf{I}_N$$
$$= \frac{(565.7\ \Omega\angle 45°)(133.5\ \text{mA}\angle 138.52°)}{2000\ \Omega + 400\ \Omega + j400\ \Omega}$$
$$= \frac{75.24\ \text{V}\angle 183.52°}{2433.1\ \Omega\angle 9.46°}$$
$$= 30.92\ \text{mA}\angle 174.06°$$

IN-PROCESS LEARNING CHECK 3

(Answers are at the end of the chapter.)

Refer to the circuit shown in Figure 20–36 (Practice Problems 4). List the steps that you would use to find the Norton equivalent circuit.

PRACTICE PROBLEMS 4

Find the Norton equivalent circuit external to R_L in the circuit of Figure 20–36. Use the equivalent circuit to find the current $\mathbf{I}_L$.

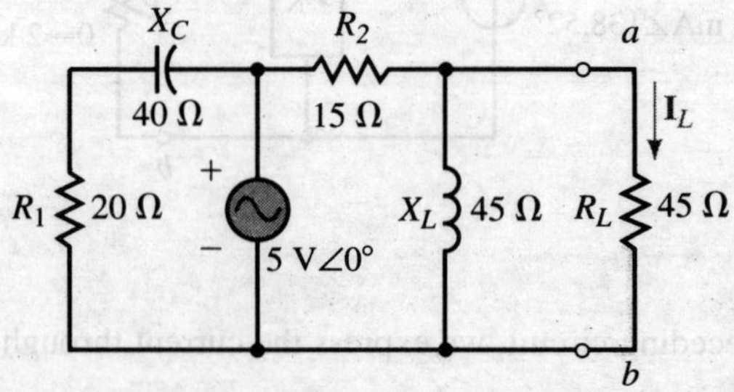

FIGURE 20–36

Answers
$\mathbf{Z}_N = 13.5\ \Omega + j4.5\ \Omega = 14.23\ \Omega\angle 18.43°$, $\mathbf{I}_N = 0.333\ \text{A}\angle 0°$,
$\mathbf{I}_L = 0.0808\ \text{A}\angle 14.03°$

20.5 Thévenin's and Norton's Theorems for Dependent Sources

If a circuit contains a dependent source that is controlled by an element outside the circuit of interest, the methods outlined in Sections 20.2 and 20.3 are used to find either the Thévenin or Norton equivalent circuit.

EXAMPLE 20–9

Given the circuit of Figure 20–37, find the Thévenin equivalent circuit external to R_L. If the voltage applied to the resistor R_1 is 10 mV, use the Thévenin equivalent circuit to calculate the minimum and maximum voltage across R_L.

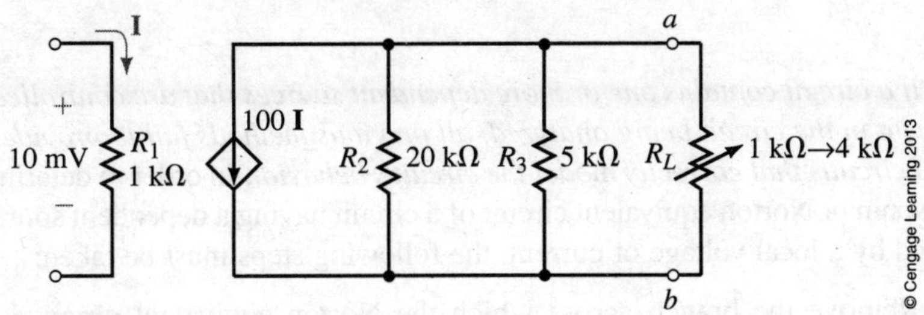

FIGURE 20–37

CircuitSim 20-2

Solution

Step 1: Removing the load resistor from the circuit and labelling the remaining terminals *a* and *b*, we have the circuit shown in Figure 20–38.

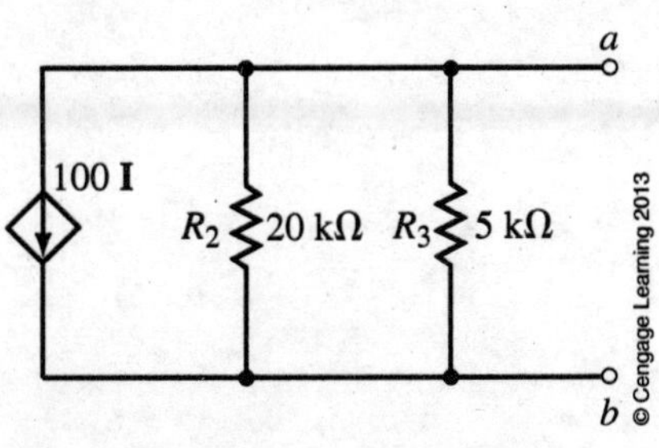

FIGURE 20–38

Steps 2 and 3: The Thévenin resistance is found by open circuiting the current source and calculating the impedance observed between the terminals *a* and *b*. Since the circuit is purely resistive, we have

$$\begin{aligned} R_{Th} &= 20\ \text{k}\Omega \| 5\ \text{k}\Omega \\ &= \frac{(20\ \text{k}\Omega)(5\ \text{k}\Omega)}{20\ \text{k}\Omega + 5\ \text{k}\Omega} \\ &= 4\ \text{k}\Omega \end{aligned}$$

Step 4: The open-circuit voltage between the terminals is found to be

$$\begin{aligned} \mathbf{V}_{ab} &= -(100\mathbf{I})(4\ \text{k}\Omega) \\ &= -(4 \times 10^5\ \Omega)\mathbf{I} \end{aligned}$$

As expected, the Thévenin voltage source is dependent upon the current **I**.

Step 5: Because the Thévenin voltage is a dependent voltage source we use the appropriate symbol when sketching the equivalent circuit, as shown in Figure 20–39.

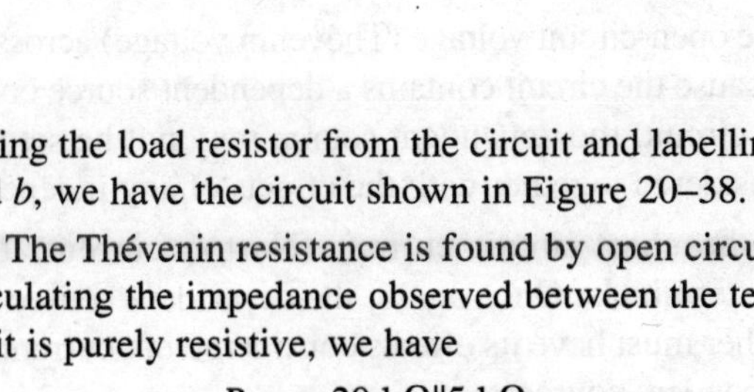

FIGURE 20–39

For the given conditions, we have

$$I = \frac{10\text{ mV}}{1\text{ k}\Omega} = 10\ \mu\text{A}$$

The voltage across the load is now determined as follows:

$$R_L = 1\text{ k}\Omega\text{:}\qquad V_{ab} = -\frac{1\text{ k}\Omega}{1\text{ k}\Omega + 4\text{ k}\Omega}(4 \times 10^5\ \Omega)(10\ \mu\text{A}) = -0.8\text{ V}$$

$$R_L = 4\text{ k}\Omega\text{:}\qquad V_{ab} = -\frac{4\text{ k}\Omega}{4\text{ k}\Omega + 4\text{ k}\Omega}(4 \times 10^5\ \Omega)(10\ \mu\text{A}) = -2.0\text{ V}$$

For an applied voltage of 10 mV, the voltage across the load resistance will vary between −0.8 V and −2.0 V as R_L is adjusted between 1 kΩ and 4 kΩ.

If a circuit contains one or more dependent sources that are controlled by an element in the circuit being analyzed, all previous methods fail to provide equivalent circuits that correctly model the circuit's behavior. In order to determine the Thévenin or Norton equivalent circuit of a circuit having a dependent source controlled by a local voltage or current, the following steps must be taken:

1. Remove the branch across which the Norton equivalent circuit is to be found. Label the resulting two terminals *a* and *b*.
2. Calculate the open-circuit voltage (Thévenin voltage) across the two terminals *a* and *b*. Because the circuit contains a dependent source controlled by an element in the circuit, the dependent source may not be set to zero. Its effects must be considered together with the effects of any independent source(s).
3. Determine the short-circuit current (Norton current) that would occur between the terminals. Once again, the dependent source may not be set to zero, but rather must have its effects considered concurrently with the effects of any independent source(s).
4. Determine the Thévenin or Norton impedance by applying Equations 20–1 and 20–2 as follows:

$$\mathbf{Z}_N = \mathbf{Z}_{Th} = \frac{\mathbf{E}_{Th}}{\mathbf{I}_N} \qquad \textbf{(20–3)}$$

5. Sketch the Thévenin or Norton equivalent circuit, as shown previously in Figure 20–26. Ensure that the portion of the network that was removed in Step 1 is reinserted as part of the equivalent circuit.

EXAMPLE 20–10

For the circuit of Figure 20–40, find the Norton equivalent circuit external to the load resistor, R_L.

Solution

Step 1: After removing the load resistor from the circuit, we have the network shown in Figure 20–41.

Step 2: At first glance, we might look into the open terminals and say that the Norton (or Thévenin) impedance appears to be 60 kΩ‖30 kΩ = 20 kΩ.

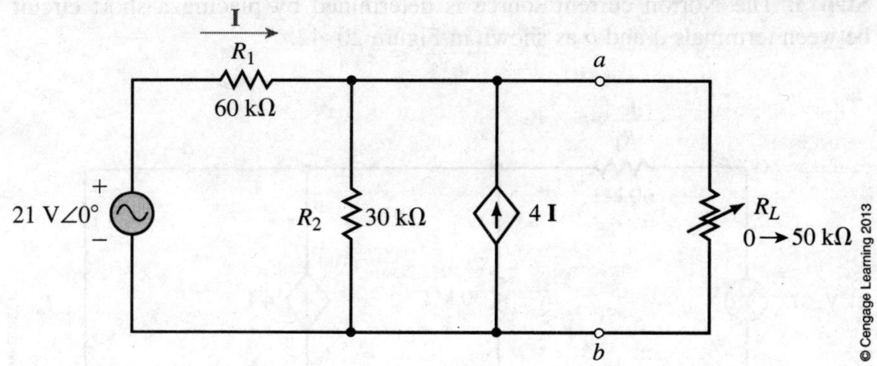

CircuitSim 20-3

FIGURE 20–40

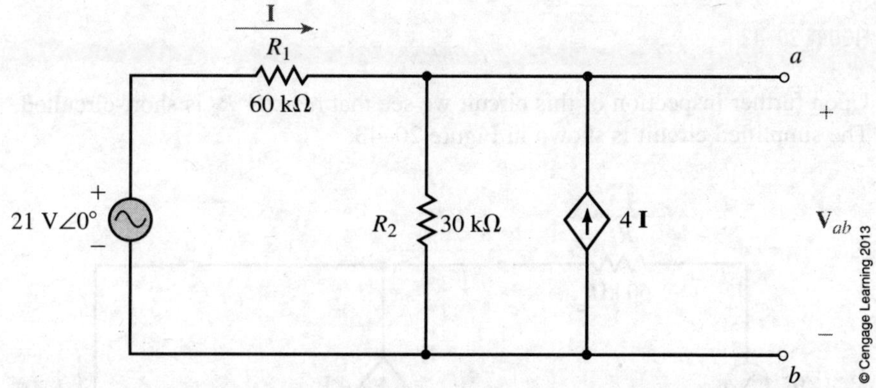

FIGURE 20–41

However, we will find that this result is incorrect. The presence of the locally controlled dependent current source makes the analysis of this circuit slightly more complicated than a circuit that contains only an independent source. We know, however, that the basic laws of circuit analysis must apply to all circuits, regardless of the complexity. Applying Kirchhoff's current law at node *a* gives the current through R_2 as

$$\mathbf{I}_{R_2} = \mathbf{I} + 4\mathbf{I} = 5\mathbf{I}$$

Now, applying Kirchhoff's voltage law around the closed loop containing the voltage source and the two resistors, gives

$$21\text{ V}\angle 0^\circ = (60\text{ k}\Omega)\mathbf{I} + (30\text{ k}\Omega)(5\mathbf{I}) = 210\text{ k}\Omega\mathbf{I}$$

which allows us to solve for the current **I** as

$$\mathbf{I} = \frac{21\text{ V}\angle 0^\circ}{210\text{ k}\Omega} = 0.100\text{ mA}\angle 0^\circ$$

Since the open-circuit voltage, $\mathbf{V}_{ab}$ is the same as the voltage across R_2, we have

$$\mathbf{E}_{\text{Th}} = \mathbf{V}_{ab} = (30\text{ k}\Omega)(5)(0.1\text{ mA}\angle 0^\circ) = 15\text{ V}\angle 0^\circ$$

Step 3: The Norton current source is determined by placing a short circuit between terminals *a* and *b* as shown in Figure 20–42.

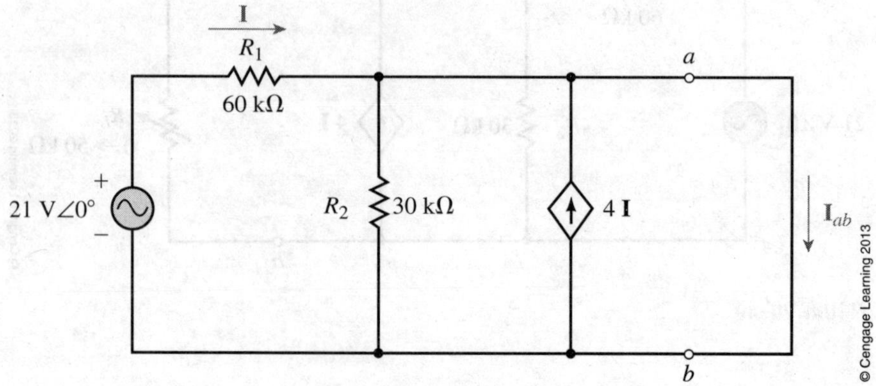

FIGURE 20–42

Upon further inspection of this circuit we see that resistor R_2 is short-circuited. The simplified circuit is shown in Figure 20–43.

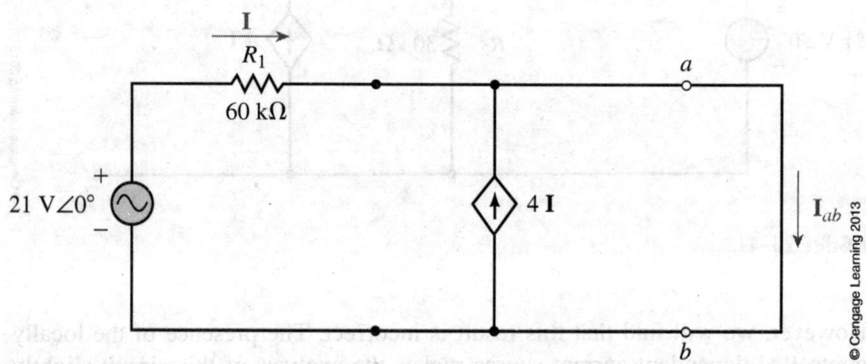

FIGURE 20–43

The short-circuit current $\mathbf{I}_{ab}$ is now easily determined by using Kirchhoff's current law at node *a*, and so we have

$$\mathbf{I}_N = \mathbf{I}_{ab} = 5\mathbf{I}$$

From Ohm's law, we have

$$\mathbf{I} = \frac{21\ \text{V}\angle 0^\circ}{60\ \text{k}\Omega} = 0.35\ \text{mA}\angle 0^\circ$$

and so

$$\mathbf{I}_N = 5(0.35\ \text{mA}\angle 0^\circ) = 1.75\ \text{mA}\angle 0^\circ$$

Step 4: The Norton (or Thévenin) impedance is now determined from Ohm's law as

$$\mathbf{Z}_N = \frac{\mathbf{E}_{Th}}{\mathbf{I}_N} = \frac{15\ \text{V}\angle 0^\circ}{1.75\ \text{mA}\angle 0^\circ} = 8.57\ \text{k}\Omega$$

Notice that this impedance is different from the originally assumed 20 kΩ. In general, this condition will occur for most circuits that contain a locally controlled voltage or current source.

Step 5: The Norton equivalent circuit is shown in Figure 20–44.

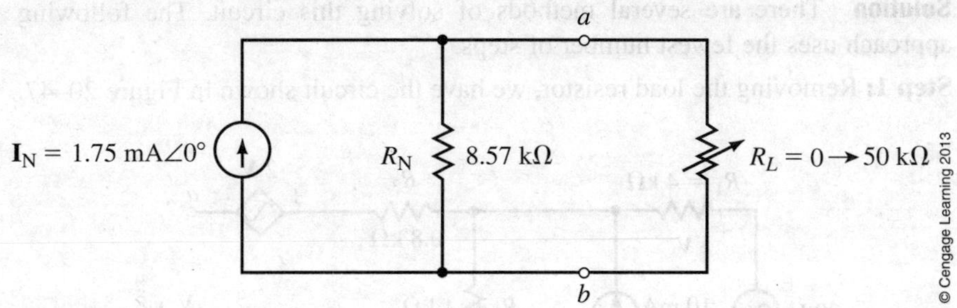

FIGURE 20–44

PRACTICE PROBLEMS 5

a. Find the Thévenin equivalent circuit external to R_L in the circuit of Figure 20–45.
b. Determine the current I_L when $R_L = 0$ and when $R_L = 20\ \Omega$.

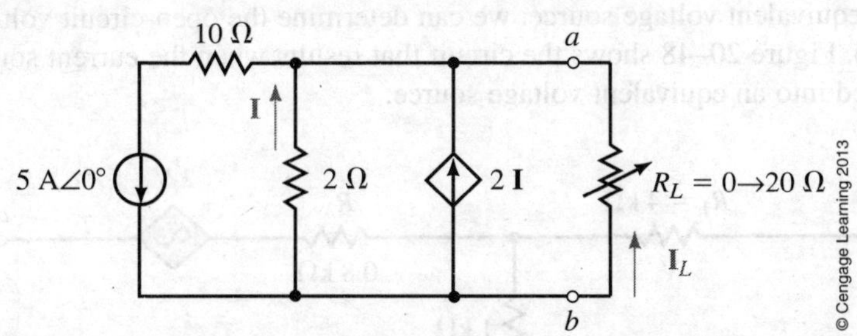

FIGURE 20–45

Answers

a. $\mathbf{E}_{Th} = \mathbf{V}_{ab} = -3.33\ V$, $\mathbf{Z}_{Th} = 0.667\ \Omega$
b. For $R_L = 0$: $\mathbf{I}_L = 5.00\ A$ (upward); for $R_L = 20\ \Omega$: $\mathbf{I}_L = 0.161\ A$ (upward)

If a circuit has more than one independent source, it is necessary to determine the open-circuit voltage and short-circuit current due to each independent source while simultaneously considering the effects of the dependent source. The following example illustrates the principle.

EXAMPLE 20–11

Find the Thévenin and Norton equivalent circuits external to the load resistor in the circuit of Figure 20–46.

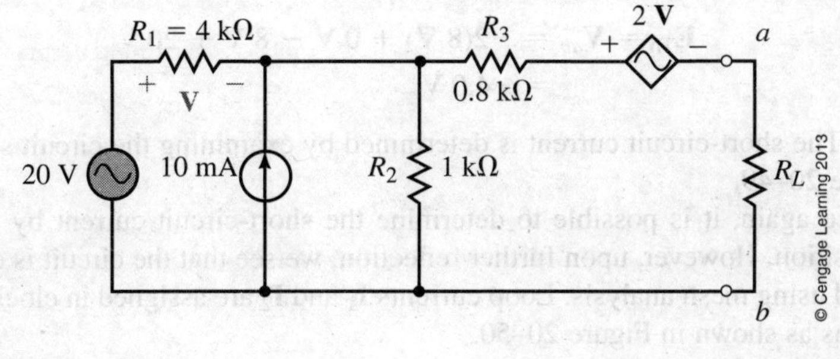

FIGURE 20–46

Solution There are several methods of solving this circuit. The following approach uses the fewest number of steps.

Step 1: Removing the load resistor, we have the circuit shown in Figure 20–47.

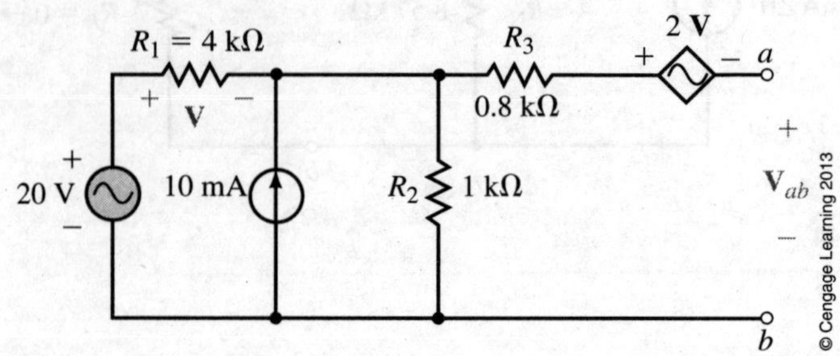

FIGURE 20–47

Step 2: In order to find the open-circuit voltage, V_{ab} of Figure 20–47, we may isolate the effects due to each independent source and then apply superposition to determine the combined result. However, by converting the current source into an equivalent voltage source, we can determine the open-circuit voltage in one step. Figure 20–48 shows the circuit that results when the current source is converted into an equivalent voltage source.

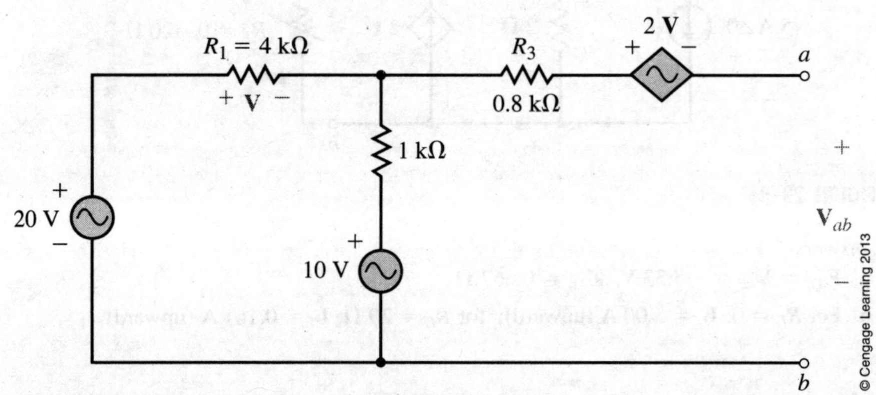

FIGURE 20–48

The controlling element (R_1) has a voltage, **V**, determined as

$$\mathbf{V} = \left(\frac{4\text{ k}\Omega}{4\text{ k}\Omega + 1\text{ k}\Omega}\right)(20\text{ V} - 10\text{ V})$$
$$= 8\text{ V}$$

which gives a Thévenin (open-circuit) voltage of

$$\mathbf{E}_{\text{Th}} = \mathbf{V}_{ab} = -2(8\text{ V}) + 0\text{ V} - 8\text{ V} + 20\text{ V}$$
$$= -4.0\text{ V}$$

Step 3: The short-circuit current is determined by examining the circuit shown in Figure 20–49.

Once again, it is possible to determine the short-circuit current by using superposition. However, upon further reflection, we see that the circuit is easily analyzed using mesh analysis. Loop currents $\mathbf{I}_1$ and $\mathbf{I}_2$ are assigned in clockwise directions as shown in Figure 20–50.

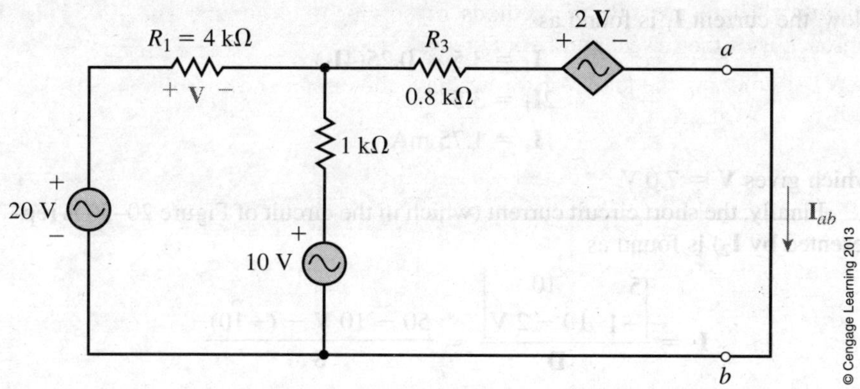

FIGURE 20–49

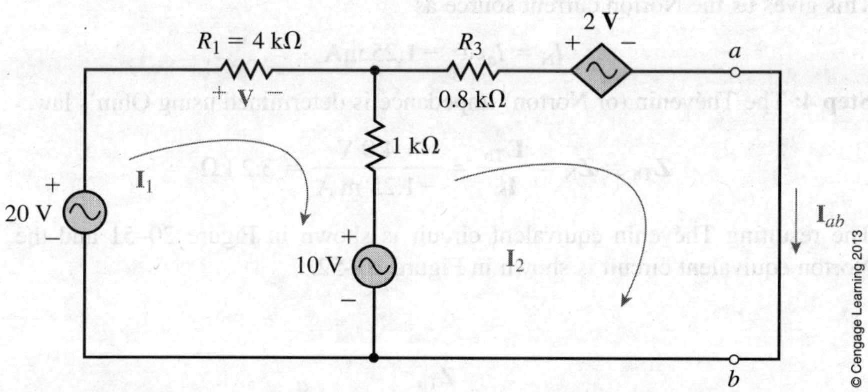

FIGURE 20–50

The loop equations are as follows:

Loop 1: $(5\ k\Omega)\mathbf{I}_1 - (1\ k\Omega)\mathbf{I}_2 = 10\ V$

Loop 2: $-(1\ k\Omega)\mathbf{I}_1 + (1.8\ k\Omega)\mathbf{I}_2 = 10\ V - 2\ \mathbf{V}$

Notice that the voltage in the second loop equation is expressed in terms of the controlling voltage across R_1. We will not worry about this right now. We may simplify our calculations by ignoring the units in the preceding equations. It is obvious that if all impedances are expressed in kΩ and all voltages are in volts, then the currents $\mathbf{I}_1$ and $\mathbf{I}_2$ must be in mA. The determinant for the denominator is found to be

$$\mathbf{D} = \begin{vmatrix} 5 & -1 \\ -1 & 1.8 \end{vmatrix} = 9 - 1 = 8$$

The current $\mathbf{I}_1$ is solved by using determinants as follows:

$$\mathbf{I}_1 = \frac{\begin{vmatrix} 10 & -1 \\ 10 - 2\mathbf{V} & 1.8 \end{vmatrix}}{\mathbf{D}} = \frac{18 - (-1)(10 - 2\mathbf{V})}{8}$$

$$= 3.5 - 0.25\ \mathbf{V}$$

The preceding result illustrates that the current $\mathbf{I}_1$ is dependent on the controlling voltage. However, by examining the circuit of Figure 20–50, we see that the controlling voltage depends on the current $\mathbf{I}_1$, and is determined from Ohm's law as

$$\mathbf{V} = (4\ k\Omega)\mathbf{I}_1$$

or more simply as

$$\mathbf{V} = 4\mathbf{I}_1$$

Now, the current $\mathbf{I}_1$ is found as

$$\mathbf{I}_1 = 3.5 - 0.25(4\mathbf{I}_1)$$
$$2\mathbf{I}_1 = 3.5$$
$$\mathbf{I}_1 = 1.75\text{ mA}$$

which gives $\mathbf{V} = 7.0$ V.

Finally, the short circuit current (which in the circuit of Figure 20–50 is represented by $\mathbf{I}_2$) is found as

$$\mathbf{I}_2 = \frac{\begin{vmatrix} 5 & 10 \\ -1 & 10 - 2\mathbf{V} \end{vmatrix}}{\mathbf{D}} = \frac{50 - 10\mathbf{V} - (-10)}{8}$$
$$= 7.5 - 1.25\mathbf{V} = 7.5 - 1.25(7.0\text{ V})$$
$$= -1.25\text{ mA}$$

This gives us the Norton current source as

$$I_N = I_{ab} = -1.25\text{ mA}$$

Step 4: The Thévenin (or Norton) impedance is determined using Ohm's law.

$$\mathbf{Z}_{Th} = \mathbf{Z}_N = \frac{\mathbf{E}_{Th}}{\mathbf{I}_N} = \frac{-4.0\text{ V}}{-1.25\text{ m A}} = 3.2\text{ k}\Omega$$

The resulting Thévenin equivalent circuit is shown in Figure 20–51 and the Norton equivalent circuit is shown in Figure 20–52.

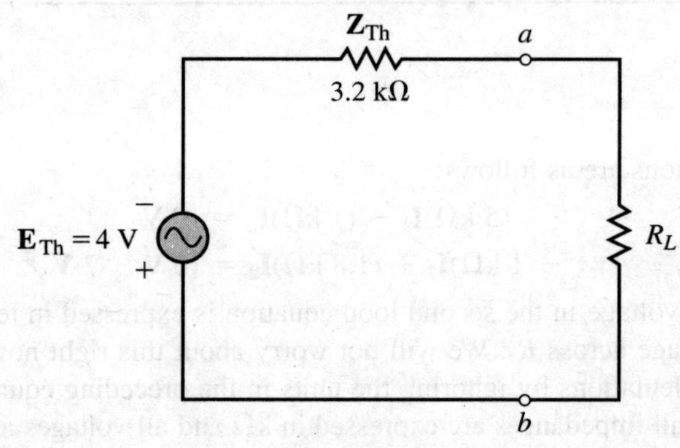

FIGURE 20–51

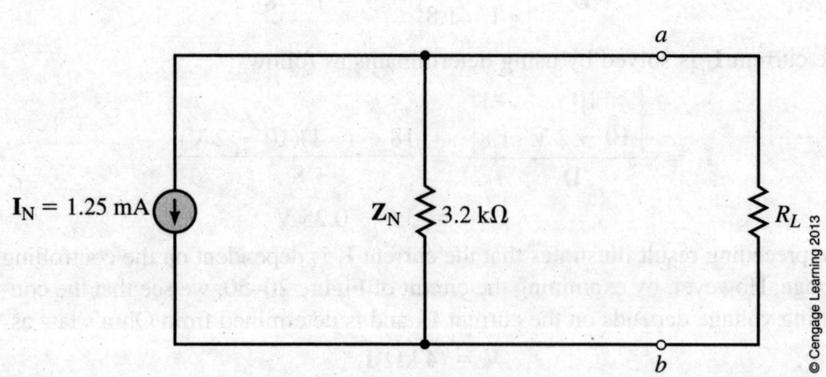

FIGURE 20–52

PRACTICE PROBLEMS 6

Find the Thévenin and Norton equivalent circuits external to the load resistor in the circuit of Figure 20–53.

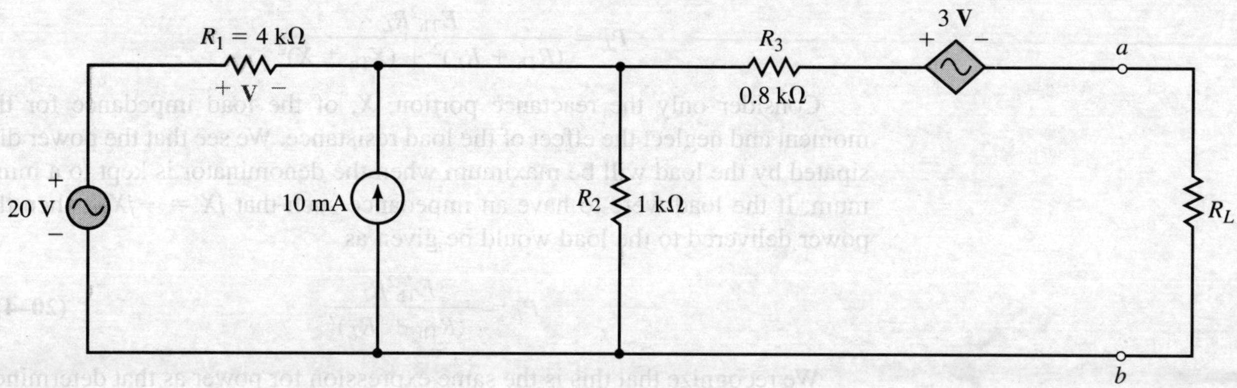

FIGURE 20–53

Answers

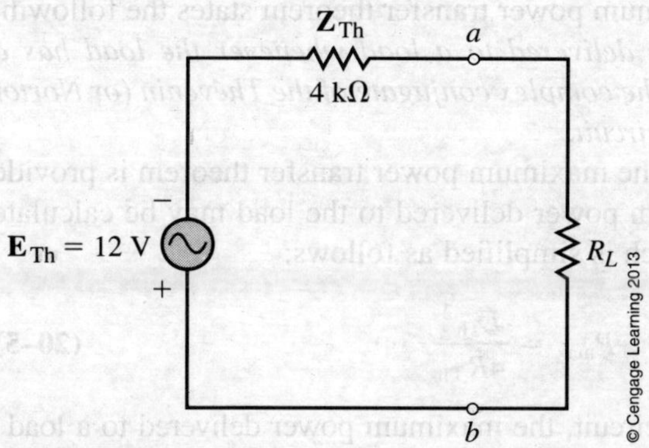

FIGURE 20–54

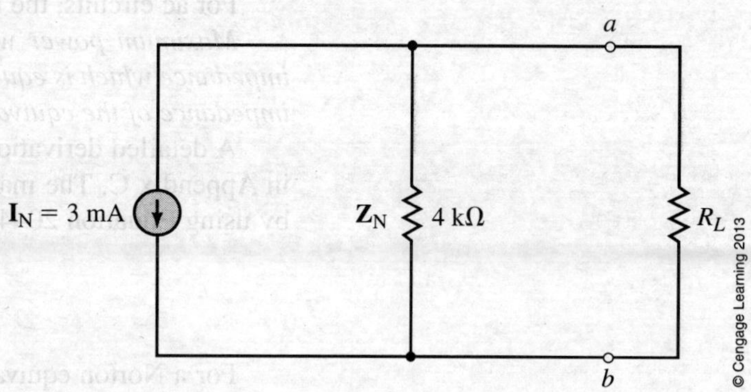

FIGURE 20–55

20.6 Maximum Power Transfer Theorem

The maximum power transfer theorem is used to determine the value of load impedance required so that the load receives the maximum amount of power from the circuit. Consider the Thévenin equivalent circuit shown in Figure 20–56.

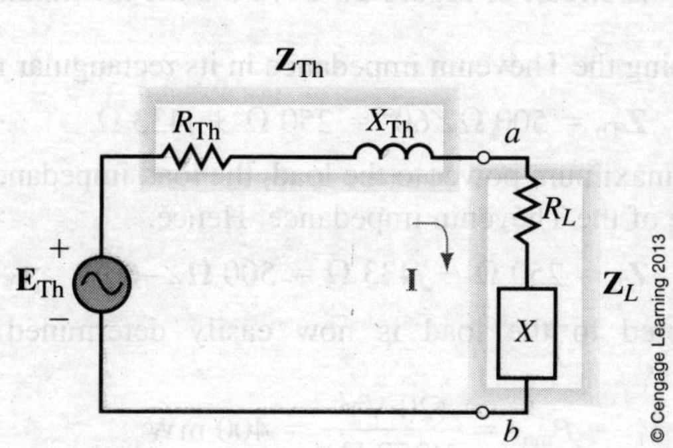

FIGURE 20–56

For any load impedance $\mathbf{Z}_L$ consisting of a resistance and a reactance such that $\mathbf{Z}_L = R_L \pm jX$, the power dissipated by the load will be determined as follows:

$$P_L = I^2 R_L$$

$$I = \frac{E_{Th}}{\sqrt{(R_{Th} + R_L)^2 + (X_{Th} \pm X)^2}}$$

$$P_L = \frac{E_{Th}{}^2 R_L}{(R_{Th} + R_L)^2 + (X_{Th} \pm X)^2}$$

Consider only the reactance portion, X, of the load impedance for the moment and neglect the effect of the load resistance. We see that the power dissipated by the load will be maximum when the denominator is kept to a minimum. If the load were to have an impedance such that $jX = -jX_{Th}$, then the power delivered to the load would be given as

$$P_L = \frac{E_{Th}{}^2 R_L}{(R_{Th} + R_L)^2} \tag{20–4}$$

We recognize that this is the same expression for power as that determined for the Thévenin equivalent of dc circuits in Chapter 9. Recall that maximum power was delivered to the load when

$$R_L = R_{Th}$$

For ac circuits, the maximum power transfer theorem states the following: *Maximum power will be delivered to a load whenever the load has an impedance which is equal to the complex conjugate of the Thévenin (or Norton) impedance of the equivalent circuit.*

A detailed derivation of the maximum power transfer theorem is provided in Appendix C. The maximum power delivered to the load may be calculated by using Equation 20–4, which is simplified as follows:

$$P_{max} = \frac{E_{Th}{}^2}{4R_{Th}} \tag{20–5}$$

For a Norton equivalent circuit, the maximum power delivered to a load is determined by substituting $E_{Th} = I_N Z_N$ into the preceding expression as follows:

$$P_{max} = \frac{I_N{}^2 Z_N{}^2}{4R_N} \tag{20–6}$$

EXAMPLE 20–12

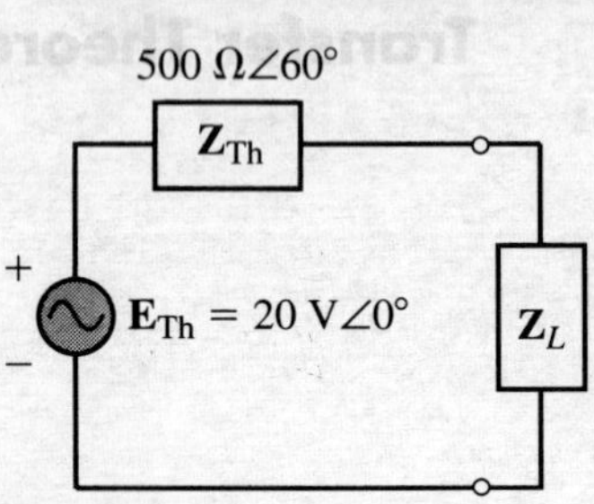

FIGURE 20–57

Determine the load impedance $\mathbf{Z}_L$ that will allow maximum power to be delivered to the load in the circuit of Figure 20–57. Find the maximum power.

Solution Expressing the Thévenin impedance in its rectangular form, we have

$$\mathbf{Z}_{Th} = 500\ \Omega\angle 60° = 250\ \Omega + j433\ \Omega$$

In order to deliver maximum power to the load, the load impedance must be the complex conjugate of the Thévenin impedance. Hence,

$$\mathbf{Z}_L = 250\ \Omega - j433\ \Omega = 500\ \Omega\angle -60°$$

The power delivered to the load is now easily determined by applying Equation 20–5:

$$P_{max} = \frac{(20\ \text{V})^2}{4(250\ \Omega)} = 400\ \text{mW}$$

PRACTICE PROBLEMS 7

Given the circuit of Figure 20–57, determine the power dissipated by the load if the load impedance is equal to the Thévenin impedance, $\mathbf{Z}_L = 500\ \Omega\angle 60°$. Compare your answer to that obtained in Example 20–12.

Answer
$P = 100$ mW, which is less than P_{max}.

Occasionally it is not possible to adjust the reactance portion of a load. In such cases, a **relative maximum power** will be delivered to the load when the load resistance has a value determined as

$$R_L = \sqrt{R_{\text{Th}}^2 + (X \pm X_{\text{Th}})^2} \qquad \text{(20–7)}$$

If the reactance of the Thévenin impedance is of the same type (both capacitive or both inductive) as the reactance in the load, then the reactances are added.

If one reactance is capacitive and the other is inductive, however, then the reactances are subtracted.

To determine the power delivered to the load in such cases, the power will need to be calculated by finding either the voltage across the load or the current through the load. Equations 20–5 and 20–6 will no longer apply, since these equations were based on the premise that the load impedance is the complex conjugate of the Thévenin impedance.

EXAMPLE 20–13

For the circuit of Figure 20–58, determine the value of the load resistor, R_L, such that maximum power will be delivered to the load.

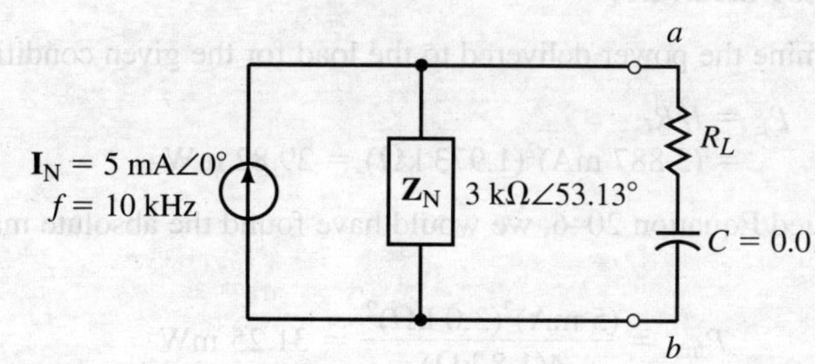

FIGURE 20–58

Solution Notice that the load impedance consists of a resistor in series with a capacitance of 0.010 μF. Since the capacitive reactance is determined by the frequency, it is quite likely that the maximum power for this circuit may only be a relative maximum, rather than the absolute maximum. For the **absolute maximum power** to be delivered to the load, the load impedance would need to be

$$\mathbf{Z}_L = 3\ \text{k}\Omega\angle -53.13° = 1.80\ \text{k}\Omega - j2.40\ \text{k}\Omega$$

The reactance of the capacitor at a frequency of 10 kHz is determined to be

$$X_C = \frac{1}{2\pi(10\ \text{kHz})(0.010\ \mu\text{F})} = 1.592\ \text{k}\Omega$$

Because the capacitive reactance is not equal to the inductive reactance of the Norton impedance, the circuit will not deliver the absolute maximum power to the load. However, relative maximum power will be delivered to the load when

$$\begin{aligned} R_L &= \sqrt{R_{\text{Th}}^2 + (X - X_{\text{Th}})^2} \\ &= \sqrt{(1.800\ \text{k}\Omega)^2 + (1.592\ \text{k}\Omega - 2.4\ \text{k}\Omega)^2} \\ &= 1.973\ \text{k}\Omega \end{aligned}$$

Figure 20–59 shows the circuit with all impedance values.

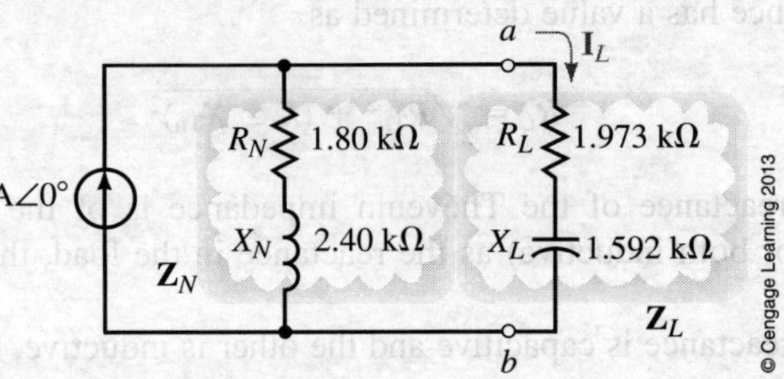

FIGURE 20–59

The load current will be

$$\begin{aligned} \mathbf{I}_L &= \frac{\mathbf{Z}_N}{\mathbf{Z}_N + \mathbf{Z}_L}\mathbf{I}_N \\ &= \frac{1.80\ \text{k}\Omega + j2.40\ \text{k}\Omega}{(1.80\ \text{k}\Omega + j2.40\ \text{k}\Omega) + (1.973\ \text{k}\Omega - j1.592\ \text{k}\Omega)}(5\ \text{mA}\angle 0°) \\ &= \frac{3\ \text{k}\Omega\angle 53.13°}{3.773\ \text{k}\Omega + j0.808\ \text{k}\Omega}(5\ \text{mA}\angle 0°) \\ &= \frac{15.0\ \text{V}\angle 53.13°}{3.859\ \text{k}\Omega\angle 12.09°} = 3.887\ \text{mA}\angle 41.04° \end{aligned}$$

We now determine the power delivered to the load for the given conditions as

$$\begin{aligned} P_{\text{L}} &= I_L^2 R_L \\ &= (3.887\ \text{mA})^2(1.973\ \text{k}\Omega) = 29.82\ \text{mW} \end{aligned}$$

If we had applied Equation 20–6, we would have found the absolute maximum power to be

$$P_{\max} = \frac{(5\ \text{mA})^2(3.0\ \text{k}\Omega)^2}{4(1.8\ \text{k}\Omega)} = 31.25\ \text{mW}$$

PRACTICE PROBLEMS 8

Refer to the Norton equivalent circuit of Figure 20–60:

a. Find the value of load resistance, R_L, such that the load receives maximum power.

b. Determine the maximum power received by the load for the given conditions.

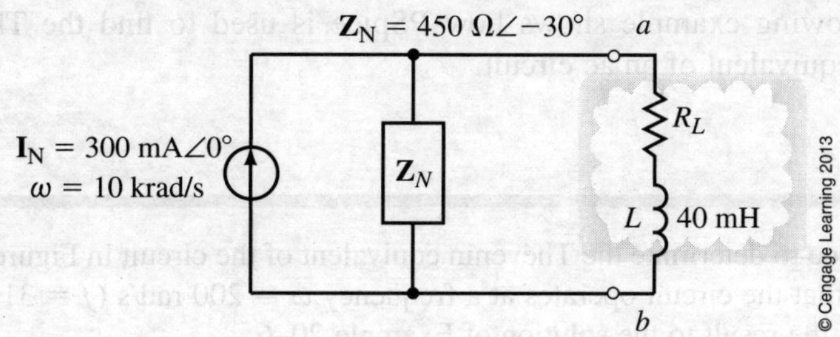

FIGURE 20–60

Answers

a. $R_L = 427\ \Omega$; b. $P_L = 11.2$ W

IN-PROCESS LEARNING CHECK 4

(Answers are at the end of the chapter.)

Refer to the Thévenin equivalent circuit of Figure 20–61.

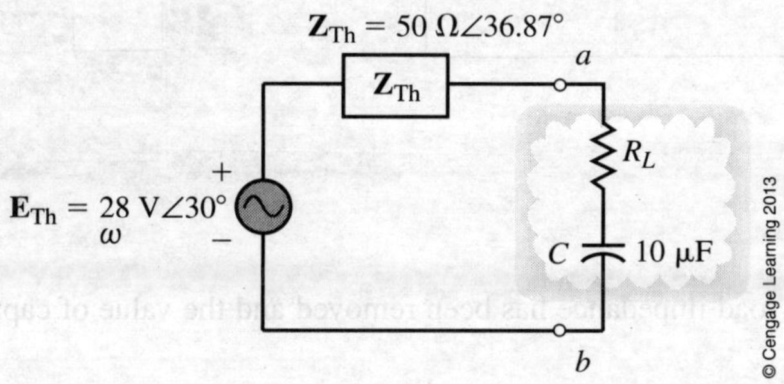

FIGURE 20–61

a. Determine the value of the unknown load resistance R_L that will result in a relative maximum power at an angular frequency of 1 krad/s.
b. Solve for the power dissipated by the load at $\omega_1 = 1$ krad/s.
c. Assuming that the Thévenin impedance remains constant at all frequencies, at what angular frequency, ω_2, will the circuit provide absolute maximum power?
d. Solve for the power dissipated by the load at ω_2.

CircuitSim 20-4

20.7 Circuit Analysis Using Computers

As demonstrated in Chapter 9, circuit analysis programs are very useful in determining the equivalent circuit between specified terminals of a dc circuit. We will use similar methods to obtain the Thévenin and Norton equivalents of ac circuits. Both Multisim and PSpice are useful in analyzing circuits with dependent sources. As we have already seen, the work in analyzing such a circuit manually is lengthy and very time consuming. In this section, you will learn how to use PSpice to find the Thévenin equivalent of a simple ac circuit. We will also use both programs to analyze circuits with dependent sources.

MULTISIM

PSPICE

PSpice

The following example shows how PSpice is used to find the Thévenin or Norton equivalent of an ac circuit.

EXAMPLE 20–14

Use PSpice to determine the Thévenin equivalent of the circuit in Figure 20–18. Assume that the circuit operates at a frequency $\omega = 200$ rad/s ($f = 31.83$ Hz). Compare the result to the solution of Example 20–6.

NOTES...

PSpice Note

The two symbols below are equivalent. The symbol on the left is the Multisim notation for an ac current source, while that on the right is the symbol used in PSpice. Notice that in PSpice, the ac current source (shown on the right) does not use the same notation as that used in the textbook, but rather appears similar to an ac voltage source. More importantly, notice the polarity of the PSpice symbol. When using an ac current source in PSpice, you may need to rotate the symbol to correctly place it into the circuit.

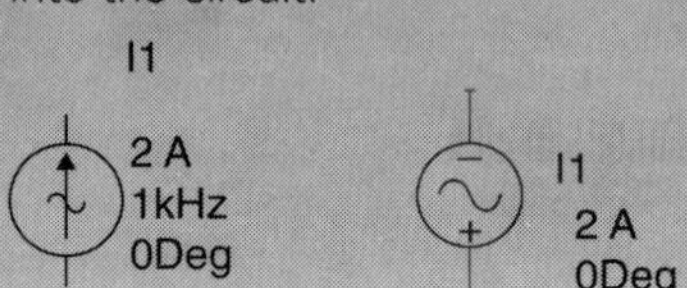

Solution We begin by using OrCAD Capture to input the circuit as shown in Figure 20–62.

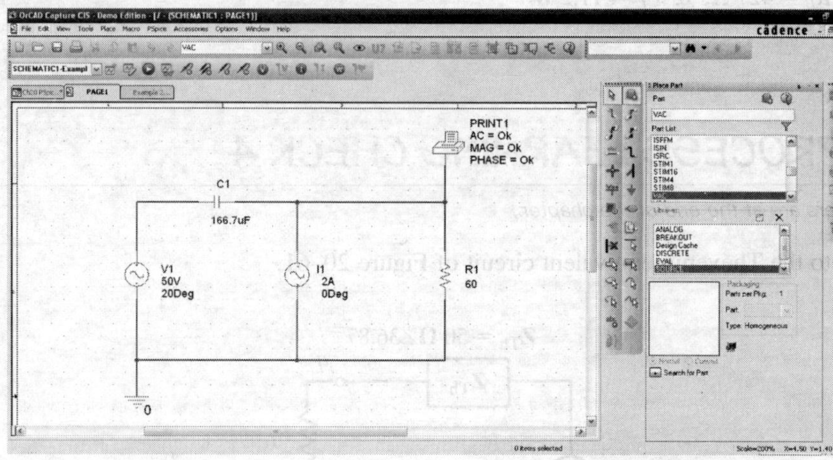

FIGURE 20–62

Notice that the load impedance has been removed and the value of capacitance is shown as

$$C = \frac{1}{\omega X_C} = \frac{1}{(200 \text{ rad/s})(30\ \Omega)} = 166.7\ \mu\text{F}$$

In order to adjust for the correct source voltage and current, we select VAC and ISRC from the source library SOURCE.slb of PSpice. The properties of the voltage source are set to ACMAG=**50 V** and ACPHASE=**20 Deg.** Similarly, the parameters of the current source are set to ACMAG=**2 A** and ACPHASE=**0 Deg.** (Ensure that you observe the correct polarity for the current source.) The open-circuit output voltage is displayed using VPRINT1, which can be set to measure magnitude and phase of an ac voltage as follows: Similarly, the properties of VPRINT1 are changed by double-clicking on the part. Use the horizontal scrollbar to find the AC, MAG, and PHASE cells. Once you have entered **OK** in each of the cells, click on *Apply*. Next, click on *Display* and select *Name* and *Value* from the display properties.

Once the circuit is entered, click on the *New Simulation Profile* tool, give the simulation a name such as **Example 20–14,** and set the analysis for *AC Sweep/Noise* with a linear sweep beginning and ending at a frequency of **31.83 Hz** (1 point). As before, it is convenient to disable the Probe postprocessor from the *Probe Window* tab in the simulation settings box.

After simulating the design, the open-circuit voltage is determined by viewing the *Output File* (found in the PSpice menu). The pertinent data from the output file is given as

```
FREQ          VM(a)         VP(a)
3.183E+01     5.690E+01     -1.583E+01
```

The preceding result gives $\mathbf{E}_{Th} = 56.90\ V\angle-15.83°$. This is the same as the value determined in Example 20–6. Recall that one way of determining the Thevenin (or Norton) impedance is to use Ohm's law, namely

$$\mathbf{Z}_{Th} = \mathbf{Z}_N = \frac{\mathbf{E}_{Th}}{\mathbf{I}_N}$$

The Norton current is found by removing the VPRINT1 device from the circuit of Figure 20–62 and inserting an IPRINT device (ammeter) between terminal *a* and ground. You will need to change the properties of IPRINT in a manner similar to that described previously for the VPRINT1 part. The result is shown in Figure 20–63.

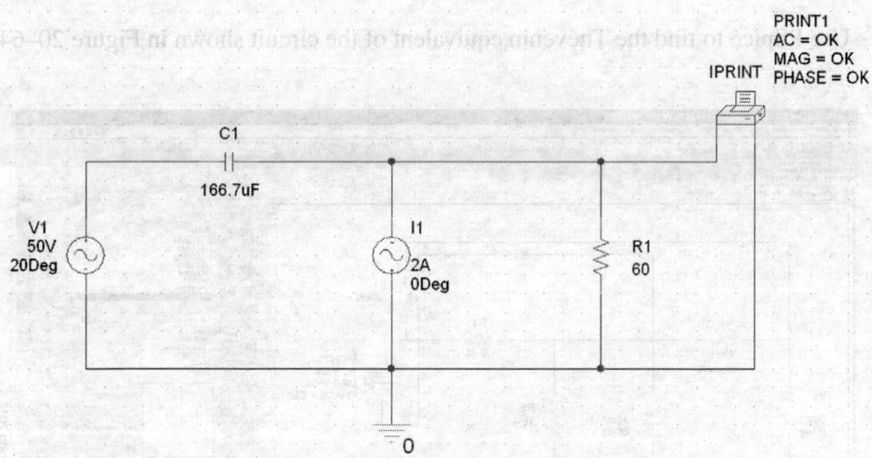

FIGURE 20–63

After simulating the design, the short-circuit current is determined by examining the output file of PSpice. The pertinent data from the output file is given as

```
FREQ          IM(V_PRINT2)IP(V_PRINT2)
3.183E+01     2.121E+00    4.761E+01
```

This result gives $\mathbf{I}_N = 21.21\ A\angle 47.61°$, and so we calculate the Thévenin impedance as

$$\mathbf{Z}_{Th} = \frac{56.90\ V\angle-15.83°}{2.121\ A\angle 47.61°} = 26.83\Omega\angle-63.44°$$

which is the same value as that obtained in Example 20–6.

In the previous example, it was necessary to determine the Thévenin impedance in two steps, by first solving for the Thévenin voltage and then solving for the Norton current. It is possible to determine the value in one step by using an additional step in the analysis. The following example shows how to determine the Thévenin impedance for a circuit having a dependent source. The same step may also be used for a circuit having independent sources.

The following parts in OrCAD Capture are used to represent dependent sources:

Voltage-controlled voltage source:	**E**
Current-controlled current source:	**F**
Voltage-controlled current source:	**G**
Current-controlled voltage source:	**H**

When using dependent sources, it is necessary to ensure that any voltage-controlled source is placed *across* the controlling voltage and any current-controlled source is placed *in series with* the controlling current. Additionally, each dependent source must have a specified **gain.** This value simply provides the ratio between the output value and the controlling voltage or current. Although the following example shows how to use only one type of dependent source, you will find many similarities between the various sources.

EXAMPLE 20–15

Use PSpice to find the Thévenin equivalent of the circuit shown in Figure 20–64.

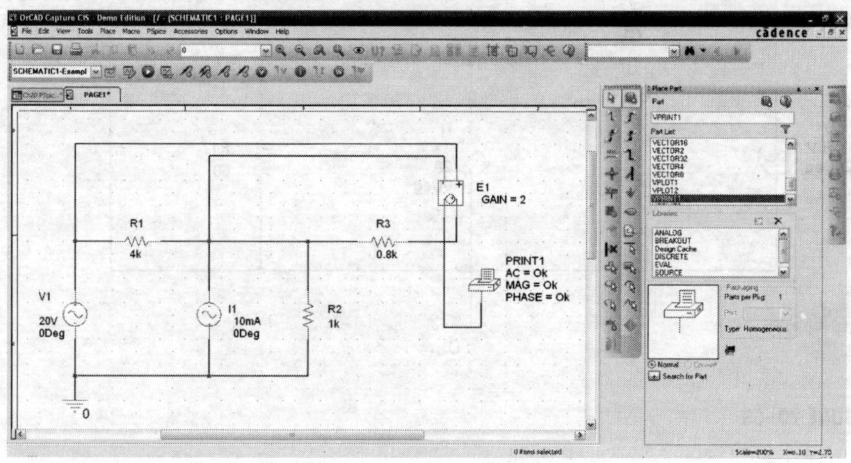

FIGURE 20–64

Solution OrCAD Capture is used to input the circuit shown in Figure 20–64. The voltage-controlled voltage source is obtained by clicking on the *Place part* tool and selecting *E* from the ANALOG library. Notice the placement of the source. Since R_1 is the controlling element, the controlling terminals are placed across this resistor. To adjust the gain of the voltage source, double-click on the symbol. Use the scrollbar to find the GAIN cell and type **2.** In order to display this value on the schematic, you will need to click on *Display* and select *Name* and *Value* from the display properties.

As in the previous circuit, you will need to first use a VPRINT1 part to measure the open circuit voltage at terminal *a*. The properties of VPRINT1 are changed by typing **OK** in the AC, MAG, and PHASE cells. Click on *Display* and select *Name* and *Value* for each of the cells. Give the simulation profile a name and run the simulation at a frequency of 1 kHz. The pertinent data in the PSpice output file provides the open-circuit (Thévenin) voltage as follows.

```
FREQ          VM(N00431)    VP(N00431)
1.000E+03     4.000E+00     1.800E+02
```

The VPRINT1 part is then replaced with IPRINT, which is connected between terminal *a* and ground to provide the short-circuit current. Remember to change the appropriate cells using the properties editor. The PSpice output file gives the short-circuit (Norton) current as:

```
FREQ          IM(V_PRINT2) IP(V_PRINT2)
1.000E+03     1.250E-03    1.800E+02
```

The Thévenin impedance is now easily determined as

$$\mathbf{Z}_{\text{Th}} = \frac{4\ \text{V}}{1.25\ \text{mA}} = 3.2\ \text{k}\Omega$$

and the resulting circuit is shown in Figure 20–65.

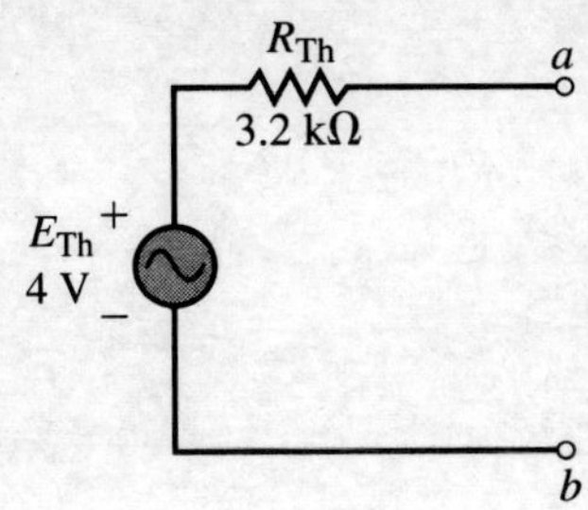

FIGURE 20–65

Multisim

Multisim has many similarities to PSpice in its analysis of ac circuits. The following example shows that the results obtained by Multisim are precisely the same as those obtained from PSpice.

EXAMPLE 20–16

Use Multisim to find the Thévenin equivalent of the circuit shown in Figure 20–47. Compare the results to those obtained in Example 20–15.

Solution The circuit is entered as shown in Figure 20–66.

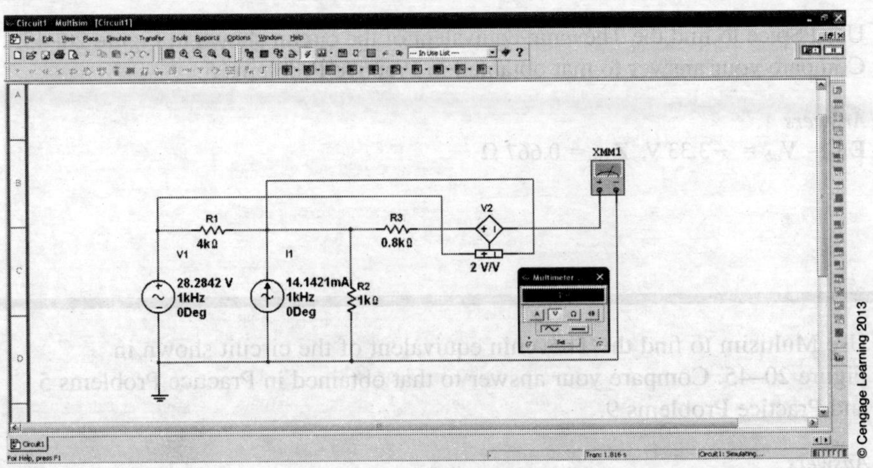

FIGURE 20–66

All components are selected from the *Virtual Components* toolbar. Since we will use an ac multimeter to measure RMS quantities, we set the amplitudes of the voltage and current sources to **28.28 V** and **14.14 mA,** respectively. We use 1 kHz as the frequency of operation, although any frequency may be used. The *Voltage-controlled* voltage source is selected from the *Sources* parts bin and the gain is adjusted by double-clicking on the symbol. The *Value* tab is selected and the *Voltage gain (E):* is set to **2 V/V.** The multimeter must be set to measure ac volts. As expected, the reading on the multimeter is 4 V. Current is easily measured by setting the multimeter onto its ac ammeter range as shown in Figure 20–67. The current reading is 1.25 mA. A limitation of Multisim modelling is that the model does not indicate the phase angle of the voltage or current.

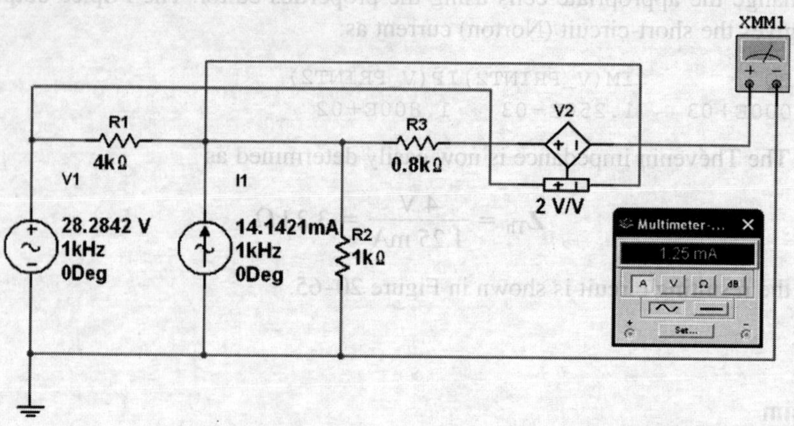

FIGURE 20–67

Now, the Thévenin impedance of the circuit is determined using Ohm's law, namely

$$\mathbf{Z}_{\text{Th}} = \frac{4\text{V}}{1.25\text{ mA}} = 3.2\text{ k}\Omega$$

These results are consistent with the calculations of Example 20–11 and the PSpice results of Example 20–15.

PRACTICE PROBLEMS 9

Use PSpice to find the Thévenin equivalent of the circuit shown in Figure 20–45. Compare your answer to that obtained in Practice Problems 5.

Answers
$\mathbf{E}_{\text{Th}} = \mathbf{V}_{ab} = -3.33\text{ V}$, $\mathbf{Z}_{\text{Th}} = 0.667\ \Omega$

PRACTICE PROBLEMS 10

Use Multisim to find the Thévenin equivalent of the circuit shown in Figure 20–45. Compare your answer to that obtained in Practice Problems 5 and Practice Problems 9.

Answers
$\mathbf{E}_{\text{Th}} = \mathbf{V}_{ab} = -3.33\text{ V}$, $\mathbf{Z}_{\text{Th}} = 0.667\ \Omega$

PRACTICE PROBLEMS 11

Use PSpice to find the Norton equivalent circuit external to the load resistor in the circuit of Figure 20–31. Assume that the circuit operates at a frequency of 20 kHz. Compare your results to those obtained in Example 20–8. Hint: You will need to place a small resistor (e.g., 1 mΩ) in series with the inductor.

Answers
$\mathbf{E}_{\text{Th}} = 7.75\text{ V}\angle -176.5°$, $\mathbf{I}_{\text{N}} = 0.1335\text{ A}\angle 138.5°$, $\mathbf{Z}_{\text{N}} = 566\Omega\angle 45°$
The results are consistent.

Putting It into Practice

In this chapter, you learned how to solve for the required load impedance to enable maximum power transfer to the load. In all cases, you worked with load impedances that were in series with the output terminals. This will not always be the case. The circuit shown in the accompanying figure shows a load that consists of a resistor in parallel with an inductor.

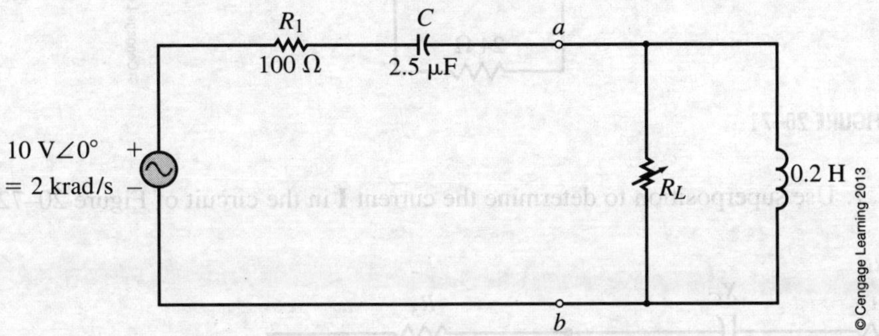

Determine the value of the resistor R_L needed to result in maximum power delivered to the load. Although several methods are possible, you may find that this example lends itself to being solved by using calculus.

Problems

20.1 Superposition Theorem—Independent Sources

1. Use superposition to determine the current in the indicated branch of the circuit in Figure 20–68.
2. Repeat Problem 1 for the circuit of Figure 20–69.
3. Use superposition to determine the voltage $\mathbf{V}_{ab}$ for the circuit of Figure 20–68.
4. Repeat Problem 3 for the circuit of Figure 20–69.
5. Consider the circuit of Figure 20–70.
   a. Use superposition to determine the indicated voltage, **V**.
   b. Show that the power dissipated by the indicated resistor cannot be determined by superposition.

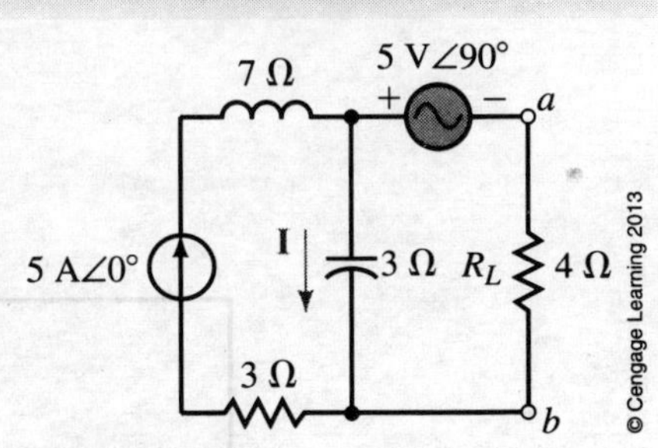

FIGURE 20–68

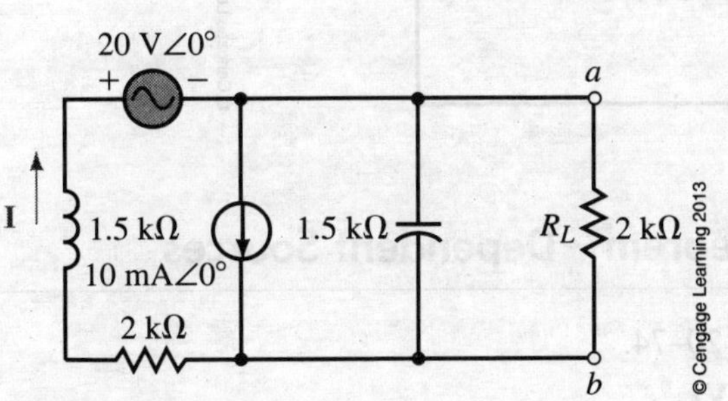

FIGURE 20–69

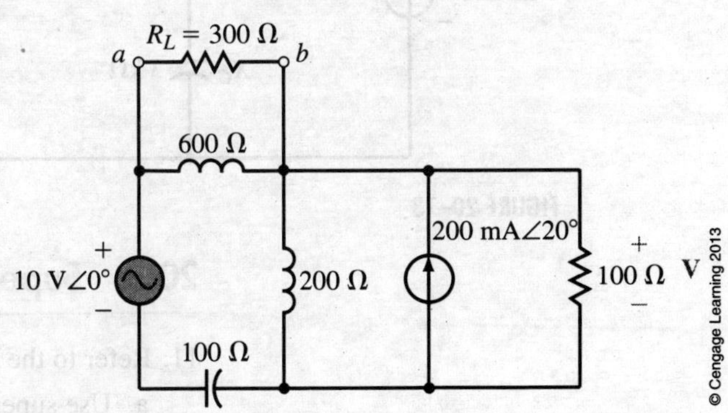

FIGURE 20–70

6. Repeat Problem 5 for the circuit of Figure 20–71.

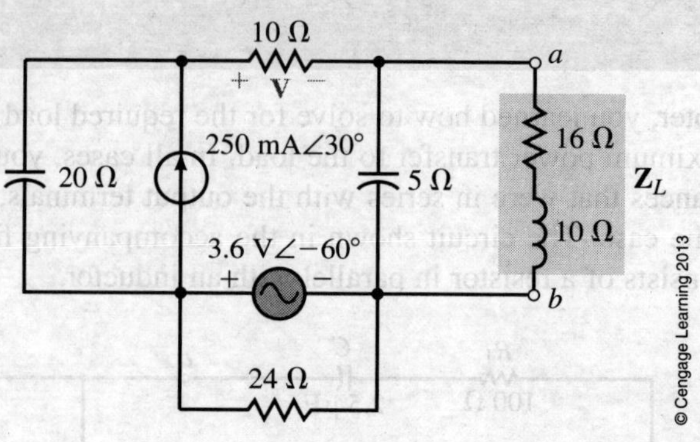

FIGURE 20–71

7. Use superposition to determine the current **I** in the circuit of Figure 20–72.

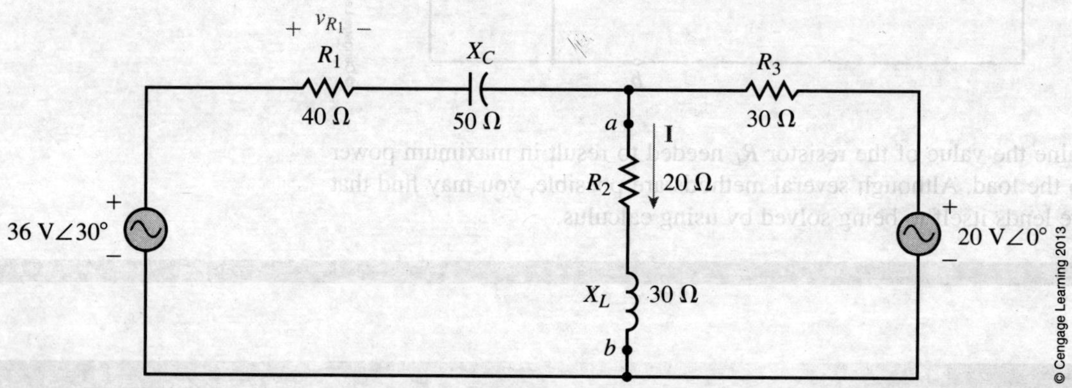

FIGURE 20–72

8. Repeat Problem 7 for the circuit of Figure 20–73.
9. Use superposition to determine the sinusoidal voltage, v_{R_1} for the circuit of Figure 20–72.
10. Repeat Problem 9 for the circuit of Figure 20–73.

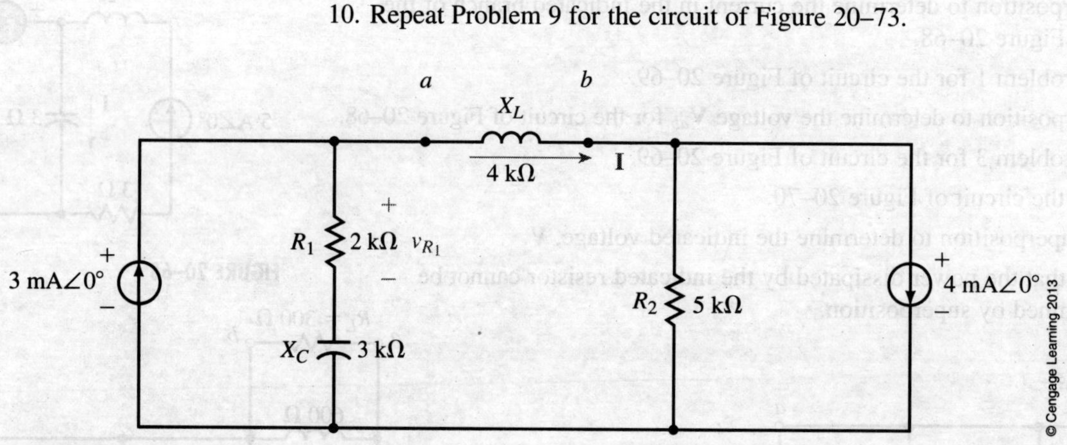

FIGURE 20–73

20.2 Superposition Theorem—Dependent Sources

11. Refer to the circuit of Figure 20–74.
    a. Use superposition to find $\mathbf{V}_L$.
    b. If the magnitude of the applied voltage **V** is increased to 200 mV, solve for the resulting $\mathbf{V}_L$.

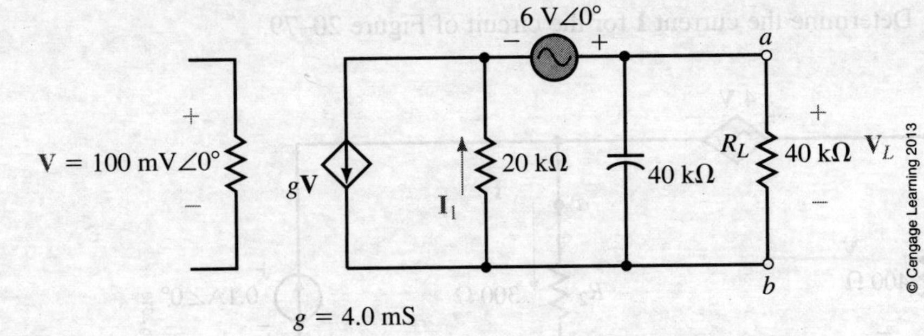

FIGURE 20–74

12. Consider the circuit of Figure 20–75.
    a. Use superposition to find $\mathbf{V}_L$.
    b. If the magnitude of the applied current **I** is decreased to 2 mA, solve for the resulting $\mathbf{V}_L$.
13. Use superposition to find the current $\mathbf{I}_1$ in the circuit of Figure 20–74.
14. Repeat Problem 13 for the circuit of Figure 20–75.
15. Find $\mathbf{V}_L$ in the circuit of Figure 20–76.

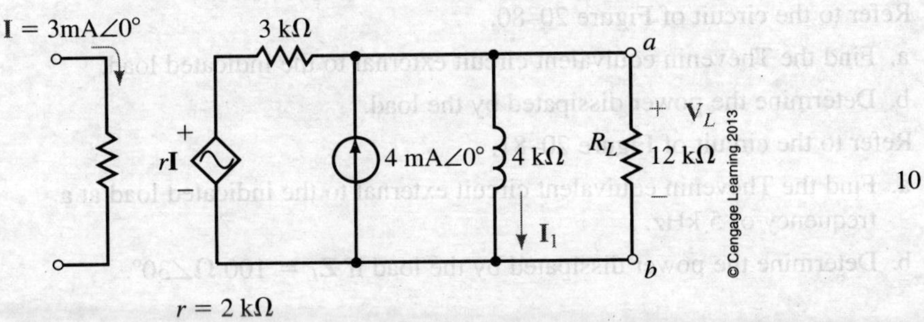

FIGURE 20–75

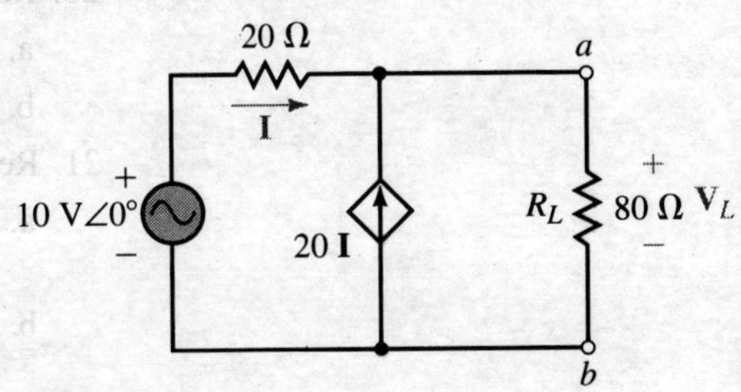

FIGURE 20–76

16. Find $\mathbf{V}_L$ in the circuit of Figure 20–77.
17. Determine the voltage $\mathbf{V}_{ab}$ for the circuit of Figure 20–78.

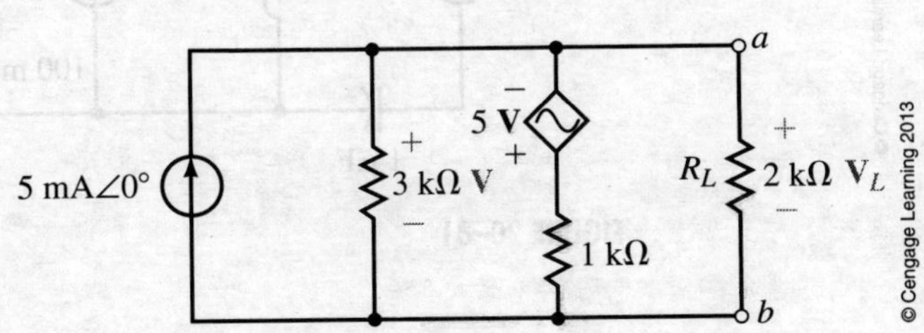

FIGURE 20–77

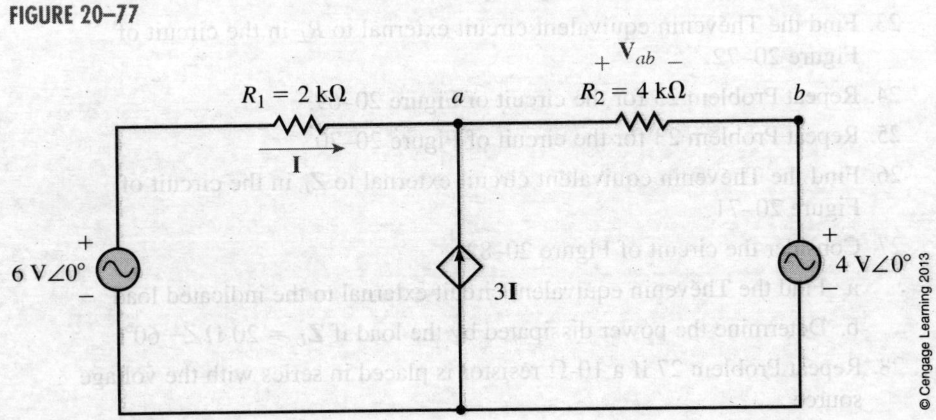

FIGURE 20–78

18. Determine the current **I** for the circuit of Figure 20–79.

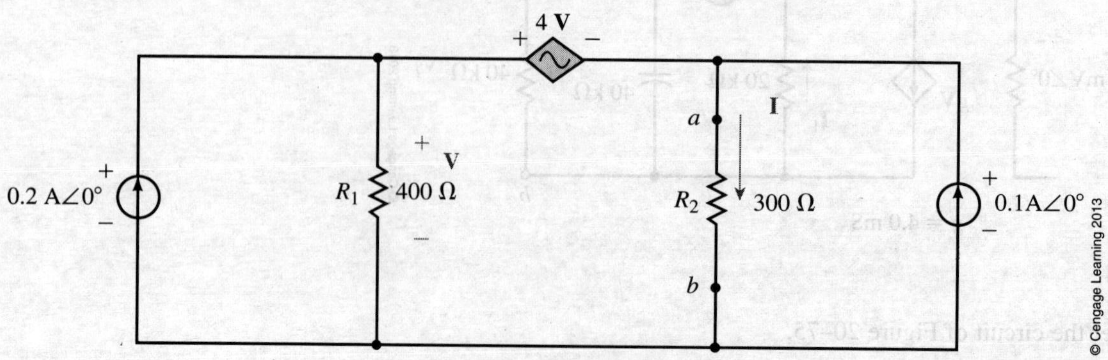

FIGURE 20–79

20.3 Thévenin's Theorem—Independent Sources

19. Find the Thévenin equivalent circuit external to the load impedance of Figure 20–68.
20. Refer to the circuit of Figure 20–80.
    a. Find the Thévenin equivalent circuit external to the indicated load.
    b. Determine the power dissipated by the load.
21. Refer to the circuit of Figure 20–81.
    a. Find the Thévenin equivalent circuit external to the indicated load at a frequency of 5 kHz.
    b. Determine the power dissipated by the load if $\mathbf{Z}_L = 100\ \Omega\angle 30°$.

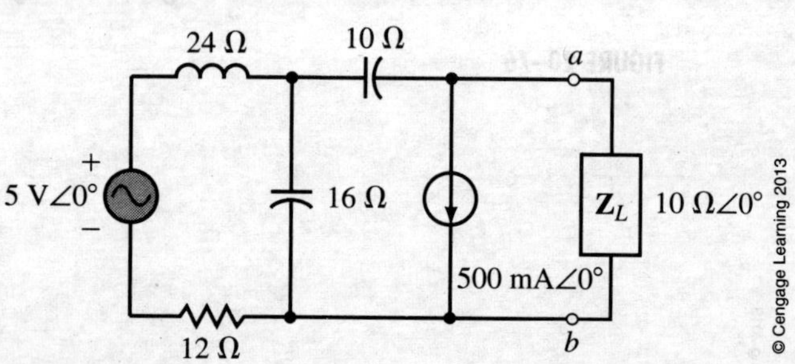

FIGURE 20–80

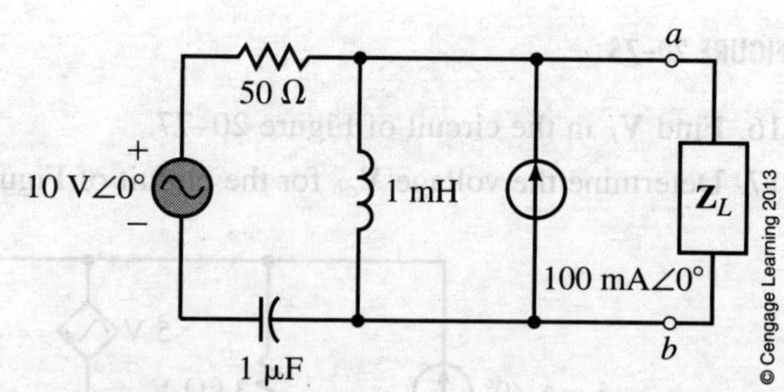

FIGURE 20–81

22. Repeat Problem 21 for a frequency of 1 kHz.
23. Find the Thévenin equivalent circuit external to R_L in the circuit of Figure 20–72.
24. Repeat Problem 23 for the circuit of Figure 20–69.
25. Repeat Problem 23 for the circuit of Figure 20–70.
26. Find the Thévenin equivalent circuit external to Z_L in the circuit of Figure 20–71.
27. Consider the circuit of Figure 20–82.
    a. Find the Thévenin equivalent circuit external to the indicated load.
    b. Determine the power dissipated by the load if $\mathbf{Z}_L = 20\ \Omega\angle -60°$.
28. Repeat Problem 27 if a 10-Ω resistor is placed in series with the voltage source.

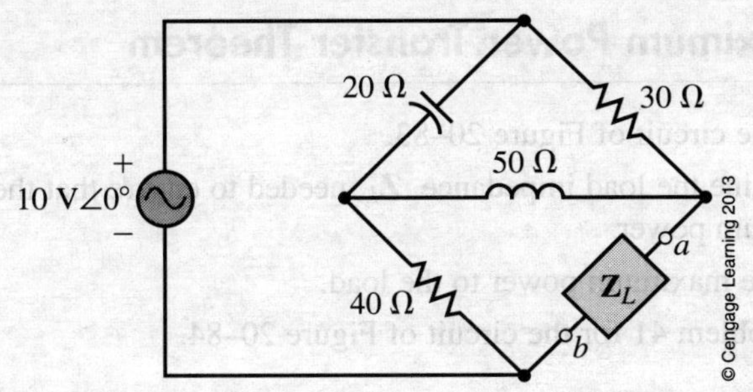

FIGURE 20–82

20.4 Norton's Theorem—Independent Sources

29. Find the Norton equivalent circuit external to the load impedance of Figure 20–68.
30. Repeat Problem 29 for the circuit of Figure 20–69.
31. a. Using the outlined procedure, find the Norton equivalent circuit external to terminals *a* and *b* in Figure 20–72.
    b. Determine the current through the indicated load.
    c. Find the power dissipated by the load.
32. Repeat Problem 31 for the circuit of Figure 20–73.
33. a. Using the outlined procedure, find the Norton equivalent circuit external to the indicated load impedance (located between terminals *a* and *b*) in Figure 20–70.
    b. Determine the current through the indicated load.
    c. Find the power dissipated by the load.
34. Repeat Problem 33 for the circuit of Figure 20–71.
35. Suppose that the circuit of Figure 20–81 operates at a frequency of 2 kHz.
    a. Find the Norton equivalent circuit external to the load impedance.
    b. If a 30-Ω load resistor is connected between terminals *a* and *b*, find the current through the load.
36. Repeat Problem 35 for a frequency of 8 kHz.

20.5 Thévenin's and Norton's Theorems for Dependent Sources

37. a. Find the Thévenin equivalent circuit external to the load impedance in Figure 20–74.
    b. Solve for the current through R_L.
    c. Determine the power dissipated by R_L.
38. a. Find the Norton equivalent circuit external to the load impedance in Figure 20–75.
    b. Solve for the current through R_L.
    c. Determine the power dissipated by R_L.
39. Find the Thévenin and Norton equivalent circuits external to the load impedance of Figure 20–76.
40. Find the Thévenin equivalent circuit external to the load impedance of Figure 20–77.

20.6 Maximum Power Transfer Theorem

41. Refer to the circuit of Figure 20–83.
    a. Determine the load impedance, $\mathbf{Z}_L$, needed to ensure that the load receives maximum power.
    b. Find the maximum power to the load.
42. Repeat Problem 41 for the circuit of Figure 20–84.

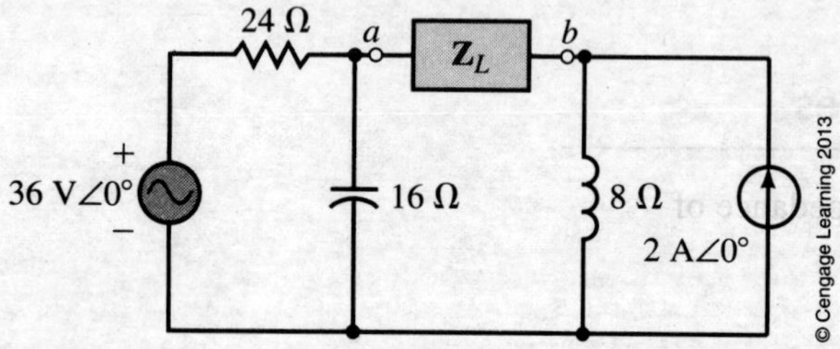

FIGURE 20–83

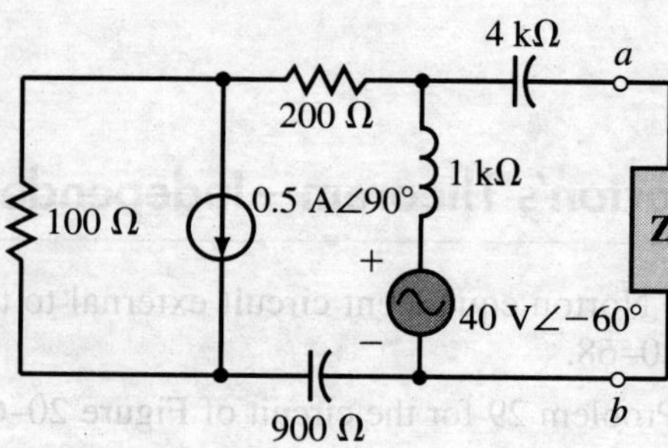

FIGURE 20–84

43. Repeat Problem 41 for the circuit of Figure 20–85.
44. Repeat Problem 41 for the circuit of Figure 20–86.

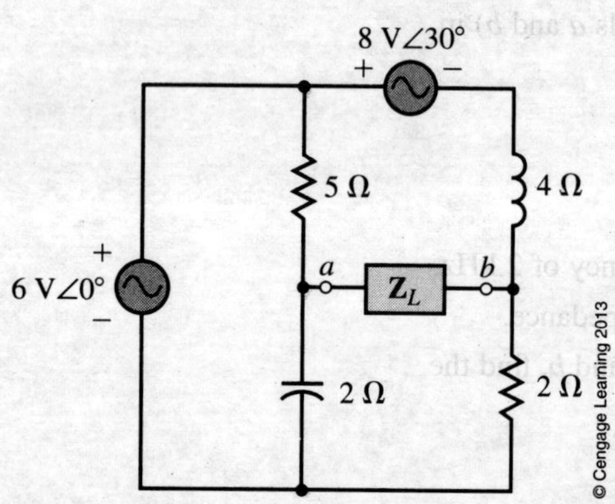

FIGURE 20–85

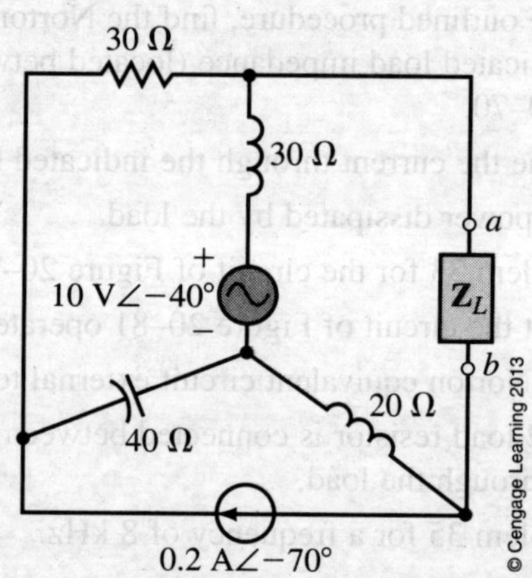

FIGURE 20–86

45. What load impedance is required for the circuit of Figure 20–71 to ensure that the load receives maximum power from the circuit?
46. Determine the load impedance required for the circuit of Figure 20–82 to ensure that the load receives maximum power from the circuit.
47. a. Determine the required load impedance, $\mathbf{Z}_L$, for the circuit of Figure 20–81 to deliver maximum power to the load at a frequency of 5 kHz.
    b. If the load impedance contains a resistor and a 1-μF capacitor, determine the value of the resistor to result in a relative maximum power transfer.
    c. Solve for the power delivered to the load in (b).
48. a. Determine the required load impedance, $\mathbf{Z}_L$, for the circuit of Figure 20–81 to deliver maximum power to the load at a frequency of 1 kHz.
    b. If the load impedance contains a resistor and a 1-μF capacitor, determine the value of the resistor to result in a relative maximum power transfer.
    c. Solve for the power delivered to the load in (b).

20.7 Circuit Analysis Using Computers

49. Use PSpice to find the Thévenin equivalent circuit external to R_L in the circuit of Figure 20–68. Assume that the circuit operates at a frequency of $\omega = 2000$ rad/s. PSPICE

    Note: PSpice does not permit a voltage source to have floating terminals. Therefore, a large resistance (e.g., 10 GΩ) must be placed across the output.
50. Repeat Problem 49 for the circuit of Figure 20–69. PSPICE
51. Use PSpice to find the Norton equivalent circuit external to R_L in the circuit of Figure 20–70. Assume that the circuit operates at a frequency of $\omega = 5000$ rad/s. PSPICE

    Note: PSpice cannot analyze a circuit with a short-circuited inductor. Consequently, it is necessary to place a small resistance (e.g., 1 nΩ) in series with an inductor.
52. Repeat Problem 51 for the circuit of Figure 20–71. PSPICE

    Note: PSpice cannot analyze a circuit with an open-circuited capacitor. Consequently, it is necessary to place a large resistance (e.g., 10 GΩ) in parallel with a capacitor.
53. Use PSpice to find the Thévenin equivalent circuit external to R_L in the circuit of Figure 20–76. Assume that the circuit operates at a frequency of $f = 1000$ Hz. PSPICE
54. Repeat Problem 53 for the circuit of Figure 20–77. PSPICE
55. Use PSpice to find the Norton equivalent circuit external to $\mathbf{V}_{ab}$ in the circuit of Figure 20–78. Assume that the circuit operates at a frequency of $f = 1000$ Hz. PSPICE
56. Repeat Problem 55 for the circuit of Figure 20–79. PSPICE
57. Use Multisim to find the Thévenin equivalent circuit external to R_L in the circuit of Figure 20–76. Assume that the circuit operates at a frequency of $f = 1000$ Hz. Multisim
58. Repeat Problem 53 for the circuit of Figure 20–77. Multisim
59. Use Multisim to find the Norton equivalent circuit external to $\mathbf{V}_{ab}$ in the circuit of Figure 20–78. Assume that the circuit operates at a frequency of $f = 1000$ Hz. Multisim
60. Repeat Problem 55 for the circuit of Figure 20–79. Multisim

ANSWERS TO IN-PROCESS LEARNING CHECKS

IN-PROCESS LEARNING CHECK 1

a. $\mathbf{I} = 3.39\text{ A}\angle 25.34°$

b. $P_T = 230.4$ W

c. $P_1 + P_2 + P_3 = 145\text{ W} \neq P_T = 230.4$ W
Superposition does not apply for power.

IN-PROCESS LEARNING CHECK 2

1. Remove the inductor from the circuit. Label the remaining terminals as *a* and *b*.
2. Set the voltage source to zero by removing it from the circuit and replacing it with a short circuit.
3. Determine the values of the impedance using the given frequency. Calculate the Thévenin impedance between terminals *a* and *b*.
4. Convert the voltage source into its equivalent phasor form. Solve for the open-circuit voltage between terminals *a* and *b*.
5. Sketch the resulting Thévenin equivalent circuit.

IN-PROCESS LEARNING CHECK 3

1. Remove the resistor from the circuit. Label the remaining terminals as *a* and *b*.
2. Set the voltage source to zero by removing it from the circuit and replacing it with a short circuit.
3. Calculate the Norton impedance between terminals *a* and *b*.
4. Solve for the short-circuit current between terminals *a* and *b*.
5. Sketch the resulting Norton equivalent circuit.

IN-PROCESS LEARNING CHECK 4

a. $R_L = 80.6\ \Omega$

b. $P_L = 3.25$ W

c. $\omega = 3333$ rad/s

d. $P_L = 4.90$ W

■ KEY TERMS

Bandwidth
Damped Oscillations
Half-Power Frequencies
Parallel Resonant Circuit
Quality Factor
Resonant Circuit
Selectivity Curve
Series Resonant Circuit

■ OUTLINE

Series Resonance
Quality Factor, Q
Impedance of a Series Resonant Circuit
Power, Bandwidth, and Selectivity of a Series Resonant Circuit
Series-to-Parallel RL and RC Conversion
Parallel Resonance
Circuit Analysis Using Computers

■ OBJECTIVES

After studying this chapter, you will be able to

- determine the resonant frequency and bandwidth of a simple series or parallel circuit,
- determine the voltages, currents, and power of elements in a resonant circuit,
- sketch the impedance, current, and power response curves of a series resonant circuit,
- find the quality factor, Q, of a resonant circuit and use Q to determine the bandwidth for a given set of conditions,
- explain the dependence of bandwidth on the L/C ratio and on R for both a series and a parallel resonant circuit,
- design a resonant circuit for a given set of parameters,
- convert a series RL network into an equivalent parallel network for a given frequency.

21

RESONANCE

CHAPTER PREVIEW

In this chapter, we build upon the knowledge obtained in previous chapters to observe how resonant circuits are able to pass a desired range of frequencies from a signal source to a load. In its most simple form, the **resonant circuit** consists of an inductor and a capacitor together with a voltage or current source. Although the circuit is simple, it is one of the most important circuits used in electronics. As an example, the resonant circuit, in one of its many forms, allows us to select a desired radio or television signal from the vast number of signals that are around us at any time. In order to obtain all the transmitted energy for a given radio station or television channel, we would like a circuit to have the frequency response shown in Figure 21–1(a). A circuit having an ideal frequency response would pass all frequency components in a band between f_1 and f_2, while rejecting all other frequencies. For a radio transmitter, the center frequency, f_r, would correspond to the *carrier frequency* of the station. The difference between the upper and lower frequencies that we would like to pass is called the **bandwidth.**

Whereas there are various configurations of resonant circuits, they all have several common characteristics. Resonant electronic circuits contain at least one inductor and one capacitor and have a bell-shaped response curve centered at some resonant frequency, f_r, as illustrated in Figure 21–1(b).

The response curve of Figure 21–1(b) indicates that power will be at a maximum at the resonant frequency, f_r. Varying the frequency in either direction results in a reduction of the power. The bandwidth of the resonant circuit is taken to be the difference between the half power points on the response curve of the filter.

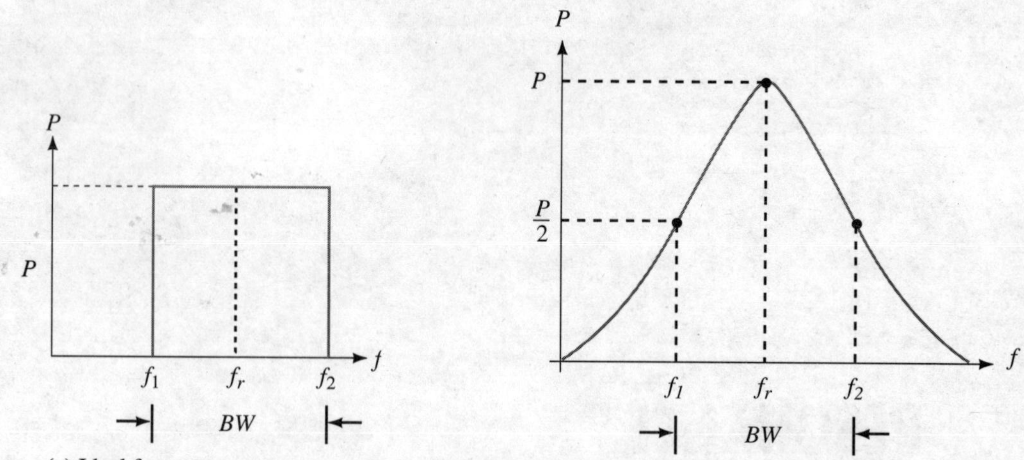

(a) Ideal frequency response curve

(b) Actual response curve of a resonant circuit

FIGURE 21–1

If we were to apply variable-frequency sinusoidal signals to a circuit consisting of an inductor and capacitor, we would find that maximum energy will transfer back and forth between the two elements at the resonant frequency. In an ideal *LC* circuit (one containing no resistance), these oscillations would continue unabated even if the signal source were turned off. However, in the practical situation, all circuits have some resistance. As a result, the stored energy will eventually be dissipated by the resistance, resulting in **damped oscillations.** In a manner similar to pushing a child on a swing, the oscillations will continue indefinitely if a small amount of energy is applied to the circuit at exactly the right moment. This phenomenon illustrates the basis of how oscillator circuits operate and therefore provides us with another application of the resonant circuit.

In this chapter, we examine in detail the two main types of resonant circuits: the **series resonant circuit** and the **parallel resonant circuit.** ■

Putting It in Perspective

Edwin Howard Armstrong—Radio Reception

© Bettmann/CORBIS

EDWIN ARMSTRONG WAS BORN in New York City on December 18, 1890. As a young man, he was keenly interested in experiments involving radio transmission and reception.

After earning a degree in electrical engineering at Columbia University, Armstrong used his theoretical background to explain and improve the operation of the triode vacuum tube, which had been invented by Lee de Forest. Edwin Armstrong was able to improve the sensitivity of receivers by using feedback to amplify a signal many times. By increasing the amount of signal feedback, Armstrong also designed and patented a circuit that used the vacuum tube as an oscillator.

Armstrong is best known for conceiving the concept of superheterodyning, in which a high frequency is lowered to a more usable intermediate frequency. Superheterodyning is still used in modern AM and FM receivers and in numerous other electronic circuits such as radar and communication equipment.

Edwin Armstrong was the inventor of FM transmission, which led to greatly improved fidelity in radio transmission.

Although Armstrong was a brilliant engineer, he was an uncompromising person who was involved in numerous lawsuits with Lee de Forest and the communications giant, RCA.

After spending nearly two million dollars in legal battles, Edwin Armstrong jumped to his death from his thirteenth-floor apartment window on January 31, 1954. ■

21.1 Series Resonance

A simple series resonant circuit is constructed by combining an ac source with an inductor, a capacitor, and optionally, a resistor as shown in Figure 21–2(a). By combining the generator resistance, R_G, with the series resistance, R_S, and the resistance of the inductor coil, R_{coil}, the circuit may be simplified as illustrated in Figure 21–2(b).

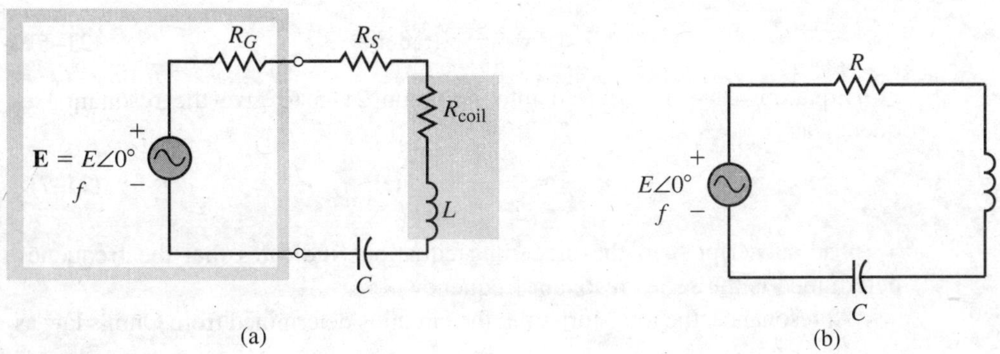

FIGURE 21–2

In this circuit, the total resistance is expressed as

$$R = R_G + R_S + R_{\text{coil}}$$

Because the circuit of Figure 21–2 is a series circuit, we calculate the total impedance as follows:

$$\begin{aligned}\mathbf{Z}_\text{T} &= R + jX_L - jX_C \\ &= R + j(X_L - X_C)\end{aligned} \qquad \textbf{(21–1)}$$

Resonance occurs when the reactance of the circuit is effectively eliminated, resulting in a total impedance that is purely resistive. We know that the reactances of the inductor and capacitor are given as follows:

$$X_L = \omega L = 2\pi f L \qquad \textbf{(21–2)}$$

$$X_C = \frac{1}{\omega C} = \frac{1}{2\pi f C} \qquad \textbf{(21–3)}$$

Examining Equation 21–1, we see that by setting the reactances of the capacitor and inductor equal to one another, the total impedance, $\mathbf{Z}_\text{T}$, is purely resistive since the inductive reactance which is on the positive j axis cancels the capacitive reactance on the negative j axis. The total impedance of the series circuit at resonance is equal to the total circuit resistance, R. Hence, at resonance,

$$Z_\text{T} = R \qquad \textbf{(21–4)}$$

By letting the reactances be equal we are able to determine the series resonance frequency ω_S (in radians per second) as follows:

$$\begin{aligned}\omega L &= \frac{1}{\omega C} \\ \omega^2 &= \frac{1}{LC} \\ \omega_\text{S} &= \frac{1}{\sqrt{LC}} \quad \text{(rad/s)}\end{aligned} \qquad \textbf{(21–5)}$$

Since the calculation of the angular frequency, ω, in radians per second is easier than solving for frequency, f, in hertz, we generally express our resonant frequencies in the simpler form. Further calculations of voltage and current will usually be much easier by using ω rather than f. If, however, it becomes necessary to determine a frequency in hertz, recall that the relationship between ω and f is as follows:

$$\omega = 2\pi f \quad \text{(rad/s)} \qquad \textbf{(21–6)}$$

Equation 21–6 is inserted into Equation 21–5 to give the resonant frequency as

$$f_\text{S} = \frac{1}{2\pi\sqrt{LC}} \quad \text{(Hz)} \qquad \textbf{(21–7)}$$

The subscript S in the preceding equations indicates that the frequency determined is the series resonant frequency.

At resonance, the total current in the circuit is determined from Ohm's law as

$$\mathbf{I} = \frac{\mathbf{E}}{\mathbf{Z}_\text{T}} = \frac{E\angle 0^\circ}{R\angle 0^\circ} = \frac{E}{R}\angle 0^\circ \qquad \textbf{(21–8)}$$

By again applying Ohm's law, we find the voltage across each of the elements in the circuit as follows:

$$\mathbf{V}_R = IR\angle 0^\circ \qquad (21\text{–}9)$$

$$\mathbf{V}_L = IX_L\angle 90^\circ \qquad (21\text{–}10)$$

$$\mathbf{V}_C = IX_C\angle -90^\circ \qquad (21\text{–}11)$$

The phasor form of the voltages and current is shown in Figure 21–3.

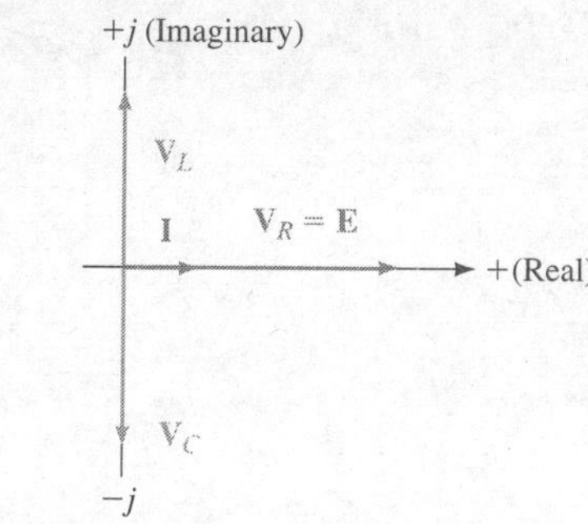

FIGURE 21–3

Notice that since the inductive and capacitive reactances have the same magnitude, the voltages across the elements must have the same magnitude but be 180° out of phase.

We determine the average power dissipated by the resistor and the reactive powers of the inductor and capacitor as follows:

$$P_R = I^2R \quad \text{(W)}$$

$$Q_L = I^2X_L \quad \text{(VAR)}$$

$$Q_C = I^2X_C \quad \text{(VAR)}$$

These powers are illustrated graphically in Figure 21–4.

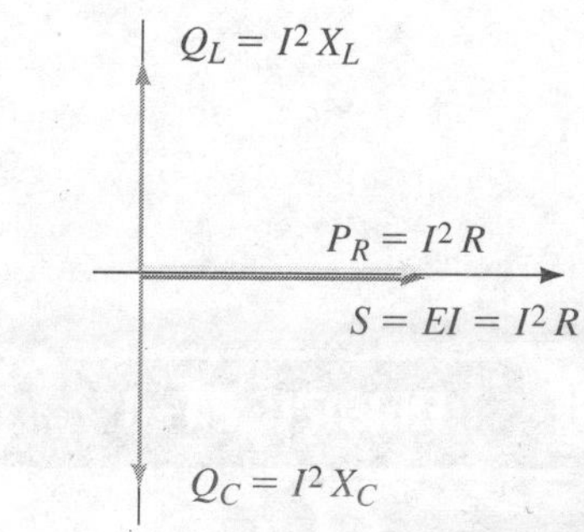

FIGURE 21–4

21.2 Quality Factor, Q

For any resonant circuit, we define the **quality factor,** Q, as the ratio of reactive power to average power, namely,

$$Q = \frac{\text{reactive power}}{\text{average power}} \qquad (21\text{–}12)$$

Because the reactive power of the inductor is equal to the reactive power of the capacitor at resonance, we may express Q in terms of either reactive power. Consequently, the preceding expression is written as follows:

$$Q_S = \frac{I^2X_L}{I^2R}$$

and so we have

$$Q_S = \frac{X_L}{R} = \frac{\omega L}{R} \qquad (21\text{–}13)$$

Quite often, the inductor of a given circuit will have a Q expressed in terms of its reactance and internal resistance, as follows:

$$Q_{\text{coil}} = \frac{X_L}{R_{\text{coil}}}$$

If an inductor with a specified Q_{coil} is included in a circuit, it is necessary to include its effects in the overall calculation of the total circuit Q.

We now examine how the Q of a circuit is used in determining other quantities of the circuit. By multiplying both the numerator and denominator of Equation 21–13 by the current, I, we have the following:

$$Q_S = \frac{IX_L}{IR} = \frac{V_L}{E} \quad \textbf{(21–14)}$$

Now, since the magnitude of the voltage across the capacitor is equal to the magnitude of the voltage across the inductor at resonance, we see that the voltages across the inductor and capacitor are related to the Q by the following expression:

$$V_C = V_L = Q_S E \quad \text{at resonance} \quad \textbf{(21–15)}$$

Note: Since the Q of a resonant circuit is generally significantly larger than 1, we see that the voltage across reactive elements can be many times greater than the applied source voltage. Therefore, it is always necessary to ensure that the reactive elements used in a resonant circuit are able to handle the expected voltages and currents.

EXAMPLE 21–1

Find the indicated quantities for the circuit of Figure 21–5.

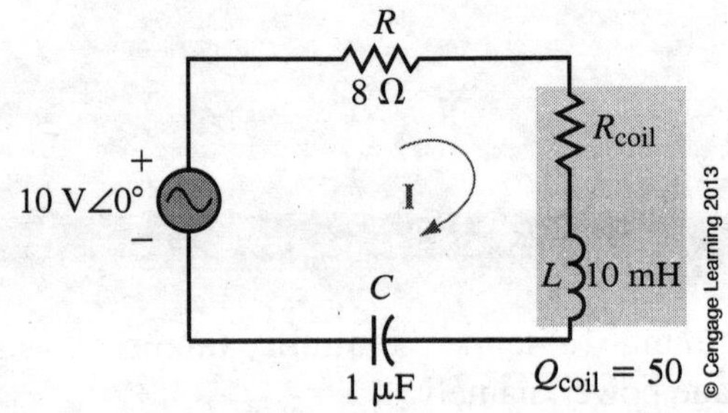

FIGURE 21–5

a. Resonant frequency expressed as ω(rad/s) and f(Hz).
b. Total impedance at resonance.
c. Current at resonance.
d. $\mathbf{V}_L$ and $\mathbf{V}_C$.
e. Reactive powers, Q_C and Q_L.
f. Quality factor of the circuit, Q_S.

Solution

a.
$$\omega_S = \frac{1}{\sqrt{LC}} = \frac{1}{\sqrt{(10\text{ mH})(1\mu\text{F})}} = 10\,000\text{ rad/s}$$

$$f_S = \frac{\omega}{2\pi} = 1592\text{ Hz}$$

b.
$$X_L = \omega L = (10\,000\text{ rad/s})(10\text{ mH}) = 100\ \Omega$$

$$R_{coil} = \frac{X_L}{Q_{coil}} = \frac{100\ \Omega}{50} = 2.00\ \Omega$$

$$R_T = R + R_{coil} = 10.0\ \Omega$$

$$\mathbf{Z}_T = 10\ \Omega\angle 0°$$

c. $$\mathbf{I} = \frac{\mathbf{E}}{\mathbf{Z}_T} = \frac{10\ \text{V}\angle 0°}{10\ \Omega\angle 0°} = 1.0\ \text{A}\angle 0°$$

d. $$\mathbf{V}_L = (100\ \Omega\angle 90°)(1.0\ \text{A}\angle 0°) = 100\ \text{V}\angle 90°$$

$$\mathbf{V}_C = (100\ \Omega\angle -90°)(1.0\ \text{A}\angle 0°) = 100\ \text{V}\angle -90°$$

Notice that the voltage across the reactive elements is 10 times greater than the applied signal voltage.

e. Although we use the symbol Q to designate both reactive power and the quality factor, the context of the question generally provides us with a clue as to which meaning to use.

$$Q_L = (1.0\ \text{A})^2(100\ \Omega) = 100\ \text{VAR}$$

$$Q_C = (1.0\ \text{A})^2(100\ \Omega) = 100\ \text{VAR}$$

f. $$Q_S = \frac{Q_L}{P} = \frac{100\ \text{VAR}}{10\ \text{W}} = 10$$

PRACTICE PROBLEMS 1

Consider the circuit of Figure 21–6:

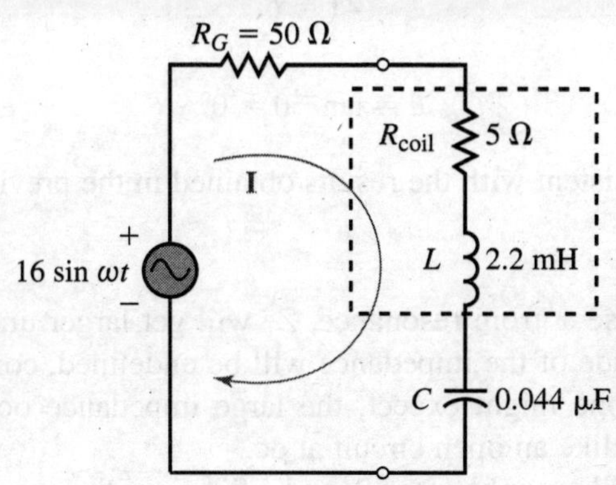

FIGURE 21–6

CircuitSim 21-1

a. Find the resonant frequency expressed as ω(rad/s) and f(Hz).

b. Determine the total impedance at resonance.

c. Solve for $\mathbf{I}$, $\mathbf{V}_L$, and $\mathbf{V}_C$ at resonance.

d. Calculate reactive powers Q_C and Q_L at resonance.

e. Find the quality factor, Q_s, of the circuit.

Answers

a. 102 krad/s, 16.2 kHz; b. 55.0 Ω∠0°

c. 0.206 A∠0°, 46.0 V∠90°, 46.0 V∠−90°; d. 9.46 VAR; e. 4.07

21.3 Impedance of a Series Resonant Circuit

In this section, we examine how the impedance of a series resonant circuit varies as a function of frequency. Because the impedances of inductors and capacitors are dependent upon frequency, the total impedance of a series resonant circuit must similarly vary with frequency. For algebraic simplicity, we use frequency expressed as ω in radians per second. If it becomes necessary to express the frequency in hertz, the conversion of Equation 21–6 is used.

The total impedance of a simple series resonant circuit is written as

$$\mathbf{Z}_\mathrm{T} = R + j\omega L - j\frac{1}{\omega C}$$

$$= R + j\left(\frac{\omega^2 LC - 1}{\omega C}\right)$$

The magnitude and phase angle of the impedance vector, $\mathbf{Z}_\mathrm{T}$, are expressed as follows:

$$Z_\mathrm{T} = \sqrt{R^2 + \left(\frac{\omega^2 LC - 1}{\omega C}\right)^2} \tag{21–16}$$

$$\theta = \tan^{-1}\left(\frac{\omega^2 LC - 1}{\omega RC}\right) \tag{21–17}$$

Examining these equations for various values of frequency, we note that the following conditions will apply:

When $\omega = \omega_\mathrm{S}$:

$$Z_\mathrm{T} = R$$

and

$$\theta = \tan^{-1}0 = 0°$$

This result is consistent with the results obtained in the previous section.

When $\omega < \omega_\mathrm{S}$:

As we decrease ω from resonance, Z_T will get larger until $\omega = 0$. At this point, the magnitude of the impedance will be undefined, corresponding to an open circuit. As one might expect, the large impedance occurs because the capacitor behaves like an open circuit at dc.

The angle θ will occur between 0° and −90° since the numerator of the argument of the arctangent function will always be negative, corresponding to an angle in the fourth quadrant. Because the angle of the impedance has a negative sign, we conclude that the impedance must appear capacitive in this region.

When $\omega > \omega_\mathrm{S}$:

As ω is made larger than resonance, the impedance Z_T will increase due to the increasing reactance of the inductor.

For these values of ω, the angle θ will always be within 0° and +90° because both the numerator and the denominator of the arctangent function are positive. Because the angle of $\mathbf{Z}_\mathrm{T}$ occurs in the first quadrant, the impedance must be inductive.

Sketching the magnitude and phase angle of the impedance $\mathbf{Z}_\mathrm{T}$ as a function of angular frequency, we have the curves shown in Figure 21–7.

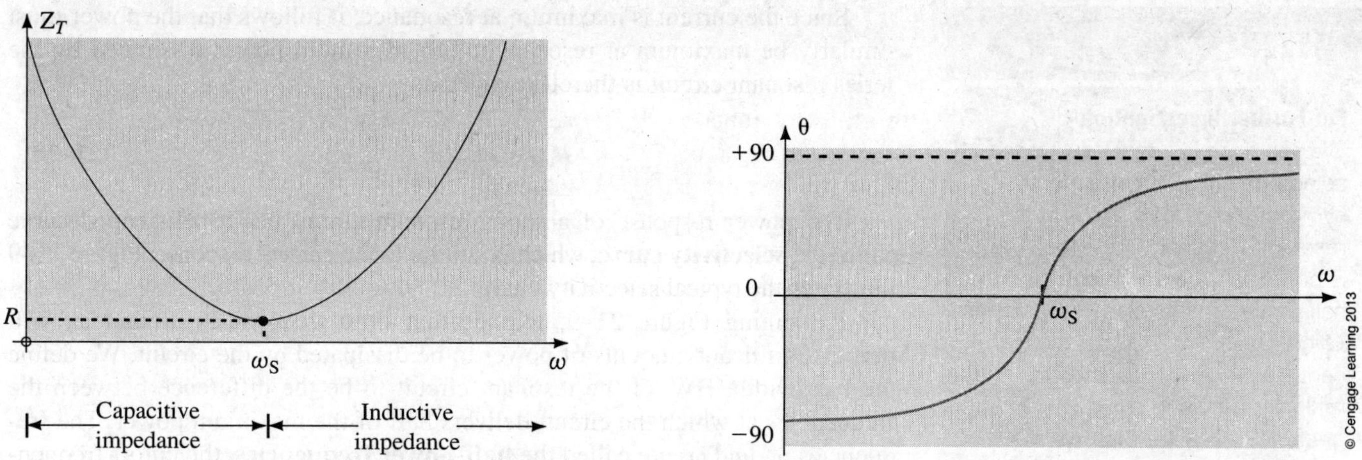

FIGURE 21–7 Impedance (magnitude and phase angle) versus angular frequency for a series resonant circuit.

21.4 Power, Bandwidth, and Selectivity of a Series Resonant Circuit

Due to the changing impedance of the circuit, we conclude that if a constant-amplitude voltage is applied to the series resonant circuit, the current and power of the circuit will not be constant at all frequencies. In this section, we examine how current and power are affected by changing the frequency of the voltage source.

Applying Ohm's law gives the magnitude of the current at resonance as follows:

$$I_{max} = \frac{E}{R} \tag{21–18}$$

For all other frequencies, the magnitude of the current will be less than I_{max} because the impedance is greater than at resonance. Indeed, when the frequency is zero (dc), the current will be zero since the capacitor is effectively an open circuit. On the other hand, at increasingly higher frequencies, the inductor begins to approximate an open circuit, once again causing the current in the circuit to approach zero. The current response curve for a typical series resonant circuit is shown in Figure 21–8.

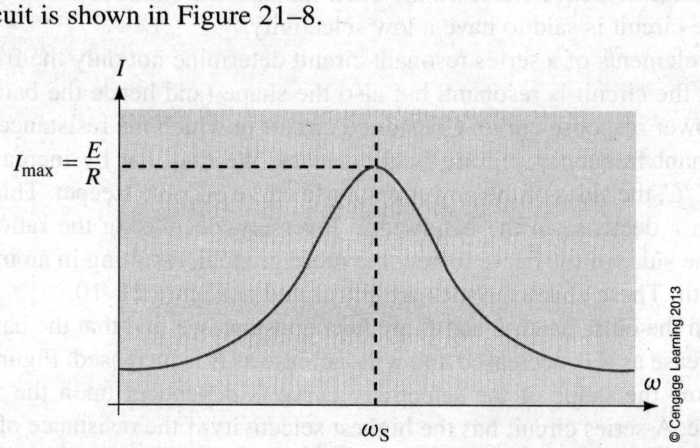

FIGURE 21–8 Current versus angular frequency for a series resonant circuit.

The total power dissipated by the circuit at any frequency is given as

$$P = I^2R \tag{21–19}$$

NOTES...

For Further Investigation

We see from Figure 21–9 that the selectivity curve is not perfectly symmetrical on both sides of the resonant frequency. As a result, ω_s is not exactly centered between the half-power frequencies. However, as Q increases, we find that the resonant frequency approaches the midpoint between ω_1 and ω_2. *In general, if $Q > 10$, then we assume that the resonant frequency is at the midpoint of half-power frequencies.* For additional information, refer to the Web site for this textbook. Go to www.cengagebrain.com and log in by entering your user name and password. Follow the links to *For Further Investigation* and select *The Importance of Selectivity in Broadcast Radio.*

Since the current is maximum at resonance, it follows that the power must similarly be maximum at resonance. The maximum power dissipated by the series resonant circuit is therefore given as

$$P_{\max} = I^2_{\max}R = \frac{E^2}{R} \tag{21–20}$$

The power response of a series resonant circuit has a bell-shaped curve called the **selectivity curve,** which is similar to the current response. Figure 21–9 illustrates the typical selectivity curve.

Examining Figure 21–9, we see that only frequencies around ω_s will permit significant amounts of power to be dissipated by the circuit. We define the bandwidth, BW, of the resonant circuit to be the difference between the frequencies at which the circuit delivers half of the maximum power. The frequencies ω_1 and ω_2 are called the **half-power frequencies,** the cutoff frequencies, or the band frequencies.

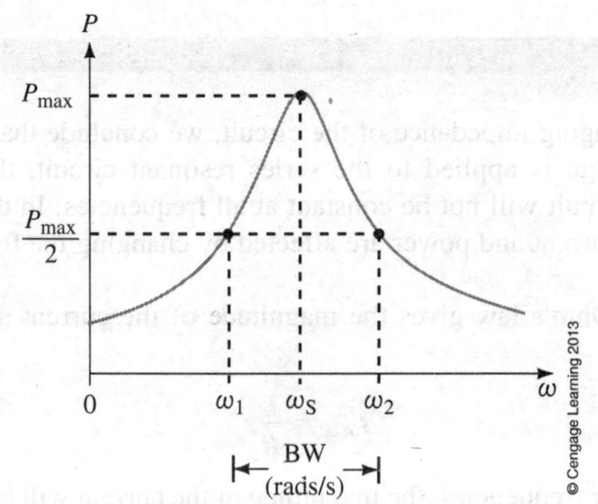

FIGURE 21–9 Selectivity curve.

If the bandwidth of a circuit is kept very narrow, the circuit is said to have a high selectivity, since it is highly selective to signals occurring within a very narrow range of frequencies. On the other hand, if the bandwidth of a circuit is large, the circuit is said to have a low selectivity.

The elements of a series resonant circuit determine not only the frequency at which the circuit is resonant, but also the shape (and hence the bandwidth) of the power response curve. Consider a circuit in which the resistance, R, and the resonant frequency, ω_s, are held constant. We find that by increasing the ratio of L/C, the sides of the power response curve become steeper. This in turn results in a decrease in the bandwidth. Inversely, decreasing the ratio of L/C causes the sides of the curve to become more gradual, resulting in an increased bandwidth. These characteristics are illustrated in Figure 21–10.

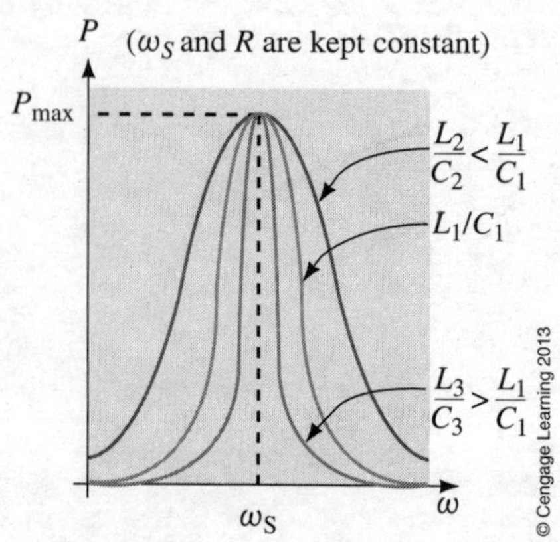

FIGURE 21–10

If, on the other hand, L and C are kept constant, we find that the bandwidth will decrease as R is decreased and will increase as R is increased. Figure 21–11 shows how the shape of the selectivity curve is dependent upon the value of resistance. A series circuit has the highest selectivity if the resistance of the circuit is kept to a minimum.

For the series resonant circuit the power at any frequency is determined as

$$\begin{aligned} P &= I^2R \\ &= \left(\frac{E}{Z_T}\right)^2 R \end{aligned}$$

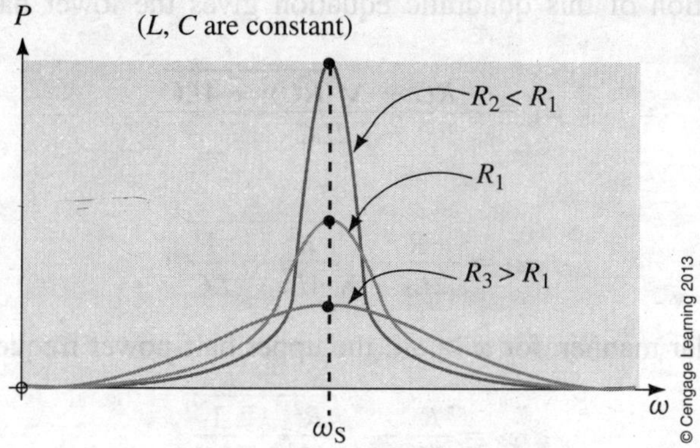

FIGURE 21–11

By substituting Equation 21–16 into the preceding expression, we arrive at the general expression for power as a function of frequency, ω:

$$P = \frac{E^2R}{R^2 + \left(\frac{\omega^2LC - 1}{\omega C}\right)^2} \tag{21–21}$$

At the half-power frequencies, the power must be

$$P_{\text{hpf}} = \frac{E^2}{2R} \tag{21–22}$$

Since the maximum current in the circuit is given as $I_{\text{max}} = E/R$, we see that by manipulating the preceding expression, the magnitude of current at the half-power frequencies is

$$I_{\text{hpf}} = \sqrt{\frac{P_{\text{hpf}}}{R}} = \sqrt{\frac{E^2}{2R^2}} = \sqrt{\frac{I_{\text{max}}^2}{2}}$$

$$I_{\text{hpf}} = \frac{I_{\text{max}}}{\sqrt{2}} \tag{21–23}$$

The cutoff frequencies are found by evaluating the frequencies at which the power dissipated by the circuit is half of the maximum power. Combining Equations 21–21 and 21–22, we have the following:

$$\frac{E^2}{2R} = \frac{E^2R}{R^2 + \left(\frac{\omega^2LC - 1}{\omega C}\right)^2}$$

$$2R^2 = R^2 + \left(\frac{\omega^2LC - 1}{\omega C}\right)^2 \tag{21–24}$$

$$\frac{\omega^2LC - 1}{\omega C} = \pm R$$

$$\omega^2LC - 1 = \pm\omega RC \quad \text{(at half-power)}$$

From the selectivity curve for a series circuit, we see that the two half-power points occur on both sides of the resonant angular frequency, ω_s.

When $\omega < \omega_S$, the term ω^2LC must be less than 1. In this case, the solution is determined as follows:

$$\omega^2LC - 1 = -\omega RC$$

$$\omega^2LC + \omega RC - 1 = 0$$

The solution of this quadratic equation gives the lower half-power frequency as

$$\omega_1 = \frac{-RC + \sqrt{(RC)^2 + 4LC}}{2LC}$$

or

$$\omega_1 = \frac{-R}{2L} + \sqrt{\frac{R^2}{4L^2} + \frac{1}{LC}} \quad \text{(21–25)}$$

In a similar manner, for $\omega > \omega_S$, the upper half-power frequency is

$$\omega_2 = \frac{R}{2L} + \sqrt{\frac{R^2}{4L^2} + \frac{1}{LC}} \quad \text{(21–26)}$$

Taking the difference between Equations 21–26 and 21–25, we find the bandwidth of the circuit as

$$\begin{aligned} \text{BW} &= \omega_2 - \omega_1 \\ &= \frac{R}{2L} + \sqrt{\frac{R^2}{4L^2} + \frac{1}{LC}} - \left(-\frac{R}{2L} + \sqrt{\frac{R^2}{4L^2} + \frac{1}{LC}}\right) \end{aligned}$$

which gives

$$\text{BW} = \frac{R}{L} \text{ (rad/s)} \quad \text{(21–27)}$$

If the preceding expression is multiplied by ω_S/ω_S we obtain

$$\text{BW} = \frac{\omega_S R}{\omega_S L}$$

and since $Q_S = \omega_S L/R$ we further simplify the bandwidth as

$$\text{BW} = \frac{\omega_S}{Q_S} \text{ (rad/s)} \quad \text{(21–28)}$$

Because the bandwidth may alternately be expressed in hertz, the preceding expression is equivalent to having

$$\text{BW} = \frac{f_S}{Q_S} \text{ (Hz)} \quad \text{(21–29)}$$

EXAMPLE 21–2

Refer to the circuit of Figure 21–12.

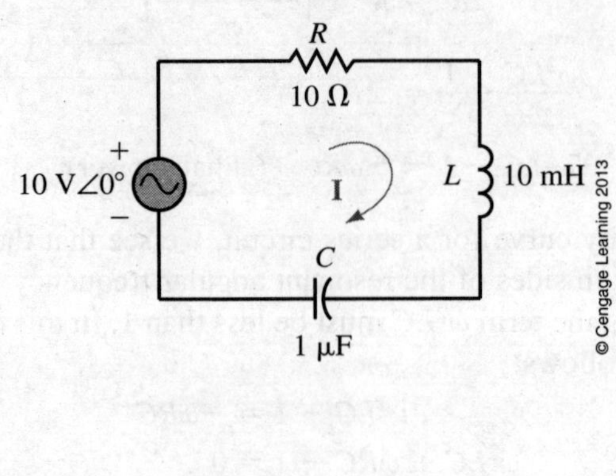

FIGURE 21–12

a. Determine the maximum power dissipated by the circuit.
b. Use the results obtained from Example 21–1 to determine the bandwidth of the resonant circuit and to arrive at the approximate half-power frequencies, ω_1 and ω_2.
c. Calculate the actual half-power frequencies, ω_1 and ω_2, from the given component values. Show two decimal places of precision.
d. Solve for the circuit current, I, and power dissipated at the lower half-power frequency, ω_1, found in part (c).

Solution

a. $$P_{max} = \frac{E^2}{R} = 10.0 \text{ W}$$

b. From Example 21–1, we had the following circuit characteristics:

$$Q_S = 10, \ \omega_S = 10 \text{ krad/s}$$

The bandwidth of the circuit is determined to be

$$\text{BW} = \omega_S Q_S = 1.0 \text{ krad/s}$$

If the resonant frequency were centered in the bandwidth, then the half-power frequencies occur at approximately

$$\omega_1 = 9.50 \text{ krad/s}$$

and

$$\omega_2 = 10.50 \text{ krad/s}$$

c. $$\omega_1 = -\frac{R}{2L} + \sqrt{\frac{R^2}{4L^2} + \frac{1}{LC}}$$

$$= -\frac{10\ \Omega}{(2)(10\text{ mH})} + \sqrt{\frac{(10\ \Omega)^2}{(4)(10\text{ mH})^2} + \frac{1}{(10\text{ mH})(1\ \mu\text{F})}}$$

$$= -500 + 10\,012.49 = 9512.49 \text{ rad/s} \quad (f_1 = 1514.0 \text{ Hz})$$

$$\omega_2 = \frac{R}{2L} + \sqrt{\frac{R^2}{4L^2} + \frac{1}{LC}}$$

$$= 500 + 10\,012.49 = 10\,512.49 \text{ rad/s} \quad (f_2 = 1673.1 \text{ Hz})$$

Notice that the actual half-power frequencies are very nearly equal to the approximate values. For this reason, if $Q \geq 10$, it is often sufficient to calculate the cutoff frequencies by using the easier approach of part (b).

d. At $\omega_1 = 9.51249$ krad/s, the reactances are as follows:

$$X_L = \omega L = (9.51249 \text{ krad/s})(10 \text{ mH}) = 95.12\ \Omega$$

$$X_C = \frac{1}{\omega C} = \frac{1}{(9.51249 \text{ krad/s})(1\ \mu\text{F})} = 105.12\ \Omega$$

The current is now determined to be

$$\mathbf{I} = \frac{10\text{ V}\angle 0^\circ}{10\ \Omega + j95.12\ \Omega - j105.12\ \Omega}$$

$$= \frac{10\text{ V}\angle 0^\circ}{14.14\ \Omega\angle -45^\circ}$$

$$= 0.707\text{ A}\angle 45^\circ$$

and the power is given as

$$P = I^2R = (0.707\text{ A})^2(10\text{ V}) = 5.0\text{ W}$$

As expected, we see that the power at the frequency ω_1 is indeed equal to half of the power dissipated by the circuit at resonance.

EXAMPLE 21–3

Refer to the circuit of Figure 21–13.

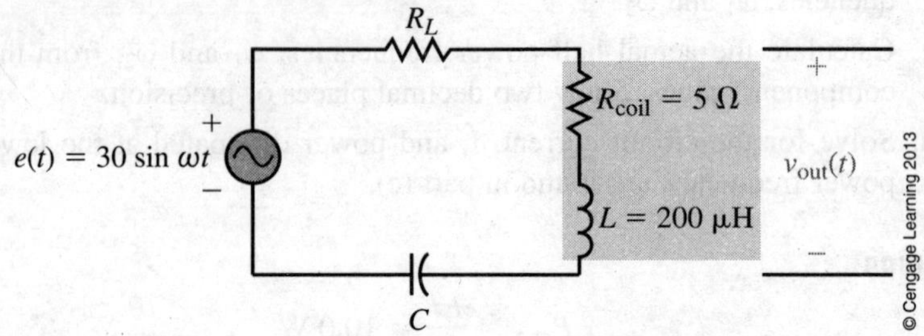

FIGURE 21–13

a. Calculate the values of R_L and C for the circuit to have a resonant frequency of 200 kHz and a bandwidth of 16 kHz.

b. Use the designed component values to determine the power dissipated by the circuit at resonance.

c. Solve for $v_{out}(t)$ at resonance.

Solution

a. Because the circuit is at resonance, we must have the following conditions:

$$Q_S = \frac{f_S}{\text{BW}} = \frac{200\text{ kHz}}{16\text{ kHz}} = 12.5$$

$$X_L = 2\pi fL = 2\pi(200\text{ kHz})(200\ \mu\text{H}) = 251.3\ \Omega$$

$$R = R_L + R_{coil} = \frac{X_L}{Q_s} = 20.1\ \Omega$$

and so R_L must be

$$R_L = 20.1\ \Omega - 5\ \Omega = 15.1\ \Omega$$

Since $X_C = X_L$, we determine the capacitance as

$$C = \frac{1}{2\pi fX_C} = \frac{1}{2\pi(200\text{ kHz})(251.3\ \Omega)} = 3.17\text{ nF } (\equiv 0.00317\ \mu\text{F})$$

b. The power at resonance is found from Equation 21–20 as

$$P_{max} = \frac{E^2}{R} = \frac{\left(\frac{30\text{ V}}{\sqrt{2}}\right)^2}{20.1\ \Omega} = 22.4\text{ W}$$

c. We see from the circuit of Figure 21–13 that the voltage $v_{out}(t)$ may be determined by applying the voltage divider rule to the circuit. However, we must first convert the source voltage from time domain into phasor domain as follows:

$$e(t) = 30 \sin \omega t \Leftrightarrow \mathbf{E} = 21.21 \text{ V}\angle 0^\circ$$

Now, applying the voltage divider rule to the circuit, we have

$$\begin{aligned}\mathbf{V}_{out} &= \frac{(R_1 + j\omega L)}{R}\mathbf{E} \\ &= \frac{(5\ \Omega + j251.3\ \Omega)}{20.1\ \Omega} 21.21 \text{ V}\angle 0^\circ \\ &= (251.4\ \Omega\angle 88.86^\circ)(1.056 \text{ A}\angle 0^\circ) \\ &= 265.5 \text{ V}\angle 88.86^\circ\end{aligned}$$

which in time domain is given as

$$v_{out}(t) = 375 \sin(\omega t + 88.86^\circ)$$

PRACTICE PROBLEMS 2

Refer to the circuit of Figure 21–14.

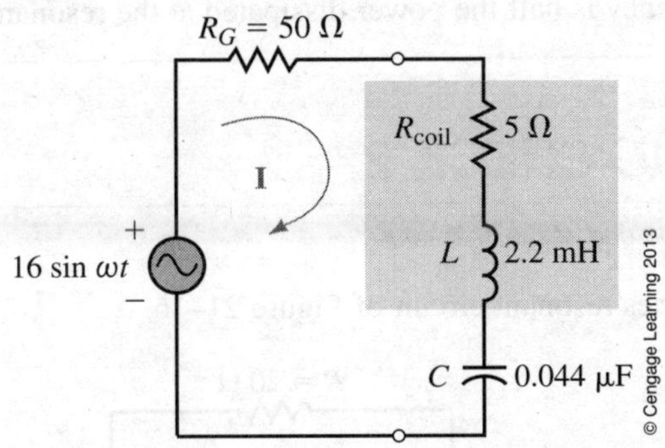

FIGURE 21–14

CircuitSim 21-2

a. Determine the maximum power dissipated by the circuit.
b. Use the results obtained from Practice Problems 1 to determine the bandwidth of the resonant circuit. Solve for the approximate values of the half-power frequencies, ω_1 and ω_2.
c. Calculate the actual half-power frequencies, ω_1 and ω_2, from the given component values. Compare your results to those obtained in part (b). Briefly explain why there is a discrepancy between the results.
d. Solve for the circuit current, **I**, and power dissipated at the lower half-power frequency, ω_1, found in part (c).

Answers

a. 2.33 W

b. BW = 25.0 krad/s (3.98 kHz), $\omega_1 \cong 89.1$ krad/s, $\omega_2 \cong 114.1$ krad/s

c. $\omega_1 = 89.9$ krad/s, $\omega_2 \cong 114.9$ krad/s. The approximation assumes that the power-frequency curve is symmetrical around ω_S, which is not quite true.

d. $\mathbf{I} = 0.145\text{ A}\angle 45^\circ$, $P = 1.16$ W

IN-PROCESS LEARNING CHECK 1

(Answers are at the end of the chapter.)

Refer to the series resonant circuit of Figure 21–15.

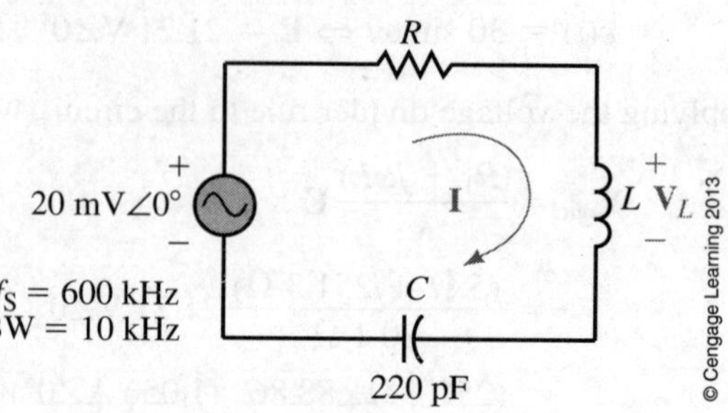

FIGURE 21–15

Suppose the circuit has a resonant frequency of 600 kHz and a bandwidth of 10 kHz.

a. Determine the value of inductor L in henries.

b. Calculate the value of resistor R in ohms.

c. Find $\mathbf{I}$, $\mathbf{V}_L$, and power, P, at resonance.

d. Find the approximate values of the half-power frequencies, f_1 and f_2.

e. Using the results of part (d), determine the current in the circuit at the lower half-power frequency, f_1, and show that the power dissipated by the resistor at this frequency is half the power dissipated at the resonant frequency.

IN-PROCESS LEARNING CHECK 2

(Answers are at the end of the chapter.)

Consider the series resonant circuit of Figure 21–16.

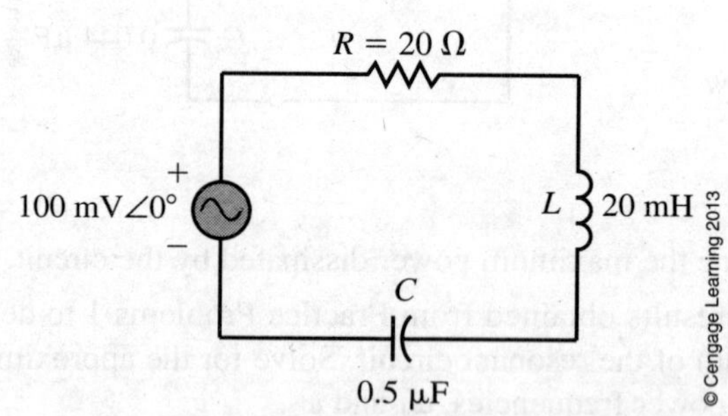

FIGURE 21–16

a. Solve for the resonant frequency of the circuit, ω_S, and calculate the power dissipated by the circuit at resonance.

b. Determine Q, BW, and the half-power frequencies, ω_1 and ω_2, in radians per second.

c. Sketch the selectivity curve of the circuit, showing P (in watts) versus ω (in radians per second).

d. Repeat parts (a) through (c) if the value of resistance is reduced to 10 Ω.

e. Explain briefly how selectivity depends upon the value of resistance in a series resonant circuit.

21.5 Series-to-Parallel *RL* and *RC* Conversion

As we have already seen, an inductor will always have some series resistance due to the length of wire used in the coil winding. Even though the resistance of the wire is generally small in comparison with the reactances in the circuit, this resistance may occasionally contribute tremendously to the overall circuit response of a parallel resonant circuit. We begin by converting the series *RL* network as shown in Figure 21–17 into an equivalent parallel *RL* network. It must be emphasized, however, that *the equivalence is only valid at a single frequency, ω.*

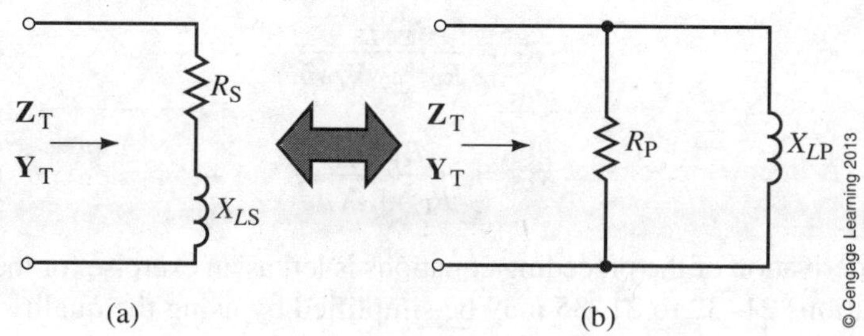

FIGURE 21–17

The networks of Figure 21–17 can be equivalent only if they each have the same input impedance, $\mathbf{Z}_T$ (and also the same input admittance, $\mathbf{Y}_T$).

The input impedance of the series network of Figure 21–17(a) is given as

$$\mathbf{Z}_T = R_S + jX_{LS}$$

which gives the input admittance as

$$\mathbf{Y}_T = \frac{1}{\mathbf{Z}_T} = \frac{1}{R_S + jX_{LS}}$$

Multiplying numerator and denominator by the complex conjugate, we have

$$\begin{aligned}\mathbf{Y}_T &= \frac{R_S - jX_{LS}}{(R_S + jX_{LS})(R_S - jX_{LS})}\\ &= \frac{R_S - jX_{LS}}{R_S^2 + X_{LS}^2} \qquad (21\text{–}30)\\ &= \frac{R_S}{R_S^2 + X_{LS}^2} - j\frac{X_{LS}}{R_S^2 + X_{LS}^2}\end{aligned}$$

From Figure 21–17(b), we see that the input admittance of the parallel network must be

$$\mathbf{Y}_T = G_P - jB_{LP}$$

which may also be written as

$$\mathbf{Y}_T = \frac{1}{R_P} - j\frac{1}{X_{LP}} \qquad (21\text{–}31)$$

The admittances of Equations 21–30 and 21–31 can only be equal if the real and the imaginary components are equal. As a result, we see that for a given frequency, the following equations enable us to convert a series RL network into its equivalent parallel network:

$$R_P = \frac{R_S^2 + X_{LS}^2}{R_S} \quad \text{(21–32)}$$

$$X_{LP} = \frac{R_S^2 + X_{LS}^2}{X_{LS}} \quad \text{(21–33)}$$

If we are given a parallel RL network, it is possible to show that the conversion to an equivalent series network is accomplished by applying the following equations:

$$R_S = \frac{R_P X_{LP}^2}{R_P^2 + X_{LP}^2} \quad \text{(21–34)}$$

$$X_{LS} = \frac{R_P^2 X_{LP}}{R_P^2 + X_{LP}^2} \quad \text{(21–35)}$$

The derivation of the preceding equations is left as an exercise for the student.

Equations 21–32 to 21–35 may be simplified by using the quality factor of the coil. Multiplying Equation 21–32 by R_S/R_S and then using Equation 21–13, we have

$$R_P = R_S \frac{R_S^2 + X_{LS}^2}{R_S^2}$$

$$R_P = R_S(1 + Q^2) \quad \text{(21–36)}$$

Similarly, Equation 21–33 is simplified as

$$X_{LP} = X_{LS} \frac{R_S^2 + X_{LS}^2}{X_{LS}^2}$$

$$X_{LP} = X_{LS}\left(1 + \frac{1}{Q^2}\right) \quad \text{(21–37)}$$

The quality factor of the resulting parallel network must be the same as for the original series network because the reactive and the average powers must be the same. Using the parallel elements, the quality factor is expressed as

$$Q = \frac{X_{LS}}{R_S} = \frac{\left(\dfrac{R_P^2 X_{LP}}{R_P^2 + X_{LP}^2}\right)}{\left(\dfrac{R_P X_{LP}^2}{R_P^2 + X_{LP}^2}\right)}$$

$$= \frac{R_P^2 X_{LP}}{R_P X_{LP}^2}$$

$$Q = \frac{R_P}{X_{LP}} \quad \text{(21–38)}$$

EXAMPLE 21–4

For the series network of Figure 21–18, find the Q of the coil at $\omega = 1000$ rad/s and convert the series RL network into its equivalent parallel network. Repeat these steps for $\omega = 10$ krad/s.

Solution

For $\omega = 1000$ rad/s,

$$X_L = \omega L = 20\ \Omega$$
$$Q = \frac{X_L}{R} = 2.0$$
$$R_P = R(1 + Q^2) = 50\ \Omega$$
$$X_{LP} = X_L\left(1 + \frac{1}{Q^2}\right) = 25\ \Omega$$

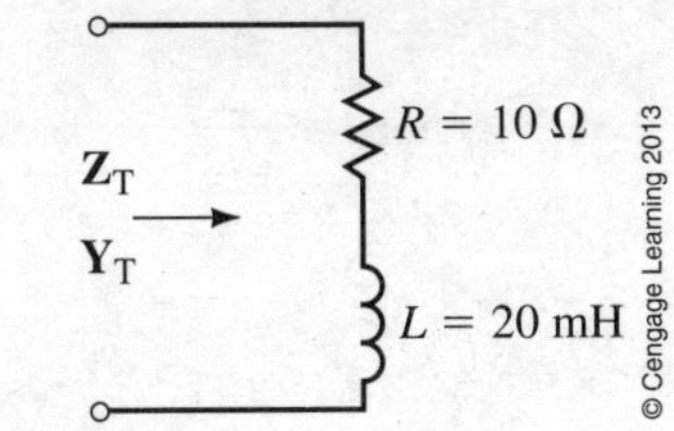

FIGURE 21–18

The resulting parallel network for $\omega = 1000$ rad/s is shown in Figure 21–19.

For $\omega = 10$ krad/s,

$$X_L = \omega L = 200\ \Omega$$
$$Q = \frac{X_L}{R} = 20$$
$$R_P = R(1 + Q^2) = 4010\ \Omega$$
$$X_{LP} = X_L\left(1 + \frac{1}{Q^2}\right) = 200.5\ \Omega$$

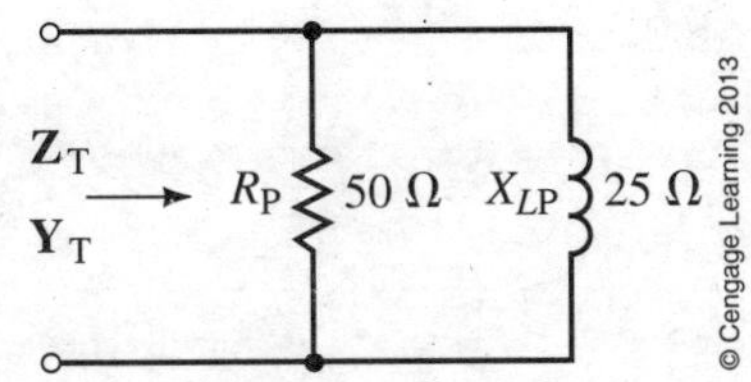

FIGURE 21–19

The resulting parallel network for $\omega = 10$ krad/s is shown in Figure 21–20.

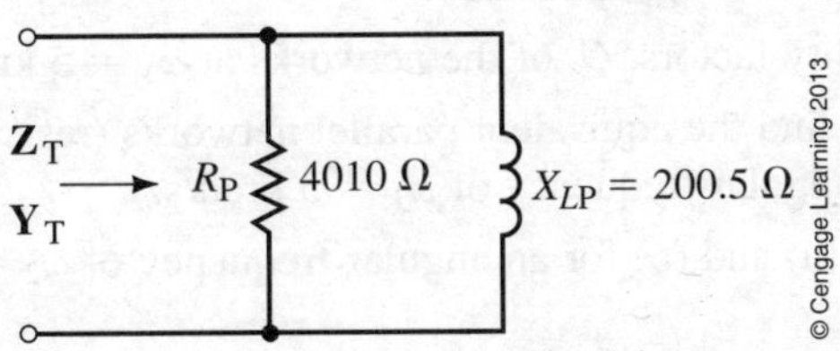

FIGURE 21–20

EXAMPLE 21–5

Find the Q of each of the networks of Figure 21–21 and determine the series equivalent for each.

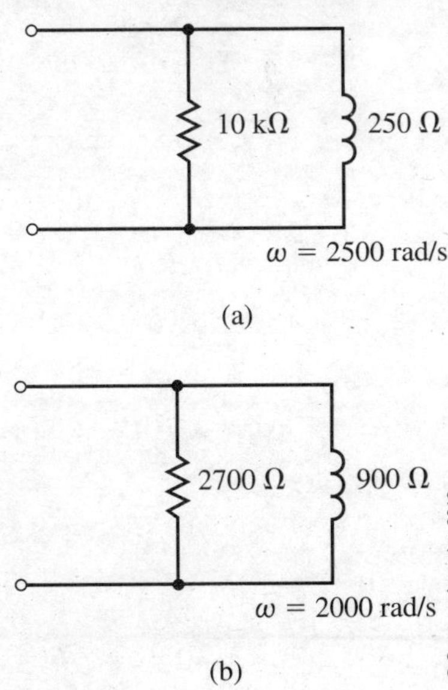

FIGURE 21–21

Solution For the network of Figure 21–21(a),

$$Q = \frac{R_P}{X_{LP}} = \frac{10\ \text{k}\Omega}{250\ \Omega} = 40$$
$$R_S = \frac{R_P}{1 + Q^2} = \frac{10\ \text{k}\Omega}{1 + 40^2} = 6.25\ \Omega$$
$$X_{LS} = QR_S = (40)(6.25\ \Omega) = 250\ \Omega$$
$$L = \frac{X_L}{\omega} = \frac{250\ \Omega}{2500\ \text{rad/s}} = 0.1\ \text{H}$$

For the network of Figure 21–21(b),

$$Q = \frac{R_P}{X_{LP}} = \frac{2700\ \Omega}{900\ \Omega} = 3$$
$$R_S = \frac{R_P}{1 + Q^2} = \frac{2700\ \Omega}{1 + 3^2} = 270\ \Omega$$

$$X_{LS} = QR_S = (3)(270\ \Omega) = 810\ \Omega$$

$$L = \frac{X_L}{\omega} = \frac{810\ \Omega}{2000\ \text{rad/s}} = 0.405\ \text{H}$$

The resulting equivalent series networks are shown in Figure 21–22.

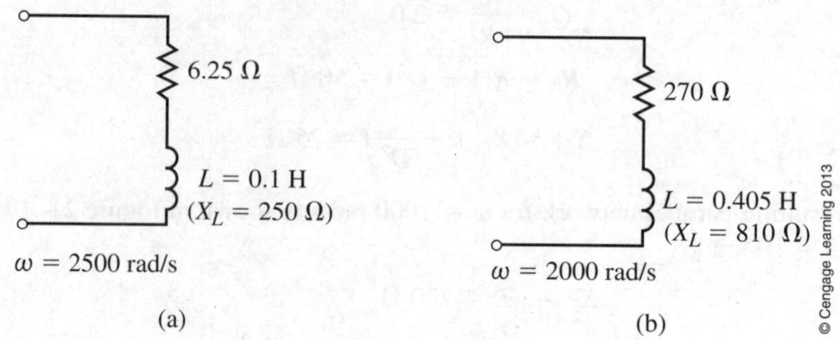

FIGURE 21–22

PRACTICE PROBLEMS 3

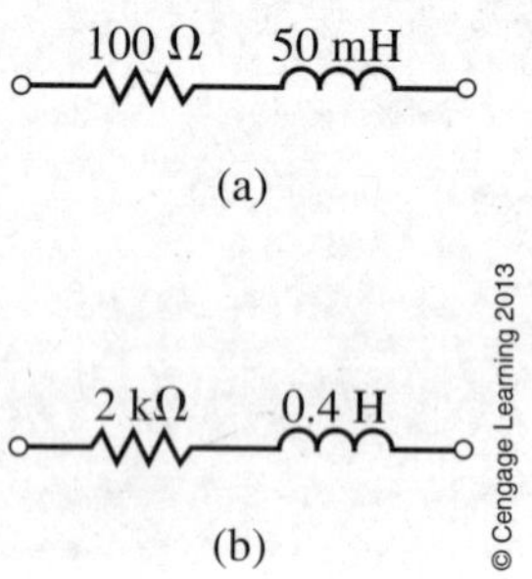

FIGURE 21–23

Refer to the networks of Figure 21–23.

a. Find the quality factors, Q, of the networks at $\omega_1 = 5$ krad/s.

b. Use the Q to find the equivalent parallel networks (resistance and reactance) at an angular frequency of $\omega_1 = 5$ krad/s.

c. Repeat parts (a) and (b) for an angular frequency of $\omega_2 = 25$ krad/s.

Answers

a. $Q_a = 2.5$	$Q_b = 1.0$		
b. Network a:	$R_P = 725\ \Omega$	$X_{LP} = 290\ \Omega$	
Network b:	$R_P = 4\ \text{k}\Omega$	$X_{LP} = 4\ \text{k}\Omega$	
c. Network a:	$Q_a = 12.5$	$R_P = 15.725\ \text{k}\Omega$	$X_{LP} = 1.258\ \text{k}\Omega$
Network b:	$Q_b = 5$	$R_P = 52\ \text{k}\Omega$	$X_{LP} = 10.4\ \text{k}\Omega$

The previous examples illustrate two important points which are valid if the Q of the network is large ($Q \geq 10$).

1. The resistance of the parallel network is approximately Q^2 larger than the resistance of the series network.
2. The inductive reactances of the series and parallel networks are approximately equal. Hence

$$R_P \cong Q^2 R_S \quad (Q \geq 10) \tag{21–39}$$

$$X_{LP} \cong X_{LS} \quad (Q \geq 10) \tag{21–40}$$

Although we have performed conversions between series and parallel *RL* circuits, it is easily shown that if the reactive element is a capacitor, the conversions apply equally well. In all cases, the equations are simply changed by replacing the terms X_{LS} and X_{LP} with X_{CS} and X_{CP}, respectively. The Q of the network is determined by the ratios

$$Q = \frac{X_{CS}}{R_S} = \frac{R_P}{X_{CP}} \tag{21–41}$$

PRACTICE PROBLEMS 4

Consider the networks of Figure 21–24.

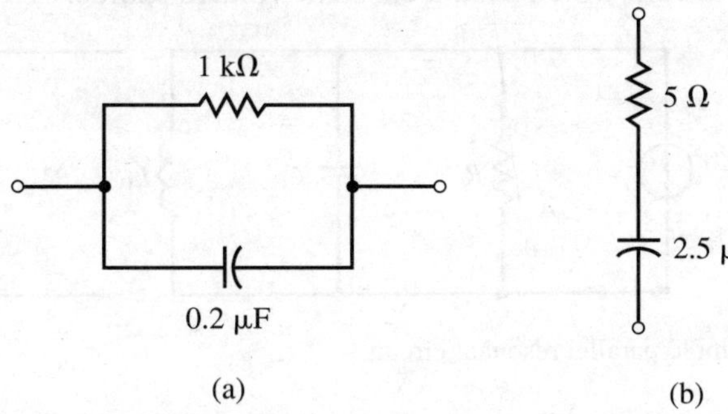

FIGURE 21–24

a. Find the Q of each network at a frequency of $f_1 = 1$ kHz.

b. Determine the series equivalent of the network in Figure 21–24(a) and the parallel equivalent of the network in Figure 21–24(b).

c. Repeat parts (a) and (b) for a frequency of $f_2 = 200$ kHz.

Answers

a. $Q_a = 1.26$ $Q_b = 12.7$

b. Network a: $R_S = 388\ \Omega$ $X_{CS} = 487\ \Omega$
Network b: $R_P = 816\ \Omega$ $X_{CP} = 64.1\ \Omega$

c. Network a: $Q_a = 251$ $R_S = 0.0158\ \Omega$ $X_{CS} = 3.98\ \Omega$
Network b: $Q_b = 0.0637$ $R_P = 5.02\ \Omega$ $X_{CP} = 78.9\ \Omega$

IN-PROCESS LEARNING CHECK 3

(Answers are at the end of the chapter.)

Refer to the networks of Figure 21–25.

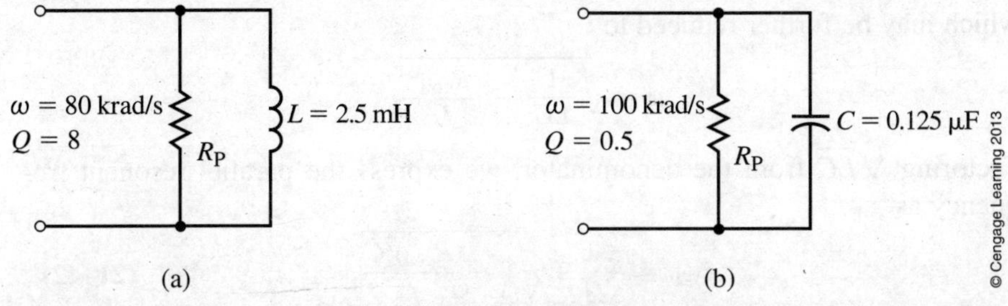

FIGURE 21–25

a. Determine the resistance, R_P, for each network.

b. Find the equivalent series network by using the quality factor for the given networks.

21.6 Parallel Resonance

A simple parallel resonant circuit is illustrated in Figure 21–26. The parallel resonant circuit is best analyzed using a constant-current source, unlike the series resonant circuit, which used a constant-voltage source.

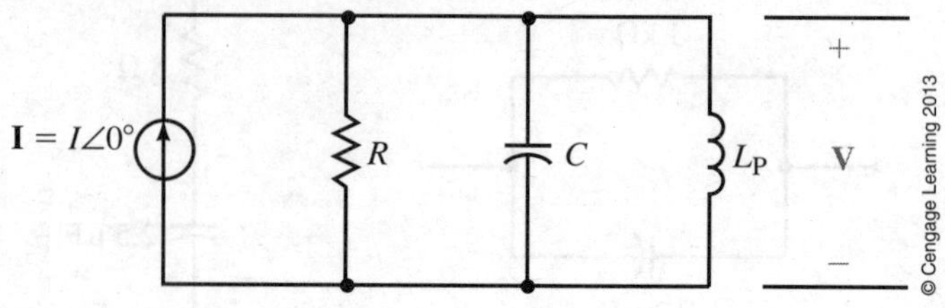

FIGURE 21–26 Simple parallel resonant circuit.

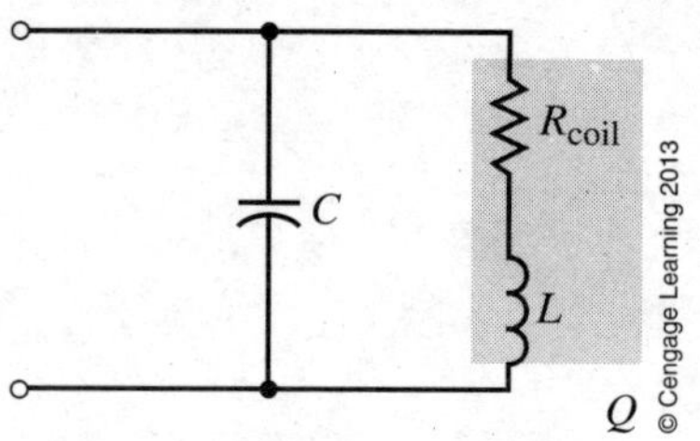

FIGURE 21–27

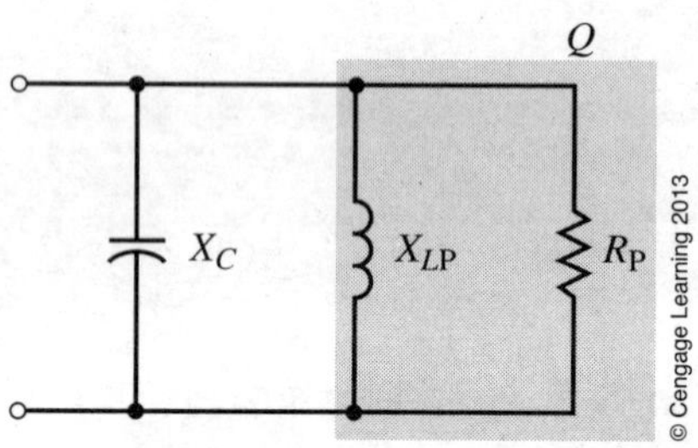

FIGURE 21–28

Consider the LC "tank" circuit shown in Figure 21–27. The tank circuit consists of a capacitor in parallel with an inductor. Due to its high Q and frequency response, the tank circuit is used extensively in communications equipment such as AM, FM, and television transmitters and receivers.

The circuit of Figure 21–27 is not exactly a parallel resonant circuit, since the resistance of the coil is in series with the inductance. In order to determine the frequency at which the circuit is purely resistive, we must first convert the series combination of resistance and inductance into an equivalent parallel network. The resulting circuit is shown in Figure 21–28.

At resonance, the capacitive and inductive reactances in the circuit of Figure 21–28 are equal. As we have observed previously, placing equal inductive and capacitive reactances in parallel effectively results in an open circuit at the given frequency. The input impedance of this network at resonance is therefore purely resistive and given as $Z_T = R_P$. We determine the resonant frequency of a tank circuit by first letting the reactances of the equivalent parallel circuit be equal:

$$X_C = X_{LP}$$

Now, using the component values of the tank circuit, we have

$$X_C = \frac{(R_{coil})^2 + X_L^2}{X_L}$$

$$\frac{1}{\omega C} = \frac{(R_{coil})^2 + (\omega L)^2}{\omega L}$$

$$\frac{L}{C} = (R_{coil})^2 + (\omega L)^2$$

which may be further reduced to

$$\omega = \sqrt{\frac{1}{LC} - \frac{R_{coil}^2}{L^2}}$$

Factoring $\sqrt{LC}$ from the denominator, we express the parallel resonant frequency as

$$\omega_P = \frac{1}{\sqrt{LC}}\sqrt{1 - \frac{(R_{coil})^2 C}{L}} \qquad \textbf{(21–42)}$$

Notice that if $R_{coil}^2 < L/C$, then the term under the radical is approximately equal to 1.

Consequently, if $L/C \geq 100R_{coil}$, the parallel resonant frequency may be simplified as

$$\omega_P \cong \frac{1}{\sqrt{LC}} \quad (\text{for } L/C \geq 100R_{coil}) \qquad \textbf{(21–43)}$$

NOTES...

For a high-Q circuit, ω_P can be approximated as $\omega_P = \frac{1}{\sqrt{LC}}$.

Recall that the quality factor, Q, of a circuit is defined as the ratio of reactive power to average power for a circuit at resonance. If we consider the parallel resonant circuit of Figure 21–29, we make several important observations.

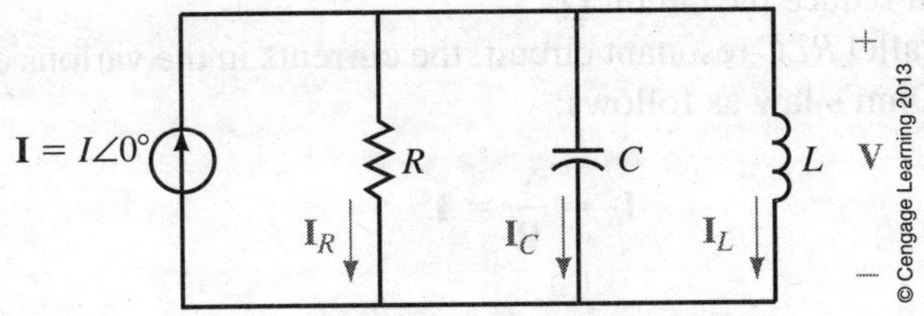

FIGURE 21–29

The inductor and capacitor reactances cancel, resulting in a circuit voltage simply determined by Ohm's law as

$$\mathbf{V} = \mathbf{I}R = IR\angle 0°$$

The frequency response of the impedance of the parallel circuit is shown in Figure 21–30.

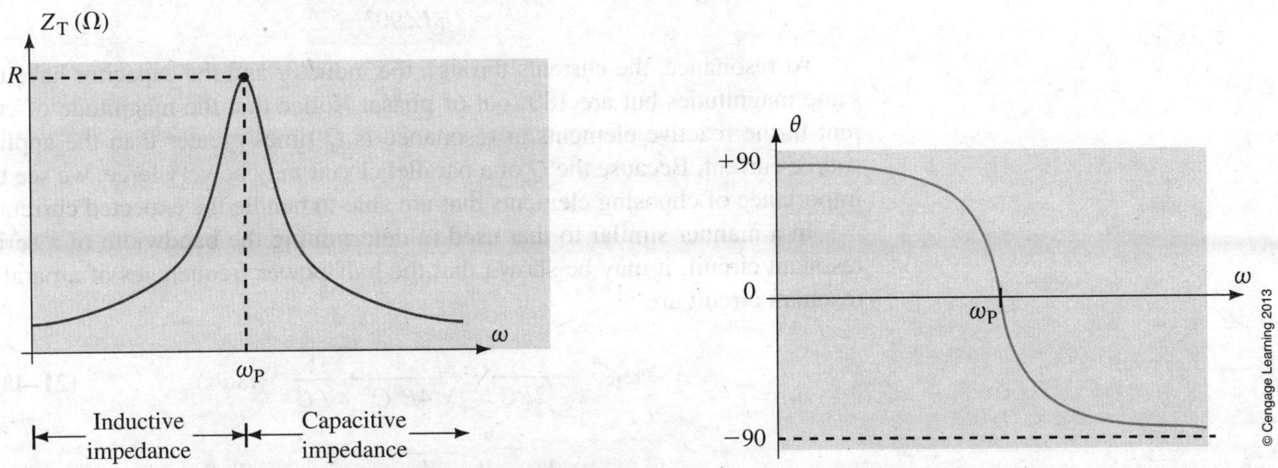

FIGURE 21–30 Impedance (magnitude and phase angle) versus angular frequency for a parallel resonant circuit.

Notice that the impedance of the entire circuit is maximum at resonance and minimum at the boundary conditions ($\omega = 0$ rad/s and $\omega \to \infty$). This result is exactly opposite to that observed in series resonant circuits, which have minimum impedance at resonance. We also see that for parallel circuits, the impedance will appear inductive for frequencies less than the resonant frequency, ω_P. Inversely, the impedance is capacitive for frequencies greater than ω_P.

The Q of the parallel circuit is determined from the definition as

$$Q_P = \frac{\text{reactive power}}{\text{average power}} = \frac{V^2/X_L}{V^2/R} \qquad \textbf{(21–44)}$$

$$Q_P = \frac{R}{X_{LP}} = \frac{R}{X_C}$$

This is precisely the same result as that obtained when we converted an *RL* series network into its equivalent parallel network. If the resistance of the coil is the only resistance within a circuit, then the circuit *Q* will be equal to the *Q* of the coil. However, if the circuit has other sources of resistance, then the additional resistance will reduce the circuit *Q*.

For a parallel *RLC* resonant circuit, the currents in the various elements are found from Ohm's law as follows:

$$\mathbf{I}_R = \frac{\mathbf{V}}{\mathbf{R}} = \mathbf{I} \tag{21–45}$$

$$\begin{aligned} \mathbf{I}_L &= \frac{\mathbf{V}}{X_L\angle 90^\circ} \\ &= \frac{V}{R/Q_P}\angle -90^\circ \\ &= Q_P I\angle -90^\circ \end{aligned} \tag{21–46}$$

$$\begin{aligned} \mathbf{I}_C &= \frac{\mathbf{V}}{X_C\angle -90^\circ} \\ &= \frac{V}{R/Q_P}\angle 90^\circ \\ &= Q_P I\angle 90^\circ \end{aligned} \tag{21–47}$$

At resonance, the currents through the inductor and the capacitor have the same magnitudes but are 180° out of phase. Notice that the magnitude of current in the reactive elements at resonance is *Q* times greater than the applied source current. Because the *Q* of a parallel circuit may be very large, we see the importance of choosing elements that are able to handle the expected currents.

In a manner similar to that used in determining the bandwidth of a series resonant circuit, it may be shown that the half-power frequencies of a parallel resonant circuit are

$$\omega_1 = -\frac{1}{2RC} + \sqrt{\frac{1}{4R^2C^2} + \frac{1}{LC}} \quad \text{(rad/s)} \tag{21–48}$$

$$\omega_2 = \frac{1}{2RC} + \sqrt{\frac{1}{4R^2C^2} + \frac{1}{LC}} \quad \text{(rad/s)} \tag{21–49}$$

The bandwidth is therefore

$$\text{BW} = \omega_2 - \omega_1 = \frac{1}{RC} \quad \text{(rad/s)} \tag{21–50}$$

If the quality factor of the circuit $Q \geq 10$, then the selectivity curve is very nearly symmetrical around ω_P, resulting in half-power frequencies which are located at $\omega_P \pm \text{BW}/2$.

Multiplying Equation 21–50 by ω_P/ω_P results in the following:

$$\begin{aligned} \text{BW} &= \frac{\omega_P}{R(\omega_P C)} = \frac{X_C}{R}\omega_P \\ \text{BW} &= \frac{\omega_P}{Q_P} \quad \text{(rad/s)} \end{aligned} \tag{21–51}$$

Notice that Equation 21–51 is the same for both series and parallel resonant circuits.

EXAMPLE 21–6

Consider the circuit shown in Figure 21–31.

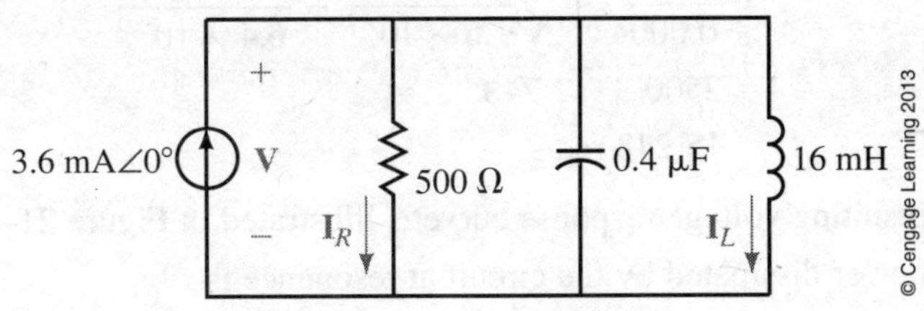

FIGURE 21–31

CircuitSim 21-3

a. Determine the resonant frequencies, ω_P(rad/s) and f_P(Hz) of the tank circuit.
b. Find the Q of the circuit at resonance.
c. Calculate the voltage across the circuit at resonance.
d. Solve for currents through the inductor and the resistor at resonance.
e. Determine the bandwidth of the circuit in both radians per second and hertz.
f. Sketch the voltage response of the circuit, showing the voltage at the half-power frequencies.
g. Sketch the selectivity curve of the circuit showing P(watts) versus ω(rad/s).

Solution

a. $$\omega_P = \frac{1}{\sqrt{LC}} = \frac{1}{\sqrt{(16\text{ mH})(0.4\ \mu\text{F})}} = 12.5\text{ krad/s}$$

$$f_P = \frac{\omega}{2\pi} = \frac{12.5\text{ krad/s}}{2\pi} = 1989\text{ Hz}$$

b. $$Q_P = \frac{R_P}{\omega L} = \frac{500\ \Omega}{(12.5\text{ krad/s})(16\text{ mH})} = \frac{500\ \Omega}{200\ \Omega} = 2.5$$

c. At resonance, $\mathbf{V}_C = \mathbf{V}_L = \mathbf{V}_R$, and so

$$\mathbf{V} = \mathbf{IR} = (3.6\text{ mA}\angle 0^\circ)(500\ \Omega\angle 0^\circ) = 1.8\text{ V}\angle 0^\circ$$

d. $$\mathbf{I}_L = \frac{\mathbf{V}_L}{\mathbf{Z}_L} = \frac{1.8\text{ V}\angle 0^\circ}{200\ \Omega\angle 90^\circ} = 9.0\text{ mA}\angle -90^\circ$$

$$\mathbf{I}_R = \mathbf{I} = 3.6\text{ mA}\angle 0^\circ$$

e. $$\text{BW(rad/s)} = \frac{\omega_P}{Q_P} = \frac{12.5\text{ krad/s}}{2.5} = 5\text{ krad/s}$$

$$\text{BW(Hz)} = \frac{\text{BW(rad/s)}}{2\pi} = \frac{5\text{ krad/s}}{2\pi} = 795.8\text{ Hz}$$

f. The half-power frequencies are calculated from Equations 21–48 and 21–49 since the Q of the circuit is less than 10.

$$\omega_1 = -\frac{1}{2RC} + \sqrt{\frac{1}{4R^2C^2} + \frac{1}{LC}}$$

$$= -\frac{1}{0.0004} + \sqrt{\frac{1}{1.6\times 10^{-7}} + \frac{1}{6.4\times 10^{-9}}}$$

$$= -2500 + 12\,748$$

$$= 10\,248\text{ rad/s}$$

$$\begin{aligned}\omega_2 &= \frac{1}{2RC} + \sqrt{\frac{1}{4R^2C^2} + \frac{1}{LC}}\\ &= \frac{1}{0.0004} + \sqrt{\frac{1}{1.6 \times 10^{-7}} + \frac{1}{6.4 \times 10^{-9}}}\\ &= 2500 + 12\,748\\ &= 15\,248 \text{ rad/s}\end{aligned}$$

The resulting voltage response curve is illustrated in Figure 21–32.

g. The power dissipated by the circuit at resonance is

$$P = \frac{V^2}{R} = \frac{(1.8 \text{ V})^2}{500\ \Omega} = 6.48 \text{ mW}$$

The selectivity curve is now easily sketched as shown in Figure 21–33.

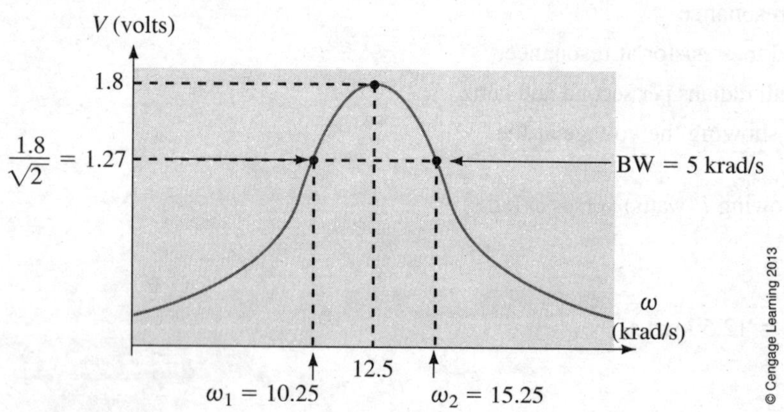

FIGURE 21–32

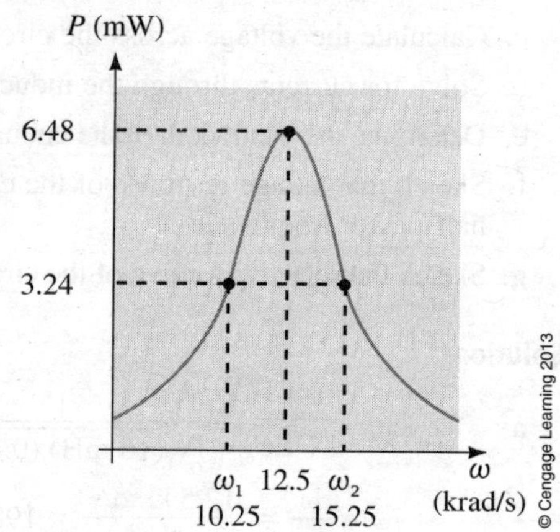

FIGURE 21–33

EXAMPLE 21–7

Consider the circuit of Figure 21–34.

CircuitSim 21-4

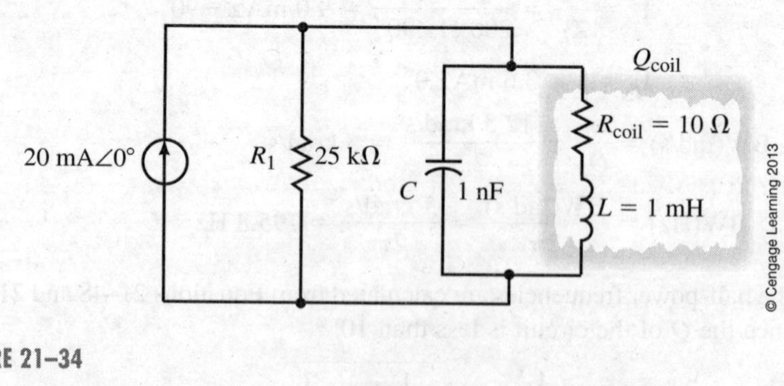

FIGURE 21–34

a. Calculate the resonant frequency, ω_P, of the tank circuit.

b. Find the Q of the coil at resonance.

c. Sketch the equivalent parallel circuit.

d. Determine the Q of the entire circuit at resonance.

e. Solve for the voltage across the capacitor at resonance.

f. Find the bandwidth of the circuit in radians per second.

g. Sketch the voltage response of the circuit showing the voltage at the half-power frequencies.

Solution

a. Since the ratio $L/C = 1000 \geq 100R_{coil}$, we use the approximation:

$$\omega_P = \frac{1}{\sqrt{LC}} = \frac{1}{\sqrt{(1\text{ mH})\,(1\text{ nF})}} = 1\text{ Mrad/s}$$

b. $$Q_{coil} = \frac{\omega L}{R_{coil}} = \frac{(1\text{ Mrad/s})\,(1\text{ mH})}{10\ \Omega} = 100$$

c. $$R_P \cong Q_{coil}{}^2 R_{coil} = (100)^2(10\ \Omega) = 100\text{ k}\Omega$$

$$X_{LP} \cong X_L = \omega L = (1\text{ Mrad/s})\,(1\text{ mH}) = 1\text{ k}\Omega$$

The circuit of Figure 21–35 shows the circuit with the parallel equivalent of the inductor.

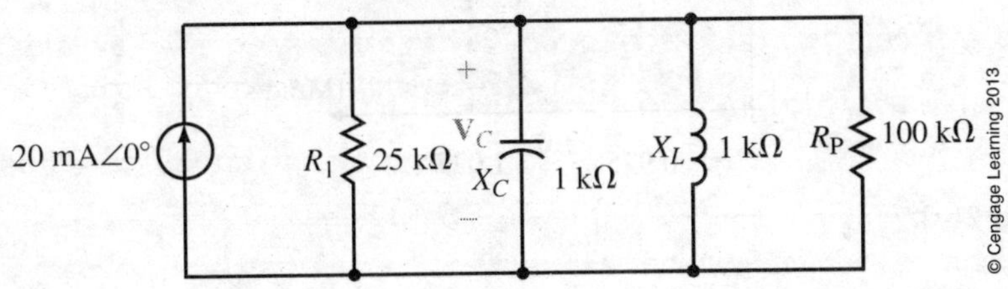

FIGURE 21–35

We see that the previous circuit may be further simplified by combining the parallel resistances:

$$R_{eq} = R_1 \| R_P = \frac{(25\text{ k}\Omega)(100\text{ k}\Omega)}{25\text{ k}\Omega + 100\text{ k}\Omega} = 20\text{ k}\Omega$$

The simplified equivalent circuit is shown in Figure 21–36.

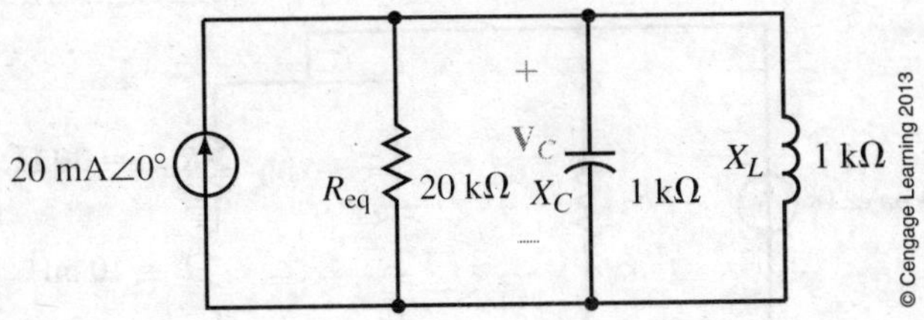

FIGURE 21–36

d. $$Q_P = \frac{R_{eq}}{X_L} = \frac{20\text{ k}\Omega}{1\text{ k}\Omega} = 20$$

e. At resonance,

$$\mathbf{V}_C = \mathbf{I}R_{eq} = (20\text{ mA}\angle 0°)\,(20\text{ k}\Omega) = 400\text{ V}\angle 0°$$

f. $$\text{BW} = \frac{\omega_P}{Q} = \frac{1\text{ Mrad/s}}{20} = 50\text{ krad/s}$$

g. The voltage response curve is shown in Figure 21–37. Since the circuit $Q \geq 10$, the half-power frequencies will occur at the following angular frequencies:

$$\omega_1 \cong \omega_P - \frac{BW}{2} = 1.0\ \text{Mrad/s} - \frac{50\ \text{krad/s}}{2} = 0.975\ \text{Mrad/s}$$

and

$$\omega_2 \cong \omega_P + \frac{BW}{2} = 1.0\ \text{Mrad/s} + \frac{50\ \text{krad/s}}{2} = 1.025\ \text{Mrad/s}$$

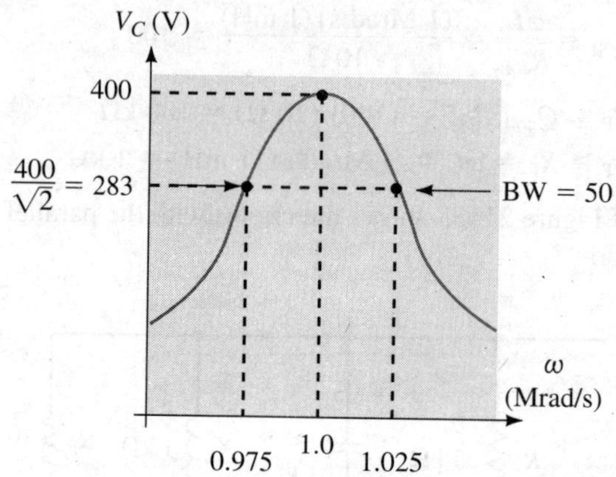

FIGURE 21–37

EXAMPLE 21–8

Determine the values of R_1 and C for the resonant tank circuit of Figure 21–38 so that the given conditions are met.

$L = 10$ mH, $R_{coil} = 30\ \Omega$

$f_P = 58$ kHz

BW = 1 kHz

Solve for the current, $\mathbf{I}_L$, through the inductor.

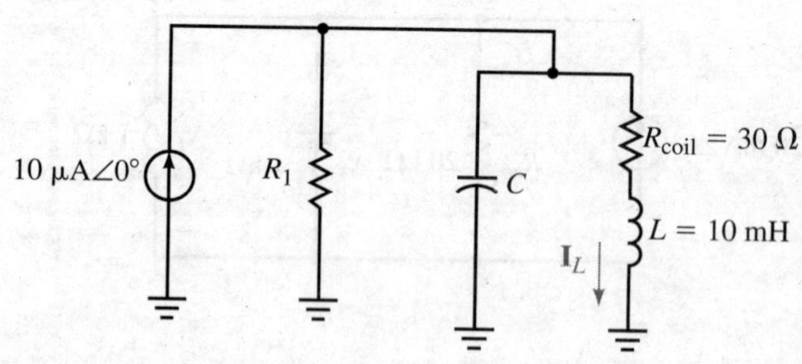

FIGURE 21–38

Solution

$$Q_P = \frac{f_P}{BW(Hz)} = \frac{58\ \text{kHz}}{1\ \text{kHz}} = 58$$

Now, because the frequency expressed in radians per second is more useful than hertz, we convert f_P to ω_P:

$$\omega_P = 2\pi f_P = (2\pi)(58\ \text{kHz}) = 364.4\ \text{krad/s}$$

The capacitance is determined from Equation 21–43 as

$$C = \frac{1}{\omega_P^2 L} = \frac{1}{(364.4\ \text{krad/s})^2(10\ \text{mH})} = 753\ \text{pF}$$

Solving for the Q of the coil permits us to easily convert the series RL network into its equivalent parallel network.

$$Q_{coil} = \frac{\omega_P L}{R_{coil}}$$

$$= \frac{(364.4\ \text{krad/s})(10\ \text{mH})}{30\ \Omega}$$

$$= \frac{3.644\ \text{k}\Omega}{30\ \Omega} = 121.5$$

$$R_P \cong Q_{coil}^2 R_{coil} = (121.5)^2(30\ \Omega) = 443\ \text{k}\Omega$$

$$X_{LP} \cong X_L = 3644\ \Omega$$

The resulting equivalent parallel circuit is shown in Figure 21–39.

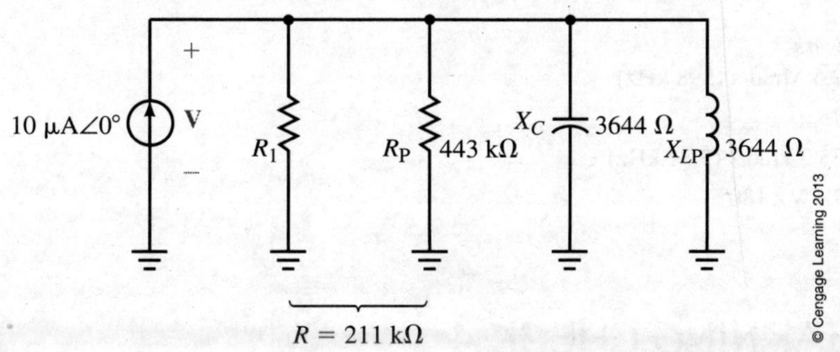

FIGURE 21–39

The quality factor Q_P is used to determine the total resistance of the circuit as

$$R = Q_P X_C = (58)(3.644\ \text{k}\Omega) = 211\ \text{k}\Omega$$

But

$$\frac{1}{R} = \frac{1}{R_1} + \frac{1}{R_P}$$

$$\frac{1}{R_1} = \frac{1}{R} - \frac{1}{R_P} = \frac{1}{211\ \text{k}\Omega} - \frac{1}{443\ \text{k}\Omega} = 2.47\ \mu\text{S}$$

And so

$$R_1 = 405\ \text{k}\Omega$$

The voltage across the circuit is determined to be

$$\mathbf{V} = \mathbf{I}R = (10\ \mu\text{A}\angle 0°)(211\ \text{k}\Omega) = 2.11\ \text{V}\angle 0°$$

and the current through the inductor is

$$\mathbf{I}_L = \frac{\mathbf{V}}{R_{coil} + jX_L}$$

$$= \frac{2.11\ \text{V}\angle 0°}{30 + j3644\ \Omega} = \frac{2.11\ \text{V}\angle 0°}{3644\ \Omega\angle 89.95°} = 579\ \mu\text{A}\angle -89.95°$$

PRACTICE PROBLEMS 5

Refer to the circuit of Figure 21–40.

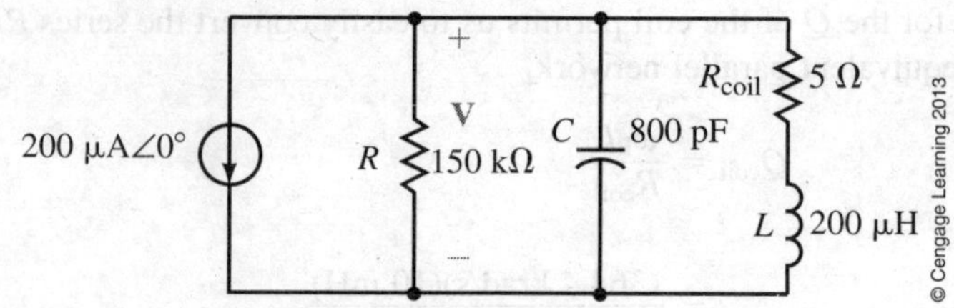

FIGURE 21–40

a. Determine the resonant frequency and express it in radians per second and in hertz.
b. Calculate the quality factor of the circuit.
c. Solve for the bandwidth.
d. Determine the voltage **V** at resonance.

Answers

a. 2.5 Mrad/s (398 kHz)
b. 75
c. 33.3 krad/s (5.31 kHz)
d. 7.5 V∠180°

IN-PROCESS LEARNING CHECK 4

(Answers are at the end of the chapter.)

Refer to the parallel resonant circuit of Figure 21–41.

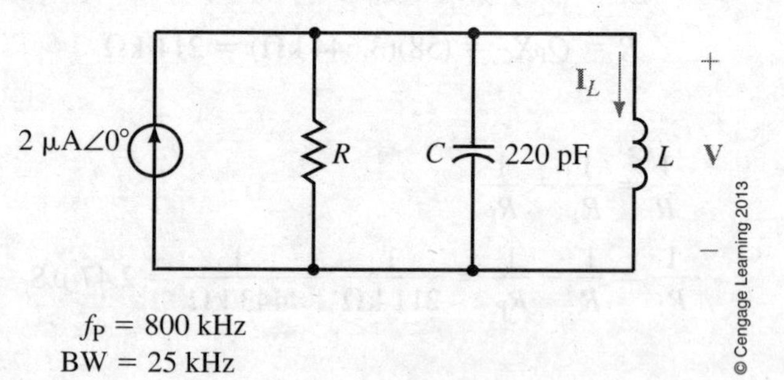

FIGURE 21–41

Suppose the circuit has a resonant frequency of 800 kHz and a bandwidth of 25 kHz.

a. Determine the value of the inductor, L, in henries.
b. Calculate the value of the resistance, R, in ohms.
c. Find **V**, $\mathbf{I}_L$, and power, P, at resonance.
d. Find the approximate values of the half-power frequencies, f_1 and f_2.
e. Determine the voltage across the circuit at the lower half-power frequency, f_1, and show that the power dissipated by the resistor at this frequency is half the power dissipated at the resonant frequency.

21.7 Circuit Analysis Using Computers

PSpice is particularly useful in examining the operation of resonant circuits. The ability of the software to provide a visual display of the frequency response is used to evaluate the resonant frequency, maximum current, and bandwidth of a circuit. The Q of the given circuit is then easily determined.

PSpice

EXAMPLE 21–9

PSPICE

Use PSpice to obtain the frequency response for current in the circuit of Figure 21–12. Use cursors to find the resonant frequency and the bandwidth of the circuit from the observed response. Compare the results to those obtained in Example 21–2.

Solution

OrCAD Capture CIS is used to input the circuit as shown in Figure 21–42. For this example, the project is titled **Ch21 PSpice1.** The voltage source used in this example is VAC and the properties are changed to **AC = 10V** and **DEG = 0Deg.** In order to obtain a plot of the circuit current, place a Current Pin tool as shown.

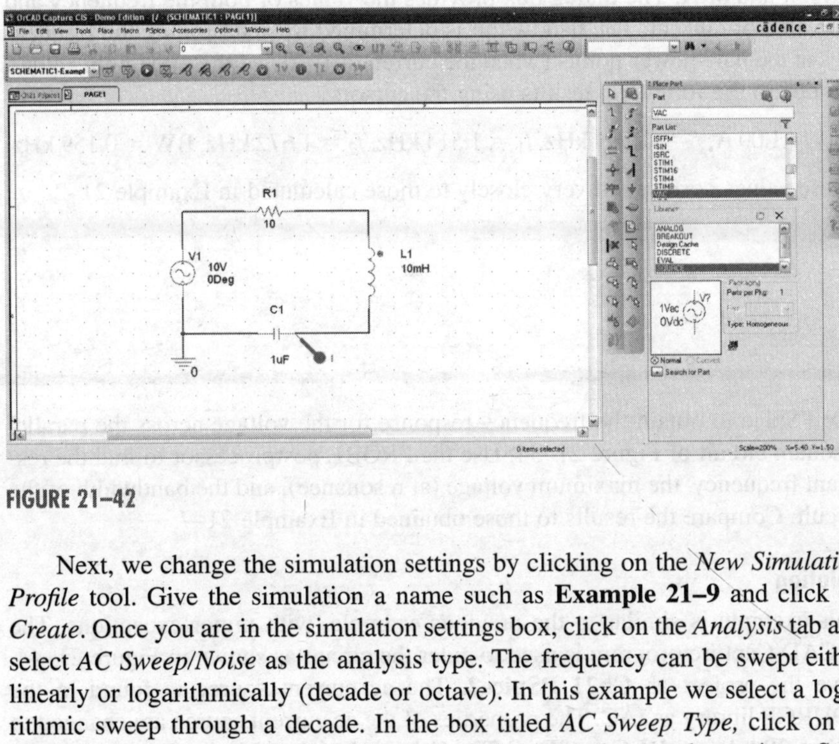

FIGURE 21–42

Next, we change the simulation settings by clicking on the *New Simulation Profile* tool. Give the simulation a name such as **Example 21–9** and click on *Create*. Once you are in the simulation settings box, click on the *Analysis* tab and select *AC Sweep/Noise* as the analysis type. The frequency can be swept either linearly or logarithmically (decade or octave). In this example we select a logarithmic sweep through a decade. In the box titled *AC Sweep Type,* click on ⊙ *Logarithmic* and select *Decade*. Enter the following values as the settings. Start Frequency: **1kHz,** End Frequency: **10kHz,** and Points/Decade: **10001.** Click *OK*.

Click on the *Run* tool. If there are no errors, the PROBE postprocessor will run automatically and display I(C1) as a function of frequency. You will notice that the selectivity curve is largely contained within a narrow range of frequencies. We may zoom into this region as follows: Select the *Plot* menu, and click on *Axis Settings* menu item. Click on the *X Axis* tab and select *User Defined Data Range*. Change the values to **1kHz** to **3kHz.** Click *OK*. The resulting display is shown in Figure 21–43.

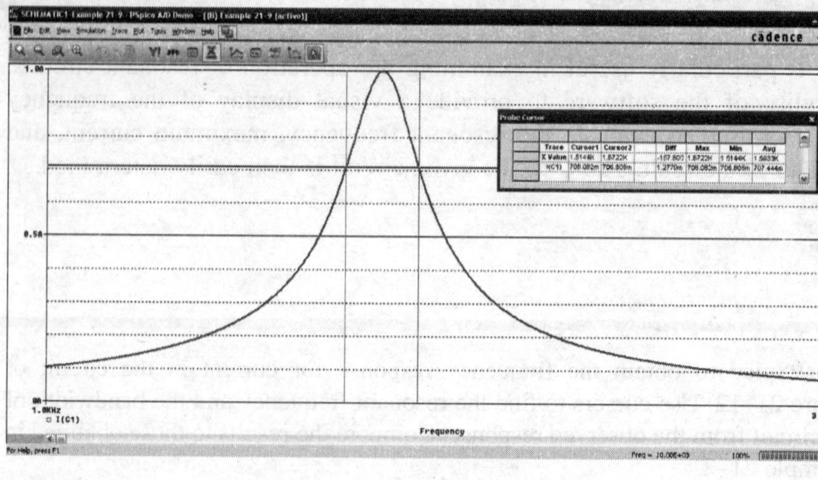

FIGURE 21–43

Finally, we may use cursors to provide us with the actual resonant frequency, the maximum current, and the half-power frequencies. Cursors are obtained as follows: Click on *Trace, Cursor,* and *Display*. The positions of the cursors are adjusted by using either the mouse or the arrow and <Shift> keys. The current at the maximum point of the curve is obtained by clicking on *Trace, Cursor,* and *Max*. The dialog box provides the values of both the frequency and the value of current. The bandwidth is determined by determining the frequencies at the half-power points (when the current is 0.707 of the maximum value). We obtain the following results using the cursors:

$I_{max} = 1.00\text{ A}, f_S = 1.591\text{ kHz}, f_1 = 1.514\text{ kHz}, f_2 = 1.672\text{ kHz}, \text{BW} = 0.159\text{ kHz}$

These values correspond very closely to those calculated in Example 21–2.

EXAMPLE 21–10

Use PSpice to obtain the frequency response for the voltage across the parallel resonant circuit of Figure 21–34. Use the PROBE postprocessor to find the resonant frequency, the maximum voltage (at resonance), and the bandwidth of the circuit. Compare the results to those obtained in Example 21–7.

Solution

This example is similar to the previous example, with minor exceptions. The OrCAD Capture program is used to enter the circuit as shown in Figure 21–44. Save the project as **Ch21 PSpice2.** The ac current source is found in the SOURCE library as IAC. The properties of the ac current source are changed to **AC = 20mA** and **DEG = 0Deg.** The *Voltage Level* tool is used to provide the voltage simulation for the circuit.

Use the *New Simulation Profile* tool to give the simulation a name such as **Example 21–10,** and to set the simulation for a logarithmic ac sweep from 100kHz to 300kHz with a total of 10001 points per decade. After running the simulation, click on the *Plot* menu to select the *Axis Settings*. Change the x-axis to indicate a *User Defined range* from **100kHz** to **300kHz** and change the y-axis to indicate a *User Defined range* from **0V** to **400V.** The resulting display is shown in Figure 21–45.

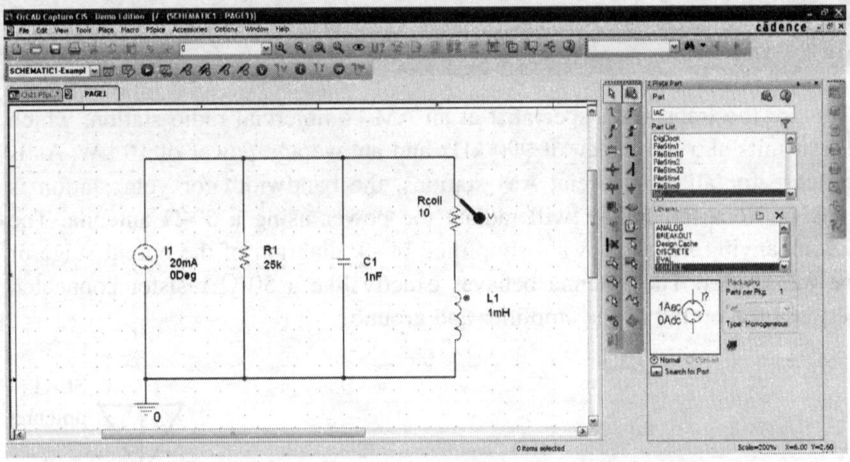

FIGURE 21–44

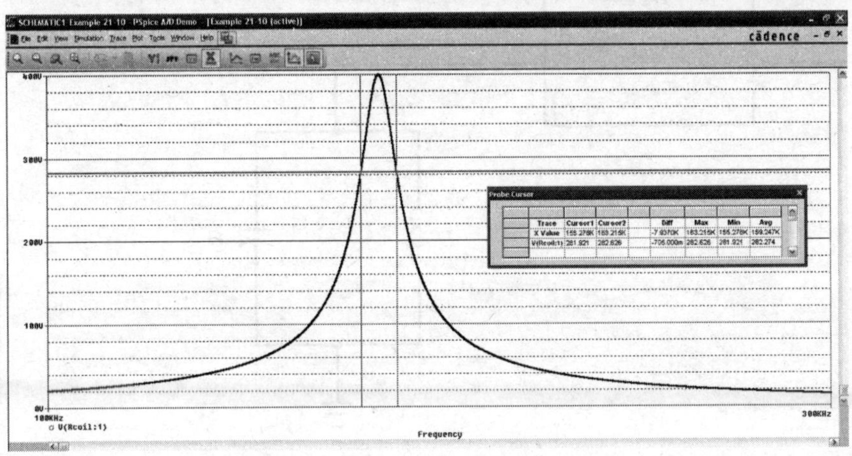

FIGURE 21–45

As in the previous example, we use cursors to observe that the maximum circuit voltage, V_{max} = 400 V, occurs at the resonance frequency as f_P = 159.2 kHz (1.00 Mrad/s). The half-power frequencies are determined when the output voltage is at 0.707 of the maximum value, namely at f_1 = 155.28 kHz (0.796 Mrad/s) and f_2 = 163.22 kHz (1.026 Mrad/s). These frequencies give a bandwidth of BW = 7.94 kHz (49.9 krad/s). The preceding results are consistent with those found in Example 21–7.

PRACTICE PROBLEMS 6

Use PSpice to obtain the frequency response of voltage, V versus f, for the circuit of Figure 21–40. Use cursors to determine the approximate values of the half-power frequencies and the bandwidth of the circuit. Compare the results to those obtained in Practice Problems 5.

Answers
V_{max} = 7.50 V, f_P = 398 kHz, f_1 = 395.3 kHz, f_2 = 400.7 kHz, BW = 5.31 kHz

Putting It into Practice

You are the transmitter specialist at an AM commercial radio station, which transmits at a frequency of 990 kHz and an average power of 10 kW. As is the case for all commercial AM stations, the bandwidth for your station is 10 kHz. Your transmitter will radiate the power using a 50-Ω antenna. The accompanying figure shows a simplified block diagram of the output stage of the transmitter. The antenna behaves exactly like a 50-Ω resistor connected between the output of the amplifier and ground.

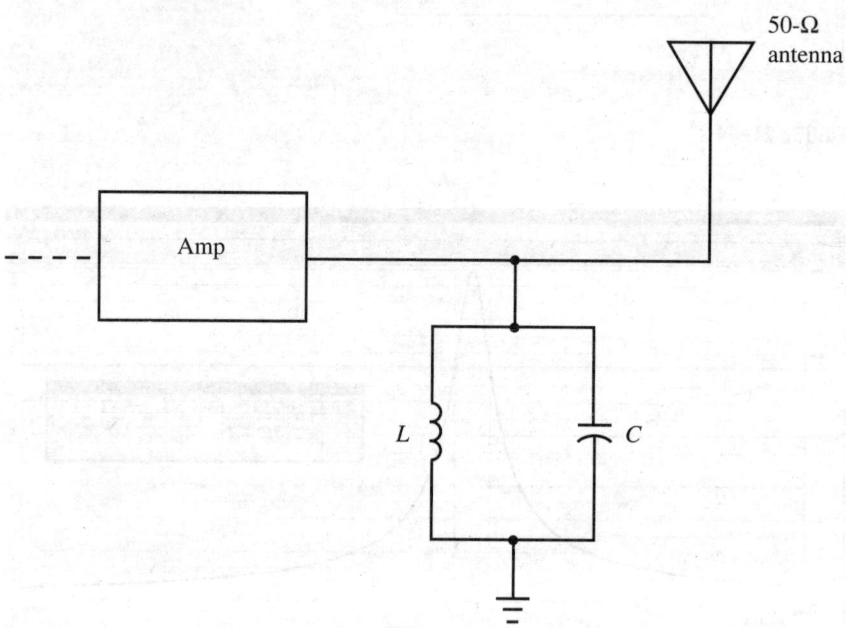

Transmitter stage of a commercial AM radio station.

You have been asked to determine the values of L and C so that the transmitter operates with the given specifications. As part of the calculations, determine the peak current that the inductor must handle and solve for the peak voltage across the capacitor. For your calculations, assume that the transmitted signal is a sinusoidal wave.

Problems

21.1 Series Resonance

1. Consider the circuit of Figure 21–46.

   a. Determine the resonant frequency of the circuit in both radians per second and hertz.

   b. Calculate the current, **I**, at resonance.

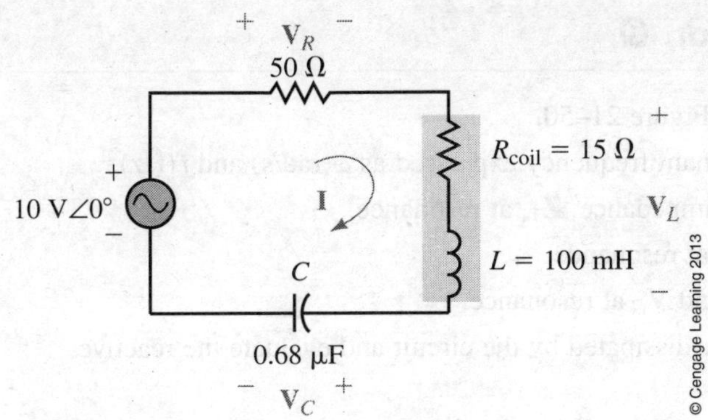

FIGURE 21–46

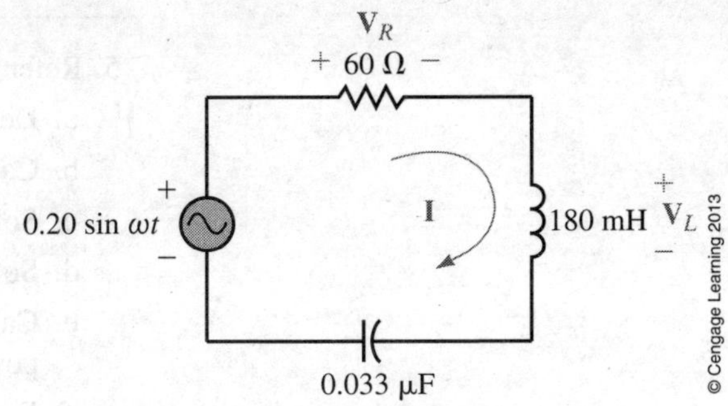

FIGURE 21–47

c. Solve for the voltages $\mathbf{V}_R$, $\mathbf{V}_L$, and $\mathbf{V}_C$. (Notice that the voltage $\mathbf{V}_L$ includes the voltage dropped across the internal resistance of the coil.)

d. Determine the power (in watts) dissipated by the inductor. (Hint: The power will not be zero.)

2. Refer to the circuit of Figure 21–47.
   a. Determine the resonant frequency of the circuit in both radians per second and hertz.
   b. Calculate the phasor current **I** at resonance.
   c. Determine the power dissipated by the circuit at resonance.
   d. Calculate the phasor voltages $\mathbf{V}_L$ and $\mathbf{V}_R$.
   e. Write the sinusoidal form of the voltages v_L and v_R.
3. Consider the circuit of Figure 21–48.
   a. Determine the values of R and C such that the circuit has a resonant frequency of 25 kHz and an rms current of 25 mA at resonance.
   b. Calculate the power dissipated by the circuit at resonance.
   c. Determine the phasor voltages, $\mathbf{V}_C$, $\mathbf{V}_L$, and $\mathbf{V}_R$ at resonance.
   d. Write the sinusoidal expressions for the voltages v_C, v_L, and v_R.

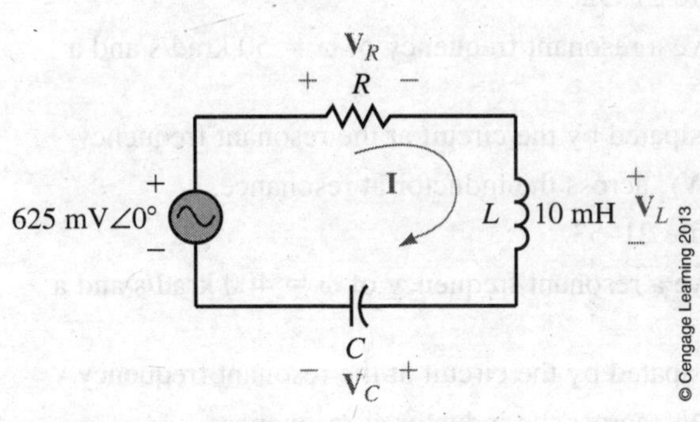

FIGURE 21–48

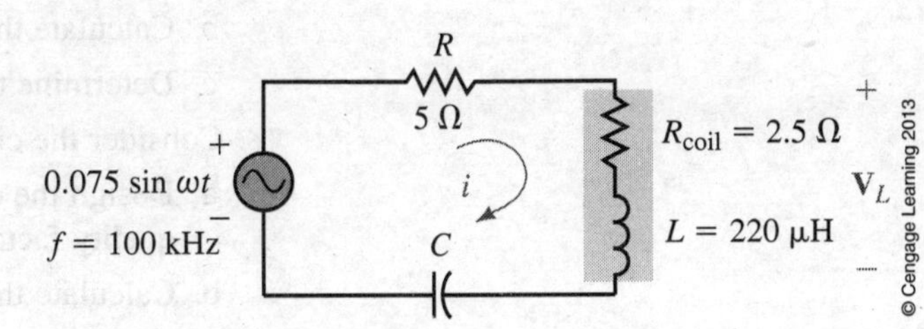

FIGURE 21–49

4. Refer to the circuit of Figure 21–49.
   a. Determine the capacitance required so that the circuit has a resonant frequency of 100 kHz.
   b. Solve for the phasor quantities **I**, $\mathbf{V}_L$, and $\mathbf{V}_R$.
   c. Find the sinusoidal expressions for i, v_L, and v_R.
   d. Determine the power dissipated by each element in the circuit.

21.2 Quality Factor, Q

5. Refer to the circuit of Figure 21–50.
   a. Determine the resonant frequency expressed as ω(rad/s) and f(Hz).
   b. Calculate the total impedance, $\mathbf{Z}_T$, at resonance.
   c. Solve for current **I** at resonance.
   d. Solve for $\mathbf{V}_R$, $\mathbf{V}_L$, and $\mathbf{V}_C$ at resonance.
   e. Calculate the power dissipated by the circuit and evaluate the reactive powers, Q_C and Q_L.
   f. Find the quality factor, Q_S, of the circuit.

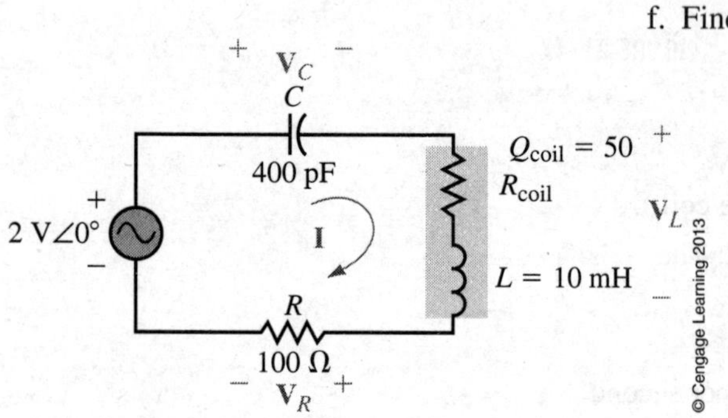

FIGURE 21–50

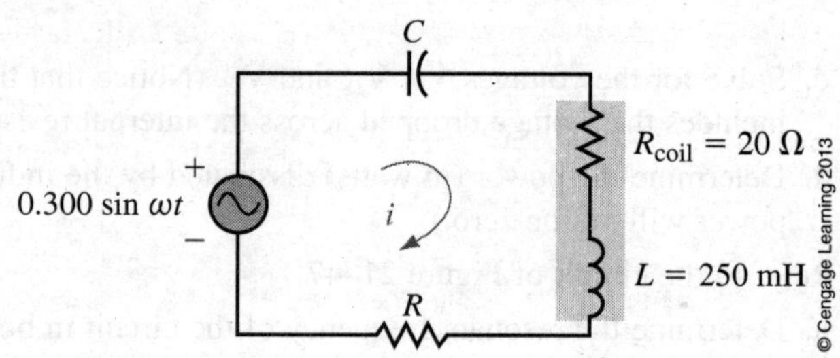

FIGURE 21–51

6. Suppose that the circuit of Figure 21–51 has a resonant frequency of $f_S = 2.5$ kHz and a quality factor of $Q_S = 10$.
   a. Determine the values of R and C.
   b. Solve for the quality factor of inductor Q_{coil}.
   c. Find $\mathbf{Z}_T$, **I**, $\mathbf{V}_C$, and $\mathbf{V}_R$ at resonance.
   d. Solve for the sinusoidal expression of current i at resonance.
   e. Calculate the sinusoidal expressions v_C and v_R at resonance.
   f. Calculate the power dissipated by the circuit and determine the reactive powers, Q_C and Q_L.
7. Refer to the circuit of Figure 21–52.
   a. Design the circuit to have a resonant frequency of $\omega = 50$ krad/s and a quality factor $Q_S = 25$.
   b. Calculate the power dissipated by the circuit at the resonant frequency.
   c. Determine the voltage, $\mathbf{V}_L$, across the inductor at resonance.
8. Consider the circuit of Figure 21–53.
   a. Design the circuit to have a resonant frequency of $\omega = 400$ krad/s and a quality factor $Q_S = 10$.
   b. Calculate the power dissipated by the circuit at the resonant frequency.
   c. Determine the voltage, $\mathbf{V}_L$, across the inductor at resonance.

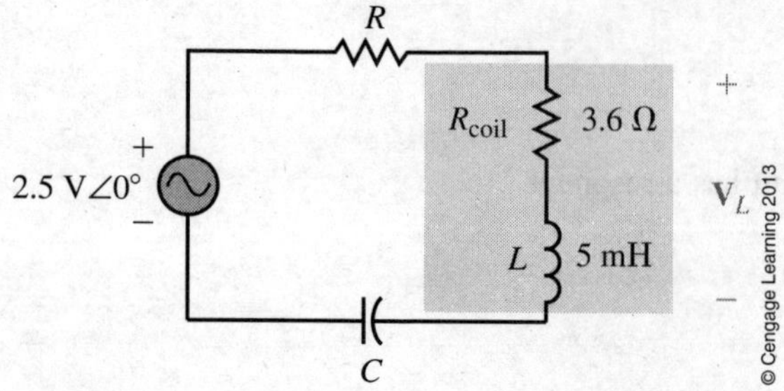

FIGURE 21–52

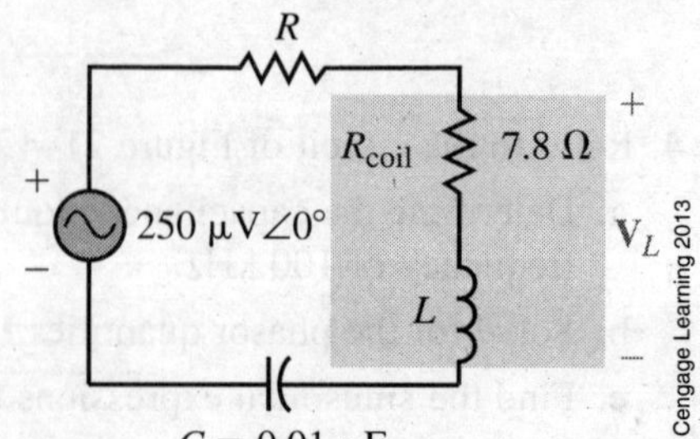

FIGURE 21–53

21.3 Impedance of a Series Resonant Circuit

9. Refer to the series resonant circuit of Figure 21–54.
   a. Determine the resonant frequency, ω_S.
   b. Solve for the input impedance, $\mathbf{Z}_T = Z\angle\theta$, of the circuit at frequencies of $0.1\omega_S$, $0.2\omega_S$, $0.5\omega_S$, ω_S, $2\omega_S$, $5\omega_S$, and $10\omega_S$.
   c. Using the results from (b), sketch a graph of Z (magnitude in ohms) versus ω (in radians per second) and a graph of θ (in degrees) versus ω (in radians per second). If possible, use log-log graph paper for the former and semilog graph paper for the latter.
   d. Using your results from (b), determine the magnitude of current at each of the given frequencies.
   e. Use the results from (d) to plot a graph of I (magnitude in amps) versus ω (in radians per second) on log-log graph paper.

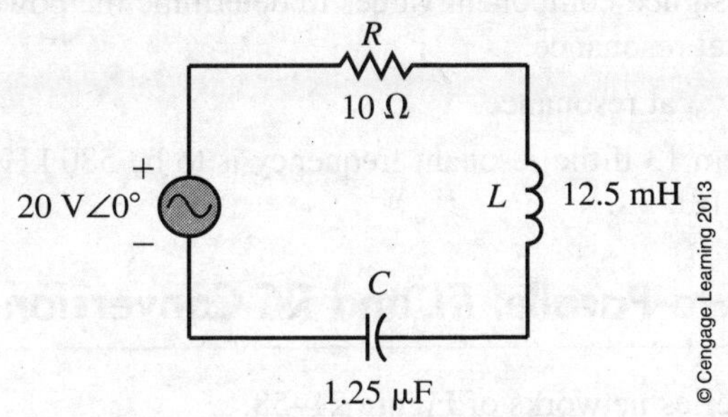

FIGURE 21–54

10. Repeat Problem 9 if the 10-Ω resistor is replaced with a 50-Ω resistor.

21.4 Power, Bandwidth, and Selectivity of a Series Resonant Circuit

11. Refer to the circuit of Figure 21–55.
    a. Find ω_S, Q, and BW (in radians per second).
    b. Calculate the maximum power dissipated by the circuit.
    c. From the results obtained in (a) solve for the approximate half-power frequencies, ω_1 and ω_2.
    d. Calculate the actual half-power frequencies, ω_1 and ω_2, using the component values and the appropriate equations.
    e. Are the results obtained in (c) and (d) comparable? Explain.
    f. Solve for the circuit current, I, and power dissipated at the lower half-power frequency, ω_1, determined in (d).

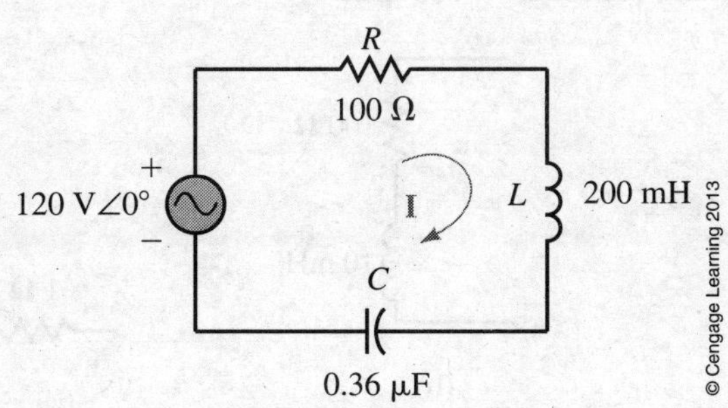

FIGURE 21–55

12. Repeat Problem 11 for the circuit of Figure 21–56.

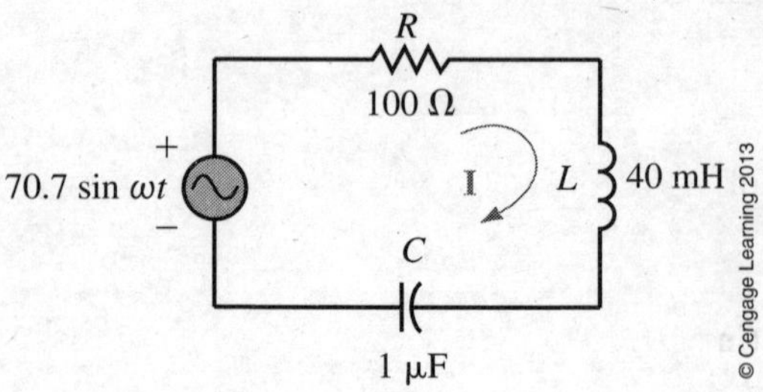

FIGURE 21–56

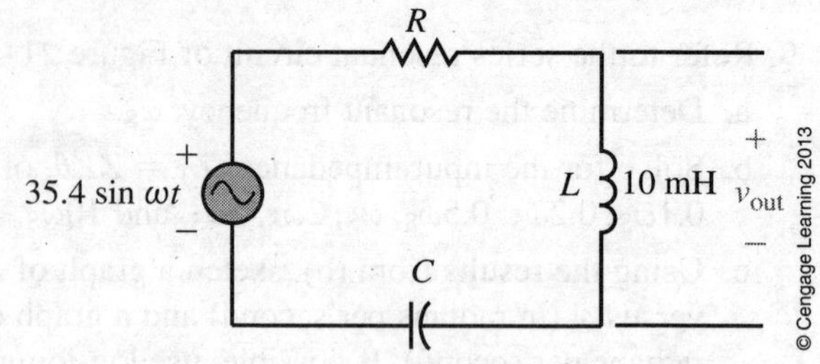

FIGURE 21–57

13. Consider the circuit of Figure 21–57.
    a. Calculate the values of R and C for the circuit to have a resonant frequency of 200 kHz and a bandwidth of 16 kHz.
    b. Use the designed component values to determine the power dissipated by the circuit at resonance.
    c. Solve for v_{out} at resonance.
14. Repeat Problem 13 if the resonant frequency is to be 580 kHz and the bandwidth is 10 kHz.

21.5 Series-to-Parallel *RL* and *RC* Conversion

15. Refer to the series networks of Figure 21–58.
    a. Find the Q of each network at $\omega = 1000$ rad/s.
    b. Convert each series RL network into an equivalent parallel network, having R_P and X_{LP} in ohms.
    c. Repeat (a) and (b) for $\omega = 10$ krad/s.

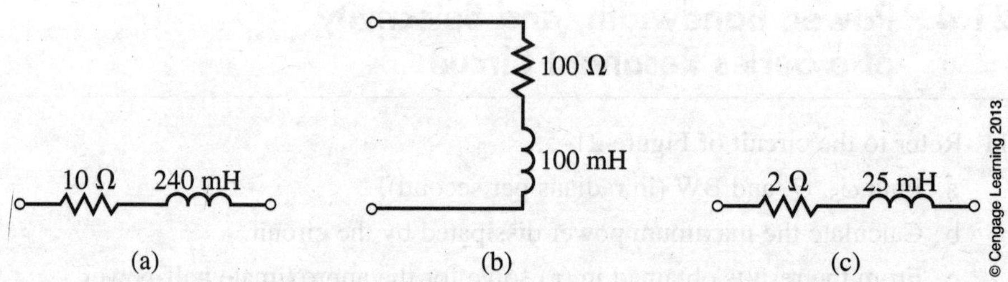

FIGURE 21–58

16. Consider the series networks of Figure 21–59.
    a. Find the Q of each coil at $\omega = 20$ krad/s.
    b. Convert each series RL network into an equivalent parallel network consisting of R_P and X_{LP} in ohms.
    c. Repeat (a) and (b) for $\omega = 100$ krad/s.

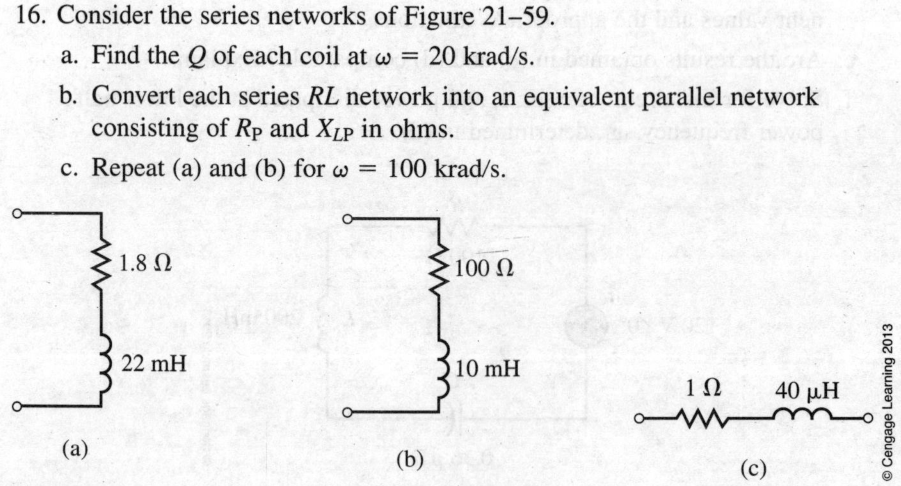

FIGURE 21–59

17. For the series networks of Figure 21–60, find the Q and convert each network into its parallel equivalent.

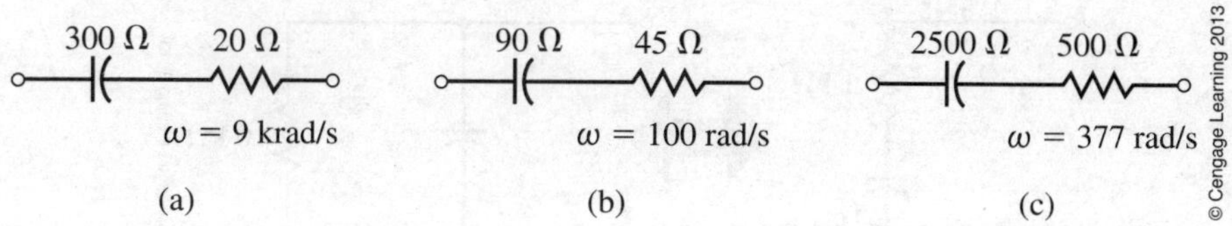

FIGURE 21–60

18. Derive Equations 21–34 and 21–35 that enable us to convert a parallel RL network into its series equivalent. (Hint: Begin by determining the expression for the input impedance of the parallel network.)
19. Find the Q of each of the networks of Figure 21–61 and determine the series equivalent of each. Express all component values in ohms.

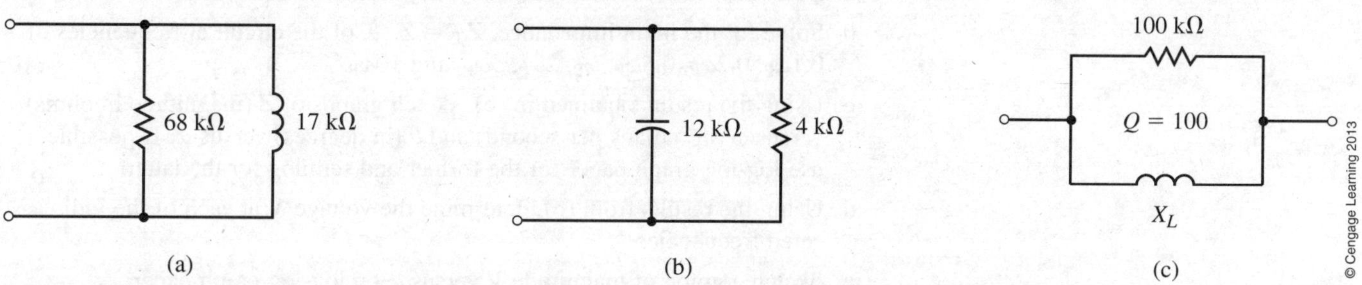

FIGURE 21–61

20. Repeat Problem 19 for the networks of Figure 21–62.

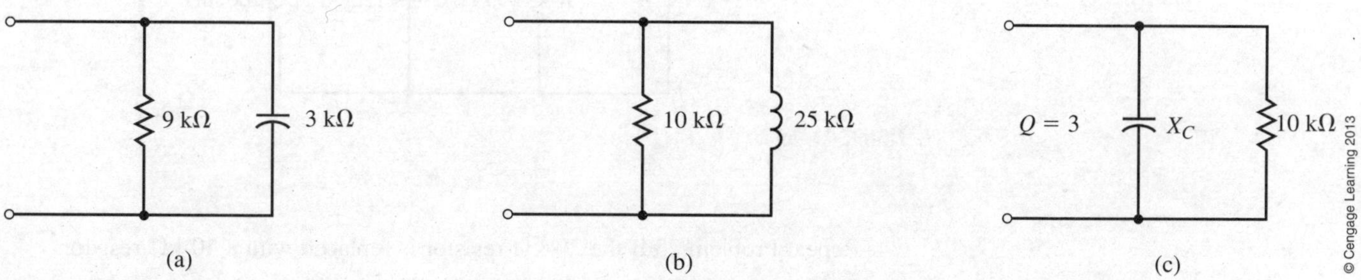

FIGURE 21–62

21. Determine the values of L_S and L_P in henries, given that the networks of Figure 21–63 are equivalent at a frequency of 250 krad/s.

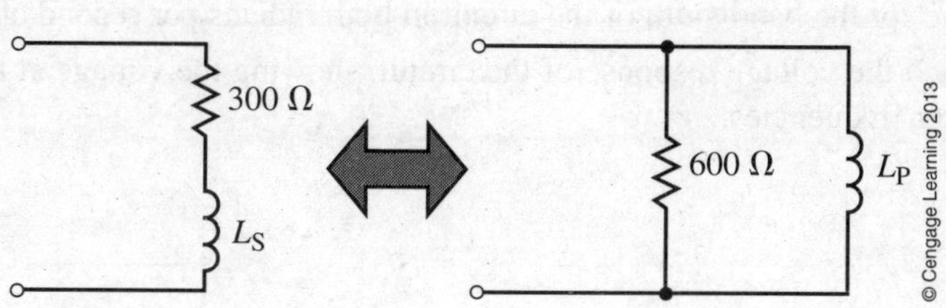

FIGURE 21–63

22. Determine the values of C_S and C_P in farads, given that the networks of Figure 21–64 are equivalent at a frequency of 48 krad/s.

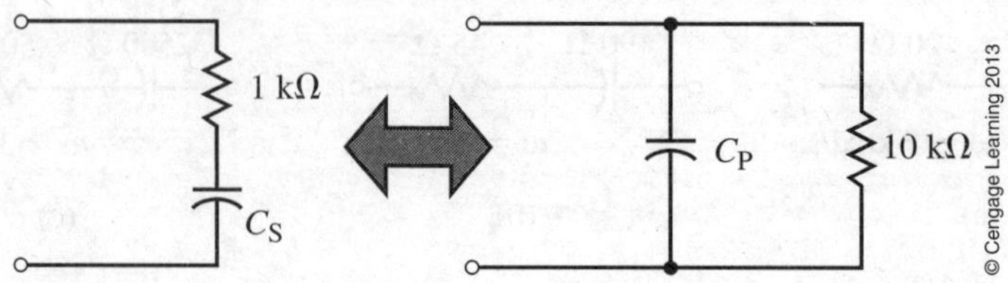

FIGURE 21–64

21.6 Parallel Resonance

23. Consider the circuit of Figure 21–65.
    a. Determine the resonant frequency, ω_P, in radians per second.
    b. Solve for the input impedance, $\mathbf{Z}_T = Z\angle\theta$, of the circuit at frequencies of $0.1\omega_P$, $0.2\omega_P$, $0.5\omega_P$, ω_P, $2\omega_P$, $5\omega_P$, and $10\omega_P$.
    c. Using the results obtained in (b), sketch graphs of Z (magnitude in ohms) versus ω (in radians per second) and θ (in degrees) versus ω. If possible, use log-log graph paper for the former and semilog for the latter.
    d. Using the results from (b), determine the voltage $\mathbf{V}$ at each of the indicated frequencies.
    e. Sketch a graph of magnitude V versus ω on log-log graph paper.

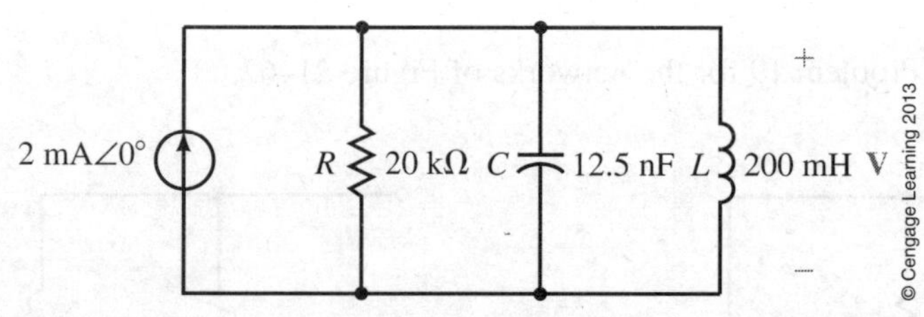

FIGURE 21–65

24. Repeat Problem 23 if the 20-kΩ resistor is replaced with a 40-kΩ resistor.
25. Refer to the circuit shown in Figure 21–66.
    a. Determine the resonant frequencies, ω_P(rad/s) and f_P(Hz).
    b. Find the Q of the circuit.
    c. Calculate $\mathbf{V}$, $\mathbf{I}_R$, $\mathbf{I}_L$, and $\mathbf{I}_C$ at resonance.
    d. Determine the power dissipated by the circuit at resonance.
    e. Solve for the bandwidth of the circuit in both radians per second and hertz.
    f. Sketch the voltage response of the circuit, showing the voltage at the half-power frequencies.

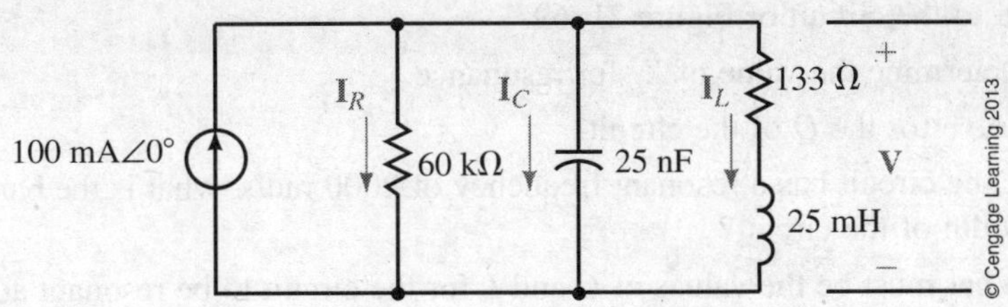

FIGURE 21–66

26. Repeat Problem 25 for the circuit of Figure 21–67.

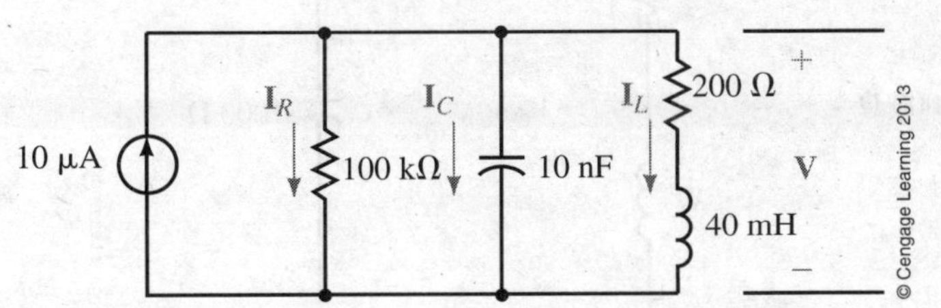

FIGURE 21–67

27. Determine the values of R_1 and C for the resonant tank circuit of Figure 21–68 so that the given conditions are met. Solve for current $\mathbf{I}_L$ through the inductor.

$$L = 25 \text{ mH}, R_{\text{coil}} = 100\ \Omega$$
$$f_P = 50 \text{ kHz}$$
$$\text{BW} = 10 \text{ kHz}$$

28. Determine the values of R_1 and C for the resonant circuit of Figure 21–68 so that the given conditions are met. Solve for the voltage, **V**, across the circuit.

$$L = 50 \text{ mH}, R_{\text{coil}} = 50\ \Omega$$
$$\omega_P = 100 \text{ krad/s}$$
$$\text{BW} = 10 \text{ krad/s}$$

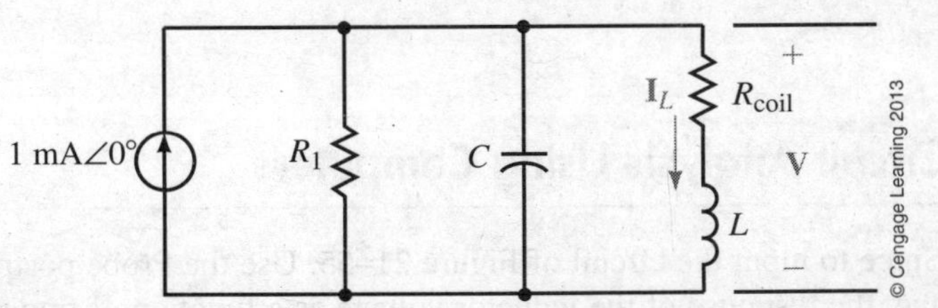

FIGURE 21–68

29. Refer to the circuit of Figure 21–69.
    a. Determine the value of X_L for resonance.
    b. Solve for the Q of the circuit.
    c. If the circuit has a resonant frequency of 2000 rad/s, what is the bandwidth of the circuit?
    d. What must be the values of C and L for the circuit to be resonant at 2000 rad/s?
    e. Calculate the voltage $\mathbf{V}_C$ at resonance.

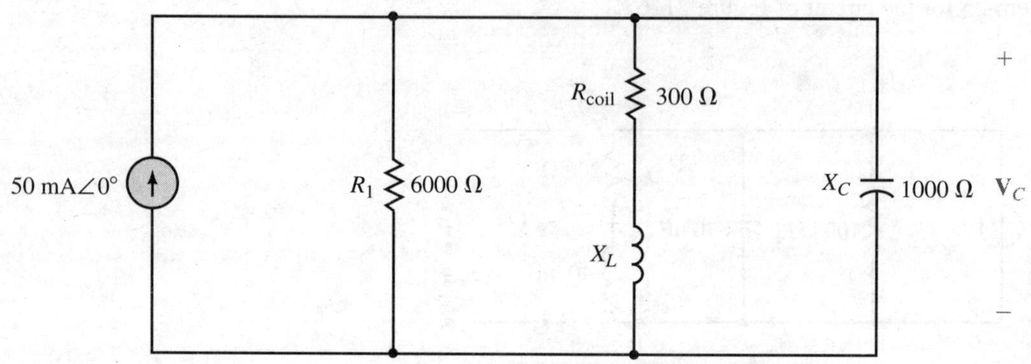

FIGURE 21–69

30. Repeat Problem 29 for the circuit of Figure 21–70.

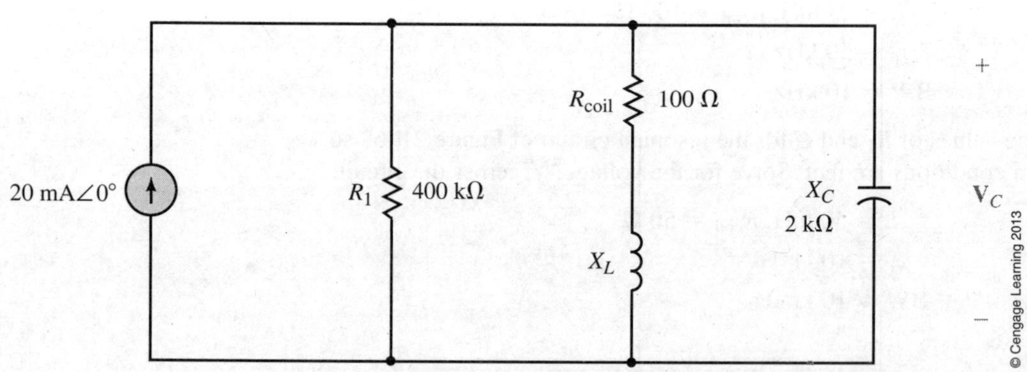

FIGURE 21–70

21.7 Circuit Analysis Using Computers

PSPICE

31. Use PSpice to input the circuit of Figure 21–55. Use the Probe postprocessor to display the response of the inductor voltage as a function of frequency. From the display, determine the maximum rms voltage, the resonant frequency, the half-power frequencies, and the bandwidth. Use the results to determine the quality factor of the circuit.

PSPICE

32. Repeat Problem 31 for the circuit of Figure 21–56.

33. Use PSpice to input the circuit of Figure 21–66. Use the Probe postprocessor to display the response of the capacitor voltage as a function of frequency. From the display, determine the maximum rms voltage, the resonant frequency, the half-power frequencies, and the bandwidth. Use the results to determine the quality factor of the circuit. PSPICE

34. Repeat Problem 33 for the circuit of Figure 21–67. PSPICE

35. Use the values of C and L determined in Problem 29 to input the circuit of Figure 21–69. Use the Probe postprocessor to display the response of the capacitor voltage as a function of frequency. From the display, determine the maximum rms voltage, the resonant frequency, the half-power frequencies, and the bandwidth. Use the results to determine the quality factor of the circuit. PSPICE

36. Use the values of C and L determined in Problem 30 to input the circuit of Figure 21–70. Repeat the measurements of Problem 35. PSPICE

ANSWERS TO IN-PROCESS LEARNING CHECKS

IN-PROCESS LEARNING CHECK 1

a. $L = 320\ \mu\text{H}$

b. $R = 20.1\ \Omega$

c. $\mathbf{I} = 0.995\ \text{mA}\angle 0°$ $\mathbf{V}_L = 1.20\ \text{V}\angle 90°$ $P = 20.0\ \mu\text{W}$

d. $f_1 = 595$ kHz $f_2 = 605$ kHz

e. $\mathbf{I} = 0.700\ \text{mA}\angle 45.28°$ $P_1 = 9.85\ \mu\text{W}$
$P_1/P = 0.492 \cong 0.5$

IN-PROCESS LEARNING CHECK 2

a. $\omega_S = 10$ krad/s $P = 500\ \mu\text{W}$

b. $Q = 10$ BW = 1 krad/s $\omega_1 = 9.5$ krad/s
$\omega_2 = 10.5$ krad/s

d. $\omega_S = 10$ krad/s $P = 1000\ \mu\text{W}$ $Q = 20$
BW = 0.5 krad/s $\omega_1 = 9.75$ krad/s
$\omega_2 = 10.25$ krad/s

e. As resistance decreases, selectivity increases.

IN-PROCESS LEARNING CHECK 3

a. $X_L = 200\ \Omega$ $R_P = 1600\ \Omega$ $X_C = 80\ \Omega$ $R_P = 40\ \Omega$

b. $X_{LS} = 197\ \Omega$ $R_S = 24.6\ \Omega$ $X_{CS} = 16\ \Omega$ $R_S = 32\ \Omega$

IN-PROCESS LEARNING CHECK 4

a. $L = 180\ \mu\text{H}$

b. $R_p = 28.9\ \text{k}\Omega$

c. $\mathbf{V} = 57.9\ \text{mV}\angle 0°$ $\mathbf{I}_L = 64.0\ \mu\text{A}\angle -90°$ $P = 115$ nW

d. $f_1 = 788$ kHz $f_2 = 813$ kHz

e. $\mathbf{V} = 41.1\ \text{mV}\angle 44.72°$ $P = 58$ nW

KEY TERMS

Amplifier
Attenuator
Bel
Bode Plots
Cutoff Frequency
Decade
Decibels (dB)
Filter Circuit
Notch Filter
Octave
Power Gain
Transfer Functions
Voltage Gain

OUTLINE

OBJECTIVES

After studying this chapter you will be able to

- evaluate the power gain and voltage gain of a given system,
- express power gain and voltage gain in decibels,
- express power levels in dBm and voltage levels in dBV and use these levels to determine power gain and voltage gain,
- identify and design simple (first-order) *RL* and *RC* low-pass and high-pass filters and explain the principles of operation of each type of filter,
- write the standard form of a transfer function for a given filter. The circuits that are studied will include band-pass and band-stop as well as low- and high-pass circuits,
- compute τ_c and use the time constant to determine the cutoff frequency(ies) in both radians per second and hertz for the transfer function of any first-order filter,
- sketch the Bode plots showing the frequency response of voltage gain and phase shift of any first-order filter,
- use PSpice to verify the operation of any first-order filter circuit.

22

FILTERS AND THE BODE PLOT

CHAPTER PREVIEW

In the previous chapter we examined how *LRC* resonant circuits react to changes in frequency. In this chapter we will continue to study how changes in frequency affect the behavior of other simple circuits. We will analyze simple low-pass, high-pass, band-pass, and band-reject filter circuits. The analysis will compare the amplitude and phase shift of the output signal with respect to the input signal.

As their names imply, low-pass and high-pass filter circuits are able to pass low frequencies and high frequencies while blocking other frequency components. A good understanding of these filters provides a basis for understanding why circuits such as amplifiers and oscilloscopes are not able to pass all signals from their input to their output.

Band-pass filters are designed to pass a range of frequencies from the input to the output. In the previous chapter, we saw that *L–C* circuits could be used to selectively pass a desired frequency range around a resonant frequency. In this chapter, we will observe that similar effects are possible by using only *R–C* or *R–L* components. Band-reject filters, on the other hand, are used to selectively prevent certain frequencies from appearing at the output, while allowing both higher and lower frequencies to pass relatively unaffected.

The analysis of all filters may be simplified by plotting the output/input voltage relationship on a semilogarithmic graph called a *Bode plot.* ■

Putting It in Perspective

Alexander Graham Bell

ALEXANDER GRAHAM BELL WAS BORN in Edinburgh, Scotland on March 3, 1847. As a young man, Bell followed his father and grandfather into research dealing with deafness, a condition that affected both his mother and his wife.

In 1873, Bell was appointed professor of vocal physiology at Boston University. His research was primarily involved in converting sound waves into electrical fluctuations. With the encouragement of Joseph Henry, who had done a great deal of work with inductors, Bell eventually developed the telephone.

In his now-famous accident in which Bell spilled acid on himself, Bell uttered the words "Watson, please come here. I want you." Watson, who was on another floor, ran to Bell's assistance.

Although others had worked on the principle of the telephone, Alexander Graham Bell was awarded the patent for the telephone in 1876. The telephone he constructed was a simple device that passed a current through carbon powder. The density of the carbon powder was determined by air fluctuations due to the sound of a person's voice. When the carbon was compressed, resistivity would decrease, allowing more current.

Bell's name has been adopted for the decibel, which is the unit used to describe sound intensities and power gain.

Although the invention of the telephone made Bell wealthy, he continued experimenting in electronics, air conditioning, and animal breeding. Bell died at the age of 75 in Baddeck, Nova Scotia, on August 2, 1922. ■

22.1 The Decibel

In electronics, we often wish to consider the effects of a circuit without examining the actual operation of the circuit itself. This black-box approach is a common technique used to simplify transistor circuits and to depict integrated circuits that may contain hundreds or even thousands of elements. Consider the system shown in Figure 22–1.

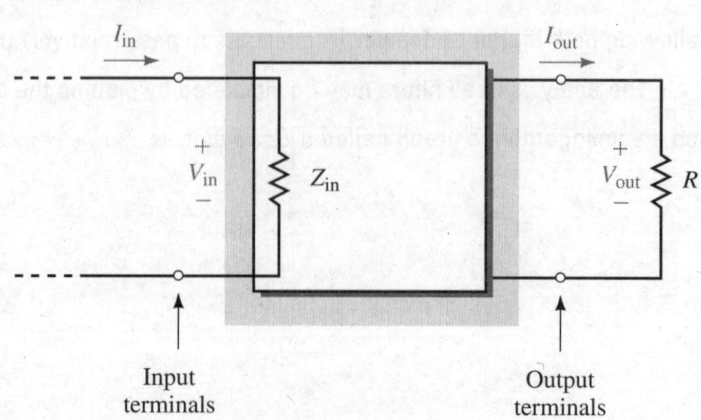

FIGURE 22–1

Although the circuit within the box may contain many elements, any source connected to the input terminals will effectively see only the input impedance, Z_{in}. Similarly, any load impedance, R_L, connected to the output terminals will have voltage and current determined by certain parameters in the circuit. These parameters usually result in an output at the load that is easily predicted for certain conditions.

We now define several terms that are used to analyze any system having two input terminals and two output terminals.

The **power gain,** A_P, is defined as the ratio of output signal power to the input signal power:

$$A_P = \frac{P_{out}}{P_{in}} \tag{22–1}$$

We must emphasize that the total output power delivered to any load can never exceed the total input power to a circuit. When we refer to the power gain of a system, we are interested in only the power contained in the ac signal, so we neglect any power due to dc. In many circuits, the ac power will be significantly less than the dc power. However, ac power gains in the order of tens of thousands are quite possible.

The **voltage gain,** A_v, is defined as the ratio of output signal voltage to the input signal voltage:

$$A_v = \frac{V_{out}}{V_{in}} \tag{22–2}$$

As mentioned, the power gain of a system may be a very large. For other applications the output power may be much smaller than the input power, resulting in a loss, or attenuation. Any circuit in which the output signal power is greater than the input signal power is referred to as an **amplifier.** Conversely, any circuit in which the output signal power is less than the input signal power is referred to as an **attenuator.**

The ratios expressing power gain or voltage gain may be either very large or very small, making it inconvenient to express the power gain as a simple ratio of two numbers. The **bel,** which is a logarithmic unit named after Alexander Graham Bell, was selected to represent a tenfold increase or decrease in power. Stated mathematically, the power gain in bels is given as

$$A_{P(\text{bels})} = \log_{10}\frac{P_{out}}{P_{in}}$$

Because the bel is an awkwardly large unit, the **decibel (dB),** which is one tenth of a bel, has been adopted as a more acceptable unit for describing the logarithmic change in power levels. One bel contains 10 decibels and so the power gain in decibels is given as

$$A_{P(\text{dB})} = 10\log_{10}\frac{P_{out}}{P_{in}} \tag{22–3}$$

If the power level in the system increases from the input to the output, then the gain in dB will be positive. If the power at the output is less than the power at the input, then the power gain will be negative. Notice that if the input and output have the same power levels, then the power gain will be 0 dB, since $\log 1 = 0$.

EXAMPLE 22–1

An amplifier has the indicated input and output power levels. Determine the power gain both as a ratio and in dB for each of the conditions:

a. $P_{in} = 1$ mW, $P_{out} = 100$ W
b. $P_{in} = 4\ \mu$W, $P_{out} = 2\ \mu$W
c. $P_{in} = 6$ mW, $P_{out} = 12$ mW
d. $P_{in} = 25$ mW, $P_{out} = 2.5$ mW

Solution

a.
$$A_P = \frac{P_{out}}{P_{in}} = \frac{100\text{ W}}{1\text{ mW}} = 100\,000$$
$$A_{P(\text{dB})} = 10\log_{10}(100\,000) = (10)(5) = 50\text{ dB}$$

b.
$$A_P = \frac{P_{out}}{P_{in}} = \frac{2\ \mu\text{W}}{4\ \mu\text{W}} = 0.5$$
$$A_{P(\text{dB})} = 10\log_{10}(0.5) = (10)(-0.30) = -3.0\text{ dB}$$

c.
$$A_P = \frac{P_{out}}{P_{in}} = \frac{12\text{ mW}}{6\text{ mW}} = 2$$
$$A_{P(\text{dB})} = 10\log_{10}2 = (10)(0.30) = 3.0\text{ dB}$$

d.
$$A_P = \frac{P_{out}}{P_{in}} = \frac{2.5\text{ mW}}{25\text{ mW}} = 0.10$$
$$A_{P(\text{dB})} = 10\log_{10}0.10 = (10)(-1) = -10\text{ dB}$$

The previous example illustrates that if the power is increased or decreased by a factor of two, the resultant power gain will be either +3 dB or −3 dB, respectively. You will recall that in the previous chapter a similar reference was made when the half-power frequencies of a resonant circuit were referred to as the 3-dB down frequencies.

The voltage gain of a system may also be expressed in dB. In order to derive the expression for voltage gain, we first assume that the input resistance and the load resistance are the same value. Then, using the definition of power gain as given by Equation 22–3, we have the following:

$$A_{P(\text{dB})} = 10\log_{10}\frac{P_{out}}{P_{in}}$$
$$= 10\log_{10}\frac{V_{out}^2/R}{V_{in}^2/R}$$
$$= 10\log_{10}\left(\frac{V_{out}}{V_{in}}\right)^2$$

which gives

$$A_{P(\text{dB})} = 20\log_{10}\frac{V_{out}}{V_{in}}$$

Because the preceding expression represents a decibel equivalent of the voltage gain, we simply write the voltage gain in dB as follows:

$$A_{v(\text{dB})} = 20\log_{10}\frac{V_{out}}{V_{in}} \qquad (22\text{–}4)$$

EXAMPLE 22–2

The amplifier circuit of Figure 22–2 has the given conditions. Calculate the voltage gain and power gain in dB.

$Z_{in} = 10\ \text{k}\Omega$
$R_L = 600\ \Omega$
$V_{in} = 20\ \text{mV}_{rms}$
$V_{out} = 500\ \text{mV}_{rms}$

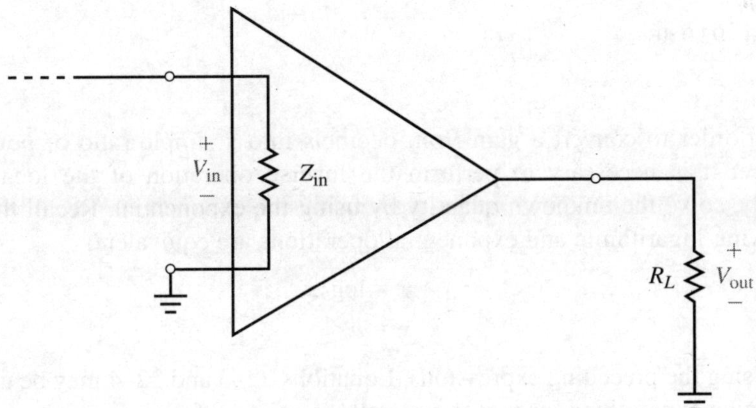

FIGURE 22–2

Solution The voltage gain of the amplifier is

$$A_v = 20 \log_{10}\frac{V_{out}}{V_{in}}$$

$$= 20 \log_{10}\frac{500\ \text{mV}}{20\ \text{mV}} = 20 \log_{10}25 = 28.0\ \text{dB}$$

The signal power available at the input of the amplifier is

$$P_{in} = \frac{V_{in}^2}{Z_{in}} = \frac{(20\ \text{mV})^2}{10\ \text{k}\Omega} = 0.040\ \mu\text{W}$$

The signal at the output of the amplifier has a power of

$$P_{out} = \frac{V_{out}^2}{R_L} = \frac{(500\ \text{mV})^2}{600\ \Omega} = 416.7\ \mu\text{W}$$

The power gain of the amplifier is

$$A_P = 10 \log_{10}\frac{P_{out}}{P_{in}}$$

$$= 10 \log_{10}\frac{416.7\ \mu\text{W}}{0.040\ \mu\text{W}}$$

$$= 10 \log_{10}(10\,417) = 40.2\ \text{dB}$$

PRACTICE PROBLEMS 1

Calculate the voltage gain and power gain (in dB) for the amplifier of Figure 22–2, given the following conditions:

$Z_{in} = 2\ k\Omega$
$R_L = 5\ \Omega$
$V_{in} = 16\ \mu V_{rms}$
$V_{out} = 32\ \mu V_{rms}$

Answers
66.0 dB, 92.0 dB

In order to convert a gain from decibels into a simple ratio of power or voltage, it is necessary to perform the inverse operation of the logarithm; namely, solve the unknown quantity by using the exponential. Recall that the following logarithmic and exponential operations are equivalent:

$$y = \log_b x$$
$$x = b^y$$

Using the preceding expressions, Equations 22–3 and 22–4 may be used to determine expressions for power and voltage gains as follows:

$$\frac{A_{P(\text{dB})}}{10} = \log_{10}\frac{P_{out}}{P_{in}}$$
$$\frac{P_{out}}{P_{in}} = 10^{A_{P(\text{dB})}/10} \qquad \textbf{(22–5)}$$

$$\frac{A_{v(\text{dB})}}{20} = \log_{10}\frac{V_{out}}{V_{in}}$$
$$\frac{V_{out}}{V_{in}} = 10^{A_{v(\text{dB})}/20} \qquad \textbf{(22–6)}$$

EXAMPLE 22–3

Convert the following from decibels to ratios:

a. $A_P = 25$ dB
b. $A_P = -6$ dB
c. $A_v = 10$ dB
d. $A_v = -6$ dB

Solution

a.
$$A_P = \frac{P_{out}}{P_{in}} = 10^{A_{P(\text{dB})}/10}$$
$$= 10^{25/10} = 316$$

b.
$$A_P = 10^{A_{P(\text{dB})}/10}$$
$$= 10^{-6/10} = 0.251$$

c.

$$A_v = \frac{V_{out}}{V_{in}} = 10^{A_{v(dB)}/20}$$
$$= 10^{10/20} = 3.16$$

d.

$$A_v = 10^{A_{v(dB)}/20}$$
$$= 10^{-6/20} = 0.501$$

Applications of Decibels

Decibels were originally intended as a measure of changes in acoustical levels. The human ear is not a linear instrument; rather, it responds to sounds in a logarithmic fashion. Because of this peculiar phenomenon, a tenfold increase in sound intensity results in a perceived doubling of sound. This means that if we wish to double the sound heard from a 10-W power amplifier, we must increase the output power to 100 W.

The minimum sound level which may be detected by the human ear is called the threshold of hearing and is usually taken to be $I_0 = 1 \times 10^{-12}$ W/m². Table 22–1 lists approximate sound intensities of several common sounds. The decibel levels are determined from the expression

$$\beta_{(dB)} = 10 \log_{10}\frac{I}{I_0}$$

Some electronic circuits operate with very small power levels. These power levels may be referenced to some arbitrary level and then expressed in decibels in a manner similar to how sound intensities are represented. For instance, power levels may be referenced to a standard power of 1 mW. In such cases, the power level is expressed in dBm and is determined as

$$P_{dBm} = 10 \log_{10}\frac{P}{1\text{ mW}} \tag{22–7}$$

If a power level is referenced to a standard of 1 W, then we have

$$P_{dBW} = 10 \log_{10}\frac{P}{1\text{ W}} \tag{22–8}$$

TABLE 22–1 Intensity Levels of Common Sounds

Sound	Intensity Level (dB)	Intensity (W/m²)
threshold of hearing, I_0	0	10^{-12}
virtual silence	10	10^{-11}
quiet room	20	10^{-10}
watch ticking at 1 m	30	10^{-9}
quiet street	40	10^{-8}
quiet conversation	50	10^{-7}
quiet motor at 1 m	60	10^{-6}
busy traffic	70	10^{-5}
door slamming	80	10^{-4}
busy office room	90	10^{-3}
jackhammer	100	10^{-2}
motorcycle	110	10^{-1}
loud indoor rock concert	120	1
threshold of pain	130	10

EXAMPLE 22–4

Express the following powers in dBm and in dBW:

a. $P_1 = 0.35\ \mu\text{W}$

b. $P_2 = 20\ \text{mW}$

c. $P_3 = 1000\ \text{W}$

d. $P_4 = 1\ \text{pW}$

Solution

a. $$P_{1(\text{dBm})} = 10 \log_{10}\frac{0.35\ \mu\text{W}}{1\ \text{mW}} = -34.6\ \text{dBm}$$

$$P_{1(\text{dBW})} = 10 \log_{10}\frac{0.35\ \mu\text{W}}{1\ \text{W}} = -64.6\ \text{dBW}$$

b. $$P_{2(\text{dBm})} = 10 \log_{10}\frac{20\ \text{mW}}{1\ \text{mW}} = 13.0\ \text{dBm}$$

$$P_{2(\text{dBW})} = 10 \log_{10}\frac{20\ \text{mW}}{1\ \text{W}} = -17.0\ \text{dBW}$$

c. $$P_{3(\text{dBm})} = 10 \log_{10}\frac{1000\ \text{W}}{1\ \text{mW}} = 60\ \text{dBm}$$

$$P_{3(\text{dBW})} = 10 \log_{10}\frac{1000\ \text{W}}{1\ \text{W}} = 30\ \text{dBW}$$

d. $$P_{4(\text{dBm})} = 10 \log_{10}\frac{1\ \text{pW}}{1\ \text{mW}} = -90\ \text{dBm}$$

$$P_{4(\text{dBW})} = 10 \log_{10}\frac{1\ \text{pW}}{1\ \text{W}} = -120\ \text{dBW}$$

Many voltmeters have a separate scale calibrated in decibels. In such cases, the voltage expressed in dBV uses 1 V_{rms} as the reference voltage. In general, any voltage reading may be expressed in dBV as follows:

$$V_{\text{dBV}} = 20 \log_{10}\frac{V_{\text{out}}}{1\ \text{V}} \tag{22–9}$$

PRACTICE PROBLEMS 2

Consider the resistors of Figure 22–3.

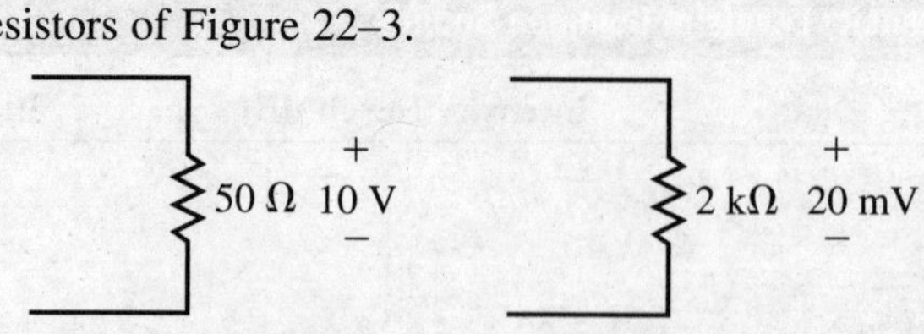

FIGURE 22–3

a. Determine the power levels in dBm and in dBW.

b. Express the voltages in dBV.

Answers

a. 33.0 dBm (3.0 dBW), −37.0 dBm (−67.0 dBW)

b. 20.0 dBV, −34.0 dBV

22.2 Multistage Systems

Quite often a system consists of several stages. In order to find the total voltage gain or power gain of the system, we would need to solve for the product of the individual gains. The use of decibels makes the solution of a multistage system easy to find. If the gain of each stage is given in decibels, then the resultant gain is simply determined as the summation of the individual gains.

Consider the system of Figure 22–4, which represents a three-stage system.

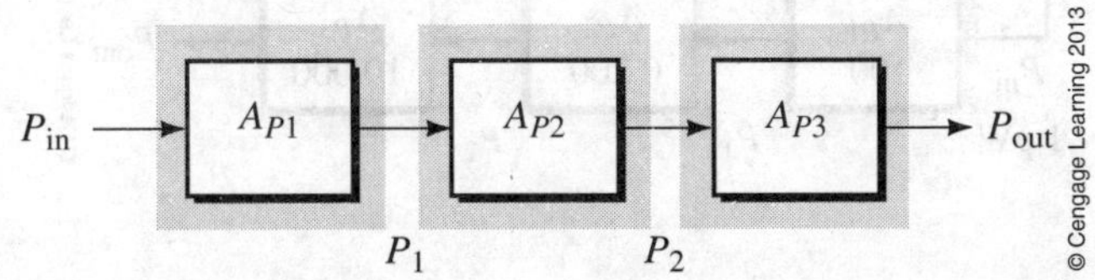

FIGURE 22–4

The power at the output of each stage is determined as follows:

$$P_1 = A_{P1}\, P_{\text{in}}$$
$$P_2 = A_{P2}\, P_1$$
$$P_{\text{out}} = A_{P3}\, P_2$$

The total power gain of the system is found as

$$\begin{aligned} A_{PT} &= \frac{P_{\text{out}}}{P_{\text{in}}} = \frac{A_{P3}\, P_2}{P_{\text{in}}} \\ &= \frac{A_{P3}\,(A_{P2}\, P_1)}{P_{\text{in}}} \\ &= \frac{A_{P3}\, A_{P2}\,(A_{P1} P_{\text{in}})}{P_{\text{in}}} \\ &= A_{P1}\, A_{P2}\, A_{P3} \end{aligned}$$

In general, for n stages the total power gain is found as the product:

$$A_{PT} = A_{P1}\, A_{P2} \cdots A_{Pn} \tag{22–10}$$

However, if we use logarithms to solve for the gain in decibels, we have the following:

$$\begin{aligned} A_{PT}(\text{dB}) &= 10 \log_{10} A_{PT} \\ &= 10 \log_{10}\,(A_{P1}\, A_{P2} \cdots A_{Pn}) \\ A_{PT}(\text{dB}) &= 10 \log_{10} A_{P1} + 10 \log_{10} A_{P2} + \cdots + 10 \log_{10} A_{Pn} \end{aligned}$$

The total power gain in decibels for n stages is determined as the summation of the individual decibel power gains:

$$A_{P(\text{dB})} = A_{P1(\text{dB})} + A_{P2(\text{dB})} + \cdots + A_{Pn(\text{dB})} \tag{22–11}$$

The advantage of using decibels in solving for power gains and power levels is illustrated in the following example.

EXAMPLE 22–5

The block diagram of Figure 22–5 represents the first three stages of a typical AM or FM receiver.

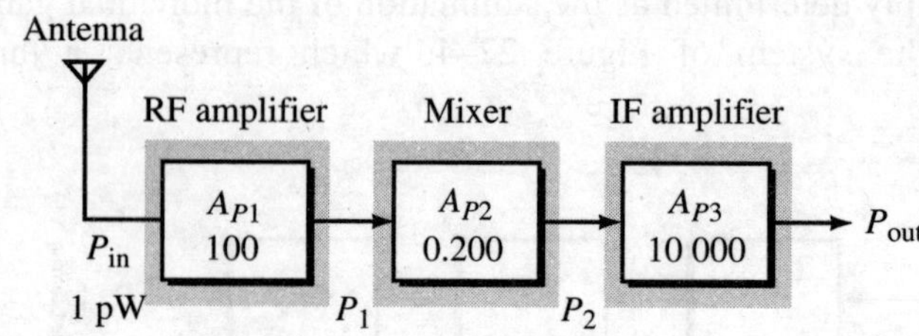

FIGURE 22–5

Find the following quantities:

a. $A_{P1(dB)}$, $A_{P2(dB)}$, and $A_{P3(dB)}$
b. $A_{PT(dB)}$
c. P_1, P_2, and P_{out}
d. $P_{in(dBm)}$, $P_{1(dBm)}$, $P_{2(dBm)}$, and $P_{out(dBm)}$

Solution

a.
$$A_{P1(dB)} = 10\log_{10}A_{P1} = 10\log_{10}100 = 20\text{ dB}$$
$$A_{P2(dB)} = \log_{10}A_{P2} = 10\log_{10}0.2 = -7.0\text{ dB}$$
$$A_{P3(dB)} = 10\log_{10}A_{P3} = 10\log_{10}(10\,000) = 40\text{ dB}$$

b.
$$A_{PT(dB)} = A_{P1(dB)} + A_{P2(dB)} + A_{P3(dB)}$$
$$= 20\text{ dB} - 7.0\text{ dB} + 40\text{ dB}$$
$$= 53.0\text{ dB}$$

c.
$$P_1 = A_{P1}P_{in} = (100)(1\text{ pW}) = 100\text{ pW}$$
$$P_2 = A_{P2}P_1 = (100\text{ pW})(0.2) = 20\text{ pW}$$
$$P_{out} = A_{P3}P_2 = (10\,000)(20\text{ pW}) = 0.20\ \mu\text{W}$$

d.
$$P_{in(dBm)} = 10\log_{10}\frac{P_{in}}{1\text{ mW}} = 10\log_{10}\frac{1\text{ pW}}{1\text{ mW}} = -90\text{ dBm}$$
$$P_{1(dBm)} = 10\log_{10}\frac{P_1}{1\text{ mW}} = 10\log_{10}\frac{100\text{ pW}}{1\text{ mW}} = -70\text{ dBm}$$
$$P_{2(dBm)} = 10\log_{10}\frac{P_2}{1\text{ mW}} = 10\log_{10}\frac{20\text{ pW}}{1\text{ mW}} = -77.0\text{ dBm}$$
$$P_{out(dBm)} = 10\log_{10}\frac{P_{out}}{1\text{ mW}} = 10\log_{10}\frac{0.20\ \mu\text{W}}{1\text{ mW}} = -37.0\text{ dBm}$$

Notice that the power level (in dBm) at the output of any stage is easily determined as the sum of the input power level (in dBm) and the gain of the stage (in dB). It is for this reason that many communication circuits express power levels in decibels rather than in watts.

PRACTICE PROBLEMS 3

Calculate the power level at the output of each of the stages in Figure 22–6.

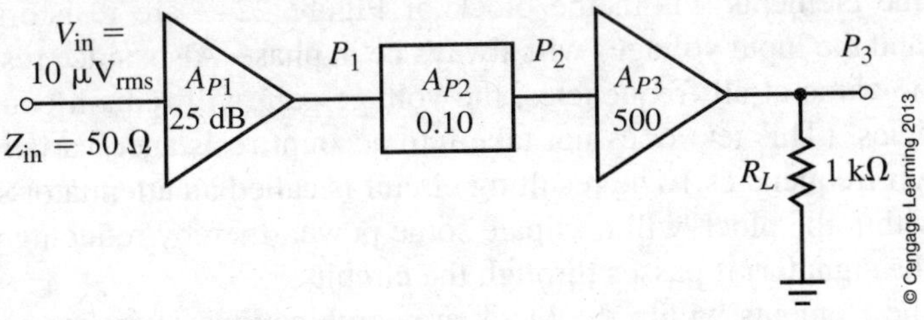

FIGURE 22–6

Answers
$P_1 = -62$ dBm, $P_2 = -72$ dBm, $P_3 = -45$ dBm

IN-PROCESS LEARNING CHECK 1

(Answers are at the end of the chapter.)

1. Given that an amplifier has a power gain of 25 dB, calculate the output power level (in dBm) for the following input characteristics:
   a. $V_{in} = 10\ mV_{rms}$, $Z_{in} = 50\ \Omega$
   b. $V_{in} = 10\ mV_{rms}$, $Z_{in} = 1\ k\Omega$
   c. $V_{in} = 400\ \mu V_{rms}$, $Z_{in} = 200\ \Omega$
2. Given amplifiers with the following output characteristics, determine the output voltage (in volts rms).
   a. $P_{out} = 8.0$ dBm, $R_L = 50\ \Omega$
   b. $P_{out} = -16.0$ dBm, $R_L = 2\ k\Omega$
   c. $P_{out} = -16.0$ dBm, $R_L = 5\ k\Omega$

22.3 Simple *RC* and *RL* Transfer Functions

Electronic circuits usually operate in a highly predictable fashion. If a certain signal is applied at the input of a system, the output will be determined by the physical characteristics of the circuit. The frequency of the incoming signal is one of the many physical conditions that determine the relationship between a given input signal and the resulting output. Although outside the scope of this text, other conditions that may determine the relationship between the input and output signals of a given circuit are temperature, light, and radiation.

For any system subjected to a sinusoidal input voltage, as shown in Figure 22–7, we define the **transfer function** as the ratio of the output voltage phasor to the input voltage phasor for any frequency ω (in radians per second).

$$\mathbf{TF}(\omega) = \frac{\mathbf{V}_{out}}{\mathbf{V}_{in}} = A_v\angle\theta \tag{22–12}$$

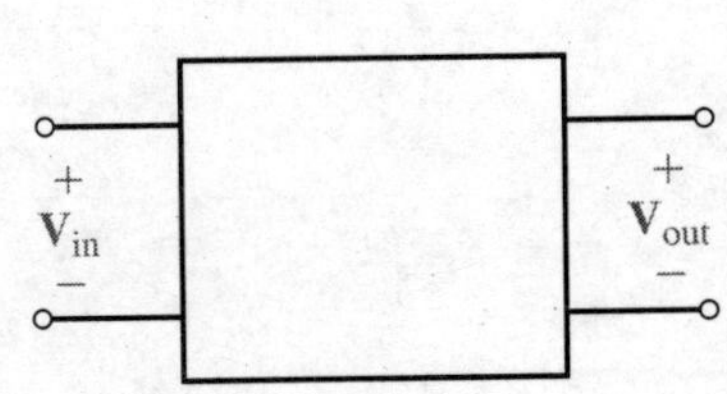

$$\mathbf{TF} = \frac{\mathbf{V}_{out}}{\mathbf{V}_{in}} = A_v\angle\,\theta$$

FIGURE 22–7

Notice that the definition of transfer function is almost the same as voltage gain. The difference is that the transfer function takes into account both amplitude and phase shift of the voltages, whereas voltage gain is a comparison of only the amplitudes.

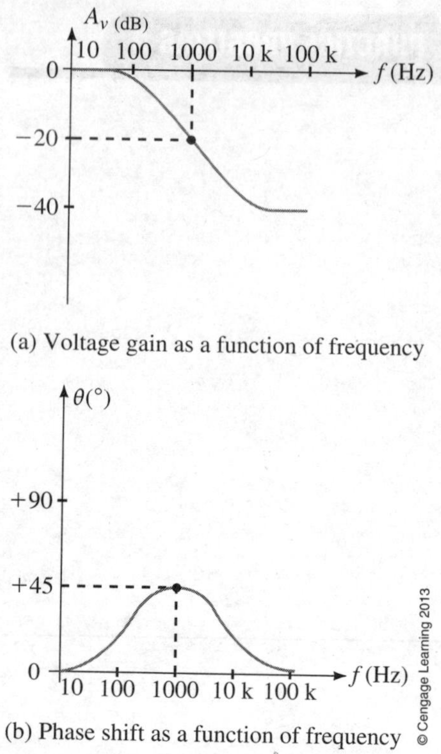

(a) Voltage gain as a function of frequency

(b) Phase shift as a function of frequency

FIGURE 22–8 Frequency response of a circuit.

From Equation 22–12 we see that the amplitude of the transfer function is in fact the voltage gain, a scalar quantity. The phase angle, θ, represents the phase shift between the input and output voltage phasors. The angle θ will be positive if the output leads the input and negative if the output lags the input waveform.

If the elements within the block of Figure 22–7 are resistors, then the output and the input voltages will always be in phase. Also, since resistors have the same value at all frequencies, the voltage gain will remain constant at all frequencies. (This text does not take into account resistance variations due to very high frequencies.) The resulting circuit is called an attenuator since resistance within the block will dissipate some power, thereby reducing (or attenuating) the signal as it passes through the circuit.

If the elements within the block are combinations of resistors, inductors, and capacitors, then the output voltage and phase will depend on frequency since the impedances of inductors and capacitors are frequency dependent. Figure 22–8 illustrates an example of how voltage gain and phase shift of a circuit may change as a function of frequency.

In order to examine the operation of the circuit over a wide range of frequencies, the abscissa (horizontal axis) is usually shown as a logarithmic scale. The ordinate (vertical axis) is generally shown as a linear scale in decibels or degrees. The semilogarithmic graphs showing the frequency response of filters are referred to as **Bode plots** and are extremely useful in predicting and understanding the operation of filters. Similar graphs are used to predict the operation of amplifiers and many other electronic components.

By examining the frequency response of a circuit, we are able to determine at a glance the voltage gain (in dB) and phase shift (in degrees) for any sinusoidal input at a given frequency. For example, at a frequency of 1000 Hz, the voltage gain is −20 dB and the phase shift is 45°. This means that the output signal is one-tenth as large as the input signal and the output leads the input by 45°.

From the frequency response of Figure 22–8, we see that the circuit that corresponds to this response is able to pass low-frequency signals, while at the same time partially attenuating high-frequency signals. Any circuit that passes a particular range of frequencies while blocking others is referred to as a **filter circuit.** Filter circuits are usually named according to their function, although certain filters are named after their inventors. The filter having the response of Figure 22–8 is referred to as a stepfilter, since the voltage gain occurs between two limits (steps). Other types of filters that are used regularly in electrical and electronic circuits include low-pass, high-pass, band-pass, and band-reject filters.

Although the design of filters is a topic in itself, we will examine Bode plots of a few common types of filters that are used in electronic circuits.

Sketching Bode Plots

In order to understand the operation of a filter (or any other frequency-dependent system) it is helpful to first sketch the Bode plots. All first-order systems (namely those consisting of *R-C* or *R-L* combinations) have transfer functions that are constructed from only four possible forms (and perhaps a constant). If voltage gain is expressed in decibels, we find that the Bode plots of more complex expressions are simply determined as the arithmetic sum of these simple forms. We begin by examining the four possible forms, and then we will combine these forms to determine the frequency responses of more complicated transfer functions. In each of the forms, the value of tau (τ) is simply a time constant determined by the components of the circuit.

One important consideration of transfer functions is that the boundary conditions must always be satisfied. The boundary conditions of a transfer function or filter circuit are determined by examining the characteristics when $\omega = 0$ rad/s and as $\omega \to \infty$.

1. $\mathbf{TF} = j\omega\tau = \omega\tau\angle 90°$

   The voltage gain of the preceding transfer function is $A_V = \omega\tau$. By examining the boundary conditions, we see that the voltage gain is $A_v = 0$ when $\omega = 0$ and $A_v = \infty$ as $\omega \to \infty$. (Although a transfer function consisting of only this term cannot occur for real systems, it is an important component when combined with other terms.) In decibels, the voltage gain becomes $[A_V]_{dB} = 20 \log \omega\tau$. Notice that the voltage gain of this expression increases as the frequency is increased.

   At a frequency of $\omega = 1/\tau$, the transfer function is evaluated as $\mathbf{TF} = j1 = 1\angle 90°$. The voltage gain is $A_V = 1$, which is equivalent to $[A_V]_{dB} = 20 \log 1 = 0$ dB.

   If the frequency is increased by a factor of 10 (increasing by one **decade**) to $\omega = 10/\tau$, the transfer function becomes $\mathbf{TF} = j10 = 10\angle 90°$. The voltage gain of $A_V = 10$ is equivalent to $[A_V]_{dB} = 20 \log 10 = 20$ dB. This illustrates that the gain of the transfer function increases at a rate of 20 dB/decade. Had the frequency been doubled (increasing by one **octave**), the transfer function would have been $\mathbf{TF} = j2 = 2\angle 90°$. The voltage gain $A_V = 2$ is equivalent to $[A_V]_{dB} = 20 \log 2 = 6$ dB. This illustrates that the gain of the transfer function increases at a rate of 6 dB/octave. (Hence, a slope of 20 dB/decade is equivalent to 6 dB/octave.)

   For the given transfer function, the phase angle (of the output voltage relative to the input voltage) is a constant 90°. Consequently, a sinusoidal signal applied at the input will result in an output signal that always leads the input by 90°.

   The Bode plots for the transfer function appear as shown in Figure 22–9.

2. $\mathbf{TF} = 1 + j\omega\tau$

   Examining the preceding expression, we determine that the magnitude of this transfer function also increases as ω is increased. However, unlike the previous transfer function form, the magnitude of this transfer function can never be less than 1. For small values of frequency, ω the real term "swamps" the imaginary term, resulting in $\mathbf{TF} \approx 1 + j0 = 1\angle 0°$. For large values of frequency the opposite happens; the imaginary term "swamps" the real term, resulting in $\mathbf{TF} \approx j\omega\tau = \omega\tau\angle 90°$.

   Now, the question is "*What do we mean by a small frequency and large frequency?*" Let's examine what happens at a frequency of $\omega_c = 1/\tau$.

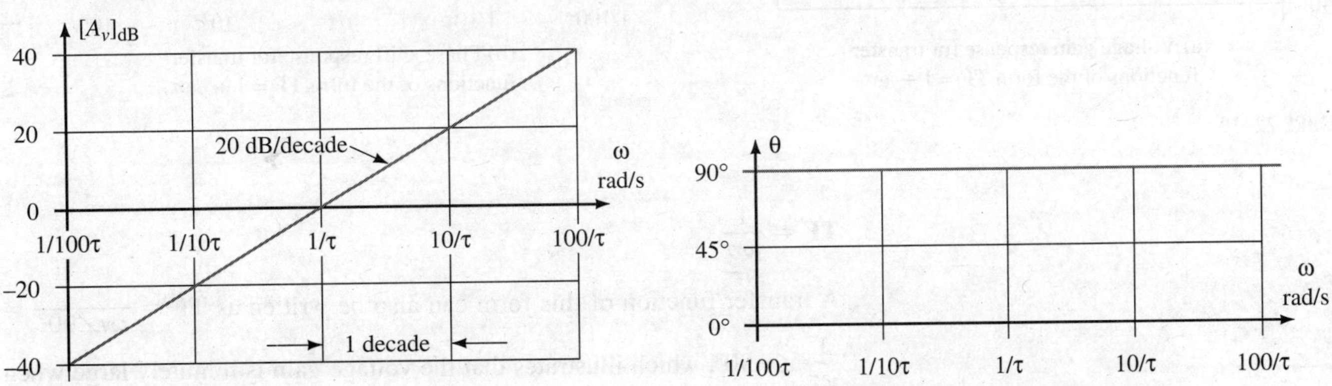

(a) Voltage gain response for transfer functions of the form $\mathbf{TF} = j\omega\tau$

(b) Phase shift response for transfer functions of the form $\mathbf{TF} = j\omega\tau$

FIGURE 22–9

At this frequency, the transfer function simply becomes **TF** $= 1 + j1 = \sqrt{2}\angle 45°$. The voltage gain is equivalent to $A_v = 20 \log\sqrt{2} = 3.0$ dB. The frequency $\omega_c = 1/\tau$ is often referred to as the **cutoff frequency,** the **critical frequency,** or the **break frequency.** If we were to decrease the frequency so that it was one-tenth of the cutoff frequency we would have **TF** $= 1 + j0.1 = 1.005\angle 5.71°$. Two very important characteristics become evident for frequencies below the cutoff frequency:

i. The voltage gain is essentially constant at $A_v = 1 \equiv 0$ dB and

ii. The phase shift below $0.1\omega_c$ is essentially constant at 0°.

Next, let's examine what happens at a frequency $\omega = 10\omega_c$. At this higher frequency, the transfer function becomes **TF** $= 1 + j10 = 10.05\angle 84.29°$. The voltage gain is equivalent to $A_v = 20 \log 10.05 = 20.04$ dB. Further analysis would show that at $\omega = 100\ \omega_c$, the voltage gain is 40 dB. Two very important characteristics become evident for frequencies above the cutoff frequency:

i. The voltage gain rises at a rate of 20 dB/decade $\equiv$ 6 dB/octave and

ii. The phase shift above $10\omega_c$ is essentially constant at 90°.

When sketching the Bode plots for this transfer function, we approximate the voltage gain response as 0 dB for all frequencies less than $\omega_c = 1/\tau$ and rising at a rate of 20 dB/decade for all frequencies above $\omega_c = 1/\tau$. The phase shift response is a bit more complicated. The phase shift is approximated as 0° for $\omega < 0.1\omega_c$ and as 90° for $\omega > 10\omega_c$. In the region between $0.1\omega_c$ and $10\omega_c$, the phase shift is approximated as 45°/decade. Figure 22–10 illustrates both the straight-line approximations and the actual curves for a transfer function of the form **TF** $= 1 + j\omega\tau$.

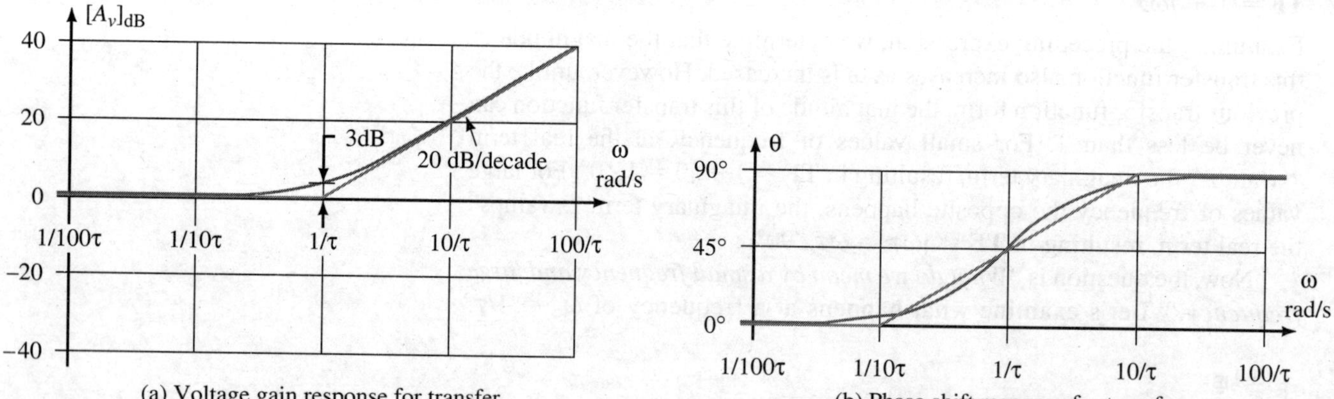

(a) Voltage gain response for transfer functions of the form **TF** = 1 + $j\omega\tau$

(b) Phase shift response for transfer functions of the form **TF** = 1 + $j\omega\tau$

FIGURE 22–10

3. $\mathbf{TF} = \dfrac{1}{j\omega\tau}$

A transfer function of this form can also be written as $\mathbf{TF} = \dfrac{1}{\omega\tau\angle 90°} = \dfrac{1}{\omega\tau}\angle -90°$, which illustrates that the voltage gain is infinitely large when $\omega = 0$ and decreases as the frequency increases. Due to the j-operator in the denominator, it is also apparent that the phase shift is −90° for all frequencies. (Once again, no real system of electronic components will have a transfer function consisting of only this term.) The Bode plots for transfer functions of the type $\mathbf{TF} = \frac{1}{j\omega\tau}$ are shown in Figure 22–11.

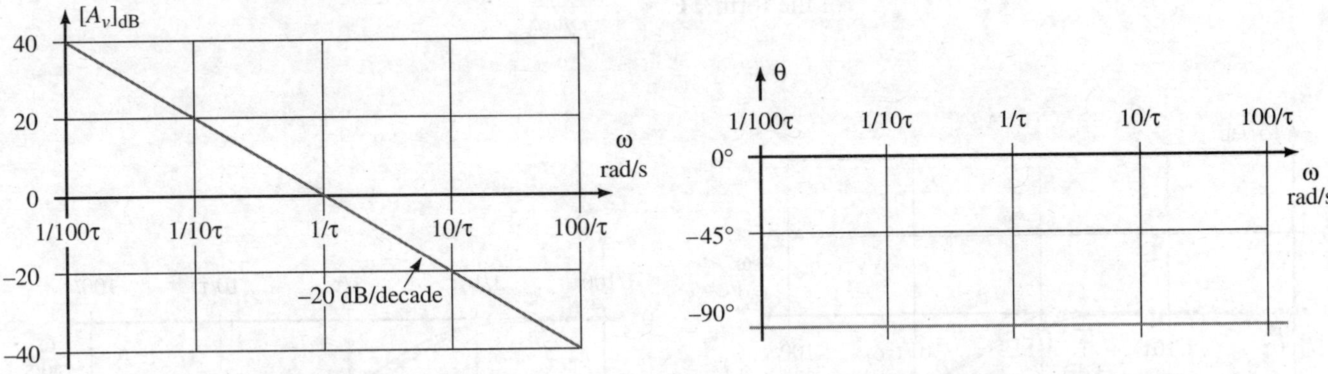

(a) Voltage gain response for transfer functions of the form $\mathbf{TF} = \frac{1}{j\omega\tau}$

(b) Phase shift response for transfer functions of the form $\mathbf{TF} = \frac{1}{j\omega\tau}$

FIGURE 22–11

4. $\mathbf{TF} = \frac{1}{1 + j\omega\tau}$

The magnitude of this transfer function decreases as ω is increased. However, unlike the previous transfer function form, the voltage gain of this transfer function can never be greater than 1. For small values of ω the real term "swamps" the imaginary term, resulting in $\mathbf{TF} \approx \frac{1}{1 + j0} = \frac{1}{1\angle 0^\circ} = 1\angle 0^\circ$. For large values of frequency the opposite happens; the imaginary term "swamps" the real term, resulting in $\mathbf{TF} \approx \frac{1}{j\omega\tau} = \frac{1}{\omega\tau\angle 90^\circ} = \frac{1}{\omega\tau}\angle -90^\circ$.

Once again, the cutoff frequency occurs at $\omega_c = 1/\tau$. At this frequency, the transfer function results in a value of $\mathbf{TF} = \frac{1}{1 + j1} = \frac{1}{\sqrt{2}\angle 45^\circ} = 0.7071\angle -45^\circ$. At the cutoff frequency, the voltage gain (in decibels) is found to be $A_v = 20 \log 0.7071 = -3.0$ dB. For a filter having this transfer function, we see that the output power at the cutoff frequency is 3 dB down from its maximum value. You will recall that a "gain" of -3 dB results in an output power that is half of the maximum power.

For a filter having a transfer function of the form $\mathbf{TF} = \frac{1}{1 + j\omega\tau}$, two characteristics are evident for frequencies below the cutoff frequency:

i. The voltage gain is essentially constant at $A_v = 1 \equiv 0$ dB and
ii. The phase shift below $0.1\omega_c$ is essentially constant at 0°.

For frequencies above the cutoff frequency:

i. The voltage gain drops at a rate of 20 dB/decade $\equiv$ 6 dB/octave and
ii. The phase shift above $10\omega_c$ is essentially constant at $-90°$.

When sketching the Bode plots for this transfer function, we approximate the voltage gain response as 0 dB for all frequencies less than $\omega_c = 1/\tau$ and decreasing at a rate of 20 dB/decade for all frequencies above $\omega_c = 1/\tau$. The phase shift is approximated as 0° for $\omega < 0.1\omega_c$ and as $-90°$

for $\omega > 10\omega_c$. In the region between $0.1\omega_c$ and $10\omega_c$, the phase shift is approximated as having a slope of $-45°$/decade. Figure 22–12 shows the straight-line approximations and the actual curves for a transfer function of the form $\mathbf{TF} = \dfrac{1}{1 + j\omega\tau}$.

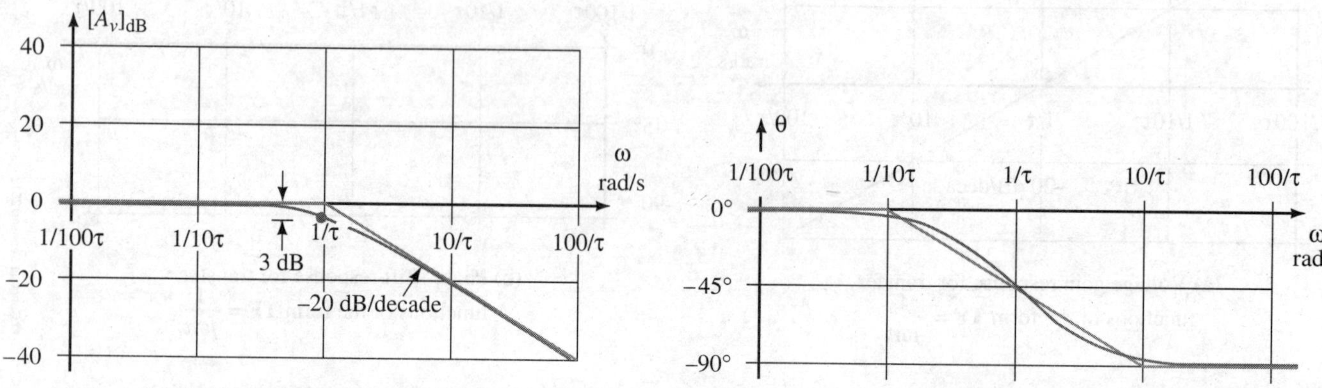

(a) Voltage gain response for transfer functions of the form $\mathbf{TF} = \dfrac{1}{1 + j\omega\tau}$

(b) Phase shift response for transfer functions of the form $\mathbf{TF} = \dfrac{1}{1 + j\omega\tau}$

FIGURE 22–12

Now we are ready to see how these various transfer function forms can be combined to sketch transfer functions of real circuits.

EXAMPLE 22–6

Given $\mathbf{TF} = \dfrac{10}{1 + j0.001\omega}$

a. Determine the boundary conditions.
b. Calculate the cutoff frequency in rad/s and in Hz.
c. Sketch the straight-line approximation of the voltage gain response.
d. Sketch the straight-line approximation of the phase shift response.

Solution

a. When $\omega = 0$, the transfer function simplifies to $\mathbf{TF} = 10 = 10\angle 0°$. The voltage gain will be $A_v = 20 \log 10 = 20$ dB and the phase shift is 0°. As $\omega \to \infty$, the magnitude of the transfer function approaches zero and the phase shift will be $-90°$ (since the real term in the denominator will be swamped by the imaginary term).

b. The cutoff frequency is determined as

$$\omega_c = \frac{1}{\tau} = \frac{1}{0.001} = 1000 \text{ rad/s} \quad (f_c = \omega_c/2\pi = 159.15 \text{ Hz})$$

c. The voltage gain response is shown in Figure 22–13.

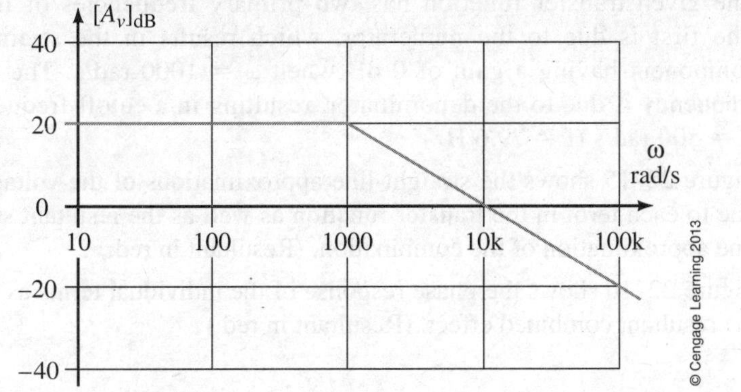

FIGURE 22–13

d. The phase shift response is shown in Figure 22–14.

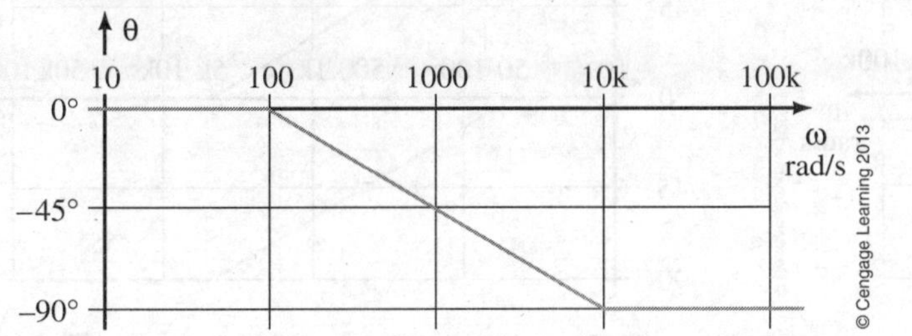

FIGURE 22–14

Note: By examining the response curves for the given transfer function, we see that this represents the characteristics of a low-pass filter. The notable exception is that a passive filter (consisting of only resistors, capacitors, or inductors) cannot have a voltage gain larger than unity ($A_v = 1$). Consequently, we must conclude that the filter includes an active component such as a transistor or an operational amplifier.

EXAMPLE 22–7

Given $\mathbf{TF} = \dfrac{j0.001\omega}{1 + j0.002\omega}$

a. Determine the boundary conditions.
b. Calculate the cutoff frequency in rad/s and in Hz.
c. Sketch the straight-line approximation of the voltage gain response.
d. Sketch the straight-line approximation of the phase shift response.

Solution

a. When $\omega = 0$, the numerator of the transfer function is $j0$ and the denominator is 1. Therefore the entire transfer function simplies to $\mathbf{TF} = j0 = 0\angle 90°$. This indicates that for low frequencies, the voltage gain is very low, and that there is a 90° phase shift. As $\omega \to \infty$, the numerator will be $j0.001\omega$, while the denominator simplifies to $j0.002\omega$ (since the real term is swamped). The reluctant transfer function will be simplified as $\mathbf{TF} = 0.5 = 0.5\angle 0°$. This corresponds to a gain $A_v = 20 \log 0.5 = -6.0$ dB and a phase shift of 0°.

b. The given transfer function has two primary frequencies of interest. The first is due to the numerator, which results in the straight-line component having a gain of 0 dB when $\omega = 1000$ rad/s. The second frequency is due to the denominator, resulting in a cutoff frequency of $\omega = 500$ rad/s (f = 79.6 Hz).

c. Figure 22–15 shows the straight-line approximations of the voltage gain due to each term in the transfer function as well as the resultant straight-line approximation of the combination. (Resultant in red.)

d. Figure 22–16 shows the phase response of the individual terms as well as the resultant combined effect. (Resultant in red.)

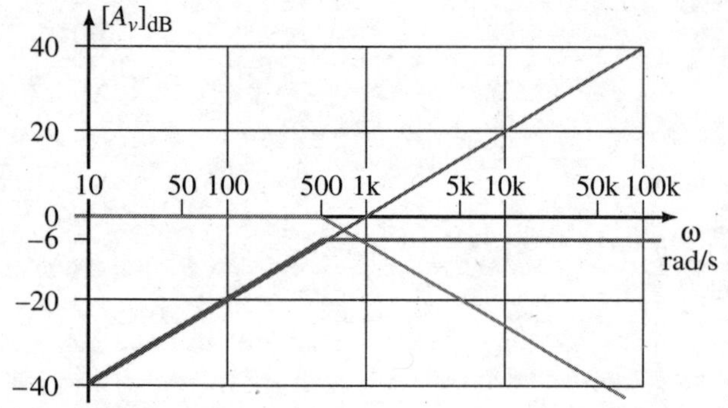

FIGURE 22–15

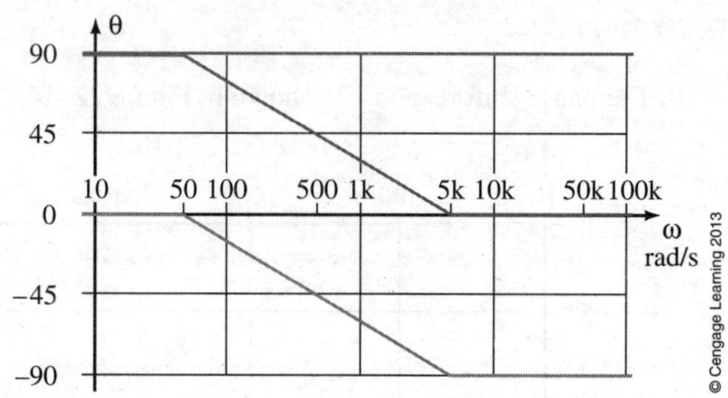

FIGURE 22–16

Note: The results of this example illustrate that frequency response plots of transfer functions are determined by arithmetically combining the results due to each component of the transfer function. Although the resultant plots may occasionally be complicated, it is important to realize that simple addition is used in all cases.

Another important consideration when sketching Bode plots is to remember that the boundary conditions of the frequency response curves must satisfy the calculated boundary conditions for the transfer function. Upon inspection of Figure 22–15, we see that for high frequencies (those above 500 rad/s) the voltage gain is −6 dB. This is precisely the value calculated from the transfer function. Similarly, we see that the phase response curve of Figure 22–16 satisfies the boundary conditions of 90° for $\omega = 0$ and 0° as $\omega \rightarrow \infty$, respectively.

PRACTICE PROBLEMS 4

Sketch the straight-line approximation for each of the transfer functions given.

a. $\mathbf{TF} = 1 + j0.02\omega$

b. $\mathbf{TF} = \dfrac{100}{1 + j0.005\omega}$

c. $\mathbf{TF} = \dfrac{1 + j0.02\omega}{20(1 + j0.002\omega)}$

Answers

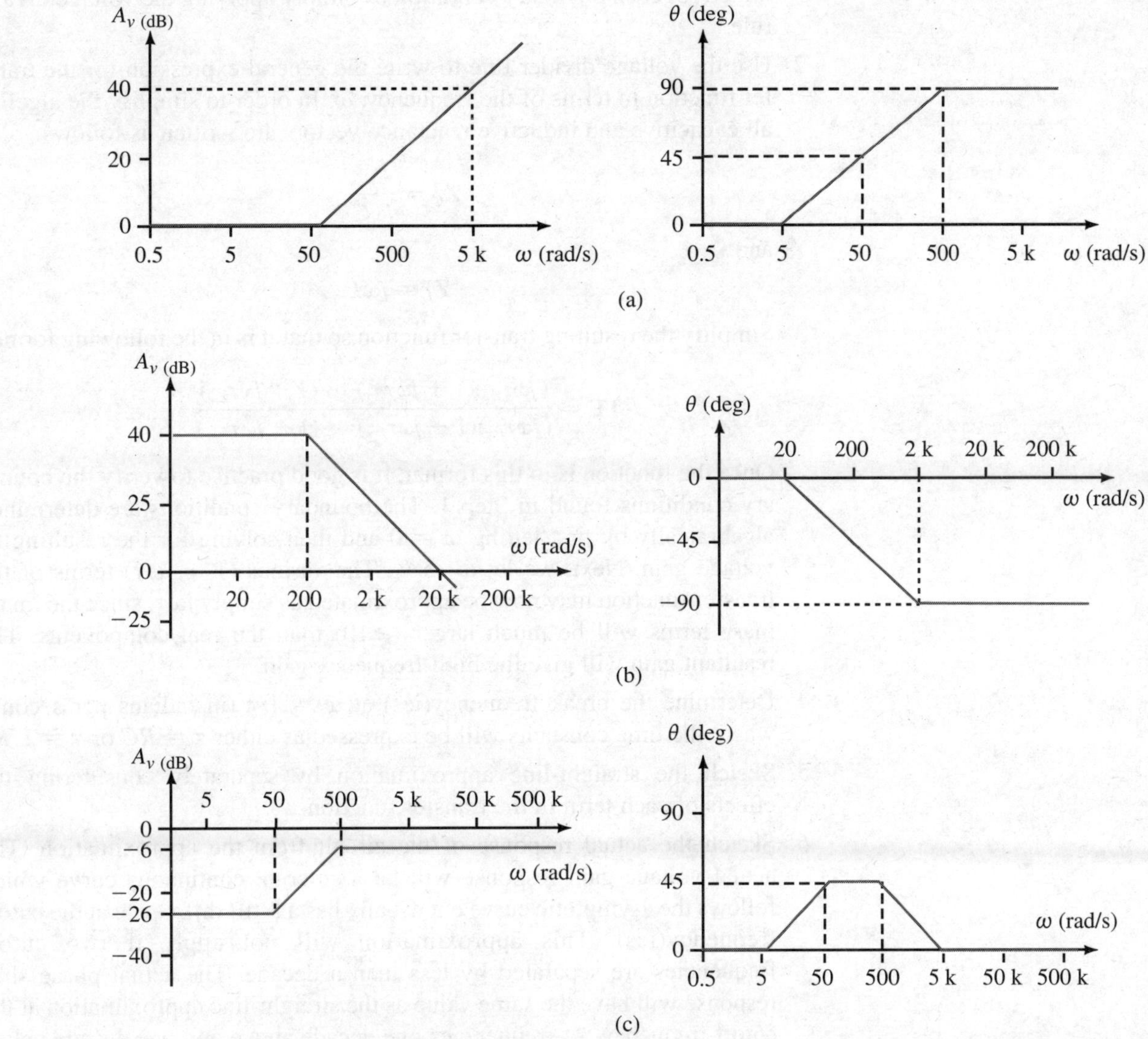

FIGURE 22–17

Writing Transfer Functions

The transfer function of any circuit is found by following a few simple steps. As we have already seen, a properly written transfer function allows us to easily calculate the cutoff frequencies and quickly sketch the corresponding Bode plot for the circuit. The steps are as follows:

1. Determine the boundary conditions for the given circuit by solving for the voltage gain when the frequency is zero (dc) and when the frequency approaches infinity. The boundary conditions are found by using the following approximations:

At $\omega = 0$,	inductors are short circuits, capacitors are open circuits.
At $\omega \to \infty$,	inductors are open circuits, capacitors are short circuits.

By using the preceding approximations, all capacitors and inductors are easily removed from the circuit. The resulting voltage gain is determined for each boundary condition by simply applying the voltage divider rule.

2. Use the voltage divider rule to write the general expression for the transfer function in terms of the frequency, ω. In order to simplify the algebra, all capacitive and inductive reactance vectors are written as follows:

$$\mathbf{Z}_C = \frac{1}{j\omega C}$$

and

$$\mathbf{Z}_L = j\omega L$$

3. Simplify the resulting transfer function so that it is in the following format:

$$\mathbf{TF} = \frac{(j\omega\tau_{Z_1})(1 + j\omega\tau_{Z_2}) \cdots (1 + j\omega\tau_{Z_n})}{(j\omega\tau_{P_1})(1 + j\omega\tau_{P_2}) \cdots (1 + j\omega\tau_{P_m})}$$

Once the function is in this format, it is good practice to verify the boundary conditions found in Step 1. The boundary conditions are determined algebraically by first letting $\omega = 0$ and then solving for the resulting dc voltage gain. Next, we let $\omega \to \infty$. The various $(1 + j\omega\tau)$ terms of the transfer function may now be approximated as simply $j\omega\tau$, since the imaginary terms will be much larger (≥ 10) than the real components. The resultant gain will give the high-frequency gain.

4. Determine the break frequency(ies) at $\omega = 1/\tau$ (in radians per second) where the time constants will be expressed as either $\tau = RC$ or $\tau = L/R$.
5. Sketch the straight-line approximation by separately considering the effects of each term in the transfer function.
6. Sketch the actual response of the circuit from the approximation. The actual voltage gain response will be a smooth, continuous curve which follows the asymptotic curve but usually has a 3-dB difference at the cutoff frequency(ies). This approximation will not apply if two cutoff frequencies are separated by less than a decade. The actual phase shift response will have the same value as the straight-line approximation at the cutoff frequency. At frequencies one decade above and one decade below the cutoff frequency, the actual phase shift will be 5.71° from the straight-line approximation.

These steps will now be used in analyzing several important types of filters.

22.4 The Low-Pass Filter

The *RC* Low-Pass Filter

The circuit of Figure 22–18 is referred to as a low-pass *RC* filter circuit since it permits low-frequency signals to pass from the input to the output while attenuating high-frequency signals.

At low frequencies, the capacitor has a very large reactance. Consequently, at low frequencies the capacitor is essentially an open circuit resulting in the voltage across the capacitor, $\mathbf{V}_{\text{out}}$, to be essentially equal to the applied voltage, $\mathbf{V}_{\text{in}}$.

At high frequencies, the capacitor has a very small reactance, which essentially short-circuits the output terminals. The voltage at the output will therefore approach zero as the frequency increases. Although we are able to easily predict what happens at the two extremes of frequency, called the *boundary conditions,* we do not yet know what occurs between the two extremes.

The circuit of Figure 22–18 is easily analyzed by applying the voltage divider rule. Namely,

$$\mathbf{V}_{\text{out}} = \frac{\mathbf{Z}_C}{\mathbf{R} + \mathbf{Z}_C}\mathbf{V}_{\text{in}}$$

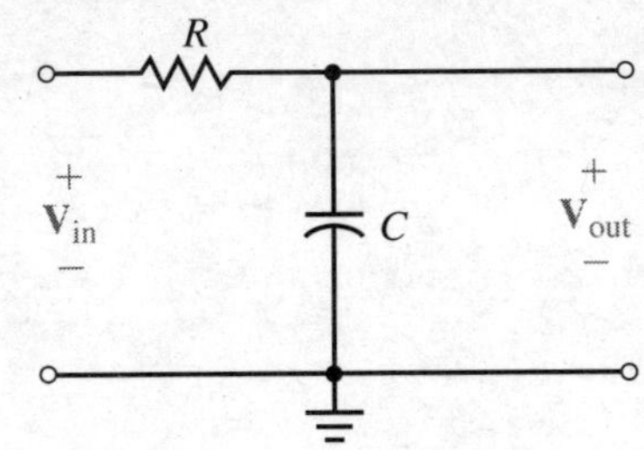

FIGURE 22–18 *RC* low-pass filter.

In order to simplify the algebra, the reactance of a capacitor is expressed as follows:

$$\mathbf{Z}_C = -j\frac{1}{\omega C} = -j\frac{j}{j\omega C} = \frac{1}{j\omega C} \qquad \textbf{(22–13)}$$

The transfer function for the circuit of Figure 22–18 is now evaluated as follows:

$$\mathbf{TF}(\omega) = \frac{\mathbf{V}_{\text{out}}}{\mathbf{V}_{\text{in}}} = \frac{\dfrac{1}{j\omega C}}{R + \dfrac{1}{j\omega C}} = \frac{\dfrac{1}{j\omega C}}{\dfrac{1 + j\omega RC}{j\omega C}}$$

$$= \frac{1}{1 + j\omega RC}$$

We define the cutoff frequency, ω_c, as the frequency at which the output power is equal to half of the maximum output power (3 dB down from the maximum). This frequency occurs when the output voltage has an amplitude which is 0.7071 of the input voltage. For the *RC* circuit, the cutoff frequency occurs at

$$\omega_c = \frac{1}{\tau} = \frac{1}{RC} \qquad \textbf{(22–14)}$$

Then the transfer function is written as

$$\mathbf{TF}(\omega) = \frac{1}{1 + j\dfrac{\omega}{\omega_c}} \qquad \textbf{(22–15)}$$

The previous transfer function results in the Bode plot shown in Figure 22–19.

Notice that the abscissas (horizontal axes) of the graphs in Figure 22–19 are shown as a ratio of ω/ω_c. Such a graph is called a normalized plot, and it eliminates the need to determine the actual cutoff frequency, ω_c. The normalized plot will have the same values for all low-pass *RC* filters. The actual frequency response of the *RC* low-pass filter can be approximated from the straight-line approximation by using the following guidelines:

1. At low frequencies ($\omega/\omega_c \leq 0.1$) the voltage gain is approximately 0 dB with a phase shift of about 0°. This means that the output signal of the filter is very nearly equal to the input signal. The phase shift at $\omega = 0.1\omega_c$ will be 5.71° less than the straight-line approximation.
2. At the cutoff frequency, $\omega_c = 1/RC$ ($f_c = 1/2\pi RC$), the gain of the filter is −3 dB. This means that at the cutoff frequency, the circuit will deliver half the power that it would deliver at very low frequencies. At the cutoff frequency, the output voltage will lag the input voltage by 45°.

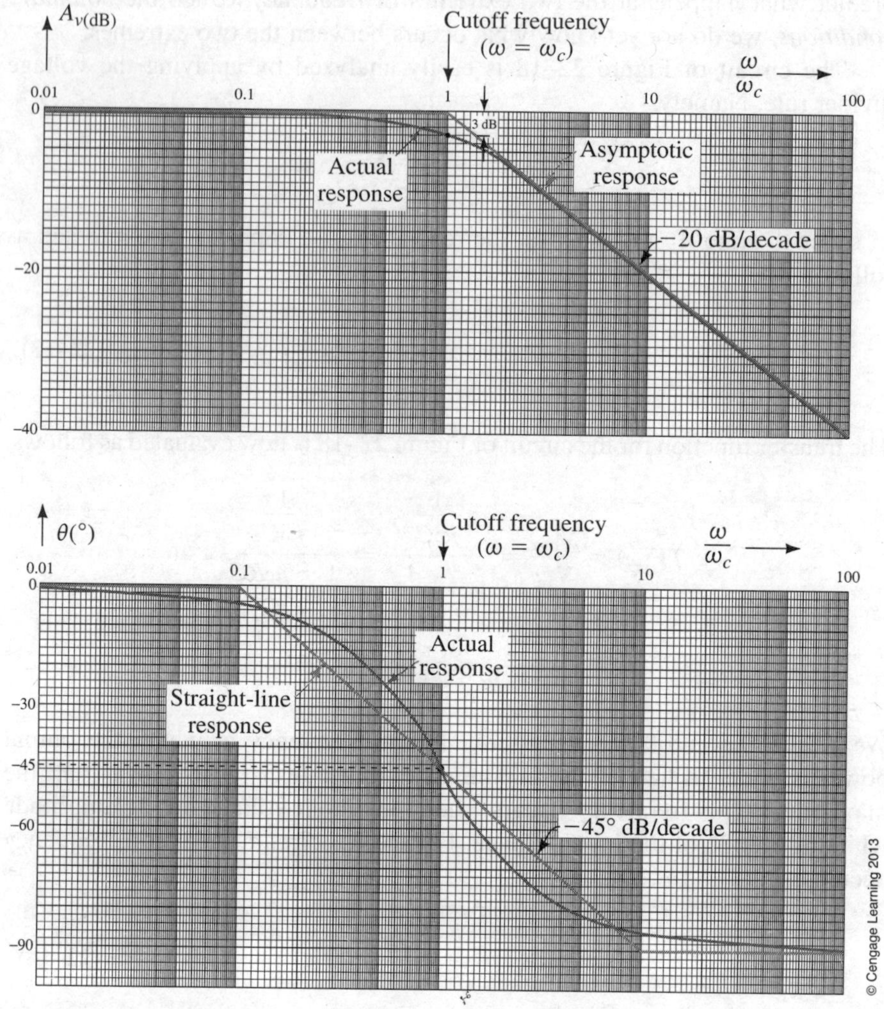

FIGURE 22–19 Normalized frequency response for an *RC* low-pass filter.

3. As the frequency increases beyond the cutoff frequency, the amplitude of the output signal decreases by a factor of approximately 10 for each tenfold increase in frequency; namely the voltage gain is −20 dB per decade. The phase shift at $\omega = 10\omega_c$ will be 5.71° greater than the straight-line approximation, namely at $\theta = -84.29°$. For high frequencies ($\omega/\omega_c \geq 10$), the phase shift between the input and output voltage approaches −90°.

The *RL* Low-Pass Filter

A low-pass filter circuit may be made up of a resistor and an inductor as illustrated in Figure 22–20.

In a manner similar to that used for the *RC* low-pass filter, we may write the transfer function for the circuit of Figure 22–20 as follows:

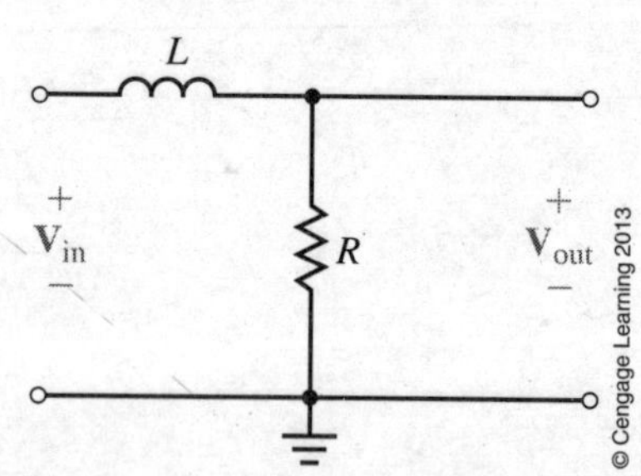

FIGURE 22–20 *RL* low-pass filter.

$$\mathbf{TF} = \frac{\mathbf{V}_{\text{out}}}{\mathbf{V}_{\text{in}}}$$

$$= \frac{\mathbf{R}}{\mathbf{R} + \mathbf{Z}_L} = \frac{R}{R + j\omega L}$$

Now, dividing the numerator and the denominator by R, we have the transfer function expressed as

$$\mathbf{TF} = \frac{1}{1 + j\omega\dfrac{L}{R}}$$

Since the cutoff frequency is found as $\omega_c = 1/\tau$, we have

$$\omega_c = \frac{1}{\tau} = \frac{1}{\dfrac{L}{R}} = \frac{R}{L}$$

and so

$$\mathbf{TF} = \frac{1}{1 + j\dfrac{\omega}{\omega_c}} \qquad \textbf{(22–16)}$$

Notice that the transfer function for the *RL* low-pass circuit in Equation 22–16 is identical to the transfer function of an *RC* circuit in Equation 22–15. In each case, the cutoff frequency is determined as the reciprocal of the time constant.

EXAMPLE 22–8

Sketch the Bode plot showing both the straight-line approximation and the actual response curves for the circuit of Figure 22–21. Show frequencies in hertz.

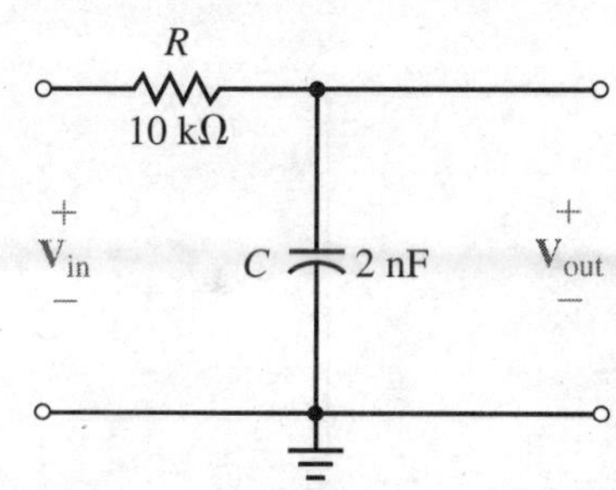

FIGURE 22–21

CircuitSim 22-1

Solution The cutoff frequency (in radians per second) for the circuit occurs at

$$\omega_c = \frac{1}{\tau} = \frac{1}{RC}$$

$$= \frac{1}{(10\text{ k}\Omega)(2\text{ nF})} = 50\text{ krad/s}$$

which gives

$$f_c = \frac{\omega_c}{2\pi} = \frac{50\text{ krad/s}}{2\pi} = 7.96\text{ kHz}$$

In order to sketch the Bode plot, we begin with the asymptotes for the voltage gain response. The circuit will have a flat response until $f_c = 7.96$ kHz. Then the gain will drop at a rate of 20 dB for each decade increase in frequency. Therefore, the voltage gain at 79.6 kHz will be −20 dB, and at 796 kHz the voltage gain will be −40 dB. At the cutoff frequency for the filter, the actual voltage gain response will pass through a point which is 3 dB down from the intersection of the two asymptotes. The frequency response of the voltage gain is shown in Figure 22–22(a).

Next, we sketch the approximate phase shift response. The phase shift at 7.96 kHz will be −45°. At a frequency one decade below the cutoff frequency (at 796 Hz) the phase shift will be approximately equal to zero, while at a frequency one decade above the cutoff frequency (at 79.6 kHz) the phase shift will be near the maximum of −90°. The actual phase shift response will be a curve which varies slightly from the asymptotic response, as shown in Figure 22–22(b).

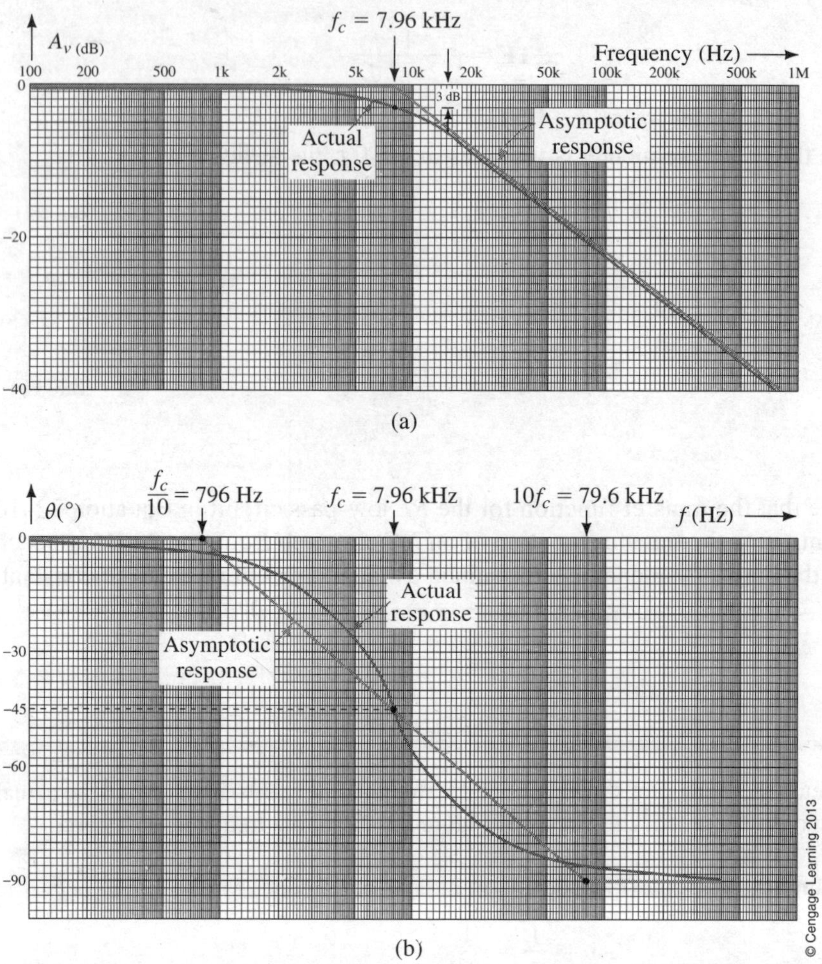

FIGURE 22–22

EXAMPLE 22–9

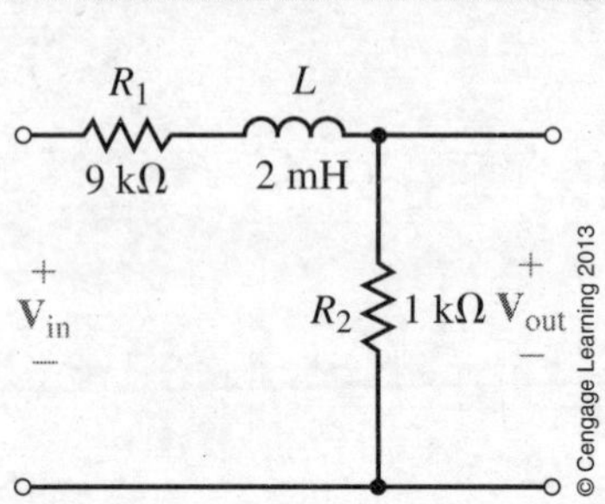

FIGURE 22–23

Consider the low-pass circuit of Figure 22–23.

a. Write the transfer function for the circuit.

b. Sketch the frequency response.

Solution

a. The transfer function of the circuit is found as

$$\mathbf{TF} = \frac{\mathbf{V}_{out}}{\mathbf{V}_{in}} = \frac{R_2}{R_2 + R_1 + j\omega L}$$

which becomes

CircuitSim 22-2

$$\mathbf{TF} = \frac{R_2}{R_1 + R_2}\left(\frac{1}{1 + j\omega\dfrac{L}{R_1 + R_2}}\right)$$

b. From the transfer function of part (a), we see that the dc gain will no longer be 1 (0 dB) but rather is found as

$$A_{v(\text{dc})} = 20\log\left(\frac{R_2}{R_1 + R_2}\right)$$
$$= 20\log\left(\frac{1}{10}\right)$$
$$= -20\text{ dB}$$

The cutoff frequency occurs at

$$\omega_c = \frac{1}{\tau} = \frac{1}{\dfrac{L}{R_1 + R_2}}$$

$$\omega_c = \frac{R_1 + R_2}{L} = \frac{10\text{ k}\Omega}{2\text{ mH}}$$
$$= 5.0\text{ Mrad/s}$$

The resulting Bode plot is shown in Figure 22–24. Notice that the frequency response of the phase shift is precisely the same as for other low-pass filters. However, the response of the voltage gain now starts at −20 dB and then drops at a rate of −20 dB/decade above the cutoff frequency, $\omega_c = 5$ Mrad/s.

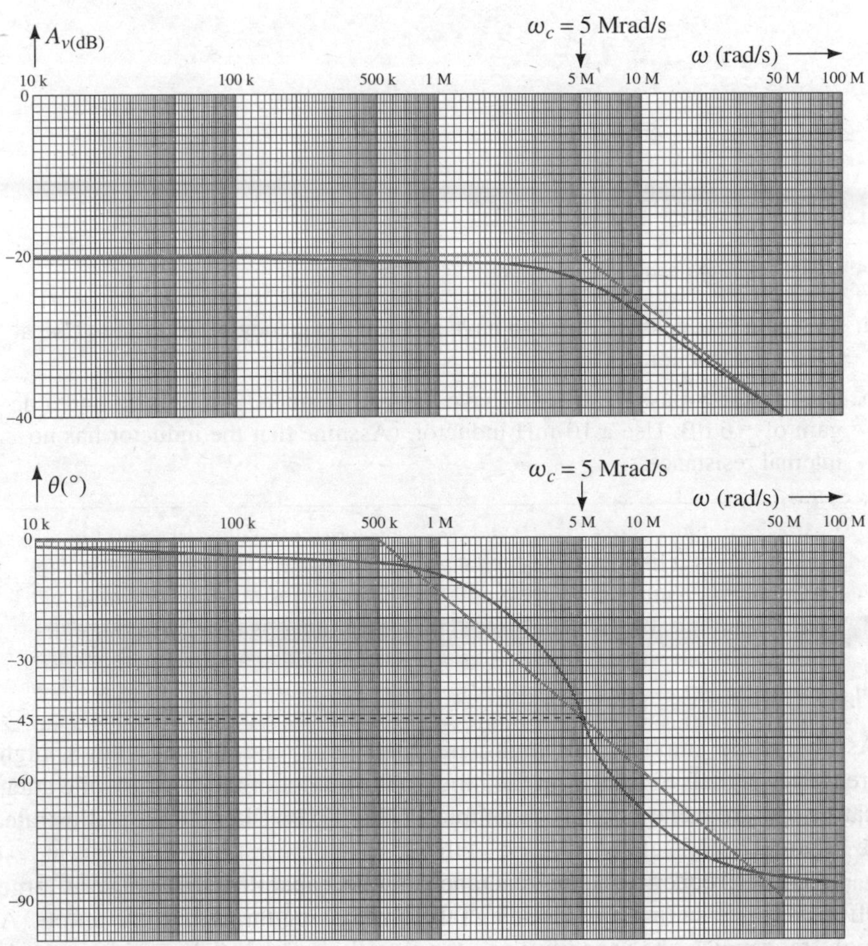

FIGURE 22–24

PRACTICE PROBLEMS 5

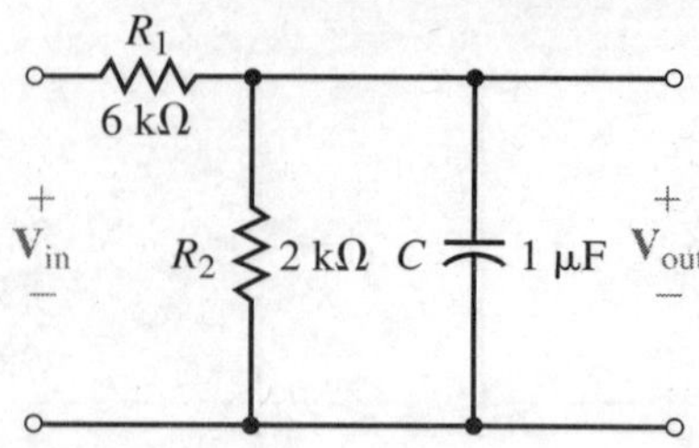

FIGURE 22–25

Refer to the low-pass circuit of Figure 22–25.

a. Write the transfer function for the circuit.

b. Sketch the frequency response.

Answers

a. $\mathbf{TF}(\omega) = \dfrac{0.25}{1 + j\omega 0.0015}$

b.

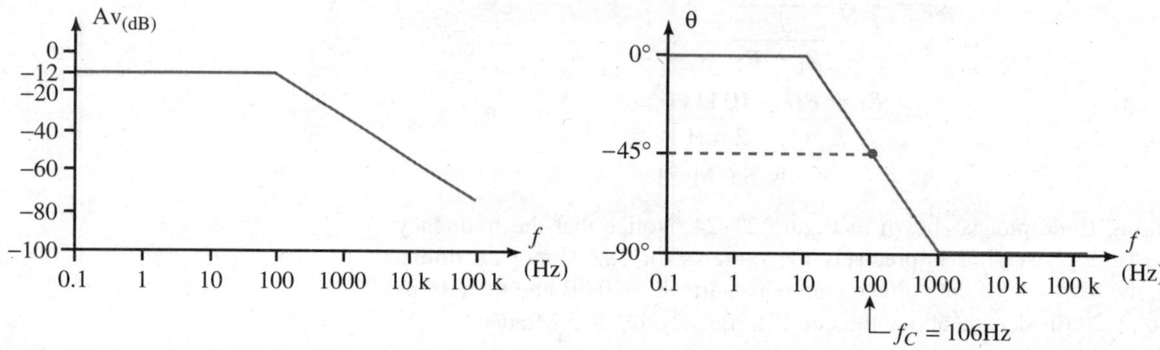

FIGURE 22–26

IN-PROCESS LEARNING CHECK 2

(Answers are at the end of the chapter.)

a. Design a low-pass *RC* filter to have a cutoff frequency of 30 krad/s. Use a 0.01-μF capacitor.

b. Design a low-pass *RL* filter to have a cutoff frequency of 20 kHz and a dc gain of −6 dB. Use a 10-mH inductor. (Assume that the inductor has no internal resistance.)

22.5 The High-Pass Filter

The *RC* High-Pass Filter

As the name implies, the high-pass filter is a circuit that allows high-frequency signals to pass from the input to the output of the circuit while attenuating low-frequency signals. A simple *RC* high-pass filter circuit is illustrated in Figure 22–27.

At low frequencies, the reactance of the capacitor will be very large, effectively preventing any input signal from passing through to the output. At high frequencies, the capacitive reactance will approach a short-circuit condition, providing a very low impedance path for the signal from the input to the output.

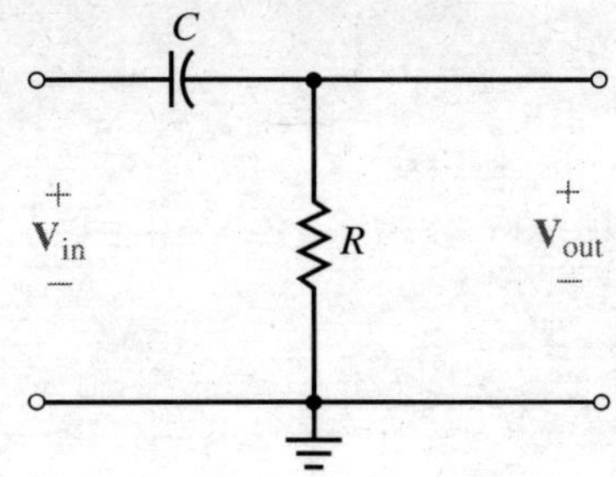

FIGURE 22–27 *RC* high-pass filter.

The transfer function of the low-pass filter is determined as follows:

$$\mathbf{TF} = \frac{\mathbf{V}_{\text{out}}}{\mathbf{V}_{\text{in}}} = \frac{\mathbf{R}}{\mathbf{R} + \mathbf{Z}_C}$$

$$= \frac{R}{R + \dfrac{1}{j\omega C}} = \frac{R}{\dfrac{j\omega RC + 1}{j\omega C}} = \frac{j\omega RC}{1 + j\omega RC}$$

Now, if we let $\omega_c = 1/\tau = 1/RC$, we have

$$\mathbf{TF} = \frac{j\dfrac{\omega}{\omega_c}}{1 + j\dfrac{\omega}{\omega_c}} \qquad \textbf{(22–17)}$$

Notice that the expression of Equation 22–17 is very similar to the expression for a low-pass filter, with the exception that there is an additional term in the numerator. Since the transfer function is a complex number which is dependent upon frequency, we may once again find the general expressions for voltage gain and phase shift as functions of frequency, ω.

The voltage gain is found as

$$A_v = \frac{\dfrac{\omega}{\omega_c}}{\sqrt{1 + \left(\dfrac{\omega}{\omega_c}\right)^2}}$$

which, when expressed in decibels, becomes

$$A_{v(\text{dB})} = 20 \log \frac{\omega}{\omega_c} - 10 \log\left[1 + \left(\frac{\omega}{\omega_c}\right)^2\right] \qquad \textbf{(22–18)}$$

The phase shift of the numerator will be a constant 90°, since the term has only an imaginary component. The overall phase shift of the transfer function is then found as follows:

$$\theta = 90° - \arctan \frac{\omega}{\omega_c} \qquad \textbf{(22–19)}$$

In order to sketch the asymptotic response of the voltage gain we need to examine the effect of Equation 22–18 on frequencies around the cutoff frequency, ω_c.

For frequencies $\omega \leq 0.1\omega_c$, the second term of the expression will be essentially equal to zero, and so the voltage gain at low frequencies is approximated as

$$A_{v(\text{dB})} \cong 20 \log \frac{\omega}{\omega_c}$$

If we substitute some arbitrary values of ω into the preceding approximation, we arrive at a general statement. For example, by letting $\omega = 0.01\omega_c$, we have the voltage gain as

$$A_v = 20 \log(0.01) = -40 \text{ dB}$$

and by letting $\omega = 0.1\omega_c$, we have

$$A_v = 20 \log(0.1) = -20 \text{ dB}$$

In general, we see that the expression $A_v = 20 \log (\omega/\omega_c)$ may be represented as a straight line on a semilogarithmic graph. The straight line will intersect the 0-dB axis at the cutoff frequency, ω_c, and have a slope of +20 dB/decade.

For frequencies $\omega >> \omega_c$, Equation 22–18 may be expressed as

$$A_{v(\text{dB})} \cong 20 \log\frac{\omega}{\omega_c} - 10 \log\left[\left(\frac{\omega}{\omega_c}\right)^2\right]$$

$$= 20 \log\frac{\omega}{\omega_c} - 20 \log\frac{\omega}{\omega_c}$$

$$= 0 \text{ dB}$$

For the particular case when $\omega = \omega_c$, we have

$$A_{v(\text{dB})} = 20 \log 1 - 10 \log 2 = -3.0 \text{ dB}$$

which is exactly the same result that we would expect, since the actual response will be 3 dB down from the asymptotic response.

Examining Equation 22–19 for frequencies $\omega \leq 0.1\omega_c$, we see that the phase shift for the transfer function will be essentially constant at 90°, while for frequencies $\omega \geq 10\omega_c$ the phase shift will be approximately constant at 0°. At $\omega = \omega_c$ we have $\theta = 90° - 45° = 45°$.

Figure 22–28 shows the normalized Bode plot of the high-pass circuit of Figure 22–27.

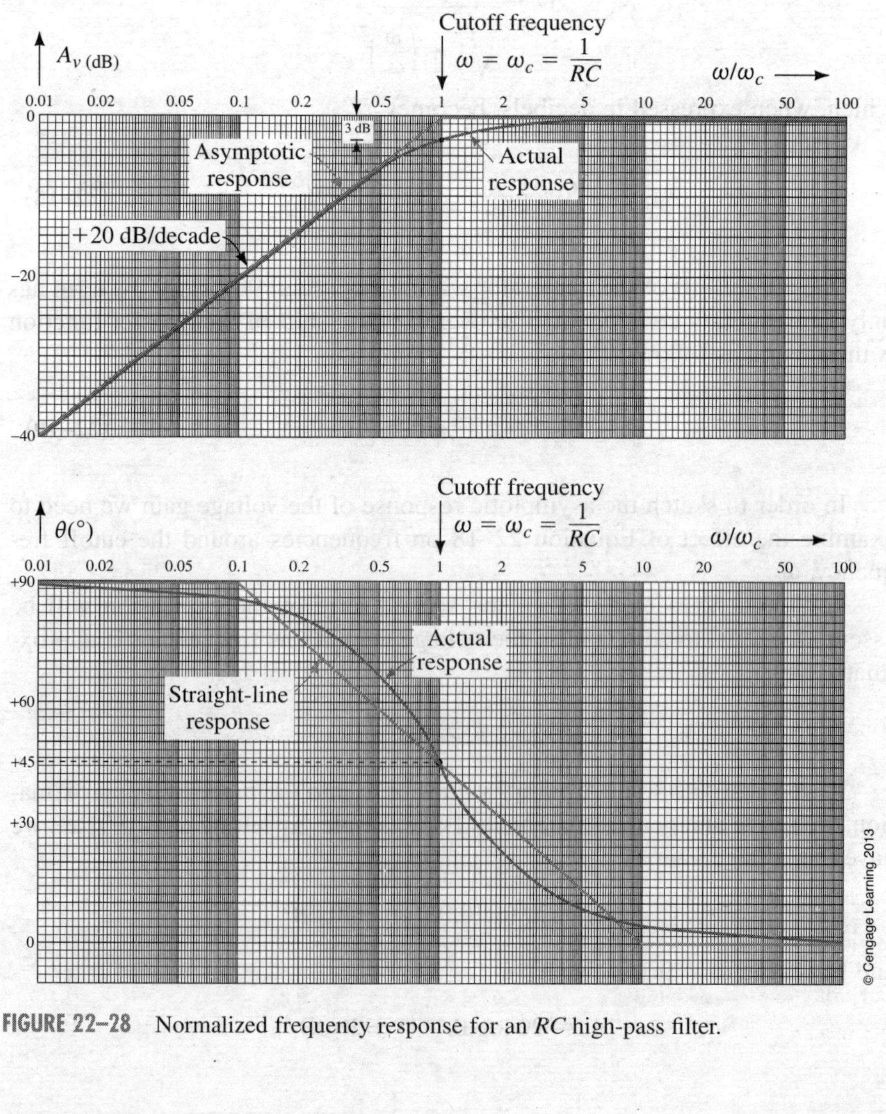

FIGURE 22–28 Normalized frequency response for an *RC* high-pass filter.

The *RL* High-Pass Filter

A typical *RL* high-pass filter circuit is shown in Figure 22–29.

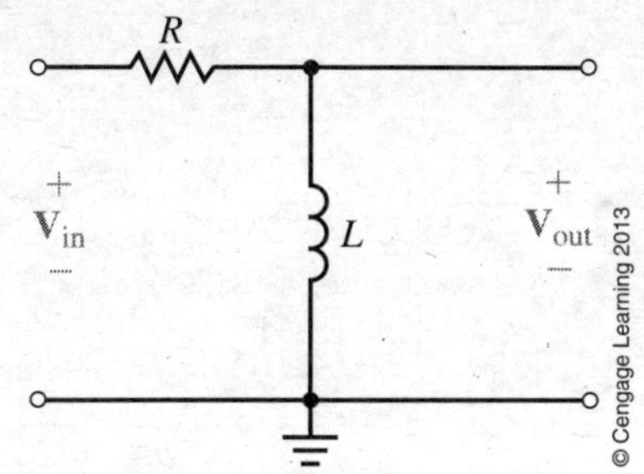

FIGURE 22–29 *RL* high-pass filter.

At low frequencies, the inductor is effectively a short circuit, which means that the output of the circuit is essentially zero at low frequencies. Inversely, at high frequencies, the reactance of the inductor approaches infinity and greatly exceeds the resistance, effectively preventing current. The voltage across the inductor is therefore very nearly equal to the applied input voltage signal. The transfer function for the high-pass *RL* circuit is derived as follows:

$$\mathbf{TF} = \frac{\mathbf{Z}_L}{\mathbf{R} + \mathbf{X}_L}$$

$$= \frac{j\omega L}{R + j\omega L} = \frac{j\omega \frac{L}{R}}{1 + j\omega \frac{L}{R}}$$

Now, letting $\omega_c = 1/\tau = R/L$, we simplify the expression as

$$\mathbf{TF} = \frac{j\omega\tau}{1 + j\omega\tau}$$

The preceding expression is identical to the transfer function for a high-pass *RC* filter, with the exception that in this case we have $\tau = L/R$.

EXAMPLE 22–10

Design the *RL* high-pass filter circuit of Figure 22–30 to have a cutoff frequency of 40 kHz. (Assume that the inductor has no internal resistance.) Sketch the frequency response of the circuit expressing the frequencies in kilohertz.

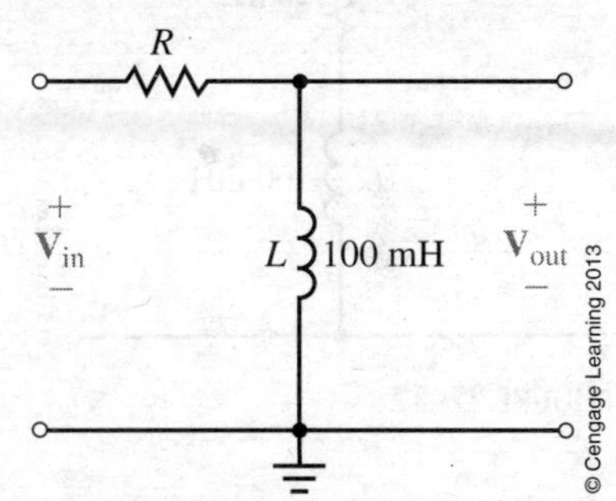

FIGURE 22–30

Solution The cutoff frequency, ω_c, in radians per second is

$$\omega_c = 2\pi f_c = 2\pi(40 \text{ kHz}) = 251.33 \text{ krad/s}$$

Now, since $\omega_c = R/L$, we have

$$R = \omega_c L = (251.33 \text{ krad/s})(100 \text{ mH}) = 25.133 \text{ k}\Omega$$

The resulting Bode plot is shown in Figure 22–31.

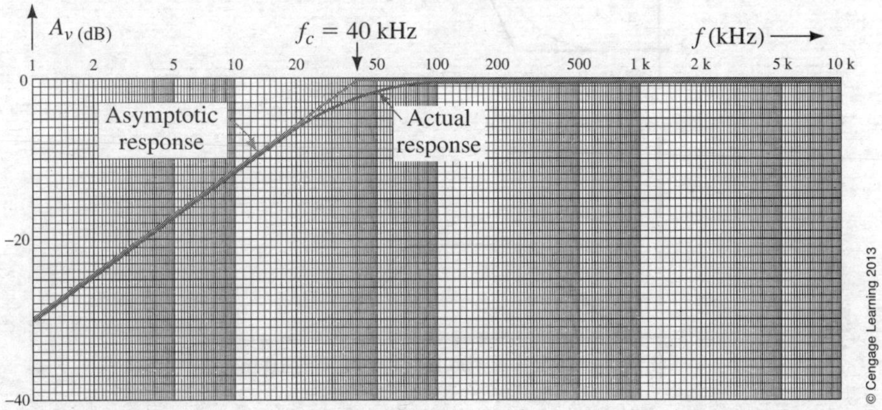

FIGURE 22–31 (Continued)

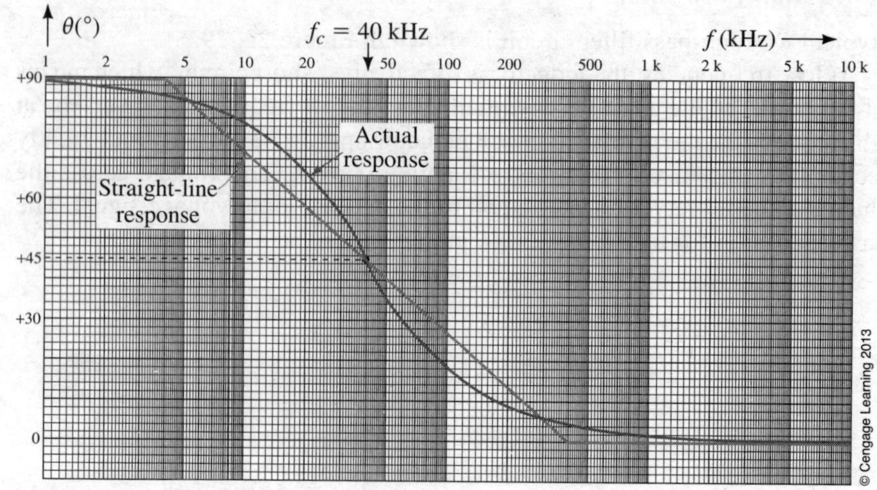

FIGURE 22–31 (Continued)

PRACTICE PROBLEMS 6

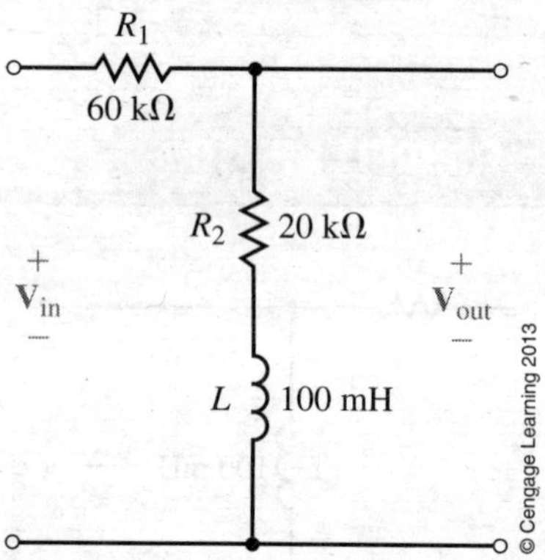

FIGURE 22–32

Consider the high-pass circuit of Figure 22–32.

a. Write the transfer function of the filter.

b. Sketch the frequency response of the filter. Show the frequency in radians per second. (Hint: The frequency response of the filter is a step response.)

Answers

a. $$\mathbf{TF}(\omega) = \left(\frac{R_2}{R_1 + R_2}\right)\left(\frac{1 + j\omega\dfrac{L}{R_2}}{1 + j\omega\dfrac{L}{R_1 + R_2}}\right)$$

f_1 = 31.8 kHz (200 krad/s), f_2 = 127 kHz (800 krad/s)

b.

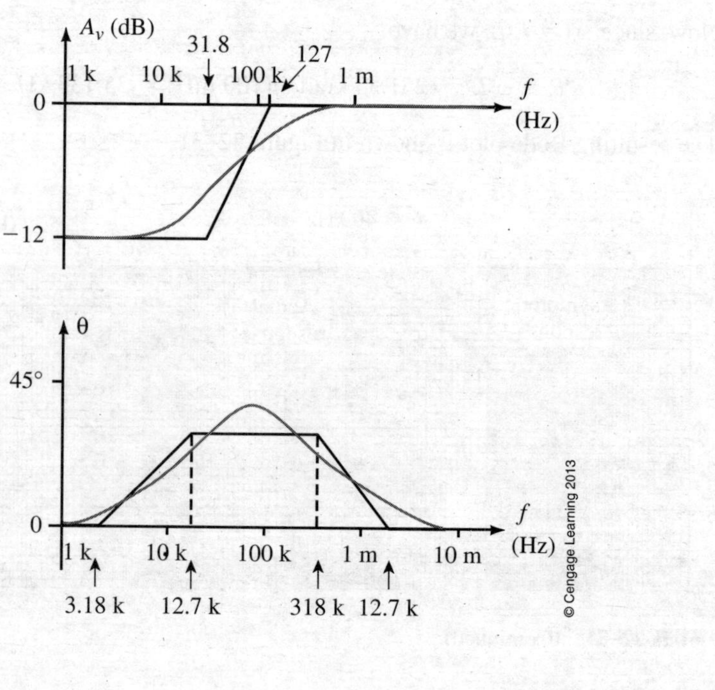

FIGURE 22–33

IN-PROCESS LEARNING CHECK 3

(Answers are at the end of the chapter.)

1. Use a 0.05-μF capacitor to design a high-pass filter having a cutoff frequency of 25 kHz. Sketch the frequency response of the filter.
2. Use a 25-mH inductor to design a high-pass filter circuit having a cutoff frequency of 80 krad/s and a high-frequency gain of −12 dB. Sketch the frequency response of the filter.

22.6 The Band-Pass Filter

A band-pass filter will permit frequencies within a certain range to pass from the input of a circuit to the output. All frequencies which fall outside the desired range will be attenuated and so will not appear with appreciable power at the output. Such a filter circuit is easily constructed by using a low-pass filter cascaded with a high-pass circuit as illustrated in Figure 22–34.

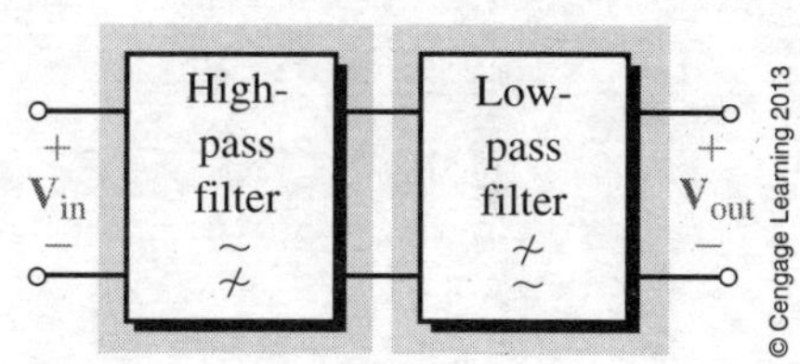

FIGURE 22–34 Block diagram of a band-pass filter.

Although the low-pass and high-pass blocks may consist of various combinations of elements, one possibility is to construct the entire filter network from resistors and capacitors as shown in Figure 22–35.

The bandwidth of the resulting band-pass filter will be approximately equal to the difference between the two cutoff frequencies, namely,

$$\text{BW} \cong \omega_2 - \omega_1 \text{ (rad/s)} \qquad \textbf{(22–20)}$$

The preceding approximation will be most valid if the cutoff frequencies of the individual stages are separated by at least one decade.

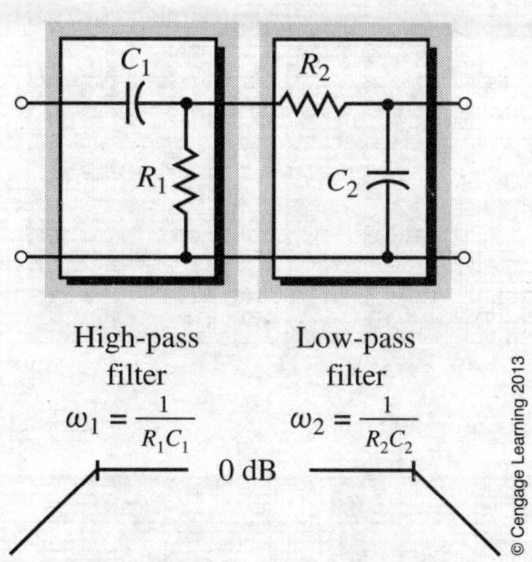

FIGURE 22–35

EXAMPLE 22–11

Write the transfer function for the circuit of Figure 22–36. Sketch the resulting Bode plot and determine the expected bandwidth for the band-pass filter.

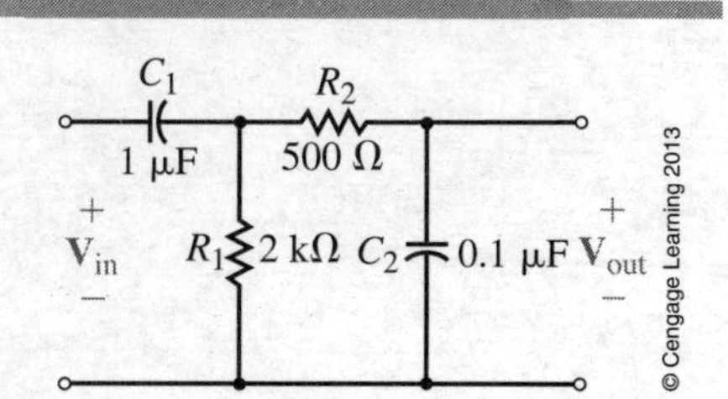

FIGURE 22–36

Solution While the transfer function for the circuit may be written by using circuit theory, it is easier to recognize that the circuit consists of two stages: one a low-pass stage and the other a high-pass stage. If the cutoff frequencies of each stage are separated by more than one decade, then we may assume that the impedance of one stage will not adversely affect the operation of the other stage. (If this is not the case, the analysis is complicated and is outside the scope of this textbook.) Based on the previous assumption, the transfer function of the first stage is determined as

$$\mathbf{TF}_1 = \frac{\mathbf{V}_1}{\mathbf{V}_{in}} = \frac{j\omega R_1 C_1}{1 + j\omega R_1 C_1}$$

and for the second stage as

$$\mathbf{TF}_2 = \frac{\mathbf{V}_{out}}{\mathbf{V}_1} = \frac{1}{1 + j\omega R_2 C_2}$$

Combining the preceding results, we have

$$\mathbf{TF} = \frac{\mathbf{V}_{out}}{\mathbf{V}_{in}} = \frac{(\mathbf{TF}_2)(\mathbf{V}_1)}{\dfrac{\mathbf{V}_1}{\mathbf{TF}_1}} = \mathbf{TF}_1\mathbf{TF}_2$$

which, when simplified, becomes

$$\mathbf{TF} = \frac{\mathbf{V}_{out}}{\mathbf{V}_{in}} = \frac{j\omega\tau_1}{(1 + j\omega\tau_1)(1 + j\omega\tau_2)} \qquad \textbf{(22–21)}$$

where $\tau_1 = R_1C_1 = 2.0$ ms and $\tau_2 = R_2C_2 = 50$ μs. The corresponding cutoff frequencies are $\omega_1 = 500$ rad/s and $\omega_2 = 20$ krad/s. The transfer function of Equation 22–21 has three separate terms which, when taken separately, result in the approximate responses illustrated in Figure 22–37.

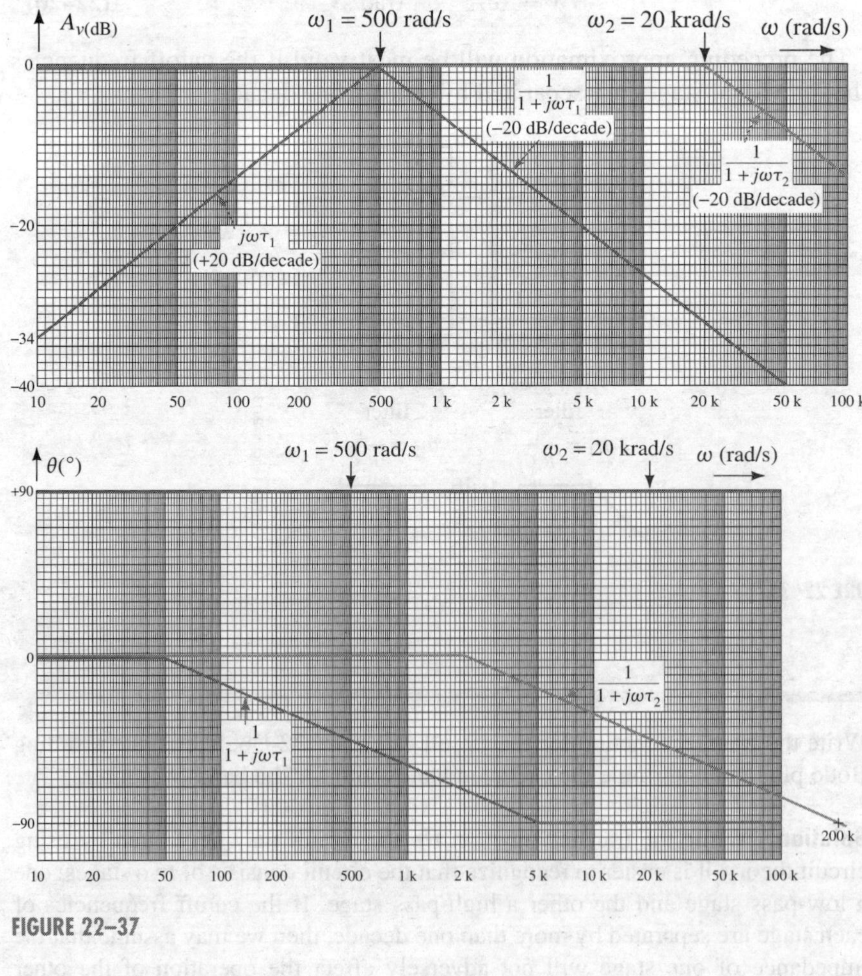

FIGURE 22–37

The resulting frequency response is determined by the summation of the individual responses as shown in Figure 22–38.

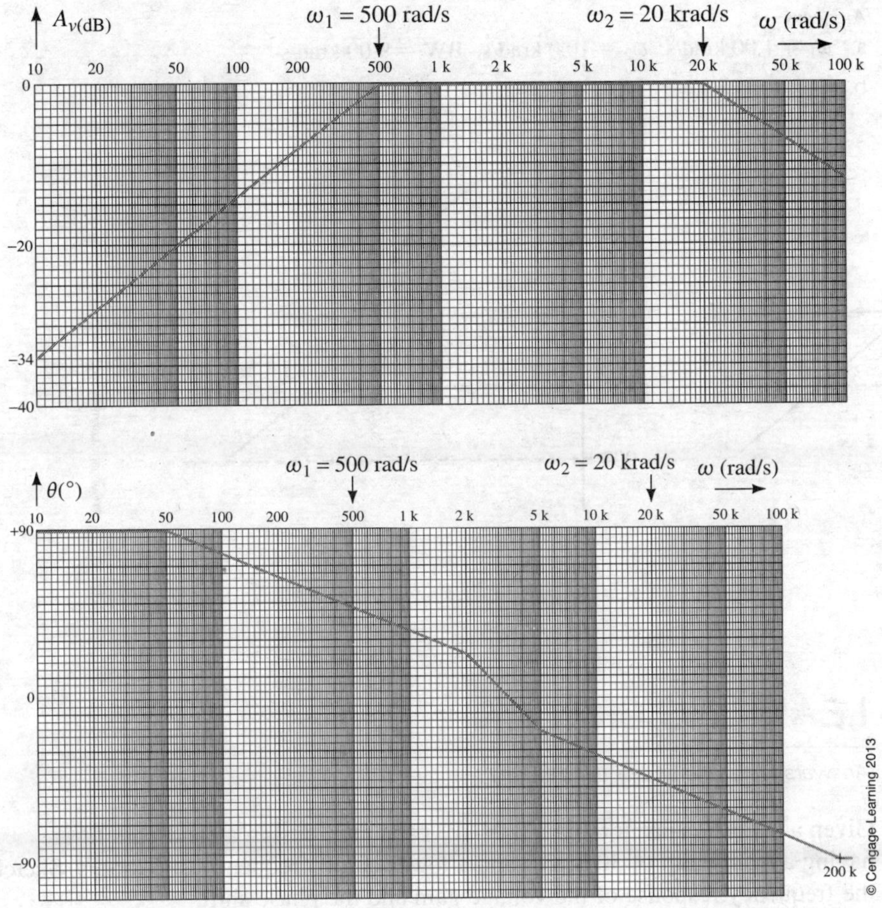

FIGURE 22–38

From the Bode plot, we determine that the bandwidth of the resulting filter is

$$\text{BW(rad/s)} = \omega_2 - \omega_1 = 20 \text{ krad/s} - 0.5 \text{ krad/s} = 19.5 \text{ krad/s}$$

PRACTICE PROBLEMS 7

Refer to the band-pass filter of Figure 22–39.

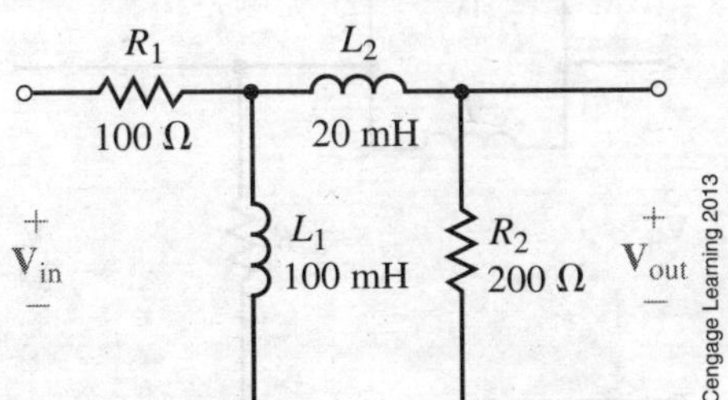

FIGURE 22–39

CircuitSim 22-3

a. Calculate the cutoff frequencies in rad/s and the approximate bandwidth.
b. Sketch the frequency response of the filter.

Answers

a. $\omega_1 = 1.00$ krad/s, $\omega_2 = 10.0$ krad/s, BW = 9.00 krad/s

b.

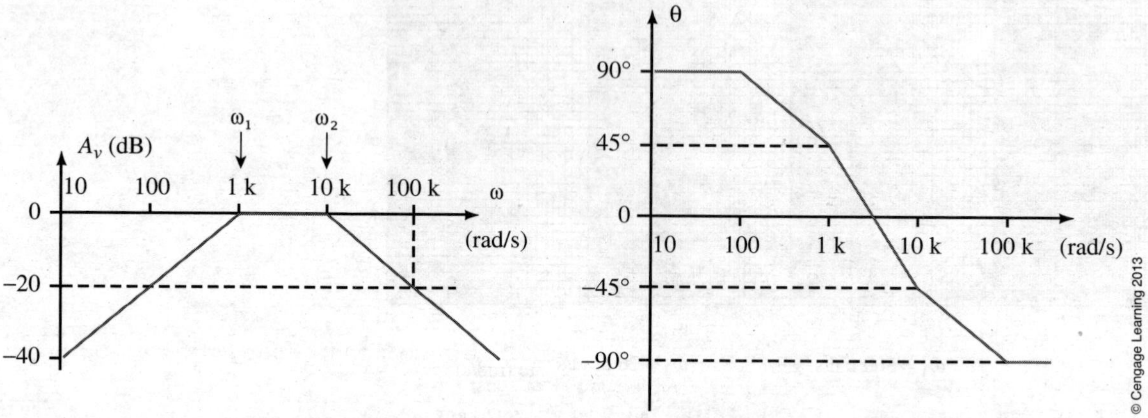

FIGURE 22–40

IN-PROCESS LEARNING CHECK 4

(Answers are at the end of the chapter.)

Given a 0.1-μF capacitor and a 0.04-μF capacitor, design a band-pass filter having a bandwidth of 30 krad/s and a lower cutoff frequency of 5 krad/s. Sketch the frequency response of the voltage gain and the phase shift.

22.7 The Band-Reject Filter

The band-reject filter has a response that is opposite to that of the band-pass filter. This filter passes all frequencies with the exception of a narrow band which is greatly attenuated. A band-reject filter constructed of a resistor, inductor, and capacitor is shown in Figure 22–41.

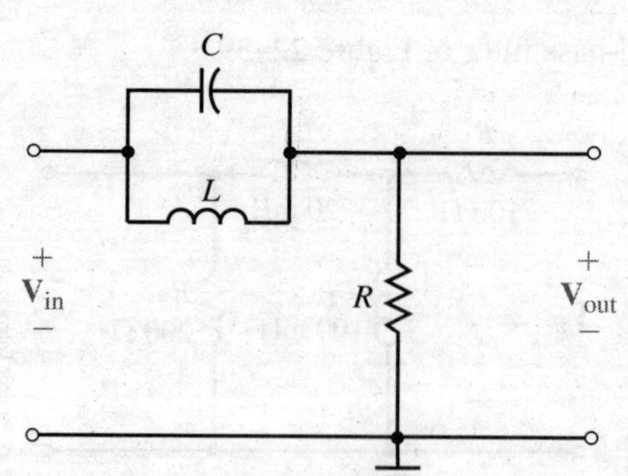

FIGURE 22–41 Notch filter.

Notice that the circuit uses a resonant tank circuit as part of the overall design. As we saw in the previous chapter, the combination of the inductor and the capacitor results in a very high tank impedance at the resonant frequency. Therefore, for any signals occurring at the resonant frequency, the output voltage is effectively zero. Because the filter circuit effectively removes any signal occurring at the resonant frequency, the circuit is often referred to as a **notch filter.** The voltage gain response of the notch filter is shown in Figure 22–42.

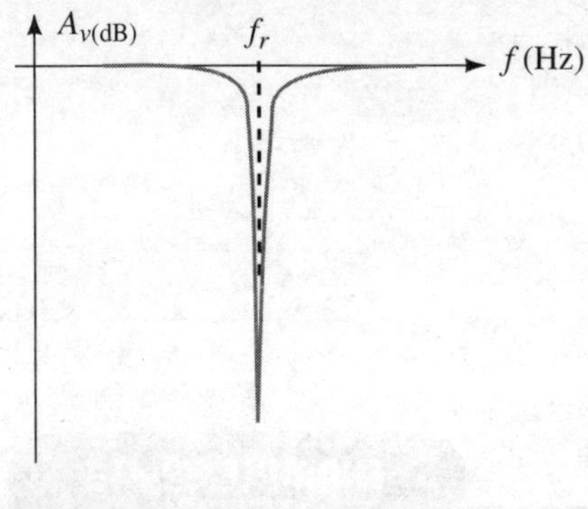

FIGURE 22–42 Voltage gain response for a notch filter.

For low-frequency signals, the inductor provides a low-impedance path from the input to the output, allowing these signals to pass from the input and appear across the resistor with minimal attenuation. Conversely, at high frequencies the capacitor provides a low-impedance path from the input to the output. Although the complete analysis of the notch filter is outside the scope of this textbook, the transfer function of the filter is determined by employing the same techniques as those previously developed.

$$\mathbf{TF} = \frac{\mathbf{R}}{\mathbf{R} + \mathbf{Z}_L \| \mathbf{Z}_C}$$

$$= \frac{R}{R + \dfrac{(j\omega L)\left(\dfrac{1}{j\omega C}\right)}{j\omega L + \dfrac{1}{j\omega C}}}$$

$$= \frac{R}{R + \dfrac{j\omega L}{1 - \omega^2 LC}}$$

$$= \frac{R(1 - \omega^2 LC)}{R - \omega^2 RLC + j\omega L}$$

$$\mathbf{TF} = \frac{1 - \omega^2 LC}{1 - \omega^2 LC + j\omega \dfrac{L}{R}} \qquad (22\text{–}22)$$

Notice that the transfer function for the notch filter is significantly more complicated than for the previous filter circuits. Due to the presence of the complex quadratic in the denominator of the transfer function, this type of filter circuit is called a second-order filter. The design of such filters is a separate field in electronics engineering.

Actual filter design often involves using operational amplifiers to provide significant voltage gain in the passband of the filter. In addition, such active filters have the advantage of providing very high input impedance to prevent loading effects. Many excellent textbooks are available for assistance in filter design.

22.8 Circuit Analysis Using Computers

Multisim

PSPICE

Despite the complexity of designing a filter for a particular application, the analysis of the filter is a relatively simple process when done by computer. We have already seen the ease with which PSpice may be used to examine the frequency response of resonant circuits. In this chapter we will again use the Probe postprocessor of PSpice to plot the frequency characteristics of a particular circuit. We find that with some minor adjustments, the program is able to simultaneously plot both the voltage gain (in decibels) and the phase shift (in degrees) of any filter circuit.

Multisim provides displays that are similar to those obtained in PSpice. The method, though, is somewhat different. Since Multisim simulates actual lab measurements, an instrument called a Bode plotter is used by the software. As one might expect, the Bode plotter provides a graph of the frequency response (showing both the gain and phase shift) of a circuit even though no such instrument is found in a real electronics lab.

PSpice

EXAMPLE 22–12

Use the PROBE postprocessor of PSpice to view the frequency response from 1 Hz to 100 kHz for the circuit of Figure 22–36. Determine the cutoff frequencies and the bandwidth of the circuit. Compare the results to those obtained in Example 22–11.

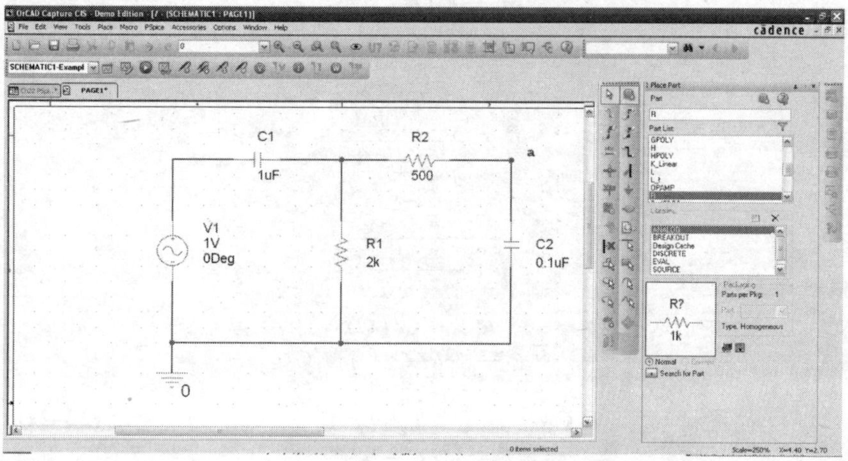

FIGURE 22–43

Solution The OrCAD Capture CIS program is used to enter the circuit as shown in Figure 22–43. You will need to set the properties of the signal generator for a magnitude of 1 V and a phase shift of 0°. Once the circuit has been entered as shown, set the analysis to perform an ac sweep from 1 Hz to 100 kHz using 1001 points per decade.

In this example, we show both the voltage gain and the phase shift on the same display. Once the PROBE screen is activated, two simultaneous displays are obtained by clicking on *Plot/Add Plot to Window*. We will use the top display to show voltage gain and the bottom to show the phase shift.

PSpice does not actually calculate the voltage gain in decibels, but rather determines the output voltage level in dBV, referenced to 1 V_{rms}. (It is for this reason that we used a supply voltage of 1 V.) The voltage at point *a* of the circuit is obtained by clicking on *Trace/Add Trace* and then selecting **DB(V(C2:1))** as the *Trace Expression*. Cursors are obtained by clicking on the *Toggle Cursor* tool. The cursor indicates that the maximum gain for the circuit is −1.02 dB. We find the bandwidth of the circuit (at the −3 dB frequencies) by moving the cursors (arrow keys and <Ctrl> arrow keys) to the frequencies at which the output of the circuit is at −4.02 dB. We determine $f_1 = 0.069$ kHz, $f_2 = 3.64$ kHz, and BW = 3.57 kHz. These results are consistent with those found in Example 22–11.

Finally, we obtain a trace of the phase shift for the circuit as follows. Click anywhere on the bottom plot. Click on *Trace/Add Trace* and then select

P(V(C2:1)) as the *Trace Expression*. The range of the ordinate (vertical axis) is changed by clicking on the axis, selecting the *Y Axis* tab, and setting the *User defined range* for a value of **−90d** to **90d**. The resulting display is shown in Figure 22–44.

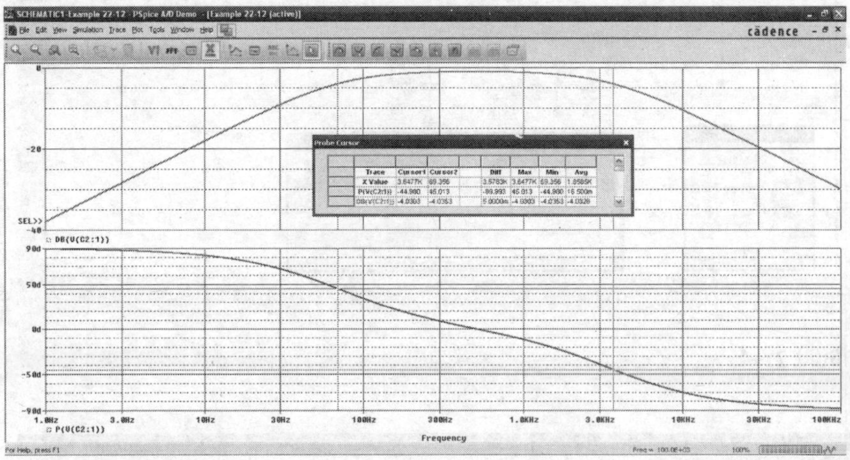

FIGURE 22–44

PRACTICE PROBLEMS 8

a. Use PSpice to input the circuit of Figure 22–39.

b. Use the Probe postprocessor to observe the frequency response from 1 Hz to 100 kHz.

c. From the display, determine the cutoff frequencies and use the cursors to determine the bandwidth.

d. Compare the results to those obtained in Practice Problems 7.

EXAMPLE 22–13

Use Multisim to obtain the frequency response for the circuit of Figure 22–36. Compare the results to those obtained in Example 22–12.

Solution In order to perform the required measurements, we need to use the function generator and the Bode plotter, both located in the Instruments parts bin. The circuit is constructed as shown in Figure 22–45.

The Bode plotter is adjusted to provide the desired frequency response by first double-clicking on the instrument. Next, we click on the *Magnitude* button. The *Vertical scale* is set to log with values between **−40** dB and **0** dB. The *Horizontal scale* is set to log with values between **1** Hz and **100** kHz. Similarly, the *Phase* is set to have a *Vertical range* of **−90°** to **90°**. After clicking the *Run* button, the Bode plotter provides a display of either the voltage gain response or the phase response. However, both displays are shown simultaneously by clicking on *Grapher* in the *View* menu. By using the cursor feature, we obtain the same results as those found in Example 22–12. Figure 22–46 shows the frequency response as viewed using the Grapher window.

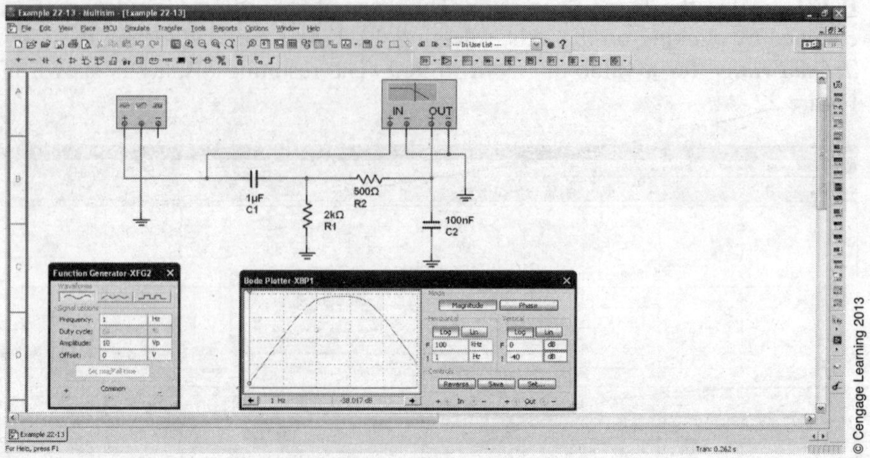

FIGURE 22–45

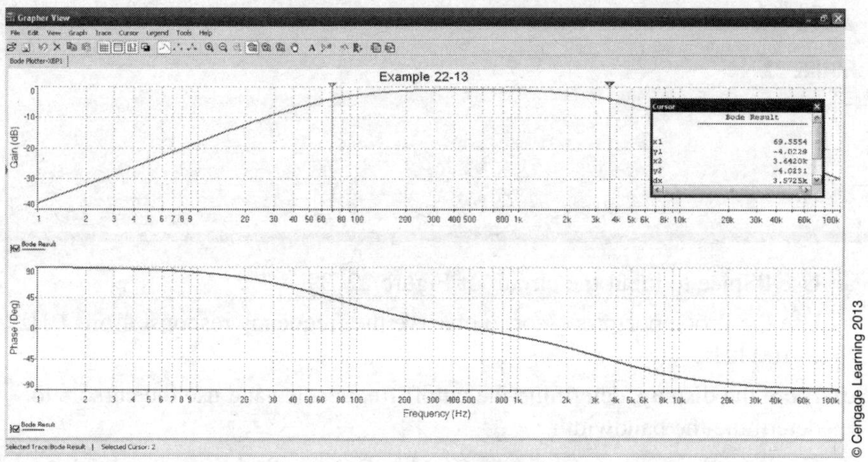

FIGURE 22–46

PRACTICE PROBLEMS 9

Use Multisim to obtain the frequency response for the circuit of Figure 22–25. Compare the results to those obtained in Practice Problems 5.

Putting It into Practice

As a designer for a sound studio, you have been asked to design band-pass filters for a color organ that will be used to provide lighting for a rock concert. The color organ will provide stage lighting that will correspond to the sound level and frequency of the music. You remember from your physics class that the human ear can perceive sounds from 20 Hz to 20 kHz.

The specifications say that the audio frequency spectrum is to be divided into three ranges. Passive *RC* filters will be used to isolate signals for each range. These signals will then be amplified and used to control lights of a particular color. The low-frequency components (20 to 200 Hz) will control blue lights, the mid-frequency components (200 Hz to 2 kHz) will control green lights, and the high-frequency components (2 kHz to 20 kHz) will control red lights.

Although the specifications call for three band-pass filters, you realize that you can simplify the design by using a low-pass filter with a break frequency of 200 Hz for the low frequencies and a high-pass filter with a break frequency of 2 kHz for the high frequencies. To simplify your work, you decide to use only 0.5-μF capacitors for all filters.

Show the design for each of the filters.

Problems

22.1 The Decibel

1. Refer to the amplifier shown in Figure 22–47. Determine the power gain both as a ratio and in decibels for the following power values:
   a. $P_{in} = 1.2$ mW, $P_{out} = 2.4$ W
   b. $P_{in} = 3.5$ μW, $P_{out} = 700$ mW
   c. $P_{in} = 6.0$ pW, $P_{out} = 12$ μW
   d. $P_{in} = 2.5$ mW, $P_{out} = 1.0$ W

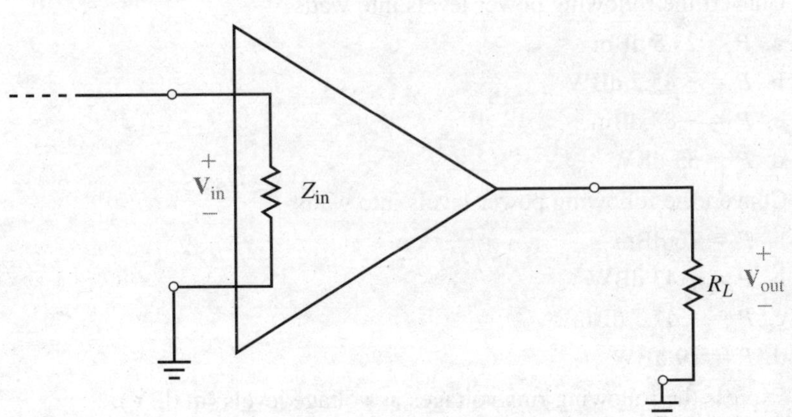

FIGURE 22–47

2. If the amplifier of Figure 22–47 has $Z_{in} = 600\ \Omega$ and $R_L = 2\ \text{k}\Omega$, find P_{in}, P_{out}, and A_P(dB) for the following voltage levels:
   a. $V_{in} = 20$ mV, $V_{out} = 100$ mV
   b. $V_{in} = 100$ μV, $V_{out} = 400$ μV
   c. $V_{in} = 320$ mV, $V_{out} = 600$ mV
   d. $V_{in} = 2$ μV, $V_{out} = 8$ V

3. The amplifier of Figure 22–47 has $Z_{in} = 2\ \text{k}\Omega$ and $R_L = 10\ \Omega$. Find the voltage gain and power gain both as a ratio and in dB for the following conditions:
   a. $V_{in} = 2$ mV, $P_{out} = 100$ mW
   b. $P_{in} = 16\ \mu$W, $V_{out} = 40$ mV
   c. $V_{in} = 3$ mV, $P_{out} = 60$ mW
   d. $P_{in} = 2$ pW, $V_{out} = 80$ mV
4. The amplifier of Figure 22–47 has an input voltage of $V_{in} = 2$ mV and an output power of $P_{out} = 200$ mW. Find the voltage gain and power gain both as a ratio and in dB for the following conditions:
   a. $Z_{in} = 5\ \text{k}\Omega$, $R_L = 2\ \text{k}\Omega$
   b. $Z_{in} = 2\ \text{k}\Omega$, $R_L = 10\ \text{k}\Omega$
   c. $Z_{in} = 300\ \text{k}\Omega$, $R_L = 1\ \text{k}\Omega$
   d. $Z_{in} = 1\ \text{k}\Omega$ $R_L = 1\ \text{k}\Omega$
5. The amplifier of Figure 22–47 has an input impedance of 5 kΩ and a load resistance of 250 Ω. If the power gain of the amplifier is 35 dB, and the input voltage is 250 mV, find P_{in}, P_{out}, V_{out}, A_v, and A_v(dB).
6. Repeat Problem 5 if the input impedance is increased to 10 kΩ. (All other quantities remain unchanged.)
7. Express the following powers in dBm and in dBW:
   a. $P = 50$ mW
   b. $P = 1$ W
   c. $P = 400$ nW
   d. $P = 250$ pW
8. Express the following powers in dBm and in dBW:
   a. $P = 250$ W
   b. $P = 250$ kW
   c. $P = 540$ nW
   d. $P = 27$ mW
9. Convert the following power levels into watts:
   a. $P = 23.5$ dBm
   b. $P = -45.2$ dBW
   c. $P = -83$ dBm
   d. $P = 33$ dBW
10. Convert the following power levels into watts:
    a. $P = 16$ dBm
    b. $P = -43$ dBW
    c. $P = -47.3$ dBm
    d. $P = 29$ dBW
11. Express the following rms voltages as voltage levels (in dBV):
    a. 2.00 V
    b. 34.0 mV
    c. 24.0 V
    d. 58.2 μV
12. Express the following rms voltages as voltage levels (in dBV):
    a. 25 μV
    b. 90 V
    c. 72.5 mV
    d. 0.84 V

13. Convert the following voltage levels from dBV to rms voltages:
    a. −2.5 dBV
    b. 6.0 dBV
    c. −22.4 dBV
    d. 10.0 dBV
14. Convert the following voltage levels from dBV to rms voltages:
    a. 20.0 dBV
    b. −42.0 dBV
    c. −6.0 dBV
    d. 3.0 dBV
15. A sinusoidal waveform is measured as 30.0 V_{p-p} with an oscilloscope. If this waveform were applied to a voltmeter calibrated to express readings in dBV, what would the voltmeter indicate?
16. A voltmeter shows a reading of 9.20 dBV. What peak-to-peak voltage would be observed on an oscilloscope?

22.2 Multistage Systems

17. Calculate the power levels (in dBm) at the output of each of the stages of the system shown in Figure 22–48. Solve for the output power (in watts).

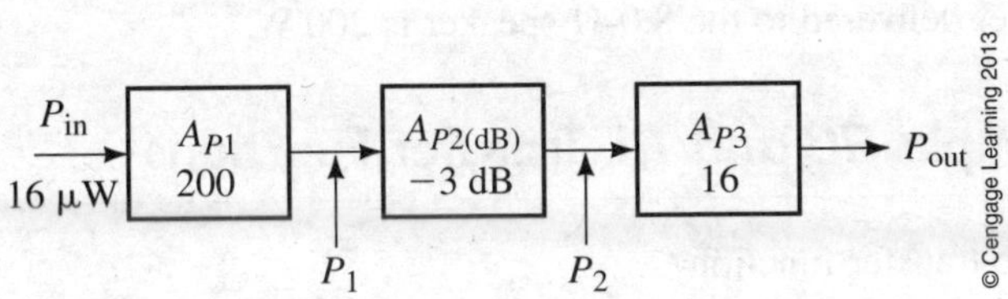

FIGURE 22–48

18. Calculate the power levels (in dBm) at the indicated locations of the system shown in Figure 22–49. Solve for the input and output powers (in watts).

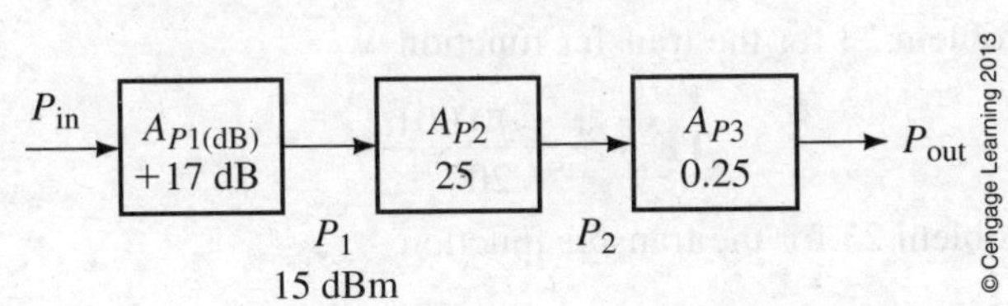

FIGURE 22–49

19. Given that power P_2 = 140 mW as shown in Figure 22–50. Calculate the power levels (in dBm) at each of the indicated locations. Solve for the voltage across the load resistor, R_L.

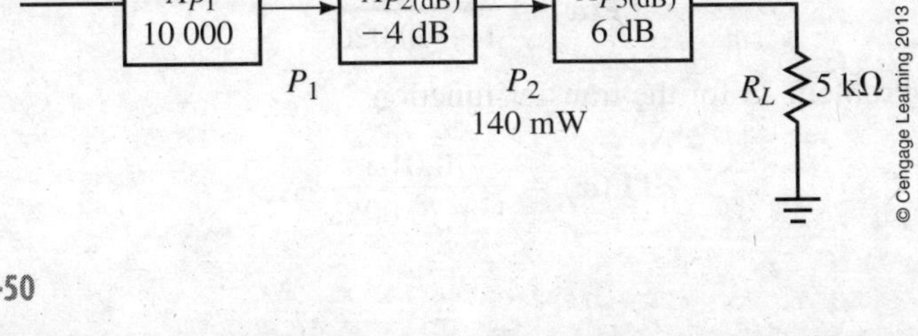

FIGURE 22–50

20. Suppose that the system of Figure 22–51 has an output voltage of 2 V.
    a. Determine the power (in watts) at each of the indicated locations.
    b. Solve for the voltage, V_{in}, if the input impedance of the first stage is 1.5 kΩ.
    c. Convert V_{in} and V_L into voltage levels (in dBV).
    d. Solve for the voltage gain, A_v (in dB).

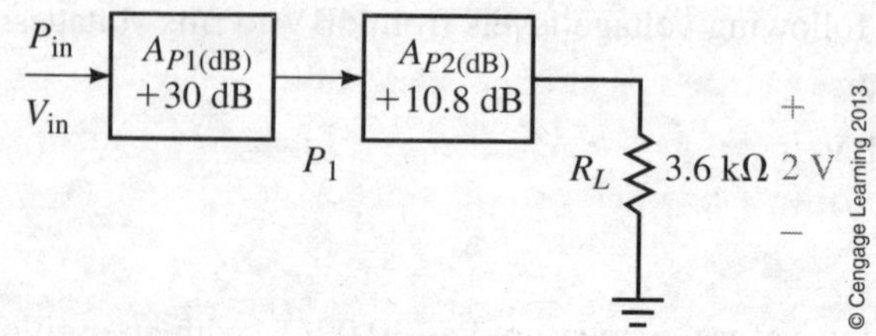

FIGURE 22–51

21. A power amplifier (P.A.) with a power gain of 250 has an input impedance of 2.0 kΩ and is used to drive a stereo speaker (output impedance of 8.0 Ω). If the output power is 100 W, determine the following:
    a. Output power level (dBm), input power level (dBm)
    b. Output voltage (rms), input voltage (rms)
    c. Output voltage level (dBV), input voltage level (dBV)
    d. Voltage gain in dB
22. Repeat Problem 21 if the amplifier has a power gain of 400 and Z_{in} = 1.0 kΩ. The power delivered to the 8.0-Ω speaker is 200 W.

22.3 Simple *RC* and *RL* Transfer Functions

23. Given the transfer function

$$\mathbf{TF} = \frac{200}{1 + j0.001\omega}$$

    a. Determine the cutoff frequency in radians per second and in hertz.
    b. Sketch the frequency response of the voltage gain and the phase shift responses. Label the abscissa in radians per second.
24. Repeat Problem 23 for the transfer function

$$\mathbf{TF} = \frac{1 + j0.001\omega}{200}$$

25. Repeat Problem 23 for the transfer function

$$\mathbf{TF} = \frac{1 + j0.02\omega}{1 + j0.001\omega}$$

26. Repeat Problem 23 for the transfer function

$$\mathbf{TF} = \frac{1 + j0.04\omega}{(1 + j0.004\omega)(1 + j0.001\omega)}$$

27. Repeat Problem 23 for the transfer function

$$\mathbf{TF}(\omega) = \frac{j0.02\omega}{1 + j0.02\omega}$$

28. Repeat Problem 23 for the transfer function

$$\mathbf{TF}(\omega) = \frac{j0.01\omega}{1 + j0.005\omega}$$

22.4 The Low-Pass Filter

29. Use a 4.0-μF capacitor to design a low-pass filter circuit having a cutoff frequency of 5 krad/s. Draw a schematic of your design and sketch the frequency response of the voltage gain and the phase shift.
30. Use a 1.0-μF capacitor to design a low-pass filter with a cutoff frequency of 2500 Hz. Draw a schematic of your design and sketch the frequency response of the voltage gain and the phase shift.
31. Use a 25-mH inductor to design a low-pass filter with a cutoff frequency of 50 krad/s. Draw a schematic of your design and sketch the frequency response of the voltage gain and the phase shift.
32. Use a 100-mH inductor (assume $R_{coil} = 0\ \Omega$) to design a low-pass filter circuit having a cutoff frequency of 15 kHz. Draw a schematic of your design and sketch the frequency response of the voltage gain and the phase shift.
33. Use a 36-mH inductor to design a low-pass filter having a cutoff frequency of 36 kHz. Draw a schematic of your design and sketch the frequency response of the voltage gain and the phase shift.
34. Use a 5-μF capacitor to design a low-pass filter circuit having a cutoff frequency of 100 krad/s. Draw a schematic of your design and sketch the frequency response of the voltage gain and the phase shift.
35. Refer to the low-pass circuit of Figure 22–52.
    a. Write the transfer function for the circuit.
    b. Sketch the frequency response of the voltage gain and phase shift.
36. Repeat Problem 35 for the circuit of Figure 22–53.

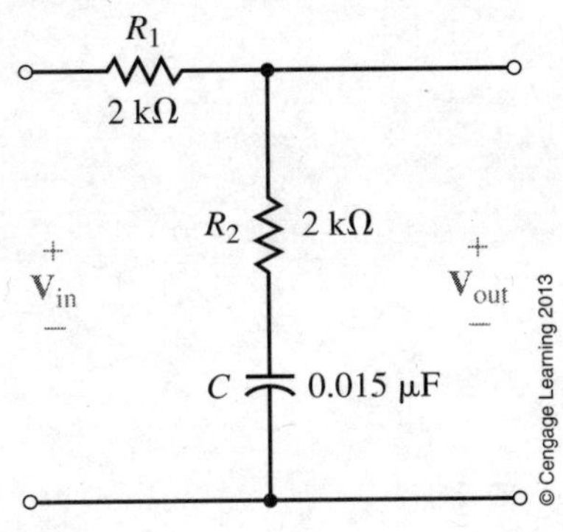

FIGURE 22–52

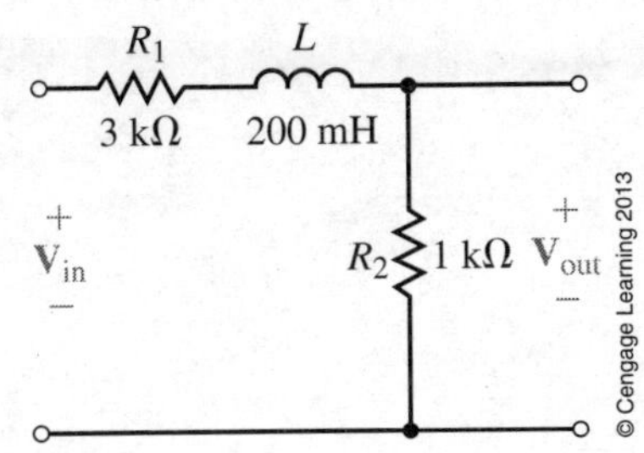

FIGURE 22–53

22.5 The High-Pass Filter

37. Use a 0.05-μF capacitor to design a high-pass filter to have cutoff frequency of 100 krad/s. Draw a schematic of your design and sketch the frequency response of the voltage gain and the phase shift.
38. Use a 2.2-nF capacitor to design a high-pass filter to have a cutoff frequency of 5 kHz. Draw a schematic of your design and sketch the frequency response of the voltage gain and phase shift.
39. Use a 2-mH inductor to design a high-pass filter to have a cutoff frequency of 36 krad/s. Draw a schematic of your design and sketch the frequency response of the voltage gain and phase shift.
40. Use a 16-mH inductor to design a high-pass filter circuit having a cutoff frequency of 250 kHz. Draw a schematic of your design and sketch the frequency response of the voltage gain and the phase shift.
41. Refer to the high-pass circuit of Figure 22–54.
    a. Write the transfer function for the circuit.
    b. Sketch the frequency response of the voltage gain and phase shift.

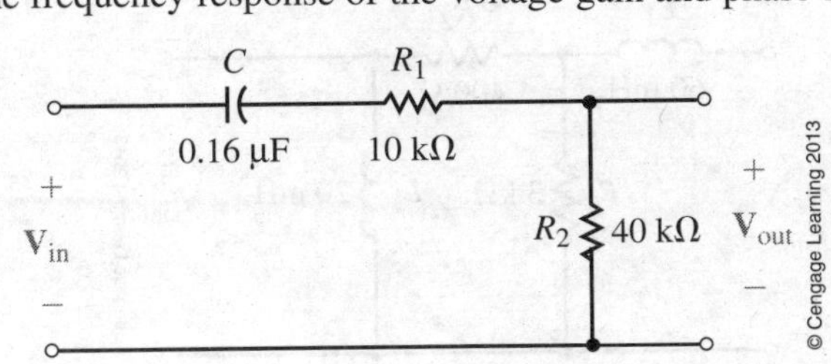

FIGURE 22–54

42. Repeat Problem 41 for the high-pass circuit of Figure 22–55.

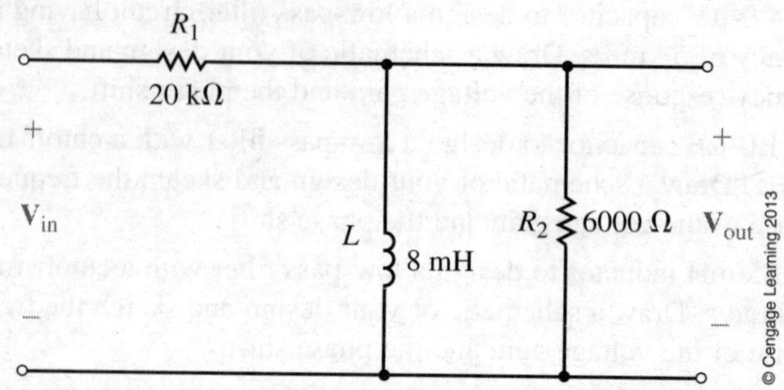

FIGURE 22–55

22.6 The Band-Pass Filter

43. Refer to the filter of Figure 22–56.
   a. Determine the approximate cutoff frequencies and bandwidth of the filter. (Assume that the two stages of the filter operate independently.)
   b. Sketch the frequency response of the voltage gain and the phase shift.

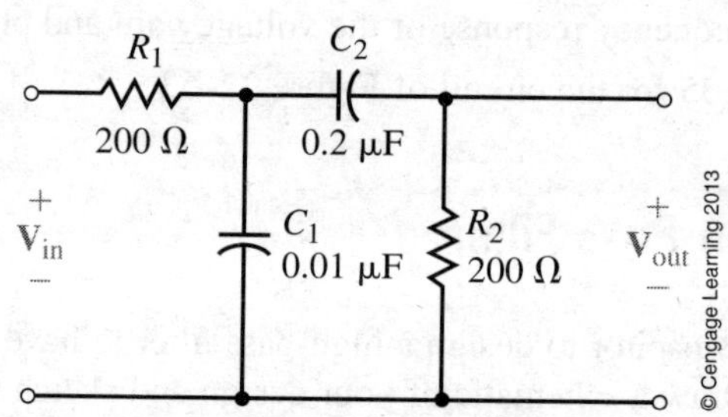

FIGURE 22–56

44. Repeat Problem 43 for the circuit of Figure 22–57.
45. a. Use two 0.01-μF capacitors to design a band-pass filter to have cutoff frequencies of 2 krad/s and 20 krad/s.
   b. Draw your schematic and sketch the frequency response of the voltage gain and the phase shift.
   c. Do you expect that the actual cutoff frequencies will occur at the designed cutoff frequencies? Explain.

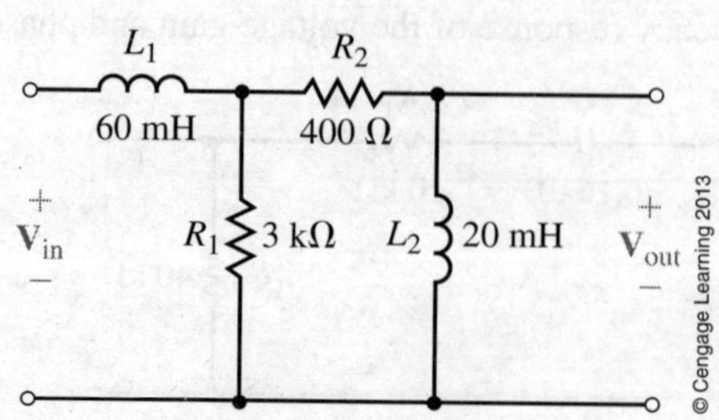

FIGURE 22–57

46. a. Use two 10-mH inductors to design a band-pass filter to have cutoff frequencies of 25 krad/s and 40 krad/s.
    b. Draw your schematic and sketch the frequency response of the voltage gain and the phase shift.
    c. Do you expect that the actual cutoff frequencies will occur at the designed cutoff frequencies? Explain.

22.7 The Band-Reject Filter

47. Given the filter circuit of Figure 22–58:
    a. Determine the "notch" frequency.
    b. Calculate the Q of the circuit.
    c. Solve for the bandwidth and determine the half-power frequencies.
    d. Sketch the voltage gain response of the circuit, showing the level (in dB) at the "notch" frequency.

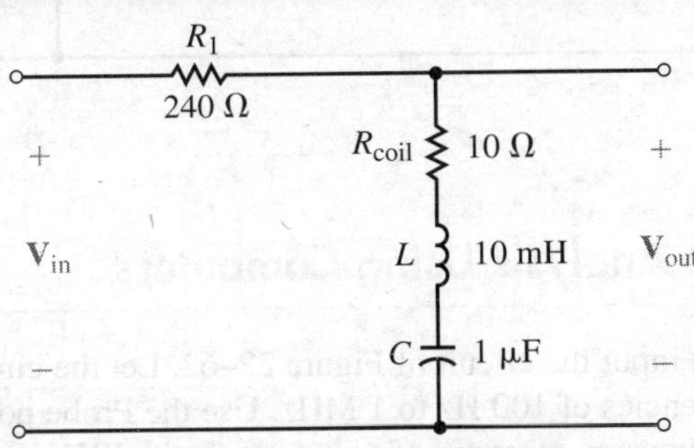

FIGURE 22–58

48. Repeat Problem 47 for the circuit of Figure 22–59.

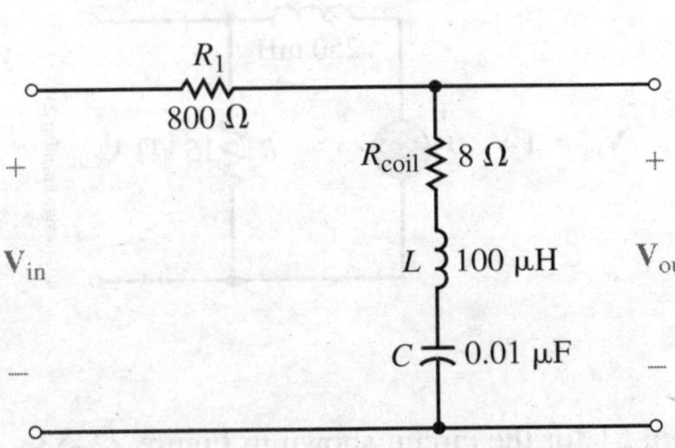

FIGURE 22–59

49. Given the circuit of Figure 22–60:

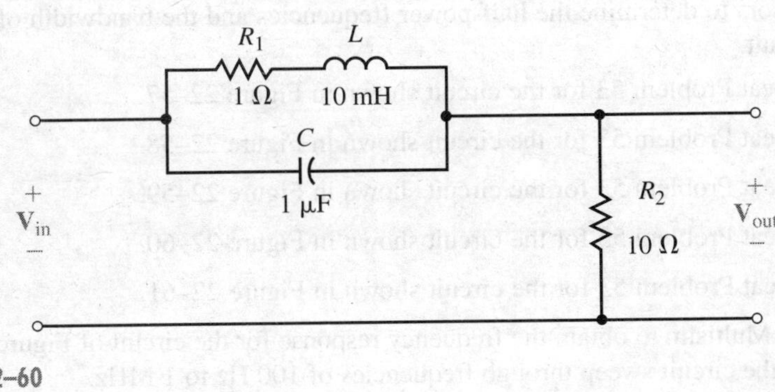

FIGURE 22–60

a. Determine the "notch" frequency in both rad/s and Hz.
b. Solve for the Q of the circuit.
c. Calculate the voltage gain (in dB) at the notch frequency.
d. Determine the voltage gain at the boundary conditions.
e. Solve for the bandwidth of the notch filter and determine the frequencies at which the voltage gain is 3 dB higher than at the notch frequency.

50. Repeat Problem 49 for the circuit of Figure 22–61.

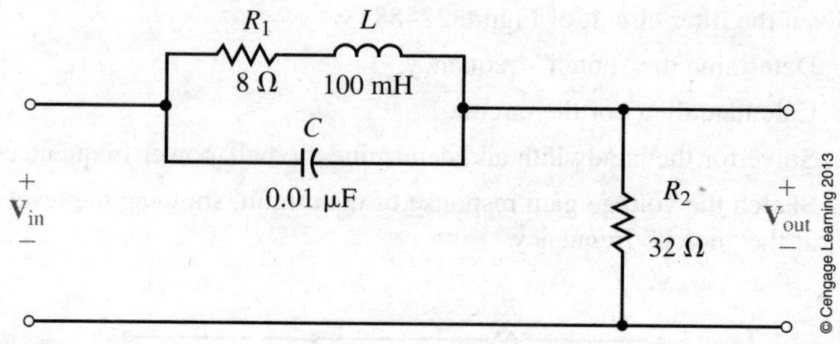

FIGURE 22–61

22.8 Circuit Analysis Using Computers

PSPICE 51. Use PSpice to input the circuit of Figure 22–62. Let the circuit sweep through frequencies of 100 Hz to 1 MHz. Use the Probe postprocessor to display the frequency response of voltage gain (in dBV) and phase shift of the circuit.

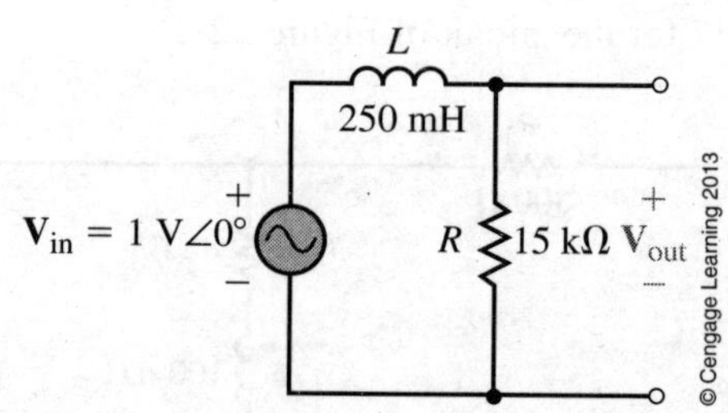

FIGURE 22–62

PSPICE 52. Repeat Problem 51 for the circuit shown in Figure 22–53.

PSPICE 53. Use PSpice to input the circuit of Figure 22–56. Use the Probe postprocessor to display the frequency response of voltage gain (in dBV) and phase shift of the circuit. Select a suitable range for the frequency sweep and use the cursors to determine the half-power frequencies and the bandwidth of the circuit.

PSPICE 54. Repeat Problem 53 for the circuit shown in Figure 22–57.

PSPICE 55. Repeat Problem 53 for the circuit shown in Figure 22–58.

PSPICE 56. Repeat Problem 53 for the circuit shown in Figure 22–59.

PSPICE 57. Repeat Problem 53 for the circuit shown in Figure 22–60.

PSPICE 58. Repeat Problem 53 for the circuit shown in Figure 22–61.

Multisim 59. Use Multisim to obtain the frequency response for the circuit of Figure 22–62. Let the circuit sweep through frequencies of 100 Hz to 1 MHz.

60. Repeat Problem 59 for the circuit of Figure 22–53. MULTISIM

61. Use Multisim to obtain the frequency response for the circuit shown in Figure 22–58. Select a suitable frequency range and use cursors to determine the "notch" frequency and the bandwidth of the circuit. MULTISIM

62. Repeat Problem 61 for the circuit shown in Figure 22–59. MULTISIM

63. Repeat Problem 61 for the circuit shown in Figure 22–60. MULTISIM

64. Repeat Problem 61 for the circuit shown in Figure 22–61. MULTISIM

ANSWERS TO IN-PROCESS LEARNING CHECKS

IN-PROCESS LEARNING CHECK 1

1. a. −1.99 dBm b. −15.0 dBm c. −66.0 dBm
2. a. 0.561 V_{rms} b. 0.224 V_{rms} c. 0.354 V_{rms}

IN-PROCESS LEARNING CHECK 2

a. $R = 3333\ \Omega$ in series with $C = 0.01\ \mu F$ (output across C)

b. $R_1 = 630\ \Omega$ in series with $L = 10$ mH and $R_2 = 630\ \Omega$ (output across R_2)

IN-PROCESS LEARNING CHECK 3

1. $\mathbf{TF} = \dfrac{j\omega(6.36 \times 10^{-6})}{1 + j\omega(6.36 \times 10^{-6})}$

   $C = 0.05\ \mu F$ is in series with $R = 127.3\ \Omega$ (output across R)

2. $\mathbf{TF} = \dfrac{j\omega(3.125 \times 10^{-6})}{1 + j\omega(12.5 \times 10^{-6})}$

   $R_1 = 8\ k\Omega$ is in series with $L = 25\ mH \| R_2 = 2.67\ k\Omega$ (output across the parallel combination)

IN-PROCESS LEARNING CHECK 4

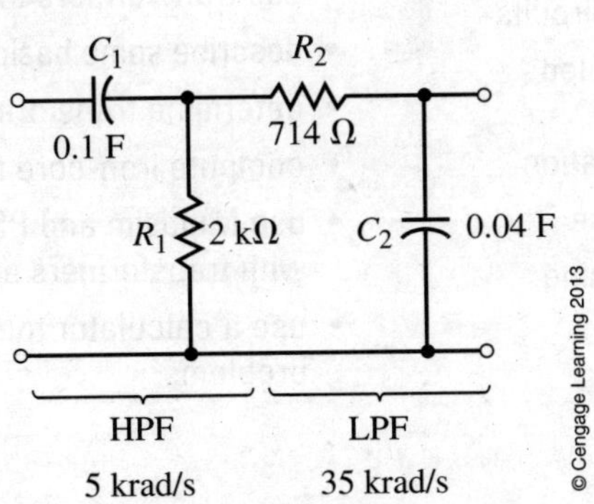

FIGURE 22–63

KEY TERMS

Autotransformer
Coefficient of Coupling
Copper Loss
Core Loss
Corresponding Terminals
Coupled Circuit
Coupled Impedance
Current Ratio
Dot Convention
Ideal Transformer
Impedance Matching
Leakage Flux
Loosely Coupled
Mutual Inductance
Mutually Induced Voltage
Open-Circuit Test
Primary
Reflected Impedance
Secondary
Self-Induced Voltage
Short-Circuit Test
Step-Up/Step-Down
Tightly Coupled
Transformer
Turns Ratio
Voltage Ratio

OUTLINE

Introduction
Iron-Core Transformers: The Ideal Model
Reflected Impedance
Power Transformer Ratings
Transformer Applications
Practical Iron-Core Transformers
Transformer Tests
Voltage and Frequency Effects
Loosely Coupled Circuits
Magnetically Coupled Circuits with Sinusoidal Excitation
Coupled Impedance
Circuit Analysis Using Computers

OBJECTIVES

After studying this chapter you will be able to

- describe how a transformer couples energy from its primary to its secondary via a changing magnetic field,
- describe basic transformer construction,
- use the dot convention to determine transformer phasing,
- determine voltage and current ratios from the turns ratio for iron-core transformers,
- compute voltage and currents in circuits containing iron-core and air-core transformers,
- use transformers to impedance match loads,
- describe some basic transformer applications,
- determine transformer equivalent circuits,
- compute iron-core transformer efficiency,
- use Multisim and PSpice to solve circuits with transformers and coupled circuits,
- use a calculator to solve coupled-circuit problems.

23

TRANSFORMERS AND COUPLED CIRCUITS

CHAPTER PREVIEW

In our study of induced voltage in Chapter 13 we found that the changing magnetic field produced by current in one coil induced a voltage in a second coil wound on the same core. A device built to utilize this effect is the **transformer.**

Transformers have many applications. They are used in electrical power systems to step up voltage for long-distance transmission and then to step it down again to a safe level for use in our homes and offices. They are used in electronic equipment power supplies to raise or lower voltages, in audio systems to match speaker loads to amplifiers, in telephone, radio, and TV systems, and so on.

In this chapter, we look at transformer fundamentals and the analysis of circuits containing transformers. We discuss transformer action, types of transformers, voltage and current ratios, and applications. Both iron-core and air-core transformers are covered along with general circuit coupling. ■

Putting It in Perspective

George Westinghouse

Omikron/Photo Researchers/Getty Images

ONE OF THE DEVICES that made possible the commercial ac power system as we know it today is the transformer. Although Westinghouse did not invent the transformer, his acquisition of the transformer patent rights and his manufacturing business helped make him an important player in the battle of dc versus ac in the emerging electrical power industry (see Chapter 15). In concert with Tesla (see Chapter 24), Westinghouse fought vigorously for ac against Edison, who favored dc. In 1893, Westinghouse's company built the Niagara Falls power system using ac, and the battle was over with ac being the clear winner. (Ironically, in recent years dc has been resurrected for use in commercial electrical power systems because at extremely high voltages, it is able to transmit power over longer distances than ac. However, this was not possible in Edison's day, and ac was and still is the correct choice for the commercial electrical power system.)

George Westinghouse was born in 1846 in Central Bridge, New York. He made his fortune with the invention of the railway air brake system. He died in 1914 and was elected to the Hall of Fame for Great Americans in 1955. ■

23.1 Introduction

A transformer is a magnetically **coupled circuit,** that is, a circuit in which the magnetic field produced by time-varying current in one circuit induces voltage in another. To illustrate, a basic iron-core transformer is shown in Figure 23–1. It consists of two coils wound on a common core. Alternating

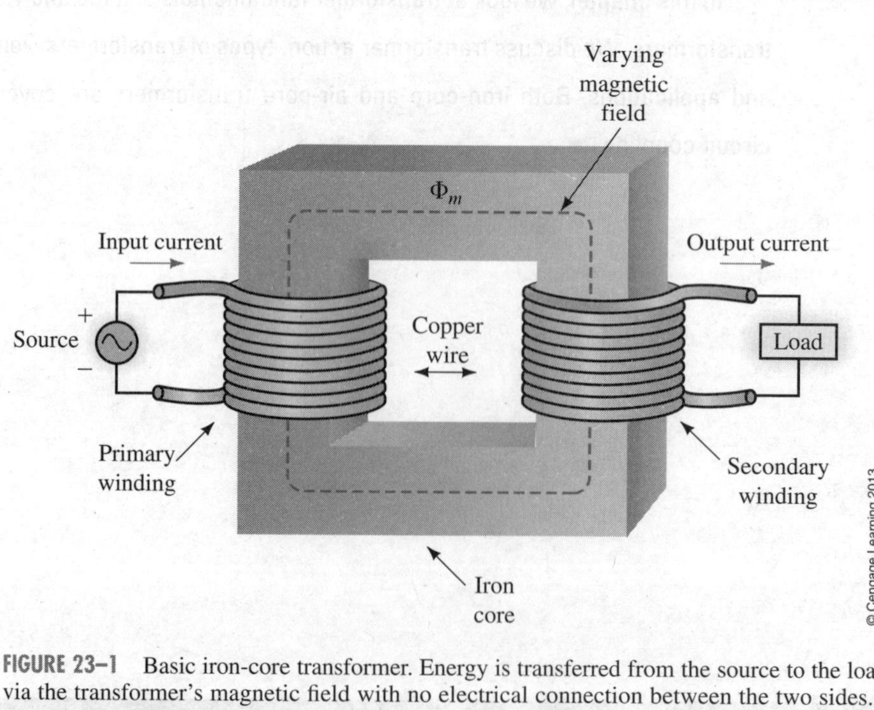

FIGURE 23–1 Basic iron-core transformer. Energy is transferred from the source to the load via the transformer's magnetic field with no electrical connection between the two sides.

current in one winding establishes a flux that links the other winding and induces a voltage in it. Power thus flows from one circuit to the other via the medium of the magnetic field, with no electrical connection between the two sides. The winding to which we supply power is called the **primary,** while the winding from which we take power is called the **secondary.** Power can flow in either direction, as either winding can be used as the primary or the secondary.

Transformer Construction

Transformers fall into two broad categories, iron core and air core. We begin with iron-core types. Iron-core transformers are generally used for low-frequency applications such as audio- and power-frequency applications. Figures 23–2 and 23–3 show a few examples of iron-core transformers.

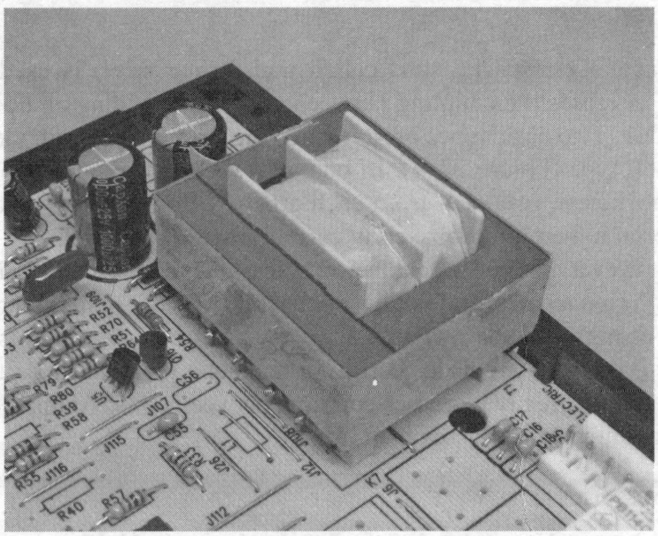

FIGURE 23–2 Iron-core transformer on the control board of an electric range. The primary and secondary windings are wound on bobbins that are then mounted on the core center leg. This transformer converts 120 VAC to the lower voltage required by the stove control circuitry.

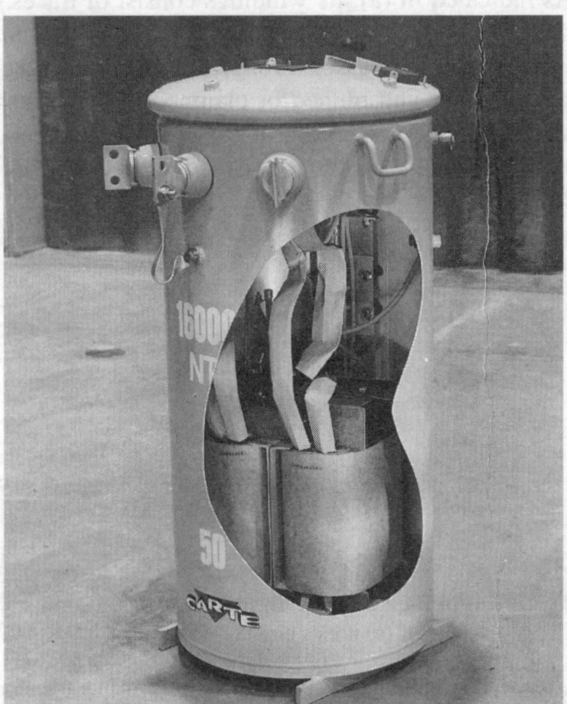

FIGURE 23–3 Distribution transformer (cutaway view) of the type used by electric utilities to distribute power to residential and commercial users. The tank is filled with oil to improve insulation and to remove heat from the core and windings.

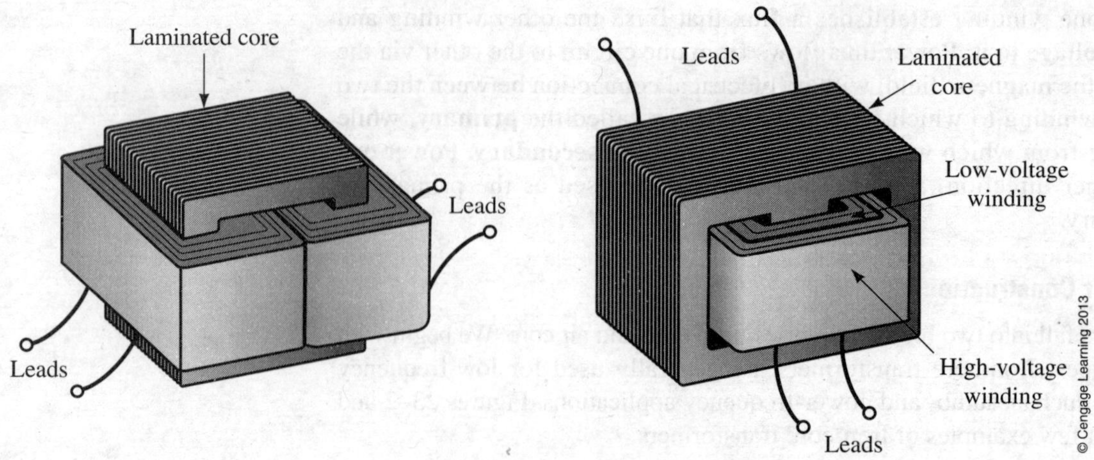

FIGURE 23–4 For the core type (left), windings are on separate legs, while for the shell type both windings are on the same leg.

Iron (actually a special steel called transformer steel) is used for cores because it increases the coupling between coils by providing an easy path for magnetic flux. Two basic types of iron-core construction are used, the core type and the shell type (Figure 23–4). In both cases, cores are made from laminations of sheet steel, insulated from each other by thin coatings of ceramic or other material to help minimize eddy current losses (Section 23.6).

Iron, however, has considerable power loss due to hysteresis and eddy currents at high frequencies, and is thus not useful as a core material above about 50 kHz. For high-frequency applications (such as in radio circuits), air-core and ferrite-core types are used. Ferrite (a magnetic material made from powdered iron oxide) greatly increases coupling between coils (compared with air) while maintaining lower losses than iron.

Planar Transformers

A relatively recent innovation in transformer design is the planar transformer, Figure 23–5. As indicated in (a), its windings consist of traces on printed circuit boards rather than coils of wire. The core consists of two flat pieces of ferrite material whose center leg passes through cutouts in the boards. The ferrite pieces are butted together and pinned or clamped in some fashion to form a continuous core. The transformer core of (a) uses a round center post, while that of (b) uses a rectangular center leg (called an E-Core). Both designs are used in practice.

(a) This transformer is rated for operation from 200 kHz to 500 kHz.

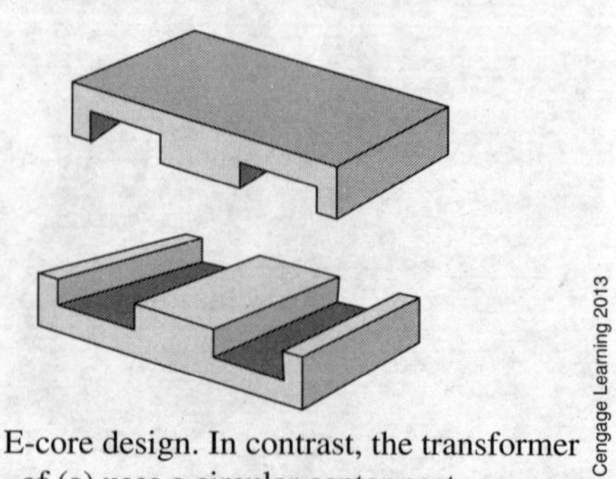

(b) An E-core design. In contrast, the transformer of (a) uses a circular center post.

FIGURE 23–5 Planar transformers have a low profile and are an excellent design for mass production. Most planar transformers use ferrite cores. After the printed circuit boards are in place, the two halves of the core are mated together and fastened securely.

While the transformer of (a) is built as a discrete component, some planar transformers are fabricated directly on the main circuit board itself. In this design, the primary winding is typically printed on one side of the board and the secondary on the other side. (Multilayer boards may have additional windings.) The core passes through cutouts in the circuit board, and the transformer thus becomes an integral part of the circuit board.

Planar transformers cost less to produce than traditional wire-wound transformers, have a very low profile, and are an excellent design for mass production. One common application is in switch mode power supplies. Planar transformers have a broad range of applications that range from telecommunications through industrial to military.

Transformer Symbols

Schematic circuit symbols for transformers are shown in Figure 23–6.

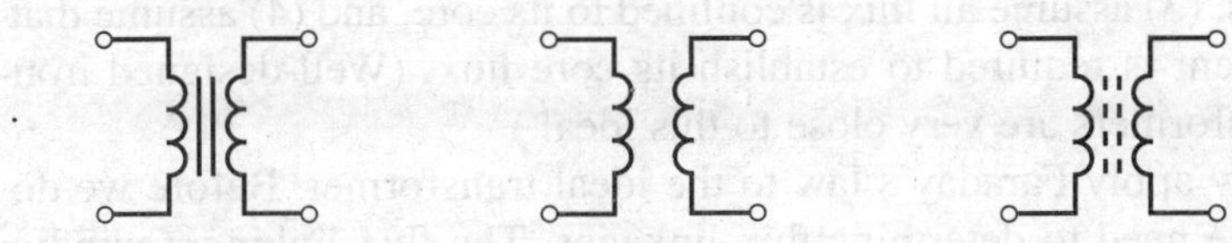

(a) Iron core (b) Air core (c) Ferrite core

FIGURE 23–6 Transformer schematic symbols.

Winding Directions

One of the advantages of a transformer is that it may be used to change the polarity of an ac voltage. This is illustrated in Figure 23–7 for a pair of iron-core transformers. For the transformer of (a) the primary and secondary voltages are in phase (for reasons to be discussed later), while for (b) they are 180° out of phase.

CircuitSim 23-1

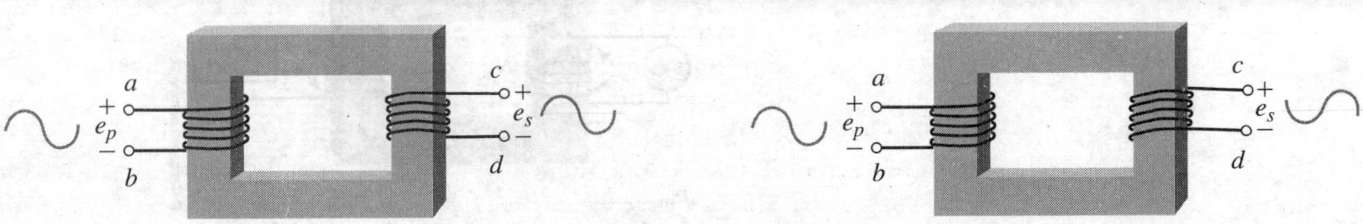

(a) 0° phase shift (b) 180° phase shift

FIGURE 23–7 The relative direction of the windings determines the phase shift.

Tightly Coupled and Loosely Coupled Circuits

If most of the flux produced by one coil links the other, the coils are said to be **tightly coupled.** Thus, iron-core transformers are tightly coupled (since close to 100% of the flux is confined to the core and thus links both windings). For air-core transformers, on the other hand, much less than 100% of the flux links both windings. They are therefore **loosely coupled.** However, regardless of whether a transformer is tightly coupled or loosely coupled, its analysis begins with Faraday's law.

Faraday's Law

Faraday's law (in SI units) states that the voltage induced in a circuit by a changing magnetic field is equal to the rate at which the flux linking the circuit is changing. When Faraday's law is applied to iron-core and air-core transformers, however, the results that emerge are quite different: Iron-core transformers are found to be characterized by their turns ratios, while air-core transformers are characterized by self- and mutual inductances. We begin with iron-core transformers.

23.2 Iron-Core Transformers: The Ideal Model

At first glance, iron-core transformers appear quite difficult to analyze because they have several characteristics such as winding resistance, core loss, and leakage flux that appear difficult to handle. Fortunately, these effects are small and often can be neglected. The result is the **ideal transformer.** Once you know how to analyze an ideal transformer, however, it is relatively easy to go back and add in the nonideal effects. This is the approach used in practice.

To idealize a transformer, (1) neglect the resistance of its coils, (2) neglect its core loss, (3) assume all flux is confined to its core, and (4) assume that negligible current is required to establish its core flux. (Well-designed iron-core power transformers are very close to this ideal.)

We now apply Faraday's law to the ideal transformer. Before we do this, however, we need to determine flux linkages. The flux linking a winding (as determined in Chapter 13) is the product of the flux that passes through the winding times the number of turns through which it passes. For flux Φ passing through N turns, flux linkage is $N\Phi$. Thus, for the ideal transformer (Figure 23–8), the primary flux linkage is $N_p\Phi_m$, while the secondary flux linkage is $N_s\Phi_m$, where the subscript "m" indicates mutual flux, that is, flux that links both windings.

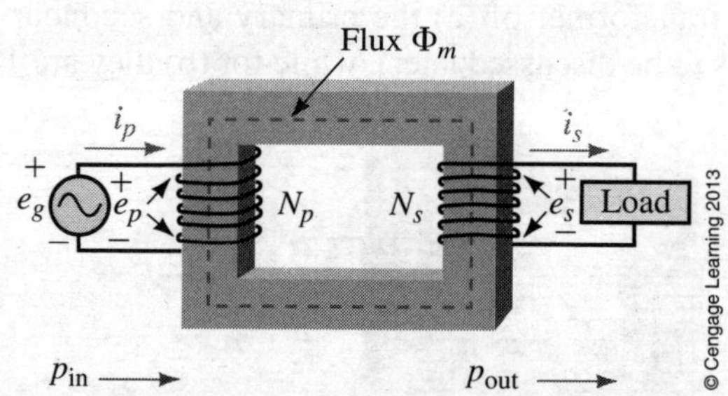

FIGURE 23–8 Ideal transformer. All flux is confined to the core and links both windings. This is a "tightly coupled" transformer.

Voltage Ratio

Now apply Faraday's law. Since the flux linkage equals $N\Phi$ and since N is constant, the induced voltage is equal to N times the rate of change of Φ, that is, $e = Nd\Phi/dt$. Thus, for the primary of an ideal transformer,

$$e_p = N_p\frac{d\Phi_m}{dt} \tag{23–1}$$

while for the secondary

$$e_s = N_s\frac{d\Phi_m}{dt} \tag{23–2}$$

Dividing Equation 23–1 by Equation 23–2 and cancelling $d\Phi_m/dt$ yields

$$\frac{e_p}{e_s} = \frac{N_p}{N_s} \tag{23–3}$$

Equation 23–3 states that *the ratio of primary voltage to secondary voltage is equal to the ratio of primary turns to secondary turns.* This ratio is called the **turns ratio** (or transformation ratio) and is given the symbol *a*. Thus,

$$a = N_p/N_s \tag{23–4}$$

and

$$e_p/e_s = a \tag{23–5}$$

For example, a transformer with 1000 turns on its primary and 250 turns on its secondary has a turns ratio of 1000/250 = 4. This is referred to as a 4:1 ratio.

Since the ratio of two instantaneous sinusoidal voltages is the same as the ratio of their effective values, Equation 23–5 can also be written as

$$E_p/E_s = a \tag{23–6}$$

As noted earlier, e_p and e_s are either in phase or 180° out of phase, depending on the relative direction of coil windings. We can therefore also express the **voltage ratio** in terms of phasors as

$$\mathbf{E}_p/\mathbf{E}_s = a \tag{23–7}$$

where the relative polarity (in phase or 180° out of phase) is determined by the direction of the coil windings (Figure 23–7).

Step-Up and Step-Down Transformers

A **step-up** transformer is one in which the secondary voltage is higher than the primary voltage, while a **step-down** transformer is one in which the secondary voltage is lower. Since $a = E_p/E_s$, a step-up transformer has $a < 1$, while for a step-down transformer, $a > 1$. If $a = 1$, the transformer's turns ratio is unity and the secondary voltage is equal to the primary voltage.

EXAMPLE 23–1

Suppose the transformer of Figure 23–7(a) has 500 turns on its primary and 1000 turns on its secondary.

a. Determine its turns ratio. Is it step-up or step-down?

b. If its primary voltage is $e_p = 25 \sin \omega t$ V, what is its secondary voltage?

c. Sketch the waveforms.

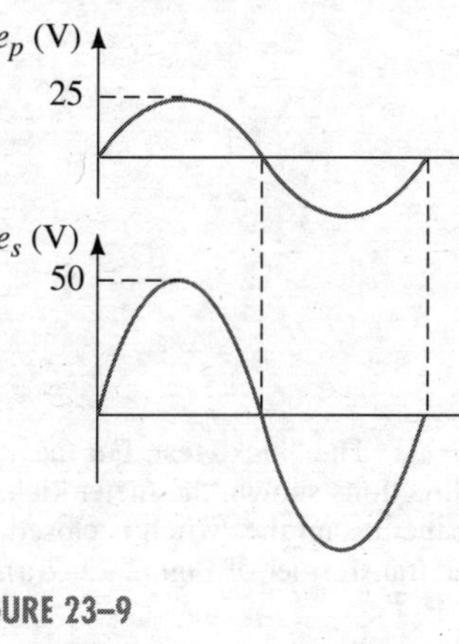

FIGURE 23–9

Solution

a. The turns ratio is $a = N_p/N_s = 500/1000 = 0.5$. This is a 1:2 step-up transformer.

b. From Equation 23–5, $e_s = e_p/a = (25 \sin \omega t)/0.5 = 50 \sin \omega t$ V.

c. Primary and secondary voltages are in phase as noted earlier. Figure 23–9 shows the waveforms.

EXAMPLE 23–2

If the transformers of Figure 23–7 have 600 turns on their primaries and 120 turns on their secondaries, and $\mathbf{E}_p = 120 \text{ V}\angle 0°$, what is $\mathbf{E}_s$ for each case?

Solution The turns ratio is $a = 600/120 = 5$. For transformer (a), $\mathbf{E}_s$ is in phase with $\mathbf{E}_p$. Therefore, $\mathbf{E}_s = \mathbf{E}_p/5 = (120 \text{ V}\angle 0°)/5 = 24 \text{ V}\angle 0°$. For transformer (b), $\mathbf{E}_s$ is 180° out of phase with $\mathbf{E}_p$. Therefore, $\mathbf{E}_s = 24 \text{ V}\angle 180°$.

PRACTICE PROBLEMS 1

Repeat Example 23–1 for the circuit of Figure 23–7(b) if $N_p = 1200$ turns and $N_s = 200$ turns.

Answer
$e_s = 4.17 \sin(\omega t + 180°)$

Current Ratio

Because an ideal transformer has no power loss, its efficiency is 100% and thus power in equals power out. Consider again Figure 23–8. At any instant, $p_{\text{in}} = e_p i_p$ and $p_{\text{out}} = e_s i_s$. Thus,

$$e_p i_p = e_s i_s \tag{23–8}$$

This yields a **current ratio** of

$$\frac{i_p}{i_s} = \frac{e_s}{e_p} = \frac{1}{a} \tag{23–9}$$

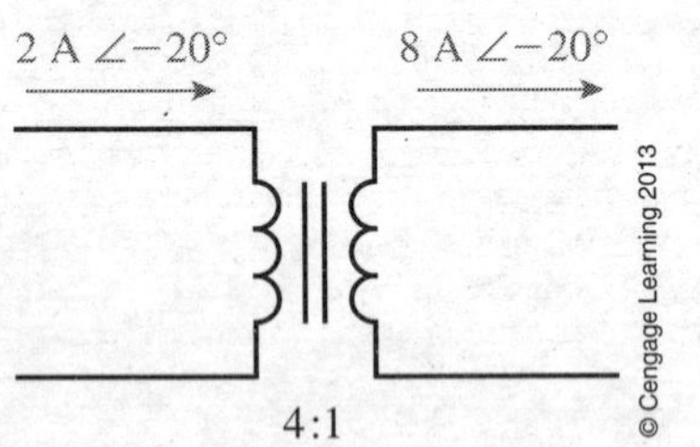

FIGURE 23–10 Currents are inverse to the turns ratio.

CircuitSim 23-2

since $e_s/e_p = 1/a$. (This means that if voltage is stepped up, current is stepped down, and vice versa.) In terms of current phasors and current magnitudes, Equation 23–9 can be written as

$$\frac{\mathbf{I}_p}{\mathbf{I}_s} = \frac{I_p}{I_s} = \frac{1}{a} \tag{23–10}$$

For example, for a transformer with $a = 4$, and $\mathbf{I}_p = 2\text{ A}\angle -20°$, $\mathbf{I}_s = a\mathbf{I}_p = 4(2\text{ A}\angle -20°) = 8\text{ A}\angle -20°$, Figure 23–10.

Polarity of Induced Voltage: The Dot Convention

As noted earlier, an iron-core transformer's secondary voltage is either in phase with its primary voltage or 180° out of phase, depending on the relative direction of its windings. We will now demonstrate why.

A simple test, called the kick test (sometimes used by electrical workers to determine transformer polarity), can help establish the idea. The basic circuit is shown in Figure 23–11. A switch is used to make and break the circuit (since voltage is induced only while flux is changing).

FIGURE 23–11 The "kick" test. For the winding directions shown, the meter kicks upscale at the instant the switch is closed. [This is the transformer of Figure 23–7(a).]

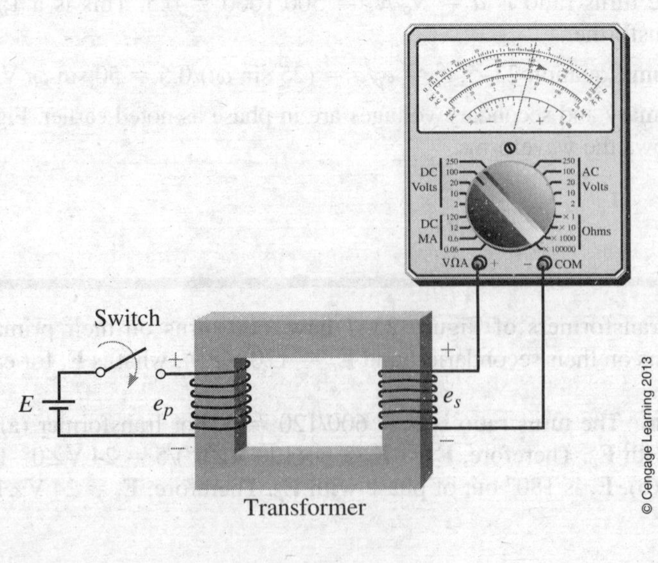

For the winding directions shown, at the instant the switch is closed, the voltmeter needle "kicks" upscale, then settles back to zero. To understand why, we need to consider magnetic fields. Before we start, however, let us place a dot on one of the primary terminals; in this case, we arbitrarily choose the top terminal. Let us also replace the voltmeter with its equivalent resistance (Figure 23–12).

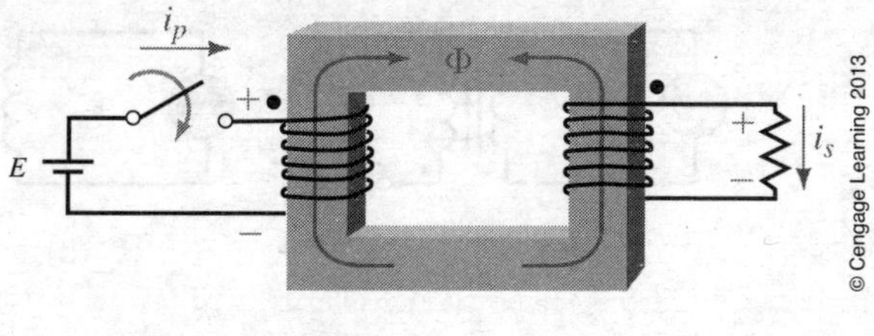

FIGURE 23–12 Determining dot positions.

At the instant the switch is closed, the polarity of the dotted primary terminal is positive with respect to the undotted primary terminal (because the + end of the source is directly connected to it). As current in the primary builds, it creates a flux in the upward direction as indicated by the blue arrow (recall the right-hand rule). According to Lenz's law, the effect that results must oppose the cause that produced it. The effect is a voltage induced in the secondary winding. The resulting current in the secondary produces a flux which, according to Lenz's law, must *oppose the buildup of the original flux*, that is, it must be in the direction of the red arrow. Applying the right-hand rule, we see that secondary current must be in the direction indicated by i_s. Placing a plus sign at the tail of this arrow shows that the top end of the resistor is positive. This means that the top end of the secondary winding is also positive. Place a dot here. Dotted terminals are called **corresponding terminals.**

As you can see, corresponding terminals are positive (with respect to their companion undotted terminals) at the instant the switch is closed. If you perform a similar analysis at the instant the switch is opened, you will find that both dotted terminals are negative. Thus, *dotted terminals have the same polarity at all instants of time.* What we have developed here is known as the **dot convention** for coupled circuits—see Notes.

PRACTICAL NOTES...

1. While we developed the dot convention using a switched dc source, it is valid for ac as well. In fact, we will use it mostly for ac.
2. The resistor of Figure 23–12 is required only to work through the physics to establish the secondary voltage polarity. You can now remove it without affecting the resulting dot position.
3. In practice, corresponding terminals may be marked by dots, by color-coded wires, or by special letter designations.
4. Much more is said about the dot convention and its use in Section 23.9.

EXAMPLE 23–3

Determine the waveform for e_s in the circuit of Figure 23–13(a).

CircuitSim 23-3

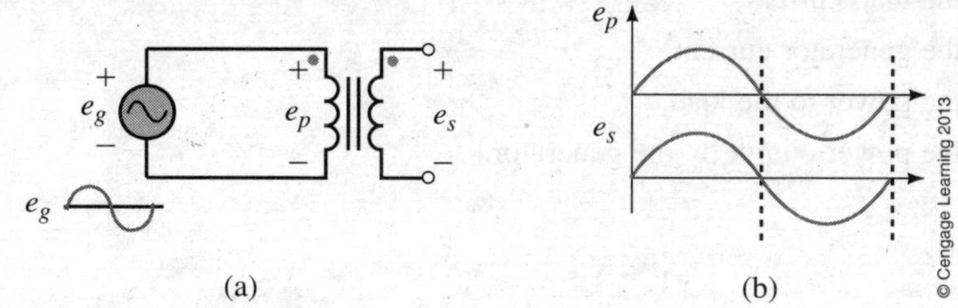

FIGURE 23–13

Solution Dotted terminals have the same polarity (with respect to their undotted terminals) at all instants. During the first half-cycle, the dotted end of the primary coil is positive. Therefore, the dotted end of the secondary is also positive. During the second half-cycle, both are negative. The polarity marks on e_s mean that we are looking at the polarity of the top end of the secondary coil with respect to its bottom end. Thus, e_s is positive during the first half-cycle and negative during the second half-cycle. It is therefore in phase with e_p as indicated in (b). Thus, if $e_p = E_{m_p} \sin \omega t$, then $e_s = E_{m_s} \sin \omega t$.

PRACTICE PROBLEMS 2

1. Determine the equation for e_s in the circuits of Figure 23–14(b), (c), and (d).

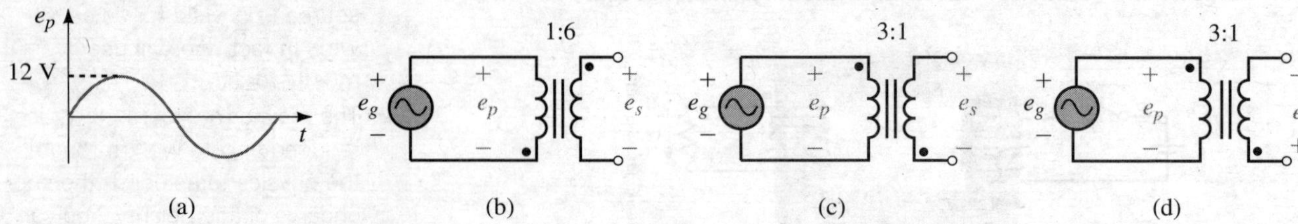

FIGURE 23–14 Waveform (a) is the applied primary voltage.

CircuitSim 23-4

2. If $\mathbf{E}_g = 120\ \text{V}\angle 30°$, determine $\mathbf{E}_s$ for each transformer of Figure 23–14.
3. Where do the dots go on the transformers of Figure 23–7?

Answers
1. $e_s = 72 \sin(\omega t + 180°)$ V; $e_s = 4 \sin(\omega t + 180°)$ V; $e_s = 4 \sin \omega t$
2. For (b), $\mathbf{E}_s = 720\ \text{V}\angle{-150°}$; for (c), $\mathbf{E}_s = 40\ \text{V}\angle{-150°}$; $40\ \text{V}\angle 30°$
3. For (a), place dots at *a* and *c*. For (b), place dots at *a* and *d*.

Analysis of Simple Transformer Circuits

Simple transformer circuits may be analyzed using the relationships described so far, namely $\mathbf{E}_p = a\mathbf{E}_s$, $\mathbf{I}_p = \mathbf{I}_s/a$, and $P_{\text{in}} = P_{\text{out}}$. This is illustrated in Example 23–4. More complex problems require some additional ideas.

EXAMPLE 23–4

For Figure 23–15(a), $\mathbf{E}_g = 120\ \text{V}\angle 0°$, the turns ratio is 5:1, and $\mathbf{Z}_L = 4\ \Omega\angle 30°$. The transformer is ideal. Find

a. the load voltage,
b. the load current,
c. the generator current,
d. the power to the load,
e. the power output by the generator.

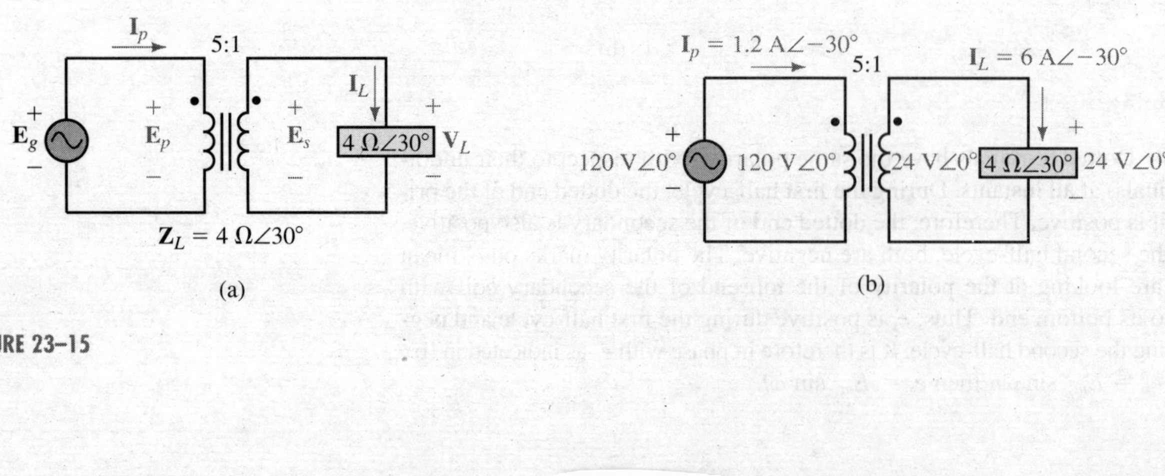

FIGURE 23–15

Solution

a. $\mathbf{E}_p = \mathbf{E}_g = 120\ \text{V}\angle 0°$

$\mathbf{V}_L = \mathbf{E}_s = \mathbf{E}_p/a = (120\ \text{V}\angle 0°)/5 = 24\ \text{V}\angle 0°$

b. $\mathbf{I}_L = \mathbf{V}_L/\mathbf{Z}_L = (24\ \text{V}\angle 0°)/(4\ \Omega\angle 30°) = 6\ \text{A}\angle -30°$

c. $\mathbf{I}_g = \mathbf{I}_p$. But $\mathbf{I}_p = \mathbf{I}_s/a = \mathbf{I}_L/a$. Thus,

$\mathbf{I}_g = (6\ \text{A}\angle -30°)/5 = 1.2\ \text{A}\angle -30°$

Values are shown on Figure 23–15(b).

d. $P_L = V_L I_L \cos\theta_L = (24)(6)\cos 30° = 124.7\ \text{W}$

CircuitSim 23-5

e. $P_g = E_g I_g \cos\theta_g$, where θ_g is the angle between $\mathbf{E}_g$ and $\mathbf{I}_g$. $\theta_g = 30°$. Thus, $P_g = (120)(1.2)\cos 30° = 124.7\ \text{W}$, which agrees with (d) as it must since the transformer is lossless.

IN-PROCESS LEARNING CHECK 1

(Answers are at the end of the chapter.)

1. A transformer has a turns ratio of 1:8. Is it step-up or step-down? If $E_p = 25$ V, what is E_s?
2. For the transformers of Figure 23–7, if $a = 0.2$ and $\mathbf{E}_s = 600\ \text{V}\angle -30°$, what is $\mathbf{E}_p$ for each case?
3. For each of the transformers of Figure 23–16, sketch the secondary voltage, showing both phase and amplitude.

CircuitSim 23-6

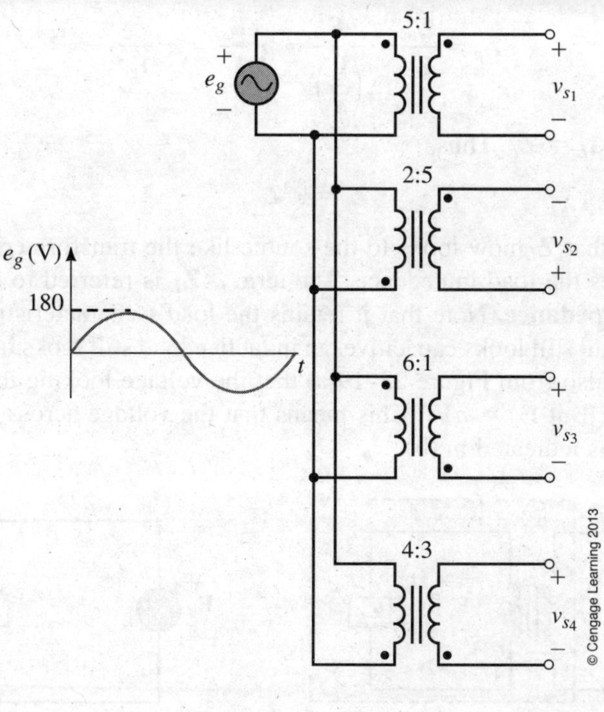

FIGURE 23–16

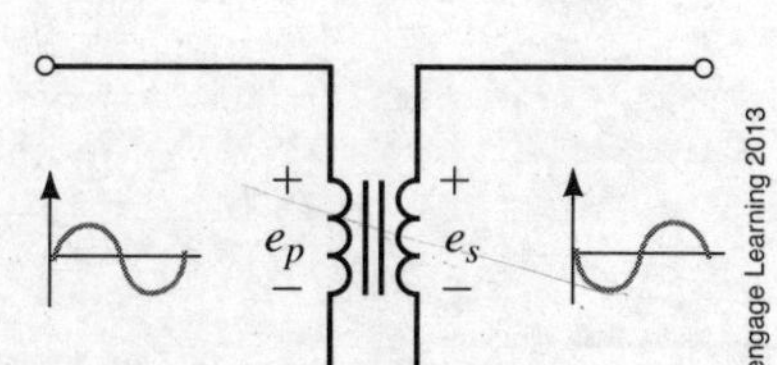

FIGURE 23–17

4. For Figure 23–17, determine the position of the missing dot.

5. Figure 23–18 shows another way to determine dotted terminals. First, arbitrarily mark one of the primary terminals with a dot. Next, connect a jumper wire and voltmeter as indicated. From the voltmeter readings, you can determine which of the secondary terminals should be dotted. For the two cases indicated, where should the secondary dot be? (*Hint:* Use KVL.)

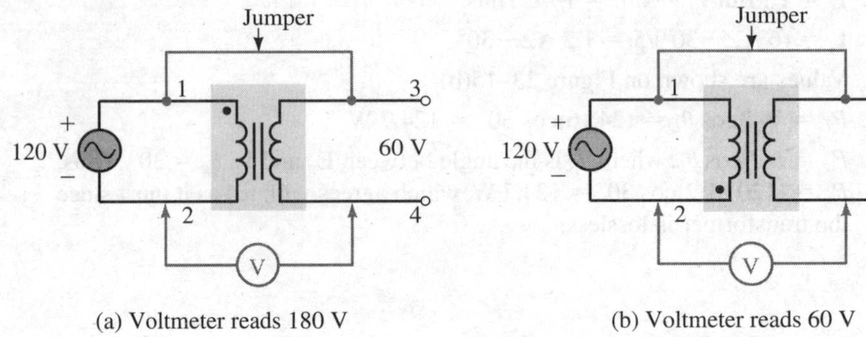

FIGURE 23–18 Each transformer is rated 120 V/60 V.

23.3 Reflected Impedance

A transformer makes a load impedance look larger or smaller, depending on its turns ratio. To illustrate, consider Figure 23–19. When connected directly to the source, the load looks like impedance $\mathbf{Z}_L$, but when connected through a transformer, it looks like $a^2\mathbf{Z}_L$. This can be illustrated as follows: First, note that $\mathbf{Z}_p = \mathbf{E}_g/\mathbf{I}_p$. But $\mathbf{E}_g = \mathbf{E}_p$, $\mathbf{E}_p = a\mathbf{E}_s$, and $\mathbf{I}_p = \mathbf{I}_s/a$. Thus,

$$\mathbf{Z}_p = \frac{\mathbf{E}_p}{\mathbf{I}_p} = \frac{a\mathbf{E}_s}{\left(\dfrac{\mathbf{I}_s}{a}\right)} = a^2\frac{\mathbf{E}_s}{\mathbf{I}_s} = a^2\frac{\mathbf{V}_L}{\mathbf{I}_L}$$

However $\mathbf{V}_L/\mathbf{I}_L = \mathbf{Z}_L$. Thus,

$$\mathbf{Z}_p = a^2\mathbf{Z}_L \tag{23–11}$$

This means that $\mathbf{Z}_L$ now looks to the source like the transformer's turns ratio squared times the load impedance. The term $a^2\mathbf{Z}_L$ is referred to as the load's **reflected impedance.** Note that it retains the load's characteristics, that is, a capacitive load still looks capacitive, an inductive load still looks inductive, and so on. Note also from Figure 23–19(a) that the voltage looking into the transformer is $\mathbf{E}_g$. But $\mathbf{E}_g = a\mathbf{V}_L$. This means that the voltage across the reflected load is $a\mathbf{V}_L$ as indicated in (b).

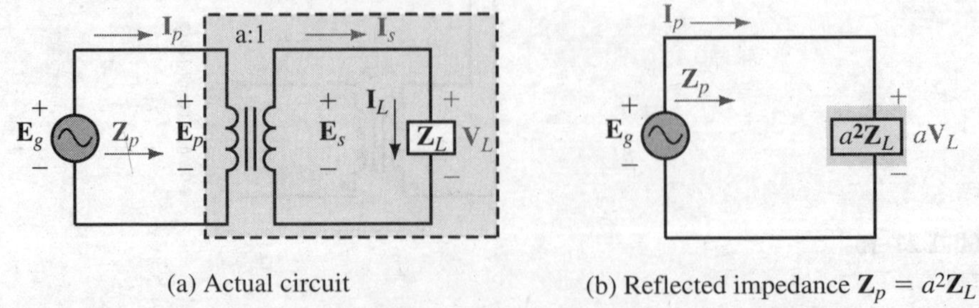

FIGURE 23–19 Concept of reflected impedance. From the primary terminals, $\mathbf{Z}_L$ looks like an impedance of $a^2\mathbf{Z}_L$ with voltage $a\mathbf{V}_L$ across it and current $\mathbf{I}_L/a$ through it.

From Equation 23–11, we see that the load will look larger if $a > 1$ and smaller if $a < 1$. To illustrate, consider Figure 23–20. If a 1-Ω resistor were connected directly to the source, it would look like a 1-Ω resistor and the generator current would be $100\text{A}\angle 0°$. However, when connected to a 10:1 transformer it looks like a $(10)^2(1\ \Omega) = 100\text{-}\Omega$ resistor and the generator current is only $1\ \text{A}\angle 0°$.

The concept of reflected impedance is useful in a number of ways. It permits us to match loads to sources (such as amplifiers) and it also provides an alternate way to solve transformer problems.

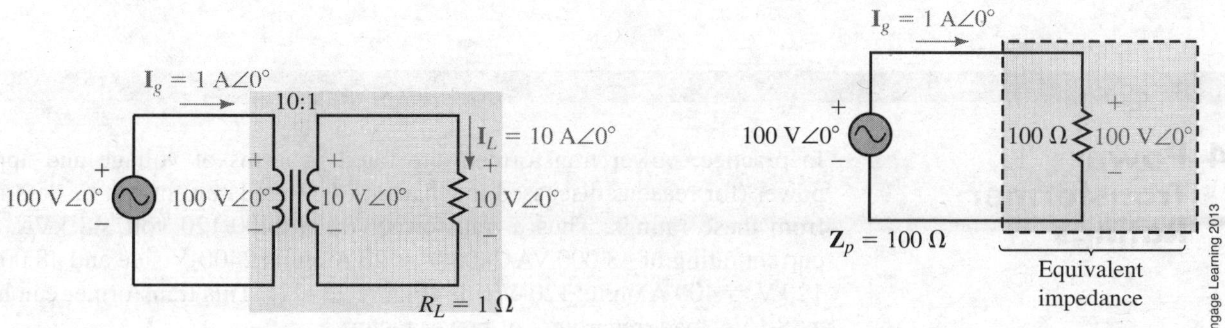

(a) Actual circuit

(b) Equivalent circuit

FIGURE 23–20 The equivalent impedance seen by the source is 100 Ω.

EXAMPLE 23–5

Use the reflected impedance idea to solve for primary and secondary currents and the load voltage for the circuit of Figure 23–21(a).

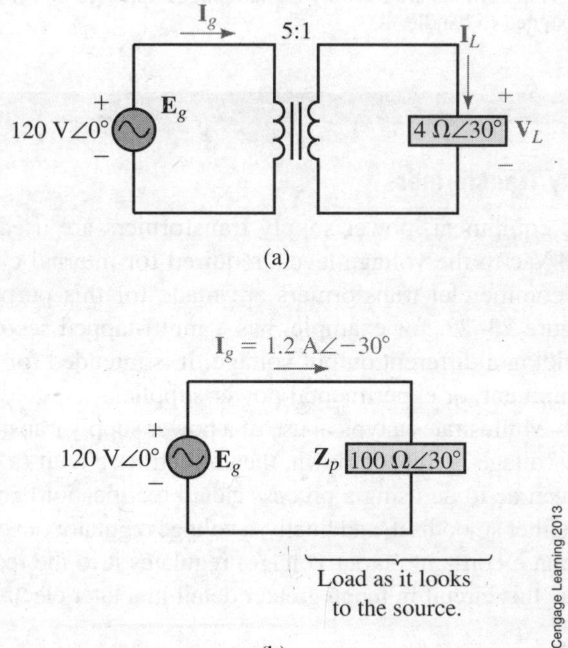

(a)

(b)

FIGURE 23–21

NOTES...

Example 23–5 is the same as Example 23–4. In Example 23–4, we analyzed the circuit in its original form, while here we used the reflected impedance method. As can be seen, the amount of work required is about the same for both solutions, illustrating that there is no particular advantage of one method over the other for a circuit as simple as this. However, for complex problems, the advantages of the reflected impedance method are considerable—you will find that it reduces the amount of work involved and greatly simplifies analysis.

Solution

$$\mathbf{Z}_p = a^2\mathbf{Z}_L = (5)^2(4\ \Omega\angle 30°) = 100\ \Omega\angle 30°$$

The equivalent circuit is shown in (b).

CircuitSim 23-7

$$\mathbf{I}_g = \mathbf{E}_g/\mathbf{Z}_p = (120\ \text{V}\angle 0°)/(100\ \Omega\angle 30°) = 1.2\ \text{A}\angle -30°$$
$$\mathbf{I}_L = a\mathbf{I}_p = a\mathbf{I}_g = 5(1.2\ \text{A}\angle -30°) = 6\ \text{A}\angle -30°$$
$$\mathbf{V}_L = \mathbf{I}_L\mathbf{Z}_L = (6\ \text{A}\angle -30°)(4\ \Omega\angle 30°) = 24\ \text{V}\angle 0°$$

The answers are the same as in Example 23–4—see Note on page 841.

23.4 Power Transformer Ratings

In practice, power transformers are rated in terms of voltage and apparent power (for reasons discussed in Chapter 17). Rated current can be determined from these ratings. Thus a transformer rated 2400/120 volt, 48 kVA, has a current rating of 48 000 VA/2400 V = 20 A on its 2400-V side and 48 000 VA/120 V = 400 A on its 120-V side (Figure 23–22). This transformer can handle a 48-kVA load, regardless of power factor.

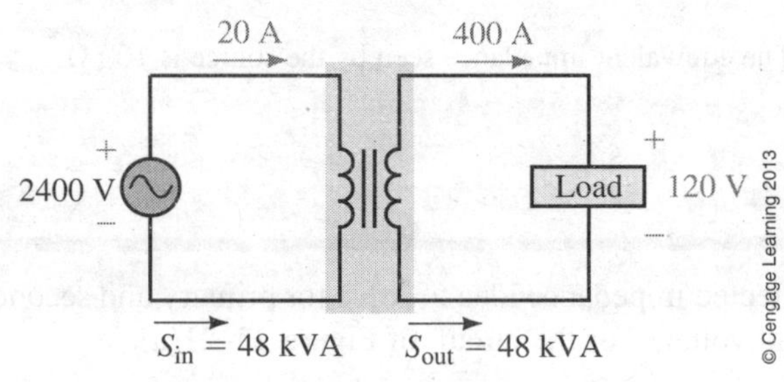

FIGURE 23–22 Transformers are rated by the amount of apparent power and the voltages that they are designed to handle.

23.5 Transformer Applications

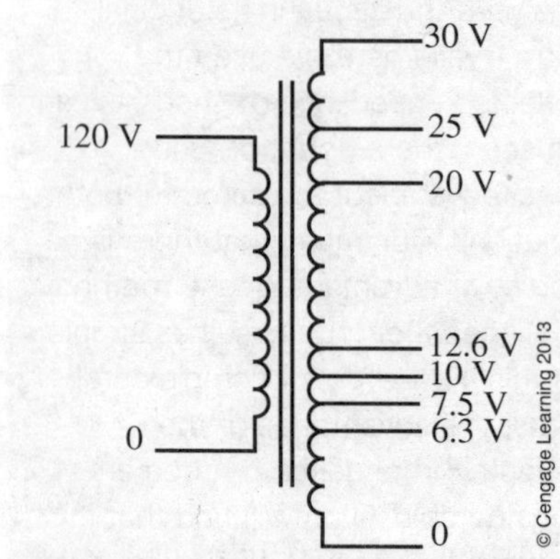

FIGURE 23–23 A multi-tapped power supply transformer. The secondary is tapped at various voltages.

Power Supply Transformers

On electronic equipment, power supply transformers are used to convert the incoming 120 Vac to the voltage levels required for internal circuit operation. A variety of commercial transformers are made for this purpose. The transformer of Figure 23–23, for example, has a multi-tapped secondary winding, each tap providing a different output voltage. It is intended for laboratory supplies, test equipment, or experimental power supplies.

Figure 23–24 illustrates a typical use of a power supply transformer. First, the incoming line voltage is stepped down, then a rectifier circuit (a circuit that uses diodes to convert ac to dc using a process called rectification) converts the ac to pulsating dc, a filter smooths it, and finally, a voltage regulator (an electronic device used to maintain a constant output voltage) regulates it to the required dc value. (You will study this circuit in much greater detail in a later electronics course.)

Transformers in Power Systems

Transformers are among the key elements that have made commercial ac power systems possible. Transformers are used at generating stations to raise

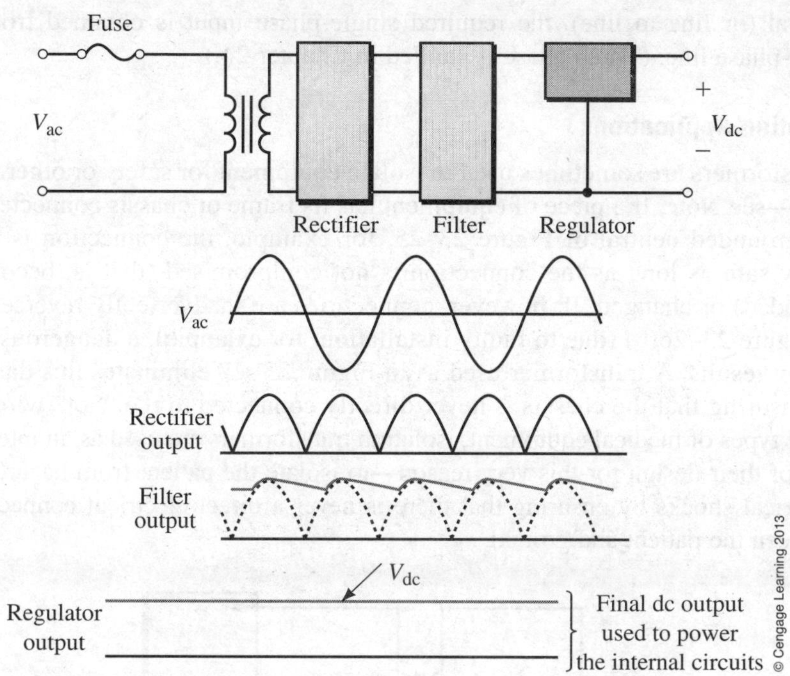

FIGURE 23–24 A transformer used in a power supply application.

voltage for long-distance transmission. This lowers the transmitted current and hence the I^2R power losses in the transmission line. At the user end, transformers reduce the voltage to a safe level for everyday use. A typical residential connection is shown in Figure 23–25. (In older neighborhoods, this is the transformer that you see hanging on power line poles. In newer neighborhoods, you may see them mounted on pads on somebody's lawn.) The taps on the primary permit the electric utility company to compensate for line voltage drops. Transformers located far from substations, for example, have lower input voltages (by a few percent) than those close to substations due to voltage drops on distribution lines. Taps permit the turns ratio to be changed to compensate. Note also the split secondary. It permits 120-V and 240-V loads to be supplied from the same transformer.

The transformer of Figure 23–25 is a single-phase unit (since residential customers require only single phase). By connecting its primary from line to

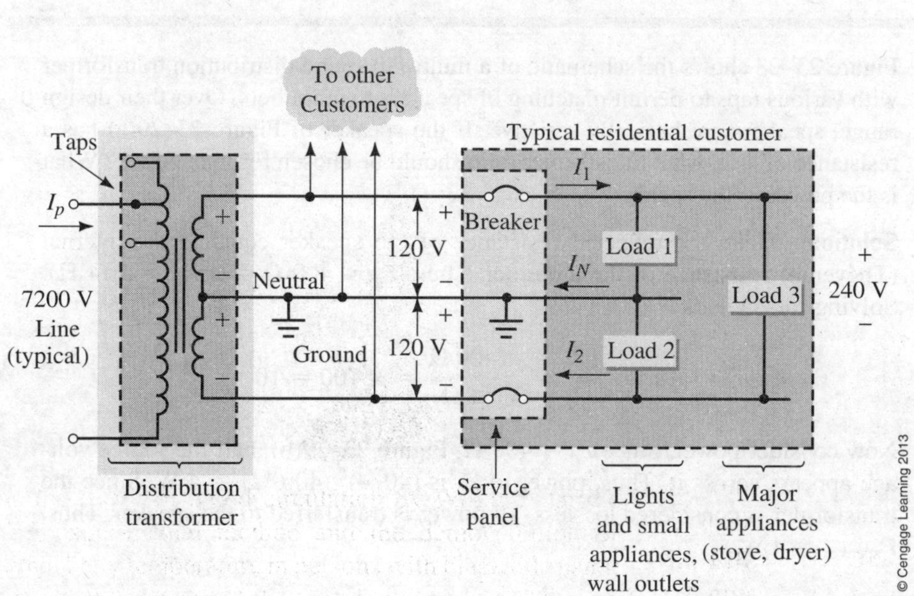

FIGURE 23–25 Typical distribution transformer. This is how power is supplied to your home.

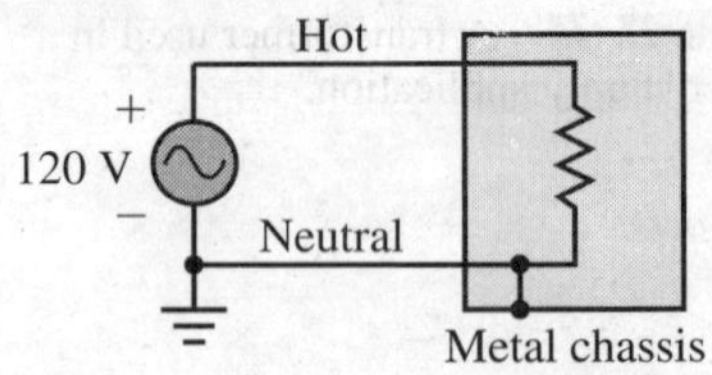

(a) Chassis is safe

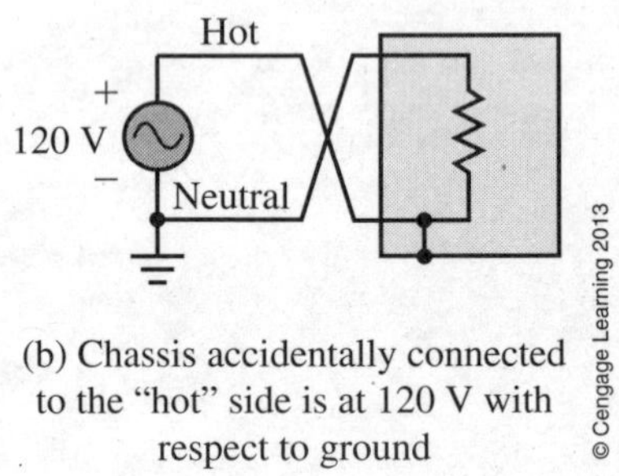

(b) Chassis accidentally connected to the "hot" side is at 120 V with respect to ground

FIGURE 23–26 If the connections are inadvertently reversed as in (b), you will get a shock if you are grounded and you touch the chassis.

neutral (or line to line), the required single-phase input is obtained from a three-phase line. (Three phase is studied in Chapter 24.)

Isolation Applications

Transformers are sometimes used to isolate equipment for safety or other reasons—see Note. If a piece of equipment has its frame or chassis connected to the grounded neutral of Figure 23–25, for example, the connection is perfectly safe as long as the connection is not compromised (that is, becomes corroded) or changed. If, however, connections are inadvertently reversed as in Figure 23–26(b) (due to faulty installation, for example), a dangerous situation results. A transformer used as in Figure 23–27 eliminates this danger by ensuring that the chassis is never directly connected to the "hot" wire. In some types of medical equipment, isolation transformers are used as an integral part of their design for this very reason—to isolate the patient from hazardous electrical shocks by ensuring that there is never a direct electrical connection between the patient and ground.

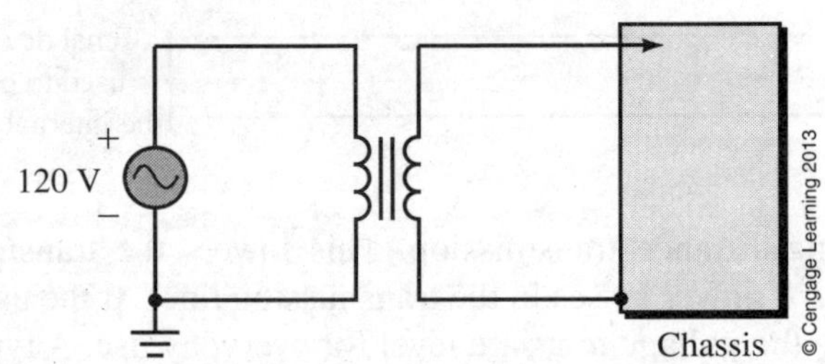

FIGURE 23–27 Using a transformer for isolation.

NOTES...

For further information, go to our Web site and follow the links to *For Further Investigation* and select *Appliance Grounding.*

Impedance Matching

As you learned earlier, a transformer can be used to raise or lower the apparent impedance of a load by choice of turns ratio. This is referred to as **impedance matching.** Impedance matching is sometimes used to match loads to amplifiers to achieve maximum power transfer. If the load and source are not matched, a transformer can be inserted between them as illustrated next.

EXAMPLE 23–6

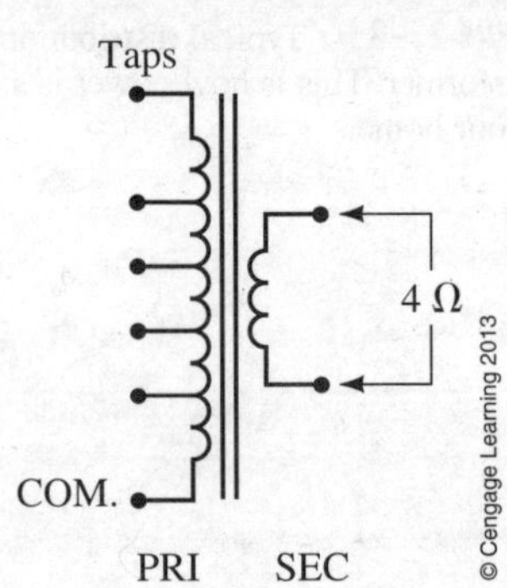

FIGURE 23–28 A tapped sound distribution transformer.

Figure 23–28 shows the schematic of a multi-tap sound distribution transformer with various taps to permit matching of speakers to amplifiers. Over their design range, speakers are basically resistive. If the speaker of Figure 23–29(a) has a resistance of 4 Ω, what transformer ratio should be chosen for max power? What is the power to the speaker?

Solution Make the reflected resistance of the speaker equal to the internal (Thévenin) resistance of the amplifier. Thus, $Z_p = 400\ \Omega = a^2 Z_L = a^2(4\ \Omega)$. Solving for a yields

$$a = \sqrt{\frac{Z_p}{Z_L}} = \sqrt{\frac{400\ \Omega}{4\ \Omega}} = \sqrt{100} = 10$$

Now consider power. Since $Z_p = 400\ \Omega$, Figure 23–29(b), half the source voltage appears across it. Thus, power to Z_p is $(40\ \text{V})^2/(400\ \Omega) = 4$ W. Since the transformer is considered lossless, all power is transferred to the speaker. Thus, $P_{\text{speaker}} = 4$ W.

CircuitSim 23-8

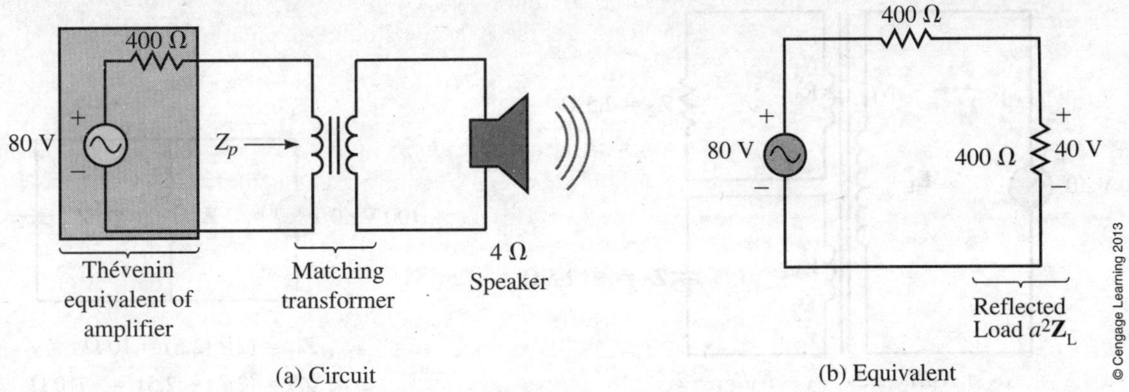

FIGURE 23–29 Matching the 4-Ω speaker to the amplifier for maximum power transfer. Power to the speaker here is 4 W. (Compare to Practice Problems 3.)

PRACTICE PROBLEMS 3

Determine the power to the speaker of Figure 23–29 if the transformer is not present (that is, the speaker is directly connected to the amplifier). Compare to Example 23–6.

Answer
0.157 W (dramatically lower)

Transformers with Multiple Secondaries

For a transformer with multiple secondaries (Figure 23–30), each secondary voltage is governed by the appropriate turns ratio, that is, $\mathbf{E}_1/\mathbf{E}_2 = N_1/N_2$ and $\mathbf{E}_1/\mathbf{E}_3 = N_1/N_3$. Loads are reflected in parallel. That is, $\mathbf{Z'}_2 = a_2^2\mathbf{Z}_2$ and $\mathbf{Z'}_3 = a_3^2\mathbf{Z}_3$ appear in parallel in the equivalent circuit, Figure 23–30(b).

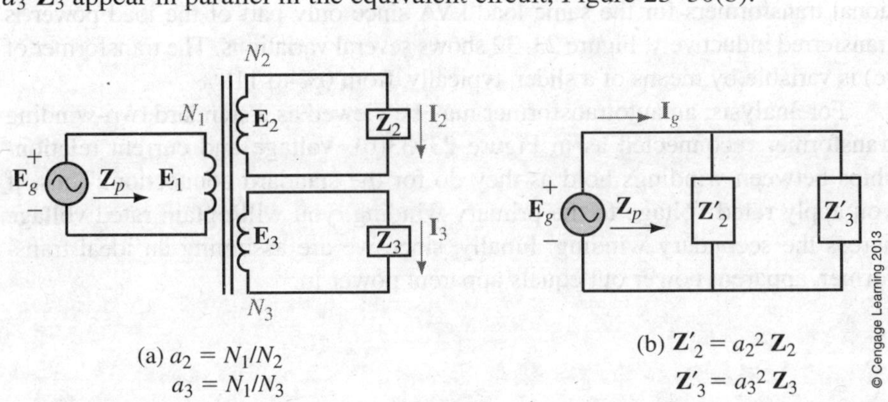

FIGURE 23–30 Loads are reflected in parallel.

EXAMPLE 23–7

For the circuit of Figure 23–31(a),

a. determine the equivalent circuit,

b. determine the generator current,

c. show that apparent power in equals apparent power out.

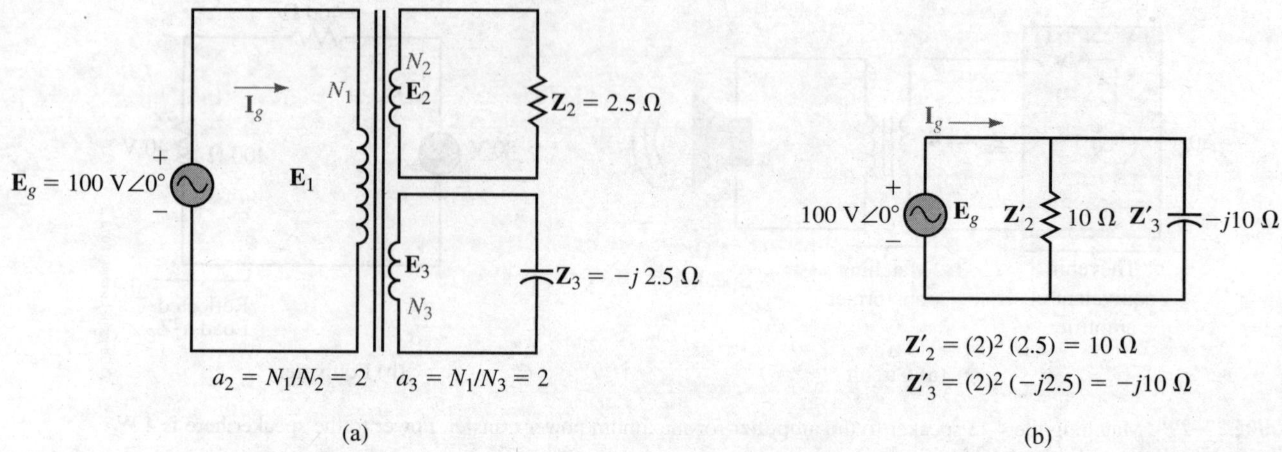

FIGURE 23–31

CircuitSim 23-9

Solution

a. See Figure 23–31(b).

b. $\mathbf{I}_g = \dfrac{\mathbf{E}_g}{\mathbf{Z}'_2} + \dfrac{\mathbf{E}_g}{\mathbf{Z}'_3} = \dfrac{100\angle 0°}{10} + \dfrac{100\angle 0°}{-j10} = 10 + j10 = 14.14\ \text{A}\angle 45°$

c. Input: $S_{in} = E_g I_g = (100\ \text{V})(14.14\ \text{A}) = 1414\ \text{VA}$
Output: From Figure 23–31(b), $P_{out} = (100\ \text{V})^2/(10\ \Omega) = 1000\ \text{W}$ and $Q_{out} = (100\ \text{V})^2/(10\ \Omega) = 1000\ \text{VAR}$. Thus, $S_{out} = \sqrt{P_{out}^2 + Q_{out}^2} = 1414\ \text{VA}$ which, as expected, is the same as S_{in}.

Autotransformers

An important variation of the transformer is the **autotransformer** (Figure 23–32). Autotransformers are unusual in that their primary circuit is not electrically isolated from their secondary. However, they are smaller and cheaper than conventional transformers for the same load kVA since only part of the load power is transferred inductively. Figure 23–32 shows several variations. The transformer of (c) is variable by means of a slider, typically, from 0% to 110%.

For analysis, an autotransformer may be viewed as a standard two-winding transformer reconnected as in Figure 23–33(b). Voltage and current relationships between windings hold as they do for the standard connection. Thus, if you apply rated voltage to the primary winding, you will obtain rated voltage across the secondary winding. Finally, since we are assuming an ideal transformer, apparent power out equals apparent power in.

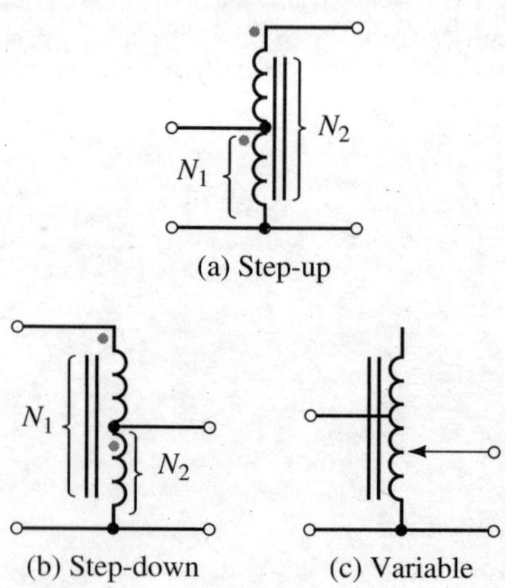

FIGURE 23–32 Autotransformers.

EXAMPLE 23–8

A 240/60-V, 3-kVA transformer [Figure 23–33(a)] is reconnected as an autotransformer to supply 300 volts to a load from a 240-V supply [Figure 23–33(b)].

a. Determine the transformer's rated primary and secondary currents.

b. Determine the maximum apparent power that can be delivered to the load.

c. Determine the supply current.

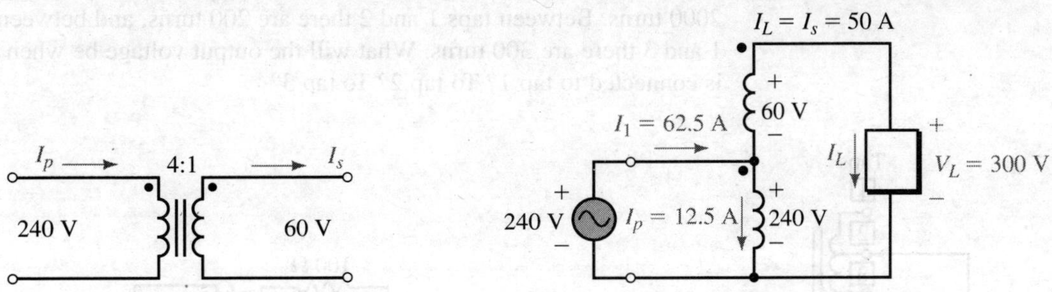

(a) 3 kVA transformer (b) Used as an autotransformer

FIGURE 23–33

CircuitSim 23-10

Solution

a. Rated current = rated kVA/rated voltage. Thus,

$$I_p = 3\text{ kVA}/240\text{ V} = 12.5\text{ A} \quad \text{and} \quad I_s = 3\text{ kVA}/60\text{ V} = 50\text{ A}$$

b. Since the 60-V winding is rated at 50 A, the transformer can deliver 50 A to the load [Figure 23–33(b)]. The load voltage is 300 V. Thus,

$$S_L = V_L I_L = (300\text{ V})(50\text{ A}) = 15\text{ kVA}$$

This is five times the rated kVA of the transformer.

c. Apparent power in = apparent power out:

$$240 I_1 = 15\text{ kVA}$$

Thus, $I_1 = 15\text{ kVA}/240\text{ V} = 62.5$ A. Current directions are as shown.

As a check, KCL at the junction of the two coils yields

$$I_1 = I_p + I_L = 12.5 + 50 = 62.5\text{ A}$$

Current magnitudes can be added here to get total current since the currents are in phase.

IN-PROCESS LEARNING CHECK 2

(Answers are at the end of the chapter.)

1. For each of the circuits of Figure 23–34, determine the required answer.

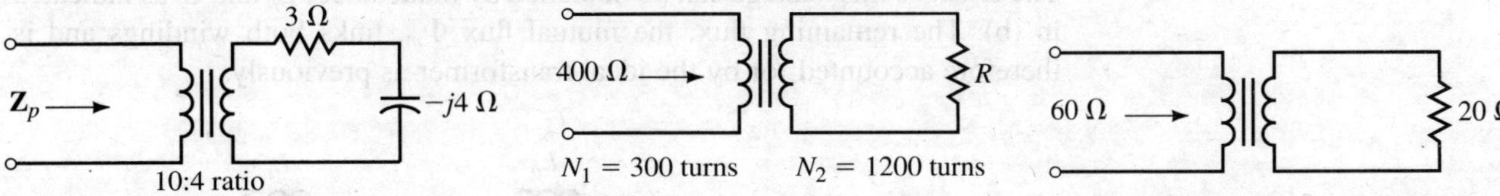

FIGURE 23–34

2. For Figure 23–35, if $a = 5$ and $\mathbf{I}_p = 5\text{ A}\angle{-60°}$, what is $\mathbf{E}_g$?
3. For Figure 23–35, if $\mathbf{I}_p = 30\text{ mA}\angle{-40°}$ and $\mathbf{E}_g = 240\text{ V}\angle 20°$, what is a?
4. a. How many amps can a 24-kVA, 7200/120-V transformer supply to a 120-V unity power factor load? To a 0.75 power factor load?

   b. How many watts can it supply to each load?

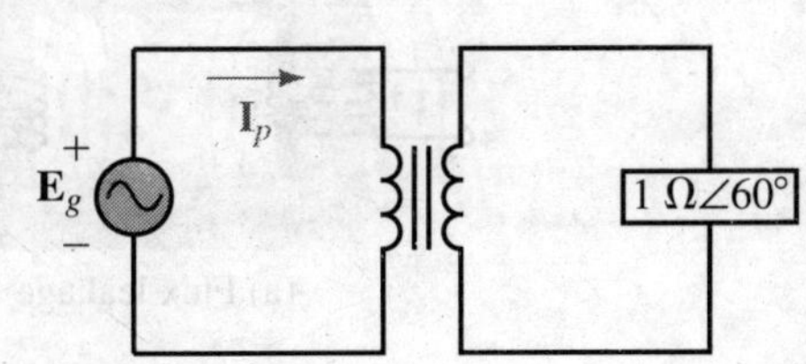

FIGURE 23–35

5. For the transformer of Figure 23–36, between position 2 and 0 there are 2000 turns. Between taps 1 and 2 there are 200 turns, and between taps 1 and 3 there are 300 turns. What will the output voltage be when the supply is connected to tap 1? To tap 2? To tap 3?

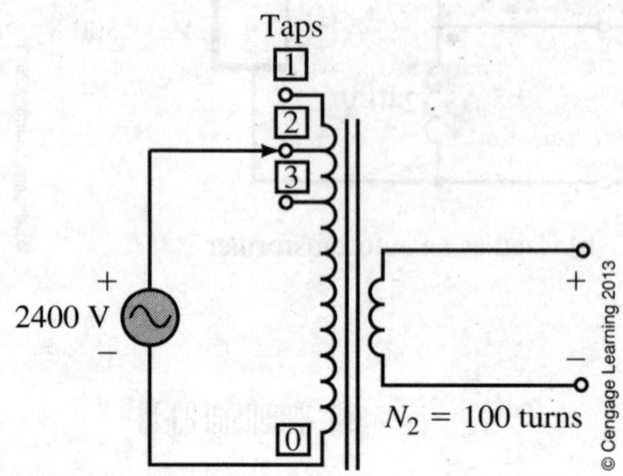

FIGURE 23–36

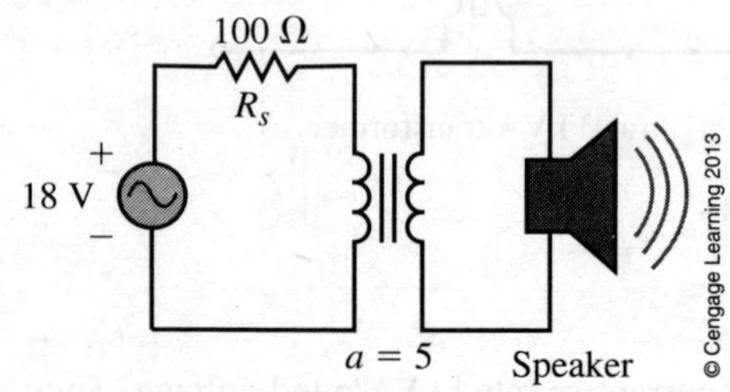

FIGURE 23–37

6. For the circuit of Figure 23–37, what is the power delivered to a 4-ohm speaker? What is the power delivered if an 8-ohm speaker is used instead? Why is the power to the 4-ohm speaker larger?

7. The autotransformer of Figure 23–38 has a 58% tap. The apparent power of the load is 7.2 kVA. Calculate the following:
   a. The load voltage and current.
   b. The source current.
   c. The current in each winding and its direction.

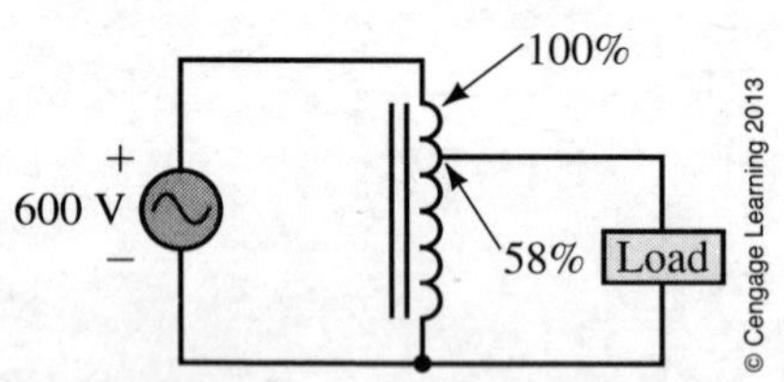

FIGURE 23–38

23.6 Practical Iron-Core Transformers

In Section 23.2, we idealized the transformer. We now add back the effects that we ignored.

Leakage Flux

While most flux is confined to the core, a small amount (called **leakage flux**) passes outside the core and through air at each winding as in Figure 23–39(a). The effect of this leakage can be modelled by inductances L_p and L_s as indicated in (b). The remaining flux, the mutual flux Φ_m, links both windings and is therefore accounted for by the ideal transformer as previously.

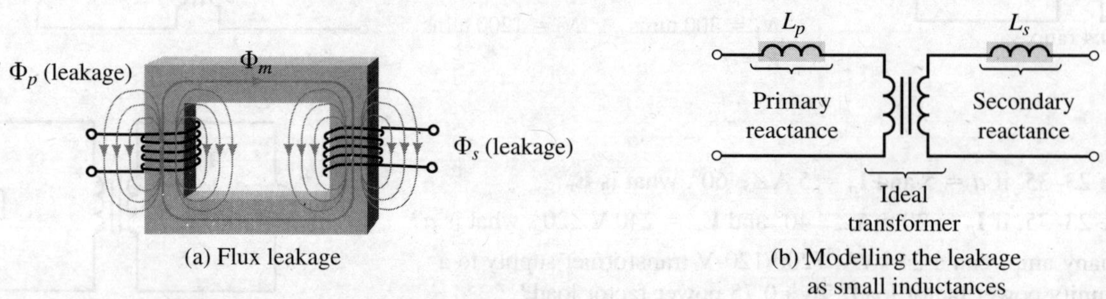

FIGURE 23–39 Adding the effect of leakage flux to the model.

Winding Resistance

The effect of coil resistance can be approximated by adding resistances R_p and R_s as shown in Figure 23–40. The effect of these resistances is to cause a slight power loss and hence a reduction in efficiency as well as a small voltage drop. [The power loss associated with coil resistance is called **copper loss** (Section 23.7) and varies as the square of the load current.]

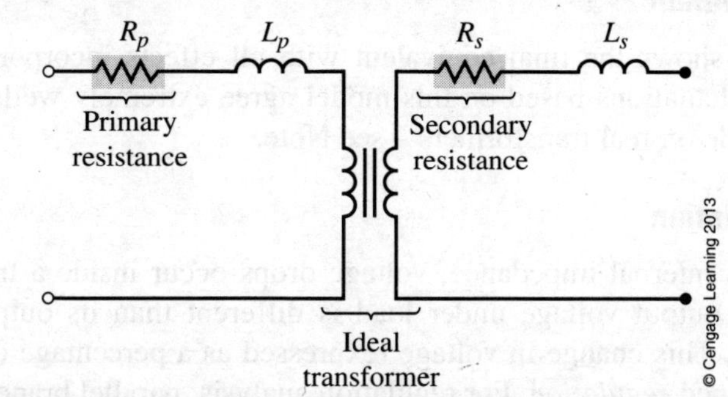

FIGURE 23–40 Adding the winding resistance to the model.

Core Loss

Losses occur in the core because of eddy currents and hysteresis. First, consider eddy currents. Since iron is a conductor, voltage is induced in the core as flux varies. This voltage creates currents that circulate as "eddies" within the core itself. One way to reduce these currents is to break their path of circulation by constructing the core from thin laminations of steel rather than using a solid block of iron. Laminations are insulated from each other by a coat of ceramic, varnish, or similar insulating material. (Although this does not eliminate eddy currents, it greatly reduces them.) Power and audio transformers are built this way (Figure 23–4). Another way to reduce eddy currents is to use powdered iron held together by an insulating binder. Ferrite and other low loss cores are made like this.

Now consider hysteresis (Chapter 12, Section 12.14). Because the flux direction constantly cycles, magnetic domains in the core steel constantly reverse as well. This takes energy. In practice, this energy is minimized by using special grain-oriented transformer steel.

The sum of hysteresis and eddy current loss is called **core loss** or iron loss. In a well-designed transformer, it is small, typically a few percent or less of the transformer rating. The effect of core loss can be modelled as a resistor, R_c in Figure 23–41. Core losses vary approximately as the square of applied voltage. As long as voltage is constant (which it normally is), core losses remain constant.

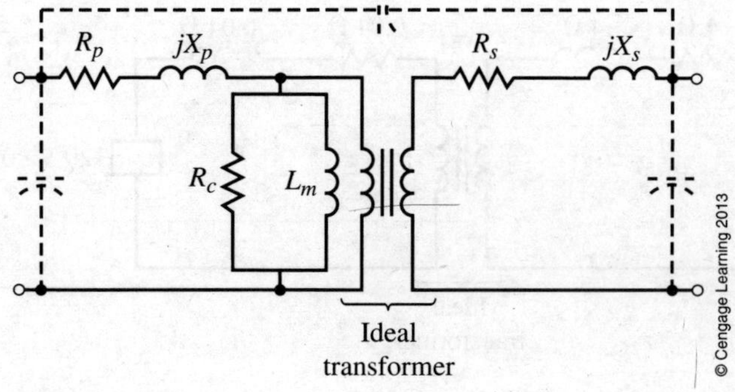

FIGURE 23–41 Final iron-core transformer equivalent circuit.

NOTES...

Although Figure 23–41 (which represents the full equivalent for iron-core transformers) yields exceptionally good results, it is complex and awkward to use. Fortunately, we can simplify the model, since certain effects are negligible for specific applications. For example, at power frequencies, the effect of capacitance is negligible, while for regulation analysis (considered next), the core branch (R_c and L_m) also has negligible effect. Thus, in practice, they can usually be omitted, leading to the simplified model illustrated next (Figure 23–42).

Other Effects

We have also neglected magnetizing current. In a real transformer, however, some current is required to magnetize the core. To account for this, add path L_m as shown in Figure 23–41. Stray capacitances also exist between various parts of the transformer. They can be approximated by lumped capacitances as indicated.

The Full Equivalent

Figure 23–41 shows the final equivalent with all effects incorporated. How good is it? Calculations based on this model agree extremely well with measurements made on real transformers—see Note.

Voltage Regulation

Because of its internal impedance, voltage drops occur inside a transformer. Therefore, its output voltage under load is different than its output voltage under no load. This change in voltage (expressed as a percentage of full-load voltage) is termed *regulation*. For regulation analysis, parallel branches R_c and L_m and stray capacitance have negligible effect and can be neglected. This yields the simplified circuit of Figure 23–42(a). Even greater simplification is achieved by reflecting secondary impedances into the primary. This yields the circuit of (b). The reflected load voltage is $a\mathbf{V}_L$ and the reflected load current is $\mathbf{I}_L/a$. The simplified equivalent of Figure 23–42(b) is the circuit that we use in practice to perform regulation analysis.

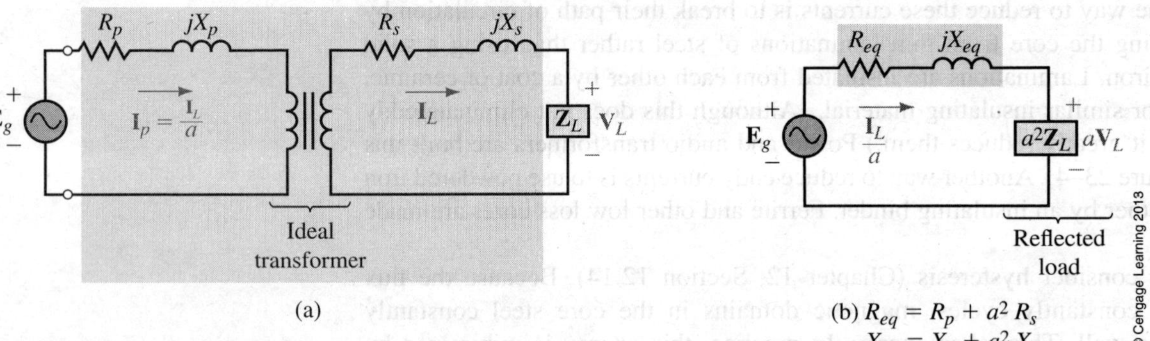

FIGURE 23–42 Simplifying the equivalent.

EXAMPLE 23–9

A 10:1 transformer has primary and secondary resistance and reactance of 4 Ω + j4 Ω and 0.04 Ω + j0.04 Ω, respectively, as in Figure 23–43.

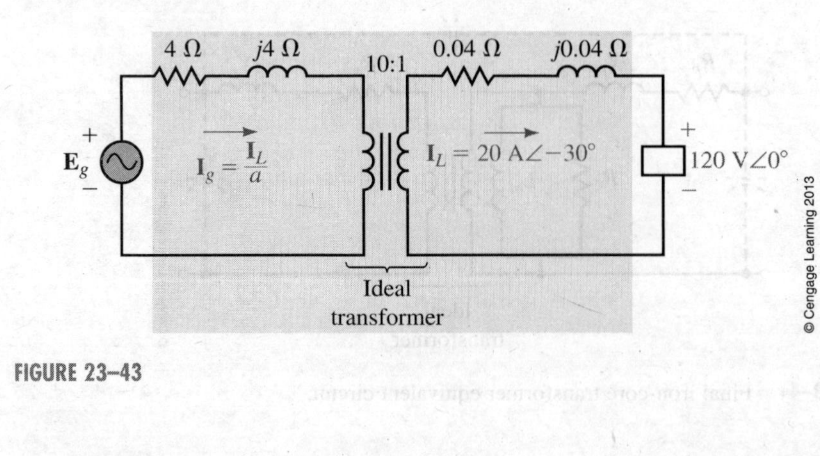

FIGURE 23–43

a. Determine its equivalent circuit.
b. If $\mathbf{V}_L = 120 \text{ V}\angle 0°$ and $\mathbf{I}_L = 20 \text{ A}\angle -30°$, what is the supply voltage, $\mathbf{E}_g$?
c. Determine the regulation.

Solution

a. $R_{eq} = R_p + a^2R_s = 4\ \Omega + (10)^2(0.04\ \Omega) = 8\ \Omega$
$X_{eq} = X_p + a^2X_s = 4\ \Omega + (10)^2(0.04\ \Omega) = 8\ \Omega$
Thus, $\mathbf{Z}_{eq} = 8\ \Omega + j8\ \Omega$ as shown in Figure 23–44.

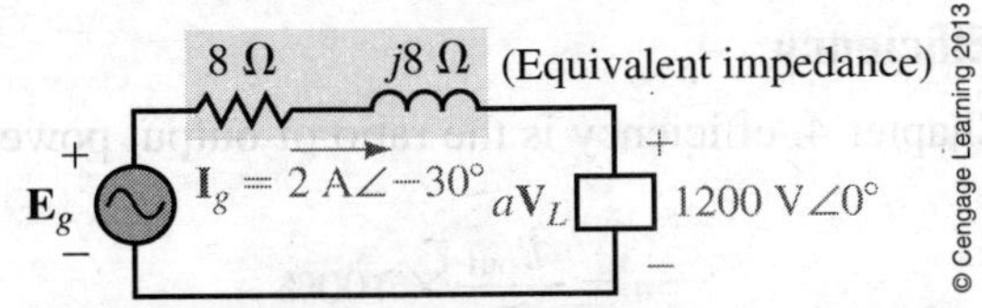

FIGURE 23–44

b. $a\mathbf{V}_L = (10)(120 \text{ V}\angle 0°) = 1200 \text{ V}\angle 0°$ and $\mathbf{I}_L/a = (20 \text{ A}\angle -30°)/10 = 2 \text{ A}\angle -30°$. From KVL, $\mathbf{E}_g = (2 \text{ A}\angle -30°)(8\ \Omega + j8) + 1200 \text{ V}\angle 0° = 1222 \text{ V}\angle 0.275°$.

Thus, there is a phase shift of 0.275° across the transformer's internal impedance and a drop of 22 V, requiring that the primary be operated slightly above its rated voltage. (This is normal.)

c. Now consider the no-load condition (Figure 23–45). Let V_{NL} be the no-load voltage. As indicated, $aV_{NL} = 1222$ V. Thus, $V_{NL} = 1222/a = 1222/10 = 122.2$ volts and

$$\text{regulation} = \frac{V_{NL} - V_{FL}}{V_{FL}} \times 100 = \frac{122.2 - 120}{120} \times 100 = 1.83\%$$

Note that only magnitudes are used in determining regulation. Note also that you can use the high-side equivalent voltages to compute regulation—see Practice Problems 4.

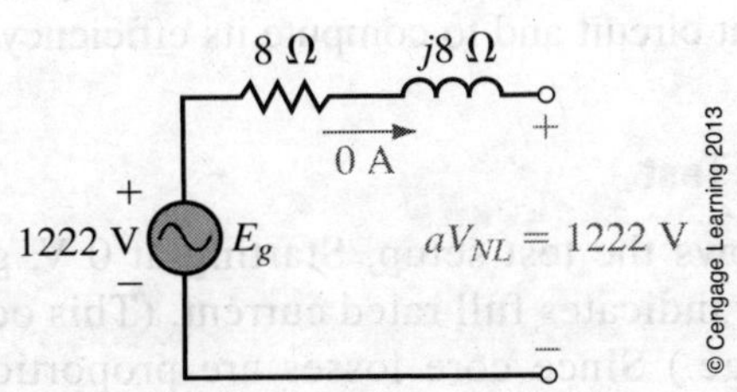

FIGURE 23–45 No-load equivalent: $aV_{NL} = E_g$.

PRACTICAL NOTES...

1. From Figure 23–45, $a = E_g/V_{NL}$. This means that the turns ratio is the ratio of input voltage to output voltage at no load.
2. The voltage rating of a transformer (such as 1200/120 V) is referred to as its *nominal rating*. The ratio of nominal voltages is the same as the turns ratio. Thus, for a nonloaded transformer, if nominal voltage is applied to the primary, nominal voltage will appear at the secondary.
3. Power transformers are normally operated close to their nominal voltages. However, depending on operating conditions, they may be a few percent above or below rated voltage at any given time.

CircuitSim 23-11

PRACTICE PROBLEMS 4

1. A transformer used in an electronic power supply has a nominal rating of 120/12 volts and is connected to a 120-Vac source. Its equivalent impedance as seen from the primary is $10\ \Omega + j10\ \Omega$. What is the magnitude of the load voltage if the load is 5 ohms resistive? Determine the regulation.

2. For Example 23–9, confirm that high-side equivalent voltages yield the same regulation as calculated using the low-side values.

Answers
1 1.8 V; 2.04%
2. regulation = (1222 V – 1200 V)/1200 V × 100 = 1.83%

Transformer Efficiency

Recall from Chapter 4, efficiency is the ratio of output power to input power.

$$\eta = \frac{P_{\text{out}}}{P_{\text{in}}} \times 100\% \tag{23–12}$$

But $P_{\text{in}} = P_{\text{out}} + P_{\text{loss}}$. For a transformer, losses are due to I^2R losses in the windings (called copper losses) and losses in the core (called core losses). Thus,

$$\eta = \frac{P_{\text{out}}}{P_{\text{out}} + P_{\text{loss}}} \times 100\% = \frac{P_{\text{out}}}{P_{\text{out}} + P_{\text{copper}} + P_{\text{core}}} \times 100\% \tag{23–13}$$

Large power transformers are exceptionally efficient, on the order of 98% to 99%. The efficiencies of smaller power transformers are around 95% or better.

23.7 Transformer Tests

The losses of Equation 23–13 may be determined experimentally using the **short-circuit test** and the **open-circuit test.** (These tests are used mainly with power transformers.) They provide the data needed to determine a transformer's equivalent circuit and to compute its efficiency.

The Short-Circuit Test

Figure 23–46 shows the test setup. Starting at 0 V, gradually increase E_g until the ammeter indicates full rated current. (This occurs at a few percent rated input voltage.) Since core losses are proportional to the square of voltage, at this low voltage, core losses are negligible. Thus, while the circuit of Figure 23–46 measures the sum of copper loss and core loss, the wattmeter reading is essentially only copper loss.

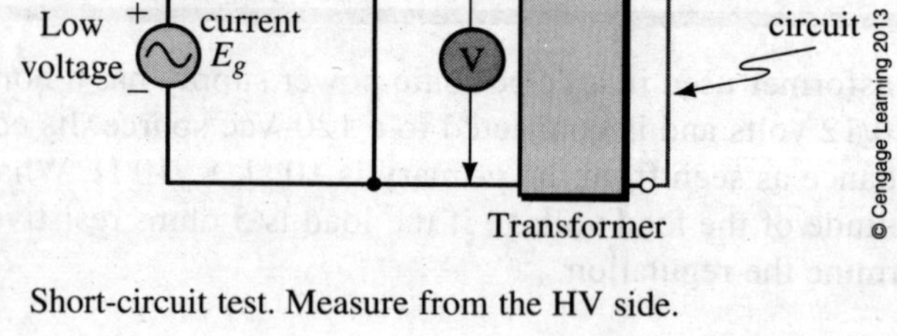

FIGURE 23–46 Short-circuit test. Measure from the HV side.

EXAMPLE 23–10

Measurements on the high side of a 240/120-volt, 4.8-kVA transformer yield $E_g = 11.5$ V and $W = 172$ W at the rated current of $I = 4.8$ kVA/240 = 20 A. Determine $\mathbf{Z}_{\text{eq}}$.

Solution See Figure 23–47. Since $Z_L = 0$, the only impedance in the circuit is Z_{eq}. Thus, $Z_{\text{eq}} = E_g/I = 11.5\text{ V}/20\text{ A} = 0.575\ \Omega$. Further, $R_{\text{eq}} = W/I^2 = 172\text{ W}/(20\text{ A})^2 = 0.43\ \Omega$. Therefore,

$$X_{\text{eq}} = \sqrt{Z_{\text{eq}}^2 - R_{\text{eq}}^2} = \sqrt{(0.575)^2 - (0.43)^2} = 0.382\ \Omega$$

and $\mathbf{Z}_{\text{eq}} = R_{\text{eq}} + jX_{\text{eq}} = 0.43\ \Omega + j0.382\ \Omega$ as shown in (b).

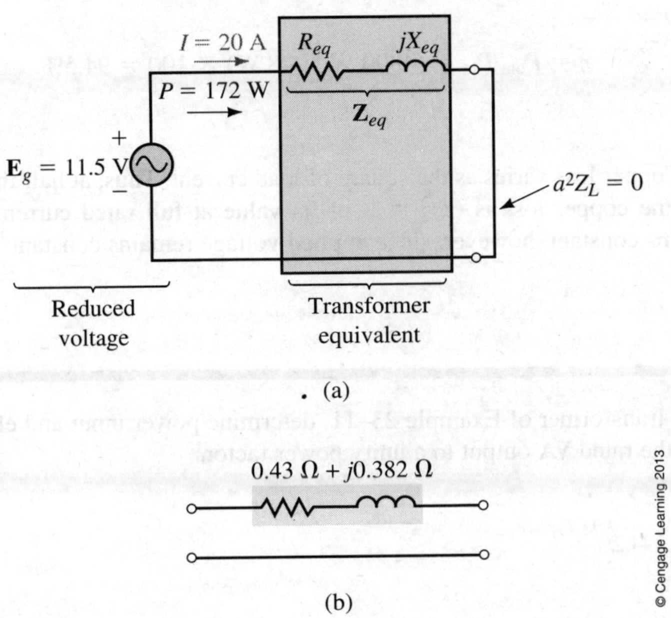

FIGURE 23–47 Determining the equivalent circuit by test.

PRACTICAL NOTES...

Since power loss is the same regardless of which side of a transformer you measure from, we usually perform the short-circuit (SC) test from the high voltage side, as currents are lower here. (For example, for a 48 kVA, 1200/120 V transformer, rated current on the HV side is 40 A, but on the LV side, it is 400 A.) Note that voltage is generally not a problem here since you need only a few percent of rated voltage to run the test. In fact, if you inadvertently applied rated voltage to a short-circuited transformer, the currents would be so high that you would probably have an explosion on your hands—thus, you must be very careful when running this test.

CircuitSim 23-12

The Open-Circuit Test

The setup for the open-circuit (OC) test is shown in Figure 23–48. Apply the full rated voltage. Since the load current is zero, only exciting current results. Since the exciting current is small, power loss in the winding resistance is negligible, and the wattmeter reading is essentially just core loss.

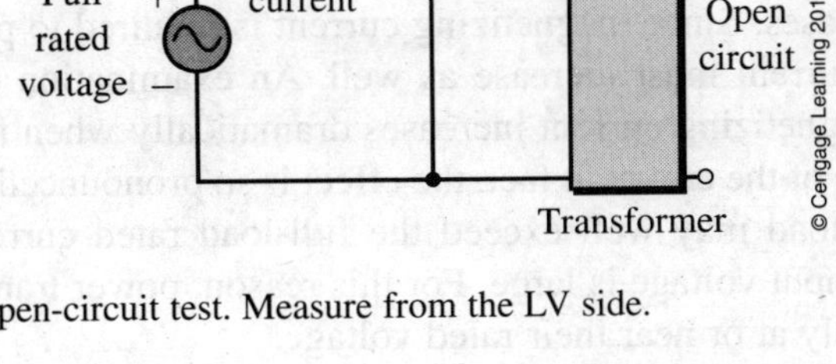

FIGURE 23–48 Open-circuit test. Measure from the LV side.

PRACTICAL NOTES...

The OC test is normally performed on the low-voltage side since this permits you to work with a lower voltage. (For example, for a 1200/120-V transformer, you would need a wattmeter capable of operation at 1200 V if measurements were made from the HV side, whereas, if the test is done from the LV side, you need only 120-V capability.)

EXAMPLE 23–11

An open-circuit test on the transformer of Example 23–10 yields a core loss of 106 W. Determine this transformer's efficiency when supplying the full, rated VA to a load at unity power factor.

Solution Since the transformer is supplying the rated VA, its current is the full, rated current. From the short-circuit test, the copper loss at full rated current is 172 W. Thus,

copper loss = 172 W (From the SC test)
core loss = 106 W (From the OC test)
output = 4800 W (Rated)
input = output + losses = 5078 W

Thus

$$\eta = P_{out}/P_{in} = (4800\text{ W}/5078\text{ W}) \times 100 = 94.5\%$$

Copper loss varies as the square of load current. Thus, at half the rated current, the copper loss is $(½)^2$ = ¼ of its value at full rated current. Core loss remains constant, however, since applied voltage remains constant.

PRACTICE PROBLEMS 5

For the transformer of Example 23–11, determine power input and efficiency at half the rated VA output to a unity power factor.

Answer
2549 W; 94.2%

23.8 Voltage and Frequency Effects

Iron-core transformer characteristics vary with frequency and voltage. To determine why, we start with Faraday's law, $e = Nd\Phi/dt$. Specializing this to the sinusoidal ac case, it can be shown that

$$E_p = 4.44 f N_p \Phi_m \quad \textbf{(23–14)}$$

where Φ_m is the mutual core flux.

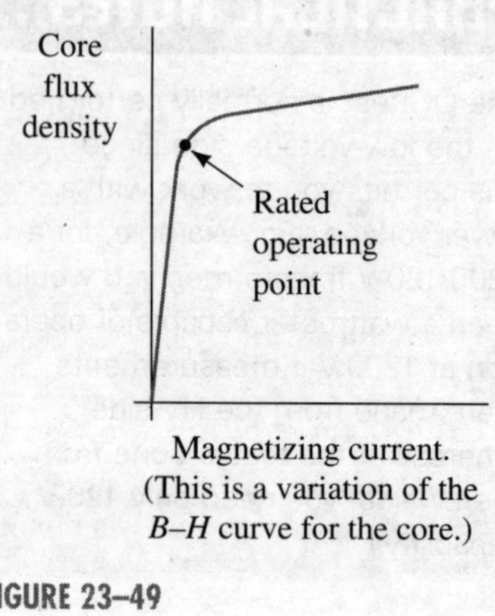

FIGURE 23–49

Effect of Voltage

First, assume constant frequency. Since $\Phi_m = E_p/4.44 f N_p$, the core flux is proportional to the applied voltage. Thus, if the applied voltage is increased, the core flux increases. Since magnetizing current is required to produce this flux, magnetizing current must increase as well. An examination of Figure 23–49 shows that magnetizing current increases dramatically when flux density rises above the knee of the curve; in fact, the effect is so pronounced that the primary current at no load may well exceed the full-load rated current of the transformer if the input voltage is large. For this reason, power transformers should be operated only at or near their rated voltage.

Effect of Frequency

Audio transformers must operate over a range of frequencies. Consider again $\Phi_m = E_p/4.44fN_p$. As this indicates, decreasing the frequency increases the core flux and hence the magnetizing current. At low frequencies, this larger current increases internal voltage drops and hence decreases the output voltage as indicated in Figure 23–50. Now consider increasing the frequency. As frequency increases, leakage inductance and shunt capacitance (recall Figure 23–41) cause the voltage to fall off. To compensate for this, audio transformers are sometimes designed so that their internal capacitances resonate with their inductances to extend the operating range. This is what causes the peaking at the high-frequency end of the curve.

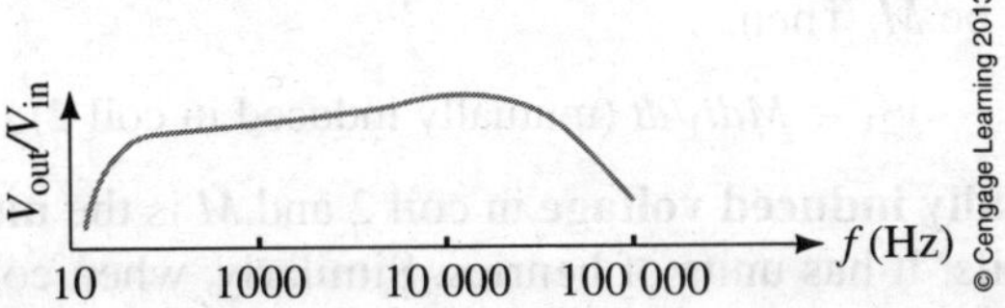

FIGURE 23–50 Frequency response curve, audio transformer.

IN-PROCESS LEARNING CHECK 3

(Answers are at the end of the chapter.)

A transformer with a nominal rating of 240/120 V, 60 Hz, has its load on the 120-V side. Suppose $R_p = 0.4\ \Omega$, $L_p = 1.061$ mH, $R_s = 0.1\ \Omega$, and $L_s = 0.2653$ mH.

a. Determine its equivalent circuit as per Figure 23–42(b).

b. If $\mathbf{E}_g = 240\ \text{V}\angle 0°$ and $\mathbf{Z}_L = 3 + j4\ \Omega$, what is $\mathbf{V}_L$?

c. Compute the regulation.

23.9 Loosely Coupled Circuits

We now turn our attention to coupled circuits that do not have iron cores. For such circuits, only a portion of the flux produced by one coil links another, and the coils are said to be loosely coupled. Loosely coupled circuits cannot be characterized by turns ratios; rather, as you will see, they are characterized by self- and mutual inductances. Air-core transformers and general inductive circuit coupling fall into this category. In this section, we develop the main ideas.

Voltages in Air-Core Coils

To begin, consider the isolated (noncoupled) coil of Figure 23–51. As shown in Chapter 13, the voltage across this coil is given by $v_L = Ldi/dt$, where i is the current through the coil and L is its inductance. Note carefully the polarity of the voltage; the plus sign goes at the tail of the current arrow. Because the coil's voltage is created by its own current, it is called a **self-induced voltage.**

Now consider a pair of coupled coils (Figure 23–52). When coil 1 alone is energized as in (a), it looks just like the isolated coil of Figure 23–51; thus its voltage is

$$v_{11} = L_1 di_1/dt \text{ (self-induced in coil 1)}$$

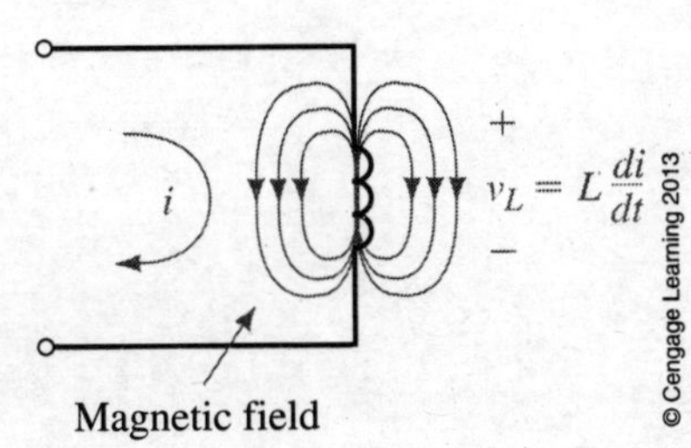

FIGURE 23–51 Place the plus sign for the self-induced voltage at the tail of the current arrow.

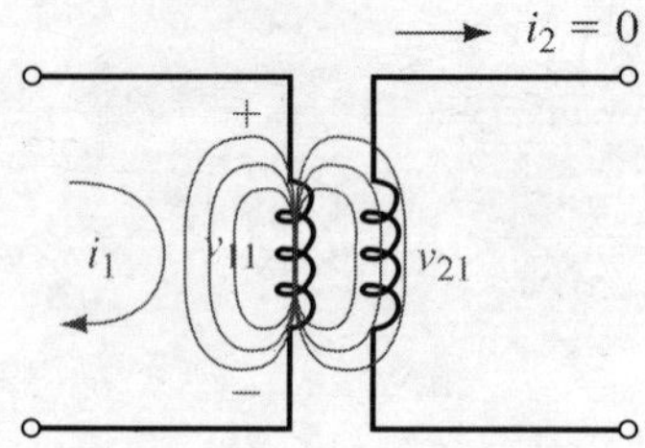

(a) v_{11} is the voltage in coil 1 due to current i_1. It is therefore a "self-induced" voltage. v_{21} is the voltage in coil 2 due to current i_1. It is therefore a "mutually-induced" voltage.

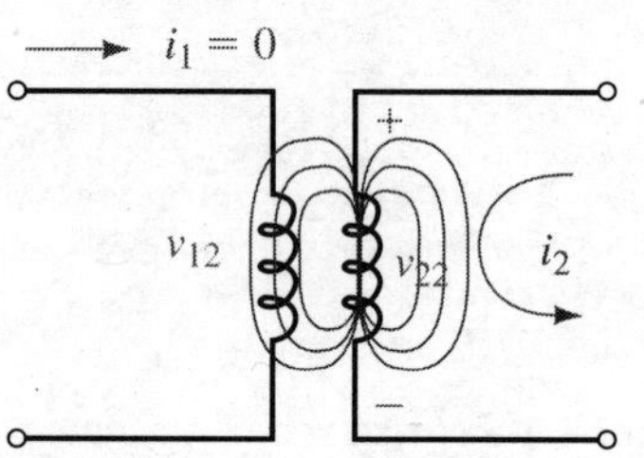

(b) v_{22} is the voltage in coil 2 due to current i_2. It is therefore a "self-induced" voltage. v_{12} is the voltage in coil 1 due to current i_2. It is therefore a "mutually-induced" voltage.

FIGURE 23–52 Self and mutual voltages. Voltages shown in (a) are due to current i_1 acting alone, while voltages shown in (b) are due to current i_2 acting alone. Note the order of subscripts on the mutual voltage.

where L_1 is the self-inductance of coil 1 and the subscripts indicate that v_{11} is the voltage across coil 1 due to its own current. Similarly, when coil 2 alone is energized as in (b), its self-induced voltage is

$$v_{22} = L_2 di_2/dt \text{ (self-induced in coil 2)}$$

For both of these self-voltages, note that the plus sign goes at the tail of their respective current arrows.

Mutual Voltages

Consider again Figure 23–52(a). When coil 1 is energized, some of the flux that it produces links coil 2, inducing voltage v_{21} in coil 2. Since the flux here is due to i_1 alone, v_{21} is proportional to the rate of change of i_1. Let the constant of proportionality be M. Then,

$$v_{21} = M di_1/dt \text{ (mutually induced in coil 2)}$$

v_{21} is the **mutually induced voltage** in coil 2 and M is the **mutual inductance** between the coils. It has units of henries. Similarly, when coil 2 alone is energized as in (b), the voltage induced in coil 1 is

$$v_{12} = M di_2/dt \text{ (mutually induced in coil 1)}$$

When both coils are energized, the voltage of each coil can be found by superposition; *in each coil, the induced voltage is the sum of its self voltage plus the voltage mutually induced due to the current in the other coil.* The sign of the self term for each coil is straightforward: It is determined by placing a plus sign at the tail of the current arrow for the coil as shown in Figures 23–52(a) and (b). The polarity of the mutual term, however, depends on whether the mutual voltage is additive or subtractive.

Additive and Subtractive Voltages

Whether self and mutual voltages add or subtract depends on the direction of currents through the coils relative to their winding directions. This can be described in terms of the dot convention. Consider Figure 23–53(a). Comparing the coils here to Figure 23–12, you can see that their top ends correspond and thus can be marked with dots. Now let currents enter both coils at dotted ends. Using the right-hand rule, you can see that their fluxes add. The total flux linking coil 1 is therefore the *sum* of that produced by i_1 and i_2; therefore, the voltage across coil 1 is the sum of that produced by i_1 and i_2. That is $v_1 = v_{11} + v_{12}$. Expanded, this is

$$v_1 = L_1\frac{di_1}{dt} + M\frac{di_2}{dt} \qquad \textbf{(23–15a)}$$

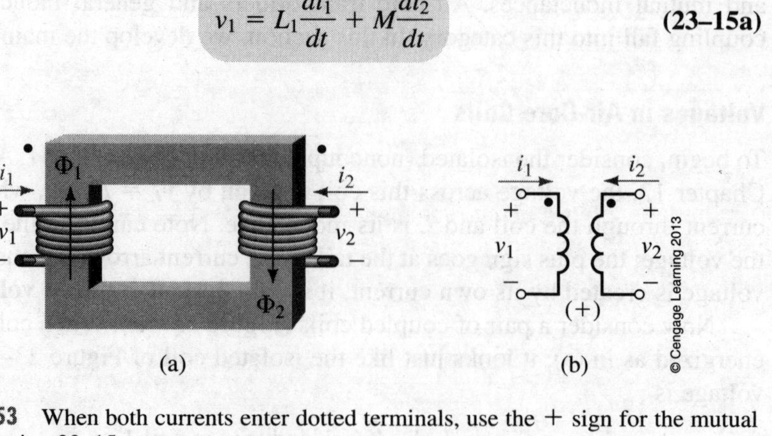

FIGURE 23–53 When both currents enter dotted terminals, use the + sign for the mutual term in Equation 23–15.

Similarly, for coil 2, $v_2 = v_{21} + v_{22}$. Thus,

$$v_2 = M\frac{di_1}{dt} + L_2\frac{di_2}{dt} \tag{23–15b}$$

Now consider Figure 23–54. Here, the fluxes oppose and the flux linking each coil is the *difference* between that produced by its own current and that produced by the current of the other coil. Thus, the sign in front of the mutual voltage terms will be negative.

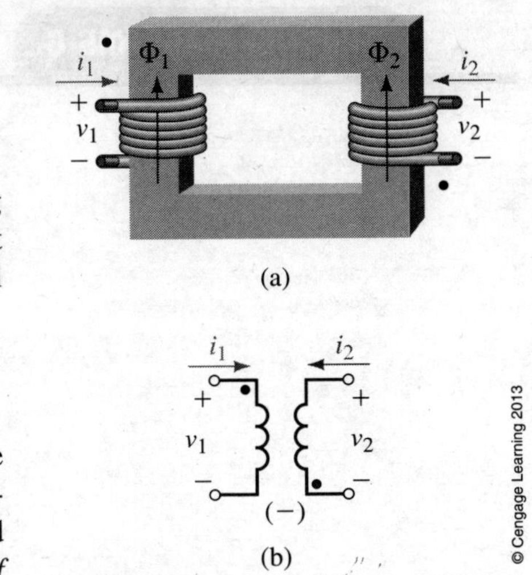

FIGURE 23–54 When one current enters a dotted terminal and the other enters an undotted terminal, use the − sign for the mutual term in Equation 23–15.

The Dot Rule

As you can see, the signs of the mutual voltage terms in Equations 23–15 are positive when both currents enter dotted terminals, but negative when one current enters a dotted terminal and the other enters an undotted terminal. Stated another way, *the sign of the mutual voltage is the same as the sign of its self voltage when both currents enter dotted (or undotted) terminals, but is opposite when one current enters a dotted terminal and the other enters an undotted terminal.* This observation provides us with a procedure for determining voltage polarities in coupled circuits.

1. Assign a direction for currents i_1 and i_2.
2. Place a plus sign at the tail of the current arrow for each coil to denote the polarity of its self-induced voltage.
3. If both currents enter (or both leave) dotted terminals, make the sign of the mutually induced voltage the same as the sign of the self-induced voltage when you write the equation.
4. If one current enters a dotted terminal and the other leaves, make the sign of the mutually induced voltage opposite to the sign of the self-induced voltage.

EXAMPLE 23–12

Write equations for v_1 and v_2 of Figure 23–55(a).

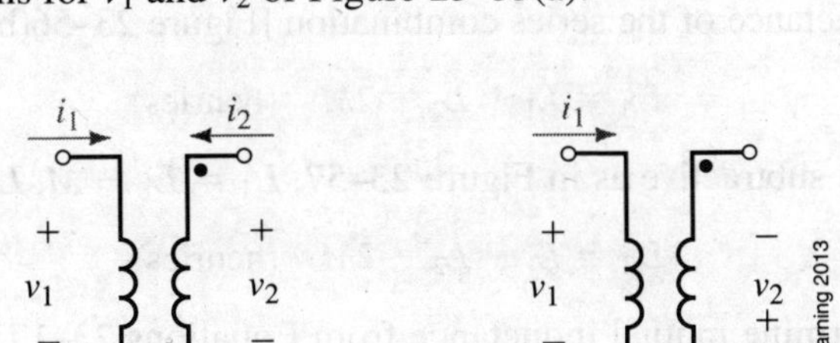

FIGURE 23–55

Solution Since one current enters an undotted terminal and the other enters a dotted terminal, place a minus sign in front of M. Thus,

$$v_1 = L_1\frac{di_1}{dt} - M\frac{di_2}{dt}$$

$$v_2 = -M\frac{di_1}{dt} + L_2\frac{di_2}{dt}$$

PRACTICE PROBLEMS 6

Write equations for v_1 and v_2 of Figure 23–55(b).

Answer
Same as Equations 23–15(a) and 23–15(b).

Coefficient of Coupling

For loosely coupled coils, not all of the flux produced by one coil links the other. To describe the degree of coupling between coils, we introduce a **coefficient of coupling,** k. Mathematically, k is defined as the ratio of the flux that links the companion coil to the total flux produced by the energized coil. For iron-core transformers, almost all the flux is confined to the core and links both coils; thus, k is very close to 1. At the other extreme (i.e., isolated coils where no flux linkage occurs), $k = 0$. Thus, $0 \leq k < 1$. Mutual inductance depends on k. It can be shown that mutual inductance, self-inductances, and the coefficient of coupling are related by the equation

$$M = k\sqrt{L_1 L_2} \qquad \textbf{(23–16)}$$

Thus, the larger the coefficient of coupling, the larger the mutual inductance.

Inductors with Mutual Coupling

If a pair of coils are in close proximity, the field of each coil couples the other, resulting in a change in the apparent inductance of each coil. To illustrate, consider Figure 23–56(a), which shows a pair of inductors with self-inductances L_1 and L_2. If coupling occurs, the effective coil inductances will no longer be L_1 and L_2. To see why, consider the voltage induced in each winding—it is the sum of the coil's own self voltage plus the voltage mutually induced from the other coil. Since current is the same for both coils, $v_1 = L_1 di/dt + M di/dt = (L_1 + M)di/dt$, which means that coil 1 has an effective inductance of $L'_1 = L_1 + M$. Similarly, $v_2 = (L_2 + M)di/dt$, giving coil 2 an effective inductance of $L'_2 = L_2 + M$. The effective inductance of the series combination [Figure 23–56(b)] is then

$$L_T^+ = L_1 + L_2 + 2M \quad \text{(henries)} \qquad \textbf{(23–17)}$$

If coupling is subtractive as in Figure 23–57, $L'_1 = L_1 - M$, $L'_2 = L_2 - M$, and

$$L_T^- = L_1 + L_2 - 2M \quad \text{(henries)} \qquad \textbf{(23–18)}$$

You can determine mutual inductance from Equations 23–17 and 23–18. It is

$$M = \frac{1}{4}(L_T^+ - L_T^-) \qquad \textbf{(23–19)}$$

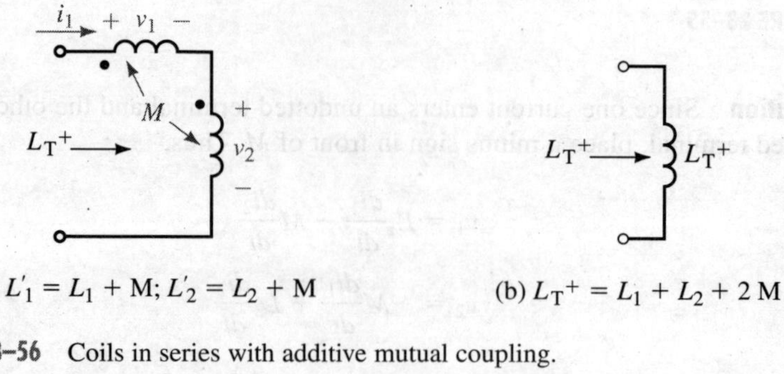

(a) $L'_1 = L_1 + \text{M}$; $L'_2 = L_2 + \text{M}$ (b) $L_T^+ = L_1 + L_2 + 2\,\text{M}$

FIGURE 23–56 Coils in series with additive mutual coupling.

EXAMPLE 23–13

Three inductors are connected in series (Figure 23–57). Coils 1 and 2 interact, but coil 3 does not.

a. Determine the effective inductance of each coil.
b. Determine the total inductance of the series connection.

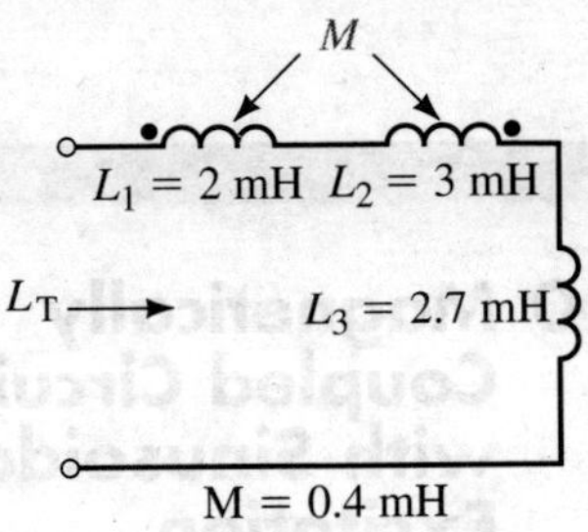

FIGURE 23–57

Solution

a. $L'_1 = L_1 - M = 2\text{ mH} - 0.4\text{ mH} = 1.6\text{ mH}$
$L'_2 = L_2 - M = 3\text{ mH} - 0.4\text{ mH} = 2.6\text{ mH}$
L'_1 and L'_2 are in series with L_3. Thus,
b. $L_T = 1.6\text{ mH} + 2.6\text{ mH} + 2.7\text{ mH} = 6.9\text{ mH}$

CircuitSim 23-13

The same principles apply when more than two coils are coupled. Thus, for the circuit of Figure 23–58, $L'_1 = L_1 - M_{12} - M_{31}$, and so on.

For two parallel inductors with mutual coupling, the equivalent inductances are

$$L'_1 = \frac{L_1L_2 - M^2}{L_2 \mp M} \tag{23–20a}$$

and

$$L'_2 = \frac{L_1L_2 - M^2}{L_1 \mp M} \tag{23–20b}$$

where, if the dots are at the same end, use the − sign. The parallel equivalent is

$$L_{eq} = \frac{L_1L_2 - M^2}{L_1 + L_2 \mp 2M} \tag{23–20c}$$

For example, if $L_1 = 20$ mH, $L_2 = 5$ mH, and $M = 2$ mH, then $L_{eq} = 4.57$ mH if the dots are both at the same ends of the coils, and $L_{eq} = 3.31$ mH when the dots are at opposite ends.

PRACTICE PROBLEMS 7

For the circuit of Figure 23–58, different "dot" symbols are used to represent coupling between sets of coils.

a. Determine the effective inductance of each coil.
b. Determine the total inductance of the series connection.

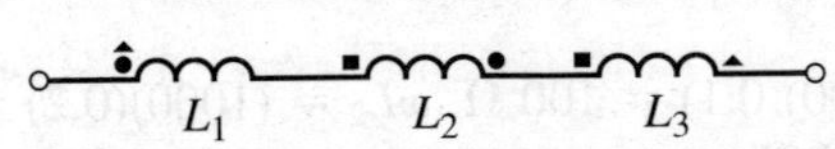

$L_1 = 10$ mH $M_{12} = 2$ mH (Mutual inductance between coils 1 and 2) (•)
$L_2 = 40$ mH $M_{23} = 1$ mH (Mutual inductance between coils 2 and 3) (▪)
$L_3 = 20$ mH $M_{31} = 0.6$ mH (Mutual inductance between coils 3 and 1) (▴)

FIGURE 23–58

Answers

a. $L'_1 = 7.4$ mH; $L'_2 = 39$ mH; $L'_3 = 20.4$ mH
b. 66.8 mH

The effect of unwanted mutual inductance can be minimized by physically separating coils or by orienting their axes at right angles. The latter technique is used where space is limited and coils cannot be spaced widely. While it does not eliminate coupling, it can help minimize its effects.

23.10 Magnetically Coupled Circuits with Sinusoidal Excitation

When coupling occurs between various parts of a circuit (whether wanted or not), the foregoing principles apply. However, since it is difficult to continue the analysis in general, we will change over to steady state ac. This will permit us to look for the main ideas. We will use the mesh approach. To use the mesh approach, (1) write mesh equations using KVL, (2) use the dot convention to determine the signs of the induced voltage components, and (3) solve the resulting equations in the usual manner.

To specialize to the sinusoidal ac case, convert voltages and currents to phasor form. To do this, recall from Chapter 16 that inductor voltage in phasor form is $\mathbf{V}_L = j\omega L\mathbf{I}$. (This is the phasor equivalent of $v_L = Ldi/dt$, Figure 23–51.) This means that Ldi/dt becomes $j\omega L\mathbf{I}$; in a similar fashion, $Mdi_1/dt \Rightarrow j\omega M\mathbf{I}_1$ and $Mdi_2/dt \Rightarrow j\omega M\mathbf{I}_2$. Thus, in phasor form Equations 23–15 become

$$\mathbf{V}_1 = j\omega L_1\mathbf{I}_1 + j\omega M\mathbf{I}_2$$
$$\mathbf{V}_2 = j\omega M\mathbf{I}_1 + j\omega L_2\mathbf{I}_2$$

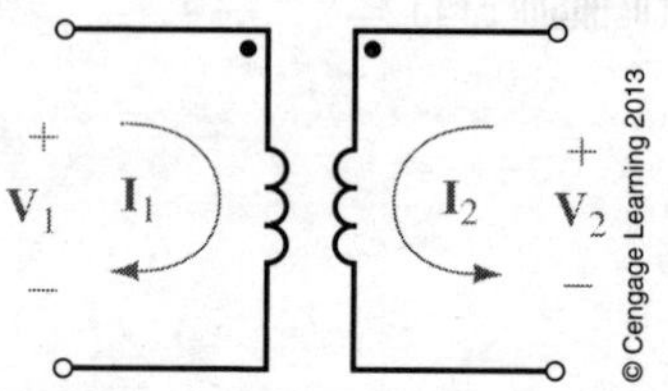

FIGURE 23–59 Coupled coils with sinusoidal ac excitation.

These equations describe the circuit of Figure 23–59, as you can see if you write KVL for each loop. (Verify this.)

EXAMPLE 23–14

For Figure 23–60, write the mesh equations and solve for $\mathbf{I}_1$ and $\mathbf{I}_2$. Let ω = 1000 rad/s, $L_1 = 0.1$ H, $L_2 = 0.2$ H, $M = 0.08$ H, $R_1 = 15\ \Omega$, and $R_2 = 20\ \Omega$.

CircuitSim 23-14

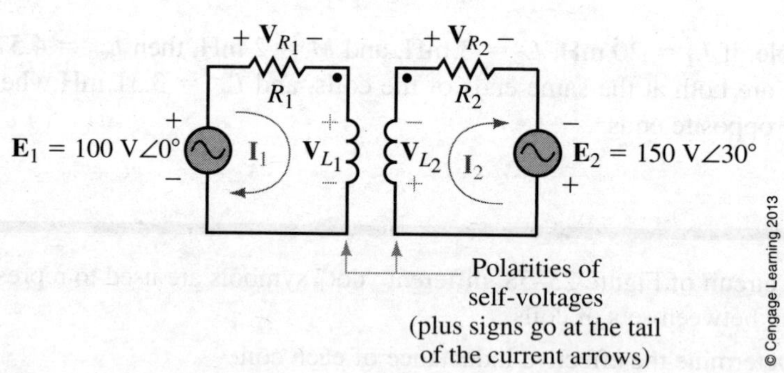

FIGURE 23–60 Air-core transformer example.

Solution $\omega L_1 = (1000)(0.1) = 100\ \Omega$, $\omega L_2 = (1000)(0.2) = 200\ \Omega$, and $\omega M = (1000)(0.08) = 80\ \Omega$. Since one current enters a dotted terminal and the other leaves, the sign of the mutual term is opposite to the sign of the self term. (To help differentiate between self and mutual terms, the self term for each equation is printed in blue and the mutual term in red.) KVL yields

opposite

Loop 1: $\mathbf{E}_1 - R_1\mathbf{I}_1 - j\omega L_1\mathbf{I}_1 + j\omega M\mathbf{I}_2 = 0$

Loop 2: $\mathbf{E}_2 - j\omega L_2\mathbf{I}_2 + j\omega M\mathbf{I}_1 - R_2\mathbf{I}_2 = 0$

opposite

Thus,

$$(15 + j100)\mathbf{I}_1 - j80\mathbf{I}_2 = 100\angle 0°$$
$$-j80\mathbf{I}_1 + (20 + j200)\mathbf{I}_2 = 150\angle 30°$$

These can be solved by computer program, by determinants, or by calculator—see Appendix B. Answers are $\mathbf{I}_1 = 2.16\angle -61.9°$ and $\mathbf{I}_2 = 1.61\angle -55.3°$.

EXAMPLE 23–15

For the circuit of Figure 23–61, determine $\mathbf{I}_1$ and $\mathbf{I}_2$.

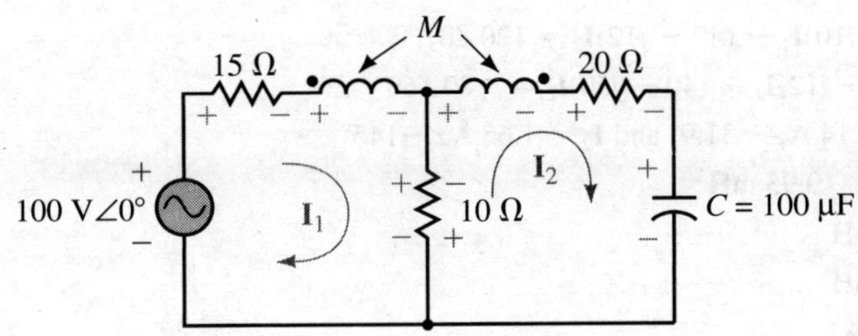

$L_1 = 0.1$ H, $L_2 = 0.2$ H, $M = 80$ mH, $\omega = 1000$ rad/s

FIGURE 23–61

Solution $\omega L_1 = 100\ \Omega$, $\omega L_2 = 200\ \Omega$, $\omega M = 80\ \Omega$, and $X_C = 10\ \Omega$.

Loop 1: $100\angle 0° - 15\mathbf{I}_1 - j100\mathbf{I}_1 + j80\mathbf{I}_2 - 10\mathbf{I}_1 + 10\mathbf{I}_2 = 0$

Loop 2: $-10\mathbf{I}_2 + 10\mathbf{I}_1 - j200\mathbf{I}_2 + j80\mathbf{I}_1 - 20\mathbf{I}_2 - (-j10)\mathbf{I}_2 = 0$

Thus,

$$(25 + j100)\mathbf{I}_1 - (10 + j80)\mathbf{I}_2 = 100\angle 0°$$
$$-(10 + j80)\mathbf{I}_1 + (30 + j190)\mathbf{I}_2 = 0$$

Solution yields $\mathbf{I}_1 = 1.43\text{ A}\angle -71.7°$ and $\mathbf{I}_2 = 0.600\text{ A}\angle -69.9°$.

PRACTICE PROBLEMS 8

1. Refer to the circuit of Figure 23–62.
   a. Determine the mesh equations.
   b. Solve for currents $\mathbf{I}_1$ and $\mathbf{I}_2$.

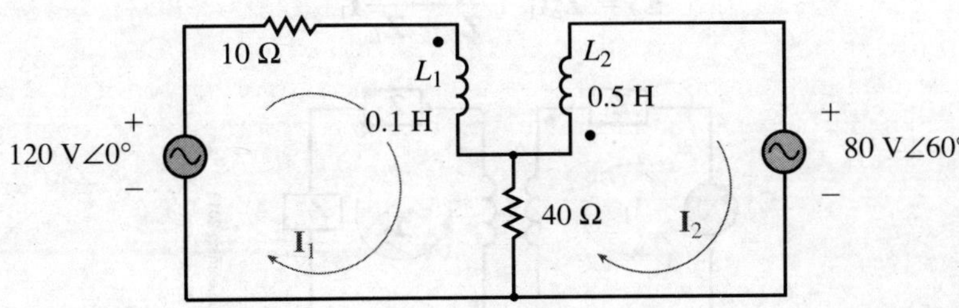

FIGURE 23–62 $M = 0.12$ H, $\omega = 100$ rad/s.

2. Consider a pair of coupled inductors (with dots at the same ends of the coils) connected in parallel. Given $f = 1$ kHz, $\mathbf{Eg} = 18\text{ V}\angle 0°$, $L_1 = 25$ mH, $L_2 = 9$ mH, and $M = 6$ mH.
   a. Determine the equivalent inductances L'_1 and L'_2.
   b. Using the computed values for L'_1 and L'_2, compute the parallel equivalent using the standard parallel inductance formula of Chapter 13.
   c. Determine the parallel equivalent inductance using Equation 23–20(c) and compare to the answer of (b).
   d. Determine source current.
   e. Determine current through each inductor.

CircuitSim 23-15

Answers

1. a. $(50 + j10)\mathbf{I}_1 - (40 - j12)\mathbf{I}_2 = 120\angle 0°$
   $-(40 - j12)\mathbf{I}_1 + (40 + j50)\mathbf{I}_2 = -80\angle 60°$
   b. $\mathbf{I}_1 = 1.14\text{ A}\angle -31.9°$ and $\mathbf{I}_2 = 1.65\text{ A}\angle -146°$
2. a. 63 mH 9.95 mH
   b. 8.59 mH
   c. 8.59 mH
   d. 0.334 A
   e. 45.5 mA 288 mA

23.11 Coupled Impedance

Earlier we found that an impedance $\mathbf{Z}_L$ on the secondary side of an iron-core transformer is reflected into the primary side as $a^2\mathbf{Z}_L$. A somewhat similar situation occurs in loosely coupled circuits. In this case, however, the impedance that you see reflected to the primary side is referred to as **coupled impedance.** To get at the idea, consider Figure 23–63. Writing KVL for each loop yields

Loop 1: $\quad \mathbf{E}_g - \mathbf{Z}_1\mathbf{I}_1 - j\omega L_1\mathbf{I}_1 - j\omega M\mathbf{I}_2 = 0$

Loop 2: $\quad -j\omega L_2\mathbf{I}_2 - j\omega M\mathbf{I}_1 - \mathbf{Z}_2\mathbf{I}_2 - \mathbf{Z}_L\mathbf{I}_2 = 0$

which reduces to

$$\mathbf{E}_g = \mathbf{Z}_p\mathbf{I}_1 + j\omega M\mathbf{I}_2 \tag{23–21a}$$

$$0 = j\omega M\mathbf{I}_1 + (\mathbf{Z}_s + \mathbf{Z}_L)\mathbf{I}_2 \tag{23–21b}$$

where $\mathbf{Z}_p = \mathbf{Z}_1 + j\omega L_1$ and $\mathbf{Z}_s = \mathbf{Z}_2 + j\omega L_2$. Solving Equation 23–21b for $\mathbf{I}_2$ and substituting this into Equation 23–21a yields, after some manipulation,

$$\mathbf{E}_g = \mathbf{Z}_p\mathbf{I}_1 + \frac{(\omega M)^2}{\mathbf{Z}_s + \mathbf{Z}_L}\mathbf{I}_1$$

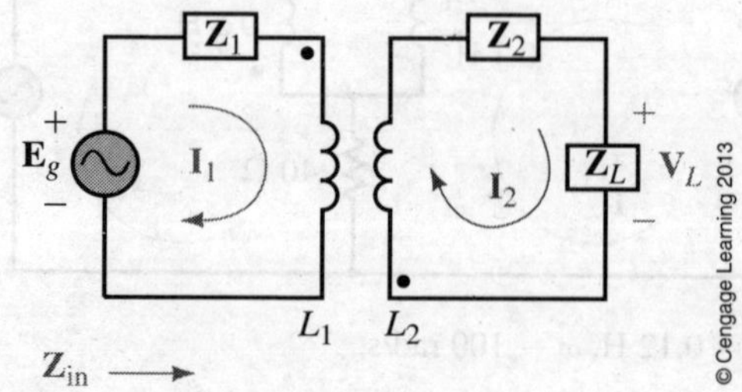

FIGURE 23–63

Now, divide both sides by $\mathbf{I}_1$, and define $\mathbf{Z}_{in} = \mathbf{E}_g/\mathbf{I}_1$. Thus,

$$\mathbf{Z}_{in} = \mathbf{Z}_p + \frac{(\omega M)^2}{\mathbf{Z}_s + \mathbf{Z}_L} \quad \text{(23–22)}$$

The term $(\omega M)^2/(\mathbf{Z}_s + \mathbf{Z}_L)$, which reflects the secondary impedances into the primary, is the coupled impedance for the circuit. Note that since secondary impedances appear in the denominator, they reflect into the primary with reversed reactive parts. Thus, capacitance in the secondary circuit looks inductive to the source and vice versa for inductance.

EXAMPLE 23–16

For Figure 23–63, let $L_1 = L_2 = 10$ mH, $M = 9$ mH, $\omega = 1000$ rad/s, $\mathbf{Z}_1 = R_1 = 5\ \Omega$, $\mathbf{Z}_2 = 1\ \Omega - j5\ \Omega$, $\mathbf{Z}_L = 1\ \Omega + j20\ \Omega$, and $\mathbf{E}_g = 100\ \text{V}\angle 0°$. Determine $\mathbf{Z}_{in}$ and $\mathbf{I}_1$.

Solution

$\omega L_1 = 10\ \Omega$. Thus, $\mathbf{Z}_p = R_1 + j\omega L_1 = 5\ \Omega + j10\ \Omega$.

$\omega L_2 = 10\ \Omega$. Thus, $\mathbf{Z}_s = \mathbf{Z}_2 + j\omega L_2 = (1\ \Omega - j5\ \Omega) + j10\ \Omega = 1\ \Omega + j5\ \Omega$.

$\omega M = 9\ \Omega$ and $\mathbf{Z}_L = 1\ \Omega + j20\ \Omega$. Thus,

$$\mathbf{Z}_{in} = \mathbf{Z}_p + \frac{(\omega M)^2}{\mathbf{Z}_s + \mathbf{Z}_L} = (5 + j10) + \frac{(9)^2}{(1 + j5) + (1 + j20)}$$

$$= 8.58\ \Omega\angle 52.2°$$

$$\mathbf{I}_1 = \mathbf{E}_g/\mathbf{Z}_{in} = (100\angle 0°)/(8.58\angle 52.2°) = 11.7\ \text{A}\angle -52.2°$$

The equivalent circuit is shown in Figure 23–64.

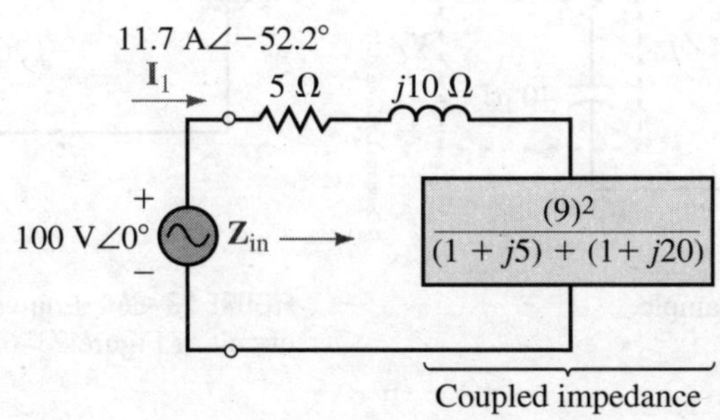

FIGURE 23–64

PRACTICE PROBLEMS 9

1. For Example 23–16, let $R_1 = 10\ \Omega$, $M = 8$ mH, and $\mathbf{Z}_L = (3 - j8)\ \Omega$. Determine $\mathbf{Z}_{in}$ and $\mathbf{I}_1$.
2. Write simultaneous equations for the circuit of Example 23–16 and solve for $\mathbf{I}_1$. Compare it to the answer obtained in the example.

Answers

1. 28.9 Ω∠41.1°; 3.72 A∠−41.1°
2. Same

23.12 Circuit Analysis Using Computers

Multisim and PSpice may be used to solve coupled circuits—see Notes.

EXAMPLE 23–17

NOTES...

1. The current version of Multisim handles both iron-core and loosely coupled circuits. However, Multisim legacy versions (e.g., Multisim 9) handle only iron core. Thus, if you are using legacy software, you will be able to do Example 23–17 but not Example 23–18.
2. Multisim and PSpice require grounds on both sides of the transformer.

As a first example, consider the iron-core transformer circuit of Figure 23–65. Solve for generator and load currents and the load voltage.

Solution

First, manually determine the answers to provide a basis for comparison. Reflecting the load impedance using $a^2\mathbf{Z}_L$ yields the equivalent circuit of Figure 23–66. From this,

$$\mathbf{I}_g = \frac{100\ \text{V}\angle 0°}{200\ \Omega + (200\ \Omega - j265.3\ \Omega)} = 208.4\ \text{mA}\angle 33.5°$$

Thus,

$$\mathbf{I}_L = a\mathbf{I}_g = 416.8\ \text{mA}\angle 33.5°$$

and

$$\mathbf{V}_L = \mathbf{I}_L\mathbf{Z}_L = 34.6\ \text{V}\angle -19.4°$$

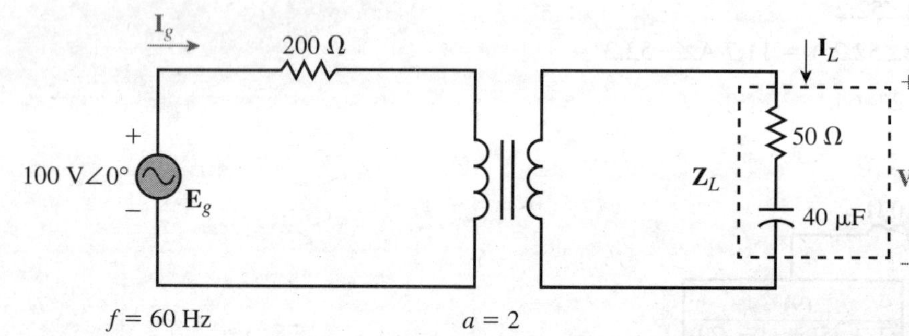

© Cengage Learning 2013

FIGURE 23–65 Iron-core circuit for first Multisim and PSpice example.

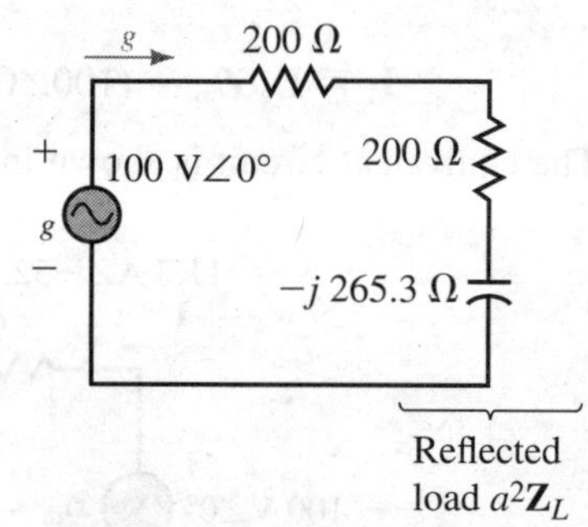

© Cengage Learning 2013

FIGURE 23–66 Equivalent of the circuit of Figure 23–65.

Multisim

Create the circuit of Figure 23–67 on your screen using the Virtual Transformer and the Power AC Source. (To access the virtual transformer, right-click *View* and ensure that the *Basic* toolbar is selected—that is, open on your screen—see Notes. Use your mouse to locate the *Place Virtual Transformer* icon, click, then position the transformer on your screen.) Double-click the transformer symbol and set its turns ratio to **2**, its primary and secondary winding resistances to **0.000001**, and its leakage inductance to **1μH**. (The values aren't critical; they just have to be small enough to be negligible.) Next, set the magnetizing inductance to **10000 H**. (This is L_m of Figure 23–41. Theoretically, for an ideal transformer it is infinity. All we need to do is make it very large.) Set all meters to AC, then click the simulation button to ON. Meter readings agree exactly with the analytic solution as determined previously.

NOTES...

Be sure that you use the Virtual Transformer, not the Virtual Audio, Virtual Misc, or Virtual Power transformer.

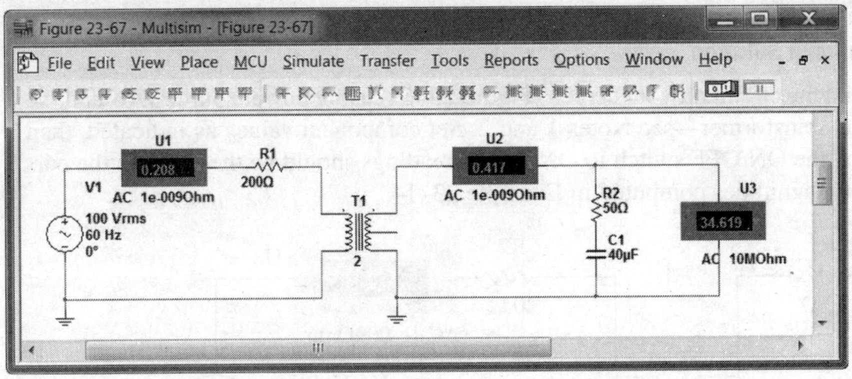

FIGURE 23–67 For Multisim iron-core transformer problems, use the virtual transformer *TS_VIRTUAL.*

PSpice Solution

Read "PSpice Notes for Coupled Circuits," page 866. As indicated, transformer element XFRM_LINEAR may be used to model iron-core transformers based solely on their turns ratios. To do this, set coupling $k = 1$, choose an arbitrarily large value for L_1, then make $L_2 = L_1/a^2$ where a is the desired turns ratio. (The actual values for L_1 and L_2 are not critical; they simply must be very large.) For example, arbitrarily choose $L_1 = 100000$ H, then compute $L_2 = 100000/(2^2) = 25000$ H. This sets $a = 2$. Now proceed as follows: Using source VAC and transformer XFMR_LINEAR, create the circuit of Figure 23–68 on your screen. Double-click the transformer and set L1 to **100000H,** L2 to **25000H,** and k to **1** as noted previously. Double-click *VPRINT1* and set *AC, MAG,* and *PHASE* to **yes** in its Properties Editor. Repeat for the *IPRINT* devices. In the simulations settings dialog box, select *AC Sweep* and set *Start* and *End* frequencies to **60Hz** and Points to **1**. Run the simulation, then scroll through the *Output File*. You should find $\mathbf{I}g = 208.3$ mA$\angle 33.6°$, $\mathbf{I}_L = 416.7$ mA$\angle 33.6°$, and $\mathbf{V}_L = 34.6$ V$\angle -19.4°$ as determined in Example 23–17.

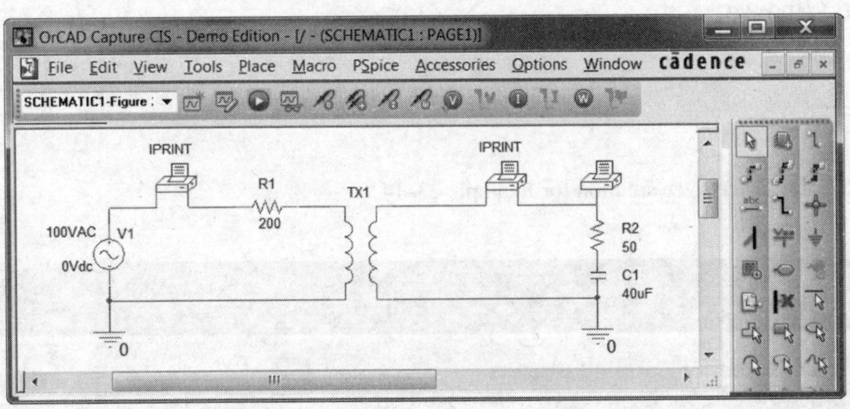

FIGURE 23–68 Using PSpice to model an iron-core transformer.

EXAMPLE 23–18

As a final example, consider the loosely coupled circuit of Figure 23–60. Convert frequency to Hz, then determine the coefficient of coupling using Equation 23–16. When you build the circuit, do not forget to rotate Source 2 as indicated.

Multisim Solution

Build the circuit on your screen (Figure 23–69) using power source *VAC* and the ideal transformer—see Notes 1 and 2. Set component values as indicated, then click the ON/OFF switch to ON. Meter readings should be the same as the current magnitudes computed in Example 23–14.

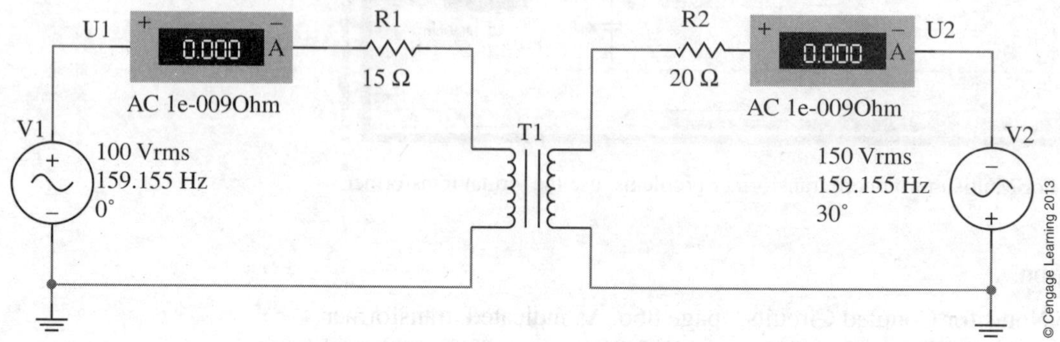

FIGURE 23–69 For Multisim air-core transformer problems, use the ideal transformer *TS_IDEAL*.

PSpice Solution

Using *VAC* for the sources and *XFMR_LINEAR* for the transformer, create the circuit of Figure 23–70 on your screen. Double-click the transformer symbol and set L_1 = **0.1H**, L_2 = **0.2H**, and k = **0.5657**. Select *AC Sweep* and set *Start* and *End* frequencies to **159.155Hz** and Points to **1**. Run the simulation. When you scroll the *Output File*, you will find that the answers match those computed in Example 23–14.

NOTES...

1. You cannot do this problem with legacy versions of Multisim.
2. With the current version of Multisim, you can incorporate coupling using either the ideal transformer (*TS_IDEAL*) or general inductor coupling (*INDUCTOR_COUPLING*). In this example, we will use the ideal transformer.
3. To access the ideal transformer, click *Place, Component, Basic,* then select *TS_IDEAL*.

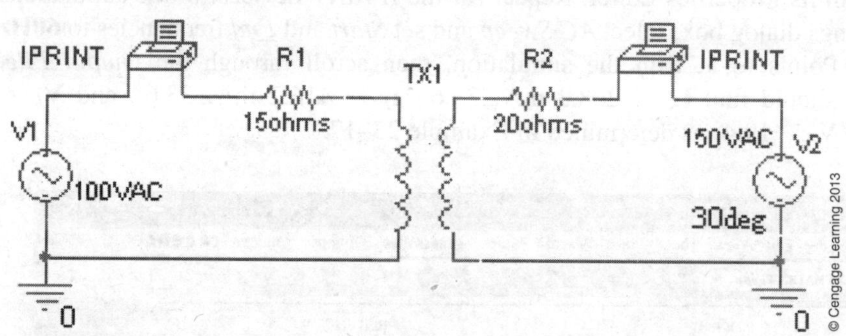

FIGURE 23–70 PSpice solution for Example 23–14.

NOTES...

PSpice Notes for Coupled Circuits

1. The PSpice transformer model XFRM_LINEAR is based on self-inductances and the coefficient of coupling and is thus able to handle loosely coupled circuits directly. It is also able to model tightly coupled circuits (such as iron-core transformers). To see how, note that basic theory shows that for an ideal iron-core transformer, $k = 1$ and L_1 and L_2 are infinite, but their ratio is $L_1/L_2 = a^2$. Thus, to represent an iron-core transformer, set L_1 to an arbitrary very large value, then compute $L_2 = L_1/a^2$. This fixes a, permitting you to model iron-core transformers based solely on their turns ratio.
2. The sign of the coefficient of coupling to use with PSpice depends on the dot locations. If dots are on adjacent coil ends (Figure 23–60), make k positive; if dots are on opposite ends (Figure 23–63), make k negative.

Putting It into Practice

A circuit you are building calls for a 3.6-mH inductor. In your parts bin, you find a 1.2-mH and a 2.4-mH inductor. You reason that if you connect them in series, the total inductance will be 3.6 mH. After you build and test the circuit, you find that it is out of spec. After careful reasoning, you become suspicious that mutual coupling between the coils is upsetting the operation. You therefore set out to measure this mutual inductance. However, you have a meter that measures only self-inductance. Then an idea hits you. You de-energize the circuit, unsolder the end of one of the inductors and measure total inductance. You get 6.32 mH. What is the mutual inductance?

Problems

23.1 Introduction

1. For the transformers of Figure 23–71, sketch the missing waveforms.

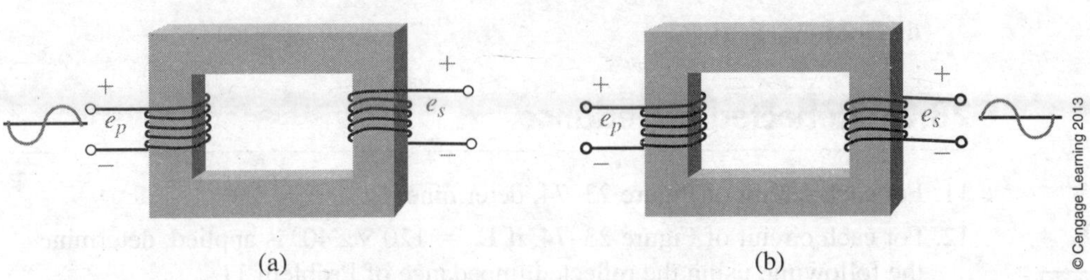

FIGURE 23–71

23.2 Iron-Core Transformers: The Ideal Model

2. List the four things that you neglect when you idealize an iron-core transformer.
3. An ideal transformer has $N_p = 1000$ turns and $N_s = 4000$ turns.
   a. Is it step-up or step-down voltage?
   b. If $e_s = 100 \sin \omega t$, what is e_p when wound as in Figure 23–7(a)?
   c. If $E_s = 24$ volts, what is E_p?
   d. If $\mathbf{E}_p = 24\text{ V}\angle 0°$, what is $\mathbf{E}_s$ when wound as in Figure 23–7(a)?
   e. If $\mathbf{E}_p = 800\text{ V}\angle 0°$, what is $\mathbf{E}_s$ when wound as in Figure 23–7(b)?
4. A 3:1 step-down voltage transformer has a secondary current of 6 A. What is its primary current?

5. For Figure 23–72, determine the expressions for v_1, v_2, and v_3.

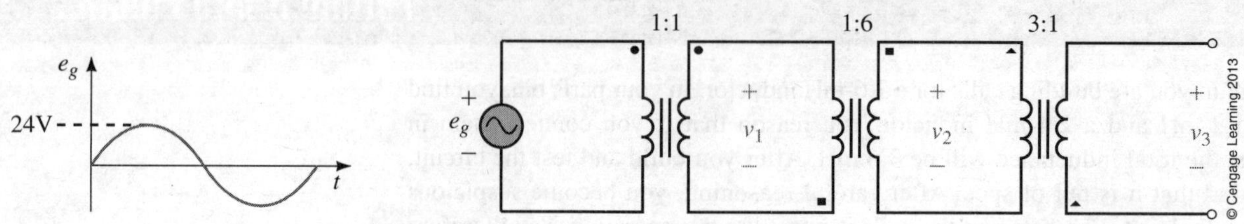

FIGURE 23–72

6. If, for Figure 23–73, $\mathbf{E}_g = 240\ \text{V}\angle 0°$, $a = 2$, and $\mathbf{Z}_L = 8\ \Omega - j6\ \Omega$, determine the following:

   a. $\mathbf{V}_L$ b. $\mathbf{I}_L$ c. $\mathbf{I}_g$

7. If, for Figure 23–73, $\mathbf{E}_g = 240\ \text{V}\angle 0°$, $a = 0.5$, and $\mathbf{I}_g = 2\ \text{A}\angle 20°$, determine the following:

   a. $\mathbf{I}_L$ b. $\mathbf{V}_L$ c. $\mathbf{Z}_L$

8. If, for Figure 23–73, $a = 2$, $\mathbf{V}_L = 40\ \text{V}\angle 0°$, and $\mathbf{I}_g = 0.5\ \text{A}\angle 10°$, determine $\mathbf{Z}_L$.

9. If, for Figure 23–73, $a = 4$, $\mathbf{I}_g = 4\ \text{A}\angle 30°$, and $\mathbf{Z}_L = 6\ \Omega - j8\ \Omega$, determine the following:

   a. $\mathbf{V}_L$ b. $\mathbf{E}_g$

10. If, for the circuit of Figure 23–73, $a = 3$, $\mathbf{I}_L = 4\ \text{A}\angle 25°$, and $\mathbf{Z}_L = 10\ \Omega\angle -5°$, determine the following:

    a. Generator current and voltage.
    b. Power to the load.
    c. Power output by the generator.
    d. Does $P_{out} = P_{in}$?

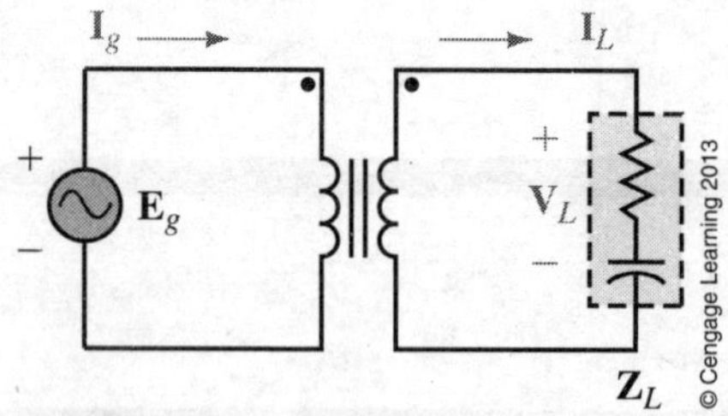

FIGURE 23–73

23.3 Reflected Impedance

11. For each circuit of Figure 23–74, determine $\mathbf{Z}_p$.
12. For each circuit of Figure 23–74, if $\mathbf{E}_g = 120\ \text{V}\angle 40°$ is applied, determine the following, using the reflected impedance of Problem 11.

    a. $\mathbf{I}_g$ b. $\mathbf{I}_L$ c. $\mathbf{V}_L$

13. For Figure 23–74(a), what turns ratio is required to make $\mathbf{Z}_p = (62.5 - j125)\ \Omega$?
14. For Figure 23–74(b), what turns ratio is required to make $\mathbf{Z}_p = 84.9\angle 58.0°\ \Omega$?
15. For each circuit of Figure 23–75, determine $\mathbf{Z}_T$.
16. For each circuit of Figure 23–75, if a generator with $\mathbf{E}_g = 120\ \text{V}\angle -40°$ is applied, determine the following:

    a. $\mathbf{I}_g$ b. $\mathbf{I}_L$ c. $\mathbf{V}_L$

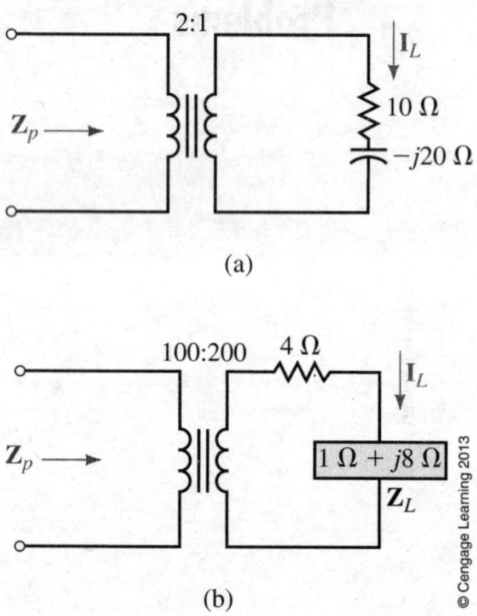

FIGURE 23–74

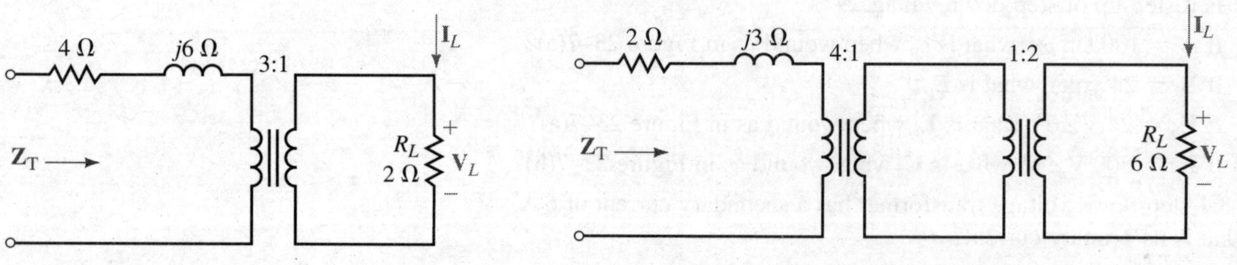

FIGURE 23–75

23.4 Power Transformer Ratings

17. A transformer has a rated primary voltage of 7.2 kV, $a = 0.2$, and a secondary rated current of 3 A. What is its kVA rating?

18. Consider a 48 kVA, 1200/120-V transformer.
    a. What is the maximum kVA load that it can handle at $F_p = 0.8$?
    b. What is the maximum power that it can supply to a 0.75 power factor load?
    c. If the transformer supplies 45 kW to a load at 0.6 power factor, is it overloaded? Justify your answer.

23.5 Transformer Applications

19. The transformer of Figure 23–25 has a 7200-V primary and center-tapped 240-V secondary. If Load 1 consists of twelve 100-W lamps, Load 2 is a 1500-W heater, and Load 3 is a 2400-W stove with $F_p = 1.0$, determine:
    a. I_1 b. I_2 c. I_N d. I_p

20. An amplifier with a Thévenin voltage of 10 V and Thévenin resistance of 128 Ω is connected to an 8-Ω speaker through a 4:1 transformer. Is the load matched? How much power is delivered to the speaker?

21. An amplifier with a Thévenin equivalent of 10 V and R_{Th} of 25 Ω drives a 4-Ω speaker through a transformer with a turns ratio of $a = 5$. How much power is delivered to the speaker? What turns ratio yields 1 W?

22. For Figure 23–76, there are 100 turns between taps 1 and 2 and 120 between taps 2 and 3. What voltage at tap 1 yields 120 V out? At tap 3?

23. For Figure 23–30(a), $a_2 = 2$ and $a_3 = 5$, $\mathbf{Z}_2 = 20\ \Omega\angle 50°$, $\mathbf{Z}_3 = (12 + j4)\ \Omega$, and $\mathbf{E}_g = 120\ \text{V}\angle 0°$. Find each load current and the generator current.

24. We wish to connect a 5-kVA, 120/240-V transformer as an autotransformer to a 120-V source to supply 360 V to a load.
    a. Draw the circuit.
    b. What is the maximum current that the load can draw?
    c. What is the maximum load kVA that can be supplied?
    d. How much current is drawn from the source?

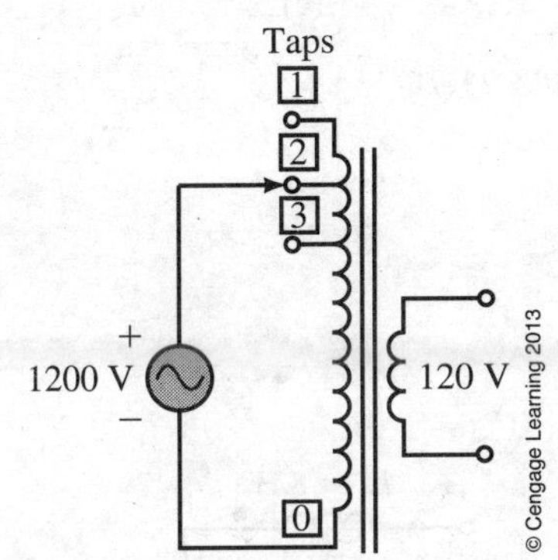

FIGURE 23–76 $N_s = 200$ turns.

23.6 Practical Iron-Core Transformers

25. For Figure 23–77, $\mathbf{E}_g = 1220\ \text{V}\angle 0°$.
    a. Draw the equivalent circuit.
    b. Determine $\mathbf{I}_g$, $\mathbf{I}_L$, and $\mathbf{V}_L$.

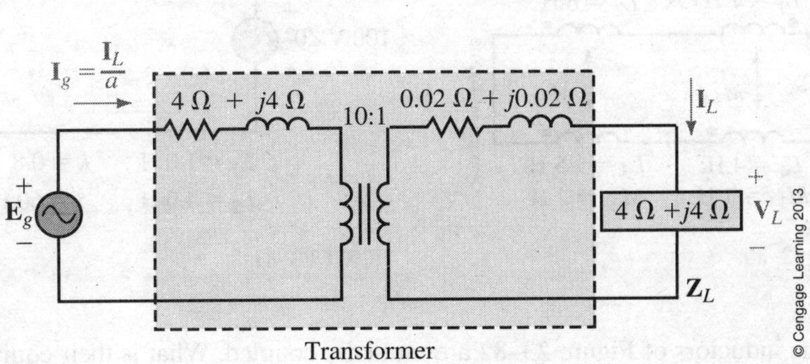

FIGURE 23–77

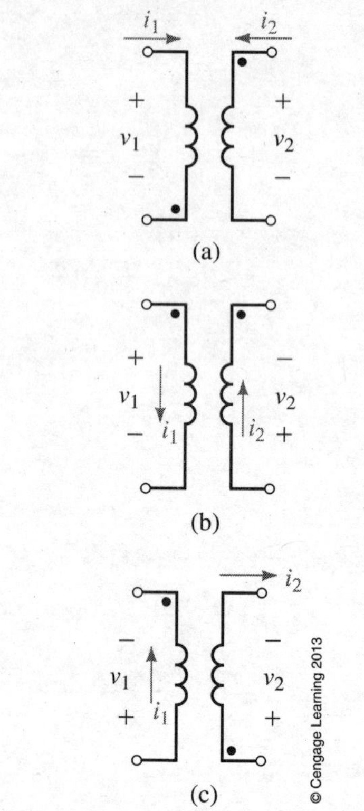

FIGURE 23–78

26. For Figure 23–77, if $\mathbf{V}_L = 118\ \text{V}\angle 0°$, draw the equivalent circuit and determine:

    a. $\mathbf{I}_L$ b. $\mathbf{I}_g$ c. $\mathbf{E}_g$
    d. no-load voltage e. regulation

27. A transformer delivering $P_{\text{out}} = 48$ kW has a core loss of 280 W and a copper loss of 450 W. What is its efficiency at this load?

23.7 Transformer Tests

28. A short-circuit test (Figure 23–46) at rated current yields a wattmeter reading of 96 W, and an open-circuit test (Figure 23–48) yields a core loss of 24 W.

    a. What is the transformer's efficiency when delivering the full, rated output of 5 kVA at unity F_p?
    b. What is its efficiency when delivering one-quarter the rated kVA at 0.8 F_p?

23.9 Loosely Coupled Circuits

29. For Figure 23–78,

$$v_1 = L_1\frac{di_1}{dt} \pm M\frac{di_2}{dt}, \qquad v_2 = \pm M\frac{di_1}{dt} + L_2\frac{di_2}{dt}$$

For each circuit, indicate whether the sign to use with M is plus or minus.

30. For a set of coils, $L_1 = 250$ mH, $L_2 = 0.4$ H, and $k = 0.85$. What is M?
31. For a set of coupled coils, $L_1 = 2$ H, $M = 0.8$ H, and the coefficient of coupling is 0.6. Determine L_2.
32. For Figure 23–52(a), $L_1 = 25$ mH, $L_2 = 4$ mH, and $M = 0.8$ mH. If i_1 changes at a rate of 1200 A/s, what are the primary and secondary induced voltages?
33. Everything the same as Problem 32 except that $i_1 = 10\,e^{-500t}$ A. Find the equations for the primary and secondary voltages. Compute them at $t = 1$ ms.
34. For each circuit of Figure 23–79, determine L_T.
35. For Figure 23–80, determine L_T.
36. For the circuit of Figure 23–81, determine **I**.

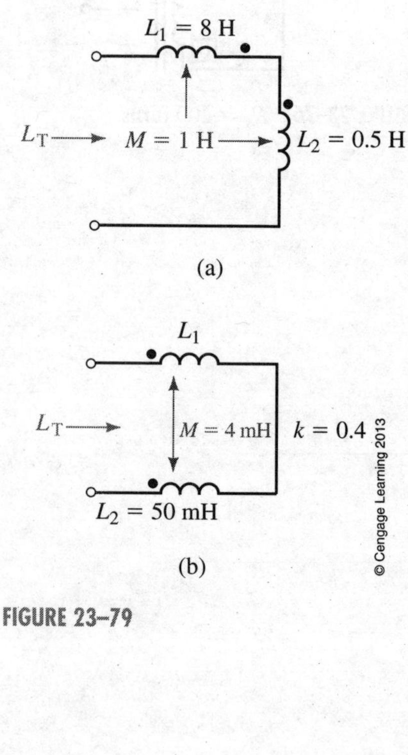

FIGURE 23–79

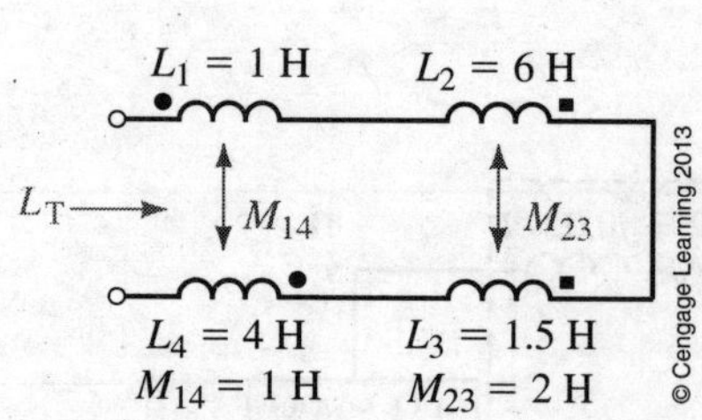

FIGURE 23–80

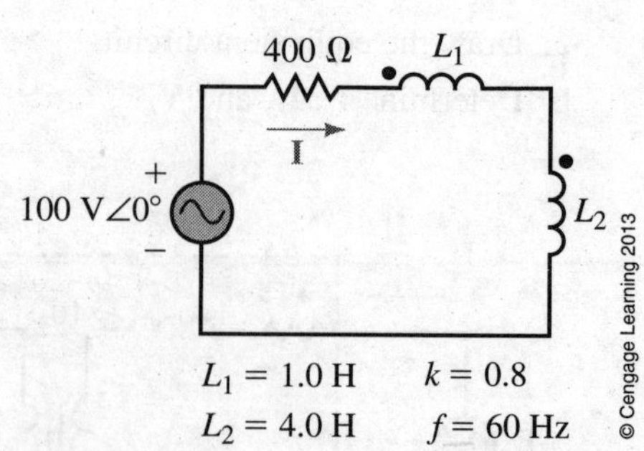

FIGURE 23–81

37. The inductors of Figure 23–82 are mutually coupled. What is their equivalent inductance? If $f = 60$ Hz, what is the source current?

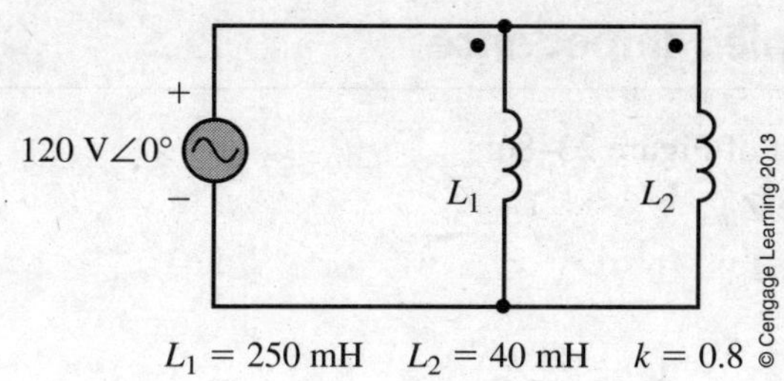

FIGURE 23–82

23.10 Magnetically Coupled Circuits with Sinusoidal Excitation

38. For Figure 23–60, $R_1 = 10\ \Omega$, $R_2 = 30\ \Omega$, $L_1 = 100$ mH, $L_2 = 200$ mH, $M = 25$ mH, and $f = 31.83$ Hz. Write the mesh equations.
39. For the circuit of Figure 23–83, write mesh equations.

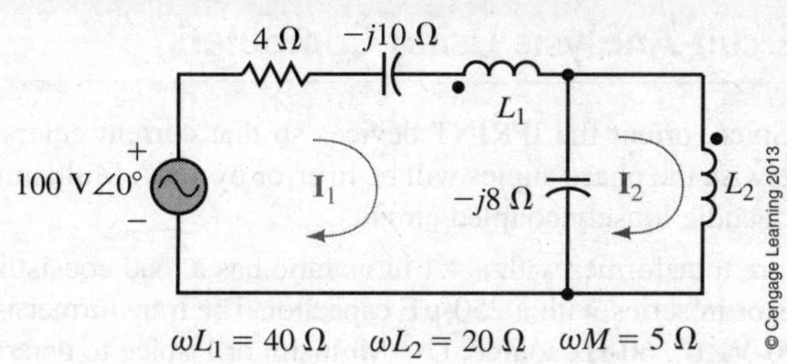

FIGURE 23–83

40. Write mesh equations for the circuit of Figure 23–84.

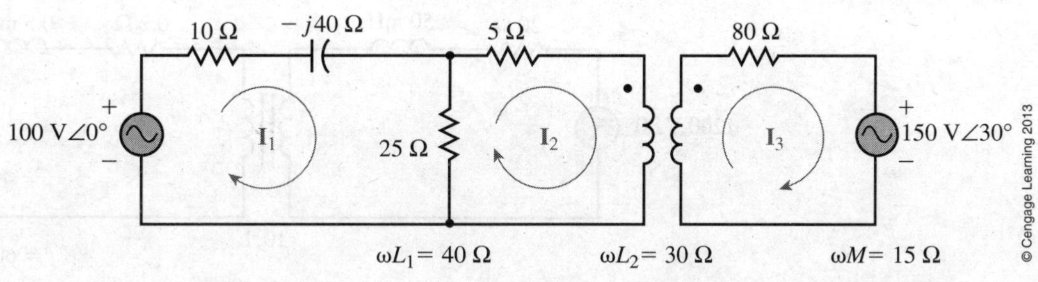

FIGURE 23–84

41. Write mesh equations for the circuit of Figure 23–85. (This is a very challenging problem.)

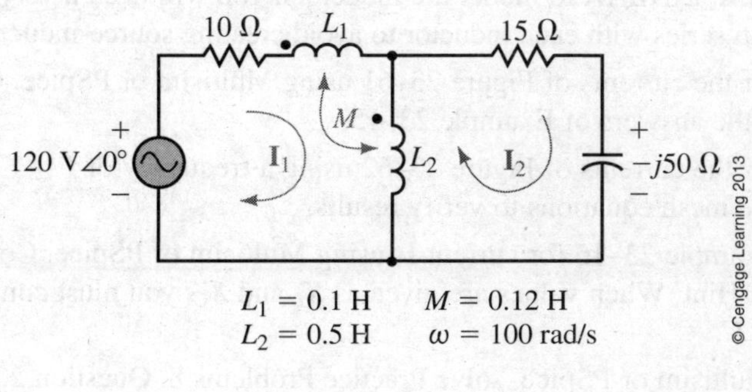

FIGURE 23–85

23.11 Coupled Impedance

42. For the circuit of Figure 23–86:
    a. Determine $\mathbf{Z}_{in}$.
    b. Determine $\mathbf{I}_g$.

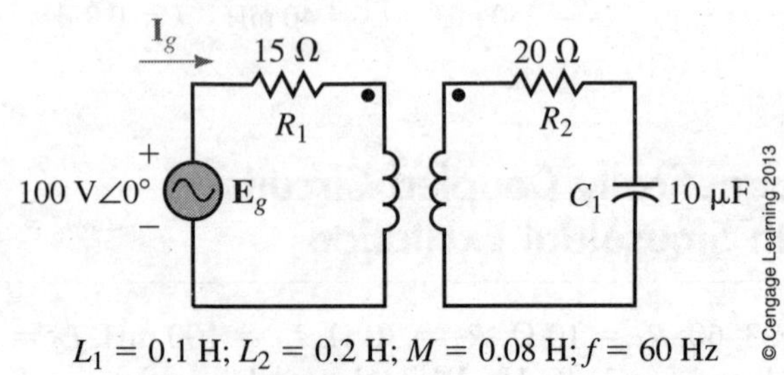

FIGURE 23–86

23.12 Circuit Analysis Using Computers

Note: With PSpice, orient the IPRINT devices so that current enters the positive terminal. Otherwise the phase angles will be in error by 180°. Multisim legacy software does not handle loosely coupled circuits.

Multisim PSPICE 43. An iron-core transformer with a 4:1 turns ratio has a load consisting of a 12-Ω resistor in series with a 250-μF capacitor. The transformer is driven from a 120-V∠0°, 60-Hz source. Use Multisim or PSpice to determine the source and load currents. Verify the answers by manual computation.

Multisim PSPICE 44. Using Multisim or PSpice, solve for the primary and secondary currents and the load voltage for Figure 23–87. Verify by manual computation.

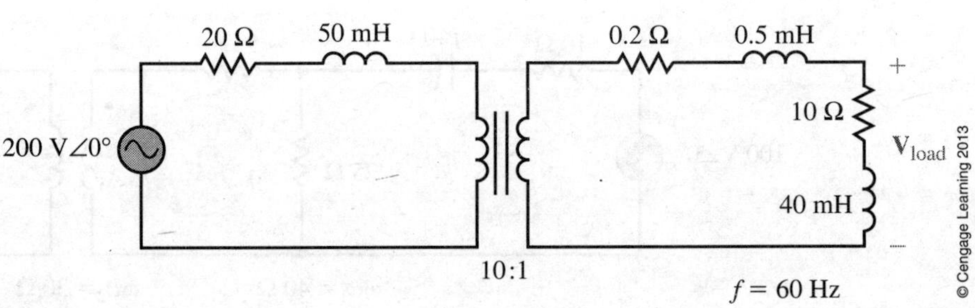

FIGURE 23–87

Multisim PSPICE 45. Using Multisim or PSpice, solve for the source current for the coupled parallel inductors of Figure 23–82. Hint: For Multisim, use TS_IDEAL; for PSpice, use XFRM_LINEAR to model the inductors. You will need a very low-value resistor in series with each inductor to avoid creating source-inductor loops.

Multisim PSPICE 46. Solve for the currents of Figure 23–61 using Multisim or PSpice. Compare these to the answers of Example 23–15.

Multisim PSPICE 47. Solve for the currents of Figure 23–62 using a frequency of f = 1 kHz. Write and solve mesh equations to verify results.

Multisim PSPICE 48. Solve Example 23–16 for current $\mathbf{I}_1$ using Multisim or PSpice. Compare answers. Hint: When values are given as X_L and X_C, you must convert them to L and C.

Multisim PSPICE 49. Using Multisim or PSpice, solve Practice Problems 8, Question 2, and compare the answers to its analytic solution.

ANSWERS TO IN-PROCESS LEARNING CHECKS

IN-PROCESS LEARNING CHECK 1

1. Step-up; 200 V
2. a. 120 V$\angle$–30°; b. 120 V$\angle$150°
3. 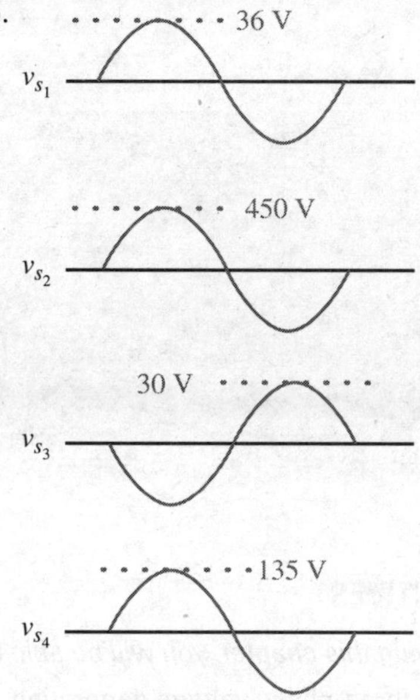

4. Secondary, upper terminal.
5. a. Terminal 4; b. Terminal 4

IN-PROCESS LEARNING CHECK 2

1. $\mathbf{Z}_p = 18.75\ \Omega - j25\ \Omega$; $R = 6400\ \Omega$; $a = 1.73$
2. 125 V$\angle$0°
3. 89.4
4. a. 200 A; 200 A; b. 24 kW; 18 kW
5. Tap 1: 109.1 V; Tap 2: 120 V; Tap 3: 126.3 V
6. 0.81 W; 0.72 W; Maximum power is delivered when $R_s = a^2R_L$.
7. a. 348 V; 20.7 A; b. 12 A; c. 12 A↓ 8.69 A↑

IN-PROCESS LEARNING CHECK 3

1. a. $\mathbf{Z}_{eq} = 0.8\ \Omega + j0.8\ \Omega$
   b. 113.6 V $\angle$0.434°
   c. 5.63%

KEY TERMS

Balanced Systems
Four-Wire System
Line Current
Line or Phase Conductor
Line-to-Line Voltage
Line-to-Neutral Voltage
Line Voltage
Load Phase Impedances
Neutral
Phase Current
Phase Sequence
Phase Voltage
Single-Phase Equivalent
Three-Wire Y-Y System
Two-Wattmeter Method
Unbalanced Systems
Y-Y System
Zero-Impedance Conductor

OUTLINE

Three-Phase Voltage Generation
Basic Three-Phase Circuit Connections
Basic Three-Phase Relationships
Examples
Power in a Balanced System
Measuring Power in Three-Phase Circuits
Unbalanced Loads
Power System Loads
Circuit Analysis Using Computers

OBJECTIVES

After studying this chapter, you will be able to

- describe three-phase voltage generation,
- represent three-phase voltages and currents in phasor form,
- describe standard three-phase load connections,
- analyze balanced three-phase circuits,
- compute active power, reactive power, and apparent power in a three-phase system,
- measure power using the two-wattmeter method and the three-wattmeter method,
- analyze simple, unbalanced three-phase circuits,
- apply Multisim and PSpice to three-phase problems.

24

THREE-PHASE SYSTEMS

CHAPTER PREVIEW

So far, we have looked only at single-phase systems. In this chapter, we consider three-phase systems. (Three-phase systems differ from single-phase systems in that they use a set of three sinusoidal ac voltages instead of one.) Three-phase systems are used for the generation and transmission of bulk electrical power. All commercial ac power systems, for example, are three-phase systems. Not all loads connected to a three-phase system need be three-phase, however—for example, the electric lights and appliances used in our homes require only single-phase ac. To get single-phase ac from a three-phase system, we simply tap off one of its phases, as you will see later in this chapter.

Three-phase systems may be **balanced** or **unbalanced.** If a system is balanced, it can be analyzed by considering just one of its phases. (This is because, once you know the solution for one phase, you can write down the solutions to the other two phases with no further computation other than the addition or subtraction of an angle.) This is significant because it makes the analysis of balanced systems only slightly more complex than the analysis of single-phase systems. Since most systems operate close to balance, many practical problems can be dealt with by assuming balance. This is the approach used in practice.

Three-phase power systems possess economic and operating advantages over single-phase power systems. For example, for the same power output, three-phase generators cost less than single-phase generators, produce uniform power rather than pulsating power, and operate with less vibration and noise.

We begin the chapter with a look at three-phase voltage generation. ■

Putting It in Perspective

Nikola Tesla

AS NOTED IN CHAPTER 15, the advent of the commercial electrical power age began with a fierce battle between Thomas A. Edison and George Westinghouse over the use of dc versus ac for the infant electrical power industry. Edison vigorously promoted dc while Westinghouse promoted ac. Tesla settled the argument in favor of ac with his development of the three-phase power system, the induction motor, and other ac devices. Coupled with the creation of a practical power transformer (Chapter 23), these developments made long-distance transmission of electrical energy possible, and ac became the clear winner.

Tesla was born in Smiljan, Croatia in 1856 and emigrated to the United States in 1884. During part of his career, he was associated with Edison, but the two had a falling out and became bitter rivals. Tesla made many important contributions in the fields of electricity and magnetism (he held over 700 patents), and the SI unit of magnetic flux density (the "tesla") is named after him. Tesla was also primarily responsible for the selection of 60 Hz as the standard power system frequency in North America and much of the world. ■

24.1 Three-Phase Voltage Generation

Three-phase generators have three sets of windings and thus produce three ac voltages instead of one. To get at the idea, consider first the elementary single-phase generator of Figure 24–1. (This is the same generator that you studied in Chapter 15, Figure 15–6.) As coil AA' rotates, it produces a sinusoidal waveform $e_{AA'}$ as indicated in (b). This voltage can be represented by phasor $\mathbf{E}_{AA'}$ as shown in (c).

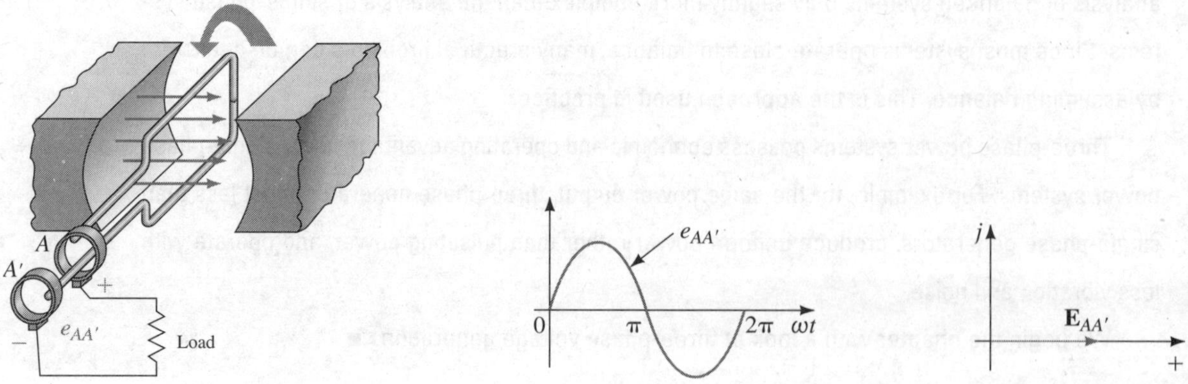

FIGURE 24–1 A basic single-phase generator.

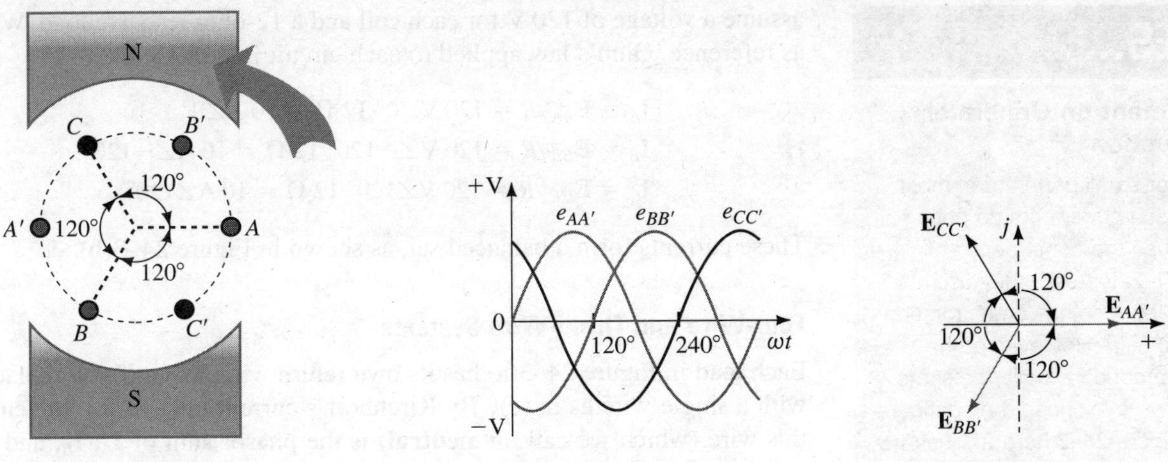

(a) Basic 3-phase generator (b) Voltage waveforms (c) Phasors

FIGURE 24–2 Generating three-phase voltages. Three sets of coils are used to produce three balanced voltages.

If two more windings are added as in Figure 24–2, two additional voltages are generated. Since these windings are identical with AA' (except for their position on the rotor), they produce identical voltages. However, since coil BB' is placed 120° behind coil AA', voltage $e_{BB'}$ lags $e_{AA'}$ by 120′; similarly, coil CC', which is placed ahead of coil AA' by 120°, produces voltage $e_{CC'}$ that leads by 120°. Waveforms are shown in (b) and phasors in (c). As indicated, the generated voltages are equal in magnitude and phase displaced by 120°. Thus, if $\mathbf{E}_{AA'}$ is at 0°, then $\mathbf{E}_{BB'}$ will be at −120° and $\mathbf{E}_{CC'}$ will be at +120°. Assuming an rms value of 120 V and a reference position of 0° for phasor $\mathbf{E}_{AA'}$ for example, yields $\mathbf{E}_{AA'} = 120\text{ V}\angle 0°$, $\mathbf{E}_{BB'} = 120\text{ V}\angle{-120°}$, and $\mathbf{E}_{CC'} = 120\text{ V}\angle 120°$. Such a set of voltages is said to be balanced. Because of this fixed relationship between balanced voltages, you can, if you know one voltage, easily determine the other two.

PRACTICE PROBLEMS 1

a. If $\mathbf{E}_{AA'} = 277\text{ V}\angle 0°$, what are $\mathbf{E}_{BB'}$ and $\mathbf{E}_{CC'}$?
b. If $\mathbf{E}_{BB'} = 347\text{ V}\angle{-120°}$, what are $\mathbf{E}_{AA'}$ and $\mathbf{E}_{CC'}$?
c. If $\mathbf{E}_{CC'} = 120\text{ V}\angle 150°$, what are $\mathbf{E}_{AA'}$ and $\mathbf{E}_{BB'}$?

Sketch the phasors for each set.

CircuitSim 24-1

Answers

a. $\mathbf{E}_{BB'} = 277\text{ V}\angle{-120°}$; $\mathbf{E}_{CC'} = 277\text{ V}\angle 120°$
b. $\mathbf{E}_{AA'} = 347\text{ V}\angle 0°$; $\mathbf{E}_{CC'} = 347\text{ V}\angle 120°$
c. $\mathbf{E}_{AA'} = 120\text{ V}\angle 30°$; $\mathbf{E}_{BB'} = 120\text{ V}\angle{-90°}$

24.2 Basic Three-Phase Circuit Connections

The generator of Figure 24–2 has three independent windings: AA', BB', and CC'. As a first thought, you might try connecting loads using six wires as in Figure 24–3(a). This will work, although it is not a scheme that is used in practice. Nonetheless, some useful insights can be gained from it. To illustrate,

> **NOTES...**
>
> **A Comment on Generator Construction**
>
> Except for small generators, most three-phase generators do not actually use the construction of Figure 24–2. Instead, they use a fixed set of windings and a rotating magnetic field. However, both designs produce exactly the same waveforms. We chose the configuration of Figure 24–2 here to illustrate because its operation is easier to visualize.

assume a voltage of 120 V for each coil and a 12-ohm resistive load. With $\mathbf{E}_{AA'}$ as reference, Ohm's law applied to each circuit in turn yields

$$\mathbf{I}_A = \mathbf{E}_{AA'}/R = 120\text{ V}\angle 0°/12\ \Omega = 10\text{ A}\angle 0°$$
$$\mathbf{I}_B = \mathbf{E}_{BB'}/R = 120\text{ V}\angle -120°/12\ \Omega = 10\text{ A}\angle -120°$$
$$\mathbf{I}_C = \mathbf{E}_{CC'}/R = 120\text{ V}\angle 120°/12\ \Omega = 10\text{ A}\angle 120°$$

These currents form a balanced set, as shown in Figure 24–3(b).

Four-Wire and Three-Wire Systems

Each load in Figure 24–3(a) has its own return wire. What if you replace them with a single wire as in (c)? By Kirchhoff's current law (KCL), the current in this wire (which we call the **neutral**) is the phasor sum of $\mathbf{I}_A$, $\mathbf{I}_B$, and $\mathbf{I}_C$. For the balanced 12-ohm load,

$$\mathbf{I}_N = \mathbf{I}_A + \mathbf{I}_B + \mathbf{I}_C = 10\text{ A}\angle 0° + 10\text{ A}\angle -120° + 10\text{ A}\angle 120°$$
$$= (10\text{ A} + j0) + (-5\text{ A} - j8.66\text{ A}) + (-5\text{ A} + j8.66\text{ A}) = 0\text{ amps}$$

Thus, the return wire carries no current at all! (This result is always true regardless of load impedance, provided the load is balanced, i.e., all phase impedances are the same.) In practice, power systems are normally operated close to balance. Thus, the return current, while not necessarily zero,

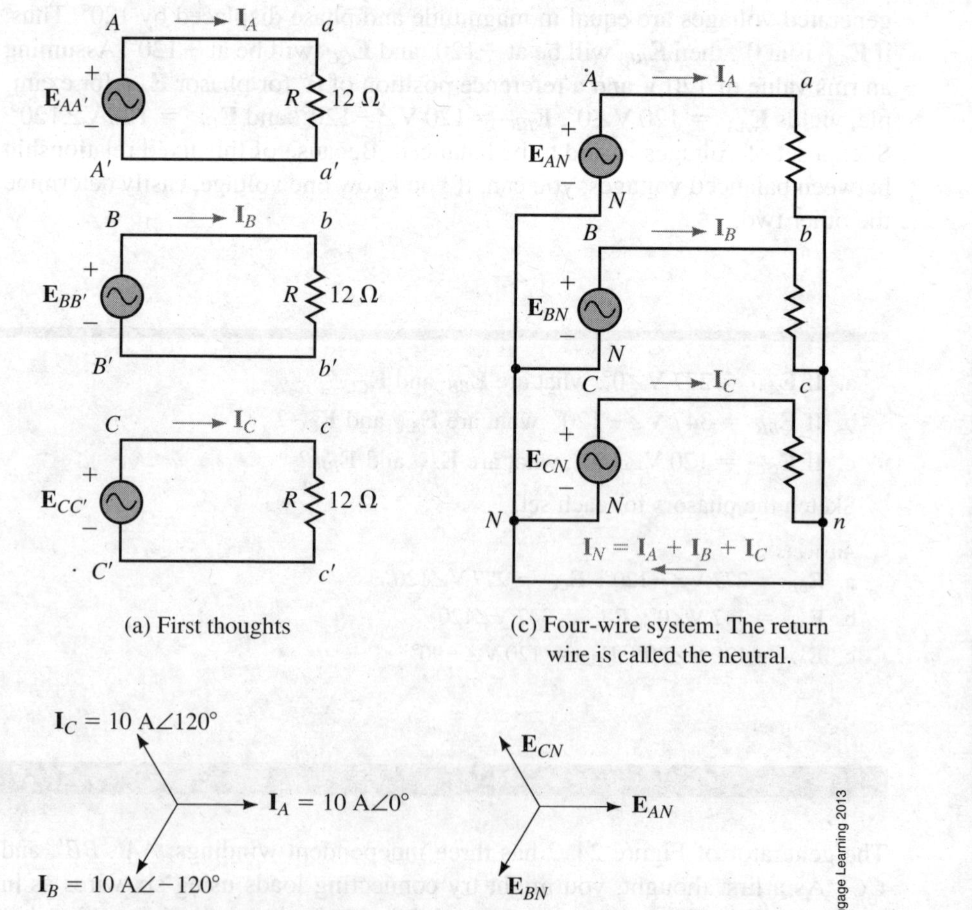

(a) First thoughts

(c) Four-wire system. The return wire is called the neutral.

(b) Currents form a balanced set

(d) Renamed phasors

CircuitSim 24-2

FIGURE 24–3 Evolution of three-phase connections.

will be quite small, and the neutral wire can be made smaller than the other three conductors. This configuration is called a **four-wire system** and is one of the configurations used in practice.

The outgoing lines of Figure 24–3(c) are called **line or phase conductors.** They are the conductors that you see suspended by insulators on transmission line towers.

Symbology

Having joined points A', B', and C' in Figure 24–3(c), we now drop the A', B', and C' notation and simply call the common point N. The voltages are then renamed $\mathbf{E}_{AN}$, $\mathbf{E}_{BN}$, and $\mathbf{E}_{CN}$ as shown in Figure 24–3(d). They are known as **line-to-neutral voltages.**

Standard Representation

Three-phase circuits are not usually drawn as in Figure 24–3. Rather, they are usually drawn as in Figure 24–4. [Figure 24–4(a), for example, shows Figure 24–3(c) redrawn in standard form.] Note that coil symbols are used to represent generator windings rather than the circle symbol that we use for single-phase ac.

As Figure 24–4(a) shows, the circuit that we have been looking at is a Y-Y circuit, it is commonly known as a four-wire **Y-Y system.** A variation, the **three-wire Y-Y system,** is shown in (b). Three-wire Y-Y circuits may be used if the load can be guaranteed to remain balanced, since under balanced conditions the neutral conductor carries no current. However, for practical reasons (discussed in Section 24.7), most Y-Y systems use four wires.

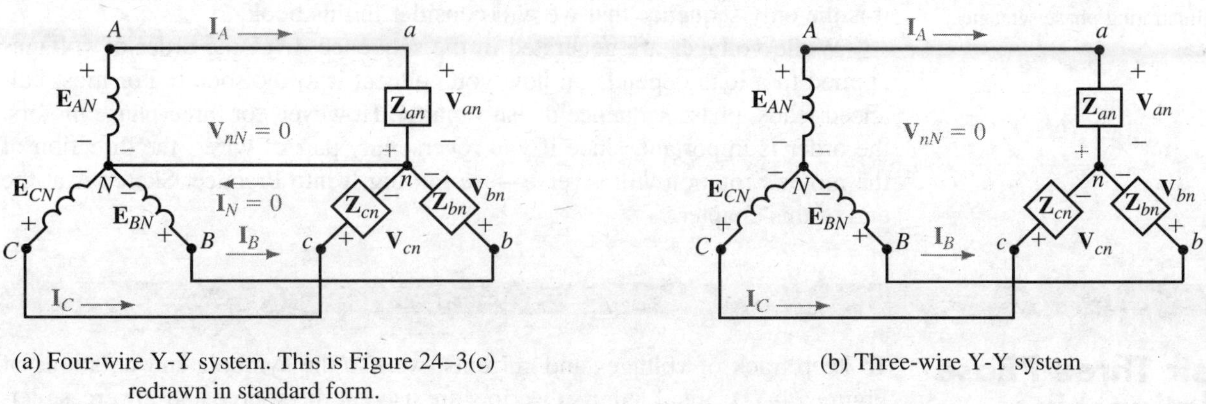

(a) Four-wire Y-Y system. This is Figure 24–3(c) redrawn in standard form.

(b) Three-wire Y-Y system

FIGURE 24–4 Standard representation of three-phase circuits.

Delta-Connected Generators

Now consider Δ connection of the generator windings. Theoretically, this is possible, as indicated in Figure 24–5. However, there are practical difficulties. For example, when generators are loaded, distortions occur in the coil voltages due to magnetic fluxes produced by load currents. In Y-connected generators, these distortions cancel, but in Δ-connected generators, they do not. These distortions create a third harmonic current that circulates within the windings of the Δ-connected generator, lowering its efficiency. (You will learn about third harmonics in Chapter 25.) For this and other reasons, Δ-connected generators are seldom used in power systems and will not be discussed in this book.

CircuitSim 24-3

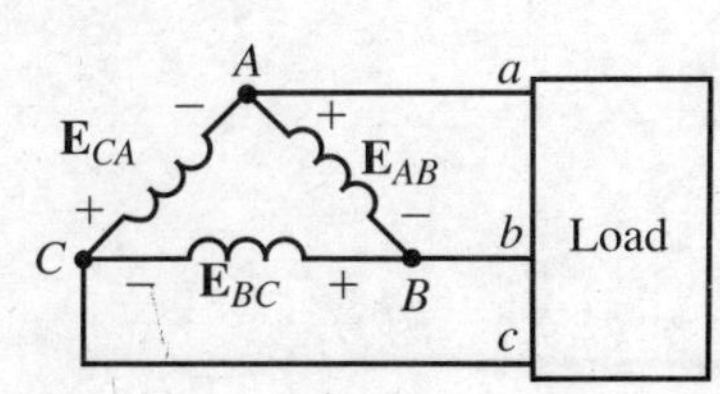

FIGURE 24–5 A delta-connected generator. For practical reasons, delta generators are seldom used in power systems.

Neutral-Neutral Voltage in a Wye-Wye Circuit

In a balanced Y-Y system, neutral current is zero because line currents sum to zero. As a consequence, the voltage between neutral points is zero. To see why, consider again Figure 24–4(a). Assume that the wire joining points *n* and *N* has impedance $\mathbf{Z}_{nN}$. This yields voltage $\mathbf{V}_{nN} = \mathbf{I}_N \times \mathbf{Z}_{nN}$. But since $\mathbf{I}_N = 0$, $\mathbf{V}_{nN} = 0$, regardless of the value of $\mathbf{Z}_{nN}$. Even if the neutral conductor is absent as in (b), $\mathbf{V}_{nN}$ is still zero. Thus, *in a balanced Y-Y system, the voltage between neutral points is zero.*

CircuitSim 24-4

EXAMPLE 24–1

Assume the circuits of Figure 24–4(a) and (b) are balanced. If $\mathbf{E}_{AN} = 247\text{ V}\angle 0°$, what are $\mathbf{V}_{an}$, $\mathbf{V}_{bn}$, and $\mathbf{V}_{cn}$?

Solution In both cases, the voltage V_{nN} between neutral points is zero. Thus, by Kirchhoff's voltage law (KVL), $\mathbf{V}_{an} = \mathbf{E}_{AN} = 247\text{ V}\angle 0°$. Since the system is balanced, $\mathbf{V}_{bn} = 247\text{ V}\angle -120$ and $\mathbf{V}_{cn} = 247\text{ V}\angle 120°$.

Phase Sequence

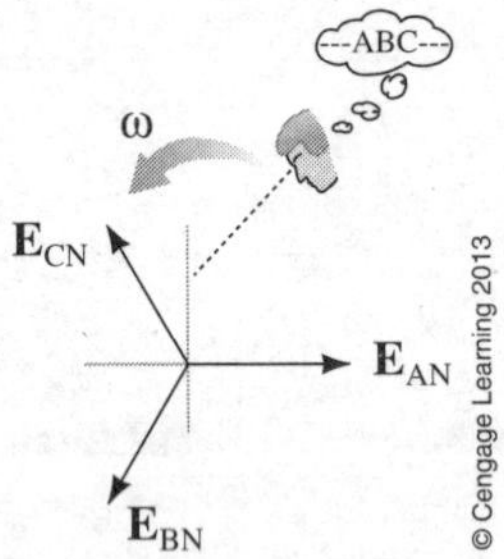

FIGURE 24–6 Illustrating phase sequence.

Phase sequence refers to the order in which three-phase voltages occur. This can be illustrated in terms of phasors. If you (conceptually) view the rotation of the phasor set of Figure 24–6, for example, you see phasors rotating by in the order . . . *ABCABC*. . . . This sequence is referred to as the ABC phase sequence or the positive phase sequence. On the other hand, if the direction of rotation were reversed, the sequence would be *ACB*. (This is called the negative phase sequence.) Since power systems generate the *ABC* sequence (recall Figure 24–2), it is the only sequence that we will consider in this book.

While voltages are generated in the sequence *ABC*, the order of voltages applied to a load depends on how you connect it to the source. For most balanced loads, phase sequence doesn't matter. However, for three-phase motors, the order is important, since if you reverse any pair of wires, the direction of the motor's rotation will reverse—see Putting It into Practice, Sketch A at the end of this chapter.

24.3 Basic Three-Phase Relationships

To keep track of voltages and currents, we use the symbols and notations of Figure 24–7. Capital letter subscripts are used at the source and lowercase letters at the load. As usual, *E* is used for source voltage and *V* for voltage drops. Note carefully the use of double subscripts, the placement of the voltage reference polarity signs, and the cyclic order of the subscripts—see Note.

Definitions

Line (also called **line-to-line**) **voltages** are voltages between lines. Thus, $\mathbf{E}_{AB}$, $\mathbf{E}_{BC}$, and $\mathbf{E}_{CA}$ are line-to-line voltages at the generator, while $\mathbf{V}_{ab}$, $\mathbf{V}_{bc}$, and $\mathbf{V}_{ca}$ are line-to-line voltages at the load.

Phase voltages are voltages across phases. For a Y load, phases are defined from line to neutral as indicated in Figure 24–7(a); thus, $\mathbf{V}_{an}$, $\mathbf{V}_{bn}$, and $\mathbf{V}_{cn}$ are phase voltages for a Y load. For a Δ load, phases are defined from line to line as shown in Figure 24–7(b); thus, $\mathbf{V}_{ab}$, $\mathbf{V}_{bc}$, and $\mathbf{V}_{ca}$ are phase voltages for a Δ. As you can see, for a Δ load, phase voltages and line voltages are the same thing. For the generator of (a), $\mathbf{E}_{AN}$, $\mathbf{E}_{BN}$, and $\mathbf{E}_{CN}$ are phase voltages.

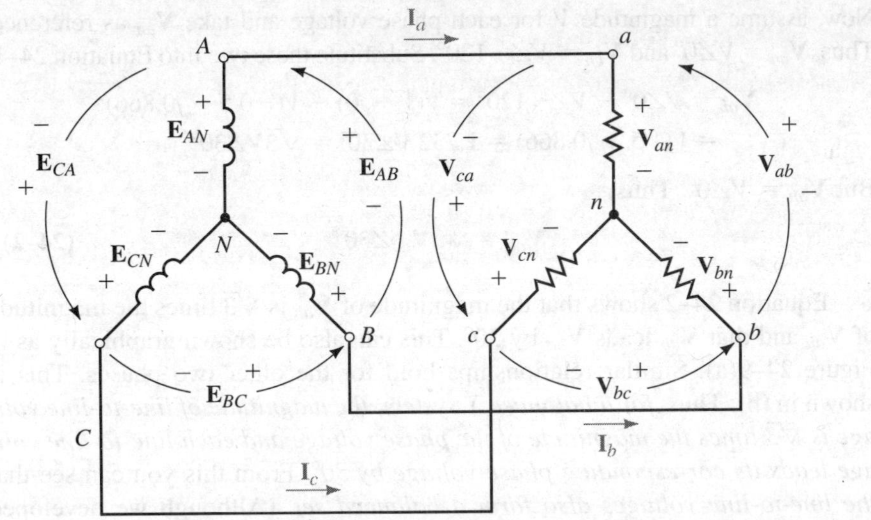

(a) For a Y, phases are defined from line to neutral

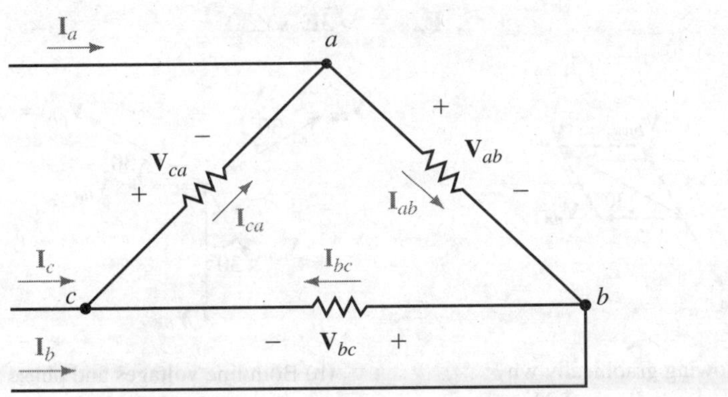

(b) For a Δ, phases are defined from line to line

FIGURE 24–7 Symbols and notation for three-phase voltages and currents.

NOTES...

Make sure you adhere scrupulously to the double subscript notation as set out in Section 5.8, Chapter 5. As always, the + sign goes at the first subscript location. Note also the cyclical order of the subscripts—specifically, the voltages defined as $\mathbf{V}_{ab}$, $\mathbf{V}_{bc}$, and $\mathbf{V}_{ca}$. As you can see, the subscript order here is *ab, bc,* and *ca.* A common mistake is to use *ac* instead of *ca.* Don't do it, since $\mathbf{V}_{ac} = -\mathbf{V}_{ca}$. Similarly, $\mathbf{I}_{ac} = -\mathbf{I}_{ca}$. Thus, if you use the wrong subscript order, you introduce a 180° phase shift error into your notation.

Line currents are the currents in the line conductors. Only a single subscript is needed. You can use either $\mathbf{I}_a$, $\mathbf{I}_b$, and $\mathbf{I}_c$ as in Figure 24–7 or $\mathbf{I}_A$, $\mathbf{I}_B$, and $\mathbf{I}_C$ as in Figure 24–4. (Some authors use double subscripts such as $\mathbf{I}_{Aa}$.)

Phase currents are currents through phases. For the Y load Figure 24–7(a), $\mathbf{I}_a$, $\mathbf{I}_b$, and $\mathbf{I}_c$ pass through phase impedances and are therefore phase currents. For the Δ load (b), $\mathbf{I}_{ab}$, $\mathbf{I}_{bc}$, and $\mathbf{I}_{ca}$ are phase currents. As you can see, for a Y load, phase currents and line currents are the same thing.

Load phase impedances for a Y load are the impedances from *a-n*, *b-n*, and *c-n* [Figure 24–7(a)] and are denoted by the symbols $\mathbf{Z}_{an}$, $\mathbf{Z}_{bn}$, and $\mathbf{Z}_{cn}$. For a Δ load (b), phase impedances are $\mathbf{Z}_{ab}$, $\mathbf{Z}_{bc}$, and $\mathbf{Z}_{ca}$. In a balanced load, impedances for all phases are the same, that is, $\mathbf{Z}_{an} = \mathbf{Z}_{bn} = \mathbf{Z}_{cn}$, and so on.

Line and Phase Voltages for a Y Circuit

We now need the relationship between line and phase voltages for a Y circuit. Consider Figure 24–8. By KVL, $\mathbf{V}_{ab} - \mathbf{V}_{an} + \mathbf{V}_{bn} = 0$. Thus,

$$\mathbf{V}_{ab} = \mathbf{V}_{an} - \mathbf{V}_{bn} \qquad (24\text{–}1)$$

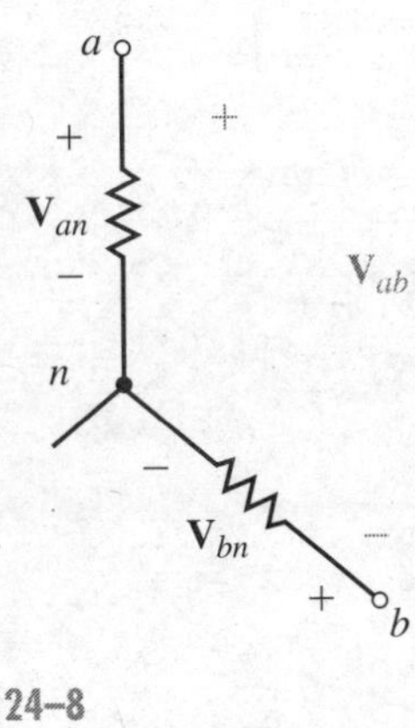

FIGURE 24–8

Now, assume a magnitude V for each phase voltage and take $\mathbf{V}_{an}$ as reference. Thus, $\mathbf{V}_{an} = V\angle 0°$ and $\mathbf{V}_{bn} = V\angle -120°$. Substitute these two into Equation 24–1:

$$\mathbf{V}_{ab} = V\angle 0° - V\angle -120° = V(1 + j0) - V(-0.5 - j0.866)$$
$$= V(1.5 + j0.866) = 1.732\,V\angle 30° = \sqrt{3}V\angle 30°$$

But $\mathbf{V}_{an} = V\angle 0°$. Thus,

$$\mathbf{V}_{ab} = \sqrt{3}\mathbf{V}_{an}\angle 30° \qquad \textbf{(24–2)}$$

Equation 24–2 shows that the magnitude of $\mathbf{V}_{ab}$ is $\sqrt{3}$ times the magnitude of $\mathbf{V}_{an}$ and that $\mathbf{V}_{ab}$ leads $\mathbf{V}_{an}$ by 30°. This can also be shown graphically as in Figure 24–9(a). Similar relationships hold for the other two phases. This is shown in (b). Thus, *for a balanced Y system, the magnitude of line-to-line voltage is $\sqrt{3}$ times the magnitude of the phase voltage and each line-to-line voltage leads its corresponding phase voltage by 30°.* From this you can see that *the line-to-line voltages also form a balanced set.* (Although we developed these relationships with $\mathbf{V}_{an}$ in the 0° reference position, they are true regardless of the choice of reference.) They also hold at the source. Thus,

$$\mathbf{E}_{AB} = \sqrt{3}\mathbf{E}_{AN}\angle 30° \qquad \textbf{(24–3)}$$

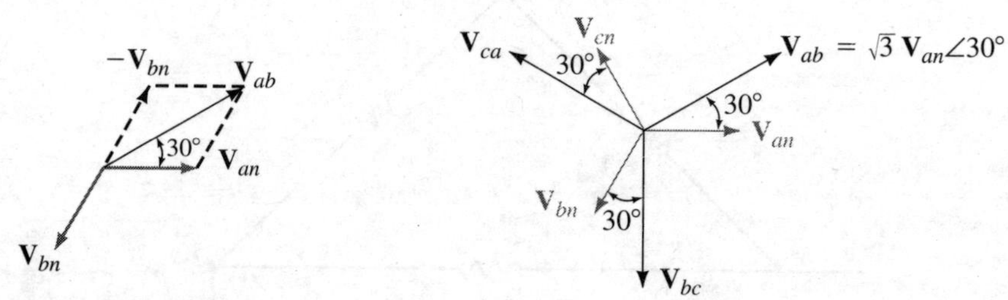

(a) Showing graphically why $\mathbf{V}_{ab}$ leads $\mathbf{V}_{an}$ by 30°

(b) Both line voltages and phase voltages form balanced sets

FIGURE 24–9 Voltages for a balanced Y load. If you know one voltage, you can determine the other five by inspection.

EXAMPLE 24–2

Given $\mathbf{V}_{an} = 120\text{ V}\angle -45°$,

a. Determine $\mathbf{V}_{ab}$ using Equation 24–2.

b. Verify $\mathbf{V}_{ab}$ by direct substitution of $\mathbf{V}_{an}$ and $\mathbf{V}_{bn}$ into Equation 24–1.

Solution

a. $\mathbf{V}_{ab} = \sqrt{3}\mathbf{V}_{an}\angle 30° = \sqrt{3}(120\text{ V}\angle -45°)(1\angle 30°) = 207.8\angle -15°$

b. $\mathbf{V}_{an} = 120\text{ V}\angle -45°$. Thus, $\mathbf{V}_{bn} = 120\text{ V}\angle -165°$.

$\mathbf{V}_{ab} = \mathbf{V}_{an} - \mathbf{V}_{bn} = (120\text{ V}\angle -45°) - (120\text{ V}\angle -165°)$
$= 207.8\text{ V}\angle -15°$ as before.

Nominal Voltages

While Example 24–2 yields 207.8 V for line-to-line voltage, we generally round this to 208 V and refer to the system as a 120/208-V system. These are nominal values. Other sets of nominal voltages used in practice are 277/480 V and 347/600 V.

EXAMPLE 24–3

For the circuits of Figure 24–4, suppose $\mathbf{E}_{AN} = 120\text{ V}\angle 0°$.

a. Determine the phase voltages at the load.
b. Determine the line voltages at the load.
c. Show all voltages on a phasor diagram.

Solution

a. $\mathbf{V}_{an} = \mathbf{E}_{AN}$. Thus, $\mathbf{V}_{an} = 120\text{ V}\angle 0°$. Since the system is balanced, $\mathbf{V}_{bn} = 120\text{ V}\angle -120°$ and $\mathbf{V}_{cn} = 120\text{ V}\angle 120°$.

b. $\mathbf{V}_{ab} = \sqrt{3}\mathbf{V}_{an}\angle 30° = \sqrt{3} \times 120\text{ V}\angle(0° + 30°) = 208\text{ V}\angle 30°$. Since line voltages form a balanced set, $\mathbf{V}_{bc} = 208\text{ V}\angle -90°$ and $\mathbf{V}_{ca} = 208\text{ V}\angle 150°$.

c. The phasors are shown in Figure 24–10.

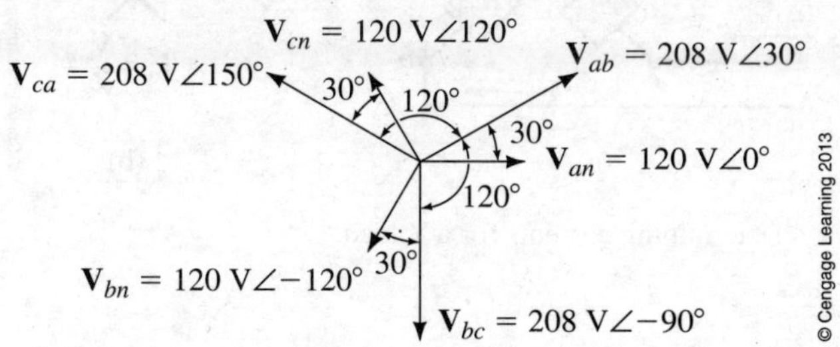

FIGURE 24–10

CircuitSim 24-5

Equations 24–2 and 24–3 permit you to calculate line voltages from phase voltages. Rearranging them yields Equation 24–4, which permits you to calculate phase voltage from line voltage.

$$\mathbf{V}_{an} = \frac{\mathbf{V}_{ab}}{\sqrt{3}\angle 30°} \qquad \mathbf{E}_{AN} = \frac{\mathbf{E}_{AB}}{\sqrt{3}\angle 30°} \tag{24–4}$$

For example, if $\mathbf{E}_{AB} = 480\text{ V}\angle 45°$, then

$$\mathbf{E}_{AN} = \frac{\mathbf{E}_{AB}}{\sqrt{3}\angle 30°} = \frac{480\text{ V}\angle 45°}{\sqrt{3}\angle 30°} = 277\text{ V}\angle 15°$$

A Milestone

You have now reached an important milestone. *Given any voltage at a point in a balanced, three-phase Y system, you can, with the aid of Equation 24–2 or 24–4, determine the remaining five voltages by inspection,* that is, by simply shifting their angles and multiplying or dividing magnitude by $\sqrt{3}$ as appropriate.

PRACTICE PROBLEMS 2

For a balanced Y generator, $\mathbf{E}_{AB} = 480\text{ V}\angle 20°$.

a. Determine the other two generator line voltages.
b. Determine the generator phase voltages.
c. Sketch the phasors.

Answers

a. $\mathbf{E}_{BC} = 480\text{ V}\angle -100°$; $\mathbf{E}_{CA} = 480\text{ V}\angle 140°$

b. $\mathbf{E}_{AN} = 277\text{ V}\angle -10°$; $\mathbf{E}_{BN} = 277\text{ V}\angle -130°$; $\mathbf{E}_{CN} = 277\text{ V}\angle 110°$

Currents for a Wye Circuit

As you saw earlier, for a Y load, line currents are the same as phase currents. Consider Figure 24–11. As indicated in (b),

$$\mathbf{I}_a = \mathbf{V}_{an}/\mathbf{Z}_{an} \qquad \textbf{(24–5)}$$

Similarly for $\mathbf{I}_b$ and $\mathbf{I}_c$. Since $\mathbf{V}_{an}$, $\mathbf{V}_{bn}$, and $\mathbf{V}_{cn}$ form a balanced set, *line currents $\mathbf{I}_a$, $\mathbf{I}_b$, and $\mathbf{I}_c$ also form a balanced set.* Thus, if you know one, you can determine the other two by inspection.

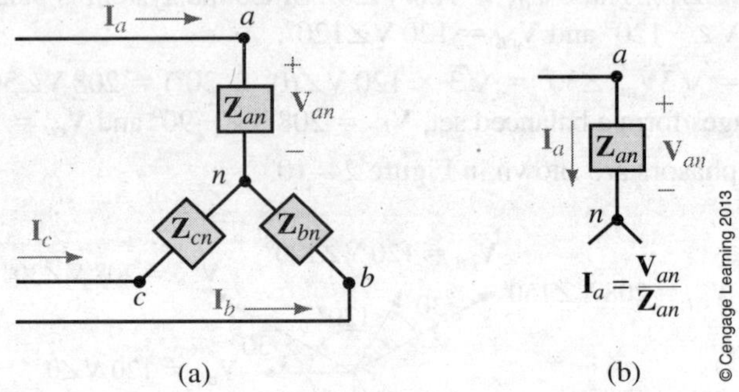

FIGURE 24–11 Determining currents for a Y load.

EXAMPLE 24–4

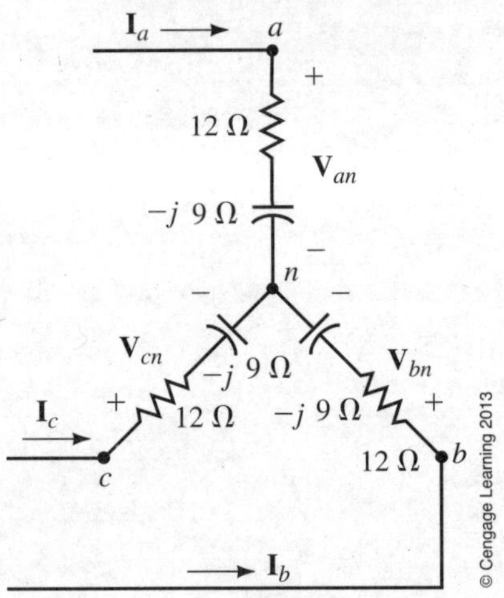

FIGURE 24–12

For Figure 24–12, suppose $\mathbf{V}_{an} = 120\ \text{V}\angle 0°$.

a. Compute $\mathbf{I}_a$, then determine $\mathbf{I}_b$ and $\mathbf{I}_c$ by inspection.
b. Verify by direct computation.

Solution

a. $\mathbf{I}_a = \dfrac{\mathbf{V}_{an}}{\mathbf{Z}_{an}} = \dfrac{120\angle 0°}{12 - j9} = \dfrac{120\angle 0°}{15\angle -36.87°} = 8.0\ \text{A}\angle 36.87°$

$\mathbf{I}_b$ lags $\mathbf{I}_a$ by 120°. Thus, $\mathbf{I}_b = 8\ \text{A}\angle -83.13°$.

$\mathbf{I}_c$ leads $\mathbf{I}_a$ by 120°. Thus, $\mathbf{I}_c = 8\ \text{A}\angle 156.87°$.

b. Since $\mathbf{V}_{an} = 120\ \text{V}\angle 0°$, $\mathbf{V}_{bn} = 120\ \text{V}\angle -120°$, and $\mathbf{V}_{cn} = 120\ \text{V}\angle 120°$. Thus,

$$\mathbf{I}_b = \frac{\mathbf{V}_{bn}}{\mathbf{Z}_{bn}} = \frac{120\angle -120°}{15\angle -36.87°} = 8.0\ \text{A}\angle -83.13°$$

$$\mathbf{I}_c = \frac{\mathbf{V}_{cn}}{\mathbf{Z}_{cn}} = \frac{120\angle 120°}{15\angle -36.87°} = 8.0\ \text{A}\angle 156.87°$$

These confirm the results obtained in (a).

CircuitSim 24-6

PRACTICE PROBLEMS 3

1. If $\mathbf{V}_{ab} = 600\ \text{V}\angle 0°$ for the circuit of Figure 24–12, what are $\mathbf{I}_a$, $\mathbf{I}_b$, and $\mathbf{I}_c$?
2. If $\mathbf{V}_{bc} = 600\ \text{V}\angle -90°$ for the circuit of Figure 24–12, what are $\mathbf{I}_a$, $\mathbf{I}_b$, and $\mathbf{I}_c$?

Answers

1. $\mathbf{I}_a = 23.1\ \text{A}\angle 6.9°$; $\mathbf{I}_b = 23.1\ \text{A}\angle -113.1°$; $\mathbf{I}_c = 23.1\ \text{A}\angle 126.9°$
2. $\mathbf{I}_a = 23.1\ \text{A}\angle 36.9°$; $\mathbf{I}_b = 23.1\ \text{A}\angle -83.1°$; $\mathbf{I}_c = 23.1\ \text{A}\angle 156.9°$

Line and Phase Currents for a Delta Circuit

Consider the delta load of Figure 24–13. Phase current $\mathbf{I}_{ab}$ can be found as in (b).

$$\mathbf{I}_{ab} = \mathbf{V}_{ab}/\mathbf{Z}_{ab} \quad (24\text{–}6)$$

Similar relationships hold for $\mathbf{I}_{bc}$ and $\mathbf{I}_{ca}$. Since line voltages are balanced, phase currents are also balanced. Now consider again Figure 23–13(a). KCL at node a yields

$$\mathbf{I}_a = \mathbf{I}_{ab} - \mathbf{I}_{ca} \quad (24\text{–}7)$$

After some manipulation, this reduces to

$$\mathbf{I}_a = \sqrt{3}\mathbf{I}_{ab}\angle{-30°} \quad (24\text{–}8)$$

Thus, the magnitude of $\mathbf{I}_a$ is $\sqrt{3}$ times the magnitude of $\mathbf{I}_{ab}$, and $\mathbf{I}_a$ lags $\mathbf{I}_{ab}$ by 30°. Similarly, for the other two phases. Thus, *in a balanced Δ, the magnitude of the line current is $\sqrt{3}$ times the magnitude of the phase current, and each line current lags its corresponding phase current by 30°*. Since phase currents are balanced, line currents are also balanced. This is shown in Figure 24–13(c). To find phase currents from line currents, use

$$\mathbf{I}_{ab} = \frac{\mathbf{I}_a}{\sqrt{3}\angle{-30°}} \quad (24\text{–}9)$$

NOTES...

Compare Equation 24–8 to Equation 24–2. No one ever forgets the $\sqrt{3}$ relationship, but students often mix up which one has the 30° shift and which one has the −30° shift. One way to deal with this is to sketch Figure 24–8, write KVL as $V_{ab} = V_{an} - V_{bn}$, substitute $V_{an} = 120\angle 0°$ and $V_{bn} = 120\angle{-120°}$, and use your calculator to obtain $V_{ab} = 207.8\angle 30°$. This establishes that voltage has the positive angle.

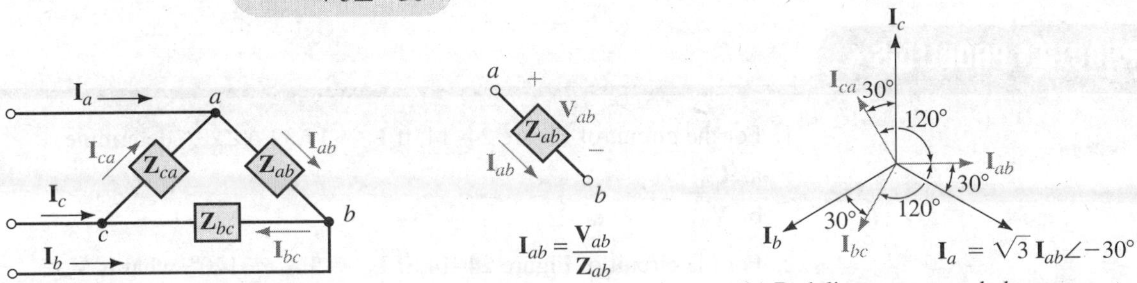

(a) $\mathbf{I}_a = \mathbf{I}_{ab} - \mathbf{I}_{ca}$ (b) (c) Both line currents and phase currents form balanced sets

FIGURE 24–13 Currents for a balanced Δ load. If you know one current, you can determine the other five by inspection.

A Second Milestone

You have reached a second milestone. *Given any current in a balanced, three-phase Δ load, you can, with the aid of Equations 24–8 or 24–9, determine all remaining currents by inspection*—see Note.

EXAMPLE 24–5

Suppose $\mathbf{V}_{ab} = 240\ \text{V}\angle 15°$ for the circuit of Figure 24–14.

a. Determine the phase currents.
b. Determine the line currents.
c. Sketch the phasor diagram.

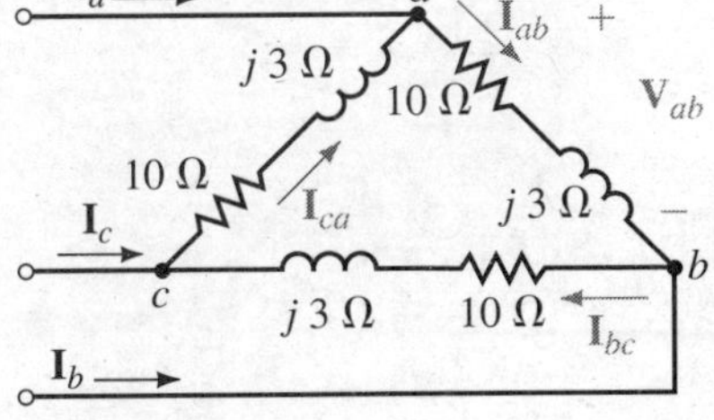

FIGURE 24–14

Solution

a. $\mathbf{I}_{ab} = \dfrac{\mathbf{V}_{ab}}{\mathbf{Z}_{ab}} = \dfrac{240\angle 15°}{10 + j3} = 23.0\ \text{A}\angle{-1.70°}$

Thus,

$$\mathbf{I}_{bc} = 23.0 \text{ A}\angle -121.7° \text{ and } \mathbf{I}_{ca} = 23.0 \text{ A}\angle 118.3°$$

b. $\mathbf{I}_a = \sqrt{3}\mathbf{I}_{ab}\angle -30° = 39.8 \text{ A}\angle -31.7°$

Thus,

$$\mathbf{I}_b = 39.8 \text{ A}\angle -151.7° \text{ and } \mathbf{I}_c = 39.8 \text{ A}\angle 88.3°$$

c. Phasors are shown in Figure 24–15.

CircuitSim 24-7

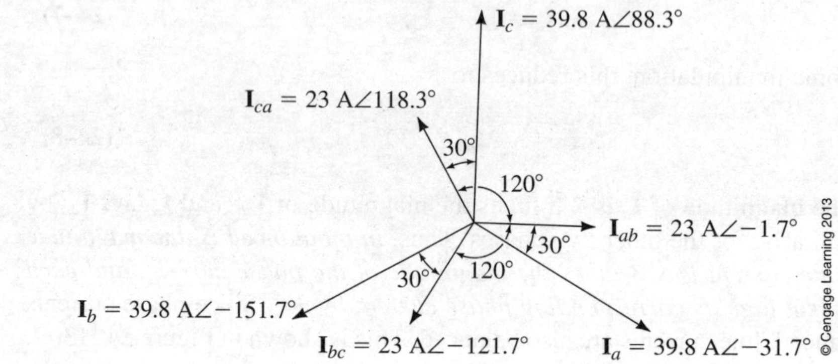

FIGURE 24–15

PRACTICE PROBLEMS 4

1. For the circuit of Figure 24–14, if $\mathbf{I}_a = 17.32 \text{ A}\angle 20°$, determine
   a. $\mathbf{I}_{ab}$
   b. $\mathbf{V}_{ab}$
2. For the circuit of Figure 24–14, if $\mathbf{I}_{bc} = 5 \text{ A}\angle -140°$, what is $\mathbf{V}_{ab}$?

Answers

1. a. 10 A∠50°; b. 104 V∠66.7°
2. 52.2 V∠−3.30°

The Single-Phase Equivalent

By now it should be apparent that if you know the solution for one phase of a balanced system, you effectively know the solution for all three phases. We will now formalize this viewpoint by developing the **single-phase equivalent** approach to solving balanced systems. Consider a Y-Y system with line impedance. The system may be either a three-wire system or a four-wire system with neutral conductor impedance. In either case, since the voltage between neutral points is zero, you can join points n and N with a **zero-impedance conductor** without disturbing voltages or currents elsewhere in the circuit. This is illustrated in Figure 24–16(a). Phase a can now be isolated as in (b). Since $V_{nN} \equiv 0$ as before, the equation describing phase a in circuit (b) is the same as that describing phase a in the original circuit; thus, circuit (b) can be used to solve the original problem. If there are Δ loads present, convert them to Y loads using the Δ-Y conversion formula for balanced loads: $\mathbf{Z}_Y = \mathbf{Z}_\Delta/3$, from Chapter 19. This procedure is valid regardless of the configuration or complexity of the circuit, as long as it is balanced. We will look at its use in Section 24.4.

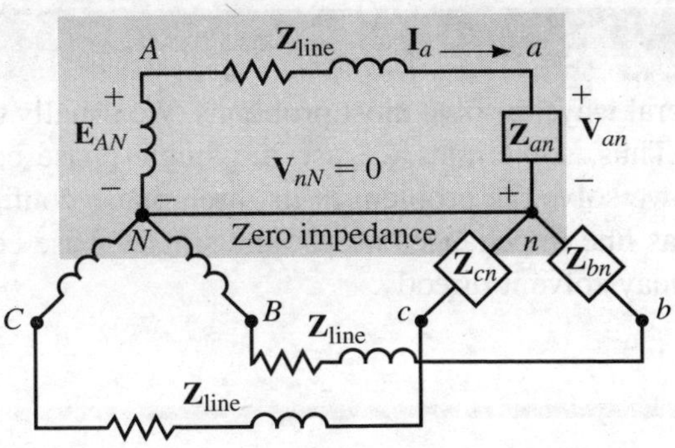

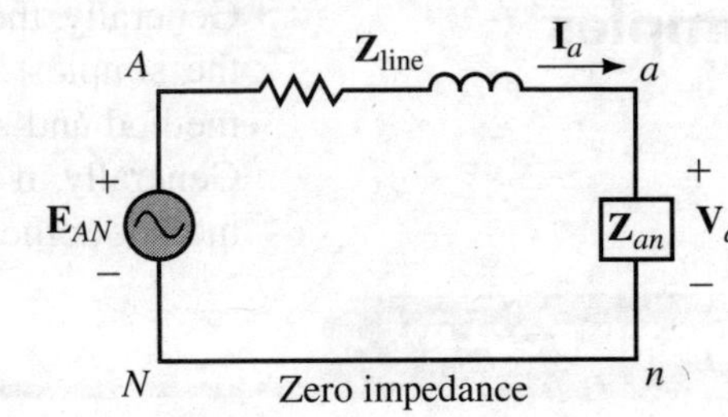

(a) Original circuit: $\mathbf{E}_{AN} = \mathbf{I}_a \mathbf{Z}_{line} + \mathbf{V}_{an}$ (b) Single-phase equivalent: $\mathbf{E}_{AN} = \mathbf{I}_a \mathbf{Z}_{line} + \mathbf{V}_{an}$

FIGURE 24–16 Reducing a circuit to its single-phase equivalent. Since both configurations are described by the same set of equations, they are equivalent.

Selecting a Reference

Before you solve a three-phase problem, you need to select a reference. For Y circuits, we normally choose $\mathbf{E}_{AN}$ or $\mathbf{V}_{an}$; for Δ circuits, we normally choose $\mathbf{E}_{AB}$ or $\mathbf{V}_{ab}$. (These are arbitrary but convenient choices.)

Summary of Basic Three-Phase Relationships

Table 24–1 summarizes the relationships developed so far. Note that in balanced systems (Y or Δ), *all* voltages and *all* currents are balanced.

TABLE 24–1 Summary of Relationships (Balanced System). All Voltages and Currents Are Balanced

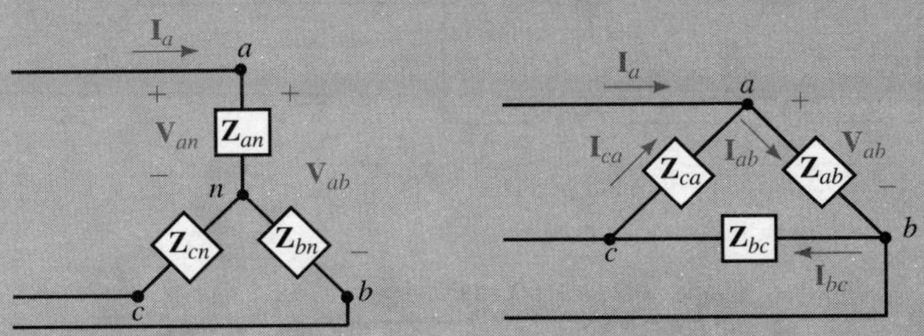

(a) Y connection	(b) Δ connection
$\mathbf{V}_{ab} = \sqrt{3}\mathbf{V}_{an}\angle 30°$	$\mathbf{I}_a = \sqrt{3}\mathbf{I}_{ab}\angle -30°$
$\mathbf{I}_a = \mathbf{V}_{an}/\mathbf{Z}_{an}$	$\mathbf{I}_{ab} = \mathbf{V}_{ab}/\mathbf{Z}_{ab}$
$\mathbf{Z}_{an} = \mathbf{Z}_{bn} = \mathbf{Z}_{cn}$	$\mathbf{Z}_{ab} = \mathbf{Z}_{bc} = \mathbf{Z}_{ca}$

Generator $\mathbf{E}_{AB} = \sqrt{3}\mathbf{E}_{AN}\angle 30°$

IN-PROCESS LEARNING CHECK 1

(Answers are at the end of the chapter.)

1. For Figure 24–4(a), if $\mathbf{E}_{AN} = 277\text{ V}\angle -20°$, determine all line and phase voltages, source, and load.
2. For Figure 24–4(a), if $\mathbf{V}_{bc} = 208\text{ V}\angle -40°$, determine all line and phase voltages, source, and load.
3. For Figure 24–12, if $\mathbf{I}_a = 8.25\text{ A}\angle 35°$, determine $\mathbf{V}_{an}$ and $\mathbf{V}_{ab}$.
4. For Figure 24–14, if $\mathbf{I}_b = 17.32\text{ A}\angle -85°$, determine all voltages.

24.4 Examples

Generally, there are several ways to solve most problems. We usually try to use the simplest approach. Thus, sometimes we use the single-phase equivalent method and sometimes we solve the problem in its three-phase configuration. Generally, if a circuit has line impedance, we use the single-phase equivalent method; otherwise, we may solve it directly.

EXAMPLE 24–6

NOTES...

For a number of the examples of this section, we use 120/207.8 volts instead of the nominal values of 120/208 volts and we carry extra digits so that the analytic solutions can be compared to the Multisim solutions.

For Figure 24–17, $\mathbf{E}_{AN} = 120\ \text{V}\angle 0°$.

a. Solve for the line currents—see Note.
b. Solve for the phase voltages at the load.
c. Solve for the line voltages at the load.

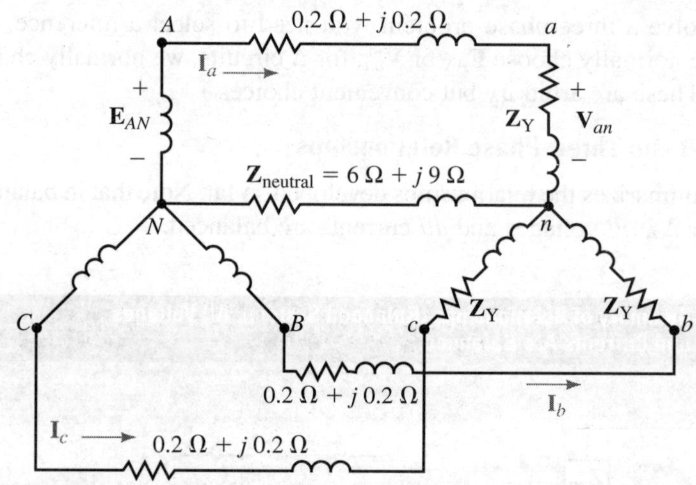

(a) $\mathbf{Z}_Y = 6\ \Omega + j\,8\ \Omega$

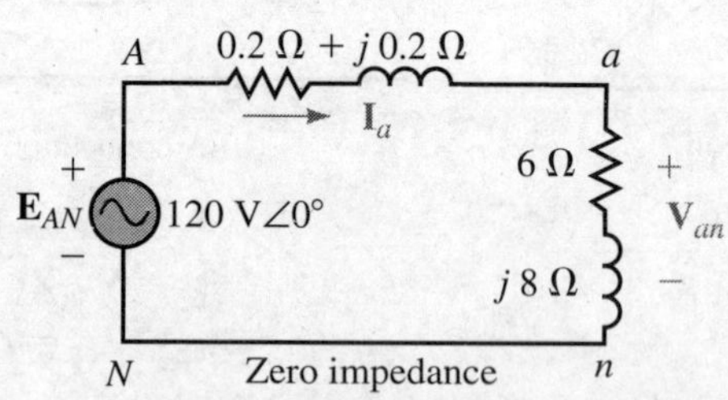

(b) Single-phase equivalent. Since the neutral conductor in (a) carries no current, its impedance has no effect on the solution.

CircuitSim 24-8

FIGURE 24–17 A Y-Y problem.

Solution

a. Reduce the circuit to its single-phase equivalent as shown in (b).

$$\mathbf{I}_a = \frac{\mathbf{E}_{AN}}{\mathbf{Z}_T} = \frac{120\angle 0°}{(0.2 + j0.2) + (6 + j8)} = 11.67\ \text{A}\angle -52.91°$$

Therefore,

$$\mathbf{I}_b = 11.67\ \text{A}\angle -172.91° \quad \text{and} \quad \mathbf{I}_c = 11.67\ \text{A}\angle 67.09°$$

b. $\mathbf{V}_{an} = \mathbf{I}_a \times \mathbf{Z}_{an} = (11.67\angle -52.91°)(6 + j8) = 116.7\ \text{V}\angle 0.223°$

Thus,

$$\mathbf{V}_{bn} = 116.7\ \text{V}\angle -119.777° \quad \text{and} \quad \mathbf{V}_{cn} = 116.7\ \text{V}\angle 120.223°$$

c. $\mathbf{V}_{ab} = \sqrt{3}\mathbf{V}_{an}\angle 30° = \sqrt{3} \times 116.7\angle(0.223° + 30°) = 202.2\ \text{V}\angle 30.223°$

Thus,

$$\mathbf{V}_{bc} = 202.2\ \text{V}\angle -89.777° \quad \text{and} \quad \mathbf{V}_{ca} = 202.2\ \text{V}\angle 150.223°$$

Note the phase shift and voltage drop across the line impedance. Note also that the impedance of the neutral conductor plays no part in the solution, since no current passes through it because the system is balanced.

PRACTICE PROBLEMS 5

For the circuit of Figure 24–18, $\mathbf{V}_{an} = 120\ \text{V}\angle 0°$.

a. Find the line currents.
b. Verify that the neutral current is zero.
c. Determine generator voltages $\mathbf{E}_{AN}$ and $\mathbf{E}_{AB}$.

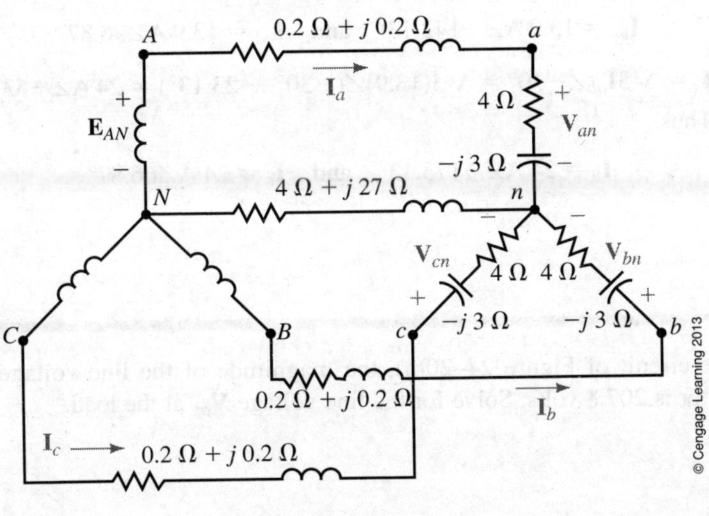

FIGURE 24–18

Answers

a. $\mathbf{I}_a = 24\ \text{A}\angle 36.9°$; $\mathbf{I}_b = 24\ \text{A}\angle -83.1°$; $\mathbf{I}_c = 24\ \text{A}\angle 156.9°$

b. $24\ \text{A}\angle 36.9° + 24\ \text{A}\angle -83.1° + 24\ \text{A}\angle 156.9° = 0$

c. $\mathbf{E}_{AN} = 121\ \text{V}\angle 3.18°$; $\mathbf{E}_{AB} = 210\ \text{V}\angle 33.18°$

EXAMPLE 24–7

For the circuit of Figure 24–19, $\mathbf{E}_{AB} = 208\ \text{V}\angle 30°$.

a. Determine the phase currents.
b. Determine the line currents.

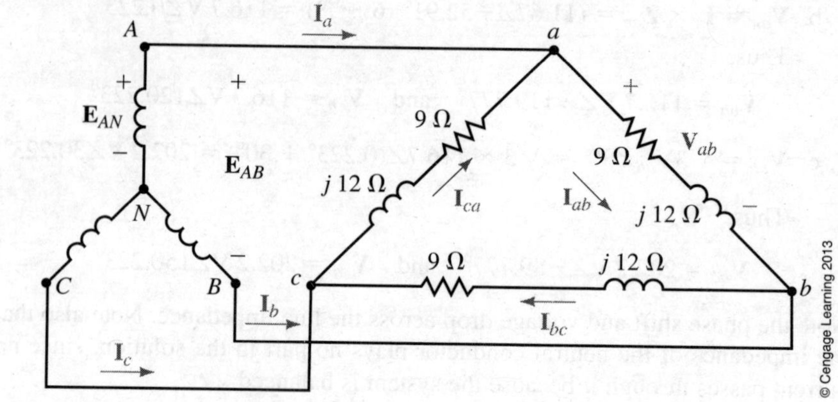

FIGURE 24–19 A Y-Δ problem.

Solution

a. Since this circuit has no line impedance, the load connects directly to the source and $\mathbf{V}_{ab} = \mathbf{E}_{AB} = 208\ \text{V}\angle 30°$. Current $\mathbf{I}_{ab}$ can be found as

$$\mathbf{I}_{ab} = \frac{\mathbf{V}_{ab}}{\mathbf{Z}_{ab}} = \frac{208\angle 30°}{9 + j12} = \frac{208\angle 30°}{15\angle 53.13°} = 13.9\ \text{A}\angle -23.13°$$

Thus,

$$\mathbf{I}_{bc} = 13.9\ \text{A}\angle -143.13° \quad \text{and} \quad \mathbf{I}_{ca} = 13.9\ \text{A}\angle 96.87°$$

b. $\mathbf{I}_a = \sqrt{3}\mathbf{I}_{ab}\angle -30° = \sqrt{3}(13.9)\angle(-30° - 23.13°) = 24\ \text{A}\angle -53.13°$

Thus,

$$\mathbf{I}_b = 24\ \text{A}\angle -173.13° \quad \text{and} \quad \mathbf{I}_c = 24\ \text{A}\angle 66.87°$$

EXAMPLE 24–8

CircuitSim 24-9

For the circuit of Figure 24–20(a), the magnitude of the line voltage at the generator is 207.8 volts. Solve for the line voltage $\mathbf{V}_{ab}$ at the load.

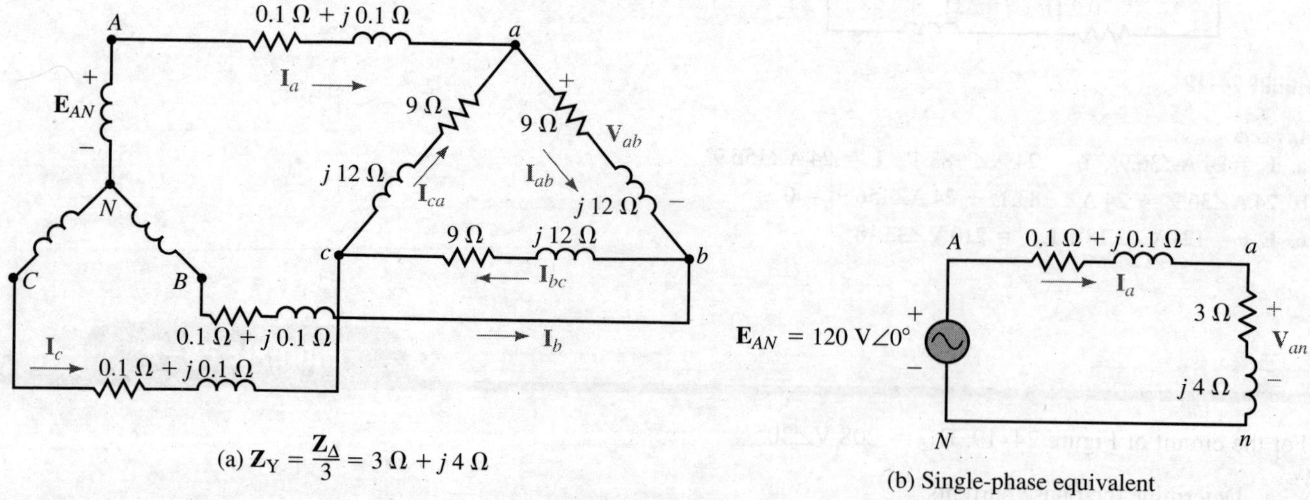

FIGURE 24–20 A circuit with line impedances.

Solution Since points *A-a* and *B-b* are not directly joined, $\mathbf{V}_{ab} \neq \mathbf{E}_{AB}$ and we cannot solve the circuit as in Example 24–7. Use the single-phase equivalent.

Phase voltage at the source is $207.8/\sqrt{3} = 120$ V. Choose $\mathbf{E}_{AN}$ as reference: $\mathbf{E}_{AN} = 120 \text{ V}\angle 0°$.

$$\mathbf{Z}_{Y} = \mathbf{Z}_{\Delta}/3 = (9 + j12)/3 = 3\ \Omega + j4\ \Omega.$$

The single-phase equivalent is shown in (b). Now use the voltage divider rule to find $\mathbf{V}_{an}$:

$$\mathbf{V}_{an} = \left(\frac{3 + j4}{3.1 + j4.1}\right) \times 120\angle 0° = 116.7 \text{ V}\angle 0.223°$$

Thus,

$$\mathbf{V}_{ab} = \sqrt{3}\mathbf{V}_{an}\angle 30° = \sqrt{3}(116.7 \text{ V})\angle 30.223° = 202.2 \text{ V}\angle 30.223°$$

EXAMPLE 24–9

For the circuit of Figure 24–20(a), the generator phase voltage is 120 volts. Find the Δ currents.

Solution Since the source voltage here is the same as in Example 24–8, load voltage $\mathbf{V}_{ab}$ will also be the same. Thus,

$$\mathbf{I}_{ab} = \frac{\mathbf{V}_{ab}}{\mathbf{Z}_{ab}} = \frac{202.2 \text{ V}\angle 30.223°}{(9 + j12)\Omega} = 13.5 \text{ A}\angle -22.9°$$

and,

$$\mathbf{I}_{bc} = 13.5 \text{ A}\angle -142.9° \quad \text{and} \quad \mathbf{I}_{ca} = 13.5 \text{ A}\angle 97.1°$$

EXAMPLE 24–10

A Y load and a Δ load are connected in parallel as in Figure 24–21(a). $\mathbf{Z}_{Y} = 20\ \Omega\angle 0°$, $\mathbf{Z}_{\Delta} = 90\ \Omega\angle 0°$, and $\mathbf{Z}_{\text{line}} = 0.5 + j\,0.5\ \Omega$. Line voltage magnitude at the generator is 207.8 V.

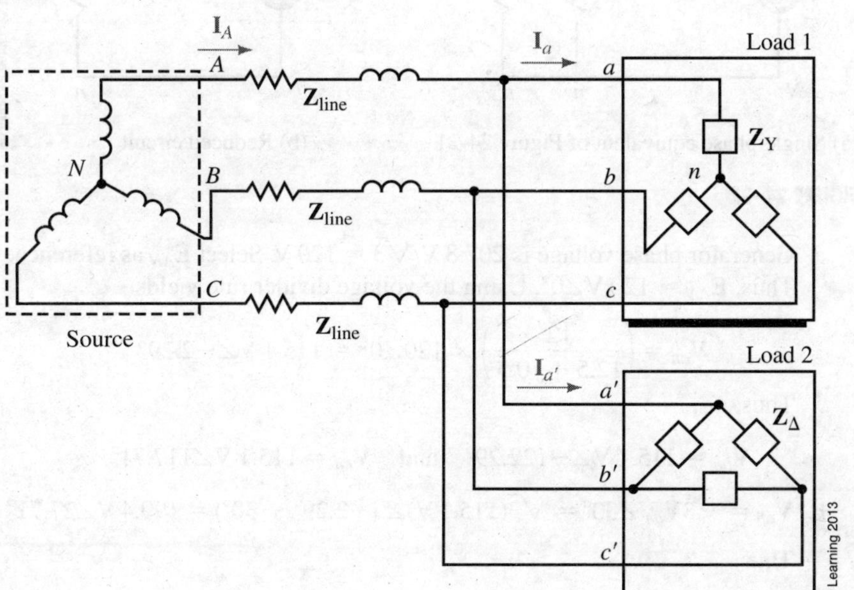

(a)

FIGURE 24–21 Parallel three-phase loads.

(Continued)

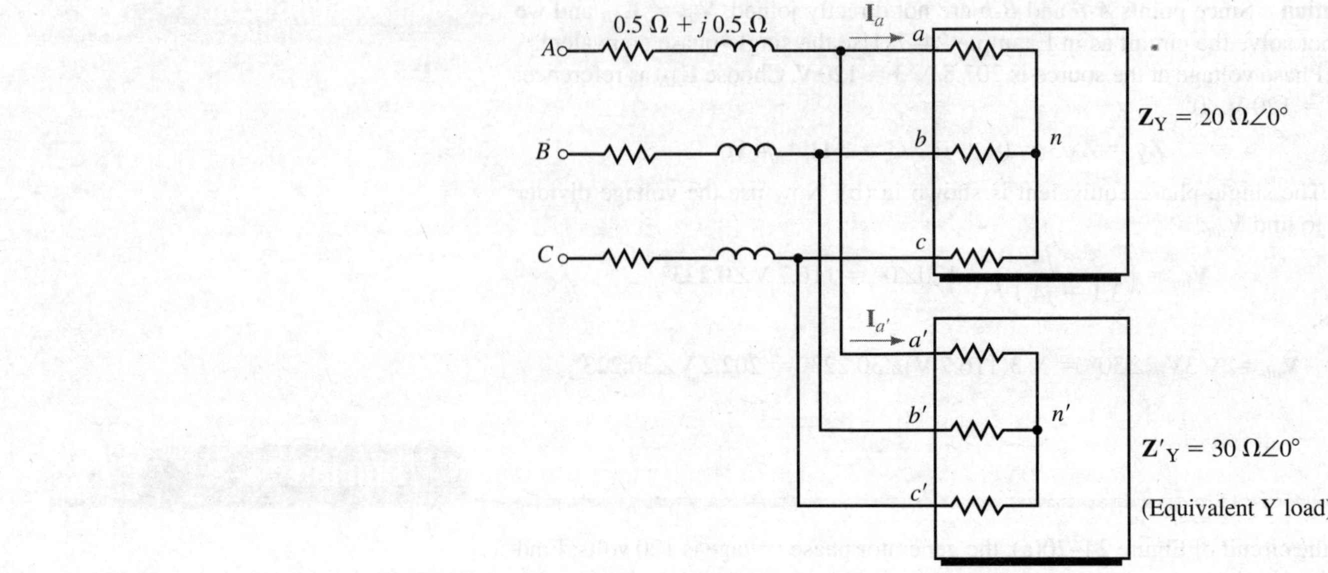

(b) Δ load converted to equivalent Y

CircuitSim 24-10

FIGURE 24–21 (Continued)

a. Find the phase voltages at the loads.
b. Find the line voltages at the loads.

Solution Convert the Δ load to a Y load. Thus, $Z'_Y = \frac{1}{3}Z_\Delta = 30\ \Omega\angle 0°$ as in (b). Now join the neutral points *N*, *n*, and *n′* with a zero-impedance conductor to get the single-phase equivalent, which is shown in Figure 24–22(a). The parallel load resistors can be combined as in (b).

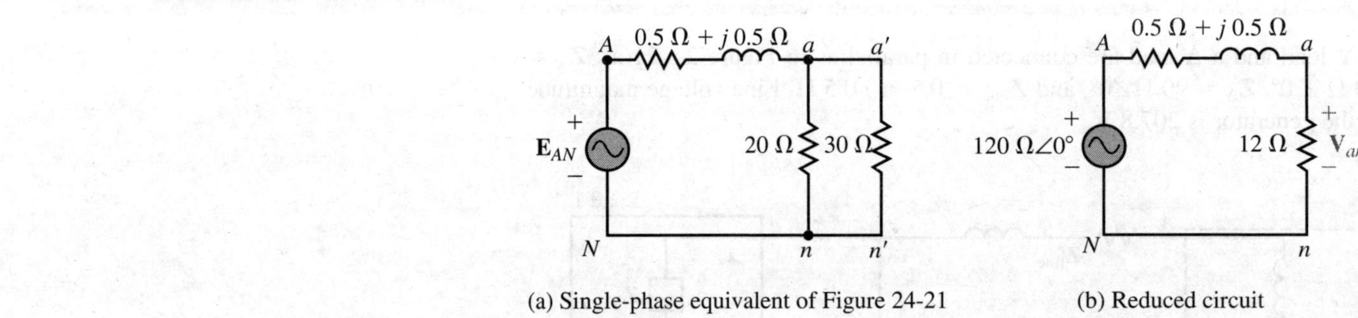

(a) Single-phase equivalent of Figure 24-21 (b) Reduced circuit

FIGURE 24–22

a. Generator phase voltage is $207.8\ \text{V}/\sqrt{3} = 120$ V. Select $\mathbf{E}_{AN}$ as reference. Thus, $\mathbf{E}_{AN} = 120\ \text{V}\angle 0°$. Using the voltage divider rule yields

$$\mathbf{V}_{an} = \left(\frac{12}{12.5 + j0.5}\right) \times 120\angle 0° = 115.1\ \text{V}\angle -2.29°$$

Thus,

$$\mathbf{V}_{bn} = 115.1\ \text{V}\angle -122.29° \quad \text{and} \quad \mathbf{V}_{cn} = 115.1\ \text{V}\angle 117.71°$$

b. $\mathbf{V}_{ab} = \sqrt{3}\mathbf{V}_{an}\angle 30° = \sqrt{3}(115.1\ \text{V})\angle(-2.29° + 30°) = 199.4\ \text{V}\angle 27.71°$

Thus,

$$\mathbf{V}_{bc} = 199.4\ \text{V}\angle -92.29° \quad \text{and} \quad \mathbf{V}_{ca} = 199.4\ \text{V}\angle 147.71°$$

These are the line voltages for both the Y and Δ loads.

PRACTICE PROBLEMS 6

1. Repeat Example 24–7 using the single-phase equivalent.
2. Determine Δ phase currents for the circuit of Figure 24–21(a).

Answers

2. $\mathbf{I}_{a'b'} = 2.22\text{ A}\angle 27.7°$; $\mathbf{I}_{b'c'} = 2.22\text{ A}\angle -92.3°$; $\mathbf{I}_{c'a'} = 2.22\text{ A}\angle 147.7°$

24.5 Power in a Balanced System

To find total power in a balanced system, determine power to one phase, then multiply by three. Per-phase quantities can be found using the formulas of Chapter 17. Since only magnitudes are involved in many power formulas and calculations and since magnitudes are the same for all three phases, we can use a simplified notation. We will use V_ϕ for magnitude of phase voltage, I_ϕ for magnitude of phase current, V_L and I_L for magnitude of line voltage and line current respectively, and Z_ϕ for phase impedance magnitude.

Active Power to a Balanced Wye Load

First, consider a Y load (Figure 24–23). The power to any phase as indicated in (b) is the product of the magnitude of the phase voltage V_ϕ times the magnitude of the phase current I_ϕ times the cosine of the angle θ_ϕ between them. Since the angle between phase voltage and current is always the angle of the load impedance, the power per phase is

$$P_\phi = V_\phi I_\phi \cos\theta_\phi \quad \text{(W)} \tag{24–10}$$

where θ_ϕ is the angle of $\mathbf{Z}_\phi$. Total power is

$$P_T = 3P_\phi = 3V_\phi I_\phi \cos\theta_\phi \quad \text{(W)} \tag{24–11}$$

It is also handy to have a formula for power in terms of line quantities. For a Y load, $I_\phi = I_L$ and $V_\phi = V_L/\sqrt{3}$, where I_L is the magnitude of the line current and V_L is the magnitude of the line-to-line voltage. Substituting these relations into Equation 24–11 and noting that $3/\sqrt{3} = \sqrt{3}$ yields

$$P_T = \sqrt{3}\, V_L I_L \cos\theta_\phi \quad \text{(W)} \tag{24–12}$$

This is a very important formula and one that is widely used. Note carefully, however, that θ_ϕ is the angle of the load impedance and not the angle between V_L and I_L.

Power per phase can also be expressed as

$$P_\phi = I_\phi^2 R_\phi = V_R^2/R_\phi \quad \text{(W)} \tag{24–13}$$

where R_ϕ is the resistive component of the phase impedance and V_R is the voltage across it. Total power is thus

$$P_T = 3I_\phi^2 R_\phi = 3V_R^2/R_\phi \quad \text{(W)} \tag{24–14}$$

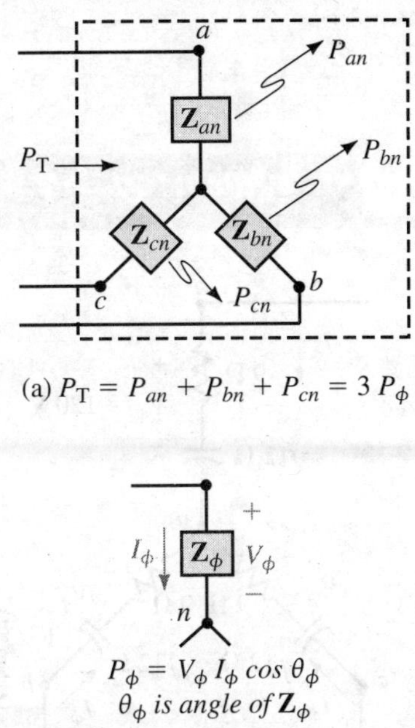

(a) $P_T = P_{an} + P_{bn} + P_{cn} = 3\,P_\phi$

$P_\phi = V_\phi\, I_\phi \cos\theta_\phi$
θ_ϕ *is angle of* $\mathbf{Z}_\phi$

(b)

FIGURE 24–23 For a balanced Y, $P_\phi = P_{an} = P_{bn} = P_{cn}$.

Reactive Power to a Balanced Wye Load

Equivalent expressions for reactive power are

$$Q_\phi = V_\phi I_\phi \sin\theta_\phi \quad \text{(VAR)} \tag{24–15}$$

$$= I_\phi^2 X_\phi = V_X^2/X_\phi \quad \text{(VAR)} \tag{24–16}$$

$$Q_T = \sqrt{3}V_L I_L \sin\theta_\phi \quad \text{(VAR)} \tag{24–17}$$

where X_ϕ is the reactive component of $\mathbf{Z}_\phi$ and V_X is the voltage across it.

Apparent Power

$$S_\phi = V_\phi I_\phi = I_\phi^2 Z_\phi = \frac{V_\phi^2}{Z_\phi} \quad \text{(VA)} \tag{24–18}$$

$$S_T = \sqrt{3}V_L I_L \quad \text{(VA)} \tag{24–19}$$

Power Factor

$$F_p = \cos\theta_\phi = P_T/S_T = P_\phi/S_\phi \tag{24–20}$$

EXAMPLE 24–11

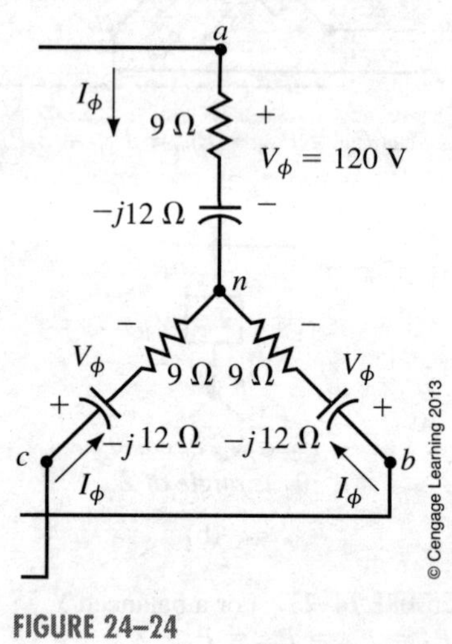

FIGURE 24–24

For Figure 24–24, the phase voltage is 120 V.

a. Compute active power to each phase and total power using each equation of this section.
b. Repeat (a) for reactive power.
c. Repeat (a) for apparent power.
d. Find the power factor.

Solution Since we want to compare answers for the various methods, we will use 207.8 V for the line voltage rather than the nominal value of 208 V.

$\mathbf{Z}_\phi = 9 - j12 = 15\ \Omega\angle -53.13°$. Thus, $\theta_\phi = -53.13°$.

$V_\phi = 120$ V and $I_\phi = V_\phi/Z_\phi = 120\text{ V}/15\ \Omega = 8.0$ A.

$V_R = (8\text{ A})(9\ \Omega) = 72$ V and $V_X = (8\text{ A})(12\ \Omega) = 96$ V

CircuitSim 24-11

a. $P_\phi = V_\phi I_\phi \cos\theta_\phi = (120)(8)\cos(-53.13°) = 576$ W

$P_\phi = I_\phi^2 R_\phi = (8^2)(9) = 576$ W

$P_\phi = V_R^2/R_\phi = (72)^2/9 = 576$ W

$P_T = 3P_\phi = 3(576) = 1728$ W

$P_T = \sqrt{3}V_L I_L \cos\theta_\phi = \sqrt{3}(207.8)(8)\cos(-53.13°) = 1728$ W

b. $Q_\phi = V_\phi I_\phi \sin\theta_\phi = (120)(8)\sin(-53.13°) = -768$ VAR

$= 768$ VAR (cap.)

$Q_\phi = I_\phi^2 X_\phi = (8)^2(12) = 768$ VAR (cap.)

$Q_\phi = V_X^2/X_\phi = (96)^2/12 = 768$ VAR (cap.)

$Q_T = 3Q_\phi = 3(768) = 2304$ VAR (cap.)

$Q_T = \sqrt{3}V_L I_L \sin\theta_\phi = \sqrt{3}(207.8)(8)\sin(-53.13°) = -2304$ VAR

$= 2304$ VAR (cap.)

c. $S_\phi = V_\phi I_\phi = (120)(8) = 960$ VA
$S_T = 3S_\phi = 3(960) = 2880$ VA
$S_T = \sqrt{3}V_L I_L = \sqrt{3}(207.8)(8) = 2880$ VA
Thus, all approaches yield the same answers.

d. The power factor is $F_p = \cos\theta_\phi = \cos 53.13° = 0.6$. (Since the load is balanced, the power factor is the same for all phases, as well as for the total load.)

Power to a Balanced Delta Load

For a Δ load [Figure 24–25(a)],

$$P_\phi = V_\phi I_\phi \cos\theta_\phi \quad \text{(W)} \tag{24–21}$$

where θ_ϕ is the angle of the Δ impedance. Note that this formula is identical to Equation 24–10 for the Y load. This is also true for reactive power, apparent power, and power factor. Thus, all power formulas are the same. Results are tabulated in Table 24–2. *Note: In all of these formulas, θ_ϕ is the angle of the load impedance, that is, the angle of $\mathbf{Z}_{an}$ for Y loads and $\mathbf{Z}_{ab}$ for Δ loads.*

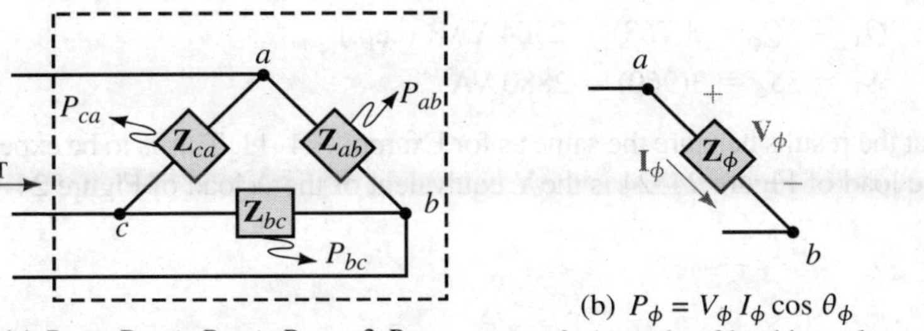

(a) $P_T = P_{ab} + P_{bc} + P_{ca} = 3\,P_\phi$

(b) $P_\phi = V_\phi I_\phi \cos\theta_\phi$
θ_ϕ is angle of load impedance

FIGURE 24–25 For a balanced Δ, $P_\phi = P_{ab} = P_{bc} = P_{ca}$.

TABLE 24–2 Summary of Power Formulas for Balanced Wye and Delta Circuits

Active power	$P_\phi = V_\phi I_\phi \cos\theta_\phi = I_\phi^2 R_\phi = \dfrac{V_R^2}{R_\phi}$
	$P_T = \sqrt{3}V_L I_L \cos\theta_\phi$
Reactive power	$Q_\phi = V_\phi I_\phi \sin\theta_\phi = I_\phi^2 X_\phi = \dfrac{V_x^2}{X_\phi}$
	$Q_T = \sqrt{3}V_L I_L \sin\theta_\phi$
Apparent power	$S_\phi = V_\phi I_\phi = I_\phi^2 Z_\phi = \dfrac{V_\phi^2}{Z_\phi}$
	$S_T = \sqrt{3}V_L I_L$
Power factor	$F_p = \cos\theta_\phi = \dfrac{P_T}{S_T} = \dfrac{P_\phi}{S_\phi}$
Power triangle	$\mathbf{S}_T = P_T + jQ_T$

EXAMPLE 24–12

Determine per phase and total power (active, reactive, and apparent) for Figure 24–26. Use $V_\phi = 207.8$ V in order to compare results.

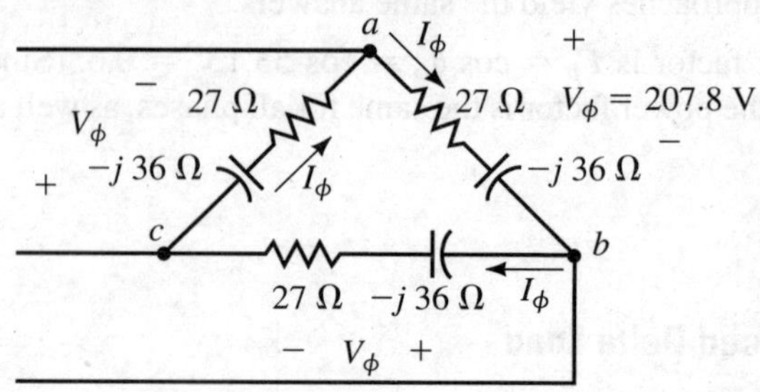

FIGURE 24–26

Solution

$$\mathbf{Z}_\phi = 27 - j36 = 45\ \Omega\angle-53.13°, \text{ so } \theta_\phi = -53.13°$$

$$V_\phi = 207.8 \text{ V and } I_\phi = V_\phi/Z_\phi = 207.8 \text{ V}/45\ \Omega = 4.62 \text{ A}$$

$$P_\phi = V_\phi I_\phi \cos\theta_\phi = (207.8)(4.62)\cos(-53.13°) = 576 \text{ W}$$

$$Q_\phi = V_\phi I_\phi \sin\theta_\phi = (207.8)(4.62)\sin(-53.13°) = -768 \text{ VAR}$$

$$= 768 \text{ VAR (cap.)}$$

$$S_\phi = V_\phi I_\phi = (207.8)(4.62) = 960 \text{ VA}$$

$$P_T = 3P_\phi = 3(576) = 1728 \text{ W}$$

$$Q_T = 3Q_\phi = 3(768) = 2304 \text{ VAR (cap.)}$$

$$S_T = 3S_\phi = 3(960) = 2880 \text{ VA}$$

Note that the results here are the same as for Example 24–11. This is to be expected since the load of Figure 24–24 is the Y equivalent of the Δ load of Figure 24–26.

PRACTICE PROBLEMS 7

Verify total active, reactive, and apparent power for the circuit of Example 24–12 using the formulas for P_T, Q_T, and S_T from Table 24–2.

Power and the Single-Phase Equivalent

You can also use the single-phase equivalent in power calculations. All single-phase active, reactive, and apparent power formulas that you learned in previous chapters apply. The equivalent yields power for one phase.

EXAMPLE 24–13

The total power to the balanced load of Figure 24–27 is 6912 W. The phase voltage at the load is 120 V. Compute the generator voltage $\mathbf{E}_{AB}$, magnitude and angle.

Solution Consider the single-phase equivalent in Figure 24–28. I_a is computed as follows:

$$P_{an} = \frac{P_T}{3} = \frac{1}{3}(6912) = 2304 \text{ W}$$

$$V_{an} = 120 \text{ V}$$

$$\theta_{an} = \cos^{-1}(0.8) = 36.87°$$

$$P_{an} = V_{an} I_a \cos\theta_{an}$$

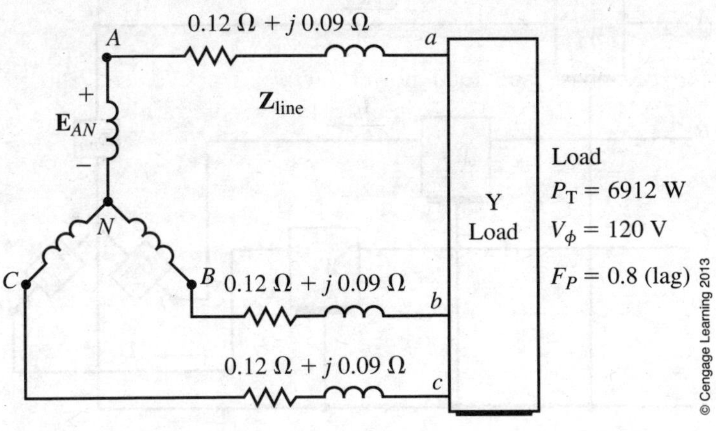

FIGURE 24–27

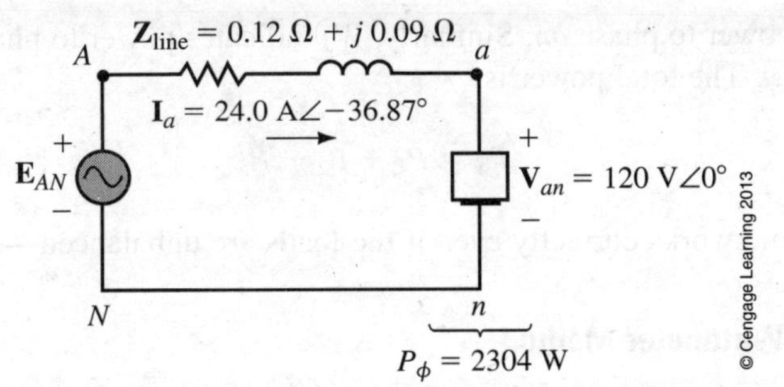

FIGURE 24–28

Therefore,

$$I_a = \frac{P_{an}}{V_{an}\cos\theta_{an}} = \frac{2304}{(120)(0.8)} = 24.0\text{ A}$$

Select $\mathbf{V}_{an}$ as reference. $\mathbf{V}_{an} = 120\text{ V}\angle 0°$. Thus, $\mathbf{I}_a = 24\text{ A}\angle -36.87°$ (since power factor was given as lagging).

$$\begin{aligned}\mathbf{E}_{AN} &= \mathbf{I}_a \times \mathbf{Z}_{\text{line}} + \mathbf{V}_{an}\\ &= (24\angle -36.87)(0.12 + j0.09) + 120\angle 0° = 123.6\text{ V}\angle 0°\\ \mathbf{E}_{AB} &= \sqrt{3}\mathbf{E}_{AN}\angle 30° = 214.1\text{ V}\angle 30°\end{aligned}$$

24.6 Measuring Power in Three-Phase Circuits

The number of wattmeters needed to measure power in a three-phase system depends on whether it's a three- or four-wire system—see Notes on page 898. For a four-wire system, you need three wattmeters, while, for a three-wire system, you need only two. Consider first, a four-wire circuit, Figure 24–29. Here, power to each phase can be measured using the same technique that you used for single-phase circuits—you connect each wattmeter so that its current coil carries the current for the phase of interest, and you connect the voltage coil of the wattmeter across the corresponding phase. For example, for wattmeter W_1, voltage is V_{an} and current is I_a. Thus, its reading is

$$P_1 = V_{an}I_a\cos\theta_{an}$$

NOTES...

1. If you could guarantee absolute balance at all times, you could use 1 wattmeter to measure power to one phase, then multiply by 3. However, balance cannot be assured, so you need more meters—as we have described in this section.
2. Although we have shown three individual wattmeters in Figure 24–29, many modern power meters incorporate this complete measurement capability in a single device. Consider for example, Figure 17–23. The meter shown is a complete system—it incorporates power measurement elements, associated electronic circuitry, and a digital readout that is designed to measure power (real, reactive, and apparent) to individual phases, as well as total power, power factor, line voltages, and line currents. The desired measurement is selected from a menu.

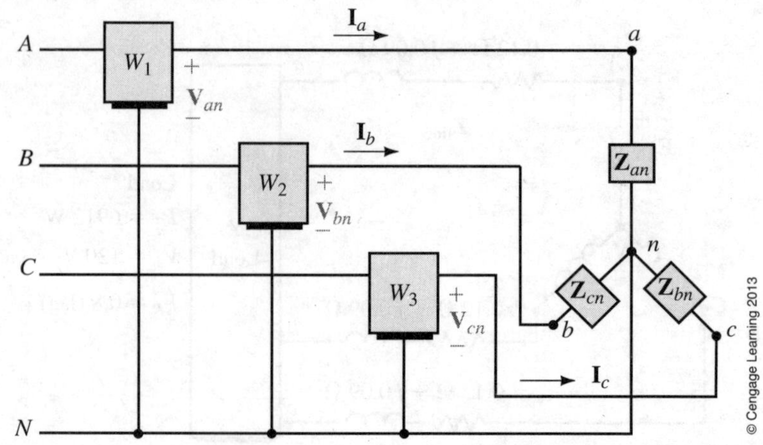

FIGURE 24–29 Three-wattmeter connection for a four-wire load.

which is power to phase *an*. Similarly, W_2 indicates power to phase *bn* and W_3 to phase *cn*. The total power is

$$P_T = P_1 + P_2 + P_3 \tag{24–22}$$

This scheme works correctly even if the loads are unbalanced—see Note 2.

CircuitSim 24-12

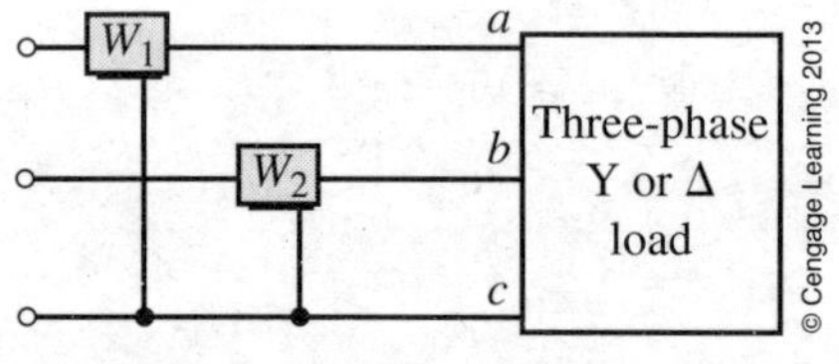

FIGURE 24–30 Two-wattmeter connection. Load may be balanced or unbalanced.

The Two-Wattmeter Method

For a three-wire system, only two wattmeters are needed. The connection is shown in Figure 24–30. Loads may be Y or Δ, balanced or unbalanced. The meters may be connected in any pair of lines with the voltage terminals connected to the third line. The total power is the algebraic sum of the meter readings. This scheme is known as the **two-wattmeter method.**

Determining Wattmeter Readings

Recall from Chapter 17, the reading of a wattmeter is equal to the product of the magnitude of its voltage, the magnitude of its current, and the cosine of the angle between them. For each meter, you must carefully determine what this angle is. This is illustrated next.

EXAMPLE 24–14

For Figure 24–31, $\mathbf{V}_{an} = 120\ \text{V}\angle 0°$. Compute the readings of each meter, then sum to determine total power. Compare P_T to the P_T found in Example 24–11.

NOTES...

We use 207.8 V here instead of the nominal 208 V so that we can compare answers to CircuitSim 24–13.

Solution $\mathbf{V}_{an} = 120\ \text{V}\angle 0°$. Thus, $\mathbf{V}_{ab} = 207.8\ \text{V}\angle 30°$ and $\mathbf{V}_{bc} = 207.8\ \text{V}\angle{-90°}$—see Note. $\mathbf{I}_a = \mathbf{V}_{an}/\mathbf{Z}_{an} = 120\ \text{V}\angle 0°/(9 - j12)\ \Omega = 8\ \text{A}\angle 53.13°$. Thus, $\mathbf{I}_c = 8\ \text{A}\angle 173.13°$.

First consider wattmeter 1, Figure 24–32. Note that W_1 is connected to terminals *a-b*; thus it has voltage $\mathbf{V}_{ab}$ across it and current $\mathbf{I}_a$ through it. Its reading is therefore $P_1 = V_{ab}I_a\cos\theta_1$, where θ_1 is the angle between $\mathbf{V}_{ab}$ and $\mathbf{I}_a$. $\mathbf{V}_{ab}$ has an angle of 30° and $\mathbf{I}_a$ has an angle of 53.13°. Thus, $\theta_1 = 53.13° - 30° = 23.13°$ and $P_1 = (207.8)(8)\cos 23.13° = 1529\ \text{W}$.

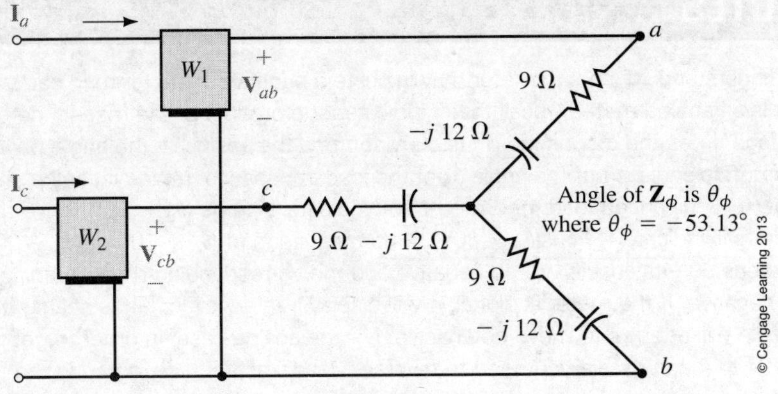

FIGURE 24–31

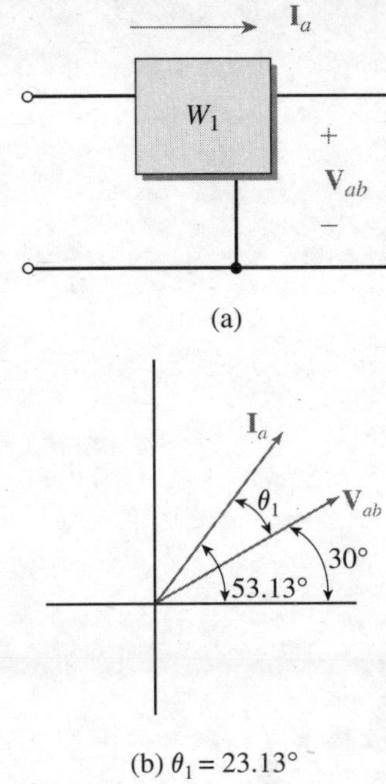

FIGURE 24–32 $P_1 = V_{ab}I_a \cos\theta_1$ where θ_1 is the angle between $\mathbf{V}_{ab}$ and $\mathbf{I}_a$.

Now consider wattmeter 2, Figure 24–33. Since W_2 is connected to terminals *c-b*, the voltage across it is $\mathbf{V}_{cb}$ and the current through it is $\mathbf{I}_c$. But $\mathbf{V}_{cb} = -\mathbf{V}_{bc} = 207.8\ \text{V}\angle 90°$ and $\mathbf{I}_c = 8\ \text{A}\angle 173.13°$. The angle between $\mathbf{V}_{cb}$ and $\mathbf{I}_c$ is thus $173.13° - 90° = 83.13°$. Therefore, $P_2 = V_{cb}I_c\cos\theta_2 = (207.8)(8)\cos 83.13° = 199$ W and $P_T = P_1 + P_2 = 1529 + 199 = 1728$ W. (This agrees well with the answer we obtained in Example 24–11.) Note that one of the wattmeters reads lower than the other. (This is generally the case for the two-wattmeter method.)

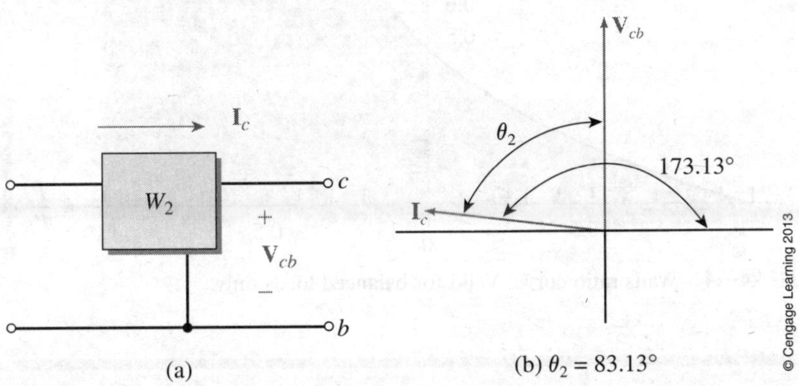

FIGURE 24–33 $P_2 = V_{cb}I_c \cos\theta_2$ where θ_2 is the angle between $\mathbf{V}_{cb}$ and $\mathbf{I}_c$.

PRACTICE PROBLEMS 8

Change the load impedances of Figure 24–31 to $15\ \Omega\angle 70°$. Repeat Example 24–14. (As a check, total power to the load is 985 W. *Hint:* One of the meters has a negative reading.)

Answers
$P_\ell = -289$ W; $P_2 = 1275$ W; $P_1 + P_2 = 986$ W

CircuitSim 24-13

If you are using analog wattmeters in Practice Problems 8, the low-reading meter (W_1) will read backward—see Note on page 900. To get an upscale reading, reverse either its voltage or current connection, then subtract. Thus, for cases like this, if P_h and P_ℓ are the high-reading and low-reading meters, respectively,

$$P_T = P_h - P_\ell \qquad (24\text{–}23)$$

NOTES...

To understand why a wattmeter may indicate a negative value (or read backward if it is an analog meter), recall that it indicates the product of the magnitude of its voltage times the magnitude of its current times the cosine of the angle between them. This angle is not the angle θ_ϕ of the load impedance; rather, it can be shown that, for a balanced load, one meter will indicate $V_L I_L \cos(\theta_\phi - 30°)$ while the other will indicate $V_L I_L \cos(\theta_\phi + 30°)$. If the magnitude of $(\theta_\phi + 30°)$ or $(\theta_\phi - 30°)$ exceeds 90°, its cosine will be negative and the corresponding meter reading will be negative. If the meter is digital, it will display a minus sign; if it is analog, the needle will attempt to move downscale. For the analog case, in order to get the meter to read upscale you need to reverse a lead—as we have noted in the text.

Watts Ratio Curve

The power factor for a balanced load can be obtained from the wattmeter readings using a simple curve called the watts ratio curve, shown in Figure 24–34.

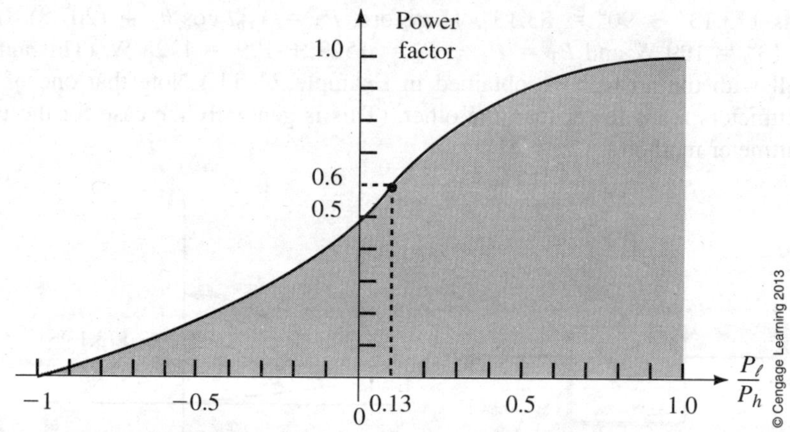

FIGURE 24–34 Watts ratio curve. Valid for balanced loads only.

EXAMPLE 24–15

Consider again Figure 24–31.

a. Determine the power factor from the load impedance.

b. Using the meter readings of Example 24–14, determine the power factor from the watts ratio curve.

Solution

a. $F_p = \cos\theta_\phi = \cos 53.13° = 0.6$

b. $P_\ell = 199$ W and $P_h = 1530$ W. Therefore, $P_\ell/P_h = 0.13$. From Figure 24–34, $F_p = 0.6$.

The problem with the watts ratio curve is that values are difficult to determine accurately from the graph. However, it can be shown that

$$\tan\theta_\phi = \sqrt{3}\left(\frac{P_h - P_\ell}{P_h + P_\ell}\right) \qquad (24\text{–}24)$$

From this, you can determine θ_ϕ and then you can compute the power factor from $F_p = \cos\theta_\phi$. Note, however, with many modern metering systems this is unnecessary as such meters automatically determine and display power factor.

24.7 Unbalanced Loads

For unbalanced loads, none of the balanced circuit relationships apply. Each problem must therefore be treated as an individual three-phase problem. We will look at a few examples that can be handled by fundamental circuit techniques such as Kirchhoff's laws and mesh analysis. Source voltages are always balanced (unless they are faulty, which we don't consider here). In practice, safety may be an issue—see Note.

NOTES...

Voltage at a neutral point can be dangerous. For example, in Figure 24–35, if the neutral is grounded at the source, the voltage at the load neutral is *floating* at some potential relative to ground. Since we are accustomed to thinking that neutrals are at ground potential and hence safe to touch, there is a potential safety hazard here—see Practice Problems 9, Question 1.

Unbalanced Wye Loads

Unbalanced four-wire Y systems without line impedance are easily handled using Ohm's law. However, for three-wire systems or four-wire systems with line and neutral impedance, you generally have to use mesh equations or computer methods.

EXAMPLE 24–16

For Figure 24–35, the generator is balanced with line-to-line voltage of 208 V. Select $\mathbf{E}_{AB}$ as reference and determine line currents and load voltages.

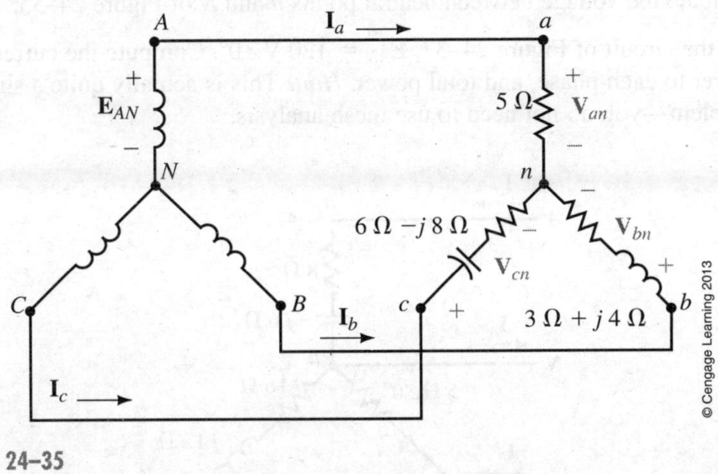

FIGURE 24–35

CircuitSim 24-14

Solution Redraw the circuit as shown in Figure 24–36, then use mesh analysis. $\mathbf{E}_{AB} = 208\text{ V}\angle 0°$ and $\mathbf{E}_{BC} = 208\text{ V}\angle -120°$.

Loop 1: $(8 + j4)\mathbf{I}_1 - (3 + j4)\mathbf{I}_2 = 208\text{ V}\angle 0°$

Loop 2: $-(3 + j4)\mathbf{I}_1 + (9 - j4)\mathbf{I}_2 = 208\text{ V}\angle -120°$

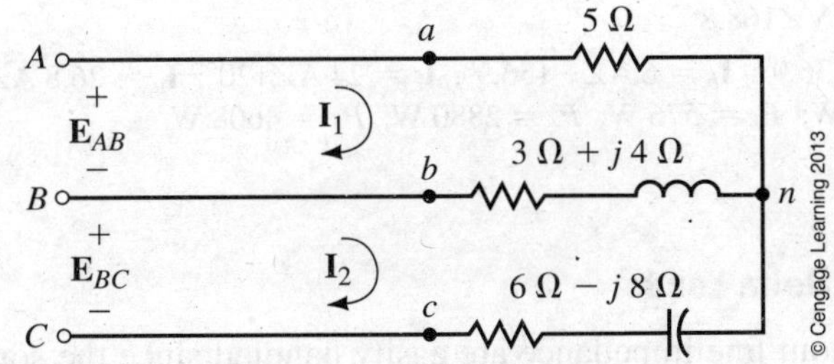

FIGURE 24–36

NOTES...

You will likely have to write down some intermediate answers and re-enter them in your calculator as you solve this example. You should probably write down four- or five-digit intermediate answers so as not to introduce round-off and truncation error in your calculations. Final answers should really be rounded to three digits, but in order to permit you to check your work, we have shown four.

These equations can be solved using standard techniques such as determinants or a calculator. Since most calculators cannot solve simultaneous equations with complex coefficients, many of you will be faced with solving by determinants—see Note. Solutions are

$$\mathbf{I}_1 = 29.85 \text{ A}\angle{-26.21°} \quad \text{and} \quad \mathbf{I}_2 = 11.81 \text{ A}\angle{-51.50°}$$

$$\text{KCL: } \mathbf{I}_a = \mathbf{I}_1 = 29.85 \text{ A}\angle{-26.21°}$$

$$\mathbf{I}_b = \mathbf{I}_2 - \mathbf{I}_1 = 19.84 \text{ A}\angle 168.50°$$

$$\mathbf{I}_c = -\mathbf{I}_2 = 11.81 \text{ A}\angle 128.50°$$

$$\mathbf{V}_{an} = \mathbf{I}_a\mathbf{Z}_{an} = (29.85 \text{ A}\angle{-26.21°})\,(5) = 149.3 \text{ V}\angle{-26.2°}$$

$$\mathbf{V}_{bn} = \mathbf{I}_b\mathbf{Z}_{bn} \quad \text{and} \quad \mathbf{V}_{cn} = \mathbf{I}_c\mathbf{Z}_{cn}$$

Thus,

$$\mathbf{V}_{bn} = 99.2 \text{ V}\angle{-138.4°} \quad \text{and} \quad \mathbf{V}_{cn} = 118.1 \text{ V}\angle 75.4°$$

PRACTICE PROBLEMS 9

1. One of the problems with unbalanced three-wire Y systems is that, because you get different voltages across each phase of the load, you get a voltage between neutral points. To illustrate, using KVL and the results from Example 24–16, compute the voltage between neutral points n and N of Figure 24–35.
2. For the circuit of Figure 24–37, $\mathbf{E}_{AN} = 120 \text{ V}\angle 0°$. Compute the currents, power to each phase, and total power. *Hint:* This is actually quite a simple problem—you do not need to use mesh analysis.

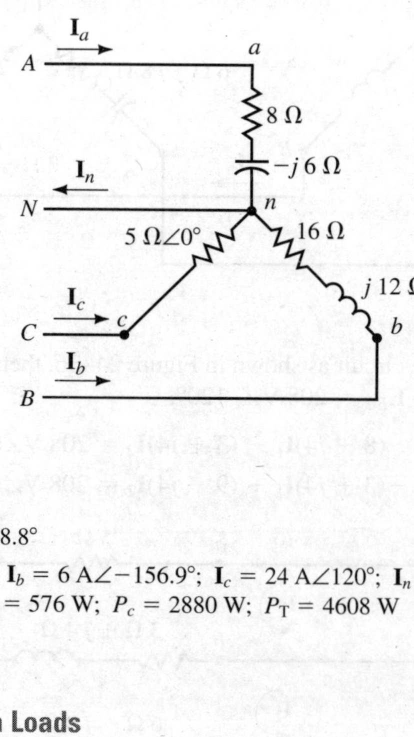

FIGURE 24–37

Answers

1. $\mathbf{V}_{nN} = 30.8 \text{ V}\angle 168.8°$
2. $\mathbf{I}_a = 12 \text{ A}\angle 36.9°$; $\mathbf{I}_b = 6 \text{ A}\angle{-156.9°}$; $\mathbf{I}_c = 24 \text{ A}\angle 120°$; $\mathbf{I}_n = 26.8 \text{ A}\angle 107.2°$
   $P_a = 1152$ W; $P_b = 576$ W; $P_c = 2880$ W; $P_T = 4608$ W

Unbalanced Delta Loads

Systems without line impedance are easily handled since the source voltage is applied directly to the load. However, for systems with line impedance, use mesh equations.

EXAMPLE 24–17

For the circuit of Figure 24–38, the line voltage is 240 V. Take $\mathbf{V}_{ab}$ as reference and do the following:

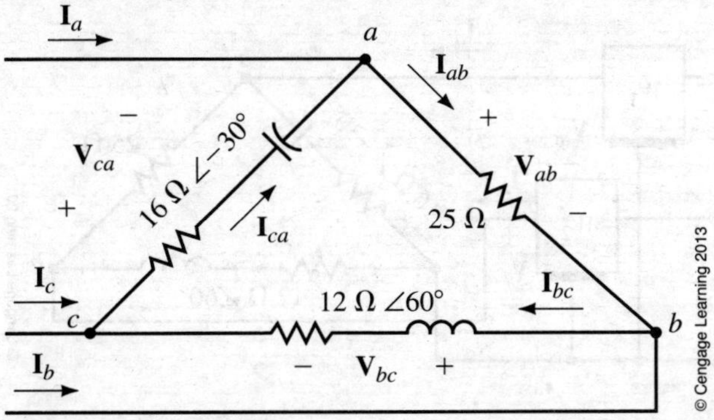

FIGURE 24–38

a. Determine the phase currents and sketch their phasor diagram.
b. Determine the line currents.
c. Determine the total power to the load.

Solution

a. $\mathbf{I}_{ab} = \mathbf{V}_{ab}/\mathbf{Z}_{ab} = (240\ \text{V}\angle 0°)/25\ \Omega = 9.6\ \text{A}\angle 0°$

$\mathbf{I}_{bc} = \mathbf{V}_{bc}/\mathbf{Z}_{bc} = (240\ \text{V}\angle -120°)/(12\ \Omega\angle 60°) = 20\ \text{A}\angle -180°$

$\mathbf{I}_{ca} = \mathbf{V}_{ca}/\mathbf{Z}_{ca} = (240\ \text{V}\angle 120°)/(16\ \Omega\angle -30°) = 15\ \text{A}\angle 150°$

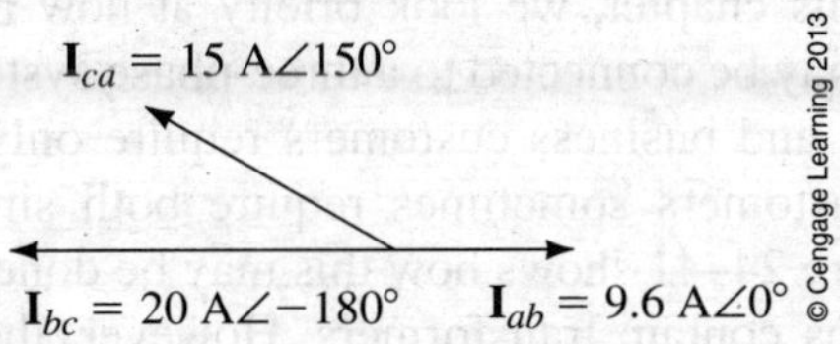

FIGURE 24–39 Phase currents for the circuit of Figure 24–38.

b. $\mathbf{I}_a = \mathbf{I}_{ab} - \mathbf{I}_{ca} = 9.6\ \text{A}\angle 0° - 15\ \text{A}\angle 150° = 23.8\ \text{A}\angle -18.4°$

$\mathbf{I}_b = \mathbf{I}_{bc} - \mathbf{I}_{ab} = 20\ \text{A}\angle -180° - 9.6\ \text{A}\angle 0° = 29.6\ \text{A}\angle 180°$

$\mathbf{I}_c = \mathbf{I}_{ca} - \mathbf{I}_{bc} = 15\ \text{A}\angle 150° - 20\ \text{A}\angle -180° = 10.3\ \text{A}\angle 46.9°$

c. $P_{ab} = V_{ab}I_{ab}\cos\theta_{ab} = (240)(9.6)\cos 0° = 2304\ \text{W}$

$P_{bc} = V_{bc}I_{bc}\cos\theta_{bc} = (240)(20)\cos 60° = 2400\ \text{W}$

$P_{ca} = V_{ca}I_{ca}\cos\theta_{ca} = (240)(15)\cos 30° = 3118\ \text{W}$

$P_{\text{T}} = P_{ab} + P_{bc} + P_{ca} = 7822\ \text{W}$

EXAMPLE 24–18

A pair of wattmeters are added to the circuit of Figure 24–38 as illustrated in Figure 24–40. Determine wattmeter readings and compare them to the total power calculated in Example 24–17.

Solution $P_1 = V_{ac}I_a\cos\theta_1$, where θ_1 is the angle between $\mathbf{V}_{ac}$ and $\mathbf{I}_a$. From Example 24–17, $\mathbf{I}_a = 23.8\text{ A}\angle{-18.4°}$ and $\mathbf{V}_{ac} = -\mathbf{V}_{ca} = 240\text{ V}\angle{-60°}$. Thus, $\theta_1 = 60° - 18.4° = 41.6°$. Therefore,

$$P_1 = (240)(23.8)\cos 41.6° = 4271\text{ W}$$

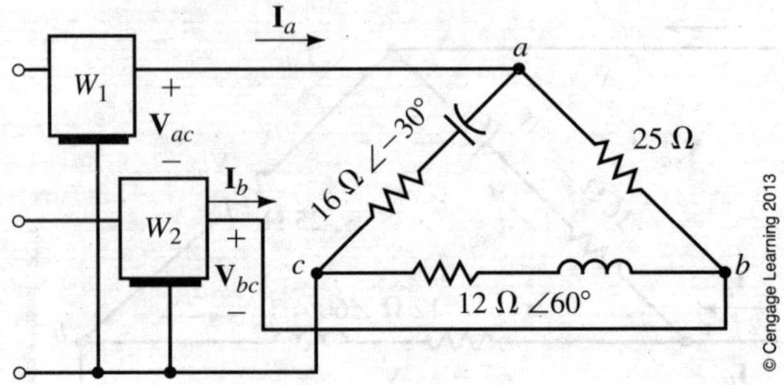

CircuitSim 24-15

FIGURE 24–40

$P_2 = V_{bc}I_b\cos\theta_2$, where θ_2 is the angle between $\mathbf{V}_{bc}$ and $\mathbf{I}_b$. $\mathbf{V}_{bc} = 240\text{ V}\angle{-120°}$ and $\mathbf{I}_b = 29.6\text{ A}\angle 180°$. Thus, $\theta_2 = 60°$. So

$$P_2 = (240)(29.6)\cos 60° = 3552\text{ W}$$

$P_T = P_1 + P_2 = 7823$ W (compared to our earlier solution of 7822 W).

24.8 Power System Loads

Before we leave this chapter, we look briefly at how both single-phase and three-phase loads may be connected to a three-phase system. (This is necessary because residential and business customers require only single-phase power, while industrial customers sometimes require both single-phase and three-phase power.) Figure 24–41 shows how this may be done. (Even this is simplified, as real systems contain transformers. However, the basic principles are correct.) Two points should be noted here.

1. In order to approximately balance the system, the utility tries to connect one-third of its single-phase loads to each phase. This yields a nearly balanced system.
2. Real loads are seldom expressed in terms of resistance, capacitance, and inductance. Instead, they are described in terms of power, power factor, and so on. This is because most loads consist of electric lights, motors, and the like, which are never described in terms of impedance. (For example, you purchase light bulbs as 60-W bulbs, 100-W bulbs, etc., and electric motors as ½ horsepower, etc. You never ask for a 240-ohm light!)

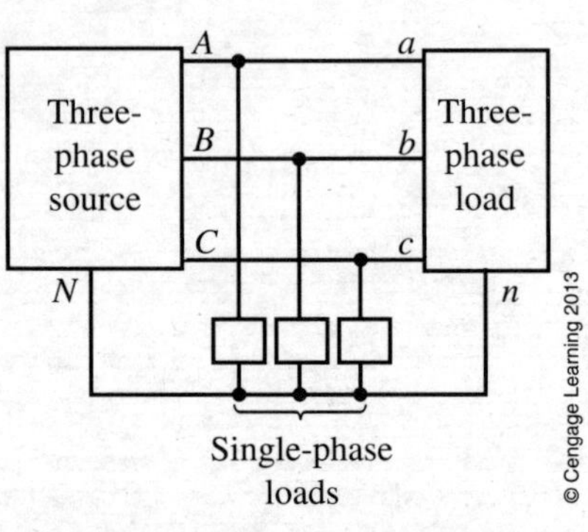

FIGURE 24–41 Single-phase loads are tapped off the three-phase lines.

24.9 Circuit Analysis Using Computers

PSpice and Multisim may be used to analyze three-phase systems (balanced or unbalanced, Y or Δ connected). As usual, PSpice yields full phasor solutions, but Multisim yields magnitudes only. Because neither software package allows component placement at an angle, Y and Δ circuits must be drawn with components placed either horizontally or vertically as in Figures 24–43 through 24–46, rather than in their traditional three-phase form. To begin, consider the balanced four-wire Y circuit of Figure 24–42. First, manually compute the currents so that you have a basis for comparison. Note that, at 60 Hz,

$$X_C = 53.05\ \Omega$$

Thus,

$$\mathbf{I}_a = \frac{120\ \text{V}\angle 0°}{(30 - j53.05)\ \Omega} = 1.969\ \text{A}\angle 60.51°$$

and

$$\mathbf{I}_b = 1.969\ \text{A}\angle -59.49°$$

and

$$\mathbf{I}_c = 1.969\ \text{A}\angle -179.49°.$$

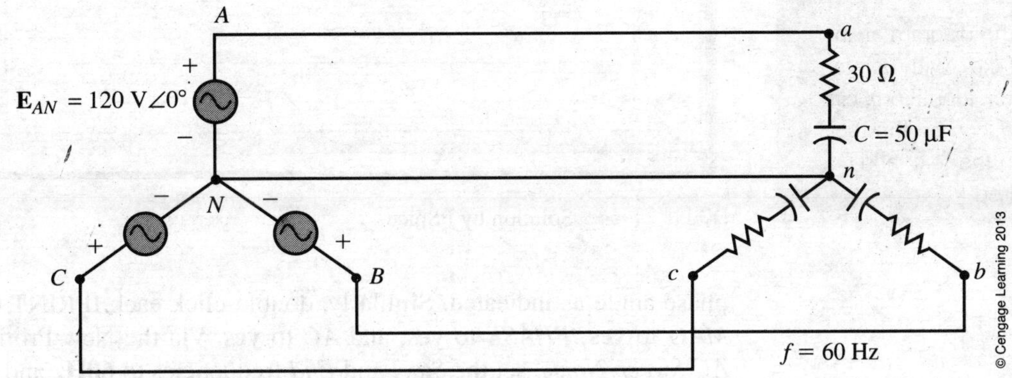

FIGURE 24–42 Balanced system for computer analysis.

Multisim

Draw the circuit on the screen as in Figure 24–43(a). Make sure the voltage sources are oriented with their + ends as shown and that their phase angles are set appropriately—see Note. Double-click the ammeters and set them to AC. Activate the circuit by clicking the power switch. Compare answers to the manual solution. Now replace the individual sources with the three-phase source of (b) and repeat.

NOTES...

Ensure that the positive phase direction for sinusoidal waveforms is set to "shift right" as detailed earlier in Chapter 15.

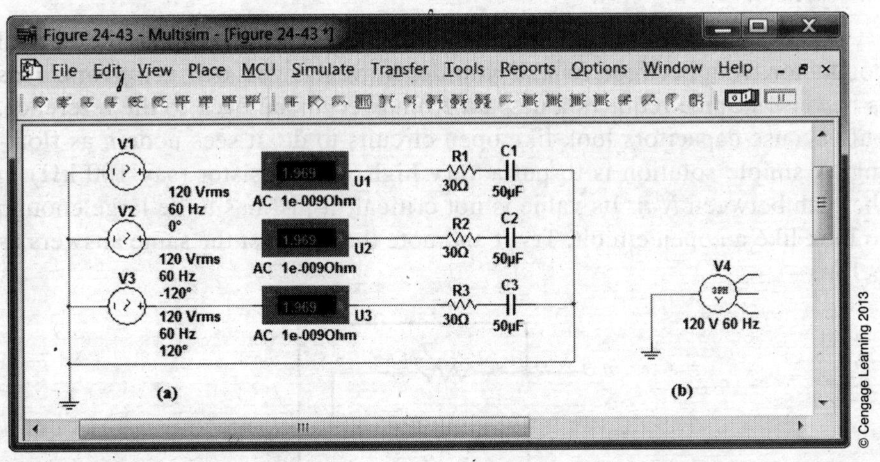

FIGURE 24–43 Solution by Multisim.

PSpice

Draw the circuit on the screen as in Figure 24–44 using source *VAC*. (Make sure all source + terminals are oriented to the right as shown.) Double-click each source in turn and, in its Property Editor, set its magnitude to **120V** with

NOTES...

1. Multisim and PSpice make no distinction between balanced circuits and unbalanced circuits; to the computer, they are simply circuits. Thus, no special considerations are necessary for unbalanced conditions.
2. Delta loads may be drawn as in Figure 24–46.
3. To annotate the diagram as in Figure 24–44 (i.e., with points A, B, C, N, etc., identified), click the *Place text* icon on the tool palette, key in the text, and place as desired.

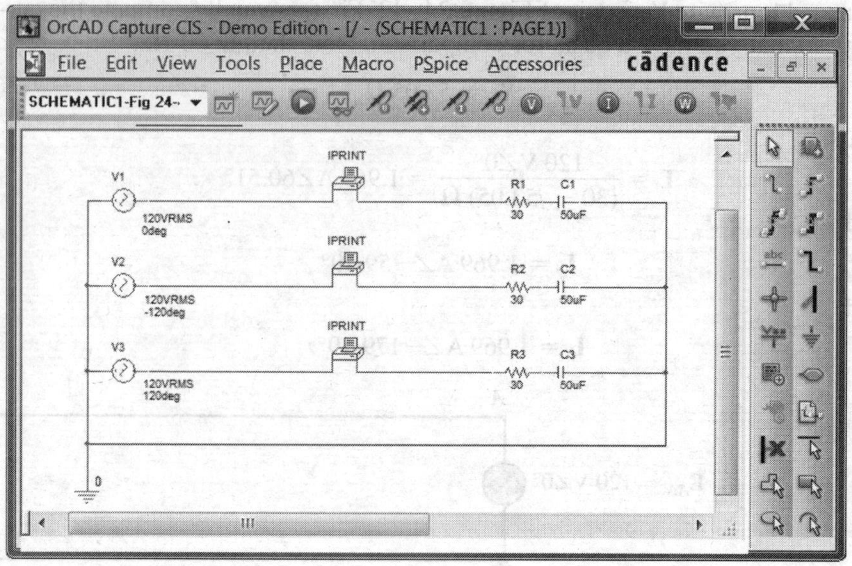

FIGURE 24–44 Solution by PSpice.

phase angle as indicated. Similarly, double-click each IPRINT device and set *MAG* to **yes,** *PHASE* to **yes,** and *AC* to **yes.** Via the New Profile icon, select *AC Sweep/Noise,* set the Start and End frequencies to **60Hz** and the number of points to **1**. Run the simulation, open the *Output File* (found under *View* in the results window), then scroll down until you find the answers. (Figure 24–45 shows the answer for current $\mathbf{I}_a$.) Examining the results, you will see that they agree exactly with what we calculated earlier.

```
FREQ          IM(V_PRINT1)   IP(V_PRINT1)

6.000E+01     1.969E+00      6.051E+01
```


FIGURE 24–45 Sample computed current: $\mathbf{I}_a = 1.969 \text{ A} \angle 60.51°$.

Since the circuit is balanced, you should be able to remove the neutral conductor from between *N-n.* If you do, however, you will get errors. This is because PSpice requires a dc path from every node back to the reference, but because capacitors look like open circuits to dc, it sees node *n* as floating. A simple solution is to put a very high-value resistor (say 100 kΩ) in the path between *N-n.* Its value is not critical; it just has to be large enough to look like an open circuit. Try it and note that you get the same answers as before.

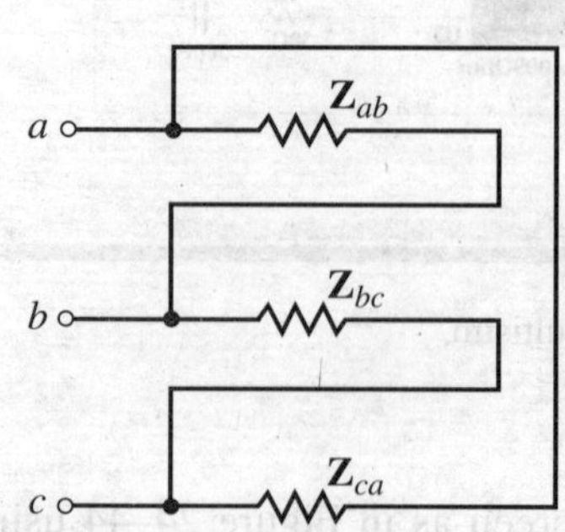

FIGURE 24–46 Representing a Δ load.

Putting It into Practice

You have been sent to a job site to supervise the installation of a 208-V, three-phase motor. The motor drives a machine, and it is essential that the motor rotate in the correct direction (in this case, clockwise), otherwise the machine may be damaged. You have a drawing that tells you to connect Line *a* of the motor to Line *A* of the three-phase system, Line *b* to Line *B*, and so on. However, you find that the three-phase lines have no markings, and you don't know which line is which. Unfortunately, you cannot simply connect the motor and determine which direction it turns for fear of damage.

You think about this for a while, then you come up with a plan. You know that the direction of rotation of a three-phase motor depends on the phase sequence of the applied voltage, so you make a sketch (sketch A). As indicated in part (a) of that drawing, the motor rotates in the correct direction when *a* is connected to *A*, *b* to *B*, and so on. You reason that the phase sequence you need is . . . *A-B-C-A-B-C* . . . (since the direction of rotation depends only on phase sequence) and that it really doesn't matter to which line *a* is connected as long as the other two are connected such that you provide sequencing in this order to the motor. To convince yourself, you make several more sketches (parts b and c). As indicated, in (b) the sequence is . . . *B-C-A-B-C-A* . . . (which fits the preceding pattern) and the motor rotates in the correct direction, but for (c) the motor rotates in the reverse direction. (Show why.)

You also remember reading about a device called a *phase sequence indicator* that allows you to determine phase sequence. It uses lights and a capacitor as in sketch B. To use the device, connect terminal *a* to the three-phase line that you have designated *A* and connect terminals *b* and *c* to the other two lines. The lamp that lights brightly is the one attached to line *B*. After a few calculations, you ask the plant electrician for some standard 120-V 60-W light bulbs and sockets. From your tool kit, you retrieve a 10-μF capacitor (one that is rated for ac operation). You make a slight modification to the schematic of sketch B, solder the parts together, wrap exposed wires with electrical tape (for safety), then connect the device and identify the three-phase lines. You then connect the motor and it rotates in the correct direction. Prepare an analysis to show why the lamp connected to *B* is much brighter than the lamp connected to line *C*. Assume each lamp resistance is 240 Ω. (Note: Don't forget that you made a slight change to the schematic. Hint: If you use single 60-W lamps in each leg as shown, they will burn out.)

CircuitSim 24-16

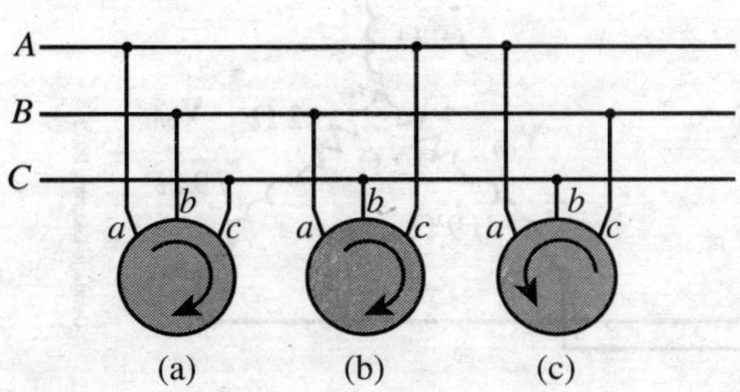

Sketch A

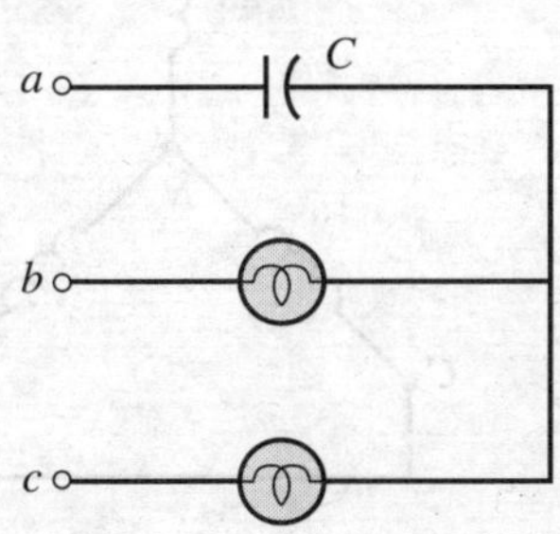

Sketch B

Problems

24.2 Basic Three-Phase Circuit Connections

1. As long as the loads and voltages of Figure 24–3 are balanced (regardless of their actual value), currents $\mathbf{I}_A$, $\mathbf{I}_B$, and $\mathbf{I}_C$ will sum to zero. To illustrate, change the load impedance from 12 Ω to 15 Ω∠30° and for $\mathbf{E}_{AA'} = 120\ \text{V}\angle 0°$, do the following:
   a. Compute currents $\mathbf{I}_A$, $\mathbf{I}_B$, and $\mathbf{I}_C$.
   b. Sum the currents. Does $\mathbf{I}_A + \mathbf{I}_B + \mathbf{I}_C = 0$?
2. For Figure 24–3(c), $\mathbf{E}_{AN} = 277\ \text{V}\angle -15°$.
   a. What are $\mathbf{E}_{BN}$ and $\mathbf{E}_{CN}$?
   b. If each resistance is 5.54 Ω, compute $\mathbf{I}_A$, $\mathbf{I}_B$, and $\mathbf{I}_C$.
   c. Sum the currents of part (b) to show that $\mathbf{I}_N = 0$.

CircuitSim 24-17

3. Sketch the set of current waveforms i_A, i_B, and i_C for the circuit of Figure 24–3(c). Assume balanced, purely resistive load resistors.

24.3 Basic Three-Phase Relationships

4. For the generators of Figure 24–4, $\mathbf{E}_{AN} = 7620\ \text{V}\angle -18°$.
   a. What are the phase voltages $\mathbf{E}_{BN}$ and $\mathbf{E}_{CN}$?
   b. Determine the line-to-line voltages.
   c. Sketch their phasor diagram.
5. For the loads of Figure 24–4, $\mathbf{V}_{bc} = 208\ \text{V}\angle -75°$.
   a. Determine the line-to-line voltages $\mathbf{V}_{ab}$ and $\mathbf{V}_{ca}$.
   b. Determine the phase voltages.
   c. Sketch their phasor diagram.
6. Repeat Problem 5 if $\mathbf{V}_{ca} = 208\ \text{V}\angle 90°$.
7. For the load of Figure 24–47, $\mathbf{V}_{an} = 347\ \text{V}\angle 15°$. Determine all line currents. Sketch their phasor diagram.

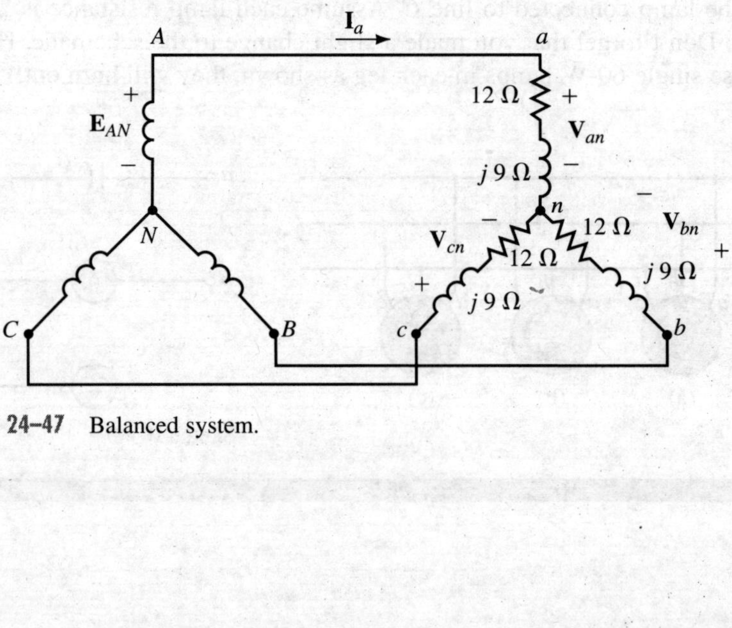

FIGURE 24–47 Balanced system.

8. For the load of Figure 24–47, if $\mathbf{I}_a = 7.8\ \text{A}\angle -10°$, determine the phase voltages and line voltages. Sketch their phasor diagram.
9. A balanced Y load has impedance $\mathbf{Z}_{an} = 14.7\ \Omega\angle 16°$. If $\mathbf{V}_{cn} = 120\ \text{V}\angle 160°$, determine all line currents.
10. For a balanced delta load, $\mathbf{I}_{ab} = 29.3\ \text{A}\angle 43°$. What is $\mathbf{I}_a$?
11. For the circuit of Figure 24–48, $\mathbf{V}_{ab} = 480\ \text{V}\angle 0°$. Find the phase and line currents.

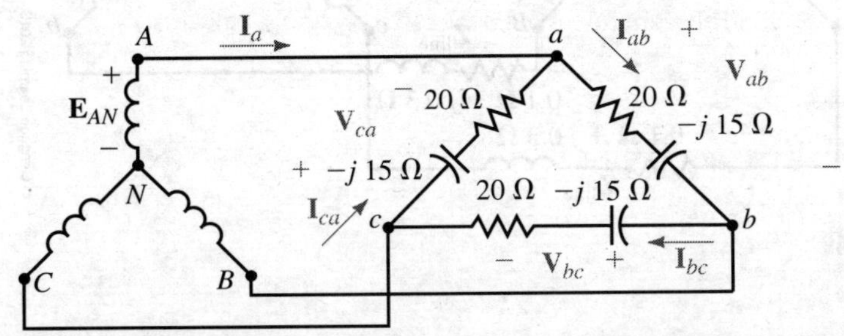

FIGURE 24–48 Balanced system.

12. For the circuit of Figure 24–48, if $\mathbf{I}_a = 41.0\ \text{A}\angle -46.7°$, find all the phase currents.
13. For the circuit of Figure 24–48, if $\mathbf{I}_{ab} = 10\ \text{A}\angle -21°$, determine all the line voltages.
14. For the circuit of Figure 24–48, if line current $\mathbf{I}_a = 11.0\ \text{A}\angle 30°$, find all the phase voltages.
15. A balanced Y load has phase impedance of $24\ \Omega\angle 33°$ and line-to-line voltage of 600 V. Take $\mathbf{V}_{an}$ as reference and determine all line currents.
16. A balanced Δ load has phase impedance of $27\ \Omega\angle -57°$ and phase voltage of 208 V. Take $\mathbf{V}_{ab}$ as reference and determine the following:
    a. Phase currents.
    b. Line currents.
17. a. For a certain balanced Y load, $\mathbf{V}_{ab} = 208\ \text{V}\angle 30°$, $\mathbf{I}_a = 24\ \text{A}\angle 40°$ and $f = 60$ Hz. Determine the load (R and L or C).
    b. Repeat (a) if $\mathbf{V}_{bc} = 208\ \text{V}\angle -30°$ and $\mathbf{I}_c = 12\ \text{A}\angle 140°$.
18. Consider Figure 24–13(a). Show that $\mathbf{I}_a = \sqrt{3}\mathbf{I}_{ab}\angle -30°$.
19. At 60 Hz, a balanced Δ load has current $\mathbf{I}_{bc} = 4.5\ \text{A}\angle -85°$. The line voltage is 240 volts and $\mathbf{V}_{ab}$ is taken as reference.
    a. Find the other phase currents.
    b. Find the line currents.
    c. Find the resistance R and capacitance C of the load.
20. A Y generator with $\mathbf{E}_{AN} = 120\ \text{V}\angle 0°$ drives a balanced Δ load. If $\mathbf{I}_a = 43.6\ \text{A}\angle -37.5°$, what are the load impedances?

24.4 Examples

21. For Figure 24–49, $\mathbf{V}_{an} = 120\ \text{V}\angle 0°$. Draw the single-phase equivalent and:
    a. Find phase voltage $\mathbf{E}_{AN}$, magnitude and angle.
    b. Find line voltage $\mathbf{E}_{AB}$, magnitude and angle.

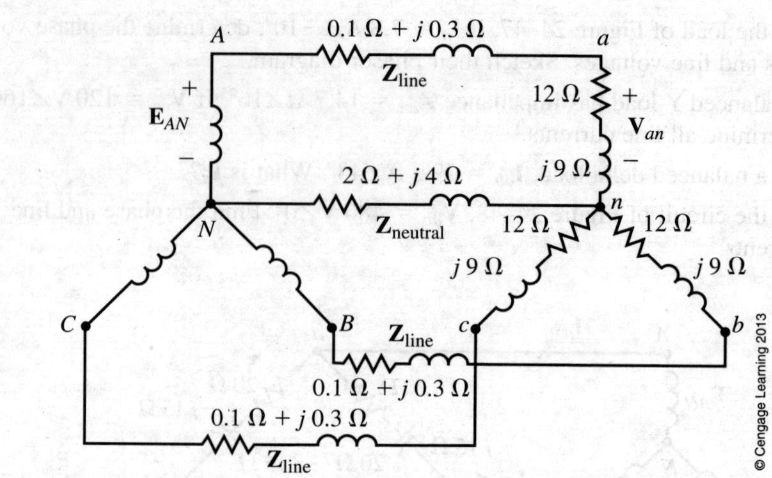

FIGURE 24–49

22. For Figure 24–49, $\mathbf{E}_{AN} = 120\ \text{V}\angle 20°$. Draw the single-phase equivalent and:
    a. Find phase voltage $\mathbf{V}_{an}$, magnitude and angle.
    b. Find line voltage $\mathbf{V}_{ab}$, magnitude and angle.
23. For Figure 24–48, $\mathbf{E}_{AN} = 120\ \text{V}\angle -10°$. Find the line currents using the single-phase equivalent method.
24. Repeat Problem 23 if $\mathbf{E}_{BN} = 120\ \text{V}\angle -100°$.
25. For Figure 24–48, assume the lines have impedance $\mathbf{Z}_{\text{line}}$ of 0.15 Ω + j0.25 Ω and $\mathbf{E}_{AN} = 120\ \text{V}\angle 0°$. Convert the Δ load to a Y and use the single-phase equivalent to find the line currents.
26. For Problem 25, find the phase currents in the Δ network.
27. For the circuit of Figure 24–49, assume $\mathbf{Z}_{\text{line}} = 0.15\ \Omega + j0.25\ \Omega$ and $\mathbf{V}_{ab} = 600\ \text{V}\angle 30°$. Determine $\mathbf{E}_{AB}$.
28. For Figure 24–21(a), $\mathbf{Z}_Y = 12\ \Omega + j9\ \Omega$, $\mathbf{Z}_\Delta = 27\ \Omega + j36\ \Omega$ and $\mathbf{Z}_{\text{line}} = 0.1\ \Omega + j0.1\ \Omega$. At the Y load, $\mathbf{V}_{an} = 120\ \text{V}\angle 0°$.
    a. Draw the single-phase equivalent.
    b. Find generator voltage $\mathbf{E}_{AN}$.
29. Same as Problem 28 except that the phase voltage at the Δ load is $\mathbf{V}_{a'b'} = 480\ \text{V}\angle 30°$. Find generator voltage $\mathbf{E}_{AB}$, magnitude and angle.
30. For Figure 24–21(a), $\mathbf{Z}_Y = 12\ \Omega + j9\ \Omega$, $\mathbf{Z}_\Delta = 36\ \Omega + j27\ \Omega$, and $\mathbf{Z}_{\text{line}} = 0.1\ \Omega + j0.1\ \Omega$. Line current $\mathbf{I}_A$ is $46.2\ \text{A}\angle -36.87°$. Find the phase currents for both loads.
31. For Figure 24–21(a), $\mathbf{Z}_Y = 15\ \Omega + j20\ \Omega$, $\mathbf{Z}_\Delta = 9\ \Omega - j12\ \Omega$, $\mathbf{Z}_{\text{line}} = 0.1\ \Omega + j0.1\ \Omega$, and $\mathbf{I}_{a'b'} = 40\ \text{A}\angle 73.13°$. Find Y phase voltage $\mathbf{V}_{an}$, magnitude and angle.

24.5 Power in a Balanced System

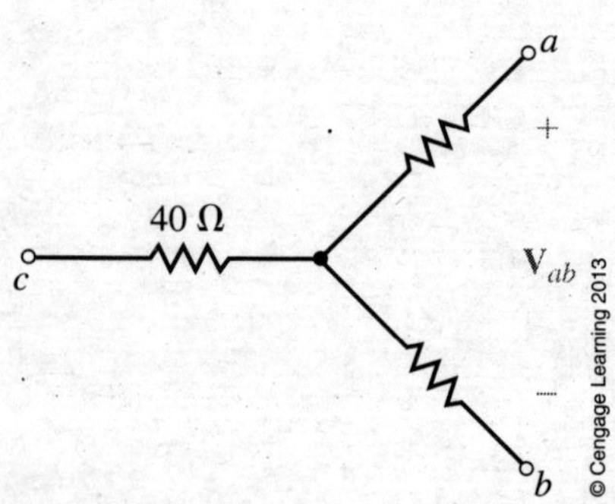

FIGURE 24–50

32. For the balanced load of Figure 24–50, $V_{ab} = 600$ V. Determine per-phase and total active, reactive, and apparent power.
33. Repeat Problem 32 for the balanced load of Figure 24–51, given $E_{AN} = 120$ V.
34. For Figure 24–47, $E_{AN} = 120$ volts.
    a. Determine per-phase real, reactive, and apparent power.
    b. Multiply per-phase quantities by 3 to get total quantities.

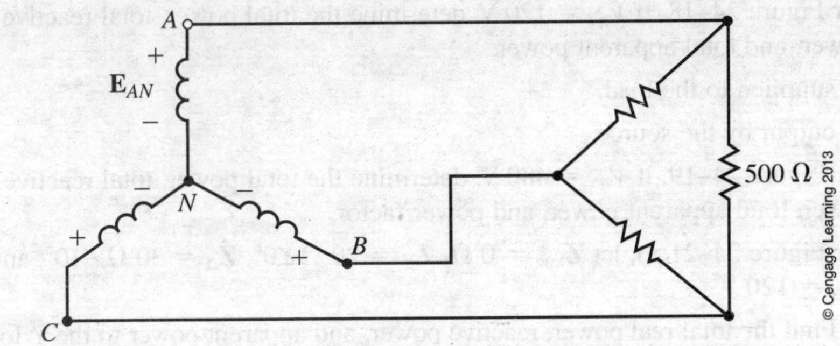

FIGURE 24–51

35. For Figure 24–47, compute the total real, reactive, and apparent power using the P_T, Q_T, and S_T formulas of Table 24–2. (Use $V_L = 207.8$ V rather than the nominal value of 208 V here.) Compare these to the results of Problem 34.

36. For Figure 24–48, $E_{AB} = 208$ volts.
   a. Determine per-phase real, reactive, and apparent power.
   b. Multiply per-phase quantities by 3 to get total quantities.

37. For Figure 24–48, $E_{AB} = 208$ V. Compute the total real, reactive, and apparent power using the P_T, Q_T, and S_T formulas of Table 24–2. Compare to the results of Problem 36.

38. For Figure 24–52, if $V_{an} = 277$ V, determine the total active power, total reactive power, total apparent power, and power factor.

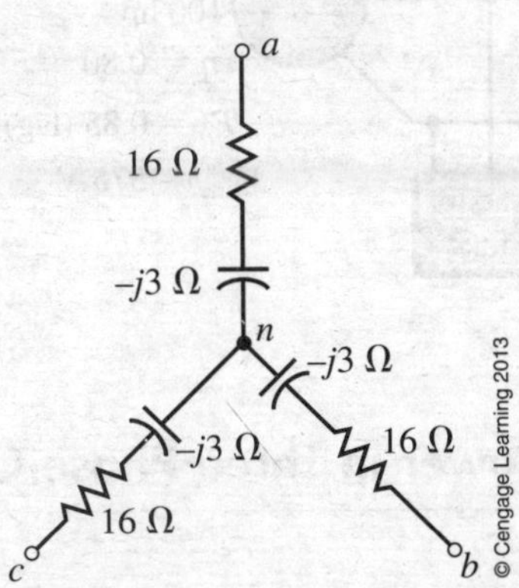

FIGURE 24–52

39. For Figure 24–53, if $V_{ab} = 600$ V, determine the total power, total reactive power, total apparent power, and power factor.

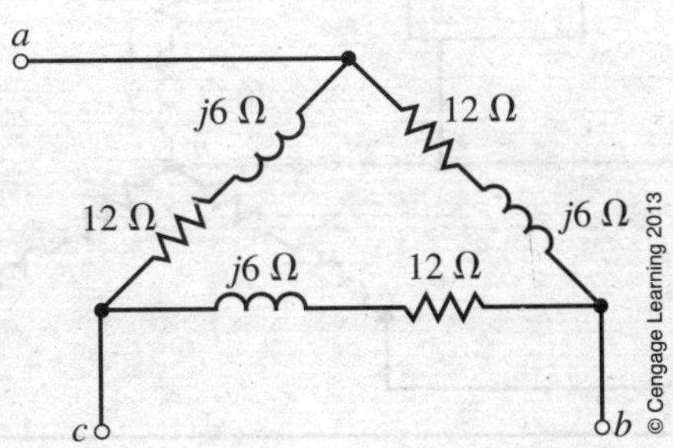

FIGURE 24–53

40. For Figure 24–18, if $V_{an} = 120$ V, determine the total power, total reactive power, and total apparent power,
    a. supplied to the load.
    b. output by the source.
41. For Figure 24–19, if $V_{ab} = 480$ V, determine the total power, total reactive power, total apparent power, and power factor.
42. For Figure 24–21(a), let $\mathbf{Z}_{\text{line}} = 0\ \Omega$, $\mathbf{Z}_{\text{Y}} = 20\ \Omega\angle 0°$, $\mathbf{Z}_{\Delta} = 30\ \Omega\angle 10°$, and $E_{AN} = 120$ V.
    a. Find the total real power, reactive power, and apparent power to the Y load.
    b. Repeat (a) for the Δ load.
    c. Determine total watts, VARs, and VA using the results of (a) and (b).
43. $V_{ab} = 208$ V for a balanced Y load, $P_{\text{T}} = 1200$ W, and $Q_{\text{T}} = 750$ VAR (ind.). Choose $\mathbf{V}_{an}$ as reference and determine $\mathbf{I}_a$. (Use power triangle.)
44. A motor whose efficiency is 80% delivers 100 hp to a load. A bank of power factor capacitors is connected as in Figure 24–54. The capacitor bank (which is balanced) is rated $Q_C = 45$ kVAR (total). Reduce the problem to its single-phase equivalent, then compute the resultant power factor of the system.
45. The capacitors of Figure 24–54 are connected in Y and each has the value $C = 120$ μF. Compute the resultant power factor. Frequency is 60 Hz.

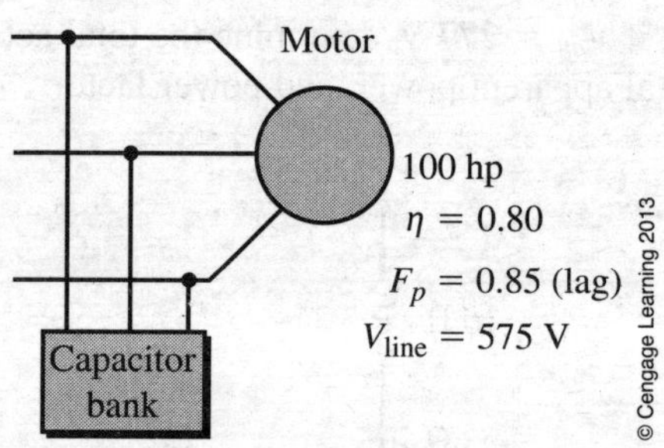

FIGURE 24–54

24.6 Measuring Power in Three-Phase Circuits

46. For Figure 24–55:
    a. Determine the wattmeter reading.
    b. If the load is balanced, what is P_{T}?

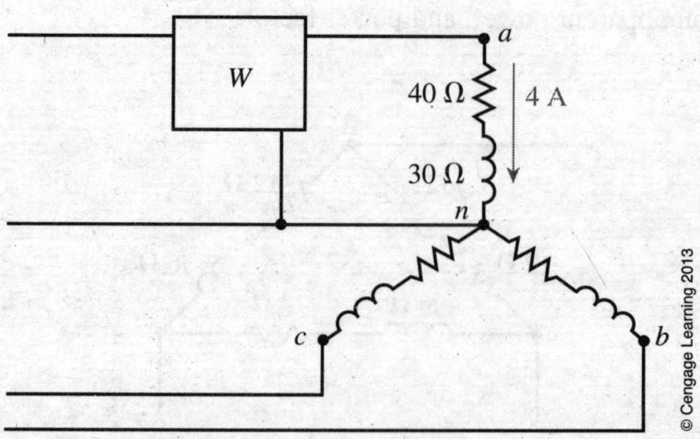

FIGURE 24–55

47. For Figure 24–47, the generator phase voltage is 120 volts.
    a. Sketch three wattmeters correctly into the circuit.
    b. Compute the reading of each wattmeter.
    c. Sum the readings and compare this to the result of 2304 W obtained in Problem 34.
48. Figures 24–30 and 24–31 show two ways that two wattmeters may be connected to measure power in a three-phase, three-wire circuit. There is one more way. Sketch it.
49. For the circuit of Figure 24–56, $\mathbf{V}_{ab} = 208\ \text{V}\angle 30°$.
    a. Determine the magnitude and angle of the currents.
    b. Determine power per phase and total power, P_T.
    c. Compute the reading of each wattmeter.
    d. Sum the meter readings and compare the result to P_T of (b).
50. Two wattmeters measure power to a balanced load. Readings are $P_h = 1000$ W and $P_\ell = -400$ W. Determine the power factor of the load from Equation 24–24 and from Figure 24–34. How do they compare?
51. Consider the circuit of Figure 24–56.
    a. Compute the power factor from the angle of the phase impedances.
    b. In Problem 49, wattmeter readings $P_h = 1164$ W and $P_\ell = 870$ W were determined. Substitute these values into Equation 24–24 and compute the power factor of the load. Compare the results to (a).

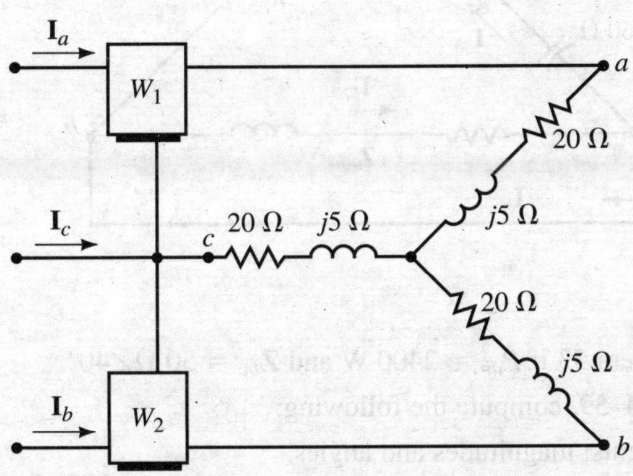

FIGURE 24–56

52. For the balanced load of Figure 24–57, $\mathbf{V}_{ab} = 208\ \text{V}\angle 0°$.
    a. Compute phase currents and line currents.
    b. Determine power per phase and total power P_T.
    c. Compute the reading of each wattmeter, then sum the wattmeter readings and compare to P_T from (b).

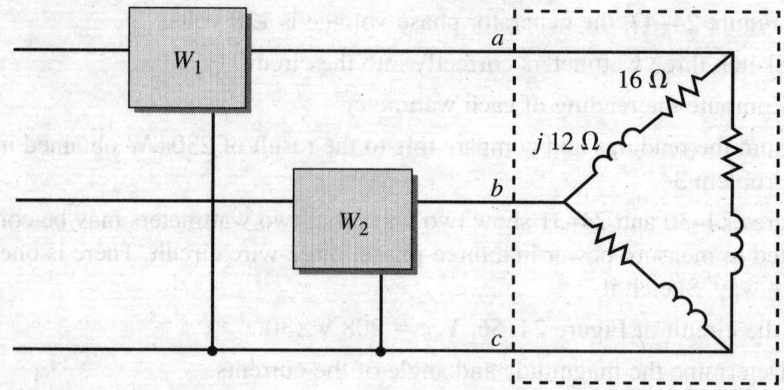

FIGURE 24–57

24.7 Unbalanced Loads

53. For Figure 24–58, $R_{ab} = 60\ \Omega$, $\mathbf{Z}_{bc} = 80\ \Omega + j60\ \Omega$. Compute the following:
    a. Phase and line currents.
    b. Power to each phase and total power.

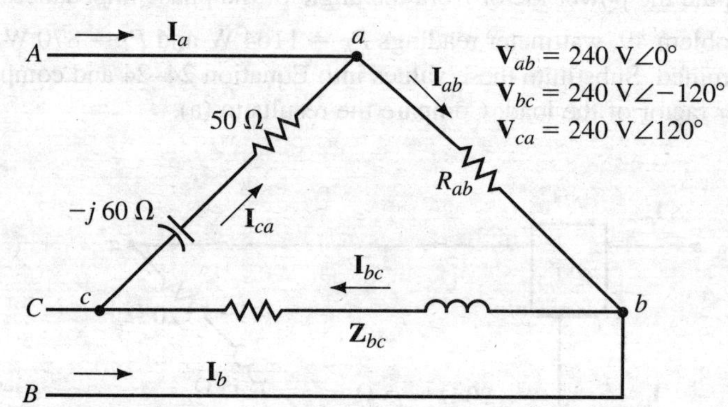

FIGURE 24–58

54. Repeat Problem 53 if $P_{ab} = 2400$ W and $\mathbf{Z}_{bc} = 50\ \Omega\angle 40°$.
55. For Figure 24–59, compute the following:
    a. Line currents, magnitudes and angles.
    b. The neutral current.
    c. Power to each phase.
    d. Total power to the load.

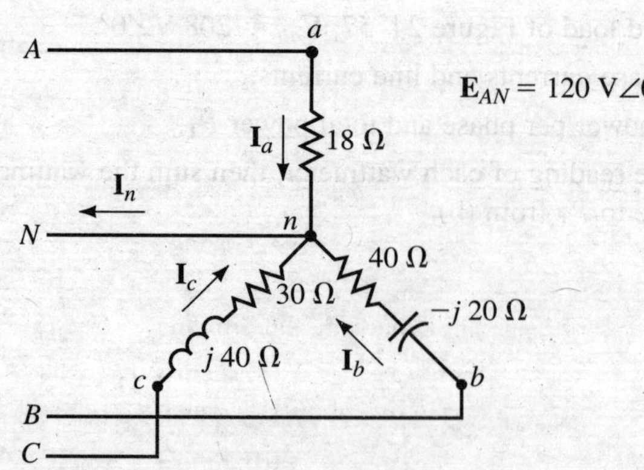

FIGURE 24–59

56. Remove the neutral conductor from the circuit of Figure 24–59 and compute the line currents. *Hint:* Use mesh equations.
57. From Problem 56, $\mathbf{I}_a = 1.94\ \text{A}\angle -0.737°$, $\mathbf{I}_b = 4.0\ \text{A}\angle -117.7°$, and $\mathbf{I}_c = 3.57\ \text{A}\angle 91.4°$. Compute the following:
    a. Voltages across each phase of the load.
    b. Voltage between the neutral of the load and the neutral of the generator.

24.9 Circuit Analysis Using Computers

For each of the following problems, use either Multisim or PSpice. With Multisim, you get only magnitude; with PSpice, solve for magnitude and angle. Caution: For PSpice, insert the IPRINT devices so that current enters the positive terminal. Otherwise, the phase angle will be in error by 180°. (See Figure 24–44.)

58. For the balanced system of Figure 24–47, let $\mathbf{E}_{AN} = 347\ \text{V}\angle 15°$, $L = 8.95$ mH, and $f = 160$ Hz. Solve for line currents. Multisim PSPICE
59. For the balanced system of Figure 24–48, let $\mathbf{E}_{AN} = 277\ \text{V}\angle -30°$, $C = 50\ \mu\text{F}$, and $f = 212$ Hz. Solve for phase currents and line currents. Multisim PSPICE
60. Repeat Problem 59 with C replaced with $L = 11.26$ mH. Multisim PSPICE
61. For Figure 24–59, let $L = 40$ mH, $C = 50\ \mu\text{F}$, and $\omega = 1000$ rad/s. Solve for line and neutral currents. Multisim PSPICE

ANSWERS TO IN-PROCESS LEARNING CHECKS

IN-PROCESS LEARNING CHECK 1

1. $\mathbf{V}_{an} = \mathbf{E}_{AN} = 277\ \text{V}\angle -20°$; $\mathbf{V}_{bn} = \mathbf{E}_{BN} = 277\ \text{V}\angle -140°$;
   $\mathbf{V}_{cn} = \mathbf{E}_{CN} = 277\ \text{V}\angle 100°$; $\mathbf{V}_{ab} = \mathbf{E}_{AB} = 480\ \text{V}\angle 10°$;
   $\mathbf{V}_{bc} = \mathbf{E}_{BC} = 480\ \text{V}\angle -110°$; $\mathbf{V}_{ca} = \mathbf{E}_{CA} = 480\ \text{V}\angle 130°$
2. $\mathbf{V}_{an} = \mathbf{E}_{AN} = 120\ \text{V}\angle 50°$; $\mathbf{V}_{bn} = \mathbf{E}_{BN} = 120\ \text{V}\angle -70°$;
   $\mathbf{V}_{cn} = \mathbf{E}_{CN} = 120\ \text{V}\angle 170°$; $\mathbf{V}_{ab} = \mathbf{E}_{AB} = 208\ \text{V}\angle 80°$;
   $\mathbf{V}_{bc} = \mathbf{E}_{BC} = 208\ \text{V}\angle -40°$; $\mathbf{V}_{ca} = \mathbf{E}_{CA} = 208\ \text{V}\angle -160°$
3. $\mathbf{V}_{an} = 124\ \text{V}\angle -1.87°$; $\mathbf{V}_{ab} = 214\ \text{V}\angle 28.13°$
4. $\mathbf{V}_{ab} = 104\ \text{V}\angle 81.7°$; $\mathbf{V}_{bc} = 104\ \text{V}\angle -38.3°$;
   $\mathbf{V}_{ca} = 104\ \text{V}\angle -158.3°$

■ KEY TERMS

Fourier Series
Frequency Spectrum
Fundamental Frequency
Harmonic Frequency
Spectrum Analyzer

■ OUTLINE

Composite Waveforms
Fourier Series
Fourier Series of Common Waveforms
Frequency Spectrum
Circuit Response to a Nonsinusoidal Waveform
Circuit Analysis Using Computers

■ OBJECTIVES

After studying this chapter, you will be able to

- solve for the coefficients of the Fourier series of a simple periodic waveform, using integration,
- use tables to write the Fourier equivalent of any simple periodic waveform,
- sketch the frequency spectrum of a periodic waveform, giving the amplitudes of various harmonics in either volts, watts, or dBm,
- calculate the power dissipated when a complex waveform is applied to a resistive load,
- determine the output of a filter circuit given the frequency spectrum of the input signal and the frequency response of the filter,
- use PSpice to observe the actual response of a filter circuit to a nonsinusoidal input signal.

25

NONSINUSOIDAL WAVEFORMS

CHAPTER PREVIEW

In our analysis of ac circuits, we have dealt primarily with sinusoidal waveforms. Although the sinusoidal wave is the most common waveform in electronic circuits, it is by no means the only type of signal used in electronics. In previous chapters we observed how sinusoidal signals were affected by the characteristics of components within a circuit. For instance, if a 1-kHz sinusoid is applied to a low-pass filter circuit having a cutoff frequency of 3 kHz, we know that the signal appearing at the output of the filter will be essentially the same as the signal applied to the input. This effect is illustrated in Figure 25–1.

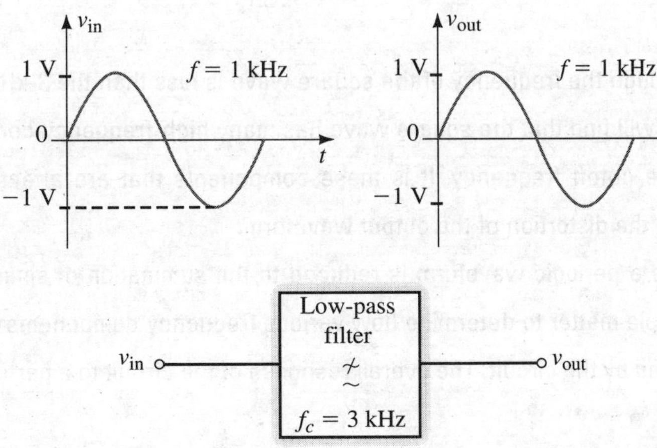

FIGURE 25–1

One would naturally expect that the 1-kHz low-pass filter would allow any other 1-kHz signal to pass from input to the output without being distorted. Unfortunately, this is not the case.

In this chapter we will find that any periodic waveform is composed of numerous sinusoidal waveforms, each of which has a unique amplitude and frequency. As we have already seen, circuits such as the low-pass filter and the resonant tank circuit do not allow all sinusoidal frequencies to pass from the input to the output in the same manner. As a result, the output signal may be dramatically different from the signal applied at the input. For example, if we were to apply a 1-kHz square wave to a low-pass filter having a 3-kHz cutoff, the output would appear as shown in Figure 25–2.

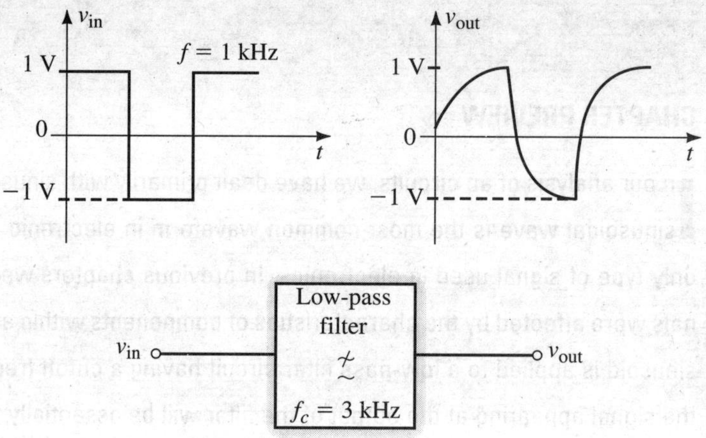

FIGURE 25–2

Although the frequency of the square wave is less than the 3-kHz cutoff frequency of the filter, we will find that the square wave has many high-frequency components which are well above the cutoff frequency. It is these components that are affected in the filter, and this results in the distortion of the output waveform.

Once a periodic waveform is reduced to the summation of sinusoidal waveforms, it is a fairly simple matter to determine how various frequency components of the original signal will be affected by the circuit. The overall response of the circuit to a particular waveform can then be found. ■

Putting It in Perspective

Jean Baptiste Joseph Fourier

FOURIER WAS BORN IN AUXERRE, Yonne, France on March 21, 1768. As a youth, Fourier reluctantly studied for the priesthood in the monastery of Saint-Benoit-sur Loire. However, his interest was in mathematics. In 1798, Fourier accompanied Napoleon to Egypt, where he was made a governor. After returning to France, Fourier was particularly interested in the study of heat transfer between two points of different temperature. He was appointed joint secretary of the Academy of Sciences in 1822.

In 1807, Fourier announced the discovery of a theorem, which made him famous. Fourier's theorem states that any periodic waveform can be written as the summation of a series of simple sinusoidal functions.

Using his theorem, Fourier was able to develop important theories in heat transfer, which were published in 1822 in a book titled *Analytic Theory of Heat.*

Although still used to describe heat transfer, Fourier's theorem is used today to predict how filters and various other electronic circuits operate when subjected to a nonsinusoidal periodic function.

Fourier died in Paris on May 16, 1830 as the result of a fall down the stairs. ■

25.1 Composite Waveforms

Any waveform that is made up of two or more separate waveforms is called a composite waveform. Most signals appearing in electronic circuits are comprised of complicated combinations of dc levels and sinusoidal waves. Consider the circuit and signal shown in Figure 25–3.

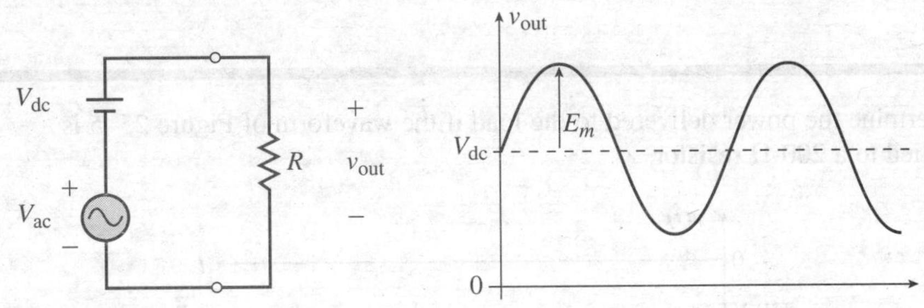

FIGURE 25–3

The voltage appearing across the load is determined by superposition as the combination of the ac source in series with a dc source. The result is a sine wave with a dc offset. As one might expect, when a composite wave is applied to a load resistor, the resulting power is determined by considering the effects of both signals. The rms voltage of the composite waveform is determined as

$$V_{rms} = \sqrt{V_{dc}^2 + V_{ac}^2} \qquad (25\text{–}1)$$

where V_{ac} is the rms value of the ac component of waveform and is found as $V_{ac} = \frac{E_m}{\sqrt{2}}$. The power delivered to a load will be determined simply as

$$P_{load} = \frac{V_{rms}^2}{R_{load}}$$

The following example illustrates this principle.

EXAMPLE 25–1

Determine the power delivered to the load if the waveform of Figure 25–4 is applied to a 500-Ω resistor.

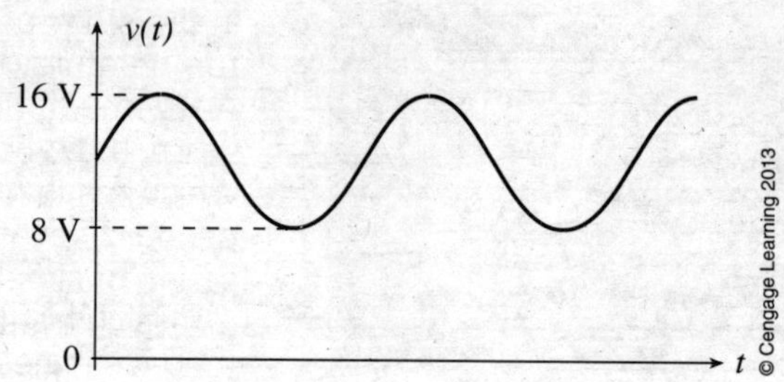

FIGURE 25–4

Solution By examining the waveform, we see that the average value is $V_{dc} =$ 12 V and the peak value of the sinusoidal is $V_m = 16\text{ V} - 12\text{ V} = 4\text{ V}$. The rms value of the sinusoidal waveform is determined to be $V_{ac} = (0.707)(4\text{ V}) =$ 2.83 V. Now we find the rms value of the composite wave as

$$\begin{aligned} V_{rms} &= \sqrt{(12\text{ V})^2 + (2.83\text{ V})^2} \\ &= \sqrt{152\text{ V}^2} \\ &= 12.3\text{ V} \end{aligned}$$

and so the power delivered to the load is

$$P_{load} = \frac{(12.3\text{ V})^2}{500\ \Omega} = 0.304\text{ W}$$

PRACTICE PROBLEMS 1

Determine the power delivered to the load if the waveform of Figure 25–5 is applied to a 200-Ω resistor.

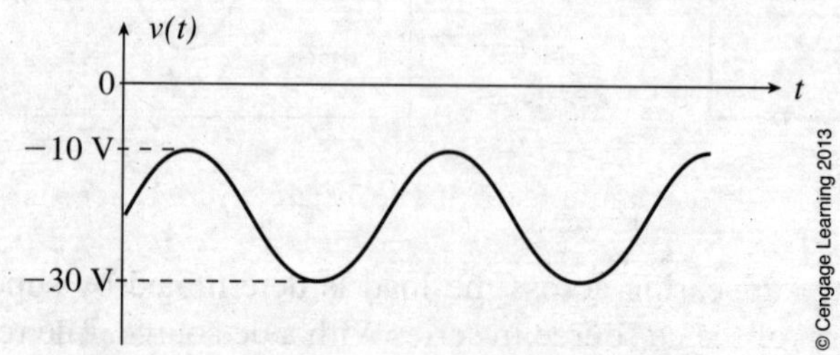

FIGURE 25–5

Answer
2.25 W

25.2 Fourier Series

In 1826, Baron Jean Baptiste Joseph Fourier developed a branch of mathematics that is used to express any periodic waveform as an infinite series of sinusoidal waveforms. Although it seems as though we are turning a simple waveform into a more complicated form, you will find that the resulting expression actually simplifies the analysis of many circuits which respond differently to signals of various frequencies. Using Fourier analysis, any periodic waveform can be written as a summation of sinusoidal waveforms as follows:

$$f(t) = a_0 + a_1 \cos \omega t + a_2 \cos 2\omega t + \cdots + a_n \cos n\omega t + \cdots + b_1 \sin \omega t + b_2 \sin 2\omega t + \cdots + b_n \sin n\omega t + \cdots \quad \text{(25–2)}$$

The coefficients of the individual terms of the **Fourier series** are found by integrating the original function over one complete period. The coefficients are determined as

$$a_0 = \frac{1}{T}\int_{t_1}^{t_1 + T} f(t)\, dt \quad \text{(25–3)}$$

$$a_n = \frac{2}{T}\int_{t_1}^{t_1 + T} f(t)\cos n\omega t\, dt \quad \text{(25–4)}$$

$$b_n = \frac{2}{T}\int_{t_1}^{t_1 + T} f(t)\sin n\omega t\, dt \quad \text{(25–5)}$$

Notice that Equation 25–2 indicates that the Fourier series of a periodic function may contain both a sine and a cosine component at each frequency. These individual components can be combined to give a single sinusoidal expression as follows:

$$a_n \cos nx + b_n \sin nx = a_n \sin(nx + 90°) + b_n \sin nx = c_n \sin(nx + \theta)$$

where

$$c_n = \sqrt{a_n^2 + b_n^2} \quad \text{(25–6)}$$

and

$$\theta = \tan^{-1}\left(\frac{a_n}{b_n}\right) \quad \text{(25–7)}$$

Therefore, the Fourier equivalent of any periodic waveform may be simplified as follows:

$$f(t) = a_0 + c_1 \sin(\omega t + \theta_1) + c_2 \sin(2\omega t + \theta_2) + \cdots$$

The a_0 term is a constant that corresponds to the average value of the periodic waveform, and the c_n coefficients give the amplitudes of the various sinusoidal terms. Notice that the first sinusoidal term ($n = 1$) has the same frequency as the original waveform. This component is referred to as the **fundamental frequency** of the given waveform. All other frequencies are integer multiples of the fundamental frequency and are called the **harmonic frequencies.** When $n = 2$, the resulting term is called the second harmonic; when $n = 3$, we have the third harmonic, and so on. Using Equation 25–3 to Equation 25–7, it is possible to derive the Fourier series for any periodic function.

NOTES...

For Further Investigation

For additional information, go to cengagebrain.com and log in by entering your user name and password. Follow the links to *For Further Investigation* and select *Fourier Series.*

EXAMPLE 25–2

Write the Fourier series for the pulse waveform shown in Figure 25–6.

$$v(t) = \begin{cases} 1: 0 < t < \dfrac{T}{2} \\ 0: \dfrac{T}{2} < t < T \end{cases}$$

FIGURE 25–6

Solution The various coefficients are calculated by integrating as follows:

$$a_0 = \frac{1}{T}\int_0^{T/2} (1)\,dt + \frac{1}{T}\int_{T/2}^{T} (0)\,dt = \frac{1}{2}$$

$$a_n = \frac{2}{T}\int_0^{T/2} (1)\cos n\omega t\,dt + \frac{2}{T}\int_{T/2}^{T} (0)\,dt$$

$$= \frac{2}{T}\left[\left(\frac{1}{n\omega}\right)\sin n\omega t\right]_0^{T/2} = \frac{1}{n\pi}\sin n\pi = 0$$

Notice that all $a_n = 0$ since $\sin n\pi = 0$ for all n.

$$b_1 = \frac{2}{T}\int_0^{T/2} (1)\sin \omega t\,dt + \frac{2}{T}\int_{T/2}^{T} (0)\,dt$$

$$= \frac{2}{T}\left[-\left(\frac{1}{\omega}\right)\cos \omega t\right]_0^{T/2} = -\frac{1}{\pi}\left[\cos\left(\frac{2\pi t}{T}\right)\right]_0^{T/2}$$

$$= -\frac{1}{\pi}[(-1) - (1)] = \frac{2}{\pi}$$

$$b_2 = \frac{2}{T}\int_0^{T/2} (1)\sin 2\omega t\,dt + \frac{2}{T}\int_{T/2}^{T} (0)\,dt$$

$$= \frac{2}{T}\left[-\left(\frac{1}{2\omega}\right)\cos 2\omega t\right]_0^{T/2}$$

$$= -\frac{1}{2\pi}\left[\cos\left(\frac{4\pi t}{T}\right)\right]_0^{T/2}$$

$$= -\frac{1}{2\pi}[(1) - (1)] = 0$$

$$b_3 = \frac{2}{T}\int_0^{T/2} (1)\sin 3\omega t\,dt + \frac{2}{T}\int_{T/2}^{T} (0)\,dt$$

$$= \frac{2}{T}\left[-\left(\frac{1}{3\omega}\right)\cos 3\omega t\right]_0^{T/2}$$

$$= -\frac{1}{3\pi}\left[\cos\left(\frac{6\pi t}{T}\right)\right]_0^{T/2}$$

$$= -\frac{1}{3\pi}[(-1) - (1)] = \frac{2}{3\pi}$$

For all odd values of n, we have $b_n = 2/n\pi$ since $\cos n\pi = -1$. Even values of n give $b_n = 0$ since $\cos n\pi = 1$.

The general expression of the Fourier series for the given pulse wave is therefore written as

$$v(t) = \frac{1}{2} + \frac{2}{\pi}\sum_{1}^{\infty}\frac{\sin n\omega t}{n} \qquad n = 1, 3, 5, \ldots \tag{25–8}$$

By examining the general expression for the pulse wave of Figure 25–6, we can observe several important characteristics. The Fourier series confirms that the wave shown has an average value of $a_0 = 0.5$. In addition, the pulse wave has only odd harmonics. In other words, a pulse wave having a frequency of 1 kHz would have harmonic components occurring at 3 kHz, 5 kHz, and so on. Although the given wave consists of an infinite number of sinusoidal components, the amplitudes of successive terms decrease as *n* increases.

If we were to consider only the first four nonzero frequency components of the pulse wave, we would have the following expression:

$$v(t) = 0.5 + \frac{2}{\pi}\sin \omega t + \frac{2}{3\pi}\sin 3\omega t + \frac{2}{5\pi}\sin 5\omega t + \frac{2}{7\pi}\sin 7\omega t \tag{25–9}$$

The graphical representation of the preceding expression is shown in Figure 25–7.

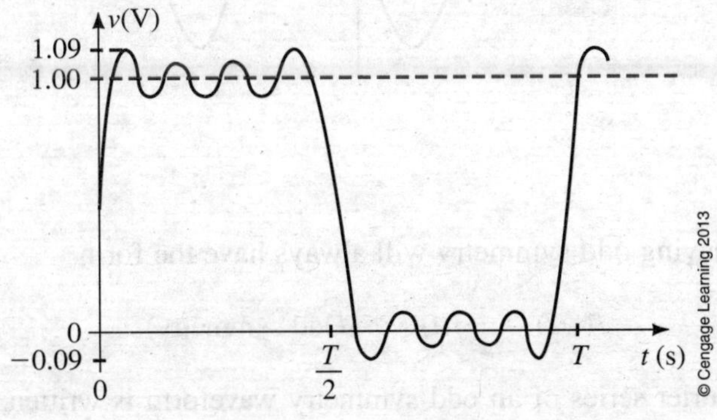

FIGURE 25–7

Although this waveform is not identical to the given pulse wave, we see that the first four nonzero harmonics provide a reasonable approximation of the original waveform.

Derivations of Fourier series for certain waveforms are simplified due to symmetry that occurs in the waveform. We will examine three types of symmetry: even, odd, and half-wave symmetry. Each type of symmetry results in consistent patterns in the Fourier series. The waveforms of Figure 25–8 are symmetrical around the vertical axis and are said to have *even symmetry* (or *cosine symmetry*).

Waveforms having even symmetry will always have the form

$$f(-t) = f(t) \qquad \text{(Even symmetry)} \tag{25–10}$$

When the Fourier series of an even symmetry waveform is written, it will only contain cosine (a_n) terms and possibly an a_0 term. All sine (b_n) terms will be zero.

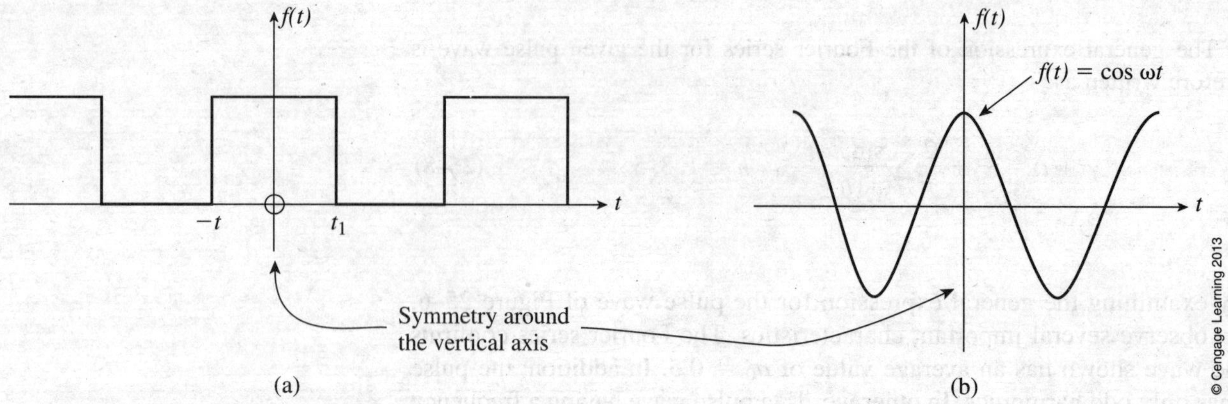

FIGURE 25–8 Even symmetry (cosine symmetry).

If the portion of the waveform to the right of the vertical axis in each signal of Figure 25–9 is rotated 180°, we find that it will exactly overlap the portion of the waveform to the left of the axis. Such waveforms are said to have *odd symmetry.*

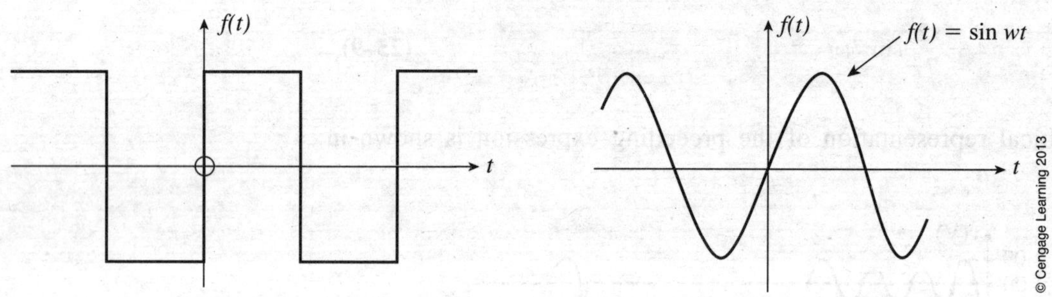

FIGURE 25–9 Odd symmetry (sine symmetry).

Waveforms having odd symmetry will always have the form

$$f(-t) = -f(t) \qquad \text{(Odd symmetry)} \tag{25–11}$$

When the Fourier series of an odd symmetry waveform is written, it will only contain sine (b_n) terms and possibly an a_0 term. All cosine (a_n) terms will be zero.

If the portion of the waveform below the horizontal axis in Figure 25–10 is the mirror image of the portion above the axis, the waveform is said to have *half-wave symmetry.*

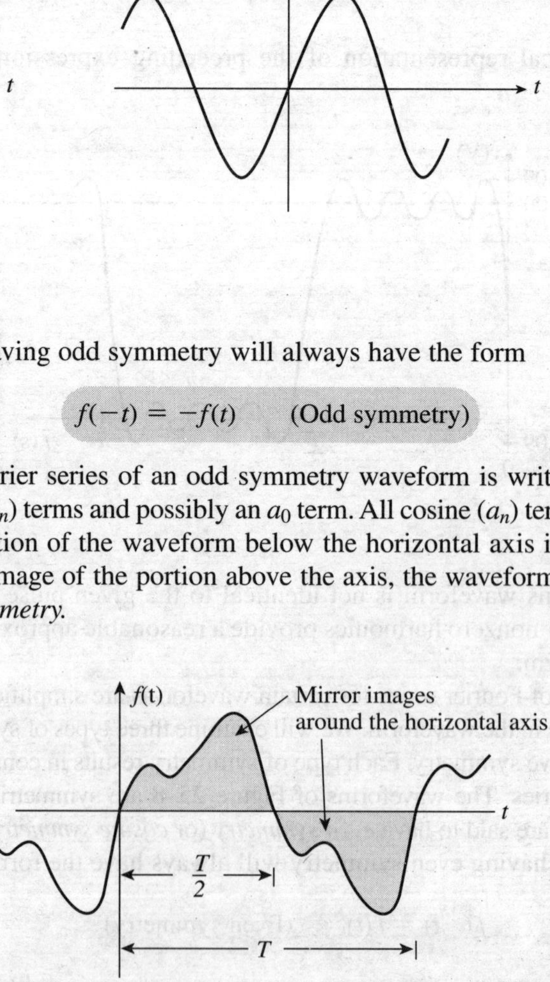

FIGURE 25–10 Half-wave symmetry.

Waveforms having half-wave symmetry will always have the form

$$f(t + T) = -f(t) \qquad \text{(Half-wave symmetry)} \qquad (25\text{–}12)$$

When the Fourier series of a half-wave symmetry waveform is written, it will have only odd harmonics and possibly an a_0 term. All even harmonic terms will be zero.

If we refer back to the waveform of Figure 25–6, we see that it has both odd symmetry and half-wave symmetry. Using the preceding rules, we would expect to find only sine terms and odd harmonics. Indeed, we see that Equation 25–9 has these conditions.

PRACTICE PROBLEMS 2

Consider the ramp function shown in Figure 25–11.

a. Does this waveform show symmetry?
b. Use calculus to determine the Fourier expression for $v(t)$.
c. Verify that the a_0 term of the Fourier series is equal to the average value of the waveform.
d. From the Fourier expression, is the ramp function made up of odd harmonics, even harmonics, or all harmonic components? Briefly justify your answer.

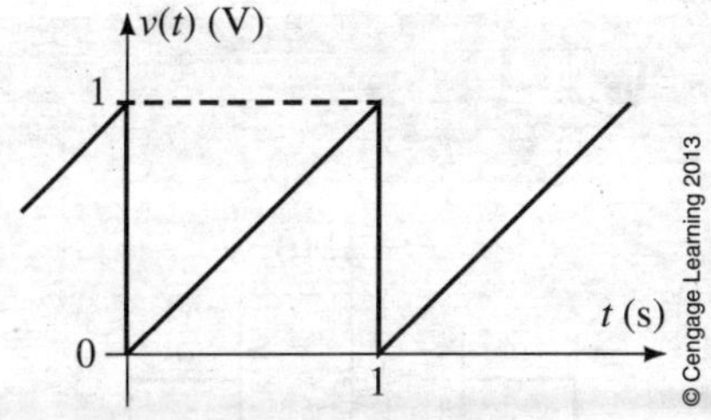

FIGURE 25–11

Answers

a. Odd symmetry
b. $v(t) = 0.5 - \dfrac{1}{\pi}\sin(2\pi t) - \dfrac{1}{2\pi}\sin(4\pi t) - \dfrac{1}{3\pi}\sin(6\pi t) \cdots$
c. $a_0 = 0.5$ V
d. All harmonic components are present since the function does not have half-wave symmetry.

IN-PROCESS LEARNING CHECK 1

(Answers are at the end of the chapter.)

Without using calculus, determine a method of rewriting the expression of Equation 25–9 to represent a square wave having an amplitude of 1 V as illustrated in Figure 25–12. *Hint:* Notice that the square wave is similar to the pulse waveform with the exception that its average value is zero and that the peak-to-peak value is twice that of the pulse wave.

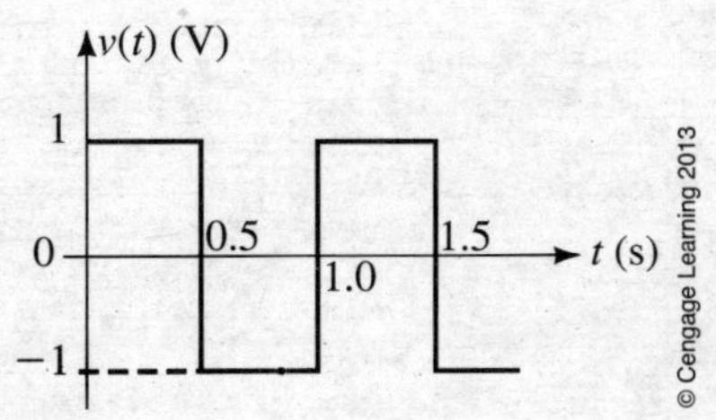

FIGURE 25–12

25.3 Fourier Series of Common Waveforms

All periodic waveforms can be converted into their Fourier equivalent series by using integration as shown in Section 25.2. Integration of common waveforms is time-consuming and prone to error. A simpler approach is to use tables such as Table 25–1, which gives the Fourier series of several common waveforms encountered in electrical circuits.

TABLE 25–1 Fourier Equivalents of Common Waveforms ($\omega = 2\pi/T$)

Waveform	Fourier Equivalent
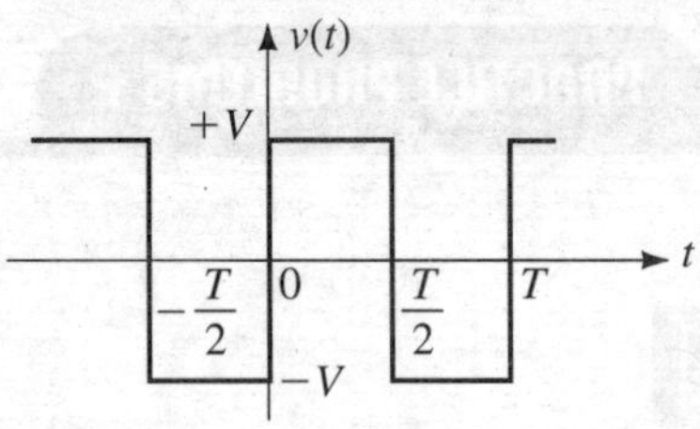 $V(t) = \begin{cases} -V, -\frac{T}{2} < t < 0 \\ +V, 0 < t < \frac{T}{2} \end{cases}$ FIGURE 25–13	$v(t) = \frac{4V}{\pi}\left(\sum_{1}^{\infty} \frac{\sin n\omega t}{n}\right) \quad n = 1, 3, 5, \ldots$
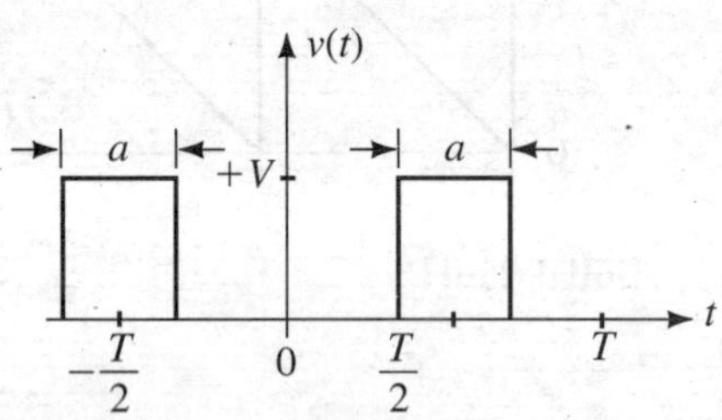 $V(t) = \begin{cases} 0, -\frac{T}{2} + a/2 < t < \frac{T}{2} - a/2 \\ \text{V}, \frac{T}{2} - a/2 < t < \frac{T}{2} + a/2 \end{cases}$ FIGURE 25–14	$v(t) = \frac{aV}{T} + \frac{2V}{\pi}\left(\sum_{1}^{\infty} (-1)^n \frac{\sin\left(\frac{\pi n a}{T}\right)\cos n\omega t}{n}\right) \quad n = 1, 2, 3, \ldots$
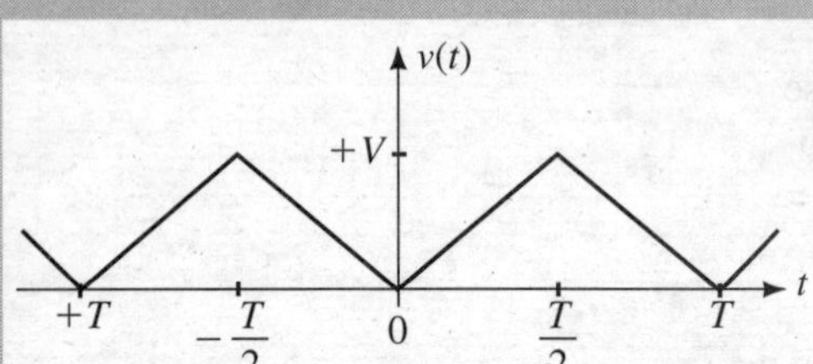 $V(t) = \begin{cases} -\frac{2Vt}{T}, -\frac{T}{2} < t < 0 \\ \frac{2Vt}{T}, 0 < t < \frac{T}{2} \end{cases}$ FIGURE 25–15	$v(t) = \frac{V}{2} - \frac{4V}{\pi^2}\left(\sum_{1}^{\infty} \frac{\cos n\omega t}{n^2}\right) \quad n = 1, 3, 5, \ldots$

TABLE 25–1 Fourier Equivalents of Common Waveforms ($\omega = 2\pi/T$) *(continued)*

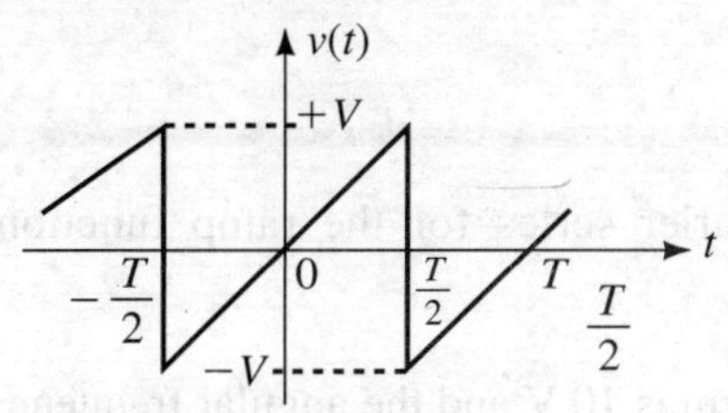

$$V(t) = \frac{2Vt}{T},\ -\frac{T}{2} < t < \frac{T}{2}$$

FIGURE 25–16

$$v(t) = -\frac{2V}{\pi}\left(\sum_{1}^{\infty} (-1)^n \frac{\sin n\omega t}{n}\right) \qquad n = 1, 2, 3, \ldots$$

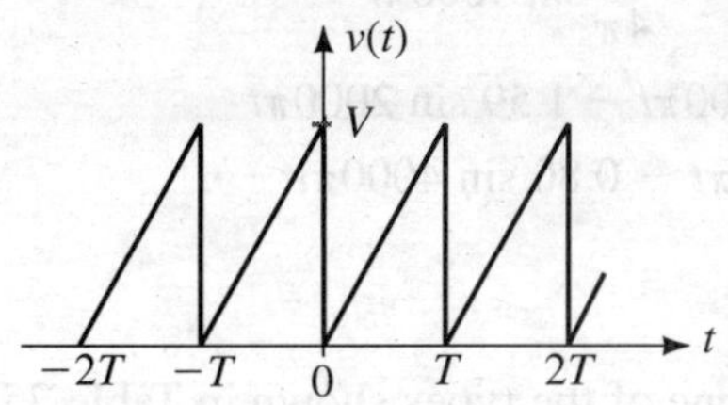

$$V(t) = \frac{Vt}{T},\ 0 < t < T$$

FIGURE 25–17

$$v(t) = \frac{V}{2} - \frac{V}{\pi}\left(\sum_{1}^{\infty} \frac{\sin n\omega t}{n}\right) \qquad n = 1, 2, 3, \ldots$$

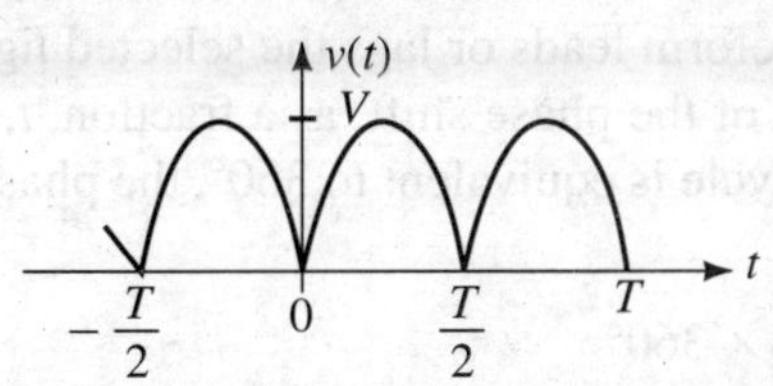

$$V(t) = \left|V \sin \omega t\right|,\ -\frac{T}{2} < t < \frac{T}{2}$$

FIGURE 25–18

$$v(t) = \frac{2V}{\pi} - \frac{4V}{\pi}\left(\frac{\cos 2\omega t}{1 \cdot 3} + \frac{\cos 4\omega t}{3 \cdot 5} + \frac{\cos 6\omega t}{5 \cdot 7} + \cdots\right)$$

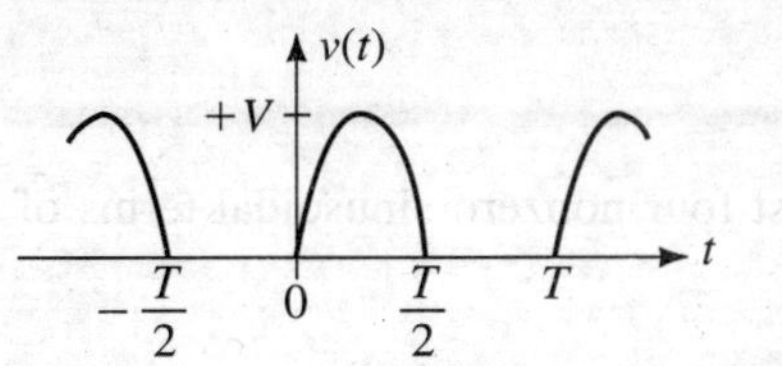

$$v(t) = \begin{cases} 0,\ -\dfrac{T}{2} < t < 0 \\ V \sin \omega t,\ 0 < t < \dfrac{T}{2} \end{cases}$$

FIGURE 25–19

$$v(t) = \frac{V}{\pi} + \frac{V}{2}\sin \omega t - \frac{2V}{\pi}\left(\frac{\cos 2\omega t}{1 \cdot 3} + \frac{\cos 4\omega t}{3 \cdot 5} + \frac{\cos 6\omega t}{5 \cdot 7} + \cdots\right)$$

The following example illustrates how a given waveform is converted into its Fourier series equivalent.

EXAMPLE 25–3

Use Table 25–1 to determine the Fourier series for the ramp function of Figure 25–20.

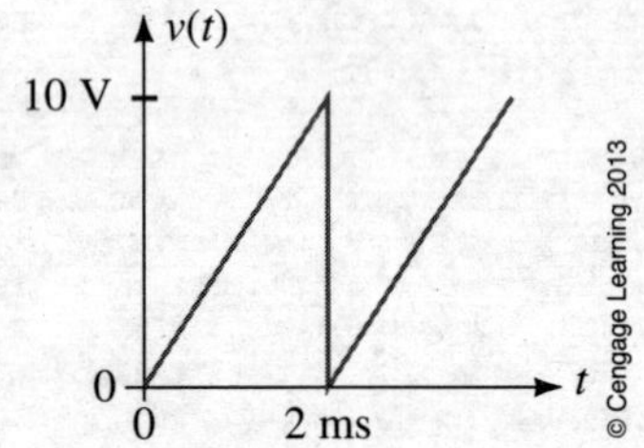

FIGURE 25–20

Solution The amplitude of the waveform is 10 V and the angular frequency of the fundamental is $\omega = 2\pi/(2 \text{ ms}) = 1000\pi$ rad/s. The resulting series is determined from Table 25–1 as

$$\begin{aligned} v(t) &= \frac{10}{2} - \frac{10}{\pi}\sin 1000\pi t - \frac{10}{2\pi}\sin 2000\pi t \\ &\quad - \frac{10}{3\pi}\sin 3000\pi t - \frac{10}{4\pi}\sin 4000\pi t - \cdots \\ &= 5 - 3.18 \sin 1000\pi t - 1.59 \sin 2000\pi t \\ &\quad - 1.06 \sin 3000\pi t - 0.80 \sin 4000\pi t - \cdots \end{aligned}$$

If a given waveform is similar to one of the types shown in Table 25–1 but is shifted along the time axis, it is necessary to include a phase shift with each of the sinusoidal terms. The phase shift is determined as follows:

1. Determine the period of the given waveform.
2. Compare the given waveform with the figures appearing in Table 25–1 and select which of the waveforms in the table best describes the given wave.
3. Determine whether the given waveform leads or lags the selected figure of Table 25–1. Calculate the amount of the phase shift, as a fraction, t, of the total period. Since one complete cycle is equivalent to 360°, the phase shift is determined as

$$\phi = \frac{t}{T} \times 360^\circ$$

4. Write the resulting Fourier expression for the given waveform. If the given waveform leads the selected figure of Table 25–1, then add the angle ϕ to each term. If the given waveform lags the selected figure, then subtract the angle ϕ from each term.

EXAMPLE 25–4

Write the Fourier expression for the first four nonzero sinusoidal terms of the waveform shown in Figure 25–21.

Solution

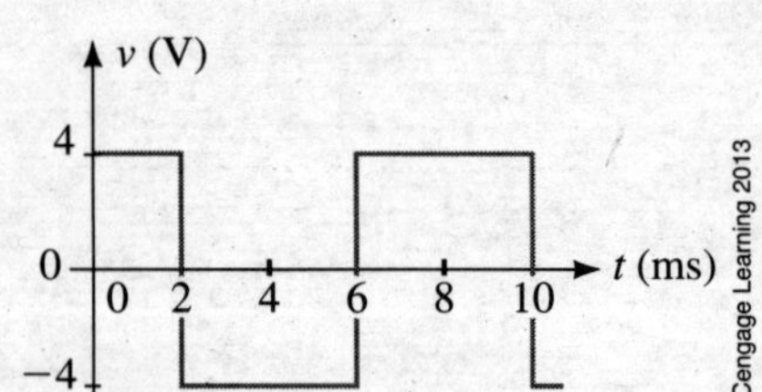

FIGURE 25–21

Step 1: The period of the given waveform is $T = 8.0$ ms, which gives a frequency of $f = 125$ Hz or an angular frequency of $\omega = 250\pi$ rad/s.

Step 2: From Table 25–1, we see that the given waveform is similar to the square wave of Figure 25–13.

Step 3: The waveform of Figure 25–21 leads the square wave of Figure 25–13 by an amount equivalent to $t = 2$ ms. This corresponds to a phase shift of

$$\phi = \frac{2 \text{ ms}}{8 \text{ ms}} \times 360^\circ = 90^\circ$$

Step 4: The Fourier expression for the first four terms of the waveform of Figure 25–21 is now written as follows:

$$v(t) = \frac{4(4)}{\pi}\sin(250\pi t + 90°) + \frac{4(4)}{3\pi}\sin[3(250\pi t + 90°)]$$
$$+ \frac{4(4)}{5\pi}\sin[5(250\pi t + 90°)] + \frac{4(4)}{7\pi}\sin[7(250\pi t + 90°)]$$

The preceding expression may be left as the summation of sine waves. However, since the cosine wave leads the sine wave by 90°, the expression may be simplified as a summation of cosine waves without any phase shift. This results in the following:

$$v(t) = 5.09 \cos 250\pi t - 1.70 \cos 750\pi t + 1.02 \cos 1250\pi t - 0.73 \cos 1750\,\pi t$$

PRACTICE PROBLEMS 3

Write the Fourier expression for the first four nonzero sinusoidal terms of the waveform shown in Figure 25–22. Express each term as a sine wave rather than as a cosine wave.

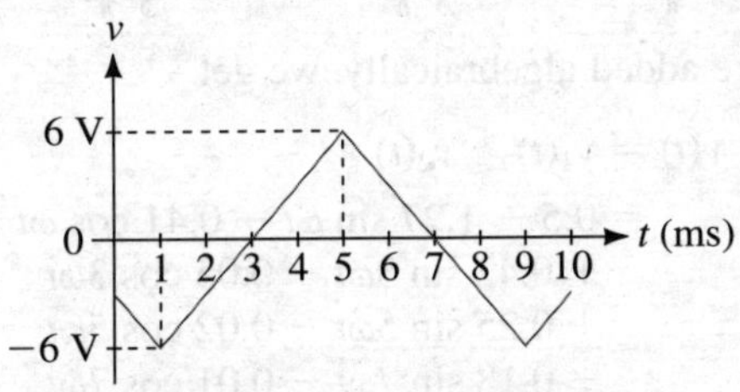

FIGURE 25–22

Answer

$$v(t) = \frac{48}{\pi^2}\sin(250\pi t - 135°) + \frac{48}{3^2\pi^2}\sin(750\pi t + 135°)$$
$$+ \frac{48}{5^2\pi^2}\sin(1250\pi t + 45°) + \frac{48}{7^2\pi^2}\sin(1750\pi t - 45°)$$

The waveforms of Table 25–1 provide most of the commonly observed waveforms. Occasionally, however, a particular waveform consists of a combination of several simple waveforms. In such a case, it is generally easiest if we first redraw the original waveform as the summation of two or more recognizable waveforms. Then the Fourier series of each of the individual component waves is determined. Finally, the resultant is expressed as the summation of the two series.

EXAMPLE 25–5

Write the first four nonzero sinusoidal terms of the Fourier series for the waveform of Figure 25–23.

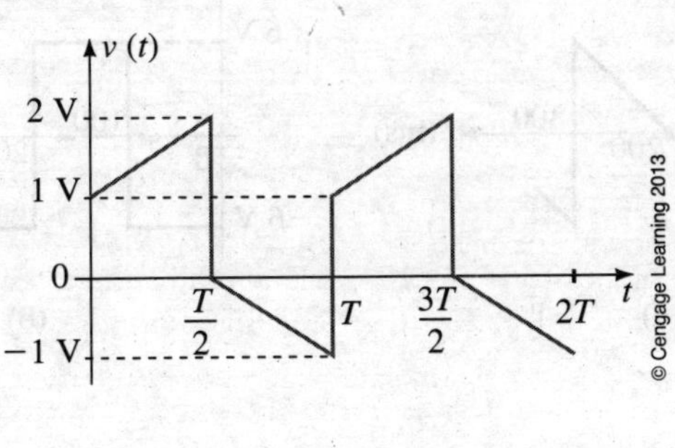

FIGURE 25–23

Solution The waveform of Figure 25–23 is made up of a combination of waves as illustrated in Figure 25–24.

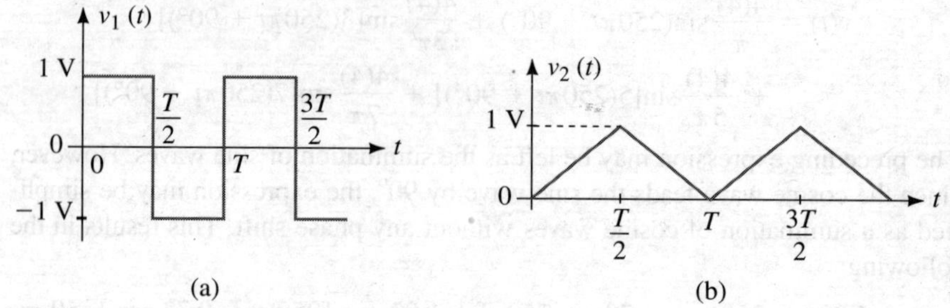

FIGURE 25–24

The Fourier series of each of the waveforms is determined from Table 25–1 as

$$v_1(t) = \frac{4}{\pi}\sin \omega t + \frac{4}{3\pi}\sin 3\omega t + \frac{4}{5\pi}\sin 5\omega t + \cdots$$

and

$$v_2(t) = \frac{1}{2} - \frac{4}{\pi^2}\cos \omega t - \frac{4}{3^2\pi^2}\cos 3\omega t - \frac{4}{5^2\pi^2}\cos 5\omega t - \cdots$$

When these series are added algebraically, we get

$$\begin{aligned} v(t) &= v_1(t) + v_2(t) \\ &= 0.5 + 1.27 \sin \omega t - 0.41 \cos \omega t \\ &\quad + 0.42 \sin 3\omega t - 0.05 \cos 3\omega t \\ &\quad + 0.25 \sin 5\omega t - 0.02 \cos 5\omega t \\ &\quad + 0.18 \sin 7\omega t - 0.01 \cos 7\omega t \\ &\quad \vdots \end{aligned}$$

The preceding series may be further simplified by using Equations 25–6 and 25–7 to provide a single coefficient and phase shift for each frequency. The resultant waveform is written as

$$\begin{aligned} v(t) &= 0.5 + 1.34 \sin(\omega t - 17.7°) + 0.43 \sin(3\omega t - 6.1°) \\ &\quad + 0.26 \sin(5\omega t - 3.6°) + 0.18 \sin(7\omega t - 2.6°) \end{aligned}$$

PRACTICE PROBLEMS 4

A composite waveform is made up of the summation of the waveforms illustrated in Figure 25–25.

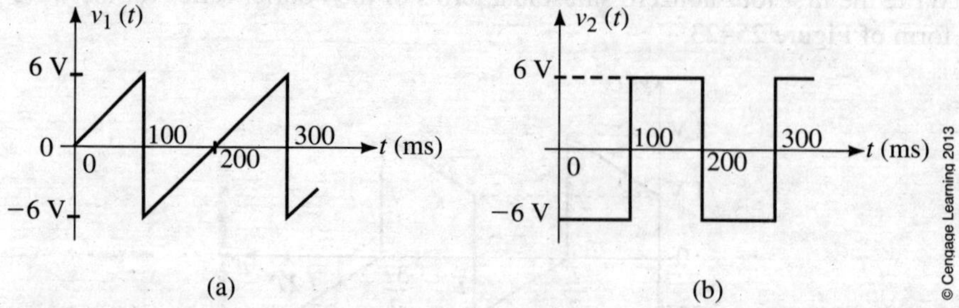

FIGURE 25–25

a. Sketch the composite waveform showing all voltage levels and time values.
b. Write the Fourier expression for the resultant waveform, $v(t)$.

Answers

a.

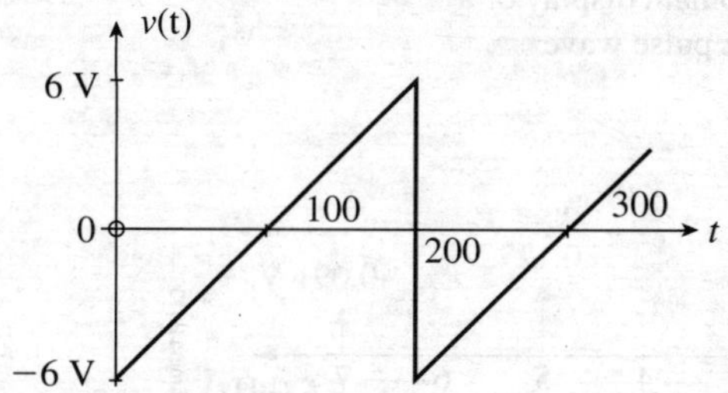

FIGURE 25–26

b. $v(t) = \frac{12}{\pi}\sin(10\pi t) - \frac{36}{2\pi}\sin(20\pi t) + \frac{12}{3\pi}\sin(30\pi t) - \cdots$

IN-PROCESS LEARNING CHECK 2

(Answers are at the end of the chapter.)

Consider the composite waveform of Figure 25–27.

a. Separate the given waveform into two recognizable waveforms appearing in Table 25–1.
b. Use Table 25–1 to write the Fourier series for each of the component waveforms of (a).
c. Combine the results to determine the Fourier series for $v(t)$.

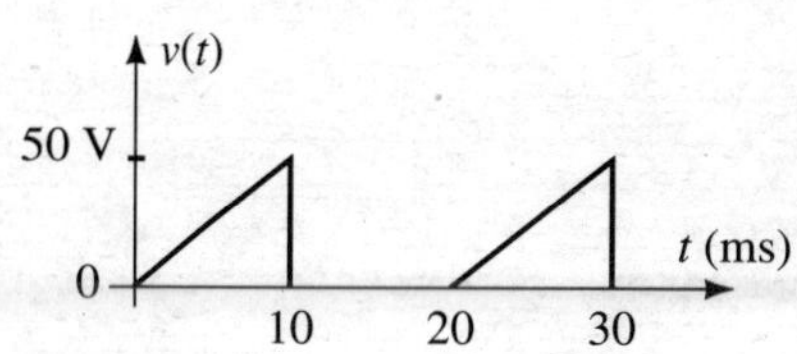

FIGURE 25–27

25.4 Frequency Spectrum

Most waveforms that we have observed were generally shown as a function of time. However, they may also be shown as a function of frequency. In such cases, the amplitude of each harmonic is indicated at the appropriate frequency. Figure 25–28 shows the display of a 1-kHz sine wave in both the time domain and the frequency domain, while Figure 25–29 shows the corresponding displays for a pulse waveform.

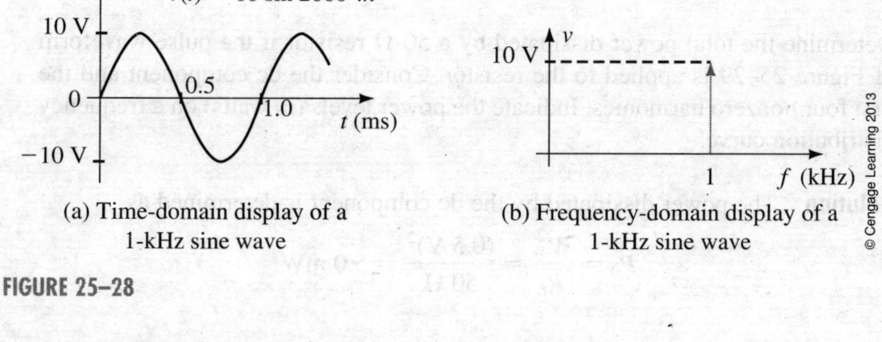

(a) Time-domain display of a 1-kHz sine wave

(b) Frequency-domain display of a 1-kHz sine wave

FIGURE 25–28

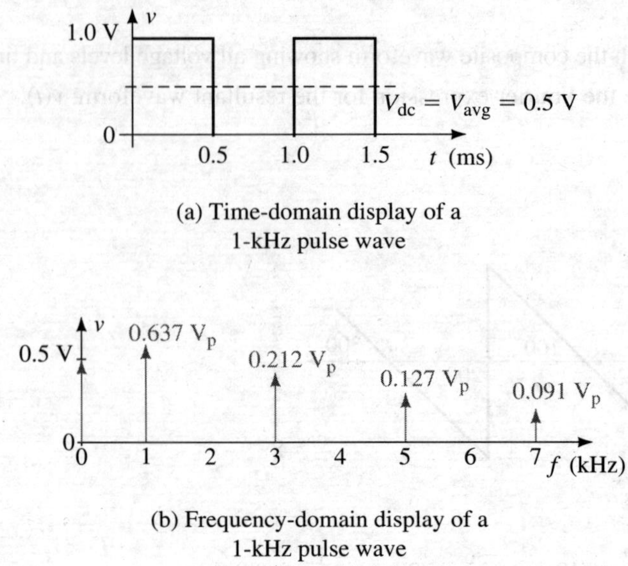

FIGURE 25–29

The **frequency spectrum** of the pulse wave shows the average value (or dc value) of the wave at a frequency of 0 kHz and illustrates the absence of even harmonics. Notice that the amplitude of successive harmonic components decreases fairly quickly.

The rms voltage of the composite waveform of Figure 25–29 is determined by considering the rms value of each frequency. The resultant rms voltage is found as

$$V_{rms} = \sqrt{V_{dc}^2 + V_1^2 + V_2^2 + V_3^2 + \cdots}$$

where each voltage, V_1, V_2, and so on, represents the rms value of the corresponding harmonic component. Using the first five nonzero terms, we would find the rms value of the pulse wave of Figure 25–29 to be

$$V_{rms} = \sqrt{(0.500)^2 + \left(\frac{0.637}{\sqrt{2}}\right)^2 + \left(\frac{0.212}{\sqrt{2}}\right)^2 + \left(\frac{0.127}{\sqrt{2}}\right)^2 + \left(\frac{0.091}{\sqrt{2}}\right)^2}$$
$$= 0.698 \text{ V}$$

This value is only slightly smaller than the actual value of $V_{rms} = 0.707$ V.

If the pulse wave were applied to a resistive element, power would be dissipated as if each frequency component had been applied independently. The total power can be determined by the summation of the individual contributions of each frequency. In order to calculate the power dissipated at each sinusoidal frequency, we need to first convert the voltages into rms values. The frequency spectrum may then be represented in terms of power rather than as voltage.

EXAMPLE 25–6

Determine the total power dissipated by a 50-Ω resistor if the pulse waveform of Figure 25–29 is applied to the resistor. Consider the dc component and the first four nonzero harmonics. Indicate the power levels (in watts) on a frequency distribution curve.

Solution The power dissipated by the dc component is determined as

$$P_0 = \frac{V_0^2}{R_L} = \frac{(0.5 \text{ V})^2}{50\ \Omega} = 5.0 \text{ mW}$$

The power dissipated by any resistor subjected to a sinusoidal frequency is determined as

$$P = \frac{V_{rms}^2}{R_L} = \frac{\left(\frac{V_P}{\sqrt{2}}\right)^2}{R_L} = \frac{V_P^2}{2R_L}$$

For the pulse wave of Figure 25–29, the power due to each of the first four nonzero sinusoidal components is found as follows:

$$P_1 = \frac{\left(\frac{2\text{ V}}{\pi}\right)^2}{(2)(50\ \Omega)} = 4.05\text{ mW}$$

$$P_3 = \frac{\left(\frac{2\text{ V}}{3\pi}\right)^2}{(2)(50\ \Omega)} = 0.45\text{ mW}$$

$$P_5 = \frac{\left(\frac{2\text{ V}}{5\pi}\right)^2}{(2)(50\ \Omega)} = 0.16\text{ mW}$$

$$P_7 = \frac{\left(\frac{2\text{ V}}{7\pi}\right)^2}{(2)(50\ \Omega)} = 0.08\text{ mW}$$

Figure 25–30 shows the power levels (in milliwatts) as a function of frequency.

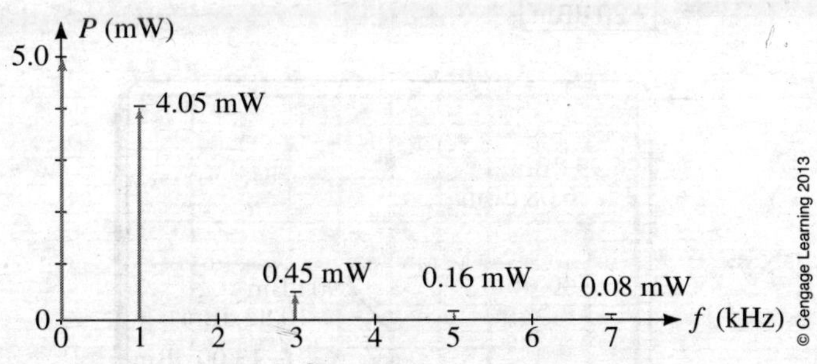

FIGURE 25–30

Using only the dc component and the first four nonzero harmonics, the total power dissipated by the resistor is $P_T = 9.74$ mW. From Chapter 15, the actual rms voltage of the pulse waveform is found to be

$$V_{rms} = \sqrt{\frac{(1\text{ V})^2(0.5\text{ ms})}{1.0\text{ ms}}} = 0.707\text{ V}$$

Therefore, using the rms voltage, the power dissipated by the resistor is found as

$$P = \frac{(0.707\text{ V})^2}{50\ \Omega} = 10.0\text{ mW}$$

Although the pulse waveform has power contained in components with frequencies above the seventh harmonic, we see that more than 97% of the total power of a pulse waveform is contained in only the first seven harmonics.

Power levels and frequencies of the various harmonics of a periodic waveform may be measured with an instrument called a **spectrum analyzer,** shown in Figure 25–31.

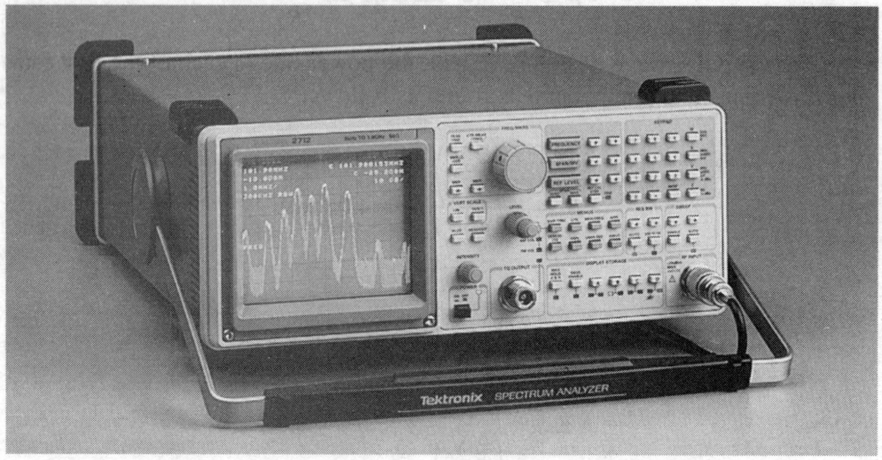

FIGURE 25–31 Spectrum analyzer.

Some spectrum analyzers are able to display either voltage levels or power levels in the frequency domain, while most others display only power levels (in dBm). When displaying power levels, the spectrum analyzer usually uses a reference 50-Ω load. Figure 25–32 shows the display of a 1.0-V pulse waveform as it would appear on a typical spectrum analyzer.

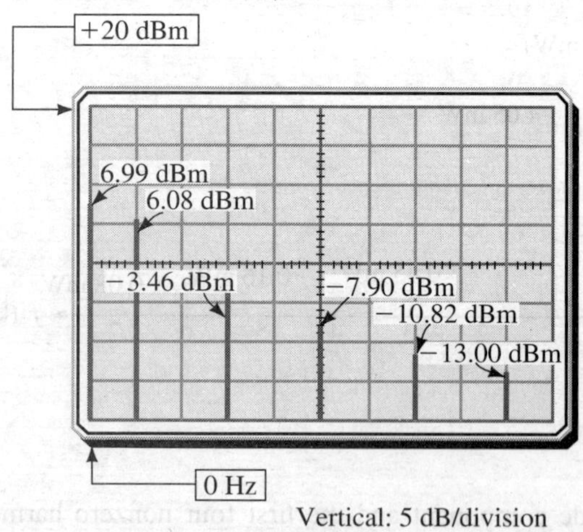

Vertical: 5 dB/division
Horizontal: 1 kHz/division
Reference: +20 dBm

FIGURE 25–32 Spectrum analyzer display of a 1-kHz pulse wave having a peak value of 1 V.

PRACTICAL NOTES...

Spectrum analyzers are very sensitive instruments. As a result, great care must be taken to ensure that the input power never exceeds the rated maximum. When in doubt, it is best to insert an extra attenuator to lower the amount of power entering the spectrum analyzer.

Notice that the spectrum analyzer has a reference of +20 dBm and that this reference is shown at the top of the display rather than at the bottom. The vertical scale of the instrument is measured in decibels, where each vertical division corresponds to 5 dB. The horizontal axis for the spectrum analyzer is scaled in hertz, where each division in Figure 25–32 corresponds to 1 kHz.

EXAMPLE 25–7

A spectrum analyzer with a 50-Ω input is used to display the power levels in dBm of the Fourier series components of the ramp waveform shown in Figure 25–33.

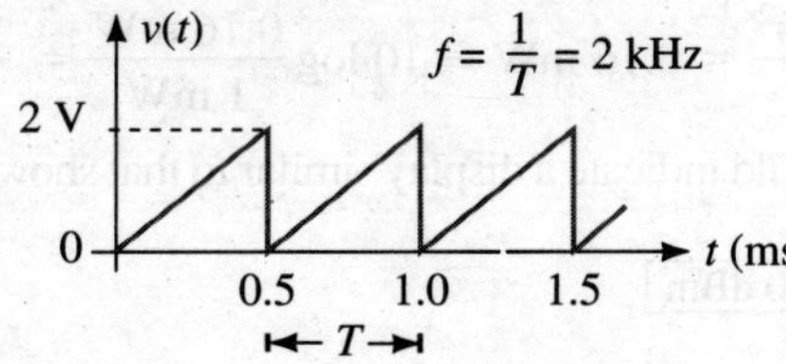

FIGURE 25–33

Determine the voltage and power levels of the various components and sketch the resultant display as it would appear on a spectrum analyzer. Assume that the spectrum analyzer has the same vertical and horizontal settings as those shown in Figure 25–32.

Solution The Fourier series of the given waveform is determined from Table 25–1 as

$$v(t) = \frac{2}{2} - \frac{2}{\pi}\sin \omega t - \frac{2}{2\pi}\sin 2\omega t - \frac{2}{3\pi}\sin 3\omega t - \cdots$$

Because the fundamental frequency occurs at $f = 2$ kHz, we see that harmonic frequencies will occur at 4 kHz, 6 kHz, and so on. However, because the spectrum analyzer is able to display only up to 10 kHz, we need go no further.

The dc component will have an average value of $v_0 = 1.0$ V, as expected. The rms values of the harmonic sinusoidal waveforms are determined as

$$V_{\text{rms}} = \frac{V_p}{\sqrt{2}}$$

which gives the following:

$$V_{1(\text{rms})} = \frac{2}{\pi\sqrt{2}} = 0.450 \text{ V}$$

$$V_{2(\text{rms})} = \frac{2}{2\pi\sqrt{2}} = 0.225 \text{ V}$$

$$V_{3(\text{rms})} = \frac{2}{3\pi\sqrt{2}} = 0.150 \text{ V}$$

$$V_{4(\text{rms})} = \frac{2}{4\pi\sqrt{2}} = 0.113 \text{ V}$$

$$V_{5(\text{rms})} = \frac{2}{5\pi\sqrt{2}} = 0.090 \text{ V}$$

The preceding rms voltages are used to calculate the powers (and power levels in dBm) of the various harmonic components.

$$P_0 = \frac{(1.0 \text{ V})^2}{50 \text{ }\Omega} = 20.0 \text{ mW} \equiv 10 \log \frac{20 \text{ mW}}{1 \text{ mW}} = 13.0 \text{ dBm}$$

$$P_1 = \frac{(0.450 \text{ V})^2}{50 \text{ }\Omega} = 4.05 \text{ mW} \equiv 10 \log \frac{4.05 \text{ mW}}{1 \text{ mW}} = 6.08 \text{ dBm}$$

$$P_2 = \frac{(0.225 \text{ V})^2}{50 \text{ }\Omega} = 1.01 \text{ mW} \equiv 10 \log \frac{1.01 \text{ mW}}{1 \text{ mW}} = 0.04 \text{ dBm}$$

$$P_3 = \frac{(0.150\text{ V})^2}{50\ \Omega} = 0.45\text{ mW} \equiv 10\log\frac{0.45\text{ mW}}{1\text{ mW}} = -3.5\text{ dBm}$$

$$P_4 = \frac{(0.113\text{ V})^2}{50\ \Omega} = 0.25\text{ mW} \equiv 10\log\frac{0.25\text{ mW}}{1\text{ mW}} = -6.0\text{ dBm}$$

$$P_5 = \frac{(0.090\text{ V})^2}{50\ \Omega} = 0.16\text{ mW} \equiv 10\log\frac{0.16\text{ mW}}{1\text{ mW}} = -7.9\text{ dBm}$$

A spectrum analyzer would indicate a display similar to that shown in Figure 25–34.

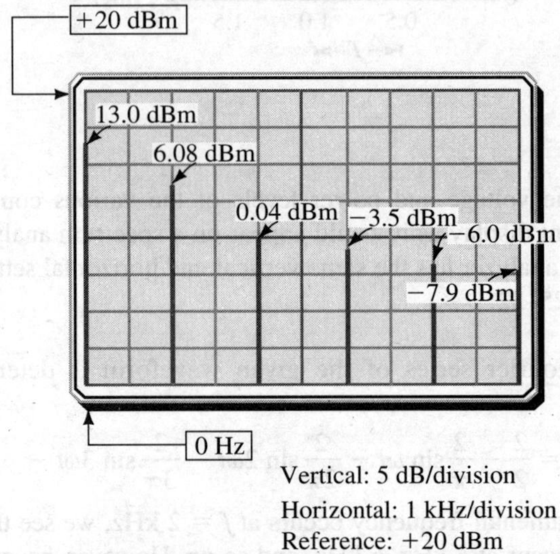

FIGURE 25–34

PRACTICE PROBLEMS 5

The sawtooth waveform of Figure 25–35 is applied to a 50-Ω spectrum analyzer. Sketch the display that would be observed, assuming that the spectrum analyzer has the same vertical and horizontal settings as those shown in Figure 25–32.

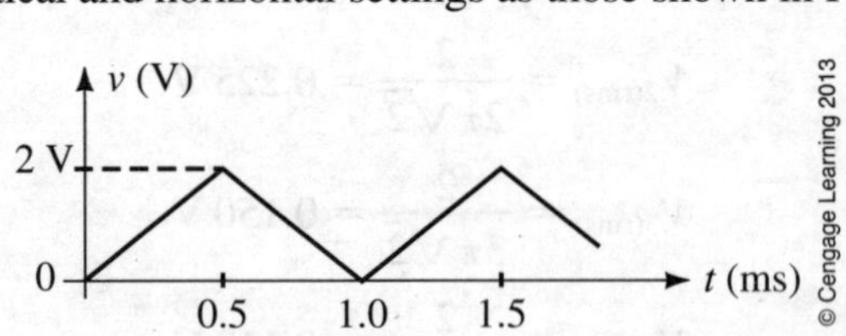

FIGURE 25–35

Answers

$P_{dc} = +13.0$ dBm, $P_{1kHz} = +8.2$ dBm, $P_{3kHz} = -10.9$ dBm, $P_{5kHz} = -19.8$ dBm. (All other components are less than −20 dBm and so will not appear.)

25.5 Circuit Response to a Nonsinusoidal Waveform

We have determined that all periodic, nonsinusoidal waveforms are comprised of numerous sinusoidal components together with a dc component. In Chapter 22 we observed how various frequencies were affected when they were applied to a given filter. We will now examine how the frequency components of a waveform will be modified when applied to the input of a given filter.

Consider what happens when a pulse waveform is applied to a band-pass filter tuned to the third harmonic, as shown in Figure 25–36.

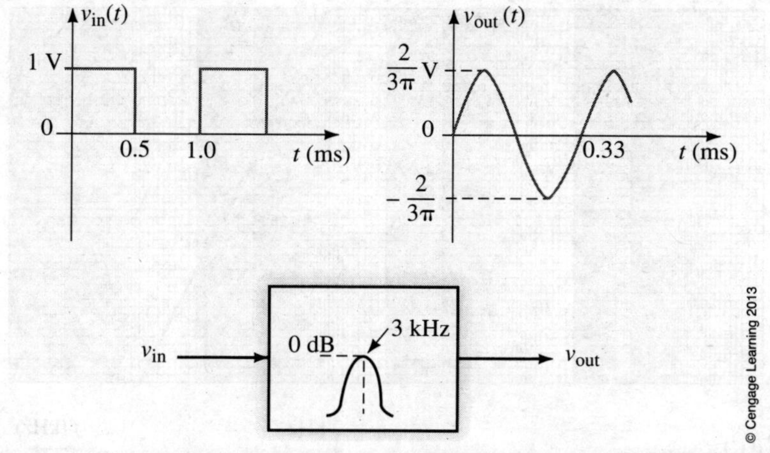

FIGURE 25–36

Since the band-pass filter is tuned to the third harmonic, only the frequency component corresponding to this harmonic will be passed from the input of the filter to the output. Further, because the filter has a voltage gain of 0 dB at the center frequency, the amplitude of the resulting sinusoidal output will be the same voltage level as the amplitude of the original third harmonic. All other frequencies, including the dc component, will be attenuated by the filter, and so are effectively eliminated from the output.

This method is used extensively in electronic circuits to provide frequency multiplication, since any distorted waveform will be rich in harmonics. The desired frequency component is easily extracted by using a tuned filter circuit. Although any integer multiplication is theoretically possible, most frequency multiplier circuits are either frequency doublers or frequency triplers since higher-order harmonics have much lower amplitudes.

In order to determine the resulting waveform after it passes through any other filter, it is necessary to determine the amplitude and phase shift of numerous harmonic components.

EXAMPLE 25–8

The circuit of Figure 25–37 has the frequency response shown in Figure 25–38.

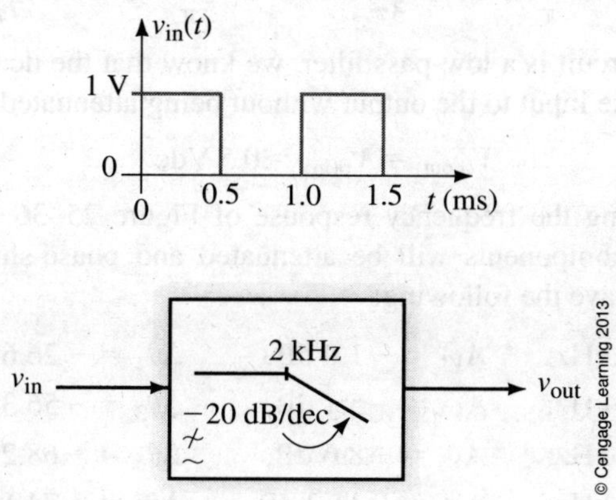

FIGURE 25–37

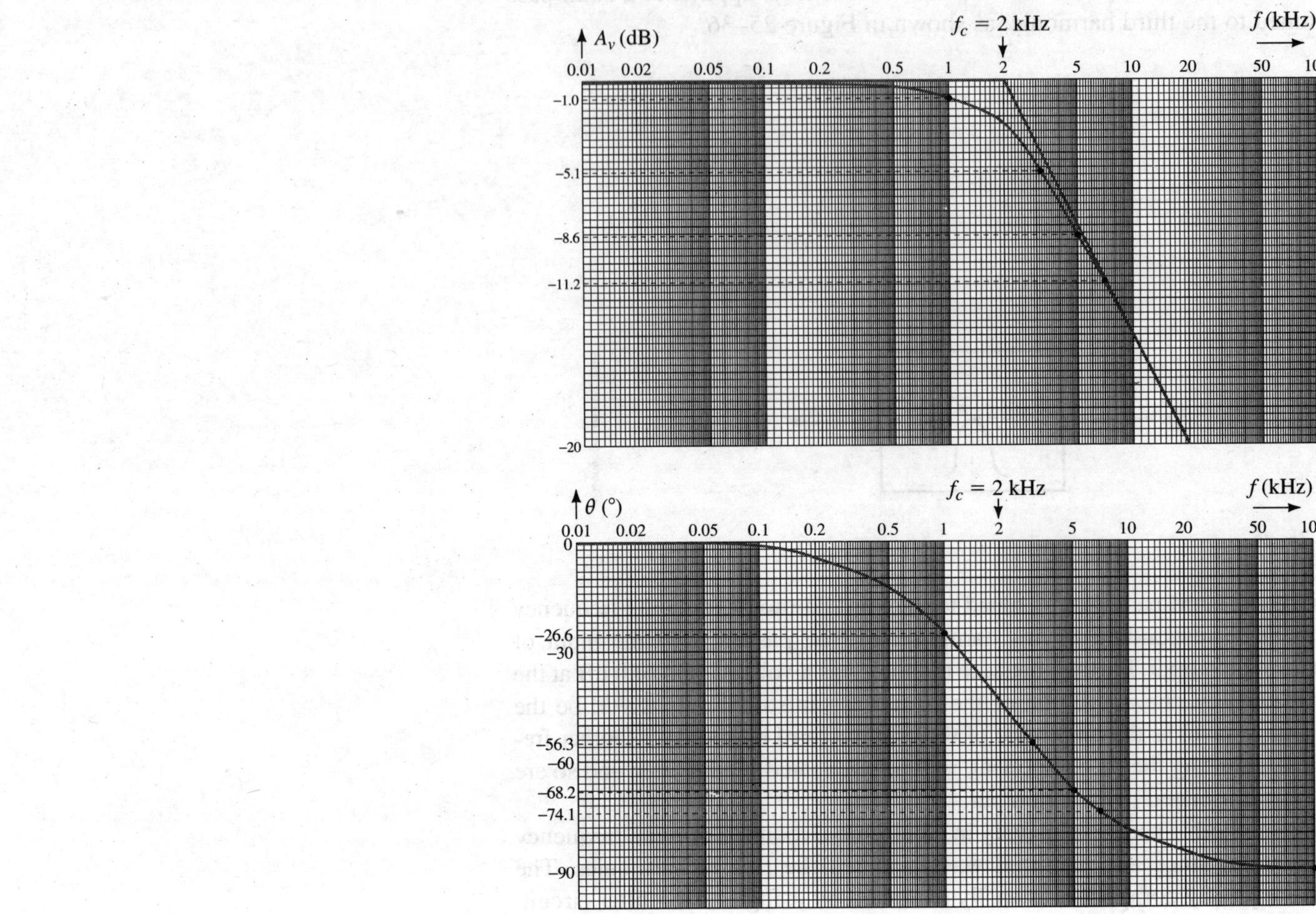

FIGURE 25–38

a. Determine the dc component at the output of the low-pass filter.
b. Calculate the amplitude and corresponding phase shift of the first four nonzero sinusoidal output components.

Solution

a. From previous examples, we have determined that the given waveform is expressed by the following Fourier series:

$$v(t) = 0.5 + \frac{2}{\pi}\sin \omega t + \frac{2}{3\pi}\sin 3\omega t + \frac{2}{5\pi}\sin 5\omega t + \frac{2}{7\pi}\sin 7\omega t$$

Since the circuit is a low-pass filter, we know that the dc component will pass from the input to the output without being attenuated. Therefore,

$$V_{0(\text{out})} = V_{0(\text{in})} = 0.5 \text{ Vdc}$$

b. By examining the frequency response of Figure 25–36, we see that all sinusoidal components will be attenuated and phase-shifted. From the graphs we have the following:

$$\begin{aligned}
&1 \text{ kHz:} && A_{V1} = -1.0 \text{ dB}, && \Delta\theta_1 = -26.6^\circ \\
&3 \text{ kHz:} && A_{V3} = -5.1 \text{ dB}, && \Delta\theta_3 = -56.3^\circ \\
&5 \text{ kHz:} && A_{V5} = -8.6 \text{ dB}, && \Delta\theta_5 = -68.2^\circ \\
&7 \text{ kHz:} && A_{V7} = -11.2 \text{ dB}, && \Delta\theta_7 = -74.1^\circ
\end{aligned}$$

The amplitudes of the various harmonics at the output of the filter are determined as follows:

$$V_{1(\text{out})} = \left(\frac{2}{\pi}\right)10^{-1.0/20} = 0.567\ \text{V}_\text{p}$$

$$V_{3(\text{out})} = \left(\frac{2}{3\pi}\right)10^{-5.1/20} = 0.118\ \text{V}_\text{p}$$

$$V_{5(\text{out})} = \left(\frac{2}{5\pi}\right)10^{-8.6/20} = 0.047\ \text{V}_\text{p}$$

$$V_{7(\text{out})} = \left(\frac{2}{7\pi}\right)10^{-11.2/20} = 0.025\ \text{V}_\text{p}$$

The Fourier series of the output waveform is now approximated as

$$v(t) = 0.5 + 0.567 \sin(\omega t - 26.6°) + 0.118 \sin(3\,\omega t - 56.3°) + 0.047 \sin(5\,\omega t - 68.2°) + 0.025 \sin(7\,\omega t - 74.1°)$$

Various computer-aided design (CAD) and mathematical applications programs are able to generate a time-domain display of a waveform from a mathematical expression. When the preceding waveform is plotted in the time domain, it appears as illustrated in Figure 25–39.

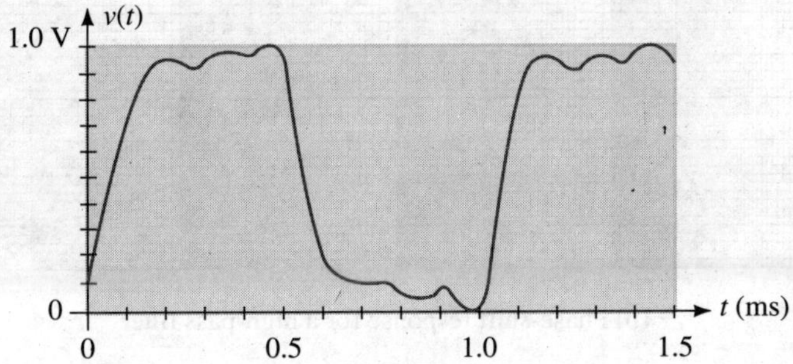

FIGURE 25–39

PRACTICE PROBLEMS 6

The voltage waveform of Figure 25–40 is applied to a high-pass filter circuit having the frequency response shown in Figure 25–41.

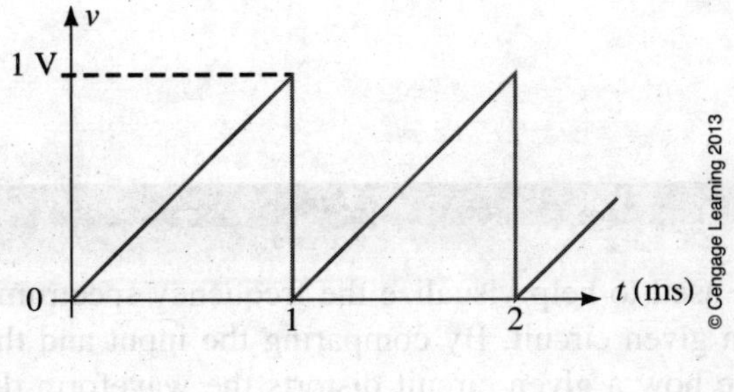

FIGURE 25–40

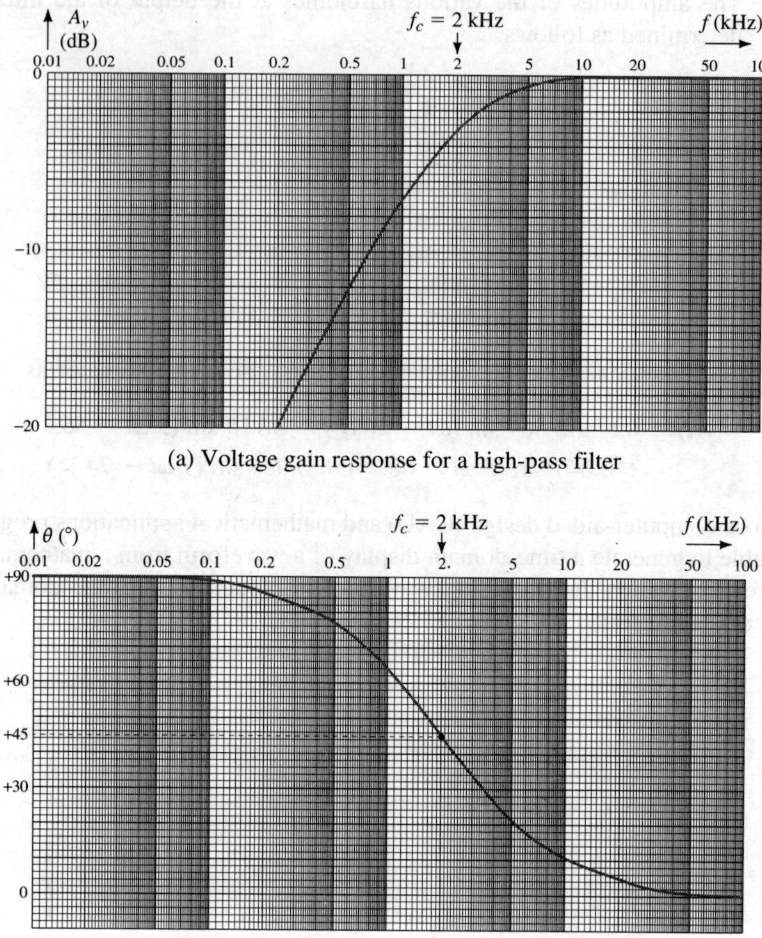

(a) Voltage gain response for a high-pass filter

(b) Phase-shift response for a high-pass filter

FIGURE 25–41

a. Determine the dc component at the output of the high-pass filter.

b. Calculate the amplitude and corresponding phase shift of the first four nonzero sinusoidal output components.

Answers

a. zero

b. 1 kHz: 0.142 V_p, −116°
   2 kHz: 0.113 V_p, −135°
   3 kHz: 0.088 V_p, −147°
   4 kHz: 0.071 V_p, −154°

25.6 Circuit Analysis Using Computers

PSPICE

PSpice can be used to help visualize the frequency spectrum at the input and the output of a given circuit. By comparing the input and the output, we are able to observe how a given circuit distorts the waveform due to attenuation and phase shift of the various frequency components.

In the following example, we use a low-pass filter that has a cutoff frequency of 3 kHz. We will observe the effects of the filter on a 1-V pulse wave. In order to complete the required analysis, it is necessary to set up PSpice correctly.

EXAMPLE 25–9

Use PSpice to find the Fourier series for both the input and the output waveforms for the circuit shown in Figure 25–42. Use the Probe postprocessor to obtain both a time-domain and a frequency-domain display of the input and output waveforms.

CircuitSim 25-1

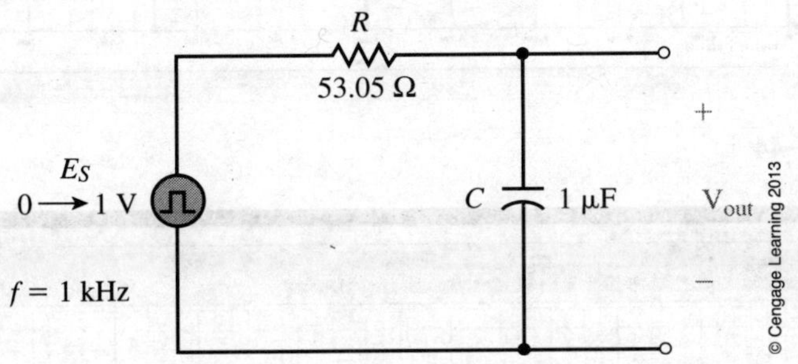

FIGURE 25–42

Solution The circuit is input as shown in Figure 25–43.

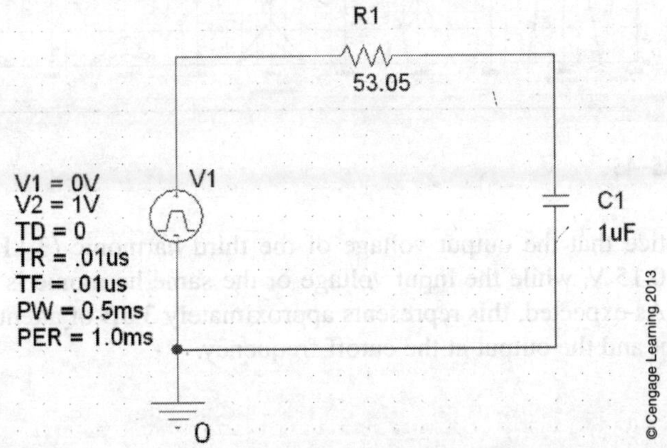

FIGURE 25–43

The voltage source is a pulse generator and is obtained from the SOURCE library by calling for VPULSE. The properties for the pulse generator are set as follows: V1 = **0V**, V2 = **1V**, TD = **0**, TR = **0.01us**, TF = **0.01us**, PW = **0.5ms**, PER = **1.0ms**.

We begin the analysis by first setting the simulation settings for *Time Domain* (*Transient*) analysis. *Run to* time is set for **2.0ms** and the *Maximum step size* is set for **2us.** Now the analysis can be run.

Once in the *Probe* window, we will simultaneously display both the input and the output waveforms. Click on *Trace* and *Add Trace*. Enter **V(V1:+), V(C1:1)** in the *Trace Expression* dialog box. You will see the time-domain output as shown in Figure 25–44.

In order to obtain the frequency-domain display, simply click on *Trace* and *Fourier*. You will need to adjust the range of the abscissa by clicking on *Plot* and *Axis Settings*. Click on the *X Axis* tab and change the range from **0Hz** to **10kHz**. The display will appear as shown in Figure 25–45.

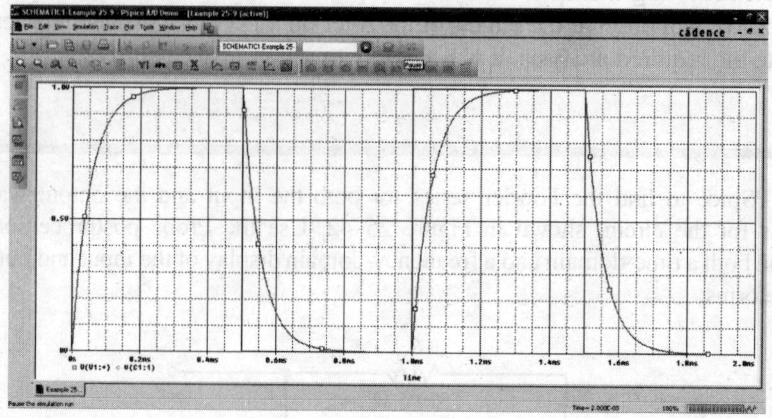

FIGURE 25–44

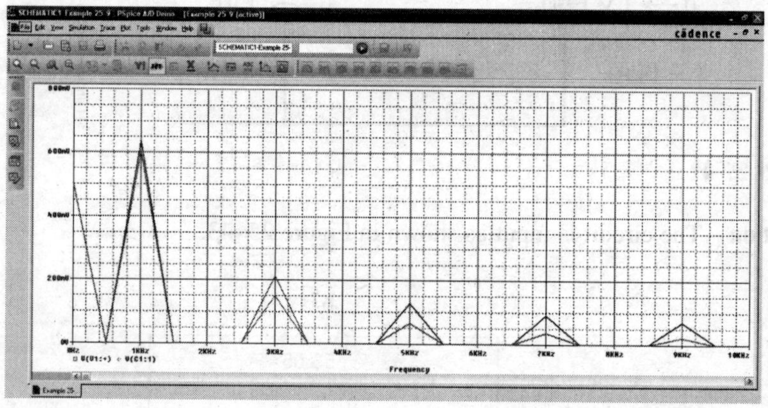

FIGURE 25–45

Notice that the output voltage of the third harmonic (3 kHz) is approximately 0.15 V, while the input voltage of the same harmonic is approximately 0.21 V. As expected, this represents approximately 3 dB of attenuation between the input and the output at the cutoff frequency.

Putting It into Practice

One method of building a frequency multiplier circuit is to generate a signal which is "rich" in harmonics. A full-wave rectifier is a circuit that converts a sine wave (which consists of only one frequency component) into one which appears as shown in Figure 25–18. As we see, the waveform at the output of the full-wave rectifier is composed of an infinite number of harmonic components. By applying this signal to a narrow-band-pass filter, it is possible to select any one of the components. The resulting output will be a pure sine wave at the desired frequency.

If a sine wave with an amplitude of 10 V is applied to a full-wave rectifier, what will be the amplitude and frequency at the output of a passive filter tuned to the third harmonic? Assume that there are no losses in the full-wave rectifier or in the filter circuit.

Problems

25.1 Composite Waveforms

1. a. Determine the rms voltage of the waveform shown in Figure 25–46.
   b. If this waveform is applied to a 50-Ω resistor, how much power will be dissipated by the resistor?

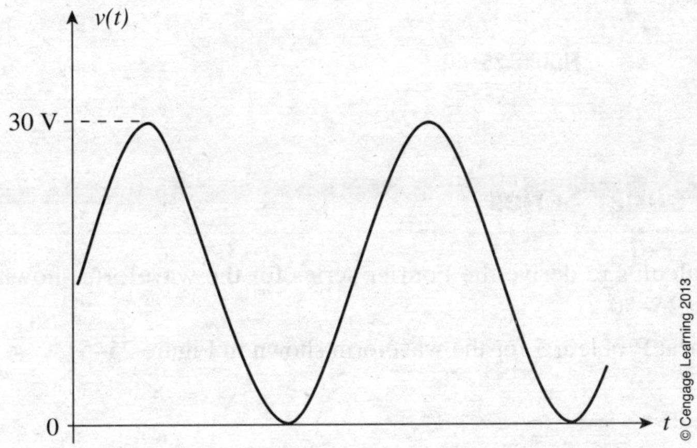

FIGURE 25–46

2. Repeat Problem 1 if the waveform of Figure 25–47 is applied to a 250-Ω resistor.

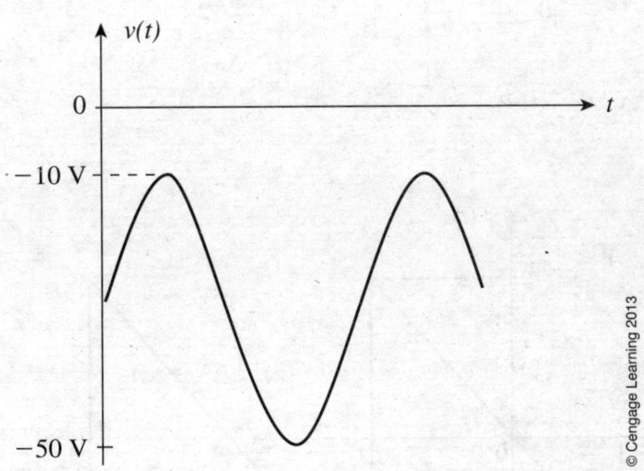

FIGURE 25–47

3. Repeat Problem 1 if the waveform of Figure 25–48 is applied to a 2.5-kΩ resistor.
4. Repeat Problem 1 if the waveform of Figure 25–49 is applied to a 10-kΩ resistor.

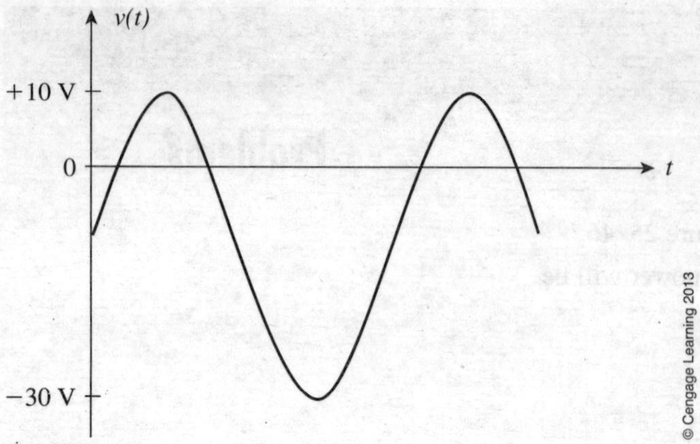

FIGURE 25–48

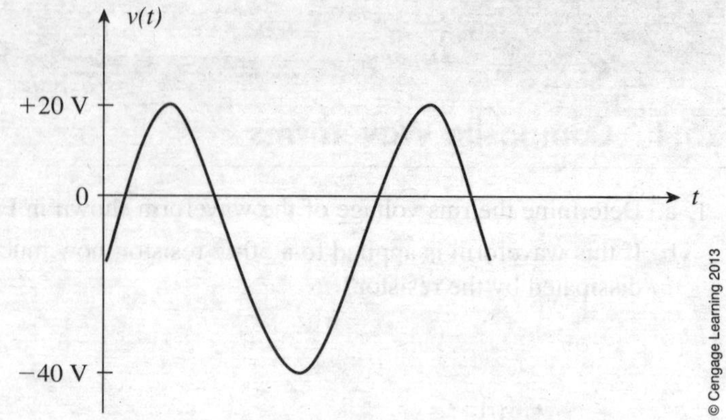

FIGURE 25–49

25.2 Fourier Series

5. Use calculus to derive the Fourier series for the waveform shown in Figure 25–50.
6. Repeat Problem 5 for the waveform shown in Figure 25–51.

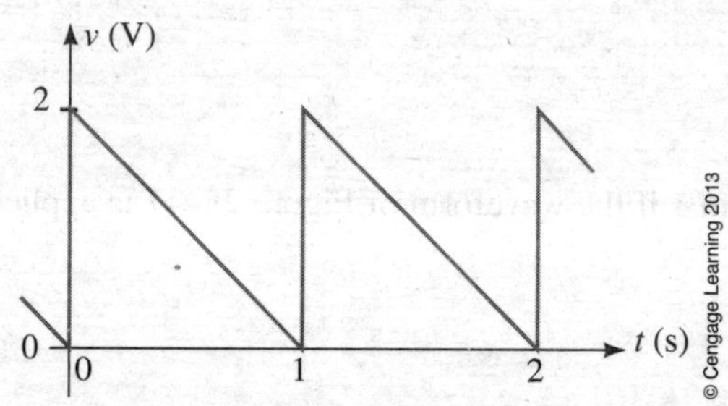

FIGURE 25–50

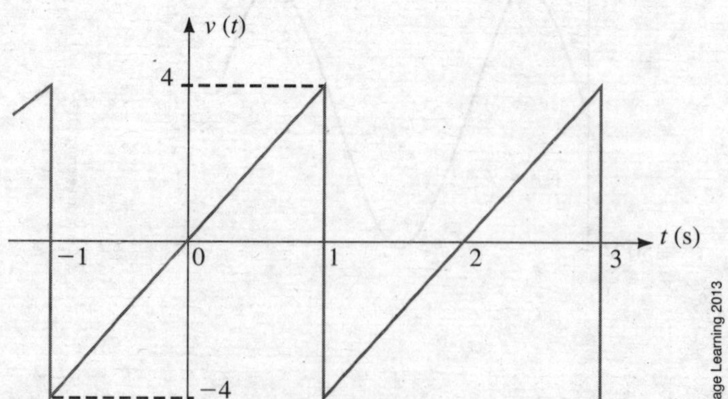

FIGURE 25–51

25.3 Fourier Series of Common Waveforms

7. Use Table 25–1 to determine the Fourier series for the waveform of Figure 25–52.
8. Repeat Problem 7 for the waveform of Figure 25–53.

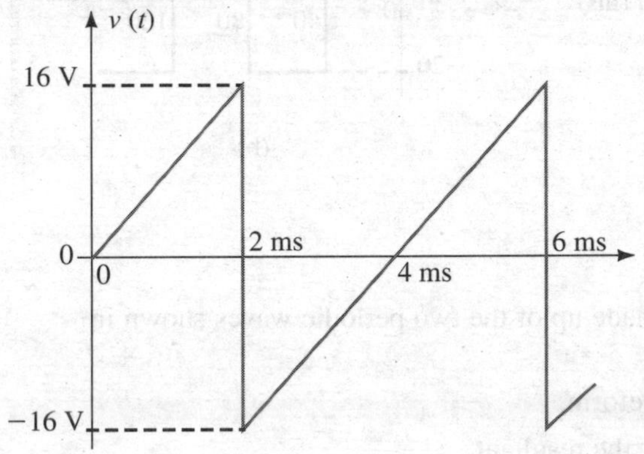

FIGURE 25–52

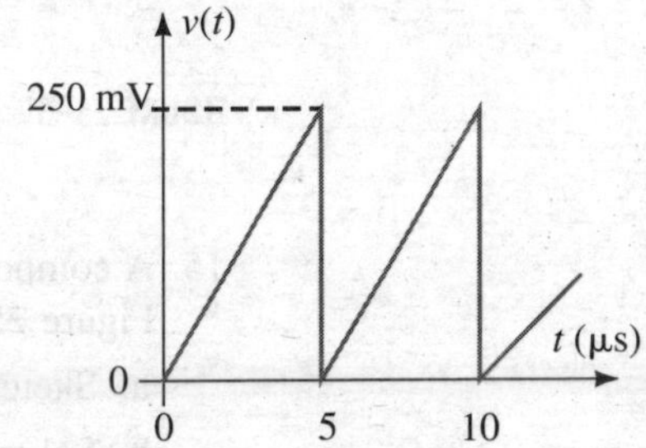

FIGURE 25–53

9. Repeat Problem 7 for the waveform of Figure 25–54.
10. Repeat Problem 7 for the waveform of Figure 25–55.
11. Write the expression including the first four sinusoidal terms of the Fourier series for the waveform of Figure 25–56.

$$v(t) = \begin{cases} 10 \sin \omega t, & 0 < t < 10 \text{ ms} \\ 0, & 10 \text{ ms} < t < 20 \text{ ms} \end{cases}$$

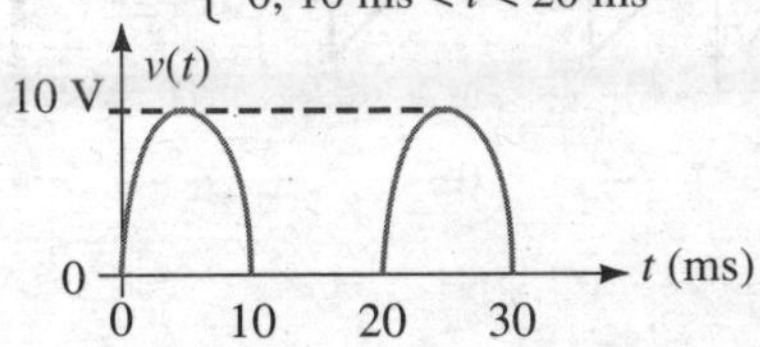

FIGURE 25–54

$$v(t) = |10 \sin \omega t|$$

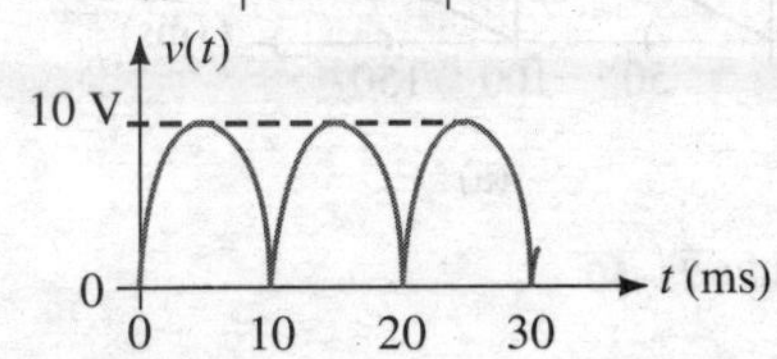

FIGURE 25–55

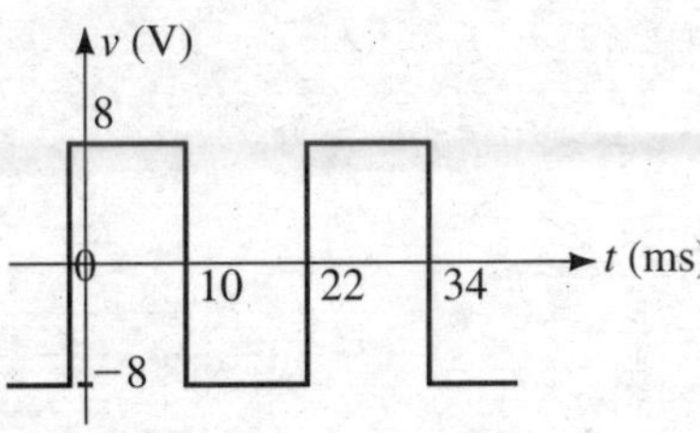

FIGURE 25–56

12. Repeat Problem 11 for the waveform of Figure 25–57.

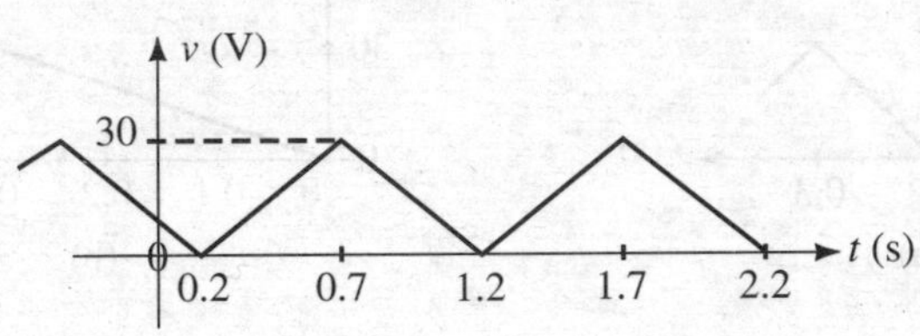

FIGURE 25–57

13. A composite waveform is made up of the two periodic waves shown in Figure 25–58.
    a. Sketch the resulting waveform.
    b. Write the Fourier series of the given waveforms.
    c. Determine the Fourier series of the resultant.

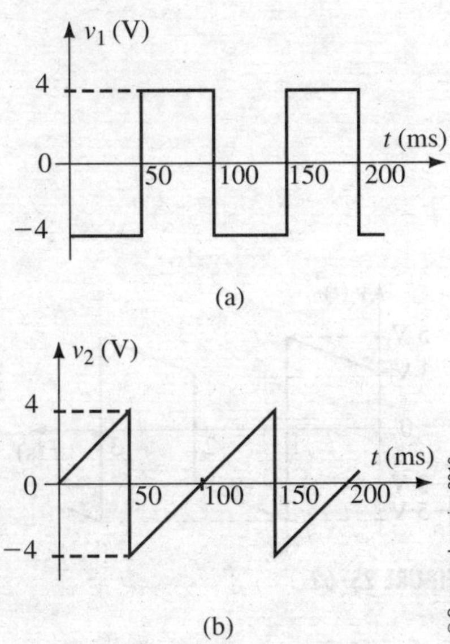

FIGURE 25–58

14. Repeat Problem 13 for the periodic waveforms shown in Figure 25–59.

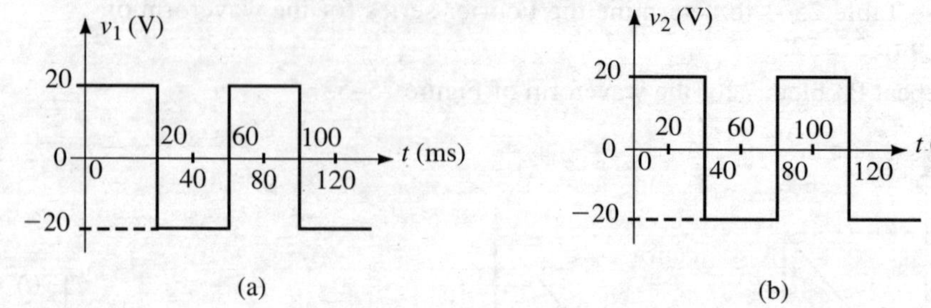

FIGURE 25–59

15. A composite waveform is made up of the two periodic waves shown in Figure 25–60.
    a. Sketch the resulting waveform.
    b. Solve for the dc value of the resultant.
    c. Write the Fourier series of the given waveforms.
    d. Determine the Fourier series of the resultant.

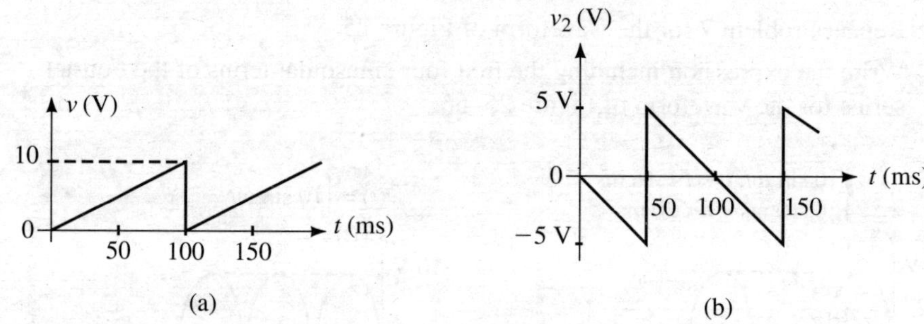

FIGURE 25–60

16. Repeat Problem 15 for the waveforms shown in Figure 25–61.

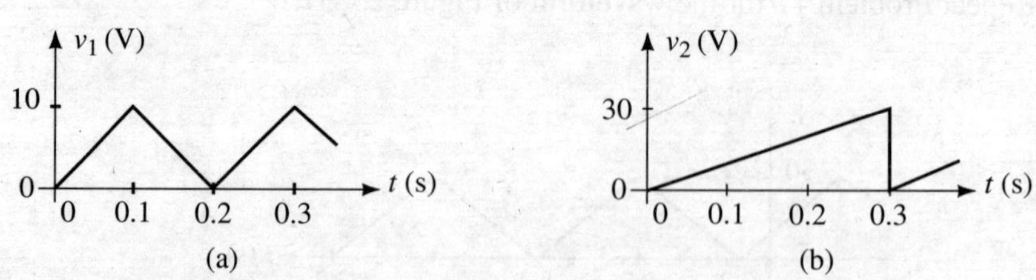

FIGURE 25–61

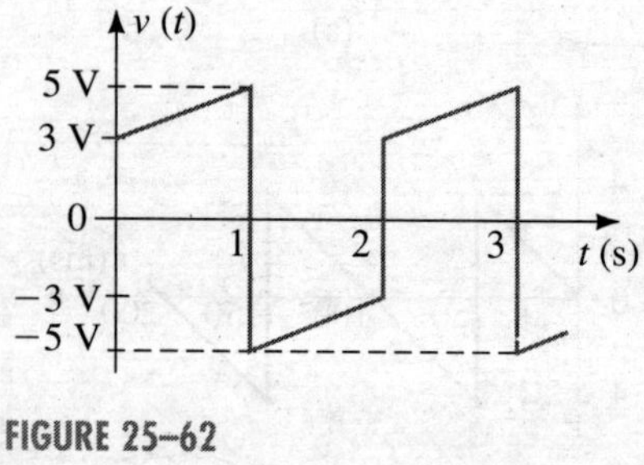

FIGURE 25–62

17. The waveform of Figure 25–62 is made up of two fundamental waveforms from Table 25–1. Sketch the two waveforms and determine the Fourier series of the composite wave.

18. The waveform of Figure 25–63 is made up of a dc voltage combined with two fundamental waveforms from Table 25–1. Determine the dc voltage and sketch the two waveforms. Determine the Fourier series of the composite wave.

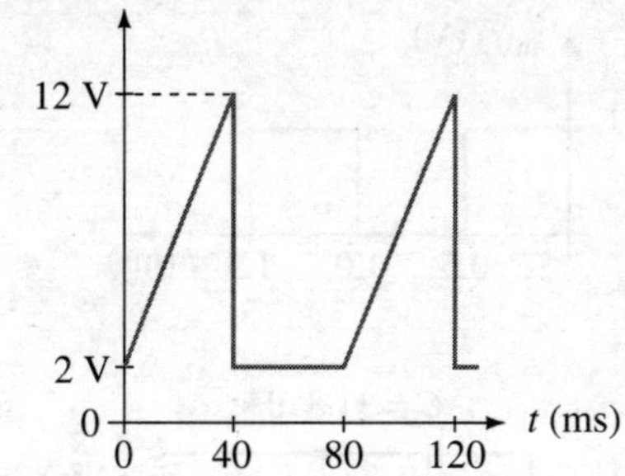

FIGURE 25–63

25.4 Frequency Spectrum

19. Determine the total power dissipated by a 50-Ω resistor, if the voltage waveform of Figure 25–52 is applied to the resistor. Consider the dc component and the first four nonzero harmonics. Indicate the power levels (in watts) on a frequency distribution curve.
20. Repeat Problem 19 for the waveform of Figure 25–53.
21. A spectrum analyzer with a 50-Ω input is used to measure the power levels in dBm of the Fourier series components of the waveform shown in Figure 25–54. Determine the power levels (in dBm) of the dc component and the first four nonzero harmonic components. Sketch the resultant display as it would appear on a spectrum analyzer.
22. Repeat Problem 21 if the waveform of Figure 25–55 is applied to the input of the spectrum analyzer.

25.5 Circuit Response to a Nonsinusoidal Waveform

23. The circuit of Figure 25–64 has the frequency response shown in Figure 25–65.
    a. Determine the dc component at the output of the filter.
    b. Calculate the amplitude and corresponding phase shift of the first four nonzero sinusoidal output components.

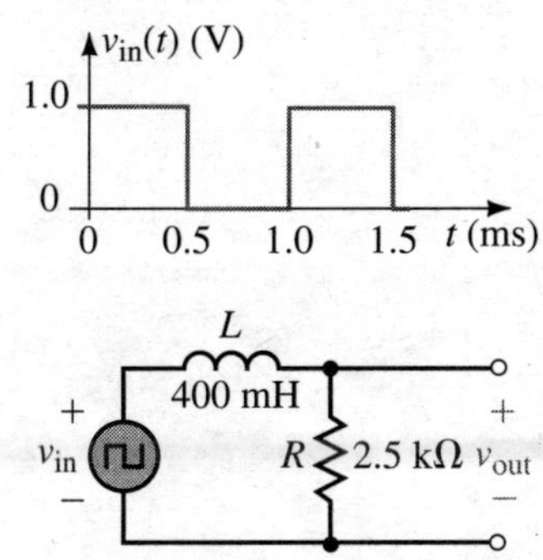

FIGURE 25–64

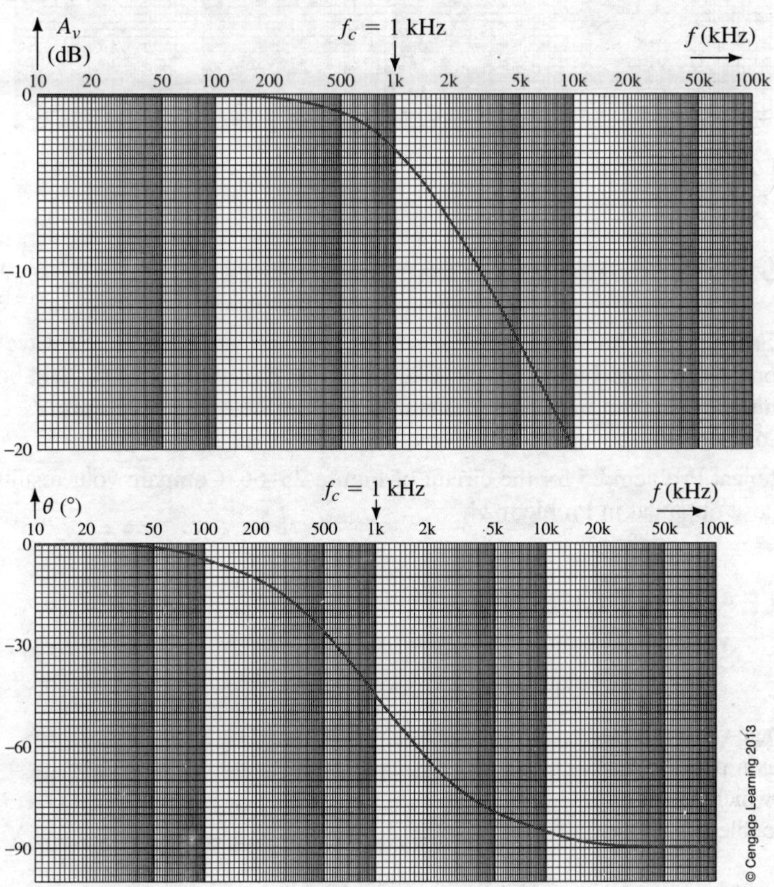

FIGURE 25–65

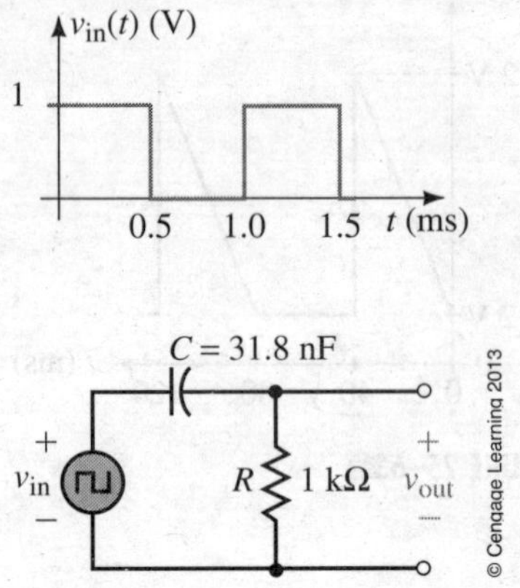

FIGURE 25–66

24. Repeat Problem 23 for Figures 25–66 and 25–67.

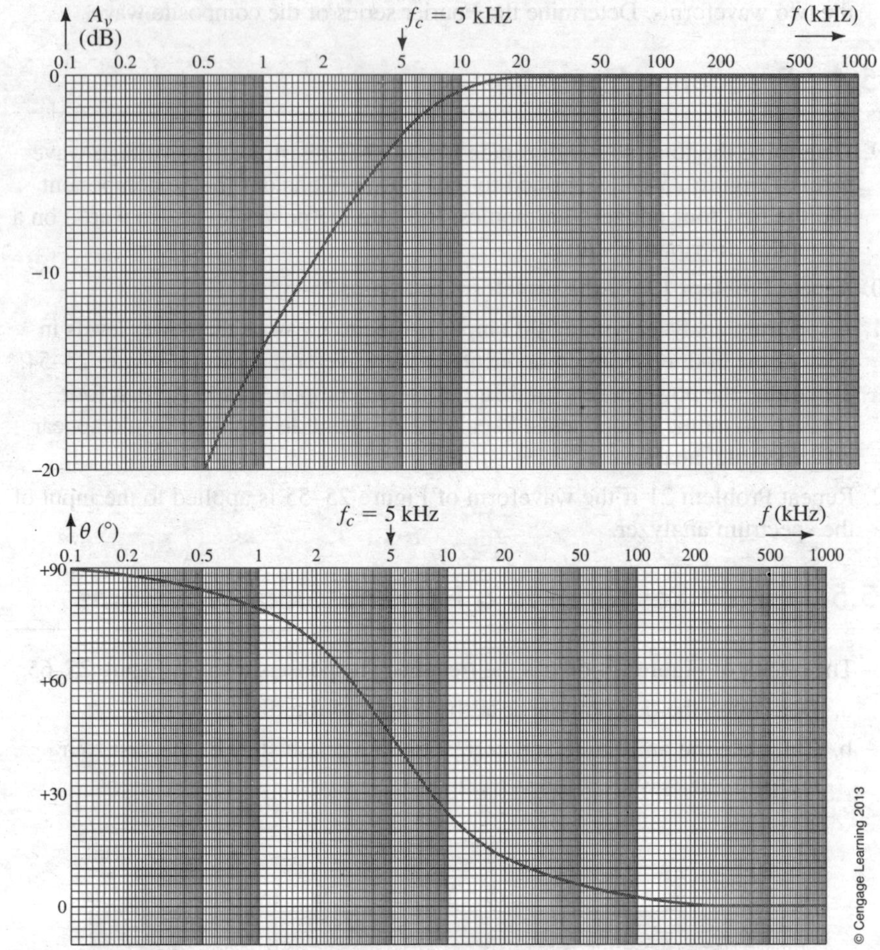

FIGURE 25–67

25.6 Circuit Analysis Using Computers

PSPICE

25. Use PSpice to find the Fourier series for both the input and output waveforms for the circuit of Figure 25–64. Use the Probe postprocessor to obtain both a time-domain and frequency-domain display of the output waveform. Compare your results to those obtained in Problem 23.

PSPICE

26. Repeat Problem 25 for the circuit of Figure 25–66. Compare your results to those obtained in Problem 24.

ANSWERS TO IN-PROCESS LEARNING CHECKS

IN-PROCESS LEARNING CHECK 1

The waveform of Figure 25–12 has an average value of zero. Therefore, $a_0 = 0$. The waveform also has a peak-to-peak value which is double that of Figure 25–6, which means that the amplitude of each harmonic must be doubled.

$$v(t) = \frac{4}{\pi}\sin \omega t + \frac{4}{3\pi}\sin 3\omega t + \frac{4}{5\pi}\sin 5\omega t + \cdots$$

IN-PROCESS LEARNING CHECK 2

a. Figure 25–15 with $V_m = 25$ V and $T = 20$ ms

b. Figure 25–16 with $V_m = 25$ V and $T = 20$ ms

c. $v(t) = 12.5 + 18.9 \sin(100\pi t - 32.48°) - 7.96 \sin(200\pi t) + 5.42 \sin(300\pi t - 11.98°) - 3.98 \sin(400\pi t)$

Appendix A

MULTISIM AND PSPICE

A.1 Introduction

In this book, we use Multisim 11 and PSpice Demo 16.3, the software versions current at the time of writing. To help get you started (and to provide an ongoing reference), we have put together this appendix. (If you are a legacy Multisim user, you can still use this appendix, but there may be some differences—see Note 1—and follow the links to *Special Instructions for Multisim 9 and 10 Users.*

This appendix assumes that you are using your own personal computer. If you are using a college or university network, your professor may have different instructions for you to follow.

Organizing Your Simulation Files

Since you will create a lot of simulation files while studying Multisim or PSpice, you should set up a simple file structure to organize them. The Windows *Documents* folder (or *My Documents* on older versions of Windows) is the standard location for saving user-created files. We will start there. (This is a Windows file creation process, and if you already know how to do it, you can skip this section.)

Multisim Users: Click the Windows *Start* button, then *Documents*. Right-click the screen that opens, and from the menu select *New* then click *Folder*. For the new folder name, type **Multisim Files**—see Note 2. This is your main folder. You need to create individual chapter subfolders. To do this, double-click the *Multisim Files* folder that you have just created, and when it opens, right-click the screen, select *New*, click *Folder*, and type in **Ch 4**. (You should save every simulation file that you create in Chapter 4 in this subfolder.) In a similar manner, create a subfolder for Ch 5. [You should now have the beginnings of the file structure shown in Figure A–1(a). Additional chapter folders can be added as you progress through the book.]

PSpice Users: Click the Windows *Start* button, then *Documents*. Right-click the screen that opens, and from the menu select *New* then click *Folder*. For the new folder name, type **PSpice Files**—see Note 2. This is your main folder. You need to create individual chapter subfolders. To do this, double-click the *PSpice Files* folder that you have just created, and when it opens, right-click

NOTES...

1. To access our Web site, please visit www.cengagebrain.com. In the search box at the top of the page, key in **Robbins and Miller** (or the book title or ISBN number). This will take you to the product page where your resources can be found.
2. You can use any name you want, but this is the name we use in this appendix and throughout the book.

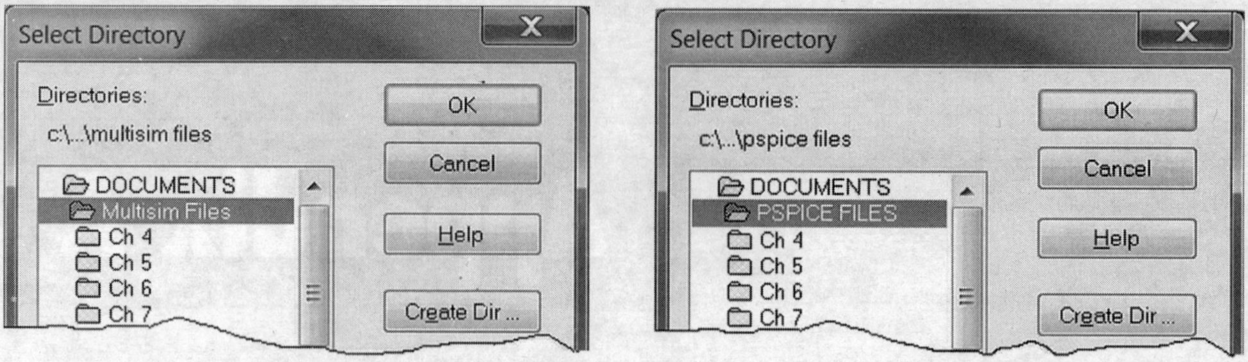

(a) Multisim directory structure (b) PSpice directory structure

FIGURE A–1 Suggested file structures for user-created simulation files. (On older versions of Windows, your files will be organized under *My Documents* instead of *Documents*.)

the screen, select *New*, click *Folder*, and type in **Ch 4**. (You should save every simulation file that you create in Chapter 4 in this subfolder.) In a similar manner, create a subfolder for Ch 5. [You should now have the beginnings of the file structure shown in Figure A–1(b). Additional chapter folders can be added as you progress through the book.]

A.2 Multisim

Multisim is a trademark of National Instruments Corporation. For the most part, its usage is straightforward. Section 4.8 of Chapter 4 describes how to get started. Alternatively, you can utilize the *Multisim Beginner's Tutorial* that we have put together—see Notes 1 and 3.

When you install Multisim on your computer, be sure to place its shortcut icon on your screen. Normally, you launch Multisim from this icon, but if it is missing, you can launch from the Windows *Start* menu—see Note 2. This opens the screen shown in Figure 4–28 of Chapter 4.

NOTES...

1. The *Multisim Beginner's Tutorial* shows you how to get started. Using a sequence of annotated screen shots and sample runs, it walks you step by step through the process of building and running your first simulation. To use the tutorial, go to our Web site, download either the Multisim 11 or Multisim 9 version, unzip, launch, then follow the instructions on your screen. (Note that this is a Multisim file and you must have Multisim loaded on you computer to use it.)
2. To launch Multisim from the *Start* menu—click the *Start* button and double-click the Multisim icon on the *Start* menu. If there is no Multisim icon on the *Start* menu, you should see the *National Instruments* icon instead. Click and you should see *Circuit Design Suite 11*, then below that, the Multisim icon. Double-click it.

Some Key Points for Multisim Users

1. All Multisim circuits require a ground. (Without a ground, Multisim refuses to work.)
2. To select a component already on the screen, click it. A dashed box appears, indicating selection. To deselect, place the cursor outside the dashed box and click.
3. To reposition a component, select it, hold down the left button, and drag it to the new position.
4. To delete a component, select it, then press the Delete key.
5. To rotate a component, select it, then, while holding down the Ctrl key, press the R key. (We denote this as *Ctrl/R* or *Ctrl + R*. Each *Ctrl/R* rotates the component by 90°.) Alternatively, right-click and select the desired rotation from the menu.
6. To place a connector dot, select *Place* on the menu bar, then click *Junction*. Position the dot on the screen and click to place it. A maximum of four wires may be connected to any junction point.
7. We sometimes use shorthand designations for mouse operations. For example, if you are told to click *View/Toolbars*, this is a shorthand way of saying, "Click *View* on the menu bar, then select *Toolbars* on the drop-down list that appears."

3. All Multisim examples in this book were prepared using the Multisim Educational version.

Component Orientation

Two-terminal components have an implied "1" end and a "2" end. When you place such a component, it takes up a default position. For example, for resistors, capacitors, and inductors, the default position is with the "1" end to the left, while for sources, it is at the top. Components may be rotated as in point number 5 under "Some Key Points for Multisim Users." To get the "1" end of a capacitor up, for example, rotate it right once from its default position. Knowing about orientation is important when you are setting initial conditions. For example, if you set a capacitor's initial voltage to 10 V, Multisim places 10 volts across the capacitor with its "1" end positive. If you have inserted the capacitor into your circuit upside down, its polarity will be reversed from what you expect. If this happens, disconnect its wiring, rotate it to its desired orientation, rewire, then set its initial condition again. For components like resistors, you don't care which end is which.

Cleaning Up Grapher View Displays

Click *View* and ensure that *Toolbars* is selected. Use your mouse pointer and locate the *Black Background* icon and click. Similarly, locate *Show Grid* and click and *Show Cursors* and click. Move your mouse pointer to an empty spot on the display area and right-click, then left-click *Show Select Marks.*

Adding Expressions to Graphic Displays

When you use *Grapher View* to display results, you can combine variables to form expressions. For example, suppose you want to multiply voltage V(1) times current I(L1) to display power. Proceed as follows: In the *Output* tab dialog box, click *Add Expression* to open the box of Figure A–2. Double-click V(1), and it appears in the *Expression* box. Then double-click *, double-click *I(L1)*, then click *OK*. The resulting expression V(1)*I(L1)

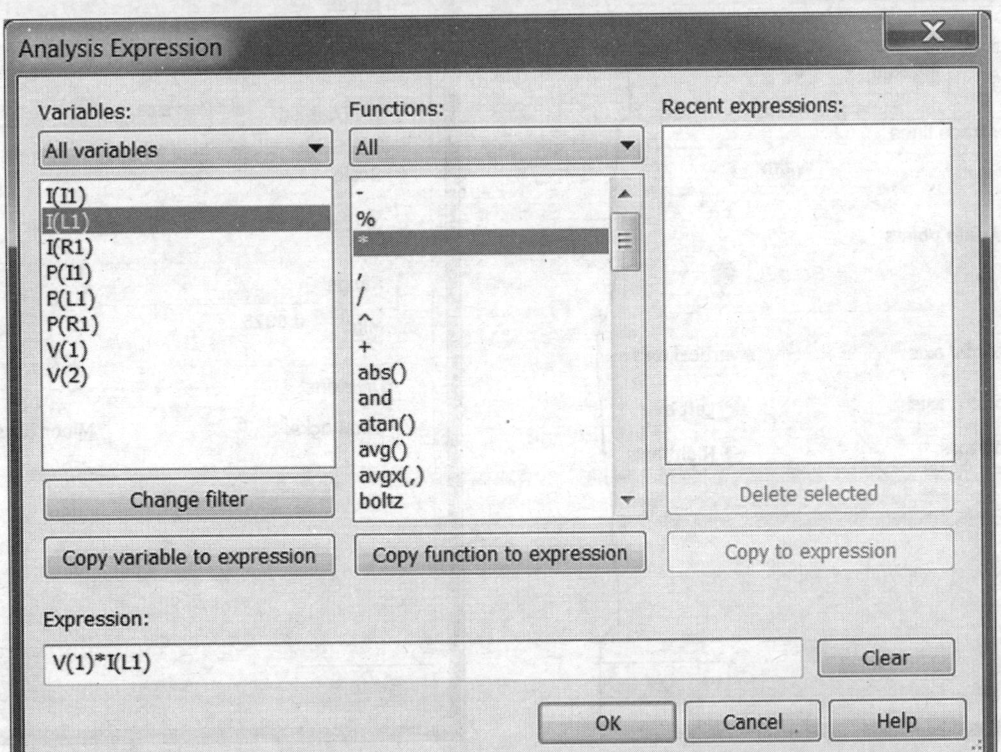

FIGURE A–2 Using the Multisim *Add expression* facility to create a power waveform by multiplying voltage times current.

appears in the *Output* display list. Now click *Simulate* and the result is graphed on your screen. (An example waveform created in this fashion is shown in Figure 17–29, Chapter 17.)

Displaying Multiple Traces

When graphing multiple traces, you can add a second vertical scale if you need one. The new scale is added to the right-hand side of the graph. To illustrate, suppose you compute and display capacitor voltage V(2) and current I(C1), but because the current (which is a few mA) is so small relative to the voltage, it displays as a flat line along the zero axis. You need an expanded current scale to display the result. Proceed as follows:

- On *Grapher View*, click *Trace/Trace Properties*. In the dialog box, select the *Traces* tab at the top of the box. The item marked *Trace Label* should indicate I(C1). If it doesn't, use the *Trace ID* up/down arrows to change it, Figure A–3(a).
- Near the bottom of the box (under *Y-vertical axis*), click the *Right axis* button, then click *Apply*. Select the *Right Axis* tab at the top of the box, click the *Axis Enabled box*, click *Auto-range*, click *Apply* then *OK*, Figure A–3(b)
- The new scale appears on the right-hand side of the graph, and the expanded waveform appears. (An example scale and waveform created in this fashion are shown in Figure 11–42.)

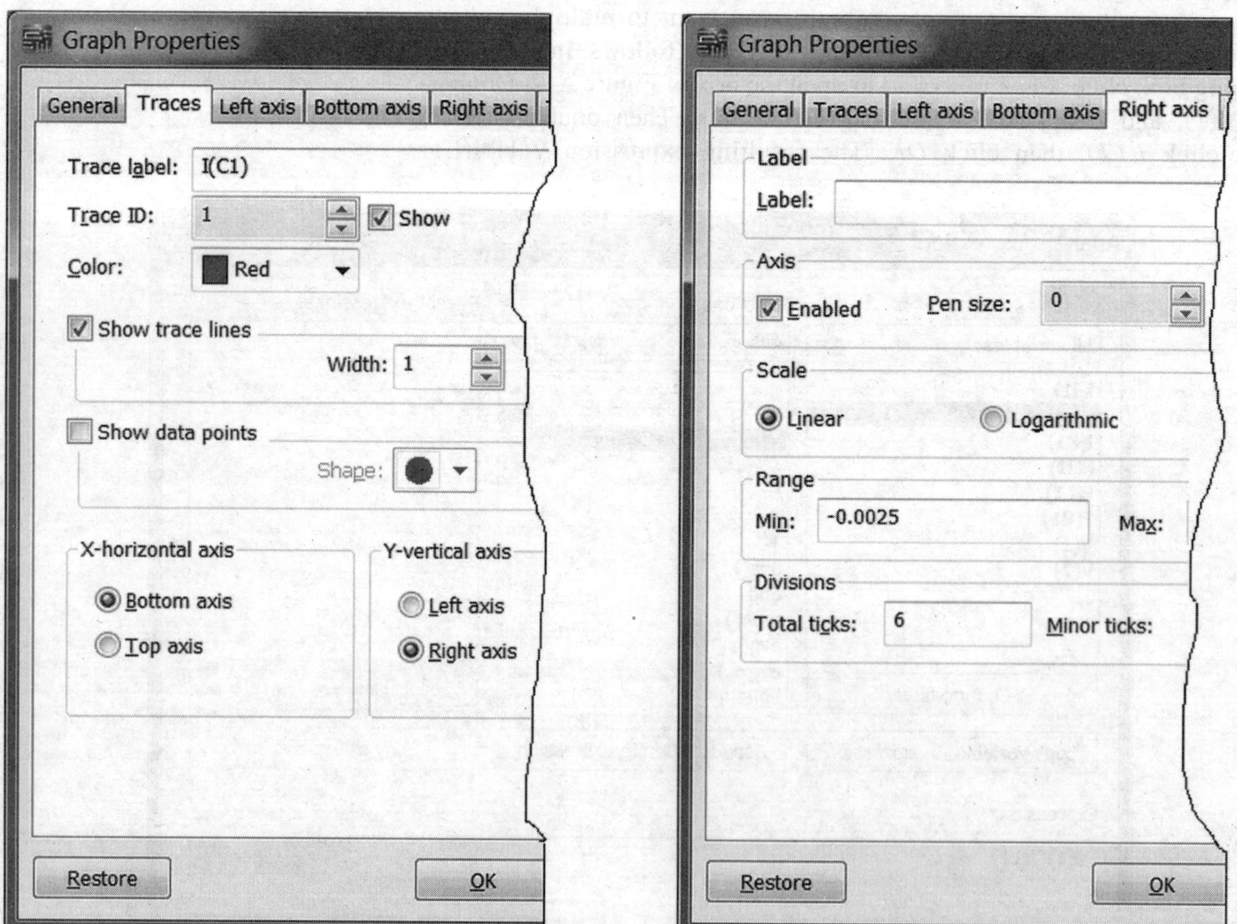

(a) (b)

FIGURE A–3 Illustrating how to create a new scale for the current trace.

Using the Cursors

Cursors are used to read values from graphs created by Multisim (for example, Figure 11–42). To activate the cursors, click the *Show Cursors* icon (or click *Cursors* on the Grapher View window, then *Show Cursors*). Two cursors appear. To move a cursor, position the mouse pointer on it, then press the left button and drag it to the display area. The x-y coordinates of the cursors appear in a display box. To move the cursor to an exact time point, right-click the cursor, click *Set X Value*, type the time value (e.g., **20m**) into the data entry box, then click *OK*. (Note: You must enter the value without a space. Thus, 20 m is wrong.)

Setting the Phase Angle for ac

In its default setting, Multisim does not generate the angle that you expect. To set the angle correctly, click *Options, Global Preferences,* select the *Simulation* tab, look for the red sine wave, and ensure that *Shift right* is selected. (Legacy users, select the *Parts* tab instead.)

A.3 PSpice

PSpice is a product of Cadence Design Systems Inc.—see Note 1. When you install it, be sure to install its shortcut icon—it will show on your screen as *ORCAD Capture CIS*. Normally, you launch from this icon, but if it is missing, you can launch from the Windows *Start* menu—see Note 2. The screen that opens is that shown in Figure 4–29, Chapter 4 (except that it is blank). Chapter 4, Section 4.8 describes how to get started using PSpice. (Alternatively, you can use the *PSpice Beginner's Tutorial* found on our Web site.)

Some Key Points for PSpice Users

1. All PSpice circuits require a ground. If you don't use a ground, you will get an error message signifying "floating nodes."
2. All circuit components have default values. To change these, double-click and type in the new values. You can include a unit if you wish, but if you don't, basic units (volts, amps, ohms) are assumed.
3. If you do include a unit, be sure that you do not leave a space. Thus, 25 and 25V are acceptable, but 25 V is not.
4. You can use either basic notation, scientific notation, or engineering prefixes. For example, you can specify a 12000-Ω resistor as 12000, 1.2E04, 12k, or 0.012Meg. Table A–1 shows PSpice prefixes.
5. For transient analysis, capacitors and inductors always need initial conditions. In some cases, PSpice can determine these for you, while in others, you will have to enter them yourself. Note the following:
   a. PSpice can automatically determine initial conditions only if it is possible to compute them by performing a pre-transient steady state analysis of the circuit. For circuits containing isolated capacitors, this is not possible, since the voltage on the capacitor is totally indeterminate—that is, it can be anything—thus, for isolated capacitors, you must always manually enter (i.e., type in) the values yourself.
   b. You can always manually enter an initial condition, even if it is possible for PSpice to compute it for you. Thus, if in doubt, do it yourself.

NOTES...

1. The PSpice Demo version is available as a free download from Cadence. To obtain a copy, go to the Internet and search for *Cadence OrCAD Downloads* and click on the Cadence Web site. Next, set up a Cadence account as noted when the Web site opens. Scroll the list of downloads until you find *OrCAD Demo (Capture and PSpice only)*. Download it and install.

2. To launch PSpice from the Windows *Start* menu, click the *Start* button and double-click the ORCAD icon that you see on the *Start* menu. If there is no ORCAD icon on your *Start* menu, you might see *Cadence* instead. Click and you should find the *ORCAD Capture* icon that you are looking for.

TABLE A–1 Standard Prefixes Used by PSpice

Symbol	Scale	Name
T or t	10^{12}	tera
G or g	10^{9}	giga
MEG or meg	10^{6}	mega
K or k	10^{3}	kilo
M or m	10^{-3}	milli
U or u	10^{-6}	micro
N or n	10^{-9}	nano
P or p	10^{-12}	pico

6. PSpice does not permit source/inductor loops—that is, you cannot have a source and an inductor alone in a loop. If you do, PSpice will display an error message. If you get such a message, add a small valued resistor to the loop. Make its resistance much smaller than any other impedance in the circuit. The process is similar for source/capacitor loops.
7. PSpice requires a dc path from all nodes to the ground. If no path exists from a particular node to ground, add a large valued resistor from that node to ground. Make the value of this resistance much larger than the impedance of any other component in the circuit.
8. To select a component that has already been placed on your screen, place your cursor over it and left-click. To deselect it, move your cursor off the component and click left.

PSpice Component Orientation

Resistors, capacitors, and inductors have an implied "1" end, a "2" with a default position, and rotational sequence as depicted in Figure A–4. For capacitors and inductors in transient analysis problems, a basic understanding of orientation is important to you. For example, one of the capacitor symbols that PSpice uses is that of Figure 10–1 in Chapter 10. When you double-click such a capacitor and type an initial condition such as IC = 10 volts, this places 10 volts across the capacitor with the flat plate positive with respect to the curved plate. However, on some versions of PSpice, the capacitor symbol has no curved plate and it is not obvious which end is which. A problem occurs if you insert such a capacitor upside down. In this case, its polarity will be reversed from what you expect and when you run the simulation, you will find that your initial conditions are reversed. (If this happens, disconnect the wiring from the capacitor, rotate it to its appropriate orientation, rewire, then set its initial condition again.) If you have no way of knowing how your capacitor is connected, you may have to experiment to find out. Similarly, you may have to do some simulation testing to find out how to handle nonzero initial inductor currents in *RL* circuits.

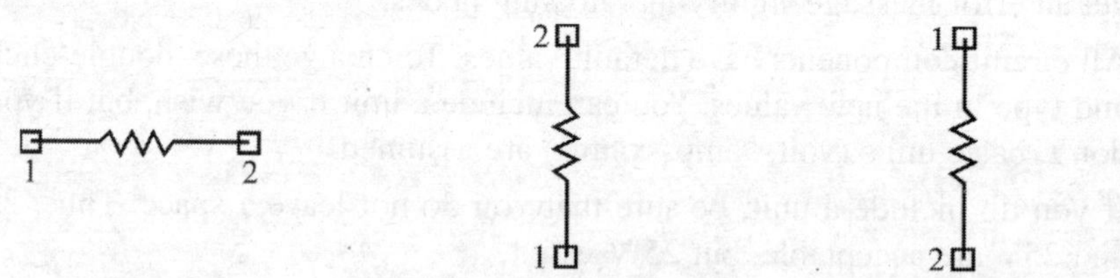

(a) Default (b) First rotation (90°) (c) Third rotation (270°)

FIGURE A–4 Rotating a component. (Second rotation not shown.)

Labeling Graph Traces and Naming Axes

To add text to a plot as in Figure A–5, select *Plot, Label,* and then *Text.* Type your text, then position it as desired. To add a Y-axis title, select *Plot, Axis Setting,* click the *Y-Axis* tab, select the *Y-Axis Number,* then in the *Axis Title* data entry box, type the desired title (e.g., **Voltage** as in Figure A–5), then click *OK.*

Displaying Multiple Traces

Multiple traces may be displayed on the same vertical axis or you may use independent axes as in Figure A–5. Often the first trace (assume voltage here)

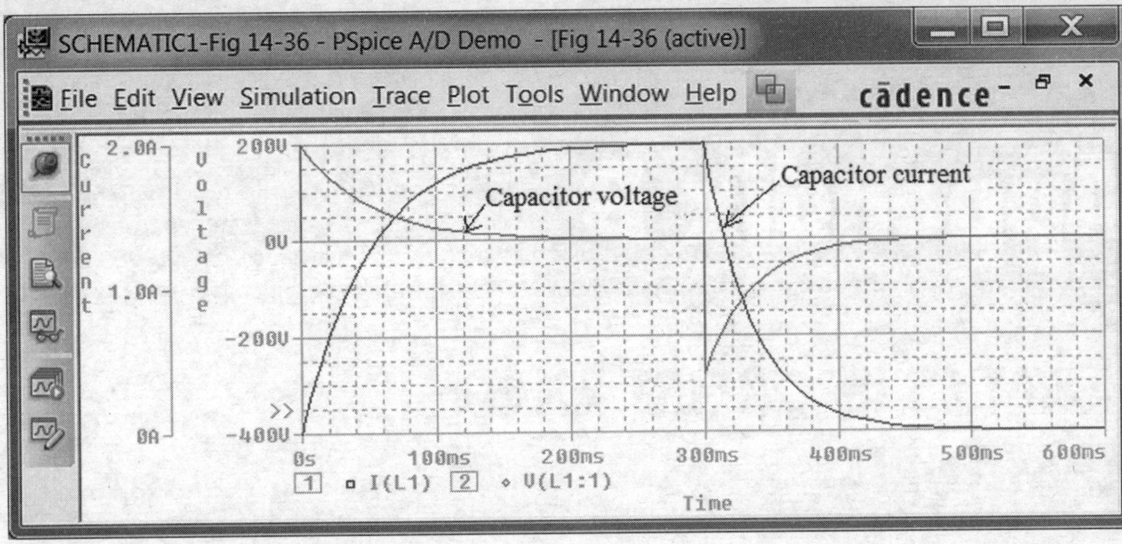

FIGURE A–5 Displaying and labelling multiple traces.

is created using a marker probe. This automatically creates the first Y-axis and scales it. To add a second axis and trace as in Figure A–5, do the following:

- Click *Plot,* then *Add Y Axis*. The second axis appears.
- Click *Trace, Add Trace,* then double-click the variable whose trace you want to add—see Figure A–5.

Using the Cursors

Two cursors are provided for reading values from graphs. To activate the first cursor, click the toggle cursor icon (or click *Trace, Cursor, Display*), then click anywhere on the screen. To move the cursor, hold the left button down and drag to the desired position. (You can also move it using the left and right arrow keys on your keyboard.) The x-y coordinates of the cursor appear in a display box. To utilize the second cursor, use the right mouse button to activate and position it.

Unidentified Subcircuit Error Message

If you get an error message such as *Subcircuit SW_tClose used by X_U1 is unidentified,* you can correct the problem as follows:

- Click the *Edit Simulation Profile* button and click the *Configuration Files* tab.
- Under *Category,* select *Library*. Click the upper *Browse* button.
- You need to find your way to where you placed OrCAD. It is probably on drive C.
- Double-click *OrCAD* and *OrCAD_16.3_Demo* (or whatever your actual version is) should appear. Double-click it.
- Double-click *tools,* double-click *pspice,* and double-click *library*.
- Click *nomd.lib,* then click *Open*. This returns you to *Simulations Settings*.
- Click *Add to Profile,* click *Apply,* then click *OK*.
- Click *Run*.

It should work now.

Note: If all of this looks scary, get a knowledgeable friend to help you.

Appendix B

MATHEMATICS IN CIRCUIT ANALYSIS: A BRIEF TUTORIAL

This tutorial looks at some of the mathematical techniques that you will need to be able to use in order to solve the problems and examples in this book. Among the topics that we cover are methods for solving simultaneous equations that have real and complex coefficients—including calculator methods and the traditional determinant approach. While many of the ideas and techniques discussed here have already been described briefly in the main text, there are some new ideas presented here, as well as some ideas that you may have already encountered in your various mathematics courses. We bring them together in this appendix to provide a handy reference and a focal point for study.

We begin with a discussion of calculator techniques.

B.1 Some Basic Mathematics and Calculator Ideas

EXAMPLE B–1

In mathematics, a common error is to write $\dfrac{1}{a+b}$ as $\dfrac{1}{a}+\dfrac{1}{b}$. The fact that this is incorrect can be seen easily if we use numbers. For example, $\dfrac{1}{2+8}=\dfrac{1}{10}=0.1$. Clearly, this does not equal $\dfrac{1}{2}+\dfrac{1}{8}=0.5+0.125=0.625$. Thus, $\dfrac{1}{a+b}\neq\dfrac{1}{a}+\dfrac{1}{b}$. Note that this is a mathematics mistake and not a calculator mistake. However, a somewhat similar error can result from incorrect calculator usage. This is illustrated next.

EXAMPLE B–2

A problem can sometimes inadvertently arise when calculators are used to perform the type of computation shown in Example B–1. To illustrate, suppose you wish to determine $\frac{1}{2+8}$. If you unthinkingly key in 1 divided by 2 plus 8, you get an answer of 8.5. Why? Well, the answer has to do with the order of operations implemented by most calculators—there is an order of precedence to consider. For example, in general, multiplication and division are performed before addition and subtraction. Thus, in this example, the calculator evaluates 1 divided by 2 first (to yield an answer of 0.5), then adds 8 to this intermediate answer in order to yield a final answer of 8.5. A way around this problem is demonstrated in Example B–3.

EXAMPLE B–3

One way to work around precedence issues is to use parentheses. Parentheses force the calculator to evaluate the parenthetical expression first. Thus, if you key in 1 divided by (2 + 8), your calculator evaluates the expression within parentheses as 10, then performs the division to yield the correct answer of 0.1. Here is a typical calculator sequence that illustrates this process:

[1] [÷] [(] [2] [+] [8] [)] [=]

Order of Precedence: We can now summarize the order of precedence used by calculators and computer programs: (1) Operations within parentheses are performed first, (2) exponentiation (roots and powers) is performed next, then (3) negation and single-key operations, (4) multiplication and division, and (5) addition and subtraction. As a recommendation, always use parentheses to remove any ambiguity about the order of evaluation.

EXAMPLE B–4

A companion problem exists when dealing with complex numbers. Suppose, for example, that you want the reciprocal of a complex number **C** where **C** = a + *j*b. A common mistake is to write $\frac{1}{a+jb}=\frac{1}{a}+\frac{1}{jb}$. The fact that this is clearly wrong can be demonstrated by using numbers. For example, suppose **C** = 1.5 – *j*2. You can find the reciprocal of this by using the inverse key. (Be sure you are in complex mode.) Enter (1.5 – *j*2) using whatever procedure your calculator requires, then use the inverse key (sometimes denoted as x^{-1} on calculators) to find $\frac{1}{1.5-j2}$. You should get an answer of 0.24 + *j* 0.32. Clearly, this does not equal $\frac{1}{1.5}+\frac{1}{(-j2)}=0.667+j\,0.5$. Remember that two complex numbers are only equal if the real components are equal and the imaginary components are equal. Thus, $\frac{1}{a+jb}\neq\frac{1}{a}+\frac{1}{jb}$.

EXAMPLE B–5

NOTES...

Depending on your calculator, the details of the procedure may be different than this one, so consult your manual. For example, on some calculators, you have to press the second function key in order to access the inverse key. The order of steps may also be different. However, the basic principle is the same—all you do is enter a number, inverse it, enter the next number, inverse it, then add the new inversed number to the first inversed number, and so on. Finally, you inverse the result.

Simplifying the Computation of Parallel Resistance, Parallel Inductance, and Series Capacitance

The inverse key can also be used to good effect in order to speed solutions and reduce errors in the computation of parallel resistance, parallel inductance, and series capacitance. This key also reduces the need to rely on memorizing numerous formulas. To illustrate, recall that the basic formula for determining parallel resistance is

$$\frac{1}{R} = \frac{1}{R_1} + \frac{1}{R_2} + \cdots + \frac{1}{R_N}$$

Suppose you have resistances of 60 Ω, 30 Ω, and 12 Ω in parallel. Following is the general procedure you can use—see Note.

Procedure:

1. Key in 60, press the inverse key, then press the + key.
2. Key in 30, press the inverse key, then press the + key.
3. Key in 12, press the inverse key, then press the Enter key.

At this point, the calculator will have computed and stored the sum of the inverses.

4. To inverse the stored result and get the final answer, press the inverse key, then press Enter. You should get an answer of 7.5 ohms.

B.2 Logarithms, Natural Logs, and Exponential Functions

Logarithms provide a way to reverse the operation of exponentiation because, technically, the logarithm function is the inverse of the exponentiation function. Stripped to its essentials, a logarithm is just an exponent.

Basics Review

If you are given 10^3, you can easily produce the answer $10^3 = 10 \times 10 \times 10 = 1000$. However, if you are given the expression $10^y = 1000$ and asked to find y, how do you do it? Note that the question that you are really asking yourself is, "Given the base 10 and its power 1000, what is the exponent that will produce it?" The exponent is called a "logarithm"—thus, in the preceding example, the exponent 3 is the logarithm. Usually, you see it written as,

$$y = \log_{10}(1000) = 3$$

This is an example of a "base 10" log. In circuit theory, you encounter base 10 logs when you study decibels (Chapter 22).

In general, if

$$b^y = x \quad \text{(B–1)}$$

then

$$y = \log_b x \quad \text{(B–2)}$$

Natural Logs and Exponentials

In transient analysis, you sometimes come across natural logs. A natural log uses the transcendental number $e = 2.71828\ldots$ as its base. Base e logs are written as *ln*. Thus, if

$$e^y = x \quad \text{(B–3)}$$

then

$$y = \ln x \quad \text{(B–4)}$$

In Chapter 11, you encountered, $e^{-t/RC} = 0.2$ where t and C were known and you were asked to solve for R. Using the relationships of Equations B–3 and B–4, you can see that $-t/RC = \ln(0.2) = -1.6094$ where $\ln(0.2) = -1.6094$ is obtained using your calculator. Now finding R is straightforward.

B.3 Solving Simultaneous Equations

Simultaneous linear equations appear often in the solution of problems in electrical/electronics technology. Solving these equations has been simplified by a branch of mathematics called linear algebra. Although the actual theorems and proofs are well outside the scope of this textbook, we will use some of the principles of linear algebra to solve simple linear equations. The following is a set of n simultaneous linear equations in n unknowns:

$$a_{11}x_1 + a_{12}x_2 + \cdots + a_{1n}x_n = b_1$$
$$a_{21}x_1 + a_{22}x_2 + \cdots + a_{2n}x_n = b_2$$
$$\vdots$$
$$a_{n1}x_1 + a_{n2}x_2 + \cdots + a_{nn}x_n = b_n$$

When analyzing electric circuits, the resulting simultaneous linear equations may have either real or complex coefficients. In either case, we can solve for the unknowns using determinants, a computer program such as Mathcad, or a calculator. Note however that, while most calculators used in circuit analysis can solve simultaneous equations with real coefficients, few (other than high-end calculators such as the TI-89) can handle simultaneous equations with complex coefficients—see Note.

NOTES...

Solving Simultaneous Equations with Complex Coefficients Using Any Calculator

As an alternative, note that it is relatively easy to convert a set of n complex equations into a set of $2n$ real equations, and once you have real equations, you can use any scientific calculator that is capable of solving simultaneous equations with real coefficients. Although there is some additional work required to arrange and input the data beyond that required for calculators like the TI-89 (which can solve the complex equations directly), it may save you time and result in fewer errors than manual solution by determinants. To learn how to do it, go to our Web site and follow the links to *For Further Investigation* and select *Solving Simultaneous Equations with Complex Coefficients.*

The Determinant Approach

Consider a set of simultaneous linear equations in two variables (shown below):

$$a_{11}x_1 + a_{12}x_2 = b_1 \quad \text{(B–5)}$$

$$a_{21}x_1 + a_{22}x_2 = b_2 \quad \text{(B–6)}$$

The unknown variables x_1 and x_2 can be determined by using *determinants*. To get at the idea, let us gather up their coefficients and arrange them in an array in exactly the same order as they appear in these equations, then enclose the array within vertical lines. The resulting structure is known as a determinant. Let us call it D. Thus,

$$D = \begin{vmatrix} a_{11} & a_{12} \\ a_{21} & a_{22} \end{vmatrix}$$

The number of rows (or columns) defines the order of the determinant. Thus, this is a second-order determinant.

A determinant has a value. By definition, the value of a second-order determinant is found by taking the product of the upper-left term and the lower-right term (elements of the principal diagonal) and then subtracting the product of the lower-left term and the upper-right term (elements of the secondary diagonal). The result is given as

$$D = a_{11}a_{22} - a_{12}a_{21}$$

The unknowns of the simultaneous linear equations are found by using a technique called *Cramer's rule*. In applying this rule, we need to solve the following determinants:

$$x_1 = \frac{\begin{vmatrix} b_1 & a_{12} \\ b_2 & a_{22} \end{vmatrix}}{\begin{vmatrix} a_{11} & a_{12} \\ a_{21} & a_{22} \end{vmatrix}} = \frac{a_{22}b_1 - a_{12}b_2}{a_{11}a_{22} - a_{21}a_{12}} \qquad x_2 = \frac{\begin{vmatrix} a_{11} & b_1 \\ a_{21} & b_2 \end{vmatrix}}{\begin{vmatrix} a_{11} & a_{12} \\ a_{21} & a_{22} \end{vmatrix}} = \frac{a_{11}b_2 - a_{21}b_1}{a_{11}a_{22} - a_{21}a_{12}}$$

The application of Cramer's rule gives the solution for each unknown by first placing the determinant of the coefficients in the denominator. The numerator is then developed by using the same determinant with the exception that the coefficients of the variable to be found are replaced by the coefficients found on the right-hand side of the equal signs.

EXAMPLE B–6

Use determinants to find solutions for the following linear equations:

$$2x_1 + 8x_2 = -2$$

$$x_1 + 2x_2 = 5$$

Solution The determinant of the denominator is found as

$$D = \begin{vmatrix} 2 & 8 \\ 1 & 2 \end{vmatrix} = (2)(2) - (1)(8) = -4$$

The variables are now calculated as

$$x_1 = \frac{\begin{vmatrix} -2 & 8 \\ 5 & 2 \end{vmatrix}}{-4} = \frac{(-2)(2) - (5)(8)}{-4} = \frac{-44}{-4} = 11$$

and

$$x_2 = \frac{\begin{vmatrix} 2 & -2 \\ 1 & 5 \end{vmatrix}}{-4} = \frac{(2)(5) - (1)(-2)}{-4} = \frac{12}{-4} = -3$$

The solution of third-order simultaneous linear equations is similar to the method used for solving second-order equations. Consider the following third-order simultaneous linear equation set:

$$a_{11}x_1 + a_{12}x_2 + a_{13}x_3 = b_1$$

$$a_{21}x_1 + a_{22}x_2 + a_{23}x_3 = b_2$$

$$a_{31}x_1 + a_{32}x_2 + a_{33}x_3 = b_3$$

Although the value of the third-order determinant may be found in one of several ways, we will examine the more general approach.

This evaluation of determinants is achieved by expansion by minors. The *minor* of an element is the determinant which remains after deleting the row and the column in which the element lies. The value of any *n*th-order determinant is found as follows:

1. For any row or column, find the product of each element and the determinant of its minor.

2. A product is given a positive sign if the sum of the row and the column of the element is even. The product is given a negative sign if the sum is odd.
3. The value of the determinant is the sum of the resulting terms.

As before, Cramer's rule is used to solve for the unknowns, x_1, x_2, and x_3, by using determinants and replacing the appropriate terms of the numerator with the terms from the right-hand side of the equal sign. The resulting determinants and solutions are given as follows:

$$x_1 = \frac{\begin{vmatrix} b_1 & a_{12} & a_{13} \\ b_2 & a_{22} & a_{23} \\ b_3 & a_{32} & a_{33} \end{vmatrix}}{\begin{vmatrix} a_{11} & a_{12} & a_{13} \\ a_{21} & a_{22} & a_{23} \\ a_{31} & a_{32} & a_{33} \end{vmatrix}} \qquad x_2 = \frac{\begin{vmatrix} a_{11} & b_1 & a_{13} \\ a_{21} & b_2 & a_{23} \\ a_{31} & b_3 & a_{33} \end{vmatrix}}{\begin{vmatrix} a_{11} & a_{12} & a_{13} \\ a_{21} & a_{22} & a_{23} \\ a_{31} & a_{32} & a_{33} \end{vmatrix}} \qquad x_3 = \frac{\begin{vmatrix} a_{11} & a_{12} & b_1 \\ a_{21} & a_{22} & b_2 \\ a_{31} & a_{32} & b_3 \end{vmatrix}}{\begin{vmatrix} a_{11} & a_{12} & a_{13} \\ a_{21} & a_{22} & a_{23} \\ a_{31} & a_{32} & a_{33} \end{vmatrix}}$$

By expansion of minors, the determinant of the denominator is found as

$$D = +a_{11}\begin{vmatrix} a_{22} & a_{23} \\ a_{32} & a_{33} \end{vmatrix} - a_{21}\begin{vmatrix} a_{12} & a_{13} \\ a_{32} & a_{33} \end{vmatrix} + a_{31}\begin{vmatrix} a_{12} & a_{13} \\ a_{22} & a_{23} \end{vmatrix}$$

$$= a_{11}(a_{22}a_{33} - a_{23}a_{32}) - a_{21}(a_{12}a_{33} - a_{13}a_{32}) + a_{31}(a_{12}a_{23} - a_{13}a_{22})$$

The solution for x_1 is now found to be

$$x_1 = \frac{+b_1\begin{vmatrix} a_{22} & a_{23} \\ a_{32} & a_{33} \end{vmatrix} - b_2\begin{vmatrix} a_{12} & a_{13} \\ a_{32} & a_{33} \end{vmatrix} + b_3\begin{vmatrix} a_{12} & a_{13} \\ a_{22} & a_{23} \end{vmatrix}}{D}$$

$$= \frac{b_1(a_{22}a_{33} - a_{23}a_{32}) - b_2(a_{12}a_{33} - a_{13}a_{32}) + b_3(a_{12}a_{23} - a_{13}a_{22})}{D}$$

Similarly, for x_2, we get

$$x_2 = \frac{-b_1(a_{21}a_{33} - a_{23}a_{31}) + b_2\,(a_{11}a_{33} - a_{13}a_{31}) + b_3\,(a_{11}a_{23} - a_{13}a_{21})}{D}$$

and for x_3 we have

$$x_3 = \frac{b_1(a_{21}a_{32} - a_{22}a_{31}) - b_2(a_{11}a_{32} - a_{12}a_{31}) + b_3\,(a_{11}a_{22} - a_{12}a_{21})}{D}$$

EXAMPLE B–7

Solve for x_1 in the following system of linear equations using minors.

$$3x_1 + x_2 - 2x_3 = 1$$
$$x_1 - 2x_2 + 3x_3 = 11$$
$$2x_1 + 3x_2 + 2x_3 = -3$$

Solution The determinant of the denominator is evaluated as follows:

$$D = \begin{vmatrix} 3 & 1 & -2 \\ 1 & -2 & 3 \\ 2 & 3 & 2 \end{vmatrix}$$

$$= +(3)\begin{vmatrix} -2 & 3 \\ 3 & 2 \end{vmatrix} - (1)\begin{vmatrix} 1 & -2 \\ 3 & 2 \end{vmatrix} + (2)\begin{vmatrix} 1 & -2 \\ -2 & 3 \end{vmatrix}$$

$$= (3)(-4 - 9) - (2 + 6) + (2)\,(3 - 4)$$

$$= -49$$

and so the unknown x_1 is calculated to be

$$x_1 = \frac{\begin{vmatrix} 1 & 1 & -2 \\ 11 & -2 & 3 \\ -3 & 3 & 2 \end{vmatrix}}{-49} = \frac{+(1)\begin{vmatrix} -2 & 3 \\ 3 & 2 \end{vmatrix} - (11)\begin{vmatrix} 1 & -2 \\ 3 & 2 \end{vmatrix} + (-3)\begin{vmatrix} 1 & -2 \\ -2 & 3 \end{vmatrix}}{-49}$$

$$= \frac{(-4 - 9) - (11)(2 + 6) - (3)(3 - 4)}{-4} = 2$$

PRACTICE PROBLEM

Use expansion by minors to solve for x_2 and x_3 in Example B–7.

Answers
$x_2 = -3,\ x_3 = 1$

The Calculator Approach

As noted previously, many calculators have built-in routines for solving simultaneous equations that have real coefficients. The Sharp EL-506W/EL-546W calculator can solve for both second-order and third-order simultaneous linear equations (having real coefficients).

EXAMPLE B–8

Equations with Real Coefficients

Consider the system of linear equations shown in the previous example:

$$3x_1 + x_2 - 2x_3 = 1$$
$$x_1 - 2x_2 + 3x_3 = 11$$
$$2x_1 + 3x_2 + 2x_3 = -3$$

Solve for the unknowns, x_1, x_2, and x_3.

Solution We begin by setting the calculator into the correct mode to solve the simultaneous linear equations. For the Sharp EL-506W/EL-546W, this is done by keying [MODE] [2]. You will see that you have the option of entering the coefficients for either a second-order (2-VLE) or a third-order (3-VLE) linear system of equations. Since we wish to solve a third-order system, we key in [1]. You will observe that you are prompted with a display showing *a1?*. This corresponds to the first coefficient of the first equation. You will need to complete the entries as follows:

a1?	[3]	[=]	a2?	[1]	[=]	a3?	[2]	[=]
b1?	[1]	[=]	b2?	[–2]	[=]	b3?	[3]	[=]
c1?	[–2]	[=]	c2?	[3]	[=]	c3?	[2]	[=]
d1?	[1]	[=]	d2?	[11]	[=]	d3?	[–3]	[=]

The results are displayed as x = 2, y = −3, and z = 1. Notice that these are the same results that were determined in Example B–7 and in the corresponding Practice Problem.

EXAMPLE B–9

Equations with Complex Coefficients

Most calculators (including the Sharp EL-506W/EL-546W) cannot directly solve simultaneous linear equations having complex coefficients. Instead, you will need to use determinants (or some other method) to solve for the unknowns in a linear equation. This example shows how to solve a 2 × 2 set of simultaneous linear equations.

Complex Linear Equation

Given the following simultaneous linear equations, solve for $\mathbf{I}_1$ and $\mathbf{I}_2$:

$$(25 + j100)\mathbf{I}_1 - (10 + j80)\mathbf{I}_2 = 100\angle 0°$$
$$-(10 + j80)\mathbf{I}_1 + (30 + j190)\mathbf{I}_2 = 0$$

Solution Similarly to Example B–6, create the denominator,

$$\mathbf{D} = \begin{vmatrix} 25 + j100 & -(10 + j80) \\ -(10 + j80) & 30 + j190 \end{vmatrix}$$

Ensure that your calculator is set for complex mode, then evaluate this determinant using a procedure similar to that of Example B–6 as shown below. (Coefficients can be entered as rectangular or polar. Depending on how your calculator is set, your answers will be either in rectangular or polar form.)

$$\begin{aligned} \mathbf{D} &= (25 + j100)(30 + j190) - [-(10 + j80)][-(10 + j80)] \\ &= (25 + j100)(30 + j190) - (10 + j80)(10 + j80) \\ &= -11{,}950 + j6150 \\ &= 13{,}439.7\angle 152.77° \end{aligned}$$

Next, as we did in Example B–6, solve for the unknowns, $\mathbf{I}_1$ and $\mathbf{I}_2$. Thus,

$$\mathbf{I}_1 = \frac{\begin{vmatrix} 100 & -(10 + j80) \\ 0 & 30 + j190 \end{vmatrix}}{\mathbf{D}} = \frac{100(30 + j190) - 0}{\mathbf{D}}$$

$$= \frac{3000 + j19{,}000}{13{,}439.7\angle 152.77°} = 1.43\text{ A}\angle -71.74°$$

and

$$\mathbf{I}_2 = \frac{\begin{vmatrix} 25 + j100 & 100 \\ -(10 + j80) & 0 \end{vmatrix}}{\mathbf{D}} = \frac{0 - 100[-(30 + j190)]}{\mathbf{D}}$$

$$= \frac{1000 + j8000}{13{,}439.7\angle 152.77°} = 0.600\text{ A}\angle -69.89°$$

NOTES...

For additional information on solving linear equations having complex coefficients, you may wish to use the procedure detailed on our Web site. See Note on page 959.

Appendix C

MAXIMUM POWER TRANSFER THEOREM

Figure C–1 shows the Thévenin equivalent of a dc circuit.

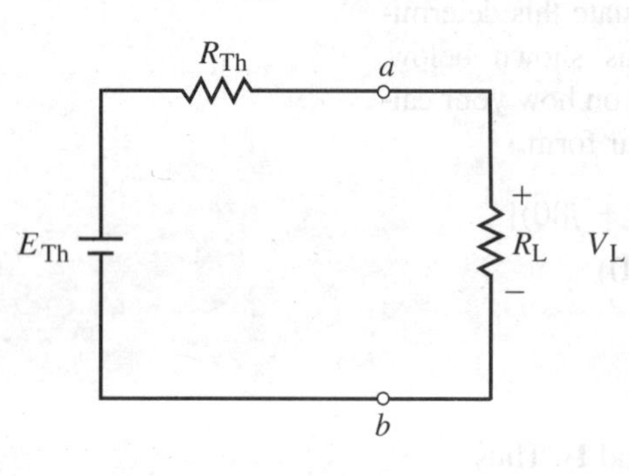

FIGURE C–1

For the preceding circuit, the values of E_{Th} and R_{Th} are constant. Therefore, power delivered to the load is determined as a function of the load resistance and is given as

$$P_L = \frac{V_L^2}{R_L} = \frac{\left(\frac{R_L E_{Th}}{R_L + R_{Th}}\right)^2}{R_L} = \frac{E_{Th}^2 R_L}{(R_L + R_{Th})^2} \tag{C–1}$$

Maximum power will be delivered to R_L when the first derivative, $\frac{dP_L}{dR_L} = 0$.

Applying the quotient rule, $\frac{d}{dx}\left(\frac{u}{v}\right) = \frac{v\frac{du}{dx} - u\frac{dv}{dx}}{v^2}$, we find the derivative of power with respect to load resistance as

$$\begin{aligned} \frac{dP_L}{dR_L} &= \frac{(R_L + R_{Th})^2(E_{Th})^2 - (E_{Th}{}^2 R_L)(2)(R_L + R_{Th})}{(R_L + R_{Th})^4} \\ &= \frac{E_{Th}{}^2[(R_L + R_{Th})^2 - 2R_L(R_L + R_{Th})]}{(R_L + R_{Th})^4} \end{aligned} \tag{C–2}$$

Now, since the first derivative can only be zero if the numerator of the preceding expression is zero, and since E_{Th} is a constant, we have

$$(R_L + R_{Th})^2 - 2R_L(R_L + R_{Th}) = 0 \qquad \textbf{(C–3)}$$

And so,

$$\begin{aligned} R_L^2 + 2R_LR_{Th} + R_{Th}^2 - 2R_LR_L - 2R_LR_{Th} &= 0 \\ R_{Th}^2 - R_L^2 &= 0 \\ R_L &= R_{Th} \end{aligned} \qquad \textbf{(C–4)}$$

Figure C–2 shows the Thévenin equivalent of an ac circuit.

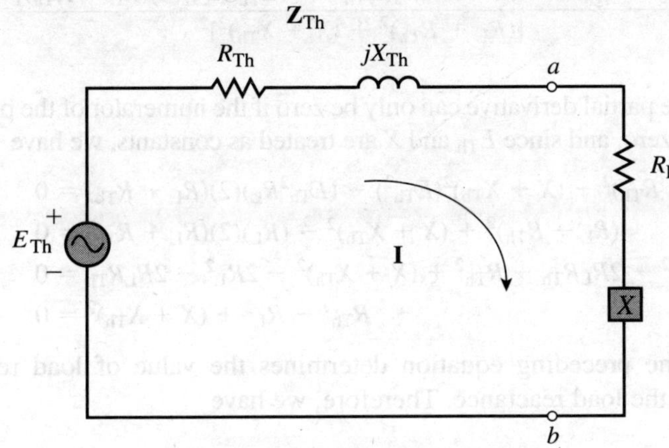

FIGURE C–2

For the preceding circuit, the values of E_{Th}, R_{Th}, and X_{Th} are constant. Although X_{Th} is shown as an inductor, it could just as easily be a capacitor. The power delivered to the load is determined as a function of the load impedance as

$$P_L = I^2R_L = \frac{E_{Th}^2R_L}{(R_L + R_{Th})^2 + (X + X_{Th})^2} \qquad \textbf{(C–5)}$$

Maximum power is dependent on two variables, R_L and X. Therefore, we will need to solve for partial derivatives. Maximum power will be transferred to the load when $\dfrac{\partial P_L}{\partial R_L} = 0$ and $\dfrac{\partial P_L}{\partial X} = 0$.

We begin by finding $\dfrac{\partial P_L}{\partial X}$. Using $\dfrac{d}{dx}\left(\dfrac{1}{v}\right) = -\dfrac{1}{v^2}\dfrac{dv}{dx}$, we get

$$\frac{\partial P_L}{\partial X} = -\frac{(E_{Th}^2R_L)(2)(X + X_{Th})}{[(R_L + R_{Th})^2 + (X + X_{Th})^2]^2} \qquad \textbf{(C–6)}$$

Now, since the partial derivative can only be zero if the numerator of the preceding expression is zero, and since E_{Th} and R_L are treated as constants, we have

$$X + X_{Th} = 0$$

or

$$X = -X_{Th} \qquad \textbf{(C–7)}$$

This result implies that if the Thévenin impedance contains an inductor of magnitude X, the load must contain a capacitor with the same magnitude.

(Conversely, if the Thévenin impedance were to contain a capacitor, then the load impedance would need to have an inductor of the same magnitude.)

Next, we determine the partial derivative, $\frac{\partial P_L}{\partial R_L}$ of Equation C–5. Applying the quotient rule,

$$\frac{d}{dx}\left(\frac{u}{v}\right) = \frac{v\frac{du}{dx} - u\frac{dv}{dx}}{v^2}$$

we get

$$\frac{\partial P_L}{\partial R_L} = \frac{[(R_L + R_{Th})^2 + (X + X_{Th})^2](E_{Th}^2) - (E_{Th}^2 R_L)[(2)(R_L + R_{Th})]}{[(R_L + R_{Th})^2 + (X + X_{Th})^2]^2} \qquad \textbf{(C–8)}$$

Now, since the partial derivative can only be zero if the numerator of the preceding expression is zero, and since E_{Th} and X are treated as constants, we have

$$(R_L + R_{Th})^2 + (X + X_{Th})^2(E_{Th}^2) - (E_{Th}^2 R_L)(2)(R_L + R_{Th}) = 0$$
$$(R_L + R_{Th})^2 + (X + X_{Th})^2 - (R_L)(2)(R_L + R_{Th}) = 0$$
$$R_L^2 + 2R_L R_{Th} + R_{Th}^2 + (X + X_{Th})^2 - 2R_L^2 - 2R_L R_{Th} = 0$$
$$R_{Th}^2 - R_L^2 + (X + X_{Th})^2 = 0$$

In general, the preceding equation determines the value of load resistance regardless of the load reactance. Therefore, we have

$$R_L = \sqrt{R_{Th}^2 + (X + X_{Th})^2} \qquad \textbf{(C–9)}$$

This result shows that the load resistance is dependent upon the load reactance, X. If the reactance of the load is the same type as the Thévenin reactive component (both inductive or both capacitive), then the reactances are added. If the load reactance is the opposite type of the Thévenin reactive component, then the reactances are subtracted. If the reactance of the load can be adjusted to result in maximum power transfer ($X = -X_{Th}$), then Equation C–9 is simplified to the expected result, namely

$$R_L = X_{Th} \qquad \textbf{(C–10)}$$

Appendix D

ANSWERS TO ODD-NUMBERED PROBLEMS

CHAPTER 1

1. a. 1620 s c. 7427 s e. 2.45 hp
   b. 2880 s d. 26 110 W f. 8280°
3. a. 0.84 m^2 c. 0.02 m^3
   b. 0.0625 m^2 d. 0.0686 m^3
5. 4500 parts/h
7. 11.5 km/l
9. 150 rpm
11. 8.33 mi
13. 0.508 m/s
15. 7.45 km
17. 20.4 min
19. Machine 1: \$25.80/h; Machine 2: \$25.00/h; Machine 2
21. a. 8.675×10^3 d. 3.72×10^{-1} g. 1.47×10^1
    b. 8.72×10^{-3} e. 3.48×10^2
    c. 1.24×10^3 f. 2.15×10^{-7}
23. a. 1.25×10^{-1} c. 2.0×10^{-2}
    b. 8×10^7 d. 2.05×10^4
25. a. 10 c. 3.6×10^3 e. −12.0
    b. 10 d. 15×10^4
27. 1.179; 4.450; Direct computation is less work for these examples.
29. 6.24×10^{18}
31. 62.6×10^{21}
33. 1.16 s
35. 13.4×10^{10} l/h
37. a. kilo, k c. giga, G e. milli, m
    b. mega, M d. micro, μ f. pico, p
39. a. 1.5 ms b. 27 μs c. 350 ns
41. a. 150, 0.15 b. 0.33, 33
43. a. 680 V b. 162.7 W
45. 1.5 kW
47. 187 A
49. 39 pF
51. Radio signal, 16.68 ms; Telephone signal, 33.33 ms; The radio signal by 16.65 ms.
53. a. 2.083 kΩ
    b. The actual value of R lies somewhere between 2.07 kΩ and 2.10 kΩ.

CHAPTER 2

1. a. 10^{29} b. 10.4×10^{23}
3. Increases by a factor of 24.
5. a. Material with many free electrons (i.e., material with 1 electron in the valence shell).
   b. Inexpensive and easily formed into wires.
   c. Full valence shell. Therefore, no free electrons.
   d. The large electrical force tears electrons out of orbit.
7. a. 630 N (repel) d. 8.22×10^{-8} N (attract)
   b. 20 N (attract) e. Neutron is uncharged, thus $F = 0$
   c. 1.6×10^{-14} N (repel)
9. 2 μC; (Attraction)
11. 0.333 μC, 1.67 μC; both (+) or both (−)
13. 30.4 μC
15. 27.7 μC (+)
17. 24 V
19. 2400 V
21. 4.25 mJ
23. 4.75 C
25. 50 mA
27. 334 μC

29. 3 mA
31. 80 A
33. 18 V, 0.966 A
35. a. 4.66 V b. 1.50 V
37. 50 h
39. 11.7 h
41. (c) Both
43. Meter 1, 7 A; Meter 2, −7 A
45. The voltmeter and ammeter are interchanged.
47. If you exceed a fuse's voltage rating, it may arc over when it "blows."

CHAPTER 3

1. a. 3.6 Ω c. 36.0 kΩ
   b. 0.90 Ω d. 36.0 mΩ
3. 0.407 inch
5. 300 m = 986 feet
7. 982×10^{-8} Ω · m (Resistivity is less than for carbon.)
9. 2.26×10^{-8} Ω · m (This alloy is not as good a conductor as copper.)
11. AWG 22: 4.86 Ω

    AWG 19: 2.42 Ω

    Diameter of AWG 19 is 1.42 times the diameter of AWG 22. The resistance of AWG 19 is half the resistance of an equal length of AWG 22.
13. AWG 19 should be able to handle 4 A.

    AWG 30 can handle about 0.30 A.
15. 405 meters
17. a. 256 CM b. 6200 CM c. 1910 MCM
19. a. 16.2 Ω b. 0.668 Ω c. 2.17×10^{-3} Ω
21. a. 4148 CM = 3260 sq mil b. 0.0644 inch
23. a. 1600 CM = 1260 sq mil b. 1930 feet
25. $R_{-30°C} = 40.2$ Ω $R_{0°C} = 46.1$ Ω $R_{200°C} = 85.2$ Ω
27. a. Positive temperature coefficient
    b. $0.00385\ (°C)^{-1}$
    c. $R_{0°C} = 18.5$ Ω $R_{100°C} = 26.2$ Ω
29. 16.8 Ω
31. $T = -260°C$
33. a. $R_{ab} = 10$ kΩ $R_{bc} = 0$ Ω
    b. $R_{ab} = 8$ kΩ $R_{bc} = 2$ kΩ
    c. $R_{ab} = 2$ kΩ $R_{bc} = 8$ kΩ
    d. $R_{ab} = 0$ kΩ $R_{bc} = 10$ kΩ
35. a. 150 kΩ ± 10%
    b. 2.8 Ω ± 5% with a reliability of 0.001%
    c. 47 MΩ ± 5%
    d. 39 Ω ± 5% with a reliability of 0.1%
37. Connect the ohmmeter between the two terminals of the light bulb. If the ohmmeter indicates an open circuit, the light bulb is burned out.
39. AWG 24 has a resistance of 25.7 Ω/1000 ft. Measure the resistance between the two ends and calculate the length as

$$\ell = \frac{R}{0.0257\ \Omega/\text{ft}}$$

41. a. 380 Ω
    b. 180 Ω
    c. Negative temperature coefficient. Resistance decreases as temperature increases.
43. a. 4.0 S c. 4.0 μS
    b. 2.0 mS d. 0.08 μS
45. 2.93 mS

CHAPTER 4

1. a. 2 A c. 5 mA e. 3 mA
   b. 7.0 A d. 4 μA f. 6 mA
3. a. 40 V c. 400 V
   b. 0.3 V d. 0.36 V
5. 96 Ω
7. 28 V
9. 6 A
11. Red, Red, Red
13. 22 V
15. a. 2.31 A b. 2.14 A
17. 2.88 V
19. 4 Ω
21. 400 V
23. 3.78 mA
25. a. + 45 V − c. − 90 V +
    b. 4 A (→) d. 7 A (←)
27. 3.19 J/s; 3.19 W
29. 36 W
31. 14.1 A
33. 47.5 V
35. 50 V, 5 mA
37. 37.9 A
39. 2656 W
41. 23.2 V; 86.1 mA
43. 361 W → 441 W
45. a. 48 W (→) c. 128 W (←)
    b. 30 W (←) d. 240 W (→)
47. a. 1.296×10^{6} J b. 360 Wh c. 2.88 cents
49. 26 cents
51. $5256
53. 5 cents
55. 51.5 kW
57. 82.7%
59. 2.15 hp
61. 8.8 hp
63. 1.97 hp
65. $137.45
67. a. 10 Ω b. 13.3 Ω

CHAPTER 5

1. a. +30 V b. −90 V
3. a. +45 V b. −60 V
   c. +90 V d. −105 V
5. a. 7 V b. $V_2 = 4$ V $V_1 = 4$ V
7. $V_3 = 12$ V $V_4 = 2$ V
9. a. 10 kΩ b. 2.94 MΩ c. 23.4 kΩ
11. Circuit 1: 1650 Ω, 6.06 mA Circuit 2: 18.15 kΩ, 16.5 mA
13. a. 10 mA
   b. 13 kΩ
   c. 5 kΩ
   d. $V_{3\text{-k}\Omega} = 30$ V $V_{4\text{-k}\Omega} = 40$ V $V_{1\text{-k}\Omega} = 10$ V $V_R = 50$ V
   e. $P_{1\text{-k}\Omega} = 100$ mW $P_{3\text{-k}\Omega} = 300$ mW $P_{4\text{-k}\Omega} = 400$ mW
   $P_R = 500$ mW
15. a. 40 mA b. $V_{R_1} = 12$ V $V_{R_3} = 10$ V c. 26 V
17. a. $V_{R_2} = 4.81$ V $V_{R_3} = 3.69$ V
   b. 1.02 mA c. 7.32 kΩ
19. a. 457 Ω
   b. 78.8 mA
   c. $V_1 = 9.45$ V $V_2 = 3.07$ V $V_3 = 6.14$ V
   $V_4 = 17.33$ V
   d. $V_T = 36$ V
   e. $P_1 = 0.745$ W $P_2 = 0.242$ W $P_3 = 0.484$ W
   $P_4 = 1.365$ W
   f. R_1: 1 W R_2: 1/4 W R_3: 1/2 W R_4: 2 W
   g. 2.836 W
21. $R = 2.02$ kΩ
23. a. 0.15 A b. 0.115 mA
25. Circuit 1: $V_{6\text{-}\Omega} = 6$ V $V_{3\text{-}\Omega} = 3$ V $V_{5\text{-}\Omega} = 5$ V $V_{8\text{-}\Omega} = 8$ V
   $V_{2\text{-}\Omega} = 2$ V $V_T = 24$ V
   Circuit 2: $V_{4.3\text{-k}\Omega} = 21.6$ V $V_{2.7\text{-k}\Omega} = 13.6$ V
   $V_{7.8\text{-k}\Omega} = 39.2$ kΩ $V_{9.1\text{-k}\Omega} = 45.7$ V $V_T = 120$ V
27. Circuit 1:
   a. $R_1 = 0.104$ kΩ $R_2 = 0.365$ kΩ $R_3 = 0.730$ kΩ
   b. $V_1 = 2.09$ V $V_2 = 7.30$ V $V_3 = 14.61$ V
   c. $P_1 = 41.7$ mW $P_2 = 146.1$ mW $P_3 = 292.2$ mW
   Circuit 2:
   a. $R_1 = 977$ Ω $R_2 = 244$ Ω $R_3 = 732$ Ω
   b. $V_1 = 25.0$ V $V_2 = 6.25$ V $V_3 = 18.75$ V
   c. $P_1 = 640$ mW $P_2 = 160$ mW $P_3 = 480$ mW
29. a. 0.2 A
   b. 5.0 V
   c. 1 W
   d. $R_T = 550$ Ω $I = 0.218$ A $V = 5.45$ V $P = 1.19$ W
   e. Life expectancy decreases.
31. Circuit 1: $V_{ab} = 9.39$ V $V_{bc} = 14.61$ V
   Circuit 2: $V_{ab} = +25.0$ V $V_{bc} = +6.25$ V
33. Circuit 1:
   $V_{3\text{k-}\Omega} = 9$ V $V_{9\text{k-}\Omega} = 27$ V $V_{6\text{k-}\Omega} = 18$ V $V_a = 45$ V
   Circuit 2: $V_{330\text{-}\Omega} = 2.97$ V $V_{670\text{-}\Omega} = 6.03$ V $V_a = 3.03$ V
35. a. 109 V b. 9.20 Ω
37. Circuit 1:
   $I_{actual} = 0.375$ mA $I_{measured} = 0.3745$ mA
   loading error = 0.125%
   Circuit 2:
   $I_{actual} = 0.375$ mA $I_{measured} = 0.3333$ mA
   loading error = 11.1%
39. Circuit 1:
   a. 1 A
   b. $V_{6\text{-}\Omega} = 6$ V $V_{3\text{-}\Omega} = 3$ V $V_{5\text{-}\Omega} = 5$ V $V_{8\text{-}\Omega} = 8$ V
   $V_{2\text{-}\Omega} = 2$ V
   Circuit 2:
   a. 5.06 mA
   b. $V_{4.3\text{-k}\Omega} = 21.7$ V $V_{2.7\text{-k}\Omega} = 13.6$ V $V_{7.8\text{-k}\Omega} = 39.1$ V
   $V_{9.1\text{-k}\Omega} = 45.6$ V
41. a. 78.8 mA
   b. $V_1 = 9.45$ V $V_2 = 3.07$ V $V_3 = 6.14$ V
   $V_4 = 17.33$ V

CHAPTER 6

1. a. A and B are in series; D and E are in series;
   C and F are parallel
   b. B, C, and D are parallel
   c. A and B are parallel; D and F are parallel;
   C and E are in series
   d. A, B, C, and D are parallel
5. a. $I_1 = 3$ A $I_2 = -1$ A
   b. $I_1 = 7$ A $I_2 = 2$ A $I_3 = -7$ A
   c. $I_1 = 4$ mA $I_2 = 20$ mA
7. a. $I_1 = 1.25$ A $I_2 = 0.0833$ A $I_3 = 1.167$ A
   $I_4 = 1.25$ A
   b. $R_3 = 4.29$ Ω
9. a. $I_1 = 200$ mA $I_2 = 500$ mA $I_3 = 150$ mA
   $I_4 = 200$ mA
   b. 2.5 V
   c. $R_1 = 12.5$ Ω $R_3 = 16.7$ Ω $R_4 = 50$ Ω
11. a. $R_T = 2.4$ Ω $G_T = 0.417$ S
   b. $R_T = 32$ kΩ $G_T = 31.25$ μS
   c. $R_T = 4.04$ kΩ $G_T = 247.6$ μS
13. a. 2.0 MΩ b. 450 Ω
15. a. $R_1 = 1250$ Ω $R_2 = 5$ kΩ $R_3 = 250$ Ω
   b. $I_{R1} = 0.40$ A $I_{R2} = 0.10$ A
   c. 2.5 A
17. a. 900 mV b. 4.5 mA
19. a. 240 Ω b. 9.392 kΩ c. 1.2 kΩ
21. $I_1 = 0.235$ mA $I_2 = 0.706$ mA $I_3 = 1.059$ mA
   $R_1 = 136$ kΩ $R_2 = 45.3$ kΩ $R_3 = 30.2$ kΩ
23. a. 12.5 kΩ b. 0 c. 75 Ω
25. $R_T \cong 15$ Ω
27. $I = 0.2$ A $I_1 = 0.1$ A $= I_2$

29. a. $I_1 = 2$ A $I_2 = 8$ A
 b. $I_1 = 4$ mA $I_2 = 12$ mA
31. a. $I_1 = 6.48$ mA $I_2 = 9.23$ mA $I_3 = 30.45$ mA
 $I_4 = 13.84$ mA
 b. $I_1 = 60$ mA $I_2 = 30$ mA $I_3 = 20$ mA $I_4 = 40$ mA
 $I_5 = 110$ mA
33. 12 Ω
35. a. 8 Ω
 b. 1.50 A
 c. $I_1 = 0.50$ A $I_2 = 0.25$ A $I_3 = 0.75$ A
 d. $\Sigma I_{in} = \Sigma I_{out} = 1.50$ A
37. a. 25 Ω $I = 9.60$ A
 b. $I_1 = 4.0$ A $I_2 = 2.40$ A $I_3 = 3.20$ A $I_4 = 5.60$ A
 c. $\Sigma I_{in} = \Sigma I_{out} = 9.60$ A
 d. $P_1 = 960$ W $P_2 = 576$ W $P_3 = 768$ W
 $P_T = 2304$ W $= P_1 + P_2 + P_3$
39. a. $I_1 = 1.00$ A $I_2 = 2.00$ A $I_3 = 5.00$ A $I_4 = 4.00$ A
 b. 12.00 A
 c. $P_1 = 20$ W $P_2 = 40$ W $P_3 = 100$ W $P_4 = 80$ W
41. a. $R_1 = 2$ kΩ $R_2 = 8$ kΩ $R_3 = 4$ kΩ $R_4 = 6$ kΩ
 b. $I_{R_1} = 24$ mA $I_{R_2} = 6$ mA $I_{R_4} = 8$ mA
 c. $I_1 = 20$ mA $I_2 = 50$ mA
 d. $P_2 = 288$ mW $P_3 = 576$ mW $P_4 = 384$ mW
43. $I_1 = 8.33$ A $I_2 = 5.00$ A $I_3 = 2.50$ A $I_4 = 7.50$ A
 $I_T = 15.83$ A
 The rated current of the fuse will be exceeded; the fuse will "blow."
45. a. $V_{measured} = 20$ V b. loading effect = 33.3%
47. 25.2 V
49. $I_1 = 4.0$ A $I_2 = 2.4$ A $I_3 = 3.2$ A
51. 20 V
53. $I_1 = 4.0$ A $I_2 = 2.4$ A $I_3 = 3.2$ A

CHAPTER 7

1. a. $R_T = R_1 + R_5 + [(R_2 + R_3) \| R_4]$
 b. $R_T = (R_1 \| R_2) + (R_3 \| R_4)$
3. a. $R_{T1} = R_1 + [(R_3 + R_4) \| R_2] + R_5$, $R_{T2} = R_5$
 b. $R_{T1} = R_1 + (R_2 \| R_3 \| R_5)$, $R_{T_2} = R_5 \| R_3 \| R_2$
7. a. 1500 Ω b. 2.33 kΩ
9. $R_{ab} = 140$ Ω $R_{cd} = 8.89$ Ω
11. a. $R_T = 314$ Ω
 b. $I_T = 63.7$ mA $I_1 = 19.2$ mA $I_2 = 44.5$ mA
 $I_3 = 34.1$ mA $I_4 = 10.4$ mA
 c. $V_{ab} = 13.6$ V $V_{bc} = -2.9$ V
13. a. $I_1 = 5.19$ mA $I_2 = 2.70$ mA $I_3 = 1.081$ mA
 $I_4 = 2.49$ mA $I_5 = 1.621$ mA $I_6 = 2.70$ mA
 b. $V_{ab} = 12.43$ V $V_{cd} = 9.73$ V
 c. $P_T = 145.3$ mW $P_1 = 26.9$ mW $P_2 = 7.3$ mW
 $P_3 = 3.5$ mW $P_4 = 30.9$ mW $P_5 = 15.8$ mW
 $P_6 = 7.0$ mW $P_7 = 53.9$ mW
15. Circuit (a):
 a. $I_1 = 4.5$ mA $I_2 = 4.5$ mA $I_3 = 1.5$ mA
 b. $V_{ab} = -9.0$ V
 c. $P_T = 162$ mW $P_{6\text{-k}\Omega} = 13.5$ mW $P_{3\text{-k}\Omega} = 27.0$ mW
 $P_{2\text{-k}\Omega} = 40.5$ mW $P_{4\text{-k}\Omega} = 81.0$ mW
 Circuit (b):
 a. $I_1 = 0.571$ A $I_2 = 0.365$ A $I_3 = 0.122$ A $I_4 = 0.449$ A
 b. $V_{ab} = -1.827$ V
 c. $P_T = 5.14$ W $P_{10\text{-}\Omega} = 3.26$ W $P_{16\text{-}\Omega} = 0.68$ W
 $P_{5\text{-}\Omega} = 0.67$ W $P_{6\text{-}\Omega} = 0.36$ W $P_{8\text{-}\Omega} = 0.12$ W
 $P_{4\text{-}\Omega} = 0.06$ W
17. $I_1 = 93.3$ mA $I_2 = 52.9$ mA $I_Z = 40.4$ mA $V_1 = 14$ V
 $V_2 = 2.06$ V $V_3 = 7.94$ V $P_T = 2240$ mW
 $P_1 = 1307$ mW $P_2 = 109$ mW $P_3 = 420$ mW
 $P_Z = 404$ mW
19. $R = 31.1\ \Omega \rightarrow 3900\ \Omega$
21. $I_C = 1.70$ mA $V_B = -1.97$ V $V_{CE} = -8.10$ V
23. a. $I_D = 3.6$ mA b. $R_S = 556$ Ω c. $V_{DS} = 7.6$ V
25. $I_C \cong 3.25$ mA $V_{CE} \cong -8.90$ mV
27. a. $V_L = 0 \rightarrow 7.2$ V b. $V_L = 2.44$ V c. $V_{ab} = 9.0$ V
29. $V_{bc} = 7.45$ V $V_{ab} = 16.55$ V
31. a. $V_{out(min)} = 0$ V $V_{out(max)} = 40$ V
 b. $R_2 = 3.82$ kΩ
33. 0 V, 8.33 V, 9.09 V
35. a. 11.33 V c. 44.1%
 b. 8.95 V d. 1.333 V
37. a. Break the circuit between the 5.6-Ω resistor and the voltage source. Insert the ammeter at the break, connecting the red (+) lead of the ammeter to the positive terminal of the voltage source and the black (−) lead to the 5.6-Ω resistor.
 b. $I_{1(loaded)} = 19.84$ mA $I_{2(loaded)} = 7.40$ mA
 $I_{3(loaded)} = 12.22$ mA
 c. loading effect (I_1) = 19.9%
 loading effect (I_2) = 18.0%
 loading effect (I_3) = 22.3%
39. 12.0 V, 30.0 V, 5.00 A, 3.00 A, 2.00 A
41. 14.1 V
43. 12.0 V, 30.0 V, 5.00 A, 3.00 A, 2.00 A

CHAPTER 8

1. 38 V
3. a. 12 mA b. $V_S = 4.4$ V $V_1 = 2.0$ V
5. $I_1 = 400\ \mu$A $I_2 = 500\ \mu$A
7. $P_T = 7.5$ mW $P_{50\text{-k}\Omega} = 4.5$ mW $P_{150\text{-k}\Omega} = 1.5$ mW
 $P_{current\ source} = 1.5$ mW
 Note: The current source is absorbing energy from the circuit rather than providing energy.
9. Circuit (a):
 0.25 A-source in parallel with a 20-Ω resistor
 Circuit (b):
 12.5-mA source in parallel with a 2-kΩ resistor

11. a. 7.2 A b. $E = 3600$ V $I_L = 7.2$ A
13. a. 21.45 V b. 6.06 mA c. 0.606 V
15. $V_2 = -80$ V $I_1 = -26.7$ mA
17. $V_{ab} = -7.52$ V $I_3 = 0.133$ mA
19. $I_1 = 0.467$ A $I_2 = 0.167$ A $I_3 = 0.300$ A
21. $I_2 = -0.931$ A
23. a. $(8\ \Omega)I_1 + 0\ I_2 - (10\ \Omega)I_3 = 24$ V
    $0\ I_1 + (4\ \Omega)I_2 + (10\ \Omega)I_3 = 16$ V
    $I_1 - I_2 + I_3 = 0$
    b. $I = 3.26$ A
    c. $V_{ab} = -13.89$ V
25. $I_1 = 0.467$ A $I_2 = 0.300$ A
27. $I_2 = -0.931$ A
29. $I_1 = -19.23$ mA $V_{ab} = 2.77$ V
31. $I_1 = 0.495$ A $I_2 = 1.879$ A $I_3 = 1.512$ A
33. $V_1 = -6.73$ V $V_2 = 1.45$ V
35. $V_1 = -6$ V $V_2 = 20$ V
37. $V_{6\Omega} = 6.10$ V
39. Network (a): $R_1 = 6.92\ \Omega$ $R_2 = 20.77\ \Omega$ $R_3 = 62.33\ \Omega$
    Network (b): $R_1 = 1.45$ kΩ $R_2 = 2.41$ kΩ $R_3 = 2.03$ kΩ
41. Network (a): $R_A = 110\ \Omega$ $R_B = 36.7\ \Omega$ $R_C = 55\ \Omega$
    Network (b): $R_A = 793$ kΩ $R_B = 1693$ kΩ $R_C = 955$ kΩ
43. $I = 6.67$ mA
45. $I = 0.149$ A
47. a. The bridge is not balanced.
    b. $(18\ \Omega)I_1 - (12\ \Omega)I_2 - (6\ \Omega)I_3 = 15$ V
    $-(12\ \Omega)I_1 + (54\ \Omega)I_2 - (24\ \Omega)I_3 = 0$
    $-(6\ \Omega)I_1 - (24\ \Omega)I_2 + (36\ \Omega)I_3 = 0$
    c. $I = 38.5$ mA
    d. $V_{R_5} = 0.923$ V
49. $I_{R_5} = 0$ $I_{RS} = 60$ mA $I_{R_1} = I_{R_3} = 45$ mA
    $I_{R_2} = I_{R_4} = 15$ mA
51. $I_{R_1} = 0.495$ A $I_{R_2} = 1.384$ A $I_{R_3} = 1.879$ A
    $I_{R_4} = 1.017$ A $I_{R_5} = 0.367$ A
53. $I_{R_1} = 6.67$ mA $I_{R_2} = 0$ $I_{R_3} = 6.67$ mA
    $I_{R_4} = 6.67$ mA $I_{R_5} = 6.67$ mA $I_{R_6} = 13.33$ mA

CHAPTER 9

1. $I_{R_1} = 75$ mA (up) $I_{R_2} = 75$ mA (to the right)
   $I_{R_3} = 87.5$ mA (down) $I_{R_4} = 12.5$ mA (to the right)
3. $V_a = -3.11$ V $I_1 = 0.1889$ A
5. $E = 30$ V $I_L(1) = 2.18$ mA $I_L(2) = 2.82$ mA
7. $R_{Th} = 20\ \Omega$ $E_{Th} = 10$ V $V_{ab} = 6.0$ V
9. $R_{Th} = 2.02$ kΩ $E_{Th} = 1.20$ V $V_{ab} = -0.511$ V
11. a. $R_{Th} = 16\ \Omega$ $E_{Th} = 5.6$ V
    b. When $R_L = 20\ \Omega$: $V_{ab} = 3.11$ V
    When $R_L = 50\ \Omega$: $V_{ab} = 4.24$ V
13. a. $E_{Th} = 75$ V $R_{Th} = 50\ \Omega$
    b. $I = 0.75$ A
15. a. $E_{Th} = 50$ V $R_{Th} = 3.8$ kΩ
    b. $I = 13.21$ mA
17. a. $R_{Th} = 60$ kΩ $E_{Th} = 25$ V
    b. $R_L = 0$: $I = -0.417$ mA
    $R_L = 10$ kΩ: $I = -0.357$ mA
    $R_L = 50$ kΩ: $I = -0.227$ mA
19. a. $E_{Th} = 28.8$ V, $R_{Th} = 16$ kΩ
    b. $R_L = 0$: $I = 1.800$ mA
    $R_L = 10$ kΩ: $I = 1.108$ mA
    $R_L = 50$ kΩ: $I = 0.436$ mA
21. $E_{Th} = 4.56$ V $R_{Th} = 7.2\ \Omega$
23. a. $E_{Th} = 8$ V $R_{Th} = 200\ \Omega$
    b. $I = 22.2$ mA (upward)
25. $I_N = 0.5$ A, $R_N = 20\ \Omega$, $I_L = 0.2$ A
27. $I_N = 0.594$ mA, $R_N = 2.02$ kΩ, $I_L = 0.341$ mA
29. a. $I_N = 0.35$ A, $R_N = 16\ \Omega$
    b. $R_L = 20\ \Omega$: $I_L = 0.156$ A
    $R_L = 50\ \Omega$: $I_L = 0.085$ A
31. a. $I_N = 1.50$ A, $R_N = 50\ \Omega$
    b. $I_N = 1.50$ A, $R_N = 50\ \Omega$
33. a. $I_N = 0.417$ mA, $R_N = 60$ kΩ
    b. $I_N = 0.417$ mA, $R_N = 60$ kΩ
35. a. $I_N = 0.633$ A, $R_N = 7.2\ \Omega$
    b. $I_N = 0.633$ A, $R_N = 7.2\ \Omega$
37. a. 60 kΩ b. 2.60 mW
39. a. 31.58 Ω b. 7.81 mW
41. a. $R_1 = 0\ \Omega$ b. 19.5 mW
43. $E = 1.5625$ V
45. $I = 0.054$ A, $P_L = 0.073$ W
47. $I = 0.284$ mA, $P_L = 0.807$ W
49. a. $I = 0.24$ A
    b. $I = 0.24$ A
    c. Reciprocity does apply.
51. a. $V = 22.5$ V
    b. Reciprocity does apply.
53. $E_{Th} = 10$ V, $R_{Th} = 20\ \Omega$
    $I_N = 0.5$ A, $R_N = 20\ \Omega$
55. $R_L = 2.02$ kΩ for maximum power.

CHAPTER 10

1. a. 800 μC b. 2 μF c. 100 μC d. 30 V e. 150 V f. 1.5 μF
3. 200 V
5. 420 μC
7. 73 pF
9. 5.65×10^{-4} m^2
11. 117 V
13. a. 2.25×10^{12} N/C
    b. 0.562×10^{12} N/C

15. 4.5 kV

17. 3.33 kV

19. a. points b. spheres c. points

21. 24.8 μF

23. 77 μF

25. 3.86 μF

27. a. 9.6 μF c. 3.6 μF
 b. 13 μF d. 0.5 μF

29. 9 μF

31. 60 μF; 30 μF

33. 81.2 μF; 1.61 μF

35. The 10-μF capacitor is in parallel with the series combination of the 1-μF and 1.5-μF capacitors.

37. a. $V_1 = 60$ V; $V_2 = V_3 = 40$ V
 b. $V_1 = 50$ V; $V_2 = V_3 = 25$ V; $V_4 = 25$ V; $V_5 = 8.3$ V; $V_6 = 16.7$ V

39. 14.4 V; 36 V; 9.6 V

41. 800 μF

43. −50 mA from 0 to 1 ms; 50 mA from 1 ms to 4 ms; 0 mA from 4 ms to 6 ms; 50 mA from 6 ms to 7 ms; −75 mA from 7 ms to 9 ms.

45. $-23.5\,e^{-0.05t}$ μA

47. 0 mJ, 0.25 mJ, 1.0 mJ, 1.0 mJ, 2.25 mJ, 0 mJ

49. a. C_3 shorted
 b. C_2 open
 c. C_2 shorted

CHAPTER 11

1. a. 0 V; 5 A b. 20 V; 0 A

3. a. Short circuit b. Voltage source c. Open circuit
 d. $i(0^-)$ = current just before $t = 0$ s; $i(0^+)$ = current just after $t = 0$ s

5. 15.1 V

7. a. $45(1 - e^{-80t})$ V b. $90e^{-80t}$ mA
 c.

t (ms)	v_C (V)	i_C (mA)
0	0	90
20	35.9	18.2
40	43.2	3.67
60	44.6	0.741
80	44.93	0.150
100	44.98	0.030

9. $40(1 - e^{-t/39\text{ ms}})$ V $10.3e^{-t/39\text{ ms}}$ mA 28.9 V 2.86 mA

11. 40 μs; 200 μs

13. v_C: 0, 12.6, 17.3, 19.0, 19.6, 19.9 (all V)
 i_C: 5, 1.84, 0.675, 0.249, 0.092, 0.034 (all A)

15. 25 kΩ; 8 μF

17. 45 V; 4.5 kΩ; 0.222 μF

19. 2.5 A

21. a. $20 + 10e^{-25\,000t}$ V
 b. $-2.5e^{-25\,000t}$ A
 c. v_C starts at 30 V and decays exponentially to 20 V in 200 μs, i_C is 0 A at $t = 0^-$, −2.5 A at $t = 0^+$, and decays exponentially to zero in 200 μs.

23. a. $50e^{-2t}$ V
 b. $-2e^{-2t}$ mA
 c. 0.5 s
 d. v_C: 50 V, 18.4 V, 6.77 V, 2.49 V, 0.916 V, 0.337 V
 i_C: −2 mA, −0.736 mA, −0.271 mA, −0.0996 mA, −0.0366 mA, −0.0135 mA

25. 14.4 V

27. a. 200 V; −12.5 mA
 b. 8 ms
 c. $200e^{-125t}$ V, $-12.5e^{-125t}$ mA

29. $45(1 - e^{-t/0.1857})$ V, 28.4 V (same)

31. a. $60(1 - e^{-500t})$ V b. $1.5e^{-500t}$ A

33. 90 V; 15 kΩ; 100 μF

35. $V_{C1} = 65$ V; $V_{C2} = 10$ V; $V_{C3} = 55$ V; $I_T = 0.5$ A

37. 14.0 μF

39. a. 5 μs b. 40% c. 200 000 pulses/s

41. 0.8 μs; 0.8 μs; 4 μs

43. 6.6 ns

45. −17.8 V, −179 mA

47. a. 51.9 V b. 261 mA

49. Verification point: At t = 20 ms, −17.8 V and −0.179 A

51. 29.3 V, 0.227 mA

CHAPTER 12

1. a. A_1
 b. 1.4 T

3. 0.50 T

5. 1.23×10^{-3} Wb

7. 1 T;

9. 264 μWb; 738 μWb; 807 μWb

11. 1061 At/m

13. $N_1I_1 = H_1\ell_1 + H_2\ell_2$; $N_2I_2 = H_2\ell_2 - H_3\ell_3$

15. 0.47 A

17. 0.88 A

19. 0.58 A

21. 0.53 A

23. 0.86 A

25. 3.7 A

27. 4.4×10^{-4} Wb

29. 1.06×10^{-4} Wb

CHAPTER 13

1. 225 V

3. 6.0 V

5. 150 mV

7. 0.111 s

9. 79.0 μH

11. $L = \frac{N\Phi}{I} = \frac{N(B_g A_g)}{I} = \frac{N(\mu_0 H_g)A_g}{I}$

$= \frac{N\mu_0\left(\frac{NI}{\ell_g}\right)A_g}{I} = \frac{\mu_0 N^2 A_g}{\ell_g}$

13. 4 H

15. a. 4 H c. 5 A

17. 84 mH

19. 4.39 mH

21. a. 21 H d. 4 H
b. 2 H e. 4 mH
c. 20 H

23. 9 μH

25. Circuit (a): 6 H; 1.5 H
Circuit (b): 2 H; 8 H

27. 1.6 H, in series with 6 H‖4 H

29. 1.2 H in series with 8 H‖12 H

31. a. 1 H in series with 3 μF
b. 2 H in series with 10 μF
c. 10 Ω, 10 H, and 25 μF in series
d. 10 Ω in series with 40 Ω‖(50 H in series with 20 μF)

33. 0.32 J

35. The path containing L_1 and L_2 is open.

CHAPTER 14

1. a. open circuit
b. Circuit (a): 1.6 A
Circuit (b): 6 A; 60 V
Circuit (c): 0 A; *E*
Circuit (d): 2 A; 30 V

3. $v_{R_1} = 180$ V; $v_{R_2} = 120$ V; $v_{R_3} = 60$ V; $v_{R_4} = 32$ V
$v_{R_5} = 28$ V; $v_{R_6} = 0$ V; $i_T = 21$ A; $i_1 = 18$ A
$i_2 = 3$ A; $i_3 = 1$ A; $i_4 = i_5 = 2$A; $i_6 = 0$ A

5. a. 50 ms
b. 250 ms
c. $3(1 - e^{-20t})$ A; $180e^{-20t}$ V
d.

t	i_L (A)	v_L (V)
0	0	180
τ	1.90	66.2
2τ	2.59	24.4
3τ	2.85	8.96
4τ	2.95	3.30
5τ	2.98	1.21

7. a. 0.2 s
b. 1 s
c. $20\,e^{-5t}$ V; $(1 - e^{-5t})$ A
d. v_L: 20, 7.36, 2.71, 0.996, 0.366, 0.135 (all V)
i_L: 0, 0.632, 0.865, 0.950, 0.982, 0.993 (all A)

9. $-182\,(1 - e^{-393t})$ mA; $-40e^{-393t}$ V; −134 mA; −10.5 V

11. 80 V; 20 Ω; 2 H

13. 40 V; 4 kΩ; 2 H

15. a. 2 ms
b. $5e^{-500t}$ A; $-1250e^{-500t}$ V
c.

t	i_L (A)	v_L (V)
0	5	−1250
τ	1.84	−460
2τ	0.677	−169
3τ	0.249	−62.2
4τ	0.092	−22.9
5τ	0.034	−8.42

17. −365 V

19. $R_1 = 20$ Ω; $R_2 = 30$ Ω

21. 5.19 A

23. a. 8.89 μs
b. $-203e^{-t/8.89\mu s}$ V; $5e^{-t/8.89\mu s}$ mA
c. −27.3 V; 0.675 mA

25. a. 10 ms
b. $90\,(1 - e^{-t/10\text{ ms}})$ mA; $36e^{-t/10\text{ ms}}$ V
c. 2.96 V 82.6 mA

27. 103.3 mA; 3.69 V

29. 33.1 V

31. 4.61 A

33. $i_L(25\text{ ms}) = 126$ mA; $i_L(50\text{ ms}) = 173$ mA

CHAPTER 15

1. AC voltage is voltage whose polarity cycles periodically between positive and negative. AC current is current whose direction cycles periodically.

3. a. The magnitude of a waveform (such as a voltage or current) at any instant of time.
b. 0, 10, 20, 20, 20, 0, −20, −20, 0 (all V)

5. 0 mA, 2.5 mA, 5 mA, 5 mA, 5 mA, 0 mA, −5 mA, −5 mA, 0 mA

7. a. 2 Hz b. 10 Hz c. 62.5 kHz

9. 7 ms; 142.9 Hz

11. 15 V; 6 mA

13. 149 919 cycles

15. 100 μs; 400 μs; 2500 Hz

17. 144.4 V

19. 100 V

21. a. 0.1745 c. 1.3963 e. 6.1087
b. 0.4363 d. 2.618 f. 10.821

23. 43.3 A; 25 A; −49.2 A; −50 A

25. $V_m = 85.1$ V. Waveform is like Figure 15–25 except $T = 50$ μs.

27. a. 62.83×10^6 rad/s c. 157.1 rad/s e. 1571 rad/s
b. 188.5 rad/s d. 314.2 rad/s

29. a. $v = 170 \sin 377t$ V
b. $i = 40 \sin 628t$ μA c. $v = 17 \sin 52.4 \times 10^3 t$ V

31. $v = 20 \sin 125.7t$ V

33. 0, 28.8, −11.4, −22 (all mA)

35. a. $5 \sin(1000\,t + 36°)$ mA
    b. $10 \sin(40\pi\,t + 120°)$ A
    c. $4 \sin(1800\pi\,t - 45°)$ V
37. 4.46 V; −3.54 V; 0.782 V
39. $v = 100 \sin(3491t + 36°)$ V
41. 6.25 ms; 13.2 ms; 38.2 ms
43. a. 20°; *i* leads c. 50°; i_1 leads
    b. in phase d. 60°; *i* leads
45. a. *A* leads by 90° b. *A* leads by 150°
47. Zero for each
49. a. 1.1 A b. −5 V c. 1.36 A
51. a. Similar to Figure 15–61(b), except positive peak is 40 V, negative peak is −10 V, and $V_{avg} = 15$ V. $T = 120$ ms.
    b. 40 V; −10 V
    c. 27.5 V; 36.7 V; 2.5 V; −6.65 V
    d. 15 V
53. 2.80 V
55. a. 12 V c. 19.7 V
    b. 17.0 mA d. 48.9 V
57. a. 5 mA b. 23.2 V
59. a. 8.94 A b. 16.8 A
61. 24 V; Its magnitude is always 24 V; therefore, it produces the same average power to a resistor as a 24-V battery.
63. 26.5 mA
65. a. 108 V c. 14.6 V
    b. 120 V d. 422 V
67. 6.25 ms; 13.2 ms; 38.2 ms

CHAPTER 16

1. a. $13\angle 67.4°$ c. $17\angle 118.1°$
   b. $10.8\angle -33.7°$ d. $10.8\angle -158.2°$
5. a. $7 + j6$ d. $-4.64 + j1.86$
   b. $1 + j10$ e. $-7.2 - j12.1$
   c. $15.4 - j3.50$ f. $0.2 - j0.1$
7. a. $14.2\angle -23.8°$ b. $1.35\angle -69.5°$ c. $5.31\angle 167.7°$
9. a. $10 \sin(\omega t + 30°)$ V b. $15 \sin(\omega t - 10°)$ V
11. a. $10\text{ V}\angle 30°$, $15\text{ V}\angle -20°$
    b. $11.5\text{ V}\angle 118.2°$
    c. $11.5 \sin(\omega t + 118.2°)$ V
13. a. $17.7\text{ mA}\angle 36°$; $28.3\text{ mA}\angle 80°$; $42.8\text{ mA}\angle 63.3°$
    b. $60.5 \sin(\omega t + 63.3°)$ mA
15. a. $4.95\text{ mA}\angle 0°$; $2.83\text{ mA}\angle -90°$; $4.2\text{ mA}\angle 90°$; $5.15\text{ mA}\angle 15.9°$
    b. $7.28 \sin(\omega t + 15.9°)$ mA
17. a. $10 \sin \omega t$ A c. $204 \sin(\omega t - 56°)$ mV
    b. $10 \sin(\omega t + 27°)$ A d. $204 \sin(\omega t - 157°)$ μV
19. a. $147 \sin(\omega t + 39°)$ V; $183.8 \sin(\omega t + 39°)$ V
    b. $330.8 \sin(\omega t + 39°)$ V; Identical
21. a. $5 \sin(\omega t - 30°)$ A
    b. $5 \sin(\omega t - 105°)$ A
    c. $10 \sin(\omega t + 120°)$ V
    d. $10 \sin(\omega t + 100°)$ V
23. a. $0.531 \sin(377t - 90°)$ A
    b. $31.8 \sin(6283t - 90°)$ mA
    c. $0.4 \sin(500t - 90°)$ A
25. a. 8.74 kHz; b. 50.9 mH; 130°
27. a. 530.5 Ω b. 31.83 Ω c. 400 Ω
29. a. $1.89 \sin(377t + 90°)$ A
    b. $79.6 \sin(2\pi \times 400t - 150°)$ mV
31. a. $48\ \Omega\angle 0°$ b. $j37.7\ \Omega$ c. $-j50\ \Omega$
33. a. $\mathbf{V}_R = 25\text{ V}\angle 0°$; $\mathbf{V}_L = 12.5\text{ V}\angle 90°$; $\mathbf{V}_C = 5\text{ V}\angle -90°$
    b. $v_R = 35.4 \sin\omega t$ V; $v_L = 17.7 \sin(\omega t + 90°)$ V; $v_C = 7.07 \sin(\omega t - 90°)$ V
35. a. 39.8 Hz b. 6.37 μF
37. Theoretical: 45.2 mA
39. $1.59\text{ A}\angle -90°$
41. 7.96 A

CHAPTER 17

1. When *p* is +, power flows from source to load. When *p* is −, power flows out of load. Out of the load.
3. a. 1000 W and 0 VAR; 500 W and 0 VAR
   b. 1500 W and 0 VAR
5. 151 VAR (cap.)
7. a. 10 Ω b. 6 Ω c. 265 μF
9. 30 Ω
11. 160 W; 400 VAR (ind.)
13. 900 W; 300 VAR (ind.)
15. a. 20 Ω c. 8 Ω
    b. 6 Ω d. 1.2 H
17. 125 VA
19. 1150 W; 70 VAR (cap.); 1152 VA; $\theta = -3.48°$
21. 2.36 Ω
23. 120 Ω
25. a. 721 W; 82.3 VAR (cap.); 726 VA
    b. $I = 6.05$ A; No
27. a. Across the load b. 73.9 μF
29. 57.3 kW
31. 2598 W
33. 160 Ω
35. Same as Figure 17–5 with peak value of $p = 3.14$ W.
37. A sawtooth wave oscillating between 8 W and −8 W.

CHAPTER 18

1. a. $0.125 \sin\omega t$
3. a. $1.87 \times 10^{-3} \sin(\omega t + 30°)$
5. a. $1.36 \sin(\omega t - 90°)$
7. a. $1333 \sin(2000\pi t + 30°)$
9. a. $62.5 \sin(10000t - 90°)$
11. a. $67.5 \sin(20000t - 160°)$

13. Network (a): $31.6\ \Omega\angle 18.43°$
    Network (b): $8.29\ k\Omega\angle -29.66°$
15. a. $42.0\ \Omega\angle 19.47° = 39.6\ \Omega + j14.0\ \Omega$
17. $R = 1.93\ k\Omega$, $L = 4.58$ mH
19. $R = 15\ \Omega$, $C = 1.93\ \mu F$
21. a. $\mathbf{Z}_T = 50\ \Omega\angle -36.87°$, $\mathbf{I} = 2.4\ A\angle 36.87°$,
    $\mathbf{V}_R = 96\ V\angle 36.87°$, $\mathbf{V}_L = 48\ V\angle 126.87°$,
    $\mathbf{V}_C = 120\ V\angle -53.13°$
    c. 230.4 W
    d. 230.4 W
23. a. $\mathbf{Z}_T = 45\ \Omega\angle -36.87°$
    b. $i = 0.533 \sin(\omega t + 36.87°), v_R = 19.20 \sin(\omega t + 36.87°)$,
    $v_C = 25.1 \sin(\omega t - 53.13°)$, $v_L = 10.7 \sin(\omega t + 126.87°)$
    e. 5.12 W f. 5.12 W
25. a. $\mathbf{V}_R = 9.49\ V\angle -18.43°$, $\mathbf{V}_L = 11.07\ V\angle 71.57°$,
    $\mathbf{V}_C = 7.91\ V\angle -108.43°$
    b. $\Sigma\mathbf{V} = 10.00\ V\angle 0°$
27. a. $\mathbf{V}_C = 317\ V\angle -30°$, $\mathbf{V}_L = 99.8\ V\angle 150°$
    b. $25\ \Omega$
29. a. $\mathbf{V}_C = 6.0\ V\angle -110°$
    b. $\mathbf{V}_Z = 13.87\ V\angle 59.92°$
    c. $69.4\ \Omega\angle 79.92°$
    d. 1.286 W
31. Network (a): $199.9\ \Omega\angle -1.99°$,
    Network (b): $485\ \Omega\angle -14.04°$
33. a. $\mathbf{Z}_T = 3.92\ k\Omega\angle -78.79°$, $\mathbf{I}_T = 2.55\ mA\angle 78.69°$,
    $\mathbf{I}_1 = 0.5\ mA\angle 0°$, $\mathbf{I}_2 = 10.0\ mA\angle -90°$,
    $\mathbf{I}_3 = 12.5\ mA\angle 90°$
    d. 5.00 mW
35. a. $5.92\ k\Omega\angle 17.4°$ b. $177.6\ V\angle 17.4°$
37. $2.55\ \Omega\angle 81.80°$
39. Network (a): $\mathbf{I}_R = 10.00\ mA\angle -31.99°$,
    $\mathbf{I}_L = 4.00\ mA\angle -121.99°$, $\mathbf{I}_C = 4.35\ mA\angle 58.01°$
    Network (b): $\mathbf{I}_R = 9.70\ mA\angle -44.04°$,
    $\mathbf{I}_{C1} = 1.62\ mA\angle 45.96°$, $\mathbf{I}_{C2} = 0.81\ mA\angle 45.96°$
41. $\mathbf{I}_L = 2.83\ mA\angle -135°$, $\mathbf{I}_C = 3.54\ mA\angle 45°$,
    $\mathbf{I}_R = 0.71\ mA{-45°}$, $\Sigma\mathbf{I}_{out} = \Sigma_{in} = 1.00\ mA\angle 0°$
43. a. $6.245\ A\angle 90°$ b. $40.0\ \Omega$ c. $8.00\ A\angle 51.32°$
45. a. $\mathbf{Z}_T = 22.5\ \Omega\angle -57.72°$
    $\mathbf{I}_L = 5.34\ A\angle 57.72°$
    $\mathbf{I}_C = 4.78\ A\angle 84.29°$
    $\mathbf{I}_R = 2.39\ A\angle -5.71°$
    c. $P_R = 342$ W
    d. $P_T = 342$ W
47. a. $\mathbf{Z}_T = 10.53\ \Omega\angle 10.95°$, $\mathbf{I}_T = 1.90\ A\angle -10.95°$,
    $\mathbf{I}_1 = 2.28\ A\angle -67.26°$, $\mathbf{I}_2 = 2.00\ A\angle 60.61°$
    b. $\mathbf{V}_{ab} = 8.87\ V\angle 169.06°$
49. a. $\mathbf{Z}_T = 7.5\ k\Omega\angle 0°$, $\mathbf{I}_1 = 0.75\ mA\angle 0°$,
    $\mathbf{I}_2 = 0.75\ mA\angle 90°$, $\mathbf{I}_3 = 0.79A\angle -71.57°$
    b. $\mathbf{V}_{ab} = 7.12\ V\angle 18.43°$
51. $\omega_C = 2000$ rad/s
53. $f_C = 3.39$ Hz
55. Network (a): 5.5-kΩ resistor in series with a 9.0-kΩ inductive reactance
    Network (b): 207.7-Ω resistor in series with a 138.5-Ω inductive reactance
57. $\omega = 1$ krad/s: $\mathbf{Y}_T = 0.01\ S + j0$, $\mathbf{Z}_T = 100\ \Omega$
    $\omega = 10$ krad/s: $\mathbf{Y}_T = 0.01\ S + j0$, $\mathbf{Z}_T = 100\ \Omega$

CHAPTER 19

1. a. $5.00\ V\angle 180°$ b. $12.50\ V\angle 0°$ c. $15.00\ V\angle -120°$
3. a. $3.20\ mV\angle 180°$ b. $8.00\ mV\angle 0°$ c. $9.60\ mV\angle -120°$
5. $7.80\ V\angle -150°$
7. Circuit (a): $\mathbf{E} = 54\ V\angle 0°$, $\mathbf{V}_L = 13.5\ V\angle 0°$
   Circuit (b): $\mathbf{E} = 450\ mV\angle -60°$, $\mathbf{V}_L = 439\ mV\angle -47.32°$
9. a. $4.69\ V\angle 180°$ b. $\mathbf{E} = (7.5\ M\Omega)\mathbf{I}$, $\mathbf{V} = 4.69\ V\angle 180°$
11. a. $(4\ \Omega + j2\ \Omega)\mathbf{I}_1 - (4\ \Omega)\mathbf{I}_2 = 20\ V\angle 0°$
    $-(4\ \Omega)\mathbf{I}_1 + (6\ \Omega + j4\ \Omega)\mathbf{I}_2 = 48.4\ V\angle -161.93°$
    b. $\mathbf{I}_1 = 2.39\ A\angle 72.63°$, $\mathbf{I}_2 = 6.04\ A\angle 154.06°$
    c. $\mathbf{I} = 6.15\ A\angle -3.33°$
13. a. $(12\ \Omega - j16\ \Omega)\mathbf{I}_1 + (j15\ \Omega)\mathbf{I}_2 = 13.23\ V\angle -79.11°$
    $(j15\ \Omega)\mathbf{I}_1 + 0\mathbf{I}_2 = 10.27\ V\angle -43.06°$
    b. $\mathbf{I}_1 = 0.684\ A\angle -133.06°$, $\mathbf{I}_2 = 1.443\ A\angle -131.93°$
    c. $\mathbf{V} = 11.39\ V\angle -40.91°$
15. $27.8\ V\angle 6.79°$ $\mathbf{I} = 6.95\ mA\angle 6.79°$
17. a. $(0.417\ S\angle 36.87°)\mathbf{V}_1 - (0.25\ S\angle 90°)\mathbf{V}_2 = 3.61\ A\angle -56.31°$
    $-(0.25\ S\angle 90°)\mathbf{V}_1 + (0.083\ S\angle 90°)\mathbf{V}_2 = 7.00\ A\angle 90°$
    b. $\mathbf{V}_1 = 30.1\ V\angle 139.97°$, $\mathbf{V}_2 = 60.0\ V\angle 75.75°$
    c. $\mathbf{I} = 13.5\ V\angle -44.31°$
19. a. $(0.0893\ S\angle 22.08°)\mathbf{V}_1 + (0.04\ S\angle 90°)\mathbf{V}_2 = 0.570\ A\angle 93.86°$
    $+(0.04\ S\angle 90°)\mathbf{V}_1 + (0.06\ S\angle 90°)\mathbf{V}_2 = 2.00\ A\angle 180°$
    b. $\mathbf{V}_1 = 17.03\ V\angle 18.95°$, $\mathbf{V}_2 = 31.5\ V\angle 109.91°$
    c. $\mathbf{V} = 11.39\ V\angle -40.91°$
21. $(0.372\ \mu S\angle -5.40°)$, $\mathbf{V} = 10.33\ mA\angle 1.39°$
    $27.8\ V\angle 6.79°$, $\mathbf{I} = 6.95\ mA\angle 6.79°$
    As expected, the answers are the same as those in Problem 15.
23. Network (a):
    $\mathbf{Z}_1 = 284.4\ \Omega\angle -20.56°$, $\mathbf{Z}_2 = 94.8\ \Omega\angle 69.44°$
    $\mathbf{Z}_3 = 31.6\ \Omega\angle 159.44°$
    Network (b):
    $\mathbf{Z}_1 = 11.84\ k\Omega\angle 9.46°$, $\mathbf{Z}_2 = 5.92\ k\Omega\angle -80.54°$
    $\mathbf{Z}_3 = 2.96\ k\Omega\angle -80.54°$
25. $\mathbf{I}_T = 0.337\ A\angle -2.82°$
27. a. $\mathbf{Z}_T = 3.03\ \Omega\angle -76.02°$
    b. $\mathbf{I} = 5.28\ A\angle 76.02°$, $\mathbf{I}_1 = 0.887\ A\angle -15.42°$
29. a. $\mathbf{Z}_2 = 1\ \Omega - j7\ \Omega = 7.07\ \Omega\angle -81.87°$
    b. $\mathbf{I} = 142.5\ mA\angle 52.13°$
31. $\mathbf{Z}_1\mathbf{Z}_4 = \mathbf{Z}_2\mathbf{Z}_3$ as required.
35. $R_3 = 50.01\ \Omega$, $R_1 = 253.3\ \Omega$
39. Same as Figure 19–21.
41. Same as Problem 26.

CHAPTER 20

1. $\mathbf{I} = 4.12\text{ A}\angle 50.91°$
3. $16\text{ V}\angle -53.13°$
5. a. $\mathbf{V} = 15.77\text{ V}\angle 36.52°$
   b. $P_{(1)} + P_{(2)} = 1.826\text{ W} \neq P_{100\text{-}\Omega} = 2.49\text{ W}$
7. $0.436\text{ A}\angle -9.27°$
9. $19.0\sin(\omega t + 68.96°)$
11. a. $\mathbf{V}_L = 1.26\text{ V}\angle 161.57°$ b. $\mathbf{V}_L = 6.32\text{ V}\angle 161.57°$
13. $0.361\text{ mA}\angle -3.18°$
15. $\mathbf{V}_L = 9.88\text{ V}\angle 0°$
17. 1.78 V
19. $\mathbf{Z}_{\text{Th}} = 3\ \Omega\angle -90°$, $\mathbf{E}_{\text{Th}} = 20\text{ V}\angle -90°$
21. a. $\mathbf{Z}_{\text{Th}} = 37.2\ \Omega\angle 57.99°$ $\mathbf{E}_{\text{Th}} = 9.63\text{ V}\angle 78.49°$
    b. 0.447 W
23. $\mathbf{Z}_{\text{Th}} = 22.3\ \Omega\angle -15.80°$, $\mathbf{E}_{\text{Th}} = 20.9\text{ V}\angle 20.69°$
25. $\mathbf{Z}_{\text{Th}} = 109.9\ \Omega\angle -28.44°$, $\mathbf{E}_{\text{Th}} = 14.5\text{ V}\angle -91.61°$
27. a. $\mathbf{Z}_{\text{Th}} = 20.6\ \Omega\angle 34.94°$, $\mathbf{E}_{\text{Th}} = 10.99\text{ V}\angle 13.36°$
    b. $P_L = 1.61\text{ W}$
29. $\mathbf{Z}_{\text{N}} = -j3\ \Omega$, $\mathbf{I}_{\text{N}} = 6.67\text{ A}\angle 0°$
31. a. $\mathbf{Z}_{\text{N}} = 22.3\ \Omega\angle -15.80°$, $\mathbf{I}_{\text{N}} = 0.935\text{ A}\angle 36.49°$
    b. $0.436\text{ A}\angle -9.27°$
    c. 3.80 W
33. a. $\mathbf{Z}_{\text{N}} = 109.9\ \Omega\angle -28.44°$, $\mathbf{I}_{\text{N}} = 0.131\text{ A}\angle -63.17°$
    b. $0.0362\text{ A}\angle -84.09°$
    c. 0.394 W
35. a. $\mathbf{Z}_{\text{N}} = 14.1\ \Omega\angle 85.41°$, $\mathbf{I}_{\text{N}} = 0.181\text{ A}\angle 29.91°$
    b. $0.0747\text{ A}\angle 90.99°$
37. a. $\mathbf{Z}_{\text{Th}} = 17.9\text{ k}\Omega\angle -26.56°$, $\mathbf{E}_{\text{Th}} = 1.79\text{ V}\angle 153.43°$
    b. $31.6\ \mu\text{A}\angle 161.56°$
    c. $40.0\ \mu\text{W}$
39. $\mathbf{E}_{\text{Th}} = 10\text{ V}\angle 0°$, $\mathbf{I}_{\text{N}} = 10.5\text{ A}\angle 0°$, $\mathbf{Z}_{\text{Th}} = 0.952\ \Omega\angle 0°$
41. a. $\mathbf{Z}_L = 8\ \Omega\angle 22.62°$ b. 40.2 W
43. a. $\mathbf{Z}_L = 2.47\ \Omega\angle 21.98°$ b. 1.04 W
45. $4.15\ \Omega\angle 85.24°$
47. a. $\mathbf{Z}_L = 37.2\ \Omega\angle -57.99°$ b. $19.74\ \Omega$ c. 1.18 W
49. $\mathbf{Z}_{\text{Th}} = 3\ \Omega\angle -90°$, $\mathbf{E}_{\text{Th}} = 20\ \Omega\angle -90°$
51. $\mathbf{Z}_{\text{Th}} = 109.9\ \Omega\angle -28.44°$, $\mathbf{E}_{\text{Th}} = 14.5\text{ V}\angle -91.61°$
53. $\mathbf{E}_{\text{Th}} = 10\text{ V}\angle 0°$, $\mathbf{I}_{\text{N}} = 10.5\text{ A}\angle 0°$, $\mathbf{Z}_{\text{Th}} = 0.952\ \Omega\angle 0°$
55. $\mathbf{Z}_{\text{N}} = 0.5\text{ k}\Omega\angle 0°$, $\mathbf{I}_{\text{N}} = 4.0\text{ mA}\angle 0°$
57. $\mathbf{E}_{\text{Th}} = 10\text{ V}\angle 0°$, $\mathbf{I}_{\text{N}} = 10.5\text{ A}\angle 0°$, $\mathbf{Z}_{\text{N}} = 0.952\ \Omega\angle 0°$
59. $\mathbf{Z}_{\text{N}} = 0.5\text{ k}\Omega\angle 0°$, $\mathbf{I}_{\text{N}} = 4.0\text{ mA}\angle 0°$

CHAPTER 21

1. a. $\omega_s = 3835\text{ rad/s}$ $f_s = 610.3\text{ Hz}$
   b. $\mathbf{I} = 153.8\text{ mA}\angle 0°$
   c. $\mathbf{V}_C = 59.0\text{ V}\angle -90°$ $\mathbf{V}_L = 59.03\text{ V}\angle 87.76°$
      $\mathbf{V}_R = 7.69\text{ V}\angle 0°$
   d. $P_L = 0.355\text{ W}$
3. a. $R = 25.0\ \Omega$ $C = 4.05\text{ nF}$
   b. $P = 15.6\text{ mW}$
   c. $X_C = 1.57\text{ k}\Omega$ $\mathbf{V}_C = 39.3\text{ V}\angle -90°$
      $\mathbf{V}_L = 39.3\text{ V}\angle 90°$
      $\mathbf{V}_R = \mathbf{E} = 0.625\text{ V}\angle -90°$
   d. $v_C = 55.5\sin(50{,}000\pi t - 90°)$
      $v_L = 55.5\sin(50{,}000\pi t + 90°)$
      $v_R = 0.884\sin(50{,}000\pi t)$
5. a. $\omega_s = 500\text{ rad/s}$ $f_s = 79.6\text{ kHz}$
   b. $\mathbf{Z}_T = 200\ \Omega\angle 0°$
   c. $\mathbf{I} = 10\text{ mA}\angle 0°$
   d. $\mathbf{V}_R = 1\text{ V}\angle 0°$ $\mathbf{V}_L = 50.01\text{ V}\angle 88.85°$
      $\mathbf{V}_C = 50\text{ V}\angle -90°$
   e. $P_T = 20\text{ mW}$ $Q_C = 0.5\text{ VAR (cap.)}$
      $Q_L = 0.5\text{ VAR (ind.)}$
   f. $Q_s = 25$
7. a. $C = 0.08\ \mu\text{F}$ $R = 6.4\ \Omega$
   b. $P_T = 0.625\text{ W}$
   c. $\mathbf{V}_L = 62.5\text{ V}\angle 90°$
9. $\omega_s = 8000\text{ rad/s}$
11. a. $\omega_s = 3727\text{ rad/s}$, $Q = 7.45$, $\text{BW} = 500\text{ rad/s}$
    b. $P_{\text{max}} = 144\text{ W}$
    c. $\omega_1 \approx 3477\text{ rad/s}$ $\omega_2 \approx 3977\text{ rad/s}$
    d. $\omega_1 = 3485.16\text{ rad/s}$ $\omega_2 = 3985.16\text{ rad/s}$
    e. The results are close, although the approximation will yield some error if used in further calculations. The error would be less if Q were larger.
13. a. $R = 1005\ \Omega$, $C = 63.325\text{ pF}$
    b. $P = 0.625\text{ W}$, $\mathbf{V}_L = 312.5\text{ V}\angle 90°$
    c. $v_{\text{out}} = 442\sin(400\pi \times 10^3 t + 90°)$
15. Network (a):
    a. $Q = 24$
    b. $R_P = 5770\ \Omega$, $X_{LP} = 240\ \Omega$
    c. $Q = 240$, $R_P = 576\ \Omega$
       $X_{LP} = 2400\ \Omega$
    Network (b):
    a. $Q = 1$
    b. $R_P = 200\ \Omega$, $X_{LP} = 200\ \Omega$
    c. $Q = 10$, $R_P = 10.1\text{ k}\Omega$
       $X_{LP} = 1.01\text{ k}\Omega$
    Network (c):
    a. $Q = 12.5$
    b. $R_P = 314.5\ \Omega$, $X_{LP} = 25.16\ \Omega$
    c. $Q = 125$, $R_P = 31.25\ \Omega$, $X_{LP} = 250\ \Omega$
17. Network (a):
    $Q = 15$, $R_P = 4500\ \Omega$, $X_{CP} = 300\ \Omega$
    Network (b):
    $Q = 2$, $R_P = 225\ \Omega$, $X_{CP} = 112.5\ \Omega$
    Network (c):
    $Q = 5$, $R_P = 13\text{ k}\Omega$, $X_{CP} = 2600\ \Omega$
19. Network (a):
    $Q = 4$, $R_s = 4\text{ k}\Omega$, $X_{LS} = 16\text{ k}\Omega$

Network (b):

$Q = 0.333$, $R_s = 3.6\ \text{k}\Omega$, $X_{CS} = 1.2\ \text{k}\Omega$

Network (c):

$Q = 100$, $R_s = 10\ \Omega$, $X_{LS} = 1\ \text{k}\Omega$

21. $L_s = 1.2$ mH, $L_P = 2.4$ mH
23. a. $\omega_P = 20$ krad/s
25. a. $\omega_P = 39.6$ krad/s, $f_P = 6310$ Hz
    b. $Q = 6.622$
    c. $\mathbf{V} = 668.2\ \text{V}\angle 0°$, $\mathbf{I}_R = 11.14\ \text{mA}\angle 0°$
       $\mathbf{I}_L = 668.2\ \text{mA}\angle -82.36°$
       $\mathbf{I}_C = 662.2\ \text{mA}\angle 90°$
    d. $P_T = 66.82$ W
    e. BW = 5.98 krad/s, BW = 952 Hz
27. $R_1 = 4194\ \text{k}\Omega$, $C = 405$ pF, $\mathbf{I}_L = 5.0\ \text{mA}\angle -89.27°$
29. a. 900 Ω d. $C = 500$ nF, $L = 0.450$ H
    b. 1.862 e. $93.1\ \text{V}\angle 0°$
    c. 1074 rad/s
33. $V = 668.2$ V, $f_P = 6310$ Hz, BW = 952 Hz, $Q = 6.622$
35. $V = 93.1$ V, $f_P = 318.3$ Hz, BW = 170.9 Hz, $Q = 1.86$

CHAPTER 22

1. a. 2000 (33.0 dB) c. 2×10^6 (63.0 dB)
   b. 200,000 (53.0 dB) d. 400 (26.0 dB)
3. a. $A_P = 50 \times 10^6$ (77.0 dB), $A_V = 500$ (54.0 dB)
   b. $A_P = 10$ (10.0 dB), $A_V = 0.224$ (−13.0 dB)
   c. $A_P = 13.3 \times 10^6$ (71.3 dB), $A_V = 258$ (48.2 dB)
   d. $A_P = 320 \times 10^6$ (85.1dB), $A_V = 1265$ (62.0 dB)
5. $P_{in} = 12.5\mu$W, $P_{out} = 39.5$ mW
   $V_{out} = 3.14$ V, $A_V = 12.6$, $[A_V]_{dB} = 22.0$ dB
7. a. 17.0 dBm (−13.0 dBW)
   b. 30.0 dBm (0 dBW)
   c. −34.0 dBm (−64.0 dBW)
   d. −66.0 dBm (−96.0 dBW)
9. a. 0.224 W c. 5.01 pW
   b. 30.2 μW d. 1995 W
17. $P_1 = 5.05$ dBm, $P_2 = 2.05$ dBm, $P_3 = 14.09$ dBm
    $P_O = 25.6$ dBm
19. $P_1 = 25.5$ dBm, $P_{in} = -14.5$ dBm, $V_{out} = 52.8$ V
23. a. $\omega_C = 1000$ rad/s, $f_C = 159.2$ Hz
25. a. $\omega_1 = 50$ rad/s $\omega_2 = 1000$ rad/s
    $f_1 = 7.96$ Hz $f_2 = 159.15$ Hz
35. a. $\text{TF} = \dfrac{1 + j0.00003}{1 + 0.00006}$
43. a. Low-pass filter: $\omega_C = 500$ rad/s
    High-pass filter: $\omega_C = 25$ krad/s
    BW = 475 krad/s
    c. The actual cutoff frequencies will be close to the designed values since the break frequencies are separated by more than one decade.
45. a. $R_1 = 5\ \text{k}\Omega$, $R_2 = 50\ \text{k}\Omega$
    c. The actual frequencies will not occur at the designed values since they are one decade apart.
47. a. 10 krad/s
    b. 10
    c. BW = 1 krad/s, $\omega_1 = 9.5$ krad/s, $\omega_2 = 10.5$ krad/s
    d. At resonance, $[A_V]_{dB} = -28.0$ dB
49. a. 10 krad/s, 1592 Hz
    b. 100
    c. −60 dB
    d. −0.8 dB, 0 dB
    e. 100 rad/s, 9.95 krad/s, 10.05 krad/s

CHAPTER 23

1. a. e_s is in phase with e_p.
   b. e_p is 180° out of phase with e_s.
3. a. Step-up c. 6 V e. $3200\ \text{V}\angle 180°$
   b. $25 \sin\omega t$ V d. $96\ \text{V}\angle 0°$
5. $v_1 = 24 \sin\omega t$ V; $v_2 = 144 \sin(\omega t + 180°)$ V;
   $v_3 = 48 \sin\omega t$ V
7. a. $1\ \text{A}\angle 20°$ b. $480\ \text{V}\angle 0°$ c. $480\ \Omega\angle -20°$
9. a. $160\ \text{V}\angle -23.1°$ b. $640\ \text{V}\angle -23.1°$
11. a. $40\ \Omega - j80\ \Omega$ b. $1.25\ \Omega + j2\ \Omega$
13. 2.5
15. a. $22\ \Omega + j6\ \Omega$ b. $26\ \Omega + j3\ \Omega$
17. 108 kVA
19. a. 20 A c. 2.5 A
    b. 22.5 A d. 0.708 A
21. 0.64 W
23. $3\ \text{A}\angle -50°$; $1.90\ \text{A}\angle -18.4°$; $1.83\ \text{A}\angle -43.8°$
25. b. $2.12\ \text{A}\angle -45°$; $21.2\ \text{A}\angle -45°$; $120.2\ \text{V}\angle 0°$
27. 98.5%
29. All are minus.
31. 0.889 H
33. $-125\, e^{-500t}$ V; $-4\, e^{-500t}$ V; −75.8 V; −2.43 V
35. 10.5 H
37. 27.69 mH; $11.5\ \text{A}\angle -90°$
39. $(4 + j22)\ \mathbf{I}_1 + j13\ \mathbf{I}_2 = 100\angle 0°$
    $j13\ \mathbf{I}_1 + j12\ \mathbf{I}_2 = 0$
41. $(10 + j84)\ \mathbf{I}_1 - j62\ \mathbf{I}_2 = 120\ \text{V}\angle 0°$
    $-j62\ \mathbf{I}_1 + 15\ \mathbf{I}_2 = 0$
43. $0.644\ \text{A}\angle -56.1°$; $6.44\ \text{A}\angle -56.1°$; $117\ \text{V}\angle 0.385°$
47. $\mathbf{I}_1 = 0.286\ \text{A}\ \angle -73.3°$; $\mathbf{I}_2 = 89.6\ \text{mA}\ \angle 121.1°$
    Simulation answers may not agree exactly.

CHAPTER 24

1. a. $8\ \text{A}\angle -30°$; $8\ \text{A}\angle -150°$; $8\ \text{A}\angle 90°$ b. yes
3. i_A is in phase with voltage e_{AN}, i_B is in phase with e_{BN}, and i_C is in phase with e_{CN}. Waveforms look identical to those of Figure 24–2(b).
5. a. $\mathbf{V}_{ab} = 208\ \text{V}\angle 45°$; $\mathbf{V}_{ca} = 208\ \text{V}\ \angle 165°$
   b. $\mathbf{V}_{an} = 120\ \text{V}\angle 15°$; $\mathbf{V}_{bn} = 120\ \text{V}\angle -105°$;
   $\mathbf{V}_{cn} = 120\ \text{V}\ \angle 135°$

7. $\mathbf{I}_a = 23.1\ \text{A}\angle{-21.9°}$; $\mathbf{I}_b = 23.1\ \text{A}\angle{-141.9°}$;
   $\mathbf{I}_c = 23.1\ \text{A}\angle 98.1°$
9. $8.16\ \text{A}\angle 24°$; $8.16\ \text{A}\angle{-96°}$; $8.16\ \text{A}\angle 144°$
11. $\mathbf{I}_{ab} = 19.2\ \text{A}\angle 36.9°$; $\mathbf{I}_{bc} = 19.2\ \text{A}\angle{-83.1°}$;
   $\mathbf{I}_{ca} = 19.2\ \text{A}\angle 156.9°$; $\mathbf{I}_a = 33.3\ \text{A}\angle 6.9°$;
   $\mathbf{I}_b = 33.3\ \text{A}\angle{-113.1°}$; $\mathbf{I}_c = 33.3\ \text{A}\angle 126.9°$
13. $\mathbf{V}_{ab} = 250\ \text{V}\angle{-57.9°}$; $\mathbf{V}_{bc} = 250\ \text{V}\angle{-177.9°}$;
   $\mathbf{V}_{ca} = 250\ \text{V}\angle 62.1°$
15. $14.4\ \text{A}\angle{-33°}$; $14.4\ \text{A}\angle{-153°}$; $14.4\ \text{A}\angle 87°$
17. a. $R = 3.83\ \Omega$; $C = 826\ \mu\text{F}$
   b. $R = 7.66\ \Omega$; $L = 17.1\ \text{mH}$
19. a. $\mathbf{I}_{ab} = 4.5\ \text{A}\angle 35°$; $\mathbf{I}_{ca} = 4.5\ \text{A}\angle 155°$
   b. $\mathbf{I}_a = 7.79\ \text{A}\angle 5°$; $\mathbf{I}_b = 7.79\ \text{A}\angle{-115°}$;
   $\mathbf{I}_c = 7.79\ \text{A}\angle 125°$
   c. $R = 43.7\ \Omega$; $C = 86.7\ \mu\text{F}$
21. a. $122.1\ \text{V}\ \angle 0.676°$
   b. $212\ \text{V}\angle 30.676°$
23. $\mathbf{I}_a = 14.4\ \text{A}\angle 26.9°$; $\mathbf{I}_b = 14.4\ \text{A}\angle{-93.1°}$;
   $\mathbf{I}_c = 14.4\ \text{A}\angle 146.9°$
25. $\mathbf{I}_a = 14.4\ \text{A}\angle 34.9°$
27. $611\ \text{V}\ \angle 30.4°$
29. b. $489\ \text{V}\angle 30°$
31. $346\ \text{V}\ \angle{-10°}$
33. $P_\phi = 86.4\ \text{W}$; $Q_\phi = 0\ \text{VAR}$; $S_\phi = 86.4\ \text{VA}$;
   For totals, multiply by 3.
35. 2303 W; 1728 VAR (ind.); 2879 VA
37. 4153 W; 3115 VAR (cap.); 5191 VA
39. 72 kW; 36 kVAR (ind.); 80.5 kVA; 0.894
41. 27.6 kW; 36.9 kVAR (ind.); 46.1 kVA; 0.60
43. $3.93\ \text{A}\angle{-32°}$
45. 0.909
47. a. Same as Figure 23–28.
   b. 768 W
   c. 2304 W
49. a. $\mathbf{I}_a = 5.82\ \text{A}\angle{-14.0°}$; $\mathbf{I}_b = 5.82\ \text{A}\angle{-134.0°}$;
   $\mathbf{I}_c = 5.82\ \text{A}\angle 106°$
   b. $P_\phi = 678\ \text{W}$; $P_T = 2034\ \text{W}$
   c. $W_1 = 1164\ \text{W}$; $W_2 = 870\ \text{W}$
   d. 2034 W
51. a. 0.970 b. 0.970
53. a. $\mathbf{I}_{ab} = 4\ \text{A}\angle 0°$; $\mathbf{I}_{bc} = 2.4\ \text{A}\angle{-156.9°}$;
   $\mathbf{I}_{ca} = 3.07\ \text{A}\angle 170.2°$; $\mathbf{I}_a = 7.04\ \text{A}\angle{-4.25°}$;
   $\mathbf{I}_b = 6.28\ \text{A}\angle{-171.4°}$; $\mathbf{I}_c = 1.68\ \text{A}\angle 119.2°$
   b. $P_{ab} = 960\ \text{W}$; $P_{bc} = 461\ \text{W}$;
   $P_{ca} = 472\ \text{W}$; 1893 W
55. a. $\mathbf{I}_a = 6.67\ \text{A}\angle 0°$; $\mathbf{I}_b = 2.68\ \text{A}\angle{-93.4°}$;
   $\mathbf{I}_c = 2.4\ \text{A}\angle 66.9°$
   b. $7.47\ \text{A}\angle{-3.62°}$
   c. $P_{an} = 800\ \text{W}$; $P_{bn} = 288\ \text{W}$; $P_{cn} = 173\ \text{W}$
   d. 1261 W
57. a. $\mathbf{V}_{an} = 34.9\ \text{V}\angle{-0.737°}$;
   $\mathbf{V}_{bn} = 179\ \text{V}\angle{-144°}$;
   $\mathbf{V}_{cn} = 178\ \text{V}\angle 145°$
   b. $\mathbf{V}_{nN} = 85.0\ \text{V}\angle 0.302°$
59. $\mathbf{I}_{ab} = 19.2\ \text{A}\angle 36.87°$; $\mathbf{I}_a = 33.2\ \text{A}\angle 6.87°$
61. $\mathbf{I}_a = 6.67\ \text{A}\angle 0°$; $\mathbf{I}_b = 2.68\ \text{A}\angle{-93.4°}$;
   $\mathbf{I}_c = 2.40\ \text{A}\angle 66.9°$; $\mathbf{I}_N = 7.64\ \text{A}\angle{-3.62°}$

CHAPTER 25

1. a. 18.37 V b. 6.75 W
3. a. 17.32 V b. 0.12 W
5. $v(t) = 1 + \frac{2}{\pi}\sin\omega t + \frac{2}{2\pi}\sin(2\omega t) + \frac{2}{3\pi}\sin(3\omega t) + \ldots$
7. $v(t) = \frac{32}{\pi}\sin 500\ \pi t - \frac{32}{2\pi}\sin(1000\pi t) + \frac{32}{3\pi}\sin(1500\pi t) - \ldots$
9. $v(t) = \frac{10}{\pi} + 5\sin\omega t - \frac{20}{\pi}\left[\frac{\cos(2\omega t)}{3} + \frac{\cos(4\omega t)}{15}\ldots\right]$
11. $v(t) = \frac{32}{\pi}\sin(\omega t + 30°) + \frac{32}{3\pi}\sin[3(\omega t + 30°)] + \frac{32}{5\pi}\sin[5(\omega t + 30°)] + \frac{32}{7\pi}\sin[7(\omega t + 30°)]$
13. b. $v_1 = -\frac{16}{\pi}\sin\omega t - \frac{16}{3\pi}\sin 3\omega t - \frac{16}{5\pi}\sin 5\omega t - \ldots$
   $v_2 = \frac{8}{\pi}\sin\omega t - \frac{8}{2\pi}\sin 2\omega t + \frac{8}{3\pi}\sin 3\omega t - \ldots$
   c. $v = -\frac{8}{\pi}\sin\omega t - \frac{8}{2\pi}\sin 2\omega t - \frac{8}{3\pi}\sin 3\omega t - \ldots$
15. b. $V_{\text{avg}} = 5\ \text{V}$
   c. $v_1 = 5 - \frac{10}{\pi}\sin\omega t - \frac{10}{2\pi}\sin 2\omega t - \frac{10}{3\pi}\sin 3\omega t - \ldots$
   $v_2 = -\frac{10}{\pi}\sin\omega t + \frac{10}{2\pi}\sin 2\omega t - \frac{10}{3\pi}\sin 3\omega t + \ldots$
   d. $v_1 + v_2 = 5 - \frac{20}{\pi}\sin\omega t - \frac{20}{3\pi}\sin 3\omega t - \frac{10}{5\pi}\sin 5\omega t - \ldots$
17. $v_1 + v_2 = \frac{16}{\pi}\sin\omega t - \frac{4}{2\pi}\sin 2\omega t + \frac{16}{3\pi}\sin 3\omega t - \ldots$
19. $P = 1.477\ \text{W}$
21. $P_0 = 23.1\ \text{dBm}$ $P_1 = 24.0\ \text{dBm}$ $P_2 = 26.1\ \text{dBm}$
   $P_3 = 2.56\ \text{dBm}$ $P_4 = -4.80\ \text{dBm}$
23. a. $V_0 = 0.5\ \text{V}$
   b. $V_1 = 0.90\ V_P$ $\theta_1 = -45°$
   $V_3 = 0.14\ V_P$ $\theta_2 = -63°$
   $V_5 = 0.05\ V_P$ $\theta_5 = -79°$
   $V_7 = 0.03\ V_P$ $\theta_7 = -82°$

Glossary

absolute maximum power In order for the absolute maximum power to be delivered to the load of an ac circuit, the load impedance must be equal to the complex conjugate of the Thévenin or Norton equivalent impedance of the circuit. Refer to *maximum power transfer theorem*.

absolute zero T = −273.15 °C is the theoretical temperature at which all molecular motion stops.

ac Although ac technically is an abreviation for alternating current, by convention, it is used to denote periodically varying voltage as well—thus, the voltage at your wall outlet is designated 120 volts ac. Standard practice is that ac is not capitalized.

ac current/alternating current Current that periodically reverses in direction.

ac voltage Voltage that periodically changes in polarity. The most common ac voltage is the sine wave.

active power In an ac system, this is the average power delivered from a source to a load. Other terms used for active power are *real power* and *average power.*

admittance (Y) A vector quantity (measured in siemens, S) that is the reciprocal of impedance. **Y = 1/Z.**

admittance diagram A two-dimensional diagram on the complex plane that is used to show all circuit admittances, typically for parallel ac circuits.

air-core coil/air-core transformer An air-core coil is a coil wound on a nonferromagnetic material. An air-core transformer has air-core coils. Magnetic flux is proportional to current for air-core coils and transformers.

alternating current Current that periodically reverses in direction, commonly called an ac current.

alternating voltage Voltage that periodically changes in polarity, commonly called an ac voltage. The most common ac voltage is the sine wave.

American Wire Gauge (AWG) An American standard for classifying wire and cable.

ammeter An instrument that measures current. Most commonly met as one of the functions of a DMM.

ampere (A or amp) The SI unit of electrical current, equal to a rate of flow of 1 coulomb of charge per second.

ampere-hour (Ah) A measure of the storage capacity of a battery.

Ampere's circuital law States that the algebraic sum of mmfs around a closed loop is zero.

ampere-turns See *mmf.*

amplifier Any active component such as a transistor that results in more ac output power applied to a load than input power applied to the circuit.

amplitude For a symmetric, periodic waveform such as dealt with in this book, amplitude is defined as the distance from its average value to its peak value, thus it is a measure of how big the wave is. Specifically, for a sine wave $v = V_m \sin \omega t$, or for a sine wave superimposed on dc such as $v = E + V_m \sin \omega t$, V_m is the amplitude.

angular frequency (ω) Frequency of an ac waveform in radians/s where $\omega = 2\pi f$ and f is frequency in Hz.

apparent power The power that apparently flows in an ac circuit. It is the combination of real power P and reactive power Q. The magnitude of S is equal to the product of effective voltage times effective current—that is, $S = VI$ and its unit is the VA (volt-amp). The relationship between P, Q, and S is represented by the power triangle—see *power triangle*.

application software Software (e.g., Multisim and PSpice) designed for a specific purpose such as analyzing electrical and electronic circuits.

atom The basic building block of matter. In the Bohr model, an atom consists of a nucleus of positively charged protons and uncharged neutrons, surrounded by negatively charged orbiting electrons. An atom normally consists of equal numbers of electrons and protons and is thus uncharged.

attenuation The amount that a signal decreases as it passes through a system. The attenuation is usually measured in decibels, dB.

attenuator Any component (active or passive) that results in less ac output power applied to the load than input power applied to the circuit.

audio frequency A frequency in the range of human hearing, which is typically from about 15 Hz to 20 kHz.

autopolarity For a voltmeter (or multimeter set to volts), it denotes a meter that automatically determines and displays the polarity of the applied voltage. For an ammeter (or multimeter set to current), it denotes a meter that automatically determines and displays the direction of measured current. Most DMMs are autopolarity devices.

autoranging A meter with autoranging automatically determines the range which yields the most significant digits and sets itself up accordingly. In contrast, for meters without autoranging, the user must select the desired measuring range manually. Most DMMs are autoranging.

autotransformer A type of transformer with partially common primary and secondary windings. Part of its energy is transferred magnetically and part conductively.

average of a waveform The mean value of a waveform, obtained by algebraically summing the areas above and below the zero axis of the waveform, divided by the cycle length of the waveform. It is equal to the dc value of the waveform as measured by an ammeter or a voltmeter.

average power An alternate name for active power—see *active power*.

***B-H* curve** The relationship between magnetic flux density B and magnetic field intensity H in ferromagnetic materials is usually shown graphically as a set of curves known as *B-H* curves. These curves, obtained by test, are available in handbooks and on the Internet for important magnetic materials such as cast iron, cast steel, sheet steel, and a few others.

balanced bridge When the voltage between midpoints on the arms of a bridge circuit is zero.

balanced system A three-phase system where the line and load impedances for each phase are identical, and the three source voltage magnitudes are identical, but each is phase displaced from the other by 120°.

band-pass filter A circuit that permits signals within a range of frequencies to pass through a circuit. Signals of all other frequencies are prevented from passing through the circuit.

band-stop filter (or notch filter) A circuit designed to prevent signals within a range of frequencies from passing through a circuit. Signals of all other frequencies freely pass through the circuit.

bandwidth (BW) The difference between the upper and lower half-power frequencies for any resonant, band-pass, or band-stop filter. The bandwidth may be expressed in either hertz or radians per second.

base In number systems, the base is the number on which the counting system is built. Our common decimal system uses base 10 numbers.

battery The most commonly used dc source. Batteries are used to power portable electronic devices such as cell phones and calculators and cordless tools such as drills. They derive their energy from electrochemical action. Primary batteries cannot be charged but secondary batteries can.

bel Named after Alexander Graham Bell, this logarithmic unit represents a tenfold increase or decrease in power.

block diagram In a block diagram, major system components are shown as blocks (usually with a short descriptor describing the function of each). Blocks are interconnected with lines, and signal flow is usually from left to right.

Bode plot (1) A graph of voltage gain (usually given in dB) and phase shift (in degrees) of a circuit as a function of frequency. (2) A graph of the transfer function of a circuit.

branch A portion of a circuit that can be simplified as having two terminals. The components between the two terminals can be any combination of resistors, voltage sources, or other elements.

branch current The current through a branch of a circuit.

branch-current analysis Circuit analysis that uses determinants to directly calculate the current in each branch of a circuit.

break frequency See *cutoff frequency*.

bridge circuit A circuit that is used to very accurately measure the values of unknown components.

bridge networks See *bridge circuit*.

capacitance A measure of charge storage capacity—see *capacitor*. Capacitance opposes a change in voltage. Unit is the farad (F).

capacitive impedance Any circuit having a net (or total) impedance that is capacitive at the frequency of operation.

capacitive loading When signals are degraded by stray capacitance, this is referred to as capacitive loading. (There are other instances, but this is the only situation considered in the book.) See also *stray capacitance*.

capacitive reactance (X_C) The opposition that capacitance presents to sinusoidal ac current, denoted by X_C where $X_C = V_C/I_C = 1/\omega C$ [ohms].

capacitive transients Also called *RC* transients. These are transients that occur when a circuit containing capacitance is disturbed. Mathematically, *RC* transients in dc circuits are always of the form of $e^{-t/RC}$ or $(1 - e^{-t/RC})$.

capacitor A device built to possess capacitance. It stores electrical charge on conductive "plates" separated by an insulating material called a dielectric. The capacitance of most capacitors typically lies in the range between nFs and a few thousand μF, except see *supercapacitor*.

cascade Two stages of a circuit are said to be in a cascade connection when the output of one stage is connected to drive the input of the following stage.

CGS system A system of units based on centimeters, grams, and seconds.

charge (1) The inherent electrical property of electrons and protons that causes a force to exist between them. Electrons are negatively charged while protons are positively charged. Charge is denoted by Q and is defined by Coulomb's law. (2) An excess or deficiency of electrons on a body. (3) To store electric charge as in to charge a capacitor or charge a battery.

charged When used in reference to a capacitor, it means that the capacitor has a voltage across it—that is, it is *charged*. See *charge*.

chassis ground When the metal chassis of an electronic circuit is used as the reference point for the circuit. When a circuit is plugged into commercial as power, the chassis will generally be connected to the ground connection of the building, which in turn is connected to earth ground.

choke Another name for an inductor.

circuit A combination of any number of interconnected components such as resistors, capacitors, inductors, and voltages sources that allows charges to flow (i.e., allows current).

circuit breaker A resettable circuit protection device that trips a set of contacts to open the circuit when current reaches a preset value.

circuit common The reference point in an electrical circuit from which voltages are measured.

circuit theory Electric circuit theory is the body of knowledge that describes (in an organized manner) what we know about electrical circuits. It encompasses the electrical behavior of circuit components, the mathematical relationships between variables such as voltage, current, and power, and the methods that we use to analyze circuits.

circular mil (CM) A unit used to specify the cross-sectional area of a conductor such as a cable or wire. The circular mil is defined as the area contained in a circle having a diameter of 1 mil (0.001 inch). When expressed in terms of square mils, we have 1 CM = $\pi/4$ sq.mil.

closed loop Any path within a circuit that starts and ends at the same point. (See also *Kirchhoff's voltage law*.)

coefficient of coupling (k) A measure of the flux linkage between circuits such as coils. If $k = 0$, there is no linkage; if $k = 1$, all of the flux produced by one coil links another. The mutual inductance M between coils is related to k by the relationship $M = k\sqrt{L_1 L_2}$, where L_1 and L_2 are the self-inductances of the coils.

coil A term commonly used to denote inductors or windings on transformers or inductors.

common-mode signal A signal that appears at both inputs of a differential amplifier.

complex number A complex number is a number with a real part and an imaginary part and may be expressed in *rectangular form* as $a + j\,b$ (where a and b are the real and imaginary parts, respectively, and $j = \sqrt{-1}$), or in *polar form* as $r\,\angle\theta$ where r is the magnitude of the number and θ is its angle.

conductance (G) The reciprocal of resistance, namely $G = 1/\text{R}$. Unit is the siemens (S).

conductor A material through which charges move easily. Copper is the most common metallic conductor.

constant-current source A source that provides the same current to a circuit regardless of the load connected to it (providing the load is not an open circuit).

continuity of current Inductor current cannot change instantaneously from one value to another but must be continuous—i.e., $i_L(0^+) = i_L(0^-)$. If switching is at time $t = t_1$, this becomes $i_L(t_1{}^+) = i_L(t_1{}^-)$. (This statement applies to isolated inductors. For coupled inductors, the statement is incomplete, but that case is beyond the scope of this textbook.)

continuity of voltage Capacitor voltage cannot change instantaneously from one value to another but must be continuous—i.e., $v_C(0^+) = v_C(0^-)$. If switching is at time $t = t_1$, this becomes $v_C(t_1{}^+) = v_C(t_1{}^-)$.

continuous In transient analysis, we are concerned about continuous and discontinuous voltages and currents. A voltage or current is continuous if it forms a perfectly smooth curve as a function of time—that is, it does not change instantaneously from one value to another. If a voltage or current does change instantaneously from one value to another, it is *discontinuous*. See *continuity of voltage*.

conventional current Current in the direction opposite to the flow of electrons in a metallic conductor. In a single-source dc circuit, conventional current emanates from the + terminal of the source.

conversion factor A multiplying factor that permits you to convert from one unit to another, say from minutes to seconds, without changing the magnitude of the quantity. To achieve this, conversion factors must have a value of 1.

copper loss The I^2R power loss in a conductor due to its resistance, such as the power loss in the windings of a transformer.

core The form or structure around which an inductor or the coils of a transformer are wound. The core material affects the magnetic properties of the device.

core loss/iron loss Power loss in the core of a transformer or inductor due to hysteresis and eddy currents. See *hysteresis loss* and *eddy currents*.

core/core type Core refers to the form or structure of a magnetic circuit such as the core of a transformer. Core type usually refers to the material from which the core is made. The core material affects the magnetic properties of the device.

corresponding terminals See *dot convention*.

coulomb (C) The SI unit of electrical charge, equal to the charge carried by 6.24×10^{18} electrons.

Coulomb's law An experimental law that states that the force (in Newtons) between charged particles is $F = Q_1 Q_2/4\pi\epsilon r^2$, where Q_1 and Q_2 are the charges (in coulombs), r is the distance between their centers in meters, and ϵ is the permittivity of the medium. For air, $\epsilon_o = 8.854 \times 10^{-12}$ F/m. (In modern terminology, ϵ_o is also called the *electric constant*.)

counter EMF (back voltage) When current changes in a coil, its changing magnetic field induces a voltage. The polarity of this voltage is always such as to oppose the change in current, hence the terminology "counter" voltage.

coupled circuit A circuit (such as a transformer) in which the magnetic field due to time varying currents in one circuit induces voltage in another. (1) Tightly coupled: Refers to circuits where almost all the flux produced by one circuit links the other. (2) Loosely coupled: Refers to circuits where only a portion of the flux produced by one circuit links the other.

coupled impedance Impedance as "seen" through the windings of a loosely coupled circuit. Note that capacitive elements on the secondary side look inductive when viewed from the primary side and vice versa. (This is quite different than what you see with an iron-core transformer—see *reflected impedance* for comparison.)

critical frequency See *cutoff frequency*.

critical temperature The temperature below which a material becomes a superconductor.

current (*I* or *i*) The rate of flow of electrical charges, measured in amperes. One amp = 1 coulomb per second.

current divider rule The current through any resistor part of a group of parallel resistors is given as $I_x = (R_T/R_x)I_T$.

current ratio In an iron-core transformer, $\mathbf{I}_p/\mathbf{I}_s = N_s/N_p = 1/a$. See *turns ratio*.

current reference arrow Arrow placed on circuit diagrams to denote the direction of conventional current. When you solve for (or measure) current, if its value is positive, it is in the same direction as the reference. If it is negative, its direction is opposite to the reference.

current source A practical current source can be modelled as an ideal current source in parallel with an internal impedance. See also *constant current source*.

cutoff frequency, f_c or ω_c The frequency at which the output power of a circuit is reduced to half of the maximum output power. The cutoff frequency may be measured in either hertz (Hz) or radians per second (rad/s).

cycle One complete variation of a periodic, time varying waveform.

damped oscillations The output of any oscillating circuit will reduce to zero when the power source is removed from the circuit, resulting in damped oscillations.

dc Although dc technically is an abreviation for direct current, by convention, it is used to denote fixed voltage as well—thus, a car battery is a 12-V dc source. Standard practice is that dc is not capitalized.

dc current Direct current—that is, unidirectional current such as that from a battery.

dc resistance See *resistance (R)*.

Δ–Δ system A three-phase system with a Δ generator and a Δ load. This is a three-wire system.

decade As used in circuit theory, a decade is a change in frequency by a factor of 10, either up or down—for example, from 10 Hz to 100 Hz.

decibel (dB) A logarithmic unit used to represent an increase (or decrease) in power levels or sound intensity.

de-energizing transient The current turn-off transient in an *RL* circuit. Such transients may result in huge voltage spikes.

delta Δ (in mathematics) A small change (increment or decrement) in a variable. For example, if current changes a small amount from i_1 to i_2, its increment is $\Delta i = i_2 - i_1$, while if time changes a small amount from t_1 to t_2, its increment is $\Delta t = t_2 - t_1$.

delta load A configuration of circuit components connected in the shape of a Δ (Greek letter delta). Sometimes called a pi (π) load.

delta-wye conversions The mathematical method of converting a delta load into an equivalent wye load.

dependent sources Voltage or current sources that have values that are dependent upon some internal or external voltage or current. These sources are used primarily to simulate the operation of active devices such as transistors.

derivative The instantaneous rate of change of a function. It is the slope of the tangent to the curve at the point of interest.

dielectric An insulating material. The term is commonly used with reference to the insulating material between the plates of a capacitor.

dielectric absorption Due to the polarization of atoms in a dielectric, you may find that after you discharge a capacitor, considerable voltage "springs back" within seconds after the discharge resistor is removed. The cause of this is the residual polarization of the dialectric atoms, and the effect is known as dielectric absorption—see *polarized*.

dielectric breakdown When stressed by high enough voltage, insulation may break down and conduct as electrons are literally torn from their orbits. In solids, this is destructive. In air, the resulting spark goes away if the voltage is dropped below its critical value. See *dielectric strength*.

dielectric constant (∊) A former term for relative permittivity, still used sometimes. See *relative permittivity*.

dielectric strength The voltage gradient at which an insulator breaks down—see *dielectric breakdown*. The dielectric strength of air is approximately 3kV/mm.

differentiator circuit A circuit that performs the mathematical operation of differentiation—that is, a circuit whose output is the derivative of its input. (The differentiator circuits of Chapter 11 are approximate differentiators.)

diode A two-terminal component made of semiconductor material, which permits current in one direction while preventing current in the opposite direction.

discontinuous See *continuous*.

DMM (digital multimeter) A digital multimeter that displays results on a numeric readout. In addition to voltage, current, and resistance, some DMMs measure other quantities such as frequency and capacitance.

domains A region within a ferromagnetic material that has uniform magnetization—that is, it is composed of magnetically aligned atoms.

dot convention A terminal marking convention that indicates *corresponding terminals* on coupled circuits by means of dots. (Corresponding terminals have the same relative polarity at all instants of time.) Variations include using letters such as X1, X2, and so on, instead of dots.

duty cycle The ratio of ON time to the duration of a pulse waveform, expressed in percent.

Δ-Y system A three-phase system with a Δ generator and a Y load. This is a three-wire system.

dynamic resistance Resistance whose *I-V* plot is not a straight line—that is, resistance whose value changes as voltage or current changes. However, at any point on the curve, Ohm's law applies.

earth ground When ac power enters a home or a commercial building, one or more ground rods are placed into the ground to provide a return path for the electrical service. Additionally, metal water pipes within the building are electrically bonded to the earth ground to provide additonal safety for inhabitants of the building.

eddy current (1) A small circulating current induced in conductive material when exposed to a time varying magnetic field. Often refers to the unwanted current that is induced in nearby metal cases, transformer cores, etc. (2) A small circulating current. Usually refers to the unwanted current that is induced in the core of an inductor or transformer by changing core flux.

eddy current loss Eddy currents result in an i^2R power loss in transformer cores (called eddy current loss, core loss, or iron loss) where *R* is the resistance of the path of the current in the iron core. See also *eddy currents*.

effective resistance Resistance defined by $R = P/I^2$. For ac, effective resistance is larger than dc resistance due to skin effect and other effects such as eddy current power losses.

effective value An equivalent dc value of a time varying waveform, hence, that value of dc that has the same heating effect as the given waveform. Also called *rms* (root mean square) value. For sinusoidal current, $I_{\text{eff}} = 0.707\, I_m$, where I_m is the amplitude of the ac waveform. Similarly, for sinusoidal voltage $V_{\text{eff}} = 0.707\, V_m$.

efficiency (η) The ratio of output power to input power, usually expressed as a percentage. $\eta = P_{\text{out}}/P_{\text{in}} \times 100\%$. Can also be expressed in terms of energy.

electric circuit Any circuit consisting of any number of sources and loads connected in any manner that allows charge to flow.

electric constant (ϵ_o) The modern name for the absolute permittivity of free space (i.e., a vacuum). It has a value of 8.8542×10^{-12} farads/meter.

electric field A region in space where an electric force is felt, hence a force field. For example, electric fields exist around electric charges.

electric field intensity (ξ) Also called electric field strength. (1) The force on a unit charge. (2) For the field between parallel capacitor plates, this becomes $\xi = V/d$ volts/meter.

electrical charge See *charge*.

electrolyte An ionic conducting fluid used in electrolytic capacitors. When a voltage of proper polarity is applied to the capacitor, the electrolyte reacts chemically to form a very thin insulating coat (the dielectric) on one of the plates. Since the dielectric is so thin, capacitance is quite large.

electrolytic capacitor A type of capacitor that uses an electrolyte—see *electrolyte*. Electrolytic capacitors are polarized, have relatively low breakdown voltages and other nonideal effects, but are nonetheless widely used in certain types of applications such as power supply filtering.

electron A negatively charged atomic particle. See *atom.*

energy (*W*) (1) The ability to do work. Its SI unit is the joule. (2) Although the SI unit of energy is the joule, in practice, it is often measured in kilowatt-hours (kWh). For example, many of us are billed for electrical usage in kWh.

engineering notation A method of representing certain common powers of 10 via standard prefixes such as milli, micro, or their symbols m and μ. For example, 0.000125 A may be expressed as 125 μA or 0.125 mA.

equivalent series resistance (ESR) The sum of all the internal series resistances of a capacitor.

exponent Consider $n^3 = n \times n \times n$. Here, 3 is the exponent and represents how many times the number n is multiplied by itself. In a more general sense, for a^x, x is the exponent.

exponential function (e^{-x}) A mathematical function of the form e^{-x} where e is the base of the natural logarithm. May be evaluated easily on your calculator using its e^x key. Encountered regularly in transient analysis—see *capacitive transients* and *inductive transients.*

fall time (t_f) The time it takes for a pulse or other abruptly changing signal to change from its 90% value to its 10% value.

farad (F) The basic unit of capacitance, named in honor of Michael Faraday. Farad is an SI unit, but it is a large unit and capacitors typically have values in the μF range.

Faraday's law An experimental law which states that voltage induced is proportional to the rate of change of flux linkage producing the voltage. Lenz's law can be used to determine its polarity.

ferrite A ceramic compound made with powdered iron oxide. Provides a good path for magnetic flux and has low enough eddy current losses that it is useful as a core material for high-frequency inductors and transformers.

ferrite core Some high-frequency transformers use ferrite cores to provide high inductance while keeping losses low. See also *ferrite.*

ferromagnetic Materials like iron, steel, nickel, cobalt (and a few others) that are attracted by magnets.

field A region in space where a force is felt, hence a force field. For example, magnetic fields exist around magnets and electric fields exist around electric charges.

field intensity The strength of a field.

filter circuit A circuit that passes certain frequencies while rejecting all other frequencies.

final condition circuit (dc) In transient analysis, this refers to a circuit drawn as it looks after all transients have subsided—that is, the circuit is in steady state. In such a circuit with dc excitation, ideal capacitors are replaced by open circuits and ideal inductors by short circuits. (We consider only dc transients in this book.)

fixed resistor A resistor for which the resistance is essentially constant at a given temperature.

flux A way of representing and visualizing force fields by drawing lines that show the strength and direction of a field at all points in space. Commonly used to depict electric or magnetic fields.

flux density (B) The number of magnetic flux lines per unit area, measured in the SI system in tesla (T), where 1 T = 1 Wb/m^2. See *flux lines.*

flux lines Also called lines of force. A conceptual representation used with both electric and magnetic fields.

flux linkage (1) For an iron-core coil of N turns with flux ϕ passing through it, *flux linkage* = $N\phi$. (2) For an air-core coil, flux linkage is proportional to its current i. Mathematically, *flux linkage* = Li where L is the coil's self-inductance.

forward biased or forward region A diode is said to be forward biased when conventional current is in the direction of the arrow. In its forward region, a diode has very low resistance and is often approximated as a short circuit.

four-wire system A three phase system with a neutral conductor is a four-wire system. Used in Y-Y connections.

Fourier series Any periodic waveform having a frequency f can be written as an infinite series of sinusoidal waveforms having frequencies $f, 2f, 3f, 4f, \ldots$.

free electron An electron that is weakly bound to its parent atom and is thus easily broken free. For materials like copper, there are billions of free electrons per cubic centimeter at room temperature. Since these electrons can break free and wander from atom to atom, they form the basis of an electric current.

frequency The number of times that a cycle repeats itself each second. Its SI unit is the hertz (Hz)—see *hertz.*

frequency spectrum A graph showing the relative amplitudes (in voltage or power) of the various components of a Fourier series.

fringing Field lines spread out when passing though a non-ferromagnetic material such as an air gap. This is called *fringing.*

fundamental frequency The frequency f of any periodic waveform. See *Fourier series.*

fuse A circuit protection device that utilizes a metallic link that melts when current exceeds the fuse rating. This opens the circuit.

gain The ratio of output voltage, current, or power to the input. Power gain for an amplifier is defined as the ratio of ac output power to ac input power, $A_p = P_{out}/P_{in}$. Gain may also be expressed in decibels. In the case of power gain, A_p (dB) = 10 log P_{out}/P_{in}.

galvanometer An instrument that is used to detect very small currents.

gauss The unit of magnetic flux density in the CGS system of units, an obsolete system. However the word *gauss* is still seen—for example, meters used to measure magnetic fields may be called "gaussmeters."

giga (G) A prefix with a value of 10^9.

ground (1) A circuit common. (2) An electrical connection to earth. (3) A short to ground, such as a ground fault.

half-power frequencies The frequencies at which the output power will be half of the output power that occurs at resonance.

Hall effect When a strip of current-carrying semiconductor is placed in a magnetic field, a small voltage transverse to the current direction is produced. This voltage is proportional to the magnetic field strength. This effect is used in gaussmeters to measure magnetic flux density.

harmonic frequency Integer multiple of the fundamental frequency f of any periodic waveform. See *Fourier series.*

harmonics See *harmonic frequency.*

Hay bridge A bridge circuit used to measure the inductance and series resistance of an inductor having a small series resistance.

henry (H) The SI unit of inductance, named in honor of Joseph Henry.

hertz (Hz) The SI unit of frequency, named in honor of Heinrich Hertz. 1 Hz equals 1 cycle per second.

high-pass filter A circuit that readily permits frequencies above the cutoff frequency to pass from the input to the output of the circuit, while attenuating frequencies below the cutoff frequency. See *cutoff frequency.*

horsepower Horsepower (hp) is a non-SI unit for measuring power, still used in everyday life but largely superseded in scientific disciplines. One hp = 746 watts.

hysteresis loop The path traced when you plot magnetic flux density versus magnetizing force H for ferromagnetic material. The path forms a loop rather than a straight line because the material saturates when H becomes large—recall the saturation of the B-H curve.

hysteresis loss (1) Power loss in a ferromagnetic material caused by the reversal of magnetic domains in a time varying magnetic field. (2) Power loss in a ferromagnetic material such as a transformer core caused by the reversal of magnetic domains. The combination of hysteresis and eddy current loss in a transformer is called core loss—see *core loss.*

ideal current source A current source having an infinite shunt (parallel) impedance. An ideal current source is able to provide the same current to all loads (except an open circuit). The voltage across the current source is determined by the value of the load impedance.

ideal transformer A transformer having no losses and characterized by its turns ratio $a = N_p/N_s$. For voltage, $\mathbf{E}_p/\mathbf{E}_s = a$, while for current $\mathbf{I}_p/\mathbf{I}_s = 1/a$.

ideal voltage source A voltage source having zero series impedance. An ideal voltage source is able to provide the same voltage across all loads (except a short circuit). The current through the voltage source is determined by the connected circuit.

impedance (Z) Total opposition that a circuit element presents to sinusoidal ac in the phasor domain. $\mathbf{Z} = \mathbf{V}/\mathbf{I}$ ohms, where $\mathbf{V}$ and $\mathbf{I}$ are voltage and current phasors, respectively. Impedance is a complex quantity with magnitude and angle. (Although it is a complex number, impedance is not a phasor.)

impedance diagram A two-dimensional diagram on the complex plane that is used to show all circuit impedances, typically for series ac circuits.

impedance matching This is the practice of matching the impedance of a load to the Thévenin impedance of the source so as to achieve maximum power transfer. For complex source impedances, make the load impedance equal to the conjugate of the source impedance. Transformers are sometimes used to achieve impedance matching.

in phase If two phasors representing waveforms coincide, their corresponding waves are in phase. See also *phase shift.*

independent sources Voltage or current sources that have values that are not dependent upon any outside factors.

induced voltage Voltage produced by changing magnetic flux linkages. The *self-induced voltage* of a coil is the voltage created by its own changing current—thus, self-voltage $v_L = L\, di/dt$ where L is self-inductance. See also *mutually induced voltage* and *self-inductance.*

inductance (L) That property of a coil (or other current-carrying conductor) that opposes a change in current. The SI unit of inductance is the henry.

inductive impedance Any circuit having a net (or total) impedance that is inductive at the frequency of operation.

inductive kick When current through an inductor is switched off, a voltage spike (the inductive kick) is induced by its rapidly falling current. This voltage can be very large, hundreds or even thousands of volts, depending on how large the inductance is and how fast the current changes.

inductive reactance (X_L) The opposition that inductance presents to sinusoidal ac current, denoted by X_L where $X_L = V_L/I_L = \omega L$ [ohms].

inductive transients Also called *RL* transients. These are transients that occur when a circuit containing inductance is disturbed. Mathematically, *RL* transients in dc circuits are always of the form of $e^{-Rt/L}$ or $(1 - e^{-Rt/L})$.

inductor A circuit element designed to possess inductance, for example, a coil of wire wound to increase its inductance.

inferred absolute temperature Is also called the *temperature intercept* of a material and is the temperature at which the linear portion of the resistance vs. temperature crosses the temperature axis.

initial condition See *continuity of voltage* and *initial condition circuit.*

initial condition circuit In transient analysis, this refers to a circuit drawn as it looks immediately after a disturbance, for example, at $t = 0^+$, with capacitors and inductors replaced by their $t = 0^+$ equivalents. In such a circuit, uncharged capacitors are replaced by short circuits, charged capacitors by voltage sources with value V_0, noncurrent-carrying inductors by open circuits, and current-carrying inductors by current sources of value I_0.

instantaneous power The product of instantaneous voltage and instantaneous current, given by $p = vi$ watts. See also *instantaneous value.*

instantaneous value The value of a quantity (such as voltage, current, or power) at some instant of time. Thus, voltages that vary with time may be referred to as instantaneous voltages.

insulator A material such as glass, rubber, or Mylar that does not conduct electricity.

integrator circuit A circuit that performs the mathematical operation of integration—that is, a circuit whose output is the integral of its input. (The integrator circuits of Chapter 11 are only approximate integrators.)

internal impedance The impedance that exists internally in a device such as a voltage source.

ion An atom that has become charged. If it has an excess of electrons, it is a negative ion, while if it has a deficiency, it is a positive ion.

iron-core coil Iron core generally denotes a core of any ferromagnetic material. With iron cores, magnetic flux is not proportional to current. A coil that uses a ferromagnetic core to achieve high inductance is called an iron core coil.

iron-core transformer Iron core generally denotes a core of any ferromagnetic material. A transformer that uses a ferromagnetic core to achieve tight coupling between windings is called an iron-core transformer.

***I-V* plot** Denotes a graph of current versus voltage for a circuit element. For a pure resistance, it is a straight line.

joule (J) The SI unit of energy, equal to 1 newton-meter.

kick test Used to determine corresponding terminals of a transformer. A momentary dc is applied to one winding, and the direction of deflection of a voltmeter on the other windings establishes corresponding terminals. See *dot convention.*

kilo (k) A prefix with the value of 10^3.

kilowatt-hour (kWh) A unit of energy equal to 1000 times 1 watt-hour and commonly used by electrical utilities. It is not an SI unit.

Kirchhoff's current law An experimental law that states that the sum of the currents entering a junction is equal to the sum leaving.

Kirchhoff's voltage law An experimental law that states that the algebraic sum of voltages around a closed path in a circuit is zero.

lag For two phasors representing waveforms, the waveform corresponding to the lagging phasor is the lagging waveform. See also *phase shift.*

lagging load A load in which current lags voltage (e.g., an inductive load). Applies to ac circuits only.

lagging power factor The power factor of an ac circuit in which the current lags the voltage—namely, an inductive circuit.

laser A light source that emits very intense monochromatic (single color) coherent (in phase) light. The term is an acronym for Light Amplification through Stimulated Emission of Radiation.

lead For two phasors representing waveforms, the waveform corresponding to the leading phasor is the leading waveform. See also *phase shift.*

leading load A load in which current leads voltage (e.g., a capacitive load). Applies to ac circuits only.

leading power factor The power factor of an ac circuit in which the current leads the voltage—namely, a capacitive circuit.

leakage current In dielectrics, the very small current that passes through the dielectric of a capacitor.

leakage flux (1) In a magnetic circuit, the small amount of flux that passes through air instead of through the ferromagnetic material. (2) For an iron-core transformer, the flux (usually very small) produced by one coil that does not link the other but instead, leaks through the air.

Lenz's law An experimental law that states that the polarity of induced voltage is such as to oppose the cause producing it.

line conductor/phase conductor Line conductors are also called phase conductors. In a three-phase Δ system, these are the wires that connect the source to the load. In a three-phase Y-Y system, you also have three line conductors, but you may also have a fourth (neutral) conductor. When you see an overhead transmission line, the three large wires are line conductors and the fourth, usually much smaller wire is the neutral.

line current In a three-phase system, line current refers to the currents in the lines. For a Y-load (or Y-generator), line current is the same as phase current. For a delta load, line current $\mathbf{I}_a = \sqrt{3}\,\mathbf{I}_{ab}\angle{-30^\circ}$ where $\mathbf{I}_{ab}$ is phase current.

line voltage/line-to-line voltage In a three-phase system, line voltage refers to the voltage from line to line. For a Δ-load, line voltage is the same as phase voltage. For a Y-load, line voltage $\mathbf{V}_{ab} = \sqrt{3}\,\mathbf{V}_{an}\angle 30^\circ$, where $\mathbf{V}_{an}$ is phase voltage.

linear bilateral network Any network that has a linear voltage-current characteristic and demonstrates the same characteristics regardless of the direction of current through the network.

linear circuit A circuit in which relationships are proportional. In a linear circuit, current is proportional to voltage.

linear resistance Resistance whose *I-V* plot is a straight line—that is, its resistance is constant regardless of voltage or current.

load (1) The device that is being driven by a circuit. Thus, the lamp in a flashlight is the load. (2) The current drawn by a load.

load phase impedances For a Y-load, refers to load impedance from line to neutral, for example, $\mathbf{Z}_{an}$. For a Δ-load, refers to load impedance from line to line, for example, $\mathbf{Z}_{ab}$.

loaded output Any circuit having output terminals across which a load is connected. The circuit could be as simple as a voltage source or as complex as an amplifier.

loading effect The degree to which the connection of a load affects a circuit's operation.

loading effect (ammeter) When an ammeter is placed into a circuit, it places a very small resistance in series with the circuit under test. The degree to which this series resistance affects the circuit operation is called the ammeter loading effect.

loading effect (voltmeter) When a voltmeter is placed into a circuit, it places a very large resistance in parallel with the circuit under test. The degree to which this parallel resistance affects the circuit operation is called the voltmeter loading effect.

loosely coupled Coupled circuits with a low coefficient of coupling. See *coupled circuit* and *coefficient of coupling.*

low-pass filter A circuit that permits frequencies below the cutoff frequency to pass through from the input to the output of the circuit, while attenuating frequencies above the cutoff frequency. See *cutoff frequency*.

magnetic circuit A ferromagnetic structure containing one or more loops of ferromagnetic material that guide and shape magnetic fields for useful purposes. Magnetic circuits are a key part of motors, generators, recording heads, and transformers.

magnetic field A region in space where a magnetic force is felt, hence, a magnetic force field.

magnetic field intensity (H) Also called magnetizing force. It is the mmf per unit length of a magnetic circuit.

magnetomotive force (mmf) The flux producing ability of a coil. In the SI system, the mmf of a coil of *N* turns with current *I* is *NI* ampere-turns.

maximum power transfer theorem For dc circuits, a load resistance will receive maximum power from a circuit when the resistance of the load is exactly equal to the Thévenin (or Norton) resistance looking back at the circuit.

maxwell (Mx) The CGS unit of magnetic flux Φ.

Maxwell bridge A bridge circuit is used to determine the inductance and series resistance of an inductor having a relatively large series resistance (in comparison to $X_L = \omega L$).

mega (M) A prefix with the value of 10^6.

mesh analysis Circuit analysis that uses determinants to determine the current in each loop (or mesh) of a circuit.

meter loading effect The degree to which the meter affects a circuit's operation since all meters use some energy from the circuit under test, resulting in measurement error.

micro (μ) A prefix with the value of 10^{-6}.

milli (m) A prefix with the value of 10^{-3}.

Millman's theorem Any circuit consisting of a number of parallel branches with each branch having a voltage source and a series resistor can be simplified as a single equivalent voltage source in series with a resistor.

mmf (magnetomotive force) A measure of the flux producing ability of a current-carrying coil. $mmf = \mathcal{F} = NI$ ampere-turns.

molded carbon composition resistor A fixed resistor having a core composed of carbon mixed with an insulating filler.

multimeter A multifunction test meter used to measure a variety of electrical quantities such as voltage, current, and resistance. Its function is selected by a switch. See also *DMM*.

Multisim Electronic schematic capture and circuit simulation software, widely used for circuit analysis and design. It is a trademark of the National Instruments Corporation.

mutual conductance Conductance that occurs between two nodes when applying nodal analysis.

mutual inductance (*M*) The inductance between circuits (such as coils) measured in henries. The voltage induced in one circuit by changing current in another circuit is equal to *M* times the rate of change of current in the first circuit—see *mutually induced voltage*.

mutual resistance Resistance that occurs in a branch that is common to two loops when applying mesh analysis.

mutually induced voltage When the changing magnetic flux created by current in a coil induces voltage in a nearby coil, this voltage is referred to as mutually induced voltage. For example, the mutually induced voltage in coil 2 due to changing current in coil 1 is given by $v_{21} = Mdi_1/dt$ where *M* is the mutual inductance.

nano (n) A prefix with the value of 10^{-9}.

negative temperature coefficient See *temperature coefficient*.

neutral (1) The common junction point in a three-phase Y-generator (or Y-load) is called its neutral point. Neutral points are often grounded to earth for safety reasons. (2) The wire joining the neutral point of a generator to the neutral point of a load is referred to as the neutral or neutral conductor. Under balanced conditions, the current in the neutral is zero.

neutron An atomic particle with no charge. See *atom*.

newton The SI unit of force. It is the force required to accelerate a mass of 1 kg at the rate of 1 meter per second per second. In more common terms, it is equal to about 0.2248 pounds of force.

nodal analysis Circuit analysis that uses determinants to determine the voltage of each node (relative to a given reference point) of a circuit.

node A junction where two or more components connect in an electric circuit.

nonlinear resistance See *dynamic resistance*.

non-ohmic device Any device for which the current-voltage relation is not linear. See *ohmic device*.

Norton's theorem Any linear bilateral network may be reduced to a simplified two-terminal circuit consisting of a single current source and a single shunt resistor.

notch filter A circuit that removes a narrow range of signals between the input and output of a circuit.

octave As used in circuit theory, an octave is a doubling (or halving) of frequency—for example, from 1 kHz to 2 kHz.

ohm (Ω) The SI unit of resistance. May be measured using an ohmmeter. Also used as the unit for reactance and impedance.

ohmic Refers to a resistive device whose *I-V* characteristic (plot) is a straight line.

ohmic device Any device that has a linear current-voltage relation. For example, if current through the device doubles, this implies that voltage across the device will also double.

ohmmeter A test instrument for measuring resistance, most commonly implemented as part of a DMM.

Ohm's law $I = E/R$. This is the basic relationship between current and voltage for resistance and was determined experimentally by Ohm, circa 1827.

open circuit A discontinuous circuit, hence, one that does not provide a continuous (complete) path for current.

open-circuit test See *transformer tests*.

oscilloscope Also called a scope. A test and measurement instrument that electronically displays voltage waveforms on a screen. The screen is ruled with a scaled grid to permit you to visually study waveform characteristics. However, to make measurements, you should use the scope's cursors or its automatic measurement capabilities as these yield better accuracy than scaling values from the grid. Most modern scopes are digital.

parallel Elements or branches are said to be in a parallel connection when they have exactly two nodes in common. The voltage across all parallel elements or branches is exactly the same.

parallel circuit Any circuit or branch having exactly two nodes in common.

parallel resonanant circuit A circuit in which the inductor is in parallel with the capacitor. The voltage across the circuit will be maximum at the resonant frequency.

peak value The maximum instantaneous value of a waveform.

peak-to-peak The magnitude of the difference between a waveform's peak and trough values.

period (*T*) The time *T* for a waveform to go through one cycle. $T = 1/f$ where *f* is frequency in Hz.

periodic Repeating at regular intervals.

periodic waveform A time varying waveform that repeats itself at regular intervals, as for example, the 60 Hz voltage at your wall outlet is a sine wave that repeats its cycle 60 times per second.

permeability A measure of how easy it is to magnetize a material, represented by the Greek letter μ. $B = \mu H$, where *B* is the resulting flux density and *H* is the magnetizing force that creates the flux. For ferromagnetic materials, μ is nonlinear and *B-H* curves obtained by test are usually required to relate *B* to *H*.

permittivity (∊) A measure of how easily a dielectric material permits the establishment of an electric field. See also *relative permittivity* and *electric constant*.

phase current Current in the phase of a three-phase load or generator. See also *line current*.

phase difference See *phase shift*.

phase sequence Refers to the order in which three-phase voltages reach their respective peak values. Positive phase sequence is *ABC*, while negative phase sequence is *ACB*. Electric power systems are based on the *ABC* sequence.

phase shift (1) The angular difference between two waveforms of the same frequency—that is, the angular difference by which one waveform leads or lags another, hence the relative displacement between time varying waveforms. (2) Mathematically, consider a voltage $v = V_m \sin(\omega t \pm \theta)$ relative to a reference $v_{ref} = V_{ref} \sin \omega t$. If $\theta = 0°$, the waveforms are in phase; for the case of $+\theta$, v leads; for the case $-\theta$, v lags.

phase voltage For a Y-load or generator, refers to the line to neutral voltage. For a Δ load (or generator), phase voltage is the same as line voltage. See also *line voltage.*

phasor A way of representing the magnitude and angle of a sine wave by a complex number. The magnitude of the phasor represents the rms value of the ac quantity, and its angle represents the waveform's phase.

phasor domain Quantities expressed as phasors are said to be in the *phasor domain.*

photocell or photoconductive cell A two-terminal transducer whose resistance is determined by the amount of light falling on the cell.

pico (p) A prefix with the value of 10^{-12}.

pictorial diagram A pictorial diagram shows an image of a circuit component as it actually looks physically.

point source A voltage source that has only one subscript indicates that this is the potential at that point with respect to the circuit reference (or ground) point. For example, V_a implies that this is the voltage at point a with respect to ground in the circuit.

polar form See *complex number.*

polarity Polarity can be defined with reference to a battery. For a battery, one terminal is always positive with respect to the other, and we define this terminal as having a positive polarity and mark it with a plus (+) sign. We then define the other terminal as having a negative polarity and mark it with a minus (−) sign. This gives us our interpretation of polarity. For voltage drops across components, we use the same +, − marking scheme.

polarized When voltage is applied to a capacitor, electron orbits in dielectrics become distorted, making atoms appear to have + and − ends. This is referred to as *polarization.* When voltage is removed, the orbits relax, that is, return to their originally shaped orbits. If the relaxation takes too long, a condition referred to as dielectric absorption remains.

positive temperature coefficient See *temperature coefficient.*

potential difference This is an alternate name for voltage difference.

potentiometer A three-terminal resistor consisting of a fixed resistance between two end terminals and a third terminal that is connected to a movable wiper arm. When the end terminals are connected to a voltage source, the voltage between the wiper and either of the other terminals is adjustable.

power (*P*, *p*) (1) The rate of doing work, with units of watts, where 1 watt equals 1 joule per second. Also called real or active power. (2) In a purely resistive circuit, power to a resistor is given by $P = VI = I^2R = V^2/R$ watts. (3) In a circuit with time varying voltage and current, the instantaneous value of power is given by $p = vi$. The average value of this is P. P may be measured using a wattmeter.

power factor The ratio of active power P to apparent power S. Maximum power factor is one (also referred to as unity). Note that $F_p = \cos \theta$ where θ is the angle between voltage and current.

power factor correction The larger the power factor, the more of the source power that is transferred to the load. Since most practical power system loads are inductive, power factor correction generally consists of adding capacitors in parallel to bring load power factor closer to unity.

power gain The ratio of output power to input power, usually for an amplifier or attenuator.

power measurement (three phase) Wattmeters are used to measure power in three-phase systems. Many power measurement instruments in use today are multifunction power/energy meters that combine many functions in one. With these, you can measure total power, power to individual phases, reactive power, apparent power, and power factor. In older setups, two wattmeters were used to measure power using the "two-wattmeter method."

power of ten notation A notation that expresses numbers as a multiple of a power of 10. Thus, the number 3251 may be written as the power of ten number 3.251×10^3.

power triangle A useful computational and visualization tool. The relationship between power P, reactive power Q, and apparent power S, given by $S^2 = P^2 + Q^2$, is shown on the *power triangle* with S the hypotenuse and P and Q its two sides.

prefixes A prefix used to denote a power of ten—for example, the prefix name for 10^3 is kilo.

primary The winding of a transformer to which we connect the source.

primary battery See *battery.*

proton A positively charged atomic particle. See *atom.*

PSpice Electronic schematic capture and circuit simulation software, widely used for circuit analysis and design. It is a trademark of Cadence Design Systems, Inc.

pulse A short duration voltage or current that abruptly changes from one value to another, then back again.

pulse repetition rate (PRR) The number of pulses per second. Also called *pulse repetition frequency.*

pulse train A repetitive stream of pulses.

pulse width The duration of a pulse. For nonideal pulses, it is measured at the 50% amplitude point.

quality factor (*Q*) (1) A figure of merit. Q for a coil is the ratio of its reactive power to its real power. The higher the Q, the more closely the coil approaches the ideal. (2) A measure of the selectivity of a resonant circuit. The higher the Q, the narrower the bandwidth.

radian frequency (ω) See *angular frequency.*

reactance (*X*) The opposition that a reactive element (capacitance or inductance) presents to sinusoidal ac, measured in ohms.

reactive power In an ac system that contains a reactive element L or C, energy gets stored, then released from the magnetic field of L or the electric field of C back into the system. This power is termed *reactive power.* Since reactive power shuttles back and forth in a system, it contributes nothing to the average flow of power from the source to the load. Symbol is Q. For inductance, $Q_L = V_L I_L = I^2 X_L$. For capacitance, $Q_C = V_C I_C = I^2 X_C$. Unit is the VAR.

reactor Another name for an inductor.

real power Another name for active power. See *active power.*

reciprocity theorem A voltage source causing a current I in any branch of a circuit may be removed from the original location and placed into that branch having the current I. The voltage source in the new location will produce a current in the original source location that is exactly equal to the originally calculated current, I.

rectangular form See *complex number.*

rectifier A circuit, generally consisting of a least one diode, which permits current in only one direction.

reflected impedance Impedance as "seen" through the windings of a tightly coupled transformer. It has a value of $a^2\mathbf{Z}_L$ where a is the turns ratio. Note that an inductive load still looks inductive from the primary, and a capacitive load looks capacitive.

relative maximum power In those circuits where it is not possible to adjust the value of the reactive component contained in a load, the relative maximum power occurs when $R_L = \sqrt{R_{Th}^2 + (X \pm X_{Th})^2}$.

relative permittivity The relative permittivity of a material is its actual permittivity relative to that of a vacuum. For example, the relative permittivity of Mylar is 3. See also *permittivity*.

relay A switching device that is opened or closed by an electrical signal. May be electromechanical or electronic.

reluctance ($\Re$) The opposition of a magnetic circuit to the establishment of flux.

residual magnetism Magnetism remaining in ferromagnetic material after its magnetizing current has been turned off.

resistance (R) (1) The opposition to current presented by a resistor—see *Ohm's law*. Current through resistance results in a power loss. (2) The definition of resistance as $R = V/I$ is inadequate for ac circuits. For ac circuits, we define resistance in terms of power dissipation as $R = P/I^2$ ohms. See *effective resistance.*

resistive circuit Any ac circuit having a net impedance that has only a real (resistive) component at the frequency of operation.

resistivity, *r* A physical property of a material that is used to calculate the resistance of a sample. In the SI system, r is normally measured at 20° C and is given in W × m. Resistivity can also be measured in CM × W/ft.

resistor A circuit component designed to possess resistance.

resonance, resonant frequency The frequency at which the output power of an *L-R-C* circuit is at a maximum. $f = \dfrac{1}{2\pi\sqrt{LC}}$.

resonant circuit In its simplest form, a resonant circuit consists of an inductor and capacitor and either a voltage or current source. At some frequency, called the resonant frequency, the output voltage or current will be at a maximum.

reverse biased or reverse region A diode is said to be reverse biased when conventional current is opposite to the direction of the arrow. In its reverse region, a diode has very high resistance and is often approximated as an open circuit.

rheostat A variable resistor connected so that current through the circuit is controlled by the position of the wiper.

rise time (t_r) The time that it takes for a pulse or or other abruptly changing signal to change from its 10% value to its 90% value.

rms value The root mean square value of a time varying waveform. See also *effective value*.

saturation The condition of a ferromagnetic material where it is fully magnetized. At saturation, if the magnetizing force (current in a coil for example) is increased, no significant increase in flux results.

schematic capture The process of drawing a schematic diagram on a computer screen using schematic editor software such as Multisim.

schematic diagram A circuit diagram that uses symbols to represent physical circuit components.

Schering bridge A bridge circuit used to determine the value of an unknown capacitance.

secondary battery See *battery.*

secondary winding The output winding of a transformer.

selectivity The ability of a filter circuit to pass a particular frequency, while rejecting all other frequency components.

selectivity curve A graph of power as a function of frequency for a resonant circuit. The curve will generally show the half-power frequencies and the bandwidth.

self-induced voltage See *induced voltage.*

self-inductance (*L*) Self-inductance of a coil is the property of a coil that causes it to oppose change in its own current. Self-inductance is denoted by the symbol *L*. See also *induced voltage* and *flux linkage.*

semiconductor A material such as silicon from which transistors, diodes, and the like, are made.

series circuit A closed loop of elements where two elements have no more than one common terminal. In a series circuit, there is only one current path, and all series elements have the same current.

series connection Any branch or circuit that is connected so that the components are in series.

series resonant circuit A circuit in which the inductor is in series with the capacitor. The current through the circuit will be maximum at the resonant frequency.

series-parallel connection A circuit that consists of components in series and in parallel.

short circuit A short circuit occurs when two terminals of an element or branch are connected by a low-resistance conductor. When a short circuit occurs, very large currents may result in sparks or a fire, particularly when the circuit is not protected by a fuse or circuit breaker.

short-circuit test See *transformer tests.*

SI system The international system of units used in science and engineering. It is an extended metric system and includes the standard units for length, mass, and time (e.g., meters, kilograms, and seconds), plus the electrical units (e.g., volts, amperes, ohms, and so on).

siemens (S) A unit of measure for conductance, admittance, and susceptance. The siemens is the reciprocal of ohm.

sine wave A periodic waveform that is described by the trigonometric sine function. It is the principal waveform used in ac systems.

single-phase equivalent A representation used to greatly simplify the analysis of a balanced three-phase system by considering only one of its phases. It is the standard method used in practice for manually analyzing three-phase systems.

skin effect At high frequencies, the tendency of current to travel in a thin layer near the surface of a conductor. The attendant reduction in conduction area results in an increase in resistance.

spectrum analyzer An instrument that displays the amplitude of a signal as the function of frequency.

square mil (sq. mil) A unit of measurement used to specify the cross-sectional area of a solid conductor. One square mil is defined as the area contained in a square having sides of length 1 mil. When expressed in terms of circular mils, we have 1 sq. mil = 4/π CM.

square wave A rectangular waveform whose high time and low time durations are the same.

stacking factor In a laminated core, the ratio of net ferromagnetic cross-sectional area to the physical cross-sectional area.

steady state The condition of operation of a circuit after all transients have subsided.

step An abrupt change in voltage or current, as for example, when a switch is closed to connect a battery to a resistor.

step up/step down A step-up transformer is one in which the secondary voltage is larger than the primary. A step-down transformer is the converse. For a step up, $a < 1$. For a step down, $a > 1$. $a = 1$ is unity turns ratio. See *turns ratio.*

stray (parasitic) capacitance The (usually unwanted) capacitance that exists between conductive parts of an electronic circuit such as that between traces on a printed circuit board or between coil turns on an inductor. In high-speed circuits, this capacitance can cause degradation of signals.

stray inductance Because all current-carrying conductors create small magnetic fields, they all have small amounts of inductance. This inductance is referred to as *stray inductance.*

substitution theorem Any branch within a circuit may be replaced by an equivalent branch, provided the replacement branch has the same current through it and voltage across it as the original branch.

supercapacitor Also called ultracapacitors or double-layer capacitors, these devices have enormously large capacitances, reaching up into the hundreds of farads (compared to a few thousand μF for the largest electrolytic capacitors), with packages no larger than ordinary capacitors. However, their working voltage is quite low.

superconductor A conductor that has no internal resistance. Current will continue unimpeded through a superconductor even though there is no externally applied voltage or current source.

superposition theorem The total current through or voltage across a resistor or branch may be determined by summing the effects due to each independent source.

surface mount device (SMD) A circuit component that is mounted directly onto the surface of a printed circuit board. SMDs (which have either very short leads or no leads at all) mount directly on solder pads, and assembly is generally done by automated machine. SMD components are usually smaller than traditional components.

susceptance The reciprocal of reactance. Unit is the siemens.

tank circuit A circuit consisting of an inductor and capacitor connected in parallel. Such an *L-C* circuit is used in oscillators and receivers to provide maximum signal at the resonant frequency. See *selectivity.*

temperature coefficient The rate at which the properties of a component change with temperature. (1) A resistor has a positive temperature coefficient if its resistance increases with an increase in temperature. Conversely, a negative temperature coefficient means that resistance decreases as temperature is increased. (2) A capacitor has a positive temperature coefficient if its capacitance increases with an increase in temperature and a negative coefficient if its capacitance decreases. The change in capacitance is due to changes in the characteristics of its dielectric with temperature.

temperature intercept Is also called the *inferred absolute temperature* of a material and is the temperature at which the linear portion of the resistance versus temperature crosses the temperature axis.

terminal voltage The voltage that appears across the terminals of a voltage source. If a load is connected across a voltage source, the terminal voltage is dependent on both the internal resistance of the voltage source and the load resistance.

tesla (T) The SI unit of magnetic flux density. One tesla = one Wb/m^2.

thermistor A two-terminal transducer in which resistance changes significantly with changes in temperature.

Thévenin's theorem Any linear bilateral network may be reduced to a simplified two-terminal circuit consisting of a single voltage source in series with a single resistor.

three-wire Y-Y system A three-phase configuration that uses three phase conductors. All Δ systems are three-wire, but Y systems are usually four-wire.

three-phase generators (3ϕ) Three-phase generators have three sets of windings that produce balanced voltages—that is, voltages that are equal in magnitude and phase displaced from each other by 120 degrees.

three-phase system This is the basis of the modern ac electrical utility system. Voltage is generated and transmitted as three phase, but is usually distributed as single phase to residential and business customers. Some industries require three phase for running large motors. See also *three-phase generators.*

tightly coupled Coupled circuits with a high coefficient of coupling. See *coupled circuit* and *coefficient of coupling.*

time constant A measure of how fast a transient changes. For example, capacitor voltage (and inductor current) change by 63.2% in the first time constant, and if $\tau = 1$ ms, this means the change has occurred in 1 ms. In contrast, if $\tau =$ 1s, the change would take 1 second to accomplish, a 1000-fold difference. For an *RC* circuit, $\tau = RC$ seconds. For an *RL* circuit, $\tau = L/R$ seconds. See also *transient duration.*

time domain Quantities expressed as functions of time are said to be in the *time domain.*

total conductance When several resistors are connected in parallel, the total conductance is given as $G_T = G_1 + G_2 + G_3 + \ldots$.

total equivalent resistance Any combination of series and parallel resistors can be replaced with a single equivalent resistance, often designated as R_T.

transducer Any device or component that causes an electrical change due to a physical change.

transfer function A vector quantity that is expressed as the ratio of output voltage to input voltage.

transformer A device with two or more coils in which energy is transferred from one winding to the other by electromagnetic action—that is, a coupled circuit. Transformers generally provide energy transfer with no direct conductive connection between the two sides.

transformer tests Refers to the *open-circuit test* and the *short-circuit test.* These are a pair of standardized tests used to determine the parameters of a transformer. Generally used with power transformers.

transient A temporary or transitional voltage or current—it is here, then it is gone!

transient duration The length of time a transient lasts. For practical purposes, transients are considered to last five time constants.

transistor A three-terminal active device that results in signal amplification.

transistor bias The connection of a dc source(s) that enables a transistor to operate as an amplifier.

transistor model An electric circuit that simulates the operation of a transistor amplifier.

trough value A minimum point in a waveform. For example, for a sine wave, the point where it reaches its most negative value.

turns ratio (*a*) Also called *transformation ratio*. The ratio of primary turns to secondary turns; $a = N_p/N_s$. (Some authors use the inverse definition.)

two-wattmeter method A connection whereby two wattmeters may be used to measure total power to a three-wire, three-phase (balanced or unbalanced) load.

unbalanced bridge When the voltage between midpoints on the arms of a bridge circuit is not zero. In such cases, there will be a current through any component or instrument connected between these nodes.

unbalanced load A load in which the impedances of each phase are not identical. This results in unbalanced currents and for Y-loads, unbalanced phase voltages at the load.

unbalanced system In the context of this book, refers to a system with an unbalanced load, since, under normal operating conditions, the generator and lines will always be balanced. (The only time that generators or lines become unbalanced in practice is when a fault occurs such as a line being broken by a falling tree.)

universal bias The most common trasnsistor bias circuit. See *transistor bias*.

universal time constant curve A graphical plot showing percent transient rise and fall times versus time constants. May be used to solve problems involving exponential functions, although, with modern calculators, it is recommended that you use your calculator instead because it yields better accuracy.

unloaded output See *loaded output*.

valence electrons and valence shells The outermost (last occupied) shell of an atom is called its valence shell, and electrons in this shell are called valence electrons.

VAR (volt-amps reactive) The SI unit of reactive power. See *reactive power*.

varactor diode (or varicap, epicap, and tuning diode) A diode that behaves as a voltage-variable capacitor.

varistor A two-terminal device that has very high resistance when the voltage applied to the device is below the breakdown value. When the applied voltage is increased beyond the breakdown value, the resistance drops to almost zero, allowing a large current through the varistor.

volt (V) The unit of voltage in the SI system.

voltage (*V, v, E, e*) Potential difference created when charges are separated, as for example by chemical means in a battery. If 1 joule of work is required to move a charge of 1 coulomb from one point to another, the potential difference between the points is 1 volt.

voltage divider rule The voltage dropped across any resistor that is part of a group of series resistors is given as $V_x = (R_x/R_T)E$.

voltage gain The ratio of output voltage to input voltage, usually for an amplifier or attenuator.

voltage ratio In an iron-core transformer, $\mathbf{V}_p/\mathbf{V}_s = N_p/N_s = a$. See *turns ratio*.

voltage reference convention The practice of placing the + sign for the voltage drop across a component at the tail of the current reference arrow.

voltage regulator A device that maintains constant output voltage to a load regardless of the input voltage or the amount of output current.

voltage source A practical voltage source can be modelled as an ideal voltage source in series with an internal impedance.

voltage subscripts Voltage subscripts are used to indicate the voltage at one point in a circuit with respect to a second point. For example, V_{ab} implies that this is the voltage at point *a* with respect to point *b* in the circuit.

volt-ampere (VA) The SI unit of apparent power. See *apparent power*.

watt (W) The SI unit of active power. Power is the rate at which work is done; 1 watt equals 1 joule/s.

watthour (Wh) A unit of energy, equal to 1 watt times 1 hour. One Wh = 3600 joules. Watthour is not an SI unit.

wattmeter An instrument for measuring power *P*, normally used in ac circuits.

waveform The variation versus time of a time varying signal, hence, the shape of the signal.

weber (Wb) The SI unit of magnetic flux.

work (*W*) The product of force times distance, measured in joules in the SI system, where 1 joule equals 1 newton-meter.

working voltage Term used with capacitors. Represents the maximum dc voltage that you should apply for continuous operation at rated temperature.

wye load A configuration of circuit components connected in the shape of a Y. Sometimes called a star or T load.

wye-delta conversions The mathematical method of converting a wye load into an equivalent delta load.

Y-Δ system A three-phase system with a Y generator and a Δ load. This is a three-wire system.

Y-Y system A three-phase system with a Y generator and a Y load. In theory, this may be a three-wire system, but in practice, it is usually a four-wire system.

zener diode A diode that normally operates in its reverse region and is used to maintain a constant output voltage.

zero-impedance conductor A hypothetical conductor of zero impedance needed to create the single-phase equivalent of a balanced three-phase system.

Index